TANGENTS AND SECANTS

30. $\int \tan u \, du = \ln |\sec u| + C$

31. $\int \sec u \, du = \ln |\sec u + \tan u| + C$

32. $\int \tan^2 u \, du = \tan u - u + C$

33. $\int \sec^2 u \, du = \tan u + C$

34. $\int \sec u \tan u \, du = \sec u + C$

35. $\int \tan^3 u \, du = \frac{1}{2} \tan^2 u + \ln |\cos u| + C$

36. $\int \sec^3 u \, du = \frac{1}{2} \sec u \tan u + \frac{1}{2} \ln |\sec u + \tan u| + C$

37. $\int \tan^n u \, du = \dfrac{\tan^{n-1} u}{n-1} - \int \tan^{n-2} u \, du$

38. $\int \sec^n u \, du = \dfrac{\sec^{n-2} u \tan u}{n-1} + \dfrac{n-2}{n-1} \int \sec^{n-2} u \, du$

COTANGENTS AND COSECANTS

39. $\int \cot u \, du = \ln |\sin u| + C$

40. $\int \csc u \, du = \ln |\csc u - \cot u| + C$

41. $\int \cot^2 u \, du = -\cot u - u + C$

42. $\int \csc^2 u \, du = -\cot u + C$

43. $\int \csc u \cot u \, du = -\csc u + C$

44. $\int \cot^3 u \, du = -\frac{1}{2} \cot^2 u - \ln |\sin u| + C$

45. $\int \csc^3 u \, du = -\frac{1}{2} \csc u \cot u + \frac{1}{2} \ln |\csc u - \cot u| + C$

46. $\int \cot^n u \, du = -\dfrac{\cot^{n-1} u}{n-1} - \int \cot^{n-2} u \, du$

47. $\int \csc^n u \, du = -\dfrac{\csc^{n-2} u \cot u}{n-1} + \dfrac{n-2}{n-1} \int \csc^{n-2} u \, du + C$

HYPERBOLIC FUNCTIONS

48. $\int \sinh u \, du = \cosh u + C$

49. $\int \cosh u \, du = \sinh u + C$

50. $\int \tanh u \, du = \ln (\cosh u) + C$

51. $\int \coth u \, du = \ln |\sinh u| + C$

52. $\int \operatorname{sech} u \, du = \tan^{-1}(\sinh u) + C$

53. $\int \operatorname{csch} u \, du = \ln \left|\tanh \frac{1}{2} u\right| + C$

54. $\int \operatorname{sech}^2 u \, du = \tanh u + C$

55. $\int \operatorname{csch}^2 u \, du = -\coth u + C$

56. $\int \operatorname{sech} u \tanh u \, du = -\operatorname{sech} u + C$

57. $\int \operatorname{csch} u \coth u \, du = -\operatorname{csch} u + C$

58. $\int \sinh^2 u \, du = \frac{1}{4} \sinh 2u - \frac{1}{2} u + C$

59. $\int \coth^2 u \, du = \frac{1}{4} \sinh 2u + \frac{1}{2} u + C$

60. $\int \tanh^2 u \, du = u - \tanh u + C$

61. $\int \coth^2 u \, du = u - \coth u - C$

62. $\int u \sinh u \, du = u \cosh u - \sinh u + C$

63. $\int u \cosh u \, du = u \sinh u - \cosh u + C$

(*table continued at the back*)

THE GREEK ALPHABET

A	α	alpha
B	β	beta
Γ	γ	gamma
Δ	δ	delta
E	ϵ	epsilon
Z	ζ	zeta
H	η	eta
Θ	θ	theta
I	ι	iota
K	κ	kappa
Λ	λ	lambda
M	μ	mu
N	ν	nu
Ξ	ξ	xi
O	o	omicron
Π	π	pi
P	ρ	rho
Σ	σ	sigma
T	τ	tau
Υ	υ	upsilon
Φ	ϕ	phi
X	χ	chi
Ψ	ψ	psi
Ω	ω	omega

CALCULUS

ONE AND SEVERAL VARIABLES

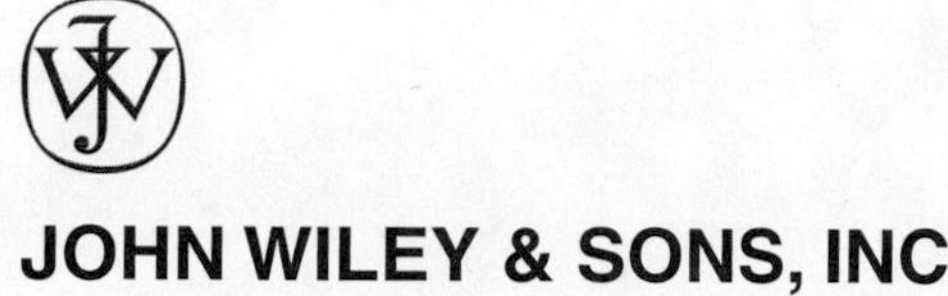

JOHN WILEY & SONS, INC

In fond remembrance of
EINAR HILLE

ACQUISITIONS EDITOR	Michael Boezi
ASSOCIATE PUBLISHER	Laurie Rosatone
ASSISTANT EDITOR	Jennifer Battista
MARKETING MANAGER	Julie Z. Lindstrom
SENIOR PRODUCTION EDITOR	Norine M. Pigliucci
COVER DESIGNER	Madelyn Lesure
COVER AND CHAPTER OPENING PHOTO	© Antonio M. Rosario/The Image Bank
PRODUCTION MANAGEMENT SERVICES	Hermitage Publishing Services

This book was set in New Times Roman by Hermitage Publishing Services and printed and bound by Von Hoffmann Corporation. The cover was printed by Brady Palmer.

This book is printed on acid-free paper. ♾

Library of Congress Cataloging-in-Publication Data

Salas, Saturnino L.
Calculus.– 9th ed. / Saturnino Salas, Einar Hille, Garrett Etgen.
p. cm.
At head of title: Salas, Hille, Etgen.
Rev. ed. of: Salas and Hille's calculus: one and several variables. 8th ed. / rev. by Garret J. Etgen. c1999.
Includes index.
ISBN 0-471-23120-7 (cloth : acid-free paper) – ISBN 0-471-38375-9 (WIE : acid-free paper)
1. Calculus. I. Hille, Einar, 1894– II. Etgen, Garret J., 1937– Salas and Hille's calculus: one and several variables. III. Title.
QA303.2 .S35 2002
515–dc21

2002190823

Printed in the United States of America
10 9 8 7 6 5

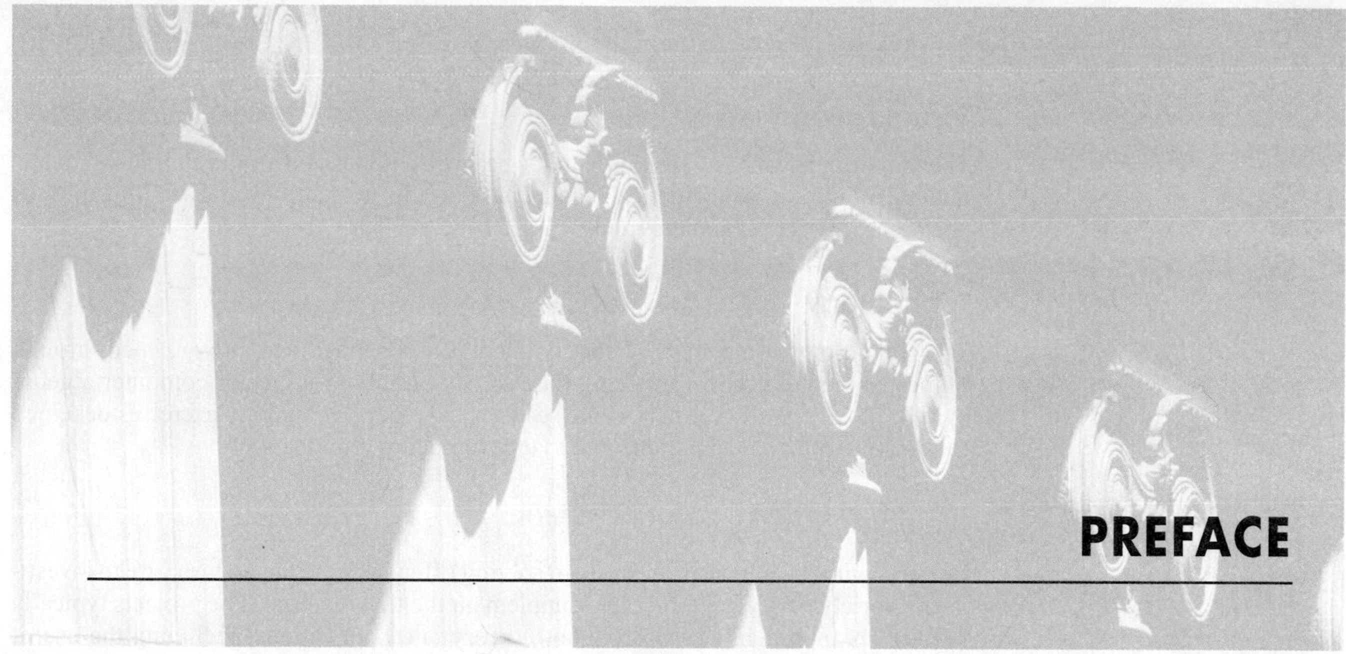

PREFACE

This text is designed for a single and multivariable calculus sequence. While applications from the sciences, engineering, business and economics are often used to motivate or illustrate mathematical ideas, the underlying emphasis throughout is on the three basic concepts of calculus: limit, derivative, integral.

This edition is the result of a substantial joint effort with S.L. Salas. He scrutinized every sentence in every chapter, seeking improved precision and readability. His gift for writing, together with his uncompromising standards for mathematical accuracy and clarity, illuminates the beauty of the subject while increasing its accessibility to students of all levels.

FEATURES OF THE NINTH EDITION

Precision and Clarity

The emphasis is on mathematical exposition, and topics are treated in a clear and understandable manner. Mathematical statements are careful and precise; the basic concepts and important points are not obscured by excess verbiage.

Balance of Theory and Applications

Problems drawn from the physical sciences are often used to introduce basic concepts in calculus. In turn, the concepts and methods of calculus are applied to a variety of problems in the sciences, engineering, business and the social sciences. Because the presentation is flexible, instructors can vary the balance of theory and applications according to the needs of their students.

Accessibility

This text is designed to be completely accessible to the beginning calculus student without sacrificing appropriate mathematical rigor. The important theorems are explained and proved, and the mathematical techniques are justified. These may be covered or omitted according to the theoretical level desired in the course.

Visualization

The importance of visualization cannot be over-emphasized in developing students' understanding of mathematical concepts. For this reason, over 1200 illustrations accompany the text examples and exercise sets.

Technology

The technology component of the text has been expanded and strengthened through numerous new exercises involving the use of a graphing utility or computer algebra system (CAS). Well over half of the sections have new technology exercises designed to illustrate or expand upon the material developed within the sections.

Projects

Projects with an emphasis on problem solving offer students the opportunity to investigate a variety of special topics that supplement the text material. The projects typically require an approach that involves both theory and applications, including the use of technology. Many of the projects are suitable for group-learning activities.

Early Coverage of Differential Equations

Differential equations are introduced informally in Chapter 7 in connection with applications to exponential growth and decay. First-order linear equations and separable equations are covered in optional sections in Chapter 8, taking advantage of their natural association with the techniques of integration.

CONTENT AND ORGANIZATION CHANGES IN THE NINTH EDITION

In our effort to produce an even more effective text, we consulted users of the Eighth Edition and other calculus instructors. Our primary goals in preparing the Ninth Edition were the following:

1. ***Improve the exposition.*** As noted above, every topic was examined for possible improvement in the clarity and accuracy of presentation. Essentially every section in the text underwent some revision and rewriting; a number of sections and subsections were completely rewritten.
2. ***Improve the illustrative examples.*** Many of the existing examples were modified to enhance students' understanding of the material. New examples were added to sections that were rewritten or substantially revised.
3. ***Revise the exercise sets.*** Every exercise set was examined for balance between drill problems, mid-level problems, and more challenging applications and conceptual problems. In many instances, the number of routine problems was reduced and new mid-level to challenging problems were added. Technology-based problems were added to more than half of the sections.

Specific changes in content and organization made to achieve these goals and meet the needs of today's students and instructors include:

Comprehensive Review Exercise Sets

Six comprehensive review exercise sets, called Skill Mastery Reviews, have been designed to test and to reinforce students' understanding of basic concepts and methods. These new exercise sets are placed at strategic points in the text and average between

70 and 80 problems per set. The Skill Mastery Reviews follow Chapters 4, 6, 8, 11, 15, and 17.

Precalculus Review (Chapter 1)

Some of the material on solving inequalities has been moved to Chapter 2 where it is covered with the intermediate-value theorem. The coverage of functions in Sections 1.5 through 1.7 has been expanded slightly.

Limits (Chapter 2)

The approach to limits is unchanged, but many of the explanations have been revised. The illustrative examples throughout the chapter have been modified, and new examples have been added.

Differentiation and Applications (Chapters 3 and 4)

The initial approach to differentiation has been modified slightly. The derivative is defined first at a fixed point c and then extended to the definition of the "derivative function." The designation of endpoints of an interval as critical points of a function has been dropped. Otherwise, the content and organization of the chapters have not been changed. Text examples and the exercise sets have been modified and improved. The Skill Mastery Review at the end of Chapter 4 covers the material in Chapters 2, 3, and 4.

Integration and Applications (Chapters 5 and 6)

A brief introductory section that motivates the definition of the definite integral has been added to the beginning of Chapter 5. The motivating problems are area and distance. Explanations, examples and exercises have been modified, but the content and organization remain as in the Eighth Edition. The Skill Mastery Review at the end of Chapter 6 surveys the basic concepts and methods of these two chapters.

The Transcendental Functions, Techniques of Integration, Polar Coordinates and Parametric Equations (Chapters 7, 8, and 9)

Coverage of the inverse trigonometric functions (Chapter 7) has been reduced slightly. The treatments of trigonometric substitutions, partial fractions, and differential equations (Chapter 8) have been revised. Minor changes in the exposition and text examples occur throughout these chapters, and many of the exercise sets have been modified. The Skill Mastery Review at the end of Chapter 8 covers the material in Chapters 7 and 8.

Sequence and Series (Chapters 10 and 11)

The length of these chapters has been reduced through re-writing and the elimination of peripheral material. Some notations and terminology have been modified to be consistent with common usage. The Skill Mastery Review at the end of Chapter 11 covers the material in Chapters 9 through 11.

Vectors and Vector Calculus (Chapters 12 and 13)

The treatment of vectors in the plane that previously paralleled the discussion of three-dimensional vectors has been rewritten. The organization of the material in Chapter 13 has been changed: curvilinear motion and curvature are treated together in a new section (13.5), and vector calculus in mechanics and the optional section on planetary motion are now the final two sections of the chapter.

Functions of Several Variables, Gradients, Extreme Values (Chapters 14 and 15)

Except for improvements in the exposition, Chapter 14 is unchanged. There are some changes in the organization and content of Chapter 15. The former section on local and absolute extrema and the second partials test has been separated into two sections: one on local extrema and the second partials test, the other on absolute extrema. This organization parallels the single variable treatment of these topics in Chapter 4. The optional section on exact differential equations has been moved to Chapter 18. The Skill Mastery Review at the end of Chapter 15 includes exercises from Chapters 12 through Chapter 15.

Multiple Integrals; Line and Surface Integrals (Chapters 16 and 17)

The basic content and organization of these two chapters are largely unchanged, but there have been improvements in the exposition, illustrative examples and exercise sets. The Skill Mastery Review at the end of Chapter 17 covers integration concepts for functions of several variables and for vector-valued functions.

Differential Equations (Chapter 18)

Each of the sections in this chapter has been completely rewritten, and the text examples and exercise sets have been modified accordingly. The material on exact differential equations from Chapter 15 has been added.

SUPPLEMENTS

Student Aids

Answers to Odd-Numbered Exercises Answers to all the odd-numbered exercises are included at the back of the text.

Student Solutions Manual This manual contains detailed solutions to all the odd-numbered exercises in the text. ISBN: 0-471-27521-2

eGrade An online assessment system that contains a large bank of skill-building problems and solutions. Instructors can now automate the process of assigning, delivering, grading, and routing all kinds of homework, quizzes, and tests while providing students with immediate scoring and feedback on their work. Wiley *eGrade* "does the math" ... and much more. For more information, visit www.wiley.com/college/egrade

Calculus Machina A web-based, intelligent software package that solves and documents calculus problems in real time. For students, Calculus Machina is a step-by-step electronic tutor. As the student works through a particular problem online, Calculus Machina provides customized feedback from the text, allowing students to identify and learn from their mistakes more efficiently. For more information, visit www.wiley.com/college/machina

Instructor Aids

Instructor's Solutions Manual This manual contains complete solutions to all the problems in the text. ISBN: 0-471-27522-0

Test Bank A wide range of problems and their solutions are keyed to the exercise sets in the text. ISBN: 0-471-27523-9

Computerized Test Bank The Computerized Test Bank allows instructors to create, customize, and print a test containing any combination of questions from the test bank. Instructors can edit existing questions from the test bank or create new ones as needed.

PowerPoint slides Key figures and examples from each chapter are supplied in PowerPoint format for use in classroom presentation and discussion. These may easily be printed onto transparencies for use with an overhead projector.

eGrade An online assessment system that contains a large bank of skill-building problems and solutions. Instructors can now automate the process of assigning, delivering, grading, and routing all kinds of homework, quizzes, and tests while providing students with immediate scoring and feedback on their work. Wiley *eGrade* "does the math" ... and much more. For more information, visit www.wiley.com/college/egrade

Calculus Machina Instructors can use Calculus Machina to preview problems and explore functions graphically and analytically. For more information, visit www.wiley.com/college/machina

ACKNOWLEDGEMENTS

The revision of a text of this magnitude and stature requires a lot of encouragement and help. I was fortunate to have an ample supply of both from many sources.

Each edition of this text was developed from those that preceded it. The present book owes much to the people who contributed to the first eight editions, most recently: Mihaly Bakonyi, Georgia State University; Edward B. Curtis, University of Washington; Kathy Davis, University of Texas-Austin; Dennis DeTurck, University of Pennsylvania; John R. Durbin, University of Texas-Austin; Charles H. Giffen, University of Virginia-Charlottesville; Michael Kinyon, Indiana University-South Bend; Nicholas Macri, Temple University; James R. McKinney, California State Polytechnic University-Pomona; Jeff Morgan, Texas A & M University; Clifford S. Queen, Lehigh University; and Yang Wang, Georgia Institute of Technology. I am deeply indebted to all of them.

The reviewers of the Ninth Edition supplied detailed criticisms and valuable suggestions. I offer my sincere appreciation to the following individuals:

Omar Adawi	Parkland College
Boris A. Datskovsky	Temple University
Ronald Gentle	Eastern Washington University
Robert W. Ghrist	Georgia Institute of Technology
Susan J. Lamon	Marquette University
Peter A. Lappan	Michigan State University
Dean Larson	Gonzaga University
James Martino	Johns Hopkins University
Peter J. Mucha	Georgia Institute of Technology
Elvira Munoz-Garcia	University of California, Los Angeles
Ralph W. Oberste-Vorth	University of South Florida
Charles Odion	Houston Community College
Charles Peters	University of Houston
J. Terry Wilson	San Jacinto College Central

I am especially grateful to Paul Lorczak, Neil Wigley, and J. Terry Wilson, who carefully read the revised material. They provided many corrections and helpful comments.

I would also like to thank Terry Wilson for his advice and guidance in the creation of the new technology exercises in the text.

I am deeply indebted to the editorial staff at John Wiley & Sons. Everyone involved in this project has been encouraging, helpful, and thoroughly professional at every stage. In particular, Laurie Rosatone, Associate Publisher, Michael Boezi, Editor, and Jennifer Battista, Assistant Editor, provided organization and support when I needed it, and prodding when prodding was required. Special thanks go to Norine Pigliucci, Production Editor, who was patient and understanding as she guided the project through the production stages; Sigmund Malinowski, Illustration Editor, who skillfully directed the art program; and Madelyn Lesure, Design Director, whose creativity produced the attractive interior design as well as the cover.

Finally, I want to acknowledge the contributions of my wife, Charlotte; without her continued support, I could not have completed this work.

Garret J. Etgen

CONTENTS

CHAPTER 1 PRECALCULUS REVIEW 1

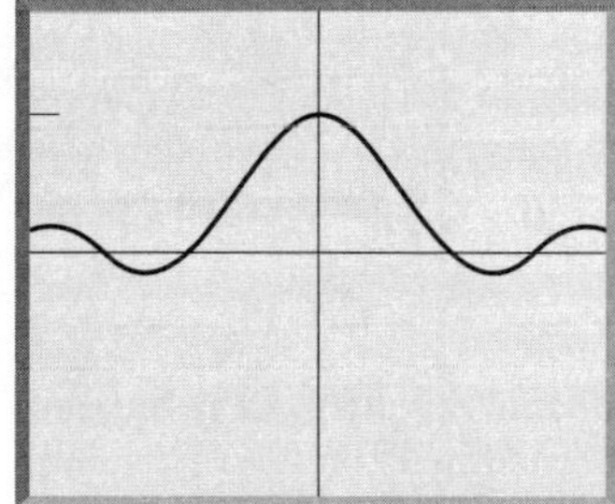

CHAPTER 2 LIMITS AND CONTINUITY 59

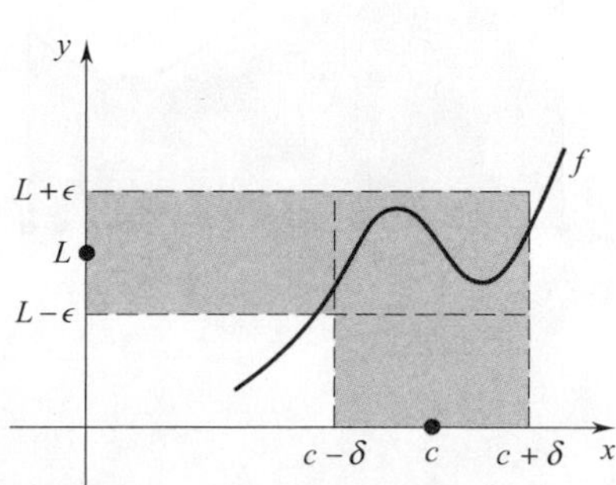

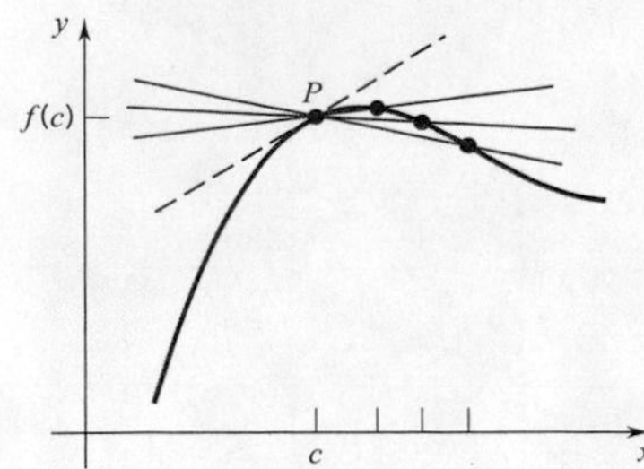

CHAPTER 3 DIFFERENTIATION 119

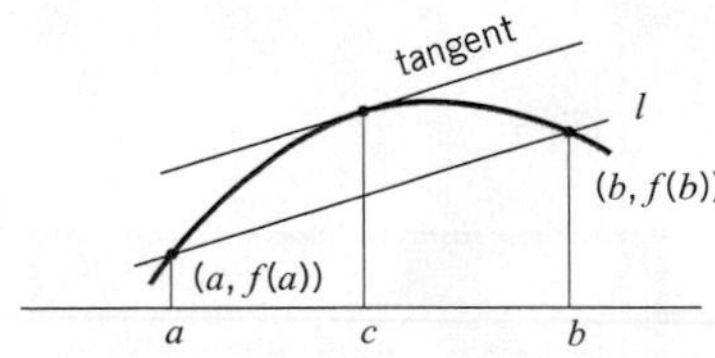

CHAPTER 4 THE MEAN-VALUE THEOREM AND APPLICATIONS 197

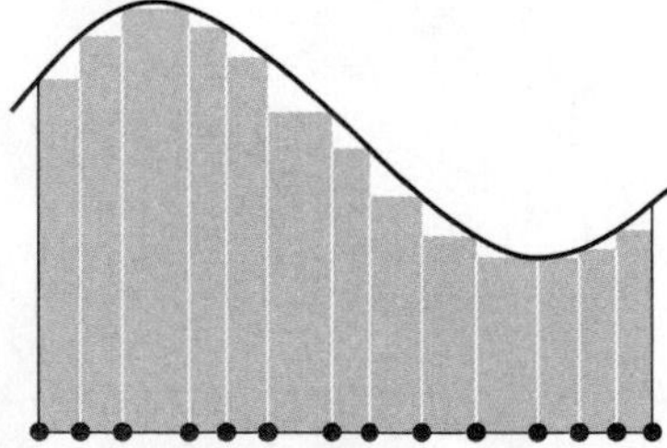

CHAPTER 5 INTEGRATION 262

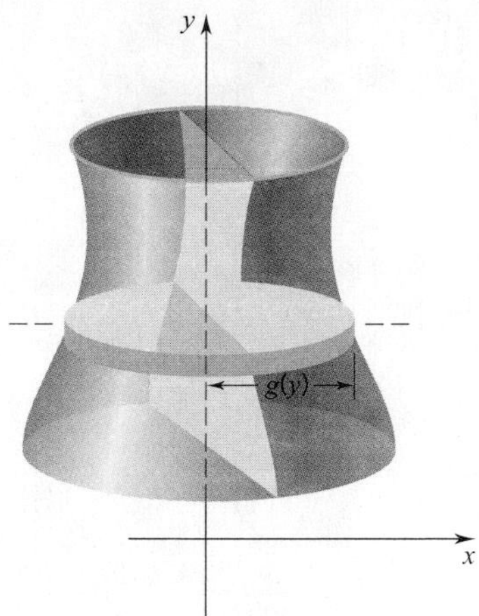

y
g(y)
x

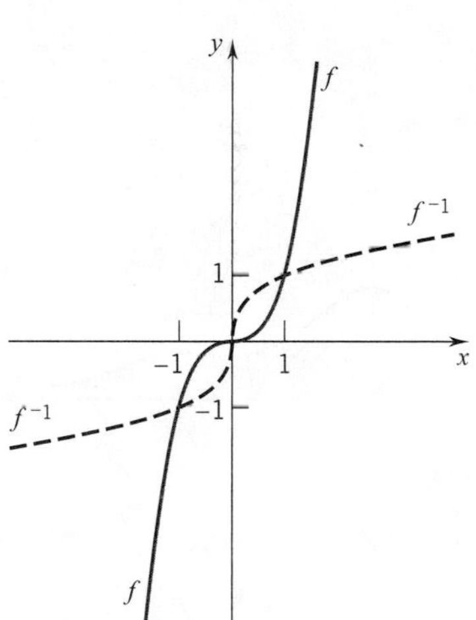

y
f
f^{-1}
1
−1
1
x
f^{-1}
−1
f

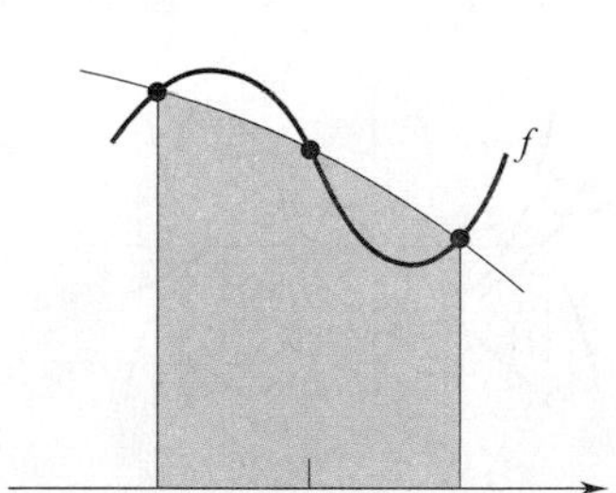

f
x

CHAPTER 9 THE CONIC SECTIONS; POLAR COORDINATES; PARAMETRIC EQUATIONS 514

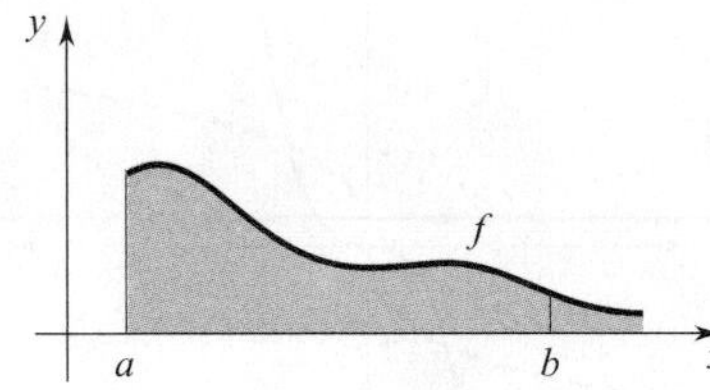

CHAPTER 10 SEQUENCES; INDETERMINATE FORMS; IMPROPER INTEGRALS 585

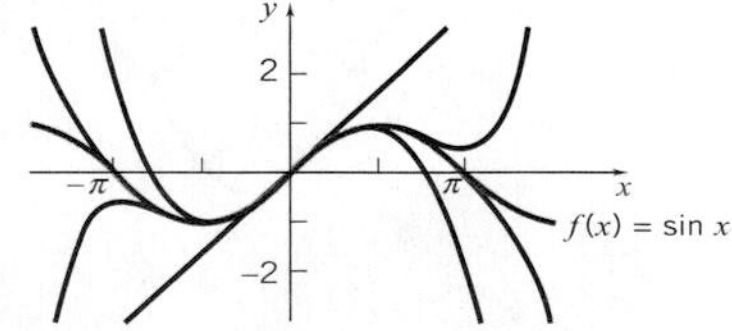

CHAPTER 11 INFINITE SERIES 632

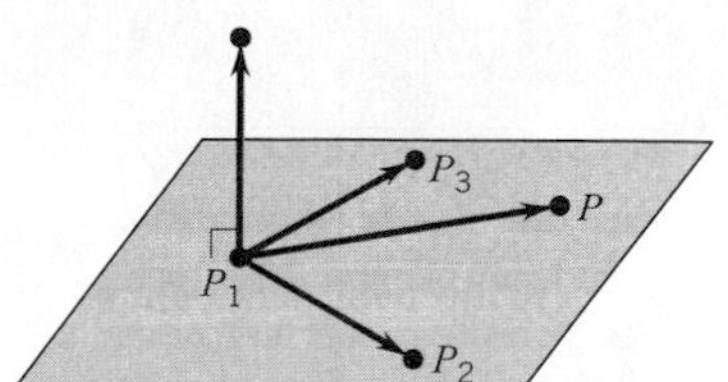

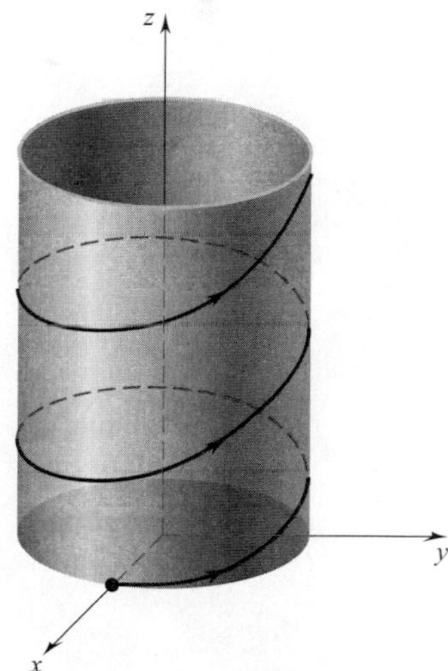

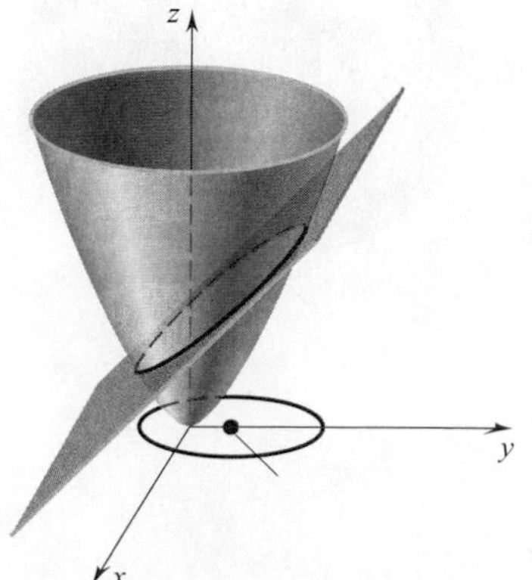

* Denote optional section.

CHAPTER 15 GRADIENTS; EXTREME VALUES; DIFFERENTIALS 861

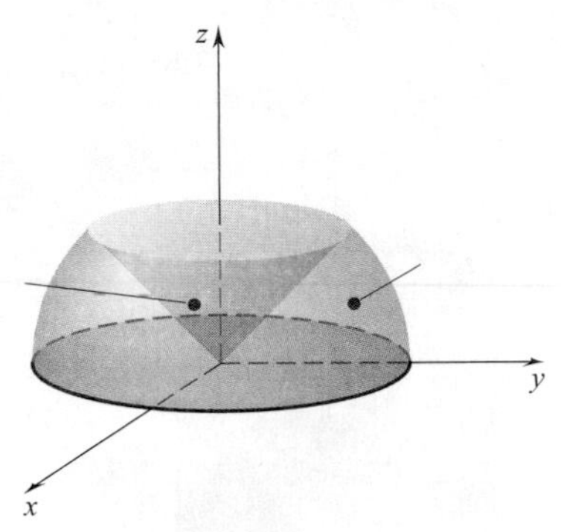

CHAPTER 16 DOUBLE AND TRIPLE INTEGRALS 942

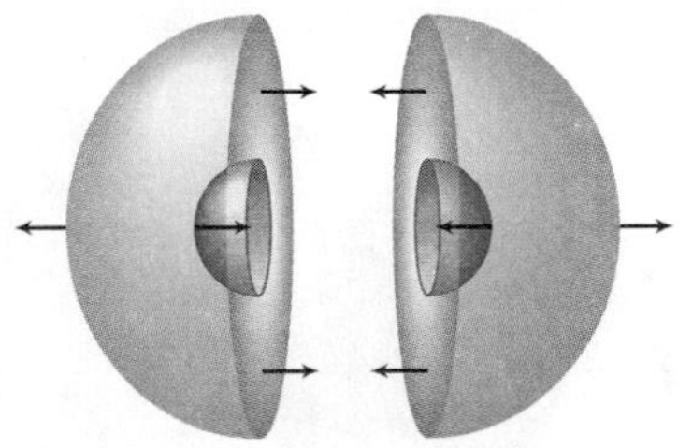

CHAPTER 17 LINE INTEGRALS AND SURFACE INTEGRALS 1019

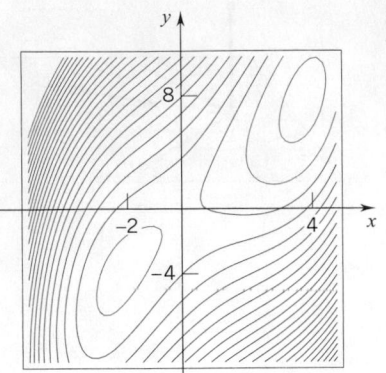

CHAPTER 1

PRECALCULUS REVIEW

This chapter is a review of topics that typically are covered in courses that serve as preparations for calculus. This is not intended to be a comprehensive review of precalculus mathematics. The focus is on those topics that are needed specifically in our study of calculus.

■ 1.1 WHAT IS CALCULUS?

To a Roman in the days of the empire, a "calculus" was a pebble used in counting and in gambling. Centuries later "calculare" came to mean "to compute," "to reckon," "to figure out." To the mathematician, physical scientist, and social scientist of today, calculus is elementary mathematics (algebra, geometry, trigonometry) enhanced by *the limit process.*

Calculus takes ideas from elementary mathematics and extends them to a more general situation. On pages 1 and 2 are some examples. On the left-hand side you will find an idea from elementary mathematics; on the right, this same idea as extended by calculus.

It seems fitting to say something about the history of calculus. The origins can be traced back to ancient Greece. The ancient Greeks raised many questions (often paradoxical) about tangents, motion, area, the infinitely small, the infinitely large—questions that today are clarified and answered by calculus. Here and there the Greeks themselves provided answers (some very elegant), but mostly they provided only questions.

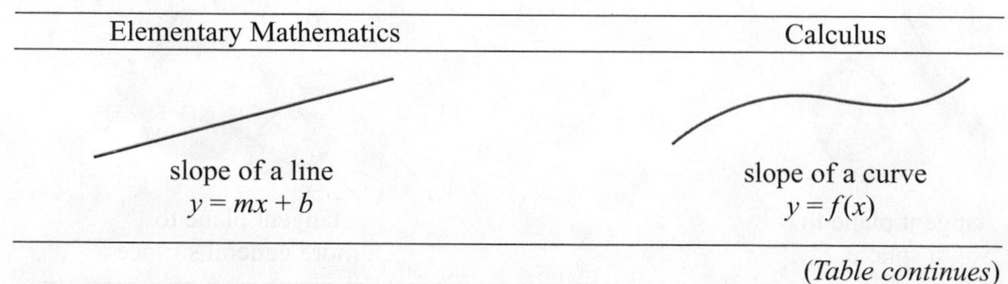

Elementary Mathematics	Calculus
slope of a line $y = mx + b$	slope of a curve $y = f(x)$

(*Table continues*)

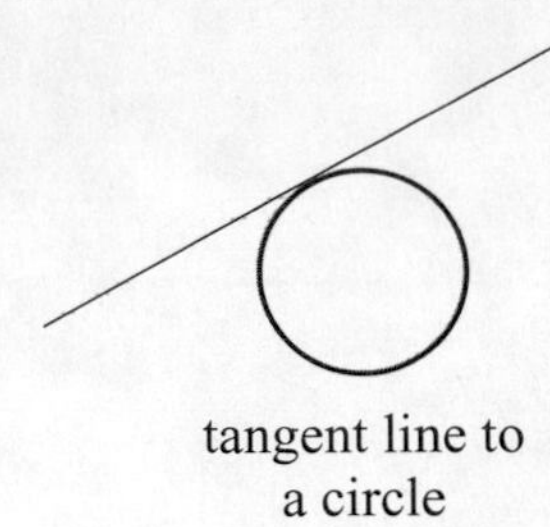
tangent line to a circle

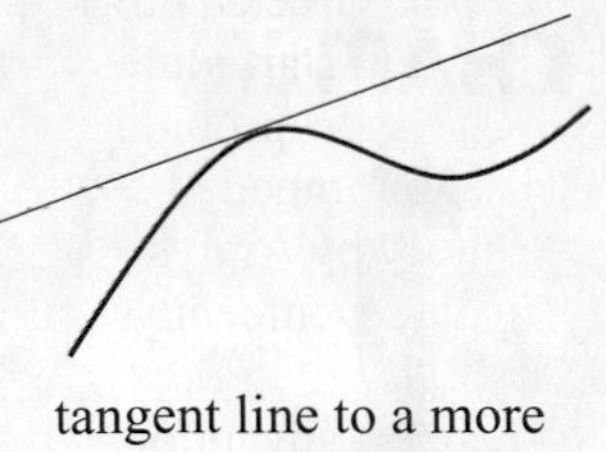
tangent line to a more general curve

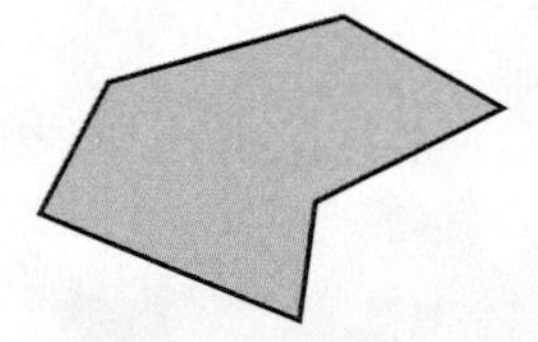
area of a region bounded by line segments

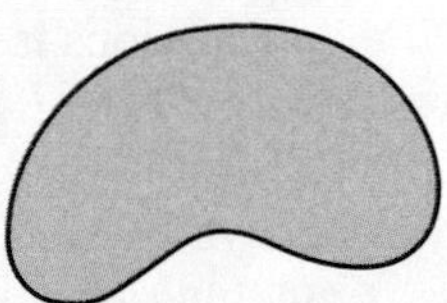
area of a region bounded by curves

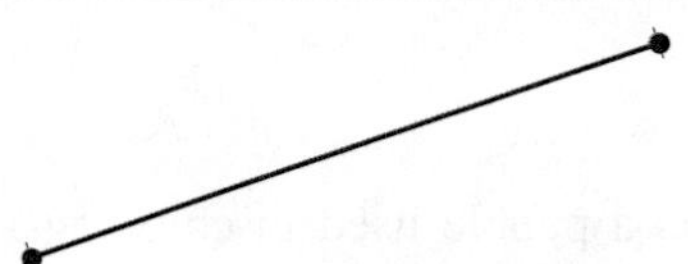
length of a line segment

length of a curve

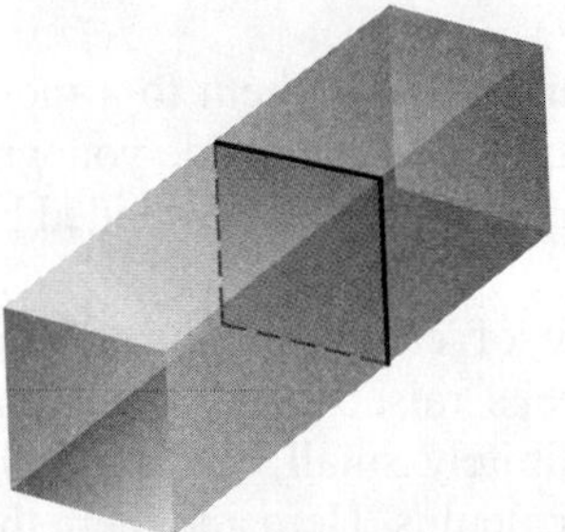
volume of a rectangular solid

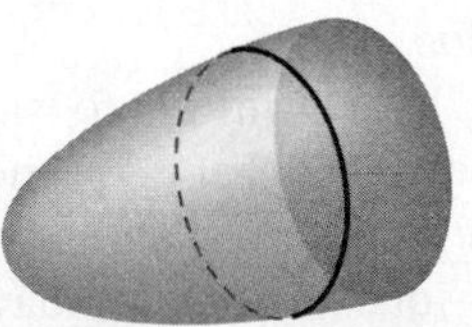
volume of a solid with a curved boundary

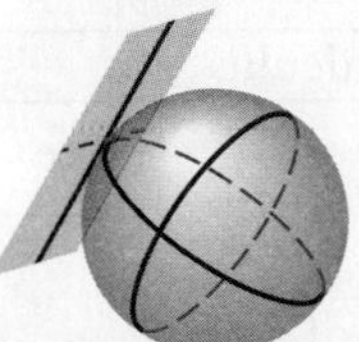
tangent plane to a sphere

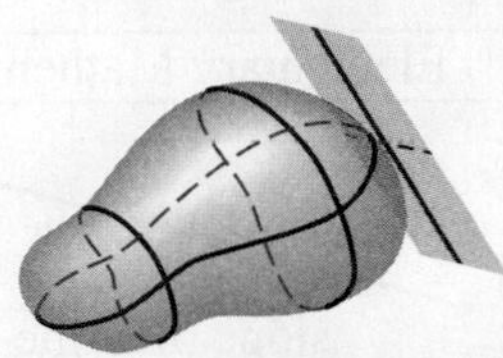
tangent plane to a more general surface

After the Greeks, progress was slow. Communication was limited, and each scholar was obliged to start almost from scratch. Over the centuries, some ingenious solutions to calculus-type problems were devised, but no general techniques were put forth. Progress was impeded by the lack of a convenient notation. Algebra, founded in the ninth century by Arab scholars, was not fully systematized until the sixteenth century. Then, in the seventeenth century, Descartes established analytic geometry, and the stage was set.

The actual invention of calculus is credited to Sir Isaac Newton (1642–1727), an Englishman, and to Gottfried Wilhelm Leibniz (1646–1716), a German. Newton's invention is one of the few good turns that the great plague did mankind. The plague forced the closing of Cambridge University in 1665, and young Isaac Newton of Trinity College returned to his home in Lincolnshire for eighteen months of meditation, out of which grew his *method of fluxions*, his *theory of gravitation*, and *his theory of light*. The method of fluxions is what concerns us here. A treatise with this title was written by Newton in 1672, but it remained unpublished until 1736, nine years after his death. The new method (calculus to us) was first announced in 1687, but in vague general terms without symbolism, formulas, or applications. Newton himself seemed reluctant to publish anything tangible about his new method, and it is not surprising that the development on the Continent, in spite of a late start, soon overtook Newton and went beyond him.

Leibniz started his work in 1673, eight years after Newton. In 1675 he initiated the basic modern notation: dx and $\int$. His first publications appeared in 1684 and 1686. These made little stir in Germany, but the two brothers Bernoulli of Basel (Switzerland) took up the ideas and added profusely to them. From 1690 onward, calculus grew rapidly and reached roughly its present state in about a hundred years. Certain theoretical subtleties were not fully resolved until the twentieth century.

■ 1.2 REVIEW OF ELEMENTARY MATHEMATICS

In this section we outline the basic terminology, notations, and formulas that are used throughout the text.

Sets

A set is a well-defined collection of distinct objects. The objects in a set are called the *elements* or *members* of the set. Capital letters $A, B, C, \ldots$ are usually used to denote sets and lowercase letters $a, b, c, \ldots,$ to denote the elements of a set.

For a collection S of objects to be a set it must be *well-defined* in the sense that given any object x whatsoever, it must be possible to determine whether or not x is an element of S. For example, the collection of letters of the alphabet, the collection of states in the United States of America, and the collection of solutions of the equation $x^2 = 9$ are all examples of sets. On the other hand, suppose we wanted to discuss the collection of beautiful cities in Europe. Different people might have different collections. Thus, the collection of beautiful cities in Europe cannot be regarded as a set. Collections based on subjective judgments, such as "all good football players" or "all intelligent adults," are not sets.

Notation

The object x is in the set A	$x \in A.$
The object x is not in the set A	$x \notin A$
The set of all x for which property P holds	$\{x : P\}$
(For example, $A = \{x : x \text{ is a vowel}\} = \{a, e, i, o, u\}$)	
A is a subset of B (A is contained in B)	$A \subseteq B$

The union of A and B ($A \cup B = \{x : x \in A \text{ or } x \in B\}$)	$A \cup B$
The intersection of A and B ($A \cap B = \{x : x \in A \text{ and } x \in B\}$)	$A \cap B$
The empty set	$\emptyset$

These are the basic notions from set theory that you will need for this text.

Real Numbers

Our study of calculus is based on the real number system. The real number system consists of the set of real numbers together with the familiar arithmetic operations—addition, subtraction, multiplication, division—and certain other properties which are reviewed briefly here.

Classification

Natural numbers (or positive integers)	$1, 2, 3, \ldots$
Integers	$0, 1, -1, 2, -2, 3, -3, \ldots$
Rational numbers	$\{x : x = p/q$, where p, q are intergers and $q \neq 0\}$. †
Examples	$\frac{2}{3}, \frac{-19}{7}, \frac{4}{1} (= 4)$.
Irrational numbers	real numbers that are not rational numbers.
Examples	$\sqrt{2}, \sqrt[3]{7}, \pi$, the solutions of the equation $x^2 - 5 = 0$.

Decimal Representation

Each real number can be represented by a decimal. If $r = p/q$ is a rational number, then its decimal representation is found by dividing the denominator q into the numerator p. The resulting decimal expansion will either *terminate* or *repeat*. For example,

$$\frac{3}{5} = 0.6, \quad \frac{27}{20} = 1.35, \quad \text{and} \quad \frac{43}{8} = 5.375$$

are terminating decimals and

$$\frac{2}{3} = 0.6666\cdots = 0.\overline{6}, \quad \frac{15}{11} = 1.363636\cdots = 1.\overline{36}, \quad \text{and} \quad \frac{116}{37} = 3.135135\cdots = 3.\overline{135}$$

are repeating decimals (the bar over the sequence of digits indicates that the sequence repeats indefinitely). The converse is also true; every terminating or repeating decimal represents a rational number.

The decimal expansion of an irrational number neither terminates nor repeats. For example, the numbers

$$\sqrt{2} = 1.414213562\cdots, \quad \pi = 3.141592653\cdots \text{ and}$$
$$r = 0.10110111011110\cdots$$

are irrational numbers.

If we stop the decimal expansion of a given number at a certain decimal place, then the result is a rational number which approximates the given number. For instance,

† Recall that division by 0 is undefined.

$1.414 = 1414/1000$ and $3.14 = 314/100$ are common rational number approximations for $\sqrt{2}$ and π, respectively. More accurate approximations can be obtained by taking more decimal places in the expansions.

Geometric Representation

A fundamental concept in mathematics which connects the abstract notion of *real number* with the geometric notion of *point* is the representation of the real numbers as points on a straight line. This is done by choosing an arbitrary point O on a horizontal line to represent the number 0, and another point U (usually taken to the right of O) to represent the number 1. See Figure 1.2.1. The point O is called the *origin*, and the distance between O and U determines a scale (a unit length). With O and U specified, each real number can be represented as a point on the line and, conversely, each point on the line represents a real number. A line representing the real numbers is called a *number line* or *real line;* the number associated with a point P on the line is called the *coordinate* of P. The *positive* numbers are identified with the points on the right side of O and the *negative* numbers with the points to the left of O. The point representing the number a ($a \neq 0$) is a units from O if a is positive and $-a$ units from O if a is negative. In this context, we will frequently refer to real numbers as "points," and by this we will mean "points on a number line." Figure 1.2.2 shows some numbers plotted as points on a number line.

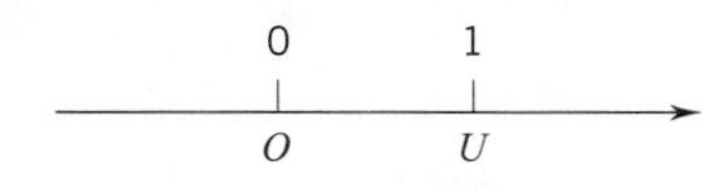

Figure 1.2.1

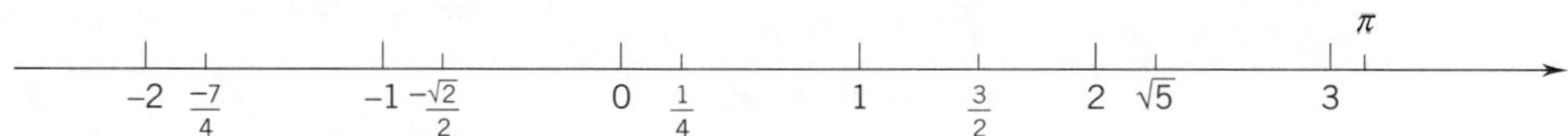

Figure 1.2.2

Order Properties

If a and b are real numbers, then *a is less than b*, denoted $a < b$, if $b - a$ is a positive number. This is equivalent to saying that *b is greater than a*, which is denoted $b > a$. Geometrically, $a < b$ if the point a lies to the left of the point b on a number line. The notation $a \leq b$ means that either $a < b$ or $a = b$ (equivalently, $b \geq a$).

The real numbers are *ordered* in the sense that if a and b are real numbers, then exactly one of the following holds:

$$a < b, \qquad a = b, \qquad a > b. \qquad \text{(trichotomy)}$$

The symbols $<, >, \leq, \geq$ are called *inequalities*. Inequalities satisfy the following properties:

(i) If $a < b$ and $b < c$, then $a < c$. (transitive property)

(ii) If $a < b$, then $a + c < b + c$ for all real numbers c.

(iii) If $a < b$ and $c < d$, then $a + c < b + d$.

(iv) If $a < b$ and $c > 0$, then $ac < bc$.

(v) If $a < b$ and $c < 0$, then $ac > bc$.

The corresponding properties hold for $>, \leq, \geq$. A crucial point is Property (v): if an inequality is multiplied by a negative quantity, then the "direction" of the inequality is reversed. Techniques for solving inequalities use these properties and are reviewed in Section 1.3.

Density

Between any two real numbers there are infinitely many rational numbers and infinitely many irrational numbers. In particular, *there is no smallest positive real number*.

Absolute Value

Two important properties of a real number a are its *sign* and its *size*, or *magnitude*. Geometrically, the sign of a tells us whether the point a is on the right or left of 0 on a number line. The magnitude of a is the distance between the point a and 0; 0 itself does not have a sign and its magnitude is 0. The magnitude of a is more commonly called *absolute value of a*, denoted $|a|$. The absolute value of a is also given by

$$|a| = \begin{cases} a, & \text{if } a \geq 0 \\ -a, & \text{if } a < 0. \end{cases}$$

Other characterizations $|a| = \max\{a, -a\}$; $|a| = \sqrt{a^2}$.

Geometric interpretations $|a| =$ distance between a and 0;

$|a - c| =$ distance between a and c.

Other properties

(i) $|a| = 0$ iff $a = 0$.†

(ii) $|-a| = |a|$.

(iii) $|ab| = |a||b|$.

(iv) $|a + b| \leq |a| + |b|$ (the triangle inequality).††

(v) $\big||a| - |b|\big| \leq |a - b|$ (a variant of the triangle inequality).

(vi) $|a|^2 = |a^2| = a^2$

Techniques for solving inequalities involving absolute values are also reviewed in Section 1.3.

Intervals

Suppose that $a < b$. The *open interval* (a, b) is the set of all numbers between a and b:

$$(a, b) = \{x : a < x < b\}.$$

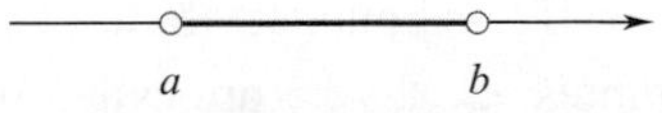

The *closed interval* $[a, b]$ is the open interval (a, b) together with the endpoints a and b:

$$[a, b] = \{x : a \leq x \leq b\}.$$

† By "iff" we mean "if and only if." This expression is used so often in mathematics that it's convenient to have an abbreviation for it.

†† The absolute value of the sum of two numbers cannot exceed the sum of their absolute values. This is analogous to the fact that in a triangle the length of one side cannot exceed the sum of the lengths of the other two sides.

There are seven other types of intervals:

$$(a, b] = \{x : a < x \le b\},$$

$$[a, b) = \{x : a \le x < b\},$$

$$(a, \infty) = \{x : a < x\},$$

$$[a, \infty) = \{x : a \le x\},$$

$$(-\infty, b) = \{x : x < b\},$$

$$(-\infty, b] = \{x : x \le b\},$$

$$(-\infty, \infty) = \text{the set of real numbers.}$$

Interval notation is easy to remember: we use a square bracket to include an endpoint and a parenthesis when the endpoint is not included. On a number line, the inclusion or exclusion of an endpoint is signified with a solid "dot" or an open "dot," respectively. The symbols ∞ and $-\infty$, read "infinity" and "negative infinity" (or "minus infinity"), are not real numbers. The ∞ symbol in the intervals given above indicates that the interval extends indefinitely in the positive direction; the $-\infty$ indicates that the interval extends indefinitely in the negative direction. Since ∞ and $-\infty$ are not real numbers, we do not have intervals of the form $[a, \infty]$ or $[-\infty, b)$, and so forth.

Boundedness

A set S of real numbers is said to be:

(i) *Bounded above* if there exists a real number M such that

$$x \le M \qquad \text{for all } x \in S;$$

M is called an *upper bound* for S.

(ii) *Bounded below* if there exists a real number m such that

$$x \ge m \qquad \text{for all } x \in S;$$

m is called a *lower bound* for S.

(iii) *Bounded* if it is bounded above and below.

For example, the intervals $(-\infty, 2]$ and $(-\infty, 2)$ are bounded above but not below; 2 is an upper bound for each of these sets. Note that any number greater than 2 is also an upper bound. The set of positive integers $N = \{1, 2, 3, 4, \dots\}$ is bounded below but not above; 0 is a lower bound for N, as are $\frac{1}{2}$ and 1. The set $\{-1, 0, 1, 2, 3\}$ and the interval $(-1, 3)$ are bounded (these sets are bounded above and below); -1 or any number less than -1 is a lower bound for each of these sets, and 3 or any number greater than 3 is an upper bound.

Factorials

Let n be a positive integer. Then n *factorial*, denoted $n!$, is the product of the integers from 1 to n. Thus,

$$1! = 1, \quad 2! = 1 \cdot 2, \quad 3! = 1 \cdot 2 \cdot 3, \quad \dots, \quad n! = 1 \cdot 2 \cdot 3 \cdot \cdots \cdot (n-1) \cdot n.$$

Note that

$$1! = 1, \quad 2! = 2, \quad 3! = 6.$$

You can verify that $4! = 24$, $5! = 120$, and so on. Finally, to ensure that certain formulas will hold for all *nonnegative* integers, we define $0! = 1$

Algebra

Basic Factoring Formulas

$$\begin{aligned}
(a+b)^2 &= a^2 + 2ab + b^2 \\
(a-b)^2 &= a^2 - 2ab + b^2 \\
(a+b)^3 &= a^3 + 3a^2b + 3ab^2 + b^3 \\
(a-b)^3 &= a^3 - 3a^2b + 3ab^2 - b^3 \\
a^2 - b^2 &= (a-b)(a+b) \\
a^3 - b^3 &= (a-b)(a^2 + ab + b^2) \\
a^3 + b^3 &= (a+b)(a^2 - ab + b^2)
\end{aligned}$$

Quadratic Equations

The roots of a quadratic equation

$$ax^2 + bx + c = 0 \qquad \text{with } a \neq 0$$

are given by the quadratic formula

$$r = \frac{-b \pm \sqrt{b^2 - 4ac}}{2a}.$$

If $b^2 - 4ac > 0$, the equation has two real roots:

$$r_1 = \frac{-b + \sqrt{b^2 - 4ac}}{2a} \quad \text{and} \quad r_2 = \frac{-b - \sqrt{b^2 - 4ac}}{2a}.$$

In this case, the quadratic expression $ax^2 + bx + c$ can be written in the factored form:

$$ax^2 + bx + c = a(x - r_1)(x - r_2).$$

If $b^2 - 4ac = 0$, then $r_1 = r_2$, and the equation has only one real root $r = -b/2a$. Here, the quadratic can be written in the factored form:

$$ax^2 + bx + c = a(x - r)^2.$$

If $b^2 - 4ac < 0$, the equation has no real roots.

To find the real roots of a quadratic equation, we first try to factor it. If this proves difficult to do, we resort to the general quadratic formula.

Geometry

Elementary Figures

Triangle

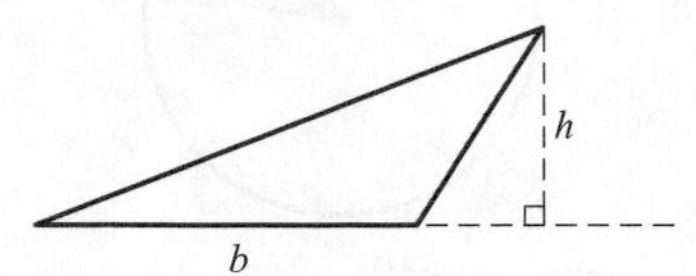

area = $\frac{1}{2}bh$

Equilateral Triangle

area = $\frac{1}{4}\sqrt{3}\,s^2$

Rectangle

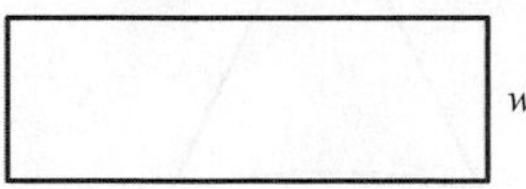

area = lw
perimeter = $2l + 2w$
diagonal = $\sqrt{l^2 + w^2}$

Rectangular Solid

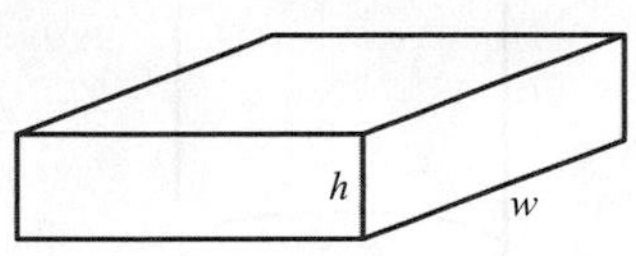

volume = lwh
surface area = $2lw + 2lh + 2wh$

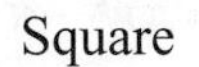

Square

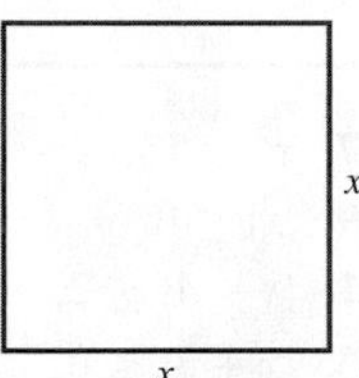

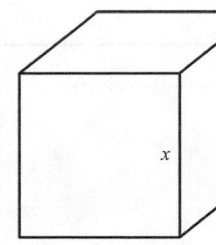

area = x^2
perimeter = $4x$
diagonal = $x\sqrt{2}$

Cube

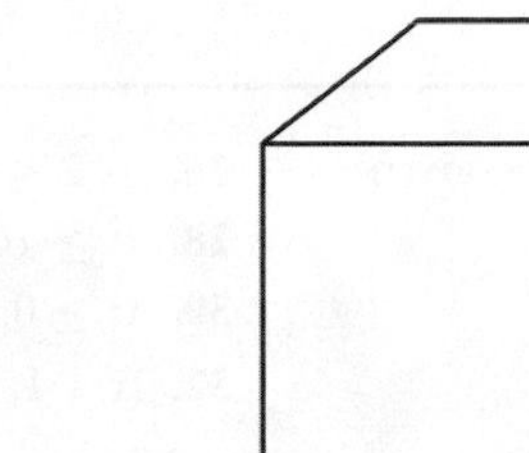

volume = x^3
surface area = $6x^2$

Circle

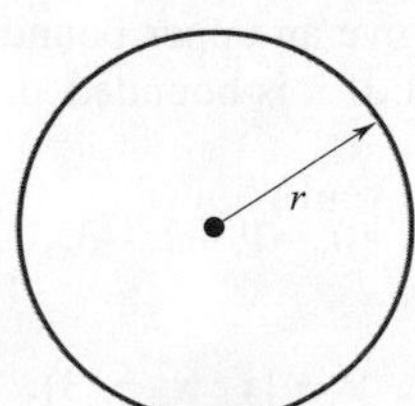

area = πr^2
circumference = $2\pi r$

Sphere

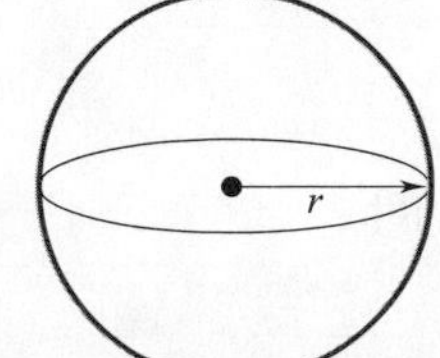

volume = $\frac{4}{3}\pi r^3$
surface area = $4\pi r^2$

Sector of a Circle: radius r, central angle θ measured in radians (see Section 1.6).

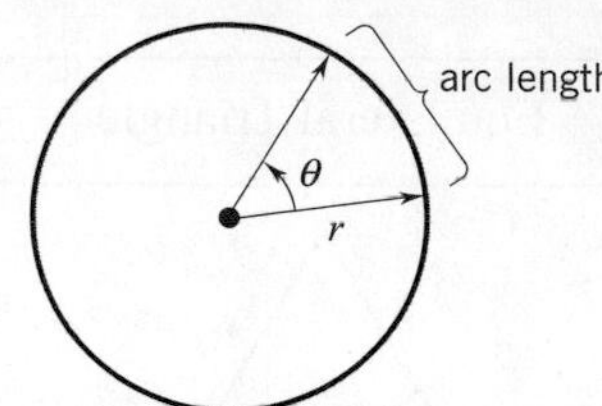

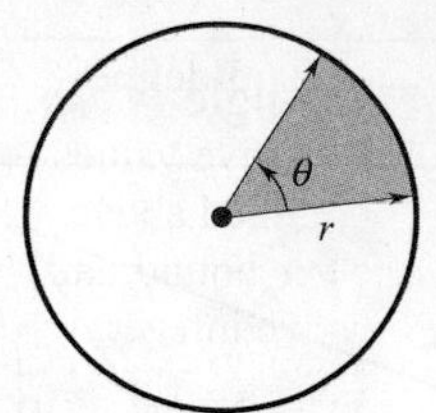

arc length $= r\theta$ area $= \frac{1}{2}r^2\theta$

Right Circular Cylinder

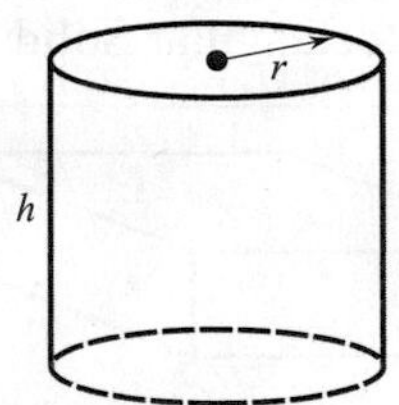

volume $= \pi r^2 h$
lateral area $= 2\pi r h$
total surface area $= 2\pi r^2 + 2\pi r h$

Right Circular Cone

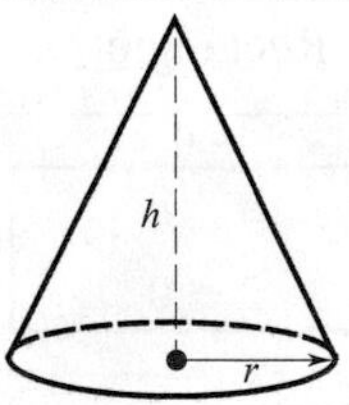

volume $= \frac{1}{3}\pi r^2 h$
slant height $= \sqrt{r^2 + h^2}$
lateral area $= \pi r\sqrt{r^2 + h^2}$
total surface area $= \pi r^2 + \pi r\sqrt{r^2 + h^2}$

EXERCISES 1.2

Use the terms *integer, rational, irrational* to classify the given number.

1. $\frac{17}{7}$. **2.** -6.
3. $2.131313\cdots = 2.\overline{13}$. **4.** $\sqrt{2} - 3$.
5. 0. **6.** $\pi - 2$.
7. $\sqrt[3]{8}$. **8.** 0.125.
9. $-\sqrt{9}$. **10.** $0.21211211121111\cdots$.

Replace the symbol $*$ by $<, >$, or $=$ to make the statement true.

11. $\frac{3}{4} * 0.75$. **12.** $0.33 * \frac{1}{3}$.
13. $\sqrt{2} * 1.414$. **14.** $4 * \sqrt{16}$.
15. $-\frac{2}{7} * -0.285714$. **16.** $\pi * \frac{22}{7}$.

Evaluate.

17. $|6|$. **18.** $|-4|$.
19. $|3 - 7|$. **20.** $|-5| - |8|$.
21. $|-5| + |-8|$. **22.** $|2 - \pi|$.
23. $|5 - \sqrt{5}|$.

Indicate on a number line all the numbers x that satisfy the given condition.

24. $x \geq 3$ **25.** $x \leq -\frac{3}{2}$.
26. $-2 \leq x \leq 3$. **27.** $x^2 < 16$.
28. $x^2 \geq 16$. **29.** $|x| \leq 0$.
30. $x^2 \geq 0$. **31.** $|x - 4| \leq 2$.
32. $|x + 1| > 3$. **33.** $|x + 3| \leq 0$.

Sketch the given set on a number line.

34. $[3, \infty)$. **35.** $(-\infty, 2)$.
36. $(-4, 3]$. **37.** $[-2, 3] \cup [1, 5]$.
38. $\left[-3, \frac{3}{2}\right) \cap \left(\frac{3}{2}, \frac{5}{2}\right]$. **39.** $(-\infty, -1) \cup (-2, \infty)$.
40. $(-\infty, 2) \cap [3, \infty)$.

Determine whether the set is bounded above, bounded below, or bounded. If a set is bounded above, give an upper bound; if it is bounded below, give a lower bound; if it is bounded, give an upper bound and a lower bound.

41. $\{0, 1, 2, 3, 4\}$. **42.** $\{0, -1, -2, -3, \ldots\}$.
43. S is the set of even integers.
44. $S = \{x : x \leq 4\}$. **45.** $S = \{x : x^2 > 3\}$.
46. $S = \left\{\frac{n-1}{n} : n = 1, 2, 3, \ldots\right\}$.
47. S is the set of rational numbers less than $\sqrt{2}$.

48. Order the following numbers from the smallest to the largest and draw them on a number line: $\sqrt[3]{\pi}, 2^{\sqrt{\pi}}, \sqrt{2}, 3^{\pi}, \pi^3$.

49. Let $x_0 = 2$ and define $x_n = \frac{17+2x_{n-1}^3}{3x_{n-1}^2}$ for $n = 1, 2, 3, 4, \ldots$. Find at least five values for x_n. Is the set $S = \{x_0, x_1, x_2, \ldots, x_n, \ldots\}$ bounded above, bounded below, or bounded? If so, give a lower bound and/or an upper bound for S. If n is a large positive integer, what is x_n (approximately)?

50. Rework Exercise 49 with $x_0 = 3$ and $x_n = \dfrac{231 + 4x_{n-1}^5}{5x_{n-1}^4}$.

Write the expression in factored form.

51. $x^2 - 10x + 25$.

52. $9x^2 - 4$.

53. $8x^6 + 64$.

54. $27x^3 - 8$.

55. $4x^2 + 12x + 9$.

56. $4x^4 + 4x^2 + 1$.

Find the real roots of the quadratic equation.

57. $x^2 - x - 2 = 0$

58. $x^2 - 9 = 0$

59. $x^2 - 6x + 9 = 0$

60. $2x^2 - 5x - 3 = 0$

61. $x^2 - 2x + 2 = 0$

62. $x^2 + 8x + 16 = 0$

63. $x^2 + 4x + 13 = 0$

64. $x^2 - 2x + 5 = 0$

Evaluate.

65. $5!$.

66. $\dfrac{5!}{8!}$.

67. $\dfrac{8!}{3!5!}$.

68. $\dfrac{9!}{3!6!}$.

69. $\dfrac{7!}{0!7!}$.

In Exercises 70–78, use the fact that the sum, difference, and product of any two integers is an integer.

70. Prove that the sum of two rational numbers is a rational number.

71. Prove that the sum of a rational number and an irrational number is irrational.

72. Prove that the product of two rational numbers is a rational number.

73. What can you say about the product of a rational number and an irrational number?

74. Show by examples that the sum or product of two irrational numbers can be either a rational number or an irrational number.

75. Prove that $\sqrt{2}$ is an irrational number. HINT: Assume that $\sqrt{2} = p/q$ and that p and q have no common factors (other than 1 and -1). Square both sides of this equation and argue that both p and q must be divisible by 2. This contradicts the assumption.

76. Prove that $\sqrt{3}$ is irrational. HINT: Use the method of Exercise 75.

77. Let R be the set of all rectangles that have the same perimeter P. Show that the square (whose side necessarily has length $P/4$) is the element of R that has the largest area.

78. Show that if a circle and a square have the same perimeter, then the circle has a larger area than the square. Suppose a circle and an arbitrary rectangle have equal perimeters. Which one has the larger area?

■ PROJECT 1.2 Decimal Expansions of Rational Numbers

The decimal form of a rational number p/q can be obtained simply by dividing the denominator into the numerator. As indicated in the text, the result will be either a *terminating decimal* or a *repeating decimal*. For example, $3/4 = 0.75$ and $24/11 = 2.181818\ldots = 2.\overline{18}$.

Problem 1. Under what conditions will the decimal expansion of p/q terminate? Repeat?
Hint: Since $p/q = p(1/q)$, it is sufficient to investigate the decimal expansions of $1/q$.
Calculate $1/q$ for enough positive integers q to form a conjecture as to whether the decimal expansion will terminate or repeat. What is your conjecture?

Problem 2. Suppose that we are given the decimal expansion of a rational number. How can we represent the decimal in the rational number form p/q?

It easy to represent a terminating decimal in the form p/q. For example, $3.74 = 374/100 = 187/50$ and $1.2516 = 12{,}516/10{,}000 = 3129/2500$. Therefore the question reduces to: How can we represent a repeating decimal in the rational number form p/q?

Example Represent the decimal $3.135135\cdots = 3.\overline{135}$ in the form p/q.

SOLUTION Set

$$r = 3.135135\cdots$$

Then

$$1000r = 3135.135135\cdots$$

and

$$1000r - r = 3132$$

so

$$999r = 3132 \quad \text{and} \quad r = \tfrac{3132}{999} = \tfrac{116}{37}$$

Problem 3. Express each of the given repeating decimals in the rational number form p/q.

(a) $13.\overline{201}$. (b) $0.\overline{27}\cdots$ (c) $0.\overline{23}$ (d) $4.16\overline{3}$.

Show that the repeating decimal $0.9999\cdots = 0.\overline{9}$. represents the number 1. Also, note that $1 = 1.0$. Thus, it follows that a rational number may have more than one decimal representation. Can you find any others?

■ 1.3 REVIEW OF INEQUALITIES

All our work with inequalities is based on the order properties of the real numbers given in Section 1.2. In this section we work with the type of inequalities that abound in calculus, namely inequalities that involve a variable.

To solve an equation in x is to find the set of numbers x for which the equation holds. To solve an inequality in x is to find the set of numbers x for which the inequality holds.

The way we solve an inequality is very similar to the way we solve an equation, but there is one important difference. We can maintain an inequality by adding the same number to both sides, or by subtracting the same number from both sides, or by multiplying or dividing both sides by the same *positive* number. But if we multiply or divide by a *negative* number, then the inequality is *reversed*:

$$x - 2 < 4 \quad \text{gives} \quad x < 6, \qquad x + 2 < 4 \quad \text{gives} \quad x < 2,$$

$$\tfrac{1}{2}x < 4 \quad \text{gives} \quad x < 8,$$

but

$$-\tfrac{1}{2}x < 4 \quad \text{gives} \quad x > -8.$$

↑ note, the inequality is reversed

Example 1 Solve the inequality

$$-3(4 - x) \leq 12.$$

SOLUTION Multiplying both sides of the inequality by $-\frac{1}{3}$, we have

$$4 - x \geq -4. \qquad \text{(the inequality has been reversed)}$$

Subtracting 4, gives

$$-x \geq -8.$$

To isolate x, we multiply by -1. This gives

$$x \leq 8. \qquad \text{(the inequality has been reversed again)}$$

The solution set is the interval $(-\infty, 8]$. □

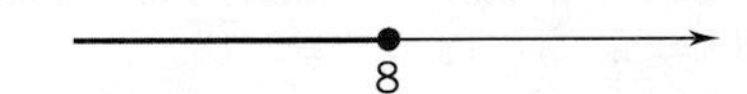

There are generally several ways to solve a given inequality. For example, the last inequality could have been solved this way:

$$-3(4 - x) \leq 12,$$

$$-12 + 3x \leq 12,$$

$$3x \leq 24, \qquad \text{(we added 12)}$$

$$x \leq 8. \qquad \text{(we divided by 3)}$$

An approach to solving a quadratic inequality is to factor the quadratic, if possible.

Example 2 Solve the inequality

$$x^2 - 4x + 3 > 0.$$

SOLUTION Factoring the quadratic, we obtain

$$(x-1)(x-3) > 0.$$

The product $(x-1)(x-3)$ is zero at 1 and 3. Mark these points on a number line (Figure 1.3.1). The points 1 and 3 separate three intervals:

$$(-\infty, 1), \qquad (1,3), \qquad (3,\infty).$$

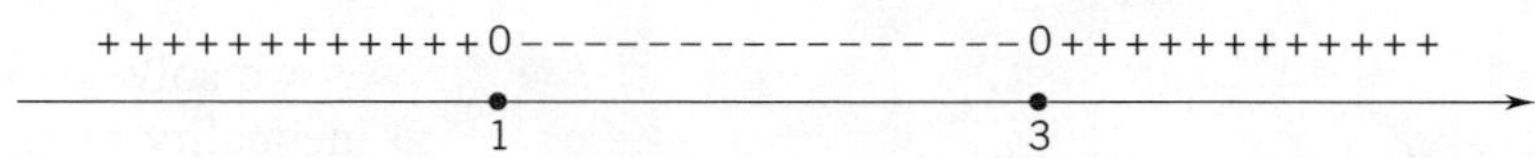

Figure 1.3.1

On each of these intervals the product $(x-1)(x-3)$ keeps a constant sign:

$$\begin{aligned}
&\text{on } (-\infty, 1) \quad [\text{to the left of } 1] \quad \text{sign of } (x-1)(x-3) = (-)(-) = +;\\
&\text{on } (1,3) \quad [\text{between 1 and } 3] \quad \text{sign of } (x-1)(x-3) = (+)(-) = -;\\
&\text{on } (3,\infty) \quad [\text{to the right of } 3] \quad \text{sign of } (x-1)(x-3) = (+)(+) = +.
\end{aligned}$$

The product $(x-1)(x-3)$ is positive on the open intervals $(-\infty, 1)$ and $(3, \infty)$. The solution set is the union $(-\infty, 1) \cup (3, \infty)$. ☐

In contrast to Example 2, consider the quadratic inequality

$$x^2 - 2x + 5 \le 0.$$

Since $b^2 - 4ac = 4 - 4(1)(5) = -16 < 0$, the quadratic $x^2 - 2x + 5$ cannot be factored. However, if we *complete the square*, we get

$$x^2 - 2x + 5 = x^2 - 2x + 1 + 4 = (x-1)^2 + 4 \ge 4 > 0 \text{ for all } x.$$

Thus, we can conclude that the given inequality has no solutions. On the other hand, note that every real number satisfies the inequality

$$x^2 - 2x + 5 > 0.$$

Returning to the approach illustrated in Example 2, consider an expression of the form

$$(x-a_1)^{k_1}(x-a_2)^{k_2}\cdots(x-a_n)^{k_n}$$

where $k_1, k_2, \ldots, k_n$ are positive integers and $a_1 < a_2 < \cdots < a_n$. Such an expression is zero at $a_1, a_2, \ldots, a_n$. It is positive on those intervals where an even number of factors are negative, and it is negative on those intervals where an odd number of factors are negative.

As an example, take the expression

$$(x+2)(x-1)(x-3).$$

This product is zero at $-2, 1, 3$. It is

negative on $(-\infty, -2)$, (3 negative terms)

positive on $(-2, 1)$, (2 negative terms)

negative on $(1, 3)$, (1 negative term)

positive on $(3, \infty)$. (0 negative terms)

See Figure 1.3.2

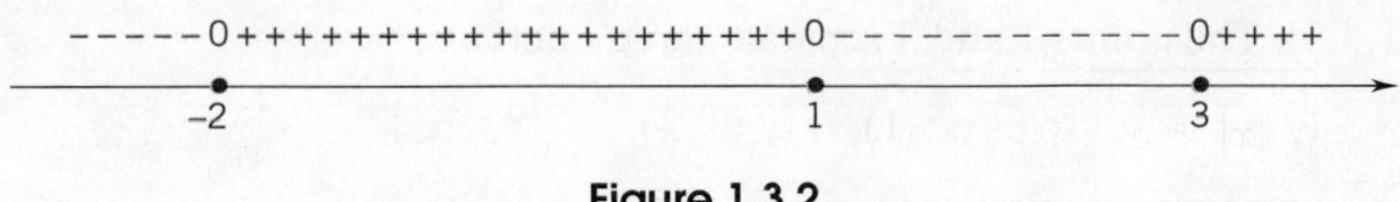

Figure 1.3.2

Example 3 Solve the inequality

$$(x+3)^5(x-1)(x-4)^2 < 0.$$

SOLUTION We view $(x+3)^5(x-1)(x-4)^2$ as the product of three factors: $(x+3)^5$, $(x-1)$, $(x-4)^2$. The product is zero at -3, 1, and 4. These points separate the intervals

$$(-\infty, -3), \quad (-3, 1), \quad (1, 4), \quad (4, \infty).$$

On each of these intervals the product keeps a constant sign. It is

positive on $(-\infty, -3)$,	(2 negative factors)
negative on $(-3, 1)$,	(1 negative factor)
positive on $(1, 4)$,	(0 negative factors)
positive on $(4, \infty)$.	(0 negative factors)

See Figure 1.3.3.

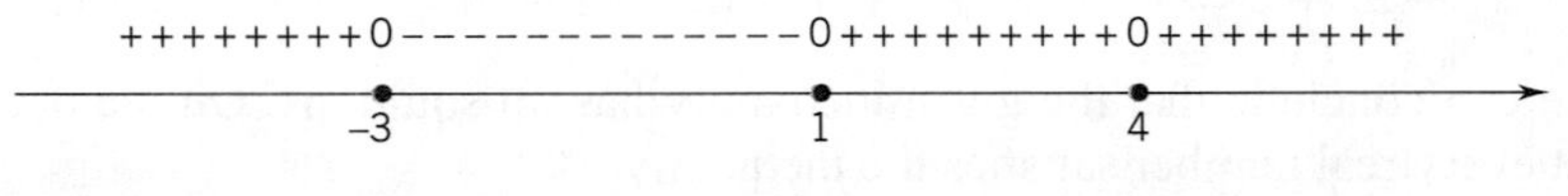

Figure 1.3.3

The solution set is the open interval $(-3, 1)$. ☐

This approach to solving inequalities will be justified in Section 2.6

Inequalities and Absolute Value

Now we take up inequalities that involve absolute values. With an eye toward developing the concept of limits in Chapter 2 we introduce two Greek letters: δ (delta) and ϵ (epsilon).

Recall that for each real number a

(1.3.1)

$$|a| = \begin{cases} a & \text{if } a \geq 0, \\ -a, & \text{if } a < 0, \end{cases} \qquad |a| = \max\{a, -a\}, \qquad |a| = \sqrt{a^2}$$

We begin with the inequality

$$|x| < \delta,$$

where δ is some positive number. To say that $|x| < \delta$ is to say that x lies within δ units of 0 or, equivalently, that x lies between $-\delta$ and δ. Thus

(1.3.2) $$|x| < \delta \quad \text{iff} \quad -\delta < x < \delta.$$

To say that $|x - c| < \delta$ is to say that x lies within δ units of c or, equivalently, that x lies between $c - \delta$ and $c + \delta$. Thus

(1.3.3) $$|x - c| < \delta \quad \text{iff} \quad c - \delta < x < c + \delta.$$

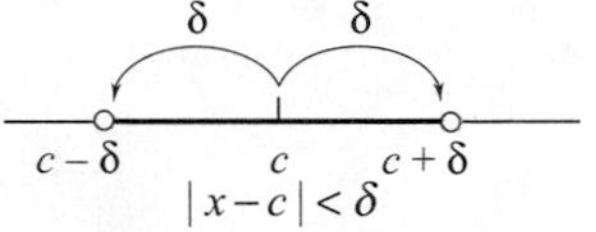

Somewhat more delicate is the inequality

$$0 < |x - c| < \delta.$$

Here we have $|x - c| < \delta$ with the additional requirement that $x \neq c$. Consequently

(1.3.4) $$0 < |x - c| < \delta \quad \text{iff} \quad c - \delta < x < c \quad \text{or} \quad c < x < c + \delta.$$

Thus, for example,

$$|x| < \tfrac{1}{2} \quad \text{iff} \quad -\tfrac{1}{2} < x < \tfrac{1}{2}; \qquad \textit{Solution}: (-\tfrac{1}{2}, \tfrac{1}{2}).$$

$$|x - 5| < 1 \quad \text{iff} \quad 4 < x < 6; \qquad \textit{Solution}: (4,6).$$

$$0 < |x - 5| < 1 \quad \text{iff} \quad 4 < x < 5 \quad \text{or} \quad 5 < x < 6; \qquad \textit{Solution}: (4,5) \cup (5,6).$$

Example 4 Solve the inequality

$$|x + 2| < 3.$$

SOLUTION The inequality $|x + 2| < 3$ holds iff

$$|x - (-2)| < 3 \quad \text{iff} \quad -2 - 3 < x < -2 + 3 \quad \text{iff} \quad -5 < x < 1.$$

The solution set is the open interval $(-5, 1)$. □

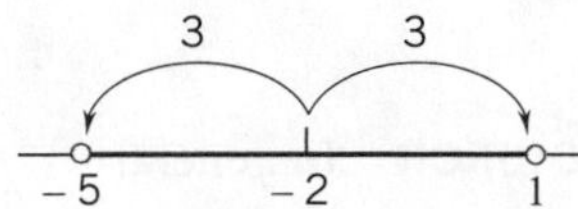

Example 5 Solve the inequality

$$|3x - 4| < 2.$$

SOLUTION Since

$$|3x - 4| = \left|3\left(x - \tfrac{4}{3}\right)\right| = |3|\left|x - \tfrac{4}{3}\right| = 3\left|x - \tfrac{4}{3}\right|,$$

the inequality can be rewritten

$$3\left|x - \tfrac{4}{3}\right| < 2.$$

This gives $\left|x - \frac{4}{3}\right| < \frac{2}{3}$. Therefore

$$\tfrac{4}{3} - \tfrac{2}{3} < x < \tfrac{4}{3} + \tfrac{2}{3},$$
$$\tfrac{2}{3} < x < 2.$$

The solution set is the open interval $(\frac{2}{3}, 2)$.

ALTERNATIVE SOLUTION The inequality

$$|3x - 4| < 2$$

is equivalent to

$$-2 < 3x - 4 < 2$$

by (1.3.2). Therefore

$$2 < 3x < 6 \qquad \text{(add 4 to both inequalities)}$$

and

$$\tfrac{2}{3} < x < 2 \qquad \text{(divide through by 3)}$$

as before. □

Let $\epsilon > 0$. If you think of $|a|$ as the distance between a and 0, then

(1.3.5) $|a| > \epsilon \quad \text{iff} \quad a > \epsilon \quad \text{or} \quad a < -\epsilon.$

$-\epsilon$ 0 ϵ

$|a| > \epsilon$

Example 6 Solve the inequality

$$|2x + 3| > 5.$$

SOLUTION In general

$$|a| > \epsilon \quad \text{iff} \quad a > \epsilon, \text{ or } a < -\epsilon.$$

So here

$$2x + 3 > 5 \quad \text{or} \quad 2x + 3 < -5.$$

The first possibility gives $2x > 2$ and thus

$$x > 1.$$

The second possibility gives $2x < -8$ and thus

$$x < -4$$

The total solution is therefore the union

$$(-\infty, -4) \cup (1, \infty). \quad \square$$

We come now to one of the fundamental inequalities of calculus: for all real numbers a and b,

(1.3.6)
$$|a + b| \leq |a| + |b|.$$

This is called the *triangle inequality* in analogy with the geometric maxim "in a triangle the length of each side is less than or equal to the sum of the lengths of the other two sides."

PROOF OF THE TRIANGLE INEQUALITY The key here is to think of $|x|$ as $\sqrt{x^2}$. Note first that

$$(a + b)^2 = a^2 + 2ab + b^2 \leq |a|^2 + 2|a||b| + |b|^2 = (|a| + |b|)^2.$$

Comparing the extremes of the inequality and taking square roots, we have

$$\sqrt{(a + b)^2} \leq |a| + |b|. \qquad \text{(Exercise 51)}$$

The result follows from observing that

$$\sqrt{(a + b)^2} = |a + b|. \quad \square$$

Here is a variant of the triangle inequality that also comes up in calculus: for all real numbers a and b,

(1.3.7)
$$\big||a| - |b|\big| \leq |a - b|.$$

The proof is left to you as an exercise.

EXERCISES 1.3

Solve the inequality and graph the solution set on a number line.

1. $2 + 3x < 5$.
2. $\frac{1}{2}(2x + 3) < 6$.
3. $16x + 64 \leq 16$.
4. $3x + 5 > \frac{1}{4}(x - 2)$.
5. $\frac{1}{2}(1 + x) < \frac{1}{3}(1 - x)$.
6. $3x - 2 \leq 1 + 6x$.
7. $x^2 - 1 < 0$.
8. $x^2 + 9x + 20 < 0$.
9. $x^2 - x - 6 \geq 0$.
10. $x^2 - 4x - 5 > 0$.
11. $2x^2 + x - 1 \leq 0$.
12. $3x^2 + 4x - 4 \geq 0$.
13. $x(x - 1)(x - 2) > 0$.
14. $x(2x - 1)(3x - 5) \leq 0$.
15. $x^3 - 2x^2 + x \geq 0$.
16. $x^2 - 4x + 4 \leq 0$.
17. $x^3(x - 2)(x + 3)^2 < 0$.
18. $x^2(x - 3)(x + 4)^2 > 0$.
19. $x^2(x - 2)(x + 6) > 0$.
20. $7x(x - 4)^2 < 0$.

Solve the inequality and express the solution set in terms of intervals.

21. $|x| < 2$.
22. $|x| \geq 1$.
23. $|x| > 3$.
24. $|x - 1| < 1$.
25. $|x - 2| < \frac{1}{2}$.
26. $|x - \frac{1}{2}| < 2$.
27. $0 < |x| < 1$.
28. $0 < |x| < \frac{1}{2}$.
29. $0 < |x - 2| < \frac{1}{2}$.
30. $0 < |x - \frac{1}{2}| < 2$.
31. $0 < |x - 3| < 8$.
32. $|3x - 5| < 3$.
33. $|2x + 1| < \frac{1}{4}$.
34. $|5x - 3| < \frac{1}{2}$.
35. $|2x + 5| > 3$.
36. $|3x + 1| > 5$.

Find an inequality of the form $|x - c| < \delta$ the solution of which is the given open interval.

37. $(-3, 3)$.
38. $(-2, 2)$.
39. $(-3, 7)$.
40. $(0, 4)$.
41. $(-7, 3)$.
42. $(-4, 0)$.

In Exercises 43–46, determine all values of $A > 0$ for which the statement is true.

43. If $|x - 2| < 1$, then $|2x - 4| < A$.

44. If $|x - 2| < A$, then $|2x - 4| < 3$.

45. If $|x + 1| < A$, then $|3x + 3| < 4$.

46. If $|x + 1| < 2$, then $|3x + 3| < A$.

47. Arrange the following in order :$1, x, \sqrt{x}, 1/x, 1/\sqrt{x}$, given that: (*a*) $x > 1$, (*b*) $0 < x < 1$.

48. Compare

$$\sqrt{\frac{x}{x+1}} \quad \text{and} \quad \sqrt{\frac{x+1}{x+2}}$$

given that $x > 0$.

49. Let a and b have the same sign. If $a < b$, show that $(1/b) < (1/a)$.

50. Let a and b be nonnegative numbers. Show that if

$$a^2 \leq b^2, \quad \text{then} \quad a \leq b.$$

51. Let a and b be nonnegative numbers. Show that if

$$a \leq b, \quad \text{then} \quad \sqrt{a} \leq \sqrt{b}.$$

52. Prove that for all real numbers a and b

$$|a - b| \leq |a| + |b|.$$

53. Prove that for all real numbers a and b

$$\big||a| - |b|\big| \leq |a - b|.$$

HINT: Calculate $\big||a| - |b|\big|^2$ and use the fact that $ab \leq |a||b|$.

54. Show that $|a + b| = |a| + |b|$ iff $ab \geq 0$.

55. Show that if

$$0 \leq a \leq b, \quad \text{then} \quad \frac{a}{1+a} \leq \frac{b}{1+b}.$$

56. Let a, b, c be nonnegative numbers. Show that if

$$a \leq b + c, \quad \text{then} \quad \frac{a}{1+a} \leq \frac{b}{1+b} + \frac{c}{1+c}.$$

57. Prove that if a and b are real numbers and $a < b$, then $a < (a + b)/2 < b$. The number $(a + b)/2$ is called the *arithmetic mean* of a and b. How are the three numbers $a, (a + b)/2, b$, related on the number line ?

58. Let a and b be nonnegative numbers with $a \leq b$. Prove that

$$a \leq \sqrt{ab} \leq \frac{a+b}{2} \leq b.$$

The number $\sqrt{ab}$ is called the *geometric mean* of a and b.

■ 1.4 COORDINATE PLANE; ANALYTIC GEOMETRY

Rectangular Coordinates

The correspondence that we have discussed between real numbers and points on a number line can be used to construct a coordinate system for the plane. In the plane, draw two number lines that are mutually perpendicular and intersect at their origins. Let O be the point of intersection. It is customary to have one of the lines be horizontal, with the positive numbers to the right of O and the other vertical with the positive numbers above O. The point O is called the *origin*, and the number lines are called the *coordinate axes*. The horizontal axis is usually labeled the *x-axis* and the vertical axis is usually labeled the *y-axis*. The coordinate axes separate the plane into four regions called *quadrants*. The quadrants are numbered in the counterclockwise direction starting with the upper right. See Figure 1.4.1.

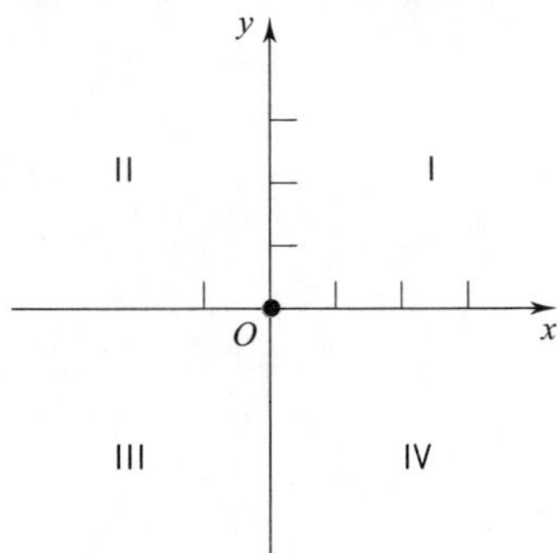

Figure 1.4.1

Rectangular coordinates are assigned to points of the plane as follows. (See Figure 1.4.2) The point on the x-axis with line coordinate a is assigned rectangular coordinates $(a, 0)$. The point on the y-axis with line coordinate b is assigned rectangular coordinates $(0, b)$. Thus the origin is assigned coordinates $(0, 0)$. A point P not on one of the coordinate axes is assigned coordinates (a, b) provided that the line l_1 that passes through P and is parallel to the y-axis intersects the x-axis at the point with coordinates $(a, 0)$, and the line l_2 that passes through P and is parallel to the x-axis intersects the y-axis at the point with coordinates $(0, b)$.

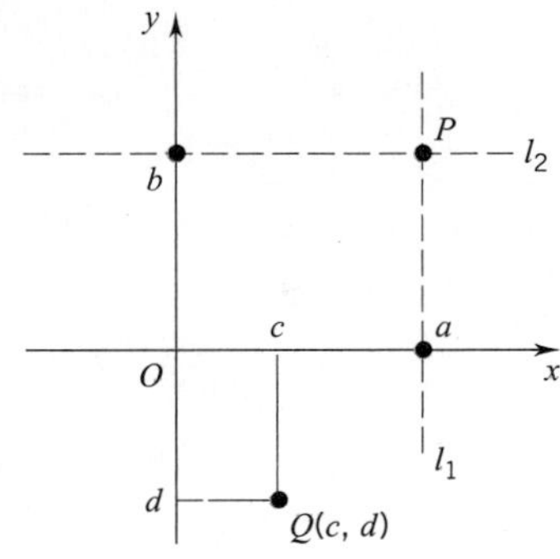

Figure 1.4.2

This procedure assigns an ordered pair of real numbers to each point of the plane. Moreover, the procedure is reversible. Given any ordered pair (c, d) of real numbers, there is a unique point Q in the plane with coordinates (c, d).

To indicate that P has coordinates (a, b), we write $P(a, b)$. The number a is called the *x-coordinate* (the *abscissa*); the number b is called the *y-coordinate* (the *ordinate*). The coordinate system that we have outlined is called the *rectangular coordinate system*, also called the *Cartesian coordinate system* after the French mathematician René Descartes (1596–1650).

Distance and Midpoint Formulas

Let $P_0(x_0, y_0)$ and $P_1(x_1, y_1)$ be points in the plane. The formula for the distance $d(P_0, P_1)$ between P_0 and P_1 follows from the *Pythagorean theorem*:

$$d(P_0, P_1) = \sqrt{|x_1 - x_0|^2 + |y_1 - y_0|^2} = \sqrt{(x_1 - x_0)^2 + (y_1 - y_0)^2}. \qquad \text{(Figure 1.4.3)}$$

$$\uparrow\!\_\_\, |a|^2 = a^2$$

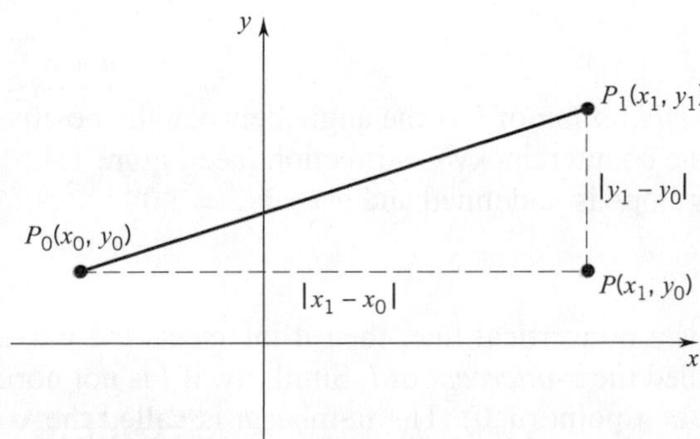

Distance : $d(P_0, P_1) = \sqrt{(x_1 - x_0)^2 + (y_1 - y_0)^2}$

Figure 1.4.3

Let $M(x, y)$ be the midpoint of the line segment joining P_0 and P_1. Using the fact that the two triangles in Figure 1.4.4 are congruent, it can be shown that

$$x = \frac{x_0 + x_1}{2} \quad \text{and} \quad y = \frac{y_0 + y_1}{2}.$$

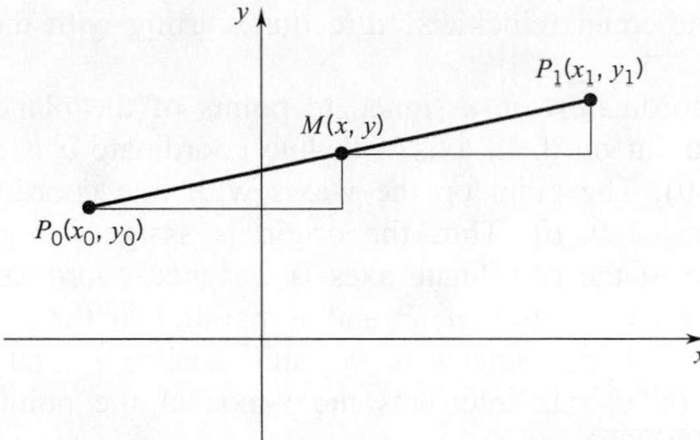

$$\text{Midpoint: } M = \left(\frac{x_0 + x_1}{2}, \frac{y_0 + y_1}{2}\right)$$

Figure 1.4.4

Lines

Let l be a straight line in the plane, and let $P_0(x_0, y_0)$ and $P_1(x_1, y_1)$ be any two distinct points on l.

(i) Slope The *slope* of l is given by

$$m = \frac{y_1 - y_0}{x_1 - x_0} \qquad \text{provided } x_1 \neq x_0.$$

Also,

$$m = \tan\theta, \dagger$$

where θ, called the *inclination* of l, is the angle between the positively directed x-axis and l measured in the counterclockwise direction (see Figure 1.4.5). If $x_1 = x_0$, then l is a vertical line; its slope is undefined and $\theta = \pi/2 = 90°$.

(ii) Intercepts If l is a nonvertical line, then it intersects the y-axis at a point $(0, b)$. The number b is called the *y-intercept* of l. Similarly, if l is not horizontal, then it will intersect the x-axis at a point $(a, 0)$. The number a is called the *x-intercept* of l. See Figure 1.4.6.

† The trigonometric functions are reviewed in Section 1.6.

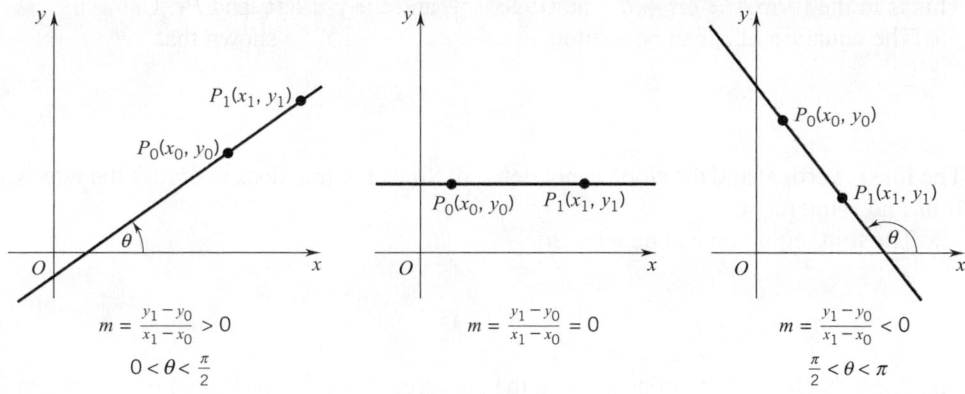

Figure 1.4.5

(iii) Equations An equation for l is an equation which involves two variables, say x and y, and has the property that a point $P(x_0, y_0)$ lies on l iff its coordinates, x_0 and y_0, satisfy the equation. The standard forms are as follows:

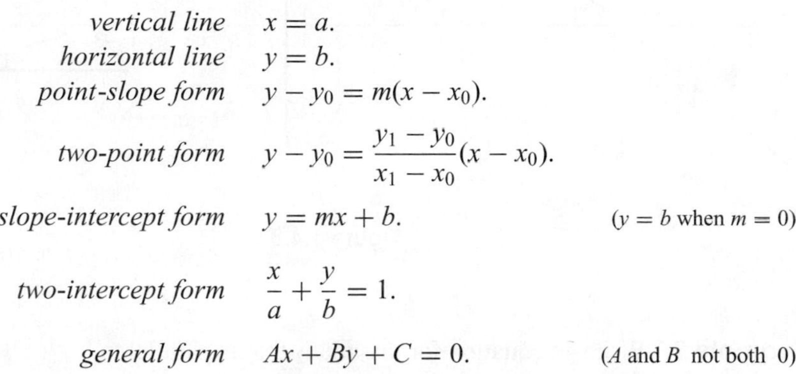

vertical line	$x = a.$	
horizontal line	$y = b.$	
point-slope form	$y - y_0 = m(x - x_0).$	
two-point form	$y - y_0 = \dfrac{y_1 - y_0}{x_1 - x_0}(x - x_0).$	
slope-intercept form	$y = mx + b.$	($y = b$ when $m = 0$)
two-intercept form	$\dfrac{x}{a} + \dfrac{y}{b} = 1.$	
general form	$Ax + By + C = 0.$	(A and B not both 0)

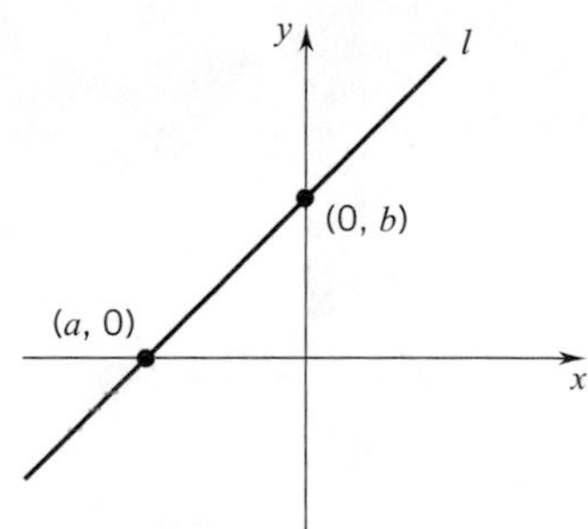

Figure 1.4.6

(iv) Parallel and Perpendicular Lines Given two nonvertical lines l_1 and l_2 with slopes m_1 and m_2, respectively, then l_1 and l_2 are

parallel iff $m_1 = m_2$,
perpendicular iff $m_1 m_2 = -1$.

(v) The Angle Between Two Lines Figure 1.4.7 shows two intersecting lines l_1 and l_2 with inclinations θ_1 and θ_2. These lines form two angles: $\alpha = \theta_2 - \theta_1$ and $\pi - \alpha$. If these two angles are equal, then $\alpha = \pi/2$ and the lines are perpendicular. If α and $\pi - \alpha$ are not equal, then the smaller of the two (the one less than $\pi/2$) is called the *angle between l_1 and l_2*. In Exercises 1.6 you will be asked to show that

$$\tan \alpha = \left| \frac{m_2 - m_1}{1 + m_1 m_2} \right|$$

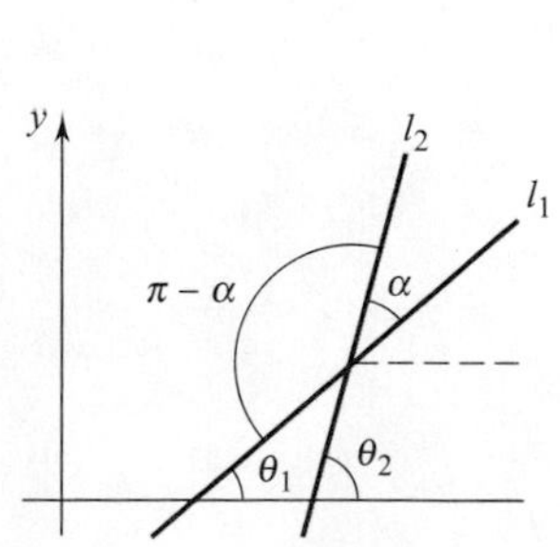

Figure 1.4.7

Example 1 Find the slope and y-intercept of each of the following lines:

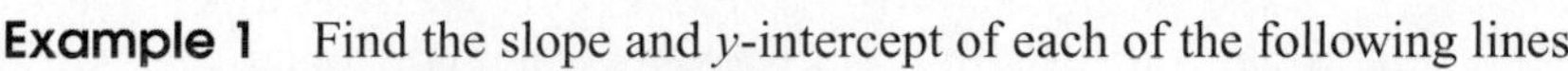

$$l_1 : 20x - 24y - 30 = 0, \quad l_2 : 2x - 3 = 0, \quad l_3 : 4y + 5 = 0.$$

SOLUTION The equation of l_1 can be written

$$y = \tfrac{5}{6}x - \tfrac{5}{4}.$$

This is in the form $y = mx + b$. The slope is $\frac{5}{6}$, and the y-intercept is $-\frac{5}{4}$.

The equation of l_2 can be written

$$x = \tfrac{3}{2}.$$

The line is vertical and the slope is not defined. Since the line does not cross the y-axis, it has no y-intercept.

The third equation can be written

$$y = -\tfrac{5}{4}.$$

The line is horizontal. Its slope is 0 and the y-intercept is $-\frac{5}{4}$. The three lines are drawn in Figure 1.4.8. ❑

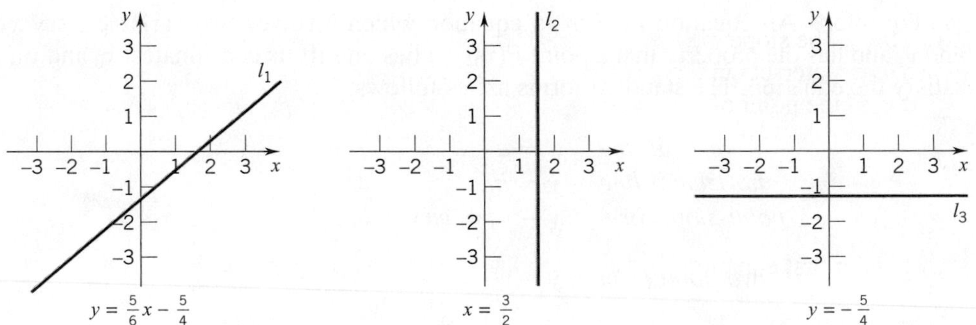

Figure 1.4.8

Example 2 Write an equation for the line l_2 that is parallel to

$$l_1 : 3x - 5y + 8 = 0$$

and passes through the point $P(-3, 2)$.

SOLUTION We can rewrite the equation of l_1 as

$$y = \tfrac{3}{5}x + \tfrac{8}{5}.$$

The slope of l_1 is $\frac{3}{5}$. The slope of l_2 must also be $\frac{3}{5}$. (*Remember:* For nonvertical parallel lines, $m_1 = m_2$.)

Since l_2 passes through (−3,2) with slope $\frac{3}{5}$, we can use the point-slope formula and write the equation as

$$y - 2 = \tfrac{3}{5}(x + 3). \quad ❑$$

Example 3 Write an equation for the line that is perpendicular to

$$l_1 : x - 4y + 8 = 0$$

and passes through the point $P(2, -4)$.

SOLUTION The equation for l_1 can be written

$$y = \tfrac{1}{4}x + 2.$$

The slope of l_1 is $\frac{1}{4}$. The slope of l_2 is therefore -4. (*Remember:* For nonvertical perpendicular lines, $m_1 m_2 = -1$.)

Since l_2 passes through $(2, -4)$ with slope -4, we can use the point-slope formula and write the equation as

$$y + 4 = -4(x - 2). \quad \square$$

Example 4 Show that the lines

$$l_1 : 3x - 4y + 8 = 0 \quad \text{and} \quad l_2 : 12x - 5y - 12 = 0$$

intersect and find their point of intersection.

SOLUTION The slope of l_1 is $\frac{3}{4}$ and the slope of l_2 is $\frac{12}{5}$. Since l_1 and l_2 have different slopes, they intersect in one point.

To find the point of intersection, we solve the two equations simultaneously:

$$\begin{aligned} 3x - 4y + 8 &= 0 \\ 12x - 5y - 12 &= 0. \end{aligned}$$

Multiply the first equation by -4 and add it to the second equation to obtain

$$\begin{aligned} 11y - 44 &= 0 \\ y &= 4. \end{aligned}$$

Substituting $y = 4$ into either of the two given equations yields $x = \frac{8}{3}$. Thus, l_1 and l_2 intersect at the point $(\frac{8}{3}, 4)$. $\square$

Conic Sections

We now consider equations of the form

$$Ax^2 + By^2 + Cx + Dy + E = 0,$$

where A, B, C, D, E are constants, and A and B are not both 0. The graph of such an equation is, in general, a conic section; that is, a circle, ellipse, parabola, or hyperbola. We say "in general" because there are degenerate cases. For example, the graph of $x^2 + y^2 = 0$ consists of a single point, (0,0). The equation $x^2 + y^2 + 1 = 0$ has no graph; there are no ordered pairs (x, y) of real numbers that satisfy this equation.

A more formal treatment of the conic sections is given in Sections 9.1 and 9.2. For purposes of review here, we list the standard forms for the equations of the conics and indicate their graphs.

Circle $A = B$; $\quad Ax^2 + Ay^2 + Cx + Dy + E = 0$ or

$$x^2 + y^2 + C_1x + D_1y + E_1 = 0 \quad \text{after dividing by } A$$

Standard forms: $x^2 + y^2 = r^2$; *circle of radius r centered at the origin* (Figure 1.4.9*a*)

$(x - h)^2 + (y - k)^2 = r^2$. *circle of radius r centered at* $P(h, k)$ (Figure 1.4.9*b*)

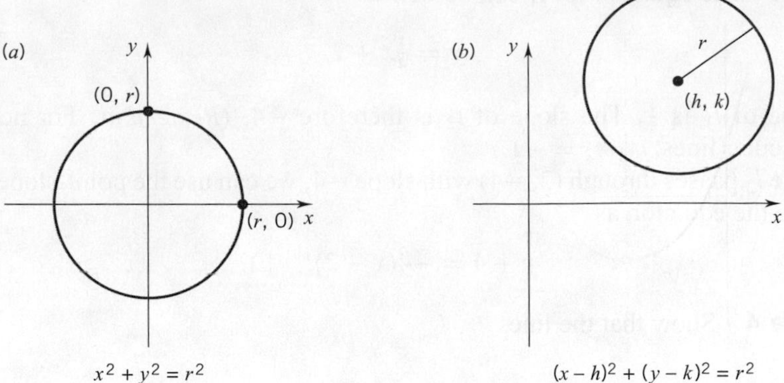

Figure 1.4.9

Ellipse A, B have the same sign, $A \neq B$; $\quad Ax^2 + By^2 + Cx + Dy + E = 0.$

Standard form: $\dfrac{(x-h)^2}{a^2} + \dfrac{(y-k)^2}{b^2} = 1$ *ellipse centered at the point* $P(h, k)$ (Figure 1.4.10).

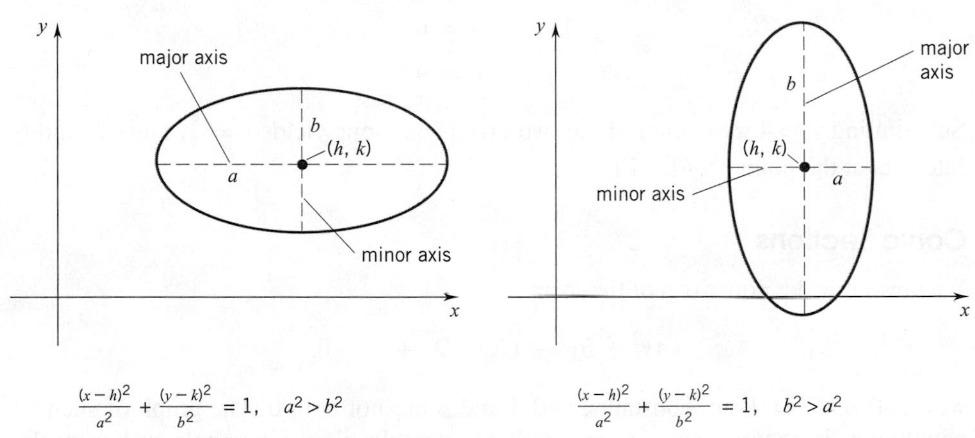

Figure 1.4.10

Parabola Either $A = 0$ or $B = 0$; $\quad Ax^2 + Cx + Dy + E = 0, \quad A \neq 0,$ or
$By^2 + Cx + Dy + E = 0, \quad B \neq 0.$

Standard forms:

$(x - h)^2 = 4c(y - k);$ *parabola, vertex at the point* $P(h, k)$ *and axis vertical* (Figure 1.4.11*a*)

$(y - k)^2 = 4c(x - h).$ *parabola, vertex at the point* $P(h, k)$ *and axis horizontal* (Figure 1.4.11*b*)

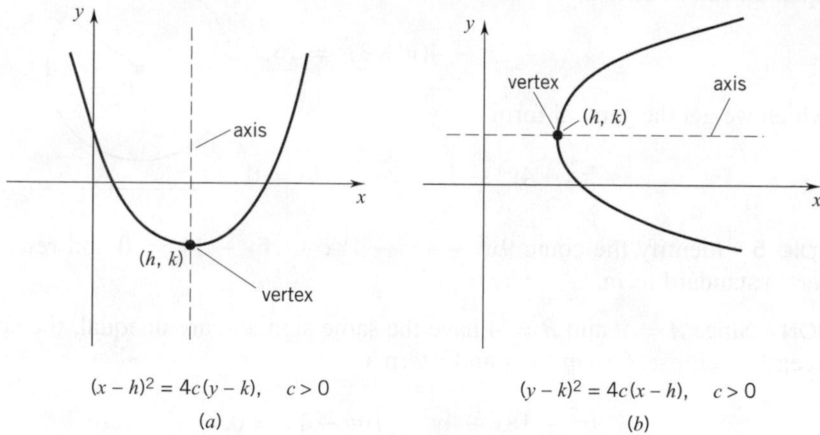

Figure 1.4.11

Hyperbola A, B have opposite sign: $\quad Ax^2 + By^2 + Cx + Dy + E = 0.$

Standard forms:

$$\frac{(x-h)^2}{a^2} - \frac{(y-k)^2}{b^2} = 1;$$

hyperbola, center at the point $P(h, k)$ and axis horizontal (Figure 1.4.12*a*);

$$\frac{(y-k)^2}{b^2} - \frac{(x-h)^2}{a^2} = 1.$$

hyperbola, center at the point $P(h, k)$ and axis vertical (Figure 1.4.12*b*).

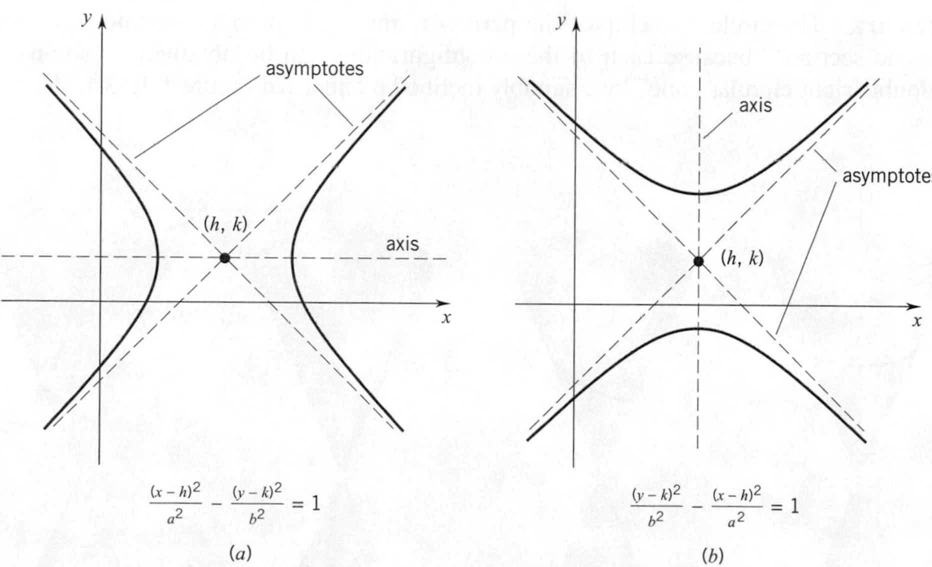

Figure 1.4.12

Standard forms are obtained from the general form $Ax^2 + By^2 + Cx + Dy + E = 0$ by completing the square; the general form is obtained from the standard forms by expanding the $(x - h^2)$ and $(y - k)^2$ terms and "clearing fractions."

For example, consider the hyperbola

$$\frac{(x-2)^2}{16} - \frac{(y+3)^2}{4} = 1.$$

This equation can be written

$$(x-2)^2 - 4(y+3)^2 = 16,$$

from which we get the general form

$$x^2 - 4y^2 - 4x - 24y - 48 = 0.$$

Example 5 Identify the conic $9x^2 + 4y^2 - 18x + 16y - 11 = 0$ and rewrite the equation in standard form.

SOLUTION Since $A = 9$ and $B = 4$ have the same sign and are unequal, the conic is (we except) an ellipse. Group the x and y terms

$$9x^2 - 18x + 4y^2 + 16y - 11 = 0,$$

complete the square

$$9(x^2 - 2x + 1) + 4(y^2 + 4y + 4) - 11 - 9 - 16 = 0$$
$$9(x-1)^2 + 4(y+2)^2 = 36,$$

and get the standard form

$$\frac{(x-1)^2}{4} + \frac{(y+2)^2}{9} = 1.$$

This is an ellipse with center at $(1, -2)$. ☐

Remark The circle, the ellipse, the parabola, and the hyperbola are known as the "conic sections" because each of these configurations can be obtained by slicing a "double right circular cone" by a suitably inclined plane. (See Figure 1.4.13.) ☐

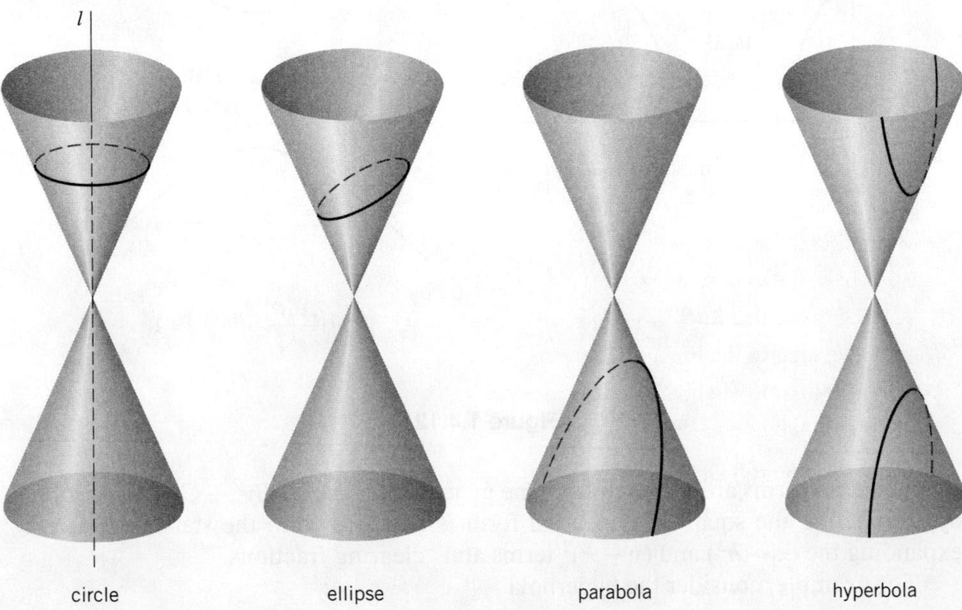

Figure 1.4.13

EXERCISES 1.4

Find the distance between the points.

1. $P_0(0, 5)$, $P_1(6, -3)$. **2.** $P_0(2, 2)$, $P_1(5, 5)$.
3. $P_0(5, -2)$, $P_1(-3, 2)$. **4.** $P_0(2, 7)$, $P_1(-4, 7)$.

Find the midpoint of the line segment P_0P_1.

5. $P_0(2, 4)$, $P_1(6, 8)$. **6.** $P_0(3, -1)$, $P_1(-1, 5)$.
7. $P_0(2, -3)$, $P_1(7, -3)$. **8.** $P_0(a, 3)$, $P_1(3, a)$.

Find the slope of the line through the given points.

9. $P_0(-2, 5)$, $P_1(4, 1)$. **10.** $P_0(4, -3)$, $P_1(-2, -7)$.
11. $P(a, b)$, $Q(b, a)$. **12.** $P(4, -1)$, $Q(-3, -1)$.
13. $P(x_0, 0)$, $Q(0, y_0)$. **14.** $O(0, 0)$, $P(x_0, y_0)$.

Find the slope and y-intercept.

15. $y = 2x - 4$. **16.** $6 - 5x = 0$.
17. $3y = x + 6$. **18.** $6y - 3x + 8 = 0$.
19. $7x - 3y + 4 = 0$. **20.** $4y = 3$.

Write an equation for the line with

21. slope 5 and y-intercept 2.
22. slope 5 and y-intercept -2.
23. slope -5 and y-intercept 2.
24. slope -5 and y-intercept -2.

In Exercises 25 and 26, write an equation for the horizontal line 3 units

25. above the x-axis.
26. below the x-axis.

In Exercises 27 and 28, write an equation for the vertical line 3 units

27. to the left of the y-axis.
28. to the right of the y-axis.

Find an equation for the line that passes through the point $P(2, 7)$ and is

29. parallel to the x-axis.
30. parallel to the y-axis.
31. parallel to the line $3y - 2x + 6 = 0$.
32. perpendicular to the line $y - 2x + 5 = 0$.
33. perpendicular to the line $3y - 2x + 6 = 0$.
34. parallel to the line $y - 2x + 5 = 0$.

Determine the point(s) where the line intersects the circle.

35. $y = x$, $x^2 + y^2 = 1$.
36. $y = mx$, $x^2 + y^2 = 4$.
37. $4x + 3y = 24$, $x^2 + y^2 = 25$.
38. $y = mx + b$, $x^2 + y^2 = b^2$.

Find the point where the lines intersect.

39. $l_1 : 4x - y - 3 = 0$, $l_2 : 3x - 4y + 1 = 0$.
40. $l_1 : 3x + y - 5 = 0$, $l_2 : 7x - 10y + 27 = 0$.
41. $l_1 : 4x - y + 2 = 0$, $l_2 : 19x + y = 0$.
42. $l_1 : 5x - 6y + 1 = 0$, $l_2 : 8x + 5y + 2 = 0$.

43. Find the area of the triangle with vertices $(1, -2), (-1, 3), (2, 4)$.

44. Find the area of the triangle with vertices $(-1, 1)$, $(3, \sqrt{2})$, $(\sqrt{2}, -1)$.

Identify the given equation as a conic section. If it is a circle, give the center and radius; if it is an ellipse, give the center; if it is a parabola, give the vertex; if it is a hyperbola, give the center.

45. $y^2 + 2y - x = 0$.
46. $x^2 + y^2 - 4x + 6y - 3 = 0$.
47. $2x^2 + 3y^2 - 8x + 6y + 5 = 0$.
48. $2x^2 + 2y^2 - 4x + 8y = -10$.
49. $y^2 - 4y - 4x^2 + 8x = 4$.
50. $4x^2 + 16x - 16y = 32$.
51. $4x^2 - y^2 - 24x - 4y + 16 = 0$.
52. $4x^2 + y^2 - 8x + 4y + 4 = 0$.

53. Determine the slope of the line that intersects the circle $x^2 + y^2 = 169$ only at the point (5, 12).

54. Find an equation for the line which is tangent to the circle $x^2 + y^2 - 2x + 6y - 15 = 0$ at the point (4, 1) on the circle. HINT: A line is tangent to a circle at a point P iff it is perpendicular to the radius at P.

55. The point $P(1, -1)$ is on the circle whose center is at $C(-1, 3)$. Find an equation for the tangent line to the circle at P.

In Exercises 56–59, use a graphing utility to estimate the point(s) of intersection.

56. $l_1 : 3x - 4y = 7$, $l_2 : -5x + 2y = 11$.
57. $l_1 : 2.41x + 3.29y = 5$, $l_2 : 5.13x - 4.27y = 13$.
58. $l_1 : 2x - 3y = 5$, circle: $x^2 + y^2 = 4$.
59. circle: $x^2 + y^2 = 9$, parabola: $y = x^2 - 4x + 5$.

The *perpendicular bisector* of the line segment $\overline{PQ}$ is the line which is perpendicular to $\overline{PQ}$ and passes through the midpoint of $\overline{PQ}$. In Exercises 60 and 61, find an equation for the perpendicular bisector of the line segment joining the given points.

60. $P(-1, 3)$, $Q(3, -4)$.
61. $P(1, -4)$, $Q(4, 9)$.

In Exercises 62–65, the given points are the vertices of a triangle. State whether the triangle is *isosceles* (two sides of equal length) a right triangle, both of these, or neither of these.

62. $P_0(-4, 3)$, $P_1(-4, -1)$, $P_2(2, 1)$.
63. $P_0(-2, 5)$, $P_1(1, 3)$, $P_2(-1, 0)$.

64. $P_0(-2, -1)$, $P_1(0, 7)$, $P_2(3, 2)$.

65. $P_0(3, 4)$, $P_1(1, 1)$, $P_2(-2, 3)$.

66. An *equilateral triangle* is a triangle in which the three sides have equal lengths. Given that two of the vertices of an equilateral triangle are (0, 0) and (4, 3), find a third vertex. How many such triangles are there?

67. Show that the midpoint M of the hypotenuse of a right triangle is equidistant from the three vertices of the triangle. HINT: Introduce a coordinate system in which the sides of the triangle are on the coordinate axes; see the figure.

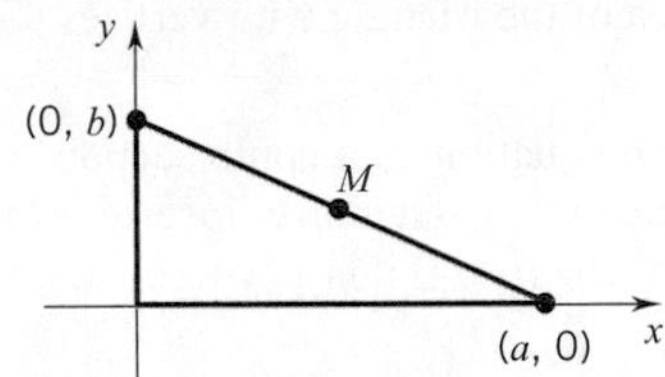

68. A *median* of a triangle is a line segment from a vertex to the midpoint of the opposite side. Find the lengths of the medians of the triangle with vertices $(-1, -2)$, $(2, 1)$, $(4, -3)$.

69. The vertices of a triangle are $(1, 0)$, $(3, 4)$, and $(-1, 6)$. Find the point(s) where the medians of this triangle intersect.

70. Show that the medians of a triangle intersect in a single point (called the *centroid* of the triangle). HINT: Introduce a coordinate system such that one vertex is at the origin and one side is on the positive x-axis; see the figure.

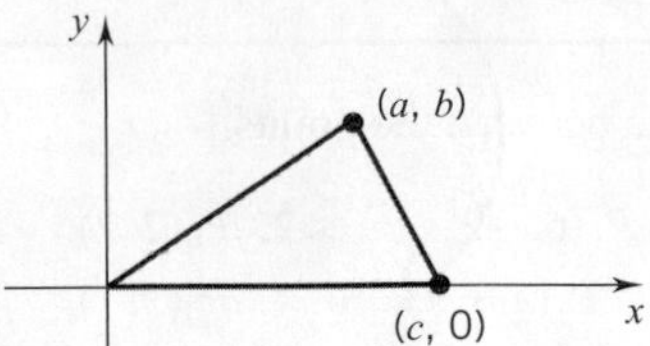

71. Prove that the diagonals of a parallelogram bisect each other. HINT: Introduce a coordinate system such that one vertex of the parallelogram is at the origin and one side is on the positive x-axis.

72. Let $P_1(x_1, y_1)$, $P_2(x_2, y_2)$, $P_3(x_3, y_3)$, $P_4(x_4, y_4)$ be the vertices of a quadrilateral. Show that the quadrilateral formed by joining the midpoints of adjacent sides is a parallelogram.

73. Temperature is usually measured either in degrees Fahrenheit (F) or in degrees Celsius (C). The relation between F and C is linear. The freezing point of water is 0°C or 32°F, and the boiling point of water is 100°C or 212°F. Find an equation giving the Fahrenheit temperature F in terms of the Celsius temperature C. Is there a temperature at which the Fahrenheit and Celsius temperatures are equal? If so, find it.

74. The relation between Fahrenheit temperature F and absolute (or Kelvin) temperature K is linear. If $K = 273°$ when $F = 32°$, and $K = 373°$ when $F = 212°$, express K in terms of F. Also, use Exercise 73 to determine the relation between Celsius temperature and absolute temperature. Is this relation linear?

■ 1.5 FUNCTIONS

The fundamental processes of calculus (called *differentiation* and *integration*) are processes applied to functions. To understand these processes and to be able to carry them out, you have to be thoroughly familiar with the function concept. Here we review some of the basic ideas.

Function; Domain and Range

First we need a working definition for the term *function*. Let D and R be sets. A *function from D into R* is a rule that assigns to each element $x \in D$ a unique element $y \in R$. The set D is called the *domain* of the function. Functions are usually denoted by letters such as f, g, F, G, and so on. If f is a function from D into R, then the element $y \in R$ that is assigned to element $x \in D$ by f is denoted $f(x)$ (read "f of x") and is called the *value of f at x*, or the *image of x under f*. See Figure 1.5.1. The elements $x \in D$ and $y \in R$ are called *variables*, with x the *independent variable* and y the *dependent variable*. The set of values of f— that is, the set $f(D) = \{y \in R : y = f(x) \text{ for some } x \in D\}$ — is called the *range of f*, or the *image of D under f*.

Our study of calculus is based on functions whose domain D is a set of real numbers, usually an interval or union of intervals, and whose range is a set of real numbers. In this context, a function f from D into R (R being the set of real numbers) is called a *real-valued function of a real variable*. For the remainder of this chapter, and throughout Chapters 2–13, the term "function" will be interpreted this way.

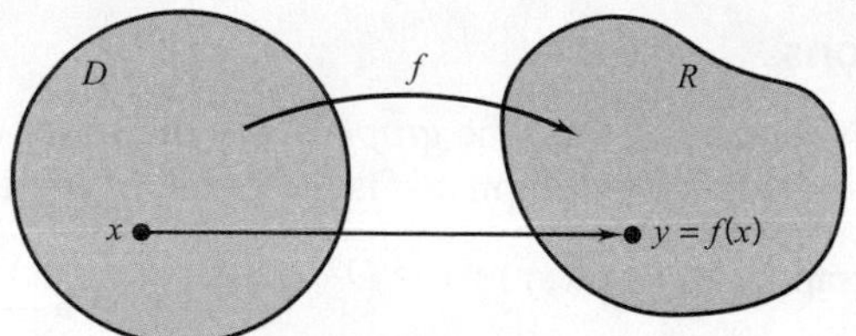

Figure 1.5.1

Here are some examples. We begin with the squaring function

$$f(x) = x^2, \quad \text{for all real numbers } x.$$

The domain of f, denoted dom (f), is given explicitly as the set of real numbers. Numerical values of f can be found by assigning numerical values to x. For example,

$$f(4) = 4^2 = 16, \quad f(-3) = (-3)^2 = 9, \quad f(0) = 0^2 = 0.$$

As x runs through the real numbers, x^2 runs through all the nonnegative numbers. Therefore, the range of f is $[0, \infty)$. In abbreviated form, we can write

$$\text{dom}(f) = (-\infty, \infty) \quad \text{and} \quad \text{range}\,(f) = [0, \infty),$$

and say that f *maps* $(-\infty, \infty)$ *onto* $[0, \infty)$.

Now consider the function

$$g(x) = \sqrt{2x + 4}, \quad x \in [0, 6].$$

The domain of g is given as the closed interval [0,6]. At $x = 0$, g takes on the value 2:

$$g(0) = \sqrt{2 \cdot 0 + 4} = \sqrt{4} = 2,$$

and at $x = 6$, g has the value 4:

$$g(6) = \sqrt{2 \cdot 6 + 4} = \sqrt{16} = 4.$$

As x runs through the numbers from [0, 6], $g(x)$ runs through the numbers from 2 to 4. Thus, the range of g is the closed interval [2, 4]. The function maps [0, 6] onto [2, 4].

A function may be defined *piecewise*. For example, take

$$h(x) = \begin{cases} 2x + 1, & \text{if } x < 0 \\ x^2, & \text{if } x \geq 0. \end{cases}$$

As explicitly stated, the domain of h is the set of real numbers. As you can verify that the range of h is also the set of real numbers. Thus the function h maps $(-\infty, \infty)$ onto $(-\infty, \infty)$.

In most cases, piecewise-defined functions will be used to illustrate mathematical concepts, but they occur in a natural way, too. For example, the *absolute value function* defined by

$$|x| = \begin{cases} x, & \text{if } \ x \geq 0 \\ -x, & \text{if } \ x < 0 \end{cases}$$

is a familiar example of a piecewise-defined function. The domain of the function is $(-\infty, \infty)$ and the range is $[0, \infty)$.

Graphs of Functions

If f is a function with domain D, then the *graph of f* is the set of all points $P\,(x, f(x))$ in the plane where $x \in D$. Thus the graph of f is the graph of the equation $y = f(x)$:

$$\text{graph of } f = \{(x, y) : x \in D \quad \text{and} \quad y = f(x)\}.$$

The most elementary way to sketch the graph of a function is to plot points. We plot enough points so that we can "see" what the graph may look like and then connect the points with a "curve." Of course, if we can identify the curve in advance (for example, a straight line, a parabola, and so on), then it is much easier to draw the graph.

The graph of the squaring function

$$f(x) = x^2, \quad x \in (-\infty, \infty)$$

is the parabola shown in Figure 1.5.2. The points that we plotted are indicated in the table and on the graph. The graph of the function

$$g(x) = \sqrt{2x + 4}, \quad x \in [0, 6]$$

is the arc shown in Figure 1.5.3.

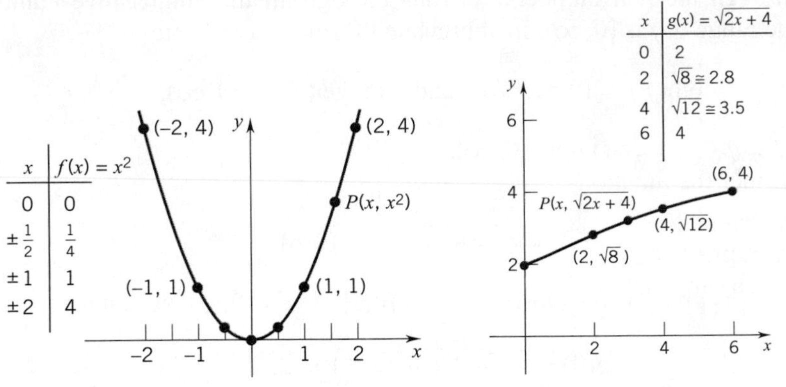

Figure 1.5.2

Figure 1.5.3

The graph of the piecewise-defined function

$$h(x) = \begin{cases} x^2, & x \geq 0 \\ 2x + 1, & x < 0 \end{cases}$$

and the graph of the absolute value function are given in Figures 1.5.4 and 1.5.5.

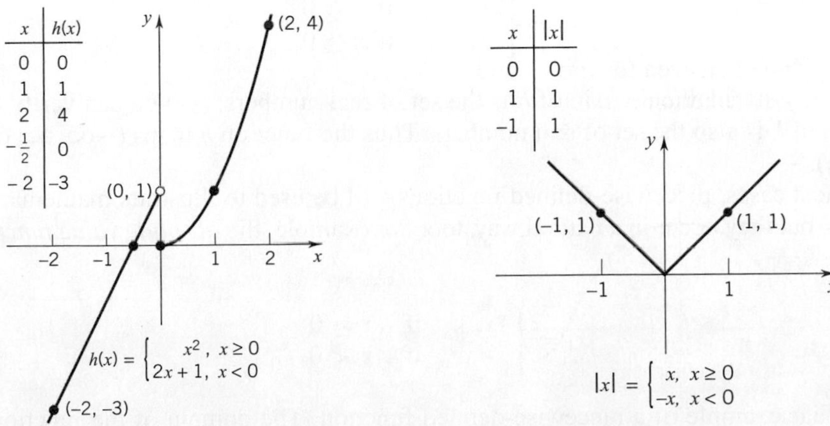

Figure 1.5.4

Figure 1.5.5

In general, the graph of a function is a "curve" in the plane; it is a geometric representation of the function. A graph provides a useful picture of the behavior of a function.

This geometric representation of a function raises the converse question: Under what conditions is a given curve in the plane the graph of a function? Since for each x in the domain of a function f there corresponds one and only one element $y = f(x)$ in the range of f, it follows that no vertical line can intersect the graph of f in more than one point. This observation leads to the so-called *vertical line test:* A curve in the plane is the graph of a function iff no vertical line intersects the curve in more than one point. Thus, the graphs of the circles, ellipses, hyperbolas, and the parabolas $(y - k)^2 = 4p\,(x - c)$ given in Section 1.4 are not the graphs of functions. The curve shown in Figure 1.5.6 is the graph of a function, while the curve shown in Figure 1.5.7 is not.

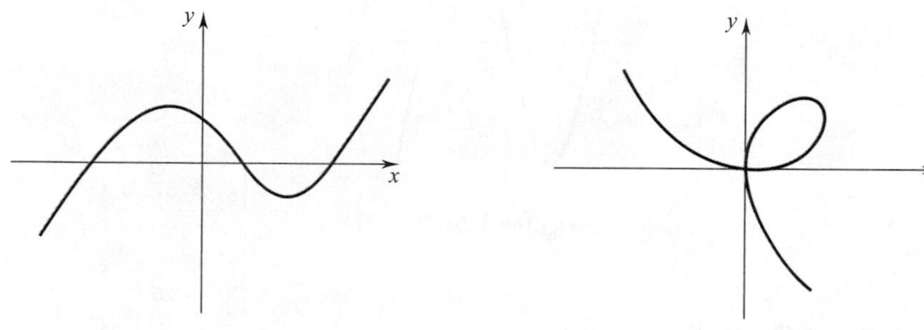

Figure 1.5.6 **Figure 1.5.7**

Graphing calculators and computer algebra systems (CAS) can be valuable aids to sketching the graphs of functions, provided they are used properly. However, these technologies are not a substitute for learning the basic methods. We will not attempt to teach you how to use a graphing calculator or computer software in this text, but technology-based exercises will occur throughout. You are encouraged to use these to enhance your understanding of the mathematical concepts and methods.

Even and Odd Functions; Symmetry

A function f is said to be *even* if

$$f(-x) = f(x) \quad \text{for all} \quad x \in \text{dom}(f).$$

A function f is said to be *odd* if

$$f(-x) = -f(x) \quad \text{for all} \quad x \in \text{dom}(f).$$

The graph of an even function is *symmetric about the y-axis* (Figure 1.5.8), and the graph of an odd function is *symmetric about the origin* (Figure 1.5.9).

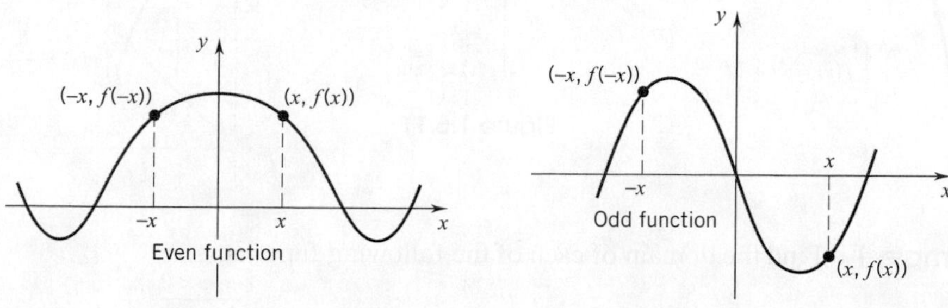

Figure 1.5.8 **Figure 1.5.9**

The squaring function $f(x) = x^2$ is even since

$$f(-x) = (-x)^2 = x^2 = f(x);$$

its graph is symmetric about the y-axis (see Figure 1.5.2). The function $f(x) = 4x - x^3$ is odd since

$$f(-x) = 4(-x) - (-x)^3 = -4x + x^3 = -(4x - x^3) = -f(x).$$

The graph of f is shown in Figure 1.5.10. Note its symmetry with respect to the origin.

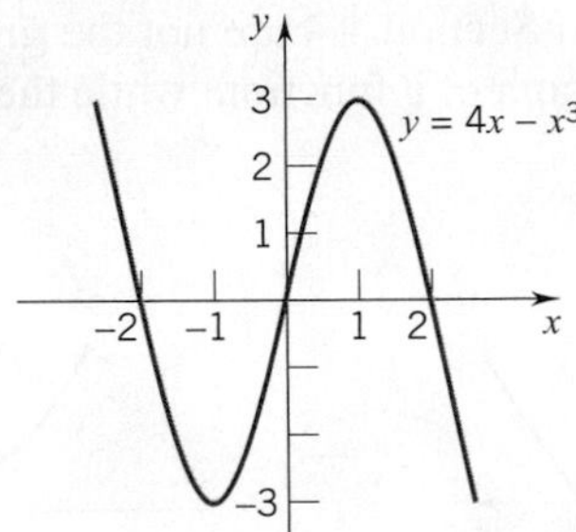

Figure 1.5.10

Convention on Domains

If the domain of a function f is not given explicitly, then by convention we take as domain the set of all real numbers x for which $f(x)$ is a real number. For the function $f(x) = x^3 + 1$, we take as domain the set of real numbers. For $g(x) = \sqrt{x}$, we take as domain the set of nonnegative numbers. For

$$h(x) = \frac{1}{x-2}$$

we take as domain the set of all real numbers $x \neq 2$. In interval notation, we write

$$\text{dom}(f) = (-\infty, \infty), \quad \text{dom}(g) = [0, \infty), \quad \text{and} \quad \text{dom}(h) = (-\infty, 2) \cup (2, \infty).$$

The graphs of the three functions are sketched in Figure 1.5.11.

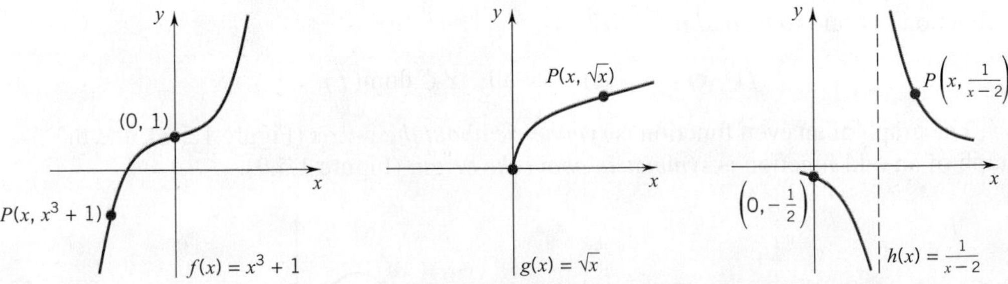

Figure 1.5.11

Example 1 Find the domain of each of the following functions:

$$\textbf{(a)}\ \ f(x) = \frac{x+1}{x^2 + x - 6}, \qquad \textbf{(b)}\ \ g(x) = \frac{\sqrt{4 - x^2}}{x - 1}.$$

SOLUTION **(a)** You can see that $f(x)$ is a real number iff $x^2 + x - 6 \neq 0$. Since

$$x^2 + x - 6 = (x + 3)(x - 2),$$

the domain of f is the set of real numbers other than -3 and 2. This set can be expressed as

$$(-\infty, -3) \cup (-3, 2) \cup (2, \infty).$$

(b) For $g(x)$ to be a real number, we need

$$4 - x^2 \geq 0 \quad \text{and} \quad x \neq 1.$$

Since $4 - x^2 \geq 0$ iff $x^2 \leq 4$ iff $-2 \leq x \leq 2$, the domain of g is the set of all numbers x in $[-2, 2]$ other than $x = 1$. This set can be expressed as the union of two half-open intervals

$$[-2, 1) \cup (1, 2]. \quad \square$$

Example 2 Find the domain and range of the function:

$$f(x) = \frac{1}{\sqrt{2 - x}} + 5.$$

SOLUTION First we look for the domain. Since $\sqrt{2 - x}$ is a real number iff $2 - x \geq 0$, we need $x \leq 2$. But at $x = 2$, $\sqrt{2 - x} = 0$ and its reciprocal is not defined. We must therefore restrict x to $x < 2$. The domain is $(-\infty, 2)$.

Now we look for the range. As x runs through $(-\infty, 2)$, $\sqrt{2 - x}$ takes on all positive values and so does its reciprocal. The range of f is therefore $(5, \infty)$.

The function f maps the interval $(-\infty, 2)$ onto the interval $(5, \infty)$. $\square$

Examples of Applications

Functions are used in a wide variety of disciplines as mathematical models which describe how variable quantities are related. In such cases, the application itself will usually dictate a domain for the function serving as the model.

Example 3 U.S. Postal Service regulations require that the length plus the girth (the perimeter of a cross section) of a package for mailing cannot exceed 108 inches. A rectangular box with a square end is designed to meet the regulation exactly (see Figure 1.5.12). Express the volume V of the box as a function of the edge length of the square end and give the domain of the function.

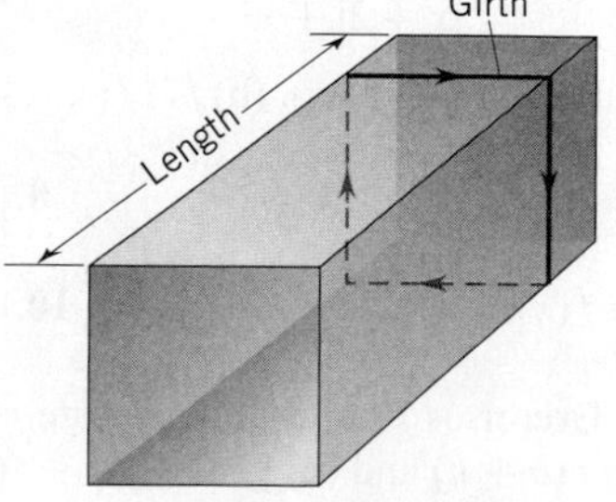

Figure 1.5.12

SOLUTION Let x denote the edge length of the square end and let h denote the length of the box. The girth is the perimeter of the square, or $4x$. Since the box meets the regulations exactly,

$$4x + h = 108 \quad \text{and therefore} \quad h = 108 - 4x.$$

The volume of the box is given by $V = x^2 h$ and so it follows that

$$V(x) = x^2(108 - 4x) = 108x^2 - 4x^3.$$

Since neither the edge length of the square end nor the length of the box can be negative, we have

$$x \geq 0 \quad \text{and} \quad h = 108 - 4x \geq 0.$$

The second condition requires $x \leq 27$. The full requirement on x, $0 \leq x \leq 27$, gives $\text{dom}(V) = [0, 27]$. ❑

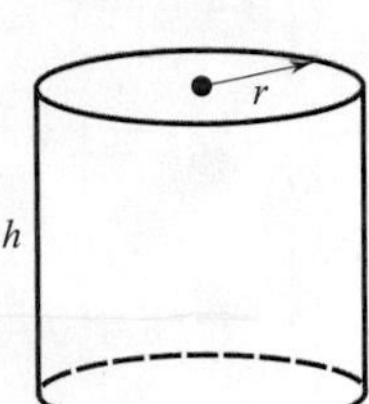

Figure 1.5.13

Example 4 A soft-drink manufacturer wants to fabricate cylindrical cans for its product (see Figure 1.5.13). The can is to have a volume of 12 fluid ounces, which is approximately 22 cubic inches. Express the total surface area S of the can as a function of its radius and give the domain of the function.

SOLUTION Let r be the radius of the can and h the height. The total surface area (top, bottom, and lateral area) of a right circular cylinder is given by

$$S = 2\pi r^2 + 2\pi rh.$$

Since the volume $V = \pi r^2 h$ must be 22 cubic inches, we have

$$\pi r^2 h = 22 \quad \text{and} \quad h = \left(\frac{22}{\pi r^2}\right)$$

and therefore

$$S(r) = 2\pi r^2 + 2\pi r\left(\frac{22}{\pi r^2}\right) = 2\pi r^2 + \frac{44}{r}.$$

Since r can take on any positive value, dom $(S) = (0, \infty)$ ❑

EXERCISES 1.5

Calculate (a) $f(0)$, (b) $f(1)$, (c) $f(-2)$, (d) $f(3/2)$.

1. $f(x) = 2x^2 - 3x + 2$.

2. $f(x) = \dfrac{2x - 1}{x^2 + 4}$.

3. $f(x) = \sqrt{x^2 + 2x}$.

4. $f(x) = |x + 3| - 5x$.

5. $f(x) = \dfrac{2x}{|x + 2| + x^2}$.

6. $f(x) = 1 - \dfrac{1}{(x + 1)^2}$.

Calculate (a) $f(-x)$, (b) $f(1/x)$, (c) $f(a + b)$.

7. $f(x) = x^2 - 2x$.

8. $f(x) = \dfrac{x}{x^2 + 1}$.

9. $f(x) = \sqrt{1 + x^2}$.

10. $f(x) = \dfrac{x}{|x^2 - 1|}$.

In Exercises 11 and 12, calculate (a) $f(a + h)$ and (b) $[f(a + h) - f(a)]/h$, $h \neq 0$.

11. $f(x) = 2x^2 - 3x$.

12. $f(x) = \dfrac{1}{x - 2}$.

Find the number(s) x, if any, where f takes on the value 1.

13. $f(x) = |2 - x|$.

14. $f(x) = \sqrt{1 + x}$.

15. $f(x) = x^2 + 4x + 5$.

16. $f(x) = 4 + 10x - x^2$.

17. $f(x) = \dfrac{2}{\sqrt{x^2 - 5}}$.

18. $f(x) = \dfrac{x}{|x|}$.

Find the domain and range of the function.

19. $f(x) = |x|$.

20. $g(x) = x^2 - 1$.

21. $f(x) = 2x - 3$.

22. $g(x) = \sqrt{x} + 5$.

23. $f(x) = \dfrac{1}{x^2}$.

24. $g(x) = \dfrac{4}{x}$.

25. $f(x) = \sqrt{1 - x}$.

26. $g(x) = \sqrt{x - 3}$.

27. $f(x) = \sqrt{7 - x} - 1$.

28. $g(x) = \sqrt{x - 1} - 1$.

29. $f(x) = \dfrac{1}{\sqrt{2 - x}}$.

30. $g(x) = \dfrac{1}{\sqrt{4 - x^2}}$.

Find the domain and sketch the graph of the function.

31. $f(x) = 1$.

32. $f(x) = -1$.

33. $f(x) = 2x$.

34. $f(x) = 2x + 1$.

35. $f(x) = \frac{1}{2}x + 2$.

36. $f(x) = -\frac{1}{2}x - 3$.

37. $f(x) = \sqrt{4 - x^2}$.

38. $f(x) = \sqrt{9 - x^2}$.

39. $f(x) = x^2 - x - 6$.

40. $f(x) = |x - 1|$.

Sketch the graph and specify the domain and range of the given function.

41. $f(x) = \begin{cases} -1, & x < 0 \\ 1, & x > 0. \end{cases}$

42. $f(x) = \begin{cases} x^2, & x \leq 0 \\ 1 - x, & x > 0. \end{cases}$

43. $f(x) = \begin{cases} 1 + x, & 0 \leq x \leq 1 \\ x, & 1 < x < 2 \\ \frac{1}{2}x + 1, & 2 \leq x. \end{cases}$

44. $f(x) = \begin{cases} x^2, & x < 0 \\ -1, & 0 < x < 2 \\ x, & 2 < x. \end{cases}$

In Exercises 45–48, state whether the curve is the graph of a function. If it is, find the domain and the range.

45.

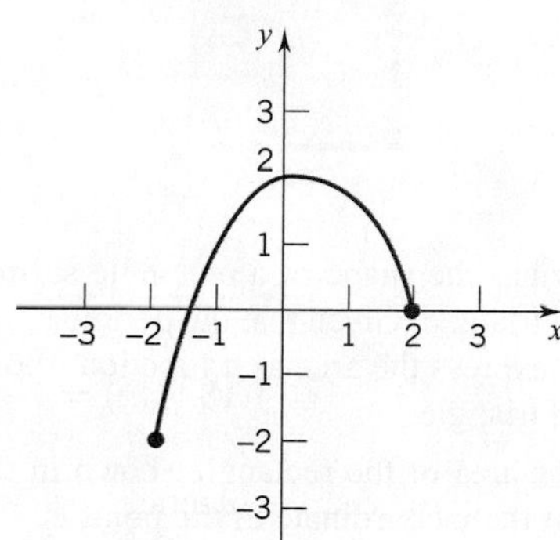

46.

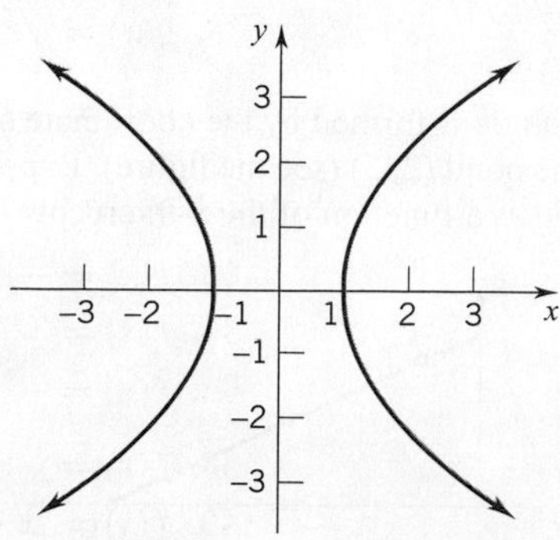

47.

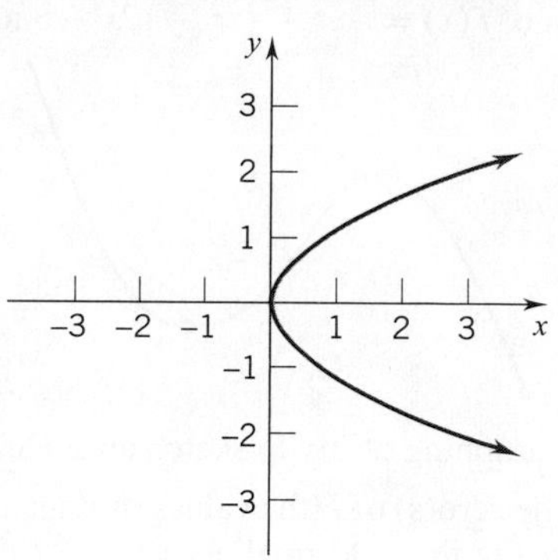

48.

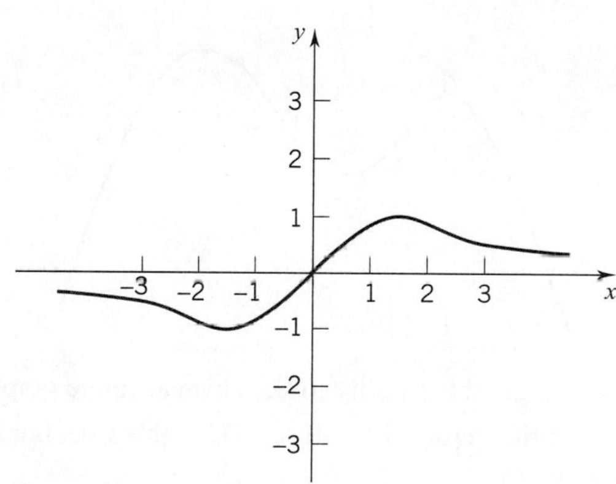

State whether the function is odd, even, or neither.

49. $f(x) = x^3$.

50. $f(x) = x^2 + 1$.

51. $g(x) = x(x - 1)$.

52. $g(x) = x(x^2 + 1)$.

53. $f(x) = \dfrac{x^2}{1 - |x|}$.

54. $F(x) = x + \dfrac{1}{x}$.

C 55. Use a CAS to find $f(-x) - f(x)$ and $f(-x) + f(x)$ for each of the following functions:

(a) $f(x) = x^4 - 4x^2 + 4$;

(b) $f(x) = \dfrac{x}{x^2 - 9}$;

(c) $f(x) = -x^6 + x^3 + 1$.

Identify these functions as even, odd, or neither. Does CAS provide *all* the simplification necessary to make these decisions easy?

C 56. Repeat Exercise 55 for the functions:

(a) $f(x) = \sqrt[5]{x - x^3}$;

(b) $\dfrac{x^4 - 16}{x^2}$;

(c) $f(x) = 3 - 6x^3 + x^5$.

57. The graph of $f(x) = \frac{1}{3}x^3 + \frac{1}{2}x^2 - 12x - 6$ looks like

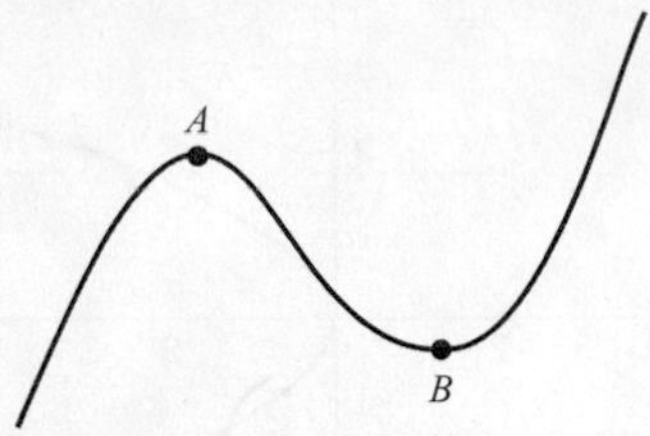

(a) Use a graphing utility to sketch an accurate graph of f.

(b) Find the zero(s) of f (the values of x such that $f(x) = 0$) accurate to three decimal places.

(c) Find the coordinates of the points marked A and B, accurate to three decimal places.

58. The graph of $f(x) = -x^4 + 8x^2 + x - 1$ looks like

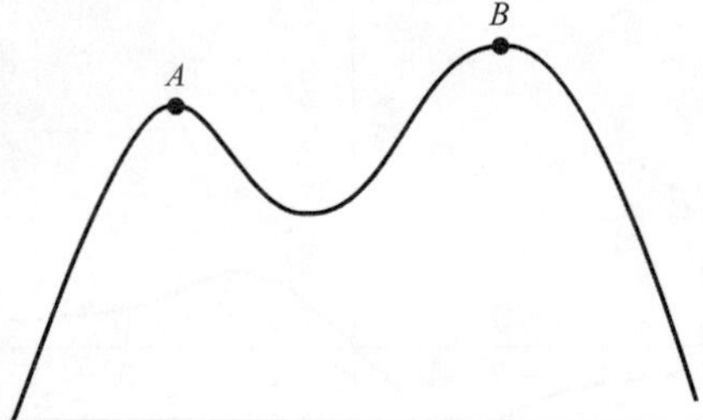

(a) Use a graphing utility to sketch an accurate graph of f.

(b) Find the zero(s) of f, if any. Use three decimal place accuracy.

(c) Find the coordinates of the points marked A and B, accurate to three decimal places.

In Exercises 59 and 60, use a graphing utility to draw several views of the graph of the function f. Select the one that most accurately shows the important features of the graph and include the ranges on the variables x and y in your answer.

59. $f(x) = |x^3 - 3x^2 - 24x + 4|$.

60. $f(x) = \sqrt{x^3 - 8}$.

61. Use a CAS to solve $y = x^2 - 4x - 5$ for x. Use the result to determine the range of this function.

62. Repeat Exercise 60 for the function $y = \dfrac{2x}{4 - x}$.

In Exercises 63–69, determine a formula for the function that is described.

63. Express the area of a circle as a function of its circumference.

64. Express the volume of a sphere as a function of its surface area.

65. Express the volume of a cube as a function of the area of one of its faces.

66. Express the volume of a cube as a function of its total surface area.

67. Express the surface area of a cube as a function of the length of a diagonal of one face.

68. Express the volume of a cube as a function of one of its diagonals.

69. Express the area of an equilateral triangle as a function of the length of a side.

70. A right triangle with hypotenuse c is rotated about one of its legs to form a cone (see the figure). Given that x is the length of the other leg, express the volume of the cone as a function of x.

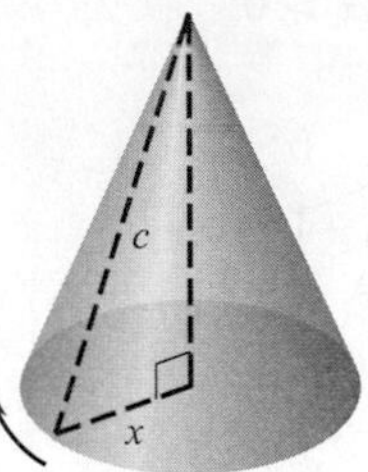

71. A Norman window is a window in the shape of a rectangle surmounted by a semicircle (see the figure). Given that the perimeter of the window is 15 feet, express the area as a function of its width x.

72. A window has the shape of a rectangle surmounted by an equilateral triangle. Given that the perimeter of the window is 15 feet, express the area as a function of one side of the equilateral triangle.

73. Express the area of the rectangle shown in the figure as a function of the x-coordinate of the point P.

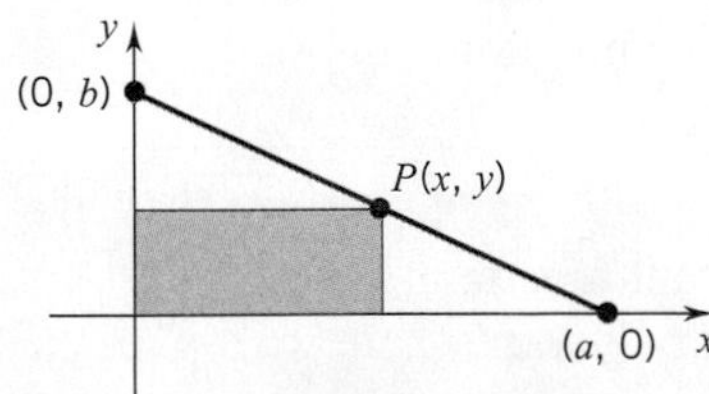

74. A right triangle is formed by the coordinate axes and a line through the point (2, 5) (see the figure). Express the area of the triangle as a function of the x-intercept.

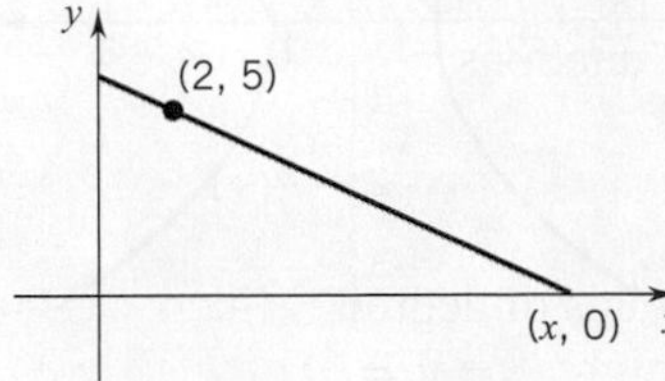

75. A string 28 inches long is to be cut into two pieces, one piece to form a square and the other to form a circle. Express the total area enclosed by the square and circle as a function of the perimeter of the square.

76. A tank in the shape of an inverted cone is being filled with water (see the figure). Express the volume of water in the tank as a function of its height.

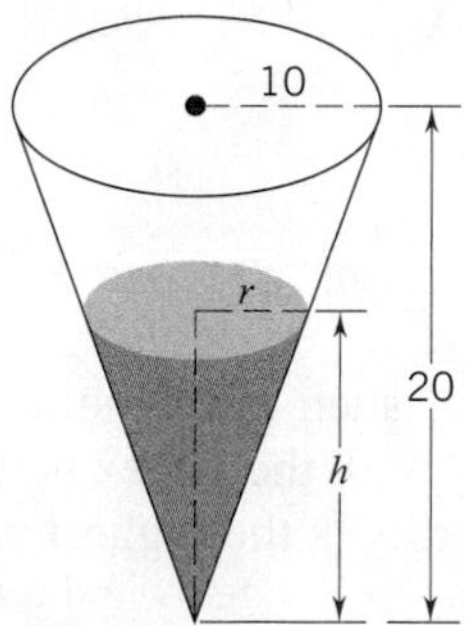

77. Suppose that a cylindrical mailing container exactly meets the U.S. Postal Service regulations given in Example 3 (see the figure). Express the volume of the container as a function of the radius of the end.

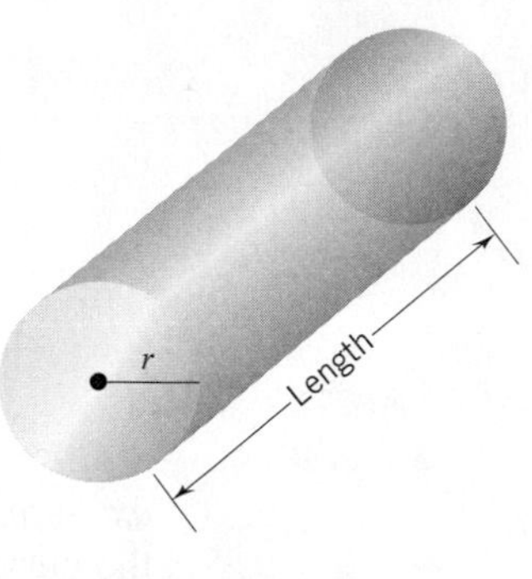

■ 1.6 THE ELEMENTARY FUNCTIONS

The functions that figure most prominently in single-variable calculus are the polynomials, the rational functions, the trigonometric functions, the exponential functions, and the logarithm functions. These functions are generally known as the *elementary functions*. Here we review polynomials, rational functions, and trigonometric functions. Exponential and logarithm functions are introduced in Chapter 7.

Polynomials

We begin with a nonnegative integer n. A function of the form

$$P(x) = a_n x^n + a_{n-1}x^{n-1} + \cdots + a_1 x + a_0 \quad \text{for all real } x,$$

where the *coefficients* $a_n, a_{n-1}, \ldots, a_1, a_0$ are real numbers and $a_n \neq 0$, is called a (*real*) *polynomial* of *degree n*.

If $n = 0$, the polynomial is simply a constant function:

$$P(x) = a_0 \quad \text{for all real } x.$$

Nonzero constant functions are polynomials of degree 0. The function $P(x) = 0$ for all real x is also a polynomial, but we assign no degree to it.

Polynomials satisfy a condition known as the *factor theorem*: If P is a polynomial and r is a real number, then

$$x - r \quad \text{is a factor of} \quad P(x) \quad \text{iff} \quad P(r) = 0.$$

The real numbers r at which $P(r) = 0$ are called the *zeros* of the polynomial.

The linear functions

$$P(x) = ax + b, \quad a \neq 0$$

are the polynomials of degree 1. Such a polynomial has only one zero: $r = -b/a$. The graph is the straight line $y = ax + b$. (See Figure 1.6.1.)

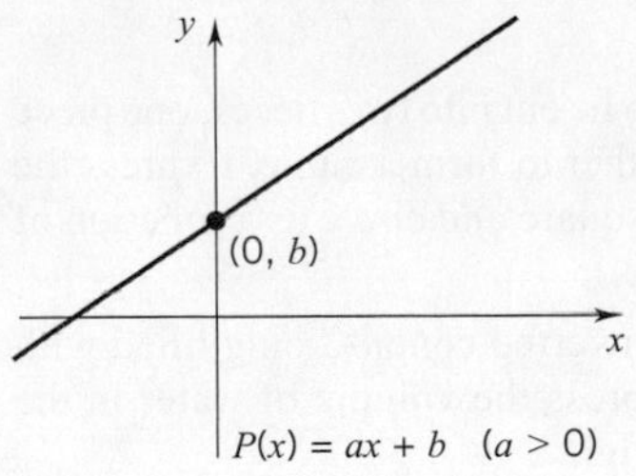

Figure 1.6.1

The quadratic functions

$$P(x) = ax^2 + bx + c, \quad a \neq 0$$

are the polynomials of degree 2. The graph of such a polynomial is a curve $y = ax^2 + bx + c$, a parabola with vertical axis. If $a > 0$, the vertex is the lowest point on the curve; the curve opens up. If $a < 0$, the vertex is the highest point on the curve; the curve opens down. (See Figure 1.6.2.)

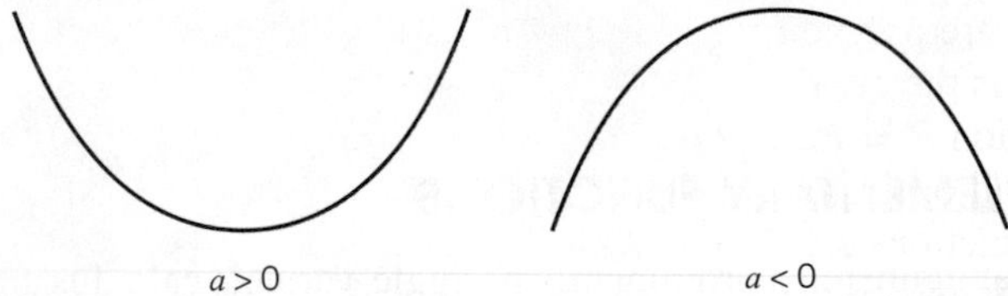

Figure 1.6.2

The zeros of the quadratic function $P(x) = ax^2 + bx + c$ are the roots of the quadratic equation

$$ax^2 + bx + c = 0.$$

The three possibilities for the roots are depicted in Figure 1.6.3 for the case $a > 0$.

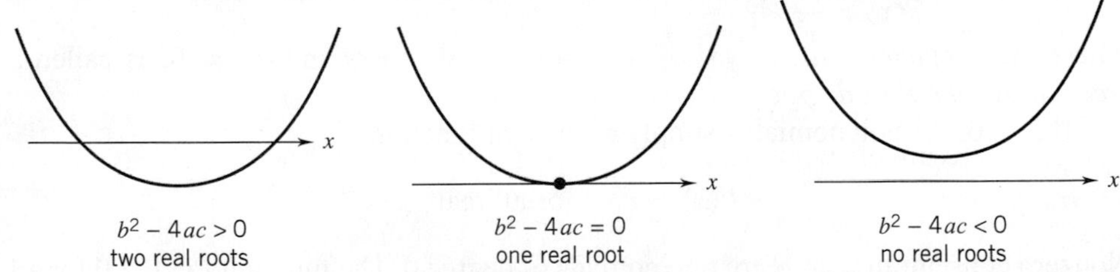

Figure 1.6.3

Polynomials of degree 3 have the form $P(x) = ax^3 + bx^2 + cx + d, a \neq 0$. These functions are called *cubics*. In general, the graph of a cubic has one of the two following shapes, again determined by the sign of a (Figure 1.6.4). Note that we have not tried to locate these graphs with respect to the coordinate axes. Our purpose here is simply to indicate the two typical shapes. You can see, however, that for a cubic there are three possibilities: 3 real roots, 2 real roots, or 1 real root.

Polynomials become more complicated as the degree increases. In Chapter 4 we show how to use calculus to analyze polynomials of higher degree and to sketch their graphs.

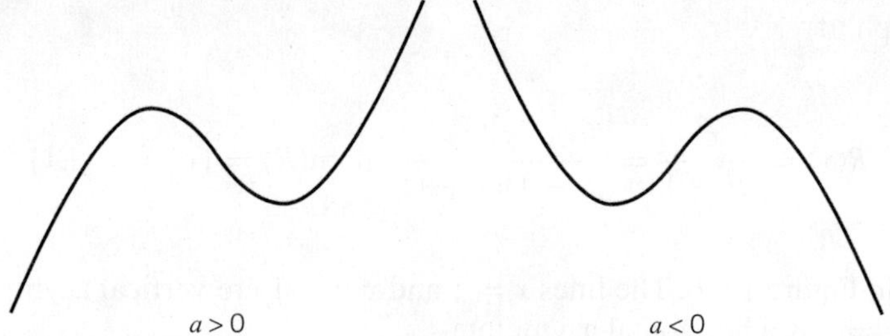

Figure 1.6.4

Rational Functions

A *rational function* is a function of the form

$$R(x) = \frac{P(x)}{Q(x)},$$

where P and Q are polynomials. Note that every polynomial P is a rational function: $P(x) = P(x)/1$ is the quotient of two polynomials. Since division by 0 is not allowed, a rational function $R = P/Q$ is not defined at those points x (if any) where $Q(x) = 0$; R is defined at all other points. Thus, $\text{dom}(R) = \{x : Q(x) \neq 0\}$.

Rational functions are more difficult to analyze and more difficult to graph than polynomials. For example, it is necessary to analyze the behavior of a rational function $R = P/Q$ in the vicinity of a zero of Q, as well as the behavior of R for large values of x, both positive and negative. If P and Q have no common factors (P/Q is in "lowest terms"), then the zeros of Q correspond to *vertical asymptotes* of the graph of R; the existence of *horizontal asymptotes* depends on the behavior of R for large positive and large negative values of x (that is, as $x \to \pm\infty$). Vertical and horizontal asymptotes will be studied in detail in Chapter 4.

A sketch of the graph of

$$R(x) = \frac{1}{x^2 - 4x + 4} = \frac{1}{(x-2)^2}; \ \text{dom}(R) = \{x : x \neq 2\}$$

is shown in Figure 1.6.5. Note that $x = 2$ is a vertical asymptote and the x-axis is a horizontal asymptote.

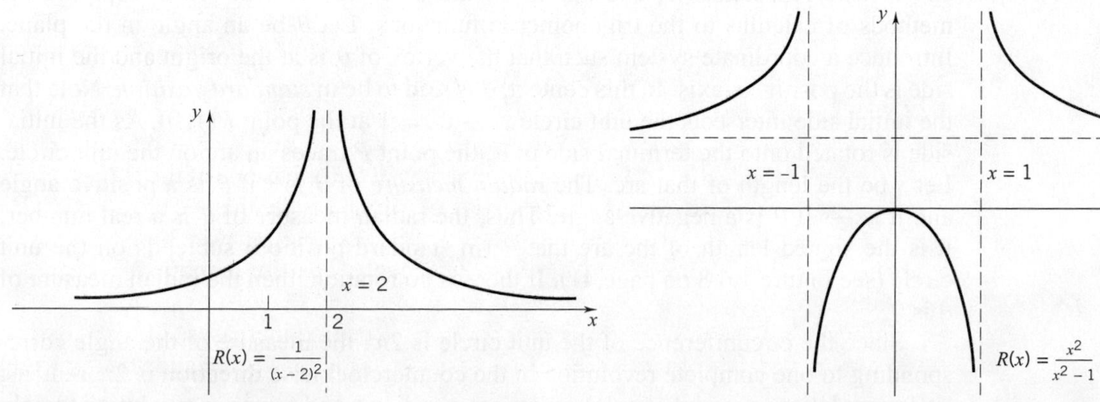

Figure 1.6.5

Figure 1.6.6

The graph of

$$R(x) = \frac{x^2}{x^2 - 1} = \frac{x^2}{(x-1)(x+1)}; \quad \text{dom}(R) = \{x : x \neq \pm 1\}$$

is sketched in Figure 1.6.6. The lines $x = 1$ and $x = -1$ are vertical asymptotes of the graph, and $y = 1$ is a horizontal asymptote.

Trigonometric Functions

We assume that the reader is already familiar with trigonometry, and so we present here a brief review of the definitions, basic properties, and graphs of the trigonometric functions.

Angles An angle in the plane is generated by rotating a ray (half-line) about its endpoint. The starting position of the ray is called *initial side* of the angle, and the final position of the ray is called the *terminal side*. The point of intersection of the initial and terminal sides is called the *vertex*. An angle is said to be *positive* if it is generated by a counterclockwise rotation and *negative* if the rotation is clockwise (see Figure 1.6.7).

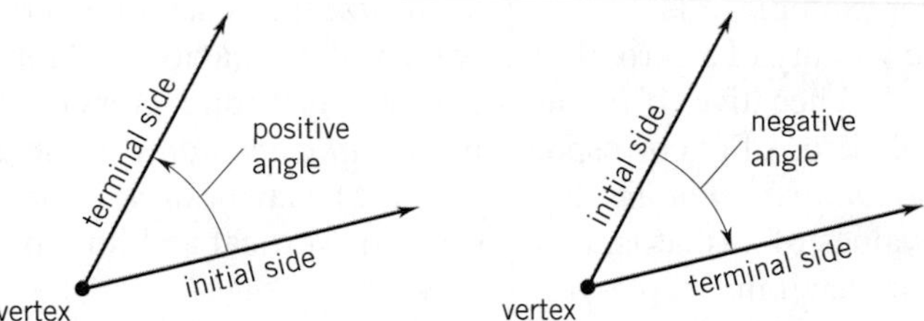

Figure 1.6.7

Degree measure, traditionally used to measure angles, has a serious drawback. It is artificial; there is no intrinsic connection between a degree and the geometry of a circle. Why choose 360° for one complete revolution? Why not 100°? or 400°?

There is another way to measure angles, called *radian measure*, that associates angles with real numbers, and this is the measure that we will use in applying the methods of calculus to the trigonometric functions. Let θ be an angle in the plane. Introduce a coordinate system such that the vertex of θ is at the origin and the initial side is the positive x-axis. In this context, θ is said to be in *standard position*. Note that the initial side intersects the unit circle $x^2 + y^2 = 1$ at the point $P(1, 0)$. As the initial side is rotated onto the terminal side of θ, the point P traces an arc on the unit circle. Let s be the length of that arc. The *radian measure* of θ is s if θ is a positive angle and it is $-s$ if θ is a negative angle. Thus, the radian measure of θ is a real number; it is the signed length of the arc that θ (in standard position) subtends on the unit circle (see Figure 1.6.8 on page 41). If there is no rotation, then the radian measure of θ is 0.

Since the circumference of the unit circle is 2π, the measure of the angle corresponding to one complete revolution in the counterclockwise direction is 2π radians; half a revolution (a straight angle) comprises π radians; and a quarter revolution (a right angle) comprises $\pi/2$ radians (see Figure 1.6.9 on page 41).

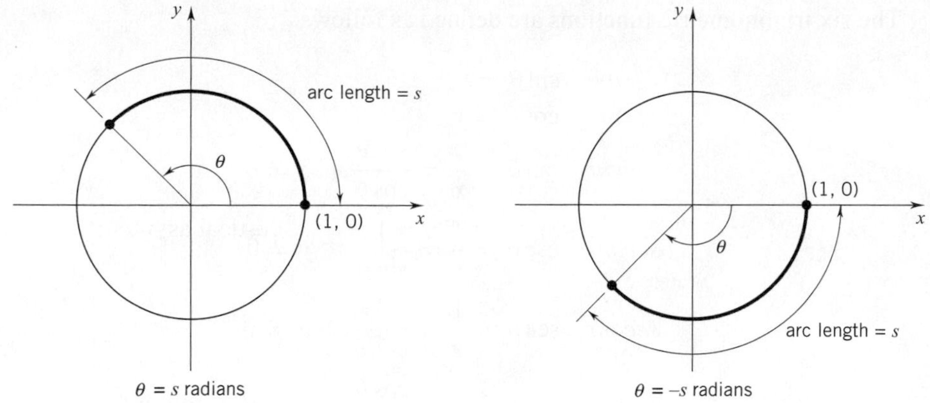

Figure 1.6.8

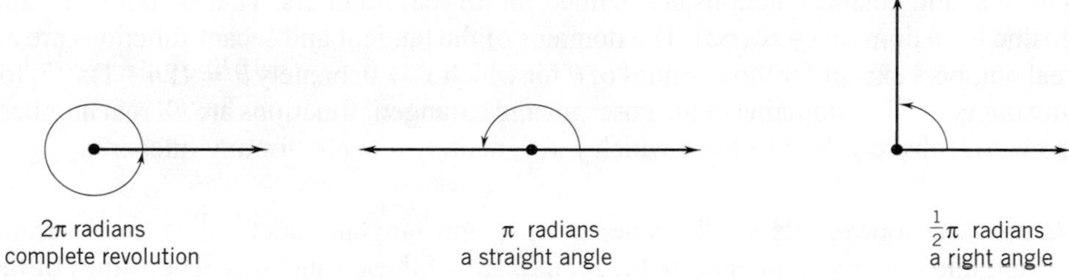

Figure 1.6.9

The conversion of degrees to radians and vice versa is made by noting that if A is the measure of an angle in degrees and x is the measure of the same angle in radians, then

$$\frac{A}{360} = \frac{x}{2\pi}.$$

In particular,

$$1 \text{ radian} = \frac{360}{2\pi}\text{degrees} \cong 57.30°$$

and
$$1° = \frac{2\pi}{360}\text{radian} \cong 0.0175 \text{ radian}.$$

The following table gives some common angles measured both in degrees and in radians.

degrees	0°	30°	45°	60°	90°	120°	135°	150°	180°	270°	360°
radians	0	$\pi/6$	$\pi/4$	$\pi/3$	$\pi/2$	$2\pi/3$	$3\pi/4$	$5\pi/6$	π	$3\pi/2$	2π

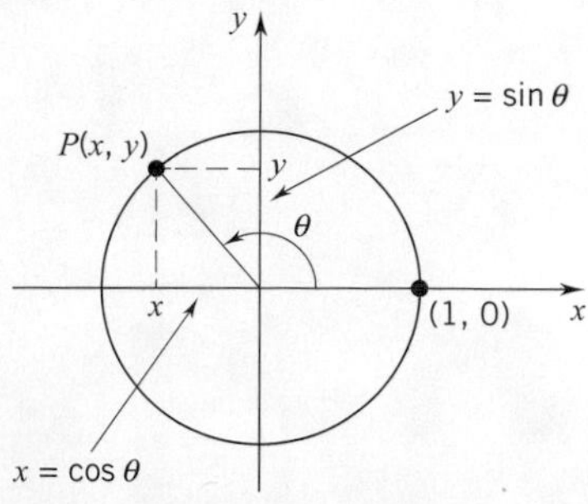

Figure 1.6.10

Definition of the Trigonometric Functions We start with the unit circle and use radian measure.

Let θ be an angle in standard position and let $P(x, y)$ be the point of intersection of the terminal side of θ and the unit circle (see Figure1.6.10).

The six trigonometric functions are defined as follows:

$$
\begin{aligned}
&\textit{sine} && \sin\theta = y \\
&\textit{cosine} && \cos\theta = x \\
&\textit{tangent} && \tan\theta = \frac{y}{x} = \frac{\sin\theta}{\cos\theta}, \quad x \neq 0 \\
&\textit{cosecant} && \csc\theta = \frac{1}{y} = \frac{1}{\sin\theta}, \quad y \neq 0 \\
&\textit{secant} && \sec\theta = \frac{1}{x} = \frac{1}{\cos\theta}, \quad x \neq 0 \\
&\textit{cotangent} && \cot\theta = \frac{x}{y} = \frac{\cos\theta}{\sin\theta}, \quad y \neq 0
\end{aligned}
$$

The sine and cosine functions are defined for all real numbers. That is, both sine and cosine have domain $(-\infty, \infty)$. The domains of the tangent and secant functions are all real numbers except for those values of θ for which $x = 0$, namely $\theta = (2n+1)\pi/2$, for any integer n. The domains of the cosecant and cotangent functions are all real numbers except for those values of θ for which $y = 0$, namely $\theta = n\pi$, for any integer n.

Values We concentrate on the sine, cosine, and tangent functions since cosecant, secant, and cotangent are simply the reciprocals of these functions. The values of the trigonometric functions for the angles $0\,(=0°)$, $\pi/6\,(=30°)$, $\pi/4\,(=45°)$, $\pi/3\,(=60°)$, $\pi/2\,(=90°)$, and selected multiples up to 2π, are given in the following table. You should already have these values memorized.

	0	$\pi/6$	$\pi/4$	$\pi/3$	$\pi/2$	$2\pi/3$	$3\pi/4$	$5\pi/6$	π	$3\pi/2$	2π
$\sin\theta$	0	$\frac{1}{2}$	$\frac{\sqrt{2}}{2}$	$\frac{\sqrt{3}}{2}$	1	$\frac{\sqrt{3}}{2}$	$\frac{\sqrt{2}}{2}$	$\frac{1}{2}$	0	-1	0
$\cos\theta$	1	$\frac{\sqrt{3}}{2}$	$\frac{\sqrt{2}}{2}$	$\frac{1}{2}$	0	$-\frac{1}{2}$	$-\frac{\sqrt{2}}{2}$	$-\frac{\sqrt{3}}{2}$	-1	0	1
$\tan\theta$	0	$\frac{\sqrt{3}}{3}$	1	$\sqrt{3}$	Undefined	$-\sqrt{3}$	-1	$-\frac{\sqrt{3}}{3}$	0	Undefined	0

In general, the (approximate) values of the trigonometric functions for any angle θ can be obtained with a hand calculator or from a table of values.

Periodicity A function f is *periodic* if there is a number $p, p \neq 0$, such that $f(x+p) = f(x)$ whenever x and $x+p$ are in the domain of f. The smallest positive number with this property is called the *period* of f. The definitions of the sine and cosine functions imply that they are periodic with period 2π. That is,

$$\sin(\theta + 2\pi) = \sin\theta \quad \text{and} \quad \cos(\theta + 2\pi) = \cos\theta \quad \text{for all } \theta.$$

It follows that the secant and cosecant functions are also periodic with period 2π. It can be verified that the tangent and cotangent functions have period π.

Graphs The graphs of the six trigonometric functions are shown in Figure 1.6.11. Since we normally use x and y to represent the independent and dependent variables when drawing a graph, we follow that convention here, replacing θ by x and writing $y = \sin x, y = \cos x$, and so forth.

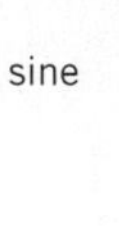

sine

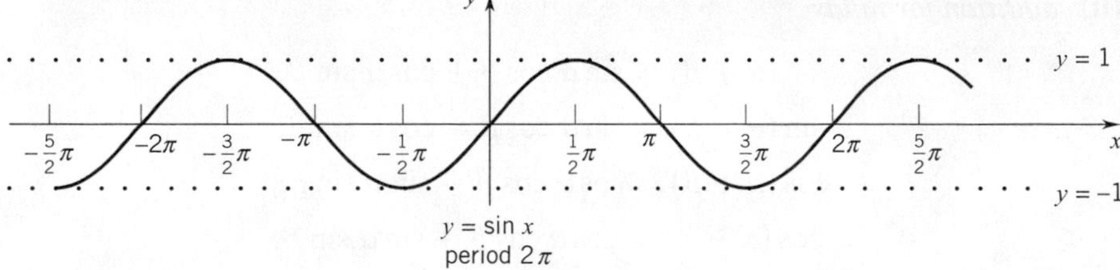

cosine

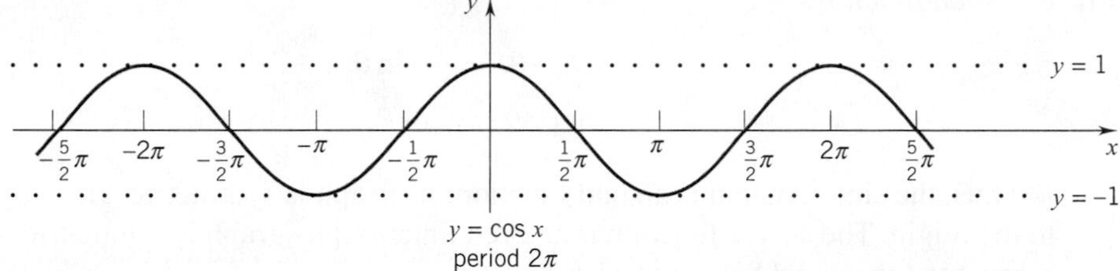

tangent

cotangent

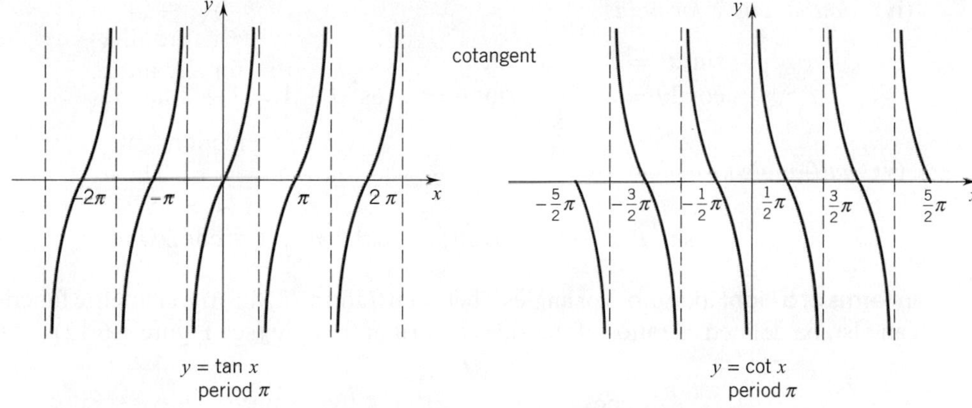

secant

cosecant

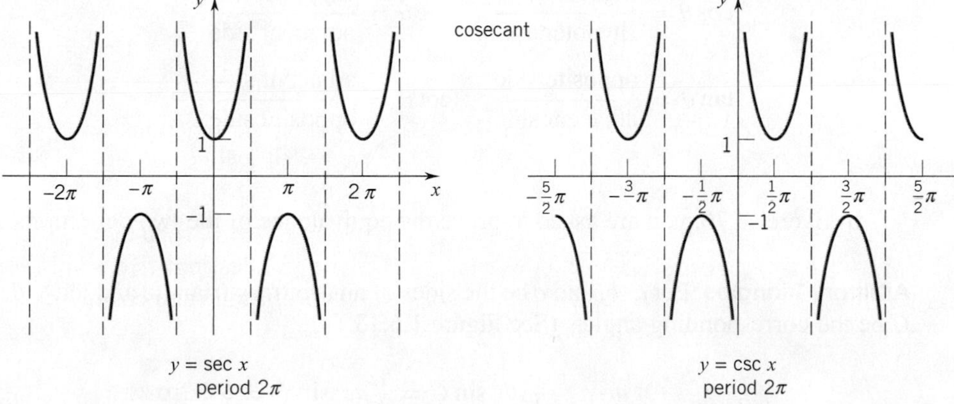

Figure 1.6.11

Important Identities

(i) *unit circle*

$$\sin^2\theta + \cos^2\theta = 1, \quad \tan^2\theta + 1 = \sec^2\theta, \quad 1 + \cot^2\theta = \csc^2\theta.$$

(ii) *addition formulas*

$$\begin{aligned}\sin(\alpha+\beta) &= \sin\alpha\cos\beta + \cos\alpha\sin\beta,\\ \sin(\alpha-\beta) &= \sin\alpha\cos\beta - \cos\alpha\sin\beta,\\ \cos(\alpha+\beta) &= \cos\alpha\cos\beta - \sin\alpha\sin\beta,\\ \cos(\alpha-\beta) &= \cos\alpha\cos\beta + \sin\alpha\sin\beta.\end{aligned}$$

(iii) *even/odd functions*

$$\begin{aligned}\sin(-\theta) &= -\sin\theta,\\ \cos(-\theta) &= \cos\theta.\end{aligned}$$

Thus, the sine function is an odd function; its graph is symmetric with respect to the origin. The cosine function is an even function; its graph is symmetric with respect to the y-axis. See Figure 1.6.11.

(iv) *double-angle formulas*

$$\begin{aligned}&\sin 2\theta = 2\sin\theta\cos\theta,\\ &\cos 2\theta = \cos^2\theta - \sin^2\theta = 2\cos^2\theta - 1 = 1 - 2\sin^2\theta.\end{aligned}$$

(v) *half-angle formulas*

$$\sin^2\theta = \tfrac{1}{2}(1-\cos 2\theta), \quad \cos^2\theta = \tfrac{1}{2}(1+\cos 2\theta).$$

In Terms of a Right Triangle For angles θ between 0 and $\pi/2$, the trigonometric functions can also be defined as ratios of the sides of a right triangle (see Figure 1.6.12).

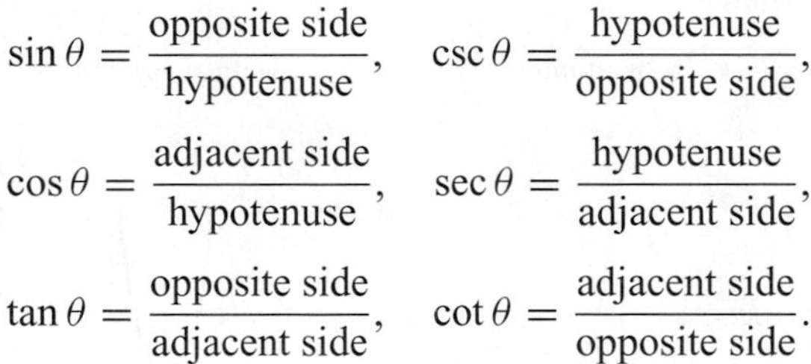

$$\begin{aligned}\sin\theta &= \frac{\text{opposite side}}{\text{hypotenuse}}, & \csc\theta &= \frac{\text{hypotenuse}}{\text{opposite side}},\\ \cos\theta &= \frac{\text{adjacent side}}{\text{hypotenuse}}, & \sec\theta &= \frac{\text{hypotenuse}}{\text{adjacent side}},\\ \tan\theta &= \frac{\text{opposite side}}{\text{adjacent side}}, & \cot\theta &= \frac{\text{adjacent side}}{\text{opposite side}}.\end{aligned}$$

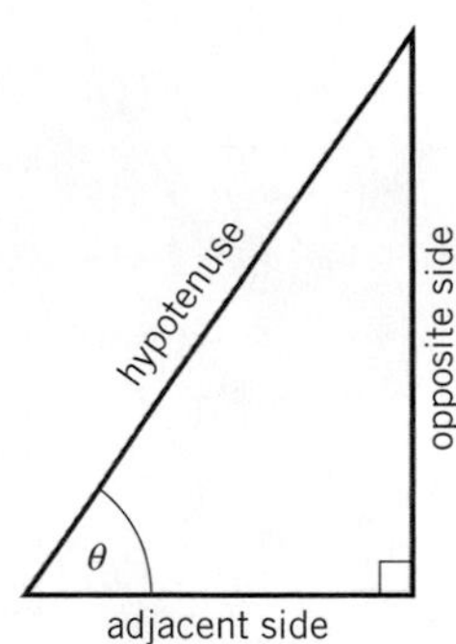

Figure 1.6.12

In Exercise 76, you are asked to prove the equivalence of the two definitions.

Arbitrary Triangles Let a, b, and c be the sides of an arbitrary triangle and let A, B, and C be the corresponding angles. (See Figure 1.6.13.)

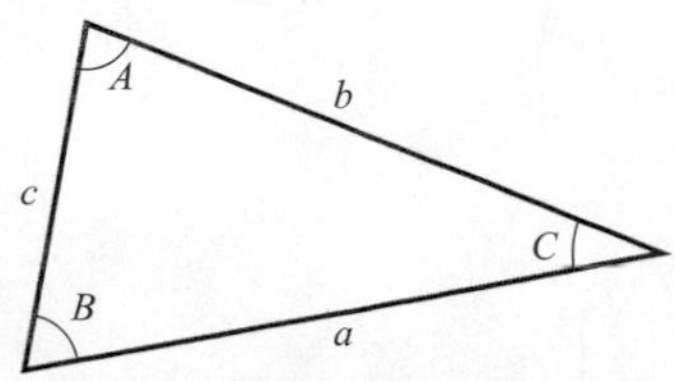

Figure 1.6.13

area $\quad \tfrac{1}{2}\,ab\,\sin C = \tfrac{1}{2}\,ac\,\sin B = \tfrac{1}{2}\,bc\sin A.$

law of sines $\quad \dfrac{\sin A}{a} = \dfrac{\sin B}{b} = \dfrac{\sin C}{c}.$

law of cosines

$$\begin{aligned}a^2 &= b^2 + c^2 - 2bc\,\cos A,\\ b^2 &= a^2 + c^2 - 2ac\,\cos B,\\ c^2 &= a^2 + b^2 - 2ab\,\cos C.\end{aligned}$$

EXERCISES 1.6

State whether the function is a polynomial, a rational function (but not a polynomial), or neither a polynomial nor a rational function. If the function is a polynomial, give the degree.

1. $f(x) = 3.$

2. $f(x) = 1 + \frac{1}{2}x.$

3. $g(x) = \dfrac{1}{x}.$

4. $h(x) = \dfrac{x^2 - 4}{\sqrt{2}}.$

5. $F(x) = \dfrac{x^3 - 3x^{3/2} + 2x}{x^2 - 1}.$

6. $f(x) = 5x^4 - \pi x^2 + \frac{1}{2}.$

7. $f(x) = \sqrt{x}(\sqrt{x} + 1).$

8. $g(x) = \dfrac{x^2 - 2x - 8}{x + 2}.$

9. $f(x) = \dfrac{\sqrt{x^2 + 1}}{x^2 - 1}.$

10. $h(x) = \dfrac{\sin^2 x + \cos^2 x}{x^3 + 8}.$

Determine the domain of the function and sketch the graph.

11. $f(x) = 3x - \frac{1}{2}.$

12. $f(x) = \dfrac{1}{x + 1}.$

13. $g(x) = x^2 - x - 6.$

14. $F(x) = x^3 - x.$

15. $f(x) = \dfrac{1}{x^2 - 4}.$

16. $g(x) = x + \dfrac{1}{x}.$

Convert the degree measure into radian measure.

17. $225°.$

18. $-210°.$

19. $-300°.$

20. $450°.$

21. $15°.$

22. $3°.$

Convert the radian measure into degree measure.

23. $-\dfrac{3\pi}{2}.$

24. $\dfrac{5\pi}{4}.$

25. $\dfrac{5\pi}{3}.$

26. $-\dfrac{11\pi}{6}.$

27. $2.$

28. $-\sqrt{3}.$

Find the number(s) x in the interval $[0, 2\pi]$ which satisfy the equation.

29. $\sin x = \frac{1}{2}.$

30. $\cos x = -\frac{1}{2}.$

31. $\tan \frac{1}{2}x = 1.$

32. $\sqrt{\sin^2 x} = 1.$

33. $\cos x = \dfrac{\sqrt{2}}{2}.$

34. $\sin 2x = -\dfrac{\sqrt{3}}{2}.$

35. $\cos 2x = 0.$

36. $\tan x = -\sqrt{3}.$

Evaluate to four decimal place accuracy.

37. $\sin 51°.$

38. $\cos 17°.$

39. $\sin (2.352).$

40. $\cos (-13.461).$

41. $\tan 72.4°.$

42. $\cot (7.311).$

43. $\sec (4.360).$

44. $\csc (-9.725).$

Find the solutions of the equation that are in the interval $[0, 2\pi)$. Express your answers in radians and use four decimal place accuracy.

45. $\sin x = 0.5231.$

46. $\cos x = -0.8243.$

47. $\tan x = 6.7192.$

48. $\cot x = -3.0649.$

49. $\sec x = -4.4073.$

50. $\csc x = 10.260.$

In Exercises 51 and 52, solve the equation $f(x) = y_0$ on $[0, 2\pi]$ by using a graphing utility to draw $f(x)$ and $y = y_0$ together and then use the trace function to find the points(s) of intersection.

51. $f(x) = \sin 3x;\quad y_0 = -1/\sqrt{2}.$

52. $f(x) = \cos\left(\frac{x}{2}\right);\quad y_0 = \frac{3}{4}.$

Give the domain and range of the function.

53. $f(x) = |\sin x|.$

54. $g(x) = \sin^2 x + \cos^2 x.$

55. $f(x) = 2\cos 3x.$

56. $F(x) = 1 + \sin x.$

57. $f(x) = 1 + \tan^2 x.$

58. $h(x) = \sqrt{\cos^2 x}.$

Sketch the graph of the function.

59. $f(x) = 3\sin 2x.$

60. $f(x) = 1 + \sin x.$

61. $g(x) = 1 - \cos x.$

62. $F(x) = \tan \frac{1}{2}x.$

63. $f(x) = \sqrt{\sin^2 x}.$

64. $g(x) = -2\cos x.$

State whether the function is odd, even, or neither.

65. $f(x) = \sin 3x.$

66. $g(x) = \tan x.$

67. $f(x) = 1 + \cos 2x.$

68. $g(x) = \sec x.$

69. $f(x) = x^3 + \sin x.$

70. $h(x) = \dfrac{\cos x}{x^2 + 1}.$

71. Suppose that l_1 and l_2 are two nonvertical lines. If $m_1 m_2 = -1$, then l_1 and l_2 intersect at right angles. Show that if l_1 and l_2 do not intersect at right angles, then the angle α between l_1 and l_2 (see Section 1.4) is given by the formula

$$\tan \alpha = \left| \frac{m_2 - m_1}{1 + m_1 m_2} \right|.$$

HINT: Use the identity

$$\tan(\theta - \phi) = \frac{\tan\theta - \tan\phi}{1 + \tan\theta\tan\phi}.$$

Find the point where the lines intersect and determine the angle between the lines.

72. $l_1: 4x - y - 3 = 0, \quad l_2: 3x - 4y + 1 = 0.$

73. $l_1: 3x + y - 5 = 0, \quad l_2: 7x - 10y + 27 = 0.$

74. $l_1: 4x - y + 2 = 0, \quad l_2: 19x + y = 0.$

75. $l_1: 5x - 6y + 1 = 0, \quad l_2: 8x + 5y + 2 = 0.$

76. For angles θ between 0 and $\pi/2$, verify that the definition of the trigonometric functions in terms of the unit circle and the definition in terms of a right triangle are equivalent.

HINT: Introduce a coordinate system with one of the angles of a right triangle in standard position. See the figure.

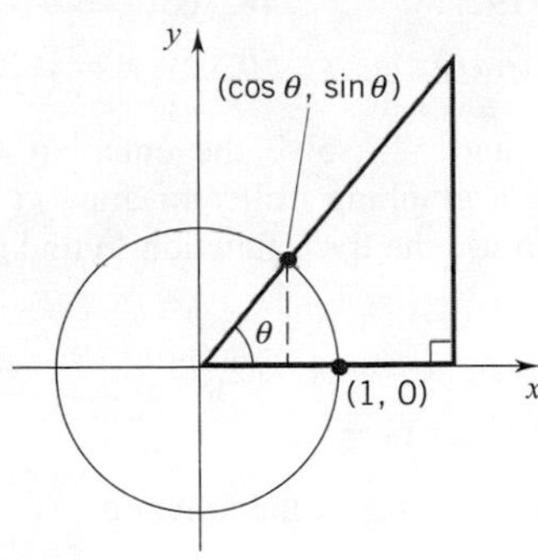

In Exercises 77 and 78, consider a general triangle with angles A, B, C and corresponding sides a, b, c. See Figure 1.6.13.

77. Prove the law of sines:

$$\frac{\sin A}{a} = \frac{\sin B}{b} = \frac{\sin C}{c}.$$

HINT: Drop a perpendicular from one vertex to the opposite side and use the two right triangles that are formed.

78. Prove the law of cosines:

$$a^2 = b^2 + c^2 - 2bc \cos A.$$

HINT: Drop a perpendicular from angle B to side b, and use trigonometry and the Pythagorean theorem.

79. Prove that the area A of an isosceles triangle is given by $A = \frac{1}{2}a^2 \sin\theta$, where a is the length of the equal sides and θ is the angle included between them.

80. (a) Use a graphing utility to graph the polynomials.

$$f(x) = x^4 + 2x^3 - 5x^2 - 3x + 1,$$
$$g(x) = -x^4 + x^3 + 4x^2 - 3x + 2.$$

(b) Based on your graphs in part (a), make a conjecture on the general shape of the graph of a polynomial of degree 4 (a *quartic* polynomial).

(c) Now graph the quartic polynomials

$$f(x) = x^4 - 4x^2 + 4x + 2 \quad \text{and} \quad g(x) = -x^4$$

and compare these graphs with your conjecture in part (b). The methods of calculus will help us distinguish between the various possibilities for the graph of a quartic polynomial. Determine one property shared by the graphs of all quartic polynomials of the form

$$P(x) = x^4 + ax^3 + bx^2 + cx + d.$$

What about quartics of the form

$$Q(x) = -x^4 + ax^3 + bx^2 + cx + d?$$

81. (a) Use a graphing utility to graph the polynomials.

$$f(x) = x^5 - 7x^3 + 6x + 2,$$
$$g(x) = -x^5 + 5x^3 - 3x - 3.$$

(b) Based on your graphs in part (a), make a conjecture on the general shape of the graph of a polynomial of degree 5 (a *quintic* polynomial).

(c) Now graph

$$P(x) = x^5 + ax^4 + bx^3 + cx^2 + dx + e$$

for several choices of a, b, c, d, and e. (For example, try $a = b = c = d = e = 0$.) How do these graphs compare with your graph of f from part (a)?

82. (a) Use a graphing utility to graph $f_A(x) = A\cos x$ for several values of A; use both positive and negative values. Compare your graphs with the graph of $f(x) = \cos x$.

(b) Now graph $f_B(x) = \cos Bx$ for several values of B. Since the cosine function is even, it is sufficient to use only positive values for B. Use some values between 0 and 1 and some values greater than 1. Again, compare your graphs with the graph of $f(x) = \cos x$.

(c) Describe the effects that the coefficients A and B have on the graph of the cosine function.

83. Let $f_n(x) = x^n, n = 1, 2, 3 \ldots.$

(a) Using a graphing utility, draw the graphs of f_n for $n = 2, 4, 6$ in one coordinate system, and in another coordinate system draw the graphs of f_n for $n = 1, 3, 5$.

(b) Based on your results in part (a), make a sketch of the graph of f_n when n is even, and when n is odd.

(c) For a given positive integer k, compare the graphs of f_k and f_{k+t} on $[0, 1]$ and on $(1, \infty)$.

84. (a) Use a graphing utility to draw the graphs of

$$f(x) = \frac{1}{x}\sin x \quad \text{and} \quad g(x) = x\sin\left(\frac{1}{x}\right)$$

$$\text{for} \quad -\frac{\pi}{2} \leq x \leq \frac{\pi}{2},\ x \neq 0.$$

(b) Use the "zoom in" feature to examine the behavior of the graphs of f and g at the origin. What can you say about the values of $f(x)$ and $g(x)$ when x is close to 0?

85. Given that $f(x) = x^2$ and $g(x) = 2^x$, calculate $f(x)/g(x)$ for $x = 10, 20, 30, 40, 50$. Conjecture a relationship between $f(x)$ and $g(x)$ as x gets "large."

86. Given that $f(x) = 10^x$ and $g(x) = x^{10}$, calculate $f(x)/g(x)$ for $x = 10, 20, 30, 40, 50$. Conjecture a relationship between $f(x)$ and $g(x)$ as x gets "large."

■ 1.7 COMBINATIONS OF FUNCTIONS

In this section we review the basic methods for combining two or more functions.

Algebraic Combinations of Functions

On the intersection of their domains, functions can be added and subtracted:

$$(f+g)(x)=f(x)+g(x), \quad (f-g)(x)=f(x)-g(x);$$

they can be multiplied:

$$(fg)(x)=f(x)g(x);$$

and, at the points x where $g(x)\neq 0$, we can form the quotient:

$$\left(\frac{f}{g}\right)(x)=\frac{f(x)}{g(x)}.$$

Example 1 Let

$$f(x)=4+\sqrt{x+3} \quad \text{and} \quad g(x)=\sqrt{5-x}-2.$$

(a) Give the domain of f and the domain of g.
(b) Determine the domain of $f+g$ and specify $(f+g)(x)$.
(c) Determine the domain of f/g and specify $(f/g)(x)$.

SOLUTION

(a) We can form $1+\sqrt{x+3}$ iff $x+3\geq 0$, which holds iff $x\geq -3$. Thus dom $(f)=[-3,\infty)$. We can form $\sqrt{5-x}-2$ iff $5-x\geq 0$, which holds iff $x\leq 5$. Thus dom $(g)=(-\infty,5]$.
(b) dom $(f+g)=$ dom $(f)\cap$ dom $(g)=[-3,\infty)\cap(-\infty,5]=[-3,5]$,
$(f+g)(x)=f(x)+g(x)=2+\sqrt{x+3}+\sqrt{5-x}$.
(c) To obtain the domain of the quotient, we must exclude from $[-3,5]$ the numbers x at which $g(x)=0$. There is only one such number: $x=1$. Therefore

$$\text{dom}\left(\frac{f}{g}\right)=\{x\in[-3,5]: x\neq 1\}=[-3,1)\cup(1,5].$$

$$\left(\frac{f}{g}\right)(x)=\frac{f(x)}{g(x)}=\frac{4+\sqrt{x+3}}{\sqrt{5-x}-2} \quad \text{for all} \quad x\in[-3,1)\cup(1,5]. \quad \square$$

With α and β real numbers, we can form *scalar multiples*

$$(\alpha f)(x)=\alpha\cdot f(x)$$

and *linear combinations*

$$(\alpha f+\beta g)=\alpha\cdot f(x)+\beta\cdot g(x).$$

If α and β are nonzero, then dom$(\alpha f)=$ dom (f) and dom $(\alpha f+\beta g)=$ dom$(f)\cap$ dom(g).

Example 2 With f and g as in Example 1,

$$(3f)(x) = 3 \cdot f(x) = 3(4 + \sqrt{x+3}) = 12 + 3\sqrt{x+3} \quad \text{for all } x \in [-3, \infty),$$

and

$$\begin{aligned}(4f - 3g)(x) &= 4(4 + \sqrt{x+3}) - 3(\sqrt{5-x} - 2) \\ &= 22 + 4\sqrt{x+3} - 3\sqrt{5-x} \quad \text{for all} \quad x \in [-3, 5]. \quad \square\end{aligned}$$

There is a special case of addition of functions that is worth noting here. For a function f and a number c, let $F(x) = f(x) + c$ for all $x \in \text{dom}(f)$. The function F is called a *vertical translation* (or *vertical shift*) of f because the graph of F is simply the graph of f shifted up $|c|$ units if $c > 0$, or shifted down $|c|$ units if $c < 0$. See Figure 1.7.1.

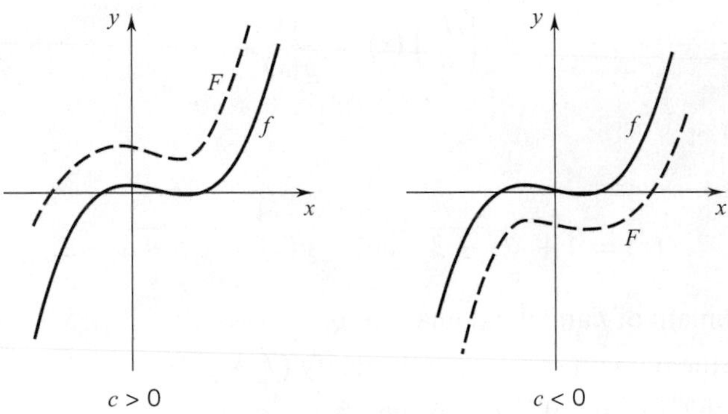

Figure 1.7.1

Example 3 Let $f(x) = x^3$ and $g(x) = \cos x$. The graphs of $F(x) = x^3 + 2$ and $G(x) = \cos x - 1$, along with the graphs of f and g, are shown in Figure 1.7.2. $\square$

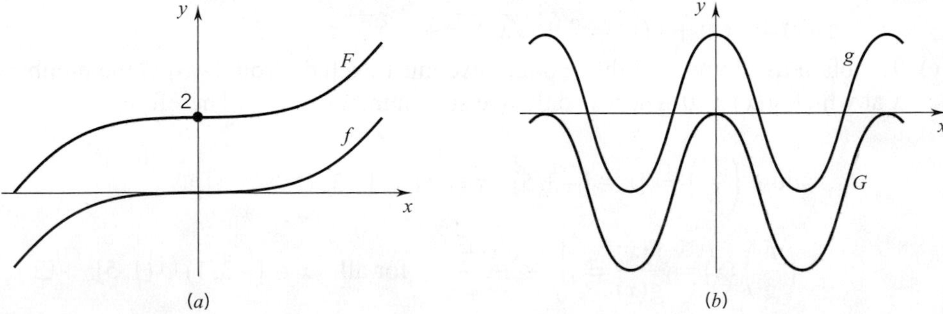

Figure 1.7.2

Composition of Functions

We have seen how to combine functions algebraically. There is another way to combine functions called *composition*. To describe it, we begin with two functions f and g, and a number x in the domain of g. By applying g to x, we get the number $g(x)$. If $g(x)$ is in the domain of f, then we can apply f to $g(x)$ and thereby obtain the number $f(g(x))$.

What is $f(g(x))$? It is the result of first applying g to x and then applying f to $g(x)$. The idea is illustrated in Figure 1.7.3. This new function — it takes x in the domain of g to $g(x)$ in the domain of f, and assigns to it the value $f(g(x))$ — is called the *composition of f with g* and is denoted by the symbol $f \circ g$ (see Figure 1.7.4).

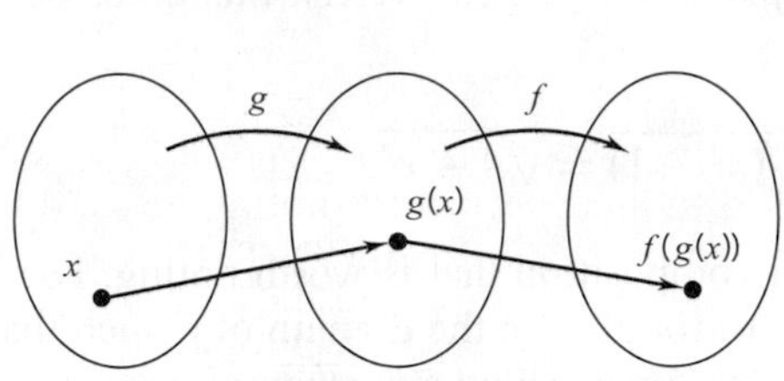

Figure 1.7.3

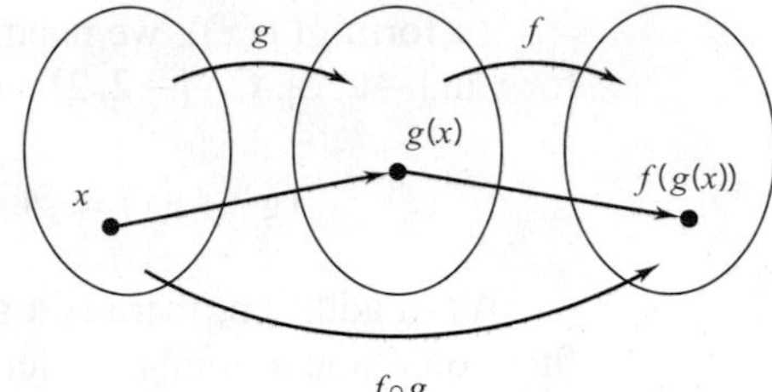

Figure 1.7.4

DEFINITION 1.7.1 COMPOSITION

Let f and g be functions. For those x in the domain of g for which $g(x)$ is in the domain of f, we define the *composition of f with g*, denoted $f \circ g$†, by

$$(f \circ g)(x) = f(g(x)).$$

For those who prefer to write domains in set notation,

$$\text{dom}(f \circ g) = \{x \in \text{dom}(g) : g(x) \in \text{dom}(f)\}$$

Example 4 Suppose that

$$g(x) = x^2 \qquad \text{(the squaring function)}$$

and

$$f(x) = x + 3. \qquad \text{(the function that adds 3)}$$

Then

$$(f \circ g)(x) = f(g(x)) = g(x) + 3 = x^2 + 3.$$

Thus, $f \circ g$ is the function that *first* squares and *then* adds 3.

On the other hand, the composition of g with f is

$$(g \circ f)(x) = g(f(x)) = (x + 3)^2.$$

Thus, $g \circ f$ is the function that *first* adds 3 and *then* squares.

Since f and g are everywhere defined, both $f \circ g$ and $g \circ f$ are also everywhere defined. Note that $g \circ f$ is *not* the same as $f \circ g$. ☐

Example 5 Let $f(x) = x^2 - 1$ and $g(x) = \sqrt{3 - x}$.

† Read "f composed with g" or "f circle g".

The domain of g is $(-\infty, 3]$. Since f is everywhere defined, the domain of $f \circ g$ is also $(-\infty, 3]$. On that interval

$$(f \circ g)(x) = f(g(x)) = \left(\sqrt{3-x}\right)^2 - 1 = (3-x) - 1 = 2 - x.$$

To form $g(f(x))$, we need $f(x) \leq 3$ (explain). As you can verify, this holds only for x in $[-2, 2]$. On $[-2, 2]$

$$(g \circ f)(x) = g(f(x)) = \sqrt{3 - (x^2 - 1)} = \sqrt{4 - x^2}. \quad \square$$

As in addition, there is a special case of composition that is worth noting. For a function f and a number c, let $g(x) = f(x - c)$ for all x in the domain of f such that $x - c$ is also in the domain of f. Then the function g is called a *horizontal translation* (or *horizontal shift*) of f because the graph of g is simply the graph of f shifted c units to the right if $c > 0$, or c units to the left if $c < 0$. See Figure 1.7.5.

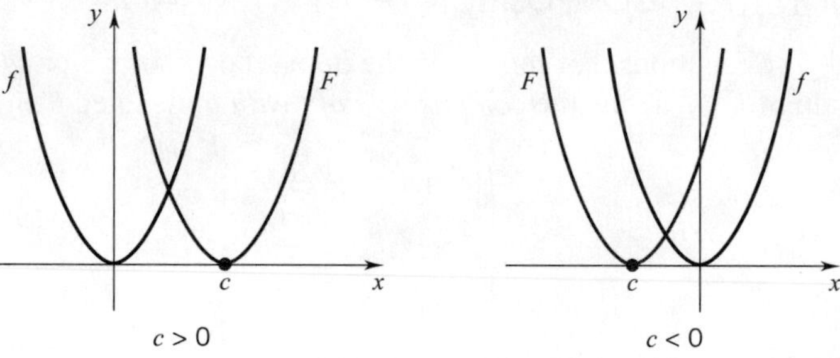

Figure 1.7.5

Example 6 Let $f(x) = x^2 + 1$ and $g(x) = \sin x$. The graphs of

$$F(x) = f(x-2) = (x-2)^2 + 1 \quad \text{and} \quad G(x) = g(x + \pi/4) = \sin(x + \pi/4),$$

along with graphs of f and g, are shown in Figure 1.7.6 (a) and (b). $\square$

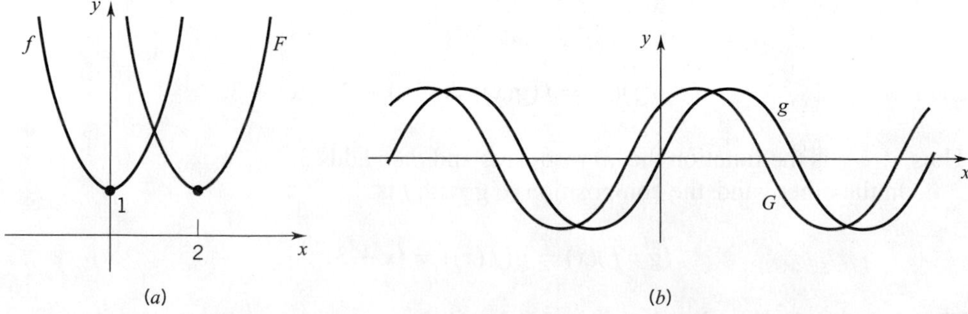

Figure 1.7.6

It is possible to form the composition of more than two functions. For example, the triple composition $f \circ g \circ h$ consists of first h, then g, and then f:

$$(f \circ g \circ h)(x) = f[g(h(x))].$$

We can go on in this manner with as many functions as we like.

Example 7 If $f(x) = \dfrac{1}{x}$, $g(x) = x^2 + 1$, $h(x) = \cos x$,

then
$$(f \circ g \circ h)(x) = f[g(h(x))] = \frac{1}{g(h(x))} = \frac{1}{[h(x)]^2 + 1}$$
$$= \frac{1}{\cos^2 x + 1}. \quad \square$$

Example 8 Find functions f and g such that $f \circ g = F$ if
$$F(x) = (x+1)^5.$$

A SOLUTION The function consists of first adding 1 and then taking the fifth power. We can therefore set
$$g(x) = x + 1 \qquad \text{(adding 1)}$$
and
$$f(x) = x^5 \qquad \text{(taking the fifth power)}$$
As you can see,
$$(f \circ g)(x) = f(g(x)) = [g(x)]^5 = (x+1)^5. \quad \square$$

Example 9 Find three functions f, g, h such that $f \circ g \circ h = F$ if
$$F(x) = \frac{1}{|x| + 3}.$$

A SOLUTION F takes the absolute value, adds 3, and then inverts. Let h take the absolute value:
$$\text{set} \quad h(x) = |x|.$$
Let g add 3:
$$\text{set} \quad g(x) = x + 3.$$
Let f do the inverting:
$$\text{set} \quad f(x) = \frac{1}{x}.$$
With this choice of f, g, h, we have
$$(f \circ g \circ h)(x) = f[g(h(x))] = \frac{1}{g(h(x))} = \frac{1}{h(x) + 3} = \frac{1}{|x| + 3}. \quad \square$$

EXERCISES 1.7

In Exercises 1–8, let $f(x) = 2x^2 - 3x + 1$ and $g(x) = x^2 + 1/x$. Determine the indicated values.

1. $(f+g)(2)$.
2. $(f-g)(-1)$.
3. $(f \cdot g)(-2)$.
4. $\left(\frac{f}{g}\right)(1)$.
5. $(2f - 3g)\left(\frac{1}{2}\right)$.
6. $\left(\frac{f+2g}{f}\right)(-1)$.
7. $(f \circ g)(1)$.
8. $(g \circ f)(1)$.

Determine $f+g, f-g, f \cdot g$, and f/g, and give the domain of each.

9. $f(x) = 2x - 3, \quad g(x) = 2 - x$.

10. $f(x) = x^2 - 1, \quad g(x) = x + \frac{1}{x}$.

11. $f(x) = \sqrt{x-1}, \quad g(x) = x - \sqrt{x+1}$.

12. $f(x) = \sin^2 x, \quad g(x) = \cos 2x$.

13. Given that $f(x) = x + 1/\sqrt{x}$ and $g(x) = \sqrt{x} - 2\sqrt{x}$, find (a) $6f + 3g$, (b) $f - g$, (c) f/g.

14. Given that

$$f(x) = \begin{cases} 1 - x, & x \le 1 \\ 2x - 1, & x > 1, \end{cases} \qquad g(x) = \begin{cases} 0, & x < 2 \\ -1, & x \ge 2, \end{cases}$$

find $f+g, f-g$, and $f \cdot g$. HINT: Break up the domains of the two functions in the same manner.

Sketch the graph with f and g as shown in the figure.

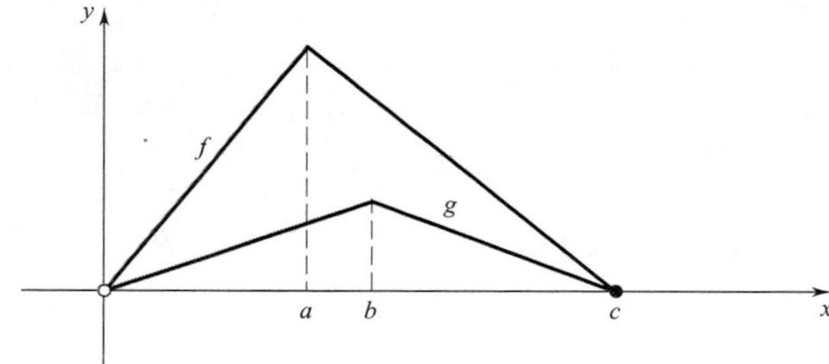

15. $2g$.
16. $\frac{1}{2}f$.
17. $-f$.
18. $0 \cdot g$.
19. $-2g$.
20. $f+g$.
21. $f-g$.
22. $f+2g$.

Form the composition $f \circ g$ and give the domain.

23. $f(x) = 2x + 5, \quad g(x) = x^2$.

24. $f(x) = x^2, \quad g(x) = 2x + 5$.

25. $f(x) = \sqrt{x}, \quad g(x) = x^2 + 5$.

26. $f(x) = x^2 + x, \quad g(x) = \sqrt{x}$.

27. $f(x) = \frac{1}{x}, \quad g(x) = \frac{x-2}{x}$.

28. $f(x) = \frac{1}{x-1}, \quad g(x) = x^2$.

29. $f(x) = \sqrt{1-x^2}, \quad g(x) = \cos 2x$.

30. $f(x) = \sqrt{1-x}, \quad g(x) = 2\cos x$ for $x \in [0, 2\pi]$.

Form the composition $f \circ g \circ h$ and give the domain.

31. $f(x) = 4x, \quad g(x) = x - 1, \quad h(x) = x^2$.

32. $f(x) = x - 1, \quad g(x) = 4x, \quad h(x) = x^2$.

33. $f(x) = \frac{1}{x}, \quad g(x) = \frac{1}{2x+1}, \quad h(x) = x^2$.

34. $f(x) = \frac{x+1}{x}, \quad g(x) = \frac{1}{2x+1}, \quad h(x) = x^2$.

Find f such that $f \circ g = F$ given that

35. $g(x) = \frac{1+x^2}{1+x^4}, \quad F(x) = \frac{1+x^4}{1+x^2}$.

36. $g(x) = x^2, \quad F(x) = ax^2 + b$.

37. $g(x) = 3x, \quad F(x) = 2\sin 3x$.

38. $g(x) = -x^2, \quad F(x) = \sqrt{a^2 + x^2}$

Find g such that $f \circ g = F$ given that

39. $f(x) = x^3, \quad F(x) = \left(1 - \frac{1}{x^4}\right)^2$.

40. $f(x) = x + \frac{1}{x}, \quad F(x) = a^2x^2 + \frac{1}{a^2x^2}$.

41. $f(x) = x^2 + 1, \quad F(x) = (2x^3 - 1)^2 + 1$.

42. $f(x) = \sin x, \quad F(x) = \sin \frac{1}{x}$.

Find $f \circ g$ and $g \circ f$.

43. $f(x) = \sqrt{x}, \quad g(x) = x^2$.

44. $f(x) = 3x + 1, \quad g(x) = x^2$.

45. $f(x) = 1 - x^2, \quad g(x) = \sin x$.

46. $f(x) = 2x, \quad g(x) = \frac{1}{2}$.

47. $f(x) = x^3 + 1, \quad g(x) = \sqrt[3]{x-1}$.

48. Suppose that f and g are odd functions. What can you conclude about $f \cdot g$? Justify your answer.

49. Suppose that f and g are even functions. What can you conclude about $f \cdot g$? Justify your answer.

50. Suppose that f is an even function and g is an odd function. What can you conclude about $f \cdot g$? Justify your answer.

51. For $x > 0$, f is defined as follows:

$$f(x) = \begin{cases} x, & 0 \le x \le 1 \\ 1, & x > 1. \end{cases}$$

How is f defined for $x < 0$ if (a) f is even? (b) f is odd?

52. For $x \ge 0$, $f(x) = x^2 - x$. How is f defined for $x < 0$ if (a) f is even? (b) f is odd?

53. Given that f is defined for all real numbers, show that the function $g(x) = f(x) + f(-x)$ is an even function.

54. Given that f is defined for all real numbers, show that the function $h(x) = f(x) - f(-x)$ is an odd function.

55. Show that every function that is defined for all real numbers can be written as the sum of an even function and an odd function.

56. For $x \neq 0, 1$, define

$$f_1(x) = x, \quad f_2(x) = \frac{1}{x}, \quad f_3(x) = 1 - x,$$

$$f_4(x) = \frac{1}{1-x}, \ f_5(x) = \frac{x-1}{x}, \ f_6(x) = \frac{x}{x-1}.$$

This family of functions is *closed* under composition; that is, the composition of any two of these functions is again one of these functions. Tabulate the results of composing these functions one with the other by filling in the table shown in the figure. To indicate that $f_i \circ f_j = f_k$, write "f_k" in the ith row, jth column. We have already made two entries in the table. Check out these two entries and then fill in the rest of the table.

	f_1	f_2	f_3	f_4	f_5	f_6
f_1						
f_2						
f_3				f_6		
f_4			f_2			
f_5						
f_6						

C In Exercises 57 and 58, let $f(x) = x^2 - 4$, $g(x) = \dfrac{3x}{2-x}$, $h(x) = \sqrt{x+4}$, and $k(x) = \dfrac{2x}{3+x}$. Use a CAS to find the indicated composition.

57. *(a)* $f \circ g$; *(b)* $g \circ k$; *(c)* $f \circ k \circ g$.

58. *(a)* $g \circ f$; *(b)* $k \circ g$; *(c)* $g \circ f \circ k$.

C In Exercises 59 and 60, let $f(x) = x^2$ and $F(x) = (x-a)^2 + b$.

59. (a) Choose a value for a and, using a graphing utility, graph F for several different values of b. Be sure to choose both positive and negative values. Compare your graphs with the graph of f, and describe the effect that varying b has on the graph of F.

(b) Now fix a value of b and graph F for several values of a; again, use both positive and negative values. Compare your graphs with the graph of f, and describe the effect that varying a has on the graph of F.

(c) Choose values for a and b, and graph $-F$. What effect does changing the sign of F have on the graph?

60. For any values of a and b, the graph of F is a parabola which opens up. Find values for a and b such that the parabola will have x-intercepts at $-\frac{3}{2}$ and 2. Verify your result algebraically.

C For Exercises 61 and 62, let $f(x) = x^3 - 3x + 1$.

61. (a) Graph $f(x) + c$ for $c = -2, -1, 0, 1, 2$. Describe the effect that varying c has on the graph of f.

(b) Using a graphing utility, graph $f(x-a)$ for $a = -2, -1, 0, 1, 2$. Describe the effect that varying a has on the graph of f.

(c) Graph $f(bx)$ for $b = -2, -1, 0, 1, 2, 3$. Describe the effect that varying b has on the graph of f.

62. Use your graphing utility to find values for a, b, c such that the graph of $f(b[x-a]) + c$ has x-intercepts at $-2, \frac{1}{2}$, and 3. Verify your result algebraically.

C For Exercises 63 and 64, let $f(x) = \sin x$.

63. (a) Using a graphing utility, graph Af for $A = -3\ -2, -1, 2, 3$. Compare your graphs with the graph of f.

(b) Now graph $f(Bx)$ for $B = -3, -2, -\frac{1}{2}, \frac{1}{3}, \frac{1}{2}, 2, 3$. Compare your graphs with the graph of f.

64. (a) Using a graphing utility, graph $f(x - C)$ for $C = -\frac{\pi}{2}, -\frac{\pi}{4}\frac{\pi}{3}, \frac{\pi}{2}, \pi, 2\pi$. Compare your graphs wth the graph of f.

(b) Now graph $Af(Bx - C)$ for several values of A, B, C. Describe the effect of A, the effect of B, the effect of C.

■ 1.8 A NOTE ON MATHEMATICAL PROOF; MATHEMATICAL INDUCTION

The notion of proof goes back to Euclid's *Elements*, and the rules of proof have changed little since they were formulated by Aristotle. We work in a deductive system where truth is argued on the basis of assumptions, definitions, and previously proved results. We cannot claim that such and such is true without clearly stating the basis on which we make that claim.

A theorem is an implication; it consists of a hypothesis and a conclusion:

$$\text{if (hypothesis)} \ldots, \text{then (conclusion)} \ldots.$$

Here is an example:

If a and b are positive numbers, then ab is positive.

A common mistake is to ignore the hypothesis and persist with the conclusion: to insist, for example, that $ab > 0$ just because a and b are numbers.

Another common mistake is to confuse a theorem

$$\text{if } A, \text{ then } B$$

with its converse

$$\text{if } B, \text{ then } A.$$

The fact that a theorem is true does not mean that its converse is true: While it is true that

if a and b are positive numbers, then ab is positive,

it is *not* true that

if ab is positive, then a and b are positive numbers;

$[(-2)(-3)$ is positive but -2 and -3 are not positive$]$.

A third, more subtle mistake is to presume that the hypothesis of a theorem represents the only condition under which the conclusion is true. There may well be other conditions under which the conclusion is true. Thus, for example, not only is it true that

if a and b are positive numbers, then ab is positive

but it is also true that

if a and b are negative numbers, then ab is positive.

In the event that a theorem

$$\text{if } A, \text{ then } B$$

and its converse

$$\text{if } B, \text{ then } A$$

are both true, then we can write

$$A \text{ if and only if } B \qquad \text{or more briefly} \qquad A \text{ iff } B.$$

We know, for example, that

$$\text{if} \quad x \geq 0, \qquad \text{then} \quad |x| = x;$$

we also know that

$$\text{if} \quad |x| = x, \qquad \text{then} \quad x \geq 0.$$

We can summarize this by writing

$$x \geq 0 \quad \text{iff} \quad |x| = x.$$

A final point. One way of proving

$$\text{if } A, \text{then } B$$

is to assume that

(1) $$A \text{ holds and } B \text{ does not hold}$$

and then arrive at a contradiction. The contradiction is taken to indicate that (1) is false and therefore

$$\text{if } A \text{ holds, then } B \text{ must hold.}$$

Some of the theorems of calculus are proved by this method.

Calculus provides procedures for solving a wide range of problems in the physical and social sciences. The fact that these procedures give us answers that seem to make sense is comforting, but it is only because we can prove our theorems that we can be confident in the results. Accordingly, your study of calculus should include the study of some proofs.

Mathematical Induction

Mathematical induction is a method of proof which can be used to show that certain propositions are true for all positive integers n. The method is based on the following axiom:

1.8.1 AXIOM OF INDUCTION

Let S be a set of positive integers. If

(A) $1 \in S$, and

(B) $k \in S$ implies that $k + 1 \in S$,

then all the positive integers are in S.

You can think of the axiom of induction as a kind of "domino theory." If the first domino falls (Figure 1.8.1), and if each domino that falls causes the next one to fall, then, according to the axiom of induction, each domino will fall.

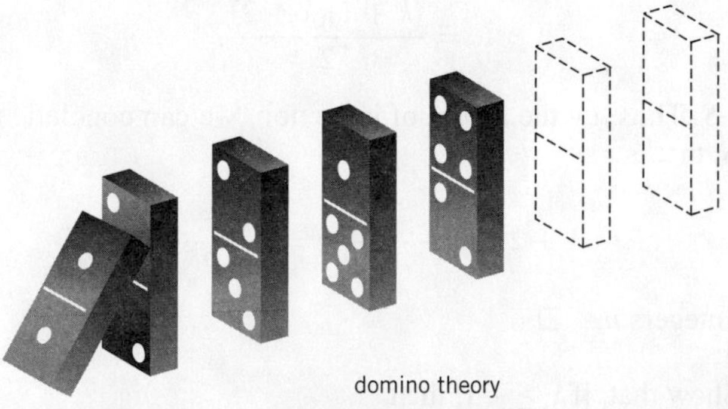

Figure 1.8.1

While we cannot prove that this axiom is valid (axioms are by their very nature assumptions and therefore not subject to proof), we can argue that it is *plausible*.

Let's assume that we have a set S that satisfies conditions (A) and (B). Now let's choose a positive integer m and "argue" that $m \in S$.

From (A) we know that $1 \in S$. Since $1 \in S$, we know from (B) that $1 + 1 \in S$, and thus that $(1 + 1) + 1 \in S$, and so on. Since m can be obtained from 1 by adding 1 successively $(m - 1)$ times, it *seems clear* that $m \in S$.

To prove that a given proposition is true for *all* positive integers n, we let S be the set of positive integers for which the proposition is true. We prove first that $1 \in S$, that is, that the proposition is true for $n = 1$. Next we assume that the proposition is true for some positive integer k, and prove that it is also true for $k + 1$, that is, $k \in S$ implies that $k + 1 \in S$. Then, by the axiom of induction, it follows that S contains the set of positive integers and so the proposition is true for all positive integers.

Example 1 Show that

$$1 + 2 + 3 + \cdots + n = \frac{n(n+1)}{2}$$

for all positive integers n.

SOLUTION Let S be the set of positive integers n for which

$$1 + 2 + 3 + \cdots + n = \frac{n(n+1)}{2}.$$

Then $1 \in S$ since

$$1 = \frac{1(1+1)}{2}.$$

Next, assume that the positive integer $k \in S$, that is, assume

$$1 + 2 + 3 + \cdots + k = \frac{k(k+1)}{2}.$$

Now consider the sum of the first $k + 1$ integers:

$$\begin{aligned}
1 + 2 + 3 + \cdots + k + (k+1) &= [1 + 2 + 3 + \cdots + k] + (k+1) \\
&= \frac{k(k+1)}{2} + (k+1) \qquad \text{(by the induction hypothesis)} \\
&= \frac{k(k+1) + 2(k+1)}{2} \\
&= \frac{(k+1)(k+2)}{2},
\end{aligned}$$

and so $k + 1 \in S$. Thus, by the axiom of induction, we can conclude that all positive integers are in S, that is

$$1 + 2 + 3 + \cdots + n = \frac{n(n+1)}{2}$$

for all positive integers n. ❑

Example 2 Show that, if $x \geq -1$, then

$$(1 + x)^n \geq 1 + nx \qquad \text{for all positive integers } n.$$

SOLUTION Take $x \geq -1$ and let S be the set of positive integers n for which

$$(1+x)^n \geq 1+nx.$$

Since
$$(1+x)^1 \geq 1+1\cdot x,$$

we have $1 \in S$.

Assume now that $k \in S$. By the definition of S,

$$(1+x)^k \geq 1+kx.$$

Since
$$(1+x)^{k+1} = (1+x)^k(1+x) \geq (1+kx)(1+x) \qquad \text{(explain)}$$

and

$$(1+kx)(1+x) = 1+(k+1)x+kx^2 \geq 1+(k+1)x,$$

if follows that

$$(1+x)^{k+1} \geq 1+(k+1)x$$

and thus that $k+1 \in S$.

We have shown that

$$1 \in S \quad \text{and that} \quad k \in S \quad \text{implies} \quad k+1 \in S.$$

By the axiom of induction, all the positive integers are in S ☐.

Remark An induction does not have to begin with the integer 1. If, for example, you want to show that some proposition is true for all integers $n \geq 3$, all you have to do is show that: (A) it is true for $n = 3$, and that (B), if it is true for $n = k$, it is then true for $n = k+1$. (Now you are starting the chain reaction by pushing on the third domino.) ☐

EXERCISES 1.8

Show that the statement holds for all positive integers n.

1. $2n \leq 2^n$.

2. $1+2n \leq 3^n$.

3. $2^0+2^1+2^2+2^3+\cdots+2^{n-1} = 2^n-1$.

4. $1+3+5+\cdots+(2n-1) = n^2$.

5. $1^2+2^2+3^2+\cdots+n^2 = \frac{1}{6}n(n+1)(2n+1)$.

6. $1^3+2^3+3^3+\cdots+n^3 = (1+2+3+\cdots+n)^2$.
HINT: Use Example 1.

7. $1^3+2^3+\cdots+(n-1)^3 < \frac{1}{4}n^4 < 1^3+2^3+\cdots+n^3$.

8. $1^2+2^2+\cdots+(n-1)^2 < \frac{1}{3}n^3 < 1^2+2^2+\cdots+n^2$.

9. $\dfrac{1}{\sqrt{1}}+\dfrac{1}{\sqrt{2}}+\dfrac{1}{\sqrt{3}}+\cdots+\dfrac{1}{\sqrt{n}} > \sqrt{n}$.

10. $\dfrac{1}{1\cdot 2}+\dfrac{1}{2\cdot 3}+\dfrac{1}{3\cdot 4}+\cdots+\dfrac{1}{n(n+1)} = \dfrac{n}{n+1}$.

11. For what integers n is $3^{2n+1}+2^{n+2}$ divisible by 7? Prove that your answer is correct.

12. For what integers n is $9^n - 8n - 1$ divisible by 64? Prove that your answer is correct.

13. Find a simplifying expression for the product

$$\left(1-\frac{1}{2}\right)\left(1-\frac{1}{3}\right)\cdots\left(1-\frac{1}{n}\right)$$

and verify its validity for all integers $n \geq 2$.

14. Find a simplifying expression for the product

$$\left(1-\frac{1}{2^2}\right)\left(1-\frac{1}{3^2}\right)\cdots\left(1-\frac{1}{n^2}\right)$$

and verify its validity for all integers $n \geq 2$.

15. Prove that an N-sided convex polygon has $\frac{1}{2}N(N-3)$ diagonals, $N > 3$.

16. Prove that the sum of the interior angles in an N-sided convex polygon is $(N-2)180°$, $N > 2$.

17. Prove that all sets with n elements have 2^n subsets. Count the empty set $\emptyset$ and the whole set as subsets.

C **18.** Find the first integer n for which $n^2 - n + 41$ is *not* a prime number.

■ CHAPTER HIGHLIGHTS

1.1 What is Calculus?

Calculus is elementary mathematics (algebra, geometry, trigonometry enhanced by the limit process).

1.2 Review of Elementary Mathematics

sets (p. 3)
real numbers: (p. 4)
  real line (p. 5) order properties (p. 5)
  absolute value (p. 6) intervals (p. 6)
  boundedness (p. 7) iff (p. 6)
algebra:
  factoring formulas (p. 8)
geometry:
  elementary figures (p. 9)

1.3 Review of Inequalities

To solve an inequality is to find the set of numbers for which the inequality holds. (p. 12)
Inequalities and Absolute Value (p. 14)

$|a+b| \leq |a| + |b|, \quad \big||a| - |b|\big| \leq |a-b|$ (p. 17)

1.4 Coordinate Plane; Analytic Geometry

rectangular coordinates (p. 19)
distance and midpoint formulas (p. 19)
lines : slope (p. 20) Intercepts (p. 20) equations (p. 21)
  parallel and perpendicular lines (p. 21)
conic sections: circle (p. 23) ellipse (p. 24)
  parabola (p. 24) hyperbola (p. 25)

1.5 Functions

domain and range (p. 28)
graphs (p. 30)
even and odd functions; symmetry (p. 31)
convention on domains (p. 32)

1.6 The Elementary Functions

polynomials (p. 37)
rational functions (p. 39)
trigonometric functions (p. 40)
  angles (p. 40) definitions of the trigonometric functions (p. 41)
  graphs (p. 43) important identities (p. 43)
  triangles (p. 44)

1.7 Combinations of Functions

algebraic combinations of functions (p. 47)
compositions of functions (p. 48)

1.8 A Note on Mathematical Proof; Mathematical Induction

axiom of induction (p. 55)

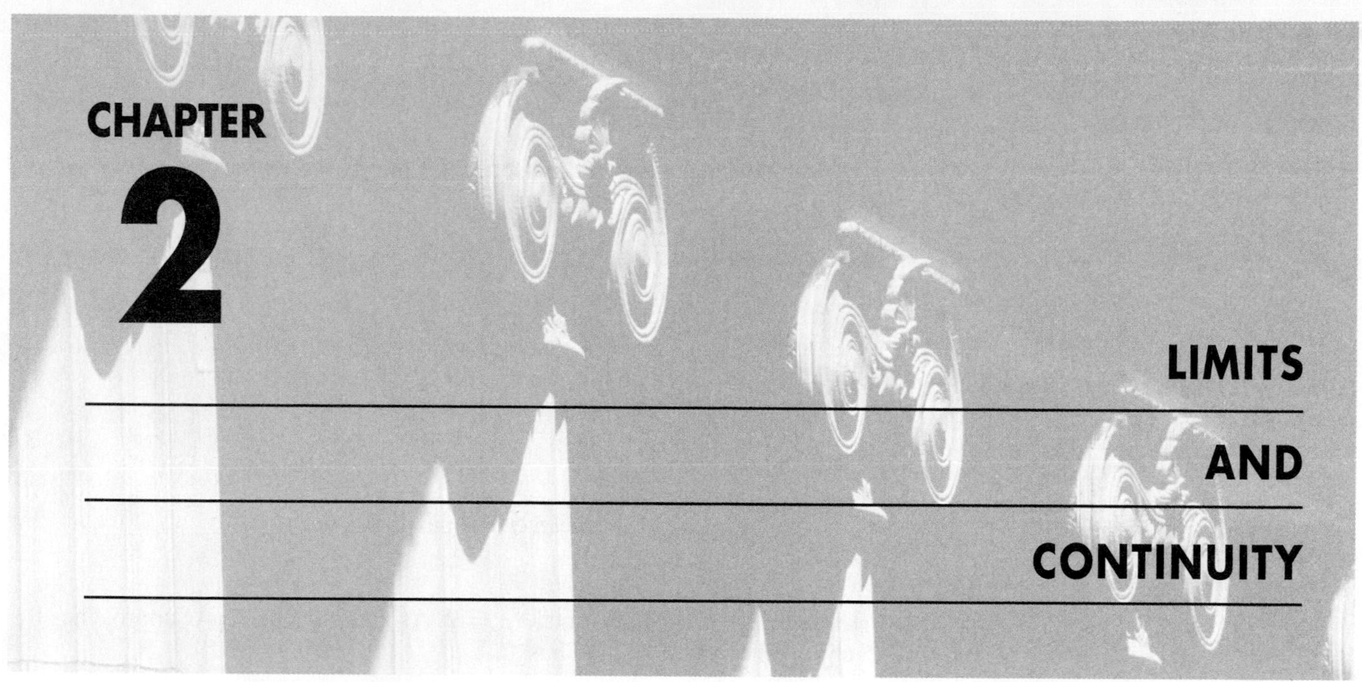

■ 2.1 THE IDEA OF LIMIT

We could begin by saying that limits are important in calculus, but that would be a major understatement. *Without limits calculus simply does not exist. Every single notion of calculus is a limit in one sense or another*. For example,

What is the slope of a curve? It is the limit of slopes of secant lines. See Figure 2.1.1.

What is the length of a curve? It is the limit of the lengths of polygonal paths. See Figure 2.1.2.

What is the area of a region bounded by a curve? It is the limit of the sum of areas of approximating rectangles. See Figure 2.1.3.

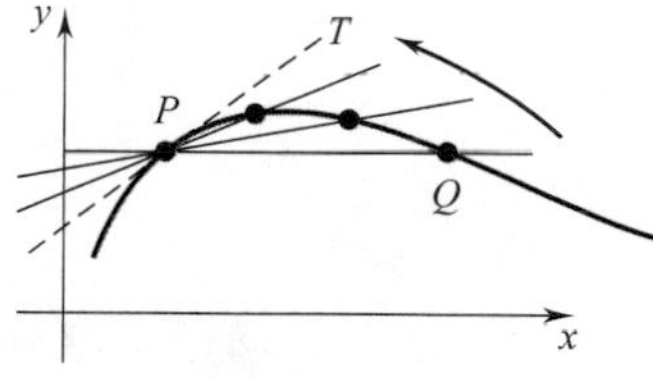

Figure 2.1.1

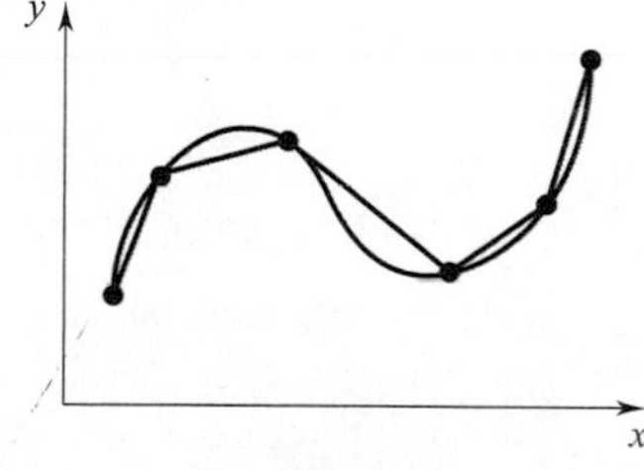

Figure 2.1.2

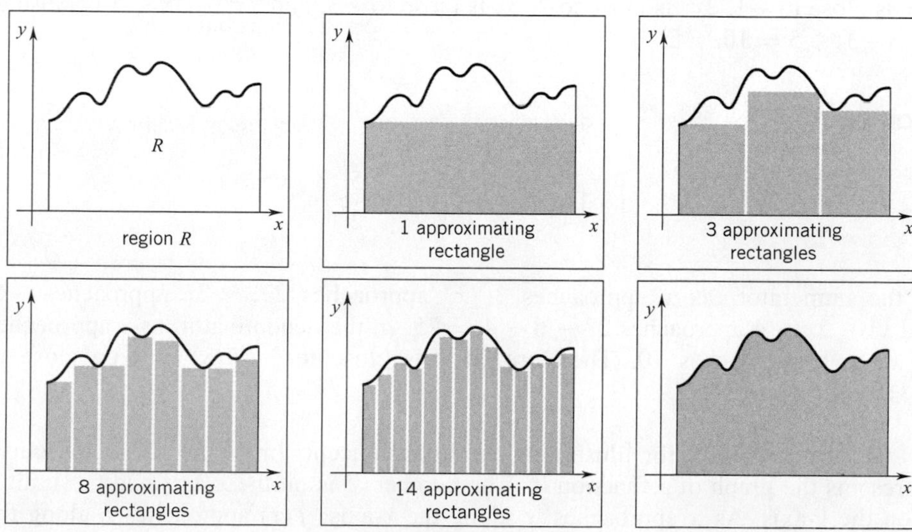

Figure 2.1.3

The Idea of Limit

We begin with a function f and a number c such that f is defined on some interval containing c but not necessarily at c itself. The number L is the *limit of* f *as* x *approaches* c, denoted

$$\lim_{x \to c} f(x) = L,$$

means that

the function values $f(x)$ *approach* L *as* x *approaches* c,

or, equivalently,

$f(x)$ *is close to* L *whenever* x *is close to* c *but* $x \neq c$.

Example 1 Let $f(x) = 4x + 5$ and take $c = 2$. As x approaches 2, $4x$ approaches 8 and $4x + 5$ approaches $8 + 5 = 13$. Thus

$$\lim_{x \to 2} f(x) = 13.$$

For the same function, suppose we let $c = -3$. If x is close to -3, but $x \neq -3$, then $4x$ is close to -12 and $4x + 5$ is close to $-12 + 5 = -7$. Therefore,

$$\lim_{x \to -3} f(x) = -7. \quad \square$$

Example 2

$$\lim_{x \to -1} (2x^2 - 3x + 5) = 10.$$

If x is close to -1, $2x^2$ is close to 2, $3x$ is close to -3, and $2x^2 - 3x + 5$ is close to $2 - (-3) + 5 = 10$. $\square$

Example 3

$$\lim_{x \to 3} \frac{x^3 - 2x + 4}{x^2 + 1} = \frac{5}{2}.$$

In the numerator, as x approaches 3, x^3 approaches 27, $-2x$ approaches -6, and $x^3 - 2x + 4$ approaches $27 - 6 + 4 = 25$. In the denominator, as x approaches 3, $x^2 + 1$ approaches 10. Therefore, if x is close to 3, $F(x)$ will be close to $25/10 = 5/2$. $\square$

Graphs are useful for illustrating the limit concept. In Figure 2.1.4, the curve represents the graph of a function f. The number c is on the x-axis and the limit L is on the y-axis. As x approaches c along the x-axis, $f(x)$ approaches L along the y-axis.

Figure 2.1.4

In taking the limit of a function f as x approaches c, it does not matter whether f is defined at c and, if so, how it is defined there. The only thing that matters is the values that f takes on when x is *near* c. For example, see Figure 2.1.5. In Figure 2.1.5*a*, $L = f(c)$; in Figure 2.1.5*b*, $f(c)$ is not defined; and in Figure 2.1.5*c*, $f(c)$ is defined but $f(c) \neq L$. However, in each case,

$$\lim_{x \to c} f(x) = L$$

because, as suggested in the figures,

as x approaches c, $f(x)$ approaches L.

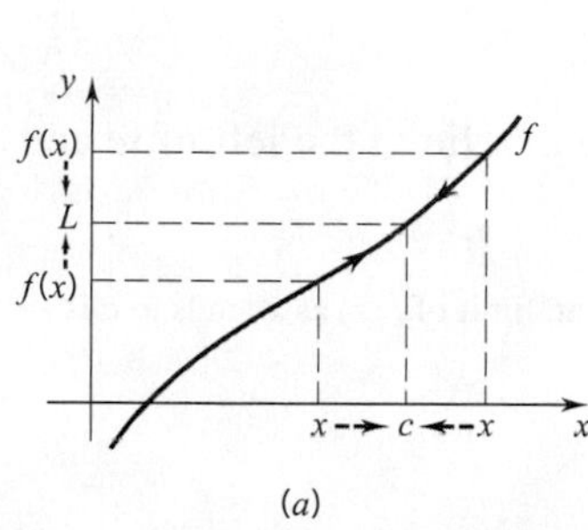

(*a*)

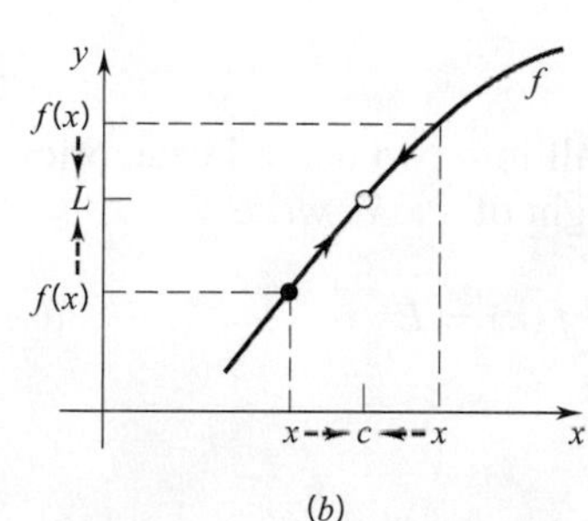

(*b*)

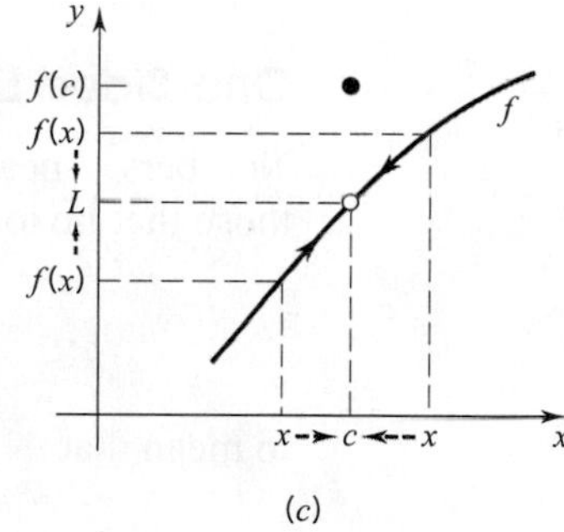

(*c*)

Figure 2.1.5

Example 4 Consider the function $f(x) = \dfrac{x^2 - 9}{x - 3}$ and let $c = 3$. Note that f is not defined at 3: both numerator and denominator are 0. But that doesn't matter. For all $x \neq 3$, and therefore *for all x near* 3,

$$\frac{x^2 - 9}{x - 3} = \frac{(x - 3)(x + 3)}{x - 3} = x + 3.$$

Therefore, if x is close to 3, then $\dfrac{x^2 - 9}{x - 3} = x + 3$ will be close to $3 + 3 = 6$. Thus,

$$\lim_{x \to 3} \frac{x^2 - 9}{x - 3} = \lim_{x \to 3} (x + 3) = 6.$$

The graph of f is shown in Figure 2.1.6. ❑

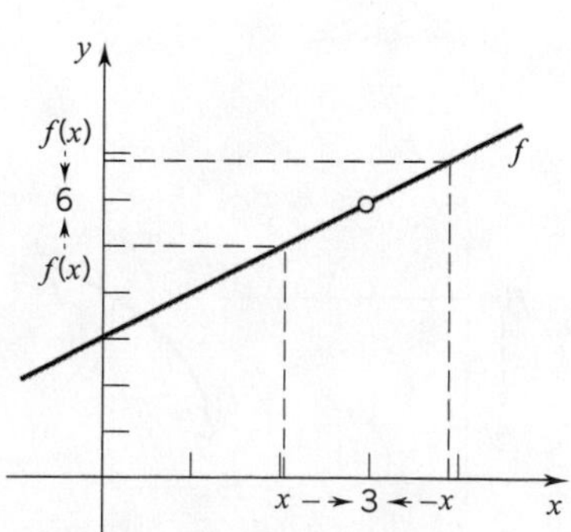

Figure 2.1.6

Example 5

$$\lim_{x\to 2} \frac{x^3 - 8}{x - 2} = 12.$$

The function $f(x) = \dfrac{x^3 - 8}{x - 2}$ is undefined at $x = 2$. But, as we said before, that doesn't matter. For all $x \neq 2$,

$$\frac{x^3 - 8}{x - 2} = \frac{(x-2)(x^2 + 2x + 4)}{x - 2} = x^2 + 2x + 4.$$

Therefore,

$$\lim_{x\to 2} \frac{x^3 - 8}{x - 2} = \lim_{x\to 2} (x^2 + 2x + 4) = 12. \quad \square$$

Example 6 If $f(x) = \begin{cases} 3x - 4, & x \neq 0 \\ 10, & x = 0, \end{cases}$ then $\lim_{x\to 0} f(x) = -4$.

It does not matter that $f(0) = 10$. For $x \neq 0$, and thus for all x near 0,

$$f(x) = 3x - 4 \quad \text{so that} \quad \lim_{x\to 0} f(x) = \lim_{x\to 0} (3x - 4) = -4. \quad \square$$

One-Sided Limits

Numbers x near c fall into two natural categories: those that lie to the left of c and those that lie to the right of c. We write

$$\lim_{x\to c^-} f(x) = L \qquad \text{[the left-hand limit of } f(x) \text{ as } x \text{ tends to } c \text{ is } L]$$

to mean that

as x approaches c from the left, $f(x)$ approaches L.

Similarly,

$$\lim_{x\to c^+} f(x) = L \qquad \text{[the right-hand limit of } f(x) \text{ as } x \text{ tends to } c \text{ is } L]$$

means that

as x approaches c from the right, $f(x)$ approaches L.†

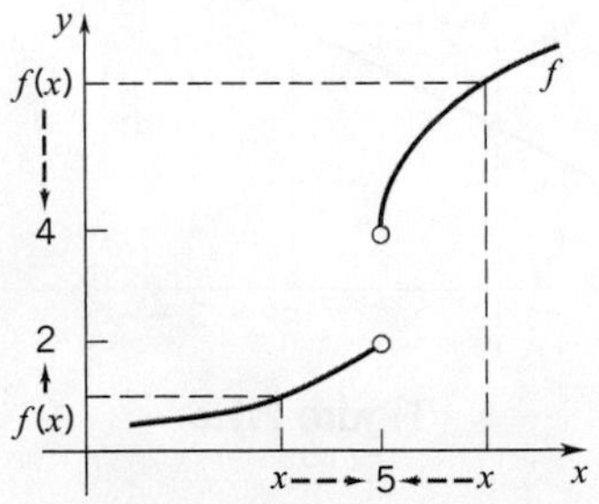

For an example, see the graph of the function f in Figure 2.1.7. As x approaches 5 from the left, $f(x)$ approaches 2. Thus,

$$\lim_{x\to 5^-} f(x) = 2.$$

As x approaches 5 from the right, $f(x)$ approaches 4, so

Figure 2.1.7

† The left-hand limit is sometimes written $\lim_{x\uparrow c} f(x)$ and the right-hand limit, $\lim_{x\downarrow c} f(x)$.

$$\lim_{x\to 5^+} f(x) = 4.$$

Note that $\lim_{x\to 5} f(x)$ does not exist, since it is not true that there is a single number L with the property that $f(x)$ is close to L *whenever* x is close to 5 [$f(x)$ is close to 2 when x is close to 5 and is less than 5; $f(x)$ is close to 4 when x is close to 5 and is greater than 5].

The left- and right-hand limits are called *one-sided limits*. The relationship between the *limit of f as x approaches c* and the two *one-sided limits of f as x approaches c* is as follows:

$$\lim_{x\to c} f(x) = L$$

if and only if (iff)† both

$$\lim_{x\to c^-} f(x) = L \quad \textit{and} \quad \lim_{x\to c^+} f(x) = L.$$

Example 7 For the function f graphed in Figure 2.1.8,

$$\lim_{x\to(-2)^-} f(x) = 5 \quad \text{and} \quad \lim_{x\to(-2)^+} f(x) = 5,$$

and therefore

$$\lim_{x\to -2} f(x) = 5.$$

For the limit, it does not matter that $f(-2) = 3$.

Examining the graph of f near $x = 4$, we find that

$$\lim_{x\to 4^-} f(x) = 7 \quad \text{whereas} \quad \lim_{x\to 4^+} f(x) = 2.$$

Since these one-sided limits are different,

$$\lim_{x\to 4} f(x) \quad \text{does not exist.} \quad \square$$

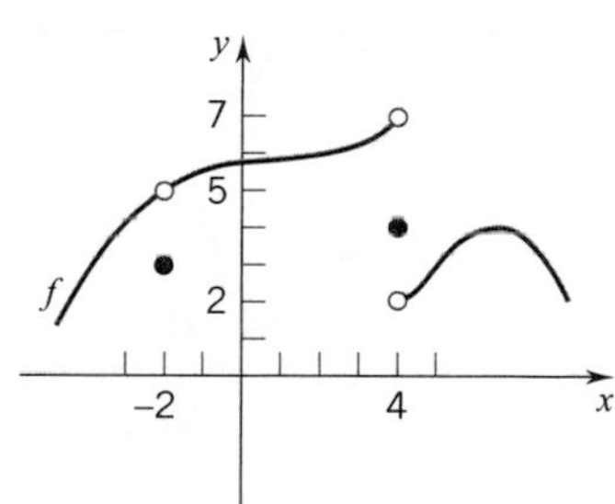

Figure 2.1.8

Example 8 Let f be the function defined by $f(x) = \dfrac{|x|}{x}$ for $x \neq 0$. This function can also be written equivalently as

$$f(x) = \begin{cases} 1, & \text{if } x > 0 \\ -1, & \text{if } x < 0. \end{cases} \qquad \text{(Figure 2.1.9)}$$

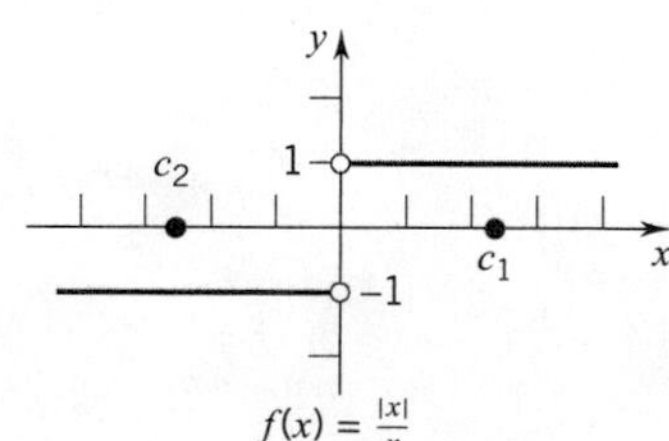

Figure 2.1.9

We consider $\lim_{x\to c} f(x)$ for several values of c.

For $c = 0$,

$$\lim_{x\to 0^-} f(x) = \lim_{x\to 0^-} (-1) = -1$$

and

$$\lim_{x\to 0^+} f(x) = \lim_{x\to 0^+} (1) = 1.$$

† Recall that the phrase "if and only if" occurs so frequently that we are abbreviating it to "iff".

Thus, $$\lim_{x\to 0} \frac{|x|}{x} \quad \text{does not exist.}$$

If $c = c_1$ is any positive number, then $f(x)$ "is close to 1" [in fact, $f(x) = 1$] when x is close to c_1. Therefore, when $c_1 > 0$

$$\lim_{x\to c_1} \frac{|x|}{x} = 1.$$

A similar analysis shows that if $c = c_2$ is a negative number, then

$$\lim_{x\to c_2} \frac{|x|}{x} = -1. \quad \square$$

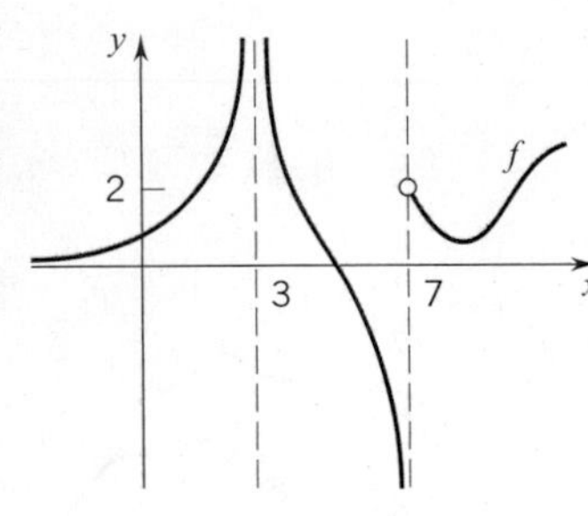

Figure 2.1.10

Example 9 Consider the function f graphed in Figure 2.1.10. As x approaches 3 from either side, $f(x)$ becomes arbitrarily large. Becoming arbitrarily large, $f(x)$ does not approach any fixed number L. Therefore

$$\lim_{x\to 3} f(x) \qquad \text{does not exist.}$$

Now let's focus on what happens near 7. As x approaches 7 from the left, $f(x)$ becomes arbitrarily large negative (i.e., less than any preassigned number). Under these circumstances $f(x)$ can't possibly approach a fixed number. Thus,

$$\lim_{x\to 7^-} f(x) \qquad \text{does not exist.}$$

Since the left-hand limit at $x = 7$ does not exist,

$$\lim_{x\to 7} f(x) \qquad \text{does not exist.}$$

[Note that the right-hand limit of f at $x = 7$ does exist; $\lim_{x\to 7^+} f(x) = 2$.] $\square$

Remark on "Infinite Limits" In Example 9 the function values $f(x)$ become arbitrarily large as x approaches 3. This particular behavior can be symbolized by writing

$$f(x) \to \infty \quad \text{as} \quad x \to 3.$$

The vertical line $x = 3$ is called a *vertical asymptote* of the graph of f. Similarly, the function values $f(x)$ become arbitrarily large negative as x approaches 7 from the left. This behavior is symbolized by writing

$$f(x) \to -\infty \quad \text{as} \quad x \to 7^-,$$

and the line $x = 7$ is a vertical asymptote of the graph. Asymptotes of the graphs of functions are discussed in detail in Section 4.7.

In general, the notations

$$f(x) \to \infty \quad \text{as} \quad x \to c, \quad f(x) \to \infty \quad \text{as} \quad x \to c^+, \quad \text{and} \quad f(x) \to \infty \quad \text{as} \quad x \to c^-$$

mean that the function values $f(x)$ become arbitrarily large as x approaches c in the indicated manner (i.e., from either side, from the right, and from the left, respectively). Similarly, the notations

$$f(x) \to -\infty \quad \text{as} \quad x \to c, \quad f(x) \to -\infty \quad \text{as} \quad x \to c^+, \quad \text{and} \quad f(x) \to -\infty \quad \textit{as} \quad x \to c^-$$

mean that the function values $f(x)$ become arbitrarily large negative as x approaches c in the indicated manner. It is important to understand that the underlying limits involved here *do not exist; ∞ is not a real number*. This is why we use the "arrow" notation rather than $\lim\limits_{x\to c}$ in these situations; this notation indicates the particular manner in which a certain limit *fails* to exist. □

Example 10 Let $f(x) = \dfrac{1}{x-2}$.

(a) $\lim\limits_{x\to 4} \dfrac{1}{x-2} = \dfrac{1}{2}$. **(b)** $\lim\limits_{x\to 2} \dfrac{1}{x-2}$ does not exist.

(a) As x approaches 4, $x - 2$ approaches 2, and $1/(x-2)$ approaches $\frac{1}{2}$.

(b) The function f is not defined at $x = 2$. However, if x is close to 2 and is greater than 2, then $1/(x-2)$ is a large positive number [$f(x) \to \infty$ as $x \to 2^+$]; if x is close to 2 and is less than 2, then $1/(x-2)$ is a large negative number [$f(x) \to -\infty$ as $x \to 2^-$]. See the table of values and Figure 2.1.11.

x	1.5	1.9	1.99	1.999	2	2.001	2.01	2.1	2.5
$f(x)$	−2	−10	−100	−1000		1000	100	10	2

□

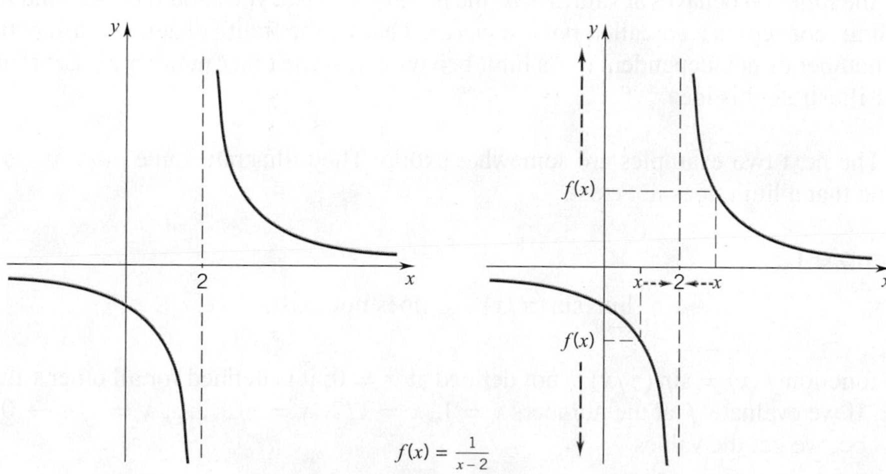

Figure 2.1.11

Example 11 Let $f(x) = \begin{cases} 1 - x^2; & x < 1 \\ 1/(x-1), & x > 1. \end{cases}$

Then

$$\lim_{x\to 1} f(x) \qquad \text{does not exist}$$

and

$$\lim_{x\to 1.5} f(x) = 2.$$

The first limit *does not exist* because

$$\lim_{x\to 1^-} f(x) = \lim_{x\to 1^-} (1 - x^2) = 0 \quad \text{and} \quad f(x) \to \infty \quad \text{as} \quad x \to 1^+.$$

The second limit *does exist* because for values of x sufficiently close to 1.5, the values of f are computed using the rule $f(x) = 1/(x - 1)$ independent of whether x is to the left or right of 1.5, and

$$\lim_{x\to 1.5} \left(\frac{1}{x-1}\right) = \frac{1}{0.5} = \frac{1}{\frac{1}{2}} = 2.$$

See Figure 2.1.12. □

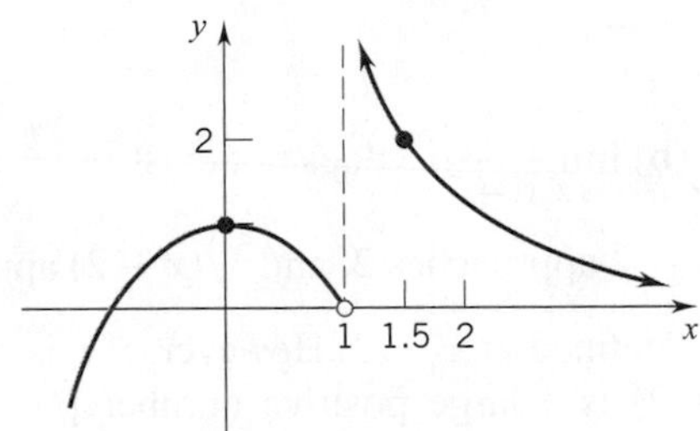

Figure 2.1.12

Remark Before going on to more examples, remember that, in taking the limit of a function as x approaches a given number c, it does not matter whether the function is defined *at the number c* and, if so, *how* it is defined there. The only thing that matters is how the function behaves at values *near* the number c. Also, you should understand that the limit concept is a so-called *point concept*. That is, the limit behavior of a function at a number c_1 is independent of its limit behavior at some other number c_2. Examples 7–11 illustrate this idea. □

The next two examples are somewhat exotic. They illustrate some other ways of seeing that a limit does not exist.

Example 12

$$\lim_{x\to 0} \sin(\pi/x) \qquad \text{does not exist.}$$

The function $f(x) = \sin(\pi/x)$ is not defined at $x = 0$; it is defined for all other values of x. If we evaluate f at the numbers $x = 1$, $x = 1/2$, $x = 1/3, \ldots$, $x = 1/n \to 0$ as $n \to \infty$, we get the values

$$\sin\frac{\pi}{1} = \sin\pi = 0, \sin\frac{\pi}{(1/2)} = \sin 2\pi = 0, \sin\frac{\pi}{(1/3)} = \sin 3\pi = 0, \quad \text{and so on.}$$

Based on these values, we might be tempted to conclude that $\lim_{x\to 0} \sin(\pi/x) = 0$. However, if we calculate the values of f at the numbers $x = 2$, $x = 2/5$, $x = 2/9$, $x = 2/13, \ldots$, $x = 2/(4n + 1)$. Then $2/(4n + 1) \to 0$ as $n \to \infty$, and

$$\sin(\pi/2) = 1$$

$$\sin\left(\pi/\tfrac{2}{5}\right) = \sin(5\pi/2) = 1,$$

$$\sin\left(\pi/\tfrac{2}{9}\right) = \sin(9\pi/2) = 1,$$

and, in general,

$$\sin\left(\pi / \tfrac{2}{4n+1}\right) = \sin\left[(4n+1)\pi/2\right] = 1.$$

This function is examined in more detail in Exercises 2.1. The graph of f is given in Figure 2.1.13. It shows that the values of f oscillate between 1 and -1 infinitely often as x approaches 0. Since the values do not approach a fixed number L, the limit does not exist. □

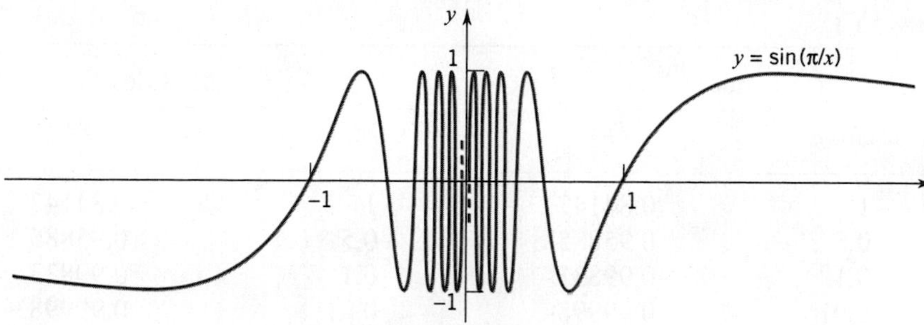

Figure 2.1.13

Our next example is an extreme illustration of the oscillatory behavior shown in Example 12. It is an adaptation of an example known as the Dirichlet function (P. G. Lejeune Dirichlet, 1805–1859).

Example 13 Let f be the function defined as follows:

$$f(x) = \begin{cases} 7, & x \text{ rational} \\ 4, & x \text{ irrational.} \end{cases} \qquad \text{(Figure 2.1.14)}$$

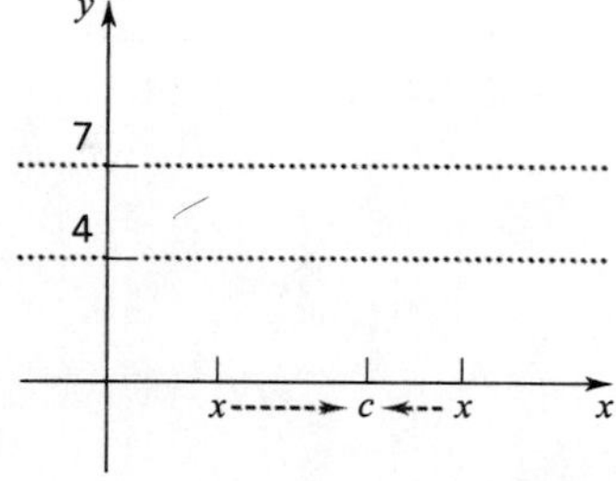

Figure 2.1.14

Now let c be any real number. As x approaches c, x passes through both rational and irrational numbers. As this happens, $f(x)$ jumps wildly back and forth between 7 and 4, and thus cannot stay close to any fixed number L. Therefore, $\lim_{x\to c} f(x)$ does not exist. □

Remark on Limits That Fail to Exist Examples 7–13 illustrate various ways in which the limit of a function f at a number c may fail to exist. We summarize the typical cases here:

(i) $\lim_{x\to c^-} f(x) = L_1$, $\lim_{x\to c^+} f(x) = L_2$ and $L_1 \neq L_2$ (Examples 7, 8).
(The left-hand and right-hand limits of f at c each exist, but they are not equal.)

(ii) $f(x) \to \pm\infty$ as $x \to c^-$, or $f(x) \to \pm\infty$ as $x \to c^+$, or both (Examples 9,10,11). (The function f is unbounded as x approaches c from the left, or from the right, or both.)

(iii) $f(x)$ "oscillates" as $x \to c^-, c^+$ or c (Examples 12, 13). □

In the preceding examples we used analytical and graphical methods to investigate the limit behavior of a function f at a number c. In the next example, we rely on a calculator and solve a limit problem using a numerical method. A calculator with graphing capabilities can be used to support numerical results. This numerical approach works well in many cases, but, as you will come to understand, there are limit problems which cannot be analyzed numerically.

Example 14 Let $f(x) = (\sin x)/x$. If we try to evaluate f at 0, we get the meaningless ratio $0/0$; $f(0)$ is not defined. However, f is defined for all $x \neq 0$, and so we can consider

$$\lim_{x \to 0} \frac{\sin x}{x}.$$

We select a sequence of numbers that approach 0 from the left and a sequence of numbers that approach 0 from the right. Using a calculator, we evaluate f at these numbers. The results are tabulated in Table 2.1.1.

■ **Table 2.1.1**

(Left side)		(Right side)	
x (radians)	$\frac{\sin x}{x}$	x (radians)	$\frac{\sin x}{x}$
−1	0.84147	1	0.84147
−0.5	0.95885	0.5	0.95885
−0.1	0.99833	0.1	0.99833
−0.01	0.99998	0.01	0.99998
−0.001	0.99999	0.001	0.99999

Based on these calculations, it appears that

$$\lim_{x \to 0^-} \frac{\sin x}{x} = 1 \quad \text{and} \quad \lim_{x \to 0^+} \frac{\sin x}{x} = 1.$$

Thus, we are led to conjecture that

$$\lim_{x \to 0} \frac{\sin x}{x} = 1.$$

The graphs shown in Figure 2.1.15 support this conjecture. First, we graphed the function on the interval $[-10, 10]$ (see Figure 2.1.15*a*). Then we zoomed in to obtain the graphs in Figures 2.1.15*b* and 2.1.15*c*. The fact that this limit *is* 1 is established rigorously in Section 2.5. □

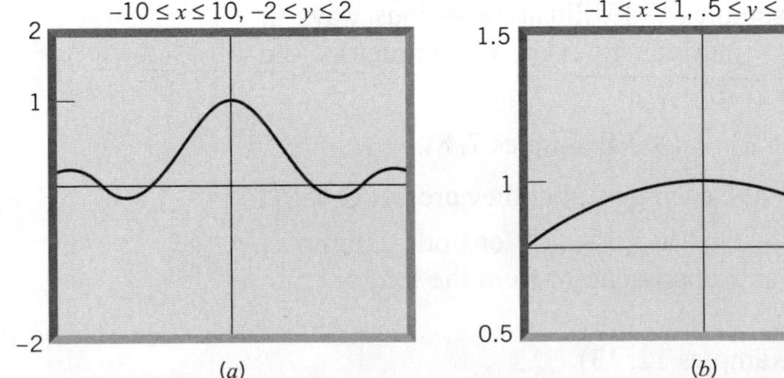

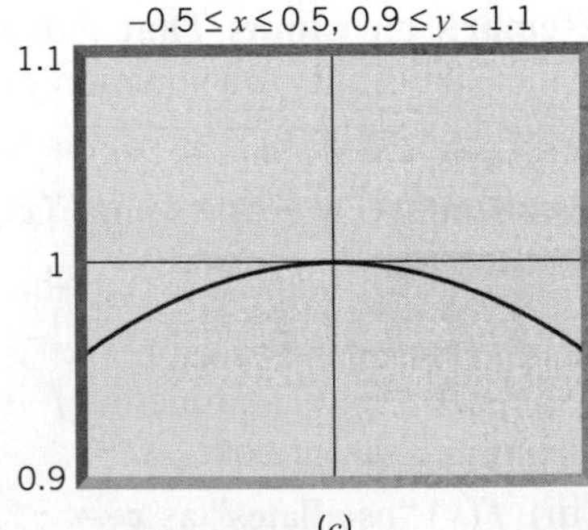

Figure 2.1.15

If you have found the treatment in this section to be imprecise, you are absolutely right. Our work so far has been imprecise, but it need not remain so. One of the great triumphs of calculus has been its capacity to formulate limit statements with precision, but for this you will have to wait until Section 2.2.

EXERCISES 2.1

In Exercises 1–10 you are given a number c and the graph of a function f. Use the graph of f to find

(a) $\lim_{x\to c^-} f(x)$ (b) $\lim_{x\to c^+} f(x)$ (c) $\lim_{x\to c} f(x)$ (d) $f(c)$.

1. $c = 2$.

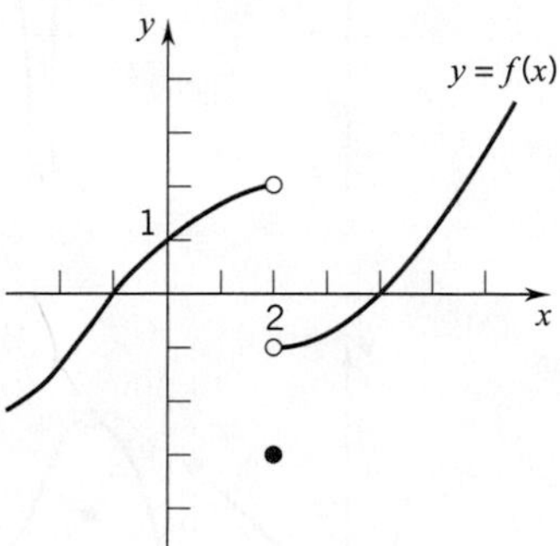

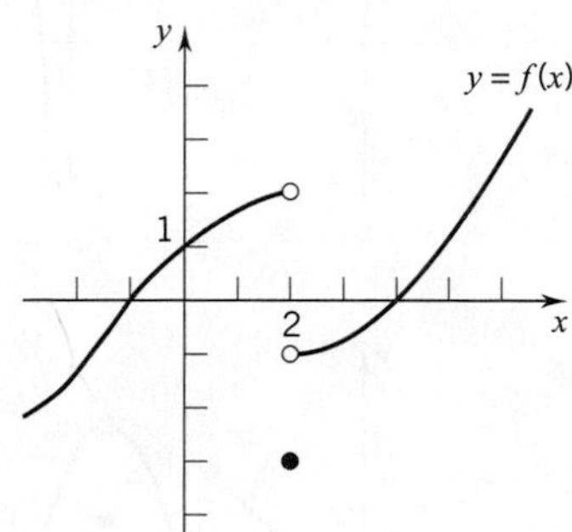

2. $c = 3$.

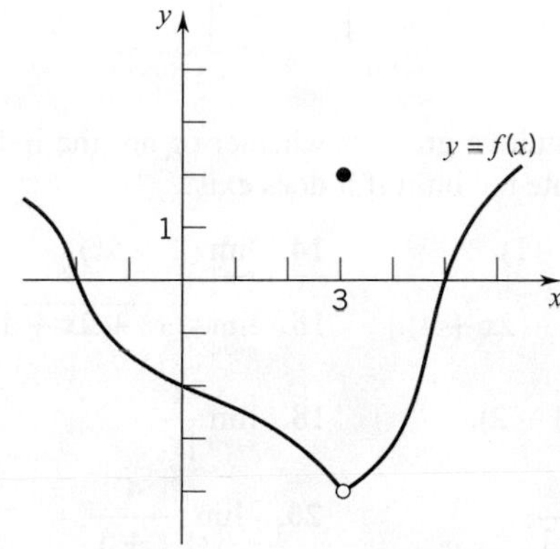

3. $c = 3$.

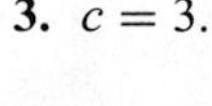

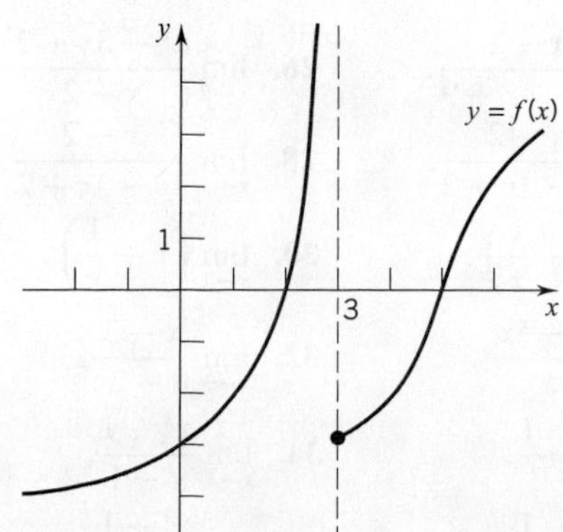

4. $c = 4$.

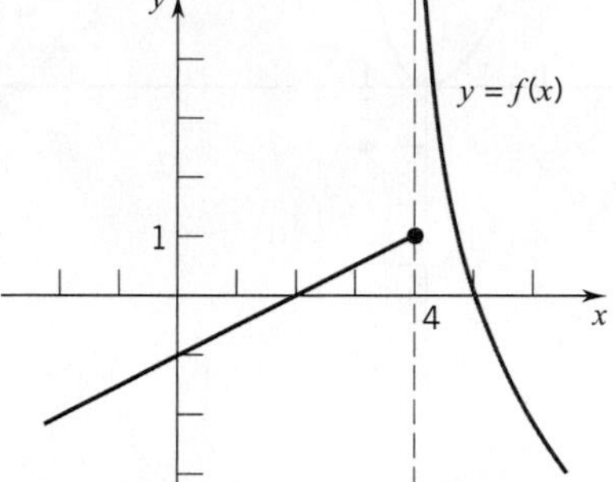

5. $c = -2$.

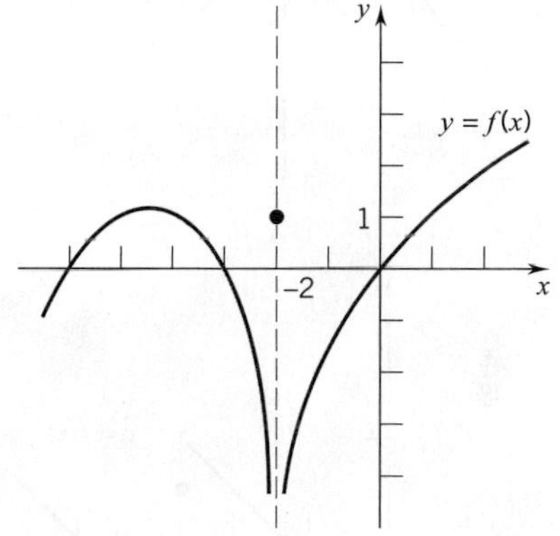

6. $c = 1$.

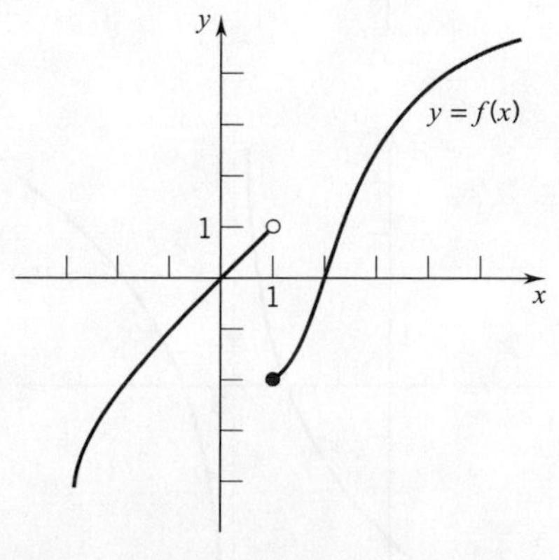

7. $c = 1$.

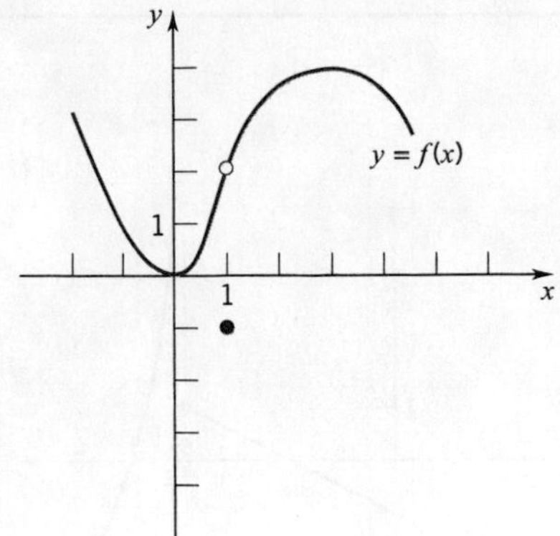

8. $c = -1$.

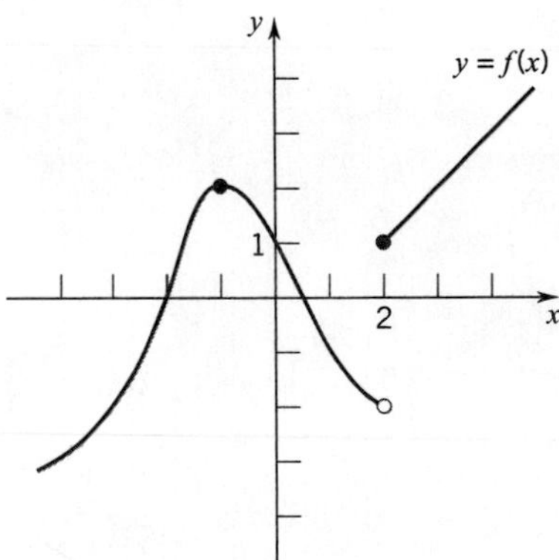

9. $c = 2$.

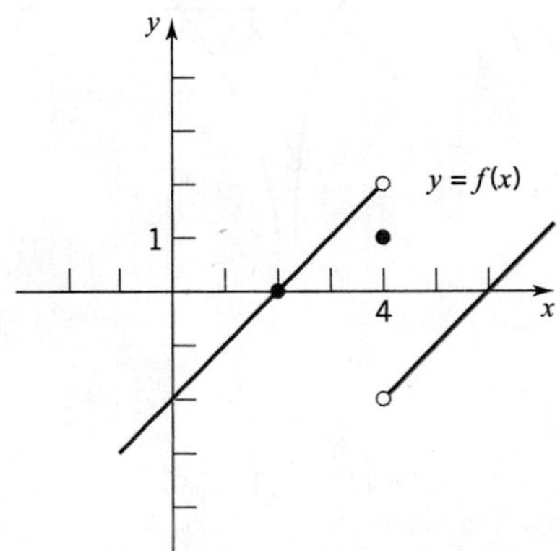

10. $c = 3$.

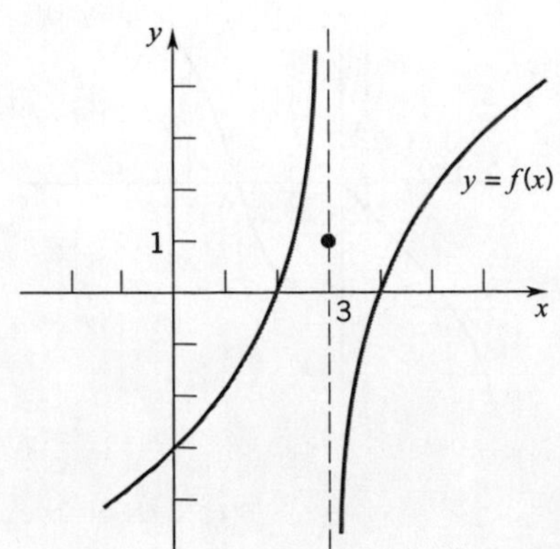

In Exercises 11 and 12, give the values of c for which $\lim_{x \to c} f(x)$ does not exist.

11.

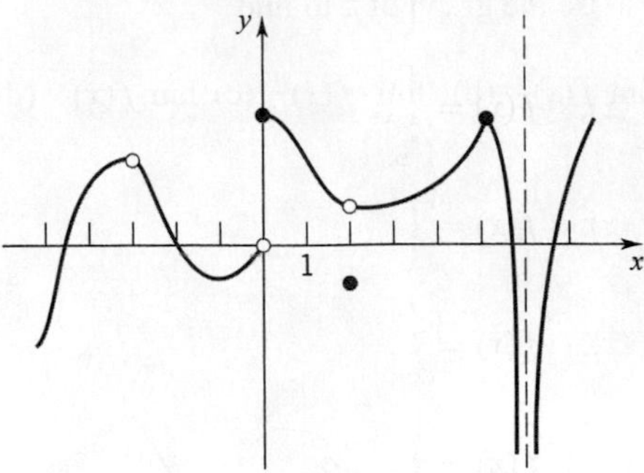

12.

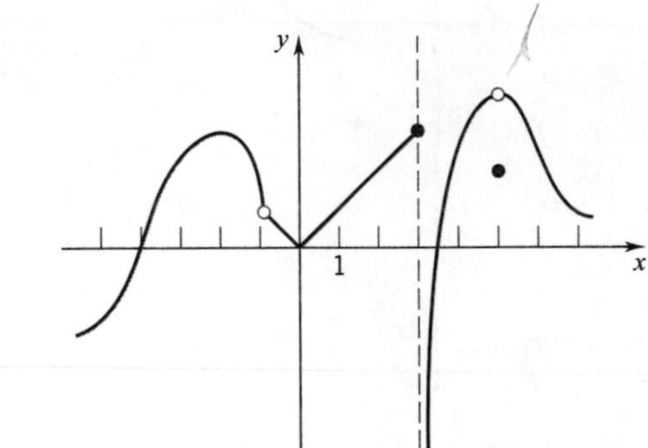

Decide on intuitive grounds whether or not the indicated limit exists; evaluate the limit if it does exist.

13. $\lim_{x \to 0} (2x - 1)$.

14. $\lim_{x \to 1} (2 - 5x)$.

15. $\lim_{x \to -2} (x^2 - 2x + 4)$.

16. $\lim_{x \to 4} \sqrt{x^2 + 2x + 1}$.

17. $\lim_{x \to -3} (|x| - 2)$.

18. $\lim_{x \to 0} \dfrac{1}{|x|}$.

19. $\lim_{x \to 1} \dfrac{3}{x + 1}$.

20. $\lim_{x \to -1} \dfrac{4}{x + 1}$.

21. $\lim_{x \to -1} \dfrac{-2}{x + 1}$.

22. $\lim_{x \to 2} \dfrac{1}{3x - 6}$.

23. $\lim_{x \to 3} \dfrac{2x - 6}{x - 3}$.

24. $\lim_{x \to 3} \dfrac{x^2 - 6x + 9}{x - 3}$.

25. $\lim_{x \to 3} \dfrac{x - 3}{x^2 - 6x + 9}$.

26. $\lim_{x \to 2} \dfrac{x^2 - 3x + 2}{x - 2}$.

27. $\lim_{x \to 2} \dfrac{x - 2}{x^2 - 3x + 2}$.

28. $\lim_{x \to 1} \dfrac{x - 2}{x^2 - 3x + 2}$.

29. $\lim_{x \to 0} \left(x + \dfrac{1}{x}\right)$.

30. $\lim_{x \to 1} \left(x + \dfrac{1}{x}\right)$.

31. $\lim_{x \to 0} \dfrac{2x - 5x^2}{x}$.

32. $\lim_{x \to 3} \dfrac{x - 3}{6 - 2x}$.

33. $\lim_{x \to 1} \dfrac{x^2 - 1}{x - 1}$.

34. $\lim_{x \to 1} \dfrac{x^3 - 1}{x - 1}$.

35. $\lim_{x \to 1} \dfrac{x^3 - 1}{x + 1}$.

36. $\lim_{x \to 1} \dfrac{x^2 + 1}{x^2 - 1}$.

37. $\lim_{x\to 0} f(x)$; $f(x) = \begin{cases} 1, & x \neq 0 \\ 3, & x = 0. \end{cases}$

38. $\lim_{x\to 1} f(x)$; $f(x) = \begin{cases} 3x, & x < 1 \\ 3, & x > 1. \end{cases}$

39. $\lim_{x\to 4} f(x)$; $f(x) = \begin{cases} x^2, & x \neq 4 \\ 0, & x = 4. \end{cases}$

40. $\lim_{x\to 0} f(x)$; $f(x) = \begin{cases} -x^2, & x < 0 \\ x^2, & x > 0. \end{cases}$

41. $\lim_{x\to 0} f(x)$; $f(x) = \begin{cases} x^2, & x < 0 \\ 1+x, & x > 0. \end{cases}$

42. $\lim_{x\to 1} f(x)$; $f(x) = \begin{cases} 2x, & x < 1 \\ x^2+1, & x > 1. \end{cases}$

43. $\lim_{x\to 2} f(x)$; $f(x) = \begin{cases} 3x, & x < 1 \\ x+2, & x \geq 1. \end{cases}$

44. $\lim_{x\to 0} f(x)$; $f(x) = \begin{cases} 2x, & x \leq 1 \\ x+1, & x > 1. \end{cases}$

45. $\lim_{x\to 0} f(x)$; $f(x) = \begin{cases} 2, & x \text{ rational} \\ -2, & x \text{ irrational}. \end{cases}$

46. $\lim_{x\to 1} f(x)$; $f(x) = \begin{cases} 2x, & x \text{ rational} \\ 2, & x \text{ irrational}. \end{cases}$

47. $\lim_{x\to 1} \dfrac{\sqrt{x^2+1}-\sqrt{2}}{x-1}$.

48. $\lim_{x\to 5} \dfrac{\sqrt{x^2+5}-\sqrt{30}}{x-5}$.

49. $\lim_{x\to 1} \dfrac{x^2-1}{\sqrt{2x+2}-2}$.

For Exercises 50 and 51 use $f(x) = \sin(\pi/x)$.

50. (a) Show that $f(1/n) = 0$ for every nonzero integer n.

(b) Show that $f(1/[2n+\frac{1}{2}]) = 1$ for every integer n.

(c) Show that $f(1/[2n+\frac{3}{2}]) = -1$ for every integer n.

51. (a) Plot the points $(x, f(x))$ on the graph of f with $x = 1, 1/2, 1/3, 1/4, 2, 2/5, 2/9, 2/3, 2/7, 2/11$.

(b) Does $\lim_{x\to 0^+} f(x)$ exist? What do you conjecture about $\lim_{x\to 0^-} f(x)$?

(c) Use a graphing utility to sketch the graph of f for $-\pi \leq x \leq \pi$.

In Exercises 52–56, estimate $\lim_{x\to c} f(x)$ by creating a table of values. Then use a graphing utility to zoom in on the graph of f near $x = c$ to justify or improve your guess.

52. Estimate

$$\lim_{x\to 0} \frac{1-\cos x}{x} \qquad \text{(radian measure)}$$

after evaluating the quotient at $x = \pm 1, \pm 0.1, \pm 0.01, \pm 0.001$.

53. Estimate

$$\lim_{x\to 0} \frac{\tan 2x}{x} \qquad \text{(radian measure)}$$

after evaluating the quotient at $x = \pm 1, \pm 0.1, \pm 0.01, \pm 0.001$.

54. Estimate

$$\lim_{x\to 0} \frac{x-\sin x}{x^3} \qquad \text{(radian measure)}$$

after evaluating the quotient at $\pm 1, \pm 0.1, \pm 0.01, \pm 0.001, \pm 0.0001$.

55. Estimate

$$\lim_{x\to 1} \frac{x^{3/2}-1}{x-1}$$

after evaluating the quotient at $x = 0.9, 0.99, 0.999, 0.9999$ and $x = 1.1, 1.01, 1.001, 1.0001$.

56. Estimate

$$\lim_{x\to 0} \frac{2^x-1}{x}$$

after evaluating the quotient at $x = \pm 1, \pm 0.1, \pm 0.01, \pm 0.001, \pm 0.0001$.

57. (a) Use a graphing utility to estimate $\lim_{x\to 4} f(x)$ for each of the following functions:

$$(i)\ f(x) = \frac{2x^2-11x+12}{x-4}; \quad (ii)\ f(x) = \frac{2x^2-11x+12}{x^2-8x+16}.$$

(b) Use a CAS to find each of the limits in part (a).

58. (a) Use a graphing utility to estimate $\lim_{x\to 4} f(x)$ for each of the following functions:

$$(i)\ f(x) = \frac{3x^2-10x-8}{5x^2+16x-16}; \quad (ii)\ f(x) = \frac{5x^2-26x+24}{4x^2-11x-20}.$$

(b) Use a CAS to find each of the limits in part (a).

59. (a) Use a graphing utility to estimate $\lim_{x\to 2} f(x)$ for each of the following functions:

$$(i)\ f(x) = \frac{\sqrt{6-x}-x}{x-2}; \quad (ii)\ f(x) = \frac{x^2-4x+4}{x-\sqrt{6-x}}.$$

(b) Use a CAS to find each of the limits in part (a).

60. (a) Use a graphing utility to estimate $\lim_{x\to 2} f(x)$ for each of the following functions:

$$(i)\ f(x) = \frac{2x-\sqrt{18-x}}{4-x^2}; \quad (ii)\ f(x) = \frac{2-\sqrt{2x}}{\sqrt{8x}-4}.$$

(b) Use a CAS to find each of the limits in part (a).

Use a graphing utility to find at least one number c such that $\lim_{x\to c} f(x)$ does not exist.

61. $f(x) = \dfrac{x+1}{|x^3+1|}$.

62. $f(x) = \dfrac{|6x^2 - x - 35|}{2x-5}$.

63. $f(x) = \dfrac{|x|}{x^5 + 2x^4 + 13x^3 + 26x^2 + 36x + 72}$.

64. $f(x) = \dfrac{5x^3 - 22x^2 + 15x + 18}{x^3 - 9x^2 + 27x - 27}$.

■ PROJECT 2.1 Tangent Lines

Let f be a function defined on an interval I and let $c \in I$. Choose a number $x \in I$, $x \neq c$. The line determined by the points $(c, f(c))$ and $(x, f(x))$ is called a *secant line* (see Figure A).

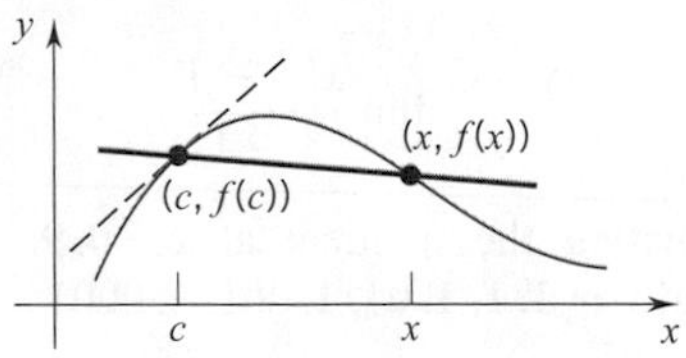

Figure A

The slope of the secant line is given by

$$m_{sec} = \frac{f(x) - f(c)}{x - c}.$$

If we let x approach c, then point $(x, f(x))$ will move along the graph of f toward the point $(c, f(c))$ and the secant line will approach the dashed line as a limiting position. The *line tangent* to the graph of f at the point $(c, f(c))$ is defined to be the line through $(c, f(c))$ with slope

$$m = \lim_{x\to c} \frac{f(x) - f(c)}{x - c}. \quad \text{(provided this limit exists)}$$

For example, let $f(x) = x^2 - x$ and $c = 2$. Then $(2, f(2)) = (2, 2)$ is the corresponding point on the graph of f. The slope of the line tangent to the graph of f at the point (2,2) is

$$m = \lim_{x\to 2} \frac{f(x) - f(2)}{x - 2} = \lim_{x\to 2} \frac{x^2 - x - 2}{x - 2}$$
$$= \lim_{x\to 2} \frac{(x-2)(x+1)}{x - 2} = \lim_{x\to 2} (x+1) = 3.$$

As an equation for the tangent line we have:

$$y - 2 = 3(x - 2) \quad \text{which can be written} \quad y = 3x - 4.$$

Figure B shows the graph of f and the line tangent to the graph at (2,2).

Problem 1. Let $f(x) = x^2 - x$ and let $c = 2$.

a. Use a graphing utility to draw the graph of f and the secant lines through $(2, f(2))$ and the points $(5, f(5))$, $(4, f(4))$, and $(3, f(3))$ together.

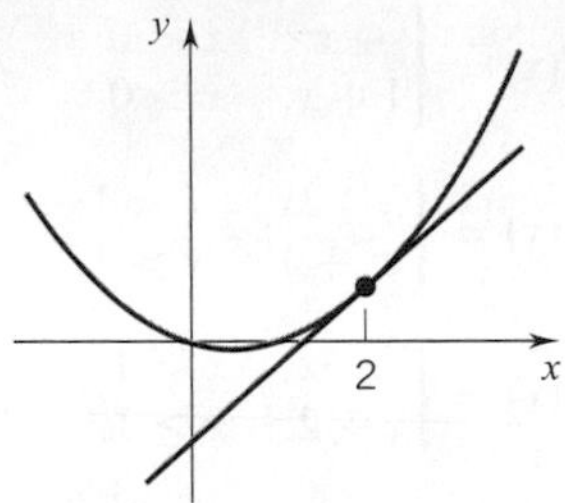

Figure B

b. Add the graph of the tangent line $y - 2 = 3(x - 2)$ to your figure in (a).

c. Experiment with other points $(a, f(a))$ approaching $(2, f(2))$; choose x-values on both sides of $x = 2$.

Problem 2. Calculate $\dfrac{f(x) - f(c)}{x - c}$, determine whether $\lim_{x\to c} \dfrac{f(x) - f(c)}{x - c}$ exists, and give an equation for the line tangent to the graph of f at $(c, f(c))$ if the limit does exist. Use a graphing utility to draw the graph of f and the tangent line together.

a. $f(x) = x^2$; $c = 2$.

b. $f(x) = x^3 + 1$; $c = 1$.

c. $f(x) = 1 - 2x + x^2$; $c = -1$.

Problem 3. Repeat Problem 2 with the functions:

a. $f(x) = 1/x$; $c = 2$.

b. $f(x) = \sqrt{x}$; $c = 1$. HINT: Rationalize the numerator.

Problem 4. Calculate $\dfrac{f(x) - f(c)}{x - c}$ and show that $\lim_{x\to c} \dfrac{f(x) - f(c)}{x - c}$ does not exist. Investigate the one-sided limits at $x = c$. Use a graphing utility to draw the graph of f. Does there appear to be a tangent line to the graph of f at $(c, f(c))$ even though the limit does not exist?

a. $f(x) = \sqrt[3]{x}$; $c = 0$.

b. $f(x) = \sqrt[3]{(x-1)^2}$; $c = 1$.

■ 2.2 DEFINITION OF LIMIT

Our work with limits in Section 2.1 was intuitive and informal. It is time to be more precise.

Let f be a function and let c be a real number. We do not require that f be defined at c, but we do require that f be defined at least on some set of the form $(c-p, c) \cup (c, c+p)$ with $p > 0$. This guarantees that we can form $f(x)$ for all $x \neq c$ that are "sufficiently close" to c.

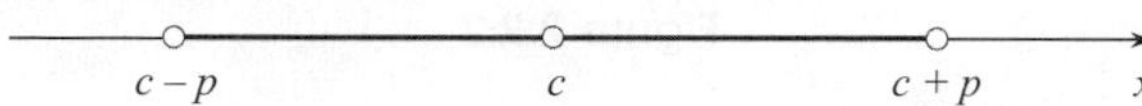

To say that

$$\lim_{x \to c} f(x) = L$$

means that we can make $f(x)$ be as close to L as we want simply by choosing x close enough to c, but $x \neq c$. That is,

Informal Statement	**ϵ, δ statement**
$\|f(x) - L\|$ can be made arbitrarily small simply by requiring that $\|x - c\|$ be sufficiently small but different from zero.	given any $\epsilon > 0$, $\|f(x) - L\|$ can be made less than ϵ simply by requiring that $\|x - c\|$ satisfy an inequality of the form $0 < \|x - c\| < \delta$ for δ sufficiently small.

Putting the various pieces together in compact form, we have the following precise definition.

DEFINITION 2.2.1 THE LIMIT OF A FUNCTION

Let f be a function defined at least on some set of the form

$$(c-p, c) \cup (c, c+p), \text{ with } p > 0.$$

We say that

$$\lim_{x \to c} f(x) = L$$

if for each $\epsilon > 0$, there exists a $\delta > 0$ such that

$$\text{if} \quad 0 < |x - c| < \delta, \quad \text{then} \quad |f(x) - L| < \epsilon$$

Figures 2.2.1 and 2.2.2 illustrate this definition.

In general, the choice of δ depends on the previous choice of ϵ. We do not require that there exists a number δ which "works" for *all* ϵ, but rather, that for each ϵ there exists a δ which "works" for it.

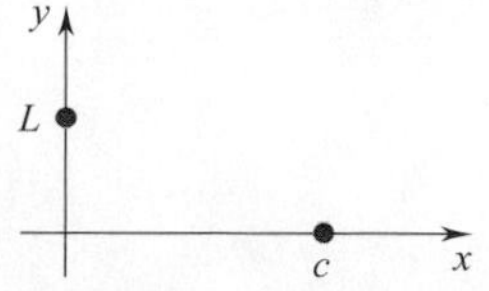

Figure 2.2.1

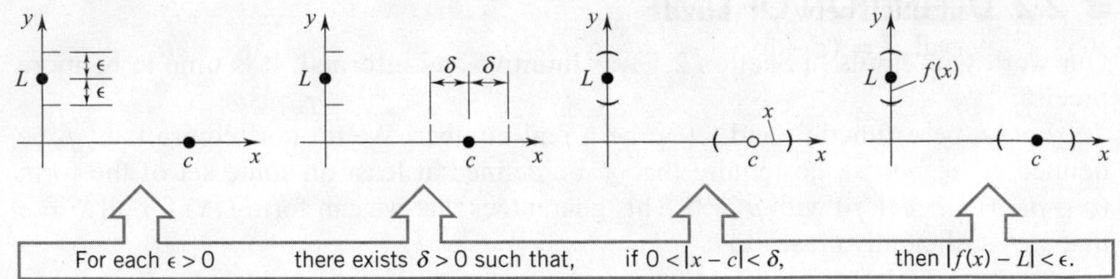

Figure 2.2.2

In Figure 2.2.3, we give two choices of ϵ and for each we display a suitable δ. For a δ to be suitable, all the points within δ of c (with the possible exception of c itself) must be taken by the function f to within ϵ of L. In Figure 2.2.3b we began with a smaller ϵ and had to use a smaller δ.

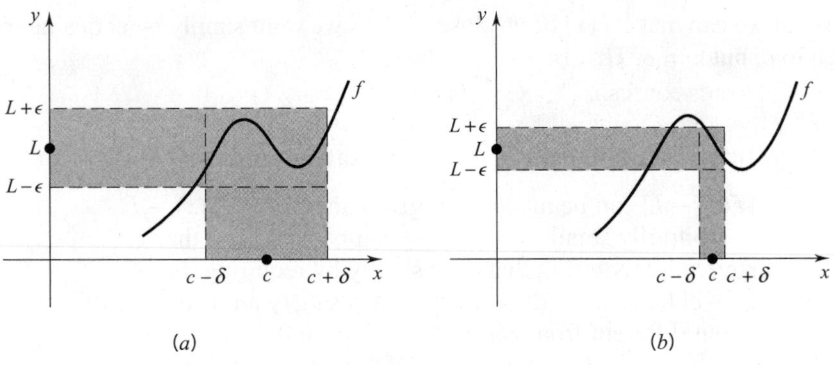

Figure 2.2.3

The δ of Figure 2.2.4 is too large for the given ϵ. In particular, the points marked x_1 and x_2 in the figure are not taken by the function f to within ϵ of L.

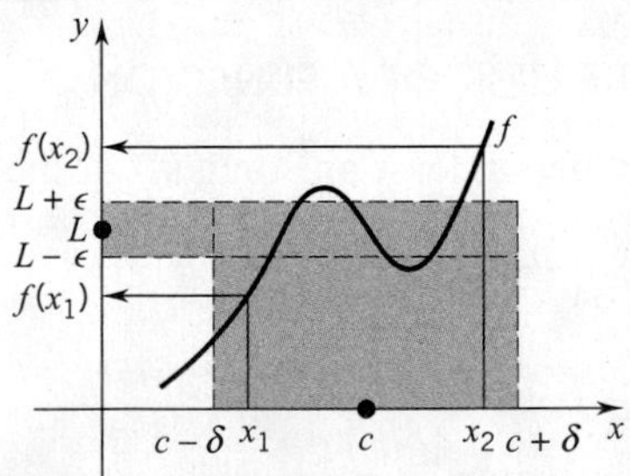

Figure 2.2.4

These illustrations suggest an equivalent definition of "limit" in terms of intervals.

Let f be defined on some open interval $(c - p, c + p)$, except, possibly, at c itself. Then

(2.2.2) $$\lim_{x \to c} f(x) = L$$

if for each open interval I centered at L, there exists an open interval J centered at c such that if $x \in J$, $x \neq c$, then $f(x) \in I$.

The connection between the two definitions is this: the interval $I = (L-\epsilon, L+\epsilon)$ and the interval $J = (c-\delta, c+\delta)$. See Figure 2.2.5.

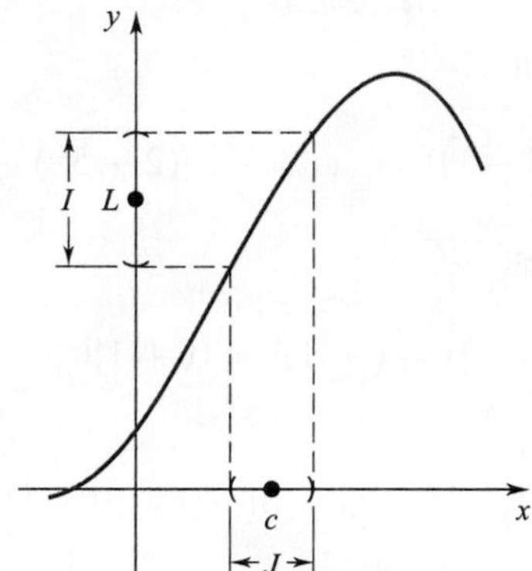

Figure 2.2.5

Next we apply the ϵ, δ definition of limit to a variety of functions. You may find the ϵ, δ arguments confusing at first, but do not be discouraged; it usually takes a little while for the ϵ, δ idea to take hold.

Example 1 Show that

$$\lim_{x\to 2}(2x-1) = 3. \qquad \text{(Figure 2.2.6)}$$

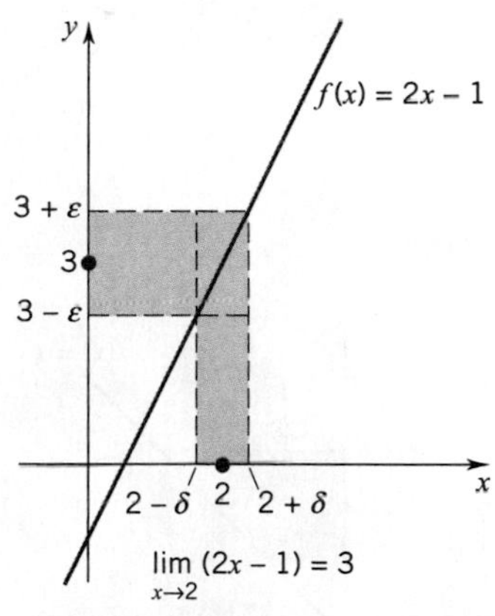

Figure 2.2.6

Finding a δ. Let $\epsilon > 0$. We seek a number $\delta > 0$ such that

$$\text{if} \quad 0 < |x-2| < \delta, \qquad \text{then} \qquad |(2x-1)-3| < \epsilon.$$

What we have to do first is establish a connection between

$$|(2x-1)-3| \qquad \text{and} \qquad |x-2|.$$

The connection is:

$$|(2x-1)-3| = |2x-4|$$

so that

$$(*) \qquad |(2x-1)-3| = 2|x-2|.$$

To make $|(2x-1)-3|$ less than ϵ, we need to make $2|x-2| < \epsilon$, which can be accomplished by making $|x-2| < \epsilon/2$. This suggests that we choose $\delta = \frac{1}{2}\epsilon$.

Showing that the δ "works." If $0 < |x-2| < \frac{1}{2}\epsilon$, then $2|x-2| < \epsilon$ and, by $(*)$, $|(2x-1)-3| < \epsilon$. □

Remark In Example 1 we chose $\delta = \frac{1}{2}\epsilon$, but we could have chosen *any* positive number δ such that $\delta \leq \frac{1}{2}\epsilon$. In general, if a certain δ^* "works" for a given ϵ, then any $\delta < \delta^*$ will also work. □

Example 2 Show that

$$\lim_{x\to -1}(2-3x) = 5. \qquad \text{(Figure 2.2.7)}$$

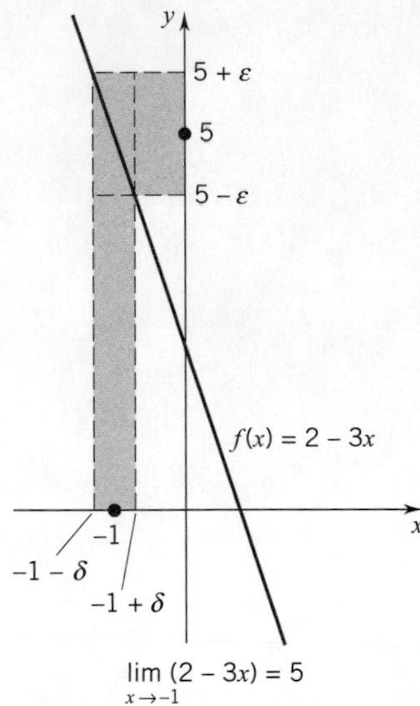

Figure 2.2.7

Finding a δ. Let $\epsilon > 0$. We seek a number $\delta > 0$ such that

$$\text{if} \quad 0 < |x - (-1)| < \delta, \qquad \text{then} \qquad |(2 - 3x) - 5| < \epsilon.$$

To find a connection between

$$|x - (-1)| \qquad \text{and} \qquad |(2 - 3x) - 5|,$$

we simplify both expressions:

$$|x - (-1)| = |x + 1|$$

and
$$|(2 - 3x) - 5| = |-3x - 3| = |-3||x + 1| = 3|x + 1|.$$

We can conclude that

$$(**) \qquad |(2 - 3x) - 5| = 3|x - (-1)|.$$

We can make the expression on the left less than ϵ by making $|x - (-1)|$ less than $\epsilon/3$. This suggests that we set $\delta = \frac{1}{3}\epsilon$.

Showing that the δ *"works."* If $0 < |x - (-1)| < \frac{1}{3}\epsilon$, then $3|x - (-1)| < \epsilon$ and, by (**), $|(2 - 3x) - 5| < \epsilon$. ❑

Three Basic Limits

We now establish three basic limits that will be used in the sections that follow.

Example 3 Let c be any number. Then

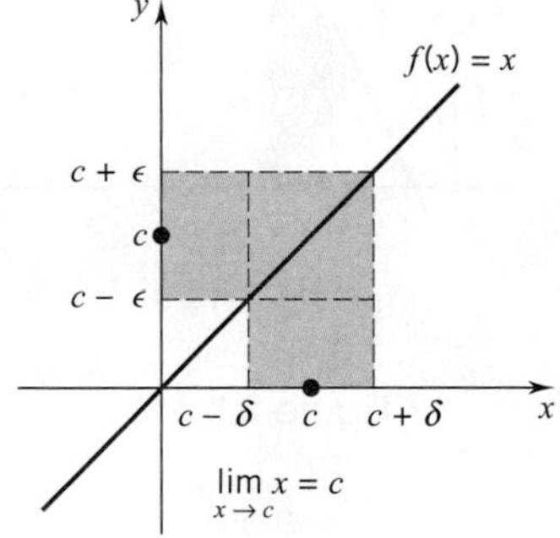

Figure 2.2.8

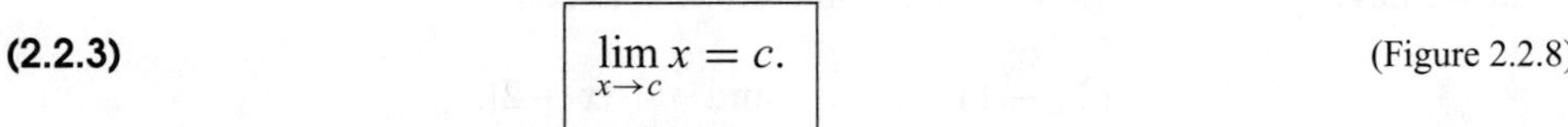

(2.2.3) $$\boxed{\lim_{x \to c} x = c.}$$ (Figure 2.2.8)

PROOF Let $\epsilon > 0$. We must find a $\delta > 0$ such that

$$\text{if} \quad 0 < |x - c| < \delta, \qquad \text{then} \quad |x - c| < \epsilon.$$

Obviously we can choose $\delta = \epsilon$. ❑

Example 4 Let c be any number. Then

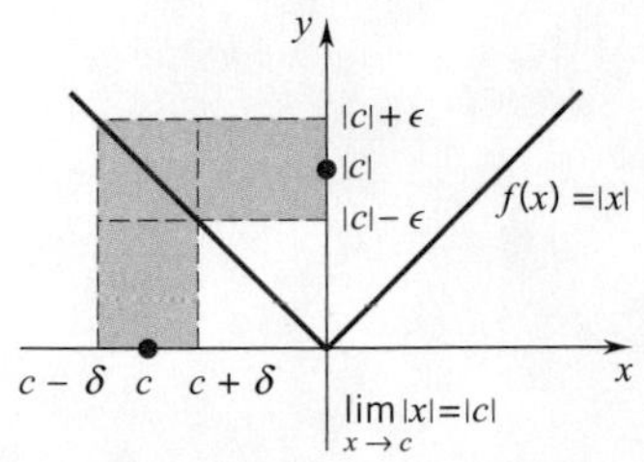

Figure 2.2.9

(2.2.4) $$\boxed{\lim_{x \to c} |x| = |c|.}$$ (Figure 2.2.9)

PROOF Let $\epsilon > 0$. We seek a $\delta > 0$ such that

$$\text{if} \quad 0 < |x - c| < \delta, \qquad \text{then} \qquad \big||x| - |c|\big| < \epsilon.$$

Since
$$\big||x| - |c|\big| \leq |x - c|,$$
(Recall: $\big||a| - |b|\big| \leq |a - b|$)

we can choose $\delta = \epsilon$; for

$$\text{if} \quad 0 < |x - c| < \epsilon, \qquad \text{then} \quad \big||x| - |c|\big| < \epsilon. \quad \square$$

Example 5

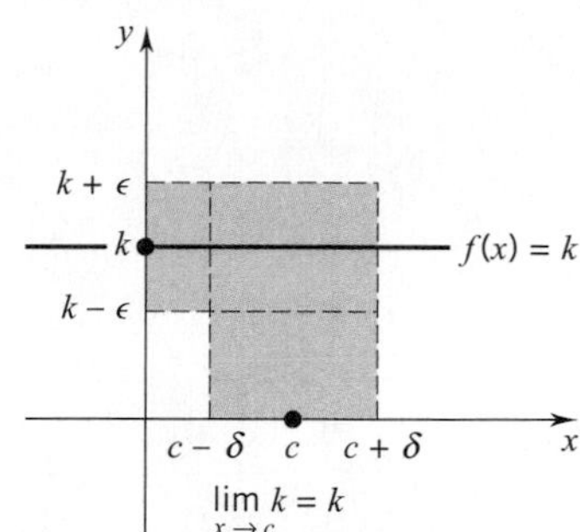

Figure 2.2.10

(2.2.5) $$\boxed{\lim_{x\to c} k = k, \quad k \text{ a constant.}}$$ (Figure 2.2.10)

PROOF Here we are dealing with the constant function

$$f(x) = k.$$

Let $\epsilon > 0$. We must find a $\delta > 0$ such that

$$\text{if} \quad 0 < |x - c| < \delta, \qquad \text{then} \quad |k - k| < \epsilon.$$

Since $|k - k| = 0$, we always have

$$|k - k| < \epsilon$$

no matter how δ is chosen; in short, any positive number will do for δ. $\square$

Usually ϵ, δ arguments are carried out in two stages. First we do a little algebraic scratch work, labeled "finding a δ" in Examples 1 and 2. This scratch work involves working backward from $|f(x) - L| < \epsilon$ to find a $\delta > 0$ sufficiently small so that we can begin with the inequality $0 < |x - c| < \delta$ to arrive at $|f(x) - L| < \epsilon$. This first stage is just preliminary, but it shows us how to proceed in the second stage. The second stage consists of showing that the δ "works" by verifying that for our choice of δ, the implication

$$\text{if} \quad 0 < |x - c| < \delta, \qquad \text{then} \quad |f(x) - L| < \epsilon$$

is true. The next two examples are more complicated and therefore may give you a better feeling for this idea of working backward to find a δ.

Example 6

$$\lim_{x\to 3} x^2 = 9 \qquad \text{(Figure 2.2.11)}$$

Finding a δ. Let $\epsilon > 0$. We seek a $\delta > 0$ such that

$$\text{if} \quad 0 < |x - 3| < \delta, \qquad \text{then} \quad |x^2 - 9| < \epsilon.$$

The needed connection between $|x - 3|$ and $|x^2 - 9|$ is found by factoring:

$$x^2 - 9 = (x + 3)(x - 3);$$

and thus,

$$(*) \qquad |x^2 - 9| = |x + 3||x - 3|.$$

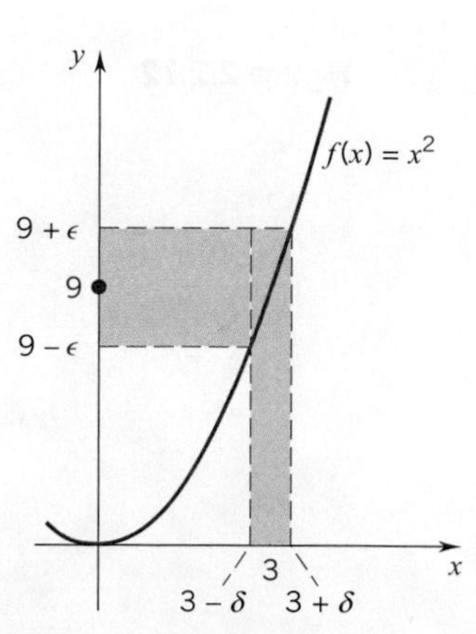

Figure 2.2.11

At this point, we need to get an estimate for the size of $|x+3|$ for x close to 3. For convenience, we'll take x within one unit of 3.

If $|x-3|<1$, then $2<x<4$ and

$$|x+3| \le |x| + |3| = x+3 < 7.$$

Therefore, by $(*)$,

$$(**) \qquad \text{if} \quad |x-3|<1, \qquad \text{then} \qquad |x^2-9| < 7|x-3|.$$

If, in addition, $|x-3| < \epsilon/7$, then it will follow that

$$|x^2-9| < 7(\epsilon/7) = \epsilon.$$

This means that we can take $\delta =$ the minimum of 1 and $\epsilon/7$.

Showing that the δ "works." Let $\epsilon > 0$. Choose $\delta = \min\{1, \epsilon/7\}$ and assume that

$$0 < |x-3| < \delta.$$

Then $\qquad |x-3| < 1 \qquad \text{and} \qquad |x-3| < \epsilon/7.$

By $(**)$,

$$|x^2-9| < 7|x-3|,$$

and since $|x-3| < \epsilon/7$, we have

$$|x^2-9| < 7(\epsilon/7) = \epsilon. \quad \square$$

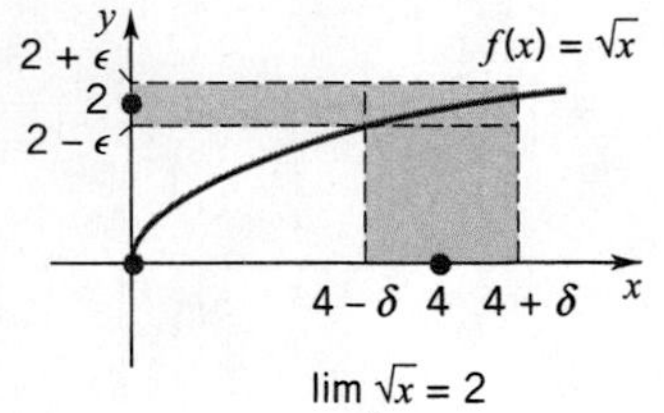

Figure 2.2.12

Example 7

$$\lim_{x\to 4} \sqrt{x} = 2. \qquad \text{(Figure 2.2.12)}$$

Finding a δ. Let $\epsilon > 0$. We seek a $\delta > 0$ such that

$$\text{if} \quad 0 < |x-4| < \delta, \qquad \text{then} \qquad |\sqrt{x}-2| < \epsilon.$$

First we want a relationship between $|x-4|$ and $|\sqrt{x}-2|$. To be able to form $\sqrt{x}$ at all, we need to have $x \ge 0$. To ensure this, we must have $\delta \le 4$. (Why?)

Remembering that we must have $\delta \le 4$, let's go on. With $x \ge 0$, we can form $\sqrt{x}$ and write

$$x - 4 = (\sqrt{x})^2 - 2^2 = (\sqrt{x}+2)(\sqrt{x}-2).$$

Taking absolute values, we have

$$|x-4| = |\sqrt{x}+2||\sqrt{x}-2|$$

Since $|\sqrt{x}+2| \ge 2 > 1$, it follows that

$$|\sqrt{x}-2| < |x-4|.$$

This last inequality suggests that we can simply set $\delta = \epsilon$. But remember the requirement $\delta \leq 4$. We can meet all requirements by setting $\delta =$ the minimum of 4, and ϵ.

Showing that the δ *"works."* Let $\epsilon > 0$. Choose $\delta = \min\{4, \epsilon\}$ and assume that

$$0 < |x - 4| < \delta.$$

Since $\delta \leq 4$, we have $x \geq 0$ and so $\sqrt{x}$ is defined. Now, as shown above,

$$|x - 4| = |\sqrt{x} + 2||\sqrt{x} - 2|$$

Since $|\sqrt{x} + 2| \geq 2 > 1$, we can conclude that

$$|\sqrt{x} - 2| < |x - 4|.$$

Since $|x - 4| < \delta$ and $\delta \leq \epsilon$, it does follow that $|\sqrt{x} - 2| < \epsilon$. ❑

There are many different ways of formulating the same limit statement. Sometimes one formulation is more convenient, sometimes another. In any case, it is useful to recognize that the following are equivalent:

(2.2.6)

$$\text{(i)} \lim_{x \to c} f(x) = L \qquad \text{(ii)} \lim_{h \to 0} f(c + h) = L$$
$$\text{(iii)} \lim_{x \to c} (f(x) - L) = 0 \qquad \text{(iv)} \lim_{x \to c} |f(x) - L| = 0.$$

For example, the equivalence of (i) and (ii) can be seen in Figure 2.2.13: simply think of h as being the signed distance from c to x. Then $x = c + h$, and x approaches c iff h approaches 0. Also, it is a good exercise in ϵ, δ technique to prove that (i) is equivalent to (ii).

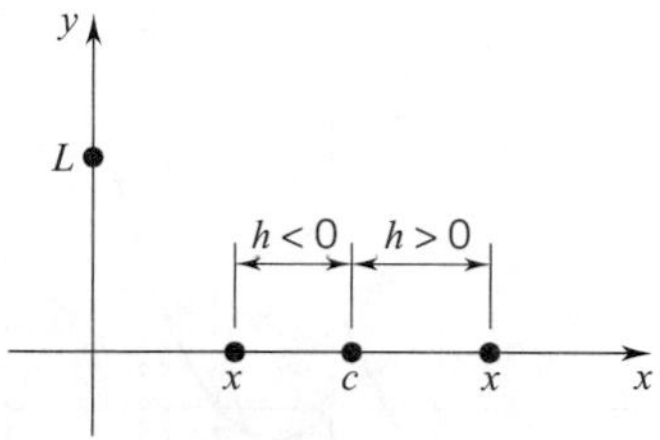

Figure 2.2.13

Example 8 For $f(x) = x^3$, we have

$$\lim_{x \to 2} x^3 = 8, \qquad \lim_{h \to 0} (2 + h)^3 = 8,$$
$$\lim_{x \to 2} (x^3 - 8) = 0, \qquad \lim_{x \to 2} |x^3 - 8| = 0. \quad ❑$$

We come now to the ϵ, δ definitions of one-sided limits. These are just the usual ϵ, δ statements, except that for a left-hand limit, the δ has to "work" only for x to the left of c, and for a right-hand limit, the δ has to "work" only for x to the right of c.

DEFINITION 2.2.7 LEFT-HAND LIMIT

Let f be a function defined at least on an interval of the form $(c - p, c)$, with $p > 0$. Then

$$\lim_{x \to c^-} f(x) = L$$

if for each $\epsilon > 0$ there exists a $\delta > 0$ such that

$$\text{if} \quad c - \delta < x < c, \qquad \text{then} \quad |f(x) - L| < \epsilon.$$

DEFINITION 2.2.8 RIGHT-HAND LIMIT

Let f be a function defined at least on an interval of the form $(c, c+p)$, with $p > 0$. Then

$$\lim_{x \to c^+} f(x) = L$$

if for each $\epsilon > 0$ there exists a $\delta > 0$ such that

$$\text{if} \quad c < x < c + \delta, \qquad \text{then} \quad |f(x) - L| < \epsilon.$$

As indicated in Section 2.1, one-sided limits give us a simple way of determining whether or not a (two-sided) limit exists:

(2.2.9) $$\lim_{x \to c} f(x) = L \quad \text{iff} \quad \lim_{x \to c^-} f(x) = L \quad \text{and} \quad \lim_{x \to c^+} f(x) = L.$$

The proof follows from the fact that any δ that "works" for the limit will work for both one-sided limits, and any δ that "works" for both one-sided limits will work for the limit. The details of the proof are left as an exercise.

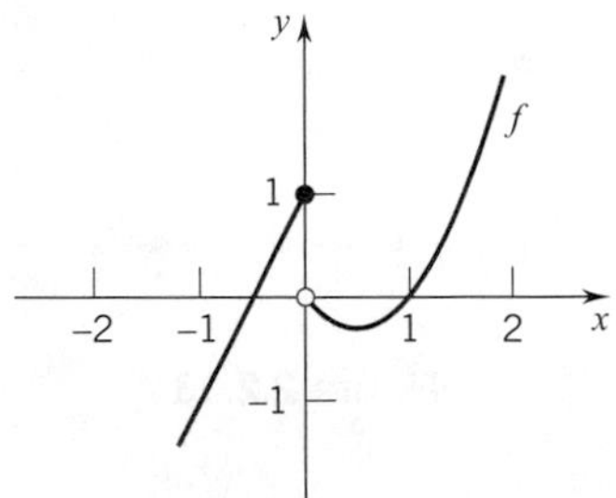

Figure 2.2.14

Example 9 If f is the function defined by

$$f(x) = \begin{cases} 2x + 1, & x \le 0 \\ x^2 - x, & x > 0, \end{cases} \qquad \text{(Figure 2.2.14)}$$

then $\lim_{x \to 0} f(x)$ does not exist.

PROOF The left- and right-hand limits at 0 are

$$\lim_{x \to 0^-} f(x) = \lim_{x \to 0^-} (2x + 1) = 1 \qquad \text{and} \qquad \lim_{x \to 0^+} f(x) = \lim_{x \to 0^+} (x^2 - x) = 0.$$

Since these one-sided limits are different, $\lim_{x \to 0} f(x)$ does not exist. □

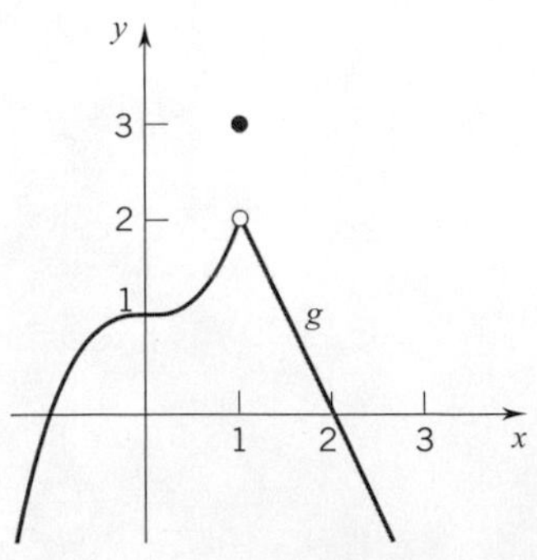

Figure 2.2.15

Example 10 If g is the function defined by

$$g(x) = \begin{cases} 1 + x^3, & x < 1 \\ 3, & x = 1 \\ 4 - 2x, & x > 1, \end{cases} \qquad \text{(Figure 2.2.15)}$$

then $\lim_{x \to 1} g(x) = 2$.

PROOF The left- and right-hand limits at 1 are

$$\lim_{x \to 1^-} g(x) = \lim_{x \to 1^-} (1 + x^3) = 2 \quad \text{and} \quad \lim_{x \to 1^+} g(x) = \lim_{x \to 1^+} (4 - 2x) = 2.$$

Thus, $\lim_{x\to 1} g(x) = 2$. NOTE: It does not matter that $g(1) \neq 2$. □

In conclusion, you should understand that one-sided limits also arise naturally, that is, independently of (two-sided) limits. For example, the domain of the function $f(x) = \sqrt{x-4}$ is the interval $[4, \infty)$, so the only limit that makes sense at $x = 4$ is the right-hand limit, and

$$\lim_{x\to 4^+} \sqrt{x-4} = 0.$$

Similarly, if the domain of a function g is the interval (a, b), then the only limit we can consider at $x = a$ is the right-hand limit, $\lim_{x\to a^+} g(x)$, and the only limit we can consider at $x = b$ is the left-hand limit, $\lim_{x\to b^-} g(x)$.

EXERCISES 2.2

Decide in the manner of Section 2.1 whether or not the indicated limit exists. Evaluate the limits that do exist.

1. $\lim_{x\to 1} \dfrac{x}{x+1}$.
2. $\lim_{x\to 0} \dfrac{x^2(1+x)}{2x}$.
3. $\lim_{x\to 0} \dfrac{x(1+x)}{2x^2}$.
4. $\lim_{x\to 4} \dfrac{x}{\sqrt{x}+1}$.
5. $\lim_{x\to 1} \dfrac{x^4-1}{x-1}$.
6. $\lim_{x\to -1} \dfrac{1-x}{x+1}$.
7. $\lim_{x\to 0} \dfrac{x}{|x|}$.
8. $\lim_{x\to 1} \dfrac{x^2-1}{x^2-2x+1}$.
9. $\lim_{x\to -2} \dfrac{|x|}{x}$.
10. $\lim_{x\to 9} \dfrac{x-3}{\sqrt{x}-3}$.
11. $\lim_{x\to 3^+} \dfrac{x+3}{x^2-7x+12}$.
12. $\lim_{x\to 0^-} \dfrac{x}{|x|}$.
13. $\lim_{x\to 1^+} \dfrac{\sqrt{x-1}}{x}$.
14. $\lim_{x\to 3^-} \sqrt{9-x^2}$.
15. $\lim_{x\to 2^+} f(x)$ if $f(x) = \begin{cases} 2x-1, & x \le 2 \\ x^2 - x, & x > 2. \end{cases}$
16. $\lim_{x\to -1^-} f(x)$ if $f(x) = \begin{cases} 1, & x \le -1 \\ x+2, & x > -1. \end{cases}$
17. $\lim_{x\to 2} f(x)$ if $f(x) = \begin{cases} 3, & x \text{ an integer} \\ 1, & \text{otherwise.} \end{cases}$
18. $\lim_{x\to 3} f(x)$ if $f(x) = \begin{cases} x^2, & x < 3 \\ 7, & x = 3 \\ 2x+3, & x > 3. \end{cases}$
19. $\lim_{x\to \sqrt{2}} f(x)$ if $f(x) = \begin{cases} 3, & x \text{ an integer} \\ 1, & \text{otherwise.} \end{cases}$
20. $\lim_{x\to 2} f(x)$ if $f(x) = \begin{cases} x^2, & x \le 1 \\ 5x, & x > 1. \end{cases}$

Finding a δ for a given ϵ. In Exercises 21–34 you are asked to find a δ corresponding to a given ϵ.

21. Which of the δ's displayed in the figure "works" for the given ϵ?

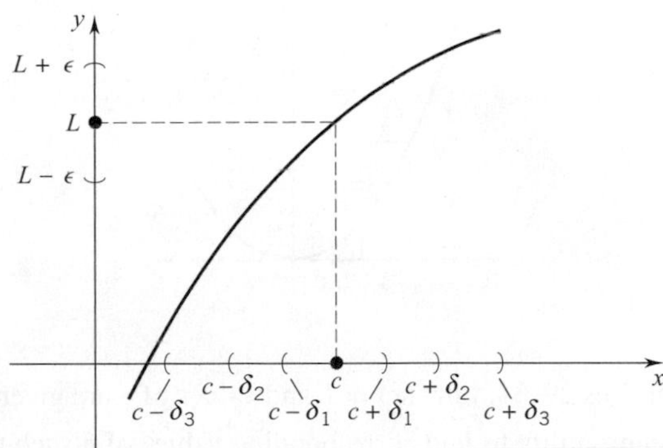

22. For which of the ϵ's given in the figure does the specified δ work?

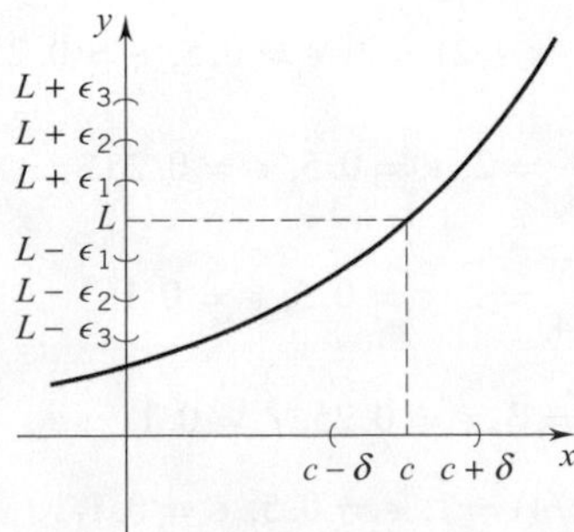

For the limits in Exercises 23–26, find the largest δ that "works" for the given ϵ.

23. $\lim_{x\to 1} 2x = 2$; $\epsilon = 0.1$.
24. $\lim_{x\to 4} 5x = 20$; $\epsilon = 0.5$.
25. $\lim_{x\to 2} \frac{1}{2}x = 1$; $\epsilon = 0.01$.
26. $\lim_{x\to 2} \frac{1}{5}x = \frac{2}{5}$; $\epsilon = 0.1$.

27. The graphs of $f(x) = \sqrt{x}$ and the horizontal lines $y = 1.5$ and $y = 2.5$ are shown in the figure. Use a graphing utility to find a value of δ such that

$$\text{if} \quad 0 < |x - 4| < \delta, \quad \text{then} \quad |\sqrt{x} - 2| < 0.5.$$

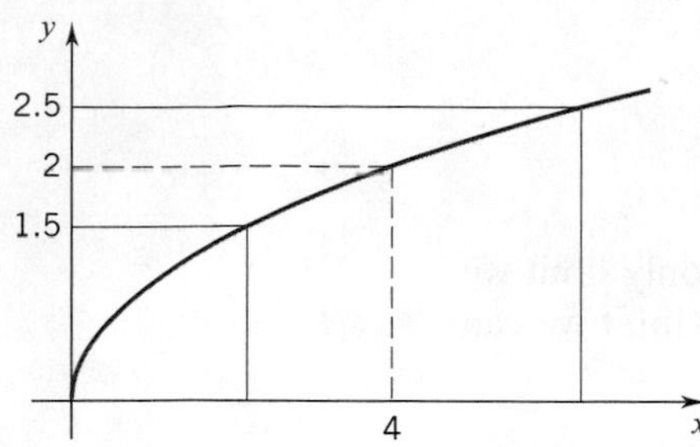

28. The graphs of $f(x) = 2x^2$ and the horizontal lines $y = 1$ and $y = 3$ are shown in the figure. Use a graphing utility to find a value of δ such that

$$\text{if} \quad 0 < |x + 1| < \delta, \quad \text{then} \quad |2x^2 - 2| < 1.$$

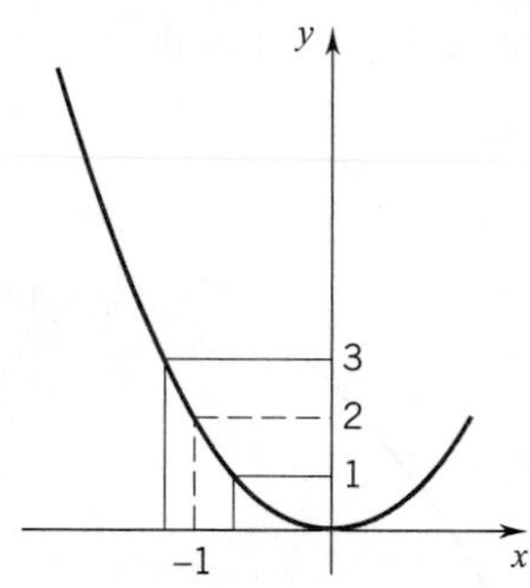

In Exercises 29–34, $\lim_{x\to c} f(x) = L$ and values of ϵ are given. Use a graphing utility to find corresponding values of δ such that if $0 < |x - c| < \delta$, then $|f(x) - L| < \epsilon$.

29. $\lim_{x\to 2} \left(\frac{1}{4}x^2 + x + 1\right) = 4$; $\epsilon = 0.5$, $\epsilon = 0.25$.

30. $\lim_{x\to -2} (x^3 - 4x + 2) = 2$; $\epsilon = 0.5$, $\epsilon = 0.25$.

31. $\lim_{x\to 1} \dfrac{x-1}{\sqrt{x}-1} = 2$; $\epsilon = 0.5$, $\epsilon = 0.25$.

32. $\lim_{x\to -1} \dfrac{1-3x}{2x+4} = 2$, $\epsilon = 0.5$, $\epsilon = 0.1$.

33. $\lim_{x\to 0} \dfrac{\sin 3x}{x} = 3$, $\epsilon = 0.25$, $\epsilon = 0.1$.

34. $\lim_{x\to 1} \tan(\pi x/4) = 1$; $\epsilon = 0.5$, $\epsilon = 0.1$.

Give an ϵ, δ proof for the following limits.

35. $\lim_{x\to 4} (2x - 5) = 3$.

36. $\lim_{x\to 2} (3x - 1) = 5$.

37. $\lim_{x\to 3} (6x - 7) = 11$.

38. $\lim_{x\to 0} (2 - 5x) = 2$.

39. $\lim_{x\to 2} |1 - 3x| = 5$.

40. $\lim_{x\to 2} |x - 2| = 0$.

41. Let f be some function for which you know only that

$$\text{if} \quad 0 < |x - 3| < 1, \quad \text{then} \quad |f(x) - 5| < 0.1.$$

Which of the following statements are necessarily true?

(a) If $|x - 3| < 1$, then $|f(x) - 5| < 0.1$.
(b) If $|x - 2.5| < 0.3$, then $|f(x) - 5| < 0.1$.
(c) $\lim_{x\to 3} f(x) = 5$.
(d) If $0 < |x - 3| < 2$, then $|f(x) - 5| < 0.1$.
(e) If $0 < |x - 3| < 0.5$, then $|f(x) - 5| < 0.1$.
(f) If $0 < |x - 3| < \frac{1}{4}$, then $|f(x) - 5| < \frac{1}{4}(0.1)$.
(g) If $0 < |x - 3| < 1$, then $|f(x) - 5| < 0.2$.
(h) If $0 < |x - 3| < 1$, then $|f(x) - 4.95| < 0.05$.
(i) If $\lim_{x\to 3} f(x) = L$, then $4.9 \le L \le 5.1$.

42. Suppose that $|A - B| < \epsilon$ for each $\epsilon > 0$. Prove that $A = B$. HINT: Consider what happens if $A \neq B$ and $\epsilon = \frac{1}{2}|A - B|$.

In Exercises 43 and 44, give the four limit statements displayed in (2.2.6), taking

43. $f(x) = \dfrac{1}{x-1}$, $c = 3$.

44. $f(x) = \dfrac{x}{x^2+2}$, $c = 1$.

45. Prove that

(2.2.10) $\lim_{x\to c} f(x) = 0, \quad \text{iff} \quad \lim_{x\to c} |f(x)| = 0.$

46. (a) Prove that

$$\text{if} \quad \lim_{x\to c} f(x) = L, \quad \text{then} \quad \lim_{x\to c} |f(x)| = |L|.$$

(b) Show that the converse is false. Give an example where

$$\lim_{x\to c} |f(x)| = |L| \quad \text{and} \quad \lim_{x\to c} f(x) = M \neq L,$$

and then give an example where

$$\lim_{x\to c} |f(x)| \text{ exists} \quad \text{but} \quad \lim_{x\to c} f(x) \quad \text{does not exist.}$$

47. Give an ϵ, δ proof that statement (i) in (2.2.6) is equivalent to (ii).

48. Give an ϵ, δ proof of (2.2.9).

49. (a) Show that $\lim_{x\to c} \sqrt{x} = \sqrt{c}$ for $c > 0$.

HINT: If x and c are positive, then

$$0 \le |\sqrt{x} - \sqrt{c}| = \frac{|x - c|}{\sqrt{x} + \sqrt{c}} < \frac{1}{\sqrt{c}}|x - c|.$$

(b) Show that $\lim_{x\to 0^+} \sqrt{x} = 0$.

Give an ϵ, δ proof for the following limits.

50. $\lim_{x\to 2} x^2 = 4.$

51. $\lim_{x\to 1} x^3 = 1.$

52. $\lim_{x\to 3} \sqrt{x+1} = 2.$

53. $\lim_{x\to 3^-} \sqrt{3-x} = 0.$

54. Prove that if

$$g(x) = \begin{cases} x, & x \text{ rational} \\ 0, & x \text{ irrational,} \end{cases}$$

then $\lim_{x\to 0} g(x) = 0$.

55. The function f defined by

$$f(x) = \begin{cases} 1, & x \text{ rational} \\ 0, & x \text{ irrational,} \end{cases}$$

is called the *Dirichlet function*. Prove that for no number c does $\lim_{x\to c} f(x)$ exist.

In Exercises 56–58, prove the given limit statement.

56. $\lim_{x\to c^-} f(x) = L$ iff $\lim_{h\to 0} f(c - |h|) = L.$

57. $\lim_{x\to c^+} f(x) = L$ iff $\lim_{h\to 0} f(c + |h|) = L.$

58. $\lim_{x\to c} f(x) = L$ iff $\lim_{x\to c} [f(x) - L] = 0.$

59. Suppose that $\lim_{x\to c} f(x) = L$.

(a) Prove that if $L > 0$, then $f(x) > 0$ for all $x \neq c$ in an interval of the form $(c - \gamma, c + \gamma), \gamma > 0$.

HINT: Use an ϵ, δ argument, setting $\epsilon = L$.

(b) Prove that if $L < 0$, then $f(x) < 0$ for all $x \neq c$ in an interval of the form $(c - \gamma, c + \gamma), \gamma > 0$.

60. Prove or give a counterexample: if $f(c) > 0$ and $\lim_{x\to c} f(x)$ exists, then $f(x) > 0$ for all x in an interval of the form $(c - \gamma, c + \gamma), \gamma > 0$.

61. Suppose that $f(x) \leq g(x)$ for all $x \in (c - p, c + p), p > 0$, except possibly at c itself.

(a) Prove that $\lim_{x\to c} f(x) \leq \lim_{x\to c} g(x)$, provided each of these limits exist.

(b) If $f(x) < g(x)$ on $(c - p, c + p)$, except possibly at c itself, does it follow that $\lim_{x\to c} f(x) < \lim_{x\to c} g(x)$?

62. Prove that if $\lim_{x\to c} f(x) = L$, then there is a number $\delta > 0$ and a number B such that $|f(x)| < B$ if $0 < |x - c| < \delta$.
HINT: It suffices to prove that $L - 1 < f(x) < L + 1$ for $0 < |x - c| < \delta$.

■ 2.3 SOME LIMIT THEOREMS

As you saw in the last section, it can be rather tedious to apply the ϵ, δ definition of the limit to test individual functions. By proving some general theorems about limits we can avoid some of this repetitive work. Of course, the theorems themselves (at least the first ones) will have to be proved by ϵ, δ methods.

We begin by showing that if a limit exists, it is unique.

THEOREM 2.3.1 THE UNIQUENESS OF A LIMIT

$$\text{If} \quad \lim_{x\to c} f(x) = L \quad \text{and} \quad \lim_{x\to c} f(x) = M, \quad \text{then} \quad L = M.$$

PROOF We show $L = M$ by proving that the assumption $L \neq M$ leads to the false conclusion

$$|L - M| < |L - M|.$$

Assume that $L \neq M$. Then $|L - M|/2 > 0$. Since $\lim_{x\to c} f(x) = L$, we know that there exists a $\delta_1 > 0$ such that

$$\text{(1)} \qquad \text{if } 0 < |x - c| < \delta_1, \quad \text{then} \quad |f(x) - L| < |L - M|/2.$$

(Here we are using $|L - M|/2$ as ϵ.)

Also, since $\lim_{x\to c} f(x) = M$, we know that there exists a $\delta_2 > 0$ such that

$$\text{(2)} \qquad \text{if } 0 < |x - c| < \delta_2, \quad \text{then} \quad |f(x) - L| < |L - M|/2.$$

(Again, using $|L - M|/2$ as ϵ.)

Now let x_1 be a number that satisfies the inequality

$$0 < |x_1 - c| < \text{ minimum of } \delta_1 \text{ and } \delta_2.$$

Then, by (1) and (2),

$$|f(x_1) - L| < \frac{|L-M|}{2} \quad \text{and} \quad |f(x_1) - M| < \frac{|L-M|}{2}.$$

It follows that

$$\begin{aligned} |L - M| &= |[L - f(x_1)] + [f(x_1) - M]| \\ &\leq |L - f(x_1)| + |f(x_1) - M| \quad \text{(by the triangle inequality)} \\ &= |f(x_1) - L| + |f(x_1) - M| < \frac{|L-M|}{2} + \frac{|L-M|}{2} = |L - M|. \quad \text{(} |a| = |-a| \text{)} \end{aligned}$$ □

THEOREM 2.3.2

If $\lim_{x\to c} f(x) = L$ and $\lim_{x\to c} g(x) = M$, then

(i) $\lim_{x\to c} [f(x) + g(x)] = L + M$,

(ii) $\lim_{x\to c} [\alpha f(x)] = \alpha L$ for each number α,

(iii) $\lim_{x\to c} [f(x)g(x)] = LM$.

PROOF Let $\epsilon > 0$. To prove (i), we must show that there exists a $\delta > 0$ such that

$$\text{if} \quad 0 < |x - c| < \delta, \quad \text{then} \quad |[f(x) + g(x)] - [L + M]| < \epsilon.$$

Note that

$$(*) \qquad \begin{aligned} |[f(x) + g(x)] - [L + M]| &= |[f(x) - L] + [g(x) - M]| \\ &\leq |f(x) - L| + |g(x) - M|. \end{aligned}$$

We make $|[f(x) + g(x)] - [L + M]|$ less than ϵ by making $|f(x) - L|$ and $|g(x) - M|$ each less that $\frac{1}{2}\epsilon$. Since $\epsilon > 0$, we know that $\frac{1}{2}\epsilon > 0$. Since

$$\lim_{x\to c} f(x) = L \quad \text{and} \quad \lim_{x\to c} g(x) = M,$$

we know that there exist positive numbers δ_1 and δ_2 such that

$$\text{if } 0 < |x - c| < \delta_1, \qquad \text{then} \quad |f(x) - L| < \tfrac{1}{2}\epsilon$$

and

$$\text{if } 0 < |x - c| < \delta_2, \qquad \text{then} \quad |g(x) - M| < \tfrac{1}{2}\epsilon.$$

Now we set $\delta =$ the minimum of δ_1 and δ_2 and note that, if $0 < |x - c| < \delta$, then

$$|f(x) - L| < \tfrac{1}{2}\epsilon \quad \text{and} \quad |g(x) - M| < \tfrac{1}{2}\epsilon,$$

and thus by (∗),

$$|[f(x)+g(x)]-[L+M]| < \epsilon.$$

In summary, by setting $\delta = \min\{\delta_1, \delta_2\}$, we find that

$$\text{if } 0 < |x-c| < \delta, \quad \text{then } |[f(x)+g(x)]-[L+M]| < \epsilon.$$

Thus (i) is proved. For proofs of (ii) and (iii), see the supplement at the end of this section. ❑

If you are wondering about $\lim_{x\to c}[f(x) - g(x)]$, note that $f(x) - g(x) = f(x) + (-1)\,g(x)$, and so the result

(2.3.3)
$$\boxed{\lim_{x\to c}[f(x)-g(x)] = L - M}$$

follows from (i) and (ii).

The results of Theorem 2.3.2 can be extended (by mathematical induction) to any finite collection of functions; namely, if

$$\lim_{x\to c} f_1(x) = L_1, \quad \lim_{x\to c} f_2(x) = L_2, \quad \dots, \quad \lim_{x\to c} f_n(x) = L_n,$$

and if $\alpha_1, \alpha_2, \dots, \alpha_n$ are numbers, then

(2.3.4)
$$\boxed{\begin{aligned}\lim_{x\to c}[\alpha_1 f_1(x) + \alpha_2 f_2(x) + \cdots + \alpha_n f_n(x)]\\ = \alpha_1 L_1 + \alpha_2 L_2 + \cdots + \alpha_n L_n\end{aligned}}$$

and

(2.3.5)
$$\boxed{\lim_{x\to c}[f_1(x) f_2(x) \cdots f_n(x)] = L_1 L_2 \cdots L_n.}$$

Let $P(x) = a_n x^n + \cdots + a_1 x + a_0$ be a polynomial and let c be any real number. Then it follows that

(2.3.6)
$$\boxed{\lim_{x\to c} P(x) = P(c).}$$

PROOF We already know that

$$\lim_{x\to c} x = c. \qquad \text{(see (2.2.3) on page 76)}$$

Applying (2.3.5) to $f(x) = x$ multiplied k times by itself, we have

$$\lim_{x\to c} x^k = c^k \quad \text{for each positive integer } k.$$

We also know from (2.2.5) on page 77 that $\lim_{x\to c} a_0 = a_0$. It follows now from (2.3.4) that

$$\lim_{x\to c}(a_n x^n + \cdots + a_1 x + a_0) = a_n c^n + \cdots + a_1 c + a_0;$$

that is,

$$\lim_{x \to c} P(x) = P(c). \quad \square$$

In words, this result says that the limit of a polynomial $P(x)$ as x approaches c not only exists, but is actually the value of P at c. Functions that have this property are said to be *continuous at c*. Thus, polynomials are continuous at $x = c$ for each number c. Continuity will be treated in detail in the next section.

Examples

$$\lim_{x \to 1} (5x^2 - 12x + 2) = 5(1)^2 - 12(1) + 2 = -5,$$

$$\lim_{x \to 0} (14x^5 - 7x^2 + 2x + 8) = 14(0)^5 - 7(0)^2 + 2(0) + 8 = 8,$$

$$\lim_{x \to -1} (2x^3 + x^2 - 2x - 3) = 2(-1)^3 + (-1)^2 - 2(-1) - 3 = -2. \quad \square$$

We come now to reciprocals and quotients.

THEOREM 2.3.7

If $\lim_{x \to c} g(x) = M$ with $M \neq 0$, then $\lim_{x \to c} \dfrac{1}{g(x)} = \dfrac{1}{M}$.

PROOF See the supplement at the end of this section. $\square$

Examples

$$\lim_{x \to 4} \frac{1}{x^2} = \frac{1}{16}, \quad \lim_{x \to 2} \frac{1}{x^3 - 1} = \frac{1}{7}, \quad \lim_{x \to -3} \frac{1}{|x|} = \frac{1}{|-3|} = \frac{1}{3}. \quad \square$$

Once you know that reciprocals present no trouble, quotients become easy to handle.

THEOREM 2.3.8

If $\lim_{x \to c} f(x) = L$ and $\lim_{x \to c} g(x) = M$ with $M \neq 0$, then $\lim_{x \to c} \dfrac{f(x)}{g(x)} = \dfrac{L}{M}$.

PROOF The key here is to observe that the quotient can be written as a product:

$$\frac{f(x)}{g(x)} = f(x)\frac{1}{g(x)}.$$

With $\lim_{x \to c} f(x) = L$ and $\lim_{x \to c} \dfrac{1}{g(x)} = \dfrac{1}{M}$, the product rule [part (iii) of Theorem 2.3.2] gives

$$\lim_{x\to c}\frac{f(x)}{g(x)} = L\frac{1}{M} = \frac{L}{M}. \quad \square$$

We can now calculate limits of rational functions by combining this theorem on quotients with the result (2.3.6) on limits of polynomials. If $R(x) = P(x)/Q(x)$ is a rational function (a quotient of two polynomials), and c is a number, then

(2.3.9) $$\lim_{x\to c} R(x) = \lim_{x\to c}\frac{P(x)}{Q(x)} = \frac{P(c)}{Q(c)} = R(c), \quad \text{provided } Q(c) \neq 0.$$

This result says that a rational function is *continuous* at the numbers c where $Q(c) \neq 0$.

Examples

$$\lim_{x\to 2}\frac{3x-5}{x^2+1} = \frac{6-5}{4+1} = \frac{1}{5}, \qquad \lim_{x\to 3}\frac{x^3-3x^2}{1-x^2} = \frac{27-27}{1-9} = 0. \quad \square$$

There is no point looking for a limit that does not exist. The next theorem gives a condition under which a quotient does not have a limit.

THEOREM 2.3.10

If $\lim_{x\to c} f(x) = L$ with $L \neq 0$ and $\lim_{x\to c} g(x) = 0$, then $\lim_{x\to c}\frac{f(x)}{g(x)}$ does not exist.

PROOF Suppose, on the contrary, that there exists a real number K such that

$$\lim_{x\to c}\frac{f(x)}{g(x)} = K.$$

Then $$L = \lim_{x\to c} f(x) = \lim_{x\to c}\left[g(x)\cdot\frac{f(x)}{g(x)}\right] = \lim_{x\to c} g(x)\cdot\lim_{x\to c}\frac{f(x)}{g(x)} = 0\cdot K = 0,$$

which contradicts our assumption that $L \neq 0$. $\square$

Examples From Theorem 2.3.10 you can see that

$$\lim_{x\to 1}\frac{x^2}{x-1}, \qquad \lim_{x\to 2}\frac{3x-7}{x^2-4}, \qquad \text{and} \qquad \lim_{x\to 0}\frac{5}{x}$$

all fail to exist. $\square$

Suppose $\lim_{x\to c} f(x) = L$ and $\lim_{x\to c} g(x) = M$. Combining Theorems 2.3.8 and 2.3.10, we have

$$\lim_{x\to c}\frac{f(x)}{g(x)} = \begin{cases} L/M & \text{if } M \neq 0 \\ \text{does not exist} & \text{if } M = 0 \text{ and } L \neq 0. \end{cases}$$

This raises the question: What happens if $L = M = 0$? This is the most difficult and the most interesting case. A limit of a quotient in which both the numerator and the denominator approach 0 is called an *indeterminate form of type* 0/0; as you will see in the examples below, the limit may or may not exist. At this stage, we will use algebraic methods to analyze limits of this type. We will learn other, more powerful, methods later.

Examples 1 Evaluate the limits that exist:

$$\textbf{(a)}\quad \lim_{x\to 3} \frac{x^2 - x - 6}{x - 3}, \qquad \textbf{(b)}\quad \lim_{x\to 4} \frac{(x^2 - 3x - 4)^2}{x - 4}, \qquad \textbf{(c)}\quad \lim_{x\to -1} \frac{x + 1}{(2x^2 + 7x + 5)^2}.$$

SOLUTION In each case both numerator and denominator tend to zero, and so we have to be careful.

(a) First we factor the numerator:

$$\frac{x^2 - x - 6}{x - 3} = \frac{(x + 2)(x - 3)}{x - 3}.$$

For $x \neq 3$,

$$\frac{x^2 - x - 6}{x - 3} = x + 2.$$

Thus,

$$\lim_{x\to 3} \frac{x^2 - x - 6}{x - 3} = \lim_{x\to 3} (x + 2) = 5.$$

(b) Note that

$$\frac{(x^2 - 3x - 4)^2}{x - 4} = \frac{[(x + 1)(x - 4)]^2}{x - 4} = \frac{(x + 1)^2(x - 4)^2}{x - 4};$$

so that, for $x \neq 4$,

$$\frac{(x^2 - 3x - 4)^2}{x - 4} = (x + 1)^2(x - 4).$$

It follows then that

$$\lim_{x\to 4} \frac{(x^2 - 3x - 4)^2}{x - 4} = \lim_{x\to 4} (x + 1)^2(x - 4) = 0.$$

(c) Since

$$\frac{x + 1}{(2x^2 + 7x + 5)^2} = \frac{x + 1}{[(2x + 5)(x + 1)]^2} = \frac{x + 1}{(2x + 5)^2(x + 1)^2},$$

you can see that, for $x \neq -1$,

$$\frac{x + 1}{(2x^2 + 7x + 5)^2} = \frac{1}{(2x + 5)^2(x + 1)}.$$

By Theorem 2.3.10,

$$\lim_{x\to -1} \frac{1}{(2x+5)^2(x+1)} \quad \text{does not exist,}$$

and therefore

$$\lim_{x\to -1} \frac{x+1}{(2x^2+7x+5)^2} \quad \text{does not exist either.} \quad \square$$

Example 2 Evaluate the following limits:

(a) $\displaystyle\lim_{x\to 2} \frac{1/x - 1/2}{x-2}$ **(b)** $\displaystyle\lim_{x\to 9} \frac{x-9}{\sqrt{x}-3}.$

SOLUTION

(a) For $x \neq 2$,

$$\frac{1/x - 1/2}{x-2} = \frac{\dfrac{2-x}{2x}}{x-2} = \frac{-(x-2)}{2x(x-2)} = \frac{-1}{2x}.$$

Thus

$$\lim_{x\to 2} \frac{1/x - 1/2}{x-2} = \lim_{x\to 2} \left[\frac{-1}{2x}\right] = -\frac{1}{4}.$$

(b) First, we "rationalize" the denominator:

$$\frac{x-9}{\sqrt{x}-3} = \frac{x-9}{\sqrt{x}-3} \cdot \frac{\sqrt{x}+3}{\sqrt{x}+3} = \frac{(x-9)(\sqrt{x}+3)}{x-9} = \sqrt{x}+3 \qquad (x \neq 9).$$

Alternatively, we could have factored the numerator:

$$\frac{x-9}{\sqrt{x}-3} = \frac{(\sqrt{x}-3)(\sqrt{x}+3)}{\sqrt{x}-3} = \sqrt{x}+3 \qquad (x \neq 9)$$

and arrived at the same result. Either way,

$$\lim_{x\to 9} \frac{x-9}{\sqrt{x}-3} = \lim_{x\to 9} [\sqrt{x}+3] = 6. \quad \square$$

Remark In this section we have phrased everything in terms of two-sided limits. Although we won't stop here to prove it, *all these results carry over to one-sided limits*. □

EXERCISES 2.3

1. Given that

$$\lim_{x\to c} f(x) = 2, \qquad \lim_{x\to c} g(x) = -1, \qquad \lim_{x\to c} h(x) = 0,$$

evaluate the limits that exist. If the limit does not exist, explain why.

(a) $\lim_{x\to c} [f(x) - g(x)]$. (b) $\lim_{x\to c} [f(x)]^2$.

(c) $\lim_{x\to c} \dfrac{f(x)}{g(x)}$. (d) $\lim_{x\to c} \dfrac{h(x)}{f(x)}$.

(e) $\lim_{x\to c} \dfrac{f(x)}{h(x)}$. (f) $\lim_{x\to c} \dfrac{1}{f(x) - g(x)}$.

2. Given that

$$\lim_{x\to c} f(x) = 3, \quad \lim_{x\to c} g(x) = 0, \quad \lim_{x\to c} h(x) = -2,$$

evaluate the limits that exist. If the limit does not exist, explain why.

(a) $\lim_{x\to c} [3f(x) + 2h(x)]$. (b) $\lim_{x\to c} [h(x)]^3$.

(c) $\lim_{x\to c} \dfrac{h(x)}{x - c}$. (d) $\lim_{x\to c} \dfrac{g(x)}{h(x)}$.

(e) $\lim_{x\to c} \dfrac{4}{f(x) - h(x)}$. (f) $\lim_{x\to c} [3 + g(x)]^2$.

3. When asked to evaluate

$$\lim_{x\to 4} \left(\frac{1}{x} - \frac{1}{4}\right)\left(\frac{1}{x-4}\right),$$

I reply that the limit is zero since $\lim_{x\to 4} [\frac{1}{x} - \frac{1}{4}] = 0$ and cite Theorem 2.3.2 (limit of a product) as justification. Verify that the limit is actually $-\frac{1}{16}$ and identify my error.

4. When asked to evaluate

$$\lim_{x\to 3} \frac{x^2 + x - 12}{x - 3},$$

I say that the limit does not exist since $\lim_{x\to 3} (x - 3) = 0$ and cite Theorem 2.3.10 (limit of a quotient) as justification. Verify that the limit is actually 7 and identify my error.

Evaluate the limits that exist.

5. $\lim_{x\to 2} 3$.

6. $\lim_{x\to 3} (5 - 4x)^2$.

7. $\lim_{x\to -4} (x^2 + 3x - 7)$.

8. $\lim_{x\to -2} 3|x - 1|$.

9. $\lim_{x\to \sqrt{3}} |x^2 - 8|$.

10. $\lim_{x\to -1} \dfrac{x^2 + 1}{3x^5 + 4}$.

11. $\lim_{x\to 0} \left(x - \dfrac{4}{x}\right)$.

12. $\lim_{x\to 5} \dfrac{2 - x^2}{4x}$.

13. $\lim_{x\to 0} \dfrac{x^2 + 1}{x - 1}$.

14. $\lim_{x\to 0} \dfrac{x^2}{x^2 + 1}$.

15. $\lim_{x\to 2} \dfrac{x}{x^2 - 4}$.

16. $\lim_{h\to 0} h\left(1 - \dfrac{1}{h}\right)$.

17. $\lim_{h\to 0} h\left(1 + \dfrac{1}{h}\right)$.

18. $\lim_{x\to 2} \dfrac{x - 2}{x^2 - 4}$.

19. $\lim_{x\to 2} \dfrac{x^2 - 4}{x - 2}$.

20. $\lim_{x\to -2} \dfrac{(x^2 - x - 6)^2}{x + 2}$.

21. $\lim_{x\to 4} \dfrac{\sqrt{x} - 2}{x - 4}$.

22. $\lim_{x\to 1} \dfrac{x - 1}{\sqrt{x} - 1}$.

23. $\lim_{x\to 1} \dfrac{x^2 - x - 6}{(x + 2)^2}$.

24. $\lim_{x\to -2} \dfrac{x^2 - x - 6}{(x + 2)^2}$.

25. $\lim_{h\to 0} \dfrac{1 - 1/h^2}{1 - 1/h}$.

26. $\lim_{h\to 0} \dfrac{1 - 1/h^2}{1 + 1/h^2}$.

27. $\lim_{h\to 0} \dfrac{1 - 1/h}{1 + 1/h}$.

28. $\lim_{h\to 0} \dfrac{1 + 1/h}{1 + 1/h^2}$.

29. $\lim_{t\to -1} \dfrac{t^2 + 6t + 5}{t^2 + 3t + 2}$.

30. $\lim_{x\to 2^+} \dfrac{\sqrt{x^2 - 4}}{x - 2}$.

31. $\lim_{t\to 0} \dfrac{t + a/t}{t + b/t}$.

32. $\lim_{x\to 1} \dfrac{x^2 - 1}{x^3 - 1}$.

33. $\lim_{x\to 1} \dfrac{x^5 - 1}{x^4 - 1}$.

34. $\lim_{h\to 0} h^2\left(1 + \dfrac{1}{h}\right)$.

35. $\lim_{h\to 0} h\left(1 + \dfrac{1}{h^2}\right)$.

36. $\lim_{x\to -4} \left(\dfrac{3x}{x + 4} + \dfrac{8}{x + 4}\right)$.

37. $\lim_{x\to -4} \left(\dfrac{2x}{x + 4} + \dfrac{8}{x + 4}\right)$.

38. $\lim_{x\to -4} \left(\dfrac{2x}{x + 4} - \dfrac{8}{x + 4}\right)$.

39. Evaluate the limits that exist.

(a) $\lim_{x\to 4} \left(\dfrac{1}{x} - \dfrac{1}{4}\right)$.

(b) $\lim_{x\to 4} \left[\left(\dfrac{1}{x} - \dfrac{1}{4}\right)\left(\dfrac{1}{x - 4}\right)\right]$.

(c) $\lim_{x\to 4} \left[\left(\dfrac{1}{x} - \dfrac{1}{4}\right)(x - 2)\right]$.

(d) $\lim_{x\to 4} \left[\left(\dfrac{1}{x} - \dfrac{1}{4}\right)\left(\dfrac{1}{x - 4}\right)^2\right]$.

40. Evaluate the limits that exist.

(a) $\lim_{x\to 3} \dfrac{x^2 + x + 12}{x - 3}$.

(b) $\lim_{x\to 3} \dfrac{x^2 + x - 12}{x - 3}$.

(c) $\lim_{x\to 3} \dfrac{(x^2 + x - 12)^2}{x - 3}$.

(d) $\lim_{x\to 3} \dfrac{x^2+x-12}{(x-3)^2}$.

41. Given that $f(x) = x^2 - 4x$, evaluate the limits that exist.

(a) $\lim_{x\to 4} \dfrac{f(x)-f(4)}{x-4}$. (b) $\lim_{x\to 1} \dfrac{f(x)-f(1)}{x-1}$.

(c) $\lim_{x\to 3} \dfrac{f(x)-f(1)}{x-3}$. (d) $\lim_{x\to 3} \dfrac{f(x)-f(2)}{x-3}$.

42. Given that $f(x) = x^3$, evaluate the limits that exist.

(a) $\lim_{x\to 3} \dfrac{f(x)-f(3)}{x-3}$. (b) $\lim_{x\to 3} \dfrac{f(x)-f(2)}{x-3}$.

(c) $\lim_{x\to 3} \dfrac{f(x)-f(3)}{x-2}$. (d) $\lim_{x\to 1} \dfrac{f(x)-f(1)}{x-1}$.

43. Show by example that $\lim_{x\to c}[f(x)+g(x)]$ can exist even if $\lim_{x\to c} f(x)$ and $\lim_{x\to c} g(x)$ do not exist.

44. Show by example that $\lim_{x\to c}[f(x)g(x)]$ can exist even if $\lim_{x\to c} f(x)$ and $\lim_{x\to c} g(x)$ do not exist.

Exercises 45–51: True or false? Justify your answers.

45. If $\lim_{x\to c}[f(x)+g(x)]$ exists but $\lim_{x\to c} f(x)$ does not exist, then $\lim_{x\to c} g(x)$ does not exist.

46. If $\lim_{x\to c}[f(x)+g(x)]$ and $\lim_{x\to c} f(x)$ exist, then it can happen that $\lim_{x\to c} g(x)$ does not exist.

47. If $\lim_{x\to c}\sqrt{f(x)}$ exists, then $\lim_{x\to c} f(x)$ exists.

48. If $\lim_{x\to c} f(x)$ exists, then $\lim_{x\to c}\sqrt{f(x)}$ exists.

49. If $\lim_{x\to c} f(x)$ exists, then $\lim_{x\to c}\dfrac{1}{f(x)}$ exists.

50. If $f(x) \le g(x)$ for all $x \ne c$, then $\lim_{x\to c} f(x) \le \lim_{x\to c} g(x)$.

51. If $f(x) < g(x)$ for all $x \ne c$, then $\lim_{x\to c} f(x) < \lim_{x\to c} g(x)$.

52. (a) Verify that

$$\max\{f(x), g(x)\} = \tfrac{1}{2}\{[f(x)+g(x)] + |f(x)-g(x)|\}.$$

(b) Find a similar expression for $\min\{f(x), g(x)\}$.

53. Let $h(x) = \min\{f(x), g(x)\}$ and $H(x) = \max\{f(x), g(x)\}$. Show that

$$\text{if } \lim_{x\to c} f(x) = L \quad \text{and} \quad \lim_{x\to c} g(x) = L,$$

$$\text{then } \lim_{x\to c} h(x) = L \quad \text{and} \quad \lim_{x\to c} H(x) = L.$$

HINT: Use Exercise 52.

54. (*The stability of limit*) Given a function f defined on an interval $I = (c-p, c+p)$, $p > 0$. Suppose the function g is also defined on I and $g(x) = f(x)$ except, possibly, at a finite set of points $x_1, x_2, \ldots, x_n$ in I.

(a) Show that if $\lim_{x\to c} f(x) = L$, then $\lim_{x\to c} g(x) = L$.

(b) Show that if $\lim_{x\to c} f(x)$ does not exist, then $\lim_{x\to c} g(x)$ does not exist.

55. (a) Suppose that $\lim_{x\to c} f(x) = 0$ and $\lim_{x\to c}[f(x)g(x)] = 1$. Prove that $\lim_{x\to c} g(x)$ does not exist.

(b) Suppose that $\lim_{x\to c} f(x) = L \ne 0$ and $\lim_{x\to c}[f(x)g(x)] = 1$. Does $\lim_{x\to c} g(x)$ exist, and if so, what is it?

56. Suppose f is a function with the following property: $\lim_{x\to c}[f(x)+g(x)]$ does not exist whenever $\lim_{x\to c} g(x)$ does not exist. Prove that $\lim_{x\to c} f(x)$ does exist.

(*Difference quotients*) For a given function f and a number c in the domain of f, the expression

$$\frac{f(c+h)-f(c)}{h}$$

is called the *difference quotient* for f at the number c. Limits of difference quotients as $h \to 0$ are the central topic of Chapter 3. In Exercises 57–60, calculate

$$\lim_{h\to 0}\frac{f(c+h)-f(c)}{h}$$

for f at the number c. See Project 2.1 for a geometric interpretation of this limit.

57. $f(x) = 2x^2 - 3x$; $c = 2$.

58. $f(x) = x^3 + 1$; $c = -1$.

59. $f(x) = \sqrt{x}$; $c = 4$.

60. $f(x) = 1/(x+1)$; $c = 1$.

61. Fix a number x and evaluate

$$\lim_{h\to 0}\frac{f(x+h)-f(x)}{h}$$

for

(a) $f(x) = x$.

(b) $f(x) = x^2$.

(c) $f(x) = x^3$.

(d) $f(x) = x^4$.

(e) Guess the limit for $f(x) = x^n$, where n is any positive integer.

62. Fix a number $x \ne 0$ and repeat Exercise 61 for

(a) $f(x) = \dfrac{1}{x}$. (b) $f(x) = \dfrac{1}{x^2}$. (c) $f(x) = \dfrac{1}{x^3}$.

(d) Guess the limit for $f(x) = 1/x^n$, where n is any positive integer.

(e) What will this limit be for $f(x) = x^n$, where n is any nonzero integer? Does your "formula" also hold for n=0?

*SUPPLEMENT TO SECTION 2.3

PROOF OF THEOREM 2.3.2 (ii)

We consider two cases: $\alpha \neq 0$ and $\alpha = 0$. If $\alpha \neq 0$, then $\epsilon/|\alpha| > 0$ and, since

$$\lim_{x \to c} f(x) = L,$$

we know that there exists $\delta > 0$ such that,

$$\text{if } \ 0 < |x - c| < \delta, \quad \text{then } \ |f(x) - L| < \frac{\epsilon}{|\alpha|}.$$

From the last inequality we obtain

$$|\alpha||f(x) - L| < \epsilon \quad \text{and thus} \quad |\alpha f(x) - \alpha L| < \epsilon.$$

The case $\alpha = 0$ was treated before in (2.2.5). □

PROOF OF THEOREM 2.3.2 (iii)

We begin with a little algebra:

$$\begin{aligned} |f(x)g(x) - LM| &= |[f(x)g(x) - f(x)M] + [f(x)M - LM]| \\ &\leq |f(x)g(x) - f(x)M| + |f(x)M - LM| \\ &= |f(x)||g(x) - M| + |M||f(x) - L| \\ &\leq |f(x)||g(x) - M| + (1 + |M|)|f(x) - L|. \end{aligned}$$

Now let $\epsilon > 0$. Since $\lim_{x \to c} f(x) = L$ and $\lim_{x \to c} g(x) = M$, we know:

(1) That there exists $\delta_1 > 0$ such that, if $0 < |x - c| < \delta_1$, then

$$|f(x) - L| < 1 \qquad \text{and thus} \quad |f(x)| < 1 + |L|.$$

(2) That there exists $\delta_2 > 0$ such that

$$\text{if } \ 0 < |x - c| < \delta_2, \qquad \text{then } \ |g(x) - M| < \left(\frac{\frac{1}{2}\epsilon}{1 + |L|}\right).$$

(3) That there exists $\delta_3 > 0$ such that

$$\text{if } \ 0 < |x - c| < \delta_3, \qquad \text{then } \ |f(x) - L| < \left(\frac{\frac{1}{2}\epsilon}{1 + |M|}\right).$$

We now set $\delta = \min\{\delta_1, \delta_2, \delta_3\}$ and observe that, if $0 < |x - c| < \delta$, then

$$\begin{aligned} |f(x) - LM| &\leq |f(x)||g(x) - M| + (1 + |M|)|f(x) - L| \\ &< \underbrace{(1 + |L|)}_{\text{by (1)}} \underbrace{\left(\frac{\frac{1}{2}\epsilon}{1 + |L|}\right)}_{\text{by (2)}} + (1 + |M|) \underbrace{\left(\frac{\frac{1}{2}\epsilon}{1 + |M|}\right)}_{\text{by (3)}} = \epsilon. \end{aligned}$$ □

PROOF OF THEOREM 2.3.7

For $g(x) \neq 0$

$$\left|\frac{1}{g(x)} - \frac{1}{M}\right| = \frac{|g(x) - M|}{|g(x)||M|}.$$

Choose $\delta_1 > 0$ such that

$$\text{if } 0 < |x - c| < \delta_1, \quad \text{then } |g(x) - M| < \frac{|M|}{2}.$$

For such x,

$$|g(x)| > \frac{|M|}{2} \quad \text{so that} \quad \frac{1}{|g(x)|} < \frac{2}{|M|}$$

and thus

$$\left|\frac{1}{g(x)} - \frac{1}{M}\right| = \frac{|g(x) - M|}{|g(x)||M|} \leq \frac{2}{|M|^2}|g(x) - M| = \frac{2}{M^2}|g(x) - M|.$$

Now let $\epsilon > 0$ and choose $\delta_2 > 0$ such that

$$\text{if } 0 < |x - c| < \delta_2, \quad \text{then } |g(x) - M| < \frac{M^2}{2}\epsilon.$$

Setting $\delta = \min\{\delta_1, \delta_2\}$, we find that

$$\text{if } 0 < |x - c| < \delta, \quad \text{then } \left|\frac{1}{g(x)} - \frac{1}{M}\right| < \epsilon. \quad \square$$

■ 2.4 CONTINUITY

In ordinary language, to say that a certain process is "continuous" is to say that it goes on without interruption and without abrupt changes. In mathematics the word "continuous" has much the same meaning.

The concept of continuity is so important in calculus and its applications that we discuss it with some care. First we treat *continuity at a point* (or number) c, and then we discuss *continuity on an interval.*

Continuity at a Point

The basic idea is this: We are given a function f and a number c. We calculate (if possible) the values of $\lim_{x \to c} f(x)$ and $f(c)$, and compare the results. The function f is continuous at c if the two values are equal. The following definition is a formal statement of this idea.

DEFINITION 2.4.1

Let f be a function defined at least on an open interval $(c - p, c + p)$ with $p > 0$. We say that *f is continuous at c* if

$$\lim_{x \to c} f(x) = f(c).$$

Remark Recall that in the definition of "limit of f at c" (Definition 2.2.1) we did not require that f be defined at c itself. In contrast, the definition of "continuity at c" requires that f be defined at c. Thus, according to this definition, a function f is continuous at a point c if:

(i) f is defined at c,

(ii) $\lim_{x\to c} f(x)$ exists, and

(iii) $\lim_{x\to c} f(x) = f(c)$.

The function f is said to be *discontinuous* at c if it is not continuous there. □

If the domain of f contains an interval $(c - p, c + p)$, $p > 0$ (so that f is defined at c), then f can fail to be continuous at c for only one of two reasons: either

(i) *f(x) does not have a limit* as x approaches c, or

(ii) *f(x) has a limit as* x approaches c, but $\lim_{x\to c} f(x) \neq f(c)$.

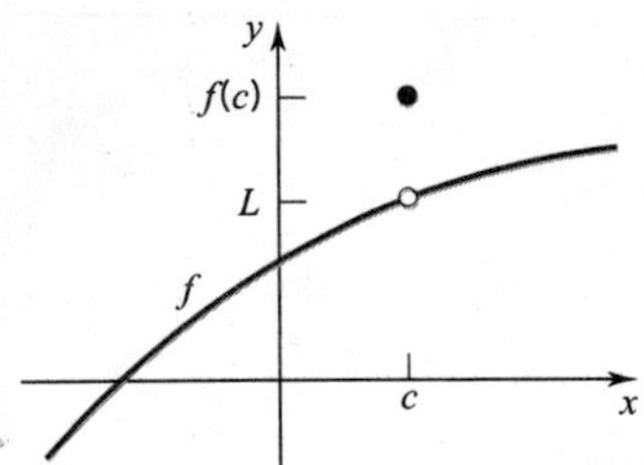

Figure 2.4.1

In case (ii), f is said to have a *removable discontinuity at c*; the discontinuity can be removed by redefining f at c. If the limit is L, redefine f at c to be L.

The function depicted in Figure 2.4.1 does have a limit at c. It is discontinuous at c because its limit at c is not its value at c. The discontinuity is removable; it can be removed by lowering the dot into place (i.e, by redefining f at c to be L).

The function depicted in Figure 2.4.2 is discontinuous at c because it does not have a limit at c. Note that $\lim_{x\to c^-} f(x)$ and $\lim_{x\to c^-} f(x)$ each exist, but they are not equal. A discontinuity of this particular type is called a *jump discontinuity*. Jump discontinuities are not removable.

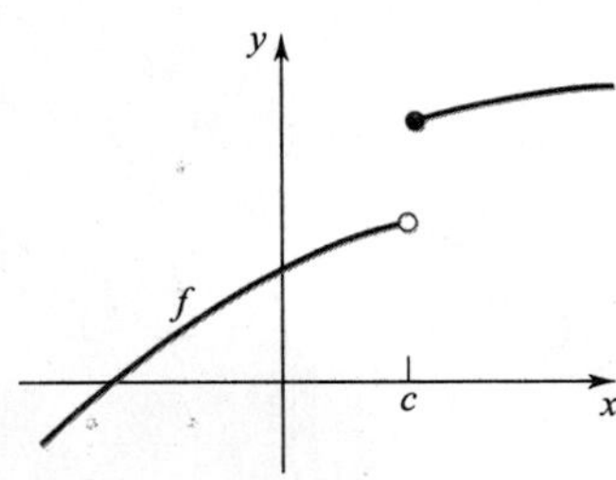

Figure 2.4.2

Additional examples of functions that fail to be continuous at a point c are shown in Figure 2.4.3. In each of these examples, at least one of the one-sided limits

$$\lim_{x\to c^-} f(x), \qquad \lim_{x\to c^+} f(x)$$

fails to exist; $f(x) \to \infty$ or $f(x) \to -\infty$ as x approaches c from either the left or right side. Discontinuities of this type are called *infinite discontinuities*. Infinite discontinuities are not removable.

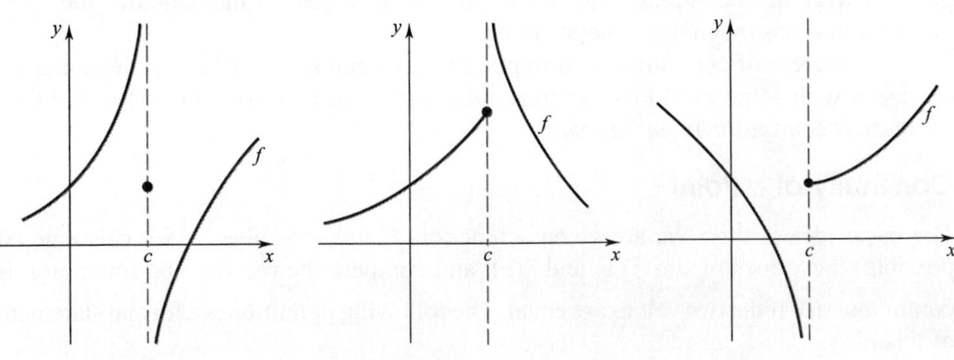

Figure 2.4.3

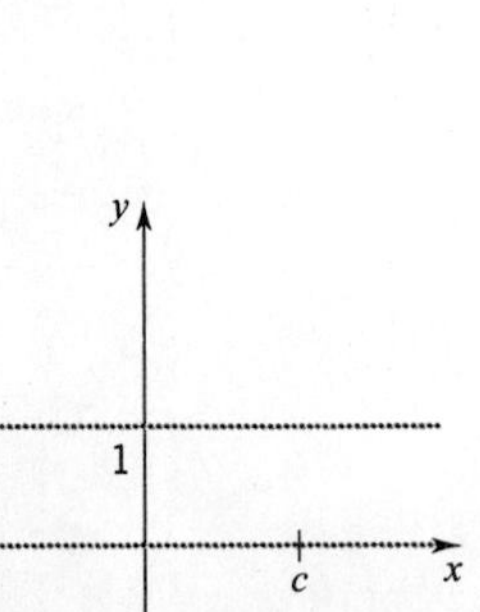

Figure 2.4.4

In Figure 2.4.4, we have tried to suggest the Dirichlet function

$$f(x) = \begin{cases} 1, & x \text{ rational} \\ 0, & x \text{ irrational.} \end{cases}$$

At no point c does f have a limit. It is therefore everywhere discontinuous.

Most of the functions that you have encountered so far are continuous at each point of their domains. In particular, this is true for polynomials P,

$$\lim_{x \to c} P(x) = P(c), \qquad \text{[see (2.3.6)]}$$

for rational functions (quotients of polynomials) $R = P/Q$,

$$\lim_{x \to c} R(x) = \lim_{x \to c} \frac{P(x)}{Q(x)} = \frac{P(c)}{Q(c)} = R(c) \qquad \text{provided} \quad Q(c) \neq 0, \qquad \text{[see (2.3.9)]}$$

and for the absolute value function,

$$\lim_{x \to c} |x| = |c|. \qquad \text{[see (2.2.4)]}$$

In Exercise 49, Section 2.2, you were asked to show that

$$\lim_{x \to c} \sqrt{x} = \sqrt{c} \qquad \text{for each } c > 0.$$

This makes the square-root function continuous at each positive number. We discuss later what happens for $c = 0$.

With f and g continuous at c, we have

$$\lim_{x \to c} f(x) = f(c), \qquad \lim_{x \to c} g(x) = g(c)$$

and thus, by the limit theorems,

$$\lim_{x \to c} [f(x) + g(x)] = f(c) + g(c), \qquad \lim_{x \to c} [f(x) - g(x)] = f(c) - g(c)$$

$$\lim_{x \to c} [\alpha f(x)] = \alpha f(c) \quad \text{for each real } \alpha \qquad \lim_{x \to c} [f(x)g(x)] = f(c)g(c)$$

and, if $g(c) \neq 0$, $\lim_{x \to c} [f(x)/g(x)] = f(c)/g(c)$.

In summary, we have the following theorem:

THEOREM 2.4.2

If f and g are continuous at c, then

(i) $f + g$ is continuous at c

(ii) $f - g$ is continuous at c,

(iii) αf is continuous at c for each real α

(iv) $f \cdot g$ is continuous at c,

(v) f/g is continuous at c provided $g(c) \neq 0$.

Parts (i)–(iv) can be combined and extended to any finite number of functions.

Example 1 The function $F(x) = 3|x| + \dfrac{x^3 - x}{x^2 - 5x + 6} + 4$ is continuous at all real numbers other than 2 and 3, You can see this by noting that

$$F = 3f + g/h + k$$

where $\quad f(x) = |x|, \quad g(x) = x^3 - x, \quad h(x) = x^2 - 5x + 6, \quad k(x) = 4.$

Since f, g, h, and k are everywhere continuous, F is continuous except at 2 and 3, the numbers where h takes on the value 0 (that is, the numbers where F is not defined). ❑

Since continuity at a point c is defined in terms of a limit, there is an ϵ, δ version of the definition. A direct translation of

$$\lim_{x\to c} f(x) = f(c)$$

into ϵ, δ terms reads like this: for each $\epsilon > 0$, there exists a $\delta > 0$ such that

$$\text{if } \; 0 < |x - c| < \delta, \qquad \text{then } \; |f(x) - f(c)| < \epsilon.$$

In the case of continuity at c, the restriction $0 < |x - c|$ is unnecessary. We can allow $|x - c| = 0$ because then $x = c, f(x) = f(c)$, and thus $|f(x) - f(c)| = 0$. Being 0, $|f(x) - f(c)|$ is certainly less than ϵ.

Thus, an ϵ, δ characterization of continuity at c reads as follows:

(2.4.3) $\quad f$ is continuous at c if $\begin{cases} \text{for each } \epsilon > 0 \text{ there exists a } \delta > 0 \text{ such that} \\ \text{if } \; |x - c| < \delta, \quad \text{then } \; |f(x) - f(c)| < \epsilon. \end{cases}$

In simple intuitive language

f is continuous at c if for x close to c, $f(x)$ is close to $f(c)$.

We are now ready to take up the continuity of composite functions. Remember the defining formula: $(f \circ g)(x) = f(g(x))$. You may wish to review Section 1.7.

THEOREM 2.4.4

If g is continuous at c and f is continuous at $g(c)$, then the composition $f \circ g$ is continuous at c.

The idea here is simple: with g continuous at c, we know that

for x close to c, $\quad g(x)$ is close to $g(c)$;

from the continuity of f at $g(c)$, we know that

with $g(x)$ close to $g(c)$, $\quad f(g(x))$ is close to $f(g(c))$.

In summary,

with x close to c, $\quad f(g(x))$ is close to $f(g(c))$.

The argument we just gave is too vague to be a proof. Here in contrast is a proof. We begin with $\epsilon > 0$. We must show that there exists a number $\delta > 0$ such that

$$\text{if } \; |x - c| < \delta, \qquad \text{then } \; |f(g(x)) - f(g(c))| < \epsilon.$$

In the first place, we observe that, since f is continuous at $g(c)$, there does exist a number $\delta_1 > 0$ such that

(1) $\qquad \text{if } \; |t - g(c)| < \delta_1, \qquad \text{then } \; |f(t) - f(g(c))| < \epsilon.$

With $\delta_1 > 0$, we know from the continuity of g at c that there exists a number $\delta > 0$ such that

$$\text{(2)} \qquad \text{if } |x - c| < \delta, \qquad \text{then } |g(x) - g(c)| < \delta_1.$$

Combining (2) and (1), we have what we want: by (2)

$$\text{if } |x - c| < \delta, \qquad \text{then } |g(x) - g(c)| < \delta_1$$

so that by (1)

$$|f(g(x)) - f(g(c))| < \epsilon. \quad \square$$

This proof is illustrated in Figure 2.4.5. The numbers within δ of c are taken by g to within δ_1 of $g(c)$, and then by f to within ϵ of $f(g(c))$.

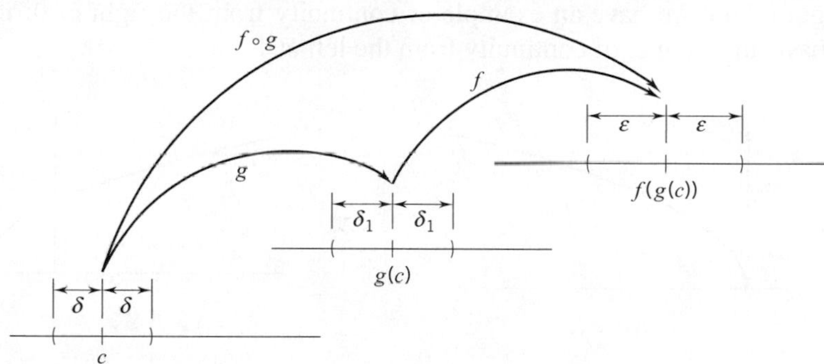

Figure 2.4.5

It is time to look at some examples

Example 2 The function $F(x) = \sqrt{\dfrac{x^2+1}{x-3}}$ is continuous at all numbers greater than 3. To see this, note that $F = f \circ g$, where

$$f(x) = \sqrt{x} \quad \text{and} \quad g(x) = \frac{x^2+1}{x-3}.$$

Now, take any $c > 3$. Since g is a rational function and g is defined at c, g is continuous at c. Also, since $g(c)$ is positive and f is continuous at each positive number, f is continuous at $g(c)$. By Theorem 2.4.4, F is continuous at c. $\square$

The continuity of composites holds for any finite number of functions. The only requirement is that each function be continuous *where it is applied.*

Example 3 The function $F(x) = \dfrac{1}{5 - \sqrt{x^2+16}}$ is continuous everywhere except at $x = \pm 3$, where it is not defined. To see this, note that $F = f \circ g \circ k \circ h$, where

$$f(x) = \frac{1}{x}, \quad g(x) = 5 - x, \quad k(x) = \sqrt{x}, \quad h(x) = x^2 + 16,$$

and observe that each of these functions is being evaluated only where it is continuous. In particular, g and h are continuous everywhere, f is being evaluated only at nonzero numbers, and k is being evaluated only at positive numbers. □

Just as we considered one-sided limits, we can consider one-sided continuity.

DEFINITION 2.4.5 ONE-SIDED CONTINUITY

A function f is called

$$\textit{continuous from the left at } c \quad \text{if} \quad \lim_{x\to c^-} f(x) = f(c).$$

It is called

$$\textit{continuous from the right at } c \quad \text{if} \quad \lim_{x\to c^+} f(x) = f(c).$$

In Figure 2.4.6 we have an example of continuity from the right at 0; in Figure 2.4.7 we have an example of continuity from the left at 1.

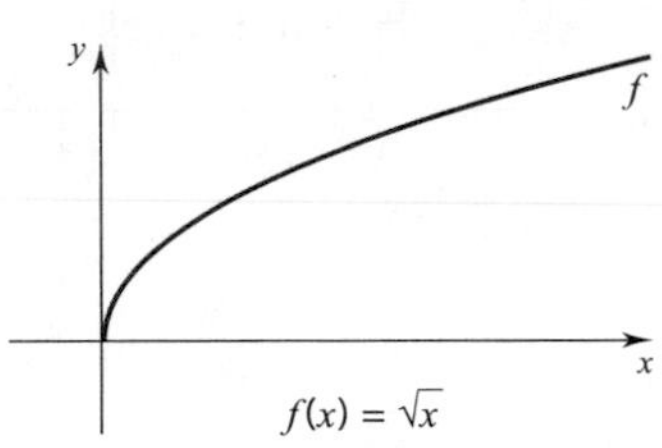

Figure 2.4.6

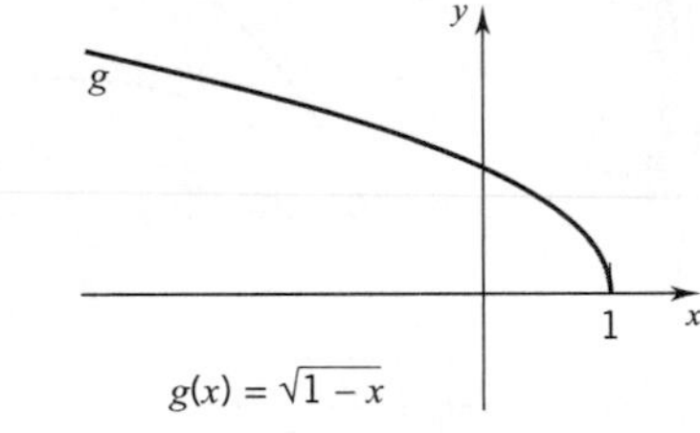

Figure 2.4.7

It follows from (2.2.9) that a function f is continuous at c iff it is continuous from both sides at c. Thus

(2.4.6) f is continuous at c iff $f(c)$, $\lim_{x\to c^-} f(x)$, and $\lim_{x\to c^+} f(x)$ all exist and are equal.

We apply this result to piecewise-defined functions in the following examples.

Example 4 Determine the discontinuities, if any, of the following function:

$$f(x) = \begin{cases} 2x+1, & x \le 0 \\ 1, & 0 < x \le 1 \\ x^2+1, & x > 1. \end{cases} \qquad \text{(Figure 2.4.8)}$$

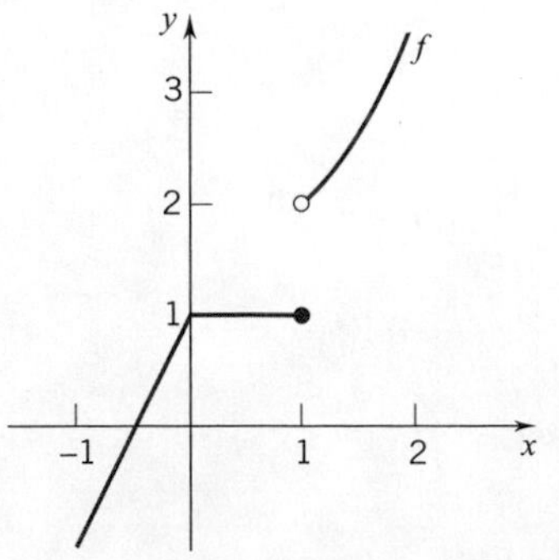

Figure 2.4.8

SOLUTION Clearly, f is continuous at each point in the open intervals $(-\infty, 0)$, $(0, 1)$, and $(1, \infty)$, since f is a polynomial in each of these intervals. Thus, we only have to check the behavior of f at $x = 0$ and $x = 1$. The figure suggests that f is continuous at 0 and discontinuous at 1. Indeed, at $x = 0$, $f(0) = 1$,

$$\lim_{x\to 0^-} f(x) = \lim_{x\to 0^-} (2x+1) = 1 \qquad \text{and} \qquad \lim_{x\to 0^+} f(x) = \lim_{x\to 0^+} (x^2+1) = 1,$$

so f is continuous at 0. At $x = 1$,

$$\lim_{x\to 1^-} f(x) = \lim_{x\to 1^-} (1) = 1 \quad \text{and} \quad \lim_{x\to 1^+} f(x) = \lim_{x\to 1^+} (x^2+1) = 2.$$

Thus, f is discontinuous at 1; in fact, f has a jump discontinuity at this point. □

Example 5 Determine the discontinuities of the following function:

$$f(x) = \begin{cases} x^3, & x \le -1 \\ x^2 - 2, & -1 < x < 1 \\ 6 - x, & 1 \le x < 4 \\ \dfrac{6}{7-x}, & 4 < x < 7 \\ 5x + 2, & x \ge 7. \end{cases}$$

SOLUTION Since the "pieces" of f consist of polynomials and rational functions, f is continuous at each point in the open intervals $(-\infty, -1), (-1, 1), (1, 4), (4, 7)$, and $(7, \infty)$. All we have to check is the behavior of f at $x = -1, 1, 4$, and 7. To do so, we apply (2.4.6).

The function f is continuous at $x = -1$ since $f(-1) = (-1)^3 = -1$,

$$\lim_{x\to -1^-} f(x) = \lim_{x\to -1^-} (x^3) = -1, \quad \text{and} \quad \lim_{x\to -1^+} f(x) = \lim_{x\to -1^+} (x^2 - 2) = -1.$$

Our findings at the other three points are displayed in the chart below. Try to verify each entry.

c	$f(c)$	$\lim_{x\to c^-} f(x)$	$\lim_{x\to c^+} f(x)$	Conclusion
1	5	-1	5	Discontinuous
4	Not defined	2	2	Discontinuous
7	37	Does not exist	37	Discontinuous

Note that the discontinuity at $x = 4$ is removable; if we define $f(4) = 2$, then f will be continuous at 4. The discontinuity at $x = 1$ is a jump discontinuity, and the discontinuity at $x = 7$ is an infinite discontinuity; $f(x) \to \infty$ as $x \to 7^-$. □

Continuity on intervals

Let (a, b) be an open interval. A function f is *continuous on* (a, b) if it is continuous at each point $c \in (a, b)$.

For a function defined on a closed interval $[a, b]$, the most continuity that we can expect is:

1. continuity on the open interval (a, b),
2. continuity from the right at a,
3. continuity from the left at b.

A function f that fulfills these requirements is said to be *continuous on* $[a, b]$. These definitions can be extended in an obvious way to functions defined on half-open intervals and on infinite intervals.

Functions that are continuous on a closed interval $[a, b]$ have some important special properties not shared by functions in general. We discuss two such properties in Section 2.6.

The concept of *continuity on an interval* also has a geometric interpretation; namely, a function f defined on an interval I is continuous on that interval if its graph has no "holes" or "jumps." That is, f is continuous on I if its graph is a "connected" curve. In practical terms, this means that you can draw the graph of f without lifting your pencil from the paper. Look back over the examples in this section to verify this.

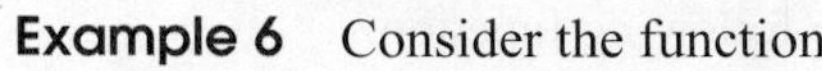

Example 6 Consider the function

$$f(x) = \sqrt{1 - x^2}.$$

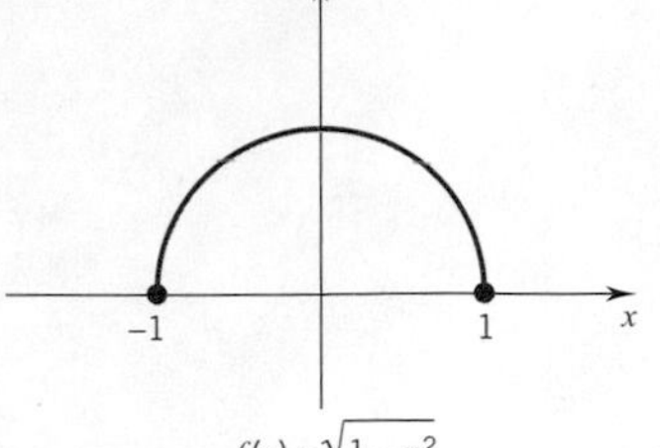

Figure 2.4.9

The graph of f is the semicircle shown in Figure 2.4.9. The function is continuous on $[-1, 1]$ because it is continuous on the open interval $(-1, 1)$, continuous from the right at -1, and continuous from the left at 1. The function maps the interval $[-1, 1]$ onto the interval $[0,1]$. ☐

Example 7 Let g be the function defined by

$$g(x) = \sqrt{x^2 - 4}.$$

Note that dom $(g) = \{x : x^2 \geq 4\}$. You can verify that this is the same as the set $(-\infty, -2] \cup [2, \infty)$. The function g is continuous on each of these intervals since it is continuous on each of the open intervals $(-\infty, -2)$ and $(2, \infty)$, continuous from the left at -2, and continuous from the right at 2. The function maps $(-\infty, -2] \cup [2, \infty)$ onto $[0, \infty)$. ☐

EXERCISES 2.4

1. The graph of a function f is given in the figure.
   (a) At which points is f discontinuous?
   (b) For each point of discontinuity found in (a), determine whether f is continuous from the right, from the left, or neither.
   (c) Which, if any, of the points of discontinuity found in (a) is removable? Which, if any, is a jump discontinuity?

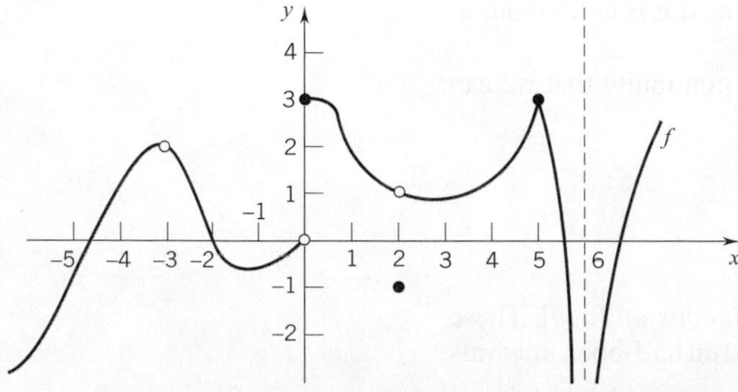

2. The graph of a function g is given in the figure. Determine the intervals on which g is continuous.

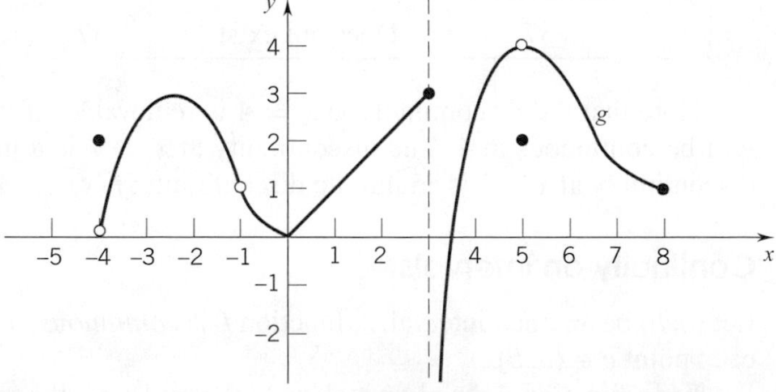

Determine whether or not the function is continuous at the indicated point. If not, determine whether the discontinuity is a removable discontinuity, a jump discontinuity, or an infinite discontinuity.

3. $f(x) = x^3 - 5x + 1; \quad x = 2.$
4. $g(x) = \sqrt{(x-1)^2 + 5}; \quad x = 1.$
5. $f(x) = \sqrt{x^2 + 9}; \quad x = 3.$
6. $f(x) = |4 - x^2|; \quad x = 2.$
7. $f(x) = \begin{cases} x^2 + 4, & x < 2 \\ x^3, & x \geq 2; \end{cases} \quad x = 2.$

8. $h(x) = \begin{cases} x^2+5, & x<2 \\ x^3, & x \geq 2; \end{cases} \quad x=2.$

9. $g(x) = \begin{cases} x^2+4, & x<2 \\ 5, & x=2 \\ x^3, & x>2; \end{cases} \quad x=2.$

10. $g(x) = \begin{cases} x^2+5, & x<2 \\ 10, & x=2 \\ 1+x^3, & x>2; \end{cases} \quad x=2.$

11. $f(x) = \begin{cases} \dfrac{|x-1|}{x-1}, & x \neq 1 \\ 0, & x=1; \end{cases} \quad x=1.$

12. $f(x) = \begin{cases} 1-x, & x<1 \\ 1, & x=1 \\ x^2-1, & x>1; \end{cases} \quad x=1.$

13. $h(x) = \begin{cases} \dfrac{x^2-1}{x+1}, & x \neq -1, \\ -2, & x=-1; \end{cases} \quad x=-1.$

14. $g(x) = \begin{cases} \dfrac{1}{x+1}, & x \neq -1 \\ 0, & x=-1; \end{cases} \quad x=-1.$

15. $f(x) = \begin{cases} \dfrac{x+2}{x^2-4}, & x \neq 2 \\ 4, & x=2; \end{cases} \quad x=2.$

16. $f(x) = \begin{cases} -x^2, & x<0 \\ 0, & x=0 \\ \dfrac{1}{x^2}, & x>0; \end{cases} \quad x=0.$

Sketch the graph and classify the discontinuities (if any) as being a removable discontinuity, a jump discontinuity, or an infinite discontinuity.

17. $f(x) = |x-1|$. **18.** $h(x) = |x^2-1|$.

19. $f(x) = \begin{cases} \dfrac{x^2-4}{x-2}, & x \neq 2 \\ 4, & x=2. \end{cases}$

20. $f(x) = \begin{cases} \dfrac{x-3}{x^2-9}, & x \neq 3, -3 \\ \frac{1}{6}, & x=3, -3 \end{cases}$

21. $f(x) = \begin{cases} \dfrac{x+2}{x^2-x-6}, & x \neq -2, 3 \\ -\frac{1}{5}, & x=-2, 3. \end{cases}$

22. $g(x) = \begin{cases} 2x-1, & x<1 \\ 0, & x=1 \\ 1/x^2, & x>1. \end{cases}$

23. $f(x) = \begin{cases} -1, & x<-1 \\ x^3, & -1 \leq x \leq 1 \\ 1, & 1<x. \end{cases}$

24. $g(x) = \begin{cases} 1, & x \leq -2 \\ \frac{1}{2}x, & -2<x<4 \\ \sqrt{x}, & 4 \leq x. \end{cases}$

25. $h(x) = \begin{cases} 1, & x \leq 0 \\ x^2, & 0<x<1 \\ 1, & 1 \leq x<2 \\ x, & 2 \leq x. \end{cases}$

26. $g(x) = \begin{cases} -x^2, & x<-1 \\ 3, & x=-1 \\ 2-x, & -1<x \leq 1 \\ 1/x^2, & 1<x. \end{cases}$

27. $f(x) = \begin{cases} 2x+9, & x<-2 \\ x^2+1, & -2<x \leq 1 \\ 3x-1, & 1<x<3 \\ x+6, & 3<x. \end{cases}$

28. $g(x) = \begin{cases} x+7, & x<-3 \\ |x-2|, & -3<x<-1 \\ x^2-2x, & -1<x<3 \\ 2x-3, & 3 \leq x. \end{cases}$

29. Sketch a graph of a function f that satisfies the following conditions:

(1) $\text{dom}(f) = [-3, 3]$.

(2) $f(-3) = f(-1) = 1; f(2) = f(3) = 2$.

(3) f has an infinite discontinuity at -1 and a jump discontinuity at 2.

(4) f is right continuous at -1 and left continuous at 2.

30. Sketch a graph of a function f that satisfies the following conditions:

(1) $\text{dom}(f) = [-2, 2]$.

(2) $f(-2) = f(-1) = f(1) = f(2) = 0$.

(3) f has an infinite discontinuity at -2, a jump discontinuity at -1, a jump discontinuity at 1, and an infinite discontinuity at 2.

(4) f is left continuous at -1 and right continuous at 1.

Each of the functions f in Exercises 31–34 is defined every where except at $x=1$. Where possible, define f at 1 so that it becomes continuous at 1.

31. $f(x) = \dfrac{x^2-1}{x-1}$. **32.** $f(x) = \dfrac{1}{x-1}$.

33. $f(x) = \dfrac{x-1}{|x-1|}$. **34.** $f(x) = \dfrac{(x-1)^2}{|x-1|}$.

35. Let $f(x) = \begin{cases} x^2, & x<1 \\ Ax-3, & x \geq 1. \end{cases}$ Find A given that f is continuous at 1.

36. Let $f(x) = \begin{cases} A^2x^2, & x \leq 2 \\ (1-A)x, & x>2. \end{cases}$ For what values of A is f continuous at 2?

37. Give necessary and sufficient conditions on A and B for the function

$$f(x) = \begin{cases} Ax-B, & x \leq 1 \\ 3x, & 1<x<2 \\ Bx^2-A, & 2 \leq x \end{cases}$$

to be continuous at $x=1$ but discontinuous at $x=2$.

38. Give necessary and sufficient conditions on A and B for the function in Exercise 37 to be continuous at $x = 2$ but discontinuous at $x = 1$.

39. Given that $f(x) = \begin{cases} 1 + cx, & x < 2 \\ c - x, & x \geq 2. \end{cases}$ Find a value of c to make f continuous on $(-\infty, \infty)$. Use a graphing utility to verify your result.

40. Given that $f(x) = \begin{cases} 1 - cx + dx^2, & x \leq -1 \\ x^2 + x, & -1 < x < 2 \\ cx^2 + dx + 4, & x \geq 2. \end{cases}$ Find a value of c and d to make f continuous on $(-\infty, \infty)$. Use a graphing utility to verify your result.

In Exercises 41–44, define the function at 5 so that it becomes continuous at 5.

41. $f(x) = \dfrac{\sqrt{x+4} - 3}{x - 5}$. **42.** $f(x) = \dfrac{\sqrt{x+4} - 3}{\sqrt{x - 5}}$.

43. $f(x) = \dfrac{\sqrt{2x - 1} - 3}{x - 5}$.

44. $f(x) = \dfrac{\sqrt{x^2 - 7x + 16} - \sqrt{6}}{(x - 5)\sqrt{x + 1}}$.

In Exercises 45–47, at what points (if any) is the given function continuous?

45. $f(x) = \begin{cases} 1, & x \text{ rational} \\ 0, & x \text{ irrational.} \end{cases}$

46. $g(x) = \begin{cases} x, & x \text{ rational} \\ 0, & x \text{ irrational.} \end{cases}$

47. $g(x) = \begin{cases} 2x, & x \text{ an integer} \\ x^2, & \text{otherwise.} \end{cases}$

48. The following functions are important in science and engineering:

1. The *Heaviside function* $H_c(x) = \begin{cases} 0, & x < c \\ 1, & x \geq c, \end{cases}$

 where c is a given number.

2. The *unit pulse function*

$$P_{\epsilon,c}(x) = \frac{1}{\epsilon}[H_c(x) - H_{c+\epsilon}(x)].$$

(a) Graph the functions H_c and $P_{\epsilon,c}$.

(b) Determine where each of the functions is continuous.

(c) Find $\lim_{x \to c^-} H_c(x)$ and $\lim_{x \to c^+} H_c(x)$. What can you say about $\lim_{x \to c} H(x)$?

49. (*Important*) Prove that

$$f \text{ is continuous at } c \quad \text{iff} \quad \lim_{h \to 0} f(c + h) = f(c).$$

50. (*Important*) Let f and g be continuous at c. Prove that if:

(a) $f(c) > 0$, then there exists $\delta > 0$ such that $f(x) > 0$ for all $x \in (c - \delta, c + \delta)$.

(b) $f(c) < 0$, then there exists $\delta > 0$ such that $f(x) < 0$ for all $x \in (c - \delta, c + \delta)$.

(c) $f(c) < g(c)$, then there exists $\delta > 0$ such that $f(x) < g(x)$ for all $x \in (c - \delta, c + \delta)$.

51. Suppose that f has a discontinuity at c which is not removable. Let g be a function such that $g(x) = f(x)$ except, possibly, at a finite set of points $x_1, x_2, \ldots, x_n$. Show that g also has a nonremovable discontinuity at c.

52. (a) Prove that if f is continuous everywhere, then $|f|$ is continuous everywhere.

(b) Give an example to show that the continuity of $|f|$ does not imply the continuity of f.

(c) Give an example of a function f such that f is continuous nowhere, but $|f|$ is continuous everywhere.

53. Suppose the function f has the property that there exists a number B such that

$$|f(x) - f(c)| \leq B|x - c|$$

for all x in the interval $(c - p, c + p), p > 0$. Prove that f is continuous at c.

54. Suppose the function f has the property that

$$|f(x) - f(t)| \leq |x - t|$$

for each pair of points x, t in the interval (a, b). Prove that f is continuous on (a, b).

55. Prove that if

$$\lim_{h \to 0} \frac{f(c + h) - f(c)}{h}$$

exists, then f is continuous at c.

56. Suppose that the function f is continuous on $(-\infty, \infty)$. Show that f can be written

$$f = f_e + f_0,$$

where f_e is an even function which is continuous on $(-\infty, \infty)$ and f_0 is an odd function which is continuous on $(-\infty, \infty)$.

In Exercises 57–60, the given function f is not defined at $x = 0$. Use a graphing utility to graph f. Use the zoom function to determine whether there is a value k such that the extended function

$$F(x) = \begin{cases} f(x), & x \neq 0 \\ k, & x = 0 \end{cases}$$

is continuous at $x = 0$. If so, what is k?

57. $f(x) = \dfrac{5^x - 1}{x}$.

58. $f(x) = \dfrac{(0.5)^x - 1}{x}$.

59. $f(x) = \dfrac{\sin x}{|x|}$.

60. $f(x) = (1 + x)^{1/x}, x > -1$.

■ 2.5 THE PINCHING THEOREM; TRIGONOMETRIC LIMITS

Figure 2.5.1 shows the graphs of three functions f, g, h. Suppose that, as suggested by the figure, for x close to c, f is trapped between g and h. (The values of these functions at c itself are irrelevant.) If, as x tends to c, both $g(x)$ and $h(x)$ tend to the same limit L, then $f(x)$ also tends to L. This idea is made precise in what we call *the pinching theorem.*

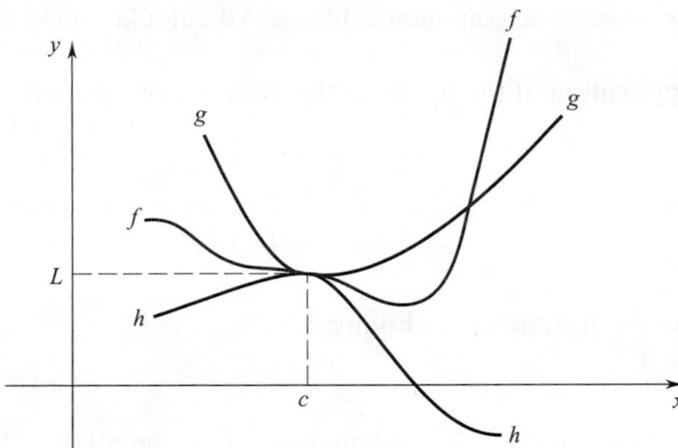

Figure 2.5.1

THEOREM 2.5.1 THE PINCHING THEOREM

Let $p > 0$. Suppose that, for all x such that $0 < |x - c| < p$,

$$h(x) \leq f(x) \leq g(x).$$

If

$$\lim_{x \to c} h(x) = L \quad \text{and} \quad \lim_{x \to c} g(x) = L,$$

then

$$\lim_{x \to c} f(x) = L.$$

PROOF Let $\epsilon > 0$. Let $p > 0$ be such that

$$\text{if } 0 < |x - c| < p, \quad \text{then} \quad h(x) \leq f(x) \leq g(x).$$

Choose $\delta_1 > 0$ such that

$$\text{if } 0 < |x - c| < \delta_1, \quad \text{then} \quad L - \epsilon < h(x) < L + \epsilon.$$

Choose $\delta_2 > 0$ such that

$$\text{if } 0 < |x - c| < \delta_2, \quad \text{then} \quad L - \epsilon < g(x) < L + \epsilon.$$

Let $\delta = \min\{p, \delta_1, \delta_2\}$. For x satisfying $0 < |x - c| < \delta$, we have

$$L - \epsilon < h(x) \leq f(x) \leq g(x) < L + \epsilon,$$

and thus

$$|f(x) - L| < \epsilon. \quad \square$$

Remark With straightforward modifications, the pinching theorem also holds for one-sided limits. We do not spell out the details here because we will be working with two-sided limits throughout this section. $\square$

We come now to some trigonometric limits. All calculations are based on radian measure.

As our first application of the pinching theorem, we prove that

(2.5.2)
$$\boxed{\lim_{x \to 0} \sin x = 0.}$$

PROOF To follow the argument, see Figure 2.5.2.†

For small $x \neq 0$

$$0 < |\sin x| = \text{length of } \overline{BP} < \text{length of } \overline{AP} < \text{length of } \widehat{AP} = |x|.$$

Thus, for such x

$$0 < |\sin x| < |x|.$$

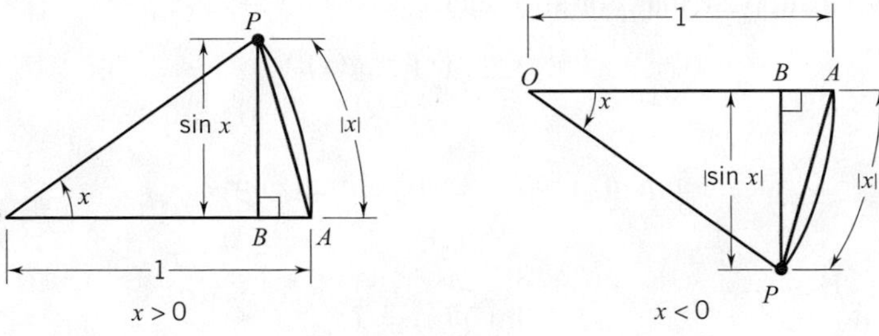

Figure 2.5.2

Since
$$\lim_{x \to 0} 0 = 0 \quad \text{and} \quad \lim_{x \to 0} |x| = 0,$$

we know from the pinching theorem that

$$\lim_{x \to 0} |\sin x| = 0 \quad \text{and therefore} \quad \lim_{x \to 0} \sin x = 0. \quad \square$$

From this it follows readily that

(2.5.3)
$$\boxed{\lim_{x \to 0} \cos x = 1.}$$

---

† Recall that in a circle of radius 1, a central angle of x radians subtends an arc of length $|x|$.

PROOF In general, $\cos^2 x + \sin^2 x = 1$. For x close to 0, the cosine is positive and we have

$$\cos x = \sqrt{1 - \sin^2 x}.$$

As x tends to 0, $\sin x$ tends to 0, $\sin^2 x$ tends to 0, and therefore $\cos x$ tends to 1. □

Next we show that the sine and cosine functions are everywhere continuous; which is to say, for all real numbers c,

(2.5.4)
$$\lim_{x\to c} \sin x = \sin c \quad \text{and} \quad \lim_{x\to c} \cos x = \cos c.$$

PROOF Take any real number c. By (2.2.6) we can write

$$\lim_{x\to c} \sin x \quad \text{as} \quad \lim_{h\to 0} \sin\,(c + h).$$

This form of the limit suggests that we use the addition formula

$$\sin\,(c + h) = \sin c \cos h + \cos c \sin h.$$

Since $\sin c$ and $\cos c$ are constants, we have

$$\begin{aligned}\lim_{h\to 0} \sin\,(c + h) &= (\sin c)(\lim_{h\to 0} \cos h) + (\cos c)(\lim_{h\to 0} \sin h) \\ &= (\sin c)(1) + (\cos c)(0) = \sin c.\end{aligned}$$

The proof that $\lim_{x\to c} \cos x = \cos c$ is left to you. □

The remaining trigonometric functions

$$\tan x = \frac{\sin x}{\cos x}, \qquad \cot x = \frac{\cos x}{\sin x}, \qquad \sec x = \frac{1}{\cos x}, \qquad \csc x = \frac{1}{\sin x}$$

are all continuous where defined. Justification? They are all quotients of continuous functions.

We turn now to two limits, the importance of which will become clear in Chapter 3:

(2.5.5)
$$\lim_{x\to 0} \frac{\sin x}{x} = 1 \qquad \text{and} \qquad \lim_{x\to 0} \frac{1 - \cos x}{x} = 0.$$

Remark These results were suggested in Section 2.1 using a numerical approach. See Example 14 and Exercise 52 in Section 2.1. □

PROOF We show that

$$\lim_{x\to 0} \frac{\sin x}{x} = 1$$

by using some simple geometry and the pinching theorem. For any x satisfying $0 < x \leq \pi/2$ (see Figure 2.5.3), length of $\overline{PB} = \sin x$, length of $\overline{OB} = \cos x$, and length $\overline{OA} = 1$. Since triangle OAQ is a right triangle, $\tan x = \overline{QA}/1 = \overline{QA}$.

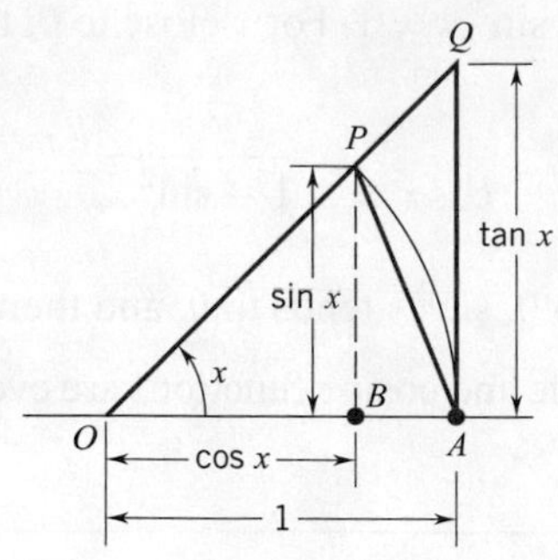

Figure 2.5.3

Now

$$\text{Area of triangle } OAP = \frac{1}{2}(1)\ \sin x = \frac{1}{2}\sin x$$

$$\text{Area of sector } OAP = \frac{1}{2}(1)^2 x = \frac{1}{2}x$$

$$\text{Area of triangle } OAQ = \frac{1}{2}(1)\ \tan x = \frac{1}{2}\frac{\sin x}{\cos x}.$$

Since triangle $OAP \subseteq$ sector $OAP \subseteq$ triangle OAQ (and these are all proper containments), we have

$$\frac{1}{2}\sin x < \frac{1}{2}x < \frac{1}{2}\frac{\sin x}{\cos x}$$

$$1 < \frac{x}{\sin x} < \frac{1}{\cos x}$$

$$\cos x < \frac{\sin x}{x} < 1.$$

(taking reciprocals, see Exercise 49, section 1.3)

This inequality was derived for $x > 0$, but since

$$\cos(-x) = \cos x \qquad \text{and} \qquad \frac{\sin(-x)}{-x} = \frac{-\sin x}{-x} = \frac{\sin x}{x},$$

this inequality also holds for $x < 0$.

We can now apply the pinching theorem. Since

$$\lim_{x\to 0} \cos x = 1 \qquad \text{and} \qquad \lim_{x\to 0} 1 = 1,$$

we can conclude that

$$\lim_{x\to 0} \frac{\sin x}{x} = 1.$$

We can now show that

$$\lim_{x\to 0} \frac{1-\cos x}{x} = 0.$$

For small $x \neq 0$, $\cos x$ is close to 1 and so $\cos x \neq -1$. Therefore, we can write

$$\frac{1-\cos x}{x} = \left(\frac{1-\cos x}{x}\right)\left(\frac{1+\cos x}{1+\cos x}\right)^{\dagger}$$

$$= \frac{1-\cos^2 x}{x(1+\cos x)}$$

$$= \frac{\sin^2 x}{x(1+\cos x)}$$

$$= \left(\frac{\sin x}{x}\right)\left(\frac{\sin x}{1+\cos x}\right)$$

Now, since

$$\lim_{x\to 0} \frac{\sin x}{x} = 1 \qquad \text{and} \qquad \lim_{x\to 0} \frac{\sin x}{1+\cos x} = \frac{0}{2} = 0,$$

it follows that

$$\lim_{x\to 0} \frac{1-\cos x}{x} = 0. \quad \square$$

Remark The limits in (2.5.5) can be extended as follows:

(2.5.6) For any number $a \neq 0$,

$$\lim_{x\to 0} \frac{\sin ax}{ax} = 1 \qquad \text{and} \qquad \lim_{x\to 0} \frac{1-\cos ax}{ax} = 0$$

See Exercise 38. □

We are now in a position to evaluate a variety of trigonometric limits.

Example 1 Find (1) $\lim_{x\to 0} \dfrac{\sin 4x}{3x}$ and (2) $\lim_{x\to 0} \dfrac{1-\cos 2x}{5x}$.

SOLUTION (1) We want to "pair off" $\sin 4x$ with $4x$ in order to use the result in (2.5.6):

$$\frac{\sin 4x}{3x} = \frac{4}{4}\cdot\frac{\sin 4x}{3x} = \frac{4}{3}\cdot\frac{\sin 4x}{4x}.$$

Therefore,

$$\lim_{x\to 0} \frac{\sin 4x}{3x} = \lim_{x\to 0}\left[\frac{4}{3}\cdot\frac{\sin 4x}{4x}\right] = \frac{4}{3}\lim_{x\to 0}\frac{\sin 4x}{4x} = \frac{4}{3}(1) = \frac{4}{3}.$$

† This "trick" is a fairly common procedure with trigonometric expressions. It is much like using "conjugates" to revise algebraic expressions:

$$\frac{3}{4+\sqrt{2}} = \frac{3}{4+\sqrt{2}}\cdot\frac{4-\sqrt{2}}{4-\sqrt{2}} = \frac{3(4-\sqrt{2})}{14}.$$

The second limit can be done the same way:

$$\lim_{x\to 0}\frac{1-\cos 2x}{5x}=\lim_{x\to 0}\frac{2}{5}\cdot\frac{1-\cos 2x}{2x}=\frac{2}{5}\lim_{x\to 0}\frac{1-\cos 2x}{2x}=\frac{2}{5}(0)=0. \quad \square$$

Example 2 Find $\lim_{x\to 0} x\cot 3x$.

SOLUTION We begin by writing

$$x\cot 3x = x\frac{\cos 3x}{\sin 3x}=\frac{1}{3}\left(\frac{3x}{\sin 3x}\right)(\cos 3x).$$

Since

$$\lim_{x\to 0}\frac{\sin 3x}{3x}=1,$$

it follows that

$$\lim_{x\to 0}\frac{3x}{\sin 3x}=1.$$

We can now conclude that

$$\lim_{x\to 0} x\cot 3x=\frac{1}{3}\lim_{x\to 0}\left(\frac{3x}{\sin 3x}\right)\lim_{x\to 0}(\cos 3x)=\frac{1}{3}(1)(1)=\frac{1}{3} \quad \square$$

Example 3 Find $\lim_{x\to \pi/4}\dfrac{\sin(x-\frac{1}{4}\pi)}{(x-\frac{1}{4}\pi)^2}$.

SOLUTION

$$\frac{\sin(x-\frac{1}{4}\pi)}{(x-\frac{1}{4}\pi)^2}=\left[\frac{\sin(x-\frac{1}{4}\pi)}{(x-\frac{1}{4}\pi)}\right]\cdot\frac{1}{x-\frac{1}{4}\pi}.$$

We know that

$$\lim_{x\to \pi/4}\frac{\sin(x-\frac{1}{4}\pi)}{x-\frac{1}{4}\pi}=1.$$

Since $\lim_{x\to \pi/4}(x-\frac{1}{4}\pi)=0$, you can see by Theorem 2.3.10 that

$$\lim_{x\to \pi/4}\frac{\sin(x-\frac{1}{4}\pi)}{(x-\frac{1}{4}\pi)^2}\quad\text{does not exist.} \quad \square$$

Example 4 Find $\lim_{x\to 0}\dfrac{x^2}{\sec x-1}$.

SOLUTION The evaluation of this limit requires a little imagination. Since both the numerator and denominator approach zero as x approaches zero, it is not clear what happens to the fraction; the limit is an indeterminate form of type 0/0. However, we can rewrite the fraction in a more amenable form by using the trick just introduced:

$$\begin{aligned}\frac{x^2}{\sec x - 1} &= \frac{x^2}{\sec x - 1}\left(\frac{\sec x + 1}{\sec x + 1}\right)\\ &= \frac{x^2(\sec x + 1)}{\sec^2 x - 1} = \frac{x^2(\sec x + 1)}{\tan^2 x}\\ &= \frac{x^2\cos^2 x(\sec x + 1)}{\sin^2 x}\\ &= \left(\frac{x}{\sin x}\right)^2(\cos^2 x)(\sec x + 1).\end{aligned}$$

Since each of these expressions has a limit as x tends to 0, the fraction we began with has a limit:

$$\begin{aligned}\lim_{x\to 0}\frac{x^2}{\sec x - 1} &= \lim_{x\to 0}\left(\frac{x}{\sin x}\right)^2 \cdot \lim_{x\to 0}\cos^2 x \cdot \lim_{x\to 0}(\sec x + 1)\\ &= (1)(1)(2) = 2. \quad \square\end{aligned}$$

EXERCISES 2.5

Evaluate the limits that exist.

1. $\lim_{x\to 0}\frac{\sin 3x}{x}$.

2. $\lim_{x\to 0}\frac{2x}{\sin x}$.

3. $\lim_{x\to 0}\frac{3x}{\sin 5x}$.

4. $\lim_{x\to 0}\frac{\sin 3x}{2x}$.

5. $\lim_{x\to 0}\frac{\sin 4x}{\sin 2x}$.

6. $\lim_{x\to 0}\frac{\sin 3x}{5x}$.

7. $\lim_{x\to 0}\frac{\sin x^2}{x}$.

8. $\lim_{x\to 0}\frac{\sin x^2}{x^2}$.

9. $\lim_{x\to 0}\frac{\sin x}{x^2}$.

10. $\lim_{x\to 0}\frac{\sin^2 x^2}{x^2}$.

11. $\lim_{x\to 0}\frac{\sin^2 3x}{5x^2}$.

12. $\lim_{x\to 0}\frac{\tan^2 3x}{4x^2}$.

13. $\lim_{x\to 0}\frac{2x}{\tan 3x}$.

14. $\lim_{x\to 0}\frac{4x}{\cot 3x}$.

15. $\lim_{x\to 0} x\csc x$.

16. $\lim_{x\to 0}\frac{\cos x - 1}{2x}$.

17. $\lim_{x\to 0}\frac{x^2}{1-\cos 2x}$.

18. $\lim_{x\to 0}\frac{x^2 - 2x}{\sin 3x}$.

19. $\lim_{x\to 0}\frac{1-\sec^2 2x}{x^2}$.

20. $\lim_{x\to 0}\frac{1}{2x\csc x}$.

21. $\lim_{x\to 0}\frac{2x^2 + x}{\sin x}$.

22. $\lim_{x\to 0}\frac{1-\cos 4x}{9x^2}$.

23. $\lim_{x\to 0}\frac{\tan 3x}{2x^2 + 5x}$.

24. $\lim_{x\to 0} x^2(1+\cot^2 3x)$.

25. $\lim_{x\to 0}\frac{\sec x - 1}{x\sec x}$.

26. $\lim_{x\to \pi/4}\frac{1-\cos x}{x}$.

27. $\lim_{x\to \pi/4}\frac{\sin x}{x}$.

28. $\lim_{x\to 0}\frac{\sin^2 x}{x(1-\cos x)}$.

29. $\lim_{x\to \pi/2}\frac{\cos x}{x - \frac{1}{2}\pi}$.

30. $\lim_{x\to \pi}\frac{\sin x}{x - \pi}$.

31. $\lim_{x\to \pi/4}\frac{\sin(x + \frac{1}{4}\pi) - 1}{x - \frac{1}{4}\pi}$. (HINT: $x + \frac{1}{4}\pi = x - \frac{1}{4}\pi + \frac{\pi}{2}$)

32. $\lim_{x\to \pi/6}\frac{\sin(x + \frac{1}{3}\pi) - 1}{x - \frac{1}{6}\pi}$.

33. Show that $\lim_{x\to c}\cos x = \cos c$ for all real numbers c.

In Exercises 34–37, a and b are nonzero constants. Evaluate the limits.

34. $\lim_{x\to 0}\frac{\sin ax}{bx}$.

35. $\lim_{x\to 0}\frac{1-\cos ax}{bx}$.

36. $\lim_{x\to 0}\frac{\sin ax}{\sin bx}$.

37. $\lim_{x\to 0}\frac{\cos ax}{\cos bx}$.

38. Show that

if $\lim_{x\to 0} f(x) = L$, then $\lim_{x\to 0} f(ax) = L$ for all $a \neq 0$.

HINT: Let $\epsilon > 0$. If $\delta_1 > 0$ "works" for the first limit, then $\delta = \delta_1/|a|$ "works" for the second limit.

In Exercises 39–42 determine whether $\lim_{h\to 0} [f(c+h)-f(c)]/h$ exists, and give the value when the limit does exist.

39. $f(x) = \sin x,\ c = \pi/4$. HINT: Use the addition formula for the sine function.

40. $f(x) = \cos x,\ c = \pi/3$.

41. $f(x) = \cos 2x,\ c = \pi/6$.

42. $f(x) = \sin 3x,\ c = \pi/2$.

43. Prove that $\lim_{x\to 0} x \sin(1/x) = 0$. HINT: Use the pinching theorem.

44. Prove that $\lim_{x\to\pi} (x-\pi)\cos^2[1/(x-\pi)] = 0$.

45. Prove that $\lim_{x\to 1} |x-1| \sin x = 0$.

46. Let f be the Dirichlet function

$$f(x) = \begin{cases} 1, & x \text{ rational} \\ 0, & x \text{ irrational.} \end{cases}$$

Prove that $\lim_{x\to 0} xf(x) = 0$.

47. Prove that if there is a number B such that $|f(x)| \leq B$ for all $x \neq 0$, then $\lim_{x\to 0} x f(x) = 0$. NOTE: Exercises 43–46 are special cases of this general result.

48. Prove that if there is a number B such that $|f(x)/x| \leq B$ for all $x \neq 0$, then $\lim_{x\to 0} f(x) = 0$.

49. Prove that if there is a number B such that $|f(x)-L|/|x-c| \leq B$ for all $x \neq c$, then $\lim_{x\to c} f(x) = L$.

50. Given that $\lim_{x\to c} f(x) = 0$ and that $|g(x)| \leq B$ on $(c-p, c+p)$, $p > 0$, prove that

$$\lim_{x\to c} f(x)g(x) = 0.$$

C 51. Assume that $f(x) = 1 - x^2 \leq g(x) \leq h(x) = 1 + x^2$ on $[-1,1]$. Use a graphing utility to draw the graphs of f and h. What do these graphs tell you about $\lim_{x\to 0} g(x)$?

C In Exercises 52–54, use the limit utility in a CAS to evaluate the limits.

52. $\lim_{x\to 0} \dfrac{20x - 15x^2}{\sin 2x}$.

53. $\lim_{x\to 0} \dfrac{\tan x}{x^2}$.

54. $\lim_{x\to 0} \dfrac{1 + \cos 3x - \cos^2 3x - \cos^3 3x}{x^2}$.

C 55. Use a graphing utility to plot $f(x) = \dfrac{x}{\tan 3x}$ on $[-0.2, 0.2]$. Estimate $\lim_{x\to 0} f(x)$; use the zoom function if necessary. Verify your result analytically.

C 56. Use a graphing utility to plot $f(x) = \dfrac{\tan x}{\tan x + x}$ on $[-0.2,0.2]$. Estimate $\lim_{x\to 0} f(x)$; use the zoom function if necessary. Verify your result analytically.

■ 2.6 TWO BASIC PROPERTIES OF CONTINUOUS FUNCTIONS

A function that is continuous on an interval does not "skip" any values, and thus its graph is an "unbroken curve". There are no "holes" in it and no "jumps." This idea is expressed in the *intermediate-value theorem*.

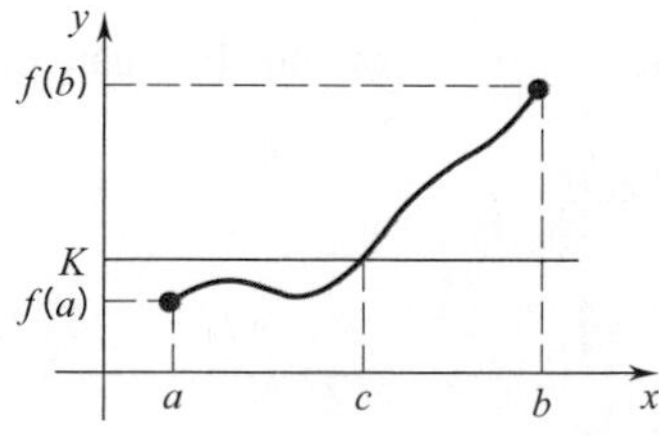

Figure 2.6.1

THEOREM 2.6.1 THE INTERMEDIATE-VALUE THEOREM

If f is continuous on $[a, b]$ and K is any number between $f(a)$ and $f(b)$, then there is at least one number c in the interval (a, b) such that $f(c) = K$.

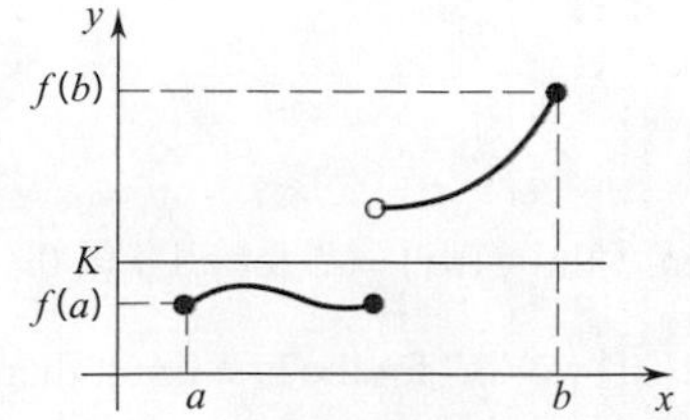

We illustrate the theorem in Figure 2.6.1. What can happen in the discontinuous case is illustrated in Figure 2.6.2. There the number K has been "skipped."

It is a small step from the intermediate-value theorem to the following observation:

continuous functions map intervals onto intervals.

A proof of the intermediate-value theorem is given in Appendix B. We will assume the result here and illustrate its usefulness.

First we apply the theorem to the important problem of locating the zeros of a function. In particular, suppose that the function f is continuous on $[a, b]$, and that either

Figure 2.6.2

$$f(a) < 0 < f(b) \quad \text{or} \quad f(b) < 0 < f(a). \qquad \text{(Figure 2.6.3)}$$

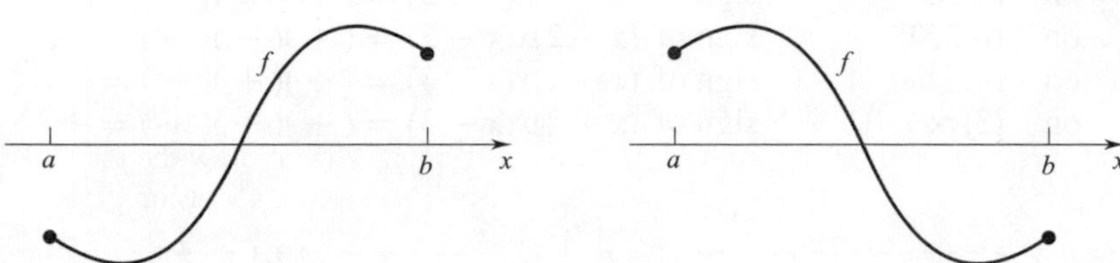

Figure 2.6.3

Then, by the intermediate-value theorem, we know that the equation $f(x) = 0$ has at least one root in (a, b).

Example 1 Let $f(x) = \cos\left(\frac{\pi}{2}x\right) - x^2$ on $[0, 1]$. Show that $f(x) = 0$ has a root in the interval $(0, 1)$.

SOLUTION Since f is the difference of two continuous functions, it is continuous. Next, we evaluate f at the two endpoints to see if there is a change in sign:

$$f(0) = \cos(0) - 0^2 = 1 > 0 \quad \text{and} \quad f(1) = \cos\left(\frac{\pi}{2}\right) - 1^2 = -1 < 0.$$

Therefore, by the intermediate-value theorem, there is at least one number $c \in (0, 1)$ such that $f(c) = 0$.

We can attempt to locate the root more precisely by repeating the process, this time evaluating f at the midpoint of the interval:

$$f\left(\frac{1}{2}\right) = \cos\left(\frac{\pi}{4}\right) - \left(\frac{1}{2}\right)^2 = \frac{\sqrt{2}}{2} - \frac{1}{4} \cong 0.7071 - 0.25 > 0.$$

Now we can conclude that $c \in (\frac{1}{2}, 1)$. Clearly, we could continue this process, "halving" the interval at each step to obtain better and better approximations of the root. This procedure is called the *bisection method*. We used a graphing utility to find that $f(x) = 0$ has exactly one root $c \in (0, 1)$, and $c \cong 0.6870$. ☐

The intermediate-value theorem also serves as the basis of a method for solving inequalities. (The method was used in Examples 2 and 3 in Section 1.3.) Suppose f is a continuous function on an interval I. As illustrated above, if f has both positive and negative values on I, then $f(x) = 0$ for some $x \in I$. Equivalently, if f has no zeros in I, then either $f(x) > 0$ or $f(x) < 0$ for all $x \in I$.

Example 2 Solve the inequality

$$x^3 - x^2 - 6x > 0.$$

SOLUTION Let $p(x) = x^3 - x^2 - 6x$. Factoring p, we have

$$p(x) = x^3 - x^2 - 6x = x(x^2 - x - 6) = x(x + 2)(x - 3).$$

Since p is a polynomial, p is continuous on $(-\infty, \infty)$. Also, $p(x) = 0$ only at $x = -2$, $x = 0$, and $x = 3$. These numbers partition the real line into four intervals: $(-\infty,-2)$, $(-2,0)$, $(0,3)$, and $(3,\infty)$, and p is nonzero on each of these intervals. Now,

$$\begin{array}{lll} \text{on} & (-\infty,-2) & \text{sign of } (x+2)x(x-3) = (-)(-)(-) = -, \\ \text{on} & (-2,0) & \text{sign of } (x+2)x(x-3) = (+)(-)(-) = +, \\ \text{on} & (0,3) & \text{sign of } (x+2)x(x-3) = (+)(+)(-) = -, \\ \text{on} & (3,\infty) & \text{sign of } (x+2)x(x-3) = (+)(+)(+) = +. \end{array}$$

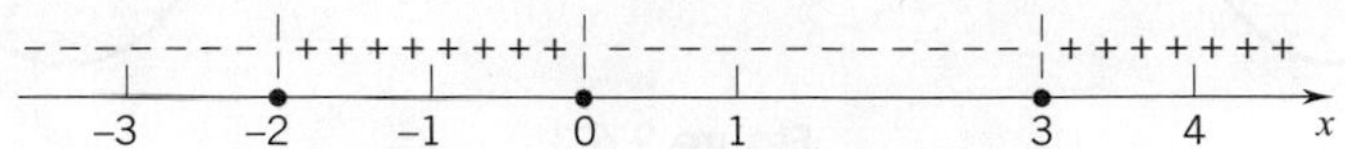

Figure 2.6.4

Therefore, the solution set of the inequality is $(-2,0)\cup(3,\infty)$.

Alternatively, the sign of p on each of the four intervals can be determined simply by evaluating p at a convenient number a on each interval. For example:

$$\begin{aligned} p(-3) = (-1)(-3)(-6) = -18 < 0, \quad &\text{so} \quad p(x) < 0 \text{ on } (-\infty,-2); \\ p(-1) = (1)(-1)(-4) = 4 > 0, \quad &\text{so} \quad p(x) > 0 \text{ on } (-2,0); \\ p(1) = (3)(1)(-2) = -6 < 0 \quad &\text{so} \quad p(x) < 0 \text{ on } (0,3); \\ p(4) = (6)(4)(1) = 24 > 0 \quad &\text{so} \quad p(x) > 0 \text{ on } (3,\infty). \end{aligned}$$

Again, $p(x) > 0$ on $(-2,0)\cup(3,\infty)$. (See Figure 2.6.4.) □

Example 3 Solve the inequality

$$(x+3)^3(2x-1)(x-4)^2 \le 0.$$

SOLUTION Since p is a polynomial, p is continuous on $(-\infty,\infty)$, and $p(x)=0$ only at $x=-3, x=1/2$, and $x=4$. These numbers partition the real line into four intervals: $(-\infty,-3), (-3,1/2), (1/2,4)$, and $(4,\infty)$, and p is nonzero on each of these intervals. Now:

$$\begin{array}{lll} \text{on} & (-\infty,-3) & \text{sign of } (x+3)^3(2x-1)(x-4)^2 = (-)(-)(+) = +, \\ \text{on} & (-3,1/2) & \text{sign of } (x+3)^3(2x-1)(x-4)^2 = (+)(-)(+) = -, \\ \text{on} & (1/2,4) & \text{sign of } (x+3)^3(2x-1)(x-4)^2 = (+)(+)(+) = +, \\ \text{on} & (3,\infty) & \text{sign of } (x+3)^3(2x-1)(x-4)^2 = (+)(+)(+) = +. \end{array}$$

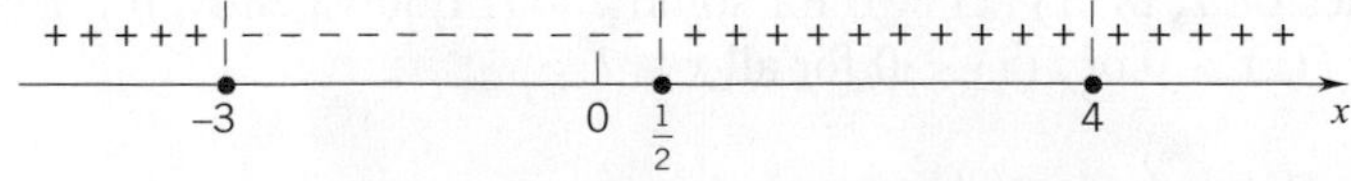

Figure 2.6.5

Therefore, $p(x) < 0$ on $(-3,1/2)$, and $p(x) = 0$ at the endpoints: $x = -3$ and $x = 1/2$. Thus, the solution set of the inequality is the closed interval $[-3,1/2]$. (See Figure 2.6.5.) □

Example 4 Solve the inequality

$$\frac{x^2-3x-10}{x-3} < 0.$$

SOLUTION Let $R(x) = \dfrac{x^2 - 3x - 10}{x - 3}$; R is a rational function. Since

$$R(x) = \frac{x^2 - 3x - 10}{x - 3} = \frac{(x+2)(x-5)}{x-3},$$

we see that $R(x) = 0$ at $x = -2$ and $x = 5$, and $R(x)$ is undefined at $x = 3$. The numbers $x = -2, x = 3$, and $x = 5$ partition the real line into four intervals $(-\infty, -2)$, $(-2, 3)$, $(3, 5)$ and $(5, \infty)$, and R is continuous and nonzero on each of these intervals. Therefore, on each interval, either $R(x) > 0$ or $R(x) < 0$. We will evaluate $R(a)$ at a conveniently chosen a on each interval to determine the sign of R (see Figure 2.6.6):

$$\begin{aligned}
R(-3) &= \frac{(-1)(-8)}{-}6 = -\tfrac{4}{3} < 0, &&\text{so} \quad R(x) < 0 \text{ on } (-\infty, -2);\\
R(0) &= \frac{2(-5)}{(-3)} = \tfrac{10}{3} > 0, &&\text{so} \quad R(x) > 0 \text{ on } (-2, 3);\\
R(4) &= \frac{6(-1)}{1} = -6 < 0, &&\text{so} \quad R(x) < 0 \text{ on } (3, 5);\\
R(6) &= \frac{8(1)}{3} = \tfrac{8}{3} > 0, &&\text{so} \quad R(x) > 0 \text{ on } (5, \infty).
\end{aligned}$$

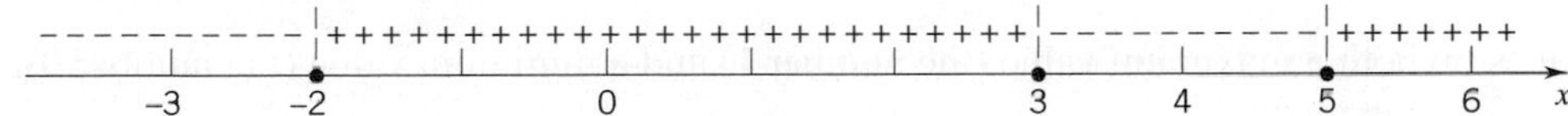

Figure 2.6.6

Thus, the solution set of the inequality is $(-\infty, -2) \cup (3, 5)$. ☐

Boundedness; Extreme Values

We start with a function f defined on some interval I. We say that f is *bounded* on I if it maps I onto a bounded set: namely, if there are numbers k and K such that

$$k \le f(x) \le K \quad \text{for all } x \in I.$$

A function which is not bounded is said to be *unbounded.*

The sine and cosine function are bounded on $(-\infty, \infty)$:

$$-1 \le \sin x \le 1 \qquad \text{and} \qquad -1 \le \cos x \le 1 \qquad \text{for all } x \in (-\infty, \infty).$$

Both functions map $(-\infty, \infty)$ onto $[-1, 1]$.

The tangent function is unbounded on $(-\pi/2, \pi/2)$ but bounded on $(-\pi/4, \pi/4)$ (see Figure 2.6.8 on page 114). The tangent function maps $(-\pi/2, \pi/2)$ onto $(-\infty, \infty)$; it maps $(-\pi/4, \pi/4)$ onto $(-1, 1)$.

Example 5 Let

$$g(x) = \begin{cases} 1/x^2, & x > 0 \\ 0, & x = 0. \end{cases} \qquad \text{(Figure 2.6.7)}$$

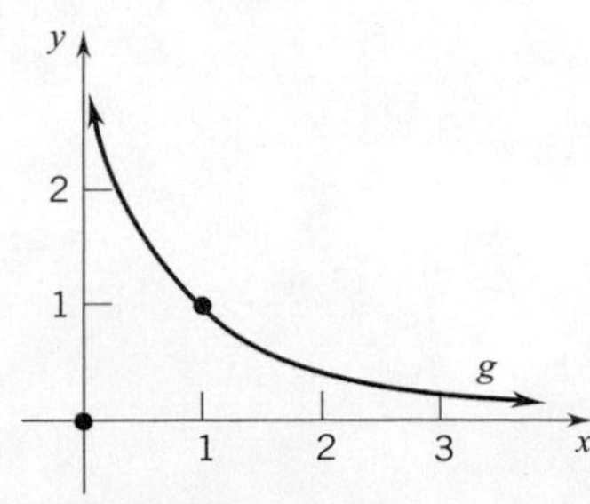

Figure 2.6.7

It is clear that g is unbounded on $[0, \infty)$. On the other hand, g is bounded on $[1, \infty)$. The function maps $[0, \infty)$ onto $[0, \infty)$, and it maps $[1, \infty)$ onto $(0, 1]$. ☐

A function may take on a maximum value; it may take on a minimum value; it may take on both a maximum value and a minimum value; it may take on neither.

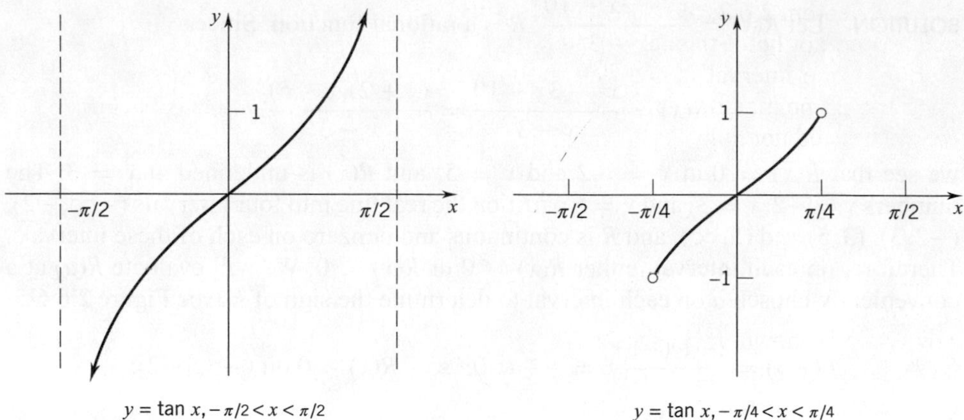

Figure 2.6.8

Here are some simple examples:

$$f(x) = \begin{cases} 1, & x \text{ rational} \\ 0, & x \text{ irrational}. \end{cases}$$

takes on both a maximum value (the number 1) and a minimum value (the number 0).

The function

$$f(x) = x^2, \quad x \in (0, 5]$$

takes on a maximum value (25), but it has no minimum value.

The function

$$f(x) = \frac{1}{x}, \quad x \in (0, \infty)$$

has no maximum value and no minimum value.

For a function continuous on a bounded closed interval, the situation is more predictable. The following theorem is fundamental.

THEOREM 2.6.2 THE EXTREME-VALUE THEOREM

A function f continuous on a bounded closed interval $[a, b]$ takes on both a maximum value M and a minimum value m.

For obvious reasons, M and m are called the *extreme values* of the function.

The result is illustrated in Figure 2.6.9. The maximum value M is taken on at the point marked d, and the minimum value m is taken on at the point marked c.

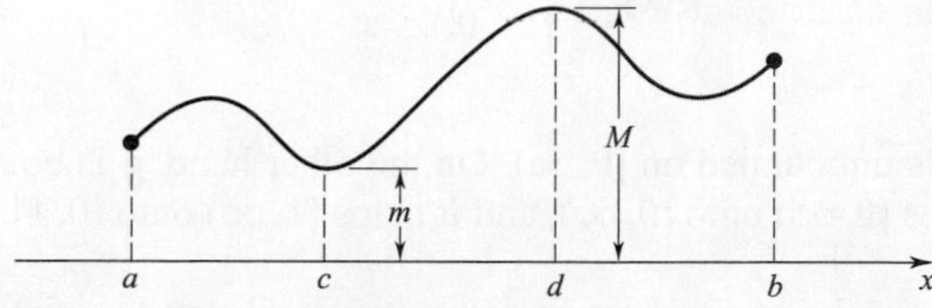

Figure 2.6.9

In Theorem 2.6.2, the full hypothesis is needed. If the interval is not bounded, the result need not hold: the cubing function $f(x) = x^3$ has no maximum on the interval $[0, \infty)$. If the interval is not closed, the result need not hold: the identity function $f(x) = x$ has no maximum and no minimum on $(0, 2)$. If the function is not continuous, the result need not hold. As an example take the function

$$f(x) = \begin{cases} 3, & x = 1 \\ x, & 1 < x < 5 \\ 3, & x = 5. \end{cases}$$

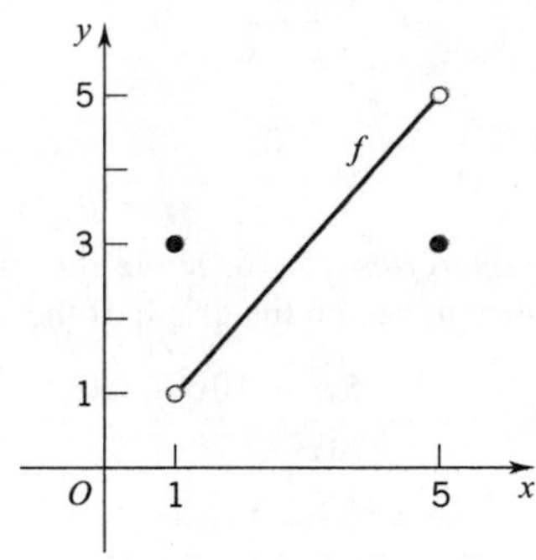

Figure 2.6.10

The graph is shown in Figure 2.6.10. The function is defined on $[1, 5]$, but it takes on neither a maximum value nor a minimum value. The function maps the closed interval $[1, 5]$ onto the open interval $(1, 5)$.

We need to make a final observation. From the intermediate-value theorem we know that

continuous functions map intervals onto intervals.

Now that we have the extreme-value theorem, we know that

continuous functions map bounded closed intervals $[a, b]$ *onto bounded closed intervals* $[m, M]$. (See Figure 2.6.11).

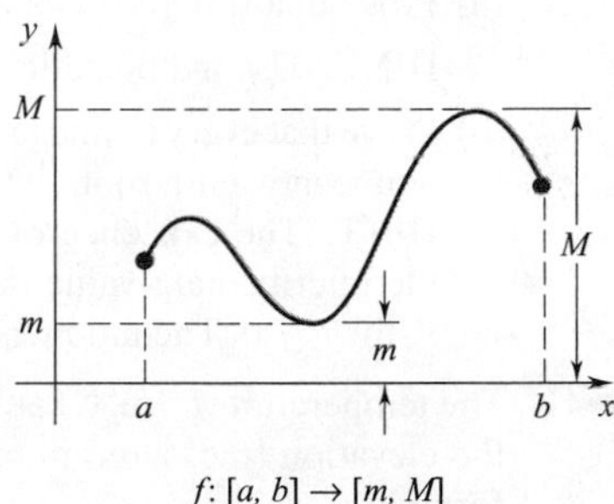

Figure 2.6.11

A proof of the extreme-value theorem is given in Appendix B. Techniques for determining the maximum and minimum values of functions are developed in Chapter 4. These techniques require an understanding of "differentiation," the subject to which we devote Chapter 3.

EXERCISES 2.6

Use the intermediate-value theorem to show that there is a solution of the given equation in the indicated interval.

1. $2x^3 - 4x^2 + 5x - 4 = 0$; $[1, 2]$.

2. $x^4 - x - 1 = 0$; $[-1, 1]$.

3. $\sin x + 2\cos x - x^2 = 0$; $[0, \pi/2]$.

4. $2\tan x - x = 1$; $[0, \pi/4]$.

5. $x^2 - 2 + \dfrac{1}{2x} = 0$; $[\frac{1}{4}, 1]$.

6. $x^{5/3} + x^{1/3} = 1$; $[-1, 1]$.

7. $x^3 = \sqrt{x + 2}$; $[1, 2]$.

8. $\sqrt{x^2 - 3x} - 2 = 0$; $[3, 5]$.

Solve the inequality for x.

9. $(x - 2)^2(10 - 2x) > 0$.

10. $x(2x - 1)(3x - 5) \geq 0$.

11. $x^3 - 2x^2 + x \leq 0$.

12. $\dfrac{2x-6}{x^2-6x+5} < 0.$

13. $\dfrac{1}{x-1} + \dfrac{4}{x-6} > 0.$

14. $\dfrac{x^2-4x}{(x+2)^2} < 0.$

C In Exercises 15–18, solve the inequality by using a graphing utility to sketch the graph of the left-hand side.

15. $\dfrac{-x^4+5x^3-10x^2+20x-24}{3x^2-5x-12} \leq 0.$

16. $\dfrac{6x^3-7x^2-11x+12}{6x^2+x-12} > 0.$

17. $\dfrac{2x^4+5x^3-9x^2-30x-18}{2x^3+11x^2+12x-9} > 0.$

18. $\dfrac{3x^3+7x^2+14x+4}{x^3-3x^2+3x-1} \leq 0.$

19. Let $f(x) = x^5 - 2x^2 + 5x$. Show that there is a number c such that $f(c) = 1$.

20. Let $f(x) = \dfrac{1}{x-1} + \dfrac{1}{x-4}$. Show that there is a number $c \in (1, 4)$ such that $f(c) = 0$.

21. Show that the equation $x^3 - 4x + 2 = 0$ has three distinct roots in $[-3, 3]$ and locate the roots between consecutive integers.

22. Use the intermediate-value theorem to prove that there exists a positive number c such that $c^2 = 2$.

In Exercises 23–34, sketch the graph of a function f that is defined on $[0, 1]$ and meets the given conditions (if possible).

23. f is continuous on $[0, 1]$, minimum value 0, maximum value $\frac{1}{2}$.

24. f is continuous on $[0, 1)$, minimum value 0, no maximum value.

25. f is continuous on $(0, 1)$, takes on the values 0 and 1, but does not take on the value $\frac{1}{2}$.

26. f is continuous on $[0, 1]$, takes on the values -1 and 1, but does not take on the value 0.

27. f is continuous on $[0, 1]$, maximum value 1, minimum value 1.

28. f is continuous on $[0, 1]$ and nonconstant, takes on no integer values.

29. f is continuous on $[0, 1]$, takes on no rational values.

30. f is not continuous on $[0, 1]$, takes on both a maximum value and a minimum value and every value in between.

31. f is continuous on $[0, 1]$, takes on only two distinct values.

32. f is continuous on $(0, 1)$, takes on only three distinct values.

33. f is continuous on $(0, 1)$, and the range of f is an unbounded interval.

34. f is continuous on $[0, 1]$, and the range of f is an unbounded interval.

35. (*Fixed-point property*) Show that if f is continuous and $0 \leq f(x) \leq 1$ for all $x \in [0, 1]$, then there exists at least one point c in $[0, 1]$ for which $f(c) = c$. HINT: Apply the intermediate-value theorem to the function $g(x) = x - f(x)$.

36. Given that f and g are continuous functions on $[a, b]$, and that $f(a) < g(a)$ and $g(b) < f(b)$, show that there exists at least one number $c \in (a, b)$ such that $f(c) = g(c)$. HINT: Consider $f(x) - g(x)$.

37. Use the intermediate-value theorem to prove that every positive number has a square root. That is, prove that if $a > 0$, then there exists a number c such that $c^2 = a$.

38. Use the intermediate-value theorem to prove that every real number has a cube root. That is, prove that for any real number a there exists a number c such that $c^3 = a$.

39. The intermediate-value theorem can be used to prove that each polynomial equation of odd degree

$$x^n + a_{n-1}x^{n-1} + \cdots + a_1x + a_o \qquad \text{with } n \text{ odd}$$

has at least one real root. Prove that the cubic equation

$$x^3 + ax^2 + bx + c = 0$$

has at least one real root.

40. Let n be a positive integer.
(a) Prove that if $0 \leq a < b$, then $a^n < b^n$.
HINT: Use mathematical induction.
(b) Prove that every nonnegative real number x has a unique nonnegative nth root $x^{1/n}$.
HINT: The existence of $x^{1/n}$ can be seen by applying the intermediate-value theorem to the function $f(t) = t^n$ for $t \geq 0$. The uniqueness follows from part (a).

C 41. The temperature T (in°C) at which water boils depends on the elevation (measured in meters h above sea level). The formula

$$T(h) = 100.862 - 0.0415\sqrt{h + 431.03}$$

gives the approximate value of T as a function of h. Use the intermediate-value theorem to show that water boils at 98°C at an elevation between 4000 and 4500 meters.

42. Assume that at any given instant, the temperature on the earth's surface varies continuously with position. Prove that there is at least one pair of points diametrically opposite each other on the equator where the temperature is the same. HINT: Form a function that relates the temperature at diametrically opposite points of the equator.

43. Let $\mathscr{C}$ denote the set of all circles with radius less than or equal to 10 inches. Prove that there is at least one member of $\mathscr{C}$ with area exactly 250 square inches.

44. Fix a positive number P. Let $\mathscr{R}$ denote the set of all rectangles with perimeter P. Prove that there is a member of $\mathscr{R}$ that has maximum area. What are the dimensions of the rectangle that has maximum area? HINT: Express the area of an arbitrary element of $\mathscr{R}$ as a function of the length of one of its sides.

45. Given a circle C of radius R. Let $\mathcal{F}$ denote the set of all rectangles that can be inscribed in C. Prove that there is a member of $\mathcal{F}$ that has maximum area.

In Exercises 46–49 use the intermediate-value theorem to locate the zeros of the given function. Then use a graphing utility to approximate the zeros to within 0.001.

46. $f(x) = 2x^3 + 4x - 4$.

47. $f(x) = x^3 - 5x + 3$.

48. $f(x) = x^5 - 3x + 1$.

49. $f(x) = x^3 - 2\sin x + \frac{1}{2}$.

In Exercises 50–53, determine whether the function f satisfies the hypothesis of the intermediate-value theorem on the interval $[a, b]$. If it does, use a graphing utility or a CAS to find a value $c \in (a, b)$ such that $f(c) = \frac{1}{2}[f(a) + f(b)]$. If not, determine if a value $c \in (a, b)$ still exists.

50. $f(x) = \dfrac{x+1}{x^2+1}$; $[-2, 3]$.

51. $f(x) = \dfrac{4x+3}{(x-1)}$; $[-3, 2]$.

52. $f(x) = \sec x$; $[-\pi, 2\pi]$.

53. $f(x) = \sin x - 3\cos 2x$; $\left[\dfrac{\pi}{2}, 2\pi\right]$.

In Exercises 54-57, use a graphing utility to graph f on the given interval. Is f bounded? If so, estimate its maximum and minimum values.

54. $f(x) = \dfrac{x^3 - 8x + 6}{4x + 1}$; $[0, 5]$.

55. $f(x) = \dfrac{2x}{1 + x^2}$; $[-2, 2]$.

56. $f(x) = \dfrac{\sin 2x}{x^2}$; $[-\pi/2, \pi/2]$.

57. $f(x) = \dfrac{1 - \cos x}{x^2}$; $[-2, 2]$.

In Exercises 58–60, use a graphing utility to find an interval $[a, b]$ on which the function f satisfies the hypothesis of the extreme-value theorem. If possible, find an interval $[c, d]$ such that

(a) $f(c) = m$, $f(d) = M$; (b) $f(c) = M$, $f(d) = m$; (c) $f(c) \neq m, M$, $f(d) \neq m, M$.

58. $f(x) = 1 - x + x^2 - x^3$.

59. $f(x) = x\sqrt{4 - x^2}$.

60. $f(x) = \dfrac{x^6 - 1}{x^2 - 1}$.

■ PROJECT 2.6 The Bisection Method

If the function f is continuous on $[a, b]$, and if $f(a)$ and $f(b)$ have opposite signs, then, by the intermediate-value theorem, the equation $f(x) = 0$ has at least one root in (a, b). For simplicity, let's assume that there is only one such root and call it c. How can we estimate the location of c? The intermediate-value theorem itself gives us no clue. The simplest method for approximating c is called the *bisection method.* It is an iterative process—a basic step is iterated (carried out repeatedly) until c is approximated with as much accuracy as we wish.

It is standard practice to label the elements of successive approximations by using subscripts $n = 1, 2, 3$, and so forth. We begin by setting $u_1 = a$ and $v_1 = b$. Now bisect $[u_1, v_1]$. If c is the midpoint of $[u_1, v_1]$, then we are done. If not, then c lies in one of the halves of $[u_1, v_1]$. Call it $[u_2, v_2]$. If c is the midpoint of $[u_2, v_2]$, then we are done. If not, then c lies in one of the halves of $[u_2, v_2]$. Call that half $[u_3, v_3]$ and continue. The first three iterations for a particular function are shown in the figure.

After n bisections, we are examining the midpoint m_n of the interval $[u_n, v_n]$. Therefore, we can be certain that

$$|c - m_n| \leq \frac{1}{2}(v_n - u_n) = \frac{1}{2}\left(\frac{v_{n-1} - u_{n-1}}{2}\right) = \cdots = \frac{b-a}{2^n}.$$

Thus, m_n approximates c to within $(b - a)/2^n$. If we want m_n to approximate c to within a given number ϵ, then we must carry out the iteration to the point where

$$\frac{b-a}{2^n} < \epsilon.$$

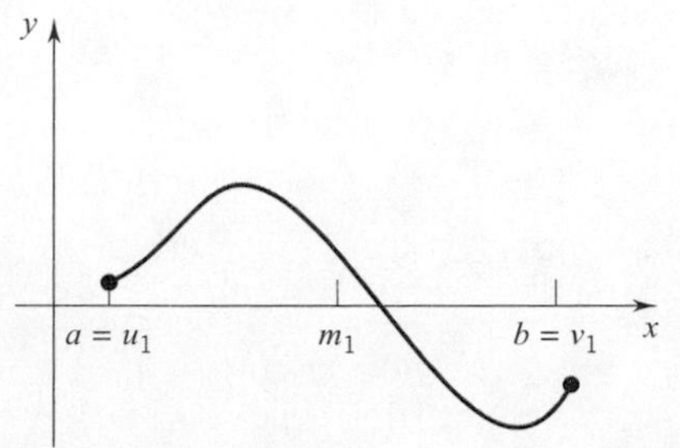

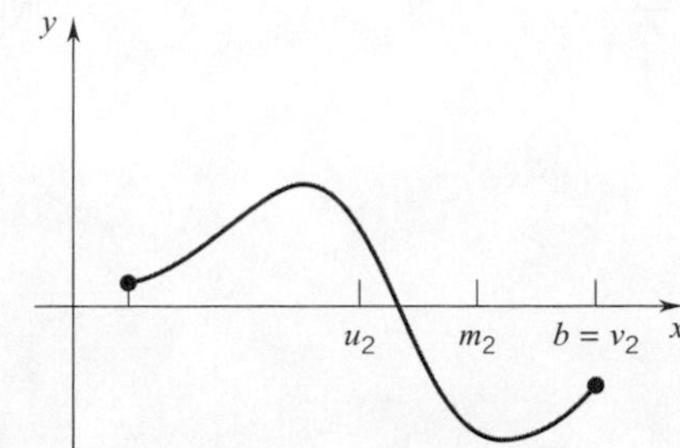

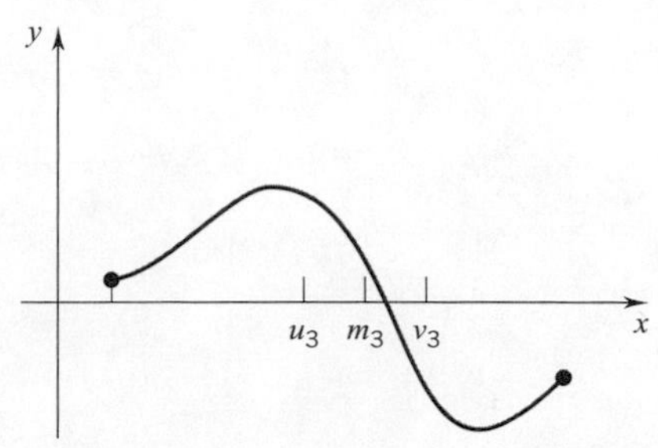

Problem 1. The function $f(x) = x^2 - 2$ is zero at $c = \sqrt{2}$. Since $f(1) < 0$ and $f(2) > 0$, we can estimate $\sqrt{2}$ by applying the bisection method to f on the interval $[1, 2]$. Suppose we want a numerical estimate accurate to within 0.01. Determine the minimum number of iterations that will be required and then do the calculations. Suppose we want a numerical estimate accurate to within 0.0001. How many iterations would be required to achieve this accuracy?

Problem 2. The function $f(x) = x^3 + x - 9$ has one zero c. Locate c between two consecutive integers and then use the bisection method to approximate c to within 0.01.

The following modification of the bisection method is sometimes used. Suppose that the continuous function f has exactly one zero c in the interval (a, b). The line connecting the points $(a, f(a))$ and $(b, f(b))$ is calculated and its x-intercept is used as the first approximation for c instead of the midpoint of the interval. The process is then repeated until the desired accuracy is obtained.

Problem 3. Carry out three iterations of the modified bisection method for the functions given in Problems 1 and 2. How does this method compare with the bisection method in terms of the rate at which the approximations converge to the zero c.

■ CHAPTER HIGHLIGHTS

2.1 The Idea of Limit

Intuitive interpretation of $\lim_{x \to c} f(x) = L$ (p. 59)
One-sided limits (p. 62)

2.2 Definition of Limit

limit of a function (p. 73) left-hand limit (p. 79)
right-hand limit (p. 80)
four formulations of the limit statement (p. 79)

2.3 Some Limit Theorems

uniqueness of limit (p. 83) stability of limit (p. 91)
The "arithmetic of limits": if $\lim_{x \to c} f(x) = L$ and $\lim_{x \to c} g(x) = M$, then

$$\lim_{x \to c} [f(x) + g(x)] = L + M, \quad \lim_{x \to c} [\alpha f(x)] = \alpha L \text{ for each real } \alpha,$$

$$\lim_{x \to c} [f(x)g(x)] = LM, \quad \lim_{x \to c} \frac{f(x)}{g(x)} = \frac{L}{M} \text{ provided } M \neq 0.$$

If $M = 0$ and $L \neq 0$, then $\lim_{x \to c} f(x)/g(x)$ does not exist.
Similar results hold for one-sided limits.

2.4 Continuity

continuity at c (p. 93) removable discontinuity (p. 94)
jump discontinuity (p. 94) infinite discontinuity (p. 94)
one-sided continuity (p. 98)
continuity on intervals (p. 99)

The sum, difference, product, quotient, and composition of continuous functions are continuous where defined. Polynomials and the absolute value function are everywhere continuous. Rational functions are continuous on their domains. The square-root function is continuous at all positive numbers and continuous from the right at 0.

2.5 The Pinching Theorem; Trigonometric Limits

pinching theorem (p. 103)

$$\lim_{x \to 0} \frac{\sin x}{x} = 1, \quad \lim_{x \to 0} \frac{1 - \cos x}{x} = 0.$$

Each trigonometric function is continuous on its domain.

2.6 Two Basic Properties of Continuous Functions

intermediate-value theorem (p. 110)
bounded functions (p. 113)
maximum value (p. 113)
minimum value (p. 113)
extreme values (p. 114)
extreme-value theorem (p. 114)
bisection method (p. 117)

■ 3.1 THE DERIVATIVE

Introduction: The Tangent to a Curve

We begin with a function f, and on its graph we choose a point $P : (c, f(c))$. See Figure 3.1.1. What line, if any, should be called tangent to the graph at that point?

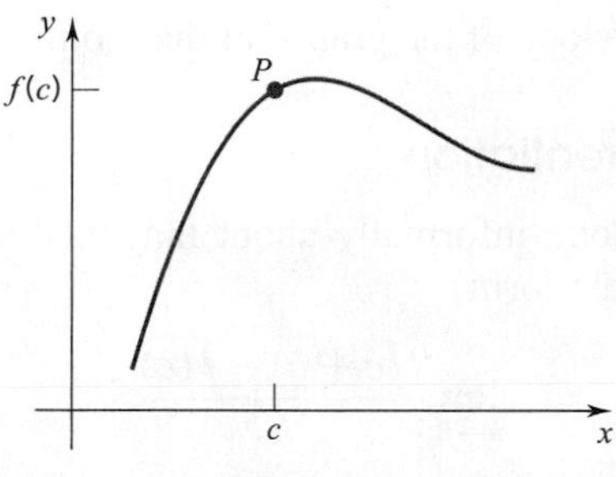

Figure 3.1.1

To answer this question, we choose a small number $h \neq 0$ and on the graph mark the point $Q : (c + h, f(c + h))$. Now we draw the secant line that passes through these two points. The situation is pictured in Figure 3.1.2, first with $h > 0$ and then with $h < 0$.

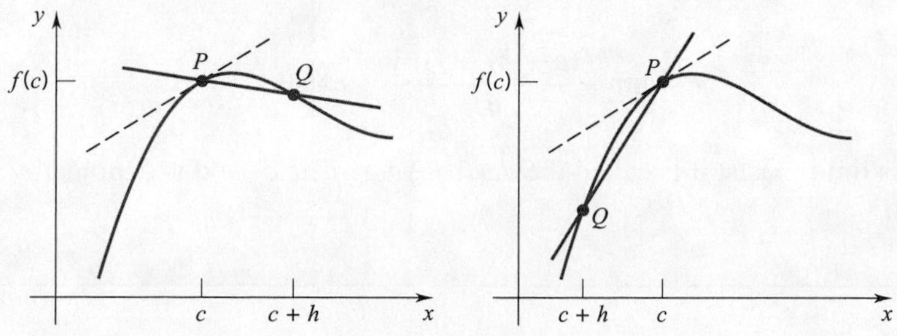

Figure 3.1.2

As h tends to zero from the right, the secant line tends to the limiting position indicated by the dashed line, and it tends to the same limiting position as h tends to zero from the left. The line at this limiting position is what we call "the tangent to the graph at the point $(c, f(c))$." Figure 3.1.3 illustrates this idea.

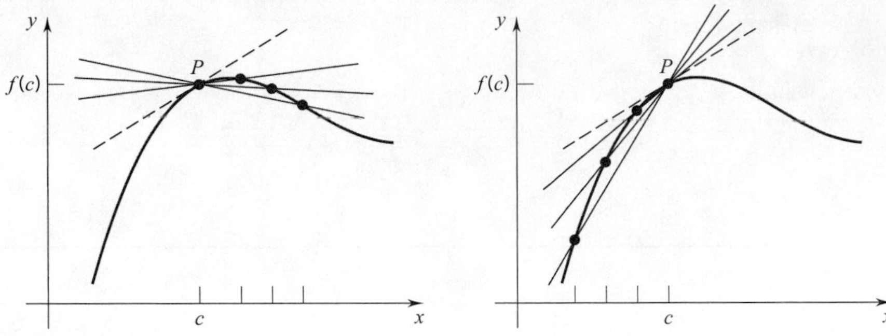

Figure 3.1.3

Since the approximating secant lines have slopes.

$$(*)\qquad \frac{f(c+h)-f(c)}{h}, \qquad \text{(verify this)}$$

you can expect the tangent line, the limiting position of these secants, to have slope

$$(**)\qquad \lim_{h\to 0}\frac{f(c+h)-f(c)}{h},$$

provided the limit exists. While $(*)$ measures the steepness of the line that passes through $(c, f(c))$ and $(c+h, f(c+h))$, $(**)$ measures the steepness of the graph at $(c, f(c))$ and is called the "slope of the graph" at that point.

Derivatives and Differentiation

In the introduction we spoke informally about tangent lines and gave a geometric interpretation to limits of the form

$$\lim_{h\to 0}\frac{f(c+h)-f(c)}{h}.$$

Here we begin the systematic study of such limits, what mathematicians call the theory of *differentiation*.

DEFINITION 3.1.1

A function f is said to be *differentiable at c* if

$$\lim_{h\to 0}\frac{f(c+h)-f(c)}{h} \quad \text{exists.}$$

If this limit exists, it is called the *derivative of f at c* and is denoted by $f'(c)$.†

† This prime notation goes back to the French mathematician Joseph-Louis Lagrange (1736–1813). Other notations are introduced later.

To calculate $f'(c)$ we must calculate

$$\lim_{h\to 0}\frac{f(c+h)-f(c)}{h}.$$

To do so, both c and the $c+h$ under consideration must lie in the domain of f. Since the derivative is a two-sided limit, f must be defined at least on some open interval that contains c.

Geometrically, $f'(c)$ is the *slope of the graph at the point* $(c,f(c))$. The line that passes through the point $(c,f(c))$ with slope $f'(c)$ is called the *tangent line at the point* $(c,f(c))$. By the point-slope formula, the tangent line has equation

(3.1.2)
$$\boxed{y-f(c)=f'(c)(x-c).}$$

This is the line that best approximates the graph of f near the point $(c,f(c))$.

It is time to consider some examples.

Example 1 We begin with the squaring function $f(x)=x^2$, and take $c=2$. To find $f'(2)$, we form the *difference quotient*

$$\frac{f(2+h)-f(2)}{h}=\frac{(2+h)^2-2^2}{h},\qquad h\neq 0.$$

Since
$$\frac{(2+h)^2-2^2}{h}=\frac{4+4h+h^2-4}{h}=\frac{4h+h^2}{h}=4+h,$$

we have

$$\frac{f(2+h)-f(2)}{h}=4+h,$$

and therefore

$$f'(2)=\lim_{h\to 0}\frac{f(2+h)-f(2)}{h}=\lim_{h\to 0}(4+h)=4.$$

The tangent line to the graph of $f(x)=x^2$ at the point (2,4) has equation

$$y-4=4(x-2)$$ (see Figure 3.1.4). ❑

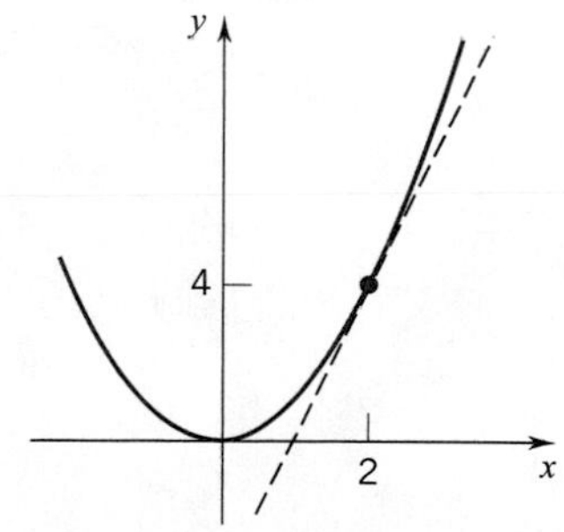

Figure 3.1.4

Example 2 For a linear function $f(x)=mx+b$ and any number c, the derivative $f'(c)$ is the number m: for $h\neq 0$,

$$\frac{f(c+h)-f(c)}{h}=\frac{[m(c+h)+b]-[mc+b]}{h}=\frac{mh}{h}=m,$$

and therefore

$$f'(c)=\lim_{h\to 0}\frac{f(c+h)-f(c)}{h}=\lim_{h\to 0}m=m.$$

This result is hardly surprising, since, as you know, the graph of the function is the line $y=mx+b$ and this line has slope m. ❑

Example 3 Find $g'(-2)$ given that

$$g(x) = 2x^2 + 3x + 1.$$

SOLUTION We form the difference quotient and simplify:

$$\begin{aligned}\frac{g(-2+h)-g(-2)}{h} &= \frac{2[-2+h]^2 + 3[-2+h] + 1 - [2(-2)^2 + 3(-2) + 1]}{h} \\ &= \frac{2[4-4h+h^2] - 6 + 3h + 1 - [3]}{h} = \frac{-5h+2h^2}{h} \\ &= -5 + 2h.\end{aligned}$$

Now, $$\lim_{h\to 0} \frac{g(-2+h)-g(-2)}{h} = \lim_{h\to 0}(-5+2h) = -5.$$

Thus, $g'(-2) = -5.$ ❑

The Derivative as a Function

In Example 1, suppose that we had wanted to find $f'(-2), f'(-1), f'(0)$, and $f'(1)$, in addition to $f'(2)$. Rather than forming each of the difference quotients and calculating each of the limits one-by-one, we'll choose an arbitrary number c and calculate $f'(c)$ in general. That is, let $f(x) = x^2$ and fix *any* number c. Then, for $h \neq 0$,

$$\frac{f(c+h)-f(c)}{h} = \frac{(c+h)^2 - c^2}{h} = \frac{c^2 + 2ch + h^2 - c^2}{h} = \frac{2ch + h^2}{h} = 2c + h,$$

and $$f'(c) = \lim_{h\to 0}\frac{f(c+h)-f(c)}{h} = \lim_{h\to 0}(2c+h) = 2c.$$

It now follows that $f'(-2) = -4, \quad f'(-1) = -2, \quad f'(0) = 0,$ and $f'(1) = 2.$

We had a similar result in Example 2, where we found that if $f(x) = mx + b$, then $f'(c) = m$ for any number c.

The point of these remarks is this: associated with the function $f(x) = x^2$ is the "derivative function" $f'(c) = 2c$, and associated with the function $f(x) = mx + b$ is the derivative function $f'(c) = m$. Following the conventional representation of functions, we replace c by x and write $f'(x) = 2x$ in the case of $f(x) = x^2$ and $f'(x) = m$ in the case of $f(x) = mx + b$.

DEFINITION 3.1.3

The *derivative* of a function f is the function f' with value at x given by:

$$f'(x) = \lim_{h\to 0}\frac{f(x+h)-f(x)}{h}, \quad \text{provided the limit exists.}$$

To *differentiate* a function f is to find its derivative.

Remark The domain of f' is necessarily a subset of the domain of f. In particular,

$$\text{dom}\,(f') = \left\{x \in \text{dom}\,(f) : \lim_{h\to 0}\frac{f(x+h)-f(x)}{h} \quad \text{exists}\right\}.$$

To apply the definition, f must be defined on some open interval containing x, and in calculating the limit, x is fixed; the "variable" is h. ❑

Example 4 Find $f'(x)$ given that

$$f(x) = x^3 - 12x.$$

SOLUTION Fix any number x, form the difference quotient, and simplify: for $h \neq 0$,

$$\begin{aligned}\frac{f(x+h)-f(x)}{h} &= \frac{[(x+h)^3 - 12(x+h)] - [x^3 - 12x]}{h}\\ &= \frac{x^3 + 3x^2h + 3xh^2 + h^3 - 12x - 12h - x^3 + 12x}{h}\\ &= \frac{3x^2h + 3xh^2 + h^3 - 12h}{h} = 3x^2 + 3xh + h^2 - 12.\end{aligned}$$

Now, $$\lim_{h\to 0}\frac{f(x+h)-f(x)}{h} = \lim_{h\to 0}(3x^2 + 3xh + h^2 - 12) = 3x^2 - 12.$$

Therefore, $f'(x) = 3x^2 - 12$. ❑

In our next example we will be dealing with the square-root function. Although this function is defined for all $x \geq 0$, you can expect a derivative only for $x > 0$.

Example 5 The square-root function

$$f(x) = \sqrt{x}, \quad x \geq 0 \qquad \text{(Figure 3.1.5)}$$

has derivative

$$f'(x) = \frac{1}{2\sqrt{x}}, \qquad \text{for } x > 0.$$

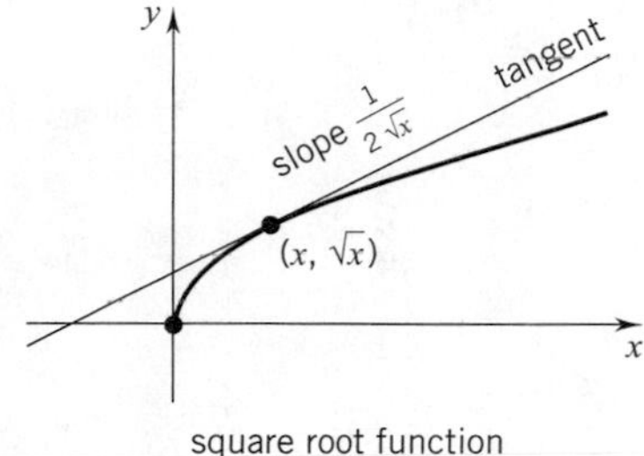

Figure 3.1.5

To verify this, fix an $x > 0$ and form the difference quotient

$$\frac{f(x+h)-f(x)}{h} = \frac{\sqrt{x+h}-\sqrt{x}}{h}, \qquad h \neq 0.$$

To simplify this expression, we multiply both the numerator and the denominator by $\sqrt{x+h}+\sqrt{x}$. This gives

$$\begin{aligned}\frac{f(x+h)-f(x)}{h} &= \left(\frac{\sqrt{x+h}-\sqrt{x}}{h}\right)\left(\frac{\sqrt{x+h}+\sqrt{x}}{\sqrt{x+h}+\sqrt{x}}\right)\\ &= \frac{(x+h)-x}{h(\sqrt{x+h}+\sqrt{x})} = \frac{1}{\sqrt{x+h}+\sqrt{x}}.\end{aligned}$$

Thus,

$$f'(x) = \lim_{h\to 0}\frac{f(x+h)-f(x)}{h} = \lim_{h\to 0}\frac{1}{\sqrt{x+h}+\sqrt{x}} = \frac{1}{2\sqrt{x}} \qquad \text{for } x > 0. \ ❑$$

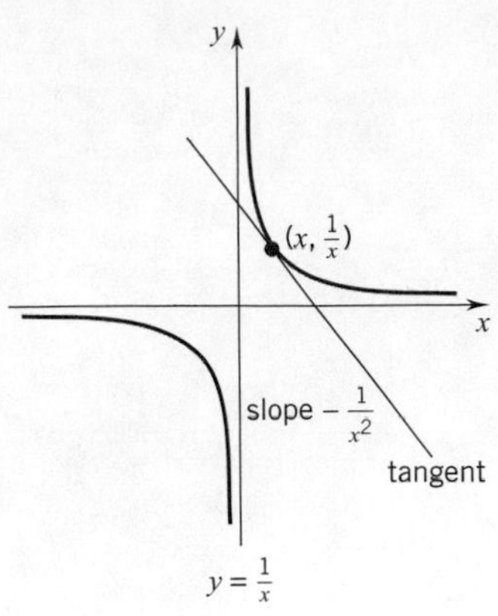

Figure 3.1.6

Example 6 Here we differentiate the function

$$f(x) = \frac{1}{x}, \quad x \neq 0.$$

(See Figure 3.1.6.) We form the difference quotient

$$\frac{f(x+h) - f(x)}{h} = \frac{\dfrac{1}{x+h} - \dfrac{1}{x}}{h}$$

and simplify:

$$\frac{\dfrac{1}{x+h} - \dfrac{1}{x}}{h} = \frac{\dfrac{x}{x(x+h)} - \dfrac{x+h}{x(x+h)}}{h} = \frac{\dfrac{-h}{x(x+h)}}{h} = \frac{-1}{x(x+h)}.$$

Thus $\qquad f'(x) = \lim_{h \to 0} \frac{f(x+h) - f(x)}{h} = \lim_{h \to 0} \frac{-1}{x(x+h)} = -\frac{1}{x^2}.$ □

Example 7 Find $f'(-3)$ and $f'(1)$ given that

$$f(x) = \begin{cases} x^2, & x \leq 1 \\ 2x - 1, & x > 1. \end{cases}$$

(see Figure 3.1.7)

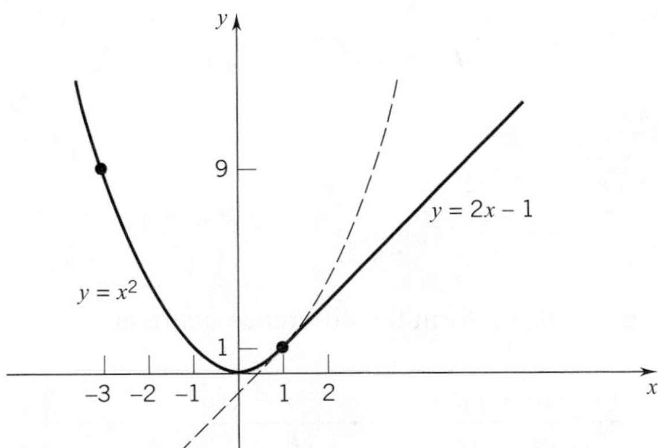

Figure 3.1.7

SOLUTION By definition $f'(-3) = \lim_{h \to 0} \frac{f(-3+h) - f(-3)}{h}$. However, for all x sufficiently close to -3, $f(x) = x^2$, and, from our discussion above, we know that $f'(x) = 2x$. Thus, $f'(-3) = 2(-3) = -6$.

Now let's find

$$f'(1) = \lim_{h \to 0} \frac{f(1+h) - f(1)}{h}.$$

Since f is not defined by the same formula on both sides of 1, we will evaluate this limit by taking one-sided limits. Note that $f(1) = 1^2 = 1$.

To the left of 1, $f(x) = x^2$. Thus

$$\begin{aligned}\lim_{h\to 0^-} \frac{f(1+h)-f(1)}{h} &= \lim_{h\to 0^-} \frac{(1+h)^2-1}{h}\\ &= \lim_{h\to 0^-} \frac{(1+2h+h^2)-1}{h} = \lim_{h\to 0^-}(2+h) = 2.\end{aligned}$$

To the right of 1, $f(x) = 2x - 1$. Thus

$$\lim_{h\to 0^+} \frac{f(1+h)-f(1)}{h} = \lim_{h\to 0^+} \frac{[2(1+h)-1]-1}{h} = \lim_{h\to 0^+} 2 = 2.$$

The limit of the difference quotient exists and is 2:

$$f'(1) = \lim_{h\to 0} \frac{f(1+h)-f(1)}{h} = 2. \quad \square$$

Tangent Lines and Normal Lines

We begin with a function f and choose a point $(c, f(c))$ on the graph. If f is differentiable at c, then, as you know, the equation for the tangent line is

$$y - f(c) = f'(c)(x - c) \qquad \text{(Equation 3.1.2)}$$

The line through $(c, f(c))$ that is perpendicular to the tangent is called the *normal line*. Since the slope of the tangent is $f'(c)$, the slope of the normal is $-1/f'(c)$, provided, of course, that $f'(c) \neq 0$. [*Remember:* For perpendicular lines, neither of which is vertical, $m_1 m_2 = -1$, (see Section 1.4).] As an equation for this normal line, we have:

(3.1.4)

$$y - f(c) = -\frac{1}{f'(c)}(x - c), \quad \text{provided that} \quad f'(c) \neq 0.$$

Some tangents and normal lines are pictured in Figure 3.1.8.

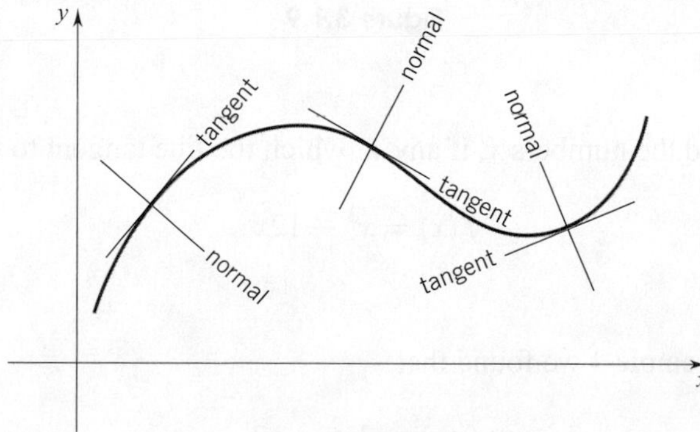

Figure 3.1.8

Example 8 Find equations for the tangent and normal lines to the graph of

$$f(x) = \sqrt{x}$$

at the point $(4, f(4)) = (4, 2)$.

SOLUTION From Example 5, we know that

$$f'(x) = \frac{1}{2\sqrt{x}}.$$

At the point (4, 2), the slope is $f'(4) = \frac{1}{4}$. The equation for the tangent line can be written

$$y - 2 = \tfrac{1}{4}(x - 4)$$

Since the slope of the normal line is the negative reciprocal of the slope of the tangent line, we can write the equation for the normal line as:

$$y - 2 = -4(x - 4). \quad \square$$

If $f'(c) = 0$, Equation 3.1.4 does not apply. If $f'(c) = 0$, the tangent line at $(c, f(c))$ is horizontal (with equation $y = f(c)$) and the normal line is vertical (with equation $x = c$)). In Figure 3.1.9 you can see several instances of a horizontal tangent. In the first two cases, the horizontal tangent occurs at the origin; the tangent line is the x-axis ($y = 0$) and the normal line is the y-axis ($x = 0$).

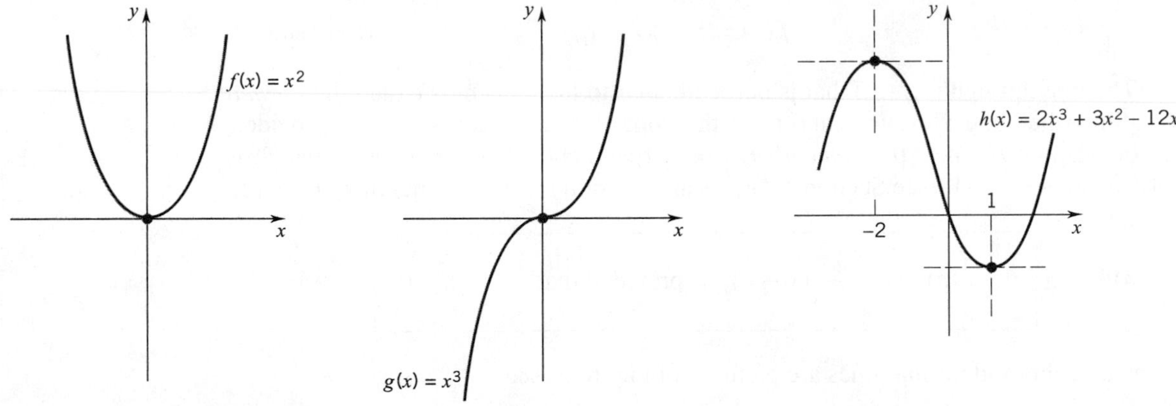

Figure 3.1.9

Example 9 Find the numbers x, if any, at which the line tangent to the graph of

$$f(x) = x^3 - 12x$$

is horizontal.

SOLUTION In Example 4 we found that

$$f'(x) = 3x^2 - 12.$$

The graph of f will have a horizontal tangent line at those points $(x, f(x))$ for which $f'(x) = 0$:

$$3x^2 - 12 = 0 \quad \text{iff} \quad x^2 = 4 \quad \text{iff} \quad x = 2, -2.$$

Thus, the graph has horizontal tangents at the points $(2, f(2)) = (2, -16)$ and $(-2, f(-2)) = (-2, 16)$. See Figure 3.1.10. $\square$

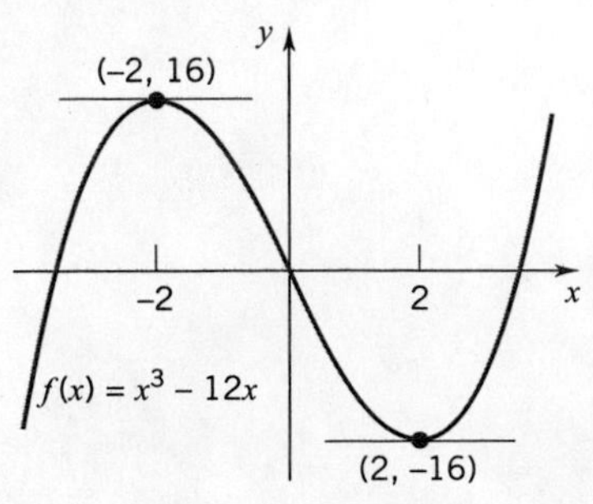

Figure 3.1.10

It is possible for the graph of a function to have a vertical tangent line as illustrated in the next example.

Example 10 Consider the cube-root function $f(x) = x^{1/3}$. The difference quotient at $x = 0$,

$$\frac{f(0+h) - f(0)}{h} = \frac{h^{1/3} - 0}{h} = \frac{1}{h^{2/3}}, \qquad h \neq 0,$$

increases without bound as h tends to zero ($1/h^{2/3} \to \infty$ as $h \to 0$). The graph has a vertical tangent (the line $x = 0$) and a horizontal normal (the line $y = 0$) at the origin. See Figure 3.1.11. □

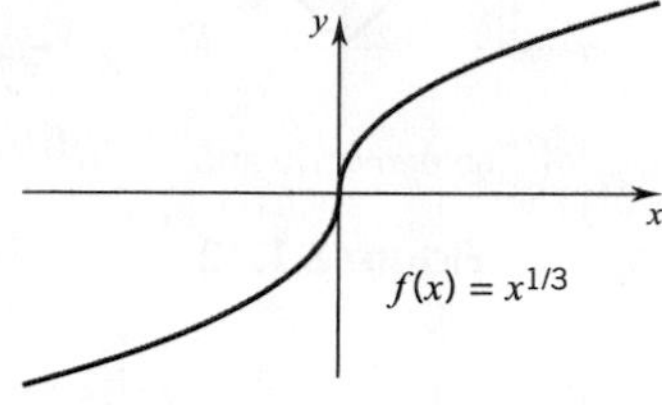

Figure 3.1.11

Vertical tangents are mentioned here only in passing. They are treated in detail in Section 4.7.

Differentiability and Continuity

Like continuity, differentiability is introduced as a point concept: a function f is differentiable at the point c if

$$\lim_{h\to 0} \frac{f(c+h) - f(c)}{h} \quad \text{exists.}$$

As with continuity, if f is differentiable at each point of an open interval I, then we shall say that f is differentiable on I. For example, $f(x) = x^2$ and $f(x) = mx + b$ are both differentiable on $(-\infty, \infty)$. The function $f(x) = 1/x$ is differentiable on $(-\infty, 0)$ and on $(0, \infty)$.

We can extend the definition of differentiability to intervals that contain endpoints by introducing one-sided derivatives. The *left-hand derivative* f'_- and the *right-hand derivative* f'_+ are defined by

$$f'_-(x) = \lim_{h\to 0^-} \frac{f(x+h) - f(x)}{h}$$

and

$$f'_+(x) = \lim_{h\to 0^+} \frac{f(x+h) - f(x)}{h},$$

provided, of course, that the limits exist. Thus, for example, a function f is differentiable on $[a, b]$ iff it is differentiable on the open interval (a, b), and has a left-hand derivative at b and a right-hand derivative at a; f is differentiable on $[a, b)$ iff it is differentiable on (a, b) and has a right-hand derivative at a, and so on. One-sided derivatives are taken up in the Exercises.

A function can be continuous at some number x without being differentiable there. For example, the absolute value function $f(x) = |x|$ is continuous at 0 (it is everywhere continuous), but it is not differentiable at 0:

$$\frac{f(0+h) - f(0)}{h} = \frac{|0+h| - |0|}{h} = \frac{|h|}{h} = \begin{cases} -1, & h < 0 \\ 1, & h > 0 \end{cases}$$

so that

$$\lim_{h\to 0^-} \frac{f(0+h) - f(0)}{h} = -1, \qquad \lim_{h\to 0^+} \frac{f(0+h) - f(0)}{h} = 1$$

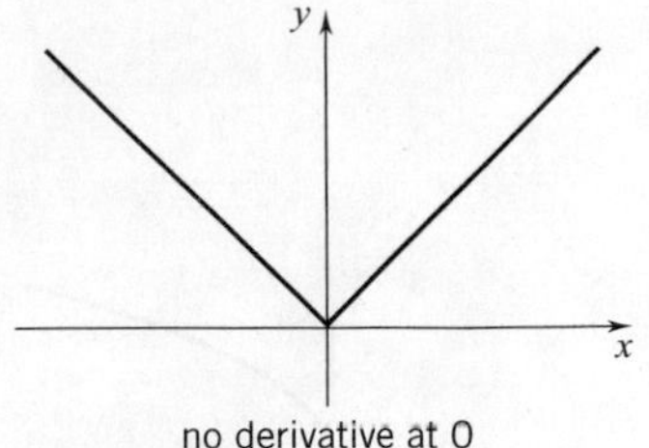

Figure 3.1.12

and thus

$$\lim_{h\to 0}\frac{f(0+h)-f(0)}{h}\quad\text{does not exist.}$$

In Figure 3.1.12 we have drawn the graph of the absolute value function. The lack of differentiability at 0 is evident geometrically. At the point $(0,0)=(0,f(0))$ the graph changes direction abruptly and there is no single tangent line. Note, however, that the absolute value function *is* differentiable on $(-\infty,0)$, as $f'(x)=-1$ on this interval; and it *is* differentiable on $(0,\infty)$ with $f'(x)=1$.

You can see a similar change of direction in the graph of

$$f(x)=\begin{cases} x^2, & x\le 1\\ \frac{1}{2}x+\frac{1}{2}, & x>1\end{cases}\qquad\text{(Figure 3.1.13)}$$

at the point (1,1). Once again, f is everywhere continuous (verify this), but it is not differentiable at 1:

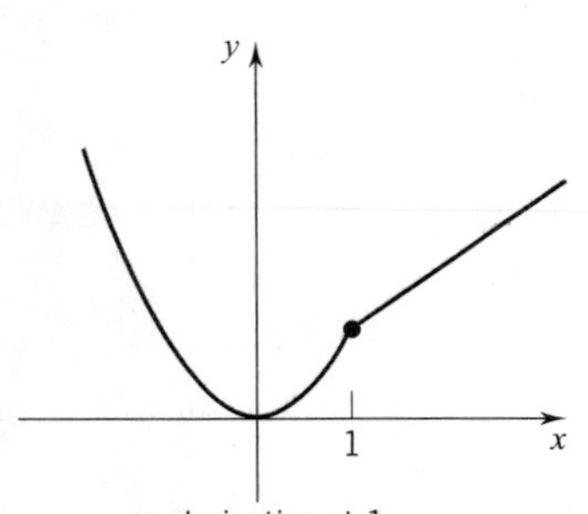

Figure 3.1.13

$$\lim_{h\to 0^-}\frac{f(1+h)-f(1)}{h}=\lim_{h\to 0^-}\frac{(1+h)^2-1}{h}=\lim_{h\to 0^-}\frac{h^2+2h}{h}=\lim_{h\to 0^-}(h+2)=2,$$

$$\lim_{h\to 0^+}\frac{f(1+h)-f(1)}{h}=\lim_{h\to 0^+}\frac{\frac{1}{2}(1+h)+\frac{1}{2}-1}{h}=\lim_{h\to 0^+}\left(\tfrac{1}{2}\right)=\tfrac{1}{2}.$$

Since these one-sided limits are different, the two-sided limit

$$\lim_{h\to 0}\frac{f(1+h)-f(1)}{h}\quad\text{does not exist.}$$

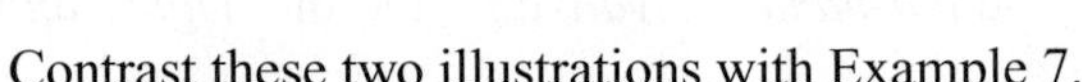

Contrast these two illustrations with Example 7.

A graphing utility was used to obtain the graph of

$$f(x)=|x^3-6x^2+8x|+3$$

shown in Figure 3.1.14. It appears that f is differentiable except, possibly, at $x=0$, $x=2$, and $x=4$, where abrupt changes in direction seem to occur. By zooming in near the point $(2,f(2))$, you can confirm that the left-hand and right-hand limits of the difference quotient both exist at $x=2$, but they are not equal. See Figure 3.1.15. A similar situation occurs at $x=0$ and $x=4$. Thus, f fails to be differentiable at $x=0,\ 2,$ and 4.

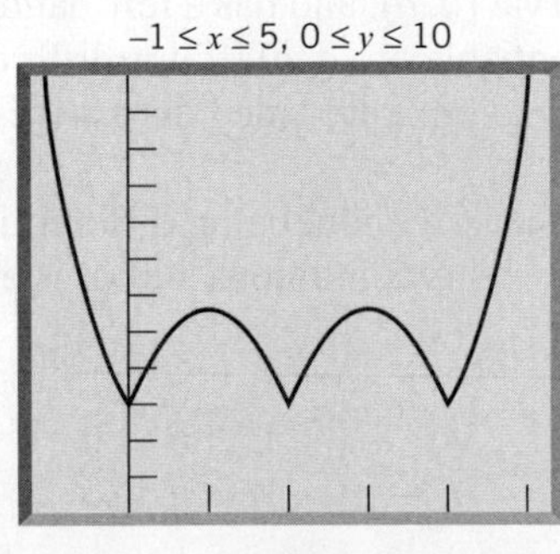

Figure 3.1.14

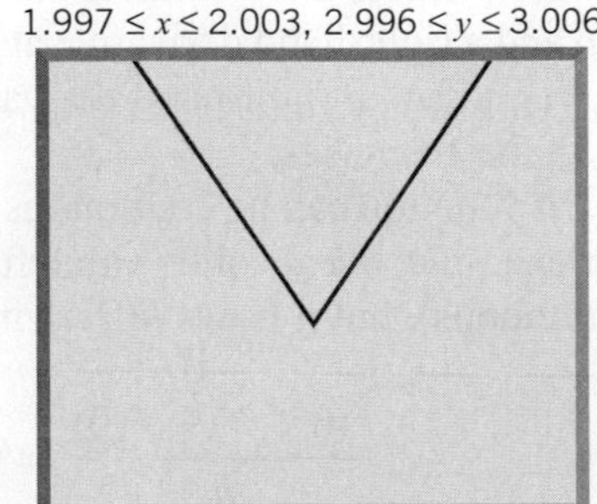

Figure 3.1.15

Although not every continuous function is differentiable, every differentiable function is continuous.

THEOREM 3.1.5

If f is differentiable at x, then f is continuous at x.

PROOF For $h \neq 0$ and $x + h$ in the domain of f,

$$f(x+h) - f(x) = \frac{f(x+h) - f(x)}{h} \cdot h.$$

With f differentiable at x,

$$\lim_{h \to 0} \frac{f(x+h) - f(x)}{h} = f'(x).$$

Since $\lim_{h \to 0} h = 0$, we have

$$\lim_{h \to 0} [f(x+h) - f(x)] = \left[\lim_{h \to 0} \frac{f(x+h) - f(x)}{h}\right] \cdot \left[\lim_{h \to 0} h\right] = f'(x) \cdot 0 = 0.$$

It follows that

$$\lim_{h \to 0} f(x+h) = f(x) \qquad \text{(explain)}$$

and thus f is continuous at x. ❑

EXERCISES 3.1

Find $f'(c)$ by forming the difference quotient

$$\frac{f(c+h) - f(c)}{h}$$

and taking the limit as $h \to 0$.

1. $f(x) = x^2 - 4x$; $c = 3$.
2. $f(x) = 7x - x^2$; $c = 2$.
3. $f(x) = 2x^3 + 1$; $c = -1$.
4. $f(x) = 5 - x^4$; $c = 1$.
5. $f(x) = \dfrac{8}{x+4}$; $c = -2$.
6. $f(x) = \sqrt{6 - x}$; $c = 2$.

Differentiate the function by forming the difference quotient

$$\frac{f(x+h) - f(x)}{h}$$

and taking the limit as h tends to 0.

7. $f(x) = 2 - 3x$.
8. $f(x) = c$, c constant.
9. $f(x) = 5x - x^2$.
10. $f(x) = 2x^3 + 1$.
11. $f(x) = x^4$.
12. $f(x) = 1/(x + 3)$.
13. $f(x) = \sqrt{x - 1}$.
14. $f(x) = x^3 - 4x$.
15. $f(x) = 1/x^2$.
16. $f(x) = 1/\sqrt{x}$.

Find equations for the tangent and normal lines at the point $(a, f(a))$.

17. $f(x) = 5x - x^2$; $a = 4$.
18. $f(x) = \sqrt{x}$; $a = 4$.
19. $f(x) = 1/x^2$; $a = -2$.
20. $f(x) = 5 - x^3$; $a = 2$.
21. The graph of a function f is shown in the figure.

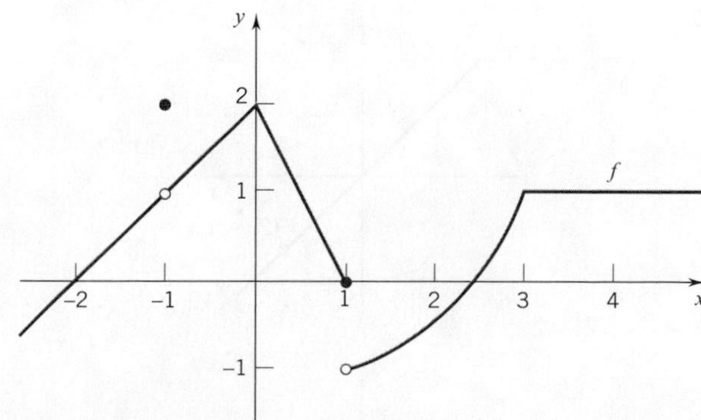

(a) For which numbers c does f fail to be continuous? For each discontinuity, state whether it is a removable discontinuity, a jump discontinuity, or neither.
(b) At which numbers c is f continuous but not differentiable?

22. Repeat Exercise 21 for the function g with graph shown in the figure below.

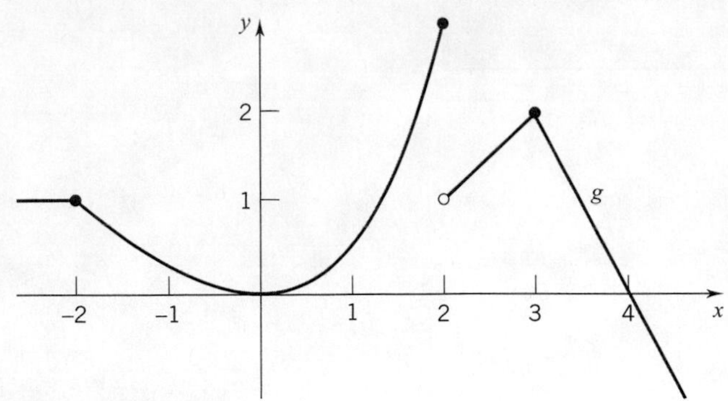

Draw the graph of the function and indicate where it is not differentiable.

23. $f(x) = |x + 1|$.

24. $f(x) = |2x - 5|$.

25. $f(x) = \sqrt{|x|}$.

26. $f(x) = |x^2 - 4|$.

27. $f(x) = \begin{cases} x^2, & x \le 1 \\ 2 - x, & x > 1. \end{cases}$

28. $f(x) = \begin{cases} x^2 - 1, & x \le 2 \\ 3, & x > 2. \end{cases}$

Find $f'(c)$ if it exists.

29. $f(x) = \begin{cases} 4x, & x < 1 \\ 2x^2 + 2, & x \ge 1; \end{cases} \quad c = 1.$

30. $f(x) = \begin{cases} 3x^2, & x \le 1 \\ 2x^3 + 1, & x > 1; \end{cases} \quad c = 1.$

31. $f(x) = \begin{cases} x + 1, & x \le -1 \\ (x+1)^2, & x > -1; \end{cases} \quad c = -1.$

32. $f(x) = \begin{cases} -\frac{1}{2}x^2, & x < 3 \\ -3x, & x \ge 3; \end{cases} \quad c = 3.$

Sketch the graph of the derivative of the function with the given graph.

33.

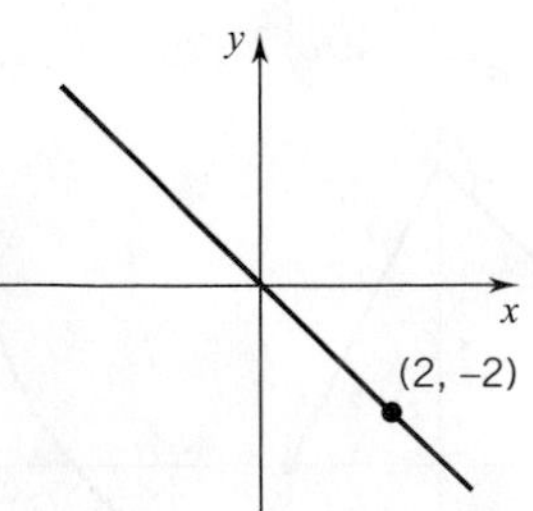

34.

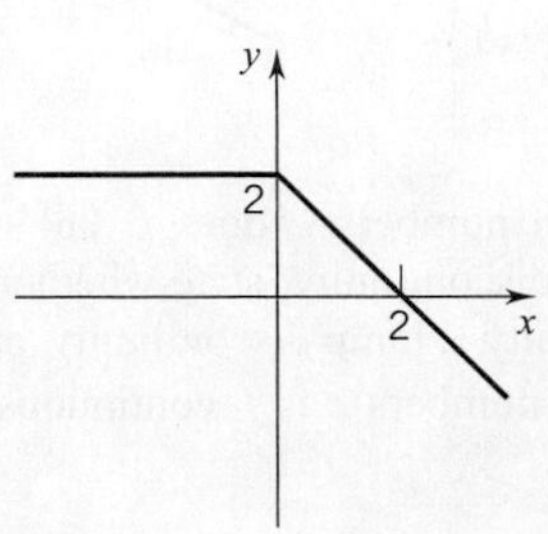

35.

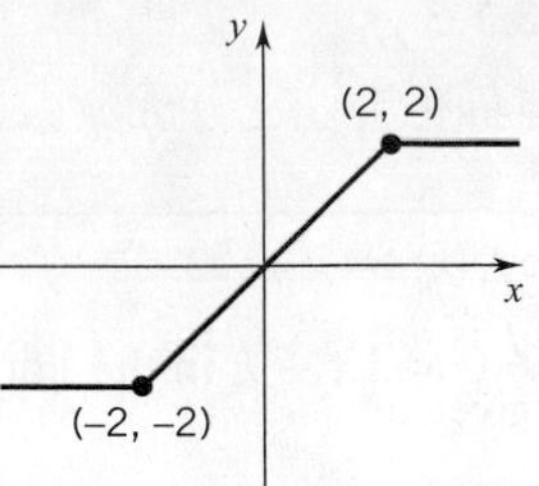

36.

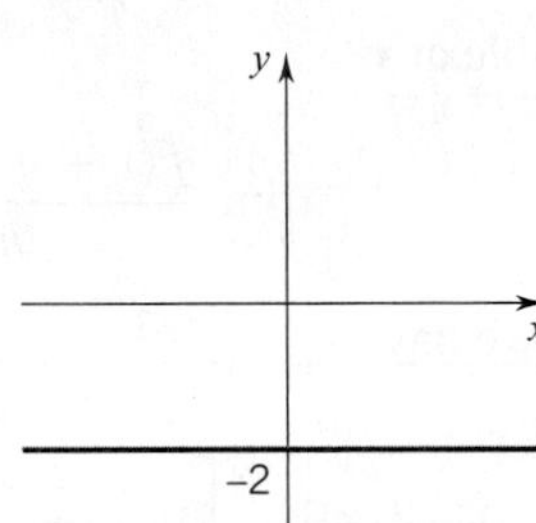

37.

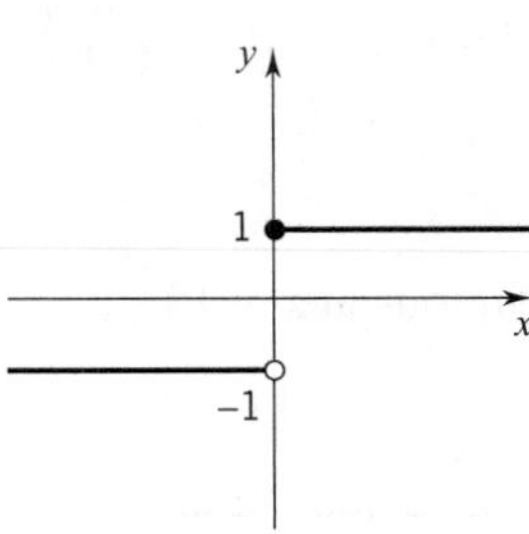

38.

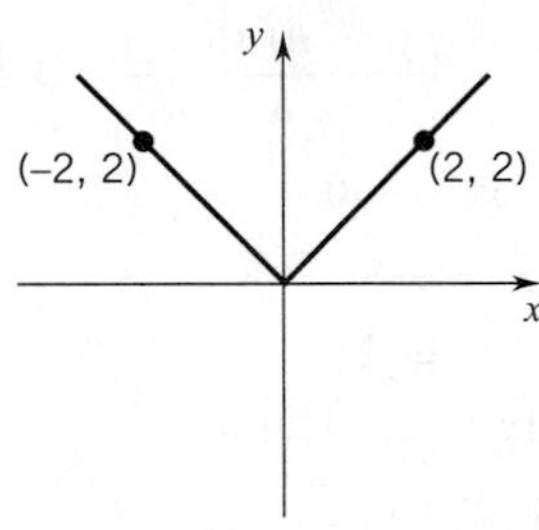

In Exercises 39–44, the limit represents the derivative of a function f at a number c. Determine f and c.

39. $\lim_{h \to 0} \dfrac{(1+h)^2 - 1}{h}$.

40. $\lim_{h \to 0} \dfrac{(-2+h)^3 + 8}{h}$.

41. $\lim_{h \to 0} \dfrac{\sqrt{4+h} - 2}{h}$.

42. $\lim_{h \to 0} \dfrac{(8+h)^{1/3} - 2}{h}$.

43. $\lim_{h \to 0} \dfrac{\cos(\pi + h) + 1}{h}$.

44. $\lim_{h \to 0} \dfrac{\sin(\frac{\pi}{6} + h) - \frac{1}{2}}{h}$.

45. Show that

$$f(x) = \begin{cases} x^2, & x \le 1 \\ 2x, & x > 1 \end{cases}$$

is not differentiable at $x = 1$.

46. Let

$$f(x) = \begin{cases} (x+1)^2, & x \le 0 \\ (x-1)^2, & x > 0. \end{cases}$$

(a) Determine $f'(x)$ for $x \neq 0$.

(b) Show that f is not differentiable at $x = 0$.

47. Find A and B given that the function

$$f(x) = \begin{cases} x^3, & x \le 1 \\ Ax + B, & x > 1. \end{cases}$$

is differentiable at $x = 1$.

48. Find A and B given that the function

$$f(x) = \begin{cases} x^2 - 2, & x \le 2 \\ Bx^2 + Ax, & x > 2 \end{cases}$$

is differentiable at $x = 2$.

In Exercises 49–54, give an example of a function f that is defined for all real numbers and satisfies the given conditions.

49. $f'(x) = 0$ for all real x.

50. $f'(x) = 0$ for all $x \neq 0$; $f'(0)$ does not exist.

51. $f'(x)$ exists for all $x \neq -1$; $f'(-1)$ does not exist.

52. $f'(x)$ exists for all $x \neq \pm 1$; neither $f'(1)$ nor $f'(-1)$ exists.

53. $f'(1) = 2$ and $f(1) = 7$.

54. $f'(x) = 1$ for $x < 0$ and $f'(x) = -1$ for $x > 0$.

Exercises 55–58 involve one-sided derivatives.

55. Let $f(x) = \begin{cases} x^2 - x, & x \le 2 \\ 2x - 2, & x > 2. \end{cases}$

(a) Show that f is continuous at 2.

(b) Find $f'_-(2)$ and $f'_+(2)$. (c) Is f differentiable at 2?

56. Let $f(x) = x\sqrt{x}, x \ge 0$.

(a) Calculate $f'(x)$ for each $x > 0$. (b) Calculate $f'_+(0)$.

57. Let $f(x) = \sqrt{1-x}$ for $0 \le x \le 1$.

(a) Calculate $f'(x)$ for each $x \in (0, 1)$.

(b) Find $f'_+(0)$, if it exists. (c) Find $f'_-(1)$, if it exists.

58. Let $f(x) = \begin{cases} 1 - x^2, & x \le 0 \\ x^2, & x > 0. \end{cases}$

(a) Find $f'_-(0)$, if it exists. (b) Find $f'_+(0)$, if it exists.

(c) Is f differentiable at 0? (d) Sketch the graph of f.

59. It is true that if f is continuous on $[a, c]$ and on $[c, b]$, it is then continuous on $[a, b]$. Does a comparable result hold for differentiability? Justify your answer.

60. Given that f is differentiable at c, let g be the function defined by

$$g(x) = \begin{cases} f(x), & x \le c \\ f'(c)(x-c) + f(c), & x > c. \end{cases}$$

(a) Show that g is differentiable at c. What is $g'(c)$?

(b) Suppose that the graph of f is as shown in the figure below. Sketch the graph of g. In what way is this exercise a generalization of Exercise 47?

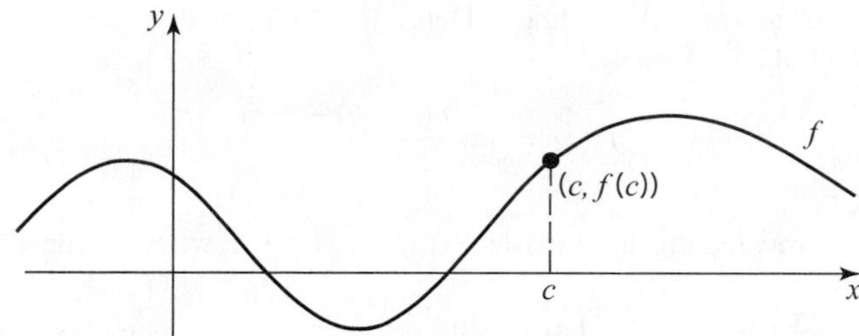

61. Given a function f and a number c in the domain of f, prove that if $f'_-(c)$ and $f'_+(c)$ both exist and are equal, then f is differentiable at c.

62. Let $f(x) = \begin{cases} x, & x \text{ rational} \\ 0, & x \text{ irrational} \end{cases}$ and

$$g(x) = \begin{cases} x^2, & x \text{ rational} \\ 0, & x \text{ irrational.} \end{cases}$$

Both functions are continuous at 0 and are discontinuous at each $x \neq 0$. (See Exercise 46, Section 2.4.)

(a) Can either function be differentiable at a point $x \neq 0$? Explain.

(b) Show that f is not differentiable at 0.

(c) Show that g is differentiable at 0 and give $g'(0)$.

63. Let $f(x) = \begin{cases} x\sin(1/x), & x \neq 0 \\ 0, & x = 0 \end{cases}$ and

$$g(x) = \begin{cases} x^2 \sin(1/x), & x \neq 0 \\ 0, & x = 0. \end{cases}$$

The graphs of f and g are indicated in the figures below.

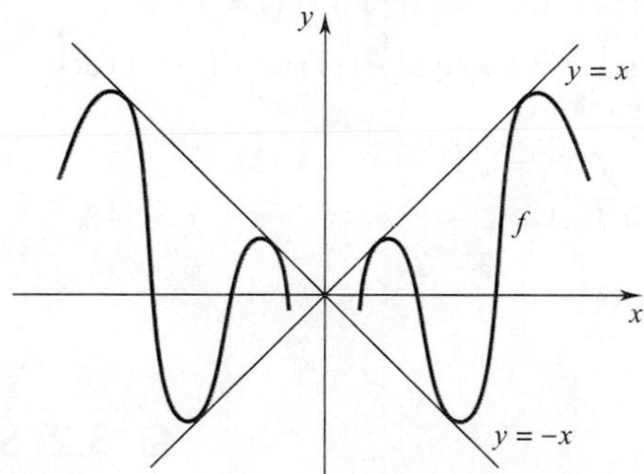

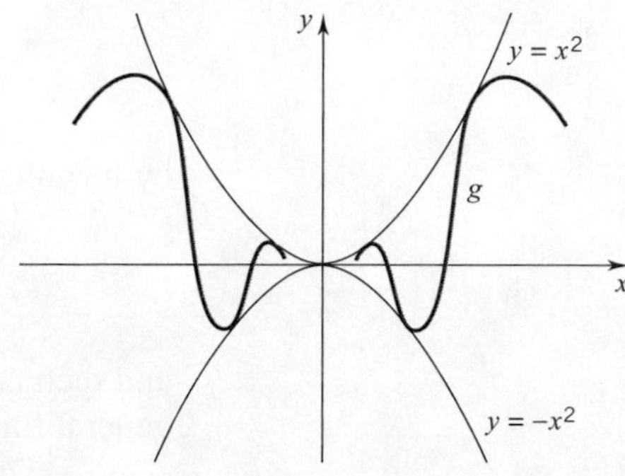

(a) Show that f and g are both continuous at 0.
(b) Show that f is not differentiable at 0.
(c) Show that g is differentiable at 0 and give $g'(0)$.

(*Important*) According to Definition 3.1.1, the derivative of f at c is given by

$$f'(c) = \lim_{h\to 0} \frac{f(c+h) - f(c)}{h}$$

provided this limit exists. Setting $x = c + h$, we can write

(3.1.6) $$f'(c) = \lim_{x\to c} \frac{f(x) - f(c)}{x - c}$$

[see (2.2.6), Section 2.2]. This is an alternative definition of derivative which has advantages in certain situations. Convince yourself of the equivalence of both definitions by calculating $f'(c)$ by both methods in Exercises 64–67.

64. $f(x) = x^3 + 1; \quad c = 2.$ **65.** $f(x) = x^2 - 3x; \quad c = 1.$

66. $f(x) = \sqrt{1+x}; \quad c = 3.$ **67.** $f(x) = x^{1/3}; \quad c = -1.$

68. $f(x) = \dfrac{1}{x+2}; \quad c = 0.$

69. Let $f(x) = x^{5/2}$ and consider the difference quotient

$$D(h) = \frac{f(2+h) - f(2)}{h}.$$

(a) Use a graphing utility to graph D for $-1 \le h \le 1$.
(b) Use the zoom feature to obtain successive magnifications of the graph of D near $h = 0$. Estimate $f'(2)$ accurate to three decimal places.
(c) Calculate D at $h = \pm 0.001$ and compare your result with your estimate in (b).

70. Repeat Exercise 69 for $f(x) = x^{2/3}$.

71. Use a CAS to find $f'(x)$ and $f'(c)$ for each of the following functions:
(a) $f(x) = \sqrt{5x - 4}; \quad c = 3.$
(b) $f(x) = 2 - x^2 + 4x^4 - x^6; \quad c = -2.$
(c) $f(x) = \dfrac{3 - 2x}{2 + 3x}; \quad c = -1.$

72. Use a CAS to evaluate

$$f'(c) = \lim_{h\to 0} \frac{f(c+h) - f(c)}{h}$$

for each of the following functions, if possible:
(a) $f(x) = |x - 1| + 2; \quad c = 1.$
(b) $f(x) = (x + 2)^{5/3} - 1; \quad c = -2.$
(c) $f(x) = (x - 3)^{2/3} + 3; \quad c = 3.$

73. Let $f(x) = 5x^2 - 7x^3$ on $[-1, 1]$.
(a) Use a graphing utility to draw the graph of f.
(b) Use the Trace Function to approximate the points on the graph where the tangent line is horizontal.
(c) Use a CAS to find $f'(x)$.
(d) Use a Solver to solve $f'(x) = 0$ and compare with your result in (b).

74. Repeat Exercise 73 with $f(x) = x^3 + x^2 - 4x + 3$ on $[-2, 2]$.

75. Let $f(x) = 4x - x^3$ and $c = \frac{3}{2}$.
(a) Use a CAS to find $f'(c)$. Then find equations for the tangent and normal lines, T and N, at $(c, f(c))$.
(b) Use a graphing utility to graph f, T and N together.
(c) Notice that the tangent line is a good approximation to the graph of f for values of x close to $x = \frac{3}{2}$. Determine an interval on which $|f(x) - T(x)| < 0.01$.

76. Repeat Exercise 75 with $f(x) = \dfrac{4x+3}{x}$ and $c = 3$.

77. Use a graphing utility to graph the function f, and determine where f is not differentiable.
(a) $f(x) = |9 - x^2|$.
(b) $f(x) = \dfrac{x+2}{|x+2|}$.
(c) $f(x) = \dfrac{x^2 - 1}{x + 1}$.

■ 3.2 SOME DIFFERENTIATION FORMULAS

Calculating the derivative of

$$f(x) = (x^3 + 2x - 3)(4x^2 + 1) \quad \text{or} \quad f(x) = \frac{6x^2 - 1}{x^4 + 5x + 1}$$

by forming the difference quotient

$$\frac{f(x+h) - f(x)}{h}$$

and then taking the limit as h tends to 0 is somewhat laborious. Here we derive some general formulas that enable us to calculate such derivatives quite quickly and easily.

We begin by pointing out that constant functions have derivative identically 0:

(3.2.1) $\boxed{\text{if } f(x) = \alpha, \quad \alpha \text{ any constant}, \quad \text{then } f'(x) = 0 \quad \text{for all } x.}$

and the identity function $f(x) = x$ has constant derivative 1:

(3.2.2) $\boxed{\text{if } f(x) = x, \quad \text{then } f'(x) = 1 \quad \text{for all } x.}$

PROOF For $f(x) = \alpha$,

$$f'(x) = \lim_{h\to 0} \frac{f(x+h) - f(x)}{h} = \lim_{h\to 0} \frac{\alpha - \alpha}{h} = \lim_{h\to 0} 0 = 0.$$

For $f(x) = x$,

$$f'(x) = \lim_{h\to 0} \frac{f(x+h) - f(x)}{h} = \lim_{h\to 0} \frac{(x+h) - x}{h} = \lim_{h\to 0} \frac{h}{h} = \lim_{h\to 0} 1 = 1. \quad \square$$

Remark These results can be verified geometrically. The graph of a constant function $f(x) = \alpha$ is a horizontal line, and the slope of a horizontal line is 0. The graph of the identity function $f(x) = x$ is the graph of the line $y = x$, which has slope 1. ☐

THEOREM 3.2.3 DERIVATIVES OF SUMS AND SCALAR MULTIPLES

Let α be a real number. If f and g are differentiable at x, then $f + g$ and αf are differentiable at x. Moreover,

$$(f+g)'(x) = f'(x) + g'(x) \quad \text{and} \quad (\alpha f)'(x) = \alpha f'(x).$$

PROOF To verify the first formula note that

$$\begin{aligned}\frac{(f+g)(x+h) - (f+g)(x)}{h} &= \frac{[f(x+h) + g(x+h)] - [f(x) + g(x)]}{h}\\ &= \frac{f(x+h) - f(x)}{h} + \frac{g(x+h) - g(x)}{h}.\end{aligned}$$

By definition

$$\lim_{h\to 0} \frac{f(x+h) - f(x)}{h} = f'(x) \quad \text{and} \quad \lim_{h\to 0} \frac{g(x+h) - g(x)}{h} = g'(x).$$

Thus
$$\lim_{h\to 0} \frac{(f+g)(x+h) - (f+g)(x)}{h} = f'(x) + g'(x),$$

which means that

$$(f+g)'(x) = f'(x) + g'(x).$$

To verify the second formula we must show that

$$\lim_{h\to 0} \frac{(\alpha f)(x+h) - (\alpha f)(x)}{h} = \alpha f'(x).$$

This follows directly from the fact that

$$\frac{(\alpha f)(x+h) - (\alpha f)(x)}{h} = \frac{\alpha f(x+h) - \alpha f(x)}{h} = \alpha\left[\frac{f(x+h) - f(x)}{h}\right]. \quad \square$$

Remark In this section and in the next few sections we will present formulas for calculating derivatives. It will be to your advantage to know these well. Some find it useful to put the formulas into words. For example, Theorem 3.2.3 says that

"the derivative of the sum of two functions is the sum of the derivatives" and

"the derivative of a constant times a function is that constant times the derivative of the function." □

Since $f - g = f + (-1)g$, it follows that if f and g are differentiable at x, then $f - g$ is differentiable at x, and

(3.2.4) $$(f - g)'(x) = f'(x) - g'(x).$$

"The derivative of the difference of two functions is the difference of the derivatives."

The results of Theorem 3.2.3 can be extended, by mathematical induction, to any finite collection of functions. That is, if $f_1, f_2 \ldots, f_n$ are each differentiable at x, and $\alpha_1, \alpha_2, \ldots, \alpha_n$ are numbers, then $\alpha_1 f_1 + \alpha_2 f_2 + \cdots + \alpha_n f_n$ is differentiable at x and

(3.2.5) $$(\alpha_1 f_1 + \alpha_2 f_2 + \cdots + \alpha_n f_n)'(x) = \alpha_1 f_1'(x) + \alpha_2 f_2'(x) + \cdots + \alpha_n f_n'(x).$$

THEOREM 3.2.6 THE PRODUCT RULE

If f and g are differentiable at x, then so is their product, and

$$(f \cdot g)'(x) = f(x)g'(x) + g(x)f'(x).$$

"The derivative of the product of two functions is the first function times the derivative of the second plus the second function times the derivative of the first."

PROOF We form the difference quotient

$$\frac{(f\cdot g)(x+h) - (f\cdot g)(x)}{h} = \frac{f(x+h)g(x+h) - f(x)g(x)}{h}$$

$$= \frac{f(x+h)g(x+h) - f(x+h)g(x) + f(x+h)g(x) - f(x)g(x)}{h}$$

and rewrite it as

$$f(x+h)\left[\frac{g(x+h)-g(x)}{h}\right]+g(x)\left[\frac{f(x+h)-f(x)}{h}\right].$$

[Here we have added and subtracted $f(x+h)g(x)$ in the numerator and then regrouped the terms so as to display the difference quotients for f and g.] Since f is differentiable at x, we know that f is continuous at x (Theorem 3.1.5) and thus

$$\lim_{h\to 0} f(x+h) = f(x). \qquad \text{(see Exercises 2.4, Problem 49)}$$

Since $\displaystyle\lim_{h\to 0}\frac{g(x+h)-g(x)}{h} = g'(x)$ and $\displaystyle\lim_{h\to 0}\frac{f(x+h)-f(x)}{h} = f'(x),$

we obtain

$$\lim_{h\to 0}\frac{(f\cdot g)(x+h)-(f\cdot g)(x)}{h} =$$

$$\lim_{h\to 0} f(x+h)\lim_{h\to 0}\left[\frac{g(x+h)-g(x)}{h}\right]+g(x)\lim_{h\to 0}\left[\frac{f(x+h)-f(x)}{h}\right]$$

$$= f(x)g'(x)+g(x)f'(x). \quad \square$$

Using the product rule, it is not hard to prove that

(3.2.7)

> for each positive integer n
>
> $p(x) = x^n$ has derivative $p'(x) = nx^{n-1}$.

In particular,

$p(x) = x$ has derivative $p'(x) = 1\cdot x^0 = 1,$

$p(x) = x^2$ has derivative $p'(x) = 2x,$

$p(x) = x^3$ has derivative $p'(x) = 3x^2,$

$p(x) = x^4$ has derivative $p'(x) = 4x^3,$

and so on.

PROOF OF (3.2.7) We proceed by mathematical induction on n. If $n = 1$, then we have the identity function

$$p(x) = x,$$

which we know satisfies

$$p'(x) = 1 = 1\cdot x^0.$$

This means that the formula holds for $n = 1$.

We suppose now that the result holds for $n = k$, that is, we assume that if $p(x) = x^k$, then $p'(x) = kx^{k-1}$, and go on to show that it holds for $n = k+1$. We let

$$p(x) = x^{k+1}$$

and note that

$$p(x) = x \cdot x^k.$$

Applying the product rule (Theorem 3.2.6) and our induction hypothesis, we obtain

$$p'(x) = x \cdot kx^{k-1} + x^k \cdot 1 = (k+1)x^k.$$

This shows that the formula holds for $n = k + 1$.

By the axiom of induction, the formula holds for all positive integers n. □

The formula for differentiating polynomials follows from (3.2.5) and (3.2.7):

(3.2.8)

If $\quad P(x) = a_n x^n + a_{n-1}x^{n-1} + \cdots + a_2 x^2 + a_1 x + a_0,$

then $\; P'(x) = na_n x^{n-1} + (n-1)a_{n-1}x^{n-2} + \cdots + 2a_2 x + a_1.$

For example,

$$P(x) = 12x^3 - 6x - 2 \quad \text{has derivative} \quad P'(x) = 36x^2 - 6$$

and $\quad Q(x) = \frac{1}{4}x^4 - 2x^2 + x + 5 \quad$ has derivative $\quad Q'(x) = x^3 - 4x + 1.$

Example 1 Differentiate $F(x) = (x^3 - 2x + 3)(4x^2 + 1)$ and find $F'(-1)$.

SOLUTION We have a product $F(x) = f(x)g(x)$ with

$$f(x) = x^3 - 2x + 3 \quad \text{and} \quad g(x) = 4x^2 + 1.$$

The product rule gives

$$\begin{aligned} F'(x) &= f(x)g'(x) + g(x)f'(x) \\ &= (x^3 - 2x + 3)(8x) + (4x^2 + 1)(3x^2 - 2) \\ &= 8x^4 - 16x^2 + 24x + 12x^4 - 5x^2 - 2 \\ &= 20x^4 - 21x^2 + 24x - 2. \end{aligned}$$

Setting $x = -1$, we have

$$F'(-1) = 20(-1)^4 - 21(-1)^2 + 24(-1) - 2 = 20 - 21 - 24 - 2 = -27. \quad □$$

Example 2 Differentiate $F(x) = (ax + b)(cx + d)$, where a, b, c, d are constants.

SOLUTION We have a product $F(x) = f(x)g(x)$ with

$$f(x) = ax + b \quad \text{and} \quad g(x) = cx + d.$$

Again we use the product rule

$$F'(x) = f(x)g'(x) + g(x)f'(x).$$

In this case

$$F'(x) = (ax + b)c + (cx + d)a = 2acx + bc + ad.$$

We can also do this problem without using the product rule by first carrying out the multiplication:

$$F(x) = acx^2 + bcx + adx + bd$$

and then differentiating:

$$F'(x) = 2acx + bc + ad.$$

The result is the same. □

Example 3 Suppose that g is differentiable at each x and that $F(x) = (x^3 - 5x)\,g(x)$. Find $F'(2)$ if it is known that $g(2) = 3$ and $g'(2) = -1$.

SOLUTION Applying the product rule, we have

$$F'(x) = [(x^3 - 5x)g(x)]' = (x^3 - 5x)g'(x) + g(x)(3x^2 - 5).$$

Therefore,

$$F'(2) = (-2)g'(2) + (7)g(2) = (-2)(-1) + (7)(3) = 23. \quad \square$$

We come now to reciprocals.

THEOREM 3.2.9 THE RECIPROCAL RULE

If g is differentiable at x and $g(x) \neq 0$, then $1/g$ is differentiable at x and

$$\left(\frac{1}{g}\right)'(x) = -\frac{g'(x)}{[g(x)]^2}.$$

PROOF Since g is differentiable at x, g is continuous at x (Theorem 3.1.5). Since $g(x) \neq 0$, we know that $1/g$ is continuous at x, and thus that

$$\lim_{h\to 0} \frac{1}{g(x+h)} = \frac{1}{g(x)}.$$

For h different from 0 and sufficiently small, $g(x+h) \neq 0$ [the continuity of g at x and the fact that $g(x) \neq 0$ guarantee this—see Exercise 50, Section 2.4]. Now the difference quotient is:

$$\frac{1}{h}\left[\frac{1}{g(x+h)} - \frac{1}{g(x)}\right] = \frac{1}{h}\left[\frac{g(x) - g(x+h)}{g(x+h)g(x)}\right]$$
$$= -\left[\frac{g(x+h) - g(x)}{h}\right]\frac{1}{g(x+h)g(x)}.$$

As h tends to zero, the right-hand side (and thus the left) tends to

$$-\frac{g'(x)}{[g(x)]^2}. \quad \square$$

From this last result we can show that the formula for the derivative of a positive integer power, x^n, also applies to negative integer powers; namely,

(3.2.10)

for each negative integer n, $p(x) = x^n$ has derivative $p'(x) = nx^{n-1}$.

This formula holds for all x except, of course, $x = 0$ where no negative power is even defined. In particular, for $x \neq 0$,

$$\begin{aligned} p(x) &= x^{-1} \quad \text{has derivative} \quad p'(x) = (-1)x^{-2} = -x^{-2}, \\ p(x) &= x^{-2} \quad \text{has derivative} \quad p'(x) = -2x^{-3}, \\ p(x) &= x^{-3} \quad \text{has derivative} \quad p'(x) = -3x^{-4}, \end{aligned}$$

and so on.

PROOF OF (3.2.10) Note that

$$p(x) = \frac{1}{g(x)} \quad \text{where} \quad g(x) = x^{-n} \text{ and } -n \text{ is a positive integer.}$$

The rule for reciprocals gives

$$p'(x) = -\frac{g'(x)}{[g(x)]^2} = -\frac{(-nx^{-n-1})}{x^{-2n}} = (nx^{-n-1})x^{2n} = nx^{n-1}. \quad \square$$

Remark A question remains. What about the derivative of $p(x) = x^0$? Since $x^0 = 1$ for all $x \neq 0$ (0^0 is another indeterminate form; it will be treated in Chapter 10), it follows that $p'(x) = 0$ for all $x \neq 0$. But,

$$0 = 0 \cdot x^{-1} \qquad \text{for all} \quad x \neq 0,$$

and so the formula also holds for $n = 0$. Thus,

for each integer n (positive, negative, or zero) $p(x) = x^n$ has derivative $p'(x) = nx^{n-1}$.

$\square$

Example 4 Differentiate $f(x) = \frac{5}{x^2} - \frac{6}{x}$ and find $f'\left(\frac{1}{2}\right)$.

SOLUTION To apply (3.2.10) we write

$$f(x) = 5x^{-2} - 6x^{-1}.$$

Differentiation gives

$$f'(x) = -10x^{-3} + 6x^{-2}.$$

Back in fractional notation

$$f'(x) = -\frac{10}{x^3} + \frac{6}{x^2}.$$

Setting $x = \frac{1}{2}$, we have

$$f'\left(\tfrac{1}{2}\right) = -\frac{10}{\left(\frac{1}{2}\right)^3} + \frac{6}{\left(\frac{1}{2}\right)^2} = -80 + 24 = -56. \quad \square$$

Example 5 Differentiate $f(x) = \dfrac{1}{ax^2 + bx + c}$, where a, b, c are constants.

SOLUTION Here we have a reciprocal $f(x) = 1/g(x)$ with

$$g(x) = ax^2 + bx + c.$$

The reciprocal rule (Theorem 3.2.9) gives

$$f'(x) = -\frac{g'(x)}{[g(x)]^2} = -\frac{2ax + b}{[ax^2 + bx + c]^2}. \quad \square$$

Finally we come to quotients in general.

THEOREM 3.2.11 THE QUOTIENT RULE

If f and g are differentiable at x and $g(x) \neq 0$, then the quotient f/g is differentiable at x and

$$\left(\frac{f}{g}\right)'(x) = \frac{g(x)f'(x) - f(x)g'(x)}{[g(x)]^2}.$$

"The derivative of a quotient is the denominator times the derivative of the numerator minus the numerator times the derivative of the denominator, all divided by the square of the denominator."

Since $f/g = f(1/g)$, the quotient rule can be obtained from the product and reciprocal rules. The proof of the quotient rule is left to you as an exercise. Finally, note that the reciprocal rule is just a special case of the quotient rule [take $f(x) = 1$].

From the quotient rule you can see that all rational functions (quotients of polynomials) are differentiable wherever they are defined.

Example 6 Differentiate $F(x) = \dfrac{6x^2 - 1}{x^4 + 5x + 1}$.

SOLUTION Here we are dealing with a quotient $F(x) = f(x)/g(x)$. The quotient rule,

$$F'(x) = \frac{g(x)f'(x) - f(x)g'(x)}{[g(x)]^2},$$

gives

$$F'(x) = \frac{(x^4 + 5x + 1)(12x) - (6x^2 - 1)(4x^3 + 5)}{(x^4 + 5x + 1)^2}$$

$$= \frac{-12x^5 + 4x^3 + 30x^2 + 12x + 5}{(x^4 + 5x + 1)^2}. \quad \square$$

Example 7 Find equations for the tangent and normal lines to the graph of

$$f(x) = \frac{3x}{1-2x}$$

at the point $(2, f(2)) = (2, -2)$.

SOLUTION We need to find $f'(2)$. Using the quotient rule, we get

$$f'(x) = \frac{(1-2x)(3) - 3x(-2)}{(1-2x)^2} = \frac{3}{(1-2x)^2}.$$

This gives

$$f'(2) = \frac{3}{[1-2(2)]^2} = \frac{3}{(-3)^2} = \frac{1}{3}.$$

Now, an equation for the tangent is

$$y - (-2) = \tfrac{1}{3}(x-2), \quad \text{which is} \quad y + 2 = \tfrac{1}{3}(x-2),$$

and the normal line has the equation: $y + 2 = -3(x-2)$. □

Example 8 Find the points on the graph of

$$f(x) = \frac{4x}{x^2+4}$$

where the tangent line is horizontal.

SOLUTION Applying the quotient rule, we get

$$f'(x) = \frac{(x^2+4)(4) - 4x(2x)}{[x^2+4]^2} = \frac{16-4x^2}{[x^2+4]^2}.$$

The tangent line will be horizontal at the points $(x, f(x))$, where $f'(x) = 0$. Therefore, we set $f'(x) = 0$ and solve for x:

$$\frac{16-4x^2}{[x^2+4]^2} = 0 \quad \text{iff} \quad 16 - 4x^2 = 0.$$

From this equation, we get

$$x^2 = 4, \quad \text{which implies} \quad x = 2 \text{ or } x = -2.$$

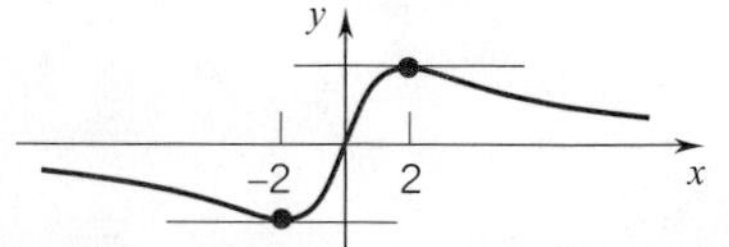

Figure 3.2.1

The points on the graph of f where the tangent line is horizontal are: $(-2, f(-2)) = (-2, -1)$ and $(2, f(2)) = (2, 1)$. See Figure 3.2.1. □

Remark Some expressions are easier to differentiate if we rewrite them in more convenient form. For example, we can differentiate

$$f(x) = \frac{x^5 - 2x}{x^2} = \frac{x^4 - 2}{x}, \quad x \neq 0,$$

by the quotient rule, or we can write

$$f(x) = (x^4 - 2)x^{-1}$$

and use the product rule; even better, we can notice that

$$f(x) = x^3 - 2x^{-1}$$

and proceed from there:

$$f'(x) = 3x^2 + 2x^{-2}. \quad \square$$

EXERCISES 3.2

Find the derivative of the function.

1. $F(x) = 1 - x.$

2. $F(x) = 2(1 + x).$

3. $F(x) = 11x^5 - 6x^3 + 8.$

4. $F(x) = \dfrac{3}{x^2}.$

5. $F(x) = ax^2 + bx + c;\quad a, b, c$ constant.

6. $F(x) = \dfrac{x^4}{4} - \dfrac{x^3}{3} + \dfrac{x^2}{2} - \dfrac{x}{1}.$

7. $F(x) = -\dfrac{1}{x^2}.$

8. $F(x) = \dfrac{(x^2 + 2)}{x^3}.$

9. $G(x) = (x^2 - 1)(x - 3).$

10. $F(x) = x - \dfrac{1}{x}.$

11. $G(x) = \dfrac{x^3}{1 - x}.$

12. $F(x) = \dfrac{ax - b}{cx - d};\quad a, b, c, d$ constant.

13. $G(x) = \dfrac{x^2 - 1}{2x + 3}.$

14. $G(x) = \dfrac{7x^4 + 11}{x + 1}.$

15. $G(x) = (x^3 - 2x)(2x + 5).$

16. $G(x) = \dfrac{x^3 + 3x}{x^2 - 1}.$

17. $G(x) = \dfrac{6 - 1/x}{x - 2}.$

18. $G(x) = \dfrac{1 + x^4}{x^2}.$

19. $G(x) = (9x^8 - 8x^9)\left(x + \dfrac{1}{x}\right).$

20. $G(x) = \left(1 + \dfrac{1}{x}\right)\left(1 + \dfrac{1}{x^2}\right).$

In Exercises 21–26, find $f'(0)$ and $f'(1)$.

21. $f(x) = \dfrac{1}{x - 2}.$

22. $f(x) = x^2(x + 1).$

23. $f(x) = \dfrac{1 - x^2}{1 + x^2}.$

24. $f(x) = \dfrac{2x^2 + x + 1}{x^2 + 2x + 1}.$

25. $f(x) = \dfrac{ax + b}{cx + d};\quad a, b, c, d$ constant.

26. $f(x) = \dfrac{ax^2 + bx + c}{cx^2 + bx + a};\quad a, b, c,$ constant.

In Exercises 27–30, find $f'(0)$ given that $h(0) = 3$ and $h'(0) = 2$.

27. $f(x) = xh(x).$

28. $f(x) = 3x^2h(x) - 5x.$

29. $f(x) = h(x) - \dfrac{1}{h(x)}.$

30. $f(x) = h(x) + \dfrac{x}{h(x)}.$

Find an equation for the line tangent to the graph of f at the point $(a, f(a))$.

31. $f(x) = \dfrac{x}{x + 2};\quad a = -4.$

32. $f(x) = (x^3 - 2x + 1)(4x - 5);\quad a = 2.$

33. $f(x) = (x^2 - 3)(5x - x^3);\quad a = 1.$

34. $f(x) = x^2 - \dfrac{10}{x};\quad a = -2.$

Find the points where the line tangent to the graph of f is horizontal.

35. $f(x) = (x - 2)(x^2 - x - 11).$

36. $f(x) = x^2 - \dfrac{16}{x}.$

37. $f(x) = \dfrac{5x}{x^2 + 1}.$

38. $f(x) = (x + 2)(x^2 - 2x - 8).$

Find x such that (a) $f'(x) = 0$, (b) $f'(x) > 0$, (c) $f'(x) < 0$.

39. $f(x) = x^4 - 8x^2 + 3.$

40. $f(x) = 3x^4 - 4x^3 - 2.$

41. $f(x) = x + \dfrac{4}{x^2}.$

42. $f(x) = \dfrac{x^2 - 2x + 4}{x^2 + 4}.$

Find the points where the line tangent to the graph of

43. $f(x) = -x^2 - 6$ is parallel to the line $y = 4x - 1$.

44. $f(x) = x^3 - 3x$ is perpendicular to the line $5y - 3x - 8 = 0$.

Find a function $f(x)$ the derivative of which is the given function $f'(x)$.

45. $f'(x) = 3x^2 + 2x + 1.$

46. $f'(x) = 4x^3 - 2x + 4.$

47. $f'(x) = 2x^2 - 3x - \dfrac{1}{x^2}.$

48. $f'(x) = x^4 + 2x^3 + \dfrac{1}{2\sqrt{x}}$.

49. Find A and B given that the derivative of

$$f(x) = \begin{cases} Ax^3 + Bx + 2, & x \le 2 \\ Bx^2 - A, & x > 2 \end{cases}$$

is everywhere continuous. HINT: First of all, f must be continuous.

50. Find A and B given that the derivative of

$$f(x) = \begin{cases} Ax^2 + B, & x < -1 \\ Bx^5 + Ax + 4, & x \ge -1 \end{cases}$$

is continuous for all x.

51. Find the area of the triangle formed by the x-axis and the lines tangent and normal to the curve $f(x) = 6x - x^2$ at the point (5, 5).

52. Find the area of the triangle formed by the x-axis and the lines tangent and normal to the curve $f(x) = 9 - x^2$ at the point (2, 5).

53. Determine the coefficients A, B, C so that the curve $f(x) = Ax^2 + Bx + C$ passes through the point (1, 3) and is tangent to the line $4x + y = 8$ at the point (2, 0).

54. Determine A, B, C, D so that the curve $f(x) = Ax^3 + Bx^2 + Cx + D$ is tangent to the line $y = 3x - 3$ at the point (1, 0) and is tangent to the line $y = 18x - 27$ at the point (2, 9).

55. Find the value(s) of x where the line tangent to the graph of the quadratic function $f(x) = ax^2 + bx + c$ is horizontal. NOTE: This gives a way to find the vertex of the parabola $y = ax^2 + bx + c$.

56. Find conditions on a, b, c and d that will guarantee that the graph of the cubic polynomial $p(x) = ax^3 + bx^2 + cx + d$ has:

(a) exactly two horizontal tangents.

(b) exactly one horizontal tangent.

(c) no horizontal tangents.

57. Find the value(s) of c, if any, where the line tangent to the graph of $f(x) = x^3 - x$ at $(c, f(c))$ is parallel to the secant line through $(-1, f(-1))$ and $(2, f(2))$.

58. Find the value(s) of c, if any, where the line tangent to the graph of $f(x) = x/(x+1)$ at $(c, f(c))$ is parallel to the secant line through $(1, f(1))$ and $(3, f(3))$.

59. Let $f(x) = 1/x, x > 0$. Show that the triangle that is formed by *any* line tangent to the graph of f and the coordinate axes has an area of 2 square units.

60. Find two lines through the point (2, 8) that are tangent to the graph of $f(x) = x^3$.

61. Find equations for all the lines tangent to the graph of $f(x) = x^3 - x$ that pass through the point $(-2, 2)$.

62. Let $f(x) = x^3$.

(a) Determine an equation for the line tangent to the graph of f at $x = c$, where $c \neq 0$.

(b) Determine whether the tangent line found in (a) intersects the graph of f at a point other than (c, c^3). If it does, determine the x-coordinate of the point of intersection.

63. Given two functions f and g, show that if f and $f + g$ are differentiable, then g is differentiable. Give an example to show that the differentiability of $f + g$ does not imply that f and g are each differentiable.

64. Given two functions f and g, if f and $f \cdot g$ are differentiable, does it follow that g is differentiable? If not, find a condition that will imply that g is differentiable if both f and $f \cdot g$ are differentiable.

65. Prove the validity of the quotient rule. HINT: $f/g = f \cdot (1/g)$.

66. Verify that, if f, g, h are differentiable, then

$$(fgh)'(x) = f'(x)g(x)h(x) + f(x)g'(x)h(x) + f(x)g(x)h'(x).$$

HINT: Apply the product rule to $[f(x)g(x)]h(x)$.

67. Use the result in Exercise 66 to find the derivative of $F(x) = (x^2 + 1)[1 + (1/x)](2x^3 - x + 1)$.

68. Use the result in Exercise 66 to find the derivative of $G(x) = \sqrt{x}\,[1/(1 + 2x)]\,(x^2 + x - 1)$.

69. Use the product rule (Theorem 3.2.6) to show that, if f is differentiable, then

$$g(x) = [f(x)]^2 \quad \text{has derivative} \quad g'(x) = 2f(x)f'(x).$$

70. Show that, if f is differentiable, then

$$g(x) = [f(x)]^n \text{ has derivative } g'(x) = n[f(x)]^{n-1}f'(x)$$

for each nonzero integer n. HINT: Mimic the inductive proof of (3.2.6).

71. Use the result in Exercise 70 to find the derivative of $g(x) = (x^3 - 2x^2 + x + 2)^3$.

72. Use the result in Exercise 70 to find the derivative of $g(x) = [x^2/(1 + 2x)]^{10}$.

C Use a CAS to find where $f'(x) = 0$, $f'(x) > 0$, $f'(x) < 0$. Verify your results with a graphing utility.

73. $f(x) = \dfrac{x^2}{x + 1}$.

74. $f(x) = 8x^5 - 60x^4 + 150x^3 - 125x^2$.

75. $f(x) = \dfrac{x^4 - 16}{x^2}$.

76. $f(x) = \dfrac{x^3 + 1}{x^4}$.

C **77.** Let $f(x) = \sin x$.

(a) Approximate $f'(x)$ at $x = 0$, $x = \pi/6$, $x = \pi/4$, $x = \pi/3$, and $x = \pi/2$ using the difference quotient

$$\frac{f(x+h) - f(x)}{h}$$

with $h = \pm 0.001$.

(b) Compare the approximate values of $f'(x)$ found in (a) with the values of $\cos x$ at each of these points.

(c) Use the results in (b) to guess a formula for the derivative of the sine function.

78. Repeat Exercise 77 with $f(x) = \cos x$, using $\sin x$ in part (b) and the cosine function in part (c).

79. Let $f(x) = 2^x$.

(a) Approximate $f'(x)$ at $x = 0, x = 1, x = 2$, and $x = 3$ using the difference quotient

$$\frac{f(x+h) - f(x)}{h}$$

with $h = \pm 0.001$.

(b) Calculate the ratio $f'(x)/f(x)$ for each of the values of x in (a).

(c) Use the results in (b) to guess a formula for the derivative of $f(x) = 2^x$.

80. Let $f(x) = x^4 + x^3 - 5x^2 + 2$.

(a) Use a graphing utility to graph f on the interval $[-4, 4]$ and estimate the x-coordinates of the points where the tangent line to the graph of f is horizontal.

(b) Use a graphing utility to graph $|f|$. Are there any points where f is not differentiable? If so, estimate where f fails to be differentiable.

■ 3.3 THE *d/dx* NOTATION; DERIVATIVES OF HIGHER ORDER

The *d/dx* Notation

So far we have indicated the derivative by a prime. There are, however, other notations that are widely used, particularly in science and engineering. The most popular of these is the "double-*d*" notation of Leibniz.† In the Leibniz notation, the derivative of a function y is indicated by writing

$$\frac{dy}{dx}, \quad \frac{dy}{dt}, \quad \text{or} \quad \frac{dy}{dz}, \quad \text{and so forth,}$$

depending on whether the letter x, t, or z, and so on, is being used for the elements of the domain of y. For instance, if y is initially defined by

$$y(x) = x^3,$$

then the Leibniz notation gives

$$\frac{d\,y(x)}{dx} = 3x^2.$$

Usually writers drop the (x) and simply write

$$y = x^3 \quad \text{and} \quad \frac{dy}{dx} = 3x^2.$$

The symbols

$$\frac{d}{dx}, \quad \frac{d}{dt}, \quad \frac{d}{dz}, \quad \text{and so forth}$$

are also used as prefixes before expressions to be differentiated. For example,

$$\frac{d}{dx}(x^3 - 4x) = 3x^2 - 4, \quad \frac{d}{dt}(t^2 + 3t + 1) = 2t + 3, \quad \frac{d}{dz}(z^5 - 1) = 5z^4.$$

† Gottfried Wilhelm Leibniz (1646–1716). The German mathematician whose role in the creation of calculus was outlined on p. 3.

In the Leibniz notation the differentiation formulas read:

$$\frac{d}{dx}[f(x)+g(x)] = \frac{d}{dx}[f(x)] + \frac{d}{dx}[g(x)], \quad \frac{d}{dx}[\alpha f(x)] = \alpha\frac{d}{dx}[f(x)],$$

$$\frac{d}{dx}[f(x)g(x)] = f(x)\frac{d}{dx}[g(x)] + g(x)\frac{d}{dx}[f(x)],$$

$$\frac{d}{dx}\left[\frac{1}{g(x)}\right] = -\frac{1}{[g(x)]^2}\frac{d}{dx}[g(x)],$$

$$\frac{d}{dx}\left[\frac{f(x)}{g(x)}\right] = \frac{g(x)\frac{d}{dx}[f(x)] - f(x)\frac{d}{dx}[g(x)]}{[g(x)]^2}.$$

Often functions f and g are replaced by u and v and the x is left out altogether. Then the formulas look like this:

$$\frac{d}{dx}(u+v) = \frac{du}{dx} + \frac{dv}{dx}, \quad \frac{d}{dx}(\alpha u) = \alpha\frac{du}{dx},$$

$$\frac{d}{dx}(uv) = u\,\frac{dv}{dx} + v\,\frac{du}{dx},$$

$$\frac{d}{dx}\left(\frac{1}{v}\right) = -\frac{1}{v^2}\frac{dv}{dx}, \quad \frac{d}{dx}\left(\frac{u}{v}\right) = \frac{v\dfrac{du}{dx} - u\dfrac{dv}{dx}}{v^2}.$$

The only way to develop a feeling for this notation is to use it. Below we work out some examples.

Example 1 Find $\dfrac{dy}{dx}$ for $y = \dfrac{3x-1}{5x+2}$.

SOLUTION We use the quotient rule:

$$\frac{dy}{dx} = \frac{(5x+2)\frac{d}{dx}(3x-1) - (3x-1)\frac{d}{dx}(5x+2)}{(5x+2)^2}$$

$$= \frac{(5x+2)(3) - (3x-1)(5)}{(5x+2)^2} = \frac{11}{(5x+2)^2}. \quad \square$$

Example 2 Find $\dfrac{dy}{dx}$ for $y = (x^3+1)(3x^5+2x-1)$.

SOLUTION Here we use the product rule:

$$\frac{dy}{dx} = (x^3+1)\frac{d}{dx}(3x^5+2x-1) + (3x^5+2x-1)\frac{d}{dx}(x^3+1)$$

$$= (x^3+1)(15x^4+2) + (3x^5+2x-1)(3x^2)$$

$$= (15x^7+15x^4+2x^3+2) + (9x^7+6x^3-3x^2)$$

$$= 24x^7+15x^4+8x^3-3x^2+2. \quad \square$$

Example 3 Find $\dfrac{d}{dt}\left(t^3 - \dfrac{t}{t^2-1}\right)$.

SOLUTION

$$\frac{d}{dt}\left(t^3 - \frac{t}{t^2-1}\right) = \frac{d}{dt}(t^3) - \frac{d}{dt}\left(\frac{t}{t^2-1}\right)$$

$$= 3t^2 - \left[\frac{(t^2-1)(1) - t(2t)}{(t^2-1)^2}\right] = 3t^2 + \frac{t^2+1}{(t^2-1)^2}. \quad \square$$

Example 4 Find $\dfrac{du}{dx}$ for $u = x(x+1)(x+2)$.

SOLUTION You can think of u as

$$[x(x+1)](x+2) \quad \text{or as} \quad x[(x+1)(x+2)].$$

From the first point of view,

$$\frac{du}{dx} = [x(x+1)](1) + (x+2)\frac{d}{dx}[x(x+1)]$$

$$= x(x+1) + (x+2)[x(1) + (x+1)(1)]$$

$$(*) \qquad = x(x+1) + (x+2)(2x+1).$$

From the second point of view,

$$\frac{du}{dx} = x\frac{d}{dx}[(x+1)(x+2)] + (x+1)(x+2)(1)$$

$$= x[(x+1)(1) + (x+2)(1)] + (x+1)(x+2)$$

$$(**) \qquad = x(2x+3) + (x+1)(x+2).$$

Both $(*)$ and $(**)$ can be multiplied out to give

$$\frac{du}{dx} = 3x^2 + 6x + 2.$$

Alternatively, this same result can be obtained by first carrying out the multiplication and then differentiating

$$u = x(x+1)(x+2) = x(x^2+3x+2) = x^3 + 3x^2 + 2x$$

so that $$\frac{du}{dx} = 3x^2 + 6x + 2. \quad \square$$

Example 5 Evaluate dy/dx at $x = 0$ and $x = 1$ given that $y = \dfrac{x^2}{x^2-4}$.

SOLUTION

$$\frac{dy}{dx} = \frac{(x^2-4)2x - x^2(2x)}{(x^2-4)^2} = -\frac{8x}{(x^2-4)^2}.$$

At $x = 0$,

$$\frac{dy}{dx} = -\frac{8\cdot 0}{(0^2-4)^2} = 0; \quad \text{at } x = 1, \quad \frac{dy}{dx} = -\frac{8\cdot 1}{(1^2-4)^2} = -\frac{8}{9}. \quad \square$$

Remark The notation

$$\left.\frac{dy}{dx}\right|_{x=a}$$

is sometimes used to emphasize the fact that we are evaluating the derivative dy/dx at $x = a$. Thus, in Example 5, we have

$$\left.\frac{dy}{dx}\right|_{x=0} = 0 \quad \text{and} \quad \left.\frac{dy}{dx}\right|_{x=1} = -\tfrac{8}{9} \quad \square$$

Derivatives of Higher Order

As we noted in Section 3.1, when we differentiate a function f we get a new function f', the derivative of f. Now suppose that f' can be differentiated. If we calculate $(f')'$, we get the *second derivative of f*, which is denoted f''. So long as we have differentiability, we can continue in this manner, forming the *third derivative of f*, denoted f''', and so on. However, the prime notation is not used beyond the third derivative. For the *fourth derivative of f*, we write $f^{(4)}$ and more generally, the nth derivative of f is denoted $f^{(n)}$. The functions f', f'', f''', $f^{(4)}, \ldots, f^{(n)}$ are called the derivatives of f of *orders* 1, 2, 3, 4, ..., n, respectively. For example, if $f(x) = x^5$, then

$$f'(x) = 5x^4, \quad f''(x) = 20x^3, \quad f'''(x) = 60x^2, \quad f^{(4)}(x) = 120x, \quad f^{(5)}(x) = 120.$$

In this case, all derivatives of order higher than five are identically zero. As a variant of this notation, you can write $y = x^5$ and then

$$y' = 5x^4, \quad y'' = 20x^3, \quad y''' = 60x^2, \quad \text{and so on.}$$

Since each polynomial P has a derivative P' that is in turn a polynomial, and each rational function Q has a derivative Q' that is in turn a rational function, polynomials and rational functions have derivatives of all orders. In the case of a polynomial of degree n, derivatives of order greater than n are all identically zero. (Explain.)

In the Leibniz notation the derivatives of higher order are written

$$\frac{d^2y}{dx^2} = \frac{d}{dx}\left(\frac{dy}{dx}\right), \quad \frac{d^3y}{dx^3} = \frac{d}{dx}\left(\frac{d^2y}{dx^2}\right), \ldots, \quad \frac{d^ny}{dx^n} = \frac{d}{dx}\left(\frac{d^{n-1}y}{dx^{n-1}}\right), \ldots$$

or

$$\frac{d^2}{dx^2}[f(x)] = \frac{d}{dx}\left[\frac{d}{dx}[f(x)]\right], \quad \frac{d^3}{dx^3}[f(x)] = \frac{d}{dx}\left[\frac{d^2}{dx^2}[f(x)]\right],$$

$$\ldots, \quad \frac{d^n}{dx^n}[f(x)] = \frac{d}{dx}\left[\frac{d^{n-1}}{dx^{n-1}}[f(x)]\right], \ldots.$$

Below we work out some examples.

Example 6 If $f(x) = x^4 - 3x^{-1} + 5$, then

$$f'(x) = 4x^3 + 3x^{-2} \quad \text{and} \quad f''(x) = 12x^2 - 6x^{-3}. \quad \square$$

Example 7

$$\frac{d}{dx}(x^5 - 4x^3 + 7x) = 5x^4 - 12x^2 + 7,$$

so that

$$\frac{d^2}{dx^2}(x^5 - 4x^3 + 7x) = \frac{d}{dx}(5x^4 - 12x^2 + 7) = 20x^3 - 24x$$

and $$\frac{d^3}{dx^3}(x^5 - 4x^3 + 7x) = \frac{d}{dx}(20x^3 - 24x) = 60x^2 - 24. \quad \square$$

Example 8 Finally we consider $y = x^{-1}$. In the Leibniz notation

$$\frac{dy}{dx} = -x^{-2}, \quad \frac{d^2y}{dx^2} = 2x^{-3}, \quad \frac{d^3y}{dx^3} = -6x^{-4}, \quad \frac{d^4y}{dx^4} = 24x^{-5}, \ldots$$

On the basis of these calculations, we are led to the general result

$$\frac{d^ny}{dx^n} = (-1)^n n! x^{-n-1}. \qquad \text{[Recall that } n! = n(n-1)(n-2)\cdots 3 \cdot 2 \cdot 1.]$$

In Exercise 61 you are asked to provide a rigorous proof of this result. In the prime notation

$$y' = -x^{-2}, \quad y'' = 2x^{-3}, \quad y''' = -6x^{-4}, \quad y^{(4)} = 24x^{-5}$$

and $$y^{(n)} = (-1)^n \, n! \, x^{-n-1}. \quad \square$$

EXERCISES 3.3

Find dy/dx.

1. $y = 3x^4 - x^2 + 1.$
2. $y = x^2 + 2x^{-4}.$
3. $y = x - \dfrac{1}{x}.$
4. $y = \dfrac{2x}{1-x}.$
5. $y = \dfrac{x}{1+x^2}.$
6. $y = x(x-2)(x+1).$
7. $y = \dfrac{x^2}{1-x}.$
8. $y = \left(\dfrac{x}{1+x}\right)\left(\dfrac{2-x}{3}\right).$
9. $y = \dfrac{x^3+1}{x^3-1}.$
10. $y = \dfrac{x^2}{(1+x)}.$

Find the indicated derivative.

11. $\dfrac{d}{dx}(2x-5).$
12. $\dfrac{d}{dx}(5x+2).$
13. $\dfrac{d}{dx}[(3x^2 - x^{-1})(2x+5)].$
14. $\dfrac{d}{dx}[(2x^2 + 3x^{-1})(2x - 3x^{-2})].$
15. $\dfrac{d}{dt}\left(\dfrac{t^4}{2t^3-1}\right).$
16. $\dfrac{d}{dt}\left(\dfrac{2t^3+1}{t^4}\right).$
17. $\dfrac{d}{du}\left(\dfrac{2u}{1-2u}\right).$
18. $\dfrac{d}{du}\left(\dfrac{u^2}{u^3+1}\right).$
19. $\dfrac{d}{du}\left(\dfrac{u}{u-1} - \dfrac{u}{u+1}\right).$
20. $\dfrac{d}{du}[u^2(1-u^2)(1-u^3)].$
21. $\dfrac{d}{dx}\left(\dfrac{x^3+x^2+x+1}{x^3-x^2+x-1}\right).$
22. $\dfrac{d}{dx}\left(\dfrac{x^3+x^2+x-1}{x^3-x^2+x+1}\right).$

Evaluate dy/dx at $x = 2$.

23. $y = (x+1)(x+2)(x+3).$
24. $y = (x+1)(x^2+2)(x^3+3).$
25. $y = \dfrac{(x-1)(x-2)}{(x+2)}.$
26. $y = \dfrac{(x^2+1)(x^2-2)}{x^2+2}.$

Find the second derivative.

27. $f(x) = 7x^3 - 6x^5.$
28. $f(x) = 2x^5 - 6x^4 + 2x - 1.$
29. $f(x) = \dfrac{x^2-3}{x}.$
30. $f(x) = x^2 - \dfrac{1}{x^2}.$
31. $f(x) = (x^2-2)(x^{-2}+2).$
32. $f(x) = (2x-3)\left(\dfrac{2x+3}{x}\right).$

Find d^3y/dx^3.

33. $y = \frac{1}{3}x^3 + \frac{1}{2}x^2 + x + 1.$
34. $y = (1+5x)^2.$
35. $y = (2x-5)^2.$
36. $y = \frac{1}{6}x^3 - \frac{1}{4}x^2 + x - 3.$
37. $y = x^3 - \dfrac{1}{x^3}.$
38. $y = \dfrac{x^4+2}{x}.$

Find the indicated derivatives.

39. $\dfrac{d}{dx}\left[x\dfrac{d}{dx}(x - x^2)\right].$
40. $\dfrac{d^2}{dx^2}\left[(x^2 - 3x)\dfrac{d}{dx}(x + x^{-1})\right].$
41. $\dfrac{d^4}{dx^4}[3x - x^4].$

42. $\dfrac{d^5}{dx^5}[ax^4 + bx^3 + cx^2 + dx + e]$,
$a, \ldots, e$ constant.

43. $\dfrac{d^2}{dx^2}\left[(1+2x)\dfrac{d^2}{dx^2}(5-x^3)\right]$.

44. $\dfrac{d^3}{dx^3}\left[\dfrac{1}{x}\dfrac{d^2}{dx^2}(x^4-5x^2)\right]$.

Find a function $y = f(x)$ for which:

45. $y' = 4x^3 - x^2 + 4x$.

46. $y' = x - \dfrac{2}{x^3} + 3$.

47. $\dfrac{dy}{dx} = 5x^4 + \dfrac{4}{x^5}$.

48. $\dfrac{dy}{dx} = 4x^5 - \dfrac{5}{x^4} - 2$.

49. Find a quadratic polynomial p such that $p(1) = 3$, $p'(1) = -2$, and $p''(1) = 4$.

50. Find a cubic polynomial p such that $p(-1) = 0$, $p'(-1) = 3$, $p''(-1) = -2$, and $p'''(-1) = 6$.

51. Let $f(x) = x^n$, where n is a positive integer.
(a) Find $f^{(k)}(x)$ for $k = n$.
(b) Find $f^{(k)}(x)$ for $k > n$.
(c) Find $f^{(k)}(x)$ for $k < n$.

52. Given the polynomial function
$p(x) = a_n x^n + a_{n-1}x^{n-1} + \cdots + a_1 x + a_0$:
(a) Find $(d^n/dx^n)[p(x)]$.
(b) What is $(d^k/dx^k)[p(x)]$ for $k > n$?

53. Let $f(x) = \begin{cases} x^2, & x \geq 0 \\ 0, & x < 0. \end{cases}$
(a) Show that f is differentiable at 0 and give $f'(0)$.
(b) Determine $f'(x)$ for all x.
(c) Show that $f''(0)$ does not exist.
(d) Sketch the graph of f and f'.

54. Let $g(x) = \begin{cases} x^3, & x \geq 0 \\ 0, & x < 0. \end{cases}$
(a) Show that $g'(0)$ and $g''(0)$ both exist and give their values.
(b) Determine $g'(x)$ and $g''(x)$ for all x.
(c) Show that $g'''(0)$ does not exist.
(d) Sketch the graph of g, g', and g''.

55. Show that in general

$$(f \cdot g)''(x) \neq f(x)g''(x) + f''(x)g(x).$$

56. Verify the identity

$$f(x)g''(x) - f''(x)g(x) = \frac{d}{dx}[f(x)g'(x) - f'(x)g(x)].$$

In Exercises 57–60, determine the values of x for which (a) $f''(x) = 0$, (b) $f''(x) > 0$, (c) $f''(x) < 0$.

57. $f(x) = x^3$.

58. $f(x) = x^4$.

59. $f(x) = x^4 + 2x^3 - 12x^2 + 1$.

60. $f(x) = x^4 + 3x^3 - 6x^2 - x$.

61. Prove by mathematical induction that

$$\text{if} \quad y = x^{-1}, \qquad \text{then} \quad \frac{d^n y}{dx^n} = (-1)^n n!\, x^{-n-1}.$$

62. Calculate y', y'', y''' for $y = 1/x^2$. Use these results to guess a formula for $y^{(n)}$ for each positive integer n, and then prove your conjecture using mathematical induction.

63. Let u, v, w be differentiable functions of x. Express the derivative of the product uvw in terms of the functions u, v, w, and their derivatives.

C 64. Use a CAS to find $\dfrac{d^n}{dx^n}[x^n]$ for $n = 1, 2, 3, 4, 5, 6, 7$. State a general result.

C 65. Use a CAS to find $\dfrac{d^{n+1}}{dx^{n+1}}[x^n]$ for $n = 1, 2, 3, 4, 5, 6, 7$. State a general result.

C 66. Let $f(x) = \dfrac{1-x}{1+x}$. Use a CAS to find a formula for $\dfrac{d^n}{dx^n}[f(x)]$.

C 67. Let $f(x) = x^3 - x$.
(a) Use a graphing utility to draw the graphs of f and the line $l : x - 2y + 12 = 0$ together.
(b) Find the points on the graph of f where its tangents are parallel to l.
(c) Verify your results in (b) by adding the graphs of these tangents to your previous drawing.

C 68. Let $f(x) = x^4 - x^2$.
(a) Use a graphing utility to draw the graphs of f and the line $l : x - 2y - 4 = 0$ together.
(b) Find the points on the graph of f where its normals are perpendicular to l.
(c) Verify your results in (b) by adding the graphs of these normals to your previous drawing.

C 69. Let $f(x) = x^3 + x^2 - 4x + 1$.
(a) Calculate $f'(x)$.
(b) Use a graphing utility to graph f and f' together.
(c) What can you say about the graph of f where $f'(x) < 0$? What can you say about the graph of f where $f'(x) > 0$?

C 70. Let $f(x) = \frac{1}{2}x^3 - 3x^2 + 3x + 3$.
(a) Calculate $f'(x)$.
(b) Use a graphing utility to graph f and f' together.
(c) Find the x-coordinates of the points where the line tangent to the graph of f is horizontal by finding the zeros of f'. Approximate the zeros of f' with three decimal place accuracy.

C 71. Let $f(x) = \frac{1}{2}x^3 - 3x^2 + 4x + 1$.
(a) Find an equation for the line tangent to the graph of f at the point $(0, 1)$.
(b) Use a graphing utility to graph f and the tangent line together. Show that the graph and the tangent line intersect at another point (a, b).
(c) Find the point (a, b).

■ PROJECT 3.3 Extending the Product Rule

In this project we will use the product rule (Theorem 3.2.6) to develop formulas for the derivative of the nth power of a function f, and for the nth-order derivative of the product of two functions f and g.

nth Powers. Suppose that f is a differentiable function and let $g(x) = [f(x)]^2 = f^2(x)$. Then $g(x) = f(x)f(x)$ and

$$g'(x) = f(x)f'(x) + f(x)f'(x) = 2f(x)f'(x).$$

Now let $g(x) = [f(x)]^3$. Then $g(x) = f(x)f^2(x)$ and

$$\begin{aligned} g'(x) &= f(x)[f^2(x)]' + f^2(x)f'(x) \\ &= f(x)[2f(x)f'(x)] + f^2(x)f'(x) = 3f^2(x)f'(x). \end{aligned}$$

Problem 1. Show that if $g(x) = [f(x)]^4 = f^4(x)$, then $g'(x) = 4f^3(x)f'(x)$. [Hint: Use either $g(x) = f(x)f^3(x)$ or $g(x) = f^2(x)f^2(x)$.]

Problem 2. Use mathematical induction to prove that if $g(x) = [f(x)]^n = f^n(x)$, then $g'(x) = nf^{n-1}(x)f'(x)$ for all positive integers n.

Problem 3. Prove that if $g(x) = [f(x)]^k = f^k(x)$, k an integer, then $g'(x) = kf^{k-1}(x)f'(x)$.

nth-Order Derivatives. Suppose that f and g are differentiable functions. Then, by the product rule,

$$(f \cdot g)'(x) = f(x)g'(x) + g(x)f'(x)$$

Now suppose that f and g are twice differentiable. Then

$$\begin{aligned} (f \cdot g)''(x) &= [(f \cdot g)'(x)]' = [f(x)g'(x) + g(x)f'(x)]' \\ &= [f(x)g'(x)]' + [g(x)f'(x)]' \\ &= f(x)g''(x) + g'(x)f'(x) + g(x)f''(x) + f'(x)g'(x) \\ &= f''(x)g(x) + 2f'(x)g'(x) + f(x)g''(x). \end{aligned}$$

Problem 4. Suppose that f and g are three-times differentiable. Show that

$$(f \cdot g)'''(x) = f'''(x)g(x) + 3f''(x)g'(x) + 3f'(x)g''(x) + f(x)g'''(x).$$

Problem 5. Suppose that f and g are n-times differentiable. The preceding results suggest that the expansion of $(f \cdot g)^{(n)}(x)$ parallels the expansion of $(a + b)^n$ by the binomial theorem. That is,

$$\begin{aligned} (f \cdot g)^{(n)} &= f^{(n)}g + nf^{(n-1)}g' + \cdots + \tbinom{n}{k} f^{(n-k)}g^{(k)} + \cdots \\ &\quad + nf'g^{(n-1)} + fg^{(n)}, \end{aligned}$$

where $\binom{n}{k} = n!/[k!\,(n-k)!]$ is the kth binomial coefficient. This formula is known as *Leibniz's rule*. Use mathematical induction to derive this formula.

■ 3.4 THE DERIVATIVE AS A RATE OF CHANGE

In the case of a linear function $y = mx + b$, the graph is a straight line and the slope m measures the steepness of the line by giving the rate of climb of the line, *the rate of change of y with respect to x*.

As x changes from x_0 to x_1, y changes m times as much:

$$y_1 - y_0 = m(x_1 - x_0). \qquad \text{(Figure 3.4.1)}$$

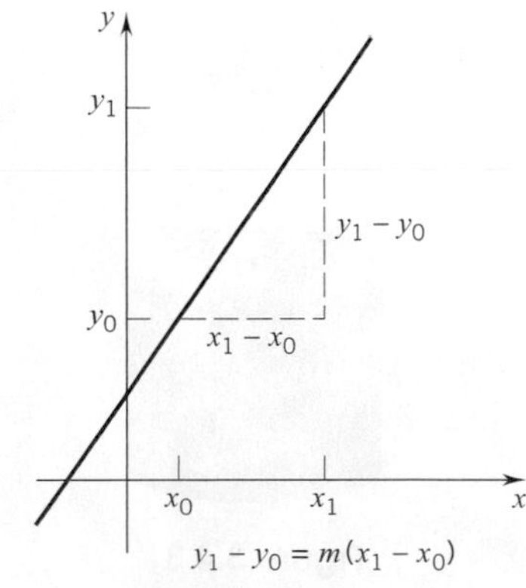

Figure 3.4.1

Thus the slope $m = (y_1 - y_0)/(x_1 - x_0)$ gives the change in y per unit change in x.

In the more general case of a differentiable function $y = f(x)$, the difference quotient

$$\frac{f(x+h) - f(x)}{(x+h) - x} = \frac{f(x+h) - f(x)}{h}, \qquad h \neq 0,$$

gives the *average rate of change of y with respect to x on the interval that joins x to x+h*. The limit as h approaches 0 is the derivative $dy/dx = f'(x)$. This can be interpreted as the *rate of change of y with respect to x at the point* $(x, f(x))$. Since the graph is a curve, the rate of change of y with respect to x can vary from point to point. At $x = x_1$ (see Figure 3.4.2) the rate of change of y with respect to x is $f'(x_1)$; the steepness of the graph is that of a line of slope $f'(x_1)$. At $x = x_2$ the rate of change of y with respect to x is $f'(x_2)$; the steepness of the graph is that of a line of slope $f'(x_2)$. At $x = x_3$ the rate of change of y with respect to x is $f'(x_3)$; the steepness of the graph is that of a line of slope $f'(x_3)$.

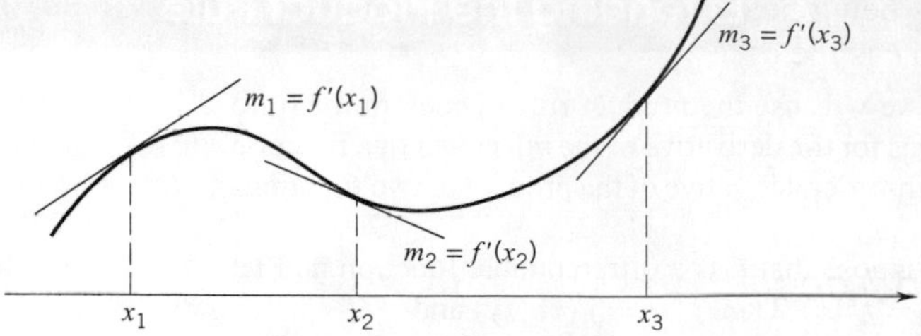

$m_3 = f'(x_3)$

$m_1 = f'(x_1)$

$m_2 = f'(x_2)$

x_1 x_2 x_3

Figure 3.4.2

From your understanding of slope, it should be apparent to you when examining Figure 3.4.2 that $f'(x_1) > 0$, $f'(x_2) < 0$, and $f'(x_3) > 0$, and that, in general, a function increases on any interval where the derivative remains positive and decreases on any interval where the derivative remains negative. We will take up this matter carefully in Chapter 4. Right now we look at rates of change in several different contexts.

Example 1 Let $y = \dfrac{x-2}{x^2}, \quad x \neq 0.$

(a) Find the rate of change of y with respect to x at $x = 2$.

(b) Find the value(s) of x, if any, at which the rate of change of y with respect to x is 0.

SOLUTION The rate of change of y with respect to x is given by the derivative, dy/dx:

$$\frac{dy}{dx} = \frac{x^2(1) - (x-2)(2x)}{x^4} = \frac{-x^2 + 4x}{x^4} = \frac{4-x}{x^3}.$$

(a) At $x = 2$,

$$\frac{dy}{dx} = \frac{4-2}{2^3} = \frac{1}{4}.$$

(b) Setting $\dfrac{dy}{dx} = 0$, we get the equation $\dfrac{4-x}{x^3} = 0$, from which it follows that $x = 4$; the rate of change of y with respect to x is 0 at $x = 4$. □

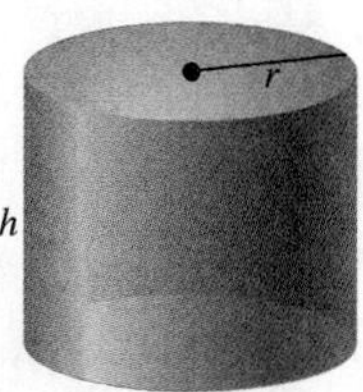

Figure 3.4.3

Example 2 Suppose that we have a right circular cylinder of changing dimensions (Figure 3.4.3). When the base radius is r and the height is h, the cylinder has volume

$$V = \pi r^2 h.$$

If r remains constant while h changes, then V can be viewed as a function of h. The rate of change of V with respect to h is the derivative

$$\frac{dV}{dh} = \pi r^2.$$

If h remains constant while r changes, then V can be viewed as a function of r. The rate of change of V with respect to r is the derivative

$$\frac{dV}{dr} = 2\pi r h.$$

Suppose now that r is changing but V is being kept constant. How is h changing with respect to r? To answer this, we express h in terms of r and V:

$$h = \frac{V}{\pi r^2} = \frac{V}{\pi} r^{-2}.$$

Since V is being held constant, h is now a function of r. The rate of change of h with respect to r is the derivative

$$\frac{dh}{dr} = -\frac{2V}{\pi}r^{-3} = -\frac{2(\pi r^2 h)}{\pi}r^{-3} = -\frac{2h}{r}. \quad \square$$

Velocity and Acceleration

Suppose that an object moves along a straight line and at each time t during a certain time interval, the object has position (coordinate) $x(t)$. Then, at time $t+h$, the position of the object is $x(t+h)$, and $x(t+h) - x(t)$ gives the change in position of the object from time t to time $t+h$. The ratio

$$\frac{x(t+h) - x(t)}{(t+h) - t} = \frac{x(t+h) - x(t)}{h}$$

gives the *average velocity* of the object during this time period. If

$$\lim_{h \to 0} \frac{x(t+h) - x(t)}{h} = x'(t)$$

exists, then $x'(t)$ gives the (*instantaneous*) *rate of change of position at time t*. This rate of change of position is called the *velocity* of the object at time t; in symbols

(3.4.1) $$\boxed{v(t) = x'(t).}$$

If the velocity function is itself differentiable, then its rate of change with respect to time is called the *acceleration*; in symbols,

(3.4.2) $$\boxed{a(t) = v'(t) = x''(t).}$$

In the Leibniz notation,

(3.4.3) $$\boxed{v = \frac{dx}{dt} \quad \text{and} \quad a = \frac{dv}{dt} = \frac{d^2x}{dt^2}.}$$

The *speed* is by definition the absolute value of the velocity:

(3.4.4) $$\boxed{\text{speed at time } t = |v(t)|.}$$

1. Positive velocity indicates motion in the positive direction (x is increasing). Negative velocity indicates motion in the negative direction (x is decreasing).
2. Positive acceleration indicates increasing velocity (increasing speed in the positive direction or decreasing speed in the negative direction). Negative acceleration indicates decreasing velocity (decreasing speed in the positive direction or increasing speed in the negative direction).
3. It follows from (2) that, if the velocity and acceleration have the same sign, the object is speeding up, but if the velocity and acceleration have opposite signs, the object is slowing down.

Example 3 An object moves along the x-axis, its position at each time t given by the function

$$x(t) = t^3 - 12t^2 + 36t - 27.$$

Let's study the motion from time $t = 0$ to time $t = 9$.

The object starts out 27 units to the left of the origin:

$$x(0) = 0^3 - 12(0)^2 + 36(0) - 27 = -27$$

and ends up 54 units to the right of the origin:

$$x(9) = 9^3 - 12(9)^2 + 36(9) - 27 = 54.$$

We can find the velocity function by differentiating the position function:

$$v(t) = x'(t) = 3t^2 - 24t + 36 = 3(t - 2)(t - 6).$$

We leave it to you to verify that

$$v(t) \text{ is } \begin{cases} \text{positive,} & \text{for } 0 \leq t < 2 \\ 0, & \text{at } t = 2 \\ \text{negative,} & \text{for } 2 < t < 6 \\ 0, & \text{at } t = 6 \\ \text{positive,} & \text{for } 6 < t \leq 9. \end{cases}$$

We can interpret all this as follows: the object begins by moving to the right [$v(t)$ is positive for $0 \leq t < 2$]; it comes to a stop at time $t = 2$ [$v(2) = 0$]; it then moves left [$v(t)$ is negative for $2 < t < 6$]; it stops at time $t = 6$ [$v(6) = 0$] ; it then moves right and keeps going right [$v(t) > 0$ for $6 < t \leq 9$].

We can find the acceleration by differentiating the velocity:

$$a(t) = v'(t) = 6t - 24 = 6(t - 4).$$

Now,

$$a(t) \text{ is } \begin{cases} \text{negative,} & \text{for } 0 \leq t < 4 \\ 0, & \text{at } t = 4 \\ \text{positive,} & \text{for } 4 < t \leq 9. \end{cases}$$

At the beginning the velocity decreases, reaching a minimum at time $t = 4$. Then the velocity starts to increase and continues to increase.

Figure 3.4.4 shows a diagram for the sign of the velocity and a corresponding diagram for the sign of the acceleration. Combining the two diagrams, we have a brief description of the motion in convenient form.

The direction of the motion at each time $t \in [0, 9]$ is represented schematically in Figure 3.4.5.

A better way to represent the motion is to graph x as a function of t, as we do in Figure 3.4.6. The velocity $v(t) = x'(t)$ then appears as the slope of the tangent to the curve. Note, for example, that at $t = 2$ and $t = 6$, the tangent is horizontal and the slope is 0. This reflects the fact that $v(2) = 0$ and $v(6) = 0$. ❑

Before going on, a few words about units. The units of velocity and acceleration depend on the units used to measure distance and the units used to measure time. The units of velocity are units of distance per unit time:

feet per second, meters per second, miles per hour, and so forth.

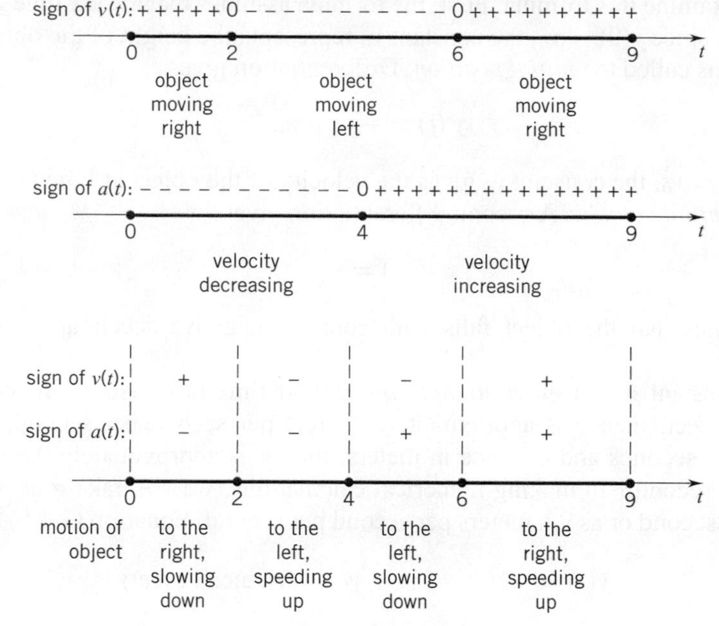

Figure 3.4.4

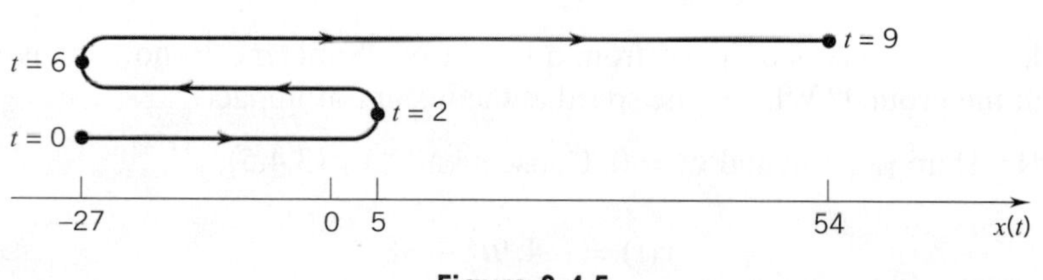

Figure 3.4.5

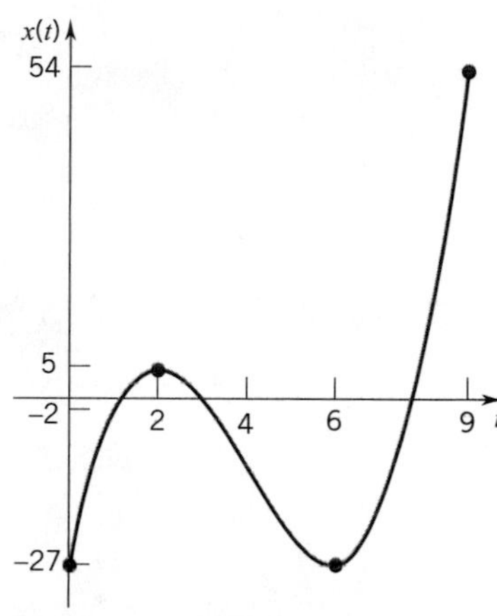

Figure 3.4.6

The units of acceleration are units of distance per unit time per unit time:

feet per second per second, meters per second per second,

miles per hour per hour, and so forth.

Free Fall (Near the Surface of the Earth)

Imagine an object (for example, a rock or an apple) falling to the ground (Figure 3.4.7). We will assume that the object is in *free fall:* namely, that the gravitational pull on the object is constant throughout the fall and that there is no air resistance.†

Galileo's formula for free fall gives the height of the object at each time t of the fall:

(3.4.5) $$y(t) = -\tfrac{1}{2}gt^2 + v_0t + y_0.\,\dagger\dagger$$

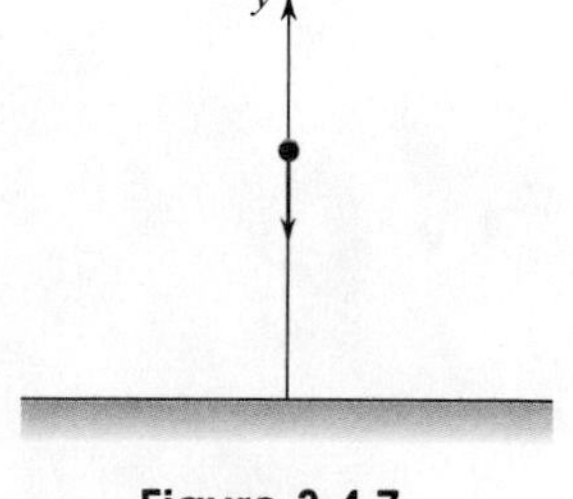

Figure 3.4.7

† In practice, neither of these conditions is ever fully met. Gravitational attraction near the surface of the earth does vary somewhat with altitude, and there is always some air resistance. Nevertheless, in the setting with which we will be working, the results that we obtain are good approximations of the actual motion.

†† Galileo Galilei (1564–1642), a great Italian astronomer and mathematician, is popularly known today for his early experiments with falling objects. His astronomical observations led him to support the Copernican view of the solar system. For this he was brought before the Inquisition.

Let's examine this formula. First, the formula assumes that the positive y direction is up. Next, since $y(0) = y_0$, the constant y_0 represents the height of the object at time $t = 0$. This is called the *initial position*. Differentiation gives

$$y'(t) = -gt + v_0.$$

Since $y'(0) = v_0$, the constant v_0 gives the velocity of the object at time $t = 0$. This is called *the initial velocity*. A second differentiation gives

$$y''(t) = -g.$$

This indicates that the object falls with constant negative acceleration $-g$. (Why negative?)

The constant g is a *gravitational constant*. If time is measured in seconds and distance in feet, then g is approximately 32 feet per second per second; if time is measured in seconds and distance in meters, then g is approximately 9.8 meters per second per second.† In making numerical calculations, we will take g as 32 feet per second per second or as 9.8 meters per second per second. Equation (3.4.5) then reads

$$y(t) = -16t^2 + v_0t + y_0 \quad \text{(distance in feet)}$$

or

$$y(t) = -4.9t^2 + v_0t + y_0 \quad \text{(distance in meters).}$$

Example 4 A stone is dropped from a height of 98 meters. In how many seconds does it hit the ground? What is the speed at the instant of impact?

SOLUTION Here $y_0 = 98$ and $v_0 = 0$. Consequently, by (3.4.5),

$$y(t) = -4.9t^2 + 98.$$

To find t at the moment of impact, we set $y(t) = 0$. This gives

$$-4.9t^2 + 98 = 0, \qquad t^2 = 20, \qquad t = \pm\sqrt{20} = \pm 2\sqrt{5}.$$

We disregard the negative value and conclude that it takes $2\sqrt{5} \cong 4.47$ seconds for the stone to hit the ground.

The velocity at impact is the velocity at time $t = 2\sqrt{5}$. Since

$$v(t) = y'(t) = -9.8t,$$

we have

$$v(2\sqrt{5}) = -(19.6)\sqrt{5} \cong -43.83.$$

The speed at impact is $|v(2\sqrt{5})| \cong 43.83$ meters per second. □

Example 5 An explosion causes debris to rise vertically with an initial velocity of 72 feet per second.

(a) In how many seconds does it attain maximum height?

(b) What is this maximum height?

† The value of this constant varies with latitude and elevation. It is approximately 32 feet per second per second at the equator at elevation zero. In Greenland it is about 32.23.

(c) What is the speed of the debris as it reaches a height of 32 feet (i) going up? (ii) coming back down?

SOLUTION The basic equation in this case is

$$y(t) = -16t^2 + v_0 t + y_0.$$

Here $y_0 = 0$ (it starts at ground level) and $v_0 = 72$ (the initial velocity is 72 feet per second). The equation of motion is therefore

$$y(t) = -16t^2 + 72t.$$

Differentiation gives

$$v(t) = y'(t) = -32t + 72.$$

The maximum height is attained when the velocity is 0. This occurs at time $t = \frac{72}{32} = \frac{9}{4}$. Since $y\left(\frac{9}{4}\right) = 81$, the maximum height attained is 81 feet.

To answer part (c), we must first find those times t for which $y(t) = 32$. Since

$$y(t) = -16t^2 + 72t,$$

the condition $y(t) = 32$ yields the equation

$$16t^2 - 72t + 32 = 0.$$

This quadratic has two solutions, $t = \frac{1}{2}$ and $t = 4$. Since $v\left(\frac{1}{2}\right) = 56$ and $v(4) = -56$, the velocity going up is 56 feet per second and the velocity coming down is -56 feet per second. In each case the speed is 56 feet per second. ☐

Economics

In business and economics one is often interested in how changes in such variables as production, supply, or price will affect other variables such as cost, revenue, or profit. If f is a function that describes the relationship between a pair of these variables, then the term *marginal* is used to specify the derivative of f.

For example, suppose $C = C(x)$ represents the cost of producing x units of a certain commodity. Although in reality x is a nonnegative integer, in theory and practice it is convenient to assume that C is defined for all x in some interval and that C is differentiable. The derivative $C'(x)$ is called the *marginal cost*.

Originally, economists defined the marginal cost at a production level x to be $C(x+1) - C(x)$, which is the cost of producing one additional unit of the commodity. Since

$$C(x+1) - C(x) = \frac{C(x+1) - C(x)}{1} \cong \lim_{h\to 0} \frac{C(x+h) - C(x)}{h} = C'(x),$$

it follows that the marginal cost $C'(x)$ at the production level x is approximately the cost of producing the $(x+1)$-st unit.

Similarly, if $R = R(x)$ is the revenue received for selling x units of the commodity, then $R'(x)$ is called the *marginal revenue*. The marginal revenue at a sales level x is approximately the revenue obtained by selling one additional unit.

If $C = C(x)$ and $R = R(x)$ are the cost and revenue functions associated with producing and selling x units of the commodity, then

$$P(x) = R(x) - C(x)$$

is called the *profit function*. The values of x (if any) at which $C(x) = R(x)$, that is, the values at which "cost"= "revenue," are called *break-even points*. The derivative,

P', is the *marginal profit*. In Chapter 4 we show that the maximum profit occurs when marginal cost equals marginal revenue.

Example 6 A manufacturer of computer components determines that the total cost C of producing x components per week is

$$C(x) = 2000 + 50x - \frac{x^2}{20} \quad \text{(dollars)}.$$

What is the marginal cost at the production level of 20 units? What is the exact cost of producing the 21st component?

SOLUTION The marginal cost at a production level x is given by the derivative

$$C'(x) = 50 - \frac{x}{10}.$$

Thus, the marginal cost at the production level of 20 components is

$$C'(20) = 50 - \tfrac{20}{10} = 48 \quad \text{(dollars)}.$$

The exact cost of producing the 21st component is

$$C(21) - C(20) = \left[2000 + 50(21) - \tfrac{(21)^2}{20}\right] - \left[2000 + 50(20) - \tfrac{(20)^2}{20}\right]$$
$$= 3027.95 - 2980 = 47.95 \quad \text{(dollars)}. \quad \square$$

Example 7 A manufacturer of digital watches determines that the cost and revenue functions involved in producing and selling x watches are

$$C(x) = 1200 + 13x$$

and

$$R(x) = 75x - \frac{x^2}{2},$$

respectively. Find the profit function and determine the break-even points. Find the marginal profit and determine the production/sales level at which the marginal profit is zero.

SOLUTION The profit function P is given by

$$P(x) = R(x) - C(x) = 75x - \frac{x^2}{2} - (1200 + 13x) = 62x - \frac{x^2}{2} - 1200.$$

To find the break-even point, set $P(x) = 0$:

$$62x - \frac{x^2}{2} - 1200 = 0$$
$$x^2 - 124x + 2400 = 0$$
$$(x - 24)(x - 100) = 0.$$

Thus, the break-even points are $x = 24$ and $x = 100$.

The marginal profit is given by derivative:

$$P'(x) = 62 - x$$

and $P'(x) = 0$ when $x = 62$.

Figure 3.4.8 gives a graphical representation of the cost and revenue functions, shows the break-even points, and indicates the regions of profit and loss. $\square$

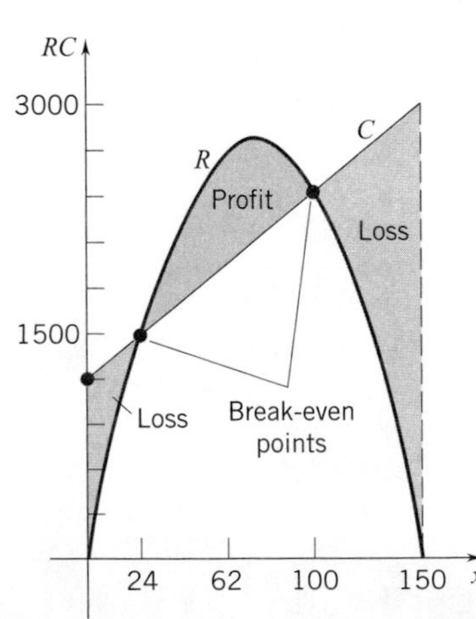

Figure 3.4.8

EXERCISES 3.4

1. Find the rate of change of the area of a circle with respect to the radius r. What is the rate when $r = 2$?

2. Find the rate of change of the volume of a cube with respect to the length s of a side. What is the rate when $s = 4$?

3. Find the rate of change of the area of a square with respect to the length z of a diagonal. What is the rate when $z = 4$?.

4. Find the rate of change of $y = 1/x$ with respect to x at $x = -1$.

5. Find the rate of change of $y = [x(x+1)]^{-1}$ with respect to x at $x = 2$.

6. Find the values of x at which the rate of change of $y = x^3 - 12x^2 + 45x - 1$ with respect to x is zero.

7. Find the rate of change of the volume of a sphere with respect to the radius r.

8. Find the rate of change of the surface area of a sphere with respect to the radius r. What is this rate of change when $r = r_0$? How must r_0 be chosen so that the rate of change is 1?

9. Find x_0 given that the rate of change of $y = 2x^2 + x - 1$ with respect to x at $x = x_0$ is 4.

10. Find the rate of change of the area A of a circle with respect to

    (a) the diameter d. (b) the circumference C.

11. Find the rate of change of the volume V of a cube with respect to

    (a) the length w of a diagonal on one of the faces.

    (b) the length z of one of the diagonals of the cube.

12. The dimensions of a rectangle are changing in such a way that the area of the rectangle remains constant. Find the rate of change of the height h with respect to the base b.

13. The area of a sector in a circle is given by the formula $A = \frac{1}{2}r^2\theta$ where r is the radius and θ is the central angle measured in radians.

    (a) Find the rate of change of A with respect to θ if r remains constant.

    (b) Find the rate of change of A with respect to r if θ remains constant.

    (c) Find the rate of change of θ with respect to r if A remains constant.

14. The total surface area of a right circular cylinder is given by the formula $A = 2\pi r(r + h)$ where r is the radius and h is the height.

    (a) Find the rate of change of A with respect to h if r remains constant.

    (b) Find the rate of change of A with respect to r if h remains constant.

    (c) Find the rate of change of h with respect to r if A remains constant.

15. For what value of x is the rate of change of

$$y = ax^2 + bx + c \quad \text{with respect to } x$$

the same as the rate of change of

$$z = bx^2 + ax + c \quad \text{with respect to } x?$$

Assume that a, b, c are constant with $a \neq b$.

16. Find the rate of change of the product $f(x)g(x)h(x)$ with respect to x at $x = 1$ given that

$$f(1) = 0, \quad g(1) = 2, \quad h(1) = -2,$$
$$f'(1) = 1, \quad g'(1) = -1, \quad h'(1) = 0.$$

In Exercises 17–22, an object moves along a coordinate line, its position at each time $t \geq 0$ given by $x(t)$. Find the position, velocity, acceleration, and speed at time t_0.

17. $x(t) = 4 + 3t - t^2; \quad t_0 = 5.$

18. $x(t) = 5t - t^3; \quad t_0 = 3.$

19. $x(t) = \dfrac{18}{t+2}; \quad t_0 = 1.$

20. $x(t) = \dfrac{2t}{t+3}; \quad t_0 = 3.$

21. $x(t) = (t^2 + 5t)(t^2 + t - 2); \quad t_0 = 1.$

22. $x(t) = (t^2 - 3t)(t^2 + 3t); \quad t_0 = 2.$

In Exercises 23–26, an object moves along a coordinate line, its position at each time $t \geq 0$ given by $x(t)$. Determine when, if ever, the object changes direction.

23. $x(t) = t^3 - 3t^2 + 3t.$

24. $x(t) = t + \dfrac{3}{t+1}.$

25. $x(t) = t + \dfrac{5}{t+2}.$

26. $x(t) = t^4 - 4t^3 + 4t^2.$

Objects A, B, C move along the x-axis. Their positions from time $t = 0$ to time $t = t_3$, have been graphed in the figure.

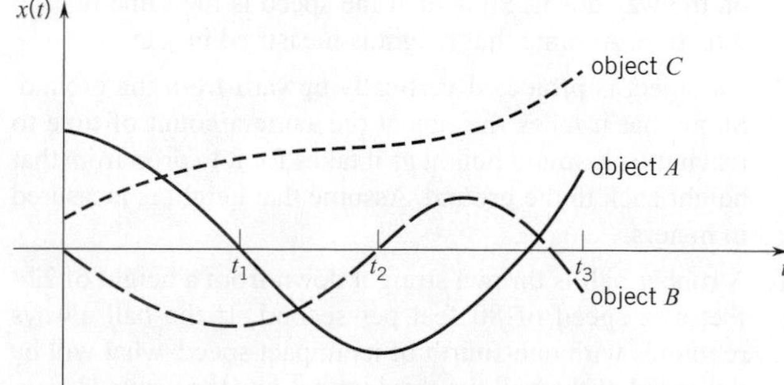

In Exercises 27–36:

27. Which object begins farthest to the right?

28. Which object finishes farthest to the right?

29. Which object has the greatest speed at time t_1?

30. Which object maintains the same direction during the time interval $[t_1, t_3]$?

31. Which object begins by moving left?

32. Which object finishes moving left?

33. Which object changes direction at time t_2?

34. Which object speeds up during the time interval $[0, t_1]$?

35. Which object slows down during the time interval $[t_1, t_2]$?

36. Which object changes direction during the time interval $[t_2, t_3]$?

An object moves along the x-axis, its position at each time $t \geq 0$ given by $x(t)$. In Exercises 37–44, determine the time interval(s), if any, during which the object satisfies the given condition.

37. $x(t) = t^4 - 12t^3 + 28t^2$; moving right.

38. $x(t) = t^3 - 12t^2 + 21t$; moving left.

39. $x(t) = 5t^4 - t^5$; speeding up.

40. $x(t) = 6t^2 - t^4$; slowing down.

41. $x(t) = t^3 - 6t^2 - 15t$; moving left and slowing down.

42. $x(t) = t^3 - 6t^2 - 15t$; moving right and slowing down.

43. $x(t) = t^4 - 8t^3 + 16t^2$; moving right and speeding up.

44. $x(t) = t^4 - 8t^3 + 16t^2$; moving left and speeding up.

In Exercises 45–58, neglect air resistance. For the numerical calculations, take g as 32 feet per second per second or as 9.8 meters per second per second.

45. An object is dropped and hits the ground 6 seconds later. From what height, in feet, was it dropped?

46. Supplies are dropped from a stationary helicopter and seconds later hit the ground at 98 meters per second. How high was the helicopter?

47. An object is projected vertically upward from ground level with velocity v_0. Find the height in meters attained by the object.

48. An object projected vertically upward from ground level returns to earth in 8 seconds. Find the initial velocity in feet per second.

49. An object projected vertically upward passes every height less than the maximum twice, once on the way up and once on the way down. Show that the speed is the same in each direction. Assume that height is measured in feet.

50. An object is projected vertically upward from the ground. Show that it takes the object the same amount of time to reach its maximum height as it takes for it to drop from that height back to the ground. Assume that height is measured in meters.

51. A rubber ball is thrown straight down from a height of 224 feet at a speed of 80 feet per second. If the ball always rebounds with one-fourth of its impact speed, what will be the speed of the ball the third time it hits the ground?

52. A ball is thrown straight up from ground level. How high will the ball go if it reaches a height of 64 feet in 2 seconds?

53. A stone is thrown upward from ground level. The initial speed is 32 feet per second. (a) In how many seconds will the stone hit the ground? (b) How high will it go? (c) With what minimum speed should the stone be thrown so as to reach a height of 36 feet?

54. To estimate the height of a bridge, a man drops a stone into the water below. How high is the bridge (a) if the stone hits the water 3 seconds later? (b) if the man hears the splash 3 seconds later? (Use 1080 feet per second as the speed of sound.)

55. A falling stone is at a certain instant 100 feet above the ground. Two seconds later it is only 16 feet above the ground. (a) From what height was it dropped? (b) If it was thrown down with an initial speed of 5 feet per second, from what height was it thrown? (c) If it was thrown upward with an initial speed of 10 feet per second, from what height was it thrown?

56. A rubber ball is thrown straight down from a height of 4 feet. If the ball rebounds with one-half of its impact speed and returns exactly to its original height before falling again, how fast was it thrown originally?

57. Ballast dropped from a balloon that was rising at the rate of 5 feet per second reached the ground in 8 seconds. How high was the balloon when the ballast was dropped?

58. Had the balloon of Exercise 57 been falling at the rate of 5 feet per second, how long would it have taken for the ballast to reach the ground?

In Exercises 59–62, a cost function for a certain commodity is given. Find the marginal cost function at a production level of 100 units and compare that with the actual cost of producing the 101st unit.

59. $C(x) = 200 + 0.02x + 0.0001x^2, \quad x \geq 0.$

60. $C(x) = 1000 + 2x + 0.02x^2 + 0.0001x^3, \quad x \geq 0.$

61. $C(x) = 200 + 0.01x + \dfrac{100}{x}, \quad x > 0.$

62. $C(x) = 2000 + 2\sqrt{x}, \quad x \geq 0.$

63. A manufacturer of electric motors estimates that the cost (in dollars) of producing x motors per day is given by

$$C(x) = 1000 + 25x - \frac{x^2}{10}, \qquad 0 \leq x \leq 200.$$

Find the marginal cost of producing 10 motors and compare it with the exact cost of producing the 11th motor.

64. The total revenue (in dollars) from the sale of x units of a certain commodity is given by

$$R(x) = 24x + 5x^2 - \frac{x^3}{3}, \quad x \geq 0.$$

(a) Find the marginal revenue function $R'(x)$ and determine the interval(s) on which $R'(x) > 0$.

(b) For what value(s) of x is the marginal revenue a maximum?

65. The cost and revenue functions for the production of x units of a certain commodity are

$$C(x) = 4x + 1400 \quad \text{and}$$

$$R(x) = 20x - \frac{x^2}{50} \quad \text{for} \quad 0 \leq x \leq 1000.$$

(a) Find the profit function and determine the break-even points.

(b) Find the marginal profit and determine the production level at which the marginal profit is zero.

(c) Sketch the cost and revenue functions in the same coordinate system and indicate the regions of profit and loss. What production level do you think will produce the maximum profit?

66. Galileo's formula for the height in feet of an object in free fall is

$$y(t) = -16t^2 + v_0 t + y_0.$$

Use a CAS to find y_0 and v_0 if it is known that $y(2) = 100$ feet and $y'(2) = 5$ feet per second.

67. The height in meters of a certain object in free fall is given by

$$y(t) = -4.9t^2 + 50t + 75.$$

Use the Trace Function on a graphing utility to approximate:

(a) the maximum height that the object attains.

(b) the time when the object hits the ground.

68. The cost and revenue functions for the production and sale of x hand calculators are

$$C(x) = 3000 + 5x \quad \text{and}$$

$$R(x) = 60x - 2x^{(3/2)}, \quad 0 \le x \le 900.$$

(a) Using a graphing utility, graph the cost and revenue functions together. Estimate the break-even points accurate to two decimal places.

(b) Graph the profit function and estimate the production level that will yield the maximum profit. Use two decimal place accuracy.

69. An object moves along the x-axis. Its position at time t is given by

$$x(t) = t^3 - 7t^2 + 10t + 5, \quad 0 \le t \le 5.$$

(a) Determine the velocity function v. Use a graphing utility to graph v.

(b) Use the graph to estimate when the object is moving to the right and when it is moving to the left.

(c) Use the graphing utility to graph the speed $|v|$ of the object. Estimate when the object stops. Estimate the maximum speed from $t = 1$ to $t = 4$.

(d) Determine the acceleration a and use the graphing utility to graph it. Estimate when the object is speeding up and when it is slowing down.

70. The cost and revenue functions for the production of a certain commodity are

$$C(x) = 4 + 0.75x \quad \text{and}$$

$$R(x) = \frac{10x}{1 + 0.25x^2} \quad \text{for } x \ge 0,$$

where x is measured in hundreds of units.

(a) Using a graphing utility, graph the cost and revenue functions together. Estimate the break-even points accurate to two decimal places.

(b) Graph the profit function and estimate the production level that will yield the maximum profit. Use two decimal place accuracy. How many units should be produced to maximize the profit?

■ 3.5 THE CHAIN RULE

In this section we take up the differentiation of composite functions. Until we get to Theorem 3.5.7, our approach is completely intuitive—no real definitions, no proofs, just informal discussion. Our purpose is to give you some experience with the standard computational procedures and some insight into why these procedures work. Theorem 3.5.7 puts it all on a sound footing.

Suppose that y is a differentiable function of u:

$$y = f(u)$$

and u in turn is a differentiable function of x:

$$u = g(x).$$

Then y is a composite function of x:

$$y = f(u) = f(g(x)) = (f \circ g)(x).$$

Does y have a derivative with respect to x? Yes it does, and dy/dx is given by a formula that is easy to remember:

(3.5.1) $$\boxed{\frac{dy}{dx} = \frac{dy}{du}\frac{du}{dx}.}$$

This formula, known as the *chain rule*, says that

> *"the rate of change of y with respect to x is the rate of change of y with respect to u times the rate of change of u with respect to x."*

Plausible as all this sounds, remember that we have proved nothing. All we have done is assert that the composition of differentiable functions is differentiable and given you a formula—a formula that needs justification and is justified at the end of this section.

Before using the chain rule in elaborate computations, let's confirm its validity in some simple instances.

If $y = 2u$ and $u = 3x$, then $y = 6x$. Clearly

$$\frac{dy}{dx} = 6 = 2 \cdot 3 = \frac{dy}{du}\frac{du}{dx},$$

and so, in this case, the chain rule is confirmed:

$$\frac{dy}{dx} = \frac{dy}{du}\frac{du}{dx}.$$

If $y = u^3$ and $u = x^2$, then $y = (x^2)^3 = x^6$. This time

$$\frac{dy}{dx} = 6x^5, \quad \frac{dy}{du} = 3u^2 = 3(x^2)^2 = 3x^4, \quad \frac{du}{dx} = 2x$$

and once again

$$\frac{dy}{dx} = 6x^5 = 3x^4 \cdot 2x = \frac{dy}{du}\frac{du}{dx}.$$

Example 1 Find dy/dx by the chain rule given that

$$y = \frac{u-1}{u+1} \quad \text{and} \quad u = x^2.$$

SOLUTION

$$\frac{dy}{du} = \frac{(u+1)(1)-(u-1)(1)}{(u+1)^2} = \frac{2}{(u+1)^2} \quad \text{and} \quad \frac{du}{dx} = 2x$$

so that

$$\frac{dy}{dx} = \frac{dy}{du}\frac{du}{dx} = \left[\frac{2}{(u+1)^2}\right]2x = \frac{4x}{(x^2+1)^2}. \quad \square$$

Remark We would have obtained the same result without the chain rule by first writing y as a function of x and then differentiating:

$$\text{with} \quad y = \frac{u-1}{u+1} \quad \text{and} \quad u = x^2 \quad \text{we have} \quad y = \frac{x^2-1}{x^2+1}$$

and therefore

$$\frac{dy}{dx} = \frac{(x^2+1)2x-(x^2-1)2x}{(x^2+1)^2} = \frac{4x}{(x^2+1)^2}. \quad \square$$

If f is a differentiable function of u and u is a differentiable function of x, then, according to (3.5.1),

(3.5.2)
$$\frac{d}{dx}[f(u(x))] = \frac{d}{du}[f(u)]\frac{du}{dx} = f'(u)\frac{du}{dx}.$$

[All we have done here is write y as $f(u)$.] This is the formulation of the chain rule that we will use most frequently in later work.

Suppose now that we were asked to calculate.

$$\frac{d}{dx}[(x^2-1)^{100}].$$

We could expand $(x^2-1)^{100}$ into polynomial form by using the binomial theorem or, if we were masochistic, by repeated multiplication, but we would have a terrible mess on our hands; for example, $(x^2-1)^{100}$ has 101 terms. With (3.5.2) we can derive a formula that will render such calculations almost trivial.

Assuming (3.5.2), we can show that, if u is a differentiable function of x and n is an integer, then

(3.5.3)
$$\frac{d}{dx}(u^n) = nu^{n-1}\frac{du}{dx}.$$

PROOF Set $f(u) = u^n$. Then

$$\frac{d}{dx}(u^n) = \underbrace{\frac{d}{du}(u^n)\frac{du}{dx}}_{(3.5.2)} = nu^{n-1}\frac{du}{dx}. \quad \square$$

To calculate

$$\frac{d}{dx}[(x^2-1)^{100}]$$

we set $u = x^2 - 1$. Then by our formula

$$\frac{d}{dx}[(x^2-1)^{100}] = 100(x^2-1)^{99}\frac{d}{dx}(x^2-1) = 100(x^2-1)^{99}2x = 200x(x^2-1)^{99}.$$

Remark While it is clear that the chain rule is the only practical way to calculate the derivative of $f(x) = (x^2-1)^{100}$, you do have a choice when differentiating a similar, but simpler, function such as $g(x) = (x^2-1)^4$. Calculating the derivative of g by the chain rule, we get

$$\frac{d}{dx}[(x^2-1)^4] = 4(x^2-1)^3\frac{d}{dx}(x^2-1) = 4(x^2-1)^3 2x = 8x(x^2-1)^3.$$

On the other hand, if we were to first expand the expression $(x^2-1)^4$, we would get

$$g(x) = x^8 - 4x^6 + 6x^4 - 4x^2 + 1$$

and then

$$g'(x) = 8x^7 - 24x^5 + 24x^3 - 8x.$$

As a final answer, this is correct, but it is somewhat unwieldy. To reconcile the two results, notice that $8x$ is a factor of g':

$$g'(x) = 8x(x^6 - 3x^4 + 3x^2 - 1),$$

and that the expression in parentheses is the binomial expansion for $(x^2 - 1)^3$. Thus,

$$g'(x) = 8x(x^2 - 1)^3,$$

as we saw above. However, the chain rule gave us the neat, compact result much more efficiently. ❑

Here are more examples of a similar sort.

Example 2

$$\frac{d}{dx}\left[\left(x+\frac{1}{x}\right)^{-3}\right] = -3\left(x+\frac{1}{x}\right)^{-4}\frac{d}{dx}\left(x+\frac{1}{x}\right) = -3\left(x+\frac{1}{x}\right)^{-4}\left(1-\frac{1}{x^2}\right). \quad ❑$$

Example 3

$$\frac{d}{dx}[1+(2+3x)^5]^3 = 3[1+(2+3x)^5]^2\frac{d}{dx}[1+(2+3x)^5].$$

Since

$$\frac{d}{dx}[1+(2+3x)^5] = 5(2+3x)^4\frac{d}{dx}(2+3x) = 5(2+3x)^4(3) = 15(2+3x)^4,$$

we have

$$\begin{aligned}\frac{d}{dx}[1+(2+3x)^5]^3 &= 3[1+(2+3x)^5]^2[15(2+3x)^4]\\ &= 45(2+3x)^4[1+(2+3x)^5]^2. \quad ❑\end{aligned}$$

Example 4 Calculate the derivative of $f(x) = 2x^3(x^2-3)^4$.

SOLUTION Here we need to use the product rule and the chain rule:

$$\begin{aligned}\frac{d}{dx}[2x^3(x^2-3)^4] &= 2x^3\frac{d}{dx}[(x^2-3)^4] + (x^2-3)^4\frac{d}{dx}(2x^3)\\ &= 2x^3[4(x^2-3)^3(2x)] + (x^2-3)^4(6x^2)\\ &= 16x^4(x^2-3)^3 + 6x^2(x^2-3)^4 = 2x^2(x^2-3)^3(11x^2-9). \quad ❑\end{aligned}$$

The formula

$$\frac{dy}{dx} = \frac{dy}{du}\frac{du}{dx}$$

can be extended to more variables. For example, if x itself depends on s, then we have

(3.5.4)
$$\boxed{\frac{dy}{ds} = \frac{dy}{du}\frac{du}{dx}\frac{dx}{ds}.}$$

If, in addition, s depends on t, then

(3.5.5)
$$\boxed{\frac{dy}{dt} = \frac{dy}{du}\frac{du}{dx}\frac{dx}{ds}\frac{ds}{dt},}$$

and so on. Each new dependence adds a new link to the chain.

Example 5 Find dy/ds given that $y = 3u + 1, \quad u = x^{-2}, \quad x = 1 - s.$

SOLUTION

$$\frac{dy}{du} = 3, \qquad \frac{du}{dx} = -2x^{-3}, \qquad \frac{dx}{ds} = -1.$$

Therefore

$$\frac{dy}{ds} = \frac{dy}{du}\frac{du}{dx}\frac{dx}{ds} = (3)(-2x^{-3})(-1) = 6x^{-3} = 6(1-s)^{-3}. \quad \square$$

Example 6 Find dy/dt at $t = 9$ given that

$$y = \frac{u+2}{u-1}, \qquad u = (3s-7)^2, \qquad s = \sqrt{t}.$$

SOLUTION As you can check,

$$\frac{dy}{du} = -\frac{3}{(u-1)^2}, \qquad \frac{du}{ds} = 6(3s-7), \qquad \frac{ds}{dt} = \frac{1}{2\sqrt{t}}.\dagger$$

At $t = 9$, we have $s = 3$ and $u = 4$, so that

$$\frac{dy}{du} = -\frac{3}{(4-1)^2} = -\frac{1}{3}, \qquad \frac{du}{ds} = 6(9-7) = 12, \qquad \frac{ds}{dt} = \frac{1}{2\sqrt{9}} = \frac{1}{6}.$$

Thus, at $t = 9$,

$$\frac{dy}{dt} = \frac{dy}{du}\frac{du}{ds}\frac{ds}{dt} = \left(-\frac{1}{3}\right)(12)\left(\frac{1}{6}\right) = -\frac{2}{3}. \quad \square$$

Note that the function $y = [1 + (2+3x)^5]^3$ in Example 3 is the composition

$$y = u^3, \; u = 1 + s^5, \; s = 2 + 3x.$$

By Formula (3.5.4),

$$\frac{dy}{dx} = \frac{dy}{du} \cdot \frac{du}{ds} \cdot \frac{ds}{dx} = 3u^2(5s^4)(3) = 45[1 + (2+3x)^5]^2(2+3x)^4,$$

as we saw above.

So far we have worked entirely in Leibniz's notation. What does the chain rule look like in prime notation?

---

† It was shown in Section 3.1 that $\frac{d}{dx}(\sqrt{x}) = \frac{1}{2\sqrt{x}}$.

Let's go back to the beginning. Once again, let y be a differentiable function of u:

$$y = f(u),$$

and let u be a differentiable function of x:

$$u = g(x).$$

Then $$y = f(u) = f(g(x)) = (f \circ g)(x)$$

and, according to the chain rule (as yet unproved),

$$\frac{dy}{dx} = \frac{dy}{du}\frac{du}{dx}.$$

Since

$$\frac{dy}{dx} = \frac{d}{dx}[(f \circ g)(x)] = (f \circ g)'(x), \quad \frac{dy}{du} = f'(u) = f'(g(x)), \quad \frac{du}{dx} = g'(x),$$

the chain rule can be written

(3.5.6) $$\boxed{(f \circ g)'(x) = f'(g(x))g'(x).}$$

The chain rule in prime notation says that

> *"the derivative of a composition $f \circ g$ at x is the derivative of f at $g(x)$ times the derivative of g at x."*

In Leibniz's notation the chain rule *appears* seductively simple, to some even obvious. "After all, to prove it, all you have to do is cancel the *du's*":

$$\frac{dy}{dx} = \frac{dy}{\cancel{du}}\frac{\cancel{du}}{dx}.$$

Of course, this is just nonsense. What would one cancel from

$$(f \circ g)'(x) = f'(g(x))g'(x)?$$

Although Leibniz's notation is useful for routine calculations, mathematicians generally turn to the prime notation when precision is required.

It is time for us to be precise. How do we know that the composition of differentiable functions is differentiable? What assumptions do we need? Under what circumstances is it true that

$$(f \circ g)'(x) = f'(g(x))g'(x)?$$

The following theorem provides the definitive answer.

THEOREM 3.5.7 THE CHAIN-RULE THEOREM

If g is differentiable at x and f is differentiable at $g(x)$, then the composition $f \circ g$ is differentiable at x and

$$(f \circ g)'(x) = f'(g(x))g'(x).$$

A proof of this theorem appears in the supplement to this section. The argument is not as easy as "canceling" the du's.

We conclude this section with an application that involves the chain rule.

Example 7 Gravel is being poured by a conveyor onto a conical pile at the constant rate of 60π cubic feet per minute. Frictional forces within the pile are such that the height is always two-thirds of the radius. How fast is the radius of the pile changing at the instant it is 5 feet?

SOLUTION The formula for the volume V of a right circular cone of radius r and height h is

$$V = \tfrac{1}{3}\pi r^2 h.$$

However, in this case we are told that $h = \frac{2}{3}r$, and so we have

$$(*) \qquad V = \tfrac{2}{9}\pi r^3.$$

Since gravel is being poured onto the pile, the volume, and hence the radius, are functions of time t. We are given that $dV/dt = 60\pi$ and we want to find dr/dt at the instant when $r = 5$. Differentiating $(*)$ with respect to t using the chain rule, we get

$$\frac{dV}{dt} = \frac{dV}{dr}\cdot\frac{dr}{dt} = \left(\tfrac{2}{3}\pi r^2\right)\frac{dr}{dt}.$$

Solving for dr/dt and using the fact that $dV/dt = 60\pi$, we find that

$$\frac{dr}{dt} = \frac{180\pi}{2\pi r^2} = \frac{90}{r^2}.$$

Now, when $r = 5$,

$$\frac{dr}{dt} = \frac{90}{(5)^2} = \frac{90}{25} = 3.6.$$

Thus, the radius is increasing at the rate of 3.6 feet per minute at the instant the radius is 5 feet. □

EXERCISES 3.5

Differentiate the function: (a) by expanding before differentiation, (b) by using the chain rule. Then reconcile your results.

1. $f(x) = (x^2 + 1)^2$.
2. $f(x) = (x^3 - 1)^2$.
3. $f(x) = (2x + 1)^3$.
4. $f(x) = (x^2 + 1)^3$.
5. $f(x) = (x + x^{-1})^2$.
6. $f(x) = (3x^2 - 2x)^2$.

Differentiate the function.

7. $f(x) = (1 - 2x)^{-1}$.
8. $f(x) = (1 + 2x)^5$.
9. $f(x) = (x^5 - x^{10})^{20}$.
10. $f(x) = \left(x^2 + \dfrac{1}{x^2}\right)^3$.
11. $f(x) = \left(x - \dfrac{1}{x}\right)^4$.
12. $f(t) = \left(\dfrac{1}{1+t}\right)^4$.
13. $f(x) = (x - x^3 + x^5)^4$.
14. $f(t) = (t - t^2)^3$.
15. $f(t) = (t^{-1} + t^{-2})^4$.
16. $f(x) = \left(\dfrac{4x+3}{5x-2}\right)^3$.
17. $f(x) = \left(\dfrac{3x}{x^2+1}\right)^4$.
18. $f(x) = [(2x + 1)^2 + (x + 1)^2]^3$.
19. $f(x) = \left(\dfrac{x^3}{3} + \dfrac{x^2}{2} + \dfrac{x}{1}\right)^{-1}$.
20. $f(x) = [(6x + x^5)^{-1} + x]^2$.

Find dy/dx at $x = 0$.

21. $y = \dfrac{1}{1 + u^2}, \quad u = 2x + 1$.
22. $y = u + \dfrac{1}{u}, \quad u = (3x + 1)^4$.

23. $y = \dfrac{2u}{1-4u}, \quad u = (5x^2+1)^4.$

24. $y = u^3 - u + 1, \quad u = \dfrac{1-x}{1+x}.$

Find dy/dt.

25. $y = \dfrac{1-7u}{1+u^2}, \quad u = 1+x^2, \quad x = 2t-5.$

26. $y = 1+u^2, \quad u = \dfrac{1-7x}{1+x^2}, \quad x = 5t+2.$

Find dy/dx at $x = 2$.

27. $y = (s+3)^2, \quad s = \sqrt{t-3}, \quad t = x^2.$

28. $y = \dfrac{1+s}{1-s}, \quad s = t - \dfrac{1}{t}, \quad t = \sqrt{x}.$

Evaluate the functions in Exercises 29–38, given that

$$f(0)=1,\ f'(0)=2,\ f(1)=0,\ f'(1)=1,\ f(2)=1,\ f'(2)=1,$$
$$g(0)=2,\ g'(0)=1,\ g(1)=1,\ g'(1)=0,\ g(2)=2,\ g'(2)=1,$$
$$h(0)=1,\ h'(0)=2,\ h(1)=2,\ h'(1)=1,\ h(2)=0,\ h'(2)=2,$$

29. $(f \circ g)'(0)$. **30.** $(f \circ g)'(1)$.

31. $(f \circ g)'(2)$. **32.** $(g \circ f)'(0)$.

33. $(g \circ f)'(1)$. **34.** $(g \circ f)'(2)$.

35. $(f \circ h)'(0)$. **36.** $(f \circ h \circ g)'(1)$.

37. $(g \circ f \circ h)'(2)$. **38.** $(g \circ h \circ f)'(0)$.

Find $f''(x)$.

39. $f(x) = (x^3+x)^4$.

40. $f(x) = (x^2-5x+2)^{10}$.

41. $f(x) = \left(\dfrac{x}{1-x}\right)^3$.

42. $f(x) = \sqrt{x^2+1}\ \left(\text{recall that } \dfrac{d}{dx}[\sqrt{x}] = \dfrac{1}{2\sqrt{x}}\right)$.

Express the derivative in terms of f'

43. $\dfrac{d}{dx}[f(x^2+1)]$. **44.** $\dfrac{d}{dx}\left[f\left(\dfrac{x-1}{x+1}\right)\right]$.

45. $\dfrac{d}{dx}[[f(x)]^2+1]$. **46.** $\dfrac{d}{dx}\left[\dfrac{f(x)-1}{f(x)+1}\right]$.

Determine the values of x for which
(a) $f'(x) = 0$. (b) $f'(x) > 0$. (c) $f'(x) < 0$.

47. $f(x) = (1+x^2)^{-2}$. **48.** $f(x) = (1-x^2)^2$.

49. $f(x) = x(1+x^2)^{-1}$. **50.** $f(x) = x(1-x^2)^3$.

An object moves along a coordinate line, its position at each time $t \geq 0$ being given by $x(t)$. Find the times at which the object changes direction.

51. $x(t) = (t+1)^2(t-9)^3$. **52.** $x(t) = t(t-8)^3$.

53. $x(t) = (t^3-12t)^4$. **54.** $x(t) = (t^2-8t+15)^3$.

Find a formula for the nth derivative.

55. $y = \dfrac{1}{1-x}$. **56.** $y = \dfrac{x}{1+x}$.

57. $y = (a+bx)^n$; n a positive integer, a, b constants.

58. $y = \dfrac{a}{bx+c}$, a, b, c constants.

Find a function $y = f(x)$ with the given derivative. Check your answer by differentiation.

59. $y' = 3(x^2+1)^2(2x)$. **60.** $y' = 2x(x^2-1)$.

61. $\dfrac{dy}{dx} = 2(x^3-2)(3x^2)$. **62.** $\dfrac{dy}{dx} = 3x^2(x^3+2)^2$.

63. A function L has the property that $L'(x) = 1/x$ for $x \neq 0$. Determine the derivative with respect to x of $L(x^2+1)$.

64. Let f and g be differentiable functions such that $f'(x) = g(x)$ and $g'(x) = f(x)$, and let

$$H(x) = [f(x)]^2 - [g(x)]^2.$$

Find $H'(x)$.

65. Let f and g be differentiable functions such that $f'(x) = g(x)$ and $g'(x) = -f(x)$, and let

$$T(x) = [f(x)]^2 + [g(x)]^2.$$

Find $T'(x)$.

66. Let f be a differentiable function. Use the chain rule to show that:

(a) if f is even, then f' is odd.

(b) if f is odd, then f' is even.

67. The number a is called a *double zero* (or a zero of *multiplicity* 2) of the polynomial function P if

$$P(x) = (x-a)^2q(x) \quad \text{and} \quad q(a) \neq 0.$$

Prove that a is a double zero of P iff a is a zero of both P and P' and $P''(a) \neq 0$.
HINT: Recall the factor theorem; see Section 1.2.

68. The number a is called a *triple zero* (or a zero of *multiplicity* 3) of the polynomial function P if

$$P(x) = (x-a)^3q(x) \quad \text{and} \quad q(a) \neq 0.$$

Prove that a is a triple zero of P iff a is a zero of P, P', and P'', and $P'''(a) \neq 0$.

69. The number a is called a *zero of multiplicity* k of the polynomial function P if

$$P(x) = (x-a)^kq(x) \quad \text{and} \quad q(a) \neq 0.$$

Use the results in Exercises 69 and 70 to state a theorem about a zero of multiplicity k.

70. An equilateral triangle of side length x and altitude h has area A given by

$$A = \frac{\sqrt{3}}{4}x^2, \quad \text{where} \quad x = \frac{2\sqrt{3}}{3}h.$$

Find the rate of change of A with respect to h and determine the rate of change of A when $h = 2\sqrt{3}$.

71. Air is being pumped into a spherical balloon in such a way that its radius is increasing at the constant rate of 2 centimeters per second. What is the rate of change of the balloon's volume at the instant the radius is 10 centimeters? (The volume V of a sphere of radius r is $V = \frac{4}{3}\pi r^3$.)

72. Air is being pumped into a spherical balloon at the constant rate of 200 cubic centimeters per second. How fast is the surface area of the balloon changing at the instant the radius is 5 centimeters? (The surface area S of a sphere of radius r is $S = 4\pi r^2$.)

73. If an object of mass m has speed v, then its *kinetic energy*, KE, is defined by

$$KE = \tfrac{1}{2}mv^2.$$

Suppose that v is a function of time. What is the rate of change of KE with respect to t?

74. Newton's Law of Gravitational Attraction states that if two bodies are at a distance r apart, then the force F exerted by one body on the other is given by

$$F(r) = -\frac{k}{r^2}$$

where k is a positive constant. Suppose that, as a function of time, the distance between the two bodies is given by

$$r(t) = 49t - 4.9t^2, \qquad 0 \le t \le 10.$$

(a) Find the rate of change of F with respect to t.

(b) Show that $(F \circ r)'(3) = -(F \circ r)'(7)$.

C 75. Use a CAS to verify Exercise 66.

(a) For f even, use the Symbolic Differentiation Utility to differentiate $f(-x) - f(x) = 0$ and verify that f' is odd.

(b) For f odd, use the Symbolic Differentiation Utility to differentiate $f(-x) - f(x) = 0$ and verify that f' is even.

C 76. Use a CAS to find $\dfrac{d}{dx}\left[x^2\dfrac{d^4}{dx^4}(x^2+1)^4\right]$.

C 77. Use a CAS to find the following derivatives in terms of $f'(x)$:

(a) $\dfrac{d}{dx}\left[f\left(\dfrac{1}{x}\right)\right]$, (b) $\dfrac{d}{dx}\left[f\left(\dfrac{x^2-1}{x^2+1}\right)\right]$,

(c) $\dfrac{d}{dx}\left[\dfrac{f(x)}{1+f(x)}\right]$.

C 78. Use a CAS to find the following derivatives:

(a) $\dfrac{d}{dx}[u_1(u_2(x))]$, (b) $\dfrac{d}{dx}[u_1(u_2(u_3(x)))]$,

(c) $\dfrac{d}{dx}[u_1(u_2(u_3(u_4(x))))]$.

C 79. Use a CAS to find a formula for $\dfrac{d^2}{dx^2}[f(g(x))]$.

*SUPPLEMENT TO SECTION 3.5

To Prove Theorem 3.5.7 it is convenient to use the formulation of the derivative given in (3.1.6), Exercises 3.1.

THEOREM 3.5.8

The function f is differentiable at x iff

$$\lim_{t\to x}\frac{f(t)-f(x)}{t-x} \text{ exists.}$$

If this limit exists, it is $f'(x)$.

PROOF For each t in the domain of f, $t \ne x$, define

$$G(t) = \frac{f(t)-f(x)}{t-x}.$$

Note that

$$G(x+h) = \frac{f(x+h)-f(x)}{h}$$

and therefore

$$f \text{ is differentiable at } x \quad \text{iff} \quad \lim_{h\to 0} G(x+h) \text{ exists.}$$

The result follows from observing that

$$\lim_{h\to 0} G(x+h) = L \quad \text{iff} \quad \lim_{t\to x} G(t) = L.$$

For the equivalence of these two limits we refer you to (2.2.6). ❑

PROOF OF THEOREM 3.5.7

By Theorem 3.5.8 it is enough to show that

$$\lim_{t\to x} \frac{f(g(t)) - f(g(x))}{t-x} = f'(g(x))g'(x).$$

We begin by defining an auxiliary function F on the domain of f by setting

$$F(y) = \begin{cases} \dfrac{f(y) - f(g(x))}{y - g(x)}, & y \neq g(x) \\ f'(g(x)), & y = g(x). \end{cases}$$

F is continuous at $g(x)$ since

$$\lim_{y\to g(x)} F(y) = \lim_{y\to g(x)} \frac{f(y) - f(g(x))}{y - g(x)},$$

and the right-hand side is (by Theorem 3.5.8) $f'(g(x))$, which is the value of F at $g(x)$. For $t \neq x$,

$$(1) \qquad \frac{f(g(t)) - f(g(x))}{t-x} = F(g(t))\left[\frac{g(t) - g(x)}{t-x}\right].$$

To see this we note that, if $g(t) = g(x)$, then both sides are 0. If $g(t) \neq g(x)$, then

$$F(g(t)) = \frac{f(g(t)) - f(g(x))}{g(t) - g(x)},$$

so that again we have equality.

Since g, being differentiable at x, is continuous at x and since F is continuous at $g(x)$, we know that the composition $F \circ g$ is continuous at x. Thus

$$\lim_{t\to x} F(g(t)) = F(g(x)) = f'(g(x)).$$

↑ by our definition of F

This, together with Equation (1), gives

$$\lim_{t\to x} \frac{f(g(t)) - f(g(x))}{t-x} = f'(g(x))\, g'(x). \quad ❑$$

■ 3.6 DIFFERENTIATING THE TRIGONOMETRIC FUNCTIONS

An outline review of trigonometry—definitions, identities, and graphs—appears in Chapter 1. As discussed there, the calculus of the trigonometric functions is simplified by the use of radian measure. Therefore, we use radian measure throughout our work and refer to degree measure only in passing.

The derivative of the sine function is the cosine function:

(3.6.1)
$$\boxed{\frac{d}{dx}(\sin x) = \cos x.}$$

PROOF Fix any number x. For $h \neq 0$,

$$\begin{aligned}\frac{\sin(x+h) - \sin x}{h} &= \frac{[\sin x \cos h + \cos x \sin h] - [\sin x]}{h}\\ &= \sin x \frac{\cos h - 1}{h} + \cos x \frac{\sin h}{h}.\end{aligned}$$

Now, as shown in Section 2.5, (2.5.5),

$$\lim_{h\to 0} \frac{\cos h - 1}{h} = 0 \quad \text{and} \quad \lim_{h\to 0} \frac{\sin h}{h} = 1.$$

Since x is fixed, $\sin x$ and $\cos x$ remain constant as h approaches zero, and it follows that

$$\begin{aligned}\lim_{h\to 0} \frac{\sin(x+h) - \sin x}{h} &= \lim_{h\to 0}\left(\sin x \frac{\cos h - 1}{h} + \cos x \frac{\sin h}{h}\right)\\ &= \sin x \left(\lim_{h\to 0} \frac{\cos h - 1}{h}\right) + \cos x \left(\lim_{h\to 0} \frac{\sin h}{h}\right).\end{aligned}$$

Thus $$\lim_{h\to 0} \frac{\sin(x+h) - \sin x}{h} = (\sin x)(0) + (\cos x)(1) = \cos x. \quad \square$$

The derivative of the cosine function is the negative of the sine function:

(3.6.2)
$$\boxed{\frac{d}{dx}(\cos x) = -\sin x.}$$

PROOF Fix any number x. For $h \neq 0$,

$$\cos(x+h) = \cos x \cos h - \sin x \sin h.$$

Therefore

$$\begin{aligned}\lim_{h\to 0} \frac{\cos(x+h) - \cos x}{h} &= \lim_{h\to 0} \frac{[\cos x \cos h - \sin x \sin h] - [\cos x]}{h}\\ &= \cos x \left(\lim_{h\to 0} \frac{\cos h - 1}{h}\right) - \sin x \left(\lim_{h\to 0} \frac{\sin h}{h}\right)\\ &= -\sin x. \quad \square\end{aligned}$$

Remark The proofs of the derivative formulas depend solely on the basic limits

$$\lim_{h\to 0} \frac{\sin h}{h} = 1 \quad \text{and} \quad \lim_{h\to 0} \frac{\cos h - 1}{h} = 0$$

that we derived in Section 2.5. Although we couldn't state it at that time, these basic limits are simply the definition of the derivative of the sine and cosine functions at $x = 0$. See Exercises 61 and 62.

Example 1 To differentiate $f(x) = \cos x \sin x$ we use the product rule:

$$\begin{aligned} f'(x) &= \cos x \frac{d}{dx}(\sin x) + \sin x \frac{d}{dx}(\cos x) \\ &= \cos x(\cos x) + \sin x(-\sin x) = \cos^2 x - \sin^2 x. \quad \square \end{aligned}$$

We come now to the tangent function. Since $\tan x = \sin x / \cos x$, we have

$$\frac{d}{dx}(\tan x) = \frac{\cos x \frac{d}{dx}(\sin x) - \sin x \frac{d}{dx}(\cos x)}{\cos^2 x} = \frac{\cos^2 x + \sin^2 x}{\cos^2 x} = \frac{1}{\cos^2 x} = \sec^2 x.$$

The derivative of the tangent function is the secant squared:

(3.6.3)
$$\boxed{\frac{d}{dx}(\tan x) = \sec^2 x.}$$

The derivatives of the other trigonometric functions are as follows:

(3.6.4)
$$\boxed{\begin{aligned} &\frac{d}{dx}(\cot x) = -\csc^2 x, \\ &\frac{d}{dx}(\sec x) = \sec x \tan x, \\ &\frac{d}{dx}(\csc x) = -\csc x \cot x. \end{aligned}}$$

The verification of these formulas is left as an exercise.

It is time for some sample problems.

Example 2 Find $f'(\pi/4)$ for $f(x) = x \cot x$.

SOLUTION We first find $f'(x)$ by the product rule:

$$f'(x) = x \frac{d}{dx}(\cot x) + \cot x \frac{d}{dx}(x) = -x \csc^2 x + \cot x.$$

Now we evaluate f' at $\pi/4$:

$$f'(\pi/4) = -\frac{\pi}{4}(\sqrt{2})^2 + 1 = 1 - \frac{\pi}{2}. \quad \square$$

Example 3 Find $\dfrac{d}{dx}\left[\dfrac{1 - \sec x}{\tan x}\right]$.

SOLUTION By the quotient rule,

$$\frac{d}{dx}\left[\frac{1-\sec x}{\tan x}\right] = \frac{\tan x \frac{d}{dx}(1-\sec x) - (1-\sec x)\frac{d}{dx}(\tan x)}{\tan^2 x}$$

$$= \frac{\tan x(-\sec x \tan x) - (1-\sec x)(\sec^2 x)}{\tan^2 x}$$

$$= \frac{\sec x(\sec^2 x - \tan^2 x) - \sec^2 x}{\tan^2 x}$$

$$= \frac{\sec x - \sec^2 x}{\tan^2 x} = \frac{\sec x(1-\sec x)}{\tan^2 x}. \quad \square$$

(recall $\sec^2 x - \tan^2 x = 1$) ↑

Example 4 Find an equation for the line tangent to the curve $y = \cos x$ at the point where $x = \pi/3$.

SOLUTION Since $\cos \pi/3 = 1/2$, the point of tangency is $(\pi/3, 1/2)$. To find the slope of the tangent line, we evaluate the derivative

$$\frac{dy}{dx} = -\sin x$$

at $x = \pi/3$. This gives $m = -\sqrt{3}/2$. An equation for the tangent line can be written

$$y - \frac{1}{2} = -\frac{\sqrt{3}}{2}\left(x - \frac{\pi}{3}\right). \quad \square$$

Example 5 An object moves along the x-axis, its position at each time t given by the function

$$x(t) = t + 2\sin t.$$

Determine those times t from 0 to 2π when the object is moving to the left.

SOLUTION The object is moving to the left only when its velocity $v(t) < 0$. Since

$$v(t) = x'(t) = 1 + 2\cos t,$$

the object is moving to the left only when $\cos t < -\frac{1}{2}$. As you can check, the only t in $[0, 2\pi]$ for which $\cos t < -\frac{1}{2}$ are those t that lie between $2\pi/3$ and $4\pi/3$. Thus from time $t = 0$ to $t = 2\pi$, the object is moving left only during the time interval $(2\pi/3, 4\pi/3)$. $\square$

The Chain Rule and the Trigonometric Functions

If f is a differentiable function of u and u is a differentiable function of x, then, as you saw in Section 3.5,

$$\frac{d}{dx}[f(u)] = \frac{d}{du}[f(u)]\frac{du}{dx} = f'(u)\frac{du}{dx}.$$

Written in this form, the derivatives of the six trigonometric functions appear as follows:

(3.6.5)
$$\frac{d}{dx}(\sin u) = \cos u\frac{du}{dx}, \qquad \frac{d}{dx}(\cos u) = -\sin u\frac{du}{dx},$$
$$\frac{d}{dx}(\tan u) = \sec^2 u\frac{du}{dx}, \qquad \frac{d}{dx}(\cot u) = -\csc^2 u\frac{du}{dx},$$
$$\frac{d}{dx}(\sec u) = \sec u \tan u\frac{du}{dx}, \qquad \frac{d}{dx}(\csc u) = -\csc u \cot u\frac{du}{dx}.$$

Example 6

$$\frac{d}{dx}(\cos 2x) = -\sin 2x\,\frac{d}{dx}(2x) = -2\sin 2x. \quad \square$$

Example 7

$$\frac{d}{dx}[\sec(x^2+1)] = \sec(x^2+1)\tan(x^2+1)\frac{d}{dx}(x^2+1)$$
$$= 2x\sec(x^2+1)\tan(x^2+1). \quad \square$$

Example 8

$$\begin{aligned}\frac{d}{dx}(\sin^3 \pi x) &= \frac{d}{dx}(\sin \pi x)^3 \\ &= 3(\sin \pi x)^2\frac{d}{dx}(\sin \pi x) \\ &= 3(\sin \pi x)^2 \cos \pi x\frac{d}{dx}(\pi x) \\ &= 3(\sin \pi x)^2 \cos \pi x\,(\pi) = 3\pi \sin^2 \pi x \cos \pi x. \quad \square\end{aligned}$$

Our treatment of the trigonometric functions has been based entirely on radian measure. When degrees are used, the derivatives of the trigonometric functions contain the extra factor $\frac{1}{180}\pi \cong 0.0175$.

Example 9 Find $\frac{d}{dx}(\sin x^\circ)$.

SOLUTION Since $x^\circ = \frac{1}{180}\pi x$ radians,

$$\frac{d}{dx}(\sin x^\circ) = \frac{d}{dx}\left(\sin \tfrac{1}{180}\pi x\right) = \tfrac{1}{180}\pi \cos \tfrac{1}{180}\pi x = \tfrac{1}{180}\pi \cos x^\circ. \quad \square$$

The extra factor $\frac{1}{180}\pi$ is a disadvantage, particularly in problems where it occurs repeatedly. This tends to discourage the use of degree measure in theoretical work.

EXERCISES 3.6

Differentiate the given function.

1. $y = 3\cos x - 4\sec x$.

2. $y = x^2 \sec x$.

3. $y = x^3 \csc x$.

4. $y = \sin^2 x$.

5. $y = \cos^2 t$.

6. $y = 3t^2 \tan t$.

7. $y = \sin^4 \sqrt{u}$.

8. $y = u \csc u^2$.

9. $y = \tan x^2$.

10. $y = \cos \sqrt{x}$.

11. $y = [x + \cot \pi x]^4$.

12. $y = [x^2 - \sec 2x]^3$.

Find the second derivative.

13. $y = \sin x$.

14. $y = \cos x$.

15. $y = \dfrac{\cos x}{1 + \sin x}$.

16. $y = \tan^3 2\pi x$.

17. $y = \cos^3 2u$.

18. $y = \sin^5 3t$.

19. $y = \tan 2t$.

20. $y = \cot 4u$.

21. $y = x^2 \sin 3x$.

22. $y = \dfrac{\sin x}{1 - \cos x}$.

23. $y = \sin^2 x + \cos^2 x$.

24. $y = \sec^2 x - \tan^2 x$.

Find the indicated derivative.

25. $\dfrac{d^4}{dx^4}(\sin x)$.

26. $\dfrac{d^4}{dx^4}(\cos x)$.

27. $\dfrac{d}{dt}\left[t^2 \dfrac{d^2}{dt^2}(t \cos 3t)\right]$.

28. $\dfrac{d}{dt}\left[t \dfrac{d}{dt}(\cos t^2)\right]$.

29. $\dfrac{d}{dx}[f(\sin 3x)]$.

30. $\dfrac{d}{dx}[\sin (f(3x))]$.

Find an equation for the line tangent to the curve at $x = a$.

31. $y = \sin x$; $a = 0$.

32. $y = \tan x$; $a = \pi/6$.

33. $y = \cot x$; $a = \pi/6$.

34. $y = \cos x$; $a = 0$.

35. $y = \sec x$; $a = \pi/4$.

36. $y = \csc x$; $a = \pi/3$.

Determine the numbers x between 0 and 2π where the line tangent to the curve is horizontal.

37. $y = \cos x$.

38. $y = \sin x$.

39. $y = \sin x + \sqrt{3} \cos x$.

40. $y = \cos x - \sqrt{3} \sin x$.

41. $y = \sin^2 x$.

42. $y = \cos^2 x$.

43. $y = \tan x - 2x$.

44. $y = 3 \cot x + 4x$.

45. $y = 2 \sec x + \tan x$.

46. $y = \cot x - 2 \csc x$.

An object moves along the x-axis, its position at each time t being given by $x(t)$. In Exercises 47–52, determine those times from $t = 0$ to $t = 2\pi$ when the object is moving to the right with increasing speed.

47. $x(t) = \sin 3t$.

48. $x(t) = \cos 2t$.

49. $x(t) = \sin t - \cos t$.

50. $x(t) = \sin t + \cos t$.

51. $x(t) = t + 2 \cos t$.

52. $x(t) = t - \sqrt{2} \sin t$.

Find dy/dt (a) by using (3.5.4) and (b) by writing y as a function of t and then differentiating.

53. $y = u^2 - 1$, $u = \sec x$, $x = \pi t$.

54. $y = [\frac{1}{2}(1 + u)]^3$, $u = \cos x$, $x = 2t$.

55. $y = [\frac{1}{2}(1 - u)]^4$, $u = \cos x$, $x = 2t$.

56. $y = 1 - u^2$, $u = \csc x$, $x = 3t$.

57. It can be shown by induction that the nth derivative of the sine function is given by the formula

$$\frac{d^n}{dx^n}(\sin x) = \begin{cases} (-1)^{(n-1)/2} \cos x, & n \text{ odd} \\ (-1)^{n/2} \sin x, & n \text{ even.} \end{cases}$$

Persuade yourself that this formula is correct and obtain a similar formula for the nth derivative of the cosine function.

58. Verify the following differentiation formulas.

(a) $\dfrac{d}{dx}(\cot x) = -\csc^2 x$.

(b) $\dfrac{d}{dx}(\sec x) = \sec x \tan x$.

(c) $\dfrac{d}{dx}(\csc x) = -\csc x \cot x$.

59. Use the identities

$$\cos x = \sin\left(\frac{\pi}{2} - x\right) \quad \text{and} \quad \sin x = \cos\left(\frac{\pi}{2} - x\right)$$

to give an alternate proof of (3.6.2).

60. The double-angle formula for the sine function is: $\sin 2x = 2 \sin x \cos x$. Differentiate this formula to obtain the double-angle formula for the cosine function.

61. Let $f(x) = \sin x$. Show that finding $f'(0)$ using the definition of the derivative amounts to finding

$$\lim_{x \to 0} \frac{\sin x}{x}. \qquad \text{(see Section 2.5)}$$

62. Let $f(x) = \cos x$. Show that finding $f'(0)$ using the definition of the derivative amounts to finding

$$\lim_{x \to 0} \frac{\cos x - 1}{x}.$$

Find a function f with the given derivative. Check your answer by differentiation.

63. $f'(x) = 2 \cos x - 3 \sin x$.

64. $f'(x) = \sec^2 x - \csc^2 x$.

65. $f'(x) = 2 \cos 2x + \sec x \tan x$.

66. $f'(x) = \sin 3x - \csc 2x \cot 2x$.

67. $f'(x) = 2x \cos (x^2) - 2 \sin 2x$.

68. $f'(x) = x^2 \sec^2 (x^3) + 2 \sec 2x \tan 2x$.

69. Let $f(x) = \begin{cases} x \sin\left(\dfrac{1}{x}\right), & x \neq 0 \\ 0, & x = 0 \end{cases}$

and $g(x) = \begin{cases} x^2 \sin\left(\dfrac{1}{x}\right), & x \neq 0 \\ 0, & x = 0. \end{cases}$

In Exercise 62, Section 3.1, you were asked to show that f is continuous at 0, but not differentiable there, and that g is differentiable at 0. Both f and g are differentiable at each $x \neq 0$.

(a) Find $f'(x)$ and $g'(x)$ for $x \neq 0$.

(b) Show that g' is not continuous at 0.

70. Let $f(x) = \begin{cases} \cos x, & x \geq 0 \\ ax + b, & x < 0. \end{cases}$

(a) Determine a and b so that f is differentiable at 0.

(b) Using the values found in part (a), sketch the graph of f.

71. Let $g(x) = \begin{cases} \sin x, & 0 \le x \le 2\pi/3 \\ ax + b, & 2\pi/3 < x \le 2\pi. \end{cases}$

(a) Determine a and b so that g is differentiable at $2\pi/3$.

(b) Using the values found in part (a), sketch the graph of g.

72. Let $f(x) = \begin{cases} 1 + a\cos x, & x \le \pi/3 \\ b + \sin(x/2), & x > \pi/3. \end{cases}$

(a) Determine a and b so that f is differentiable at $\pi/3$.

(b) Using the values found in part (a), sketch the graph of f.

73. Let $y = A\sin\omega t + B\cos\omega t$, where A, B, and ω are constants. Show that y satisfies the equation

$$\frac{d^2y}{dt^2} + \omega^2 y = 0.$$

74. A simple pendulum consists of a mass m swinging at the end of a (massless) rod or wire of length L (see the figure). The angular displacement θ of the pendulum at time t is given by

$$\theta(t) = a\sin(\omega t + \phi),$$

where a, ω, and ϕ are constants.

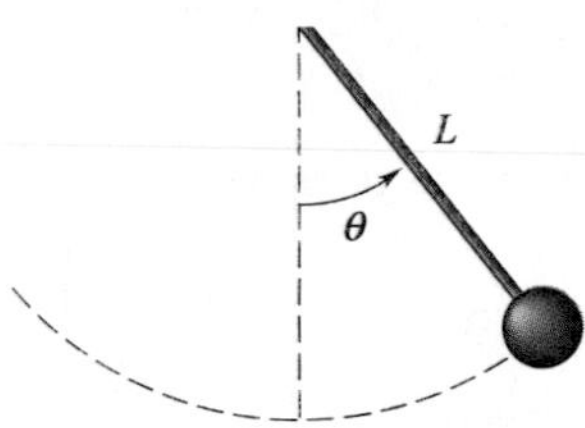

(a) Show that θ satisfies the equation $d^2\theta/dt^2 + \omega^2\theta = 0$. (Compare this result with the result in Exercise 73.)

(b) Show that θ can be written in the form

$$\theta(t) = A\sin\omega t + B\cos\omega t,$$

where A, B, and ω are constants.

75. An isosceles triangle has equal sides of length c, and the angle between them is x radians. Express the area A of the triangle as a function of x and find the rate of change of A with respect to x.

76. A triangle has sides of length a and b, and the angle between them is x radians. If a and b are constants, find the rate of change of the third side c with respect to x. HINT: Use the law of cosines.

C 77. Let θ be measured in degrees.

(a) Estimate

$$\lim_{\theta\to 0}\frac{\sin\theta}{\theta}$$

by letting $\theta = 5,\ 1,\ 0.1,\ 0.01$, and 0.001 degrees.

(b) Find the value of $\pi/180$ and compare the result with the limit found in part (a). This gives another demonstration of the reason for using radian measure rather than degree measure.

C 78. Let $f(x) = \cos kx$ where k is a positive integer. Use a CAS to:

(a) Find $\dfrac{d^n}{dx^n}[f(x)]$.

(b) Find all positive integers m for which $f(x)$ is a solution of $y'' + my = 0$ for some value of k.

C 79. Use a CAS to show that $f(x) = A\cos\sqrt{2}x + B\sin\sqrt{2}x$ is a solution of $y'' + 2y = 0$. Find A and B given that $f(0) = 2$ and $f'(0) = -3$. Verify your results analytically.

C 80. Let $f(x) = \sin x - \cos 2x$ for $0 \le x \le 2\pi$.

(a) Use a graphing utility to approximate the points on the graph where f has horizontal tangents.

(b) Use a CAS to approximate the points on the graph at which $f'(x) = 0$.

(c) Compare the results in (a) and (b).

C 81. Repeat Exercise 80 with $f(x) = \sin x - \sin^2 x$ for $0 \le x \le 2\pi$.

C 82. Let $f(x) = \sin x$, $-\pi \le x \le \pi$.

(a) Using a graphing utility to graph f and $p_1(x) = x$ together. Notice that $p_1 = x$ is a good approximation for $\sin x$ near 0.

(b) Compare the graphs of f and $p_3(x) = x - \frac{1}{6}x^3$.

(c) Compare the graphs of f and $p_5(x) = x - \frac{1}{6}x^3 + \frac{1}{120}x^5$.

C 83. Let $f(x) = \cos x$, $-\pi \le x \le \pi$. Repeat Exercise 82 for f and $p_0(x) = 1$, $p_2(x) = 1 - \frac{1}{2}x^2$, and $p_4(x) = 1 - \frac{1}{2}x^2 + \frac{1}{24}x^4$. The use of polynomials to approximate the trigonometric functions (and other functions) is one of the main topics in Chapter 11.

■ 3.7 IMPLICIT DIFFERENTIATION; RATIONAL POWERS

Up to this point we have been considering functions defined by relations in which the dependent variable is given *explicitly* in terms of the independent variable. For instance,

$$y = (x^2 - 1)^3, \quad y = x^2\sin 2x, \quad y = \begin{cases} x^2 + 1, & x \le 0 \\ 1 - x, & x > 0, \end{cases}$$

or, in general, $y = f(x)$. In this section we consider functional relationships between two variables in which neither of the variables is given explicitly in terms of the other.

Suppose we know that y is a differentiable function of x that satisfies a particular equation in x and y. If we find it difficult to obtain the derivative of y, either because the calculations are burdensome or because we are unable to express y *explicitly* in terms of x, we may still be able to obtain dy/dx by using a process called *implicit differentiation*. This process is based on differentiating both sides of the equation satisfied by x and y.

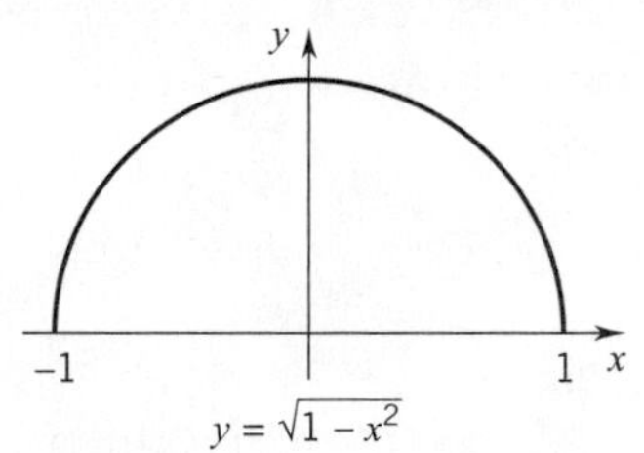

Figure 3.7.1

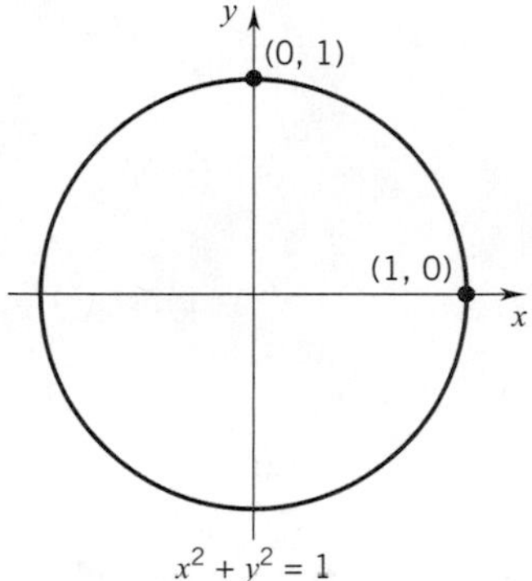

Figure 3.7.2

Example 1 We know that the function $y = \sqrt{1-x^2}$ (Figure 3.7.1) satisfies the equation

$$x^2 + y^2 = 1. \qquad \text{(Figure 3.7.2)}$$

We can obtain dy/dx by carrying out the differentiation in the usual manner, or we can do it more simply by working with the equation $x^2 + y^2 = 1$.

Differentiating both sides of the equation with respect to x (remembering that y is a function of x), we have

$$\frac{d}{dx}(x^2) + \frac{d}{dx}(y^2) = \frac{d}{dx}(1),$$

$$2x + \underbrace{2y\frac{dy}{dx}}_{\text{(by the chain rule)}} = 0,$$

$$\frac{dy}{dx} = -\frac{x}{y}.$$

We have obtained dy/dx in terms of x and y. Usually this is as far as we can go. Here we can go further since we have y explicitly in terms of x. The relation $y = \sqrt{1-x^2}$ gives

$$\frac{dy}{dx} = -\frac{x}{\sqrt{1-x^2}}.$$

Verify this result by differentiating $y = \sqrt{1-x^2}$. □

Example 2 Use implicit differentiation to express dy/dx in terms of x and y.

(a) $2x^2y - y^3 + 1 = x + 2y$. **(b)** $\cos(x - y) = (2x + 1)^3 y$.

SOLUTION

(a) Differentiating both sides of the equation with respect to x, we have

$$\underbrace{2x^2\frac{dy}{dx} + 4xy}_{\text{(by the product rule)}} - \underbrace{3y^2\frac{dy}{dx}}_{\text{(by the chain rule)}} = 1 + 2\frac{dy}{dx}$$

$$(2x^2 - 3y^2 - 2)\frac{dy}{dx} = 1 - 4xy.$$

Therefore,

$$\frac{dy}{dx} = \frac{1-4xy}{2x^2-3y^2-2}.$$

(b) We differentiate both sides of the equation with respect to x:

$$\underbrace{-\sin(x-y)\left[1-\frac{dy}{dx}\right]}_{\text{(by the chain rule)}} = (2x+1)^3\frac{dy}{dx} + 3(2x+1)^2(2)y,$$

$$[\sin(x-y)-(2x+1)^3]\frac{dy}{dx} = 6(2x+1)^2y + \sin(x-y).$$

Thus
$$\frac{dy}{dx} = \frac{6(2x+1)^2y+\sin(x-y)}{\sin(x-y)-(2x+1)^3}. \quad \square$$

Example 3 The curve $2x^3 + 2y^3 = 9xy$ is a *folium of Descartes*. Find the slope of the curve at the point (1, 2).

SOLUTION First we check that the point does lie on the curve:

$$2(1^3)+2(2)^3 = 18 = 9(1)(2). \qquad \checkmark$$

Now we differentiate both sides of the equation:

$$6x^2 + 6y^2\frac{dy}{dx} = 9x\left(\frac{dy}{dx}\right) + 9y,$$

$$2x^2 + 2y^2\frac{dy}{dx} = 3x\frac{dy}{dx} + 3y.$$

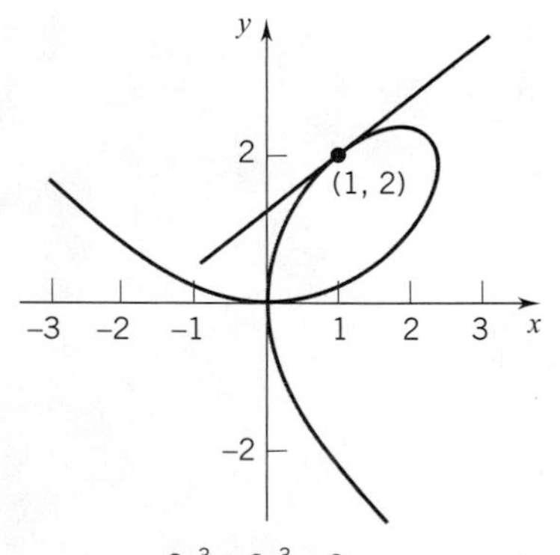

$2x^3 + 2y^3 = 9xy$

Figure 3.7.3

Solving for dy/dx, we get

$$\frac{dy}{dx} = \frac{3y-2x^2}{2y^2-3x}.$$

Setting $x = 1, y = 2$, we have

$$\frac{dy}{dx} = \frac{6-2}{8-3} = \frac{4}{5}.$$

The slope of the curve at the point (1, 2) is 4/5. See Figure 3.7.3. $\square$

We can also find higher derivatives by implicit differentiation.

Example 4 Express d^2y/dx^2 in terms of x and y given that

$$y^3 - x^2 = 4.$$

SOLUTION Differentiation with respect to x gives

$$(*) \qquad 3y^2\frac{dy}{dx} - 2x = 0.$$

Differentiating again, we have

$$\underbrace{3y^2\frac{d}{dx}\left(\frac{dy}{dx}\right)+\left(\frac{dy}{dx}\right)\frac{d}{dx}(3y^2)}-2=0$$

(by the product rule)

$$3y^2\,\frac{d^2y}{dx^2}+6y\left(\frac{dy}{dx}\right)^2-2=0.$$

Since (∗) gives

$$\frac{dy}{dx}=\frac{2x}{3y^2},$$

we have

$$3y^2\,\frac{d^2y}{dx^2}+6y\left(\frac{2x}{3y^2}\right)^2-2=0.$$

Now, as you can check

$$\frac{d^2y}{dx^2}=\frac{6y^3-8x^2}{9y^5}.\quad\square$$

Remark If we differentiate $x^2+y^2=-1$ implicitly, we find that

$$2x+2y\frac{dy}{dx}=0\quad\text{and therefore}\quad\frac{dy}{dx}=-\frac{y}{x}.$$

However, the result is meaningless. It is meaningless because there is no real-valued function y of x that satisfies the equation $x^2+y^2=-1$. Implicit differentiation can be applied meaningfully to an equation in x and y only if there is a function y of x that satisfies the equation. ☐

Rational Powers

The differentiation formula

$$\frac{d}{dx}(x^n)=nx^{n-1}$$

holds for all integers n. It also hold for all rational exponents p/q:

(3.7.1)
$$\boxed{\frac{d}{dx}(x^{p/q})=\frac{p}{q}\,x^{(p/q)-1}.}$$

PROOF The relation $y=x^{p/q}$ gives

$$y^q=x^p.$$

Implicit differentiation with respect to x yields

$$q\,y^{q-1}\frac{dy}{dx}=p\,x^{p-1}$$

and therefore

$$\frac{dy}{dx}=\frac{px^{p-1}}{qy^{q-1}}=\frac{p}{q}x^{p-1}y^{1-q}.$$

Now, replacing y by $x^{p/q}$, we have

$$\frac{dy}{dx}=\frac{p}{q}x^{p-1}[x^{p/q}]^{1-q}=\frac{p}{q}x^{p-1}x^{(p/q)-p}=\frac{p}{q}x^{(p/q)-1}. \quad \square$$

Here are some simple examples:

$$\frac{d}{dx}(x^{2/3})=\tfrac{2}{3}x^{-1/3},\qquad \frac{d}{dx}(x^{5/2})=\tfrac{5}{2}x^{3/2},\qquad \frac{d}{dx}(x^{-7/9})=-\tfrac{7}{9}x^{-16/9}.$$

If u is a differentiable function of x, then, by the chain rule

(3.7.2)
$$\boxed{\frac{d}{dx}(u^{p/q})=\frac{p}{q}\,u^{(p/q)-1}\frac{du}{dx}.}$$

The verification of this is left to you. The result holds on every open x-interval where $u^{(p/q)-1}$ is defined.

Example 4

(a) $\dfrac{d}{dx}[(1+x^2)^{1/5}]=\tfrac{1}{5}(1+x^2)^{-4/5}(2x)=\tfrac{2}{5}x(1+x^2)^{-4/5}.$

(b) $\dfrac{d}{dx}[(1-x^2)^{2/3}]=\tfrac{2}{3}(1-x^2)^{-1/3}(-2x)=-\tfrac{4}{3}x(1-x^2)^{-1/3}.$

(c) $\dfrac{d}{dx}[(1-x^2)^{1/4}]=\tfrac{1}{4}(1-x^2)^{-3/4}(-2x)=-\tfrac{1}{2}x(1-x^2)^{-3/4}.$

As you can verify, the first statement holds for all real x, the second for all $x\neq\pm1$, and the third only for $x\in(-1,1)$. $\square$

Example 5

$$\begin{aligned}
\frac{d}{dx}\left[\left(\frac{x}{1+x^2}\right)^{1/2}\right]&=\frac{1}{2}\left(\frac{x}{1+x^2}\right)^{-1/2}\frac{d}{dx}\left(\frac{x}{1+x^2}\right)\\
&=\frac{1}{2}\left(\frac{x}{1+x^2}\right)^{-1/2}\frac{(1+x^2)(1)-x(2x)}{(1+x^2)^2}\\
&=\frac{1}{2}\left(\frac{1+x^2}{x}\right)^{1/2}\frac{1-x^2}{(1+x^2)^2}\\
&=\frac{1-x^2}{2x^{1/2}(1+x^2)^{3/2}}.
\end{aligned}$$

The result holds for all $x>0$. $\square$

EXERCISES 3.7

Use implicit differentiation to obtain dy/dx in terms of x and y.

1. $x^2 + y^2 = 4$. **2.** $x^3 + y^3 - 3xy = 0$.

3. $4x^2 + 9y^2 = 36$. **4.** $\sqrt{x} + \sqrt{y} = 4$.

5. $x^4 + 4x^3y + y^4 = 1$.

6. $x^2 - x^2y + xy^2 + y^2 = 1$.

7. $(x - y)^2 - y = 0$. **8.** $(y + 3x)^2 - 4x = 0$.

9. $\sin(x + y) = xy$. **10.** $\tan xy = xy$.

Express d^2y/dx^2 in terms of x and y.

11. $y^2 + 2xy = 16$. **12.** $x^2 - 2xy + 4y^2 = 3$.

13. $y^2 + xy - x^2 = 9$. **14.** $x^2 - 3xy = 18$.

15. $4\tan y = x^3$. **16.** $\sin^2 x + \cos^2 y = 1$.

Evaluate dy/dx and d^2y/dx^2 at the point indicated.

17. $x^2 - 4y^2 = 9$; $(5, 2)$.

18. $x^2 + 4xy + y^3 + 5 = 0$; $(2, -1)$.

19. $\cos(x + 2y) = 0$; $(\pi/6, \pi/6)$.

20. $x = \sin^2 y$; $(\frac{1}{2}, \pi/4)$.

Find equations for the tangent and normal lines at the point indicated.

21. $2x + 3y = 5$; $(-2, 3)$.

22. $9x^2 + 4y^2 = 72$; $(2, 3)$.

23. $x^2 + xy + 2y^2 = 28$; $(-2, -3)$.

24. $x^3 - axy + 3ay^2 = 3a^3$; (a, a).

25. $x = \cos y$; $(\frac{1}{2}, \frac{\pi}{3})$.

26. $\tan xy = x$; $(1, \frac{\pi}{4})$.

Find dy/dx.

27. $y = (x^3 + 1)^{1/2}$. **28.** $y = (x + 1)^{1/3}$.

29. $y = \sqrt[4]{2x^2 + 1}$.

30. $y = (x + 1)^{1/3}(x + 2)^{2/3}$.

31. $y = \sqrt{2 - x^2}\sqrt{3 - x^2}$.

32. $y = \sqrt{(x^4 - x + 1)^3}$.

Carry out the differentiation.

33. $\dfrac{d}{dx}\left(\sqrt{x} + \dfrac{1}{\sqrt{x}}\right)$. **34.** $\dfrac{d}{dx}\left(\sqrt{\dfrac{3x + 1}{2x + 5}}\right)$.

35. $\dfrac{d}{dx}\left(\dfrac{x}{\sqrt{x^2 + 1}}\right)$. **36.** $\dfrac{d}{dx}\left(\dfrac{\sqrt{x^2 + 1}}{x}\right)$.

37. (*Important*) Show the general form of the graph.
(a) $f(x) = x^{1/n}$, n a positive even integer.
(b) $f(x) = x^{1/n}$, n a positive odd integer.
(c) $f(x) = x^{2/n}$, n an odd integer greater than 1.

Find the second derivative.

38. $y = \sqrt{a^2 + x^2}$. **39.** $y = \sqrt[3]{a + bx}$.

40. $y = x\sqrt{a^2 - x^2}$. **41.** $y = \sqrt{x}\tan\sqrt{x}$.

42. $y = \sqrt{x}\sin\sqrt{x}$.

43. Show that all normals to the circle $x^2 + y^2 = r^2$ pass through the center of the circle.

44. Determine the x-intercept of the tangent to the parabola $y^2 = x$ at the point where $x = a$.

The angle between two curves is the angle between their tangents at the point of intersection. If the slopes are m_1 and m_2, then the angle of intersection α can be obtained from the formula

$$\tan\alpha = \left|\frac{m_2 - m_1}{1 + m_1 m_2}\right|.$$

45. At what angles do the parabolas $y^2 = 2px + p^2$ and $y^2 = p^2 - 2px$ intersect?

46. At what angles does the line $y = 2x$ intersect the curve $x^2 - xy + 2y^2 = 28$?

47. The curves $y = x^2$ and $x = y^3$ intersect at the points $(1, 1)$ and $(0, 0)$. Find the angle between the curves at each of these points.

48. Find the angles at which the circles $(x - 1)^2 + y^2 = 10$ and $x^2 + (y - 2)^2 = 5$ intersect.

Two curves are said to be *orthogonal* iff, at each point of intersection, the angle between them is a right angle. In Exercises 49 and 50, show that the given curves are orthogonal.

49. The hyperbola $x^2 - y^2 = 5$ and the ellipse $4x^2 + 9y^2 = 72$.

50. The ellipse $3x^2 + 2y^2 = 5$ and $y^3 = x^2$.
HINT: The curves intersect at $(1,1)$ and $(-1, 1)$

Two families of curves are said to be *orthogonal trajectories* (of each other) iff each member of one family is orthogonal to each member of the other family. In Exercises 51 and 52, show that the given families of curves are orthogonal trajectories.

51. The family of circles $x^2 + y^2 = r^2$ and the family of lines $y = mx$.

52. The family of parabolas $x = ay^2$ and the family of ellipses $x^2 + \frac{1}{2}y^2 = b$.

53. Find equations for the lines tangent to the ellipse $4x^2 + y^2 = 72$ that are perpendicular to the line $x + 2y + 3 = 0$.

54. Find equations for the lines normal to the hyperbola $4x^2 - y^2 = 36$ that are parallel to the line $2x + 5y - 4 = 0$.

55. The curve defined by the equation $(x^2 + y^2)^2 = x^2 - y^2$ is called a *lemniscate*. The curve is shown in the figure. Find the four points on the curve at which the tangent line is horizontal.

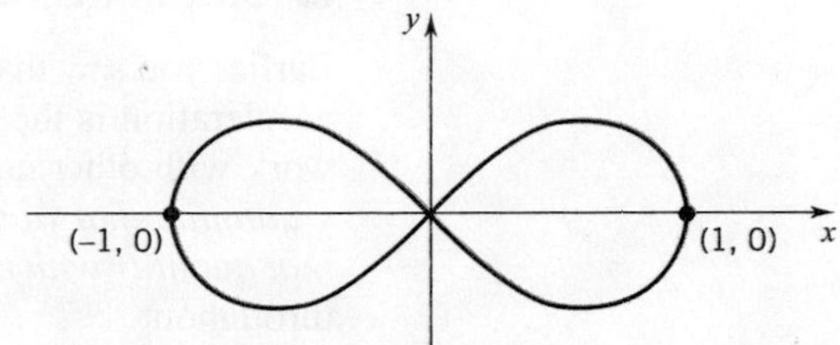

56. The curve defined by the equation $x^{2/3} + y^{2/3} = a^{2/3}$ is called an astroid. The curve is shown in the figure.

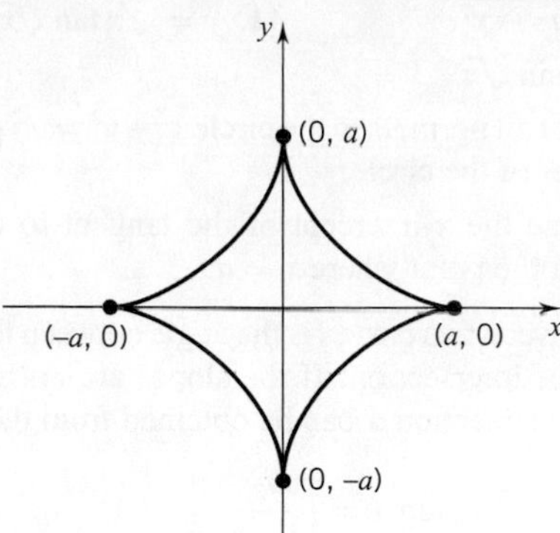

(a) Find the slope of the tangent line to the graph at an arbitrary point (x_1, y_1), $x_1 \neq 0, \pm a$.

(b) At what points on the curve is the slope of the tangent line 0, 1, or -1?

57. Prove that the sum of the x- and y-intercepts of any tangent line to the graph of

$$x^{1/2} + y^{1/2} = c^{1/2}$$

is constant and equal to c.

58. A circle of radius 1 with center on the y-axis is inscribed in the parabola $y = 2x^2$. See the figure. Find the points of intersection. HINT: The circle has equation $x^2 + (y - a)^2 = 1$, and the circle and parabola have the same tangent lines at the points of intersection.

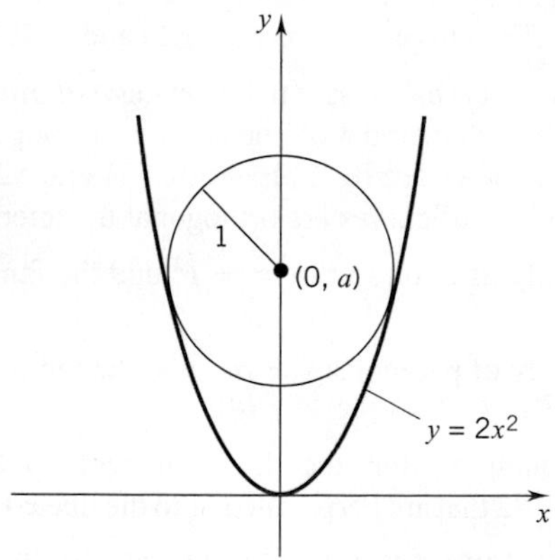

C **59.** Let $f(x) = 3\sqrt[3]{x}$. Use a CAS to:

(a) Find $d(h) = \dfrac{f(h) - f(0)}{h}$.

(b) Find $\lim_{h \to 0^-} d(h)$ and $\lim_{h \to 0^+} d(h)$.

(c) Is there a tangent line at (0, 0)? Explain.

(d) Use a graphing utility to draw the graph of f on $[-2, 2]$.

C **60.** Repeat Exercise 59 with $f(x) = 3\sqrt[3]{x^2}$.

C In Exercises 61 and 62, use a graphing utility to determine where:

(a) $f'(x) = 0$, (b) $f'(x) > 0$. (c) $f'(x) < 0$,

61. $f(x) = x\sqrt[3]{x^2 + 1}$.

62. $f(x) = \dfrac{x^2 + 1}{\sqrt{x}}$.

C **63.** A graphing utility in parametric mode can be used to graph some equations in x and y. Sketch the graph of the equation $x^2 + y^2 = 4$ by first setting $x = t, y = \sqrt{4 - t^2}$ and then setting $x = t, y = -\sqrt{4 - t^2}$.

C In Exercises 64–66, use a CAS to find the slope of the line tangent to the curve at the given point. Also, use a graphing utility to sketch the curve and the tangent line together.

64. $3x^2 + 4y^2 = 16$; $P(2, 1)$.

65. $4x^2 - y^2 = 20$; $P(3, 4)$.

66. $2\sin y - \cos x = 0$; $P(0, \frac{\pi}{6})$.

C **67.** (a) Use a graphing utility to sketch the curve $x^4 = x^2 - y^2$. This curve is called a *figure eight curve*.

(b) Find the x-coordinates of the points on the graph where the tangent line is horizontal.

C **68.** Use a graphing utility to sketch the curve $(2 - x)y^2 = x^3$. This curve is called a *cissoid*.

■ 3.8 RATES OF CHANGE PER UNIT TIME

Earlier you saw that velocity is the rate of change of position with respect to time and acceleration is the rate of change of velocity with respect to time. In this section we work with other quantities that vary with time. The fundamental point is this: *if Q is a quantity that varies with time, then the derivative dQ/dt gives the rate of change of that quantity with respect to time.* Implicit differentiation with respect to t will be used throughout.

Example 1 A spherical balloon is expanding. If the radius is increasing at the rate of 2 inches per minute, at what rate is the volume increasing when the radius is 5 inches?

SOLUTION

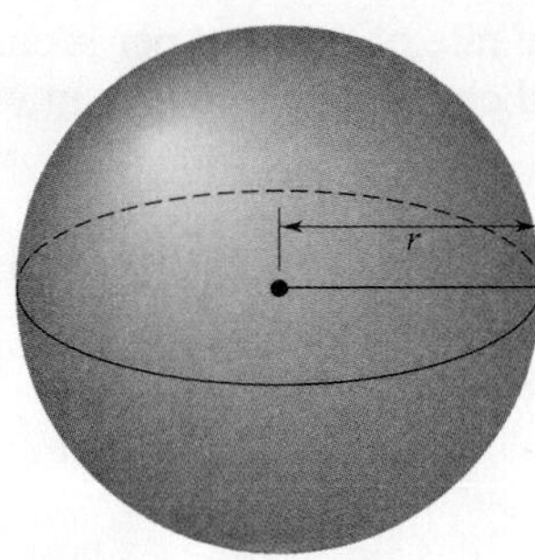

Find dV/dt when $r = 5$ inches, given that $dr/dt = 2$ in/min and

$$V = \tfrac{4}{3}\pi r^3. \quad \text{(volume of a sphere of radius } r\text{)}$$

Since r and V are functions of t, we differentiate the equation $V = \frac{4}{3}\pi r^3$ implicitly with respect to t. We have

$$\frac{dV}{dt} = 4\pi r^2 \frac{dr}{dt}.$$

Substituting $r = 5$ and $dr/dt = 2$, we find that

$$\frac{dV}{dt} = 4\pi(5)^2(2) = 200\pi.$$

When the radius is 5 inches, the volume is increasing at the rate of 200π cubic inches per minute. □

Example 2 An object is moving in the clockwise direction around the unit circle $x^2 + y^2 = 1$. As it passes through the point $(1/2, \sqrt{3}/2)$, its y-coordinate is decreasing at the rate of 3 units per second. At what rate is the x-coordinate changing at this point?

SOLUTION

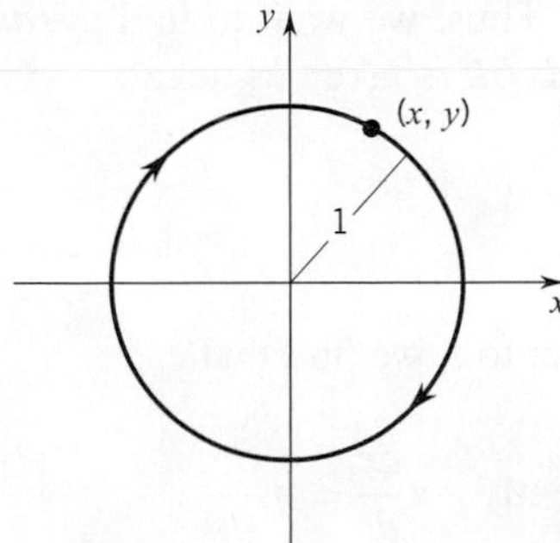

Find dx/dt when $x = \frac{1}{2}$ and $y = \frac{\sqrt{3}}{2}$, given that $dy/dt = -3$ units/sec and

$$x^2 + y^2 = 1. \quad \text{(equation of circle)}$$

Differentiating $x^2 + y^2 = 1$ implicitly with respect to t, we have

$$2x\frac{dx}{dt} + 2y\frac{dy}{dt} = 0 \quad \text{and thus} \quad x\frac{dx}{dt} + y\frac{dy}{dt} = 0.$$

Substitution of $x = \frac{1}{2}, y = \frac{\sqrt{3}}{2}$, and $dy/dt = -3$ gives

$$\frac{1}{2}\frac{dx}{dt} + \frac{\sqrt{3}}{2}(-3) = 0 \quad \text{so that} \quad \frac{dx}{dt} = 3\sqrt{3}.$$

As the object passes through the point $(\frac{1}{2}, \frac{1}{2}\sqrt{3})$, the x-coordinate is increasing at the rate of $3\sqrt{3}$ units per second. □

Example 3 A ladder 13 feet long is leaning against the side of a building. If the foot of the ladder is pulled away from the building at the rate of 0.1 foot per second, how fast is the angle formed by the ladder and the ground changing at the instant when the top of the ladder is 12 feet above the ground?

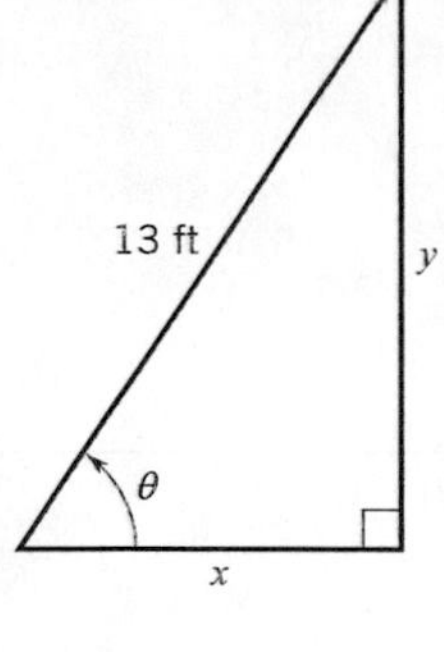

SOLUTION Find $d\theta/dt$ when $y = 12$ feet, given that $dx/dt = 0.1$ ft/ sec and

$$\cos\theta = \frac{x}{13}.$$

Implicit differentiation with respect to t gives

$$-\sin\theta\frac{d\theta}{dt} = \tfrac{1}{13}\frac{dx}{dt}.$$

When $y = 12, \sin\theta = \frac{12}{13}$. Substituting $\sin\theta = \frac{12}{13}$ and $dx/dt = 0.1$, we have

$$-\left(\frac{12}{13}\right)\frac{d\theta}{dt} = \frac{1}{13}(0.1) \quad \text{and thus} \quad \frac{d\theta}{dt} = -\frac{1}{120}.$$

At the instant when the top of the ladder is 12 feet above the ground, the angle formed by the ladder and the ground is decreasing at the rate of $\frac{1}{120}$ radians per second (about half a degree per second). □

Example 4 Two ships, one heading west and the other east, approach each other on parallel courses 8 nautical miles apart.† Given that each ship is cruising at 20 nautical miles per hour (knots), at what rate is the distance between them diminishing when they are 10 nautical miles apart?

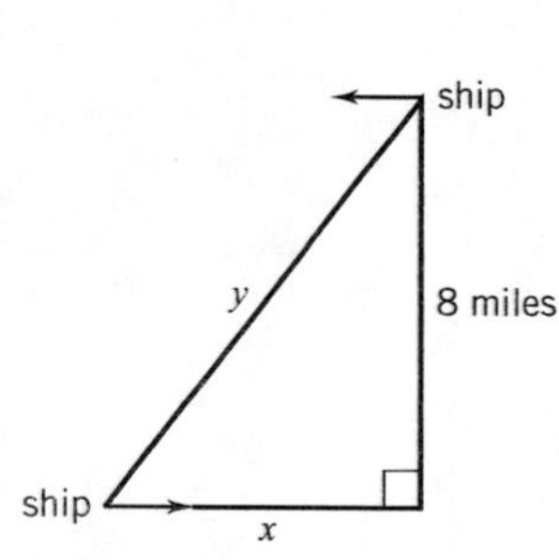

SOLUTION Let y be the distance between the ships measured in nautical miles. Since the ships are moving in opposite directions at the rate of 20 knots each, the distance x (see figure) is decreasing at the rate of 40 knots. Thus, we want to find dy/dt when $y = 10$, given that $dx/dt = -40$ knots (note that dx/dt is taken as negative since x is decreasing) and

$$x^2 + 8^2 = y^2. \qquad \text{(Pythagorean theorem)}$$

Differentiating $x^2 + 8^2 = y^2$ implicitly with respect to t, we find that

$$2x\frac{dx}{dt} + 0 = 2y\frac{dy}{dt} \quad \text{and consequently} \quad x\frac{dx}{dt} = y\frac{dy}{dt}.$$

When $y = 10, x = 6$. (Why?) Substituting $x = 6, y = 10$, and $dx/dt = -40$, we have

$$6(-40) = 10\frac{dy}{dt} \quad \text{so that} \quad \frac{dy}{dt} = -24.$$

(Note that dy/dt is negative since y is decreasing.) When the two ships are 10 miles apart, the distance between them is diminishing at the rate of 24 knots. □

† The international nautical mile measures 6080 feet.

The preceding sample problems were solved by the same general method, a method that we recommend to you for solving problems of this type:

Step 1. Draw a diagram where relevant, and indicate the quantities that vary.

Step 2. Specify in mathematical form the rate of change you are looking for, and record all given information.

Step 3. Find an equation involving the variable whose rate of change is to be found.

Step 4. Differentiate with respect to time t the equation found in step 3.

Step 5. State the final answer in coherent form, specifying the units that you are using.

Example 5 A conical paper cup 8 inches across the top and 6 inches deep is full of water. The cup springs a leak at the bottom and loses water at the rate of 2 cubic inches per minute. How fast is the water level dropping at the instant when the water is exactly 3 inches deep?

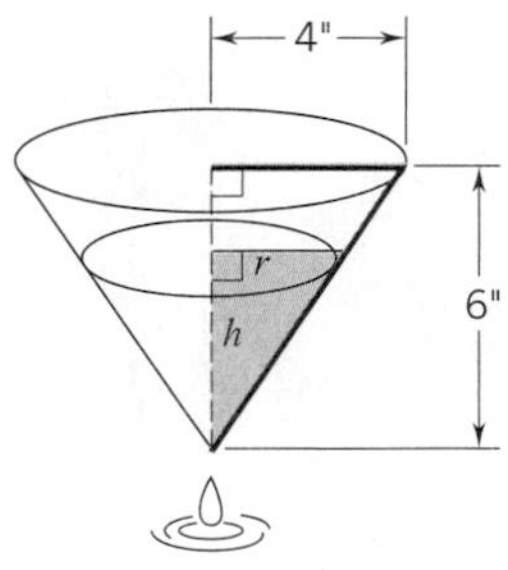

Figure 3.8.1

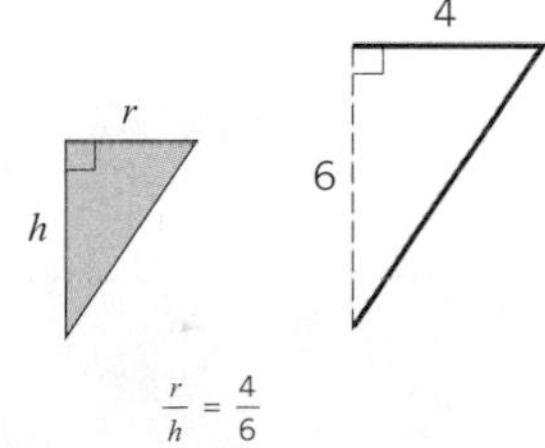

Figure 3.8.2

SOLUTION We begin with a diagram that represents the situation after the cup has been leaking for a while (Figure 3.8.1). We label the radius and height of the remaining "cone of water" r and h. We can relate r and h by similar triangles (Figure 3.8.2). Let h be the depth of the water measured in inches. Find dh/dt when $h = 3$, given that $dV/dt = -2\ \text{in}^3/\text{min}$,

$$V = \frac{1}{3}\pi r^2 h \quad \text{(volume of a cone)} \quad \text{and} \quad \frac{r}{h} = \frac{4}{6} = \frac{2}{3} \quad \text{(similar triangles).}$$

Using the second equation to eliminate r from the first equation, we have

$$V = \frac{1}{3}\pi\left(\frac{2h}{3}\right)^2 h = \frac{4}{27}\pi h^3.$$

Implicit differentiation with respect to t gives

$$\frac{dV}{dt} = \frac{4}{9}\pi h^2 \frac{dh}{dt}.$$

Substituting $h = 3$ and $dV/dt = -2$, we have

$$-2 = \frac{4}{9}\pi(3)^2 \frac{dh}{dt} \quad \text{and thus} \quad \frac{dh}{dt} = -\frac{1}{2\pi}.$$

At the instant when the water is exactly 3 inches deep, the water level is dropping at the rate of $1/2\pi$ inches per minute (about 0.16 inches per minute). □

Example 6 A balloon leaves the ground 500 feet away from an observer and rises vertically at the rate of 140 feet per minute. At what rate is the angle of inclination of the observer's line of sight increasing at the instant when the balloon is exactly 500 feet above the ground?

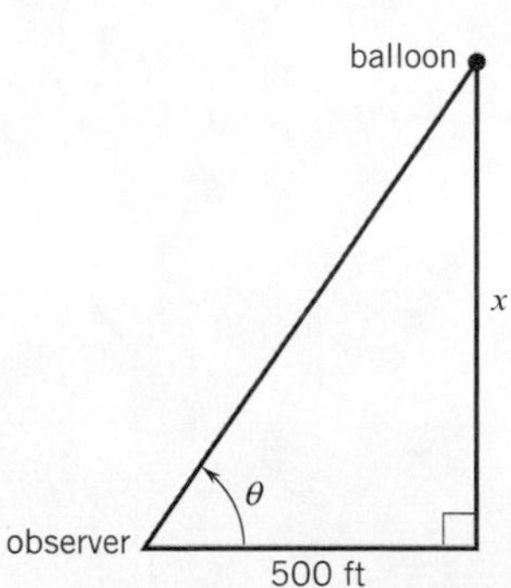

SOLUTION Let x be the altitude of the balloon and θ the angle of inclination of the observer's line of sight. Find $d\theta/dt$ when $x = 500$ feet, given that $dx/dt = 140\ \text{ft}/\text{min}$ and

$$\tan\theta = \frac{x}{500}.$$

Implicit differentiation with respect to t gives

$$\sec^2\theta\,\frac{d\theta}{dt} = \tfrac{1}{500}\,\frac{dx}{dt}.$$

When $x = 500$, the triangle is isosceles, which implies $\theta = \pi/4$ and $\sec\theta = \sqrt{2}$. Substituting $\sec\theta = \sqrt{2}$ and $dx/dt = 140$, we have

$$(\sqrt{2})^2\frac{d\theta}{dt} = \frac{1}{500}(140) \quad \text{and therefore} \quad \frac{d\theta}{dt} = 0.14.$$

At the instant when the baloon is exactly 500 feet above the ground, the inclination of the observer's line of sight is increasing at the rate of 0.14 radians per minute (about 8 degrees per minute). ❑

Example 7 A water trough with vertical cross section in the shape of an equilateral triangle is being filled at a rate of 4 cubic feet per minute. Given that the trough is 12 feet long, how fast is the level of the water rising at the instant the water reaches a depth of $1\frac{1}{2}$ feet?

SOLUTION Let x be the depth of the water in feet and V the volume of water in cubic feet. Find dx/dt when $x = 3/2$, given that $dV/dt = 4$ ft^3/ min and

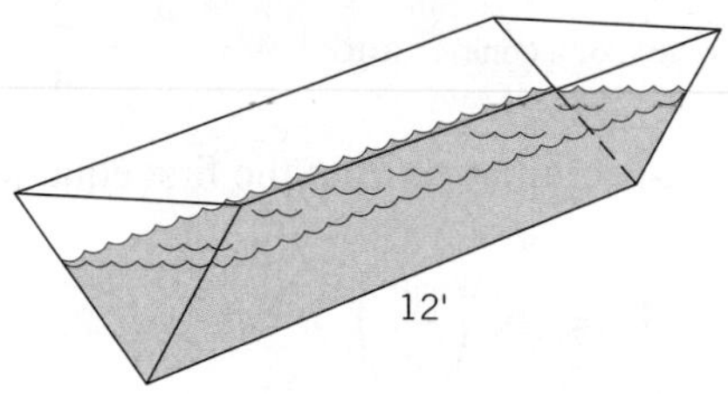

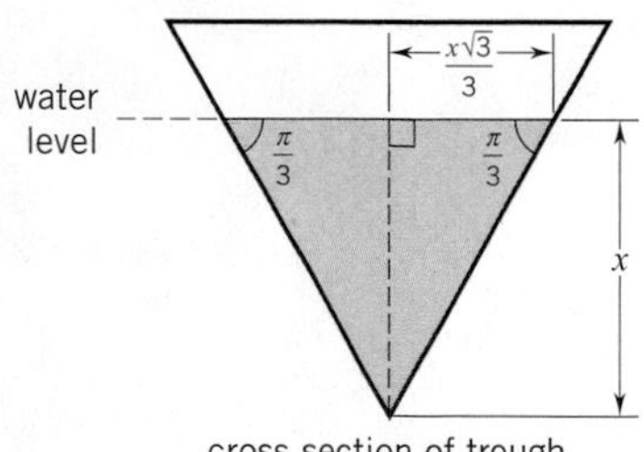

cross section of trough

$$\text{Area of cross section} = \frac{1}{2}\left(\frac{2x}{\sqrt{3}}\right)x = \frac{\sqrt{3}}{3}x^2,$$

$$\text{Volume of water} = 12\left(\frac{\sqrt{3}}{3}x^2\right) = 4\sqrt{3}x^2.$$

Implicit differentiation of $V = 4\sqrt{3}x^2$ with respect to t gives

$$\frac{dV}{dt} = 8\sqrt{3}\,x\,\frac{dx}{dt}.$$

Substituting $x = \frac{3}{2}$ and $dV/dt = 4$, we have

$$4 = 8\sqrt{3}\left(\tfrac{3}{2}\right)\frac{dx}{dt} \quad \text{and thus} \quad \frac{dx}{dt} = \frac{1}{3\sqrt{3}} = \frac{1}{9}\sqrt{3}.$$

At the instant that the water reaches a depth of $1\frac{1}{2}$ feet, the water level is rising at the rate of $\frac{1}{9}\sqrt{3}$ feet per second (about 0.19 feet per second). ❑

EXERCISES 3.8

1. A point moves along the straight line $x + 2y = 2$. Find (a) the rate of change of the y-coordinate, given that the x-coordinate is increasing at a rate of 4 units per second; (b) the rate of change of the x-coordinate, given that the y-coordinate is decreasing at a rate of 2 units per second.
2. A particle is moving in the circular orbit $x^2 + y^2 = 25$. As it passes through the point (3, 4), its y-coordinate is decreasing at the rate of 2 units per second. At what rate is the x-coordinate changing?
3. A particle is moving along the parabola $y^2 = 4(x+2)$. As it passes through the point (7, 6), its y-coordinate is increasing at the rate of 3 units per second. How fast is the x-coordinate changing at this instant?
4. A particle is moving along the parabola $4y = (x+2)^2$ in such a way that its x-coordinate is increasing at the constant rate of 2 units per second. How fast is the particle's distance to the point $(-2, 0)$ changing at the instant it is at the point (2, 4)?
5. A particle is moving along the ellipse $4x^2 + 16y^2 = 64$. At any time t its x- and y-coordinates are given by $x = 4\cos t$, $y = 2\sin t$. At what rate is the particle's distance to the origin changing at an arbitrary time t? At what rate is the distance to the origin changing when $t = \pi/4$?
6. A particle is moving along the curve $y = x\sqrt{x}, x \geq 0$. At time t its coordinates are $(x(t), y(t))$. Find the points on the curve, if any, at which both coordinates are changing at the same rate.
7. A heap of rubbish in the shape of a cube is being compacted into a smaller cube. Given that the volume decreases at a rate of 2 cubic meters per minute, find the rate of change of an edge of the cube when the volume is exactly 27 cubic meters. What is the rate of change of the surface area of the cube at that instant?
8. The volume of a spherical balloon is increasing at a constant rate of 8 cubic feet per minute. How fast is the radius increasing when the radius is exactly 10 feet? How fast is the surface area increasing at that instant?
9. An isosceles triangle has equal sides of length 10 centimeters. Let θ denote the angle included between the two equal sides.
   (a) Express the area A of the triangle as a function of θ (radians).
   (b) Suppose that θ is increasing at the rate of 10° per minute. How fast is A changing at the instant $\theta = \pi/3$?
   (c) At what value of θ will the triangle have maximum area?
10. At a certain instant the side of an equilateral triangle is α centimeters long and increasing at the rate of k centimeters per minute. How fast is the area increasing?
11. The perimeter of a rectangle is fixed at 24 centimeters. If the length l of the rectangle is increasing at the rate of 1 centimeter per second, for what value of l does the area of the rectangle start to decrease?
12. The dimensions of a rectangle are changing in such a way that the perimeter is always 24 inches. Show that, at the instant when the area is 32 square inches, the area is either increasing or decreasing 4 times as fast as the length is increasing.
13. A rectangle is inscribed in a circle of radius 5 inches. If the length of the rectangle is decreasing at the rate of 2 inches per second, how fast is the area changing at the instant when the length is 6 inches? HINT: A diagonal of the rectangle is a diameter of the circle.
14. A boat is held by a bow line that is wound about a bollard 6 feet higher than the bow of the boat. If the boat is drifting away at the rate of 8 feet per minute, how fast is the line unwinding when the bow is exactly 30 feet from the bollard?
15. Two boats are racing with constant speed toward a finish marker, boat A sailing from the south at 13 mph and boat B approaching from the east. When equidistant from the marker, the boats are 16 miles apart and the distance between them is decreasing at the rate of 17 mph. Which boat will win the race?
16. A spherical snowball is melting in such a manner that its radius is changing at a constant rate, decreasing from 16 cm to 10 cm in 30 minutes. How fast is the volume of the snowball changing at the instant the radius is 12 cm?
17. A 13-foot ladder is leaning against a vertical wall. If the bottom of the ladder is being pulled away from the wall at the rate of 2 feet per second, how fast is the area of the triangle formed by the wall, the ground, and the ladder changing at the instant the bottom of the ladder is 12 feet from the wall?
18. A ladder 13 feet long is leaning against a wall. If the foot of the ladder is pulled away from the wall at the rate of 0.5 foot per second, how fast will the top of the ladder be dropping at the instant when the base is 5 feet from the wall?
19. A tank contains 1000 cubic feet of natural gas at a pressure of 5 pounds per square inch. Find the rate of change of the volume if the pressure decreases at a rate of 0.05 pounds per square inch per hour. (Assume Boyle's law: *pressure* × *volume* = *constant*.)
20. The adiabatic law for the expansion of air is $PV^{1.4} = C$. At a given instant the volume is 10 cubic feet and the pressure is 50 pounds per square inch. At what rate is the pressure changing if the volume is decreasing at a rate of 1 cubic foot per second?
21. A man standing 3 feet from the base of a lamppost casts a shadow 4 feet long. If the man is 6 feet tall and walks away from the lamppost at a speed of 400 feet per minute, at what rate will his shadow lengthen? How fast is the tip of his shadow moving?
22. A light is attached to the wall of a building 64 feet above the ground. A ball is dropped from the same height, but 20 feet away from the side of the building. The height y of the ball at time t is given by $y(t) = 64 - 16t^2$. How fast is the shadow of the ball moving along the ground after 1 second?
23. An object that weighs 150 pounds on the surface of the earth will weigh $150\,(1+\frac{1}{4000}r)^{-2}$ pounds when it is r miles above the surface. Given that the altitude of the object is increasing

at the rate of 10 miles per second, how fast is the weight decreasing at the instant it is 400 miles above the surface?

24. In the special theory of relativity the mass of a particle moving at velocity v is

$$\frac{m}{\sqrt{1 - v^2/c^2}}$$

where m is the mass at rest and c is the speed of light. At what rate is the mass changing when the particle's velocity is $\frac{1}{2}c$ and the rate of change of the velocity is $0.01c$ per second?

25. Water is dripping through the bottom of a conical cup 4 inches across and 6 inches deep. Given that the cup loses half a cubic inch of water per minute, how fast is the water level dropping when the water is 3 inches deep?

26. Water is poured into a reservoir in the shape of a cone 6 feet tall with a radius of 4 feet. If the water level is rising at the constant rate of 0.5 feet per second, how fast is the water being poured in at the instant the depth is 2 feet?

27. At what rate is the volume of a sphere changing at the instant when the surface area is increasing at the rate of 4 square centimeters per minute and the radius is increasing at the rate of 0.1 centimeter per minute?

28. Water flows from a faucet into a hemispherical basin 14 inches in diameter at a rate of 2 cubic inches per second. How fast does the water rise (a) when the water is exactly halfway to the top? (b) just as it runs over? (The volume of a spherical segment is given by $\pi rh^2 - \frac{1}{3}\pi h^3$ where r is the radius of the sphere and h is the depth of the segment.)

29. The base of an isosceles triangle is 6 feet. If the altitude is 4 feet and increasing at the rate of 2 inches per minute, at what rate is the vertex angle changing?

30. As a boy winds up the cord, his kite is moving horizontally at a height of 60 feet with a speed of 10 feet per minute. How fast is the inclination of the cord changing when its length is 100 feet?

31. A revolving searchlight $\frac{1}{2}$ mile from shore makes 1 revolution per minute. How fast is the light traveling along the straight beach at the instant it passes over a shorepoint 1 mile away from the shorepoint nearest to the searchlight?

32. A revolving searchlight on an island 1 mile from shore turns at the rate of 2 revolutions per minute in the counterclockwise direction.

(a) How fast is the light beam traveling along the straight beach at the instant it makes an angle of 45° with the shore?

(b) How fast is the light beam traveling at the instant the angle is 90°?

33. A man starts at a point A and walks 40 feet north. He then turns and walks due east at 4 feet per second. If a searchlight placed at A follows him, at what rate is the light turning 15 seconds after he started walking east?

34. The diameter and height of a right circular cylinder are found at a certain instant to be 10 centimeters and 20 centimeters, respectively. If the diameter is increasing at the rate of 1 centimeter per second, what change in height will keep the volume constant?

35. A horizontal trough 12 feet long has a vertical cross section in the shape of a trapezoid. The bottom is 3 feet wide, and the sides are inclined to the vertical at an angle whose sine is $\frac{4}{5}$. Given that water is poured into the trough at the rate of 10 cubic feet per minute, how fast is the water level rising at the instant the water is exactly 2 feet deep?

36. Two cars, car A traveling east at 30 mph and car B traveling north at 22.5 mph, are heading toward an intersection I. At what rate is the angle IAB changing at the instant when cars A and B are 300 feet and 400 feet, respectively, from the intersection?

37. A rope 32 feet long is attached to a weight and passed over a pulley 16 feet above the ground. The other end of the rope is pulled away along the ground at the rate of 3 feet per second. At what rate is the angle between the rope and the ground changing at the instant the weight is exactly 4 feet off the ground?

38. A slingshot is made by fastening the two ends of a 10-inch rubber strip 6 inches apart. If the midpoint of the strip is drawn back at the rate of 1 inch per second, at what rate is the angle between the segments of the strip changing 8 seconds later?

39. A balloon is released 500 feet away from an observer. If the balloon rises vertically at the rate of 100 feet per minute and at the same time the wind is carrying it horizontally away from the observer at the rate of 75 feet per minute, at what rate is the angle of inclination of the observer's line of sight changing 6 minutes after the balloon has been released?

40. A searchlight is trained on a plane that flies directly above the light at an altitude of 2 miles and a speed of 400 miles per hour. How fast must the light be turning 2 seconds after the plane passes directly overhead?

C **41.** A baseball diamond is a square 90 feet on a side. A player is running from second base to third base at the rate of 15 feet per second. Find the rate of change of the distance from the player to home plate at the instant he is 10 feet from third base.

C **42.** An airplane is flying at a constant speed and altitude on a line that will take it directly over a radar station on the ground. At the instant the plane is 12 miles from the station, it is noted that the plane's angle of elevation is 30° and is increasing at the rate of 0.5° per second. Find the speed of the plane in miles per hour.

C In Exercises 43–45, the variables x and y are functions of t and are related by the given formula. Use a CAS to find the specified derivative.

43. $2xy^2 - y = 22$. Given that $\dfrac{dy}{dt} = -2$ when $x = 3$ and $y = 2$, find $\dfrac{dx}{dt}$.

44. $x - \sqrt{xy} = 4$. Given that $\dfrac{dy}{dt} = 3$ when $x = 8$ and $y = 2$, find $\dfrac{dy}{dt}$.

45. $\sin x = 4\cos y - 1$. Given that $\dfrac{dx}{dt} = -1$ when $x = \pi$ and $y = \dfrac{\pi}{3}$, find $\dfrac{dy}{dt}$.

■ PROJECT 3.8 Angular Velocity; Uniform Circular Motion

As a particle moves along a circle of radius r, it effects a change in the central angle (measured in radians) marked θ in Figure A. The *angular velocity*, ω, of the particle is the time rate of change of θ, that is, $\omega = d\theta/dt$. Circular motion with constant, positive angular velocity is called *uniform circular motion*.

Problem 1. A particle in uniform circular motion traces out a circular arc. The time rate of change of the length of that arc is called the *speed* of the particle. What is the speed of a particle that moves around a circle of radius r with constant, positive angular velocity ω?

Problem 2. The *kinetic energy*, KE, of a particle of mass m is given by

$$KE = \tfrac{1}{2}mv^2$$

where v is the speed of the particle. Suppose the particle in Problem 1 has mass m. What is the kinetic energy of the particle?

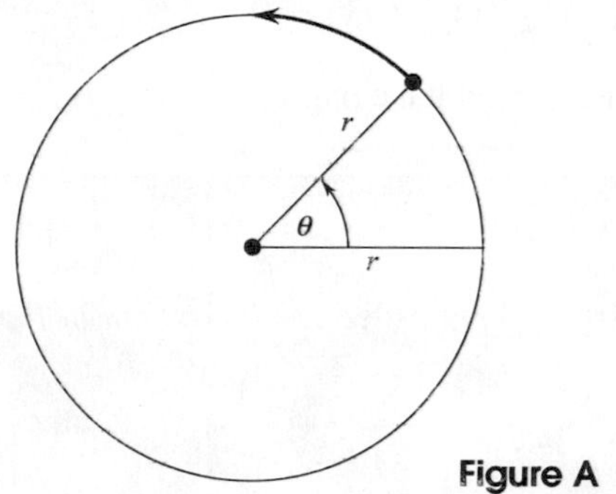

Figure A

Problem 3. A point P moves uniformly along the circle $x^2 + y^2 = r^2$ with constant angular velocity ω. Find the x- and y-coordinates of P at time t given that the motion starts at time $t = 0$ with $\theta = \theta_0$. Then find the velocity and acceleration of the projection of P onto the x-axis and onto the y-axis. [The projection of P onto the x-axis is the point $(x, 0)$; the projection of P onto the y-axis is the point $(0, y)$.]

Problem 4. Figure B shows a sector in a circle of radius r. The sector is the union of the triangle T and the segment S. Suppose that the radius vector rotates counterclockwise with a constant angular velocity of ω radians per second. Show that the area of the sector changes at a constant rate but that the area of T and the area of S do not change at a constant rate.

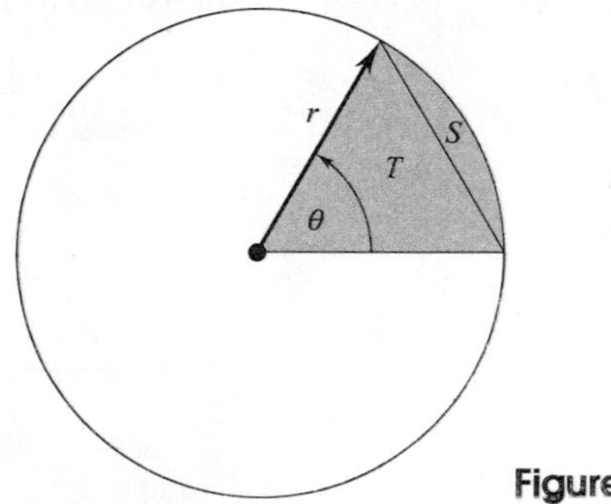

Figure B

Problem 5. Take S and T as in Problem 4. While the area of S and the area of T change at different rates, there is one value of θ between 0 and π at which both areas have the same instantaneous rate of change. Find this value of θ.

■ 3.9 DIFFERENTIALS; NEWTON-RAPHSON APPROXIMATIONS

Differentials

Let f be a differentiable function. In Figure 3.9.1 on page 188, you can see the graph of f and below it the tangent line at the point $(x, f(x))$. As the figure suggests, for small h, $h \neq 0$, $f(x+h) - f(x)$, the change in f from x to $x + h$, can be approximated by the product $f'(x)h$:

(3.9.1)
$$f(x+h) - f(x) \cong f'(x)h.$$

How good is this approximation? It is good in the sense that, for small h, the difference between the two quantities,

$$[f(x+h) - f(x)] - f'(x)h,$$

is small compared to h. How small? Small enough that the ratio

$$\frac{[f(x+h) - f(x)] - f'(x)h}{h}$$

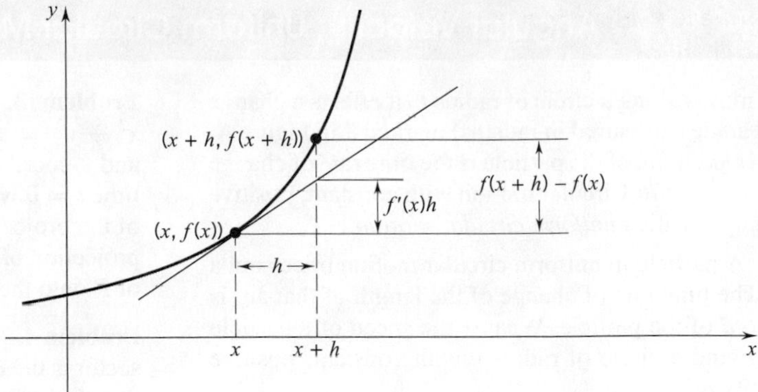

Figure 3.9.1

tends to 0 as h tends to 0:

$$\lim_{h\to 0}\frac{[f(x+h)-f(x)]-f'(x)h}{h} = \lim_{h\to 0}\frac{f(x+h)-f(x)}{h} - \lim_{h\to 0}\frac{f'(x)h}{h}$$

$$= f'(x) - f'(x) = 0.$$

The quantities $f(x+h)-f(x)$ and $f'(x)h$ have names:

> **DEFINITION 3.9.2**
>
> Let $h \neq 0$. The difference $f(x+h)-f(x)$ is called the *increment of f from x to x + h* and is denoted by Δf:
>
> $$\Delta f = f(x+h) - f(x).\dagger$$
>
> The product $f'(x)h$ is called the *differential of f at x with increment h* and is denoted by df:
>
> $$df = f'(x)h.$$

Display (3.9.1) says that, for small h, Δf and df are approximately equal:

$$\boxed{\Delta f \cong df.}$$

How approximately equal are they? Enough so that the ratio

$$\frac{\Delta f - df}{h}$$

tends to 0 as h tends to 0.

Let's see what all this amounts to in a very simple case. The area of a square of side x is given by the function

$$f(x) = x^2, \quad x > 0.$$

† The symbol Δ is the capital of the Greek letter δ, which corresponds to the English letter d; Δf is read "delta f."

If the length of each side is increased from x to $x + h$, the area increases from $f(x)$ to $f(x + h)$. The change in area is the increment Δf:

$$\begin{aligned}\Delta f &= f(x+h) - f(x)\\ &= (x+h)^2 - x^2\\ &= (x^2 + 2xh + h^2) - x^2\\ &= 2xh + h^2.\end{aligned}$$

As an estimate for this change, we can use the differential

$$df = f'(x)h = 2xh. \qquad \text{(Figure 3.9.2)}$$

The error of this estimate, the difference between the actual change Δf and the estimated change df, is the difference

$$\Delta f - df = h^2.$$

As promised, the error is small compared to h in the sense that

$$\frac{\Delta f - df}{h} = \frac{h^2}{h} = h$$

tends to 0 as h tends to 0.

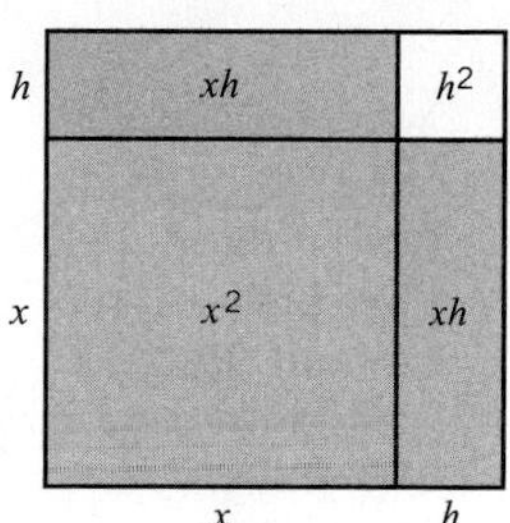

Figure 3.9.2

Example 1 Use a differential to estimate the change in $f(x) = x^{2/5}$ if:

(a) x is increased from 32 to 34. (b) x is decreased from 1 to $\frac{9}{10}$.

SOLUTION Since $f'(x) = \frac{2}{5}x^{-3/5} = 2/(5x^{3/5})$, we have

$$df = f'(x)h = \frac{2}{5x^{3/5}}h.$$

For part (a) we set $x = 32$ and $h = 2$. The differential then becomes

$$df = \frac{2}{5(32)^{3/5}}(2) = \frac{4}{40} = 0.1.$$

A change in x from 32 to 34 increases the value of f by approximately 0.1. Checking this result with a hand calculator, we have

$$\Delta f = f(34) - f(32) = (34)^{2/5} - (32)^{2/5} \cong 4.0982 - 4 = 0.0982.$$

For part (b) we set $x = 1$ and $h = -\frac{1}{10}$. In this case, the differential is

$$df = \frac{2}{5(1)^{3/5}}\left(-\frac{1}{10}\right) = -\frac{2}{50} = -0.04.$$

A change in x from 1 to $\frac{9}{10}$ decreases the value of f by approximately 0.04. Checking this result with a hand calculator, we have

$$\Delta f = f(0.9) - f(1) = f(0.9)^{2/5} - (1)^{2/5} \cong 0.9587 - 1 = -0.0413. \quad \square$$

Example 2 Use a differential to estimate:
(a) $\sqrt{104}$ (b) $\cos 40°$

SOLUTION
(a) We know $\sqrt{100} = 10$. What we need is an estimate for the increase of

$$f(x) = \sqrt{x}$$

when x increases from 100 to 104. Here,

$$f'(x) = \frac{1}{2\sqrt{x}} \quad \text{and} \quad df = f'(x)h = \frac{h}{2\sqrt{x}}.$$

With $x = 100$ and $h = 4$, df becomes

$$\frac{4}{2\sqrt{100}} = \frac{1}{5} = 0.2$$

A change in x from 100 to 104 increases the value of the square root by approximately 0.2. It follows that

$$\sqrt{104} \cong \sqrt{100} + 0.2 = 10 + 0.2 = 10.2.$$

As you can check, $(10.2)^2 = 104.04$, so the estimate is not far off.

(b) Let $f(x) = \cos x$. We know that $\cos 45° = \cos\frac{\pi}{4} = \sqrt{2}/2$. Converting 40° to radians, we have

$$40° = 45° - 5° = \frac{\pi}{4} - \left(\frac{\pi}{180}\right)5 = \frac{\pi}{4} - \frac{\pi}{36} \text{ radians.}$$

We use a differential to estimate the change in $\cos x$ when x decreases from $\pi/4$ to $(\pi/4) - (\pi/36)$:

$$f'(x) = -\sin x \quad \text{and} \quad df = f'(x)h = -h\sin x.$$

With $x = \pi/4$ and $h = -\pi/36$, df is given by

$$df = -\left(-\frac{\pi}{36}\right)\sin\left(\frac{\pi}{4}\right) = \frac{\pi}{36}\frac{\sqrt{2}}{2} = \frac{\pi\sqrt{2}}{72} \cong 0.0617.$$

A decrease in x from $\pi/4$ to $(\pi/4) - (\pi/36)$ increases the value of the cosine function by approximately 0.0617. Therefore,

$$\cos 40° \cong \cos 45° + 0.0617 \cong 0.7071 + 0.0617 = 0.7688$$

Checking this result with a hand calculator, we find that $\cos 40° \cong 0.7660$. □

Example 3 A metal sphere with a radius of 10 cm is to be covered with a 0.02-cm coating of silver. Approximately how much silver will be required?

SOLUTION We will use a differential to estimate the increase in the volume of the sphere when the radius is increased from 10 cm to 10.02 cm. The formula for the volume of a sphere of radius r is:

$$V = \tfrac{4}{3}\pi r^3 \quad \text{and so} \quad dV = 4\pi r^2 h.$$

Now, with $r = 10$ and $h = 0.02$, we have

$$dV = 4\pi(10)^2(0.02) = 8\pi \cong 25.133.$$

Thus, it will take approximately 25.133 cubic cm of silver to coat the sphere. □

Newton-Raphson Method

Figure 3.9.3 shows the graph of a function f. Since the graph of f crosses the x-axis at $x = c$, the number c is a solution (root) of the equation $f(x) = 0$. In the setup of Figure 3.9.3, we can approximate c as follows: Start at a point x_1(see the figure). The tangent line at $(x_1,f(x_1))$ intersects the x-axis at a point x_2, which is closer to c than x_1. The tangent line at $(x_2,f(x_2))$ intersects the x-axis at a point x_3, which is closer to c than x_2. Continuing in this manner, we will obtain better and better approximations $x_4, x_5, \cdots, x_n$ to the root c.

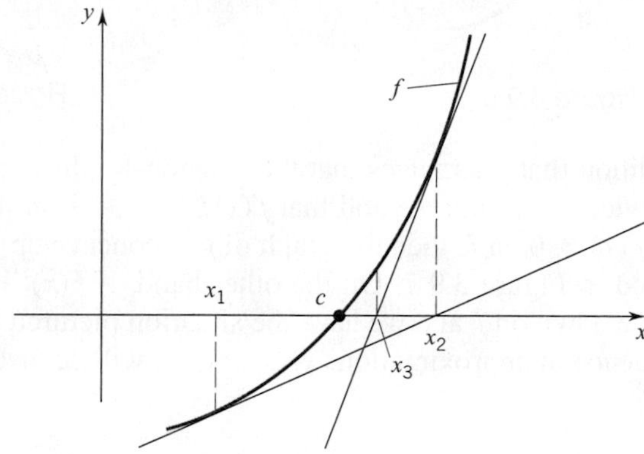

Figure 3.9.3

There is an algebraic connection between x_n and x_{n+1} that we now develop. The tangent line at $(x_n,f(x_n))$ has the equation

$$y - f(x_n) = f'(x_n)(x - x_n).$$

The x-intercept of this line, x_{n+1}, can be found by setting $y = 0$:

$$0 - f(x_n) = f'(x_n)(x_{n+1} - x_n).$$

Solving this equation for x_{n+1}, we have

(3.9.3)
$$\boxed{x_{n+1} = x_n - \frac{f(x_n)}{f'(x_n)}.}$$

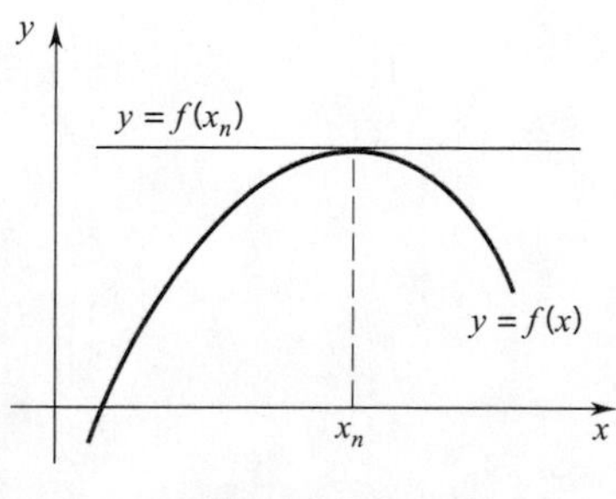

Figure 3.9.4

This method for locating a root of an equation $f(x) = 0$ is called the *Newton-Raphson method*. The method does not work in all cases. First, there are some conditions that must be placed on the function f. Clearly, f must be differentiable on some interval that contains the root c. Also, note that if $f'(x_n) = 0$ for some n, then the tangent line at $(x_n,f(x_n))$ is horizontal and the next approximation x_{n+1} cannot be calculated. See Figure 3.9.4. Thus, we assume that $f'(x) \neq 0$ in some interval containing c.

The method can also fail if proper care is not taken in choosing the first approximation x_1. For example, it can happen that the first approximation x_1 produces a second approximation x_2, which, in turn, gives the same x_1 as the third approximation, and so on—the approximations simply alternate between x_1 and x_2. See Figure 3.9.5. Another type of difficulty can arise if $f'(x_1)$ is close to zero. In this case, the second approximation x_2 could be worse than x_1, the third approximation x_3 could be worse than x_2, and so forth. See Figure 3.9.6.

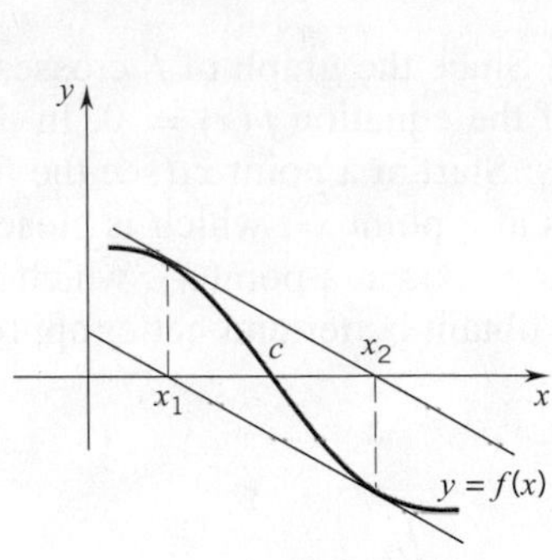

Figure 3.9.5

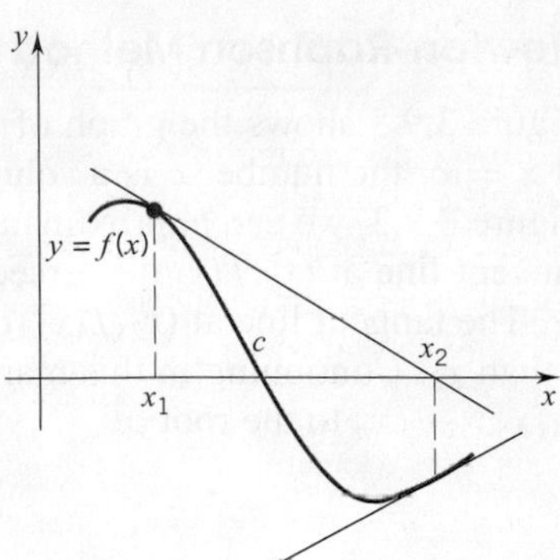

Figure 3.9.6

There is a condition that guarantees that the Newton-Raphson method will work. Suppose that f is twice differentiable and that $f(x)f''(x) > 0$ on the open interval I joining c and x_1. If $f(x) > 0$ on I, then the graph of f is concave up† on I and we have the situation pictured in Figure 3.9.7. On the other hand, if $f(x) < 0$ on I, then the graph of f is concave down on I and we have the situation pictured in Figure 3.9.8. In either case, the sequence of approximations $x_1, x_2, x_3, \ldots$ will "converge" to the root c.

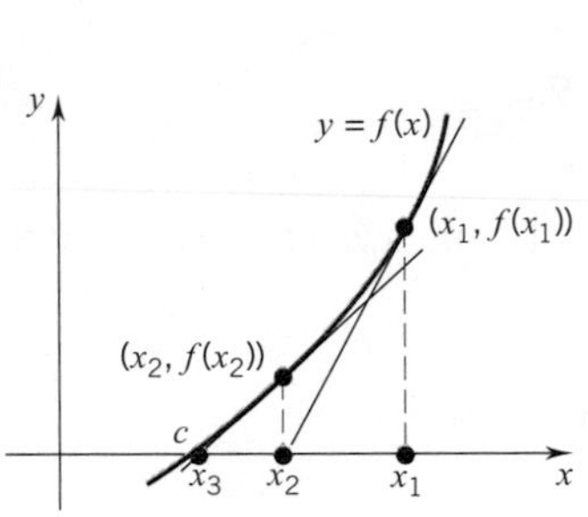

Figure 3.9.7

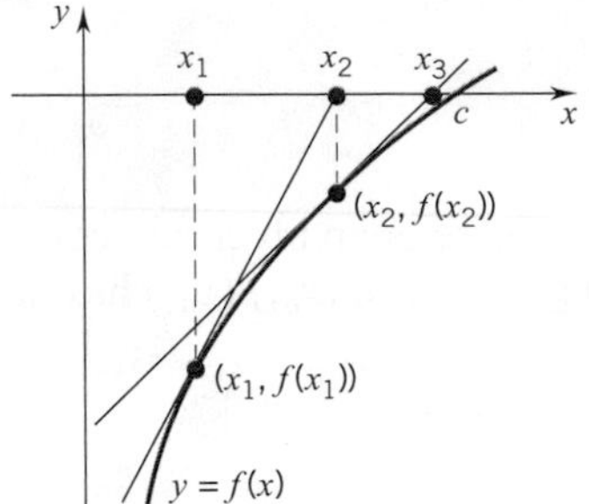

Figure 3.9.8

Example 4 The number $\sqrt{3}$ is a root of the equation $x^2 - 3 = 0$. We will estimate $\sqrt{3}$ by applying the Newton-Raphson method to the function $f(x) = x^2 - 3$ starting at $x_1 = 2$. [As you can check, $f(x)f''(x) > 0$ on $(\sqrt{3},\ 2)$ and therefore we can be sure that the method applies.] Since $f'(x) = 2x$, the Newton-Raphson formula gives

$$x_{n+1} = x_n - \left(\frac{x_n^2 - 3}{2x_n}\right) = \frac{x_n^2 + 3}{2x_n}.$$

Successive calculations with this formula (using a calculator) are given in the following table:

n	x_n	$x_{n+1} = \frac{x_n^2+3}{2x_n}$
1	2	1.75000
2	1.75000	1.73214
3	1.73214	1.73205

Since $(1.73205)^2 \cong 2.999997$, the method has generated a very accurate estimate of $\sqrt{3}$ in only three steps. ☐

† The concavity of a graph and the role of the second derivative are treated in Chapter 4

EXERCISES 3.9

1. Use a differential to estimate the change in the volume of a cube caused by an increase h in the length of each side. Interpret geometrically the error of your estimate $\Delta V - dV$.

2. Use a differential to estimate the area of a ring of inner radius r and width h. What is the exact area?

Use differentials to estimate the value of the indicated expression. Then compare your estimate with the result given by a calculator.

3. $\sqrt[3]{1002}$.
4. $1/\sqrt{24.5}$.
5. $\sqrt[4]{15.5}$.
6. $(26)^{2/3}$.
7. $(33)^{3/5}$.
8. $(33)^{-1/5}$.

Use a differential to estimate the value of the expression. (Remember to convert to radian measure.) Compare your estimate with the result given by a calculator.

9. $\sin 46°$.
10. $\cos 62°$.
11. $\tan 28°$.
12. $\sin 43°$.

13. Estimate $f(2.8)$ given that $f(3) = 2$ and $f'(x) = (x^3+5)^{1/5}$.

14. Estimate $f(5.4)$ given that $f(5) = 1$ and $f'(x) = \sqrt[3]{x^2+2}$.

15. Find the approximate volume of a thin cylindrical sheet with open ends given that the inner radius is r, the height is h, and the thickness is t.

16. The diameter of a steel ball is measured to be 16 centimeters, with a maximum error of 0.3 centimeters. Estimate by differentials the maximum error (a) in the surface area when calculated by the formula $S = 4\pi r^2$; (b) in the volume when calculated by the formula $V = \frac{4}{3}\pi r^3$.

17. A box is to be constructed in the form of a cube to hold 1000 cubic feet. Use a differential to estimate how accurately the inner edge must be made so that the volume will be correct to within 3 cubic feet.

18. Use differentials to estimate the values of x for which

    (a) $\sqrt{x+1} - \sqrt{x} < 0.01$.

    (b) $\sqrt[4]{x+1} - \sqrt[4]{x} < 0.002$.

19. A hemispherical dome with radius 50 feet will be given a coat of paint 0.01 inch thick. The contractor for the job wants to estimate the number of gallons of paint that will be needed. Use differentials to obtain an estimate (there are 231 cubic inches in a gallon). HINT: Approximate the change in the volume of the hemisphere corresponding to an increase of 0.01 inch in the radius.

20. Assume that the earth is a sphere of radius 4000 miles. The volume of ice at the north and south poles is estimated to be 8 million cubic miles. Suppose that this ice melts and the water produced distributes uniformly over the surface of the earth. Estimate the depth of the added water at any point on the earth.

21. For oscillations of small amplitude, the relationship between the period P of one complete oscillation and the length L of a simple pendulum (see Exercise 74, Section 3.6) is given by the equation

$$P = 2\pi\sqrt{\frac{L}{g}}$$

where g is the (constant) acceleration of gravity. Show that a small change dL in the length of a pendulum produces a change dP in the period that satisfies

$$\frac{dP}{P} = \frac{1}{2}\frac{dL}{L}.$$

22. Suppose that the pendulum in a pendulum clock has length 90 centimeters. Use the result in Exercise 21 to determine how the length of the pendulum should be adjusted if the clock is losing 15 seconds per hour.

23. A pendulum has length 3.26 feet and goes through one complete oscillation in 2 seconds. Use Exercise 21 to find the approximate change in P if the pendulum is lengthened by 0.01 foot.

24. As a metal cube is heated, the length of each edge increases $\frac{1}{10}$% per degree increase in temperature. Show by differentials that the surface area increases about $\frac{2}{10}$% per degree and the volume increases about $\frac{3}{10}$% per degree.

25. We are trying to determine the area of a circle by measuring the diameter. How accurately must we measure the diameter if our estimate is to be correct to within 1%?

26. Estimate by differentials how precisely x must be determined (a) if x^n is to be accurate to within 1%; (b) if $x^{1/n}$ is to be accurate to within 1%.

In Exercises 27–32, use the Newton-Raphson method to estimate a root of the equation $f(x) = 0$ starting at the indicated value of x: (a) Express x_{n+1} in terms of x_n. (b) Give x_4 rounded off to five decimal places and evaluate f at that approximation.

27. $f(x) = x^2 - 24; \quad x_1 = 5$.

28. $f(x) = x^3 - 4x + 1; \quad x_1 = 2$.

29. $f(x) = x^3 - 25; \quad x_1 = 3$.

30. $f(x) = x^5 - 30; \quad x_1 = 2$.

31. $f(x) = \cos x - x; \quad x_1 = 1$.

32. $f(x) = \sin x - x^2; \quad x_1 = 1$.

33. The function $f(x) = x^{1/3}$ is 0 at $x = 0$. Show that the Newton-Raphson method applied to f starting at *any* value $x_1 \neq 0$ fails to generate values that approach the solution $x = 0$. Describe the sequence $x_1, x_2, x_3, \ldots$ that the method generates.

34. Let $f(x) = 2x^3 - 3x^2 - 1$.

    (a) Prove that the equation $f(x) = 0$ has a root in the interval $(1, 2)$. HINT: Use the intermediate-value theorem.

    (b) Show that the Newton-Raphson method with $x_1 = 1$ fails to generate values that approach the root in $(1, 2)$.

(c) Estimate the root by starting at $x_1 = 2$. Determine x_4 rounded off to five decimal places, and evaluate $f(x_4)$.

35. The function $f(x) = x^4 - 2x^2 - \frac{17}{16}$ has two zeros, one at a, where $0 < a < 2$, and the other at $-a$ since f is an even function.

(a) Show that the Newton-Raphson method fails if $x_1 = \frac{1}{2}$ is used as the initial estimate for finding the root a. Describe the sequence $x_1, x_2, x_3, \ldots$ that the method generates.

(b) Estimate the root a by starting at $x_1 = 2$. Determine x_4 rounded off to five decimal places, and evaluate $f(x_4)$.

36. Let $f(x) = x^2 - a$, where $a > 0$. The roots of the equation $f(x) = 0$ are $\pm\sqrt{a}$.

(a) Show that if $x_1 > 0$ is any initial estimate for $\sqrt{a}$, then the Newton-Raphson method gives the iteration formula

$$x_{n-1} = \tfrac{1}{2}\left(x_n + \frac{a}{x_n}\right), \qquad n \geq 1.$$

(b) Let $a = 5$. Beginning with $x_1 = 2$, use the formula in part (a) to calculate x_4 rounded off to five decimal places and evaluate $f(x_4)$.

37. Let $f(x) = x^k - a$, where k is a positive integer and $a > 0$. The number $a^{1/k}$ is a root of the equation $f(x) = 0$.

(a) Show that if $x_1 > 0$ is any initial estimate for $a^{1/k}$, then the Newton-Raphson method gives the iteration formula

$$x_{n-1} = \frac{1}{k}\left[(k-1)x_n + \frac{a}{x_n^{k-1}}\right].$$

Note that for $k = 2$, this formula reduces to the formula given in part (a) of Exercise 36.

(b) Use the formula in part (a) to approximate $\sqrt[3]{23}$. Begin with $x_1 = 3$ and calculate x_4 rounded off to five decimal places. Evaluate $f(x_4)$.

38. Let $f(x) = \dfrac{1}{x} - a, a \neq 0$.

a. Apply the Newton-Raphson method to derive the iteration formula

$$x_{n+1} = 2x_n - ax_n^2, \qquad n \geq 1.$$

Note that this algorithm provides a method for calculating reciprocals without dividing.

b. Use the formula in part(a) to calculate 1/2.7153 rounded off to five decimal places.

Let g be a function defined at least on some open interval containing the number 0. We say that g is of *smaller order than h*, or that $g(h)$ is *little-o*(h) and write $g(h) = o(h)$ iff $g(h)$ is small enough compared with h that

$$\lim_{h\to 0} \frac{g(h)}{h} = 0, \qquad \text{or equivalently,} \qquad \lim_{h\to 0} \frac{g(h)}{|h|} = 0.$$

39. Which of the following statements is true?

(a) $h^3 = o(h)$. (b) $\dfrac{h^2}{h-1} = o(h)$. (c) $h^{1/3} = o(h)$.

40. Show that, if $g(h) = o(h)$, then $\lim_{h\to 0} g(h) = 0$.

41. Show that, if $g_1(h) = o(h)$ and $g_2(h) = o(h)$, then

$$g_1(h) + g_2(h) = o(h) \qquad \text{and} \qquad g_1(h)g_2(h) = o(h).$$

42. The figure shows the graph of a differentiable function f and a line with slope m that passes through the point $(x, f(x))$. The vertical separation at $x + h$ between the line with slope m and the graph of f has been labeled g (h).

(a) Calculate g (h)

(b) Show that, of all lines that pass through $(x, f(x))$, the tangent line is the line that best approximates the graph of f near the point $(x, f(x))$ by showing that

$$g(h) = o(h) \qquad \text{iff} \qquad m = f'(x).$$

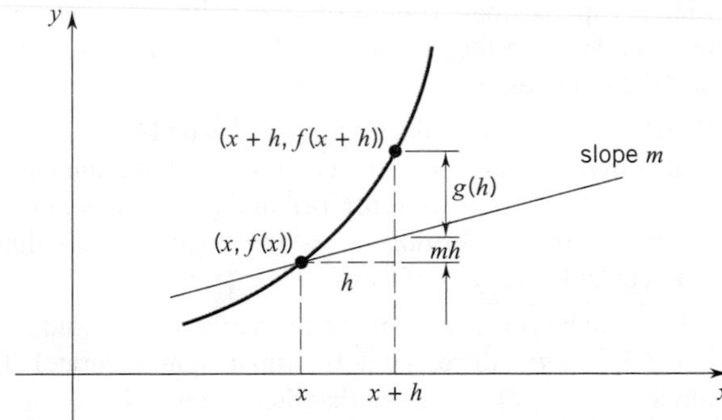

43. Let $f(x) = \cos x$ and $g(x) = x^2$.

(a) Use a graphing utility to draw their graphs together.

(b) Is the number of intersection points finite or infinite?

(c) Use the Newton-Raphson method to find a nonzero x value of an intersection point.

(d) If possible, find the x values of the remaining intersection points without using the Newton-Raphson method.

(e) If not possible, find one more x value for a remaining intersection point using the Newton-Raphson method.

44. Let $f(x) = \cos x$ and $g(x) = 2 \sin 2x$. Repeat Exercise 43.

45. Let $f(x) = \tan x$ and $g(x) = x$. Repeat Exercise 43.

■ PROJECT 3.9 Fixed Points

Suppose that the function f is defined on the interval I. A number $c \in I$ is a *fixed point* of f if $f(c) = c$. That is, c is a fixed point of f if f maps c to itself. This concept has a simple geometric interpretation: c is a fixed point of f if the graph of f intersects the line $y = x$ at the point where $x = c$. We can conclude from this that a given function f may or may not have a fixed point. For example, the function $f(x) = x^2 + 1$ does not have a fixed point, and the function $g(x) = x^2 - 2$ has two fixed points. (See the figures.)

Problem 1. Let f be a continuous function on the interval $[a, b]$, and suppose that $a \le f(x) \le b$ for all $x \in [a, b]$; that is, suppose that f maps $[a, b]$ into $[a, b]$. Prove that f has at least one fixed point $c \in [a, b]$. [HINT: Let $F(x) = f(x) - x$ and use the intermediate-value theorem.]

If f is differentiable, then the Newton-Raphson method can be applied to the function $F(x) = f(x) - x$ to approximate the fixed points of f, if any.

Problem 2. In each of the following, you are given a function f and an interval $[a, b]$. First verify that f has at least one fixed point in $[a, b]$, and then use the Newton-Raphson method to approximate the fixed point(s) with four decimal place accuracy.

a. $f(x) = 2x^3 - 4x - 3$; $[1, 2]$ (begin with $x_1 = 2$).
b. $f(x) = \frac{1}{2}\cos x$; $[0, \pi/2]$ (begin with $x_1 = 0$).
c. $f(x) = \frac{2}{3}\sin(x) + 1$; $[0, 2]$ (begin with $x_1 = 2$).

The geometric interpretation of a fixed point provides a simple graphical solution to the problem of finding the fixed point of a given function.

Problem 3. For each of the functions given in Problem 2, use a graphing utility to graph f and the line $y = x$ in the same coordinate system, and then use the features of your utility to verify the approximation(s) that you found.

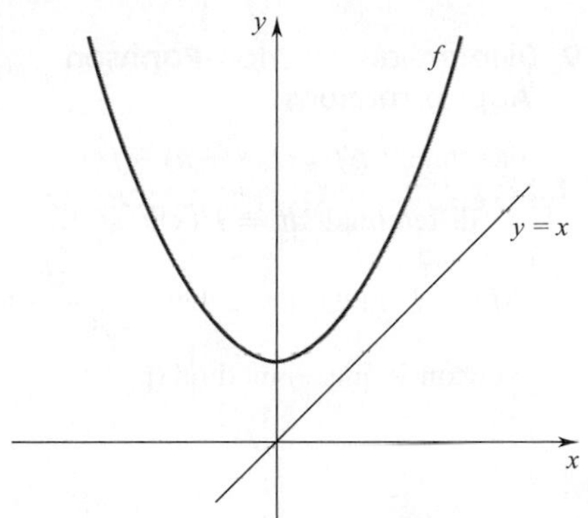

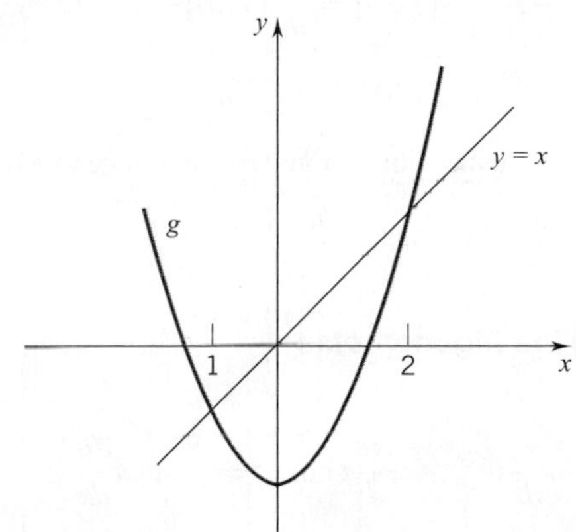

■ CHAPTER HIGHLIGHTS

3.1 The Derivative

derivative of f at c: $\lim_{h\to 0} \dfrac{f(c+h) - f(c)}{h}$ (p. 120)
tangent line (p. 121) normal line (p. 125)
vertical tangent (p. 127)

If f is differentiable at x, then f is continuous at x. The converse is false.

left-hand derivative of f at x: $\lim_{h\to 0^-} \dfrac{f(x+h) - f(x)}{h}$.

right-hand derivative of f at x: $\lim_{h\to 0^-} \dfrac{f(x+h) - f(x)}{h}$.

alternative definition of the derivative at c:

$\lim_{x\to c} \dfrac{f(x) - f(c)}{x - c}$.

3.2 Some Differentiation Formulas

If $f(x) = \alpha$, α a constant, then $f'(x) = 0$. If $f(x) = x$, then $f'(x) = 1$.

$$(f+g)'(x) = f'(x) + g'(x), \qquad (\alpha f)'(x) = \alpha f'(x),$$

$$(fg)'(x) = f(x)g'(x) + g(x)f'(x),$$

$$\left(\frac{f}{g}\right)'(x) = \frac{g(x)f'(x) - f(x)g'(x)}{[g(x)]^2}, \quad [g(x) \ne 0].$$

For any integer n, $p(x) = x^n$ has derivative $p'(x) = nx^{n-1}$.
If $p(x) = a_n x^n + a_{n-1}x^{n-1} + \cdots + a_2x^2 + a_1x + a_0$ is a polynomial, then

$$p'(x) = na_nx^{n-1} + (n-1)a_{n-1}x^{n-2} + \cdots + 2a_2x + a_1.$$

3.3 The *d/dx* Notation; Derivatives of Higher Order

Another notation for the derivative is the *double-d* notation of Leibniz: if $y = f(x)$, then

$$\frac{dy}{dx} = f'(x).$$

Higher derivatives are calculated by repeated differentiation. For example, the second derivative of $y = f(x)$ is the derivative of the derivative:

$$f''(x) = [f'(x)]' \quad \text{or} \quad \frac{d^2y}{dx^2} = \frac{d}{dx}\left(\frac{dy}{dx}\right).$$

3.4 The Derivative as a Rate of Change

dy/dx gives the rate of change of y with respect to x.

Velocity and Acceleration

velocity, acceleration, speed (p. 151)
free fall (p. 153)

Economics

marginal cost (p. 155), marginal revenue (p. 155), marginal profit (p. 155), break-even points (p. 155).

3.5 The Chain Rule

Various forms of the chain rule:

$$\frac{dy}{dx} = \frac{dy}{du}\frac{du}{dx}, \quad \frac{d}{dx}[f(u)] = \frac{d}{du}[f(u)]\frac{du}{dx},$$

$$(f \circ g)'(x) = f'(g(x))\,g'(x).$$

If u is a differentiable function of x and n is an integer, then

$$\frac{d}{dx}(u^n) = nu^{n-1}\frac{du}{dx}.$$

3.6 Differentiating the Trigonometric Functions

$$\frac{d}{dx}(\sin u) = \cos u\frac{du}{dx}, \quad \frac{d}{dx}(\cos u) = -\sin u\frac{du}{dx},$$

$$\frac{d}{dx}(\tan u) = \sec^2 u\frac{du}{dx}, \quad \frac{d}{dx}(\cot u) = -\csc^2 u\frac{du}{dx},$$

$$\frac{d}{dx}(\sec u) = \sec u \tan u\frac{du}{dx}, \quad \frac{d}{dx}(\csc u) = -\csc u \cot u\frac{du}{dx}.$$

3.7 Implicit Differentiation; Rational Powers

finding dy/dx from an equation in x and y without first solving the equation for y
If u is a differentiable function of x and r is a rational number, then

$$\frac{d}{dx}(u^r) = ru^{r-1}\frac{du}{dx}$$

3.8 Rates of Change per Unit Time

If a quantity Q varies with time t, then the derivative dQ/dt gives the rate of change of that quantity with respect to time.
A five-step procedure for solving *related rate problems* (p. 183).
angular velocity; uniform circular motion. (p. 187)

3.9 Differentials; Newton-Raphson Approximations

increment; $\Delta f = f(x+h) - f(x)$

differential: $df = f'(x)h$

$\Delta f \cong df$ in the sense that $\dfrac{\Delta f - df}{h}$ tends to 0 as $h \to 0$

Newton-Raphson method (p. 191).

■ 4.1 THE MEAN-VALUE THEOREM

In this section we prove a result known as *the mean-value theorem.* First stated by the French mathematician Joseph-Louis Lagrange† (1736–1813), this theorem has come to permeate the theoretical structure of all calculus.

THEOREM 4.1.1 THE MEAN-VALUE THEOREM

If f is differentiable on the open interval (a, b) and continuous on the closed interval $[a, b]$, then there is at least one number c in (a, b) for which

$$f'(c) = \frac{f(b) - f(a)}{b - a}$$

or, equivalently,

$$f(b) - f(a) = f'(c)(b - a).$$

The number

$$\frac{f(b) - f(a)}{b - a}$$

is the slope of the line l that passes through the points $(a, f(a))$ and $(b, f(b))$. To say that there is at least one number c for which

$$f'(c) = \frac{f(b) - f(a)}{b - a}$$

† Lagrange, whom you encountered earlier in connection with the prime notation for differentiation, was born in Turin, Italy. He spent twenty years as mathematician in residence at the court of Frederick the Great. Later he taught at the renowned École Polytechnique in France.

is to say that the graph of f has at least one point $(c, f(c))$ at which the tangent line is parallel to l. See Figure 4.1.1.

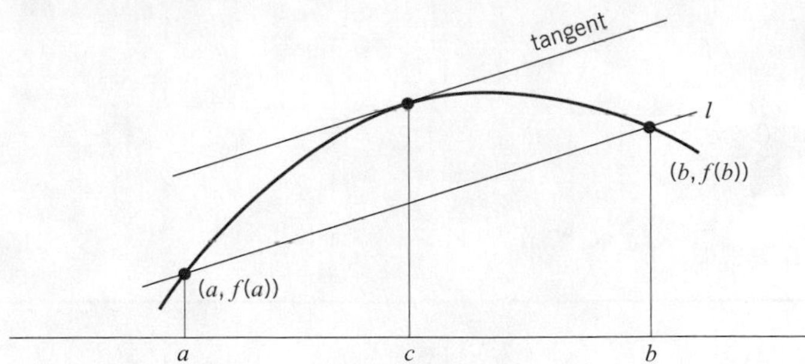

Figure 4.1.1

We can also give a "physical" interpretation of the mean-value theorem. Suppose that an object is moving in one direction along a straight line and that, during a certain time interval $t = a$ to $t = b$, its position at time t is denoted by $x(t)$ (see Section 3.4). Assume that the position function $x(t)$ is continuous on $[a, b]$ and differentiable on (a, b). Now, $x(b) - x(a)$ is the distance that the object travels during the time interval $[a, b]$ and

$$\frac{x(b) - x(a)}{b - a}$$

is the average velocity of the object over this interval. By the mean-value theorem, there is at least one time $t = c$ for which

$$x'(c) = \frac{x(b) - x(a)}{b - a}.$$

But $x'(c) = v(c)$ is the instantaneous velocity of the object at $t = c$. Thus, the mean-value theorem tells us that there is at least one time $t = c$ at which the velocity at that instant equals the average velocity over the whole interval. Think about this when you are driving your car. For example, if you drive 240 miles in 4 hours, then your average velocity is 60 miles per hour and, by the mean-value theorem, your speedometer must have registered 60 miles per hour at least once during your trip.

We will prove the mean-value theorem in steps. First we will show that if a function f has a nonzero derivative at some point x_0, then, close to x_0, $f(x)$ is greater than $f(x_0)$ on one side of x_0 and less than $f(x_0)$ on the other side of x_0.

THEOREM 4.1.2

Let f be differentiable at x_0. If $f'(x_0) > 0$, then

$$f(x_0 - h) < f(x_0) < f(x_0 + h)$$

for all positive h sufficiently small. If $f'(x_0) < 0$, then

$$f(x_0 - h) > f(x_0) > f(x_0 + h)$$

for all positive h sufficiently small.

PROOF We take the case $f'(x_0) > 0$ and leave the other case to you. By the definition of the derivative,

$$\lim_{k \to 0} \frac{f(x_0 + k) - f(x_0)}{k} = f'(x_0).$$

With $f'(x_0) > 0$ we can use $f'(x_0)$ itself as ϵ and conclude that there exists $\delta > 0$ such that

$$\text{if } 0 < |k| < \delta, \quad \text{then} \quad \left| \frac{f(x_0 + k) - f(x_0)}{k} - f'(x_0) \right| < f'(x_0).$$

For such k we have

$$-f'(x_0) < \frac{f(x_0 + k) - f(x_o)}{k} - f'(x_0) < f'(x_0)$$

and it follows that

$$0 < \frac{f(x_o + k) - f(x_0)}{k} < 2f'(x_0). \qquad \text{(Why?)}$$

In particular,

$$(*) \qquad \frac{f(x_0 + k) - f(x_0)}{k} > 0.$$

Now let $0 < h < \delta$. Replacing k with h in $(*)$, we have

$$\frac{f(x_0 + h) - f(x_0)}{h} > 0, \quad \text{which implies that} \quad f(x_0) < f(x_0 + h).$$

Replacing k with $-h$ in $(*)$, we have

$$\frac{f(x_0 - h) - f(x_0)}{-h} > 0, \quad \text{which implies that} \quad f(x_0 - h) < f(x_0). \quad \square$$

Next we prove a special case of the mean-value theorem, known as Rolle's theorem [after the French mathematician Michel Rolle (1652–1719), who first announced the result in 1691]. In Rolle's theorem we make the additional assumption that $f(a)$ and $f(b)$ are both 0. (See Figure 4.1.2.) In this case the line through $(a, f(a))$ and $(b, f(b))$ is horizontal. (It is the x-axis.) The conclusion is that there is a point $(c, f(c))$ at which the tangent line is horizontal.

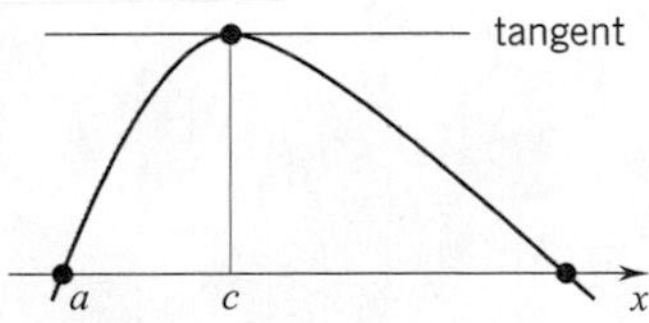

Figure 4.1.2

THEOREM 4.1.3 ROLLE'S THEOREM

Let f be differentiable on the open interval (a, b) and continuous on the closed interval $[a, b]$. If $f(a)$ and $f(b)$ are both 0, then there is at least one number c in (a, b) at which

$$f'(c) = 0.$$

PROOF If f is constantly 0 on $[a, b]$, then $f'(c) = 0$ for all $c \in (a, b)$. If f is not constantly 0 on $[a, b]$, then f takes on either some positive values or some negative values. We assume the former and leave the other case to you.

Since f is continuous on $[a, b]$, f must take on a maximum value at some point c of $[a, b]$. (See Theorem 2.6.2.) This maximum value, $f(c)$, must be positive. Since $f(a)$ and $f(b)$ are both 0, c cannot be a and it cannot be b. This means that c must lie in the open interval (a, b) and therefore $f'(c)$ exists. Now $f'(c)$ cannot be greater than 0 and it cannot be less than 0 because each of these conditions would imply that f takes on values greater than $f(c)$. (This follows from Theorem 4.1.2.) We conclude therefore that $f'(c) = 0$. ❑

Remark Rolle's theorem is sometimes stated as:

> Let g be differentiable on the open interval (a, b) and continuous on the closed interval $[a, b]$. If $g(a) = g(b)$, then there is at least one number c in (a, b) for which
> $$g'(c) = 0.$$

In Exercise 42, you are asked to show that this version of Rolle's theorem is an immediate consequence of the given version. ❑

Rolle's theorem has some useful applications independent of the mean-value theorem.

Example 1 Show that the polynomial function $p(x) = 2x^3 + 5x - 1$ has exactly one real zero.

SOLUTION From Exercise 39, Section 2.6, we know that p has at least one real zero. Suppose that p has more than one real zero. In particular, suppose that $p(a) = p(b) = 0$, where a and b are real numbers and $a \neq b$. Without loss of generality, assume that $a < b$. Since a polynomial is differentiable everywhere, p is differentiable on (a, b) and continuous on $[a, b]$. Thus, by Rolle's theorem, there is a number c in (a, b) for which $p'(c) = 0$. But

$$p'(x) = 6x^2 + 5 \geq 5 \qquad \text{for all} \quad x,$$

and so $p'(x) \neq 0$ for all x. Thus, we have a contradiction and we can conclude that p has exactly one real zero. ❑

We are now ready to give a proof of the mean-value theorem.

PROOF OF THE MEAN-VALUE THEOREM We will create a function g that satisfies the conditions of Rolle's theorem and is so related to f that the conclusion $g'(c) = 0$ leads to the conclusion

$$f'(c) = \frac{f(b) - f(a)}{b - a}.$$

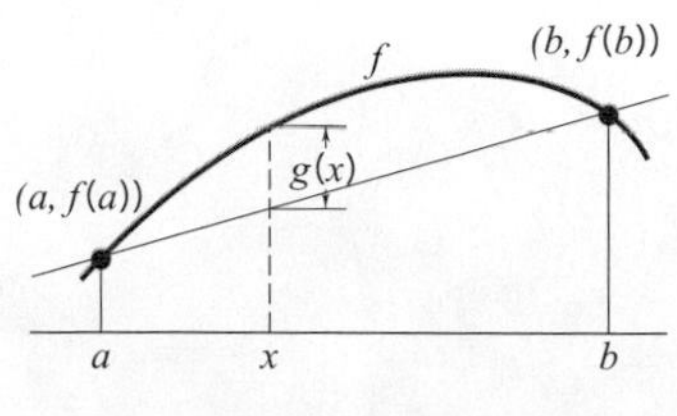

Figure 4.1.3

The function g given by

$$g(x) = f(x) - \left[\frac{f(b) - f(a)}{b - a}(x - a) + f(a)\right]$$

is exactly such a function. Geometrically $g(x)$ is represented in Figure 4.1.3. The line that passes through $(a, f(a))$ and $(b, f(b))$ has the equation

$$y = \frac{f(b) - f(a)}{b - a}(x - a) + f(a).$$

[This is not hard to verify. The slope is right, and, when $x = a, y = f(a)$.] The difference

$$g(x) = f(x) - \left[\frac{f(b) - f(a)}{b - a}(x - a) + f(a)\right]$$

is simply the vertical separation between the graph of f and the line in question.

If f is differentiable on (a, b) and continuous on $[a, b]$, then so is g. As you can check, $g(a)$ and $g(b)$ are both 0. Therefore, by Rolle's theorem, there is at least one number c in (a, b) for which $g'(c) = 0$. Since

$$g'(x) = f'(x) - \frac{f(b) - f(a)}{b - a},$$

we have

$$g'(c) = f'(c) - \frac{f(b) - f(a)}{b - a}.$$

With $g'(c) = 0$, it follows that

$$f'(c) = \frac{f(b) - f(a)}{b - a}. \quad \square$$

Example 2 Let $f(x) = x^3 - 4x$ on $[0, 3]$. Since f is a polynomial, f is continuous on $[0,3]$ and differentiable on $(0,3)$. Therefore, by the mean-value theorem, there exists at least one number $c \in (0, 3)$ such that

$$f(3) - f(0) = f'(c)(3 - 0).$$

Since $f(3) = 15$, $f(0) = 0$, and $f'(x) = 3x^2 - 4$, we get

$$15 = (3c^2 - 4)(3) = 9c^2 - 12.$$

Therefore

$$9c^2 = 27 \quad \text{which gives} \quad c^2 = 3 \quad \text{and} \quad c = \pm\sqrt{3}.$$

Since $c \in (0, 3)$, we must have $c = \sqrt{3}$. Figure 4.1.4 shows the graph of f, the secant line through $(0, 0)$ and $(3, 15)$, and the line tangent to the curve at $(\sqrt{3}, f(\sqrt{3}))$. $\square$

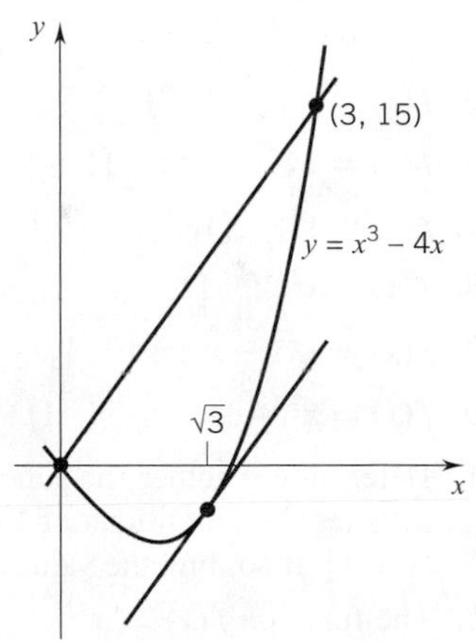

Figure 4.1.4

Example 3 Suppose that f is continuous on $[1, 4]$, differentiable on $(1, 4)$, and $f(1) = 2$. If we are given that $2 \le f'(x) \le 3$ on $(1, 4)$, what is the least possible value of $f(4)$? What is the greatest possible value of $f(4)$?

SOLUTION By the mean-value theorem, there is at least one number $c \in (1, 4)$ such that

$$f(4) - f(1) = f'(c)(4 - 1) = 3f'(c).$$

Solving this equation for $f(4)$, we obtain

$$f(4) = f(1) + 3f'(c) = 2 + 3f'(c).$$

Since $f'(x) \ge 2$ for every $x \in (1,4)$, we know that $f'(c) \ge 2$, and so

$$f(4) \ge 2 + 3(2) = 8.$$

Similarly, $f'(x) \leq 3$ for every x implies $f'(c) \leq 3$, and

$$f(4) \leq 2 + 3(3) = 11.$$

The least possible value of $f(4)$ is 8 and the greatest possible value of $f(4)$ is 11. □

The hypotheses of the mean-value theorem require f to be differentiable on the open interval (a, b) and continuous on the closed interval $[a, b]$. As you would expect, if either of these conditions fails to hold, then the conclusion of the theorem may not hold. This is illustrated in the Exercises.

EXERCISES 4.1

Show that f satisfies the conditions of Rolle's theorem on the indicated interval and find all numbers c on the interval for which $f'(c) = 0$.

1. $f(x) = x^3 - x$; $[0, 1]$.

2. $f(x) = x^4 - 2x^2 - 8$; $[-2, 2]$.

3. $f(x) = \sin 2x$; $[0, 2\pi]$.

4. $f(x) = x^{2/3} - 2x^{1/3}$; $[0, 8]$.

Verify that f satisfies the conditions of the mean-value theorem on the indicated interval and find all numbers c that satisfy the conclusion of the theorem.

5. $f(x) = x^2$; $[1, 2]$.

6. $f(x) = 3\sqrt{x} - 4x$; $[1, 4]$.

7. $f(x) = x^3$; $[1, 3]$.

8. $f(x) = x^{2/3}$; $[1, 8]$.

9. $f(x) = \sqrt{1 - x^2}$; $[0, 1]$.

10. $f(x) = x^3 - 3x$; $[-1, 1]$.

11. Determine whether the function $f(x) = \sqrt{1 - x^2}/(3 + x^2)$ satisfies the conditions of Rolle's theorem on the interval $[-1, 1]$. If so, find the values of c for which $f'(c) = 0$.

12. The function $f(x) = x^{2/3} - 1$ has zeros at $x = -1$ and at $x = 1$.

(a) Show that f' has no zeros in $(-1, 1)$.

(b) Show that this does not contradict Rolle's theorem.

13. Does there exist a differentiable function f that satisfies $f(0) = 2$, $f(2) = 5$, and $f'(x) \leq 1$ on $(0, 2)$? If not, why not?

14. Does there exist a differentiable function f that has the value 1 only at $x = 0, 2$, and 3, and $f'(x) = 0$ only at $x = -1, 3/4$, and $3/2$? If not, why not?

15. Sketch the graph of

$$f(x) = \begin{cases} 2x + 2, & x \leq -1 \\ x^3 - x, & x > -1 \end{cases}$$

and find the derivative. Determine whether f satisfies the conditions of the mean-value theorem on the interval $[-3, 2]$ and, if so, find the values of c guaranteed by the theorem.

16. Sketch the graph of

$$f(x) = \begin{cases} 2 + x^3, & x \leq 1 \\ 3x, & x > 1 \end{cases}$$

and find the derivative. Determine whether f satisfies the conditions of the mean-value theorem on the interval $[-1, 2]$ and, if so, find the values of c guaranteed by the theorem.

17. Consider the quadratic function $f(x) = Ax^2 + Bx + C$. Show that, for any interval $[a, b]$, the value of c that satisfies the conclusion of the mean-value theorem is $(a + b)/2$, the midpoint of the interval.

18. Set $f(x) = x^{-1}, a = -1, b = 1$. Verify that there is no number c for which

$$f'(c) = \frac{f(b) - f(a)}{b - a}.$$

Explain how this does not violate the mean-value theorem.

19. Repeat Exercise 18 with $f(x) = |x|$.

20. Graph the function $f(x) = |2x - 1| - 3$ and compute the derivative. Verify that $f(-1) = 0 = f(2)$ and yet $f'(x)$ is never 0. Explain how this does not violate Rolle's theorem.

21. Show that the equation $6x^4 - 7x + 1 = 0$ does not have more than two distinct real roots. (Use Rolle's theorem.)

22. Show that the equation $6x^5 + 13x + 1 = 0$ has exactly one real root. (Use Rolle's theorem and the intermediate- value theorem.)

23. Show that the equation $x^3 + 9x^2 + 33x - 8 = 0$ has exactly one real root.

24. (a) Let f be differentiable on (a, b). Prove that if $f'(x) \neq 0$ for each $x \in (a, b)$, then f has at most one zero in (a, b).

(b) Let f be twice differentiable on (a, b). Prove that if $f''(x) \neq 0$ for each $x \in (a, b)$, then f has at most two zeros in (a, b).

25. Let $P(x) = a_n x^n + \cdots + a_1 x + a_0$ be a nonconstant polynomial. Show that between any two consecutive roots of the equation $P'(x) = 0$ there is at most one root of the equation $P(x) = 0$.

26. Let f be twice differentiable. Show that, if the equation $f(x) = 0$ has n distinct real roots, then the equation $f'(x) = 0$ has at least $n - 1$ distinct real roots and the equation $f''(x) = 0$ has at least $n - 2$ distinct real roots.

27. Recall that a number c is a fixed point of a function f if $f(c) = c$ (see Project 3.9). Prove that if f is differentiable on an interval I and $f'(x) < 1$ for all $x \in I$, then f has at most one fixed point in I. HINT: Assume that f has two fixed points in I and let $g(x) = f(x) - x$.

28. Show that the equation $x^3 + ax + b = 0$ has exactly one real root if $a \geq 0$ and at most one real root between $-\frac{1}{3}\sqrt{3}|a|$ and $\frac{1}{3}\sqrt{3}|a|$ if $a < 0$.

29. Let $f(x) = x^3 - 3x + b$.
   (a) Show that $f(x) = 0$ for at most one number x in $[-1, 1]$.
   (b) Determine the values of b such that $f(x) = 0$ for some x in $[-1, 1]$.

30. Let $f(x) = x^3 - 3a^2x + b, a > 0$. Show that $f(x) = 0$ for at most one number x in $[-a, a]$.

31. Show that the equation $x^n + ax + b = 0$, n an even positive integer, has at most two distinct real roots.

32. Show that the equation $x^n + ax + b = 0$, n an odd positive integer, has at most three distinct real roots.

33. Given that $|f'(x)| \leq 1$ for all real numbers x, show that $|f(x_1) - f(x_2)| \leq |x_1 - x_2|$ for all real numbers x_1 and x_2.

34. Let f be differentiable on an open interval I. Prove that, if $f'(x) = 0$ for all x in I, then f is constant on I.

35. Let f be differentiable on (a, b) with $f(a) = f(b) = 0$ and $f'(c) = 0$ for some c in (a, b). Show by example that f need not be continuous on $[a, b]$.

36. Prove that for all real x and y
   (a) $|\cos x - \cos y| \leq |x - y|$.
   (b) $|\sin x - \sin y| \leq |x - y|$.

37. Let f be differentiable on (a, b) and continuous on $[a, b]$.
   (a) Prove that if there is a constant M such that $f'(x) \leq M$ for all $x \in (a, b)$, then
   $$f(b) \leq f(a) + M(b - a).$$
   (b) Prove that if there is a constant m such that $f'(x) \geq m$ for all $x \in (a, b)$, then
   $$f(b) \geq f(a) + m(b - a).$$
   (c) Parts (a) and (b) imply that if there exists a constant L such that $|f'(x)| \leq L$ on (a, b), then
   $$f(a) - L(b - a) \leq f(b) \leq f(a) + L(b - a).$$
   Derive this result.

38. Suppose that f and g are differentiable and that $f(x)g'(x) - g(x)f'(x)$ has no zeros on some interval I. Assume that there are numbers a, b in I with $a < b$ such that $f(a) = f(b) = 0$, and that f has no zeros in (a, b). Prove that if $g(a) \neq 0$ and $g(b) \neq 0$, then g has exactly one zero in (a, b). HINT: Suppose that g has no zeros in (a, b) and consider $h = f/g$. Then reverse the roles of f and g.

39. Let $f(x) = \cos x$ and $g(x) = \sin x$ on $I = (-\infty, \infty)$. Prove that the zeros of f and g separate each other on I; that is, prove that between two consecutive zeros of $\cos x$ there is exactly one zero of $\sin x$ and conversely. HINT: Use Exercise 38.

40. (*Important*) Use the mean-value theorem to show that if f is continuous at x and $x + h$ and differentiable in between, then
$$f(x + h) - f(x) = f'(x + \theta h)h$$
for some number θ between 0 and 1. (In some texts this is how the mean-value theorem is stated.)

41. Let $h > 0$. Suppose that f is continuous on $[x_0 - h, x_0 + h]$ and differentiable on $(x_0 - h, x_0) \cup (x_0, x_0 + h)$. Show that if
$$\lim_{x \to x_0^-} f'(x) = \lim_{x \to x_0^+} f'(x) = L,$$
then f is differentiable at x_0 and $f'(x_0) = L$. HINT: Use Exercise 40.

42. Let f be continuous on $[a, b]$ and differentiable on (a, b). Without assuming the mean-value theorem, prove that if $f(a) = f(b)$, then there is at least one number $c \in (a, b)$ for which $f'(c) = 0$. HINT: Suppose $f(a) = f(b) = k$ and let $g(x) = f(x) - k$.

43. *Generalization of the mean-value theorem*. Let f and g both satisfy the hypotheses of the mean-value theorem, Theorem 4.1.1. In addition, assume that g' has no zeros in (a, b). Prove that there is at least one number c in (a, b) such that
$$\frac{f(b) - f(a)}{g(b) - g(a)} = \frac{f'(c)}{g'(c)}.$$
This result is known as the *Cauchy mean-value theorem*; it will be used in Chapter 10. Note that this result reduces to Theorem 4.1.1 if $g(x) = x$. HINT: Consider the function $F(x) = [f(b) - f(a)]g(x) - [g(b) - g(a)]f(x)$.

44. At 1:00 P.M. a car's speedometer reads 30 miles per hour and at 1:15 P.M. it reads 60 miles per hour. Prove that the car's acceleration was exactly 120 miles per hour per hour at least once between 1:00 and 1:15.

45. Two race horses start a race at the same time and finish in a tie. Prove that there must have been at least one time t *during* the race at which the two horses had exactly the same speed. HINT: Let $f_1(t)$ and $f_2(t)$ denote the positions of the two horses at time t and consider $f(t) = f_1(t) - f_2(t)$.

46. Continuing Exercise 45, suppose that the two horses cross the finish line together at the same speed. Show that they had the same acceleration at some instant during the race.

47. A certain tollway is 120 miles long and the speed limit is 65 miles per hour. If a driver's entry ticket at one end of the tollway is stamped 12 noon and she exits at the other end at 1:40 P.M., should she be given a speeding ticket? Explain.

48. A car is stationary at a toll booth. Twenty minutes later, at a point 20 miles down the road, the car is clocked at 60 mph. Explain why the car must have exceeded the 60-mph speed limit at some time after leaving the toll booth, but before the car was clocked at 60 mph.

49. The results of an investigation of a car accident showed that the driver applied his brakes and skidded 280 feet in 6 seconds. If the speed limit on the street where the accident occurred was 30 miles per hour, was the driver exceeding the speed limit at the instant he applied his brakes? Explain. HINT: 30 miles per hour = 44 feet per second.

50. Let f be a differentiable function. Apply the mean-value theorem to f with $a = x$ and $b = x + \Delta x$, $\Delta x > 0$, to show that

$$\Delta f = f(x + \Delta x) - f(x) = f'(c)\Delta x,$$

where $x < c < x + \Delta x$. Compare this result with the result in Section 3.9. If f' is continuous and Δx is small, then $f'(c) \cong f'(x)$ and we have

$$\Delta f \cong f'(x)\Delta x,$$

which is the result in Section 3.9.

51. Use the mean-value theorem to estimate $\sqrt{65}$. HINT: Let $f(x) = \sqrt{x}$, $a = 64$, $b = 65$, and use Exercise 50.

C In Exercises 52 and 53, show that the given function satisfies the hypotheses of Rolle's theorem on the indicated interval. Use a graphing utility to graph f' and estimate the number(s) c where $f'(c) = 0$. Round off your estimates to three decimal places.

52. $f(x) = 2x^3 + 3x^2 - 3x - 2$; $[-2, 1]$.

53. $f(x) = 1 - x^3 - \cos(\pi x/2)$; $[0, 1]$.

C **54.** Let $f(x) = x^4 - x^3 + x^2 - x$. Find a number b, if possible, such that Rolle's theorem is satisfied on $[0, b]$. If b does exist, find the number c of Rolle's theorem and use a graphing utility to draw the graphs of f and $y = f(c)$ together.

C **55.** Repeat Exercise 54 with $f(x) = x^4 + x^3 + x^2 - x$.

Use a CAS in Exercises 56–58. Find the x-intercepts of the function f. Between each pair of intercepts, find the number c of Rolle's theorem, if possible.

56. $f(x) = \dfrac{x^2 - x}{x^2 + 2x + 2}$.

57. $f(x) = \dfrac{x^4 - 16}{x^2 + 4}$.

58. $f(x) = 125x^7 - 300x^6 - 760x^5 + 2336x^4 + 80x^3 - 4288x^2 + 3840x - 1024$.

Suppose that the function f satisfies the hypotheses of the mean-value theorem on an interval $[a, b]$. We can find the numbers c that satisfy the conclusion of the mean-value theorem by finding the zeros of the function g given by

$$g(x) = f'(x) - \frac{f(b) - f(a)}{b - a}.$$

C In Exercises 59–60, use a graphing utility to graph the function g corresponding to the given function f on the indicated interval. Estimate the zeros of g accurate to three decimal places. For each zero c in the interval, graph the tangent line to the graph of f at $(c, f(c))$, and graph the line through $(a, f(a))$ and $(b, f(b))$. Verify that they are parallel.

59. $f(x) = x^4 - 7x^2 + 2$; $[1, 3]$.

60. $f(x) = x\cos x + 4\sin x$; $[-\pi/2, \pi/2]$.

■ 4.2 INCREASING AND DECREASING FUNCTIONS

To place our discussion on a solid footing, we begin with a definition.

DEFINITION 4.2.1

A function f is said to:

(i) *increase* on the interval I if for every two numbers x_1, x_2 in I,

$$x_1 < x_2 \quad \text{implies that} \quad f(x_1) < f(x_2);$$

(ii) *decrease* on the interval I if for every two numbers x_1, x_2 in I,

$$x_1 < x_2 \quad \text{implies that} \quad f(x_1) > f(x_2).$$

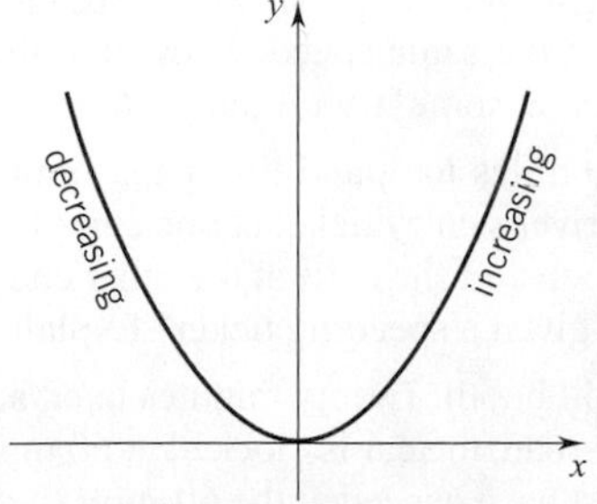

Figure 4.2.1

Preliminary Examples

(a) The quadratic function

$$f(x) = x^2 \qquad \text{(Figure 4.2.1)}$$

decreases on $(-\infty, 0]$ and increases on $[0, \infty)$.

(b) The function

$$f(x) = \begin{cases} 1, & x < 0 \\ x, & x \geq 0 \end{cases} \qquad \text{(Figure 4.2.2)}$$

is constant on $(-\infty, 0)$, it neither increases nor decreases; f increases on $[0, \infty)$.

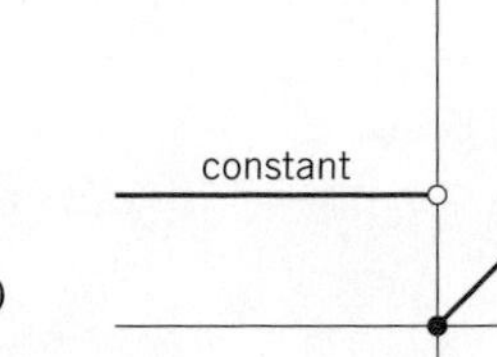

Figure 4.2.2

(c) The cubic function

$$f(x) = x^3 \qquad \text{(Figure 4.2.3)}$$

is everywhere increasing.

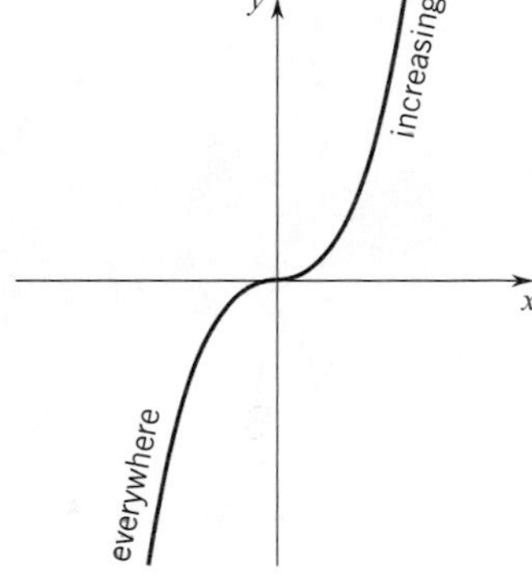

Figure 4.2.3

(d) In the case of the Dirichlet function

$$g(x) = \begin{cases} 1, & x \text{ rational} \\ 0, & x \text{ irrational}, \end{cases} \qquad \text{(Figure 4.2.4)}$$

there is no interval on which the function increases and no interval on which the function decreases. On every interval the function jumps back and forth between 0 and 1 an infinite number of times. ☐

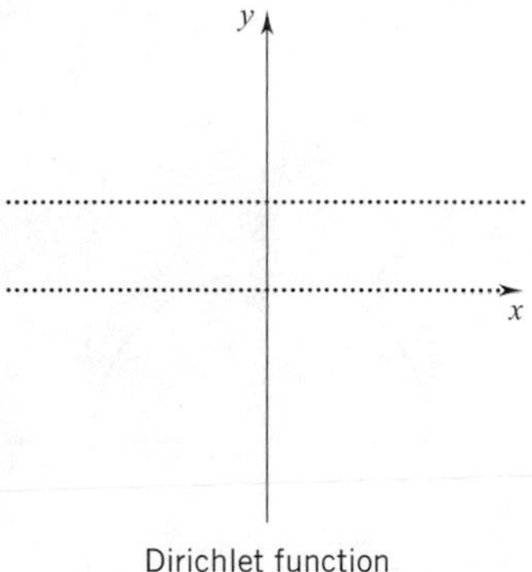

Dirichlet function

Figure 4.2.4

If f is a differentiable function, then we can determine the intervals on which f increases and the intervals on which f decreases by examining the sign of the first derivative.

THEOREM 4.2.2

Let f be differentiable on an open interval I.

(i) If $f'(x) > 0$ for all x in I, then f increases on I.
(ii) If $f'(x) < 0$ for all x in I, then f decreases on I.
(iii) If $f'(x) = 0$ for all x in I, then f is constant on I.

PROOF Choose any two numbers x_1 and x_2 in I with $x_1 < x_2$. Since f is differentiable on I, it is continuous on I. Therefore we know that f is differentiable on (x_1, x_2) and continuous on $[x_1, x_2]$. By the mean-value theorem there is a number c in (x_1, x_2) for which

$$f'(c) = \frac{f(x_2) - f(x_1)}{x_2 - x_1}.$$

In (i), $f'(x) > 0$ for all x. Therefore, $f'(c) > 0$ and we have

$$\frac{f(x_2) - f(x_1)}{x_2 - x_1} > 0, \qquad \text{which implies that } f(x_1) < f(x_2).$$

In (ii), $f'(x) < 0$ for all x. Therefore, $f'(c) < 0$ and we have

$$\frac{f(x_2) - f(x_1)}{x_2 - x_1} < 0, \qquad \text{which implies that } f(x_1) > f(x_2).$$

In (iii), $f'(x) = 0$ for all x. Therefore, $f'(c) = 0$ and we have

$$\frac{f(x_2) - f(x_1)}{x_2 - x_1} = 0, \qquad \text{which implies that } f(x_1) = f(x_2). \quad ☐$$

Remark In Section 3.2 we showed that if a function f is constant on an open interval I, then $f'(x) = 0$ for all $x \in I$. Part (iii) of Theorem 4.2.2 gives the converse: if $f'(x) = 0$ for all x in an open interval I, then f is constant on I. Combining these two statements, we have:

> Suppose that f is differentiable on an open interval I. Then
>
> $$f \text{ is constant on } I \text{ iff } f'(x) = 0 \text{ for all } x \in I.$$

□

Theorem 4.2.2 is useful, but it doesn't tell the complete story. Look, for example, at the function $f(x) = x^2$. The derivative $f'(x) = 2x$ is negative for x in $(-\infty, 0)$, zero at $x = 0$, and positive for x in $(0, \infty)$. Theorem 4.2.2 assures us that

$$f \text{ decreases on } (-\infty, 0) \text{ and increases on } (0, \infty),$$

but actually

$$f \text{ decreases on } (-\infty, 0] \text{ and increases on } [0, \infty).$$

To get these stronger results, we need a theorem that works for closed intervals, too.

To extend Theorem 4.2.2 so that it works for an arbitrary interval I, the only additional condition we need is continuity at the endpoint(s).

> **THEOREM 4.2.3**
>
> Let f be continuous on an arbitrary interval I and differentiable on the interior of I.
>
> **(i)** If $f'(x) > 0$ for all x in the interior of I, then f increases on all of I.
> **(ii)** If $f'(x) < 0$ for all x in the interior of I, then f decreases on all of I.
> **(iii)** If $f'(x) = 0$ for all x in the interior of I, then f is constant on all of I.

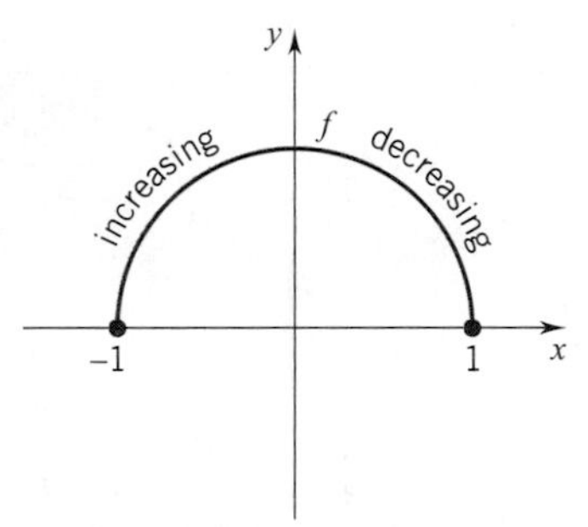

Figure 4.2.5

The method of proof of Theorem 4.2.2 can also be used to establish this result.

It is time for some examples.

Example 1 The function $f(x) = \sqrt{1 - x^2}$ has derivative $f'(x) = -\dfrac{x}{\sqrt{1 - x^2}}$.

Since $f'(x) > 0$ for all x in $(-1, 0)$ and f is continuous on $[-1, 0]$, f increases on $[-1, 0]$. Since $f'(x) < 0$ for all x in $(0, 1)$ and f is continuous on $[0, 1]$, f decreases on $[0, 1]$. The graph of f is a semicircle. (See Figure 4.2.5.) □

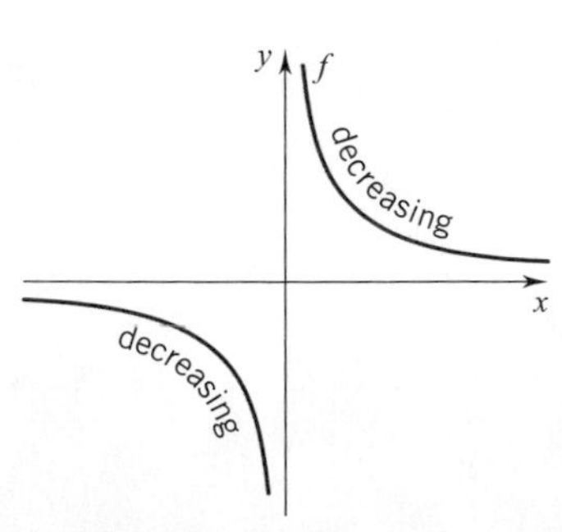

Figure 4.2.6

Example 2 The function $f(x) = \dfrac{1}{x}$ is defined for all $x \neq 0$. The derivative $f'(x) = -\dfrac{1}{x^2}$ is negative for all $x \neq 0$. Thus the function f decreases on $(-\infty, 0)$ and on $(0, \infty)$. (See Figure 4.2.6). □

Example 3 The function $g(x) = \frac{4}{5}x^5 - 3x^4 - 4x^3 + 22x^2 - 24x + 6$ is a polynomial. It is therefore everywhere continuous and everywhere differentiable.

Differentiation gives

$$\begin{aligned} g'(x) &= 4x^4 - 12x^3 - 12x^2 + 44x - 24 \\ &= 4(x^4 - 3x^3 - 3x^2 + 11x - 6) \\ &= 4(x+2)(x-1)^2(x-3). \end{aligned}$$

The derivative g' takes on the value 0 at -2, at 1, and at 3. These numbers determine four intervals on which g' keeps a constant sign (see Section 2.6):

$$(-\infty, -2), \quad (-2, 1), \quad (1, 3), \quad (3, \infty).$$

The sign of g' on these intervals and the consequences for g are as follows:

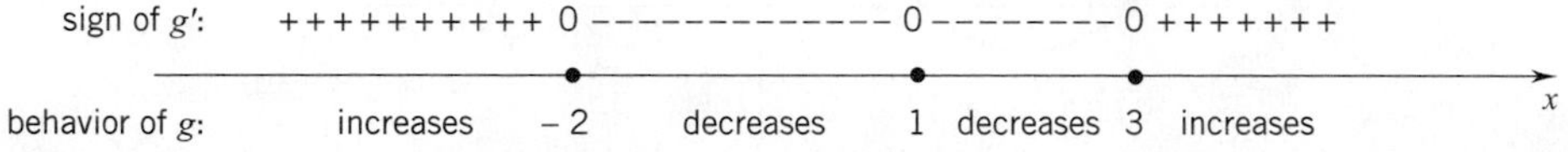

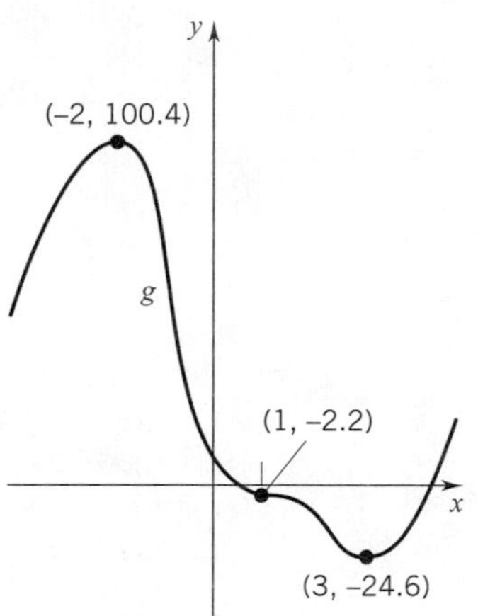

Figure 4.2.7

Since g is everywhere continuous, g increases on $(-\infty, -2]$, decreases on $[-2, 3]$, and increases on $[3, \infty)$. (See Figure 4.2.7.) □

Remark From Theorem 4.2.2 we know that if $f'(x) > 0\,(f'(x) < 0)$ for all x in an open interval I, then f increases (decreases) on I. The cubing function $f(x) = x^3$ [Preliminary Example (c)] and the function g of Example 3 illustrate that the converse does not hold. That is, with f differentiable, f increasing (decreasing) on I *does not imply* that $f'(x) > 0\,(f'(x) < 0)$ for all $x \in I$: the cubing function is increasing on $(-\infty, \infty)$, yet $f'(x) = 3x^2$ and $f'(0) = 0$; the function g of Example 3 is decreasing on $(-2, 3)$ yet $g'(1) = 0$. You are asked to prove a partial converse of Theorem 4.2.2 in Exercise 53. An extension of Theorem 4.2.2 that covers examples such as those just cited is given by

(4.2.4)

> Let f be differentiable on an open interval I.
>
> **(i)** If $f'(x) \geq 0$ for all $x \in I$ with $f'(x) = 0$ for at most finitely many numbers $x \in I$, then f is increasing on I.
>
> **(ii)** If $f'(x) \leq 0$ for all $x \in I$ with $f'(x) = 0$ for at most finitely many numbers $x \in I$, then f is decreasing on I.

You are asked to prove this result in Exercise 54. □

Example 4 Let $f(x) = x - 2\sin x,\ 0 \leq x \leq 2\pi$. Find the intervals on which f increases and the intervals on which f decreases.

SOLUTION The derivative of f is $f'(x) = 1 - 2\cos x$. Setting $f'(x) = 0$, we have

$$\begin{aligned} 1 - 2\cos x &= 0 \\ \cos x &= \tfrac{1}{2}. \end{aligned}$$

The solutions are $x = \pi/3$ and $x = 5\pi/3$. Thus, f' has constant sign on the intervals $(0, \pi/3)$, $(\pi/3, 5\pi/3)$, and $(5\pi/3, 2\pi)$. The sign of f' and the behavior of f are recorded below.

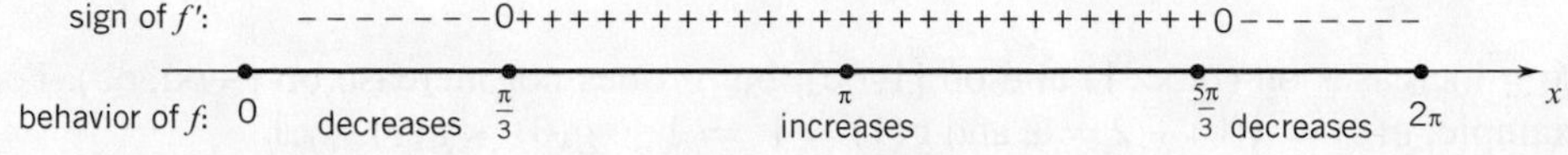

Since f is continuous on $[0, 2\pi]$, f decreases on $[0, \pi/3]$, increases on $[\pi/3, 5\pi/3]$, and decreases on $[5\pi/3, 2\pi]$. (See Figure 4.2.8.) □

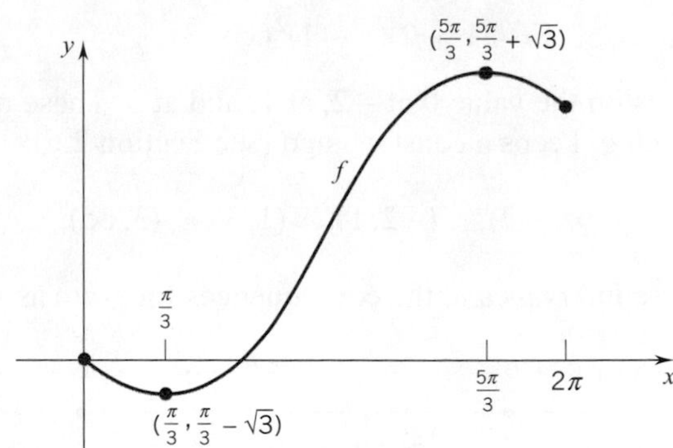

Figure 4.2.8

While Theorems 4.2.2 and 4.2.3 have wide applicability, there are some limitations. If, for example, f is discontinuous at some point in its domain, then Theorems 4.2.2 and 4.2.3 do not tell the whole story.

Example 5 The function

$$f(x) = \begin{cases} x^3, & x < 1 \\ \frac{1}{2}x + 2, & x \geq 1 \end{cases}$$

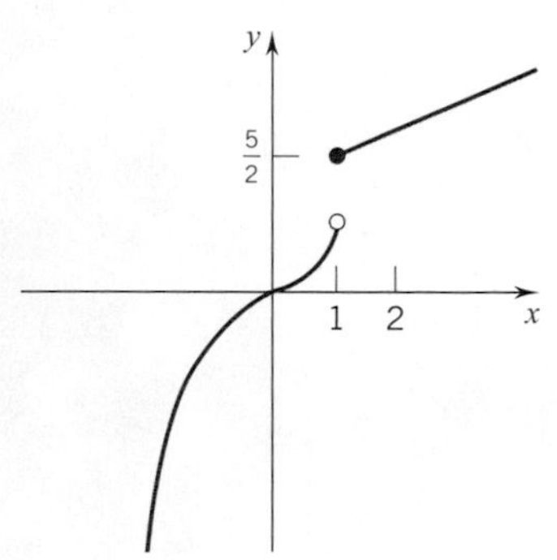

Figure 4.2.9

is graphed in Figure 4.2.9. Obviously there is a discontinuity at $x = 1$. Differentiation gives

$$f'(x) = \begin{cases} 3x^2, & x < 1 \\ \text{does not exist}, & x = 1 \\ \frac{1}{2}, & x > 1. \end{cases}$$

Since $f'(x) > 0$ on $(-\infty, 0) \cup (0, 1)$, we know from Theorem 4.2.2 that f increases on $(-\infty, 1)$. Since $f'(x) > 0$ on $(1, \infty)$ and f is continuous on $[1, \infty)$, we know from Theorem 4.2.3 that f increases on $[1, \infty)$. The fact that f increases on all of $(-\infty, \infty)$ is not derivable from the theorems.

Now consider the function

$$g(x) = \begin{cases} \frac{1}{2}x + 2, & x < 1 \\ x^3, & x \geq 1, \end{cases}$$

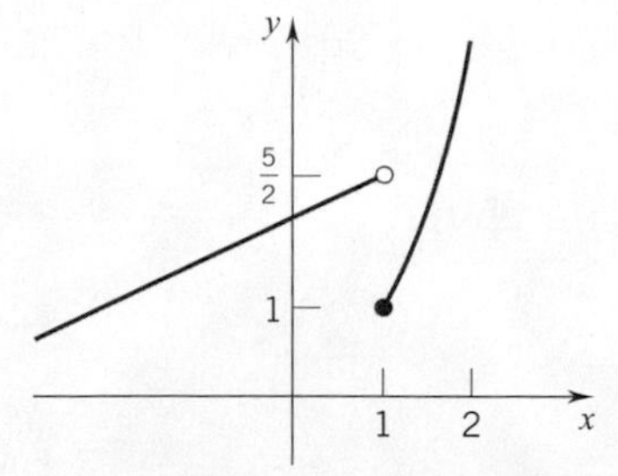

Figure 4.2.10

which is graphed in Figure 4.2.10. Again, there is a discontinuity at $x = 1$. The derivative of g is

$$g'(x) = \begin{cases} \frac{1}{2}, & x < 1 \\ \text{does not exist}, & x = 1 \\ 3x^2, & x > 1, \end{cases}$$

so g increases on $(-\infty, 1)$ and on $[1, \infty)$, but g does not increase on $(-\infty, \infty)$. For example, $g(0) = \frac{1}{2}(0) + 2 = 2$ and $g(1) = 1^3 = 1$ so $g(0) > g(1)$. □

Equality of Derivatives

If two differentiable functions differ by a constant,

$$f(x) = g(x) + C,$$

then their derivatives are equal:

$$f'(x) = g'(x).$$

The converse is also true. In fact, we have the following theorem.

THEOREM 4.2.5

(i) Let I be an open interval. If $f'(x) = g'(x)$ for all x in I, then f and g differ by a constant on I.

(ii) Let I be an arbitrary interval. If $f'(x) = g'(x)$ for all x in the interior of I, and f and g are continuous on I, then f and g differ by a constant on I.

PROOF Set $H = f - g$. For the first assertion apply (iii) of Theorem 4.2.2 to H. For the second assertion apply (iii) of Theorem 4.2.3 to H. We leave the details as an exercise. ☐

We illustrate the theorem in Figure 4.2.11. At points with the same x-coordinate the slopes are equal, and thus the curves have the same steepness. The separation between the curves remains constant; the curves are "parallel."

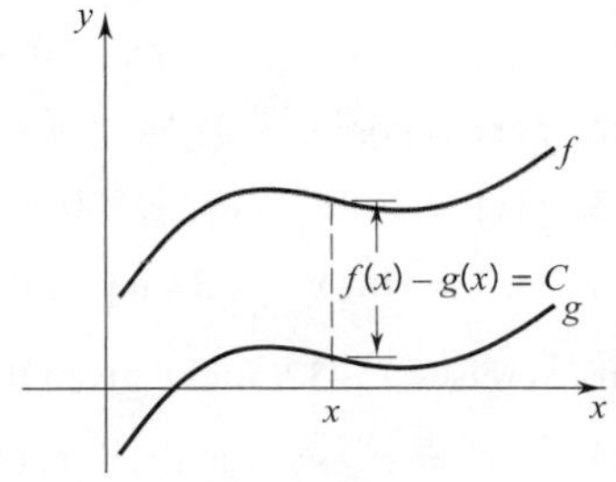

Figure 4.2.11

Example 6 Find f given that $f'(x) = 6x^2 - 7x - 5$ for all x and $f(2) = 1$.

SOLUTION It is not hard to find a function that has $6x^2 - 7x - 5$ as its derivative:

$$\frac{d}{dx}\left(2x^3 - \frac{7}{2}x^2 - 5x\right) = 6x^2 - 7x - 5.$$

By Theorem 4.2.4 we know that $f(x)$ differs from $2x^3 - \frac{7}{2}x^2 - 5x$ only by some constant C. Thus we can write

$$f(x) = 2x^3 - \tfrac{7}{2}x^2 - 5x + C.$$

To evaluate C we use the fact that $f(2) = 1$. Since $f(2) = 1$ and also

$$f(2) = 2(2)^3 - \tfrac{7}{2}(2)^2 - 5(2) + C = 16 - 14 - 10 + C = -8 + C,$$

we have $-8 + C = 1$, and therefore $C = 9$. Thus

$$f(x) = 2x^3 - \tfrac{7}{2}x^2 - 5x + 9$$

has the specified properties. ☐

EXERCISES 4.2

Find the intervals on which f increases and the intervals on which f decreases.

1. $f(x) = x^3 - 3x + 2$.
2. $f(x) = x^3 - 3x^2 + 6$.
3. $f(x) = x + \dfrac{1}{x}$.
4. $f(x) = (x - 3)^3$.
5. $f(x) = x^3(1 + x)$.
6. $f(x) = x(x + 1)(x + 2)$.
7. $f(x) = (x + 1)^4$.
8. $f(x) = 2x - \dfrac{1}{x^2}$.
9. $f(x) = \dfrac{1}{|x - 2|}$.
10. $f(x) = \dfrac{x}{1 + x^2}$.
11. $f(x) = \dfrac{x^2 + 1}{x^2 - 1}$.
12. $f(x) = \dfrac{x^2}{x^2 + 1}$.
13. $f(x) = |x^2 - 5|$.
14. $f(x) = x^2(1 + x)^2$.
15. $f(x) = \dfrac{x - 1}{x + 1}$.
16. $f(x) = x^2 + \dfrac{16}{x^2}$.
17. $f(x) = \sqrt{\dfrac{1 + x^2}{2 + x^2}}$.
18. $f(x) = |x + 1||x - 2|$.
19. $f(x) = x - \cos x, \quad 0 \le x \le 2\pi$.
20. $f(x) = x + \sin x, \quad 0 \le x \le 2\pi$.
21. $f(x) = \cos 2x + 2\cos x, \quad 0 \le x \le \pi$.
22. $f(x) = \cos^2 x, \quad 0 \le x \le \pi$.
23. $f(x) = \sqrt{3}\,x - \cos 2x, \quad 0 \le x \le \pi$.
24. $f(x) = \sin^2 x - \sqrt{3}\sin x, \quad 0 \le x \le \pi$.

In Exercises 25–32, find f given the following information.

25. $f'(x) = x^2 - 1$ for all x, $f(1) = 2$.
26. $f'(x) = 2x - 5$ for all x, $f(2) = 4$.
27. $f'(x) = 5x^4 + 4x^3 + 3x^2 + 2x + 1$ for all x, $f(0) = 5$.
28. $f'(x) = 4x^{-3}$ for $x > 0$, $f(1) = 0$.
29. $f'(x) = x^{1/3} - x^{1/2}$ for $x > 0$, $f(0) = 1$.
30. $f'(x) = x^{-5} - 5x^{-1/5}$ for $x > 0$, $f(1) = 0$.
31. $f'(x) = 2 + \sin x$ for all x, $f(0) = 3$.
32. $f'(x) = 4x + \cos x$ for all x, $f(0) = 1$.

Find the intervals on which f increases and the intervals on which f decreases.

33. $f(x) = \begin{cases} x + 7, & x < -3 \\ |x + 1|, & -3 \le x < 1 \\ 5 - 2x, & 1 \le x. \end{cases}$

34. $f(x) = \begin{cases} (x - 1)^2, & x < 1 \\ 5 - x, & 1 \le x < 3 \\ 7 - 2x, & 3 \le x. \end{cases}$

35. $f(x) = \begin{cases} 4 - x^2, & x < 1 \\ 7 - 2x, & 1 \le x < 3 \\ 3x - 10, & 3 \le x. \end{cases}$

36. $f(x) = \begin{cases} x + 2, & x < 0 \\ (x - 1)^2, & 0 < x < 3 \\ 8 - x, & 3 < x < 7 \\ 2x - 9, & 7 < x \\ 6, & x = 0, 3, 7. \end{cases}$

In Exercises 37–40, f is a differentiable function and the graph of its derivative f' is given. If $f(0) = 1$, give a rough sketch of the graph of f.

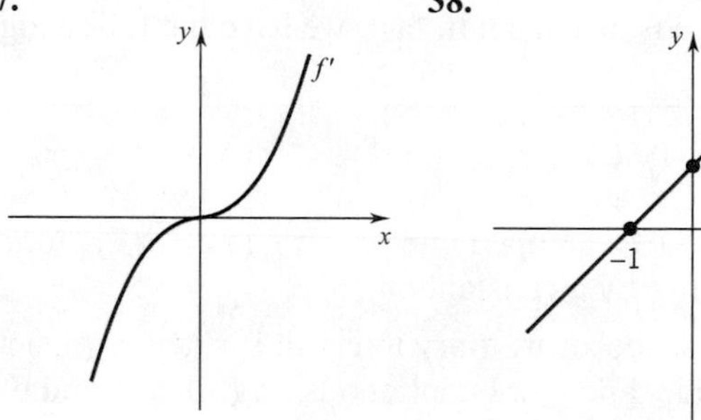

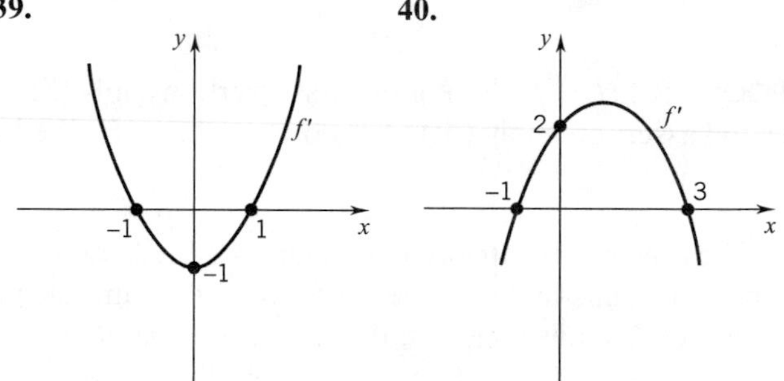

In Exercises 41 and 42 the graph of a continuous function f is given. Sketch the graph f'. Give the intervals on which $f'(x) > 0$ and the intervals on which $f'(x) < 0$.

41.

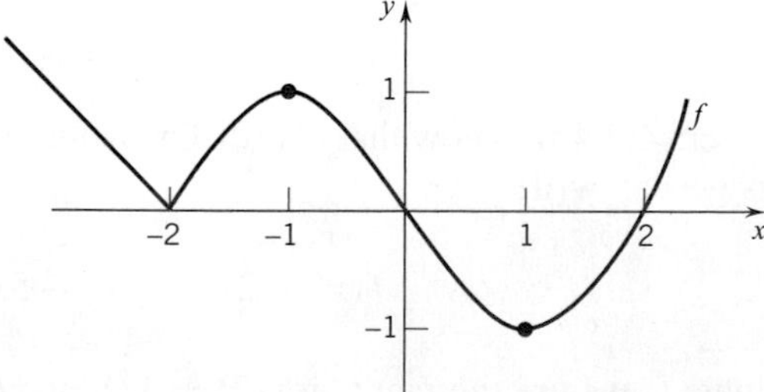

42.

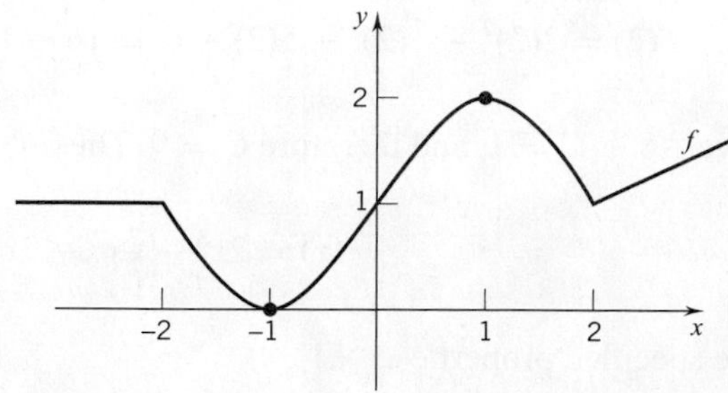

In Exercises 43–46, sketch the graph of a differentiable function f that satisfies the given conditions, if possible. If it is not possible, explain why.

43. $f(x) > 0$ for all x, $f(0) = 1$, and $f'(x) < 0$ for all x.

44. $f(1) = -1$, $f'(x) < 0$ for all $x \neq 1$, and $f'(1) = 0$.

45. $f(-1) = 4$, $f(2) = 2$, and $f'(x) > 0$ for all x.

46. f has x-intercepts only at $x = 1$ and $x = 2$, $f(3) = 4$, and $f(5) = -1$.

In Exercises 47–50, an object is moving along a straight line with its position at time t given by $x(t)$. Determine the time interval(s) when the object is moving to the right, determine the time interval(s) when it is moving to the left, and find the time(s) when it changes direction. Also, find the time interval(s) on which the velocity of the object is increasing and those on which it is decreasing. Draw a figure that illustrates the motion of the object. See Example 3, Section 3.4.

47. $x(t) = t^3 - 6t^2 + 9t + 2$.

48. $x(t) = (2t - 1)(t - 1)^2$.

49. $x(t) = 2\sin 3t, t \in [0, \pi]$.

50. $x(t) = 3\cos(2t + \frac{\pi}{4}), t \in [0, 2\pi]$.

51. Suppose that the function f increases on (a, b) and on (b, c). Suppose further that $\lim_{x\to b^-} f(x) = M$, $\lim_{x\to b^+} f(x) = N$, and $f(b) = L$. For which values of L, if any, can you conclude that f increases on (a, c) if (a) $M < N$? (b) $M > N$? (c) $M = N$?

52. Suppose that the function f increases on (a, b) and decreases on (b, c). Under what conditions can you conclude that f increases on $(a, b]$ and decreases on $[b, c)$?

53. Prove that if f is differentiable and increasing on the interval (a, b), then $f'(x) \geq 0$ for each x in (a, b). This is a partial converse of Theorem 4.2.2(i). HINT: Assume there is a number c in (a, b) for which $f'(c) < 0$ and apply Theorem 4.1.2.

54. Prove (4.2.4).

55. Let $f(x) = x - \sin x$.

(a) Prove that f increases on $(-\infty, \infty)$.

(b) Use the result in part (a) to show that $\sin x < x$ on $(0, \infty)$ and $\sin x > x$ on $(-\infty, 0)$.

56. Prove Theorem 4.2.5.

57. Let $f(x) = \sec^2 x$ and $g(x) = \tan^2 x$ on the interval $I = (-\pi/2, \pi/2)$.

(a) Show that $f'(x) = g'(x)$ for all $x \in I$.

(b) The result in part (a) implies, by Theorem 4.2.5, that $f - g = C$, a constant, on $(-\pi/2, \pi/2)$. Find the value of C.

58. Let L be a differentiable function on $(0, \infty)$ such that $L'(x) = 1/x$ and $L(1) = 0$. This function is studied in detail in Chapter 7. Prove that for any two positive numbers a and b, $L(ab) = L(a) + L(b)$. HINT: Begin by showing that, for each $x > 0$, $L(ax)$ and $L(x)$ have the same derivative.

59. Let f and g be differentiable functions such that $f'(x) = -g(x)$ and $g'(x) = f(x)$ for all x.

(a) Prove that $f^2(x) + g^2(x) = C$, a constant.

(b) Suppose there is a number a such that $f(a) = 1$ and $g(a) = 0$. Use this information to find the value of C. Give a pair of functions f, g that have these properties.

60. Let f and g be differentiable functions on the interval $(-c, c)$ such that $f(0) = g(0)$.

(a) Prove that if $f'(x) > g'(x)$ for all $x \in (0, c)$, then $f(x) > g(x)$ for all $x \in (0, c)$. HINT: Consider $F = f - g$.

(b) Prove that if $f'(x) > g'(x)$ for all $x \in (-c, 0)$, then $f(x) < g(x)$ for all $x \in (-c, 0)$.

61. Prove that $\tan x > x$ for all $x \in (0, \pi/2)$. HINT: Use Exercise 60(a).

62. Prove that $1 - x^2/2 < \cos x$ for all $x \in (0, \infty)$. HINT: Use Exercises 59 and 60.

63. Let $n > 1$ be an integer. Prove that $(1 + x)^n > 1 + nx$ for all $x \in (0, \infty)$.

64. Prove that $x - x^3/6 < \sin x$ for all $x \in (0, \infty)$.

C 65. It follows from Exercises 55 and 64 that

$$x - \tfrac{1}{6}x^3 < \sin x < x \quad \text{for } x > 0.$$

Use this result to estimate $\sin 4°$.

C 66. (a) Prove that $\cos x < 1 - \frac{1}{2}x^2 + \frac{1}{24}x^4$ for all $x \in (0, \infty)$. HINT: Use Exercise 64.

(b) It follows from part (a) and Exercise 62 that

$$1 - \tfrac{1}{2}x^2 < \cos x < 1 - \tfrac{1}{2}x^2 + \tfrac{1}{24}x^4 \quad \text{for } x > 0.$$

Use this result to estimate $\cos 6°$.

C In Exercises 67–70, use a graphing utility to graph the function f and its derivative f' on the indicated interval. Estimate the zeros of f' correct to three decimal places. Then estimate the subintervals on which f increases and the subintervals on which f decreases.

67. $f(x) = 3x^4 - 10x^3 - 4x^2 + 10x + 9$; $[-2, 5]$.

68. $f(x) = 2x^3 - x^2 - 13x - 6$; $[-3, 4]$.

69. $f(x) = x\cos x - 3\sin 2x$; $[0, 6]$.

70. $f(x) = x^4 + 3x^3 - 2x^2 + 4x + 4$; $[-5, 3]$.

C In Exercises 71–74, use a CAS to find where:

(a) $f'(x) = 0$, (b) $f'(x) > 0$. (c) $f'(x) < 0$,

71. $f(x) = \cos^3 x$, $0 \leq x \leq 2\pi$.

72. $f(x) = \dfrac{x}{\sqrt{x^2 + 4}}$.

73. $f(x) = \dfrac{x^2 - 1}{x^2 + 1}$.

74. $f(x) = 8x^5 - 36x^4 + 6x^3 + 73x^2 + 48x + 9$.

C 75. Use a graphing utility to draw the graph of $f(x) = \sin x \sin(x + 2) - \sin^2(x + 1)$. From the graph, what do you conclude about f? Calculate f' and verify the implication of the graph of f.

■ PROJECT 4.2 Energy of a Falling Body (Near the Surface of the Earth)

If we lift an object, we counteract the force of gravity. In so doing, we increase what physicists call the *gravitational potential energy* of the object. The gravitational potential energy, GPE, of an object is defined by

$$\text{GPE} = \text{weight} \times \text{height}.$$

Since the weight of an object of mass m is mg (we take this from physics), where g is the gravitational constant, we can write

$$\text{GPE} = mgy,$$

where y is the height.

If we lift an object and release it, the object drops. As it drops, it loses height and therefore loses gravitational potential energy, but its speed increases. The speed with which the object falls gives the object another form of energy called *kinetic energy,* the energy of motion. The kinetic energy, KE, of an object in straight-line motion is given by

$$\text{KE} = \tfrac{1}{2}mv^2,$$

where v is the velocity of the object.

Problem 1. Prove the *Law of Conservation of Energy*: $\text{GPE} + \text{KE} = C$, constant. HINT: Differentiate the expression GPE + KE and use the fact that $dv/dt = -g$.

Problem 2. An object initially at rest falls freely from height y_0. Show that the speed of the object at height y is given by

$$|v| = \sqrt{2g(y_0 - y)}.$$

Problem 3. According to the results in Section 3.4, the position of an object that falls from rest from a height y_0 is given by

$$y(t) = -\tfrac{1}{2}gt^2 + y_0.$$

Derive the speed of the object from this equation and show that the result is equivalent to the result found in Problem 2.

Problem 4. A bobsled run descends from a height of 150 meters from its starting point to the bottom of the run. Suppose that the bobsled starts from rest at the top and slides down the run without friction. Find the speed of the bobsled at the bottom of the run (use $g = 9.8$ meters per second per second). NOTE: To solve this problem by calculating the forces and acceleration would require a detailed knowledge of the path of the run down the hill. The Law of Conservation of Energy provides a simple method for finding the speed at the bottom.

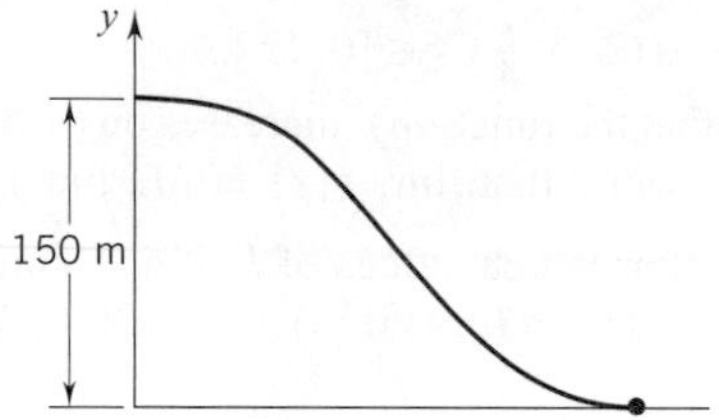

■ 4.3 LOCAL EXTREME VALUES

In many problems of economics, engineering, and physics it is important to determine how large or how small a certain quantity can be. If the problem admits a mathematical formulation, it is often reducible to the problem of finding the maximum or minimum value of some function.

In this section we consider maximum and minimum values for functions defined on an *open interval* or on a *union of open intervals*. We will take up functions defined on closed or half-closed intervals in the next section. We begin with a definition.

DEFINITION 4.3.1 LOCAL EXTREME VALUES

A function f is said to take on a *local maximum at c* if

$$f(c) \geq f(x) \qquad \text{for all } x \text{ sufficiently close to } c.\dagger$$

A function f is said to take on a *local minimum at c* if

$$f(c) \leq f(x) \qquad \text{for all } x \text{ sufficiently close to } c.$$

The local maxima and local minima of the f are called *local extreme values of f.*

† What do we mean by saying that "such and such is true *for all x sufficiently close to c*"? We mean that it is true for all x in some open interval $(c - \delta,\ c + \delta)$ centered at c.

We illustrate these notions in Figure 4.3.1. A careful look at the figure suggests that local maxima and minima occur only at points where the tangent is horizontal [$f'(c) = 0$] or where there is no tangent line [$f'(c)$ does not exist]. This is indeed the case.

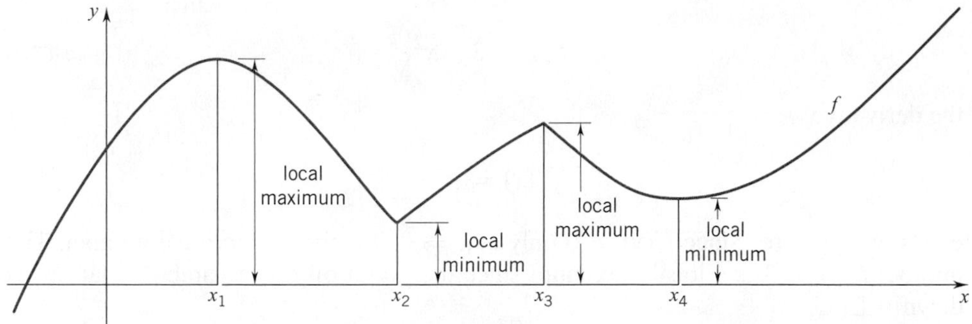

Figure 4.3.1

THEOREM 4.3.2

If f takes on a local maximum or minimum at c, then either

$$f'(c) = 0 \qquad \text{or} \qquad f'(c) \text{ does not exist.}$$

PROOF We are given that f has a local extreme value at c. Suppose that $f'(c)$ exists. If $f'(c) > 0$ or $f'(c) < 0$, then, by Theorem 4.1.2, there must be numbers x_1 and x_2 that are arbitrarily close to c which satisfy

$$f(x_1) < f(c) < f(x_2).$$

This makes it impossible for a local maximum or a local minimum to occur at c. Therefore, if $f'(c)$ exists, then it must have the value 0. The only other possibility is that $f'(c)$ does not exist. □

On the basis of this result, we have the following important definition:

DEFINITION 4.3.3 CRITICAL NUMBER

The numbers c in the domain of a function f for which either

$$f'(c) = 0 \qquad \text{or} \qquad f'(c) \text{ does not exist,}$$

are called the *critical numbers of f*.†

In view of Theorem 4.3.2, when searching for the local maxima and minima of a function f, the only numbers that we need to consider are the critical numbers of f.

†The terms *critical values* and *critical points* are also used.

We illustrate the technique for finding local maxima and minima by some examples. In each example the first step will be to find the critical numbers. We begin with very simple cases.

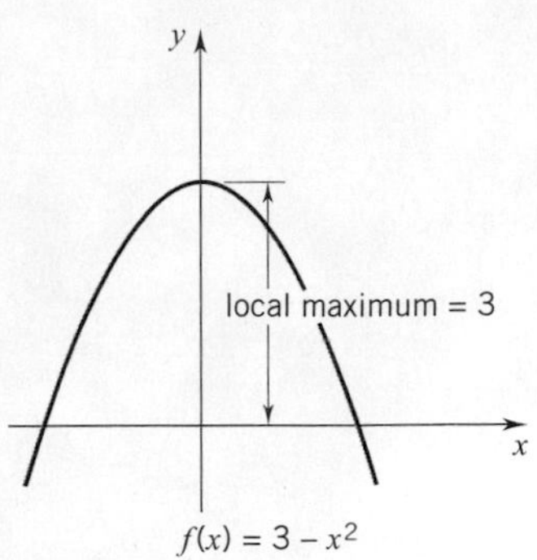

$f(x) = 3 - x^2$

Figure 4.3.2

Example 1 For

$$f(x) = 3 - x^2, \qquad \text{(Figure 4.3.2)}$$

the derivative

$$f'(x) = -2x$$

exists everywhere. Since $f'(x) = 0$ only at $x = 0, 0$ is the only critical number. The number $f(0) = 3$ is a local maximum since the graph of f is a parabola that opens down. ❑

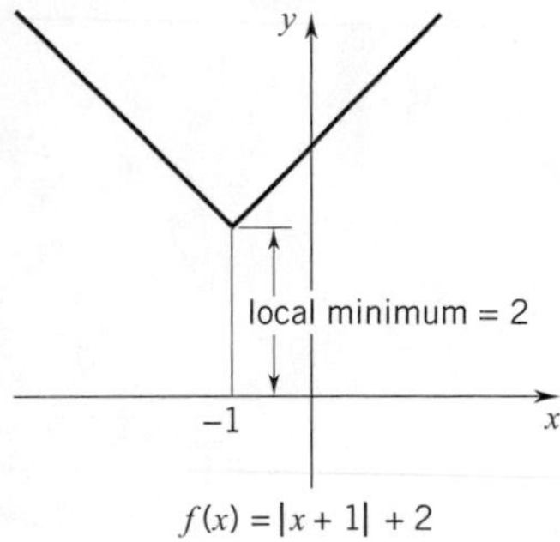

$f(x) = |x + 1| + 2$

Figure 4.3.3

Example 2 In the case of

$$f(x) = |x + 1| + 2 = \begin{cases} -x + 1, & x < -1 \\ x + 3, & x \geq -1, \end{cases} \qquad \text{(Figure 4.3.3)}$$

differentiation gives

$$f'(x) = \begin{cases} -1, & x < -1 \\ \text{does not exist}, & x = -1 \\ 1, & x > -1. \end{cases}$$

This derivative is never 0. It fails to exist only at -1. The number -1 is the only critical number. The value $f(-1) = 2$ is a local minimum. ❑

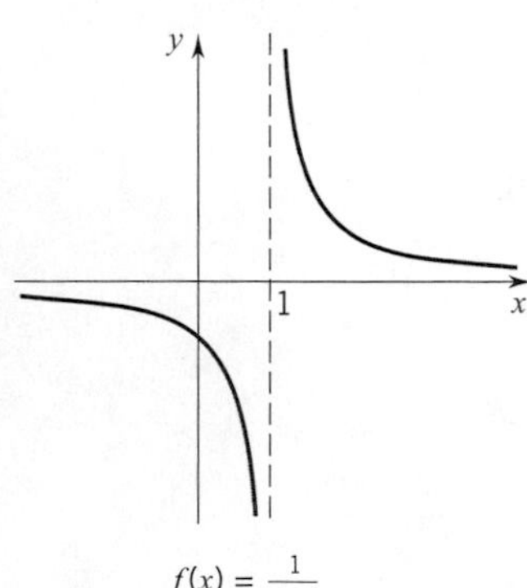

$f(x) = \frac{1}{x - 1}$

Figure 4.3.4

EXAMPLE 3 Figure 4.3.4 shows the graph of the function $f(x) = \dfrac{1}{x - 1}$. The domain is $(-\infty, 1) \cup (1, \infty)$. The derivative

$$f'(x) = -\frac{1}{(x - 1)^2}$$

exists throughout the domain of f and is never 0. Thus there are no critical numbers. In particular, 1 is not a critical number of f because 1 is not in the domain of f. Since f has no critical numbers, there are no local extreme values. ❑

CAUTION The fact that c is a critical number of f does not guarantee that $f(c)$ is a local extreme value. This is illustrated in the next two examples. ❑

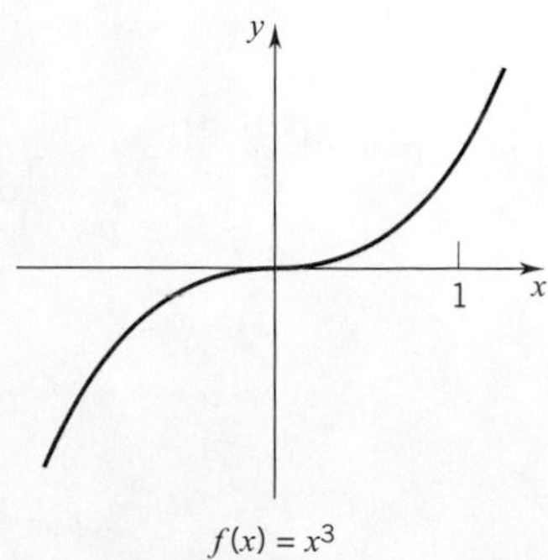

$f(x) = x^3$

Figure 4.3.5

Example 4 In the case of the function

$$f(x) = x^3, \qquad \text{(Figure 4.3.5)}$$

the derivative $f'(x) = 3x^2$ is 0 at 0, but $f(0) = 0$ is not a local extreme value. This function is everywhere increasing. ❑

Example 5 The function

$$f(x) = \begin{cases} -2x + 5, & x < 2 \\ -\frac{1}{2}x + 2, & x \geq 2 \end{cases} \qquad \text{(Figure 4.3.6)}$$

is everywhere decreasing. Although 2 is a critical number [$f'(2)$ does not exist], $f(2) = 1$ is not a local extreme value. □

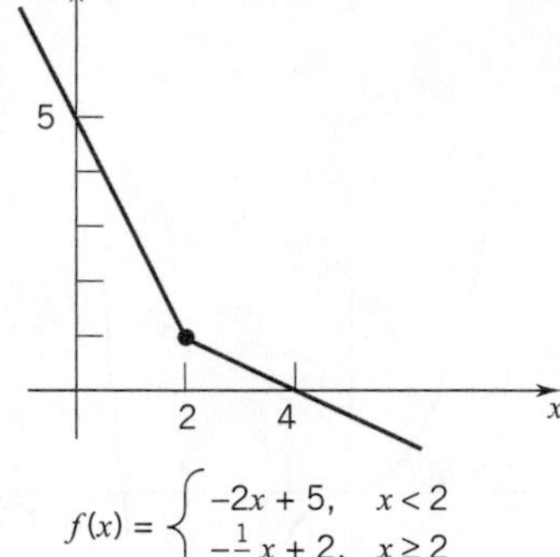

Figure 4.3.6

There are two widely used tests for determining the behavior of a function at a critical number. The first test (given in Theorem 4.3.4) requires that we examine the sign of the first derivative on both sides of the critical number. The second test (given in Theorem 4.3.5) requires that we examine the sign of the second derivative at the critical number itself.

> **THEOREM 4.3.4 THE FIRST-DERIVATIVE TEST**
>
> Suppose that c is a critical number of f and f is continuous at c. If there is a positive number δ such that:
>
> **(i)** $f'(x) > 0$ for all x in $(c - \delta, c)$ and $f'(x) < 0$ for all x in $(c, c + \delta)$, then $f(c)$ is a local maximum. (Figures 4.3.7 and 4.3.8)
>
> **(ii)** $f'(x) < 0$ for all x in $(c - \delta, c)$ and $f'(x) > 0$ for all x in $(c, c + \delta)$, then $f(c)$ is a local minimum. (Figures 4.3.9 and 4.3.10)
>
> **(iii)** $f'(x)$ keeps constant sign on $(c - \delta, c) \cup (c, c + \delta)$, then $f(c)$ is not a local extreme value.

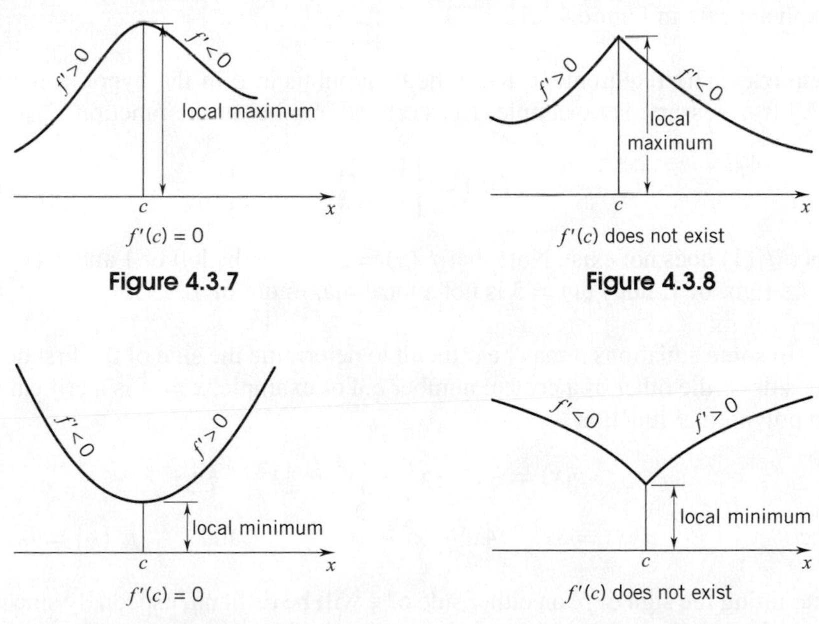

Figure 4.3.7 **Figure 4.3.8** **Figure 4.3.9** **Figure 4.3.10**

PROOF The result is a direct consequence of Theorem 4.2.3. The details of the proof are left to you as Exercise 29. □

Example 6 The function $f(x) = x^4 - 2x^3$ has derivative

$$f'(x) = 4x^3 - 6x^2 = 2x^2(2x - 3).$$

The only critical numbers are 0 and $\frac{3}{2}$. The sign of f' is recorded below.

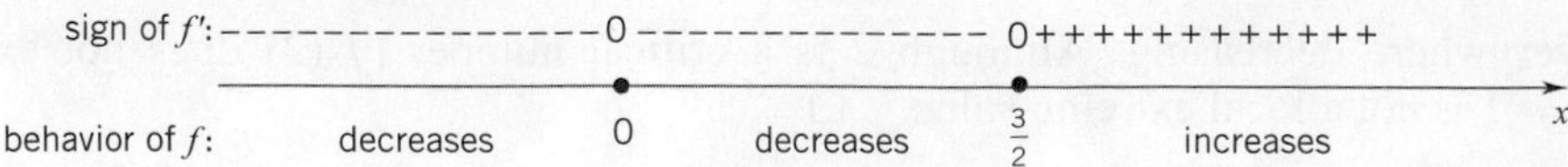

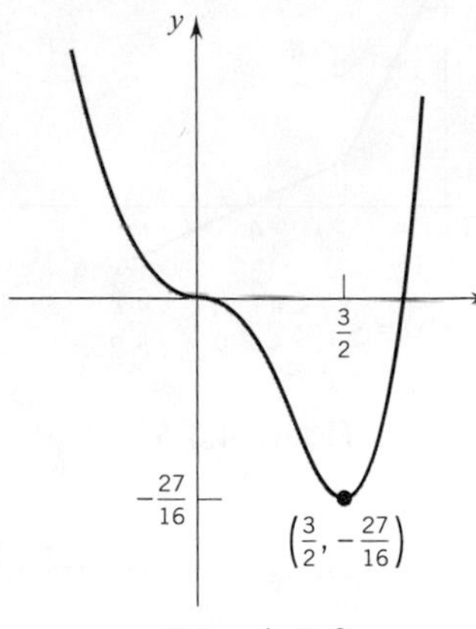

Figure 4.3.11

Since f' keeps the same sign on both sides of $0, f(0) = 0$ is not local extreme value. However, $f\left(\frac{3}{2}\right) = -\frac{27}{16}$ is a local minimum. The graph of f appears in Figure 4.3.11. ❑

Example 7 The function $f(x) = 2x^{5/3} + 5x^{2/3}$ is defined for all real x. The derivative of f is given by

$$f'(x) = \tfrac{10}{3}x^{2/3} + \tfrac{10}{3}x^{-1/3} = \tfrac{10}{3}x^{-1/3}(x+1), \qquad x \neq 0.$$

Since $f'(-1) = 0$ and $f'(0)$ does not exist, the critical numbers are -1 and 0. The sign of f' is recorded below. (To save space in the diagram, we write "dne" for "does not exist.")

sign of f': ++++++++++++ 0 ------------ dne ++++++++++++

behavior of f: increases -1 decreases 0 increases x

(−1, 3)

$f(x) = 2x^{5/3} + 5x^{2/3}$

Figure 4.3.12

In this case $f(-1) = 3$ is a local maximum and $f(0) = 0$ is a local minimum. The graph appears in Figure 4.3.12. ❑

Remark The requirement that f be continuous at c in the hypothesis of Theorem 4.3.4 is necessary. For example, 1 is a critical number of the function

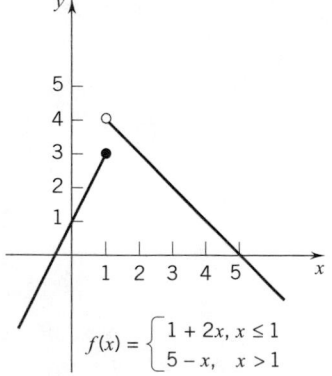

$$f(x) = \begin{cases} 1 + 2x, & x \leq 1 \\ 5 - x, & x > 1 \end{cases} \qquad \text{(Figure 4.3.13)}$$

since $f'(1)$ does not exist. Note that $f'(x) = 2 > 0$ to the left of 1 and $f'(x) = -1 < 0$ to the right of 1, but $f(1) = 3$ is not a local maximum of f. ❑

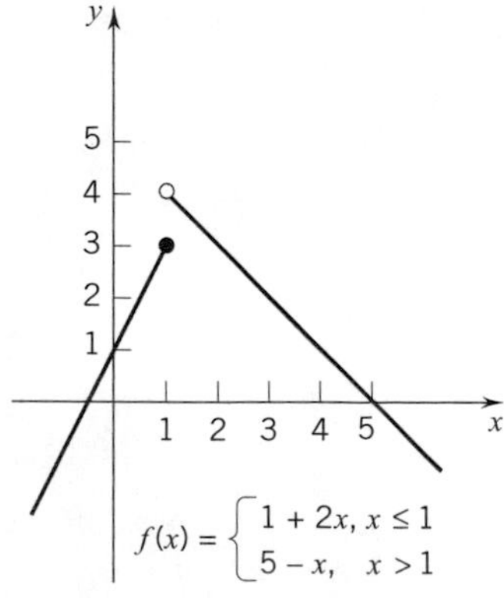

Figure 4.3.13

In some situations it may be difficult to determine the sign of the first derivative on one side or the other of a critical number c. For example, $x = \frac{2}{3}$ is a critical number of the polynomial function

$$p(x) = \tfrac{3}{5}x^5 + x^4 - \tfrac{1}{3}x^3 - \tfrac{1}{2}x^2 - \tfrac{2}{3}x + 1$$

since $\qquad p'(x) = 3x^4 + 4x^3 - x^2 - x - \frac{2}{3} \qquad$ and $\qquad p'\left(\frac{2}{3}\right) = 0.$

Examining the sign of p' on either side of $\frac{2}{3}$ will be difficult, especially since we do not know the location of the other critical numbers of p.

If c is a critical number for a function f and f is twice differentiable at c, then the following test is sometimes easier to apply.

THEOREM 4.3.5 THE SECOND-DERIVATIVE TEST

Suppose that $f'(c) = 0$ and $f''(c)$ exists.

(i) If $f''(c) > 0$, then $f(c)$ is a local minimum.

(ii) If $f''(c) < 0$, then $f(c)$ is a local maximum.

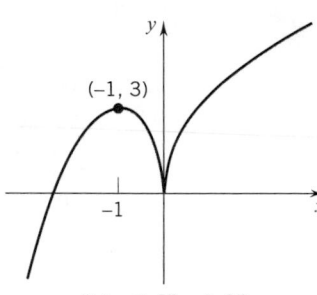

PROOF We handle the case $f''(c) > 0$. The other is left as an exercise. Since f'' is the derivative of f', we see from Theorem 4.1.2 that there exists a $\delta > 0$ such that, if

$$c - \delta < x_1 < c < x_2 < c + \delta,$$

then
$$f'(x_1) < f'(c) < f'(x_2).$$

Since $f'(c) = 0$, we have

$$f'(x) < 0 \quad \text{for } x \text{ in } (c - \delta, c) \quad \text{and} \quad f'(x) > 0 \quad \text{for } x \text{ in } (c, c + \delta).$$

By the first-derivative test, $f(c)$ is a local minimum. ❑

Continuing with the illustration begun earlier,

$$p''(x) = 12x^3 + 12x^2 - 2x - 1$$

and
$$p''\left(\tfrac{2}{3}\right) = 12\left(\tfrac{2}{3}\right)^3 + 12\left(\tfrac{2}{3}\right)^2 - 2\left(\tfrac{2}{3}\right) - 1 = \tfrac{59}{9} > 0.$$

Thus p has a local minimum at $x = \frac{2}{3}$.

Example 8 For $f(x) = 2x^3 - 3x^2 - 12x + 5$ we have

$$f'(x) = 6x^2 - 6x - 12 = 6(x^2 - x - 2) = 6(x - 2)(x + 1)$$

and
$$f''(x) = 12x - 6.$$

The critical numbers are 2 and -1; the first derivative is 0 at each of these points. Since $f''(2) = 18 > 0$ and $f''(-1) = -18 < 0$, we can conclude from the second-derivative test that $f(2) = -15$ is a local minimum and $f(-1) = 12$ is a local maximum. ❑

Comparing the First- and Second-Derivative Tests

The first-derivative test is more general than the second-derivative test. For example, we can apply the first-derivative test even at critical numbers where the function is not differentiable (provided, of course, that it is continuous there). In contrast, the second-derivative test has certain limitations:

(a) It can be applied only at critical numbers where the function is *twice* differentiable; and if $f''(c)$ does exist, then

(b) information is obtained only if $f''(c) \neq 0$.

The following examples illustrate these limitations.

Example 9 Consider the function $f(x) = x^{4/3}$. We have $f'(x) = \frac{4}{3}x^{1/3}$ so that

$$f'(0) = 0, \quad f'(x) < 0 \quad \text{for} \quad x < 0, \quad f'(x) > 0 \quad \text{for} \quad x > 0.$$

By the first-derivative test, $f(0) = 0$ is a local minimum. We cannot get this information from the second derivative because $f''(x) = \frac{4}{9}x^{-2/3}$ is not defined at $x = 0$. ❑

y
$f(x) = x^3$
x

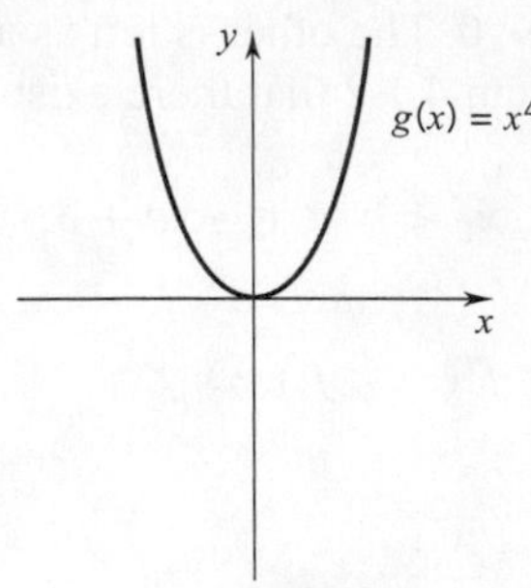

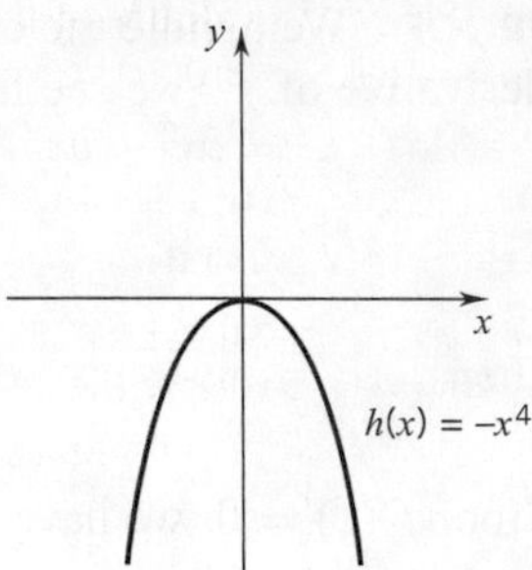

Figure 4.3.14

Example 10 To show what happen if the second derivative is zero at a critical number c, consider the functions

$$f(x) = x^3, \qquad g(x) = x^4, \qquad h(x) = -x^4. \qquad \text{(See Figure 4.3.14)}$$

In each case $x = 0$ is a critical number.

$$f'(x) = 3x^2; \quad g'(x) = 4x^3; \quad h'(x) = -4x^3;$$
$$f'(0) = 0; \quad g'(0) = 0; \quad h'(0) = 0;$$

and in each case the second derivative is zero at $x = 0$:

$$f''(x) = 6x; \quad g''(x) = 12x^2; \quad h''(x) = -12x^2;$$
$$f''(0) = 0; \quad g''(0) = 0; \quad h''(0) = 0.$$

In the first case, we already know that $f(x) = x^3$ has neither a local maximum nor a local minimum at 0 (see Example 4). In the case of $g(x) = x^4$, we have $g'(x) = 4x^3$ and

$$g'(x) < 0 \quad \text{for} \quad x < 0, \qquad g'(x) > 0 \quad \text{for} \quad x > 0.$$

so that $g(0) = 0$ is a local minimum by the first-derivative test. Finally, since the function h is simply the negative of the function g, it follows that h has a local maximum at $x = 0$. □

EXERCISES 4.3

Find the critical numbers of f and the local extreme values.

1. $f(x) = x^3 + 3x - 2$.

2. $f(x) = 2x^4 - 4x^2 + 6$.

3. $f(x) = x + \dfrac{1}{x}$.

4. $f(x) = x^2 - \dfrac{3}{x^2}$.

5. $f(x) = x^2(1 - x)$.

6. $f(x) = (1 - x)^2(1 + x)$.

7. $f(x) = \dfrac{1 + x}{1 - x}$.

8. $f(x) = \dfrac{2 - 3x}{2 + x}$.

9. $f(x) = \dfrac{2}{x(x + 1)}$.

10. $f(x) = |x^2 - 16|$.

11. $f(x) = x^3(1 - x)^2$.

12. $f(x) = \left(\dfrac{x - 2}{x + 2}\right)^3$.

13. $f(x) = (1 - 2x)(x - 1)^3$.

14. $f(x) = (1 - x)(1 + x)^3$.

15. $f(x) = \dfrac{x^2}{1 + x}$.

16. $f(x) = x\sqrt[3]{1 - x}$.

17. $f(x) = x^2\sqrt[3]{2 + x}$.

18. $f(x) = \dfrac{1}{x + 1} - \dfrac{1}{x - 2}$.

19. $f(x) = |x - 3| + |2x + 1|$.

20. $f(x) = x^{7/3} - 7x^{1/3}$.

21. $f(x) = x^{2/3} + 2x^{-1/3}$.

22. $f(x) = \dfrac{x^3}{x + 1}$.

23. $f(x) = \sin x + \cos x, \quad 0 < x < 2\pi$.

24. $f(x) = x + \cos 2x, \quad 0 < x < \pi$.

25. $f(x) = \sin^2 x - \sqrt{3}\sin x, \quad 0 < x < \pi$.

26. $f(x) = \sin^2 x, \quad 0 < x < 2\pi$.

27. $f(x) = \sin x \cos x - 3\sin x + 2x, \quad 0 < x < 2\pi$.

28. $f(x) = 2\sin^3 x - 3\sin x, \quad 0 < x < \pi$.

29. Prove Theorem 4.3.4 by applying Theorem 4.2.3.

30. Prove the validity of the second-derivative test in the case that $f''(c) < 0$.

31. Let $f(x) = ax^2 + bx + c, a \neq 0$. Prove that f has a local maximum at $x = -b/(2a)$ if $a < 0$ and a local minimum at $x = -b/(2a)$ if $a > 0$.

32. Let $f(x) = ax^3+bx^2+cx+d, a \neq 0$. Under what conditions on a, b, and c will f have: (1) two local exterma, (2) only one local extremum, (3) no local etrema?

33. Find the critical numbers and the local extreme values of the polynomial

$$P(x) = x^4 - 8x^3 + 22x^2 - 24x + 4.$$

Then show that the equation $P(x) = 0$ has exactly two real roots, both positive.

34. A function f has derivative f' given by

$$f'(x) = x^3(x-1)^2(x+1)(x-2).$$

At what numbers x, if any, does f have a local maximum? A local minimum?

35. Suppose that $p(x) = a_nx^n + a_{n-1}x^{n-1} + \cdots + a_1x + a_0$ has critical numbers at $x = -1, 1, 2, 3$, and corresponding values $p(-1) = 6$, $p(1) = 1$, $p(2) = 3$, $p(3) = 1$. Sketch a possible graph for p if: (a) n is odd, (b) n is even.

36. Suppose that $f(x) = Ax^2 + Bx + C$ has a local minimum at $x = 2$ and passes through the points $(-1, 3)$ and $(3, -1)$. Find A, B, and C.

37. Determine the numbers a and b such that the function $f(x) = ax/(x^2+b^2)$ has a local minimum at $x = -2$ and $f'(0) = 1$.

38. Let $f(x) = x^p(1-x)^q$, where $p \geq 2$ and $q \geq 2$ are integers.

(a) Show that the critical numbers of f are $x = 0, p/(p+q)$, and 1.

(b) Show that if p is even, then f has a local minimum at 0.

(c) Show that if q is even, then f has a local minimum at 1.

(d) Show that f has a local maximum at $p/(p+q)$ for all p and q.

39. Let

$$f(x) = \begin{cases} x^2 \sin(1/x), & x \neq 0. \\ 0, & x = 0. \end{cases}$$

In Exercise 63, Section 3.1, we saw that f is differentiable at 0 and that $f'(0) = 0$. Show that f has neither a local maximum nor a local minimum at 0.

40. Suppose that $C(x)$, $R(x)$, and $P(x)$ are the cost, revenue, and profit functions corresponding to the production and sale of x units of a certain item. Suppose, also, that C and R are differentiable functions. Then, since $P = R - C$, it follows that P is differentiable. Prove that if it is possible to maximize the profit by producing and selling x_0 items, then $C'(x_0) = R'(x_0)$. That is, the marginal cost equals the marginal revenue when the profit is maximized.

41. Let $y = f(x)$ be differentiable and suppose that the graph of f does not pass through the origin. Then the distance D from the origin to a point $P(x, f(x))$ on the graph is given by

$$D = \sqrt{x^2 + [f(x)]^2}.$$

Show that if D has a local extreme value at c, then the line through $(0, 0)$ and $(c, f(c))$ is perpendicular to the line tangent the graph of f at $(c, f(c))$

42. Prove that a polynomial of degree n has at most $n - 1$ local extreme values.

C 43. Let $f(x) = x^4 - 7x^2 - 8x - 3$.

(a) Show that f has exactly one critical number c in the interval (2,3).

(b) Use the Newton-Raphson method to approximate c; calculate x_3 and round off your answer to four decimal places. Does f have a local maximum, a local minimum, or neither a maximum nor a minimum at c?

C 44. Let $f(x) = \sin x + (x^2/2) - 2x$.

(a) Show that f has exactly one critical number in the interval [2, 3].

(b) Use the Newton-Raphson method to approximate c; calculate x_3 and round off your answer to four decimal places. Does f have a local maximum, a local minimum, or neither a maximum nor a minimum at c?

C 45. Let $f(x) = \dfrac{ax^2 + b}{cx^2 + d}$ with $d \neq 0$. Use a CAS to show that f has a minimum at $x = 0$ if $ad - bc > 0$ and a maximum at $x = 0$ if $ad - bc < 0$. Verify this result by finding $ad - bc$ for the function f and then using a graphing utility to draw the graph.

(a) $f(x) = \dfrac{2x^2 + 3}{4 - x^2}$. (b) $f(x) = \dfrac{3 - 2x^2}{x^2 + 2}$.

C In Exercises 46–48, use a graphing utility to graph the function f on the indicated interval. (a) Use the graph to estimate the critical numbers and the local extreme values; and (b) estimate the intervals on which f increases and the intervals on which f decreases. Round off your estimates to three decimal places.

46. $f(x) = 3x^3 - 7x^2 - 14x + 24$; $[-3, 4]$.

47. $f(x) = |3x^3 + x^2 - 10x + 2| + 3x$; $[-4, 4]$.

48. $f(x) = \dfrac{8 \sin 2x}{1 + \frac{1}{2}x^2}$; $[-3, 3]$.

C In Exercises 49–52, find the extreme values of f by using a graphing utility to draw the graphs of f and f' together and noting the values of x at which $f'(x) = 0$.

49. $f(x) = -x^5 + 13x^4 - 67x^3 + 171x^2 - 216x + 108$.

50. $f(x) = \dfrac{x^2 - 9}{x^2 + 9}$.

51. $f(x) = x^2\sqrt{3x - 2}$.

52. $f(x) = \cos^2 2x$.

C In Exercises 53 and 54, the derivative f' of a function f is given. Use a graphing utility to graph f' on the indicated interval. Estimate the critical numbers of f and determine whether f has a local maximum, a local minimum, or neither a maximum nor a minimum at each. Round off your estimates to three decimal places.

53. $f'(x) = \sin^2 x + 2 \sin 2x$; $[-2, 2]$.

54. $f'(x) = 2x^3 + x^2 - 4x + 3$; $[-4, 4]$.

■ 4.4 ENDPOINT AND ABSOLUTE EXTREME VALUES

Endpoint Maxima and Minima

For functions defined on an open interval or on a union of open intervals, the critical numbers are the numbers where the derivative is 0 or where the derivative does not exist. Here we consider functions defined on a closed or half-closed interval,

$$[a, b], \quad [a, b), \quad (a, b], \quad [a, \infty), \quad \text{or} \quad (-\infty, b],$$

or on a union of such intervals.

Endpoints can give rise to what are called *endpoint maxima* and *endpoint minima*. See, for example, Figures 4.4.1, 4.4.2, 4.4.3, and 4.4.4.

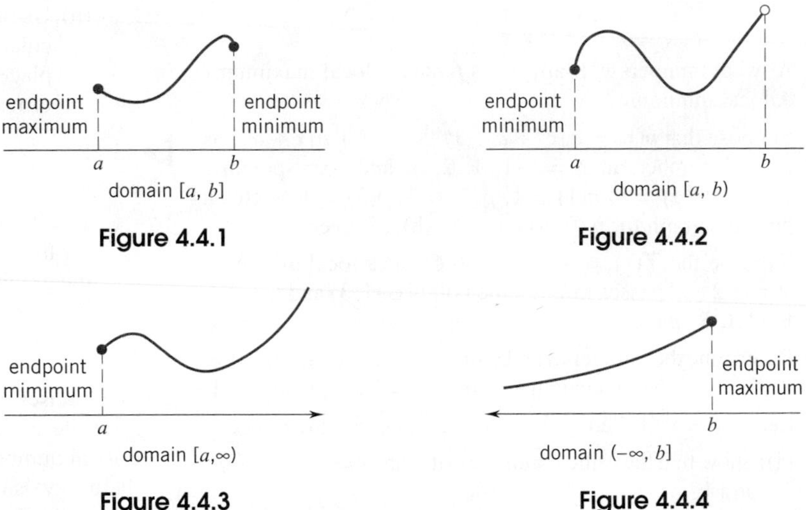

Figure 4.4.1

Figure 4.4.2

Figure 4.4.3

Figure 4.4.4

> **DEFINITION 4.4.1 ENDPOINT EXTREME VALUES**
>
> If c is an endpoint of the domain of f, then f is said to have an *endpoint maximum* at c if
>
> $$f(c) \geq f(x) \quad \text{for all } x \text{ in the domain of } f \text{ sufficiently close to } c.$$
>
> It is said to have an *endpoint minimum* at c if
>
> $$f(c) \leq f(x) \quad \text{for all } x \text{ in the domain of } f \text{ sufficiently close to } c.$$

Endpoints are usually tested either by calculating a one-sided derivative at the endpoint or by examining the sign of the derivative at nearby points. Suppose, for example, that a is the left endpoint and f is continuous from the right at a. If $f'_+(a) < 0$, or if $f'(x) < 0$ for all x sufficiently close to a, then f decreases on an interval of the form $[a, a + \delta]$ and therefore $f(a)$ must be an endpoint maximum. (See Figure 4.4.1.) On the other hand, if $f'_+(a) > 0$, or if $f'(x) > 0$ for all x sufficiently close to a, then f increases on an interval of the form $[a, a+\delta)$ and so $f(a)$ must be an endpoint minimum. (See Figure 4.4.2.) Similar reasoning can be applied to right endpoints.

Absolute Maxima and Minima

Whether or not a function f has a local or endpoint extreme value at some point depends entirely on the behavior of f for x close to that point. Absolute extreme values, which we define below, depend on the behavior of the function on its entire domain.

We begin with a number d in domain of f. Here d can be an interior point or an endpoint.

> **DEFINITION 4.4.2 ABSOLUTE EXTREME VALUES**
>
> A function f is said to have an *absolute maximum at d* if
>
> $$f(d) \geq f(x) \qquad \text{for all } x \text{ in the domain of } f,$$
>
> f is said to have an *absolute minimum* at d if
>
> $$f(d) \leq f(x) \qquad \text{for all } x \text{ in the domain of } f.$$

A function can be continuous on an interval (or even differentiable there) without taking on an absolute maximum or an absolute minimum. (The functions depicted in Figures 4.4.2 and 4.4.3 have no absolute maximum. The function depicted in Figure 4.4.4 has no absolute minimum.) All we can say in general is that if f takes on an absolute extreme value, then it does so at a critical number or at an endpoint.

There are, however, special conditions that guarantee the existence of absolute extreme values. From Section 2.6 we know that continuous functions map bounded closed intervals $[a, b]$ onto bounded closed intervals $[m, M]$; M is the maximum value taken on by f on $[a, b]$ and m is the minimum value. If $[a, b]$ constitutes the entire domain of f, then, clearly, M is the absolute maximum value of f and m is the absolute minimum. Thus, the absolute extreme values of a continuous function f on a bounded closed interval $[a, b]$ can be obtained as follows:

Step 1. Find the critical numbers $c_1, c_2, \cdots$ of f in the open interval (a, b).

Step 2. Calculate $f(a)$, $f(c_1)$, $f(c_2)$, $\cdots$, $f(b)$.

Step 3. The largest of the numbers found in step 2 is the absolute maximum value of f and the smallest is the absolute minimum.

Example 1 Find the critical numbers and classify all extreme values of

$$f(x) = 1 + 4x^2 - \tfrac{1}{2}x^4, \quad x \in [-1, 3].$$

SOLUTION Since f is continuous on the bounded closed interval $[-1, 3]$, we know f has an absolute maximum and an absolute minimum. To find the critical numbers of f in $(-1, 3)$, we differentiate:

$$f'(x) = 8x - 2x^3 = 2x(4 - x^2) = 2x(2 - x)(2 + x).$$

Now, $f'(x)$ is defined for all $x \in (-1, 3)$ and $f'(x) = 0$ at $x = 0$ and $x = 2$. Thus, 0 and 2 are the critical numbers.

The sign of f' and the behavior of f are as follows:

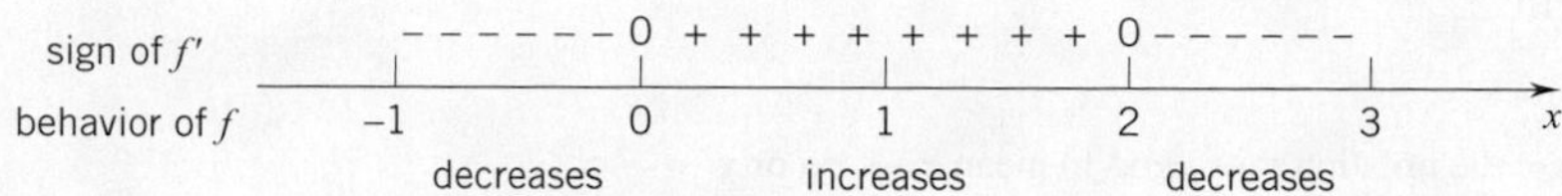

Therefore

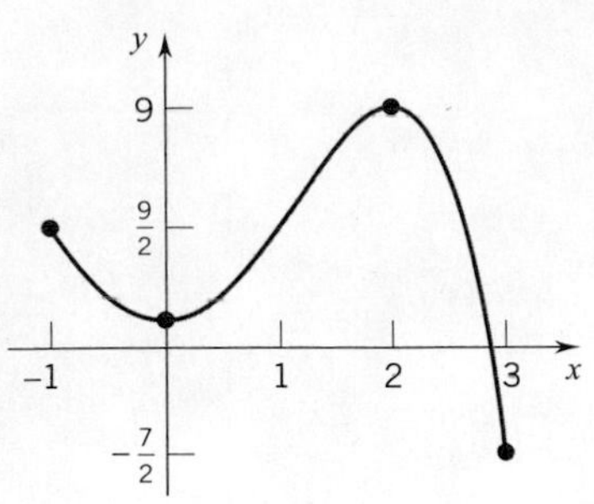

Figure 4.4.5

$$f(-1) = 1 + 4(-1)^2 - \tfrac{1}{2}(-1)^4 = \tfrac{9}{2} \quad \text{is an endpoint maximum;}$$
$$f(0) = 1 \quad \text{is a local minimum;}$$
$$f(2) = 1 + 4(2)^2 - \tfrac{1}{2}(2^4) = 9 \quad \text{is a local maximum;}$$
$$f(3) = 1 + 4(3)^2 - \tfrac{1}{2}(3)^4 = -\tfrac{7}{2} \quad \text{is an endpoint minimum.}$$

The smallest of these extrema, $f(3) = -\frac{7}{2}$, is the absolute minimum; the largest of these extrema, $f(2) = 9$, is the absolute maximum. The graph of the function is shown in Figure 4.4.5. ☐

Example 2 Find the critical numbers and classify all extreme values of the function

$$f(x) = x^2 - 2|x| + 2 = \begin{cases} x^2 + 2x + 2, & -\frac{1}{2} \le x < 0 \\ x^2 - 2x + 2, & 0 \le x \le 2. \end{cases}$$

SOLUTION Since f is continuous on its entire domain, which is the bounded closed interval $[-\frac{1}{2}, 2]$, We know that f has an absolute maximum and an absolute minimum. This function is differentiable on the open interval $(-\frac{1}{2}, 2)$, except at $x = 0$:

$$f'(x) = \begin{cases} 2x + 2, & -\frac{1}{2} < x < 0 \\ \text{does not exist}, & x = 0 \\ 2x - 2, & 0 < x < 2. \end{cases} \qquad \text{(verify this)}$$

This makes $x = 0$ a critical number. Since $f'(x) = 0$ at $x = 1$, 1 is a critical number.

The sign of f' and the behavior of f are as follows:

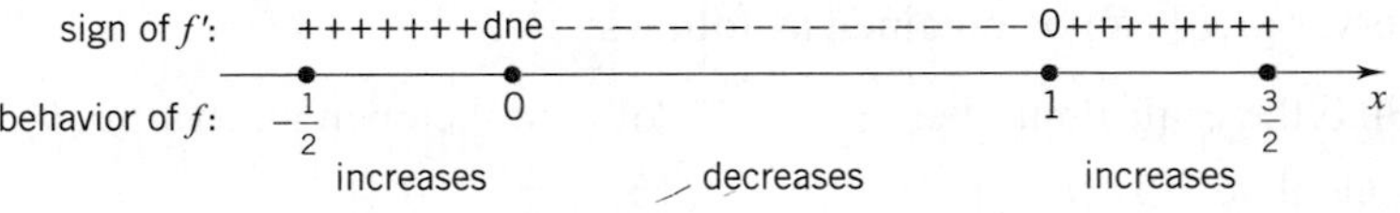

Therefore

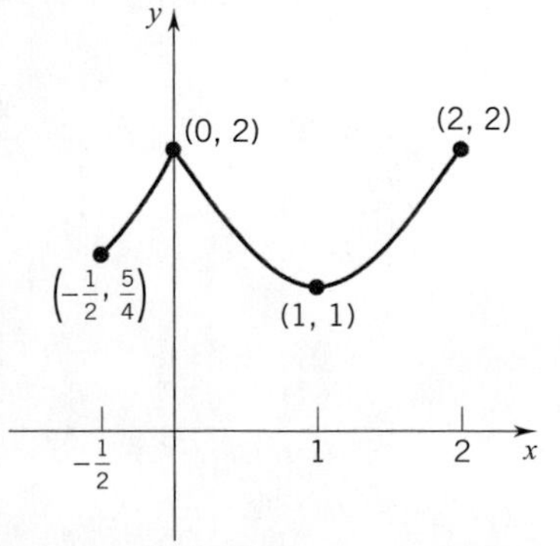

Figure 4.4.6

$$f\left(-\tfrac{1}{2}\right) = \tfrac{1}{4} - 1 + 2 = \tfrac{5}{4} \quad \text{is an endpoint minimum;}$$
$$f(0) = 2 \quad \text{is a local maximum;}$$
$$f(1) = 1 - 2 + 2 = 1 \quad \text{is a local minimum;}$$
$$f(2) = 2 \quad \text{is an endpoint maximum.}$$

The least of these extrema, $f(1) = 1$, is the absolute minimum; the greatest of these extrema, $f(0) = f(2) = 2$, is the absolute maximum. The graph of the function is shown in Figure 4.4.6. ☐

Behavior of f as $x \to \pm\infty$†

We now state four definitions. Once you grasp the first one, the others become transparent.

† We will use the notation $x \to \pm\infty$ to mean $x \to \infty$ or $x \to -\infty$.

To say that

$$\text{as } x \to \infty, \quad f(x) \to \infty$$

means that, *as x increases without bound, f(x) becomes arbitrarily large.* In formal terms, the definition is as follows: given any positive number M, there exists a positive number K such that

$$\text{if } x \geq K, \quad \text{then} \quad f(x) \geq M.$$

Thus, for example, as $x \to \infty$,

$$x^2 \to \infty, \quad \sqrt{1+x^2} \to \infty, \quad \tan\left(\frac{\pi}{2} - \frac{1}{x^2}\right) \to \infty$$

To say that

$$\text{as } x \to \infty, \quad f(x) \to -\infty$$

is to say that, *as x increases without bound, f(x) becomes arbitrary large negative.* Again in formal terms, the definition is as follows: given any negative number M, there exists a positive number K such that

$$\text{if } \quad x \geq K, \quad \text{then} \quad f(x) \leq M.$$

Thus, for example, as $x \to \infty$,

$$-x^4 \to -\infty, \quad 1 - \sqrt{x} \to -\infty, \quad \tan\left(\frac{1}{x^2} - \frac{\pi}{2}\right) \to -\infty.$$

To say that

$$\text{as } \quad x \to -\infty, \quad f(x) \to \infty$$

is to say that, *as x decreases without bound, f(x) becomes arbitrarily large*: given any positive number M, there exists a negative number K such that

$$\text{if } \quad x \leq k, \quad \text{then} \quad f(x) \geq M.$$

Thus, for example, as $x \to -\infty$,

$$x^2 \to \infty, \quad \sqrt{1-x} \to \infty, \quad \tan\left(\frac{\pi}{2} - \frac{1}{x^2}\right) \to \infty.$$

Finally, to say that

$$\text{as } \quad x \to -\infty, \quad f(x) \to -\infty$$

is to say that, *as x decreases without bound, f(x) becomes arbitrarily large negative*: given any negative number M, there exists a negative number K such that,

$$\text{if } \quad x \leq K, \quad \text{then} \quad f(x) \leq M.$$

Thus, for example, as $x \to -\infty$,

$$x^3 \to -\infty, \quad -\sqrt{1-x} \to -\infty, \quad \tan\left(\frac{1}{x^2} - \frac{\pi}{2}\right) \to -\infty.$$

Remark As you can readily see, $f(x) \to -\infty$ iff $-f(x) \to \infty$. □

Suppose now that P is a nonconstant polynomial:

$$P(x) = a_n x^n + a_{n-1}x^{n-1} + \cdots + a_1 x + a_0 \qquad (a_n \neq 0, n \geq 1).$$

For large $|x|$—that is, for large positive x or for large negative x— the leading term $a_n x^n$ dominates. Thus, what happens to $P(x)$ as $x \to \pm\infty$ depends entirely on what happens to $a_n x^n$. (For confirmation, see Exercise 43).

Example 3

(a) As $x \to \infty$, $3x^4 \to \infty$, and therefore $3x^4 - 100x^3 + 2x - 5 \to \infty$.

(b) As $x \to -\infty$, $5x^3 \to -\infty$, and therefore $5x^3 + 12x^2 + 80 \to -\infty$. □

Finally, we point out that if $f(x) \to \infty$, then f cannot have an absolute maximum value, and if $f(x) \to -\infty$, then f cannot have an absolute minimum value.

Example 4 Find the critical numbers and classify all the extreme values of

$$f(x) = 6\sqrt{x} - x\sqrt{x}.$$

SOLUTION The domain is $[0, \infty)$. To differentiate, we rewrite f using fractional exponents:

$$f(x) = 6x^{1/2} - x^{3/2}.$$

Now, on $(0, \infty)$,

$$f'(x) = 3x^{-1/2} - \frac{3}{2}x^{1/2} = \frac{3(2-x)}{2\sqrt{x}}. \qquad \text{(verify this)}$$

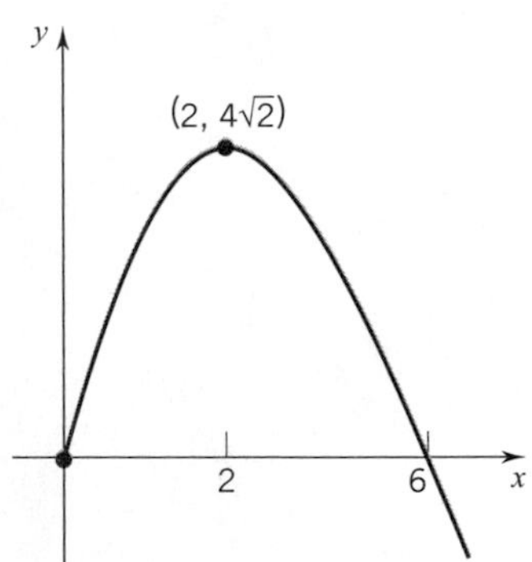

Figure 4.4.7

Since $f'(x) = 0$ at $x = 2$, we see that 2 is an critical number.

The sign of f' and the behavior of f are as follows:

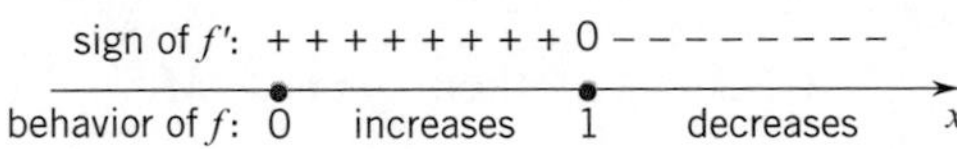

Therefore,

$$f(0) = 0 \text{ is an endpoint minimum;}$$
$$f(2) = 6\sqrt{2} - 2\sqrt{2} = 4\sqrt{2} \text{ is a local maximum.}$$

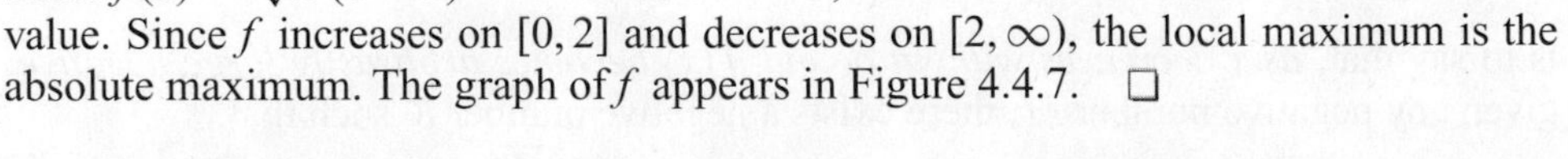

Since $f(x) = \sqrt{x}\,(6 - x) \to -\infty$ as $x \to \infty$, the function has no absolute minimum value. Since f increases on $[0, 2]$ and decreases on $[2, \infty)$, the local maximum is the absolute maximum. The graph of f appears in Figure 4.4.7. □

A Summary for Finding All the Extreme Values (Local, Endpoint, and Absolute) of a Continuous Function f

Step 1. Find the critical numbers — the interior numbers c at which $f'(c) = 0$ or $f'(c)$ does not exist.

Step 2. Test each endpoint of the domain by examining the sign of the first derivative nearby.

Step 3. Test each critical number c by examining the sign of the first derivative on both sides of c (first-derivative test) or by checking the sign of the second derivative at c itself (second derivative test).

Step 4. If the domain of f is unbounded, determine the behavior of f as $x \to \infty$ or as $x \to -\infty$.

Step 5. Determine whether any of the endpoint extremes and local extremes are absolute extremes.

Example 5 Find the critical numbers and classify all the extreme values of

$$f(x) = \tfrac{1}{4}(x^3 - \tfrac{3}{2}x^2 - 6x + 2), \quad x \in [-2, \infty).$$

SOLUTION To find the critical numbers, we differentiate:

$$f'(x) = \tfrac{1}{4}(3x^2 - 3x - 6) = \tfrac{3}{4}(x+1)(x-2).$$

Since $f'(x) = 0$ at $x = -1$ and $x = 2$, the numbers -1 and 2 are critical numbers.

The sign of f' and the behavior of f are as follows:

sign of f': + + + + + + + + 0 - - - - - - - - - - - - - - - - 0 + + + + + +

behavior of f: -2 increases -1 decreases 2 increases x

We can see from the sign of f' that

$$f(-2) = \tfrac{1}{4}(-8 - 6 + 12 + 2) = 0 \quad \text{is an endpoint minimum;}$$
$$f(-1) = \tfrac{1}{4}(-1 - \tfrac{3}{2} + 6 + 2) = \tfrac{11}{8} \quad \text{is a local maximum;}$$
$$f(2) = \tfrac{1}{4}(8 - 6 - 12 + 2) = -2 \quad \text{is a local minimum.}$$

The function takes on no absolute maximum value since $f(x) \to \infty$ as $x \to \infty$; $f(2) = -2$ is the absolute minimum value of f. The graph of f is shown in Figure 4.4.8. ☐

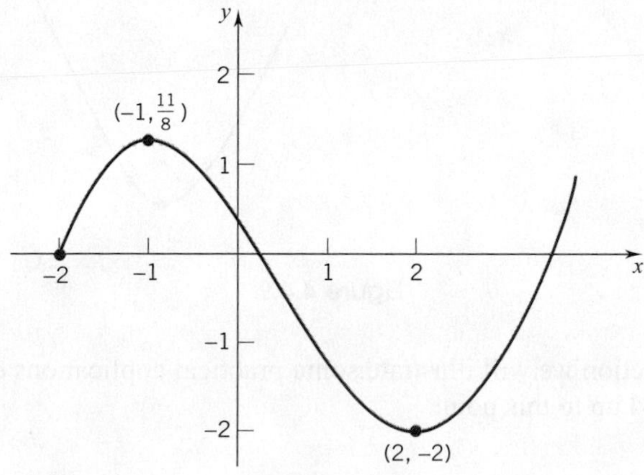

Figure 4.4.8

Example 6 Find the critical numbers and classify all the extreme values of

$$f(x) = \sin x - \sin^2 x, \quad x \in [0, 2\pi].$$

SOLUTION On $(0, 2\pi)$

$$f'(x) = \cos x - 2\sin x \cos x$$
$$= \cos x(1 - 2\sin x).$$

Setting $f'(x) = 0$, we have

$$\cos x(1 - 2\sin x) = 0.$$

This equation is satisfied when $\cos x = 0$, which gives $x = \pi/2$ and $x = 3\pi/2$, and when $\sin x = \frac{1}{2}$, which gives $x = \pi/6$ and $x = 5\pi/6$. Thus, the numbers $\pi/6$, $\pi/2$, $5\pi/6$, and $3\pi/2$ are critical numbers.

The sign of f' and the behavior of f are as follows:

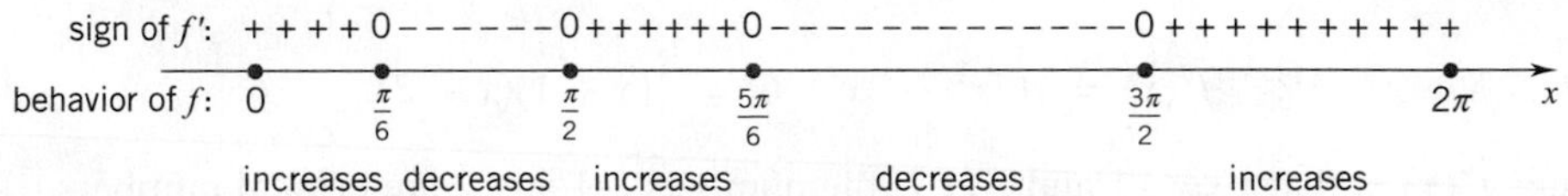

Therefore,

$f(0) = 0$ is an endpoint minimum; $f(\pi/6) = \frac{1}{4}$ is a local maximum;

$f(\pi/2) = 0$ is a local minimum; $f(5\pi/6) = \frac{1}{4}$ is a local maximum;

$f(3\pi/2) = -2$ is a local minimum; $f(2\pi) = 0$ is an endpoint maximum.

Also, $f(\pi/6) = f(5\pi/6) = \frac{1}{4}$ is the absolute maximum of f, and $f(3\pi/2) = -2$ is the absolute minimum. The graph of the function is shown in Figure 4.4.9. □

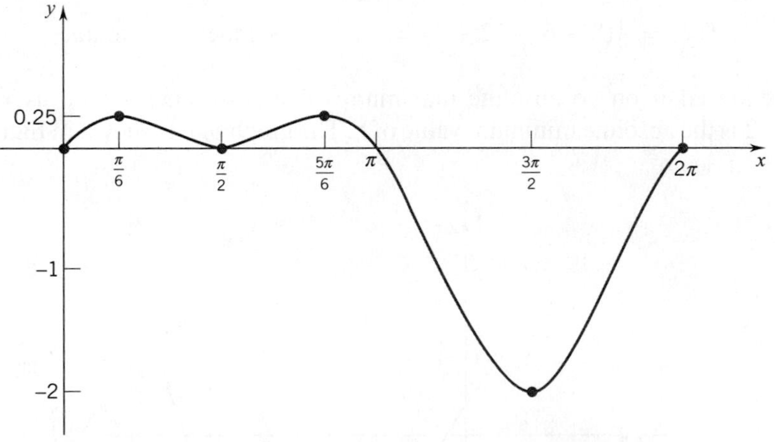

Figure 4.4.9

In the next section we will illustrate some practical applications of the theory that we have developed up to this point.

EXERCISES 4.4

Find the critical numbers and classify the extreme values.

1. $f(x) = \sqrt{x+2}$.

2. $f(x) = (x-1)(x-2)$.

3. $f(x) = x^2 - 4x + 1, \quad x \in [0, 3]$.

4. $f(x) = 2x^2 + 5x - 1, \quad x \in [-2, 0]$.

5. $f(x) = x^2 + \dfrac{1}{x}$.

6. $f(x) = x + \dfrac{1}{x^2}$.

7. $f(x) = x^2 + \dfrac{1}{x}, \quad x \in [\frac{1}{10}, 2]$.

8. $f(x) = x + \dfrac{1}{x^2}, \quad x \in [1, \sqrt{2}]$.

9. $f(x) = (x-1)(x-2), \quad x \in [0, 2]$.

10. $f(x) = (x-1)^2(x-2)^2, \quad x \in [0, 4]$.

11. $f(x) = \dfrac{x}{4+x^2}, \quad x \in [-3, 1]$.

12. $f(x) = \dfrac{x^2}{1+x^2}, \quad x \in [-1, 2]$.

13. $f(x) = (x - \sqrt{x})^2$.

14. $f(x) = x\sqrt{4-x^2}$.

15. $f(x) = x\sqrt{3-x}$.

16. $f(x) = \sqrt{x} - \dfrac{1}{\sqrt{x}}$.

17. $f(x) = 1 - \sqrt[3]{x-1}$.

18. $f(x) = (4x-1)^{1/3}(2x-1)^{2/3}$.

19. $f(x) = \sin^2 x - \sqrt{3}\cos x, \quad 0 \le x \le \pi$.

20. $f(x) = \cot x + x, \quad 0 < x \le \dfrac{2\pi}{3}$.

21. $f(x) = 2\cos^3 x + 3\cos x, \quad 0 \le x \le \pi$.

22. $f(x) = \sin 2x - x, \quad 0 \le x \le \pi$.

23. $f(x) = \tan x - x, \quad -\dfrac{\pi}{3} \le x < \dfrac{\pi}{2}$.

24. $f(x) = \sin^4 x - \sin^2 x, \quad 0 \le x \le \dfrac{2\pi}{3}$.

25. $f(x) = \begin{cases} -2x, & 0 \le x < 1 \\ x-3, & 1 \le x \le 4 \\ 5-x, & 4 < x \le 7. \end{cases}$

26. $f(x) = \begin{cases} x+9, & -8 \le x < -3 \\ x^2+x, & -3 \le x \le 2 \\ 5x-4, & 2 < x < 5. \end{cases}$

27. $f(x) = \begin{cases} x^2+1, & -2 \le x < -1 \\ 5+2x-x^2, & -1 \le x \le 3 \\ x-1, & 3 < x < 6. \end{cases}$

28. $f(x) = \begin{cases} 2-2x-x^2, & -2 \le x \le 0 \\ |x-2|, & 0 < x < 3 \\ \frac{1}{3}(x-2)^3, & 3 \le x \le 4. \end{cases}$

29. $f(x) = \begin{cases} |x+1|, & -3 \le x < 0 \\ x^2-4x+2, & 0 \le x < 3 \\ 2x-7, & 3 \le x < 4. \end{cases}$

30. $f(x) = \begin{cases} -x^2, & 0 \le x < 1 \\ -2x, & 1 < x < 2 \\ -\frac{1}{2}x^2, & 2 \le x \le 3. \end{cases}$

In Exercises 31–34, sketch the graph of a differentiable function that satisfies the given conditions, if possible. If it is not possible, state why.

31. Domain is $(-\infty, \infty)$, local maximum at -1, local minimum at 1, $f(3)=6$ is the absolute maximum, no absolute minimum.

32. Domain is $[0, \infty), f(0) = 1$ is the absolute minimum, local maximum at 4, local minimum at 7, no absolute maximum.

33. Domain is $[-3, 3]$, $f(-3) = f(3) = 0$, $f(x) \ne 0$ for all $x \in (-3, 3)$, $f'(-1) = f'(1) = 0$, $f'(x) > 0$ if $|x| > 1$, $f'(x) < 0$ if $|x| < 1$.

34. Domain is $(-1, 3]$, local minimum at $2, f(1) = 5$ is the absolute maximum, no absolute minimum.

35. Show that the cubic polynomial $p(x) = x^3 + ax^2 + bx + c$ has no extreme values iff $a^2 \le 3b$.

36. Let r be a rational number, $r > 1$, and let f be the function defined by

$$f(x) = (1+x)^r - (1+rx) \quad \text{for} \quad x \ge -1.$$

Show that 0 is a critical number of f and show that $f(0) = 0$ is the absolute minimum of f.

37. Suppose that c is a critical number for f and $f'(x) > 0$ for $x \ne c$. Prove that if $f(c)$ is a local maximum, then f is not continuous at c.

38. What can be said about the function f continuous on $[a, b]$, if for some c in $(a, b), f(c)$ is both a local maximum and a local minimum?

39. Suppose that f is continuous on its domain $[a, b]$ and $f(a) = f(b)$. Prove that f has at least one critical number in (a, b).

40. Suppose that $c_1 < c_2$ and that f takes on local maxima at c_1 and c_2. Prove that if f is continuous on $[c_1, c_2]$, then there is at least one c in (c_1, c_2) at which f takes on a local minimum.

41. Give an example of a nonconstant function that takes on both its absolute maximum and absolute minimum on every interval.

42. Give an example of a nonconstant function that has an infinite number of local maxima and an infinite number of local minima.

43. Let P be a polynomial with positive leading coefficient:

$$P(x) = a_n x^n + a_{n-1}x^{n-1} + \cdots + a_1 x + a_0 \qquad (n \ge 1, a_n > 0).$$

Clearly, as $x \to \infty, a_n x^n \to \infty$. Show that, as $x \to \infty$, $P(x) \to \infty$ by showing that, given any positive number M, there exists a positive number K such that, if $x \ge K$, then $P(x) \ge M$.

44. An object which moves along a straight line with position $x(t)$ given by a function of the form

$$x(t) = A\sin(\omega t + \phi_0),$$

with A, ω, and ϕ_0 as constants, is said to be in *simple harmonic motion*. Such is the case for a mass m which, suspended by a spring, oscillates up and down about the equilibrium position.

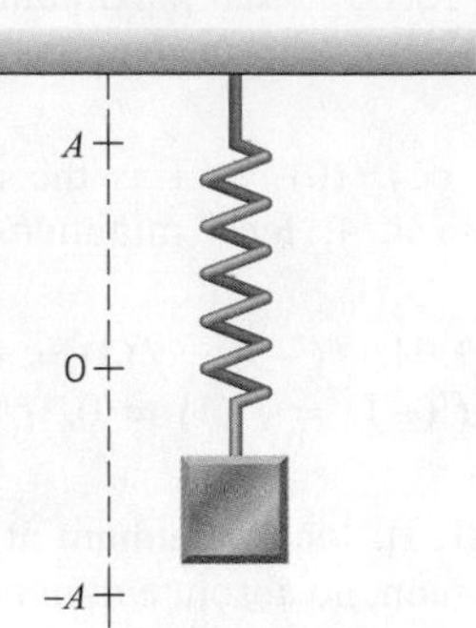

(a) Show that the position function $x(t)$ satisfies
$$x''(t) + \omega^2 x(t) = 0.$$
(b) Find the absolute maximum and the absolute minimum values of $x(t)$.

45. Of all rectangles with a given diagonal of length c, prove that the square has the largest area.

46. The sum of two numbers is 16. Find the numbers if the sum of their cubes is an absolute minimum.

47. If the angle of elevation of a cannon is x and a projectile is fired with muzzle velocity v ft/sec, then the range is given by
$$R = \frac{v^2 \sin 2x}{32} \text{ feet.}$$
What angle of elevation maximizes the range?

48. A piece of wire of length L will be cut into two pieces, one piece to form a square and the other piece to form an equilateral triangle. How should the wire be cut so as to:
(a) maximize the sum of the areas of the square and the triangle?
(b) minimize the sum of the areas of the square and the triangle?

C In Exercises 49-52,use a graphing utility to graph the function f on the indicated interval. Estimate the critical numbers of f and classify all the extreme values. Round off your estimates to three decimal places.

49. $f(x) = x^3 - 4x + 2x \sin x; [-2.5, 3]$.

50. $f(x) = x^4 - 7x^2 + 10x + 3; [-3, 3]$.

51. $f(x) = x \cos 2x - \cos^2 x; [-\pi, \pi]$.

52. $f(x) = 5x^{2/3} + 3x^{5/3} + 1; [-3, 1]$.

C In Exercises 53–55, use a graphing utility to determine whether f satisfies the extreme-value theorem on $[a, b]$. If the theorem is satisfied, find the absolute maximum value M and the absolute minimum value m. If the theorem is not satisfied, determine whether M or m exist anyway.

53. $f(x) = \begin{cases} 1 - \sqrt{2 - x}, & \text{if } 1 \le x \le 2 \\ 1 - \sqrt{x - 2}, & \text{if } 2 < x \le 3; \end{cases} \quad [a, b] = [1, 3].$

54. $f(x) = \begin{cases} \frac{11}{4}x - \frac{19}{4}, & \text{if } 0 \le x \le 3 \\ \sqrt{x - 3} + 2, & \text{if } 3 < x \le 4; \end{cases} \quad [a, b] = [0, 4].$

55. $f(x) = \begin{cases} \frac{1}{2}x + 1, & \text{if } 1 \le x < 4 \\ \sqrt{x - 3} + 2, & \text{if } 4 \le x \le 6; \end{cases} \quad [a, b] = [1, 6].$

C In Exercises 56 and 57, use a CAS to find the absolute extreme values of f.

56. $f(x) = \sin x - \cos 2x, \quad 0 \le x \le 2\pi.$

57. $f(x) = \dfrac{x^2 - x}{x^2 + 4}, \quad -2 \le x \le 3.$

C In Exercises 58–60, use a CAS in conjunction with the second-derivative test to find the absolute extreme values of f.

58. $f(x) = \sin^2 x - 2\cos^2 x, \quad -\pi \le x \le \pi.$

59. $f(x) = x^4 + 2x^3 + 5x^2, \quad -2 \le x \le 2.$

60. $f(x) = \dfrac{x^2 + 2x + 2}{x^2}, \quad -4 \le x \le -1.$

■ 4.5 SOME MAX-MIN PROBLEMS

The techniques of the preceding two sections can be brought to bear on a wide variety of optimization problems. The *key* idea in solving such problems is to express the quantity to be maximized or minimized as a function of one variable. If the function is differentiable, we can differentiate and analyze the results. We begin with a geometrical example.

Example 1 An isosceles triangle has a base of 6 units and a height of 12 units. Find the maximum possible area of a rectangle that can be placed inside the triangle with one side resting on the base of the triangle. What are the dimensions of the rectangle(s) of maximum area?

SOLUTION, Figure 4.5.1 shows the isosceles triangle with a typical rectangle placed inside. In Figure 4.5.2, we have introduced a rectangular coordinate system. With x and y as indicated in Figure 4.5.2, we want to maximize the area A, which is given by
$$A = 2xy.$$

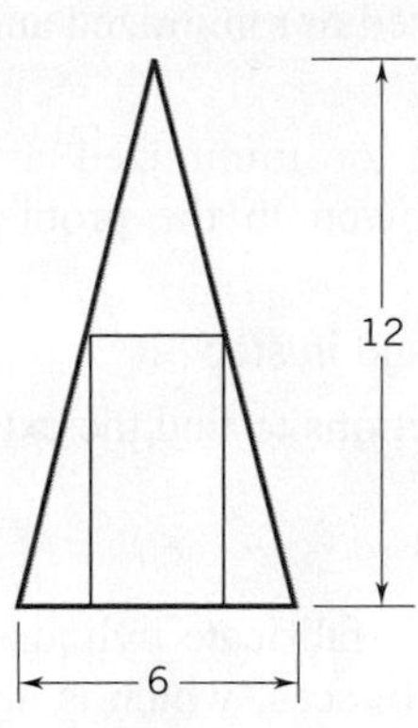

Figure 4.5.1

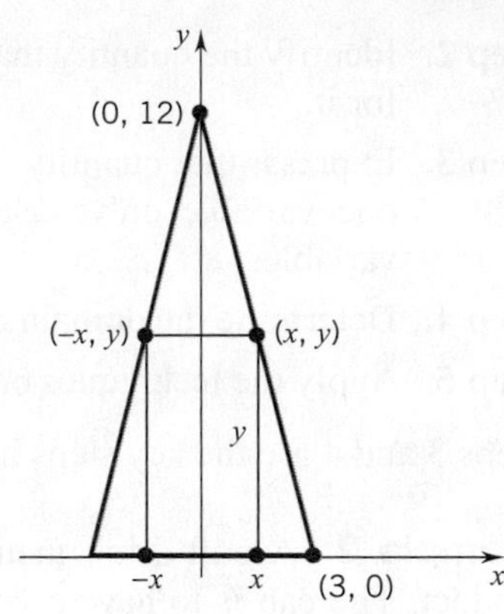

Figure 4.5.2

Since the point (x, y) lies on the line through $(0, 12)$ and $(3, 0)$, it can be shown that

$$y = 12 - 4x.$$

Now the area can be expressed in terms of x alone:

$$A(x) = 2x(12 - 4x) = 24x - 8x^2.$$

Also, since x and y represent lengths and A is an area, these variables cannot be negative. It follows that $0 \leq x \leq 3$.

Our problem can now be formulated as follows: find the absolute maximum of the function

$$A(x) = 24x - 8x^2 \quad x \in [0, 3].$$

The derivative

$$A'(x) = 24 - 16x$$

is defined for all $x \in (0, 3)$. Setting $A'(x) = 0$ yields

$$24 - 16x = 0, \qquad \text{which implies} \quad x = \tfrac{3}{2}.$$

The only critical number is $x = \frac{3}{2}$. Evaluating A at the two endpoints and at the critical number, we have:

$$\begin{aligned}
A(0) &= 24(0) - 8(0)^2 = 0,\\
A\left(\tfrac{3}{2}\right) &= 24\left(\tfrac{3}{2}\right) - 8\left(\tfrac{3}{2}\right)^2 = 18,\\
A(3) &= 24(3) - 8(3)^2 = 0.
\end{aligned}$$

The maximum possible area is $A(\frac{3}{2}) = 18$ square units. The dimensions of the rectangle of maximum possible area are: base $= 3$ units, height $= 6$ units. □

This example illustrates a basic strategy for solving max-min problems.

Strategy

Step 1. Draw a representative figure and assign labels to the relevant quantities involved in the problem.

Step 2. Identify the quantity that is to be maximized or minimized and find a formula for it.

Step 3. Express the quantity to be maximized or minimized as a function of one variable only; use the conditions given in the problem to eliminate variables.

Step 4. Determine the domain of the function found in step 3.

Step 5. Apply the techniques of the preceding sections to find the extreme value(s).

Steps 3 and 4 are the key steps in this procedure.

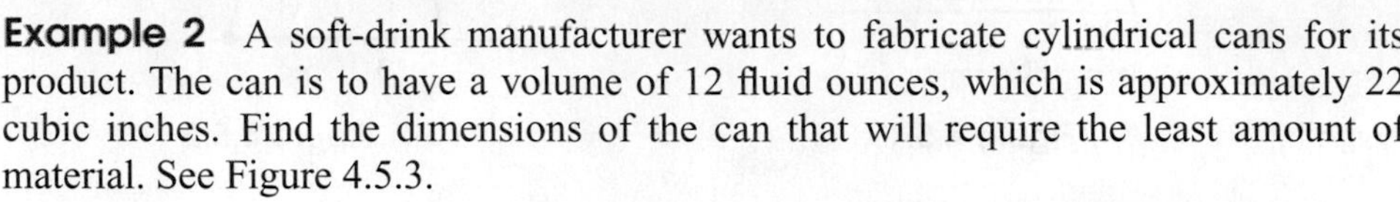

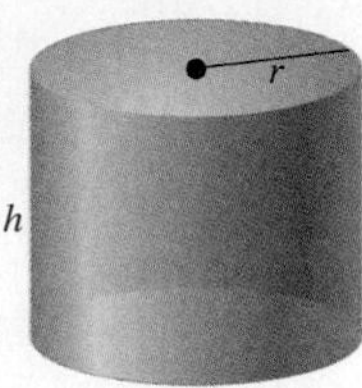

Figure 4.5.3

Example 2 A soft-drink manufacturer wants to fabricate cylindrical cans for its product. The can is to have a volume of 12 fluid ounces, which is approximately 22 cubic inches. Find the dimensions of the can that will require the least amount of material. See Figure 4.5.3.

SOLUTION Let r be the radius of the can and h the height. The total surface area (top, bottom, lateral area) of a circular cylinder of radius r and height h is given by the formula

$$S = 2\pi r^2 + 2\pi rh.$$

This is the quantity that we want to minimize.

Now, since the volume $V = \pi r^2 h$ has to be 22 cubic inches, we have

$$\pi r^2 h = 22$$

and

$$h = \frac{22}{\pi r^2}.$$

It follows from these equations that r and h must both be positive.

Thus, we want to minimize the function

$$S(r) = 2\pi r^2 + 2\pi r\left(\frac{22}{\pi r^2}\right)$$

$$= 2\pi r^2 + \frac{44}{r}, \qquad r \in (0, \infty).$$

(*Key steps completed.*)
Differentiation gives

$$\frac{dS}{dr} = 4\pi r - \frac{44}{r^2} = \frac{4\pi r^3 - 44}{r^2} = \frac{4(\pi r^3 - 11)}{r^2}.$$

The derivative is 0 where $\pi r^3 - 11 = 0$, which is the point $r_0 = (11/\pi)^{1/3}$. Since

$$\frac{dS}{dr} \text{ is } \begin{cases} \text{negative for } & r < r_0 \\ 0 \text{ for } & r = r_0 \\ \text{positive for } & r > r_0, \end{cases}$$

S decreases on $(0, r_0]$ and increases on $[r_0, \infty)$. Therefore, the function S is minimized by setting $r = r_0 = (11/\pi)^{1/3}$.

The dimensions of the can that will require the least amount of material are as follows:

$$\text{radius } r = \sqrt[3]{\frac{11}{\pi}} \cong 1.5 \text{ inches}, \qquad \text{height } h = \frac{22}{\pi(11/\pi)^{2/3}} = 2\sqrt[3]{\frac{11}{\pi}} \cong 3 \text{ inches};$$

the height of the cylindrical can should be twice the radius. □

Example 3 A window in the shape of a rectangle capped by a semicircle is to have perimeter p. Choose the radius of the semicircular part so that the window admits the greatest amount of light.

SOLUTION The window of maximum area will admit the greatest amount of light. As in Figure 4.5.4, we let x be the radius of the semicircular part and y be the height of the rectangular part. We want to express the area

$$A = \tfrac{1}{2}\pi x^2 + 2xy$$

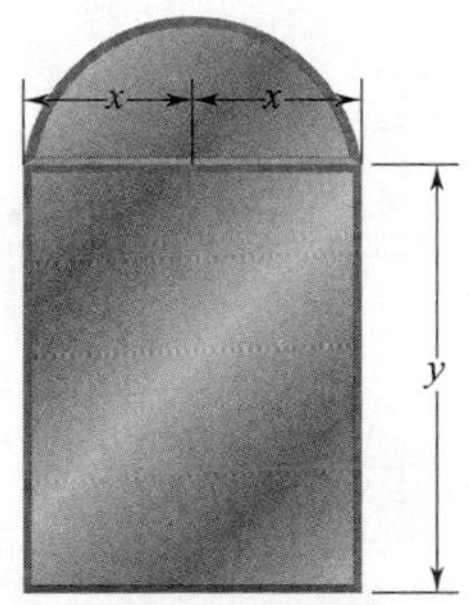

Figure 4.5.4

as a function of x alone. To do this, we must express y in terms of x.

Since the perimeter is p, we have

$$p = 2x + 2y + \pi x$$

and thus

$$y = \tfrac{1}{2}[p - (2 + \pi)x].$$

Since x and y represent lengths, these variables must be nonnegative, and so it follows that $0 \le x \le p/(2+\pi)$.

The area can now be expressed in terms of x alone:

$$\begin{aligned} A(x) &= \tfrac{1}{2}\pi x^2 + 2xy \\ &= \tfrac{1}{2}\pi x^2 + 2x\left\{\tfrac{1}{2}[p - (2+\pi)x]\right\} \\ &= \tfrac{1}{2}\pi x^2 + px - (2+\pi)x^2 = px - (2 + \tfrac{1}{2}\pi)x^2. \end{aligned}$$

We want to maximize the function

$$A(x) = px - \left(2 + \tfrac{1}{2}\pi\right)x^2, \quad x \in \left[0, \frac{p}{2+\pi}\right].$$

(Key steps completed.)

The derivative

$$A'(x) = p - (4+\pi)x$$

is 0 only at $x = p/(4+\pi)$. Since $A(0) = A(p/(2+\pi)) = 0$, and since $A'(x) > 0$ for $0 < x < p/(4+\pi)$ and $A'(x) < 0$ for $p/(4+\pi) < x < p/(2+\pi)$, the function A is maximized by setting $x = p/(4+\pi)$. Thus the window has maximum area when the radius of the semicircular part is $p/(4+\pi)$. □

Example 4 A state highway department plans to construct a new road between towns A and B. Town A lies on an abandoned road that runs east-west. Town B is 3 miles north

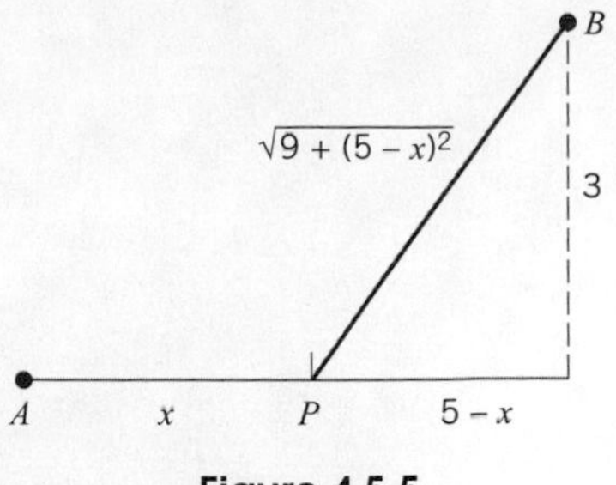

Figure 4.5.5

of the point on that road that is 5 miles east of A. The engineering division proposes that the road be constructed by restoring a section of the old road from A up to a point P and joining it to a new road that connects P and B. If the cost of restoring the old road is \$200,000 per mile and the cost of the new road is \$400,000 per mile, how much of the old road should be restored in order to minimize the department's costs?

SOLUTION Figure 4.5.5 shows the geometry of the problem. Notice that we have chosen a straight line joining P and B rather than some curved path (the shortest connection between two points is provided by the straight-line path). We let x be the amount of old road that will be restored. Then

$$\sqrt{9+(5-x)^2} = \sqrt{34-10x+x^2}$$

is the length of the new road. The total cost of constructing the two sections of the road is

$$C(x) = 200,000x + 400,000[34-10x+x^2]^{1/2}, \quad 0 \le x \le 5.$$

We want to find the value of x that minimizes this function.
(Key steps completed.)
Differentiation gives

$$\begin{aligned} C'(x) &= 200,000 + 400,000\left(\tfrac{1}{2}\right)[34-10x+x^2]^{-1/2}(2x-10) \\ &= 200,000 + \frac{400,000(x-5)}{[34-10x+x^2]^{1/2}}, \qquad 0 < x < 5. \end{aligned}$$

Setting $C'(x) = 0$, we find that

$$\begin{aligned} 1 + \frac{2(x-5)}{[34-10x+x^2]^{1/2}} &= 0 \\ 2(x-5) &= -[34-10x+x^2]^{1/2} \\ 4(x^2-10x+25) &= 34-10x+x^2 \\ 3x^2-30x+66 &= 0 \\ x^2-10x+22 &= 0. \end{aligned}$$

Now, by the quadratic formula, we have

$$x = \frac{10 \pm \sqrt{100-4(22)}}{2} = 5 \pm \sqrt{3}.$$

The value $x = 5+\sqrt{3}$ is not in the domain of our function; the value we want is $x = 5-\sqrt{3}$. We analyze the sign of C':

sign of C': 0 – – – – – – – – 0 + + + + + 0
behavior of C: 0 .... $5-\sqrt{3}$ .... 5 x
decreases .... increases

Thus $x = 5-\sqrt{3} \cong 3.27$ gives the absolute minimum value for C. The highway department will minimize its costs by restoring 3.27 miles of the old road.

As you can check, the minimum cost is $\$200,000(5+3\sqrt{3}) \cong \$2,039,230$. Using all of the old road, the cost would be exactly \$2,200,000, while using none it would cost $\$400,000\sqrt{34} \cong \$2,332,400$. The most expensive option costs almost \$300,000 more than the cheapest option. ❑

Example 5 (*The angle of incidence equals the angle of reflection.*) Figure 4.5.6 depicts light from a point A reflected to a point B by a mirror. Two angles have been marked: the *angle of incidence*, θ_i, and the *angle of reflection*, θ_r. Experiment shows that $\theta_i = \theta_r$. Derive this result by postulating that the light that travels from A to the mirror and then to B follows the shortest possible path.†

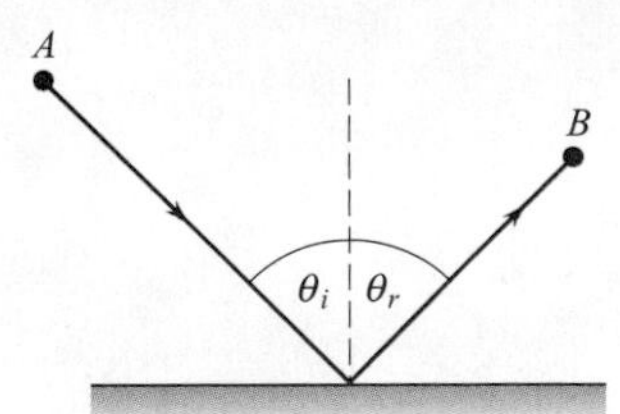

Figure 4.5.6

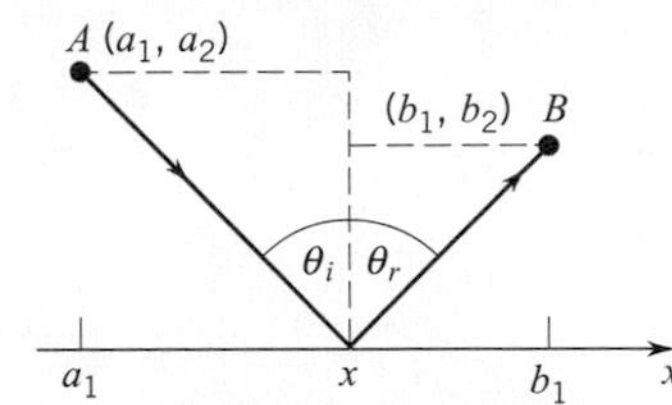

Figure 4.5.7

SOLUTION We can write the length of the path as a function of x. In the setup of Figure 4.5.7,

$$L(x) = \sqrt{(x - a_1)^2 + a_2^2} + \sqrt{(x - b_1)^2 + b_2^2}, \quad x \in [a_1, b_1].$$

Differentiation gives

$$L'(x) = \frac{x - a_1}{\sqrt{(x - a_1)^2 + a_2^2}} + \frac{x - b_1}{\sqrt{(x - b_1)^2 + b_2^2}}.$$

Therefore

$$\begin{aligned} L'(x) = 0 \quad \text{iff} \quad & \frac{x - a_1}{\sqrt{(x - a_1)^2 + a_2^2}} = \frac{b_1 - x}{\sqrt{(x - b_1)^2 + b_2^2}} \\ \text{iff} \quad & \sin\theta_i = \sin\theta_r \qquad \text{(see the figure)} \\ \text{iff} \quad & \theta_i = \theta_r. \end{aligned}$$

That $L(x)$ is minimal when $\theta_i = \theta_r$ can be seen by noting that $L''(x)$ is always positive,

$$L''(x) = \frac{a_2^2}{[(x - a_1)^2 + a_2^2]^{3/2}} + \frac{b_2^2}{[(x - b_1)^2 + b_2^2]^{3/2}} > 0,$$

and applying the second-derivative test. □

We must admit that there is a simpler way to do Example 5 that requires no calculus. Can you find it?

Now we will work out a problem in which the function to be optimized is defined not on an interval or on a union of intervals, but on a discrete set of points, in this case a finite collection of integers.

Example 6 A manufacturing plant has a capacity of 25 articles per week. Experience has shown that n articles per week can be sold at a price of p dollars each where $p = 110 - 2n$ and the cost of producing n articles is $600 + 10n + n^2$ dollars. How many articles should be made each week to give the largest profit?

SOLUTION The profit (P dollars) on the sale of n articles is

$$P = \text{revenue} - \text{cost} = np - (600 + 10n + n^2).$$

With $p = 110 - 2n$, this simplifies to

$$P = 100n - 600 - 3n^2.$$

† This is a special case of Fermat's *Principle of Least Time,* which says that, of all (neighboring) paths, light chooses the one that demands the least time. If light passes from one medium to another, the geometrically shortest path is not necessarily the path of least time.

In this problem n must be an integer, and thus it makes no sense to differentiate P with respect to n. The formula shows that P is negative if n is less than 8 or greater than 25. By direct calculation we construct Table 4.5.1. The table shows that the largest profit comes from setting production at 17 articles per week.

■ **Table 4.5.1**

n	P	n	P	n	P
8	8	14	212	20	200
9	57	15	225	21	177
10	100	16	232	22	148
11	137	17	233	23	113
12	168	18	228	24	72
13	193	19	217	25	25

We can avoid such massive computation by considering the function

$$f(x) = 100x - 600 - 3x^2, \quad 8 \leq x \leq 25.$$

This function is differentiable with respect to x, and for integral values of x it agrees with P. Differentiation of f gives

$$f'(x) = 100 - 6x.$$

Obviously, $f'(x) = 0$ at $x = \frac{100}{6} = 16\frac{2}{3}$. Since $f'(x) > 0$ on $(8, 16\frac{2}{3})$, f increases on $[8, 16\frac{2}{3}]$. Since $f'(x) < 0$ on $(16\frac{2}{3}, 25)$, f decreases on $[16\frac{2}{3}, 25]$. The largest value of f corresponding to an integer value of x will therefore occur at $x = 16$ or $x = 17$. Direct calculation of $f(16)$ and $f(17)$ shows that the choice of $x = 17$ is correct. ☐

EXERCISES 4.5

1. Find the greatest possible value of xy given that x and y are both positive and $x + y = 40$.
2. Find the dimensions of the rectangle of perimeter 24 that has the largest area.
3. A rectangular garden 200 square feet in area is to be fenced off against rabbits. Find the dimensions that will require the least amount of fencing if one side of the garden is already protected by a barn.
4. Find the largest possible area for a rectangle with base on the x-axis and upper vertices on the curve $y = 4 - x^2$.
5. Find the largest possible area for a rectangle inscribed in a circle of radius 4.
6. Find the dimensions of the rectangle of area A square units that has the smallest perimeter.
7. A rectangular playground is to be fenced off and divided into two parts by a fence parallel to one side of the playground. Six hundred feet of fencing is used. Find the dimensions of the playground that will enclose the greatest total area.
8. A rectangular warehouse will have 5000 square feet of floor space and will be separated into two rectangular rooms by an interior wall. The cost of the exterior walls is \$150 per linear foot and the cost of the interior wall is \$100 per linear foot. Find the dimensions that will minimize the cost of building the warehouse.
9. Rework Example 3, this time assuming that the semicircular portion of the window admits only one-third as much light per square foot as does the rectangular portion of the window.
10. One side of a rectangular field is bounded by a straight river. The other three sides are bounded by straight fences. The total length of the fence is 800 feet. Determine the dimensions of the field given that its area is a maximum.
11. Find the coordinates of P that maximize the area of the rectangle shown in the figure.

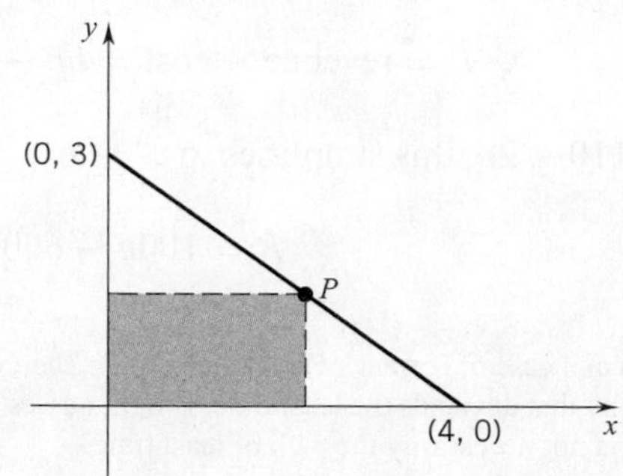

12. The base of a triangle is on the x-axis, one side lies along the line $y = 3x$, and the third side passes through the point $(1, 1)$. What is the slope of the third side if the area of the triangle is to be a minimum?

13. A triangle is formed by the coordinate axes and a line through the point $(2, 5)$ as in the figure. Determine the slope of this line if the area of the triangle is to be a minimum.

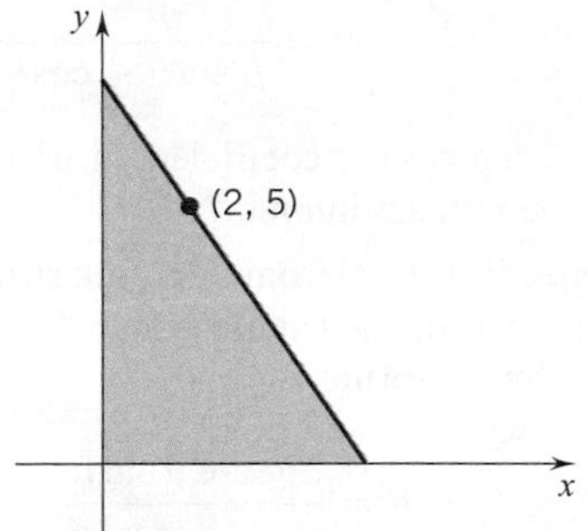

14. In the setting of Exercise 13, determine the slope of the line if the area is to be a maximum.

15. What are the dimensions of the base of the rectangular box of greatest volume that can be constructed from 100 square inches of cardboard if the base is to be twice as long as it is wide? Assume that the box has top.

16. Repeat Exercise 15 under the assumption that the box has no top.

17. Find the dimensions of the isosceles triangle of largest area with perimeter 12.

18. Find the point(s) on the parabola $y = \frac{1}{8}x^2$ closest to the point $(0, 6)$.

19. Find the point(s) on the parabola $x = y^2$ closest to the point $(0,3)$.

20. Find A and B given that the function $y = Ax^{-1/2} + Bx^{1/2}$ has a minimum value of 6 at $x = 9$.

21. A pentagon with a perimeter of 30 inches is to be constructed by adjoining an equilateral triangle to a rectangle. Find the dimensions of the rectangle and triangle that will maximize the area of the pentagon.

22. A 10-foot section of gutter is made from a 12-inch-wide strip of sheet metal by folding up 4-inch strips on each side so that they make the same angle with the bottom of the gutter. Determine the depth of the gutter that has the greatest carrying capacity. *Caution*: There are two ways to sketch the trapezoidal cross section, as shown in the Figure.

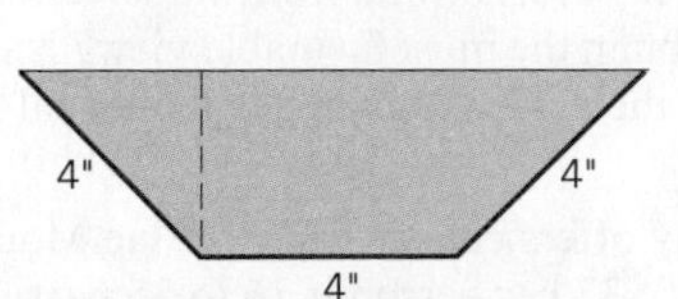

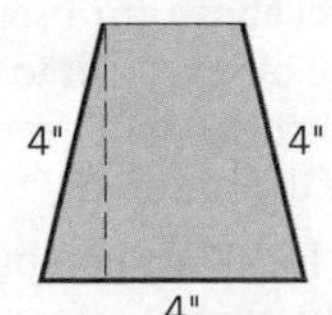

23. From a rectangular piece of cardboard of dimensions 8×15, four congruent squares are to be cut out, one at each corner. (See the figure.) The remaining crosslike piece is then to be folded into an open box. What size squares should be cut out if the volume of the resulting box is to be a maximum?

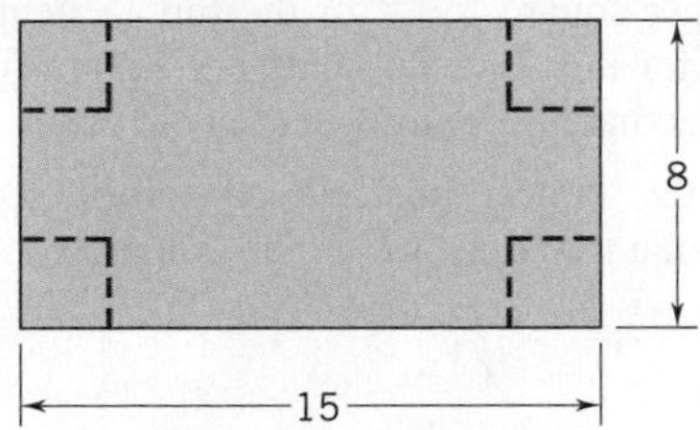

24. A page is to contain 81 square centimeters of print. The margins at the top and bottom are to be 3 centimeters each and, at the sides, 2 centimeters each. Find the most economical dimensions given that the cost of a page varies directly with the perimeter of the page.

25. Let ABC be a triangle with vertices $A = (-3, 0)$, $B = (0,6)$, $C = (3, 0)$. Let P be a point on the line segment that joins B to the origin. Find the position of P that minimizes the sum of the distances between P and the vertices.

26. Solve Exercise 25 with $A = (-6, 0), B = (0, 3), C = (6, 0)$.

27. An 8-foot-high fence is located 1 foot from a building. Determine the length of the shortest ladder that can be leaned against the building and touch the top of the fence.

28. Two hallways, one 8 feet wide and the other 6 feet wide, meet at right angles. Determine the length of the longest ladder that can be carried horizontally from one hallway into the other.

29. A rectangular banner has a red border and a rectangular white center. The width of the border at top and bottom is 8 inches, and along the sides it is 6 inches. The total area is 27 square feet. What should be the dimensions of the banner if the area of the white center is to be a maximum?

30. Conical paper cups are usually made so that the depth is $\sqrt{2}$ times the radius of the rim. Show that this design requires the least amount of paper per unit volume.

31. A string 28 inches long is to be cut into two pieces, one piece to form a square and the other to form a circle. How should the string be cut so as to (a) maximize the sum of the two areas? (b) minimize the sum of the two areas?

32. What is the maximum volume for a rectangular box (square base, no top) made from 12 square feet of cardboard?

33. The figure shows a cylinder inscribed in a right circular cone of height 8 and base radius 5. Find the dimensions of the cylinder if its volume is to be a maximum.

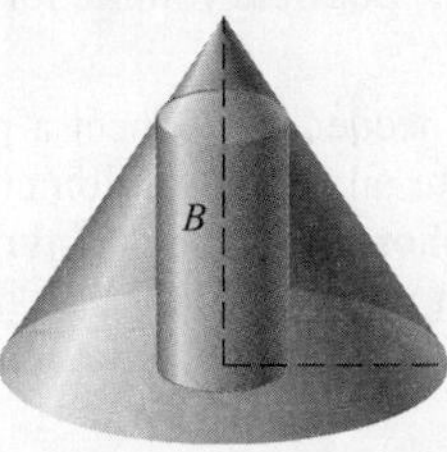

34. As a variant of Exercise 33, this time find the dimensions of the cylinder if the area of its curved surface is to be a maximum.

35. A rectangular box with square base and top is to be made to contain 1250 cubic feet. The material for the base costs 35 cents per square foot, for the top 15 cents per square foot, and for the sides 20 cents per square foot. Find the dimensions that will minimize the cost of the box.

36. What is the largest possible area for a parallelogram inscribed in a triangle ABC in the manner of the figure.

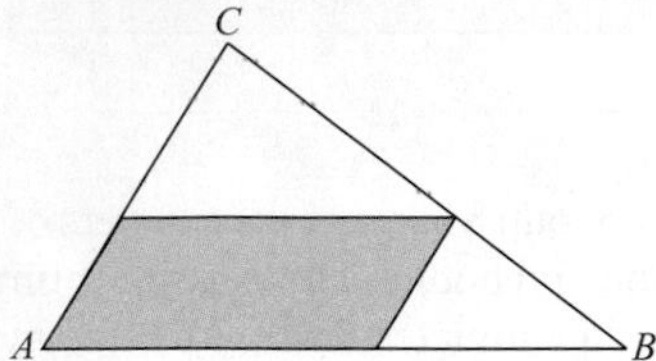

37. Find the dimensions of the isosceles triangle of least area that circumscribes a circle of radius r.

38. What is the maximum possible area for a triangle inscribed in a circle of radius r?

39. The figure shows a right circular cylinder inscribed in a sphere of radius r. Find the dimensions of the cylinder if its volume is to be a maximum.

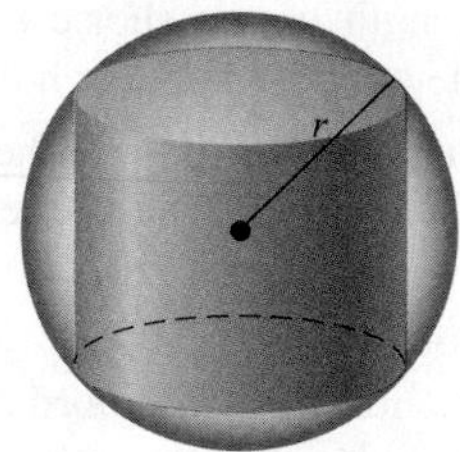

40. As a variant of Exercise 39, this time find the dimensions of the right circular cylinder if the area of its curved surface is to be a maximum.

41. A right circular cone is inscribed in a sphere of radius r as in the figure. Find the dimensions of the cone if its volume is to be a maximum.

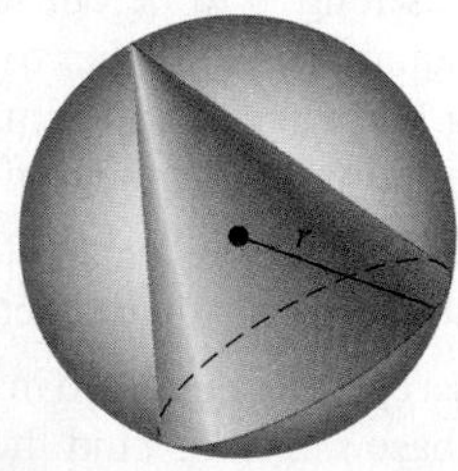

42. What is the largest possible volume for a right circular cone of slant height a?

43. A power line is needed to connect a power station on the shore of a river to an island 4 kilometers downstream and 1 kilometer offshore. Find the minimum cost for such a line given that it costs \$50,000 per kilometer to lay wire under water and \$30,000 per kilometer to lay wire under ground.

44. A tapestry 7 feet high hangs on a wall. The lower edge is 9 feet above an observer's eye. How far from the wall should the observer stand to obtain the most favorable view? Namely, what distance from the wall maximizes the visual angle of the observer? HINT: Use the formula for $\tan(A - B)$.

45. An object of weight W is dragged along a horizontal plane by means of a force P whose line of action makes an angle θ with the plane. The magnitude of the force is given by the equation

$$P = \frac{\mu W}{\mu \sin\theta + \cos\theta}$$

where μ denotes the coefficient of friction. For what value of θ is the pull a minimum?

46. If a projectile is fired from O so as to strike an inclined plane that makes a constant angle α with the horizontal, its range is given by the formula

$$R = \frac{2v^2 \cos\theta \sin(\theta - \alpha)}{g \cos^2\alpha},$$

where v and g are constants and θ is the angle of elevation. Calculate θ for maximum range.

47. Two sources of heat are placed s meters apart—a source of intensity a at A and a source of intensity b at B. The intensity of heat at a point P on the line segment between A and B is give by the formula

$$I = \frac{a}{x^2} + \frac{b}{(s - x)^2},$$

where x is the distance between P and A measured in meters. At what point between A and B will the temperature be lowest?

48. The distance from a point to a line is the distance from that point to the closest point of the line. What point of the line $ax + by + c = 0$ is closest to the point (x_1, y_1)? What is the distance from (x_1, y_1) to the line?

49. Given that $\overline{PQ}$ is the longest or shortest line segment that can be drawn from the point $P(a, b)$ to the point Q on the graph of the differentiable function $y = f(x)$, show that $\overline{PQ}$ is perpendicular to the graph at Q.

50. Draw the parabola $y = x^2$. On the parabola mark a point $P \neq O$. Through P draw the normal line. The normal line intersects the parabola at another point Q. Show that the distance between P and Q is minimized by setting $P = \left(\pm\frac{\sqrt{2}}{2}, \frac{1}{2}\right)$.

51. The lower edge of a movie theater screen 30 feet high is 6 feet above an observer's eye. How far from the screen should the observer sit to obtain the most favorable view? Namely, what distance from the screen maximizes the visual angle of the observer?

52. A local bus company offers charter trips to Blue Mountain Museum at a fare of \$37 per person if 16 to 35 passengers sign up for the trip. The company does not charter trips for fewer than 16 passengers. The bus has 48 seats. If more than 35 passengers sign up, then the fare for every passenger is reduced by 50 cents for each passenger in excess of 35 that signs up. Determine the number of passengers that generates the greatest revenue for the bus company.

53. The Hotwheels Rent-A-Car Company derives an average net profit of $12 per customer if it services 50 customers or less. If it services more than 50 customers, then the average net profit is decreased by 6 cents for each customer over 50. What number of customers produces the greatest total net profit for the company?

54. A steel plant is capable of producing x tons per day of low-grade steel and y tons per day of high-grade steel, where

$$y = \frac{40 - 5x}{10 - x}.$$

If the market price of low-grade steel is half that of high-grade steel, show that about $5\frac{1}{2}$ tons of low-grade steel should be produced per day for maximum receipts.

55. Let $C(x)$ denote the cost of producing x units of a certain commodity. Then the derivative dC/dx is called the marginal cost (see Section 3.4). The *average cost* of producing x units is $A(x) = C(x)/x$. Show that the average cost is a minimum at the production levels x where the marginal cost equals the average cost.

56. An oil drum is to be made in the form of a right circular cylinder to contain 16π cubic feet. The upright drum is to be taller than it is wide, but not more than 6 feet tall. Determine the dimensions of the drum with minimal surface area.

57. The equation of the path of a ball is $y = mx - \frac{1}{800}(m^2 + 1)x^2$, where the origin is taken as the point from which the ball is thrown and m is the initial slope of the trajectory. For what value of m will the ball strike at the greatest distance at the same horizontal level?

58. In the setting of Exercise 57, for what value of m will the ball strike at the greatest height on a vertical wall 300 feet away?

59. A truck is to be driven 300 miles on a freeway at a constant speed of x miles per hour. Speed laws require that $35 \le x \le 55$. Assume that fuel costs $1.35 per gallon and is consumed at the rate of $2 + (\frac{1}{600})x^2$ gallons per hour. Given that the driver's wages are $13 per hour, at what speed should the truck be driven to minimize the truck owner's expenses?

60. The cost of erecting an office building is $1,000,000 for the first story, $1,100,000 for the second, $1,200,000 for the third, and so on. Other expenses (lot, basement, etc.) are $5,000,000. Assume that the net annual income is $200,000 per story. How many stories will provide the greatest return on investment?

C 61. Let $f(x) = x^2 - x$ and let P be the point (4, 3).

(a) Use a graphing utility to draw f and P.

(b) Use a CAS to find the point(s) on the graph of f that are closest to P.

(c) Let Q be a point which satisfies (b). Determine the equation for the line l_{PQ} through P and Q and draw f, P, and l_{PQ} together.

(d) Determine the equation of the line l_N normal to the graph of f at $(Q, f(Q))$.

(e) Compare l_{PQ} and l_N.

C 62. Repeat Exercise 61 with $f(x) = x - x^3$ and $P(1,8)$.

C 63. Use the distance formula to find a formula $D(x)$ for the distance from a point (x, y) on the line $y + 3x = 7$ to the origin. Use a graphing utility to draw the graph of $D(x)$ and then use the Trace Function to estimate the point on the line whose distance to the origin is a minimum.

C 64. Use the distance formula to find a formula $D(x)$ for the distance from a point (x, y) on the graph of $f(x) = 4 - x^2$ to the point $P(4, 3)$. Use a graphing utility to draw the graph of $D(x)$ and then use the Trace Function to estimate the point on the graph of f whose distance to P is a minimum.

C 65. Find a formula for the area $A(x)$ of a rectangle inscribed in the ellipse $16x^2 + 9y^2 = 144$. Use a graphing utility to draw the graph of $A(x)$ and then use the Trace Function to estimate the value $x \ge 0$ at which $A(x)$ takes on its absolute maximum value. Also, estimate this maximum area.

C 66. Use a CAS to obtain a formula for the absolute maximum area of a rectangle inscribed in the ellipse $b^2x^2 + a^2y^2 = a^2b^2$. Does your formula verify the result in Exercise 65?

■ PROJECT 4.5 Flight Paths of Birds

Ornithologists studying the flight behavior of birds have determined that certain species tend to avoid flying over large bodies of water during daylight hours. A possible explanation for this is based on the belief that it takes more energy to fly over water than over land because air typically rises over land and falls over water during the day. Suppose that a bird with this tendency is released from an island that is 6 miles from the nearest point A on a straight shoreline. It flies to a point B on the shore and then flies along the shore to its nesting area C, which is 12 miles from A (see the Figure).

Let W denote the energy required to fly over water, and let L denote the energy required to fly over land (assume that W and L are measured in energy units per mile).

Problem 1. Show that the total energy E expended by the bird in flying from the island to its nesting area is given by

$$E(x) = W\sqrt{36 + x^2} + L(12 - x), \quad 0 \le x \le 12,$$

where x is the distance from A to B.

Problem 2. Suppose that $W = 1.5L$; that is, suppose it takes 50% more energy to fly over water than over land.

(a) Use the methods of this section to find the point B to which the bird should fly to minimize the total energy expended.

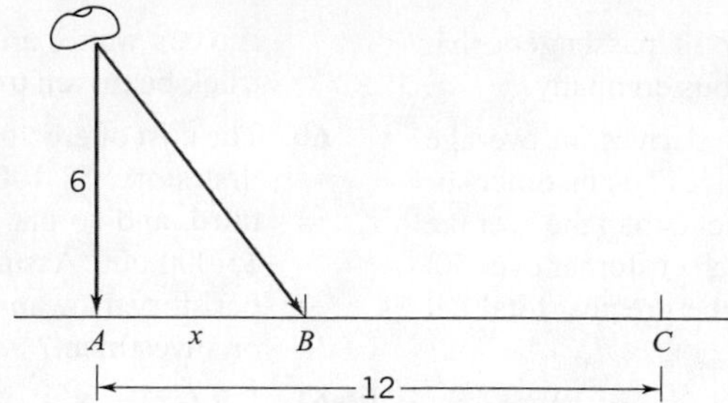

(b) Use a graphing utility to graph E, and then find the minimum value to confirm your result in part (a).

Problem 3. In general, suppose $W = kL, k > 1$.

(a) Find the point B (as a function of k) to which the bird should fly to minimize the total energy.

(b) Use a graphing utility to experiment with different values of k to find out how the point B moves as k increases/ decreases.

(c) Find the value(s) of k such that the bird will minimize the total energy expended by flying directly to its nest.

(d) Are there any values of k such that the bird will minimize the total energy expended by flying directly to the point A and then along the shore to C?

■ 4.6 CONCAVITY AND POINTS OF INFLECTION

We begin with a picture of the graph of a function f, Figure 4.6.1. To the left of c_1 and between c_2 and c_3, the graph "curves up" (we call it *concave up*); between c_1 and c_2, and to the right of c_3, the graph "curves down"(we call it *concave down*). These terms deserve a precise definition.

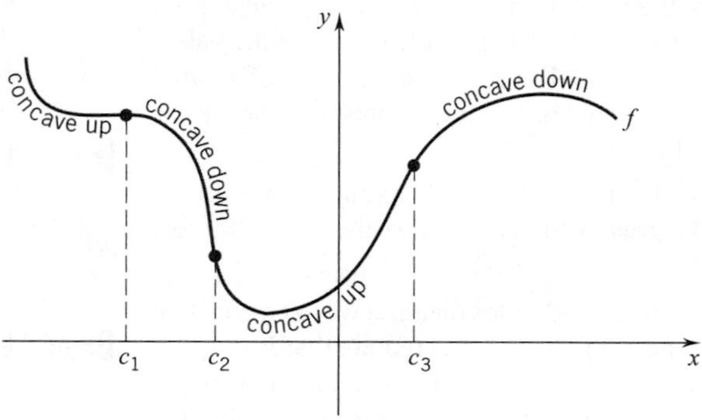

Figure 4.6.1

DEFINITION 4.6.1 CONCAVITY

Let the function f be differentiable on an open interval I. The graph of f is said to be *concave up* on I if f' increases on I; it is said to be *concave down* on I if f' decreases on I.

In other words, the graph is concave up on an open interval where the slope increases and concave down on an open interval where the slope decreases.

There is also a geometric interpretation of concavity: the graph of f is concave up on an interval I if the tangent lines to the graph are *below* the curve; the graph of f is concave down on I if the tangent lines to the graph are *above* the curve. You can verify this interpretation using Figure 4.6.1.

Points that join arcs of opposite concavity are called *points of inflection*. The graph in Figure 4.6.1 has three of them: $(c_1,f(c_1)), (c_2,f(c_2)), (c_3,f(c_3))$. Here is a formal definition.

DEFINITION 4.6.2 POINT OF INFLECTION

Let the function f be continuous at $x = c$. The point $(c,f(c))$ is called a *point of inflection* if there exists a $\delta > 0$ such that the graph of f is concave in one sense on $(c - \delta, c)$ and concave in the opposite sense on $(c, c + \delta)$.

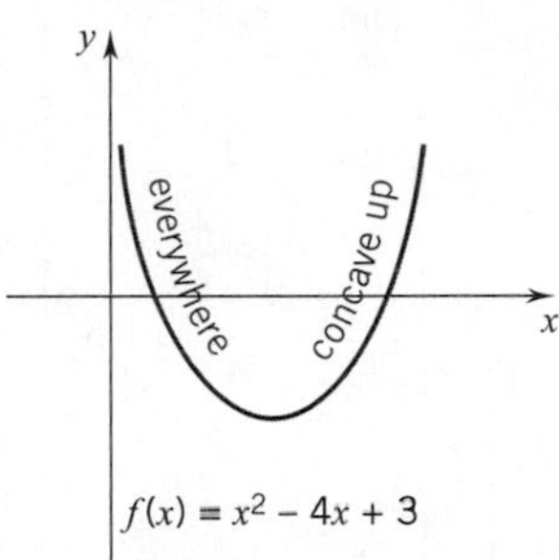

Figure 4.6.2

Example 1 The graph of the quadratic function $f(x) = x^2 - 4x + 3$ is concave up everywhere since the derivative $f'(x) = 2x - 4$ is everywhere increasing. (See Figure 4.6.2.) ❑

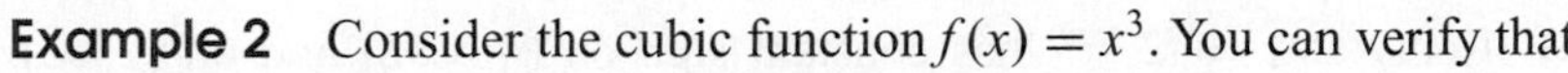

Example 2 Consider the cubic function $f(x) = x^3$. You can verify that

$$f'(x) = 3x^2 \qquad \text{decreases on } (-\infty, 0] \text{ and increases on } [0, \infty).$$

Thus, the graph of f is concave down on $(-\infty, 0)$ and concave up on $(0, \infty)$. The origin, $(0, f(0)) = (0, 0)$, is a point of inflection. (See Figure 4.6.3) ❑

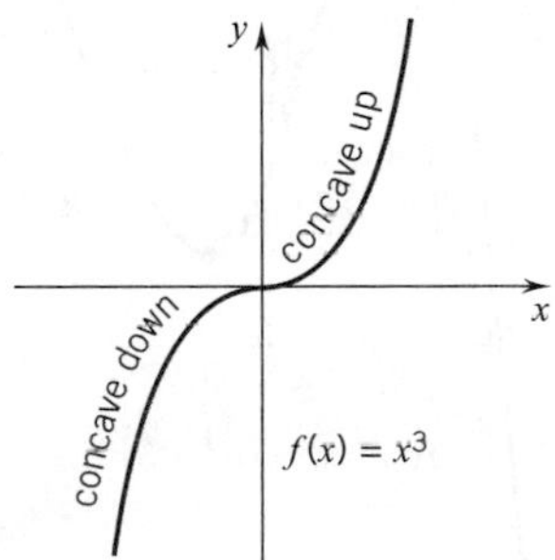

Figure 4.6.3

If f is twice differentiable, then, remembering that $f'' = (f')'$, we can determine the concavity of the graph by looking at the sign of the second derivative.

THEOREM 4.6.3

Let f be twice differentiable on an open interval I.

(i) If $f''(x) > 0$ for all x in I, then f' increases on I, and the graph of f is concave up.

(ii) If $f''(x) < 0$ for all x in I, then f' decreases on I and the graph of f is concave down.

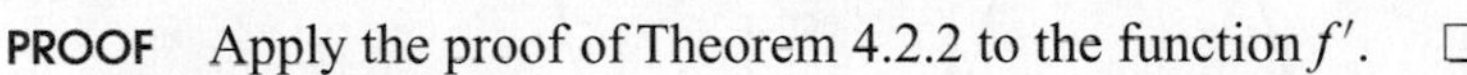

PROOF Apply the proof of Theorem 4.2.2 to the function f'. ❑

The following result gives us a way of identifying possible points of inflection.

THEOREM 4.6.4

If the point $(c,f(c))$ is a point of inflection, then either

$$f''(c) = 0 \quad \text{or} \quad f''(c) \quad \text{does not exist.}$$

PROOF Suppose that $(c,f(c))$ is a point of inflection. Let's assume that the graph of f is concave up to the left of c and concave down to the right of c. The other case can be handled in a similar manner.

In this situation f' increases on an interval $(c - \delta, c)$ and decreases on an interval $(c, c + \delta)$.

Suppose now that $f''(c)$ exists. Then f' is continuous at c. It follows that f' increases on the half-open interval $(c-\delta, c]$ and decreases on the half-open interval $[c, c+\delta)$†. This says that f' has a local maximum at c. Since, by assumption, $f''(c)$ exists, $f''(c) = 0$. (Theorem 4.3.2 applied to f'.)

We have shown that if $f''(c)$ exists, then $f''(c) = 0$. The only other possibility is that $f''(c)$ does not exist (see Example 4 below). ❑

Example 3 For the function

$$f(x) = x^3 - 6x^2 + 9x + 1 \qquad \text{(Figure 4.6.4)}$$

we have

$$f'(x) = 3x^2 - 12x + 9 = 3(x^2 - 4x + 3)$$

and

$$f''(x) = 6x - 12 = 6(x - 2).$$

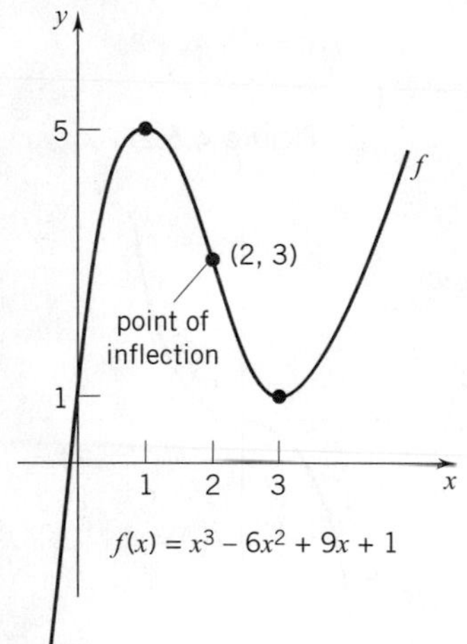

Figure 4.6.4

Since $f''(x) = 0$ only at $x = 2$, f'' has constant sign on $(-\infty, 2)$ and on $(2, \infty)$. The sign of f'' on these intervals and the consequences for the graph of f are as follows:

sign of f'': - 0 + + + + + + + + + + + + + + + + + + +

graph of f: concave down 2 (point of inflection) concave up x

The point $(2, f(2)) = (2, 3)$ is a point of inflection. ❑

Example 4 For

$$f(x) = 3x^{5/3} - 5x \qquad \text{(Figure 4.6.5)}$$

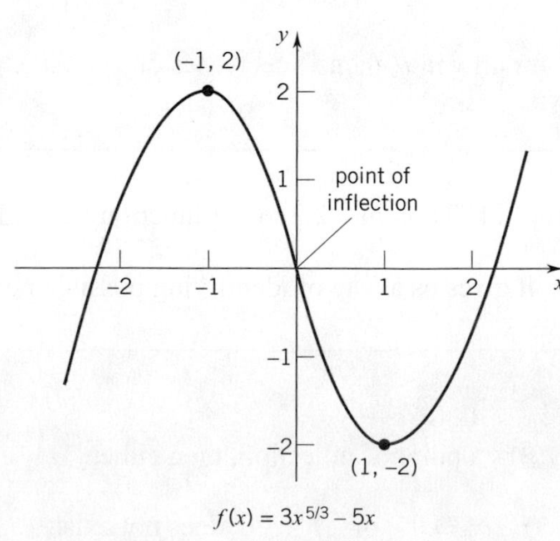

Figure 4.6.5

we have

$$f'(x) = 5x^{2/3} - 5 \quad \text{and} \quad f''(x) = \tfrac{10}{3}x^{-1/3}.$$

† See Exercise 52, Section 4.2.

The second derivative does not exist at $x = 0$. Since

$$f''(x) \text{ is } \begin{cases} \text{negative}, & \text{for } x < 0 \\ \text{positive}, & \text{for } x > 0, \end{cases}$$

the graph is concave down on $(-\infty, 0)$ and concave up on $(0, \infty)$. Since f is continuous at 0, the point $(0, f(0)) = (0, 0)$ is a point of inflection. □

CAUTION The fact that $f''(c) = 0$ or $f''(c)$ does not exist does not guarantee that $(c, f(c))$ is a point of inflection. As you can verify, the function $f(x) = x^4$ satisfies $f''(0) = 0$, but the graph is always concave up and there are no points of inflection. If f is discontinuous at c, then $f''(c)$ does not exist, but $(c, f(c))$ cannot be a point of inflection. A point of inflection occurs at c iff f is continuous at c and the point $(c, f(c))$ joins arcs of opposite concavity. □

Example 5 Determine the concavity and find the points of inflection (if any) of the graph of

$$f(x) = x + \cos x, \qquad x \in [0, 2\pi].$$

SOLUTION For $x \in (0, 2\pi)$, we have

$$f'(x) = 1 - \sin x$$

and

$$f''(x) = -\cos x.$$

Since $f''(x) = 0$ at $x = \pi/2$ and $3\pi/2$, f'' has constant sign on $(0, \pi/2)$, $(\pi/2, 3\pi/2)$, and $(3\pi/2, 2\pi)$. The sign of f'' on these intervals and the consequences for f are as follows:

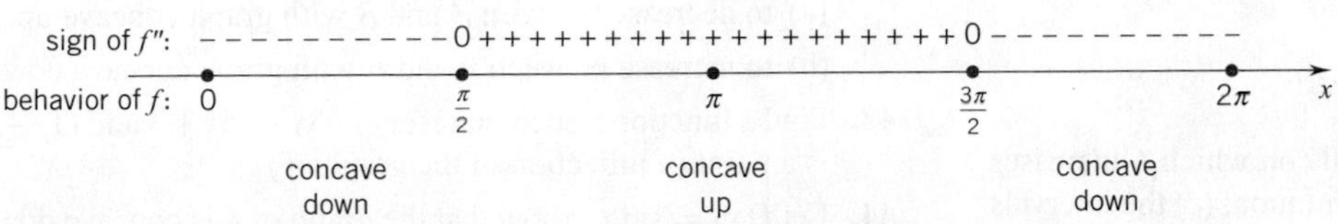

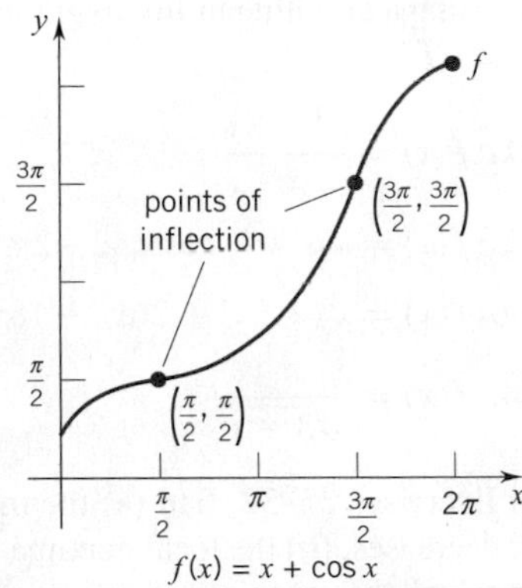

Figure 4.6.6

The points $(\pi/2, f(\pi/2)) = (\pi/2, \pi/2)$ and $(3\pi/2, f(3\pi/2)) = (3\pi/2, 3\pi/2)$ are points of inflection. The graph of f is shown in Figure 4.6.6. □

EXERCISES 4.6

1. The graph of a function f is given in the figure. (a) Determine the intervals on which f increases and the intervals on which f decreases; (b) determine the intervals on which the graph of f is concave up, the intervals on which the graph is concave down, and give the x-coordinates of the points of inflection.

2. Repeat Exercise 1 for the function g whose graph is given in the figure.

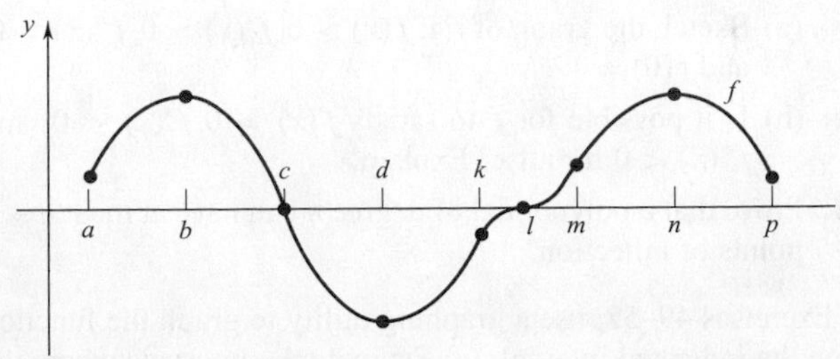

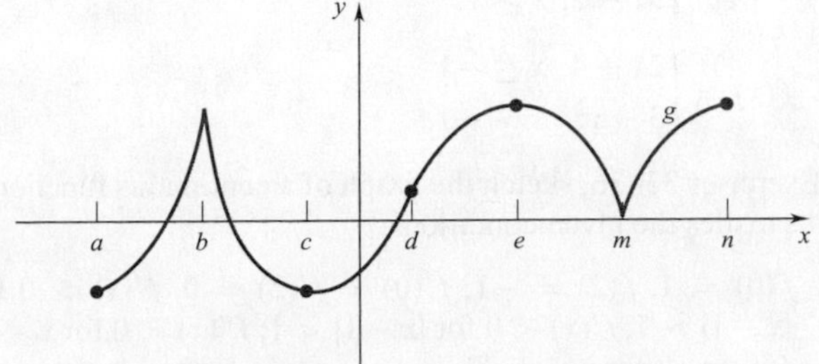

Describe the concavity of the graph of f and find the points of inflection (if any).

3. $f(x) = \dfrac{1}{x}$.

4. $f(x) = x + \dfrac{1}{x}$.

5. $f(x) = x^3 - 3x + 2$.

6. $f(x) = 2x^2 - 5x + 2$.

7. $f(x) = \frac{1}{4}x^4 - \frac{1}{2}x^2$.

8. $f(x) = x^3(1 - x)$.

9. $f(x) = \dfrac{x}{x^2 - 1}$.

10. $f(x) = \dfrac{x+2}{x-2}$.

11. $f(x) = (1 - x)^2(1 + x)^2$.

12. $f(x) = \dfrac{6x}{x^2 + 1}$.

13. $f(x) = \dfrac{1 - \sqrt{x}}{1 + \sqrt{x}}$.

14. $f(x) = (x - 3)^{1/5}$.

15. $f(x) = (x + 2)^{5/3}$.

16. $f(x) = x\sqrt{4 - x^2}$.

17. $f(x) = \sin^2 x, \quad x \in [0, \pi]$.

18. $f(x) = 2\cos^2 x - x^2, \quad x \in [0, \pi]$.

19. $f(x) = x^2 + \sin 2x, \quad x \in [0, \pi]$.

20. $f(x) = \sin^4 x, \quad x \in [0, \pi]$.

In Exercises 21–24, find the points of inflection of the graph of f by using a graphing utility to graph f and f'' together and noting where $f''(x) = 0$.

21. $f(x) = \dfrac{x^4 - 81}{x^2}$.

22. $f(x) = \sin^2 x - \cos x, \quad -2\pi \le x \le 2\pi$.

23. $f(x) = x^5 + 9x^4 + 26x^3 + 18x^2 - 27x - 27$.

24. $f(x) = \dfrac{x}{\sqrt{3 - x}}$.

In Exercises 25–32, find (a) the intervals on which f increases or decreases; (b) the local maxima and minima; (c) the intervals on which the graph of f is concave up and the intervals on which it is concave down; and (d) the points of inflection. Use this information to sketch the graph of f.

25. $f(x) = x^3 - 9x$.

26. $f(x) = 3x^4 + 4x^3 + 1$.

27. $f(x) = \dfrac{2x}{x^2 + 1}$.

28. $f(x) = x^{1/3}(x - 6)^{2/3}$.

29. $f(x) = x + \sin x, \quad x \in [-\pi, \pi]$.

30. $f(x) = \sin x + \cos x, \quad x \in [0, 2\pi]$.

31. $f(x) = \begin{cases} x^3, & x < 1 \\ 3x - 2, & x \ge 1. \end{cases}$

32. $f(x) = \begin{cases} 2x + 4, & x \le -1 \\ 3 - x^2, & x > -1. \end{cases}$

In Exercises 33–36, sketch the graph of a continuous function f that satisfies the given conditions.

33. $f(0) = 1, f(2) = -1; f'(0) = f'(2) = 0, f'(x) > 0$ for $|x - 1| > 1, f'(x) < 0$ for $|x - 1| < 1; f''(x) < 0$ for $x < 1$, $f''(x) > 0$ for $x > 1$.

34. $f''(x) > 0$ if $|x| > 2, f''(x) < 0$ if $|x| < 2$; $f'(0) = 0$, $f'(x) > 0$ if $x < 0, f'(x) < 0$ if $x > 0$; $f(0) = 1, f(-2) = f(2) = \frac{1}{2}, f(x) > 0$ for all x, f is an even function.

35. $f''(x) < 0$ if $x < 0$, $f''(x) > 0$ if $x > 0$; $f'(-1) = f'(1) = 0, f'(0)$ does not exist, $f'(x) > 0$ if $|x| > 1, f'(x) < 0$ if $|x| < 1 (x \ne 0)$; $f(-1) = 2, f(1) = -2$, f is an odd function.

36. $f(-2) = 6$, $f(1) = 2$, $f(3) = 4$; $f'(1) = f'(3) = 0$, $f'(x) < 0$ for $|x - 2| > 1, f'(x) > 0$ for $|x - 2| < 1$; $f''(x) < 0$ for $|x + 1| < 1$ or $x > 2, f''(x) > 0$ for $|x - 1| < 1$ or $x < -2$.

37. Find d given that $(d, f(d))$ is a point of inflection of the graph of

$$f(x) = (x - a)(x - b)(x - c).$$

38. Find c given that the graph of $f(x) = cx^2 + x^{-2}$ has a point of inflection at $(1, f(1))$.

39. Find a and b given that the graph of $f(x) = ax^3 + bx^2$ passes through $(-1, 1)$ and has a point of inflection at $x = \frac{1}{3}$.

40. Determine A and B so that the curve

$$y = Ax^{1/2} + Bx^{-1/2}$$

will have a point of inflection at $(1, 4)$.

41. Determine A and B so that the curve

$$y = A\cos 2x + B\sin 3x$$

will have a point of inflection at $(\pi/6, 5)$.

42. Find necessary and sufficient conditions on A and B for $f(x) = Ax^2 + Bx + C$:

(a) to decrease between A and B with graph concave up.

(b) to increase between A and B with graph concave down.

43. Find a function f such that $f'(x) = 3x^2 - 6x + 3$ and $(1, -2)$ is a point of inflection of the graph of f.

44. Let $f(x) = \sin x$. Show that the graph of f is concave down when it is above the x-axis and concave up when it is below the x-axis. Does $g(x) = \cos x$ have the same property?

45. Given the cubic polynomial $p(x) = x^3 + ax^2 + bx + c$.

(a) Prove that the graph of p has exactly one point of inflection.

(b) Show that p has local extreme values iff $a^2 > 3b$.

46. Prove that if the cubic polynomial $p(x) = x^3 + ax^2 + bx + c$ has a local maximum and a local minimum, then the point of inflection is the midpoint of the line segment connecting the local extrema on the graph.

47. Suppose that the function f is defined for all x.

(a) Sketch the graph of f if $f(x) > 0, f'(x) > 0, f''(x) > 0$, and $f(0) = 1$

(b) Is it possible for f to satisfy $f(x) > 0, f'(x) < 0$, and $f''(x) < 0$ for all x? Explain.

48. Prove that a polynomial of degree n can have at most $n - 2$ points of inflection.

In Exercises 49–52, use a graphing utility to graph the function f on the indicated interval. (a) Estimate the intervals where the graph of f is concave up and the intervals where it is concave down. (b) Estimate the x-coordinate of each point of inflection. Round off your estimates to three decimal places.

49. $f(x) = x^4 - 5x^2 + 3;\ [-4, 4].$

50. $f(x) = x \sin x;\ [-2\pi, 2\pi].$

51. $f(x) = 1 + x^2 - 2x \cos x;\ [-\pi, \pi].$

52. $f(x) = x^{2/3}(x^2 - 4);\ [-5, 5].$

C In Exercises 53–56, use a CAS to determine where: (a) $f''(x) = 0$, (b) $f''(x) > 0$ (c) $f''(x) < 0$.

53. $f(x) = 2\cos^2 x - \cos x, \quad 0 \le x \le 2\pi.$

54. $f(x) = \dfrac{x^2}{x^4 - 1}.$

55. $f(x) = x^{11} - 4x^9 + 6x^7 - 4x^5 + x^3.$

56. $f(x) = x\sqrt{16 - x^2}.$

■ 4.7 VERTICAL AND HORIZONTAL ASYMPTOTES; VERTICAL TANGENTS AND CUSPS

Vertical and Horizontal Asymptotes

In Figure 4.7.1 you can see the graph of the function

$$f(x) = \frac{1}{|x - c|} \quad \text{for } x \text{ close to } c.$$

As $x \to c$, $f(x) \to \infty$; that is, given any positive number M, there exists a positive number δ such that

$$\text{if } 0 < |x - c| < \delta, \qquad \text{then} \quad f(x) \ge M.$$

The line $x = c$ is called a *vertical asymptote*. Figure 4.7.2 shows the graph of

$$g(x) = -\frac{1}{|x - c|} \quad \text{for } x \text{ close to } c.$$

In this case, we see that $g(x) \to -\infty$ as $x \to c$. Again the line $x = c$ is called a *vertical asymptote*.

Vertical asymptotes can also arise from one-sided behavior. With f and g as in Figure 4.7.3, we write

$$\text{as } x \to c^-, \ f(x) \to \infty \quad \text{and} \quad g(x) \to -\infty.$$

With f and g as in Figure 4.7.4, on page 244, we write

$$\text{as } x \to c^+, \ f(x) \to \infty \quad \text{and} \quad g(x) \to -\infty.$$

In each case the line $x = c$ is a vertical asymptote for both functions.

This discussion can be summarized as follows:

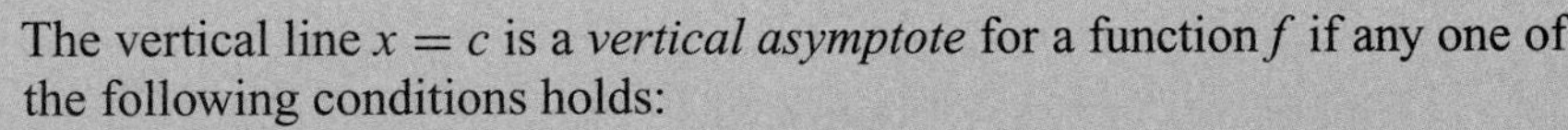

The vertical line $x = c$ is a *vertical asymptote* for a function f if any one of the following conditions holds:

$f(x) \to \infty$ or $-\infty$ as $x \to c$;
$f(x) \to \infty$ or $-\infty$ as $x \to c^-$;
$f(x) \to \infty$ or $-\infty$ as $x \to c^+$.

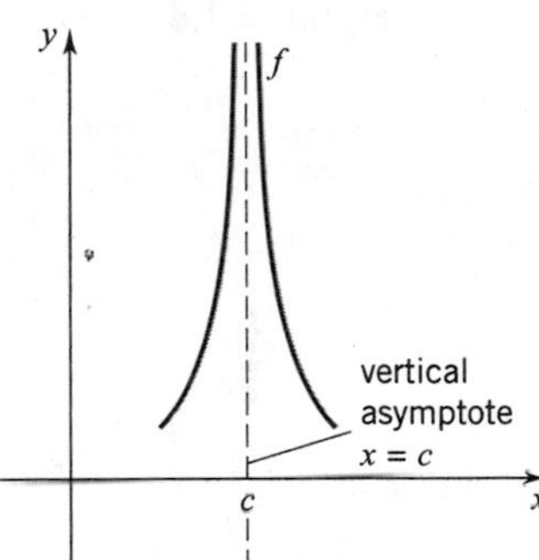

Figure 4.7.1

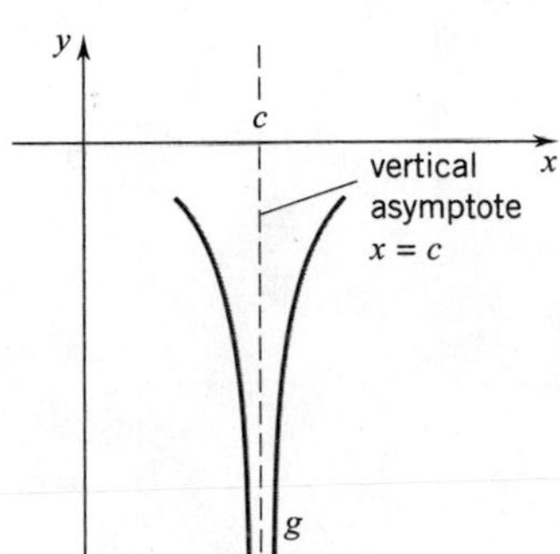

Figure 4.7.2

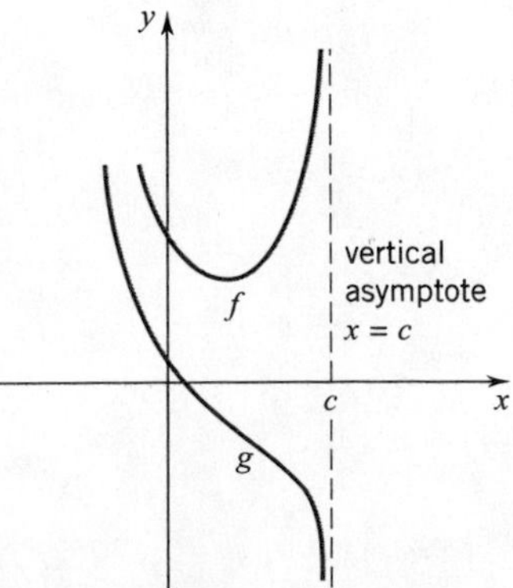

Figure 4.7.3

Typically, to locate the vertical asymptotes for a function f, find the values $x = c$ at which f is discontinuous and examine the behavior of f as x approaches c.

Example 1 The function

$$f(x) = \frac{3x + 6}{x^2 - 2x - 8} = \frac{3(x + 2)}{(x + 2)(x - 4)}$$

is defined and continuous everywhere except at $x = 4$ and $x = -2$. As $x \to 4^+$, $f(x) \to \infty$; as $x \to 4^-, f(x) \to -\infty$. Thus, the line $x = 4$ is a vertical asymptote. Since $\lim_{x \to -2} f(x) = -\frac{1}{2}$ exists, f does not have a vertical asymptote at $x = -2$. See Figure 4.7.5. □

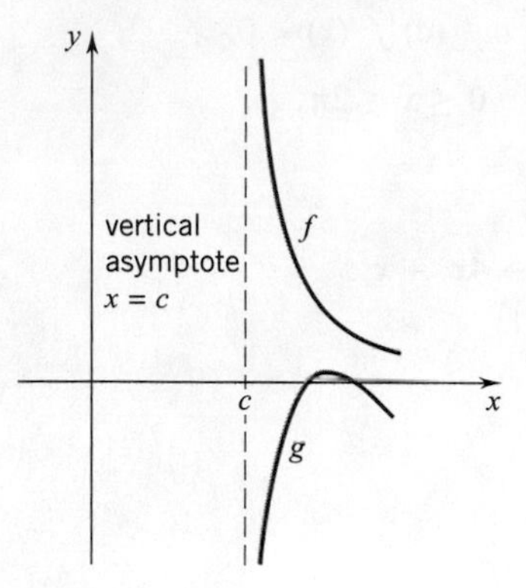

Figure 4.7.4

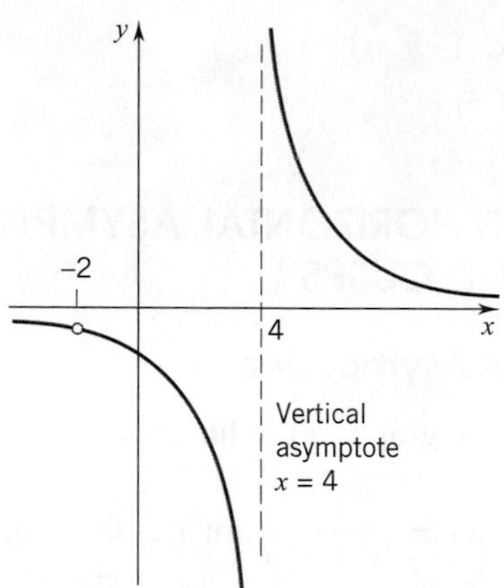

Figure 4.7.5

Recall that $\tan x \to \infty$ as $x \to (\pi/2)^-$ and $\tan x \to -\infty$ as $x \to (\pi/2)^+$. Thus the line $x = \pi/2$ is a vertical asymptote for the tangent function. Indeed, the lines $x = (2n + 1)\,\pi/2, n = 0, \pm 1, \pm 2, \ldots$, are all vertical asymptotes for the tangent function (Figure 4.7.6)

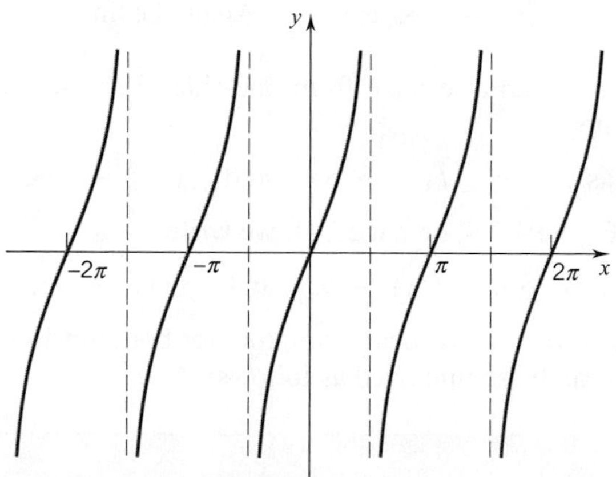

Figure 4.7.6

It is also possible for a function to have a *horizontal asymptote*:

> If there exists a number L such that $f(x) \to L$ as $x \to \infty$, or if $f(x) \to L$ as $x \to -\infty$, then the horizontal line $y = L$ is called a *horizontal asymptote*. See Figures 4.7.7 and 4.7.8.

To be precise, $f(x) \to L$ as $x \to \infty$ means that given a positive number ϵ, there is a positive number K such that if $x \geq K$, then $|f(x) - L| < \epsilon$. Similarly, $f(x) \to L$ as

$x \to -\infty$ means that given a positive number ϵ, there is a negative number M such that if $x \leq M$, then $|f(x) - L| < \epsilon$.

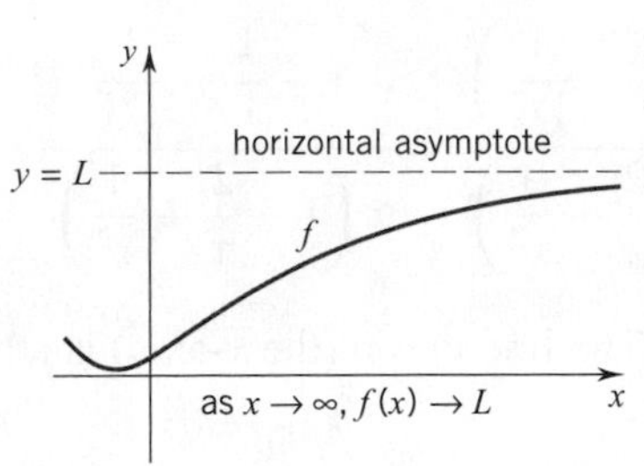

Figure 4.7.7

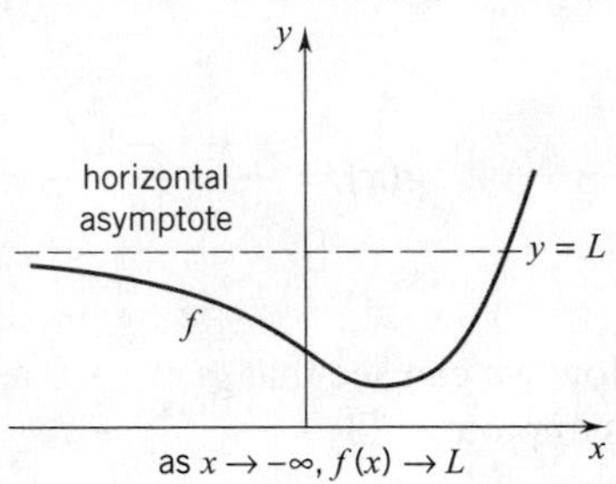

Figure 4.7.8

Example 2 Figure 4.7.9 shows the graph of the function

$$f(x) = \frac{x}{x-2}.$$

As $x \to 2^-$, $f(x) \to -\infty$; as $x \to 2^+$, $f(x) \to \infty$. The line $x = 2$ is a vertical asymptote.

As $x \to \pm\infty$,†

$$f(x) = \frac{x}{x-2} = \frac{x}{x\left(1 - \dfrac{2}{x}\right)} = \frac{1}{1 - \dfrac{2}{x}} \to 1.$$

The line $y = 1$ is a horizontal asymptote. □

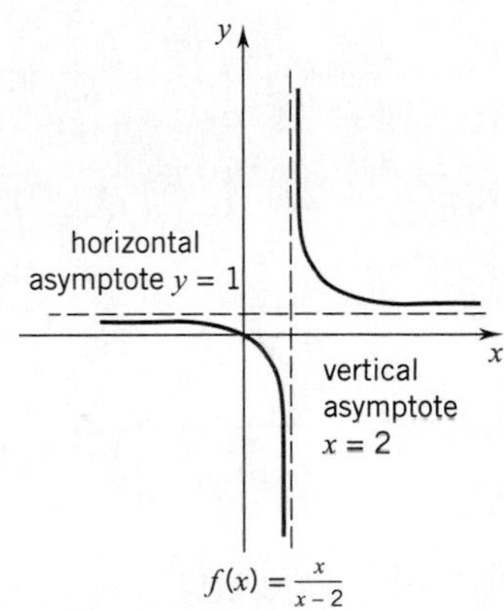

Figure 4.7.9

Example 3 Figure 4.7.10 shows the graph of the function

$$f(x) = \frac{\cos x}{x}.$$

As $x \to 0^-$, $f(x) \to -\infty$; as $x \to 0^+$, $f(x) \to \infty$. The line $x = 0$ (the y-axis) is a vertical asymptote.

As $x \to \pm\infty$,

$$f(x) = \frac{\cos x}{x} \to 0.$$

This follows from the fact that

$$\left|\frac{\cos x}{x}\right| \leq \frac{1}{|x|} \quad \text{for all } x$$

and $1/|x| \to 0$ as $x \to \pm\infty$. Thus, the line $y = 0$ (the x-axis) is a horizontal asymptote, Note that f is an odd function $[f(-x) = -f(x)]$, so its graph is symmetric with respect to the origin. This example also illustrates that a horizontal asymptote does not have to be approached by the graph from one side only. Here the graph of f crosses its horizontal asymptote infinitely many times. □

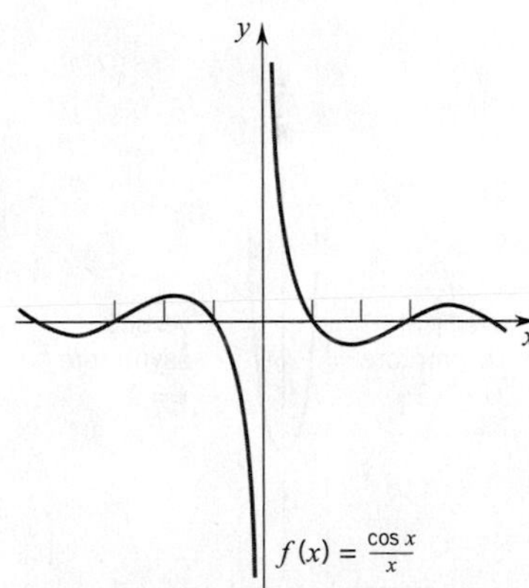

Figure 4.7.10

Example 4 Find the vertical and horizontal asymptotes, if any, of the function

$$g(x) = \frac{x + 1 - \sqrt{x}}{x^2 - 2x + 1} = \frac{x + 1 - \sqrt{x}}{(x-1)^2}.$$

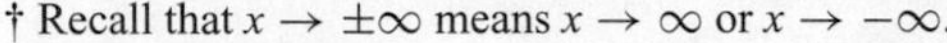

† Recall that $x \to \pm\infty$ means $x \to \infty$ or $x \to -\infty$.

SOLUTION The domain of g is $0 \le x < \infty$, $x \ne 1$. As $x \to 1$, $g(x) \to \infty$. Thus, the line $x = 1$ is a vertical asymptote. The behavior of g as $x \to \infty$ can be made more apparent by writing

$$g(x) = \frac{x + 1 - \sqrt{x}}{x^2 - 2x + 1} = \frac{x\left(1 + \dfrac{1}{x} - \dfrac{1}{\sqrt{x}}\right)}{x^2\left(1 - \dfrac{2}{x} + \dfrac{1}{x^2}\right)} = \frac{1 + \dfrac{1}{x} - \dfrac{1}{\sqrt{x}}}{x\left(1 - \dfrac{2}{x} + \dfrac{1}{x^2}\right)}.$$

Now we can see that $g(x) \to 0$ as $x \to \infty$. The line $y = 0$ (the x-axis) is a horizontal asymptote. □

The technique suggested in Examples 2 and 4, factoring out the highest power of x from the numerator and the denominator, can be used to prove the following general result concerning the behavior of a rational function as $x \to \pm\infty$. Let

$$R(x) = \frac{p(x)}{q(x)} = \frac{a_n x^n + \cdots + a_1 x + a_0}{b_k x^k + \cdots + b_1 x + b_0}, \qquad a_n \ne 0, b_k \ne 0,$$

be a rational function. Then,

(4.7.1) $$\text{as } x \to \pm\infty, \begin{cases} R(x) \to 0, & \text{if } n < k \\ R(x) \to \dfrac{a_n}{b_n}, & \text{if } n = k \\ R(x) \to \pm\infty, & \text{if } n > k. \end{cases}$$

Example 5 The function $f(x) = \dfrac{5 - 3x^2}{1 - x^2}$ is continuous everywhere except $x = \pm 1$.

The line $x = 1$ is a vertical asymptote:

$$\text{as } x \to 1, \quad 5 - 3x^2 \to 2, \quad \text{and} \quad 1 - x^2 \to 0.$$

In particular,

$$\text{as } x \to 1^-, \quad 1 - x^2 \text{ is positive so } f(x) \to \infty;$$
$$\text{as } x \to 1^+, \quad 1 - x^2 \text{ is negative so } f(x) \to -\infty.$$

The line $x = -1$ is a vertical asymptote:

$$\text{as } x \to -1, \quad 5 - 3x^2 \to 2, \quad \text{and} \quad 1 - x^2 \to 0.$$

In particular,

$$\text{as } x \to -1^+, \quad 1 - x^2 \text{ is positive so } f(x) \to \infty;$$
$$\text{as } x \to -1^-, \quad 1 - x^2 \text{ is negative so } f(x) \to -\infty.$$

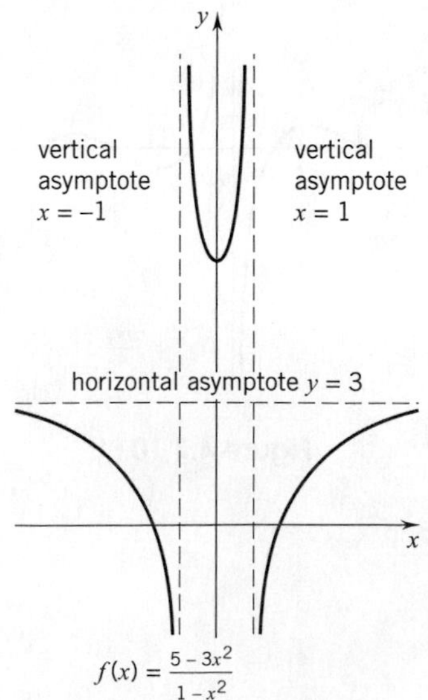

Figure 4.7.11

Since f is a rational function, we can use (4.7.1) to investigate the behavior as $x \to \pm\infty$. We have $n = k = 2$, and so $f(x) \to -3/-1 = 3$ as $x \to \pm\infty$. Thus, the line $y = 3$ is a horizontal asymptote. The graph of f is shown in Figure 4.7.11. Note that f is an even function $[f(-x) = f(x)]$; its graph is symmetric with respect to the y-axis. □

Vertical Tangents; Vertical Cusps

(For the remainder of this section, assume that f is continuous at $x = c$ and differentiable for $x \neq c$.)

We say that the graph of f has a *vertical tangent* at the point $(c, f(c))$ if

$$\text{as} \quad x \to c, \quad f'(x) \to \infty \quad \text{or} \quad f'(x) \to -\infty.$$

Examples

(a) The graph of the function $f(x) = x^{1/3}$ has a vertical tangent at the point $(0, 0)$ since

$$f'(x) = \tfrac{1}{3}x^{-2/3} \to \infty \quad \text{as} \quad x \to 0 \qquad \text{(Figure 4.7.12}a\text{).}$$

(b) The graph of the function $g(x) = (2-x)^{1/5}$ has a vertical tangent at the point $(2, 0)$ since

$$g'(x) = -\tfrac{1}{5}(2-x)^{-4/5} \to -\infty \quad \text{as} \quad x \to 2 \qquad \text{(Figure 4.7.12}b\text{).} \quad \square$$

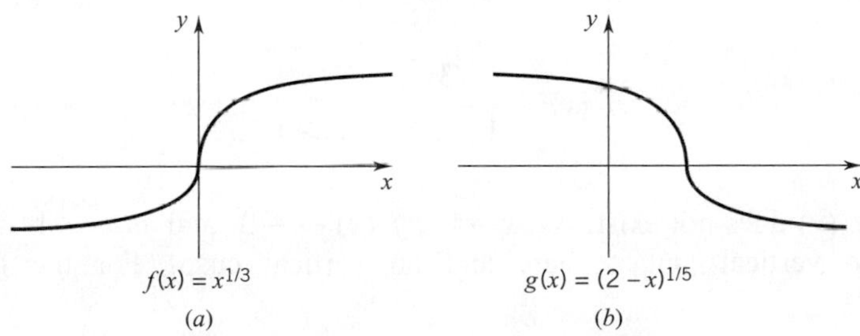

Figure 4.7.12

On occasion you will see a graph become almost vertical and then virtually double back on itself. Such a pattern signals the presence of what is known as a "vertical cusp." We say that the graph of f has a *vertical cusp* at $(c, f(c))$ if

$$\text{as } x \to c^-, \quad f'(x) \to -\infty \quad \text{and} \quad \text{as} \quad x \to c^+, \quad f'(x) \to \infty,$$

or

$$\text{as } x \to c^-, \quad f'(x) \to \infty \quad \text{and} \quad \text{as} \quad x \to c^+, \quad f'(x) \to -\infty.$$

Examples

(a) The graph of the function $f(x) = x^{2/3}$ has a vertical cusp at $(0, 0)$:

$f'(x) = \tfrac{2}{3}x^{-1/3}$ and

$$\text{as } x \to 0^-, \quad f'(x) \to -\infty, \quad \text{and} \quad \text{as} \quad x \to 0^+, \quad f'(x) \to \infty$$

(Figure 4.7.13a).

(b) The graph of the function $g(x) = 2 - (x-1)^{2/5}$ has a vertical cusp at $(1, 2)$:

$g'(x) = -\tfrac{2}{5}(x-1)^{-3/5}$ and

$$\text{as} \quad x \to 1^-, \quad g'(x) \to \infty, \quad \text{and} \quad \text{as} \quad x \to 1^+, \quad g'(x) \to -\infty.$$

(Figure 4.7.13b). $\square$

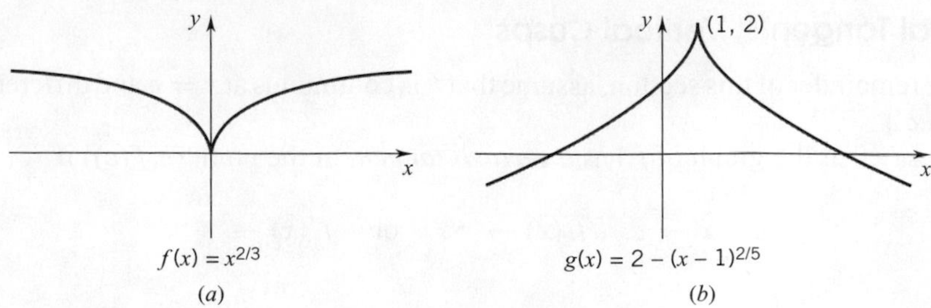

Figure 4.7.13

The fact that $f'(c)$ does not exist does *not* mean that the graph of f has either a vertical tangent or a vertical cusp at $(c, f(c))$. Unless the conditions spelled out earlier are met, the graph of f can simply be making a "corner" at $(c, f(c))$. For example, the function

$$f(x) = |x^3 - 1|$$

has derivative

$$f'(x) = \begin{cases} -3x^2, & x < 1 \\ 3x^2, & x > 1. \end{cases}$$

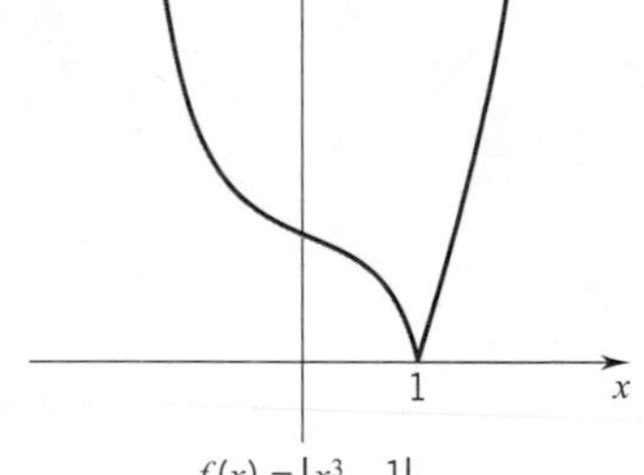

Figure 4.7.14

At $x = 1$, $f'(x)$ does not exist. As $x \to 1^-$, $f'(x) \to -3$, and as $x \to 1^+$, $f'(x) \to 3$. There is no vertical tangent here and no vertical cusp. For the graph, see Figure 4.7.14.

EXERCISES 4.7

1. The graph of a function f is given in the figure.

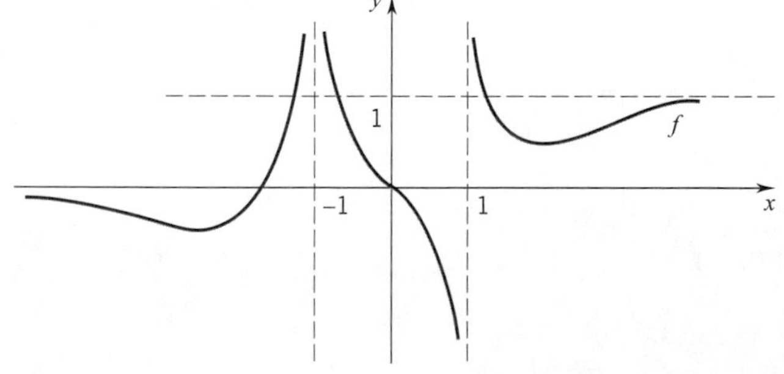

(a) As $x \to -1$, $f(x) \to ?$

(b) As $x \to 1^-$, $f(x) \to ?$

(c) As $x \to 1^+$, $f(x) \to ?$

(d) As $x \to \infty$, $f(x) \to ?$

(e) As $x \to -\infty$, $f(x) \to ?$

(f) Give the equations of the vertical asymptotes, if any.

(g) Give the equations of the horizontal asymptotes, if any.

2. The graph of a function g is given in the figure.

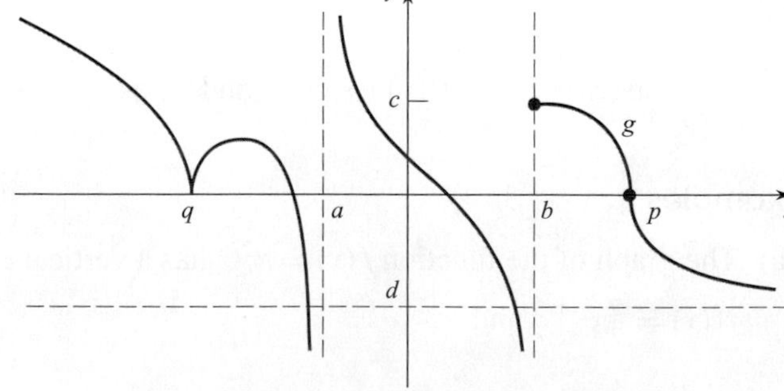

(a) As $x \to \infty$, $g(x) \to ?$ (b) As $x \to b^+$, $g(x) \to ?$

(c) Give the equations of the vertical asymptotes, if any.

(d) Give the equations of the horizontal asymptotes, if any.

(e) Give the numbers c, if any, at which the graph of g has a vertical tangent line.

(f) Give the numbers c, if any, at which the graph of g has a vertical cusp.

Find the vertical and horizontal asymptotes

3. $f(x) = \dfrac{x}{3x-1}$. **4.** $f(x) = \dfrac{x^3}{x+2}$.

5. $f(x) = \dfrac{x^2}{x-2}$. **6.** $f(x) = \dfrac{4x}{x^2+1}$.

7. $f(x) = \dfrac{2x}{x^2-9}$. **8.** $f(x) = \dfrac{\sqrt{x}}{4\sqrt{x}-x}$.

9. $f(x) = \left(\dfrac{2x-1}{4+3x}\right)^2$. **10.** $f(x) = \dfrac{4x^2}{(3x-1)^2}$.

11. $f(x) = \dfrac{3x}{(2x-5)^2}$. **12.** $f(x) = \left(\dfrac{x}{1-2x}\right)^3$.

13. $f(x) = \dfrac{3x}{\sqrt{4x^2+1}}$. **14.** $f(x) = \dfrac{x^{1/3}}{x^{2/3}-4}$.

15. $f(x) = \dfrac{\sqrt{x}}{2\sqrt{x}-x-1}$. **16.** $f(x) = \dfrac{2x}{\sqrt{x^2-1}}$.

17. $f(x) = \sqrt{x+4} - \sqrt{x}$. **18.** $f(x) = \sqrt{x} - \sqrt{x-2}$.

19. $f(x) = \dfrac{\sin x}{\sin x - 1}$. **20.** $f(x) = \dfrac{1}{\sec x - 1}$.

Determine whether or not the graph of f has a vertical tangent or a vertical cusp at c.

21. $f(x) = (x+3)^{4/3}$; $c = -3$.

22. $f(x) = 3 + x^{2/5}$; $c = 0$.

23. $f(x) = (2-x)^{4/5}$; $c = 2$.

24. $f(x) = (x+1)^{-1/3}$; $c = -1$.

25. $f(x) = 2x^{3/5} - x^{6/5}$; $c = 0$.

26. $f(x) = (x-5)^{7/5}$; $c = 5$.

27. $f(x) = (x+2)^{-2/3}$; $c = -2$.

28. $f(x) = 4 - (2-x)^{3/7}$; $c = 2$.

29. $f(x) = \sqrt{|x-1|}$; $c = 1$.

30. $f(x) = x(x-1)^{1/3}$; $c = 1$.

31. $f(x) = |(x+8)^{1/3}|$; $c = -8$.

32. $f(x) = \sqrt{4-x^2}$; $c = 2$.

33. $f(x) = \begin{cases} x^{1/3}+2, & x \le 0 \\ 1 - x^{1/5}, & x > 0; \end{cases}$ $c = 0$.

34. $f(x) = \begin{cases} 1+\sqrt{-x}, & x \le 0 \\ (4x-x^2)^{1/3}, & x > 0; \end{cases}$ $c = 0$.

Sketch the graph of the function, showing all asymptotes.

35. $f(x) = \dfrac{x+1}{x-2}$. **36.** $f(x) = \dfrac{1}{(x+1)^2}$.

37. $f(x) = \dfrac{x}{1+x^2}$. **38.** $f(x) = \dfrac{x-2}{x^2-5x+6}$.

C In Exercises 39–42, find (a) the intervals on which f increases or decreases, and (b) the intervals on which the graph of f is concave up and the intervals on which it is concave down. Also, determine whether the graph of f has vertical tangents or vertical cusps. Confirm your results using a graphing utility. Then sketch the graph of f.

39. $f(x) = x - 3x^{1/3}$. **40.** $f(x) = x^{2/3} - x^{1/3}$.

41. $f(x) = \frac{3}{5}x^{5/3} - 3x^{2/3}$. **42.** $f(x) = \sqrt{|x|}$.

In Exercises 43–48, sketch the graph of a continuous function f that satisfies the given conditions. Indicate whether the graph of f has any horizontal or vertical asymptotes, and whether the graph has any vertical tangent lines or vertical cusps. If no such function exists, explain why.

43. $f(3) = 0, f(0) = 4, f(-1) = 0, f(-2) = -3, f(x) \to \infty$ as $x \to 1^-, f(x) \to -\infty$ as $x \to 1^+, f(x) \to 2$ as $x \to \infty$, $f(x) \to 0$ as $x \to -\infty$; $f'(x) < 0$ if $x < -2, f'(x) > 0$ if $x > -2, x \ne 1$; $f''(x) < 0$ if $x > 1$ or if $x < -4, f''(x) > 0$ if $-4 < x < 1$.

44. $f(0) = 0, f(3) = f(-3) = 0, f(x) \to -\infty$ as $x \to 1$, $f(x) \to -\infty$ as $x \to -1, f(x) \to 1$ as $x \to \infty, f(x) \to 1$ as $x \to -\infty$; $f''(x) < 0$ for all $x, x \ne \pm 1$.

45. $f(x) \ge 1$ for all x, $f(0) = 1$; $f''(x) < 0$ for all $x \ne 0$; $f'(x) \to \infty$ as $x \to 0^+, f'(x) \to -\infty$ as $x \to 0^-$.

46. $f(0) = 1, f(x) \to 4$ as $x \to \infty, f(x) \to -\infty$ as $x \to -\infty$; $f'(x) \to \infty$ as $x \to 0$; $f''(x) > 0$ if $x < 0, f''(x) < 0$ if $x > 0$.

47. $f(0) = 0, f(x) \to -1$ as $x \to \infty, f(x) \to 1$ as $x \to -\infty$; $f'(x) \to -\infty$ as $x \to 0$; $f''(x) < 0$ if $x < 0, f''(x) > 0$ if $x > 0$; f is an odd function.

48. $f(0) = 1, f(2) = 0; f'(x) \to -\infty$ as $x \to 0^-, f'(x) \to 0$ as $x \to 0^+$; $f''(x) < 0$ for all $x \ne 0$.

49. Let p and q be positive integers with q odd and $p < q$. Let $f(x) = x^{p/q}$. Specify conditions on p and q so that:

(a) The graph of f will have a vertical tangent line at $x = 0$.

(b) The graph of f will have a vertical cusp at $x = 0$.

50. *Oblique Asymptotes* Let $r(x) = p(x)/q(x)$ be a rational function and suppose that (degree of p) = (degree of q)+1.

(a) Show that r can be written in the form

$$r(x) = ax + b + \frac{Q(x)}{q(x)},$$

where (degree Q) < (degree q). HINT: divide q into p.

(b) Prove that $[r(x) - (ax+b)] \to 0$ as $x \to \infty$ and $[r(x) - (ax+b)] \to 0$ as $x \to -\infty$. Thus the graph of f "approaches" the line $y = ax + b$ as $x \to \pm\infty$. The line $y = ax + b$ is called an *oblique asymptote* of the graph of r.

In Exercise 51–54, sketch the graph of the given fuction showing all vertical and oblique asymptotes.

51. $f(x) = \dfrac{x^2-4}{x}$. **52.** $f(x) = \dfrac{2x^2+3x-2}{x+1}$.

53. $f(x) = \dfrac{x^3}{(x-1)^2}$. **54.** $f(x) = \dfrac{1+x-3x^2}{x}$.

C **55.** Let $f(x) = \sqrt{x^4-x^2} - x^2$. Use a graphing utility to determine $\lim_{x\to\infty} f(x)$. Verify your result analytically.

56. Let $f(x) = (1 - x)x^{1/3}$. Use a graphing utility of find the horizontal and vertical lines tangent to the graph of f. Verify your results analytically.

In Exercises 57–58, use a CAS to find the oblique asymptotes to the graph of f. Then use a graphing utility to draw the graph of f and its asymptotes to verify your result.

57. $f(x) = \dfrac{3x^4 - 4x^3 - 2x^2 + 2x + 2}{x^3 - x}$.

58. $f(x) = \dfrac{5x^3 - 3x^2 + 4x - 4}{x^2 + 1}$.

59. Use a graphing utility to find any horizontal asymptotes that the function f may have.

(a) $f(x) = \dfrac{\sin 3x}{x^3}$. (b) $f(x) = x\cos(1/x)$. (c) $f(x) = \dfrac{\tan 2x}{x^2}$.

60. Verify that $\lim_{x\to 2} f(x) = +\infty$ where $f(x) = \dfrac{x+4}{(x-2)^2}$. Recall that $\lim_{x\to c} f(x) = +\infty$ means that for every $M > 0$, there is a $\delta > 0$ such that if $0 < |x - c| < \delta$, then $f(x) \geq M$. With $M = 100$, use a graphing utility to draw the graphs of f and $y = M$ together. Use the Zoom Function to estimate the maximum δ that satisfies the definition.

61. Verify that $\lim_{x\to\infty} f(x) = +\infty$ where $f(x) = \dfrac{x^2}{x+2}$. Recall that $\lim_{x\to\infty} f(x) = +\infty$. means that for every $M > 0$, there is a $K > 0$ such that if $x \geq K$, then $f(x) \geq M$. With $M = 250$, use a graphing utility to draw the graphs of f and $y = M$ together. Then find a number K that satisfies the definition.

62. Verify that $\lim_{x\to\infty} f(x) = 2$ where $f(x) = \dfrac{2x+1}{x+2}$. Recall that $\lim_{x\to\infty} f(x) = L$ means that for every $\epsilon > 0$, there is a $K > 0$ such that if $x \geq K$, then $|f(x) - L| < \epsilon$. With $\epsilon = 0.05$, use a graphing utility to draw the graphs of $f, y = L - \epsilon$, and $y = L + \epsilon$ together. Then find a number K that satisfies the definition.

■ 4.8 SOME CURVE SKETCHING

During the course of the last few sections you have learned how to find the extreme values of a function, the intervals where it increases, and the intervals where it decreases; you have seen how to determine the concavity of the graph and how to find the points of inflection, and in the preceding section you learned how to investigate the asymptotic properties of a function. This knowledge makes it possible to sketch an accurate graph of a somewhat complicated function without having to plot point, after point, after point.

Before attempting to sketch the graph of a function f, we try to gather the necessary information and record it in an organized form. Here is an outline of the procedure we will follow.

1. Domain of f: Determine the domain of f; identify endpoints; find the vertical asymptotes of f; determine the behavior of f as $x \to \pm\infty$; find the horizontal asymptotes.
2. Intercepts: Determine the x- and y- intercepts of the graph; the y- intercept is the value $f(0)$; the x-intercepts are the solutions of the equation $f(x) = 0$.
3. Symmetry and periodicity: If f is an even function $[f(-x) = f(x)]$, then the graph of f is symmetric with respect to the y-axis; if f is an odd function $[f(-x) = -f(x)]$, then the graph is symmetric with respect to the origin. If f is periodic with period p, then the graph replicates itself on intervals of length p.
4. Calculate f': Determine the critical numbers of f; examine the sign of f' to determine the intervals on which f increases and the intervals on which f decreases; determine vertical tangents and cusps.
5. Calculate f'': Examine the sign of f'' to determine the intervals on which the graph is concave up and the intervals on which the graph is concave down; determine the points of inflection.
6. Plot the points of interest in a preliminary sketch: intercept points, extreme points (local extreme points, absolute extreme points, endpoint extreme points), and points of inflection.

7. Sketch the graph by connecting the points in the preliminary sketch; make sure the curve "rises," "falls," and "bends" in the proper way. It will be a worthwhile exercise to verify your sketches by using a graphing utility to plot the graph.

Figure 4.8.1 shows some of the elements that we might include in a preliminary sketch.

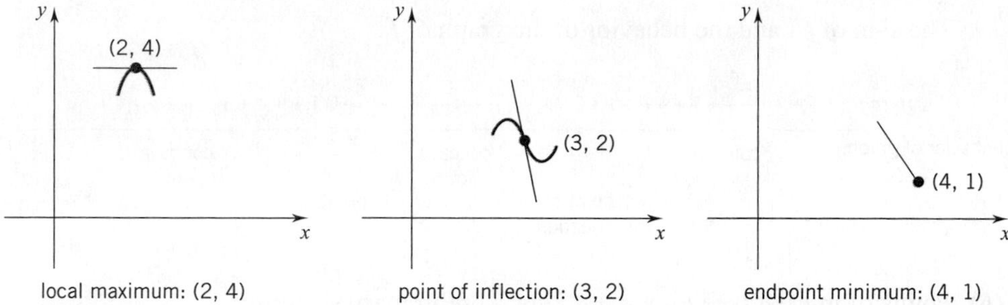

Figure 4.8.1

Example 1 Sketch the graph of the function $f(x) = \frac{1}{4}x^4 - 2x^2 + \frac{7}{4}$.

SOLUTION

(1) Domain: This is a polynomial function, so its domain is the set of all real numbers. Since the leading term is $\frac{1}{4}x^4$, $f(x) \to \infty$ as $x \to \pm\infty$. There are no asymptotes.

(2) Intercepts: The y-intercept is $f(0) = \frac{7}{4}$. To find the x-intercepts, we solve the equation

$$\begin{aligned} f(x) = \tfrac{1}{4}x^4 - 2x^2 + \tfrac{7}{4} &= 0. \\ x^4 - 8x^2 + 7 &= 0, \\ (x^2 - 1)(x^2 - 7) &= 0, \\ (x+1)(x-1)(x+\sqrt{7})(x-\sqrt{7}) &= 0, \end{aligned}$$

Thus, the x-intercepts are $x = \pm 1$ and $x = \pm\sqrt{7}$.

(3) Symmetry/periodicity: Since

$$f(-x) = \tfrac{1}{4}(-x)^4 - 2(-x)^2 + \tfrac{7}{4} = \tfrac{1}{4}x^4 - 2x^2 + \tfrac{7}{4} = f(x),$$

f is an even function and its graph is symmetric with respect to the y-axis; f is not a periodic function.

(4) First derivative:

$$f'(x) = x^3 - 4x = x(x^2 - 4) = x(x+2)(x-2).$$

The critical numbers of f are $x = 0$, $x = \pm 2$. The sign of f' and behavior of f:

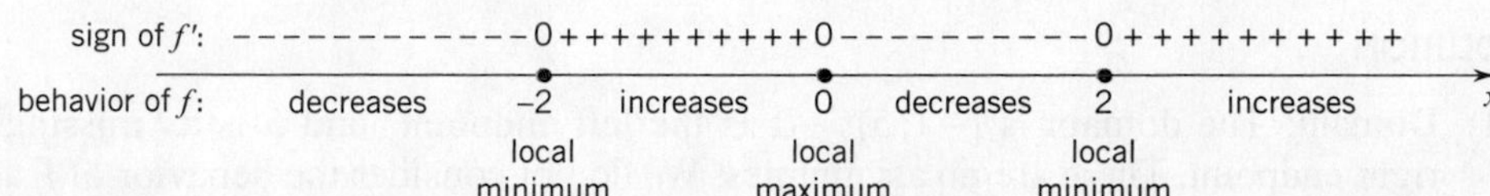

(5) Second derivative

$$f''(x) = 3x^2 - 4 = 3\left(x - \frac{2}{\sqrt{3}}\right)\left(x + \frac{2}{\sqrt{3}}\right).$$

The sign of f'' and the behavior of the graph of f:

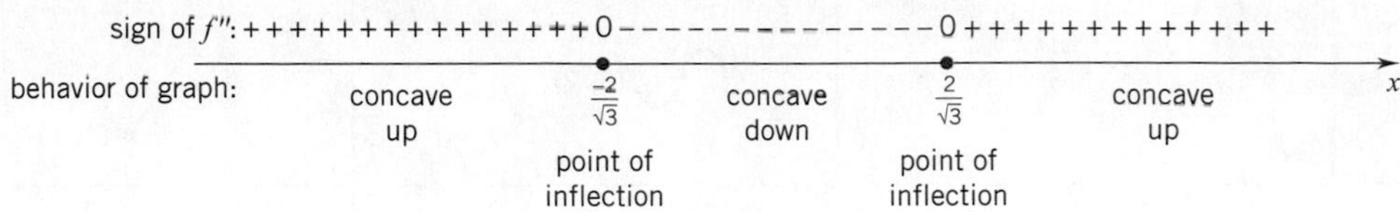

(6) Points of interest (see Figure 4.8.2 for a preliminary sketch):

$$\left(0, \tfrac{7}{4}\right): \text{ y-intercept point.}$$

$$(-1,0), (1,0), (-\sqrt{7},0), (\sqrt{7},0): \text{ x-intercept points.}$$

$$\left(0, \tfrac{7}{4}\right): \text{ local maximum point.}$$

$$\left(-2, -\tfrac{9}{4}\right), \left(2, -\tfrac{9}{4}\right): \text{ local and absolute minimum points.}$$

$$(-2/\sqrt{3}, -17/36), (2/\sqrt{3}, -17/36): \text{ points of inflection.}$$

(7) Sketch the graph: Since the graph is symmetric with respect to the y-axis, we can sketch the graph for $x \geq 0$, and then obtain the graph for $x \leq 0$ by a reflection in the y-axis. See Figure 4.8.3. ☐

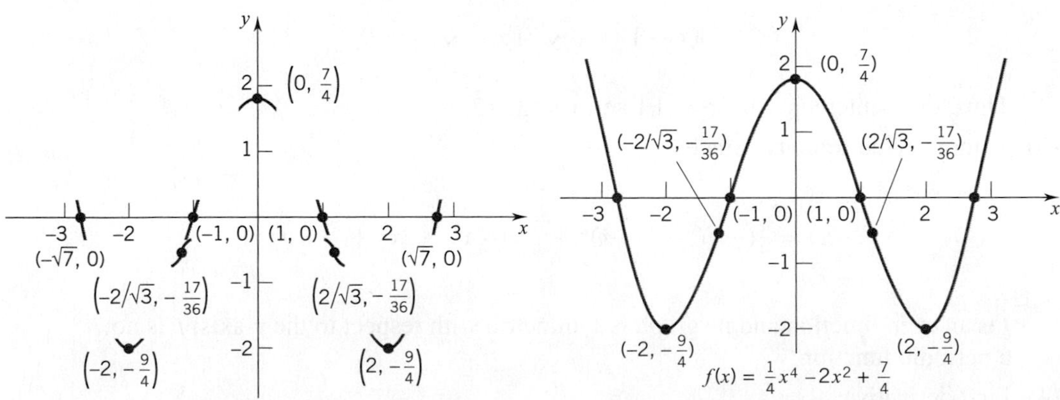

Figure 4.8.2 **Figure 4.8.3**

Example 2 Sketch the graph of the function $f(x) = x^4 - 4x^3 + 1, \quad -1 \leq x < 5$.

SOLUTION

(1) Domain: The domain is $[-1, 5)$; -1 is the left endpoint, and 5 is a "missing" right endpoint. There are no asymptotes. We do not consider the behavior of f as $x \to \pm\infty$ since f is defined only on $[-1, 5)$.

(2) Intercepts: The y-intercept is $f(0) = 1$. To find the x-intercepts, we must solve the equation

$$x^4 - 4x^3 + 1 = 0.$$

We cannot do this exactly, but we can verify that $f(0) > 0$ and $f(1) < 0$, and that $f(3) < 0$ and $f(4) > 0$. Thus there are x-intercepts in the interval $(0, 1)$ and in the interval $(3, 4)$. We could find approximate values for these intercepts by the Newton-Raphson method, but we won't stop to do this since our aim here is a sketch of the graph, not a detailed drawing.

(3) Symmetry/periodicity: The graph is not symmetric with respect to the y-axis, nor with respect to the origin $[f(-x) \neq \pm f(x)]$; f is not periodic.

(4) First derivative: For $x \in (-1, 5)$

$$f'(x) = 4x^3 - 12x^2 = 4x^2(x - 3).$$

The critical numbers of f are $x = 0$ and $x = 3$.

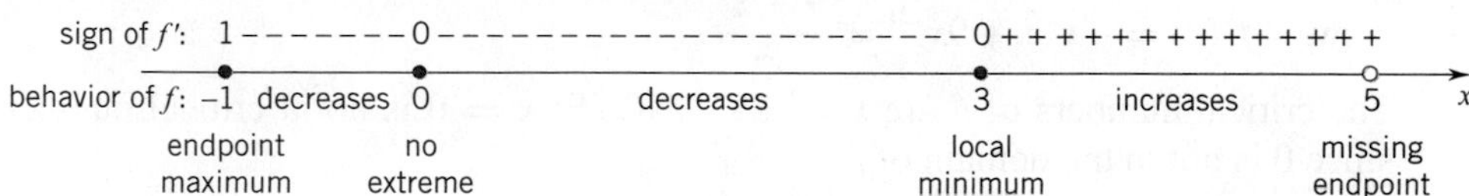

(5) Second derivative:

$$f''(x) = 12x^2 - 24x = 12x(x - 2).$$

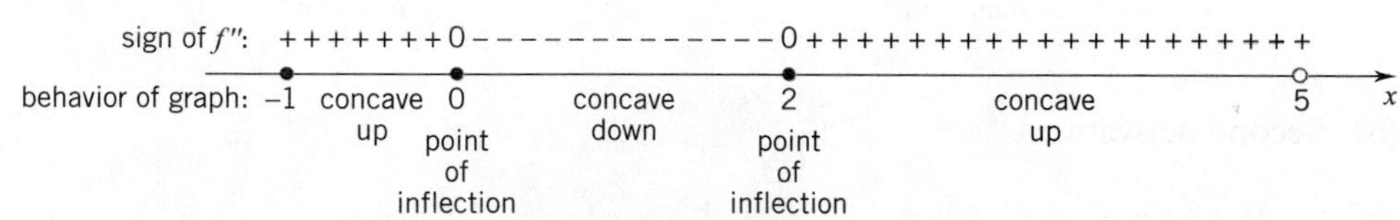

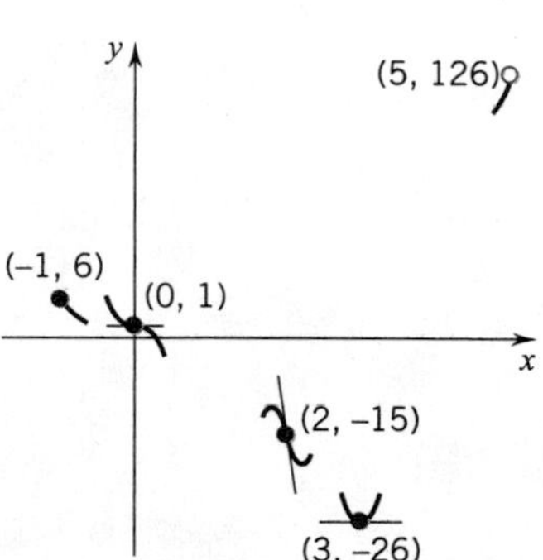

Figure 4.8.4

(6) Points of interest (see Figure 4.8.4 for a preliminary sketch):

$(0, 1)$: y-intercept point; point of inflection with horizontal tangent.
$(-1, 6)$: endpoint maximum point.
$(2, -15)$: point of inflection.
$(3, -26)$: local and absolute minimum point.

As x approaches the missing endpoint 5 from the left, $f(x)$ increases toward a value of 126.

(7) Sketch the graph. Since the range of f makes a scale drawing impractical, we must be content with a rough sketch as in Figure 4.8.5. In cases like this, it is particularly important to give the coordinates of the points of interest. ❑

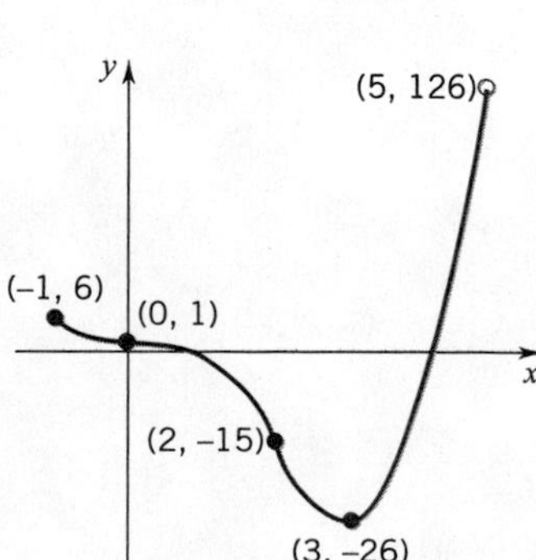

Figure 4.8.5

Example 3 Sketch the graph of the function $f(x) = \dfrac{x^2 - 3}{x^3}$.

SOLUTION

(1) Domain: The domain of f is $\{x : x \neq 0\} = (-\infty, 0) \cup (0, \infty)$. The y-axis (the line $x = 0$) is a vertical asymptote since $f(x) \to \infty$ as $x \to 0^-$ and $f(x) \to -\infty$

as $x \to 0^+$. The x-axis (the line $y = 0$) is a horizontal asymptote: $f(x) \to 0$ as $x \to \pm\infty$.

(2) Intercepts: There is no y-intercept since f is not defined at $x = 0$. The x-intercepts are $x = \pm\sqrt{3}$.

(3) Symmetry: Since

$$f(-x) = \frac{(-x)^2 - 3}{(-x)^3} = -\frac{x^2 - 3}{x^3} = -f(x),$$

the graph is symmetric with respect to the origin; f is not periodic.

(4) First derivative: It is easier to calculate f' if we first rewrite f as

$$f(x) = \frac{x^2 - 3}{x^3} = x^{-1} - 3x^{-3}.$$

Now, $f'(x) = -x^{-2} + 9x^{-4} = \dfrac{9 - x^2}{x^4}.$

The critical numbers of f are $x = \pm 3$. NOTE: $x = 0$ is not a critical number since 0 is not in the domain of f.

The sign of f' and the behavior of f:

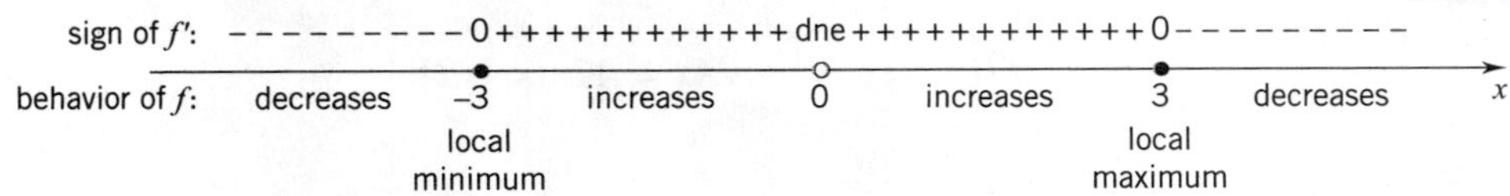

(5) Second derivative:

$$f''(x) = 2x^{-3} - 36x^{-5} = \frac{2(x^2 - 18)}{x^5} = \frac{2(x - 3\sqrt{2})(x + 3\sqrt{2})}{x^5}.$$

The sign of f'' and the behavior of the graph of f:

sign of f'': - - - - - - 0 + + + + + + + + + + + + dne - - - - - - - - - - - - 0 + + + + + + + +

behavior of graph: concave down | $-3\sqrt{2}$ point of inflection | concave up | 0 | concave down | $3\sqrt{2}$ point of inflection | concave up | x

(6) Points of interest (see Figure 4.8.6 for a preliminary sketch):

$$(-\sqrt{3}, 0), (\sqrt{3}, 0): \text{ } x\text{-intercept points.}$$

$$(-3, -2/9): \text{ local minimum point.}$$

$$(3, 2/9): \text{ local maximum point.}$$

$$(-3\sqrt{2}, -5\sqrt{2}/36), (3\sqrt{2}, 5\sqrt{2}/36): \text{ points of inflection.}$$

(7) Sketch the graph: See Figure 4.8.7. □

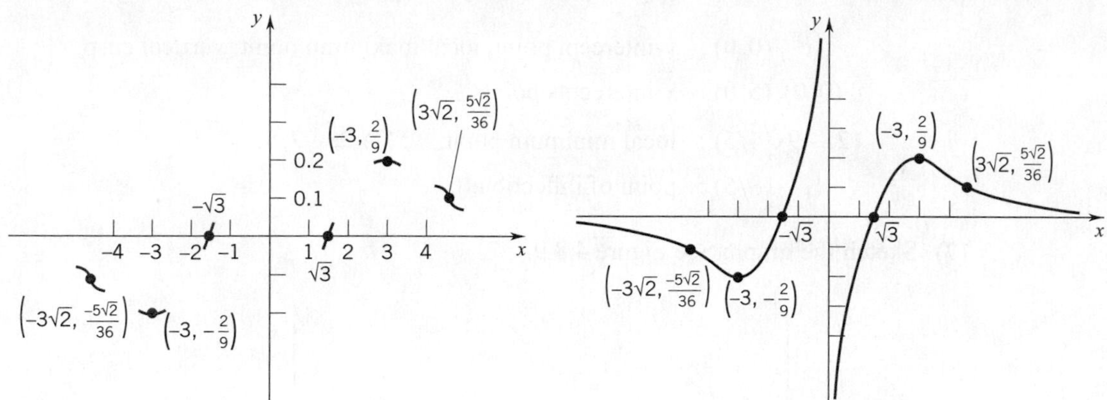

Figure 4.8.6 **Figure 4.8.7**

Example 4 Sketch the graph of the function $f(x) = \frac{3}{5}x^{5/3} - 3x^{2/3}$.

SOLUTION

(1) Domain: The domain of f is the set of real numbers. Since we can express f as $\frac{3}{5}x^{2/3}(x-5)$, we see that, as $x \to \infty$, $f(x) \to \infty$, and as $x \to -\infty$, $f(x) \to -\infty$. There are no asymptotes.

(2) Intercepts: $f(0) = 0$ is both the y-intercept and an x-intercept; $x = 5$ is also an x-intercept.

(3) Symmetry/periodicity: There is no symmetry; f is not periodic.

(4) First derivative:

$$f'(x) = x^{2/3} - 2x^{-1/3} = \frac{x-2}{x^{1/3}}.$$

The critical numbers of f are $x = 0$ and $x = 2$. The sign of f' and the behavior of f:

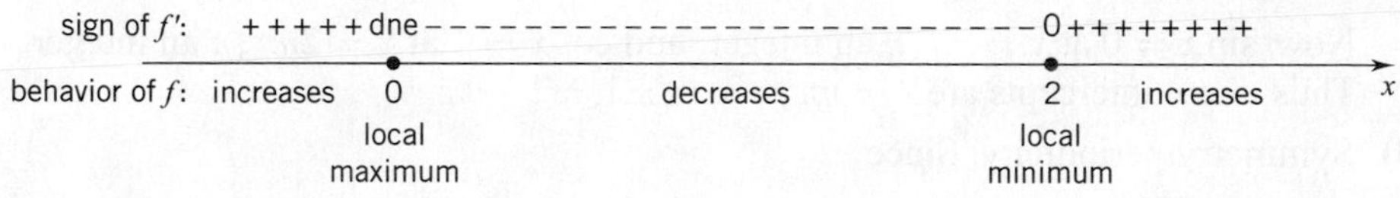

Note that, as $x \to 0^-$, $f'(x) \to \infty$, and as $x \to 0^+$, $f'(x) \to -\infty$. Thus the graph of f has a vertical cusp at $x = 0$.

(5) Second derivative:

$$f''(x) = \tfrac{2}{3}x^{-1/3} + \tfrac{2}{3}x^{-4/3} = \tfrac{2}{3}x^{-4/3}(x+1).$$

The sign of f'' and the behavior of the graph of f:

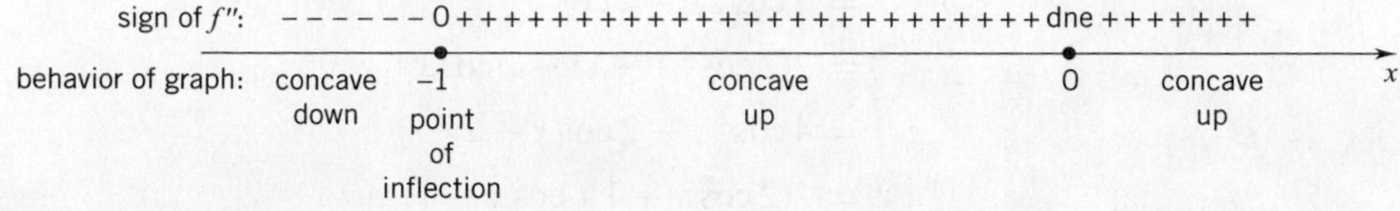

(6) Points of interest (see Figure 4.8.8 for a preliminary sketch):

$$\begin{aligned}(0,0):&\ \ y\text{-intercept point, local maximum point; vertical cusp.}\\(0,0),(5,0):&\ \ x\text{-intercepts points.}\\(2,-9\sqrt[3]{4}/5):&\ \ \text{local minimum point,}\quad f(2)\cong -2.9.\\(-1,-18/5):&\ \ \text{point of inflection.}\end{aligned}$$

(7) Sketch the graph: See Figure 4.8.9. ❑

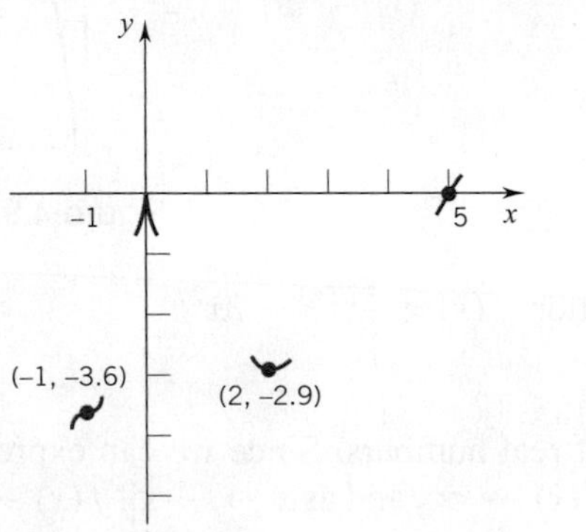

Figure 4.8.8

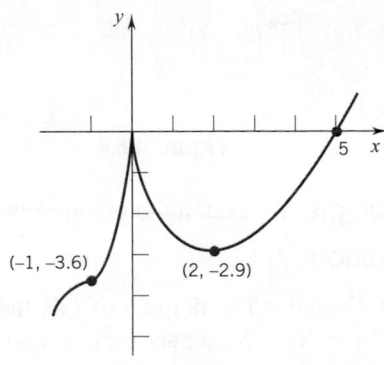

Figure 4.8.9

Example 5 Sketch the graph of $f(x) = \sin 2x - 2\sin x$.

SOLUTION

(1) Domain: The domain of f is the set of all real numbers. There are no asymptotes and, as you can verify, the graph of f oscillates between $\frac{3}{2}\sqrt{3}$ and $-\frac{3}{2}\sqrt{3}$ as $x \to \pm\infty$.

(2) Intercepts: The y-intercept is $f(0) = 0$. To find the x-intercepts, we set $f(x) = 0$:

$$\begin{aligned}\sin 2x - 2\sin x &= 2\sin x\cos x - 2\sin x\\&= 2\sin x(\cos x - 1) = 0.\end{aligned}$$

Now, $\sin x = 0$ at $x = n\pi$, n an integer; and $\cos x = 1$ at $x = 2n\pi$, n an integer. Thus, the x-intercepts are $x = n\pi, n = 0, \pm 1, \pm 2, \cdots$

(3) Symmetry/periodicity: Since

$$f(-x) = \sin 2[(-x)] - 2\sin(-x) = -\sin 2x + 2\sin x = -f(x),$$

f is an odd function and thus the graph is symmetric with respect to the origin. Also, f is periodic with period 2π. On the basis of these two properties, it would be sufficient to sketch the graph of f on the interval $[0, \pi]$. The result could then be extended to the interval $[-\pi, 0]$ using the symmetry, and then to $(-\infty, \infty)$ using the periodicity. However, for purposes of illustration here, we will consider f and its derivatives on $[-\pi, \pi]$.

(4) First derivative:

$$\begin{aligned}f'(x) &= 2\cos 2x - 2\cos x\\&= 2(2\cos^2 x - 1) - 2\cos x\\&= 4\cos^2 x - 2\cos x - 2\\&= 2(2\cos x + 1)(\cos x - 1).\end{aligned}$$

The critical points of f (on $[-\pi, \pi]$) are $x = -2\pi/3, 2\pi/3$, and $x = 0$.

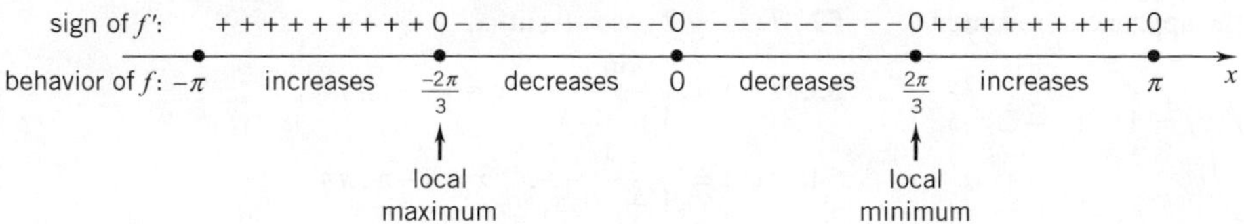

(5) Second derivative:

$$\begin{aligned} f''(x) &= -4\sin 2x + 2\sin x \\ &= -8\sin x\cos x + 2\sin x \\ &= 2\sin x(-4\cos x + 1). \end{aligned}$$

Now $f''(x) = 0$ at $x = -\pi, 0, \pi$, and at the value of x where $\cos x = \frac{1}{4}$, which yields $x \cong \pm 1.3$. The sign of f'' and the behavior of the graph (on $[-\pi, \pi]$):

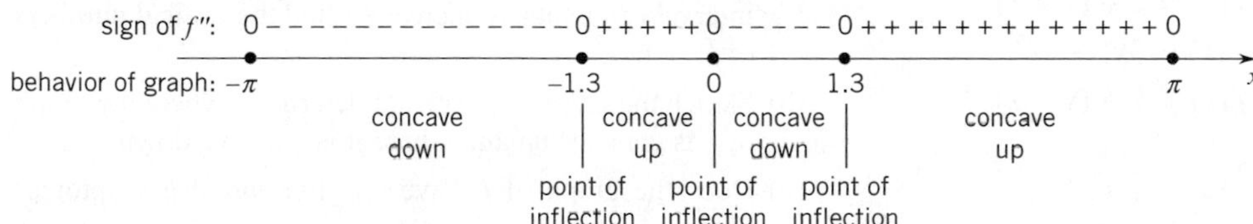

(6) Points of interest (see Figure 4.8.10 for a preliminary sketch):

$(0, 0)$: y-intercept point.

$(-\pi, 0), (0, 0), (\pi, 0)$: x-intercept points, (0,0) is a point of inflection.

$(-\frac{2}{3}\pi, \frac{3}{2}\sqrt{3})$: local and absolute maximum point; $\frac{3}{2}\sqrt{3} \cong 2.6$.

$(\frac{2}{3}\pi, -\frac{3}{2}\sqrt{3})$: local and absolute minimum point; $-\frac{3}{2}\sqrt{3} \cong -2.6$.

$(-1.3, 1.4), (1.3, -1.4)$: points of inflection (approximately).

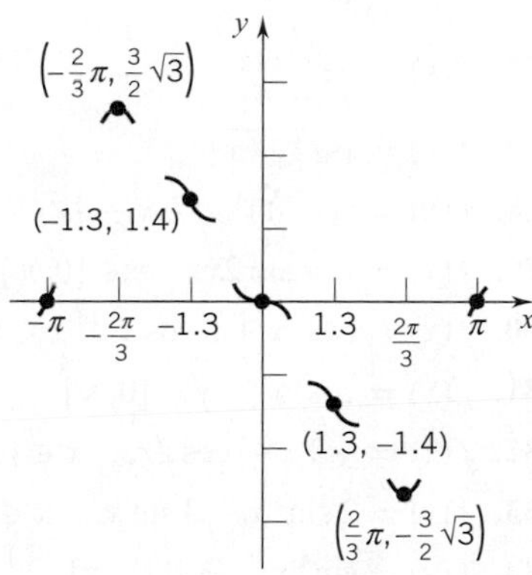

Figure 4.8.10

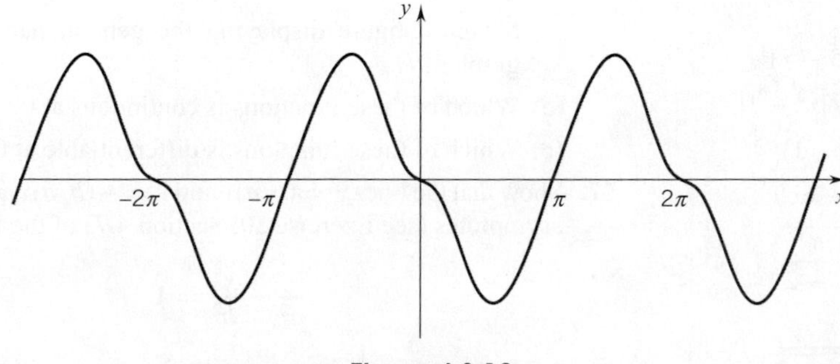

Figure 4.8.12

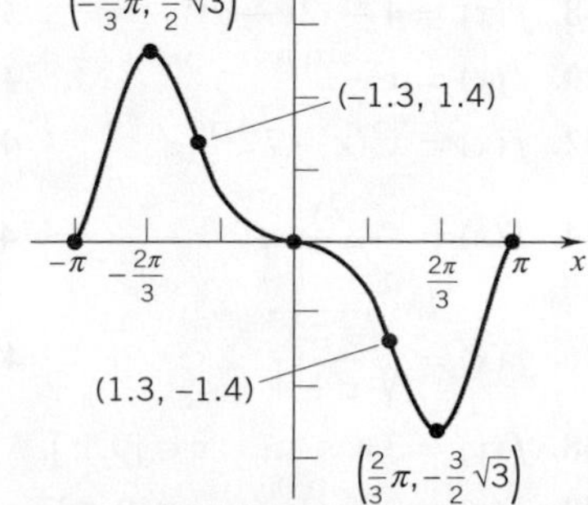

Figure 4.8.11

(7) Sketch the graph: The graph of f on the interval $[-\pi, \pi]$ is shown in Figure 4.8.11. An indication of the complete graph is shown in Figure 4.8.12. □

EXERCISES 4.8

Sketch the graph of the function f using the approach presented in this section.

1. $f(x) = (x-2)^2$.

2. $f(x) = 1-(x-2)^2$.

3. $f(x) = x^3 - 2x^2 + x + 1$.

4. $f(x) = x^3 - 9x^2 + 24x - 7$.

5. $f(x) = x^3 + 6x^2, \quad x \in [-4, 4]$.

6. $f(x) = x^4 - 8x^2, \quad x \in (0, \infty)$.

7. $f(x) = \frac{2}{3}x^3 - \frac{1}{2}x^2 - 10x - 1$.

8. $f(x) = x(x^2+4)^2$.

9. $f(x) = x^2 + \dfrac{2}{x}$.

10. $f(x) = x - \dfrac{1}{x}$.

11. $f(x) = \dfrac{x-4}{x^2}$.

12. $f(x) = \dfrac{x+2}{x^3}$.

13. $f(x) = 2\sqrt{x} - x, \quad x \in [0, 4]$.

14. $f(x) = \frac{1}{4}x - \sqrt{x}, \quad x \in [0, 9]$.

15. $f(x) = 2 + (x+1)^{6/5}$.

16. $f(x) = 2 + (x+1)^{7/5}$.

17. $f(x) = 3x^5 + 5x^3$.

18. $f(x) = 3x^4 + 4x^3$.

19. $f(x) = 1 + (x-2)^{5/3}$.

20. $f(x) = 1 + (x-2)^{4/3}$.

21. $f(x) = \dfrac{x^2}{x^2+4}$.

22. $f(x) = \dfrac{2x^2}{x+1}$.

23. $f(x) = \dfrac{x}{(x+3)^2}$.

24. $f(x) = \dfrac{x}{x^2+1}$.

25. $f(x) = \dfrac{x^2}{x^2-4}$.

26. $f(x) = \dfrac{1}{x^3-x}$.

27. $f(x) = x\sqrt{1-x}$.

28. $f(x) = (x-1)^4 - 2(x-1)^2$.

29. $f(x) = x + \sin 2x, \quad x \in [0, \pi]$.

30. $f(x) = \cos^3 x + 6\cos x, \quad x \in [0, \pi]$.

31. $f(x) = \cos^4 x, \quad x \in [0, \pi]$.

32. $f(x) = \sqrt{3}\,x - \cos 2x, \quad x \in [0, \pi]$.

33. $f(x) = 2\sin^3 x + 3\sin x, \quad x \in [0, \pi]$.

34. $f(x) = \sin^4 x, \quad x \in [0, \pi]$.

35. $f(x) = (x+1)^3 - 3(x+1)^2 + 3(x+1)$.

36. $f(x) = x^3(x+5)^2$.

37. $f(x) = x^2(5-x)^3$.

38. $f(x) = 4 - |2x - x^2|$.

39. $f(x) = 3 - |x^2 - 1|$.

40. $f(x) = x - x^{1/3}$.

41. $f(x) = x(x-1)^{1/5}$.

42. $f(x) = x^2(x-7)^{1/3}$.

43. $f(x) = x^2 - 6x^{1/3}$.

44. $f(x) = \dfrac{2x}{\sqrt{x^2+1}}$.

45. $f(x) = \sqrt{\dfrac{x}{x-2}}$.

46. $f(x) = \sqrt{\dfrac{x}{x+4}}$.

47. $f(x) = \dfrac{x^2}{\sqrt{x^2-2}}$.

48. $f(x) = 3\cos 4x, \quad x \in [0, \pi]$.

49. $f(x) = 2\sin 3x, \quad x \in [0, \pi]$.

50. $f(x) = 3 + 2\cot x + \csc^2 x, \quad x \in (0, \frac{1}{2}\pi)$.

51. $f(x) = 2\tan x - \sec^2 x, \quad x \in (0, \frac{1}{2}\pi)$.

52. $f(x) = 2\cos x + \sin^2 x$.

53. $f(x) = \dfrac{\sin x}{1 - \sin x}, \quad x \in (-\pi, \pi)$.

54. $f(x) = \dfrac{1}{1-\cos x}, \quad x \in (-\pi, \pi)$.

55. A function f is continuous on $(-\infty, \infty)$, differentiable for all $x \neq 0$, and $f(0) = 0$. The graph of the derivative of f is shown in the figure.

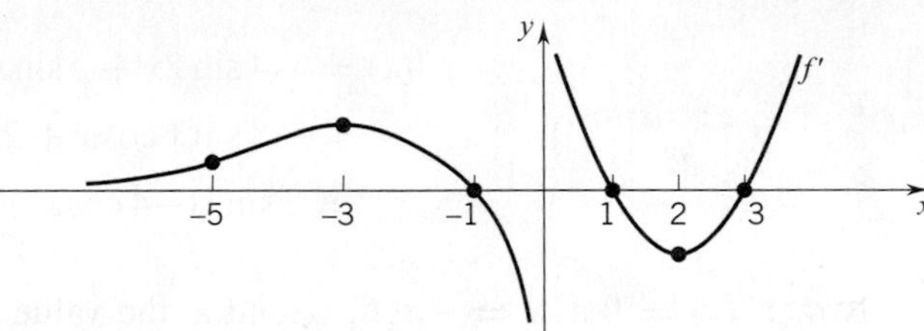

(a) Determine the intervals on which f increases and the intervals on which it decreases; find the critical numbers of f.

(b) Sketch the graph of $f''(x)$ and determine where the graph of f is concave up and where it is concave down.

(c) Does the graph of f have any horizontal asymptotes? Sketch the graph of f.

56. Set

$$F(x) = \begin{cases} \sin(1/x), & x \neq 0 \\ 0, & x = 0 \end{cases}$$

$$G(x) = \begin{cases} x\sin(1/x), & x \neq 0 \\ 0, & x = 0, \end{cases}$$

$$H(x) = \begin{cases} x^2\sin(1/x), & x \neq 0 \\ 0, & x = 0. \end{cases}$$

(a) Sketch a figure displaying the general nature of the graph of F.

(b) Sketch a figure displaying the general nature of the graph of G.

(c) Sketch a figure displaying the general nature of the graph of H.

(d) Which of these functions is continuous at 0?

(e) Which of these functions is differentiable at 0?

57. Show that the lines $y = (b/a)x$ and $y = -(b/a)x$ are oblique asymptotes (see Exercise 50, Section 4.7) of the hyperbola

$$\frac{x^2}{a^2} - \frac{y^2}{b^2} = 1.$$

58. Let $f(x) = \dfrac{x^3-1}{x}$. Show that

$$f(x) - x^2 \to 0 \quad \text{as} \quad x \to \pm\infty.$$

This implies that the graph of f is asymptotic to the graph of the parabola $y = x^2$. Use this fact to sketch the graph of f.

■ CHAPTER HIGHLIGHTS

4.1 The Mean-Value Theorem

mean-value theorem (p. 197) Rolle's theorem (p. 199)

4.2 Increasing and Decreasing Functions

f increases on an interval I (p. 204)
f decreases on an interval I (p. 204)

One can find the intervals on which a differentiable function increases or decreases or is constant by examining the sign of the derivative.

If two functions have the same derivative on an interval, then they differ by a constant on that interval (p. 209).

4.3 Local Extreme Values

local extrema: local maximum, local minimum (p. 212)
critical number (p. 213) first-derivative test (p. 215).
second-derivative test (p. 216)

If f has a local extreme value at c, then either $f'(c) = 0$ or $f'(c)$ does not exist; the converse is false.

4.4 Endpoint and Absolute Extreme Values

endpoint maximum and endpoint minimum (p. 220)
absolute maximum and absolute minimum (p. 221)
$f(x) \to \pm\infty$ as $x \to \pm\infty$ (p. 223)

4.5 Some Max-Min Problems

strategy for solving max-min problems (p. 229)

The *key steps* are to express the quantity to be maximized or minimized as a function of one variable and to specify the domain of the function.

4.6 Concavity and Points of Inflection

concave up and concave down (p. 238)
point of inflection (p. 239)

If $(c, f(c))$ is a point of inflection, then either $f''(c) = 0$ or $f''(c)$ does not exist; the converse is false.

4.7 Vertical and Horizontal Asymptotes; Vertical Tangents and Cusps

$f(x) \to \pm\infty$ as $x \to c$ (p. 243)
$f(x) \to L$ as $x \to \pm\infty$ (p. 244)
vertical asymptote (p. 243)
horizontal asymptote (p. 244)
vertical tangent (p. 247)
vertical cusp (p. 248)

4.8 Some Curve Sketching

systematic procedure for sketching graphs (pp. 250–251)

Solve the inequality. Express the solution set in terms of intervals and graph the solution set on a number line.

1. $x^2 - x - 6 \geq 0$ **2.** $x^3 - 2x^2 \leq 8x$

3. $\dfrac{x+1}{x^2-4} > 0$ **4.** $\dfrac{x^2-4x+4}{x^2-x-12} \leq 0$

Decide whether the limit exists, and evaluate the limit if it does exist.

5. $\lim_{x\to -2}\left(x^3 - 2x + \dfrac{2}{x}\right)$ **6.** $\lim_{x\to 3}\dfrac{x^2-3}{x+3}$

7. $\lim_{x\to 2}\dfrac{x^2+4}{x^2+2x+1}$ **8.** $\lim_{x\to 3}\dfrac{x^2-9}{x^2-5x+6}$

9. $\lim_{x\to 3^-}\dfrac{x-3}{|x-3|}$ **10.** $\lim_{x\to 1}\dfrac{1-1/x}{1-1/x^2}$

11. $\lim_{x\to 3^+}\dfrac{x^2-2x-3}{\sqrt{x-3}}$ **12.** $\lim_{x\to 3^+}\dfrac{\sqrt{x^2-2x-3}}{x-3}$

13. $\lim_{x\to 2}\dfrac{x^3-8}{x^4-3x^2-4}$ **14.** $\lim_{x\to 0}\dfrac{\sin 5x}{6x}$

15. $\lim_{x\to 0}\dfrac{1-\cos 3x}{4x}$ **16.** $\lim_{x\to 0}\dfrac{\sin ax}{\sin bx}, a, b \neq 0$

17. $\lim_{x\to 0}\dfrac{\tan^2 2x}{3x^2}$ **18.** $\lim_{x\to 0}\dfrac{\sin 3x}{5x^2-4x}$

19. $\lim_{x\to \pi/2}\dfrac{\cos 2x}{\sin x}$ **20.** $\lim_{x\to 0}\dfrac{5x^2}{1-\cos 2x}$

21. $\lim_{x\to 2} f(x)$ if $f(x) = \begin{cases} x+1, & x < 2 \\ 3x - x^2, & x > 2 \end{cases}$

22. $\lim_{x\to -2} f(x)$ if $f(x) = \begin{cases} 3+x, & x < -2 \\ 5, & x = -2 \\ x^2 - 3, & x > -2 \end{cases}$

23. (a) Sketch the graph of

$$f(x) = \begin{cases} 3x+4, & x < -1 \\ -2x-2, & -1 < x < 2 \\ 2x, & x > 2 \\ x^2, & x = -1, 2 \end{cases}$$

(b) Evaluate the limits that exist.

(*i*) $\lim_{x\to -1^-} f(x)$ (*ii*) $\lim_{x\to -1^+} f(x)$ (*iii*) $\lim_{x\to -1} f(x)$

(*iv*) $\lim_{x\to 2^-} f(x)$ (*v*) $\lim_{x\to 2^+} f(x)$ (*vi*) $\lim_{x\to 2} f(x)$

(c) (*i*) Is f continuous the left at -1? Is f continuous from the right at -1?

(*ii*) Is f continuous from the left at 2? Is f continuous from the right at 2?

In Exercises 24–26 the function f is continuous everywhere except at the given point a. If possible, define f at a so that it becomes continuous at a.

24. $f(x) = \dfrac{x^2-2x-15}{x+3}$, $a = -3$

25. $f(x) = \dfrac{\sqrt{x+2}-2}{x-2}$, $a = 2$

26. $f(x) = \dfrac{\sin \pi x}{x}$, $a = 0$

27. (a) Find $\lim_{x\to c} f(x)$ given that $|f(x)| \leq M|x-c|$ for all $x \neq c$.

(b) Given that $\lim_{x\to 0} f(x)/x$ exists, prove that $\lim_{x\to 0} f(x) = 0$.

Use Definition 3.1.3 to find the derivative of f.

28. $f(x) = x^3 - 4x + 3$ **29.** $f(x) = \dfrac{1}{x-2}$

Find the derivative of the function.

30. $f(x) = 2x^4 + 4x^{1/4}$ **31.** $f(x) = \dfrac{ax^2+bx+c}{x}$

32. $f(t) = (2-3t^2)^2$ **33.** $y = \left(a + \dfrac{b}{x^2}\right)^3$

34. $y = x^2\sqrt{a+bx}$ **35.** $g(x) = \dfrac{4+x^2}{2x-x^2}$

36. $y = (3x+1)^2\sqrt{x^2+2}$ **37.** $f(x) = \sec(x^2+1) + \tan(\sqrt{x})$

38. $y = \dfrac{\cos(x^3)+1}{1-\sin 2x}$ **39.** $s = \sqrt[3]{\dfrac{2+3t}{2-3t}}$

40. $f(x) = \csc^2 2x - \cot^2 2x$

Find equations for the tangent and normal lines to the graph of f at the given point.

41. $f(x) = 2x^3 - x^2 + 3$; $(1, 4)$

42. $f(x) = \cos 2x$; $(\frac{\pi}{6}, \frac{1}{2})$

Find the second derivative of the function.

43. $y = (x^2+4)^{3/2}$ **44.** $g(x) = x\sin x$

Find a formula for the nth derivative.

45. $y = (a+bx)^n$

46. $y = \dfrac{a}{bx+c}$

Use implicit differentiation to find dy/dx.

47. $x^3y + xy^3 = 2$

48. $x^2 + 3x\cos y = \dfrac{x}{y}$

49. Find the points on the curve $y = \frac{2}{3}x^{3/2}$ where the inclination of the tangent line is (a) $\pi/4$, (b) $60°$, (c) $\pi/6$.

50. Find equations for all tangents to the curve $y = x^3 - x$ that pass through the point $(-2, 2)$.

Find the following limits. [They are derivatives set in the form of Definition 3.1.1.]

51. $\lim_{h\to 0} \dfrac{\sqrt{9+h}-3}{h}$

52. $\lim_{h\to 0} \dfrac{\sin\left(\frac{\pi}{2}+h\right)-1}{h}$

53. Use differentials to estimate the value of (a) $\sqrt{63.5}$, (b) $\sqrt[3]{(8.7)^2}$

54. Use differentials to estimate the value of (a) $\tan 32°$, (b) $\cos 42°$

55. Verify that $f(x) = \sin x + \cos x - 1$, $0 \le x \le 2\pi$, satisfies the hypotheses of Rolle's theorem and find the value(s) of c specified by the conclusion of the theorem.

56. Verify that $f(x) = x^3 - 2x + 1$, $-2 \le x \le 3$, satisfies the hypotheses of the mean-value theorem and find the value(s) of c specified by the conclusion of the theorem.

57. If $f(x) = (x-1)^{4/5} - 1$ on $[0, 2]$, then $f(0) = f(2) = 0$. Verify that there does not exist a number $c \in (0, 2)$ such that $f'(c) = 0$. Explain why this does not violate Rolle's theorem.

Find the absolute extreme values of f on the indicated interval

58. $f(x) = x^4 - 8x^2 + 2;\ [-2, 3]$

59. $f(x) = \cos^2 x + \sin x;\ [0, 2\pi]$

Sketch the graph of f using the procedures presented in Section 4.8.

60. $f(x) = (x+2)^2(x-1)^3$

61. $f(x) = x\sqrt{4-x}$

62. $f(x) = \sin^2 x - \cos x,\ x \in [0, 2\pi]$

63. $f(x) = x^{2/3}(x-10)$

64. $f(x) = 6 + 4x^3 - 3x^4$

65. $f(x) = \dfrac{2x}{x^2+4}$

66. $f(x) = 2 + (4-x)^{1/3}$

67. $f(x) = \sin x + \sqrt{3}\cos x,\ x \in [0, 2\pi]$

68. $f(x) = 3x^5 - 5x^3 + 1$

69. Sketch the graph of a function f that satisfies the following conditions: $f(-1) = 3$, $f(0) = 0$, $f(2) = -4$; $f'(-1) = f'(2) = 0$, $f'(x) > 0$ if $x < -1$ or if $x > 2$, $f'(x) < 0$ if $-1 < x < 2$; $f''(x) < 0$ if $x < \frac{1}{2}$, $f''(x) > 0$ if $x > \frac{1}{2}$.

70. A car rental agency rents 100 cars per day at a rate of \$30 per car. For each \$1 increase in the rental rate, two fewer cars are rented. Let x denote the amount of increase above \$30. Express the daily rental income as a function of x and give the domain. Is there a value of x that will maximize the daily income? If so, what is it?

71. The radius of a cone is increasing at the rate of 0.3 inches per minute but the volume remains constant. At what rate is the height of the cone changing at the instant the radius is 4 inches and the height is 15 inches?

72. An object moves along a coordinate line, its position at time t given by the function $x(t) = t + 2\cos t$. Find those times t from 0 to 2π when the object is slowing down.

73. An object moves along a coordinate line, its position at time t given by the function $x(t) = \sqrt{t+1}$. (a) Show that the acceleration is negative and proportional to the cube of the velocity. (b) Use differentials to find numerical estimates for the position, velocity, and acceleration of the object at time $t = 17$. Base your estimate on $t = 15$.

74. A boy walks on a straight, horizontal path away from a light that hangs 12 feet above the path. How fast does his shadow lengthen if he is 5 feet tall and walks at the rate of 168 feet per minute?

75. A rocket is fired from the ground straight up into the air with an initial velocity of 128 feet per second. (a) When does the rocket reach its maximum height and what is the maximum height? (b) When does the rocket hit the ground and what is its velocity when it hits?

76. A closed rectangular box with a square base is to be built subject to the following conditions; the volume is to be 27 cubic feet, the area of the base may not exceed 18 square feet, the height of the box may not exceed 4 feet. Determine the dimensions of the box for (a) minimal surface area; (b) maximal surface area.

77. Draw a line through $P(1, 2)$ that intersects the positive x-axis at $A(a, 0)$ and the positive y-axis at $B(0, b)$. Choose A and B so that the area of the right triangle OAB is a minimum.

78. A right circular cylinder is generated by rotating a rectangle of given perimeter P about one of its sides. What dimensions of the rectangle will generate the cylinder of maximum volume?

■ 5.1 AN AREA PROBLEM; A SPEED-DISTANCE PROBLEM

An Area Problem

In Figure 5.1.1 you can see a region Ω bounded above by the graph of a continuous function f, bounded below by the x-axis, bounded on the left by the line $x = a$, and bounded on the right by the line $x = b$. The question before us is this: What number, if any, should be called the area of Ω?

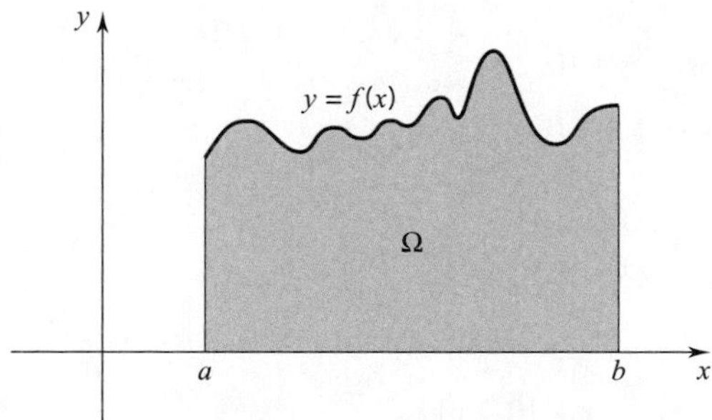

Figure 5.1.1

To begin to answer this question, we split up the interval $[a, b]$ into a finite number of subintervals

$$[x_0, x_1], [x_1, x_2], \ldots, [x_{n-1}, x_n] \quad \text{with} \quad a = x_0 < x_1 < \cdots < x_n = b.$$

This breaks up the region Ω into n subregions:

$$\Omega_1, \Omega_2, \ldots, \Omega_n. \qquad \text{(Figure 5.1.2)}$$

We can estimate the total area of Ω by estimating the area of each subregion Ω_i and adding up the results. Let's denote by M_i the maximum value of f on $[x_{i-1}, x_i]$ and

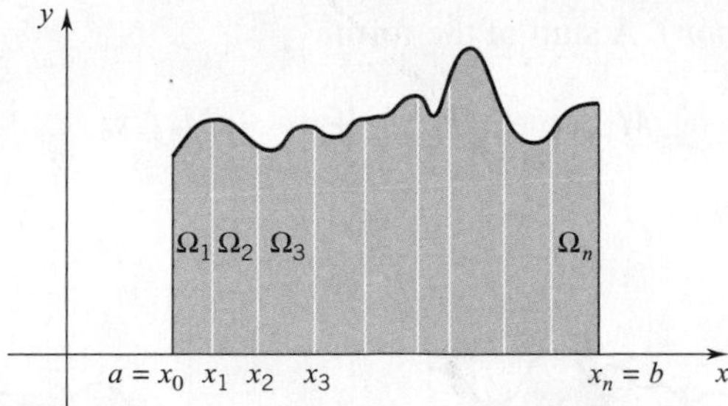

Figure 5.1.2

by m_i the minimum value. (Recall Theorem 2.6.2 and Section 4.4.) Consider now the rectangles r_i and R_i of Figure 5.1.3. Since

$$r_i \subseteq \Omega_i \subseteq R_i,$$

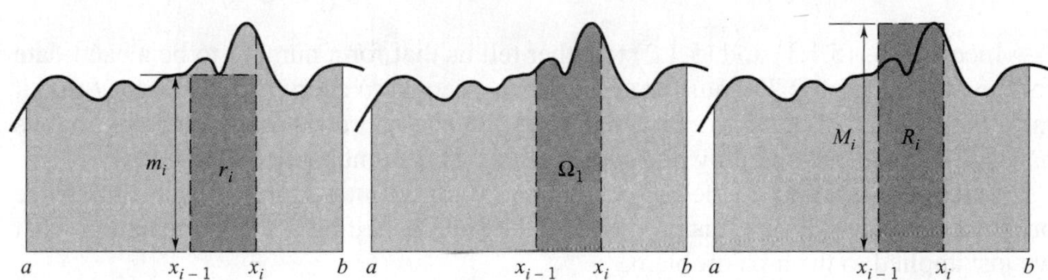

Figure 5.1.3

we must have

$$\text{area of } r_i \leq \text{area of } \Omega_i \leq \text{area of } R_i.$$

Since the area of a rectangle is the length times the width,

$$m_i(x_i - x_{i-1}) \leq \text{area of } \Omega_i \leq M_i(x_i - x_{i-1}).$$

Setting $\Delta x_i = x_i - x_{i-1}$ we have

$$m_i \Delta x_i \leq \text{area of } \Omega_i \leq M_i \Delta x_i.$$

This inequality holds for $i = 1,\ i = 2, \ldots,\ i = n$. Adding up these inequalities we get on the one hand

(5.1.1) $$m_1 \Delta x_1 + m_2 \Delta x_2 + \cdots + m_n \Delta x_n \leq \text{area of } \Omega,$$

and on the other hand

(5.1.2) $$\text{area of } \Omega \leq M_1 \Delta x_1 + M_2 \Delta x_2 + \cdots + M_n \Delta x_n.$$

A sum of the form

$$m_1 \Delta x_1 + m_2 \Delta x_2 + \cdots + m_n \Delta x_n \qquad \text{(Figure 5.1.4)}$$

is called a *lower sum for f*. A sum of the form

$$M_1\Delta x_1 + M_2\Delta x_2 + \cdots + M_n\Delta x_n \qquad \text{(Figure 5.1.5)}$$

is called an *upper sum for f*.

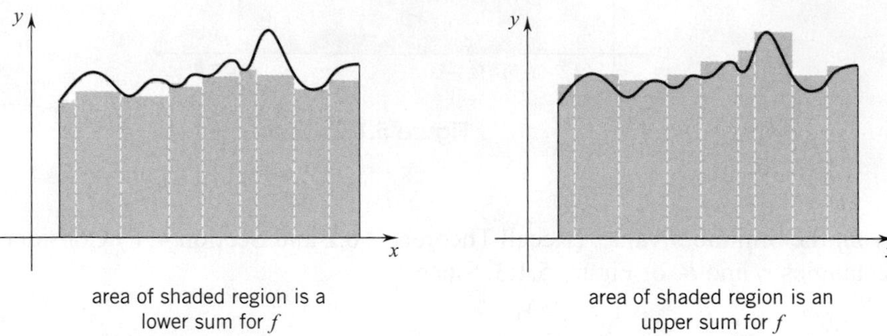

area of shaded region is a lower sum for f

Figure 5.1.4

area of shaded region is an upper sum for f

Figure 5.1.5

Inequalities (5.1.1) and (5.1.2) together tell us that for a number to be a candidate for the title "area of Ω," it must be greater than or equal to every lower sum for f and it must be less than or equal to every upper sum. It can be proved that with f continuous on $[a, b]$ there is one and only one such number. This number we call *the area of* Ω.

Later we will return to the subject of area. At this point we turn to a speed-distance problem. As you will see, this problem can be approached by the same technique that we just applied to the area problem.

A Speed-Distance Problem

If an object moves at a constant speed for a given period of time, then the total distance traveled is given by the familiar formula

$$\text{distance} = \text{speed} \times \text{time}.$$

Suppose now that during the course of the motion the speed does not remain constant but instead varies continuously. How can the total distance traveled be computed then?

To answer this question, we suppose that the motion begins at time a, ends at time b, and that during the time interval $[a, b]$ the speed varies continuously.

As in the case of the area problem we begin by breaking up the interval $[a, b]$ into a finite number of subintervals:

$$[t_0, t_1], [t_1, t_2], \ldots, [t_{n-1}, t_n], \quad \text{with} \quad a = t_0 < t_1 < \cdots < t_n = b.$$

On each subinterval $[t_{i-1}, t_i]$ the object attains a certain maximum speed M_i and a certain minimum speed m_i. (How do we know this?) If throughout the time interval $[t_{i-1}, t_i]$ the object were to move constantly at its minimum speed, m_i, then it would cover a distance of $m_i\Delta t_i$ units. If instead it were to move constantly at its maximum speed, M_i, then it would cover a distance of $M_i\Delta t_i$ units. As it is, the actual distance traveled, call it s_i, must lie somewhere in between; namely, we must have

$$m_i\Delta t_i \leq s_i \leq M_i\Delta t_i.$$

The total distance traveled during the time interval $[a, b]$, call it s, must be the sum of the distances traveled during the subintervals $[t_{i-1}, t_i]$. In other words we must have

$$s = s_1 + s_2 + \cdots + s_n.$$

Since

$$\begin{aligned} m_1\,\Delta t_1 &\le s_1 \le M_1\,\Delta t_1 \\ m_2\,\Delta t_2 &\le s_2 \le M_2\,\Delta t_2 \\ &\vdots \\ m_n\Delta t_n &\le s_n \le M_n\Delta t_n, \end{aligned}$$

it follows by the addition of these inequalities that

$$m_1\Delta t_1 + m_2\Delta t_2 + \cdots + m_n\Delta t_n \le s \le M_1\Delta t_1 + M_2\Delta t_2 + \cdots + M_n\Delta t_n.$$

A sum of the form

$$m_1\Delta t_1 + m_2\Delta t_2 + \cdots + m_n\Delta t_n$$

is called a *lower sum* for the speed function. A sum of the form

$$M_1\Delta t_1 + M_2\Delta t_2 + \cdots + M_n\Delta t_n$$

is called an *upper sum* for the speed function. The inequality we just obtained for s tells us that s must be greater than or equal to every lower sum for the speed function, and it must be less than or equal to every upper sum. As in the case of the area problem, it turns out that there is one and only one such number, and this is the total distance traveled.

■ 5.2 THE DEFINITE INTEGRAL OF A CONTINUOUS FUNCTION

The process we used to solve the two problems of Section 5.1 is called *integration*, and the end results of this process are called *definite integrals*. Our purpose here is to establish these notions more precisely.

(5.2.1) By a *partition* of the closed interval $[a, b]$ we mean a finite subset of $[a, b]$ which contains the points a and b.

We index the elements of a partition according to their natural order. Thus, if we write

$$P = \{x_0, x_1, x_2, \cdots, x_{n-1}, x_n\} \quad \text{is a partition of } [a, b],$$

you can conclude that

$$a = x_0 < x_1 < \cdots < x_n = b.$$

Example 1 The sets

$$\{0, 1\}, \quad \{0, \tfrac{1}{2}, 1\}, \quad \{0, \tfrac{1}{4}, \tfrac{1}{2}, 1\}, \quad \{0, \tfrac{1}{4}, \tfrac{1}{3}, \tfrac{1}{2}, \tfrac{5}{8}, 1\}$$

are all partitions of the interval [0,1]. □

If $P = \{x_0, x_1, x_2, \cdots, x_{n-1}, x_n\}$ is a partition of $[a, b]$, then P breaks up $[a, b]$ into n subintervals

$$[x_0, x_1], [x_1, x_2], \ldots, [x_{n-1}, x_n]$$

of lengths $\Delta x_1, \Delta x_2, \ldots, \Delta x_n$, respectively.

Suppose now that f is continuous on $[a, b]$. Then on each interval $[x_{i-1}, x_i]$ the function f takes on a maximum value, M_i, and a minimum value, m_i.

(5.2.2) The number

$$U_f(P) = M_1\Delta x_1 + M_2\Delta x_2 + \cdots + M_n\Delta x_n$$

is called *the P upper sum for f*, and the number

$$L_f(P) = m_1\Delta x_1 + m_2\Delta x_2 + \cdots + m_n\Delta x_n$$

is called *the P lower sum for f*.

Example 2 The quadratic function $f(x) = x^2$ is continuous on $[1, 3]$. The partition $P = \{1, \frac{3}{2}, 2, 3\}$ breaks up $[1, 3]$ into three subintervals:

$$[x_0, x_1] = [1, \tfrac{3}{2}], \quad [x_1, x_2] = [\tfrac{3}{2}, 2], \quad [x_2, x_3] = [2, 3]$$

of lengths

$$\Delta x_1 = \tfrac{3}{2} - 1 = \tfrac{1}{2}, \qquad \Delta x_2 = 2 - \tfrac{3}{2} = \tfrac{1}{2}, \qquad \Delta x_3 = 3 - 2 = 1,$$

respectively. Since f is increasing on $[1, 3]$, it takes on its maximum value at the right endpoint of each subinterval:

$$M_1 = f(\tfrac{3}{2}) = \tfrac{9}{4}, \quad M_2 = f(2) = 4, \quad M_3 = f(3) = 9.$$

The minimum values of f are taken on at the left endpoints of each subinterval:

$$m_1 = f(1) = 1, \quad m_2 = f(\tfrac{3}{2}) = \tfrac{9}{4}, \quad m_3 = f(2) = 4.$$

Thus,

$$U_f(P) = M_1\Delta x_1 + M_2\Delta x_2 + M_3\Delta x_3 = \tfrac{9}{4}(\tfrac{1}{2}) + 4(\tfrac{1}{2}) + 9(1) = \tfrac{97}{8} = 12.125$$

(see Figure 5.2.1a)

and $$L_f(P) = m_1\Delta x_1 + m_2\Delta x_2 + m_3\Delta x_3 = 1(\tfrac{1}{2}) + \tfrac{9}{4}(\tfrac{1}{2}) + 4(1) = \tfrac{45}{8} = 5.625.$$

(see Figure 5.2.1b) □

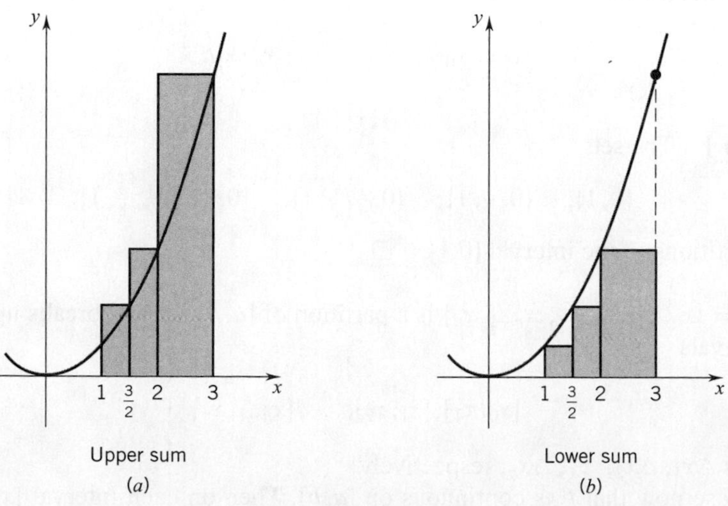

Figure 5.2.1

Example 3 The function $f(x) = 5 - x^2$ is continuous on the interval $[-1, 3]$. The partition $P = \{-1, \frac{1}{2}, \frac{3}{2}, 2, 3\}$ breaks up $[-1, 3]$ into four subintervals

$$[x_0, x_1] = [-1, \tfrac{1}{2}], \quad [x_1, x_2] = [\tfrac{1}{2}, \tfrac{3}{2}], \quad [x_2, x_3] = [\tfrac{3}{2}, 2], \quad [x_3, x_4] = [2, 3]$$

of lengths

$$\Delta x_1 = \tfrac{1}{2} - (-1) = \tfrac{3}{2}, \quad \Delta x_2 = \tfrac{3}{2} - \tfrac{1}{2} = 1, \quad \Delta x_3 = 2 - \tfrac{3}{2} = \tfrac{1}{2}, \quad \Delta x_4 = 3 - 2 = 1,$$

respectively. See Figure 5.2.2.

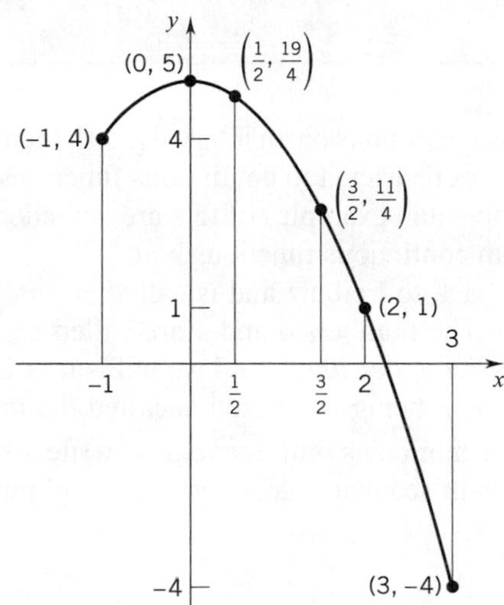

Figure 5.2.2

As you can check, the maximum value of f on each subinterval is given by:

$$M_1 = f(0) = 5 \text{ on } \left[-1, \tfrac{1}{2}\right], \qquad M_2 = f\left(\tfrac{1}{2}\right) = \tfrac{19}{4} \text{ on } \left[\tfrac{1}{2}, \tfrac{3}{2}\right]$$
$$M_3 = f\left(\tfrac{3}{2}\right) = \tfrac{11}{4} \text{ on } \left[\tfrac{3}{2}, 2\right], \qquad M_4 = f(2) = 1 \text{ on } [2, 3].$$

The minimum value of f on each subinterval is given by

$$m_1 = f(-1) = 4 \text{ on } \left[-1, \tfrac{1}{2}\right], \quad m_2 = f(\tfrac{3}{2}) = \tfrac{11}{4} \text{ on } [\tfrac{1}{2}, \tfrac{3}{2}]$$
$$m_3 = f(2) = 1 \text{ on } \left[\tfrac{3}{2}, 2\right], \quad m_4 = f(3) = -4 \text{ on } [2, 3].$$

Therefore,

$$U_f(P) = 5\left(\tfrac{3}{2}\right) + \tfrac{19}{4}(1) + \tfrac{11}{4}\left(\tfrac{1}{2}\right) + 1(1) = \tfrac{117}{8} = 14.625$$

and

$$L_f(P) = 4\left(\tfrac{3}{2}\right) + \tfrac{11}{4}(1) + 1\left(\tfrac{1}{2}\right) + (-4)(1) = \tfrac{21}{4} = 5.25 \quad \square$$

By an argument that we omit here (it appears in Appendix B.4), it can be proved that, with f continuous on $[a, b]$, there is one and only one number I that satisfies the inequality

$$L_f(P) \leq I < U_f(P) \qquad \text{for } all \text{ partitions } P \text{ of } [a, b].$$

This is the number we want.

DEFINITION 5.2.3 THE DEFINITE INTEGRAL
A function f defined on an interval $[a, b]$ is *integrable on* $[a, b]$ if there is one and only one number I that satisfies the inequality

$$L_f(P) \leq I < U_f(P) \quad \text{for } all \text{ partitions } P \text{ of } [a, b].$$

This unique number I is called the *definite integral* (or more simply *the integral*) of f from a to b and is denoted by

$$\int_a^b f(x)\,dx$$

As noted above, if f is continuous on an interval $[a, b]$, then it is integrable on $[a, b]$. However, integrability is not restricted to continuous functions. Certain discontinuous functions are also integrable and examples of this are considered in the Exercises. We will continue to work with continuous functions here.

The symbol $\int$ dates back to Leibniz and is called an *integral sign*. It is really an elongated S—as in *Sum*. The numbers a and b are called *the limits of integration* (a is the *lower limit* and b is the *upper limit*), and we will speak of *integrating* a function f from a to b. The function f being integrated is called the *integrand*. This is not the only notation. Some mathematicians omit the dx and write simply $\int_a^b f$. We will keep the dx. As we go on, you will see that it does serve a useful purpose.

In the expression

$$\int_a^b f(x)\,dx$$

the letter x is a "dummy variable"; in other words, it can be replaced by any other letter not already in use. Thus, for example,

$$\int_a^b f(x)\,dx, \qquad \int_a^b f(t)\,dt, \qquad \int_a^b f(z)\,dz$$

all denote exactly the same quantity, the definite integral of f from a to b.

An immediate application of the definite integral was given in the introduction: If f is nonnegative on $[a, b]$, then

$$A = \int_a^b f(x)\,dx$$

gives the area below the graph of f. See Figure 5.1.1. We will come back to this and other applications later. Right now we do some simple computations.

Example 4 If $f(x) = k$, constant, for all x in $[a, b]$, then

(5.2.4) $$\boxed{\int_a^b f(x)\,dx = k(b-a).}$$

To see this, take $P = \{x_0, x_1, \ldots, x_n\}$ as an arbitrary partition of $[a, b]$. Since f is constantly k on $[a, b]$, it is constantly k on each subinterval $[x_{i-1}, x_i]$. Thus, M_i and m_i

are both k. It follows that

$$\begin{aligned} U_f(P) &= k\Delta x_1 + k\Delta x_2 + \cdots + k\Delta x_n \\ &= k(\Delta x_1 + \Delta x_2 + \cdots + \Delta x_n) \\ &= k[(x_1 - x_0) + (x_2 - x_1) + (x_3 - x_2) + \cdots + (x_{n-1} - x_{n-2}) + (x_n - x_{n-1})] \\ &= k(b-a) \quad (\text{recall } x_0 = a \text{ and } x_n = b) \end{aligned}$$

and

$$\begin{aligned} L_f(P) &= k\Delta x_1 + k\Delta x_2 + \cdots + k\Delta x_n \\ &= k(\Delta x_1 + \Delta x_2 + \cdots + \Delta x_n) = k(b-a). \end{aligned}$$

Obviously then

$$L_f(P) \le k(b-a) \le U_f(P)$$

Since this inequality holds for all partitions P of $[a, b]$, we can conclude that

$$\int_a^b f(x)\,dx = k(b-a). \quad \square$$

For example

$$\int_{-1}^{1} 3\,dx = 3[1-(-1)] = 3(2) = 6 \quad \text{and}$$

$$\int_4^{10} -2\,dx = -2(10-4) = -2(6) = -12.$$

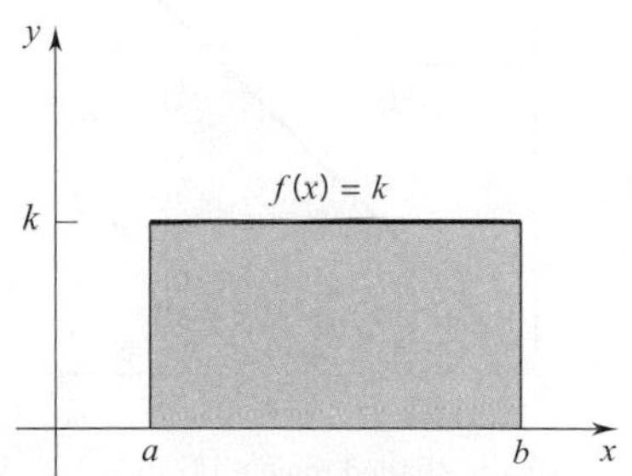

Figure 5.2.3

If $k > 0$, the region between the graph and the x-axis is a rectangle of height k erected on the interval $[a, b]$ (Figure 5.2.3). The integral gives the area of the rectangle.

Example 5 Let $f(x) = x$ for all $x \in [a, b]$ (see Figure 5.2.4). Then

(5.2.5)
$$\boxed{\int_a^b x\,dx = \tfrac{1}{2}(b^2 - a^2).}$$

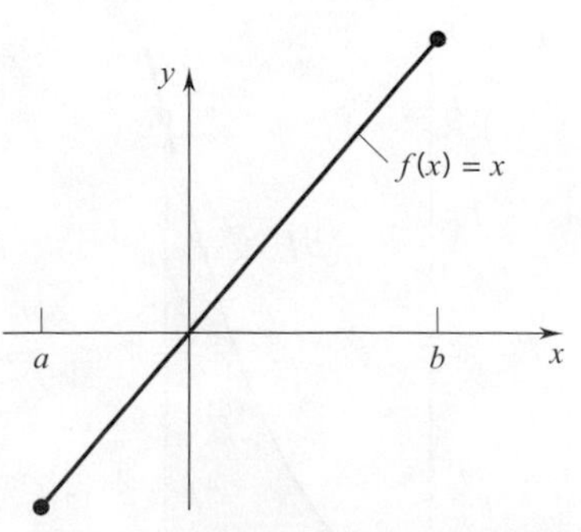

Figure 5.2.4

To see this, take $P = \{x_0, x_1, \ldots, x_n\}$ as an arbitrary partition of $[a, b]$. On each subinterval $[x_{i-1}, x_i]$, the function $f(x) = x$ has a maximum value M_i and a minimum value m_i. Since f is an increasing function, the maximum value occurs at the right endpoint of the subinterval and the minimum value occurs at the left endpoint. Thus $M_i = x_i$ and $m_i = x_{i-1}$. It follows that

$$U_f(P) = x_1\Delta x_1 + x_2\Delta x_2 + \cdots + x_n\Delta x_n$$

and

$$L_f(P) = x_0\Delta x_1 + x_1\Delta x_2 + \cdots + x_{n-1}\Delta x_n.$$

For each index i

$$(*) \qquad x_{i-1} \le \tfrac{1}{2}(x_i + x_{i-1}) \le x_i. \qquad \text{(explain)}$$

Multiplication by $\Delta x_i = x_i - x_{i-1}$ gives

$$x_{i-1}\Delta x_i \leq \tfrac{1}{2}(x_i + x_{i-1})(x_i - x_{i-1}) \leq x_i \Delta x_i,$$

which we write as

$$x_{i-1}\Delta x_i \leq \tfrac{1}{2}(x_i^2 - x_{i-1}^2) \leq x_i \Delta x_i.$$

Summing from $i = 1$ to $i = n$, we find that

$$(**) \qquad L_f(P) \leq \tfrac{1}{2}(x_1^2 - x_0^2) + \tfrac{1}{2}(x_2^2 - x_1^2) + \cdots + \tfrac{1}{2}(x_n^2 - x_{n-1}^2) \leq U_f(P).$$

The sum in the middle collapses to

$$\tfrac{1}{2}(x_n^2 - x_0^2) = \tfrac{1}{2}(b^2 - a^2).$$

Consequently we have

$$L_f(P) \leq \tfrac{1}{2}(b^2 - a^2) \leq U_f(P).$$

Since P was chosen arbitrarily, we can conclude that this inequality holds for all partitions P of $[a, b]$. It follows then that

$$\int_a^b x\,dx = \tfrac{1}{2}(b^2 - a^2). \quad \square$$

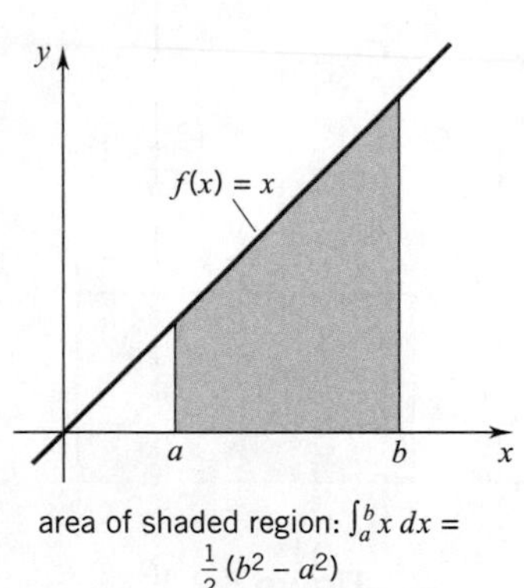

area of shaded region: $\int_a^b x\,dx = \frac{1}{2}(b^2 - a^2)$

Figure 5.2.5

For example

$$\int_{-1}^{3} x\,dx = \tfrac{1}{2}[3^2 - (-1)^2] = \tfrac{1}{2}(8) = 4 \quad \text{and} \quad \int_{-2}^{2} x\,dx = \tfrac{1}{2}[2^2 - (-2)^2] = 0.$$

If the interval $[a, b]$ lies to the right of the origin, then the region below the graph of

$$f(x) = x, \quad x \in [a, b]$$

is the trapezoid of Figure 5.2.5. The integral

$$\int_a^b x\,dx$$

gives the area of this trapezoid: $A = (b - a)[\frac{1}{2}(a + b)] = \frac{1}{2}(b^2 - a^2)$.

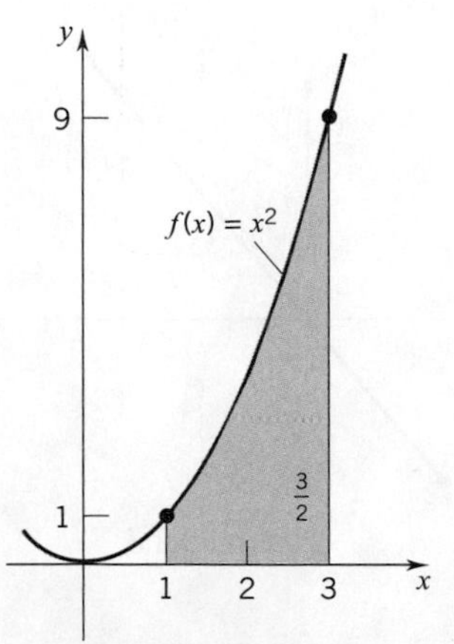

Area of the shaded region: $\int_1^3 x^2\,dx = \frac{26}{3}$

Figure 5.2.6

Example 6

$$\int_1^3 x^2\,dx = \tfrac{26}{3}. \qquad \text{(Figure 5.2.6)}$$

See Example 2. This time take $P = \{x_0, x_1, \ldots, x_n\}$ as an arbitrary partition of $[1, 3]$. On each subinterval $[x_{i-1}, x_i]$ the increasing function $f(x) = x^2$ has a maximum $M_i = x_i^2$ and a minimum $m_i = x_{i-1}^2$. It follows that

$$U_f(P) = x_1^2 \Delta x_1 + \cdots + x_n^2 \Delta x_n$$

and

$$L_f(P) = x_0^2 \Delta x_1 + \cdots + x_{n-1}^2 \Delta x_n.$$

For each index i,

$$x_{i-1}^2 \le \tfrac{1}{3}(x_{i-1}^2 + x_{i-1}x_i + x_i^2) \le x_i^2. \qquad \text{(verify this)}$$

(As we will see later in this chapter, $\frac{1}{3}(x_{i-1}^2 + x_{i-1}x_i + x_i^2)$ is the average value of x^2 on $[x_{i-1}, x_i]$.) We now multiply this inequality by $\Delta x_i = x_i - x_{i-1}$. The middle term is

$$\tfrac{1}{3}(x_{i-1}^2 + x_{i-1}x_i + x_i^2)(x_i - x_{i-1}) = \tfrac{1}{3}(x_i^3 - x_{i-1}^3),$$

showing us that

$$x_{i-1}^2 \Delta x_i \le \tfrac{1}{3}(x_i^3 - x_{i-1}^3) \le x_i^2 \Delta x_i.$$

The sum of the terms on the left is $L_f(P)$. The sum of all the middle terms collapses to $\frac{26}{3}$:

$$\tfrac{1}{3}(x_1^3 - x_0^3 + x_2^3 - x_1^3 + \cdots + x_n^3 - x_{n-1}^3) = \tfrac{1}{3}(x_n^3 - x_0^3) = \tfrac{1}{3}(3^3 - 1^3) = \tfrac{26}{3}.$$

The sum of the terms on the right is $U_f(P)$. Clearly then

$$L_f(P) \le \tfrac{26}{3} \le U_f(P).$$

Since P was chosen arbitrarily, we can conclude that this inequality holds for all partitions P of $[1, 3]$. It follows therefore that

$$\int_1^3 x^2\, dx = \tfrac{26}{3}. \quad \square$$

Riemann Sums

Let f be continuous on the interval $[a, b]$. With our approach to integration, the definite integral

$$\int_a^b f(x)\, dx$$

is the unique number that satisfies the inequality

$$L_f(P) \le \int_a^b f(x)\, dx \le U_f(P)$$

for all partitions P of $[a, b]$. This method of obtaining the definite integral by *squeezing* toward it with upper and lower sums is called the *Darboux method.*†

There is another way to obtain the integral that is frequently used. Take a partition $P = \{x_0, x_1, \ldots, x_n\}$ of $[a, b]$. Then P breaks up $[a, b]$ into n subintervals

$$[x_0, x_1], [x_1, x_2], \ldots, [x_{n-1}, x_n]$$

of lengths

$$\Delta x_1, \Delta x_2, \ldots, \Delta x_n.$$

† After the French mathematician J.-G. Darboux (1842–1917).

Now pick a point x_1^* from $[x_0, x_1]$ and form the product $f(x_1^*)\Delta x_1$; pick a point x_2^* from $[x_1, x_2]$ and form the product $f(x_2^*)\Delta x_2$; go on in this manner until you have formed the products

$$f(x_1^*)\,\Delta x_1,\quad f(x_2^*)\,\Delta x_2,\quad \ldots,\quad f(x_n^*)\,\Delta x_n.$$

The sum of these products

$$S^*(P) = f(x_1^*)\Delta x_1 + f(x_2^*)\Delta x_2 + \cdots + f(x_n^*)\Delta x_n$$

is called a *Riemann sum*.† As an exercise, you are asked to show that if P is any partition of the interval $[a, b]$ and $S^*(P)$ is any corresponding Riemann sum, then

$$L_f(P) \leq S^*(P) \leq U_f(P).$$

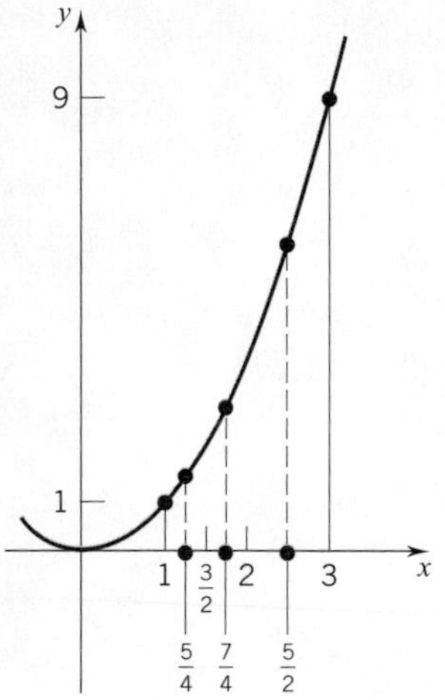

Example 7 Let $f(x) = x^2,\ x \in [1, 3]$. Take $P = \{1, \frac{3}{2}, 2, 3\}$ and let $x_1^* = \frac{5}{4}, x_2^* = \frac{7}{4}$, and $x_3^* = \frac{5}{2}$. See Figure 5.2.7. Then $\Delta x_1 = \Delta x_2 = \frac{1}{2}$, and $\Delta x_3 = 1$. Now

$$S^*(P) = f\left(\tfrac{5}{4}\right)\cdot\tfrac{1}{2} + f\left(\tfrac{7}{4}\right)\cdot\tfrac{1}{2} + f\left(\tfrac{5}{2}\right)\cdot 1 = \tfrac{25}{16}\left(\tfrac{1}{2}\right) + \tfrac{49}{16}\left(\tfrac{1}{2}\right) + \tfrac{25}{4}(1) = \tfrac{137}{16} = 8.5625.$$

As we saw in Example 2,

$$U_f(P) = \tfrac{9}{4}\left(\tfrac{1}{2}\right) + 4\left(\tfrac{1}{2}\right) + 9(1) = \tfrac{97}{8} = 12.125,$$

and
$$L_f(P) = 1\left(\tfrac{1}{2}\right) + \tfrac{9}{4}\left(\tfrac{1}{2}\right) + 4(1) = \tfrac{45}{8} = 5.625.$$

From Example 6, the exact value is: $\displaystyle\int_1^3 x^2 dx = \left[\tfrac{1}{3}x^3\right]_1^3 = \tfrac{27}{3} - \tfrac{1}{3} = \tfrac{26}{3} \cong 8.667.$ □

Figure 5.2.7

The definite integral of f can be viewed as the *limit* of Riemann sums in the following sense. For any partition P of $[a, b]$, define $||P||$, the *norm* of P, by setting

$$||P|| = \max \Delta x_i, \qquad i = 1, 2, \ldots, n.$$

Then, given any $\epsilon > 0$, there exists a $\delta > 0$ such that

$$\text{if}\quad ||P|| < \delta, \quad\text{then}\quad \left|S^*(P) - \int_a^b f(x)\,dx\right| < \epsilon,$$

no matter how the x_i^* are chosen within $[x_{i-1}, x_i]$.

In symbols we write

(5.2.6)
$$\int_a^b f(x)\,dx = \lim_{||P||\to 0}\left[f(x_1^*)\Delta x_1 + f(x_2^*)\Delta x_2 + \cdots + f(x_n^*)\Delta x_n\right].$$

A proof of this assertion appears in Appendix B.5. Figure 5.2.8 illustrates the idea. Here the base interval is broken up into 8 subintervals. The point x_1^* is chosen from $[x_0, x_1]$, x_2^* from $[x_1, x_2]$, and so on. While the integral represents the area under the

† After the German mathematician G. F. B. Riemann (1826–1866).

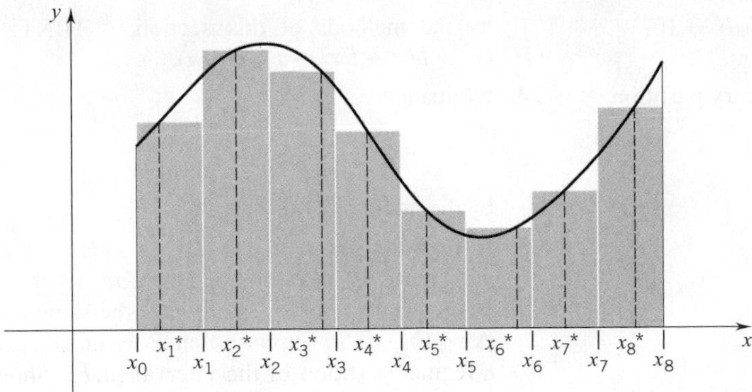

Figure 5.2.8

curve, the Riemann sum represents the sum of the areas of the shaded rectangles. The difference between the two can be made as small as we wish (less than ϵ) simply by making the maximum length of the base subintervals sufficiently small—that is, by making $||P||$ sufficiently small. □

You have now seen two approaches to the definite integral of a continuous function, through upper and lower sums, and through Riemann sums. Since

$$L_f(P) \leq S^*(P) \leq U_f(P)$$

for any partition P, it follows that the definite integral of f exists in the sense of upper and lower sums iff it exists in the sense of Riemann sums.

In the Exercises we show that the definite integrals of certain discontinuous functions also exist. See Exercises 41 and 42. In general, a function f defined on the interval $[a, b]$ is *integrable on* $[a, b]$ if its definite integral

$$\int_a^b f(x)\,dx$$

exists either in the sense of upper and lower sums, or in the Riemann sum sense. Determining whether a given function is integrable is generally a very difficult problem. In addition to continuous functions, however, a function f is integrable on an interval $[a, b]$ if:

1. f is increasing on $[a, b]$, or f is decreasing on $[a, b]$.

2. f is bounded on $[a, b]$ and has at most a finite number of discontinuities.

EXERCISES 5.2

Calculate $L_f(P)$ and $U_f(P)$.

1. $f(x) = 2x, \quad x \in [0, 1]; \quad P = \{0, \frac{1}{4}, \frac{1}{2}, 1\}$.

2. $f(x) = 1 - x, \quad x \in [0, 2]; \quad P = \{0, \frac{1}{3}, \frac{3}{4}, 1, 2\}$.

3. $f(x) = x^2, \quad x \in [-1, 0]; \quad P = \{-1, -\frac{1}{2}, -\frac{1}{4}, 0\}$.

4. $f(x) = 1 - x^2, \quad x \in [0, 1]; \quad P = \{0, \frac{1}{4}, \frac{1}{2}, 1\}$.

5. $f(x) = 1 + x^3, \quad x \in [0, 1]; \quad P = \{0, \frac{1}{2}, 1\}$.

6. $f(x) = \sqrt{x}, \quad x \in [0, 1]; \quad P = \{0, \frac{1}{25}, \frac{4}{25}, \frac{9}{25}, \frac{16}{25}, 1\}$.

7. $f(x) = x^2, \quad x \in [-1, 1]; \quad P = \{-1, -\frac{1}{4}, \frac{1}{4}, \frac{1}{2}, 1\}$.

8. $f(x) = x^2, \quad x \in [-1, 1]; \quad P = \{-1, -\frac{3}{4}, -\frac{1}{4}, \frac{1}{4}, \frac{1}{2}, 1\}$.

9. $f(x) = \sin x, \quad x \in [0, \pi]; \quad P = \{0, \frac{1}{6}\pi, \frac{1}{2}\pi, \pi\}$.

10. $f(x) = \cos x, \quad x \in [0, \pi]; \quad P = \{0, \frac{1}{3}\pi, \frac{1}{2}\pi, \pi\}$.

11. Let f be a function continuous on $[-1, 1]$ and take P as a partition of $[-1, 1]$. Show that each of the following three statements is false.

(a) $L_f(P) = 3$ and $U_f(P) = 2$.

(b) $L_f(P) = 3, \quad U_f(P) = 6,$ and $\int_{-1}^{1} f(x)\,dx = 2$.

(c) $L_f(P) = 3, \quad U_f(P) = 6, \quad \text{and} \quad \int_{-1}^{1} f(x)\,dx = 10.$

12. (a) Given that $P = \{x_0, x_1, \ldots, x_n\}$ is an arbitrary partition of $[a, b]$, find $L_f(P)$ and $U_f(P)$ for $f(x) = x + 3$.

(b) Use your answers to part (a) to evaluate

$$\int_a^b f(x)\,dx.$$

13. Repeat Exercise 12 with $f(x) = -3x$.

14. Repeat Exercise 12 with $f(x) = 1 + 2x$.

In Exercises 15–18, express each limit as a definite integral on the indicated interval $[a, b]$.

15. $\lim_{\|P\|\to 0} \left[(x_1^2 + 2x_1 - 3)\Delta x_1 + (x_2^2 + 2x_2 - 3)\Delta x_2 + \cdots + (x_n^2 + 2x_n - 3)\Delta x_n\right]; \; [-1, 2].$

16. $\lim_{\|P\|\to 0} \left[(x_0^3 - 3x_0)\Delta x_1 + (x_1^3 - 3x_1)\Delta x_2 + \cdots + (x_{n-1}^3 - 3x_{n-1})\Delta x_n\right]; \; [0, 3].$

17. $\lim_{\|P\|\to 0} \left[t_1^2 \sin(2t_1 + 1)\Delta t_1 + t_2^2 \sin(2t_2 + 1)\Delta t_2 + \cdots + t_n^2 \sin(2t_n + 1)\Delta t_n\right]$ where $t_i \in (t_{i-1}, t_i)$, $i = 1, 2, \ldots, n$; $[0, 2\pi]$.

18. $\lim_{\|P\|\to 0} \left[\frac{\sqrt{t_1}}{t_1^2 + 1}\Delta t_1 + \frac{\sqrt{t_2}}{t_2^2 + 1}\Delta t_2 + \cdots + \frac{\sqrt{t_n}}{t_n^2 + 1}\Delta t_n\right]$ where $t_i \in (t_{i-1}, t_i)$, $i = 1, 2, \ldots, n$; $[1, 4]$.

19. Let Ω be the region below the graph of $f(x) = x^2$, $x \in [0, 1]$. Draw a figure showing the Riemann sum $S^*(P)$ as an estimate for this area. Take

$$P = \{0, \tfrac{1}{4}, \tfrac{1}{2}, \tfrac{3}{4}, 1\} \quad \text{and}$$
$$x_1^* = \tfrac{1}{8}, \quad x_2^* = \tfrac{3}{8}, \quad x_3^* = \tfrac{5}{8}, \quad x_4^* = \tfrac{7}{8}.$$

20. Let Ω be the region below the graph of $f(x) = \frac{3}{2}x + 1$, $x \in [0, 2]$. Draw a figure showing the Riemann sum $S^*(P)$ as an estimate for this area. Take $P = \{0, \frac{1}{4}, \frac{3}{4}, 1, \frac{3}{2}, 2\}$ and let the x_i^* be the midpoints of the subintervals.

21. Let $f(x) = 2x$, $x \in [0, 1]$. Take $P = \{0, \frac{1}{8}, \frac{1}{4}, \frac{1}{2}, \frac{3}{4}, 1\}$ and set

$$x_1^* = \tfrac{1}{16}, \quad x_2^* = \tfrac{3}{16}, \quad x_3^* = \tfrac{3}{8}, \quad x_4^* = \tfrac{5}{8}, \quad x_5^* = \tfrac{3}{4}.$$

Calculate the following:

(a) $L_f(P)$. (b) $S^*(P)$.

(c) $U_f(P)$. (d) $\int_0^1 f(x)\,dx$.

22. Let f be continuous on $[a, b]$, let $P = \{x_0, x_1, \ldots, x_n\}$ be a partition of $[a, b]$, and let $S^*(P)$ be any Riemann sum generated by P. Show that

$$L_f(P) \le S^*(P) \le U_f(P).$$

23. Evaluate

$$\int_0^1 x^3\,dx$$

by the methods of this section. HINT: $b^4 - a^4 = (b^3 + b^2a + ba^2 + a^3)(b - a)$.

24. Evaluate

$$\int_0^1 x^4\,dx$$

by the methods of this section.

25. A partition $P = \{x_0, x_1, x_2, \ldots, x_{n-1}, x_n\}$ of the interval $[a, b]$ is said to be *regular* if the subintervals $[x_{i-1}, x_i]$, $i = 1, 2, \ldots, n$, all have the same length, namely, $\Delta x = (b - a)/n$. Suppose that $P = \{x_0, x_1, \ldots, x_{n-1}, x_n\}$ is a regular partition of the interval $[a, b]$. Show that if f is continuous and increasing on $[a, b]$, then

$$U_f(P) - L_f(P) = [f(b) - f(a)]\;\Delta x.$$

26. Let $P = \{x_0, x_1, x_2, \ldots, x_{n-1}, x_n\}$ be a regular partition of the interval $[a, b]$. Show that if f is continuous and decreasing on $[a, b]$, then

$$U_f(P) - L_f(P) = [f(a) - f(b)]\;\Delta x.$$

In Exercises 27–32, assume that f and g are continuous, that $a < b$, and that $\int_a^b f(x)\,dx > \int_a^b g(x)\,dx$. Which of the statements necessarily hold for all partitions P of $[a, b]$? Explain your answers.

27. $L_g(P) < U_f(P)$.

28. $L_g(P) < L_f(P)$.

29. $L_g(P) < \int_a^b f(x)\,dx$.

30. $U_g(P) < U_f(P)$.

31. $U_f(P) > \int_a^b g(x)\,dx$.

32. $U_g(P) < \int_a^b f(x)\,dx$.

33. Let f be continuous on $[a, b]$ and set $I = \int_a^b f(x)\,dx$. Let $P = \{x_0, x_1, x_2, \ldots, x_{n-1}, x_n\}$ be an arbitrary partition of $[a, b]$.

(a) Show that $I - L_f(P) \le U_f(P) - L_f(p)$. The difference $I - L_f(P)$ is called the *error* in using $L_f(P)$ to approximate I.

(b) Show that $U_f(P) - I \le U_f(P) - L_f(P)$. The difference $U_f(P) - I$ is the error in using $U_f(P)$ to approximate I.

34. Let $P = \{x_0, x_1, x_2, \ldots, x_{n-1}, x_n\}$ be a regular partition of the interval $[a, b]$, let f be continuous and set $I = \int_a^b f(x)\,dx$. Show that if f is increasing on $[a, b]$, or if f is decreasing on $[a, b]$, then

$$I - L_f(P) \le |f(b) - f(a)|\Delta x \quad \text{and}$$
$$U_f(P) - I \le |f(b) - f(a)|\Delta x.$$

C 35. Let $f(x) = \sqrt{1 + x^2}$.

(a) Verify that f is increasing on $[0, 2]$.

(b) Use Exercise 34 to determine a value of n such that, if $P = \{x_0, x_1, x_2, \ldots, x_n\}$ is a regular partition of $[0, 2]$, then

$$\int_0^2 f(x)\,dx - L_f(P) \le 0.1.$$

(c) Use a programmable calculator or computer to calculate $\int_0^2 f(x)\,dx$ with an error of less than 0.1.

36. Let $f(x) = 1/(1+x^2)$.

(a) Verify that f is decreasing on $[0, 1]$.

(b) Use Exercise 34 to determine a value of n such that, if $P = \{x_o, x_1, x_2, \ldots, x_n\}$ is a regular partition of $[0, 1]$, then

$$U_f(P) - \int_0^1 f(x)\,dx \le 0.05.$$

(c) Use a programmable calculator or computer to calculate $\int_0^1 f(x)\,dx$ with an error of less than 0.05. NOTE: You will see in Chapter 7 that the exact value of this integral is $\pi/4$.

37. Use mathematical induction to show that the sum of the first k positive integers is $\frac{1}{2}k(k+1)$:

$$1+2+3+\cdots+k = \frac{k(k+1)}{2}$$

for each positive integer k.

38. Use mathematical induction to show that the sum of the squares of the first k positive integers is $\frac{1}{6}k(k+1)(2k+1)$:

$$1^2+2^2+3^2+\cdots+k^2 = \frac{k(k+1)(2k+1)}{6},$$

for each positive integer k.

39. Let $P = \{x_0, x_1, x_2, \ldots, x_{n-1}, x_n\}$ be a regular partition of the interval $[0, b]$, and let $f(x) = x$.

(a) Show that

$$L_f(P) = \frac{b^2}{n^2}[0+1+2+3+\cdots+(n-1)].$$

(b) Show that

$$U_f(P) = \frac{b^2}{n^2}[1+2+3+\cdots+n].$$

(c) Use Exercise 37 to show that $L_f(P), U_f(P) \to b^2/2$ as $||P|| \to 0$. Thus, $\int_0^b x\,dx = \frac{1}{2}b^2$. Compare with Example 5.

40. Let $P = \{x_0, x_1, x_2, \ldots, x_{n-1}, x_n\}$ be a regular partition of $[0, b]$, and let $f(x) = x^2$.

(a) Show that

$$L_f(P) = \frac{b^3}{n^3}[0^2+1^2+2^2+\cdots+(n-1)^2].$$

(b) Show that

$$U_f(P) = \frac{b^3}{n^3}[1^2+2^2+3^2+\cdots+n^2].$$

(c) Use Exercise 38 to show that $L_f(P)$, $U_f(P) \to b^3/3$ as $||P|| \to 0$. Thus, $\int_0^b x^2 dx = \frac{1}{3}b^3$. Compare with Example 6.

If $P = \{x_0, x_1, x_2, \cdots x_n\}$ is a regular partition of $[a, b]$, then the Riemann sums have the form

$$S_n^* = \frac{b-a}{n}[f(x_1^*) + f(x_2^*) + \cdots + f(x_n^*)]$$

In Exercises 41–43, choose $\epsilon > 0$. Break up the interval $[0, 1]$ into n subintervals each of length $1/n$. Take $x_1^* = 1/n$, $x_2^* = 2/n, \ldots, x_n^* = n/n$.

41. (a) Determine the Riemann sum S_n^* for

$$\int_0^1 x\,dx. \qquad \text{(see Exercise 39)}$$

(b) Show that

$$\left|S_n^* - \int_0^1 x\,dx\right| < \epsilon \quad \text{if } n > 1/\epsilon.$$

HINT: See Exercise 37.

42. (a) Determine the Riemann sum S_n^* for

$$\int_0^1 x^2 dx. \qquad \text{(See Exercise 40)}$$

(b) Show that

$$\left|S_n^* - \int_0^1 x^2 dx\right| < \epsilon \quad \text{if} \quad n > 1/\epsilon.$$

HINT: See Exercise 38.

43. (a) Determine the Riemann sum S_n^* for

$$\int_0^1 x^3 dx.$$

(b) Show that

$$\left|s_n^* - \int_0^1 x^3 dx\right| < \epsilon \quad \text{if } n > 1/\epsilon.$$

HINT: $1^3 + 2^3 + \cdots + n^3 = (1+2+\cdots+n)^2$.

44. Suppose the speed of an object is given by $s(t) = t^2 + t + 1$ units per second for $0 \le t \le 3$. Find the total distance traveled by the object.

45. Suppose the speed of an object is given by $s(t) = t^3 + 2t$ units per second for $0 \le t \le 1$. Find the total distance traveled by the object.

46. Let f be a continuous function on $[a, b]$. Show that if P is a partition of $[a, b]$, then $L_f(P)$, $U_f(P)$, and $\frac{1}{2}[L_f(P) + U_f(P)]$ are all Riemann sums.

47. (*Important*) The definition of integral that we have given (Definition 5.2.5) can be applied to functions with a finite number of discontinuities. Suppose that f is continuous on $[a, b]$. If g is defined on $[a, b]$ and differs from f only at a finite number of points, then g can be integrated on $[a, b]$ and

$$\int_a^b g(x)\,dx = \int_a^b f(x)\,dx.$$

For example, the function

$$g(x) = \begin{cases} 2, & x \in [0,3) \cup (3,4] \\ 7, & x = 3 \end{cases}$$

differs from the constant function

$$f(x) = 2, \quad x \in [0, 4]$$

only at $x = 3$.

Clearly $\int_0^4 f(x)\,dx = 8$. Show that $\int_0^4 g(x)\,dx = 8$ by showing that 8 is the unique number I that satisfies the inequality

$$L_g(P) \le I \le U_g(P) \quad \text{for all partitions } P \text{ of } [0, 4].$$

48. (*Important*) A function f is said to be *piecewise continuous* on $[a, b]$ if it is continuous except, possibly, for a finite set of points at which it has a jump discontinuity. It can be shown that if f is piecewise continuous on $[a, b]$, then $\int_a^b f(x)\,dx$ exists.

Let

$$f(x) = \begin{cases} x^2 & 0 \le x \le 2 \\ x & 2 < x \le 5. \end{cases}$$

(a) Sketch the graph of f.

(b) Find the area of the region bounded by the graph of f and the x-axis between $x = 0$ and $x = 5$. HINT: $\int_0^5 f(x)\,dx = \int_0^2 f(x)\,dx + \int_2^5 f(x)\,dx$, then use Exercise 40 and Example 5.

49. Consider the function f on $[2, 10]$:

$$f(x) = \begin{cases} 7 & \text{for } x \text{ rational} \\ 4 & \text{for } x \text{ irrational.} \end{cases}$$

(a) Show that for all partitions P of $[2, 10]$, $L_f(P) \le 40 \le U_f(P)$.

(b) Explain why you can *not* conclude that

$$\int_2^{10} f(x)\,dx = 40.$$

(c) Show that, for all partitions P of $[2, 10]$, $L_f(P) = 32$ and $U_f(P) = 56$. Does $\int_2^{10} f(x)\,dx$ exist?

C In Exercises 50–53, approximate the value of the given integral as follows. Use a regular partition P with 10 subintervals and

(a) Calculate $L_f(P)$ and $U_f(P)$.

(b) Calculate $\frac{1}{2}[L_f(P) + U_f(P)]$.

(c) Calculate $S^*(P)$ using the midpoints of the subintervals. How does this result compare with the result in part (b)?

50. $\int_0^2 (x^3 + 2)\,dx$.

51. $\int_0^1 \sqrt{x}\,dx$.

52. $\int_0^2 \dfrac{1}{1 + x^2}\,dx$.

53. $\int_0^1 \sin(\pi x)\,dx$.

■ 5.3 THE FUNCTION $F(x) = \int_a^x f(t)\,dt$

The evaluation of a definite integral

$$\int_a^x f(x)\,dx$$

directly from its definition as the unique number I satisfying the inequality $L_f(P) \le I \le U_f(P)$ for all partitions P of $[a, b]$ is usually a laborious and difficult process. Try, for example, to evaluate

$$\int_2^5 \left(x^3 + x^{52} - \frac{2x}{1 - x^2}\right) dx \quad \text{or} \quad \int_{-1/2}^{1/4} \frac{x}{1 - x^2}\,dx$$

by this method. Theorem 5.4.2. called *the fundamental theorem of integral calculus*, gives us another way to evaluate such integrals. This other way depends on a connection between integration and differentiation described in Theorem 5.3.5. The main purpose of this section is to prove Theorem 5.3.5. Along the way we will pick up some information that is of interest in itself.

THEOREM 5.3.1

Let f be continuous on $[a, b]$ and let P and Q be partitions of the interval $[a, b]$. If $P \subseteq Q$, then

$$L_f(P) \le L_f(Q) \quad \text{and} \quad U_f(Q) \le U_f(P).$$

This result can be justified as follows: By adding points to a partition we make the subintervals $[x_{i-1}, x_i]$ smaller. This tends to make the minima, m_i, larger and the maxima, M_i, smaller. Thus the lower sums are made bigger, and the upper sums are made smaller. The idea is illustrated (for a positive function) in Figures 5.3.1 and 5.3.2.

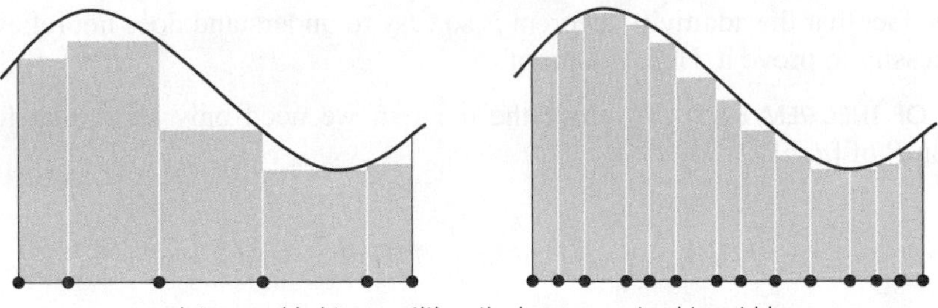

as points are added to a partition, the lower sums tend to get bigger

Figure 5.3.1

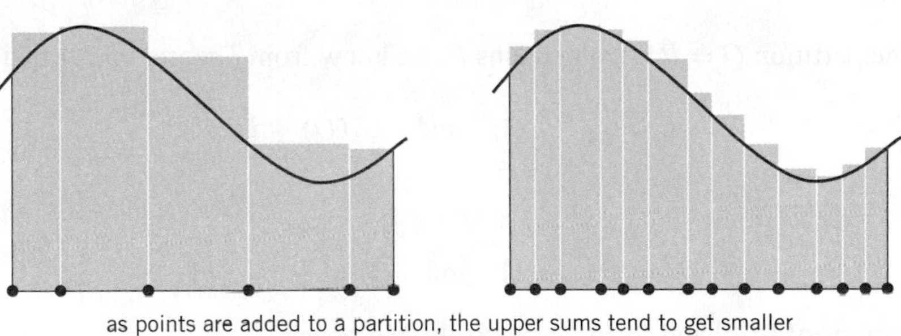

as points are added to a partition, the upper sums tend to get smaller

Figure 5.3.2

The next theorem says that the integral is *additive* on intervals.

THEOREM 5.3.2

If f is continuous on $[a, b]$ and $a < c < b$, then

$$\int_a^c f(t)\,dt + \int_c^b f(t)\,dt = \int_a^b f(t)\,dt.$$

For nonnegative functions f this theorem is easily understood in terms of area. The area of part I in Figure 5.3.3 is given by

$$\int_a^c f(t)\,dt;$$

the area of part II by

$$\int_c^b f(t)\,dt;$$

and the area of the entire region by

$$\int_a^b f(t)\,dt.$$

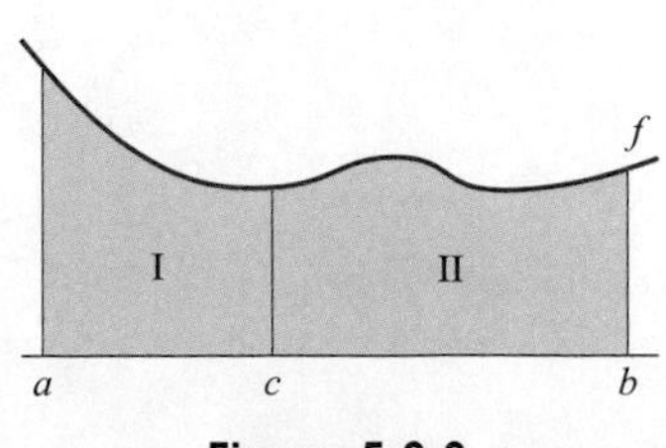

Figure 5.3.3

The theorem says that

the area of part I $+$ the area of part II $=$ the area of the entire region.

The fact that the additivity theorem is so easy to understand does not relieve us of the necessity to prove it. Here is a proof.

PROOF OF THEOREM 5.3.2 To prove the theorem we need only show that for each partition P of $[a, b]$

$$L_f(P) \leq \int_a^c f(t)\,dt + \int_c^b f(t)\,dt \leq U_f(P). \qquad \text{(Why?)}$$

We begin with an arbitrary partition of $[a, b]$:

$$P = \{x_0, x_1, \ldots, x_n\}.$$

Since the partition $Q = P \cup \{c\}$ contains P, we know from Theorem 5.3.1 that

$$\text{(1)} \qquad L_f(P) \leq L_f(Q) \quad \text{and} \quad U_f(Q) \leq U_f(P).$$

The sets

$$Q_1 = Q \cap [a, c] \quad \text{and} \quad Q_2 = Q \cap [c, b]$$

are partitions of $[a, c]$ and $[c, b]$, respectively. Moreover

$$\text{(2)} \qquad L_f(Q_1) + L_f(Q_2) = L_f(Q) \quad \text{and} \quad U_f(Q_1) + U_f(Q_2) = U_f(Q).$$

Since

$$L_f(Q_1) \leq \int_a^c f(t)\,dt \leq U_f(Q_1) \quad \text{and} \quad L_f(Q_2) \leq \int_c^b f(t)\,dt \leq U_f(Q_2),$$

we have

$$L_f(Q_1) + L_f(Q_2) \leq \int_a^c f(t)\,dt + \int_c^b f(t)\,dt \leq U_f(Q_1) + U_f(Q_2),$$

and thus by (2),

$$L_f(Q) \leq \int_a^c f(t)\,dt + \int_c^b f(t)\,dt \leq U_f(Q).$$

Therefore, by (1),

$$L_f(P) \leq \int_a^c f(t)\,dt + \int_c^b f(t)\,dt \leq U_f(P). \quad \square$$

Until now we have integrated only from left to right: from a number a to a number b greater than a. We integrate in the other direction by defining

(5.3.3)

$$\boxed{\int_b^a f(t)\,dt = -\int_a^b f(t)\,dt.}$$

The integral from any number to itself is defined to be zero:

(5.3.4)
$$\boxed{\int_c^c f(t)\,dt = 0.}$$

With these additional conventions, the additivity condition

$$\int_a^c f(t)\,dt + \int_c^b f(t)\,dt = \int_a^b f(t)\,dt$$

holds for all choices of a, b, c from an interval on which f is continuous, no matter what the order of a, b, c happens to be. We have left the proof of this to you as an exercise.

We are now ready to state the all-important connection that exists between integration and differentiation. Our first step is to point out that if f is continuous on $[a, b]$, then for each x in $[a, b]$, the integral

$$\int_a^x f(t)\,dt$$

is a number, and consequently we can define a function F on $[a, b]$ by setting

$$F(x) = \int_a^x f(t)\,dt.$$

THEOREM 5.3.5

Let f be continuous on $[a, b]$. The function F defined on $[a, b]$ by

$$F(x) = \int_a^x f(t)\,dt$$

is continuous on $[a, b]$, differentiable on (a, b), and has derivative

$$F'(x) = f(x) \qquad \text{for all } x \text{ in } (a, b).$$

PROOF We begin with x in the half-open interval $[a, b)$ and show that

$$\lim_{h\to 0^+} \frac{F(x+h) - F(x)}{h} = f(x).$$

(For a pictorial outline of the proof for f nonnegative, see Figure 5.3.4.)

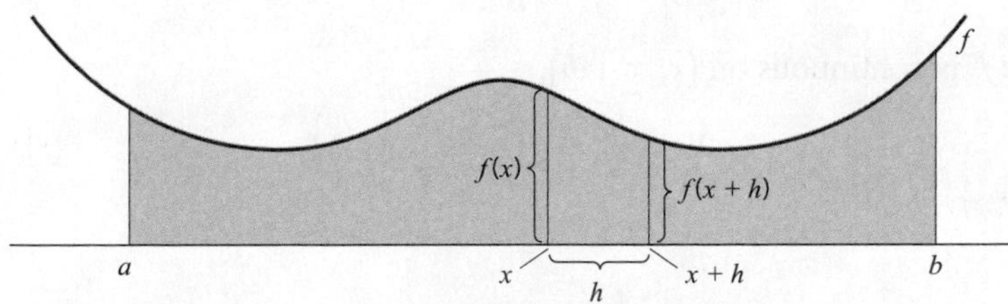

$F(x)$ = area from a to x and $F(x+h)$ = area from a to $x+h$.
Therefore $F(x+h) - F(x)$ = area from x to $x+h \cong f(x)\,h$ if h is small and

$$\frac{F(x+h) - F(x)}{h} \cong \frac{f(x)\,h}{h} = f(x).$$

Figure 5.3.4

If $x < x + h \leq b$, then

$$F(x+h) - F(x) = \int_a^{x+h} f(t)\,dt - \int_a^x f(t)\,dt.$$

It follows that

$$(1) \qquad F(x+h) - F(x) = \int_x^{x+h} f(t)\,dt. \qquad \text{(justify this step)}$$

Now set

$$M_h = \text{ maximum value of } f \text{ on } [x,\ x+h]$$

and
$$m_h = \text{minimum value of } f \text{ on } [x,\ x+h].$$

See Figure 5.3.5. Since

$$M_h[(x+h) - x] = M_h \cdot h$$

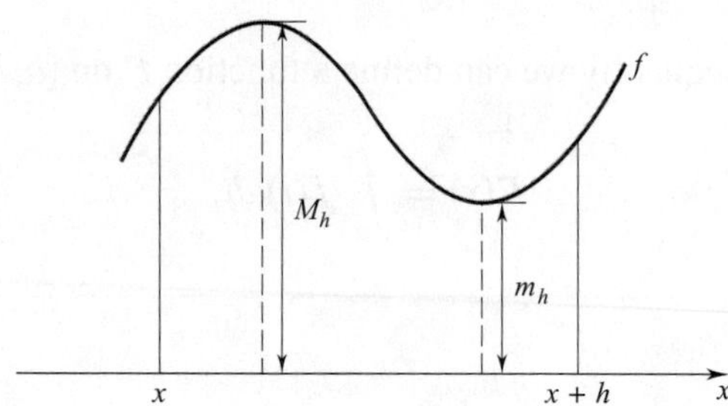

Figure 5.3.5

is an upper sum for f on $[x,\ x+h]$ and

$$m_h[(x+h) - x] = m_h \cdot h$$

is a lower sum for f on $[x,\ x+h]$, we have

$$m_h \cdot h \leq \int_x^{x+h} f(t)\,dt \leq M_h \cdot h.$$

Now, using (1) and the fact that $h > 0$, it follows that

$$m_h \leq \frac{F(x+h) - F(x)}{h} \leq M_h.$$

Also, since f is continuous on $[x,\ x+h]$,

$$\lim_{h\to 0^+} m_h = f(x) = \lim_{h\to 0^+} M_h$$

and thus

$$(2) \qquad \lim_{h\to 0^+} \frac{F(x+h) - F(x)}{h} = f(x)$$

by the "pinching theorem," Theorem 2.5.1, which, as we remarked in Section 2.5, applies also to one-sided limits. In a similar manner we can prove that, for x in the half-open interval $(a, b]$,

$$(3) \qquad \lim_{h\to 0^-} \frac{F(x+h) - F(x)}{h} = f(x).$$

For x in the open interval (a, b), both (2) and (3) hold, and we have

$$F'(x) = \lim_{h\to 0} \frac{F(x+h) - F(x)}{h} = f(x).$$

This proves that F is differentiable on (a, b) and has derivative $F'(x) = f(x)$.

All that remains to be shown is that F is continuous from the right at a and continuous from the left at b. Limit (2) at $x = a$ gives

$$\lim_{h\to 0^+} \frac{F(a+h) - F(a)}{h} = f(a).$$

Now, for $h > 0$,

$$F(a+h) - F(a) = \frac{F(a+h) - F(a)}{h} \cdot h,$$

and so

$$\lim_{h\to 0^+} [F(a+h) - F(a)] = \lim_{h\to 0^+} \left(\frac{F(a+h) - F(a)}{h} \cdot h\right) = f(a) \cdot \lim_{h\to 0^+} h = 0.$$

Therefore $$\lim_{h\to 0^+} F(a+h) = F(a).$$

This shows that F is continuous from the right at a. The continuity of F from the left at b can be shown by applying limit (3) at $x = b$. □

Example 1 If F is defined by $F(x) = \int_{-1}^{x} (2t + t^2)\,dt$ for $-1 \le x \le 5$, then

$$F'(x) = 2x + x^2 \quad \text{on } (-1, 5). \quad \square$$

Example 2 Given that F is defined by

$$F(x) = \int_0^x \sin \pi t\,dt$$

for all real numbers x, determine $F'(3/4)$.

SOLUTION By Theorem 5.3.5.

$$F'(x) = \sin \pi x.$$

Thus, $F'(3/4) = \sin(3\pi/4) = \sqrt{2}/2$. □

Example 3 Let F be defined by $F(x) = \int_0^x \frac{1}{1+t^2}\,dt$, for all real numbers x.

(a) Find the critical numbers of F and determine the intervals on which F is increasing and the intervals on which F is decreasing.

(b) Determine the concavity of the graph of F and find the points of inflection (if any).

(c) Sketch the graph of F.

SOLUTION

(a) To find where F is increasing or decreasing, we need to examine the first derivative. By Theorem 5.3.5,

$$F'(x) = \frac{1}{1+x^2}.$$

Since $F'(x) > 0$ for all x, F is increasing on $(-\infty, \infty)$; there are no critical numbers.

(b) We use the second derivative to determine the concavity of the graph and to find the points of inflection:

$$F''(x) = \frac{-2x}{(1+x^2)^2}.$$

The sign of F'' and the behavior of the graph of F are as follows:

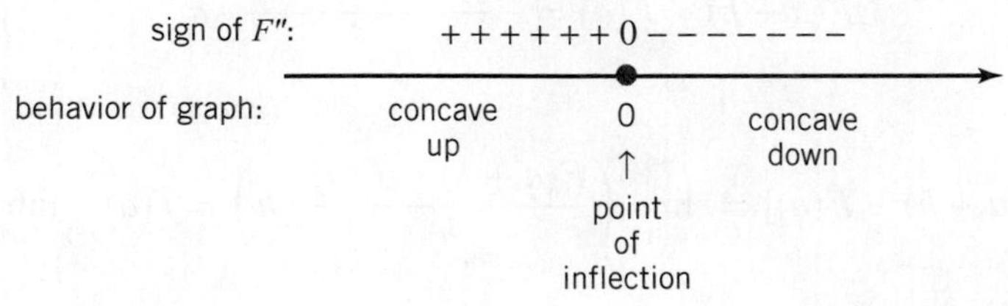

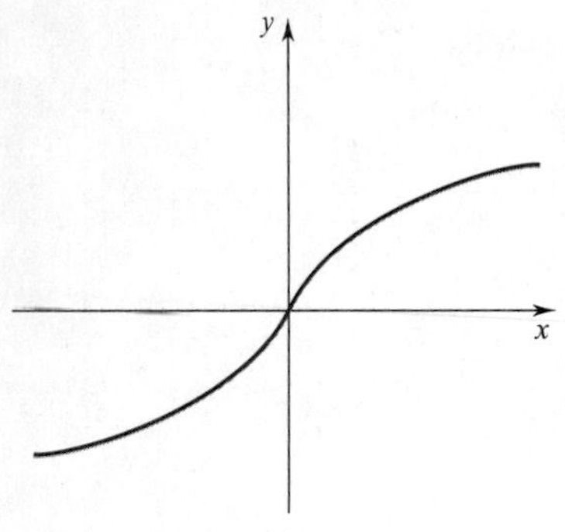

Figure 5.3.6

(c) Since $F(0) = 0$ and $F'(0) = 1$, the graph goes through the origin with slope 1. A sketch of the graph is shown in Figure 5.3.6. Although it is not at all clear now, you will see in Chapter 7 that $y = \pi/2$ and $y = -\pi/2$ are horizontal asymptotes of the graph. ❑

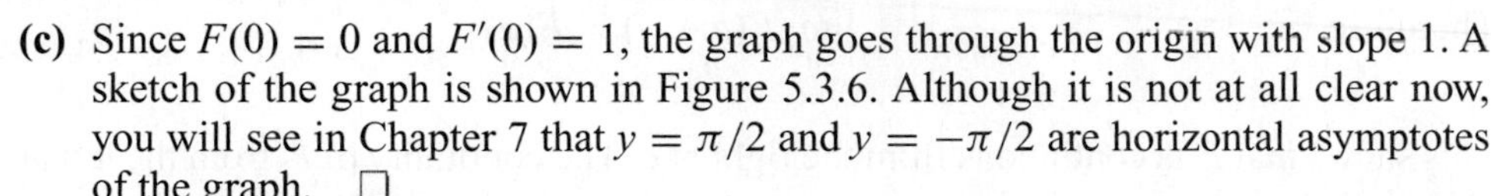

EXERCISES 5.3

1. Given that

$$\int_0^1 f(x)\,dx = 6, \quad \int_0^2 f(x)\,dx = 4, \quad \int_2^5 f(x)\,dx = 1,$$

find each of the following:

(a) $\int_0^5 f(x)\,dx.$ (b) $\int_1^2 f(x)\,dx.$ (c) $\int_1^5 f(x)\,dx.$

(d) $\int_0^0 f(x)\,dx.$ (e) $\int_2^0 f(x)\,dx.$ (f) $\int_5^1 f(x)\,dx.$

2. Given that

$$\int_1^4 f(x)\,dx = 5, \quad \int_3^4 f(x)\,dx = 7, \quad \int_1^8 f(x)\,dx = 11,$$

find each of the following:

(a) $\int_4^8 f(x)\,dx.$ (b) $\int_4^3 f(x)\,dx.$ (c) $\int_1^3 f(x)\,dx.$

(d) $\int_3^8 f(x)\,dx.$ (e) $\int_8^4 f(x)\,dx.$ (f) $\int_4^4 f(x)\,dx.$

3. Use upper and lower sums to show that

$$0.5 < \int_1^2 \frac{dx}{x} < 1.$$

4. Use upper and lower sums to show that

$$0.6 < \int_0^1 \frac{dx}{1+x^2} < 1.$$

5. For $x > -1$, set $F(x) = \int_0^x t\sqrt{t+1}\,dt$.

(a) Find $F(0)$. (b) Find $F'(x)$. (c) Find $F'(2)$.

(d) Express $F(2)$ as an integral of $t\sqrt{t+1}$.

(e) Express $-F(x)$ as an integral of $t\sqrt{t+1}$.

6. Let $F(x) = \int_\pi^x t \sin t\,dt$.

(a) Find $F(\pi)$. (b) Find $F'(x)$. (c) Find $F'(\pi/2)$.

(d) Express $F(2\pi)$ as an integral of $t \sin t$.

(e) Express $-F(x)$ as an integral of $t \sin t$.

Calculate the following for each function F:

(a) $F'(-1)$. (b) $F'(0)$. (c) $F'(\frac{1}{2})$. (d) $F''(x)$.

7. $F(x) = \int_0^x \frac{dt}{t^2+9}$. **8.** $F(x) = \int_x^0 \sqrt{t^2+1}\,dt$.

9. $F(x) = \int_x^1 t\sqrt{t^2+1}\,dt$. **10.** $F(x) = \int_1^x \sin \pi t\,dt$.

11. $F(x) = \int_1^x \cos \pi t\,dt$. **12.** $F(x) = \int_2^x (t+1)^3\,dt$.

13. Explain why each of the following statements must be false.

(a) $U_f(P_1) = 4$ for the partition $P_1 = \{0, 1, \frac{3}{2}, 2\}$ and
$U_f(P_2) = 5$ for the partition $P_2 = \{0, \frac{1}{4}, 1, \frac{3}{2}, 2\}$.

(b) $L_f(P_1) = 5$ for the partition $P_1 = \{0, 1, \frac{3}{2}, 2\}$ and
$L_f(P_2) = 4$ for the partition $P_2 = \{0, \frac{1}{4}, 1, \frac{3}{2}, 2\}$.

14. Show that if f is continuous on an interval I, then

$$\int_a^c f(t)\,dt + \int_c^b f(t)\,dt = \int_a^b f(t)\,dt$$

for *every* choice of a, b, c from I. HINT: Assume $a < b$ and consider the four cases: $c = a, c = b, c < a, b < c$. Then consider what happens if $a > b$ or $a = b$.

In Exercises 15 and 16, a function F is given. Find the critical numbers of F and, at each critical number, determine whether F has a local maximum, a local minimum, or neither.

15. $F(x) = \int_0^x \frac{t-1}{1+t^2}\,dt$. **16.** $F(x) = \int_0^x \frac{t-4}{1+t^2}\,dt$.

17. For $x > 0$, set $F(x) = \int_1^x (1/t)\,dt$.

(a) Find the critical numbers of F, if any, and determine the intervals on which F is increasing and the intervals on which F is decreasing.

(b) Determine the concavity of the graph of F and find the points of inflection, if any.

(c) Sketch the graph of F.

18. Let $F(x) = \int_0^x t(t-3)^2\,dt$.

(a) Find the critical numbers of F and determine the intervals on which F is increasing and the intervals on which F is decreasing.

(b) Determine the concavity of the graph of F and find the points of inflection, if any.

(c) Sketch the graph of F.

19. Suppose that f is differentiable with $f'(x) > 0$ for all x, and suppose that $f(1) = 0$. Set

$$F(x) = \int_0^x f(t)\,dt.$$

Explain why each of the following statements must be true.

(a) F is continuous.

(b) F is twice differentiable.

(c) $x = 1$ is a critical number of F.

(d) F takes on a local minimum at $x = 1$.

(e) $F(1) < 0$.

Make a rough sketch of the graph of F.

20. Suppose that g is differentiable with $g'(x) < 0$ for all $x < 1$, $g'(1) = 0$, and $g'(x) > 0$ for all $x > 1$, and suppose that $g(1) = 0$. Set

$$G(x) = \int_0^x g(t)\,dt.$$

Explain why each of the following statements must be true.

(a) G is continuous.

(b) G is twice differentiable.

(c) $x = 1$ is a critical number of G.

(d) The graph of G is concave down for $x < 1$ and concave up for $x > 1$.

(e) G is an increasing function.

Make a rough sketch of the graph of G.

21. (a) Sketch the graph of the function

$$f(x) = \begin{cases} 2 - x, & 0 \le x \le 1 \\ 2 + x, & 1 < x \le 3. \end{cases}$$

(b) Find the function $F(x) = \int_0^x f(t)\,dt,\ 0 \le x \le 3$, and sketch its graph.

(c) What can you conclude about f and F at $x = 1$?

22. (a) Sketch the graph of the function

$$f(x) = \begin{cases} x^2 + x, & 0 \le x \le 1 \\ 2x, & 1 < x \le 3. \end{cases}$$

(b) Find the function $F(x) = \int_0^x f(t)\,dt,\ 0 \le x \le 3$, and sketch its graph.

(c) What can you conclude about f and F at $x = 1$?

In Exercises 23–26, calculate the derivative of function F.

23. $F(x) = \int_0^{x^3} t\cos t\,dt$. HINT: Let $u = x^3$ and use the chain rule.

24. $F(x) = \int_1^{\cos x} \sqrt{1-t^2}\,dt$.

25. $F(x) = \int_{x^2}^1 (t - \sin^2 t)\,dt$.

26. $F(x) = \int_0^{\sqrt{x}} \frac{t^2}{1+t^4}\,dt$.

27. Let $F(x) = 2x + \int_0^x \frac{\sin 2t}{1+t^2}\,dt$. Determine

(a) $F(0)$. (b) $F'(0)$. (c) $F''(0)$.

28. Let $F(x) = 2x + \int_0^{x^2} \frac{\sin 2t}{1+t^2}\,dt$. Determine

(a) $F(0)$. (b) $F'(x)$.

29. Assume that f is a continuous function and that

$$\int_0^x f(t)\,dt = \frac{2x}{4+x^2}.$$

(a) Determine $f(0)$.

(b) Find the zeros of f, if any.

30. Assume that f is a continuous function and that

$$\int_0^x t f(t)\,dt = \sin x - x\cos x.$$

(a) Determine $f(\pi/2)$.

(b) Find $f'(x)$.

31. (*A mean-value theorem for integrals*) Show that if f is continuous on $[a, b]$, then there is a least one number c in (a, b) for which

$$\int_a^b f(x)\,dx = f(c)(b-a).$$

32. Show the validity of equation (2) in the proof of Theorem 5.3.5.

33. Extend Theorem 5.3.5 by showing that, if f is continuous on $[a, b]$ and c is *any* number in $[a, b]$, then the function

$$F(x) = \int_c^x f(t)\,dt$$

is continuous on $[a, b]$, differentiable on (a, b), and satisfies

$$F'(x) = f(x) \quad \text{for all } x \text{ in } (a, b).$$

34. Complete the proof of Theorem 5.3.5 by showing that

$$\lim_{h\to 0^-} \frac{F(x+h) - F(x)}{h} = f(x) \quad \text{for all } x \text{ in } (a, b].$$

35. Let f be continuous on $[a, b]$ and let the functions F and G be defined by

$$F(x) = \int_c^x f(t)\,dt, \qquad \text{and} \qquad G(x) = \int_d^x f(t)\,dt$$

where $c, d \in [a, b]$.

(a) Show that F and G differ by a constant.

(b) Show that $F(x) - G(x) = \int_c^d f(t)\,dt$.

36. Let f be continuous and define F by

$$F(x) = \int_0^x \left[t \int_1^t f(u)\,du \right] dt.$$

Find (a) $F'(x)$. (b) $F'(1)$. (c) $F''(x)$. (d) $F''(1)$.

C In Exercises 37–40, $F(x) = \int f(t)\,dt$ is given. Use a CAS to carry out the following steps:

(a) Solve the equation $F'(x) = 0$. Determine the intervals on which F is increasing and the intervals on which F is decreasing. Graph F and F' together.

(b) Solve the equation $F''(x) = 0$. Determine the intervals on which the graph of F is concave up and the intervals on which the graph of F is concave down. Graph F and F'' together.

37. $F(x) = \int_0^x (t^2 - 3t - 4)\,dt$.

38. $F(x) = \int_0^x (2 - 3\cos t)\,dt, \quad x \in [0, 2\pi]$

39. $F(x) = \int_x^0 \sin 2t\,dt, \quad x \in [0, 2\pi]$

40. $F(x) = \int_x^0 (2-t)^2\,dt$.

■ PROJECT 5.3 Functions Represented by an Integral

There is no simple way to represent the function

$$F(x) = \int_0^x \sin^2 t^2\,dt.$$

For example, we cannot represent F in terms of sums, differences, products, quotients, or compositions of elementary functions. However, we can still obtain a lot of information about F by using the results of Section 5.2 and this section, and then applying methods developed in Chapter 4.

Problem 1.

(a) Give the domain of F.

(b) Explain why F is continuous.

(c) Discuss the symmetry of the graph of F.

Problem 2.

(a) Explain why F is differentiable for all x.

(b) Find $F'(x)$.

(c) Find the critical numbers of F.

(d) Find the intervals on which F is increasing (if any) and the intervals on which F is decreasing (if any).

Problem 3.

(a) Find the intervals on which the graph of F is concave up (if any) and the intervals on which the graph is concave down (if any).

(b) Find the points of inflection of the graph of F.

Problem 4.

(a) Based on the information found in Problems 1–3, sketch the graph of F.

(b) Use a graphing utility to draw the graph of F to verify the sketch you made in part (a).

■ 5.4 THE FUNDAMENTAL THEOREM OF INTEGRAL CALCULUS

DEFINITION 5.4.1 ANTIDERIVATIVE

Let f be continuous on $[a, b]$. A function G is called an *antiderivative* for f on $[a, b]$ if:

$$G \text{ is continuous on } [a, b] \quad \text{and} \quad G'(x) = f(x) \text{ for all } x \in (a, b).$$

Theorem 5.3.5 says that if f is continuous on $[a, b]$, then

$$F(x) = \int_a^x f(t)\, dt$$

is an antiderivative for f on $[a, b]$. This gives us a prescription for constructing antiderivatives. It tells us that we can construct an antiderivative for f by integrating f.

The "fundamental theorem" goes the other way. It gives us a prescription, not for finding antiderivatives, but for evaluating integrals. It tells us that we can evaluate

$$\int_a^b f(t)\, dt$$

by finding an antiderivative for f and evaluating it at b and at a.

THEOREM 5.4.2 THE FUNDAMENTAL THEOREM OF INTEGRAL CALCULUS

Let f be continuous on $[a, b]$. If G is any antiderivative for f on $[a, b]$, then

$$\int_a^b f(t)\, dt = G(b) - G(a).$$

PROOF From Theorem 5.3.5 we know that the function

$$F(x) = \int_a^x f(t)\, dt$$

is an antiderivative for f on $[a, b]$. If G is also an antiderivative for f on $[a, b]$, then both F and G are continuous on $[a, b]$ and satisfy $F'(x) = G'(x)$ for all x in (a, b). From Theorem 4.2.5 we know that there exists a constant C such that

$$F(x) = G(x) + C \qquad \text{for all } x \text{ in } [a, b].$$

Since $F(a) = 0$,

$$G(a) + C = 0 \quad \text{and thus} \quad C = -G(a).$$

It follows that

$$F(x) = G(x) - G(a) \quad \text{for all } x \text{ in } [a, b].$$

In particular,

$$\int_a^b f(t)\,dt = F(b) = G(b) - G(a). \quad \square$$

We now evaluate some integrals by applying the fundamental theorem. In each case we use the simplest antiderivative we can think of.

Example 1 Evaluate $\displaystyle\int_1^4 x^2\,dx$.

SOLUTION As an antiderivative for $f(x) = x^2$, we can use the function

$$G(x) = \tfrac{1}{3}x^3. \qquad \text{(verify this)}$$

By the fundamental theorem

$$\int_1^4 x^2\,dx = G(4) - G(1) = \tfrac{1}{3}(4)^3 - \tfrac{1}{3}(1)^3 = \tfrac{64}{3} - \tfrac{1}{3} = 21.$$

NOTE: Any other antiderivative of $f(x) = x^2$ has the form $H(x) = \frac{1}{3}x^3 + C$ for some constant C. Had we chosen such an H instead of G, then we would have had

$$\int_1^4 x^2\,dx = H(4) - H(1) = \left[\tfrac{1}{3}(4)^3 + C\right] - \left[\tfrac{1}{3}(1)^3 + C\right] = \tfrac{64}{3} + C - \tfrac{1}{3} - C = 21;$$

the C "cancels out." $\square$

Example 2 Evaluate $\displaystyle\int_0^{\pi/2} \sin x\,dx$.

SOLUTION Here we use the antiderivative $G(x) = -\cos x$:

$$\begin{aligned}\int_0^{\pi/2} \sin x\,dx &= G(\pi/2) - G(0)\\ &= -\cos(\pi/2) - [-\cos(0)] = 0 - (-1) = 1. \quad \square\end{aligned}$$

Notation Expressions of the form $G(b) - G(a)$ are conveniently written

$$\Big[G(x)\Big]_a^b.$$

In this notation

$$\int_1^4 x^2\,dx = \left[\tfrac{1}{3}x^3\right]_1^4 = \tfrac{1}{3}(4)^3 - \tfrac{1}{3}(1)^3 = 21$$

and

$$\int_0^{\pi/2} \sin x\,dx = \Big[-\cos x\Big]_0^{\pi/2} = -\cos(\pi/2) - [-\cos(0)] = 1. \quad \square$$

The essential step in applying the fundamental theorem is the determination of an antiderivative G for the integrand f. Thus, we need to have a supply of antiderivatives of functions. From our study of differentiation, we know that

$$\frac{d}{dx}(x^r) = rx^{r-1}, \quad r \text{ any rational number.}$$

It follows immediately from this formula that

$$G(x) = \frac{x^{r+1}}{r+1}, \qquad (r \neq -1)$$

is an antiderivative of $f(x) = x^r$. Similarly, the differentiation formula

$$\frac{d}{dx}(\sin x) = \cos x$$

implies that $G(x) = \sin x$ is an antiderivative of $f(x) = \cos x$. In general, each differentiation formula yields a corresponding antidifferentiation formula. Table 5.4.1 lists some particular antiderivatives for our use in this section

■ Table 5.4.1

Function	Antiderivative
x^r	$\dfrac{x^{r+1}}{r+1}$ $(r$ a rational number $\neq -1)$
$\sin x$	$-\cos x$
$\cos x$	$\sin x$
$\sec^2 x$	$\tan x$
$\sec x \tan x$	$\sec x$
$\csc^2 x$	$-\cot x$
$\csc x \cot x$	$-\csc x$

You can verify that the derivative of each function in the right column is the corresponding function in the left column.

The following examples make use of the entries in Table 5.4.1.

$$\int_1^2 \frac{dx}{x^3} = \int_1^2 x^{-3}\,dx = \left[\frac{x^{-2}}{-2}\right]_1^2 = \left[-\frac{1}{2x^2}\right]_1^2 = -\tfrac{1}{8} - (-\tfrac{1}{2}) = \tfrac{3}{8},$$

$$\int_0^1 t^{5/3}\,dt = \left[\tfrac{3}{8}t^{8/3}\right]_0^1 = \tfrac{3}{8}(1)^{8/3} - \tfrac{3}{8}(0)^{8/3} = \tfrac{3}{8}.$$

$$\int_{-\pi/4}^{\pi/3} \sec^2 x\,dx = \Big[\tan x\Big]_{-\pi/4}^{\pi/3} = \tan\frac{\pi}{3} - \tan\frac{-\pi}{4} = \sqrt{3} - (-1) = \sqrt{3} + 1.$$

$$\int_{\pi/6}^{\pi/2} \csc x \cot x\,dx = \Big[-\csc x\Big]_{\pi/6}^{\pi/2} = -\csc\frac{\pi}{2} - \left[-\csc\frac{\pi}{6}\right] = -1 - (-2) = 1.$$

Example 3 Evaluate $\displaystyle\int_0^1 (2x - 6x^4 + 5)\,dx.$

SOLUTION As an antiderivative we use $G(x) = x^2 - \frac{6}{5}x^5 + 5x$:

$$\int_0^1 (2x - 6x^4 + 5)\,dx = \left[x^2 - \tfrac{6}{5}x^5 + 5x\right]_0^1 = 1 - \tfrac{6}{5} + 5 = \tfrac{24}{5}. \quad \square$$

Example 4 Evaluate $\displaystyle\int_{-1}^{1} (x-1)(x+2)\,dx$.

SOLUTION First we carry out the indicated multiplication:

$$(x-1)(x+2) = x^2 + x - 2.$$

As an antiderivative we use $G(x) = \frac{1}{3}x^3 + \frac{1}{2}x^2 - 2x$:

$$\int_{-1}^{1} (x-1)(x+2)\,dx = \left[\tfrac{1}{3}x^3 + \tfrac{1}{2}x^2 - 2x\right]_{-1}^{1} = -\tfrac{10}{3}. \quad \square$$

We now give some slightly more complicated examples. The essential step in each case is the determination of an antiderivative. Check each computation in detail.

$$\int_1^2 \frac{x^4+1}{x^2}\,dx = \int_1^2 (x^2 + x^{-2})\,dx = \left[\tfrac{1}{3}x^3 - x^{-1}\right]_1^2 = \tfrac{17}{6}.$$

$$\int_1^5 \sqrt{x-1}\,dx = \int_1^5 (x-1)^{1/2}dx = \left[\tfrac{2}{3}(x-1)^{3/2}\right]_1^5 = \tfrac{16}{3}.$$

$$\int_0^1 (4-\sqrt{x})^2\,dx = \int_0^1 (16 - 8\sqrt{x} + x)\,dx = \left[16x - \tfrac{16}{3}x^{3/2} + \tfrac{1}{2}x^2\right]_0^1 = \tfrac{67}{6}.$$

$$\int_1^2 -\frac{dt}{(t+2)^2} = \int_1^2 -(t+2)^{-2}\,dt = \left[(t+2)^{-1}\right]_1^2 = -\tfrac{1}{12}.$$

The Linearity of the Integral

The preceding examples suggest some simple properties of the integral that are used regularly in computations. Throughout, take f and g as continuous functions and α and β as constants.

I. Constants may be factored through the integral sign:

(5.4.3)
$$\int_a^b \alpha f(x)\,dx = \alpha \int_a^b f(x)\,dx.$$

For example,

$$\int_1^4 \tfrac{3}{7}\sqrt{x}\,dx = \tfrac{3}{7}\int_1^4 x^{1/2}\,dx = \tfrac{3}{7}\left[\tfrac{x^{3/2}}{3/2}\right]_1^4 = \tfrac{2}{7}[(4)^{3/2} - (1)^{3/2}] = \tfrac{2}{7}[8-1]$$
$$= 2.$$

$$\int_0^{\pi/4} 2\cos x\,dx = 2\int_0^{\pi/4} \cos x\,dx = 2\Big[\sin x\Big]_0^{\pi/4} = 2\left[\sin\frac{\pi}{4} - \sin 0\right]$$
$$= 2\frac{\sqrt{2}}{2} = \sqrt{2}.$$

Remark We want to emphasize that only constants can be factored through the integral sign; an expression containing a variable cannot be factored through the integral sign! For example,

$$\int_0^{\pi/2} x \cos x \, dx \neq x \int_0^{\pi/2} \cos x \, dx.$$

You can verify that an antiderivative for $h(x) = x \cos x$ is $H(x) = x \sin x + \cos x$, and so

$$\int_0^{\pi/2} x \cos x \, dx = \Big[x \sin x + \cos x \Big]_0^{\pi/2} = \tfrac{\pi}{2} - 1.$$

On the other hand,

$$x \int_0^{\pi/2} \cos x \, dx = x \Big[\sin x \Big]_0^{\pi/2} = x. \quad \square$$

II. The integral of a sum is the sum of the integrals:

(5.4.4)
$$\int_a^b [f(x) + g(x)] \, dx = \int_a^b f(x) \, dx + \int_a^b g(x) \, dx.$$

For example,

$$\begin{aligned}\int_1^2 \left[(x-1)^2 + \frac{1}{(x+2)^2} \right] dx &= \int_1^2 (x-1)^2 \, dx + \int_1^2 \frac{dx}{(x+2)^2} \\ &= \Big[\tfrac{1}{3}(x-1)^3 \Big]_1^2 + \Big[-(x+2)^{-1} \Big]_1^2 \\ &= \tfrac{1}{3} - \tfrac{1}{4} + \tfrac{1}{3} = \tfrac{5}{12}.\end{aligned}$$

III. The integral of a linear combination is the linear combination of the integrals:

(5.4.5)
$$\int_a^b [\alpha f(x) + \beta g(x)] \, dx = \alpha \int_a^b f(x) \, dx + \beta \int_a^b g(x) \, dx.$$

For example,

$$\begin{aligned}\int_0^1 (2x - 6x^4 + 5) \, dx &= 2 \int_0^1 x \, dx - 6 \int_0^1 x^4 \, dx + \int_0^1 5 \, dx \\ &= 2\left[\frac{x^2}{2} \right]_0^1 - 6\left[\frac{x^5}{5} \right]_0^1 + \Big[5x \Big]_0^1 = 1 - \tfrac{6}{5} + 5 = \tfrac{24}{5},\end{aligned}$$

as we saw in Example 3.

Properties I and II are particular instances of Property III. To prove III, Let F be an antiderivative for f and let G be an antiderivative for g. Then, since

$$[\alpha F(x) + \beta G(x)]' = \alpha F'(x) + \beta G'(x) = \alpha f(x) + \beta g(x),$$

it follows that $\alpha F + \beta G$ is an antiderivative for $\alpha f + \beta g$. Therefore,

$$\begin{aligned}\int_a^b [\alpha f(x) + \beta g(x)]\,dx &= \Big[\alpha F(x) + \beta G(x)\Big]_a^b \\ &= [\alpha F(b) + \beta G(b)] - [\alpha F(a) + \beta G(a)] \\ &= \alpha [F(b) - F(a)] + \beta [G(b) - G(a)] \\ &= \alpha \int_a^b f(x)\,dx + \beta \int_a^b g(x)\,dx.\end{aligned}$$

Example 5 Evaluate $\displaystyle\int_0^{\pi/4} \sec x\,[2\tan x - 5\sec x]\,dx.$

SOLUTION

$$\begin{aligned}\int_0^{\pi/4} \sec x\,[2\tan x - 5\sec x]\,dx &= \int_0^{\pi/4} [2\sec x \tan x - 5\sec^2 x]\,dx \\ &= 2\int_0^{\pi/4} \sec x \tan x\,dx - 5\int_0^{\pi/4} \sec^2 x\,dx \\ &= 2\Big[\sec x\Big]_0^{\pi/4} - 5\Big[\tan x\Big]_0^{\pi/4} \\ &= 2\left[\sec\frac{\pi}{4} - \sec 0\right] - 5\left[\tan\frac{\pi}{4} - \tan 0\right] \\ &= 2[\sqrt{2} - 1] - 5[1 - 0] = 2\sqrt{2} - 7. \quad \square\end{aligned}$$

EXERCISES 5.4

Evaluate the definite integral.

1. $\displaystyle\int_0^1 (2x - 3)\,dx.$
2. $\displaystyle\int_0^1 (3x + 2)\,dx.$
3. $\displaystyle\int_{-1}^0 5x^4\,dx.$
4. $\displaystyle\int_1^2 (2x + x^2)\,dx.$
5. $\displaystyle\int_1^4 2\sqrt{x}\,dx.$
6. $\displaystyle\int_0^4 \sqrt[3]{x}\,dx.$
7. $\displaystyle\int_1^5 2\sqrt{x - 1}\,dx.$
8. $\displaystyle\int_1^2 \left(\frac{3}{x^3} + 5x\right)dx.$
9. $\displaystyle\int_{-2}^0 (x + 1)(x - 2)\,dx.$
10. $\displaystyle\int_1^0 (t^3 + t^2)\,dt.$
11. $\displaystyle\int_1^2 \left(3t + \frac{4}{t^2}\right)dt.$
12. $\displaystyle\int_{-1}^{-1} 7x^6\,dx.$
13. $\displaystyle\int_0^1 (x^{3/2} - x^{1/2})\,dx.$
14. $\displaystyle\int_0^1 (x^{3/4} - 2x^{1/2})\,dx.$
15. $\displaystyle\int_0^1 (x + 1)^{17}\,dx.$
16. $\displaystyle\int_0^a (a^2x - x^3)\,dx.$
17. $\displaystyle\int_0^a (\sqrt{a} - \sqrt{x})^2\,dx.$
18. $\displaystyle\int_{-1}^1 (x - 2)^2\,dx.$
19. $\displaystyle\int_1^2 \frac{6 - t}{t^3}\,dt.$
20. $\displaystyle\int_1^3 \left(x^2 - \frac{1}{x^2}\right)dx.$
21. $\displaystyle\int_1^2 2x(x^2 + 1)\,dx.$
22. $\displaystyle\int_0^1 3x^2(x^3 + 1)\,dx.$
23. $\displaystyle\int_0^{\pi/2} \cos x\,dx.$
24. $\displaystyle\int_0^{\pi} 3\sin x\,dx.$
25. $\displaystyle\int_0^{\pi/4} 2\sec^2 x\,dx.$
26. $\displaystyle\int_{\pi/6}^{\pi/3} \sec x \tan x\,dx.$
27. $\displaystyle\int_{\pi/6}^{\pi/4} \csc u \cot u\,du.$
28. $\displaystyle\int_{\pi/4}^{\pi/3} -\csc^2 u\,du.$
29. $\displaystyle\int_0^{2\pi} \sin x\,dx.$
30. $\displaystyle\int_0^{\pi} \tfrac{1}{2}\cos x\,dx.$

31. $\int_0^{\pi/3} \left(\frac{2}{\pi}x - 2\sec^2 x\right) dx.$

32. $\int_{\pi/4}^{\pi/2} \csc x(\cot x - 3\csc x)\, dx.$

33. $\int_0^3 \left[\frac{d}{dx}(\sqrt{4+x^2})\right] dx.$ **34.** $\int_0^{\pi/2} \left[\frac{d}{dx}(\sin^3 x)\right] dx.$

In Exercises 35–38, calculate the derivative of the given function (a) without integrating; that is, using the results of Section 5.2; (b) by integrating and then differentiating the result.

35. $\int_1^x (t+2)^2\, dt.$ **36.** $\int_0^x (\cos t - \sin t)\, dt.$

37. $\int_1^{2x+1} \frac{1}{2}\sec u \tan u\, du.$ **38.** $\int_{x^2}^2 t(t-1)\, dt.$

39. Define a function F on $[1, 8]$ such that $F'(x) = 1/x$ and (a) $F(2) = 0$; (b) $F(2) = -3$.

40. Define a function F on $[0, 4]$ such that $F'(x) = \sqrt{1+x^2}$ and (a) $F(3) = 0$; (b) $F(3) = 1$.

In Exercises 41–44, verify that the function f is nonnegative on the given interval, and then calculate the area between the graph of f and the x-axis on that interval.

41. $f(x) = 4x - x^2;\quad [0, 4].$

42. $f(x) = x\sqrt{x} + 1;\quad [1, 9].$

43. $f(x) = 2\cos x;\quad [-\pi/2, \pi/4].$

44. $f(x) = \sec x \tan x;\quad [0, \pi/3].$

Evaluate the given integrals.

45. (a) $\int_2^5 (x-3)\, dx.$ (b) $\int_2^5 |x-3|\, dx.$

46. (a) $\int_{-4}^2 (2x+3)\, dx.$ (b) $\int_{-4}^2 |2x+3|\, dx.$

47. (a) $\int_{-2}^2 (x^2-1)\, dx.$ (b) $\int_{-2}^2 |x^2-1|\, dx.$

48. (a) $\int_{-\pi/2}^{\pi} \cos x\, dx.$ (b) $\int_{-\pi/2}^{\pi} |\cos x|\, dx.$

In Exercises 49–52, determine whether the given calculation is valid. If it is not valid, explain why.

49. $\int_0^{2\pi} x\cos x\, dx = \Big[x\sin x + \cos x\Big]_0^{2\pi} = 1 - 1 = 0.$

50. $\int_0^{2\pi} \sec^2 x\, dx = \tan x\Big|_0^{2\pi} = 0 - 0 = 0.$

51. $\int_{-2}^2 \frac{1}{x^3} dx = \frac{-1}{2x^2}\Big|_{-2}^2 = -\frac{1}{8} - (-\frac{1}{8}) = 0.$

52. $\int_{-2}^2 |x|\, dx = \frac{1}{2}x|x|\Big|_{-2}^2 = 2 - (-2) = 4.$

53. An object starts at the origin and moves along the x-axis with velocity

$$v(t) = 10t - t^2,\quad 0 \le t \le 10.$$

(a) What is the position of the object at any time t, $0 \le t \le 10$?

(b) When is the object's velocity a maximum, and what is its position at that time?

54. The velocity of a weight suspended on a spring is given

$$v(t) = 3\sin t + 4\cos t,\quad t \ge 0.$$

At time $t = 0$, the weight is one unit below the equilibrium position (see the figure).

(a) Determine the position of the weight at any time $t \ge 0$.

(b) What is the weight's maximum displacement from the equilibrium position?

(NOTE: The motion of the weight is called *simple harmonic motion*. See Exercises 73 and 74, Section 3.6.)

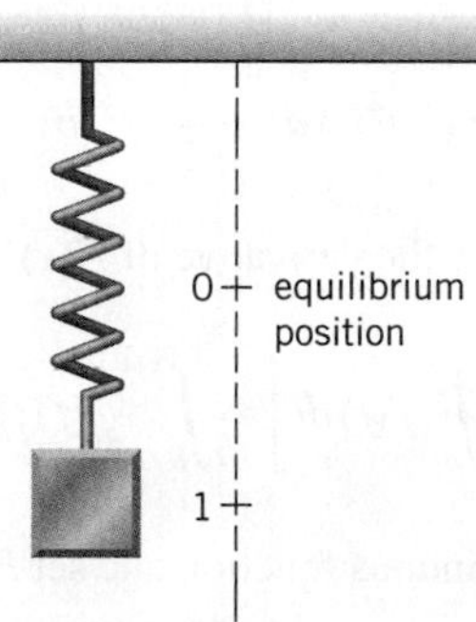

Evaluate the definite integral.

55. $\int_0^4 f(x)\, dx;\quad f(x) = \begin{cases} 2x+1, & 0 \le x \le 1 \\ 4-x, & 1 < x \le 4. \end{cases}$

56. $\int_{-2}^4 f(x)\, dx;\quad f(x) = \begin{cases} x^2, & -2 \le x < 0 \\ \frac{1}{2}x + 2, & 0 \le x \le 4. \end{cases}$

57. $\int_{-\pi/2}^{\pi} f(x)\, dx;\quad f(x) = \begin{cases} \cos x, & -\pi/2 \le x \le \pi/3 \\ (3/\pi)x + 1, & \pi/3 < x \le \pi. \end{cases}$

58. $\int_0^{3\pi/2} f(x)\, dx;\quad f(x) = \begin{cases} 2\sin x, & 0 \le x \le \pi/2 \\ \frac{1}{2}\cos x, & \pi/2 < x \le 3\pi/2. \end{cases}$

59. Let $f(x) = \begin{cases} x+2, & -2 \le x \le 0 \\ 2, & 0 < x \le 1 \\ 4-2x, & 1 < x \le 2, \end{cases}$

and let $g(x) = \int_{-2}^x f(t)\, dt.$

(a) Carry out the integration.

(b) Sketch the graphs of f and g.

(c) Where is f continuous? Where is f differentiable? Where is g differentiable?

60. Let $f(x) = \begin{cases} 1 - x^2, & -1 \le x \le 1 \\ 1, & 1 < x < 3 \\ 2x - 5, & 3 \le x \le 5 \end{cases}$

and let $g(x) = \int_{-1}^{x} f(t)\, dt$.

(a) Carry out the integration.

(b) Sketch the graphs of f and g.

(c) Where is f continuous? Where is f differentiable? Where is g differentiable?

61. (*Important*) Let f be a function such that f' is continuous on $[a, b]$. Prove that

$$\int_a^b f'(t)\, dt = f(b) - f(a).$$

This is a useful special case of Theorem 5.4.2.

62. Let f be a function such that f' is continuous on $[a, b]$. Show that

$$\int_a^b f(t)f'(t)\, dt = \frac{1}{2}[f^2(b) - f^2(a)].$$

HINT: Calculate the derivative of $F(x) = f^2(x)$.

63. Compare $\dfrac{d}{dx}\left[\int_a^x f(t)\, dt\right]$ to $\int_a^x \dfrac{d}{dt}[f(t)]\, dt$.

64. Let f be a continuous function and set $F(x) = \int_0^x x f(t)\, dt$. Find $F'(x)$. HINT: The answer is not $x f(x)$.

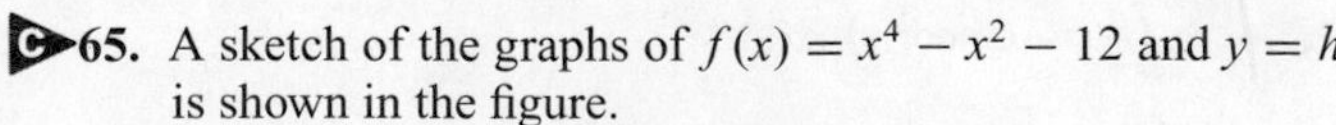

65. A sketch of the graphs of $f(x) = x^4 - x^2 - 12$ and $y = h$ is shown in the figure.

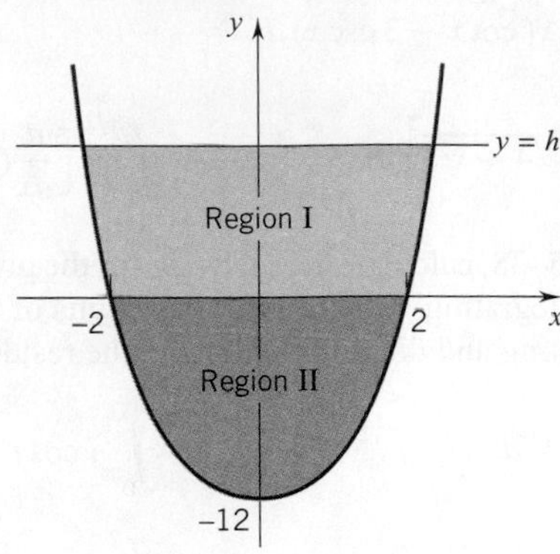

(a) Use a graphing utility to get an accurate drawing of these curves.

(b) Estimate h so that region I and region II have equal areas.

(c) Use a CAS to find the area of region I.

(d) Use a CAS to find h exactly.

66. Repeat Exercise 65 with $f(x) = x^3 - x^4$ using the figure.

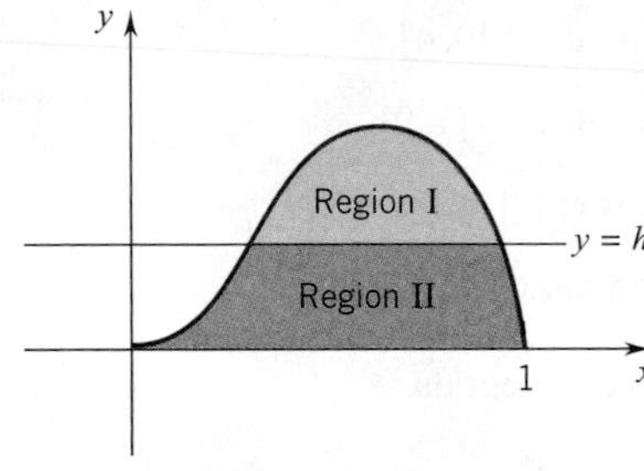

■ 5.5 SOME AREA PROBLEMS

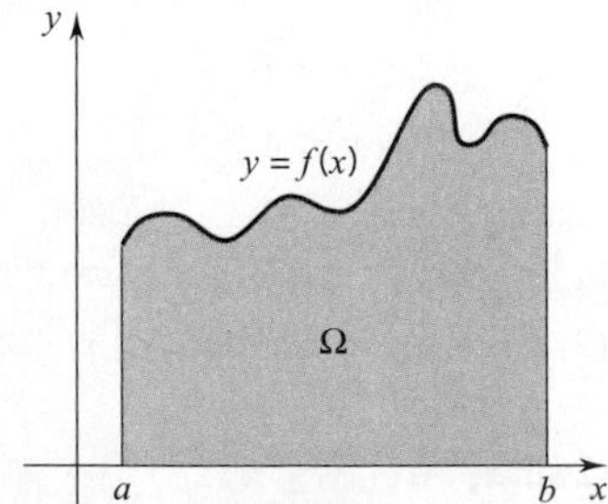

Figure 5.5.1

In Section 5.2 we noted that if f is nonnegative and continuous on $[a, b]$, then the integral of f from a to b gives the area of the region below the graph of f, that is, the area of the region bounded by the graph of f and the x-axis between the lines $x = a$ and $x = b$. With Ω as in Figure 5.5.1,

(5.5.1)
$$\text{area of } \Omega = \int_a^b f(x)\, dx.$$

Example 1 Find the area below the graph of the square-root function from $x = 0$ to $x = 1$.

SOLUTION The graph is pictured in Figure 5.5.2. The area below the graph is $\frac{2}{3}$.

$$\int_0^1 \sqrt{x}\, dx = \int_0^1 x^{1/2}\, dx = \left[\tfrac{2}{3}x^{3/2}\right]_0^1 = \tfrac{2}{3}. \quad \square$$

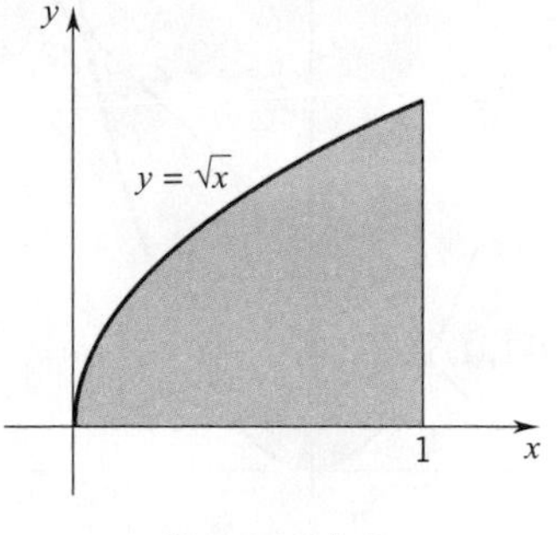

Figure 5.5.2

Example 2 Find the area of the region bounded above by the curve $y = 4 - x^2$ and below by the x-axis.

SOLUTION The curve intersects the x-axis at $x = -2$ and $x = 2$. See Figure 5.5.3. The area of the region is $\frac{32}{3}$.

$$\int_{-2}^{2} (4 - x^2)\,dx = \left[4x - \tfrac{1}{3}x^3\right]_{-2}^{2} = \tfrac{32}{3}.$$

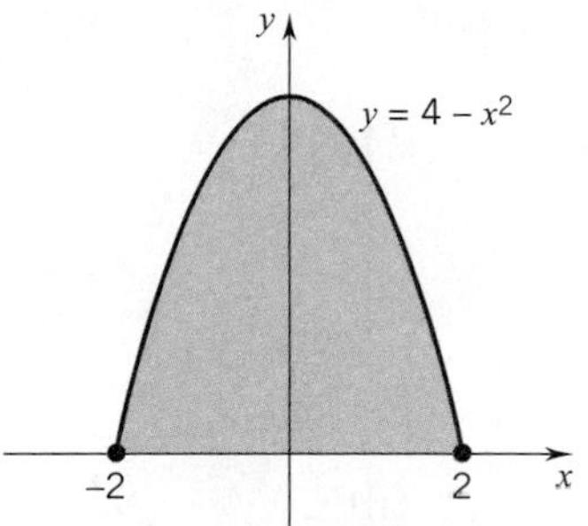

Figure 5.5.3

NOTE: The region is symmetric with respect to the y-axis. Therefore, the area of the region is also $2\int_0^2 (4 - x^2)\,dx$, and this integral is a little easier to calculate:

$$2\int_{0}^{2} (4 - x^2)\,dx = 2\left[4x - \tfrac{1}{3}x^3\right]_{0}^{2} = 2\left(8 - \tfrac{8}{3}\right) = 2\left(\tfrac{16}{3}\right) = \tfrac{32}{3}.$$

We will have more to say about symmetry in integration in Section 5.8. □

Now we calculate the areas of somewhat more complicated regions, such as region Ω shown in Figure 5.5.4. To avoid excessive repetition, let's agree at the outset that throughout this section the symbols f, g, h represent continuous functions.

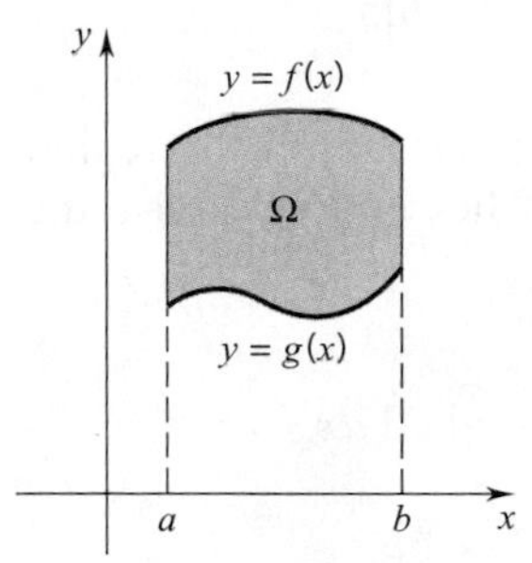

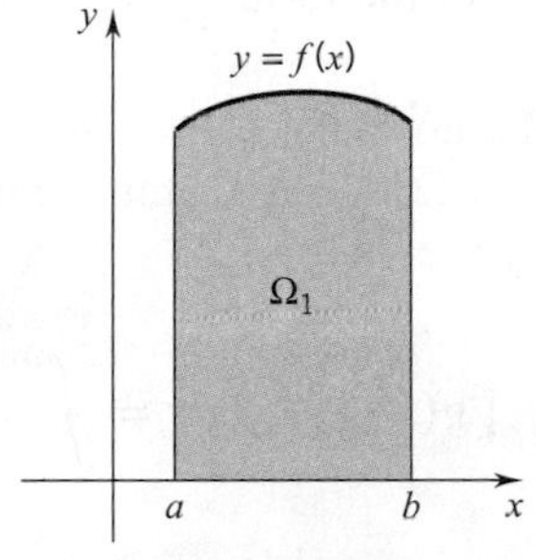

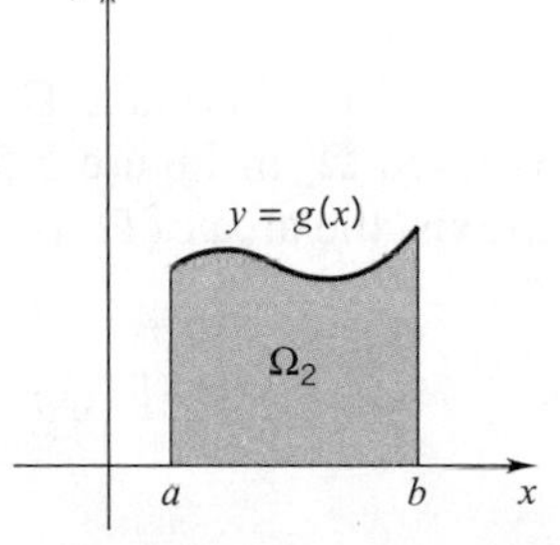

area of Ω = area of Ω_1 − area of Ω_2

Figure 5.5.4

Look at the region Ω shown in Figure 5.5.4. The upper boundary of Ω is the graph of a nonnegative function f and the lower boundary is the graph of a nonnegative function g. We can obtain the area of Ω by calculating the area of Ω_1 and subtracting off the area of Ω_2. Since

$$\text{area of } \Omega_1 = \int_a^b f(x)\,dx \quad \text{and} \quad \text{area of } \Omega_2 = \int_a^b g(x)\,dx,$$

we have

$$\text{area of } \Omega = \int_a^b f(x)\,dx - \int_a^b g(x)\,dx.$$

We can combine the two integrals [by (5.4.5)] and write

(5.5.2)
$$\text{area of } \Omega = \int_a^b [f(x) - g(x)]\,dx.$$

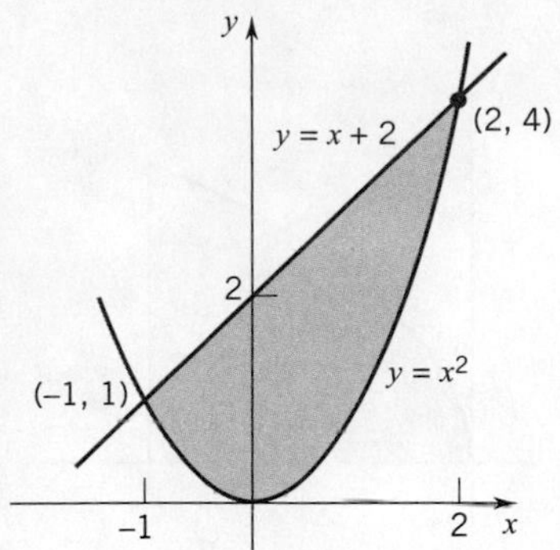

Figure 5.5.5

Example 3 Find the area of the region bounded above by $y = x + 2$ and below by $y = x^2$.

SOLUTION The region bounded by the graphs of the two equations is shown in Figure 5.5.5. Out first step is to find the points of intersection of the two curves:

$$x + 2 = x^2 \quad \text{iff} \quad x^2 - x - 2 = 0$$
$$\text{iff} \quad (x + 1)(x - 2) = 0,$$

and so $x = -1$ or $x = 2$. The curves intersect at $(-1, 1)$ and $(2,4)$.

The area of the region is given by:

$$\int_{-1}^{2} \left[(x+2) - x^2\right] dx = \left[\tfrac{1}{2}x^2 + 2x - \tfrac{1}{3}x^3\right]_{-1}^{2}$$
$$= \left(2 + 4 - \tfrac{8}{3}\right) - \left(\tfrac{1}{2} - 2 + \tfrac{1}{3}\right)$$
$$= \tfrac{9}{2}. \quad \square$$

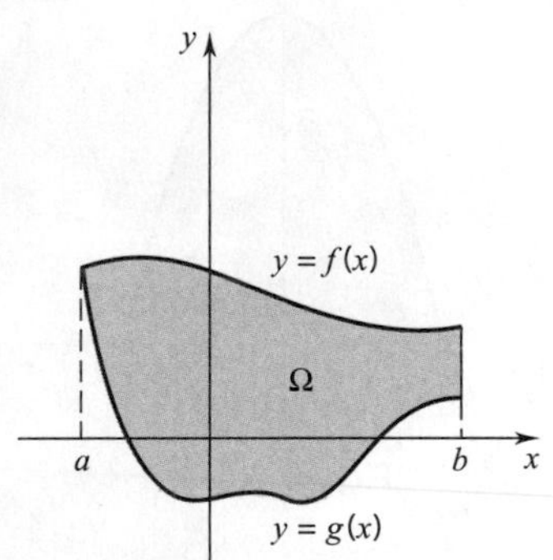

Figure 5.5.6

We derived Formula 5.5.2 under the assumption that f and g were both nonnegative, but that assumption is unnecessary. The formula holds for any region Ω that has

an upper boundary of the form $y = f(x), \quad x \in [a, b]$

and a lower boundary of the form $y = g(x), \quad x \in [a, b]$.

To see this, take Ω as in Figure 5.5.6. Obviously, Ω is congruent to the region marked Ω' in Figure 5.5.7; Ω' is Ω raised C units. Since Ω' lies entirely above the x-axis, the area of Ω' is given by the integral

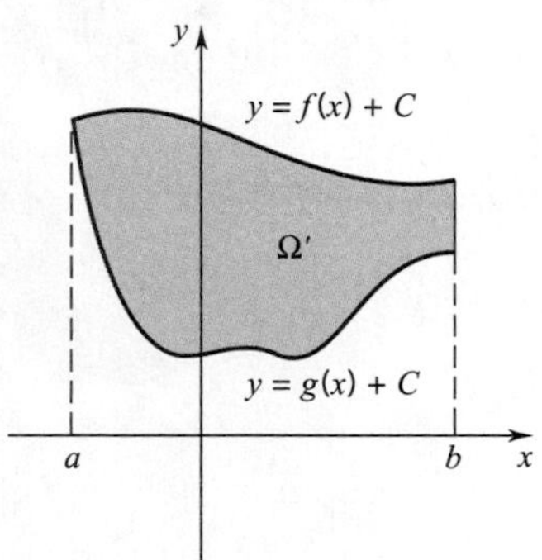

Figure 5.5.7

$$\int_a^b \{[f(x) + C] - [g(x) + C]\}\, dx = \int_a^b [f(x) - g(x)]\, dx.$$

Since area of Ω = area of Ω',

$$\text{area of } \Omega = \int_a^b [f(x) - g(x)]\, dx$$

as asserted.

Example 4 Find the area of the region shown in Figure 5.5.8.

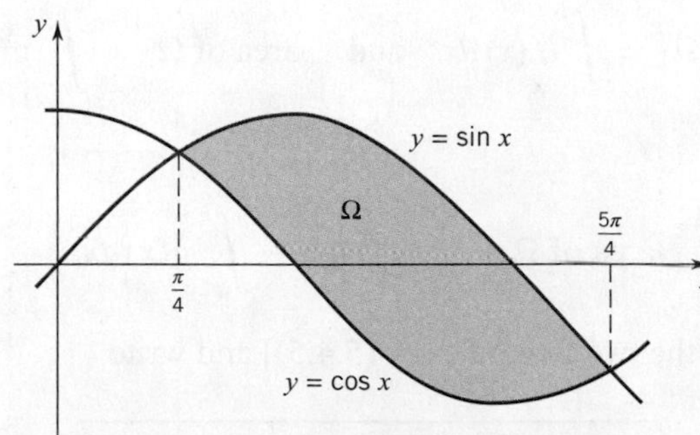

Figure 5.5.8

SOLUTION

$$\text{Area of } \Omega = \int_{\pi/4}^{5\pi/4} [\sin x - \cos x]\, dx$$

$$= \Big[-\cos x - \sin x\Big]_{\pi/4}^{5\pi/4} = 2\sqrt{2}. \quad \square$$

Example 5 Find the area between the curves

$$y = 4x \quad \text{and} \quad y = x^3$$

from $x = -2$ to $x = 2$. See Figure 5.5.9.

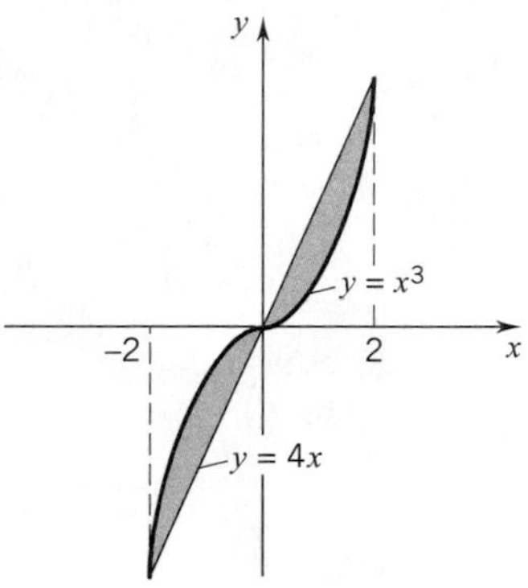

Figure 5.5.9

SOLUTION Notice that $y = x^3$ is the upper boundary from $x = -2$ to $x = 0$, but it is the lower boundary from $x = 0$ to $x = 2$. Thus

$$\text{area} = \int_{-2}^{0} [x^3 - 4x]\, dx + \int_{0}^{2} [4x - x^3]\, dx$$

$$= \Big[\tfrac{1}{4}x^4 - 2x^2\Big]_{-2}^{0} + \Big[2x^2 - \tfrac{1}{4}x^4\Big]_{0}^{2}$$

$$= [0 - (-4)] + [4 - 0] = 8. \quad \square$$

Example 6 Use integrals to represent the area of the region $\Omega = \Omega_1 \cup \Omega_2$ shaded in Figure 5.5.10.

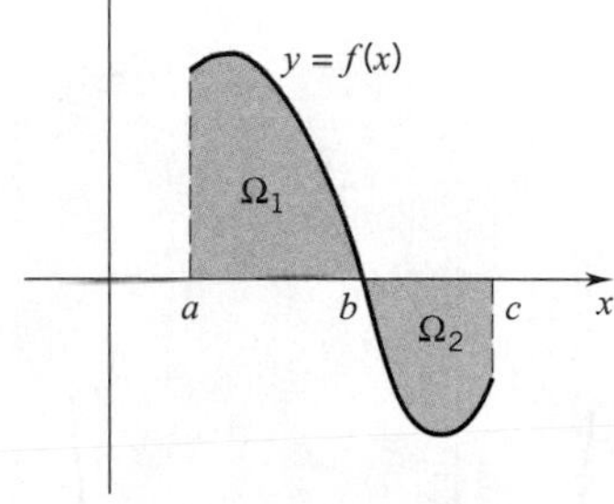

Figure 5.5.10

SOLUTION From $x = a$ to $x = b$, the curve $y = f(x)$ is above the x-axis. Therefore

$$\text{area of } \Omega_1 = \int_a^b f(x)\, dx.$$

From $x = b$ to $x = c$, the curve $y = f(x)$ is below the x-axis. The upper boundary for Ω_2 is the curve $y = 0$ (the x-axis) and the lower boundary is the curve $y = f(x)$. Thus

$$\text{area of } \Omega_2 = \int_b^c [0 - f(x)]\, dx = -\int_b^c f(x)\, dx.$$

The area of Ω is the sum of these two areas:

$$\text{area of } \Omega = \int_a^b f(x)\, dx - \int_b^c f(x)\, dx. \quad \square$$

Signed Area

The solution of Example 6 suggests a general geometric interpretation of the definite integral of a continuous function f. For example, if f is as denoted in Figure 5.5.10, then $\int_a^c f(x)\, dx$ is a number. But what does this number represent in terms of area?

Since $$\int_a^c f(x)\, dx = \int_a^b f(x)\, dx + \int_b^c f(x)\, dx = \text{area } \Omega_1 - \text{area } \Omega_2,$$

it follows that $\int_a^c f(x)dx$ is the area of the region above the x-axis *minus* the area of the region below the x-axis. This result is true in general: If f changes sign on the interval $[a, b]$, then $\int_a^b f(x)\,dx$ can be interpreted as the *difference* of two areas—the total area of the regions that lie above the x-axis *minus* the total area of the regions that lie below the x-axis. Consider the function f shown in Figure 5.5.11:

$$\begin{aligned}\int_a^b f(x)\,dx &= \int_a^c f(x)\,dx + \int_c^d f(x)\,dx + \int_d^e f(x)\,dx + \int_e^b f(x)\,dx \\ &= \text{area of } \Omega_1 - \text{ area of } \Omega_2 + \text{area of } \Omega_3 - \text{ area of} \Omega_4 \\ &= [\text{area of } \Omega_1 + \text{ area of } \Omega_3] - [\text{ area of } \Omega_2 + \text{ area of } \Omega_4].\end{aligned}$$

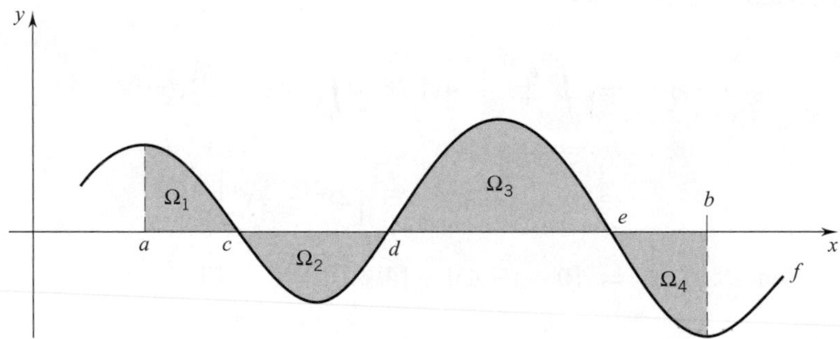

Figure 5.5.11

This value is sometimes referred to as the *signed area*; $\int_a^b f(x)\,dx$ can be interpreted geometrically as the signed area of the region bounded by the graph of f and the x-axis between $x = a$ and $x = b$.

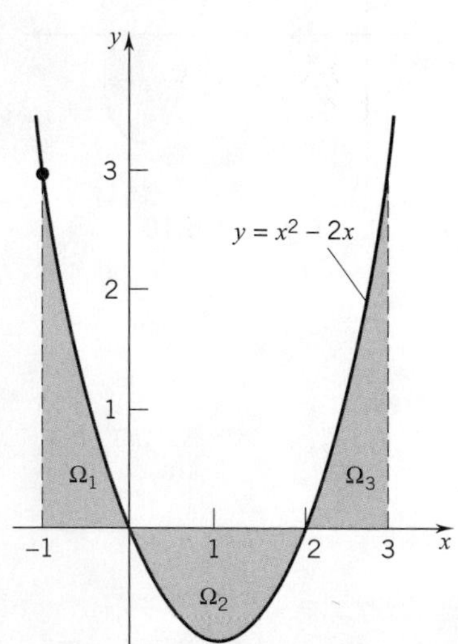

Figure 5.5.12

Example 7 Evaluate $\int_{-1}^{3} (x^2 - 2x)\,dx$ and interpret the result in terms of areas. Also, determine the area of the region bounded by the graph of $f(x) = x^2 - 2x$ and the x-axis between $x = -1$ and $x = 3$.

SOLUTION The graph of $f(x) = x^2 - 2x$ is shown in Figure 5.5.12.

$$\begin{aligned}\int_{-1}^{3} (x^2 - 2x)\,dx &= \left[\tfrac{1}{3}x^3 - x^2\right]_{-1}^{3} \\ &= \left[\tfrac{1}{3}(3)^3 - (3)^2\right] - \left[\tfrac{1}{3}(-1)^3 - (-1)^2\right] = \tfrac{4}{3}.\end{aligned}$$

This integral represents a signed area: area of Ω_1–area of Ω_2 + area of Ω_3.

The actual area of the region bounded by the graph of f and the x-axis from $x = -1$ to $x = 3$ is given by A = area of Ω_1 + area of Ω_2 + area of Ω_3.

$$\text{area of } \Omega_1 = \int_{-1}^{0} (x^2 - 2x)\,dx, \quad \text{area of } \Omega_2 = -\int_0^2 (x^2 - 2x)\,dx \quad \text{(see Example 6)}$$

and

$$\text{area of } \Omega_3 = \int_2^3 (x^2 - 2x)\,dx.$$

Thus,

$$A = \int_{-1}^{0} (x^2 - 2x)\,dx + \left[-\int_{0}^{2} (x^2 - 2x)\,dx\right] + \int_{2}^{3} (x^2 - 2x)\,dx$$

$$= \int_{-1}^{0} (x^2 - 2x)\,dx + \int_{0}^{2} (2x - x^2)\,dx + \int_{2}^{3} (x^2 - 2x)\,dx$$

$$= \left[\tfrac{1}{3}x^3 - x^2\right]_{-1}^{0} + \left[x^2 - \tfrac{1}{3}x^3\right]_{0}^{2} + \left[\tfrac{1}{3}x^3 - x^2\right]_{2}^{3} = \tfrac{4}{3} + \tfrac{4}{3} + \tfrac{4}{3} = 4. \quad \square$$

Example 8 Use integrals to represent the area of the region shaded in Figure 5.5.13.

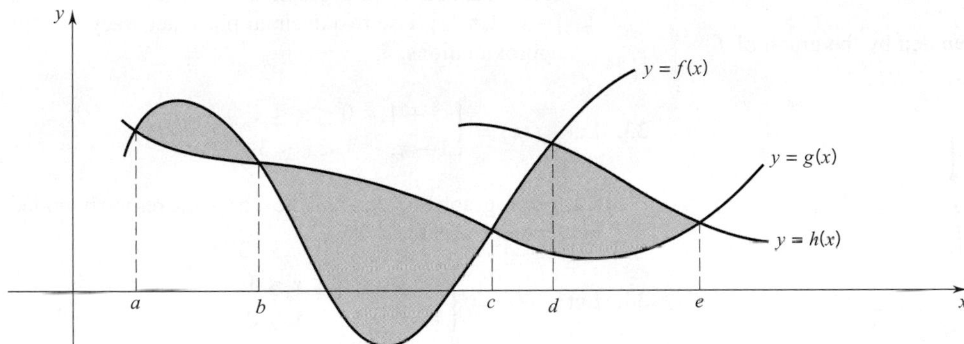

Figure 5.5.13

SOLUTION

$$\text{Area} = \int_{a}^{b} [f(x) - g(x)]\,dx + \int_{b}^{c} [g(x) - f(x)]\,dx$$

$$+ \int_{c}^{d} [f(x) - g(x)]\,dx + \int_{d}^{e} [h(x) - g(x)]\,dx. \quad \square$$

Remark The examples in this section demonstrate the importance of having an accurate figure to illustrate the problem being considered. □

EXERCISES 5.5

Find the area between the graph of f and the x-axis.

1. $f(x) = 2 + x^3, \quad x \in [0, 1].$
2. $f(x) = (x + 2)^{-2}, \quad x \in [0, 2].$
3. $f(x) = \sqrt{x+1}, \quad x \in [3, 8].$
4. $f(x) = x^2(3 + x), \quad x \in [0, 8].$
5. $f(x) = (2x^2 + 1)^2, \quad x \in [0, 1].$
6. $f(x) = \frac{1}{2}(x + 1)^{-1/2}, \quad x \in [0, 8].$
7. $f(x) = x^2 - 4, \quad x \in [1, 2].$
8. $f(x) = \cos x, \quad x \in [\frac{1}{6}\pi, \frac{1}{3}\pi].$
9. $f(x) = \sin x, \quad x \in [\frac{1}{3}\pi, \frac{1}{2}\pi].$
10. $f(x) = x^3 + 1, \quad x \in [-2, -1].$

Sketch the region bounded by the curves and find its area.

11. $y = \sqrt{x}, \quad y = x^2.$
12. $y = 6x - x^2, \quad y = 2x.$
13. $y = 5 - x^2, \quad y = 3 - x.$
14. $y = 8, \quad y = x^2 + 2x.$
15. $y = 8 - x^2, \quad y = x^2.$
16. $y = \sqrt{x}, \quad y = \frac{1}{4}x.$
17. $x^3 - 10y^2 = 0, \quad x - y = 0.$
18. $y^2 - 27x = 0, \quad x + y = 0.$

19. $x - y^2 + 3 = 0, \quad x - 2y = 0.$

20. $y^2 = 2x, \quad x - y = 4.$

21. $y = x, \quad y = 2x, \quad y = 4.$

22. $y = x^2, \quad y = -\sqrt{x}, \quad x = 4.$

23. $y = \cos x, \quad y = 4x^2 - \pi^2.$

24. $y = \sin x, \quad y = \pi x - x^2.$

25. $y = x, \quad y = \sin x, \quad x = \pi/2.$

26. $y = x + 1, \quad y = \cos x, \quad x = \pi.$

27. The graph of $f(x) = x^2 - x - 6$ is shown in the figure.

(a) Evaluate $\int_{-3}^{4} f(x)\, dx$ and interpret the result in terms of areas.

(b) Find the area of the region bounded by the graph of f and the x-axis for $x \in [-3, 4]$.

(c) Find the area of the region bounded by the graph of f and the x-axis for $x \in [-2, 3]$.

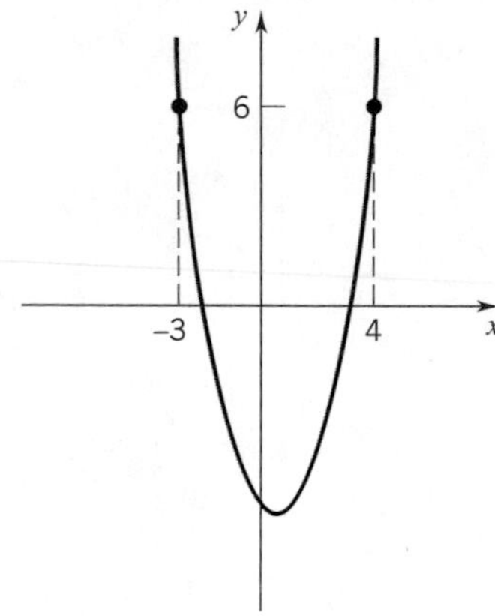

28. The graph of $f(x) = 2\sin x$, $x \in [-\pi/2, 3\pi/4]$ is shown in the figure.

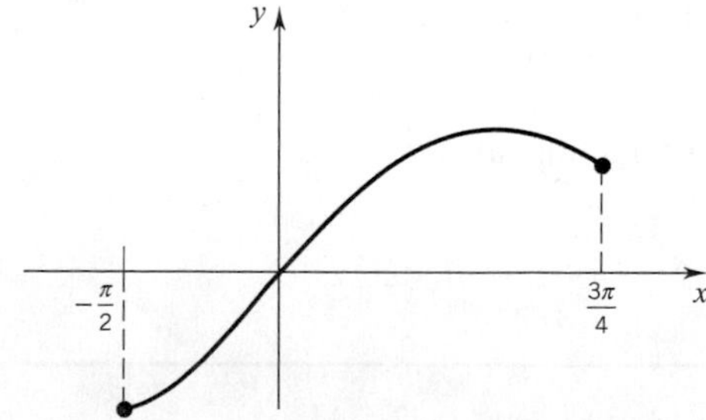

(a) Evaluate $\int_{-\pi/2}^{3\pi/4} f(x)\, dx$ and interpret the result in terms of areas.

(b) Find the area of the region bounded by the graph of f and the x-axis for $x \in [-\pi/2, 3\pi/4]$.

(c) Find the area of the region bounded by the graph of f and the x-axis for $x \in [-\pi/2, 0]$.

29. Let $f(x) = x^3 - x$.

(a) Evaluate $\int_{-2}^{2} f(x)\, dx$.

(b) Sketch the graph of f and find the area of the region bounded by the graph and the x-axis for $x \in [-2, 2]$.

30. Let $f(x) = \cos x + \sin x$, $x \in [-\pi, \pi]$.

(a) Evaluate $\int_{-\pi}^{\pi} f(x)\, dx$.

(b) Sketch the graph of f and find the area of the region bounded by the graph and the x-axis for $x \in [-\pi, \pi]$.

C 31. Let $f(x) = x^3 - 4x + 2$.

(a) Evaluate $\int_{-2}^{3} f(x)\, dx$.

(b) Use a graphing utility to graph f and estimate the area bounded by the graph and the x-axis for $x \in [-2, 3]$. Use two decimal place accuracy in your approximations.

C 32. Let $f(x) = 3x^2 - 2\cos x$.

(a) Evaluate $\int_{-\pi/2}^{\pi/2} f(x)\, dx$.

(b) Use a graphing utility to graph f and estimate the area bounded by the graph and the x-axis for $x \in [-\pi/2, \pi/2]$. Use two decimal place accuracy in your approximations.

33. Let $f(x) = \begin{cases} x^2 + 1, & 0 \le x \le 1 \\ 3 - x, & 1 < x \le 3. \end{cases}$

Sketch the graph of f and find the area of the region bounded by the graph and the x-axis.

34. Let $f(x) = \begin{cases} 3\sqrt{x}, & 0 \le x \le 1 \\ 4 - x^2, & 1 < x \le 2. \end{cases}$

Sketch the graph of f and find the area of the region bounded by the graph and the x-axis.

35. Sketch the region bounded by the x-axis and the curves $y = \sin x$ and $y = \cos x$, $x \in [0, \pi/2]$, and find its area.

36. Sketch the region bounded by the curves $y = 1$ and $y = 1 + \cos x$ for $x \in [0, \pi]$, and find its area.

C 37. Use a graphing utility to sketch the region bounded by the curves $y = x^3 + 2x$ and $y = 3x + 1$ for $x \in [0, 2]$, and estimate its area. Use two decimal place accuracy in your approximations.

C 38. Use a graphing utility to sketch the region bounded by the curves $y = x^4 - 2x^2$ and $y = 4 - x^2$ for $x \in [-2, 2]$, and estimate its area. Use two decimal place accuracy in your approximations.

C 39. Let $f(x) = |x - 1|$ on $[-2, 3]$.

(a) Use a graphing utility to obtain the graph of f.

(b) Find $\int_{-2}^{3} f(x)\, dx$ using the graph found in part (a).

(c) Find a piecewise defined formula for f.

(d) Find $\int_{-2}^{3} f(x)\, dx$ using the fundamental theorem of calculus.

C 40. Let $f(x) = 3|x + 2| - |x - 3|$ on $[-5, 5]$.

(a) Use a graphing utility to obtain the graph of f.

(b) Find $\int_{-5}^{5} f(x)\, dx$ using the graph found in part (a).

(c) Find a piecewise defined formula for f.

(d) Find $\int_{-5}^{5} f(x)\, dx$ using the fundamental theorem of calculus.

■ 5.6 INDEFINITE INTEGRALS

We begin with a continuous function f. If F is an antiderivative for f on $[a, b]$, then

(1) $$\int_a^b f(x)\,dx = \Big[F(x)\Big]_a^b.$$

If C is a constant, then

$$\Big[F(x) + C\Big]_a^b = [F(b) + C] - [F(a) + C] = F(b) - F(a) = \Big[F(x)\Big]_a^b.$$

Thus we can replace (1) by writing

$$\int_a^b f(x)\,dx = \Big[F(x) + C\Big]_a^b.$$

If we have no particular interest in the interval $[a, b]$ but wish instead to emphasize that F is an antiderivative for f on *some* interval, then we can omit the a and the b and simply write

$$\int f(x)\,dx = F(x) + C.$$

Antiderivatives expressed in this manner are called *indefinite integrals*. The constant C is called the *constant of integration*; it is an *arbitrary* constant since it can be assigned any real value.

The indefinite integral of a function f is actually a *family* of functions; a specific member of the family is determined by assigning a particular value to the constant of integration. This family has the property that each of its members is an antiderivative of f and, conversely, every antiderivative of f is a member of the family. The latter fact follows from Theorem 4.2.5.

Thus, for example,

$$\int x^2\,dx = \tfrac{1}{3}x^3 + C \quad \text{and} \quad \int \sqrt{s}\,ds = \tfrac{2}{3}s^{3/2} + C.$$

The graphs of some specific members of these families are sketched in Figures 5.6.1 and 5.6.2, respectively.

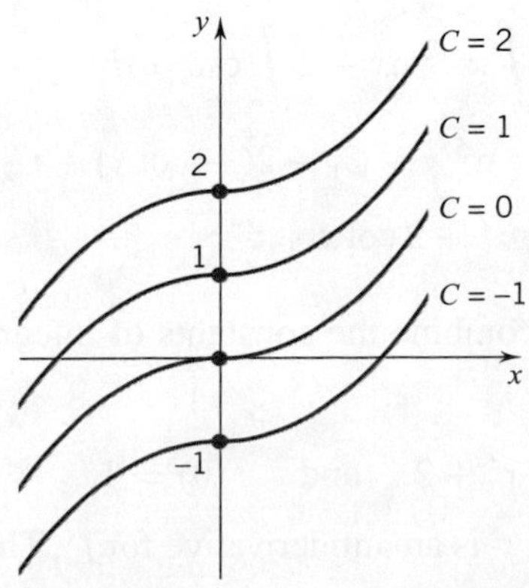

Figure 5.6.1

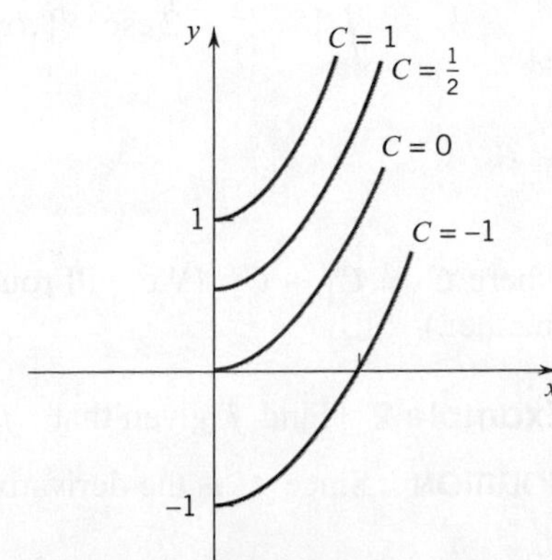

Figure 5.6.2

The Table of Antiderivatives (5.4.1) can be restated in terms of indefinite integrals:

■ Table 5.6.1

$$\int x^r\,dx = \frac{x^{r+1}}{r+1} + C \qquad (r \text{ rational, } r \neq -1)$$

$$\int \sin x\,dx = -\cos x + C \qquad \int \cos x\,dx = \sin x + C$$

$$\int \sec^2 x\,dx = \tan x + C \qquad \int \sec x \tan x\,dx = \sec x + C$$

$$\int \csc^2 x\,dx = -\cot x + C \qquad \int \csc x \cot x\,dx = -\csc x + C$$

The linearity properties of definite integrals (5.4.3), (5.4.4), and (5.4.5) also hold for indefinite integrals.

(5.6.1)
$$\int \alpha f(x)\,dx = \alpha \int f(x)\,dx, \quad \alpha \text{ a constant,}$$

(5.6.2)
$$\int [f(x) + g(x)]\,dx = \int f(x)\,dx + \int g(x)\,dx,$$

and, in general,

(5.6.3)
$$\int [\alpha f(x) + \beta g(x)]\,dx = \alpha \int f(x)\,dx + \beta \int g(x)\,dx,$$
where α and β are constants.

Example 1 Calculate $\int [5x^{3/2} - 2\csc^2 x]\,dx$.

SOLUTION

$$\begin{aligned}\int [5x^{3/2} - 2\csc^2 x]\,dx &= 5\int x^{3/2}\,dx - 2\int \csc^2 x\,dx \\ &= 5(\tfrac{2}{5})x^{5/2} + C_1 - 2(-\cot x) + C_2 \\ &= 2x^{5/2} + 2\cot x + C,\end{aligned}$$

where $C = C_1 + C_2$. (We will routinely combine the constants of integration in this manner.) □

Example 2 Find f given that $f'(x) = x^3 + 2$ and $f(0) = 1$.

SOLUTION Since f' is the derivative of f, f is an antiderivative for f'. Thus

$$f(x) = \int (x^3 + 2)\,dx = \tfrac{1}{4}x^4 + 2x + C$$

for some value of the constant C. To evaluate C we use the fact that $f(0) = 1$. Since

$$f(0) = 1 \quad \text{and} \quad f(0) = \tfrac{1}{4}(0)^4 + 2(0) + C = C,$$

we see that $C = 1$. Therefore

$$f(x) = \tfrac{1}{4}x^4 + 2x + 1.$$

Some of the members of the family of curves $F(x) = \frac{1}{4}x^4 + 2x + C$ are shown in Figure 5.6.3. The graph of f is appears in black. □

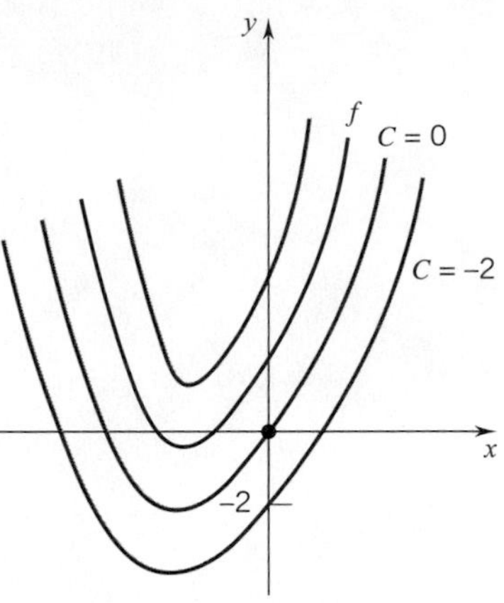

Figure 5.6.3

Example 3 Find f given that

$$f''(x) = 6x - 2, \quad f'(1) = -5, \quad \text{and} \quad f(1) = 3.$$

SOLUTION First we get f' by integrating f'':

$$f'(x) = \int (6x - 2)\,dx = 3x^2 - 2x + C.$$

Since $\qquad f'(1) = -5 \quad \text{and} \quad f'(1) = 3(1)^2 - 2(1) + C = 1 + C,$

we have $\qquad -5 = 1 + C \quad \text{and thus} \quad C = -6.$

Therefore $\qquad f'(x) = 3x^2 - 2x - 6.$

Now we get to f by integrating f':

$$f(x) = \int (3x^2 - 2x - 6)\,dx = x^3 - x^2 - 6x + K.$$

(We are writing the constant of integration as K because we used C before and it would be confusing to have C with two different values in the same problem.) Since

$$f(1) = 3 \quad \text{and} \quad f(1) = (1)^3 - (1)^2 - 6(1) + K = -6 + K,$$

we have $\qquad 3 = -6 + K \quad \text{and thus} \quad K = 9.$

Therefore $\qquad f(x) = x^3 - x^2 - 6x + 9.$ □

Some Motion Examples

Example 4 An object moves along a coordinate line with velocity

$$v(t) = 2 - 3t + t^2 \quad \text{units per second.}$$

Its initial position (position at time $t = 0$) is 2 units to the right of the origin. Find the position of the object 4 seconds later.

SOLUTION Let $x(t)$ be the position (coordinate) of the object at time t. We are given that $x(0) = 2$. Since $x'(t) = v(t)$,

$$x(t) = \int v(t)\,dt = \int (2 - 3t + t^2)\,dt = 2t - \tfrac{3}{2}t^2 + \tfrac{1}{3}t^3 + C.$$

Since $x(0) = 2$ and $x(0) = 2(0) - \frac{3}{2}(0)^2 + \frac{1}{3}(0)^3 + C = C$, we have $C = 2$ and

$$x(t) = 2t - \tfrac{3}{2}t^2 + \tfrac{1}{3}t^3 + 2.$$

The position of the object at time $t = 4$ is the value of this function at $t = 4$:

$$x(4) = 2(4) - \tfrac{3}{2}(4)^2 + \tfrac{1}{3}(4)^3 + 2 = 7\tfrac{1}{3}.$$

At the end of 4 seconds the object is $7\frac{1}{3}$ units to the right of the origin.

The motion of the object is represented schematically in Figure 5.6.4. □

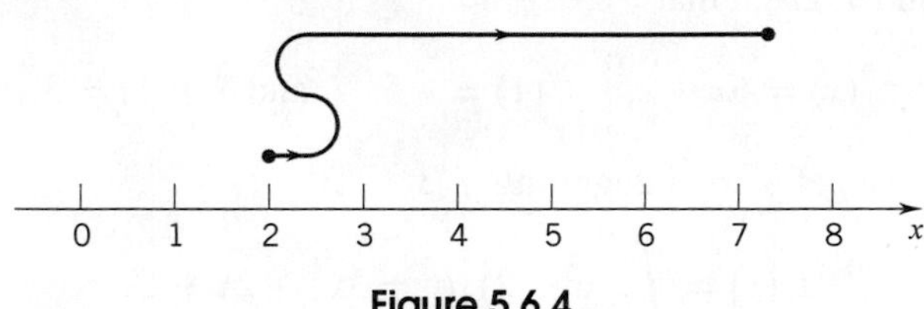

Figure 5.6.4

Recall that speed is the absolute value of velocity (see Section 3.4):

$$\text{Speed at time } t = |v(t)|,$$

and the integral of the speed function gives the distance traveled (see Section 5.1):

(5.6.4)
$$\int_a^b |v(t)|\, dt = \text{distance traveled from time } t = a \text{ to time } t = b.$$

Example 5 An object moves along the x-axis with acceleration $a(t) = 2t - 2$ units per second per second. Its initial position (position at time $t = 0$) is 5 units to the right of the origin. One second later the object is moving left at the rate of 4 units per second.

(a) Find the position of the object at time $t = 4$ seconds.
(b) How far does the object travel during these 4 seconds?

SOLUTION **(a)** Let $x(t)$ and $v(t)$ denote the position and velocity of the object at time t. We are given that $x(0) = 5$ and $v(1) = -4$. Since $v'(t) = a(t)$,

$$v(t) = \int a(t)\, dt = \int (2t - 2)\, dt = t^2 - 2t + C.$$

Since $\quad v(1) = -4 \quad \text{and} \quad v(1) = (1)^2 - 2(1) + C = -1 + C,$

we have $C = -3$ and therefore

$$v(t) = t^2 - 2t - 3.$$

Since $x'(t) = v(t)$,

$$x(t) = \int v(t)\, dt = \int (t^2 - 2t - 3)\, dt = \tfrac{1}{3}t^3 - t^2 - 3t + K.$$

Since $\quad x(0) = 5 \quad$ and $\quad x(0) = \frac{1}{3}(0)^3 - (0)^2 - 3(0) + K = K,$

we have $K = 5$. Therefore

$$x(t) = \tfrac{1}{3}t^3 - t^2 - 3t + 5.$$

As you can check, $x(4) = -\frac{5}{3}$. At time $t = 4$ the object is $\frac{5}{3}$ units to the left of the origin.

(b) The distance traveled from time $t = 0$ to $t = 4$ is given by the integral

$$s = \int_0^4 |v(t)|\,dt = \int_0^4 |t^2 - 2t - 3|\,dt.$$

To evaluate this integral we first remove the absolute value sign. As you can verify,

$$|t^2 - 2t - 3| = \begin{cases} -(t^2 - 2t - 3), & 0 \le t < 3 \\ \quad t^2 - 2t - 3, & 3 \le t \le 4. \end{cases}$$

Thus
$$s = \int_0^3 (3 + 2t - t^2)\,dt + \int_3^4 (t^2 - 2t - 3)\,dt$$

$$= \left[3t + t^2 - \tfrac{1}{3}t^3\right]_0^3 + \left[\tfrac{1}{3}t^3 - t^2 - 3t\right]_3^4 = \tfrac{34}{3}.$$

During the 4 seconds the object travels a distance of $\frac{34}{3}$ units.

The motion of the object is represented schematically in Figure 5.6.5. ☐

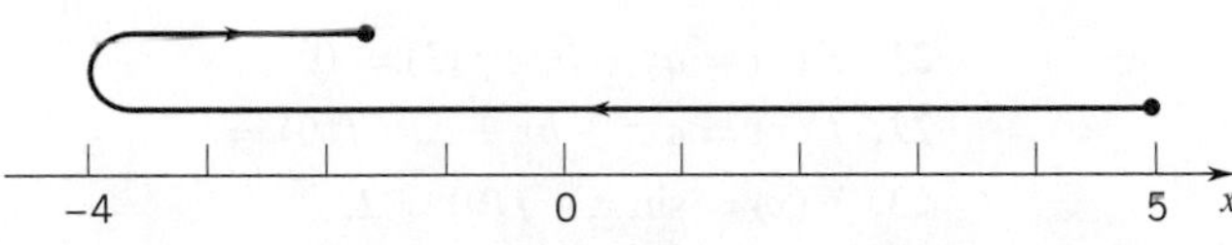

Figure 5.6.5

QUESTION The object in Example 5 leaves $x = 5$ at time $t = 0$ and arrives at $x = -\frac{5}{3}$ at time $t = 4$. The separation between $x = 5$ and $x = -\frac{5}{3}$ is only $|5 - (-\frac{5}{3})| = \frac{20}{3}$. How is it possible that the object travels a distance of $\frac{34}{3}$ units?

ANSWER The object does not maintain a fixed direction. It changes direction at time $t = 3$. You can see this by noting that the velocity function

$$v(t) = t^2 - 2t - 3 = (t - 3)(t + 1)$$

changes signs at $t = 3$. ☐

Example 6 Find the equation of motion for an object that moves along a straight line with constant acceleration a from an initial position x_0 with initial velocity v_0.

SOLUTION Call the line of the motion the x-axis. Here $a(t) = a$ at all time t. To find the velocity we integrate the acceleration

$$v(t) = \int a\,dt = at + C.$$

The constant C is the initial velocity v_0:

$$v_0 = v(0) = a \cdot 0 + C = C.$$

We see therefore that

$$v(t) = at + v_0.$$

To find the position function we integrate the velocity:

$$x(t) = \int v(t)\,dt = \int (at + v_0)\,dt = \tfrac{1}{2}at^2 + v_0 t + K.$$

The constant K is the initial position x_0:

$$x_0 = x(0) = \tfrac{1}{2}a \cdot 0^2 + v_0 \cdot 0 + K = K.$$

The equation of motion can be written

(5.6.5) $$\boxed{x(t) = \tfrac{1}{2}at^2 + v_0 t + x_0.}\ \dagger \quad \square$$

† In the case of a free-falling body, $a = -g$ and we have Galileo's equation for free fall. See (3.4.5).

EXERCISES 5.6

Calculate the indefinite integral.

1. $\int \frac{dx}{x^4}.$

2. $\int (x-1)^2\,dx.$

3. $\int (ax+b)\,dx.$

4. $\int (ax^2+b)\,dx.$

5. $\int \frac{dx}{\sqrt{1+x}}.$

6. $\int \left(\frac{x^3+1}{x^5}\right) dx.$

7. $\int \left(\frac{x^3-1}{x^2}\right) dx.$

8. $\int \left(\sqrt{x} - \frac{1}{\sqrt{x}}\right) dx.$

9. $\int (t-a)(t-b)\,dt.$

10. $\int (t^2-a)(t^2-b)\,dt.$

11. $\int \frac{(t^2-a)(t^2-b)}{\sqrt{t}}\,dt.$

12. $\int (2-\sqrt{x})(2+\sqrt{x})\,dx.$

13. $\int g(x)g'(x)\,dx.$

14. $\int \sin x \cos x\,dx.$

15. $\int \tan x \sec^2 x\,dx.$

16. $\int \frac{g'(x)}{[g(x)]^2}\,dx.$

17. $\int \frac{4}{(4x+1)^2}\,dx.$

18. $\int \frac{3x^2}{(x^3+1)^2}\,dx.$

Find f from the information given.

19. $f'(x) = 2x - 1, \quad f(3) = 4.$

20. $f'(x) = 3 - 4x, \quad f(1) = 6.$

21. $f'(x) = ax + b, \quad f(2) = 0.$

22. $f'(x) = ax^2 + bx + c, \quad f(0) = 0.$

23. $f'(x) = \sin x, \quad f(0) = 2.$

24. $f'(x) = \cos x, \quad f(\pi) = 3.$

25. $f''(x) = 6x - 2, \quad f'(0) = 1, \quad f(0) = 2.$

26. $f''(x) = -12x^2, \quad f'(0) = 1, \quad f(0) = 2.$

27. $f''(x) = x^2 - x, \quad f'(1) = 0, \quad f(1) = 2.$

28. $f''(x) = 1 - x, \quad f'(2) = 1, \quad f(2) = 0.$

29. $f''(x) = \cos x, \quad f'(0) = 1, \quad f(0) = 2.$

30. $f''(x) = \sin x, \quad f'(0) = -2, \quad f(0) = 1.$

31. $f''(x) = 2x - 3, \quad f(2) = -1, \quad f(0) = 3.$

32. $f''(x) = 5 - 4x, \quad f(1) = 1, \quad f(0) = -2.$

33. Compare $\frac{d}{dx}\left[\int f(x)\,dx\right]$ to $\int \frac{d}{dx}[f(x)]\,dx.$

34. Calculate

$$\int [f(x)g''(x) - g(x)f''(x)]\,dx.$$

35. An object moves along a coordinate line with velocity $v(t) = 6t^2 - 6$ units per second. Its initial position (position at time $t = 0$) is 2 units to the left of the origin. (a) Find the position of the object 3 seconds later. (b) Find the total distance traveled by the object during those 3 seconds.

36. An object moves along a coordinate line with acceleration $a(t) = (t+2)^3$ units per second per second. (a) Find the velocity function given that the initial velocity is 3 units per second. (b) Find the position function given that the initial velocity is 3 units per second and the initial position is the origin.

37. An object moves along a coordinate line with acceleration $a(t) = (t+1)^{-1/2}$ units per second per second. (a) Find the velocity function given that the initial velocity is 1 unit per second. (b) Find the position function given that the initial velocity is 1 unit per second and the initial position is the origin.

38. An object moves along a coordinate line with velocity $v(t) = t(1-t)$ units per second. Its initial position is 2 units to the left of the origin. (a) Find the position of the object 10 seconds later. (b) Find the total distance traveled by the object during those 10 seconds.

39. A car traveling at 60 mph decelerates at 20 feet per second per second. (a) How long does it take for the car to come to a complete stop? (b) What distance is required to bring the car to a complete stop?

40. An object moves along the x-axis with constant acceleration. Express the position $x(t)$ in terms of the initial position x_0, the initial velocity v_0, the velocity $v(t)$, and the elapsed time t.

41. An object moves along the x-axis with constant acceleration a. Verify that

$$[v(t)]^2 = v_0^2 + 2a[x(t) - x_0].$$

42. A bobsled moving at 60 mph decelerates at a constant rate to 40 mph over a distance of 264 feet and continues to decelerate at that same rate until it comes to a full stop. (a) What is the acceleration of the sled in feet per second per second? (b) How long does it take to reduce the speed to 40 mph? (c) How long does it take to bring the sled to a complete stop from 60 mph? (d) Over what distance does the sled come to a complete stop from 60 mph?

43. In the AB-run, minicars start from a standstill at point A, race along a straight track, and come to a full stop at point B one-half mile away. Given that the cars can accelerate uniformly to a maximum speed of 60 mph in 20 seconds and can brake at a maximum rate of 22 feet per second per second , what is the best possible time for the completion of the AB-run?

In Exercises 44–46, find the general law of motion of an object that moves in a straight line with acceleration $a(t)$. Write x_0 for initial position and v_0 for initial velocity.

44. $a(t) = \sin t$.

45. $a(t) = 2A + 6Bt$.

46. $a(t) = \cos t$.

47. As a particle moves about the plane, its x-coordinate changes at the rate of $t^2 - 5$ units per second and its y-coordinate changes at the rate of $3t$ units per second. If the particle is at the point (4, 2) when $t = 2$ seconds, where is the particle 4 seconds later?

48. As a particle moves about the plane, its x-coordinate changes at the rate of $t - 2$ units per second and its y-coordinate changes at the rate of $\sqrt{t}$ units per second. If the particle is at the point (3, 1) when $t = 4$ seconds, where is the particle 5 second later?

49. A particle moves along the x-axis with velocity $v(t) = At + B$. Determine A and B given that the initial velocity of the particle is 2 units per second and the position of the particle after 2 seconds of motion is 1 unit to the left of the initial position.

50. A particle moves along the x-axis with velocity $v(t) = At^2 + 1$. Determine A given that $x(1) = x(0)$. Compute the total distance traveled by the particle during the first second.

51. An object moves along a coordinate line with velocity $v(t) = \sin t$ units per second. The object passes through the origin at time $t = \pi/6$ seconds. When is the next time: (a) that the object passes through the origin? (b) that the object passes through the origin moving from left to right?

52. Repeat Exercise 51 with $v(t) = \cos t$.

53. An automobile with varying velocity $v(t)$ moves in a fixed direction for 5 minutes and covers a distance of 4 miles. What theorem would you invoke to argue that for at least one instant the speedometer must have read 48 miles per hour?

54. A speeding motorcyclist sees his way blocked by a haywagon some distance s ahead and slams on his brakes. Given that the brakes impart to the motorcycle a constant negative acceleration a and that the haywagon is moving with speed v_1 in the same direction as the motorcycle, show that the motorcyclist can avoid collision only if he is traveling at a speed less than $v_1 + \sqrt{2|a|s}$.

55. Find the velocity $v(t)$ given that $a(t) = 2[v(t)]^2$ and $v_0 \neq 0$.

(C) In Exercises 56 and 57, use a CAS to find and compare

$$\frac{d}{dx}\left(\int f(x)\,dx\right) \quad \text{and} \quad \int \frac{d}{dx}[f(x)]\,dx.$$

56. $f(x) = \dfrac{x^2 - x^3 + x^4}{\sqrt{x}}$.

57. $f(x) = \cos x - 2\sin x$.

(C) Use a CAS to find f from the information given.

58. $f'(x) = \dfrac{\sqrt{x}+1}{\sqrt{x}}; \quad f(4) = 2.$

59. $f'(x) = \cos x - 2\sin x; \quad f(\pi/2) = 2.$

60. $f''(x) = 3\sin x + 2\cos x; \quad f(0) = 0,\ f'(0) = 0.$

61. $f''(x) = 5 - 3x + x^2; \quad f(0) = -3,\ f'(0) = 4.$

■ 5.7 THE u-SUBSTITUTION; CHANGE OF VARIABLES

When we differentiate a composite function, we do so by the chain rule. In trying to calculate an indefinite integral, we are often called on to apply the chain rule in reverse. This process can be facilitated by making what we call a "u-substitution."

An integral of the form

$$\int f(g(x))\, g'(x)\, dx$$

can be written

$$\int f(u)\, du$$

by setting

$$u = g(x), \quad du = g'(x)\, dx.\dagger$$

If F is an antiderivative for f, then

$$\frac{d}{dx}[F(g(x))] = F'(g(x))\, g'(x) = f(g(x))\, g'(x)$$

(by the chain rule; $F' = f$)

and so

$$\int f(g(x))\, g'(x)\, dx = \int \frac{d}{dx}[F(g(x))]\, dx = F(g(x)) + C$$

Note that we can obtain the same result by calculating

$$\int f(u)\, du$$

and then substituting $g(x)$ back in for u:

$$\int f(u)\, du = F(u) + C = F(g(x)) + C.$$

Example 1 Calculate $\int (x^2 - 1)^4 2x\, dx$ and then check the result by differentiation.

SOLUTION Since $x^2 - 1$ is raised to a power and $2x$ is the derivative of $x^2 - 1$, set

$$u = x^2 - 1 \quad \text{so that} \quad du = 2x\, dx.$$

Then

$$\int (x^2 - 1)^4\, 2x\, dx = \int u^4\, du = \tfrac{1}{5}u^5 + C = \tfrac{1}{5}(x^2 - 1)^5 + C.$$

† Think of $du = g'(x)\, dx$ as a "formal differential," writing dx for h. See Section 3.9.

CHECKING

$$\frac{d}{dx}\left[\tfrac{1}{5}(x^2-1)^5+C\right]=\tfrac{5}{5}(x^2-1)^4\frac{d}{dx}(x^2-1)\overset{\checkmark}{=}(x^2-1)^4\,2x. \quad \square$$

Example 2 Calculate $\int 3x^2\cos(x^3+2)\,dx$, and then check the result by differentiation.

SOLUTION Since $3x^2$ is the derivative of x^3+2, set

$$u=x^3+2 \quad \text{so that} \quad du=3x^2\,dx.$$

Then $$\int 3x^2\cos(x^3+2)\,dx=\int \cos u\,du=\sin u+C=\sin(x^3+2)+C.$$

CHECKING

$$\frac{d}{dx}\left[\sin(x^3+2)+C\right]=\cos(x^3+2)\frac{d}{dx}[x^3+2]\overset{\checkmark}{=}3x^2\cos(x^3+2). \quad \square$$

Example 3 Calculate $\int \sin x\cos x\,dx$, and then check the result by differentiation.

SOLUTION Set $u=\sin x, \quad du=\cos x\,dx.$

Then $$\int \sin x\cos x\,dx=\int u\,du=\tfrac{1}{2}u^2+C=\tfrac{1}{2}\sin^2 x+C.$$

CHECKING

$$\frac{d}{dx}\left[\tfrac{1}{2}\sin^2 x+C\right]=\tfrac{1}{2}(2)\sin x\frac{d}{dx}[\sin x]\overset{\checkmark}{=}\sin x\cos x.$$

ALTERNATE SOLUTION Since $\sin x\cos x=\frac{1}{2}\sin 2x$,

$$\int \sin x\cos x\,dx=\tfrac{1}{2}\int \sin 2x\,dx.$$

Set $$u=2x, \quad du=2\,dx \quad \text{or} \quad dx=\tfrac{1}{2}\,du.$$

Then $$\tfrac{1}{2}\int \sin 2x\,dx=\tfrac{1}{2}\int \sin u\left(\tfrac{1}{2}\,du\right)=\tfrac{1}{4}\int \sin u\,du=-\tfrac{1}{4}\cos u+C$$
$$=-\tfrac{1}{4}\cos 2x+C.$$

We leave it to you to check this result and to reconcile it with the result given in the first solution. □

Example 4 Calculate $\int \frac{dx}{(3+5x)^2}$, and then check the result by differentiation.

SOLUTION Set $u=3+5x$ so that $du=5\,dx$.

Then

$$\frac{dx}{(3+5x)^2} = \frac{\frac{1}{5}\,du}{u^2} = \frac{1}{5}\frac{du}{u^2}$$

and

$$\int \frac{dx}{(3+5x^2)} = \tfrac{1}{5}\int \frac{du}{u^2} = \tfrac{1}{5}\int u^{-2}\,du = \frac{1}{5}\frac{u^{-1}}{-1} + C\dagger$$

$$= -\frac{1}{5u} + C = -\frac{1}{5(3+5x)} + C.$$

CHECKING

$$\frac{d}{dx}\left[-\frac{1}{5(3+5x)} + C\right] = \frac{d}{dx}\left[-\tfrac{1}{5}(3+5x)^{-1}\right]$$

$$= \left(-\tfrac{1}{5}\right)(-1)(3+5x)^{-2}(5) \overset{\checkmark}{=} \frac{1}{(3+5x)^2}.\quad \square$$

In the remaining examples we leave the checking to you.

Example 5 Calculate $\int x^2\sqrt{4+x^3}\,dx$.

SOLUTION Set $u = 4 + x^3$, $du = 3x^2\,dx$.
Then

$$x^2\sqrt{4+x^3}\,dx = \underbrace{(4+x^3)^{1/2}}_{u^{1/2}}\underbrace{x^2\,dx}_{\frac{1}{3}\,du} = \tfrac{1}{3}u^{1/2}du$$

and

$$\int x^2\sqrt{4+x^3}\,dx = \tfrac{1}{3}\int u^{1/2}\,du = \tfrac{2}{9}u^{3/2} + C = \tfrac{2}{9}(4+x^3)^{3/2} + C.\quad \square$$

As suggested by the preceding examples, the key step in making a u-substitution is to find a substitution $u = g(x)$ such that the expression $du = g'(x)\,dx$ appears in the original integral (at least up to a constant factor) and the new integral

$$\int f(u)\,du$$

is easier to calculate than the original integral. Often the form of the original integral will suggest a good choice for u. The following table is the analog of Table 5.6.1. It is based on the chain rule and it indicates the transformation of the original integral into the form $\int f(u)\,du$ of 5.6.1.

† We could write

$$\tfrac{1}{5}\int u^{-2}\,du = \tfrac{1}{5}[u^{-1} + C] = -\tfrac{1}{5u} - \tfrac{C}{5}.$$

But, since C is arbitrary, $-C/5$ is arbitrary, and we can therefore write C instead.

■ **Table 5.7.1**

Original Integral	$u = g(x)$, $du = g'(x)dx$	New Integral
$\int [g(x)]^r g'(x)\,dx$	$\to$	$\int u^r\,du = \dfrac{u^{r+1}}{r+1} + C = \dfrac{[g(x)]^{r+1}}{r+1} + C \quad (r \neq -1)$
$\int \sin\,[g(x)]\,g'(x)\,dx$	$\to$	$\int \sin u\,du = -\cos u + C = -\cos\,[g(x)] + C$
$\int \cos\,[g(x)]\,g'(x)\,dx$	$\to$	$\int \cos u\,du = \sin u + C = \sin\,[g(x)] + C$
$\int \sec^2\,[g(x)]\,g'(x)\,dx$	$\to$	$\int \sec^2 u\,du = \tan u + C = \tan\,[g(x)] + C$
$\int \sec\,[g(x)]\tan\,[g(x)]g'(x)\,dx$	$\to$	$\int \sec u\,\tan u\,du = \sec u + C = \sec\,[g(x)] + C$
$\int \csc^2\,[g(x)]\,g'(x)\,dx$	$\to$	$\int \csc^2 u\,du = -\cot u + C = -\cot\,[g(x)] + C$
$\int \csc\,[g(x)]\cot\,[g(x)]\,g'(x)\,dx$	$\to$	$\int \csc u \cot u\,du = -\csc u + C = -\csc\,[g(x)] + C$

Here are more examples.

Example 6 Find $\displaystyle\int 2x^3 \sec^2 (x^4 + 1)\,dx.$

SOLUTION Set $u = x^4 + 1, \quad du = 4x^3\,dx.$
Then

$$2x^3 \sec^2 (x^4 + 1)\,dx = 2\,\underbrace{\sec^2 (x^4 + 1)}_{\sec^2 u}\underbrace{x^3\,dx}_{\frac{1}{4}du} = \tfrac{1}{2}\sec^2 u\,du$$

and

$$\int 2x^3 \sec^2 (x^4 + 1)\,dx = \tfrac{1}{2}\int \sec^2 u\,du = \tfrac{1}{2}\tan u + C = \tfrac{1}{2}\tan (x^4 + 1) + C. \quad \square$$

Example 7 Find $\displaystyle\int \sec^3 x \tan x\,dx.$

SOLUTION We can write $\sec^3 x \tan x\,dx$ as $\sec^2 x \sec x \tan x\,dx$. Setting

$$u = \sec x, \quad du = \sec x \tan x\,dx,$$

we have

$$\sec^3 x \tan x\,dx = \underbrace{\sec^2 x}_{u^2}\underbrace{(\sec x \tan x)\,dx}_{du} = u^2\,du.$$

Therefore $$\int \sec^3 x \tan x\,dx = \int u^2\,du = \tfrac{1}{3}u^3 + C = \tfrac{1}{3}\sec^3 x + C. \quad \square$$

Example 8 Evaluate the definite integral $\displaystyle\int_0^2 (x^2 - 1)(x^3 - 3x + 2)^3\,dx.$

SOLUTION We need to find an antiderivative for the integrand. The indefinite integral

$$\int (x^2-1)(x^3-3x+2)^3\,dx$$

gives the set of all antiderivatives, and so we will calculate this first. Set

$$u = x^3 - 3x + 2, \quad du = (3x^2-3)\,dx = 3(x^2-1)\,dx.$$

Then

$$(x^2-1)(x^3-3x+2)^3 dx = \underbrace{(x^3-3x+2)^3}_{u^3}\,\underbrace{(x^2-1)\,dx}_{\frac{1}{3}du} = \tfrac{1}{3}u^3\,du.$$

It follows that

$$\int (x^2-1)(x^3-3x+2)^3\,dx = \tfrac{1}{3}\int u^3\,du = \tfrac{1}{12}u^4 + C = \tfrac{1}{12}(x^3-3x+2)^4 + C.$$

To evaluate the given definite integral, we need only one antiderivative, and so we will choose the one with $C = 0$. This gives

$$\int_0^2 (x^2-1)(x^3-3x+2)^3\,dx = \left[\tfrac{1}{12}(x^3-3x+2)^4\right]_0^2 = 20. \quad \square$$

Remark Thus far, all of the integrals that we have calculated using a u-substitution can be calculated without substitution. All that is required is a good sense of the chain rule. For example:

$\int (x^2-1)^4\,2x\,dx$ (Example 1). The derivative of x^2-1 is $2x$. Therefore,

$$\int (x^2-1)^4\,2x\,dx = \int (x^2-1)^4 \frac{d}{dx}(x^2-1)\,dx = \tfrac{1}{5}(x^2-1)^5 + C.$$

$\int 3x^2\cos(x^3+2)\,dx$ (Example 2). The derivative of x^3+2 is $3x^2$. Thus,

$$\int 3x^2\cos(x^3+2)\,dx = \int \cos(x^3+2)\frac{d}{dx}(x^3+2)\,dx = \sin(x^3+2) + C.$$

$\int x^2\sqrt{4+x^3}\,dx$ (Example 5). The derivative of $4+x^3$ is $3x^2$. Therefore,

$$\frac{d}{dx}(4+x^3) = 3x^2 \quad \text{and} \quad x^2 = \tfrac{1}{3}\frac{d}{dx}(4+x^3).$$

Thus,

$$\begin{aligned}\int x^2\sqrt{4+x^3}\,dx &= \tfrac{1}{3}\int (4+x^3)^{1/2}\frac{d}{dx}(4+x^3)\,dx \\ &= \tfrac{1}{3}\cdot\tfrac{2}{3}(4+x^3)^{3/2} + C = \tfrac{2}{9}(4+x^3)^{3/2} + C.\end{aligned}$$

$\int \sec^3 x \tan x\, dx$ (Example 7). Write the integrand as

$$\sec^2 x(\sec x \tan x) = \sec^2 x \frac{d}{dx}(\sec x).$$

Then $$\int \sec^2 x \tan x\, dx = \int \sec^2 x \frac{d}{dx}(\sec x)\, dx = \tfrac{1}{3}\sec^3 x + C.$$

There is nothing wrong with calculating integrals by substitution. All we are saying is that, with some experience, you will be able to calculate many integrals without formally using it. □

Substitution in Definite Integrals

You can calculate a definite integral that entails a change of variables by the method we used in Example 8 or, if you prefer, by employing the following formula:

(5.7.1) $$\int_a^b f(g(x))\, g'(x)\, dx = \int_{g(a)}^{g(b)} f(u)\, du.$$

This formula, called the *change of variables formula,* holds provided that f and g' are both continuous. More precisely, g' must be continuous on $[a, b]$ and f must be continuous on the set of values taken on by g.

PROOF Let F be an antiderivative for f. Then $F' = f$ and

$$\begin{aligned}\int_a^b f(g(x))\, g'(x)\, dx &= \int_a^b F'(g(x))\, g'(x)\, dx \\ &= \Big[F(g(x))\Big]_a^b = F(g(b)) - F(g(a)) = \int_{g(a)}^{g(b)} f(u)\, du.\end{aligned}$$ □

We redo Example 8, this time using the change of variables formula.

Example 9 Evaluate $\displaystyle\int_0^2 (x^2 - 1)(x^3 - 3x + 2)^3 dx.$

SOLUTION As before, set $u = x^3 - 3x + 2, \quad du = 3(x^2 - 1)\, dx.$

Then

$$(x^2 - 1)(x^3 - 3x + 2)^3 dx = \tfrac{1}{3}u^3 du.$$

At $x = 0,\ u = 2$. At $x = 2,\ u = 4$. Therefore,

$$\begin{aligned}\int_0^2 (x^2 - 1)(x^3 - 3x + 2)^3 dx &= \tfrac{1}{3}\int_2^4 u^3\, du \\ &= \Big[\tfrac{1}{12}u^4\Big]_2^4 = \tfrac{1}{12}(4)^4 - \tfrac{1}{12}(2)^4 = 20.\end{aligned}$$ □

Example 10 Evaluate $\displaystyle\int_0^{1/2} \cos^3 \pi x \sin \pi x\, dx.$

SOLUTION Set $u = \cos \pi x, \quad du = -\pi \sin \pi x \, dx.$

Then

$$\cos^3 \pi x \sin \pi x \, dx = \underbrace{\cos^3 \pi x}_{u^3} \underbrace{\sin \pi x \, dx}_{-\frac{1}{\pi} du} = -\tfrac{1}{\pi} u^3 \, du.$$

At $x = 0$, $u = 1$. At $x = 1/2$, $u = 0$. Therefore,

$$\int_0^{1/2} \cos^3 \pi x \sin \pi x \, dx = -\tfrac{1}{\pi} \int_1^0 u^3 \, du = \tfrac{1}{\pi} \int_0^1 u^3 \, du = \tfrac{1}{\pi} \left[\tfrac{1}{4} u^4 \right]_0^1 = \tfrac{1}{4\pi}. \quad \square$$

Up to this point our focus has been on substitutions connected with the chain rule. However, as we shall illustrate here and later in Chapter 8, "substitution" is a very general method of integration. Choosing the proper substitution is a matter of practice and experience. At this early stage, we will provide some hints and a little guidance.

Example 11 Calculate $\int x(x-3)^5 \, dx$.

SOLUTION Set $u = x - 3$. Then $du = dx$ and $x = u + 3$.

Now

$$x(x-3)^5 \, dx = (u+3)u^5 \, du = (u^6 + 3u^5) \, du$$

and

$$\begin{aligned} \int x(x-3)^5 \, dx &= \int (u^6 + 3u^5) \, du \\ &= \tfrac{1}{7} u^7 + \tfrac{1}{2} u^6 + C = \tfrac{1}{7}(x-3)^7 + \tfrac{1}{2}(x-3)^6 + C. \quad \square \end{aligned}$$

Example 12 Evaluate $\int_0^{\sqrt{3}} x^5 \sqrt{x^2+1} \, dx$.

SOLUTION Set $u = x^2 + 1$. Then $du = 2x \, dx$ and $x^2 = u - 1$.

Now

$$x^5 \sqrt{x^2+1} \, dx = \underbrace{x^4}_{(u-1)^2} \underbrace{\sqrt{x^2+1}}_{\sqrt{u}} \underbrace{x \, dx}_{\frac{1}{2} du} = \tfrac{1}{2}(u-1)^2 \sqrt{u} \, du.$$

At $x = 0$, $u = 1$. At $x = \sqrt{3}$, $u = 4$. Thus

$$\begin{aligned} \int_0^{\sqrt{3}} x^5 \sqrt{x^2+1} \, dx &= \tfrac{1}{2} \int_1^4 (u-1)^2 \sqrt{u} \, du \\ &= \tfrac{1}{2} \int_1^4 (u^{5/2} - 2u^{3/2} + u^{1/2}) \, du \\ &= \tfrac{1}{2} \left[\tfrac{2}{7} u^{7/2} - \tfrac{4}{5} u^{5/2} + \tfrac{2}{3} u^{3/2} \right]_1^4 \\ &= \left[u^{3/2} \left(\tfrac{1}{7} u^2 - \tfrac{2}{5} u + \tfrac{1}{3} \right) \right]_1^4 = \tfrac{848}{105}. \quad \square \end{aligned}$$

EXERCISES 5.7

Calculate the integral by a u-substitution.

1. $\displaystyle\int \frac{dx}{(2-3x)^2}$.

2. $\displaystyle\int \frac{dx}{\sqrt{2x+1}}$.

3. $\displaystyle\int \sqrt{2x+1}\,dx$.

4. $\displaystyle\int \sqrt{ax+b}\,dx$.

5. $\displaystyle\int (ax+b)^{3/4}\,dx$.

6. $\displaystyle\int 2ax(ax^2+b)^4\,dx$.

7. $\displaystyle\int \frac{t}{(4t^2+9)^2}\,dt$.

8. $\displaystyle\int \frac{3t}{(t^2+1)^2}\,dt$.

9. $\displaystyle\int x^2(1+x^3)^{1/4}dx$.

10. $\displaystyle\int x^{n-1}\sqrt{a+bx^n}\,dx$.

11. $\displaystyle\int \frac{s}{(1+s^2)^3}\,ds$.

12. $\displaystyle\int \frac{2s}{\sqrt[3]{6-5s^2}}ds$.

13. $\displaystyle\int \frac{x}{\sqrt{x^2+1}}\,dx$.

14. $\displaystyle\int \frac{x^2}{(1-x^3)^{2/3}}\,dx$.

15. $\displaystyle\int 5x(x^2+1)^{-3}\,dx$.

16. $\displaystyle\int 2x^3(1-x^4)^{-1/4}\,dx$.

17. $\displaystyle\int x^{-3/4}(x^{1/4}+1)^{-2}\,dx$.

18. $\displaystyle\int \frac{4x+6}{\sqrt{x^2+3x+1}}\,dx$.

19. $\displaystyle\int \frac{b^3x^3}{\sqrt{1-a^4x^4}}\,dx$.

20. $\displaystyle\int \frac{x^{n-1}}{\sqrt{a+bx^n}}\,dx$.

Evaluate by a u-substitution.

21. $\displaystyle\int_0^1 x(x^2+1)^3dx$.

22. $\displaystyle\int_{-1}^0 3x^2(4+2x^3)^2\,dx$.

23. $\displaystyle\int_{-1}^1 \frac{r}{(1+r^2)^4}\,dr$.

24. $\displaystyle\int_0^3 \frac{r}{\sqrt{r^2+16}}\,dr$.

25. $\displaystyle\int_0^a y\sqrt{a^2-y^2}dy$.

26. $\displaystyle\int_{-a}^0 y^2\left(1-\frac{y^3}{a^2}\right)^{-2} dy$.

In Exercises 27–30, f is nonnegative on the indicated interval. Find the area of the region bounded by the graph of f and the x-axis.

27. $f(x)=x\sqrt{2x^2+1},\quad x\in[0,2]$.

28. $f(x)=\dfrac{x}{(2x^2+1)^2},\quad x\in[0,2]$.

29. $f(x)=x^{-3}(1+x^{-2})^{-3},\quad x\in[1,2]$.

30. $f(x)=\dfrac{2x+5}{(x+2)^2(x+3)^2},\quad x\in[0,1]$.

Calculate by a substitution.

31. $\displaystyle\int x\sqrt{x+1}\,dx$. [set $u=x+1$]

32. $\displaystyle\int 2x\sqrt{x-1}\,dx$. [set $u=x-1$]

33. $\displaystyle\int x\sqrt{2x-1}\,dx$.

34. $\displaystyle\int t(2t+3)^8\,dt$.

35. $\displaystyle\int_0^1 \frac{x+3}{\sqrt{x+1}}\,dx$.

36. $\displaystyle\int_{-1}^0 x^3(x^2+1)^6\,dx$.

37. Find an equation $y=f(x)$ for the curve that passes through the point (0, 1) and has slope

$$\frac{dy}{dx}=x\sqrt{x^2+1}.$$

38. Find an equation $y=f(x)$ for the curve that passes through the point $(4,\frac{1}{3})$ and has slope

$$\frac{dy}{dx}=-\frac{1}{2\sqrt{x}(1+\sqrt{x})^2}.$$

Calculate the indefinite integral.

39. $\displaystyle\int \cos(3x+1)\,dx$.

40. $\displaystyle\int \sin 2\pi x\,dx$.

41. $\displaystyle\int \csc^2 \pi x\,dx$.

42. $\displaystyle\int \sec 2x\tan 2x\,dx$.

43. $\displaystyle\int \sin(3-2x)\,dx$.

44. $\displaystyle\int \sin^2 x\cos x\,dx$.

45. $\displaystyle\int \cos^4 x\sin x\,dx$.

46. $\displaystyle\int x\sec^2 x^2\,dx$.

47. $\displaystyle\int \frac{\sin\sqrt{x}}{\sqrt{x}}\,dx$

48. $\displaystyle\int \csc(1-2x)\cot(1-2x)\,dx$.

49. $\displaystyle\int \sqrt{1+\sin x}\cos x\,dx$.

50. $\displaystyle\int \frac{\sin x}{\sqrt{1+\cos x}}\,dx$.

51. $\displaystyle\int \frac{1}{\cos^2 x}\,dx$.

52. $\displaystyle\int (1+\tan^2 x)\sec^2 x\,dx$.

53. $\displaystyle\int x\sin^3 x^2\cos x^2\,dx$.

54. $\displaystyle\int \frac{1}{\sin^2 x}\,dx$.

55. $\displaystyle\int \frac{\sec^2 x}{\sqrt{1+\tan x}}\,dx$.

56. $\displaystyle\int x^2\sin(4x^3-7)\,dx$.

Evaluate the integral.

57. $\displaystyle\int_{-\pi}^{\pi} \sin x\,dx$.

58. $\displaystyle\int_{-\pi/3}^{\pi/3} \sec x\tan x\,dx$.

59. $\displaystyle\int_{1/4}^{1/3} \sec^2 \pi x\,dx$.

60. $\displaystyle\int_0^1 \cos^2\frac{\pi}{2}x\sin\frac{\pi}{2}x\,dx$.

61. $\displaystyle\int_0^{\pi/2} \sin^3 x\cos x\,dx$.

62. $\displaystyle\int_0^{\pi} x\cos x^2\,dx$.

63. Derive the formula

$$\int \sin^2 x\,dx=\tfrac{1}{2}x-\tfrac{1}{4}\sin 2x+C.$$

HINT: Recall the half-angle formula $\sin^2\theta=\frac{1}{2}(1-\cos 2\theta)$.

64. Derive the formula

$$\int \cos^2 x\,dx = \tfrac{1}{2}x + \tfrac{1}{4}\sin 2x + C.$$

HINT: Recall the half-angle formula for cosine.

Evaluate the integrals.

65. $\int \cos^2 5x\,dx.$ **66.** $\int \sin^2 3x\,dx.$

67. $\int_0^{\pi/2} \cos^2 2x\,dx.$ **68.** $\int_0^{2\pi} \sin^2 x\,dx.$

Find the area bounded by the following curves.

69. $y = \cos x, \quad y = -\sin x, \quad x = 0, \quad x = \frac{\pi}{2}.$

70. $y = \cos \pi x, \quad y = \sin \pi x, \quad x = 0, \quad x = \frac{1}{4}.$

71. $y = \cos^2 \pi x, \quad y = \sin^2 \pi x, \quad x = 0, \quad x = \frac{1}{4}.$

72. $y = \cos^2 \pi x, \quad y = -\sin^2 \pi x, \quad x = 0, \quad x = \frac{1}{4}.$

73. $y = \csc^2 \pi x, \quad y = \sec^2 \pi x, \quad x = \frac{1}{6}, \quad x = \frac{1}{4}.$

74. In Example 3 we found that

$$\int \sin x \cos x\,dx = \tfrac{1}{2}\sin^2 x + C$$

by setting $u = \sin x$. Calculate the integral by setting $u = \cos x$ and then reconcile the two answers.

75. Calculate

$$\int \sec^2 x \tan x\,dx$$

(a) Setting $u = \sec x$. (b) Setting $u = \tan x$.

(c) Reconcile your answers to (a) and (b).

76. (*The area of a circular region*) The circle $x^2 + y^2 = r^2$ encloses a circular disc of radius r. Justify the familiar formula $A = \pi r^2$ by integration. HINT: The quarter disc in the first quadrant is the region below the curve $y = \sqrt{r^2 - x^2}, x \in [0, r]$. Therefore

$$A = 4\int_0^r \sqrt{r^2 - x^2}\,dx.$$

Evaluate the integral by setting $x = r\sin u,\ dx = r\cos u\,du$.

77. Find the area enclosed by the ellipse $b^2x^2 + a^2y^2 = a^2b^2$.

78. Let f be a continuous function, c a real number. Show that

(a) $$\int_{a+c}^{b+c} f(x - c)\,dx = \int_a^b f(x)\,dx,$$

and if $c \neq 0$

(b) $$\frac{1}{c}\int_{ac}^{bc} f(x/c)\,dx = \int_a^b f(x)dx.$$

Ⓒ Use a CAS to calculate the integrals. Use the methods of this section to verify your results.

79. $\displaystyle\int \frac{1}{\sqrt{x\sqrt{x} + x}}\,dx.$

80. $\displaystyle\int \frac{\csc^2 2x}{\sqrt{2 + \cot 2x}}\,dx.$

81. $\displaystyle\int_1^4 \frac{x^3 + 1}{(x^4 + 4x)^2}\,dx.$

82. $\displaystyle\int_{\pi/6}^{\pi/3} \cos^3 2x \sin 2x\,dx.$

Ⓒ Use a CAS to evaluate the pairs of definite integrals. Explain why each member of the pair has the same value.

83. $\displaystyle\int_3^8 x\sqrt{x+1}\,dx; \qquad \int_4^9 (u^{3/2} - u^{1/2})\,du.$

84. $\displaystyle\int_{\pi/6}^{\pi/4} \frac{\sin x}{\cos^2 x}\,dx; \qquad -\int_{\sqrt{3}/2}^{\sqrt{2}/2} u^{-2}\,du.$

■ 5.8 ADDITIONAL PROPERTIES OF THE DEFINITE INTEGRAL

We now present some important general properties of the definite integral. The proofs are left mostly to you. You can assume throughout that the functions involved are continuous and that $a < b$.

I. The integral of a nonnegative function is nonnegative:

(5.8.1) if $f(x) \geq 0$ for all $x \in [a, b]$, then $\displaystyle\int_a^b f(x)\,dx \geq 0.$

The integral of a positive function is positive:

(5.8.2) if $f(x) > 0$ for all $x \in [a, b]$, then $\displaystyle\int_a^b f(x)\,dx > 0.$

The next property is an immediate consequence of Property I and linearity (5.4.5).

II. The integral is order-preserving:

(5.8.3) if $f(x) \leq g(x)$ for all $x \in [a, b]$, then $$\int_a^b f(x)\,dx \leq \int_a^b g(x)\,dx$$

and

(5.8.4) if $f(x) < g(x)$ for all $x \in [a, b]$, then $$\int_c^b f(x)\,dx < \int_a^b g(x)\,dx.$$

PROOF OF (5.8.3) Since $f(x) \leq g(x)$, it follows that $g(x) - f(x) \geq 0$. Therefore, using the linearity of the integral and (5.7.1), we have

$$\int_a^b g(x)\,dx - \int_a^b f(x)\,dx = \int_a^b [g(x) - f(x)]\,dx \geq 0,$$

and so $$\int_a^b f(x)\,dx \leq \int_a^b g(x)\,dx.$$

The proof of (5.8.4) can be carried out in the same manner. ☐

III. Just as the absolute value of a sum is less than or equal to the sum of the absolute values,

$$|x_1 + x_2 + \cdots + x_n| \leq |x_1| + |x_2| + \cdots + |x_n|,$$

the absolute value of an integral is less than or equal to the integral of the absolute value:

(5.8.5) $$\left|\int_a^b f(x)\,dx\right| \leq \int_a^b |f(x)|\,dx.$$

PROOF OF (5.8.5) Since $-|f(x)| \leq f(x) \leq |f(x)|$, it follows that

$$-\int_a^b |f(x)|\,dx \leq \int_a^b f(x)\,dx \leq \int_a^b |f(x)|\,dx$$

by (5.8.3). This pair of inequalities is equivalent to (5.8.5). ☐

You are already familiar with the next property.

IV. If m is the minimum value of f on $[a, b]$ and M is the maximum value, then

(5.8.6) $$m(b - a) \leq \int_a^b f(x)\,dx \leq M(b - a).$$

Recall Theorem 5.3.5, which states that

$$\frac{d}{dx}\left(\int_a^x f(t)\,dt\right) = f(x).$$

The next property is a generalization of this theorem. Some specific examples of this result were given in Exercises 23–26, Section 5.3. Its importance will become apparent in Chapter 7.

V. If u is a differentiable function of x and f is continuous, then

(5.8.7)
$$\boxed{\frac{d}{dx}\left(\int_a^u f(t)\,dt\right) = f(u)\frac{du}{dx}.}$$

PROOF The function

$$F(u) = \int_a^u f(t)\,dt$$

is differentiable with respect to u and

$$\frac{d}{du}[F(u)] = f(u). \qquad \text{(Theorem 5.3.5)}$$

Therefore

$$\frac{d}{dx}\left(\int_a^u f(t)\,dt\right) = \frac{d}{dx}[F(u)] = \frac{d}{du}[F(u)]\frac{du}{dx} = f(u)\,\frac{du}{dx}. \quad \square$$

by the chain rule ↑

Example 1 Find $\dfrac{d}{dx}\left(\displaystyle\int_0^{x^3} \frac{dt}{1+t}\right)$.

SOLUTION At this stage you probably cannot carry out the integration; it requires the natural logarithm function which is not introduced in this text until Chapter 7. But, for our purposes, that doesn't matter. We have $f(t) = 1/(1+t)$ and $u(x) = x^3$. By (5.8.7), you can see that

$$\frac{d}{dx}\left(\int_0^{x^3} \frac{dt}{1+t}\right) = f(u)\frac{du}{dx} = \frac{1}{1+x^3}\,3x^2 = \frac{3x^2}{1+x^3}$$

without carrying out the integration. $\square$

Example 2 Find $\dfrac{d}{dx}\left(\displaystyle\int_x^{2x} \frac{dt}{1+t^2}\right)$.

SOLUTION The idea is to express the integral in terms of integrals that have constant lower limits of integration. Once we have done that, we can apply (5.8.7). In this case, we choose 0 as a convenient lower limit. Then, by the additivity of the integral,

$$\int_0^x \frac{dt}{1+t^2} + \int_x^{2x} \frac{dt}{1+t^2} = \int_0^{2x} \frac{dt}{1+t^2}.$$

Thus
$$\int_x^{2x} \frac{dt}{1+t^2} = \int_0^{2x} \frac{dt}{1+t^2} - \int_0^x \frac{dt}{1+t^2}.$$

Differentiation gives

$$\begin{aligned}\frac{d}{dx}\left(\int_x^{2x} \frac{dt}{1+t^2}\right) &= \frac{d}{dx}\left(\int_0^{2x} \frac{dt}{1+t^2}\right) - \frac{d}{dx}\left(\int_0^x \frac{dt}{1+t^2}\right)\\ &= \frac{1}{1+(2x)^2}(2) - \frac{1}{1+x^2}(1) = \frac{2}{1+4x^2} - \frac{1}{1+x^2}. \quad \square\end{aligned}$$

(5.8.7)⟶↑

VI. Now a few words about the role of symmetry in integration. Suppose that f is a continuous function defined on some interval of the form $[-a, a]$, a closed interval that is symmetric about the origin.

(5.8.8)

(a) if f is odd on $[-a, a]$, then
$$\int_{-a}^{a} f(x)\,dx = 0.$$
(b) If f is even on $[-a, a]$, then
$$\int_{-a}^{a} f(x)\,dx = 2\int_0^a f(x)\,dx.$$

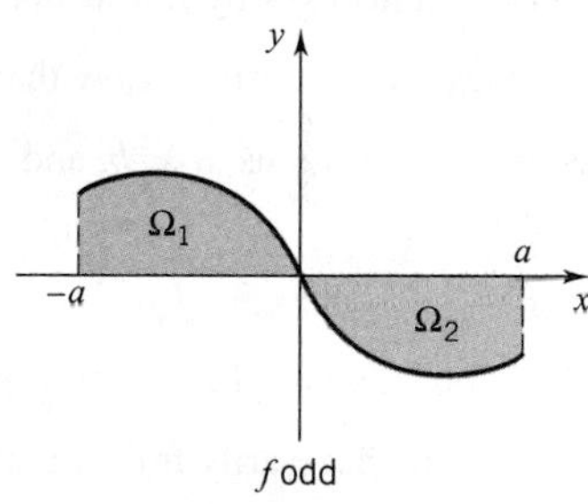

Figure 5.8.1

These assertions can be verified by a simple change of variables (Exercise 36). Here we look at these assertions from the standpoint of area, referring to Figures 5.8.1 and 5.8.2.

For the odd function,

$$\int_{-a}^{a} f(x)\,dx = \int_{-a}^{0} f(x)\,dx + \int_0^a f(x)\,dx = \text{ area of } \Omega_1 - \text{ area of } \Omega_2 = 0.$$

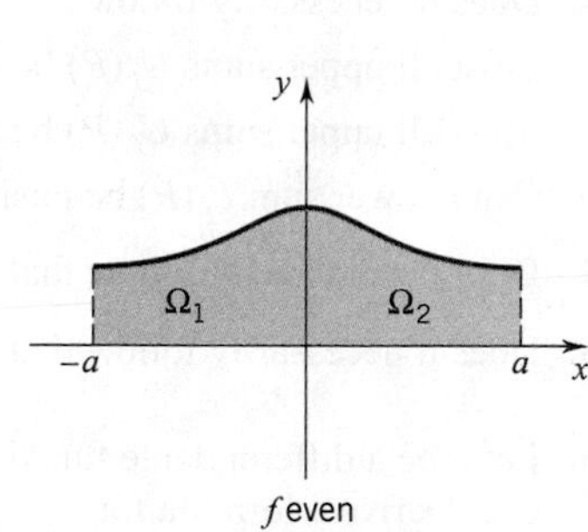

Figure 5.8.2

For the even function,

$$\int_{-a}^{a} f(x)\,dx = \text{ area of } \Omega_1 + \text{ area of } \Omega_2 = 2\,(\text{area of } \Omega_2) = 2\int_0^a f(x)\,dx.$$

Suppose we were asked to evaluate

$$\int_{-\pi}^{\pi} (\sin x - x\cos x)^3\,dx.$$

A laborious calculation would show that this integral is zero. We don't have to carry out that calculation. The integrand is an odd function, and the interval of integration is symmetric about the origin. Thus we can tell immediately that the integral is zero.

Exercises 5.8

Assume: f and g continuous, $a < b$, and

$$\int_a^b f(x)\,dx > \int_a^b g(x)\,dx.$$

Answer questions 1–6, giving supporting reasons.

1. Does it necessarily follow that $\int_a^b [f(x) - g(x)]\,dx > 0$?
2. Does it necessarily follow that $f(x) > g(x)$ for all $x \in [a, b]$?
3. Does it necessarily follow that $f(x) > g(x)$ for at least some $x \in [a, b]$?
4. Does it necessarily follow that

$$\left|\int_a^b f(x)\,dx\right| > \left|\int_a^b g(x)\,dx\right|?$$

5. Does it necessarily follow that $\int_a^b |f(x)|\,dx > \int_a^b |g(x)|\,dx$?
6. Does it necessarily follow that $\int_a^b |f(x)|\,dx > \int_a^b g(x)\,dx$?

Assume f continuous, $a < b$, and

$$\int_a^b f(x)\,dx = 0.$$

Answer questions 7–15, giving supporting reasons.

7. Does it necessarily follow that $f(x) = 0$ for all $x \in [a, b]$?
8. Does it necessarily follow that $f(x) = 0$ for at least some $x \in [a, b]$?
9. Does it necessarily follow that $\int_a^b |f(x)|\,dx = 0$?
10. Does it necessarily follow that $|\int_a^b f(x)\,dx| = 0$?
11. Must all upper sums $U_f(P)$ be nonnegative?
12. Must all upper sums $U_f(P)$ be positive?
13. Can a lower sum $L_f(P)$ be positive?
14. Does it necessarily follow that $\int_a^b [f(x)]^2\,dx = 0$?
15. Does it necessarily follow that $\int_a^b [f(x) + 1]\,dx = b - a$?

16. Let u be a differentiable function of x and let f be continuous. Derive a formula for

$$\frac{d}{dx}\left(\int_{u(x)}^b f(t)\,dt\right).$$

Calculate the derivative.

17. $\frac{d}{dx}\left(\int_0^{1+x^2} \frac{dt}{\sqrt{2t+5}}\right)$.

18. $\frac{d}{dx}\left(\int_1^{x^2} \frac{dt}{t}\right)$.

19. $\frac{d}{dx}\left(\int_x^a f(t)\,dt\right)$.

20. $\frac{d}{dx}\left(\int_0^{x^3} \frac{dt}{\sqrt{1+t^2}}\right)$.

21. $\frac{d}{dx}\left(\int_{x^2}^3 \frac{\sin t}{t}\,dt\right)$.

22. $\frac{d}{dx}\left(\int_{\tan x}^4 \sin(t^2)\,dt\right)$.

23. $\frac{d}{dx}\left(\int_1^{\sqrt{x}} \frac{t^2}{1+t^2}\,dt\right)$.

24. Show that, if u and v are differentiable functions of x and f is continuous, then

$$\frac{d}{dx}\left(\int_{u(x)}^{v(x)} f(t)\,dt\right) = f(v)\frac{dv}{dx} - f(u)\frac{du}{dx}.$$

HINT: Take a number a from the domain of f. Express the integral as the difference of two integrals, each with lower limit a.

Calculate the derivative using Exercise 24.

25. $\frac{d}{dx}\left(\int_x^{x^2} \frac{dt}{t}\right)$.

26. $\frac{d}{dx}\left(\int_{\sqrt{x}}^{x^2+x} \frac{dt}{2+\sqrt{t}}\right)$.

27. $\frac{d}{dx}\left(\int_{\tan x}^{2x} t\sqrt{1+t^2}\,dt\right)$.

28. $\frac{d}{dx}\left(\int_{3x}^{1/x} \cos 2t\,dt\right)$.

29. Verify (5.8.1): (a) by considering lower sums $L_f(P)$; (b) by using an antiderivative; (c) by using Riemann sums.

30. Verify (5.8.4).

31. Verify (5.8.6).

32. (*Important*) Prove that, if f is continuous on $[a, b]$ and

$$\int_a^b |f(x)|\,dx = 0,$$

then $f(x) = 0$ for all x in $[a, b]$. HINT: See Exercise 50, Section 2.4.

33. Find $H'(2)$ given that

$$H(x) = \int_{2x}^{x^3-4} \frac{x}{1+\sqrt{t}}\,dt.$$

34. Find $H'(3)$ given that

$$H(x) = \frac{1}{x}\int_3^x [2t - 3H'(t)]\,dt.$$

35. (a) Let f be continuous on $[-a, 0]$. Use a change of variable to prove that

$$\int_{-a}^0 f(x)\,dx = \int_0^a f(-x)\,dx.$$

(b) Let f be continuous on $[-a, a]$. Prove that

$$\int_{-a}^a f(x)\,dx = \int_0^a [f(x) + f(-x)]\,dx.$$

36. Let f be a continuous function on $[-a, a]$. Use Exercise 35 to prove that:

(a) $\int_{-a}^a f(x)\,dx = 0$ if f is odd.

(b) $\int_{-a}^a f(x)\,dx = 2\int_0^a f(x)\,dx$ if f is even.

Use Exercise 36 to evaluate the integral.

37. $\int_{-\pi/4}^{\pi/4} (x + \sin 2x)\,dx$.

38. $\int_{-3}^3 \frac{t^3}{1+t^2}\,dt$.

39. $\displaystyle\int_{-\pi/3}^{\pi/3} (1 + x^2 - \cos x)\, dx.$

40. $\displaystyle\int_{-\pi/4}^{\pi/4} (x^2 - 2x + \sin x + \cos 2x)\, dx.$

C 41. Let $f(x) = x^3 - 6x^2 + 12x - 9$ on $[3, 5]$.

(a) Use a graphing utility to sketch the graph of f. Note that $f(x) \geq 0$ on $[3, 5]$.

(b) Use a CAS to verify that $\int_3^5 f(x)\, dx \geq 0$.

(c) Find $\int_3^5 f(x)\, dx$ by evaluating the definite integral.

C 42. Let $f(x) = \cos x$ and $g(x) = \sin 2x$ on $[\pi/6, \pi/2]$.

(a) Use a graphing utility to sketch the graphs of f and g together. Note that $f(x) \leq g(x)$ on $[\pi/6, \pi/2]$.

(b) Use a CAS to verify that $\int_{\pi/6}^{\pi/2} f(x)\, dx \leq \int_{\pi/6}^{\pi/2} g(x)\, dx$.

(c) Find $\int_{\pi/6}^{\pi/2} f(x)\, dx$ and $\int_{\pi/6}^{\pi/2} g(x)\, dx$ by evaluating the definite integrals.

C 43. Let $f(x) = x^3 - 6x^2 + 3x - 1$ on $[3, 6]$.

(a) Use a CAS to verify that $|\int_3^6 f(x)\, dx| \leq \int_3^6 |f(x)|\, dx$.

(b) Find $\int_3^6 f(x)\, dx$ and $\int_3^6 |f(x)|\, dx$ by evaluating the definite integrals.

C 44. Let $f(x) = \sin x - 3\cos x$ on $[0, 3\pi/2]$.

(a) Use the trace function on a graphing utility to determine the minimum m and the maximum M of f on $[0, 3\pi/2]$.

(b) Evaluate the definite integral $\int_0^{3\pi/2} f(x)\, dx$ and verify that

$$m\left(\tfrac{3\pi}{2}\right) \leq \int_0^{3\pi/2} f(x)\, dx \leq M\left(\tfrac{3\pi}{2}\right).$$

C 45. (a) Show that the function $f(x) = 2x - \sin x$ is an odd function.

(b) Use a CAS to verify that $\int_{-5}^5 f(x)\, dx = 0$.

(c) Find $\int_{-5}^5 f(x)\, dx$ by evaluating the definite integral.

C 46. (a) Show that the function $f(x) = x^2 + \cos x$ is an even function.

(b) Use a CAS to verify that $\int_{-3}^3 f(x)\, dx = 2\int_0^3 f(x)\, dx$.

(c) Find $\int_0^3 f(x)\, dx$ by evaluating the definite integral.

■ 5.9 MEAN-VALUE THEOREMS FOR INTEGRALS; AVERAGE VALUES

We begin with a result that we asked you to prove in Exercise 31, Section 5.3.

THEOREM 5.9.1 THE FIRST MEAN-VALUE THEOREM FOR INTEGRALS

If f is continuous on $[a, b]$, then there is at least one number c in (a, b) for which

$$\int_a^b f(x)\, dx = f(c)(b - a).$$

This number $f(c)$ is called *the average* (or *mean*) *value of f on* $[a, b]$.

We then have the following identity:

(5.9.2) $$\int_a^b f(x)\, dx = (\textit{the average value of } f \textit{ on } [a, b]) \cdot (b - a).$$

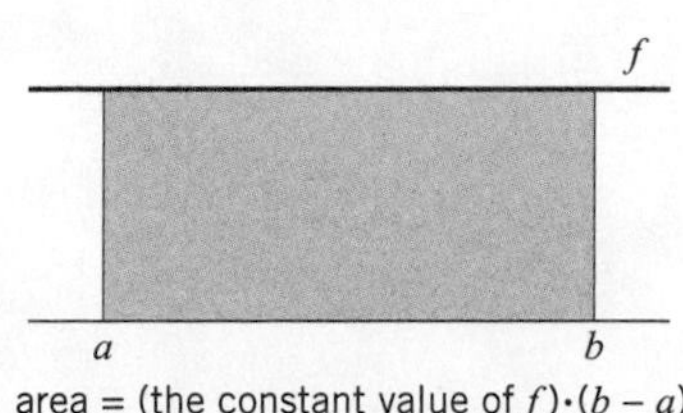

area = (the constant value of f)·$(b - a)$

Figure 5.9.1

This identity provides a powerful, intuitive way of viewing the definite integral. Think for a moment about area. If f is constant and positive on $[a, b]$, then Ω, the region below the graph, is a rectangle. Its area is given by the formula

$$\text{area of } \Omega = (\text{the constant value of } f \text{ on } [a, b]) \cdot (b - a). \qquad \text{(Figure 5.9.1)}$$

If f is now allowed to vary continuously on $[a, b]$, then we have

$$\text{area of } \Omega = \int_a^b f(x)\,dx,$$

and the area formula reads

$$\boxed{\textit{area of } \Omega = (\textit{the average value of } f \textit{ on } [a, b]) \cdot (b - a).} \qquad \text{(Figure 5.9.2)}$$

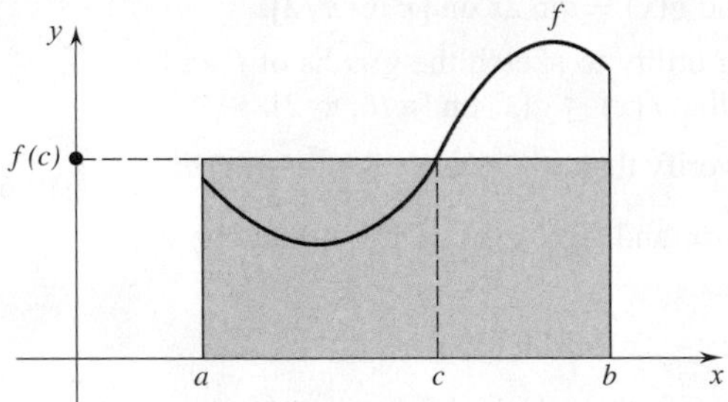

Figure 5.9.2

Think now about motion. If an object moves along a line with constant speed $|v|$ during the time interval $[a, b]$, then

$$\text{distance traveled} = (\text{the constant value of } |v| \text{ on } [a, b]) \cdot (b - a).$$

If the speed $|v|$ varies, then we have

$$\text{distance traveled} = \int_a^b |v(t)|\,dt,$$

and the formula reads

$$\boxed{\textit{distance traveled} = (\textit{the average speed on } [a, b]) \cdot (b - a).}$$

We take an interval $[a, b]$ and calculate the average value of the simplest possible functions on that interval. For convenience, let f_{avg} denote the average value of f on $[a, b]$. Solving identity (5.9.2) for f_{avg}, we have

$$f_{\text{avg}} = \frac{1}{b-a}\int_a^b f(x)\,dx.$$

The average value of a constant function $f(x) = k$ is, of course, k:

$$f_{\text{avg}} = \frac{1}{b-a}\int_a^b k\,dx = \frac{k}{b-a}\Big[x\Big]_a^b = \frac{k}{b-a}(b-a) = k.$$

The average value of the identity function $f(x) = x$ on $[a, b]$ is $\frac{1}{2}(b + a)$:

$$f_{\text{avg}} = \frac{1}{b-a}\int_a^b x\,dx = \frac{1}{b-a}\left[\frac{x^2}{2}\right]_a^b = \frac{1}{b-a}\left(\frac{b^2-a^2}{2}\right) = \tfrac{1}{2}(b+a).$$

What is the average value of $f(x) = x^2$?

$$f_{\text{avg}} = \frac{1}{b-a}\int_a^b x^2\,dx = \frac{1}{b-a}\left[\frac{x^3}{3}\right]_a^b = \frac{1}{b-a}\left(\frac{b^3-a^3}{3}\right)$$

$$= \frac{1}{b-a}\left[\frac{(b^2+ab+a^2)(b-a)}{3}\right] = \tfrac{1}{3}(b^2+ab+a^2).$$

Thus, the average value of $f(x) = x^2$ on $[a, b]$ is $\frac{1}{3}(b^2+ab+a^2)$ (see Example 6 in Section 5.2.). On $[1, 3]$ the values of x^2 range from 1 to 9; the average value is $\frac{13}{3}$.

There is an extension of Theorem 5.9.1 that, as you will see, is useful in applications.

THEOREM 5.9.3 THE SECOND MEAN-VALUE THEOREM FOR INTEGRALS

If f and g are continuous on $[a, b]$ and g is nonnegative, then there is a number c in (a, b) for which

$$\int_a^b f(x)g(x)\,dx = f(c)\int_a^b g(x)\,dx.$$

This number $f(c)$ is called the *g-weighted average of f on* $[a, b]$.

We will prove this theorem (and thereby have a proof of Theorem 5.9.1) at the end of the section. First, a brief excursion into physics for some motivation.

The Mass of a Rod Imagine a thin rod (a straight material wire of negligible thickness) lying on the x-axis from $x = a$ to $x = b$. If the *mass density* of the rod (the mass per unit length) is constant, then the total mass M of the rod is simply the density λ times the length of the rod: $M = \lambda(b - a)$.† If the density λ varies continuously from point to point, say $\lambda = \lambda(x)$, then the mass of the rod is the average density of the rod times the length of the rod:

$$M = (\text{average density}) \cdot (\text{length}).$$

This is an integral:

(5.9.4) $$\boxed{M = \int_a^b \lambda(x)\,dx.}$$

The Center of Mass of a Rod Continue with that same rod. If the rod is homogeneous (constant density), then the center of mass x_M of the rod is simply the midpoint of the rod:

$$x_M = \tfrac{1}{2}(a+b). \qquad \text{(the average of } x \text{ from } a \text{ to } b)$$

† The symbol λ is the Greek letter "lambda"

If the rod is not homogeneous, the center of mass is still an average, but now a weighted average, *the density-weighted average of x from a to b;* namely, x_M is the point for which

$$x_M \int_a^b \lambda(x)\,dx = \int_a^b x\,\lambda(x)\,dx.$$

Since the integral on the left is simply M, we have

(5.9.5)
$$x_M M = \int_a^b x\,\lambda(x)\,dx.$$

Example 1 A rod of length L is placed on the x-axis from $x = 0$ to $x = L$. Find the mass of the rod and the center of mass if the density of the rod varies directly as the distance from the $x = 0$ endpoint of the rod.

SOLUTION Here $\lambda(x) = kx$ where k is some positive constant. Therefore

$$M = \int_0^L kx\,dx = \left[\tfrac{1}{2}kx^2\right]_0^L = \tfrac{1}{2}kL^2$$

and

$$x_M M = \int_0^L x(kx)\,dx = \int_0^L kx^2\,dx = \left[\tfrac{1}{3}kx^3\right]_0^L = \tfrac{1}{3}kL^3.$$

Division by M gives $x_M = \frac{2}{3}L$.

0 $\quad x_M = \frac{2}{3}L \quad L$

In this instance the center of mass is to the right of the midpoint. This makes sense. After all, the density increases from left to right. Thus mass accumulates near the right tip of the rod. ☐

We return to the center of mass in the Exercises. Below we try to provide some physical insight.

We know from physics that, close to the surface of the earth, where the force of gravity is given by the familiar formula $W = mg$, the center of mass is the *center of gravity*. This is the balance point. For the rod of Example 1, the balance point is at $x = \frac{2}{3}L$. Supported at that point, the rod will be in balance.

Later (in Project 9.6) you will see that a projectile fired at an angle follows a parabolic path. (Here we are disregarding air resistance.) Suppose that a rod is hurled into the air end over end. Certainly not every point of the rod can follow a parabolic path. What moves in a parabolic arc is the center of mass of the rod.

We go back now to Theorem 5.9.3 and prove it. [There is no reason to worry about Theorem 5.9.1. That is just Theorem 5.9.3 with $g(x)$ identically 1.]

PROOF OF THEOREM 5.9.3 Since f is continuous on $[a, b]$, f takes on a minimum value m on $[a, b]$ and a maximum value M. Since g is nonnegative on $[a, b]$,

$$mg(x) \le f(x)g(x) \le Mg(x) \quad \text{for all } x \text{ in } [a, b].$$

Therefore
$$\int_a^b m\,g(x)\,dx \le \int_a^b f(x)g(x)\,dx \le \int_a^b M\,g(x)\,dx.$$

and
$$m\int_a^b g(x)\,dx \le \int_a^b f(x)g(x)\,dx \le M\int_a^b g(x)\,dx.$$

We know that $\int_a^b g(x)\,dx \geq 0$. If $\int_a^b g(x)\,dx = 0$, then $\int_a^b f(x)\,g(x)\,dx = 0$ and the theorem holds for all choices of c in (a, b). If $\int_a^b g(x)\,dx > 0$, then

$$m \leq \frac{\displaystyle\int_a^b f(x)\,g(x)\,dx}{\displaystyle\int_a^b g(x)\,dx} \leq M$$

and by the intermediate-value theorem (Theorem 2.6.1) there exists a number c in (a, b) for which

$$f(c) = \frac{\displaystyle\int_a^b f(x)\,g(x)\,dx}{\displaystyle\int_a^b g(x)\,dx}.$$

Obviously, then,

$$f(c)\int_a^b g(x)\,dx = \int_a^b f(x)\,g(x)\,dx. \quad \square$$

EXERCISES 5.9

Determine the average value on the indicated interval and find a point in this interval at which the function takes on this average value.

1. $f(x) = mx + b, \quad x \in [0, c]$.
2. $f(x) = x^2, \quad x \in [-1, 1]$.
3. $f(x) = x^3, \quad x \in [-1, 1]$.
4. $f(x) = x^{-2}, \quad x \in [1, 4]$.
5. $f(x) = |x|, \quad x \in [-2, 2]$.
6. $f(x) = x^{1/3}, \quad x \in [-8, 8]$.
7. $f(x) = 2x - x^2, \quad x \in [0, 2]$.
8. $f(x) = 3 - 2x, \quad x \in [0, 3]$.
9. $f(x) = \sqrt{x}, \quad x \in [0, 9]$.
10. $f(x) = 4 - x^2, \quad x \in [-2, 2]$.
11. $f(x) = \sin x, \quad x \in [0, 2\pi]$.
12. $f(x) = \cos x, \quad x \in [0, \pi]$.
13. Let $f(x) = x^n$, n a positive integer. Determine the average value of f on the interval $[a, b]$.
14. Given that f is continuous on $[a, b]$, compare

$$f(b)(b-a) \quad \text{and} \quad \int_a^b f(x)\,dx$$

if f is (a) constant on $[a, b]$; (b) increasing on $[a, b]$; (c) decreasing on $[a, b]$.

15. In Chapter 3, we viewed $[f(b)-f(a)]/(b-a)$ as the average rate of change of f on $[a, b]$ and $f'(t)$ as the instantaneous rate of change at time t. If our new sense of average is to be consistent with the old one, we must have

$$\frac{f(b)-f(a)}{b-a} = \text{average of } f' \text{ on } [a, b].$$

Prove that this is the case.

16. Determine whether the assertion is true or false.

(a) $$\begin{pmatrix}\text{the average of } f+g \\ \text{on } [a, b]\end{pmatrix} = \begin{pmatrix}\text{the average of } f \\ \text{on } [a, b]\end{pmatrix} + \begin{pmatrix}\text{the average of } g \\ \text{on } [a, b]\end{pmatrix}.$$

(b) $$\begin{pmatrix}\text{the average of } \alpha f \\ \text{on } [a, b]\end{pmatrix} = \alpha\begin{pmatrix}\text{the average of } f \\ \text{on } [a, b]\end{pmatrix}, \quad \alpha \text{ a constant.}$$

(c) $$\begin{pmatrix}\text{the average of } fg \\ \text{on } [a, b]\end{pmatrix} = \begin{pmatrix}\text{the average of } f \\ \text{on } [a, b]\end{pmatrix}\begin{pmatrix}\text{the average of } g \\ \text{on } [a, b]\end{pmatrix}.$$

17. Let $P(x, y)$ be an arbitrary point on the curve $y = x^2$. Express as a function of x the distance from P to the origin and calculate the average of this distance as x ranges from 0 to $\sqrt{3}$.
18. Let $P(x, y)$ be an arbitrary point on the line $y = mx$. Express as a function of x the distance from P to the

origin and calculate the average of this distance as x ranges from 0 to 1.

19. A stone falls from rest in a vacuum for t seconds (see Section 3.4). (a) Compare its terminal velocity to its average velocity; (b) compare its average velocity during the first $\frac{1}{2}t$ seconds to its average velocity during the next $\frac{1}{2}t$ seconds.

20. Let f be continuous. Show that, if f is an odd function, then its average value on every interval of the form $[-a, a]$ is zero.

21. Suppose that f is continuous on $[a, b]$ and $\int_a^b f(x)\,dx = 0$. Prove that there is at least one number $c \in (a, b)$ such that $f(c) = 0$.

22. Find the average value of the linear function $f(x) = mx + k$ on the interval $[a, b]$. At what point (or points) does f take on its average value?

23. An object starts from rest at the point x_0 and moves along the x-axis with constant acceleration a.

(a) Derive formulas for the velocity and position of the object at each time $t \geq 0$.

(b) Show that the average velocity over any time interval $[t_1, t_2]$ is the average of the initial and final velocities on that interval.

24. Find the point on the rod of Example 1 that breaks up that rod into two pieces of equal mass. (Observe that this point is not the center of mass.)

25. A rod 6 meters long is placed on the x-axis from $x = 0$ to $x = 6$. The mass density is $12/\sqrt{x+1}$ kilograms per meter.

(a) Find the mass of the rod and the center of mass.

(b) What is the average mass density of the rod?

26. For the rod of (5.9.5) the integral

$$\int_a^b (x - c)\lambda(x)\,dx$$

gives what is called the *mass moment* of the rod about the point $x = c$. Show that the mass moment about the center of mass is zero. (The center of mass can be defined as the point about which the mass moment is zero.)

27. A rod of length L is placed on the x-axis from $x = 0$ to $x = L$. Find the mass of the rod and the center of mass if the mass density of the rod varies directly: (a) as the square root of the distance from $x = 0$; (b) as the square of the distance from $x = L$.

28. A rod of varying density, mass M, and center of mass x_M, extends from $x = a$ to $x = b$. A partition $P = \{x_0, x_1, \ldots, x_n\}$ of $[a, b]$ decomposes the rod into n pieces in the obvious way. Show that, if the n pieces have masses $M_1, M_2, \ldots, M_n$ and centers of mass $x_{M_1}, x_{M_2}, \ldots, x_{M_n}$, then

$$x_M M = x_{M_1} M_1 + x_{m_2} M_2 + \cdots + x_{M_n} M_n.$$

29. A rod that has mass M and extends from $x = 0$ to $x = L$ consists of two pieces with masses M_1, M_2. Given that the center of mass of the entire rod is at $x = \frac{1}{4}L$ and the center of mass of the first piece is at $x = \frac{1}{8}L$, determine the center of mass of the second piece.

30. A rod that has mass M and extends from $x = 0$ to $x = L$ consists of two pieces. Find the mass of each piece given that the center of mass of the entire rod is at $x = \frac{2}{3}L$, the center of mass of the first piece is at $x = \frac{1}{4}L$, and the center of mass of the second piece is at $x = \frac{7}{8}L$.

31. A rod of mass M and length L is to be cut from a long piece that extends to the right from $x = 0$. Where should the cuts be made if the density of the long piece varies directly as the distance from $x = 0$? (Assume that $M \geq \frac{1}{2}kL^2$ where k is the constant of proportionality in the density function.)

32. Let f be a function such that f' is continuous on $[a, b]$. Use the definition of average value of a function to find the *average slope* of the graph of f on $[a, b]$. What is the geometric interpretation of this result?

33. Prove Theorem 5.9.1.

34. Let f be continuous on $[a, b]$. Let $a < c < b$. Prove that

$$f(c) = \lim_{h \to 0^+} (\text{average value of } f \text{ on } [c - h, c + h]).$$

35. Prove that two distinct continuous functions cannot have the same average on every interval.

36. The *arithmetic average* of n numbers is the sum of the numbers divided by n. Let f be a continuous function on $[a, b]$. Show that the average value of f is the limit of arithmetic averages of values of f on $[a, b]$. HINT: Partition $[a, b]$ into n subintervals of equal length $(b - a)/n$ and let $S^*(P)$ be a corresponding Riemann sum. Show that $S^*(P)/(b - a)$ is an arithmetic average of n values of f. Then take the limit as $||P|| \to 0$.

37. A partition $P = \{x_0, x_1, x_2, \ldots, x_n\}$ of $[a, b]$ breaks up $[a, b]$ into n subintervals

$$[x_0, x_1], [x_1, x_2], \ldots, [x_{n-1}, x_n].$$

Show that if f is continuous on $[a, b]$, then there are n numbers $x_i^* \in [x_{i-1}, x_i]$ such that

$$\int_a^b f(x)\,dx = f(x_1^*)\Delta x_1 + f(x_2^*)\Delta x_2 + \cdots + f(x_n^*)\Delta x_n.$$

(Thus each partition P of $[a, b]$ gives rise to a Riemann sum which is exactly equal to the definite integral.)

38. (*The mass of a rod*) The total mass M of a system of point masses $m_1, m_2, \ldots, m_n$ is the sum

$$M = m_1 + m_2 + \cdots + m_n.$$

Use this to derive the formula for the mass of a rod given in (5.9.4). HINT: Break up the interval $[a, b]$ into n subintervals and build a Riemann sum.

39. (*Center of mass of a rod*)

(a) Start with a system of point masses $m_1, m_2, \ldots, m_n$ located at points $x_1, x_2, \ldots, x_n$ of the x-axis. The *mass moment of the system about the origin* is by definition the sum

$$m_1x_1 + m_2x_2 + \cdots + m_nx_n.$$

Let M be the total mass of the system:

$$M = m_1 + m_2 + \cdots + m_n.$$

The point x_M for which

$$x_M M = m_1x_1 + m_2x_2 + \cdots + m_nx_n$$

is called the *center of mass* of the system. Use this to derive formula (5.9.5). (HINT: Break up the interval $[a, b]$ into n subintervals and build a Riemann sum.)

(b) Show that the mass moment of a rod about its center of mass is zero. (HINT: The locations $x_1, x_2, \ldots, x_n$ with respect to the origin become $x_1 - x_M, x_2 - x_M, \ldots, x_n - x_M$ with respect to the center of mass.)

40. A system consists of three point masses: $m_1 = m, m_2 = \frac{1}{2}m, m_3 = \frac{1}{3}m$. Given that m_1 is located at $x = -1$ and m_2 is located at $x = 4$, place m_3 so that the origin becomes the center of mass of the system.

41. A homogeneous seesaw laid out from $x = -L$ to $x = L$ is balanced at $x = 0$. Sit a 40 lb. child at $x = L$ and a 100 lb. youngster at $x = \frac{1}{2}L$. Where should a 150 lb. man sit so as to keep the seesaw in balance?

42. Let $f(x) = x^3 - x + 1$ for $x \in [-1, 2]$.

(a) Find the average value of f on this interval.

(b) Estimate with three decimal place accuracy a value c in the interval at which f takes on its average value.

(c) Use a graphing utility to illustrate your results with a figure similar to Figure 5.9.2.

43. Repeat Exercise 42 with $f(x) = \sin x$ for $x \in [0, \pi]$.

44. Repeat Exercise 42 with $f(x) = 2\cos 2x$ for $x \in [-\pi/4, \pi/6]$.

45. Use a graphing utility to draw the graph of $f(x) = x^2+2x+6$ on $[a, b] = [-5, 5]$.

(a) Estimate h so that $h(b - a)$ equals the area under the graph of f.

(b) Draw the graphs of $y = h$ and f together.

(c) Estimate $d \in [a, b]$ so that $f(d) = h$.

(d) Find $\int_a^b f(x)\,dx$ exactly.

(e) Solve

$$\frac{\int_a^b f(x)\,dx}{b - a} = f(c) \quad \text{for} \quad c \in (a, b).$$

and compare your result with the number d you found in (c).

46. Repeat Exercise 45 with $f(x) = x^3 - x + 10$ on $[a, b] = [-2, 4]$.

47. Use a graphing utility to draw the graph of $f(x) = -x^4 + 10x^2 + 25$ on $[a, b]$ where $f(a) = f(b) = 0, a < b$.

(a) Find $c \in (a, b)$ such that $\int_a^b f(x)\,dx = f(c)(b - a)$.

(b) Draw the graphs of $y = f(c)$ and f together.

(c) Calculate $\int_a^b [f(x) - f(c)]\,dx$.

48. Repeat Exercise 47 with $f(x) = 8 + x^2 - x^4$.

■ PROJECT 5.9 Moving Averages

When economists study systems such as the stock market or the seasonal sales of some product and wish to determine long-term trends, they try to smooth out short-term fluctuations. To do this, they employ what are called *moving averages*, averages that reflect not only what is happening now, but also what happened before. If, for example, $f(x)$ denotes sales during the current period x, then the function

$$g_f(x) = \frac{f(x-1) + f(x)}{2}$$

is a moving average that takes into account what happened during the previous period $x - 1$.

Problem 1. Let $f(x) = x^4 - 3x^2 + 0.1x + 1$.

(a) Use a graphing utility to draw the graphs of f and g_f in the same coordinate system.

(b) Find and compare the domains and ranges of f and g_f.

(c) Find the points of intersection of the graphs of f and g_f. What characteristic do the points of intersection have?

Sometimes it is preferable to weight a moving average more heavily in favor of recent values. For example, the function

$$h_f(x) = \frac{f(x-1) + 2f(x)}{3}$$

gives twice as much weight to the current value than to the value for the preceding period.

Problem 2. Continue to let $f(x) = x^4 - 3x^2 + 0.1x + 1$.

(a) Use a graphing utility to draw the graphs of f and h_f in the same coordinate system.

(b) Find the domain and range of h_f.

(c) Find the points of intersection of the graphs of f and h_f and compare with the intersections of the graphs of f and g_f.

(d) Make and prove a conjecture about the points of intersection of the graphs of f and a moving average for f.

Instead of averaging just two values of a function, we can average n values:

$$\begin{aligned} A(x) &= \frac{1}{n}[f(x_1) + f(x_2) + \cdots + f(x_n)] \\ &= \frac{1}{a}\left\{\frac{a}{n}[f(x_1) + f(x_2) + \cdots + f(x_n)]\right\} \end{aligned}$$

where $a > 0$ and $x_i = x - a + ia/n, i = 1, 2, \ldots, n$. Since the expression on the right is a Riemann sum for A, we can take the limit as $n \to \infty$ to obtain the "continuous" moving average for f given by

$$F(x) = \frac{1}{a}\int_{x-a}^{x} f(t)\,dt.$$

Problem 3. Let $f(x) = \sin(3x + 2)$ and $a = \pi$.

(a) Use a graphing utility to graph F and

$$g_f = \frac{f(x-\pi) + f(x)}{2}$$

in the same coordinate system. Which of these moving averages seems to do the best job of "smoothing" the data?

(b) Let G be the moving average of f' : $G(x) = \dfrac{1}{\pi}\displaystyle\int_{x-\pi}^{x} f'(t)\,dt$. Find F' and compare it with G. Does this result hold in general?

■ CHAPTER HIGHLIGHTS

5.2 The Definite Integral of a Continuous Function

partition (p. 265) upper sum; lower sum (p. 266)
integrable (p. 268) definite integral (p. 268)
limits of integration; integrand (p. 268)
Darboux method (p. 271) Riemann sum (p. 272)

$$\int_a^b f(x)\,dx = \lim_{||P||\to 0} [f(x_1^*)\,\Delta x_1 + f(x_2^*)\,\Delta x_2 + \cdots + f(x_n^*)\,\Delta x_n].$$

5.3 The Function $F(x) = \int_a^x f(t)\,dt$

additivity of the integral (p. 277)

$\dfrac{d}{dx}\left[\displaystyle\int_a^x f(t)\,dt\right] = f(x)$ provided f is continuous (p. 279)

5.4 The Fundamental Theorem of Integral Calculus

antiderivative (p. 285) fundamental theorem (p. 285)
Table of Antiderivatives (p. 287)
linearity of the integral (p. 288)

5.5 Some Area Problems

If f and g are continuous and $f(x) \geq g(x)$ for all x in $[a, b]$, then

$$\int_a^b [f(x) - g(x)]\,dx$$

gives the area between the graph of f and the graph of g over $[a, b]$.
signed area (p. 296)

5.6 Indefinite Integrals

indefinite integral; constant of integration (p. 299)
Table of Indefinite Integrals (p. 300)
some motion problems (p. 301)

$\displaystyle\int_a^b |v(t)|\,dt$ = distance traveled from time $t = a$ to time $t = b$.
equation for linear motion with constant acceleration (p. 304)

5.7 The *u*-Substitution; Change of Variables

An integral of the form $\int f(g(x))\,g'(x)\,dx$ can be written $\int f(u)\,du$ by setting

$$u = g(x), \quad du = g'(x)\,dx.$$

For definite integrals

$$\int_a^b f(g(x))\,g'(x)\,dx = \int_{b(a)}^{g(b)} f(u)\,du.$$

5.8 Additional Properties of the Definite Integral

The integral of a nonnegative function is nonnegative; the integral of a positive function is positive; the integral is order-preserving (pp. 314–315)

$$\left|\int_a^b f(x)\,dx\right| \leq \int_a^b |f(x)|\,dx, \quad \frac{d}{dx}\left(\int_a^u f(t)\,dt\right) = f(u)\frac{du}{dx}.$$

5.9 Mean-Value Theorems for Integrals; Average Values

first mean-value theorem for integrals (p. 319)
average value (p. 319)
second mean-value theorem for integrals (p. 321)
weighted average (p. 321)
mass of a rod (p. 321)
center of mass of a rod (p. 321)

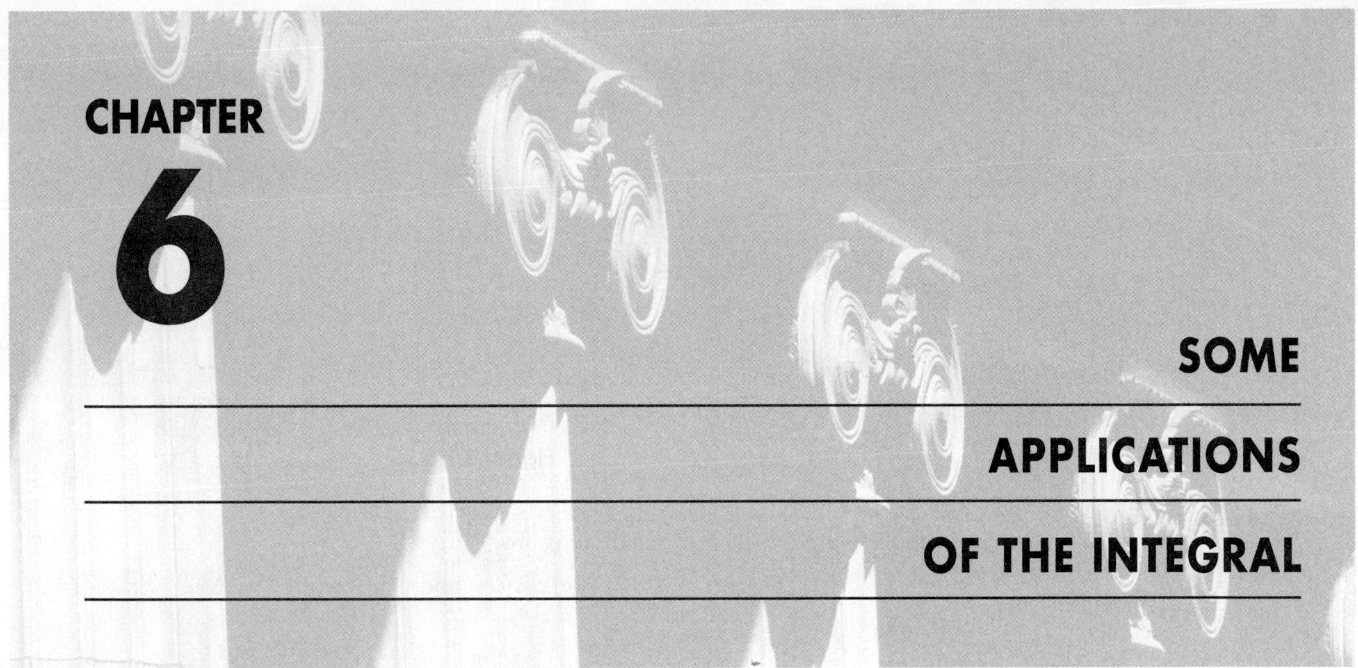

■ 6.1 MORE ON AREA

Representative Rectangles

You have seen that the definite integral can be viewed as the limit of Riemann sums:

$$(1)\qquad \int_a^b f(x)dx = \lim_{\|P\|\to 0} [f(x_1^*)\,\Delta x_1 + f(x_2^*)\,\Delta x_2 + \cdots + f(x_n^*)\,\Delta x_n].$$

With x_i^* chosen arbitrarily from $[x_{i-1}, x_i]$, you can think of $f(x_i^*)$ as a *representative* value of f for that interval. If f is positive, then the product

$$f(x_i^*)\,\Delta x_i$$

gives the area of the *representative rectangle* shown in Figure 6.1.1. Formula (1) tells us that we can approximate the area under the curve as closely as we wish by adding up the areas of representative rectangles (Figure 6.1.2).

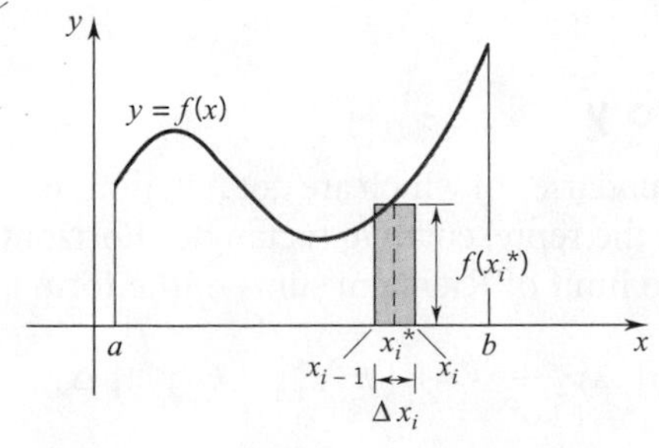

Figure 6.1.1

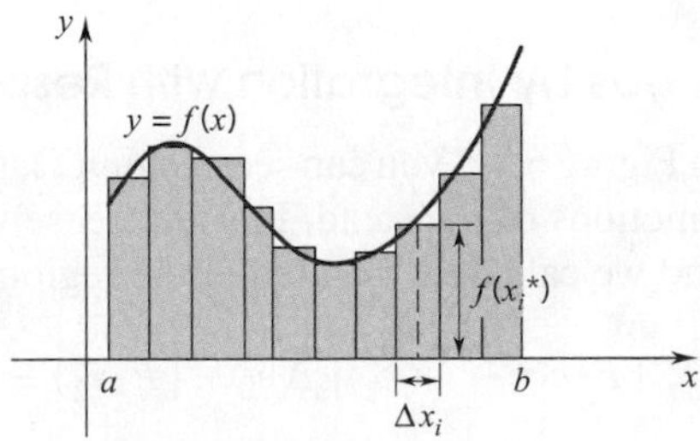

Figure 6.1.2

Figure 6.1.3 shows a region Ω bounded above by the graph of a function f and bounded below by the graph of a function g. In this case, approximating Riemann sums for the area of Ω are of the form

$$[f(x_1^*) - g(x_1^*)]\,\Delta x_1 + [f(x_2^*) - g(x_2^*)]\,\Delta x_2 + \cdots + [f(x_n^*) - g(x_n^*)]\,\Delta x_n.$$

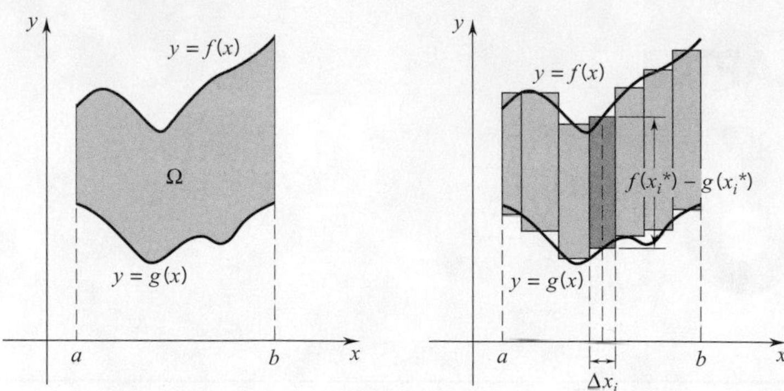

Figure 6.1.3

A representative rectangle has dimensions:

$$\text{“height”} = f(x_i^*) - g(x_i^*) \qquad \text{and} \qquad \text{“width”} = \Delta x_i,$$

and area:

$$[f(x_i^*) - g(x_i^*)]\ \Delta x_i.$$

Thus we can calculate the area of Ω by integrating with respect to x the *vertical separation*

$$f(x) - g(x)$$

from $x = a$ to $x = b$:

$$\text{area}\ (\Omega) = \int_a^b [f(x) - g(x)]\ dx.$$

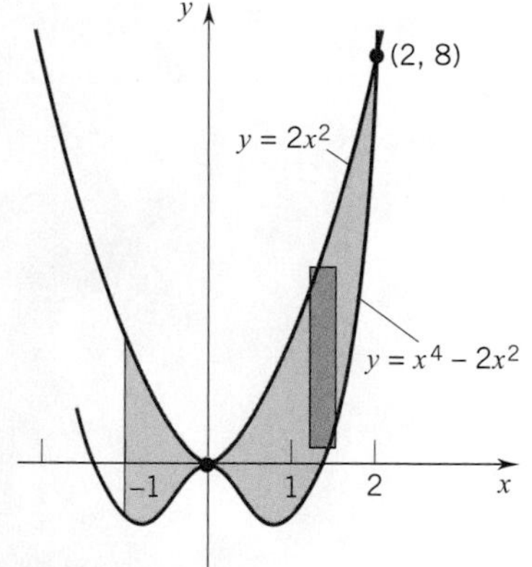

Figure 6.1.4

Example 1 Find the area of the shaded region shown in Figure 6.1.4.

SOLUTION The vertical separation of the curves is : $2x^2 - (x^4 - 2x^2), -1 \leq x \leq 2$. Thus the area of the shaded region is given by

$$A = \int_{-1}^{2} [2x^2 - (x^4 - 2x^2)]\ dx = \int_{-1}^{2} (4x^2 - x^4)\ dx$$

$$= \left[\tfrac{4}{3}x^3 - \tfrac{1}{5}x^5\right]_{-1}^{2} = \left[\tfrac{32}{3} - \tfrac{32}{5}\right] - \left[-\tfrac{4}{3} + \tfrac{1}{5}\right] = \tfrac{27}{5}. \quad \square$$

Areas by Integration with Respect to y

In Figure 6.1.5 you can see a region Ω, the boundaries of which are not functions of x but functions of y instead. In such a case we set the representative rectangles horizontally and we calculate the area of the region as the limit of Riemann sums of the form

$$[F(y_1^*) - G(y_1^*)]\ \Delta y_1 + [F(y_2^*) - G(y_2^*)]\ \Delta y_2 + \cdots + [F(y_n^*) - G(y_n^*)]\ \Delta y_n.$$

The dimensions of a representative rectangle are now

$$\text{“width”} = F(y_i^*) - G(y_i^*) \qquad \text{and} \qquad \text{“height”} = \Delta y_i,$$

The area of Ω is thus given by the integral

$$\int_c^d [F(y) - G(y)]\ dy.$$

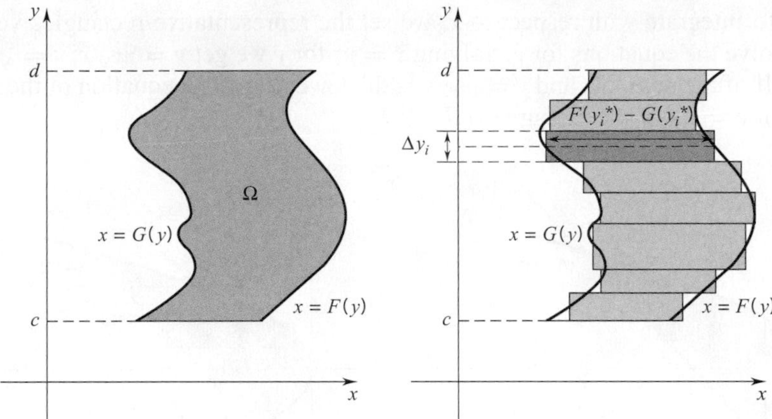

Figure 6.1.5

Here we are integrating the *horizontal separation*

$$F(y) - G(y)$$

with respect to y from $y = c$ to $y = d$.

Example 2 Find the area of the region bounded on the left by $x = y^2$ and on the right by $x = 3 - 2y^2$.

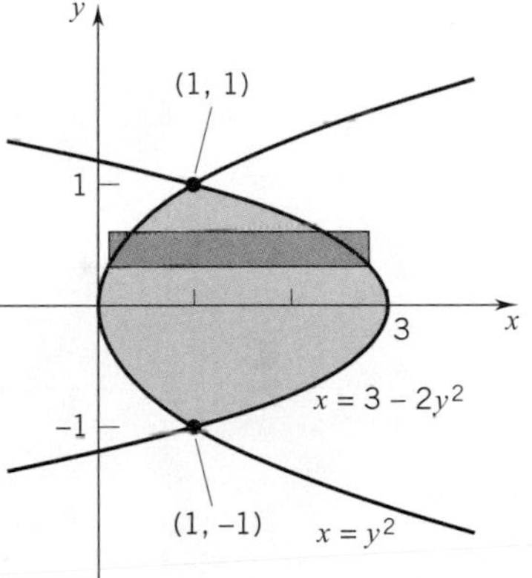

Figure 6.1.6

SOLUTION The region is sketched in Figure 6.1.6. The points of intersection are found by solving the two equations simultaneously:

$$x = y^2 \qquad \text{and} \qquad x = 3 - 2y^2$$

together imply

$$y = \pm 1.$$

Thus the points of intersection are $(1, 1)$ and $(1, -1)$. The easiest way to calculate the area is to set our representative rectangles horizontally and integrate with respect to y. We find the area of the region by integrating the horizontal separation

$$(3 - 2y^2) - y^2 = 3 - 3y^2$$

from $y = -1$ to $y = 1$:

$$A = \int_{-1}^{1} (3 - 3y^2)\, dy = \Big[3y - y^3\Big]_{-1}^{1} = 4.$$

NOTE: In our solution we did not take advantage of the symmetry. The region is symmetric with respect to the x-axis (the integrand is an even function of y,) and so the area is also given by

$$A = 2\int_{0}^{1} (3 - 3y^2)\, dy = 2\Big[3y - y^3\Big]_{0}^{1} = 4. \quad \square$$

Example 3 Calculate the area of the region bounded by the curves $x = y^2$ and $x - y = 2$ by integrating (**a**) with respect to x, (**b**) with respect to y.

SOLUTION You can verify that the two curves intersect at the points $(1, -1)$ and $(4, 2)$.

(a) To integrate with respect to x, we set the representative rectangles vertically, and we solve the equations for y. Solving $x = y^2$ for y we get $y = \pm\sqrt{x}$; $y = \sqrt{x}$ is the upper half of the parabola and $y = -\sqrt{x}$ is the lower half. The equation of the line can be written $y = x - 2$. See Figure 6.1.7.

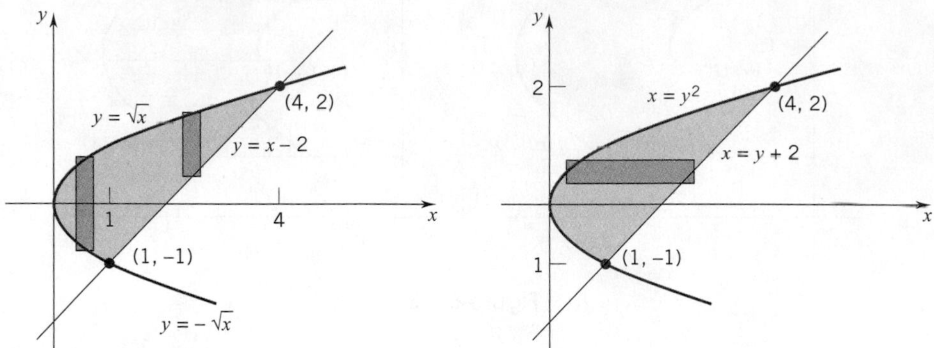

Figure 6.1.7

Figure 6.1.8

The upper boundary of the region is the curve $y = \sqrt{x}$. However, the lower boundary is described by two different equations: $y = -\sqrt{x}$ from $x = 0$ to $x = 1$, and $y = x - 2$ from $x = 1$ to $x = 4$. Thus, we need two integrals:

$$\begin{aligned} A &= \int_0^1 [\sqrt{x} - (-\sqrt{x})]\, dx + \int_1^4 [\sqrt{x} - (x - 2)]\, dx \\ &= 2\int_0^1 \sqrt{x}\, dx + \int_1^4 (\sqrt{x} - x + 2)\, dx = \left[\tfrac{4}{3}x^{3/2}\right]_0^1 + \left[\tfrac{2}{3}x^{3/2} - \tfrac{1}{2}x^2 + 2x\right]_1^4 = \tfrac{9}{2}. \end{aligned}$$

(b) To integrate with respect to y, we set the representative rectangles horizontally. See Figure 6.1.8. Now the right boundary is the line $x = y + 2$ and the left boundary is the curve $x = y^2$. Since y ranges from -1 to 2,

$$A = \int_{-1}^2 [(y + 2) - y^2]\, dy = \left[\tfrac{1}{2}y^2 + 2y - \tfrac{1}{3}y^3\right]_{-1}^2 = \tfrac{9}{2}. \quad \square$$

EXERCISES 6.1

Sketch the region bounded by the curves. Represent the area of the region by one or more integrals (a) in terms of x; (b) in terms of y.

1. $y = x^2, \quad y = x + 2.$

2. $y = x^2, \quad y = -4x.$

3. $y = x^3, \quad y = 2x^2.$

4. $y = \sqrt{x}, \quad y = x^3.$

5. $y = -\sqrt{x}, \quad y = x - 6, \quad y = 0.$

6. $x = y^3, \quad x = 3y + 2.$

7. $y = |x|, \quad 3y - x = 8.$

8. $y = x, \quad y = 2x, \quad y = 3.$

9. $x + 4 = y^2, \quad x = 5.$

10. $x = |y|, \quad x = 2.$

11. $y = 2x, \quad x + y = 9, \quad y = x - 1.$

12. $y = x^3, \quad y = x^2 + x - 1.$

13. $y = x^{1/3}, \quad y = x^2 + x - 1.$

14. $y = x + 1, \quad y + 3x = 13, \quad 3y + x + 1 = 0.$

Sketch the region bounded by the curves and find its area.

15. $4x = 4y - y^2, \quad 4x - y = 0.$

16. $x + y^2 - 4 = 0, \quad x + y = 2.$

17. $x = y^2, \quad x = 12 - 2y^2.$

18. $x + y = 2y^2, \quad y = x^3.$

19. $x + y - y^3 = 0, \quad x - y + y^2 = 0.$

20. $8x = y^3, \quad 8x = 2y^3 + y^2 - 2y.$

21. $y = \cos x, \quad y = \sec^2 x, \quad x \in [-\pi/4, \pi/4].$

22. $y = \sin^2 x, \quad y = \tan^2 x, \quad x \in [-\pi/4, \pi/4].$

HINT: See Exercise 63, Section 5.7.

23. $y = 2\cos x, \quad y = \sin 2x, \quad x \in [-\pi, \pi].$

24. $y = \sin x, \quad y = \sin 2x, \quad x \in [0, \pi/2].$

25. $y = \sin^4 x \cos x, \quad x \in [0, \pi/2].$

26. $y = \sin 2x, \quad y = \cos 2x, \quad x \in [0, \pi/4]$.

In Exercises 27 and 28, use integration to find the area of the triangle with the given vertices.

27. $(0, 0), (1, 3), (3, 1)$.

28. $(0, 1), (2, 0), (3, 4)$.

29. Use integration to find the area of the trapezoid with vertices $(-2, -2), (1, 1), (5, 1), (7, -2)$.

30. Sketch the region bounded by the three curves $y = x^3$, $y = -x, y = 1$, and find its area.

31. Sketch the region bounded by the three curves $y = 6 - x^2$, $y = x \ (x < 0)$ and $y = -x \ (x > 0)$, and find its area.

32. Find the area of the region bounded by the parabolas $x^2 = 4py$ and $y^2 = 4px$, p a positive constant.

33. Sketch the region bounded by $y = x^2$ and $y = 4$. This region is divided into two subregions of equal area by the line $y = c$. Find c.

34. The region between $y = \cos x$ and the x-axis for $x \in [0, \pi/2]$ is divided into two subregions of equal area by the line $x = c$. Find c.

In Exercise 35–38, *set up* a definite integral (or integrals) the value of which gives the area of the region.

35. The region in the first quadrant bounded by the x-axis, the line $y = \sqrt{3}\,x$, and the circle $x^2 + y^2 = 4$.

36. The region in the first quadrant bounded by the y-axis, the line $y = \sqrt{3}\,x$, and the circle $x^2 + y^2 = 4$.

37. The region determined by the intersection of the circles $x^2 + y^2 = 4$ and $(x - 2)^2 + (y - 2)^2 = 4$.

38. The region in the first quadrant bounded by the x-axis, the parabola $y = x^2/3$ and the circle $x^2 + y^2 = 4$.

39. A rectangle has one vertex at the origin, opposite vertex on the curve $y = bx^n$ at the point where $x = a \ (a > 0, b > 0, n > 0)$, and sides parallel to the coordinate axes. Show that the ratio of the area of that part of the rectangle that lies below the curve to the total area of the of the rectangle depends only on n; the ratio is independent of a and b.

40. (a) Calculate the area of the region in the first quadrant bounded by the coordinate axes and the parabola $y = 1 + a - ax^2, a > 0$.

(b) Determine a so that the area found in part (a) is a minimum.

41. Use a graphing utility to sketch the region bounded by the curves $y = x^4 - 2x^2$ and $y = x + 2$. Then find (approximately) the area of the region.

42. Use a graphing utility to sketch the region bounded by the curves $y = \sin x$ and $y = |x - 1|$. Then find (approximately) the area of the region.

43. Let $f(x) = 2x^3 + 27x^2$ and $g(x) = x^5 + 54$.

(a) Use a graphing utility to draw the graphs of f and g together.

(b) Use a CAS to find the points of intersection of the graphs.

(c) Use a CAS to find the area of the region bounded by the graphs.

44. Repeat Exercise 43 with $f(x) = \cos x$ and $g(x) = \sin(x/2)$ on $[0, 2\pi]$.

45. A section of rain gutter is 8 feet long. Vertical cross sections of the gutter are in the shape of the parabolic region bounded by $y = \frac{4}{9}x^2$ and $y = 4$, with x and y measured in inches. What is the volume of the rain gutter?

HINT: $V =$ (cross-sectional area) $\times$ length.

46. (a) Calculate the area A of the region bounded by the graph of $f(x) = 1/x^2$ and x-axis with $x \in [1, b]$.

(b) The result in part (a) depends on b. Calculate the limit of $A(b)$ as $b \to \infty$. What can you conclude about the region?

47. (a) Calculate the area A of the region bounded by the graph of $f(x) = 1/\sqrt{x}$ and x-axis with $x \in [1, b]$.

(b) The result in part (a) depends on b. Calculate the limit of $A(b)$ as $b \to \infty$. What can you conclude about the region?

48. (a) Let $p > 1$. Calculate the area A of the region bounded by the graph of $f(x) = 1/x^p$ and the x-axis with $x \in [1, b]$, and then calculate $\lim_{b\to\infty} A(b)$.

(b) Let $0 < p < 1$. Calculate the area A of the region bounded by the graph of $f(x) = 1/x^p$ and the x-axis with $x \in [1, b]$. and then calculate $\lim_{b\to\infty} A(b)$.

■ PROJECT 6.1 Distribution of Income

Economists use a cumulative distribution called a *Lorenz curve* to measure the distribution of income among households in a given country. Typically, a Lorenz curve begins at (0, 0) and ends at (1, 1), is continuous, increasing and concave up. See the figure. The points on the curve are determined by ranking all households by income and then comparing the percentage of households whose total income is less than or equal to a given percentage of the total income of the country. For example, a point (a, b) on the curve indicates that a% of the households receive b% of the total income. *Absolute equality* of income distribution would occur if the bottom a% of the households received a% of the total income. The Lorenz curve in this case would be the line $y = x$. The area of the region between the line $y = x$ and the Lorenz curve measures the extent to which the income distribution differs from absolute equality. The *coefficient of inequality*, also called the *index of income concentration*, is the ratio of the area of the region between $y = x$ and the Lorenz curve to the area under $y = x$.

Problem 1. Let $y = L(x)$ be a Lorenz curve.

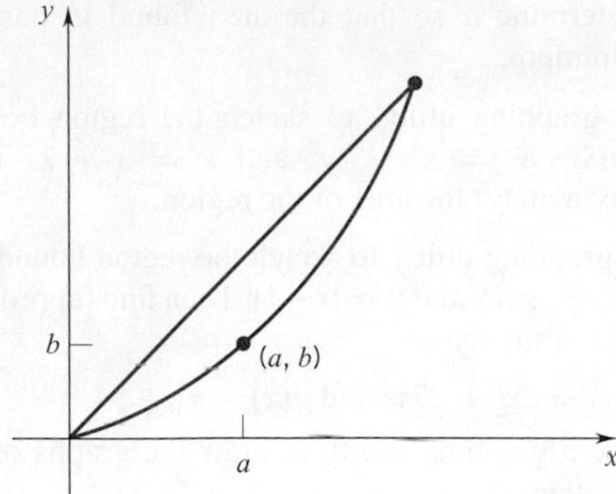

(a) Show that the coefficient of inequality C is given by

$$C = 2\int_0^1 [x - L(x)]\, dx.$$

(b) Show that C is always a number between 0 and 1. What does a coefficient close to 0 represent? What does a coefficient close to 1 represent?

Problem 2. The income distribution of a certain country is represented by the Lorenz curve

$$L(x) = \tfrac{7}{12}x^2 + \tfrac{5}{12}x, \quad \text{for } x \in [0, 1]$$

(a) What is the percentage of total income received by the bottom 50% of the households?

(b) Determine the coefficient of inequality.

Problem 3. An economist studying the effects of World War II on the U.S. economy used data from the Bureau of the Census to develop Lorenz curves for the distribution of income in 1935; $L(x) = x^{2.4}$, and 1947: $L(x) = x^{1.6}$. Find the coefficient of inequality correct to three decimal places for each Lorenz curve and interpret the results.

Problem 4. Let $L(x) = \dfrac{5x^3}{4 + x^2}$, for $x \in [0, 1]$

(a) Show that L has the characteristics of a Lorenz curve: continuous, increasing, concave up.

(b) Estimate the coefficient of inequality by calculating the Riemann sum for $f(x) = x - L(x)$ with the partition $P = \{0,\ 0.2,\ 0.4,\ 0.6,\ 0.8,\ 1\}$ taking $x_1^* = 0.1$, $x_2^* = 0.3, \cdots, x_5^* = 0.9$.

(c) Use a graphing utility to find the coefficient of inequality and compare the result with your estimate in part (b).

■ 6.2 VOLUME BY PARALLEL CROSS SECTION; DISCS AND WASHERS

In this section we use define integrals to find the volumes of certain three-dimensional solids. To begin, let Ω be a plane region. A *right cylinder with cross section* Ω is a solid formed by translating Ω along a line which is perpendicular to the plane of Ω. Such a line is called an *axis* for the cylinder. See Figure 6.2.1. Let A be the area of Ω. If a right cylinder is formed by translating the region Ω through a distance h, then the volume of the cylinder is given by the product

$$V = A \cdot h.$$

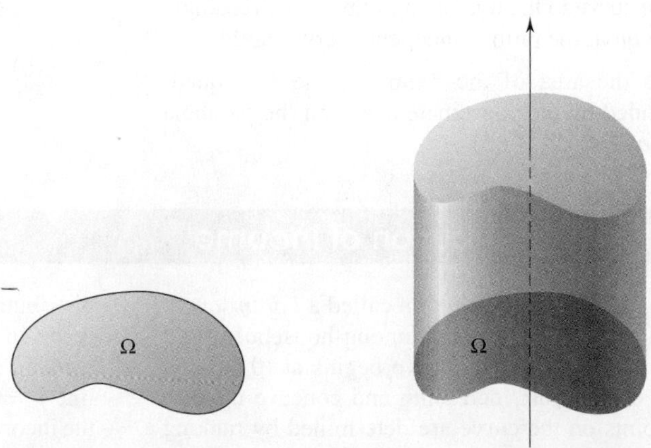

Figure 6.2.1

Two familiar examples are shown in Figure 6.2.2: a right circular cylinder of radius r and height h, and a rectangular box of length l, width w, and height h.

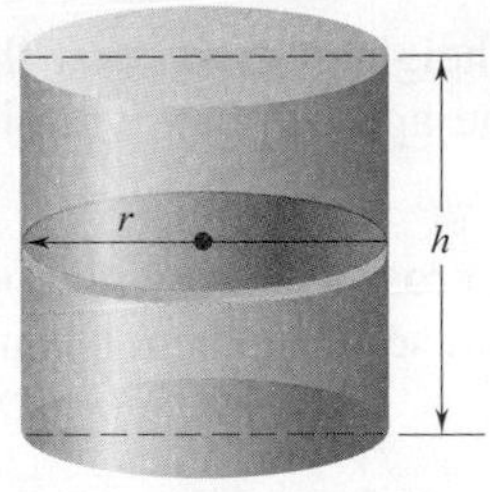

$V = \pi r^2 h =$ (cross-sectional area) · height

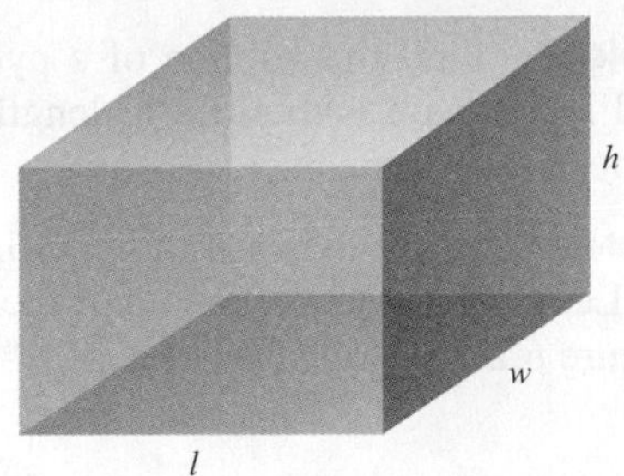

$V = l \cdot w \cdot h =$ (cross-sectional area) · height

Figure 6.2.2

To calculate the volume of a more general solid, we introduce a coordinate axis and then examine the cross sections of the solid that are perpendicular to that axis. In Figure 6.2.3 we picture a solid and a coordinate axis that we label the x-axis. As in the figure, we suppose that the solid lies entirely between $x = a$ and $x = b$. The figure shows an arbitrary cross section perpendicular to the x-axis. We denote by $A(x)$ the area of the cross section that has coordinate x.

If the cross-sectional area, $A(x)$, varies continuously with x, then we can find the volume V of the solid by integrating $A(x)$ from a to b:

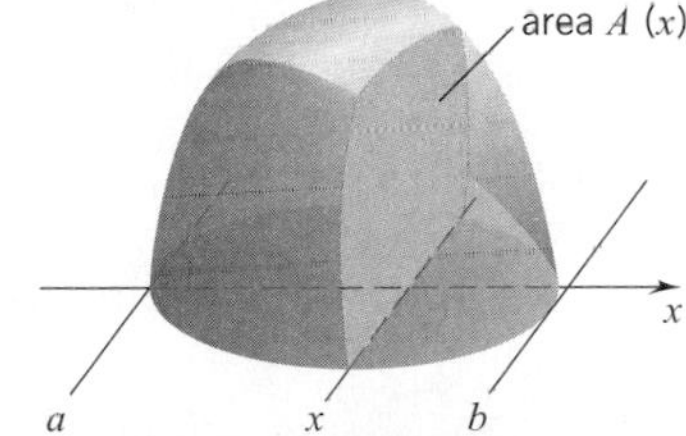

Figure 6.2.3

(6.2.1)

$$V = \int_a^b A(x)\,dx.$$

PROOF Let $P = \{x_0, x_1 \cdots, x_n\}$ be a partition of $[a, b]$. For each subinterval $[x_{i-1}, x_i]$ choose a point x_i^*. The solid from x_{i-1} to x_i can be approximated by a cylinder of cross-sectional area $A(x_i^*)$ and thickness Δx_i. The volume of this approximating cylinder is the product

$$A(x_i^*)\,\Delta x_i \qquad \text{(Figure 6.2.4)}$$

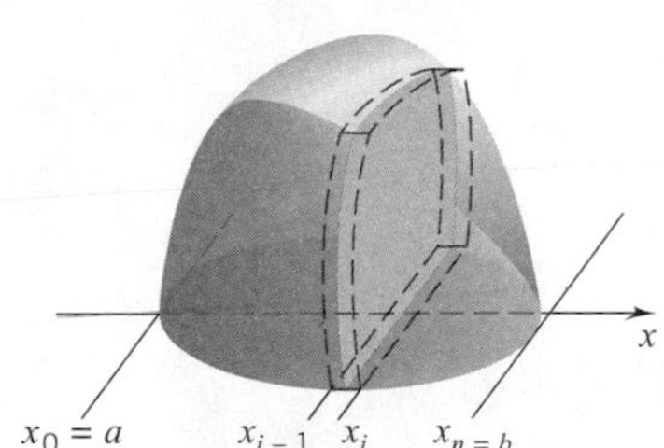

Figure 6.2.4

The sum of these products,

$$A(x_1^*)\,\Delta x_1 + A(x_2^*)\,\Delta x_2 + \cdots A(x_n^*)\,\Delta x_n,$$

is a Riemann sum which approximates the volume of the entire solid. As $||P|| \to 0$, the Riemann sum converges to

$$\int_a^b A(x)\,dx. \quad \square$$

Remark By the mean-value theorem for integrals [see (5.9.1)], the formula for the volume of a solid with varying cross-sectional area $A = A(x)$ can be written

(6.2.2)

$$V = (\text{the average cross-sectional area}) \cdot (b - a) = A_{\text{avg}}\,(b - a). \quad \square$$

Finding the volume of a solid with a more general shape is left to Chapter 16. Here we restrict our attention to solids with simple cross sections.

Example 1 Find the volume of a pyramid of height h given that the base of the pyramid is a square with sides of length r and the apex of the pyramid lies directly above the center of the base.

SOLUTION Set the x-axis as in Figure 6.2.5a. The cross section with coordinate x is a square. Let $s = s(x)$ denote length of the side of that square. Then, by similar triangles (see Figure 6.2.5b), we have

$$\frac{\frac{1}{2}s}{h-x} = \frac{\frac{1}{2}r}{h},$$

and so

$$s(x) = \frac{r}{h}(h-x).$$

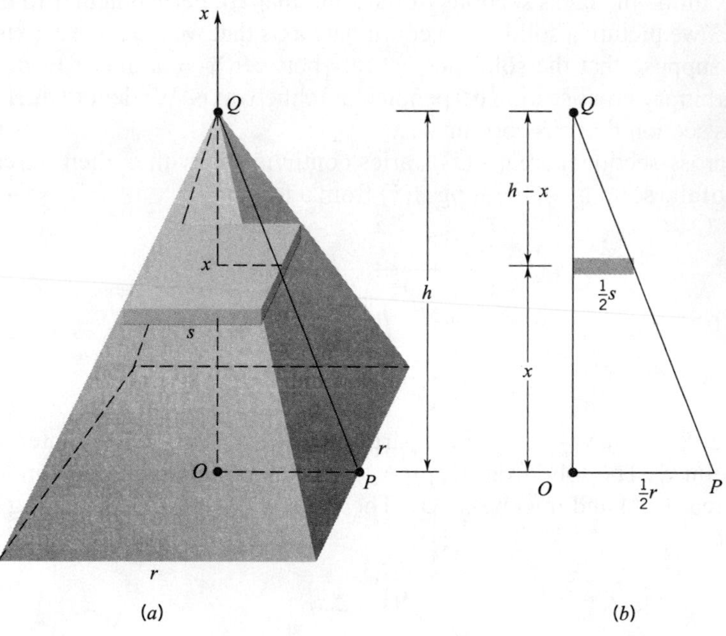

Figure 6.2.5

The area of the square at coordinate x is $A(x) = s^2 = (r^2/h^2)(h-x)^2$. Thus

$$V = \int_0^h A(x)\,dx = \frac{r^2}{h^2}\int_0^h (h-x)^2\,dx$$

$$= \frac{r^2}{h^2}\left[-\frac{(h-x)^3}{3}\right]_0^h = \tfrac{1}{3}r^2h. \quad \square$$

Example 2 The base of a solid is the region bounded by the ellipse

$$\frac{x^2}{a^2} + \frac{y^2}{b^2} = 1.$$

Find the volume of the solid given that each cross section perpendicular to the x-axis is an isosceles triangle with base in the region and altitude equal to one-half the base.

SOLUTION Set the x-axis as in Figure 6.2.6. The cross section with coordinate x is an isosceles triangle with base $\overline{PQ}$ and altitude $\frac{1}{2}\overline{PQ}$. The equation of the ellipse can be written

$$y^2 = \frac{b^2}{a^2}(a^2 - x^2).$$

Since
$$\text{length of } \overline{PQ} = 2y = \frac{2b}{a}\sqrt{a^2 - x^2},$$

the isosceles triangle has area

$$A(x) = \tfrac{1}{2}bh = \tfrac{1}{2}\left(\frac{2b}{a}\sqrt{a^2 - x^2}\right)\left(\frac{b}{a}\sqrt{a^2 - x^2}\right) = \frac{b^2}{a^2}(a^2 - x^2).$$

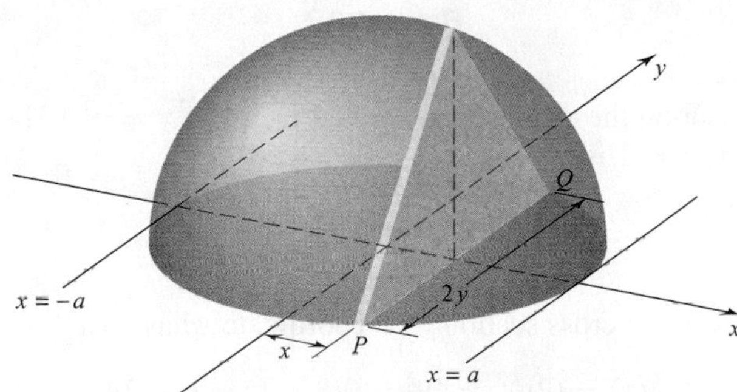

Figure 6.2.6

We can find the volume of the solid by integrating $A(x)$ from $x = -a$ to $x = a$:

$$V = \int_{-a}^{a} A(x)\,dx = 2\int_{0}^{a} A(x)\,dx$$

by symmetry

$$= \frac{2b^2}{a^2}\int_{0}^{a}(a^2 - x^2)\,dx$$

$$= \frac{2b^2}{a^2}\left[a^2x - \frac{x^3}{3}\right]_0^a = \tfrac{4}{3}ab^2. \quad \square$$

Example 3 The base of a solid is the region between the parabolas

$$x = y^2 \quad \text{and} \quad x = 3 - 2y^2 \qquad \text{(See Example 2, Section 6.1)}$$

Find the volume of the solid given that the cross sections perpendicular to the x-axis are squares.

SOLUTION The two parabolas intersect at $(1,1)$ and $(1,-1)$ (Figure 6.2.7). From $x = 0$ to $x = 1$, the cross section with coordinate x has area

$$A(x) = (2y)^2 = 4y^2 = 4x.$$

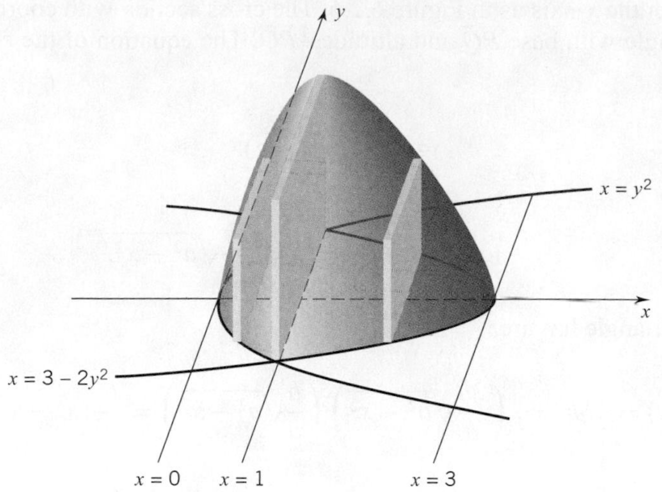

Figure 6.2.7

(Here we are measuring the span across the first parabola $x = y^2$.) The volume of the solid from $x = 0$ to $x = 1$ is

$$V_1 = \int_0^1 4x\,dx = \left[2x^2\right]_0^1 = 2.$$

From $x = 1$ to $x = 3$, the cross section with coordinate x has area

$$A(x) = (2y)^2 = 4y^2 = 2(3 - x) = 6 - 2x.$$

(Here we are measuring the span across the second parabola $x = 3 - 2y^2$.) The volume of the solid from $x = 1$ to $x = 3$ is

$$V_2 = \int_1^3 (6 - 2x)\,dx = \left[6x - x^2\right]_1^3 = 4.$$

The total volume is

$$V_1 + V_2 = 6. \quad \square$$

Solids of Revolution: Disc Method

Suppose that f is nonnegative and continuous on $[a, b]$. (See Figure 6.2.8.) If we revolve the region bounded by the graph of f and the x-axis about the x-axis, we obtain a solid.

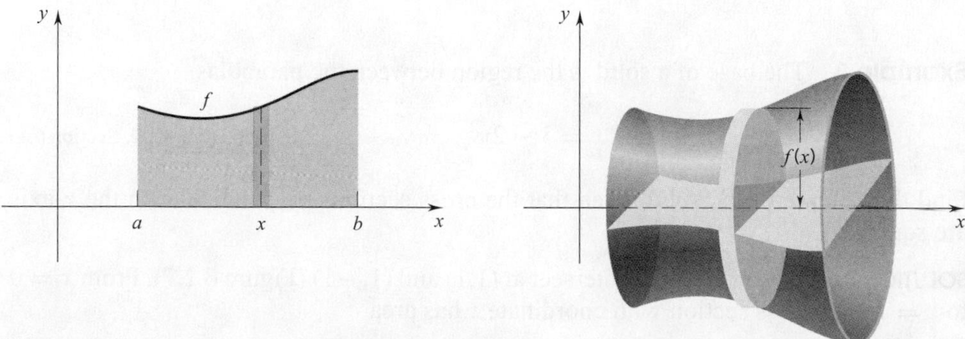

Figure 6.2.8

The volume of this solid is given by the formula

(6.2.3)
$$V = \int_a^b \pi[f(x)]^2\,dx.$$

VERIFICATION The cross section with coordinate x is a circular *disc* of radius $f(x)$. The cross-sectional area is thus $\pi[f(x)]^2$ ❑

Among the simplest solids of revolution are the cone and the sphere.

Example 4 We can generate a cone of base radius r and height h by revolving about the x-axis the region below the graph of

$$f(x) = \frac{r}{h}x, \qquad 0 \le x \le h. \qquad \text{(Figure 6.2.9)}$$

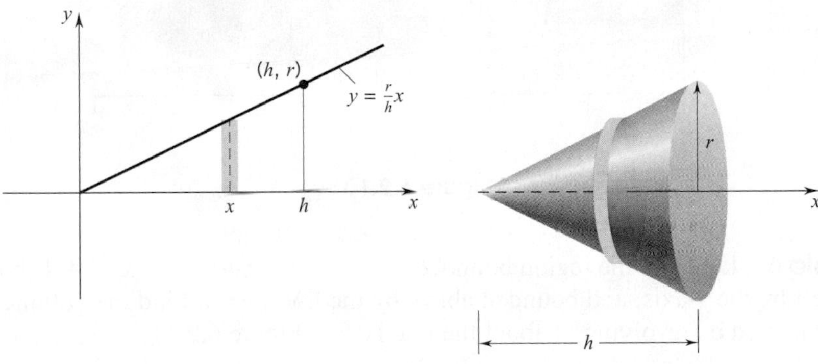

Figure 6.2.9

By Formula 6.2.3,

$$\text{volume of cone} = \int_0^h \pi\left[\frac{r}{h}x\right]^2 dx = \frac{\pi r^2}{h^2}\int_0^h x^2\,dx$$
$$= \frac{\pi r^2}{h^2}\left[\frac{x^3}{3}\right]_0^h = \tfrac{1}{3}\pi r^2 h. \quad ❑$$

Example 5 A sphere of radius r can be obtained by revolving about the x-axis the region below the graph of

$$f(x) = \sqrt{r^2 - x^2}, \qquad -r \le x \le r. \qquad \text{(draw a figure)}$$

Therefore

$$\text{volume of sphere} = \int_{-r}^{r} \pi(r^2 - x^2)\,dx = \pi\left[r^2x - \tfrac{1}{3}x^3\right]_{-r}^{r} = \tfrac{4}{3}\pi r^3$$

NOTE: Archimedes derived the formula for the volume of a sphere in the third century B.C. ❑

Now suppose that $x = g(y)$ is continuous and nonnegative for $c \le y \le d$ (see Figure 6.2.10). If we revolve the region bounded by the graph of g and the y-axis about the y-axis, we obtain a solid with each cross section perpendicular to the y-axis a disc of area $A(y) = \pi[g(y)]^2$. Thus the volume of this solid is given by

(6.2.4)
$$V = \int_c^d \pi[g(y)]^2\,dy.$$

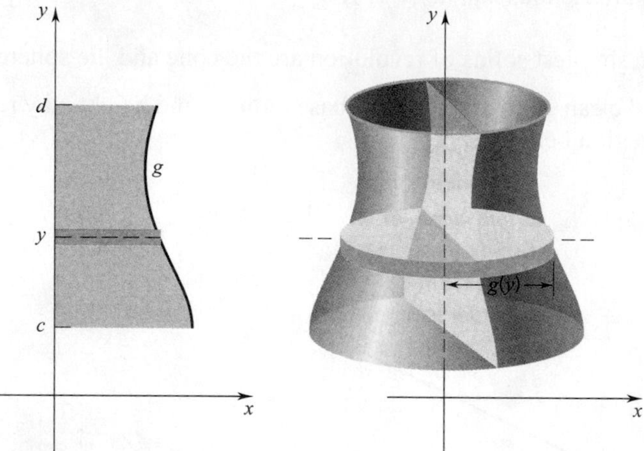

Figure 6.2.10

Example 6 Let Ω be the region bounded below by the curve $y = x^{2/3} + 1$, bounded to the left by the y-axis, and bounded above by the line $y = 5$. Find the volume of the solid generated by revolving Ω about the y-axis. See Figure 6.2.11.

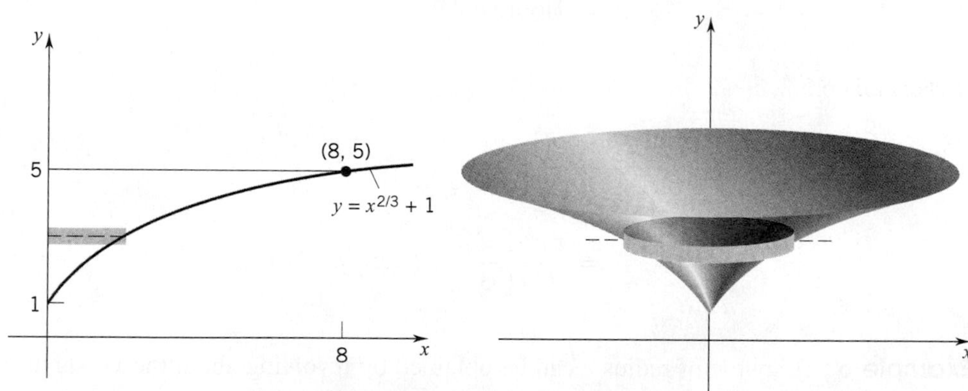

Figure 6.2.11

SOLUTION We need to express the right-hand boundary of Ω as a function of y:

$$y = x^{2/3} + 1 \quad \text{gives} \quad x^{2/3} = y - 1 \quad \text{and thus} \quad x = (y-1)^{3/2}$$

The volume of the solid obtained by revolving Ω about the y-axis is given by the integral

$$V = \int_1^5 \pi[g(y)]^2\,dy = \pi\int_1^5 [(y-1)^{3/2}]^2\,dy$$
$$= \pi\int_1^5 (y-1)^3\,dy = \pi\left[\frac{(y-1)^4}{4}\right]_1^5 = 64\pi \quad \square$$

Solids of Revolution: Washer Method

The washer method is a slight generalization of the disc method. Suppose that f and g are nonnegative continuous functions with $g(x) \leq f(x)$ for all x in [a,b] (see Figure 6.2.12). If we revolve the region Ω about the x-axis, we obtain a solid. The volume of this solid is given by the formula

(6.2.5)
$$V = \int_a^b \pi([f(x)]^2 - [g(x)]^2)\,dx.$$
(washer method about x-axis)

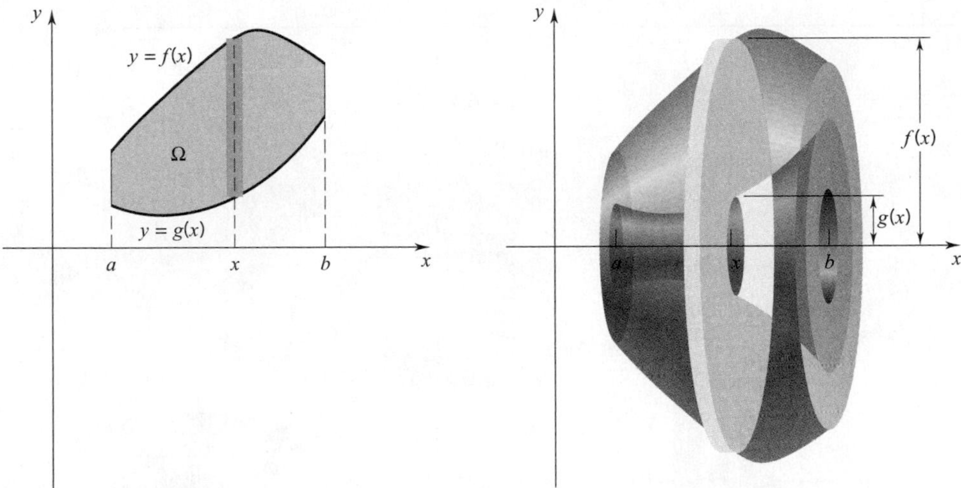

Figure 6.2.12

VERIFICATION The cross section with coordinate x is a *washer* of outer radius $f(x)$, inner radius $g(x)$ and area

$$A(x) = \pi[f(x)]^2 - \pi[g(x)]^2 = \pi\left([f(x)]^2 - [g(x)]^2\right).$$

We can get the volume of the solid by integrating this function from a to b. □

Suppose now that the boundaries are functions of y rather than x (see Figure 6.2.13 on page 340). By revolving Ω *about the y-axis*, we obtain a solid. It follows from (6.2.4) that in this case

(6.2.6)
$$V = \int_c^d \pi\left([F(y)]^2 - [G(y)]^2\right)dy.$$
(washer method about y-axis)

Example 7 Find the volume of the solid generated by revolving the region between $y = x^2$ and $y = 2x$ about the x-axis.

SOLUTION Setting $x^2 = 2x$, we get $x = 0, 2$. Thus, the curves intersect at the points (0, 0) and (2, 4). See Figure 6.2.14. For each x from 0 to 2 , the x cross section is a washer of outer radius $2x$ and inner radius x^2. By (6.2.5)

$$V = \int_0^2 \pi[(2x)^2 - (x^2)^2]\,dx = \pi\int_0^2 (4x^2 - x^4)dx$$
$$= \pi\left[\tfrac{4}{3}x^3 - \tfrac{1}{5}x^5\right]_0^2 = \tfrac{64}{15}\pi. \quad □$$

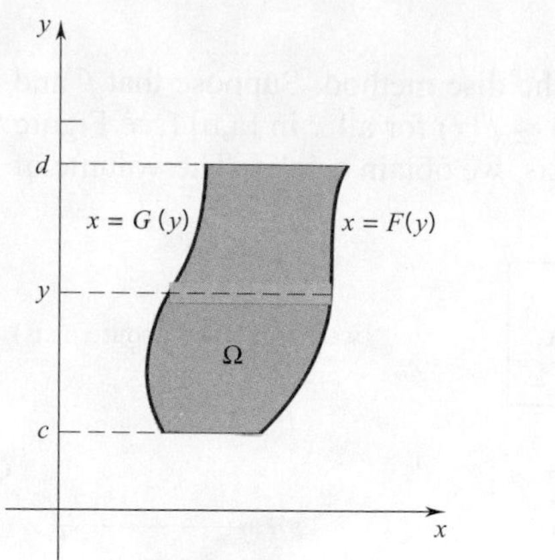

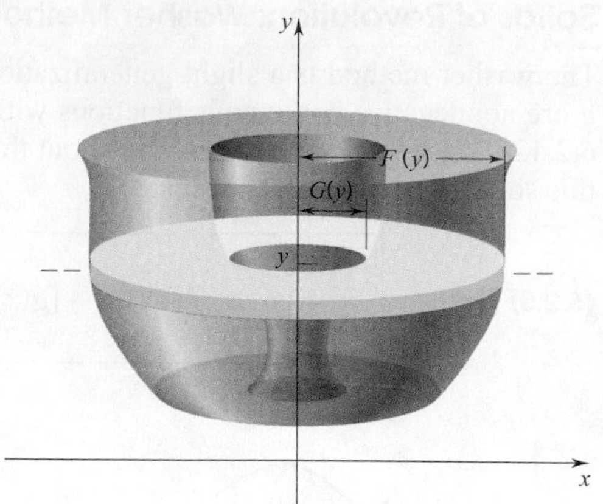

Figure 6.2.13

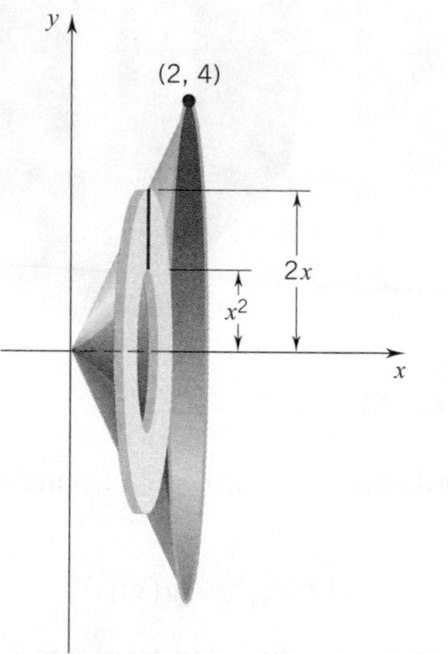

Figure 6.2.14

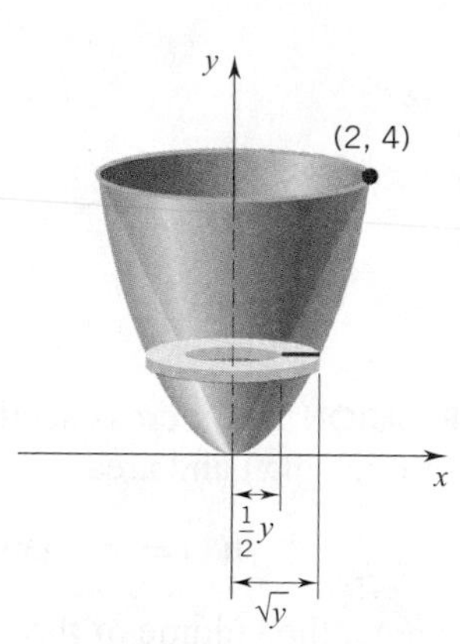

Figure 6.2.15

Example 8 Find the volume of the solid generated by revolving the region between $y = x^2$ and $y = 2x$ about the y-axis.

SOLUTION The solid is depicted in Figure 6.2.15. For each y from 0 to 4, the y cross section is a washer of outer radius $\sqrt{y}$ and inner radius $\frac{1}{2}y$. By (6.2.6)

$$V = \int_0^4 \pi\left[(\sqrt{y})^2 - \left(\tfrac{1}{2}y\right)^2\right] dy = \pi \int_0^4 \left(y - \tfrac{1}{4}y^2\right) dy$$

$$= \pi\left[\tfrac{1}{2}y^2 - \tfrac{1}{12}y^3\right]_0^4 = \tfrac{8}{3}\pi. \quad \square$$

Remark These last two examples concerned solids generated by revolving the *same* region about *different* axes. Notice that the solids are different and have different volumes. □

EXERCISES 6.2

Sketch the region Ω bounded by the curves and find the volume of the solid generated by revolving this region about the x-axis.

1. $y = x, \quad y = 0, \quad x = 1.$

2. $x + y = 3, \quad y = 0, \quad x = 0.$

3. $y = x^2, \quad y = 9.$

4. $y = x^3, \quad y = 8, \quad x = 0.$

5. $y = \sqrt{x}, \quad y = x^3.$

6. $y = x^2, \quad y = x^{1/3}.$

7. $y = x^3, \quad x + y = 10, \quad y = 1.$

8. $y = \sqrt{x}, \quad x + y = 6, \quad y = 1.$

9. $y = x^2, \quad y = x + 2.$

10. $y = x^2, \quad y = 2 - x.$

11. $y = \sqrt{4 - x^2}, \quad y = 0.$

12. $y = 1 - |x|, \quad y = 0.$

13. $y = \sec x, \quad x = 0, \quad x = \frac{1}{4}\pi, \quad y = 0.$

14. $y = \csc x, \quad x = \frac{1}{4}\pi, \quad x = \frac{3}{4}\pi, \quad y = 0.$

15. $y = \cos x, \quad y = x + 1, \quad x = \frac{1}{2}\pi.$

16. $y = \sin x, \quad x = \frac{1}{4}\pi, \quad x = \frac{1}{2}\pi, \quad y = 0.$

Sketch the region Ω bounded by the curves and find the volume of the solid generated by revolving this region about the y-axis.

17. $y = 2x, \quad y = 4, \quad x = 0.$

18. $x + 3y = 6, \quad x = 0, \quad y = 0.$

19. $x = y^3, \quad x = 8, \quad y = 0.$

20. $x = y^2, \quad x = 4.$

21. $y = \sqrt{x}, \quad y = x^3.$

22. $y = x^2, \quad y = x^{1/3}.$

23. $y = x, \quad y = 2x, \quad x = 4.$

24. $x + y = 3, \quad 2x + y = 6, \quad x = 0.$

25. $x = y^2, \quad x = 2 - y^2.$

26. $x = \sqrt{9 - y^2}, \quad x = 0.$

27. The base of a solid is the circle $x^2 + y^2 = r^2$. Find the volume of the solid given that the cross sections perpendicular to the x-axis are: (a) squares; (b) equilateral triangles.

28. The base of a solid is the region bounded by the ellipse $4x^2 + 9y^2 = 36$. Find the volume of the solid given that cross sections perpendicular to the x-axis are: (a) equilateral triangles; (b) squares.

29. The base of a solid is the region bounded by $y = x^2$ and $y = 4$. Find the volume of the solid given that the cross sections perpendicular to the x-axis are : (a) squares; (b) semicircles; (c) equilateral triangles.

30. The base of a solid is the region between the parabolas $x = y^2$ and $x = 3 - 2y^2$. Find the volume of the solid given that the cross sections perpendicular to the x-axis are:

(a) rectangles of height h;

(b) equilateral triangles;

(c) isosceles right triangles each with hypotenuse on the xy-plane.

31. Repeat Exercise 29 with the cross sections perpendicular to the y-axis.

32. Repeat Exercise 30 with the cross sections perpendicular to the y-axis.

33. The base of a solid is the triangular region bounded by the y-axis and the lines $x + 2y = 4$, $x - 2y = 4$. Find the volume of the solid given that the cross sections perpendicular to the x-axis are: (a) squares; (b) isosceles right triangles with hypotenuse on the xy-plane.

34. The base of a solid is the region bounded by the ellipse $b^2x^2 + a^2y^2 = a^2b^2$. Find the volume of the solid given that the cross sections perpendicular to the x-axis are: (a) isosceles right triangles each with hypotenuse on the xy-plane; (b) squares; (c) isosceles triangles of height 2.

35. The base of a solid is the region bounded by $y = 2\sqrt{\sin x}$ and the x-axis with $x \in [0, \pi/2]$. Find the volume of the solid given that cross sections perpendicular to the x-axis are: (a) equilateral triangles; (b) squares.

36. The base of a solid is the region bounded by $y = \sec x$ and $y = \tan x$ with $x \in [0, \pi/4]$. Find the volume of the solid given that cross sections perpendicular to the x-axis are: (a) semicircles; (b) squares.

37. Find the volume of the solid generated by revolving the ellipse $b^2x^2 + a^2y^2 = a^2b^2$ about the x-axis.

38. Repeat Exercise 37 with the ellipse revolved about the y-axis.

39. Derive a formula for the volume of the frustum of a right circular cone in terms of the height h, the lower base radius R, and the upper base radius r. (See the figure.)

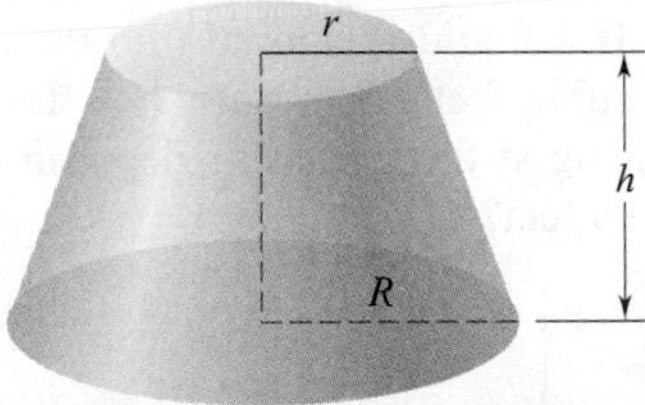

40. Find the volume enclosed by the surface that is generated by revolving the equilateral triangle with vertices (0, 0), $(a, 0)$, $(\frac{1}{2}a, \frac{1}{2}\sqrt{3}a)$ about the x-axis.

41. A hemispherical basin of radius r feet is being used to store water. To what percent of capacity is it filled when the water is:

(a) $\frac{1}{2}r$ feet deep?

(b) $\frac{1}{3}r$ feet deep?

42. A sphere of radius r is cut by two parallel planes: one is a units above the equator; the other, b units above the equator. Find the volume of the portion of the sphere that lies between the two planes. Assume that $a < b$.

43. A sphere of radius r is cut by a plane h units above the equator $(0 < h < r)$. The top portion is called a *cap*. Derive the formula for the volume of a cap.

44. A hemispherical punch bowl 2 feet in diameter is filled to within 1 inch of the top. Thirty minutes after the party starts, there are only 2 inches of punch left at the bottom of the bowl.

(a) How many gallons of punch were in the bowl at the beginning?

(b) How many gallons of punch were consumed?

HINT: See Exercises 42 and 43.

45. Let f be defined by $f(x) = x^{-2/3}$ for $x > 0$.

(a) Sketch the graph of f.

(b) Show that the area of the region bounded by the graph of f and the x-axis between $x = 1$ and $x = b, b > 1$, is given by $A(b) = 3(b^{1/3} - 1)$.

(c) The region in part (b) is rotated about the x-axis. Show that the volume of the resulting solid is given by $V(b) = 3\pi(1 - b^{1/3})$.

(d) Show that $A(b) \to \infty$ and $V(b) \to 3\pi$ as $b \to \infty$. Note that a region with "infinite" area when rotated about the x-axis generates a solid with a finite volume.

46. This is a continuation of Exercise 45.

(a) Show that the area of the region bounded by the graph of f and the x-axis between $x = c$ and $x = 1, 0 < c < 1$, is given by $A(c) = 3(1 - c^{1/3})$.

(b) The region in part (a) is rotated about the x-axis. Show that the volume of the resulting solid is given by $V(c) = 3\pi(c^{1/3} - 1)$.

(c) Show that $A(c) \to 3$ and $V(c) \to \infty$ as $c \to 0$. In this case a region with finite area generates a solid with infinite volume.

47. The parabolic arc shown in the figure is revolved about the y-axis to form a parabolic container. The indicated units are in feet. If a liquid is poured into the container at the rate of two cubic feet per minute, how fast is the level of the liquid rising at the instant its depth in the container is one foot? Two feet?

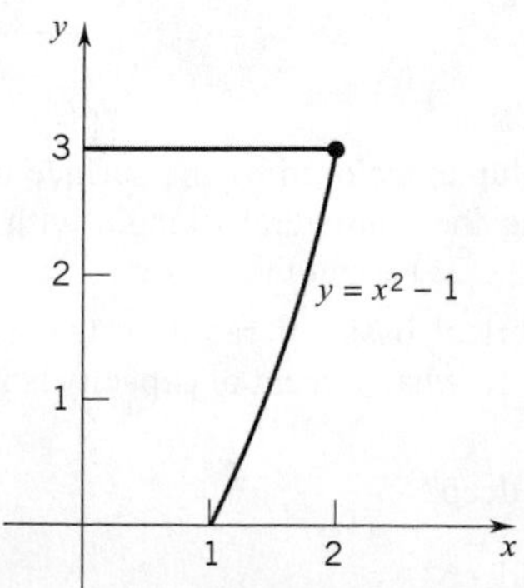

48. Let $f(x) = x^3$ and $g(x) = x$.

(a) Use a graphing utility to draw the graphs of f and g together.

(b) Use a CAS to find the points of intersection of the two graphs.

(c) Use a CAS to find the area of the region bounded by the two curves.

(d) The region in part (c) is rotated about the x-axis. Use a CAS to find the volume of the solid that is generated.

(e) (Optional) Use a CAS to draw the solid of revolution.

49. Repeat Exercise 48 with $f(x) = \sqrt{2x - 1}$ and $g(x) = x^2 - 4x + 4$.

Rotations About Lines Parallel to a Coordinate Axis

For Exercises 50 and 51, let f be a continuous, nonnegative function on $[a, b]$ and let Ω be the region between the graph of f and the x-axis.

50. Let k be a positive constant. Show that if the region Ω is revolved around the line $y = -k$ (Figure A), then the volume of the solid of revolution (Figure B) is given by

$$V = \int_a^b \pi\left([f(x)]^2 + 2kf(x)\right)dx.$$

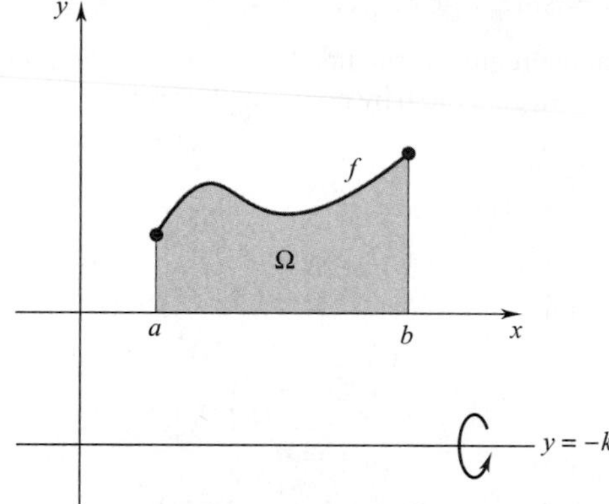

Figure A

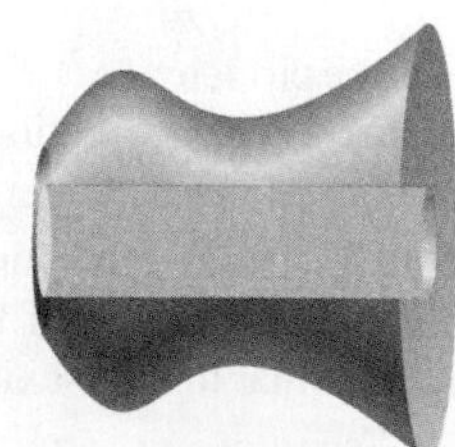

Figure B

51. Let k be a positive constant satisfying $k - f(x) \geq 0$ on $[a, b]$. Show that if the region Ω is revolved about the line $y = k$ (Figure C), then the volume of the solid of revolution (Figure D) is given by

$$V = \int_a^b \pi\left(2kf(x) - [f(x)]^2\right)dx.$$

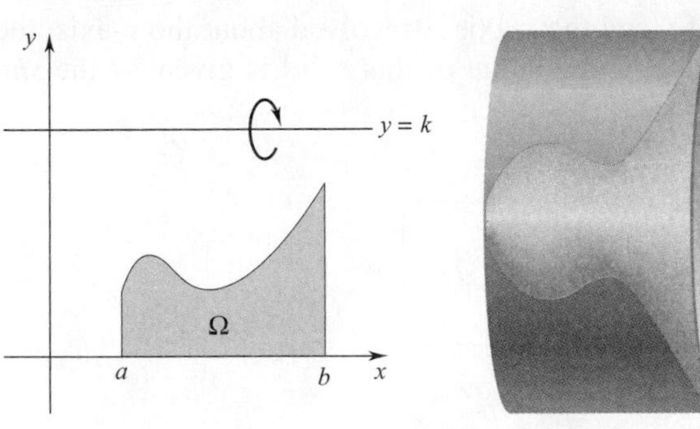

Figure C Figure D

52. Let f and g be functions continuous on $[a, b]$ and let k be a constant.

(a) Suppose that $f(x) \geq g(x) \geq k$ on $[a, b]$. Let Ω be the region between the graphs of f and g. Show that if Ω is revolved about the line $y = k$, then the volume of the resulting solid of revolution is given by

$$V = \int_a^b \pi([f(x) - k]^2 - [g(x) - k]^2)\, dx.$$

(b) Suppose that $k \geq f(x) \geq g(x)$ on $[a, b]$, and let Ω be the region between the graphs of f and g. Derive a formula for the volume of the solid which is generated by revolving Ω about the line $y = k$.

NOTE: Formulas analogous to those of Exercises 50–52 hold for functions $x = h(y)$ and regions revolved about vertical lines $x = l$.

53. The region between the graph of $f(x) = \sqrt{x}$ and the x-axis for $0 \leq x \leq 4$ is revolved about the line $y = 2$. Find the volume of the solid that is generated.

54. The region bounded by the curves $y = (x-1)^2$ and $y = x+1$ is revolved about the line $y = -1$. Find the volume of the solid that is generated.

55. The region between the graph of $y = \sin x$ and the x-axis, $0 \leq x \leq \pi$, is revolved about the line $y = 1$. Find the volume of the solid that is generated. HINT: See Exercise 63, Section 5.7.

56. The region bounded by $y = \sin x$ and $y = \cos x, \pi/4 \leq x \leq \pi$, is revolved about the line $y = 1$. Find the volume of the solid that is generated. HINT: Use a trigonometric identity.

57. The region bounded by the curves $y = x^2 - 2x$ and $y = 3x$ is revolved about the line $y = -1$. Find the volume of the solid that is generated.

58. Find the volume of the solid generated by revolving the region bounded by $y = x^2$ and $x = y^2$ about: (a) the line $x = -2$; (b) the line $x = 3$.

59. Find the volume of the solid generated by revolving the region bounded by $y^2 = 4x$ and $y = x$ about: (a) the x-axis; (b) the line $x = 4$.

60. Find the volume of the solid generated by revolving the region bounded by $y = x^2$ and $y = 4x$ about: (a) the line $x = 5$; (b) the line $x = -1$.

61. Find the volume if the region OAB in Figure E is revolved about: (a) the x-axis; (b) the line AB; (c) the line CA; (d) the y-axis.

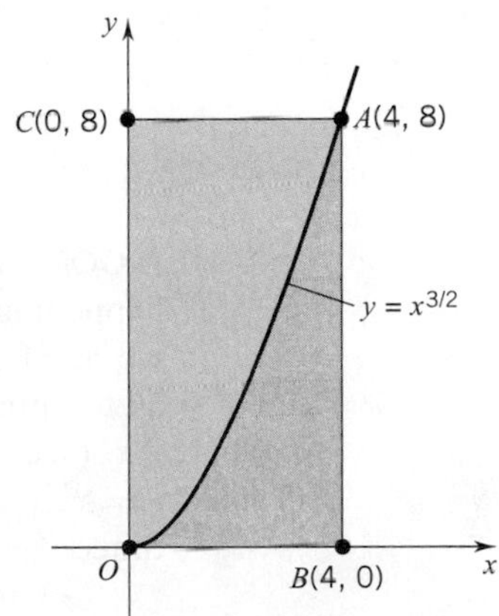

Figure E

62. Find the volume if the region OAC in Figure E is revolved about: (a) the y-axis; (b) the line CA; (c) the line AB; (d) the x-axis.

■ 6.3 VOLUME BY THE SHELL METHOD

To describe the shell method of calculating volumes, we begin with a solid cylinder of radius R and height h, and from it we cut out a cylindrical core of radius r (Figure 6.3.1).

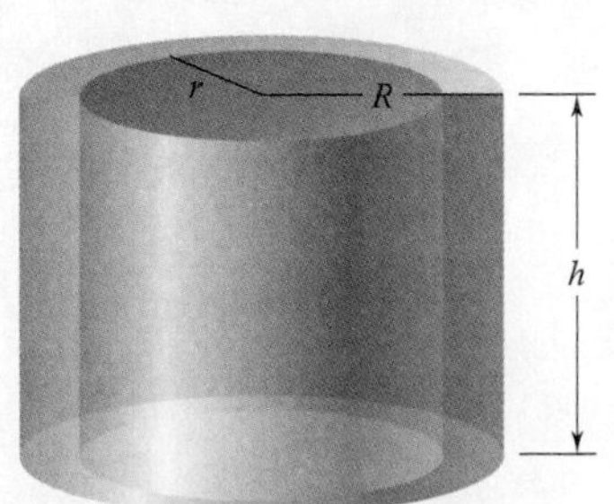

Figure 6.3.1

Since the original cylinder had volume $\pi R^2 h$ and the piece removed had volume $\pi r^2 h$, the solid cylindrical shell that remains has volume

(6.3.1) $$\pi R^2 h - \pi r^2 h = \pi h(R + r)(R - r).$$

We will use this shortly.

Now let $[a, b]$ be an interval that does not contain 0 in its interior, and let f be a nonnegative continuous function on $[a, b]$. For convenience, we assume that $a \geq 0$. If the region bounded by the graph of f and the x-axis is revolved about the y-axis, then a solid T is generated (Figure 6.3.2). The volume of this solid is given by the *shell method formula:*

(6.3.2)
$$V = \int_a^b 2\pi x f(x)\, dx.$$

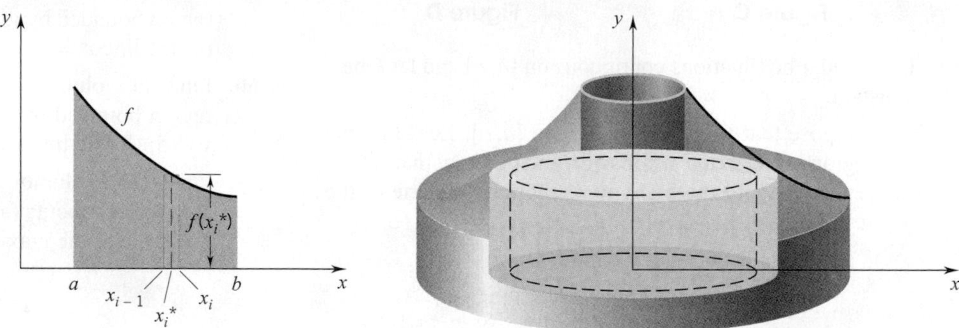

Figure 6.3.2

PROOF We take a partition $p = \{x_0, x_1, \ldots, x_n\}$ of $[a, b]$ and concentrate on what's happening on the ith subinterval $[x_{i-1}, x_i]$. Recall that, when we form a Riemann sum, we are free to choose x_i^* as any point from $[x_{i-1}, x_i]$. For convenience we take x_i^* as the midpoint $\frac{1}{2}(x_{i-1} + x_i)$. The representative rectangle of height $f(x_i^*)$ and base Δx_i (see Figure 6.3.2) generates a cylindrical shell of height $h = f(x_i^*)$, inner radius $r = x_{i-1}$, and outer radius $R = x_i$. We can calculate the volume of this shell by (6.3.1). Since

$$h = f(x_i^*) \quad \text{and} \quad R + r = x_i + x_{i-1} = 2x_i^* \quad \text{and} \quad R - r = \Delta x_i,$$

the volume of this shell is

$$\pi h(R + r)(R - r) = 2\pi x_i^* f(x_i^*)\Delta x_i.$$

The volume of the entire solid can be approximated by adding up the volumes of these shells:

$$V \cong 2\pi x_1^* f(x_1^*)\,\Delta x_1 + 2\pi x_2^* f(x_2^*)\,\Delta x_2 + \cdots + 2\pi x_n^* f(x_n^*)\,\Delta x_n.$$

The sum on the right is a Riemann sum, which, as $||P|| \to 0$, converges to

$$\int_a^b 2\pi x f(x)\, dx. \quad \square$$

For a simple interpretation of Formula 6.3.2, we refer to Figure 6.3.3. As the region below the graph of f is revolved about the y-axis, the line segment x units from the y-axis generates a cylinder of radius x, height $f(x)$, and lateral area $2\pi x f(x)$. The shell method formula expresses the volume of a solid of revolution as the definite integral from a to b of the lateral areas of these cylinders.

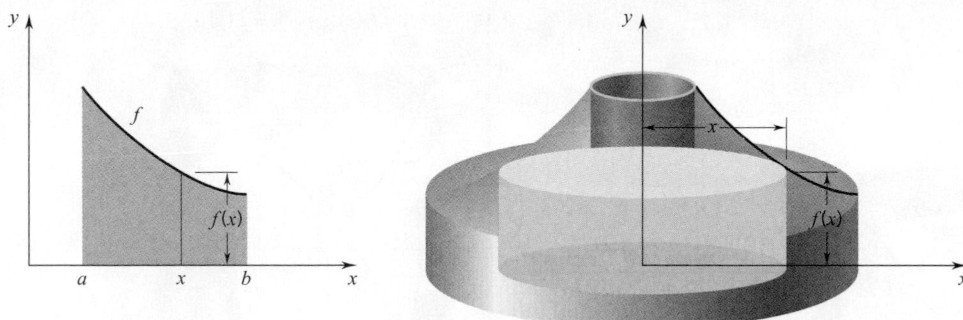

Figure 6.3.3

Example 1 The region bounded by $f(x) = 4x - x^2$ and the x-axis between $x = 1$ and $x = 4$ is rotated about the y-axis. Find the volume of the solid that is generated.

SOLUTION See Figure 6.3.4. The line segment x units from the y-axis, $1 \le x \le 4$, generates a cylinder of radius x, height $f(x)$, and lateral area $2\pi x f(x)$. Thus by (6.3.2)

$$V = \int_1^4 2\pi x(4x - x^2)\,dx = 2\pi \int_1^4 (4x^2 - x^3)\,dx = 2\pi\left[\tfrac{4}{3}x^3 - \tfrac{1}{4}x^4\right]_1^4 = \tfrac{81}{2}\pi. \quad \square$$

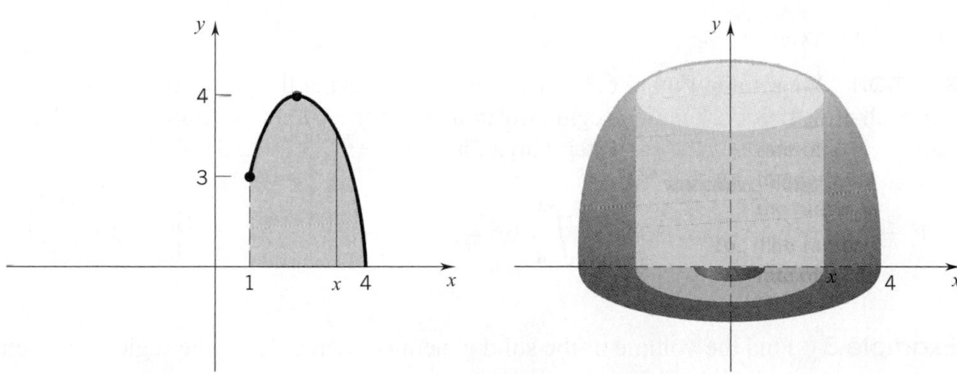

Figure 6.3.4

The shell method formula can be generalized. With Ω as in Figure 6.3.5, the volume generated by revolving Ω about the y-axis is given by the formula

(6.3.3)
$$V = \int_a^b 2\pi x[f(x) - g(x)]\,dx. \qquad \text{(shell method about } y\text{-axis)}$$

The integrand $2\pi x[f(x) - g(x)]$ is the lateral area of the cylinder in Figure 6.3.5.

We can also apply the shell method to solids generated by revolving a region about the x-axis. See Figure 6.3.6. In this instance the curved boundaries are functions of y rather than x, and we have

(6.3.4)
$$V = \int_c^d 2\pi y[F(y) - G(y)]\,dy. \qquad \text{(shell method about } x\text{-axis)}$$

The integrand $2\pi y[F(y) - G(y)]$ is the lateral area of the cylinder in Figure 6.3.6.

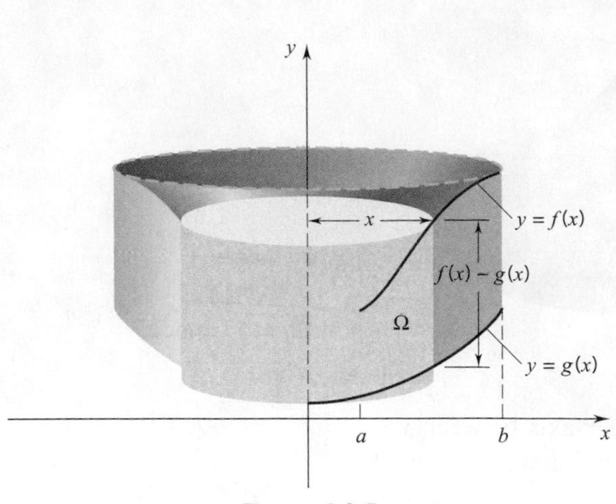

Figure 6.3.5

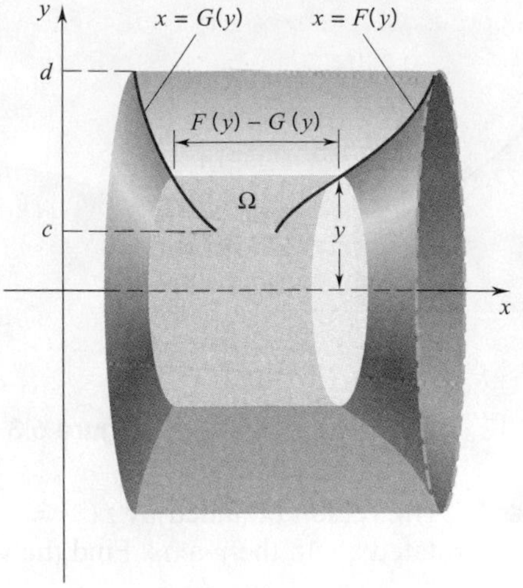

Figure 6.3.6

Example 2 Find the volume of the solid generated by revolving the region between

$$y = x^2 \quad \text{and} \quad y = 2x$$

about the y-axis.

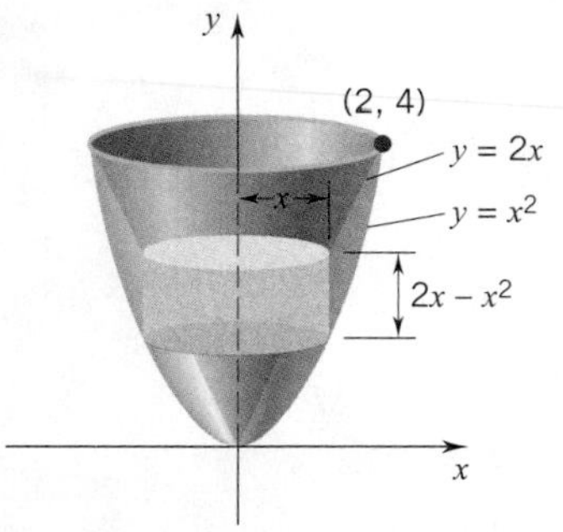

Figure 6.3.7

SOLUTION We refer to Figure 6.3.7. The curves intersect at the points (0, 0) and (2, 4). For each x from 0 to 2 the line segment x units from the y-axis generates a cylinder of radius x, height $(2x - x^2)$, and lateral area $2\pi x(2x - x^2)$. By (6.3.3)

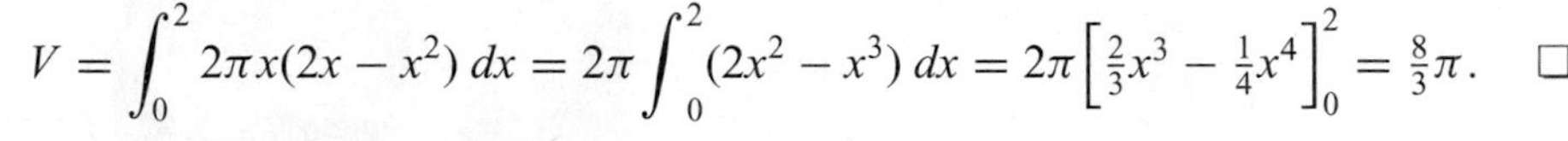

$$V = \int_0^2 2\pi x(2x - x^2)\,dx = 2\pi \int_0^2 (2x^2 - x^3)\,dx = 2\pi\left[\tfrac{2}{3}x^3 - \tfrac{1}{4}x^4\right]_0^2 = \tfrac{8}{3}\pi. \quad \square$$

Example 3 Find the volume of the solid generated by revolving the region between

$$y = x^2 \quad \text{and} \quad y = 2x$$

about the x-axis.

SOLUTION We begin by expressing these boundaries as functions of y. We write $x = \sqrt{y}$ for the right boundary and $x = \frac{1}{2}y$ for the left boundary (see Figure 6.3.8). The shell of radius y has height $(\sqrt{y} - \frac{1}{2}y)$. Thus, by (6.3.4)

$$V = \int_0^4 2\pi y(\sqrt{y} - \tfrac{1}{2}y)\,dy = \pi \int_0^4 (2y^{3/2} - y^2)\,dy$$

$$= \pi\left[\tfrac{4}{5}y^{5/2} - \tfrac{1}{3}y^3\right]_0^4 = \tfrac{64}{15}\pi. \quad \square$$

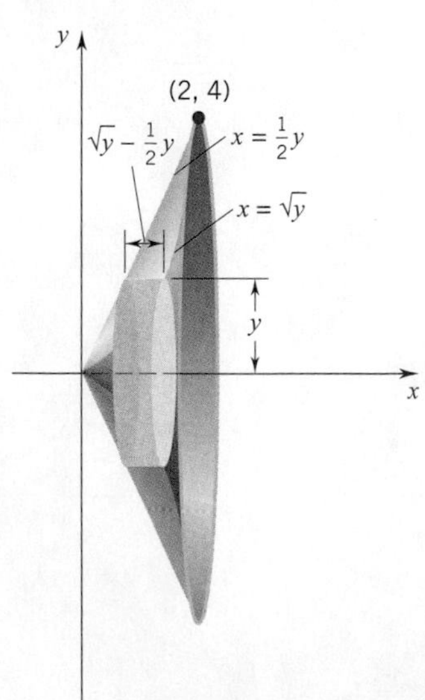

Figure 6.3.8

Remark In section 6.2 we calculated the volumes of the same solids (and got the same answers) by the washer method. These examples raise an obvious question: Which of the three methods—disc, washer, or shell—should you use to solve a particular problem? As you may have guessed, the "best" method depends on the problem. Since a disc is a special case of a washer, you will typically have to decide between the shell method and the washer method. For the solids in Examples 2 and 3, the shell and washer methods were essentially equivalent. On the other hand, if you try to use the washer method to calculate the volume of the solid in Example 1, you will be led to the expression

$$V = \int_0^3 (\pi[g_1(y)]^2 - \pi[1]^2)\,dy + \int_3^4 (\pi[g_1(y)]^2 - \pi[g_2(y)]^2)\,dy, \quad \text{(see Figure 6.3.4)}$$

where $x = g_1(y)$ and $x = g_2(y)$ are the functions obtained by solving $y = 4x - x^2$ for x. While the two functions g_1 and g_2 can be determined and the integrals can be evaluated, it is not easy to carry out the calculations. Try it! Moreover, it may not be possible to solve a given equation $y = f(x)$ for x, nor an equation $x = g(y)$ for y. The selection of the method to use on a particular problem will depend on the region that is being revolved, the axis of revolution, and the flexibility you have in expressing the functions involved in terms of either x or y, whichever is required. □

Example 4 A round hole of radius r is drilled through the center of a hemisphere of radius a $(r < a)$. Find the volume of the portion of the hemisphere that remains.

SOLUTION A hemisphere of radius a can be generated by revolving the region in the first quadrant bounded by $x^2 + y^2 = a^2$ around the y-axis. See Figure 6.3.9*a*. The portion of the hemisphere that remains after the hole is drilled is the solid generated by revolving the shaded region around the y-axis (Figure 6.3.9*b*).

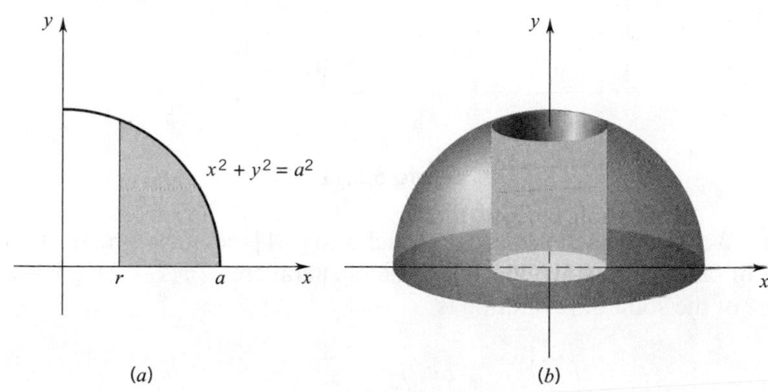

Figure 6.3.9

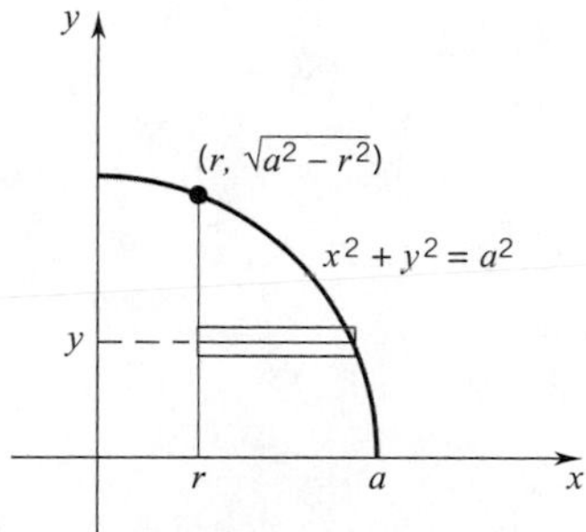

Figure 6.3.10

(a) Washer method. Refer to Figure 6.3.10.

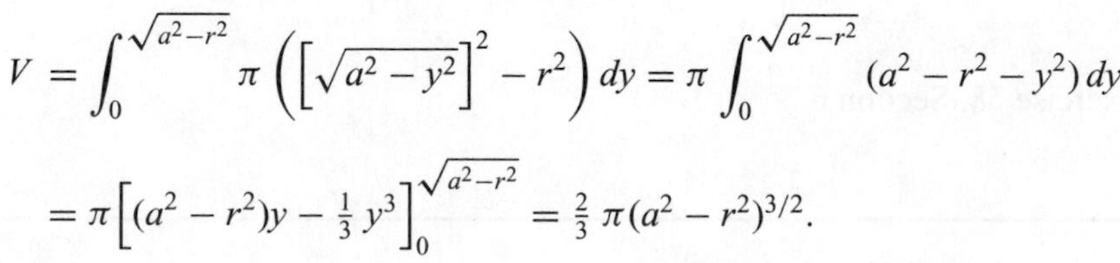

$$V = \int_0^{\sqrt{a^2-r^2}} \pi\left(\left[\sqrt{a^2-y^2}\right]^2 - r^2\right)dy = \pi\int_0^{\sqrt{a^2-r^2}} (a^2 - r^2 - y^2)\,dy$$

$$= \pi\left[(a^2 - r^2)y - \tfrac{1}{3}y^3\right]_0^{\sqrt{a^2-r^2}} = \tfrac{2}{3}\pi(a^2 - r^2)^{3/2}.$$

(b) Shell method. Refer to Figure 6.3.11.

$$V = \int_r^a 2\pi x\sqrt{a^2 - x^2}\,dx.$$

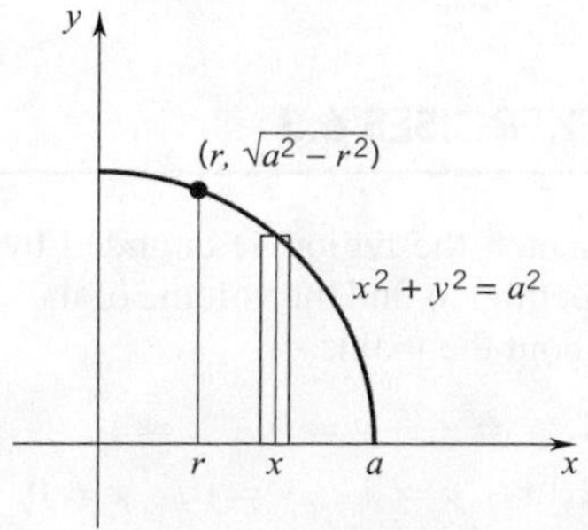

Figure 6.3.11

Let $u = a^2 - x^2$. Then $du = -2x\,dx$, and $u = a^2 - r^2$ at $x = r$, $u = 0$ at $x = a$.

Thus

$$V = \int_r^a 2\pi x\sqrt{a^2 - x^2}\,dx = -\pi\int_{a^2-r^2}^0 u^{1/2}\,du = \pi\int_0^{a^2-r^2} u^{1/2}\,du$$

$$= \pi\left[\tfrac{2}{3}u^{3/2}\right]_0^{a^2-r^2} = \tfrac{2}{3}\pi(a^2 - r^2)^{3/2}.$$

Note that if we let $r = 0$, then $V = \frac{2}{3}\pi a^3$, the volume of the hemisphere. □

We conclude this section with an example in which we revolve a region about a line parallel to the y-axis.

Example 5 The region Ω between $y = \sqrt{x}$ and $y = x^2, 0 \leq x \leq 1$, is revolved about the line $x = -2$ (see Figure 6.3.12). Find the volume of the solid which is generated.

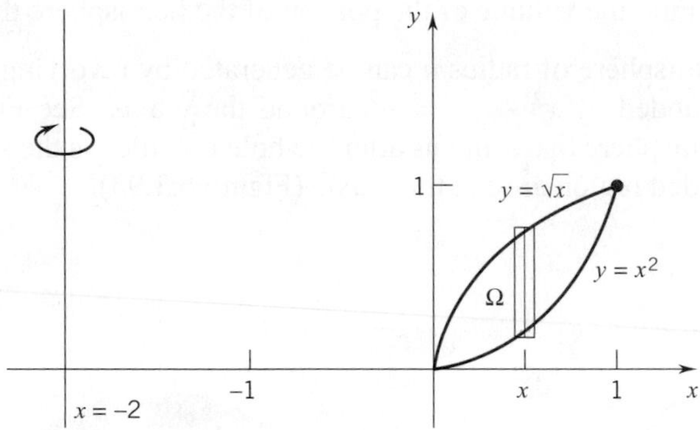

Figure 6.3.12

SOLUTION We use the shell method. For each x in [0,1] the line segment at x generates a cylinder of radius $x + 2$, height $\sqrt{x} - x^2$, and lateral area $2\pi(x + 2)(\sqrt{x} - x^2)$. Thus, the volume of the solid of revolution is

$$V = \int_0^1 2\pi(x + 2)(\sqrt{x} - x^2)\,dx$$

$$= 2\pi\int_0^1 (x^{3/2} + 2x^{1/2} - x^3 - 2x^2)\,dx$$

$$= 2\pi\left[\tfrac{2}{5}x^{5/2} + \tfrac{4}{3}x^{3/2} - \tfrac{1}{4}x^4 - \tfrac{2}{3}x^3\right]_0^1 = \tfrac{49}{30}\pi.$$

See Exercise 58, Section 6.2. □

EXERCISES 6.3

Sketch the region Ω bounded by the curves and use the shell method to find the volume of the solid generated by revolving Ω about the y-axis.

1. $y = x, \quad y = 0, \quad x = 1.$

2. $x + y = 3, \quad y = 0, \quad x = 0.$

3. $y = \sqrt{x}, \quad x = 4, \quad y = 0.$

4. $y = x^3, \quad x = 2, \quad y = 0.$

5. $y = \sqrt{x}, \quad y = x^3.$

6. $y = x^2, \quad y = x^{1/3}.$

7. $y = x, \quad y = 2x, \quad y = 4.$

8. $y = x, \quad y = 1, \quad x + y = 6.$

9. $x = y^2, \quad x = y + 2.$

10. $x = y^2, \quad x = 2 - y.$

11. $x = \sqrt{9 - y^2}, \quad x = 0.$

12. $x = |y|, \quad x = 2 - y^2.$

Sketch the region Ω bounded by the curves and use the shell method to find the volume of the solid generated by revolving Ω about the x-axis.

13. $x+3y=6, \quad y=0, \quad x=0.$

14. $y=x, \quad y=5, \quad x=0.$

15. $y=x^2, \quad y=9.$

16. $y=x^3, \quad y=8, \quad x=0.$

17. $y=\sqrt{x}, \quad y=x^3.$

18. $y=x^2, \quad y=x^{1/3}.$

19. $y=x^2, \quad y=x+2.$

20. $y=x^2, \quad y=2-x.$

21. $y=x, \quad y=2x, \quad x=4.$

22. $y=x, \quad x+y=8, \quad x=1.$

23. $y=\sqrt{1-x^2}, \quad x+y=1.$

24. $y=x^2, \quad y=2-|x|.$

The curves $y=x^2, y=\sqrt{x}, x=0, x=1, y=0$ and $y=1$ determine the three regions indicated in the figure. In Exercises 25–30, express the volume of the solid that is generated when the indicated region is revolved about the given line by: (a) a definite integral in which you will be integrating with respect to x; (b) a definite integral in which you will be integrating with respect to y. Then choose one of the two integrals and calculate the volume of the solid.

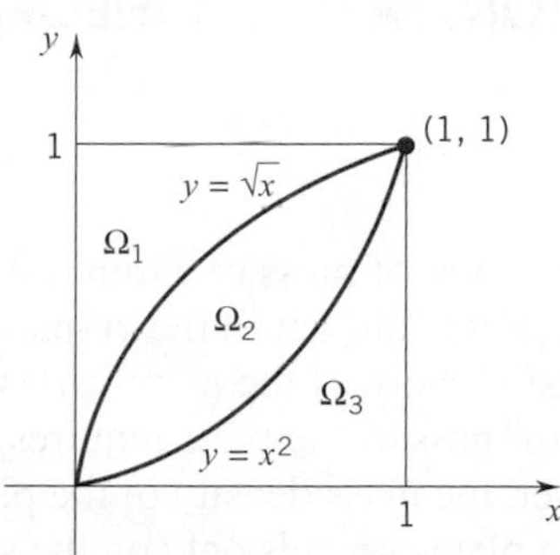

25. Ω_1 is revolved about the y-axis.

26. Ω_1 is revolved about the line $y=2$.

27. Ω_2 is revolved about the x-axis.

28. Ω_2 is revolved about the line $x=-3$.

29. Ω_3 is revolved about the y-axis.

30. Ω_3 is revolved about the line $y=-1$.

31. Use the shell method to find the volume enclosed by the surface obtained by revolving the ellipse $b^2x^2+a^2y^2=a^2b^2$ about the y-axis.

32. Repeat Exercise 31 with the ellipse revolved about the x-axis.

33. Find the volume enclosed by the surface generated by revolving the equilateral triangle with vertices $(0,0)$, $(a,0)$, $(\frac{1}{2}a, \frac{1}{2}\sqrt{3}a)$ about the y-axis.

34. A ball of radius r is cut into two pieces by a horizontal plane a units above the center of the ball. Determine the volume of the upper piece by using the shell method.

35. Carry out Exercise 61 of Section 6.2, this time using the shell method.

36. Carry out Exercise 62 of Section 6.2, this time using the shell method.

37. (a) Verify that $F(x)=x\sin x+\cos x$ is an antiderivative of $f(x)=x\cos x$.

(b) Use the shell method to find the volume of the solid that is generated by revolving the region between $y=\cos x$ and the x-axis, $0\le x\le \pi/2$, about the y-axis.

38. (a) Sketch the region in the right half-plane that is outside the parabola $y=x^2$ and is between the lines $y=x+2$ and $y=2x-2$.

(b) The region in part (a) is revolved about the y-axis. Use the method that you find most practical to calculate the volume of the solid that is generated.

In Exercises 39–42, let

$$f(x)=\begin{cases}\sqrt{3}\,x, & 0\le x<1\\ \sqrt{4-x^2}, & 1\le x\le 2,\end{cases}$$

and let Ω be the region between the graph of f and the x-axis (see the figure).

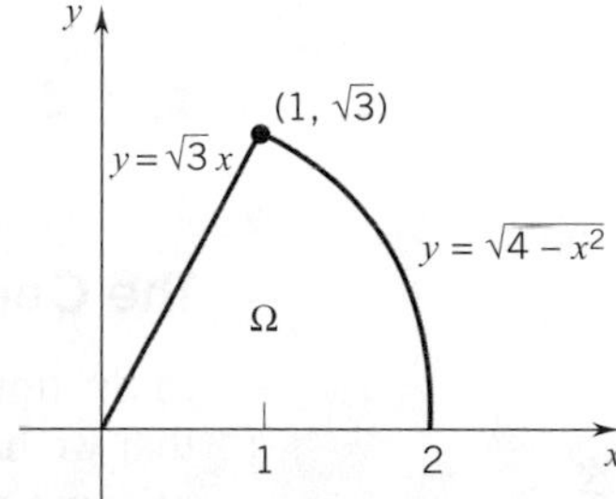

39. Revolve Ω about the y-axis.

(a) Express the volume of the solid that is generated as a definite integral in which you will be integrating with respect to x.

(b) Express the volume of the solid that is generated as a definite integral in which you will be integrating with respect to y.

(c) Choose one of the two integrals and determine the volume of the solid.

40. Revolve Ω about the x-axis and repeat Exercise 39.

41. Revolve Ω about the line $x=2$ and repeat parts (a) and (b) of Exercise 39.

42. Revolve Ω about the line $y=-1$ and repeat parts (a) and (b) of Exercise 39.

43. Let Ω be the circular disc $(x-b)^2+y^2\le a^2$, where $b>a>0$. The doughnut-shaped region generated by revolving Ω about the y-axis is called a *torus*. Express the volume V of the torus as:

(a) A definite integral in which you will be integrating with respect to x.

(b) A definite integral in which you will be integrating with respect to y.

44. The circular disc $x^2 + y^2 \leq a^2, a > 0$, is revolved about the line $x = a$. Find the volume of the solid that is generated.

45. Let r and h be positive numbers. The region in the first quadrant bounded by the line $x/r + y/h = 1$ and the coordinate axes is rotated about the y-axis. Use the shell method to derive the formula for the volume of a cone of radius r and height h.

46. A hole is drilled through the center of a ball of radius r, leaving a solid with a hollow cylindrical core of height h. Show that the volume of this solid is independent of the radius of the ball.

47. (a) Use a graphing utility to draw the graph $f(x) = \sin(\pi x^2)$ where $x \in [-3, 3]$.

(b) Let Ω be the region bounded by the graph of f and the x-axis with $x \in [0, 1]$. Show that if Ω is rotated about the x-axis and the disc method is used to calculate the volume of the solid of revolution, then the resulting definite integral *cannot* be evaluated by using the fundamental theorem of calculus. Use a CAS to find an approximate value for the volume.

(c) Show that if Ω is rotated about the y-axis and the shell method is used to calculate the volume of that solid, then the definite integral *can* be evaluated by using the fundamental theorem of calculus. Calculate the volume.

48. Let $f(x) = x^3$ and $g(x) = x$.

(a) Use a graphing utility to draw the graphs of f and g together.

(b) Use a CAS to find the points of intersection of the two graphs.

(c) Use a CAS to find the area of the region bounded by the two curves.

(d) The region in part (c) is rotated about the y-axis. Use a CAS to find the volume of the solid that is generated.

(e) (Optional) Use a CAS to draw the solid of revolution.

49. Repeat Exercise 48 with $f(x) = \sqrt{2x - 1}$ and $g(x) = x^2 - 4x + 4$.

■ 6.4 THE CENTROID OF A REGION; PAPPUS'S THEOREM ON VOLUMES

The Centroid of a Region

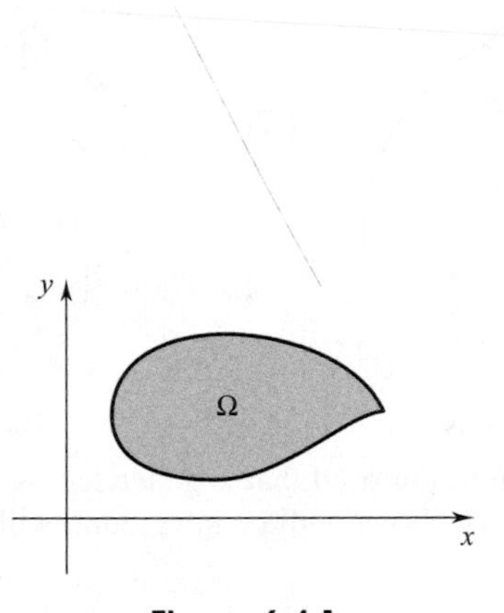

Figure 6.4.1

In Section 5.9 you saw how to calculate the center of mass of a thin rod. Suppose now that we have a thin distribution of matter, a *plate*, laid out in the xy-plane in the shape of some region Ω (Figure 6.4.1). If the mass density of the plate varies from point to point, then the determination of the center of mass of the plate requires the evaluation of a double integral (Chapter 16). If, however, the mass density of the plate is constant throughout Ω, then the center of mass of the plate depends only on the shape of Ω and falls on a point $(\overline{x}, \overline{y})$ that we call the *centroid*. Unless Ω has a very complicated shape, we can calculate its centroid by ordinary one-variable integration.

We will use two guiding principles to find the centroid of a plane region Ω. The first is obvious. The second we take from physics; this result is easily justified by double integration.

Principle 1: Symmetry If the region has an axis of symmetry, then the centroid $(\overline{x}, \overline{y})$ lies somewhere along that axis. In particular, if the region has a center, then the center is the centroid.

Principle 2: Additivity If the region, having area A, consists of a finite number of pieces with areas $A_1, \ldots, A_n$ and centroids $(\overline{x}_1, \overline{y}_1), \ldots, (\overline{x}_n, \overline{y}_n)$, then

(6.4.1) $$\overline{x}A = \overline{x}_1A_1 + \cdots + \overline{x}_nA_n \quad \text{and} \quad \overline{y}A = \overline{y}_1A_1 + \cdots + \overline{y}_nA_n.$$

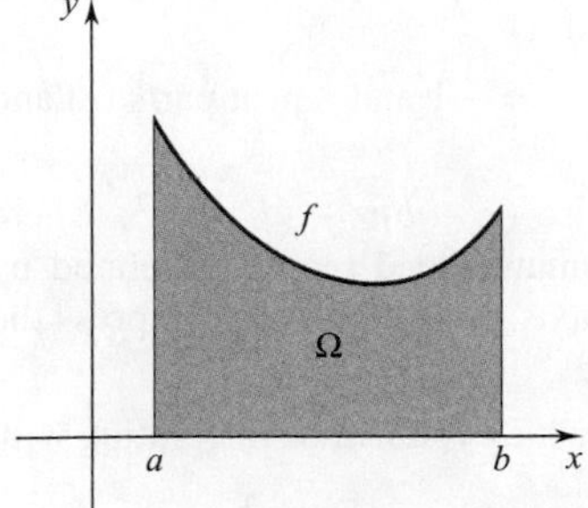

Figure 6.4.2

We are now ready to bring the techniques of calculus into play. Figure 6.4.2 shows the region Ω under the graph of a continuous function f. Denote the area of Ω by A. The centroid $(\bar{x}, \bar{y})$ of Ω can be obtained from the following formulas:

(6.4.2)
$$\bar{x}A = \int_a^b x f(x)\,dx, \quad \bar{y}A = \int_a^b \tfrac{1}{2}[f(x)]^2\,dx.$$

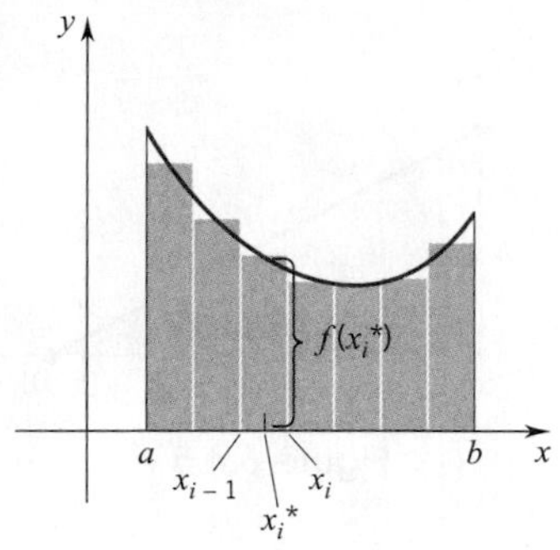

Figure 6.4.3

PROOF To derive these formulas we choose a partition $P = \{x_0, x_1, \ldots, x_n\}$ of $[a, b]$. This breaks up $[a, b]$ into n subintervals $[x_{i-1}, x_i]$. Choosing x_i^* as the midpoint of $[x_{i-1}, x_i]$, we form the midpoint rectangles R_i shown in Figure 6.4.3. The area of R_i is $f(x_i^*)\,\Delta x_i$, and the centroid of R_i is its center $(x_i^*, \frac{1}{2}f(x_i^*))$. By (6.4.1), the centroid $(\bar{x}_p, \bar{y}_p)$ of the union of all these rectangles satisfies the following equations:

$$\bar{x}_p A_p = x_1^* f(x_1^*)\,\Delta x_1 + \cdots + x_n^* f(x_n^*)\,\Delta x_n,$$
$$\bar{y}_p A_p = \tfrac{1}{2}\left[f(x_1^*)\right]^2 \Delta x_1 + \cdots + \tfrac{1}{2}\left[f(x_n^*)\right]^2 \Delta x_n$$

(Here A_P represents the area of the union of the n rectangles.) As $\|P\| \to 0$, the union of rectangles tends to the shape of Ω and the equations we just derived tend to the formulas given in (6.4.2). ☐

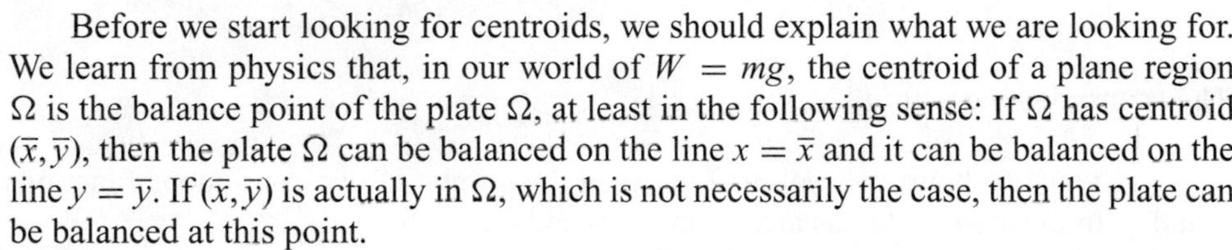

Before we start looking for centroids, we should explain what we are looking for. We learn from physics that, in our world of $W = mg$, the centroid of a plane region Ω is the balance point of the plate Ω, at least in the following sense: If Ω has centroid $(\bar{x}, \bar{y})$, then the plate Ω can be balanced on the line $x = \bar{x}$ and it can be balanced on the line $y = \bar{y}$. If $(\bar{x}, \bar{y})$ is actually in Ω, which is not necessarily the case, then the plate can be balanced at this point.

Example 1 Find the centroid of the quarter-disc shown in Figure 6.4.4.

SOLUTION The quarter-disc is symmetric about the line $y = x$. We therefore know that $\bar{x} = \bar{y}$. Here

$$\bar{y}A = \int_0^r \tfrac{1}{2}[f(x)]^2\,dx = \int_0^r \tfrac{1}{2}(r^2 - x^2)\,dx = \tfrac{1}{2}\left[r^2 x - \tfrac{1}{3}x^3\right]_0^r = \tfrac{1}{3}r^3.$$

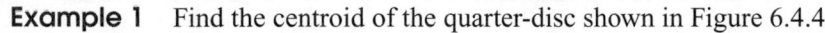

$f(x) = \sqrt{r^2 - x^2}$ ↑

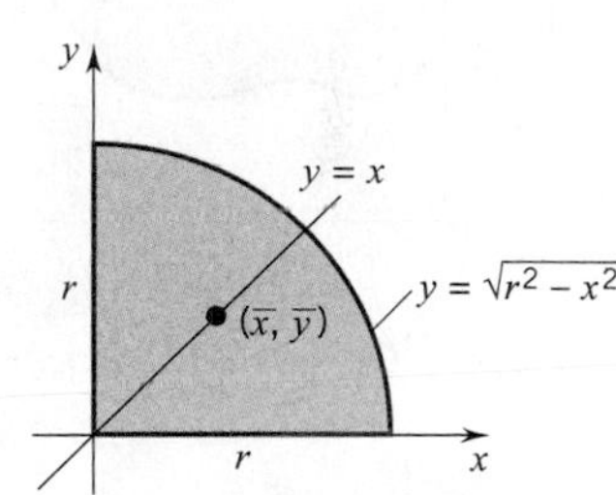

Figure 6.4.4

Since $A = \frac{1}{4}\pi r^2$,

$$\bar{y} = \frac{\frac{1}{3}r^3}{\frac{1}{4}\pi r^2} = \frac{4r}{3\pi}.$$

The centroid of the quarter-disc is the point

$$\left(\frac{4r}{3\pi}, \frac{4r}{3\pi}\right).$$

NOTE: It is almost as easy to calculate $\bar{x}A$,

$$\bar{x}A = \int_0^r x f(x)\,dx = \int_0^r x\sqrt{r^2 - x^2}\,dx$$
$$= -\tfrac{1}{2}\int_{r^2}^0 u^{1/2}\,du \qquad [u = (r^2 - x^2),\ du = -2x\,dx]$$
$$= -\tfrac{1}{2}\left[\tfrac{2}{3}u^{3/2}\right]_{r^2}^0 = \tfrac{1}{3}r^3,$$

and proceed from there. ☐

Example 2 Find the centroid of the right triangle shown in Figure 6.4.5.

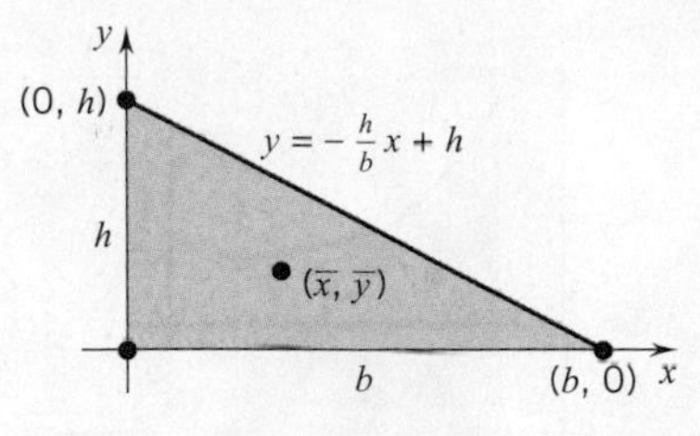

Figure 6.4.5

SOLUTION The hypotenuse lies on the line

$$y = -\frac{h}{b}x + h.$$

Hence
$$\overline{x}A = \int_0^b x f(x)\,dx = \int_0^b \left(-\frac{h}{b}x^2 + hx\right) dx = \tfrac{1}{6}b^2h$$

and
$$\overline{y}A = \int_0^b \tfrac{1}{2}[f(x)]^2\,dx = \tfrac{1}{2}\int_0^b \left(\frac{h^2}{b^2}x^2 - \frac{2h^2}{b}x + h^2\right) dx = \tfrac{1}{6}bh^2.$$

Since $A = \frac{1}{2}bh$, we have

$$\overline{x} = \frac{\frac{1}{6}b^2h}{\frac{1}{2}bh} = \tfrac{1}{3}b \qquad \text{and} \qquad \overline{y} = \frac{\frac{1}{6}bh^2}{\frac{1}{2}bh} = \tfrac{1}{3}h.$$

The centroid is the point $\left(\frac{1}{3}b, \frac{1}{3}h\right)$. □

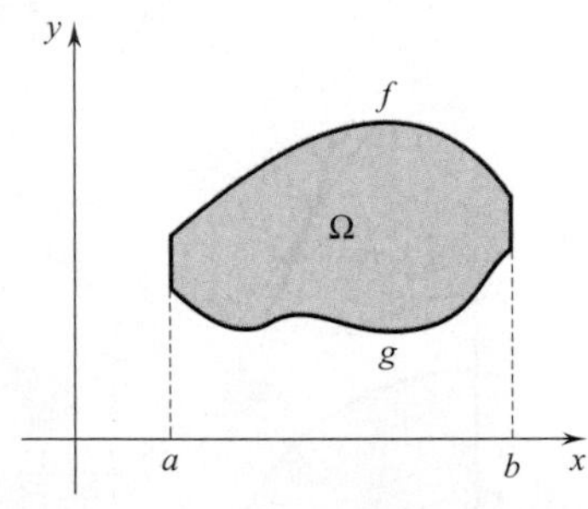

Figure 6.4.6

Figure 6.4.6 shows the region Ω between the graphs of two continuous functions f and g. In this case, if Ω has area A and centroid $(\overline{x}, \overline{y})$, then

(6.4.3)
$$\overline{x}A = \int_a^b x[f(x) - g(x)]\,dx, \quad \overline{y}A = \int_a^b \tfrac{1}{2}\left([f(x)]^2 - [g(x)]^2\right) dx.$$

PROOF Let A_f be the area below the graph of f and let A_g be the area below the graph of g. Then, in obvious notation,

$$\overline{x}A + \overline{x}_g A_g = \overline{x}_f A_f \qquad \text{and} \qquad \overline{y}A + \overline{y}_g A_g = \overline{y}_f A_f \tag{6.4.1}$$

Therefore

$$\overline{x}A = \overline{x}_f A_f - \overline{x}_g A_g = \int_a^b x f(x)\,dx - \int_a^b x g(x)\,dx = \int_a^b x[f(x) - g(x)]\,dx$$

and
$$\overline{y}A = \overline{y}_f A_f - \overline{y}_g A_g = \int_a^b \tfrac{1}{2}[f(x)]^2\,dx - \int_a^b \tfrac{1}{2}[g(x)]^2\,dx$$
$$= \int_a^b \tfrac{1}{2}\left([f(x)]^2 - [g(x)]^2\right) dx. \quad □$$

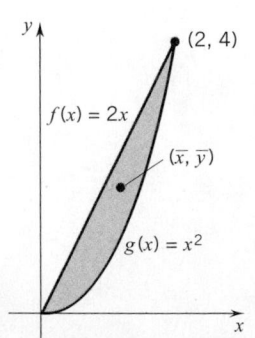

Figure 6.4.7

Example 3 Find the centroid of the region shown in Figure 6.4.7.

SOLUTION Here there is no symmetry we can appeal to. We must carry out the calculations.

$$A = \int_0^2 [f(x) - g(x)]\,dx = \int_0^2 (2x - x^2)\,dx = \left[x^2 - \tfrac{1}{3}x^3\right]_0^2 = \tfrac{4}{3},$$

$$\bar{x}A = \int_0^2 x[f(x) - g(x)]\,dx = \int_0^2 (2x^2 - x^3)\,dx = \left[\tfrac{2}{3}x^3 - \tfrac{1}{4}x^4\right]_0^2 = \tfrac{4}{3},$$

$$\bar{y}A = \int_0^2 \tfrac{1}{2}([f(x)]^2 - [g(x)]^2)\,dx = \tfrac{1}{2}\int_0^2 (4x^2 - x^4)\,dx = \tfrac{1}{2}\left[\tfrac{4}{3}x^3 - \tfrac{1}{5}x^5\right]_0^2 = \tfrac{32}{15}.$$

Therefore $\bar{x} = \frac{4}{3}/\frac{4}{3} = 1$ and $\bar{y} = \frac{32}{15}/\frac{4}{3} = \frac{8}{5}$. □

Pappus's Theorem on Volumes

All the formulas that we have derived for volumes of solids of revolution are simple corollaries to an observation made by a brilliant ancient Greek, Pappus of Alexandria (circa A.D. 300).

THEOREM 6.4.4 PAPPUS'S THEOREM ON VOLUMES†

A plane region is revolved about an axis that lies in its plane. If the region does not cross the axis, then the volume of the resulting solid of revolution is the area of the region multiplied by the circumference of the circle described by the centroid of the region:

$$V = 2\pi \bar{R} A$$

where A is the area of the region and $\bar{R}$ is the distance from the axis to the centroid of the region. See Figure 6.4.8.

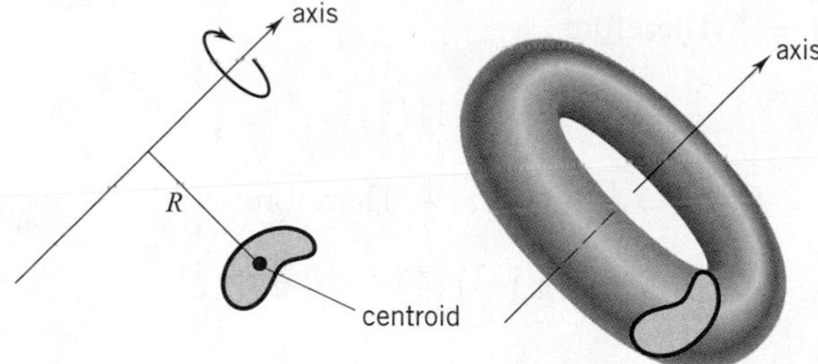

Figure 6.4.8

Basically we have derived only two formulas for the volumes of solids of revolution:

1. *The Washer Method Formula.* If the region Ω of Figure 6.4.6 is revolved about the x-axis, the resulting solid has volume

$$V_x = \int_a^b \pi\left([f(x)]^2 - [g(x)]^2\right) dx.$$

---

† This theorem is found in Book VII of Pappus's "Mathematical Collection," largely a survey of ancient geometry to which Pappus made many original contributions (among them this theorem). Much of what we know today of Greek geometry we owe to Pappus.

2. *The Shell Method Formula.* If the region Ω of Figure 6.4.6 is revolved about the y-axis, the resulting solid has volume

$$V_y = \int_a^b 2\pi x[f(x) - g(x)]\,dx.$$

Note that

$$\begin{aligned} V_x &= \int_a^b \pi\left([f(x)]^2 - [g(x)]^2\right) dx \\ &= 2\pi \int_a^b \tfrac{1}{2}\left([f(x)]^2 - [g(x)]^2\right) dx = 2\pi\bar{y}A = 2\pi\bar{R}A \end{aligned}$$

and $\qquad V_y = \displaystyle\int_a^b 2\pi x[f(x) - g(x)]\,dx = 2\pi\bar{x}A = 2\pi\bar{R}A,$

exactly as predicted by Pappus.

Remark In stating Pappus's theorem, we assumed a complete revolution. If Ω is only partially revolved about the given axis, then the volume of the resulting solid is simply the area of Ω multiplied by the length of the circular arc described by the centroid of Ω. □

Applications of Pappus's Theorem

Example 4 Earlier we saw that the region in Figure 6.4.7 has area $\frac{4}{3}$ and centroid $\left(1, \frac{8}{5}\right)$. Find the volumes of the solids formed by revolving this region **(a)** about the y-axis, **(b)** about the line $y = 5$.

SOLUTION **(a)** We have already calculated this volume by two methods: by the washer method and by the shell method. (Example 8, Section 6.2, and Example 2, Section 6.3.) The result was $V = \frac{8}{3}\pi$. Now we calculate the volume by Pappus's theorem. Here we have $\bar{R} = 1$ and $A = \frac{4}{3}$. Therefore

$$V = 2\pi(1)\left(\tfrac{4}{3}\right) = \tfrac{8}{3}\pi.$$

(b) In this case $\bar{R} = 5 - \frac{8}{5} = \frac{17}{5}$ and $A = \frac{4}{3}$. Therefore

$$V = 2\pi\left(\tfrac{17}{5}\right)\left(\tfrac{4}{3}\right) = \tfrac{136}{15}\pi. \quad \square$$

Example 5 Find the volume of the torus generated by revolving the circular disc

$$(x - h)^2 + (y - k)^2 \leq c^2, \quad h, k \geq c > 0 \qquad \text{(Figure 6.4.9)}$$

about **(a)** the x-axis, **(b)** the y-axis.

SOLUTION The centroid of the disc is the center (h, k). This lies k units from the x-axis and h units from the y-axis. The area of the disc is πc^2. Therefore **(a)** $V_x = 2\pi(k)(\pi c^2) = 2\pi^2 kc^2$, **(b)** $V_y = 2\pi(h)(\pi c^2) = 2\pi^2 hc^2$. □

Example 6 Find the centroid of the half-disc

$$x^2 + y^2 \leq r^2, \qquad y \geq 0$$

by appealing to Pappus's theorem.

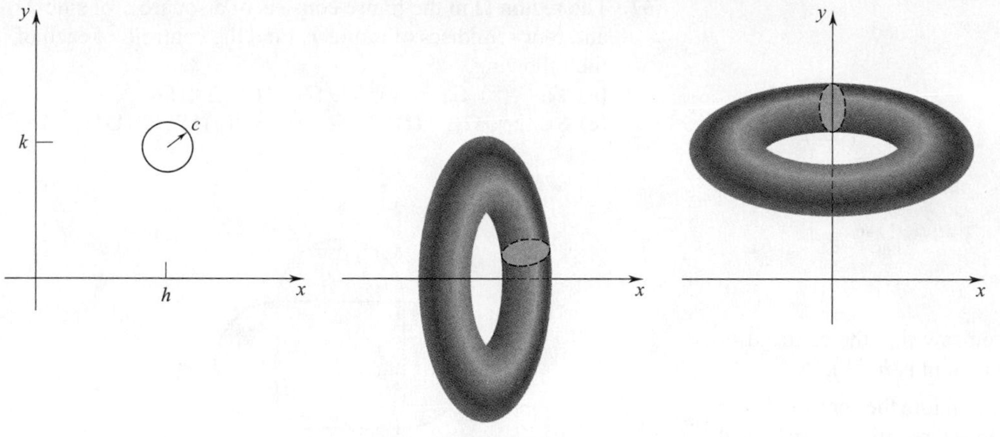

Figure 6.4.9

SOLUTION Since the half-disc is symmetric about the y-axis, we know that $\bar{x} = 0$. All we need is $\bar{y}$.

If we revolve the half-disc about the x-axis, we obtain a solid ball of volume $\frac{4}{3}\pi r^3$. The area of the half-disc is $\frac{1}{2}\pi r^2$. By Pappus's theorem

$$\tfrac{4}{3}\pi r^3 = 2\pi\bar{y}\left(\tfrac{1}{2}\pi r^2\right).$$

Simple division gives $\bar{y} = 4r/3\pi$. ☐

Remark Centroids of solids of revolution are discussed in Project 6.4. ☐

EXERCISES 6.4

Sketch the region bounded by the curves. Determine the centroid of the region and the volume generated by revolving the region about each of the coordinate axes.

1. $y = \sqrt{x}, \quad y = 0, \quad x = 4.$

2. $y = x^3, \quad y = 0, \quad x = 2.$

3. $y = x^2, \quad y = x^{1/3}.$

4. $y = x^3, \quad y = \sqrt{x}.$

5. $y = 2x, \quad y = 2, \quad x = 3.$

6. $y = 3x, \quad y = 6, \quad x = 1.$

7. $y = x^2 + 2, \quad y = 6, \quad x = 0.$

8. $y = x^2 + 1, \quad y = 1, \quad x = 3.$

9. $\sqrt{x} + \sqrt{y} = 1, \quad x + y = 1.$

10. $y = \sqrt{1 - x^2}, \quad x + y = 1.$

11. $y = x^2, \quad y = 0, \quad x = 1, \quad x = 2.$

12. $y = x^{1/3}, \quad y = 1, \quad x = 8.$

13. $y = x, \quad x + y = 6, \quad y = 1.$

14. $y = x, \quad y = 2x, \quad x = 3.$

Find the centroid of the bounded region determined by the following curves.

15. $y = 6x - x^2, \quad y = x.$

16. $y = 4x - x^2, \quad y = 2x - 3.$

17. $x^2 = 4y, \quad x - 2y + 4 = 0.$

18. $y = x^2, \quad 2x - y + 3 = 0.$

19. $y^3 = x^2, \quad 2y = x.$

20. $y^2 = 2x, \quad y = x - x^2.$

21. $y = x^2 - 2x, \quad y = 6x - x^2.$

22. $y = 6x - x^2, \quad x + y = 6.$

23. $x + 1 = 0, \quad x + y^2 = 0.$

24. $\sqrt{x} + \sqrt{y} = \sqrt{a}, \quad x = 0, \quad y = 0.$

25. Let Ω be the annular region (ring) formed by the circles

$$x^2 + y^2 = \tfrac{1}{4} \qquad x^2 + y^2 = 4.$$

(a) Locate the centroid of Ω. (b) Locate the centroid of the first-quadrant part of Ω. (c) Locate the centroid of the upper half of Ω.

26. The ellipse $b^2x^2 + a^2y^2 = a^2b^2$ encloses a region of area πab. Locate the centroid of the upper half of the region.

27. The rectangle in the figure is revolved about the line marked l. Find the volume of the resulting solid.

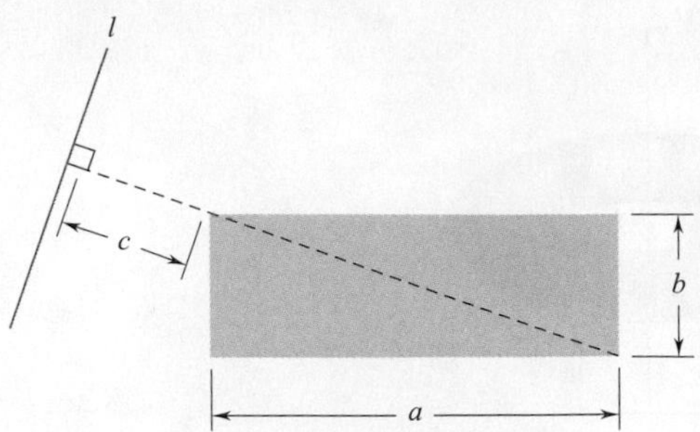

28. In Example 2 of this section you saw that the centroid of the triangle in Figure 6.4.5 is the point $(\frac{1}{3}b, \frac{1}{3}h)$.

(a) Verify that the line segments that join the centroid to the vertices divide the triangle into three triangles of equal area.

(b) Find the distance d from the centroid of the triangle to the hypotenuse.

(c) Find the volume generated by revolving the triangle about the hypotenuse.

29. The triangular region in the figure is the union of two right triangles Ω_1, Ω_2. Find the centroid: (a) of Ω_1; (b) of Ω_2; (c) of the entire region.

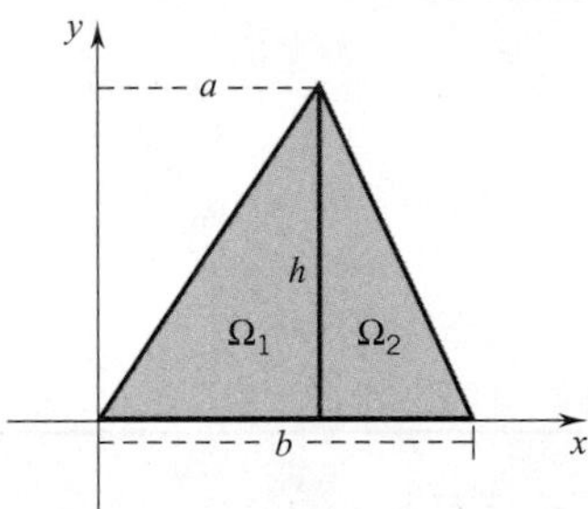

30. Find the volume of the solid generated by revolving the entire triangular region in the figure in Exercise 29 about: (a) the x-axis; (b) the y-axis.

31. (a) Find the volume of the ice-cream cone shown in figure A (a right circular cone topped by a solid hemisphere).

(b) Find $\bar{x}$ for the region Ω in figure B

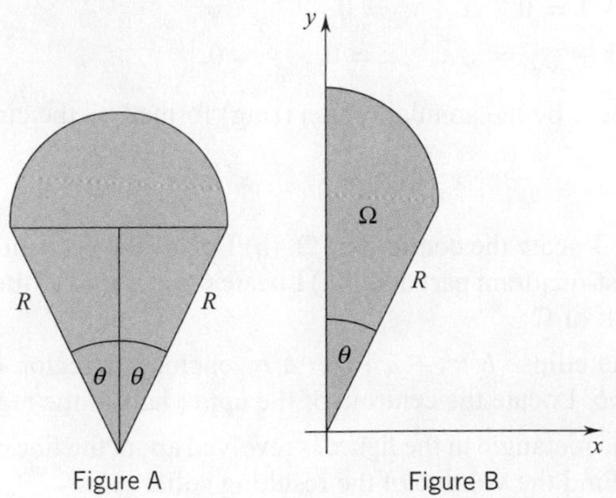

32. The region Ω in the figure consists of a square S of side $2r$ and four semidiscs of radius r. Find the centroid of each of the following.

(a) Ω. (b) Ω_1. (c) $S \cup \Omega_1$. (d) $S \cup \Omega_3$.
(e) $S \cup \Omega_1 \cup \Omega_3$. (f) $S \cup \Omega_1 \cup \Omega_2$. (g) $S \cup \Omega_1 \cup \Omega_2 \cup \Omega_3$.

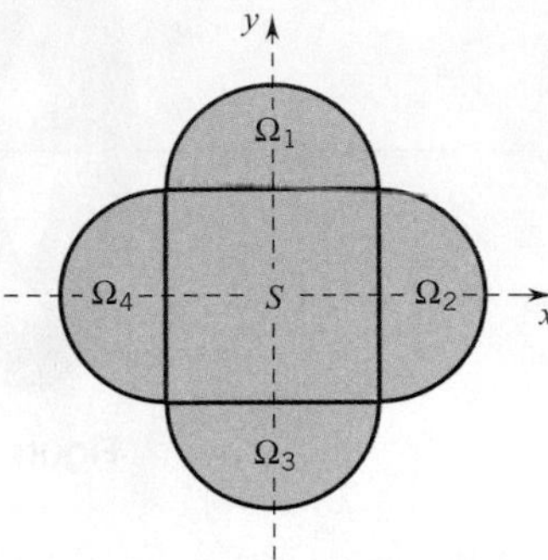

33. Give an example of a region that does not contain its centroid.

34. The centroid of a triangular region can be located without integration. Find the centroid of the region shown in the figure by applying Principles 1 and 2. Then verify that all three medians meet at this point.

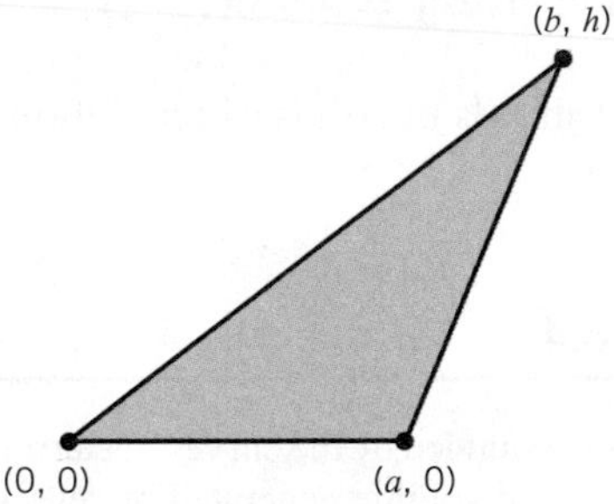

C 35. Use a graphing utility to draw the graphs of $y = \sqrt[3]{x}$ and $y = x^3$ for $x \geq 0$. Let Ω be the region bounded by the two curves. Use a CAS to find:

(a) the area of Ω.

(b) the centroid of Ω.

(c) the volume of the solid generated by revolving Ω about the x-axis.

(d) the volume of the solid generated by revolving Ω about the y-axis.

C 36. Repeat Exercise 35 with $y = x^2 - 2x + 4$ and $y = 2x + 1$.

C 37. Use a graphing utility to draw the graphs of $y = 16 - 8x$ and $y = x^4 - 5x^2 + 4$. Let Ω be the region bounded by the two curves. Use a CAS to find:

(a) the area of Ω.

(b) the centroid of Ω.

C 38. Repeat Exercise 37 with $y = 2 + \sqrt{x+2}$ and $y = \frac{1}{6}(5x^2 + 3x - 2)$.

■ PROJECT 6.4 Centroid of a Solid of Revolution

If a solid is *homogeneous* (constant mass density), then the center of mass depends only on the shape of the solid and is called the *centroid*. The determination of the centroid of a solid requires triple integration (Chapter 16). However, if the solid is a solid of revolution, then the centroid can be found by ordinary one-variable integration.

Let Ω be the region shown in the figure, and let T be the solid generated by revolving Ω about the x-axis. By symmetry, the centroid of T is on the x-axis. Thus the centroid of T is determined solely by its x-coordinate $\bar{x}$.

Problem 1. Show that $\bar{x}V = \int_a^b \pi x[f(x)]^2\,dx$, where V is the volume of T.

HINT: Use the following principle: If a solid of volume V consists of a finite number of pieces with volume $V_1, V_2, \ldots, V_n$ and the centroids of the pieces have x-coordinates $\bar{x}_1, \bar{x}_2, \ldots, \bar{x}_n$, then $\bar{x}V = \bar{x}_1V_1 + \bar{x}_2V_2 + \cdots + \bar{x}_nV_n$.

Now revolve Ω about the y-axis and let S be the solid generated. By symmetry, the centroid of S lies on the y-axis and is determined solely by its y-coordinate $\bar{y}$.

Problem 2. Show that $\bar{y}V = \int_a^b \pi x[f(x)]^2\,dx$, where V is the volume of S.

Problem 3. Use the results in Problem 1 and 2 to locate the centroids of each of the following solids:

a. A solid cone of base radius r and height h.

b. A solid hemisphere of radius r.

c. The solid generated by revolving about the x-axis the first-quadrant part of the ellipse $b^2x^2 + a^2y^2 = a^2b^2$.

d. The solid generated by revolving the region below the graph of $f(x) = \sqrt{x}, x \in [0, 1]$, about (i) the x-axis; (ii) the y-axis.

e. The solid generated by revolving the region below the graph of $f(x) = 4 - x^2, x \in [0, 2]$, about (i) the x-axis; (ii) the y-axis.

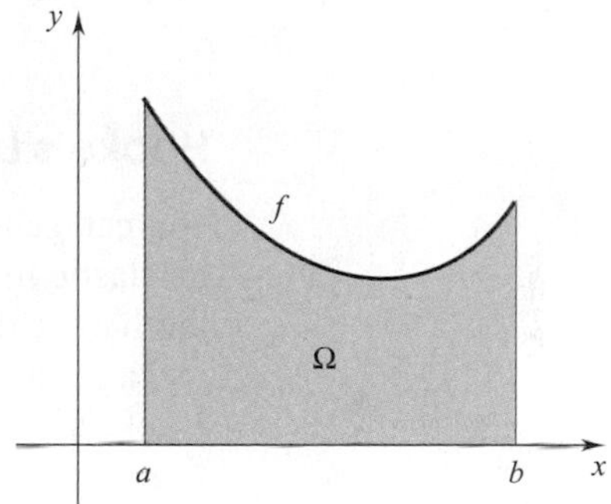

■ 6.5 THE NOTION OF WORK

We begin with a constant force F that acts along some line that we call the x-axis. By convention F is positive if it acts in the direction of increasing x and negative if it acts in the direction of decreasing x (Figure 6.5.1).

Figure 6.5.1

Suppose now that an object moves along the x-axis from $x = a$ to $x = b$ subject to this constant force F. The *work* done by F during the displacement is by definition the *force times the displacement.*

(6.5.1) $$W = F \cdot (b - a).$$

It is not hard to see that, if F acts in the direction of the motion, then $W > 0$, but if F acts against the motion, then $W < 0$. Thus, for example, if an object slides off a table and falls to the floor, then the work done by gravity is positive (after all, earth's gravity points down). But if an object is lifted from the floor and raised to tabletop level, then the work done by gravity is by definition negative. However, the work done by the hand that lifts the object is positive.

To repeat, if an object moves from $x = a$ to $x = b$ subject to a constant force F, then the work done by F is the constant value of F times $b - a$. What is the work done by F if F does not remain constant but instead varies continuously as a function of x?

As you would expect, we then define the work done by F as the *average value* of F times $b-a$:

(6.5.2) $$W = \int_a^b F(x)\,dx.$$ (Figure 6.5.2)

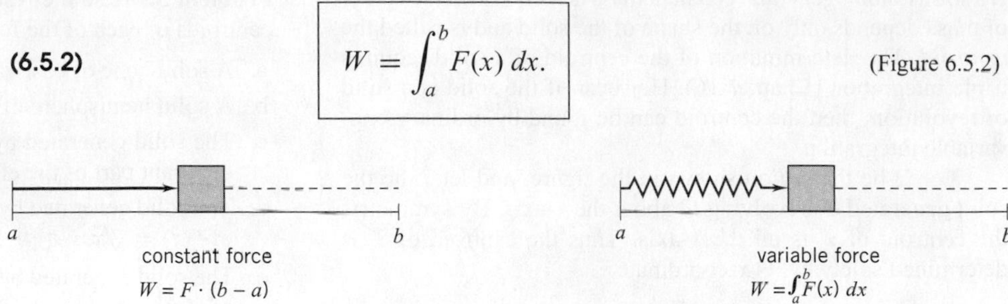

Figure 6.5.2

Hooke's Law

You can sense a variable force in the action of a steel spring. Stretch a spring within its elastic limit and you feel a pull in the opposite direction. The greater the stretching, the harder the pull of the spring. Compress a spring within its elastic limit and you feel a push against you. The greater the compression, the harder the push. According to Hooke's law (Robert Hooke, 1635–1703), the force exerted by the spring can be written

$$F(x) = -k\,x,$$

where k is a positive number, called *the spring constant,* and x is the displacement from the equilibrium position. The minus sign indicates that the spring force always acts in the direction opposite to the direction in which the spring has been deformed (the force always acts so as to restore the spring to its equilibrium state.)

Remark Hooke's law is only an approximation, but it is a good approximation for small displacements. In the problems that follow, we assume that the restoring force of the spring is given by Hooke's law. ☐

Example 1 A spring of natural length L compressed to length $\frac{7}{8}L$ exerts a force F_0.

(a) Find the work done by the spring in restoring itself to natural length.

(b) What work must be done to stretch the spring to length $\frac{11}{10}L$?

SOLUTION Place the spring on the x-axis so that the equilibrium point falls at the origin. View compression as a move to the left. See Figure 6.5.3.

Compressed $\frac{1}{8}L$ units to the left, the spring exerts a force F_0. Thus by Hooke's law

$$F_0 = F(-\tfrac{1}{8}L) = -k(-\tfrac{1}{8}L) = \tfrac{1}{8}kL.$$

This tells us that $k = 8F_0/L$. Therefore the force law for this spring reads

$$F(x) = -\left(\frac{8F_0}{L}\right)x.$$

(a) To find the work done by this spring in restoring itself to equilibrium, we integrate $F(x)$ from $x = -\frac{1}{8}L$ to $x = 0$:

$$W = \int_{-L/8}^{0} F(x)\,dx = \int_{-L/8}^{0} -\left(\frac{8F_0}{L}\right)x\,dx = -\frac{8F_0}{L}\left[\frac{x^2}{2}\right]_{-L/8}^{0} = \frac{LF_0}{16}. \quad ☐$$

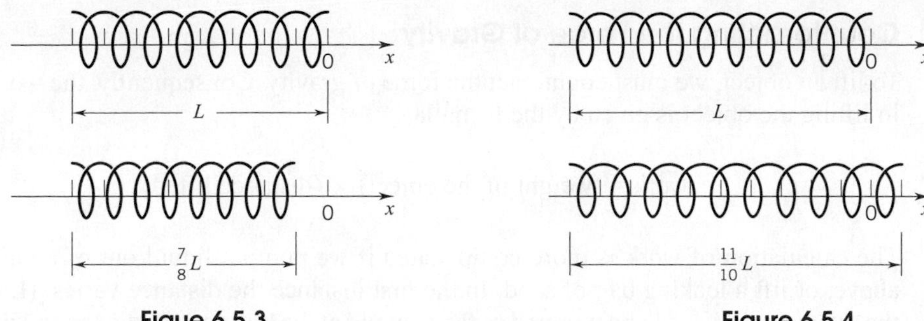

Figue 6.5.3

Figure 6.5.4

(b) Refer to Figure 6.5.4. To stretch the spring, we must counteract the force of the spring. The force exerted by the spring when stretched x units is

$$F(x) = -\left(\frac{8F_0}{L}\right)x.$$

To counter this force, we must apply the opposite force

$$-F(x) = \left(\frac{8F_0}{L}\right)x.$$

The work we must do to stretch the spring to length $\frac{11}{10}L$ can be found by integrating $-F(x)$ from $x = 0$ to $x = \frac{1}{10}L$:

$$W = \int_0^{L/10} -F(x)\,dx = \int_0^{L/10} \left(\frac{8F_0}{L}\right) x\,dx = \frac{8F_0}{L}\left[\frac{x^2}{2}\right]_0^{L/10} = \frac{LF_0}{25}. \quad \square$$

Remark on Units One unit of work is the work done by a unit force in moving an object a unit distance. If force is measured in pounds and distance in feet, then the units of work are *foot -pounds*. In the metric system, the standard units of force are *newtons* (1 newton is the force required to give a mass of 1 kilogram an acceleration of 1 meter per second per second) and *dynes* (1 dyne is the force required to give a mass of 1 gram an acceleration of 1 centimeter per second per second). The units of work used in the metric system are *newton-meters* (called *joules*) and *dyne-centimeters* (called *ergs*). □

Example 2 Stretched $\frac{1}{3}$ meter beyond its natural length, a certain spring exerts a restoring force of 10 newtons. What work must we do to stretch the spring another $\frac{1}{3}$ meter?

SOLUTION Place the spring on the x-axis so that the equilibrium point falls at the origin. View stretching as moving to the right and assume Hooke's law: $F(x) = -kx$.

When the spring is stretched $\frac{1}{3}$ meter, it exerts a force of -10 newtons (10 newtons to the left). Therefore $-10 = -k(\frac{1}{3})$ and $k = 30$ (that is, 30 newtons per meter).

To find the work necessary to stretch the spring another $\frac{1}{3}$ meter, we integrate the opposite force $-F(x) = 30x$ from $x = \frac{1}{3}$ to $x = \frac{2}{3}$.

$$W = \int_{1/3}^{2/3} 30x\,dx = 30\left[\tfrac{1}{2}x^2\right]_{1/3}^{2/3} = 5 \text{ joules.} \quad \square$$

Counteracting the Force of Gravity

To lift an object, we must counteract the force of gravity. Consequently, the work done in lifting the object is given by the formula

$$\text{work} = (\text{weight of the object}) \times (\text{distance lifted}).$$

The calculation of work is more complicated if we pump a liquid out of a tank from above, or lift a leaking bag of sand. In the first instance the distance varies. (Liquid at the top does not have to be pumped as far as liquid at the bottom). In the second instance the weight varies during the motion. (There is less sand in the bag as we keep lifting.)

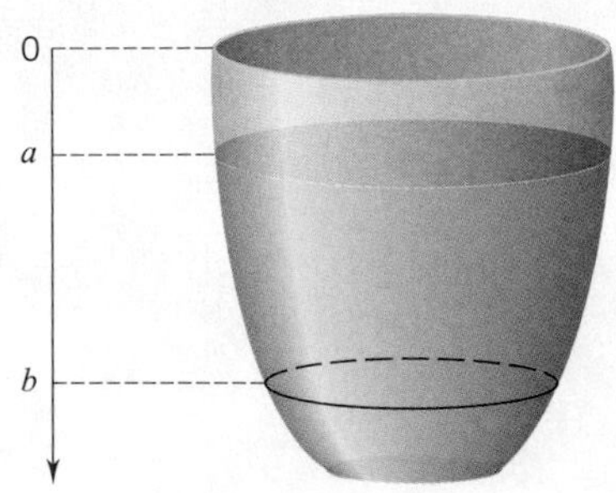

Figure 6.5.5

Pumping Out a Tank: Figure 6.5.5 depicts a storage tank filled to within a feet of the top with some liquid. Assume that the liquid is homogeneous and weights σ† pounds per cubic foot. Suppose now that this storage tank is pumped out from above until the level of the liquid drops to b feet below the top of the tank. How much work has been done?

We can answer this question by the methods of integral calculus. For each $x \in [a, b]$, we let

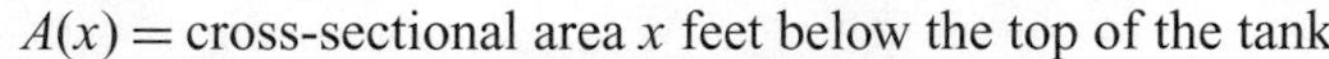

$$A(x) = \text{cross-sectional area } x \text{ feet below the top of the tank,}$$
$$s(x) = \text{distance that the } x\text{-level must be lifted.}$$

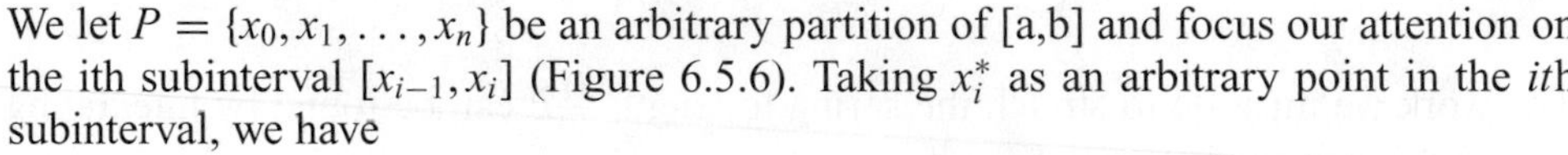

We let $P = \{x_0, x_1, \ldots, x_n\}$ be an arbitrary partition of [a,b] and focus our attention on the ith subinterval $[x_{i-1}, x_i]$ (Figure 6.5.6). Taking x_i^* as an arbitrary point in the ith subinterval, we have

$$A(x_i^*)\Delta x_i = \text{approximate volume of the } i\text{th layer of liquid,}$$
$$\sigma A(x_i^*)\Delta x_i = \text{approximate weight of this volume,}$$
$$s(x_i^*) = \text{approximate distance this weight is to be moved.}$$

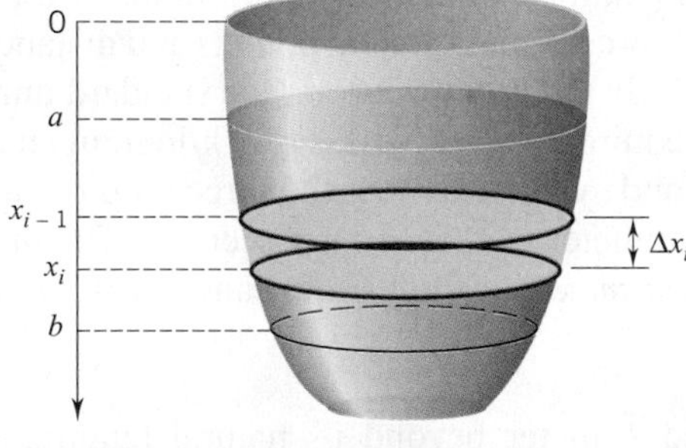

Figure 6.5.6

Therefore

$$\sigma s(x_i^*)A(x_i^*)\Delta x_i = \text{approximate work (weight} \times \text{distance) required to pump this layer of liquid to the top of the tank.}$$

The work required to pump out all the liquid can be approximated by adding up all these terms:

$$W \cong \sigma s(x_1^*)A(x_1^*)\Delta x_1 + \sigma s(x_2^*)A(x_2^*)\Delta x_2 + \cdots + \sigma s(x_n^*)A(x_n^*)\Delta x_n.$$

† The symbol σ is the lowercase Greek letter "sigma."

The sum on the right is a Riemann sum, which as $||P|| \to 0$, converges to give

(6.5.3)
$$W = \int_a^b \sigma s(x)\, A(x)\, dx.$$

We use this formula in the next problem.

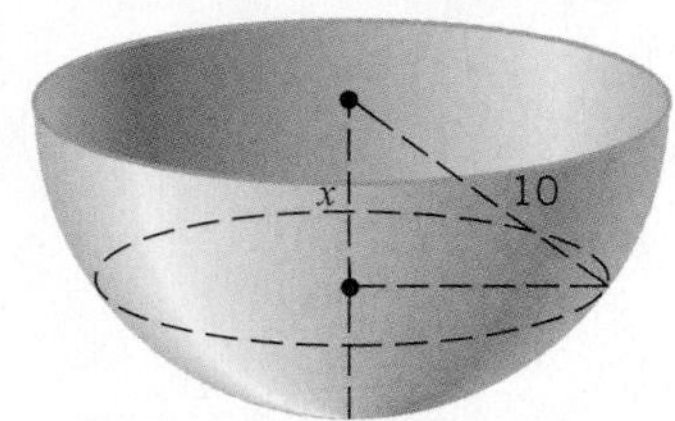

Figure 6.5.7

Example 3 A hemispherical water tank of radius 10 feet is being pumped out (Figure 6.5.7). Find the work done in lowering the water level from 2 feet below the top of the tank to 4 feet below the top of the tank given that the pump is placed **(a)** at the top of the tank, **(b)** 3 feet above the top of the tank.

SOLUTION Take 62.5 pounds per cubic foot as the weight of water. From the figure, you can see that the cross section x feet from the top of the tank is a disc of radius $\sqrt{100 - x^2}$. Its area is therefore

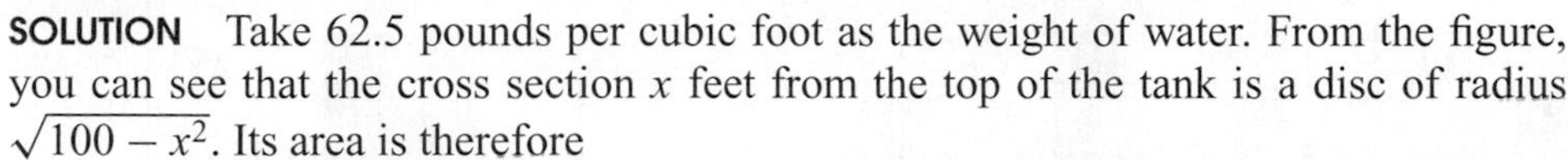

$$A(x) = \pi\left(100 - x^2\right).$$

For part **(a)** we have $s(x) = x$, so that

$$W = \int_2^4 62.5\pi x\left(100 - x^2\right) dx = 33,750\,\pi \cong 106,029 \text{ foot-pounds.}$$

For part **(b)** we have $s(x) = x + 3$, so that

$$W = \int_2^4 62.5\pi(x+3)\left(100 - x^2\right) dx = 67,750\,\pi \cong 212,843 \text{ foot-pounds.} \quad \square$$

Lifting a Leaky Bag of Sand: A bag of sand weighing M pounds is hoisted by a cable to the top of a building H feet high. The bag has a hole in it and sand leaks out at the rate of p pounds for each foot that the bag is raised. (Assume $pH < M$ so that there is some sand in the bag when it reaches the top.) Neglecting the weight of the cable, how much work is done in lifting the bag of sand to the top of the building?

In hoisting the bag of sand to the top of the building we are counteracting the force of gravity which equals the weight of the sand. At a point x feet above the ground, the weight $F(x)$ is given by

$$F(x) = M - px. \qquad \text{(Figure 6.5.8)}$$

Therefore, the work done in lifting the bag of sand to the top of the building is

$$W = \int_0^H F(x)\, dx = \int_0^H (M - px)\, dx$$

$$= \left[Mx - \tfrac{1}{2}px^2\right]_0^H = MH - \tfrac{1}{2}pH^2 \quad \text{foot-pounds.}$$

Example 4 A bag of sand weighing 150 pounds is hoisted by a cable from the ground to the top of a 50-foot building. Sand leaks out of the bag at the rate of 0.75 pounds for each foot that the bag is raised.

(a) How much work is required to lift the bag of sand to the top of the building if the weight of the cable is negligible?

(b) How much work is required to lift the bag of sand if the cable weighs 1.5 pounds per foot?

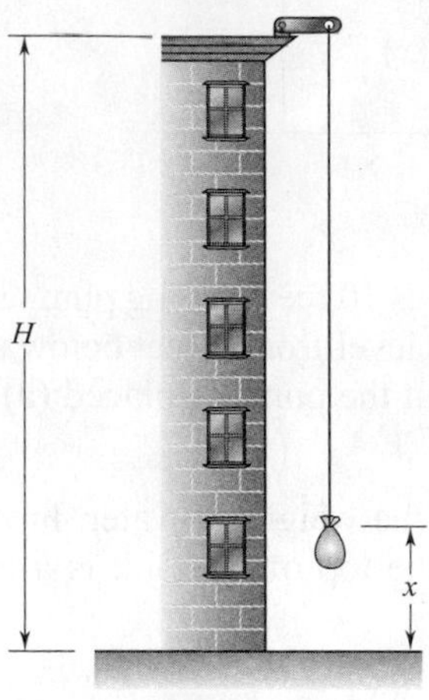

Figure 6.5.8

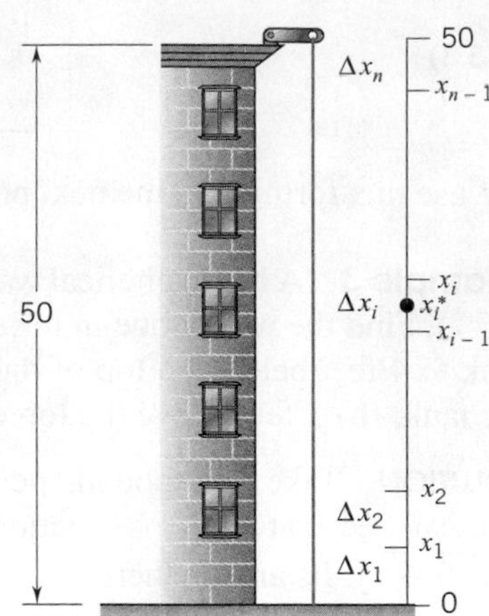

Figure 6.5.9

SOLUTION (a) By our result above, the work required to lift the bag of sand is given by

$$W_1 = (150)(50) - \tfrac{1}{2}(0.75)(50)^2 = 6562.5 \text{ foot-pounds.}$$

(b) The work required to lift the bag of sand and the cable to the top of the building is equal to the work needed to lift the sand plus the work needed to lift the cable. Thus, we simply have to add the work required to lift the cable to our result in part(a).

The approach here is similar to pumping out a tank. A partition $P = \{x_0, x_1, x_2 \ldots, x_n\}$ of the interval [0, 50] will partition the chain into n pieces of length $\Delta x_i, i = 1, 2, \ldots, n$. For each integer i, $1 \leq i \leq n$, let x_i^* be an arbitrary point in the ith subinterval (Figure 6.5.9).

Now, the ith piece of cable weighs $1.5\Delta x_i$ and is approximately $50 - x_i$ feet from the top of the building. Thus, the work required to lift this piece to the top is $x_i^*(1.5\Delta x_i) = 1.5x_i^*\Delta x_i$ foot-pounds. Adding the work required to lift each piece to the top, we have

$$1.5x_1^*\Delta x_1 + 1.5x_2^*\Delta x_2 + \cdots + 1.5x_n^*\Delta x_n \quad \text{foot-pounds.}$$

This is a Riemann sum which, as $||P|| \to 0$, converges to the definite integral $\int_0^{50} 1.5x\, dx$. Thus, the work required to lift the cable to the top of the building is

$$W_2 = \int_0^{50} 1.5x\, dx = 1.5\left[\tfrac{1}{2}x^2\right]_0^{50} = 1875 \quad \text{foot-pounds.}$$

[Note: This problem can be solved without integration by viewing the weight of the entire cable as concentrated at the center of mass of the cable. The cable weighs $1.5 \times 50 = 75$ pounds. Since the cable is homogeneous, the center of mass is at the center of the cable, 25 feet above the ground. The work required to lift 75 pounds a distance of 25 feet is

$$75 \text{ pounds} \times 25 \text{ feet} = 1875 \text{ foot-pounds.}$$

In Exercise 29 you are asked to show that this center of mass argument also applies to the nonhomogeneous case.]

Finally, the total work required to lift the bag of sand and the cable to the top of the building is given by

$$W = W_1 + W_2 = 8437.5 \text{ foot-pounds.} \quad \square$$

EXERCISES 6.5

In Exercises 1 and 2, an object is moving along the x-axis under the action of a force of $F(x)$ pounds when it is x feet from the origin. Find the work done by F in moving the object from $x = a$ to $x = b$.

1. $F(x) = x(x^2+1)^2$; $a = 1, b = 4$.

2. $F(x) = 2x\sqrt{x+1}$; $a = 3, b = 8$.

In Exercises 3–6, an object is moving along the x-axis under the action of force of $F(x)$ newtons when it is x meters from the origin. Find the work done by F in moving the object from $x = a$ to $x = b$.

3. $F(x) = x\sqrt{x^2+7}$; $a = 0, b = 3$.

4. $F(x) = x^2 + \cos 2x$; $a = 0$, $b = \frac{1}{4}\pi$.

5. $F(x) = x + \sin 2x$; $a = \frac{1}{6}\pi, b = \pi$.

6. $F(x) = \dfrac{\cos 2x}{\sqrt{2+\sin 2x}}$; $a = 0, b = \frac{1}{2}\pi$.

7. A 600-pound force will compress a 10-inch automobile coil spring 1 inch. How much work must be done to compress the spring to 5 inches?

8. Five foot-pounds of work are needed to stretch a certain spring from 1 foot to 3 feet beyond its natural length. How far beyond its natural length will a 6-pound weight stretch the spring?

9. Stretched 4 feet beyond its natural length, a certain spring exerts a restoring force of 200 pounds. What work must we do to stretch the spring: (a) 1 foot beyond its natural length? (b) $1\frac{1}{2}$ feet beyond its natural length?

10. A certain spring has natural length L. Given that W is the work done in stretching the spring from L feet to $L+a$ feet, find the work done in stretching the spring: (a) from L feet to $L+2a$ feet; (b) from L feet to $L+na$ feet; (c) from $L+a$ feet to $L+2a$ feet; (d) from $L+a$ feet to $L+na$ feet.

11. Find the natural length of a heavy metal spring, given that the work done in stretching it from a length of 2 feet to a length to 2.1 feet is one-half of the work done in stretching it from a length of 2.1 feet to a length of 2.2 feet.

12. A vertical cylindrical tank of radius 2 feet and height 6 feet is full of water. Find the work done in pumping out the water: (a) to an outlet at the top of the tank; (b) to a level 5 feet above the top of the tank. (Assume that the water weighs 62.5 pounds per cubic foot.)

13. A horizontal cylindrical tank of radius 3 feet and length 8 feet is half full of oil weighing 60 pounds per cubic foot.

(a) Show that the work done in pumping out the oil to the top of the tank is given by the integral

$$960\int_0^3 (x+3)\sqrt{9-x^2}\,dx.$$

Evaluate this integral by evaluating the integrals

$$\int_0^3 x\sqrt{9-x^2}\,dx \quad \text{and} \quad \int_0^3 \sqrt{9-x^2}\,dx$$

separately. HINT: Verify that the second integral represents one-quarter of the area of a circle of radius 3.

(b) What is the work done in pumping out the oil to a level 4 feet above the top of the tank?

14. What is the work done by gravity if the tank of Exercise 13 is completely drained through an opening at the bottom?

15. A conical container (vertex down) of radius r feet and height h feet is full of a liquid weighing σ pounds per cubic foot. Find the work done in pumping out the top $\frac{1}{2}h$ feet of liquid: (a) to the top of the tank; (b) to a level k feet above the top of the tank.

16. What is the work done by gravity if the tank of Exercise 15 is completely drained through an opening at the bottom?

17. A tank filled with water is in the shape of the solid generated by revolving the parabola $y = \frac{3}{4}x^2, 0 \le x \le 4$, around the y-axis. The dimensions of the tank are measured in meters. Find the work done in pumping the water: (a) to an outlet at the top of the tank; (b) to an outlet 1 meter above the top of the tank. Take $\sigma = 9800$.

18. The force of gravity exerted by the earth on a mass m at a distance r from the center of the earth is given by Newton's formula

$$F = -G\frac{mM}{r^2},$$

where M is the mass of the earth and G is the gravitational constant. Find the work done by gravity in pulling a mass m from $r = r_1$ to $r = r_2$.

19. A chain that weighs 15 pounds per foot is hanging from the top of an 80-foot building to the ground. How much work is done in pulling the chain to the top of the building?

20. A box that weighs w pounds is dropped to the floor from a height of d feet. (a) What is the work done by gravity? (b) Show that the work is the same if the box slides to the floor along a smooth inclined plane.

21. A bucket of sand weighing 200 pounds is hoisted at a constant rate by a chain from the ground to the top of a building 100 feet high.

(a) How much work is required to lift the bucket if the weight of the chain is negligible?

(b) How much work is required to lift the bucket if the chain weighs 2 pounds per foot?

22. Suppose that the bucket in Exercise 21 has a hole in the bottom and the sand leaks out at a constant rate so that only 150 pounds of sand are left when the bucket reaches the top.

(a) How much work is required to lift the bucket if the weight of the chain is negligible?

(b) How much work is required to lift the bucket if the chain weighs 2 pounds per foot?

23. A 100-pound bag of sand is lifted for 2 seconds at the rate of 4 feet per second. Find the work done in lifting the bag if the sand leaks out at the rate of $1\frac{1}{2}$ pounds per second.

24. A bucket of water is to be raised by a rope from the bottom of a 40-foot well. When the bucket is full of water it weighs 40 pounds; however, there is a hole in the bottom, and the water leaks out at the constant rate of $\frac{1}{2}$ gallon for each 10 feet that the bucket is raised. If the weight of the rope is negligible, how much work is done in raising the bucket to the top of the well? (Assume that water weighs 8.3 pounds per gallon.)

25. A rope of length l feet that weighs σ pounds per foot is lying on the ground. What is the work done in lifting the rope so that it hangs from a beam: (a) l feet high; (b) $2l$ feet high?

26. A load of weight w is lifted from the bottom of a shaft h feet deep. Find the work done given that the rope used to hoist the load weighs σ pounds per foot.

27. An 800-pound steel beam hangs from a 50-foot cable which weighs 6 pounds per foot. Find the work done in winding 20 feet of the cable about a steel drum.

28. A water container initially weighing w pounds is hoisted by a crane at the rate of n feet per second. What is the work done if the tank is raised m feet and the water leaks out constantly at the rate of p gallons per second? (Assume that the water weighs 8.3 pounds per gallon.)

29. A chain of variable mass density hangs to the ground from the top of a building of height H. Show that the work required to pull the chain to the top of the building can be obtained by assuming that the weight of the entire chain is concentrated at the center of mass of the chain.

30. Rework Exercise 19 using the result of Exercise 29.

31. Rework Exercise 21(b) using the result of Exercise 29.

32. Rework Exercise 26 using the result of Exercise 29.

33. An object of mass m is moving along the x-axis. At $x = a$, it has velocity v_a and at $x = b$ it has velocity v_b. Use Newton's second law of motion, $F = ma = m(dv/dt)$, to show that

$$W = \int_a^b F(x)\,dx = \tfrac{1}{2}mv_b^2 - \tfrac{1}{2}mv_a^2.$$

HINT: $\dfrac{dv}{dt} = \dfrac{dv}{dx}\dfrac{dx}{dt} = v\dfrac{dv}{dx}.$

Use the result of Exercise 33 in Exercises 34–37.

34. If the same amount of work done on two objects results in the speed of one of them being three times that of the other, how are the masses of the two objects related?

35. A major league baseball weighs 5 oz. How much work is required to throw a baseball at a speed of 95 mph? (The ball's mass is its weight in pounds divided by 32 ft/ sec^2, the acceleration of gravity.)

36. How much work is required to increase the speed of a 2000-lb vehicle from 30 mph to 55 mph?

37. The speed of an earth satellite at an altitude of 100 miles is approximately 17,000 mph. How much work is required to launch a 1000-lb satellite into a 100-mile orbit?

Power. Power measures the rate at which work is done. Suppose an object starts at $x = a$ and moves along the x-axis under the action of a force F. Then the work done by F in moving the object to the position x is

$$W(x) = \int_a^x F(u)du.$$

If the position of the object is a function of time, that is, if $x = x(t)$, then

$$p = \frac{dW}{dt} = F[x(t)]\frac{dx}{dt} = F[x(t)]v(t)$$

is called the *power input* of the force F. The units of measurement of power are foot-pounds per second if force is measured in pounds, distance in feet, and time in seconds; or joules per second, or *watts*, if force is measured in newtons, distance in meters, and time in seconds. In addition,

$$\begin{aligned}1 \text{ horsepower} &= 550 \text{ foot-pounds per second}\\ &= 746 \text{ watts.}\end{aligned}$$

38. (a) Assuming constant acceleration, what horsepower must an engine produce to accelerate a 3000-pound truck from 0 to 60 miles per hour (88 feet per second) in 15 seconds along a level road?
(b) What horsepower must the engine produce if the road rises 4 feet for every 100 feet of road? HINT: Integration is not required to answer these questions.

39. A vertical cylindrical tank with a height of 10 feet and a radius of 5 feet is half filled with water. If a 1 horsepower pump can do 550 foot-pounds of work per second, how long will it take a $\frac{1}{2}$-horsepower pump to:
(a) pump the water out of the top of the tank?
(b) pump the water to a point 5 feet above the top of the tank?

40. A tank filled with oil weighing 60 pounds per cubic foot is in the shape of a hemisphere surmounted by a cylinder. The hemisphere and cylinder each have a radius of 4 feet, and the height of the cylinder is 8 feet.
(a) What is the work done in pumping the oil to the top of the tank?
(b) How long will it take a $\frac{1}{2}$-horsepower motor to empty the tank?

■ *6.6 FLUID FORCE

If you pour oil into a container of water, you'll see that the oil soon rises to the top. Oil weighs less than water.

For any fluid, the weight per unit volume is called the *weight density* of the fluid. We'll denote this by the Greek letter σ.

An object submerged in a fluid experiences a compressive force that acts at right angles to each surface exposed to the fluid. (It is to counter these compressive forces that submarines have to be built so structurally strong.)

Fluid in a container exerts a downward force on the base of the container. What is the magnitude of this force? It is the weight in the column of fluid directly above it. (Figure 6.6.1). If a container with base area A is filled to a depth h by a fluid of weight density σ, the downward force on the base of the container is given by the product

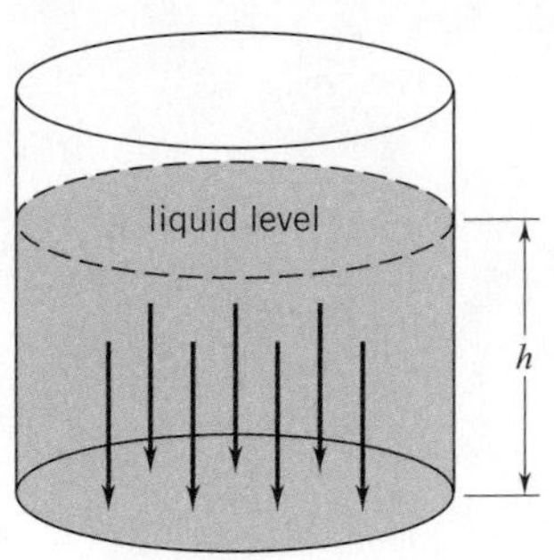

Figure 6.6.1

(6.6.1) $$F = \sigma hA.$$

A Word about Units. If area is measured in square feet, depth in feet, and weight density in pounds per cubic foot, then the fluid force is given in pounds. If area is measured in square meters, depth in meters, and weight density in newtons per cubic meter, the fluid force is given in newtons.

For example, the weight density of water is approximately 62.5 pounds per cubic foot. A plane surface with an area of 4 square feet submerged horizontally in water at a depth of 6 feet experiences a downward force of 1500 pounds:

$$(62.5)(6)(4) = 1500.$$

Fluid force acts not only on the base of the container but also on the walls of the container. In Figure 6.6.2, we have depicted a vertical wall standing against a body of liquid. (Think of it as the wall of a container or as a dam at the end of a lake.) We want to calculate the force exerted by the liquid on this wall.

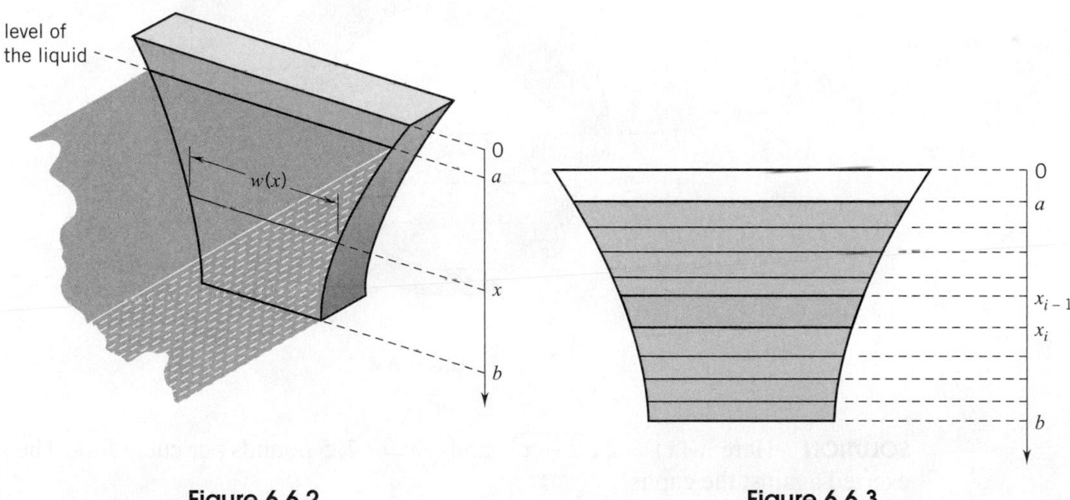

Figure 6.6.2

Figure 6.6.3

As in the figure, we assume that the liquid extends from depth a to depth b, and we let $w(x)$ denote the width of the wall at depth x. A partition $P = \{x_0, x_1, \ldots, x_n\}$ of $[a, b]$ of small norm subdivides the wall into n narrow horizontal strips (Figure 6.6.3).

We can estimate the force on the ith strip by taking x_i^* as the midpoint of $[x_{i-1}, x_i]$. Then

$$w(x_i^*) = \text{the approximate width of the } i\text{th strip}$$

and $$w(x_i^*)\Delta x_i = \text{the approximate area of the } i\text{th strip}.$$

Since this strip is narrow, all the points of the strip are approximately at depth x_i^*. Thus, using (6.6.1), we can estimate the force on the ith strip by the product

$$\sigma x_i^* w(x_i^*)\Delta x_i.$$

Adding up all these estimates, we have an estimate for the force on the entire wall:

$$F \cong \sigma x_i^* w(x_i^*)\Delta x_1 + \sigma x_2^* w(x_2^*)\Delta x_2 + \cdots + \sigma x_n^* w(x_n^*)\Delta x_n.$$

The sum on the right is a Riemann sum for the integral

$$\int_a^b \sigma x w(x)\, dx,$$

and as such converges to that integral as $||P|| \to 0$. Thus we have

(6.6.2) $$\boxed{\text{force against wall} = \int_a^b \sigma x w(x)\, dx.}$$

Example 1 A horizontal circular water main (Figure 6.6.4) 6 feet in diameter is capped when half full of water. What is the force of the water against the cap?

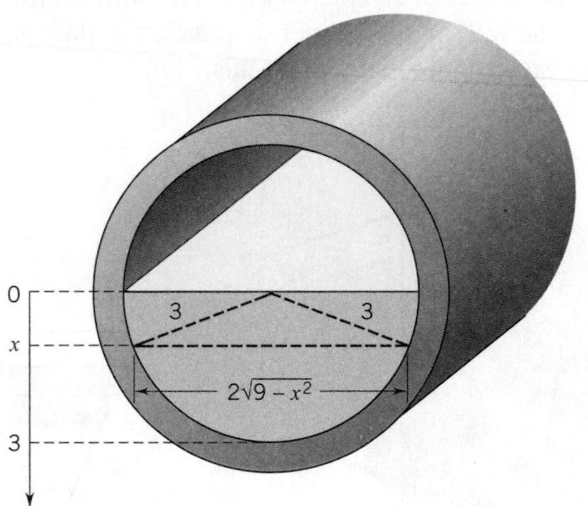

Figure 6.6.4

SOLUTION Here $w(x) = 2\sqrt{9 - x^2}$ and $\sigma = 62.5$ pounds per cubic foot. The force exerted against the cap is

$$F = \int_0^3 (62.5)x(2\sqrt{9 - x^2})\, dx = 62.5 \int_0^3 2x\sqrt{9 - x^2}\, dx = 1125 \text{ pounds.} \quad \square$$

Example 2 A metal plate in the form of a trapezoid is affixed to a vertical dam as in Figure 6.6.5. The dimensions shown are given in meters; the weight density of water in the metric system is approximately 9800 newtons per cubic meter. Find the force on the plate.

SOLUTION First we find the width of the plate x meters below the water level. By similar triangles (see Figure 6.6.6),

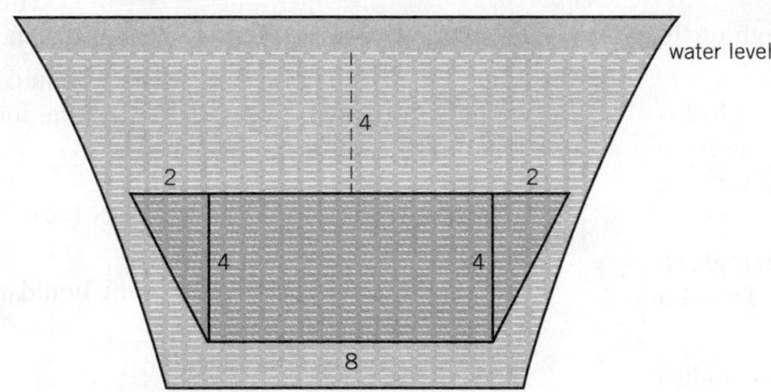

Figure 6.6.5

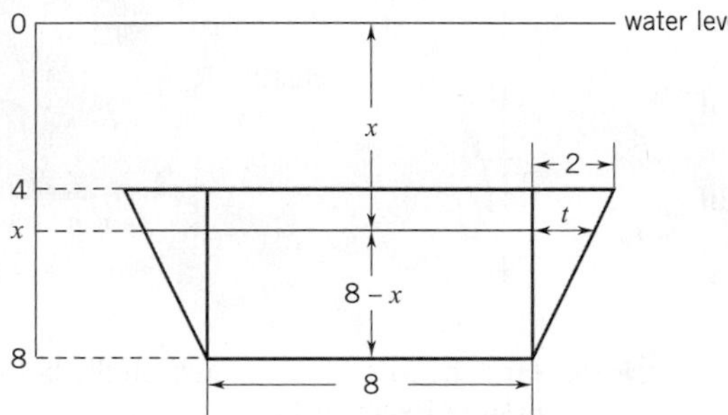

Figure 6.6.6

$$t = \tfrac{1}{2}(8 - x) \qquad \text{so that} \qquad w(x) = 8 + 2t = 16 - x.$$

The force against the plate is

$$\int_4^8 9800x(16 - x)\,dx = 9800 \int_4^8 (16x - x^2)\,dx$$

$$= 9800\Big[8x^2 - \tfrac{1}{3}x^3\Big]_4^8 \cong 2,300,000 \text{ newtons.} \quad \square$$

EXERCISES *6.6

1. A rectangular plate 8 feet wide by 6 feet deep is submerged vertically in a tank of water with the upper edge (that is, the 8-foot edge) at the surface of the water. Find the force of the water on one side of the plate.

2. A 6-foot by 6-foot square plate is submerged vertically in a tank of water with upper edge parallel to the surface. The center of the plate is 4 feet below the surface. Find the force of the water on one side of the plate.

3. A vertical dam in a river is in the shape of an isosceles trapezoid 100 meters across at the surface of the river, and 60 meters across at the bottom. If the river is 20 meters deep, find the force of the water on the dam.

4. A 5-meter by 5-meter square gate is at the bottom of the dam in Exercise 3. What is the force of the water on the gate?

5. A gate in the shape of an isosceles trapezoid 4 meters at the top, 6 meters at the bottom, and 3 meters high has its upper edge 10 meters below the top of the dam in Exercise 3. Find the force of the water on this gate.

6. A vertical dam in the shape of a rectangle is 1000 feet wide and 100 feet high. Calculate the force on the dam when:

(a) the water level is 75 feet above the bottom.

(b) the water level is 50 feet above the bottom.

7. Each end of a horizontal oil tank is an ellipse with horizontal axis 12 feet along and vertical axis 6 feet long. Calculate the force on one end when the tank is half full of oil weighing 60 pounds per cubic foot.

8. Each vertical end of a vat is a segment of a parabola (vertex down) 8 feet across the top and 16 feet deep. Calculate the force on an end when the vat is full of liquid weighing 70 pounds per cubic foot.

9. The vertical ends of a water trough are isosceles right triangles with the 90° angle at the bottom. Calculate the force on each triangle when the trough is full of water given that the legs of the triangle are 8 feet long.

10. The vertical ends of a water trough are isosceles triangles 5 feet across the top and 5 feet deep. Calculate the force on an end when the trough is full of water.

11. The ends of a water trough are semicircular discs with radius 2 feet. Find the force of the water on one end if the trough is full of water.

12. The ends of a water trough have the shape of the parabolic region bounded by $y = x^2 - 4$ and $y = 0$, where the measurements are in feet. Assuming that the trough is full of water, set up a definite integral whose value is the force of the water on one end.

13. A horizontal cylindrical tank of diameter 8 feet is half full of oil weighing 60 pounds per cubic foot. Calculate the force on one end.

14. Calculate the force on one end if the tank of Exercise 13 is full.

15. A rectangular metal plate 10 feet by 6 feet is affixed to a vertical dam with the center of the plate 11 feet below water level. Find the force on the plate if: (a) the 10-foot sides are horizontal; (b) the 6-foot sides are horizontal.

16. A vertical cylindrical tank of diameter 30 feet and height 50 feet is full of oil weighing 60 pounds per cubic foot. Find the force on the curved surface.

17. A swimming pool is 8 meters wide and 14 meters long. The pool is 1 meter deep at the shallow end and 3 meters deep at the other end; the depth increases linearly from the shallow to the deep end. If the pool is full of water, find:

(a) the force of the water on each of the sides.

(b) the force of the water on each of the ends.

18. Prove that if a plate is submerged in a liquid at an angle of θ with the vertical, $0 < \theta < \pi/2$, then the force on the plate is given by

$$F = \int_a^b \sigma x w(x) \sec\theta \, dx,$$

where σ is the weight density of the liquid, and $x, w(x), a, b$ are as shown in the figure below.

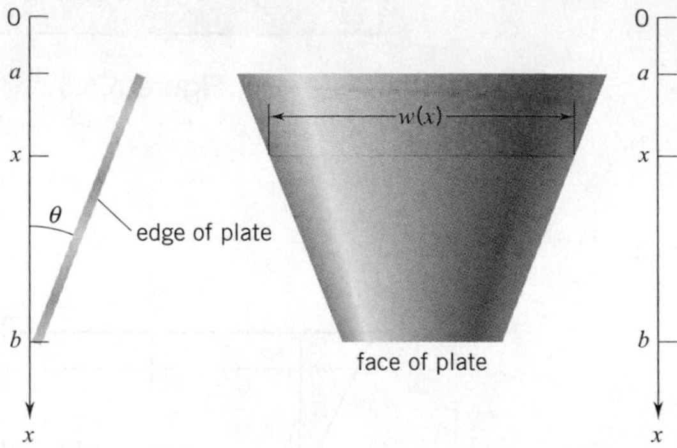

19. Find the force of the water on the bottom of the swimming pool in Exercise 17.

20. The face of a rectangular dam in a river is 1000 feet wide by 100 feet deep, and it makes an angle of 30° with the vertical. Find the force of the water on the face of the dam if:

(a) the water level is at the top of the damp.

(b) the water level is 75 feet from the bottom of the river.

21. Relate the force on a dam to the centroid of the submerged surface of the dam.

22. Two identical metal plates are affixed to a vertical dam. The centroid of the first plate is at depth h_1, and the centroid of the second plate is a depth h_2. Compare the forces on the two plates. HINT: See Exercise 21.

■ CHAPTER HIGHLIGHTS

6.1 More on Area

representative rectangle (p. 327)
area by integration with respect to y (p. 328)

6.2 Volume by Parallel Cross Sections; Discs and Washers

volume by parallel cross sections (p. 333)
disc method (p. 336)
washer method x-axis (p. 339)
washer method y-axis (p. 339)

6.3 Volume by the Shell Method

shell method y-axis (p. 344) shell method (x-axis) (p. 345)

6.4 The Centroid of a Region; Pappus's Theorem on Volumes

principles for finding the centroid of a region (p. 350)

$$\bar{x}A = \int_a^b x[f(x) - g(x)]\, dx,$$

$$\bar{x}A = \int_a^b \tfrac{1}{2}([f(x)]^2 - [g(x)]^2)\, dx$$

Pappus's theorem on volumes (p. 353)
centroid of a solid of revolution:

$$\bar{x}V_x = \int_a^b \pi x[f(x)]^2\, dx, \bar{y}V_y = \int_a^b \pi x[f(x)]^2\, dx$$

6.5 The Notion of Work

Hooke's law (p. 358)
counteracting the force of gravity (p. 360)

6.6 Fluid Force

force against a vertical wall $= \int_a^b \sigma x w(x)\, dx$ (p. 366)

SKILL MASTERY REVIEW Integration and Applications

Evaluate the integral.

1. $\displaystyle\int \frac{x^3 - 2x + 1}{\sqrt{x}}\, dx$

2. $\displaystyle\int (x^{3/5} - 3x^{5/3})\, dx$

3. $\displaystyle\int t^2(1 + t^3)^{10}\, dt$

4. $\displaystyle\int (1 + 2\sqrt{x})^2\, dx$

5. $\displaystyle\int \frac{(t^{2/3} - 1)^2}{t^{1/3}}\, dt$

6. $\displaystyle\int x\sqrt{x^2 - 2}\, dx$

7. $\displaystyle\int x\sqrt{2 - x}\, dx$

8. $\displaystyle\int x^2(2 + 2x^3)^4\, dx$

9. $\displaystyle\int \frac{(1 + \sqrt{x})^5}{\sqrt{x}}\, dx$

10. $\displaystyle\int \frac{\sin(1/x)}{x^2}\, dx$

11. $\displaystyle\int \frac{\cos x}{\sqrt{1 + \sin x}}\, dx$

12. $\displaystyle\int (\sec\theta - \tan\theta)^2\, d\theta$

13. $\displaystyle\int (\tan 3\theta - \cot 3\theta)^2\, d\theta$

14. $\displaystyle\int x\sin^3(x^2)\cos(x^2)\, dx$

15. $\displaystyle\int \frac{1}{1 + \cos 2x}\, dx$

16. $\displaystyle\int \frac{1}{1 - \sin 2x}\, dx$

17. $\displaystyle\int \sec^3 \pi x \tan \pi x\, dx$

18. $\displaystyle\int ax\sqrt{1 + bx^2}\, dx$

19. $\displaystyle\int ax\sqrt{1 + bx}\, dx$

20. $\displaystyle\int ax^2\sqrt{1 + bx}\, dx$

21. $\displaystyle\int \frac{g(x)g'(x)}{\sqrt{1 + g^2(x)}}\, dx$

22. $\displaystyle\int \frac{g'(x)}{g^3(x)}\, dx$

Evaluate the definite integral.

23. $\displaystyle\int_{-1}^{2} (x^2 - 2x + 3)\, dx$

24. $\displaystyle\int_0^1 \frac{x}{(x^2 + 1)^3}\, dx$

25. $\displaystyle\int_0^{\pi/4} \sin^3 2x \cos 2x\, dx$

26. $\displaystyle\int_0^{\pi/8} (\tan^2 2x + \sec^2 2x)\, dx$

27. $\displaystyle\int_0^2 (x^2 + 1)(x^3 + 3x - 6)^{1/3}\, dx$

28. $\displaystyle\int_{-\pi/2}^{\pi/4} (\sin x + \cos x)^2\, dx$

29. $\displaystyle\int_1^8 \frac{(1 + x^{1/3})^2}{x^{2/3}}\, dx$

30. At every point of a curve C the slope is given by $y' = x\sqrt{x^2 + 1}$. Find an equation $y = f(x)$ for C given that the curve passes through the point $(0, 1)$.

Sketch the region bounded by the curves and find its area.

31. $y = 4 - x^2, \quad y = x + 2$
32. $y = 4 - x^2, \quad x + y + 2 = 0$
33. $y^2 = x, \quad x^2 = 3y$
34. $y = \sqrt{x}$, the x-axis, $y = 6 - x$
35. $y = x^3$, the x-axis, $x + y = 2$
36. $4y = x^2 - x^4, \quad x + y + 1 = 0$

Sketch the region bounded by the curves. Represent the area of the region by one or more definite integrals (a) in terms of x; (b) in terms of y. Find the area of the region using the most convenient representation.

37. $y = 2 - x^2, \quad y = -x$
38. $y = x^3, \quad y = -x, \quad y = 1$
39. $y^2 = 2(x - 1), \quad x - y = 5$
40. $y^3 = x^2, \quad x - 3y + 4 = 0$

Carry out the differentiation.

41. $\displaystyle\frac{d}{dx}\left(\int_0^x \frac{dt}{1 + t^2}\right)$

42. $\displaystyle\frac{d}{dx}\left(\int_0^{x^2} \frac{dt}{1 + t^2}\right)$

43. $\dfrac{d}{dx}\left(\displaystyle\int_x^{x^2} \frac{dt}{1+t^2}\right)$

44. Let $F(x) = \displaystyle\int_0^x \frac{1}{t^2+t+1}\,dt$, x any real number.

(a) Does F take on the value 0? If so, where?

(b) Show that F is an increasing function on $(-\infty, \infty)$.

(c) Determine the concavity of the graph of F.

(d) Sketch the graph of F.

45. Assume that f is a continuous function and that

$$\int_0^x tf(t)\,dt = x \sin x + \cos x - 1.$$

(a) Determine $f(\pi)$. (b) Find $f'(x)$.

Find the average value of f on the indicated interval.

46. $f(x) = \dfrac{x}{\sqrt{x^2+9}}$; $[0, 4]$

47. $f(x) = x + 2\sin x$; $[0, \pi]$

48. A rod extends from $x = 0$ to $x = a$. Find the center of mass if the density of the rod varies directly as the distance from $x = 2a$.

Sketch the region Ω bounded by the given curves and find the volume of the solid generated by revolving Ω about the indicated axis.

49. $x^2 = 4y$, $y = \frac{1}{2}x$; x-axis

50. $x^2 = 4y$, $y = \frac{1}{2}x$; y-axis

51. $y = x^3$, $y = 1$, $x = 0$; x-axis

52. $y = x^3$, $y = 1$, $x = 0$; y-axis

53. $y = \tan x$, $0 \le x \le \pi/4$; x-axis

54. $y = \cos x$, $-\pi/2 \le x \le \pi/2$; x-axis

55. $y = \sin(x^2)$, $0 \le x \le \sqrt{\pi}$; y-axis

56. $y = 3x - x^2$, $y = x^2 - 3x$, y-axis

57. The base of a solid is the disc $x^2 + y^2 \le r^2$. Find the volume of the solid if cross-sections perpendicular to the x-axis are (a) semicircles; (b) isosceles right triangles with hypotenuse on xy-plane.

58. The base of a solid is an equilateral triangle with one vertex at the origin and an altitude along the positive x-axis. Find the volume of the solid if cross-sections perpendicular to the x-axis are squares with one side in the base of the solid.

59. A solid in the shape of a right circular cylinder of radius 3 has its base on the xy-plane. A wedge is cut from the cylinder by a plane through a diameter of the base and inclined to the xy-plane at an angle of 30°. Find the volume of the wedge.

Find the centroid of the bounded region determined by the given curves.

60. $y = 4 - x^2$, $y = 0$

61. $y = x^3$, $y = 4x$

62. $y = x^2 - 4$, $y = 2x - x^2$

63. $y = \cos x$, $-\pi/2 \le x \le \pi/2$

Sketch the region bounded by the given curves. Determine the centroid of the region and the volume of the solid generated by revolving the region about each of the coordinate axes.

64. $y = x$, $y = 2 - x^2$, $0 \le x \le 1$

65. $y = x^3$, $x = y^3$, $0 \le x \le 1$.

66. One of the springs used to support a truck has a natural length of 12 inches and a force of 8000 pounds compresses it $\frac{1}{2}$ inch. Find the work done in compressing the spring from 12 to 9 inches.

67. The work required to stretch a spring from 9 inches to 10 inches is 1.5 times the work needed to stretch the spring from 8 inches to 9 inches. What is the natural length of the spring?

68. A conical tank, 10 meters deep and 8 meters across the top is filled with water to a depth of 5 meters. Find the work done in pumping the water: (a) to an outlet at the top of the tank; (b) to an outlet 1 meter below the top of the tank. Take $\sigma = 9800$.

69. A bucket, weighing 5 pounds when empty, is filled with sand weighing 60 pounds, and then lifted to the top of a 20 foot building at a constant rate. The sand leaks out of the bucket at a uniform rate and two-thirds of the sand remains when the bucket reaches the top. Find the work done in lifting the bucket of sand to the top of the building.

70. A spherical oil tank of radius 10 feet is half full of oil weighing 60 pounds per cubic foot. Find the work done in pumping the oil to an outlet at the top of the tank.

CHAPTER 7

THE TRANSCENDENTAL FUNCTIONS

Some real numbers satisfy polynomial equations with integer coefficients:

$$\tfrac{3}{5} \text{ satisfies the equation } 5x - 3 = 0,$$
$$\sqrt{2} \text{ satisfies the equation } x^2 - 2 = 0.$$

Such numbers are called *algebraic*. There are, however, numbers that are not algebraic, among them π. Such numbers are called *transcendental*.

Some functions f satisfy polynomial equations with polynomial coefficients:

$$f(x) = \frac{x}{\pi x + \sqrt{2}} \quad \text{satisfies the equation} \quad (\pi x + \sqrt{2})f(x) - x = 0,$$
$$f(x) = 2\sqrt{x} - 3x^2 \quad \text{satisfies the equation} \quad [f(x)]^2 + 6x^2 f(x) + (9x^4 - 4x) = 0.$$

Such functions are called *algebraic*. There are, however, functions that are not algebraic; these are called *transcendental*. For example, the trigonometric functions are transcendental functions. In this chapter we introduce other transcendental functions: the logarithm function, the exponential function, and the inverse trigonometric functions. But first, a little more on functions in general.

■ 7.1 ONE-TO-ONE FUNCTIONS; INVERSES

One-to-One Functions

A function can take on the same value at different points of its domain. Constant functions, for example, take on the same value at all points of their domains. The quadratic function $f(x) = x^2$ takes on the same value at $-c$ as it does at c; so does the absolute-value function $g(x) = |x|$. The function

$$f(x) = 1 + (x - 3)(x - 5)$$

takes on the same value at $x = 5$ as it does at $x = 3$:

$$f(3) = 1, \qquad f(5) = 1.$$

Functions for which this kind of repetition *does not* occur are called *one-to-one functions*.

DEFINITION 7.1.1

A function f is said to be *one-to-one* if there are no two distinct numbers in the domain of f at which f takes on the same value:

$$f(x_1) = f(x_2) \quad \text{implies} \quad x_1 = x_2.$$

Thus, if f is one-to-one and x_1, x_2 are different points of the domain, then

$$f(x_1) \neq f(x_2).$$

The functions

$$f(x) = x^3 \qquad \text{and} \qquad f(x) = \sqrt{x}$$

are both one-to-one. The cubing function is one-to-one because no two distinct numbers have the same cube. The square-root function is one-to-one because no two distinct nonnegative numbers have the same square root.

There is a simple geometric test, called the *horizontal line test,* which can be used to determine whether a function is one-to-one. Look at the graph of the function. If some horizontal line intersects the graph more than once, then the function is not one-to-one (Figure 7.1.1). If, on the other hand, no horizontal line intersects the graph more than once, then the function is one-to-one (Figure 7.1.2).

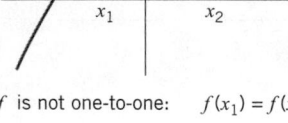

f is not one-to-one: $f(x_1) = f(x_2)$

Figure 7.1.1

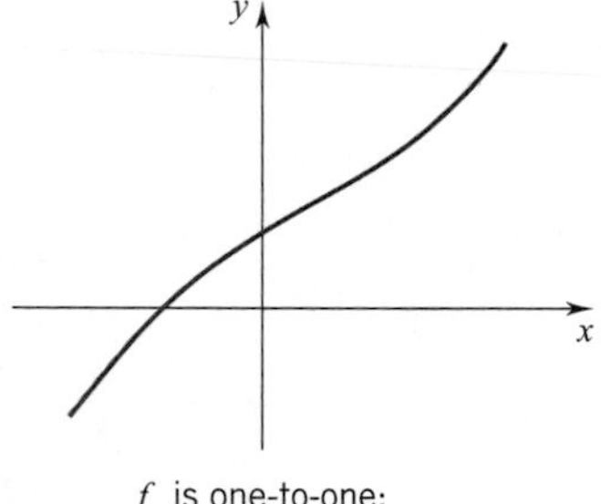

f is one-to-one:

Figure 7.1.2

Inverses

We begin with a theorem about one-to-one functions.

THEOREM 7.1.2

If f is a one-to-one function, then there is one and only one function g with domain equal to the range of f that satisfies the equation

$$f(g(x)) = x \quad \text{for all } x \text{ in the range of } f.$$

PROOF The proof is straightforward. If x is in the range of f, then f must take on the value x at some number. Since f is one-to-one, there can be only one such number. We have called that number $g(x)$. □

The function that we have named g in the theorem is called the *inverse* of f and is usually denoted by the symbol f^{-1}.

DEFINITION 7.1.3 INVERSE FUNCTION

Let f be a one-to-one function. The *inverse* of f, denoted by f^{-1}, is the unique function with domain equal to the range of f that satisfies the equation

$$f(f^{-1}(x)) = x \quad \text{for all } x \text{ in the range of } f. \qquad \text{(Figure 7.1.3)}$$

f $f^{-1}(x)$ $x = f(f^{-1}(x))$ f^{-1}

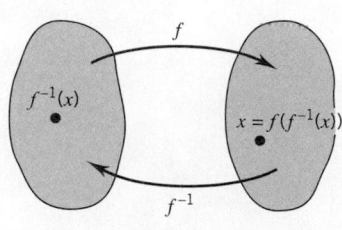

Figure 7.1.3

Remark The notation f^{-1} for the inverse function is standard. Unfortunately, there is the danger of confusing f^{-1} with the reciprocal of f, that is, with $1/f(x)$. The "-1" in the notation for the inverse of f is *not an exponent*; $f^{-1}(x)$ *does not mean* $1/f(x)$. On those occasions when we want to express $1/f(x)$ using the exponent -1, we will write $[f(x)]^{-1}$. ❑

Example 1 You have seen that the cubing function

$$f(x) = x^3$$

is one-to-one. Find the inverse.

SOLUTION We set $y = f^{-1}(x)$ and solve the equation $f(y) = x$ for y:

$$\begin{aligned} f(y) &= x \\ y^3 &= x \qquad (f \text{ is the cubing function}) \\ y &= x^{1/3}. \end{aligned}$$

Substituting $f^{-1}(x)$ back in for y, we have

$$f^{-1}(x) = x^{1/3}.$$

The inverse of the cubing function is the cube-root function. The graphs of $f(x) = x^3$ and $f^{-1}(x) = x^{1/3}$ are shown in Figure 7.1.4. ❑

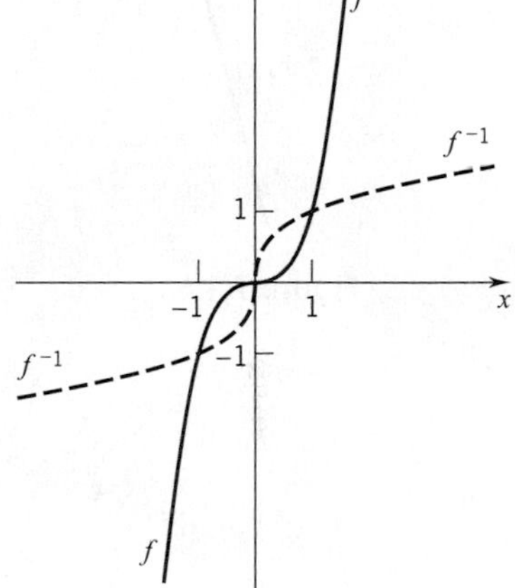

Figure 7.1.4

Remark We substitute y for $f^{-1}(x)$ merely to simplify the calculations. It is easier to work with the symbol y than with the string of symbols $f^{-1}(x)$. ❑

Example 2 Show that the linear function

$$f(x) = 3x - 5$$

is one-to-one. Then find the inverse.

SOLUTION To show that f is one-to-one , let's suppose that

$$f(x_1) = f(x_2).$$

Then

$$\begin{aligned} 3x_1 - 5 &= 3x_2 - 5 \\ 3x_1 &= 3x_2 \\ x_1 &= x_2. \end{aligned}$$

The function is one-to-one since

$$f(x_1) = f(x_2) \quad \text{implies} \quad x_1 = x_2.$$

(Viewed geometrically, the result is obvious. The graph is a line with slope 3 and as such cannot be intersected by a horizontal line more than once.)

Now let's find the inverse. To do this, we set $y = f^{-1}(x)$ and solve the equation $f(y) = x$ for y:

$$\begin{aligned} f(y) &= x \\ 3y - 5 &= x \\ 3y &= x + 5 \\ y &= \tfrac{1}{3}x + \tfrac{5}{3}. \end{aligned}$$

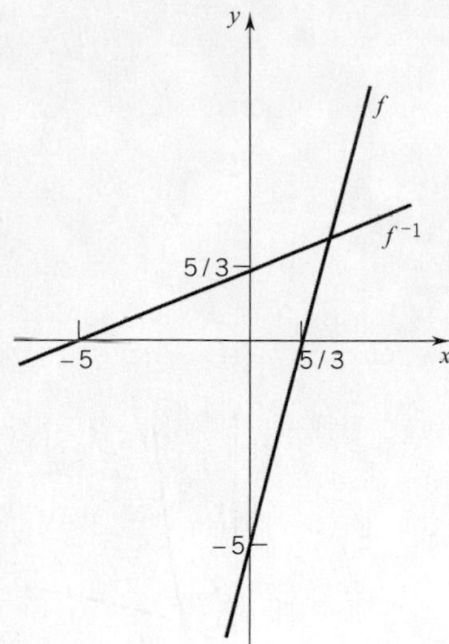

Figure 7.1.5

Substituting $f^{-1}(x)$ back in for y, we have

$$f^{-1}(x) = \tfrac{1}{3}x + \tfrac{5}{3}.$$

The graphs of f and f^{-1} are shown in Figure 7.1.5. □

Example 3 Find the inverse of the function

$$f(x) = (1 - x^3)^{1/5} + 2.$$

SOLUTION We set $y = f^{-1}(x)$ and solve the equation $f(y) = x$ for y:

$$\begin{aligned} f(y) &= x \\ (1 - y^3)^{1/5} + 2 &= x \\ (1 - y^3)^{1/5} &= x - 2 \\ 1 - y^3 &= (x - 2)^5 \\ y^3 &= 1 - (x - 2)^5 \\ y &= [1 - (x - 2)^5]^{1/3}. \end{aligned}$$

Substituting $f^{-1}(x)$ back in for y, we have

$$f^{-1}(x) = [1 - (x - 2)^5]^{1/3}. \quad \square$$

Example 4 Show that the function

$$F(x) = x^5 + 2x^3 + 3x - 4$$

is one-to-one.

SOLUTION Setting $F(x_1) = F(x_2)$, we have

$$\begin{aligned} x_1^5 + 2x_1^3 + 3x_1 - 4 &= x_2^5 + 2x_2^3 + 3x_2 - 4 \\ x_1^5 + 2x_1^3 + 3x_1 &= x_2^5 + 2x_2^3 + 3x_2. \end{aligned}$$

How to go on from here is far from clear. The algebra is very complicated.

Here is another approach. Differentiating F, we get

$$F'(x) = 5x^4 + 6x^2 + 3.$$

Note that $F'(x) > 0$ for all x and therefore F is an increasing function. Increasing functions are clearly one-to-one: $x_1 < x_2$ implies $F(x_1) < F(x_2)$, and so $F(x_1)$ cannot possibly equal $F(x_2)$. □

Remark In Example 4 we used the sign of the derivative to test for one-to-oneness. The connection between the sign of the derivative and one-to-oneness can be summarized as follows; functions with positive derivatives are increasing functions and therefore one-to-one; functions with negative derivatives are decreasing functions and therefore also one-to-one. □

Suppose that the function f has an inverse. Then, by definition, f^{-1} satisfies the equation

(7.1.4) $\quad f(f^{-1}(x)) = x \quad$ for all x in the range of f.

It is also true that

(7.1.5) $f^{-1}(f(x)) = x$ for all x in the domain of f.

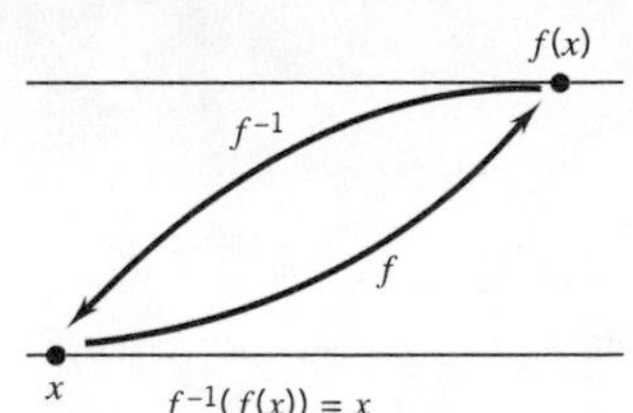

Figure 7.1.6

PROOF Take x in the domain of f and set $y = f(x)$. Since y is the range of f,

$$f(f^{-1}(y)) = y.$$

This means that

$$f(f^{-1}(f(x))) = f(x)$$

and tells us that f takes on the same value at $f^{-1}(f(x))$ as it does at x. With f one-to-one, this can only happen if

$$f^{-1}(f(x)) = x. \quad \square$$

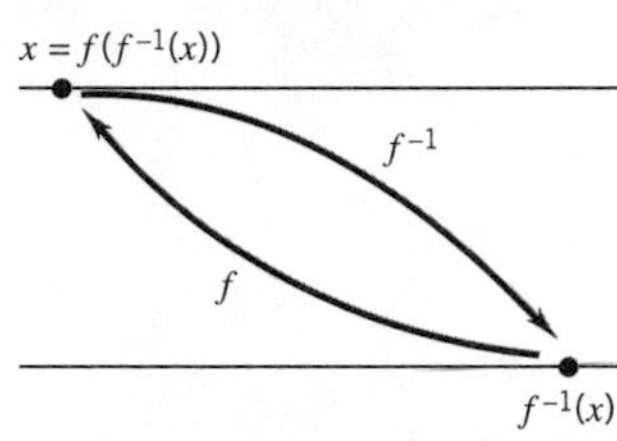

Figure 7.1.7

Equation (7.1.5) tells us that f^{-1} undoes what is done by f:

f takes x to $f(x)$; f^{-1} takes $f(x)$ back to x (Figure 7.1.6)

Equation (7.1.4) tells us that f undoes what is done by f^{-1}:

f^{-1} takes x to $f^{-1}(x)$; f takes $f^{-1}(x)$ back to x (Figure 7.1.7)

It is evident from this that

domain of f^{-1} = range of f and range of f^{-1} = domain of f.

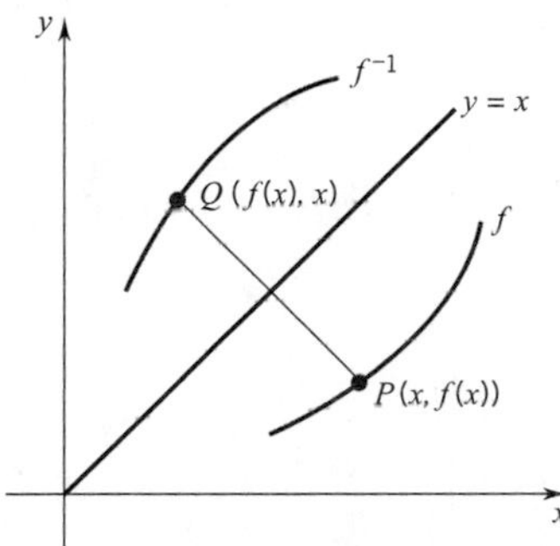

Figure 7.1.8

The Graphs of f and f^{-1}

The graph of f consists of the points $(x, f(x))$. Since f^{-1} takes on the value x at $f(x)$, the graph of f^{-1} consists of points of the form $(f(x), x)$. If, as usual, we use the same scale on the y-axis as we do on the x-axis, then the points $(x, f(x))$ and $(f(x), x)$ are symmetric with respect to the line $y = x$ (Figure 7.1.8). Hence

the graph of f^{-1} is the graph of f reflected in the line $y = x$.

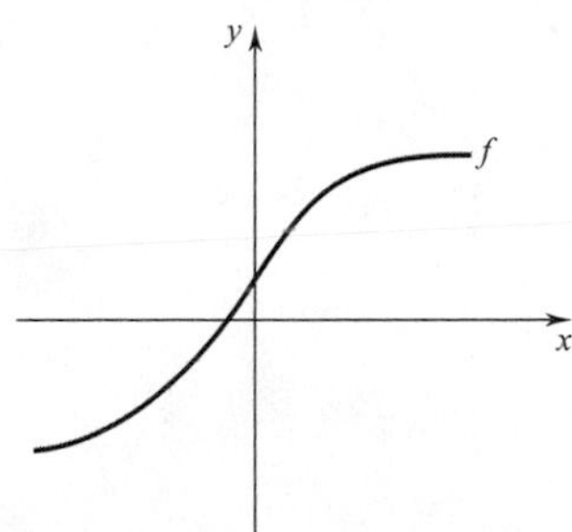

Figure 7.1.9

This idea pervades all that follows.

Example 5 Given the graph of f as in Figure 7.1.9, sketch the graph of f^{-1}.

SOLUTION First we draw the line $y = x$. Then we reflect the graph of f in that line. The result is shown in Figure 7.1.10. $\square$

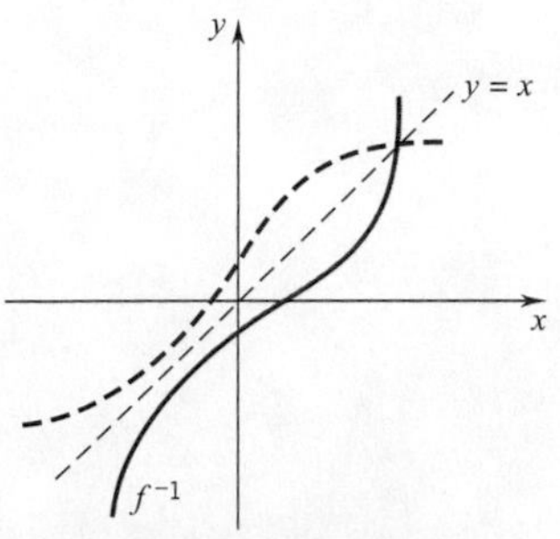

Figure 7.1.10

Continuity and Differentiability of Inverses

Let f be a one-to-one function. Then f has an inverse, f^{-1}. Suppose, in addition, that f is continuous. Since the graph of f has no "holes" or "gaps", and since the graph of f^{-1} is simply the reflection of the graph of f in the line $y = x$, we can conclude that the graph of f^{-1} has no holes or gaps. That is, f^{-1} must also be continuous. We state this result formally; a proof is given in Appendix B3.

THEOREM 7.1.6

Let f be a one-to-one function defined on an open interval I. If f is continuous, then its inverse f^{-1} is also continuous.

Now suppose that f is differentiable. Is f^{-1} necessarily differentiable? Let's assume so for the moment.

From the definition of inverse, we know that

$$f(f^{-1}(x)) = x \qquad \text{for all } x \text{ in the range of } f.$$

Differentiation gives

$$\frac{d}{dx}[f(f^{-1}(x))] = 1.$$

However, by the chain rule

$$\frac{d}{dx}[f(f^{-1}(x))] = f'(f^{-1}(x))(f^{-1})'(x),$$

and so

$$f'(f^{-1}(x))(f^{-1})'(x) = 1.$$

Therefore, if $f'(f^{-1}(x)) \neq 0$, then

(7.1.7)
$$(f^{-1})'(x) = \frac{1}{f'(f^{-1}(x))}.$$

For a geometric understanding of this relation, we refer you to Figure 7.1.11.

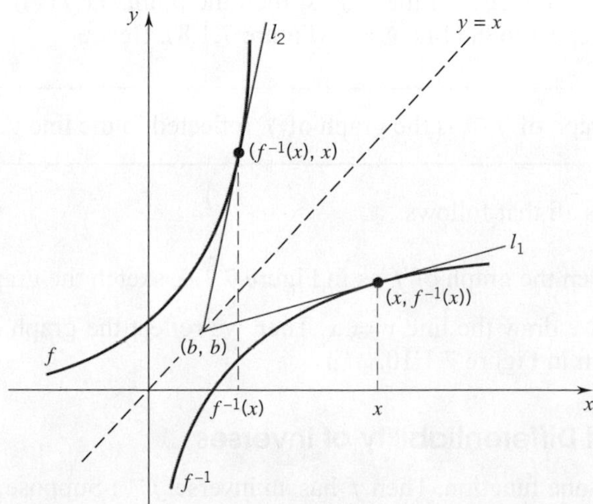

Figure 7.1.11

The graphs of f and f^{-1} are reflections of each other in the line $y = x$. The tangent lines l_1 and l_2 are also reflections of one another. From the figure,

$$(f^{-1})'(x) = \text{slope of } l_1 = \frac{f^{-1}(x) - b}{x - b}, \qquad f'(f^{-1}(x)) = \text{slope of } l_2 = \frac{x - b}{f^{-1}(x) - b},$$

so that $(f^{-1})'(x)$ and $f'(f^{-1}(x))$ are indeed reciprocals.

The figure shows the two tangent lines intersecting the line $y = x$ at a common point. If the tangent lines have slope 1, they do not intersect that line at all. However, in that case, both slopes are 1, the derivatives are also 1, and the relation holds. One more observation. We have assumed that $f'(f^{-1}(x)) \neq 0$. If $f'(f^{-1}(x)) = 0$, then the tangent line to the graph of f at $(f^{-1}(x), x)$ is horizontal, and the tangent line to the graph of f^{-1} at $(x, f^{-1}(x))$ is vertical, in which case f^{-1} is not differentiable.

Formula (7.1.7) has an unwieldy look about it; too many fussy little symbols. The following characterization of $(f^{-1})'$ may be easier to understand:

THEOREM 7.1.8

Let f be a one-to-one function differentiable on an open interval I. Let a be a point of I and let $f(a) = b$. If $f'(a) \neq 0$, then f^{-1} is differentiable at b and

$$(f^{-1})'(b) = \frac{1}{f'(a)}.$$

This theorem, proven in Appendix B3, places our discussion on a firm footing.

Remark Note that $a = f^{-1}(b)$, and therefore

$$(f^{-1})'(b) = \frac{1}{f'(a)} = \frac{1}{f'(f^{-1}(b))}.$$

This is simply (7.1.7) with $x = b$. ❑

We rely on Theorem 7.1.8 when we cannot solve for f^{-1} explicitly and yet we want to evaluate $(f^{-1})'$ at a particular number.

Example 6 The function $f(x) = x^3 + \frac{1}{2}x$ is differentiable and has range $(-\infty, \infty)$.

(a) Show that f is one-to-one.

(b) Calculate $(f^{-1})'(9)$.

SOLUTION

(a) To show that f is one-to-one, we note that

$$f'(x) = 3x^2 + \tfrac{1}{2} > 0 \qquad \text{for all real } x.$$

Thus f is an increasing function and therefore one-to-one.

(b) To calculate $(f^{-1})'(9)$, we want to find a number a for which $f(a) = 9$. Then $(f^{-1})'(9)$ is simply $1/f'(a)$.

The assumption $f(a) = 9$ gives

$$a^3 + \tfrac{1}{2}a = 9$$

and tells us $a = 2$. (We must admit that this example was contrived so that the algebra would be easy to carry out.) Since $f'(2) = 3(2)^2 + \frac{1}{2} = \frac{25}{2}$, we conclude that

$$(f^{-1})'(9) = \frac{1}{f'(2)} = \frac{1}{\frac{25}{2}} = \frac{2}{25}. \quad \square$$

Finally, a few words about differentiating inverses in the Leibniz notation. Suppose that y is a one-to-one function of x:

$$y = y(x).$$

Then x is a one-to-one function of y:

$$x = x(y).$$

Moreover, $\qquad y(x(y)) = y \qquad$ for all y in the domain of x.

Assuming that y is a differentiable function of x and x is a differentiable function of y, we have

$$y'(x(y))\,x'(y) = 1,$$

which, when $y'(x(y)) \neq 0$, gives

$$x'(y) = \frac{1}{y'(x(y))}.$$

In the Leibniz notation, we have

(7.1.9) $$\boxed{\frac{dx}{dy} = \frac{1}{dy/dx}.}$$

The rate of change of x with respect to y is the reciprocal of the rate of change of y with respect to x.

Where are the rates of change to be evaluated? Given that $y(a) = b$, the left side is to be evaluated at b and the right side at a.

EXERCISES 7.1

Determine whether or not the given function is one-to-one and, if so, find the inverse. If f has an inverse, give the domain of f^{-1}.

1. $f(x) = 5x + 3$. **2.** $f(x) = 3x + 5$.

3. $f(x) = 1 - x^2$. **4.** $f(x) = x^5$.

5. $f(x) = x^5 + 1$. **6.** $f(x) = x^2 - 3x + 2$.

7. $f(x) = 1 + 3x^3$. **8.** $f(x) = x^3 - 1$.

9. $f(x) = (1 - x)^3$. **10.** $f(x) = (1 - x)^4$.

11. $f(x) = (x + 1)^3 + 2$. **12.** $f(x) = (4x - 1)^3$.

13. $f(x) = x^{3/5}$. **14.** $f(x) = 1 - (x - 2)^{1/3}$.

15. $f(x) = (2 - 3x)^3$. **16.** $f(x) = (2 - 3x^2)^3$.

17. $f(x) = \sin x, \quad -\frac{\pi}{2} \leq x \leq \frac{\pi}{2}$.

18. $f(x) = \cos x, \quad -\frac{\pi}{2} \leq x \leq \frac{\pi}{2}$.

19. $f(x) = \dfrac{1}{x}$. **20.** $f(x) = \dfrac{1}{1 - x}$.

21. $f(x) = x + \dfrac{1}{x}$. **22.** $f(x) = \dfrac{x}{|x|}$.

23. $f(x) = \dfrac{1}{x^3 + 1}$. **24.** $f(x) = \dfrac{1}{1 - x} - 1$.

25. $f(x) = \dfrac{x + 2}{x + 1}$. **26.** $f(x) = \dfrac{1}{(x + 1)^{2/3}}$.

27. What relation is there between f and $(f^{-1})^{-1}$?

In Exercises 28–31, sketch the graph of f^{-1} given the graph of f.

28.

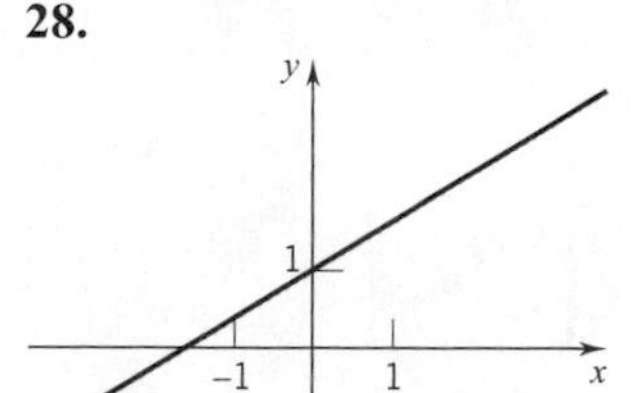

29.

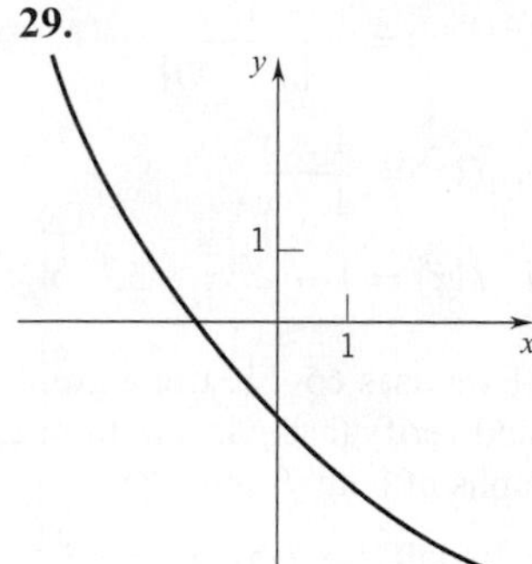

30.

31.

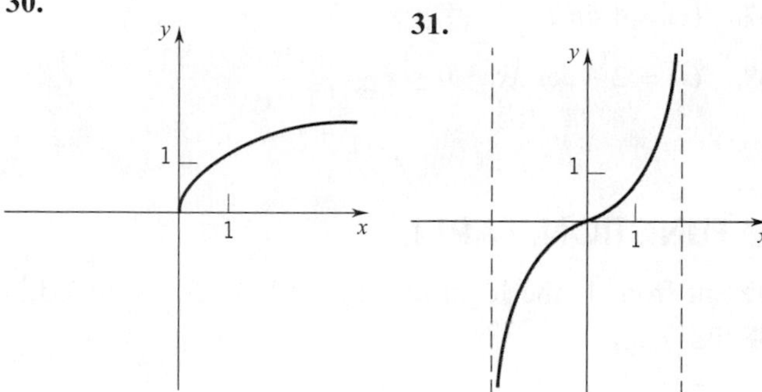

32. (a) Show that the composition of two one-to-one functions, f and g, is one-to-one.
(b) Express $(f \circ g)^{-1}$ in terms of f^{-1} and g^{-1}.

33. a. Let $f(x) = \frac{1}{3}x^3 + x^2 + kx$, k a constant. For what values of k is f one-to-one?
b. Let $g(x) = x^3 + kx^2 + x$, k a constant. For what values of k is g one-to-one?

34. a. Suppose that f has an inverse and $f(2) = 5$, $f'(2) = -\frac{3}{4}$. What is $(f^{-1})'(5)$?
b. Suppose that f has an inverse and $f(2) = -3$, $f'(2) = \frac{2}{3}$. If $g = 1/f^{-1}$, what is $g'(-3)$?

In Exercises 35–44, the given function f is differentiable. Verify that f has an inverse and find $(f^{-1})'(c)$.

35. $f(x) = x^3 + 1$; $\quad c = 9$.

36. $f(x) = 1 - 2x - x^3$; $\quad c = 4$.

37. $f(x) = x + 2\sqrt{x}$, $\quad x > 0$; $\quad c = 8$.

38. $f(x) = \sin x$, $-\frac{\pi}{2} < x < \frac{\pi}{2}$; $\quad c = -\frac{1}{2}$.

39. $f(x) = 2x + \cos x$; $\quad c = \pi$.

40. $f(x) = \dfrac{x+3}{x-1}$, $\quad x > 1$; $\quad c = 3$.

41. $f(x) = \tan x$, $\quad -\frac{\pi}{2} < x < \frac{\pi}{2}$; $\quad c = \sqrt{3}$.

42. $f(x) = x^5 + 2x^3 + 2x$; $\quad c = -5$.

43. $f(x) = 3x - \dfrac{1}{x^3}$; $\quad x > 0$, $\quad c = 2$.

44. $f(x) = x - \pi + \cos x$, $\quad 0 < x < 2\pi$, $\quad c = -1$.

In Exercise 45–47, find a formula for $(f^{-1})'(x)$ given that f is one-to-one and its derivative satisfies the indicated equation.

45. $f'(x) = f(x)$. $\qquad$ **46.** $f'(x) = 1 + [f(x)]^2$.

47. $f'(x) = \sqrt{1 - [f(x)]^2}$.

48. Let

$$f(x) = \begin{cases} x^3 - 1, & x < 0. \\ x^2, & x \geq 0. \end{cases}$$

(a) Sketch the graph of f and verify that f is one-to-one.
(b) Find f^{-1}.

In Exercises 49 and 50, let $f(x) = \dfrac{ax+b}{cx+d}$, $\quad x \neq -\dfrac{d}{c}$.

49. (a) Show that f is one-to-one iff $ad - bc \neq 0$.
(b) Suppose that $ad - bc \neq 0$. Find f^{-1}.

50. Determine the constants a, b, c, d so that $f = f^{-1}$.

51. Let

$$f(x) = \int_2^x \sqrt{1+t^2}\, dt.$$

(a) Prove that f has an inverse.
(b) Find $(f^{-1})'(0)$.

52. Let

$$f(x) = \int_1^{2x} \sqrt{16+t^4}\, dt.$$

(a) Prove that f has an inverse.
(b) Find $(f^{-1})'(0)$.

53. Let f be defined by $f(x) = \int_3^x g(t)\, dt$, where g is a function continuous on $[a, b]$.
(a) What conditions on g will imply that f has an inverse?
(b) Given that f has an inverse, what conditions on g will imply that f is differentiable?
(c) Given that f^{-1} is differentiable, find $(f^{-1})'$.

54. Suppose that the function f has an inverse and fix a number $c \neq 0$.
(a) Let $g(x) = f(x+c)$ for all x such that $x + c \in \text{dom}(f)$. Prove that g has an inverse and find it.
(b) Let $h(x) = f(cx)$ for all x such that $cx \in \text{dom}(f)$. Prove that h has an inverse and find it.

55. Let f be a one-to-one, twice differentiable function and let $g = f^{-1}$.
(a) Show that

$$g''(x) = -\frac{f''[g(x)]}{(f'[g(x)]^3)}.$$

(b) Suppose that the graph of f is concave up (down) on an interval I. What can you say about the graph of f^{-1}?

56. Let $P(x) = a_n x^n + a_{n-1}x^{n-1} + \cdots + a_1 x + a_0$ be a polynomial.
(a) Can P have an inverse if its degree n is even? Justify your answer.

(b) Can P have an inverse if n is odd? If so, give an example. Also, give an example of a polynomial of odd degree that does not have an inverse.

57. Suppose that f is differentiable and has an inverse. If we let $y = f^{-1}(x)$, then $f(y) = x$ and we can find dy/dx by differentiating implicitly. The function $f(x) = \sin x$, $-\frac{\pi}{2} < x < \frac{\pi}{2}$, is one-to-one and differentiable. Let $y = f^{-1}(x)$ and find dy/dx. Express your result as a function of x.

58. Repeat Exercise 57 for the function $f(x) = \tan x$, $-\frac{\pi}{2} < x < \frac{\pi}{2}$.

In Exercises 59–62, use a CAS to find f^{-1}. Verify your result by showing that $f[f^{-1}(x)] = x$ and $f^{-1}[f(x)] = x$.

59. $f(x) = 2 - (x+1)^3$.

60. $f(x) = \dfrac{3x}{2x+5}, \quad x \neq -\frac{5}{2}$.

61. $f(x) = 4 + 3\sqrt{x-1}, \quad x \geq 1$.

62. $f(x) = \sqrt[3]{8-x} + 2$.

Use a CAS in Exercises 63–64 to find f^{-1} and then show that $(f^{-1})'(x) = \dfrac{1}{f'[f^{-1}(x)]}$.

63. $f(x) = \dfrac{1-x}{1+x}$.

64. $f(x) = 1 - \sqrt{3x+5}, \quad x \geq -\frac{5}{3}$.

In Exercises 65–68, use a graphing utility to draw the graph of f and verify that f is one-to-one. Draw a figure that displays the graphs of both f and f^{-1}.

65. $f(x) = x^3 + 3x + 2$.

66. $f(x) = x^{3/5} - 1$.

67. $f(x) = 4\sin 2x, \quad -\frac{\pi}{4} \leq x \leq \frac{\pi}{4}$.

68. $f(x) = 2 - \cos 3x, \quad 0 \leq x \leq \frac{\pi}{3}$.

■ 7.2 THE LOGARITHM FUNCTION, PART I

If B is a positive number different from 1, the logarithm to the base B is defined in elementary mathematics by setting

$$C = \log_B A \quad \text{iff} \quad B^C = A.$$

Historically, the base 10 was chosen because our number system is based on the powers of 10. The defining relation then becomes:

$$C = \log_{10} A \quad \text{iff} \quad 10^C = A.$$

The basic properties of $\log_{10}$ can then be summarized as follows: with $A,\ B > 0$,

$$\log_{10}(AB) = \log_{10}A + \log_{10}B, \qquad \log_{10}1 = 0,$$
$$\log_{10}(1/B) = -\log_{10}B, \qquad \log_{10}(A/B) = \log_{10}A - \log_{10}B,$$
$$\log_{10}A^B = B\log_{10}A, \qquad \log_{10}10 = 1.$$

This elementary notion of logarithm is inadequate for calculus. It is unclear: What is meant by 10^C if C is irrational? It does not lend itself well to the methods of calculus: How would you differentiate $B = \log_{10}A$ knowing only that $10^B = A$?

Here we take an entirely different approach to logarithms. Instead of trying to tamper with the elementary definition, we discard it altogether. From our point of view the fundamental property of logarithms is that they transform multiplication into addition:

the log of a product = the sum of the logs.

Taking this as the central idea, we are led to a general notion of logarithm that encompasses the elementary notion, lends itself well to the methods of calculus, and leads us naturally to a choice of base that simplifies many calculations.

DEFINITION 7.2.1

A *logarithm* function is a nonconstant differentiable function f defined on the set of positive numbers such that for all $a > 0$ and $b > 0$

$$f(ab) = f(a) + f(b).$$

Let's assume for the time being that such logarithm functions exist, and let's see what we can find out about them. In the first place, if f is such a function, then

$$f(1) = f(1 \cdot 1) = f(1) + f(1) = 2f(1) \quad \text{and so} \quad f(1) = 0.$$

Taking $b > 0$, we have

$$0 = f(1) = f(b \cdot 1/b) = f(b) + f(1/b),$$

and therefore

$$f(1/b) = -f(b).$$

Taking $a > 0$ and $b > 0$, we have

$$f(a/b) = f(a \cdot 1/b) = f(a) + f(1/b),$$

which, in view of the previous result, means that

$$f(a/b) = f(a) - f(b).$$

(Thus, f shares many of the properties of $\log_{10}$.)

We are now ready to look for the derivative. (Remember, we are *assuming* that f is differentiable). We begin by forming the difference quotient

$$\frac{f(x+h) - f(x)}{h},$$

where $x > 0$ is fixed. From what we have discovered about f,

$$f(x+h) - f(x) = f\left(\frac{x+h}{x}\right) = f(1 + h/x),$$

and therefore

$$\frac{f(x+h) - f(x)}{h} = \frac{f(1 + h/x)}{h}.$$

Multiplying the denominator by x/x and using the fact that $f(1) = 0$, we can write the difference quotient as

$$\frac{f(x+h) - f(x)}{h} = \frac{1}{x}\left[\frac{f(1 + h/x) - f(1)}{h/x}\right].$$

Now let $k = h/x$. Then we have

$$\frac{f(x+h) - f(x)}{h} = \frac{1}{x}\left[\frac{f(1 + k) - f(1)}{k}\right].$$

Since $k \to 0$ iff $h \to 0$ (remember, x is fixed), it follows that

$$f'(x) = \lim_{h\to 0} \frac{f(x+h) - f(x)}{h} = \lim_{k\to 0} \frac{1}{x}\left[\frac{f(1+k) - f(1)}{k}\right]$$
$$= \frac{1}{x} \lim_{k\to 0}\left[\frac{f(1+k) - f(1)}{k}\right] = \frac{1}{x} f'(1).$$

In short,

(7.2.2) $$\boxed{f'(x) = \frac{1}{x} f'(1).}$$

Thus we have shown that if f is a logarithm function and x is any positive number, then

$$f(1) = 0 \qquad \text{and} \qquad f'(x) = \frac{1}{x} f'(1).$$

We can't have $f'(1) = 0$, for that would make f constant. (Explain.) The most natural choice, the one that will keep calculations as simple as possible, is to set $f'(1) = 1$.† The derivative is then $1/x$.

This function, which takes on the value 0 at 1 and has derivative $1/x$ for $x > 0$, must, by Theorem 5.3.5, take the form

$$\int_1^x \frac{dt}{t}. \qquad \text{(verify this)}$$

DEFINITION 7.2.3

The function

$$L(x) = \int_1^x \frac{dt}{t}, \quad x > 0,$$

is called the *(natural) logarithm function.*

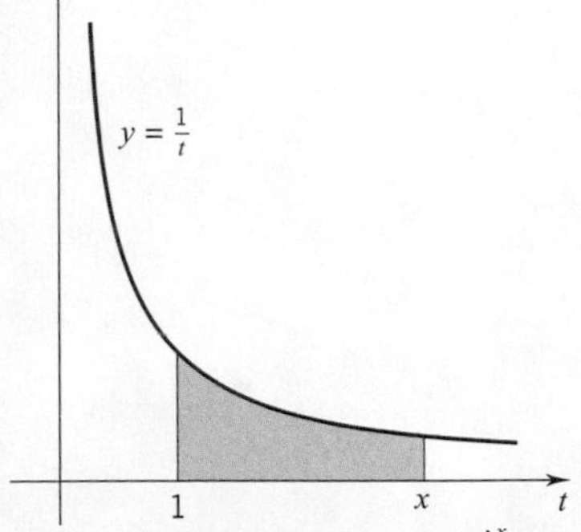

area of shaded region $= L(x) = \int_1^x \frac{dt}{t}$

Figure 7.2.1

Here are some properties of L:

(1) L is defined on $(0, \infty)$ with derivative

$$L'(x) = \frac{1}{x} \quad \text{for all } x > 0.$$

L' is positive on $(0, \infty)$, and therefore L is an increasing function.

(2) L is continuous on $(0, \infty)$, since it is differentiable there.

(3) For $x > 1$, $L(x)$ gives the area of the region shaded in Figure 7.2.1.

(4) $L(x)$ is negative if $0 < x < 1$, $L(1) = 0$, $L(x)$ is positive if $x > 1$.

† This, as you will see later, is tantamount to a choice of base.

The following result is fundamental; it establishes that L *is* a logarithm function by showing that it satisfies the equation in Definition 7.2.1.

THEOREM 7.2.4

If a and b are positive, then

$$L(ab) = L(a) + L(b).$$

PROOF Fix any positive number a. Then, since

$$\frac{d}{dx}[L(x)] = \frac{1}{x} \quad \text{and} \quad \frac{d}{dx}[L(ax)] \underset{\text{chain rule}}{=} \frac{1}{ax} \cdot a = \frac{1}{x},$$

$L(x)$ and $L(ax)$ have the same derivative, and so we know that

$$L(ax) = L(x) + C$$

for some constant C (Theorem 4.2.5). We can evaluate the constant by taking $x = 1$:

$$L(a) = L(1 \cdot a) = L(1) + C \underset{L(1)=0}{=} C.$$

Therefore, $L(ax) = L(a) + L(x)$ for all $x > 0$. Now let $x = b$ to get the statement in the theorem. □

We come now to another important result.

THEOREM 7.2.5

If a is positive and p/q is rational, then

$$L(a^{p/q}) = \frac{p}{q}L(a).$$

PROOF You have seen that $d[L(x)]/dx = 1/x$. By the chain rule

$$\frac{d}{dx}[L(x^{p/q})] = \frac{1}{x^{p/q}}\frac{d}{dx}(x^{p/q}) \underset{(3.7.1)}{=} \frac{1}{x^{p/q}}\left(\frac{p}{q}\right)x^{p/q-1} = \frac{p}{q}\left(\frac{1}{x}\right) = \frac{d}{dx}\left[\frac{p}{q}L(x)\right].$$

Since $L(x^{p/q})$ and $\frac{p}{q}L(x)$ have the same derivative, they differ by a constant:

$$L(x^{p/q}) = \frac{p}{q}L(x) + C.$$

Since both functions are zero at $x = 1$, $C = 0$. Therefore $L(x^{p/q}) = \frac{p}{q}L(x)$ for all $x > 0$, and we get the statement in the theorem by letting $x = a$. □

The domain of L is $(0, \infty)$. What is the range of L?

THEOREM 7.2.6

The range of L is $(-\infty, \infty)$.

PROOF Since L is continuous on $(0, \infty)$, we know from the intermediate-value theorem that it "skips" no values. Thus, its range is an interval. To show that the interval is $(-\infty, \infty)$, we need only show that it is unbounded above and unbounded below. We can do this by taking M as an arbitrary positive number and showing that L takes on values greater than M and values less than $-M$.

Let M be an arbitrary positive number. Since

$$L(2) = \int_1^2 \frac{dt}{t}$$

is positive (explain), we know that some positive multiple of $L(2)$ must be greater than M; namely, we know that there exists a positive integer n such that

$$nL(2) > M.$$

Multiplying this equation by -1 we have

$$-nL(2) < -M.$$

Since $\quad nL(2) = L(2^n) \quad$ and $\quad -nL(2) = L(2^{-n}),$ (Theorem 7.2.5)

we have $\quad L(2^n) > M \quad$ and $\quad L(2^{-n}) < -M.$

This proves the unboundedness. ❑

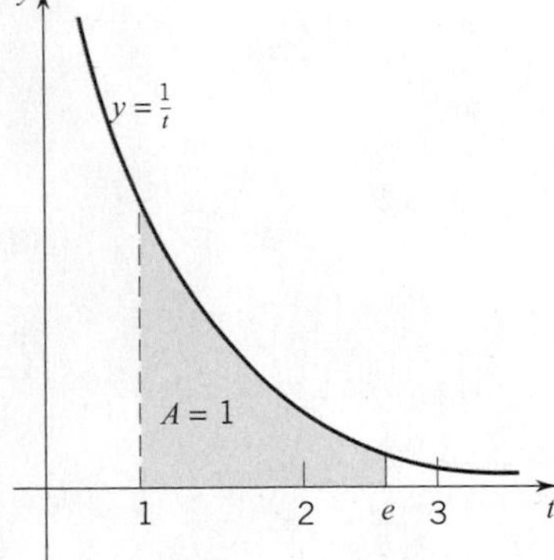

Figure 7.2.2

The Number e

Since the range of L is $(-\infty, \infty)$ and L is an increasing function, we know that L takes on every value and does so only once. In particular, there is one and only one number at which the function L takes on the value 1. *This unique number is denoted by the letter e.* †

The number e can also be defined geometrically: e is the unique number with the property that the area under the graph of $f(t) = 1/t$ from $t = 1$ to $t = e$ is 1. See Figure 7.2.2.

Since

(7.2.7) $$L(e) = \int_1^e \frac{dt}{t} = 1,$$

it follows from Theorem 7.2.5 that

(7.2.8) $$L(e^{p/q}) = \frac{p}{q} \quad \text{for all rational numbers } \frac{p}{q}.$$

† After the Swiss mathematician Leonhard Euler (1707–1783), considered by many the greatest mathematician of the eighteenth century.

Because of this relation, we call L *the logarithm to the base e* and sometimes write

$$L(x) = \log_e x.$$

The number e arises naturally in many settings. Accordingly, we call $L(x)$ the *natural logarithm* and write

(7.2.9) $$\boxed{L(x) = \ln x.}\ \dagger$$

Here are the basic properties we have established for $\ln x$:

(7.2.10)
$$\begin{array}{|ll|}\hline \ln 1 = 0, \quad \ln e = 1. & \\ \ln ab = \ln a + \ln b & (a > 0, b > 0). \\ \ln 1/b = -\ln b & (b > 0. \\ \ln a/b = \ln a - \ln b & (a > 0,\ b > 0). \\ \ln a^r = r \ln a & (a > 0,\ r \text{ rational}). \\ \hline \end{array}$$

Notice how closely these rules parallel the familiar rules for common logarithms (base 10). Later we will show that the last of these rules also holds for irrational exponents.

The Graph of the Logarithm Function

You know that the logarithm function

$$\ln x = \int_1^x \frac{dt}{t}$$

has domain $(0, \infty)$, range $(-\infty, \infty)$, and derivative

$$\frac{d}{dx}(\ln x) = \frac{1}{x} > 0.$$

For small x the derivative is large (near 0 the curve is steep); for large x the derivative is small (far out the curve flattens out). At $x = 1$ the logarithm is 0 and its derivative $1/x$ is 1. [The graph crosses the x-axis at the point $(1, 0)$, and the tangent line at that point is parallel to the line $y = x$.] The second derivative

$$\frac{d^2}{dx^2}(\ln x) = -\frac{1}{x^2}$$

is negative on $(0, \infty)$. (The graph is concave down throughout.) We have sketched the graph in Figure 7.2.3. The y-axis is a vertical asymptote:

$$\text{as } x \to 0^+, \qquad \ln x \to -\infty.$$

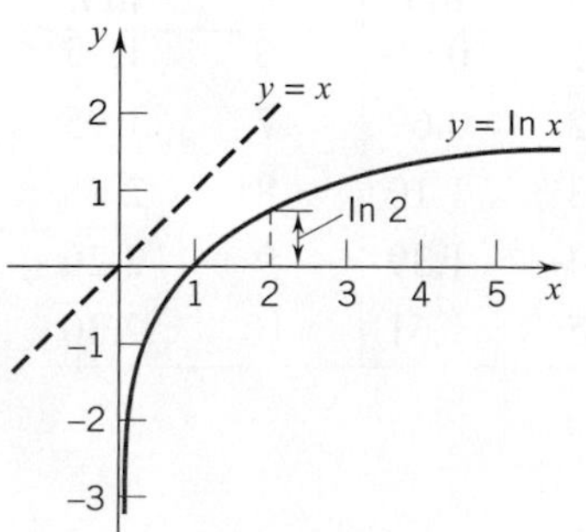

Figure 7.2.3

† Logarithms to bases other than e will be taken up later [they arise by other choices of $f'(1)$], but by far the most important logarithm in calculus is the logarithm to the base e. So much so, that when we speak of the logarithm of a number x and don't specify the base, you can be sure that we are talking about the *natural logarithm* $\ln x$.

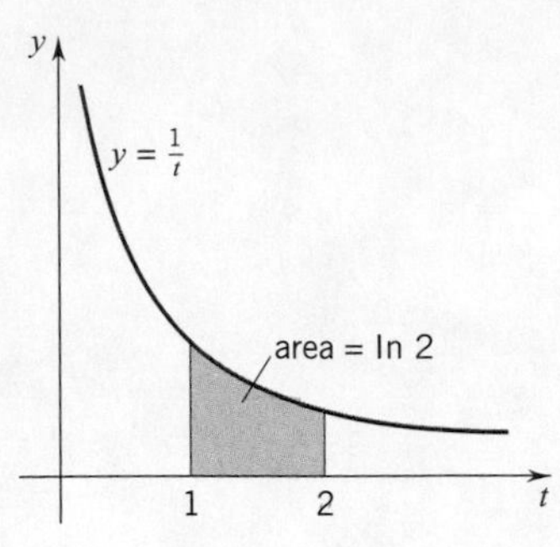

Figure 7.2.4

Example 1 Using upper and lower sums, estimate

$$\ln 2 = \int_1^2 \frac{dt}{t} \qquad \text{(Figure 7.2.4)}$$

from the partition

$$P = \{1 = \tfrac{10}{10}, \tfrac{11}{10}, \tfrac{12}{10}, \tfrac{13}{10}, \tfrac{14}{10}, \tfrac{15}{10}, \tfrac{16}{10}, \tfrac{17}{10}, \tfrac{18}{10}, \tfrac{19}{10}, \tfrac{20}{10} = 2\}.$$

SOLUTION Using a calculator, we find that

$$\begin{aligned} L_f(P) &= \tfrac{1}{10}\left(\tfrac{10}{11} + \tfrac{10}{12} + \tfrac{10}{13} + \tfrac{10}{14} + \tfrac{10}{15} + \tfrac{10}{16} + \tfrac{10}{17} + \tfrac{10}{18} + \tfrac{10}{19} + \tfrac{10}{20}\right) \\ &= \tfrac{1}{11} + \tfrac{1}{12} + \tfrac{1}{13} + \tfrac{1}{14} + \tfrac{1}{15} + \tfrac{1}{16} + \tfrac{1}{17} + \tfrac{1}{18} + \tfrac{1}{19} + \tfrac{1}{20} > 0.668 \end{aligned}$$

and

$$\begin{aligned} U_f(P) &= \tfrac{1}{10}\left(\tfrac{10}{10} + \tfrac{10}{11} + \tfrac{10}{12} + \tfrac{10}{13} + \tfrac{10}{14} + \tfrac{10}{15} + \tfrac{10}{16} + \tfrac{10}{17} + \tfrac{10}{18} + \tfrac{10}{19}\right) \\ &= \tfrac{1}{10} + \tfrac{1}{11} + \tfrac{1}{12} + \tfrac{1}{13} + \tfrac{1}{14} + \tfrac{1}{15} + \tfrac{1}{16} + \tfrac{1}{17} + \tfrac{1}{18} + \tfrac{1}{19} < 0.719. \end{aligned}$$

Thus we have

$$0.668 < L_f(P) \le \ln 2 \le U_f(P) < 0.719.$$

The average of these two estimates is

$$\tfrac{1}{2}(0.668 + 0.179) = 0.6935.$$

Rounded off to four decimal places, the value of ln 2 given on a calculator is 0.6931, so we are not far off. ❑

Table 7.2.1 gives the natural logarithms of the integers 1 through 10 rounded off to the nearest hundredth.

■ **Table 7.2.1**

n	$\ln n$	n	$\ln n$
1	0.00	6	1.79
2	0.69	7	1.95
3	1.10	8	2.08
4	1.39	9	2.20
5	1.61	10	2.30

Example 2 Use the properties of logarithms and Table 7.2.1 to estimate the following logarithms.

(a) ln 0.2. **(b)** ln 0.25. **(c)** ln 2.4. **(d)** ln 90.

SOLUTION

(a) $\ln 0.2 = \ln \frac{1}{5} = -\ln 5 \cong -1.61$. **(b)** $\ln 0.25 = \ln \frac{1}{4} = -\ln 4 \cong -1.39$.

(c) $\ln 2.4 = \ln \frac{12}{5} = \ln \frac{(3)(4)}{5} = \ln 3 + \ln 4 - \ln 5 \cong 0.88$.

(d) $\ln 90 = \ln[(9)(10)] = \ln 9 + \ln 10 \cong 4.50$. ❑

Example 3 Estimate e on the basis of Table 7.2.1.

SOLUTION So far all we know is that $\ln e = 1$. From the table you can see that

$$3 \ln 3 - \ln 10 \cong 1.$$

The expression on the left can be written

$$\ln 3^3 - \ln 10 = \ln 27 - \ln 10 = \ln \tfrac{27}{10} = \ln 2.7.$$

Thus $\ln 2.7 \cong 1$, and so $e \cong 2.7$. □

Remark It can be shown that e is an irrational number; in fact a transcendental number. The decimal expansion of e to twelve decimal places reads

$$e \cong 2.718281828459.\dagger \quad \square$$

---

† Exercise 68 in Section 11.5 guides you through a proof of the irrationality of e. A proof that e is transcendental is beyond the reach of this text.

EXERCISES 7.2

Estimate the given natural logarithm on the basis of Table 7.2.1; check your results on a calculator.

1. $\ln 20$. **2.** $\ln 16$.

3. $\ln 1.6$. **4.** $\ln 3^4$.

5. $\ln 0.1$. **6.** $\ln 2.5$.

7. $\ln 7.2$. **8.** $\ln \sqrt{630}$.

9. $\ln \sqrt{2}$. **10.** $\ln 0.4$.

11. Interpret the equation $\ln n = \ln mn - \ln m$ in terms of area under the curve $y = 1/x$. Draw a figure.

12. Given that $0 < x < 1$, express as a logarithm the area under the curve $y = 1/t$ from $t = x$ to $t = 1$.

13. Estimate

$$\ln 1.5 = \int_1^{1.5} \frac{dt}{t}.$$

using the approximation $\frac{1}{2}[L_f(P) + U_f(P)]$ with $P = \{1 = \frac{8}{8}, \frac{9}{8}, \frac{10}{8}, \frac{11}{8}, \frac{12}{8} = 1.5\}$.

14. Estimate

$$\ln 2.5 = \int_1^{2.5} \frac{dt}{t}$$

using the approximation $\frac{1}{2}[L_f(P) + U_f(P)]$ with $P = \{1 = \frac{4}{4}, \frac{5}{4}, \frac{6}{4}, \frac{7}{4}, \frac{8}{4}, \frac{9}{4}, \frac{10}{4} = \frac{5}{2}\}$.

15. Taking $\ln 5 \cong 1.61$, use differentials to estimate;

(a) $\ln 5.2$, (b) $\ln 4.8$, (c) $\ln 5.5$,

16. Taking $\ln 10 \cong 2.30$, use differentials to estimate:

(a) $\ln 10.3$, (b) $\ln 9.6$, (c) $\ln 11$,

In Exercises 17–22, solve the given equation for x.

17. $\ln x = 2$. **18.** $\ln x = -1$.

19. $(2 - \ln x) \ln x = 0$. **20.** $\frac{1}{2} \ln x = \ln (2x - 1)$.

21. $\ln [(2x + 1)(x + 2)] = 2 \ln (x + 2)$.

22. $2 \ln (x + 2) - \frac{1}{2} \ln x^4 = 1$.

23. Show that

$$\lim_{x\to 1} \frac{\ln x}{x - 1} = 1.$$

HINT: $\dfrac{\ln x}{x-1} = \dfrac{\ln x - \ln 1}{x - 1}$; interpret the limit in terms of the derivative of $\ln x$.

24. (a) Use the mean-value theorem to show that

$$\frac{x - 1}{x} \le \ln x \le x - 1 \quad \text{for all } x > 0.$$

HINT: Consider the cases $x \ge 1$ and $0 < x < 1$ separately.

(b) Use the result in part (a) to show that

$$\lim_{x\to 1} \frac{\ln x}{x - 1} = 1.$$

25. (a) Show that for $n \ge 2$,

$$\frac{1}{2} + \frac{1}{3} + \cdots + \frac{1}{n} < \ln n < 1 + \frac{1}{2} + \frac{1}{3} + \cdots + \frac{1}{n-1}.$$

HINT: See the figure.

(b) Show that the area of the shaded part is given by

$$1 + \frac{1}{2} + \frac{1}{3} + \cdots + \frac{1}{n-1} - \ln n.$$

As $n \to \infty$, this area approaches the number γ known as *Euler's constant.*

(c) Use geometric reasoning to show that $\frac{1}{2} < \gamma < 1$. (To three decimal places, $\gamma \cong 0.577$).

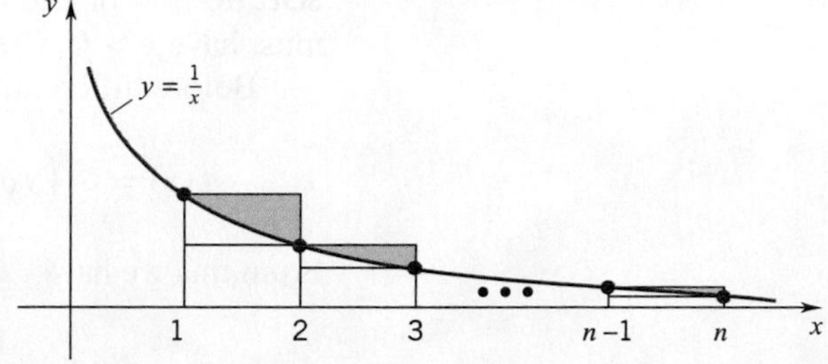

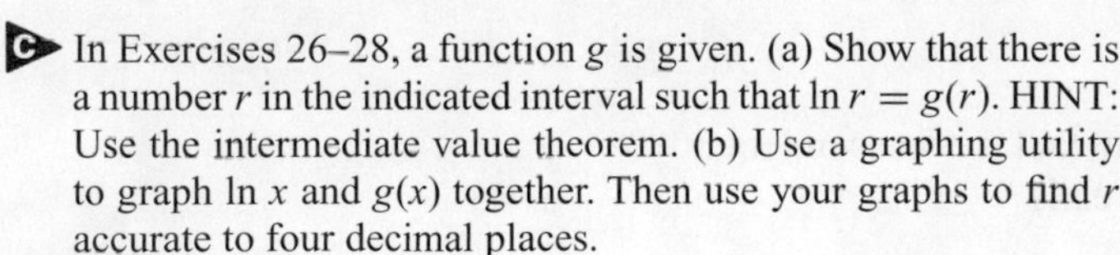

In Exercises 26–28, a function g is given. (a) Show that there is a number r in the indicated interval such that $\ln r = g(r)$. HINT: Use the intermediate value theorem. (b) Use a graphing utility to graph $\ln x$ and $g(x)$ together. Then use your graphs to find r accurate to four decimal places.

26. $g(x) = 2x - 3$; $[1, 2]$.

27. $g(x) = \sin x$; $[2, 3]$.

28. $g(x) = \dfrac{1}{x^2}$; $[1, 2]$

In Exercises 29 and 30, estimate $\lim_{x\to a} f(x)$ numerically by evaluating f at the indicated values. Then use a graphing utility to zoom in on the graph near $x = a$ to justify your estimate.

29. $\lim_{x\to 1} \dfrac{\ln x}{x-1}$; $x = 1 \pm 0.5, 1 \pm 0.1, 1 \pm 0.01, 1 \pm 0.001, 1 \pm 0.0001$.

30. $\lim_{x\to 0^+} \sqrt{x} \ln x$; $x = 0.5, 0.1, 0.01, 0.001, 0.0001$.

31. (a) Use a graphing utility to draw the graph of $f(x) = \cos(\ln x)$.
(b) Estimate some of the zeros of f.
(c) Use a CAS to find a general formula for the zeros of f.

32. Repeat Exercise 31 with $f(x) = \ln(\cos x)$.

■ 7.3 THE LOGARITHM FUNCTION, PART II

Differentiation and Graphing

We know that for $x > 0$,

$$\frac{d}{dx}(\ln x) = \frac{1}{x}.$$

Now suppose that u is a positive, differentiable function of x. Then

(7.3.1)
$$\frac{d}{dx}(\ln u) = \frac{1}{u}\frac{du}{dx}.$$

PROOF By the chain rule,

$$\frac{d}{dx}(\ln u) = \frac{d}{du}(\ln u)\frac{du}{dx} = \frac{1}{u}\frac{du}{dx}. \quad \square$$

For example,

$$\frac{d}{dx}[\ln(1+x^2)] = \frac{1}{1+x^2}\cdot 2x = \frac{2x}{1+x^2} \qquad \text{for all } x,$$

and

$$\frac{d}{dx}[\ln(1+3x)] = \frac{1}{1+3x}\cdot 3 = \frac{3}{1+3x} \qquad \text{for all } x > -\tfrac{1}{3}.$$

Example 1 Find the domain of f and find $f'(x)$ if

$$f(x) = \ln(x\sqrt{4+x^2}).$$

SOLUTION For x to be in the domain of f, we must have $x\sqrt{4+x^2} > 0$, and thus we must have $x > 0$. The domain of f is the set of positive numbers.

Before differentiating f, we make use of the special properties of the logarithm:

$$f(x) = \ln\left(x\sqrt{4+x^2}\right) = \ln x + \ln[(4+x^2)^{1/2}] = \ln x + \tfrac{1}{2}\ln(4+x^2).$$

From this we have

$$f'(x) = \frac{1}{x} + \frac{1}{2}\cdot\frac{1}{4+x^2}\cdot 2x = \frac{1}{x} + \frac{x}{4+x^2} = \frac{4+2x^2}{x(4+x^2)}. \quad \square$$

Example 2 Sketch the graph of

$$f(x) = \ln |x|.$$

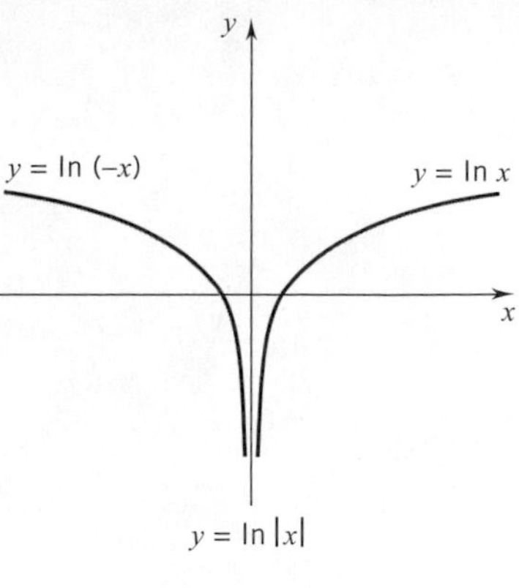

Figure 7.3.1

SOLUTION The function, defined at all $x \neq 0$, is an even function: $f(-x) = f(x)$ for all $x \neq 0$. The graph has two branches:

$$y = \ln(-x), \qquad x < 0 \qquad \text{and} \qquad y = \ln x, \quad x > 0.$$

Each branch is the mirror image of the other. (Figure 7.3.1) □

Example 3 (*Important*) Show that

(7.3.2)
$$\boxed{\frac{d}{dx}(\ln |x|) = \frac{1}{x} \quad \text{for all} \quad x \neq 0.}$$

SOLUTION For $x > 0$,

$$\frac{d}{dx}(\ln |x|) = \frac{d}{dx}(\ln x) = \frac{1}{x}.$$

For $x < 0$, we have $|x| = -x > 0$, so that

$$\frac{d}{dx}(\ln |x|) = \frac{d}{dx}[\ln(-x)] = \frac{1}{-x}\frac{d}{dx}(-x) = \left(\frac{1}{-x}\right)(-1) = \frac{1}{x}. \quad \square$$

It follows that if u is a differentiable function of x, then where $u(x) \neq 0$,

(7.3.3)
$$\boxed{\frac{d}{dx}(\ln |u|) = \frac{1}{u}\frac{du}{dx}.}$$

PROOF

$$\frac{d}{dx}(\ln |u|) = \frac{d}{du}(\ln |u|)\frac{du}{dx} = \frac{1}{u}\frac{du}{dx}. \quad \square$$

Here are two examples:

$$\frac{d}{dx}(\ln |1 - x^3|) = \frac{1}{1 - x^3}\frac{d}{dx}(1 - x^3) = \frac{-3x^2}{1 - x^3} = \frac{3x^2}{x^3 - 1}.$$

$$\frac{d}{dx}\left(\ln \left|\frac{x-1}{x-2}\right|\right) = \frac{d}{dx}(\ln |x - 1|) - \frac{d}{dx}(\ln |x - 2|) = \frac{1}{x - 1} - \frac{1}{x - 2}.$$

Example 4 Let $f(x) = x \ln x$.
(a) Specify the domain of f and find the intercepts, if any. **(b)** On what intervals does f increase? Decrease? **(c)** Find the extreme values of f. **(d)** Determine the concavity of the graph and find the points of inflection. **(e)** Sketch the graph of f.

SOLUTION Since the logarithm function is defined only for positive numbers, the domain of f is $(0, \infty)$ and there is no y-intercept. Since $f(1) = 1 \cdot \ln 1 = 0$, 1 is an x-intercept.

Differentiating f, we have

$$f'(x) = x \cdot \frac{1}{x} + \ln x = 1 + \ln x.$$

To find the critical numbers of f, we set $f'(x) = 0$:

$$1 + \ln x = 0, \qquad \ln x = -1, \qquad x = \frac{1}{e} \qquad \text{(verify this).}$$

Recalling that the logarithm function is increasing on $(0, \infty)$, and that $\ln x \to -\infty$ as $x \to 0^+$ and $\ln x \to \infty$ as $x \to \infty$, we have

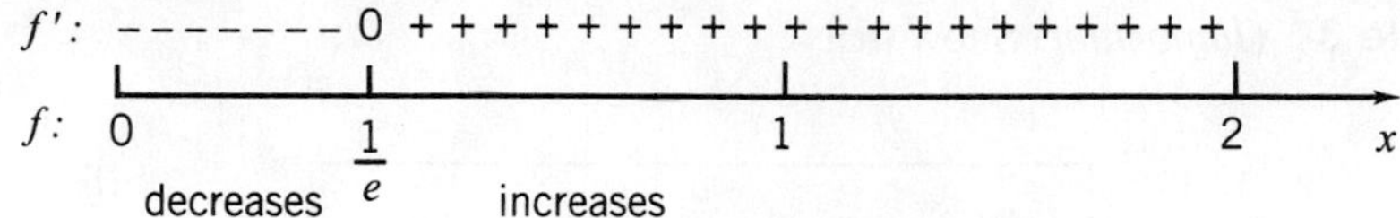

Therefore, f decreases on $(0, 1/e]$ and increases on $[1/e, \infty)$. By the first derivative test,

$$f(1/e) = \frac{1}{e} \ln\left(\frac{1}{e}\right) = -\frac{1}{e} \cong -0.368$$

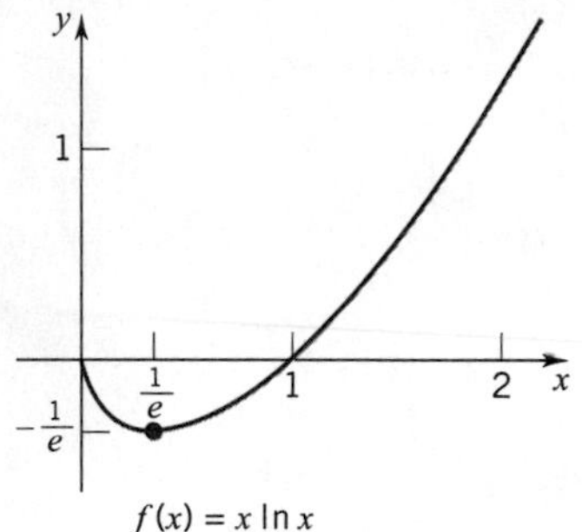

Figure 7.3.2

is a local and absolute minimum of f.

Since $f''(x) = 1/x > 0$ for $x > 0$, the graph of f is concave up on $(0, \infty)$; there are no points of inflection.

You can verify numerically that $\lim_{x \to 0^+} x \ln x = 0$ (see Exercise 30, Section 7.2) and that $x \ln x \to \infty$ as $x \to \infty$. Finally, note that $\lim_{x \to 0^+} f'(x) = \lim_{x \to 0^+} (1 + \ln x) = -\infty$.

A sketch of the graph of f is shown in Figure 7.3.2. □

Example 5 Let $f(x) = \ln\left(\dfrac{x^4}{x-1}\right)$.

(a) Specify the domain of f. **(b)** On what intervals does f increase? Decrease? **(c)** Find the extreme values of f. **(d)** Determine the concavity of the graph and find the points of inflection. **(e)** Sketch the graph, specifying the asymptotes if any.

SOLUTION Since the logarithm function is defined only for positive numbers, the domain of f is the open interval $(1, \infty)$.

Making use of the special properties of the logarithm, we write

$$f(x) = \ln x^4 - \ln (x-1) = 4 \ln x - \ln (x-1).$$

Differentiation gives

$$f'(x) = \frac{4}{x} - \frac{1}{x-1} = \frac{3x-4}{x(x-1)}$$

and

$$f''(x) = -\frac{4}{x^2} + \frac{1}{(x-1)^2} = -\frac{(x-2)(3x-2)}{x^2(x-1)^2}.$$

Since the domain of f is $(1, \infty)$, we consider only the values of x greater than 1. Note that $f'(x) = 0$ at $x = 4/3$ (critical number) and we have:

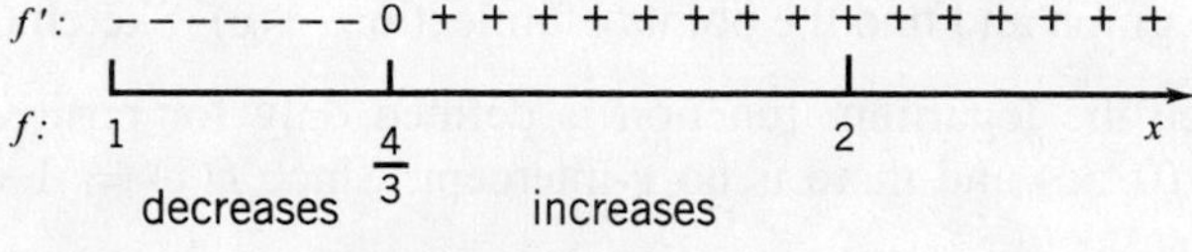

Thus f decreases on $(1, \frac{4}{3}]$, increases on $[\frac{4}{3}, \infty)$, and

$$f(\tfrac{4}{3}) = 4 \ln 4 - 3 \ln 3 \cong 2.25$$

is a local and absolute minimum. There are no other extreme values.

Testing for concavity, we see that $f''(x) = 0$ at $x = 2$ (we ignore $x = 2/3$ since $2/3$ is not part of the domain of f). The sign chart for f'' looks like this:

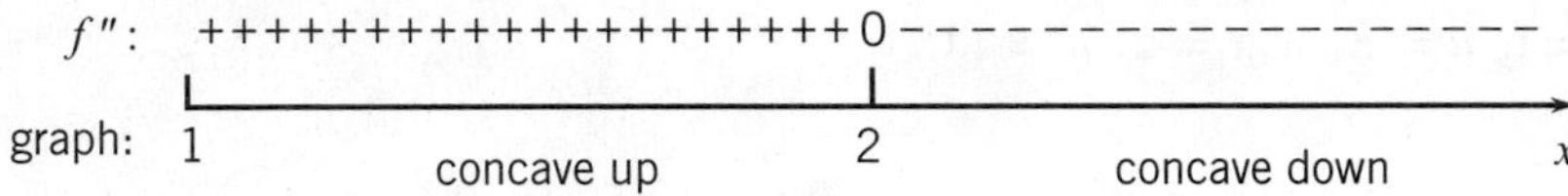

Therefore, the graph is concave up on $(1, 2)$, concave down on $(2, \infty)$, and

$$(2, f(2)) = (2, 4 \ln 2) \cong (2, 2.77)$$

is the only point of inflection.

Before sketching the graph, note that the derivative

$$f'(x) = \frac{4}{x} - \frac{1}{x-1}$$

is very large negative for x close to 1 and very close to 0 for x large. This means that the graph is very steep for x close to 1 and very flat for x large. See Figure 7.3.3.The line $x = 1$ is a vertical asymptote: as $x \to 1^+, f(x) \to \infty$. □

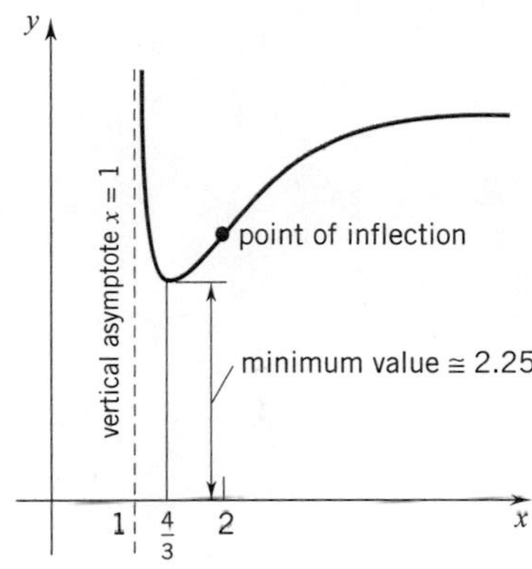

Figure 7.3.3

Integration

The integral counterpart of (7.3.2) takes the form

(7.3.4)
$$\int \frac{dx}{x} = \ln |x| + C, \qquad x \neq 0.$$

Integrals of the form

$$\int \frac{g'(x)}{g(x)} dx \qquad \text{can be reduced to} \qquad \int \frac{du}{u}$$

by setting
$$u = g(x), \qquad du = g'(x)\, dx.$$

This substitution gives

(7.3.5)
$$\int \frac{g'(x)}{g(x)} dx = \ln |g(x)| + C, \qquad g(x) \neq 0.$$

Example 6 Calculate $\displaystyle\int \frac{x^2}{1 - 4x^3} dx.$

SOLUTION Note that x^2 is the derivative of $1 - 4x^3$, except for a constant factor. Therefore, set

$$u = 1 - 4x^3, \qquad du = -12x^2 dx.$$

$$\int \frac{x^2}{1-4x^3}\,dx = -\tfrac{1}{12}\int \frac{du}{u} = -\tfrac{1}{12}\ln|u| + C = -\tfrac{1}{12}\ln|1-4x^3| + C. \quad \square$$

Example 7 Evaluate $\displaystyle\int_1^2 \frac{6x^2+2}{x^3+x+1}\,dx.$

SOLUTION Set $u = x^3 + x + 1, \quad du = 3x^2 + 1.$
At $x = 1,\ u = 3$; at $x = 2,\ u = 11$.

$$\begin{aligned}\int_1^2 \frac{6x^2+2}{x^3+x+1}\,dx &= 2\int_3^{11} \frac{du}{u} = 2\Big[\ln|u|\Big]_3^{11}\\ &= 2(\ln 11 - \ln 3) = 2\ln\left(\tfrac{11}{3}\right). \quad \square\end{aligned}$$

Here is an example of a different sort.

Example 8 Calculate $\displaystyle\int \frac{\ln x}{x}\,dx.$

SOLUTION Note that $1/x$ is the derivative of $\ln x$. Thus, we set

$$u = \ln x, \qquad du = \frac{1}{x}dx,$$

and
$$\int \frac{\ln x}{x}\,dx = \int u\,du = \tfrac{1}{2}u^2 + C = \tfrac{1}{2}(\ln x)^2 + C. \quad \square$$

Integration of the Trigonometric Functions

In Section 5.5, you saw that

$$\int \cos x\,dx = \sin x + C, \qquad \int \sin x\,dx = -\cos x + C,$$

$$\int \sec^2 x\,dx = \tan x + C, \qquad \int \csc^2 x\,dx = -\cot x + C,$$

$$\int \sec x\tan x\,dx = \sec x + C, \qquad \int \csc x\cot x\,dx = -\csc x + C.$$

Now that you are familiar with the natural logarithm function, we can add four more basic formulas to the list:

(7.3.6)

$$\int \tan x\,dx = -\ln|\cos x| + C = \ln|\sec x| + C,$$

$$\int \cot x\,dx = \ln|\sin x| + C,$$

$$\int \sec x\,dx = \ln|\sec x + \tan x| + C,$$

$$\int \csc x\,dx = \ln|\csc x - \cot x| + C.$$

The derivation of these formulas is based on (7.3.5).

(i) $$\int \tan x\,dx = \int \frac{\sin x}{\cos x}\,dx \qquad (\text{set } u = \cos x,\ du = -\sin x\,dx)$$
$$= -\int \frac{du}{u} = -\ln|u| + C$$
$$= -\ln|\cos x| + C = \ln\left|\frac{1}{\cos x}\right| + C$$
$$= \ln|\sec x| + C.$$

(ii) $$\int \cot x\,dx = \int \frac{\cos x}{\sin x}\,dx \qquad (\text{set } u = \sin x,\quad du = \cos x\,dx)$$
$$= \int \frac{du}{u} = \ln|u| + C = \ln|\sin x| + C.$$

(iii) $$\int \sec x\,dx \overset{\dagger}{=} \int \sec x\,\frac{\sec x + \tan x}{\sec x + \tan x}\,dx$$
$$= \int \frac{\sec x \tan x + \sec^2 x}{\sec x + \tan x}\,dx$$
$$[\text{set } u = \sec x + \tan x,\quad du = (\sec x \tan x + \sec^2 x)\,dx]$$
$$= \int \frac{du}{u} = \ln|u| + C = \ln|\sec x + \tan x| + C.$$

The derivation of the formula for $\int \csc x\,dx$ is left to you.

Example 9 Calculate $\int \cot \pi x\,dx$.

SOLUTION Set $u = \pi x, \quad du = \pi\,dx.$

$$\int \cot \pi x\,dx = \frac{1}{\pi}\int \cot u\,du = \frac{1}{\pi}\ln|\sin u| + C = \frac{1}{\pi}\ln|\sin \pi x| + C. \quad \square$$

Remark The u-substitution simplifies many calculations, but you will find with experience that you can carry out many integrations without it. □

Example 10 Evaluate $\int_0^{\pi/8} \sec 2x\,dx$.

SOLUTION It is easy to check that $\frac{1}{2}\ln|\sec 2x + \tan 2x|$ is an antiderivative of $\sec 2x$. Thus

$$\int_0^{\pi/8} \sec 2x\,dx = \tfrac{1}{2}\Big[\ln|\sec 2x + \tan 2x|\Big]_0^{\pi/8}$$
$$= \tfrac{1}{2}\,[\ln(\sqrt{2}+1) - \ln 1] = \tfrac{1}{2}\ln(\sqrt{2}+1) \cong 0.44 \quad \square$$

When the integrand is a quotient, it is worth checking to see if the integral can be written in the form

$$\int \frac{du}{u}.$$

† Only experience prompts us to multiply numerator and denominator by $\sec x + \tan x$.

Example 11 Calculate $\displaystyle\int \frac{\sec^2 3x}{1+\tan 3x}\,dx.$

SOLUTION Set $u = 1 + \tan 3x, \qquad du = 3\sec^2 3x\,dx.$

$$\int \frac{\sec^2 3x}{1+\tan 3x}\,dx = \tfrac{1}{3}\int \frac{du}{u} = \tfrac{1}{3}\ln|u| + C = \tfrac{1}{3}\ln|1+\tan 3x| + C. \quad \square$$

Logarithmic Differentiation

We can differentiate a lengthy product

$$g(x) = g_1(x)g_2(x)\cdots g_n(x)$$

by first writing

$$\begin{aligned}\ln|g(x)| &= \ln(|g_1(x)||g_2(x)|\cdots|g_n(x)|)\\ &= \ln|g_1(x)| + \ln|g_2(x)| + \cdots + \ln|g_n(x)|\end{aligned}$$

and then differentiating:

$$\frac{g'(x)}{g(x)} = \frac{g_1'(x)}{g_1(x)} + \frac{g_2'(x)}{g_2(x)} + \cdots + \frac{g_n'(x)}{g_n(x)}.$$

Multiplication by $g(x)$ gives

(7.3.7)
$$g'(x) = g(x)\left(\frac{g_1'(x)}{g_1(x)} + \frac{g_2'(x)}{g_2(x)} + \cdots + \frac{g_n'(x)}{g_n(x)}\right).$$

The process by which $g'(x)$ was obtained is called *logarithmic differentiation*. Logarithmic differentiation is valid at all points x where $g(x) \neq 0$. At points x where $g(x) = 0$, none of it makes sense.

A product of n factors

$$g(x) = g_1(x)g_2(x)\cdots g_n(x)$$

can, of course, also be differentiated by repeated applications of the product rule, Theorem 3.2.6 The great advantage of logarithmic differentiation is that it gives us an explicit formula for the derivative, a formula that's easy to remember and easy to work with.

Example 12 Given that

$$g(x) = x(x-1)(x-2)(x-3)$$

find $g'(x)$ for $x \neq 0,\ 1,\ 2,\ 3$.

SOLUTION We can write down $g'(x)$ directly from Formula (7.3.7),

$$g'(x) = x(x-1)(x-2)(x-3)\left(\frac{1}{x} + \frac{1}{x-1} + \frac{1}{x-2} + \frac{1}{x-3}\right),$$

or we can go through the process by which we derived Formula (7.3.7):

$$\begin{aligned}\ln|g(x)| &= \ln|x| + \ln|x-1| + \ln|x-2| + \ln|x-3|,\\ \frac{g'(x)}{g(x)} &= \frac{1}{x} + \frac{1}{x-1} + \frac{1}{x-2} + \frac{1}{x-3},\\ g'(x) &= x(x-1)(x-2)(x-3)\left(\frac{1}{x} + \frac{1}{x-1} + \frac{1}{x-2} + \frac{1}{x-3}\right). \quad \square\end{aligned}$$

Logarithmic differentiation can also be used with quotients.

Example 13 Given that

$$g(x) = \frac{(x^2+1)^3(2x-5)^2}{(x^2+5)^2},$$

find $g'(x)$ for $x \neq \frac{5}{2}$.

SOLUTION Our first step is to write

$$g(x) = (x^2+1)^3(2x-5)^2(x^2+5)^{-2}.$$

Then according to Formula 7.3.7

$$\begin{aligned} g'(x) &= \frac{(x^2+1)^3(2x-5)^2}{(x^2+5)^2}\left[\frac{3(x^2+1)^2(2x)}{(x^2+1)^3} + \frac{2(2x-5)(2)}{(2x-5)^2} + \frac{(-2)(x^2+5)^{-3}(2x)}{(x^2+5)^{-2}}\right] \\ &= \frac{(x^2+1)^3(2x-5)^2}{(x^2+5)^2}\left(\frac{6x}{x^2+1} + \frac{4}{2x-5} - \frac{4x}{x^2+5}\right). \end{aligned}$$

Equivalently, using the basic properties of the logarithm function, we have

$$\begin{aligned} \ln|g(x)| &= \ln(x^2+1)^3 + \ln(2x-5)^2 - \ln(x^2+5)^2 \\ &= 3\ln(x^2+1) + 2\ln|2x-5| - 2\ln(x^2+5). \end{aligned}$$

(We have omitted absolute value in the first and third terms since x^2+1 and x^2+5 are positive for all x.) Differentiation gives

$$\frac{g'(x)}{g(x)} = \frac{3(2x)}{x^2+1} + \frac{2(2)}{2x-5} - \frac{2(2x)}{x^2+5},$$

and so

$$g'(x) = g(x)\left(\frac{6x}{x^2+1} + \frac{4}{2x-5} - \frac{4}{x^2+5}\right),$$

as we saw above. □

EXERCISES 7.3

Determine the domain and find the derivative.

1. $f(x) = \ln 4x$.

2. $f(x) = \ln(2x+1)$.

3. $f(x) = \ln(x^3+1)$.

4. $f(x) = \ln\left[(x+1)^3\right]$.

5. $f(x) = \ln\sqrt{1+x^2}$.

6. $f(x) = (\ln x)^3$.

7. $f(x) = \ln|x^4-1|$.

8. $f(x) = \ln(\ln x)$.

9. $f(x) = (2x+1)^2\ln(2x+1)$.

10. $f(x) = \ln\left|\dfrac{x+2}{x^3-1}\right|$.

11. $f(x) = \dfrac{1}{\ln x}$.

12. $f(x) = \ln\sqrt[4]{x^2+1}$.

13. $f(x) = \sin(\ln x)$.

14. $f(x) = \cos(\ln x)$.

Calculate the integral.

15. $\displaystyle\int \frac{dx}{x+1}$.

16. $\displaystyle\int \frac{dx}{3-x}$.

17. $\displaystyle\int \frac{x}{3-x^2}\,dx$.

18. $\displaystyle\int \frac{x+1}{x^2}\,dx$.

19. $\displaystyle\int \tan 3x\,dx$.

20. $\displaystyle\int \sec\tfrac{1}{2}\pi x\,dx$.

21. $\displaystyle\int x\sec x^2\,dx$.

22. $\displaystyle\int \frac{\csc^2 x}{2+\cot x}\,dx$.

23. $\displaystyle\int \frac{x}{(3-x^2)^2}\,dx.$ **24.** $\displaystyle\int \frac{\ln(x+a)}{x+a}\,dx.$

25. $\displaystyle\int \frac{\sin x}{2+\cos x}\,dx.$ **26.** $\displaystyle\int \frac{\sec^2 2x}{4-\tan 2x}\,dx.$

27. $\displaystyle\int \frac{1}{x\ \ln x}\,dx.$ **28.** $\displaystyle\int \frac{x^2}{2x^3-1}\,dx.$

29. $\displaystyle\int \frac{dx}{x\,(\ln x)^2}.$ **30.** $\displaystyle\int \frac{\sec 2x\tan 2x}{1+\sec 2x}\,dx.$

31. $\displaystyle\int \frac{\sin x-\cos x}{\sin x+\cos x}\,dx.$

32. $\displaystyle\int \frac{1}{\sqrt{x}\,(1+\sqrt{x})}\,dx.$ HINT: Let $u = 1+\sqrt{x}$.

33. $\displaystyle\int \frac{\sqrt{x}}{1+x\sqrt{x}}\,dx.$ **34.** $\displaystyle\int \frac{\tan(\ln x)}{x}\,dx.$

35. $\displaystyle\int (1+\sec x)^2\,dx.$ **36.** $\displaystyle\int (3-\csc x)^2\,dx.$

Evaluate the definite integral.

37. $\displaystyle\int_1^e \frac{dx}{x}.$ **38.** $\displaystyle\int_1^{e^2} \frac{dx}{x}.$

39. $\displaystyle\int_e^{e^2} \frac{dx}{x}.$ **40.** $\displaystyle\int_0^1 \left(\frac{1}{x+1}-\frac{1}{x+2}\right)dx.$

41. $\displaystyle\int_4^5 \frac{x}{x^2-1}\,dx.$ **42.** $\displaystyle\int_{1/4}^{1/3} \tan \pi x\,dx.$

43. $\displaystyle\int_{\pi/6}^{\pi/2} \frac{\cos x}{1+\sin x}\,dx.$ **44.** $\displaystyle\int_{\pi/4}^{\pi/2} (1+\csc x)^2\,dx.$

45. $\displaystyle\int_{\pi/4}^{\pi/2} \cot x\,dx.$ **46.** $\displaystyle\int_1^e \frac{\ln x}{x}\,dx.$

47. Explain why the formula $\displaystyle\int_1^4 \frac{1}{x-2} = \Big[\ln|x-2|\Big]_1^5 = \ln 3$ is not correct.

48. Show that $\displaystyle\lim_{x\to 0} \frac{\ln(1+x)}{x} = 1$ by using the definition of the derivative.

Calculate the derivative by logarithmic differentiation.

49. $g(x) = (x^2+1)^2(x-1)^5x^3.$

50. $g(x) = x(x+a)(x+b)(x+c).$

51. $g(x) = \dfrac{x^4(x-1)}{(x+2)(x^2+1)}.$

52. $g(x) = \sqrt{\dfrac{(x-1)(x-2)}{(x-3)(x-4)}}.$

Sketch the region bounded by the curves and find its area.

53. $y = \sec x,\quad y = 2,\quad x = 0,\quad x = \pi/6.$

54. $y = \csc \tfrac{1}{2}\pi x,\quad y = x,\quad x = \tfrac{1}{2}.$

55. $y = \tan x,\quad y = 1,\quad x = 0.$

56. $y = \sec x,\quad y = \cos x,\quad x = 0,\quad x = \dfrac{\pi}{4}.$

In Exercises 57 and 58, find the area of the part of the first quadrant that lies between

57. $x + 4y - 5 = 0$ and $xy = 1.$

58. $x + y - 3 = 0$ and $xy = 2.$

59. The region bounded by the graph of $f(x) = 1/\sqrt{1+x}$ and the x-axis for $0 \le x \le 8$ is revolved about the x-axis. Find the volume of the solid that is generated.

60. The region bounded by the graph of $f(x) = 3/(1+x^2)$ and the x-axis for $0 \le x \le 3$ is revolved about the y-axis. Find the volume of the solid that is generated.

61. The region bounded by the graph of $f(x) = \sqrt{\sec x}$ and the x-axis for $-\pi/3 \le x \le \pi/3$ is revolved about the x-axis. Find the volume of the solid that is generated.

62. The region bounded by the graph of $f(x) = \tan x$ and the x-axis for $0 \le x \le \pi/4$ is revolved about the x-axis. Find the volume of the solid that is generated.

63. A particle moves along a line with acceleration $a(t) = -(t+1)^{-2}$ feet per second per second. Find the distance traveled by the particle during the time interval $[0, 4]$, given that the initial velocity $v(0)$ is 1 foot per second.

64. Repeat Exercise 63 with $v(0)$ as 2 feet per second.

In Exercises 65 and 66, find a formula for the nth derivative.

65. $\dfrac{d^n}{dx^n}(\ln x).$ **66.** $\dfrac{d^n}{dx^n}[\ln(1-x)].$

67. Show that $\int \csc x\,dx = \ln|\csc x - \cot x| + C$ using the methods of this section.

68. (a) Show that for $n = 2$, Formula (7.3.7) reduces to the product rule (3.2.6).

(b) Show that Formula (7.3.7) applied to

$$g(x) = \frac{g_1(x)}{g_2(x)}$$

reduces to the quotient rule (3.2.10).

In Exercises 69–74, (i) find the domain of f, (ii) find the intervals where the function increases and the intervals where it decreases, (iii) find the extreme values, (iv) determine the concavity of the graph and find the points of inflection, and, finally, (v) sketch the graph, indicating asymptotes.

69. $f(x) = \ln(4-x).$ **70.** $f(x) = x - \ln x.$

71. $f(x) = x^2 \ln x.$ **72.** $f(x) = \ln(4-x^2).$

73. $f(x) = \ln\left[\dfrac{x}{1+x^2}\right].$ **74.** $f(x) = \ln\left[\dfrac{x^3}{x-1}\right].$

75. Show that the average slope of the logarithm curve from $x = a$ to $x = b$ is

$$\frac{1}{b-a}\ln\left(\frac{b}{a}\right).$$

76. (a) Show that $f(x) = \ln 2x$ and $g(x) = \ln 3x$ have the same derivative.
(b) Calculate the derivative of $F(x) = \ln kx$, where k is any positive number.
(c) Explain these results in terms of the properties of logarithms.

In Exercises 77–80, use a graphing utility to graph f on the indicated interval. Estimate the x-intercepts of the graph of f and the values of x where f has either a local or absolute extremum (if any). Use four decimal place accuracy in your answers.

77. $f(x) = \sqrt{x}\ln x$; $[0, 10]$.

78. $f(x) = x^3 \ln x$; $[0, 2]$.

79. $f(x) = \sin(\ln x)$; $[1, 100]$.

80. $f(x) = x^2 \ln(\sin x)$; $[0, 2]$.

81. A particle moves along a line with acceleration $a(t) = 4 - 2(t+1) + 3/(t+1)$ feet per second per second from $t = 0$ to $t = 3$.
(a) Find the velocity v of the particle at each time t during the time interval $[0, 3]$, given that $v(0) = 2$.
(b) Use a graphing utility to graph v and a together.
(c) Estimate the time $t \in [0, 3]$ at which the particle has maximum velocity and the time at which it has minimum velocity. Use four decimal place accuracy.

82. Repeat Exercise 81 for a particle whose acceleration is $a(t) = 2\cos 2(t+1) + 2/(t+1)$ feet per second per second from $t = 0$ to $t = 7$.

83. Let $f(x) = \dfrac{1}{x}$ and $g(x) = -x^2 + 4x - 2$.
(a) Use a graphing utility to graph f and g together.
(b) Use a CAS to find the points of intersection of the graphs.
(c) Use a CAS to find the area of the region bounded by the graphs.

84. Repeat Exercise 83 with $f(x) = \dfrac{x-1}{x}$ and $g(x) = |x - 2|$.

85. Use a CAS to find a one-term expression for f in each of the following:
(a) $\int f(x)\,dx = x\ln x - x$.
(b) $\int f(x)\,dx = \frac{1}{2}x^2 \ln x - \frac{1}{4}x^2$.
(c) $\int f(x)\,dx = \frac{1}{3}x^3 \ln x - \frac{1}{9}x^3$.
Verify your results by differentiating the expressions on the right-hand side.

In Exercise 86–88, use a CAS to find (a) f' and f''; (b) where each of f, f', and f'' is zero; and (c) the intervals on which each of f, f', and f'' is positive, negative.

86. $f(x) = x\ln x$.

87. $f(x) = \dfrac{\ln x}{x^2}$.

88. $f(x) = \dfrac{1 + 2\ln x}{2\sqrt{\ln x}}$.

■ 7.4 THE EXPONENTIAL FUNCTION

Rational powers of e already have an established meaning: by $e^{p/q}$ we mean the qth root of e raised to the pth power. But what is meant by $e^{\sqrt{2}}$ or e^{π}?

Earlier we proved that each rational power $e^{p/q}$ has logarithm p/q:

(7.4.1) $$\ln e^{p/q} = \frac{p}{q}.$$

The definition of e^z for z irrational is patterned after this relation.

DEFINITION 7.4.2

If z is irrational, then by e^z we mean the unique number that has logarithm z:

$$\ln e^z = z.$$

What is $e^{\sqrt{2}}$? It is the unique number that has logarithm $\sqrt{2}$. What is e^{π} ? It is the unique number that has logarithm π. Note that e^x now has meaning for every real value of x: it is the unique number that has logarithm x.

DEFINITION 7.4.3

The function

$$E(x) = e^x \quad \text{for all real } x$$

is called the *exponential function.*

Some properties of the exponential function are listed below.

(1) In the first place,

(7.4.4) $$\ln e^x = x \qquad \text{for all real } x$$

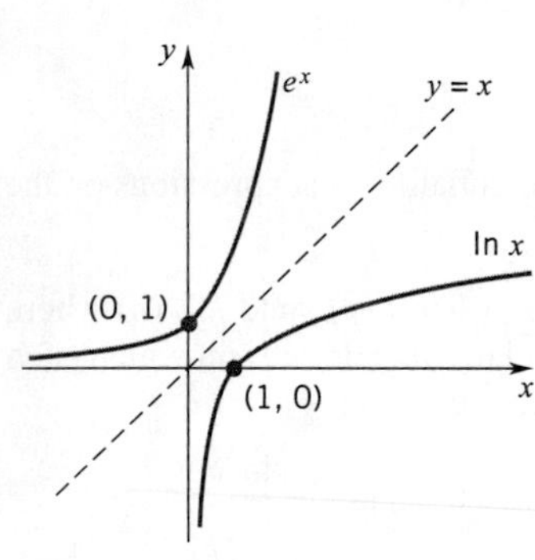

Figure 7.4.1

Writing $L(x) = \ln x$ and $E(x) = e^x$, we have

$$L(E(x)) = x \quad \text{for all real } x.$$

This says that the *exponential function is the inverse of the logarithm function.*

(2) The graph of the exponential function appears in Figure 7.4.1. It can be obtained from the graph of the logarithm by reflection in the line $y = x$.

(3) Since the graph of the logarithm lies to the right of the y-axis, the graph of the exponential function lies above the x-axis:

(7.4.5) $$e^x > 0 \quad \text{for all real } x.$$

(4) Since the graph of the logarithm crosses the x-axis at (1, 0), the graph of the exponential function crosses the y-axis at (0,1):

$$\ln 1 = 0 \quad \text{gives} \quad e^0 = 1.$$

(5) Since the y-axis is a vertical asymptote for the graph of the logarithm function, the x-axis is a horizontal asymptote for the graph of the exponential function:

$$\text{as } x \to -\infty, \quad e^x \to 0.$$

(6) Since the exponential function is the inverse of the logarithm function, the logarithm function is the inverse of the exponential function; thus

(7.4.6) $$e^{\ln x} = x \qquad \text{for all } x > 0.$$

You can verify this equation directly by observing that both sides have the same logarithm:

$$\ln (e^{\ln x}) = \ln x$$

since, for all real t, $\ln e^t = t$.

You know that for rational exponents

$$e^{(p/q + r/s)} = e^{p/q} \cdot e^{r/s}.$$

This property holds for all exponents, including irrational exponents.

THEOREM 7.4.7

$$e^{a+b} = e^a \cdot e^b \quad \text{for all real } a \text{ and } b.$$

PROOF

$$\ln(e^a \cdot e^b) = \ln e^a + \ln e^b = a + b = \ln e^{a+b}.$$

Since the logarithm function is one-to-one, we must have

$$e^{a+b} = e^a \cdot e^b. \quad \square$$

We leave it to you to verify that

(7.4.8)
$$e^{-b} = \frac{1}{e^b} \quad \text{and} \quad e^{a-b} = \frac{e^a}{e^b}.$$

We come now to one of the most important results in calculus. It is marvelously simple.

THEOREM 7.4.9

The exponential function is its own derivative: for all real x,

$$\frac{d}{dx}(e^x) = e^x.$$

PROOF The logarithm function is differentiable, and its derivative is never 0. It follows from Section 7.1 that its inverse, the exponential function, is also differentiable. Knowing this, we can show that

$$\frac{d}{dx}(e^x) = e^x$$

by differentiating the identity

$$\ln e^x = x.$$

On the left-hand side, the chain rule gives

$$\frac{d}{dx}(\ln e^x) = \frac{1}{e^x}\frac{d}{dx}(e^x).$$

On the right-hand side, the derivative is 1:

$$\frac{d}{dx}(x) = 1.$$

Equating these derivatives, we have

$$\frac{1}{e^x}\frac{d}{dx}(e^x) = 1 \quad \text{and thus} \quad \frac{d}{dx}(e^x) = e^x. \quad \square$$

We frequently run across expressions of the form e^u, where u is a function of x. If u is differentiable, then the chain rule gives

(7.4.10)
$$\frac{d}{dx}(e^u) = e^u \frac{du}{dx}.$$

PROOF

$$\frac{d}{dx}(e^u) = \frac{d}{du}(e^u)\frac{du}{dx} = e^u \frac{du}{dx}. \quad \square$$

Example 1

(a) $\frac{d}{dx}(e^{kx}) = e^{kx} \cdot k = ke^{kx}.$

(b) $\frac{d}{dx}(e^{\sqrt{x}}) = e^{\sqrt{x}} \frac{d}{dx}(\sqrt{x}) = e^{\sqrt{x}}\left(\frac{1}{2\sqrt{x}}\right) = \frac{1}{2\sqrt{x}}e^{\sqrt{x}}.$

(c) $\frac{d}{dx}(e^{-x^2}) = e^{-x^2}\frac{d}{dx}(-x^2) = e^{-x^2}(-2x) = -2x\,e^{-x^2}. \quad \square$

The relation

$$\frac{d}{dx}(e^x) = e^x \quad \text{and its corollary} \quad \frac{d}{dx}(e^{kx}) = k\,e^{kx}$$

have important applications to engineering, physics, chemistry, biology, and economics. We discuss some of these applications in Section 7.6.

Example 2 Let $f(x) = xe^{-x}$ for all real x.

(a) On what intervals does f increase? Decrease? **(b)** Find the extreme values of f.

(c) Determine the concavity of the graph and find the points of inflection.

(d) Sketch the graph indicating the asymptotes if any.

SOLUTION We have

$$f(x) = xe^{-x},$$
$$f'(x) = xe^{-x}(-1) + e^{-x} = (1-x)e^{-x},$$
$$f''(x) = (1-x)e^{-x}(-1) - e^{-x} = (x-2)e^{-x}.$$

Since $e^{-x} > 0$ for all x, we have $f'(x) = 0$ only at $x = 1$ (critical number). The sign of f' and the behavior of f are as follows:

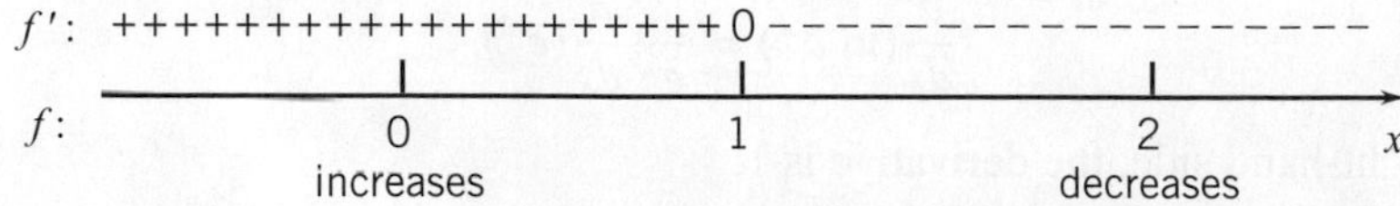

Thus, f increases on $(-\infty, 1]$ and decreases on $[1, \infty)$. The number

$$f(1) = \frac{1}{e} \cong \frac{1}{2.72} \cong 0.368$$

is a local and absolute maximum. There are no other extreme values.

Now consider f''. The sign of f'' and the behavior of the graph of f are as follows:

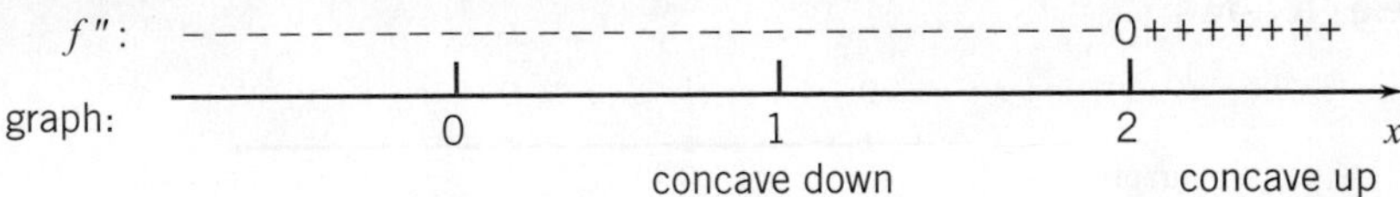

Thus, the graph is concave down on $(-\infty, 2)$ and concave up on $(2, \infty)$. The point

$$(2, f(2)) = (2, 2\,e^{-2}) \cong \left(2, \frac{2}{(2.72)^2}\right) \cong (2, 0.27)$$

is the only point of inflection. In Section 10.6, we prove that $f(x) = x/e^x \to 0$ as $x \to \infty$. Accepting this result for now, it follows that the x-axis is a horizontal asymptote. The graph is given in Figure 7.4.2. ☐

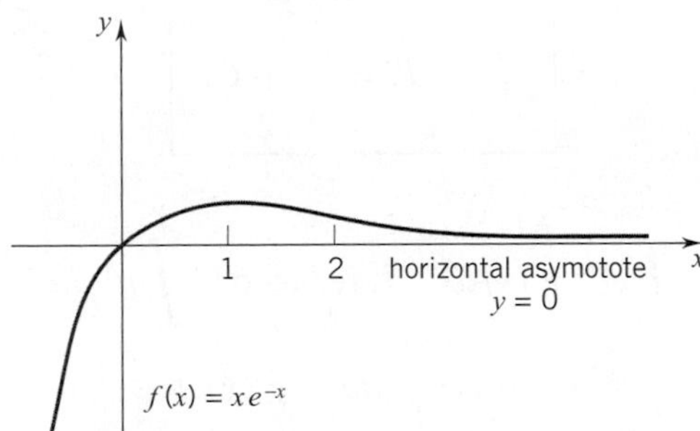

Figure 7.4.2

Example 3 Let $f(x) = e^{-x^2/2}$ for all real x.
(a) Determine the symmetry of the graph and the asymptotes, if any. **(b)** On what intervals does f increase? Decrease? **(c)** Find the extreme values. **(d)** Determine the concavity of the graph and find the points of inflection. **(e)** Sketch the graph.

NOTE: This function plays a very important role in the mathematical fields of probability and statistics. As you will see after we complete (a)–(e), the graph is the familiar bell-shaped curve.

SOLUTION Since $f(-x) = e^{-(-x)^2/2} = e^{-x^2/2} = f(x)$, f is an even function. Thus the graph is symmetric with respect to the y-axis. As $x \to \pm\infty, e^{-x^2/2} \to 0$. Therefore, the x-axis is a horizontal asymptote. There are no vertical asymptotes.

Differentiating f, we have

$$f'(x) = e^{-x^2/2}(-x) = -xe^{-x^2/2}$$
$$f''(x) = -x(-xe^{-x^2/2}) - e^{-x^2/2} = (x^2 - 1)e^{-x^2/2}.$$

Since $e^{-x^2/2} > 0$ for all x, we have $f'(x) = 0$ only at $x = 0$ (critical number). The sign of f' and the behavior of f are as follows:

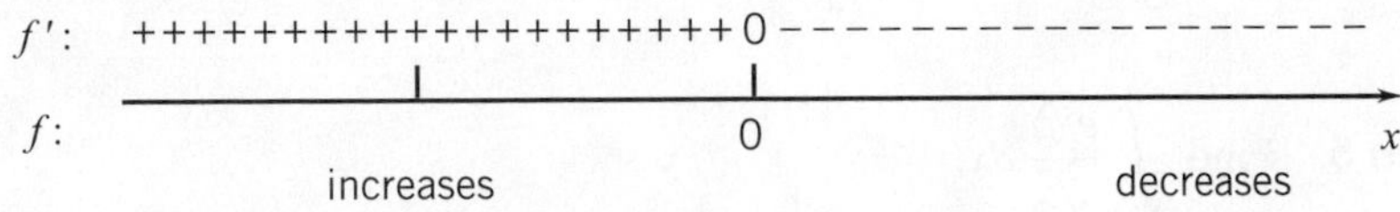

Thus, f is increasing on $(-\infty, 0]$ and decreasing on $[0, \infty)$. The number

$$f(0) = e^0 = 1$$

is a local and absolute maximum. There are no other extreme values.

Now consider $f''(x) = (x^2 - 1)e^{-x^2/2}$. The sign of f'' and the behavior of the graph of f are as follows:

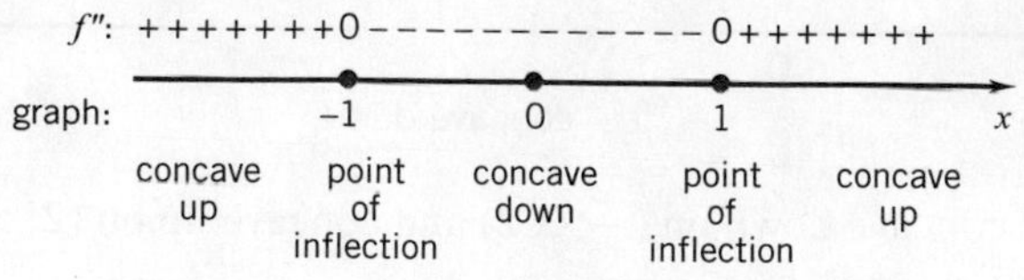

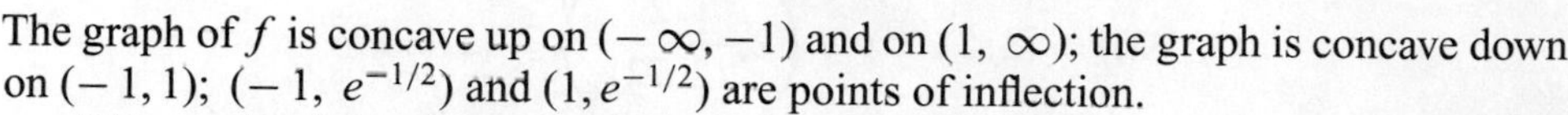

The graph of f is concave up on $(-\infty, -1)$ and on $(1, \infty)$; the graph is concave down on $(-1, 1)$; $(-1, e^{-1/2})$ and $(1, e^{-1/2})$ are points of inflection.

The graph of f is shown in Figure 7.4.3. □

$f(x) = e^{-x^2/2}$

Figure 7.4.3

The integral counterpart of Theorem 7.4.9 takes the form

(7.4.11)
$$\int e^x\,dx = e^x + C.$$

In practice

$$\int e^{g(x)}g'(x)\,dx \quad \text{is reduced to} \quad \int e^u\,du$$

by setting

$$u = g(x), \quad du = g'(x)\,dx.$$

This substitution gives

(7.4.12)
$$\int e^{g(x)}g'(x)\,dx = e^{g(x)} + C.$$

Example 4 Find $\int 9\,e^{3x}\,dx$.

SOLUTION Set $u = 3x, \quad du = 3\,dx$.

$$\int 9\,e^{3x}\,dx = 3\int e^u\,du = 3\,e^u + C = 3\,e^{3x} + C.$$

If you recognize at the very beginning that

$$3\,e^{3x} = \frac{d}{dx}(e^{3x}),$$

then you can dispense with the u-substitution and simply write

$$\int 9\,e^{3x}\,dx = 3\int 3\,e^{3x}\,dx = 3\,e^{3x} + C. \quad \square$$

Example 5 Find $\int \frac{e^{\sqrt{x}}}{\sqrt{x}}\,dx$.

SOLUTION Set $u = \sqrt{x}, \quad du = \frac{1}{2\sqrt{x}}\,dx$.

$$\int \frac{e^{\sqrt{x}}}{\sqrt{x}}\,dx = 2\int e^u\,du = 2\,e^u + C = 2\,e^{\sqrt{x}} + C.$$

If you recognize that

$$\tfrac{1}{2}\left(\frac{e^{\sqrt{x}}}{\sqrt{x}}\right) = \frac{d}{dx}(e^{\sqrt{x}}),$$

then you can dispense with the u-substitution and integrate directly:

$$\int \frac{e^{\sqrt{x}}}{\sqrt{x}}\,dx = 2\int \frac{1}{2}\left(\frac{e^{\sqrt{x}}}{\sqrt{x}}\right) dx = 2\,e^{\sqrt{x}} + C. \quad \square$$

Example 6 Find $\displaystyle\int \frac{e^{3x}}{e^{3x}+1}\,dx$

SOLUTION We can put this integral in the form

$$\int \frac{du}{u}$$

by setting $\qquad u = e^{3x} + 1, \quad du = 3\,e^{3x}\,dx.$

Then $\displaystyle\int \frac{e^{3x}}{e^{3x}+1}\,dx = \tfrac{1}{3}\int \frac{du}{u} = \tfrac{1}{3}\ln|u| + C = \tfrac{1}{3}\ln(e^{3x}+1) + C. \quad \square$

Example 7 Intervals involving $x\,e^{-x^2/2}$ play an important role in probability and statistics. Evaluate

$$\int_0^{\sqrt{2\ln 3}} x\,e^{-x^2/2}\,dx.$$

SOLUTION Set $u = -\tfrac{1}{2}x^2, \quad du = -x\,dx.$
At $x = 0,\ u = 0;$ at $x = \sqrt{2\ln 3},\ u = -\ln 3.$ Thus

$$\int_0^{\sqrt{2\ln 3}} xe^{-x^2/2}dx = -\int_0^{-\ln 3} e^u\,du = -\Big[e^u\Big]_0^{-\ln 3} = 1 - e^{-\ln 3} = 1 - \tfrac{1}{3} = \tfrac{2}{3}. \quad \square$$

Example 8 Evaluate $\displaystyle\int_0^1 e^x(e^x+1)^{1/5}\,dx.$

SOLUTION Set $u = e^x + 1, \quad du = e^x dx.$
At $x = 0,\ u = 2;$ at $x = 1,\ u = e + 1.$ Thus

$$\int_0^1 e^x(e^x+1)^{1/5}dx = \int_2^{e+1} u^{1/5}\,du = \Big[\tfrac{5}{6}u^{6/5}\Big]_2^{e+1} = \tfrac{5}{6}[(e+1)^{6/5} - 2^{6/5}]. \quad \square$$

Remark The u-substitution simplifies many calculations, but you will find that in many cases you can carry out the integration more quickly without it. $\square$

EXERCISES 7.4

Differentiate the given function.

1. $y = e^{-2x}$.
2. $y = 3e^{2x+1}$.
3. $y = e^{x^2-1}$.
4. $y = 2e^{-4x}$.
5. $y = e^x \ln x$.
6. $y = x^2 e^x$.
7. $y = x^{-1} e^{-x}$.
8. $y = e^{\sqrt{x}+1}$.
9. $y = \frac{1}{2}(e^x + e^{-x})$.
10. $y = \frac{1}{2}(e^x - e^{-x})$.
11. $y = e^{\sqrt{x}} \ln \sqrt{x}$.
12. $y = (3 - 2e^{-x})^3$.
13. $y = (e^{x^2} + 1)^2$.
14. $y = (e^{2x} - e^{-2x})^2$.
15. $y = (x^2 - 2x + 2)e^x$.
16. $y = x^2 e^x - xe^{x^2}$.
17. $y = \dfrac{e^x - 1}{e^x + 1}$.
18. $y = \dfrac{e^{2x} - 1}{e^{2x} + 1}$.
19. $y = e^{4 \ln x}$.
20. $y = \ln e^{3x}$.
21. $f(x) = \sin(e^{2x})$.
22. $f(x) = e^{\sin 2x}$.
23. $f(x) = e^{-2x} \cos x$.
24. $f(x) = \ln(\cos e^{2x})$.

Calculate the following indefinite integrals.

25. $\int e^{2x}\,dx$.
26. $\int e^{-2x}\,dx$.
27. $\int e^{kx}\,dx$.
28. $\int e^{ax+b}\,dx$.
29. $\int xe^{x^2}\,dx$.
30. $\int xe^{-x^2}\,dx$.
31. $\int \dfrac{e^{1/x}}{x^2}\,dx$.
32. $\int \dfrac{e^{2\sqrt{x}}}{\sqrt{x}}\,dx$.
33. $\int \ln e^x\,dx$.
34. $\int e^{\ln x}\,dx$.
35. $\int \dfrac{4}{\sqrt{e^x}}\,dx$.
36. $\int \dfrac{e^x}{e^x + 1}\,dx$.
37. $\int \dfrac{e^x}{\sqrt{e^x + 1}}\,dx$.
38. $\int \dfrac{xe^{ax^2}}{e^{ax^2} + 1}\,dx$.
39. $\int \dfrac{e^{2x}}{2e^{2x} + 3}\,dx$.
40. $\int \dfrac{\sin(e^{-2x})}{e^{2x}}\,dx$.
41. $\int \cos x\, e^{\sin x}\,dx$.
42. $\int e^{-x}[1 + \cos(e^{-x})]\,dx$.

Evaluate the following definite integrals.

43. $\int_0^1 e^x\,dx$.
44. $\int_0^1 e^{-kx}\,dx$.
45. $\int_0^{\ln \pi} e^{-6x}\,dx$.
46. $\int_0^1 xe^{-x^2}\,dx$.
47. $\int_0^1 \dfrac{e^x + 1}{e^x}\,dx$.
48. $\int_0^1 \dfrac{4 - e^x}{e^x}\,dx$.
49. $\int_0^{\ln 2} \dfrac{e^x}{e^x + 1}\,dx$.
50. $\int_0^1 \dfrac{e^x}{4 - e^x}\,dx$.
51. $\int_0^1 x(e^{x^2} + 2)\,dx$.
52. $\int_0^{\ln \pi/4} e^x \sec e^x\,dx$.

53. Let a be a positive constant.
    (a) Find a formula for the nth derivative of $f(x) = e^{ax}$.
    (b) Find a formula for the nth derivative of $f(x) = e^{-ax}$.

54. A particle moves on a coordinate line with its position at time t given by the function $x(t) = Ae^{kt} + Be^{-kt}$, $A > 0, B > 0, k > 0$.
    (a) Find the time t at which the particle is closest to the origin.
    (b) Show that the acceleration of the particle is proportional to its position. What is the constant of proportionality?

55. A rectangle has one side on the x-axis and the upper two vertices on the graph of $y = e^{-x^2}$. Where should the vertices be placed so as to maximize the area of the rectangle?

56. A rectangle has two sides on the positive x- and y-axes and one vertex at a point P that moves along the curve $y = e^x$ in such a way that y increases at the rate of $\frac{1}{2}$ unit per minute. How fast is the area of the rectangle changing at the instant when $y = 3$?

57. The function $f(x) = e^{-x^2}$ is very important in statistics. (a) What is the symmetry of the graph? (b) On what intervals does the function increase? Decrease? (c) What are the extreme values of the function? (d) Determine the concavity of the graph and find the points of inflection. (e) The graph has a horizontal asymptote. What is it? (f) Sketch the graph.

58. Let R be the region bounded by the graph of $y = e^x$ and the x-axis from $x = 0$ to $x = 1$.
    (a) Find the volume of the solid generated by revolving R about the x-axis.
    (b) Set up the definite integral that gives the volume of the solid generated by revolving R about the y-axis using the shell method. (You will learn how to evaluate this integral in Section 8.2.)

59. Let Ω be the region bounded by the graph of $y = e^{-x^2}$ and the x-axis from $x = 0$ to $x = 1$.
    (a) Find the volume of the solid generated by revolving Ω about the y-axis.
    (b) Set up the definite integral that gives the volume of the solid generated by revolving Ω about the x-axis using the disc method. Note that you cannot evaluate this integral by the fundamental theorem of calculus.

Sketch the region bounded by the curves and find its area.

60. $x = e^{2y}$, $x = e^{-y}$, $x = 4$.
61. $y = e^x$, $y = e^{2x}$, $y = e^4$.
62. $y = e^x$, $y = e$, $y = x$, $x = 0$.
63. $x = e^y$, $y = 1$, $y = 2$, $x = 2$.

For each of the functions f in Exercises 64–68, find: (a) the domain; (b) the intervals where f increases, decreases;

(c) the extreme values; (d) the concavity of the graph of f and the points of inflection; (e) Sketch the graph, indicating all asymptotes.

64. $f(x) = (1 - x)e^x$.

65. $f(x) = e^{(1/x^2)}$.

66. $f(x) = x^2 e^{-x}$.

67. $f(x) = x^2 \ln x$.

68. $f(x) = (x - x^2)e^{-x}$

69. Draw the graphs of $f(x) = x^k \ln x$ for various positive values of k.

(a) What is $\lim_{x \to 0^+} f(x)$ when $0 < k < 1$? When $k \geq 1$?

(b) f has an absolute minimum if $k \geq 1$. What is the x-coordinate of the point where f takes on its absolute minimum?

70. Draw the graphs of $f(x) = x^k e^{-x}$ for various positive values of k.

(a) As $x \to \infty$, $f(x) \to$?

(b) f has an absolute maximum on $[0, \infty)$. What is the x-coordinate of the point where f takes on its absolute maximum?

71. Take $a > 0$ and refer to the figure.

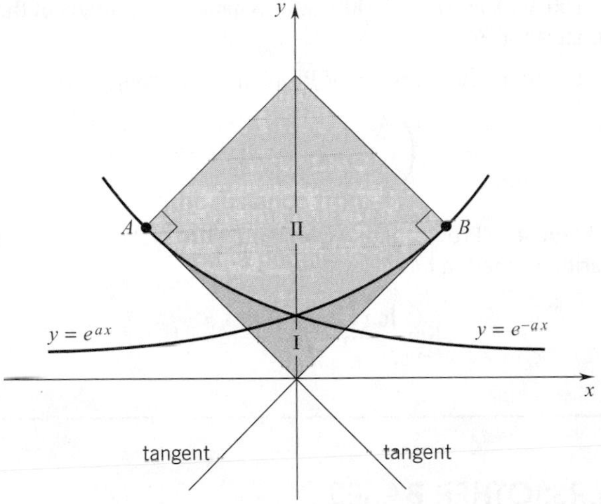

(a) Find the points of tangency, marked A and B.

(b) Find the area of region I.

(c) Find the area of region II.

72. Prove that for all $x > 0$ and all positive integers n

$$e^x > 1 + x + \frac{x^2}{2!} + \frac{x^3}{3!} + \cdots + \frac{x^n}{n!}.$$

Recall, $n! = n(n-1)(n-2)\cdots 3 \cdot 2 \cdot 1$.

HINT: $$e^x = 1 + \int_0^x e^t\,dt > 1 + \int_0^x dt = 1 + x$$

$$e^x = 1 + \int_0^x e^t\,dt > 1 + \int_0^x (1+t)\,dt = 1 + x + \frac{x^2}{2}, \quad \text{and so on.}$$

73. Prove that, if n is a positive integer, then

$$e^x > x^n \quad \text{for all } x \text{ sufficiently large.}$$

HINT: Use Exercise 72.

74. Let $f(x) = e^{-x^2}$ and $g(x) = x^2$.

(a) Use a graphing utility to sketch the graphs of f and g.

(b) Estimate the x-coordinates of the points of intersection of the two curves. Use four decimal place accuracy.

(c) Estimate the area of the region bounded by the two curves.

75. Repeat Exercise 74 for the functions $f(x) = e^x$ and $g(x) = 4 - x^2$.

In Exercises 76–78 use a graphing utility to graph the functions f and g. Your graphs should suggest that f and g are inverses of each other. Confirm this by using the methods in Section 7.1.

76. $f(x) = e^{2x}$; $\quad g(x) = \ln\sqrt{x}$.

77. $f(x) = e^{x^2}$; $\quad g(x) = \sqrt{\ln x}$.

78. $f(x) = e^{x-2}$; $\quad g(x) = 2 + \ln x$.

79. Use a graphing utility to graph $f(x) = \sin(e^x)$.

(a) Estimate the zeros of f.

(b) Use a CAS to obtain a general formula for the zeros of f.

80. Repeat Exercise 79 with $f(x) = e^{\sin x} - 1$.

81. Let $f(x) = e^{-x}$ and $g(x) = \ln x$.

(a) Use a graphing utility to draw the graphs of f and g together.

(b) Estimate the x-coordinate of a point of intersection of the two graphs.

(c) Find the slope of the tangent lines for f and g at the point found in part (b).

(d) Are these tangent lines perpendicular to each other?

82. (a) Use a graphing utility to draw the graphs of $f(x) = 10e^{-x}$ and $g(x) = 7 - e^x$ together.

(b) Use a CAS to find the points of intersection of the two graphs.

(c) Use a CAS to find the area of the region between the graphs.

83. Use a CAS to find the following indefinite integrals.

(a) $\displaystyle\int \frac{1}{1 - e^x}\,dx.$

(b) $\displaystyle\int e^{-x}\left(\frac{1 - e^x}{e^x}\right)^4 dx.$

(c) $\displaystyle\int \frac{e^{\tan x}}{\cos^2 x}\,dx.$

84. Use a CAS to solve the equation $\int_0^{x_n} e^u\,du = n$ for x_n where n is a positive integer. How much area is in the region bounded by the x-axis and $y = e^x$ from $x = 0$ to $x = x_n$?

■ PROJECT 7.4 Estimating the Number *e*

Since *e* is an irrational number, we cannot hope to express *e* as a repeating or terminating decimal. In this project, we derive a numerical estimate for *e* from the numerical representation of the natural logarithm function:

$$\ln x = \int_1^x \frac{1}{t}\,dt, \quad x > 0.$$

Let n be a positive integer. Then

$$\ln\left(1+\frac{1}{n}\right) = \int_1^{1+1/n} \frac{1}{t}\,dt$$

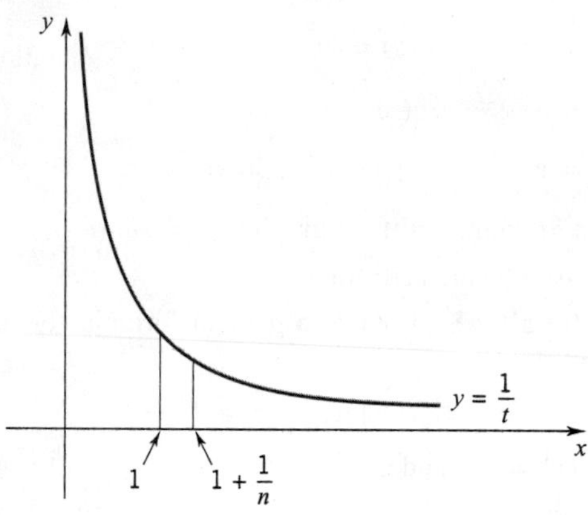

Problem 1. Show that

$$\frac{1}{n+1} \le \ln\left(1+\frac{1}{n}\right) \le \frac{1}{n}$$

by showing:

a. $\ln\left(1+\frac{1}{n}\right) \le \frac{1}{n}$ (HINT : $\frac{1}{t} \le 1$ on$[1, 1+1/n]$.)

b. $\ln\left(1+\frac{1}{n}\right) \ge \frac{1}{n+1}$

(HINT : $\frac{1}{t} \ge \frac{1}{1+1/n}$ on $[1, 1+1/n]$.)

Problem 2. Show that

$$\textbf{(1)} \qquad \left(1+\frac{1}{n}\right)^n \le e \le \left(1+\frac{1}{n}\right)^{n+1}.$$

The result in Problem 2 is an elegant characterization of *e* but not a very efficient tool for calculating *e*. For example,

$$\left(1+\tfrac{1}{100}\right)^{100} \approx 2.7048138 \quad \text{and} \quad \left(1+\tfrac{1}{100}\right)^{101} \approx 2.7318619$$

gives *e* rounded off to one decimal place: $e \approx 2.7$.

Problem 3. Evaluate (1) for $n = 1{,}000$, $n = 10{,}000$, $n = 100{,}000$, and $n = 1{,}000{,}000$. What is the accuracy of these estimates for *e*?

Based on these results of Problem 2, we conjecture that

$$\textbf{(2)} \qquad \left(1+\frac{1}{x}\right)^x \to e \text{ as } x \to \infty.$$

Problem 4. Prove that (2) holds. (HINT: At $x = 1$, the logarithm function has derivative

$$\lim_{h\to 0} \frac{\ln(1+h) - \ln 1}{h} = 1$$

.

■ 7.5 ARBITRARY POWERS; OTHER BASES

Arbitrary Powers: The Function $f(x) = x^r$

The elementary notion of exponent applies only to rational numbers. Expressions such as

$$10^5, \quad 2^{1/3}, \quad 7^{-4/5}, \quad \pi^{-1/2}$$

make sense, but so far we have attached no meaning to expressions such as

$$10^{\sqrt{2}}, \quad 2^{\pi}, \quad 7^{-\sqrt{3}}, \quad \pi^{e}.$$

The extension of our sense of exponent to allow for irrational exponents is conveniently done by making use of the logarithm function and the exponential function. The heart of the matter is to observe that for $x > 0$ and p/q rational

$$x^{p/q} = e^{(p/q)\ln x}.$$

(To verify this, take the natural log of both sides.) We *define* x^z for irrational z by setting

$$x^z = e^{z \ln x}.$$

We then have the following result:

(7.5.1)

$$\text{if } x > 0, \text{ then} \quad x^r = e^{r \ln x} \quad \text{for all real numbers } r.$$

In particular

$$10^{\sqrt{2}} = e^{\sqrt{2} \ln 10}, \; 2^{\pi} = e^{\pi \ln 2}, \; 7^{-\sqrt{3}} = e^{-\sqrt{3} \ln 7}, \; \pi^e = e^{e \ln \pi}.$$

With this extended sense of exponent, the usual laws of exponents

(7.5.2)

$$x^{r+s} = x^r x^s, \quad x^{r-s} = \frac{x^r}{x^s}, \quad (x^r)^s = x^{rs}$$

still hold:

$$\begin{aligned} x^{r+s} &= e^{(r+s)\ln x} = e^{r \ln x} \cdot e^{s \ln x} = x^r x^s, \\ x^{r-s} &= e^{(r-s)\ln x} = e^{r \ln x} \cdot e^{-s \ln x} = \frac{e^{r \ln x}}{e^{s \ln x}} = \frac{x^r}{x^s}, \\ (x^r)^s &= e^{s \ln x^r} = e^{rs \ln x} = x^{rs}. \quad \square \end{aligned}$$

In Chapter 3 we proved that $d(x^p)/dx = px^{p-1}$ for any rational number p. We can now extend this result to arbitrary powers. For *any real number r:*

(7.5.3)

$$\frac{d}{dx}(x^r) = r x^{r-1} \quad \text{for all } x > 0.$$

PROOF

$$\frac{d}{dx}(x^r) = \frac{d}{dx}(e^{r \ln x}) = e^{r \ln x}\frac{d}{dx}(r \ln x) = x^r \frac{r}{x} = r x^{r-1}.$$

You can also write $f(x) = x^r$ and use logarithmic differentiation:

$$\begin{aligned} \ln f(x) &= r \ln x \\ \frac{f'(x)}{f(x)} &= \frac{r}{x} \\ f'(x) &= \frac{r f(x)}{x} = \frac{r x^r}{x} = r x^{r-1}. \quad \square \end{aligned}$$

Thus

$$\frac{d}{dx}(x^{\sqrt{2}}) = \sqrt{2}\, x^{\sqrt{2}-1}, \quad \frac{d}{dx}(x^{\pi}) = \pi x^{\pi-1}.$$

If u is a positive differentiable function of x and r is any real number, then, by the chain rule,

(7.5.4)
$$\frac{d}{dx}(u^r) = r\,u^{r-1}\frac{du}{dx}.$$

PROOF

$$\frac{d}{dx}(u^r) = \frac{d}{du}(u^r)\frac{du}{dx} = r\,u^{r-1}\frac{du}{dx}. \quad \square$$

For example,

$$\frac{d}{dx}[(x^2+5)^{\sqrt{3}}] = \sqrt{3}(x^2+5)^{\sqrt{3}-1}(2x) = 2\sqrt{3}\,x(x^2+5)^{\sqrt{3}-1}.$$

Example 1 Find $\dfrac{d}{dx}[(x^2+1)^{3x}]$.

SOLUTION One way to find this derivative is to observe that $(x^2+1)^{3x} = e^{3x\ln(x^2+1)}$ and then differentiate:

$$\begin{aligned}\frac{d}{dx}[(x^2+1)^{3x}] &= \frac{d}{dx}[e^{3x\ln(x^2+1)}] = e^{3x\ln(x^2+1)}\left[3x\cdot\frac{2x}{x^2+1} + 3\ln(x^2+1)\right]\\ &= (x^2+1)^{3x}\left[\frac{6x^2}{x^2+1} + 3\ln(x^2+1)\right].\end{aligned}$$

Another way to find the derivative of $(x^2+1)^{3x}$ is to set $f(x) = (x^2+1)^{3x}$ and use logarithmic differentiation:

$$\ln f(x) = 3x\cdot\ln(x^2+1)$$

$$\frac{f'(x)}{f(x)} = 3x\cdot\frac{2x}{x^2+1} + [\ln(x^2+1)](3) = \frac{6x^2}{x^2+1} + 3\ln(x^2+1)$$

$$\begin{aligned}f'(x) &= f(x)\left[\frac{6x^2}{x^2+1} + 3\ln(x^2+1)\right]\\ &= (x^2+1)^{3x}\left[\frac{6x^2}{x^2+1} + 3\ln(x^2+1)\right]. \quad \square\end{aligned}$$

Each derivative formula gives rise to a companion integral formula. The integral version of (7.5.3) takes the form

(7.5.5)
$$\int x^r\,dx = \frac{x^{r+1}}{r+1} + C, \quad \text{for} \quad r \neq -1.$$

Note the exclusion of $r = -1$. What is the integral if $r = -1$?

Example 2 Find $\displaystyle\int \frac{x^3}{(2x^4+1)^{\pi}}\,dx$.

SOLUTION Set $u = 2x^4 + 1, \quad du = 8x^3 dx.$

$$\int \frac{x^3}{(2x^4+1)^\pi}\,dx = \frac{1}{8}\int u^{-\pi}\,du = \frac{1}{8}\left(\frac{u^{1-\pi}}{1-\pi}\right) + C = \frac{(2x^4+1)^{1-\pi}}{8(1-\pi)} + C. \quad \square$$

Base p: The Function $f(x) = p^x$

To form the function $f(x) = x^r$ we take a positive variable x and raise it to a constant power r. To form the function $f(x) = p^x$ we take a positive constant p and raise it to a variable power x. Since $1^x = 1$ for all x, the function is of interest only if $p \neq 1$.

Functions of the form $f(x) = p^x$ are called *exponential functions with base p.* The high status enjoyed by Euler's number e comes from the fact that

$$\frac{d}{dx}(e^x) = e^x.$$

For other bases the derivative has an extra factor:

(7.5.6)
$$\frac{d}{dx}(p^x) = p^x \ln p.$$

PROOF

$$\frac{d}{dx}(p^x) = \frac{d}{dx}(e^{x \ln p}) = e^{x \ln p} \ln p = p^x \ln p. \quad \square$$

For example,

$$\frac{d}{dx}(2^x) = 2^x \ln 2 \qquad \text{and} \qquad \frac{d}{dx}(10^x) = 10^x \ln 10.$$

If u is a differentiable function of x, then, by the chain rule,

(7.5.7)
$$\frac{d}{dx}(p^u) = p^u \ln p \frac{du}{dx}.$$

PROOF

$$\frac{d}{dx}(p^u) = \frac{d}{du}(p^u)\frac{du}{dx} = p^u \ln p \frac{du}{dx}. \quad \square$$

For example,

$$\frac{d}{dx}(2^{3x^2}) = 2^{3x^2}(\ln 2)(6x) = 6x\, 2^{3x^2} \ln 2.$$

The integral version of (7.5.6) reads

(7.5.8)
$$\int p^x\,dx = \frac{1}{\ln p}p^x + C, \quad p > 0, \quad p \neq 1.$$

For example,

$$\int 2^x\,dx = \frac{1}{\ln 2}2^x + C.$$

Example 3 Find $\int x5^{-x^2}\,dx$.

SOLUTION Set $u = -x^2, \quad du = -2x\,dx$.

$$\int x5^{-x^2}\,dx = -\tfrac{1}{2}\int 5^u du = -\tfrac{1}{2}\left(\frac{1}{\ln 5}\right)5^u + C$$
$$= \frac{-1}{2\ln 5}5^{-x^2} + C. \quad \square$$

Example 4 Evaluate $\int_1^2 3^{2x-1}\,dx$.

SOLUTION Set $u = 2x - 1, \quad du = 2\,dx$.
At $x = 1,\ u = 1$; at $x = 2,\ u = 3$. Thus

$$\int_1^2 3^{2x-1}\,dx = \tfrac{1}{2}\int_1^3 3^u\,du = \tfrac{1}{2}\left[\frac{1}{\ln 3}\cdot 3^u\right]_1^3 = \frac{12}{\ln 3} \cong 10.923. \quad \square$$

Base p: The Function $f(x) = \log_p x$

If $p > 0$, then

$$\ln p^t = t \ln p \qquad \text{for all } t.$$

If p is also different from 1, then $\ln p \neq 0$, and we have

$$\frac{\ln p^t}{\ln p} = t.$$

This indicates that the function

$$f(x) = \frac{\ln x}{\ln p}, \qquad x > 0,$$

satisfies the relation

$$f(p^t) = t \quad \text{for all real } t.$$

In view of this we call

$$\frac{\ln x}{\ln p}$$

the logarithm of x to the base p and write:

(7.5.9) $$\boxed{\log_p x = \frac{\ln x}{\ln p}.}$$

The relation holds for all $x > 0$ and assumes that p is a positive number different from 1.

For example,

$$\log_2 32 = \frac{\ln 32}{\ln 2} = \frac{\ln 2^5}{\ln 2} = \frac{5 \ln 2}{\ln 2} = 5$$

and $$\log_{100}\left(\tfrac{1}{10}\right) = \frac{\ln(\frac{1}{10})}{\ln 100} = \frac{\ln 10^{-1}}{\ln 10^2} = \frac{-\ln 10}{2 \ln 10} = -\frac{1}{2}. \quad \square$$

We can obtain these same results more directly from the relation

(7.5.10) $$\boxed{\log_p p^t = t.}$$

Accordingly

$$\log_2 32 = \log_2 2^5 = 5 \quad \text{and} \quad \log_{100}\left(\tfrac{1}{10}\right) = \log_{100}(100^{-1/2}) = -\tfrac{1}{2}.$$

Since $\log_p x$ and $\ln x$ differ only by a constant factor, there is no reason to introduce new differentiation and integration formulas. For the record, we simply point out that

$$\frac{d}{dx}(\log_p x) = \frac{d}{dx}\left(\frac{\ln x}{\ln p}\right) = \frac{1}{x \ln p}$$

and $$\frac{d}{dx}(\log_p u) = \frac{d}{dx}\left(\frac{\ln u}{\ln p}\right) = \frac{1}{u \ln p} \cdot \frac{du}{dx}$$

if u is a positive, differentiable function of x.

If p is e, the factor $\ln p$ is 1 and we have

$$\frac{d}{dx}(\log_e x) = \frac{1}{x}.$$

The logarithm to the base e, $\ln = \log_e$, is called the *natural logarithm* because it is the logarithm with the simplest derivative.

Example 5 Find

(a) $\frac{d}{dx}(\log_5 |x|)$ **(b)** $\frac{d}{dx}[\log_2(3x^2 + 1)]$ **(c)** $\int \frac{1}{x \ln 10} dx$

SOLUTION

(a) $$\frac{d}{dx}(\log_5 |x|) = \frac{d}{dx}\left[\frac{\ln |x|}{\ln 5}\right] = \frac{1}{x \ln 5}.$$

(b) $$\frac{d}{dx}[\log_2 (3x^2 + 1)] = \frac{d}{dx}\left[\frac{\ln (3x^2 + 1)}{\ln 2}\right] = \frac{1}{(3x^2 + 1) \ln 2} \cdot 6x = \frac{6x}{(3x^2 + 1) \ln 2}.$$

(c) $$\int \frac{1}{x \ln 10} dx = \frac{1}{\ln 10} \int \frac{1}{x} dx = \frac{\ln |x|}{\ln 10} + C = \log_{10} |x| + C. \quad \square$$

EXERCISES 7.5

Evaluate.

1. $\log_2 64$.

2. $\log_2 \frac{1}{64}$.

3. $\log_{64} \frac{1}{2}$.

4. $\log_{10} 0.01$.

5. $\log_5 1$.

6. $\log_5 0.2$.

7. $\log_5 125$.

8. $\log_2 4^3$.

Show that the given identity holds.

9. $\log_p xy = \log_p x + \log_p y$.

10. $\log_p \frac{1}{x} = -\log_p x$.

11. $\log_p x^y = y \log_p x$.

12. $\log_p \frac{x}{y} = \log_p x - \log_p y$.

Find the numbers x, if any, which satisfy the given equation.

13. $10^x = e^x$.

14. $\log_5 x = 0.04$.

15. $\log_x 10 = \log_4 100$.

16. $\log_x 2 = \log_3 x$.

17. Estimate $\ln a$ given that $e^{t_1} < a < e^{t_2}$.

18. Estimate e^b given that $\ln x_1 < b < \ln x_2$.

Differentiate the function.

19. $f(x) = 3^{2x}$.

20. $g(x) = 4^{3x^2}$.

21. $f(x) = 2^{5x}\, 3^{\ln x}$.

22. $F(x) = 5^{-2x^2+x}$.

23. $g(x) = \sqrt{\log_3 x}$.

24. $h(x) = 7^{\sin x^2}$.

25. $f(x) = \tan(\log_5 x)$.

26. $g(x) = \frac{\log_{10} x}{x^2}$.

27. $F(x) = \cos(2^x + 2^{-x})$.

28. $h(x) = a^{-x} \cos bx$.

Calculate the integral.

29. $\int 3^x dx$.

30. $\int 2^{-x} dx$.

31. $\int (x^3 + 3^{-x})\, dx$.

32. $\int x10^{-x^2} dx$.

33. $\int \frac{dx}{x \ln 5}$.

34. $\int \frac{\log_5 x}{x} dx$.

35. $\int \frac{\log_2 x^3}{x} dx$.

36. Show that, if a, b, c are positive, then

$$\log_a c = \log_a b \log_b c$$

provided that a and b are both different from 1.

Find $f'(e)$ in Exercises 37–40.

37. $f(x) = \log_3 x$.

38. $f(x) = x \log_3 x$.

39. $f(x) = \ln(\ln x)$.

40. $f(x) = \log_3(\log_2 x)$.

In Exercises 41 and 42, derive the formula for $f'(x)$ by logarithmic differentiation.

41. $f(x) = p^x$.

42. $f(x) = p^{g(x)}$.

Find the derivative by logarithmic differentiation.

43. $\frac{d}{dx}[(x+1)^x]$.

44. $\frac{d}{dx}[(\ln x)^x]$.

45. $\frac{d}{dx}[(\ln x)^{\ln x}]$.

46. $\frac{d}{dx}\left[\left(\frac{1}{x}\right)^x\right]$.

47. $\frac{d}{dx}[x^{\sin x}]$.

48. $\frac{d}{dx}[(\cos x)^{(x^2+1)}]$.

49. $\frac{d}{dx}[(\sin x)^{\cos x}]$.

50. $\frac{d}{dx}[x^{(x^2)}]$.

51. $\frac{d}{dx}[x^{(2^x)}]$.

52. $\frac{d}{dx}[(\tan x)^{\sec x}]$.

Sketch figures in which you compare the following pairs of graphs.

53. $f(x) = e^x$ and $g(x) = 2^x$.

54. $f(x) = e^x$ and $g(x) = 3^x$.

55. $f(x) = \ln x$ and $g(x) = \log_3 x$.

56. $f(x) = 2^x$ and $g(x) = 2^{-x}$.

57. $f(x) = 2^x$ and $g(x) = \log_2 x$.

58. $f(x) = \ln x$ and $g(x) = \log_2 x$.

Evaluate the integral.

59. $\int_1^2 2^{-x} dx$.

60. $\int_0^1 4^x dx$.

61. $\int_1^4 \frac{dx}{x \ln 2}$.

62. $\int_0^2 p^{x/2} dx$.

63. $\int_0^1 x10^{1+x^2}\, dx$.

64. $\int_0^1 \frac{5p^{\sqrt{x+1}}}{\sqrt{x+1}} dx$.

65. $\int_0^1 (2^x + x^2)\, dx$.

Evaluate numerically and then give the exact value.

66. $7^{1/\ln 7}$.

67. $5^{(\ln 17)/(\ln 5)}$.

68. $(16)^{1/\ln 2}$.

69. (a) Use a graphing utility to graph $f(x) = 2^x$ and $g(x) = x^2 - 1$ together.

(b) Use a CAS to find the points of intersection of the two graphs.

(c) Use a CAS to find the area of the region bounded by the two graphs.

70. Repeat Exercise 69 with the functions $f(x) = 2^{-x}$ and $g(x) = 1/x^2$, $x > 0$.

■ 7.6 EXPONENTIAL GROWTH AND DECAY

We begin by comparing exponential change to linear change. Let $y = y(t)$ be a function of time t.

If y is a linear function

$$y(t) = kt + C, \qquad k, C \text{ constants},$$

then y changes by the *same additive amount during all periods of the same duration*:

$$y(t + \Delta t) = k(t + \Delta t) + C = (kt + C) + k\,\Delta t = y(t) + k\Delta t.$$

During every period of length Δt, y changes by the amount $k\Delta t$.

If y is an exponential function

$$y(t) = C\,e^{kt}, \qquad k, C \quad \text{constants},$$

then y changes by the *same multiplicative factor during all periods of the same duration*:

$$y(t + \Delta t) = C\,e^{k(t+\Delta t)} = C\,e^{kt + k\Delta t} = C\,e^{kt} \cdot e^{k\Delta t} = e^{k\Delta t} y(t).$$

During every period of length Δt, y changes by the factor $e^{k\Delta t}$.

An exponential function

$$f(t) = C\,e^{kt}$$

has the property that its derivative $f'(t)$ is proportional to $f(t)$:

$$f'(t) = C\,ke^{kt} = k\,C\,e^{kt} = kf(t).$$

Moreover, it is the only such function:

THEOREM 7.6.1

If

$$f'(t) = k\,f(t) \qquad \text{for all } t \text{ in some interval}$$

then f is an exponential function

$$f(t) = C\,e^{kt}$$

where C is an arbitrary constant.

PROOF We assume that

$$f'(t) = k\,f(t)$$

and write

$$f'(t) - k\,f(t) = 0.$$

Multiplying this equation by e^{-kt}, we have

$$(*) \qquad e^{-kt}f'(t) - k\,e^{-kt}f(t) = 0.$$

Observe now that the left side of this equation is the derivative

$$\frac{d}{dt}\left[e^{-kt}f(t)\right]. \qquad \text{(verify this)}$$

Equation $(*)$ can therefore be written

$$\frac{d}{dt}\left[e^{-kt}f(t)\right] = 0.$$

It follows that

$$e^{-kt}f(t) = C \qquad \text{for some constant } C.$$

Multiplication by e^{kt} gives

$$f(t) = C\,e^{kt}. \quad \square$$

Remark In the study of exponential growth or decay, time is usually measured from time $t = 0$. The constant C is the value of f at time $t = 0$:

$$f(0) = C\,e^0 = C.$$

This is called the *initial value of* f. The exponential $f(t) = C\,e^{kt}$ is conveniently written

$$f(t) = f(0)\,e^{kt}. \quad \square$$

Example 1 Find $f(t)$ given that $f'(t) = 2f(t)$ for all t and $f(0) = 5$.

SOLUTION The fact that $f'(t) = 2f(t)$ tells us that $f(t) = C\,e^{2t}$ where C is some constant. Since $f(0) = C = 5$, we have $f(t) = 5e^{2t}$. $\square$

Population Growth

Under ideal conditions (unlimited space, adequate food supply, immunity to disease, and so on), the rate of increase of a population P at time t is proportional to the size of the population at time t. That is,

$$P'(t) = kP(t),$$

where $k > 0$ is a constant, called the *growth constant*.† Thus, by our theorem, the size of the population at any time t is given by

$$P(t) = P(0)\,e^{kt},$$

and the population is said to grow *exponentially*. This is a model of uninhibited growth. In reality, the rate of increase of a population does not continue to be proportional to the

† The growth constant of a population is often expressed as a percentage. In such instances, the percentage must be converted to its decimal equivalent for use in the formula. For example, if the growth constant of a certain population is given as 1.7%, then $k = 0.017$.

size of the population. After some time has passed, factors such as limitations on space or food supply, diseases, and so forth, set in and affect the growth rate of the population.

Example 2 The number of bacteria in a certain culture is increasing at a rate proportional to the number present. Suppose that there are 1000 bacteria present initially, and that 1500 are present after 2 hours.

(a) Determine the number of bacteria in the culture at any time t. How many bacteria are in the culture after 5 hours?

(b) How long will it take for the number of bacteria in the culture to double?

SOLUTION Let $P(t)$ be the number of bacteria in the culture at time t. The basic equation $P'(t) = kP(t)$ gives

$$P(t) = P(0)\, e^{kt}.$$

Since $P(0) = 1000$, we have

$$(1) \qquad P(t) = 1000\, e^{kt}.$$

We can use the fact that $P(2) = 1500$ to eliminate k from (1):

$$1500 = 1000\, e^{2k}, \qquad e^{2k} = \tfrac{1500}{1000} = \tfrac{3}{2}, \quad 2k = \ln\left[\tfrac{3}{2}\right],$$

and so

$$(2) \qquad k = \tfrac{1}{2}\ln\left[\tfrac{3}{2}\right] \cong 0.203.$$

(a) The number of bacteria in the culture at any time t is (approximately)

$$P(t) = 1000\, e^{0.203t}.$$

After 5 hours, the number of bacteria in the culture is

$$P(5) = 1000\, e^{0.203(5)} = 1000\, e^{1.015} \cong 2759$$

(b) To find out how long it will take for the number of bacteria to double, we need to find the value of t for which $P(t) = 2000$.† Thus we set

$$1000\, e^{0.203t} = 2000$$

and solve for t:

$$e^{0.203\,t} = 2, \qquad 0.203t = \ln 2, \qquad \text{and} \qquad t = \frac{\ln 2}{0.203} \cong 3.4 \text{ hours.} \quad \square$$

Remark There is another way of expressing P that uses the exact value of k. From (2), we have $k = \frac{1}{2}\ln\left[\frac{3}{2}\right]$. Thus

$$P(t) = 1000\, e^{(t/2)\ln[3/2]} = 1000\, e^{\ln[3/2]^{t/2}} = 1000\left(\tfrac{3}{2}\right)^{t/2}.$$

† The length of time that it takes for a population to double in size is called the *doubling time*.

Now, for example, to find the doubling time, we solve

$$1000\left(\tfrac{3}{2}\right)^{t/2} = 2000$$

for t:

$$\left(\tfrac{3}{2}\right)^{t/2} = 2, \qquad \left(\tfrac{t}{2}\right)\ln\left(\tfrac{3}{2}\right) = \ln 2, \qquad t = \frac{2\ln 2}{\ln(3/2)} \cong 3.4. \quad \square$$

Example 3 The world population in 1980 was approximately 4.5 billion, and in 2000 approximately 6 billion. Assume that the world population at time t is increasing at a rate proportional to the size of the population at time t.

(a) Determine the growth constant and the (approximate) population at any time t.

(b) How long will it take for the world's population to reach 9 billion (double the 1980 population)?

(c) The current (2002) world population is estimated to be 6.2 billion. What does the result determined in part (a) predict for the population in 2002?

SOLUTION Let $P(t)$ be the world population (in billions) t years after 1980. The basic equation $P'(t) = kP(t)$, with $P(0) = 4.5$, gives

$$P(t) = 4.5e^{kt}.$$

(a) Since $P(20) = 6$, we have

$$4.5e^{20k} = 6, \qquad 20k = \ln\left(\frac{6}{4.5}\right), \qquad \text{and} \qquad k = \frac{\ln 1.3333}{20} \cong 0.0143.$$

Therefore the growth constant is approximately 0.0143, or 1.43%. The population at any time t is given (approximately) by

$$P(t) = 4.5e^{0.0143t}.$$

(b) To find the value of t for which $P(t) = 9$, we set $4.5e^{0.0143t} = 9$:

$$e^{0.0143t} = 2, \qquad 0.0143t = \ln 2, \qquad \text{and} \qquad t = \frac{\ln 2}{0.0143} \cong 48.47.$$

Thus, based on the given data, the world population will reach 9 billion in approximately $48\frac{1}{2}$ years (from 1980); that is, midyear in the year 2028. (As of January 1, 2002, demographers were predicting that the world population will peak at 9 billion in the year 2070 and will begin to decline thereafter.)

(c) According to the result in part (a), $P(22) = 4.5e^{0.0143(22)} = 4.5e^{0.3146} \cong 6.164$ billion. This is very close to the January 1, 2002, estimate of 6.196 billion. $\square$

Radioactive Decay

Experimental evidence shows that radioactive substances decay at a rate which is proportional to the amount of such substance present. Therefore, if $A(t)$ is the amount of a radioactive substance present at time t, then

$$A'(t) = kA(t)$$

where $k < 0$ is a constant, called the *decay constant*. Thus

$$A(t) = A(0)\,e^{kt}$$

where $A(0)$ is the amount of the substance present initially.

The *half-life* of a radioactive substance is the length of time it takes for half of the initial amount to decay. If we let T denote the half-life and k the decay constant, then the relation between T and k is given by

(7.6.2)
$$\boxed{kT = -\ln 2.}$$

PROOF We have $A(T) = \frac{1}{2}A(0)$. Thus,

$$\begin{aligned}\tfrac{1}{2}A(0) &= A(0)\,e^{kT}\\ e^{kT} &= \tfrac{1}{2}\\ kT &= \ln\left(\tfrac{1}{2}\right) = -\ln 2. \quad \square\end{aligned}$$

Example 4 One-third of a radioactive substance decays every 5 years. Today we have $A(0) = A_0$ grams of the substance.

(a) Find the decay constant and determine how much will be left t years from now.

(b) What is the half-life of the substance?

SOLUTION Let $A(t)$ be the amount of the radioactive substance present at time t. Since $A'(t) = kA(t)$, we know that $A(t) = A_0\,e^{kt}$.

(a) At the end of 5 years, one-third of the substance will have decayed and therefore two-thirds of A_0 will remain:

$$A(5) = \tfrac{2}{3}A_0.$$

Now
$$\begin{aligned}\tfrac{2}{3}A_0 &= A_0\,e^{5k}\\ e^{5k} &= \tfrac{2}{3}\\ 5k &= \ln\left(\tfrac{2}{3}\right)\end{aligned}$$

and therefore

$$k = \frac{\ln\left(\frac{2}{3}\right)}{5} \cong -0.081.$$

The amount of the substance that remains after t years is (approximately):

$$A(t) = A_0 e^{-0.081t}.$$

(b) We use Equation (7.6.2) to find the half-life of the substance:

$$T = \frac{-\ln 2}{k} = \frac{-\ln 2}{-0.081} \cong 8.56.$$

Thus the half-life of the substance is approximately 8.56 years. $\square$

Remark As you can verify, the exact value of A in Example 4 is $A(t) = A_0(\frac{2}{3})^{t/5}$, and the exact value of the half-life is $T = (5 \ln 2)/(\ln 1.5)$. ❑

Example 5 Cobalt-60 is a radioactive substance that is used extensively in medical radiology. It has a half-life of 5.3 years. Suppose that an initial sample of cobalt-60 has a mass of 100 grams.

(a) Find the decay constant and determine an expression for the amount of the sample that will remain t years from now.

(b) How long will it take for 90% of the sample to decay?

SOLUTION **(a)** The decay constant k is given by

$$k = \frac{-\ln 2}{T} = \frac{-\ln 2}{5.3} \cong -0.131.$$

With $A(0) = 100$, the amount of material that will remain after t years is

$$A(t) = 100\, e^{-0.131t}.$$

(b) If 90% of the sample decays, then 10%, which is 10 grams, remains. We seek the time t at which

$$100\, e^{-0.131t} = 10.$$

We solve this equation for t:

$$e^{-0.131t} = 0.1, \quad -0.131t = \ln (0.1), \quad t = \frac{\ln (0.1)}{-0.131} \cong 17.6$$

It will take approximately 17.6 years for 90% of the sample to decay. ❑

Compound Interest

Consider money invested at interest rate r. If the accumulated interest is credited once a year, then the interest is said to be compounded annually; if twice a year, then semiannually; if four times a year, then quarterly. The idea can be pursued further. Interest can be credited every day, every hour, every second, every half-second, and so on. In the limiting case, interest is credited instantaneously. Economists call this *continuous compounding*.

The economists' formula for continuous compounding is a simple exponential:

(7.6.3) $$\boxed{A(t) = A_0\, e^{rt}.}$$

Here t is measured in years,

$$A(t) = \text{the principal in dollars at time } t,$$
$$A_0 = A(0) = \text{the initial investment},$$
$$r = \text{the annual interest rate.}$$

The rate r is also called the *nominal* interest rate, and it is conventional to express this rate as a percentage. However, as in the case of the growth and decay constants, the decimal equivalent of the percentage must be used in all calculations.

A DERIVATION OF THE COMPOUND INTEREST FORMULA Fix t and take h as a small time increment. Then

$$A(t+h) - A(t) = \text{interest earned from time } t \text{ to time } t+h.$$

Had the principal remained $A(t)$ from time t to time $t+h$, the interest earned during this time period would have been

$$rhA(t).$$

Had the principal been $A(t+h)$ throughout the time interval, the interest earned would have been

$$rhA(t+h).$$

The actual interest earned must be somewhere in between:

$$rhA(t) \leq A(t+h) - A(t) \leq rhA(t+h).$$

Dividing by h, we get

$$rA(t) \leq \frac{A(t+h) - A(t)}{h} \leq rA(t+h).$$

If A varies continuously, then, as h tends to zero, $rA(t+h)$ tends to $rA(t)$ and (by the pinching theorem) the difference quotient in the middle must also tend to $rA(t)$:

$$\lim_{h \to 0} \frac{A(t+h) - A(t)}{h} = rA(t).$$

This says that

$$A'(t) = rA(t).$$

Thus, with continuous compounding, the principal increases at a rate proportional to the amount present and the growth constant is the nominal rate r. Now, it follows that

$$A(t) = C\,e^{rt}.$$

If A_0 is the initial investment, we have $C = A_0$ and therefore

$$A(t) = A_0\,e^{rt}.$$

Remark It is interesting to compare the accumulation of principal at a fixed nominal rate r using different compounding periods. For example, a principal of \$1000 invested at 6% compounded:

(a) annually (once per year) will have the value $A(1) = 1000(1 + 0.06) = \$1060$ at the end of one year.

(b) quarterly (4 times per year) will have the value $A(1) = 1000[1 + (0.06/4)]^4 \cong \$1061.36$ at the end of one year.

(c) monthly (12 times per year) will have the value $A(1) = 1000[1 + (0.06/12)]^{12} \cong \$1061.67$ at the end of one year.

(d) continuously will have the value $A(1) = 1000\,e^{0.06} \cong \$1061.84$ at the end of one year. □

Example 6 Suppose that \$1000 is deposited in a bank that pays 5% compounded continuously. How much money will be in the account after 5 years, and what is the interest earned during this period?

SOLUTION Here we have $A_0 = 1000$ and $r = 0.05$. Thus, the amount of money in the account at any time t is given by

$$A(t) = 1000\, e^{0.05t}.$$

The amount of money in the account after 5 years is

$$A(5) = 1000\, e^{0.05(5)} = 1000\, e^{0.25} \cong \$1284.03,$$

and the interest earned during this period is \$284.03 □

Example 7 How long does it take for an investment to double at an interest rate r compounded continuously?

SOLUTION If an amount A_0 is invested at an interest rate r compounded continuously, then the value of the investment at time t is

$$A(t) = A_0\, e^{rt}.$$

To find the "doubling time", we solve

$$2A_0 = A_0\, e^{rt}$$

for t: $$e^{rt} = 2, \quad rt = \ln 2, \quad t = \frac{\ln 2}{r} \cong \frac{0.69}{r}. \quad \square$$

For example, if the interest rate is $8\% = 0.08$, then it will take $\frac{0.69}{0.08} = 8.625$ years for the investment to double in value.

Remark A popular estimate for the "doubling time" of an investment at an interest rate $r\%$ is the *rule of 72*:

$$\text{doubling time} = \frac{72}{r}.$$

For example, at an interest of 8%, it will take approximately $\frac{72}{8} = 9$ years for an investment to double in value. Here's how this rule originated:

$$\frac{0.69}{r\%} = \frac{69}{r} \cong \frac{72}{r}. \quad \square$$

Since we are only *estimating*, 72 is preferred to 69 because it has many more divisors.

EXERCISES 7.6

C NOTE: Some of these exercises require a calculator or graphing utility.

1. Find the amount of interest earned by \$500 compounded continuously for 10 years:
(a) at 6%, (b) at 8%, (c) at 10%.

2. How long does it take for a sum of money to double if compounded continuously:
(a) at 6%? (b) at 8%? (c) at 10%?

3. At what rate r of continuous compounding does a sum of money triple in 20 years?

4. At what rate r of continuous compounding does a sum of money double in 10 years?

5. A certain species of virulent bacteria is being grown in a culture. It is observed that the rate of growth of the bacterial population is proportional to the number present. If there were 1000 bacteria in the initial population and the number doubled after the first 30 minutes, how many bacteria will be present after 2 hours?

6. In a bacteria growing experiment, a biologist observes that the number of bacteria in a certain culture triples every

4 hours. After 12 hours, it is estimated that there are 1 million bacteria in the culture.

(a) How many bacteria were present initially?

(b) What is the doubling time for the bacteria population?

7. A population P of insects increases at a rate proportional to the current population. Suppose there are 10,000 insects initially and 20,000 insects 1 week later.

(a) Find an expression for the number $P(t)$ of insects at any time t.

(b) How many insects will there be in $\frac{1}{2}$ year? In 1 year?

8. Determine the period in which $y = Ce^{kt}$ changes by a factor of q.

9. The population of a certain country is increasing at the rate of 3.5% per year. By what factor does it increase every 10 years? What percentage increase per year will double the population every 15 years?

10. According to the Bureau of the Census, the population of the United States in 1990 was approximately 249 million and in 2000, 281 million. Use this information to estimate the population in 1980. (The actual figure was about 227 million.)

11. Use the data of Exercise 10 to predict the population for 2010. Compare the prediction for 2001 with the actual (estimated) figure of 284.8 million.

12. Use the data of Exercise 10 to estimate how long it will take for the U.S population to double.

13. It is estimated that $\frac{1}{3}$ of an acre is needed to provide food for one person. It is further estimated that there are 10 billion square miles of arable land on the earth. Thus, a maximum of 30 billion people can be sustained according to current estimates. Use the data of Example 3 to find when the maximum population will be reached.

14. Water is pumped into a tank to dilute a saline solution. The volume of the solution, call it V, is kept constant by continuous outflow. The amount of salt in the tank, call it s, depends on the amount of water that has been pumped in, call this x. Given that

$$\frac{ds}{dx} = -\frac{s}{V},$$

find the amount of water that must be pumped into the tank to eliminate 50% of the salt. Take V as 10,000 gallons.

15. A 200-liter tank initially full of water develops a leak at the bottom. Given that 20% of the water leaks out in the first 5 minutes, find the amount of water left in the tank t minutes after the leak develops if the water drains off at a rate that is proportional to the amount of water present.

In Exercises 16–20, remember that the rate of decay of a radioactive substance is proportional to the amount of substance present.

16. What is the half-life of a radioactive substance if it takes 5 years for one-third of the substance to decay?

17. Two years ago there were 5 grams of a radioactive substance. Now there are 4 grams. How much will remain 3 years from now?

18. A year ago there were 4 grams of a radioactive substance. Now there are 3 grams. How much was there 10 years ago?

19. Suppose the half-life of a radioactive substance is n years. What percentage of the substance present at the start of a year will decay during the ensuing year?

20. A radioactive substance weighed n grams at time $t = 0$. Today, 5 years later, the substance weighs m grams. How much will it weigh 5 years from now?

21. The half-life of radium-226 is 1620 years. What percentage of a given amount of the radium will remain after 500 years? How long will it take for the original amount to be reduced by 75%?

22. Cobalt-60 is used extensively in medical technology. It has a half-life of 5.3 years. What percentage of a given amount of cobalt will remain after 8 years? If you have 100 grams of cobalt now, how much was there 3 years ago?

23. (*The power of exponential growth*) Imagine two racers competing on the x-axis (which has been calibrated in meters), a linear racer LIN [position function of the form $x_1(t) = kt + C$] and an exponential racer EXP [position function of the form $x_2(t) = e^{kt} + C$]. Suppose that both racers start out simultaneously from the origin, LIN at one million meters per second, EXP at only one meter per second. In the early stages of the race, fast-starting LIN will move far ahead of EXP, but in time EXP will catch up to LIN, pass her, and leave her hopelessly behind. Show that this is true as follows:

(a) Express the position of each racer as a function of time, measuring t in seconds.

(b) Show that LIN's lead over EXP starts to decline about 13.8 seconds into the race.

(c) Show that LIN is still ahead of EXP some 15 seconds into the race but far behind 3 seconds later.

(d) Show that, once EXP passes LIN, LIN can never catch up.

24. (*The weakness of logarithmic growth*) Having been soundly beaten in the race in Exercise 23, LIN finds an opponent she can beat, LOG, the logarithmic racer [position function $x_3(t) = k \ln(t+1) + C$]. Once again the racetrack is the x-axis calibrated in meters. Both racers start out at the origin, LOG at one million meters per second, LIN at only one meter per second. (LIN is tired from the previous race.) In this race LOG will shoot ahead and remain ahead for a long time, but eventually LIN will catch up to LOG, pass her, and leave her permanently behind. Show that this is true as follows:

(a) Express the position of each racer as a function of time t, measuring t in seconds.

(b) Show that LOG's lead over LIN starts to decline $10^6 - 1$ seconds into the race.

(c) Show that LOG is still ahead of LIN $10^7 - 1$ seconds into the race but behind LIN $10^8 - 1$ seconds into the race.

(d) Show that, once LIN passes LOG, LOG can never catch up.

25. Atmospheric pressure p varies with altitude h according to the equation

$$\frac{dp}{dh} = kp, \quad \text{where } k \text{ is a constant.}$$

Given that p is 15 pounds per square inch at sea level and 10 pounds per square inch at 10,000 feet, find p at: (a) 5000 feet; (b) 15,000 feet.

26. According to the compound interest formula (7.6.3), if P dollars are deposited in an account now at an interest rate r compounded continuously, then the amount of money in the account t years from now will be

$$Q = Pe^{rt}.$$

The quantity Q is sometimes called the *future value* of P (at the interest rate r). Solving this equation for P, we get

$$P = Qe^{-rt}.$$

In this formulation, the quantity P is called the *present value* of Q. Find the present value of \$20,000 at 6% compounded continuously for 4 years.

27. Find the interest rate r that is needed to have \$6000 be the present value of \$10,000 over an 8-year period.

28. You are 45 years old and are looking forward to an annual pension of \$50,000 per year at age 65. What is the present-day purchasing power (present value) of your pension if money can be invested over this period at a continuously compounded interest rate of: (a) 4%? (b) 6%? (c) 8%?

29. The cost of the tuition, fees, room, and board at XYZ College is currently \$16,000 per year. What would you expect to pay 3 years from now if the costs at XYZ are rising at the continuously compounded rate of: (a) 5%? (b) 8%? (c) 12%?

30. A boat moving in still water is subject to a retardation proportional to its velocity. Show that the velocity t seconds after the power is shut off is given by the formula $v = ce^{-kt}$ where c is the velocity at the instant the power is shut off.

31. A boat is drifting in still water at 4 miles per hour; 1 minute later, at 2 miles per hour. How far has the boat drifted in that 1 minute? (See Exercise 30.)

32. During the process of inversion, the amount A of raw sugar present decreases at a rate proportional to A. During the first 10 hours, 1000 pounds of raw sugar have been reduced to 800 pounds. How many pounds will remain after 10 more hours of inversion?

33. The method of *carbon dating* makes use of the fact that all living organisms contain two isotopes of carbon, carbon-12, denoted ^{12}C (a stable isotope), and carbon-14, denoted ^{14}C (a radioactive isotope). The ratio of the amount of ^{14}C to the amount of ^{12}C is essentially constant (approximately 1/10,000). When an organism dies, the amount of ^{12}C present remains unchanged, but the ^{14}C decays at a rate proportional to the amount present with a half-life of approximately 5700 years. This change in the amount of ^{14}C relative to the amount of ^{12}C makes it possible to estimate the time at which the organism lived. A fossil found in an archaeological dig was found to contain 25% of the original amount of ^{14}C. What is the approximate age of the fossil?

34. The Dead Sea Scrolls are approximately 2000 years old. How much of the original ^{14}C remains in them?

In Exercises 35–37, find all the functions f that satisfy the equation for all real t.

35. $f'(t) = t\,f(t)$. HINT: Write $f'(t) - t\,f(t) = 0$ and multiply the equation by $e^{-t^2/2}$.

36. $f'(t) = \sin t\,f(t)$.

37. $f'(t) = \cos t\,f(t)$.

38. Let g be a function everywhere continuous and not identically zero. Show that if $f'(t) = g(t)f(t)$ for all real t, then either f is identically zero, or f does not take on the value zero at any t.

■ 7.7 THE INVERSE TRIGONOMETRIC FUNCTIONS

Since none of the trigonometric functions are one-to-one, none of them have inverses. What, then, are the inverse trigonometric functions?

The Inverse Sine

The graph of $y = \sin x$ is shown in Figure 7.7.1. Since every horizontal line between -1 and 1 intersects the graph at infinitely many points, the sine function does not have an inverse. However, observe that if we restrict the domain to the interval $[-\frac{1}{2}\pi, \frac{1}{2}\pi]$ (the solid portion of the graph in Figure 7.7.1), then $y = \sin x$ is one-to-one, and on that interval it takes on as a value every number in $[-1,1]$. Thus, if $x \in [-1, 1]$, there is one and only one number in the interval $[-\frac{1}{2}\pi, \frac{1}{2}\pi]$ at which the sine function has the value x. This number is called the *inverse sine of* x, or *the angle whose sine is* x, and is written $\sin^{-1}x$.

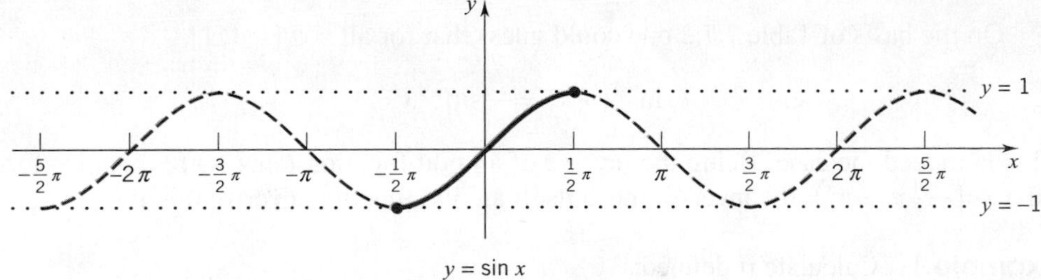

Figure 7.7.1

Remarks Another common notation for the inverse sine function is arcsin x, read "arc sine of x." For consistency in the treatment here, we use $\sin^{-1}x$ throughout, but we use both notations in the Exercises. Remember that the "−1" is not an exponent; do not confuse $\sin^{-1}x$ with the reciprocal $1/\sin x$. □

The *inverse sine function*

$$y = \sin^{-1}x, \quad \text{domain: } [-1, 1], \quad \text{range: } [-\tfrac{1}{2}\pi, \tfrac{1}{2}\pi]$$

is the inverse of the function

$$y = \sin x, \quad \text{domain: } [-\tfrac{1}{2}\pi, \tfrac{1}{2}\pi], \quad \text{range: } [-1, 1].$$

The graphs of these functions are shown in Figure 7.7.2. Each curve is the reflection of the other in the line $y = x$.

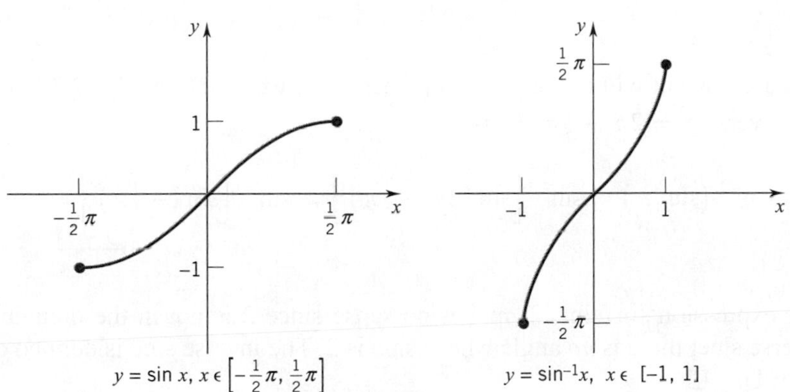

Figure 7.7.2

Because these functions are inverses,

(7.7.1) $$\text{for all } x \in [-1, 1], \qquad \sin(\sin^{-1}x) = x$$

and

(7.7.2) $$\text{for all } x \in [-\tfrac{1}{2}\pi, \tfrac{1}{2}\pi], \quad \sin^{-1}(\sin x) = x.$$

Table 7.7.1 gives some representative values of the sine function from $x = -\frac{1}{2}\pi$ to $x = \frac{1}{2}\pi$. Reversing the order of the columns, we have a table for the inverse sine (Table 7.7.2).

■ **Table 7.7.1**

x	$\sin x$
$-\frac{1}{2}\pi$	-1
$-\frac{1}{3}\pi$	$-\frac{1}{2}\sqrt{3}$
$-\frac{1}{4}\pi$	$-\frac{1}{2}\sqrt{2}$
$-\frac{1}{6}\pi$	$-\frac{1}{2}$
0	0
$\frac{1}{6}\pi$	$\frac{1}{2}$
$\frac{1}{4}\pi$	$\frac{1}{2}\sqrt{2}$
$\frac{1}{3}\pi$	$\frac{1}{2}\sqrt{3}$
$\frac{1}{2}\pi$	1

■ **Table 7.7.2**

x	$\sin^{-1}x$
-1	$-\frac{1}{2}\pi$
$-\frac{1}{2}\sqrt{3}$	$-\frac{1}{3}\pi$
$-\frac{1}{2}\sqrt{2}$	$-\frac{1}{4}\pi$
$-\frac{1}{2}$	$-\frac{1}{6}\pi$
0	0
$\frac{1}{2}$	$\frac{1}{6}\pi$
$\frac{1}{2}\sqrt{2}$	$\frac{1}{4}\pi$
$\frac{1}{2}\sqrt{3}$	$\frac{1}{3}\pi$
1	$\frac{1}{2}\pi$

On the basis of Table 7.7.2 one could guess that for all $x \in [-1, 1]$,

$$\sin^{-1}(-x) = -\sin^{-1}x.$$

This is indeed the case. Being the inverse of an odd function ($\sin(-x) = -\sin x$ for all $x \in [-\frac{1}{2}\pi, \frac{1}{2}\pi]$), the inverse sine is itself an odd function. (Verify this.)

Example 1 Calculate if defined:

(a) $\sin^{-1}(\sin\frac{1}{16}\pi)$; **(b)** $\sin^{-1}(\sin\frac{7}{3}\pi)$; **(c)** $\sin(\sin^{-1}\frac{1}{3})$;
(d) $\sin^{-1}(\sin\frac{9}{5}\pi)$; **(e)** $\sin(\sin^{-1}2)$.

SOLUTION

(a) Since $\frac{1}{16}\pi$ is within the interval $[-\frac{1}{2}\pi, \frac{1}{2}\pi]$, we know by (7.7.2) that

$$\sin^{-1}(\sin\tfrac{1}{16}\pi) = \tfrac{1}{16}\pi.$$

(b) Since $\frac{7}{3}\pi$ is not in the interval $[-\pi/2, \pi/2]$, we cannot apply (7.7.2) directly. However, $\frac{7}{3}\pi = \frac{1}{3}\pi + 2\pi$ and $\sin(\frac{1}{3}\pi + 2\pi) = \sin(\frac{1}{3}\pi)$ (recall that the sine function is periodic with period 2π). Thus

$$\sin^{-1}(\sin\tfrac{7}{3}\pi) = \sin^{-1}\left(\sin\left[\tfrac{1}{3}\pi + 2\pi\right]\right) = \sin^{-1}(\sin\tfrac{1}{3}\pi) \underset{\text{by (7.7.2)}}{=} \tfrac{1}{3}\pi.$$

(c) By (7.7.1)

$$\sin(\sin^{-1}\tfrac{1}{3}) = \tfrac{1}{3}.$$

(d) Since $\frac{9}{5}\pi$ is not within the interval $[-\frac{1}{2}\pi, \frac{1}{2}\pi]$, we cannot apply (7.7.2) directly. However, $\frac{9}{5}\pi = 2\pi - \frac{1}{5}\pi$. Thus

$$\sin^{-1}\left(\sin\tfrac{9}{5}\pi\right) = \sin^{-1}\left[\sin\left(2\pi - \tfrac{1}{5}\pi\right)\right] = \sin^{-1}\left[\sin\left(-\tfrac{1}{5}\pi\right)\right] \underset{\text{by (7.7.2)}}{=} -\tfrac{1}{5}\pi.$$

(e) The expression $\sin(\sin^{-1}2)$ makes no sense since 2 is not in the domain of the inverse sine; there is *no* angle whose sine is 2. The inverse sine is defined only on $[-1, 1]$. □

If $0 < x < 1$, then $\sin^{-1}x$ is the radian measure of the acute angle whose sine is x. We can construct an angle of radian measure $\sin^{-1}x$ by drawing a right triangle with a leg of length x and a hypotenuse of length 1. See Figure 7.7.3.

Reading from the figure we have

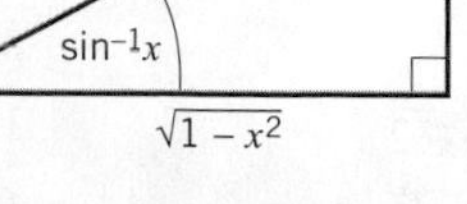

Figure 7.7.3

$$\sin(\sin^{-1}x) = x \qquad \cos(\sin^{-1}x) = \sqrt{1-x^2}$$

$$\tan(\sin^{-1}x) = \frac{x}{\sqrt{1-x^2}} \qquad \cot(\sin^{-1}x) = \frac{\sqrt{1-x^2}}{x}$$

$$\sec(\sin^{-1}x) = \frac{1}{\sqrt{1-x^2}} \qquad \csc(\sin^{-1}x) = \frac{1}{x}.$$

Since the derivative of the sine function,

$$\frac{d}{dx}(\sin x) = \cos x,$$

does not take on the value 0 on the *open* interval $(-\frac{1}{2}\pi, \frac{1}{2}\pi)$ (recall that $\cos x > 0$ for $-\frac{1}{2}\pi < x < \frac{1}{2}\pi$), the inverse sine function is differentiable on the open interval $(-1,1)$.† We can find the derivative as follows:

$$y = \sin^{-1}x,$$

$$\sin y = x,$$

$$\cos y\frac{dy}{dx} = 1,$$

$$\frac{dy}{dx} = \frac{1}{\cos y} = \frac{1}{\sqrt{1-x^2}} \quad \text{(see the figure).}$$

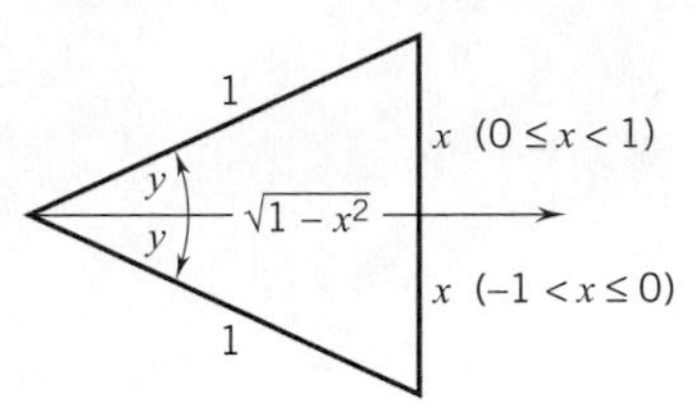

In short

(7.7.3) $$\frac{d}{dx}(\sin^{-1}x) = \frac{1}{\sqrt{1-x^2}}.$$

If u is a differentiable function of x with $|u| < 1$, then, by the chain rule,

$$\frac{d}{dx}(\sin^{-1}u) = \frac{1}{\sqrt{1-u^2}} \cdot \frac{du}{dx}.$$

Example 2

$$\frac{d}{dx}[\sin^{-1}(3x^2)] = \frac{1}{\sqrt{1-(3x^2)^2}} \cdot \frac{d}{dx}(3x^2) = \frac{6x}{\sqrt{1-9x^4}}.$$

NOTE: Although we have not made a point of it, we continue with the convention that if the domain of a function f is not specified explicitly, then it is understood to be the maximal set of real numbers x for which $f(x)$ is a real number. In this case, the domain is the set of all real numbers x such that $-1 \leq 3x^2 \leq 1$, which is the interval $[-1/\sqrt{3}, 1/\sqrt{3}]$. □

Example 3 Show that for $a > 0$

(7.7.4) $$\int \frac{dx}{\sqrt{a^2-x^2}} = \sin^{-1}\left(\frac{x}{a}\right) + C.$$

SOLUTION We change variables so that the a^2 becomes 1 and we can use (7.7.3). We set

$$au = x, \qquad a\,du = dx.$$

Then

$$\int \frac{dx}{\sqrt{a^2-x^2}} = \int \frac{a\,du}{\sqrt{a^2-a^2u^2}} = \int \frac{a\,du}{a\sqrt{1-u^2}} = \int \frac{du}{\sqrt{1-u^2}}$$

(since $a > 0$)

$$= \sin^{-1}u + C = \sin^{-1}\left(\frac{x}{a}\right) + C. \quad \square$$

† See Section 7.1.

Example 4 Evaluate $\displaystyle\int_0^{\sqrt{3}} \frac{dx}{\sqrt{4-x^2}}$.

SOLUTION By (7.7.4)

$$\int \frac{dx}{\sqrt{4-x^2}} = \sin^{-1}\left(\frac{x}{2}\right) + C.$$

It follows that

$$\int_0^{\sqrt{3}} \frac{dx}{\sqrt{4-x^2}} = \left[\sin^{-1}\left(\frac{x}{2}\right)\right]_0^{\sqrt{3}} = \sin^{-1}\left(\frac{\sqrt{3}}{2}\right) - \sin^{-1}0 = \frac{\pi}{3} - 0 = \frac{\pi}{3}. \quad \square$$

The Inverse Tangent

Although not one-to-one on its full domain, the tangent function is one-to-one on the open interval $(-\frac{1}{2}\pi, \frac{1}{2}\pi)$, and on that interval it takes on as a value every real number (see Figure 7.7.4). Thus, for any real number x, there is one and only one number in the open interval $(-\frac{1}{2}\pi, \frac{1}{2}\pi)$ at which the tangent function has the value x. This number is called the *inverse tangent of* x, or *the angle whose tangent is* x, and is written $\tan^{-1}x$ or $\arctan x$.

The *inverse tangent function*

$$y = \tan^{-1}x, \quad \text{domain: } (-\infty, \infty), \quad \text{range: } (-\tfrac{1}{2}\pi, \tfrac{1}{2}\pi)$$

is the inverse of the function

$$y = \tan x, \quad \text{domain: } (-\tfrac{1}{2}\pi, \tfrac{1}{2}\pi), \quad \text{range: } (-\infty, \infty).$$

The graphs of these two functions are given in Figure 7.7.4. Each curve is a reflection

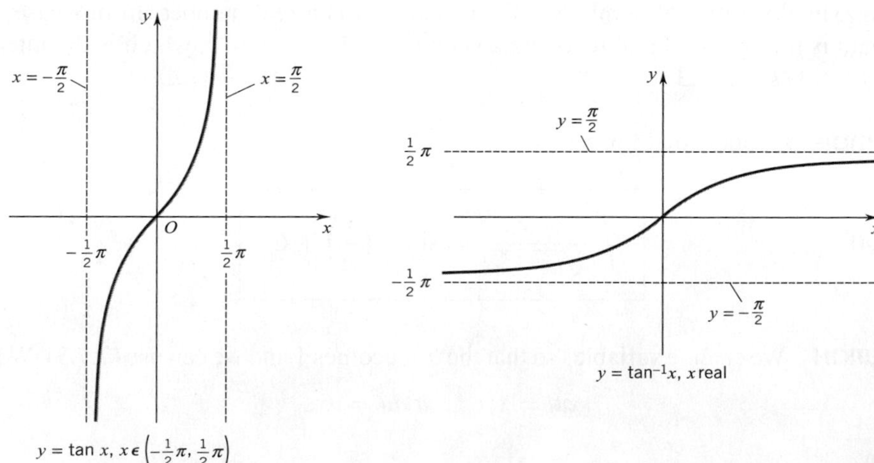

Figure 7.7.4

of the other in the line $y = x$. While the tangent has vertical asymptotes, the inverse tangent has horizontal asymptotes. Both functions are odd functions.

Because these functions are inverses,

(7.7.5) $\quad$ for all real numbers x, $\quad \tan(\tan^{-1}x) = x$

and

(7.7.6) $$\text{for all } x \in (-\tfrac{1}{2}\pi, \tfrac{1}{2}\pi), \quad \tan^{-1}(\tan x) = x.$$

It is hard to make a mistake with the first relation since it applies for all real numbers. The second relation requires the usual care:

$$\tan^{-1}(\tan \tfrac{1}{4}\pi) = \tfrac{1}{4}\pi \quad \text{but} \quad \tan^{-1}(\tan \tfrac{7}{5}\pi) \neq \tfrac{7}{5}\pi.$$

We can calculate $\tan^{-1}(\tan\frac{7}{5}\pi)$ as follows: $\frac{7}{5}\pi = \frac{2}{5}\pi + \pi$ and $\tan(\frac{2}{5}\pi + \pi) = \tan(\frac{2}{5}\pi)$ (recall that the tangent function is periodic with period π). The relation $\tan^{-1}(\tan \frac{2}{5}\pi) = \frac{2}{5}\pi$ is valid because $\frac{2}{5}\pi$ is within the interval $(-\frac{1}{2}\pi, \frac{1}{2}\pi)$.

If $x > 0$, then $\tan^{-1}x$ is the radian measure of the acute angle that has tangent x. We can construct an angle of radian measure $\tan^{-1}x$ by drawing a right triangle with legs of length x and 1 (Figure 7.7.5). Reading from the triangle, we have

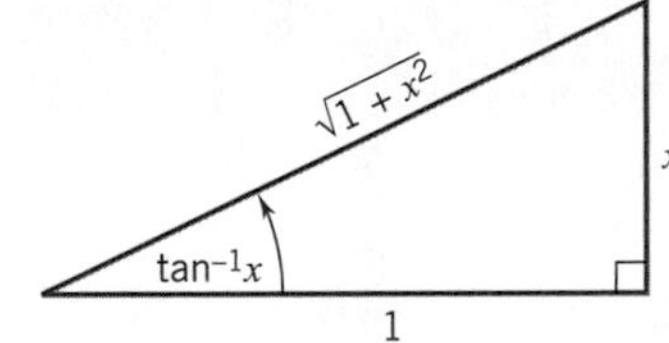

Figure 7.7.5

$$\tan(\tan^{-1}x) = x \qquad \cot(\tan^{-1}x) = \frac{1}{x}$$

$$\sin(\tan^{-1}x) = \frac{x}{\sqrt{1+x^2}} \qquad \cos(\tan^{-1}x) = \frac{1}{\sqrt{1+x^2}}$$

$$\sec(\tan^{-1}x) = \sqrt{1+x^2} \qquad \csc(\tan^{-1}x) = \frac{\sqrt{1+x^2}}{x}.$$

Since the derivative of the tangent function,

$$\frac{d}{dx}(\tan x) = \sec^2 x = \frac{1}{\cos^2 x},$$

is never 0 on $(-\frac{1}{2}\pi, \frac{1}{2}\pi)$, the inverse tangent function is everywhere differentiable (Section 7.1). We can find the derivative as we did for the inverse sine:

$$y = \tan^{-1} x,$$

$$\tan y = x,$$

$$\sec^2 y \frac{dy}{dx} = 1,$$

$$\frac{dy}{dx} = \frac{1}{\sec^2 y} = \cos^2 y = \frac{1}{1+x^2} \qquad \text{(see the figure).}$$

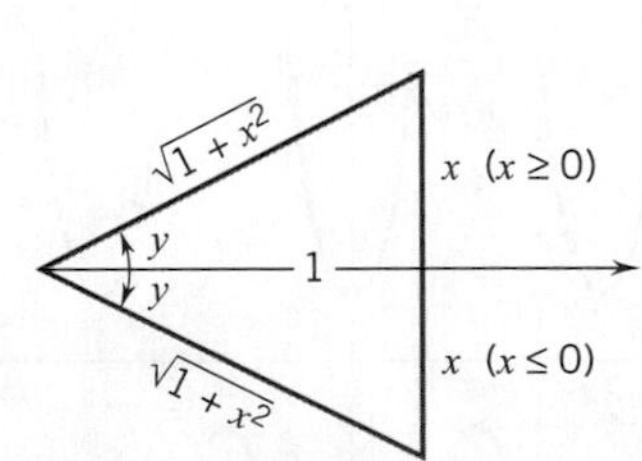

We have found that

(7.7.7) $$\frac{d}{dx}(\tan^{-1}x) = \frac{1}{1+x^2}.$$ (See Example 3, Section 5.3)

If u is a differentiable function of x, then the chain rule gives

$$\frac{d}{dx}(\tan^{-1}u) = \frac{1}{1+u^2} \cdot \frac{du}{dx}.$$

Example 5

$$\frac{d}{dx}\left[\tan^{-1}(ax^2+bx+c)\right] = \frac{1}{1+(ax^2+bx+c)^2}\cdot\frac{d}{dx}(ax^2+bx+c)$$
$$= \frac{2ax+b}{1+(ax^2+bx+c)^2}. \quad \square$$

Example 6 Show that, for $a \neq 0$,

(7.7.8)
$$\int \frac{dx}{a^2+x^2} = \frac{1}{a}\tan^{-1}\left(\frac{x}{a}\right)+C.$$

SOLUTION We change variables so that a^2 is replaced by 1 and we can use (7.7.7). We set

$$au = x, \qquad a\,du = dx.$$

Then
$$\int \frac{dx}{a^2+x^2} = \int \frac{a\,du}{a^2+a^2u^2} = \frac{1}{a}\int \frac{du}{1+u^2}$$
$$= \frac{1}{a}\tan^{-1}u + C = \frac{1}{a}\tan^{-1}\left(\frac{x}{a}\right)+C. \quad \square$$

(7.7.7) ↑

Example 7 Evaluate $\displaystyle\int_0^2 \frac{dx}{4+x^2}$.

SOLUTION By (7.7.8)

$$\int \frac{dx}{4+x^2} = \int \frac{dx}{2^2+x^2} = \frac{1}{2}\tan^{-1}\left(\frac{x}{2}\right)+C,$$

so that

$$\int_0^2 \frac{dx}{4+x^2} = \left[\frac{1}{2}\tan^{-1}\left(\frac{x}{2}\right)\right]_0^2 = \frac{1}{2}\tan^{-1}(1) - \frac{1}{2}\tan^{-1}(0) = \frac{\pi}{8}. \quad \square$$

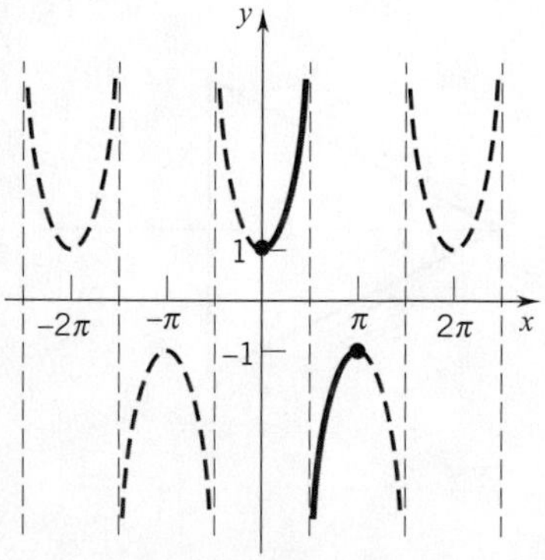

Figure 7.7.6

Inverse Secant

The graph of $y = \sec x$ is shown in Figure 7.7.6. Note that $|\sec x| \geq 1$ for all x in the domain. If we restrict the domain to the set $[0, \frac{1}{2}\pi) \cup (\frac{1}{2}\pi, \pi]$† (the solid portion of the graph in Figure 7.7.6), then the secant function is one-to-one, and for each number x such that $|x| \geq 1$ there is one and only one number in $[0, \frac{1}{2}\pi) \cup (\frac{1}{2}\pi, \pi]$ at which the secant function has the value x. This number is called the *inverse secant of x,* or *the angle whose secant is x.*

The *inverse secant function*

$$y = \sec^{-1}x, \quad \text{domain: } (-\infty, -1] \cup [1, \infty), \quad \text{range: } [0, \tfrac{1}{2}\pi) \cup (\tfrac{1}{2}\pi, \pi]$$

† Some authors use the set $[0, \frac{1}{2}\pi) \cup (\pi, \frac{3}{2}\pi]$ for the restricted domain of $y = \sec x$. There is no agreed-upon convention in this matter. Some consequences of this choice are considered in the Exercises.

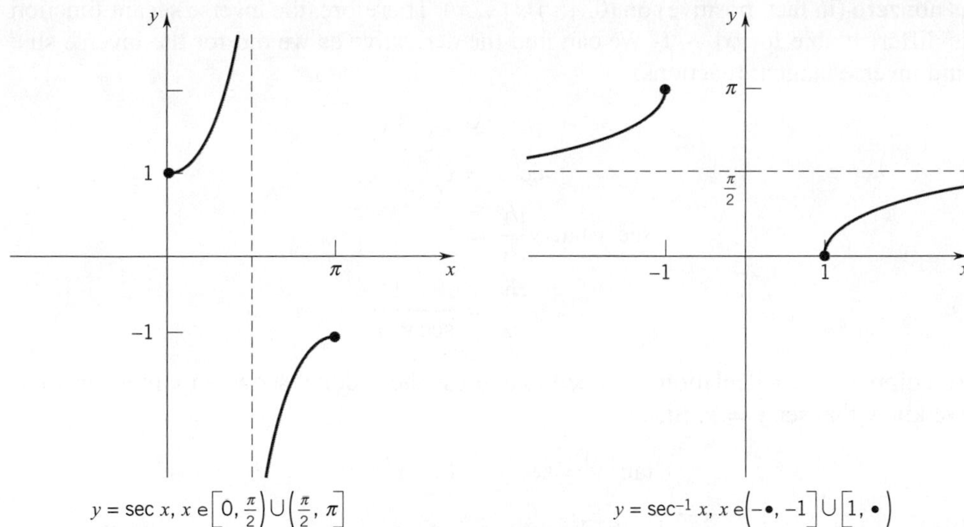

Figure 7.7.7

is the inverse of

$$y = \sec x, \quad \text{domain: } [0, \tfrac{1}{2}\pi) \cup (\tfrac{1}{2}\pi, \pi], \quad \text{range: } (-\infty, -1] \cup [1, \infty).$$

The graphs of these two functions are given in Figure 7.7.7. As the graphs indicate, each is an increasing function.

The secant function and inverse secant function are related as follows:

$$\text{for all } x \text{ such that } |x| \geq 1, \qquad \sec(\sec^{-1} x) = x,$$

$$\text{for all } x \in [0, \tfrac{1}{2}\pi) \cup (\tfrac{1}{2}\pi, \pi], \quad \sec^{-1}(\sec x) = x.$$

As you would expect, the second relation requires careful attention:

$$\sec^{-1}(\sec \tfrac{1}{3}\pi) = \tfrac{1}{3}\pi \quad \text{but} \quad \sec^{-1}(\sec \tfrac{7}{4}\pi) \neq \tfrac{7}{4}\pi.$$

We calculate $\sec^{-1}(\sec \frac{7}{4}\pi)$ as follows:

$$\sec^{-1}(\sec \tfrac{7}{4}\pi) = \sec^{-1}(\sec[2\pi - \tfrac{1}{4}\pi]) = \sec^{-1}(\sec \tfrac{1}{4}\pi) = \tfrac{1}{4}\pi.$$

If $x > 1$, then $\sec^{-1} x$ is the radian measure of the acute angle that has secant x. We can construct an angle of radian measure $\sec^{-1} x$ by drawing a right triangle with hypotenuse of length x and a side of length 1 (Figure 7.7.8). The values

$$\sec(\sec^{-1} x) = x \qquad \csc(\sec^{-1} x) = \frac{x}{\sqrt{x^2 - 1}}$$

$$\sin(\sec^{-1} x) = \frac{\sqrt{x^2 - 1}}{x} \qquad \cos(\sec^{-1} x) = \frac{1}{x}$$

$$\tan(\sec^{-1} x) = \sqrt{x^2 - 1} \qquad \cot(\sec^{-1} x) = \frac{1}{\sqrt{x^2 - 1}}$$

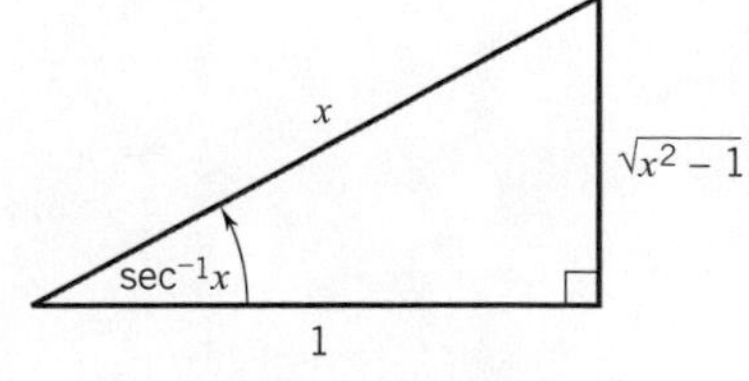

Figure 7.7.8

can all be read from the triangle.

The derivative of the secant function,

$$\frac{d}{dx}(\sec x) = \sec x \tan x,$$

is nonzero (in fact, positive) on $(0, \frac{1}{2}\pi) \cup (\frac{1}{2}, \pi)$. Therefore, the inverse secant function is differentiable for $|x| > 1$. We can find the derivative as we did for the inverse sine and inverse tangent functions:

$$\begin{aligned} y &= \sec^{-1} x \\ \sec\, y &= x \\ \sec\, y \tan\, y \frac{dy}{dx} &= 1 \\ \frac{dy}{dx} &= \frac{1}{\sec y\, \tan y}. \end{aligned}$$

To complete the calculation, we need to express the product $\sec y \tan y$ in terms of x. We know that $\sec y = x$. Since

$$\tan^2 y = \sec^2 y - 1 = x^2 - 1,$$

We have $\tan y = \pm\sqrt{x^2 - 1}$, which, with $\sec y = x$, gives

$$\sec y\, \tan y = \pm x\sqrt{x^2 - 1}.$$

If $0 < y < \frac{1}{2}\pi$, then $\sec y$ and $\tan y$ are both positive; if $\frac{1}{2}\pi < y < \pi$, then $\sec y$ and $\tan y$ are both negative. In each case, the product $\sec y \tan y$ is *positive*. It follows that $\sec y \tan y = |x|\sqrt{x^2 - 1}$ and therefore

(7.7.9)

$$\frac{d}{dx}\left(\sec^{-1} x\right) = \frac{1}{|x|\sqrt{x^2 - 1}}.$$

Note that the derivative is positive, confirming the fact that the inverse secant function is increasing.

If u is a differentiable function of x, with $|u| > 1$, then, by the chain rule,

$$\frac{d}{dx}\left(\sec^{-1} u\right) = \frac{1}{|u|\sqrt{u^2 - 1}} \cdot \frac{du}{dx}.$$

Example 8

$$\begin{aligned} \frac{d}{dx}\left[\sec^{-1}(2x^3)\right] &= \frac{1}{|2x^3|\sqrt{(2x^3)^2 - 1}} \cdot \frac{d}{dx}(2x^3) \\ &= \frac{6x^2}{|2x^3|\sqrt{4x^6 - 1}} = \frac{6x^2}{2x^2|x|\sqrt{4x^6 - 1}} = \frac{3}{|x|\sqrt{4x^6 - 1}}. \quad \square \end{aligned}$$

Example 9 Show that

$$\int \frac{dx}{x\sqrt{x^2 - 1}} = \sec^{-1}|x| + C.$$

SOLUTION If $x > 1$, $\sec^{-1}|x| = \sec^{-1} x$ and

$$\frac{d}{dx}\left(\sec^{-1} x\right) = \frac{1}{|x|\sqrt{x^2 - 1}} = \frac{1}{x\sqrt{x^2 - 1}}.$$

If $x < -1$, $\sec^{-1}|x| = \sec^{-1}(-x)$ and

$$\frac{d}{dx}[\sec^{-1}(-x)] = \frac{1}{|-x|\sqrt{x^2-1}}(-1) = \frac{-1}{-x\sqrt{x^2-1}} = \frac{1}{x\sqrt{x^2-1}}.$$

Thus, it follows that $f(x) = \sec^{-1}|x|$ is an antiderivative for $f(x) = \dfrac{1}{x\sqrt{x^2-1}}$. □

We leave it as an exercise to show that if $a > 0$ is a constant, then

(7.7.10)
$$\int \frac{1}{x\sqrt{x^2-a^2}}\,dx = \frac{1}{a}\sec^{-1}\left(\frac{|x|}{a}\right) + C.$$

The Other Trigonometric Inverses

There are three other trigonometric inverses:

the *inverse cosine,* $y = \cos^{-1}x$, is the inverse of $y = \cos x, x \in [0, \pi]$;
the *inverse cotangent,* $y = \cot^{-1}x$, is the inverse of $y = \cot x, x \in (0, \pi)$;
the *inverse cosecant,* $y = \csc^{-1}x$, is the inverse of $y = \csc x, x \in [-\frac{1}{2}\pi, 0] \cup (0, \frac{1}{2}\pi]$.

Figure 7.7.9 illustrates each of these inverses between 0 and $\frac{1}{2}\pi$ in terms of right triangles.

The differentiation formulas for these functions are as follows:

$$\frac{d}{dx}(\cos^{-1}x) = \frac{-1}{\sqrt{1-x^2}} = -\frac{d}{dx}(\sin^{-1}x)$$
$$\frac{d}{dx}(\cot^{-1}x) = \frac{-1}{1+x^2} = -\frac{d}{dx}(\tan^{-1}x)$$
$$\frac{d}{dx}(\csc^{-1}x) = \frac{-1}{|x|\sqrt{x^2-1}} = -\frac{d}{dx}(\sec^{-1}x).$$

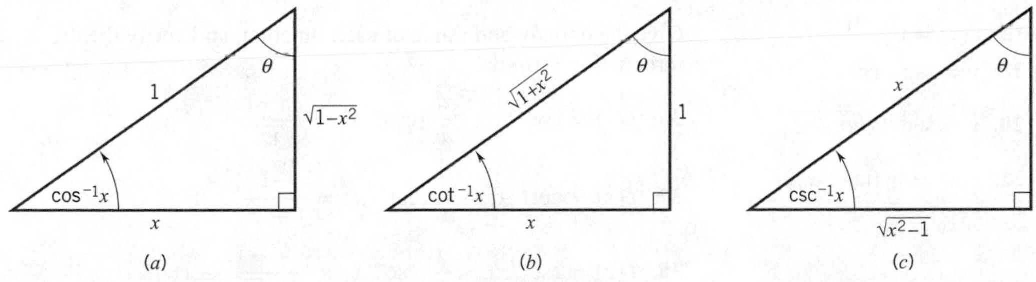

Figure 7.7.9

You are asked to verify these formulas in the Exercises. Since the derivatives of these functions differ from the derivatives of their corresponding cofunctions by a constant factor, these last three trigonometric inverses are not needed for finding antiderivatives.

Remark In Figure 7.7.9*a*, we let θ denote the other acute angle of the right triangle. Since the two acute angles in a right triangle are complementary, we have

$$\theta + \cos^{-1}x = \frac{\pi}{2}.$$

But note that $\sin\theta = x$, which means that $\theta = \sin^{-1}x$. Therefore,

$$\sin^{-1}x + \cos^{-1}x = \frac{\pi}{2} \qquad \text{or} \qquad \cos^{-1}x = \frac{\pi}{2} - \sin^{-1}x.$$

You can verify that the corresponding relations hold for the pairs $\{\tan x, \cot x\}$ and $\{\sec x, \csc x\}$:

$$\cot^{-1}x = \frac{\pi}{2} - \tan^{-1}x \quad \text{and} \quad \csc^{-1}x = \frac{\pi}{2} - \sec^{-1}x.$$

This is another reason why it is sufficient to restrict our attention to $\sin^{-1}x$, $\tan^{-1}x$, and $\sec^{-1}x$. □

EXERCISES 7.7

Determine the exact value.

1. (a) $\tan^{-1} 0$; (b) $\sin^{-1}(-\sqrt{3}/2)$.

2. (a) arc sec 2; (b) $\tan^{-1}(\sqrt{3})$.

3. (a) $\cos^{-1}\left(-\frac{1}{2}\right)$; (b) $\sec^{-1}(-\sqrt{2})$.

4. (a) $\sec\,(\sec^{-1}[-2/\sqrt{3}])$; (b) $\sec\,(\arccos[-\frac{1}{2}])$.

5. (a) $\cos\,(\sec^{-1}2)$; (b) $\arctan\,(\sec 0)$.

6. (a) $\arcsin\,(\sin[11\pi/6])$; (b) $\arctan\,(\tan[11\pi/4])$.

7. (a) $\cos^{-1}(\sec[7\pi/6])$; (b) $\sec^{-1}(\sin[13\pi/6])$.

8. (a) $\cos\left(\sin^{-1}[\frac{3}{5}]\right)$; (b) $\sec\left(\tan^{-1}[\frac{4}{3}]\right)$.

9. (a) $\sin\left(2\cos^{-1}[\frac{1}{2}]\right)$; (b) $\cos\left(2\sin^{-1}[\frac{4}{5}]\right)$.

10. (a) $\cos\left(\sin^{-1}[\frac{1}{2}] + \sin^{-1}[-1]\right)$;
(b) $\tan\left(\sin^{-1}[\sqrt{2}/2] + \cos^{-1}[\frac{1}{2}]\right)$.

Differentiate the function.

11. $y = \tan^{-1}(x+1)$.

12. $y = \tan^{-1}\sqrt{x}$.

13. $f(x) = \sec^{-1}(2x^2)$.

14. $f(x) = e^x \sin^{-1}x$.

15. $f(x) = x\sin^{-1}2x$.

16. $f(x) = e^{\tan^{-1}x}$.

17. $u = (\sin^{-1}x)^2$.

18. $v = \tan^{-1}(e^x)$.

19. $y = \dfrac{\tan^{-1}x}{x}$.

20. $y = \sec^{-1}\sqrt{x^2+2}$.

21. $f(x) = \sqrt{\tan^{-1}2x}$.

22. $f(x) = \ln(\tan^{-1}x)$.

23. $y = \tan^{-1}(\ln x)$.

24. $g(x) = \sec^{-1}(\cos x + 2)$.

25. $\theta = \sin^{-1}(\sqrt{1-r^2})$.

26. $\theta = \sin^{-1}\left(\dfrac{r}{r+1}\right)$.

27. $g(x) = x^2\sec^{-1}\left(\dfrac{1}{x}\right)$.

28. $\theta = \tan^{-1}\left(\dfrac{1}{1+r^2}\right)$.

29. $y = \sin[\sec^{-1}(\ln x)]$.

30. $f(x) = e^{\sec^{-1}x}$.

31. $f(x) = \sqrt{c^2-x^2} + c\sin^{-1}\left(\dfrac{x}{c}\right), c > 0$.

32. $y = \dfrac{x}{\sqrt{c^2-x^2}} - \sin^{-1}\left(\dfrac{x}{c}\right), c > 0$,

33. Show that for $a > 0$

$$\textbf{(7.7.11)} \qquad \int \frac{dx}{\sqrt{a^2-(x+b)^2}} = \sin^{-1}\left(\frac{x+b}{a}\right) + C.$$

34. Show that for $a \neq 0$

$$\textbf{(7.7.12)} \qquad \int \frac{dx}{a^2+(x+b)^2} = \frac{1}{a}\tan^{-1}\left(\frac{x+b}{a}\right) + C.$$

35. (a) Verify (7.7.10).
(b) Show that for $a > 0$

$$\int \frac{dx}{(x+b)\sqrt{(x+b)^2-a^2}} = \frac{1}{a}\sec^{-1}\left(\frac{|x+b|}{a}\right) + C.$$

Give the domain and range of each function, and verify the differentiation formula.

36. $f(x) = \cos^{-1}x$; $\quad \dfrac{d}{dx}(\cos^{-1}x) = \dfrac{-1}{\sqrt{1-x^2}}$.

37. $f(x) = \cot^{-1}x$; $\quad \dfrac{d}{dx}(\cot^{-1}x) = \dfrac{-1}{1+x^2}$.

38. $f(x) = \csc^{-1}x$; $\quad \dfrac{d}{dx}(\csc^{-1}x) = \dfrac{-1}{|x|\sqrt{x^2-1}}$.

Evaluate the integral.

39. $\displaystyle\int_0^1 \frac{dx}{1+x^2}$.

40. $\displaystyle\int_{-1}^1 \frac{dx}{1+x^2}$.

41. $\displaystyle\int_0^{1/\sqrt{2}} \frac{dx}{\sqrt{1-x^2}}$.

42. $\displaystyle\int_0^1 \frac{dx}{\sqrt{4-x^2}}$.

43. $\displaystyle\int_0^5 \frac{dx}{25+x^2}$.

44. $\displaystyle\int_5^8 \frac{dx}{x\sqrt{x^2-16}}$.

45. $\displaystyle\int_0^{3/2} \frac{dx}{9+4x^2}$. **46.** $\displaystyle\int_2^{5} \frac{dx}{9+(x-2)^2}$.

47. $\displaystyle\int_{3/2}^{3} \frac{dx}{x\sqrt{16x^2-9}}$.

48. $\displaystyle\int_4^{6} \frac{dx}{(x-3)\sqrt{x^2-6x+8}}$.

49. $\displaystyle\int_{-3}^{-2} \frac{dx}{\sqrt{4-(x+3)^2}}$. **50.** $\displaystyle\int_{\ln 2}^{\ln 3} \frac{e^{-x}}{\sqrt{1-e^{-2x}}}\,dx$.

51. $\displaystyle\int_0^{\ln 2} \frac{e^x}{1+e^{2x}}\,dx$. **52.** $\displaystyle\int_0^{1/2} \frac{1}{\sqrt{3-4x^2}}\,dx$.

Calculate the indefinite integral.

53. $\displaystyle\int \frac{x}{\sqrt{1-x^4}}\,dx$. **54.** $\displaystyle\int \frac{\sec^2 x}{\sqrt{9-\tan^2 x}}\,dx$.

55. $\displaystyle\int \frac{x}{1+x^4}\,dx$. **56.** $\displaystyle\int \frac{dx}{\sqrt{4x-x^2}}$.

57. $\displaystyle\int \frac{\sec^2 x}{9+\tan^2 x}\,dx$. **58.** $\displaystyle\int \frac{\cos x}{3+\sin^2 x}\,dx$.

59. $\displaystyle\int \frac{\sin^{-1} x}{\sqrt{1-x^2}}\,dx$. **60.** $\displaystyle\int \frac{\tan^{-1} x}{1+x^2}\,dx$.

61. $\displaystyle\int \frac{dx}{x\sqrt{1-(\ln x)^2}}$. **62.** $\displaystyle\int \frac{dx}{x[1+(\ln x)^2]}$.

63. Explain why the formula $\displaystyle\int_0^3 \frac{1}{\sqrt{1-x^2}}\,dx = \sin^{-1} 3$ is NOT correct.

64. Explain why the formula $\displaystyle\int_{-1}^{1} \frac{1}{x\sqrt{x^2-1}}\,dx = \sec^{-1}(1) - \sec^{-1}(-1) = -\pi$ is NOT correct.

65. Find the area of the region bounded by the graph of $y = 1/\sqrt{4-x^2}$ and the x-axis between $x=-1$ and $x=1$.

66. Find the area of the region bounded by the graph of $y = 3/(9+x^2)$ and the x-axis between $x=-3$ and $x=3$.

67. Sketch the region bounded by the graphs of $4y = x^2$ and $y = 8/(x^2+4)$ and find the area.

68. The region bounded by the graph of $y = 1/\sqrt{4+x^2}$ between $x=0$ and $x=2$ is revolved about the x-axis. Find the volume of the resulting solid.

69. The region in Exercise 68 is revolved about the y-axis. Find the volume of the resulting solid.

70. The region bounded by the graph of $y = 1/(x^2\sqrt{x^2-9})$ and the x-axis between $x = 2\sqrt{3}$ and $x=6$ is revolved about the y-axis. Find the volume of the resulting solid.

71. A billboard k feet wide is perpendicular to a straight road and is s feet from the road. At what point on the road would a motorist have the best view of the billboard; that is, at what point on the road is the angle θ subtended by the billboard a maximum (see the figure)?

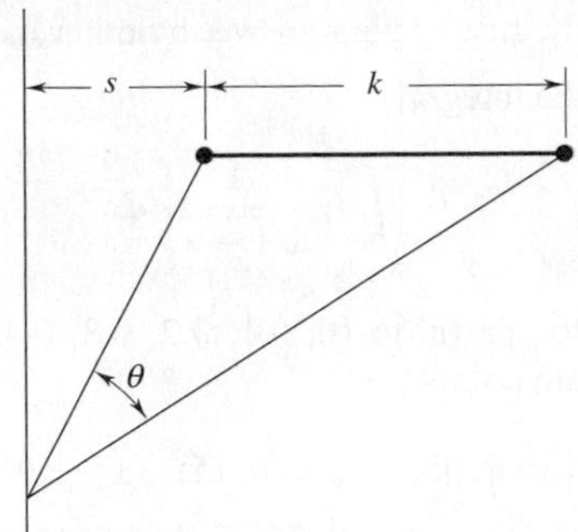

72. A person walking along a straight path at the rate of 6 feet per second is followed by a spotlight that is located 30 feet from the path. How fast is the spotlight turning at the instant the person is 50 feet past the point on the path that is closest to the spotlight?

73. (a) Show that

$$F(x) = \frac{x}{2}\sqrt{a^2-x^2} + \frac{a^2}{2}\sin^{-1}\left(\frac{x}{a}\right), \quad a>0$$

is an antiderivative of $f(x) = \sqrt{a^2-x^2}$.

(b) Use the result in part (a) to calculate $\displaystyle\int_{-a}^{a}\sqrt{a^2-x^2}\,dx$ and interpret your result as an area.

74. Let

$$f(x) = \tan^{-1}\left(\frac{a+x}{1-ax}\right), \quad x \neq 1/a.$$

(a) Show that $f'(x) = \dfrac{1}{1+x^2}$, $x \neq 1/a$.

(b) Show that there is no constant C such that $f(x) = \tan^{-1}x + C$ for all $x \neq 1/a$.

(c) Find constants C_1 and C_2 such that

$$f(x) = \tan^{-1}x + C_1 \quad \text{for } x < 1/a$$
$$f(x) = \tan^{-1}x + C_2 \quad \text{for } x > 1/a.$$

75. Let $f(x) = \tan^{-1}(1/x)$, $x \neq 0$.

(a) Use a graphing utility to graph f and $g(x) = \tan^{-1}x$. What do you notice about these two graphs?

(b) Use your graph of f to estimate the limits

$$\lim_{x\to 0^+} f(x) \quad \text{and} \quad \lim_{x\to 0^-} f(x)$$

(c) Show that $f'(x) = \dfrac{-1}{1+x^2}$, $x \neq 0$.

(d) Show that there does not exist a constant C such that $f(x) + \tan^{-1}x = C$ for all x.

(e) Find constants C_1 and C_2 such that

$$f(x) + \tan^{-1}x = C_1 \quad \text{for } x > 0$$
$$f(x) + \tan^{-1}x = C_2 \quad \text{for } x < 0.$$

76. Evaluate

$$\lim_{x\to 0}\frac{\sin^{-1}x}{x}$$

numerically. Justify your answer by other means.

77. Estimate the integral

$$\int_0^{0.5}\frac{1}{\sqrt{1-x^2}}\,dx$$

by using the partition $\{0, 0.1, 0.2, 0.3, 0.4, 0.5\}$ and the intermediate points

$$x_1^* = 0.05,\quad x_2^* = 0.15,\quad x_3^* = 0.25,$$
$$x_4^* = 0.35,\quad x_5^* = 0.45.$$

Note that the sine of your estimate is close to 0.5. Explain the reason for this.

78. Use a graphing utility to draw the graph of $f(x) = \dfrac{1}{1+x^2}$ on $[0, 10]$.

(a) Calculate $\int_0^n f(x)\,dx$ for $n =$250, 500, 750, 1000.

(b) What are the numbers in part (a) getting "close" to?

(c) Determine the value of

$$\lim_{n\to\infty}\int_0^n \frac{1}{1+x^2}\,dx.$$

79. Use a CAS to find the exact values of the following expressions:

(a) $\tan\,(\arccos[3/5] - \arctan[12/13])$.

(b) $\cos\,(\sin^{-1}[4/5] - \cos^{-1}[4/5])$.

(c) $\sin\,(\arctan[-12/5] - \arccos\,[5/13])$.

Verify your answers analytically.

80. Use a CAS to find a one-term expression for f:

(a) $\displaystyle\int f(x)\,dx = \sqrt{x^2-9} + 3\,\tan^{-1}\left(\frac{3}{\sqrt{x^2-9}}\right) + C.$

(b) $\displaystyle\int f(x)\,dx = 2\,\arcsin\left(\frac{x}{2}\right) + \left(\frac{x^3}{4} - \frac{x}{2}\right)\sqrt{4-x^2} + C.$

(c) $\displaystyle\int f(x)\,dx = x^2\,\sec x - \sqrt{x^2-1} + C.$

(d) Verify your results by differentiating the right-hand sides.

81. Use a graphing utility to graph $f(x) = \tan^{-1}x$ and $g(x) = \cos^{-1}x$ together.

(a) Use a CAS to find the point of intersection of the graphs.

(b) Use a CAS to find the area of the region bounded by graphs of f, g, and the x-axis.

82. Repeat Exercise 81 with the functions $f(x) = \sin^{-1}x$ and $g(x) = \cos^{-1}x$.

■ PROJECT 7.7 Refraction

Dip a straight stick in a pool of water and it appears to bend. Only in a vacuum does light travel at speed c (the famous c of $E = mc^2$). Light does not travel as fast through a material medium. The *index of refraction n* of a medium relates the speed of light in that medium to c:

$$n = \frac{c}{\text{speed of light in the medium}}.$$

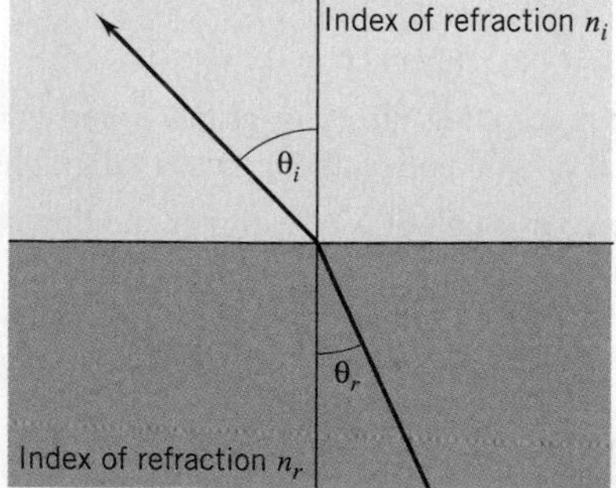

When light travels from one medium to another, it changes direction; we say that light is *refracted.* Experiment shows that the *angle of refraction* θ_r is related to the *angle of incidence* θ_i by Snell's law.

$$n_i\sin\theta_i = n_r\sin\theta_r.$$

Like the Law of Reflection (see Example 5, Section 4.5), Snell's Law of Refraction can be derived from Fermat's Principle of Least Time.

Problem 1. A light beam passes from a medium with index of refraction n_1 through a plane sheet of material whose top and bottom faces are parallel and then out into some other medium with index of refraction n_2. Show that Snell's law implies that $n_1\sin\theta_1 = n_2\sin\theta_2$ regardless of the thickness of the sheet or its index of refraction.

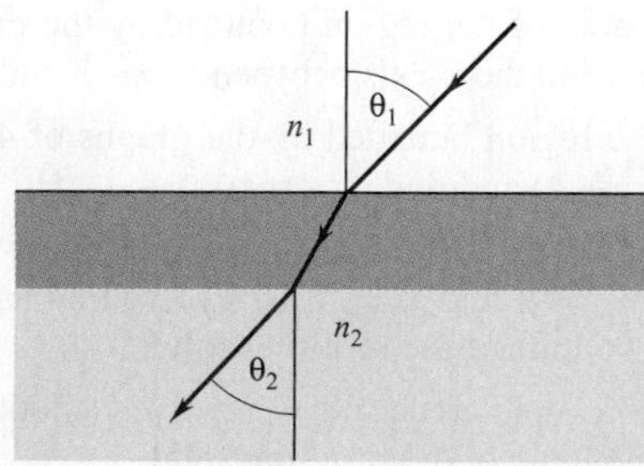

A star is not where it appears to be. Consider a beam of light traveling through an atmosphere whose index of refraction varies with height, $n = n(y)$. The light follows some curved path $y = y(x)$. Think of the atmosphere as a succession of parallel slabs. When a light ray strikes a slab at height y, it is traveling at some angle θ to the vertical; when it emerges at height $y + \Delta y$, it is traveling at a slightly different angle, $\theta + \Delta\theta$.

Problem 2.

a. Use the result in Problem 1 to show that

$$\frac{1}{n}\frac{dn}{dy} = -\cot\theta\frac{d\theta}{dy} = \frac{d^2y/dx^2}{1+(dy/dx)^2}.$$

b. Verify that the slope of the light path must vary in such a way that

$$1 + (dy/dx)^2 = (\text{constant})\,[n(y)]^2.$$

c. How must n vary with height y for the light to travel along a circular arc?

■ 7.8 THE HYPERBOLIC SINE AND COSINE

Certain combinations of the exponential functions e^x and e^{-x} occur so frequently in mathematical applications that they are given special names. The *hyperbolic sine* (sinh) and *hyperbolic cosine* (cosh) are the functions defined by

(7.8.1)
$$\sinh x = \tfrac{1}{2}(e^x - e^{-x}), \quad \cosh x = \tfrac{1}{2}(e^x + e^{-x}).$$

The reasons for these names will become apparent as we go on.

Since
$$\frac{d}{dx}(\sinh x) = \frac{d}{dx}\left[\tfrac{1}{2}(e^x - e^{-x})\right] = \tfrac{1}{2}(e^x + e^{-x})$$

and
$$\frac{d}{dx}(\cosh x) = \frac{d}{dx}[\tfrac{1}{2}(e^x + e^{-x})] = \tfrac{1}{2}(e^x - e^{-x}),$$

we have

(7.8.2)
$$\frac{d}{dx}(\sinh x) = \cosh x, \quad \frac{d}{dx}(\cosh x) = \sinh x.$$

In short, each of these functions is the derivative of the other.

The Graphs

We begin with the hyperbolic sine. Since

$$\sinh(-x) = \tfrac{1}{2}(e^{-x} - e^x) = -\tfrac{1}{2}(e^x - e^{-x}) = -\sinh x,$$

the hyperbolic sine is an odd function. The graph is therefore symmetric about the origin. Since

$$\frac{d}{dx}(\sinh x) = \cosh x = \tfrac{1}{2}(e^x + e^{-x}) > 0 \quad \text{for all real } x,$$

the hyperbolic sine increases everywhere. Since

$$\frac{d^2}{dx^2}(\sinh x) = \frac{d}{dx}(\cosh x) = \sinh x = \tfrac{1}{2}(e^x - e^{-x}),$$

you can see that

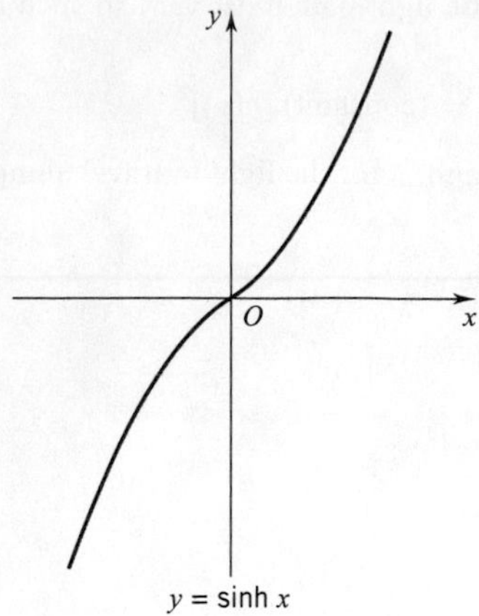

Figure 7.8.1

$$\frac{d^2}{dx^2}(\sinh x) \quad \text{is} \quad \begin{cases} \text{negative,} & \text{for} \quad x < 0 \\ 0, & \text{at} \quad x = 0 \\ \text{positive,} & \text{for} \quad x > 0. \end{cases}$$

The graph is therefore concave down on $(-\infty, 0)$ and concave up on $(0, \infty)$. The point $(0, \sinh 0) = (0, 0)$ is the only point of inflection. The slope at the origin is $\cosh 0 = 1$. A sketch of the graph appears in Figure 7.8.1.

We turn now to the hyperbolic cosine. Since

$$\cosh(-x) = \tfrac{1}{2}(e^{-x} + e^{x}) = \tfrac{1}{2}(e^{x} + e^{-x}) = \cosh x,$$

the hyperbolic cosine is an even function. The graph is therefore symmetric about the y-axis. Since

$$\frac{d}{dx}(\cosh x) = \sinh x,$$

you can see that

$$\frac{d}{dx}(\cosh x) \quad \text{is} \quad \begin{cases} \text{negative,} & \text{for} \quad x < 0 \\ 0, & \text{at} \quad x = 0 \\ \text{positive,} & \text{for} \quad x > 0. \end{cases}$$

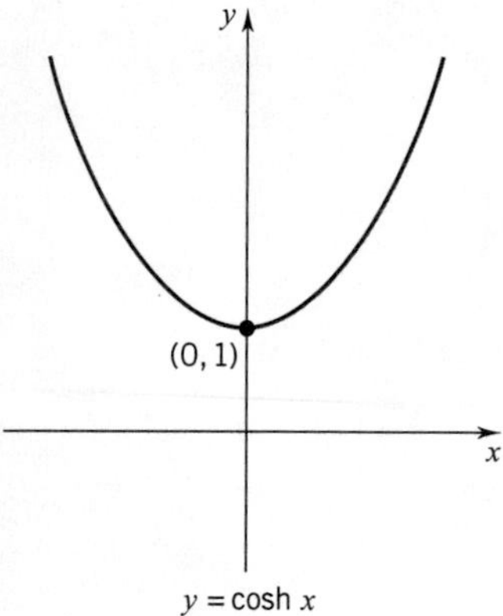

Figure 7.8.2

The function therefore decreases on $(-\infty, 0]$ and increases on $[0, \infty)$. The number

$$\cosh 0 = \tfrac{1}{2}(e^{0} + e^{-0}) = \tfrac{1}{2}(1 + 1) = 1$$

is a local and absolute minimum. There are no other extreme values. Since

$$\frac{d^2}{dx^2}(\cosh x) = \frac{d}{dx}(\sinh x) = \cosh x > 0 \quad \text{for all real } x,$$

the graph is everywhere concave up. (See Figure 7.8.2.)

Figure 7.8.3 shows the graphs of three functions

$$y = \sinh x = \tfrac{1}{2}(e^{x} - e^{-x}), \quad y = \tfrac{1}{2}e^{x}, \quad y = \cosh x = \tfrac{1}{2}(e^{x} + e^{-x}).$$

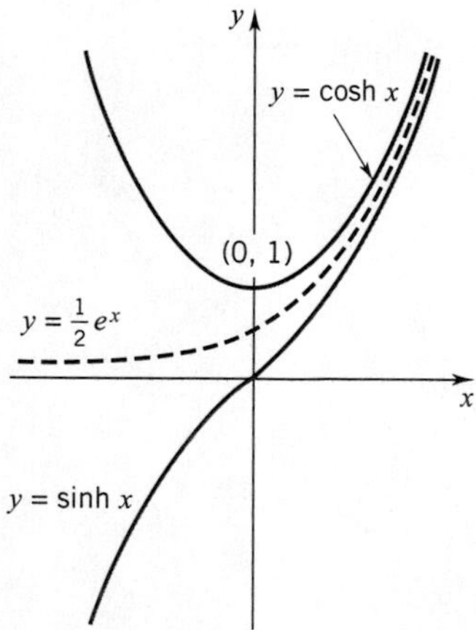

Figure 7.8.3

Since $e^{-x} > 0$, it follows that

$$\sinh x < \tfrac{1}{2}e^{x} < \cosh x \qquad \text{for all real } x.$$

Although markedly different for negative x, these functions are almost indistinguishable for large positive x. This follows from the fact that $e^{-x} \to 0$ as $x \to \infty$.

Applications

The hyperbolic functions have a variety of applications in science and engineering. Perhaps the best known application is the use of the hyperbolic cosine function to describe the shape of a flexible chain or cable that is suspended between two points. For example, think of a telephone wire or power line that sags under its own weight (Figure 7.8.4).

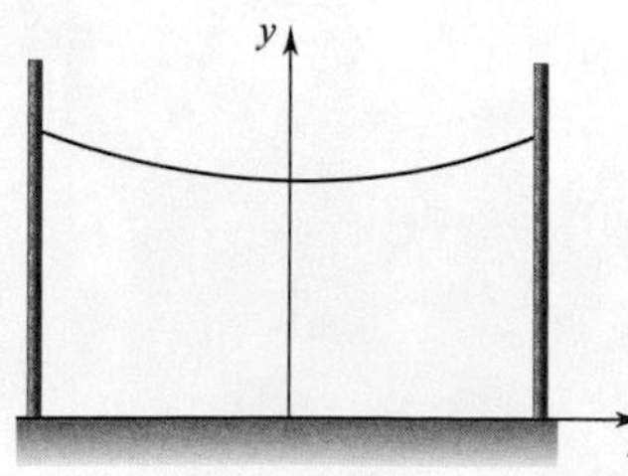

Figure 7.8.4

Suppose we have a flexible cable of uniform density that is suspended between two points of equal height. If we introduce an x, y-coordinate system so that the lowest point of the cable is on the y-axis (see Figure 7.8.5), then it can be shown that the shape of the curve $y = f(x)$ must satisfy the equation

$$\frac{d^2y}{dx^2} = \frac{1}{a}\sqrt{1 + \left(\frac{dy}{dx}\right)^2},$$

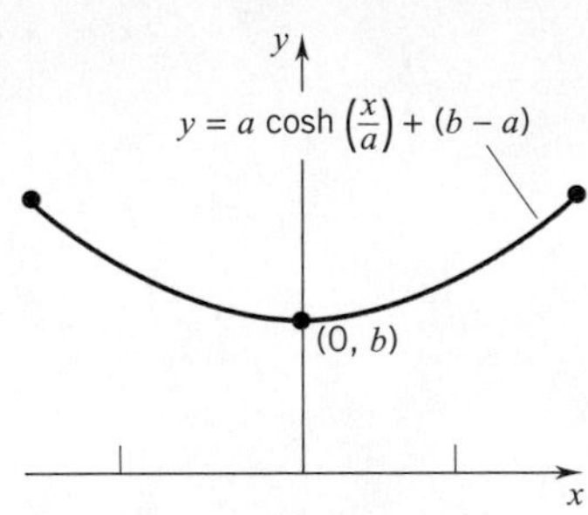

Figure 7.8.5

where a is a constant that depends on the density of the cable and on the tension (the horizontal force at the two ends). In the Exercises you are asked to show that

$$y = a \cosh\left(\frac{x}{a}\right) + C$$

is a solution of this equation. The graph of this solution is called a *catenary*.

The Gateway to the West Arch in St. Louis, Missouri, is in the shape of an inverted catenary (see Figure 7.8.6). This arch is 630 feet high at its center, and it measures 630 across the base. The value of the constant a for this arch is approximately 127.7, and its equation takes the form

$$y = -127.7 \cosh\ (x/127.7) + 757.7.$$

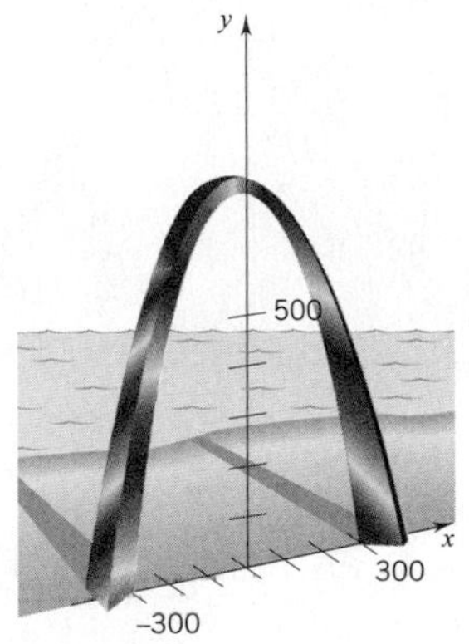

Figure 7.8.6

Identities

The hyperbolic sine and cosine functions satisfy identities similar to those satisfied by the "circular" sine and cosine.

(7.8.3)

$$\begin{aligned}
&\cosh^2 t - \sinh^2 t = 1,\\
&\sinh\ (t + s) = \sinh t \cosh s + \cosh t \sinh s,\\
&\cosh\ (t + s) = \cosh t \cosh s + \sinh t \sinh s,\\
&\sinh 2t = 2 \sinh t \cosh t,\\
&\cosh 2t = \cosh^2 t + \sinh^2 t.
\end{aligned}$$

The verification of these identities is left to you as an exercise.

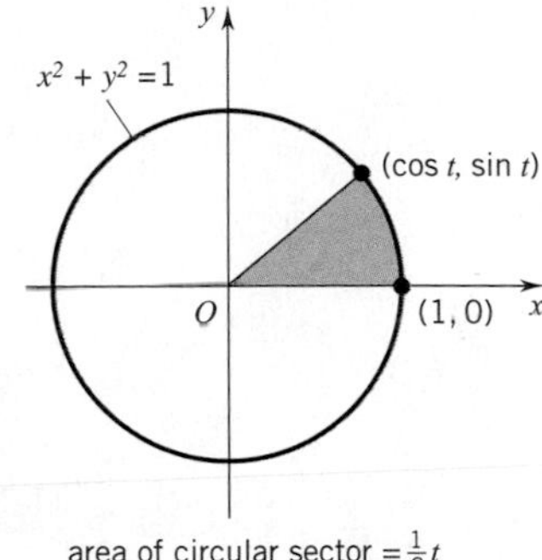

area of circular sector $= \frac{1}{2}t$

Figure 7.8.7

Relation to the Hyperbola $x^2 - y^2 = 1$

The hyperbolic sine and cosine are related to the hyperbola $x^2 - y^2 = 1$ much as the "circular" sine and cosine are related to the circle $x^2 + y^2 = 1$:

1. For each real t,

$$\cos^2 t + \sin^2 t = 1,$$

and thus the point $(\cos t, \sin t)$ lies on the circle $x^2 + y^2 = 1$. For each real t,

$$\cosh^2 t - \sinh^2 t = 1,$$

and thus the point $(\cosh t, \sinh t)$ lies on the hyperbola $x^2 - y^2 = 1$.

2. For each t in $[0, 2\pi]$ (see Figure 7.8.7), the number $\frac{1}{2}t$ gives the area of the circular sector generated by the circular arc that begins at $(1, 0)$ and ends at $(\cos t, \sin t)$. As we prove below, for each $t > 0$ (see Figure 7.8.8), the number $\frac{1}{2}t$ gives the area of

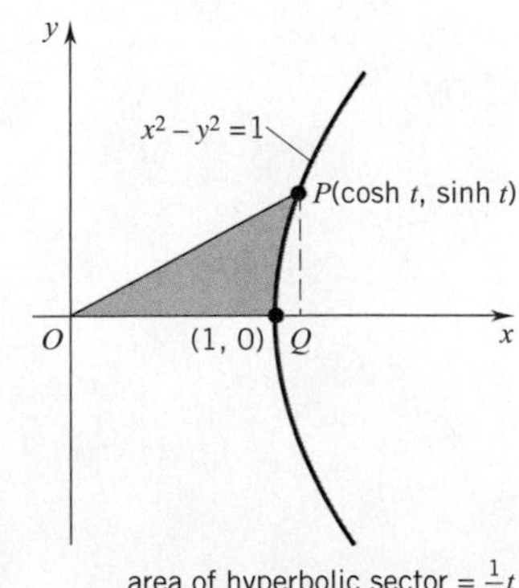

area of hyperbolic sector $= \frac{1}{2}t$

Figure 7.8.8

the hyperbolic sector generated by the hyperbolic arc that begins at (1, 0) and ends at (cosh t, sinh t).

PROOF Let $A(t)$ be the area of the hyperbolic sector. Then,

$$A(t) = \tfrac{1}{2}\cosh t \sinh t - \int_1^{\cosh t} \sqrt{x^2 - 1}\, dx.$$

The first term, $\frac{1}{2}\cosh t \sinh t$, gives the area of the triangle OPQ, and the integral

$$\int_1^{\cosh t} \sqrt{x^2 - 1}\, dx$$

gives the area of the unshaded portion of the triangle. We wish to show that

$$A(t) = \tfrac{1}{2}t \quad \text{for all } t \geq 0.$$

We will do so by showing that

$$A'(t) = \tfrac{1}{2} \quad \text{for all } t > 0 \quad \text{and} \quad A(0) = 0.$$

Differentiating $A(t)$, we have

$$A'(t) = \tfrac{1}{2}\left[\cosh t \,\frac{d}{dt}(\sinh t) + \sinh t\, \frac{d}{dt}(\cosh t)\right] - \frac{d}{dt}\left(\int_1^{\cosh t} \sqrt{x^2 - 1}\, dx\right),$$

and therefore

$$\text{(1)} \qquad A'(t) = \tfrac{1}{2}(\cosh^2 t + \sinh^2 t) - \frac{d}{dt}\left(\int_1^{\cosh t} \sqrt{x^2 - 1}\, dx\right).$$

Now we differentiate the integral:

$$\frac{d}{dt}\left(\int_1^{\cosh t} \sqrt{x^2 - 1}\, dx\right) \underset{(5.8.7)}{=} \sqrt{\cosh^2 t - 1}\,\frac{d}{dt}(\cosh t) = \sinh t \cdot \sinh t = \sinh^2 t.$$

Substituting this last expression into (1), we have

$$A'(t) = \tfrac{1}{2}(\cosh^2 t + \sinh^2 t) - \sinh^2 t = \tfrac{1}{2}(\cosh^2 t - \sinh^2 t) = \tfrac{1}{2}.$$

It is not hard to see that $A(0) = 0$:

$$A(0) = \tfrac{1}{2}\cosh 0 \sinh 0 - \int_1^{\cosh 0} \sqrt{x^2 - 1}\, dx = \tfrac{1}{2}(1)(0) - \int_1^1 \sqrt{x^2 - 1}\, dx = 0. \quad \square$$

EXERCISES 7.8

Differentiate the function.

1. $y = \sinh x^2$.
2. $y = \cosh(x + a)$.

3. $y = \sqrt{\cosh ax}$.
4. $y = (\sinh ax)(\cosh ax)$.

5. $y = \dfrac{\sinh x}{\cosh x - 1}$.
6. $y = \dfrac{\sinh x}{x}$.

7. $y = a \sinh bx - b \cosh ax$.

8. $y = e^x(\cosh x + \sinh x)$.

9. $y = \ln|\sinh ax|$.
10. $y = \ln|1 - \cosh ax|$.

11. $y = \sinh(e^{2x})$.
12. $y = \cosh(\ln x^3)$.

13. $y = e^{-x} \cosh 2x$.
14. $y = \tan^{-1}(\sinh x)$.

15. $y = \ln(\cosh x)$.
16. $y = \ln(\sinh x)$.

17. $y = (\sinh x)^x$.
18. $y = x^{\cosh x}$.

Verify the identity.

19. $\cosh^2 t - \sinh^2 t = 1$.

20. $\sinh(t + s) = \sinh t \cosh s + \cosh t \sinh s$.

21. $\cosh(t + s) = \cosh t \cosh s + \sinh t \sinh s$.

22. $\sinh 2t = 2 \sinh t \cosh t$.

23. $\cosh 2t = \cosh^2 t + \sinh^2 t = 2\cosh^2 t - 1 = 2\sinh^2 t + 1$.

24. $\cosh(-t) = \cosh t$; the hyperbolic cosine function is even.

25. $\sinh(-t) = -\sinh t$; the hyperbolic sine function is odd.

Find the absolute extreme values.

26. $y = 5\cosh x + 4\sinh x$.

27. $y = -5\cosh x + 4\sinh x$.

28. $y = 4\cosh x + 5\sinh x$.

29. Show that for each positive integer n

$$(\cosh x + \sinh x)^n = \cosh nx + \sinh nx.$$

30. Verify that $y = A \cosh cx + B \sinh cx$ satisfies the equation $y'' - c^2 y = 0$.

31. Determine A, B, and c so that $y = A \cosh cx + B \sinh cx$ satisfies the conditions $y'' - 9y = 0$, $y(0) = 2$, $y'(0) = 1$. Take $c > 0$.

32. Determine A, B, and c so that $y = A \cosh cx + B \sinh cx$ satisfies the conditions $4y'' - y = 0$, $y(0) = 1$, $y'(0) = 2$. Take $c > 0$.

Calculate the indefinite integral.

33. $\displaystyle\int \cosh ax\, dx$.
34. $\displaystyle\int \sinh ax\, dx$.

35. $\displaystyle\int \sinh^2 ax \cosh ax\, dx$.

36. $\displaystyle\int \sinh ax \cosh^2 ax\, dx$.

37. $\displaystyle\int \frac{\sinh ax}{\cosh ax}\, dx$.
38. $\displaystyle\int \frac{\cosh ax}{\sinh ax}\, dx$.

39. $\displaystyle\int \frac{\sinh ax}{\cosh^2 ax}\, dx$.
40. $\displaystyle\int \sinh^2 x\, dx$.

41. $\displaystyle\int \cosh^2 x\, dx$.
42. $\displaystyle\int \sinh 2x\, e^{\cosh 2x}\, dx$.

43. $\displaystyle\int \frac{\sinh \sqrt{x}}{\sqrt{x}}\, dx$.
44. $\displaystyle\int \frac{\sinh x}{1 + \cosh x}\, dx$.

In Exercises 45 and 46, find the average value of the function on the interval indicated.

45. $f(x) = \cosh x, \quad x \in [-1, 1]$.

46. $f(x) = \sinh 2x, \quad x \in [0, 4]$.

47. Find the area of the region bounded by the graph of $y = \sinh x$ and the x-axis from $x = 0$ to $x = \ln 10$.

48. Find the area of the region bounded by the graph of $y = \cosh 2x$ and the x-axis from $x = -\ln 5$ to $x = \ln 5$.

49. Find the volume of the solid generated by revolving the region bounded by $y = \cosh x$ and $y = \sinh x$ between $x = 0$ and $x = 1$ about the x-axis.

50. The region bounded by the graph of $y = \sinh x$ and the x-axis from $x = 0$ to $x = \ln 5$ is revolved about the x-axis. Find the volume of the resulting solid.

51. The region bounded by the graph of $y = \cosh 2x$ and the x-axis from $x = -\ln 5$ to $x = \ln 5$ is revolved about the x-axis. Find the volume of the resulting solid.

52. (a) Evaluate the limit

$$\lim_{x\to\infty} \frac{\sinh x}{e^x}.$$

(b) Evaluate the limit

$$\lim_{x\to\infty} \frac{\cosh x}{e^{ax}}$$

if $0 < a < 1$ and if $a > 1$.

C 53. Use a graphing utility to graph $f(x) = 2 - \sinh x$ and $g(x) = \cosh x$ together.

(a) Use a CAS to find the points of intersection of the two graphs.

(b) Use a CAS to find the area of the region in the first quadrant bounded by the graphs of f and g and the y-axis.

C 54. Use a graphing utility to graph $f(x) = \cosh x - 1$ and $g(x) = 1/\cosh x$ together.

(a) Use a CAS to find the points of intersection of the two graphs.

(b) Use a CAS to find the area of the region bounded by the graphs of f and g.

■ *7.9 THE OTHER HYPERBOLIC FUNCTIONS

The hyperbolic tangent is defined by setting

$$\tanh x = \frac{\sinh x}{\cosh x} = \frac{e^x - e^{-x}}{e^x + e^{-x}}.$$

There is also a *hyperbolic cotangent*, a *hyperbolic secant*, and a *hyperbolic cosecant*:

$$\coth x = \frac{\cosh x}{\sinh x}, \quad \operatorname{sech} x = \frac{1}{\cosh x}, \quad \operatorname{csch} x = \frac{1}{\sinh x}.$$

The derivatives are as follows:

(7.9.1)

$$\frac{d}{dx}(\tanh x) = \operatorname{sech}^2 x, \qquad \frac{d}{dx}(\coth x) = -\operatorname{csch}^2 x,$$
$$\frac{d}{dx}(\operatorname{sech} x) = -\operatorname{sech} x \tanh x, \qquad \frac{d}{dx}(\operatorname{csch} x) = -\operatorname{csch} x \coth x.$$

These formulas are easy to verify. For instance,

$$\frac{d}{dx}(\tanh x) = \frac{d}{dx}\left(\frac{\sinh x}{\cosh x}\right) = \frac{\cosh x \frac{d}{dx}(\sinh x) - \sinh x \frac{d}{dx}(\cosh x)}{\cosh^2 x}$$
$$= \frac{\cosh^2 x - \sinh^2 x}{\cosh^2 x} = \frac{1}{\cosh^2 x} = \operatorname{sech}^2 x.$$

We leave it to you to verify the other formulas.

Let's examine the hyperbolic tangent a little further. Since

$$\tanh(-x) = \frac{\sinh(-x)}{\cosh(-x)} = \frac{-\sinh x}{\cosh x} = -\tanh x,$$

the hyperbolic tangent is an odd function and thus the graph is symmetric about the origin. Since

$$\frac{d}{dx}(\tanh x) = \operatorname{sech}^2 x > 0 \qquad \text{for all real } x,$$

the function is everywhere increasing. From the relation

$$\tanh x = \frac{e^x - e^{-x}}{e^x + e^{-x}} = \frac{e^x - e^{-x}}{e^x + e^{-x}} \cdot \frac{e^x}{e^x} = \frac{e^{2x} - 1}{e^{2x} + 1} = \frac{e^{2x} + 1 - 2}{e^{2x} + 1} = 1 - \frac{2}{e^{2x} + 1},$$

you can see that $\tanh x$ always remains between -1 and 1. Moreover,

$$\text{as } x \to \infty, \quad \tanh x \to 1 \quad \text{and} \quad \text{as } x \to -\infty, \quad \tanh x \to -1.$$

The lines $y = 1$ and $y = -1$ are horizontal asymptotes. To check on the concavity of the graph, we take the second derivative:

$$\frac{d^2}{dx^2}(\tanh x) = \frac{d}{dx}(\operatorname{sech}^2 x) = 2 \operatorname{sech} x \frac{d}{dx}(\operatorname{sech} x)$$
$$= 2 \operatorname{sech} x(-\operatorname{sech} x \tanh x)$$
$$= -2 \operatorname{sech}^2 x \tanh x.$$

Since

$$\tanh x = \frac{e^x - e^{-x}}{e^x + e^{-x}} \quad \text{is} \quad \begin{cases} \text{negative}, & \text{for} \quad x < 0 \\ 0, & \text{at} \quad x = 0 \\ \text{positive}, & \text{for} \quad x > 0, \end{cases}$$

you can see that

$$\frac{d^2}{dx^2}(\tanh x) \quad \text{is} \quad \begin{cases} \text{positive}, & \text{for} \quad x < 0 \\ 0, & \text{at} \quad x = 0 \\ \text{negative}, & \text{for} \quad x > 0. \end{cases}$$

The graph is therefore concave up on $(-\infty, 0)$ and concave down on $(0, \infty)$. The point $(0, \tanh 0) = (0, 0)$ is a point of inflection. At the origin the slope is

$$\text{sech}^2 0 = \frac{1}{\cosh^2 0} = 1.$$

The graph is shown in Figure 7.9.1.

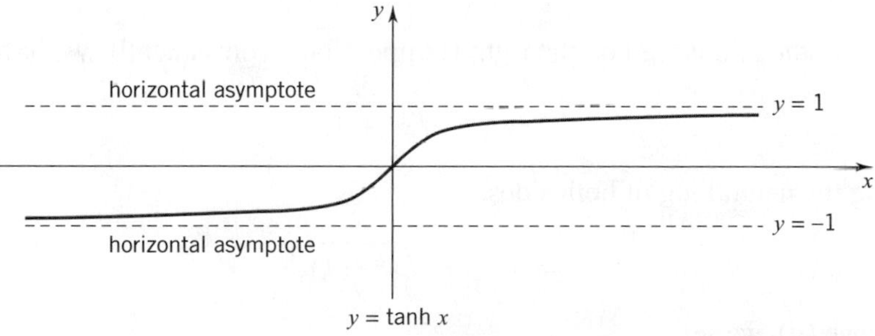

Figure 7.9.1

The Hyperbolic Inverses

Of the six hyperbolic functions, only the hyperbolic cosine and its reciprocal, hyperbolic secant, fail to be one-to-one (refer to the graphs of $y = \sinh x$, $y = \cosh x$, and $y = \tanh x$). Thus, the hyperbolic sine, hyperbolic tangent, hyperbolic cosecant, and hyperbolic cotangent functions all have inverses. If we restrict the domains of the hyperbolic cosine and hyperbolic secant functions to $x \geq 0$, then these functions will also have inverses. The hyperbolic inverses that are important to us are the *inverse hyperbolic sine*, the *inverse hyperbolic cosine*, and the *inverse hyperbolic tangent*. These functions,

$$y = \sinh^{-1} x, \quad y = \cosh^{-1} x, \quad y = \tanh^{-1} x,$$

are the inverses of

$$y = \sinh x, \quad y = \cosh x \quad (x \geq 0), \quad y = \tanh x,$$

respectively.

THEOREM 7.9.2

(i) $\sinh^{-1}x = \ln(x + \sqrt{x^2+1}),$ $\quad x$ real

(ii) $\cosh^{-1}x = \ln x + \sqrt{x^2-1},$ $\quad x \geq 1$

(iii) $\tanh^{-1}x = \frac{1}{2}\ln\left(\frac{1+x}{1-x}\right),$ $\quad -1 < x < 1.$

PROOF To prove (i), we set $y = \sinh^{-1}x$ and note that

$$\sinh y = x.$$

This gives in sequence:

$$\tfrac{1}{2}(e^y - e^{-y}) = x, \quad e^y - e^{-y} = 2x, \quad e^y - 2x - e^{-y} = 0, \quad e^{2y} - 2x\,e^y - 1 = 0.$$

This last equation is a quadratic equation in e^y. From the general quadratic formula, we find that

$$e^y = \tfrac{1}{2}(2x \pm \sqrt{4x^2+4}) = x \pm \sqrt{x^2+1}.$$

Since $e^y > 0$, the minus sign on the right is impossible. Consequently, we have

$$e^y = x + \sqrt{x^2+1},$$

and, taking the natural log of both sides,

$$y = \ln(x + \sqrt{x^2+1}).$$

To prove (ii), we set

$$y = \cosh^{-1}x, \quad x \geq 1$$

and note that

$$\cosh y = x \quad \text{and} \quad y \geq 0.$$

This gives in sequence:

$$\tfrac{1}{2}(e^y + e^{-y}) = x, \quad e^y + e^{-y} = 2x, \quad e^{2y} - 2x\,e^y + 1 = 0.$$

Again we have a quadratic in e^y. Here the general quadratic formula gives

$$e^y = \tfrac{1}{2}(2x \pm \sqrt{4x^2-4}) = x \pm \sqrt{x^2-1}.$$

Since y is nonnegative,

$$e^y = x \pm \sqrt{x^2-1}$$

cannot be less than 1. This renders the negative sign impossible (check this out) and leaves

$$e^y = x + \sqrt{x^2-1}$$

as the only possibility. Taking the natural log of both sides, we get

$$y = \ln(x + \sqrt{x^2 - 1}).$$

The proof of (iii) is left as an exercise. □

EXERCISES 7.9

Differentiate the function.

1. $y = \tanh^2 x.$
2. $y = \tanh^2 3x.$
3. $y = \ln(\tanh x).$
4. $y = \tanh(\ln x).$
5. $y = \sinh(\tan^{-1} e^{2x}).$
6. $y = \operatorname{sech}(3x^2 + 1).$
7. $y = \coth(\sqrt{x^2 + 1}).$
8. $y = \ln(\operatorname{sech} x).$
9. $y = \dfrac{\operatorname{sech} x}{1 + \cosh x}.$
10. $y = \dfrac{\cosh x}{1 + \operatorname{sech} x}.$

Verify the differentiation formula.

11. $\dfrac{d}{dx}(\coth x) = -\operatorname{csch}^2 x.$
12. $\dfrac{d}{dx}(\operatorname{sech} x) = -\operatorname{sech} x \tanh x.$
13. $\dfrac{d}{dx}(\operatorname{csch} x) = -\operatorname{csch} x \coth x.$
14. Show that

$$\tanh(t + s) = \frac{\tanh t + \tanh s}{1 + \tanh t \tanh s}.$$

15. Given that $\tanh x_0 = \frac{4}{5}$, find (a) $\operatorname{sech} x_0$. HINT: $1 - \tanh^2 x = \operatorname{sech}^2 x$. Then find (b) $\cosh x_0$, (c) $\sinh x_0$, (d) $\coth x_0$, (e) $\operatorname{csch} x_0$.
16. Given that $\tanh t_0 = -\frac{5}{12}$, evaluate the remaining hyperbolic functions at t_0.
17. Show that, if $x^2 \geq 1$, then $x - \sqrt{x^2 - 1} \leq 1$.
18. Show that

$$\tanh^{-1} x = \frac{1}{2} \ln\left(\frac{1 + x}{1 - x}\right), \quad -1 < x < 1.$$

19. Show that

$$\textbf{(7.9.3)} \qquad \frac{d}{dx}(\sinh^{-1} x) = \frac{1}{\sqrt{x^2 + 1}}, \quad x \text{ real}.$$

20. Show that

$$\textbf{(7.9.4)} \qquad \frac{d}{dx}(\cosh^{-1} x) = \frac{1}{\sqrt{x^2 - 1}}, \quad x > 1.$$

21. Show that

$$\textbf{(7.9.5)} \qquad \frac{d}{dx}(\tanh^{-1} x) = \frac{1}{1 - x^2}, \quad -1 < x < 1.$$

22. Show that

$$\frac{d}{dx}(\operatorname{sech}^{-1} x) = \frac{-1}{x\sqrt{1 - x^2}}, \quad 0 < x < 1.$$

23. Show that

$$\frac{d}{dx}(\operatorname{csch}^{-1} x) = \frac{-1}{|x|\sqrt{1 + x^2}}, \quad x \neq 0.$$

24. Show that

$$\frac{d}{dx}(\coth^{-1} x) = \frac{1}{1 - x^2}, \quad |x| > 1.$$

25. Sketch the graph of $y = \operatorname{sech} x$, giving: (a) the extreme values; (b) the points of inflection; and (c) the concavity.
26. Sketch the graphs of (a) $y = \coth x$, (b) $y = \operatorname{csch} x$.
27. Graph $y = \sinh x$ and $y = \sinh^{-1} x$ in the same coordinate system. Find all points of inflection.
28. Sketch the graphs of (a) $y = \cosh^{-1} x$, (b) $y = \tanh^{-1} x$.
29. Given that $\tan \phi = \sinh x$, show that
    (a) $\dfrac{d\phi}{dx} = \operatorname{sech} x.$
    (b) $x = \ln(\sec \phi + \tan \phi).$
    (c) $\dfrac{dx}{d\phi} = \sec \phi.$
30. The region bounded by the graph of $y = \operatorname{sech} x$ between $x = -1$ and $x = 1$ is revolved about the x-axis. Find the volume of the solid that is generated.

Calculate the integral.

31. $\int \tanh x \, dx.$
32. $\int \coth x \, dx.$
33. $\int \operatorname{sech} x \, dx.$
34. $\int \operatorname{csch} x \, dx.$
35. $\int \operatorname{sech}^3 x \tanh x \, dx.$
36. $\int x \operatorname{sech}^2 x^2 \, dx.$

37. $\displaystyle\int \tanh x \ln(\cosh x)\, dx.$ **38.** $\displaystyle\int \frac{1+\tanh x}{\cosh^2 x}\, dx.$

39. $\displaystyle\int \frac{\operatorname{sech}^2 x}{1+\tanh x}\, dx.$ **40.** $\displaystyle\int \tanh^5 x \operatorname{sech}^2 x\, dx.$

In Exercises 41–43, verify the given integration formula. In each case, take $a > 0$.

41. $\displaystyle\int \frac{1}{\sqrt{a^2+x^2}} dx = \sinh^{-1}\left(\frac{x}{a}\right) + C.$

42. $\displaystyle\int \frac{1}{\sqrt{x^2-a^2}} dx = \cosh^{-1}\left(\frac{x}{a}\right) + C.$

43. $$\int \frac{1}{a^2-x^2}\, dx = \begin{cases} \dfrac{1}{a}\tanh^{-1}\left(\dfrac{x}{a}\right) + C & \text{If } |x| < a. \\[2ex] \dfrac{1}{a}\coth^{-1}\left(\dfrac{x}{a}\right) + C & \text{If } |x| > a. \end{cases}$$

44. If a body of mass m falling from rest under the action of gravity encounters air resistance that is proportional to the square of its velocity, then the velocity $v(t)$ of the body at time t satisfies the equation

$$m\frac{dv}{dt} = mg - kv^2$$

where $k > 0$ is the constant of proportionality and g is the gravitational constant.

(a) Show that

$$v(t) = \sqrt{\frac{mg}{k}} \tanh\left(\sqrt{\frac{gk}{m}}\, t\right)$$

is a solution of the equation which satisfies $v(0) = 0$.

(b) Find

$$\lim_{t\to\infty} v(t).$$

This limit is called the *terminal velocity* of the body.

C 45. Use a CAS to find a one-term expression for f.

(a) $\displaystyle\int f(x)\, dx = -\frac{\operatorname{sech} x(2+\tanh x)}{3(1+\tanh x)^2}.$

(b) $\displaystyle\int f(x)\, dx = \frac{x^2 \tanh^{-1} x^2}{2} + \frac{\ln(x^4-1)}{4}.$

Verify your results by differentiating the right-hand sides.

C 46. Use a CAS to find a one-term expression for f.

(a) $\displaystyle\int f(x)\, dx = \frac{x}{2}\sqrt{16+x^2} - 8\sinh^{-1}\left(\frac{x}{4}\right).$

(b) $\displaystyle\int f(x)\, dx = -\frac{x^2+1}{2}\sqrt{\frac{x^2-1}{x^2+1}} + \frac{x^2\cosh^{-1} x^2}{2}.$

Verify your results by differentiating the right-hand sides.

■ CHAPTER HIGHLIGHTS

7.1 One-to-One Function; Inverses

one-to-one function; inverse function (p. 371)
one-to-one and increasing/decreasing functions; derivatives (p. 374)
relation between graph of f and the graph of f^{-1} (p. 375)
continuity and differentiability of inverse functions (p. 375)
derivative of an inverse (p. 376)

7.2 The Logarithm Function, Part 1

definition of a logarithm function (p. 381)
natural logarithm: In $x = \displaystyle\int_t^x \frac{dt}{t}, x > 0;$
domain $(0, \infty)$, range $(-\infty, \infty)$
basic properties of $\ln x$ (p. 385)
graph of $y = \ln x$ (p. 385)

7.3 The Logarithm Function, Part II

$$\frac{d}{dx}(\ln |u|) = \frac{1}{u}\frac{du}{dx},$$

$$\int \frac{g'(x)}{g(x)} dx = \ln |g(x)| + C,$$

$$\int \tan x\, dx = \ln |\sec x| + C,$$

$$\int \sec x\, dx = \ln |\sec x + \tan x| + C,$$

$$\int \cot x\, dx = \ln |\sin x| + C,$$

$$\int \csc x\, dx = \ln |\csc x - \cot x| + C.$$

logarithmic differentiation (p. 394)

7.4 The Exponential Function

The exponential function $y = e^x$ is the inverse of the logarithm function $y = \ln x$.
graph of $y = e^x$ (p. 398);
domain $(-\infty, \infty)$, range $(0, \infty)$
basic properties of the exponential function (p. 398)

$$\frac{d}{dx}(e^u) = e^u\frac{du}{dx}, \qquad \int e^{g(x)} g'(x)\, dx = e^{g(x)} + C$$

7.5 Arbitrary Powers; Other Bases

$x^r = e^{r \ln x}$ for all $x > 0$, all real r

$$\frac{d}{dx}(p^u) = p^u \ln p \frac{du}{dx} \quad (p \text{ a positive constant}), \qquad \log_p x = \frac{\ln x}{\ln p}$$

$$\frac{d}{dx}(\log_p u) = \frac{1}{u \ln p}\frac{du}{dx}$$

$$\left(1+\frac{1}{n}\right)^n \le e \le \left(1+\frac{1}{n}\right)^{n+1} \qquad e \cong 2.71828$$

7.6 Exponential Growth and Decay

All the functions that satisfy the equation $f'(t) = kf(t)$ are of the form $f(t) = C\,e^{kt}$ (p. 413)

population growth (p. 414)
radioactive decay, half-life (p. 417)
compound interest, continuous compounding (p. 418)
rule of 72 (p. 420)

7.7 The Inverse Trigonometric Functions

The inverse sine, $y = \sin^{-1} x$, is the inverse of $y = \sin x$, $x \in [-\frac{1}{2}\pi, \frac{1}{2}\pi]$.
The inverse tangent, $y = \tan^{-1} x$, is the inverse of $y = \tan x$, $x \in (-\frac{1}{2}\pi, \frac{1}{2}\pi)$.
The inverse secant, $y = \sec^{-1} x$, is the inverse of $y = \sec x$, $x \in [0, \frac{1}{2}\pi) \cup (\frac{1}{2}\pi, \pi]$.

graph of $y = \sin^{-1} x$ (p. 423) graph of $y = \tan^{-1} x$ (p. 426).
graph of $y = \sec^{-1} x$ (p. 429)

$$\frac{d}{dx}(\sin^{-1} x) = \frac{1}{\sqrt{1-x^2}}$$

$$\int \frac{dx}{\sqrt{a^2-x^2}} = \sin^{-1}\left(\frac{x}{a}\right) + C \qquad (a > 0)$$

$$\frac{d}{dx}(\tan^{-1} x) = \frac{1}{1+x^2}$$

$$\int \frac{dx}{a^2+x^2} = \frac{1}{a}\tan^{-1}\left(\frac{x}{a}\right) + C \qquad (a \neq 0)$$

$$\frac{d}{dx}(\sec^{-1} x) = \frac{1}{|x|\sqrt{x^2-1}}$$

$$\int \frac{dx}{x\sqrt{x^2-a^2}} = \frac{1}{a}\sec^{-1}\left(\frac{|x|}{a}\right) + C \quad (a > 0)$$

definition of the remaining inverse trigonometric functions (p. 431)

7.8 The Hyperbolic Sine and Cosine

$\sinh x = \frac{1}{2}(e^x - e^{-x})$, $\quad \cosh x = \frac{1}{2}(e^x + e^{-x})$,
$\frac{d}{dx}(\sinh x) = \cosh x$, $\quad \frac{d}{dx}(\cosh x) = \sinh x$.

graphs (p. 436) basic identities (p. 437)

*7.9 The Other Hyperbolic Functions

$$\tanh x = \frac{\sinh x}{\cosh x}, \quad \coth x = \frac{\cosh x}{\sinh x},$$

$$\text{sech } x = \frac{1}{\cosh x}, \quad \text{csch } x = \frac{1}{\sinh x}.$$

derivatives (p. 440) hyperbolic inverses (p. 441)
derivatives of hyperbolic inverses (p. 443).

■ 8.1 INTEGRAL TABLES AND REVIEW

We begin by listing the more important integrals with which you are already familiar.

1. $\int k\,dx = kx + C, \qquad k$ constant.

2. $\int x^r\,dx = \dfrac{1}{r+1}x^{r+1} + C, \quad r \neq -1.$

3. $\int \dfrac{dx}{x} = \ln|x| + C.$

4. $\int e^x\,dx = e^x + C.$

5. $\int p^x\,dx = \dfrac{p^x}{\ln p} + C, \quad p > 0$ constant, $p \neq 1$.

6. $\int \sin x\,dx = -\cos x + C.$

7. $\int \cos x\,dx = \sin x + C.$

8. $\int \tan x\,dx = \ln|\sec x| + C.$

9. $\int \cot x\,dx = \ln|\sin x| + C.$

10. $\int \sec x\,dx = \ln|\sec x + \tan x| + C.$

11. $\int \csc x\,dx = \ln|\csc x - \cot x| + C.$

12. $\int \sec x \tan x\,dx = \sec x + C.$

13. $\int \csc x \cot x\,dx = -\csc x + C.$

14. $\int \sec^2 x\,dx = \tan x + C.$

15. $\int \csc^2 x\,dx = -\cot x + C.$

16. $\int \dfrac{dx}{\sqrt{a^2 - x^2}} = \sin^{-1}\left(\dfrac{x}{a}\right) + C, \quad a > 0$ constant.

17. $\int \dfrac{dx}{a^2 + x^2} = \dfrac{1}{a}\tan^{-1}\left(\dfrac{x}{a}\right) + C, \quad a > 0$ constant.

18. $\displaystyle\int \frac{dx}{x\sqrt{x^2-a^2}} = \frac{1}{a}\sec^{-1}\left(\frac{|x|}{a}\right) + C, \quad a > 0$ constant.

19. $\displaystyle\int \sinh x\,dx = \cosh x + C.$ **20.** $\displaystyle\int \cosh x\,dx = \sinh x + C.$

For review we work out a few integrals involving u-substitutions.

Example 1 Find $\displaystyle\int x \tan x^2\,dx.$

SOLUTION Set $u = x^2$, $du = 2x\,dx$. Then

$$\int x\tan x^2\,dx = \tfrac{1}{2}\int \tan u\,du = \tfrac{1}{2}\ln|\sec u| + C = \tfrac{1}{2}|\sec x^2| + C. \quad \square$$

(Formula 8)

Example 2 Calculate $\displaystyle\int_0^1 \frac{e^x}{e^x+2}\,dx.$

SOLUTION Set

$$u = e^x + 2, \quad du = e^x\,dx. \quad \text{At } x = 0,\ u = 3; \text{ at } x = 1,\ u = e + 2.$$

Thus

$$\int_0^1 \frac{e^x}{e^x+2}\,dx = \int_3^{e+2}\frac{du}{u} = \Big[\ln|u|\Big]_3^{e+2}$$

$$= \ln(e+2) - \ln 3 = \ln\left[\tfrac{1}{3}(e+2)\right] \cong 0.45. \quad \square$$

(Formula 3)

Example 3 Find $\displaystyle\int \frac{\cos 2x}{(2+\sin 2x)^{1/3}}\,dx.$

SOLUTION Set $u = 2 + \sin 2x$, $du = 2\cos 2x\,dx$. Then

$$\int \frac{\cos 2x}{(2+\sin 2x)^{1/3}}\,dx = \tfrac{1}{2}\int \frac{1}{u^{1/3}}\,du = \tfrac{1}{2}\int u^{-1/3}\,du = \tfrac{1}{2}\left(\tfrac{3}{2}\right)u^{2/3} + C$$

$$= \tfrac{3}{4}(2+\sin 2x)^{2/3} + C. \quad \square$$

(Formula 2)

The final example involves in little algebra.

Example 4 Find $P = \displaystyle\int \frac{dx}{x^2+2x+5}.$

SOLUTION First we complete the square in the denominator:

$$P = \int \frac{dx}{x^2+2x+5} = \int \frac{dx}{(x^2+2x+1)+4} = \int \frac{dx}{(x+1)^2+2^2}.$$

We know that

$$\int \frac{du}{u^2+a^2} = \frac{1}{a}\tan^{-1}\left(\frac{u}{a}\right) + C.$$

Setting

$$u = x + 1, \quad du = dx, \quad a = 2,$$

we have

$$P = \int \frac{du}{u^2 + 2^2} = \tfrac{1}{2}\tan^{-1}\left(\tfrac{u}{2}\right) + C = \tfrac{1}{2}\tan^{-1}\left(\frac{x+1}{2}\right) + C. \quad \square$$

Using an Integral Table A table of over 100 integrals, including those listed at the beginning of this section, appears on the inside covers of this text. This is a relatively short list. Mathematical handbooks such as *Burington's Handbook of Mathematical Tables and Formulas* and *CRC Standard Mathematical Tables* contain extensive tables; the table in the CRC reference lists 600 integrals.

The integrals in an integral table are grouped by the form of the integrand. For example, "forms containing $a + bu$," "forms containing $\sqrt{a^2 - u^2}$," "trigonometric forms," and so forth. The table on the inside covers is arranged in this manner.

Example 5 Use the integral table given on the inside covers to calculate

$$\int \frac{dx}{\sqrt{4 + x^2}}.$$

SOLUTION Find the group containing the form $\sqrt{a^2 + u^2}$ and see Formula 77. Replacing a by 2 and u by x, we have

$$\int \frac{dx}{\sqrt{4 + x^2}} = \ln|x + \sqrt{4 + x^2}| + C. \quad \square$$

Example 6 Use the integral table to calculate $\displaystyle\int \frac{dx}{3x^2(2x - 1)}$.

SOLUTION Find the group containing $a + bu$ and see Formula 109. Replacing a by -1, b by 2, and u by x, we have

$$\int \frac{dx}{3x^2(2x-1)} = \frac{1}{3}\int \frac{dx}{x^2(2x-1)} = \frac{1}{3}\left(\frac{1}{x} + 2\ln\left|\frac{2x-1}{x}\right|\right) + C. \quad \square$$

Sometimes a preliminary substitution has to be made before using the table.

Example 7 Use the integral table to calculate $\displaystyle\int \frac{\sqrt{9 - 4x^2}}{x^2}\,dx$.

SOLUTION The integral table contains the group $\sqrt{a^2 - u^2}$, and Formula 90 is close to what we need. But first we have to rewrite the integral so that it fits the formula. Set

$$u = 2x, \quad du = 2\,dx.$$

Then

$$\int \frac{\sqrt{9-4x^2}}{x^2}\,dx = \frac{1}{2}\int \frac{\sqrt{9-u^2}}{(u^2/4)}\,du = 2\left(-\frac{1}{u}\sqrt{9-u^2} - \sin^{-1}\frac{u}{3}\right) + C$$

$$= 2\left(-\frac{1}{2x}\sqrt{9-4x^2} - \sin^{-1}\frac{2x}{3}\right) + C. \quad \square$$

EXERCISES 8.1

Calculate the integral.

1. $\int e^{2-x}\,dx.$

2. $\int \cos\frac{2}{3}x\,dx.$

3. $\int_0^1 \sin\pi x\,dx.$

4. $\int_0^1 \sec\pi x\tan\pi x\,dx.$

5. $\int \sec^2(1-x)\,dx.$

6. $\int \frac{dx}{5^x}.$

7. $\int_{\pi/6}^{\pi/3} \cot x\,dx.$

8. $\int_0^1 \frac{x^3}{1+x^4}\,dx.$

9. $\int \frac{x}{\sqrt{1-x^2}}\,dx.$

10. $\int_{-\pi/4}^{\pi/4} \frac{dx}{\cos^2 x}.$

11. $\int_{-\pi/4}^{\pi/4} \frac{\sin x}{\cos^2 x}\,dx.$

12. $\int \frac{e^{\sqrt{x}}}{\sqrt{x}}\,dx.$

13. $\int_1^2 \frac{e^{1/x}}{x^2}\,dx.$

14. $\int \frac{x^3}{\sqrt{1-x^4}}\,dx.$

15. $\int_0^c \frac{dx}{x^2+c^2}.$

16. $\int a^x e^x\,dx.$

17. $\int \frac{\sec^2\theta}{\sqrt{3\tan\theta+1}}\,d\theta.$

18. $\int \frac{\sin\phi}{3-2\cos\phi}\,d\phi.$

19. $\int \frac{e^x}{a\,e^x-b}\,dx.$

20. $\int \frac{dx}{x^2-4x+13}.$

21. $\int \frac{x}{(x+1)^2+4}\,dx.$

22. $\int \frac{\ln x}{x}\,dx.$

23. $\int \frac{x}{\sqrt{1-x^4}}\,dx.$

24. $\int \frac{e^x}{1+e^{2x}}\,dx.$

25. $\int \frac{dx}{x^2+6x+10}.$

26. $\int e^x\tan e^x\,dx.$

27. $\int x\sin x^2\,dx.$

28. $\int \frac{x}{9+x^4}\,dx.$

29. $\int \tan^2 x\,dx.$

30. $\int \cosh 2x\,\sinh^3 2x\,dx.$

31. $\int_1^e \frac{\ln x^3}{x}\,dx.$

32. $\int_0^{\pi/4} \frac{\tan^{-1}x}{1+x^2}\,dx.$

33. $\int \frac{\sin^{-1}x}{\sqrt{1-x^2}}\,dx.$

34. $\int e^x\cosh(2-e^x)\,dx.$

35. $\int \frac{1}{x\ln x}\,dx.$

36. $\int_{-1}^1 \frac{x^2}{x^2+1}\,dx.$

37. $\int_0^{\pi/4} \frac{1+\sin x}{\cos^2 x}\,dx.$

38. $\int_0^{1/2} \frac{1+x}{\sqrt{1-x^2}}\,dx.$

Calculate using an integral table.

39. $\int \sqrt{x^2-4}\,dx.$

40. $\int \sqrt{4-x^2}\,dx.$

41. $\int \cos^3 2t\,dt.$

42. $\int \sec^4 t\,dt.$

43. $\int \frac{dx}{x(2x+3)}.$

44. $\int \frac{x\,dx}{2+3x}.$

45. $\int \frac{\sqrt{x^2+9}}{x^2}\,dx.$

46. $\int \frac{dx}{x^2\sqrt{x^2-2}}.$

47. $\int x^3\ln x\,dx.$

48. $\int x^3\sin x\,dx.$

49. Evaluate $\int_0^{\pi} \sqrt{1+\cos x}\,dx.$
HINT: $\cos x = 2\cos^2(x/2)-1.$

50. Calculate $\int \sec^2 x\tan x\,dx$ in two ways.

(a) Let $u=\tan x$ and verify that the result is

$$\int \sec^2 x\tan x\,dx = \tfrac{1}{2}\tan^2 x + C_1.$$

(b) Let $u=\sec x$ and verify that the result is

$$\int \sec^2 x\tan x\,dx = \tfrac{1}{2}\sec^2 x + C_2.$$

(c) Reconcile the results in parts (a) and (b)

51. Verify that, for each positive integer n:

(a) $\int_0^{\pi} \sin^2 nx\,dx = \frac{1}{2}\pi.$
HINT: $\sin^2\theta = \frac{1}{2}(1-\cos 2\theta)$

(b) $\int_0^{\pi} \sin nx\cos nx\,dx = 0.$

(c) $\int_0^{\pi/n} \sin nx\cos nx\,dx = 0.$

52. (a) Calculate $\int \sin^3 x\,dx.$ HINT: $\sin^2 x = 1-\cos^2 x.$

(b) Calculate $\int \sin^5 x\,dx.$

(c) Explain how to calculate $\int \sin^{2k-1} x\,dx$ for each positive integer k.

53. (a) Calculate $\int \tan^3 x\,dx.$ HINT: $\tan^2 x = \sec^2 x - 1.$

(b) Calculate $\int \tan^5 x\,dx.$

(c) Calculate $\int \tan^7 x\,dx.$

(d) Explain how to calculate $\int \tan^{2k+1} x\,dx$ for each positive integer k.

54. (a) Sketch the region bounded on the left by the line $x = \pi/6$, and by the curves $y=\csc x$ and $y=\sin x$ for $\pi/6 \le x \le \pi/2$.

(b) Calculate the area of the region in part (a).

(c) The region in part (a) is revolved about the x-axis. Find the volume of the solid that is generated.

55. (a) Use a graphing utility to sketch the graph of

$$f(x) = \frac{1}{\sin x + \cos x} \quad \text{for} \quad 0 \le x \le \frac{\pi}{2}.$$

(b) Find A and B such that $\sin x + \cos x = A \sin(x + B)$.

(c) Find the area of the region bounded by the graph of f and the x-axis. HINT: Use the result from part (b).

56. (a) Use a graphing utility to sketch the graph of $f(x) = e^{-x^2}$.

(b) Fix a number $a > 0$ and revolve the region bounded by the graph of f and the y-axis on the interval $[0, a]$ about the y-axis. Find the volume of the solid that is generated.

(c) Find a such that the solid in part (b) has volume 2 cubic units.

57. (a) Use a graphing utility to sketch the graph of

$$f(x) = \frac{x^2 + 1}{x + 1}, \quad x > -1, \quad \text{and} \quad x + 2y = 16$$

in the same coordinate system.

(b) The two graphs intersect at two points and determine a bounded region R. Estimate the x-coordinates of the two points of intersection accurate to two decimal places.

(c) Determine the approximate area of the region R.

58. (a) Use a graphing utility to sketch the graph of

$$y^2 = x^2(1 - x).$$

(b) Your sketch in part (a) should show that the curve forms a loop for $0 \le x \le 1$. Calculate the area of the loop. HINT: Use the symmetry of the curve.

■ 8.2 INTEGRATION BY PARTS

We begin with the formula for the derivative of a product

$$f(x)g'(x) + f'(x)g(x) = (f \cdot g)'(x).$$

Integrating both sides, we get

$$\int f(x)g'(x)\,dx + \int f'(x)g(x)\,dx = \int (f \cdot g)'(x)\,dx.$$

Since

$$\int (f \cdot g)'(x)\,dx = f(x)g(x) + C,$$

we have

$$\int f(x)g'(x)\,dx + \int f'(x)g(x)\,dx = f(x)g(x) + C,$$

and therefore

$$\int f(x)g'(x)\,dx = f(x)g(x) - \int f'(x)g(x)\,dx + C.$$

Since the calculation of

$$\int f'(x)g(x)\,dx$$

will yield its own arbitrary constant, there is no reason to keep the constant C. We therefore drop it and write

(8.2.1) $$\int f(x)g'(x)\,dx = f(x)g(x) - \int f'(x)g(x)\,dx.$$

This formula, called the formula for *integration by parts*, enables us to find

$$\int f(x)g'(x)\,dx$$

by finding

$$\int f'(x)g(x)\,dx$$

instead. Of course, it is of practical use only if the second integral is easier to calculate than the first.

In practice we usually set

$$u = f(x), \quad dv = g'(x)\,dx.$$

Then
$$du = f'(x)\,dx, \quad v = g(x).$$

Now, with these substitutions, the formula for integration by parts can be written

(8.2.2)
$$\int u\,dv = uv - \int v\,du.$$

Success with this formula depends on choosing u and dv so that

$$\int v\,du \quad \text{is easier to calculate than} \quad \int u\,dv.$$

Example 1 Calculate $\displaystyle\int x\,e^x\,dx$.

SOLUTION We want to separate x from e^x. Setting

$$u = x, \quad dv = e^x\,dx,$$

we have

$$du = dx, \quad v = e^x.$$

Accordingly,

$$\int x\,e^x\,dx = \int u\,dv = uv - \int v\,du = xe^x - \int e^x\,dx = x\,e^x - e^x + C.$$

Our choice of u and dv worked out very well. Does the choice of u and dv make a difference? Suppose we had set

$$u = e^x, \quad dv = x\,dx;$$

then we would have had

$$du = e^x\,dx, \quad v = \tfrac{1}{2}x^2.$$

Integration by parts would then have given

$$\int xe^x\,dx = \int u\,dv = uv - \int v\,du = \tfrac{1}{2}x^2\,e^x - \tfrac{1}{2}\int x^2e^x\,dx,$$

an integral more complicated than the one we started with. Clearly, the proper selection of u and dv is important. ❑

Example 2 Calculate $\int x \sin 2x\,dx$.

SOLUTION Setting $u = x, \quad dv = \sin 2x\,dx, \quad$ we have

$$du = dx, \quad v = -\tfrac{1}{2}\cos 2x.$$

Therefore,

$$\int x \sin\,2x\,dx = -\tfrac{1}{2}x\cos 2x - \int -\tfrac{1}{2}\cos 2x\,dx = -\tfrac{1}{2}x\cos 2x\,+\tfrac{1}{4}\sin 2x\,+C.$$

You can verify that if we had set

$$u = \sin 2x, \quad dv = x\,dx,$$

then we would run into the same kind of difficulty that we saw in Example 1. ❑

In Examples 1 and 2 there was only one effective choice for u and dv. However, in some cases there may be more than one way to choose u and dv.

Example 3 Calculate $\int x \ln x\,dx$.

SOLUTION Setting $u = \ln x, \quad dv = x\,dx, \quad$ we have

$$du = \frac{1}{x}dx, \quad v = \frac{x^2}{2}.$$

Then

$$\begin{aligned}\int x \ln x\,dx &= \int u\,dv = uv - \int v\,du \\ &= \frac{x^2}{2}\ln x - \int \frac{1}{x}\frac{x^2}{2}\,dx = \tfrac{1}{2}x^2\ln x - \tfrac{1}{2}\int x\,dx = \tfrac{1}{2}x^2\ln x - \tfrac{1}{4}x^2 + C.\end{aligned}$$

ALTERNATE SOLUTION This time we set

$$u = x\ln x, \quad dv = dx,$$

so that

$$du = (1 + \ln x)\,dx, \quad v = x.$$

Substituting these selections in

$$\int u\,dv = uv - \int v\,du,$$

we find that

$$(1) \qquad \int x \ln x \, dx = x^2 \ln x - \int x(1 + \ln x)\, dx.$$

It may seem that the integral on the right is more complicated than the one we started with. However, we can rewrite equation (1) as

$$\int x \ln x \, dx = x^2 \ln x - \int x \, dx - \int x \ln x \, dx.$$

Adding $\int x \ln dx$ to both sides of this equation, we get

$$2\int x \ln x \, dx = x^2 \ln x - \int x \, dx = x^2 \ln x - \tfrac{1}{2} x^2 + C$$

and so

$$\int x \ln x \, dx = \tfrac{1}{2}x^2 \ln x - \tfrac{1}{4} x^2 + C \qquad \text{(as usual, we replace } C/2 \text{ by } C)$$

as before. ☐

Remark Given the two successful approaches in Example 3, you may also be tempted to set

$$u = x, \quad dv = \ln x \, dx.$$

This won't work, however, because you can't calculate v; that is, you don't know an antiderivative for $\ln x$. Actually, as you will see at the end of this section, $\int \ln x \, dx$ can itself be found by integration by parts. ☐

To calculate some integrals you may have to integrate by parts more than once.

Example 4 Evaluate $\displaystyle\int x^2 e^{-x} \, dx.$

SOLUTION Setting $u = x^2, \quad dv = e^{-x}\, dx,$ we have

$$du = 2x \, dx, \quad v = -e^{-x}$$

and thus

$$\int x^2 e^{-x} \, dx = \int u \, dv = uv - \int v \, du = -x^2 e^{-x} - \int -2xe^{-x} \, dx$$
$$= -x^2 e^{-x} + \int 2xe^{-x} \, dx.$$

We now calculate the integral on the right again by parts. This time we set

$$u = 2x, \quad dv = e^{-x}\, dx.$$

This gives

$$du = 2 \, dx, \quad v = -e^{-x},$$

and thus

$$\int 2xe^{-x}\,dx = \int u\,dv = uv - \int v\,du = -2xe^{-x} - \int -2e^{-x}\,dx$$
$$= -2xe^{-x} + \int 2e^{-x}\,dx = -2xe^{-x} - 2e^{-x} + C.$$

This, together with our earlier calculation, gives

$$\int x^2e^{-x}\,dx = -x^2e^{-x} - 2xe^{-x} - 2e^{-x} + C = -(x^2 + 2x + 2)e^{-x} + C. \quad \square$$

Example 5 Find $\int e^x \cos x\,dx$.

SOLUTION Here we integrate by parts twice. First we write

$$u = e^x, \quad dv = \cos x\,dx,$$
$$du = e^x\,dx, \quad v = \sin x.$$

This gives

$$\text{(1)} \qquad \int e^x \cos x\,dx = \int u\,dv = uv - \int v\,du = e^x \sin x - \int e^x \sin x\,dx.$$

Now we work with the integral on the right. Setting

$$u = e^x, \quad dv = \sin x\,dx,$$
$$du = e^x\,dx, \quad v = -\cos x,$$

we have

$$\text{(2)} \qquad \int e^x \sin x\,dx = \int u\,dv = uv - \int v\,du = -e^x \cos x + \int e^x \cos x\,dx.$$

Substituting (2) into (1), we get

$$\int e^x \cos x\,dx = e^x \sin x + e^x \cos x - \int e^x \cos x\,dx,$$

and we have a situation similar to the one we had in the alternate solution in Example 3. "Solving" this equation for $\int e^x \cos x\,dx$, we have

$$2\int e^x \cos x\,dx = e^x(\sin x + \cos x),$$
$$\int e^x \cos x\,dx = \tfrac{1}{2}e^x(\sin x + \cos x).$$

Since this integral is an indefinite integral, we add an arbitrary constant C:

$$\int e^x \cos x\,dx = \tfrac{1}{2}e^x(\sin x + \cos x) + C. \quad \square$$

Remark You should verify that if you switch the roles of u and dv in Example 5 by letting

$$u = \cos x, \qquad dv = e^x\, dx,$$
$$du = -\sin x\, dx, \qquad v = e^x,$$

then you will get precisely the same result in precisely the same way. □

Integration by parts is often used to calculate integrals where the integrand is a mixture of function types; for example, polynomials and exponentials, or polynomials and trigonometric functions, and so forth. Some integrands, however, are better left as mixtures; for example,

$$\int 2xe^{x^2}\, dx = e^{x^2} + C \quad \text{and} \quad \int 3x^2 \cos x^3\, dx = \sin x^3 + C.$$

Any attempt to separate these integrands for integration by parts is counterproductive. The mixtures in these integrands arise from the chain rule, and we need these mixtures to calculate the integrals.

The next example illustrates a clever choice of u and dv.

Example 6 Calculate $\int x^5 \cos(x^3)\, dx$.

SOLUTION To integrate $\cos(x^3)$, we need an x^2 factor. So we will keep x^2 together with $\cos(x^3)$ and set

$$u = x^3, \quad dv = x^2 \cos(x^3).$$

Then

$$du = 3x^2\, dx, \quad v = \tfrac{1}{3}\sin(x^3) \qquad \text{(verify this)}$$

and

$$\int x^5 \cos(x^3)\, dx = \tfrac{1}{3}x^3 \sin(x^3) - \int x^2 \sin(x^3)\, dx$$
$$= \tfrac{1}{3}x^3 \sin(x^3) + \tfrac{1}{3}\cos(x^3) + C. \quad □$$

Integration by parts can also be used in connection with definite integrals. Suppose that f' and g' are continuous on the interval $[a, b]$. Integrating the formula for the derivative of a product, we have

$$\int_a^b f(x)g'(x)\, dx + \int_a^b f'(x)g(x)\, dx = \int_a^b [f(x)g(x)]'\, dx.$$

However, by the fundamental theorem of calculus, Theorem 5.4.2,

$$\int_a^b [f(x) \cdot g(x)]' dx = \Big[f(x)g(x)\Big]_a^b.$$

Thus

$$\int_a^b f(x)g'(x)\, dx + \int_a^b f'(x)g(x)\, dx = \Big[f(x)g(x)\Big]_a^b$$

and

(8.2.3)
$$\int_a^b f(x)g'(x)\,dx = \Big[f(x)g(x)\Big]_a^b - \int_a^b f'(x)g(x)\,dx,$$

which is the definite integral version of Formula (8.2.1).

Example 7 Evaluate $\int_1^2 x^3 \ln x\,dx$.

SOLUTION Setting $u = \ln x,\quad dv = x^3\,dx,$ we have

$$du = \frac{1}{x}dx, \quad v = \tfrac{1}{4}x^4.$$

Substitution into (8.2.3) gives

$$\begin{aligned}\int_1^2 x^3 \ln x\,dx &= \Big[\tfrac{1}{4}x^4 \ln x\Big]_1^2 - \tfrac{1}{4}\int_1^2 x^3\,dx\\ &= 4\ln 2 - \tfrac{1}{16}\Big[x^4\Big]_1^2 = 4\ln 2 - \tfrac{15}{16}. \quad \square\end{aligned}$$

Finally, the technique of integration by parts enables us to integrate the logarithm function and the inverse trigonometric functions:

(8.2.4)
$$\int \ln x\,dx = x\ln x - x + C.$$

(8.2.5)
$$\int \sin^{-1} x\,dx = x\sin^{-1} x + \sqrt{1-x^2} + C.$$

(8.2.6)
$$\int \tan^{-1} x\,dx = x\tan^{-1} x - \tfrac{1}{2}\ln(1+x^2) + C.$$

(8.2.7)
$$\int \sec^{-1} x\,dx = x\sec^{-1} x - \ln|x + \sqrt{x^2-1}| + C.$$

We will derive (8.2.5). Formulas (8.2.4) and (8.2.6) are left to you as exercises. Formula (8.2.7) requires integration by parts *and* a trigonometric substitution. You will be asked to derive it in the Exercises for Section 8.4.

Set $\qquad u = \sin^{-1} x, \quad dv = dx.$

Then $\qquad du = \dfrac{1}{\sqrt{1-x^2}}\,dx, \quad v = x,$

and $\displaystyle\int \sin^{-1} x\,dx = x\sin^{-1} x - \int \frac{x}{\sqrt{1-x^2}}\,dx.$

The new integral is evaluated by means of the substitution

$$w = 1 - x^2, \quad dw = -2x\,dx,$$

which gives

$$\int \frac{x}{\sqrt{1-x^2}}\,dx = -\tfrac{1}{2}\int \frac{dw}{\sqrt{w}} = -\sqrt{w} = -\sqrt{1-x^2}.$$

Therefore,

$$\int \sin^{-1} x\,dx = x\sin^{-1} x + \sqrt{1-x^2} + C.$$

EXERCISES 8.2

Calculate the integral.

1. $\displaystyle\int xe^{-x}\,dx.$

2. $\displaystyle\int_0^2 x2^x\,dx.$

3. $\displaystyle\int x^2e^{-x^3}\,dx.$

4. $\displaystyle\int x\ln x^2\,dx.$

5. $\displaystyle\int_0^1 x^2e^{-x}\,dx.$

6. $\displaystyle\int x^3e^{-x^2}\,dx.$

7. $\displaystyle\int \frac{x^2}{\sqrt{1-x}}\,dx.$

8. $\displaystyle\int \frac{dx}{x(\ln x)^3}.$

9. $\displaystyle\int_1^{e^2} x\ln\sqrt{x}\,dx.$

10. $\displaystyle\int_0^3 x\sqrt{x+1}\,dx.$

11. $\displaystyle\int \frac{\ln(x+1)}{\sqrt{x+1}}\,dx.$

12. $\displaystyle\int x^2(e^x-1)\,dx.$

13. $\displaystyle\int (\ln x)^2\,dx.$

14. $\displaystyle\int x(x+5)^{-14}\,dx.$

15. $\displaystyle\int x^3 3^x\,dx.$

16. $\displaystyle\int \sqrt{x}\ln x\,dx.$

17. $\displaystyle\int x(x+5)^{14}dx.$

18. $\displaystyle\int (2^x+x^2)^2dx.$

19. $\displaystyle\int_0^{1/2} x\cos\pi x\,dx.$

20. $\displaystyle\int_0^{\pi/2} x^2\sin x\,dx.$

21. $\displaystyle\int x^2(x+1)^9\,dx.$

22. $\displaystyle\int x^2(2x-1)^{-7}\,dx.$

23. $\displaystyle\int e^x\sin x\,dx.$

24. $\displaystyle\int (e^x+2x)^2\,dx.$

25. $\displaystyle\int_0^1 \ln(1+x^2)\,dx.$

26. $\displaystyle\int x\ln(x+1)\,dx.$

27. $\displaystyle\int x^n\ln x\,dx\ (n\neq -1).$

28. $\displaystyle\int e^{3x}\cos 2x\,dx.$

29. $\displaystyle\int x^3\sin x^2\,dx.$

30. $\displaystyle\int x^3\sin x\,dx.$

31. $\displaystyle\int_0^{1/4} \sin^{-1} 2x\,dx.$

32. $\displaystyle\int \frac{\sin^{-1} 2x}{\sqrt{1-4x^2}}\,dx.$

33. $\displaystyle\int_0^1 x\tan^{-1} x^2\,dx.$

34. $\displaystyle\int \cos\sqrt{x}\,dx.$ HINT: Let $u=\sqrt{x}$.

35. $\displaystyle\int x^2\cosh 2x\,dx.$

36. $\displaystyle\int_{-1}^1 x\sinh(2x^2)\,dx.$

37. $\displaystyle\int \frac{1}{x}\sin^{-1}(\ln x)\,dx.$

38. $\displaystyle\int \cos(\ln x)\,dx.$ HINT: Integrate by parts twice.

39. $\displaystyle\int \sin(\ln x)\,dx.$

40. $\displaystyle\int_1^{2e} x^2(\ln x)^2\,dx.$

41. Derive Formula (8.2.4): $\displaystyle\int \ln x\,dx = x\ln x - x + C.$

42. Derive Formula (8.2.6):

$$\int \tan^{-1} x\,dx = x\tan^{-1} x - \tfrac{1}{2}\ln(1+x^2) + C.$$

Derive the integration formula.

43. $\displaystyle\int x^k\ln x\,dx = \frac{x^{k+1}}{k+1}\ln x - \frac{x^{k+1}}{(k+1)^2} + C\ (k\neq -1).$

44. $\displaystyle\int e^{ax}\cos bx\,dx = \frac{e^{ax}(a\cos bx + b\sin bx)}{a^2+b^2} + C.$

45. $\displaystyle\int e^{ax}\sin bx\,dx = \frac{e^{ax}(a\sin bx - b\cos bx)}{a^2+b^2} + C.$

46. What happens if you use integration by parts to evaluate $\int e^{ax}\cosh ax\,dx$? Evaluate this integral by some other method.

47. If $f(x) = x\sin x$ on $[0, \pi]$, then $f(0) = f(\pi) = 0$ and $f(x) > 0$ for $x \in (0, \pi)$. Find the area of the region bounded by the graph of f and the x-axis.

48. If $g(x) = x\cos(x/2)$ on $[0, \pi]$, then $g(0) = g(\pi) = 0$ and $g(x) > 0$ for $x \in (0, \pi)$. Find the area of the region bounded by the graph of g and the x-axis.

In Exercises 49 and 50, find the area of the region bounded by the graph of f and the x-axis.

49. $f(x) = \sin^{-1}x, \quad x \in [0, \frac{1}{2}].$

50. $f(x) = xe^{-2x}, \quad x \in [0, 2].$

51. Let Ω be the region bounded by the graph of $f(x) = \ln x$ and the x-axis between $x = 1$ and $x = e$. (a) Find the area of Ω . (b) Find the centroid of Ω . (c) Find the volume of the solids generated by revolving Ω about each of the coordinate axes.

52. Let $f(x) = \dfrac{\ln x}{x}, \ x \in [1, 2e].$

(a) Find the area of the region R bounded by the graph of f and the x-axis.

(b) Find the volume of the solid generated by revolving R about the x-axis.

In Exercises 53–56, find the centroid of the region under the graph.

53. $f(x) = e^x, \quad x \in [0, 1].$

54. $f(x) = e^{-x}, \quad x \in [0, 1].$

55. $f(x) = \sin x, \quad x \in [0, \pi].$

56. $f(x) = \cos x, \quad x \in [0, \frac{1}{2}\pi].$

57. The mass density of a rod that extends from $x = 0$ to $x = 1$ is given by the function $\lambda(x) = e^{kx}$, where k is a constant. (a) Calculate the mass of the rod. (b) Find the center of mass of the rod.

58. The mass density of a rod that extends from $x = 2$ to $x = 3$ is given by the logarithm function $f(x) = \ln x$. (a) Calculate the mass of the rod. (b) Find the center of mass of the rod.

In Exercises 59–62, find the volume generated by revolving the region under the graph about the y-axis.

59. $f(x) = \cos\frac{1}{2}\pi x, \quad x \in [0, 1].$

60. $f(x) = x\sin x, \quad x \in [0, \pi].$

61. $f(x) = x\,e^x, \quad x \in [0, 1].$

62. $f(x) = x\cos x, \quad x \in [0, \frac{1}{2}\pi].$

63. Let Ω be the region under the curve $y = e^x$, $x \in [0, 1]$. Find the centroid of the solid generated by revolving Ω about the x-axis. (See Project 6.4)

64. Let Ω be the region under the curve $y = \sin x$, $x \in [0, \frac{1}{2}\pi]$. Find the centroid of the solid geneated by revolving Ω about the x-axis. (See Project 6.4)

65. Let Ω be the region bounded by the graph of $y = \cosh x$ and the x-axis between $x = 0$ and $x = 1$. Find the area of Ω and determine its centroid.

66. Let Ω be the region given in Exercise 65. Find the centroid of the solid generated by revolving Ω about: (a) the x-axis; (b) the y-axis

67. Let n be a positive integer. Use integration by parts to derive the formula

$$\int x^n e^{ax}\,dx = \frac{x^n e^{ax}}{a} - \frac{n}{a}\int x^{n-1}e^{ax}\,dx, \quad a \neq 0.$$

NOTE: This formula is known as a *reduction formula* since the exponent n in the integrand has been reduced.

68. Let n be a positive integer. Verify the reduction formula

$$\int (\ln x)^n\,dx = x(\ln x)^n - n\int (\ln x)^{n-1}\,dx.$$

Use the reduction formulas of Exercises 67 and 68 to evaluate the integral.

69. $\displaystyle\int x^3 e^{2x}\,dx.$ **70.** $\displaystyle\int x^2 e^{-x}\,dx.$

71. $\displaystyle\int (\ln x)^3\,dx.$ **72.** $\displaystyle\int (\ln x)^4\,dx.$

73. (a) You know that if you were to calculate $\int x^3 e^x\,dx$ using integration by parts, the result would be of the form

$$\int x^3 e^x\,dx = Ax^3e^x + Bx^2e^x + Cxe^x + De^x + E.$$

Differentiate both sides of this equation and solve for the coefficients A, B, C, D. Thus, you can calculate the integral without integrating.

(b) Verify the results in part (a) by calculating the integral using the reduction formula of Exercise 67.

74. Show that if $P(x)$ is a polynomial of degree k, then

$$\int P(x)e^x\,dx = [P(x) - P'(x) + P''(x) - \cdots \pm P^{(k)}(x)]e^x + C.$$

75. Use the result of Exercise 74 to calculate:

(a) $\displaystyle\int (x^2 - 3x + 1)e^x\,dx.$ (b) $\displaystyle\int (x^3 - 2x)e^x\,dx.$

76. Use integration by parts to show that if f has an inverse, then

$$\int f^{-1}(x)\,dx = xf^{-1}(x) - \int x\,(f^{-1})'(x)\,dx.$$

77. Show that if f and g have continuous second derivatives, and $f(a) = g(a) = f(b) = g(b) = 0$, then

$$\int_a^b f(x)g''(x)\,dx = \int_a^b g(x)f''(x)\,dx.$$

78. You are familiar with the identity

$$f(b) - f(a) = \int_a^b f'(x)\,dx.$$

(a) Assume that f has a continuous second derivative. Use integration by parts to derive the identity

$$f(b) - f(a) = f'(a)(b - a) - \int_a^b f''(x)\,(x - b)\,dx.$$

(b) Assume that f has a continuous third derivative. Use the result in part (a) and integration by parts to derive the identity

$$f(b) - f(a) = f'(a)\,(b - a) + \frac{f''(a)}{2}(b - a)^2 + \int_a^b \frac{f'''(x)}{2}(x - b)^2\,dx.$$

These results generalize Exercise 61, Section 5.4. The identities begun here will be extended and used in Chapter 11.

79. Use a graphing utility to draw the graph of $f(x) = x \sin x$, $x \geq 0$. Then use a CAS to calculate the area of the region bounded by the graph of f and the x-axis for:

(a) $0 \leq x \leq \pi$.

(b) $\pi \leq x \leq 2\pi$.

(c) $2\pi \leq x \leq 3\pi$.

(d) What is the area of the region bounded by the graph of f and the x-axis for $n\pi \leq x \leq (n + 1)\pi$, n an arbitrary nonnegative integer?

80. Use a graphing utility to draw the graph of $g(x) = x \cos x$, $x \geq 0$. Then use a CAS to calculate the area of the region bounded by the graph of g and the x-axis for:

(a) $\pi/2 \leq x \leq 3\pi/2$.

(b) $3\pi/2 \leq x \leq 5\pi/2$.

(c) $5\pi/2 \leq x \leq 7\pi/2$.

(d) What is the area of the region bounded by the graph of g and the x-axis for $\frac{1}{2}(2n - 1)\pi \leq x \leq \frac{1}{2}(2n + 1)\pi$, n an arbitrary positive integer?

81. Use a graphing utility to draw the graph of $f(x) = 1 - \sin x$ on $[0, \pi]$. Then use a CAS to find:

(a) the area of the region R bounded by the graph of f and the x-axis.

(b) the volume of the solid generated by revolving R about the y-axis.

(c) the centroid of R.

82. Use a graphing utility to draw the graph of $g(x) = xe^{-x}$ on $[0,10]$. Then use a CAS to find:

(a) the centroid of the region R bounded by the graph of g and the x-axis.

(b) the volume of the solid generated by revolving R about the x-axis.

(c) the volume of the solid generated by revolving R about the y-axis.

■ PROJECT 8.2 Revenue Streams

In Section 7.6, we considered continuous compounding of interest and saw that if P dollars are invested at an interest rate r compounded continuously, then the value of the investment at some future time n is given by

$$Q = Pe^{rn} \qquad \text{(see Formula 7.6.3).}$$

The quantity Q is called the *future value* of the investment P (at the interest rate r). Solving this equation for P, we get

$$P = Qe^{-rn}.$$

In this formulation, P is called the *present value* of Q at the interest rate r.

Problem 1. Rather than having a fixed amount Q at time n, suppose that revenue flows continuously at the constant rate of R dollars per year for n years. Show that the present value of such a revenue stream is given by

$$P.V. = \int_0^n Re^{-rt}\,dt.$$

HINT: Partition $[0, n]$ into subintervals $[t_{i-1}, t_i]$ and estimate the present value of R dollars compounded continuously at the interest rate r on $[t_{i-1}, t_i]$.

Problem 2. Given a constant revenue stream of $R = \$1000$ per year, find the present value of the first four years of revenue if the annual rate of continuous compounding is: (a) 4%? (b) 8%?

Revenue streams usually do not flow at a constant rate R; rather they flow at a time-dependent rate $R(t)$. In general, $R(t)$ tends to increase when business is good and tends to decrease when business is poor. "Growth" companies are companies that have a continuously increasing flow rate, while the so-called cyclical companies owe their name to a fluctuating flow rate. If we postulate a time-dependent flow rate $R(t)$, then the present value of an n-year revenue stream is given by

$$P.V. = \int_0^n R(t)e^{-rt}\,dt.$$

In Problems 3 and 4, assume a time-dependent revenue stream of $R(t) = 1000 + 60t$ dollars per year.

Problem 3. What is the present value of the first two years of revenue if the annual rate of continuous compounding is : (a) 5%? (b) 10%?

Problem 4. What is the present value of the third year of revenue if the annual rate of continuous compounding is; (a) 5%? (b) 10%?

Problem 5. Give an example of a time-dependent revenue stream that could be attached to a cyclical company and give the formula for the present value of such an n-year stream.

■ 8.3 POWERS AND PRODUCTS OF TRIGONOMETRIC FUNCTIONS

Sines and Cosines

I. We begin by explaining how to calculate integrals of the form

$$\int \sin^m x \cos^n x\,dx \qquad \text{with } m \text{ or } n \text{ odd positive integers.}$$

Suppose that n is odd. If $n = 1$, we have

$$(1) \qquad \int \sin^m x \cos x\,dx = \frac{1}{m+1}\sin^{m+1} x + C, \quad m \neq -1.$$

If $n > 1$, write

$$\cos^n x = \cos^{n-1} x \cos x.$$

Since $n - 1$ is even, $\cos^{n-1} x$ can be expressed in powers of $\sin^2 x$ using the identity $\cos^2 x = 1 - \sin^2 x$. The integral then takes the form

$$\int (\text{sum of powers of } \sin x) \cdot \cos x\,dx,$$

which can be broken up into integrals of the form (1).

Similarly, if m is odd, write

$$\sin^m x = \sin^{m-1} x \sin x.$$

and use the sustitution $\sin^2 x = 1 - \cos^2 x$.

Example 1

$$\begin{aligned}\int \sin^2 x \cos^5 x\,dx &= \int \sin^2 x \cos^4 x \cos x\,dx\\ &= \int \sin^2 x(1 - \sin^2 x)^2 \cos x\,dx\\ &= \int (\sin^2 x - 2\sin^4 x + \sin^6 x)\cos x\,dx\\ &= \tfrac{1}{3}\sin^3 x - \tfrac{2}{5}\sin^5 x + \tfrac{1}{7}\sin^7 x + C. \quad \square\end{aligned}$$

Example 2

$$\begin{aligned}\int \sin^5 x\,dx &= \int \sin^4 x \sin x\,dx\\ &= \int (1-\cos^2 x)^2 \sin x\,dx\\ &= \int (1-2\cos^2 x+\cos^4 x)\sin x\,dx\\ &= \int \sin x\,dx - 2\int \cos^2 x \sin x\,dx + \int \cos^4 x \sin x\,dx\\ &= -\cos x + \tfrac{2}{3}\cos^3 x - \tfrac{1}{5}\cos^5 x + C. \quad \square\end{aligned}$$

II. To calculate integrals of the form

$$\boxed{\int \sin^m x \cos^n x\,dx \qquad \text{with } m \text{ and } n \text{ even positive integers.}}$$

use the following trigonometric identities:

$$\sin x \cos x = \tfrac{1}{2}\sin 2x, \qquad \sin^2 x = \tfrac{1}{2} - \tfrac{1}{2}\cos 2x, \qquad \cos^2 x = \tfrac{1}{2} + \tfrac{1}{2}\cos 2x.$$

The first of these identities is derived from the double-angle formula for the sine function. The other two are the half-angle formulas for sine and cosine.

Example 3

$$\begin{aligned}\int \cos^2 x\,dx &= \int \left(\tfrac{1}{2} + \tfrac{1}{2}\cos 2x\right) dx\\ &= \tfrac{1}{2}\int dx + \tfrac{1}{2}\int \cos 2x\,dx = \tfrac{1}{2}x + \tfrac{1}{4}\sin 2x + C. \quad \square\end{aligned}$$

Example 4

$$\begin{aligned}\int \sin^2 x \cos^2 x\,dx &= \int (\sin x \cos x)^2\,dx\\ &= \tfrac{1}{4}\int \sin^2 2x\,dx\\ &= \tfrac{1}{4}\int (\tfrac{1}{2} - \tfrac{1}{2}\cos 4x)\,dx\\ &= \tfrac{1}{8}\int dx - \tfrac{1}{8}\int \cos 4x\,dx = \tfrac{1}{8}x - \tfrac{1}{32}\sin 4x + C. \quad \square\end{aligned}$$

Example 5

$$\begin{aligned}\int \sin^4 x\,dx &= \int \left(\frac{1-\cos 2x}{2}\right)^2 dx\\ &= \tfrac{1}{4}\int (1-2\cos 2x+\cos^2 2x)\,dx\\ &= \tfrac{1}{4}\int dx - \tfrac{1}{2}\int \cos 2x\,dx + \tfrac{1}{8}\int (1+\cos 4x)\,dx\\ &= \tfrac{1}{4}x - \tfrac{1}{4}\sin 2x + \tfrac{1}{8}x + \tfrac{1}{32}\sin 4x + C\\ &= \tfrac{3}{8}x - \tfrac{1}{4}\sin 2x + \tfrac{1}{32}\sin 4x + C. \quad \square\end{aligned}$$

III. Next we derive *reduction formulas* for integrals of the form

$$\int \sin^n x\,dx \quad \text{and} \quad \int \cos^n x\,dx, \qquad \text{where } n \text{ is a positive integer.}$$

We will use integration by parts to show that

(8.3.1)
$$\int \sin^n x\,dx = -\frac{1}{n}\sin^{n-1} x\cos x + \frac{n-1}{n}\int \sin^{n-2} x\,dx.$$

The corresponding formula for $\int \cos^n x\,dx$ reads

(8.3.2)
$$\int \cos^n x\,dx = \frac{1}{n}\cos^{n-1} x\sin x + \frac{n-1}{n}\int \cos^{n-2} x\,dx.$$

The verification of this formula is left as an exercise.

Formula such as (8.3.1) and (8.3.2) are called reduction formulas because the exponent is reduced (from n to $n-2$ in each of these cases).

To establish (8.3.1), we write $\sin^n x$ as $\sin^{n-1} x \sin x$ and set

$$u = \sin^{n-1} x, \quad dv = \sin x\,dx.$$

Then
$$du = (n-1)\sin^{n-2} x\cos x\,dx, \quad v = -\cos x,$$

and
$$\begin{aligned}\int \sin^n x\,dx &= -\sin^{n-1} x\cos x + (n-1)\int \sin^{n-2} x\cos^2 x\,dx\\ &= -\sin^{n-1} x\cos x + (n-1)\int \sin^{n-2} x(1-\sin^2 x)\,dx.\end{aligned}$$

Therefore,

$$\int \sin^n x\,dx = -\sin^{n-1} x\cos x + (n-1)\int \sin^{n-2} x\,dx - (n-1)\int \sin^n x\,dx.$$

Now, adding $(n-1)\int \sin^n x\,dx$ to both sides, we get

$$n\int \sin^n x\,dx = -\sin^{n-1} x\cos x + (n-1)\int \sin^{n-2} x\,dx$$

and (8.3.1) follows. $\square$

Repeated applications of Formula (8.3.1) or (8.3.2) will reduce a power of the sine or cosine to 0 or 1, depending on whether n is even or odd.

Example 6

$$\begin{aligned}\int \sin^5 x\,dx &= -\tfrac{1}{5}\sin^4 x\cos x + \tfrac{4}{5}\int \sin^3 x\,dx\\ &= -\tfrac{1}{5}\sin^4 x\cos x + \tfrac{4}{5}\Big[-\tfrac{1}{3}\sin^2 x\cos x + \tfrac{2}{3}\int \sin x\,dx\Big]\\ &= -\tfrac{1}{5}\sin^4 x\cos x - \tfrac{4}{15}\sin^2 x\cos x - \tfrac{8}{15}\cos x + C.\end{aligned}$$

Verify that this result is equivalent to the result in Example 2. □

IV. Finally we come to integrals of the form

$$\boxed{\int \sin mx\cos nx\,dx,\quad \int \sin mx\sin nx\,dx,\quad \int \cos mx\cos nx\,dx.}$$

If $m = n$, there is no difficulty. For $m \neq n$, use the identities

$$\begin{aligned}\sin A\cos B &= \tfrac{1}{2}[\sin(A-B) + \sin(A+B)],\\ \sin A\sin B &= \tfrac{1}{2}[\cos(A-B) - \cos(A+B)],\\ \cos A\cos B &= \tfrac{1}{2}[\cos(A-B) + \cos(A+B)].\end{aligned}$$

These identities follow readily from the familiar addition formulas:

$$\begin{aligned}\sin(A+B) &= \sin A\cos B + \cos A\sin B,\\ \sin(A-B) &= \sin A\cos B - \cos A\sin B,\\ \cos(A+B) &= \cos A\cos B - \sin A\sin B,\\ \cos(A-B) &= \cos A\cos B + \sin A\sin B.\end{aligned}$$

Example 7

$$\begin{aligned}\int \sin 5x\sin 3x\,dx &= \int \tfrac{1}{2}[\cos(5x-3x) - \cos(5x+3x)]\,dx\\ &= \tfrac{1}{2}\int [\cos 2x - \cos 8x]\,dx = \tfrac{1}{4}\sin 2x - \tfrac{1}{16}\sin 8x + C.\end{aligned}$$ □

Other Trigonometric Powers

I. First we consider integrals of the form

$$\boxed{\int \tan^n x\,dx,\quad \int \cot^n x\,dx,\quad n \geq 2.}$$

To integrate $\tan^n x$, set

(1) $\tan^n x = \tan^{n-2} x\tan^2 x = (\tan^{n-2} x)(\sec^2 x - 1) = \tan^{n-2} x\sec^2 x - \tan^{n-2} x.$

If $n - 2 \geq 2$, then repeat the reduction to get

$$\begin{aligned}\tan^n x &= \tan^{n-2} x \sec^2 x - (\tan^{n-4} x \sec^2 x - \tan^{n-4} x)\\ &= \tan^{n-2} x \sec^2 x - \tan^{n-4} x \sec^2 x + \tan^{n-4} x.\end{aligned}$$

Continue until the final term is either $\pm \tan x$ (n odd) or ± 1(n even). The idea is to obtain a sum of terms of the form $\tan^k x \sec^2 x$, pairing a power of $\tan x$ with $\sec^2 x$. These terms will be easy to integrate since they have the form $\int u^k du$, where $u = \tan x$, $du = \sec^2 x\, dx$.

Example 8

$$\begin{aligned}\int \tan^6 x\, dx &= \int (\tan^4 x \sec^2 x - \tan^4 x)\, dx\\ &= \int (\tan^4 x \sec^2 x - \tan^2 x \sec^2 x + \tan^2 x)\, dx\\ &= \int (\tan^4 x \sec^2 x - \tan^2 x \sec^2 x + \sec^2 x - 1)\, dx\\ &= \int \tan^4 x \sec^2 x\, dx - \int \tan^2 x \sec^2 x\, dx + \int \sec^2 x\, dx - \int dx\\ &= \tfrac{1}{5} \tan^5 x - \tfrac{1}{3} \tan^3 x + \tan x - x + C. \quad \square\end{aligned}$$

Remark The reduction formula

(8.3.3)
$$\boxed{\int \tan^n x\, dx = \frac{1}{n-1} \tan^{n-1} x - \int \tan^{n-2} x\, dx, \quad n \geq 2.}$$

follows immediately from (1). □

Powers of the cotangent function are handled in the same manner:

$$\cot^n x = \cot^{n-2} x \cot^2 x = (\cot^{n-2} x)(\csc^2 x - 1) = \cot^{n-2} x \csc^2 x - \cot^{n-2} x.$$

The reduction formula for $\int \cot^n x\, dx$ is left as an exercise.

II. Next we consider the integrals

$$\boxed{\int \sec^n x\, dx, \quad \int \csc^n x\, dx, \quad n \geq 2.}$$

We use integration by parts to derive a reduction formula. Set

$$u = \sec^{n-2} x, \quad dv = \sec^2 x\, dx.$$

Then

$$du = (n-2) \sec^{n-3} x \sec x \tan x\, dx = (n-2) \sec^{n-2} x \tan x\, dx, \quad v = \tan x,$$

and

$$\begin{aligned}\int \sec^n x\, dx &= \sec^{n-2} x \tan x - (n-2) \int \sec^{n-2} x \tan^2 x\, dx\\ &= \sec^{n-2} x \tan x - (n-2) \int \sec^{n-2} x (\sec^2 x - 1)\, dx\\ &= \sec^{n-2} x \tan x - (n-2) \int \sec^n x\, dx + (n-2) \int \sec^{n-2} x\, dx.\end{aligned}$$

Now, solving for $\int \sec^n x\,dx$, we get

$$(n-1)\int \sec^n x\,dx = \sec^{n-2} x \tan x + (n-2)\int \sec^{n-2} x\,dx,$$

from which it follows that

(8.3.4) $$\int \sec^n x\,dx = \frac{1}{n-1}\sec^{n-2} x \tan x + \frac{n-2}{n-1}\int \sec^{n-2} x\,dx, \quad n \geq 2.$$

The corresponding reduction formula for $\int \csc^n x\,dx$ is left as an exercise.

Example 9

$$\begin{aligned}\int \sec^3 x\,dx &= \tfrac{1}{2}\sec x \tan x + \tfrac{1}{2}\int \sec x\,dx\\ &= \tfrac{1}{2}\sec x \tan x + \tfrac{1}{2}\ln|\sec x + \tan x| + C.\end{aligned}$$

This integral occurs so frequently in applications that you will find it listed on the inside covers of this text. ❑

Remark There is another approach to calculating $\int \sec^n x\,dx$ when n is small and even. It is similar to the derivation of the reduction formula for powers of the tangent function. Assume n is a positive even integer:

$$\sec^n x = \sec^{n-2} x \sec^2 x = (\sec^2 x)^{(n-2)/2}\sec^2 x = (1+\tan^2 x)^{(n-2)/2}\sec^2 x.$$

Thus, an even power of the secant can be expressed as a sum of terms of the form $\tan^k x \sec^2 x$, which are integrated as $u^k du$. For example,

$$\begin{aligned}\int \sec^6 x\,dx = \int \sec^4 x \sec^2 x\,dx &= \int (\tan^2 x + 1)^2 \sec^2 x\,dx\\ &= \int (u^2+1)^2 du \quad (u = \tan x,\ du = \sec^2 x\,dx)\\ &= \int (u^4 + 2u^2 + 1)\,du = \tfrac{1}{5}u^5 + \tfrac{2}{3}u^3 + u + C\\ &= \tfrac{1}{5}\tan^5 x + \tfrac{2}{3}\tan^3 x + \tan x + C.\end{aligned}$$ ❑

III. Finally we come to integrals of the form

$$\int \tan^m x \sec^n x\,dx, \qquad \int \cot^m x \csc^n x\,dx.$$

Case 1. If n is an even positive integer, write

$$\tan^m x \sec^n x = \tan^m x \sec^{n-2} x \sec^2 x$$

and express $\sec^{n-2} x$ entirely in terms of $\tan^2 x$ using $\sec^2 x = \tan^2 x + 1$.

Example 10

$$\begin{aligned}\int \tan^5 x \sec^4 x\, dx &= \int \tan^5 x \sec^2 x \sec^2 x\, dx\\ &= \int \tan^5 x(\tan^2 x + 1)\sec^2 x\, dx\\ &= \int \tan^7 x \sec^2 x\, dx + \int \tan^5 x \sec^2 x\, dx\\ &= \tfrac{1}{8}\tan^8 x + \tfrac{1}{6}\tan^6 x + C. \quad \Box\end{aligned}$$

Case 2. If m is an odd positive integer, write

$$\tan^m x \sec^n x = \tan^{m-1} x \sec^{n-1} x \sec x \tan x$$

and express $\tan^{m-1} x$ entirely in terms of $\sec^2 x$ using $\tan^2 x = \sec^2 x - 1$.

Example 11

$$\begin{aligned}\int \tan^5 x \sec^3 x\, dx &= \int \tan^4 x \sec^2 x \sec x \tan x\, dx\\ &= \int (\sec^2 x - 1)^2 \sec^2 x \sec x \tan x\, dx\\ &= \int (\sec^6 x - 2\sec^4 x + \sec^2 x)\sec x \tan x\, dx\\ &= \tfrac{1}{7}\sec^7 x - \tfrac{2}{5}\sec^5 x + \tfrac{1}{3}\sec^3 x + C. \quad \Box\end{aligned}$$

Case 3. Finally, if n is odd and m is even, use $\tan^2 x = \sec^2 x - 1$ to write the product as a sum of odd powers of the secant. Then you can either use integration by parts or the reduction formula (8.3.4).

Example 12

$$\begin{aligned}\int \tan^2 x \sec x\, dx &= \int (\sec^2 x - 1)\sec x\, dx\\ &= \int (\sec^3 x - \sec x)\, dx = \int \sec^3 x\, dx - \int \sec x\, dx.\end{aligned}$$

We have calculated each of these integrals before:

$$\int \sec^3 x\, dx = \tfrac{1}{2}\sec x \tan x + \tfrac{1}{2}\ln|\sec x + \tan x| + C$$

and

$$\int \sec x\, dx = \ln|\sec x + \tan x| + C.$$

It follows that

$$\int \tan^2 x \sec x\, dx = \tfrac{1}{2}\sec x \tan x - \tfrac{1}{2}\ln|\sec x + \tan x| + C. \quad \Box$$

You can handle integrals of $\cot^m x \csc^n x$ in a similar manner.

EXERCISES 8.3

Calculate the integral.

1. $\int \sin^3 x\, dx.$

2. $\int_0^{\pi/8} \cos^2 4x\, dx.$

3. $\int_0^{\pi/6} \sin^2 3x\, dx.$

4. $\int \cos^3 x\, dx.$

5. $\int \cos^4 x \sin^3 x\, dx.$

6. $\int \sin^3 x \cos^2 x\, dx.$

7. $\int \sin^3 x \cos^3 x\, dx.$

8. $\int \sin^2 x \cos^4 x\, dx.$

9. $\int \sec^2 \pi x\, dx.$

10. $\int \csc^2 2x\, dx.$

11. $\int \tan^3 x\, dx.$

12. $\int \cot^3 x\, dx.$

13. $\int_0^{\pi} \sin^4 x\, dx.$

14. $\int \cos^3 x \cos 2x\, dx.$

15. $\int \sin 2x \cos 3x\, dx.$

16. $\int_0^{\pi/2} \cos 2x \sin 3x\, dx.$

17. $\int \tan^2 x \sec^2 x\, dx.$

18. $\int \cot^2 x \csc^2 x\, dx.$

19. $\int \sin^2 x \sin 2x\, dx.$

20. $\int_0^{\pi/2} \cos^4 x\, dx.$

21. $\int \sin^6 x\, dx.$

22. $\int \cos^5 x \sin^5 x\, dx.$

23. $\int_{\pi/6}^{\pi/2} \cot^2 x\, dx.$

24. $\int \tan^4 x\, dx.$

25. $\int \cot^3 x \csc^3 x\, dx.$

26. $\int \tan^3 x \sec^3 x\, dx.$

27. $\int \sin 5x \sin 2x\, dx.$

28. $\int \sec^4 3x\, dx.$

29. $\int \sin^{5/2} x \cos^3 x\, dx.$

30. $\int \dfrac{\sin^3 x}{\cos x} dx.$

31. $\int \tan^5 3x\, dx.$

32. $\int \cot^5 2x\, dx.$

33. $\int_{-1/6}^{1/3} \sin^4 (3\pi x) \cos^3 (3\pi x)\, dx.$

34. $\int_0^{1/2} \cos (\pi x) \cos (\tfrac{\pi}{2} x)\, dx.$

35. $\int_0^{\pi/4} \cos 4x \sin 2x\, dx.$

36. $\int (\sin 3x - \sin x)^2\, dx.$

37. $\int \tan^4 x \sec^4 x\, dx.$

38. $\int \cot^4 x \csc^4 x\, dx.$

39. $\int \sin (\tfrac{1}{2} x) \cos (2x)\, dx.$

40. $\int_0^{2\pi} \sin^2 ax\, dx, \quad a \neq 0.$

41. $\int_0^{\pi/4} \tan^3 x \sec^2 x\, dx.$

42. $\int_{\pi/4}^{\pi/2} \csc^3 x \cot x\, dx.$

43. $\int_0^{\pi/6} \tan^2 2x\, dx.$

44. $\int_0^{\pi/3} \tan x \sec^{3/2} x\, dx.$

45. Find the area of the region R bounded by the graph of $f(x) = \sin^2 x$ and the x-axis, $x \in [0, \pi]$.

46. The region bounded by the graph of $f(x) = \cos x$ and the x-axis, $x \in [-\pi/2, \pi/2]$, is revolved about the x-axis. Find the volume of the resulting solid.

47. The region R in Exercise 45 is revolved about the x-axis. Find the volume of the resulting solid.

48. The region bounded by the y-axis and the graphs of $y = \sin x$ and $y = \cos x$, $0 \le x \le \pi/4$, is revolved about the x-axis. Find the volume of the resulting solid.

49. The region bounded by the y-axis, the line $y = 1$, and the graph of $y = \tan x$, $x \in [0, \pi/4]$, is revolved about the x-axis. Find the volume of the resulting solid.

50. The region bounded by the graph of $y = \tan^2 x$ and the x-axis, $x \in [0, \pi/4]$, is revolved about the x-axis. Find the volume of the resulting solid.

51. The region bounded by the graph of $y = \tan x$ and the x-axis, $x \in [0, \pi/4]$, is revolved about the line $y = -1$. Find the volume of the resulting solid.

52. The region bounded by the graph of $y = \sec^2 x$ and the x-axis, $x \in [0, \pi/4]$, is revolved about the x-axis. Find the volume of the resulting solid.

53. Let m and n be positive integers. Prove that

$$\int \sin mx \sin nx\, dx = \begin{cases} \dfrac{\sin (m-n)x}{2(m-n)} - \dfrac{\sin (m+n)x}{2(m+n)} + C, & \text{if } m \neq n \\[2mm] \dfrac{x}{2} - \dfrac{\sin 2mx}{4m} + C, & \text{if } m = n. \end{cases}$$

54. Derive formulas corresponding to the formula in Exercise 53 for

(a) $\int \sin mx \cos nx\, dx.$

(b) $\int \cos mx \cos nx\, dx.$

55. Use the result in Exercise 53 to prove that

$$\int_{-\pi}^{\pi} \sin mx \sin nx\, dx = \begin{cases} 0, & \text{if } m \neq n \\ \pi, & \text{if } m = n. \end{cases}$$

56. (a) Evaluate $\int_{-\pi}^{\pi} \sin mx \cos nx\, dx.$

(b) Evaluate $\int_{-\pi}^{\pi} \cos mx \cos nx \, dx$.

NOTE: The formulas in Exercises 55 and 56 are important in applied mathematics. They arise in Fourier analysis, which involves the approximation of functions by sums of sines and cosines.

57. Verify reduction formula (8.3.2).

58. (a) Use reduction formula (8.3.1) to show that

$$\int_0^{\pi/2} \sin^n x \, dx = \frac{n-1}{n} \int_0^{\pi/2} \sin^{n-2} x \, dx.$$

(b) Show that

$$\int_0^{\pi/2} \sin^n x \, dx =$$

$$\begin{cases} \dfrac{(n-1)\cdots 5\cdot 3\cdot 1}{n\cdots 6\cdot 4\cdot 2}\cdot\dfrac{\pi}{2}, & n \text{ even}, \ n \geq 2 \\ \dfrac{(n-1)\cdots 4\cdot 2}{n\cdots 5\cdot 3}, & n \text{ odd}, \ n \geq 3. \end{cases}$$

These formulas are known as the *Wallis sine formulas*.

(c) Show that

$$\int_0^{\pi/2} \cos^n x \, dx = \int_0^{\pi/2} \sin^n x \, dx.$$

Use the results in Exercise 58 to evaluate the integrals in Exercises 59 and 60.

59. $\displaystyle\int_0^{\pi/2} \sin^7 x \, dx.$

60. $\displaystyle\int_0^{\pi/2} \cos^6 x \, dx.$

61. Derive the reduction formula: for $n > 1$,

$$\int \cot^n x \, dx = \frac{-\cot^{n-1} x}{n-1} - \int \cot^{n-2} x \, dx.$$

62. Use Exercise 61 to calculate the integrals:

(a) $\displaystyle\int \cot^3 x \, dx$; (b) $\displaystyle\int \cot^4 x \, dx$; (c) $\displaystyle\int \cot^5 2x \, dx$.

63. Derive the reduction formula: for $n > 1$,

$$\int \csc^n x \, dx = \frac{-csc^{n-2} x \cot x}{n-1} + \frac{n-2}{n-1} \int \csc^{n-2} x \, dx.$$

64. Use Exercise 63 to calculate the integrals:

(a) $\displaystyle\int \csc^3 x \, dx$; (b) $\displaystyle\int \csc^4 x \, dx$; (c) $\displaystyle\int \csc^5 3x \, dx$.

65. Calculate $\displaystyle\int \sin x \cos x \, dx$ in three ways.

a. Let $u = \sin x$.

b. Let $u = \cos x$.

c. Set $\sin x \cos x = \frac{1}{2} \sin 2x$.

d. Reconcile the results in parts (a), (b), and (c).

C **66.** Use a graphing utility to draw the graph of $f(x) = x + \sin 2x$, $x \in [0, \pi]$. The region bounded by the graph of f and the x-axis is revolved about the x-axis.

(a) Use a CAS to calculate the volume of the resulting solid.

(b) Use the methods of this section to calculate the volume of the solid.

C **67.** Use a graphing utility to draw the graph of $g(x) = \sin^2(x^2)$, $x \in [0, \sqrt{\pi}]$. The region bounded by the graph of g and the x-axis is revolved about the y-axis.

(a) Use a CAS to calculate the volume of the resulting solid.

(b) Use the methods of this section to calculate the volume of the solid.

C **68.** Use a graphing utility to draw the graphs of $f(x) = 1 + \cos x$ and $g(x) = \sin(x/2)$, $x \in [0, 2\pi]$, together.

(a) Use a CAS to find the points of intersection of the two curves and then find the area of the region between the curves.

(c) The region between the curves is rotated about the x-axis. Use a CAS to find the volume of the solid of revolution.

■ 8.4 INTEGRALS INVOLVING $\sqrt{a^2 - x^2}$, $\sqrt{a^2 + x^2}$, $\sqrt{x^2 - a^2}$

Such integrals can often by calculated by a *trigonometric substitution*:

for $\sqrt{a^2 - x^2}$, set $a \sin u = x$;

for $\sqrt{a^2 + x^2}$, set $a \tan u = x$;

for $\sqrt{x^2 - a^2}$, set $a \sec u = x$.

In each case, take $a > 0$.

Example 1 Find $\displaystyle\int \frac{dx}{(a^2-x^2)^{3/2}}$.

SOLUTION First note that

$$\int \frac{dx}{(a^2-x^2)^{3/2}} = \int \frac{dx}{\left[\sqrt{a^2-x^2}\right]^3};$$

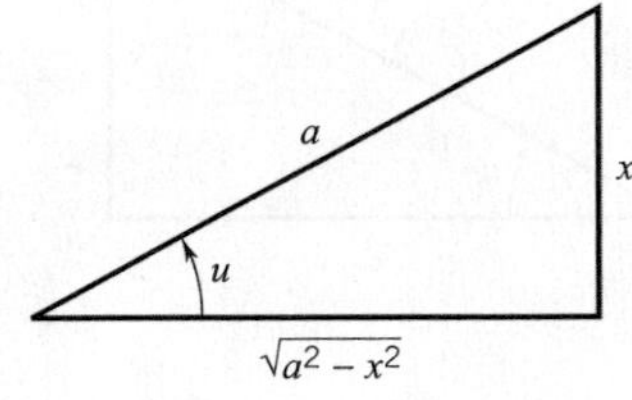

the integral involves $\sqrt{a^2-x^2}$. Therefore, we set

$$a \sin u = x, \quad a \cos u \, du = dx.$$

Then $$\sqrt{a^2-x^2} = a\cos u.$$

Now
$$\begin{aligned}
\int \frac{dx}{(a^2-x^2)^{3/2}} &= \int \frac{a\cos u}{(a\cos u)^3}\, du \\
&= \frac{1}{a^2}\int \frac{\cos u}{(\cos u)^3}\, du \\
&= \frac{1}{a^2}\int \sec^2 u \, du \\
&= \frac{1}{a^2}\tan u + C = \frac{x}{a^2\sqrt{a^2-x^2}} + C. \quad \square
\end{aligned}$$

Example 2 Find $\displaystyle\int \sqrt{a^2+x^2}\, dx$.

SOLUTION Set $a\tan u = x, \quad a\sec^2 u\, du = dx.$ Then

$$\sqrt{a^2+x^2} = a\sec u.$$

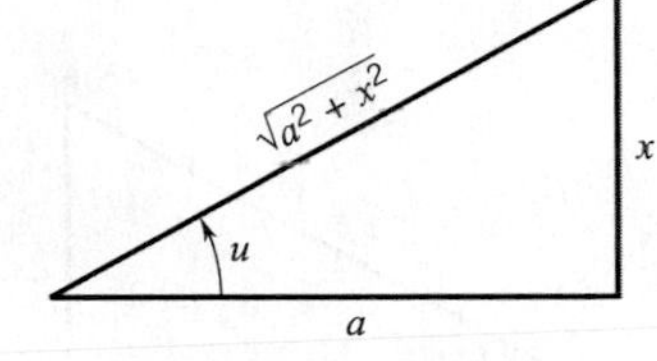

Now
$$\begin{aligned}
\int \sqrt{a^2+x^2}\, dx &= \int (a\sec u)\, a\sec^2 u\, du \\
&= a^2 \int \sec^3 u\, du \\
&= \frac{a^2}{2}(\sec u \tan u + \ln|\sec u + \tan u|) + C \\
&= \frac{a^2}{2}\left[\frac{\sqrt{a^2+x^2}}{a}\left(\frac{x}{a}\right) + \ln\left|\frac{\sqrt{a^2+x^2}}{a} + \frac{x}{a}\right|\right] + C \\
&= \tfrac{1}{2}x\sqrt{a^2+x^2} + \tfrac{1}{2}a^2 \ln(x+\sqrt{a^2+x^2}) - \tfrac{1}{2}a^2 \ln a + C.
\end{aligned}$$

Example 9, Section 8.3 ↑

We can absorb the constant $-\frac{1}{2}a^2 \ln a$ in C and write

(8.4.1) $$\int \sqrt{a^2+x^2}\, dx = \tfrac{1}{2}x\sqrt{a^2+x^2} + \tfrac{1}{2}a^2 \ln(x+\sqrt{a^2+x^2}) + C.$$

This is a standard integration formula. $\square$

Example 3 Find $\displaystyle\int \frac{dx}{x^2\sqrt{x^2-4}}$.

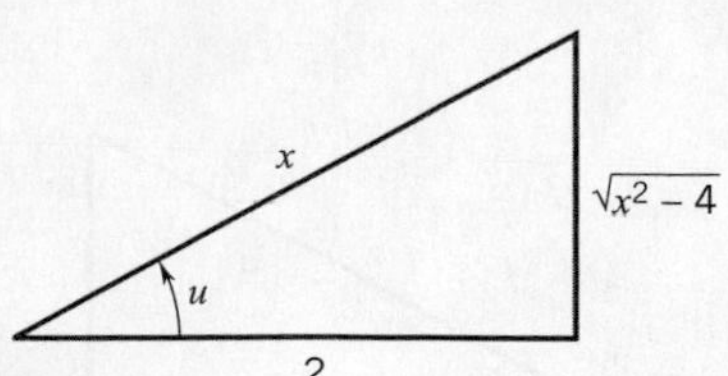

SOLUTION Set $2\sec u = x, \quad 2\sec u \tan u\, du = dx.$ Then

$$\sqrt{x^2-4} = 2\tan u.$$

Now
$$\int \frac{dx}{x^2\sqrt{x^2-4}} = \int \frac{2\sec u \tan u}{4\sec^2 u \cdot 2\tan u}\, du$$
$$= \tfrac{1}{4}\int \frac{1}{\sec u}\, du$$
$$= \tfrac{1}{4}\int \cos u\, du$$
$$= \tfrac{1}{4}\sin u + C = \frac{\sqrt{x^2-4}}{4x} + C. \quad \square$$

Remark Sometimes there is more than one way to calculate an integral. For example, to calculate

$$\int \frac{x}{\sqrt{a^2-x^2}}\, dx$$

you do not need to set $a\sin u = x$. You can carry out the integration more directly by substituting $u = a^2 - x^2$. (Try both substitutions and decide which you like better.) $\square$

Example 4 Find $\displaystyle\int \frac{dx}{x\sqrt{4x^2+9}}$.

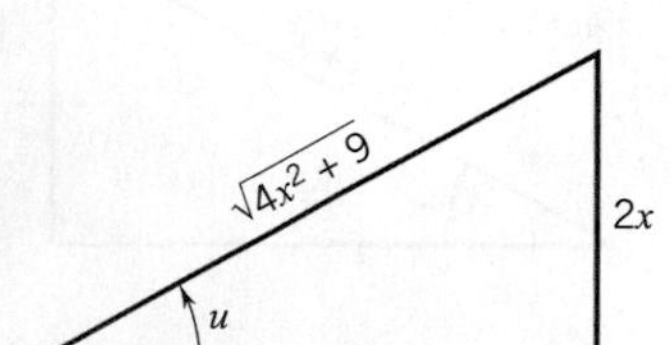

SOLUTION Set $3\tan u = 2x, \quad 3\sec^2 u\, du = 2\, dx.$ Then

$$\sqrt{4x^2+9} = 3\sec u$$

and
$$\int \frac{dx}{x\sqrt{4x^2+9}} = \int \frac{\frac{3}{2}\sec^2 u}{\frac{3}{2}\tan u \cdot 3\sec u}\, du$$
$$= \tfrac{1}{3}\int \frac{\sec u}{\tan u}\, du$$
$$= \tfrac{1}{3}\int \csc u\, du$$
$$= \tfrac{1}{3}\ln|\csc u - \cot u| + C$$
$$= \frac{1}{3}\ln\left|\frac{\sqrt{4x^2+9}-3}{2x}\right| + C. \quad \square$$

The next example requires that we first complete the square under the radical.

Example 5 Find $\displaystyle\int \frac{x}{\sqrt{x^2+2x-3}}\, dx$.

SOLUTION First note that

$$\int \frac{x}{\sqrt{x^2 + 2x - 3}}\, dx = \int \frac{x}{\sqrt{(x+1)^2 - 4}}\, dx.$$

Now set $2 \sec u = x + 1,\quad 2 \sec u \tan u \, du = dx.$ Then

$$\sqrt{(x+1)^2 - 4} = 2 \tan u$$

and

$$\begin{aligned}
\int \frac{x}{\sqrt{(x+1)^2 - 4}}\, dx &= \int \frac{(2 \sec u - 1)\, 2 \sec u \tan u}{2 \tan u}\, du \\
&= \int (2 \sec^2 u - \sec u)\, du \\
&= 2 \tan u - \ln |\sec u + \tan u| + C \\
&= \sqrt{x^2 + 2x - 3} - \ln \left| \frac{x + 1 + \sqrt{x^2 + 2x - 3}}{2} \right| + C. \quad \square
\end{aligned}$$

In our last example, we calculate a definite integral.

Example 6 Calculate $\displaystyle\int_{-r}^{r} \sqrt{r^2 - x^2}\, dx$, where r is a positive constant.

SOLUTION Set $r \sin u = x,\quad r \cos u \, du = dx.$ Then

$$\sqrt{r^2 - x^2} = r \cos u.$$

Observe that $u = -\frac{\pi}{2}$ at $x = -r$ and $u = \frac{\pi}{2}$ at $x = r$. Thus

$$\begin{aligned}
\int_{-r}^{r} \sqrt{r^2 - x^2} dx &= \int_{-\pi/2}^{\pi/2} (r \cos u) \cdot r \cos u \, du = r^2 \int_{-\pi/2}^{\pi/2} \cos^2 u \, du \\
&= r^2 \int_{-\pi/2}^{\pi/2} \left(\tfrac{1}{2} + \tfrac{1}{2} \cos 2u\right) du \\
&= \frac{r^2}{2} \Big[u + \tfrac{1}{2} \sin 2u \Big]_{-\pi/2}^{\pi/2} = \frac{\pi r^2}{2}.
\end{aligned}$$

Note that the graph of $y = \sqrt{r^2 - x^2}, -r \le x \le r$, is the semicircle of radius r shown in Figure 8.4.1. Thus we have found that the area enclosed by a semicircle of radius r is $\pi r^2/2$. The familiar formula for the area enclosed by a circle of radius r, $A = \pi r^2$, follows immediately from this result. $\square$

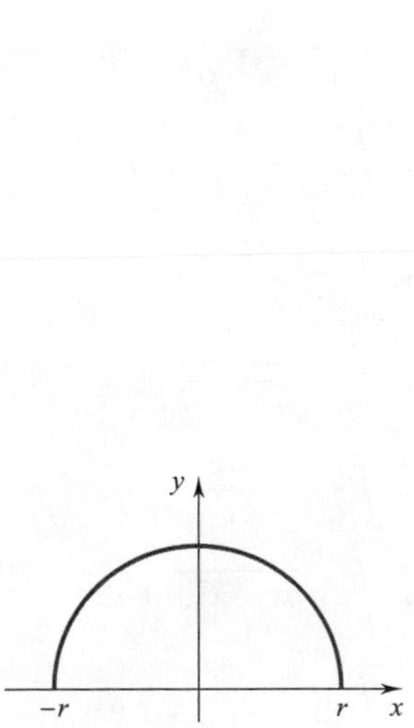

Figure 8.4.1

A trigonometric substitution may be effective even in cases where the quadratic in the integrand is not under a radical. For example, the reduction formula

(8.4.2)
$$\int \frac{dx}{(x^2 + a^2)^n} = \frac{1}{a^{2n-1}} \int \cos^{2(n-1)} u \, du$$

can be obtained by setting $a \tan u = x,\ a \sec^2 u \, du = dx$. The proof is left to you as an exercise.

EXERCISES 8.4

Calculate the integral.

1. $\displaystyle\int \frac{dx}{\sqrt{a^2 - x^2}}.$

2. $\displaystyle\int_{5/2}^{4} \frac{x}{\sqrt{x^2 - 4}}\,dx.$

3. $\displaystyle\int \sqrt{x^2 - 1}\,dx.$

4. $\displaystyle\int \frac{x}{\sqrt{4 - x^2}}\,dx.$

5. $\displaystyle\int \frac{x^2}{\sqrt{4 - x^2}}\,dx$

6. $\displaystyle\int \frac{x^2}{\sqrt{x^2 - 4}}\,dx.$

7. $\displaystyle\int \frac{x}{(1 - x^2)^{3/2}}\,dx.$

8. $\displaystyle\int \frac{x^2}{\sqrt{4 + x^2}}\,dx.$

9. $\displaystyle\int_0^{1/2} \frac{x^2}{(1 - x^2)^{3/2}}\,dx.$

10. $\displaystyle\int \frac{x}{a^2 + x^2}\,dx.$

11. $\displaystyle\int x\sqrt{4 - x^2}\,dx.$

12. $\displaystyle\int_0^{2} \frac{x^3}{\sqrt{16 - x^2}}\,dx.$

13. $\displaystyle\int_0^{5} x^2\sqrt{25 - x^2}\,dx.$

14. $\displaystyle\int \frac{\sqrt{1 - x^2}}{x^4}\,dx.$

15. $\displaystyle\int \frac{x^2}{(x^2 + 8)^{3/2}}\,dx.$

16. $\displaystyle\int_0^{a} \sqrt{a^2 - x^2}\,dx.$

17. $\displaystyle\int \frac{dx}{x\sqrt{a^2 - x^2}}.$

18. $\displaystyle\int \frac{\sqrt{x^2 - 1}}{x}\,dx.$

19. $\displaystyle\int_0^{3} \frac{x^3}{\sqrt{9 + x^2}}\,dx.$

20. $\displaystyle\int \frac{dx}{x^2\sqrt{a^2 - x^2}}.$

21. $\displaystyle\int \frac{dx}{x^2\sqrt{a^2 + x^2}}.$

22. $\displaystyle\int \frac{dx}{(x^2 + 2)^{3/2}}.$

23. $\displaystyle\int_0^{1} \frac{dx}{(5 - x^2)^{3/2}}.$

24. $\displaystyle\int \frac{dx}{e^x\sqrt{4 + e^{2x}}}.$

25. $\displaystyle\int \frac{dx}{x^2\sqrt{x^2 - a^2}}.$

26. $\displaystyle\int \frac{e^x}{\sqrt{9 - e^{2x}}}\,dx.$

27. $\displaystyle\int \frac{dx}{e^x\sqrt{e^{2x} - 9}}.$

28. $\displaystyle\int \frac{dx}{\sqrt{x^2 - 2x - 3}}.$

29. $\displaystyle\int \frac{dx}{(x^2 - 4x + 4)^{3/2}}.$

30. $\displaystyle\int \frac{x}{\sqrt{6x - x^2}}\,dx.$

31. $\displaystyle\int x\sqrt{6x - x^2 - 8}\,dx.$

32. $\displaystyle\int \frac{x + 2}{\sqrt{x^2 + 4x + 13}}\,dx.$

33. $\displaystyle\int \frac{x}{(x^2 + 2x + 5)^2}\,dx.$

34. $\displaystyle\int \frac{x}{\sqrt{x^2 - 2x + 3}}\,dx.$

35. Verify integration formula (8.2.7):

$$\int \sec^{-1} x\,dx = x\sec^{-1} x - \ln\left[x + \sqrt{x^2 - 1}\right] + C.$$

36. Calculate $\displaystyle\int \frac{\sqrt{a^2 - x^2}}{x}\,dx$ by using:

(a) The substitution $u = \sqrt{a^2 - x^2}$.

(b) A trigonometric substitution.

(c) Reconcile your two results.

37. Verify reduction formula (8.4.2).

In Exercises 38 and 39, use reduction formula (8.4.2) to calculate the integral.

38. $\displaystyle\int \frac{1}{(x^2 + 1)^2}\,dx.$

39. $\displaystyle\int \frac{1}{(x^2 + 1)^3}\,dx.$

In Exercises 40 and 41, use integration by parts and then a trigonometric substitution to calculate the integral.

40. $\displaystyle\int x\tan^{-1} x\,dx.$

41. $\displaystyle\int x\sin^{-1} x\,dx.$

42. Find the area under the graph of $y = (\sqrt{x^2 - 9})/x$ from $x = 3$ to $x = 5$.

43. The region bounded by the graph of $f(x) = 1/(1 + x^2)$ and the x-axis between $x = 0$ and $x = 1$ is revolved about the x-axis. Find the volume of the solid that is generated.

44. In a disc of radius r a chord h units from the center generates a region of the disc called a *segment* (see the figure). Find a formula for the area of the segment.

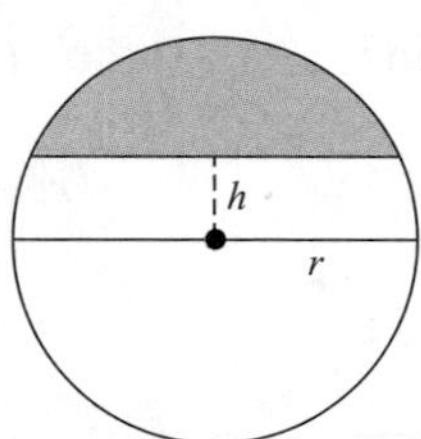

45. Derive the formula $A = \frac{1}{2}r^2\theta$ for the area of a sector of a circle of radius r and central angle θ (measured in radians). HINT: Assume first that $0 < \theta < \frac{1}{2}\pi$ and subdivide the region as indicated in the figure. Then verify that the formula holds for any sector.

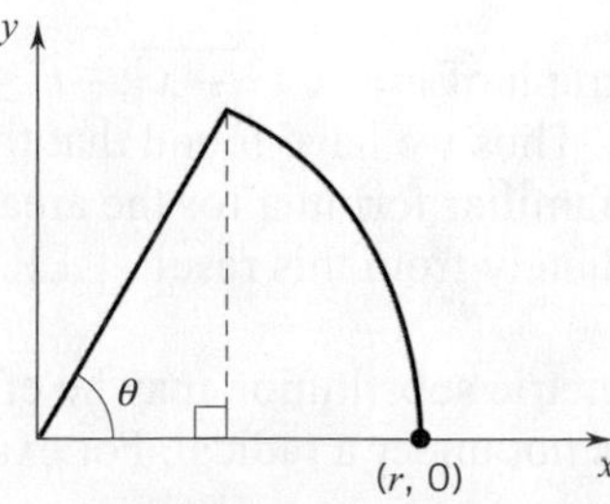

46. Find the area of the region bounded on the left and right by the two branches of the hyperbola $(x^2/a^2) - (y^2/b^2) = 1$, and above and below by the lines $y = \pm b$.

47. Find the area of the region bounded by the right branch of the hyperbola $(x^2/9) - (y^2/16) = 1$ and the line $x = 5$.

48. If the circle $(x-b)^2+y^2=a^2, b>a>0$, is revolved about the y-axis, the resulting "doughnut-shaped" solid is called a *torus*. Use the shell method to find the formula for the volume of the torus. See Exercise 43 in Section 6.3 and Example 5 in Section 6.4.

49. Calculate the mass and the center of mass of a rod that extends from $x=0$ to $x=a>0$ and has mass density $\lambda(x)=(x^2+a^2)^{-1/2}$.

50. Calculate the mass and the center of the rod in Exercise 49 if the mass density is given by $\lambda(x)=(x^2+a^2)^{-3/2}$.

For Exercises 51–53, let Ω be the region under the curve $y=\sqrt{x^2-a^2}, x\in[a,\sqrt{2}a]$.

51. Sketch Ω, find the area, and locate the centroid.

52. Find the volume of the solid generated by revolving Ω about the x-axis and determine the centroid of that solid.

53. Find the volume of the solid generated by revolving Ω about the y-axis and determine the centroid of that solid.

54. (a) Use a trigonometric substitution to show that, for $a>0$,

$$\int \frac{1}{\sqrt{a^2+x^2}}\,dx = \ln\left|x+\sqrt{a^2+x^2}\right| + C.$$

(b) Use the *hyperbolic substitution* $x=a\sinh u$ to show that

$$\int \frac{1}{\sqrt{a^2+x^2}}\,dx = \sinh^{-1}\left(\frac{x}{a}\right) + C.$$

See Exercise 41, Section 7.9.

55. (a) Use a trigonometric substitution to show that, for $a>0$,

$$\int \frac{1}{\sqrt{x^2-a^2}}\,dx = \ln\left|x+\sqrt{x^2-a^2}\right| + C.$$

(b) Use the *hyperbolic substitution* $x=a\cosh u$ with $x>a$ to show that

$$\int \frac{1}{\sqrt{x^2-a^2}}\,dx = \cosh^{-1}\left(\frac{x}{a}\right) + C.$$

See Exercise 42, Section 7.9.

C **56.** Let

$$f(x)=\frac{x^2}{\sqrt{1-x^2}}.$$

(a) Use a graphing utility to sketch the graph of f.

(b) Find the area of the region bounded by the graph of f and the x-axis from $x=0$ to $x=\frac{1}{2}$.

(c) Find the volume of the solid generated by revolving the region in part (b) about the y-axis.

C **57.** Let

$$f(x)=\frac{\sqrt{x^2-9}}{x^2}, \quad x\geq 3.$$

(a) Use a graphing utility to sketch the graph of f.

(b) Find the area of the region bounded by the graph of f and the x-axis from $x=3$ to $x=6$.

(c) Find the centroid of the region.

■ 8.5 RATIONAL FUNCTIONS; PARTIAL FRACTIONS

In this section we present a method for integrating rational functions. Recall that a rational function is, by definition, the quotient of two polynomials. For example,

$$\frac{1}{x^2-4}, \quad \frac{2x^2+3}{x(x-1)^2}, \quad \frac{3x^4-20x^2+17}{x^3+2x^2-7}$$

are rational functions, but

$$\frac{1}{\sqrt{x}}, \quad \frac{x^2+1}{\ln x}, \quad \frac{|x-2|}{x^2+1}$$

are not rational functions.

A rational function $R(x)=P(x)/Q(x)$ is said to be *proper* if the degree of the numerator is less than the degree of the denominator. If the degree of the numerator is greater than or equal to the degree of the denominator, then the rational function is *improper*.† We will focus our attention on *proper rational functions* because any

† These terms are taken from the familiar terms used to described rational numbers $\frac{p}{q}$.

improper rational function can be written as the sum of a polynomial and a proper rational function.

$$\frac{P(x)}{Q(x)} = p(x) + \frac{r(x)}{Q(x)}.\dagger$$

This is accomplished simply by dividing the denominator into the numerator (the polynomial p is called the *quotient* and the polynomial r is called the *remainder*).

It is shown in precalculus algebra courses that every proper rational function can be written in one and only one way as a sum of fractions of the form

(8.5.1)
$$\frac{A}{(x-\alpha)^k} \quad \text{and} \quad \frac{Bx+C}{(x^2+\beta x+\gamma)^k}$$

with the quadratic $x^2 + \beta x + \gamma$ irreducible (that is, not factorable into linear factors with real coefficients; $\beta^2 - 4\gamma < 0$). Such fractions are called *partial fractions*. When a rational function $R(x)$ is expressed as a sum of partial fractions, the sum is called the *partial fraction decomposition of R*.

We begin by giving some examples of partial fraction decompositions. We then illustrate how these can be used to calculate integrals involving rational functions.

Example 1 (*The denominator splits into distinct linear factors.*) For

$$\frac{2x}{x^2-x-2} = \frac{2x}{(x-2)(x+1)},$$

we write
$$\frac{2x}{x^2-x-2} = \frac{A}{x-2} + \frac{B}{x+1},$$

where A and B are constants whose values are to be determined.

Multiplication by $(x-2)(x+1)$ yields the equation

$$(1) \qquad 2x = A(x+1) + B(x-2).$$

We illustrate two methods for finding A and B.

METHOD 1 Substitute numbers for x in (1):

setting $x = 2$, we get $4 = 3A$, which gives $A = \frac{4}{3}$;

setting $x = -1$, we get $-2 = -3B$, which gives $B = \frac{2}{3}$.

The desired decomposition reads

$$\frac{2x}{x^2-x-2} = \frac{4}{3(x-2)} + \frac{2}{3(x+1)}.$$

You can verify this by carrying out the addition on the right.

METHOD 2 This method is based on the general fact that for polynomials

$$P(x) = a_n x^n + a_{n-1}x^{n-1} + \cdots + a_0 \quad \text{and} \quad Q(x) = b_k x^k + b_{k-1}x^{k-1} + \cdots + b_0;$$

† This is analogous to writing an improper fraction as a so-called *mixed number*.

$P(x) = Q(x)$ *for all* x, iff

$$n = k, \quad \text{and} \quad a_n = b_n,\ a_{n-1} = b_{n-1}, \ldots, a_0 = b_0.$$

Now, rewrite (1) as

$$2x = (A + B)x + (A - 2B)$$

and then equate the coefficients of corresponding powers of x to produce the system of equations

$$A + B = 2$$
$$A - 2B = 0.$$

We can then get A and B by solving these equations simultaneously. Again, the solutions are: $A = \frac{4}{3}, \quad B = \frac{2}{3}$. □

In general, each distinct linear factor $x - \alpha$ in the denominator gives rise to a term of the form

$$\frac{A}{x - \alpha}.$$

Example 2 (*The denominator has a repeated linear factor.*) For

$$\frac{2x^2 + 3}{x(x-1)^2},$$

we write

$$\frac{2x^2 + 3}{x(x-1)^2} = \frac{A}{x} + \frac{B}{x-1} + \frac{C}{(x-1)^2}.$$

This leads to

$$2x^2 + 3 = A(x-1)^2 + Bx(x-1) + Cx.$$

To determine the three coefficients A, B, C, we need to substitute three values for x. We select 0 and 1 because for those values of x several terms on the right side will drop out. As a third value of x, any other number will do; we select 2 just to keep the arithmetic simple.

Setting $x = 0$, we get $3 = A$.

Setting $x = 1$, we get $5 = C$.

Setting $x = 2$, we get $11 = A + 2B + 2C$

which, with $A = 3$ and $C = 5$, gives $B = -1$.

Therefore the decomposition is:

$$\frac{2x^2 + 3}{x(x-1)^2} = \frac{3}{x} - \frac{1}{x-1} + \frac{5}{(x-1)^2}. \quad □$$

In general, each factor of the form $(x-\alpha)^k$ in the denominator gives rise to an expression of the form

$$\frac{A_1}{x-\alpha}+\frac{A_2}{(x-\alpha)^2}+\cdots+\frac{A_k}{(x-\alpha)^k}.$$

Example 3 (*The denominator has an irreducible quadratic factor.*) For

$$\frac{x^2+5x+2}{(x+1)(x^2+1)},$$

we write

$$\frac{x^2+5x+2}{(x+1)(x^2+1)}=\frac{A}{x+1}+\frac{Bx+C}{x^2+1}$$

and obtain

$$(2)\ \ x^2+5x+1=A(x^2+1)+(Bx+C)(x+1)=(A+B)x^2+(B+C)x+A+C.$$

This time we equate the coefficients of corresponding powers of x to illustrate the second method for finding the unknowns A, B, C in the decomposition. We obtain the system of equations:

$$\begin{aligned} A+B &= 1,\\ B+C &= 5,\\ A\ \ +C &= 2. \end{aligned}$$

The solution of this system is $A=-1,\ B=2,\ C=3$ (verify). The decomposition reads

$$\frac{x^2+5x+2}{(x+1)(x^2+1)}=\frac{-1}{x+1}+\frac{2x+3}{x^2+1}.$$

NOTE: You could have used the first method to solve for A, B, C equally as well. For example, try setting $x=-1, x=0$, and $x=1$ in (2). □

Example 4 (*The denominator has an irreducible quadratic factor.*) For

$$\frac{1}{x(x^2+x+1)}$$

we write

$$\frac{1}{x(x^2+x+1)}=\frac{A}{x}+\frac{Bx+C}{x^2+x+1}$$

and obtain

$$1=A(x^2+x+1)+(Bx+C)x.$$

Again we select values of x that produce zeros or simple arithmetic.

$$\begin{aligned} 1 &= A & &(x = 0),\\ 1 &= 3A + B + C & &(x = 1),\\ 1 &= A + B - C & &(x = -1). \end{aligned}$$

From this we find that

$$A = 1, \qquad B = -1, \qquad C = -1,$$

and therefore

$$\frac{1}{x(x^2+x+1)} = \frac{1}{x} - \frac{x+1}{x^2+x+1}. \quad \square$$

In general, each irreducible quadratic factor $x^2 + \beta x + \gamma$ in the denominator gives rise to a term of the form

$$\frac{Ax+B}{x^2+\beta x+\gamma}.$$

Example 5 (*The denominator has a repeated irreducible quadratic factor.*) For

$$\frac{3x^4 + x^3 + 20x^2 + 3x + 31}{(x+1)(x^2+4)^2},$$

we write

$$\frac{3x^4 + x^3 + 20x^2 + 3x + 31}{(x+1)(x^2+4)^2} = \frac{A}{x+1} + \frac{Bx+C}{x^2+4} + \frac{Dx+E}{(x^2+4)^2}.$$

This gives

$$3x^4 + x^3 + 20x^2 + 3x + 31$$
$$= A(x^2+4)^2 + (Bx+C)(x+1)(x^2+4) + (Dx+E)(x+1).$$

This time we use $-1, 0, 1, 2$, and -2.

$$\begin{aligned} 50 &= 25A & &(x=-1),\\ 31 &= 16A + 4C + E & &(x=0),\\ 58 &= 25A + 10B + 10C + 2D + 2E & &(x=1),\\ 173 &= 64A + 48B + 24C + 6D + 3E & &(x=2),\\ 145 &= 64A + 16B - 8C + 2D - E & &(x=-2). \end{aligned}$$

With a little patience, you can determine that

$$A = 2, \qquad B = 1, \qquad C = 0, \qquad D = 0, \qquad \text{and} \qquad E = -1.$$

This gives the decomposition

$$\frac{3x^4 + x^3 + 20x^2 + 3x + 31}{(x+1)(x^2+4)^2} = \frac{2}{x+1} + \frac{x}{x^2+4} - \frac{1}{(x^2+4)^2}. \quad \square$$

In general, each multiple irreducible quadratic factor $(x^2 + \beta x + \gamma)^k$ in the denominator gives rise to an expression of the form

$$\frac{A_1 x + B_1}{x^2 + \beta x + \gamma} + \frac{A_2 x + B_2}{(x^2 + \beta x + \gamma)^2} + \cdots + \frac{A_k x + B_k}{(x^2 + \beta x + \gamma)^k}.$$

As indicated at the beginning of this section, if the rational function is improper, then a polynomial will appear in the decomposition.

Example 6 (*An improper rational function.*) The rational function

$$\frac{x^5 + 2}{x^2 - 1}$$

is improper. Dividing the denominator into the numerator, we find that

$$\frac{x^5 + 2}{x^2 - 1} = x^3 + x + \frac{x + 2}{x^2 - 1}. \qquad \text{(verify this)}$$

The decomposition of the remaining fraction reads

$$\frac{x + 2}{x^2 - 1} = \frac{A}{x + 1} + \frac{B}{x - 1}.$$

As you can verify, $A = -\frac{1}{2}$ and $B = \frac{3}{2}$, and we have

$$\frac{x^5 + 2}{x^2 - 1} = x^3 + x - \frac{1}{2(x + 1)} + \frac{3}{2(x - 1)}. \quad \square$$

We have been decomposing rational functions into partial fractions in order to integrate them. Here we carry out the integrations, leaving some of the details to you.

Example 1′

$$\begin{aligned} \int \frac{2x}{x^2 - x - 2}\, dx &= \int \left[\frac{4}{3(x - 2)} + \frac{2}{3(x + 1)}\right] dx \\ &= \tfrac{4}{3} \ln |x - 2| + \tfrac{2}{3} \ln |x + 1| + C \\ &= \tfrac{1}{3} \ln [(x - 2)^4 (x + 1)^2] + C. \quad \square \end{aligned}$$

Example 2′

$$\begin{aligned} \int \frac{2x^2 + 3}{x(x - 1)^2}\, dx &= \int \left[\frac{3}{x} - \frac{1}{x - 1} + \frac{5}{(x - 1)^2}\right] dx \\ &= 3 \ln |x| - \ln |x - 1| - \frac{5}{x - 1} + C \\ &= \ln \left|\frac{x^3}{x - 1}\right| - \frac{5}{x - 1} + C \quad \square \end{aligned}$$

Example 3′

$$\int \frac{x^2+5x+2}{(x+1)(x^2+1)}\,dx = \int \left(\frac{-1}{x+1}+\frac{2x+3}{x^2+1}\right)dx$$
$$= -\int \frac{1}{x+1}\,dx + \int \frac{2x+3}{x^2+1}\,dx.$$

Since

$$-\int \frac{1}{x+1}dx = -\ln|x+1| + C_1$$

and

$$\int \frac{2x+3}{x^2+1}\,dx = \int \frac{2x}{x^2+1}\,dx + 3\int \frac{1}{x^2+1}\,dx = \ln(x^2+1) + 3\tan^{-1}x + C_2,$$

We have

$$\int \frac{x^2+5x+2}{(x+1)(x^2+1)}\,dx = -\ln|x+1| + \ln(x^2+1) + 3\tan^{-1}x + C$$
$$= \ln\left|\frac{x^2+1}{x+1}\right| + 3\tan^{-1}x + C. \quad \square$$

Example 4′

$$P = \int \frac{dx}{x(x^2+x+1)} = \int \left(\frac{1}{x} - \frac{x+1}{x^2+x+1}\right)dx = \ln|x| - \int \frac{x+1}{x^2+x+1}\,dx.$$

To calculate the remaining integral, note that $(d/dx)(x^2+x+1) = 2x+1$. We will manipulate the integrand to get a term of the form du/u, where $u = x^2+x+1$ and $du = (2x+1)\,dx$:

$$\frac{x+1}{x^2+x+1} = \frac{\frac{1}{2}[2x+1]+\frac{1}{2}}{x^2+x+1} = \frac{1}{2}\,\frac{2x+1}{x^2+x+1} + \frac{1}{2}\,\frac{1}{x^2+x+1}.$$

Therefore,

$$\int \frac{x+1}{x^2+x+1}\,dx = \frac{1}{2}\int \frac{2x+1}{x^2+x+1}\,dx + \frac{1}{2}\int \frac{1}{x^2+x+1}\,dx.$$

Now

$$\frac{1}{2}\int \frac{2x+1}{x^2+x+1}\,dx = \frac{1}{2}\ln(x^2+x+1) + C_1. \qquad \text{(NOTE: } x^2+x+1>0 \text{ for all } x.)$$

You can verify that the second integral is an inverse tangent:

$$\frac{1}{2}\int \frac{dx}{x^2+x+1} = \frac{1}{2}\int \frac{dx}{(x+\frac{1}{2})^2 + (\sqrt{3}/2)^2} = \frac{1}{\sqrt{3}}\tan^{-1}\left[\frac{2}{\sqrt{3}}\left(x+\frac{1}{2}\right)\right] + C_2.$$

Combining results, we have

$$P = \ln|x| - \frac{1}{2}\ln(x^2+x+1) - \frac{1}{\sqrt{3}}\tan^{-1}\left[\frac{2}{\sqrt{3}}\left(x+\frac{1}{2}\right)\right] + C. \quad \square$$

Example 5′

$$\int \frac{3x^4 + x^3 + 20x^2 + 3x + 31}{(x+1)(x^2+4)^2}\,dx = \int \left[\frac{2}{x+1} + \frac{x}{x^2+4} - \frac{1}{(x^2+4)^2}\right] dx.$$

The first two fractions are easy to integrate:

$$\int \frac{2}{x+1}\,dx = 2\ln|x+1| + C_1,$$

$$\int \frac{x}{x^2+4}\,dx = \tfrac{1}{2}\int \frac{2x}{x^2+4}\,dx = \tfrac{1}{2}\ln(x^2+4) + C_2.$$

The integral of the last fraction is of the form

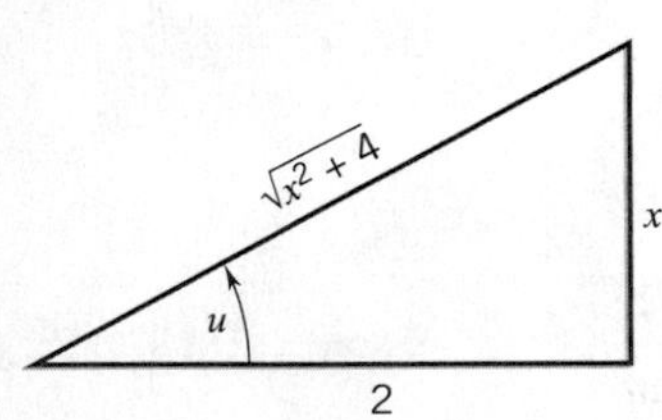

$$\int \frac{dx}{(x^2+a^2)^n}.$$

As you saw in the preceding section, such integrals can be calculated by the trigonometric substitution $a\tan u = x$ [see (8.4.2)]. Here we have

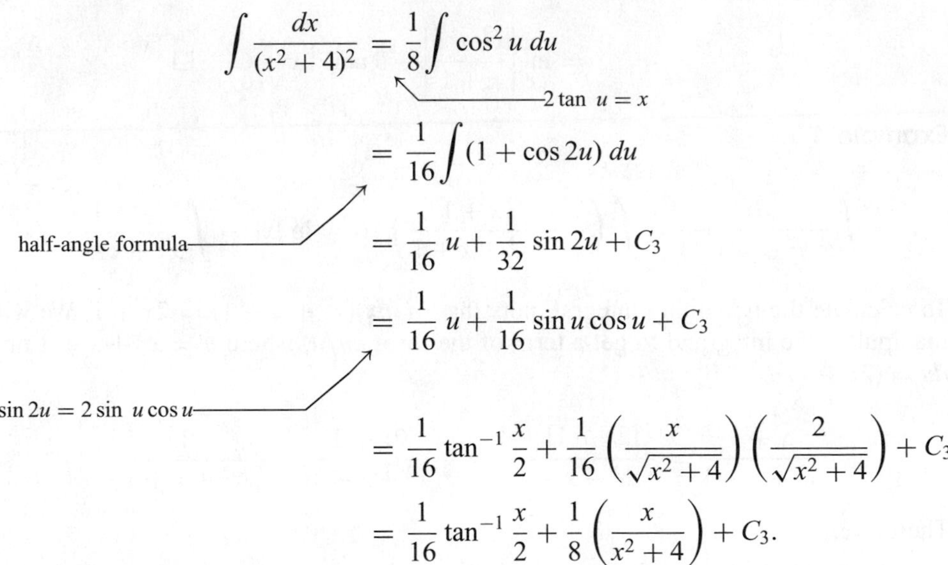

$$\int \frac{dx}{(x^2+4)^2} = \frac{1}{8}\int \cos^2 u\,du$$

$$= \frac{1}{16}\int (1+\cos 2u)\,du$$

$$= \frac{1}{16}u + \frac{1}{32}\sin 2u + C_3$$

$$= \frac{1}{16}u + \frac{1}{16}\sin u\cos u + C_3$$

$$= \frac{1}{16}\tan^{-1}\frac{x}{2} + \frac{1}{16}\left(\frac{x}{\sqrt{x^2+4}}\right)\left(\frac{2}{\sqrt{x^2+4}}\right) + C_3$$

$$= \frac{1}{16}\tan^{-1}\frac{x}{2} + \frac{1}{8}\left(\frac{x}{x^2+4}\right) + C_3.$$

The integral we want is equal to

$$2\ln|x+1| + \tfrac{1}{2}\ln(x^2+4) - \tfrac{1}{8}\left(\frac{x}{x^2+4}\right) - \tfrac{1}{16}\tan^{-1}\frac{x}{2} + C. \quad \square$$

Example 6′

$$\int \frac{x^5+2}{x^2-1}\,dx = \int \left[x^3 + x - \frac{1}{2(x+1)} + \frac{3}{2(x-1)}\right] dx$$

$$= \tfrac{1}{4}x^4 + \tfrac{1}{2}x^2 - \tfrac{1}{2}\ln|x+1| + \tfrac{3}{2}\ln|x-1| + C. \quad \square$$

Partial fractions come up in a variety of applications involving solutions of differential equations. See Section 8.9 for some examples.

EXERCISES 8.5

Find the partial fraction decomposition.

1. $r(x) = \dfrac{1}{x^2 + 7x + 6}$.

2. $R(x) = \dfrac{x^2}{(x-1)(x^2+4x+5)}$.

3. $r(x) = \dfrac{x}{x^4 - 1}$.

4. $R(x) = \dfrac{x^4}{(x-1)^3}$.

5. $r(x) = \dfrac{x^2 - 3x - 1}{x^3 + x^2 - 2x}$.

6. $R(x) = \dfrac{x^3 + x^2 + x + 2}{x^4 + 3x^2 + 2}$.

7. $r(x) = \dfrac{2x^2 + 1}{x^3 - 6x^2 + 11x - 6}$.

8. $R(x) = \dfrac{1}{x(x^2+1)^2}$.

Calculate the integral.

9. $\displaystyle\int \frac{7}{(x-2)(x+5)}\,dx$.

10. $\displaystyle\int \frac{x}{(x+1)(x+2)(x+3)}\,dx$.

11. $\displaystyle\int \frac{2x^4 - 4x^3 + 4x^2 + 3}{x^3 - x^2}\,dx$.

12. $\displaystyle\int \frac{x^2 + 1}{x(x^2-1)}\,dx$.

13. $\displaystyle\int \frac{x^5}{(x-2)^2}\,dx$.

14. $\displaystyle\int \frac{x^5}{x-2}\,dx$.

15. $\displaystyle\int \frac{x+3}{x^2 - 3x + 2}\,dx$.

16. $\displaystyle\int \frac{x^2+3}{x^2 - 3x + 2}\,dx$.

17. $\displaystyle\int \frac{dx}{(x-1)^3}$.

18. $\displaystyle\int \frac{dx}{x^2 + 2x + 2}$.

19. $\displaystyle\int \frac{x^2}{(x-1)^2(x+1)}\,dx$.

20. $\displaystyle\int \frac{2x - 1}{(x+1)^2(x-2)^2}\,dx$.

21. $\displaystyle\int \frac{dx}{x^4 - 16}$.

22. $\displaystyle\int \frac{3x^5 - 3x^2 + x}{x^3 - 1}\,dx$.

23. $\displaystyle\int \frac{x^3 + 4x^2 - 4x - 1}{(x^2+1)^2}\,dx$.

24. $\displaystyle\int \frac{dx}{(x^2+16)^2}$.

25. $\displaystyle\int \frac{dx}{x^4 + 4}$.†

26. $\displaystyle\int \frac{dx}{x^4 + 16}$.†

27. $\displaystyle\int \frac{x-3}{x^3 + x^2}\,dx$.

28. $\displaystyle\int \frac{1}{(x-1)(x^2+1)^2}\,dx$.

29. $\displaystyle\int \frac{x+1}{x^3 + x^2 - 6x}\,dx$.

30. $\displaystyle\int \frac{x^3 + x^2 + x + 3}{(x^2+1)(x^2+3)}\,dx$.

Evaluate the integral.

31. $\displaystyle\int_0^2 \frac{x}{x^2 + 5x + 6}\,dx$.

32. $\displaystyle\int_1^3 \frac{1}{x^3 + x}\,dx$.

33. $\displaystyle\int_1^3 \frac{x^2 - 4x + 3}{x^3 + 2x^2 + x}\,dx$.

34. $\displaystyle\int_0^2 \frac{x^3}{(x^2+2)^2}\,dx$.

Calculate the integral.

35. $\displaystyle\int \frac{\cos\theta}{\sin^2\theta - 2\sin\theta - 8}\,d\theta$.

36. $\displaystyle\int \frac{e^t}{e^{2t} + 5e^t + 6}\,dt$.

37. $\displaystyle\int \frac{1}{t\left([\ln t]^2 - 4\right)}\,dt$.

38. $\displaystyle\int \frac{\sec^2\theta}{\tan^3\theta - \tan^2\theta}\,d\theta$

Derive the formula.

39. Formula 106 for $\displaystyle\int \frac{u\,du}{a + bu}$.

40. Formula 108 for $\displaystyle\int \frac{du}{u(a + bu)}$.

41. Formula 109 for $\displaystyle\int \frac{du}{u^2(a + bu)}$.

42. Formula 111 for $\displaystyle\int \frac{du}{u(a + bu)^2}$.

Derive the formula.

43.
$$\int \frac{du}{(a+bu)(c+du)} = \frac{1}{ad - bc}\ln\left|\frac{c + du}{a + bu}\right| + C, \quad ad - bc \neq 0.$$

44. $\displaystyle\int \frac{du}{a^2 - u^2} = \frac{1}{2a}\ln\left|\frac{a+u}{a-u}\right| + C.$

45. $\displaystyle\int \frac{u\,du}{a^2 - u^2} = -\tfrac{1}{2}\ln|a^2 - u^2| + C.$

46. $\displaystyle\int \frac{u^2\,du}{a^2 - u^2} = -u + \frac{a}{2}\ln\left|\frac{a+u}{a-u}\right| + C.$

47. (a) Let $a > 0$. Use partial fractions to derive the formula
$$\int \frac{1}{a^2 - x^2}\,dx = \frac{1}{2a}\ln\frac{a + x}{a - x} + C.$$

(b) Derive this formula by making a trigonometric substitution.

(c) Use a *hyperbolic substitution* to obtain the formula
$$\int \frac{1}{a^2 - x^2}\,dx = \begin{cases} \dfrac{1}{a}\tanh^{-1}\left(\dfrac{x}{a}\right) + C & \text{if } |x| < a \\[2ex] \dfrac{1}{a}\coth^{-1}\left(\dfrac{x}{a}\right) + C & \text{if } |x| > a. \end{cases}$$

See Exercise 43, Section 7.9.

† HINT: With $a > 0$, $x^4 + a^2 = (x^2 + \sqrt{2a}\,x + a)(x^2 - \sqrt{2a}\,x + a)$.

48. Show that if $y = \dfrac{1}{x^2 - 1}$, then

$$\frac{d^n y}{dx^n} = \frac{(-1)^n n!}{2}\left[\frac{1}{(x-1)^{n+1}} - \frac{1}{(x+1)^{n+1}}\right].$$

49. Find the volume of the solid generated by revolving the region bounded by the graph of $y = 1/\sqrt{4 - x^2}$ and the x-axis from $x = 0$ to $x = 3/2$ about : (a) the x-axis; (b) the y-axis.

50. Calculate

$$\int x^3 \tan^{-1} x\, dx.$$

51. Find the centroid of the region under the curve $y = (x^2 + 1)^{-1}$, $x \in [0, 1]$.

52. Find the centroid of the solid generated by revolving the region of Exercise 51 about: (a) the x-axis; (b) the y-axis.

C 53. Use a CAS to find the partial fraction decomposition of f.

(a) $f(x) = \dfrac{6x^4 + 11x^3 - 2x^2 - 5x - 2}{x^2(x+1)^3}$.

(b) $f(x) = -\dfrac{x^3 + 20x^2 + 4x + 93}{(x^2 + 4)(x^2 - 9)}$.

(c) $f(x) = \dfrac{x^2 + 7x + 12}{x(x^2 + 2x + 4)}$.

C 54. Use a CAS to calculate the integrals

$$\int \frac{1}{x^2 + 2x + n}\, dx \text{ where } n = 0, 1, 2.$$

Verify your results by differentiating.

C 55. Let

$$f(x) = \frac{x}{x^2 + 5x + 6}.$$

(a) Use a graphing utility to sketch the graph of f.

(b) Calculate the area of the region bounded by the graph of f and the x-axis between $x = 0$ and $x = 4$.

56. (a) The region in Exercise 55 is revolved about the y-axis. Find the volume of the solid that is generated.

(b) Find the centroid of the solid in part (a).

C 57. Let

$$f(x) = \frac{9 - x}{(x+3)^2}.$$

(a) Use a graphing utility to sketch the graph of f.

(b) Find the area of the region bounded by the graph of f and the x-axis from $x = -2$ to $x = 9$.

58. (a) The region in Exercise 57 is revolved about the x-axis. Find the volume of the solid that is generated.

(b) Find the centroid of the solid in part (a).

■ *8.6 SOME RATIONALIZING SUBSTITUTIONS

These are integrands which are not rational functions but can be transformed into rational functions by a suitable substitution. Such substitutions are known as *rationalizing substitutions*.

First we consider integrals in which the integrand involves an expression of the form $\sqrt[n]{f(x)}$ for some function f. In such cases, the substitution $u = \sqrt[n]{f(x)}$, which is equivalent to setting $u^n = f(x)$, is sometimes effective. The idea is to replace fractional exponents by integer exponents; integer exponents are usually easier to handle.

Example 1 Find $\displaystyle\int \frac{dx}{1 + \sqrt{x}}$.

SOLUTION To rationalize the integrand, we set

$$u^2 = x, \qquad 2u\, du = dx.$$

Then $u = \sqrt{x}$ and

$$\int \frac{dx}{1+\sqrt{x}} = \int \frac{2u}{1+u}\, du = \int \left(2 - \frac{2}{1+u}\right) du$$

divide

$$= 2u - 2\ln|1+u| + C$$

$$= 2\sqrt{x} - 2\ln|1 + \sqrt{x}| + C. \quad \square$$

Example 2 Find $\displaystyle\int \frac{dx}{\sqrt[3]{x} + \sqrt{x}}$.

SOLUTION Here the integrand contains two distinct roots, $x^{1/3}$ and $x^{1/2}$. If we set $u^6 = x$, then both terms will be rationalized simultaneously:

$$u^6 = x, \quad 6u^5\,du = dx, \quad \text{and} \quad x^{1/3} = u^2, \quad x^{1/2} = u^3.$$

Thus, we have

$$\begin{aligned}\int \frac{dx}{\sqrt[3]{x}+\sqrt{x}} = \int \frac{6u^5}{u^2+u^3}\,du &= 6\int \frac{u^3}{1+u}\,du \\ &= 6\int \left(u^2 - u + 1 - \frac{1}{1+u}\right) du \\ &= 6\left(\tfrac{1}{3}u^3 - \tfrac{1}{2}u^2 + u - \ln|1+u|\right) + C \\ &= 2\sqrt{x} - 3\sqrt[3]{x} + 6\sqrt[6]{x} - 6\ln|1+\sqrt[6]{x}| + C. \quad \square\end{aligned}$$

divide

Example 3 Find $\int \sqrt{1-e^x}\,dx$

SOLUTION To rationalize the integrand, we set

$$u^2 = 1 - e^x.$$

To find dx in terms of u and du, we solve the equation for x:

$$1 - u^2 = e^x, \quad \ln(1-u^2) = x, \quad -\frac{2u}{1-u^2}\,du = dx.$$

The rest is straightforward:

$$\begin{aligned}\int \sqrt{1-e^x}\,dx &= \int u\left(-\frac{2u}{1-u^2}\right) du \\ &= \int \frac{2u^2}{u^2-1}\,du = \int \left(2 + \frac{1}{u-1} - \frac{1}{u+1}\right) du \\ &= 2u + \ln|u-1| - \ln|u+1| + C \\ &= 2u + \ln\left|\frac{u-1}{u+1}\right| + C, \\ &= 2\sqrt{1-e^x} + \ln\left|\frac{\sqrt{1-e^x}-1}{\sqrt{1-e^x}+1}\right| + C. \quad \square\end{aligned}$$

divide, then partial fractions

We now consider rational expressions in sine and cosine. Suppose, for example, that we want to calculate

$$\int \frac{1}{3\sin x - 4\cos x}\,dx.$$

To convert the integrand to a rational function in u, we set

$$u = \tan\left(\frac{x}{2}\right), \quad -\pi < u < \pi.$$

Then $$\cos\left(\frac{x}{2}\right) = \frac{1}{\sec(x/2)} = \frac{1}{\sqrt{1+\tan^2(x/2)}} = \frac{1}{\sqrt{1+u^2}},$$

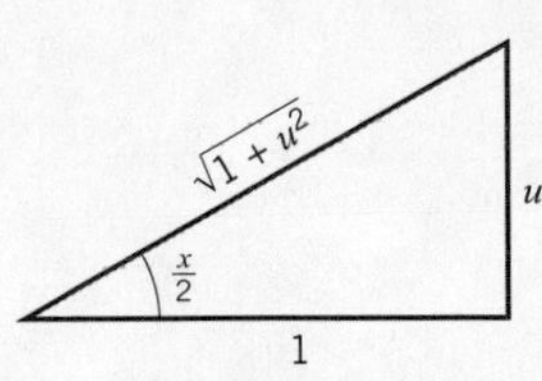

Figure 8.6.1

and $$\sin\left(\frac{x}{2}\right) = \cos\left(\frac{x}{2}\right)\tan\left(\frac{x}{2}\right) = \frac{u}{\sqrt{1+u^2}}.$$

The right triangle in Figure 8.6.1 illustrates these relationships.

Observe that

$$\sin x = 2\sin x\left(\frac{x}{2}\right)\cos\left(\frac{x}{2}\right) = \frac{2u}{1+u^2}.$$

and

$$\cos x = \cos^2\left(\frac{x}{2}\right) - \sin^2\left(\frac{x}{2}\right) = \frac{1-u^2}{1+u^2}.$$

Since $u = \tan(x/2)$, $x = 2\tan^{-1} u$, and so

$$dx = \frac{2}{1+u^2}\,du.$$

To summarize, if an integrand is a rational expression in sine and cosine, then the substitution

$$\sin x = \frac{2u}{1+u^2}, \qquad \cos x = \frac{1-u^2}{1+u^2}, \qquad dx = \frac{2}{1+u^2}\,du,$$

where $u = \tan(x/2)$, will convert the integrand into a rational function in u. The resulting integral can then be calculated by the methods of Section 8.5.

Example 4 Find $\displaystyle\int \frac{1}{3\sin x - 4\cos x}\,dx$.

SOLUTION Set $u = \tan\left(\frac{x}{2}\right)$. Then

$$\frac{1}{3\sin x - 4\cos x} = \frac{1}{[6u/(1+u^2)] - [4(1-u^2)/(1+u^2)]} = \frac{1+u^2}{4u^2+6u-4}$$

and

$$\int \frac{1}{3\sin x - 4\cos x}\,dx = \int \frac{1+u^2}{4u^2+6u-4}\cdot\frac{2}{1+u^2}\,du = \int \frac{1}{2u^2+3u-2}\,du.$$

Since

$$\frac{1}{2u^2+3u-2} = \frac{1}{(u+2)(2u-1)} = \frac{-1/5}{u+2} + \frac{2/5}{2u-1}, \qquad \text{(partial fractions)}$$

we have

$$\begin{aligned}\int \frac{1}{2u^2+3u-2}\,du &= -\tfrac{1}{5}\int \frac{1}{u+2}\,du + \tfrac{2}{5}\int \frac{1}{2u-1}\,du \\ &= -\tfrac{1}{5}\ln|u+2| + \tfrac{1}{5}\ln|2u-1| + C \\ &= \tfrac{1}{5}\ln\left|\frac{2u-1}{u+2}\right| + C \\ &= \tfrac{1}{5}\ln\left|\frac{2\tan(x/2)-1}{\tan(x/2)+2}\right| + C. \quad \square\end{aligned}$$

EXERCISES *8.6

Calculate the integral.

1. $\displaystyle\int \frac{dx}{1-\sqrt{x}}$.

2. $\displaystyle\int \frac{\sqrt{x}}{1+x}\,dx$.

3. $\displaystyle\int \sqrt{1+e^x}\,dx$.

4. $\displaystyle\int \frac{dx}{x(x^{1/3}-1)}$.

5. $\displaystyle\int x\sqrt{1+x}\,dx$. [(a) set $u^2 = 1+x$; (b) set $u = 1+x$]

6. $\displaystyle\int x^2\sqrt{1+x}\,dx$. [(a) set $u^2 = 1+x$; (b) set $u = 1+x$]

7. $\displaystyle\int (x+2)\sqrt{x-1}\,dx$.

8. $\displaystyle\int (x-1)\sqrt{x+2}\,dx$.

9. $\displaystyle\int \frac{x^3}{(1+x^2)^3}\,dx$.

10. $\displaystyle\int x(1+x)^{1/3}\,dx$.

11. $\displaystyle\int \frac{\sqrt{x}}{\sqrt{x}-1}\,dx$.

12. $\displaystyle\int \frac{x}{\sqrt{x+1}}\,dx$.

13. $\displaystyle\int \frac{\sqrt{x-1}+1}{\sqrt{x-1}-1}\,dx$.

14. $\displaystyle\int \frac{1-e^x}{1+e^x}\,dx$.

15. $\displaystyle\int \frac{dx}{\sqrt{1+e^x}}$.

16. $\displaystyle\int \frac{dx}{1+e^{-x}}$.

17. $\displaystyle\int \frac{x}{\sqrt{x+4}}\,dx$.

18. $\displaystyle\int \frac{x+1}{x\sqrt{x-2}}\,dx$.

19. $\displaystyle\int 2x^2(4x+1)^{-5/2}\,dx$.

20. $\displaystyle\int x^2\sqrt{x-1}\,dx$.

21. $\displaystyle\int \frac{x}{(ax+b)^{3/2}}\,dx$.

22. $\displaystyle\int \frac{x}{\sqrt{ax+b}}\,dx$.

23. $\displaystyle\int \frac{1}{1+\cos x-\sin x}\,dx$.

24. $\displaystyle\int \frac{1}{2+\cos x}\,dx$.

25. $\displaystyle\int \frac{1}{2+\sin x}\,dx$.

26. $\displaystyle\int \frac{\sin x}{1+\sin^2 x}\,dx$.

27. $\displaystyle\int \frac{1}{\sin x+\tan x}\,dx$.

28. $\displaystyle\int \frac{1}{1+\sin x+\cos x}\,dx$.

29. $\displaystyle\int \frac{1-\cos x}{1+\sin x}\,dx$.

30. $\displaystyle\int \frac{1}{5+3\sin x}\,dx$.

Evaluate the definite integral.

31. $\displaystyle\int_0^4 \frac{x^{3/2}}{x+1}\,dx$.

32. $\displaystyle\int_0^8 \frac{1}{1+\sqrt[3]{x}}\,dx$.

33. $\displaystyle\int_0^{\pi/2} \frac{\sin 2x}{2+\cos x}\,dx$.

34. $\displaystyle\int_0^{\pi/2} \frac{1}{1+\sin x}\,dx$.

35. $\displaystyle\int_0^{\pi/3} \frac{1}{\sin x-\cos x-1}\,dx$.

36. $\displaystyle\int_0^1 \frac{\sqrt{x}}{1+\sqrt{x}}\,dx$.

37. Use the method of this section to show that

$$\int \sec x\,dx = \int \frac{1}{\cos x}\,dx = \ln\left|\frac{1+\tan(x/2)}{1-\tan(x/2)}\right| + C.$$

38. (a) Another way to calculate $\int \sec x\,dx$ is to write

$$\int \sec x\,dx = \int \frac{\cos x}{\cos^2 x}\,dx = \int \frac{\cos x}{1-\sin^2 x}\,dx.$$

Use the method of this section to show that

$$\int \sec x\,dx = \ln\sqrt{\frac{1+\sin x}{1-\sin x}} + C.$$

(b) Show that the result of part (a) is equivalent to the familiar formula

$$\int \sec x\,dx = \ln|\sec x+\tan x| + C.$$

39. (a) Use the approach given in Exercise 38(a) to show that

$$\int \csc x\,dx = \ln\sqrt{\frac{1-\cos x}{1+\cos x}} + C.$$

(b) Show that the result of part (a) is equivalent to the formula

$$\int \csc x\,dx = \ln|\csc x-\cot x| + C.$$

40. The integral of a rational function of $\sinh x$ and $\cosh x$ can be transformed into a rational function of u by means of the substitution $u = \tanh(x/2)$. Show that this substitution, gives

$$\sinh x = \frac{2u}{1-u^2},\qquad \cosh x = \frac{1+u^2}{1-u^2},\qquad dx = \frac{2}{1-u^2}\,du.$$

In Exercises 41–44, set $u = \tanh(x/2)$ and carry out the integration

41. $\displaystyle\int \operatorname{sech} x\,dx$.

42. $\displaystyle\int \frac{1}{1+\cosh x}\,dx$.

43. $\displaystyle\int \frac{1}{\sinh x+\cosh x}\,dx$.

44. $\displaystyle\int \frac{1-e^x}{1+e^x}\,dx$.

■ 8.7 NUMERICAL INTEGRATION

To evaluate a definite integral by the formula

$$\int_a^b f(x)\,dx = F(b) - F(a),$$

we must be able to find an antiderivative F and we must to able to evaluate this antiderivative both at a and at b. When this is not possible, the method fails.

The method fails even for such simple-looking integrals as

$$\int_0^1 \sqrt{x}\sin x\,dx \quad \text{or} \quad \int_0^1 e^{-x^2}\,dx.$$

There are no *elementary functions* with derivative $\sqrt{x}\sin x$ or e^{-x^2}.

Here we take up some simple numerical methods for estimating definite integrals—methods that you can use whether or not you can find an antiderivative. All the methods we describe involve only simple arithmetic and are ideally suited to the computer.

We focus now on

$$\int_a^b f(x)\,dx.$$

We suppose that f is continuous on $[a, b]$ and, for pictorial convenience, assume that f is positive. Take a regular partition $P = \{x_0, x_1, x_2, \ldots, x_{n-1}, x_n\}$ of $[a, b]$, subdividing the interval into n subintervals each of length $(b - a)/n$:

$$[a, b] = [x_0, x_1] \cup \cdots \cup [x_{i-1}, x_i] \cup \cdots \cup [x_{n-1}, x_n]$$

with $$\Delta x_i = \frac{b-a}{n}.$$

The region Ω_i pictured in Figure 8.7.1, can be approximated in many ways.

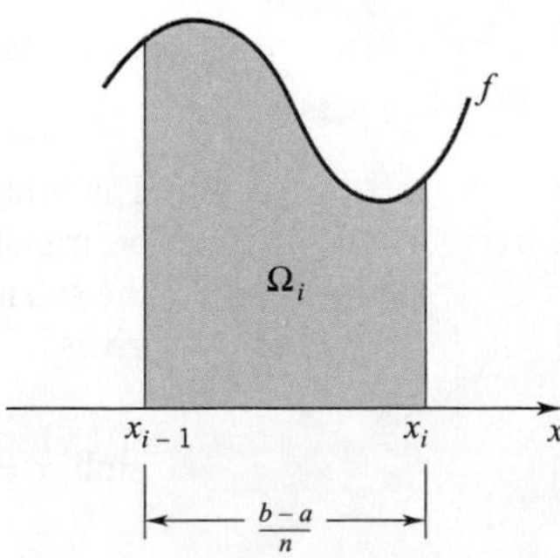

Figure 8.7.1

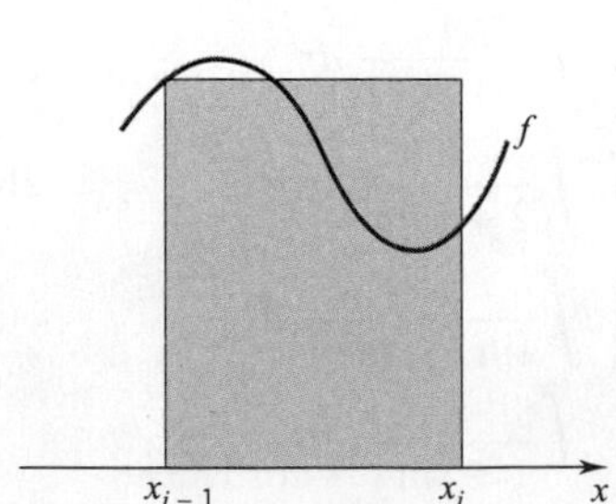

Figure 8.7.2

(1) By the left-endpoint rectangle (Figure 8.7.2):

$$\begin{aligned}\text{area} &\cong f(x_{i-1})\,\Delta x_i\\ &= f(x_{i-1})\left(\frac{b-a}{n}\right).\end{aligned}$$

(2) By the right-endpoint rectangle (Figure 8.7.3):

$$\text{area} \cong f(x_i)\,\Delta x_i = f(x_i)\left(\frac{b-a}{n}\right).$$

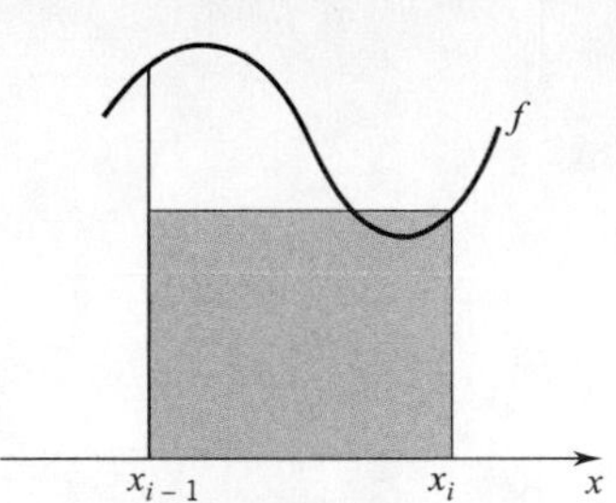

Figure 8.7.3

(3) By the midpoint rectangle (Figure 8.7.4):

$$\text{area} \cong f\left(\frac{x_{i-1}+x_i}{2}\right)\Delta x_i = f\left(\frac{x_{i-1}+x_i}{2}\right)\left(\frac{b-a}{n}\right)$$

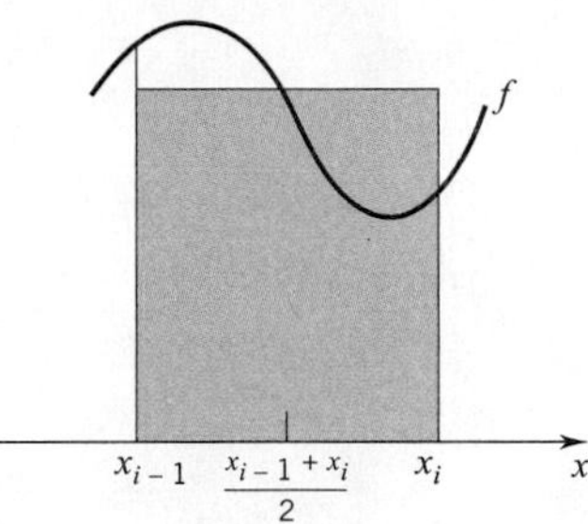

Figure 8.7.4

(4) By a trapezoid (Figure 8.7.5):

$$\text{area} \cong \frac{1}{2}[f(x_{i-1})+f(x_i)]\,\Delta x_i = \frac{1}{2}[f(x_{i-1})+f(x_i)]\left(\frac{b-a}{n}\right)$$

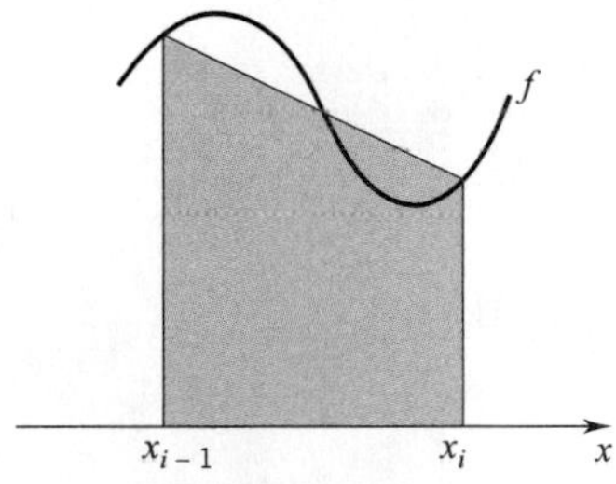

Figure 8.7.5

(5) By a parabolic region (Figure 8.7.6): take the parabola $y = Ax^2 + Bx + C$ that passes through the three points indicated.

$$\text{area} \cong \frac{1}{6}\left[f(x_{i-1})+4f\left(\frac{x_{i-1}+x_i}{2}\right)+f(x_i)\right]\Delta x_i = \left[f(x_{i-1})+4f\left(\frac{x_{i-1}+x_i}{2}\right)+f(x_i)\right]\left(\frac{b-a}{6n}\right).$$

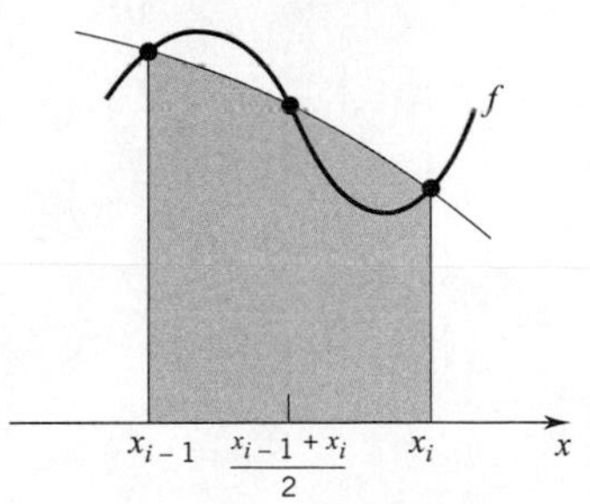

Figure 8.7.6

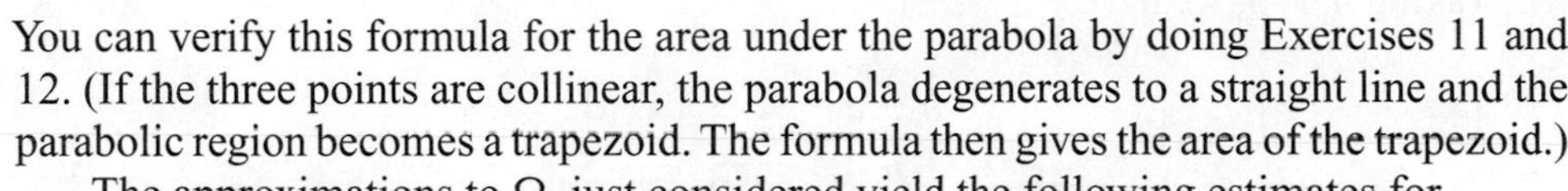

You can verify this formula for the area under the parabola by doing Exercises 11 and 12. (If the three points are collinear, the parabola degenerates to a straight line and the parabolic region becomes a trapezoid. The formula then gives the area of the trapezoid.)

The approximations to Ω_i just considered yield the following estimates for

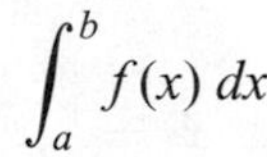

$$\int_a^b f(x)\,dx.$$

(1) The left-endpoint estimate:

$$L_n = \frac{b-a}{n}\,[f(x_0)+f(x_1)+\cdots+f(x_{n-1})].$$

(2) The right-endpoint estimate:

$$R_n = \frac{b-a}{n}\,[f(x_1)+f(x_2)+\cdots+f(x_n)].$$

(3) The midpoint estimate:

$$M_n = \frac{b-a}{n}\left[f\left(\frac{x_0+x_1}{2}\right)+\cdots+f\left(\frac{x_{n-1}+x_n}{2}\right)\right].$$

(4) The trapezoidal estimate (*trapezoidal rule*):

$$T_n = \frac{b-a}{n}\left[\frac{f(x_0)+f(x_1)}{2} + \frac{f(x_1)+f(x_2)}{2} + \cdots + \frac{f(x_{n-1})+f(x_n)}{2}\right]$$

$$= \frac{b-a}{2n}\,[f(x_0)+2f(x_1)+\cdots+2f(x_{n-1})+f(x_n)].$$

(5) The parabolic estimate (*Simpson's rule*):

$$S_n = \frac{b-a}{6n}\left\{f(x_0)+f(x_n)+2[f(x_1)+\cdots+f(x_{n-1})]\right.$$

$$\left. +4\left[f\left(\frac{x_0+x_1}{2}\right)+\cdots+f\left(\frac{x_{n-1}+x_n}{2}\right)\right]\right\}.$$

The first three estimates, L_n, R_n, M_n, are Riemann sums (Section 5.2); T_n and S_n, although not explicitly written as Riemann sums, can be written as Riemann sums. (See Exercise 26.) It follows from (5.2.6) that any one of these estimates can be used to approximate the integral as closely as we may wish. All we have to do is take n sufficiently large.

As an example, we will find the approximate value of

$$\ln 2 = \int_1^2 \frac{dx}{x}$$

by applying each of the five estimates. Here

$$f(x) = \frac{1}{x}, \quad [a, b] = [1, 2].$$

Taking $n = 5$, we have

$$\frac{b-a}{n} = \frac{2-1}{5} = \frac{1}{5}.$$

The partition points are

$$x_0 = \tfrac{5}{5}, \quad x_1 = \tfrac{6}{5}, \quad x_2 = \tfrac{7}{5}, \quad x_3 = \tfrac{8}{5}, \quad x_4 = \tfrac{9}{5}, \quad x_5 = \tfrac{10}{5}. \qquad \text{(Figure 8.7.7)}$$

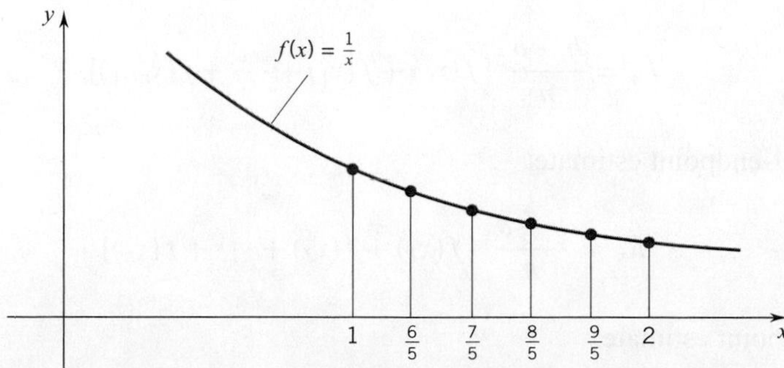

Figure 8.7.7

Using a calculator and rounding off to four decimal places, we have the following estimates:

$$L_5 = \tfrac{1}{5}\left(\tfrac{5}{5}+\tfrac{5}{6}+\tfrac{5}{7}+\tfrac{5}{8}+\tfrac{5}{9}\right) = \left(\tfrac{1}{5}+\tfrac{1}{6}+\tfrac{1}{7}+\tfrac{1}{8}+\tfrac{1}{9}\right) \cong 0.7456.$$

$$R_5 = \tfrac{1}{5}\left(\tfrac{5}{6}+\tfrac{5}{7}+\tfrac{5}{8}+\tfrac{5}{9}+\tfrac{5}{10}\right) = \left(\tfrac{1}{6}+\tfrac{1}{7}+\tfrac{1}{8}+\tfrac{1}{9}+\tfrac{1}{10}\right) \cong 0.6456.$$

$$M_5 = \tfrac{1}{5}\left(\tfrac{10}{11}+\tfrac{10}{13}+\tfrac{10}{15}+\tfrac{10}{17}+\tfrac{10}{19}\right) = 2\left(\tfrac{1}{11}+\tfrac{1}{13}+\tfrac{1}{15}+\tfrac{1}{17}+\tfrac{1}{19}\right) \cong 0.6919.$$

$$T_5 = \tfrac{1}{10}\left(\tfrac{5}{5}+\tfrac{10}{6}+\tfrac{10}{7}+\tfrac{10}{8}+\tfrac{10}{9}+\tfrac{5}{10}\right) = \left(\tfrac{1}{10}+\tfrac{1}{6}+\tfrac{1}{7}+\tfrac{1}{8}+\tfrac{1}{9}+\tfrac{1}{20}\right) \cong 0.6956.$$

$$S_5 = \tfrac{1}{30}\left[\tfrac{5}{5}+\tfrac{5}{10}+2\left(\tfrac{5}{6}+\tfrac{5}{7}+\tfrac{5}{8}+\tfrac{5}{9}\right)+4\left(\tfrac{10}{11}+\tfrac{10}{13}+\tfrac{10}{15}+\tfrac{10}{17}+\tfrac{10}{19}\right)\right] \cong 0.6932.$$

Since the integrand $1/x$ decreases throughout the interval $[1, 2]$, you can expect the left-endpoint estimate, 0.7456, to be too large and you can expect the right-endpoint estimate, 0.6456, to be too small. The other estimates should be better.

The value of ln 2 given on a calculator is $\ln 2 \cong 0.69314718$, or 0.6931 rounded to four decimal places. Thus S_5 is correct to the nearest thousandth.

Example 1 Find the approximate value of $\int_0^3 \sqrt{4+x^3}\,dx$ by the trapezoidal rule. Take $n = 6$.

SOLUTION Each subinterval has length $\dfrac{b-a}{n} = \dfrac{3-0}{6} = \dfrac{1}{2}$.

The partition points are

$$x_0 = 0, \quad x_1 = \tfrac{1}{2}, \quad x_2 = 1, \quad x_3 = \tfrac{3}{2}, \quad x_4 = 2, \quad x_5 = \tfrac{5}{2}, \quad x_6 = 3. \quad \text{(Figure 8.7.8)}$$

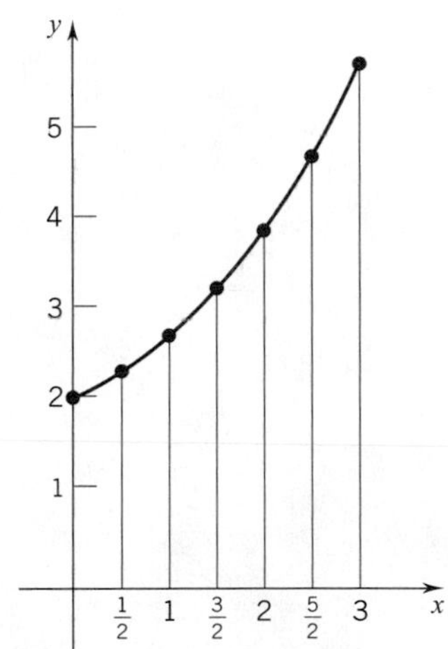

Figure 8.7.8

Now $\quad T_6 \cong \tfrac{1}{4}[f(0)+2f(\tfrac{1}{2})+2f(1)+2f(\tfrac{3}{2})+2f(2)+2f(\tfrac{5}{2})+f(3)],$

with $f(x) = \sqrt{4+x^3}$. Using a calculator and rounding off to three decimal places we have

$$f(0) = 2.000, \quad f(\tfrac{1}{2}) \cong 2.031, \quad f(1) \cong 2.236, \quad f(\tfrac{3}{2}) \cong 2.716,$$
$$f(2) \cong 3.464, \quad f(\tfrac{5}{2}) \cong 4.430, \quad f(3) \cong 5.568.$$

Thus

$$t_6 \cong \tfrac{1}{4}(2.000+4.062+4.472+5.432+6.928+8.860+5.568) \cong 9.331. \quad \square$$

Example 2 Find the approximate value of

$$\int_0^3 \sqrt{4+x^3}\,dx$$

by Simpson's rule. Take $n = 3$.

SOLUTION There are three subintervals each of length

$$\frac{b-a}{n} = \frac{3-0}{3} = 1.$$

Here
$$x_0 = 0, \quad x_1 = 1, \quad x_2 = 2, \quad x_3 = 3,$$
$$\frac{x_0 + x_1}{2} = \frac{1}{2}, \quad \frac{x_1 + x_2}{2} = \frac{3}{2}, \quad \frac{x_2 + x_3}{2} = \frac{5}{2}.$$

Simpson's rule yields

$$S_3 = \tfrac{1}{6}[f(0) + f(3) + 2f(1) + 2f(2) + 4f(\tfrac{1}{2}) + 4f(\tfrac{3}{2}) + 4f(\tfrac{5}{2})],$$

with $f(x) = \sqrt{4 + x^3}$. Using the values of f from Example 1, we have

$$S_3 = \tfrac{1}{6}(2.000 + 5.568 + 4.472 + 6.928 + 8.124 + 10.864 + 17.72) \cong 9.279.$$

For comparison, the value of this integral accurate to five decimal places is 9.27972. □

Error Estimates

A numerical estimate is useful only to the extent that we can gauge its accuracy. When we use any kind of approximation method, we face two forms of error: the error inherent in the method we use (we call this the *theoretical error*) and the error that accumulates from rounding off the decimals that arise during the course of computation (we call this the *round-off error*). The nature of round-off error is obvious. We will speak first about theoretical error.

We begin with a function f continuous and increasing on $[a, b]$. We subdivide $[a, b]$ into n nonoverlapping intervals, each of length $(b - a)/n$. We want to estimate

$$\int_a^b f(x)\, dx$$

by the left-endpoint method. What is the theoretical error? It should be clear from Figure 8.7.9 that the theoretical error does not exceed

$$[f(b) - f(a)]\left(\frac{b - a}{n}\right).$$

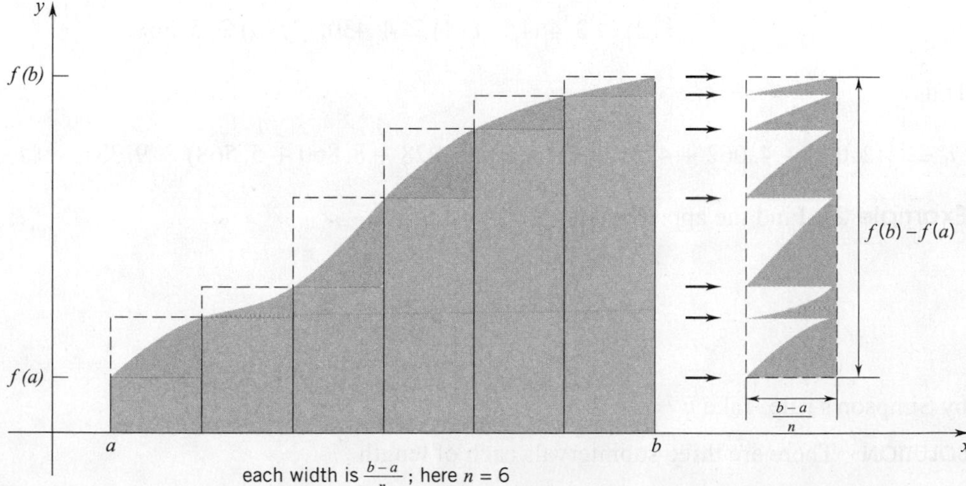

Figure 8.7.9

The error is represented by the sum of the areas of the shaded regions. These regions when shifted to the right, all fit together within a rectangle of height $f(b) - f(a)$ and base $(b - a)/n$.

Similar reasoning shows that, under the same circumstances, the theoretical error associated with the trapezoidal method does not exceed

$$(*) \qquad \tfrac{1}{2}[f(b) - f(a)]\left(\frac{b-a}{n}\right).$$

In this setting, at least, the trapezoidal estimate does a better job than the left-endpoint estimate.

The trapezoidal rule is more accurate than $(*)$ suggests. As is shown in texts on numerical analysis, if f is continuous on $[a, b]$ and twice differentiable on (a, b), then the theoretical error of the trapezoidal rule,

$$E_n^T = \int_a^b f(x)\,dx - T_n,$$

can be written

(8.7.1)
$$E_n^T = -\frac{(b-a)^3}{12\,n^2} f''(c),$$

where c is some number between a and b. Usually we cannot pinpoint c any further. However, if f'' is bounded on $[a, b]$, say $|f''(x)| \leq M$ for $a \leq x \leq b$, then

(8.7.2)
$$|E_n^T| \leq \frac{(b-a)^3}{12\,n^2} M.$$

Recall the trapezoidal-rule estimate of ln 2 derived at the beginning of this section:

$$\ln 2 = \int_1^2 \frac{1}{x}\,dx \cong 0.696.$$

To find the theoretical error, we apply (8.72). Here

$$f(x) = \frac{1}{x}, \quad f'(x) = -\frac{1}{x^2}, \quad f''(x) = \frac{2}{x^3}.$$

Since f'' is a decreasing function, it takes on its maximum value at the left endpoint of the interval. Thus $|f''(x)| \leq f''(1) = 2$ on $[1, 2]$. Therefore, with $a = 1, b = 2$, and $n = 5$, we have

$$|E_5^T| \leq \frac{(2-1)^3}{12 \cdot 5^2} \cdot 2 = \frac{1}{150} < 0.007.$$

The estimate 0.696 is in theoretical error by less than 0.007.

To get an estimate for

$$\ln 2 = \int_1^2 \frac{1}{x}\,dx$$

that is accurate to four decimal places, we need

$$(1) \qquad \frac{(b-a)^3}{12\,n^2}M < 0.00005.$$

Since
$$\frac{(b-a)^3}{12\,n^2}M \leq \frac{(2-1)^3}{12\,n^2}\cdot 2 \leq \frac{1}{6n^2},$$

we can satisfy (1) by having

$$\frac{1}{6n^2} < 0.00005,$$

which is to say, by having

$$n^2 > 3333.$$

As you can check, $n = 58$ is the smallest integer that satisfies this inequality. In this case, the trapezoidal rule requires a regular partition with at least 58 subintervals to guarantee four decimal place accuracy.

Simpson's rule is more effective than the trapezoidal rule. If f is continuous on $[a, b]$ and $f^{(4)}$ exists on (a, b), then the theoretical error for Simpson's rule,

$$E_n^S = \int_a^b f(x)\,dx - S_n,$$

can be written

(8.7.3)
$$E_n^S = -\frac{(b-a)^5}{2880n^4}f^{(4)}(c),$$

where, as before, c is some number between a and b. Whereas (8.7.1) varies as $1/n^2$, this quantity varies as $1/n^4$. Thus, for comparable n, we can expect greater accuracy from Simpson's rule. In addition, if we assume that $f^{(4)}(x)$ is bounded on $[a, b]$, say $|f^{(4)}(x)| \leq M$ for $a \leq x \leq b$, then

(8.7.4)
$$|E_n^S| \leq \frac{(b-a)^5}{2880n^4}M,$$

and this can be used to analyze the theoretical error for Simpson's rule in the same way that we used (8.7.2) to analyze the error for the trapezoidal rule.

For example, to achieve four decimal place accuracy in estimating

$$\ln 2 = \int_1^2 \frac{dx}{x}$$

using Simpson's rule, we set $b = 2$ and $a = 1$ in (8.7.4), and then try to find n so that

$$|E_n^S| \leq \frac{(2-1)^5}{2880n^4}M = \frac{1}{2880n^4}M < 0.00005,$$

where M is a bound for

$$|f^{(4)}(x)| = \frac{24}{x^5} \text{ on } [1, 2].$$

Now, $|f^{(4)}(x)| \leq 24$ on $[1, 2]$, and so we want to find n such that

$$\frac{1}{2880n^4}24 = \frac{1}{120n^4} < 0.00005.$$

This is equivalent to requiring

$$n^4 > 167.$$

You can verify that $n = 4$ is the smallest positive integer that satisfies this inequality. Obviously this is a considerable improvement in efficiency over the trapezoidal rule.

Finally, a word about round-off error. Any numerical procedure requires careful consideration of round-off error. To illustrate this point, we rework our trapezoidal estimate for

$$\int_1^2 \frac{dx}{x},$$

again taking $n = 5$, but this time assuming that our computer or calculator can store only two significant digits. As before,

$$\int_1^2 \frac{dx}{x} \cong \tfrac{1}{10}[\tfrac{1}{1} + 2(\tfrac{5}{6}) + 2(\tfrac{5}{7}) + 2(\tfrac{5}{8}) + 2(\tfrac{5}{9}) + \tfrac{1}{2}]. \tag{2}$$

Now our limited round-off machine goes to work:

$$\begin{aligned}\int_1^2 \frac{dx}{x} &\cong (0.10)[(1.0) + 2(0.83) + 2(0.71) + 2(0.62) + 2(0.44) + (0.50)]\\ &\text{“}=\text{”}\ (0.10)[(1.0) + (1.7) + (1.4) + (1.2) + (0.88) + (0.50)]\\ &= (0.10)[6.7] = 0.67.\end{aligned}$$

Earlier we used (8.7.1) to show that estimate (2) is in error by no more than 0.007 and found 0.70 as the approximation. Now with our limited round-off machine, we simplified (2) in a different way and obtained 0.67 as the approximation. The apparent error due to crude round off, 0.70–0.67 = 0.03, exceeds the error of the approximation method itself. The lesson should be clear: round-off error is important.

EXERCISES 8.7

In Exercises 1–10, round your answers to four decimal places.

1. Estimate

$$\int_0^{12} x^2\,dx$$

by: (a) the left-endpoint estimate, $n = 12$; (b) the right-endpoint estimate $n = 12$; (c) the midpoint estimate, $n = 6$; (d) the trapezoidal rule, $n = 12$; (e) Simpson's rule, $n = 6$.

Check your results by performing the integration.

2. Estimate

$$\int_0^1 \sin^2 \pi x\,dx$$

by: (a) the midpoint estimate, $n = 3$; (b) the trapezoidal rule, $n = 6$; (c) Simpson's rule, $n = 3$. Check your results by performing the integration.

3. Estimate

$$\int_0^3 \frac{dx}{1+x^3}$$

by: (a) the left-endpoint estimate, $n = 6$; (b) the right-endpoint estimate, $n = 6$; (c) the midpoint estimate, $n = 3$; (d) the trapezoidal rule, $n = 6$; (e) Simpson's rule, $n = 3$.

4. Estimate

$$\int_0^\pi \frac{\sin x}{\pi + x}\,dx$$

by: (a) the trapezoidal rule, $n = 6$; (b) Simpson's rule, $n = 3$.

5. Find the approximate value of π by estimating the integral

$$\frac{\pi}{4} = \tan^{-1} 1 = \int_0^1 \frac{dx}{1+x^2}$$

by: (a) the trapezoidal rule, $n = 4$; (b) Simpson's rule, $n = 4$.

6. Estimate

$$\int_0^2 \frac{dx}{\sqrt{4+x^3}}$$

by: (a) the trapezoidal rule, $n = 4$; (b) Simpson's rule, $n = 2$.

7. Estimate

$$\int_{-1}^1 \cos(x^2)\,dx$$

by: (a) the midpoint estimate, $n = 4$; (b) the trapezoidal rule, $n = 8$; (c) Simpson's rule, $n = 4$.

8. Estimate

$$\int_1^2 \frac{e^x}{x}\,dx$$

by: (a) the midpoint estimate, $n = 4$; (b) the trapezoidal rule, $n = 8$; (c) Simpson's rule, $n = 4$.

9. Estimate

$$\int_0^2 e^{-x^2}\,dx$$

by: (a) the trapezoidal rule, $n = 10$; (b) Simpson's rule, $n = 5$.

10. Estimate

$$\int_2^4 \frac{1}{\ln x}\,dx$$

by: (a) the midpoint estimate, $n = 4$; (b) the trapezoidal rule, $n = 8$; (c) Simpson's rule, $n = 4$.

11. Show that there is one and only one curve of the form $y = Ax^2 + Bx + C$ through three distinct points with different x-coordinates.

12. Show that the function $g(x) = Ax^2 + Bx + C$ satisfies the condition

$$\int_a^b g(x)\,dx = \frac{b-a}{6}\left[g(a) + 4g\left(\frac{a+b}{2}\right) + g(b)\right]$$

for every interval $[a, b]$.

In Exercises 13–22, determine the values of n which guarantee a theoretical error less than ϵ if the integral is estimated by: (a) the trapezoidal rule; (b) Simpson's rule.

13. $\int_1^4 \sqrt{x}\,dx$; $\epsilon = 0.01$.

14. $\int_1^3 x^5\,dx$; $\epsilon = 0.01$.

15. $\int_1^4 \sqrt{x}\,dx$; $\epsilon = 0.00001$.

16. $\int_1^3 x^5\,dx$; $\epsilon = 0.00001$.

17. $\int_0^\pi \sin x\,dx$; $\epsilon = 0.001$.

18. $\int_0^\pi \cos x\,dx$; $\epsilon = 0.001$.

19. $\int_1^3 e^x dx$; $\epsilon = 0.01$.

20. $\int_1^e \ln x\,dx$; $\epsilon = 0.01$.

21. $\int_0^2 e^{-x^2}\,dx$; $\epsilon = 0.0001$.

22. $\int_0^2 e^x\,dx$; $\epsilon = 0.00001$.

23. Show that Simpson's rule is exact (theoretical error zero) for every polynomial of degree 3 or less.

24. Show that the trapezoidal rule is exact (theoretical error zero) if f is linear.

25. (a) Let $f(x) = x^2$ on [0,1] and let $n = 2$. Show that

$$E_n^T = \frac{(b-a)^3}{12n^2}M,$$

where M is the maximum value of f'' on [a,b]. Thus, the inequality (8.7.2) can reduce to equality.

(b) Let $f(x) = x^4$ on [0,1] and let $n = 1$. Show that

$$E_n^S = \frac{(b-a)^5}{2880n^4}M,$$

where M is the maximum value of $f^{(4)}$ on $[a, b]$.

26. Show that, if f is continuous, then T_n and S_n can both be written as Riemann sums. HINT: Both

$$\tfrac{1}{2}[f(x_{i-1})+f(x_i)] \text{ and } \tfrac{1}{6}\left[f(x_{i-1})+4f\left(\frac{x_{i-1}+x_i}{2}\right)+f(x_i)\right]$$

lie between m_i and M_i, the minimum and maximum values of f on $[x_{i-1}, x_i]$.

27. Let f be a twice differentiable function. Show that if $f(x) > 0$ and $f''(x) > 0$ on [a,b], then

$$M_n \leq \int_a^b f(x)\,dx \leq T_n \quad \text{for any } n.$$

HINT: See the figure and note that the area of the rectangle $ABCD$ equals the area of the trapezoid $AEFD$.

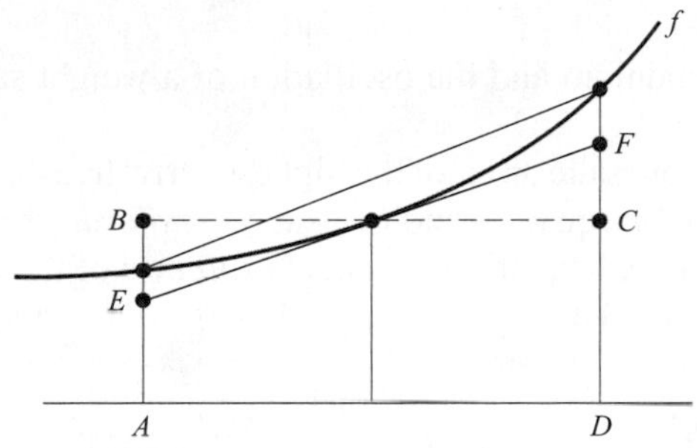

28. Show that $\frac{1}{3}T_n + \frac{2}{3}M_n = S_n$.

29. Use a CAS and the trapezoidal rule to estimate:

(a) $\displaystyle\int_0^{10} (x + \cos x)\,dx, \quad n = 50.$

(b) $\displaystyle\int_{-4}^{7} (x^5 - 5x^4 + x^3 - 3x^2 - x + 4)\,dx, \quad n = 30.$

30. Use a CAS and Simpson's rule to estimate:

(a) $\displaystyle\int_{-4}^{3} \frac{x^2}{x^2+4}\,dx, \quad n = 50.$

(b) $\displaystyle\int_0^{\pi/6} (x + \tan x)\,dx, \quad n = 25.$

31. Analyze the error if Simpson's rule with $n = 20$ is used to estimate

$$\int_1^5 \frac{x^2-4}{x^2+9}\,dx.$$

32. Analyze the error if the trapezoidal rule with $n = 30$ is used to estimate

$$\int_2^7 \frac{x^2}{x^2+1}\,dx.$$

33. Show that $\displaystyle\int_0^1 \frac{4}{1+x^2}\,dx = \pi.$

(a) Use the trapezoidal rule with $n = 30$ to approximate π.

(b) Use Simpson's rule with $n = 25$ to approximate π.

■ PROJECT 8.7 The Error Function

The function B defined by

$$B(x) = \int_0^x e^{-t^2}\,dt, \quad x \in (-\infty, \infty)$$

is of fundamental importance in science and engineering, as well as in mathematical statistics. In particular,

$$\text{erf}(x) = \frac{2}{\sqrt{\pi}}\int_0^x e^{-t^2}\,dt$$

is called the *error function* and

$$\Phi(x) = \frac{1}{\sqrt{2\pi}}\int_0^x e^{-t^2/2}\,dt$$

is called the *probability integral*. Since e^{-t^2} does not have an elementary antiderivative, we must approximate the value of B numerically. Nevertheless, we can determine many of the properties of B directly from the definition.

Problem 1.

a. Show that B is an odd function.

b. Show that B is differentiable on $(-\infty, \infty)$.

c. Calculate $B'(x)$ and show that B is increasing on $(-\infty, \infty)$.

d. Show that the graph of B has exactly one point of inflection and find it.

Problem 2.

a. Use a graphing utility to estimate $B(n)$ for $n = 1, 2, 3, 4$.

b. Estimate $\lim_{x\to\infty} B(x)$.

c. What is $\lim_{x\to\infty} \text{erf}(x)$?

d. What is $\lim_{x\to\infty} \Phi(x)$?

e. What value would you assign to $\displaystyle\frac{1}{\sqrt{2\pi}}\int_{-\infty}^{\infty} e^{-t^2/2}\,dt$

Problem 3.

a. Sketch the graph of B.

b. Show that B has an inverse and sketch the graph of B^{-1}.

Problem 4.

Show that the error function and the probability integral are related as follows:

$$\text{erf}(x) = 2\Phi(\sqrt{2}\,x).$$

■ *8.8 DIFFERENTIAL EQUATIONS; FIRST-ORDER LINEAR EQUATIONS

An equation that relates an unknown function to one or more of its derivatives is called a *differential equation*. We have already introduced some differential equations. In Chapter 7 we used the differential equation

$$\frac{dy}{dt} = ky \tag{1}$$

to model exponential growth and decay. In various exercises (Section 3.6 and 4.4) we used the differential equation

$$\frac{d^2y}{dt^2} + \omega^2 y = 0 \tag{2}$$

to model the motion of a simple pendulum and the oscillation of a weight suspended at the end of a spring.

The *order* of a differential equation is the order of the highest derivative that appears in the equation. Thus (1) is a *first-order* equation and (2) is a *second order* equation

A function that satisfies a differential equation is called a *solution* of the equation. All functions $y = Ce^{kt}$ are solutions of (1):

$$\frac{dy}{dt} = kCe^{kt} = ky.$$

All functions $y = A \sin \omega t + B \cos \omega t$ are solutions of (2):

$$\begin{aligned} y &= A \sin \omega t + B \cos \omega t \\ \frac{dy}{dt} &= \omega A \cos \omega t - \omega B \sin \omega t \\ \frac{d^2y}{dt^2} &= -\omega^2 A \sin \omega t - \omega^2 B \cos \omega t = -\omega^2 y \end{aligned}$$

and therefore

$$\frac{d^2y}{dt^2} + \omega^2 y = 0.$$

Remark Differential equations reach far outside the boundaries of pure mathematics. Countless processes in the physical sciences, in the life sciences, in engineering, and in the social sciences are modeled by differential equations.

Extensive study of differential equations is beyond the scope of this text. In this section, and in the next one, we take up some simple, yet important, differential equations. Some additional topics in differential equations appear in later chapters. □

First-Order Linear Equations

A differential equation of the form

(8.8.1) $$\boxed{y' + p(x)y = q(x)}$$

is called a *first-order linear differential equation*. Here p and q are given functions defined and continuous on some interval I. In the simplest case, $p(x) = 0$ for all x, the equation reduces to

$$y' = q(x).$$

The solutions of this equation are the antiderivatives of q.

Solution Method To solve (8.8.1), we first calculate

$$H(x) = \int p(x)\,dx$$

(omitting the constant of integration) and form $e^{H(x)}$. We then multiply the equation by $e^{H(x)}$ and obtain

$$e^{H(x)}y' + e^{H(x)}p(x)\,y = e^{H(x)}q(x).$$

The left side of this equation is the derivative of $e^{H(x)}y$ (verify this), and so we have

$$\frac{d}{dx}[e^{H(x)}y] = e^{H(x)}\,q(x).$$

Integration gives

$$e^{H(x)}y = \int e^{H(x)}\,q(x)\,dx + C$$

and yields

(8.8.2)
$$\boxed{y(x) = e^{-H(x)}\left[\int e^{H(x)}\,q(x)\,dx + C.\right]} \quad \square$$

Remark The key step in solving $y' + p(x)y = q(x)$ is the multiplication by $e^{H(x)}$ where $H(x) = \int p(x)\,dx$. It is multiplication by this factor, called an *integrating factor*, that enables us to write the left-hand side in a form that we can integrate directly. $\square$

Note that (8.8.2) contains an arbitrary constant C. A close look at the steps taken to obtain (8.8.2) makes it clear that (8.8.2) includes *all* the functions that satisfy (8.8.1). For this reason (8.8.2) is called the *general solution of* (8.8.1). If we assign a particular value to the constant C, we obtain what is called a *particular solution*.

Example 1 Find the general solution of the equation

$$y' + a\,y = b, \quad a, b \text{ constants}, \quad a \neq 0.$$

SOLUTION First we calculate an integrating factor:

$$H(x) = \int a\,dx = ax \quad \text{and therefore} \quad e^{H(x)} = e^{ax}.$$

Multiplying the differential equation by e^{ax}, we get

$$e^{ax}\,y' + a\,e^{ax}y = b\,e^{ax}.$$

Now the left-hand side is the derivative of $e^{ax}y$ (verify). Thus, we have

$$\frac{d}{dx}[e^{ax}y] = b\,e^{ax}.$$

We integrate this equation and get

$$e^{ax}y = \frac{b}{a}e^{ax} + C.$$

Therefore
$$y = \frac{b}{a} + Ce^{-ax}.$$

This is the general solution. □

Example 2 Find the general solution of

$$y' + 2xy = x$$

and find the particular solution that satisfies $y(0) = 2$.

SOLUTION This equation has the form of (8.8.1). To solve it, we calculate the integrating factor $e^{H(x)}$:

$$H(x) = \int 2x\,dx = x^2 \quad \text{and so} \quad e^{H(x)} = e^{x^2}.$$

Multiplication by e^{x^2} gives

$$e^{x^2}y' + 2xe^{x^2}y = xe^{x^2},$$
$$\frac{d}{dx}\left[e^{x^2}y\right] = xe^{x^2}.$$

Integrating this equation, we get

$$e^{x^2}y = \tfrac{1}{2}e^{x^2} + C,$$

which we write as

$$y = \tfrac{1}{2} + Ce^{-x^2}.$$

This is the general solution. To find the solution that satisfies $y(0) = 2$, we set $x = 0$ and $y = 2$ in the general solution and solve for C:

$$2 = \tfrac{1}{2} + Ce^0 = \tfrac{1}{2} + C \quad \text{and so} \quad C = \tfrac{3}{2}.$$

Therefore the function

$$y = \tfrac{1}{2} + \tfrac{3}{2}e^{-x^2}$$

is the particular solution that satisfies the given condition. (Figure 8.8.1). □

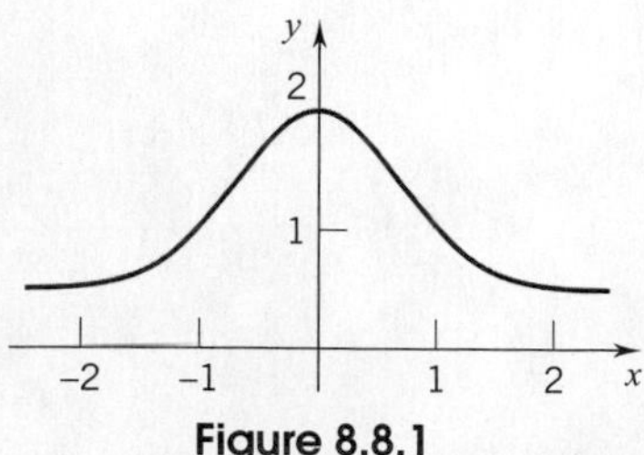

Figure 8.8.1

Remark When a differential equation is used as a mathematical model in some application, there is usually an *initial condition* $y(x_0) = y_0$ that makes it possible to evaluate the arbitrary constant that appears in the general solution. The problem of finding a particular solution of a differential equation that satisfies a given condition is called an *initial-value problem.* ☐

Example 3 Solve the initial-value problem:

$$xy' - 2y = 3x^4, \quad y(-1) = 2.$$

SOLUTION This differential equation does not have the form of (8.8.1), but we can put it in the required form by dividing the equation by x:

$$y' - \frac{2}{x}y = 3x^3.$$

Now we set

$$H(x) = \int -\frac{2}{x}dx = -2\ln x = \ln x^{-2}$$

and get the integrating factor

$$e^{H(x)} = e^{\ln x^{-2}} = x^{-2}.$$

Multiplication by x^{-2} gives

$$x^{-2}y' - 2x^{-3}y = 3x,$$

$$\frac{d}{dx}[x^{-2}y] = 3x.$$

Integrating this equation, we get

$$x^{-2}y = \tfrac{3}{2}x^2 + C,$$

which we write as

$$y = \tfrac{3}{2}x^4 + Cx^2.$$

This is the general solution. Applying the initial condition, we have

$$y(-1) = 2 = \tfrac{3}{2}(-1)^4 + C(-1)^2 = \tfrac{3}{2} + C.$$

It follows that $C = \frac{1}{2}$; the function $y = \frac{3}{2}x^4 + \frac{1}{2}x^2$ is the solution of the initial-value problem. (Figure 8.8.2) ☐

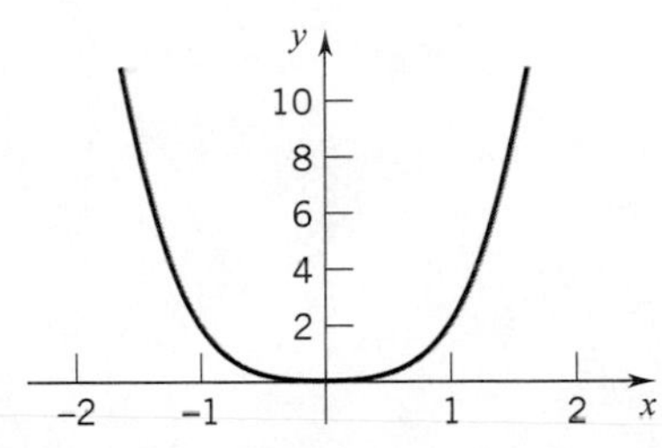

Figure 8.8.2

Applications

Newton's Law of Cooling Newton's law of cooling states that the rate of change of the temperature T of an object is proportional to the difference between T and the (constant) temperature τ of the surrounding medium, called the *ambient temperature*. If the temperature of the object is given by a differentiable function, the mathematical formulation of Newton's law takes the form

$$\frac{dT}{dt} = m(T - \tau), \quad \text{where } m \text{ is a constant.}$$

Remark The constant m in this model must be negative; for if the object is warmer than the ambient temperature ($T - \tau > 0$), then its temperature will decrease ($dT/dt < 0$), which implies $m < 0$; if the object is colder than the ambient temperature $(T - \tau < 0)$, its temperature will increase $(dT/dt > 0)$,which again implies $m < 0$. □

To emphasize that the constant of proportionality is negative, we write Newton's law of cooling as

(8.8.3)
$$\frac{dT}{dt} = -k(T - \tau), \quad k > 0 \text{ constant.}$$

This equation can be rewritten as

$$\frac{dT}{dt} + kT = k\tau,$$

a first-order linear equation with $p(t) = k$ and $q(t) = k\tau$ constant. From the result in Example 1, we see that

$$T = \frac{k\tau}{k} + Ce^{-kt} = \tau + Ce^{-kt}.$$

The constant C is determined by the initial temperature $T(0)$:

$$T(0) = \tau + Ce^{0} = \tau + C \quad \text{so that} \quad C = T(0) - \tau.$$

The temperature of the object at any time t is given by the function

(8.8.4)
$$T(t) = \tau + [T(0) - \tau]e^{-kt}.$$

Example 4 A cup of coffee is served to you at 185° F in a room where the temperature is 65° F. Two minutes later, the temperature of the coffee has dropped to 155° F. How many more minutes would you expect to wait for the coffee to cool to 105° F?

SOLUTION We are given $\tau = 65$ and $T(0) = 185$. Therefore

$$T(t) = 65 + [185 - 65]e^{-kt} = 65 + 120e^{-kt}.$$

To determine the constant of proportionality k, we use the fact that $T(2) = 155$:

$$T(2) = 65 + 120e^{-2k} = 155, \quad e^{-2k} = \tfrac{90}{120} = \tfrac{3}{4}, \quad k = -\tfrac{1}{2}\ln(3/4) \approx 0.144.$$

Therefore the function $T(t) = 65 + 120e^{-0.144t}$ gives the temperature of the coffee at any time t.

We now want to find the value of t for which $T(t) = 105°F$. To do this, we solve the equation

$$65 + 120e^{-0.144t} = 105$$

for t: $\quad 120e^{-0.144t} = 40, \quad -0.144t = \ln(1/3), \quad t \approx 7.63$ min.

Therefore you would expect to wait another 5.63 minutes. □

A Mixing Problem

Example 5 A chemical manufacturing company has a 1000-gallon holding tank which it uses to control the release of pollutants into a sewage system. Initially the tank has 360 gallons of fluid containing 2 pounds of pollutant per gallon. Fluid containing 3 pounds of pollutant per gallon enters the tank at the rate of 80 gallons per hour and is uniformly mixed with the fluid already in the tank. Simultaneously, fluid is released from the tank at the rate of 40 gallons per hour. Determine how many pounds of pollutant are in the tank at any time t. Then determine the rate (lbs/gal) at which the pollutant is being released after 10 hours.

SOLUTION Let $P(t)$ be the amount of pollutant (in pounds) in the tank at time t. The rate of change of pollutant in the tank, dP/dt, is given by

$$\frac{dP}{dt} = (\text{rate in}) - (\text{rate out}).$$

The pollutant is entering the tank at the rate of $3 \times 80 = 240$ pounds per hour (= rate in).

Fluid is entering the tank at the rate of 80 gallons per hour and is leaving at the rate of 40 gallons per hour. The amount of fluid in the tank is increasing at the rate of 40 gallons per hour, and so there are $360 + 40t$ gallons of fluid in the tank at time t. We can now conclude that the amount of pollutant per gallon in the tank at time t is given by

$$\frac{P(t)}{360 + 40t},$$

and the rate at which pollutant is leaving the tank is

$$40\frac{P(t)}{360 + 40t} = \frac{P(t)}{9 + t} \quad (= \text{rate out}).$$

Therefore, our differential equation reads

$$\frac{dP}{dt} = (\text{rate in}) - (\text{rate out}) = 240 - \frac{P}{9 + t},$$

which we can write as

$$\frac{dP}{dt} + \frac{1}{9 + t}P = 240.$$

This is a first-order linear differential equation. Here we have

$$p(t) = \frac{1}{9 + t} \quad \text{and} \quad H(t) = \int \frac{1}{9 + t}\, dt = \ln|9 + t| = \ln(9 + t) \quad (9 + t > 0).$$

As an integrating factor we use

$$e^{H(t)} = e^{\ln(9+t)} = 9 + t.$$

Multiplying the differential equation by $9 + t$, we have

$$(9+t)\frac{dP}{dt} + P = 240(9+t),$$

$$\frac{d}{dt}[(9+t)P] = 240(9+t),$$

$$(9+t)P = 120(9+t)^2 + C,$$

and
$$P(t) = 120(9+t) + \frac{C}{9+t}.$$

Since the amount of pollutant in the tank is initially $2 \times 360 = 720$ (pounds), we see that

$$P(0) = 120(9) + \frac{C}{9} = 720 \quad \text{which implies that} \quad C = -3240.$$

Thus, the function

$$P(t) = 120(9+t) - \frac{3240}{9+t} \quad \text{(pounds).}$$

gives the amount of pollutant in the tank at any time t. After 10 hours, there are $360 + 40(10) = 760$ gallons of fluid in the tank, and there are

$$P(10) = 120(19) - \tfrac{3240}{19} \cong 2109$$

pounds of pollutant. Therefore, the rate at which pollutant is being released into the sewage system after 10 hours is $\frac{2109}{760} \cong 2.78$ pounds per gallon. ❑

EXERCISES *8.8

Determine whether the functions satisfy the differential equation.

1. $2y' - y = 0; \quad y_1(x) = e^{x/2}, \quad y_2(x) = x^2 + 2e^{x/2}.$

2. $y' + xy = x; \quad y_1(x) = e^{-x^2/2}, \quad y_2(x) = 1 + Ce^{-x^2/2},$

3. $y' + y = y^2; \quad y_1(x) = \dfrac{1}{e^x + 1}, \quad y_2(x) = \dfrac{1}{C\,e^x + 1},$

4. $y'' + 4y = 0; \quad y_1(x) = 2\sin 2x, \quad y_2(x) = 2\cos x.$

5. $y'' - 4y = 0; \quad y_1(x) = e^{2x}, \quad y_2(x) = C\sinh 2x,$

6. $y'' - 2y' - 3y = 7e^{3x}; \quad y_1(x) = e^{-x} + 2e^{3x}, \quad y_2(x) = \frac{7}{4}xe^{3x}.$

Find the general solution.

7. $y' - 2y = 1.$

8. $xy' - 2y = -x.$

9. $2y' + 5y = 2.$

10. $y' - y = -2\,e^{-x}.$

11. $y' - 2y = 1 - 2x.$

12. $xy' + 2y = \dfrac{\cos x}{x}.$

13. $xy' - 4y = -2nx.$

14. $y' + y = 2 + 2x.$

15. $y' - e^x y = 0.$

16. $y' - y = e^x.$

17. $(1 + e^x)\,y' + y = 1.$

18. $xy' + y = (1 + x)\,e^x.$

19. $y' + 2xy = x\,e^{-x^2}.$

20. $xy' - y = 2x\ln x.$

21. $y' + \dfrac{2}{x+1}y = 0.$

22. $y' + \dfrac{2}{x+1}y = (x+1)^{5/2}.$

Find the particular solution determined by the initial condition.

23. $y' + y = x, \quad y(0) = 1.$

24. $y' - y = e^{2x}, \quad y(1) = 1.$

25. $y' + y = \dfrac{1}{1 + e^x}, \quad y(0) = e.$

26. $y' + y = \dfrac{1}{1 + 2\,e^x}, \quad y(0) = e.$

27. $xy' - 2y = x^3 e^x, \quad y(1) = 0.$

28. $xy' + 2y = x\,e^{-x}, \quad y(1) = -1.$

29. Find all functions that satisfy the differential equation $y' - y = y'' - y'$. HINT: Set $z = y' - y$.

30. Find the general solution of $y' + ry = 0$ on $[0, \infty)$ where r is a constant.

(a) Show that if y is a solution and $y(a) = 0$ at some number $a \geq 0$, then $y(x) = 0$ for all x. (A solution y is either identically zero or never zero.)

(b) Show that if $r < 0$, then all nonzero solutions are unbounded.

(c) Show that if $r > 0$, then all solutions have limit 0 as $x \to \infty$.

(d) Describe all solutions of the equation in the case $r = 0$.

Exercises 31 and 32 are concerned with the first-order linear differential equation

$$(*) \qquad y' + p(x)y = 0$$

where the function p is continuous on an interval I.

31. (a) Show that if y_1 and y_2 are solutions of $(*)$, then $u = y_1 + y_2$ is also a solution of $(*)$.

(b) Show that if y is a solution of $(*)$ and C is a constant, then $u = Cy$ is also a solution of $(*)$.

32. (a) Let $a \in I$. Show that the general solution of $(*)$ can be written as

$$y(x) = Ce^{-\int_a^x p(t)\,dt}.$$

(b) Show that if y is a solution of $(*)$ and $y(b) = 0$ for some $b \in I$, then $y(x) = 0$ for all $x \in I$.

(c) Show that if y_1 and y_2 are solutions of $(*)$ and $y_1(b) = y_2(b)$ for some $b \in I$, then $y_1(x) = y_2(x)$ for all $x \in I$.

Exercises 33 and 34 are concerned with the differential equation (8.8.1):

$$y' + p(x)y = q(x)$$

where p and q are functions continuous on some interval I.

33. Let $a \in I$ and let $H(x) = \int_a^x p(t)\,dt$. Show that

$$y(x) = e^{-H(x)} \int_a^x q(t)\, e^{H(t)} dt$$

is the solution of the differential equation that satisfies the initial condition $y(a) = 0$.

34. Show that if y_1 and y_2 are solutions of (8.8.1), then $z = y_1 - y_2$ is a solution of the differential equation $(*)$ above.

35. A thermometer is taken from a room where the temperature is 72°F to the outside, where the temperature is 32°F. After $\frac{1}{2}$ minute the thermometer reads 50°F. What will the thermometer read at $t = 1$ minute? How long will it take for the thermometer to read 35°?

36. A solid metal sphere at room temperature of $20°C$ is dropped into a container of boiling water $(100°C)$. If the temperature of the sphere increases 2° in 2 seconds, what will the temperature be at time $t = 6$ seconds? How long will it take for the temperature of the sphere to reach $90°C$?

37. An object falling from rest in air is subject not only to the gravitational force but also to air resistance. Assume that the air resistance is proportional to the velocity and acts in a direction opposite to the motion. Then the velocity of the object at time t satisfies

$$v' = 32 - kv(t), \quad k > 0 \text{ constant}, \quad v(0) = 0.$$

Here we are measuring distance in feet and the positive direction is down.

(a) Find $v(t)$.

(b) Show that $v(t)$ cannot exceed $32/k$ and that $v(t) \to 32/k$ as $t \to \infty$.

(c) Sketch the graph of $v(t)$.

38. Suppose that a certain population P has a known birth rate dB/dt and a known death rate dD/dt. Then the rate of change of P is given by

$$\frac{dP}{dt} = \frac{dB}{dt} - \frac{dD}{dt}.$$

(a) Assume that $dB/dt = aP$ and $dD/dt = bP$, where a and b are constants. Find $P(t)$ if $P(0) = P_0 > 0$.

(b) Analyze the cases (i) $a > b$, (ii) $a = b$, (iii) $a < b$. Find $\lim_{t\to\infty} P(t)$ in each case.

39. The current i in an electrical circuit consisting of a resistance R, inductance L, and voltage E varies with time (measured in seconds) according to the formula

$$L\frac{di}{dt} + Ri = E, \qquad R > 0, L > 0, E > 0 \text{ constant}.$$

(a) Find i if $i(0) = 0$.

(b) What upper limit does the current approach as $t \to \infty$?

(c) In how many seconds will the current reach 90% of its limit?

40. The current i in an electrical circuit consisting of a resistance R, inductance L, and a voltage $E \sin \omega t$, varies with time according to the formula

$$L\frac{di}{dt} + Ri = E \sin \omega t, \qquad R > 0, L > 0, E > 0 \text{ constant}$$

(a) Find i if $i(0) = i_0$.

(b) Does $\lim_{t\to\infty} i(t)$ exist? Compare with Exercise 39.

(c) Sketch the graph of i.

41. A 200-liter tank initially full of water develops a leak at the bottom. Given that 20% of the water leaks out in the first 5 minutes, find the amount of water left in the tank t minutes after the leak develops:

(a) if the water drains off at a rate that is proportional to the amount of water present.

(b) if the water drains off at a rate that is proportional to the product of the time elapsed and the amount of water present.

42. At a certain moment a 100-gallon mixing tank is full of brine containing 0.25 pounds of salt per gallon. Find the amount of salt present t minutes later if the brine is being continuously drawn off at the rate of 3 gallons per minute and replaced by brine containing 0.2 pounds of salt per gallon.

43. An advertising company is trying to expose a new product to a certain metropolitan area by advertising on television.

Suppose that the exposure to new people is proportional to the number of people who have not seen the product out of a total population of M viewers. Let $P(t)$ denote the number of viewers who have been exposed to the product at time t. The company has determined that no one was aware of the product at the start of the campaign $[P(0) = 0]$ and that 30% of the viewers were aware of the product after 10 days.

(a) Determine the differential equation that describes the number of viewers who are aware of the product at time t.

(b) Determine the solution of the differential equation from part (a) that satisfies the initial condition $P(0) = 0$.

(c) How long will it take for 90% of the population to be aware of the product?

44. A drug is fed intravenously into a patient's bloodstream at the constant rate r. Simultaneously, the drug diffuses into the patient's body at a rate proportional to the amount of drug present.

(a) Determine the differential equation that describes the amount $Q(t)$ of the drug in the patient's bloodstream at time t.

(b) Determine the solution $Q = Q(t)$ of the differential equation found in part (a) that satisfies the initial condition $Q(0) = 0$.

(c) Find $\lim_{t\to\infty} Q(t)$.

45. (a) The differential equation

$$\frac{dP}{dt} = (2\cos 2\pi t)P$$

describes a population that undergoes periodic fluctuations. Assume that $P(0) = 1000$ and find $P(t)$. Use a graphing utility to draw the graph of P.

(b) The differential equation

$$\frac{dP}{dt} = (2\cos 2\pi t)P + 2000\cos 2\pi t$$

describes a population that undergoes periodic fluctuations as well as periodic migration. Continue to assume that $P(0) = 1000$ and find $P(t)$ in this case. Use a graphing utility to draw the graph of P and estimate the maximum value of P.

46. The *Gompertz equation*

$$\frac{dP}{dt} = P(a - b\ln P),$$

where a and b are positive constants, is another model of population growth.

(a) Find the solution of this differential equation that satisfies the initial condition $P(0) = P_0$, where $P_0 > 0$. HINT: Define a new dependent variable Q by $Q = \ln P$.

(b) Find $\lim_{t\to\infty} P(t)$.

(c) Determine the concavity of the graph of P.

(d) Use a graphing utility to draw the graph of P in the case where $a = 4, b = 2$, and $P_0 = \frac{1}{2}e^2$. Does the graph confirm your result in part (c)?

■ *8.9 INTEGRAL CURVES; SEPARABLE EQUATIONS

Integral Curves

If a function $y = y(x)$ satisfies a first-order differential equation, then along the graph of the function, the numbers x, y, y' are related as prescribed by the equation. In the case of a nonlinear differential equation, it is usually difficult (and sometimes impossible) to find functions which are explicitly defined and satisfy the differential equation. More often, we can find plane curves which, though not the graphs of functions, do have the property that along the curve the numbers x, y, y' are related as prescribed by the differential equation. Such curves are called *integral curves* (*solution curves*) of the differential equation. To *solve* a nonlinear differential equation is to find the integral curves of the equation.

Separable Equations

A first-order differential equation is said to be *separable* if it can be written in the form

(8.9.1) $$\boxed{p(x) + q(y)y' = 0.}$$

The functions p and q are assumed to be continuous where defined.

Solution Method We expand equation (8.9.1) to read

$$p(x) + q[y(x)]\,y'(x) = 0.$$

Integrating this equation with respect to x, we find that

$$\int p(x)\,dx + \int q[y(x)]\,y'(x)\,dx = C,$$

where C is an arbitrary constant. From $y = y(x)$, we have $dy = y'(x)\,dx$. Therefore

$$\int p(x)\,dx + \int q(y)\,dy = C.$$

The variables have been separated. Now, if P is an antiderivative of p and Q is an antiderivative of q, then this last equation can be written

(8.9.2) $$\boxed{P(x) + Q(y) = C.}$$

This equation represents a family of curves in the xy-plane. As we show below, all these curves are integral curves for Equation (8.9.1). To prove this statement,we must show that, if x and y are related by the equation $P(x) + Q(y) = C$, then x, y, y' are related by the equation $p(x) + q(y)y' = 0$. The proof is straightforward: if

$$P(x) + Q(y) = C,$$

then, by implicit differentiation with respect to x, we have

$$\begin{aligned} p(x) + q(y)\,y' &= p(x) + q[y(x)]\,y'(x) \\ &= \frac{d}{dx}\Big(P(x) + Q[y(x)]\Big) \\ &= \frac{d}{dx}[P(x) + Q(y)] = \frac{d}{dx}[C] = 0. \quad \square \end{aligned}$$

Example 1 The differential equation

$$x + yy' = 0$$

is separable. We can find the integral curves by writing

$$\int x\,dx + \int y\,dy = C.$$

Carrying out the integration, we have

$$\tfrac{1}{2}x^2 + \tfrac{1}{2}y^2 = C,$$

which, since C is arbitrary, we can write as

$$x^2 + y^2 = C.$$

For this equation to give us a curve, we must take $C \geq 0$. For $C > 0$, the integral curves are circles of radius $\sqrt{C}$ centered at the origin. For $C = 0$, the integral curve degenerates to a single point, the origin. □

Example 2 Show that the differential equation

$$y' = \frac{xy - y}{y + 1}$$

is separable and find an integral curve which satisfies the initial condition: $y(2) = 1$.

SOLUTION To show that the differential equation is separable, we try to write it in the form (8.9.1):

$$\begin{aligned}(y+1)y' &= y(x-1)\\ y(1-x) + (y+1)y' &= 0.\end{aligned}$$

The next step is to divide this equation by y. As you can verify, the horizontal line $y = 0$ is an integral curve. However, we can ignore it because it doesn't satisfy the initial condition $y(2) = 1$. Therefore, with $y \neq 0$, we can write

$$\begin{aligned}1 - x + \frac{y+1}{y}y' &= 0,\\ 1 - x + \left(1 + \frac{1}{y}\right)y' &= 0.\end{aligned}$$

The equation is separable. Writing

$$\int (1-x)\,dx + \int \left(1 + \frac{1}{y}\right) dy = C,$$

we find that the integral curves take the form

$$x - \tfrac{1}{2}x^2 + y + \ln|y| = C.$$

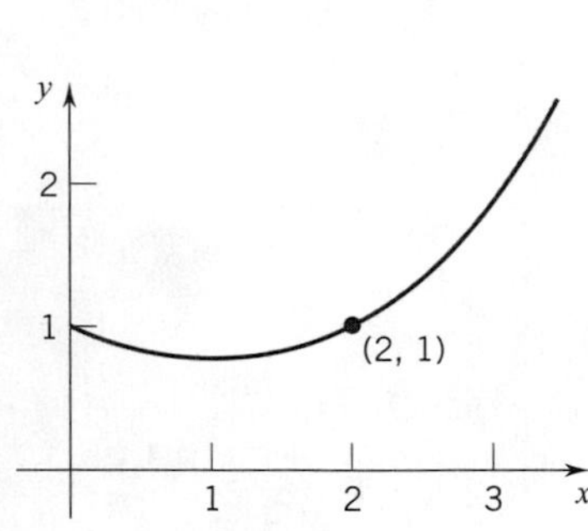

Figure 8.9.1

The condition $y(2) = 1$ forces

$$2 - \tfrac{1}{2}(2)^2 + 1 + \ln(1) = C, \quad \text{which implies} \quad C = 1.$$

The integral curve that satisfies the initial condition is the curve

$$x - \tfrac{1}{2}x^2 + y + \ln y = 1 \quad \square \qquad \text{(Figure 8.9.1)}$$

Example 3 Show that the differential equation

$$y' = xe^{y-x}$$

is separable and find the integral curves.

SOLUTION Since $e^{y-x} = e^y e^{-x}$, the equation can be written

$$xe^{-x}e^y - y' = 0,$$

and thus

$$xe^{-x} - e^{-y}y' = 0.$$

The equation is separable. Writing

$$\int xe^{-x}dx - \int e^{-y}dy = C,$$

we have

$$-xe^{-x} - e^{-x} + e^{-y} = C \quad \text{and therefore} \quad e^{-y} = xe^{-x} + e^{-x} + C.$$

These are the integral curves. In this case, we can solve for y and express the curves as functions of x:

$$y = -\ln(xe^{-x} + e^{-x} + C).$$

The graphs of several of the integral curves are shown in Figure 8.9.2. □

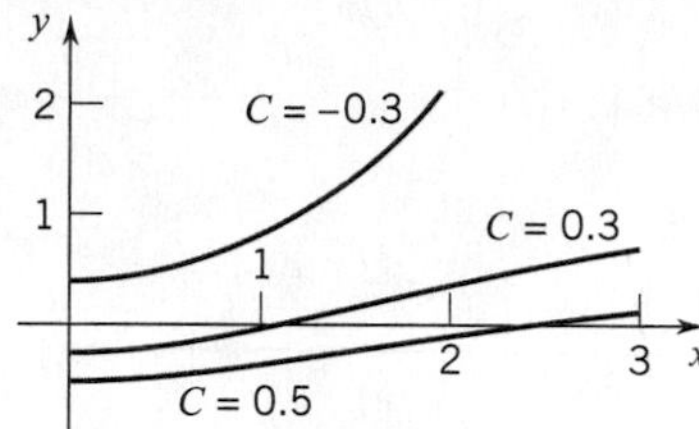

Figure 8.9.2

Applications

In the mid-nineteenth century, the Belgian biologist P. F. Verhulst used the differential equation

(8.9.3) $$\boxed{\frac{dy}{dt} = ky(M - y),}$$

where k and M are positive constants, to study the population growth of various countries. This equation is now known as the *logistic equation*, and its solutions are called *logistic functions*. Life scientists have used this equation to model the spread of a disease through a population, and social scientists have used it to study the flow of information. In the case of a disease, if M denotes the total number of people in the population and $y(t)$ is the number of infected people at time t, then the differential equation states that the rate of growth of the disease is proportional to the product of the number of people who are infected and the number who are not.

The differential equation is separable since it can be written in the form

$$k - \frac{1}{y(M-y)}y' = 0.$$

Integrating this equation, we have

$$\int k\,dt - \int \frac{1}{y(M-y)}dy = C,$$

$$kt - \int \left(\frac{1/M}{y} + \frac{1/M}{M-y}\right) dy = C, \quad \text{(partial fraction decomposition)}$$

and therefore

$$\frac{1}{M}\ln|y| - \frac{1}{M}\ln|M-y| = kt + C.$$

It is a good exercise in manipulating logarithms, exponentials, and arbitrary constants to show that this equation can be solved for y, and thus the solution can be expressed as a function of t. The result can be written as

$$y = \frac{CM}{C + e^{-Mkt}}. \qquad \text{(not the same C as above)}$$

If $y = y(t)$ satisfies the initial condition $y(0) = R, R < M$, then

$$R = \frac{CM}{C+1}, \quad \text{which implies} \quad C = \frac{R}{M-R},$$

and we have

(8.9.4)
$$y(t) = \frac{MR}{R + (M-R)e^{-Mkt}}$$

This particular solution is shown graphically in Figure 8.9.3. Note that y is increasing for all $t \geq 0$. In the Exercises you are asked to show that the graph is concave up on $[0, t_1]$ and concave down on $[t_1, \infty)$. In the case of a disease, this means that the disease is spreading at an increasing rate up to time $t = t_1$; and after t_1 the disease is still spreading, but at a decreasing rate. Note, also, that $y(t) \to M$ as $t \to \infty$.

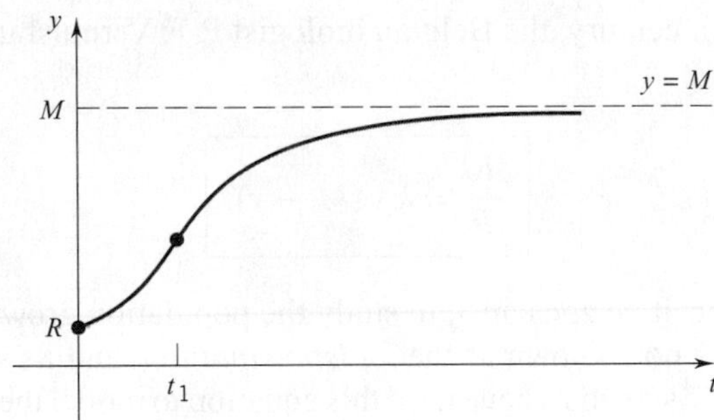

Figure 8.9.3

Example 4 A rumor spreads through a population of 5000 people at a rate proportional to the product of the number of people who have heard it and the number who have not. Suppose that 100 people initiate the rumor and that a total of 500 people have heard the rumor after two days. How long will it take for half the people to hear the rumor?

SOLUTION Let $y(t)$ denote the number of people who know the rumor at time t. Then y satisfies the logistic equation with $M = 5000$ and the initial condition $R = y(0) = 100$. Thus, by (8.9.4),

$$y(t) = \frac{100(5000)}{100 + 4900e^{-5000\,kt}} = \frac{5000}{1 + 49e^{-5000\,kt}}.$$

The constant of proportionality k can be determined from the condition that $y(2) = 500$. We have

$$\begin{aligned} 500 &= \frac{5000}{1 + 49e^{-10{,}000\,k}}, \\ 1 + 49e^{-10{,}000\,k} &= 10, \\ e^{-10{,}000\,k} &= \frac{9}{49}, \\ -10{,}000\,k &= \ln(9/49) \quad \text{and thus} \quad k \cong 0.00017. \end{aligned}$$

Therefore, $$y(t) = \frac{5000}{1 + 49e^{-0.85\,t}}.$$

To determine how long it will take for half the population to hear the rumor, we solve the equation

$$2500 = \frac{5000}{1 + 49e^{-0.85t}}$$

for t. We get

$$1 + 49e^{-0.85t} = 2, \quad e^{-0.85t} = \frac{1}{49}, \quad t = \frac{\ln(1/49)}{-0.85} \cong 4.58$$

It will take slightly more than $4\frac{1}{2}$ days for half the population to hear the rumor. □

EXERCISES *8.9

Find the integral curves

1. $y' = y\sin(2x + 3)$.
2. $y' = (x^2 + 1)(y^2 + y)$.
3. $y' = (xy)^3$.
4. $y' = 3x^2(1 + y^2)$.
5. $y' = -\dfrac{\sin 1/x}{x^2 y \cos y}$.
6. $y' = \dfrac{y^2 + 1}{y + yx}$.
7. $y' = x\,e^{x+y}$.
8. $y' = xy^2 - x - y^2 + 1$.
9. $(y \ln x)\,y' = \dfrac{(y + 1)^2}{x}$.
10. $e^y \sin 2x\,dx + \cos x\,(e^{2y} - y)\,dy = 0$.
11. $(y \ln x)\,y' = \dfrac{y^2 + 1}{x}$.
12. $y' = \dfrac{1 + 2y^2}{y \sin x}$.

Solve the initial value problem.

13. $y' = x\sqrt{\dfrac{1 - y^2}{1 - x^2}}, \quad y(0) = 0$.
14. $y' = \dfrac{e^{x-y}}{1 + e^x}, \quad y(1) = 0$.
15. $y' = \dfrac{x^2 y - y}{y + 1}, \quad y(3) = 1$.
16. $x^2 y' = y - xy, \quad y(-1) = -1$.
17. $(xy^2 + y^2 + x + 1)\,dx + (y - 1)\,dy = 0, \quad y(2) = 0$.
18. $\cos y\,dx + (1 + e^{-x})\sin y\,dy = 0, \quad y(0) = \pi/4$.
19. $y' = 6e^{2x-y}, \quad y(0) = 0$.
20. $xy' - y = 2x^2 y, \quad y(1) = 1$.

Orthogonal trajectories If two curves intersect at right angles, one with slope m_1 and the other with slope m_2, then $m_1 m_2 = -1$. A curve that intersects every member of a family of curves at right angles is called an *orthogonal trajectory* for that family of curves. A differential equation of the form

$$y' = f(x, y),$$

where $f(x, y)$ is an expression in x and y, generates a family of curves: all the integral curves of that particular equation. The orthogonal trajectories of this family are the integral curves of the differential equation

$$y' = -\frac{1}{f(x, y)}.$$

The figure shows the family of a parabolas $y = Cx^2$. The orthogonal trajectories of this family of parabolas are ellipses.

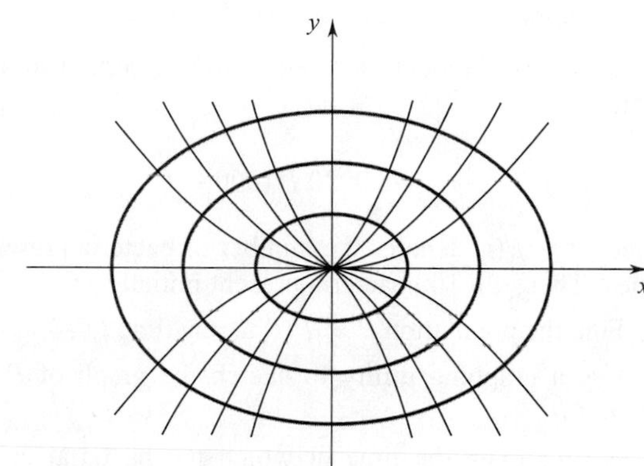

We can establish this by starting with the equation

$$y = Cx^2$$

and differentiating. Differentiation gives

$$y' = 2Cx.$$

Since $y = Cx^2$, we have $C = y/x^2$, and thus

$$y' = 2\left(\frac{y}{x^2}\right)x = \frac{2y}{x}.$$

The orthogonal trajectories are the integral curves of the differential equation

$$y' = -\frac{x}{2y}.$$

As you can check, the integral curves of this equation are all of the form

$$x^2 + 2y^2 = K.$$

In Exercises 21–26, find the orthogonal trajectories for the following families of curves. In each case draw several of the curves and several of the orthogonal trajectories.

21. $2x + 3y = C$.

22. $y = Cx$.

23. $xy = C$.

24. $y = Cx^3$.

25. $y = Ce^x$.

26. $x = Cy^4$.

C In Exercises 27 and 28, show that the given family is *self-orthogonal*. Use a graphing utility to graph at least four members of the family.

27. $y^2 = 4C(x + C)$.

28. $\dfrac{x^2}{C^2} + \dfrac{y^2}{C^2 - 4} = 1$.

29. Suppose that a chemical A combines with a chemical B to form a compound C. In addition, suppose that the rate at which C is produced at time t varies directly with the amounts of A and B present at time t. With this model, if A_0 grams of A are mixed with B_0 grams of B, then

$$\frac{dC}{dt} = k(A_0 - C)(B_0 - C).$$

(a) Find the amount of compound C present at time t if $A_0 = B_0$.

(b) Find the amount of compound C present at time t if $A_0 \neq B_0$.

C **30.** A mathematical model for the growth of a certain strain of bacteria is

$$\frac{dP}{dt} = 0.0020\, P\,(800 - P),$$

where $P = P(t)$ denotes the number of bacteria present at time t. There are 100 bacteria present initially.

(a) Find the population $P = P(t)$ at any time t.

(b) Use a graphing utility to sketch the graph of P and dP/dt.

(c) Approximate the time at which the bacterial culture experiences its most rapid growth rate. That is, locate the maximum value of dP/dt. Use three decimal place accuracy. What point on the graph of P corresponds to the maximum growth rate?

31. When an object of mass m is moving through air or a viscous medium, it is acted on by a frictional force that acts in a direction opposite to its motion. This frictional force depends on the velocity of the object and (within close approximation) is given by

$$F(v) = -\alpha v - \beta v^2,$$

where α and β are positive constants.

(a) From Newton's second law, $F = ma$, we have

$$m\frac{dv}{dt} = -\alpha v - \beta v^2.$$

Solve this differential equation to find $v = v(t)$.

(b) Find v if the object has initial velocity $v(0) = v_0$.

(c) What is $\lim_{t\to\infty} v(t)$?

32. A descending parachutist is acted on by two forces: a constant downward force mg and the upward force of air resistance, which (within close approximation) is of the form $-\beta v^2$ where β is a positive constant. (In this problem we are taking the downward direction as positive.)

(a) Express t in terms of the velocity v, the initial velocity v_0, and the constant $v_c = \sqrt{mg/\beta}$.

(b) Express v as a function of t.

(c) Express the acceleration a as a function of t. Verify that the acceleration never changes sign and in time tends to zero.

(d) Show that in time v tends to v_c. (This number v_c is called the *terminal velocity.*)

C **33.** A flu virus is spreading rapidly through a small town with a population of 25,000. The disease is spreading at a rate proportional to the product of the number of people who have it and the number who don't. Suppose that 100 people had the disease initially, and 400 had it after 10 days.

(a) How many people will have the disease at an arbitrary time t? How many will have it after 20 days?

(b) How long it will take for half the population to have the flu?

(c) Use a graphing utility to sketch the graph of your solution in part (a).

34. Consider the logistic equation (8.9.3). Show that dy/dt is increasing if $y < M/2$ and is decreasing if $y > M/2$. What can you conclude about dy/dt at the instant $y = M/2$? Explain.

C **35.** A rescue package whose mass is 100 kilograms is dropped from an airplane at a height of 4000 meters. As it falls, the air resistance is equal to twice its velocity. After 10 seconds, the package's parachute opens and the air resistance is now four times the square of its velocity.

(a) What is the package's velocity at the instant the parachute opens?

(b) Determine an expression for the velocity of the package at time t after the parachute opens.

(c) What is the terminal velocity of the package?

HINT: There are two differential equations that govern the package's velocity and position: one for the free-fall period and one for the period after the parachute opens.

36. It is known that m parts of chemical A combine with n parts of chemical B to produce a compound C. Suppose that the rate at which C is produced varies directly with the product of the amounts of A and B present at that instant. Find the amount of C produced in t minutes from an initial mixing of A_0 pounds of A with B_0 pounds of B, given that:

(a) $n = m$, $A_0 = B_0$, and A_0 pounds of C are produced in the first minute.

(b) $n = m$, $A_0 = \frac{1}{2}B_0$, and A_0 pounds of C are produced in the first minute.

(c) $n \neq m$, $A_0 = B_0$, and A_0 pounds of C are produced in the first minute.

HINT: Denote by $A(t)$, $B(t)$, and $C(t)$ the amounts of A, B, and C present at time t. Observe that $C'(t) = kA(t)B(t)$. Then note that

$$A_0 - A(t) = \frac{m}{m+n}C(t) \quad \text{and} \quad B_0 - B(t) = \frac{n}{m+n}C(t)$$

and thus

$$C'(t) = k\left[A_0 - \frac{m}{m+n}C(t)\right]\left[B_0 - \frac{n}{m+n}C(t)\right].$$

■ CHAPTER HIGHLIGHTS

8.1 Integral Tables and Review

Important integral formulas (p. 446)
A table of integral appears on the inside covers.

8.2 Integration by Parts

$$\int u\,dv = uv - \int v\,du. \text{ (p. 451)}$$

Success with the technique depends on choosing u and dv so that $\int v du$ is easier to integrate than $\int u dv$.
The integral of $\ln x$ and the integrals of the inverse trigonometric functions are calculated using integration by parts (p. 453).

8.3 Powers and Products of Trigonometric Functions

Integrals of the form

$$\int \sin^m x \cos^n x\,dx$$

can be calculated by using the basic identity $\sin^2 x \cos^2 x = 1$ and the double-angle formulas for sine and cosine
Reduction formulas:

$$\int \sin^n x\,dx = -\frac{1}{n}\sin^{n-1} x \cos x + \frac{n-1}{n}\int \sin^{n-2} x\,dx,$$

$$\int \cos^n x\,dx = \frac{1}{n}\cos^{n-1} x \sin x + \frac{n-1}{n}\int \cos^{n-2} x\,dx.$$

The main tools for calculating such integrals as $\int \tan^m x \sec^n x\,dx$ are the identities $1 + \tan^2 x = \sec^2 x$ and integration by parts.

8.4 Integrals Involving $\sqrt{a^2 - x^2}, \sqrt{a^2 + x^2}, \sqrt{x^2 - a^2}$

Such integrals can be calculated by a trigonometric substitution:

for $\sqrt{a^2 - x^2}$ set $a \sin u = x$,

for $\sqrt{a^2 + x^2}$ set $a \tan u = x$,

for $\sqrt{x^2 - a^2}$ set $a \sec u = x$.

It may be necessary to complete the square under the radical before making a trigonometric substitution (p. 470).

8.5 Partial Fractions

A proper rational function may be integrated by writing it as a sum of fractions of the form

$$\frac{A}{(x-\alpha)^k} \quad \text{and} \quad \frac{Bx + C}{(x^2 + \beta x + \gamma)k}$$

called the *partial fraction decomposition* (p. 474).
To integrate an improper rational functions, express it as a polynomial plus a proper rational function by dividing the denominator into the numerator (p. 474).

*8.6 Some Rationalizing Substitutions

For integrals involving $\sqrt[n]{f(x)}$ for some function f, let $u^n = f(x), nu^{n-1}du = f'(x)\,dx$. For rational expressions in sine and cosine, let $u = \tan(x/2)$; then

$$\sin x = \frac{2u}{1+u^2}, \quad \cos x = \frac{1-u^2}{1+u^2}, \quad dx = \frac{2}{1+u^2}du \text{ (p. 484).}$$

8.7 Numerical Integration

Left-endpoint, right-endpoint, and midpoint estimates; trapezoidal rule; Simpson's rule (p. 487)
The theoretical error in the trapezoidal rule varies as $1/n^2$ (p. 491).
The theoretical error in Simpson's rule varies as $1/n^4$ (p. 492).

*8.8 Differential Equations; First-Order Linear Equations

differential equation (p. 496)
order (p. 496)
solution (p. 496)
general solution (p. 497)
particular solution (p. 497)
initial-value problem (p. 499)
integrating factor (p. 497)
Newton's law of cooling (p. 499)

A first-order differential equation is *linear* if it can be written in the form $y' + p(x)y = q(x)$, where p and q are continuous functions on some interval I.

The general solution is $y = e^{-H(x)}\left\{\int e^{H(x)}q(x)\,dx + C\right\}$, where $H(x) = \int p(x)dx$.

*8.9 Integral Curves; Separable Equations

Integral curves (p. 504)
A first-order differential equation is separable if it can be written in the form

$$p(x) + q(y)y' = 0.$$

The integral curves are of the form

$$\int p(x)dx + \int q(y)dy = C.$$

The logistic equation (p. 507).

■ SKILL MASTERY REVIEW Transcendental Functions and Techniques of Integration

Determine whether the function f is one-to-one and, if so, find its inverse.

1. $f(x) = x^{1/3} + 2$

2. $f(x) = x^2 - x - 6$

3. $f(x) = \sin 2x + \cos x$

4. $f(x) = \dfrac{2x+1}{3-2x}$

Show that f has an inverse and find $(f^{-1})'(c)$.

5. $f(x) = \dfrac{1}{1+e^x}; \quad c = \frac{1}{2}$

6. $f(x) = 3x - \dfrac{1}{x^3}, \quad x > 0; \quad c = 2$

7. $f(x) = \displaystyle\int_0^x \sqrt{4+t^2}\,dt; \quad c = 0$

8. $f(x) = x - \pi + \cos x; \quad c = -1$

Calculate the derivative.

9. $f(x) = (\ln x^2)^3$

10. $y = 2\sin(e^{3x})$

11. $g(x) = \dfrac{e^x}{1+e^{2x}}$

12. $f(x) = (x^2+1)^{\sinh x}$

13. $y = \ln(x^3 + 3^x)$

14. $g(x) = \tan^{-1}(\cosh x)$

15. $f(x) = (\cosh x)^{1/x}$

16. $f(x) = 2x^3 \sin^{-1} x^2$

Evaluate the integral.

17. $\displaystyle\int_1^e \frac{\sqrt{\ln x}}{x}\,dx$

18. $\displaystyle\int \frac{\cos x}{4+\sin^2 x}\,dx$

19. $\displaystyle\int 2x \sinh x\,dx$

20. $\displaystyle\int (\tan x + \cot x)^2\,dx$

21. $\displaystyle\int \frac{x-3}{x^2(x+1)}\,dx$

22. $\displaystyle\int_1^8 \frac{x^{1/3}}{x^{4/3}+1}\,dx$

23. $\displaystyle\int \sin 2x \cos x\,dx$

24. $\displaystyle\int 3xe^{-3x}\,dx$

25. $\displaystyle\int \ln\sqrt{x+1}\,dx$

26. $\displaystyle\int \frac{2}{x(1+x^2)}\,dx$

27. $\displaystyle\int \frac{\sin^3 x}{\cos x}\,dx$

28. $\displaystyle\int \frac{\cos x}{\sin^3 x}\,dx$

29. $\displaystyle\int_0^1 e^{-x}\cosh x\,dx$

30. $\displaystyle\int_0^{\pi/4} \frac{x^2}{1+x^2}\,dx$

31. $\displaystyle\int \tan x \ln(\cos x)\,dx$

32. $\displaystyle\int 2^x \sinh 2^x\,dx$

33. $\displaystyle\int_2^5 \frac{1}{x^2-4x+13}\,dx$

34. $\displaystyle\int \frac{1}{x^3-1}\,dx$

35. $\displaystyle\int x\,2^x\,dx$

36. $\displaystyle\int \ln(x\sqrt{x})\,dx$

37. $\displaystyle\int \frac{\sqrt{a^2-x^2}}{x^2}\,dx$

38. $\displaystyle\int_0^2 x^2 e^{x^3}\,dx$

39. $\displaystyle\int \frac{\sin^5 x}{\cos^7 x}\,dx$

40. $\displaystyle\int \left(\frac{\sqrt{4+x^2}}{x} - \frac{x}{\sqrt{4+x^2}}\right)dx$

41. $\displaystyle\int_{\pi/6}^{\pi/3} \frac{\sin x}{\sin 2x}\,dx$

42. $\displaystyle\int \frac{x+3}{\sqrt{x^2+2x-8}}\,dx$

43. $\displaystyle\int \frac{x^2+x}{\sqrt{1-x^2}}\,dx$

44. $\displaystyle\int x\tan^2 2x\,dx$

45. Find the average value of $f(x) = \sin^2 x$ from $x = 0$ to $x = \pi$. (This average is used in the theory of alternating currents.)

46. (a) Apply the mean-value theorem to the function $f(x) = \ln(1 + x)$ to show that

$$\frac{x}{1+x} < \ln(1 + x) < x$$

for all $x > -1$.

(b) Use the result in part (a) to show that $\lim_{x\to 0} \frac{\ln(1+x)}{x} = 1$

47. Find the area of the region bounded by the curve $xy = a^2$, the x-axis, and the vertical lines $x = a, x = 2a$.

48. Let Ω be the region bounded by the graph of $y = (1 + x^2)^{-1/2}$ and the x-axis, $x \in [0, \sqrt{3}]$. Find the volume of the solid generated by revolving Ω about (a) the x-axis; (b) the y-axis.

49. Find the centroid of the region bounded by the graph of $y = (1 - x^2)^{-1/2}$ and the x-axis, $x \in [0, \frac{1}{2}]$.

50. Let $f(x) = \sin^3 x$ on $[0, \pi]$.

(a) Calculate the area of the region Ω bounded by the graph of f and the x-axis.

(b) Calculate the volume of the solid generated by revolving Ω about the x-axis.

51. Let $f(x) = x + \sin x$ and $g(x) = x$ on $[0, \pi]$.

(a) Sketch the graphs of f and g in the same coordinate system.

(b) Calculate the area of the region R between the graphs of f and g.

(c) Calculate the centroid of R.

52. (a) Find the volume of the solid generated by revolving the region R in Exercise 51 about the x-axis.

(b) Find the volume of the solid generated by revolving the region R in Exercise 51 about the y-axis.

53. The region bounded by the graph of $y = \ln 2x$ and the x-axis between $x = 1$ and $x = e$ is revolved about the y-axis. Find the volume of the solid that is generated.

54. Let $f(x) = \frac{\ln x}{x}$ on $(0, \infty)$. (a) Find the intervals where f increases and the intervals where it decreases; (b) find the extreme values; (c) determine the concavity of the graph and find the points of inflection; (d) sketch the graph, including all asymptotes.

55. Repeat Exercise 54 for the function $f(x) = x^2 e^{-x^2}$.

56. A certain bacterial culture, growing exponentially, increases from 200 grams to 400 grams in the period from 6 a.m. to 8 a.m.

(a) How many grams will be present at noon?

(b) How long will it take for the culture to reach 2000 grams?

57. A certain radioactive substance loses 20% of its mass per year. What is the half-life of the substance?

58. Polonium-210 decays exponentially with a half-life of 140 days.

(a) If a sample of polonium-210 has a mass of 100 grams initially, determine an expression for the mass at any time t.

(b) How long will it take for the 100-gram mass to decay to 75 grams?

59. Radioactive carbon has a half-life of 5700 years.

(a) What will remain of 10 grams after 3000 years?

(b) If 10% of an original sample is present today, when was the sample created?

60. In 1980 the population of the United States was 227 million and the population of Mexico was 62 million. In 1990, the respective populations were 249 million and 79 million. If the populations of the United States and Mexico continue to grow at these rates, when will the two populations be equal?

■ 9.1 TRANSLATIONS; THE PARABOLA

Translations

In Figure 9.1.1 we have drawn a rectangular coordinate system and marked a point $O'(x_0, y_0)$. Think of the Oxy system as a rigid frame and in your mind slide it along the plane, without letting it turn, so that the origin O falls on the point O' (Figure 9.1.2). Such a move, called a *translation*, produces a new coordinate system $O'XY$.

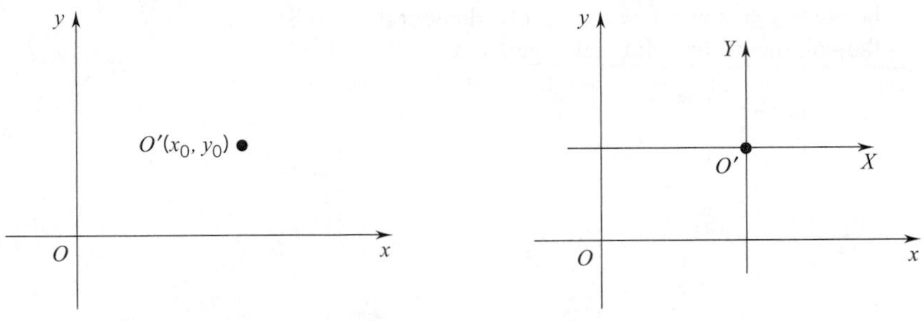

Figure 9.1.1

Figure 9.1.2

A point P now has two pairs of coordinates: a pair (x, y) with respect to the Oxy system and a pair (X, Y) with respect to the $O'XY$ system (Figure 9.1.3). To see the relation between these coordinates, note that, starting at O, we can reach P by first going to O' and then going on to P; thus

$$x = x_0 + X, \quad y = y_0 + Y.$$

Translations are often used to simplify geometric arguments. To illustrate this, we will derive a formula for the distance between a point and a line.

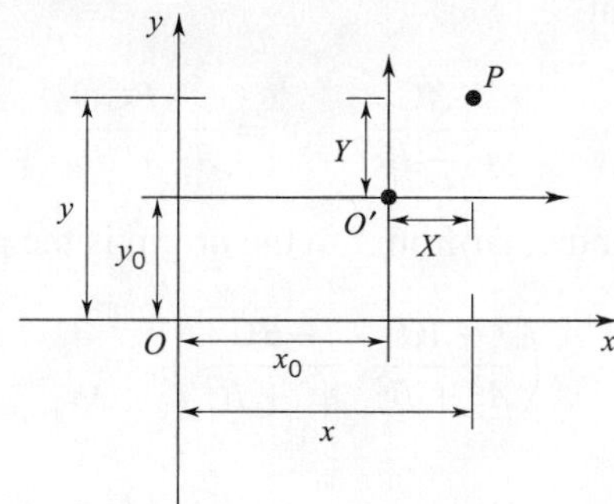

Figure 9.1.3

Let l be a line and let P be a point that is not on l. It is easy to see that the point Q on l that is closest to P is the foot of the perpendicular from P to l. See Figure 9.1.4. The *distance between P and l*, denoted $d(P, l)$, is defined to be $d(P, Q)$.

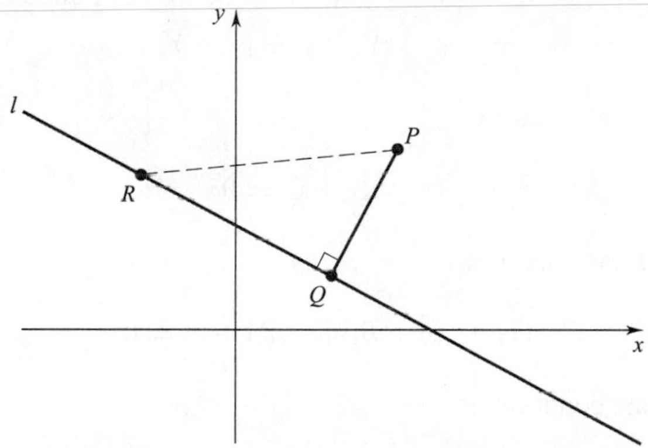

Figure 9.1.4

9.1.1 DISTANCE BETWEEN A POINT AND A LINE

The distance between the line $l : Ax + By + C = 0$ and the point $P_1(x_1, y_1)$ is given by the formula

$$d(P_1, l) = \frac{|Ax_1 + By_1 + C|}{\sqrt{A^2 + B^2}}.$$

PROOF First we find the distance between the origin O and a line $l : Ax+By+C = 0$. We assume $B \neq O$ and leave the case $B = O$ to you. Since l has slope $-A/B$, the line through the origin perpendicular to l has slope B/A and equation

$$y = \frac{B}{A}x, \qquad \text{which we write as} \qquad Bx - Ay = 0.$$

Solving the equations

$$\begin{aligned} Ax + By + C &= 0 \\ Bx - Ay \quad &= 0 \end{aligned}$$

simultaneously, we find that

$$x = \frac{-AC}{A^2+B^2}, \quad y = \frac{-BC}{A^2+B^2}.$$

Thus, the foot of the perpendicular from l to the origin is the point Q with coordinates

$$\left(\frac{-AC}{A^2+B^2}, \frac{-BC}{A^2+B^2}\right). \qquad \text{(Figure 9.1.5)}$$

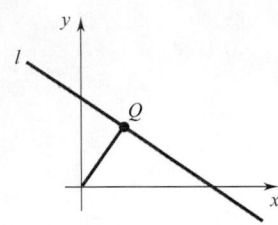

Figure 9.1.5

Therefore

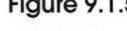

$$d(O,l) = d(O,Q) = \sqrt{\left[\frac{-AC}{A^2+B^2}\right]^2 + \left[\frac{-BC}{A^2+B^2}\right]^2} = \frac{|C|}{\sqrt{A^2+B^2}}.$$

Now let $P_1(x_1, y_1)$ be an arbitrary point in the plane. We translate the Oxy coordinate system to obtain a new coordinate system $O'XY$ with O' falling on P_1 (see Figure 9.1.6). The new coordinates are related to the old coordinates as follows:

$$x = x_1 + X, \quad y = y_1 + Y.$$

In the xy-system, l has equation

$$Ax + By + C = 0.$$

In the XY-system, l has equation

$$A(x_1 + X) + B(y_1 + Y) + C = 0.$$

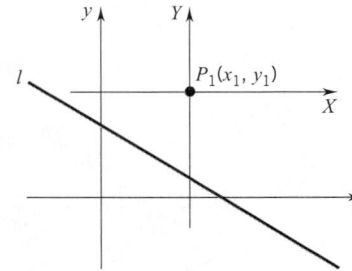

Figure 9.1.6

We can write this last equation as

$$AX + BY + K = 0 \qquad \text{with} \quad K = Ax_1 + By_1 + C.$$

The distance we want is the distance between the line $AX + BY + K = 0$ and the new origin O'. As noted above, this distance is

$$\frac{|K|}{\sqrt{A^2+B^2}}.$$

Since $K = Ax_1 + By_1 + C$, we have

$$d(P_1, l) = \frac{|Ax_1 + By_1 + C|}{\sqrt{A^2+B^2}}. \qquad \square$$

Example 1 Find the distance between the line $l : 3x + 4y - 5 = 0$ and **(a)** the origin, **(b)** the point $P(-6, 2)$.

SOLUTION

(a) $d(O,l) = \dfrac{|C|}{\sqrt{A^2+B^2}} = \dfrac{|-5|}{\sqrt{3^2+4^2}} = \dfrac{5}{\sqrt{25}} = 1.$

(b) $d(P,l) = \dfrac{|Ax_1 + By_1 + C|}{\sqrt{A^2+B^2}} = \dfrac{|3(-6) + 4(2) - 5|}{\sqrt{3^2+4^2}} = \dfrac{15}{\sqrt{25}} = 3.$ $\square$

Conic Sections

Conic sections were introduced and discussed briefly in Section 1.4. We expand on that discussion here.

If a "double right circular cone" is cut by a plane, the resulting intersection is called a *conic section* or, more briefly, a *conic*. In Figure 9.1.7, we show a double cone and depict three important cases: the parabola, ellipse, and hyperbola.

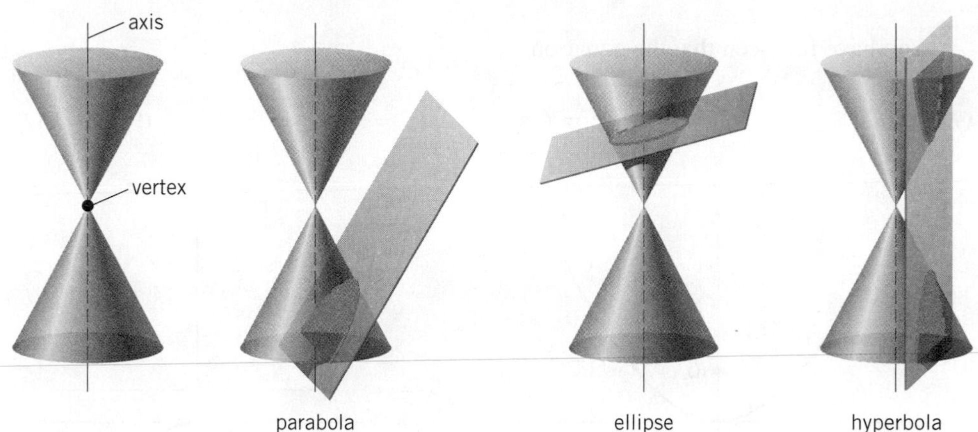

Figure 9.1.7

By choosing a plane perpendicular to the axis of the cone, we can obtain a circle. Other less interesting cases occur when the plane passes through the vertex of the cone. Depending on the orientation of the plane with respect to the axis of the cone, the intersection may either be a point, a line, or a pair of intersecting lines. Try to visualize these possibilities.

$\bullet F$

l

Figure 9.1.8

This three-dimensional approach to the conic sections goes back to Apollonius of Perga, a Greek of the third century B.C. He wrote eight books on the subject.

We will take a different approach. We will define parabola, ellipse, and hyperbola entirely in terms of plane geometry.

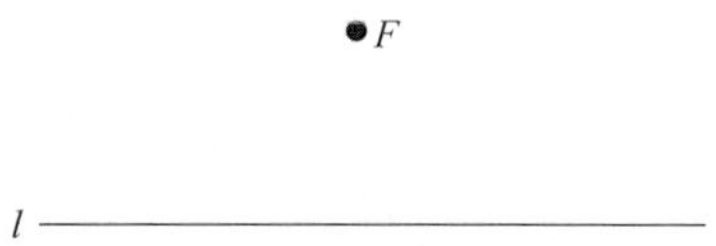

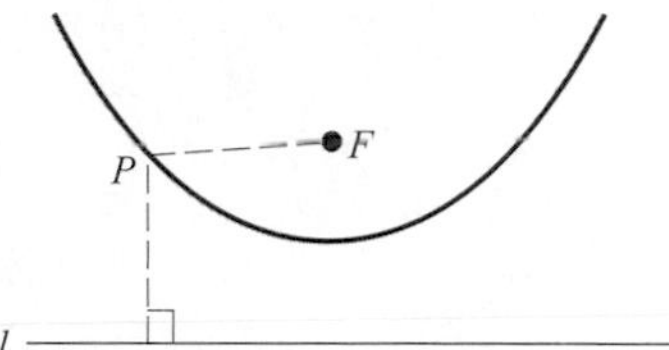

Figure 9.1.9

The Parabola

Figure 9.1.8 shows a line l and a point F not on l.

(9.1.2) The set of points P equidistant from F and l is called a *parabola*.

See Figure 9.1.9.

The line l is called the *directrix* of the parabola, and the point F is called the *focus*. (You will see why later on.) The line through F perpendicular to l is called the *axis* of the parabola. (It is the axis of symmetry.) The point at which the axis intersects the parabola is called the *vertex*. (See Figure 9.1.10.)

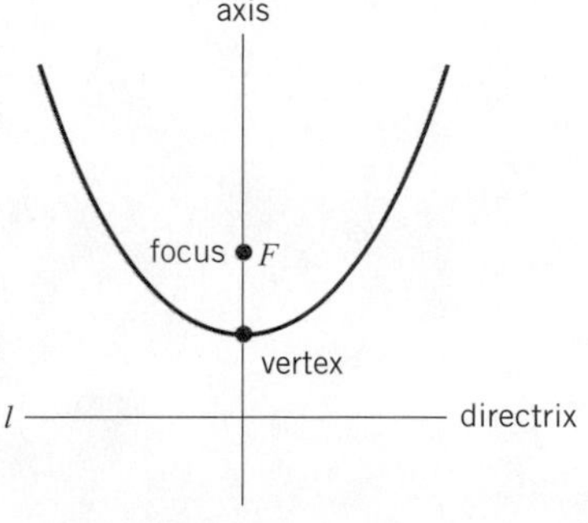

Figure 9.1.10

The equation of a parabola is particularly simple if we place the vertex at the origin and the focus on one of the coordinate axes. Suppose for the moment that the focus F is on the y-axis. Then F has coordinates of the form (0, c). With the vertex at the origin, the directrix has equation $y = -c$. See Figure 9.1.11.

Every point $P(x, y)$ that lies on this parabola has the property that

$$d(P, F) = d(P, l).$$

Since $\quad d(P, F) = \sqrt{x^2 + (y - c)^2} \quad \text{and} \quad d(P, l) = |y + c|,$

you can see that

$$\begin{aligned}\sqrt{x^2+(y-c)^2} &= |y+c|,\\ x^2+(y-c)^2 &= |y+c|^2=(y+c)^2,\\ x^2+y^2-2cy+c^2 &= y^2+2cy+c^2,\\ x^2 &= 4cy. \quad \square\end{aligned}$$

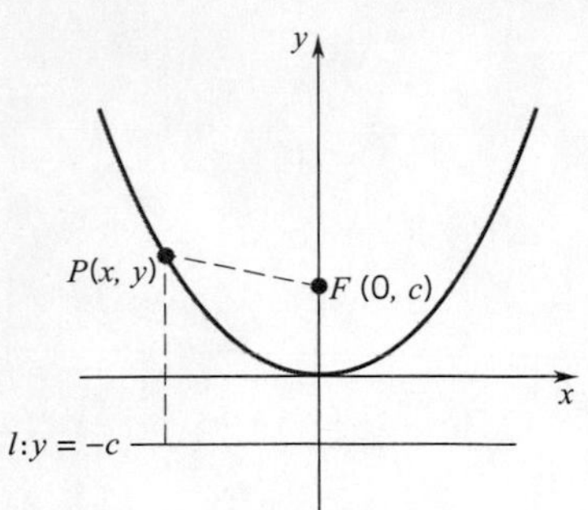

Figure 9.1.11

You have just seen that the equation

(9.1.3) $$\boxed{x^2 = 4cy}$$ (Figure 9.1.12)

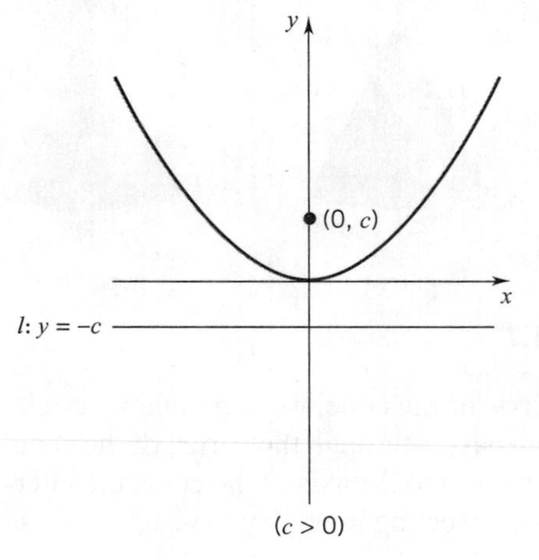

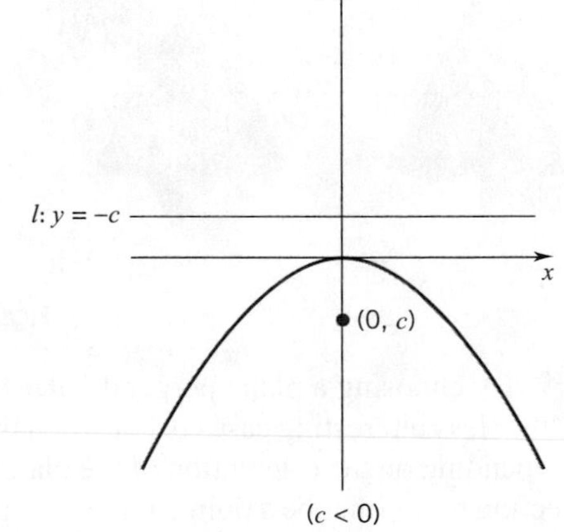

$x^2 = 4cy$

Figure 9.1.12

represents a parabola with vertex at the origin and focus at $(0, c)$. By interchanging the roles of x and y, you can see that the equation

(9.1.4) $$\boxed{y^2 = 4cx}$$ (Figure 9.1.13)

represents a parabola with vertex at the origin and focus at $(c, 0)$.

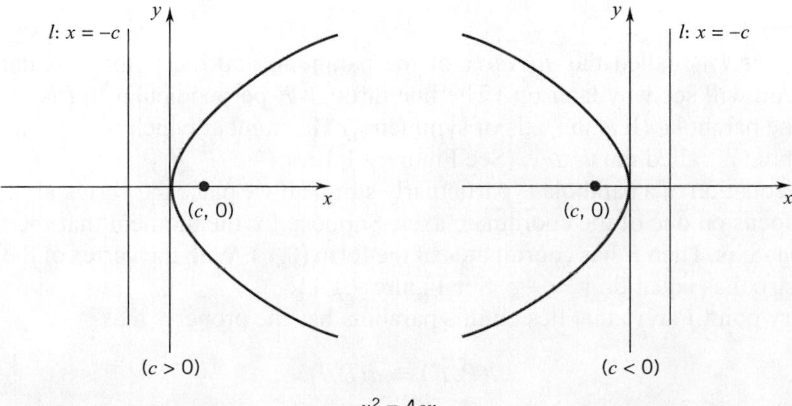

$y^2 = 4cx$

Figure 9.1.13

Example 2 Sketch the parabola, specifying the vertex, focus, directrix, and axis:
(a) $x^2 = -4y$. **(b)** $y^2 = 3x$.

SOLUTION

(a) The equation $x^2 = -4y$ has the form

$$x^2 = 4cy \quad \text{with} \quad c = -1.$$

The vertex is at the origin, and the focus is at $(0, -1)$; the directrix is the horizontal line $y = 1$; the axis of the parabola is the y-axis. We also plotted a few points to ensure the accuracy of our sketch. See Figure 9.1.14.

(b) The equation $y^2 = 3x$ has the form

$$y^2 = 4cx \quad \text{with} \quad c = \tfrac{3}{4}. \qquad \text{(Figure 9.1.14)}$$

The vertex is at the origin, and the focus is at $(\frac{3}{4}, 0)$; the directrix is the vertical line $x = -\frac{3}{4}$; the axis of the parabola is the x-axis. □

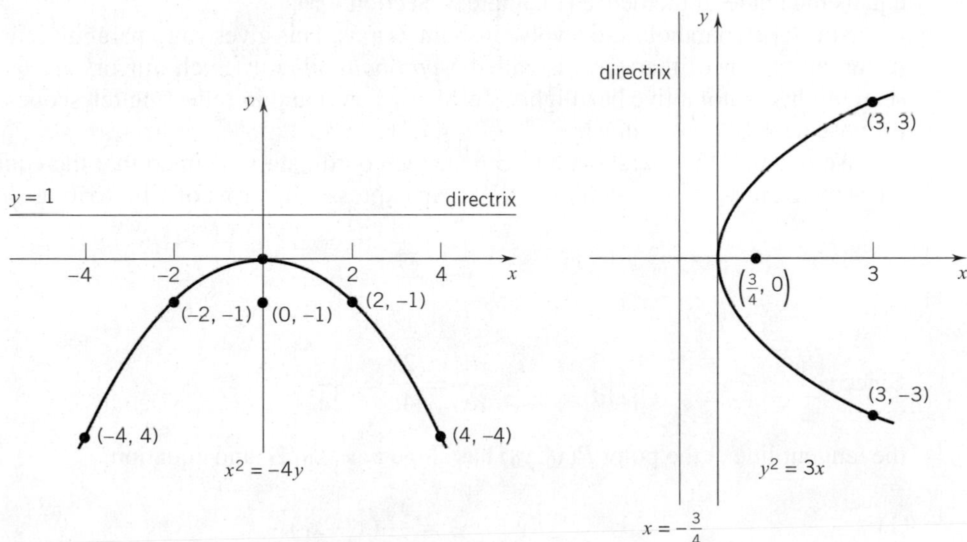

Figure 9.1.14

Every parabola with vertical axis is a translation of a parabola with equation of the form $x^2 = 4cy$, and every parabola with horizontal axis is a translation of a parabola with equation of the form $y^2 = 4cx$.

Example 3 Identify the curve $(x - 4)^2 = 8(y + 3)$.

SOLUTION The curve is the parabola

$$x^2 = 8y \qquad \text{(here } c = 2\text{)}$$

displaced 4 units right and 3 units down. The parabola $x^2 = 8y$ has vertex at the origin; the focus is at $(0, 2)$, and the directrix is the line $y = -2$. Thus, the parabola $(x - 4)^2 = 8(y + 3)$ has vertex at $(4, -3)$: the focus is at $(4, -1)$, and the directrix is the line $y = -5$. (See Figure 9.1.15.) □

Figure 9.1.15

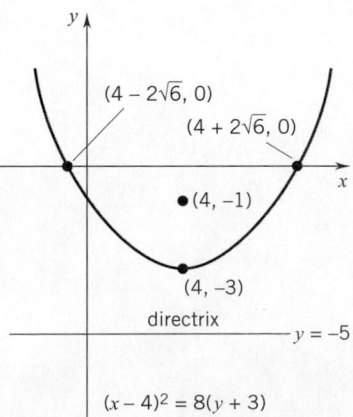

Example 4 Identify the curve $y^2 = 2y - 12x - 37$.

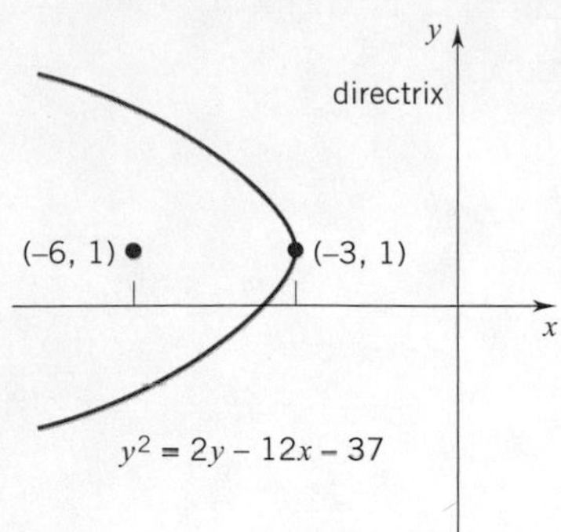

Figure 9.1.16

SOLUTION First we gather the y terms on the left-hand side of the equation and then complete the square:

$$\begin{aligned} y^2 - 2y &= -12x - 37, \\ y^2 - 2y + 1 &= -12x - 36, \\ (y-1)^2 &= -12(x+3). \end{aligned}$$

This is the parabola

$$y^2 = -12x \qquad \text{(here } c = -3\text{)}$$

displaced 3 units left, 1 unit up. The parabola $y^2 = -12x$ has vertex at the origin; the focus is at $(-3, 0)$, and the directrix is the vertical line $x = 3$. Thus, the parabola $(y-1)^2 = -12(x+3)$ has vertex at $(-3, 1)$; the focus is at $(-6, 1)$, and the directrix is the line $x = 0$ (the y-axis). See Figure 9.1.16. □

Parabolic Mirrors

You may want to review the geometric principle of reflected light: the angle of reflection equals the angle of incidence (Example 5, Section 4.5).

Now take a parabola and revolve it about its axis. This gives you a parabolic surface. A curved mirror of that form is called a *parabolic mirror*. Such mirrors are used in searchlights (automotive headlights, flashlights, etc.) and in reflecting telescopes. Our purpose here is to explain why.

We begin with a parabola and choose the coordinate system so that the equation takes the form $x^2 = 4cy$ with $c > 0$. We can express y in terms of x by writing

$$y = \frac{x^2}{4c}.$$

Since

$$\frac{dy}{dx} = \frac{2x}{4c} = \frac{x}{2c},$$

the tangent line at the point $P(x_0, y_0)$ has slope $m = x_0/2c$ and equation

$$(1) \qquad (y - y_0) = \frac{x_0}{2c}(x - x_0).$$

For the rest we refer to Figure 9.1.17.

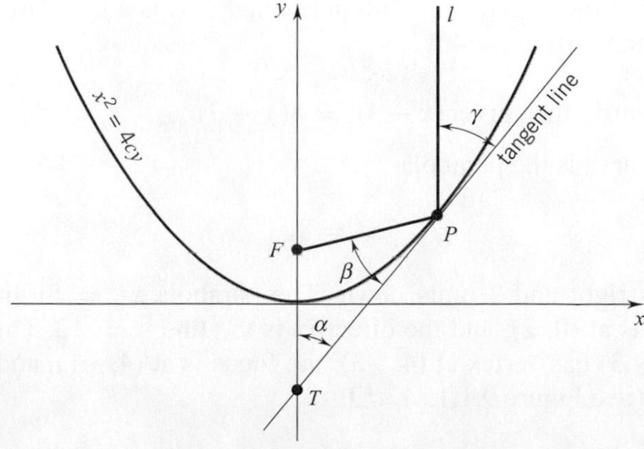

Figure 9.1.17

In the figure we have drawn a ray (a half-line) l parallel to the axis of the parabola, the y-axis. We want to show that the angles marked β and γ are equal.

First we find the y-intercept T of the tangent line. Setting $x = 0$ in Equation (1), we find that

$$y = y_0 - \frac{x_0^2}{2c}.$$

Since the point (x_0, y_0) lies on the parabola, we have $x_0^2 = 4cy_0$, and thus

$$y_0 - \frac{x_0^2}{2c} = y_0 - \frac{4cy_0}{2c} = -y_0.$$

The y-coordinate of the point marked T is $-y_0$. Since the focus F is at $(0, c)$,

$$d(F, T) = y_0 + c.$$

The distance between F and P is also $y_0 + c$:

$$d(F, P) = \sqrt{x_0^2 + (y_0 - c)^2} = \sqrt{4cy_0 + (y_0 - c)^2} = \sqrt{(y_0 + c)^2} = y_0 + c.$$

$x_0^2 = 4cy_0$ ↑ ↑ $y_0 + c > 0$

Since $d(F, T) = d(F, P)$, the triangle TFP is isosceles and the angles marked α and β are equal. Since l is parallel to the y-axis, $\alpha = \gamma$ and thus (and this is what we wanted to show) $\beta = \gamma$.

The fact that $\beta = \gamma$ has important optical consequences. It means (Figure 9.1.18) that light emitted from a source at the focus of a parabolic mirror is reflected in a beam parallel to the axis of that mirror; this is the principle of the searchlight. It also means that light coming to a parabolic mirror in a beam parallel to the axis of the mirror is reflected entirely to the focus; this is the principle of the reflecting telescope.

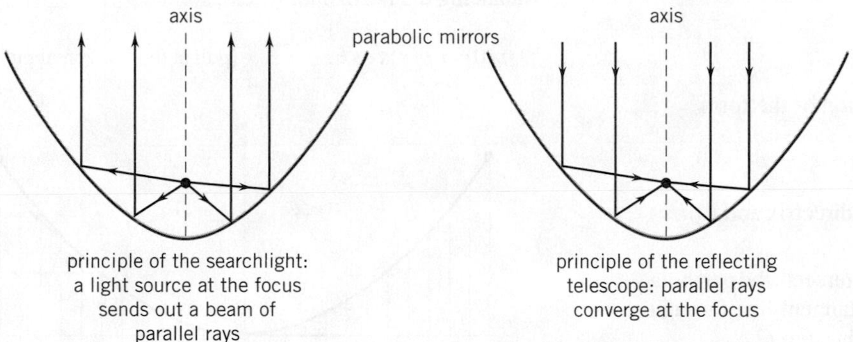
axis
parabolic mirrors
axis
principle of the searchlight: a light source at the focus sends out a beam of parallel rays
principle of the reflecting telescope: parallel rays converge at the focus

Figure 9.1.18

EXERCISES 9.1

1. Find the distance between the line $5x + 12y + 2 = 0$ and
   (a) The origin. (b) the point $P(1, -3)$.
2. Find the distance between the line $2x - 3y + 1 = 0$ and
   (a) the origin. (b) the point $P(-2, 5)$.
3. Which of the points $(0, 1), (1, 0)$, and $(-1, 1)$ is closest to $l : 8x + 7y - 6 = 0$? Which is farthest from l?
4. Consider the triangle with vertices $A(2, 0), B(4, 3), C(5, -1)$. Which of these vertices is farthest from the opposite side?
5. Find the area of the triangle with vertices $(1, -2), (-1, 3), (2, 4)$.
6. Find the area of the triangle with vertices $(-1, 1), (3, \sqrt{2}), (\sqrt{2}, -1)$.
7. Write the equation of a line $Ax + By + C = 0$ in *normal form*:

$$x \cos\alpha + y \sin\alpha = p \quad \text{with } p \geq 0.$$

What is the geometric significance of p? of α?

8. Find an expression for the distance between the parallel lines

$$Ax + By + C = 0 \quad \text{and} \quad Ax + By + C' = 0, \quad C \neq C'.$$

Sketch the parabola and give an equation for it.

9. vertex $(0, 0)$, focus $(2, 0)$.

10. vertex $(0, 0)$, focus $(-2, 0)$.

11. vertex $(-1, 3)$, focus $(-1, 0)$.

12. vertex $(1, 2)$, focus $(1, 3)$.

13. focus $(1, 1)$, directrix $y = -1$.

14. focus $(2, -2)$, directrix $x = -5$.

15. focus $(1, 1)$, directrix $x = 2$.

16. focus $(2, 0)$, directrix $y = 3$.

Find the vertex, focus, axis, and directrix; then sketch the parabola.

17. $y^2 = 2x$.

18. $x^2 = -5y$.

19. $2y = 4x^2 - 1$.

20. $y^2 = 2(x - 1)$.

21. $(x + 2)^2 = 12 - 8y$.

22. $y - 3 = 2(x - 1)^2$.

23. $x = y^2 + y + 1$.

24. $y = x^2 + x + 1$.

Find an equation for the indicated parabola.

25. focus $(1, 2)$, directrix $x + y + 1 = 0$.

26. vertex $(2, 0)$, directrix $2x - y = 0$.

27. vertex $(2, 0)$, focus $(0, 2)$.

28. vertex $(3, 0)$, focus $(0, 1)$.

29. Show that every parabola has an equation of the form

$$(\alpha x + \beta y)^2 = \gamma x + \delta y + \epsilon \quad \text{with} \quad \alpha^2 + \beta^2 \neq 0.$$

HINT: Take $l: Ax + By + C = 0$ as the directrix and $F(a, b)$ as the focus.

30. A line through the focus of a parabola intersects the parabola at two points P and Q. Show that the tangent line through P is perpendicular to the tangent line through Q.

31. A parabola intersects a rectangle of area A at two opposite vertices. Show that, if one side of the rectangle falls on the axis of the parabola, then the parabola subdivides the rectangle into two pieces, one of area $\frac{1}{3}A$, the other of area $\frac{2}{3}A$.

32. (a) Show that every parabola with axis parallel to the y-axis has an equation of the form $y = Ax^2 + Bx + C$ with $A \neq 0$. (b) Find the vertex, the focus, and the directrix of the parabola $y = Ax^2 + Bx + C$.

33. Find equations for all the parabolas that pass through the point $(5,6)$ and have directrix $y = 1$, axis $x = 2$.

34. Find an equation for the parabola that has horizontal axis, vertex $(-1, 1)$, and passes through the point $(-6, 13)$.

The line that passes through the focus of a parabola and is parallel to the directrix intersects the parabola at two points A and B. The line segment $\overline{AB}$ is called the *latus rectum* of the parabola.

In Exercises 35–38 we work with the parabola $x^2 = 4cy, c > 0$. By Ω we mean the region bounded below by the parabola and above by the latus rectum.

35. Find the length of the latus rectum.

36. What is the slope of the parabola at the endpoints of the latus rectum?

37. Determine the area of Ω and locate the centroid.

38. Find the volume of the solid generated by revolving Ω about the y-axis and determine the centroid of the solid. See Project 6.4.

39. Suppose that a flexible inelastic cable (see the figure) fixed at the ends supports a horizontal load. (Imagine a suspension bridge and think of the load on the cable as the roadway.) Show that, if the load has constant weight per unit length, then the cable hangs in the form of a parabola.
HINT: The part of the cable that supports the load from 0 to x is subject to the following forces:

(1) the weight of the load, which in this case is proportional to x.

(2) the horizontal pull at $0 : p(0)$.

(3) the tangential pull at $x : p(x)$.

Balancing the vertical forces, we have

$$kx = p(x) \sin \theta. \qquad \text{[weight = vertical pull at } x\text{]}$$

Balancing the horizontal forces, we have

$$p(0) = p(x) \cos \theta. \qquad \text{[pull at 0 = horizontal pull at } x\text{]}$$

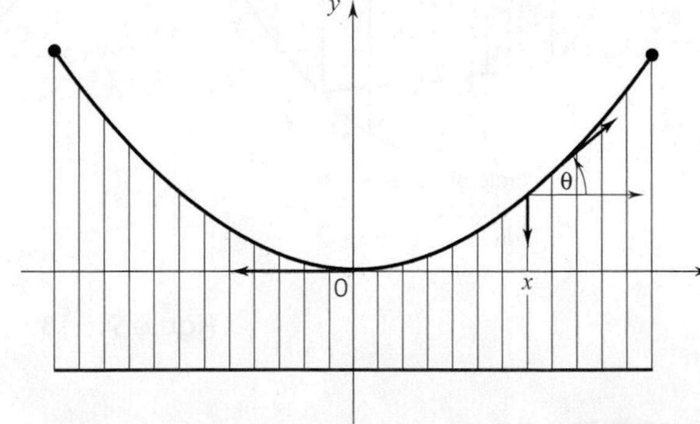

40. The parabolic mirror of a telescope gathers parallel light rays from a distant star and directs them all to the focus. Show that all the light paths to the focus are of the same length.

41. All equilateral triangles are similar; they differ only in scale. Show that the same is true of all parabolas.

42. A searchlight reflector is in the shape of a parabolic mirror. If it is 5 feet in diameter and 2 feet deep at the center, how far is the focus from the vertex of the mirror?

■ 9.2 THE ELLIPSE AND HYPERBOLA

The Ellipse

Start with two points F_1, F_2 and a number k greater than the distance between them.

(9.2.1) The set of all points P for which

$$d(P, F_1) + d(P, F_2) = k$$

is called an *ellipse*; F_1 and F_2 are called the *foci*.

The idea is illustrated in Figure 9.2.1. A string is looped over tacks placed at the foci. The pencil placed in the loop traces out an ellipse.

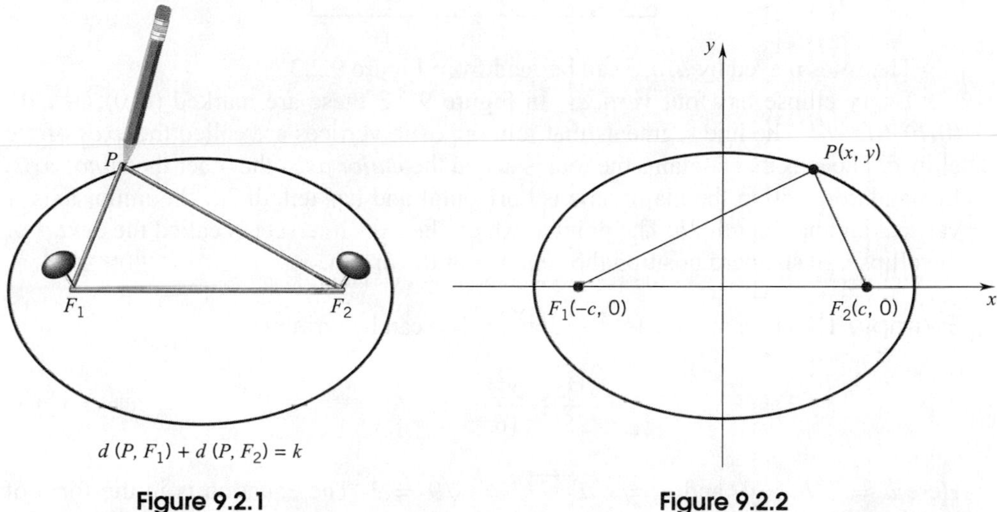

Figure 9.2.1 **Figure 9.2.2**

Figure 9.2.2 shows an ellipse in what is called *standard position:* foci along the x-axis at equal distances from the origin. We now derive an equation for this ellipse setting $a = k/2$ so that $k = 2a$. The advantage of this change of constant will become apparent.

A point $P(x, y)$ lies on the ellipse iff

$$d(P, F_1) + d(P, F_2) = 2a.$$

With F_1 at $(-c, 0)$ and F_2 at $(c, 0)$, $c > 0$, we have

$$\sqrt{(x+c)^2 + y^2} + \sqrt{(x-c)^2 + y^2} = 2a.$$

Transferring the second term to the right-hand side and squaring both sides, we get

$$(x+c)^2 + y^2 = 4a^2 + (x-c)^2 + y^2 - 4a\sqrt{(x-c)^2 + y^2}.$$

This reduces to

$$4a\sqrt{(x-c)^2 + y^2} = 4(a^2 - cx).$$

Canceling the factor 4 and squaring again, we obtain

$$a^2(x^2 - 2cx + c^2 + y^2) = a^4 - 2a^2cx + c^2x^2.$$

This in turn reduces to

$$(a^2 - c^2)x^2 + a^2y^2 = a^2(a^2 - c^2),$$

which we write as

$$\frac{x^2}{a^2} + \frac{y^2}{a^2 - c^2} = 1.$$

Usually we set $b = \sqrt{a^2 - c^2}$. The equation for an ellipse in standard position then takes the form

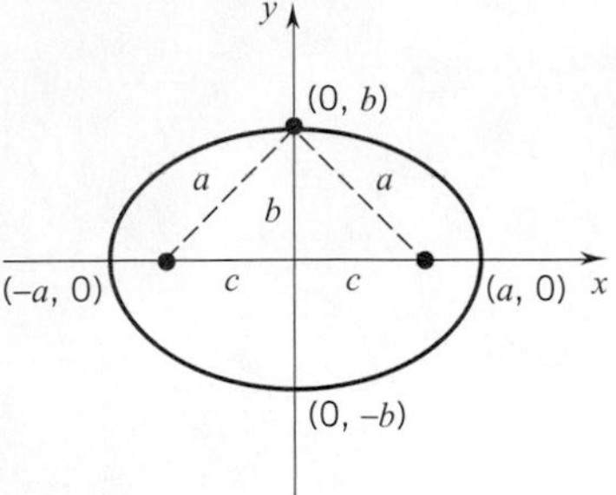

Figure 9.2.3

(9.2.2) $$\frac{x^2}{a^2} + \frac{y^2}{b^2} = 1 \quad \text{with } a > b.$$

The roles played by a, b, c can be read from Figure 9.2.3.

Every ellipse has four *vertices*. In Figure 9.2.3 these are marked $(a, 0), (-a, 0), (0, b), (0, -b)$. The line segments that join opposite vertices are called the *axes* of the ellipse. The axis that contains the foci is called the *major axis*, the other the *minor axis*. In standard position the major axis is horizontal and has length $2a$; the minor axis is vertical and has length $2b$. The point at which the axes intersect is called the *center* of the ellipse. In standard position the center is at the origin.

Example 1 The equation $16x^2 + 25y^2 = 400$ can be written

$$\frac{x^2}{25} + \frac{y^2}{16} = 1. \qquad \text{(divide by 400)}$$

Here $a = 5, b = 4$, and $c = \sqrt{a^2 - b^2} = \sqrt{9} = 3$. The equation is in the form of (9.2.2). It is an ellipse in standard position with foci at $(-3, 0)$ and $(3, 0)$. The major axis has length $2a = 10$, and the minor axis has length $2b = 8$. The center is at the origin. The ellipse is sketched in Figure 9.2.4. ☐

Example 2 The equation

$$\frac{x^2}{16} + \frac{y^2}{25} = 1$$

does not represent an ellipse in standard position because $25 > 16$. This equation is the equation of Example 1 with x with y interchanged. It represents the ellipse of Example 1 reflected in the line $y = x$. (See Figure 9.2.5.) The foci are now on the y-axis, at $(0, -3)$ and $(0, 3)$. The major axis, now vertical, has length 10, and the minor axis has length 8. The center remains at the origin. ☐

Example 3 Figure 9.2.6 shows two ellipses:

$$\frac{x^2}{25} + \frac{y^2}{9} = 1 \quad \text{and} \quad \frac{(x-1)^2}{25} + \frac{(y+4)^2}{9} = 1.$$

The first ellipse is in standard position. Here $a = 5, b = 3, c = \sqrt{a^2 - b^2} = 4$. The foci are at $(-4, 0)$ and $(4, 0)$. The major axis has length 10, and the minor axis has length 6.

The second ellipse is the first ellipse displaced 1 unit right and 4 units down. The center is now at the point $(1, -4)$. The foci are at $(-3, -4)$ and $(5, -4)$. ☐

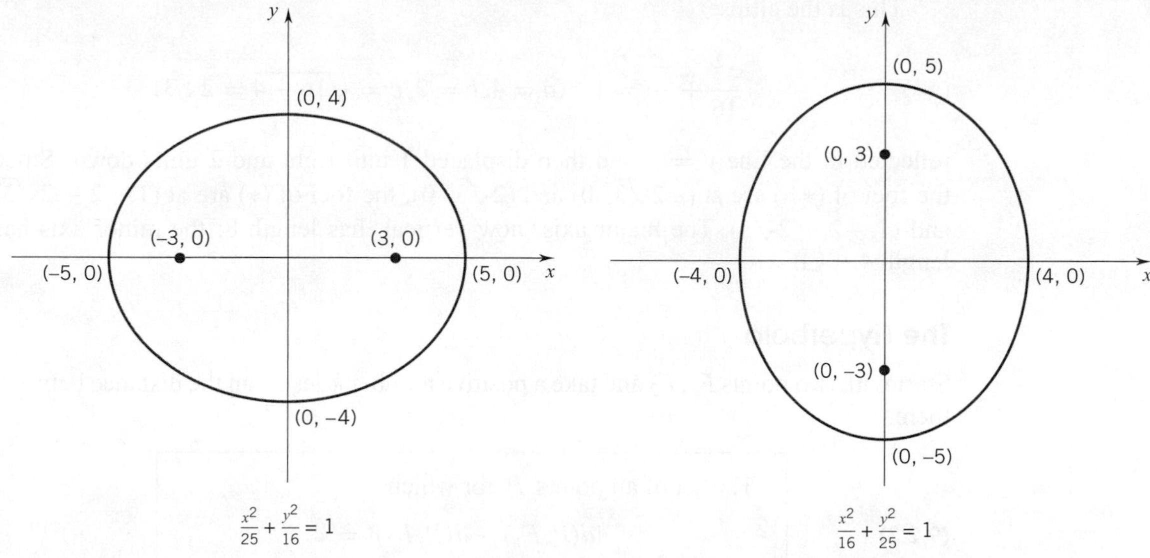

Figure 9.2.4

Figure 9.2.5

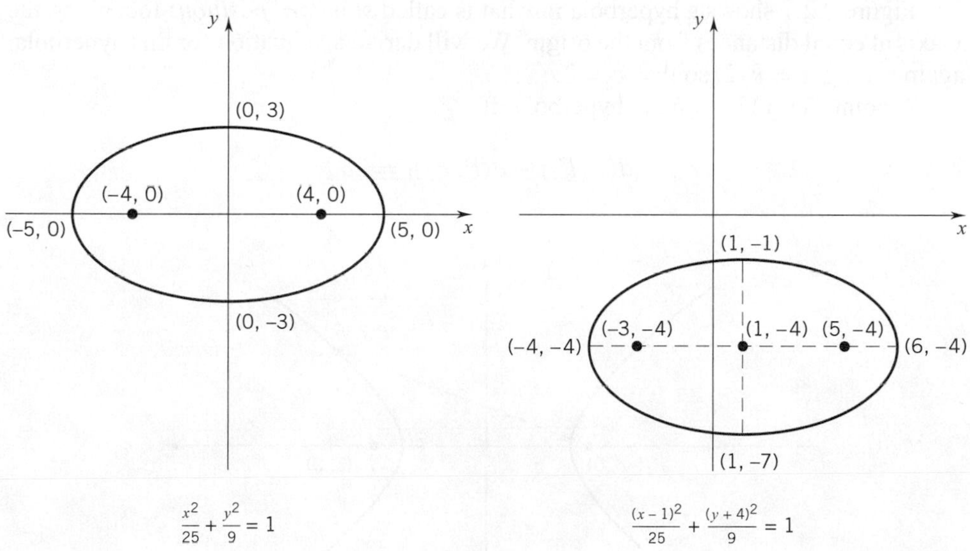

Figure 9.2.6

Example 4 To identify the curve

$$4x^2 - 8x + y^2 + 4y - 8 = 0,$$

we write

$$4(x^2 - 2x + \quad) + (y^2 + 4y + \quad) = 8,$$

and complete the squares within the parentheses. This gives

$$4(x^2 - 2x + 1) + (y^2 + 4y + 4) = 8 + 4 + 4,$$

$$4(x-1)^2 + (y+2)^2 = 16,$$

$$(*) \qquad \frac{(x-1)^2}{4} + \frac{(y+2)^2}{16} = 1.$$

This is the ellipse

$$(**) \qquad \frac{x^2}{16} + \frac{y^2}{4} = 1 \quad (a = 4, b = 2, c = \sqrt{16 - 4} = 2\sqrt{3})$$

reflected in the line $y = x$ and then displaced 1 unit right and 2 units down. Since the foci of $(**)$ are at $(-2\sqrt{3},\ 0)$ and $(2\sqrt{3},\ 0)$, the foci of $(*)$ are at $(1, -2 - 2\sqrt{3})$ and $(1, -2 + 2\sqrt{3})$. The major axis, now vertical, has length 8; the minor axis has length 4. □

The Hyperbola

Start with two points F_1, F_2 and take a positive number k less than the distance between them.

(9.2.3) The set of all points P for which

$$|d(P, F_1) - d(P, F_2)| = k$$

is called a *hyperbola*; F_1 and F_2 are called the *foci*.

Figure 9.2.7 shows a hyperbola in what is called *standard position:* foci along the x-axis at equal distances from the origin. We will derive an equation for this hyperbola, again setting $a = k/2$, so that $k = 2a$.

A point $P(x, y)$ lies on the hyperbola iff

$$|d(P, F_1) - d(P, F_2)| = 2a.$$

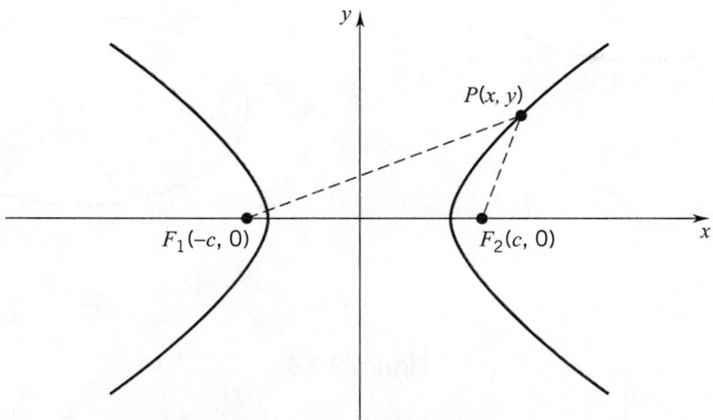

Figure 9.2.7

With F_1 at $(-c, 0)$ and F_2 at $(c, 0)$, $c > 0$, we have

$$\sqrt{(x + c)^2 + y^2} - \sqrt{(x - c)^2 + y^2} = \pm 2a. \qquad \text{(explain)}$$

Transferring the second term to the right and squaring both sides, we obtain

$$(x + c)^2 + y^2 = 4a^2 \pm 4a\sqrt{(x - c)^2 + y^2} + (x - c)^2 + y^2.$$

This equation reduces to

$$xc - a^2 = \pm a\sqrt{(x - c)^2 + y^2}.$$

Squaring once more, we find that

$$x^2c^2 - 2a^2xc + a^4 = a^2(x^2 - 2xc + c^2 + y^2),$$

which reduces to

$$(c^2 - a^2)x^2 - a^2y^2 = a^2(c^2 - a^2),$$

and thus to

$$\frac{x^2}{a^2} - \frac{y^2}{c^2 - a^2} = 1.$$

Usually we set $b = \sqrt{c^2 - a^2}$ (verify that $c > a$). The equation for a hyperbola in standard position then takes the form

(9.2.4)
$$\boxed{\frac{x^2}{a^2} - \frac{y^2}{b^2} = 1.}$$

The roles played by a, b, c can be read from Figure 9.2.8. As the figure suggests, the hyperbola remains between the lines

$$y = \frac{b}{a}x \quad \text{and} \quad y = -\frac{b}{a}x.$$

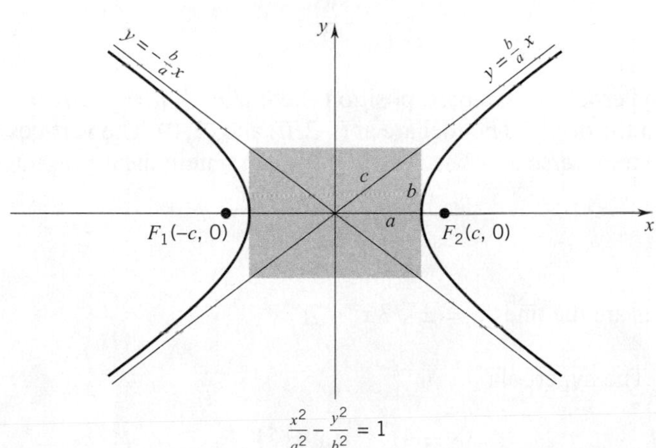

Figure 9.2.8

These lines are called the *asymptotes* of the hyperbola. They can be obtained from Equation (9.2.4) by replacing the 1 on the right-hand side by 0:

$$\frac{x^2}{a^2} - \frac{y^2}{b^2} = 0 \quad \text{gives} \quad y = \pm\frac{b}{a}x.$$

As $x \to \pm\infty$, the vertical separation between the hyperbola and the asymptotes tends to zero. To see this, solve the equation

$$\frac{x^2}{a^2} - \frac{y^2}{b^2} = 1$$

for y. This gives

$$y = \pm\sqrt{\frac{b^2}{a^2}x^2 - b^2} = \pm\frac{b}{a}\sqrt{x^2 - a^2}.$$

In the first quadrant the vertical separation between the hyperbola and the asymptote can be written

$$\frac{b}{a}x - \frac{b}{a}\sqrt{x^2 - a^2} = \frac{b}{a}\left(x - \sqrt{x^2 - a^2}\right). \qquad \text{(check this)}$$

As $x \to \infty$,

$$\frac{b}{a}\left(x - \sqrt{x^2 - a^2}\right) = \frac{b}{a}\left(x - \sqrt{x^2 - a^2}\right)\frac{x + \sqrt{x^2 - a^2}}{x + \sqrt{x^2 - a^2}}$$

$$= \frac{b}{a} \cdot \frac{x^2 - (x^2 - a^2)}{x + \sqrt{x^2 - a^2}} = \frac{ab}{x + \sqrt{x^2 - a^2}} \to 0.$$

Figure 9.2.9

Corresponding arguments hold for the other quadrants.

The line determined by the foci of a hyperbola intersects the hyperbola at two points, called the *vertices*. The line segment that joins the vertices is called the *transverse axis*. The midpoint of the transverse axis is called the *center* of the hyperbola.

In standard position (Figure 9.2.8), the vertices are $(\pm a, 0)$, the transverse axis has length $2a$, and the center is at the origin.

Example 5 The equation

$$\frac{x^2}{1} - \frac{y^2}{3} = 1 \qquad \text{(Figure 9.2.9)}$$

represents a hyperbola in standard position; here $a = 1, b = \sqrt{3}, c = \sqrt{1+3} = 2$. The center is at the origin. The foci are at $(-2, 0)$ and $(2, 0)$. The vertices are at $(-1, 0)$ and $(1, 0)$. The transverse axis has length 2. We can obtain the asymptotes by setting

$$\frac{x^2}{1} - \frac{y^2}{3} = 0.$$

The asymptotes are the lines $y = \pm\sqrt{3}\,x$. ❑

Figure 9.2.10

Example 6 The hyperbola

$$\frac{(x-4)^2}{1} - \frac{(y+5)^2}{3} = 1 \qquad \text{(Figure 9.2.10)}$$

is the hyperbola

$$\frac{x^2}{1} - \frac{y^2}{3} = 1$$

of Example 5 displaced 4 units right and 5 units down. The center of the hyperbola is now at the point $(4, -5)$. The foci are at $(2, -5)$ and $(6, -5)$. The vertices are at $(3, -5)$ and $(5, -5)$. The new asymptotes are the lines $y + 5 = \pm\sqrt{3}(x - 4)$. ❑

Example 7 The hyperbola

$$\frac{y^2}{1} - \frac{x^2}{3} = 1 \qquad \text{(Figure 9.2.11)}$$

is the hyperbola

$$\frac{x^2}{1} - \frac{y^2}{3} = 1$$

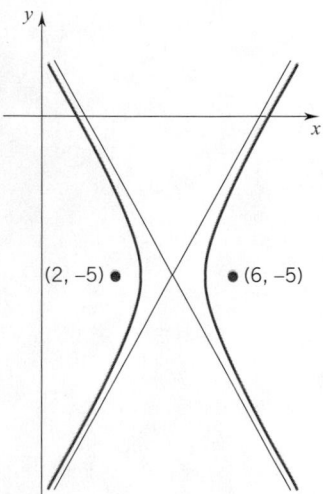

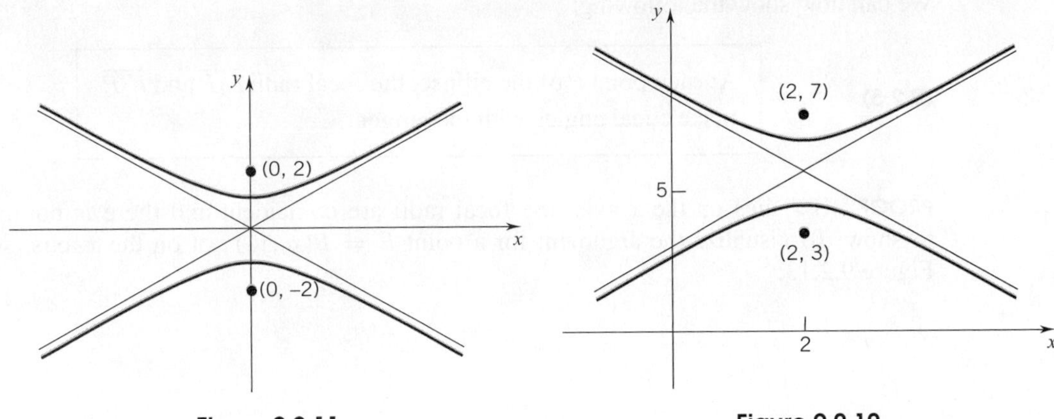

Figure 9.2.11

Figure 9.2.12

of Example 5 reflected in the line $y = x$. The center is still at the origin. The foci are now at $(0, -2)$ and $(0, 2)$. The vertices are at $(0, -1)$ and $(0, 1)$. The asymptotes are the lines $x = \pm\sqrt{3}y$. ☐

Example 8 To identify the curve $x^2 - 3y^2 - 4x + 30y = 68$ we write $(x^2 - 4x \quad) - 3(y^2 - 10y \quad) = 68$ and complete the squares within the parentheses. This gives

$$(x^2 - 4x + 4) - 3(y^2 - 10y + 25) = 68 + 4 - 75$$

$$(x - 2)^2 - 3(y - 5)^2 = -3$$

$$\frac{(y-5)^2}{1} - \frac{(x-2)^2}{3} = 1.$$

This is the hyperbola of Example 7 displaced 2 units right and 5 units up (Figure 9.2.12). ☐

Elliptical Reflectors

Like the parabola, the ellipse has an interesting reflecting property. To derive it, we consider the ellipse

$$\frac{x^2}{a^2} + \frac{y^2}{b^2} = 1.$$

Differentiating implicitly with respect to x, we get

$$\frac{2x}{a^2} + \frac{2y}{b^2}\frac{dy}{dx} = 0 \quad \text{and thus} \quad \frac{dy}{dx} = -\frac{b^2x}{a^2y}.$$

The slope at the point $P(x_0, y_0)$ is therefore

$$-\frac{b^2x_0}{a^2y_0}, \quad y_0 \neq 0,$$

and the tangent line has equation

$$y - y_0 = -\frac{b^2x_0}{a^2y_0}(x - x_0).$$

We can rewrite this last equation as

$$(b^2x_0)x + (a^2y_0)y - a^2b^2 = 0.$$

We can now show the following:

(9.2.5) At each point P of the ellipse, the focal radii $\overline{F_1P}$ and $\overline{F_2P}$ make equal angles with the tangent.

PROOF If P lies on the x-axis, the focal radii are coincident and there is nothing to show. To visualize the argument for a point $P = P(x_0, y_0)$ not on the x-axis, see Figure 9.2.13.

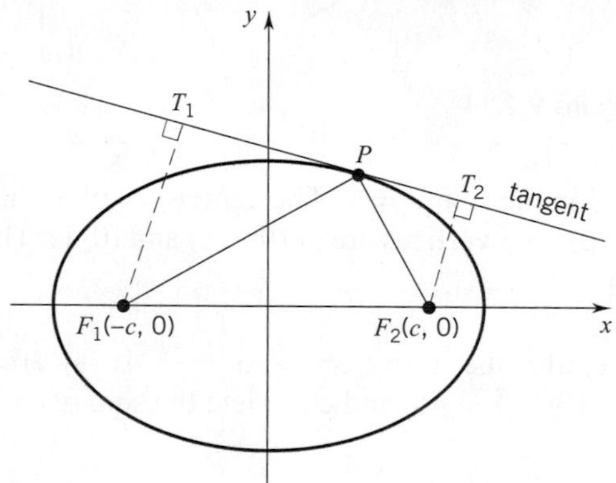

Figure 9.2.13

To show that $\overline{F_1P}$ and $\overline{F_2P}$ make equal angles with the tangent, we need only show that the triangles PT_1F_1 and PT_2F_2 are similar. We can do this by showing that

$$\frac{d(T_1, F_1)}{d(F_1, P)} = \frac{d(T_2, F_2)}{d(F_2, P)}$$

or, equivalently, by showing that

$$\frac{|-b^2x_0c - a^2b^2|}{\sqrt{(x_0 + c)^2 + y_0^2}} = \frac{|b^2x_0c - a^2b^2|}{\sqrt{(x_0 - c)^2 + y_0^2}}. \qquad \text{(verify this)}$$

The validity of this last equation can be seen by canceling the factor b^2 and then squaring. This gives

$$\frac{(x_0c + a^2)^2}{(x_0 + c)^2 + y_0^2} = \frac{(x_0c - a^2)^2}{(x_0 - c)^2 + y_0^2},$$

which can be simplified to

$$(a^2 - c^2)x_0^2 + a^2y_0^2 = a^2(a^2 - c^2) \qquad \text{and thus to} \qquad \frac{x_0^2}{a^2} + \frac{y_0^2}{b^2} = 1.$$

This last equation holds since the point $P(x_0, y_0)$ is on the ellipse. □

The result we just proved has the following physical consequence:

(9.2.6) An elliptical reflector takes light or sound originating at one focus and reflects it to the other focus.

In elliptical rooms called "whispering chambers," a whisper at one focus, inaudible nearby, is easily heard at the other focus. You will experience this phenomenon if you visit the Statuary Room in the Capitol in Washington, D.C. In many hospitals there are elliptical water tubs designed to break up kidney stones. The patient is positioned so that the stone is at one focus. Small vibrations set off at the other focus are so efficiently concentrated that the stone is shattered.

Hyperbolic Reflectors

A straightforward calculation that you are asked to carry out in the Exercises shows that

(9.2.7) at each point P of a hyperbola, the tangent line bisects the angle between the focal radii $\overline{F_1P}$ and $\overline{F_2P}$.

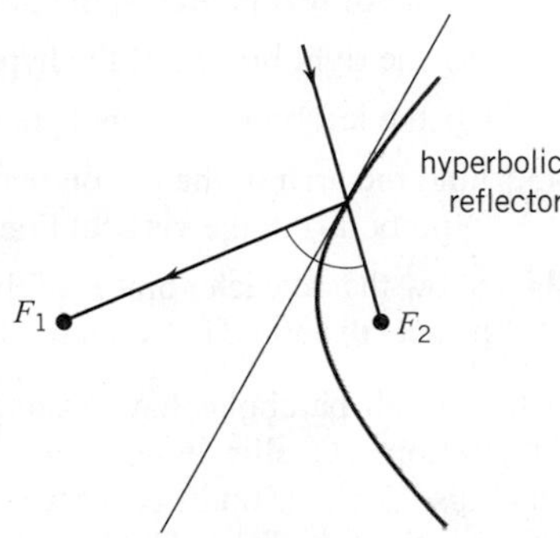

Figure 9.2.14

The optical consequences of this are illustrated in Figure 9.2.14. There you see the right branch of a hyperbola with foci F_1, F_2. Light or sound aimed at F_2 from any point to the left of the reflector is beamed to F_1.

An Application to Range Finding If observers, located at two listening posts at a known distance apart, time the firing of a cannon, the time difference multiplied by the velocity of sound gives the value of $2a$ and hence determines a hyperbola on which the cannon must be located. A third listening post gives two more hyperbolas. The cannon is found where the hyperbolas intersect. A system for long-range navigation called LORAN is based on a similar use of hyperbolas.

EXERCISES 9.2

Find (a) the center, (b) the foci, (c) the length of the major axis, and (d) the length of the minor axis of the given ellipse. Then sketch the figure.

1. $x^2/9 + y^2/4 = 1$. **2.** $x^2/4 + y^2/9 = 1$.

3. $3x^2 + 2y^2 = 12$. **4.** $3x^2 + 4y^2 - 12 = 0$.

5. $4x^2 + 9y^2 - 18y = 27$. **6.** $4x^2 + y^2 - 6y + 5 = 0$.

7. $4(x - 1)^2 + y^2 = 64$.

8. $16(x - 2)^2 + 25(y - 3)^2 = 400$.

Find an equation for the ellipse that satisfies the given conditions.

9. foci at $(-1, 0), (1, 0)$; major axis 6.

10. foci at $(0, -1), (0, 1)$; major axis 6.

11. foci at $(1, 3), (1, 9)$; minor axis 8.

12. foci at $(3, 1), (9, 1)$; minor axis 10.

13. focus at $(1, 1)$; center at $(1, 3)$; major axis 10.

14. center at $(2, 1)$; vertices at $(2, 6), (1, 1)$.

15. major axis 10; vertices at $(3, 2), (3, -4)$.

16. focus at $(6, 2)$; vertices at $(1, 7), (1, -3)$.

Find an equation for the indicated hyperbola.

17. foci at $(-5, 0), (5, 0)$; transverse axis 6.

18. foci at $(-13, 0), (13, 0)$; transverse axis 10.

19. foci at $(0, -13), (0, 13)$; transverse axis 10.

20. foci at $(0, -13), (0, 13)$; transverse axis 24.

21. foci at $(-5, 1), (5, 1)$; transverse axis 6.

22. foci at $(-3, 1), (7, 1)$; transverse axis 6.

23. foci at $(-1, -1), (-1, 1)$; transverse axis $\frac{1}{2}$.

24. foci at $(2, 1), (2, 5)$; transverse axis 3.

Find the center, the vertices, the foci, the asymptotes, and the length of the transverse axis of the given hyperbola. Then sketch the figure.

25. $x^2 - y^2 = 1$. **26.** $y^2 - x^2 = 1$.

27. $x^2/9 - y^2/16 = 1$. **28.** $x^2/16 - y^2/9 = 1$.

29. $y^2/16 - x^2/9 = 1$. **30.** $y^2/9 - x^2/16 = 1$.

31. $(x - 1)^2/9 - (y - 3)^2/16 = 1$.

32. $(x - 1)^2/16 - (y - 3)^2/9 = 1$.

33. $4x^2 - 8x - y^2 + 6y - 1 = 0$. **34.** $-3x^2 + y^2 - 6x = 0$.

35. What is the length of the string in Figure 9.2.1?

36. Show that the set of all points $(a \cos t, b \sin t)$ with t real lie on an ellipse.

37. Find the distance between the foci of an ellipse of area A if the length of the major axis is $2a$.

38. Show that in an ellipse the product of the distances between the foci and a tangent to the ellipse $[d(F_1, T_1)\, d(F_2, T_2)$ in Figure 9.2.13] is always the square of one-half the length of the minor axis.

39. Locate the foci of the ellipse given that the point $(3, 4)$ lies on the ellipse and the ends of the major axis are at $(0, 0)$ and $(10, 0)$.

40. Find the centroid of the first-quadrant portion of the elliptical region $b^2x^2 + a^2y^2 \leq a^2b^2$.

41. Find the center, the vertices, the foci, the asymptotes, and the length of the transverse axis of the hyperbola with equation $xy = 1$. HINT: Define new XY-coordinates by setting $x = X + Y$ and $y = X - Y$.

For Exercises 42–44 we refer to the hyperbola in Figure 9.2.8.

42. Find functions $x = x(t), y = y(t)$ such that, as t ranges over the set of real numbers, the points $(x(t), y(t))$ traverse

(a) the right branch of the hyperbola.

(b) the left branch of the hyperbola.

43. Find the area of the region between the right branch of the hyperbola and the vertical line $x = 2a$.

44. Show that at each point P of the hyperbola the tangent at P bisects the angle between the focal radii $\overline{F_1P}$ and $\overline{F_2P}$.

Although all parabolas have exactly the same shape (Exercise 41, Section 9.1), ellipses come in different shapes. The shape of an ellipse depends on its *eccentricity* e. This is half the distance between the foci divided by half the length of the major axis:

(9.2.8) $$e = c/a.$$

For every ellipse, $0 < e < 1$.

In Exercises 45–48, determine the eccentricity of the ellipse.

45. $x^2/25 + y^2/16 = 1$.

46. $x^2/16 + y^2/25 = 1$.

47. $(x - 1)^2/25 + (y + 2)^2/9 = 1$.

48. $(x + 1)^2/169 + (y - 1)^2/144 = 1$.

49. Suppose that E_1 and E_2 are both ellipses with the same major axis. Compare the shape of E_1 to the shape of E_2 if $e_1 < e_2$.

50. What happens to an ellipse with major axis $2a$ if e tends to 0?

51. What happens to an ellipse with major axis $2a$ if e tends to 1?

In Exercises 52 and 53, write an equation for the ellipse.

52. Major axis from $(-3, 0)$ to $(3, 0)$, eccentricity $\frac{1}{3}$.

53. Major axis from $(-3, 0)$ to $(3, 0)$, eccentricity $\frac{2}{3}\sqrt{2}$.

54. Let l be a line and let F be a point not on l. You have seen that the set of points P for which

$$d(F, P) = d(l, P)$$

is a parabola. Show that, if $0 < e < 1$, then the set of all points P for which

$$d(F, P) = e\,d(l, P)$$

is an ellipse of eccentricity e. HINT: Begin by choosing a coordinate system whereby F falls on the origin and l is a vertical line $x = k$.

The shape of a hyperbola is determined by its *eccentricity* e. This is half the distance between the foci divided by half the length of the transverse axis:

(9.2.9) $$e = c/a.$$

For all hyperbolas, $e > 1$.

In Exercises 55–58, determine the eccentricity of the hyperbola.

55. $x^2/9 - y^2/16 = 1$.

56. $x^2/16 - y^2/9 = 1$.

57. $x^2 - y^2 = 1$.

58. $x^2/25 - y^2/144 = 1$.

59. Suppose H_1 and H_2 are both hyperbolas with the same transverse axis. Compare the shape of H_1 to the shape of H_2 if $e_1 < e_2$.

60. What happens to a hyperbola if e tends to 1?

61. What happens to a hyperbola if e increases without bound?

62. (Compare to Exercise 54.) Let l be a line and let F be a point not on l. Show that, if $e > 1$, then the set of all points P for which

$$d(F, P) = e\,d(l, P)$$

is a hyperbola of eccentricity, e. HINT: Begin by choosing a coordinate system whereby F falls on the origin and l is a vertical line $x = k$.

C 63. A meteor crashes somewhere in the hills that lie north of point A. The impact is heard at point A and four seconds later it is heard at point B. Two seconds still later it is heard at point C. Locate the point of impact given that A lies two miles due east of B and two miles due west of C. (Take 0.20 miles per second as the speed of sound.)

C 64. A radio signal is received at the points marked P_1, P_2, P_3, P_4 in the figure. Suppose that the signal arrives at P_1 600 microseconds after it arrives at P_2 and arrives at P_4 800 microseconds after it arrives at P_3. Locate the source of the signal given that radio waves travel at the speed of light, 186,000 miles per second. (A microsecond is a millionth of a second.)

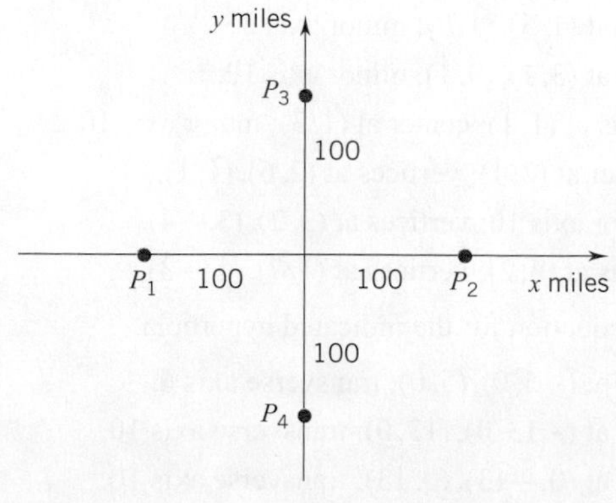

■ 9.3 POLAR COORDINATES

The purpose of coordinates is to fix position with respect to a frame of reference. When we use rectangular coordinates, our frame of reference is a pair of lines that intersect at right angles. In this section we introduce an alternative to the rectangular coordinate system called the *polar coordinate system*. This system lends itself particularly well to the representation of curves that have symmetry about a point or spiral about a point. In the polar coordinate system, the frame of reference is a point O that we call the *pole* and a ray that emanates from it that we call the *polar axis* (Figure 9.3.1).

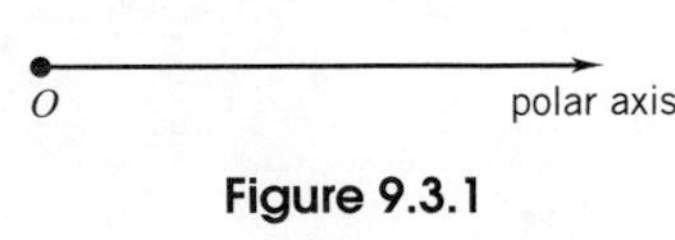

Figure 9.3.1

In Figure 9.3.2 we have drawn two more rays from the pole. One lies at an angle of θ radians from the polar axis; we call it ray θ. The opposite ray lies at an angle of $\theta + \pi$ radians; we call it ray $\theta + \pi$.

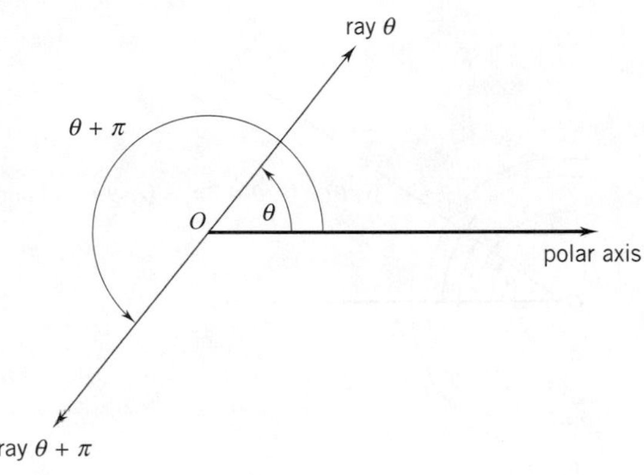

Figure 9.3.2

Figure 9.3.3 shows some points along these same rays, labeled with *polar coordinates.*

(9.3.1) In general, a point is given *polar coordinates* $[r, \theta]$ iff it lies at a distance $|r|$ from the pole

along the ray θ, if $r \geq 0$, and along the ray $\theta + \pi$, if $r < 0$.

Figure 9.3.4 shows the point $[2, \frac{2}{3}\pi]$ at a distance of 2 units from the pole along the ray $\frac{2}{3}\pi$. The point $[-2, \frac{2}{3}\pi]$ also lies 2 units from the pole, not along the ray $\frac{2}{3}\pi$, but along the opposite ray.

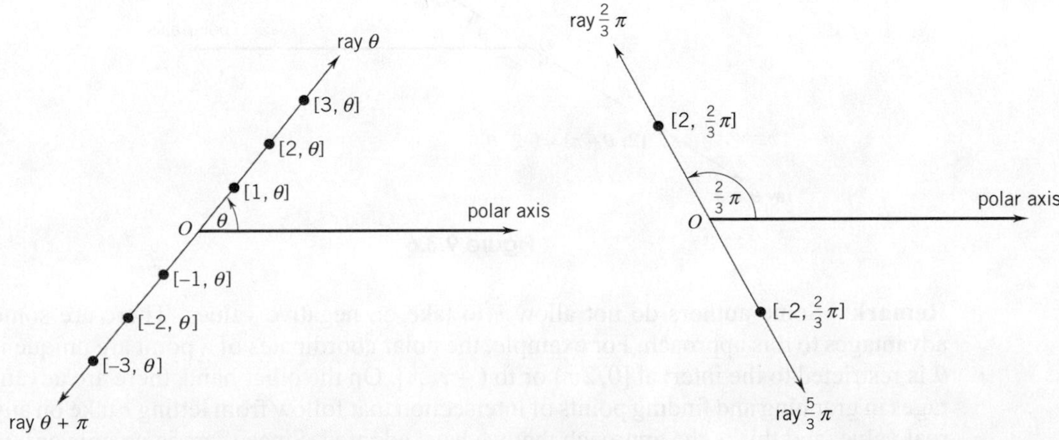

Figure 9.3.3 Figure 9.3.4

Polar coordinates are not unique. Many pairs $[r, \theta]$ can represent the same point.

1. If $r = 0$, it does not matter how we choose θ. The resulting point is still the pole:

(9.3.2) $$O = [0, \theta] \quad \text{for all } \theta.$$

2. Geometrically there is no distinction between angles that differ by an integer multiple of 2π. Consequently, as suggested in Figure 9.3.5,

(9.3.3) $$[r, \theta] = [r, \theta + 2n\pi] \quad \text{for all integers } n.$$

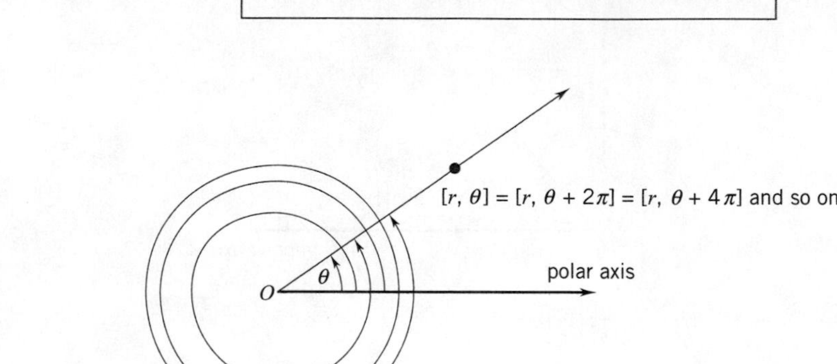

Figure 9.3.5

3. Adding π to the second coordinate is equivalent to changing the sign of the first coordinate:

(9.3.4) $$[r, \theta + \pi] = [-r, \theta].$$ (Figure 9.3.6)

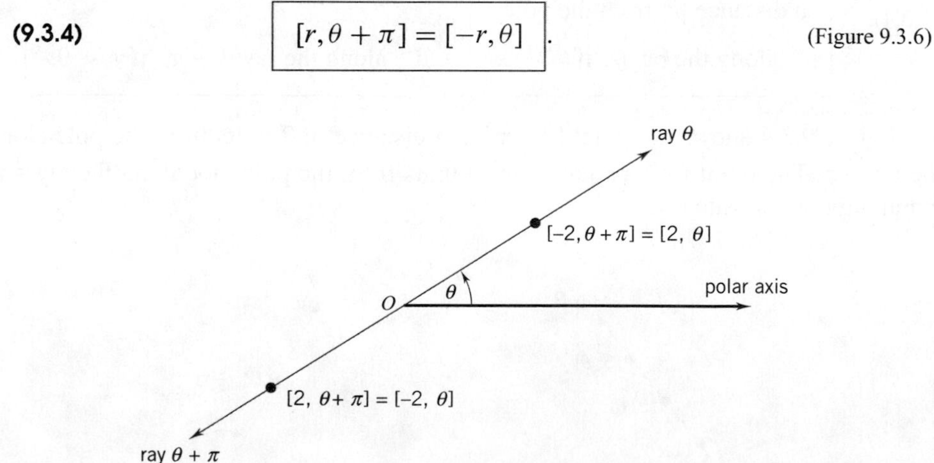

Figure 9.3.6

Remark Some authors do not allow r to take on negative values. There are some advantages to this approach. For example, the polar coordinates of a point are unique if θ is restricted to the interval $[0, 2\pi)$ or to $(-\pi, \pi]$. On the other hand, there are advantages in graphing and finding points of intersection that follow from letting r take on any real value, and this is the approach that we have adopted. Since there is no convention on this issue, you should be aware of the two approaches. □

Relation to Rectangular Coordinates

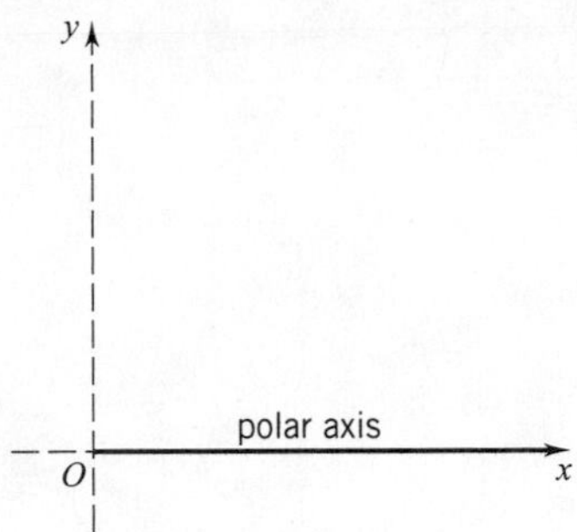

Figure 9.3.7

In Figure 9.3.7 we have superimposed a polar coordinate system on a rectangular coordinate system. We have placed the pole at the origin and the polar axis along the positive x-axis.

The relation between polar coordinates $[r, \theta]$ and rectangular coordinates (x, y) is given by the following equations:

(9.3.5) $$x = r\cos\theta, \quad y = r\sin\theta.$$

PROOF If $r = 0$, the formulas hold, since the point $[r, \theta]$ is then the origin and both x and y are 0:

$$0 = 0\cos\theta, \quad 0 = 0\sin\theta.$$

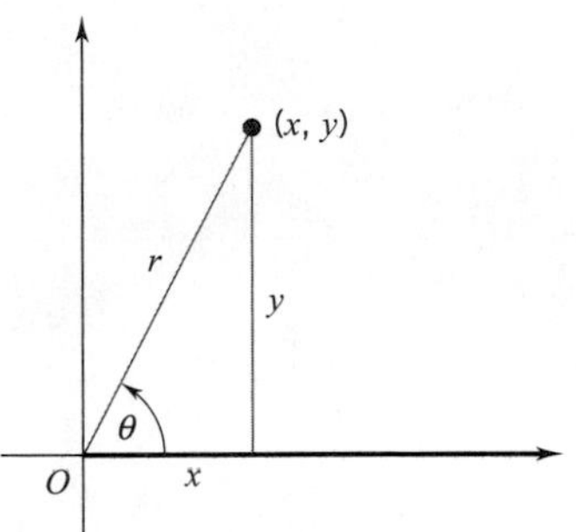

Figure 9.3.8

For $r > 0$, we refer to Figure 9.3.8.† From the figure,

$$\cos\theta = \frac{x}{r}, \quad \sin\theta = \frac{y}{r},$$

and therefore

$$x = r\cos\theta, \quad y = r\sin\theta.$$

Suppose now that $r < 0$. Since $[r, \theta] = [-r, \theta + \pi]$ and $-r > 0$, we know from the previous case that

$$x = -r\cos(\theta + \pi), \quad y = -r\sin(\theta + \pi).$$

Since $\quad \cos(\theta + \pi) = -\cos\theta \quad$ and $\quad \sin(\theta + \pi) = -\sin\theta.$

once again we have

$$x = r\cos\theta, \quad y = r\sin\theta. \quad \square$$

From the relations we just proved, you can see that, unless $x = 0$,

(9.3.6) $$\tan\theta = \frac{y}{x},$$

and, under all circumstances,

(9.3.7) $$x^2 + y^2 = r^2. \qquad \text{(check this out)}$$

Example 1 Find the rectangular coordinates of the point P with polar coordinates $[-2, \frac{1}{3}\pi]$.

SOLUTION The relations $x = r\cos\theta, \quad y = r\sin\theta$

give $\quad x = -2\cos\frac{1}{3}\pi = -2(\frac{1}{2}) = -1, \quad y = -2\sin\frac{1}{3}\pi = -2(\frac{1}{2}\sqrt{3}) = -\sqrt{3}.$

† For simplicity we have placed (x, y) in the first quadrant. A similar argument works in each of the other quadrants.

The point P has rectangular coordinates $(-1, -\sqrt{3})$. □

Example 2 Find all possible polar coordinates for the point P that has rectangular coordinates $(-2, 2\sqrt{3})$.

SOLUTION We know that $r\cos\theta = -2$, $r\sin\theta = 2\sqrt{3}$. We can get the possible values of r by squaring these expressions and then adding them:

$$r^2 = r^2\cos^2\theta + r^2\sin^2\theta = (-2)^2 + (2\sqrt{3})^2 = 16,$$

so that $r = \pm 4$.

Taking $r = 4$, we have

$$4\cos\theta = -2, \qquad 4\sin\theta = 2\sqrt{3}$$
$$\cos\theta = -\tfrac{1}{2}, \qquad \sin\theta = \tfrac{1}{2}\sqrt{3}.$$

These equations are satisfied by setting $\theta = \frac{2}{3}\pi$, or more generally by setting

$$\theta = \tfrac{2}{3}\pi + 2n\pi.$$

The polar coordinates of P with first coordinate $r = 4$ are all pairs of the form

$$[4, \tfrac{2}{3}\pi + 2n\pi]$$

where n ranges over the set of all integers.

We could go through the same process again, this time taking $r = -4$, but there is no need to do so. Since $[r, \theta] = [-r, \theta + \pi]$, we know that

$$[4, \tfrac{2}{3}\pi + 2n\pi] = [-4, (\tfrac{2}{3}\pi + \pi) + 2n\pi].$$

The polar coordinates of P with first coordinate $r = -4$ are thus all pairs of the form

$$[-4, \tfrac{5}{3}\pi + 2n\pi]$$

where n again ranges over the set of all integers. □

Let's specify some simple sets in polar coordinates.

1. In rectangular coordinates the circle of radius a centered at the origin has equation

$$x^2 + y^2 = a^2.$$

The equation for this circle in polar coordinates is simply

$$r = a.$$

The interior of the circle is given by $0 \le r < a$ and the exterior by $r > a$.

2. In rectangular coordinates the line through the origin with inclination α has equation $y = mx$, where $m = \tan\alpha$ (Figure 9.3.9). The polar equation of this line is

$$\theta = \alpha \qquad \text{(verify this)}$$

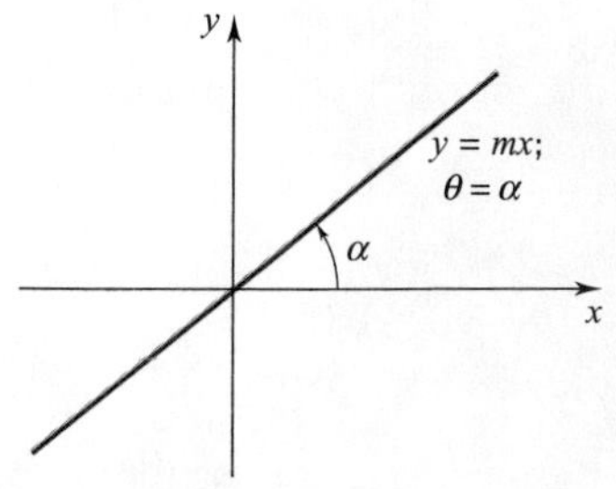

Figure 9.3.9

3. The vertical line $x = a$ becomes

$$r\cos\theta = a \qquad \text{which can be written} \qquad r = a\sec\theta.$$

and the horizontal line $y = b$ becomes

$$r\sin\theta = b \qquad \text{which can be written} \qquad r = a\csc\theta$$

4. The line $Ax + By + C = 0$ can be written

$$r(A\cos\theta + B\sin\theta) + C = 0.$$

Example 3 Find an equation in polar coordinates for the hyperbola $x^2 - y^2 = a^2$.

SOLUTION Setting $x = r\cos\theta$ and $y = r\sin\theta$, we have

$$\begin{aligned} r^2\cos^2\theta - r^2\sin^2\theta &= a^2, \\ r^2(\cos^2\theta - \sin^2\theta) &= a^2, \\ r^2\cos 2\theta &= a^2. \quad \square \end{aligned}$$

Example 4 Show that the equation $r = 2a\cos\theta, a > 0,$ represents a circle.

SOLUTION Multiplication by r gives

$$\begin{aligned} r^2 &= 2ar\cos\theta, \\ x^2 + y^2 &= 2ax, \\ x^2 - 2ax + y^2 &= 0, \\ x^2 - 2ax + a^2 + y^2 &= a^2, \\ (x-a)^2 + y^2 &= a^2. \end{aligned}$$

This is a circle of radius a centered at the point with rectangular coordinates $(a, 0)$. See Figure 9.3.10. □

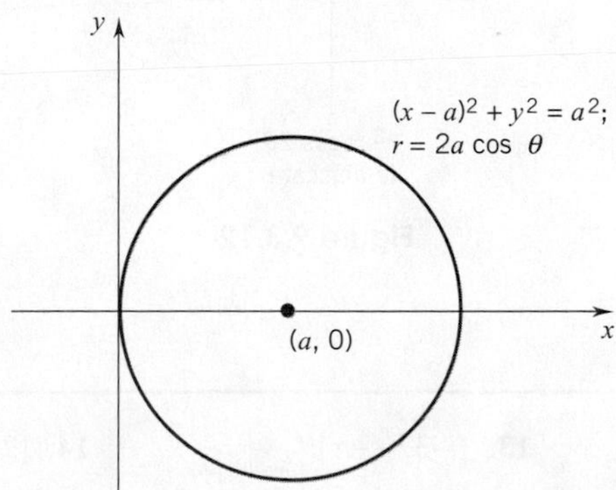

Figure 9.3.10

Symmetry

Symmetry with respect to each of the coordinate axes and with respect to the origin is illustrated in Figure 9.3.11. The coordinates marked are, of course, not the only ones possible. (The difficulties that this can cause are explored in Section 9.4.)

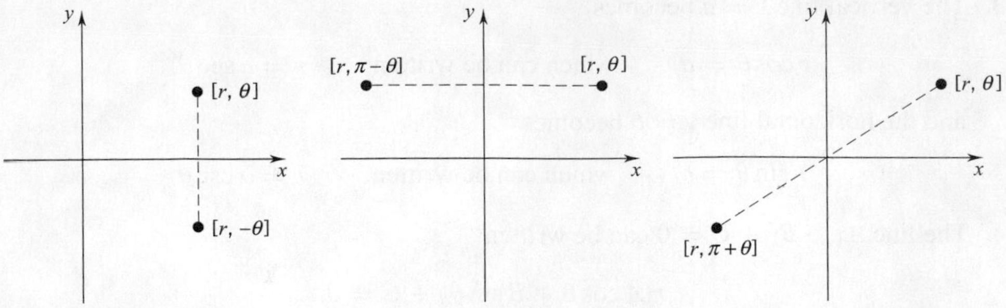

symmetry about the x-axis   symmetry about the y-axis   symmetry about the origin

Figure 9.3.11

Example 5 Test the curve $r^2 = \cos 2\theta$ for symmetry.

SOLUTION Since $\cos[2(-\theta)] = \cos(-2\theta) = \cos 2\theta$, you can see that, if $[r, \theta]$ is on the curve, then so is $[r, -\theta]$. This says that the curve is symmetric about the x-axis. Since

$$\cos[2(\pi - \theta)] = \cos(2\pi - 2\theta) = \cos(-2\theta) = \cos 2\theta,$$

you can see that, if $[r, \theta]$ is on the curve, then so is $[r, \pi - \theta]$. The curve is therefore symmetric about the y-axis.

Being symmetric about both axes, the curve must also be symmetric about the origin. You can verify this directly by noting that

$$\cos[2(\pi + \theta)] = \cos(2\pi + 2\theta) = \cos 2\theta,$$

so that, if $[r, \theta]$ lies on the curve, then so does $[r, \pi + \theta]$. A sketch of the curve, which is called a *lemniscate*, appears in Figure 9.3.12. ❑

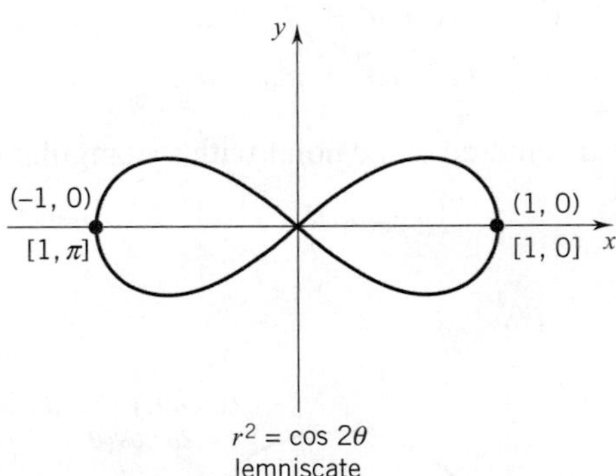

$r^2 = \cos 2\theta$
lemniscate

Figure 9.3.12

EXERCISES 9.3

Plot the point given in polar coordinates.

1. $[1, \frac{1}{3}\pi]$.
2. $[1, \frac{1}{2}\pi]$.
3. $[-1, \frac{1}{3}\pi]$.
4. $[-1, -\frac{1}{3}\pi]$.
5. $[4, \frac{5}{4}\pi]$.
6. $[-2, 0]$.
7. $[-\frac{1}{2}, \pi]$.
8. $[\frac{1}{3}, \frac{2}{3}\pi]$.

Find the rectangular coordinates of the point.

9. $[3, \frac{1}{2}\pi]$.
10. $[4, \frac{1}{6}\pi]$.
11. $[-1, -\pi]$.
12. $[-1, \frac{1}{4}\pi]$.
13. $[-3, -\frac{1}{3}\pi]$.
14. $[2, 0]$.
15. $[3, -\frac{1}{2}\pi]$.
16. $[2, 3\pi]$.

Points are specified in rectangular coordinates. Give all possible polar coordinates for each point.

17. $(0, 1)$.
18. $(1, 0)$.
19. $(-3, 0)$.
20. $(4, 4)$.
21. $(2, -2)$.
22. $(3, -3\sqrt{3})$.
23. $(4\sqrt{3}, 4)$.
24. $(\sqrt{3}, -1)$.

25. Find a formula for the distance between $[r_1, \theta_1]$ and $[r_2, \theta_2]$.

26. Show that for $r_1 > 0, r_2 > 0, |\theta_1 - \theta_2| < \pi$ the distance formula you found in Exercise 25 reduces to the law of cosines.

Find the point $[r, \theta]$ symmetric to the given point about: (a) the x-axis; (b) the y-axis; (c) the origin. Express your answer with $r > 0$ and $\theta \in [0, 2\pi)$.

27. $[\frac{1}{2}, \frac{1}{6}\pi]$.
28. $[3, -\frac{5}{4}\pi]$.

29. $[-2, \frac{1}{3}\pi]$.
30. $[-3, -\frac{7}{4}\pi]$.

Test the curve for symmetry about the coordinate axes and for symmetry about the origin.

31. $r = 2 + \cos\theta$.
32. $r = \cos 2\theta$.

33. $r(\sin\theta + \cos\theta) = 1$.
34. $r\sin\theta = 1$.

35. $r^2 \sin 2\theta = 1$.
36. $r^2 \cos 2\theta = 1$.

Write the equation in polar coordinates.

37. $x = 2$.
38. $y = 3$.

39. $2xy = 1$.
40. $x^2 + y^2 = 9$.

41. $x^2 + (y-2)^2 = 4$.
42. $(x-a)^2 + y^2 = a^2$.

43. $y = x$.
44. $x^2 - y^2 = 4$.

45. $x^2 + y^2 + x = \sqrt{x^2 + y^2}$.

46. $y = mx$.
47. $(x^2 + y^2)^2 = 2xy$.

48. $(x^2 + y^2)^2 = x^2 - y^2$.

Identify the curve and write the equation in rectangular coordinates.

49. $r\sin\theta = 4$.
50. $r\cos\theta = 4$.

51. $\theta = \frac{1}{3}\pi$.
52. $\theta^2 = \frac{1}{9}\pi^2$.

53. $r = 2(1 - \cos\theta)^{-1}$.
54. $r = 4\sin(\theta + \pi)$.

55. $r = 3\cos\theta$.
56. $\theta = -\frac{1}{2}\pi$.

57. $\tan\theta = 2$.
58. $r = 2\sin\theta$.

Write the equation in rectangular coordinates and identify the curve.

59. $r = \dfrac{4}{2 - \cos\theta}$.
60. $r = \dfrac{6}{1 + 2\sin\theta}$.

61. $r = \dfrac{4}{1 - \cos\theta}$.
62. $r = \dfrac{2}{3 + 2\sin\theta}$.

63. Show that if a and b are not both zero, then the curve

$$r = a\sin\theta + b\cos\theta$$

is a circle. Find the center and the radius.

64. Find a polar equation for the set of points $P[r, \theta]$ such that the distance from P to the pole equals the distance from P to the line $x = -d$. See the figure.

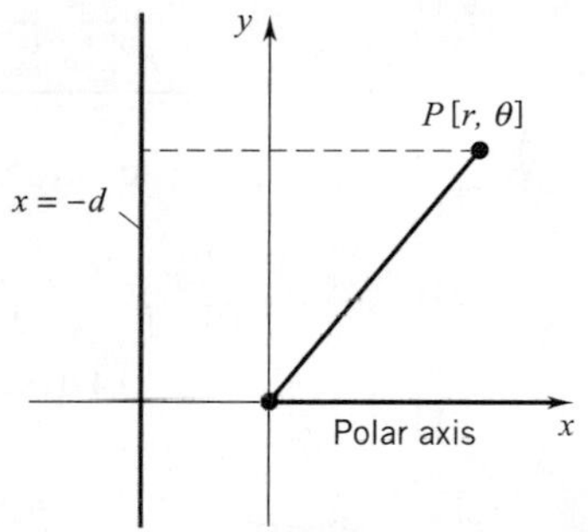

65. Find a polar equation for the set of points $P[r, \theta]$ such that the distance from P to the pole is half the distance from P to the line $x = -d$.

66. Find a polar equation for the set of points $P[r, \theta]$ such that the distance from P to the pole is twice the distance from P to the line $x = -d$.

■ 9.4 GRAPHING IN POLAR COORDINATES

We begin with the curve

$$r = \theta, \quad \theta \geq 0.$$

The curve is a nonending spiral, part of the famous *spiral of Archimedes*. The curve is shown in detail from $\theta = 0$ *to* $\theta = 2\pi$ in Figure 9.4.1. At $\theta = 0$, $r = 0$; at $\theta = \frac{1}{4}\pi, r = \frac{1}{4}\pi$; at $\theta = \frac{1}{2}\pi, r = \frac{1}{2}\pi$; and so on.

The next examples involve trigonometric functions.

Example 1 Sketch the curve $r = 1 - 2\cos\theta$.

SOLUTION Since the cosine function is periodic with period 2π, the curve $r = 1 - 2\cos\theta$ is a closed curve. We will draw it from $\theta = 0$ to $\theta = 2\pi$. The curve just repeats itself for values of θ outside the interval $[0, 2\pi]$.

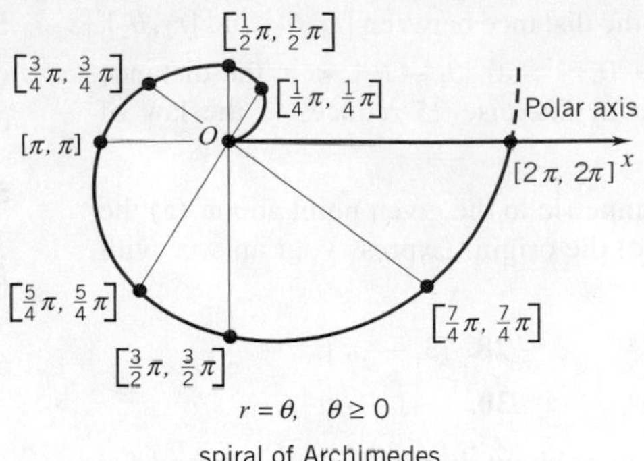

Figure 9.4.1

We begin by compiling a table of values:

θ	0	$\pi/4$	$\pi/3$	$\pi/2$	$2\pi/3$	$3\pi/4$	π	$5\pi/4$	$4\pi/3$	$3\pi/2$	$5\pi/3$	$7\pi/4$	2π
r	-1	-0.41	0	1	2	2.41	3	2.41	2	1	0	-0.41	-1

The values of θ for which $r = 0$ or $|r|$ is a local maximum are as follows:

$r = 0$ at $\theta = \frac{1}{3}\pi, \frac{5}{3}\pi$ for then $\cos\theta = \frac{1}{2}$; $|r|$ is a local maximum at $\theta = 0, \pi, 2\pi$.

These five values of θ generate four intervals:

$$[0, \tfrac{1}{3}\pi], \quad [\tfrac{1}{3}\pi, \pi], \quad [\pi, \tfrac{5}{3}\pi], \quad [\tfrac{5}{3}\pi, 2\pi].$$

We sketch the curve in four stages. These stages are shown in Figure 9.4.2.

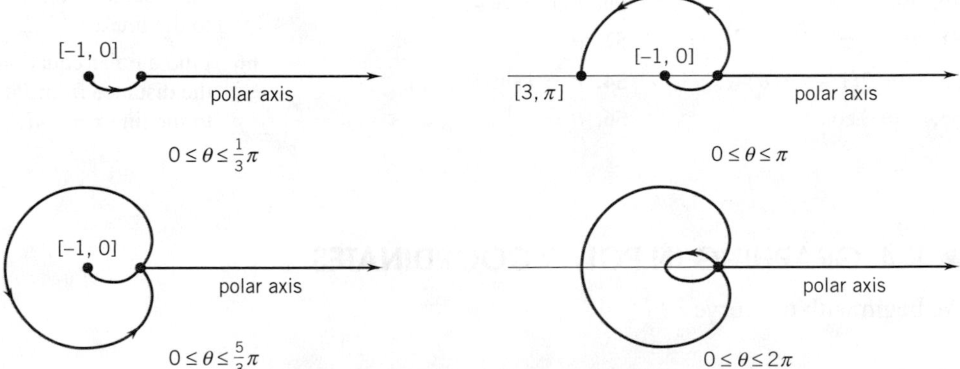

Figure 9.4.2

As θ increases from 0 to $\frac{1}{3}\pi$, $\cos\theta$ decreases from 1 to $\frac{1}{2}$ and $r = 1 - 2\cos\theta$ increases from -1 to 0.

As θ increases from $\frac{1}{3}\pi$ to π, $\cos\theta$ decreases from $\frac{1}{2}$ to -1 and r increases from 0 to 3.

As θ increases from π to $\frac{5}{3}\pi$, $\cos\theta$ increases from -1 to $\frac{1}{2}$ and r decreases from 3 to 0.

Finally, as θ increases from $\frac{5}{3}\pi$ to 2π, $\cos\theta$ increases from $\frac{1}{2}$ to 1 and r decreases from 0 to -1.

As we could have read from the equation, the curve is symmetric about the x-axis $[r(-\theta) = 1 - 2\cos(-\theta) = 1 - 2\cos\theta = r(\theta)]$. □

Example 2 Sketch the curve $r = \cos 2\theta, \quad 0 \leq \theta \leq 2\pi$.

SOLUTION As an alternative to compiling a table of values as we did in Example 1, we refer to the graph of $\cos 2\theta$ in rectangular coordinates for the values of r. See Figure 9.4.3. The values of θ for which r is zero or has an extreme value are as follows:

$$r = 0 \text{ at } \theta = \tfrac{\pi}{4}, \tfrac{3\pi}{4}, \tfrac{5\pi}{4}, \tfrac{7\pi}{4}; \qquad \text{local maxima/minima at } \theta = 0, \tfrac{\pi}{2}, \pi, \tfrac{3\pi}{2}, 2\pi.$$

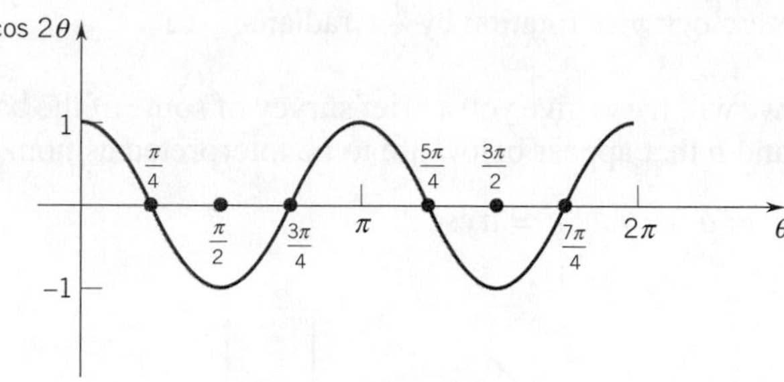

Figure 9.4.3

In Figure 9.4.4 we sketch the curve in eight stages. □

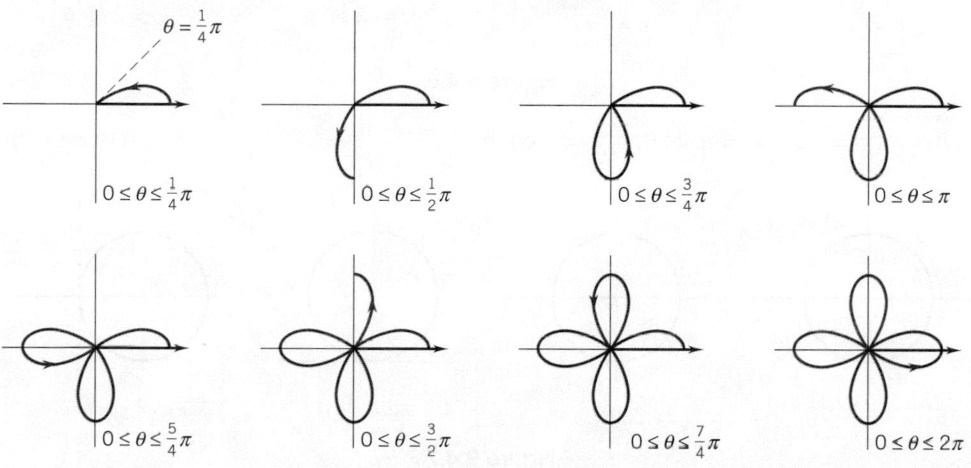

Figure 9.4.4

Example 3 Figure 9.4.5 shows four *cardioids*, heart-shaped curves. Rotation of $r = 1 + \cos\theta$ by $\frac{1}{2}\pi$ radians, measured in the counterclockwise direction, gives

$$r = 1 + \cos(\theta - \tfrac{1}{2}\pi) = 1 + \sin\theta.$$

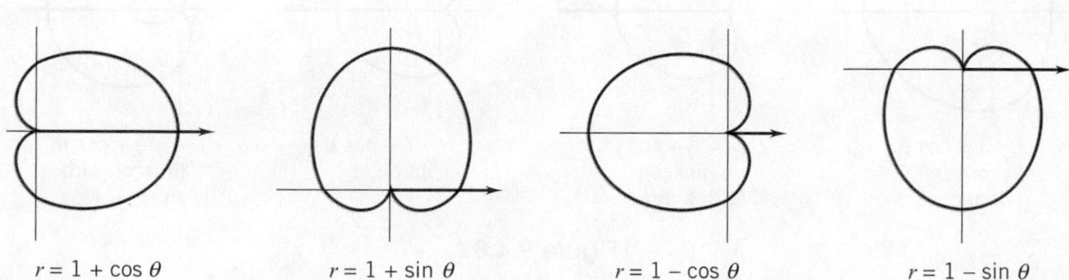

Figure 9.4.5

Rotation by another $\frac{1}{2}\pi$ radians gives

$$r = 1 + \cos(\theta - \pi) = 1 - \cos\theta.$$

Rotation by yet another $\frac{1}{2}\pi$ radians gives

$$r = 1 + \cos(\theta - \tfrac{3}{2}\pi) = 1 - \sin\theta.$$

Notice how easy it is to rotate axes in polar coordinates: each change

$$\cos\theta \to \sin\theta \to -\cos\theta \to -\sin\theta$$

represents a counterclockwise rotation by $\frac{1}{2}\pi$ radians. □

At this point we will try to give you a brief survey of some of the basic polar curves. (The numbers a and b that appear below are to be interpreted as nonzero constants.)

Lines : $\theta = a, \quad r = a\sec\theta, \quad r = a\csc\theta.$ (Figure 9.4.6)

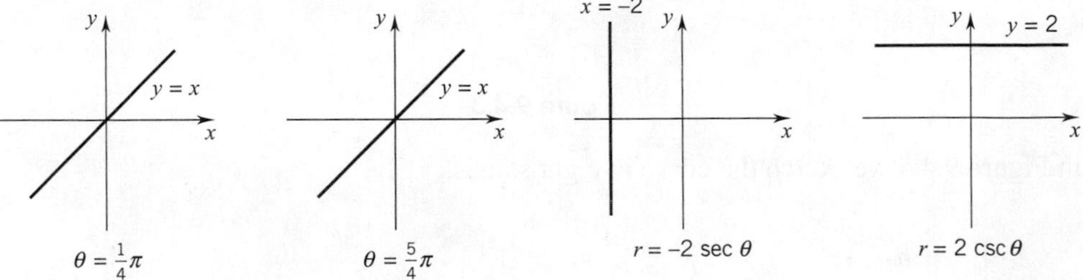

Figure 9.4.6

Circles : $r = a, \quad r = a\sin\theta, \quad r = a\cos\theta.$ (Figure 9.4.7)

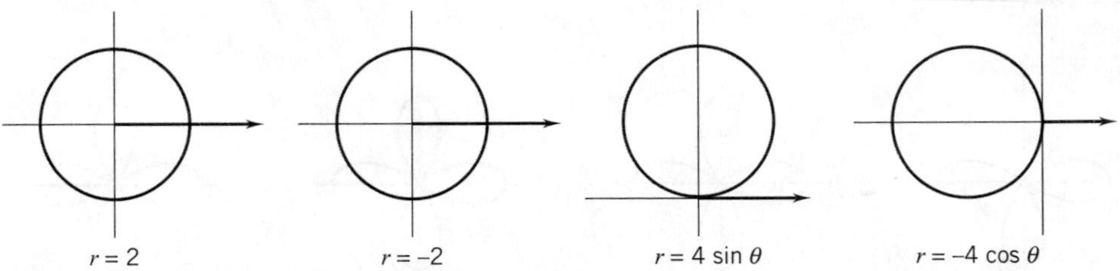

Figure 9.4.7

Limaçons :†$r = a + b\sin\theta, \quad r = a + b\cos\theta.$ (Figure 9.4.8)

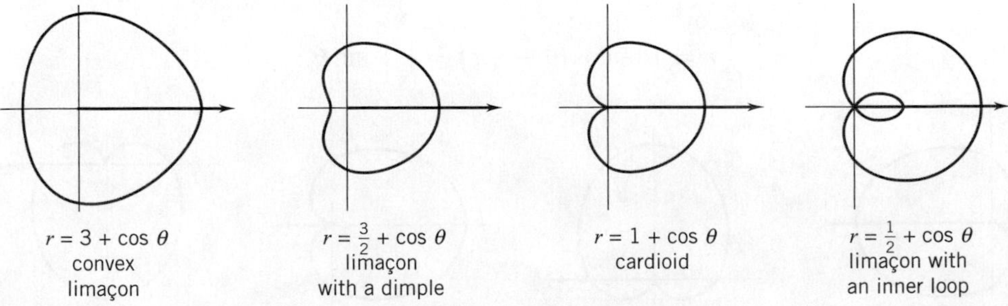

Figure 9.4.8

The general shape of the curve depends on the relative magnitudes of $|a|$ and $|b|$.

† From the French term for "snail". The word is pronounced with a soft c.

Lemniscates : †$r^2 = a \sin 2\theta$, $\quad r^2 = a \cos 2\theta$. (Figure 9.4.9)

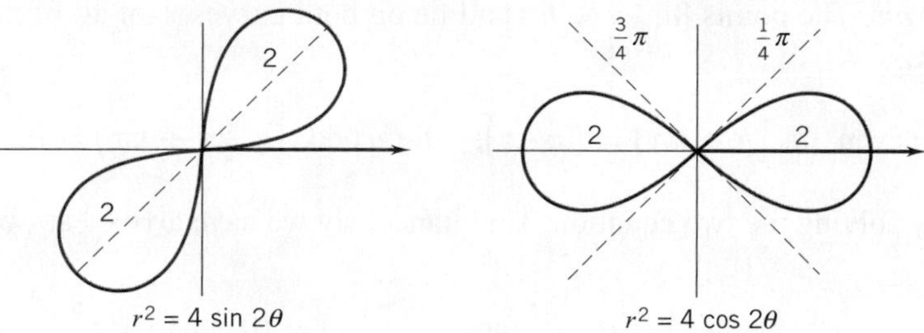

Figure 9.4.9

Petal Curves : $r = a \sin n\theta, \quad r = a \cos n\theta, \quad$ integer n. (Figure 9.4.10)

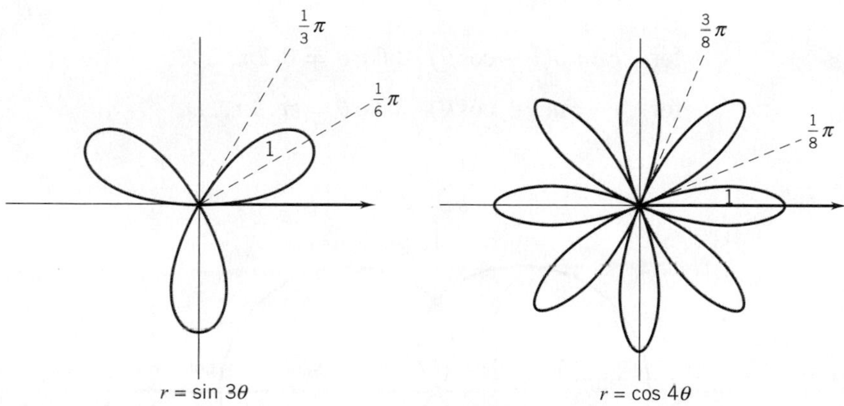

Figure 9.4.10

If n is *odd*, there are n petals. If n is *even*, there are $2n$ petals.

The Intersection of Polar Curves

The fact that a single point has many pairs of polar coordinates can cause complications. In particular, it means that a point $[r_1, \theta_1]$ can lie on a curve given by a polar equation although the coordinates r_1 and θ_1 do not satisfy the equation. For example, the coordinates of $[2, \pi]$ do not satisfy the equation $r^2 = 4 \cos \theta$:

$$r^2 = 2^2 = 4 \quad \text{but} \quad 4 \cos \theta = 4 \cos \pi = -4.$$

Nevertheless the point $[2, \pi]$ does lie on the curve $r^2 = 4 \cos \theta$. It lies on the curve because $[2, \pi] = [-2, 0]$ and the coordinates of $[-2, 0]$ do satisfy the equation:

$$r^2 = (-2)^2 = 4, \quad 4 \cos \theta = 4 \cos 0 = 4.$$

In general, a point $P[r_1, \theta_1]$ lies on a curve given by a polar equation if it has at least one polar coordinate representation $[r, \theta]$ which satisfies the equation. The difficulties are compounded when we deal with two or more curves. Here is an example.

Example 4 Find the points where the cardioids

$$r = a(1 - \cos \theta) \quad \text{and} \quad r = a(1 + \cos \theta) \qquad (a > 0)$$

intersect.

---

† From the Latin *lemniscatus,* meaning "adorned with pendant ribbons."

SOLUTION We begin by solving the two equations simultaneously. Adding these equations, we get $2r = 2a$ and thus $r = a$. This tells us that $\cos\theta = 0$ and therefore $\theta = \frac{1}{2}\pi + n\pi$. The points $[a, \frac{1}{2}\pi + n\pi]$ all lie on both curves. Not all of these points are distinct:

$$\text{for } n \text{ even, } [a, \tfrac{1}{2}\pi + n\pi] = [a, \tfrac{1}{2}\pi]; \quad \text{for } n \text{ odd, } [a, \tfrac{1}{2}\pi + n\pi] = [a, \tfrac{3}{2}\pi].$$

In short, by solving the two equations simultaneously we have arrived at two common points:

$$[a, \tfrac{1}{2}\pi] = (0, a) \quad \text{and} \quad [a, \tfrac{3}{2}\pi] = (0, -a).$$

However, by sketching the two curves (see Figure 9.4.11), we see that there is a third point at which the curves intersect; the two curves intersect at the origin, which clearly lies on both curves:

$$\text{for} \quad r = a(1 - \cos\theta) \quad \text{take } \theta = 0, 2\pi, \ldots,$$

$$\text{for} \quad r = a(1 + \cos\theta) \quad \text{take } \theta = \pi, 3\pi, \ldots.$$

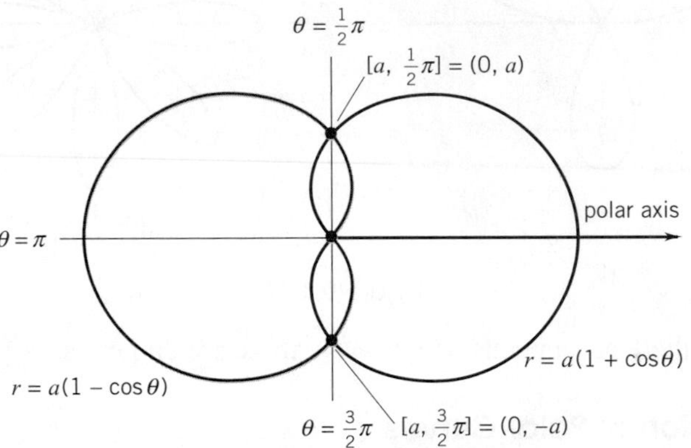

Figure 9.4.11

The reason that the origin does not appear when we solve the two equations simultaneously is that the curves do not pass through the origin "simultaneously"; that is, they do not pass through the origin for the same values of θ. Think of each of the equations

$$r = a(1 - \cos\theta) \qquad \text{and} \qquad r = a(1 + \cos\theta)$$

as giving the position of an object at time θ. At the points we found by solving the two equations simultaneously, the objects collide. (They both arrive there at the same time.) At the origin the situation is different. Both objects pass through the origin, but no collision takes place because the objects pass through the origin at *different* times. □

Example 5 Sketch the polar curves $r = 2\sin\theta$ and $r = 2\sin 2\theta$, and find their points of intersection, if any.

SOLUTION The curve $r = 2\sin\theta$ is a circle of radius 1 and center on the ray $\theta = \pi/2$. The entire circle is traced out as θ varies from 0 to π. The curve $r = 2\sin 2\theta$ is a petal curve with four petals. This curve is traced out as θ varies from 0 to 2π. The two curves are shown in Figure 9.4.12.

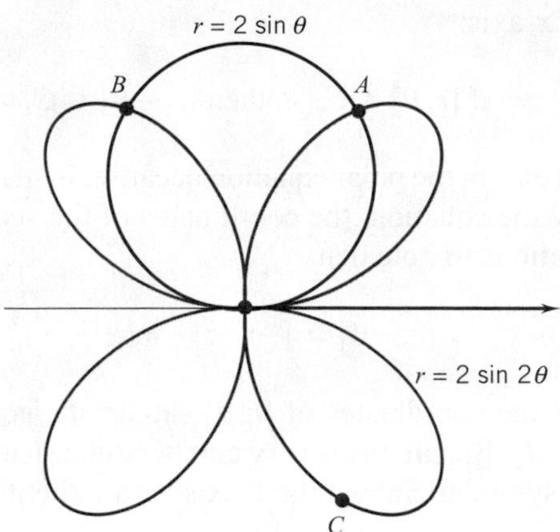

Figure 9.4.12

From the figure, we see that there are three points of intersection: the origin, the point labeled A, and the point labeled B. Solving the two equations simultaneously, we have

$$\begin{aligned} 2\sin 2\theta &= 2\sin\theta, \\ 2\sin\theta\cos\theta &= \sin\theta, \\ \sin\theta(2\cos\theta - 1) &= 0. \end{aligned}$$

Setting $\sin\theta = 0$, we get $\theta = n\pi$, n an integer. We can take $[0, 0]$ as the coordinates of this point of intersection.

Setting $2\cos\theta - 1 = 0$, we get $\theta = (\pi/3) + 2n\pi$ and $\theta = (5\pi/3) + 2n\pi$. The point A clearly has coordinates $[\sqrt{3}, \pi/3]$, but the point B involves a complication. The coordinates $[\sqrt{3}, 2\pi/3]$ satisfy the equation for the circle, but they do not satisfy the equation for the petal curve: at $\theta = 2\pi/3, r = 2\sin 2(2\pi/3) = -\sqrt{3}$ (the point labeled C in the figure).

You can verify that both of the sets of coordinates $[\sqrt{3}, (2\pi/3) + 2n\pi]$ and $[-\sqrt{3}, (5\pi/3) + 2n\pi]$ satisfy $r = 2\sin\theta$, while only $[-\sqrt{3}, (5\pi/3) + 2n\pi]$ satisfies $r = 2\sin 2\theta$. As in Example 4, if the equations describe the positions of two objects at time θ, one moving around the circle and the other around the petal curve, then at $\theta = 2\pi/3$ the object on the circle is at B and the object on the petal curve is C. However, at time $\theta = 5\pi/3$ both objects are at the point B and they collide. The point $[-\sqrt{3}, 5\pi/3]$ satisfies both equations simultaneously. ☐

Remark Problems of incidence (does such and such a point lie on the curve with the following polar equation?) and problems of intersection (where do such and such polar curves intersect?) can usually be analyzed by sketching the curves. However, there are situations where such problems can be handled more readily by first changing to rectangular coordinates. For example, the rectangular equation of the curve $r^2 = 4\cos\theta$ discussed above is

$$(x^2 + y^2)^3 = 16x^2.$$

The point $P[2, \pi]$ has rectangular coordinates $(-2, 0)$, and it is easy to verify that the pair $x = -2, y = 0$ satisfies this equation.

In a similar manner, the symmetry properties of a curve can often be analyzed more easily in rectangular coordinates. To illustrate this, the curve C given by

$$r^2 = \sin\theta$$

is symmetric about the x-axis:

$$(1) \qquad \text{if } [r, \theta] \in C, \quad \text{then } [r, -\theta] \in C.$$

But this is not easy to see from the polar equation because, in general, if the coordinates of the first point satisfy the equation, the coordinates of the second point do not. One way to see that (1) is valid is to note that

$$[r, -\theta] = [-r, \pi - \theta]$$

and then verify that, if the coordinates of $[r, \theta]$ satisfy the equation, then so do the coordinates of $[-r, \pi - \theta]$. But all this is very cumbersome. The easiest way to see that the curve $r^2 = \sin\theta$ is symmetric about the x-axis is to write it as

$$(x^2 + y^2)^3 = y^2.$$

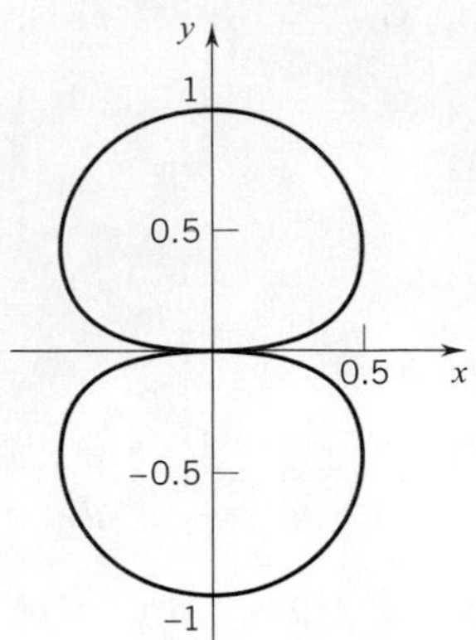

Figure 9.4.13

The other symmetries of the curve, symmetric about the y-axis and symmetric about the origin, can also be seen from this equation (see Figure 9.4.13).

EXERCISES 9.4

Sketch the polar curve.

1. $\theta = -\frac{1}{4}\pi$.
2. $r = -3$.
3. $r = 4$.
4. $r = 3\cos\theta$.
5. $r = -2\sin\theta$.
6. $\theta = \frac{2}{3}\pi$.
7. $r\csc\theta = 3$.
8. $r = 1 - \cos\theta$.
9. $r = \theta, \quad -\frac{1}{2}\pi \le \theta \le \pi$.
10. $r\sec\theta = -2$.
11. $r = \sin 3\theta$.
12. $r^2 = \cos 2\theta$.
13. $r^2 = \sin 2\theta$.
14. $r = \cos 2\theta$.
15. $r^2 = 4, \quad 0 \le \theta \le \frac{3}{4}\pi$.
16. $r = \sin\theta$.
17. $r^3 = 9r$.
18. $\theta = -\frac{1}{4}\pi, \quad 1 \le r < 2$.
19. $r = -1 + \sin\theta$.
20. $r^2 = 4r$.
21. $r = \sin 2\theta$.
22. $r = \cos 3\theta, \quad 0 \le \theta \le \frac{1}{2}\pi$.
23. $r = \cos 5\theta, \quad 0 \le \theta \le \frac{1}{2}\pi$.
24. $r = e^\theta, \quad -\pi \le \theta \le \pi$ (logarithmic spiral).
25. $r = 2 + \sin\theta$.
26. $r = \cot\theta$.
27. $r = \tan\theta$.
28. $r = 2 - \cos\theta$.
29. $r = 2 + \sec\theta$.
30. $r = 3 - \csc\theta$.
31. $r = -1 + 2\cos\theta$.
32. $r = 1 + 2\sin\theta$.

Determine whether the point lies on the curve.

33. $r^2\cos\theta = 1; \quad [1, \pi]$.
34. $r^2 = \cos 2\theta; \quad [1, \frac{1}{4}\pi]$.
35. $r = \sin\frac{1}{3}\theta; \quad [\frac{1}{2}, \frac{1}{2}\pi]$.
36. $r^2 = \sin 3\theta; \quad [1, -\frac{5}{6}\pi]$.
37. Show that the point $[2, \pi]$ lies both on $r^2 = 4\cos\theta$ and on $r = 3 + \cos\theta$.
38. Show that the point $[2, \frac{1}{2}\pi]$ lies both on $r^2\sin\theta = 4$ and on $r = 2\cos 2\theta$.

Sketch the curves and find the points at which they intersect. Express your answers in rectangular coordinates.

39. $r = \sin\theta, \quad r = -\cos\theta$.
40. $r^2 = \sin\theta, \quad r = 2 - \sin\theta$.
41. $r = \cos^2\theta, \quad r = -1$.
42. $r = 2\sin\theta, \quad r = 2\cos\theta$.
43. $r = 1 - \cos\theta, \quad r = \cos\theta$.
44. $r = 1 - \cos\theta, \quad r = \sin\theta$.
45. $r = \sin 2\theta, \quad r = \sin\theta$.
46. $r = 1 - \cos\theta, \quad r = 1 + \sin\theta$.
47. Show that the polar equation $r = 2a\sin\theta + 2b\cos\theta$ represents a circle. Find the center and radius of the circle and sketch the circle.
48. (C) a. Use a graphing utility to draw the curves
$$r = 1 + \cos(\theta - \tfrac{1}{3}\pi) \quad \text{and} \quad r = 1 + \cos(\theta + \tfrac{1}{6}\pi).$$
How do your curves compare with the graph of $r = 1 + \cos\theta$? (See Figure 9.4.5)
b. In general, what is the relationship between the graph of $r = f(\theta - \alpha)$ and the graph of $r = f(\theta)$?
49. (C) a. Use a graphing utility to draw the curves
$$r = 1 + \sin\theta \quad \text{and} \quad r^2 = 4\sin 2\theta$$
in the same coordinate system.

b. Use a CAS to find the points of intersection of the two curves.

50. Repeat Exercise 49 for the pair of equations

$$r = 1 + \cos\theta \qquad \text{and} \qquad r = 1 + \cos\left(\tfrac{1}{2}\theta\right).$$

51. Repeat Exercise 49 for the pair of equations

$$r = 2\cos\theta \qquad \text{and} \qquad r = 2\sin 2\theta.$$

52. Repeat Exercise 49 for the pair of equations

$$r = 2 \qquad \text{and} \qquad r = 2\sin 3\theta.$$

53. Repeat Exercise 49 for the pair of equations

$$r = 1 - 3\cos\theta \qquad \text{and} \qquad r = 2 - 5\sin\theta.$$

54. (a) The electrostatic charge distribution consisting of a charge $q\,(q > 0)$ at the point $[r, 0]$ and a charge $-q$ at $[r, \pi]$ is called a *dipole*. The *lines of force* for the dipole are given by the equations

$$r = k\sin^2\theta.$$

Use a graphing utility to draw the lines of force for $k = 1, 2, 3$.

(b) The *equipotential lines* (the set of points with equal electric potential) for the dipole are given by the equations

$$r^2 = m\cos\theta.$$

Use a graphing utility to draw the equipotential lines for $m = 1, 2, 3$.

(c) Draw the curves $r = 2\sin^2\theta$ and $r^2 = 2\cos\theta$ in the same coordinate system and estimate the coordinates of their points of intersection. Use four decimal place accuracy.

55. Use a graphing utility to draw the curves

$$r = 1 + \sin k\theta + \cos^2(2k\theta)$$

for $k = 1, 2, 3, 4, 5$. If you were asked to give a name to this family of curves, what would you suggest?

56. Use a graphing utility to draw the graph $r = e^{\cos\theta} - 2\cos 4\theta$. What name would you suggest for this curve?

57. The graphs of $r = A\cos k\theta$ and $r = A\sin k\theta$ are called petal curves (see Figure 9.4.10). Use a graphing utility to draw the curves

$$r = 2\cos k\theta \qquad \text{and} \qquad r = 2\sin k\theta$$

for $k = \frac{3}{2}$ and $k = \frac{5}{2}$. Conjecture what the curve will look like for $k = m/2$ for any odd integer m.

58. Use a graphing utility to draw the curves

$$r = 2\cos k\theta \qquad \text{and} \qquad r = 2\sin k\theta$$

for $k = \frac{4}{3}$ and $k = \frac{5}{3}$. Conjecture what the curve will look like for $k = m/3$ for any even integer m that is not a multiple of 3. For any odd integer m that is not a multiple of 3?

■ PROJECT 9.4 The Conic Sections in Polar Coordinates

In Section 9.1, we defined parabola in terms of a focus and a directrix, but our definitions of the ellipse and hyperbola in Section 9.2 were given in terms of two foci; there was no mention of a directrix for either of these conics. In this project, we give a unified approach to the conic sections, an approach that involves a focus and a directrix in all three cases.

In the plane, let F be a fixed point (the *focus*) and l a fixed line (the *directrix*) which does not pass through F. Let e be a positive number (the *eccentricity*) and consider the set of points P that satisfy

$$\text{(1)} \qquad \frac{\text{distance from } P \text{ to } F}{\text{distance from } P \text{ to } l} = e.$$

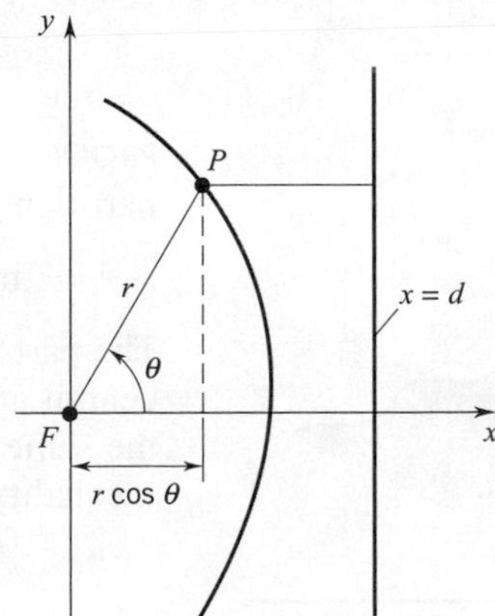

In the figure, we have superimposed a polar and rectangular coordinate system and, without loss of generality, we have taken F as the origin and l as the vertical line $x = d, d > 0$.

Problem 1. Show that the set of all points P that satisfy (1) is described by the polar equation

$$\text{(2)} \qquad r = \frac{ed}{1 + e\cos\theta}.$$

Problem 2. Show that:

a. If $0 < e < 1$, the equation is an ellipse of eccentricity e with right focus at the origin, major axis horizontal:

$$\frac{(x+c)^2}{a^2} + \frac{y^2}{a^2 - c^2} = 1 \quad \text{with} \quad a = \frac{ed}{1 - e^2}, \qquad c = ea.$$

b. If $e = 1$, the equation represents a parabola with focus at the origin and directrix $x = d$:

$$y^2 = -4\frac{d}{2}\left(x - \frac{d}{2}\right).$$

c. If $e > 1$, the equation represents a hyperbola of eccentricity e with left focus at the origin, transverse axis horizontal:

$$\frac{(x-c)^2}{a^2} - \frac{y^2}{c^2 - a^2} = 1 \quad \text{with} \quad a = \frac{ed}{e^2 - 1}, \qquad c = ea.$$

Problem 3. Identify each of the following conic sections and write the equation in rectangular coordinates.

a. $r = \dfrac{8}{4 + 3\cos\theta}$. b. $r = \dfrac{6}{1 + 2\cos\theta}$. c. $r = \dfrac{6}{2 + 2\cos\theta}$.

Consider the polar equation $r = \alpha/(1 + \beta\sin\theta), \alpha, \beta > 0$. Since

$$r = \frac{\alpha}{1 + \beta\sin\theta} = \frac{\alpha}{1 + \beta\cos(\theta - \pi/2)}$$

we can conclude that $r = \alpha/(1 + \beta\sin\theta)$ is simply the conic section $r = \alpha/(1 + \beta\cos\theta)$ rotated $\pi/2$ radians in the counterclockwise direction.

Problem 4. Relate the polar curves

$$r = \frac{\alpha}{1 - \beta\cos\theta} \quad \text{and} \quad r = \frac{\alpha}{1 - \beta\sin\theta}, \qquad \alpha, \beta > 0$$

to the conic section

$$r = \frac{\alpha}{1 + \beta\cos\theta}.$$

■ 9.5 AREA IN POLAR COORDINATES

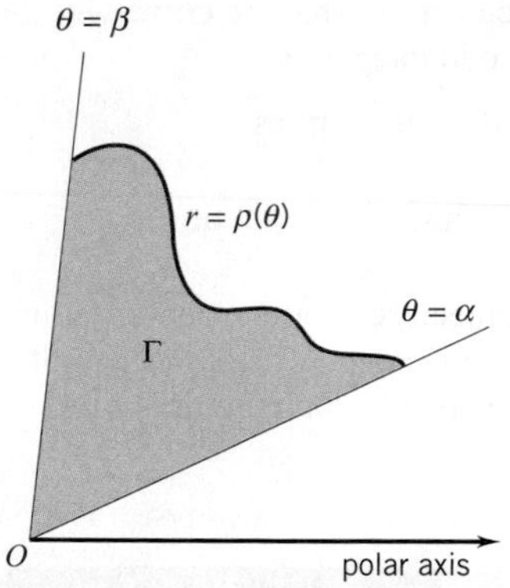

Here we develop a technique for calculating the area of a region the boundary of which is given in polar coordinates.

As a start, we suppose that α and β are two real numbers with $\alpha < \beta \leq \alpha + 2\pi$. We take ρ as a function that is continuous on $[\alpha, \beta]$ and keeps a constant sign on that interval. We want the area of the polar region Γ generated by the curve

$$r = \rho(\theta), \quad \alpha \leq \theta \leq \beta.$$

Such a region is portrayed in Figure 9.5.1.

$\theta = \beta$; $r = \rho(\theta)$; $\theta = \alpha$; Γ; O; polar axis

Figure 9.5.1

In the figure $\rho(\theta)$ remains nonnegative. If $\rho(\theta)$ were negative, the region Γ would appear on the opposite side of the pole. In either case, the area of Γ is given by the formula

(9.5.1)
$$A = \int_\alpha^\beta \tfrac{1}{2}[\rho(\theta)]^2 d\theta.$$

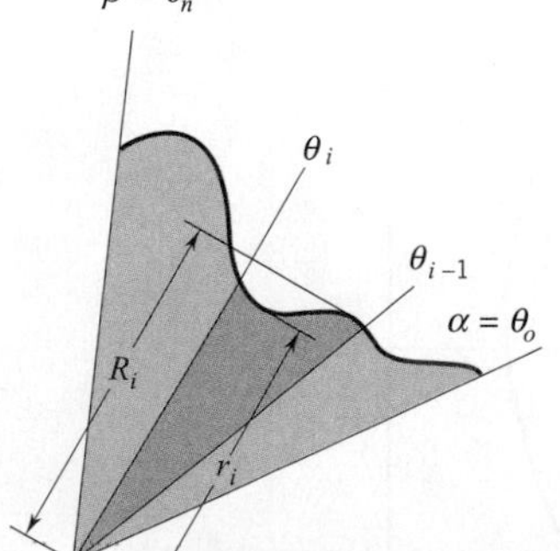

Figure 9.5.2

PROOF We consider the case where $\rho(\theta) \geq 0$. We take $P = \{\theta_0, \theta_1, \ldots, \theta_n\}$ as a partition of $[\alpha, \beta]$ and direct our attention to what happens from θ_{i-1} to θ_i. We set

$$r_i = \text{ min value of } \rho \text{ on } [\theta_{i-1}, \theta_i] \qquad \text{and} \qquad R_i = \text{ max value of } \rho \text{ on } [\theta_{i-1}, \theta_i].$$

The part of Γ that lies between θ_{i-1} and θ_i contains a circular sector of radius r_i and central angle $\Delta\theta_i = \theta_i - \theta_{i-1}$ and is contained in a circular sector of radius R_i with the same central angle $\Delta\theta_i$. (See Figure 9.5.2.) Its area A_i must therefore satisfy the inequality

$$\tfrac{1}{2}r_i^2\,\Delta\theta_i \leq A_i \leq \tfrac{1}{2}R_i^2\,\Delta\theta_i.\dagger$$

By summing these inequalities from $i = 1$ to $i = n$, we can see that the total area A must satisfy the inequality

(1) $$L_f(P) \leq A \leq U_f(P)$$

---

† The area of a circular sector of radius r and central angle α is $\frac{1}{2}r^2\alpha$.

where $$f(\theta) = \tfrac{1}{2}[\rho(\theta)]^2.$$

Since f is continuous and (1) holds for every partition P of $[\alpha, \beta]$, we can conclude that

$$A = \int_\alpha^\beta f(\theta)\, d\theta = \int_\alpha^\beta \tfrac{1}{2}[\rho(\theta)]^2 d\theta. \quad \square$$

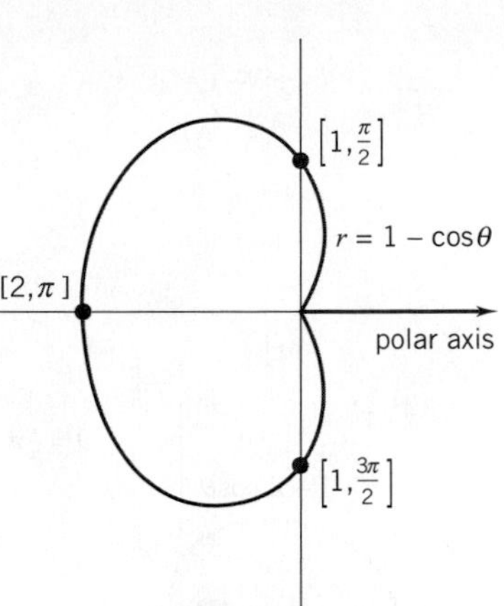

Figure 9.5.3

Example 1 Calculate the area enclosed by the cardioid

$$r = 1 - \cos\theta \qquad \text{(Figure 9.5.3)}$$

SOLUTION The entire curve is traced out as θ increases from 0 to 2π, and $1 - \cos\theta \geq 0$ on $[0, 2\pi]$. Therefore

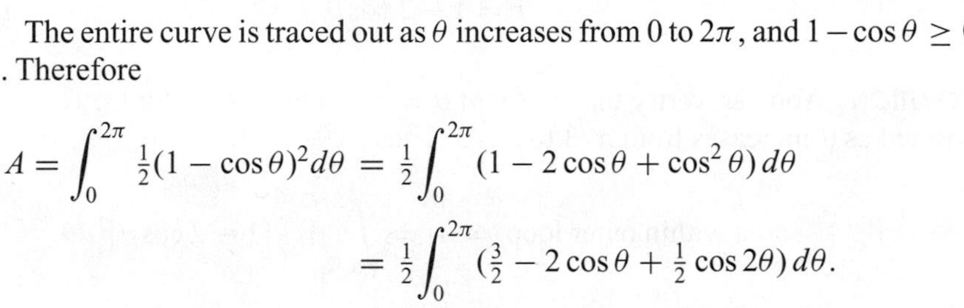

$$A = \int_0^{2\pi} \tfrac{1}{2}(1 - \cos\theta)^2 d\theta = \tfrac{1}{2}\int_0^{2\pi} (1 - 2\cos\theta + \cos^2\theta)\, d\theta$$

$$= \tfrac{1}{2}\int_0^{2\pi} (\tfrac{3}{2} - 2\cos\theta + \tfrac{1}{2}\cos 2\theta)\, d\theta.$$

↑ half-angle formula: $\cos^2\theta = \frac{1}{2} + \frac{1}{2}\cos 2\theta$

Since $\displaystyle\int_0^{2\pi} \cos\theta\, d\theta = \sin\theta \Big|_0^{2\pi} = 0$ and $\displaystyle\int_0^{2\pi} \cos 2\theta\, d\theta = \tfrac{1}{2}\sin 2\theta \Big|_0^{2\pi} = 0,$

we have $$A = \tfrac{1}{2}\int_0^{2\pi} \tfrac{3}{2}\, d\theta = \tfrac{3}{4}\int_0^{2\pi} d\theta = \tfrac{3}{2}\pi. \quad \square$$

A slightly more complicated type of region is pictured in Figure 9.5.4. We approach the problem of calculating the area of the region Ω in the same way that we calculated the area between two curves in Section 5.5; that is, we calculate the area out to $r = \rho_2(\theta)$ and subtract from it the area out to $r = \rho_1(\theta)$. This gives

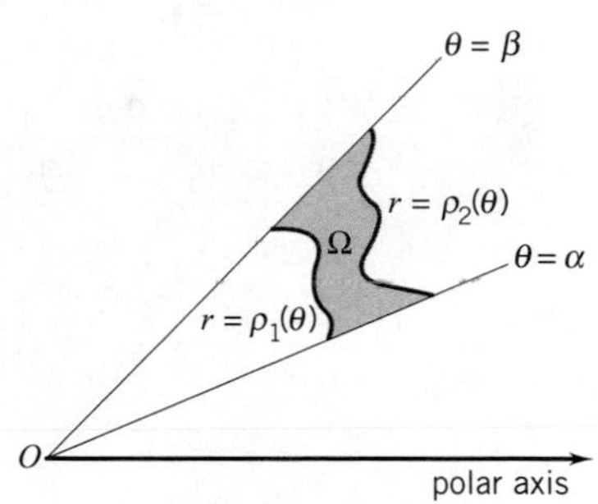

Figure 9.5.4

$$\text{area of } \Omega = \int_\alpha^\beta \tfrac{1}{2}[\rho_2(\theta)]^2 d\theta - \int_\alpha^\beta \tfrac{1}{2}[\rho_1(\theta)]^2 d\theta,$$

which can be written as

(9.5.2) $$\text{area of } \Omega = \int_\alpha^\beta \tfrac{1}{2}([\rho_2(\theta)]^2 - [\rho_1(\theta)]^2)\, d\theta.$$

To find the area between two polar curves, we first determine the curves that serve as outer and inner boundaries of the region and the intervals of θ values over which these boundaries are traced out. Since the polar coordinates of a point are not unique, extra care is needed to determine these intervals of θ values.

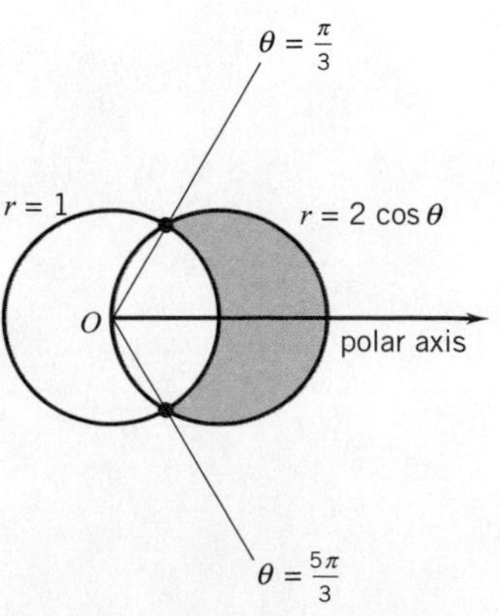

Figure 9.5.5

Example 2 Find the area of the region that consists of all points that lie within the circle $r = 2\cos\theta$ but outside the circle $r = 1$.

SOLUTION The region is shown in Figure 9.5.5. Our first step is to find values of θ for the two points where the circles intersect:

$$2\cos\theta = 1, \quad \cos\theta = \tfrac{1}{2}, \quad \theta = \tfrac{1}{3}\pi, \tfrac{5}{3}\pi.$$

Since the region is symmetric about the polar axis, the area below the polar axis equals the area above the polar axis. Thus

$$A = 2\int_0^{\pi/3} \tfrac{1}{2}([2\cos\theta]^2 - [1]^2)\,d\theta.$$

If you carry out the integration, you will see that $A = \frac{1}{3}\pi + \frac{1}{2}\sqrt{3} \cong 1.91.$ □

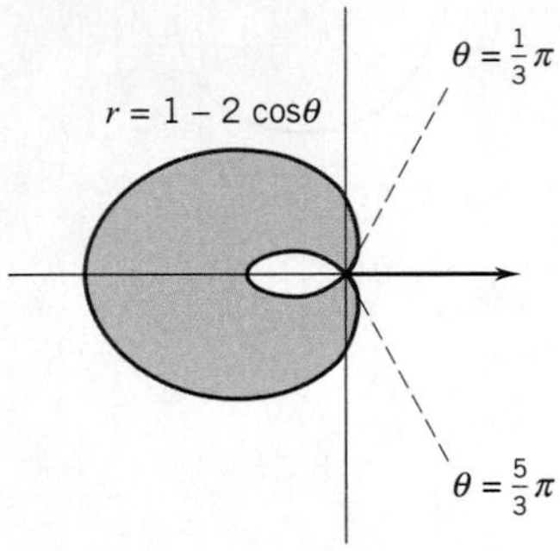

$r = 1 - 2\cos\theta$, $\theta = \frac{1}{3}\pi$, $\theta = \frac{5}{3}\pi$

Figure 9.5.6

Example 3 Find the area A of the region between the inner and outer loops of the limaçon

$$r = 1 - 2\cos\theta. \qquad \text{(Figure 9.5.6)}$$

SOLUTION You can verify that $r = 0$ at $\theta = \pi/3$ and at $\theta = 5\pi/3$. The outer loop is formed as θ increases from $\pi/3$ to $5\pi/3$. Thus

$$\text{area within outer loop } = A_1 = \int_{\pi/3}^{5\pi/3} \tfrac{1}{2}[1 - 2\cos\theta]^2 d\theta.$$

The lower half of the inner loop is formed as θ increases from 0 to $\pi/3$, and the upper half as θ increases from $5\pi/3$ to 2π (verify this). Therefore

$$\text{area within inner loop } = A_2 = \int_0^{\pi/3} \tfrac{1}{2}[1 - 2\cos\theta]^2 d\theta + \int_{5\pi/3}^{2\pi} \tfrac{1}{2}[1 - 2\cos\theta]^2 d\theta.$$

Note that

$$\begin{aligned}\int \tfrac{1}{2}[1 - 2\cos\theta]^2 d\theta &= \tfrac{1}{2}\int [1 - 4\cos\theta + 4\cos^2\theta]\,d\theta \\ &= \tfrac{1}{2}\int [1 - 4\cos\theta + 2(1 + \cos 2\theta)]\,d\theta \\ &= \tfrac{1}{2}\int [3 - 4\cos\theta + 2\cos 2\theta]\,d\theta \\ &= \tfrac{1}{2}[3\theta - 4\sin\theta + \sin 2\theta] + C.\end{aligned}$$

Therefore

$$A_1 = \tfrac{1}{2}\Big[3\theta - 4\sin\theta + \sin 2\theta\Big]_{\pi/3}^{5\pi/3} = 2\pi + \tfrac{3}{2}\sqrt{3}$$

and

$$\begin{aligned}A_2 &= \tfrac{1}{2}\Big[3\theta - 4\sin\theta + \sin 2\theta\Big]_0^{\pi/3} + \tfrac{1}{2}\Big[3\theta - 4\sin\theta + \sin 2\theta\Big]_{5\pi/3}^{2\pi} \\ &= \tfrac{1}{2}\pi - \tfrac{3}{4}\sqrt{3} + \tfrac{1}{2}\pi - \tfrac{3}{4}\sqrt{3} = \pi - \tfrac{3}{2}\sqrt{3}.\end{aligned}$$

Thus,

$$A = A_1 - A_2 = 2\pi + \tfrac{3}{2}\sqrt{3} - (\pi - \tfrac{3}{2}\sqrt{3}) = \pi + 3\sqrt{3} \cong 8.34. \quad \square$$

Remark We could have done Example 3 more efficiently by exploiting the symmetry of the region. The region is symmetric about the x-axis. Therefore

$$A = 2\int_{\pi/3}^{\pi} \tfrac{1}{2}[1 - 2\cos\theta]^2 d\theta - 2\int_0^{\pi/3} \tfrac{1}{2}[1 - 2\cos\theta]^2 d\theta. \quad \square$$

Example 4 The region Ω common to the circle $r = 2\sin\theta$ and the limaçon $r = \frac{3}{2} - \sin\theta$ is indicated in Figure 9.5.7. The θ coordinates of the points of intersection can be found by solving the two equations simultaneously:

$$2\sin\theta = \tfrac{3}{2} - \sin\theta, \quad \sin\theta = \tfrac{1}{2}, \quad \text{and} \quad \theta = \tfrac{1}{6}\pi, \tfrac{5}{6}\pi.$$

Thus, the area of Ω can be represented as follows:

$$\text{area of } \Omega = \int_0^{\pi/6} \tfrac{1}{2}[2\sin\theta]^2 d\theta + \int_{\pi/6}^{5\pi/6} \tfrac{1}{2}[\tfrac{3}{2} - \sin\theta]^2 d\theta + \int_{5\pi/6}^{\pi} \tfrac{1}{2}[2\sin\theta]^2\, d\theta;$$

or, by the symmetry of the region,

$$\text{area of } \Omega = 2\int_0^{\pi/6} \tfrac{1}{2}[2\sin\theta]^2 d\theta + 2\int_{\pi/6}^{\pi/2} \tfrac{1}{2}[\tfrac{3}{2} - \sin\theta]^2 d\theta.$$

As you can verify, the area of Ω is $\frac{5}{4}\pi - \frac{15}{8}\sqrt{3} \cong 0.68$. □

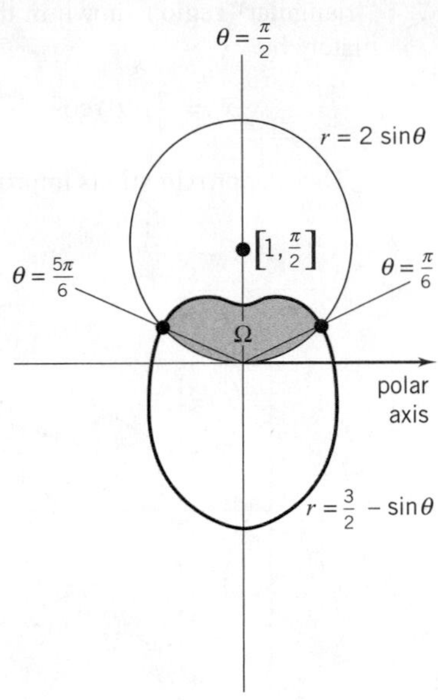

Figure 9.5.7

EXERCISES 9.5

Calculate the area enclosed by the given curve. Take $a > 0$.

1. $r = a\cos\theta$ from $\theta = -\frac{1}{2}\pi$ to $\theta = \frac{1}{2}\pi$.
2. $r = a\cos 3\theta$ from $\theta = -\frac{1}{6}\pi$ to $\theta = \frac{1}{6}\pi$.
3. $r = a\sqrt{\cos 2\theta}$ from $\theta = -\frac{1}{4}\pi$ to $\theta = \frac{1}{4}\pi$.
4. $r = a(1 + \cos 3\theta)$ from $\theta = -\frac{1}{3}\pi$ to $\theta = \frac{1}{3}\pi$.
5. $r^2 = a^2\sin^2\theta$.
6. $r^2 = a^2\sin^2 2\theta$.

Calculate the area of the given region.

7. $r = \tan 2\theta$ and the rays $\theta = 0$, $\theta = \frac{1}{8}\pi$.
8. $r = \cos\theta$, $r = \sin\theta$, and the rays $\theta = 0$, $\theta = \frac{1}{4}\pi$.
9. $r = 2\cos\theta$, $r\cos\theta$, and the rays $\theta = 0$, $\theta = \frac{1}{4}\pi$.
10. $r = 1 + \cos\theta$, $r = \cos\theta$, and the rays $\theta = 0$, $\theta = \frac{1}{2}\pi$.
11. $r = a(4\cos\theta - \sec\theta)$ and the rays $\theta = 0$, $\theta = \frac{1}{4}\pi$.
12. $r = \frac{1}{2}\sec^2 \frac{1}{2}\theta$ and the vertical line through the origin.

Find the area between the curves.

13. $r = e^\theta$, $0 \le \theta \le \pi$; $r = \theta$, $0 \le \theta \le \pi$; the rays $\theta = 0$, $\theta = \pi$.
14. $r = e^\theta$, $2\pi \le \theta \le 3\pi$; $r = \theta$, $0 \le \theta \le \pi$; the rays $\theta = 0$, $\theta = \pi$.
15. $r = e^\theta$, $0 \le \theta \le \pi$; $r = e^{\theta/2}$, $0 \le \theta \le \pi$; the rays $\theta = 2\pi$, $\theta = 3\pi$.
16. $r = e^\theta$, $0 \le \theta \le \pi$; $r = e^\theta$, $2\pi \le \theta \le 3\pi$; the rays $\theta = 0$, $\theta = \pi$.

Represent the area by one or more integrals.

17. Outside $r = 2$, but inside $r = 4\sin\theta$.
18. Outside $r = 1 - \cos\theta$, but inside $r = 1 + \cos\theta$.
19. Inside $r = 4$, and to the right of $r = 2\sec\theta$.
20. Inside $r = 2$, but outside $r = 4\cos\theta$.
21. Inside $r = 4$, and between the lines $\theta = \frac{1}{2}\pi$ and $r = 2\sec\theta$.
22. Inside the inner loop of $r = 1 - 2\sin\theta$.
23. Inside one petal of $r = 2\sin 3\theta$.
24. Outside $r = 1 + \cos\theta$, but inside $r = 2 - \cos\theta$.
25. Interior to both $r = 1 - \sin\theta$ and $r = \sin\theta$.
26. Inside one petal of $r = 5\cos 6\theta$.
27. Outside $r = \cos 2\theta$, but inside $r = 1$.
28. Interior to both $r = 2a\cos\theta$ and $r = 2a\sin\theta, a > 0$.
29. Find the area of the region that is common to the three circles: $r = 1, r = 2\cos\theta$, and $r = 2\sin\theta$.
30. Find the area of the region outside the circle $r = a$ and inside the lemniscate $r^2 = 2a^2\cos 2\theta$.
31. Fix $a > 0$ and let n be a positive integer. Prove that the petal curves $r = a\cos 2n\theta$ and $r = a\sin 2n\theta$ all enclose exactly the same area. Find the area.
32. Fix $a > 0$ and let n be a positive integer. Prove that the petal curves $r = a\cos([2n+1]\theta)$ and $r = a\sin([2n+1]\theta$ all enclose exactly the same area. Find the area.

Centroids in Polar Coordinates

Let Ω be the region bounded by the polar curve $r = \rho(\theta)$ between $\theta = \alpha$ and $\theta = \beta$. Since the centroid of a triangle lies on each median, two-thirds of the distance from the vertex to the opposite side (see Exercise 29, Section 6.4), it follows that the x and y coordinates of the centroid of the "triangular" region shown in the figure are given approximately by

$$\bar{x} = \tfrac{2}{3}\rho(\theta)\cos\theta \quad \bar{y} = \tfrac{2}{3}\rho(\theta)\sin\theta$$

These approximations improve as the triangle narrows.

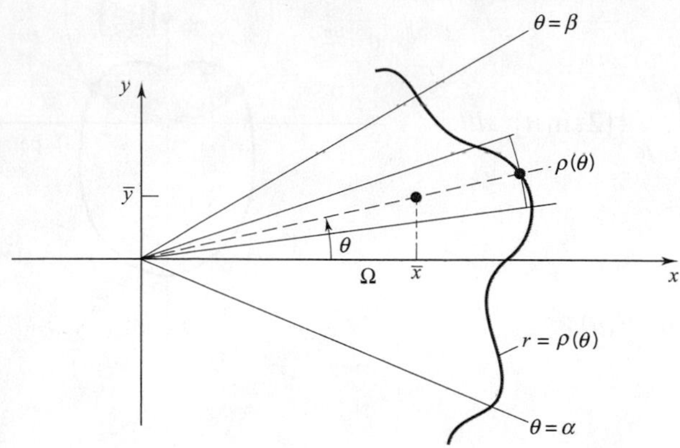

33. Following the approach used in Section 6.4, show that the rectangular coordinates of the centroid of Ω are given by

$$\bar{x}A = \tfrac{1}{3}\int_{\alpha}^{\beta} [\rho(\theta)]^3 \cos\theta \, d\theta, \quad \bar{y}A = \tfrac{1}{3}\int_{\alpha}^{\beta} [\rho(\theta)]^3 \sin\theta \, d\theta$$

where A is the area of Ω.

In Exercises 34–36, use the result of Exercise 33 to find the rectangular coordinates of the centroid of the given region.

34. The region enclosed by $r = 4$ between $\theta = -\alpha$ and $\theta = \alpha, 0 < \alpha < \pi/2$.

35. The region enclosed by the cardioid $r = 1 + \cos\theta$.

36. The region enclosed by $r = 2 + \sin\theta$.

(C) In Exercises 37 and 38 ,use a graphing utility to draw the polar curve. Then use a CAS to find the area of the region it encloses.

37. $r = 2 + \cos\theta$.

38. $r = 2\cos 3\theta$.

(C) In Exercises 39 and 40, use a graphing utility to drawn the polar curves. Then use a CAS to find the area of the region inside the first curve and outside the second curve.

39. $r = 4\cos 3\theta, \quad r = 2$.

40. $r = 2\cos\theta, \quad r = 1 - \cos\theta$.

(C) **41.** The curve

$$y^2 = x^2\left(\frac{a - x}{a + x}\right), \quad a > 0,$$

is called a *strophoid*.

(a) Show that the polar equation of this curve has the form

$$r = a\cos 2\theta \sec\theta.$$

(b) Use a graphing utility to draw the curves for $a = 1, 2,$ and 4.

(c) Let $a = 2$. Find the area inside the loop.

(C) **42.** The curve

$$(x^2 + y^2)^2 = ax^2y, \quad a > 0,$$

is called a *bifolium*.

(a) Show that the polar equation of this curve has the form

$$r = a\sin\theta\cos^2\theta.$$

(b) Use a graphing utility to draw the curves for $a = 1, 2$ and 4.

(c) Let $a = 2$. Find the area inside one of the loops.

■ 9.6 CURVES GIVEN PARAMETRICALLY

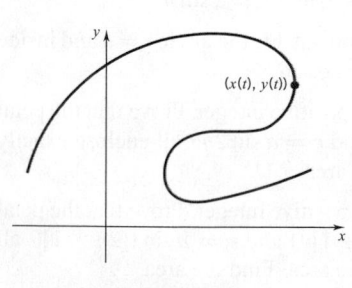

So far we have specified curves by equations in rectangular coordinates or by equations in polar coordinates. Here we introduce a more general method. We begin with a pair of functions $x = x(t), y = y(t)$ differentiable on the interior of an interval I. At the endpoints of I (if any) we require only continuity.

For each number t in I we can interpret $(x(t), y(t))$ as the point with x-coordinate $x(t)$ and y-coordinate $y(t)$. Then, as t ranges over I, the point $(x(t), y(t))$ traces out a path in the xy-plane (Figure 9.6.1). We call such a path a *parametrized curve* and refer to t as the *parameter*.

Figure 9.6.1

Example 1 Identify the curve parametrized by the functions

$$x(t) = t + 1, \quad y(t) = 2t - 5, \quad t \in (-\infty, \infty).$$

SOLUTION We can express $y(t)$ in terms of $x(t)$:

$$y(t) = 2[x(t) - 1] - 5 = 2x(t) - 7.$$

The functions parametrize the line $y = 2x - 7$: as t ranges over the set of real numbers, the point $(x(t), y(t))$ traces out the line $y = 2x - 7$. □

Example 2 Identify the curve parametrized by the functions

$$x(t) = 2t, \quad y(t) = t^2, \quad t \in [0, \infty).$$

SOLUTION From the first equation $t = \frac{1}{2}x(t)$, and so

$$y(t) = \tfrac{1}{4}[x(t)]^2.$$

The functions parametrize that part of the parabola $y = \frac{1}{4}x^2$ which lies in the right half plane: as t ranges over the interval $[0, \infty)$, the point $(x(t), y(t))$ traces out the parabolic path $y = \frac{1}{4}x^2, x \geq 0$ (Figure 9.6.2). □

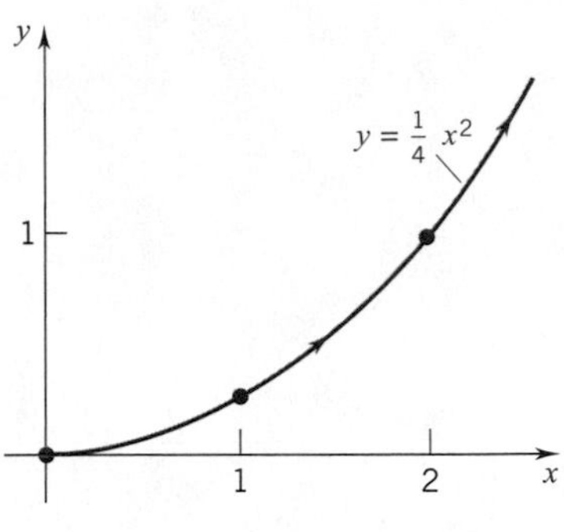

Figure 9.6.2

Example 3 Identify the curve parametrized by the functions

$$x(t) = \sin^2 t, \quad y(t) = \cos t, \quad t \in [0, \pi].$$

SOLUTION Note first that

$$x(t) = \sin^2 t = 1 - \cos^2 t = 1 - [y(t)]^2.$$

The points $(x(t), y(t))$ all lie on the parabola

$$x = 1 - y^2. \qquad \text{(Figure 9.6.3)}$$

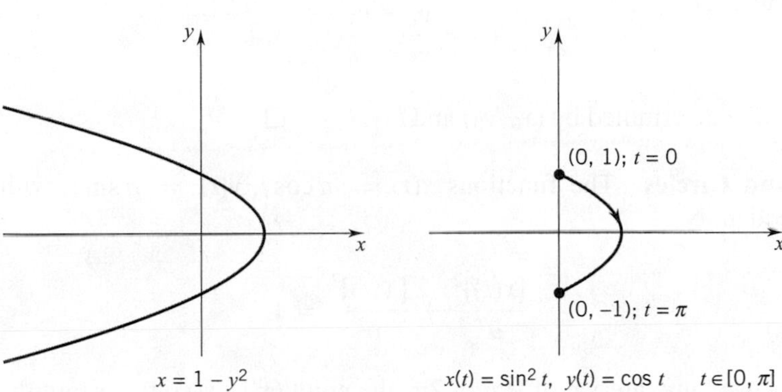

Figure 9.6.3

At $t = 0, x = 0$ and $y = 1$; at $t = \pi, x = 0$ and $y = -1$. As t ranges from 0 to π, the point $(x(t), y(t))$ traverses the parabolic arc

$$x = 1 - y^2, \quad -1 \leq y \leq 1$$

from the point (0, 1) to the point (0, −1). □

Remark Changing the domain in Example 3 to all real t does *not* give us any more of the parabola. For any given t we still have

$$0 \leq x(t) \leq 1 \qquad \text{and} \qquad -1 \leq y(t) \leq 1.$$

As t ranges over the set of real numbers, the point $(x(t), y(t))$ traces out that same parabolic arc back and forth an infinite number of times. □

Straight Lines Given that $(x_0, y_0) \neq (x_1, y_1)$, the functions

(9.6.1) $$x(t) = x_0 + t(x_1 - x_0), \quad y(t) = y_0 + t(y_1 - y_0), \quad t \in (-\infty, \infty)$$

parametrize the line that passes through (x_0, y_0) and (x_1, y_1).

PROOF If $x_1 = x_0$, then we have

$$x(t) = x_0, \quad y(t) = y_0 + t(y_1 - y_0). \qquad (y_0 \neq y_1)$$

As t ranges over the set of real numbers, $x(t)$ remains constantly x_0 and $y(t)$ ranges over the set of real numbers. The functions parametrize the vertical line $x = x_0$. Since $x_1 = x_0$, both (x_0, y_0) and (x_1, y_1) lie on this vertical line.

If $x_1 \neq x_0$, then we can solve the first equation for t:

$$t = \frac{x(t) - x_0}{x_1 - x_0}.$$

Substituting this into the second equation, we obtain the identity

$$y(t) - y_0 = \frac{y_1 - y_0}{x_1 - x_0}[x(t) - x_0].$$

The functions parametrize the line with equation

$$y - y_0 = \frac{y_1 - y_0}{x_1 - x_0}(x - x_0).$$

This is the line determined by (x_0, y_0) and (x_1, y_1). ☐

Ellipses and Circles The functions $x(t) = a\cos t$, $y(t) = b\sin t$, with $a, b > 0$, satisfy the identity

$$\frac{[x(t)]^2}{a^2} + \frac{[y(t)]^2}{b^2} = 1.$$

As t ranges over any interval of length 2π, the point $(x(t), y(t))$ traces out the ellipse

$$\frac{x^2}{a^2} + \frac{y^2}{b^2} = 1.$$

Usually we let t range from 0 to 2π and parametrize the ellipse by setting

(9.6.2) $$x(t) = a\cos t, \quad y(t) = b\sin t, \quad t \in [0, 2\pi].$$

If $b = a$, we have a circle. We can parametrize the circle

$$x^2 + y^2 = a^2$$

by setting

(9.6.3) $$x(t) = a\cos t, \quad y(t) = a\sin t, \quad t \in [0, 2\pi].$$ ☐

Hyperbolas The functions $x(t) = a\cosh t$, $y(t) = b\sinh t$, with $a, b > 0$, satisfy the identity

$$\frac{[x(t)]^2}{a^2} - \frac{[y(t)]^2}{b^2} = 1.$$

Since $x(t) = a\cosh t > 0$ for all t, as t ranges over the set of real numbers, the point $(x(t), y(t))$ traces out the right branch of the hyperbola

$$\frac{x^2}{a^2} - \frac{y^2}{b^2} = 1. \qquad \text{(Figure 9.6.4)}$$

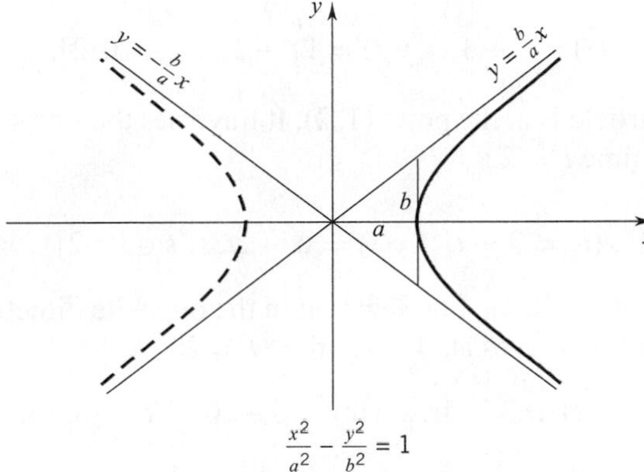

Figure 9.6.4

Thus, the right branch of the hyperbola can be parametrized by setting

$$x(t) = a\cosh t, \quad y(t) = b\sinh t, \quad t \in (-\infty, \infty).$$

The left branch can be parametrized by setting

$$x(t) = -a\cosh t, \quad y(t) = b\sinh t, \quad t \in (-\infty, \infty). \quad \square$$

The connection between curves given parametrically and the graphs of functions such as $y = f(x)$ or $r = \rho(\theta)$ is easy to illustrate. First, let $y = f(x)$ be a *continuously differentiable* function† on the interval I. Then, the graph of f is a curve C in the plane that is parametrized by the functions

$$x(t) = t, \quad y(t) = f(t), \quad t \in I.$$

[Replacing t by x in the second equation gives $y = f(x)$]. Similarly, if $r = \rho(\theta)$ is continuously differentiable on a θ-interval I, then the graph of this polar equation is a curve C that is parametrized by the functions

$$x(\theta) = r\cos\theta = \rho(\theta)\cos\theta, \quad y(\theta) = r\sin\theta = \rho(\theta)\sin\theta, \quad \theta \in I.$$

Interpreting the parameter t as time measured, say, in seconds, we can think of a pair of parametric equations, $x = x(t)$ and $y = y(t)$, as describing the motion of a particle in the xy-plane. Different parametrizations of the same curve represent different ways of traversing that curve.

† The function f is continuously differentiable on I if f' is a continuous function on I.

Example 4 The line that passes through the points (1, 2) and (3, 6) has equation $y = 2x$. The line segment that joins these same points is given by

$$y = 2x, \quad 1 \leq x \leq 3.$$

We will parametrize this line segment in different ways and interpret each parametrization as the motion of a particle.

We begin by setting

$$x(t) = t, \quad y(t) = 2t, \quad t \in [1, 3].$$

At time $t = 1$, the particle is at the point $(1, 2)$. It traverses the line segment and arrives at the point (3, 6) at time $t = 3$.

Now we set

$$x(t) = t + 1, \quad y(t) = 2t + 2, \quad t \in (0, 2].$$

At time $t = 0$, the particle is at the point $(1, 2)$. It traverses the line segment and arrives at the point (3, 6) at time $t = 2$.

The equations

$$x(t) = 3 - t, \quad y(t) = 6 - 2t, \quad t \in [0, 2]$$

represent a traversal of that same line segment in the opposite direction. At time $t = 0$, the particle is at (3, 6). It arrives at (1, 2) at time $t = 2$.

Set

$$x(t) = 3 - 4t, \quad y(t) = 6 - 8t, \quad t \in [0, \tfrac{1}{2}].$$

Now the particle traverses the same line segment in only half a second. At time $t = 0$, the particle is at (3, 6). It arrives at (1, 2) at time $t = \frac{1}{2}$.

Finally we set

$$x(t) = 2 - \cos t, \quad y(t) = 4 - 2\cos t, \quad t \in [0, 4\pi].$$

In this instance the particle begins and ends its motion at the point (1,2), having traced and retraced the line segment twice during a span of 4π seconds. See Figure 9.6.5. □

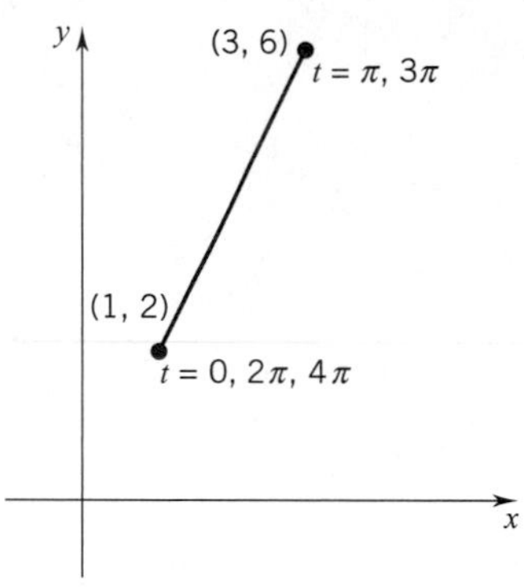

Figure 9.6.5

Remark If the path of an object is given in terms of a time parameter t and we eliminate the parameter to obtain an equation in x and y, it may be that we obtain a clearer view of the path, but we do so at considerable expense. The equation in x and y does not tell us where the particle is at any time t. The parametric equations do. □

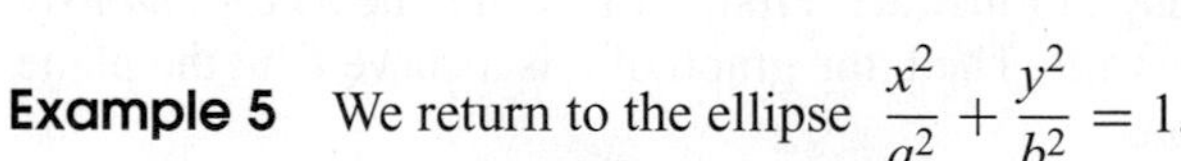

Example 5 We return to the ellipse $\dfrac{x^2}{a^2} + \dfrac{y^2}{b^2} = 1.$

A particle with position given by the equations

$$x(t) = a\cos t, \quad y(t) = b\sin t, \quad t \in [0, 2\pi]$$

traverses the ellipse in a counterclockwise manner. It begins at the point $(a, 0)$ and makes a full circuit in 2π seconds. If the equations of motion are

$$x(t) = a\cos 2\pi t, \quad y(t) = -b\sin 2\pi t, \quad t \in [0, 1],$$

the particle still travels the same ellipse, but in a different manner. Once again it starts at $(a, 0)$, but this time it moves clockwise and makes the full circuit in only 1 second. If the equations of motion are

$$x(t) = a\sin 4\pi t, \quad y(t) = b\cos 4\pi t, \quad t \in [0, \infty],$$

the motion begins at $(0, b)$ and goes on in perpetuity. The motion is clockwise, a complete circuit taking place every half second. □

Intersections and Collisions

Example 6 Two particles start at the same instant, the first along the linear path

$$x_1(t) = \tfrac{16}{3} - \tfrac{8}{3}t, \quad y_1(t) = 4t - 5, \quad t \geq 0$$

and the second along the elliptical path

$$x_2(t) = 2\sin\tfrac{1}{2}\pi t, \quad y_2(t) = -3\cos\tfrac{1}{2}\pi t, \quad t \geq 0.$$

(a) At what points, if any, do the paths intersect?

(b) At what points, if any, do the particles collide?

SOLUTION To see where the paths intersect, we find equations for them in x and y. The linear path can be written

$$3x + 2y - 6 = 0, \quad x \leq \tfrac{16}{3}$$

and the elliptical path

$$\frac{x^2}{4} + \frac{y^2}{9} = 1.$$

Solving the two equations simultaneously, we get

$$x = 2, \quad y = 0 \qquad \text{and} \qquad x = 0, \quad y = 3.$$

This means that the paths intersect at the points (2, 0) and (0, 3). This answers part (a).

Now for part (b). The first particle passes through (2, 0) only when

$$x_1(t) = \tfrac{16}{3} - \tfrac{8}{3}t = 2 \qquad \text{and} \qquad y_1(t) = 4t - 5 = 0.$$

As you can check, this happens only when $t = \frac{5}{4}$. When $t = \frac{5}{4}$, the second particle is elsewhere. Hence no collision takes place at (2, 0). There is, however, a collision at (0,3), because both particles get there at exactly the same time, $t = 2$:

$$x_1(2) = 0 = x_2(2), \qquad y_1(2) = 3 = y_2(2). \qquad \text{(See Figure 9.6.6.)} \ \square$$

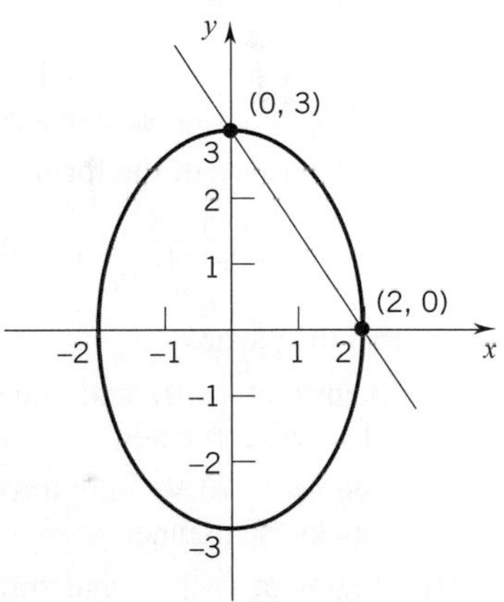

Figure 9.6.6

EXERCISES 9.6

Express the curve by an equation in x and y.

1. $x(t) = t^2, \quad y(t) = 2t + 1.$

2. $x(t) = 3t - 1, \quad y(t) = 5 - 2t.$

3. $x(t) = t^2, \quad y(t) = 4t^4 + 1.$

4. $x(t) = 2t - 1, \quad y(t) = 8t^3 - 5.$

5. $x(t) = 2\cos t, \quad y(t) = 3\sin t.$

6. $x(t) = \sec^2 t, \quad y(t) = 2 + \tan t.$

7. $x(t) = \tan t, \quad y(t) = \sec t.$

8. $x(t) = 2 - \sin t, \quad y(t) = \cos t.$

9. $x(t) = \sin t, \quad y(t) = 1 + \cos^2 t.$

10. $x(t) = e^t, \quad y(t) = 4 - e^{2t}.$

11. $x(t) = 4\sin t, \quad y(t) = 3 + 2\sin t.$

12. $x(t) = \csc t, \quad y(t) = \cot t.$

Express the curve by an equation in x and y; then sketch the curve.

13. $x(t) = e^{2t}, \quad y(t) = e^{2t} - 1, \quad t \leq 0.$

14. $x(t) = 3\cos t, \quad y(t) = 2 - \cos t, \quad 0 \leq t \leq \pi.$

15. $x(t) = \sin t, \quad y(t) = \csc t, \quad 0 < t \leq \frac{1}{4}\pi.$

16. $x(t) = 1/t, \quad y(t) = 1/t^2, \quad 0 < t < 3.$

17. $x(t) = 3 + 2t, \quad y(t) = 5 - 4t, \quad -1 \leq t \leq 2.$

18. $x(t) = \sec t, \quad y(t) = \tan t, \quad 0 \leq t \leq \frac{1}{4}\pi.$

19. $x(t) = \sin \pi t, \quad y(t) = 2t, \quad 0 \leq t \leq 4.$

20. $x(t) = 2\sin t, \quad y(t) = \cos t, \quad 0 \leq t \leq \frac{1}{2}\pi.$

21. $x(t) = \cot t, \quad y(t) = \csc t, \quad \frac{1}{4}\pi \leq t < \frac{1}{2}\pi.$

22. (*Important*) Parametrize: (a) the curve $y = f(x)$, $x \in [a, b]$; (b) the polar curve $r = f(\theta)$, $\theta \in [\alpha, \beta]$.

23. A particle with position given by the equations

$$x(t) = \sin 2\pi t, \quad y(t) = \cos 2\pi t, \quad t \in [0, 1].$$

starts at the point (0, 1) and traverses the unit circle $x^2 + y^2 = 1$ once in a clockwise manner. Write equations of the form

$$x(t) = f(t), \quad y(t) = g(t), \quad t \in [0, 1].$$

so that the particle

(a) begins at (0,1) and traverses the circle once in a counterclockwise manner;

(b) begins at (0,1) and traverses the circle twice in a clockwise manner;

(c) traverses the quarter circle from (1,0) to (0,1);

(d) traverses the three-quarter circle from (1,0) to (0,1).

24. A particle with position given by the equations

$$x(t) = 3\cos 2\pi t, \quad y(t) = 4\sin 2\pi t, \quad t \in [0, 1].$$

starts at the point (3, 0) and traverses the ellipse $16x^2 + 9y^2 = 144$ once in a counterclockwise manner. Write equations of the form

$$x(t) = f(t), \quad y(t) = g(t), \quad t \in [0, 1],$$

so that the particle

(a) begins at (3, 0) and traverses the ellipse once in a clockwise manner;

(b) begins at (0, 4) and traverses the ellipse once in a clockwise manner;

(c) begins at (−3, 0) and traverses the ellipse twice in a counterclockwise manner;

(d) traverses the upper half of the ellipse from (3, 0) to (0, 3).

25. Find a parametrization

$$x = x(t), \quad y = y(t), \quad t \in (-1, 1),$$

for the horizontal line $y = 2$.

26. Find a parametrization

$$x(t) = \sin f(t), \quad y(t) = \cos f(t), \quad t \in (0, 1),$$

which traces out the unit circle infinitely often.

Find a parametrization

$$x = x(t), \quad y = y(t), \quad t \in [0, 1].$$

for the given curve.

27. The line segment from (3, 7) to (8, 5).

28. The line segment from (2, 6) to (6, 3).

29. The parabolic arc $x = 1 - y^2$ from (0, −1), to (0, 1).

30. The parabolic arc $x = y^2$ from (4, 2) to (0, 0).

31. The curve $y^2 = x^3$ from (4, 8) to (1, 1).

32. The curve $y^3 = x^2$ from (1, 1) to (8, 4).

(*Important*) In Exercises 33–36, suppose that the curve

$$C: \quad x = x(t), \quad y = y(t), \quad t \in [c, d],$$

is the graph of a nonnegative function $y = f(x)$ over an interval $[a, b]$. Suppose that $x'(t)$ and $y(t)$ are continuous, $x(c) = a$ and $x(d) = b$.

33. (*The area under a parametrized curve*) Show that

(9.6.4)
$$\text{the area below } C = \int_c^d y(t)x'(t)\,dt.$$

HINT: Since C is the graph of f, we know that $y(t) = f(x(t))$.

34. (*The centroid of a region under a parametrized curve*) Show that, if the region under C has area A and centroid $(\bar{x}, \bar{y})$, then

(9.6.5)
$$\bar{x}A = \int_c^d x(t)y(t)x'(t)\,dt,$$
$$\bar{y}A = \int_c^d \tfrac{1}{2}[y(t)]^2 x'(t)\,dt.$$

35. (*The volume of the solid generated by revolving about a coordinate axis the region under a parametrized curve*) Show that

(9.6.6)
$$V_x = \int_c^d \pi[y(t)]^2 x'(t)\,dt,$$
$$V_y = \int_c^d 2\pi x(t)y(t)x'(t)\,dt.$$
↑ provided $x(c) \geq 0$

36. (*The centroid of the solid generated by revolving about a coordinate axis the region under a parametrized curve*) Show that

(9.6.7)
$$\bar{x}V_x = \int_c^d \pi x(t)[y(t)]^2 x'(t)\,dt,$$
$$\bar{y}V_y = \int_c^d \pi x(t)[y(t)]^2 x'(t)\,dt.$$
↑ provided $x(c) \geq 0$

37. Sketch the curve

$$x(t) = at, \quad y(t) = a(1 - \cos t), \quad a > 0, \quad t \in [0, 2\pi]$$

and find the area below it.

38. Determine the centroid of the region under the curve in Exercise 37.

39. Find the volume generated by revolving the region in Exercise 38 about: (a) the x-axis; (b) the y-axis.

40. Find the centroid of the solid generated by revolving the region of Exercise 38 about: (a) the x-axis; (b) the y-axis.

41. Give a parametrization for the upper half of the ellipse $b^2x^2 + a^2y^2 = a^2b^2$ that satisfies the requirements given for Exercises 33–36.

42. Use the parametrization you chose for Exercise 41 to find (a) the area of the region enclosed by the ellipse; (b) the centroid of the upper half of that region.

43. Two particles start at the same instant, the first along the ray

$$x(t) = 2t + 6, \quad y(t) = 5 - 4t, \quad t \geq 0,$$

and the second along the circular path

$$x(t) = 3 - 5\cos\pi t, \quad y(t) = 1 + 5\sin\pi t, \quad t \geq 0.$$

(a) At what points, if any, do these paths intersect?

(b) At what points, if any, will the particles collide?

44. Two particles start at the same instant, the first along the elliptical path

$$x_1(t) = 2 - 3\cos\pi t, \quad y_1(t) = 3 + 7\sin\pi t, \quad t \geq 0,$$

and the second along the parabolic path

$$x_2(t) = 3t + 2, \quad y_2(t) = -\tfrac{7}{15}(3t+1)^2 + \tfrac{157}{15}, \quad t \geq 0.$$

(a) At what points, if any, do these paths intersect?

(b) At what points, if any, will the particles collide?

We can determine the points where a parametrized curve

$$C: \quad x = x(t), \quad y = y(t), \quad t \in I$$

intersects itself by finding the numbers r and s in $I (r \neq s)$ for which

$$x(r) = x(s) \quad \text{and} \quad y(r) = y(s).$$

Use this method to find the point(s) of self-intersection for each of the curves in Exercises 45–48.

45. $x(t) = t^2 - 2t, \quad y(t) = t^3 - 3t^2 + 2t, \quad t$ real.

46. $x(t) = \cos t(1 - 2\sin t), \; y(t) = \sin t(1 - 2\sin t), \; t \in [0, \pi]$.

47. $x(t) = \sin 2\pi t, \quad y(t) = 2t - t^2, \quad t \in [0, 4]$.

48. $x(t) = t^3 - 4t, \quad y(t) = t^3 - 3t^2 + 2t, \quad t$ real.

C In Exercises 49 and 50, use a CAS to obtain an equation of the curve in x and y. Then use a graphing utility to draw the curve in both the parametric and the x, y-forms.

49. $x = \cosh t, \quad y = \sinh t$. **50.** $x = 2\cos t, \quad y = 5\sin t$.

C In Exercises 51–54, a particle moves along the curve described by the parametric equations $x = f(t), y = g(t)$. Use a graphing utility to draw the path of the particle and describe the motion of the particle as it moves along the curve.

51. $x = 2t, \quad y = 4t - t^2, \quad 0 \leq t \leq 6$.

52. $x = 3(t^2 - 3), \quad y = t^3 - 3t, \quad -3 \leq t \leq 3$.

53. $x = \cos(t^2 + t), \quad y = \sin(t^2 + t), \quad 0 \leq t \leq 2.1$.

54. $x = \cos(\ln t), \quad y = \sin(\ln t), \quad 1 \leq t \leq e^{2\pi}$.

C 55. As a wheel rolls along a straight line each point on the rim traces out a curve. Such a curve is called a *cycloid*. The cycloid depicted below can be parametrized as follows:

$$x(\theta) = a(\theta - \sin\theta),$$
$$y(\theta) = a(1 - \cos\theta).$$

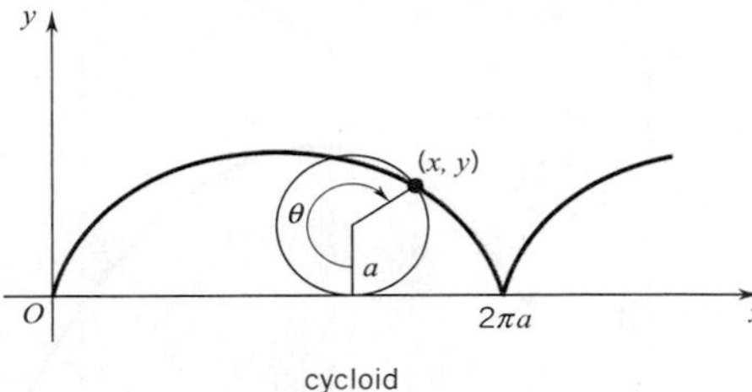

cycloid

(a) Use a graphing utility to draw the graph of the cycloid for $a = 1, 2$, and 4. What effect does the coefficient a have on the graph?

(b) The curve has a cusp at $\theta = 0, 2\pi$, and so on. Let $a = 2$. Use your graph to determine the slope of the curve for points near the cusp corresponding to $\theta = 2\pi$.

(c) Does the curve have a vertical tangent line at the cusp?

C 56. Use a graphing utility to draw the curve of

$$x(\theta) = \cos\theta\,(a - b\sin\theta),$$
$$y(\theta) = \sin\theta\,(a - b\sin\theta), \quad 0 \leq \theta \leq 2\pi,$$

for the following values of a and b:

(a) $a = 1, b = 2$

(b) $a = 2, b = 2$.

(c) $a = 2, b = 1$.

(d) In general, what can you say about the graphs when $a < b$, and $a > b$?

C Curves with parametric equations of the form $x = A\cos mt$, $y = B\sin nt$ are known as *Lissajous curves*. Use a graphing utility to draw the Lissajous curves in Exercises 57 and 58.

57. (a) $x = \cos 3t, \quad y = \sin kt$ for $k = 1, 2, 3, 4, 5$, $0 \leq t \leq 2\pi$.

(b) Let $x = \cos mt, \quad y = \sin nt$. Experiment with several combinations of m and n.

(c) Explain why the curves in parts (a) and (b) are confined to the square of side length 2 with center at the origin.

58. (a) $x = 3\cos 3t, \quad y = 2\sin kt$ for $k = 1, 2, 3, 4, 5$, $0 \leq t \leq 2\pi$.

(b) Let $x = A\cos 3t, \quad y = B\sin 4t$. Experiment with several combinations of A and B.

(c) Describe the regions that contain the curves in parts (a) and (b).

■ PROJECT 9.6 Parabolic Trajectories

In the early part of the seventeenth century Galileo Galilei observed the motion of stones projected from the tower of Pisa and concluded that their trajectory was parabolic. Using calculus, together with some simplifying assumptions, we obtain results that agree with Galileo's observations.

Consider a projectile fired at an angle $\theta, 0 < \theta < \pi/2$, from a point (x_0, y_0) with initial velocity v_0 (Figure A). The horizontal component of v_0 is $v_0 \cos\theta$, and the vertical component is $v_0 \sin\theta$ (Figure B.) Let $x = x(t), y = y(t)$ be parametric equations for the path of the projectile.

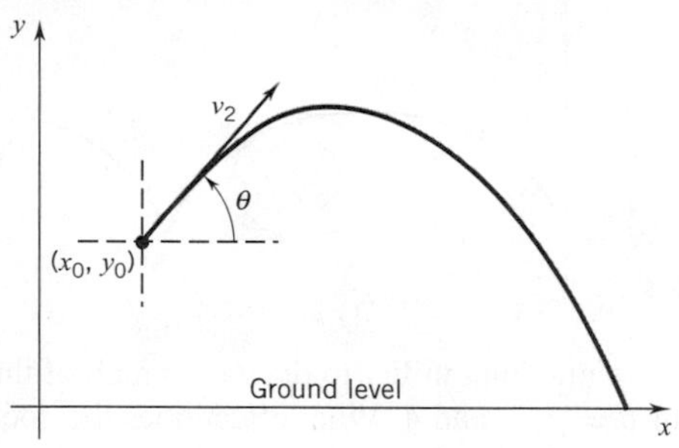

Figure A

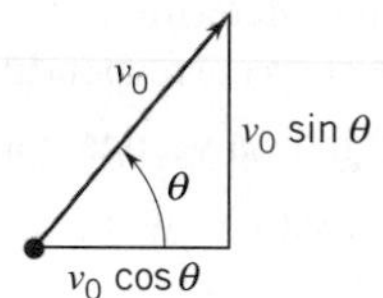

Figure B

We neglect air resistance and the curvature of the earth. Under these circumstances there is no horizontal acceleration; therefore

$$x''(t) = 0.$$

The only vertical acceleration is due to gravity; therefore

$$y''(t) = -g$$

Problem 1. Show that the path of the projectile (the trajectory) is given by:

$$x(t) = (v_0 \cos\theta)t - x_0, \quad y(t) = -\tfrac{1}{2}gt^2 + (v_0 \sin\theta)t + y_0.$$

Problem 2. Show that the rectangular equation of the trajectory can be written

$$x = -\frac{g}{2v_0^2}\sec^2\theta(x - x_0)^2 + \tan\theta(x - x_0) + y_0.$$

Problem 3. Measure distance in feet, time in seconds and set $g = 32\,(\text{ft}/\text{sec}^2)$. Take (x_0, y_0) as the origin and the x-axis as ground level. Consider a projectile fired at an angle θ with initial velocity v_0.

a. Give parametric and rectangular equations for the trajectory.
b. What is the maximum height attained by the projectile?
c. Find the range (horizontal distance) of the projectile.
d. How many seconds after firing does the impact take place?
e. How should θ be chosen to maximize the range?
f. How should θ be chosen so that the point of impact is at $x = b$?

Problem 4.

a. Use a graphing utility to draw the path of a projectile fired at an angle of 30° with an initial velocity of $v_0 = 1500$ ft/sec. Determine the height and range of the projectile and compare with the theoretical results of Problem 3.
b. Keeping $v_0 = 1500$ ft/sec, experiment with several values of θ. Confirm that $\theta = 45°$ maximizes the range. What angle maximizes the height?

■ 9.7 TANGENTS TO CURVES GIVEN PARAMETRICALLY

Let C be a curve parametrized by the functions

$$x = x(t), \qquad y = y(t)$$

where x and y are defined on some interval I. Since a curve can intersect itself, at any given point C can have

(i) one tangent, (ii) two or more tangents, or (iii) no tangent at all.

We illustrate these possibilities in Figure 9.7.1.

As before, we are assuming that $x'(t)$ and $y'(t)$ exist, at least on the interior of I. To make sure that at least one tangent line exists at each point of C, we will make the additional *assumption* that

(9.7.1) $$[x'(t)]^2 + [y'(t)]^2 \neq 0.$$

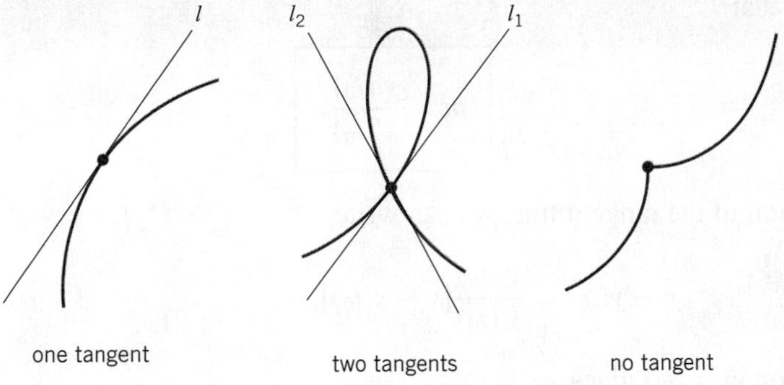

Figure 9.7.1

This assumption is equivalent to requiring that $x'(t)$ and $y'(t)$ are not simultaneously equal to 0. Without this assumption almost anything can happen. See Exercises 31–35.

Now choose a point (x_0, y_0) on the curve C and a time t_0 at which

$$x(t_0) = x_0 \quad \text{and} \quad y(t_0) = y_0.$$

We want the slope of the curve as it passes through the point (x_0, y_0) at time t_0.† To find this slope, we assume that $x'(t_0) \neq 0$. With $x'(t_0) \neq 0$, we can be sure that, for h sufficiently small, $h \neq 0$,

$$x(t_0 + h) - x(t_0) \neq 0. \qquad \text{(explain)}$$

For such h we can form the quotient

$$\frac{y(t_0 + h) - y(t_0)}{x(t_0 + h) - x(t_0)}.$$

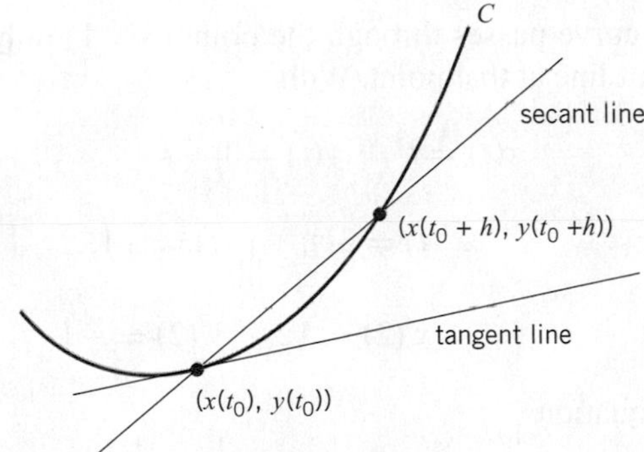

Figure 9.7.2

This quotient is the slope of the secant line pictured in Figure 9.7.2. The limit of this quotient as h tends to zero is the slope of the tangent line and thus the slope of the curve. Since

$$\frac{y(t_0 + h) - y(t_0)}{x(t_0 + h) - x(t_0)} = \frac{(1/h)[y(t_0 + h) - y(t_0)]}{(1/h)[x(t_0 + h) - x(t_0)]} \to \frac{y'(t_0)}{x'(t_0)} \quad \text{as } h \to 0,$$

† It could pass through the point (x_0, y_0) at other times also.

you can see that

(9.7.2) $$m = \frac{y'(t_0)}{x'(t_0)}.$$

As an equation of the tangent line, we can write

$$y - y(t_0) = \frac{y'(t_0)}{x'(t_0)}[x - x(t_0)]. \qquad \text{(point-slope form)}$$

Multiplication by $x'(t_0)$ gives

$$y'(t_0)[x - x(t_0)] - x'(t_0)[y - y(t_0)] = 0,$$

and thus

(9.7.3) $$y'(t_0)[x - x_0] - x'(t_0)[y - y_0] = 0.$$

We derived this equation under the assumption that $x'(t_0) \neq 0$. If $x'(t_0) = 0$, Equation (9.7.3) still makes sense. It is simply $y'(t_0)[x - x_0] = 0$, which, since $y'(t_0) \neq 0$,† can be simplified to read

(9.7.4) $$x = x_0.$$

In this instance the line is vertical, and we say that the curve has a *vertical tangent.*

Example 1 Find an equation for each tangent to the curve

$$x(t) = t^3, \quad y(t) = 1 - t, \quad t \in (-\infty, \infty)$$

at the point $(8, -1)$.

SOLUTION Since the curve passes through the point $(8, -1)$ only when $t = 2$, there can be only one tangent line at that point. With

$$x(t) = t^3, \quad y(t) = 1 - t,$$

we have

$$x'(t) = 3t^2, \quad y'(t) = -1,$$

and therefore

$$x'(2) = 12, \quad y'(2) = -1.$$

The tangent line has equation

$$(-1)[x - 8] - (12)\ [y - (-1)] = 0. \qquad \text{[by (9.7.3)]}$$

This reduces to

$$x + 12y + 4 = 0. \quad \square$$

Example 2 Find the points on the curve

$$x(t) = 3 - 4\sin t, \qquad y(t) = 4 + 3\cos t, \qquad t \in (-\infty, \infty)$$

at which there is (i) a horizontal tangent, (ii) a vertical tangent.

† We are assuming that $[x'(t)]^2 + [y'(t)]^2$ is never 0. Since $x'(t_0) = 0, y'(t_0) \neq 0$.

SOLUTION Observe first of all that the derivatives

$$x'(t) = -4\cos t \quad \text{and} \quad y'(t) = -3\sin t$$

are never 0 simultaneously.

To find the points at which there is a horizontal tangent, we set $y'(t) = 0$. This gives $t = n\pi, n = 0, \pm 1, \pm 2, \ldots$. Horizontal tangents occur at all points of the form $(x(n\pi), y(n\pi))$. Since

$$x(n\pi) = 3 - 4\sin n\pi = 3 \quad \text{and} \quad y(n\pi) = 4 + 3\cos n\pi = \begin{cases} 7, & n \text{ even} \\ 1, & n \text{ odd,} \end{cases}$$

there are horizontal tangents at (3, 7) and (3, 1).

To find the vertical tangents, we set $x'(t) = 0$. This gives $t = \frac{1}{2}\pi + n\pi$, $n = 0, \pm 1, \pm 2, \ldots$. Vertical tangents occur at all points of the form $(x(\frac{1}{2}\pi + n\pi), y(\frac{1}{2}\pi + n\pi))$. Since

$$x(\tfrac{1}{2}\pi + n\pi) = 3 - 4\sin(\tfrac{1}{2}\pi + n\pi) = \begin{cases} -1, & n \text{ even} \\ 7, & n \text{ odd} \end{cases}$$

and

$$y(\tfrac{1}{2}\pi + n\pi) = 4 + 3\cos(\tfrac{1}{2}\pi + n\pi) = 4,$$

there are vertical tangents at $(-1, 4)$ and $(7, 4)$. □

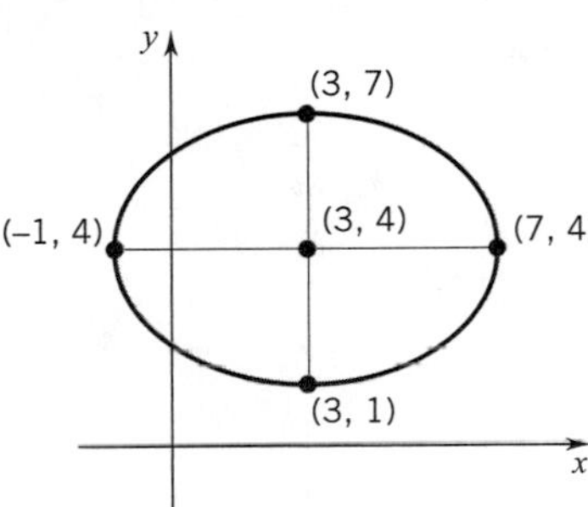

Figure 9.7.3

Remark We leave it to you to verify that the functions of Example 2 parametrize the ellipse (Figure 9.7.3)

$$\frac{(x-3)^2}{16} + \frac{(y-4)^2}{9} = 1 \quad \square$$

Example 3 The curve parametrized by the functions

$$x(t) = \frac{1-t^2}{1+t^2}, \quad y(t) = \frac{t(1-t^2)}{1+t^2}, \quad t \in (-\infty, \infty)$$

is called a *strophoid*. The curve is shown in Figure 9.7.4. Find equations for the lines tangent to the curve at the origin. Then find the points at which there is a horizontal tangent.

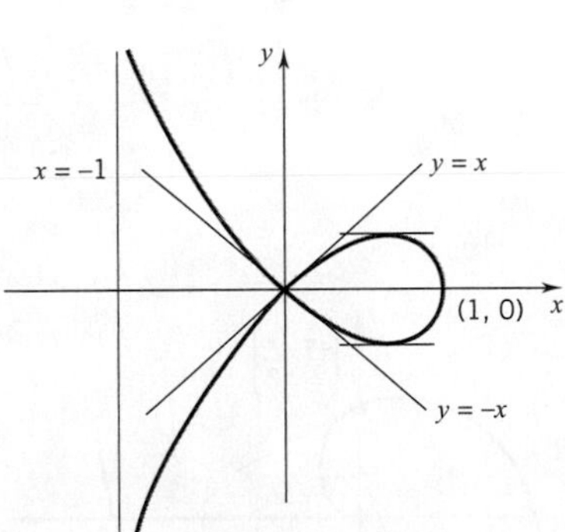

Figure 9.7.4

SOLUTION The curve passes through the origin when $t = -1$ and when $t = 1$ (verify this). Differentiating $x(t)$ and $y(t)$, we have

$$x'(t) = \frac{(1+t^2)(-2t) - (1-t^2)(2t)}{(1+t^2)^2} = \frac{-4t}{(1+t^2)^2}$$

and

$$y'(t) = \frac{(1+t^2)(1-3t^2) - t(1-t^2)(2t)}{(1+t^2)^2} = \frac{1-4t^2-t^4}{(1+t^2)^2}.$$

At time $t = -1$, the curve passes through the origin with slope

$$\frac{y'(-1)}{x'(-1)} = \frac{-1}{1} = -1.$$

Therefore the tangent line has equation $y = -x$.

At time $t = 1$, the curve passes through the origin with slope

$$\frac{y'(1)}{x'(1)} = \frac{-1}{-1} = 1.$$

Therefore the tangent line has equation $y = x$.

To find the points at which there is a horizontal tangent, we set $y'(t) = 0$. This implies that

$$1 - 4t^2 - t^4 = 0.$$

This equation is a quadratic in t^2. By the quadratic formula,

$$t^2 = \frac{-4 \pm \sqrt{16+4}}{2} = -2 \pm \sqrt{5}.$$

Since $t^2 \geq 0$, it follows that $t^2 = \sqrt{5} - 2$ and $t = \pm\sqrt{\sqrt{5} - 2}$. Therefore,

$$x(\pm\sqrt{\sqrt{5}-2}) = \frac{1 - (\sqrt{5} - 2)}{1 + (\sqrt{5} - 2)} = \frac{\sqrt{5} - 1}{2} \cong 0.62$$

and $\quad y(\pm\sqrt{\sqrt{5}-2}) = (\pm\sqrt{\sqrt{5}-2})x = (\pm\sqrt{\sqrt{5}-2})\left(\frac{\sqrt{5}-1}{2}\right) \cong \pm 0.30.$

There is a horizontal tangent line at the points $(0.62, \pm 0.30)$ (approximately). ❑

We can apply these ideas to a curve given in polar coordinates by an equation of the form $r = \rho(\theta)$. The coordinate transformations

$$x = r\cos\theta, \quad y = r\sin\theta$$

enable us to parametrize such a curve by setting

$$x(\theta) = \rho(\theta)\,\cos\theta, \quad y(\theta) = \rho(\theta)\,\sin\theta.$$

Example 4 Take $a > 0$. Find the slope of the spiral $\; r = a\theta, \quad \theta \in [0, \infty)$ at $\theta = \frac{1}{2}\pi$.

SOLUTION We write

$$x(\theta) = r\cos\theta = a\theta\cos\theta, \quad y(\theta) = r\sin\theta = a\theta\sin\theta.$$

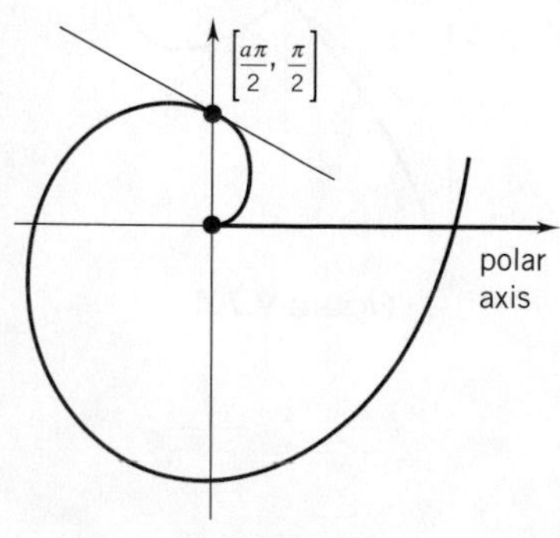

Figure 9.7.5

Now we differentiate:

$$x'(\theta) = -a\theta\sin\theta + a\cos\theta, \quad y'(\theta) = a\theta\cos\theta + a\sin\theta.$$

Since $\quad x'(\frac{1}{2}\pi) = -\frac{1}{2}\pi a \quad$ and $\quad y'(\frac{1}{2}\pi) = a,$

the slope of the curve at $\theta = \frac{1}{2}\pi$ is

$$\frac{y'(\frac{1}{2}\pi)}{x'(\frac{1}{2}\pi)} = -\frac{2}{\pi} \cong -0.64.$$

See Figure 9.7.5. ❑

Example 5 Find the points of the cardioid $r = 1 - \cos\theta$ at which the tangent line is vertical.

SOLUTION Since the cosine function has period 2π, we need only concern ourselves with θ in $[0, 2\pi)$. The curve can be parametrized by setting

$$x(\theta) = (1 - \cos\theta)\cos\theta, \quad y(\theta) = (1 - \cos\theta)\sin\theta.$$

Differentiating and simplifying, we find that

$$x'(\theta) = (2\cos\theta - 1)\sin\theta, \quad y'(\theta) = (1 - \cos\theta)(1 + 2\cos\theta).$$

The only numbers in the interval $[0, 2\pi)$ at which x' is zero and y' is not zero are $\frac{1}{3}\pi, \pi$, and $\frac{5}{3}\pi$. The tangent line is vertical at

$$[\tfrac{1}{2}, \tfrac{1}{3}\pi], \quad [2, \pi], \quad [\tfrac{1}{2}, \tfrac{5}{3}\pi].$$

These points have rectangular coordinates $(\frac{1}{4}, \frac{1}{4}\sqrt{3}), (-2, 0), (\frac{1}{4}, -\frac{1}{4}\sqrt{3})$. See Figure 9.7.6 □

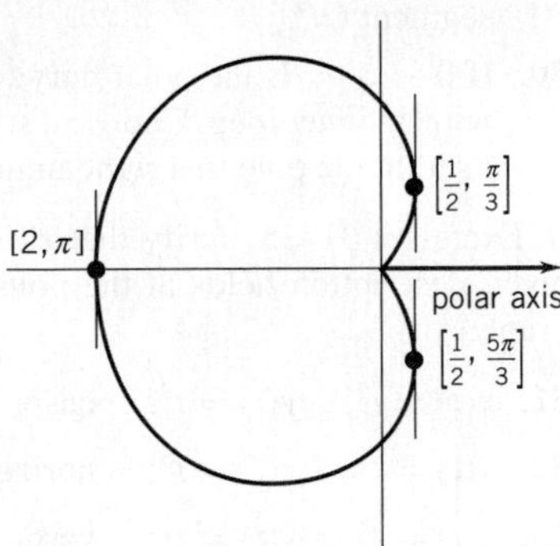

Figure 9.7.6

EXERCISES 9.7

Find an equation in x and y for the line tangent to the curve.

1. $x(t) = t, \quad y(t) = t^3 - 1; \quad t = 1.$

2. $x(t) = t^2, \quad y(t) = t + 5; \quad t = 2.$

3. $x(t) = 2t, \quad y(t) = \cos\pi t; \quad t = 0.$

4. $x(t) = 2t - 1, \quad y(t) = t^4; \quad t = 1.$

5. $x(t) = t^2, \quad y(t) = (2 - t)^2; \quad t = \frac{1}{2}.$

6. $x(t) = 1/t, \quad y(t) = t^2 + 1; \quad t = 1.$

7. $x(t) = \cos^3 t, \quad y(t) = \sin^3 t \quad t = \frac{1}{4}\pi.$

8. $x(t) = e^t, \quad y(t) = 3e^{-t}; \quad t = 0.$

Find an equation in x and y for the line tangent to the polar curve.

9. $r = 4 - 2\sin\theta, \quad \theta = 0.$

10. $r = 4\cos 2\theta, \quad \theta = \frac{1}{2}\pi.$

11. $r = \dfrac{4}{5 - \cos\theta}, \quad \theta = \frac{1}{2}\pi.$

12. $r = \dfrac{5}{4 - \cos\theta}, \quad \theta = \frac{1}{6}\pi.$

13. $r = \dfrac{\sin\theta - \cos\theta}{\sin\theta + \cos\theta}, \quad \theta = 0.$

14. $r = \dfrac{\sin\theta + \cos\theta}{\sin\theta - \cos\theta}, \quad \theta = \frac{1}{2}\pi.$

Parametrize the curve by a pair of differentiable functions

$$x = x(t), \quad y = y(t) \qquad \text{with} \qquad [x'(t)]^2 + [y'(t)]^2 \neq 0.$$

Sketch the curve and determine the tangent line at the origin by the method of this section.

15. $y = x^3.$ **16.** $x = y^3.$

17. $y^5 = x^3.$ **18.** $y^3 = x^5.$

Find the points (x, y) at which the curve has: (a) a horizontal tangent; (b) a vertical tangent. Then sketch the curve.

19. $x(t) = 3t - t^3, \quad y(t) = t + 1.$

20. $x(t) = t^2 - 2t, \quad y(t) = t^3 - 12t.$

21. $x(t) = 3 - 4\sin t, \quad y(t) = 4 + 3\cos t.$

22. $x(t) = \sin 2t, \quad y(t) = \sin t.$

23. $x(t) = t^2 - 2t, \quad y(t) = t^3 - 3t^2 + 2t.$

24. $x(t) = 2 - 5\cos t, \quad y(t) = 3 + \sin t.$

25. $x(t) = \cos t, \quad y(t) = \sin 2t.$

26. $x(t) = 3 + 2\sin t, \quad y(t) = 2 + 5\sin t.$

27. Find the tangent(s) to the curve

$$x(t) = -t + 2\cos\tfrac{1}{4}\pi t, \quad y(t) = t^4 - 4t^2$$

at the point $(2, 0)$.

28. Find the tangent(s) to the curve

$$x(t) = t^3 - t, \quad y(t) = t\sin\tfrac{1}{2}\pi t$$

at the point $(0, 1)$.

29. Let $P = [r_1, \theta]$ be a point on a polar curve $r = f(\theta)$ as in the figure. Show that, if $f'(\theta_1) = 0$ but $f(\theta_1) \neq 0$,

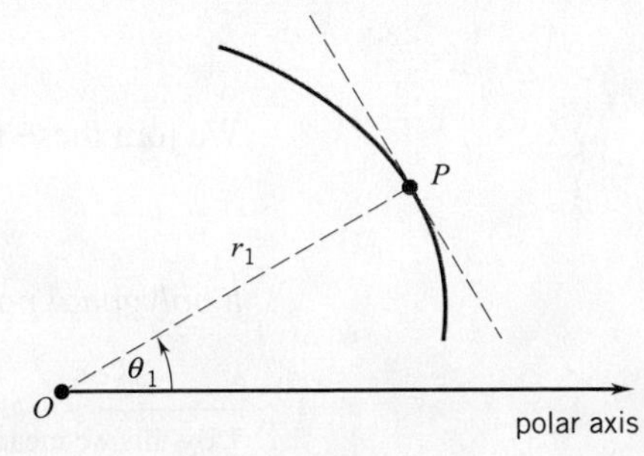

then the tangent line at P is perpendicular to the line segment $\overline{OP}$.

30. If $0 < a < 1$, the polar curve $r = a - \cos\theta$ is a limaçon with an inner loop. Choose a so that the curve will intersect itself at the pole in a right angle.

In Exercises 31–35, verify that $x'(0) = y'(0) = 0$ and that the given description holds at the point where $t = 0$. Sketch the graph.

31. $x(t) = t^3, \quad y(t) = t^2;$ cusp.

32. $x(t) = t^3, \quad y(t) = t^5;$ horizontal tangent.

33. $x(t) = t^5, \quad y(t) = t^3;$ vertical tangent.

34. $x(t) = t^3 - 1, \quad y(t) = 2t^3;$ tangent with slope 2.

35. $x(t) = t^2, \quad y(t) = t^2 + 1;$ no tangent line.

36. Suppose that $x = x(t)$, $y = y(t)$ are twice differentiable functions that parametrize a curve. Take a point on the curve at which $x'(t) \neq 0$ and d^2y/dx^2 exists. Show that

(9.7.5)
$$\frac{d^2y}{dx^2} = \frac{x'(t)y''(t) - y'(t)x''(t)}{[x'(t)]^3}$$

Calculate d^2y/dx^2 at the indicated point without eliminating the parameter.

37. $x(t) = \cos t, \quad y(t) = \sin t; \quad t = \frac{1}{6}\pi.$

38. $x(t) = t^3, \quad y(t) = t - 2; \quad t = 1.$

39. $x(t) = e^t, \quad y(t) = e^{-t}; \quad t = 0.$

40. $x(t) = \sin^2 t, \quad y(t) = \cos t; \quad t = \frac{1}{4}\pi.$

C **41.** Let $x = 2 + \cos t$, $y = 2 - \sin t$. Use a CAS to find d^2y/dx^2.

C **42.** Use a CAS to find an equation in x and y for the line tangent to the curve:

$$x = \sin^2 t, \quad y = \cos^2 t; \quad t = \tfrac{1}{4}\pi.$$

Then use a graphing utility to draw the curve and the tangent line together.

C **43.** Repeat Exercise 42 with $x = e^{-3t}, \quad y = e^t; \quad t = \ln 2.$

C **44.** Use a CAS to find an equation in x and y for the line tangent to the polar curve:

$$r = \frac{4}{2 + \sin\theta}; \quad \theta = \tfrac{1}{3}\pi.$$

Then use a graphing utility to draw the curve and the tangent line together.

■ 9.8 ARC LENGTH AND SPEED

Arc Length

In Figure 9.8.1 we have sketched the path C traced out by a pair of *continuously differentiable* functions†

$$x = x(t), \quad y = y(t), \quad t \in [a, b].$$

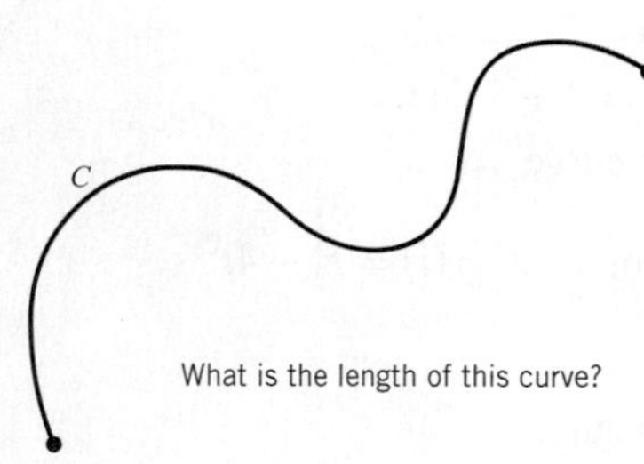

Figure 9.8.1

We want to determine the length of C.

Here our experience in Chapter 5 can be used as a model. To decide what should be meant by the area of a region Ω, we approximated Ω by the union of a finite number of rectangles. To decide what should be meant by the length of C, we approximate C by the union of a finite number of line segments.

Each number t in $[a, b]$ gives rise to a point $P = P(x(t), y(t))$ that lies on C. By choosing a finite number of points in $[a, b]$,

$$a = t_0 < t_1 < \cdots < t_{i-1} < t_i < \cdots < t_{n-1} < t_n = b,$$

we obtain a finite number of points on C,

$$P_0, P_1, \ldots, P_{i-1}, P_i, \ldots, P_{n-1}, P_n.$$

We join these points consecutively by line segments and call the resulting path,

$$\gamma = \overline{P_0P_1} \cup \cdots \cup \overline{P_{i-1}P_i} \cup \cdots \cup \overline{P_{n-1}P_n},$$

a *polygonal path* inscribed in C. (See Figure 9.8.2.)

† By this we mean functions that have continuous first derivatives.

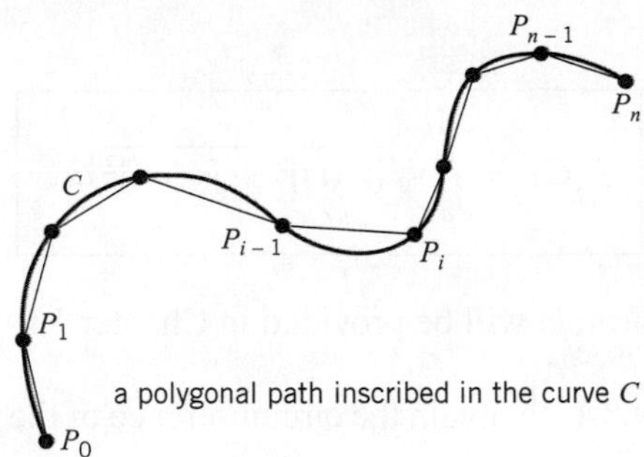

Figure 9.8.2

The length of such a polygonal path is the sum of the distances between consecutive vertices:

$$\text{length of } \gamma = L(\gamma) = d(P_0, P_1) + \cdots + d(P_{i-1}, P_i) + \cdots + d(P_{n-1}, P_n).$$

The ith line segment $\overline{P_{i-1}P_i}$ has length

$$\begin{aligned} d(P_{i-1}, P_i) &= \sqrt{[x(t_i) - x(t_{i-1})]^2 + [y(t_i) - y(t_{i-1})]^2} \\ &= \sqrt{\left[\frac{x(t_i) - x(t_{i-1})}{t_i - t_{i-1}}\right]^2 + \left[\frac{y(t_i) - y(t_{i-1})}{t_i - t_{i-1}}\right]^2}\,(t_i - t_{i-1}). \end{aligned}$$

By the mean-value theorem, there exist points t_i^* and t_i^{**}, both in the interval (t_{i-1}, t_i), such that

$$\frac{x(t_i) - x(t_{i-1})}{t_i - t_{i-1}} = x'(t_i^*) \quad \text{and} \quad \frac{y(t_i) - y(t_{i-1})}{t_i - t_{i-1}} = y'(t_i^{**}).$$

Letting $\Delta t_i = t_i - t_{i-t}$, we have

$$d(P_{i-1}, P_i) = \sqrt{[x'(t_i^*)]^2 + [y'(t_i^{**})]^2}\,\Delta t_i.$$

Adding up these terms, we obtain

$$\begin{aligned} L(\gamma) = \sqrt{[x'(t_1^*)]^2 + [y'(t_1^{**})]^2}\,\Delta t_1 + \sqrt{[x'(t_2^*)]^2 + [y'(t_2^{**})]^2}\,\Delta t_2 + \\ \cdots + \sqrt{[x'(t_n^*)]^2 + [y'(t_n^{**})]^2}\,\Delta t_n. \end{aligned}$$

As written, $L(\gamma)$ is not a Riemann sum: in general, $t_i^* \neq t_i^{**}$. It is nevertheless true (and at the moment we ask you to take on faith) that, as $\max \Delta t_i \to 0$, $L(\gamma)$ approaches the integral

$$\int_a^b \sqrt{[x'(t)]^2 + [y'(t)]^2}\,dt.$$

Arc Length Formulas

By the intuitive argument just given, we have obtained a way to calculate arc length. The length of the path C traced out by a pair of continuously differentiable functions

$$x = x(t), \quad y = y(t), \quad t \in [a, b]$$

is given by the formula

(9.8.1)
$$L(C) = \int_a^b \sqrt{[x'(t)]^2 + [y'(t)]^2}\, dt.$$

More insight into this formula will be provided in Chapter 13.

Let's use Formula (9.8.1) to obtain the circumference of the unit circle. Parametrizing the unit circle by

$$x(t) = \cos t, \qquad y(t) = \sin t, \qquad t \in [0, 2\pi],$$

we have
$$x'(t) = -\sin t, \qquad y'(t) = \cos t,$$

and thus

$$\text{circumference} = \int_0^{2\pi} \sqrt{\sin^2 t + \cos^2 t}\, dt = \int_0^{2\pi} 1\, dt = 2\pi.$$

Nothing surprising here. But, suppose we parametrize the unit circle by setting

$$x(t) = \cos 2t, \quad y(t) = \sin 2t, \quad t \in [0, 2\pi].$$

Then we have

$$x'(t) = -2\sin 2t, \quad y'(t) = 2\cos 2t,$$

and the arc length formula gives

$$L(C) = \int_0^{2\pi} \sqrt{4\sin^2 2t + 4\cos^2 2t}\, dt = \int_0^{2\pi} 2\, dt = 4\pi.$$

This is not the circumference of the unit circle. What's wrong here? There is nothing wrong here. Formula (9.8.1) gives the length of the path traced out by the parametrizing functions. The functions $x(t) = \cos 2t$, $y(t) = \sin 2t$, $t \in [0, 2\pi]$ trace out the unit circle not once, but twice. Hence the discrepancy.

When applying Formula (9.8.1) to calculate the arc length of a curve given to us geometrically, we must make sure that the functions that we use to parametrize the curve trace out each arc of the curve only once.

Suppose now that C is the graph of a continuously differentiable function

$$y = f(x), \quad x \in [a, b].$$

We can parametrize C by setting

$$x(t) = t, \quad y(t) = f(t), \quad t \in [a, b].$$

Since
$$x'(t) = 1 \quad \text{and} \quad y'(t) = f'(t),$$

Formula (9.8.1) gives

$$L(C) = \int_a^b \sqrt{1 + [f'(t)]^2}\, dt.$$

Replacing t by x we can write

(9.8.2) $$\text{the length of the graph of } f = \int_a^b \sqrt{1+[f'(x)]^2}\,dx.$$

A direct derivation of this formula is outlined in Exercise 52.

Example 1 If $f(x) = \frac{1}{6}x^3 + \frac{1}{2}x^{-1}$, then

$$f'(x) = \tfrac{1}{2}x^2 - \tfrac{1}{2}x^{-2}.$$

Therefore

$$1+[f'(x)]^2 = 1 + (\tfrac{1}{4}x^4 - \tfrac{1}{2} + \tfrac{1}{4}x^{-4}) = \tfrac{1}{4}x^4 + \tfrac{1}{2} + \tfrac{1}{4}x^{-4} = (\tfrac{1}{2}x^2 + \tfrac{1}{2}x^{-2})^2.$$

The length of the graph from $x = 1$ to $x = 3$ is

$$\int_1^3 \sqrt{1+[f'(x)]^2}\,dx = \int_1^3 \sqrt{(\tfrac{1}{2}x^2 + \tfrac{1}{2}x^{-2})^2}\,dx$$
$$= \int_1^3 (\tfrac{1}{2}x^2 + \tfrac{1}{2}x^{-2})\,dx = \left[\tfrac{1}{6}x^3 - \tfrac{1}{2}x^{-1}\right]_1^3 = \tfrac{14}{3}. \quad \square$$

Example 2 The graph of the function $f(x) = x^2$, $x \in [0, 1]$, is a parabolic arc. The length of this arc is given by

$$\int_0^1 \sqrt{1+[f'(x)]^2}\,dx = \int_0^1 \sqrt{1+4x^2}\,dx = 2\int_0^1 \sqrt{(\tfrac{1}{2})^2 + x^2}\,dx$$
$$= \left[x\sqrt{(\tfrac{1}{2})^2 + x^2} + (\tfrac{1}{2})^2 \ln\left(x + \sqrt{(\tfrac{1}{2})^2 + x^2}\right)\right]_0^1$$

by (8.4.1)

$$= \tfrac{1}{2}\sqrt{5} + \tfrac{1}{4}\ln(2+\sqrt{5}) \cong 1.48. \quad \square$$

Suppose now that C is the graph of a polar function

$$r = \rho(\theta), \quad \alpha \le \theta \le \beta,$$

where ρ is continuously differentiable. We can parametrize C by setting

$$x(\theta) = \rho(\theta)\cos\theta, \quad y(\theta) = \rho(\theta)\sin\theta, \quad \theta \in [\alpha, \beta].$$

A straightforward calculation that we leave to you shows that

$$[x'(\theta)]^2 + [y'(\theta)]^2 = [\rho(\theta)]^2 + [\rho'(\theta)]^2.$$

The arc length formula now reads

(9.8.3) $$L(C) = \int_\alpha^\beta \sqrt{[\rho(\theta)]^2 + [\rho'(\theta)]^2}\,d\theta.$$

Example 3 For fixed $a > 0$, the equation $r = a$ represents a circle of radius a. Here

$$\rho(\theta) = a \qquad \text{and} \qquad \rho'(\theta) = 0.$$

The circle is traced out once as θ ranges from 0 to 2π. Therefore the length of the curve (the circumference of the circle) is given by

$$\int_0^{2\pi} \sqrt{[\rho(\theta)]^2 + [\rho'(\theta)^2]}\, d\theta = \int_0^{2\pi} \sqrt{a^2 + 0^2}\, d\theta = \int_0^{2\pi} a\, d\theta = 2\pi a. \quad \square$$

Example 4 We calculate the arc length of the cardioid $r = a(1 - \cos\theta)$, a a positive constant. To make sure that no arc of the curve is traced out more than once, we restrict θ to the interval $[0, 2\pi]$. Here

$$\rho(\theta) = a(1 - \cos\theta) \qquad \text{and} \qquad \rho'(\theta) = a\sin\theta,$$

so that

$$[\rho(\theta)]^2 + [\rho'(\theta)]^2 = a^2(1 - 2\cos\theta + \cos^2\theta) + a^2\sin^2\theta = 2a^2(1 - \cos\theta).$$

The identity

$$\tfrac{1}{2}(1 - \cos\theta) = \sin^2 \tfrac{1}{2}\theta$$

gives

$$[\rho(\theta)]^2 + [\rho'(\theta)]^2 = 4a^2 \sin^2 \tfrac{1}{2}\theta.$$

The length of the cardioid is $8a$:

$$\int_0^{2\pi} \sqrt{[\rho(\theta)]^2 + [\rho'(\theta)^2]}\, d\theta \underset{\uparrow}{=} \int_0^{2\pi} 2a\sin\tfrac{1}{2}\theta\, d\theta = 4a\Big[-\cos\tfrac{1}{2}\theta\Big]_0^{2\pi} = 8a. \quad \square$$

$\sin\frac{1}{2}\theta \geq 0$ for $\theta \in [0, 2\pi]$

The Geometric Significance of dx/ds and dy/ds

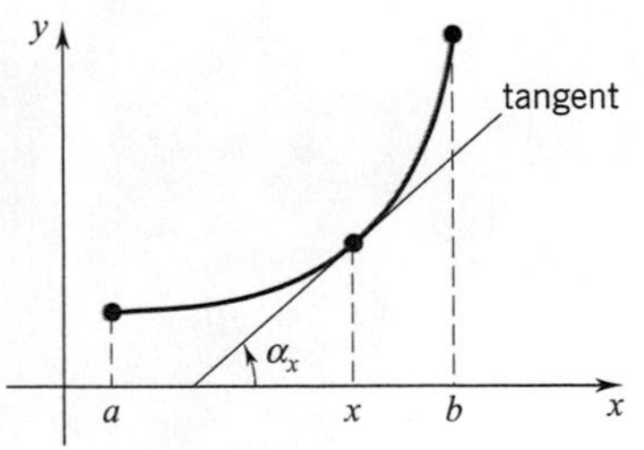

Figure 9.8.3

Figure 9.8.3 shows the graph of a function $y = f(x)$ which we assume to be continuously differentiable on (a, b). At the point (x, y) the tangent line has an inclination marked α_x.

The length of the arc on the graph from a to x can be written

$$s(x) = \int_a^x \sqrt{1 + [f'(t)]^2}\, dt.$$

Differentiation with respect to x gives $s'(x) = \sqrt{1 + [f'(x)]^2}$ (Theorem 5.3.5). Using the Leibniz notation we have

$$\frac{ds}{dx} = \sqrt{1 + \left(\frac{dy}{dx}\right)^2} = \sqrt{1 + \tan^2\alpha_x} \underset{\uparrow}{=} \sec\alpha_x.$$

$\sec\alpha_x > 0$ for $\alpha_x \in (-\frac{1}{2}\pi, \frac{1}{2}\pi)$

By (7.1.9)

$$\frac{dx}{ds} = \frac{1}{\sec\alpha_x} = \cos\alpha_x.$$

To find dy/ds we note that

$$\tan\alpha_x = \frac{dy}{dx} = \frac{dy}{ds}\frac{ds}{dx} \underset{\uparrow}{=} \frac{dy}{ds}\sec\alpha_x.$$

chain rule

Multiplication by $\cos \alpha_x$ gives

$$\frac{dy}{ds} = \sin \alpha_x.$$

For the record,

(9.8.4) $$\frac{dx}{ds} = \cos \alpha_x \quad \text{and} \quad \frac{dy}{ds} = \sin \alpha_x \quad \text{where } \alpha_x \text{ is the inclination of the tangent line at the point } (x, y).$$

Speed Along a Plane Curve

So far we have talked about speed only in connection with straight-line motion. How can we calculate the speed of an object that moves along a curve? Imagine an object moving along some curved path. Suppose that $(x(t), y(t))$ gives the position of the object at time t. The distance traveled by the object from time zero to any later time t is simply the length of the path up to time t:

$$s(t) = \int_0^t \sqrt{[x'(u)]^2 + [y'(u)]^2}\, du.$$

The time rate of change of this distance is what we call the *speed* of the object. Denoting the speed of the object at time t by $v(t)$ we have

(9.8.5) $$v(t) = s'(t) = \sqrt{[x'(t)]^2 + [y'(t)]^2}.$$

Example 5 The position of a particle at time t is given by the parametric equations:

$$x(t) = 3\cos 2t, \quad y(t) = 4\sin 2t, \quad t \in [0, 2\pi].$$

Find the speed of the particle at time t and determine when the speed is a maximum and when it is a minimum.

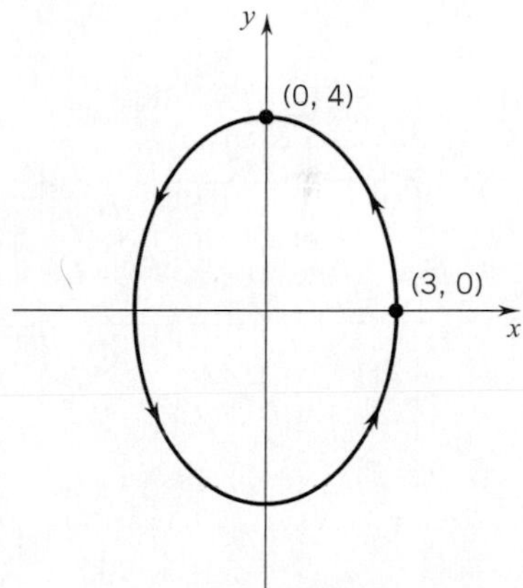

Figure 9.8.4

SOLUTION The path of the particle is the ellipse

$$\frac{x^2}{9} + \frac{y^2}{16} = 1, \qquad \text{(Figure 9.8.4)}$$

and the particle moves around the curve in the counterclockwise direction, as indicated by the arrows. The speed of the particle at time t is given by

$$v(t) = \sqrt{[x'(t)]^2 + [y'(t)]^2} = \sqrt{(-6\sin 2t)^2 + (8\cos 2t)^2} = \sqrt{36\sin^2 2t + 64\cos^2 2t}$$

$$= \sqrt{36 + 28\cos^2 2t}. \qquad (\sin^2 2t = 1 - \cos^2 2t)$$

The maximum speed is 8, and this occurs when $\cos^2 2t = 1$; that is, when $t = 0, \pi/2, \pi, 3\pi/2, 2\pi$. At these times, $\sin^2 2t = 0$ and the particle is at either end of the minor axis. The minimum speed is 6, which occurs when $\cos^2 2t = 0$ (and $\sin^2 2t = 1$), that is, when $t = \pi/4, 3\pi/4, 5\pi/4, 7\pi/4$. At these times, the particle is at either end of the major axis. ☐

In the Leibniz notation the equation for speed reads

(9.8.6) $$v = \frac{ds}{dt} = \sqrt{\left(\frac{dx}{dt}\right)^2 + \left(\frac{dy}{dt}\right)^2}.$$

If we know the speed of an object and we know its mass, then we can calculate its kinetic energy.

Example 6 A particle of mass m slides down a frictionless curve (see Figure 9.8.5) from a point (x_0, y_0) to a point (x_1, y_1) under the force of gravity. As discussed in Project 4.2, the particle has two forms of energy during the motion: gravitational potential energy mgy and kinetic energy $\frac{1}{2}mv^2$. Show that the sum of these two quantities remains constant:

$$\overset{\text{GPE}}{mgy} + \overset{\text{KE}}{\tfrac{1}{2}mv^2} = C.$$

SOLUTION The particle is subjected to a vertical force $-mg$ (a downward force of magnitude mg). Since the particle is constrained to remain on the curve, the effective force on the particle is tangential. The tangential component of the vertical force is $-mg\sin\alpha$ (see Figure 9.8.5). The speed of the particle is ds/dt and the tangential acceleration is d^2s/dt^2. (It is as if the particle were moving along the tangent line.)

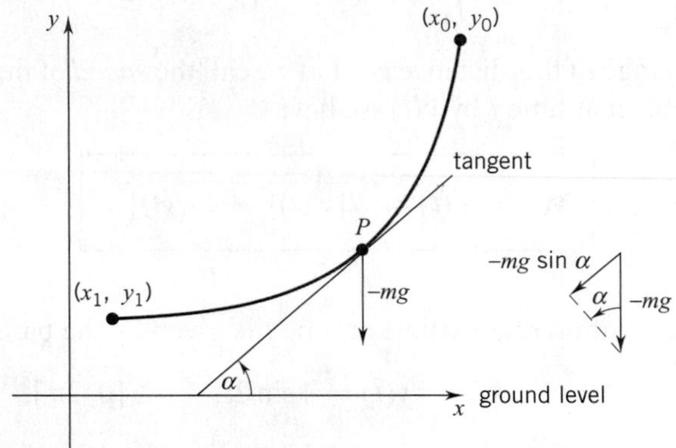

Figure 9.8.5

Therefore, by Newton's law $F = ma$, we have

$$m\frac{d^2s}{dt^2} = -mg\sin\alpha \underset{\text{by (9.8.4)}}{=} -mg\frac{dy}{ds}.$$

We can therefore write

$$mg\frac{dy}{ds} + m\frac{d^2s}{dt^2} = 0.$$

$$mg\frac{dy}{ds}\frac{ds}{dt} + m\frac{ds}{dt}\frac{d^2s}{dt^2} = 0. \qquad \text{(multiplied by } \frac{ds}{dt}\text{)}$$

$$mg\frac{dy}{dt} + mv\frac{dv}{dt} = 0. \qquad \text{(chain rule)}$$

Integrating with respect to t we have

$$mgy + \tfrac{1}{2}mv^2 = C,$$

as asserted. □

EXERCISES 9.8

Find the length of the graph and compare it to the straight-line distance between the endpoints of the graph.

1. $f(x) = 2x + 3, \quad x \in [0, 1].$

2. $f(x) = 3x + 2, \quad x \in [0, 1].$

3. $f(x) = (x - \frac{4}{9})^{3/2}, \quad x \in [1, 4].$

4. $f(x) = x^{3/2}, \quad x \in [0, 44].$

5. $f(x) = \frac{1}{3}\sqrt{x}\,(x - 3), \quad x \in [0, 3].$

6. $f(x) = \frac{2}{3}(x - 1)^{3/2}, \quad x \in [1, 2].$

7. $f(x) = \frac{1}{3}(x^2 + 2)^{3/2}, \quad x \in [0, 1].$

8. $f(x) = \frac{1}{3}(x^2 - 2)^{3/2}, \quad x \in [2, 4].$

9. $f(x) = \frac{1}{4}x^2 - \frac{1}{2}\ln x, \quad x \in [1, 5].$

10. $f(x) = \frac{1}{8}x^2 - \ln x, \quad x \in [1, 4].$

11. $f(x) = \frac{3}{8}x^{4/3} - \frac{3}{4}x^{2/3}, \quad x \in [1, 8].$

12. $f(x) = \frac{1}{10}x^5 + \frac{1}{6}x^{-3}, \quad x \in [1, 2].$

13. $f(x) = \ln(\sec x), \quad x \in [0, \frac{1}{4}\pi].$

14. $f(x) = \frac{1}{2}x^2, \quad x \in [0, 1].$

15. $f(x) = \frac{1}{2}x\sqrt{x^2 - 1} - \frac{1}{2}\ln(x + \sqrt{x^2 - 1}), \quad x \in [1, 2].$

16. $f(x) = \cosh x, \quad x \in [0, \ln 2].$

17. $f(x) = \frac{1}{2}x\sqrt{3 - x^2} + \frac{3}{2}\sin^{-1}(\frac{1}{3}\sqrt{3}x), \quad x \in [0, 1].$

18. $f(x) = \ln(\sin x), \quad x \in [\frac{1}{6}\pi, \frac{1}{2}\pi].$

The equations in Exercises 19–24 give the position of a particle at each time t during the time interval specified. Find the initial speed of the particle, the terminal speed, and the distance traveled.

19. $x(t) = t^2, \quad y(t) = 2t, \quad$ from $t = 0$ to $t = \sqrt{3}$.

20. $x(t) = t - 1, \quad y(t) = \frac{1}{2}t^2, \quad$ from $t = 0$ to $t = 1$.

21. $x(t) = t^2, \quad y(t) = t^3, \quad$ from $t = 0$ to $t = 1$.

22. $x(t) = a\cos^3 t, \quad y(t) = a\sin^3 t, \quad$ from $t = 0$ to $t = \frac{1}{2}\pi$.

23. $x(t) = e^t \sin t, \quad y(t) = e^t \cos t, \quad$ from $t = 0$ to $t = \pi$.

24. $x(t) = \cos t + t\sin t, \quad y(t) = \sin t - t\cos t, \quad$ from $t = 0$ to $t = \pi$.

25. Let $a > 0$. Find the length of the path traced out by

$$x(\theta) = a(\theta - \sin\theta), \quad y(\theta) = a(1 - \cos\theta),$$

as θ ranges from 0 to 2π.

26. Let $a > 0$. Find the length of the path traced out by

$$x(\theta) = 2a\cos\theta - a\cos 2\theta,$$
$$y(\theta) = 2a\sin\theta - a\sin 2\theta,$$

as θ ranges from 0 to 2π

27. (a) Let $a > 0$. Find the length of the path traced out by

$$x(\theta) = 3a\cos\theta + a\cos 3\theta,$$
$$y(\theta) = 3a\sin\theta - a\sin 3\theta,$$

as θ ranges from 0 to 2π.

(b) Show that this curve can also be represented by the parametric equations

$$x(\theta) = 4a\cos^3\theta, \quad y(\theta) = 4a\sin^3\theta, \quad 0 \le \theta \le 2\pi.$$

28. The curve defined parametrically by

$$x(\theta) = \theta\cos\theta, \quad y(\theta) = \theta\sin\theta,$$

is called an *Archimedean spiral*. Find the length of the arc traced out as θ ranges from 0 to 2π.

Find the length of the polar curve.

29. $r = 1 \quad$ from $\theta = 0$ to $\theta = 2\pi$.

30. $r = 3 \quad$ from $\theta = 0$ to $\theta = \pi$.

31. $r = e^\theta \quad$ from $\theta = 0$ to $\theta = 4\pi$. (logarithmic spiral)

32. $r = a\,e^\theta, a > 0, \quad$ from $\theta = -2\pi$ to $\theta = 2\pi$.

33. $r = e^{2\theta} \quad$ from $\theta = 0$ to $\theta = 2\pi$.

34. $r = 1 + \cos\theta \quad$ from $\theta = 0$ to $\theta = 2\pi$.

35. $r = 1 - \cos\theta \quad$ from $\theta = 0$ to $\theta = \frac{1}{2}\pi$.

36. $r = 2a\sec\theta, a > 0, \quad$ from $\theta = 0$ to $\theta = \frac{1}{4}\pi$.

37. At time t a particle has position

$$x(t) = 1 + \tan^{-1} t, \quad y(t) = 1 - \ln\sqrt{1 + t^2}.$$

Find the total distance traveled from time $t = 0$ to time $t = 1$. Give the initial speed and the terminal speed.

38. At time t a particle has position

$$x(t) = 1 - \cos t, \quad y(t) = t - \sin t.$$

Find the total distance traveled from time $t = 0$ to time $t = 2\pi$. Give the initial speed and the terminal speed.

39. Find c given that the length of the curve $y = \ln x$ from $x = 1$ to $x = e$ equals the length of the curve $y = e^x$ from $x = 0$ to $x = c$.

40. Find the length of the curve $y = x^{2/3}$, $x \in [1, 8]$.
HINT: Work with the mirror image $y = x^{3/2}$, $x \in [1, 4]$.

C **41.** Let $f(x) = 3x - 5$ on $[-3, 4]$.

(a) Draw the graph of f and find the coordinates of the midpoint of the line segment.

(b) Find the coordinates of the midpoint by using a CAS to solve the equation

$$\int_{-3}^{c} \sqrt{1 + [f'(x)]^2}\,dx = \int_{c}^{4} \sqrt{1 + [f'(x)]^2}\,dx$$

for c. Compare this result with your result in part (a).

C **42.** Let $f(x) = x^{3/2}$ on $[1, 5]$. Use a graphing utility to draw the graph of f and a CAS to find the coordinates of the "midpoint." HINT: Use the method of Exercise 41 (b).

C **43.** Show that the function $f(x) = \cosh x$ has the following property: for every interval $[a, b]$ the length of the graph equals the area under the graph.

44. Let $f(x) = 2 \ln x$ on $[1, e]$. Use a graphing utility to draw the graph of f and a CAS to find its arc length.

45. Let $f(x) = \sin x - x \cos x$ on $[0, \pi]$. Use a graphing utility to draw the graph of f and a CAS to find its arc length.

46. (a) Use a graphing utility to draw the curve defined parametrically by

$$x(t) = e^{2t} \cos 2t, \quad y(t) = e^{2t} \sin 2t, \quad 0 \le t \le \pi/3.$$

(b) Use a CAS to find the approximate length of the curve. Round off your answer to four decimal places.

47. (a) Use a graphing utility to draw the curve defined parametrically by

$$x(t) = t^2, \quad y(t) = t^3 - t, \quad -\infty < t < \infty.$$

(b) Your drawing should show that the curve has a loop. Use a CAS to find the approximate length of the loop. Round off your answer to four decimal places.

48. The curve

$$x(t) = \frac{3t}{t^3 + 1}, \quad y(t) = \frac{3t^2}{t^3 + 1}, \quad t \ne -1.$$

is called the *folium of Descartes*.

(a) Use a graphing utility to draw this curve.

(b) Your drawing in part (a) should show that the curve has a loop in the first quadrant. Use a CAS to estimate the length of the loop. Round off your answer to four decimal places. HINT: Use symmetry.

In Exercises 49 and 50, use a graphing utility to draw the graph of the polar curve and a CAS to find its arc length. Round off your answers to four decimal places.

49. $r = 1 - \cos\theta, \quad 0 \le \theta \le \pi.$

50. $r = \sin 5\theta, \quad 0 \le \theta \le 2\pi.$

51. (a) Let $a > b > 0$. Show that the length of the ellipse

$$x(t) = a \cos t, \ y(t) = b \sin t, \ 0 \le t \le 2\pi.$$

is given by

$$L = 4a \int_0^{\pi/2} \sqrt{1 - e^2 \cos^2 t}\, dt,$$

where $e = \sqrt{a^2 - b^2}/a$ is the eccentricity. The integrand does not have an elementary antiderivative. This integral belongs to a class of integrals known as *elliptic integrals*.

(b) Let $a = 5$ and $b = 4$, and approximate the length of the ellipse using a CAS. Round off your answer to four decimal places.

52. The figure shows the graph of a continuously differentiable function f from $x = a$ to $x = b$ together with a polygonal approximation. Show that the length of this polygonal approximation can be written as the following Riemann sum:

$$\sqrt{1 + [f'(x_1^*)]^2}\,\Delta x_1 + \sqrt{1 + [f'(x_2^*)]^2}\,\Delta x_2 + \cdots + \sqrt{1 + [f'(x_n^*)]^2}\,\Delta x_n.$$

As $\|P\| = \max \Delta x_i$ tends to 0, such Riemann sums tend to

$$\int_a^b \sqrt{1 + [f'(x)]^2}\,dx.$$

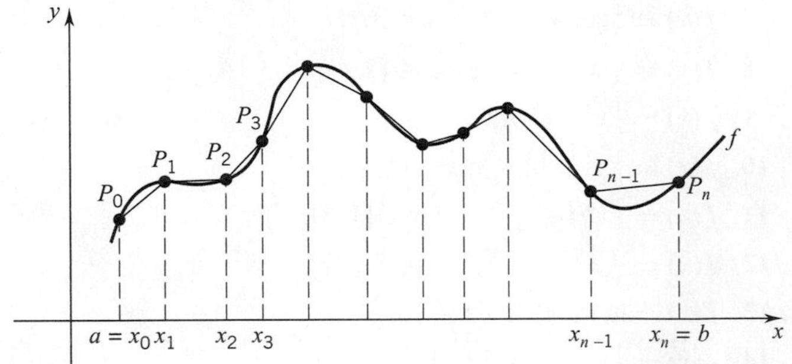

53. Suppose that f is continuously differentiable from $x = a$ to $x = b$. Show that the

(9.8.7) length of the graph of $f = \int_a^b |\sec[\alpha(x)]|\, dx$, where $\alpha(x)$ is the inclination of the tangent line at $(x, f(x))$.

54. Show that a homogeneous, flexible, inelastic rope hanging from two fixed points assumes the shape of a *catenary*:

$$f(x) = a \cosh\left(\frac{x}{a}\right) = \frac{a}{2}(e^{x/a} + e^{-x/a}), a > 0.$$

HINT: Refer to the figure. The part of the rope that corresponds to the interval $[0, x]$ is subject to the following forces:

(1) its weight, which is proportional to its length;

(2) a horizontal pull at 0, $p(0)$;

(3) a tangential pull at x, $p(x)$.

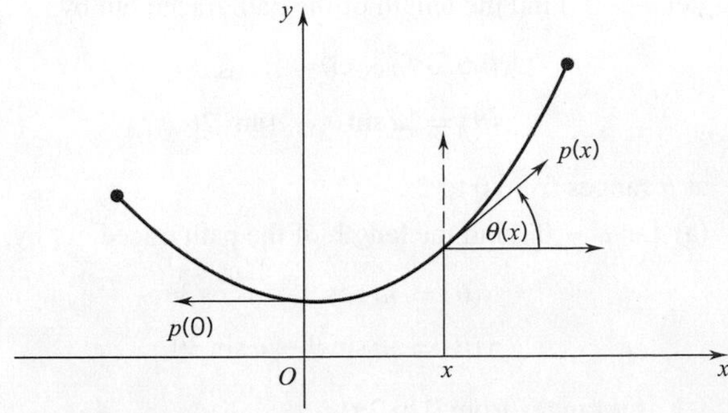

■ 9.9 THE AREA OF A SURFACE OF REVOLUTION; THE CENTROID OF A CURVE; PAPPUS'S THEOREM ON SURFACE AREA

The Area of a Surface of Revolution

In Figure 9.9.1 you can see the frustum of a cone; one radius marked r, the other R. The slant height is marked s. An interesting elementary calculation that we leave to you shows that the area of this slanted surface is given by the formula

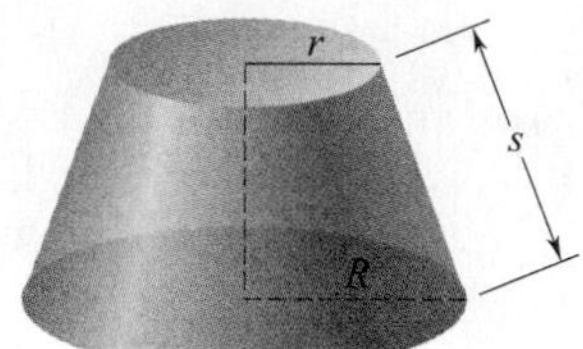

Figure 9.9.1

(9.9.1) $$A = \pi(r + R)s.$$ (Exercise 21)

This formula forms the basis for all that follows.

Let C be a curve in the upper half-plane (Figure 9.9.2). The curve can meet the x-axis, but only at a finite number of points. We will assume that C is parametrized by a pair of continuously differentiable functions

$$x = x(t), \quad y = y(t), \quad t \in [c, d].$$

Furthermore, we will assume that C is *simple*: no two values of t between c and d give rise to the same point of C; that is, the curve does not intersect itself.

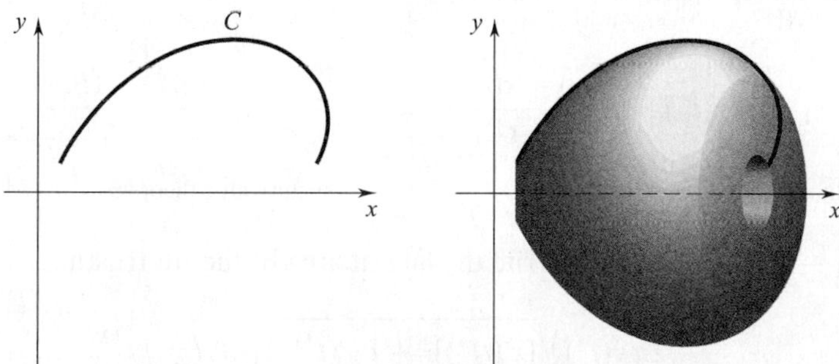

Figure 9.9.2

If we revolve C about the x-axis, we obtain a surface of revolution. The area of that surface is given by the formula

(9.9.2) $$A = \int_c^d 2\pi y(t)\sqrt{[x'(t)]^2 + [y'(t)]^2}\, dt.$$

We will try to outline how this formula comes about. The argument is similar to the one given in Section 9.8 for the length of a curve.

Each partition $P = \{c = t_0 < t_1 < \cdots < t_n = d\}$ of $[c, d]$ generates a polygonal approximation to C (Figure 9.9.3). Call this polygonal approximation C_p. By revolving C_p about the x-axis, we get a surface made up of n conical frustums.

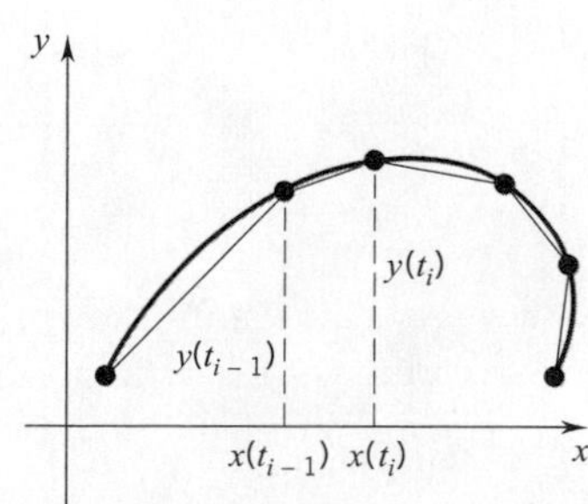

Figure 9.9.3

The ith frustum (Figure 9.9.4) has slant height

$$s_i = \sqrt{[x(t_i) - x(t_{i-1})]^2 + [y(t_i) - y(t_{i-1})]^2}$$

$$= \sqrt{\left[\frac{x(t_i) - x(t_{i-1})}{t_i - t_{i-1}}\right]^2 + \left[\frac{y(t_i) - y(t_{i-1})}{t_i - t_{i-1}}\right]^2}\,(t_i - t_{i-1}).$$

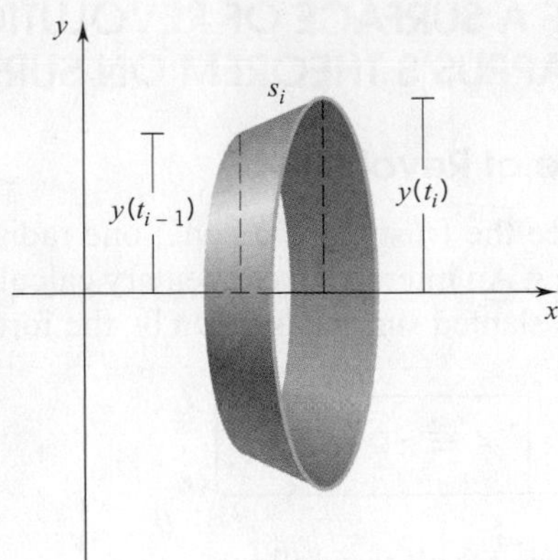

Figure 9.9.4

The lateral area $\pi[y(t_{i-1}) + y(t_i)]\, s_i$ [see Formula (9.9.1)] can be written

$$\pi[y(t_{i-1}) + y(t_i)]\sqrt{\left[\frac{x(t_i) - x(t_{i-1})}{t_i - t_{i-1}}\right]^2 + \left[\frac{y(t_i) - y(t_{i-1})}{t_i - t_{i-1}}\right]^2}\,(t_i - t_{i-1}).$$

There exist points $t_i^*, t_i^{**}, t_i^{***}$ all in $[t_{i-1}, t_i]$ such that

$$y(t_i) + y(t_{i-1}) = 2y(t_i^*), \qquad \frac{x(t_i) - x(t_{i-1})}{t_i - t_{i-1}} = x'(t_i^{**}), \qquad \frac{y(t_i) - y(t_{i-1})}{t_i - t_{i-1}} = y'(t_i^{***}).$$

↑ intermediate-value theorem; ↑ mean-value theorem ↑

Let $\Delta t_i = t_i - t_{i-1}$. We can now write the lateral area of the ith frustum as

$$2\pi y(t_i^*)\sqrt{[x'(t_i^{**})]^2 + [y'(t_i^{***})]^2}\,\Delta t_i.$$

The area generated by revolving all of C_p is the sum of these terms:

$$2\pi y(t_1^*)\sqrt{[x'(t_1^{**})]^2 + [y'(t_1^{***})]^2}\,\Delta t_1 + \cdots + 2\pi y(t_n^*)\sqrt{[x'(t_n^{**})]^2 + [y'(t_n^{***})]^2}\,\Delta t_n.$$

This is not a Riemann sum: we don't know that $t_i^* = t_i^{**} = t_i^{***}$. But it is "close" to a Riemann sum. Close enough that, as $||P|| \to 0$, this "almost" Riemann sum tends to the integral.

$$\int_c^d 2\pi y(t)\sqrt{[x'(t)]^2 + [y'(t)]^2}\,dt.$$

That this is so follows from a theorem of advanced calculus known as Duhamel's principle. We will not attempt to fill in the details. □

Example 1 Derive a formula for the surface area of a sphere from (9.9.2).

SOLUTION We can generate a sphere of radius r by revolving the arc

$$x(t) = r\cos t, \quad y(t) = r\sin t, \quad t \in [0, \pi]$$

about the x-axis. Differentiation gives

$$x'(t) = -r\sin t, \quad y'(t) = r\cos t.$$

By Formula (9.9.2)

$$A = 2\pi \int_0^{\pi} r\sin t\sqrt{r^2(\sin^2 t + \cos^2 t)}\,dt$$

$$= 2\pi r^2 \int_0^{\pi} \sin t\,dt = 2\pi r^2\Big[-\cos t\Big]_0^{\pi} = 4\pi r^2. \quad \square$$

Example 2 Find the area of the surface generated by revolving about the x-axis the curve

$$y^2 - 2\ln y = 4x \quad \text{from} \quad y = 1 \text{ to } y = 2.$$

SOLUTION We can represent the curve parametrically by setting

$$x(t) = \tfrac{1}{4}(t^2 - 2\ln t), \quad y(t) = t, \quad t \in [1, 2].$$

Here
$$x'(t) = \tfrac{1}{2}(t - t^{-1}), \quad y'(t) = 1$$

and
$$[x'(t)]^2 + [y'(t)]^2 = [\tfrac{1}{2}(t + t^{-1})]^2. \qquad \text{(check this)}$$

It follows that

$$A = \int_1^2 2\pi t[\tfrac{1}{2}(t + t^{-1})]\,dt = \int_1^2 \pi(t^2 + 1)\,dt = \pi\Big[\tfrac{1}{3}t^3 + t\Big]_1^2 = \tfrac{10}{3}\pi. \quad \square$$

Suppose now that f is a continuously differentiable nonnegative function defined from $x = a$ to $x = b$. The area of the surface generated by revolving the graph of f about the x-axis is given by the formula

(9.9.3)
$$A = \int_a^b 2\pi f(x)\sqrt{1 + [f'(x)]^2}\,dx.$$

This follows readily from (9.9.2). Set

$$x = t, \quad y(t) = f(t), \quad t \in [a, b].$$

Apply (9.9.2) and then replace the dummy variable t by x.

Example 3 Find the area of the surface generated by revolving about the x-axis the graph of the sine function from $x = 0$ to $x = \frac{1}{2}\pi$.

SOLUTION Setting $f(x) = \sin x$, we have $f'(x) = \cos x$ and therefore

$$A = \int_0^{\pi/2} 2\pi \sin x\sqrt{1 + \cos^2 x}\,dx.$$

To calculate this integral, we set

$$u = \cos x, \quad du = -\sin x\,dx.$$

At $x = 0, u = 1$; at $x = \frac{1}{2}\pi, u = 0$. Therefore

$$A = -2\pi \int_1^0 \sqrt{1+u^2}\,du = 2\pi \int_0^1 \sqrt{1+u^2}\,du$$

$$= 2\pi \left[\tfrac{1}{2}u\sqrt{1+u^2} + \tfrac{1}{2}\ln\left(u + \sqrt{1+u^2}\right)\right]_0^1$$

by (8.4.1)

$$= \pi[\sqrt{2} + \ln(1+\sqrt{2})] \cong 2.3\pi \cong 7.23. \quad \square$$

The Centroid of a Curve

The centroid of a plane region Ω is the center of mass of a homogeneous plate in the shape of Ω. Likewise, the centroid of a solid of revolution T is the center of mass of a homogeneous solid in the shape of T. All this was covered in Section 6.4.

What do we mean by the centroid of a plane curve C? Exactly what you would expect. By the *centroid* of a plane curve C, we mean the center of mass of a homogeneous wire in the shape of C. It should be noted that the centroid of a plane curve does not necessarily lie on the curve. We can calculate the centroid of a curve from the following principles, which we take from physics.

Principle 1: Symmetry If a curve has an axis of symmetry, then the centroid $(\overline{x}, \overline{y})$ lies somewhere along that axis.

Principle 2: Additivity If a curve with length L is broken up into a finite number of pieces with arc lengths $\Delta s_1, \ldots, \Delta s_n$ and centroids $(\overline{x}_1, \overline{y}_1), \ldots, (\overline{x}_n, \overline{y}_n)$, then

$$\overline{x}L = \overline{x}_1\Delta s_1 + \cdots + \overline{x}_n\Delta s_n \quad \text{and} \quad \overline{y}L = \overline{y}_1\Delta s_1 + \cdots + \overline{y}_n\Delta s_n.$$

Figure 9.9.5 shows a curve C that begins at A and ends at B. Let's suppose that the curve is continuously differentiable and that the length of the curve is L. We want a formula for the centroid $(\overline{x}, \overline{y})$.

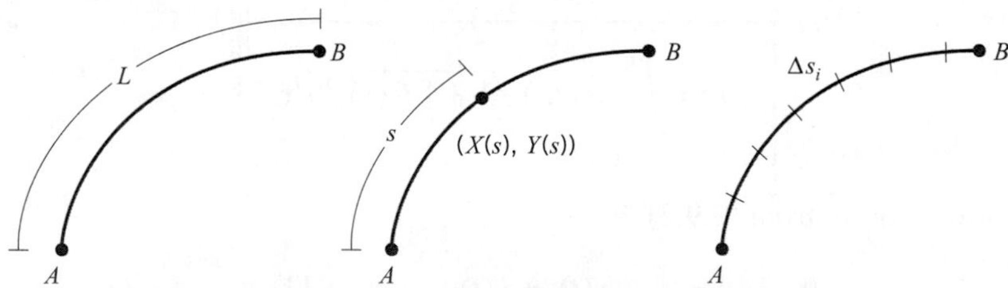

Figure 9.9.5

Let $(X(s), Y(s))$ be the point on C that is at an arc distance s from the initial point A. (What we are doing here is called *parametrizing C by arc length.*) A partition $P = \{0 = s_0 < s_1 < \cdots < s_n = L\}$ of $[0, L]$ breaks up C into n little pieces of lengths $\Delta s_1, \ldots, \Delta s_n$ and centroids $(\overline{x}_1, \overline{y}_1), \ldots, (\overline{x}_n, \overline{y}_n)$. From Principle 2 we know that

$$\overline{x}L = \overline{x}_1\Delta s_1 + \cdots + \overline{x}_n\Delta s_n \quad \text{and} \quad \overline{y}L = \overline{y}_1\Delta s_i + \cdots + \overline{y}_n\Delta s_n.$$

Now for each i there exists s_i^* in $[s_{i-1}, s_i]$ for which $\overline{x}_i = X(s_i^*)$ and s_i^{**} in $[s_{i-1}, s_i]$ for which $\overline{y}_i = Y(s_i^{**})$. We can therefore write

$$\overline{x}L = X(s_1^*)\Delta s_1 + \cdots + X(s_n^*)\Delta s_n, \quad \overline{y}L = Y(s_1^{**})\Delta s_1 + \cdots + Y(s_n^{**})\Delta s_n.$$

The sums on the right are Riemann sums tending to easily recognizable limits: letting $||P|| \to 0$ we have

(9.9.4) $$\bar{x}L = \int_0^L X(s)\,ds \quad \text{and} \quad \bar{y}L = \int_0^L Y(s)\,ds.$$

These formulas give the centroid of a curve in terms of the arc length parameter. It is but a short step from here to formulas having a more convenient form.

Suppose that the curve C is given parametrically by the functions

$$x = x(t), \quad y = y(t), \quad t \in [c, d]$$

where t is now an arbitrary parameter. Then

$$s(t) = \int_c^t \sqrt{[x'(u)]^2 + [y'(u)]^2}\,du, \quad ds = s'(t)\,dt = \sqrt{[x'(t)]^2 + [y'(t)]^2}\,dt.$$

At $s = 0, t = c$; at $s = L, t = d$. Changing variables in (9.9.4) from s to t, we have

$$\bar{x}L = \int_c^d X(s(t))\,s'(t)\,dt = \int_c^d X(s(t))\sqrt{[x'(t)]^2 + [y'(t)]^2}\,dt$$

and

$$\bar{y}L \int_c^d Y(s(t))\,s'(t)\,dt = \int_c^d Y(s(t))\sqrt{[x'(t)]^2 + [y'(t)]^2}\,dt.$$

A moment's reflection shows that

$$X(s(t)) = x(t) \quad \text{and} \quad Y(s(t)) = y(t).$$

We can then write

(9.9.5) $$\bar{x}L = \int_c^d x(t)\sqrt{[x'(t)]^2 + [y'(t)]^2}\,dt,$$
$$\bar{y}L = \int_c^d y(t)\sqrt{[x'(t)]^2 + [y'(t)]^2}\,dt.$$

These are the centroid formulas in their most useful form.

Example 4 Find the centroid of the quarter-circle shown in Figure 9.9.6.

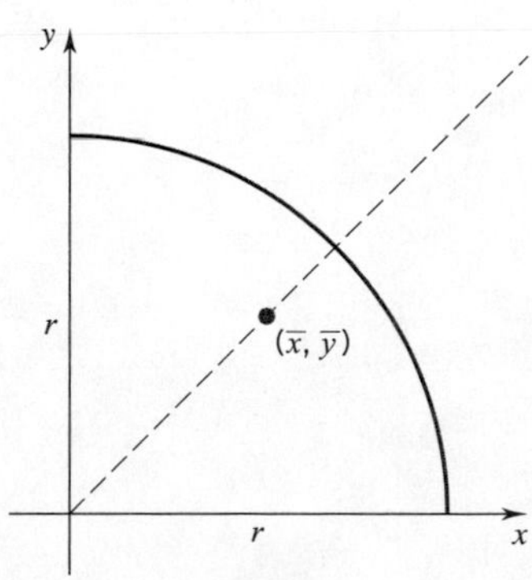

Figure 9.9.6

SOLUTION We can parametrize that quarter-circle by setting

$$x(t) = r\cos t, \quad y(t) = r\sin t, \quad t \in [0, \pi/2].$$

Since the curve is symmetric about the line $x = y$, we know that $\bar{x} = \bar{y}$. Here $x'(t) = -r\sin t$ and $y'(t) = r\cos t$. Therefore

$$\sqrt{[x'(t)]^2 + [y'(t)]^2} = \sqrt{r^2\sin^2 t + r^2\cos^2 t} = r.$$

By (9.9.5)

$$\bar{y}L = \int_0^{\pi/2} (r\sin t)r\,dt = r^2\int_0^{\pi/2} \sin t\,dt = r^2\Big[-\cos t\Big]_0^{\pi/2} = r^2.$$

Note that $L = \pi r/2$. Therefore $\overline{y} = r^2/L = 2r/\pi$. The centroid of the quarter-circle is the point $(2r/\pi, 2r/\pi)$. [Note that this point is closer to the curve than the centroid of the quarter-disc (Example 1, Section 6.4).] ❑

Example 5 Take $a > 0$. Find the centroid of the cardioid $r = a(1 - \cos\theta)$.

SOLUTION The curve (see Figure 9.4.11) is symmetric about the x-axis. Thus $\overline{y} = 0$.

To find $\overline{x}$ we parametrize the curve as follows: taking $\theta \in [0, 2\pi]$, we set

$$x(\theta) = r\cos\theta = a(1 - \cos\theta)\cos\theta,$$
$$y(\theta) = r\sin\theta = a(1 - \cos\theta)\sin\theta.$$

A straightforward calculation shows that

$$[x'(\theta)]^2 + [y'(\theta)]^2 = 4a^2\sin^2\tfrac{1}{2}\theta.$$

Applying (9.9.5) we have

$$\overline{x}L = \int_0^{2\pi} [a(1 - \cos\theta)\cos\theta][2a\sin\tfrac{1}{2}\theta]\,d\theta = -\tfrac{32}{5}a^2.$$

check this out

By Example 4 in Section 9.8, $L = 8a$. Thus $\overline{x} = (-\tfrac{32}{5}a^2)/8a = -\tfrac{4}{5}a$. The centroid of the curve is the point $(-\tfrac{4}{5}a, 0)$. ❑

If C is a curve of the form

$$y = f(x), \quad x \in [a, b].$$

where f is continuously differentiable, then the formulas in (9.9.5) give

(9.9.6)
$$\overline{x}L = \int_a^b x\sqrt{1 + [f'(x)]^2}\,dx, \quad \overline{y}L = \int_a^b f(x)\sqrt{1 + [f'(x)]^2}\,dx.$$

The details are left to you.

Pappus's Theorem on Surface Area

That same Pappus who gave us that wonderful theorem on volumes of solids of revolution (Theorem 6.4.4) gave us the following equally marvelous result on surface area:

THEOREM 9.9.7 PAPPUS'S THEOREM ON SURFACE AREA

A plane curve is revolved about an axis that lies in its plane. The curve may meet the axis but, if so, only at a finite number of points. If the curve does not cross the axis, then the area of the resulting surface of revolution is the length of the curve multiplied by the circumference of the circle described by the centroid of the curve:

$$A = 2\pi\overline{R}L,$$

where L is the length of the curve and $\overline{R}$ is the distance from the axis to the centroid of the curve.

Pappus did not have calculus to help him when he made his inspired guesses; he did his work 13 centuries before Newton or Leibniz was born. With the formulas that we have developed through calculus (through Newton and Leibniz, that is), Pappus's theorem is easily verified. Call the plane of the curve the xy-plane and call the axis of rotation the x-axis. Then $\overline{R} = \overline{y}$ and

$$\begin{aligned} A &= \int_c^d 2\pi y(t)\sqrt{[x'(t)]^2 + [y'(t)]^2}\,dt \\ &= 2\pi \int_c^d y(t)\sqrt{[x'(t)]^2 + [y'(t)]^2}\,dt = 2\pi\overline{y}L = 2\pi\overline{R}L. \end{aligned}$$

EXERCISES 9.9

Find the length of the curve, locate the centroid, and determine the area of the surface generated by revolving the curve about the x-axis.

1. $f(x) = 4, \quad x \in [0, 1]$.

2. $f(x) = 2x, \quad x \in [0, 1]$.

3. $y = \frac{4}{3}x, \quad x \in [0, 3]$.

4. $y = -\frac{12}{5}x + 12, \quad x \in [0, 5]$.

5. $x(t) = 3t, \quad y(t) = 4t; \quad t \in [0, 2]$.

6. $r = 5, \quad \theta \in [0, \frac{1}{4}\pi]$.

7. $x(t) = 2\cos t, \quad y(t) = 2\sin t; \quad t \in [0, \frac{1}{6}\pi]$.

8. $x(t) = \cos^3 t, \quad y(t) = \sin^3 t; \quad t \in [0, \frac{1}{2}\pi]$.

9. $x^2 + y^2 = a^2, \quad x \in [-\frac{1}{2}a, \frac{1}{2}a], \quad y > 0, \quad a > 0$.

10. $r = 1 + \cos\theta, \quad \theta \in [0, \pi]$.

Find the area of the surface generated by revolving the curve about the x-axis.

11. $f(x) = \frac{1}{3}x^3, \quad x \in [0, 2]$.

12. $f(x) = \sqrt{x}, \quad x \in [1, 2]$.

13. $4y = x^3, \quad x \in [0, 1]$.

14. $y^2 = 9x, \quad x \in [0, 4]$.

15. $y = \cos x, \quad x \in [0, \frac{1}{2}\pi]$.

16. $f(x) = 2\sqrt{1 - x}, \quad x \in [-1, 0]$.

17. $r = e^\theta, \quad \theta \in [0, \frac{1}{2}\pi]$.

18. $y = \cosh x, \quad x \in [0, \ln 2]$.

19. Take a > 0. One arch of a cycloid can be defined parametrically by setting

$$x(\theta) = a(\theta - \sin\theta), \quad y(\theta) = a(1 - \cos\theta),$$

letting θ range from 0 to 2π.

(a) Find the area under the curve.

(b) Find the area of the surface generated by revolving the arch about the x-axis.

C 20. Take $a > 0$. The curve

$$\begin{aligned} x(\theta) &= 3a\cos\theta + a\cos 3\theta, \\ y(\theta) &= 3a\sin\theta - a\sin 3\theta, \end{aligned}$$

with θ ranging from 0 to 2π, is called a *hypocycloid*.

(a) Use a graphing utility to draw the curve.

(b) Find the area enclosed by the curve.

(c) Set up a definite integral that gives the area of the surface generated by revolving the curve about the x-axis.

21. By cutting a cone of slant height s and base radius r along a lateral edge and laying the surface flat, we can form a sector of a circle of radius s. (See the figure.) Use this idea to verify Formula (9.9.1).

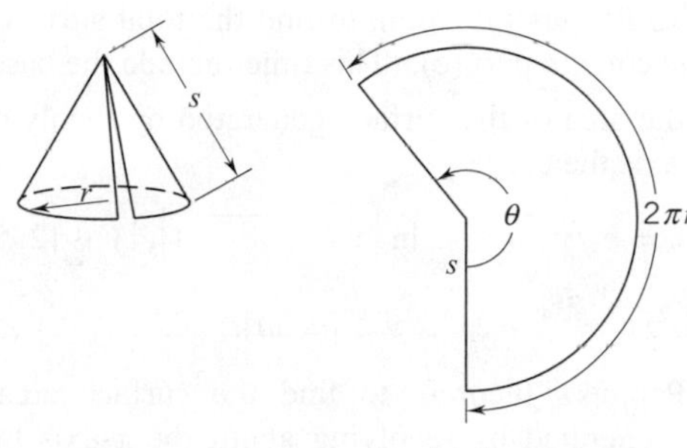

22. The figure shows a ring formed by two quarter-circles. Call the corresponding quarter-discs Ω_a and Ω_r. By Section 6.4, Ω_a has centroid $(4a/3\pi, 4a/3\pi)$ and Ω_r has centroid $(4r/3\pi, 4r/3\pi)$.

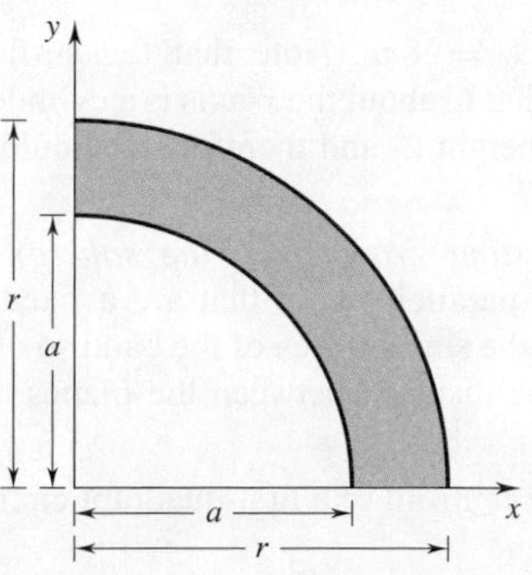

(a) Without integration, calculate the centroid of the ring.

(b) Find the centroid of the outer arc from your answer to part (a) by letting a tend to r.

23. (a) Find the centroid of each side of the triangle in the figure.

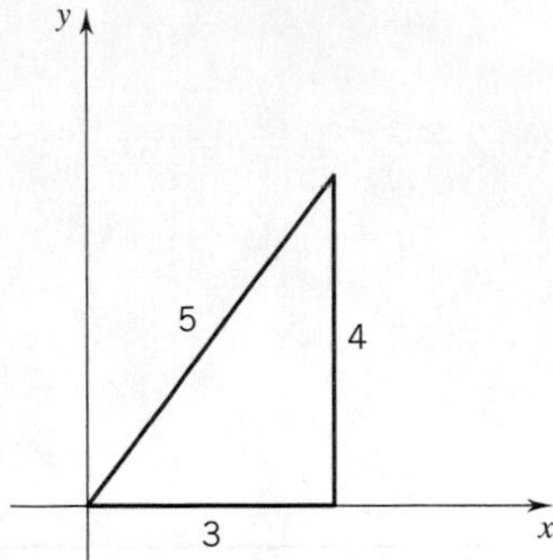

(b) Use your answers in part (a) to calculate the centroid of the triangle.

(c) What is the centroid of the triangular region?

(d) What is the centroid of the curve consisting of sides 4 and 5?

(e) Use Pappus's theorem to find the slanted surface area of a cone of base radius 4 and height 3.

(f) Use Pappus's theorem to find the total surface area of the cone in part (e). (This time include the base.)

24. Find the area of the surface generated by revolving about the x-axis the curve

(a) $2x = y\sqrt{y^2-1} + \ln\left|y - \sqrt{y^2-1}\right|, \quad y \in [2,5].$

(b) $6a^2xy = y^4 + 3a^4, \quad y \in [a, 3a].$

25. Use Pappus's theorem to find the surface area of the *torus* generated by revolving about the x-axis the circle $x^2 + (y-b)^2 = a^2 \quad (0 < a \le b)$

26. (a) We calculated the total surface area of a sphere from (9.9.2) not (9.9.3). Could we just as will have used (9.9.3)? Explain.

(b) Verify that Formula (9.9.2) applied to

$$C\colon\ x(t) = \cos t, \quad y(t) = r, \quad \text{with } t \in [0, 2\pi].$$

gives $A = 8\pi r$. Note that the surface obtained by revolving C about the x-axis is a cylinder of base radius r and height 2, and therefore A should be $4\pi r$. What's wrong?

27. (*An interesting property of the sphere*) Slice a sphere along two parallel planes that are a fixed distance apart. Show that the surface area of the band so obtained depends only on the distance between the planes and not on their location.

28. Locate the centroid of a first-quadrant circular arc

$$C\colon\ x(t) = r\cos t, \quad y(t) = r\sin t, \quad t \in [\theta_1, \theta_2].$$

29. Find the surface area of the ellipsoid obtained by revolving the ellipse

$$\frac{x^2}{a^2} + \frac{y^2}{b^2} = 1 \qquad (0 < b < a)$$

(a) about its major axis; (b) about its minor axis.

The Centroid of a Surface of Revolution

If a material surface of revolution is homogeneous (constant mass density), then the center of mass of that material surface is called the *centroid*. The determination of the centroid of a surface of arbitrary shape requires surface integration (Chapter 17). However, if the surface is a surface of revolution, then the centroid can be found by ordinary one-variable integration.

30. Let C be a simple curve in the upper half-plane parametrized by a pair of continuously differentiable functions.

$$x = x(t), \quad y = y(t). \quad t \in [c, d].$$

By revolving C about the x-axis we obtain a surface of revolution, the area of which we denote by A. By symmetry, the centroid of the surface lies on the x-axis. Thus the centroid is completely determined by its x-coordinate $\bar{x}$. Show that

(9.9.8) $$\bar{x}A = \int_c^d 2\pi x(t)y(t)\sqrt{[x'(t)]^2 + [y'(t)]^2}\,dt$$

by assuming the following additivity principle: If the surface is broken up into n surfaces of revolution with areas $A_1, \ldots, A_n$ and the centroids of the surfaces have x-coordinates $\bar{x}_1, \ldots, \bar{x}_n$, then

$$\bar{x}A = \bar{x}_1A_1 + \cdots + \bar{x}_nA_n.$$

31. Locate the centroid of a hemisphere of radius r.

32. Locate the centroid of a conical surface of base radius r and height h.

33. Where is the centroid of the lateral surface of the frustum of a cone of height h with base radii r and R?

34. (a) Show that the circle

$$(x-a)^2 + y^2 = b^2$$

can be parametrized by

$$x(t) = a + b\cos t, \quad y(t) = b\sin t, \quad 0 \le t \le 2\pi.$$

(b) Suppose that $0 < b < a$. The solid generated by revolving the circle around the y-axis is a torus. Find the volume of the torus.

(c) Find the surface area of the torus.

■ PROJECT 9.9 The Cycloid

Take a wheel (a roll of tape will do) and mark a point on the rim. Call that point P. Now roll the wheel slowly, keeping your eyes on P. The jumping-kangaroo path described by P is called a *cycloid*. To obtain a mathematical characterization of the cycloid, let the radius of the wheel be R and set the wheel on the x-axis so that the point P starts out at the origin. The figure shows P after a turn of θ radians.

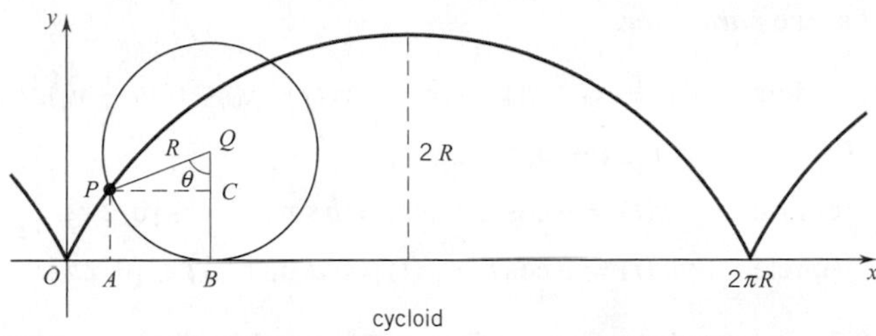

cycloid

Problem 1. Show that the cycloid can be parametrized by the functions

$$x(\theta) = R(\theta - \sin\theta), \qquad y(\theta) = R(1 - \cos\theta).$$

HINT: Length of $\overline{OB} =$ length of $PB = R\theta$.

Problem 2.

(a) At the end of each arch, the cycloid comes to a cusp. Show that x' and y' are both 0 at the end of each arch.

(b) Show that the area under an arch of the cycloid is three times the area of the rolling circle.

(c) Find the length of an arch of the cycloid.

Problem 3.

(a) Locate the centroid of the region under the first arch of the cycloid.

(b) Find the volume of the solid generated by revolving the region under an arch of the cycloid about the x-axis.

(c) Find the volume of the solid generated by revolving the region under an arch of the cycloid about the y-axis.

The curve given parametrically by

$$x(\phi) = R(\phi + \sin\phi), \quad y(\phi) = R(1 - \cos\phi), \quad \phi \in [-\pi, \pi]$$

is called the *inverted cycloid*. It is obtained by reflecting one arch of the cycloid in the x-axis, and then translating the resulting curve so that the low point is at the origin.

Problem 4. Use a graphing utility to draw the inverted cycloid.

Problem 5.

(a) Find the inclination α of the line tangent to the inverted cycloid at the point $(x(\phi), y(\phi))$.

(b) Let s be the arc distance from the low point of the inverted cycloid to the point $(x(\phi), y(\phi))$. Show that $s = 4R\sin\frac{1}{2}\phi = 4R\sin\alpha$, where α is the inclination of the tangent line at $(x(\phi), y(\phi))$.

Visualize two particles sliding without friction down an arch of the inverted cycloid. If the two particles are released at the same time from different positions, which one will reach the bottom first? Neither—they will both get there at exactly the same time. Being the only curve that has this property, the inverted arch of a cycloid is known as the *tautochrone*, the *same-time* curve.

Problem 6. Verify that the inverted arch of a cycloid has the tautochrone property by showing that:

(a) The effective gravitational force on a particle of mass m is $-mg\sin\alpha$, where α is the inclination of the tangent line at the position of the particle. From this, conclude that

$$(*) \qquad \frac{d^2s}{dt^2} = -g\sin\alpha.$$

(b) Combine ($*$) with Problem 5(b) to show that the particle is in simple harmonic motion with period

$$T = 4\pi\sqrt{R/g}.$$

Thus, while the amplitude of the motion depends on the point of release, the frequency does not. Two particles released simultaneously from different points of the curve will reach the low point of the curve in exactly the same amount of time: $T/4 = \pi\sqrt{R/g}$.

Suppose now that a particle descends without friction along a curve from point A to a point B not directly below it. What should be the shape of the curve so that the particle descends from A to B in the least possible time? This question was first formulated by Johann Bernoulli and posed by him as a challenge to the scientific community in 1696. The challenge was readily accepted and within months the answer was found—by Johann Bernoulli himself, by his brother Jacob, by Newton, by Leibniz, and by L'Hospital. The answer? Part of an inverted cycloid. Because of this, the inverted cycloid is heralded as the *brachystochrone*, the *least-time* curve.

A proof that the inverted cycloid is the least-time curve, the curve of quickest descent, is beyond our reach. The argument requires a sophisticated variant of calculus known as *the calculus of variations*. We can, however, compare the time of descent along a cycloid to the time of descent along a straight-line path.

Problem 7. You have seen that a particle descends along the inverted arch of a cycloid from $(\pi R, 2R)$ to $(0, 0)$ in time $t = T/4 = \pi\sqrt{R/g}$. What is the time of descent along a straight-line path?

■ CHAPTER HIGHLIGHTS

9.1 Translations; The Parabola

translation of axes: $x = x_0 + X, y = y_0 + Y$ (p. 514)
distance between a point and a line (p. 515)
the parabola
  directrix, focus, axis, vertex (p. 517)
  reflecting property (p. 520)

A parabola is the set of points equidistant from a fixed line and a fixed point not on that line.

9.2 The Ellipse and Hyberbola

the ellipse
  foci, standard position (p. 523)
  vertices, axes, center (p. 524)
  elliptical reflectors (p. 529)
  eccentricity (p. 532)

An ellipse is the set of points the sum of whose distances from two fixed points is constant.

the hyperbola
  foci, standard position (p. 526)
  asymptotes (p. 527)
  vertices, transverse axis, center (p. 528)
  hyperbolic reflectors (p. 531)
  eccentricity (p. 532)

A hyperbola is the set of points the difference of whose distances from two fixed points is constant.

9.3 Polar Coordinates

relation between rectangular coordinates (x, y) and polar coordinates $[r, \theta]$:

$x = r\cos\theta,\ y = r\sin\theta,\ \tan\theta = \dfrac{y}{x},\ x^2+y^2 = r^2$ (p. 535)

9.4 Graphing in Polar Coordinates

spirals (p. 539) petal curves (p. 543)
circles (p. 542)
cardioids (p. 541) lines (p. 542)
lemniscates (p. 543)
limaçons (p. 542)

If we think of each of two polar curves as giving the position of an object at time θ, then the simultaneous solution of the two equations gives the points where the objects collide. A figure will often help to identify the points where two polar curves intersect but do not collide; conversion to rectangular coordinates may also help to locate such points.

9.5 Area in Polar Coordinates

Let ρ_1 and ρ_2 be positive continuous functions defined on a closed interval $[\alpha, \beta]$ of length 2π, at most. If $\rho_1(\theta) \le \rho_2(\theta)$ for all θ in $[\alpha, \beta]$, then the area of the region between the polar curves $r = \rho_1(\theta)$ and $r = \rho_2(\theta)$ is given by the formula

$$A = \int_\alpha^\beta \tfrac{1}{2}([\rho_2(\theta)]^2 - [\rho_1(\theta)]^2)\, d\theta.$$

9.6 Curves Given Parametrically

Let $x = x(t), y = y(t)$ be a pair of functions differentiable on the interior of some interval I. At the endpoints of I (if any) we require only continuity. For each t in I we can interpret $(x(t), y(t))$ as the point with x-coordinate $x(t)$ and y-coordinate $y(t)$. Then, as t ranges over I, the point $(x(t), y(t))$ traces out a path in the xy-plane. We call such a path a *parametrized curve* and refer to t as the *parameter*.

$$\text{line}:\ x(t) = x_0 + t(x_1 - x_0),\quad y(t) = y_0 + t(y_1 - y_0),\quad t \in (-\infty, \infty).$$

$$\text{ellipse}:\ x(t) = a\cos t,\quad y(t) = b\sin t,\quad t \in [0, 2\pi].$$

$$\text{circle}:\ x(t) = a\cos t,\quad y(t) = a\sin t,\quad t \in [0, 2\pi].$$

9.7 Tangents to Curves Given Parametrically

tangent line at (x_0, y_0) : $y'(t_0)[x - x_0] - x'(t_0)[y - y_0] = 0$ [where $x(t_0) = x_0, y(t_0) = y_0$], provided that $[x'(t_0)]^2 + [y'(t_0)]^2 \ne 0$

9.8 Arc Length and Speed

polygonal path (p. 566)
significance of $dx/ds, dy/ds$ (p. 570)
length of a parametrized curve:

$$L = \int_a^b \sqrt{x'(t)]^2 + [y'(t)]^2}dt$$

length of a graph:

$$L = \int_a^b \sqrt{1 + [f'(x)]^2}\, dx$$

speed along a curve parametrized by time t : $v(t) = \sqrt{[x'(t)]^2 + [y'(t)]^2}$

9.9 The Area of a Surface of Revolution; The Centroid of a Curve; Pappus's Theorem on Surface Area

revolution of a curve about x-axis:

$$A = \int_c^d 2\pi\, y(t)\sqrt{[x'(t)]^2 + [y'(t)]^2}\, dt$$

revolution of a graph about x-axis:

$$A = \int_a^b 2\pi f(x)\sqrt{1 + [f'(x)]^2}\, dx$$

principles for finding the centroid of a plane curve (p. 578)
parametrizing a curve by arc length (p. 578)
centroid of a plane curve:

$$\overline{x}L = \int_c^d x(t)\sqrt{x'(t)]^2 + [y'(t)]^2}\, dt$$

$$\overline{y}L = \int_c^d y(t)\sqrt{x'(t)]^2 + [y'(t)]^2}\, dt$$

Pappus's theorem on surface area: $A = 2\pi\overline{R}L$

CHAPTER

10

SEQUENCES; INDETERMINATE FORMS; IMPROPER INTEGRALS

■ 10.1 THE LEAST UPPER BOUND AXIOM

So far, our approach to the real number system has been somewhat primitive. We have simply taken the point of view that there is a one-to-one correspondence between the set of points on a line and the set of real numbers, and that this enables us to measure all distances, take all roots of nonnegative numbers, and, in short, fill in all the gaps left by the set of rational numbers. This point of view is basically correct and has served us well, but it is not sufficiently sharp to put our theorems on a sound basis, nor is it sufficiently sharp for the work that lies ahead.

We begin with a nonempty set S of real numbers. As indicated in Section 1.2, a number M is an *upper bound* for S if

$$x \leq M \quad \text{for all } x \in S.$$

It follows that if M is an upper bound for S, then every number in $[M, \infty)$ is also an upper bound for S. Of course, not all sets of real numbers have upper bounds. Those that do are said to be *bounded above*.

It is clear that every set that has a largest element has an upper bound: if b is the largest element of S, then $x \leq b$ for all $x \in S$. This makes b an upper bound for S. The converse is false: the sets

$$S_1 = (-\infty, 0) \quad \text{and} \quad S_2 = \left\{\frac{1}{2}, \frac{2}{3}, \frac{3}{4}, \ldots, \frac{n}{n+1}, \ldots\right\}$$

both have upper bounds (for instance, 2 is an upper bound for each set), but neither has a largest element.

Let's return to the first set, S_1. While $(-\infty, 0)$ does not have a largest element, the set of its upper bounds, namely $[0, \infty)$, does have a smallest element, namely 0. We call 0 the *least upper bound* of $(-\infty, 0)$.

Now let's reexamine S_2. While the set of quotients

$$\frac{n}{n+1} = 1 - \frac{1}{n+1}, \; n = 1, 2, 3, \ldots,$$

does not have a greatest element, the set of its upper bounds, $[1, \infty)$, does have a least element, 1. The number 1 is the *least upper bound* of that set of quotients.

In general, if S is a nonempty set of numbers which is bounded above, then the *least upper bound* of S is an upper bound that is less than or equal to any other upper bound for S.

We now state explicitly one of the key *assumptions* that we make about the real number system. It is called the *least upper bound axiom*, and it provides the sharpness and the clarity that we require.

AXIOM 10.1.1 LEAST UPPER BOUND AXIOM

Every nonempty set of real numbers that has an upper bound has a *least* upper bound.

Some find this axiom obvious; some find it unintelligible. For those of you who find it obvious, note that the axiom is not satisfied by the rational number system; namely, it is not true that every nonempty set of rational numbers that has a rational upper bound has a least rational upper bound. (For a detailed illustration of this, we refer you to Exercise 33). Those who find the axiom unintelligible will come to understand it by working with it.

We indicate the least upper bound of a set S by writing lub S. As you will see from the examples below, the least upper bound idea has wide applicability.

1. $\text{lub}\,(-\infty, 0) = 0, \qquad \text{lub}\,(-\infty, 0] = 0.$

2. $\text{lub}\,(-4, -1) = -1 \qquad \text{lub}\,(-4, -1] = -1.$

3. $\text{lub}\left\{\frac{1}{2}, \frac{2}{3}, \frac{3}{4}, \ldots, \frac{n}{n+1}, \ldots\right\} = 1.$

4. $\text{lub}\left\{-\frac{1}{2}, -\frac{1}{8}, -\frac{1}{27}, \ldots, -\frac{1}{n^3}, \ldots\right\} = 0.$

5. $\text{lub}\,\{x : x^2 < 3\} = \text{lub}\,\{-\sqrt{3} < x < \sqrt{3}\} = \sqrt{3}.$

6. For each decimal fraction

$$a = 0.a_1a_2a_3\cdots$$

we have

$$a = \text{lub}\,\{0.a_1,\ 0.a_1a_2,\ 0.a_1a_2a_3, \ldots\}.$$

7. If S consists of the lengths of all polygonal paths inscribed in a semicircle of radius 1, then lub $S = \pi$ (half the circumference of the unit circle).

The least upper bound of a set has a special property that deserves particular attention. The idea is this: the fact that M is the least upper bound of set S does not guarantee that M is in S (indeed, it need not be, as illustrated in the preceding examples), but it guarantees that we can approximate M as closely as we wish by elements of S.

THEOREM 10.1.2

If M is the least upper bound of the set S and ϵ is a positive number, then there is at least one number s in S such that

$$M - \epsilon < s \le M.$$

PROOF Let $\epsilon > 0$. Since M is an upper bound for S, the condition $s \leq M$ is satisfied by all numbers s in S. All we have to show therefore is that there is some number s in S such that

$$M - \epsilon < s.$$

Suppose on the contrary that there is no such number in S. We then have

$$x \leq M - \epsilon \quad \text{for all } x \in S.$$

This makes $M - \epsilon$ an upper bound for S. But this cannot be, for then $M - \epsilon$ is an upper bound for S that is *less* than M, which contradicts the assumption that M is the *least* upper bound. □

The theorem we just proved is illustrated in Figure 10.1.1. Take S as the set of points marked in the figure. If $M = \text{lub } S$, then S has at least one element in every half-open interval of the form $(M - \epsilon, M]$.

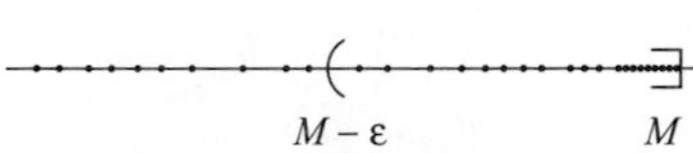

Figure 10.1.1

Example 1

(a) Let $$S = \left\{\frac{1}{2}, \frac{2}{3}, \frac{3}{4}, \ldots, \frac{n}{n+1}, \ldots\right\}$$

and take $\epsilon = 0.0001$. Since 1 is the least upper bound of S, there must be a number s in S such that

$$1 - 0.0001 < s \leq 1.$$

There is: take, for example, $s = \frac{99999}{100000}$.

(b) Let $$S = \{0, 1, 2, 3\}$$
and take $\epsilon = 0.00001$. It is clear that 3 is the least upper bound of S. Therefore, there must be a number $s \in S$ such that

$$3 - 0.00001 < s \leq 3.$$

There is: $s = 3$. □

We come now to lower bounds. Recall that a number m is called a *lower bound* for a nonempty set S if

$$m \leq x \quad \text{for all } x \in S.$$

Sets that have lower bounds are said to be *bounded below*. Not all sets have lower bounds, but those that do have *greatest lower bounds*. We don't have to assume this. We can prove it by using the least upper bound axiom.

> **THEOREM 10.1.3**
>
> Every nonempty set of real numbers that has a lower bound has a *greatest* lower bound.

PROOF Suppose that S is nonempty and that it has a lower bound k. Then

$$k \leq s \quad \text{for all } s \in S.$$

It follows that $-s \leq -k$ for all $s \in S$; that is,

$$\{-s : s \in S\} \quad \text{has an upper bound } -k.$$

From the least upper bound axiom we conclude that $\{-s : s \in S\}$ has a least upper bound; call it m. Since $-s \leq m$ for all $s \in S$, we can see that

$$-m \leq s \text{ for all } s \in S,$$

and thus $-m$ is a lower bound for S. We now assert that $-m$ is the greatest lower bound of the set S. To see this, note that, if there existed a number m_1 satisfying

$$-m < m_1 \leq s \text{ for all } s \in S,$$

then we would have

$$-s \leq -m_1 < m \text{ for all } s \in S,$$

and thus m would not be the *least* upper bound of $\{-s : s \in S\}$.† □

The greatest lower bound of a set, although not necessarily in the set, can be approximated as closely as we wish by members of the set. In short, we have the following theorem, the proof of which is left as an exercise.

THEOREM 10.1.4

If m is the greatest lower bound of the set S and ϵ is a positive number, then there is at least one number s in S such that

$$m \leq s < m + \epsilon.$$

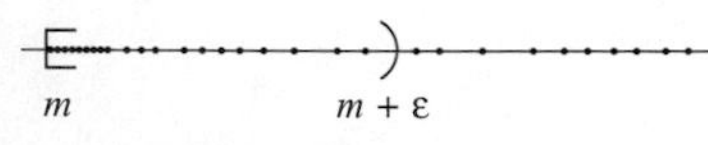

Figure 10.1.2

The theorem is illustrated in Figure 10.1.2. If $m = \text{glb}\, S$ (that is, if m is the greatest lower bound of the set S), then S has at least one element in every half-open interval of the form $[m, m + \epsilon)$.

Remark Given that a function f is defined and continuous on $[a, b]$, what do we know about f. Certainly the following:

1. We know that if f takes on two values, then it takes on every value in between; f maps intervals onto intervals (the intermediate-value theorem).
2. We know that f takes on both a maximum value and minimum value (the extreme-value theorem).

We so-call "know" this, but actually we have proven none of it. With the least upper bound axiom in hand, we can prove both theorems. (Appendix B.) □

† We proved Theorem 10.1.3 by assuming the least upper bound axiom. We could have proceeded the other way. We could have set Theorem 10.1.3 as an axiom, and then proved the least upper bound axiom as a theorem.

EXERCISES 10.1

Find the least upper bound (if it exists) and the greatest lower bound (if it exists) for the given set.

1. $(0, 2)$.

2. $[0, 2]$.

3. $(0, \infty)$.

4. $(-\infty, 1)$.

5. $\{x : x^2 < 4\}$.

6. $\{x : |x - 1| < 2\}$.

7. $\{x : x^3 \geq 8\}$.

8. $\{x : x^4 \leq 16\}$.

9. $\{2\frac{1}{2}, 2\frac{1}{3}, 2\frac{1}{4}, \ldots\}$.

10. $\{-1, -\frac{1}{2}, -\frac{1}{3}, -\frac{1}{4}, \ldots\}$.

11. $\{0.9, 0.99, 0.999, \ldots\}$.

12. $\{-2, 2, -2.1, 2.1, -2.11, 2.11, \ldots\}$.

13. $\{x : \ln x < 1\}$.

14. $\{x : \ln x > 0\}$.

15. $\{x : x^2 + x - 1 < 0\}$.

16. $\{x : x^2 + x + 2 \geq 0\}$.

17. $\{x : x^2 > 4\}$.

18. $\{x : |x - 1| > 2\}$.

19. $\{x : \sin x \geq -1\}$.

20. $\{x : e^x < 1\}$.

Illustrate the validity of Theorem 10.1.4 taking S and ϵ as given below.

21. $S = \{\frac{1}{11}, (\frac{1}{11})^2, (\frac{1}{11})^3, \ldots, (\frac{1}{11})^n, \ldots\}, \quad \epsilon = 0.001$.

22. $S = \{1, 2, 3, 4\}, \quad \epsilon = 0.0001$.

23. $S = \{\frac{1}{10}, \frac{1}{1000}, \frac{1}{100000}, \ldots, (\frac{1}{10})^{2n-1}, \ldots\}$,
$\epsilon = (\frac{1}{10})^k \ (k \geq 1)$.

24. $S = \{\frac{1}{2}, \frac{1}{4}, \frac{1}{8}, \ldots, (\frac{1}{2})^n, \ldots\}, \quad \epsilon = (\frac{1}{4})^k \ (k \geq 1)$.

25. Prove Theorem 10.1.4 by imitating the proof of Theorem 10.1.2

26. Let $S = \{a_1, a_2, a_3, \ldots, a_n\}$ be a nonempty, finite set of real numbers.
    (a) Show that S is bounded.
    (b) Show that lub S and glb S are elements of S.

27. Suppose that b is an upper bound for a set S of real numbers. Show that if $b \in S$, then $b = \text{lub}\, S$.

28. Let S be a bounded set of real numbers and suppose that $\text{lub}\, S = \text{glb}\, S$. What can you conclude about S?

29. Suppose that S is a nonempty, bounded set of real numbers and that T is a nonempty subset of S.
    (a) Show that T is bounded.
    (b) Show that $\text{glb}\, S \leq \text{glb}\, T \leq \text{lub}\, T \leq \text{lub}\, S$.

30. Let S and T be nonempty sets of real numbers such that $x \leq y$ for all $x \in S$ and all $y \in T$.
    (a) Show that $\text{lub}\, S \leq y$ for all $y \in T$.
    (b) Show that $\text{lub}\, S \leq \text{glb}\, T$.

31. Let c be a positive number. Prove that the set $S = \{c, 2c, 3c, \ldots, nc, \ldots\}$ is not bounded above.

32. Show that if a and b are any two positive numbers with $a < b$, then there exists a rational number r such that $a < r < b$. HINT: Show that there is a positive integer n such that $na > 1$ and $n(b - a) > 1$. Then show that there is a positive integer m such that $m > na$ and $m - 1 \leq na$. Let $r = m/n$.

33. The set S of rational numbers x with $x^2 < 2$ has rational upper bounds but no least rational upper bound. The argument goes like this. Suppose that S has a least rational upper bound and call it x_0. Then either

$$x_0^2 = 2, \qquad \text{or} \qquad x_0^2 > 2, \qquad \text{or} \qquad x_0^2 < 2.$$

    (a) Show that $x_0^2 = 2$ is impossible by showing that if $x_0^2 = 2$, then x_0 is not rational.
    (b) Show that $x_0^2 > 2$ is impossible by showing that if $x_0^2 > 2$, then there is a positive integer n for which $(x_0 - \frac{1}{n})^2 > 2$, which makes $x_0 - \frac{1}{n}$ a rational upper bound for S that is less than the least rational upper bound x_0.
    (c) Show that $x_0^2 < 2$ is impossible by showing that if $x_0^2 < 2$, then there is a positive integer n for which $(x_0 + \frac{1}{n})^2 < 2$. This places $x_0 + \frac{1}{n}$ in S and shows that x_0 cannot be an upper bound for S.

C 34. Recall that a *prime number* is an integer $p > 1$ that has no divisors other than 1 and p. Let S be the set of prime numbers

$$S = \{2, 3, 5, 7, 11, 13, 17, 19, \ldots, p_n, \ldots\}.$$

There are infinitely many prime numbers, a famous theorem proved by Euclid.
    (a) Use a CAS to find p_{25} and p_{50}.
    (b) Does S have a least upper bound? A greatest lower bound?

C 35. Let S be the set of numbers

$$S = \{\tfrac{1}{2}, (\tfrac{2}{3})^2, (\tfrac{3}{4})^3, (\tfrac{4}{5})^4, \ldots, (\tfrac{n}{n+1})^n, \ldots\}.$$

    (a) Use a CAS to calculate $(\frac{n}{n+1})^n$ for $n = 5, 10, 100, 1000, 10,000$.
    (b) Does S have a least upper bound? If so, what is it? Does S have a greatest lower bound? If so, what is it?

C 36. Let $S = \{a_1, a_2, a_3, \ldots, a_n, \ldots\}$ where $a_1 = 4$ and for each positive integer n, $a_{n+1} = 3 - \dfrac{3}{a_n}$.
    (a) Calculate the numbers $a_2, a_3, a_4, \ldots, a_{10}$.
    (b) Use a CAS to calculate $a_{20}, a_{30}, \ldots, a_{50}$.
    (c) Does S have a least upper bound? If so, what is it? Does S have a greatest lower bound? If so, what is it?

C 37. Let S be the set of irrational numbers

$$S = \{\sqrt{2}, \sqrt{2\sqrt{2}}, \sqrt{2\sqrt{2\sqrt{2}}}, \ldots\}.$$

That is, $S = \{a_1, a_2, a_3, \ldots, a_n \ldots\}$ where $a_1 = \sqrt{2}$, and for each positive integer n, $a_{n+1} = \sqrt{2a_n}$.
    (a) Calculate the numbers $a_1, a_2, a_3, \ldots, a_{10}$.
    (b) Use mathematical induction to show that $a_n < 2$ for all n.
    (c) Is 2 the least upper bound of S?
    (d) Choose a positive number other than 2 and repeat this exercise. What can you conclude?

C 38. Let S be the set of irrational numbers

$$S = \{\sqrt{2}, \sqrt{2 + \sqrt{2}}, \sqrt{2 + \sqrt{2 + \sqrt{2}}}, \ldots\}.$$

That is, $a_1 = \sqrt{2}$, and for each positive integer n, $a_{n+1} = \sqrt{2 + a_n}$.
    (a) Calculate the numbers $a_1, a_2, a_3, \ldots, a_{10}$.
    (b) Use mathematical induction to show that $a_n < 2$ for all n.
    (c) Is 2 the least upper bound of S?
    (d) Choose a positive number other than 2 and repeat this exercise. What can you conclude?

■ 10.2 SEQUENCES OF REAL NUMBERS

To this point we have considered sequences only in a peripheral manner. Here we focus on them.

What is a sequence of real numbers?

DEFINITION 10.2.1 SEQUENCE OF REAL NUMBERS

A *sequence of real numbers* is a real-valued function defined on the set of positive integers.

You may find this definition somewhat surprising, but in a moment you will see that it makes sense.

Suppose we have a sequence of real numbers

$$a_1,\ a_2,\ a_3, \ldots,\ a_n, \ldots.$$

What is a_1? It is the image of 1. What is a_2? It is the image of 2. What is a_3? It is the image of 3. In general, a_n is the image of n.

By convention, a_1 is called the *first term* of the sequence, a_2 the *second term*, and so on. More generally, a_n, the term with *index n*, is called the *n*th *term*.

Sequences can be defined by giving the law of formation. For example:

$$a_n = \frac{1}{n} \quad \text{is the sequence} \quad 1, \tfrac{1}{2}, \tfrac{1}{3}, \tfrac{1}{4}, \ldots;$$

$$b_n = \frac{n}{n+1} \quad \text{is the sequence} \quad \tfrac{1}{2}, \tfrac{2}{3}, \tfrac{3}{4}, \tfrac{4}{5}, \ldots;$$

$$c_n = n^2 \quad \text{is the sequence} \quad 1, 4, 9, 16, \ldots.$$

It's like defining f by giving $f(x)$.

Sequences can be multiplied by constants; they can be added, subtracted, and multiplied. From

$$a_1,\ a_2,\ a_3, \ldots,\ a_n, \ldots \quad \text{and} \quad b_1,\ b_2,\ b_3, \ldots,\ b_n, \ldots$$

we can form

$$\text{the scalar product sequence}: \ \alpha a_1,\ \alpha a_2,\ \alpha a_3, \ldots,\ \alpha a_n, \ldots,$$

$$\text{the sum sequence}: \ a_1 + b_1,\ a_2 + b_2,\ a_3 + b_3, \ldots,\ a_n + b_n, \ldots,$$

$$\text{the difference sequence}: \ a_1 - b_1,\ a_2 - b_2,\ a_3 - b_3, \ldots,\ a_n - b_n, \ldots,$$

$$\text{the product sequence}: \ a_1 b_1,\ a_2 b_2,\ a_3 b_3, \ldots,\ a_n b_n, \ldots.$$

If the b_i's are all different from zero, we can form

$$\text{the reciprocal sequence}: \ \frac{1}{b_1}, \frac{1}{b_2}, \frac{1}{b_3}, \ldots, \frac{1}{b_n}, \ldots,$$

$$\text{the quotient sequence}: \ \frac{a_1}{b_1}, \frac{a_2}{b_2}, \frac{a_3}{b_3}, \ldots, \frac{a_n}{b_n}, \ldots.$$

The *range* of a sequence is the set of values taken on by the sequence. The range of the sequence $a_1,\ a_2,\ a_3, \ldots,\ a_n, \ldots$ is the set

$$\{a_n : n = 1, 2, 3, \ldots\}.$$

Thus the range of the sequence

$$0, 1, 0, 1, 0, 1, 0, 1, \ldots$$

made up of alternating zeros and ones is the set $\{0, 1\}$. The range of the sequence

$$0, 1, -1, 2, 2, -2, 3, 3, 3, -3, 4, 4, 4, 4, -4, \ldots$$

is the set of integers.

Boundedness and unboundedness for sequences are what they are for other real-valued functions. Thus the sequence $a_n = 2^n$ is bounded below (with greatest lower bound 2) but unbounded above. The sequence $b_n = 2^{-n}$ is bounded. It is bounded below with greatest lower bound 0, and it is bounded above with least upper bound $\frac{1}{2}$.

Many of the sequences we work with have some regularity. They either have an upward tendency or they have a downward tendency. The following terminology is standard. The sequence $\{a_n\}$ is said to be

increasing	if	$a_n < a_{n+1}$	for all n,
nondecreasing	if	$a_n \leq a_{n+1}$	for all n,
decreasing	if	$a_n > a_{n+1}$	for all n,
nonincreasing	if	$a_n \geq a_{n+1}$	for all n.

A sequence that satisfies any of these conditions is called *monotonic*.

The sequences

$$1, \tfrac{1}{2}, \tfrac{1}{3}, \tfrac{1}{4}, \ldots, \tfrac{1}{n}, \ldots$$
$$2, 4, 8, 16, \ldots, 2^n$$
$$2, 2, 4, 4, 6, 6 \ldots, 2n, 2n, \ldots$$

are monotonic. The sequence

$$1, \tfrac{1}{2}, 1, \tfrac{1}{3}, 1, \tfrac{1}{4}, 1, \ldots$$

is not monotonic.

Now to some examples that are less trivial.

Example 1 The sequence $a_n = \dfrac{n}{n+1}$ is increasing. It is bounded below by $\frac{1}{2}$ (the greatest lower bound) and above by 1 (the least upper bound).

PROOF Since

$$\frac{a_{n+1}}{a_n} = \frac{(n+1)/(n+2)}{n/(n+1)} = \frac{n+1}{n+2} \cdot \frac{n+1}{n} = \frac{n^2+2n+1}{n^2+2n} > 1,$$

we have $a_n < a_{n+1}$. This confirms that the sequence is increasing. The sequence can be displayed as

$$\tfrac{1}{2}, \tfrac{2}{3}, \tfrac{3}{4}, \tfrac{4}{5}, \ldots, \tfrac{98}{99}, \tfrac{99}{100}, \ldots.$$

It is clear that $\frac{1}{2}$ is the greatest lower bound and 1 is the least upper bound (see the two representations in Figure 10.2.1).

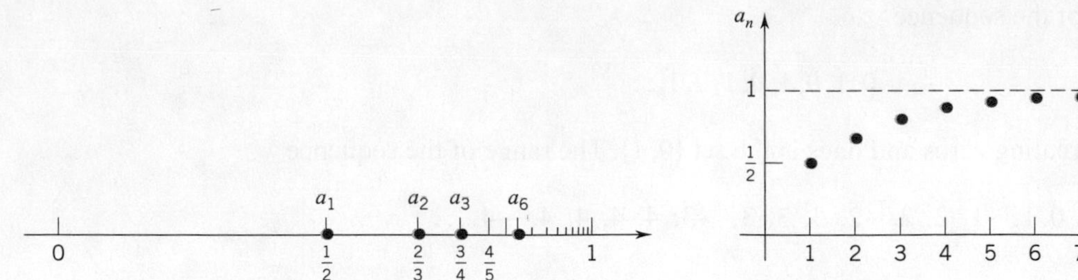

Figure 10.2.1

Example 2 The sequence $a_n = \dfrac{2^n}{n!}$ is nonincreasing and starts decreasing at $n = 2$. †

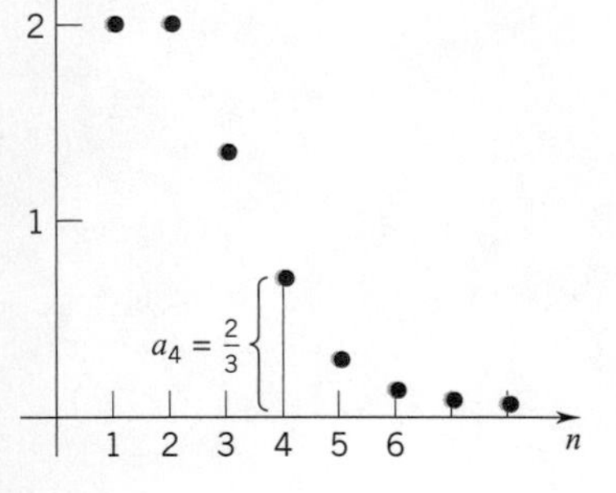

Figure 10.2.2

PROOF The first two terms are equal:

$$a_1 = \frac{2^1}{1!} = 2 = \frac{2^2}{2!} = a_2.$$

For $n \geq 2$ the sequence decreases:

$$\frac{a_{n+1}}{a_n} = \frac{2^{n+1}}{(n+1)!} \cdot \frac{n!}{2^n} = \frac{2}{n+1} < 1 \qquad \text{(see Figure 10.2.2).} \quad \square$$

Example 3 For $c > 1$, the sequence $a_n = c^n$ increases without bound.

PROOF Choose a number $c > 1$. Then

$$\frac{a_{n+1}}{a_n} = \frac{c^{n+1}}{c^n} = c > 1.$$

This shows that the sequence increases. To show the unboundedness, we take an arbitrary positive number M and show that there exists a positive integer k for which

$$c^k \geq M.$$

A suitable k is one such that

$$k \geq \frac{\ln M}{\ln c},$$

for then

$$k \ln c \geq \ln M, \quad \ln c^k \geq \ln M, \quad c^k \geq M. \quad \square$$

Since sequences are defined on the set of positive integers and not on an interval, they are not directly susceptible to the methods of calculus. Fortunately, we can sometimes circumvent this difficulty by dealing initially, not with the sequence itself, but with a function of a real variable x that agrees with the given sequence at the positive integers n.

Example 4 The sequence $a_n = \dfrac{n}{e^n}$ is decreasing. It is bounded above by $1/e$ and below by 0.

† Recall that $n! = n(n-1)(n-2)\cdots 3 \cdot 2 \cdot 1$. See Section 1.2.

PROOF We will work with the function

$$f(x) = \frac{x}{e^x}. \qquad \text{(See Example 2, Section 7.4.)}$$

Note that $f(1) = 1/e = a_1$, $f(2) = 2/e^2 = a_2$, $f(3) = 3/e^3 = a_3$, and so on.

Differentiating f, we get

$$f'(x) = \frac{e^x - x\,e^x}{e^{2x}} = \frac{1-x}{e^x}.$$

Since $f'(x) < 0$ for $x > 1$, f decreases on $[1, \infty)$. Thus $f(1) > f(2) > f(3) > \cdots$, that is, $a_1 > a_2 > a_3 > \cdots$. The sequence is decreasing.

The first term $a_1 = 1/e$ is the least upper bound of the sequence. Since all the terms of the sequence are positive, 0 is a lower bound for the sequence.

In Figure 10.2.3 we have sketched the graph of $f(x) = x/e^x$ and marked some points (n, a_n). As the figure suggests, 0 is the greatest lower bound of the sequence. □

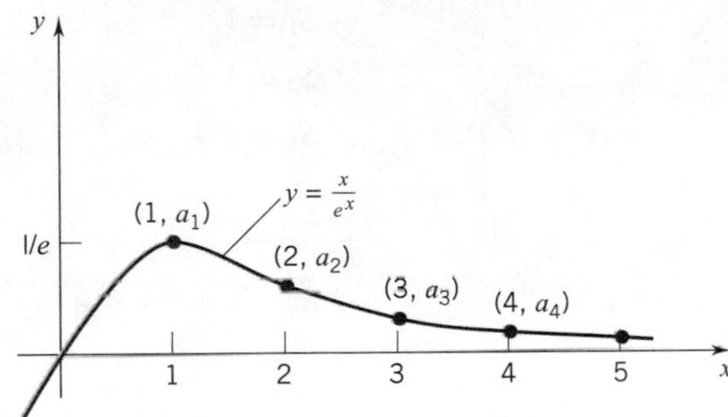

Figure 10.2.3

Example 5 The sequence $a_n = n^{1/n}$ decreases for $n \geq 3$.

PROOF We could compare a_n with a_{n+1} directly, but it is easier to consider the function

$$f(x) = x^{1/x}$$

instead. Since $f(x) = e^{(1/x)\ln x}$,

we have

$$f'(x) = e^{(1/x)\ln x}\frac{d}{dx}\left(\frac{1}{x}\ln x\right) = x^{1/x}\left(\frac{1-\ln x}{x^2}\right).$$

For $x > e$, $f'(x) < 0$. This shows that f decreases on $[e, \infty)$. Since $3 > e$, the function f decreases on $[3, \infty)$, and the sequence decreases for $n \geq 3$. □

EXERCISES 10.2

The first several terms of a sequence $\{a_n\}$ are given. Assume that the pattern continues as indicated and find an explicit formula for a_n.

1. $2, 5, 8, 11, 14, \ldots$

2. $2, 0, 2, 0, 2, \ldots$

3. $1, -\frac{1}{3}, \frac{1}{5}, -\frac{1}{7}, \frac{1}{9}, \ldots$

4. $\frac{1}{2}, \frac{3}{4}, \frac{7}{8}, \frac{15}{16}, \frac{31}{32}, \ldots$

5. $2, \frac{5}{2}, \frac{10}{3}, \frac{17}{4}, \frac{26}{5}, \ldots$

6. $-\frac{1}{4}, \frac{2}{9}, -\frac{3}{16}, \frac{4}{25}, -\frac{5}{36}, \ldots$

7. $1, \frac{1}{2}, 3, \frac{1}{4}, 5, \frac{1}{6}, \ldots$

8. $1, 2, \frac{1}{9}, 4, \frac{1}{25}, 6, \frac{1}{49}, \ldots$

Determine the boundedness and monotonicity of the sequence with a_n as indicated.

9. $\frac{2}{n}$.

10. $\frac{(-1)^n}{n}$.

11. $\frac{n + (-1)^n}{n}$.

12. $(1.001)^n$.

13. $(0.9)^n$.

14. $\frac{n-1}{n}$.

15. $\dfrac{n^2}{n+1}$.

16. $\sqrt{n^2+1}$.

17. $\dfrac{4n}{\sqrt{4n^2+1}}$.

18. $\dfrac{2^n}{4^n+1}$.

19. $\dfrac{4^n}{2^n+100}$.

20. $\dfrac{n^2}{\sqrt{n^3+1}}$.

21. $\ln\left(\dfrac{2n}{n+1}\right)$.

22. $\dfrac{n+2}{3^{10}\sqrt{n}}$.

23. $\dfrac{(n+1)^2}{n^2}$.

24. $(-1)^n\sqrt{n}$.

25. $\sqrt{4-\dfrac{1}{n}}$.

26. $\ln\left(\dfrac{n+1}{n}\right)$.

27. $(-1)^{2n+1}\sqrt{n}$.

28. $\dfrac{\sqrt{n+1}}{\sqrt{n}}$.

29. $\dfrac{2^n-1}{2^n}$.

30. $\dfrac{1}{2n}-\dfrac{1}{2n+3}$.

31. $\sin\dfrac{\pi}{n+1}$.

32. $(-\frac{1}{2})^n$.

33. $(1.2)^{-n}$.

34. $\dfrac{n+3}{\ln(n+3)}$.

35. $\dfrac{1}{n}-\dfrac{1}{n+1}$.

36. $\cos n\pi$.

37. $\dfrac{\ln(n+2)}{n+2}$.

38. $\dfrac{(-2)^n}{n^{10}}$.

39. $\dfrac{3^n}{(n+1)^2}$.

40. $\dfrac{1-(\frac{1}{2})^n}{(\frac{1}{2})^n}$.

41. Show that the sequence $a_n = 5^n/n!$ decreases for $n \geq 5$. Is the sequence nonincreasing?

42. Let M be a positive integer. Show that $a_n = M^n/n!$ decreases for $n \geq M$.

43. Show that, if $0 < c < d$, then the sequence

$$a_n = (c^n + d^n)^{1/n}$$

is bounded and monotonic.

44. Show that linear combinations and products of bounded sequences are bounded.

Sequences can be defined *recursively*: one or more terms are given explicitly; the remaining ones are then defined in terms of their predecessors. In Exercises 45–56, give the first six terms of the sequence and then give the nth term.

45. $a_1 = 1;\quad a_{n+1} = \dfrac{1}{n+1}a_n$.

46. $a_1 = 1;\quad a_{n+1} = a_n + 3n(n+1) + 1$.

47. $a_1 = 1;\quad a_{n+1} = \frac{1}{2}(a_n + 1)$.

48. $a_1 = 1;\quad a_{n+1} = \frac{1}{2}a_n + 1$.

49. $a_1 = 1;\quad a_{n+1} = a_n + 2$.

50. $a_1 = 1;\quad a_{n+1} = \dfrac{n}{n+1}a_n$.

51. $a_1 = 1;\quad a_{n+1} = a_n + 2n + 1$.

52. $a_1 = 1;\quad a_{n+1} = 2a_n + 1$.

53. $a_1 = 1;\quad a_{n+1} = a_n + \cdots + a_1$.

54. $a_1 = 3;\quad a_{n+1} = 4 - a_n$.

55. $a_1 = 1,\quad a_2 = 3;\quad a_{n+1} = 2a_n - a_{n-1},\quad n \geq 2$.

56. $a_1 = 1,\quad a_2 = 3;\quad a_{n+1} = 3a_n - 2n - 1,\quad n \geq 2$.

In Exercises 57–60, use mathematical induction to prove the following assertions for all $n \geq 1$.

57. If $a_1 = 1$ and $a_{n+1} = 2a_n + 1$, then $a_n = 2^n - 1$.

58. If $a_1 = 3$ and $a_{n+1} = a_n + 5$, then $a_n = 5n - 2$.

59. If $a_1 = 1$ and $a_{n+1} = \dfrac{n+1}{2n}a_n$, then $a_n = \dfrac{n}{2^{n-1}}$.

60. If $a_1 = 1$ and $a_{n+1} = a_n - \dfrac{1}{n(n+1)}$, then $a_n = \dfrac{1}{n}$.

61. Let r be a real number, $r \neq 0$. Define a sequence $\{S_n\}$ by

$$\begin{aligned} S_1 &= 1 \\ S_2 &= 1 + r \\ S_3 &= 1 + r + r^2 \\ &\vdots \\ S_n &= 1 + r + r^2 + \cdots + r^{n-1} \\ &\vdots \end{aligned}$$

(a) Suppose $r = 1$. What is S_n for $n = 1, 2, 3, \ldots$?

(b) Suppose $r \neq 1$. Find a formula for S_n that does not involve adding up the powers of r. HINT: Calculate $S_n - rS_n$.

62. Set $a_n = \dfrac{1}{n(n+1)}$, $n = 1, 2, 3, \ldots$, and form the sequence

$$\begin{aligned} S_1 &= a_1 \\ S_2 &= a_1 + a_2 \\ S_3 &= a_1 + a_2 + a_3 \\ &\vdots \\ S_n &= a_1 + a_2 + a_3 + \cdots + a_n \\ &\vdots \end{aligned}$$

Find a formula for $S_n, n = 1, 2, 3, \ldots$, that does not involve adding up the terms $a_1, a_2, a_3, \ldots$. HINT: Use partial fractions to write $1/[k(k+1)]$ as the sum of two fractions.

63. A ball is dropped from a height of 100 feet. Each time it hits the ground, it rebounds to 75% of its previous height.

(a) Let S_n be the distance that the ball travels between the nth and $(n+1)$st bounce, $n = 1, 2, 3, \ldots$. Find a formula for S_n.

(b) Let T_n be the time that the ball is in the air between the nth and $(n+1)$st bounce, $n = 1, 2, 3, \ldots$. Find a formula for T_n.

64. Suppose that the number of bacteria in a culture is growing exponentially (see Section 7.6) and that the number doubles every 12 hours. Find a formula for the number P_n of bacteria in the culture after n hours, given that there are 500 bacteria initially.

C 65. Let $\{a_n\}$ be the sequence defined by setting $a_n = \sqrt{n^2+n} - n$, $n = 1, 2, 3, \ldots$. Use a CAS to determine whether $\{a_n\}$ is increasing, nondecreasing, decreasing, nonincreasing, or none of these.

C 66. Repeat Exercise 65 with the sequence $\{a_n\}$ defined recursively by setting

$$a_1 = 100; \qquad a_{n+1} = \sqrt{2 + a_n}, \qquad n = 1, 2, 3, \ldots.$$

C 67. Let $\{a_n\}$ be the sequence defined recursively by setting

$$a_1 = 1; \qquad a_{n+1} = 1 + \sqrt{a_n}, \qquad n = 1, 2, 3, \ldots.$$

(a) Show by induction that $\{a_n\}$ is an increasing sequence.

(b) Show by induction that $\{a_n\}$ is bounded above.

(c) Calculate $a_2, a_3, a_4, \ldots, a_{15}$. Estimate the least upper bound of the sequence.

C 68. Let $\{a_n\}$ be the sequence defined recursively by setting

$$a_1 = 1; \qquad a_{n+1} = \sqrt{3a_n}, \qquad n = 1, 2, 3, \ldots.$$

(a) Show by induction that $\{a_n\}$ is an increasing sequence.

(b) Show by induction that $\{a_n\}$ is bounded above.

(c) Calculate $a_2, a_3, a_4, \ldots, a_{15}$. Estimate the least upper bound of the sequence.

C 69. Let p_n denote the nth prime number (see Exercise 34, Section 10.1). Define the sequence $\{a_n\}$ by setting

$$a_n = \frac{p_n}{n \ln n}, \qquad n = 2, 3, 4, \ldots.$$

(a) Use a CAS to investigate a_n for large n.

(b) Is $\{a_n\}$ a bounded sequence?

(c) Does it appear that $\lim_{n\to\infty} a_n$ exists? If so, what is the limit?

C 70. Let $\{a_n\}$ be the sequence defined recursively by setting

$$a_1 = 1, \; a_2 = 3, \; \text{and} \; a_{n+2} = \frac{1 + a_{n+1}}{a_n}, \qquad n = 1, 2, 3, \ldots.$$

(a) Use a CAS to investigate the behavior of the sequence. Is the sequence monotonic? Is it bounded? If so, give the least upper bound and the greatest lower bound.

(b) Experiment with other "initial" values a_1 and a_2.

■ 10.3 LIMIT OF A SEQUENCE

You have seen the limit process applied in various settings. The limit process applied to sequences is exactly what you would expect.

> **DEFINITION 10.3.1 LIMIT OF A SEQUENCE**
>
> $$\lim_{n\to\infty} a_n = L$$
>
> if for each $\epsilon > 0$, there exists a positive integer K such that
>
> $$\text{if} \quad n \geq K, \qquad \text{then} \qquad |a_n - L| < \epsilon.$$

Example 1 $\displaystyle\lim_{n\to\infty} \frac{4n-1}{n} = 4.$

PROOF Let $\epsilon > 0$. We must show that there exists an integer K such that

$$\text{if} \quad n \geq K, \qquad \text{then} \qquad \left|\frac{4n-1}{n} - 4\right| < \epsilon.$$

Note that

$$\left|\frac{4n-1}{n} - 4\right| = \left|\frac{4n-1-4n}{n}\right| = \left|\frac{-1}{n}\right| = \frac{1}{n}.$$

Choose an integer $K > 1/\epsilon$. Then $1/K < \epsilon$. Now, if $n \geq K$, then $1/n \leq 1/K$ and

$$\left|\frac{4n-1}{n} - 4\right| = \frac{1}{n} \leq \frac{1}{K} < \epsilon. \quad \square$$

Example 2 $\displaystyle\lim_{n\to\infty} \frac{2\sqrt{n}}{\sqrt{n}+1} = 2.$

PROOF Let $\epsilon > 0$. We want a positive integer K such that

$$\text{if} \quad n \geq K, \qquad \text{then} \qquad \left|\frac{2\sqrt{n}}{\sqrt{n}+1} - 2\right| < \epsilon.$$

Note that

$$\left|\frac{2\sqrt{n}}{\sqrt{n}+1} - 2\right| = \left|\frac{2\sqrt{n} - 2(\sqrt{n}+1)}{\sqrt{n}+1}\right| = \left|\frac{-2}{\sqrt{n}+1}\right| = \frac{2}{\sqrt{n}+1} < \frac{2}{\sqrt{n}}.$$

We want $2/\sqrt{n} < \epsilon$. We can get this by having $\sqrt{n} > 2/\epsilon$; that is, by having $n > (2/\epsilon)^2$. Choose $K > (2/\epsilon)^2$. Then

$$n \geq K \quad \text{implies} \quad n > (2/\epsilon)^2, \quad \text{which forces} \quad \left|\frac{2\sqrt{n}}{\sqrt{n}+1} - 2\right| < \epsilon. \quad \square$$

The next example justifies the familiar statement

$$\tfrac{1}{3} = 0.333\ldots.$$

Example 3 The decimal fractions $a_n = 0.\overbrace{33\ldots3}^{n}, \quad n = 1, 2, 3, \ldots$ tend to $\frac{1}{3}$ as a limit:

$$\lim_{n\to\infty} a_n = \tfrac{1}{3}.$$

PROOF Let $\epsilon > 0$. In the first place

$$(1) \qquad \left|a_n - \frac{1}{3}\right| = \left|0.\overbrace{33\cdots3}^{n} - \frac{1}{3}\right| = \left|\frac{0.\overbrace{99\cdots9}^{n} - 1}{3}\right| = \frac{1}{3}\cdot\frac{1}{10^n} < \frac{1}{10^n}.$$

Now choose K so that $1/10^K < \epsilon$. If $n \geq K$, then by (1)

$$\left|a_n - \frac{1}{3}\right| < \frac{1}{10^n} \leq \frac{1}{10^K} < \epsilon. \quad \square$$

Limit Theorems

The limit process for sequences is so similar to the limit processes you have already studied that you may find you can prove many of the limit theorems yourself. In any case, try to come up with your own proofs and refer to these only if necessary.

THEOREM 10.3.2 UNIQUENESS OF LIMIT

If $\lim_{n\to\infty} a_n = L$ and $\lim_{n\to\infty} a_n = M$, then $L = M$.

A proof, similar to the proof of Theorem 2.3.1, is given in the supplement at the end of this section.

DEFINITION 10.3.3

A sequence that has a limit is said to be *convergent*. A sequence that has no limit is said to be *divergent*.

Instead of writing

$$\lim_{n\to\infty} a_n = L,$$

we will often write

$$a_n \to L \qquad \text{(read “}a_n\text{ converges to }L\text{”)}$$

or more fully,

$$a_n \to L \quad \text{as} \quad n \to \infty.$$

THEOREM 10.3.4

Every convergent sequence is bounded.

PROOF Assume that $a_n \to L$ and choose any positive number: 1, for instance. Using 1 as ϵ, you can see that there must exist a positive integer K such that

$$|a_n - L| < 1 \quad \text{for all } n \geq K.$$

Since $|a_n| - |L| \leq \big||a_n| - |L|\big| \leq |a_n - L|$, we have

$$|a_n| < 1 + |L| \quad \text{for all } n \geq K.$$

It follows that,

$$|a_n| \leq \max\{|a_1|, |a_2|, \ldots, |a_{K-1}|, 1 + |L|\} \qquad \text{for all } n.$$

This proves that the sequence is bounded. □

Since every convergent sequence is bounded, a sequence that is not bounded cannot be convergent; namely,

(10.3.5) every unbounded sequence is divergent.

The sequences

$$a_n = \tfrac{1}{2}n, \quad b_n = \frac{n^2}{n+1}, \quad c_n = n \ln n$$

are all unbounded. Each of these sequences is therefore divergent.

Boundedness does not imply convergence. As a counterexample, consider the "oscillating" sequence

$$\{1, 0, 1, 0, 1, 0, \ldots\}$$

This sequence is certainly bounded (above by 1 and below by 0), but it does not converge: the limit would have to be arbitrarily close to both 0 and 1 simultaneously.

Boundedness together with monotonicity does imply convergence.

THEOREM 10.3.6

A bounded, nondecreasing sequence converges to its least upper bound; a bounded, nonincreasing sequence converges to its greatest lower bound.

PROOF Suppose that the sequence $\{a_n\}$ is bounded and nondecreasing. If L is the least upper bound of this sequence, then

$$a_n \leq L \quad \text{for all } n.$$

Now let ϵ be an arbitrary positive number. By Theorem 10.1.2 there exists a_k such that

$$L - \epsilon < a_k.$$

Since the sequence is nondecreasing,

$$a_k \leq a_n \quad \text{for all } n \geq k.$$

It follows that

$$L - \epsilon < a_n \leq L \quad \text{for all } n \geq k.$$

This shows that

$$|a_n - L| < \epsilon \quad \text{for all } n \geq k$$

and proves that

$$a_n \to L.$$

The nonincreasing case can be handled in a similar manner. □

Example 4 We shall show that the sequence $a_n = (3^n + 4^n)^{1/n}$ is convergent. Since

$$3 = (3^n)^{1/n} < (3^n + 4^n)^{1/n} < (4^n + 4^n)^{1/n} = (2 \cdot 4^n)^{1/n} = 2^{1/n} \cdot 4 \leq 8,$$

the sequence is bounded. Note that

$$\begin{aligned}(3^n + 4^n)^{(n+1)/n} &= (3^n + 4^n)^{1/n}\,(3^n + 4^n)\\ &= (3^n + 4^n)^{1/n}\,3^n + (3^n + 4^n)^{1/n}\,4^n.\end{aligned}$$

Since

$$(3^n + 4^n)^{1/n} > (3^n)^{1/n} = 3 \quad \text{and} \quad (3^n + 4^n)^{1/n} > (4^n)^{1/n} = 4,$$

we have

$$(3^n + 4^n)^{(n+1)/n} > 3 \cdot (3^n) + 4 \cdot (4^n) = 3^{n+1} + 4^{n+1}.$$

Taking the $(n+1)$st root of the left and right sides of this inequality, we obtain

$$(3^n + 4^n)^{1/n} > (3^{n+1} + 4^{n+1})^{1/(n+1)}.$$

The sequence is decreasing. Being also bounded, it must be convergent. (In Section 10.6, we show that the limit is 4.) □

THEOREM 10.3.7

Let α be a real number. If $a_n \to L$ and $b_n \to M$, then

$$\text{(i) } a_n + b_n \to L + M, \quad \text{(ii) } \alpha a_n \to \alpha L, \quad \text{(iii) } a_n b_n \to LM.$$

If, in addition, $M \neq 0$ and each $b_n \neq 0$, then

$$\text{(iv) } \frac{1}{b_n} \to \frac{1}{M} \qquad \text{(v) } \frac{a_n}{b_n} \to \frac{L}{M}.$$

Proofs of parts (i) and (ii) are left as exercises. For proofs of parts (iii)–(v), see the supplement at the end of this section.

We are now in a position to handle all sequences of the form

$$a_n = \frac{\alpha_k n^k + \alpha_{k-1} n^{k-1} + \cdots + \alpha_0}{\beta_j n^j + \beta_{j-1} n^{j-1} + \cdots + \beta_0}.$$

To determine the behavior of such a sequence, we need only divide both numerator and denominator by the highest power of n that occurs.

Example 5

$$\frac{3n^4 - 2n^2 + 1}{n^5 - 3n^3} = \frac{3/n - 2/n^3 + 1/n^5}{1 - 3/n^2} \to \frac{0}{1} = 0. \quad \square$$

↑ divide by n^5

Example 6

$$\frac{1 - 4n^7}{n^7 + 12n} = \frac{1/n^7 - 4}{1 + 12/n^6} \to \frac{-4}{1} = -4. \quad \square$$

↑ divide by n^7

Example 7

$$\frac{n^4 - 3n^2 + n + 2}{n^3 + 7n} = \frac{1 - 3/n^2 + 1/n^3 + 2/n^4}{1/n + 7/n^3}.$$

↑ divide by n^4

Since the numerator tends to 1 and the denominator tends to 0, the sequence is unbounded. Therefore, it cannot converge. □

THEOREM 10.3.8

$$a_n \to L \quad \text{iff} \quad a_n - L \to 0 \quad \text{iff} \quad |a_n - L| \to 0.$$

We leave the proof to you.

THEOREM 10.3.9 THE PINCHING THEOREM FOR SEQUENCES

Suppose that for all n greater than some integer K,

$$a_n \leq b_n \leq c_n.$$

If $a_n \to L$ and $c_n \to L$, then $b_n \to L$.

Once again the proof is left to you.

As an immediate and obvious consequence of the pinching theorem, we have the following corollary.

(10.3.10) Suppose that for all n greater than some integer K

$$|b_n| \leq c_n.$$

If $c_n \to 0$, then $b_n \to 0$.

Example 8

$$\frac{\cos n}{n} \to 0 \quad \text{since} \quad \left|\frac{\cos n}{n}\right| \leq \frac{1}{n} \quad \text{and} \quad \frac{1}{n} \to 0. \quad \square$$

Example 9

$$\sqrt{4 + \left(\frac{1}{n}\right)^2} \to 2$$

since

$$2 \leq \sqrt{4 + \left(\frac{1}{n}\right)^2} \leq \sqrt{4 + 4\left(\frac{1}{n}\right) + \left(\frac{1}{n}\right)^2} = 2 + \frac{1}{n}$$

and

$$2 + \frac{1}{n} \to 2 \quad \square$$

Example 10

(10.3.11)

$$\lim_{n\to\infty} \left(1 + \frac{1}{n}\right)^n = e.$$

PROOF In the Exercises for Section 7.4 you were asked to show that

$$\left(1+\frac{1}{n}\right)^n \le e \le \left(1+\frac{1}{n}\right)^{n+1} \qquad \text{for all positive integers } n.$$

Dividing the right-hand inequality by $1 + 1/n$, we have

$$\frac{e}{1+1/n} \le \left(1+\frac{1}{n}\right)^n.$$

Combining this with the left-hand inequality, we can write

$$\frac{e}{1+1/n} \le \left(1+\frac{1}{n}\right)^n \le e.$$

Since

$$\frac{e}{1+1/n} \to \frac{e}{1} = e,$$

we can conclude from the pinching theorem that

$$\left(1+\frac{1}{n}\right)^n \to e. \quad \square$$

The sequences

$$a_n = \cos\frac{\pi}{n}, \quad a_n = \ln\left(\frac{n}{n+1}\right), \quad a_n = \tan\left(\sqrt{\frac{\pi^2 n^2 - 8}{16n^2}}\right)$$

are all of the form $\{f(c_n)\}$ with f a continuous function. Such sequences are frequently easy to deal with. The basic idea is this: when a continuous function is applied to a convergent sequence, the result is itself a convergent sequence. More precisely, we have the following theorem.

THEOREM 10.3.12

Suppose that

$$c_n \to c$$

and all the c_n are in the domain of f. If f is continuous at c, then

$$f(c_n) \to f(c).$$

PROOF We assume that f is continuous at c and take $\epsilon > 0$. From the continuity of f at c we know that there exists a $\delta > 0$ such that

$$\text{if } |x - c| < \delta, \quad \text{then } |f(x) - f(c)| < \epsilon.$$

Since $c_n \to c$, we know that there exists a positive integer K such that

$$\text{if } n \ge K, \quad \text{then } |c_n - c| < \delta.$$

It follows therefore that

$$\text{if } n \geq K, \quad \text{then } |f(c_n) - f(c)| < \epsilon. \quad \square$$

Example 11 Since $\pi/n \to 0$ and the cosine function is continuous at 0,

$$\cos\left(\frac{\pi}{n}\right) \to \cos 0 = 1. \quad \square$$

Example 12 Since $\dfrac{n}{n+1} = \dfrac{1}{1+1/n} \to 1$ and the logarithm function is continuous at 1,

$$\ln\left(\frac{n}{n+1}\right) \to \ln 1 = 0. \quad \square$$

Example 13 Since $\dfrac{\pi^2 n^2 - 8}{16n^2} = \dfrac{\pi^2 - 8/n^2}{16} \to \dfrac{\pi^2}{16}$ and the function $f(x) = \tan\sqrt{x}$ is continuous at $\pi^2/16$,

$$\tan\left(\sqrt{\frac{\pi^2 n^2 - 8}{16n^2}}\right) \to \tan\left(\sqrt{\frac{\pi^2}{16}}\right) = \tan\frac{\pi}{4} = 1. \quad \square$$

Example 14 Since the absolute value function is everywhere continuous,

$$a_n \to L \quad \text{implies} \quad |a_n| \to |L|. \quad \square$$

Remark For some time now we have asked you to take on faith two fundamentals of integration: that continuous functions do have definite integrals and that these integrals can be expressed as limits of Riemann sums. We could not give you proofs of these assertions because we did not have the necessary tools. Now we do. Proofs are given in Appendix B. □

EXERCISES 10.3

State whether the sequence converges and, if it does, find the limit.

1. 2^n.

2. $\dfrac{2}{n}$.

3. $\dfrac{(-1)^n}{n}$.

4. $\sqrt{n}$.

5. $\dfrac{n-1}{n}$.

6. $\dfrac{n+(-1)^n}{n}$.

7. $\dfrac{n+1}{n^2}$.

8. $\sin\dfrac{\pi}{2n}$.

9. $\dfrac{2^n}{4^n+1}$.

10. $\dfrac{n^2}{n+1}$.

11. $(-1)^n\sqrt{n}$.

12. $\dfrac{4n}{\sqrt{n^2+1}}$.

13. $(-\frac{1}{2})^n$.

14. $\dfrac{4^n}{2^n+10^6}$.

15. $\tan\dfrac{n\pi}{4n+1}$.

16. $\dfrac{10^{10}\sqrt{n}}{n+1}$.

17. $\dfrac{(2n+1)^2}{(3n-1)^2}$.

18. $\ln\left(\dfrac{2n}{n+1}\right)$.

19. $\dfrac{n^2}{\sqrt{2n^4+1}}$.

20. $\dfrac{n^4-1}{n^4+n-6}$.

21. $\cos n\pi$.

22. $\dfrac{n^5}{17n^4+12}$.

23. $e^{1/\sqrt{n}}$.

24. $\sqrt{4-\dfrac{1}{n}}$.

25. $\ln n - \ln(n+1)$.

26. $\dfrac{2^n-1}{2^n}$.

27. $\dfrac{\sqrt{n+1}}{2\sqrt{n}}$.

28. $\dfrac{1}{n} - \dfrac{1}{n+1}$.

29. $\left(1+\dfrac{1}{n}\right)^{2n}$.

30. $\left(1+\dfrac{1}{n}\right)^{n/2}$.

31. $\dfrac{2^n}{n^2}$.

32. $2\ln 3n - \ln(n^2+1)$.

In Exercises 33 and 34, use technology (graphing utility or CAS) to determine whether the sequence converges, and if it does, give the limit.

33. (a) $\dfrac{\sin n}{\sqrt{n}}$. (b) $\tan^{-1}\left(\dfrac{n}{n+1}\right)$.
(c) $\sqrt{n^2+n}-n$.

34. (a) $\sqrt[n]{n}$. (b) $\dfrac{3^n}{n!}$
(c) $\dfrac{n}{\sqrt[n]{n!}}$.

35. Show that if $0 < a < b$, then $\sqrt[n]{a^n + b^n} \to b$.

36. (a) Determine the values of r for which r^n converges.
(b) Determine the values of r for which nr^n converges.

37. Prove that if $a_n \to L$ and $b_n \to M$, then $a_n + b_n \to L + M$.

38. Let α be a real number. Prove that if $a_n \to L$, then $\alpha a_n \to \alpha L$.

39. Given that

$$\left(1+\frac{1}{n}\right)^n \to e \quad \text{show that} \quad \left(1+\frac{1}{n}\right)^{n+1} \to e.$$

40. Determine the convergence or divergence of the sequence

$$a_n = \frac{\alpha_k n^k + \alpha_{k-1}n^{k-1} + \cdots + \alpha_0}{\beta_j n^j + \beta_{j-1}n^{j-1} + \cdots + \beta_0}$$

given that: (a) $k = j$; (b) $k < j$; (c) $k > j$. Justify your answers. Assume that $\alpha_k \neq 0$ and $\beta_j \neq 0$.

41. Prove that a bounded nonincreasing sequence converges to its greatest lower bound.

42. From a sequence with terms a_n, collect the even-numbered terms $e_n = a_{2n}$ and the odd-numbered terms $o_n = a_{2n-1}$. Show that

$$a_n \to L \quad \text{iff} \quad e_n \to L \quad \text{and} \quad o_n \to L.$$

43. Prove the pinching theorem for sequences.

44. Let $\{a_n\}$ and $\{b_n\}$ be sequences such that $a_n \to 0$ and $\{b_n\}$ is bounded. Prove that $a_n b_n \to 0$.

45. Suppose that $a_n \to L$. Show that if $a_n \leq M$ for all n, then $L \leq M$.

46. According to Example 14, if $a_n \to L$, then $|a_n| \to |L|$. Is the converse true? That is, if $|a_n| \to |L|$, does it follow that $a_n \to L$? Prove or give a counterexample.

47. Prove that $a_n \to 0$ iff $|a_n| \to 0$.

48. Suppose that $a_n \to L$ and $b_n \to L$. Show that the sequence

$$a_1,\ b_1,\ a_2,\ b_2,\ a_3,\ b_3, \ldots$$

converges to L.

49. Let f be a continuous function on $(-\infty, \infty)$ and let r be a real number. Define the sequence $\{a_n\}$ as follows:

$$a_1 = r, \quad a_2 = f(r), \quad a_3 = f[f(r)],$$
$$a_4 = f\{f[f(r)]\}, \ldots.$$

Prove that if $a_n \to L$, then $f(L) = L$; that is, L is a fixed point of f.

50. Show that

$$\frac{2^n}{n!} \to 0.$$

HINT: First show that

$$\frac{2^n}{n!} = \frac{2}{1}\cdot\frac{2}{2}\cdot\frac{2}{3}\cdot\cdots\cdot\frac{2}{n} \leq \frac{4}{n}.$$

51. Prove that $(1/n)^{1/p} \to 0$ for all positive integers p.

52. Prove Theorem 10.3.8.

The sequences in Exercises 53–58 are defined recursively.† Determine in each case whether the sequence converges and, if so, find the limit. Start each sequence with $a_1 = 1$.

53. $a_{n+1} = \dfrac{1}{e}a_n$.

54. $a_{n+1} = 2^{n+1}a_n$.

55. $a_{n+1} = \dfrac{1}{n+1}a_n$.

56. $a_{n+1} = \dfrac{n}{n+1}a_n$.

57. $a_{n+1} = 1 - a_n$.

58. $a_{n+1} = \frac{1}{2}a_n + 1$.

In Exercises 59–66, evaluate numerically the limit of each sequence as $n \to \infty$. Some of these sequences converge more rapidly than others. Determine for each sequence the least value of n for which the nth term differs from the limit by less than 0.001.

59. $\dfrac{1}{n^2}$.

60. $\dfrac{1}{\sqrt{n}}$.

61. $\dfrac{n}{10^n}$.

62. $\dfrac{n^{10}}{10^n}$.

63. $\dfrac{1}{n!}$.

64. $\dfrac{2^n}{n!}$.

65. $\dfrac{\ln n}{n^2}$.

66. $\dfrac{\ln n}{n}$.

67. (a) Find the exact value of the limit of the sequence $\{a_n\}$ given in Exercise 67, Section 10.2.
(b) Find the exact value of the limit of the sequence $\{a_n\}$ given in Exercise 68, Section 10.2.

68. Let $\{a_n\}$ be the sequence defined recursively by

$$a_1 = 1,\ a_n = \sqrt{6 + a_{n-1}},\ n = 2, 3, 4, \ldots.$$

(a) Approximate a_2, a_3, a_4, a_5, a_6. Round off your answers to six decimal places.
(b) Use mathematical induction to show that $a_n \leq 3$ for all n.
(c) Show that $\{a_n\}$ is an increasing sequence. HINT: $a_{n+1}^2 - a_n^2 = (3 - a_n)(2 + a_n)$.
(d) What is the limit of this sequence?

† The notion was introduced in Exercises 10.2.

69. Let $\{a_n\}$ be the sequence defined recursively by

$$a_1 = 1, \quad a_n = \cos a_{n-1}, \quad n = 2, 3, 4, \ldots.$$

(a) Approximate $a_2, a_3, a_4, \ldots, a_{10}$. Round off your answers to six decimal places.

(b) Assuming that $a_n \to L$, approximate L to six decimal places and interpret your result geometrically. HINT: Use Exercise 49.

70. Let $\{a_n\}$ be the sequence defined recursively by

$$a_1 = 1, \quad a_n = a_{n-1} + \cos a_{n-1}, \quad n = 2, 3, 4, \ldots.$$

(a) Approximate $a_2, a_3, a_4, \ldots, a_{10}$. Round off your answers to six decimal places.

(b) Assuming that $a_n \to L$, approximate L to six decimal places and interpret your result geometrically.

■ PROJECT 10.3 Sequences and the Newton-Raphson Method

Let R be a positive number. The sequence defined recursively by setting

$$(1) \qquad a_1 = 1, \quad a_n = \frac{1}{2}\left(a_{n-1} + \frac{R}{a_{n-1}}\right), \quad n = 2, 3, 4, \ldots.$$

can be used to approximate $\sqrt{R}$.

Problem 1. Let $R = 3$.

a. Calculate $a_2, a_3, \ldots, a_8$. Round off your answers to six decimal places.

b. Show that if $a_n \to L$, then $L = \sqrt{3}$.

The Newton–Raphson method (Sec. 3.9) applied to a differentiable function f generates a sequence which, under certain conditions, converges to a zero of f. The recurrence relation is of the form

$$x_{n+1} = x_n - \frac{f(x_n)}{f'(x_n)}, \quad n = 1, 2, 3, \ldots.$$

Problem 2. Show that recurrence relation (1) is the recurrence relation generated by the Newton–Raphson method applied to the function $f(x) = x^2 - R$.

Problem 3. Each of the following recurrence relations is based on the Newton–Raphson method. Determine whether the sequence converges and if so, give the limit.

a. $x_1 = 1, \quad x_{n+1} = x_n - \dfrac{x_n^3 - 8}{3x_n^2}.$

b. $x_1 = 0, \quad x_{n+1} = x_n - \dfrac{\sin x_n - 0.5}{\cos x_n}.$

c. $x_1 = 1, \quad x_{n+1} = x_n - \dfrac{\ln x_n - 1}{1/x_n}.$

Problem 4. For each part of Problem 3, find a function f that generates the recurrence relation. Then check each of your answers by evaluating f.

*SUPPLEMENT TO SECTION 10.3

PROOF OF THEOREM 10.3.2

If $L \neq M$, then

$$\tfrac{1}{2}|L - M| > 0.$$

The assumption that $\lim_{n\to\infty} a_n = L$ and $\lim_{n\to\infty} a_n = M$ gives the existence of K_1 such that

$$\text{if} \quad n \geq K_1, \quad \text{then} \quad |a_n - L| < \tfrac{1}{2}|L - M|$$

and the existence of K_2 such that

$$\text{if} \quad n \geq K_2, \quad \text{then} \quad |a_n - M| < \tfrac{1}{2}|L - M|.\dagger$$

For $n \geq \max\{K_1, K_2\}$, we have

$$|a_n - L| + |a_n - M| < |L - M|.$$

† We can reach these conclusions from Definition 10.3.1 by taking $\frac{1}{2}|L - M|$ as ϵ.

By the triangle inequality we have

$$|L - M| = |(L - a_n) + (a_n - M)| \le |L - a_n| + |a_n - M| = |a_n - L| + |a_n - M|.$$

Combining the last two statements, we have

$$|L - M| < |L - M|.$$

The hypothesis $L \neq M$ has led to an absurdity. We conclude that $L = M$. □

PROOF OF THEOREM 10.3.7 (iii)–(v)

To prove (iii), we set $\epsilon > 0$. For each n,

$$\begin{aligned} |a_n b_n - LM| &= |(a_n b_n - a_n M) + (a_n M - LM|) \\ &\le |a_n||b_n - M| + |M||a_n - L|. \end{aligned}$$

Since convergent sequences are bounded, there exists $Q > 0$ such that

$$|a_n| \le Q \quad \text{for all } n.$$

Since $|M| < |M| + 1$, we have

(1) $$|a_n b_n - LM \le Q|b_n - M| + (|M| + 1)|a_n - L|.\dagger$$

Since $b_n \to M$, we know that there exists K_1 such that

$$\text{if} \quad n \ge K_1, \quad \text{then} \quad |b_n - M| < \frac{\epsilon}{2Q}.$$

Since $a_n \to L$, we know that there exists K_2 such that

$$\text{if} \quad n \ge K_2, \quad \text{then} \quad |a_n - L| < \frac{\epsilon}{2(|M| + 1)}.$$

For $n \ge \max\{K_1, K_2\}$, both conditions hold, and consequently

$$Q|b_n - M| + (|M| + 1)|a_n - L| < \frac{\epsilon}{2} + \frac{\epsilon}{2} = \epsilon.$$

In view of (1), we can conclude that

$$\text{if} \quad n \ge \max\{K_1, K_2\}, \quad \text{then} \quad |a_n b_n - LM| < \epsilon.$$

This proves that

$$a_n b_n \to LM. \quad □$$

To prove (iv), once again we set $\epsilon > 0$. In the first place

$$\left| \frac{1}{b_n} - \frac{1}{M} \right| = \left| \frac{M - b_n}{b_n M} \right| = \frac{|b_n - M|}{|b_n||M|}.$$

† Soon we will want to divide by the coefficient of $|a_n - L|$. We have replaced $|M|$ by $|M| + 1$ because $|M|$ can be zero.

Since $b_n \to M$ and $|M|/2 > 0$, there exists K_1 such that

$$\text{if} \quad n \geq K_1, \quad \text{then} \quad |b_n - M| < \frac{|M|}{2}.$$

This tells us that for $n \geq K_1$ we have

$$|b_n| > \frac{|M|}{2} \quad \text{and thus} \quad \frac{1}{|b_n|} < \frac{2}{|M|}.$$

Thus for $n \geq K_1$ we have

$$\left|\frac{1}{b_n} - \frac{1}{M}\right| \leq \frac{2}{|M|^2}|b_n - M|. \tag{2}$$

Since $b_n \to M$ there exists K_2 such that

$$\text{if} \quad n \geq K_2, \quad \text{then} \quad |b_n - M| < \frac{\epsilon |M|^2}{2}.$$

Thus for $n \geq K_2$ we have

$$\frac{2}{|M|^2}|b_n - M| < \epsilon.$$

In view of (2), we can be sure that

$$\text{if} \quad n \geq \max\{K_1, K_2\}, \quad \text{then} \quad \left|\frac{1}{b_n} - \frac{1}{M}\right| < \epsilon.$$

This proves that

$$\frac{1}{b_n} \to \frac{1}{M}. \quad \square$$

The proof of (v) is now easy:

$$\frac{a_n}{b_n} = a_n \cdot \frac{1}{b_n} \to L \cdot \frac{1}{M} = \frac{L}{M}. \quad \square$$

■ 10.4 SOME IMPORTANT LIMITS

Our purpose here is to familiarize you with some limits that are particularly important in calculus and to give you more experience with limit arguments.

(10.4.1) If $x > 0$, then

$$x^{1/n} \to 1 \quad \text{as} \quad n \to \infty.$$

PROOF Fix any $x > 0$. Note that

$$\ln(x^{1/n}) = \frac{1}{n}\ln x \to 0 \quad \text{as} \quad n \to \infty.$$

Since the exponential function is continuous at 0, it follows from Theorem 10.3.12 that

$$x^{1/n} = e^{(1/n)\ln x} \to e^0 = 1. \quad \square$$

(10.4.2)

> If $|x| < 1$, then
> $$x^n \to 0 \quad \text{as} \quad n \to \infty.$$

PROOF The result clearly holds for $x = 0$. Now fix any $x \neq 0$ with $|x| < 1$ and observe that the sequence $a_n = |x|^n$ is a decreasing sequence:

$$|x|^{n+1} = |x||x|^n < |x|^n.$$

Let $\epsilon > 0$. Since

$$\epsilon^{1/n} \to 1 \quad \text{as} \quad n \to \infty, \tag{10.4.1}$$

there exists an integer $k > 0$ for which

$$|x| < \epsilon^{1/k}. \qquad \text{(explain)}$$

This implies that $|x|^k < \epsilon$. Since the $|x|^n$ form a decreasing sequence, we have

$$|x^n| = |x|^n < \epsilon \quad \text{for all } n \geq k. \quad \square$$

(10.4.3)

> For each $\alpha > 0$
> $$\frac{1}{n^\alpha} \to 0 \quad \text{as} \quad n \to \infty.$$

PROOF Since $\alpha > 0$, there exists an odd positive integer p such that $1/p < \alpha$. Then

$$0 < \frac{1}{n^\alpha} = \left(\frac{1}{n}\right)^\alpha \leq \left(\frac{1}{n}\right)^{1/p}.$$

Since $1/n \to 0$ and $f(x) = x^{1/p}$ is continuous at 0, we have

$$\left(\frac{1}{n}\right)^{1/p} \to 0 \quad \text{and thus by the pinching theorem} \quad \frac{1}{n^\alpha} \to 0. \quad \square$$

(10.4.4)

> For each real x
> $$\frac{x^n}{n!} \to 0 \quad \text{as} \quad n \to \infty.$$

PROOF Fix any real number x and choose an integer k such that $k > |x|$. For $n > k+1$,

$$\frac{k^n}{n!} = \left(\frac{k^k}{k!}\right)\left[\frac{k}{k+1}\frac{k}{k+2}\cdots\frac{k}{n-1}\right]\left(\frac{k}{n}\right) < \left(\frac{k^{k+1}}{k!}\right)\left(\frac{1}{n}\right).$$

the middle term is less than 1

Since $k > |x|$, we have

$$0 < \frac{|x|^n}{n!} < \frac{k^n}{n!} < \left(\frac{k^{k+1}}{k!}\right)\left(\frac{1}{n}\right).$$

Since k is fixed and $1/n \to 0$, it follows from the pinching theorem that

$$\frac{|x|^n}{n!} \to 0 \quad \text{and thus} \quad \frac{x^n}{n!} \to 0. \quad \square$$

(10.4.5)

$$\frac{\ln n}{n} \to 0 \quad \text{as} \quad n \to \infty.$$

PROOF A routine proof can be based on L'Hôpital's rule (Theorem 10.6.1), but that is not available to us yet. We will appeal to the pinching theorem and base our argument on the integral representation of the logarithm:

$$0 \leq \frac{\ln n}{n} = \frac{1}{n}\int_1^n \frac{dt}{t} \leq \frac{1}{n}\int_1^n \frac{dt}{\sqrt{t}} = \frac{2}{n}(\sqrt{n}-1)$$

$$= \frac{2(\sqrt{n}-1)}{n} \to 0. \quad \square$$

(10.4.6)

$$n^{1/n} \to 1 \quad \text{as} \quad n \to \infty.$$

PROOF We know that $\qquad n^{1/n} = e^{(1/n)\ln n}.$

Since $\qquad (1/n)\ln n \to 0 \qquad$ (10.4.5)

and the exponential function is continuous at 0, it follows from Theorem 10.3.12 that

$$n^{1/n} \to e^0 = 1. \quad \square$$

(10.4.7)

For each real x

$$\left(1+\frac{x}{n}\right)^n \to e^x \quad \text{as} \quad n \to \infty.$$

PROOF For $x = 0$, the result is obvious. For $x \neq 0$,

$$\ln\left(1+\frac{x}{n}\right)^n = n\ln\left(1+\frac{x}{n}\right) = x\left[\frac{\ln(1+x/n)-\ln 1}{x/n}\right].$$

The crux here is to recognize that the bracketed expression is a difference quotient for the logarithm function. Once we see this, we let $h = x/n$ and write

$$\lim_{n\to\infty}\left[\frac{\ln(1+x/n)-\ln 1}{x/n}\right] = \lim_{h\to 0}\left[\frac{\ln(1+h)-\ln 1}{h}\right] = 1.\dagger$$

† For each $t > 0$

$$\lim_{h\to 0}\frac{\ln(t+h)-\ln t}{h} = \frac{d}{dt}(\ln t) = \frac{1}{t}.$$

It follows that

$$\ln\left(1+\frac{x}{n}\right)^n \to x \quad \text{and therefore} \quad \left(1+\frac{x}{n}\right)^n = e^{\ln(1+x/n)^n} \to e^x. \quad \square$$

EXERCISES 10.4

State whether the sequence converges as $n \to \infty$; if it does, find the limit.

1. $2^{2/n}$.
2. $e^{-\alpha/n}$.
3. $\left(\frac{2}{n}\right)^n$.
4. $\frac{\log_{10} n}{n}$.
5. $\frac{\ln(n+1)}{n}$.
6. $\frac{3^n}{4^n}$.
7. $\frac{x^{100n}}{n!}$.
8. $n^{1/(n+2)}$.
9. $n^{\alpha/n}, \quad \alpha > 0$.
10. $\ln\left(\frac{n+1}{n}\right)$.
11. $\frac{3^{n+1}}{4^{n-1}}$.
12. $\int_{-n}^{0} e^{2x}\,dx$.
13. $(n+2)^{1/n}$.
14. $\left(1-\frac{1}{n}\right)^n$.
15. $\int_0^n e^{-x}\,dx$.
16. $\frac{2^{3n-1}}{7^{n+2}}$.
17. $\int_{-n}^{n} \frac{dx}{1+x^2}$.
18. $\int_0^n e^{-nx}\,dx$.
19. $(n+2)^{1/(n+2)}$.
20. $n^2 \sin n\pi$.
21. $\frac{\ln n^2}{n}$.
22. $\int_{-1+1/n}^{1+1/n} \frac{dx}{\sqrt{1-x^2}}$.
23. $n^2 \sin \frac{\pi}{n}$.
24. $\frac{n!}{2n}$.
25. $\frac{5^{n+1}}{4^{2n-1}}$.
26. $\left(1+\frac{x}{n}\right)^{3n}$.
27. $\left(\frac{n+1}{n+2}\right)^n$.
28. $\int_{1/n}^{1} \frac{dx}{\sqrt{x}}$.
29. $\int_n^{n+1} e^{-x^2}\,dx$.
30. $\left(1+\frac{1}{n^2}\right)^n$.
31. $\frac{n^n}{2^{n^2}}$.
32. $\int_0^{1/n} \cos e^x\,dx$.
33. $\left(1+\frac{x}{2n}\right)^{2n}$.
34. $\left(1+\frac{1}{n}\right)^{n^2}$.
35. $\int_{-1/n}^{1/n} \sin x^2\,dx$.
36. $\left(t+\frac{x}{n}\right)^n, \quad t > 0, \quad x > 0$.

C In Exercises 37 and 38, use technology (graphing utility or CAS) to determine whether the sequence converges, and if it does, give the limit.

37. (a) $\frac{\sin(6/n)}{\sin(3/n)}$. (b) $\frac{\arctan n}{n}$.
    (c) $n \sin(1/n)$.
38. (a) $\frac{n^2 \cos(n^2\pi)}{n^2+1}$.
    (b) $n\,(2^{1/n}-1)$.
    (c) $n^e e^{-n}$.
39. Show that $\lim_{n\to\infty} (\sqrt{n+1}-\sqrt{n}) = 0$.
40. Show that $\lim_{n\to\infty} (\sqrt{n^2+n}-n) = \frac{1}{2}$.
41. (a) Show that a regular polygon of n sides inscribed in a circle of radius r has perimeter $p_n = 2rn \sin(\pi/n)$.
    (b) Find
    $$\lim_{n\to\infty} p_n$$
    and give a geometric interpretation of your result.
42. Show that
    $$\text{if} \quad 0 < c < d, \quad \text{then} \quad (c^n + d^n)^{1/n} \to d.$$
    See Example 4, Section 10.3.

In Exercises 41–43, find the indicated limit.

43. $\lim_{n\to\infty} \frac{1+2+\cdots+n}{n^2}$.
    HINT: $1+2+\cdots+n = \frac{n(n+1)}{2}$.
44. $\lim_{n\to\infty} \frac{1^2+2^2+\cdots+n^2}{(1+n)(2+n)}$.
    HINT: $1^2+2^2+\cdots+n^2 = \frac{n(n+1)(2n+1)}{6}$.
45. $\lim_{n\to\infty} \frac{1^3+2^3+\cdots+n^3}{2n^4+n-1}$.
    HINT: $1^3+2^3+\cdots+n^3 = \frac{n^2(n+1)^2}{4}$.

46. A sequence $\{a_n\}$ is said to be a *Cauchy sequence*† if

(10.4.8) for each $\epsilon > 0$ there exists an index K such that $|a_n - a_m| < \epsilon$ for all $m, n \geq K$.

Show that

(10.4.9) every convergent sequence is a Cauchy sequence.

It is also true that every Cauchy sequence is convergent, but this is more difficult to prove.

47. (*Arithmetic means*) For a given sequence $\{a_n\}$, let

$$m_n = \frac{1}{n}(a_1 + a_2 + \cdots + a_n).$$

(a) Prove that if $\{a_n\}$ is increasing, then $\{m_n\}$ is increasing.

(b) Prove that if $a_n \to 0$, then $m_n \to 0$. HINT: Choose an integer $j > 0$ such that, if $n \geq j$, then $a_n < \epsilon/2$. Then for $n \geq j$,

$$|m_n| < \frac{|a_1 + a_2 + \cdots + a_j|}{n} + \frac{\epsilon}{2}\left(\frac{n-j}{n}\right).$$

48. (a) Let $\{a_n\}$ be a convergent sequence. Prove that

$$\lim_{n\to\infty} (a_n - a_{n-1}) = 0.$$

(b) What can you say about the converse? That is, suppose that $\{a_n\}$ is a sequence such that

$$\lim_{n\to\infty} (a_n - a_{n-1}) = 0.$$

Does $\{a_n\}$ necessarily converge? Prove or give a counterexample.

49. Let a and b be positive numbers with $b > a$. Define two sequences $\{a_n\}$ and $\{b_n\}$ as follows:

$$a_1 = \frac{a+b}{2} \text{ (the arithmetic mean of } a \text{ and } b),$$

$$b_1 = \sqrt{ab} \text{ (the geometric mean of } a \text{ and } b),$$

and

$$a_n = \frac{a_{n-1} + b_{n-1}}{2},$$

$$b_n = \sqrt{a_{n-1}\, b_{n-1}}, \quad n = 2, 3, 4, \ldots.$$

(a) Use mathematical induction to show that

$$a_{n-1} > a_n > b_n > b_{n-1} \quad \text{for } n = 2, 3, 4 \ldots.$$

(b) Prove that $\{a_n\}$ and $\{b_n\}$ are convergent sequences and that $\lim_{n\to\infty} a_n = \lim_{n\to\infty} b_n$. The value of this common limit is known as the *arithmetic-geometric mean of a* and *b*.

C 50. You have seen that for all real x

$$\lim_{n\to\infty}\left(1 + \frac{x}{n}\right)^n = e^x.$$

However, the rate of convergence is different for different x. Verify that at $n = 100$, $(1 + 1/n)^n$ is within 1% of its limit, while $(1 + 5/n)^n$ is still about 12% from its limit. Give comparable accuracy estimates for these two sequences at $n = 1000$.

C 51. Evaluate

$$\lim_{n\to\infty}\left(\sin\frac{1}{n}\right)^{1/n}$$

numerically and justify your answer by other means.

C 52. We have stated that

$$\lim_{n\to\infty} (\sqrt{n^2 + n} - n) = \tfrac{1}{2}. \qquad \text{(Exercise 40)}$$

Evaluate numerically

$$\lim_{n\to\infty} [(n^3 + n^2)^{1/3} - n].$$

Formulate a conjecture about

$$\lim_{n\to\infty} [(n^k + n^{k-1})^{1/k} - n], \quad k = 1, 2, 3, \cdots$$

and prove that your conjecture is valid.

C 53. The *Fibonacci sequence* is defined recursively by

$$a_{n+2} = a_{n+1} + a_n, \quad \text{starting with} \quad a_1 = a_2 = 1.$$

(a) Calculate $a_3, a_4, a_5, \cdots, a_{10}$.

(b) Now define the sequence $\{r_n\}$ by setting

$$r_n = \frac{a_{n+1}}{a_n}.$$

Calculate $r_1, r_2, \cdots, r_6$.

(c) Assume that $r_n \to L$, and find L.

HINT: Show that $r_n = 1 + \dfrac{1}{r_{n-1}}$.

54. Let

$$a_n = \frac{1}{n^2} + \frac{2}{n^2} + \frac{3}{n^2} + \cdots + \frac{n}{n^2}.$$

Show that a_n is a Riemann sum for $\int_0^1 x\,dx$ for each $n \geq 1$. Does the sequence $\{a_n\}$ converge? If so, to what?

---

† After the French baron Augustin Louis Cauchy (1789–1857), one of the most prolific mathematicians of all time.

■ 10.5 THE INDETERMINATE FORM (0/0)

Here we are concerned with limits of quotients where the numerator and denominator both tend to 0 and elementary methods fail or are difficult to apply. We call such limits *indeterminates of the form* $0/0$.

THEOREM 10.5.1 L'HÔPITAL'S RULE $(0/0)$†

Suppose that

$$f(x) \to 0 \quad \text{and} \quad g(x) \to 0$$

as $x \to c^+$, $x \to c^-$, $x \to c$, $x \to \infty$, or $x \to -\infty$.

$$\text{If} \quad \frac{f'(x)}{g'(x)} \to L, \quad \text{then} \quad \frac{f(x)}{g(x)} \to L.$$

NOTE: This theorem includes the possibility that $L = \infty$ or $-\infty$.

We will prove the validity of L'Hôpital's rule later in the section. First we demonstrate its usefulness.

Example 1 Find $\displaystyle\lim_{x\to\pi/2} \frac{\cos x}{\pi - 2x}$.

SOLUTION As $x \to \pi/2$, both the numerator $f(x) = \cos x$ and the denominator $g(x) = \pi - 2x$ tend to zero, but it is not at all obvious what happens to the quotient

$$\frac{f(x)}{g(x)} = \frac{\cos x}{\pi - 2x}.$$

Therefore we test the quotient of derivatives:

$$\frac{f'(x)}{g'(x)} = \frac{-\sin x}{-2} = \frac{\sin x}{2} \to \frac{1}{2} \quad \text{as} \quad x \to \frac{\pi}{2}.$$

It follows from L'Hôpital's rule that

$$\frac{\cos x}{\pi - 2x} \to \frac{1}{2} \quad \text{as} \quad x \to \frac{\pi}{2}.$$

We can express all this on just one line using $*$ to indicate the differentiation of numerator and denominator:

$$\lim_{x\to\pi/2} \frac{\cos x}{\pi - 2x} \overset{*}{=} \lim_{x\to\pi/2} \frac{-\sin x}{-2} = \lim_{x\to\pi/2} \frac{\sin x}{2} = \frac{1}{2}. \quad \square$$

Example 2 Find $\displaystyle\lim_{x\to 0^+} \frac{x}{\sin\sqrt{x}}$.

SOLUTION As $x \to 0^+$, both numerator and denominator tend to 0. Since

$$\frac{f'(x)}{g'(x)} = \frac{1}{(\cos\sqrt{x})(1/[2\sqrt{x}])} = \frac{2\sqrt{x}}{\cos\sqrt{x}} \to \frac{0}{1} = 0 \quad \text{as} \quad x \to 0^+,$$

† Named after the Frenchman G. F. A. L'Hôpital (1661–1704). The result was actually discovered by John Bernoulli (1667–1748), who communicated the result to L'Hôpital in 1694.

it follows from L'Hôpital's rule that

$$\frac{x}{\sin\sqrt{x}} \to 0 \quad \text{as} \quad x \to 0^+.$$

In short, we can write

$$\lim_{x\to 0^+} \frac{x}{\sin\sqrt{x}} \overset{*}{=} \lim_{x\to 0^+} \frac{2\sqrt{x}}{\cos\sqrt{x}} = 0. \quad \square$$

Remark There is a tendency to abuse the limit-finding technique given in Theorem 10.5.1. L'Hôpital's rule *does not apply* in cases where the numerator or the denominator has a finite non-zero limit. For example,

$$\lim_{x\to 0} \frac{x}{x+\cos x} = \frac{0}{1} = 0,$$

but a blind application of L'Hôpital's rule would lead to

$$\lim_{x\to 0} \frac{x}{x+\cos x} \overset{*}{=} \lim_{x\to 0} \frac{1}{1-\sin x} = 1.$$

This is, of course, incorrect. □

Sometimes it is necessary to differentiate numerator and denominator more than once. Such is the case in the next example.

Example 3 Find $\lim_{x\to 0} \dfrac{e^x - x - 1}{x^2}$.

SOLUTION As $x \to 0$, both numerator and denominator tend to 0. Here

$$\frac{f'(x)}{g'(x)} = \frac{e^x - 1}{2x}.$$

Since both numerator and denominator still tend to 0 as $x \to 0$, we differentiate again:

$$\frac{f''(x)}{g''(x)} = \frac{e^x}{2}.$$

Since this last quotient tends to $\frac{1}{2}$, we can conclude that

$$\frac{e^x - 1}{2x} \to \frac{1}{2} \quad \text{and therefore} \quad \frac{e^x - x - 1}{x^2} \to \frac{1}{2} \quad \text{as} \quad x \to 0.$$

In short, we can write

$$\lim_{x\to 0} \frac{e^x - x - 1}{x^2} \overset{*}{=} \lim_{x\to 0} \frac{e^x - 1}{2x} \overset{*}{=} \lim_{x\to 0} \frac{e^x}{2} = \frac{1}{2}. \quad \square$$

In the next example we use L'Hôpital's rule to find the limit of a sequence.

Example 4 Find $\lim_{n\to\infty} \dfrac{e^{2/n} - 1}{1/n}$.

SOLUTION This is an indeterminate of the form 0/0. To apply the methods of this section, we replace the integer variable n by the real variable x and examine the behavior of

$$\frac{e^{2/x} - 1}{1/x} \quad \text{as} \quad x \to \infty.$$

Applying L'Hôpital's rule, we have

$$\lim_{x\to\infty}\frac{e^{2/x}-1}{1/x} \overset{*}{=} \lim_{x\to\infty}\frac{e^{2/x}(-2/x^2)}{(-1/x^2)} = 2\lim_{x\to\infty} e^{2/x} = 2.$$

Thus, $\quad$ as $n\to\infty$, $\quad \dfrac{e^{2/n}-1}{1/n}\to 2.$ □

To derive L'Hôpital's rule, we need a generalization of the mean-value theorem.

THEOREM 10.5.2 THE CAUCHY MEAN-VALUE THEOREM†

Suppose that f and g are differentiable on (a, b) and continuous on $[a, b]$. If g' is never 0 in (a, b), then there is a number r in (a, b) for which

$$\frac{f'(r)}{g'(r)} = \frac{f(b)-f(a)}{g(b)-g(a)}.$$

PROOF We can prove this by applying Rolle's theorem (4.1.3) to the function

$$G(x) = [g(b)-g(a)][f(x)-f(a)] - [g(x)-g(a)][f(b)-f(a)].$$

Since

$$G(a) = 0 \quad \text{and} \quad G(b) = 0,$$

there exists (by Rolle's theorem) a number r in (a, b) for which $G'(r) = 0$.

Now

$$G'(x) = [g(b)-g(a)]f'(x) - g'(x)[f(b)-f(a)].$$

Setting $x = r$, we get

$$[g(b)-g(a)]f'(r) - g'(r)][f(b)-f(a)] = 0,$$

and thus

$$[g(b)-g(a)]f'(r) = g'(r)[f(b)-f(a)].$$

Since g' is never 0 in (a, b),

$$g'(r) \neq 0 \quad \text{and} \quad g(b)-g(a) \neq 0.$$

↑ explain

We can therefore divide by these numbers and obtain

$$\frac{f'(r)}{g'(r)} = \frac{f(b)-f(a)}{g(b)-g(a)}.$$ □

Now we prove L'Hôpital's rule for the case $x\to c^+$. We assume that, as $x\to c^+$,

$$f(x)\to 0, \qquad g(x)\to 0 \qquad \text{and} \qquad \frac{f'(x)}{g'(x)}\to L.$$

† Another contribution of A.-L. Cauchy, after whom Cauchy sequences were named.

We want to show that

$$\frac{f(x)}{g(x)} \to L.$$

PROOF The fact that

$$\frac{f'(x)}{g'(x)} \to L \quad \text{as} \quad x \to c^+$$

assures that both $f'(x)$ and $g'(x)$ exist on an interval of the form $[c, c+h)$ and that g' is not zero there. By setting $f(c) = 0$ and $g(c) = 0$, we ensure that f and g are both continuous on $[c, c+h]$. We can now apply the Cauchy mean-value theorem and conclude that there exists a number c_h between c and $c+h$ such that

$$\frac{f'(c_h)}{g'(c_h)} = \frac{f(c+h) - f(c)}{g(c+h) - g(c)} = \frac{f(c+h)}{g(c+h)}.$$

The result is now obtained by letting $h \to 0^+$. Since the left side tends to L, the right side tends to L. □

The case $x \to c^-$ can be handled in the same manner. The two cases together prove the rule for the case $x \to c$.

Here is a proof of L'Hôpital's rule for the case $x \to \infty$.

PROOF The key here is to set $x = 1/t$:

$$\lim_{x\to\infty} \frac{f'(x)}{g'(x)} = \lim_{t\to 0^+} \frac{[f(1/t)]'}{[g(1/t)]'} = \lim_{t\to 0^+} \frac{-t^{-2}f'(1/t)}{-t^{-2}g'(1/t)}$$

$$= \lim_{t\to 0^+} \frac{f'(1/t)}{g'(1/t)} = \lim_{t\to 0^+} \frac{f(1/t)}{g(1/t)} = \lim_{x\to\infty} \frac{f(x)}{g(x)}. \quad \square$$

(by L'Hôpital's rule for the case $t \to 0^+$)

EXERCISES 10.5

Calculate the indicated limit.

1. $\lim_{x\to 0^+} \dfrac{\sin x}{\sqrt{x}}$.
2. $\lim_{x\to 1} \dfrac{\ln x}{1-x}$.
3. $\lim_{x\to 0} \dfrac{e^x - 1}{\ln(1+x)}$.
4. $\lim_{x\to 4} \dfrac{\sqrt{x} - 2}{x - 4}$.
5. $\lim_{x\to \pi/2} \dfrac{\cos x}{\sin 2x}$.
6. $\lim_{x\to a} \dfrac{x - a}{x^n - a^n}$.
7. $\lim_{x\to 0} \dfrac{2^x - 1}{x}$.
8. $\lim_{x\to 0} \dfrac{\tan^{-1} x}{x}$.
9. $\lim_{x\to 1} \dfrac{x^{1/2} - x^{1/4}}{x - 1}$.
10. $\lim_{x\to 0} \dfrac{e^x - 1}{x(1+x)}$.
11. $\lim_{x\to 0} \dfrac{e^x - e^{-x}}{\sin x}$.
12. $\lim_{x\to 0} \dfrac{1 - \cos x}{3x}$.
13. $\lim_{x\to 0} \dfrac{x + \sin \pi x}{x - \sin \pi x}$.
14. $\lim_{x\to 0} \dfrac{a^x - (a+1)^x}{x}$.
15. $\lim_{x\to 0} \dfrac{e^x + e^{-x} - 2}{1 - \cos 2x}$.
16. $\lim_{x\to 0} \dfrac{x - \ln(x+1)}{1 - \cos 2x}$.
17. $\lim_{x\to 0} \dfrac{\tan \pi x}{e^x - 1}$.
18. $\lim_{x\to 0} \dfrac{\cos x - 1 + x^2/2}{x^4}$.
19. $\lim_{x\to 0} \dfrac{1 + x - e^x}{x(e^x - 1)}$.
20. $\lim_{x\to 0} \dfrac{\ln(\sec x)}{x^2}$.
21. $\lim_{x\to 0} \dfrac{x - \tan x}{x - \sin x}$.
22. $\lim_{x\to 0} \dfrac{x\, e^{nx} - x}{1 - \cos nx}$.
23. $\lim_{x\to 1^-} \dfrac{\sqrt{1 - x^2}}{\sqrt{1 - x^3}}$.
24. $\lim_{x\to 0} \dfrac{2x - \sin \pi x}{4x^2 - 1}$.
25. $\lim_{x\to \pi/2} \dfrac{\ln(\sin x)}{(\pi - 2x)^2}$.
26. $\lim_{x\to 0^+} \dfrac{\sqrt{x}}{\sqrt{x} + \sin\sqrt{x}}$.
27. $\lim_{x\to 0} \dfrac{\cos x - \cos 3x}{\sin(x^2)}$.
28. $\lim_{x\to 0} \dfrac{\sqrt{a + x} - \sqrt{a - x}}{x}$.
29. $\lim_{x\to \pi/4} \dfrac{\sec^2 x - 2\tan x}{1 + \cos 4x}$.
30. $\lim_{x\to 0} \dfrac{x - \sin^{-1} x}{\sin^3 x}$.
31. $\lim_{x\to 0} \dfrac{\tan^{-1} x}{\tan^{-1} 2x}$.
32. $\lim_{x\to 0} \dfrac{\sin^{-1} x}{x}$.

Find the limit of the sequence.

33. $\lim_{n\to\infty} \dfrac{(\pi/2 - \tan^{-1} n)}{1/n}$.

34. $\lim_{n\to\infty} \dfrac{\ln(1 - 1/n)}{\sin(1/n)}$.

35. $\lim_{n\to\infty} \dfrac{1}{n[\ln(n+1) - \ln n]}$.

36. $\lim_{n\to\infty} \dfrac{\sinh \pi/n - \sin \pi/n}{\sin^3 \pi/n}$.

C Use technology (graphing utility or CAS) to calculate the limits in Exercises 37–42.

37. $\lim_{x\to 0} \dfrac{x^3}{4^x - 1}$.

38. $\lim_{x\to 0} \dfrac{4x}{\sin^2 x}$.

39. $\lim_{x\to 0} \dfrac{x}{\frac{\pi}{2} - \arccos x}$.

40. $\lim_{x\to 2^+} \dfrac{\sqrt{2x} - 2}{\sqrt{x-2}}$.

41. $\lim_{x\to 0} \dfrac{\tanh x}{x}$.

42. $\lim_{x\to \frac{\pi}{2}} \dfrac{1 + \cos 2x}{1 - \sin x}$.

43. Find the fallacy:

$$\lim_{x\to 0} \frac{2 + x + \sin x}{x^3 + x - \cos x} \overset{*}{=} \lim_{x\to 0} \frac{1 + \cos x}{3x^2 + 1 + \sin x}$$

$$\overset{*}{=} \lim_{x\to 0} \frac{-\sin x}{6x + \cos x} = \frac{0}{1} = 0.$$

44. Show that, if $a > 0$, then

$$\lim_{n\to\infty} n(a^{1/n} - 1) = \ln a.$$

45. Find values for a and b for which

$$\lim_{x\to 0} \frac{\cos ax - b}{2x^2} = -4.$$

46. Find values for a and b for which

$$\lim_{x\to 0} \frac{\sin 2x + ax + bx^3}{x^3} = 0.$$

47. Calculate $\lim_{x\to 0} \dfrac{(1+x)^{1/x} - e}{x}$.

48. Let f be a twice differentiable function and fix a value of x.

(a) Show that

$$\lim_{h\to 0} \frac{f(x+h) - f(x-h)}{2h} = f'(x).$$

(b) Show that

$$\lim_{h\to 0} \frac{f(x+h) - 2f(x) + f(x-h)}{h^2} = f''(x).$$

49. Given that f is continuous, use L'Hôpital's rule to determine

$$\lim_{x\to 0} \left(\frac{1}{x} \int_0^x f(t)dt \right).$$

50. The integral $Si(x) = \displaystyle\int_0^x \frac{\sin t}{t} dt$ is a special function in applied mathematics. Calculate the limits:

(a) $\lim_{x\to 0} \dfrac{Si(x)}{x}$. (b) $\lim_{x\to 0} \dfrac{Si(x) - x}{x^3}$.

51. The *Fresnel function* $C(x) = \displaystyle\int_0^x \cos^2 t\, dt$ arises in the study of the diffraction of light. Calculate the limits:

(a) $\lim_{x\to 0} \dfrac{C(x)}{x}$. (b) $\lim_{x\to 0} \dfrac{C(x) - x}{x^3}$.

52. (a) Given that the function f is differentiable, $f(a) = 0$ and $f'(a) \neq 0$, determine

$$\lim_{x\to a} \frac{\int_a^x f(t)dt}{f(x)}$$

(b) Suppose f is k-times differentiable, $f(a) = f'(a) = \cdots = f^{k-1}(a) = 0$, and $f^k(a) \neq 0$. Determine

$$\lim_{x\to a} \frac{\int_a^x f(t)dt}{f(x)}.$$

53. Let $A(b)$ be the area of the region bounded by the parabola $y = x^2$ and the horizontal line $y = b$ $(b > 0)$, and let $T(b)$ be the area of the triangle AOB (see the figure). Find $\lim_{b\to 0^+} T(b)/A(b)$.

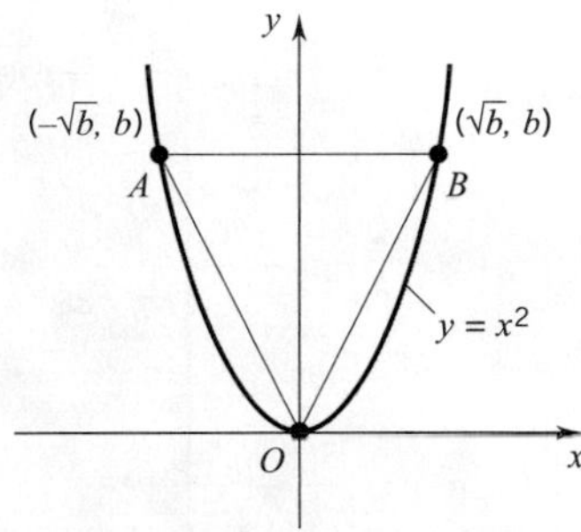

54. Choose an angle $\theta, 0 < \theta < \pi/2$, in standard position as shown in the figure. Let $T(\theta)$ be the area of the triangle ABC, and let $S(\theta)$ be the area of the segment of the circle formed by the chord AB. Find $\lim_{\theta\to 0^+} T(\theta)/S(\theta)$.

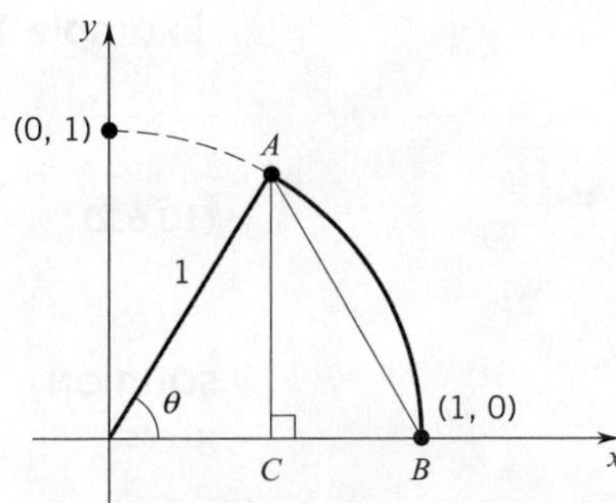

C **55.** Let

$$f(x) = \frac{x^2 - 16}{\sqrt{x^2 + 9} - 5}.$$

(a) Use a graphing utility to graph f. What is the behavior of $f(x)$ as $x \to \infty$ and as $x \to -\infty$?

(b) What is the behavior of $f(x)$ as $x \to 4$? Confirm your answer using L'Hôpital's rule.

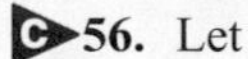**56.** Let

$$f(x) = \frac{x - \sin x}{x^3}.$$

(a) Use a graphing utility to graph f. What is the behavior of $f(x)$ as $x \to \infty$ and as $x \to -\infty$?

(b) What is the behavior of $f(x)$ as $x \to 0$? Confirm your answer using L'Hôpital's rule.

57. Let $f(x) = \dfrac{2^{\sin x} - 1}{x}$.

(a) Use a graphing utility to graph f. Estimate

$$\lim_{x\to 0} f(x).$$

(b) Use L'Hôpital's rule to confirm your estimate in part (a).

58. Let $g(x) = \dfrac{3^{\cos x} - 3}{x^2}$.

(a) Use a graphing utility to graph g. Estimate

$$\lim_{x\to 0} g(x).$$

(b) Use L'Hôpital's rule to confirm your estimate in part (a).

■ 10.6 THE INDETERMINATE FORM (∞/∞); OTHER INDETERMINATE FORMS

We come now to limits of quotients $f(x)/g(x)$ where numerator and denominator both tend to ∞. Such limits are called *indeterminates of the form* ∞/∞.

THEOREM 10.6.1 L'HÔPITAL'S RULE (∞/∞)

Suppose that

$$f(x) \to \pm\infty \quad \text{and} \quad g(x) \to \pm\infty$$

as $x \to c^+$, $x \to c^-$, $x \to c$, $x \to \infty$ or $x \to -\infty$.

$$\text{If} \quad \frac{f'(x)}{g'(x)} \to L, \quad \text{then} \quad \frac{f(x)}{g(x)} \to L.$$

NOTE: This theorem includes the possibility that $L = \infty$ or $-\infty$.

While the proof of L'Hôpital's rule in this setting is a little more complicated than it was in the (0/0) case,† the application of the rule is much the same.

Example 1 Let α be any positive number. Show that

(10.6.2)
$$\lim_{x\to\infty} \frac{\ln x}{x^\alpha} = 0.$$

SOLUTION Both numerator and denominator tend to ∞ as $x \to \infty$. L'Hôpital's rule gives

$$\lim_{x\to\infty} \frac{\ln x}{x^\alpha} \overset{*}{=} \lim_{x\to\infty} \frac{1/x}{\alpha x^{\alpha-1}} = \lim_{x\to\infty} \frac{1}{\alpha x^\alpha} = 0. \quad \square$$

† We omit the proof.

For example, as $x \to \infty$,

$$\frac{\ln x}{x^{0.01}} \to 0 \quad \text{and} \quad \frac{\ln x}{x^{0.001}} \to 0.$$

Example 2 Let k be any positive integer. Show that

$$\text{(10.6.3)} \qquad \boxed{\lim_{x\to\infty} \frac{x^k}{e^x} = 0.}$$

SOLUTION Here we differentiate numerator and denominator k times:

$$\lim_{x\to\infty} \frac{x^k}{e^x} \overset{*}{=} \lim_{x\to\infty} \frac{kx^{k-1}}{e^x} \overset{*}{=} \lim_{x\to\infty} \frac{k(k-1)x^{k-2}}{e^x} \overset{*}{=} \cdots \overset{*}{=} \lim_{x\to\infty} \frac{k!}{e^x} = 0. \quad \square$$

For example, as $x \to \infty$,

$$\frac{x^{100}}{e^x} \to 0 \quad \text{and} \quad \frac{x^{1000}}{e^x} \to 0$$

Remark The limits (10.6.2) and (10.6.3) tell us that $\ln x$ tends to infinity *more slowly than* any positive power of x and that e^x tends to infinity *faster than* any positive integral power of x. In the Exercises you are asked to show that e^x tends to infinity faster than *any* positive power of x and that *any* positive power of $\ln x$ tends to infinity more slowly than x. That is, for any positive number α,

$$\lim_{x\to\infty} \frac{x^\alpha}{e^x} = 0 \quad \text{and} \quad \lim_{x\to\infty} \frac{[\ln x]^\alpha}{x} = 0.$$

Other comparisons of logarithmic and exponential growth were given in Exercises 23 and 24 of Section 7.6. □

Example 3 Determine the behavior of $a_n = \dfrac{2^n}{n^2}$ as $n \to \infty$.

SOLUTION To use the methods of calculus, we investigate $\lim_{x\to\infty} \dfrac{2^x}{x^2}$. Since both numerator and denominator tend to ∞ with x, we try L'Hôpital's rule:

$$\lim_{x\to\infty} \frac{2^x}{x^2} \overset{*}{=} \lim_{x\to\infty} \frac{2^x \ln 2}{2x} \overset{*}{=} \lim_{x\to\infty} \frac{2^x(\ln 2)^2}{2} = \infty.$$

Therefore the sequence must also diverge to ∞. □

Other Indeterminate Forms: $0 \cdot \infty$, $\infty - \infty$, 0^0, 1^∞, ∞^0

If f tends to 0 and g tends to ∞ (or $-\infty$) as x approaches some number c (or $\pm\infty$), then it is not clear what the product $f \cdot g$ will do. For example, as $x \to 1$,

$$(x-1)^3 \to 0, \quad \frac{1}{(x-1)^2} \to \infty, \quad \text{and} \quad (x-1)^3 \cdot \frac{1}{(x-1)^2} = (x-1) \to 0.$$

On the other hand,

$$\lim_{x\to 1} \left[(x-1)^3 \cdot \frac{1}{(x-1)^4}\right] = \lim_{x\to 1} \left[\frac{1}{x-1}\right] \quad \text{does not exist.}$$

A limit of this type is called an *indeterminate of the form* $0 \cdot \infty$. We handle such indeterminates by writing the product $f \cdot g$ as a quotient

$$\frac{f}{1/g} \quad \text{or} \quad \frac{g}{1/f}.$$

In the first case, the result will be an indeterminate of the form $0/0$, and in the second it will have the form ∞/∞.

Example 4 Find $\lim_{x\to 0^+} \sqrt{x}\ln x$.

SOLUTION As $x \to 0^+$, $\sqrt{x} \to 0$ and $\ln x \to -\infty$. Thus the given limit is an indeterminate of the form $0 \cdot \infty$. Rewriting the product as a quotient, we have

$$\lim_{x\to 0^+} \sqrt{x}\ln x = \lim_{x\to 0^+} \frac{\ln x}{1/\sqrt{x}} \overset{*}{=} \lim_{x\to 0^+} \frac{1/x}{-\frac{1}{2}x^{-3/2}} = \lim_{x\to 0^+} -2\sqrt{x} = 0.$$

Of course, we could have chosen to write $\sqrt{x}\ln x$ as the quotient

$$\frac{\sqrt{x}}{1/\ln x}.$$

Try to evaluate the limit using this quotient. □

If f and g both tend to ∞, or if both tend to $-\infty$, as x tends to c (or $\pm\infty$), then $\lim (f - g)$ is called an *indeterminate of the form* $\infty - \infty$. Such indeterminates can be handled by converting them to quotients.

Example 5 Find $\lim_{x\to(\pi/2)^-} (\tan x - \sec x)$.

SOLUTION Both $\tan x$ and $\sec x$ tend to ∞ as x tends to $\pi/2$ from the left. We first rewrite the difference as a quotient:

$$\tan x - \sec x = \frac{\sin x}{\cos x} - \frac{1}{\cos x} = \frac{\sin x - 1}{\cos x}.$$

Now

$$\lim_{x\to(\pi/2)^-} \frac{\sin x - 1}{\cos x}$$

is an indeterminate of the form $0/0$, and

$$\lim_{x\to(\pi/2)^-} \frac{\sin x - 1}{\cos x} \overset{*}{=} \lim_{x\to(\pi/2)^-} \frac{\cos x}{-\sin x} = \frac{0}{-1} = 0.$$

Thus, $\lim_{x\to(\pi/2)^-} (\tan x - \sec x) = 0$. □

Limits involving exponential expressions $[f(x)]^{g(x)}$ are indeterminate: (1) if f and g both tend to 0; (2) if f tends to 1 and g tends to $\pm\infty$; and (3) if f tends to $\pm\infty$ and g tends to 0. These cases lead to *indeterminates of the form* 0^0, 1^∞, and ∞^0 respectively. We can usually gain insight into the behavior of an exponential indeterminate form by applying the natural logarithm function:

$$\text{replace} \quad y = [f(x)]^{g(x)} \qquad \text{by} \qquad \ln y = \ln\,[f(x)]^{g(x)} = g(x)\ln\,[f(x)].$$

Example 6 Show that

(10.6.4)
$$\boxed{\lim_{x\to 0^+} x^x = 1.}$$

SOLUTION Here we are dealing with an indeterminate of the form 0^0. Our first step is to take the logarithm of x^x. Then we apply L'Hôpital's rule:

$$\lim_{x\to 0^+} \ln(x^x) = \lim_{x\to 0^+} (x \ln x) = \lim_{x\to 0^+} \frac{\ln x}{1/x} \overset{*}{=} \lim_{x\to 0^+} \frac{1/x}{-1/x^2} = \lim_{x\to 0^+} (-x) = 0.$$

Thus, as $x \to 0$, $\ln(x^x) \to 0$ and $x^x = e^{\ln(x^x)} \to 1$. □

Example 7 Find $\lim_{x\to 0^+} (1+x)^{1/x}$.

SOLUTION Here we are dealing with an indeterminate of the form 1^∞: as $x \to 0^+$, $1+x \to 1$ and $1/x$ increases without bound. Taking the logarithm and then applying L'Hôpital's rule, we have

$$\lim_{x\to 0^+} \ln(1+x)^{1/x} = \lim_{x\to 0^+} \frac{\ln(1+x)}{x} \overset{*}{=} \lim_{x\to 0^+} \frac{1}{1+x} = 1.$$

Therefore, as $x \to 0^+$, $\ln(1+x)^{1/x} \to 1$ and thus $(1+x)^{1/x} = e^{\ln(1+x)^{1/x}} \to e^1 = e$. Set $x = 1/n$ and we have the familiar result: as $n \to \infty$, $[1+(1/n)]^n \to e$. □

Example 8 As asserted without proof in Example 4, Section 10.3

$$\lim_{n\to\infty} (3^n + 4^n)^{1/n} = 4.$$

SOLUTION We shall calculate $\lim_{x\to\infty} (3^x + 4^x)^{1/x}$. This is an indeterminate of the form ∞^0. Taking the logarithm and then applying L'Hôpital's rule, we find that

$$\lim_{x\to\infty} \ln(3^x + 4^x)^{1/x} = \lim_{x\to\infty} \frac{\ln(3^x + 4^x)}{x} \overset{*}{=} \lim_{x\to\infty} \frac{\dfrac{(3^x \ln 3 + 4^x \ln 4)}{(3^x + 4^x)}}{1}$$

$$= \lim_{x\to\infty} \frac{3^x \ln 3 + 4^x \ln 4}{3^x + 4^x} = \lim_{x\to\infty} \frac{(3/4)^x \ln 3 + \ln 4}{(3/4)^x + 1} = \ln 4.$$

Therefore, $\lim_{x\to\infty} (3^x + 4^x)^{1/x} = 4$ and the result follows. □

Remark Suppose that $\lim(f/g)$ is an indeterminate form. Both versions of L'Hôpital's rule (Theorems 10.5.1 and 10.6.1) tell us that

$$\text{if} \quad \lim \frac{f'}{g'} = L, \qquad \text{then} \qquad \lim \frac{f}{g} = L.$$

However, the rules do not provide any information when $\lim(f'/g')$ fails to exist. In this case $\lim(f/g)$ may or may not exist. See Project 10.6 □

Finally, remember that L'Hôpital's rule does not apply when the numerator or denominator has a finite nonzero limit. While

$$\frac{1+x}{\sin x} \to \infty \qquad \text{as} \qquad x \to 0^+,$$

a misapplication of L'Hôpital's rule leads to the *incorrect conclusion* that

$$\lim_{x\to 0^+} \frac{1+x}{\sin x} \overset{*}{=} \lim_{x\to 0^+} \frac{1}{\cos x} = 1.$$

EXERCISES 10.6

Calculate the indicated limit.

1. $\lim\limits_{x\to -\infty} \dfrac{x^2+1}{1-x}$.
2. $\lim\limits_{x\to\infty} \dfrac{20x}{x^2+1}$.
3. $\lim\limits_{x\to\infty} \dfrac{x^3}{1-x^3}$.
4. $\lim\limits_{x\to\infty} \dfrac{x^3-1}{2-x}$.
5. $\lim\limits_{x\to\infty} \left(x^2 \sin\dfrac{1}{x}\right)$.
6. $\lim\limits_{x\to\infty} \dfrac{\ln x^k}{x}$.
7. $\lim\limits_{x\to\pi/2^-} \dfrac{\tan 5x}{\tan x}$.
8. $\lim\limits_{x\to 0} (x \ln|\sin x|)$.
9. $\lim\limits_{x\to 0^+} x^{2x}$.
10. $\lim\limits_{x\to\infty} \left(x \sin\dfrac{\pi}{x}\right)$.
11. $\lim\limits_{x\to 0} [x(\ln|x|)^2]$.
12. $\lim\limits_{x\to 0^+} \dfrac{\ln x}{\cot x}$.
13. $\lim\limits_{x\to\infty} \left(\dfrac{1}{x}\displaystyle\int_0^x e^{t^2}\,dt\right)$.
14. $\lim\limits_{x\to\infty} \dfrac{\sqrt{1+x^2}}{x^2}$.
15. $\lim\limits_{x\to 0} \left[\dfrac{1}{\sin^2 x} - \dfrac{1}{x^2}\right]$.
16. $\lim\limits_{x\to 0} |\sin x|^x$.
17. $\lim\limits_{x\to 1} x^{1/(x-1)}$.
18. $\lim\limits_{x\to 0^+} x^{\sin x}$.
19. $\lim\limits_{x\to\infty} \left(\cos\dfrac{1}{x}\right)^x$.
20. $\lim\limits_{x\to\pi/2} |\sec x|^{\cos x}$.
21. $\lim\limits_{x\to 0} \left[\dfrac{1}{\ln(1+x)} - \dfrac{1}{x}\right]$.
22. $\lim\limits_{x\to\infty} (x^2+a^2)^{(1/x)^2}$.
23. $\lim\limits_{x\to 0} \left(\dfrac{1}{x} - \cot x\right)$.
24. $\lim\limits_{x\to\infty} \ln\left(\dfrac{x^2-1}{x^2+1}\right)^3$.
25. $\lim\limits_{x\to\infty} (\sqrt{x^2+2x} - x)$.
26. $\lim\limits_{x\to\infty} \dfrac{1}{x}\displaystyle\int_0^x \sin\left(\dfrac{1}{t+1}\right) dt$.
27. $\lim\limits_{x\to\infty} (x^3+1)^{1/\ln x}$.
28. $\lim\limits_{x\to\infty} (e^x+1)^{1/x}$.
29. $\lim\limits_{x\to\infty} (\cosh x)^{1/x}$.
30. $\lim\limits_{x\to\infty} \left(1+\dfrac{1}{x}\right)^{3x}$.
31. $\lim\limits_{x\to 0} \left(\dfrac{1}{\sin x} - \dfrac{1}{x}\right)$.
32. $\lim\limits_{x\to 0} (e^x+3x)^{1/x}$.
33. $\lim\limits_{x\to 1} \left(\dfrac{1}{\ln x} - \dfrac{x}{x-1}\right)$.
34. $\lim\limits_{x\to 0} \left(\dfrac{1+2^x}{2}\right)^{1/x}$.

Find the limit of the sequence.

35. $\lim\limits_{n\to\infty} \left(\dfrac{1}{n}\ln\dfrac{1}{n}\right)$.
36. $\lim\limits_{n\to\infty} \dfrac{n^k}{2^n}$.
37. $\lim\limits_{n\to\infty} (\ln n)^{1/n}$.
38. $\lim\limits_{n\to\infty} \dfrac{\ln n}{n^p}, \quad (p>0)$.
39. $\lim\limits_{n\to\infty} (n^2+n)^{1/n}$.
40. $\lim\limits_{n\to\infty} n^{\sin(\pi/n)}$.
41. $\lim\limits_{n\to\infty} \dfrac{n^2 \ln n}{e^n}$.
42. $\lim\limits_{n\to\infty} (\sqrt{n}-1)^{1/\sqrt{n}}$.

C Use technology (graphing utility or CAS) to calculate the limits in Exercises 43–46.

43. $\lim\limits_{x\to 0} (\sin x)^x$.
44. $\lim\limits_{x\to\frac{\pi}{4}} (\tan x)^{\tan 2x}$.
45. $\lim\limits_{x\to 0} \left(\dfrac{1}{\sin x} - \dfrac{1}{\tan x}\right)$.
46. $\lim\limits_{x\to 0^+} (\sinh x)^{-x}$.

Sketch the curve, specifying all vertical and horizontal asymptotes.

47. $y = x^2 - \dfrac{1}{x^3}$.
48. $y = \sqrt{\dfrac{x}{x-1}}$.
49. $y = x\,e^x$.
50. $y = x\,e^{-x}$.
51. $y = x^2 e^{-x}$.
52. $y = \dfrac{\ln x}{x}$.

The graphs of two functions $y = f(x)$ and $y = g(x)$ are said to be *asymptotic as* $x \to \infty$ if

$$\lim_{x\to\infty} [f(x) - g(x)] = 0;$$

they are said to be *asymptotic as* $x \to -\infty$ if

$$\lim_{x\to-\infty} [f(x) - g(x)] = 0.$$

53. Show that the hyperbolic arc $y = (b/a)\sqrt{x^2 - a^2}$ is asymptotic to the line $y = (b/a)x$ as $x \to \infty$.

54. Show that the graphs of $y = \cosh x$ and $y = \sinh x$ are asymptotic as $x \to \infty$.

55. Give an example of a function the graph of which is asymptotic to the parabola $y = x^2$ as $x \to \infty$ and crosses the graph of the parabola exactly twice.

56. Give an example of a function the graph of which is asymptotic to the line $y = x$ as $x \to \infty$ and crosses the line infinitely often.

57. Find the fallacy:

$$\lim_{x\to 0^+} \frac{x^2}{\sin x} \overset{*}{=} \lim_{x\to 0^+} \frac{2x}{\cos x} \overset{*}{=} \lim_{x\to 0^+} \frac{2}{-\sin x} = -\infty.$$

58. Show that the exponential forms 1^0, 0^1, and 0^∞ are not indeterminate. What are the "limits" in each of these cases? For example, if $\lim_{x\to c} f(x) = 1$ and $\lim_{x\to c} g(x) = 0$, what is

$$\lim_{x\to c} [f(x)]^{g(x)}?$$

59. (a) Show by induction that, for each positive integer k,

$$\lim_{x\to\infty} \frac{(\ln x)^k}{x} = 0.$$

(b) Show that, for each positive number α,

$$\lim_{x\to\infty} \frac{(\ln x)^\alpha}{x} = 0.$$

60. Let α be a positive number. Show that

$$\lim_{x\to\infty} \frac{x^\alpha}{e^x} = 0.$$

61. The *geometric mean* of two positive numbers a and b is $\sqrt{ab}$. Show that

$$\sqrt{ab} = \lim_{x\to\infty} \left(\frac{a^{1/x} + b^{1/x}}{2}\right)^x.$$

62. The differential equation satisfied by the velocity of an object of mass m dropped from rest under the influence of gravity with air resistance directly proportional to the velocity can be written

$$(*) \qquad m\frac{dv}{dt} + kv = mg.$$

where $k > 0$ is the constant of proportionality, g is the gravitational constant and $v(0) = 0$. See Exercises 37, Section 8.8. The velocity of the object at time t is given by

$$v(t) = (mg/k)(1 - e^{-(k/m)t}).$$

(a) Fix t and find $\lim_{k\to 0^+} v(t)$.

(b) Set $k = 0$ in equation ($*$) and solve

$$m\frac{dv}{dt} = mg, \qquad v(0) = 0.$$

Does this result agree with the result found in part (a)?

In Exercises 63 and 64, let $f(x) = xe^{-x}$. Use a graphing utility to draw the graph of f on $[0, 20]$. For the purposes of the exercises, consider f on the interval $[0, b]$.

63. Use a CAS to find:

(a) an expression for the area of the region Ω between the graph of f and the x-axis; and

(b) an expression for the centroid of Ω.

(c) Find the limit of each expression as $b \to \infty$. Interpret your results geometrically.

64. Use a CAS to find:

(a) an expression for the volume of the solid generated by revolving Ω about the x-axis; and

(b) an expression for the volume of the solid generated by revolving Ω about the y-axis.

(c) Find the limit of each expression as $b \to \infty$. Interpret your results geometrically.

65. Let $f(x) = (1 + x)^{1/x}$ and $g(x) = (1 + x^2)^{1/x}$ on $(0, \infty)$.

(a) Use a graphing utility to graph f and g in the same coordinate system. Estimate

$$\lim_{x\to 0^+} g(x).$$

(b) Use L'Hôpital's rule to confirm your estimate.

66. Let $f(x) = \sqrt{x^2 + 3x + 1} - x$.

(a) Use a graphing utility to graph f. Then use your graph to estimate

$$\lim_{x\to\infty} f(x).$$

(b) Use L'Hôpital's rule to confirm your estimate. HINT: "Rationalize."

67. Let $g(x) = \sqrt[3]{x^3 - 5x^2 + 2x + 1} - x$.

(a) Use a graphing utility to graph g. Then use your graph to estimate

$$\lim_{x\to\infty} g(x).$$

(b) Use L'Hôpital's rule to confirm your estimate.

68. Exercises 66 and 67 can be generalized as follows: Let n be a positive integer and let P be the polynomial

$$P(x) = x^n + b_1x^{n-1} + b_2x^{n-2} + \cdots + b_{n-1}x + b_n.$$

Prove that

$$\lim_{x\to\infty} ([P(x)]^{1/n} - x) = \frac{b_1}{n}.$$

■ PROJECT 10.6 Peculiar Indeterminate Forms

Sometimes the application of L'Hôpital's rule to evaluate an indeterminate form leads to peculiar situations. In this project, we investigate two such cases.

First, we let

$$f(x) = \frac{2x}{\sqrt{1+x^2}}$$

and consider $\lim_{x\to\infty} f(x)$. This limit is an indeterminate of the form ∞/∞.

Problem 1.

a. Try to evaluate $\lim_{x\to\infty} \dfrac{2x}{\sqrt{1+x^2}}$ using L'Hôpital's rule and describe what happens.

b. Use a graphing utility to graph f and determine $\lim_{x\to\infty} f(x)$.

c. Actually this limit is easy to calculate by algebraic methods. Use algebra to verify the limit that you found in part (b).

Now consider the function

$$g(x) = \frac{e^{-1/x^2}}{x}$$

and consider $\lim_{x\to 0} g(x)$. This is an indeterminate of the form $0/0$.

Problem 2.

a. Try to evaluate $\lim_{x\to 0} \dfrac{e^{-1/x^2}}{x}$ using L'Hôpital's rule and describe what happens.

b. Write g in the equivalent form

$$g(x) = \frac{1/x}{e^{1/x^2}}$$

and use L'Hôpital's rule to evalute $\lim_{x\to 0} g(x)$.

c. Use a graphing utility to graph g and verify the limit you found in part (b).

Problem 3. Define the function h by

$$h(x) = \begin{cases} e^{-1/x^2} & x \neq 0 \\ 0 & x = 0. \end{cases}$$

a. Use the definition of the derivative to show that h is differentiable at $x = 0$. What is $h'(0)$?

b. Show that $\lim_{x\to 0} \dfrac{e^{-1/x^2}}{x^n} = 0$ for every positive integer n.

c. Calculate $h'(x), h''(x), \cdots$, for $x \neq 0$.

d. Use part (b) and the results in part (c) to argue that $h^{(n)}(0) = 0$ for every positive integer n.

■ 10.7 IMPROPER INTEGRALS

So far we have worked with integrals

$$\int_a^b f(x)\,dx$$

in which the interval of integration is bounded and the function being integrated is bounded. By a limit process described below, we can extend the integration process to intervals that are unbounded and to functions that are unbounded. The resulting integrals are called *improper integrals*.

Integrals over Unbounded Intervals

We begin with a function f continuous on an unbounded interval $[a, \infty)$. For each number $b > a$ we can form the definite integral

$$\int_a^b f(x)\,dx. \qquad \text{(Figure 10.7.1)}$$

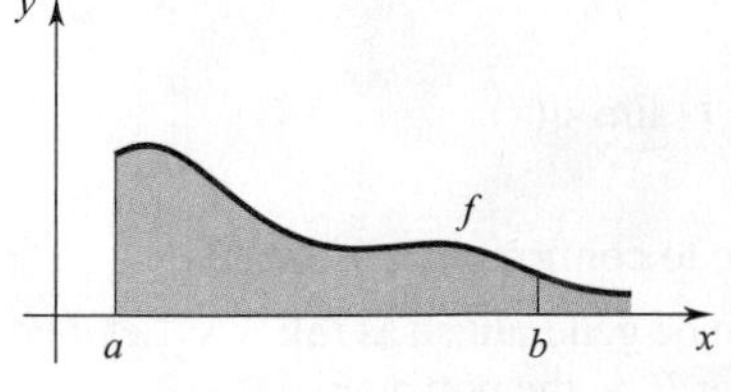

Figure 10.7.1

If, as b tends to ∞, this integral tends to a finite limit L,

$$\lim_{b\to\infty} \int_a^b f(x)\,dx = L,$$

then we write

$$\int_a^\infty f(x)\,dx = L$$

and say that

$$\textit{the improper integral} \quad \int_a^\infty f(x)\,dx \quad \textit{converges to } L.$$

Otherwise, we say that

$$\textit{the improper integral} \quad \int_a^\infty f(x)\,dx \quad \textit{diverges.}$$

In a similar manner,

$$\text{improper integrals} \quad \int_{-\infty}^b f(x)\,dx \quad \text{arise as limits of the form} \quad \lim_{a\to-\infty} \int_a^b f(x)\,dx.$$

Example 1

(a) $\displaystyle\int_0^\infty e^{-2x}dx = \tfrac{1}{2}.$ **(b)** $\displaystyle\int_1^\infty \frac{dx}{x}$ diverges.

(c) $\displaystyle\int_1^\infty \frac{dx}{x^2} = 1.$ **(d)** $\displaystyle\int_{-\infty}^1 \cos \pi x\,dx$ diverges.

VERIFICATION

(a) $$\int_0^\infty e^{-2x}\,dx = \lim_{b\to\infty}\int_0^b e^{-2x}\,dx = \lim_{b\to\infty}\left[-\frac{e^{-2x}}{2}\right]_0^b = \lim_{b\to\infty}\left(\frac{1}{2} - \frac{1}{2e^{2b}}\right) = \frac{1}{2}.$$

(b) $$\int_1^\infty \frac{dx}{x} = \lim_{b\to\infty}\int_1^b \frac{dx}{x} = \lim_{b\to\infty} \ln b = \infty.$$

(c) $$\int_1^\infty \frac{dx}{x^2} = \lim_{b\to\infty}\int_1^b \frac{dx}{x^2} = \lim_{b\to\infty}\left[-\frac{1}{x}\right]_1^b = \lim_{b\to\infty}\left(1 - \frac{1}{b}\right) = 1.$$

(d) Note first that

$$\int_a^1 \cos \pi x\,dx = \left[\frac{1}{\pi}\sin \pi x\right]_a^1 = -\frac{1}{\pi}\sin \pi a.$$

As a tends to $-\infty$, $\sin \pi a$ oscillates between -1 and 1. Therefore the integral oscillates between $1/\pi$ and $-1/\pi$ and does not converge. □

The usual formulas for area and volume are extended to the unbounded case by means of improper integrals.

Example 2 Let p be a positive number. If Ω is the region between the x-axis and the graph of

$$f(x) = \frac{1}{x^p}, \quad x \ge 1, \qquad \text{(Figure 10.7.2)}$$

then

$$\text{area of } \Omega = \begin{cases} \dfrac{1}{p-1}, & \text{if } p > 1 \\ \infty, & \text{if } p \le 1. \end{cases}$$

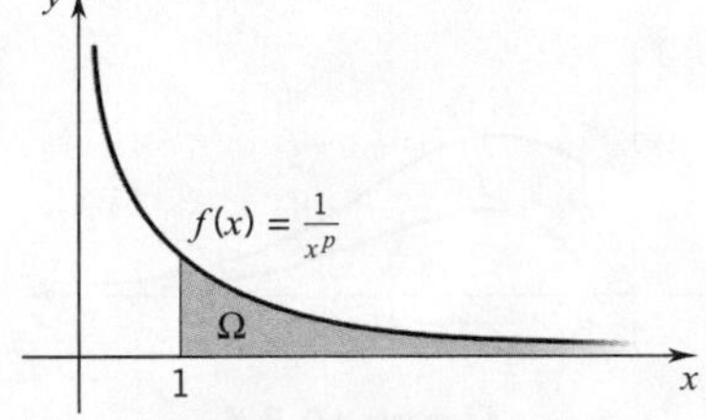

Figure 10.7.2

This comes about from setting

$$\text{area of } \Omega = \lim_{b\to\infty}\int_1^b \frac{dx}{x^p} = \int_1^\infty \frac{dx}{x^p}.$$

For $p \neq 1$,

$$\int_1^\infty \frac{dx}{x^p} = \lim_{b\to\infty}\int_1^b \frac{dx}{x^p} = \lim_{b\to\infty}\frac{1}{1-p}(b^{1-p}-1) = \begin{cases} \dfrac{1}{p-1}, & \text{if } p>1 \\ \infty, & \text{if } p<1. \end{cases}$$

For $p = 1$,

$$\int_1^\infty \frac{dx}{x^p} = \int_1^\infty \frac{dx}{x} = \infty,$$

as you have seen already. ❑

For future reference we record the following: Let $p > 0$,

(10.7.1)

$$\boxed{\int_1^\infty \frac{dx}{x^p} \text{ converges for } p>1 \text{ and diverges for } p \le 1.}$$

Example 3 We know that the region below the graph of $f(x) = 1/x$, $x \geq 1$, has infinite area. Suppose that this region with infinite area is revolved about the x-axis (see Figure 10.7.3). What is the volume V of the resulting solid? It may surprise you somewhat, but the volume is not infinite. In fact, it is π. Using the disc method to calculate the volume (Section 6.2), we have

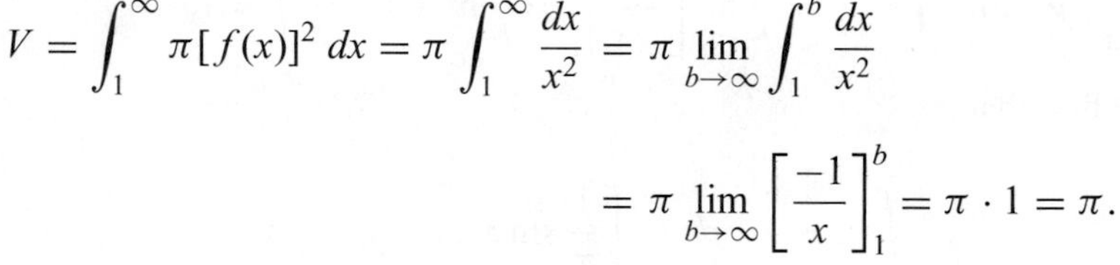

$$V = \int_1^\infty \pi[f(x)]^2\,dx = \pi\int_1^\infty \frac{dx}{x^2} = \pi\lim_{b\to\infty}\int_1^b \frac{dx}{x^2}$$

$$= \pi\lim_{b\to\infty}\left[\frac{-1}{x}\right]_1^b = \pi\cdot 1 = \pi.$$

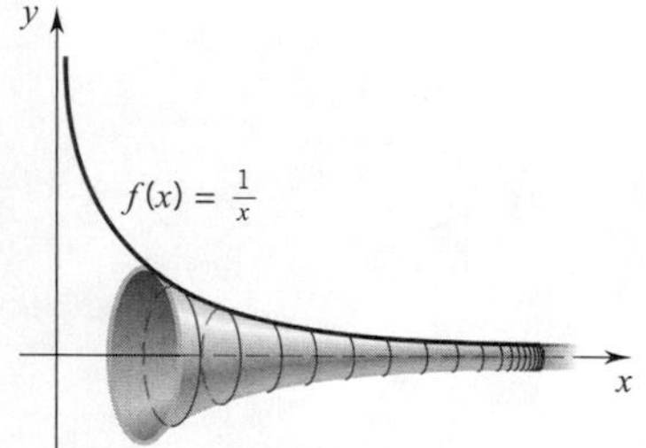

Figure 10.7.3

It may surprise you further to learn that the surface area of this solid is infinite. See Exercise 41. This surface is known as *Gabriel's horn.* ❑

It is often difficult to determine the convergence or divergence of a given improper integral by direct methods, but we can usually gain some information by comparison with integrals of known behavior.

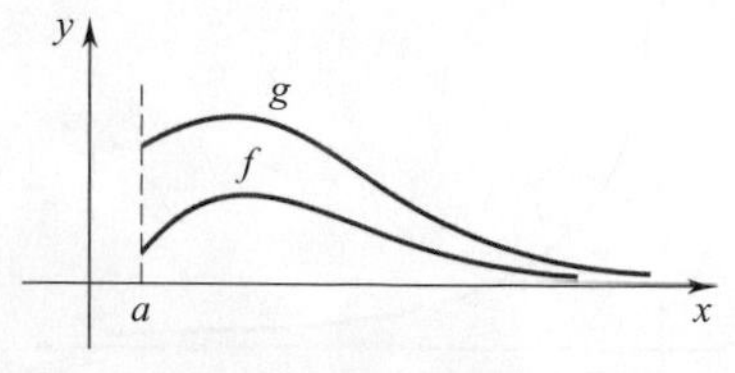

Figure 10.7.4

(10.7.2)

> (*A comparison test*) Suppose that f and g are continuous and
> $0 \le f(x) \le g(x)$ for all $x \in [a, \infty)$. (Figure 10.7.4)
>
> (i) If $\int_a^\infty g(x)\,dx$ converges, then $\int_a^\infty f(x)\,dx$ converges.
>
> (ii) If $\int_a^\infty f(x)\,dx$ diverges, then $\int_a^\infty g(x)\,dx$ diverges.

A similar result holds for integrals form $-\infty$ to b. The proof of (10.7.2) is left to you as an exercise.

Example 4 The improper integral $\displaystyle\int_1^\infty \frac{dx}{\sqrt{1+x^3}}$ converges since

$$\frac{1}{\sqrt{1+x^3}} < \frac{1}{x^{3/2}} \text{ for } x \in [1,\infty) \quad \text{and} \quad \int_1^\infty \frac{dx}{x^{3/2}} \text{ converges.}$$

In contrast, if we tried to evaluate

$$\lim_{b\to\infty} \int_1^b \frac{dx}{\sqrt{1+x^3}}$$

directly, we would first have to calculate the integral

$$\int \frac{dx}{\sqrt{1+x^3}},$$

and this cannot be done by any of the methods we have developed so far. ❑

Example 5 The improper integral $\displaystyle\int_1^\infty \frac{dx}{\sqrt{1+x^2}}$ diverges since

$$\frac{1}{1+x} \leq \frac{1}{\sqrt{1+x^2}} \quad \text{for } x \in [1,\infty) \quad \text{and} \quad \int_1^\infty \frac{dx}{1+x} \text{ diverges.}$$

This result can be obtained by evaluating

$$\int_1^b \frac{dx}{\sqrt{1+x^2}}$$

and then calculating the limit as $b \to \infty$. Try it. ❑

Suppose now that f is continuous on $(-\infty,\infty)$. The *improper integral*

$$\int_{-\infty}^\infty f(x)\,dx$$

is said to *converge* if

$$\int_{-\infty}^0 f(x)\,dx \quad \text{and} \quad \int_0^\infty f(x)\,dx$$

both converge. We then set

$$\int_{-\infty}^\infty f(x) = L + M,$$

where

$$\int_{-\infty}^0 f(x)\,dx = L \quad \text{and} \quad \int_0^\infty f(x)\,dx = M.$$

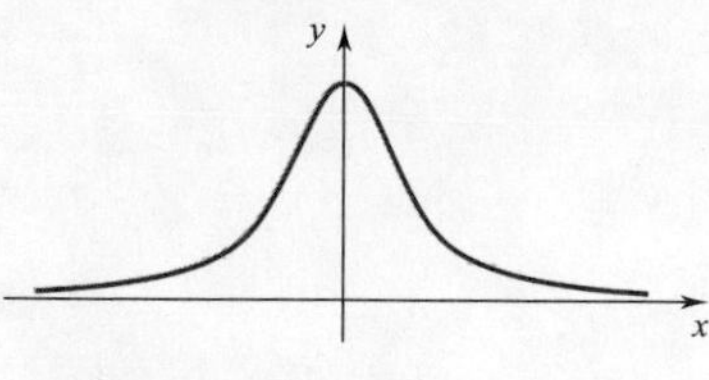

Figure 10.7.5

Example 6 Let $r > 0$. Determine whether the improper integral

$$\int_{-\infty}^{\infty} \frac{r}{r^2 + x^2}\,dx \qquad \text{(Figure 10.7.5)}$$

converges or diverges. If it converges, gives its value.

SOLUTION According to the definition, we need to determine the convergence or divergence of each of the improper integrals

$$\int_{-\infty}^{0} \frac{r}{r^2 + x^2}\,dx \quad \text{and} \quad \int_{0}^{\infty} \frac{r}{r^2 + x^2}\,dx.$$

For the first integral:

$$\int_{-\infty}^{0} \frac{r}{r^2 + x^2}\,dx = \lim_{a\to-\infty} \int_{a}^{0} \frac{r}{r^2 + x^2}\,dx = \lim_{a\to-\infty} \left[\tan^{-1}\left(\frac{x}{r}\right)\right]_a^0$$
$$= -\lim_{a\to-\infty} \tan^{-1}\left(\frac{a}{r}\right) = -\left(-\frac{\pi}{2}\right) = \frac{\pi}{2}.$$

For the second integral:

$$\int_{0}^{\infty} \frac{r}{r^2 + x^2}\,dx = \lim_{b\to\infty} \int_{0}^{b} \frac{r}{r^2 + x^2}\,dx = \lim_{b\to\infty} \left[\tan^{-1}\left(\frac{x}{r}\right)\right]_0^b$$
$$= \lim_{b\to\infty} \tan^{-1}\left(\frac{b}{r}\right) = \frac{\pi}{2}.$$

Since both of these integrals converge, the improper integral

$$\int_{-\infty}^{\infty} \frac{r}{r^2 + x^2}\,dx$$

converges. Its value is $\frac{1}{2}\pi + \frac{1}{2}\pi = \pi$. □

Remark Note that we did not define

(1) $$\int_{-\infty}^{\infty} f(x)\,dx$$

as

(2) $$\lim_{b\to\infty} \int_{-b}^{b} f(x)\,dx.$$

It is easy to show that if (1) exists, then (2) exists and (1) = (2). However, the existence of (2) does not imply the existence of (1). For example, (2) exists and is 0 for every odd function f, but this is certainly not the case for (1). You are asked to verify these statements in Exercise 57. □

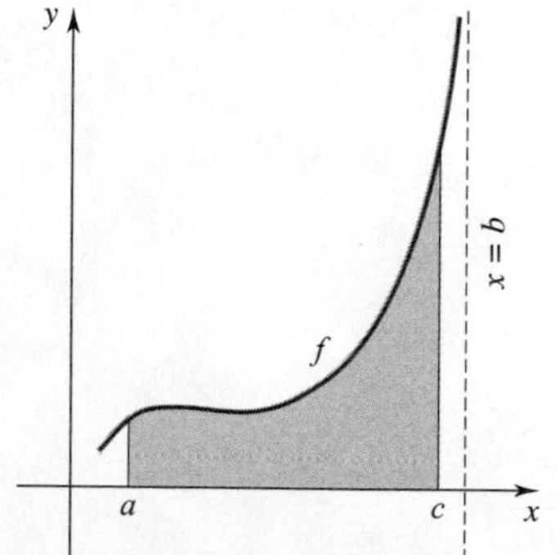

Figure 10.7.6

Integrals of Unbounded Functions

Improper integrals can also arise on bounded intervals. Suppose that f is continuous on the half-open interval $[a, b)$ but is unbounded there. See Figure 10.7.6. For each number $c < b$, we can form the definite integral

$$\int_{a}^{c} f(x)\,dx.$$

If, as $c \to b^-$, the integral tends to a finite limit L,

$$\lim_{c \to b^-} \int_a^c f(x)\, dx = L,$$

then we write

$$\int_a^b f(x)\, dx = L$$

and say that

$$\textit{the improper integral} \quad \int_a^b f(x)\, dx \quad \textit{converges to } L.$$

Otherwise, we say that *the improper integral diverges*.

Similarly, functions continuous but unbounded on half-open intervals of the form $(a, b]$ lead to limits of the form

$$\lim_{c \to a^+} \int_c^b f(x)\, dx.$$

As above,

$$\textit{the improper integral} \quad \int_a^b f(x)\, dx \quad \textit{converges to } L$$

if

$$\lim_{c \to a^+} \int_c^b f(x)\, dx = L.$$

Otherwise, *the improper integral diverges*.

Example 7

(a) $\displaystyle\int_0^1 (1 - x)^{-2/3}\, dx = 3.$ **(b)** $\displaystyle\int_0^2 \frac{dx}{x}$ diverges .

VERIFICATION

$$\textbf{(a)} \quad \int_0^1 (1 - x)^{-2/3}\, dx = \lim_{c \to 1^-} \int_0^c (1 - x)^{-2/3}\, dx$$

$$= \lim_{c \to 1^-} \Big[-3(1 - x)^{1/3} \Big]_0^c = \lim_{c \to 1^-} [-3(1 - c)^{1/3} + 3] = 3.$$

$$\textbf{(b)} \quad \int_0^2 \frac{dx}{x} = \lim_{c \to 0^+} \int_c^2 \frac{dx}{x} = \lim_{c \to 0^+} \Big[\ln x \Big]_c^2 = \lim_{c \to 0^+} [\ln 2 - \ln c] = \infty. \quad \square$$

Now suppose that f is continuous on an interval $[a, b]$ except at some point c in (a, b) where $f(x) \to \pm\infty$ as $x \to c^-$ or as $x \to c^+$. We say that the *improper integral*

$$\int_a^b f(x)\, dx$$

converges iff *both* of the integrals

$$\int_a^c f(x)\,dx \quad \text{and} \quad \int_c^b f(x)\,dx$$

converge. If

$$\int_a^c f(x)\,dx = L \quad \text{and} \quad \int_c^b f(x)\,dx = M,$$

we then set

$$\int_a^b f(x)\,dx = L + M.$$

Example 8 To evaluate

$$(*) \qquad \int_1^4 \frac{dx}{(x-2)^2},$$

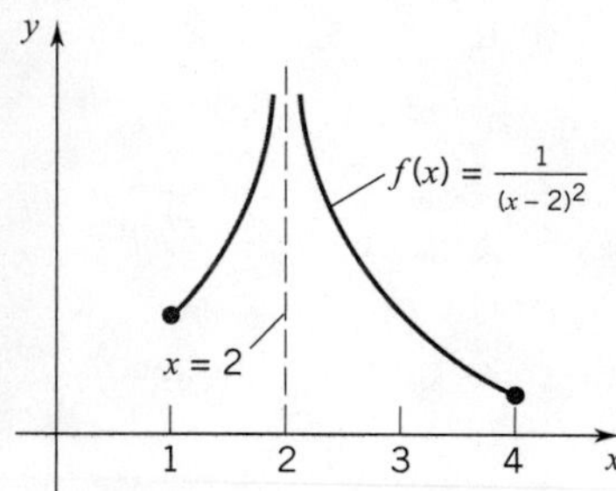

Figure 10.7.7

we need to calculate

$$\lim_{c\to 2^-} \int_1^c \frac{dx}{(x-2)^2} \quad \text{and} \quad \lim_{c\to 2^+} \int_c^4 \frac{dx}{(x-2)^2}$$

(the integrand blows up at $x = 2$; see Figure 10.7.7). For $c < 2$,

$$\int_1^c \frac{dx}{(x-2)^2} = \left[\frac{-1}{(x-2)}\right]_1^c = -\frac{1}{c-2} - 1 \to \infty \quad \text{as} \quad c \to 2^-.$$

Since

$$\int_1^c \frac{dx}{(x-2)^2}$$

does not converge, improper integral $(*)$ does not converge. (As you can verify, the second integral also fails to converge.)

Notice that if we ignore the fact that integral $(*)$ is improper, then we are led to the *incorrect conclusion* that

$$\int_1^4 \frac{dx}{(x-2)^2} = \left[\frac{-1}{(x-2)}\right]_1^4 = -\frac{3}{2}. \quad \square$$

Example 9 Evaluate $\displaystyle\int_{-2}^1 \frac{dx}{x^{4/5}}$.

SOLUTION Since $1/x^{4/5} \to \infty$ as $x \to 0^-$ and as $x \to 0^+$, the given integral is improper. We need to calculate

$$\int_{-2}^0 \frac{dx}{x^{4/5}} \quad \text{and} \quad \int_0^1 \frac{dx}{x^{4/5}}.$$

Now

$$\int_{-2}^0 \frac{dx}{x^{4/5}} = \lim_{c\to 0^-} \int_{-2}^c \frac{dx}{x^{4/5}} = \lim_{c\to 0^-} \left[5x^{1/5}\right]_{-2}^c = \lim_{c\to 0^-} [5c^{1/5} - 5(-2)^{1/5}] = 5(2^{1/5})$$

and $$\int_0^1 \frac{dx}{x^{4/5}} = \lim_{c\to 0+} \int_c^1 \frac{dx}{x^{4/5}} = \lim_{c\to 0+} \left[5x^{1/5}\right]_c^1 = \lim_{c\to 0+} [5 - 5c^{1/5}] = 5.$$

The improper integral converges and

$$\int_{-2}^1 \frac{dx}{x^{4/5}} = 5 + 5(2^{1/5}) \cong 10.74. \quad \square$$

EXERCISES 10.7

Evaluate the improper integrals that converge.

1. $\int_1^\infty \frac{dx}{x^2}$.
2. $\int_0^\infty \frac{dx}{1+x^2}$.
3. $\int_0^\infty \frac{dx}{4+x^2}$.
4. $\int_0^\infty e^{-px}\,dx, \quad p > 0.$
5. $\int_0^\infty e^{px}\,dx, \quad p > 0.$
6. $\int_0^1 \frac{dx}{\sqrt{x}}$.
7. $\int_0^8 \frac{dx}{x^{2/3}}$.
8. $\int_0^1 \frac{dx}{x^2}$.
9. $\int_0^1 \frac{dx}{\sqrt{1-x^2}}$.
10. $\int_0^1 \frac{dx}{\sqrt{1-x}}$.
11. $\int_0^2 \frac{x}{\sqrt{4-x^2}}\,dx$.
12. $\int_0^a \frac{dx}{\sqrt{a^2-x^2}}$.
13. $\int_e^\infty \frac{\ln x}{x}\,dx$.
14. $\int_e^\infty \frac{dx}{x\ln x}$.
15. $\int_0^1 x\ln x\,dx$.
16. $\int_e^\infty \frac{dx}{x(\ln x)^2}$.
17. $\int_{-\infty}^\infty \frac{dx}{1+x^2}$.
18. $\int_2^\infty \frac{dx}{x^2-1}$.
19. $\int_{-\infty}^\infty \frac{dx}{x^2}$.
20. $\int_{1/3}^3 \frac{dx}{\sqrt[3]{3x-1}}$.
21. $\int_1^\infty \frac{dx}{x(x+1)}$.
22. $\int_{-\infty}^0 x\,e^x\,dx$.
23. $\int_3^5 \frac{x}{\sqrt{x^2-9}}\,dx$.
24. $\int_1^4 \frac{dx}{x^2-4}$.
25. $\int_{-3}^3 \frac{dx}{x(x+1)}$.
26. $\int_1^\infty \frac{x}{(1+x^2)^2}\,dx$.
27. $\int_{-3}^1 \frac{dx}{x^2-4}$.
28. $\int_{-\infty}^\infty \frac{1}{e^x+e^{-x}}\,dx$.
29. $\int_0^\infty \cosh x\,dx$.
30. $\int_1^4 \frac{dx}{x^2-5x+6}$.
31. $\int_0^\infty e^{-x}\sin x\,dx$.
32. $\int_0^\infty \cos^2 x\,dx$.
33. $\int_0^1 \frac{e^{\sqrt{x}}}{\sqrt{x}}\,dx$.
34. $\int_0^{\pi/2} \frac{\cos x}{\sqrt{\sin x}}\,dx$.

C In Exercises 35–36, use a graphing utility to draw the graph of the integrand in each of the improper integrals. Then use a CAS to determine whether the integral converges or diverges.

35. (a) $\int_0^\infty \frac{x}{(16+x^2)^2}\,dx$. (b) $\int_0^\infty \frac{x^2}{(16+x^2)^2}\,dx$.
(c) $\int_0^\infty \frac{x}{16+x^4}\,dx$. (d) $\int_0^\infty \frac{x}{16+x^2}\,dx$.

36. (a) $\int_0^2 \frac{x^3}{\sqrt[3]{2-x}}\,dx$. (b) $\int_0^2 \frac{1}{\sqrt{2-x}}\,dx$.
(c) $\int_0^2 \frac{x}{\sqrt{2-x}}\,dx$. (d) $\int_0^2 \frac{1}{\sqrt{2x-x^2}}\,dx$.

37. Evaluate
$$\int_0^1 \sin^{-1} x\,dx$$
by using integration by parts even though the technique leads to an improper integral.

38. (a) For what values of r is
$$\int_0^\infty x^r e^{-x}\,dx$$
convergent?
(b) Use mathematical induction to show that
$$\int_0^\infty x^n e^{-x}\,dx = n!, \quad n = 1, 2, 3, \ldots.$$

39. The integral
$$\int_0^\infty \frac{1}{\sqrt{x}\,(1+x)}\,dx$$
is improper in two distinct ways: the interval of integration is unbounded and the integrand is unbounded. If we rewrite the integral as
$$\int_0^1 \frac{1}{\sqrt{x}\,(1+x)}\,dx + \int_1^\infty \frac{1}{\sqrt{x}(1+x)}\,dx,$$
then we have two improper integrals, the first having an unbounded integrand and the second defined on an unbounded interval. If each of these integrals converges with values L_1 and L_2, respectively, then the original integral converges and has the value $L_1 + L_2$. Evaluate the original integral.

40. Evaluate

$$\int_1^{\infty} \frac{1}{x\sqrt{x^2-1}}\,dx$$

using the method given in Exercise 39.

41. Show that if the region below the graph of $f(x) = 1/x,\ x \geq 1$, is revolved about the x-axis, then the surface area of the resulting solid is infinite (see Example 3).

42. Sketch the graphs of $y = \sec x$ and $y = \tan x$ for $0 \leq x < \pi/2$. Calculate the area of the region between the two curves.

43. Let Ω be the region bounded by the coordinate axes, the graph of $y = 1/\sqrt{x}$, and the line $x = 1$. (a) Sketch Ω. (b) Show that Ω has finite area and find it. (c) Show that if Ω is revolved about the x-axis, the solid obtained does not have finite volume.

44. Let Ω be the region between the graph of $y = 1/(1+x^2)$ and the x-axis, $x \geq 0$. (a) Sketch Ω. (b) Find the area of Ω. (c) Find the volume of the solid obtained by revolving Ω about the x-axis. (d) Find the volume of the solid obtained by revolving Ω about the y-axis.

45. Let Ω be the region bounded by the curve $y = e^{-x}$ and the x-axis, $x \geq 0$. (a) Sketch Ω. (b) Find the area of Ω. (c) Find the volume of the solid obtained by revolving Ω about the x-axis. (d) Find the volume obtained by revolving Ω about the y-axis. (e) Find the lateral surface area of the solid in part (c).

46. What point would you call the centroid of the region in Exercise 45? Does Pappus's theorem work in this instance?

47. Let Ω be the region bounded by the curve $y = e^{-x^2}$ and the x-axis, $x \geq 0$. (a) Show that Ω has finite area. (The area is actually $\frac{1}{2}\sqrt{\pi}$, as you will see in Chapter 16.) (b) Calculate the volume generated by revolving Ω about the y-axis.

48. Let Ω be the region bounded below by $y(x^2+1) = x$, above by $xy = 1$, and to the left by $x = 1$. (a) Find the area of Ω. (b) Show that the solid generated by revolving Ω about the x-axis has finite volume. (c) Calculate the volume generated by revolving Ω about the y-axis.

49. Let Ω be the region bounded by the curve $y = x^{-1/4}$ and the x-axis, $0 < x \leq 1$. (a) Sketch Ω. (b) Find the area of Ω. (c) Find the volume of the solid obtained by revolving Ω about the x-axis. (d) Find the volume of the solid obtained by revolving Ω about the y-axis.

50. Prove the validity of the comparison test (10.7.2).

In Exercises 51–56, use the comparison test (10.7.2) to determine whether the integral converges.

51. $\displaystyle\int_1^{\infty} \frac{x}{\sqrt{1+x^5}}\,dx.$

52. $\displaystyle\int_1^{\infty} 2^{-x^2}\,dx.$

53. $\displaystyle\int_0^{\infty} (1+x^5)^{-1/6}\,dx.$

54. $\displaystyle\int_{\pi}^{\infty} \frac{\sin^2 2x}{x^2}\,dx.$

55. $\displaystyle\int_1^{\infty} \frac{\ln x}{x^2}\,dx.$

56. $\displaystyle\int_e^{\infty} \frac{dx}{\sqrt{x+1}\ln x}.$

57. (a) Show that the improper integral

$$\int_0^{\infty} \frac{2x}{1+x^2}\,dx$$

diverges. Thus, the improper integral

$$\int_{-\infty}^{\infty} \frac{2x}{1+x^2}\,dx$$

diverges.

(b) Show that $\displaystyle\lim_{b\to\infty} \int_{-b}^{b} \frac{2x}{1+x^2}\,dx = 0.$

58. Show that

(a) $\displaystyle\int_{-\infty}^{\infty} \sin x\,dx$ diverges and

(b) $\displaystyle\lim_{b\to\infty} \int_{-b}^{b} \sin x\,dx = 0.$

59. Calculate the arc distance from the origin to the point $(x(\theta_1),\ y(\theta_1))$ along the exponential spiral $r = a\,e^{c\theta}$. (Take $a > 0, c > 0$.)

60. The function

$$f(x) = \frac{1}{\sqrt{2\pi}} \int_{-\infty}^{x} e^{-t^2/2}\,dt$$

is important in statistics. Prove that the integral on the right converges for all real x.

Exercises 59–62: *Laplace transforms*. Let f be continuous on $[0, \infty)$. The *Laplace transform* of f is the function F defined by

$$F(s) = \int_0^{\infty} e^{-sx} f(x)\,dx.$$

The domain of F is the set of all real numbers s such that the improper integral converges. Find the Laplace transform F of each of the following functions and give the domain of F.

61. $f(x) = 1.$

62. $f(x) = x.$

63. $f(x) = \cos 2x.$

64. $f(x) = e^{ax}.$

Exercises 63–66: *Probability density functions*. A nonnegative function f defined on $(-\infty, \infty)$ is called a *probability density function* if

$$\int_{-\infty}^{\infty} f(x)\,dx = 1.$$

65. Show that the function

$$f(x) = \begin{cases} 6x/(1+3x^2)^2 & x \geq 0 \\ 0 & x < 0 \end{cases}$$

is a probability density function.

66. Let $k > 0$. Show that the function

$$f(x) = \begin{cases} ke^{-kx} & x \geq 0 \\ 0 & x < 0, \end{cases}$$

is a probability density function. It is called the *exponential density function*.

67. The *mean* of a probability density function f is defined as the number

$$\mu = \int_{-\infty}^{\infty} x f(x)\,dx.$$

Calculate the mean of the exponential density function.

68. The *standard deviation* of a probability density function f is defined as the number

$$\sigma = \left[\int_{-\infty}^{\infty} (x-\mu)^2 f(x)\,dx\right]^{1/2}$$

where μ is the mean. Calculate the standard deviation for the exponential density function.

69. (*Useful later*) Let f be a continuous, positive, decreasing function on $[1,\infty)$. Show that

$$\int_1^{\infty} f(x)\,dx$$

converges iff the sequence

$$a_n = \int_1^n f(x)\,dx$$

converges.

■ CHAPTER HIGHLIGHTS

10.1 The Least Upper Bound Axiom

upper bound, bounded above, least upper bound (p. 585)
least upper bound axiom (p. 586)
lower bound, bounded below, greatest lower bound (p. 587)

10.2 Sequences of Real Numbers

sequence (p. 590)
bounded above, bounded below, bounded (p. 591)
increasing, nondecreasing, decreasing, nonincreasing (p. 591)
recurrence relation (p. 594)

It is sometimes possible to obtain useful information about a sequence $y_n = f(n)$ by applying the techniques of calculus to the function $y = f(x)$.

10.3 Limit of a Sequence

limit of a sequence (p. 595)
uniqueness of the limit (p. 597)
convergent, divergent (p. 597)
pinching theorem (p. 600)

Every convergent sequence is bounded (p. 597); thus, every unbounded sequence is divergent.

A bounded, monotonic sequence converges. (p. 598)

Suppose that $c_n \to c$ as $n \to \infty$, and all the c_n are in the domain of f. If f is continuous at c, then $f(c_n) \to f(c)$. (p. 601)

10.4 Some Important Limits

for $x > 0$, $\lim_{n\to\infty} x^{1/n} = 1$; for $|x| < 1$ $\lim_{n\to\infty} x^n = 0$

for each $\alpha > 0$, $\lim_{n\to\infty} \dfrac{1}{n^\alpha} = 0$

for each real x, $\lim_{n\to\infty} \dfrac{x^n}{n!} = 0$

$\lim_{n\to\infty} \dfrac{\ln n}{n} = 0$ $\lim_{n\to\infty} n^{1/n} = 1$

for each real x, $\lim_{n\to\infty} \left(1 - \dfrac{x}{n}\right)^n = e^x$

Cauchy sequence (p. 610)

10.5 The Indeterminate Form (0/0)

L'Hôpital's rule (0/0) (p. 611)
Cauchy mean-value theorem (p. 613)

10.6 The Indeterminate Form (∞/∞); other Indeterminate Forms

L'Hôpital's rule (∞/∞) (p. 616)

$\lim_{x\to\infty} \dfrac{\ln x}{x^\alpha} = 0$ $\lim_{x\to\infty} \dfrac{x^k}{e^x} = 0$ $\lim_{x\to 0+} x^x = 1$

other indeterminate forms: $0\cdot\infty, \infty-\infty, 0^0, 1^\infty, \infty^\infty$ (p. 617)

10.7 Improper Integrals

integrals over infinite intervals (p. 622)
convergent, divergent (p. 623)

$\displaystyle\int_1^{\infty} \frac{dx}{x^p}$ converges for $p > 1$ and diverges for $p \le 1$.

a comparison test (p. 624)
integrals of unbounded functions (p. 626)
convergent, divergent (p. 627)

CHAPTER

11

INFINITE SERIES

■ 11.1 INFINITE SERIES

While it is possible to add two numbers, three numbers, a hundred numbers, or even a million numbers, it is impossible to add an infinite number of numbers. The theory of *infinite series* arose from attempts to circumvent this impossibility.

But why would someone want to add an infinite number of numbers? For a simple illustration, consider the decimal expansion of a rational number. A rational number with a terminating decimal expansion can be expressed as a finite sum. For example,

$$\tfrac{3}{8} = 0.375 = 0.3 + 0.07 + 0.005.$$

On the other hand, a rational number with a repeating decimal expansion is represented by what would seem to be an infinite sum. For instance

$$\tfrac{1}{3} = 0.33333 \cdots = 0.3 + 0.03 + 0.003 + 0.0003 + 0.00003 + \cdots$$

We begin our study of infinite series with the notation that will be used throughout.

Sigma Notation

To indicate the sequence

$$1, \tfrac{1}{2}, \tfrac{1}{4}, \tfrac{1}{8}, \tfrac{1}{16}, \cdots$$

we can set $a_n = \left(\frac{1}{2}\right)^{n-1}$ and write

$$a_1,\ a_2,\ a_3,\ a_4,\ a_5, \ldots.$$

We can also set $b_n = \left(\frac{1}{2}\right)^n$ and write

$$b_0,\ b_1,\ b_2,\ b_3,\ b_4, \ldots,$$

thereby beginning the sequence with index 0. In general, we can set $c_n = \left(\frac{1}{2}\right)^{n-p}$ and write

$$c_p,\ c_{p+1},\ c_{p+2},\ c_{p+3},\ c_{p+4},\ldots$$

and thereby begin with index p. In this chapter we will often begin with an index other than 1.

The symbol Σ is the capital Greek letter "sigma." We write

$$\sum_{k=0}^{n} a_k \tag{1}$$

(read "the sum of a sub k from k equals 0 to k equals n") to indicate the sum

$$a_0 + a_1 + \cdots + a_n.$$

More generally, if $n \geq m$, we write

$$\sum_{k=m}^{n} a_k. \tag{2}$$

to indicate the sum

$$a_m + a_{m+1} + \cdots + a_n.$$

In (1) and (2) the letter "k" is being used as a "dummy" variable. That is, it can be replaced by any letter not already engaged. For instance,

$$\sum_{i=3}^{7} a_i, \qquad \sum_{j=3}^{7} a_j, \qquad \sum_{k=3}^{7} a_k$$

can all be used to denote the sum

$$a_3 + a_4 + a_5 + a_6 + a_7.$$

Translating

$$(a_0 + \cdots + a_n) + (b_0 + \cdots + b_n) = (a_0 + b_0) + \cdots + (a_n + b_n),$$

$$\alpha(a_0 + \cdots + a_n) = \alpha a_0 + \cdots + \alpha a_n,$$

$$(a_0 + \cdots + a_m) + (a_{m+1} + \cdots + a_n) = a_0 + \cdots + a_n$$

into the Σ-notation. We have

$$\sum_{k=0}^{n} a_k + \sum_{k=0}^{n} b_k = \sum_{k=0}^{n} (a_k + b_k), \quad \alpha \sum_{k=0}^{n} a_k = \sum_{k=0}^{n} \alpha a_k,$$

$$\sum_{k=0}^{m} a_k + \sum_{k=m+1}^{n} a_k = \sum_{k=0}^{n} a_k.$$

At times it is convenient to change indices. In this connection note that

$$\sum_{k=j}^{n} a_k = \sum_{i=0}^{n-j} a_{i+j}. \qquad \text{(set } i = k - j\text{)}$$

Both expressions are abbreviations for $a_j + a_{j+1} + \cdots + a_n$.

You can familiarize yourself further with this notation by doing the first 24 exercises of this section, but first one more remark. If all the a_k are equal to some fixed number r, then

$$\sum_{k=0}^{n} a_k \qquad \text{can be written} \qquad \sum_{k=0}^{n} r.$$

Then

$$\sum_{k=0}^{n} r = \overbrace{r + r + \cdots + r}^{n+1} = (n+1)r.$$

In particular,

$$\sum_{k=0}^{n} 1 = n + 1.$$

Infinite Series: Introduction

To form an infinite series, we begin with an infinite sequence of real numbers: $a_o, a_1, a_2, \ldots$. We can't form the sum of all the a_k (there are an infinite number of them), but we can form the *partial sums*

$$s_0 = a_0 = \sum_{k=0}^{0} a_k,$$

$$s_1 = a_0 + a_1 = \sum_{k=0}^{1} a_k,$$

$$s_2 = a_0 + a_1 + a_2 = \sum_{k=0}^{2} a_k,$$

$$s_3 = a_0 + a_1 + a_2 + a_3 = \sum_{k=0}^{3} a_k,$$

$$\vdots$$

$$s_n = a_0 + a_1 + a_2 + a_3 + \cdots + a_n = \sum_{k=0}^{n} a_k,$$

$$\vdots$$

Continuing in this way, we are led to consider the "infinite sum" $\sum_{k=0}^{\infty} a_k$, which is called an *infinite series*. The corresponding sequence $\{s_n\}$ is called the *sequence of partial sums* of the series.

DEFINITION 11.1.1

If the sequence $\{s_n\}$ of partial sums converges to a finite limit L, we write

$$\sum_{k=0}^{\infty} a_k = L$$

and say that

$$\text{the } \textit{series} \quad \sum_{k=0}^{\infty} a_k \quad \textit{converges} \text{ to } L.$$

We call L the *sum* of the series. If the sequence of partial sums diverges, we say that

$$\text{the } \textit{series} \quad \sum_{k=0}^{\infty} a_k \quad \textit{diverges}.$$

Remark It is important to note that the sum of a series is not a sum in the ordinary sense. It is a limit. ❑

Here are some examples.

Example 1 We begin with the series

$$\sum_{k=0}^{\infty} \frac{1}{(k+1)(k+2)}.$$

To determine whether this series converges, we must examine the partial sums.
Since

$$\frac{1}{(k+1)(k+2)} = \frac{1}{k+1} - \frac{1}{k+2}, \qquad \text{(partial fraction decomposition, Section 8.5)}$$

you can see that

$$\begin{aligned} s_n &= \frac{1}{1\cdot 2} + \frac{1}{2\cdot 3} + \cdots + \frac{1}{n(n+1)} + \frac{1}{(n+1)(n+2)} \\ &= \left(\frac{1}{1} - \frac{1}{2}\right) + \left(\frac{1}{2} - \frac{1}{3}\right) + \cdots + \left(\frac{1}{n} - \frac{1}{n+1}\right) + \left(\frac{1}{n+1} - \frac{1}{n+2}\right) \\ &= 1 - \frac{1}{2} + \frac{1}{2} - \frac{1}{3} + \cdots + \frac{1}{n} - \frac{1}{n+1} + \frac{1}{n+1} - \frac{1}{n+2}. \end{aligned}$$

Since all but the first and last terms occur in pairs with opposite signs, the sum "telescopes" to give

$$s_n = 1 - \frac{1}{n+2}.$$

Now, as $n \to \infty, s_n \to 1$. This means that the series converges to 1:

$$\sum_{k=0}^{\infty} \frac{1}{(k+1)(k+2)} = 1. \quad \square$$

Example 2 Here we examine two divergent series

$$\sum_{k=0}^{\infty} 2^k \qquad \text{and} \qquad \sum_{k=0}^{\infty} (-1)^k.$$

The partial sums of the first series take the form

$$s_n = \sum_{k=0}^{n} 2^k = 1 + 2 + 2^2 + \cdots + 2^n.$$

Since $s_n > 2^n$, the sequence $\{s_n\}$ is unbounded and therefore divergent (10.3.5). This means that the series diverges.

For the second series we have

$$s_n = \sum_{k=0}^{n} (-1)^k = 1 - 1 + 1 - 1 + \cdots + (-1)^n,$$

and it follows that

$$s_n = \begin{cases} 1 & \text{if } n \text{ is even} \\ 0 & \text{if } n \text{ is odd.} \end{cases}$$

The sequence of partial sums looks like this: $1, 0, 1, 0, \ldots$. Since this sequence diverges, the series diverges. $\square$

Remark Example 2 illustrates two types of divergence. In the first case, $s_n \to \infty$ as $n \to \infty$. The notation $\sum_{k=0}^{\infty} a_k = \infty$ is sometimes used to denote this type of divergence. In the second case, s_n "oscillates" between 0 and 1. $\square$

The Geometric Series

The sequence

$$1, x, x^2, x^3, \ldots$$

is called a *geometric progression*. If $|x| < 1$, then $x^n \to 0$ ["special limit" 10.4.2]. If $x = 1$, $\{x^n\}$ is the constant sequence $1, 1, 1, \ldots$ which clearly converges to 1; if $x = -1$, the sequence "oscillates" between -1 and 1 and hence diverges. Finally, if $|x| > 1$, $\{x^n\}$ is unbounded and therefore diverges.

The sums

$$1, \quad 1 + x, \quad 1 + x + x^2, \quad 1 + x + x^2 + x^3, \ldots$$

generated by numbers in geometric progression are the partial sums of what is known as the *geometric series*:

$$\sum_{k=0}^{\infty} x^k.$$

The geometric series arises in so many contexts that it merits special attention.

The following result is fundamental:

(11.1.2)

(i) If $\lvert x\rvert < 1$, then	$\sum_{k=0}^{\infty} x^k = \dfrac{1}{1-x}$.
(ii) If $\lvert x\rvert \geq 1$, then	$\sum_{k=0}^{\infty} x^k$ diverges.

PROOF The nth partial sum of the geometric series

$$\sum_{k=0}^{\infty} x^k$$

takes the form

(1) $$s_n = 1 + x + \cdots + x^n.$$

Multiplication by x gives

$$xs_n = x + x^2 + \cdots + x^{n+1}.$$

Subtracting the second equation from the first, we find that

$$(1 - x)s_n = 1 - x^{n+1}.$$

For $x \neq 1$, this gives

(2) $$s_n = \frac{1 - x^{n+1}}{1 - x}.$$

If $|x| < 1$, then $x^{n+1} \to 0$ as $n \to \infty$ and thus, by Equation (2),

$$s_n \to \frac{1}{1-x}.$$

This proves (i).

Now let's prove (ii). For $x = 1$, use Equation (1) and deduce that $s_n = n + 1$. Obviously $\{s_n\}$ diverges. For $x = -1$, use Equation (1) and deduce that s_n alternates between 0 and 1. Thus $\{s_n\}$ diverges. For $|x| > 1$, use Equation (2). Since $\{x^{n+1}\}$ is unbounded, $\{s_n\}$ diverges. ❑

You may have seen (11.1.2) before, written as

$$a + ar + ar^2 + \cdots + ar^n + \cdots = \begin{cases} \dfrac{a}{1-r}, & |r| < 1 \\ \text{diverges}, & |r| \geq 1. \end{cases} \qquad (a \neq 0).$$

Taking $a = 1$ and $r = \frac{1}{2}$, we have

$$\sum_{k=0}^{\infty} \frac{1}{2^k} = \frac{1}{1 - \frac{1}{2}} = 2.$$

Begin the summation at $k = 1$ instead of at $k = 0$, and you see that

(11.1.3)
$$\boxed{\sum_{k=1}^{\infty} \frac{1}{2^k} = 1,}$$

an elegant and powerful result. The partial sums of this series

$$\begin{aligned}
s_1 &= \tfrac{1}{2}, \\
s_2 &= \tfrac{1}{2} + \tfrac{1}{4} = \tfrac{3}{4}, \\
s_3 &= \tfrac{1}{2} + \tfrac{1}{4} + \tfrac{1}{8} = \tfrac{7}{8}, \\
s_4 &= \tfrac{1}{2} + \tfrac{1}{4} + \tfrac{1}{8} + \tfrac{1}{16} = \tfrac{15}{16}, \\
s_5 &= \tfrac{1}{2} + \tfrac{1}{4} + \tfrac{1}{8} + \tfrac{1}{16} + \tfrac{1}{32} = \tfrac{31}{32}, \\
&\text{etc.}
\end{aligned}$$

are illustrated in Figure 11.1.1. Each new partial sum lies halfway between the previous partial sum and the number 1.

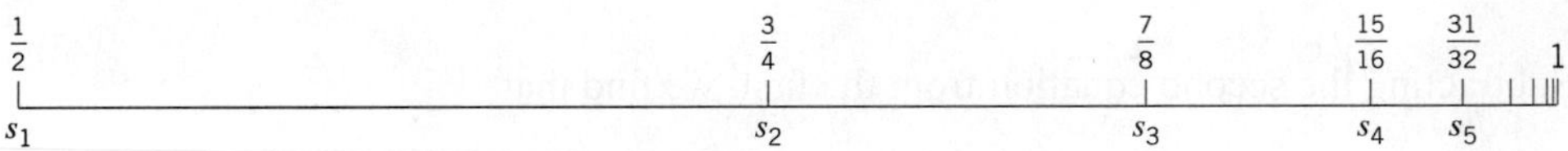

Figure 11.1.1

The convergence of the geometric series at $x = \frac{1}{10}$ enables us to assign a precise meaning to infinite decimals. Begin with the fact that

$$\sum_{k=0}^{\infty} \frac{1}{10^k} = \sum_{k=0}^{\infty} \left(\frac{1}{10}\right)^k = \frac{1}{1 - \frac{1}{10}} = \frac{10}{9}.$$

This gives

$$\sum_{k=1}^{\infty} \frac{1}{10^k} = \left(\sum_{k=0}^{\infty} \frac{1}{10^k}\right) - 1 = \frac{1}{9}$$

and shows that the partial sums

$$s_n = \frac{1}{10} + \frac{1}{10^2} + \cdots + \frac{1}{10^n}$$

are all less than $\frac{1}{9}$. Now take a series of the form

$$\sum_{k=1}^{\infty} \frac{a_k}{10^k} \qquad \text{with} \qquad a_k = 0, 1, \ldots, \text{ or } 9.$$

Its partial sums

$$t_n = \frac{a_1}{10} + \frac{a_2}{10^2} + \cdots + \frac{a_n}{10^n}$$

are all less than 1:

$$t_n = \frac{a_1}{10} + \frac{a_2}{10^2} + \cdots + \frac{a_n}{10^n} \le 9\left(\frac{1}{10} + \frac{1}{10^2} + \cdots + \frac{1}{10^n}\right) = 9s_n < 9\left(\frac{1}{9}\right) = 1.$$

Since $\{t_n\}$ is nondecreasing, as well as bounded above, $\{t_n\}$ is convergent; this means that the series

$$\sum_{k=1}^{\infty} \frac{a_k}{10^k}$$

is convergent. The sum of this series is what we mean by the infinite decimal

$$0.a_1a_2a_3 \cdots a_n \cdots.$$

Example 3 Suppose that a ball dropped from a height h hits the floor and rebounds to a height proportional to h, that is, to a height σh with $\sigma < 1$. It then falls from the height σh, hits the floor, and rebounds to the height $\sigma(\sigma h) = \sigma^2 h$, and so on. Find the total distance traveled by the ball.

SOLUTION The motion of the ball is depicted in Figure 11.1.2. The total distance D traveled by the ball is given by

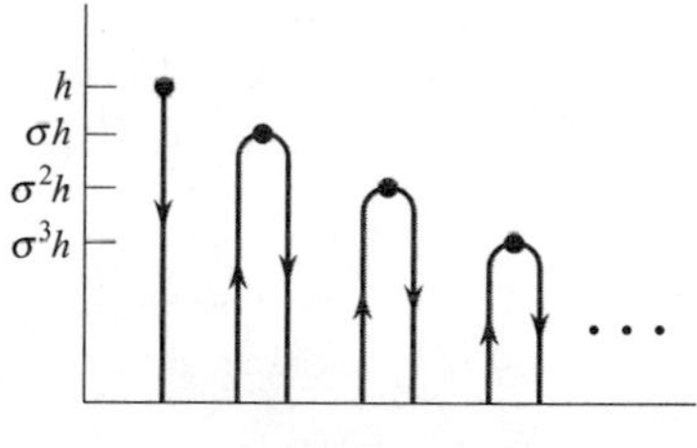

Figure 11.1.2

$$\begin{aligned} D &= h + 2\sigma h + 2\sigma^2 h + 2\sigma^3 h + \cdots \\ &= h + 2\sigma h[1 + \sigma + \sigma^2 + \cdots] = h + 2\sigma h \sum_{k=0}^{\infty} \sigma^k. \end{aligned}$$

The series in this expression is a geometric series in σ. Thus we have

$$D = h + 2\sigma h \frac{1}{1-\sigma}.$$

If h is 6 feet and $\sigma = \frac{2}{3}$, then

$$D = 6 + 2\left(\tfrac{2}{3}\right)6\frac{1}{1-\frac{2}{3}} = 6 + 24 = 30 \text{ feet.} \quad \square$$

We will return to the geometric series later. Right now we turn our attention to series in general.

Some Basic Results

THEOREM 11.1.4

1. If $\sum_{k=0}^{\infty} a_k$ converges and $\sum_{k=0}^{\infty} b_k$ converges, then $\sum_{k=0}^{\infty} (a_k + b_k)$ converges.

   Moreover, if $\sum_{k=0}^{\infty} a_k = L$ and $\sum_{k=0}^{\infty} b_k = M$, then $\sum_{k=0}^{\infty} (a_k + b_k) = L + M$.

2. If $\sum_{k=0}^{\infty} a_k$ converges, then $\sum_{k=0}^{\infty} \alpha\, a_k$ converges for each real number α.

   Moreover, if $\sum_{k=0}^{\infty} a_k = L$, then $\sum_{k=0}^{\infty} \alpha\, a_k = \alpha\, L$.

PROOF Let

$$s_n = \sum_{k=0}^{n} a_k, \quad t_n = \sum_{k=0}^{n} b_k, \quad u_n = \sum_{k=0}^{n} (a_k + b_k), \quad v_n = \sum_{k=0}^{n} \alpha\, a_k.$$

Note that

$$u_n = s_n + t_n \qquad \text{and} \qquad v_n = \alpha\, s_n.$$

If $s_n \to L$ and $t_n \to M$, then

$$u_n \to L + M \qquad \text{and} \qquad v_n \to \alpha L. \quad \square$$

THEOREM 11.1.5

The *kth term* of a convergent series tends to 0; namely,

$$\text{if} \quad \sum_{k=0}^{\infty} a_k \quad \text{converges,} \quad \text{then} \quad a_k \to 0 \quad \text{as} \quad k \to \infty.$$

PROOF To say that the series converges is to say that the sequence of partial sums converges to some number L:

$$s_n = \sum_{k=0}^{\infty} a_k \to L.$$

Therefore, it follows that $s_{n-1} \to L$ as well. Since $a_n = s_n - s_{n-1}$, we have $a_n \to L - L = 0$. A change in notation gives $a_k \to 0$. $\square$

The next result is a very important consequence of Theorem 11.1.5.

THEOREM 11.1.6 BASIC DIVERGENCE TEST

$$\text{If} \quad a_k \not\to 0 \quad \text{as } k \to \infty, \quad \text{then} \quad \sum_{k=0}^{\infty} a_k \quad \text{diverges.}$$

Example 4

(a) Since $\dfrac{k}{k+1} \to 1 \neq 0$ as $k \to \infty$, the series

$$\sum_{k=0}^{\infty} \frac{k}{k+1} = 0 + \frac{1}{2} + \frac{2}{3} + \frac{3}{4} + \frac{4}{5} + \cdots \quad \text{diverges.}$$

(b) Since $\sin k \not\to 0$ as $k \to \infty$, the series

$$\sum_{k=0}^{\infty} \sin k = \sin 0 + \sin 1 + \sin 2 + \sin 3 + \cdots \quad \text{diverges.} \quad \square$$

CAUTION Theorem 11.1.5 does *not* say that, if $a_k \to 0$, then $\sum_{k=0}^{\infty} a_k$ converges. There are divergent series for which $a_k \to 0$. □

Example 5 In the case of $\displaystyle\sum_{k=1}^{\infty} \frac{1}{\sqrt{k}} = \frac{1}{\sqrt{1}} + \frac{1}{\sqrt{2}} + \frac{1}{\sqrt{3}} + \frac{1}{\sqrt{4}} + \cdots,$

we have
$$a_k = \frac{1}{\sqrt{k}} \to 0 \quad \text{as } k \to \infty.$$

However,
$$s_n = \frac{1}{\sqrt{1}} + \frac{1}{\sqrt{2}} + \cdots + \frac{1}{\sqrt{n}} \geq \underbrace{\frac{1}{\sqrt{n}} + \frac{1}{\sqrt{n}} + \cdots + \frac{1}{\sqrt{n}}}_{n \text{ terms}} = \frac{n}{\sqrt{n}} = \sqrt{n}.$$

Thus, the sequence of partial sums is unbounded, and the series diverges. □

EXERCISES 11.1

Evaluate the given expression.

1. $\sum_{k=0}^{2} (3k+1)$. **2.** $\sum_{k=1}^{4} (3k-1)$.

3. $\sum_{k=0}^{3} 2^k$. **4.** $\sum_{k=0}^{3} (-1)^k\, 2^{k+1}$.

5. $\sum_{k=3}^{5} \frac{(-1)^k}{k!}$. **6.** $\sum_{k=2}^{4} \frac{1}{3^{k-1}}$.

Express in sigma notation.

7. $1 + 3 + 5 + 7 + \cdots + 21$.

8. $1 - 3 + 5 - 7 + \cdots - 19$.

9. $1 \cdot 2 + 2 \cdot 3 + 3 \cdot 4 + \cdots + 35 \cdot 36$.

10. The lower sum $m_1\,\Delta x_1 + m_2 \Delta x_2 + \cdots + m_n \Delta x_n$.

11. The upper sum $M_1\,\Delta x_1 + M_2 \Delta x_2 + \cdots + M_n \Delta x_n$.

12. The Riemann sum $f(x_1^*)\,\Delta x_1 + f(x_2^*)\Delta x_2 + \cdots + f(x_n^*)\Delta x_n$.

Write the given sums as $\sum_{k=3}^{10} a_k$ and $\sum_{i=0}^{7} a_{i+3}$.

13. $\frac{1}{2^3} + \frac{1}{2^4} + \cdots + \frac{1}{2^{10}}$.

14. $\frac{3^3}{3!} + \frac{4^4}{4!} + \cdots + \frac{10^{10}}{10!}$.

15. $\frac{3}{4} - \frac{4}{5} + \cdots - \frac{10}{11}$.

16. $\frac{1}{3} + \frac{1}{5} + \frac{1}{7} + \cdots + \frac{1}{17}$.

Verify by a change of indices that the two sums are identical.

17. $\sum_{k=2}^{10} \frac{k}{k^2+1}$; $\sum_{n=-1}^{7} \frac{n+3}{n^2+6n+10}$.

18. $\sum_{n=2}^{12} \frac{(-1)^n}{n-1}$; $\sum_{k=1}^{11} \frac{(-1)^{k+1}}{k}$.

19. $\sum_{k=4}^{25} \frac{1}{k^2-9}$; $\sum_{n=7}^{28} \frac{1}{n^2-6n}$.

20. $\sum_{k=0}^{15} \frac{3^{2k}}{k!}$; $(81) \sum_{n=-2}^{13} \frac{3^{2n}}{(n+2)!}$.

The following formulas can be verified by mathematical induction:

$$\sum_{k=1}^{n} k = \frac{n(n+1)}{2}, \qquad \sum_{k=1}^{n} (2k-1) = n^2,$$

$$\sum_{k=1}^{n} k^2 = \frac{n(n+1)(2n+1)}{6}, \qquad \sum_{k=1}^{n} k^3 = \left(\sum_{k=1}^{n} k\right)^2.$$

Use these formulas to evaluate the sums.

21. $\sum_{k=1}^{10} (2k+3)$. **22.** $\sum_{k=1}^{10} (2k^2+3k)$.

23. $\sum_{k=1}^{8} (2k-1)^2$. **24.** $\sum_{k=1}^{n} k(k^2-5)$.

Find the sum of the series.

25. $\sum_{k=1}^{\infty} \frac{1}{2k(k+1)}$. **26.** $\sum_{k=3}^{\infty} \frac{1}{k^2-k}$.

27. $\sum_{k=1}^{\infty} \frac{1}{k(k+3)}$. **28.** $\sum_{k=0}^{\infty} \frac{1}{(k+1)(k+3)}$.

29. $\sum_{k=0}^{\infty} \frac{3}{10^k}$. **30.** $\sum_{k=0}^{\infty} \frac{(-1)^k}{5^k}$.

31. $\sum_{k=0}^{\infty} \frac{1-2^k}{3^k}$. **32.** $\sum_{k=0}^{\infty} \frac{1}{2^{k+3}}$.

33. $\sum_{k=0}^{\infty} \frac{2^{k+3}}{3^k}$. **34.** $\sum_{k=2}^{\infty} \frac{3^{k-1}}{4^{3k+1}}$.

Write the decimal fraction as an infinite series and express the sum as the quotient of two integers.

35. $0.777\ldots$ **36.** $0.999\ldots$

37. $0.\widehat{24}\widehat{24}\ldots$ **38.** $0.\widehat{89}\widehat{89}\ldots$

39. $0.6\widehat{245}\widehat{45}\ldots$ **40.** $0.\widehat{1120}\widehat{19019}\ldots$

41. Using series, show that every repeating decimal represents a rational number (the quotient of two integers).

42. (a) Let j be a positive integer. Prove that $\sum_{k=0}^{\infty} a_k$ converges iff $\sum_{k=j}^{\infty} a_k$ converges.

(b) Show that if $\sum_{k=0}^{\infty} a_k = L$, then $\sum_{k=j}^{\infty} a_k = L - \sum_{k=0}^{j-1} a_k$.

(c) Show that if $\sum_{k=j}^{\infty} a_k = M$, then $\sum_{k=0}^{\infty} a_k = M + \sum_{k=0}^{j-1} a_k$.

In Exercises 43 and 44, derive the indicated result from the geometric series.

43. $\sum_{k=0}^{\infty} (-1)^k x^k = \frac{1}{1+x}, \quad |x| < 1.$

44. $\sum_{k=0}^{\infty} (-1)^k x^{2k} = \frac{1}{1+x^2}, \quad |x| < 1.$

Find a series expansion for the expression.

45. $\frac{x}{1-x}$ for $|x| < 1$. **46.** $\frac{x}{1+x}$ for $|x| < 1$.

47. $\frac{x}{1+x^2}$ for $|x| < 1$. **48.** $\frac{x}{1+4x^2}$ for $|x| < \frac{1}{2}$.

Show that the series diverges.

49. $1 + \frac{3}{2} + \frac{9}{4} + \frac{27}{8} + \frac{81}{16} + \cdots.$

50. $\sum_{k=0}^{\infty} \frac{(-5)^k}{4^{k+1}}.$ **51.** $\sum_{k=1}^{\infty} \left(\frac{k+1}{k}\right)^k.$

52. $\sum_{k=2}^{\infty} \frac{k^{k-2}}{3k}.$

53. Use a CAS to calculate the sum. Does the result help you to estimate the sum "to infinity"? If so, what is the sum of the infinite series?

(a) $\sum_{k=1}^{2000} \frac{1}{\sqrt{k}}.$ (b) $\sum_{k=1}^{2000} \frac{1}{k}.$

(c) $\sum_{k=1}^{2000} \frac{1}{k^2}.$ (d) $\sum_{k=1}^{2000} \frac{1}{k!}.$

54. Use a CAS to determine if the series converges or diverges.

(a) $\sum_{k=1}^{\infty} (-1)^k 5^{-k}.$ (b) $\sum_{k=1}^{\infty} (1234567)10^{-7k}.$

(c) $\sum_{k=1}^{\infty} \frac{1}{k(k+2)}.$ (d) $\sum_{k=3}^{\infty} \frac{2}{(k-1)(k-2)}.$

55. Given that a ball dropped to the floor rebounds to a height proportional to the height from which it is dropped, find the total distance traveled by a ball dropped from a height of 6 feet if it rebounds initially to a height of 3 feet.

56. In the setting of Exercise 55, to what height does the ball rebound initially if the total distance traveled by the ball is 21 feet?

57. How much money must you deposit at r% interest compounded annually to enable your descendants to withdraw n_1 dollars at the end of the first year, n_2 dollars at the end of the second year, n_3 dollars at the end of the third year, and so on in perpetuity? Assume that the sequence $\{n_k\}$ is bounded, $n_k \le N$ for all k, and express your answer as an infinite series.

58. Sum the series you obtained in Exercise 57 setting

(a) $r = 5, n_k = 5000(\frac{1}{2})^{k-1}$.

(b) $r = 6, n_k = 1000(0.8)^{k+1}$.

(c) $r = 5, n_k = N$.

59. Suppose that 90% of each dollar is recirculated into the economy. That is, suppose that when a dollar is put into circulation, 90% of it is spent; then 90% of that is spent, and so on. What is the total economic value of the dollar?

60. Consider the following sequence of steps. First, take the unit interval [0,1] and delete the open interval $(\frac{1}{3}, \frac{2}{3})$. Next delete the two open intervals $(\frac{1}{9}, \frac{2}{9})$ and $(\frac{7}{9}, \frac{8}{9})$ from the two intervals that remain after the first step. For the third step, delete the middle thirds from the four intervals that remain after the second step. Continue on in this manner. What is the sum of the lengths of the intervals that have been deleted? The set that remains after all of the "middle thirds" have been deleted is called the *Cantor middle third set*. Give some points that are in the Cantor set.

61. Start with a square that has sides four units long. Join the midpoints of the sides of the square to form a second square inside the first. Then join the midpoints of the sides of the second square to form a third square, and so on. See the figure. Find the sum of the areas of the squares.

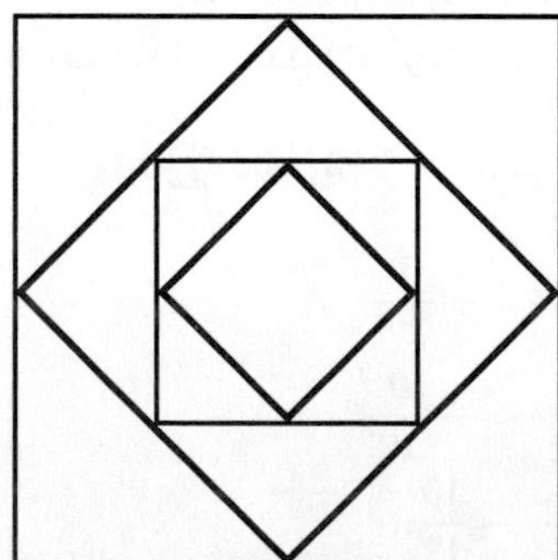

62. (a) Show that if the series $\sum a_k$ converges and the series $\sum b_k$ diverges, then the series $\sum (a_k + b_k)$ diverges.

(b) Give examples to show that if $\sum a_k$ and $\sum b_k$ both diverge, then each of the series

$$\sum (a_k + b_k) \quad \text{and} \quad \sum (a_k - b_k)$$

may either converge or diverge.

63. Let $\sum_{k=0}^{\infty} a_k$ be a convergent series and let $R_n = \sum_{k=n+1}^{\infty} a_k$. Prove that $R_n \to 0$ as $n \to \infty$. Note that if s_n is the nth partial sum of the series, then $\sum_{k=0}^{\infty} a_k = s_n + R_n$; R_n is called the *remainder*.

64. (a) Prove that if $\sum_{k=0}^{\infty} a_k$ is a convergent series and $a_k \neq 0$ for all k, then $\sum_{k=0}^{\infty} (1/a_k)$ is divergent.

(b) Suppose that $a_k > 0$ for all k and $\sum_{k=0}^{\infty} a_k$ diverges. Show by means of example that $\sum_{k=0}^{\infty} (1/a_k)$ may either converge or diverge.

65. Let $\{s_n\}$ be the sequence of partial sums of the series $\sum_{k=0}^{\infty} (-1)^k$. Find a formula for s_n.

66. Repeat Exercise 65 for the series $\sum_{k=1}^{\infty} \ln\left(\dfrac{k}{k+1}\right)$.

67. Show that

$$\sum_{k=1}^{\infty} \ln\left(\frac{k+1}{k}\right) \quad \text{diverges,}$$

$$\text{even though} \quad \lim_{k\to\infty} \ln\left(\frac{k+1}{k}\right) = 0.$$

68. Show that

$$\sum_{k=1}^{\infty} \left(\frac{k}{k+1}\right)^k \quad \text{diverges.}$$

69. (a) Let $\{d_k\}$ be a sequence of real numbers that converges to 0. Show that

$$\sum_{k=1}^{\infty} (d_k - d_{k+1}) = d_1.$$

(b) Sum the following series:

$$\text{(i)} \quad \sum_{k=1}^{\infty} \frac{\sqrt{k+1} - \sqrt{k}}{\sqrt{k(k+1)}}. \qquad \text{(ii)} \quad \sum_{k=1}^{\infty} \frac{2k+1}{2k^2(k+1)^2}.$$

70. Show that

$$\sum_{k=1}^{\infty} kx^{k-1} = \frac{1}{(1-x)^2} \quad \text{for} \quad |x| < 1.$$

HINT: Verify that s_n, the nth partial sum, satisfies the identity

$$(1-x)^2 s_n = 1 - (n+1)x^n + nx^{n+1}.$$

Ⓒ *Speed of convergence* Suppose that $\sum_{k=0}^{\infty} a_k$ is a convergent series with sum L and let $\{s_n\}$ be its sequence of partial sums. In the notation of Exercise 63, $|L - s_n| = |R_n|$. In Exercises 71–74, find the smallest integer N such that $|L - s_N| < 0.0001$.

71. $\sum_{k=0}^{\infty} \dfrac{1}{4^k}$.

72. $\sum_{k=0}^{\infty} (0.9)^k$.

73. $\sum_{k=1}^{\infty} \dfrac{1}{k(k+2)}$.

74. $\sum_{k=0}^{\infty} \left(\dfrac{2}{3}\right)^k$.

75. Start with the geometric series $\sum_{k=0}^{\infty} x^k$ with $|x| < 1$, and a positive number ϵ. Determine the smallest positive integer N such that $|L - s_N| < \epsilon$, where L is the sum of the series and s_N is the Nth partial sum.

76. Prove that the series $\sum_{k=1}^{\infty} (a_{k+1} - a_k)$ converges iff the sequence $\{a_n\}$ converges.

■ 11.2 THE INTEGRAL TEST; COMPARISON TESTS

Here and in the next section we direct our attention to *series with nonnegative terms:* $a_k \geq 0$ for all k. Note that for such series, the sequence of partial sums is nondecreasing:

$$s_{n+1} = \sum_{k=0}^{n+1} a_k = a_{n+1} + \sum_{k=0}^{n} a_k \geq \sum_{k=0}^{n} a_k = s_n.$$

The following theorem is fundamental.

THEOREM 11.2.1

A series with nonnegative terms converges iff the sequence of partial sums is bounded.

PROOF Assume that the series converges. Then the sequence of partial sums is convergent and therefore bounded (Theorem 10.3.4).

Suppose now that the sequence of partial sums is bounded. Since the terms are nonnegative, the sequence is nondecreasing. By being bounded and nondecreasing, the

the sequence of partial sums converges (Theorem 10.3.6). This means that the series converges. □

The convergence or divergence of a series can sometimes be deduced from the convergence or divergence of a closely related improper integral.

THEOREM 11.2.2 THE INTEGRAL TEST

If f is continuous, decreasing, and positive on $[1, \infty)$, then

$$\sum_{k=1}^{\infty} f(k) \quad \text{converges} \quad \text{iff} \quad \int_1^{\infty} f(x)\,dx \quad \text{converges.}$$

PROOF In Exercise 69, Section 10.7, you were asked to show that with f continuous, decreasing, and positive on $[1, \infty)$

$$\int_1^{\infty} f(x)\,dx \quad \text{converges} \quad \text{iff} \quad \text{the sequence} \quad a_n = \int_1^{n} f(x)\,dx \quad \text{converges.}$$

We assume this result and base our proof on the behavior of the sequence of integrals. To visualize our argument, see Figure 11.2.1.

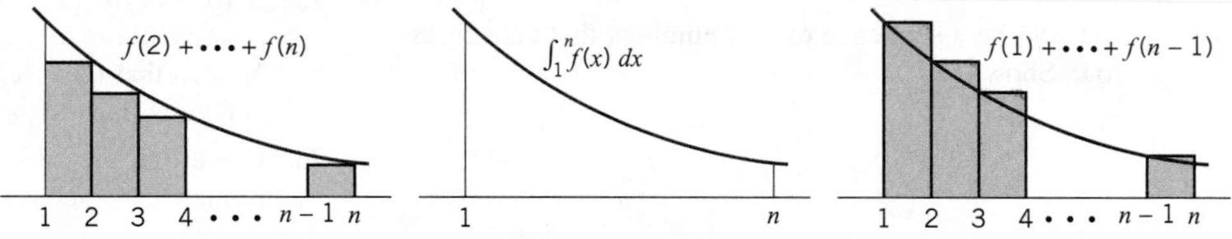

Figure 11.2.1

Since f decreases on the interval $[1, n]$

$$f(2) + \cdots + f(n) \quad \text{is a lower sum for } f \text{ on } [1, n].$$

and

$$f(1) + \cdots + f(n-1) \quad \text{is an upper sum for } f \text{ on } [1, n].$$

Consequently

$$\text{(1)} \quad f(2) + \cdots + f(n) \le \int_1^{n} f(x)dx \quad \text{and} \quad \int_1^{n} f(x)dx \le f(1) + \cdots + f(n-1).$$

If the sequence of integrals converges, it is bounded. By the first inequality the sequence of partial sums is bounded and the series is therefore convergent.

Suppose now that the sequence of integrals diverges. Since f is positive, the sequence of integrals increases:

$$\int_1^{n} f(x)\,dx < \int_1^{n+1} f(x)\,dx.$$

Since this sequence diverges, it must be unbounded. By the second inequality, the sequence of partial sums must be unbounded and the series is divergent. □

Remark The inequalities established in the proof of Theorem 11.2.2 lead to bounds on the sum of the infinite series

$$\sum_{k=1}^{\infty} f(k)$$

where f is continuous, decreasing, and positive on $[1, \infty)$. In particular, it follows from the second inequality in (1) that

$$\int_1^{\infty} f(x)\,dx \leq \sum_{k=1}^{\infty} f(k),$$

and from the first inequality in (1) that

$$\sum_{k=1}^{\infty} f(k) \leq f(1) + \int_1^{\infty} f(x)\,dx.$$

Combining these two inequalities, we have

$$\int_1^{\infty} f(x)\,dx \leq \sum_{k=1}^{\infty} f(k) \leq f(1) + \int_1^{\infty} f(x)\,dx.$$

These inequalities spell out the relation that holds between infinite series and the corresponding improper integral. □

Applying the Integral Test

Example 1 (*The harmonic series*)

(11.2.3) $$\sum_{k=1}^{\infty} \frac{1}{k} = 1 + \frac{1}{2} + \frac{1}{3} + \frac{1}{4} + \cdots \quad \text{diverges.}$$

PROOF The function $f(x) = 1/x$ is continuous, decreasing, and positive on $[1, \infty)$. We know that

$$\int_1^{\infty} \frac{dx}{x} \quad \text{diverges.} \tag{10.7.1}$$

By the integral test,

$$\sum_{k=1}^{\infty} \frac{1}{k} \quad \text{diverges. } \square$$

The next example gives a more general result.

Example 2 (*The p-series*)

$$\text{(11.2.4)} \qquad \sum_{k=1}^{\infty} \frac{1}{k^p} = 1 + \frac{1}{2^p} + \frac{1}{3^p} + \frac{1}{4^p} + \cdots \quad \text{converges} \quad \text{iff} \quad p > 1.$$

PROOF If $p \leq 0$, then each term of the series is greater than or equal to 1. Therefore the sequence of terms does not tend to zero and, by the basic divergence test (11.1.6), the series cannot converge. We assume, therefore, that $p > 0$. The function $f(x) = 1/x^p$ is then continuous, decreasing, and positive on $[1, \infty)$. Thus, by the integral test,

$$\sum_{k=1}^{\infty} \frac{1}{k^p} \quad \text{converges} \quad \text{iff} \quad \int_1^{\infty} \frac{dx}{x^p} \quad \text{converges.}$$

Earlier you saw that

$$\int_1^{\infty} \frac{dx}{x^p} \quad \text{converges} \quad \text{iff} \quad p > 1. \qquad (10.7.1)$$

It follows that

$$\sum_{k=1}^{\infty} \frac{1}{k^p} \quad \text{converges} \quad \text{iff} \quad p > 1. \quad \square$$

Example 3 Show that the series $\displaystyle\sum_{k=1}^{\infty} \frac{1}{k \ln(k+1)} = \frac{1}{\ln 2} + \frac{1}{2 \ln 3} + \frac{1}{3 \ln 4} + \cdots$ diverges.

SOLUTION We begin by setting $f(x) = \dfrac{1}{x \ln(x+1)}$. Since f is continuous, decreasing, and positive on $[1, \infty)$, we can use the integral test. Note first that

$$\frac{1}{x \ln(x+1)} > \frac{1}{(x+1) \ln(x+1)} \qquad \text{on } [1, \infty).$$

Therefore we have

$$\int_1^b \frac{1}{x \ln(x+1)}\, dx > \int_1^b \frac{1}{(x+1) \ln(x+1)}\, dx = \Big[\ln[\ln(x+1)] \Big]_1^b$$
$$= \ln[\ln(b+1)] - \ln[\ln 2].$$

As $b \to \infty$, $\ln[\ln(b+1)] \to \infty$. This shows that

$$\int_1^{\infty} \frac{1}{x \ln(x+1)}\, dx$$

diverges. Therefore the series diverges. $\square$

Remark on Notation You have seen that for each $j \geq 0$

$$\sum_{k=0}^{\infty} a_k \quad \text{converges} \quad \text{iff} \quad \sum_{k=j}^{\infty} a_k \quad \text{converges.}$$

(Exercise 42, Section 11.1). This tells you that, in determining whether a series converges, it does not matter where we begin the summation. Where detailed indexing would contribute nothing, we will omit it and write $\sum a_k$ without specifying where the summation begins. For instance, it makes sense to say that

$$\sum \frac{1}{k^2} \quad \text{converges} \quad \text{and} \quad \sum \frac{1}{k} \quad \text{diverges}$$

without specifying where we begin the summation. ❑

The convergence or divergence of a series with nonnegative terms can sometimes be determined by comparison with a series of known behavior.

THEOREM 11.2.5 THE BASIC COMPARISON TEST

Let $\sum a_k$ be a series with nonnegative terms.

(i) $\sum a_k$ converges if there exists a convergent series $\sum c_k$ with nonnegative terms such that $a_k \leq c_k$ for all k sufficiently large;

(ii) $\sum a_k$ diverges if there exists a divergent series $\sum d_k$ with nonnegative terms such that $d_k \leq a_k$ for all k sufficiently large.

PROOF The proof is just a matter of noting that, in the first instance, the partial sums of $\sum a_k$ form a bounded sequence and, in the second instance, they form an unbounded sequence. The details are left to you. ❑

Example 4

(a) $\sum \dfrac{1}{2k^3+1}$ converges by comparison with $\sum \dfrac{1}{k^3}$:

$$\frac{1}{2k^3+1} < \frac{1}{k^3} \quad \text{and} \quad \sum \frac{1}{k^3} \text{ converges.}$$

(b) $\sum \dfrac{1}{3k+1}$ diverges by comparison with $\sum \dfrac{1}{3(k+1)}$:

$$\frac{1}{3(k+1)} < \frac{1}{3k+1} \quad \text{and} \quad \sum \frac{1}{3(k+1)} = \tfrac{1}{3} \sum \frac{1}{k+1} \quad \text{diverges}$$

(it's the series $\sum 1/k$ with a change of index).

(c) $\sum \dfrac{k^3}{k^5+5k^4+7}$ converges by comparison with $\sum \dfrac{1}{k^2}$:

$$\frac{k^3}{k^5+5k^4+7} < \frac{k^3}{k^5} = \frac{1}{k^2} \quad \text{and} \quad \sum \frac{1}{k^2} \quad \text{converges.} \quad ❑$$

Example 5 Show that $\sum \dfrac{1}{\ln(k+6)}$ diverges.

SOLUTION We know that as $k \to \infty$,

$$\frac{\ln k}{k} \to 0. \qquad \text{[see (10.4.5)]}$$

It follows that

$$\frac{\ln(k+6)}{k+6} \to 0.$$

Thus, for k sufficiently large,

$$\frac{\ln(k+6)}{k+6} < 1,$$

and so

$$\ln(k+6) < k+6 \quad \text{and} \quad \frac{1}{k+6} < \frac{1}{\ln(k+6)}.$$

Since $$\sum \frac{1}{k+6} \text{ diverges,}$$

(It's the series $\sum 1/k$ with a change of index), we can conclude that

$$\sum \frac{1}{\ln(k+6)} \quad \text{diverges.} \quad \square$$

Remark Another way to show that $\ln(k+6) < k+6$ is to examine the function $f(x) = x + 6 - \ln(x+6)$. At $x = 0$ the function is positive:

$$f(0) = 6 - \ln 6 \cong 6 - 1.792 > 0.$$

Since $$f'(x) = 1 - \frac{1}{x+6} > 0 \quad \text{for all } x > 0,$$

$f(x) > 0$ for all $x \geq 0$. It follows that

$$\ln(x+6) < x+6 \quad \text{for all } x \geq 0. \quad \square$$

To apply the basic comparison test to a series $\sum a_k$, you must show that the terms a_k are smaller than the terms c_k of a known convergent series to establish convergence, or larger than the terms of d_k of a known divergent series to establish divergence. However, there are situations where a given series $\sum a_k$ has a natural comparison series $\sum b_k$ but the inequalities "go the wrong way." For example, consider the series

$$\sum_{k=2}^{\infty} \frac{1}{k^3 - 1}.$$

You would expect this series to be convergent by comparison with the convergent series

$$\sum_{k=2}^{\infty} \frac{1}{k^3},$$

but $$\frac{1}{k^3 - 1} > \frac{1}{k^3} \quad \text{for all} \quad k \geq 2,$$

and so the basic comparison test does not apply. A similar situation occurs if you try to apply the comparison test to establish the divergence of

$$\sum_{k=1}^{\infty} \frac{1}{1+\sqrt{k}} \quad \text{by comparison with} \quad \sum_{k=1}^{\infty} \frac{1}{\sqrt{k}};$$

$$\frac{1}{1+\sqrt{k}} < \frac{1}{\sqrt{k}} \quad \text{for all} \quad k \geq 1.$$

We come now to a somewhat more sophisticated comparison test, more sophisticated in that it involves the evaluation of a limit. Our proof relies on the basic comparison test

THEOREM 11.2.6 THE LIMIT COMPARISON TEST

Let $\sum a_k$ and $\sum b_k$ be series with *positive terms*. If $a_k/b_k \to L$, where L is some *positive* number, then

$$\sum a_k \quad \text{converges} \qquad \text{iff} \quad \sum b_k \quad \text{converges}.$$

PROOF Choose ϵ between 0 and L. Since $a_k/b_k \to L$, we know that for all k sufficiently large (for all k greater than some k_0)

$$\left|\frac{a_k}{b_k} - L\right| < \epsilon.$$

For such k we have

$$L - \epsilon < \frac{a_k}{b_k} < L + \epsilon,$$

and thus

$$(L - \epsilon)b_k < a_k < (L + \epsilon)b_k.$$

This last inequality is what we need:

if $\sum a_k$ converges, then $\sum (L - \epsilon)b_k$ converges, and thus $\sum b_k$ converges;

if $\sum b_k$ converges, then $\sum (L + \epsilon)b_k$ converges, and thus $\sum a_k$ converges. □

We can now complete the examples that we began above:

$$\sum_{k=2}^{\infty} \frac{1}{k^3 - 1}$$

converges since

$$\sum_{k=2}^{\infty} \frac{1}{k^3} \quad \text{converges} \quad \text{and} \quad \left(\frac{1}{k^3 - 1}\right) \div \left(\frac{1}{k^3}\right) = \frac{k^3}{k^3 - 1} \to 1 \text{ as } k \to \infty;$$

$$\sum_{k=1}^{\infty} \frac{1}{1+\sqrt{k}}$$

diverges since

$$\sum_{k=1}^{\infty} \frac{1}{\sqrt{k}} \quad \text{diverges} \quad \text{and} \quad \left(\frac{1}{1+\sqrt{k}}\right) \div \left(\frac{1}{\sqrt{k}}\right) = \frac{\sqrt{k}}{1+\sqrt{k}} \to 1 \text{ as } k \to \infty.$$

To apply the limit comparison test to a series $\sum a_k$, we must first find a series $\sum b_k$ of known behavior for which a_k/b_k converges to a positive number.

Example 6 Determine whether the series

$$\sum \frac{3k^2+2k+1}{k^3+1}$$

converges or diverges.

SOLUTION For large k, the terms with the highest powers of k dominate. Here $3k^2$ dominates the numerator and k^3 dominates the denominator. Thus, for large k,

$$\frac{3k^2+2k+1}{k^3+1} \quad \text{differs little from} \quad \frac{3k^2}{k^3} = \frac{3}{k}.$$

Since

$$\frac{3k^2+2k+1}{k^3+1} \div \frac{3}{k} = \frac{3k^3+2k^2+k}{3k^3+3} = \frac{1+2/(3k)+1/(3k^2)}{1+1/k^3} \to 1 \text{ as } k \to \infty$$

and

$$\sum \frac{3}{k} = 3\sum \frac{1}{k} \quad \text{diverges,}$$

we know that the series diverges. □

Example 7 Determine whether the series $\sum \frac{5\sqrt{k}+100}{2k^2\sqrt{k}+9\sqrt{k}}$ converges or diverges.

SOLUTION For large values of k, $5\sqrt{k}$ dominates the numerator and $2k^2\sqrt{k}$ dominates the denominator. Thus, for such k,

$$\frac{5\sqrt{k}+100}{2k^2\sqrt{k}+9\sqrt{k}} \quad \text{differs little from} \quad \frac{5\sqrt{k}}{2k^2\sqrt{k}} = \frac{5}{2k^2}.$$

Since

$$\frac{5\sqrt{k}+100}{2k^2\sqrt{k}+9\sqrt{k}} \div \frac{5}{2k^2} = \frac{10k^2\sqrt{k}+200k^2}{10k^2\sqrt{k}+45\sqrt{k}} = \frac{1+20/\sqrt{k}}{1+9/2k^2} \to 1 \quad \text{as } k \to \infty$$

and

$$\sum \frac{5}{2k^2} = \frac{5}{2}\sum \frac{1}{k^2} \quad \text{converges,}$$

the series converges. □

Example 8 Determine whether the series $\sum \sin \frac{\pi}{k}$ converges or diverges.

SOLUTION Recall that

$$\text{as } x \to 0, \quad \frac{\sin x}{x} \to 1. \qquad (2.5.5)$$

As $k \to \infty$, $\pi/k \to 0$ and thus

$$\frac{\sin(\pi/k)}{\pi/k} \to 1.$$

Since $\sum \pi/k$ diverges, $\sum \sin(\pi/k)$ diverges. □

Remark The question of what we can and cannot conclude by limit comparison if $a_k/b_k \to 0$ or if $a_k/b_k \to \infty$ is taken up in Exercises 47 and 48. □

EXERCISES 11.2

Determine whether the series converges or diverges.

1. $\sum \dfrac{k}{k^3+1}$.

2. $\sum \dfrac{1}{3k+2}$.

3. $\sum \dfrac{1}{(2k+1)^2}$.

4. $\sum \dfrac{\ln k}{k}$.

5. $\sum \dfrac{1}{\sqrt{k+1}}$.

6. $\sum \dfrac{1}{k^2+1}$.

7. $\sum \dfrac{1}{\sqrt{2k^2-k}}$.

8. $\sum \left(\dfrac{5}{2}\right)^{-k}$.

9. $\sum \dfrac{\tan^{-1} k}{1+k^2}$.

10. $\sum \dfrac{\ln k}{k^3}$.

11. $\sum \dfrac{1}{k^{2/3}}$.

12. $\sum \dfrac{1}{k(k+1)(k+2)}$.

13. $\sum \left(\dfrac{3}{4}\right)^{-k}$.

14. $\sum \dfrac{1}{1+2\ln k}$.

15. $\sum \dfrac{\ln \sqrt{k}}{k}$.

16. $\sum \dfrac{2}{k(\ln k)^2}$.

17. $\sum \dfrac{1}{2+3^{-k}}$.

18. $\sum \dfrac{7k+2}{2k^5+7}$.

19. $\sum \dfrac{2k+5}{5k^3+3k^2}$.

20. $\sum \dfrac{k^4-1}{3k^2+5}$.

21. $\sum \dfrac{1}{k\ln k}$.

22. $\sum \dfrac{1}{2^{k+1}-1}$.

23. $\sum \dfrac{1+2^k}{1+5^k}$.

24. $\sum \dfrac{k^{3/2}}{k^{5/2}+2k-1}$.

25. $\sum \dfrac{2k+1}{\sqrt{k^4+1}}$.

26. $\sum \dfrac{2k+1}{\sqrt{k^3+1}}$.

27. $\sum \dfrac{2k+1}{\sqrt{k^5+1}}$.

28. $\sum \dfrac{1}{\sqrt{2k(k+1)}}$.

29. $\sum ke^{-k^2}$.

30. $\sum k^2 2^{-k^3}$.

31. $\sum \dfrac{2+\sin k}{k^2}$.

32. $\sum \dfrac{2+\cos k}{\sqrt{k+1}}$.

33. $\sum \dfrac{1}{1+2+\cdots+k}$.

34. $\sum \dfrac{k}{1+2^2+\cdots+k^2}$.

35. (C) Use a CAS to determine the convergence or divergence of the series.

(a) $\displaystyle\sum_{k=0}^{\infty} \frac{2k!}{(2k)!}$.

(b) $\displaystyle\sum_{k=0}^{\infty} \frac{2k}{(2k)!}$.

(c) $\displaystyle\sum_{k=0}^{\infty} \frac{2k!}{(2+k)!}$.

36. (C) Use a CAS to evaluate an improper integral that will determine the convergence or divergence of the series.

(a) $\displaystyle\sum_{k=1}^{\infty} \frac{k}{\sqrt{1+k^2}}$.

(b) $\displaystyle\sum_{k=1}^{\infty} \frac{k}{1+k^2}$.

(c) $\displaystyle\sum_{k=1}^{\infty} \frac{k}{1+k^4}$.

37. Find the values of p for which the series $\displaystyle\sum_{k=2}^{\infty} \frac{1}{k(\ln k)^p}$ converges.

38. Find the values of p for which the series $\displaystyle\sum_{k=2}^{\infty} \frac{\ln k}{k^p}$ converges.

39. (a) Show that $\displaystyle\sum_{k=0}^{\infty} e^{-\alpha k}$ converges for each $\alpha > 0$.

(b) Show that $\displaystyle\sum_{k=0}^{\infty} ke^{-\alpha k}$ converges for each $\alpha > 0$.

(c) In general, show that $\displaystyle\sum_{k=0}^{\infty} k^n e^{-\alpha k}$ converges for each nonnegative integer n and each $\alpha > 0$.

40. Let $p > 1$. Use the integral test to show that

$$\frac{1}{(p-1)(n+1)^{p-1}} < \sum_{k=1}^{\infty} \frac{1}{k^p} - \sum_{k=1}^{n} \frac{1}{k^p} < \frac{1}{(p-1)n^{p-1}}.$$

This result gives bounds on the *error* (the remainder) R_n that results from using s_n to approximate the sum of the series.

(C) In Exercises 41–42, (a) compute the sum of the first four terms of the given series; use four decimal place accuracy. (b) Use the result in Exercise 40, to give upper and lower bounds on R_4. (c) Use parts (a) and (b) to estimate the sum of the series.

41. $\sum_{k=1}^{\infty} \frac{1}{k^3}$. **42.** $\sum_{k=1}^{\infty} \frac{1}{k^4}$.

In Exercises 43–46, use the error bounds given in Exercise 40.

C 43. (a) If you were to use s_{100} to approximate $\sum_{k=1}^{\infty} \frac{1}{k^2}$, what would be the bounds on your error?
(b) How large would you have to choose n to ensure that R_n is less than 0.0001?

C 44. (a) If you were to use s_{100} to approximate $\sum_{k=1}^{\infty} \frac{1}{k^3}$, what would be the bounds on your error?
(b) How large would you have to choose n to ensure that R_n is less than 0.0001?

C 45. (a) How many terms of the series $\sum_{k=1}^{\infty} \frac{1}{k^4}$ should you use to ensure that R_n is less than 0.0001?
(b) Estimate $\sum_{k=1}^{\infty} \frac{1}{k^4}$ to three decimal places.

C 46. Repeat Exercise 45, for the series $\sum_{k=1}^{\infty} \frac{1}{k^5}$.

Exercises 47 and 48 complete the limit comparison test.

47. Let $\sum a_k$ and $\sum b_k$ be series with positive terms. Suppose that $a_k/b_k \to 0$.
(a) Show that if $\sum b_k$ converges, then $\sum a_k$ converges.
(b) Show that if $\sum a_k$ diverges, then $\sum b_k$ diverges.
(c) Show by example that if $\sum a_k$ converges, then $\sum b_k$ may converge or diverge.
(d) Show by example that if $\sum b_k$ diverges, then $\sum a_k$ may converge or diverge.
[Parts (c) and (d) explain why we stipulated $L > 0$ in Theorem 11.2.6.]

48. Let $\sum a_k$ and $\sum b_k$ be series with positive terms. Suppose that $a_k/b_k \to \infty$.
(a) Show that if $\sum b_k$ diverges, then $\sum a_k$ diverges.
(b) Show that if $\sum a_k$ converges, then $\sum b_k$ converges.
(c) Show by example that if $\sum a_k$ diverges, then $\sum b_k$ may converge or diverge.
(d) Show by example that if $\sum b_k$ converges, then $\sum a_k$ may converge or diverge.

49. Let $\sum a_k$ be a series with nonnegative terms.
(a) Prove that if $\sum a_k$ converges, then $\sum a_k^2$ converges.
(b) Give an example where $\sum a_k^2$ converges and $\sum a_k$ converges; give an example where $\sum a_k^2$ converges but $\sum a_k$ diverges.

50. Let $\sum a_k$ be a series with nonnegative terms. Prove that if $\sum a_k^2$ converges, then $\sum (a_k/k)$ converges.

51. Let f be a continuous, positive, decreasing function on $[1, \infty)$ such that $\int_1^{\infty} f(x)\,dx$ converges. Then we know that the series $\sum_{k=1}^{\infty} f(k)$ also converges. Prove that

$$0 < L - s_n < \int_n^{\infty} f(x)\,dx,$$

where L is the sum of the series and s_n is the nth partial sum.

In Exercises 52 and 53, use the result of Exercise 51 to determine the smallest integer N such that the difference between the sum of the given series and the Nth partial sum is less than 0.001.

52. $\sum_{k=1}^{\infty} \frac{1}{k^2+1}$. **53.** $\sum_{k=1}^{\infty} k\,e^{-k^2}$.

54. (a) Use the method of the proof of the integral test to show that

$$\ln(n+1) < 1 + \tfrac{1}{2} + \tfrac{1}{3} + \cdots + \tfrac{1}{n} < 1 + \ln n.$$

(b) How many terms of the harmonic series are needed to ensure that $s_n > 100$?

55. This exercise demonstrates that we cannot always use the same testing series for both the basic comparison test and the limit comparison test.
(a) Show that

$$\sum \frac{\ln k}{k\sqrt{k}} \quad \text{converges by comparison with} \quad \sum \frac{1}{k^{5/4}}.$$

(b) Show that the limit comparison test does not apply.

56. Let p and q be polynomials with nonnegative coefficients. State conditions that will imply the convergence or divergence of the series

$$\sum \frac{p(k)}{q(k)}.$$

■ 11.3 THE ROOT TEST; THE RATIO TEST

We continue our study of series with nonnegative terms. Comparison with the geometric series

$$\sum x^k$$

leads to two important tests for convergence: the root test and the ratio test.

THEOREM 11.3.1 THE ROOT TEST

Let $\sum a_k$ be a series with nonnegative terms, and suppose that.

$$(a_k)^{1/k} \to \rho.$$

(a) If $\rho < 1$, then $\sum a_k$ converges.
(b) If $\rho > 1$, then $\sum a_k$ diverges.
(c) If $\rho = 1$, then the test is inconclusive. The series may converge; it may diverge.

PROOF We suppose first that $\rho < 1$ and choose μ so that

$$\rho < \mu < 1.$$

Since $(a_k)^{1/k} \to \rho$, we have

$$(a_k)^{1/k} < \mu \qquad \text{for all } k \text{ sufficiently large.} \qquad \text{(explain)}$$

Thus

$$a_k < \mu^k \qquad \text{for all } k \text{ sufficiently large.}$$

Since $\sum \mu^k$ converges (a geometric series with $0 < \mu < 1$), we know by the basic comparison test that $\sum a_k$ converges.

We suppose now that $\rho > 1$ and choose μ so that

$$\rho > \mu > 1.$$

Since $(a_k)^{1/k} \to \rho$, we have

$$(a_k)^{1/k} > \mu \qquad \text{for all } k \text{ sufficiently large.} \qquad \text{(explain)}$$

Thus

$$a_k > \mu^k \qquad \text{for all } k \text{ sufficiently large.}$$

Since $\sum \mu^k$ diverges (a geometric series with $\mu > 1$), we know by the basic comparison test that $\sum a_k$ diverges.

To see the inconclusiveness of the root test when $\rho = 1$, consider the series $\sum(1/k^2)$ and $\sum(1/k)$. The first series converges and the second series diverges. However, in each case $(a_k)^{1/k} \to 1$:

$$(a_k)^{1/k} = \left(\frac{1}{k^2}\right)^{1/k} = \left(\frac{1}{k^{1/k}}\right)^2 \to 1^2 = 1,$$

$$(a_k)^{1/k} = \left(\frac{1}{k}\right)^{1/k} = \frac{1}{k^{1/k}} \to 1.$$

[Recall that $k^{1/k} \to 1$ as $k \to \infty$, see (10.4.6).] ☐

Applying the Root Test

Example 1 For the series $\sum \dfrac{1}{(\ln k)^k}$, we have

$$(a_k)^{1/k} = \frac{1}{\ln k} \to 0.$$

The series converges. ☐

Example 2 For the series $\sum \dfrac{2^k}{k^3}$, we have

$$(a_k)^{1/k} = 2\left(\frac{1}{k}\right)^{3/k} = 2\left[\left(\frac{1}{k}\right)^{1/k}\right]^3 = 2\left[\frac{1}{k^{1/k}}\right]^3 \to 2 \cdot 1^3 = 2 \quad \text{as } k \to \infty.$$

The series diverges. □

Example 3 In the case of $\sum \left(1 - \dfrac{1}{k}\right)^k$, we have

$$(a_k)^{1/k} = 1 - \frac{1}{k} \to 1.$$

Here the root test is inconclusive. It is also unnecessary: since $a_k = (1 - 1/k)^k$ converges to $1/e$ and not to 0 (10.4.7), the series diverges (11.1.6). □

THEOREM 11.3.2 THE RATIO TEST

Let $\sum a_k$ be a series with positive terms and suppose that

$$\frac{a_{k+1}}{a_k} \to \lambda.$$

(a) If $\lambda < 1$, then $\sum a_k$ converges.
(b) If $\lambda > 1$, then $\sum a_k$ diverges.
(c) If $\lambda = 1$, then the test is inconclusive. The series may converge; it may diverge.

PROOF We suppose first that $\lambda < 1$ and choose μ so that $\lambda < \mu < 1$. Since

$$\frac{a_{k+1}}{a_k} \to \lambda,$$

we know that there exists $k_0 > 0$ such that

$$\text{if } k \geq k_0, \quad \text{then } \frac{a_{k+1}}{a_k} < \mu. \qquad \text{(explain)}$$

This gives

$$a_{k_0+1} < \mu a_{k_0}, \quad a_{k_0+2} < \mu a_{k_0+1} < \mu^2 a_{k_0},$$

and more generally,

$$a_{k_0+j} < \mu^j a_{k_0}, \quad j = 1, 2, \ldots.$$

For $k > k_0$ we have

$$(1) \qquad a_k < \mu^{k-k_0} a_{k_0} = \frac{a_{k_0}}{\mu^{k_0}} \mu^k.$$

↑ set $j = k - k_0$

Since $\mu < 1$,

$$\sum \frac{a_{k_0}}{\mu^{k_0}} \mu^k = \frac{a_{k_0}}{\mu^{k_0}} \sum \mu^k \quad \text{converges.}$$

Therefore, it follows from (1) and the basic comparison test that $\sum a_k$ converges. The proof of the rest of the theorem is left to the Exercises. □

Remark Contrary to some people's intuition, the root and ratio tests are *not* equivalent. See Exercise 50. □

Applying the Ratio Test

Example 4 The ratio test shows that the series $\sum \frac{1}{k!}$ converges:

$$\frac{a_{k+1}}{a_k} = \frac{1}{(k+1)!} \cdot \frac{k!}{1} = \frac{1}{k+1} \to 0. \quad \square$$

Example 5 For the series $\sum \frac{k}{10^k}$, we have

$$\frac{a_{k+1}}{a_k} = \frac{k+1}{10^{k+1}} \cdot \frac{10^k}{k} = \frac{1}{10}\frac{k+1}{k} \to \frac{1}{10}.$$

The series converges.† □

Example 6 For the series $\sum \frac{k^k}{k!}$, we have

$$\frac{a_{k+1}}{a_k} = \frac{(k+1)^{k+1}}{(k+1)!} \cdot \frac{k!}{k^k} = \left(\frac{k+1}{k}\right)^k = \left(1 + \frac{1}{k}\right)^k \to e.$$

Since $e > 1$, the series diverges. □

Example 7 For the series $\sum \frac{1}{2k+1}$, the ratio test is inconclusive:

$$\frac{a_{k+1}}{a_k} = \frac{1}{2(k+1)+1} \cdot \frac{2k+1}{1} = \frac{2k+1}{2k+3} = \frac{2+1/k}{2+3/k} \to 1 \text{ as } k \to \infty.$$

Therefore, we have to look further. Limit comparison with the harmonic series shows that the series diverges:

$$\frac{1}{2k+1} \div \frac{1}{k} = \frac{k}{2k+1} \to \frac{1}{2} \quad \text{and} \quad \sum \frac{1}{k} \text{ diverges.} \quad \square$$

† This series can be summed explicitly. See Exercise 43.

Summary on Convergence Tests

In general, the root test is used only if powers are involved. The ratio test is particularly effective with factorials and with combinations of powers and factorials. If the terms are rational functions of k, the ratio test is inconclusive and the root test is difficult to apply. Rational terms are most easily handled by comparison or limit comparison with a p-series, $\sum 1/k^p$. If the terms have the configuration of a derivative, you may be able to apply the integral test. Finally, keep in mind that, if $a_k \not\to 0$, then there is no reason to apply any special convergence test; the series diverges (Theorem 11.1.6).

EXERCISES 11.3

Determine whether the series converges or diverges.

1. $\sum \dfrac{10^k}{k!}$.

2. $\sum \dfrac{1}{k2^k}$.

3. $\sum \dfrac{1}{k^k}$.

4. $\sum \left(\dfrac{k}{2k+1}\right)^k$.

5. $\sum \dfrac{k!}{100^k}$.

6. $\sum \dfrac{(\ln k)^2}{k}$.

7. $\sum \dfrac{k^2+2}{k^3+6k}$.

8. $\sum \dfrac{1}{(\ln k)^k}$.

9. $\sum k\left(\dfrac{2}{3}\right)^k$.

10. $\sum \dfrac{1}{(\ln k)^{10}}$.

11. $\sum \dfrac{1}{1+\sqrt{k}}$.

12. $\sum \dfrac{2k+\sqrt{k}}{k^3+\sqrt{k}}$.

13. $\sum \dfrac{k!}{10^{4k}}$.

14. $\sum \dfrac{k^2}{e^k}$.

15. $\sum \dfrac{\sqrt{k}}{k^2+1}$.

16. $\sum \dfrac{2^k k!}{k^k}$.

17. $\sum \dfrac{k!}{(k+2)!}$.

18. $\sum \dfrac{1}{k}\left(\dfrac{1}{\ln k}\right)^{3/2}$.

19. $\sum \dfrac{1}{k}\left(\dfrac{1}{\ln k}\right)^{1/2}$.

20. $\sum \dfrac{1}{\sqrt{k^3-1}}$.

21. $\sum \left(\dfrac{k}{k+100}\right)^k$.

22. $\sum \dfrac{(k!)^2}{(2k)!}$.

23. $\sum k^{-(1+1/k)}$.

24. $\sum \dfrac{11}{1+100^{-k}}$.

25. $\sum \dfrac{\ln k}{e^k}$.

26. $\sum \dfrac{k!}{k^k}$.

27. $\sum \dfrac{\ln k}{k^2}$.

28. $\sum \dfrac{k!}{1\cdot 3\cdots(2k-1)}$.

29. $\sum \dfrac{2\cdot 4\cdots 2k}{(2k)!}$.

30. $\sum \dfrac{(2k+1)^{2k}}{(5k^2+1)^k}$.

31. $\sum \dfrac{k!(2k)!}{(3k)!}$.

32. $\sum \dfrac{\ln k}{k^{5/4}}$.

33. $\sum \dfrac{k^{k/2}}{k!}$.

34. $\sum \dfrac{k^k}{(3^k)^2}$.

35. $\sum \dfrac{k^k}{3^{(k^2)}}$.

36. $\sum \left(\sqrt{k}-\sqrt{k-1}\right)^k$.

37. $\dfrac{1}{2}+\dfrac{2}{3^2}+\dfrac{4}{4^3}+\dfrac{8}{5^4}+\cdots$.

38. $1+\dfrac{1\cdot 2}{1\cdot 3}+\dfrac{1\cdot 2\cdot 3}{1\cdot 3\cdot 5}+\dfrac{1\cdot 2\cdot 3\cdot 4}{1\cdot 3\cdot 5\cdot 7}+\cdots$.

39. $\dfrac{1}{4}+\dfrac{1\cdot 3}{4\cdot 7}+\dfrac{1\cdot 3\cdot 5}{4\cdot 7\cdot 10}+\dfrac{1\cdot 3\cdot 5\cdot 7}{4\cdot 7\cdot 10\cdot 13}+\cdots$.

40. $\dfrac{2}{3}+\dfrac{2\cdot 4}{3\cdot 7}+\dfrac{2\cdot 4\cdot 6}{3\cdot 7\cdot 11}+\dfrac{2\cdot 4\cdot 6\cdot 8}{3\cdot 7\cdot 11\cdot 15}+\cdots$.

C 41. Use a CAS to implement the ratio test to determine whether the series converges or diverges.

(a) $\sum_{k=1}^{\infty} \dfrac{2^k}{k!}$. (b) $\sum_{k=1}^{\infty} \dfrac{2^k}{k}$.

C 42. Use a CAS to implement the root test to determine whether the series converges or diverges.

(a) $\sum_{k=1}^{\infty} \dfrac{2^k}{k^{k+1}}$ (b) $\sum_{k=1}^{\infty} \left(\dfrac{k+2}{2k}\right)^k$

43. Find the sum of the series $\frac{1}{10}+\frac{2}{100}+\frac{3}{1000}+\frac{4}{10000}+\cdots$.
HINT: Exercise 70 of Section 11.1.

44. Complete the proof of the ratio test.

(a) Prove that, if $\lambda > 1$, then $\sum a_k$ diverges.

(b) Prove that, if $\lambda = 1$, the ratio test is inconclusive. HINT: Consider $\sum 1/k$ and $\sum 1/k^2$.

45. Show that the sequence $a_n = \frac{n!}{n^n} \to 0$. HINT: Consider the series $\sum \dfrac{k!}{k^k}$.

46. Let r be a positive number. Show that the sequence $a_n = \frac{r^n}{n!} \to 0$.

47. Let $p \geq 2$ be an integer. Find the values of p (if any) such that $\sum \dfrac{(k!)^2}{(pk)!}$ converges.

48. Let r be a positive number. For what values of r (if any) does $\sum \frac{r^k}{k^r}$ converge?

49. Let $\{a_k\}$ be a sequence of positive numbers and take $r > 0$. Use the root test to show that, if $(a_k)^{1/k} \to \rho$ and $\rho < 1/r$, then $\sum a_k r^k$ converges.

50. Consider the series $\frac{1}{2} + 1 + \frac{1}{8} + \frac{1}{4} + \frac{1}{32} + \frac{1}{16} + \ldots$ formed by rearranging a convergent geometric series.

(a) Use the root test to show that the series converges.

(b) Show that the ratio test does not apply.

■ 11.4 ABSOLUTE AND CONDITIONAL CONVERGENCE; ALTERNATING SERIES

In this section we consider series that have both positive and negative terms.

Absolute and Conditional Convergence

Let $\sum a_k$ be a series with positive and negative terms. One way to show that $\sum a_k$ converges is to show that the series of absolute values $\sum |a_k|$ converges.

THEOREM 11.4.1

$$\text{If} \quad \sum |a_k| \quad \text{converges,} \quad \text{then} \quad \sum a_k \quad \text{converges.}$$

PROOF For each k,

$$-|a_k| \le a_k \le |a_k| \qquad \text{and therefore} \qquad 0 \le a_k + |a_k| \le 2|a_k|.$$

If $\sum |a_k|$ converges, then $\sum 2|a_k| = 2\sum |a_k|$ converges, and therefore, by the basic comparison theorem, $\sum (a_k + |a_k|)$ converges. Since

$$a_k = (a_k + |a_k|) - |a_k|,$$

we can conclude that $\sum a_k$ converges. □

Series $\sum a_k$ for which $\sum |a_k|$ converges are called *absolutely convergent.* The theorem we have just proved says that

(11.4.2) absolutely convergent series are convergent.

As we will show presently, the converse is false. There are convergent series that are not absolutely convergent. Such series are called *conditionally convergent.*

Example 1 Consider the series

$$\sum_{k=1}^{\infty} \frac{(-1)^{k+1}}{k^2} = 1 - \frac{1}{2^2} + \frac{1}{3^2} - \frac{1}{4^2} + \frac{1}{5^2} - \cdots.$$

If we replace each term by its absolute value, we obtain the series

$$\sum_{k=1}^{\infty} \frac{1}{k^2} = 1 + \frac{1}{2^2} + \frac{1}{3^2} + \frac{1}{4^2} + \frac{1}{5^2} + \cdots.$$

This is a p-series with $p = 2$. It is therefore convergent. This means that the initial series is absolutely convergent. □

Example 2 Consider the series

$$1 - \frac{1}{2} - \frac{1}{2^2} + \frac{1}{2^3} - \frac{1}{2^4} + \frac{1}{2^5} + \frac{1}{2^6} - \frac{1}{2^7} - \frac{1}{2^8} + \cdots.$$

If we replace each term by its absolute value, we obtain the series

$$1 + \frac{1}{2} + \frac{1}{2^2} + \frac{1}{2^3} + \frac{1}{2^4} + \frac{1}{2^5} + \frac{1}{2^6} + \frac{1}{2^7} + \frac{1}{2^8} + \cdots.$$

This is a convergent geometric series. The initial series is therefore absolutely convergent. □

Example 3 The series

$$1 - \frac{1}{2} + \frac{1}{3} - \frac{1}{4} + \frac{1}{5} - \frac{1}{6} + \cdots$$

is only conditionally convergent.†

It is convergent (see the next theorem), but it is not absolutely convergent: if we replace each term by its absolute value, we get the divergent harmonic series

$$1 + \frac{1}{2} + \frac{1}{3} + \frac{1}{4} + \frac{1}{5} + \frac{1}{6} + \cdots. \quad □$$

Alternating Series

Series in which the consecutive terms have opposite sign are called *alternating series.* For example, the series

$$\sum_{k=0}^{\infty} \frac{(-1)^k}{k+1} = 1 - \frac{1}{2} + \frac{1}{3} - \frac{1}{4} + \frac{1}{5} - \frac{1}{6} + \cdots$$

and

$$\sum_{k=1}^{\infty} \frac{(-1)^k}{\sqrt{k}} = -1 + \frac{1}{\sqrt{2}} - \frac{1}{\sqrt{3}} + \frac{1}{\sqrt{4}} - \frac{1}{\sqrt{5}} + \cdots$$

are alternating series.

The series

$$1 - \frac{1}{2} - \frac{1}{3} + \frac{1}{4} - \frac{1}{5} - \frac{1}{6} + \cdots$$

is not an alternating series because there are consecutive terms with the same sign.

In general, an alternating series will either have the form

$$a_0 - a_1 + a_2 - a_3 + a_4 - \cdots = \sum_{k=0}^{\infty} (-1)^k a_k$$

or the form

$$-a_0 + a_1 - a_2 + a_3 - a_4 + \cdots = \sum_{k=0}^{\infty} (-1)^{k+1} a_k,$$

where $\{a_k\}$ is a sequence of positive numbers. Since the second form is simply the negative of the first, we will focus our attention on the first form.

---

† In Section 11.5 we show that the series $1 - \frac{1}{2} + \frac{1}{3} - \frac{1}{4} + \frac{1}{5} - \frac{1}{6} + \cdots$ converges to ln 2.

THEOREM 11.4.3 ALTERNATING SERIES TEST†

Let $\{a_k\}$ be a decreasing sequence of positive numbers.

$$\text{If} \quad a_k \to 0, \qquad \text{then} \quad \sum_{k=0}^{\infty} (-1)^k a_k \quad \text{converges.}$$

PROOF First we look at the even partial sums, s_{2m}. Since

$$s_{2m} = (a_0 - a_1) + (a_2 - a_3) + \cdots + (a_{2m-2} - a_{2m-1}) + a_{2m}$$

is the sum of positive numbers, the even partial sums are all positive. Since

$$s_{2m+2} = s_{2m} - (a_{2m+1} - a_{2m+2}) \quad \text{and} \quad a_{2m+1} - a_{2m+2} > 0,$$

we have

$$s_{2m+2} < s_{2m}.$$

This means that the sequence of even partial sums is decreasing. Being bounded below by 0, it is convergent; say,

$$s_{2m} \to L \quad \text{as } m \to \infty.$$

Now

$$s_{2m+1} = s_{2m} - a_{2m+1}.$$

Since $a_{2m+1} \to 0$ as $m \to \infty$, we also have

$$s_{2m+1} \to L.$$

Since both the even and the odd partial sums tend to L, the sequence of all partial sums tends to L (Exercise 42, Section 10.3). □

From this theorem you can see that the following series all converge:

$$1 - \frac{1}{2} + \frac{1}{3} - \frac{1}{4} + \frac{1}{5} - \frac{1}{6} + \cdots, \quad 1 - \frac{1}{\sqrt{2}} + \frac{1}{\sqrt{3}} - \frac{1}{\sqrt{4}} + \frac{1}{\sqrt{5}} - \frac{1}{\sqrt{6}} + \cdots,$$

$$1 - \frac{1}{2!} + \frac{1}{3!} - \frac{1}{4!} + \frac{1}{5!} - \frac{1}{6!} + \cdots.$$

The first two series converge only conditionally; the third is absolutely convergent.

Remark In the proof of Theorem 11.4.3, we showed that the sequence of even partial sums, $\{s_{2m}\}$, is decreasing and bounded below, and hence convergent. In Exercise 48 you are asked to show that the sequence of odd partial sums $\{s_{2m+1}\}$ is increasing and bounded above. Thus, we have an alternative way to show that the sequence of odd partial sums converges. Moreover, since

$$s_{2m+1} - s_{2m} = a_{2m+1} \to 0 \quad as \quad m \to \infty,$$

†This theorem dates back to Leibniz. He proved the result in 1705.

the even partial sums and the odd partial sums each converge to the same limit, say L. Figure 11.4.1 illustrates the monotonic behavior of the even and odd partial sums.

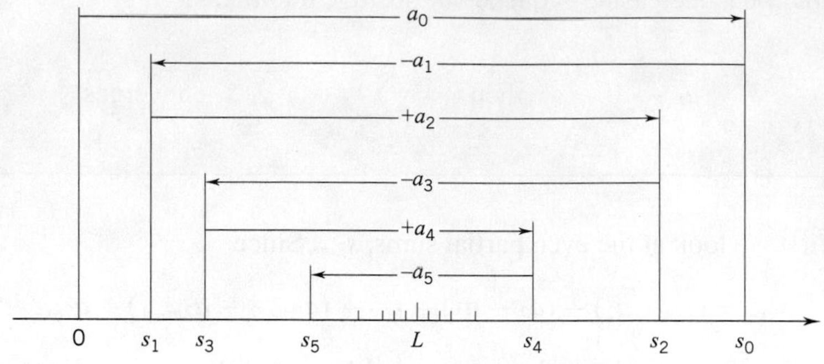

Figure 11.4.1

An Estimate for Alternating Series You have seen that if $\{a_k\}$ is a decreasing sequence of positive numbers that tends to 0, then

$$\sum_{k=0}^{\infty}(-1)^k a_k \text{ converges to a sum } L.$$

(11.4.4) The sum L of a convergent alternating series lies between consecutive partial sums s_n, s_{n+1}, and thus s_n approximates L to within a_{n+1}:

$$|s_n - L| < a_{n+1}.$$

PROOF For all n,

$$a_{n+1} > a_{n+2}.$$

If n is odd,

$$s_{n+2} = s_n + a_{n+1} - a_{n+2} > s_n;$$

if n is even,

$$s_{n+2} = s_n - a_{n+1} + a_{n+2} < s_n.$$

The odd partial sums increase toward L; the even partial sums decrease toward L.

For odd n,

$$s_n < L < s_n + a_{n+1} = s_{n+1},$$

and for even n,

$$s_n - a_{n+1} = s_{n+1} < L < s_n.$$

Thus, for all n, L lies between s_n and s_{n+1}. We can now conclude that

$$|L - s_n| < |s_{n+1} - s_n| = a_{n+1}$$

and so s_n approximates L to within a_{n+1}. □

Example 4 Both

$$1 - \frac{1}{2} + \frac{1}{3} - \frac{1}{4} + \frac{1}{5} - \frac{1}{6} + \cdots \quad \text{and} \quad 1 - \frac{1}{2^2} + \frac{1}{3^2} - \frac{1}{4^2} + \frac{1}{5^2} - \frac{1}{6^2} + \cdots$$

are convergent alternating series. The nth partial sum of the first series approximates the sum of that series within $1/(n+1)$; the nth partial sum of the second series approximates the sum of the second series within $1/(n+1)^2$. The second series converges more rapidly than the first series. □

Example 5 Give a numerical estimate for the sum of the series

$$\sum_{k=0}^{\infty} \frac{(-1)^k}{(2k+1)!} = 1 - \frac{1}{3!} + \frac{1}{5!} - \frac{1}{7!} + \cdots$$

correct to three decimal places.

SOLUTION This is a convergent alternating series. Call the sum L. For s_n to approximate L to three decimal places, we must have $|L - s_n| < 0.0005$. Writing out the first few terms of the series, we have

$$1 - \frac{1}{3!} + \frac{1}{5!} - \frac{1}{7!} + \cdots = 1 - \frac{1}{6} + \frac{1}{120} - \frac{1}{5040} + \cdots.$$

Since $a_3 = \dfrac{1}{7!} = \dfrac{1}{5040} < \dfrac{1}{5000} = 0.0002 < 0.0005$,

$$s_2 = 1 - \frac{1}{3!} + \frac{1}{5!} = 1 - \frac{1}{6} + \frac{1}{120} = \frac{101}{120}$$

approximates L to within 0.0005. As you can verify, $0.8416 < \frac{101}{120} < 0.8417$, and so the estimate 0.842 is correct to three decimal places.† □

Rearrangements

A *rearrangement* of a series $\sum a_k$ is a series that has exactly the same terms but in a different order. Thus, for example,

$$1 + \frac{1}{3^3} - \frac{1}{2^2} + \frac{1}{5^5} - \frac{1}{4^4} + \frac{1}{7^7} - \frac{1}{6^6} + \cdots$$

and

$$1 + \frac{1}{3^3} + \frac{1}{5^5} - \frac{1}{2^2} - \frac{1}{4^4} + \frac{1}{7^7} + \frac{1}{9^9} - \cdots$$

are both rearrangements of

$$1 - \frac{1}{2^2} + \frac{1}{3^3} - \frac{1}{4^4} + \frac{1}{5^5} - \frac{1}{6^6} + \frac{1}{7^7} - \cdots.$$

In 1867 Riemann published a theorem on rearrangements of series that underscores the importance of distinguishing between absolute convergence and conditional convergence. According to this theorem, all rearrangements of an absolutely convergent series converge absolutely to the same sum. In sharp contrast, a series that is only conditionally convergent can be rearranged to converge to any number we please. It can also be arranged to diverge to $+\infty$, or to diverge to $-\infty$, or even to oscillate between any two bounds we choose .†† These possibilities are explored in Project 11.4.

†In Section 11.5 you will see that the series $1 - \frac{1}{3!} + \frac{1}{5!} - \frac{1}{7!} + \ldots$ converges to sin 1.

†† For a complete proof see pp. 138-139, 318-320 in Konrad Knopp's *Theory and Applications of Infinite Series* (Second English Edition), Blackie and Son Limited, London, 1951.

EXERCISES 11.4

Test these series for (a) absolute convergence, (b) conditional convergence.

1. $1 + (-1) + 1 + \cdots + (-1)^k + \cdots$.

2. $\frac{1}{4} - \frac{1}{6} + \frac{1}{8} - \frac{1}{10} + \cdots + \frac{(-1)^k}{2k} + \cdots$.

3. $\frac{1}{2} - \frac{2}{3} + \frac{3}{4} - \frac{4}{5} + \cdots + (-1)^{k+1}\frac{k}{(k+1)} + \cdots$.

4. $\frac{1}{2\ln 2} - \frac{1}{3\ln 3} + \frac{1}{4\ln 4} - \frac{1}{5\ln 5} + \cdots + (-1)^k\frac{1}{k\ln k} + \cdots$.

5. $\sum (-1)^k \frac{\ln k}{k}$.

6. $\sum (-1)^k \frac{k}{\ln k}$.

7. $\sum \left(\frac{1}{k} - \frac{1}{k!}\right)$.

8. $\sum \frac{k^3}{2^k}$.

9. $\sum (-1)^k \frac{1}{2k+1}$.

10. $\sum (-1)^k \frac{(k!)^2}{(2k)!}$.

11. $\sum \frac{k!}{(-2)^k}$.

12. $\sum \sin\left(\frac{k\pi}{4}\right)$.

13. $\sum (-1)^k \left(\sqrt{k+1} - \sqrt{k}\right)$.

14. $\sum (-1)^k \frac{k}{k^2+1}$.

15. $\sum \sin\left(\frac{\pi}{4k^2}\right)$.

16. $\sum \frac{(-1)^k}{\sqrt{k(k+1)}}$.

17. $\sum (-1)^k \frac{k}{2^k}$.

18. $\sum \left(\frac{1}{\sqrt{k}} - \frac{1}{\sqrt{k+1}}\right)$.

19. $\sum \frac{(-1)^k}{k - 2\sqrt{k}}$.

20. $\sum (-1)^k \frac{k+2}{k^2+k}$.

21. $\sum (-1)^k \frac{4^{k-2}}{e^k}$.

22. $\sum (-1)^k \frac{k^2}{2^k}$.

23. $\sum (-1)^k k \sin(1/k)$.

24. $\sum (-1)^{k+1} \frac{k^k}{k!}$.

25. $\sum (-1)^k k e^{-k}$.

26. $\sum \frac{\cos \pi k}{k}$.

27. $\sum (-1)^k \frac{\cos \pi k}{k}$.

28. $\sum \frac{\sin(\pi k/2)}{k\sqrt{k}}$.

29. $\sum \frac{\sin(\pi k/4)}{k^2}$.

30. $\frac{1}{2} - \frac{1}{3} - \frac{1}{4} + \frac{1}{5} - \frac{1}{6} - \frac{1}{7} + \cdots + \frac{1}{3k+2} - \frac{1}{3k+3} - \frac{1}{3k+4} + \cdots$.

31. $\frac{2\cdot 3}{4\cdot 5} - \frac{5\cdot 6}{7\cdot 8} + \cdots + (-1)^k \frac{(3k+2)(3k+3)}{(3k+4)(3k+5)} + \cdots$.

Estimate the error if the partial sum s_n is used to approximate the sum of the series.

32. $\sum_{k=1}^{\infty} (-1)^{k+1} \frac{1}{k}$; s_{20}.

33. $\sum_{k=0}^{\infty} (-1)^k \frac{1}{\sqrt{k+1}}$; s_{80}.

34. $\sum_{k=0}^{\infty} (-1)^k \frac{1}{(10)^k}$; s_4.

35. $\sum_{k=1}^{\infty} (-1)^{k+1} \frac{1}{k^3}$; s_9.

36. Let s_n be the nth partial sum of the series

$$\sum_{k=0}^{\infty} (-1)^k \frac{1}{10^k}.$$

Find the least value of n for which s_n approximates the sum of the series within (a) 0.001, (b) 0.0001.

37. Find the sum of the series in Exercise 36.

In Exercises 38 and 39, find the smallest integer N such that s_N will approximate the sum of the given alternating series to within the indicated accuracy.

38. $\sum_{k=1}^{\infty} (-1)^k \frac{(0.9)^k}{k}$; 0.001.

39. $\sum_{k=0}^{\infty} (-1)^k \frac{1}{\sqrt{k+1}}$; 0.005.

C 40. Use a CAS to find the sum of the series.

(a) $\sum_{k=0}^{\infty} \frac{(-1)^k}{(2k+1)!}$; compare your result with sin 1.

(b) $\sum_{k=1}^{\infty} \frac{(-1)^{k+1}}{k}$; compare your result with ln 2.

(c) $\sum_{k=0}^{\infty} \frac{(-1)^k}{k!}$. Is this result related to e? If so, how?

C 41. The alternating series $\sum_{k=0}^{\infty} (-1)^k \frac{k}{k^2+1}$ converges to a number L. Use a CAS to find the smallest positive integer n such that

$$\left| L - \sum_{k=0}^{n} (-1)^k \frac{k}{k^2+1} \right| < 0.001.$$

42. Verify that the series

$$1 - \frac{1}{2} + \frac{1}{2} - \frac{1}{3} + \frac{1}{2} - \frac{1}{3} - \frac{1}{4} + \frac{1}{3} - \frac{1}{4} + \frac{1}{3} - \frac{1}{4} + \cdots$$

diverges and explain how this does not violate the theorem on alternating series.

43. Let L be the sum of the series

$$\sum_{k=0}^{\infty} (-1)^k \frac{1}{k!}$$

and let s_n be the nth partial sum. Find the least value of n for which s_n approximates L to within (a) 0.01, (b) 0.001.

44. Let $\{a_k\}$ be a nonincreasing sequence of positive numbers that converges to 0. Does the alternating series $\sum (-1)^k a_k$ necessarily converge?

45. Can the hypothesis of Theorem 11.4.3 be relaxed to require only that $\{a_{2k}\}$ and $\{a_{2k+1}\}$ be decreasing sequences of positive numbers with limit zero?

46. Show that if $\sum a_k$ is absolutely convergent and $|b_k| \le |a_k|$ for all k, then $\sum b_k$ is absolutely convergent.

47. (a) Show that if $\sum a_k$ is absolutely convergent then $\sum a_k^2$ is convergent.

(b) Show by means of an example that the converse of the result in part (a) is false.

48. Let $\sum_{k=0}^{\infty}(-1)^k a_k$ be an alternating series with $\{a_k\}$ a decreasing sequence. Show that the sequence of odd partial sums $\{s_{2m+1}\}$ increases and is bounded above.

49. In Section 11.7 we prove that, if $\sum a_k c^k$ converges, then $\sum a_k x^k$ converges absolutely for all x such that $|x| < |c|$. Try to prove this now.

50. Let a and b be positive numbers and form the series

$$a - \frac{b}{2} + \frac{a}{3} - \frac{b}{4} + \frac{a}{5} - \frac{b}{6} + \cdots.$$

(a) Express this series in $\sum$ notation.

(b) For what values of a and b is this series absolutely convergent? Conditionally convergent?

51. (a) Show that $\displaystyle\sum_{k=1}^{2n}\frac{(-1)^{k+1}}{k} = \frac{1}{n+1} + \frac{1}{n+2} + \cdots + \frac{1}{2n}$.

HINT: Let $S_m = \displaystyle\sum_{k=1}^{m}\frac{1}{k}$ and calculate $S_{2m} - S_m$.

(b) Show that $\displaystyle\frac{1}{n+1} + \frac{1}{n+2} + \cdots + \frac{1}{2n}$ is a Riemann sum for $\displaystyle\int_n^{2n}(1/x)\,dx$.

(c) Use the results in parts (a) and (b) to show that

$$\sum_{k=1}^{\infty}\frac{(-1)^{k+1}}{k} = \ln 2.$$

52. (a) Use a graphing utility to draw the graph of $f(x) = (\sin x)/x$ for $x \ge 0$.

(b) Use Riemann sums to show that the improper integral $\displaystyle\int_0^{\infty}\frac{\sin x}{x}\,dx$ converges. Verify the result with a CAS.

(c) Show that the improper integral $\displaystyle\int_0^{\infty}\frac{|\sin x|}{x}\,dx$ diverges.

■ PROJECT 11.4 Convergence of Alternating Series

Part I. Approximation. The alternating series

$$\sum_{k=0}^{\infty}(-1)^k a_k$$

converges if $\{a_k\}$ is a decreasing sequence with limit 0. Moreover, if the series converges, then the sum s is always between consecutive partial sums, s_n and s_{n+1}. Here we investigate procedures that provide better approximations for the sum of such series.

Consider the convergent alternating series

$$\sum_{k=0}^{\infty}(-1)^k\frac{k^2+1}{k^3+1}.$$

Denote the sum of the series by s.

Problem 1. Estimate the sum of the first 10, 20, 30,..., 100 terms of this series.

Problem 2. Determine the least integer n such that s_n approximates s with three decimal place accuracy.

For each positive integer n, let t_n and u_n be defined by

$$t_n = \frac{s_n + s_{n-1}}{2} \quad\text{and}\quad u_n = \frac{a_n s_n + a_{n-1}s_{n-1}}{a_n + a_{n-1}}.$$

For each n, the numbers t_n and u_n are averages: t_n is a simple average, u_n is a weighted average.

Problem 3. Show that $\lim_{n\to\infty} t_n = \lim_{n\to\infty} u_n = s$.

Problem 4.

a. Calculate t_n for $n = 10, 20, 30, \cdots, 100$.

b. Calculate u_n for $n = 10, 20, 30, \ldots, 100$.

c. Compare the speed of convergence of the three sequences.

Part II. Rearrangements Here we will work with the conditionally convergent series

$$\sum_{k=0}^{\infty}(-1)^k\frac{1}{k+1}$$

In Exercise 51 you were asked to show that the sum of this series is ln 2. This result will be shown by another means in the next section. Accepting this fact for now, we have

$$1 - \tfrac{1}{2} + \tfrac{1}{3} - \tfrac{1}{4} + \tfrac{1}{5} - \tfrac{1}{6} + \cdots = \ln 2$$

and

$$\tfrac{1}{2} - \tfrac{1}{4} + \tfrac{1}{6} - \tfrac{1}{8} + \tfrac{1}{10} - \tfrac{1}{12} + \cdots = \tfrac{1}{2}\ln 2 \qquad (\text{multiply by } \tfrac{1}{2})$$

Adding the two series, we get a rearrangement of the original series with a new sum:

$$1 + \tfrac{1}{3} - \tfrac{1}{2} + \tfrac{1}{5} + \tfrac{1}{7} - \tfrac{1}{4} + \cdots = \tfrac{3}{2}\ln 2.$$

Problem 5. Write out the first ten terms of this rearrangement.

Problem 6.

a. Show that $\displaystyle\sum_{k=0}^{\infty}\frac{1}{2k+1}$ (the sum of the odd-numbered terms) diverges to ∞.

b. Show that $\sum_{k=1}^{\infty} \frac{-1}{2k}$ (the sum of the even-numbered terms) diverges to $-\infty$.

We can use the results of Problem 6 to rearrange the terms to obtain a series in which the partial sums exhibit unbounded oscillation:

$$1 + \tfrac{1}{3} > 1;$$

add enough consecutive negative terms so that

$$(1) \qquad 1 + \tfrac{1}{3} - \tfrac{1}{2} - \tfrac{1}{4} - \tfrac{1}{8} - \cdots - \tfrac{1}{2n} < -2;$$

now add enough consecutive positive terms so that

$$(2) \qquad 1 + \frac{1}{3} - \frac{1}{2} - \frac{1}{4} - \cdots - \frac{1}{2n} + \frac{1}{5} + \frac{1}{7} + \cdots + \frac{1}{2m+1} > 3;$$

and so on.

Problem 7. Find the least n and m that satisfy the conditions of (1) and (2).

The terms can be rearranged so that the partial sums converge to any preassigned number.

Problem 8. Describe how to obtain a rearrangement of the series that will converge to 1.5. Write out the first ten terms of your rearranged series.

Here is a "proof" that $2 = 1$:

$$1 - \tfrac{1}{2} + \tfrac{1}{3} - \tfrac{1}{4} + \tfrac{1}{5} - \tfrac{1}{6} + \cdots = \ln 2$$

$$2\left(1 - \tfrac{1}{2} + \tfrac{1}{3} - \tfrac{1}{4} + \tfrac{1}{5} - \tfrac{1}{6} + \tfrac{1}{7} - \tfrac{1}{8} + \tfrac{1}{9} - \tfrac{1}{10} + \cdots\right) = 2\ln 2$$

$$2 - 1 + \tfrac{2}{3} - \tfrac{1}{2} + \tfrac{2}{5} - \tfrac{1}{3} + \tfrac{2}{7} - \tfrac{1}{4} + \tfrac{2}{9} - \tfrac{1}{5} + \cdots = 2\ln 2$$

Now add the terms with the same denominator: $2 - 1 = 1$, $\frac{2}{3} - \frac{1}{3} = \frac{1}{3}$, $\frac{2}{5} - \frac{1}{5} = \frac{1}{5}, \cdots$ to get

$$1 - \tfrac{1}{2} + \tfrac{1}{3} - \tfrac{1}{4} + \tfrac{1}{5} - \tfrac{1}{6} + \ldots = 2\ln 2.$$

Thus, $2\ln 2 = \ln 2$ and $2 = 1$.

Problem 9. What is wrong with the argument just given?

■ 11.5 TAYLOR POLYNOMIALS IN *x*; TAYLOR SERIES IN *x*

Taylor Polynomials in *x*

We begin with a function f continuous at 0 and set $P_0(x) = f(0)$. If f is differentiable at 0, the linear function that best approximates f at points close to 0 is the linear polynomial

$$P_1(x) = f(0) + f'(0)x;$$

P_1 has the same value as f at 0 and also the same first derivative (the same rate of change):

$$P_1(0) = f(0), \quad P_1'(0) = f'(0).$$

If f is twice differentiable at 0, then we can get a better approximation to f by using the quadratic polynomial

$$P_2(x) = f(0) + f'(0)x + \frac{f''(0)}{2!}x^2;$$

P_2 has the same value as f at 0 and the same first two derivatives:

$$P_2(0) = f(0), \quad P_2'(0) = f'(0), \quad P_2''(0) = f''(0).$$

If f has three derivatives at 0, we can form the cubic polynomial

$$P_3(x) = f(0) + f'(0)x + \frac{f''(0)}{2!}x^2 + \frac{f'''(0)}{3!}x^3;$$

P_3 has the same value as f at 0 and the same first three derivatives:

$$P_3(0) = f(0), \quad P_3'(0) = f'(0), \quad P_3''(0) = f''(0), \quad P_3'''(0) = f'''(0).$$

In general, if f has n derivatives at 0, we can form the polynomial

$$P_n(x) = f(0) + f'(0)x + \frac{f''(0)}{2!}x^2 + \cdots + \frac{f^{(n)}(0)}{n!}x^n;$$

P_n is the polynomial that has the same value as f at 0 and the same first n derivatives:

$$P_n(0) = f(0), \quad P_n'(0) = f'(0), \quad P_n''(0) = f''(0), \ldots, \quad P_n^{(n)}(0) = f^{(n)}(0).$$

These approximating polynomials $P_0(x), P_1(x), P_2(x), \cdots, P_n(x)$ are called *Taylor polynomials* after the English mathematician Brook Taylor (1685–1731). Taylor introduced these polynomials in the year 1712.

Example 1 The exponential function $f(x) = e^x$ has derivatives

$$f'(x) = e^x, f''(x) = e^x, f'''(x) = e^x, \quad \text{and so on.}$$

Thus $\quad f(0) = 1, \quad f'(0) = 1, \quad f''(0) = 1, \quad f'''(0) = 1, \cdots, \quad f^{(n)}(0) = 1.$

The nth Taylor polynomial takes the form

$$P_n(x) = 1 + x + \frac{x^2}{2!} + \frac{x^3}{3!} + \cdots + \frac{x^n}{n!}.$$

Figure 11.5.1 shows the graph of the exponential function and the graphs of the first four approximating polynomials. ❑

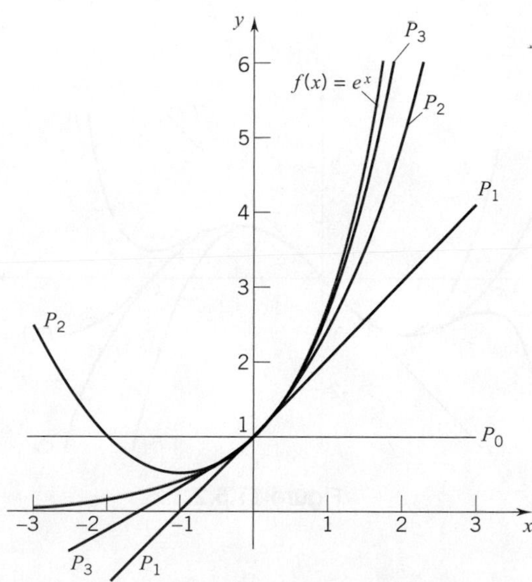

Figure 11.5.1

Example 2 To find the Taylor polynomials that approximate the sine function, we write

$$f(x) = \sin x, \quad f'(x) = \cos x, \quad f''(x) = -\sin x, \quad f'''(x) = -\cos x.$$

The pattern now repeats itself:

$$f^{(4)}(x) = \sin x, \quad f^{(5)}(x) = \cos x, \quad f^{(6)}(x) = -\sin x, \quad f^{(7)}(x) = -\cos x.$$

At $x = 0$, the sine function and all its even derivatives are 0. The odd derivatives are alternately 1 and -1:

$$f'(0) = 1, \quad f'''(0) = -1, \quad f^{(5)}(0) = 1, \quad f^{(7)}(0) = -1, \quad \text{and so on.}$$

Therefore the Taylor polynomials are as follows:

$$\begin{aligned}
&P_0(x) = 0,\\
&P_1(x) = P_2(x) = x,\\
&P_3(x) = P_4(x) = x - \frac{x^3}{3!},\\
&P_5(x) = P_6(x) = x - \frac{x^3}{3!} + \frac{x^5}{5!},\\
&P_7(x) = P_8(x) = x - \frac{x^3}{3!} + \frac{x^5}{5!} - \frac{x^7}{7!}, \quad \text{and so on .}
\end{aligned}$$

Only odd powers appear. This is not surprising since the sine function is an odd function. The graphs of the sine function and the first few polynomial approximations appear in Figure 11.5.2. □

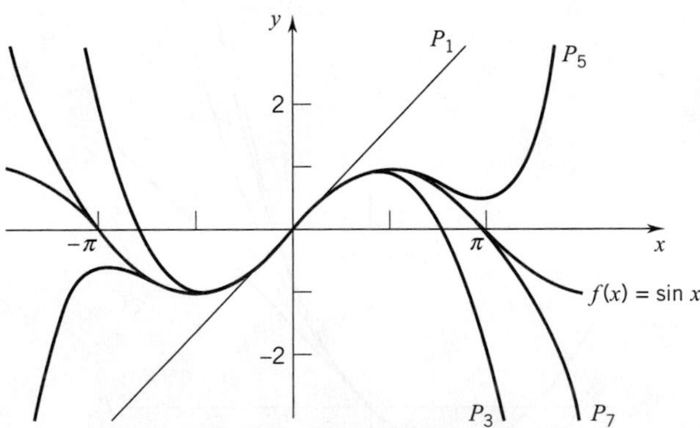

Figure 11.5.2

It is not enough to say that the Taylor polynomials

$$P_n(x) = f(0) + f'(0)x + \frac{f''(0)}{2!}x^2 + \cdots + \frac{f^{(n)}(0)}{n!}x^n$$

approximate $f(x)$. We must describe the accuracy of the approximation.

Our first step is to prove a result known as Taylor's theorem.

THEOREM 11.5.1 TAYLOR'S THEOREM

If f has $n+1$ continuous derivatives on an open interval I that contains 0, then for each $x \in I$

$$f(x) = f(0) + f'(0)x + \frac{f''(0)}{2!}x^2 + \cdots + \frac{f^{(n)}(0)}{n!}x^n + R_n(x)$$

where the *remainder* $R_n(x) = \dfrac{1}{n!}\displaystyle\int_0^x f^{(n+1)}(t)\,(x-t)^n dt.$

PROOF Fix x in the interval I. Since

$$\int_0^x f'(t)\,dt = f(x) - f(0),$$

we have

$$(1) \qquad f(x) = f(0) + \int_0^x f'(t)\,dt.$$

We now integrate by parts. We set

$$u = f'(t) \qquad \text{and} \qquad dv = dt.$$

This forces

$$du = f''(t)dt.$$

For v we may choose any function of t with derivative identically 1. To suit our purpose, we choose

$$v = -(x-t).$$

We carry out the integration by parts and get

$$\int_0^x f'(t) = \Big[-f'(t)(x-t)\Big]_0^x + \int_0^x f''(t)(x-t)\,dt$$
$$= f'(0)x + \int_0^x f''(t)(x-t)\,dt,$$

which with (1) gives

$$f(x) = f(0) + f'(0)x + \int_0^x f''(t)(x-t)\,dt.$$

This completes the first step toward the proof of the theorem.

Integrating by parts again [set $u = f''(t)$, $dv = (x-t)\,dt$, leading to $du = f'''(t)\,dt$, $v = -\frac{1}{2}(x-t)^2$], we get

$$f(x) = f(0) + f'(0)x + \frac{f''(0)}{2!}x^2 + \frac{1}{2!}\int_0^x f'''(t)(x-t)^2\,dt.$$

Continuing to integrate by parts (see Exercise 61), we get after n steps

$$f(x) = f(0) + f'(0)x + \frac{f''(0)}{2!}x^2 + \cdots + \frac{f^{(n)}(0)}{n!}x^n + \frac{1}{n!}\int_0^x f^{(n+1)}(t)(x-t)^n\,dt,$$

which is what have been trying to prove. ❑

To see how closely

$$P_n(x) = f(0) + f'(0)x + \frac{f''(0)}{2!}x^2 + \cdots + \frac{f^{(n)}(0)}{n!}x^n$$

approximates $f(x)$, we need an estimate for the remainder term $R_n(x)$. The following corollary to Taylor's theorem gives a more convenient form of the remainder. It was established by Joseph Lagrange in 1797, and it is known as the Lagrange formula for the remainder. The proof is left to you as an exercise.

COROLLARY 11.5.2 LAGRANGE FORMULA FOR THE REMAINDER

Suppose that f has $n+1$ continuous derivatives on an open interval I that contains 0. Let $x \in I$ and let $P_n(x)$ be the nth Taylor polynomial for f. Then

$$R_n(x) = \frac{f^{(n+1)}(c)}{(n+1)!}x^{n+1}$$

where c is some number between 0 and x.

Remark If we rewrite Taylor's theorem using the Lagrange formula for the remainder, we have

$$f(x) = f(0) + f'(0)x + \frac{f''(0)}{2!}x^2 + \cdots + \frac{f^{(n)}(0)}{n!}x^n + \frac{f^{(n+1)}(c)}{(n+1)!}x^{(n+1)}$$

where c is some number between 0 and x. This result is an extension of the mean-value theorem (Theorem 4.1.1), for if we let $n = 0$ and take $x > 0$, we get

$$f(x) = f(0) + f'(c)x, \qquad \text{which can be written} \qquad f(x) - f(0) = f'(c)(x-0)$$

where $c \in (0, x)$. This is the mean-value theorem for f on the interval $[0, x]$. ❑

The following estimate for $R_n(x)$ is an immediate consequence of Corollary 11.5.2:

(11.5.3)
$$|R_n(x)| \le \left(\max_{t \in J} \left|f^{(n+1)}(t)\right|\right)\frac{|x|^{n+1}}{(n+1)!}$$

where J is the interval joining 0 to x.

Example 3 The Taylor polynomials of the exponential function $f(x) = e^x$ take the form

$$P_n(x) = 1 + x + \frac{x^2}{2!} + \cdots + \frac{x^n}{n!}. \qquad \text{(Example 1)}$$

We will show with our remainder estimate that for each real x

$$R_n(x) \to 0 \quad \text{as} \quad n \to \infty,$$

and therefore we can approximate e^x as closely as we wish by Taylor polynomials.

We begin by fixing x and letting M be the maximum value of the exponential function on the interval J that joins 0 to x. (If $x > 0$, then $M = e^x$; if $x < 0$, $M = e^0 = 1$.) Since

$$f^{(n+1)}(t) = e^t \quad \text{for all } n,$$

we have

$$\max_{t \in J} |f^{(n+1)}(t)| = M \quad \text{for all } n.$$

Thus, by (11.5.3),

$$|R_n(x)| \leq M \frac{|x|^{n+1}}{(n+1)!}.$$

By (10.4.4), we know that

$$\frac{|x|^{n+1}}{(n+1)!} \to 0 \quad \text{as } n \to \infty.$$

It follows then that $R_n(x) \to 0$ as asserted. ❑

Example 4 We return to the sine function $f(x) = \sin x$ and its Taylor polynomials

$$P_1(x) = P_2(x) = x$$

$$P_3(x) = P_4(x) = x - \frac{x^3}{3!}$$

$$P_5(x) = P_6(x) = x - \frac{x^3}{3!} + \frac{x^5}{5!}, \quad \text{and so on.}$$

The pattern of derivatives was established in Example 2; namely, for all k,

$$f^{(4k)}(x) = \sin x, \quad f^{(4k+1)}(x) = \cos x,$$
$$f^{(4k+2)}(x) = -\sin x, \quad f^{(4k-3)}(x) = -\cos x.$$

Thus, for all n and all real t,

$$|f^{(n+1)}(t)| \leq 1.$$

It follows from our remainder estimate (11.5.3) that

$$|R_n(x)| \leq \frac{|x|^{n+1}}{(n+1)!}.$$

Since

$$\frac{|x|^{n+1}}{(n+1)!} \to 0 \quad \text{for all real } x,$$

we see that $R_n(x) \to 0$ for all real x. Thus the sequence of Taylor polynomials converges to the sine function and therefore can be used to approximate $\sin x$ for any real number x with as much accuracy as we wish. ❑

Taylor Series in x

By definition $0! = 1$. Adopting the convention that $f^{(0)} = f$, we can write Taylor polynomials

$$P_n(x) = f(0) + f'(0)\,x + \frac{f''(0)}{2!}x^2 + \cdots + \frac{f^{(n)}(0)}{n!}x^n$$

in $\sum$ notation:

$$P_n(x) = \sum_{k=0}^{n} \frac{f^{(k)}(0)}{k!}x^k.$$

If f is infinitely differentiable on an open interval I containing 0, then we have

$$f(x) = \sum_{k=0}^{n} \frac{f^{(k)}(0)}{k!}x^k + R_n(x), \quad x \in I,$$

for all positive integers n. If, as in the case of the exponential function and the sine function, $R_n(x) \to 0$ as $n \to \infty$ for each $x \in I$, then

$$\sum_{k=0}^{n} \frac{f^{(k)}(0)}{k!}x^k \to f(x).$$

In this case, we say that $f(x)$ can be expanded as a *Taylor series in* x and write

(11.5.4)
$$f(x) = \sum_{k=0}^{\infty} \frac{f^{(k)}(0)}{k!}x^k.$$

Taylor series in x are sometimes called Maclaurin series after Colin Maclaurin, a Scottish mathematician (1698–1746). In some circles the name Maclaurin remains attached to these series, although Taylor considered them some twenty years before Maclaurin.

It follows from Example 3 that

(11.5.5)
$$e^x = \sum_{k=0}^{\infty} \frac{x^k}{k!} = 1 + x + \frac{x^2}{2!} + \frac{x^3}{3!} + \cdots \text{ for all real } x.$$

From Example 4 we have

(11.5.6)
$$\sin x = \sum_{k=0}^{\infty} \frac{(-1)^k}{(2k+1)!}x^{2k+1} = x - \frac{x^3}{3!} + \frac{x^5}{5!} - \frac{x^7}{7!} + \cdots \text{ for all real } x.$$

Note that $\sin 1 = 1 - \frac{1}{3!} + \frac{1}{5!} - \frac{1}{7!} + \cdots$ as suggested in Section 11.4.

We leave it to you as an exercise to show that

(11.5.7)
$$\cos x = \sum_{k=0}^{\infty} \frac{(-1)^k}{(2k)!}x^{2k} = 1 - \frac{x^2}{2!} + \frac{x^4}{4!} - \frac{x^6}{6!} + \cdots \text{ for all real } x.$$

We come now to the logarithm function. Since $\ln x$ is not defined at $x = 0$, we cannot expand $\ln x$ in powers of x. We work instead with $\ln(1 + x)$.

(11.5.8) $$\ln(1+x) = \sum_{k=1}^{\infty} \frac{(-1)^{k+1}}{k} x^k = x - \frac{x^2}{2} + \frac{x^3}{3} - \cdots \quad \text{for } -1 < x \leq 1.$$

PROOF† The function

$$f(x) = \ln(1+x)$$

is defined on $(-1, \infty)$ and has derivatives

$$f'(x) = \frac{1}{1+x}, \quad f''(x) = -\frac{1}{(1+x)^2}, \quad f'''(x) = \frac{2}{(1+x)^3},$$

$$f^{(4)}(x) = -\frac{3!}{(1+x)^4}, \quad f^{(5)}(x) = \frac{4!}{(1+x)^5}, \quad \text{and so on.}$$

For $k \geq 1$,

$$f^{(k)}(x) = (-1)^{k+1}\frac{(k-1)!}{(1+x)^k}, \quad f^{(k)}(0) = (-1)^{k+1}(k-1)!, \quad \frac{f^{(k)}(0)}{k!} = \frac{(-1)^{k+1}}{k}.$$

Since $f(0) = 0$, the nth Taylor polynomial takes the form

$$P_n(x) = \sum_{k=1}^{n} (-1)^{k+1}\frac{x^k}{k} = x - \frac{x^2}{2} + \cdots + (-1)^{n+1}\frac{x^n}{n}.$$

Therefore, all we have to show is that

$$R_n(x) \to 0 \quad \text{for } -1 < x \leq 1.$$

Instead of trying to apply our usual remainder estimate [in this case, that estimate is not delicate enough to show that $R_n(x) \to 0$ for $-1 < x < -\frac{1}{2}$], we keep the remainder in its integral form. From Taylor's theorem

$$R_n(x) = \frac{1}{n!}\int_0^x f^{(n+1)}(t)(x-t)^n\,dt,$$

so that in this case

$$R_n(x) = \frac{1}{n!}\int_0^x (-1)^{n+2}\frac{n!}{(1+t)^{n+1}}(x-t)^n dt = (-1)^n \int_0^x \frac{(x-t)^n}{(1+t)^{n+1}}\,dt.$$

For $0 \leq x \leq 1$ we have

$$|R_n(x)| = \int_0^x \frac{(x-t)^n}{(1+t)^{n+1}}\,dt \underset{\uparrow\text{explain}}{\leq} \int_0^x (x-t)^n\,dt = \frac{x^{n+1}}{n+1} \to 0.$$

† The proof we give here illustrates the methods of this section. A much simpler way of obtaining this series expansion is given in Section 11.8

For $-1 < x < 0$ we have

$$|R_n(x)| = \left| \int_0^x \frac{(x-t)^n}{(1+t)^{n+1}} dt \right| = \int_x^0 \left(\frac{t-x}{1+t}\right)^n \frac{1}{1+t} dt.$$

By the first mean-value theorem for integrals (Theorem 5.9.1), there exists a number x_n between x and 0 such that

$$\int_x^0 \left(\frac{t-x}{1+t}\right)^n \frac{1}{1+t} dt = \left(\frac{x_n - x}{1+x_n}\right)^n \left(\frac{1}{1+x_n}\right)(-x).$$

Since $-x = |x|$ and $0 < 1 + x < 1 + x_n$, we can conclude that

$$|R_n(x)| < \left(\frac{x_n + |x|}{1+x_n}\right)^n \left(\frac{|x|}{1+x}\right).$$

Since $|x| < 1$ and $x_n < 0$, we have

$$x_n < |x|x_n, \qquad x_n + |x| < |x|x_n + |x| = |x|(1+x_n),$$

and thus

$$\frac{x_n + |x|}{1+x_n} < |x|.$$

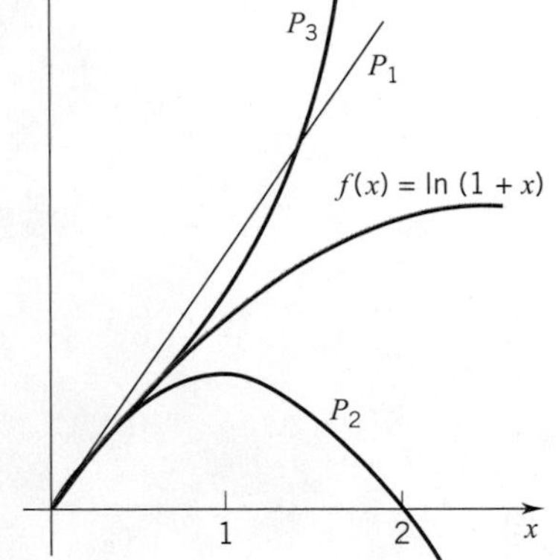

Figure 11.5.3

It follows that

$$|R_n(x)| < |x|^n \left(\frac{|x|}{1+x}\right)$$

and, since $|x| < 1$, that $R_n(x) \to 0$ as $n \to \infty$. □

Remark The series expansion for $\ln(1+x)$ that we have just verified for $-1 < x \le 1$ cannot be extended to other values of x. For $x \le -1$ neither side makes sense: $\ln(1+x)$ is not defined, and the series on the right diverges. For $x > 1$, $\ln(1+x)$ is defined, but the series on the right diverges and hence does not represent the function. At $x = 1$, the series gives the intriguing result that was mentioned in Section 11.4:

$$\ln 2 = 1 - \tfrac{1}{2} + \tfrac{1}{3} - \tfrac{1}{4} + \cdots.$$

The graph of $f(x) = \ln(1+x)$ and the graphs of the first three approximating polynomials are shown in Figure 11.5.3. □

We want to emphasize again the role played by the remainder term $R_n(x)$. We can form a Taylor series

$$\sum_{k=0}^{\infty} \frac{f^{(k)}(0)}{k!} x^k$$

for any function f with derivatives of all orders at $x = 0$, but such a series need not converge at any number $x \neq 0$. Even if it does converge, the sum need not be $f(x)$. (See Exercise 63.) The Taylor series converges to $f(x)$ iff the remainder term $R_n(x)$ tends to 0.

Some Numerical Calculations

If the Taylor series converges to $f(x)$, we can use the partial sums (the Taylor polynomials) to calculate $f(x)$ as accurately as we wish. In what follows we show some sample calculations. For ready reference we list some values of $k!$ in Table 11.5.1.

■ **Table 11.5.1**

		k!
2!	=	2
3!	=	6
4!	=	24
5!	=	120
6!	=	720
7!	=	5,040
8!	=	40,320

Example 5 Determine the maximum possible error we incur by using $P_6(x)$ to approximate $f(x) = e^x$ for x in the interval [0,1].

SOLUTION For all x,

$$e^x = 1 + x + \frac{x^2}{2!} + \frac{x^3}{3!} + \cdots + \frac{x^n}{n!} + \cdots.$$

Now fix any x in the interval [0,1]. From Example 3, we have

$$P_6(x) = 1 + x + \frac{x^2}{2!} + \frac{x^3}{3!} + \frac{x^4}{4!} + \frac{x^5}{5!} + \frac{x^6}{6!}$$

and

$$|R_6(x)| \le \max_{0\le t\le x} |f^{(7)}(t)|\frac{|x|^7}{7!} \le \max_{0\le t\le x} e^t\frac{1}{7!} \le \underbrace{e^x\frac{1}{7!} < \frac{e}{7!}}_{e^x < e} < \underbrace{\frac{3}{7!}}_{e<3} = \frac{1}{1680} < 0.0006.$$

The maximum possible error we can incur by using $P_6(x)$ to approximate e^x on [0, 1] is less than 0.0006. In particular, we can be sure that

$$P_6(1) = 1 + 1 + \frac{1}{2!} + \frac{1}{3!} + \frac{1}{4!} + \frac{1}{5!} + \frac{1}{6!} = \frac{1957}{720}$$

differs from e by less than 0.0006. Our calculator gives $\frac{1957}{720} \cong 2.7180556$ and $e \cong 2.7182818$. This difference is actually less than 0.0002. □

Example 6 Give an estimate for $e^{0.2}$ correct to three decimal places (that is, remainder less than 0.0005).

SOLUTION From Example 3 we know that the nth Taylor polynomial of e^x evaluated at $x = 0.2 = \frac{1}{5}$,

$$P_n(0.2) = 1 + 0.2 + \frac{(0.2)^2}{2!} + \frac{(0.2)^3}{3!} + \cdots + \frac{(0.2)^n}{n!},$$

approximates $e^{0.2}$ to within

$$|R_n(x)(0.2)| \le e^{0.2}\frac{|0.2|^{n+1}}{(n+1)!} < e\frac{(0.2)^{n+1}}{(n+1)!} < \underbrace{3}_{e<3}\frac{1}{5^{n+1}(n+1)!}.$$

We want

$$\frac{3}{5^{n+1}(n+1)!} < 0.0005,$$

which is equivalent to

$$5^{n+1}(n+1)! > 6000.$$

The smallest integer that satisfies this inequality is $n = 3$ $[5^4(4!) = 15,000]$. Thus,

$$P_3(0.2) = 1 + 0.2 + \frac{(0.2)^2}{2!} + \frac{(0.2)^3}{3!} = \frac{7.328}{6} \cong 1.22133$$

differs from $e^{0.2}$ by less than 0.0005. Our calculator gives $e^{0.2} \cong 1.2214028$, so in fact our estimate differs by less than 0.00008. □

Example 7 Estimate sin 0.5 to within 0.001.

SOLUTION At $x = 0.5 = \frac{1}{2}$, the sine series gives

$$\sin 0.5 = 0.5 - \frac{(0.5)^3}{3!} + \frac{(0.5)^5}{5!} - \frac{(0.5)^7}{7!} + \cdots.$$

From Example 4, we know that $P_n(0.5)$ approximates sin 0.5 to within

$$|R_n(0.5)| \le \frac{(0.5)^{n+1}}{(n+1)!} = \frac{1}{2^{n+1}(n+1)!}.$$

Now, $$\frac{1}{2^{n+1}(n+1)!} < 0.001 \quad \text{if} \quad 2^{(n+1)}(n+1)! > 1000.$$

The smallest integer that satisfies this inequality is $n = 4$ $[2^5(5!) = 3840]$. Thus,

$$\underbrace{P_4(0.5)}_{\text{the coefficient of } x^4 \text{ is } 0} = P_3(0.5) = 0.5 - \frac{(0.5)^3}{3!} = \frac{23}{48}$$

approximates sin 0.5 to within 0.001.

Our calculator gives

$$\frac{23}{48} \cong 0.4791667 \qquad \text{and} \qquad \sin 0.5 \cong 0.4794255. \quad □$$

Remark We could have solved the last problem without reference to the remainder estimate derived in Example 4. The series for sin 0.5 is a convergent alternating series with decreasing terms. By (11.4.4) we can conclude immediately that sin 0.5 lies between every two consecutive partial sums. In particular

$$0.4791667 \cong 0.5 - \frac{(0.5)^3}{3!} < \sin 0.5 < 0.5 - \frac{(0.5)^3}{3!} + \frac{(0.5)^5}{5!} \cong 0.4794271. \quad □$$

Example 8 Estimate ln 1.4 to within 0.01.

SOLUTION By (11.5.8),

$$\ln 1.4 = \ln(1 + 0.4) = 0.4 - \tfrac{1}{2}(0.4)^2 + \tfrac{1}{3}(0.4)^3 - \tfrac{1}{4}(0.4)^4 + \cdots.$$

This is a convergent alternating series with decreasing terms. Therefore ln 1.4 lies between every two consecutive partial sums.

The first term less than 0.01 is

$$\tfrac{1}{4}(0.4)^4 = \tfrac{1}{4}(0.0256) = 0.0064.$$

The relation

$$0.4 - \tfrac{1}{2}(0.4)^2 + \tfrac{1}{3}(0.4)^3 - \tfrac{1}{4}(0.4)^4 < \ln 1.4 < 0.4 - \tfrac{1}{2}(0.4)^2 + \tfrac{1}{3}(0.4)^3$$

gives $$0.335 < \ln 1.4 < 0.341.$$

Within the prescribed limits of accuracy, we can take $\ln 1.4 \cong 0.34$. □†

†A much more effective tool for computing logarithms is given in the Exercises.

EXERCISES 11.5

Find the Taylor polynomial P_4 for the given function.

1. $f(x) = x - \cos x$. **2.** $f(x) = \sqrt{1+x}$.

3. $f(x) = \ln \cos x$. **4.** $f(x) = \sec x$.

Find the Taylor polynomial P_5 for the given function.

5. $f(x) = (1+x)^{-1}$. **6.** $f(x) = e^x \sin x$.

7. $f(x) = \tan x$. **8.** $f(x) = x \cos x^2$.

9. Determine P_0, P_1, P_2, P_3 for $f(x) = 1 - x + 3x^2 + 5x^3$.

10. Determine P_0, P_1, P_2, P_3 for $f(x) = (x+1)^3$.

Determine the nth Taylor polynomial P_n for the function f.

11. $f(x) = e^{-x}$. **12.** $f(x) = \sinh x$.

13. $f(x) = \cosh x$. **14.** $f(x) = \ln(1-x)$.

15. $f(x) = e^{rx}$, r a real number.

16. $f(x) = \cos bx$, b a real number.

In Exercises 17–20, assume that f is a function such that $|f^{(n)}(x)| \le 1$ for all n and x (the sine and cosine functions have this property).

17. Estimate the error if $P_5(1/2)$ is used to approximate $f(1/2)$.

18. Estimate the error if $P_7(-2)$ is used to approximate $f(-2)$.

19. Find the least integer n for which $P_n(2)$ approximates $f(2)$ to within 0.001.

20. Find the least integer n for which $P_n(-4)$ approximates $f(-4)$ to within 0.001.

In Exercises 21–24, assume that f is a function such that $|f^{(n)}(x)| \le 3$ for all n and x.

21. Find the least integer n for which $P_n(1/2)$ approximates $f(1/2)$ with four decimal place accuracy.

22. Find the least integer n for which $P_n(2)$ approximates $f(2)$ with three decimal place accuracy.

23. Find the values of x such that the error in the approximation of f by P_5 will be less than 0.05.

24. Find the values of x such that the error in the approximation of f by P_9 will be less than 0.05.

Use Taylor polynomials to estimate the following to within 0.01.

25. $\sqrt{e}$ **26.** $\sin 0.3$.

27. $\sin 1$. **28.** $\ln 1.2$.

29. $\cos 1$. **30.** $e^{0.8}$.

31. $\sin 10°$ **32.** $\cos 6°$.

Find the Lagrange form of the remainder R_n for the given function and the indicated integer n.

33. $f(x) = e^{2x}$; $n = 4$.

34. $f(x) = \ln(1+x)$; $n = 5$.

35. $f(x) = \cos 2x$; $n = 4$. **36.** $f(x) = \sqrt{x+1}$; $n = 3$.

37. $f(x) = \tan x$; $n = 2$. **38.** $f(x) = \sin x$; $n = 5$.

39. $f(x) = \tan^{-1} x$; $n = 2$. **40.** $f(x) = \dfrac{1}{1+x}$; $n = 4$.

Write the remainder R_n in Lagrange form.

41. $f(x) = e^{-x}$. **42.** $f(x) = \sin 2x$.

43. $f(x) = \dfrac{1}{1-x}$. **44.** $f(x) = \ln(1+x)$.

45. Let P_n be the nth Taylor polynomial of the function

$$f(x) = \ln(1+x).$$

Find the least integer n for which: (a) $P_n(0.5)$ approximates $\ln 1.5$ to within 0.01; (b) $P_n(0.3)$ approximates $\ln 1.3$ to within 0.01; (c) $P_n(1)$ approximates $\ln 2$ to within 0.001.

46. Let P_n be the nth Taylor polynomial of the function

$$f(x) = \sin x.$$

Find the least integer n for which: (a) $P_n(1)$ approximates $\sin 1$ to within 0.001; (b) $P_n(2)$ approximates $\sin 2$ to within 0.001. (c) $P_n(3)$ approximate $\sin 3$ to within 0.001.

C **47.** Let $f(x) = e^x$.

(a) Find the Taylor polynomial P_n of f of least degree that approximates $\sqrt{e}$ with four decimal place accuracy. Then evaluate that polynomial to obtain an approximation of $\sqrt{e}$.

(b) Find the Taylor polynomial P_n of f of least degree that approximates $1/e$ with three decimal place accuracy. Then evaluate that polynomial to obtain an approximation of $1/e$.

48. Let $g(x) = \cos x$.

(a) Find the Taylor polynomial P_n of g of least degree that approximates $\cos(\pi/30)$ with three decimal place accuracy. Then evaluate that polynomial to obtain an approximation of $\cos(\pi/30)$.

(b) Find the Taylor polynomial P_n of g of least degree that approximates $\cos 9°$ with four decimal place accuracy. Then evaluate that polynomial to obtain an approximation of $\cos 9°$. (Remember to convert to radian measure.)

49. Show that a polynomial $P(x) = a_0 + a_1 x + \cdots + a_n x^n$ is its own Taylor series.

50. Show that

$$\cos x = \sum_{k=0}^{\infty} \frac{(-1)^k}{(2k)!} x^{2k} \quad \text{for all real } x.$$

51. Show that

$$\sinh x = \sum_{k=0}^{\infty} \frac{1}{(2k+1)!} x^{2k+1} \quad \text{for all real } x.$$

52. Show that

$$\cosh x = \sum_{k=0}^{\infty} \frac{1}{(2k)!} x^{2k} \quad \text{for all real } x.$$

In Exercises 53–57, derive a series expansion in x for the given function and specify the numbers x for which the expansion is valid. Take $a > 0$.

53. $f(x) = e^{ax}$

54. $f(x) = \sin ax$.

55. $f(x) = \cos ax$.

56. $f(x) = \ln(1 - ax)$.

57. $f(x) = \ln(a + x)$.

58. The series we derived for $\ln(1 + x)$ converges too slowly to be of much practical use. The following logarithm series converges much more quickly:

(11.5.9)
$$\ln\left(\frac{1+x}{1-x}\right) = 2\left(x + \frac{x^3}{3} + \frac{x^5}{5} + \cdots\right) \quad \text{for } -1 < x < 1.$$

Derive this series expansion.

59. Set $x = \frac{1}{3}$ and use the first three nonzero terms of (11.5.9) to estimate ln 2.

60. Use the first two nonzero terms of (11.5.9) to estimate ln 1.4.

61. Verify the identify

$$\frac{f^{(k)}(0)}{k!} x^k = \frac{1}{(k-1)!} \int_0^x f^{(k)}(t)(x-t)^{k-1}\, dt - \frac{1}{k!} \int_0^x f^{(k+1)}(t)(x-t)^k\, dt$$

by using integration by parts on the second integral.

62. Prove Corollary 11.5.2 and then derive the remainder estimate (11.5.3).

63. (a) Use a graphing utility to draw the graph of the function

$$f(x) = \begin{cases} e^{-1/x^2}, & x \neq 0. \\ 0, & x = 0. \end{cases}$$

(b) Use L' Hôpital's rule to show that for every positive integer n,

$$\lim_{x \to 0} \frac{e^{-1/x^2}}{x^n} = 0.$$

(c) Use mathematical induction to prove that $f^{(n)}(0) = 0$ for all $n \geq 1$.

(d) What is the Taylor series of f?

(e) For what values of x does the Taylor series of f actually give $f(x)$?

64. Let $f(x) = \cos x$. Use a graphing utility to graph the Taylor polynomials P_2, P_4, P_6, and P_8 of f.

65. Let $g(x) = \ln(1 + x)$. Use a graphing utility to graph the Taylor polynomials P_2, P_3, P_4, and , P_5 of g.

66. Let $f(x) = \sqrt{1 + x}$.

(a) Use a CAS to determine the Taylor series for f and give the maximal interval on which the series converges.

(b) Using a graphing utility, graph f and P_1, P_2, P_4, P_6 together. On what interval does it appear that the polynomials are converging to f?

67. Repeat Exercise 66 with $f(x) = 3^x$.

Show that e is irrational by following these steps.

68. (1) Take the expansion

$$e = \sum_{k=0}^{\infty} \frac{1}{k!}$$

and show that the qth partial sum

$$S_q = \sum_{k=0}^{q} \frac{1}{k!}$$

satisfies the inequality

$$0 < q!(e - s_q) < \frac{1}{q}.$$

(2) Show that $q!s_q$ is an integer and argue that, if e were of the form p/q, then $q!(e - s_q)$ would be a positive integer less than 1.

■ 11.6 TAYLOR POLYNOMIALS AND TAYLOR SERIES IN $x-a$

So far we have encountered series expansions only in powers of x. Here we generalize to expansions in powers of $x - a$, where a is an arbitrary real number. We begin with a more general version of Taylor's theorem.

THEOREM 11.6.1 TAYLOR'S THEOREM

If g has $n + 1$ continuous derivatives on an open interval I that contains the point a, then for each $x \in I$,

$$g(x) = g(a) + g'(a)(x-a) + \frac{g''(a)}{2!}(x-a)^2 + \cdots + \frac{g^{(n)}(a)}{n!}(x-a)^n + R_n(x)$$

where

$$R_n(x) = \frac{1}{n!}\int_a^x g^{(n+1)}(t)(x-t)^n\,dt.$$

The polynomial

$$P_n(x) = g(a) + g'(a)(x-a) + \frac{g''(a)}{2!}(x-a)^2 + \cdots + \frac{g^{(n)}(a)}{n!}(x-a)^n$$

is called the *nth Taylor polynomial for g in powers of x−a.* In this more general setting, the *Lagrange formula for the remainder,* $R_n(x)$, takes the form

(11.6.2) $$R_n(x) = \frac{g^{(n+1)}(c)}{(n+1)!}(x-a)^{n+1}$$

where c is some number between a and x.

Now let $x \in I$, $x \neq a$, and let J be the interval that joins a to x. Then

(11.6.3) $$|R_n(x)| \le \left(\max_{t\in J} |g^{(n+1)}(t)|\right)\frac{|x-a|^{n+1}}{(n+1)!}.$$

If $R_n(x) \to 0$, then we have the series representation

$$g(x) = g(a) + g'(a)(x-a) + \frac{g''(a)}{2!}(x-a)^2 + \cdots + \frac{g^{(n)}(a)}{n!}(x-a)^n + \cdots,$$

which, in sigma notation, takes the form

(11.6.4) $$g(x) = \sum_{k=0}^{\infty} \frac{g^{(k)}(a)}{k!}(x-a)^k.$$

This is known as the Taylor expansion of $g(x)$ in powers of $x - a$. The series on the right is called a *Taylor series in* $x - a$.

All this differs from what you saw before only by a translation. Define

$$f(x) = g(x+a).$$

Then $$f^{(k)}(x) = g^{(k)}(x+a) \quad \text{and} \quad f^{(k)}(0) = g^{(k)}(a).$$

The results of this section as stated for g can all be derived by applying the results of Section 11.5 to the function f.

Example 1 Expand $g(x) = 4x^3 - 3x^2 + 5x - 1$ in powers of $x - 2$.

SOLUTION We need to evaluate g and its derivatives at $x = 2$:

$$\begin{aligned} g(x) &= 4x^3 - 3x^2 + 5x - 1, \\ g'(x) &= 12x^2 - 6x + 5, \\ g''(x) &= 24x - 6, \\ g'''(x) &= 24. \end{aligned}$$

All higher derivatives are identically 0.

Substitution gives $g(2) = 29,\ g'(2) = 41,\ g''(2) = 42,\ g'''(2) = 24$, and $g^{(k)}(2) = 0$ for all $k \geq 4$. Thus, from (11.6.4),

$$\begin{aligned} g(x) &= 29 + 41(x-2) + \frac{42}{2!}(x-2)^2 + \frac{24}{3!}(x-2)^3 \\ &= 29 + 41(x-2) + 21(x-2)^2 + 4(x-2)^3. \quad \square \end{aligned}$$

Example 2 Expand $g(x) = x^2 \ln x$ in powers of $x - 1$.

SOLUTION We need to evaluate g and its derivatives at $x = 1$:

$$\begin{aligned} g(x) &= x^2 \ln x, \\ g'(x) &= x + 2x \ln x, \\ g''(x) &= 3 + 2 \ln x, \\ g'''(x) &= 2x^{-1}, \\ g^{(4)}(x) &= -2x^{-2}, \\ g^{(5)}(x) &= (2)(2)x^{-3}, \\ g^{(6)}(x) &= -(2)(2)(3)x^{-4} = -2(3!)x^{-4}, \\ g^{(7)}(x) &= (2)(2)(3)(4)x^{-5} = (2)(4!)x^{-5}, \quad \text{and so on.} \end{aligned}$$

The pattern is now clear: for $k \geq 3$

$$g^{(k)}(x) = (-1)^{k+1} 2(k-3)!\, x^{-k+2}.$$

Evaluation at $x = 1$ gives $g(1) = 0,\ g'(1) = 1,\ g''(1) = 3$ and, for $k \geq 3$,

$$g^{(k)}(1) = (-1)^{k+1}\, 2(k-3)!.$$

The expansion in powers of $x - 1$ can be written

$$\begin{aligned} g(x) &= (x-1) + \frac{3}{2!}(x-1)^2 + \sum_{k=3}^{\infty} \frac{(-1)^{k+1}(2)(k-3)!}{k!}(x-1)^k \\ &= (x-1) + \frac{3}{2}(x-1)^2 + 2\sum_{k=3}^{\infty} \frac{(-1)^{k+1}}{k(k-1)(k-2)}(x-1)^k. \quad \square \end{aligned}$$

Another way to expand $g(x)$ in powers of $x-a$ is to expand $g(t+a)$ in powers of t and then set $t = x-a$. This is the approach we take when the expansion in t is either known to us or is readily available.

Example 3 We can expand $g(x) = e^{x/2}$ in powers of $x-3$ by expanding

$$g(t+3) = e^{(t+3)/2} \qquad \text{in powers of } t.$$

and then setting $t = x-3$.

Note that

$$g(t+3) = e^{3/2}e^{t/2} = e^{3/2}\sum_{k=0}^{\infty}\frac{(t/2)^k}{k!} = e^{3/2}\sum_{k=0}^{\infty}\frac{1}{2^k k!}t^k.$$

exponential series ↑

Setting $t = x-3$, we have

$$g(x) = e^{3/2}\sum_{k=0}^{\infty}\frac{1}{2^k k!}(x-3)^k.$$

Since the expansion of $g(t+3)$ is valid for all real t, the expansion of $g(x)$ is valid for all real x.

Taking this same approach, we can prove that

(11.6.5)

$$\text{for } 0 < x \le 2a, \quad \ln x = \ln a + \frac{1}{a}(x-a) - \frac{1}{2a^2}(x-a)^2 + \frac{1}{3a^3}(x-a)^3 - \cdots.$$

PROOF We will expand $\ln(a+t)$ in powers of t and then set $t = x-a$. In the first place

$$\ln(a+t) = \ln\left[a\left(1+\frac{t}{a}\right)\right] = \ln a + \ln\left(1+\frac{t}{a}\right).$$

From (11.5.8), we know that the expansion

$$\ln\left(1+\frac{t}{a}\right) = \frac{t}{a} - \frac{1}{2}\left(\frac{t}{a}\right)^2 + \frac{1}{3}\left(\frac{t}{a}\right)^3 - \cdots$$

holds if $-1 < t/a \le 1$, that is, for $-a < t \le a$. Adding $\ln a$ to both sides, we have

$$\ln(t+a) = \ln a + \frac{1}{a}t - \frac{1}{2a^2}t^2 + \frac{1}{3a^3}t^3 - \cdots \qquad \text{for } -a < t \le a.$$

Setting $t = x-a$, we find that

$$\ln x = \ln a + \frac{1}{a}(x-a) - \frac{1}{2a^2}(x-a)^2 + \frac{1}{3a^3}(x-a)^3 - \cdots$$

for all x such that $-a < x-a \le a$, that is, for all x such that $0 < x \le 2a$. □

EXERCISES 11.6

Find the Taylor polynomial of the function f for the given values of a and n, and give the Lagrange form of the remainder.

1. $f(x) = \sqrt{x}$; $a = 4$, $n = 3$.
2. $f(x) = \cos x$; $a = \pi/3$, $n = 4$.
3. $f(x) = \sin x$; $a = \pi/4$, $n = 4$.
4. $f(x) = \ln x$; $a = 1$, $n = 5$.
5. $f(x) = \tan^{-1} x$; $a = 1$, $n = 3$.
6. $f(x) = \cos \pi x$; $a = \frac{1}{2}$, $n = 4$.

Expand $g(x)$ as indicated and specify the values of x for which the expansion is valid.

7. $g(x) = 3x^3 - 2x^2 + 4x + 1$ in powers of $x - 1$.
8. $g(x) = x^4 - x^3 + x^2 - x + 1$ in powers of $x - 2$.
9. $g(x) = 2x^5 + x^2 - 3x - 5$ in powers of $x + 1$.
10. $g(x) = x^{-1}$ in powers of $x - 1$.
11. $g(x) = (1 + x)^{-1}$ in powers of $x - 1$.
12. $g(x) = (b + x)^{-1}$ in powers of $x - a$, $a \neq -b$.
13. $g(x) = (1 - 2x)^{-1}$ in powers of $x + 2$.
14. $g(x) = e^{-4x}$ in powers of $x + 1$.
15. $g(x) = \sin x$ in powers of $x - \pi$.
16. $g(x) = \sin x$ in powers of $x - \frac{1}{2}\pi$.
17. $g(x) = \cos x$ in powers of $x - \pi$.
18. $g(x) = \cos x$ in powers of $x - \frac{1}{2}\pi$.
19. $g(x) = \sin \frac{1}{2}\pi x$ in powers of $x - 1$.
20. $g(x) = \sin \pi x$ in powers of $x - 1$.
21. $g(x) = \ln (1 + 2x)$ in powers of $x - 1$.
22. $g(x) = \ln (2 + 3x)$ in powers of $x - 4$.

Expand $g(x)$ as indicated.

23. $g(x) = x \ln x$ in powers of $x - 2$.
24. $g(x) = x^2 + e^{3x}$ in powers of $x - 2$.
25. $g(x) = x \sin x$ in powers of x.
26. $g(x) = \ln (x^2)$ in powers of $x - 1$.
27. $g(x) = (1 - 2x)^{-3}$ in powers of $x + 2$.
28. $g(x) = \sin^2 x$ in powers of $x - \frac{1}{2}\pi$.
29. $g(x) = \cos^2 x$ in powers of $x - \pi$.
30. $g(x) = (1 + 2x)^{-4}$ in powers of $x - 2$.
31. $g(x) = x^n$ in powers of $x - 1$.
32. $g(x) = (x - 1)^n$ in powers of x.
33. (a) Expand e^x in powers of $x - a$.
    (b) Use the expansion to show that $e^{x_1 + x_2} = e^{x_1} e^{x_2}$.
    (c) Expand e^{-x} in powers of $x - a$.
34. (a) Expand $\sin x$ and $\cos x$ in powers of $x - a$.
    (b) Show that both series absolutely convergent for all real x.
    (c) As noted earlier (Section 11.4), Riemann proved that the order of the terms of an absolutely convergent series may be changed without altering the sum of the series. Use Riemann's discovery and the Taylor expansions of part (a) to derive the addition formulas

$$\sin (x_1 + x_2) = \sin x_1 \cos x_2 + \cos x_1 \sin x_2,$$

$$\cos (x_1 + x_2) = \cos x_1 \cos x_2 - \sin x_1 \sin x_2.$$

35. (a) Find the Taylor polynomial P_n in powers of $(x - \pi/6)$ of least degree that approximates $\sin 35°$ with four decimal place accuracy.
    (b) Evaluate the polynomial that you found in part (a) to obtain your approximation of $\sin 35°$.
36. (a) Find the Taylor polynomial P_n in powers of $(x - \pi/3)$ of least degree that approximates $\cos 57°$ with four decimal place accuracy.
    (b) Evaluate the polynomial that you found in part (a) to obtain your approximation of $\cos 57°$.
37. Choose an appropriate Taylor polynomial of $f(x) = \sqrt{x}$ to approximate $\sqrt{38}$ with three decimal place accuracy.
38. Choose an appropriate Taylor polynomial for $f(x) = \sqrt{x}$ to approximate $\sqrt{61}$ with three decimal place accuracy.
39. Use a CAS to determine the Taylor polynomial P_6 in powers of $(x - 1)$ for $f(x) = \tan^{-1} x$.
40. Use a CAS to determine the Taylor polynomial P_8 in powers of $(x - 2)$ for $f(x) = \cosh 2x$.

■ 11.7 POWER SERIES

You have become familiar with Taylor series

$$\sum_{k=0}^{\infty} \frac{f^{(k)}(0)}{k!} x^k \quad \text{and} \quad \sum_{k=0}^{\infty} \frac{f^{(k)}(a)}{k!} (x - a)^k.$$

Here we study series of the form

$$\sum_{k=0}^{\infty} a_k x^k \quad \text{and} \quad \sum_{k=0}^{\infty} a_k (x - a)^k$$

without regard to how the coefficients a_k have been generated. Such series are called *power series*: the first is a *power series in x*; the second is a *power series in $x - a$*.

Since a simple translation converts

$$\sum_{k=0}^{\infty} a_k(x-a)^k \quad \text{into} \quad \sum_{k=0}^{\infty} a_k x^k,$$

we can focus our attention on power series of the form

$$\sum_{k=0}^{\infty} a_k x^k.$$

When detailed indexing is unnecessary, we will omit it and write

$$\sum a_k x^k.$$

DEFINITION 11.7.1

A power series $\sum a_k x^k$ is said to converge

(i) at c if $\sum a_k c^k$ converges;

(ii) on the set S if $\sum a_k x^k$ converges for each $x \in S$.

The following result is fundamental.

THEOREM 11.7.2

If $\sum a_k x^k$ converges at $c \neq 0$, then it converges absolutely for all x such that $|x| < |c|$.

If $\sum a_k x^k$ diverges at d, then it diverges for all x such that $|x| > |d|$.

PROOF If $\sum a_k c^k$ converges, then $a_k c^k \to 0$ as $k \to \infty$. In particular, for k sufficiently large,

$$|a_k c^k| \leq 1,$$

and thus
$$|a_k x^k| = |a_k c^k| \left|\frac{x}{c}\right|^k \leq \left|\frac{x}{c}\right|^k.$$

For $|x| < |c|$, we have

$$\left|\frac{x}{c}\right| < 1.$$

The convergence of $\sum |a_k x^k|$ follows by comparison with the geometric series. This proves the first statement.

Suppose now that $\sum a_k d^k$ diverges. By a similar argument, there cannot exist x with $|x| > |d|$ such that $\sum a_k x^k$ converges. The existence of such an x would imply the absolute convergence of $\sum a_k d^k$. This proves the second statement. ❑

It follows from the theorem we just proved that there are exactly three possibilities for a power series:

Case I. *The series converges only at* $x = 0$. This is what happens with

$$\sum k^k x^k.$$

For $x \neq 0$, $\lim_{k\to\infty} k^k x^k \neq 0$, and so the series cannot converge (Theorem 11.1.6).

Case II. *The series is absolutely convergent for all real numbers* x. This is what happens with the exponential series

$$\sum \frac{x^k}{k!}.$$

Case III. *There exists a positive number* r *such that the series converges for* $|x| < r$ *and diverges for* $|x| > r$. This is what happens with the geometric series

$$\sum x^k.$$

For the geometric series, there is absolute convergence for $|x| < 1$ and divergence for $|x| > 1$.

Associated with each case is a *radius of convergence:*

In Case I, we say that the radius of convergence is 0.
In Case II, we say that the radius of convergence is ∞.
In Case III, we say that the radius of convergence is r.

The three cases are pictured in Figure 11.7.1.

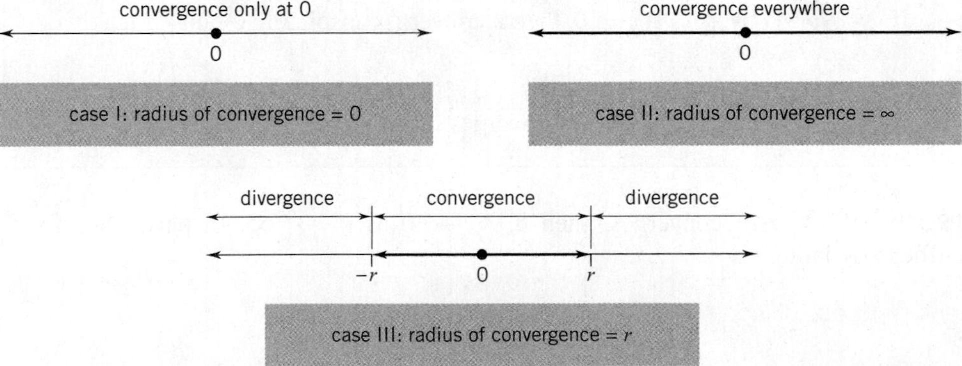

Figure 11.7.1

In general, the behavior of a power series at $-r$ and at r is not predictable. For example, the series

$$\sum x^k, \quad \sum \frac{(-1)^k}{k} x^k, \quad \sum \frac{1}{k} x^k, \quad \sum \frac{1}{k^2} x^k$$

all have radius of convergence 1, but the first series converges only on $(-1, 1)$ the second series converges on $(-1, 1]$, the third on $[-1, 1)$, and the fourth on $[-1, 1]$.

The maximal interval on which a power series converges is called the *interval of convergence.* For a series with infinite radius of convergence, the interval of convergence is $(-\infty, \infty)$. For a series with radius of convergence r, the interval of convergence can be $[-r, r], (-r, r], [-r, r)$, or $(-r, r)$. For a series with radius of convergence 0, the interval of convergence reduces to a point, $\{0\}$.

Example 1 Verify that the series

$$\sum \frac{(-1)^k}{k} x^k \tag{1}$$

has interval of convergence $(-1, 1]$.

SOLUTION First we show that the radius of convergence is 1(that is, the series converges absolutely for $|x| < 1$ and diverges for $|x| > 1$). We do this by forming the series

$$\sum \left| \frac{(-1)^k}{k} x^k \right| = \sum \frac{1}{k} |x|^k \tag{2}$$

and applying the ratio test.

We set

$$b_k = \frac{1}{k} |x|^k$$

and note that

$$\frac{b_{k+1}}{b_k} = \frac{|x|^{k+1}/(k+1)}{|x|^k/k} = \frac{k}{k+1} \frac{|x|^{k+1}}{|x|^k} = \frac{k}{k+1} |x| \to |x| \quad \text{as } k \to \infty.$$

By the ratio test, series (2) converges for $|x| < 1$ and diverges for $|x| > 1$.† It follows that series (1) converges absolutely for $|x| < 1$ and diverges for $|x| > 1$. The radius of convergence is therefore 1.

Now we test the endpoints $x = -1$ and $x = 1$. At $x = -1$,

$$\sum \frac{(-1)^k}{k} x^k \quad \text{becomes} \quad \sum \frac{(-1)^k}{k} (-1)^k = \sum \frac{1}{k}.$$

This is the harmonic series, which, as you know, diverges. At $x = 1$,

$$\sum \frac{(-1)^k}{k} x^k \quad \text{becomes} \quad \sum \frac{(-1)^k}{k}.$$

This is a (conditionally) convergent alternating series.

We have shown that series (1) converges absolutely for $|x| < 1$, diverges at -1, and converges at 1. The interval of convergence is $(-1, 1]$. □

Remark The same arguments can be used to show that the series

$$\sum \frac{1}{k} x^k$$

converges on $[-1, 1)$. □

---

† We could also have used the root test:

$$(b_k)^{1/k} = \left| \frac{1}{k} \right|^{1/k} |x| = \frac{1}{k^{1/k}} |x| \to |x|.$$

Example 2 Verify that the series

$$(1)\qquad \sum \frac{1}{k^2}x^k$$

has interval of convergence $[-1, 1]$.

SOLUTION First we examine the series

$$(2)\qquad \sum \left|\frac{1}{k^2}x^k\right| = \sum \frac{1}{k^2}|x|^k.$$

Here again we use the ratio test. We set

$$b_k = \frac{1}{k^2}|x|^k$$

and note that

$$\frac{b_{k+1}}{b_k} = \frac{k^2}{(k+1)^2}\frac{|x|^{k+1}}{|x|^k} = \left(\frac{k}{k+1}\right)^2 |x| \to |x| \quad \text{as } k \to \infty.$$

By the ratio test, (2) converges for $|x| < 1$ and diverges for $|x| > 1$.† This shows that (1) converges absolutely for $|x| < 1$ and diverges for $|x| > 1$. The radius of convergence is therefore 1.

Now for the endpoints. At $x = -1$,

$$\sum \frac{1}{k^2}x^k \quad \text{takes the form} \quad \sum \frac{(-1)^k}{k^2} = -1 + \tfrac{1}{4} - \tfrac{1}{9} + \tfrac{1}{16} - \cdots.$$

This is an absolutely convergent alternating series. At $x = 1$,

$$\sum \frac{1}{k^2}x^k \quad \text{becomes} \quad \sum \frac{1}{k^2}.$$

This is a convergent p-series. The interval of convergence is therefore the closed interval $[-1, 1]$. □

Example 3 Find the interval of convergence of the series

$$(1)\qquad \sum \frac{k}{6^k}x^k.$$

SOLUTION We begin by examining the series

$$(2)\qquad \sum \left|\frac{k}{6^k}x^k\right| = \sum \frac{k}{6^k}|x|^k.$$

We set

$$b_k = \frac{k}{6^k}|x|^k$$

---

† Once again we could have used the root test:

$$(b_k)^{1/k} = \frac{1}{k^{2/k}}|x| \to |x| \quad \text{as } k \to \infty.$$

and apply the root test. (The ratio test will also work.) Since

$$(b_k)^{1/k} = \tfrac{1}{6}k^{1/k}|x| \to \tfrac{1}{6}|x| \quad \text{as } k \to \infty, \qquad \text{(recall } k^{1/k} \to 1\text{)}$$

you can see that (2) converges

$$\text{for} \quad \tfrac{1}{6}|x| < 1, \qquad \text{that is, for} \quad |x| < 6,$$

and diverges

$$\text{for} \quad \tfrac{1}{6}|x| > 1, \qquad \text{that is, for} \quad |x| > 6.$$

Testing the endpoints, we have:

$$\text{at} \quad x = 6, \qquad \sum \frac{k}{6^k}6^k = \sum k, \quad \text{which is divergent;}$$

$$\text{at} \quad x = -6, \qquad \sum \frac{k}{6^k}(-6)^k = \sum (-1)^k k, \quad \text{which is also divergent.}$$

Thus, the interval of convergence is $(-6, 6)$. □

Example 4 Find the interval of convergence of the series

$$\text{(1)} \qquad \sum \frac{k!}{(3k)!}x^k.$$

SOLUTION Again, we begin by examining the series

$$\text{(2)} \qquad \sum \left|\frac{k!}{(3k)!}x^k\right| = \sum \frac{k!}{(3k)!}|x|^k.$$

Set

$$b_k = \frac{k!}{(3k)!}|x|^k.$$

Since factorials are involved, we will use the ratio test. Note that

$$\frac{b_{k+1}}{b_k} = \frac{(k+1)!}{[3(k+1)]!} \cdot \frac{(3k)!}{k!} \cdot \frac{|x|^{k+1}}{|x|^k} = \frac{k+1}{(3k+3)(3k+2)(3k+1)}|x|$$

$$= \frac{1}{3(3k+2)(3k+1)}|x|.$$

Since

$$\frac{1}{3(3k+2)(3k+1)} \to 0 \quad \text{as} \quad k \to \infty,$$

the ratio b_{k+1}/b_k tends to 0 no matter what x is. By the ratio test, (2) converges for all x and therefore (1) converges absolutely for all x. The radius of convergence is ∞ and the interval of convergence is $(-\infty, \infty)$. □

Example 5 Find the interval of convergence of the series $\sum \frac{k^k}{2^k}x^k$.

SOLUTION We set $b_k = \frac{k^k}{2^k}|x|^k$

and apply the root test. Since $(b_k)^{1/k} = \frac{1}{2}k|x| \to \infty$ as $k \to \infty$ for every $x \neq 0$, the series diverges for all $x \neq 0$; the series converges only at $x = 0$. ❑

Example 6 Find the interval of convergence of the series

$$\sum \frac{(-1)^k}{k^2 3^k}(x+2)^k.$$

SOLUTION We consider the series

$$\sum \left|\frac{(-1)^k}{k^2 3^k}(x+2)^k\right| = \sum \frac{1}{k^2 3^k}|x+2|^k.$$

Set
$$b_k = \frac{1}{k^2 3^k}|x+2|^k$$

and apply the ratio test (the root test will work equally well):

$$\frac{b_{k+1}}{b_k} = \frac{k^2 3^k}{(k+1)^2 3^{k+1}} \cdot \frac{|x+2|^{k+1}}{|x+2|^k} = \frac{k^2}{3(k+1)^2}|x+2| \to \tfrac{1}{3}|x+2| \text{ as } k \to \infty.$$

The series is absolutely convergent for $-5 < x < 1$:

$$\tfrac{1}{3}|x+2| < 1 \quad \text{iff} \quad |x+2| < 3 \quad \text{iff} \quad -5 < x < 1.$$

We now check the endpoints. At $x = -5$,

$$\sum \frac{(-1)^k}{k^2 3^k}(-3)^k = \sum \frac{1}{k^2}.$$

This is a convergent p-series. At $x = 1$,

$$\sum \frac{(-1)^k}{k^2 3^k}(3)^k = \sum \frac{(-1)^k}{k^2}.$$

This is a convergent alternating series. The interval of convergence is the closed interval $[-5, 1]$. ❑

EXERCISES 11.7

1. Suppose that the power series $\sum_{k=0}^{\infty} a_k x^k$ converges at $x = 3$, that is, suppose that $\sum_{k=0}^{\infty} a_k 3^k$ converges. What can you say about the convergence or divergence of the following?

(a) $\sum_{k=0}^{\infty} a_k 2^k$. (b) $\sum_{k=0}^{\infty} a_k (-2)^k$.

(c) $\sum_{k=0}^{\infty} a_k (-3)^k$. (d) $\sum_{k=0}^{\infty} a_k 4^k$.

2. Suppose that the power series $\sum_{k=0}^{\infty} a_k x^k$ converges at $x = -3$ and diverges at $x = 5$. What can you say about the convergence or divergence of the following?

(a) $\sum_{k=0}^{\infty} a_k 2^k$. (b) $\sum_{k=0}^{\infty} a_k (-6)^k$.

(c) $\sum_{k=0}^{\infty} a_k 4^k$. (d) $\sum_{k=0}^{\infty} (-1)^k a_k 3^k$.

Find the interval of convergence.

3. $\sum kx^k$. **4.** $\sum \frac{1}{k}x^k$.

5. $\sum \frac{1}{(2k)!}x^k$. **6.** $\sum \frac{2^k}{k^2}x^k$.

7. $\sum (-k)^{2k} x^{2k}$. **8.** $\sum \frac{(-1)^k}{\sqrt{k}}x^k$.

9. $\sum \frac{1}{k 2^k}x^k$. **10.** $\sum \frac{1}{k^2 2^k}x^k$.

11. $\sum \left(\frac{k}{100}\right)^k x^k$.

12. $\sum \frac{k^2}{1+k^2} x^k$.

13. $\sum \frac{2^k}{\sqrt{k}} x^k$.

14. $\sum \frac{1}{\ln k} x^k$.

15. $\sum \frac{k-1}{k} x^k$.

16. $\sum k a^k x^k$.

17. $\sum \frac{k}{10^k} x^k$.

18. $\sum \frac{3k^2}{e^k} x^k$.

19. $\sum \frac{x^k}{k^k}$.

20. $\sum \frac{7^k}{k!} x^k$.

21. $\sum \frac{(-1)^k}{k^k}(x-2)^k$.

22. $\sum k! x^k$.

23. $\sum (-1)^k \frac{2^k}{3^{k+1}} x^k$.

24. $\sum \frac{2^k}{(2k)!} x^k$.

25. $\sum (-1)^k \frac{k!}{k^3}(x-1)^k$.

26. $\sum \frac{(-e)^k}{k^2} x^k$.

27. $\sum \left(\frac{k}{k-1}\right) \frac{(x+2)^k}{2^k}$.

28. $\sum \frac{\ln k}{k}(x+1)^k$.

29. $\sum (-1)^k \frac{k^2}{(k+1)!}(x+3)^k$.

30. $\sum \frac{k^3}{e^k}(x-4)^k$.

31. $\sum \left(1+\frac{1}{k}\right)^k x^k$.

32. $\sum \frac{(-1)^k a^k}{k^2}(x-a)^k$.

33. $\sum \frac{\ln k}{2^k}(x-2)^k$.

34. $\sum \frac{1}{(\ln k)^k}(x-1)^k$.

35. $\sum (-1)^k \left(\frac{2}{3}\right)^k (x+1)^k$.

36. $\sum \frac{2^{1/k}\pi^k}{k(k+1)(k+2)}(x-2)^k$.

37. $1 - \frac{x}{2} + \frac{2x^2}{4} - \frac{3x^3}{8} + \frac{4x^4}{16} - \cdots$.

38. $\frac{(x-1)}{5^2} + \frac{4}{5^4}(x-1)^2 + \frac{9}{5^6}(x-1)^3 + \frac{16}{5^8}(x-1)^4 + \cdots$.

39. $\frac{3x^2}{4} + \frac{9x^4}{9} + \frac{27x^6}{16} + \frac{81x^8}{25} + \cdots$.

40. $\frac{1}{16}(x+1) - \frac{2}{25}(x+1)^2 + \frac{3}{36}(x+1)^3 - \frac{4}{49}(x+1)^4 + \cdots$.

41. Suppose that the power series $\sum_{k=0}^{\infty} a_k(x-1)^k$ converges at $x = 3$. What can you say about the convergence or divergence of the following?

(a) $\sum_{k=0}^{\infty} a_k$.

(b) $\sum_{k=0}^{\infty} (-1)^k a_k$.

(c) $\sum_{k=0}^{\infty} (-1)^k a_k 2^k$.

42. Suppose that the power series $\sum_{k=0}^{\infty} a_k(x+2)^k$ converges at $x = 4$. At what other values of x must $\sum_{k=0}^{\infty} a_k(x+2)^k$ converge? Does the power series converge at $x = -8$? Explain.

43. Let $\sum a_k x^k$ be a power series with radius of convergence $r > 0$.

(a) Show that if the power series is absolutely convergent at one endpoint of its interval of convergence, then it is absolutely convergent at the other endpoint.

(b) Show that if the interval of convergence is $(-r, r]$, then the power series is conditionally convergent at r.

44. Let $r > 0$ be arbitrary. Give an example of a power series $\sum a_k x^k$ with radius of convergence r.

45. The power series $\sum_{k=0}^{\infty} a_k x^k$ has the property that $a_{k+3} = a_k$ for all $k \geq 0$.

(a) Show that the power series has radius of convergence $r = 1$.

(b) Find an explicit formula in terms of a_0, a_1, a_2 for the sum of the series.

46. Find the interval of convergence of the series $\sum s_k x^k$ where s_k is the kth partial sum of the series

$$\sum_{n=1}^{\infty} \frac{1}{n}.$$

47. Let $\sum a_k x^k$ be a power series, and let r be its radius of convergence.

(a) Given that $|a_k|^{1/k} \to \rho$, show that, if $\rho \neq 0$, then $r = 1/\rho$ and, if $\rho = 0$, then $r = \infty$.

(b) Given that $|a_{k+1}/a_k| \to \lambda$, show that, if $\lambda \neq 0$, then $r = 1/\lambda$ and, if $\lambda = 0$, then $r = \infty$.

48. Let $\sum a_k x^k$ be a power series and let r, $0 < r < \infty$, be its radius of convergence. Prove that the power series $\sum a_k x^{2k}$ has radius of convergence $\sqrt{r}$.

C 49. Use a CAS to implement the ratio test to find the radius of convergence of the power series.

(a) $\sum_{k=0}^{\infty} \frac{2^k}{k!} x^k$.

(b) $\sum_{k=0}^{\infty} \frac{2^k}{k} x^k$.

(c) $\sum_{k=0}^{\infty} (-1)^k \left(\frac{x}{2}\right)^k$.

C 50. Consider the power series $\sum_{k=1}^{\infty} \frac{(-1)^{k+1}}{k} x^k$.

(a) Graph the partial sums $s_2(x), s_4(x), s_6(x), s_8(x), s_{10}(x)$ together.

(b) On the same screen, graph $f(x) = \ln(1+x)$.

(c) On what interval does it appear that $s_n(x)$ may be converging to $f(x)$?

C 51. Repeat Exercise 50 with $\sum_{k=0}^{\infty} x^k$ and $f(x) = \frac{1}{1-x}$.

C 52. Repeat Exercise 50 with $\sum_{k=1}^{\infty} \frac{(-1)^{k+1}}{2k-1} x^{2k-1}$ and $f(x) = \tan^{-1} x$.

■ 11.8 DIFFERENTIATION AND INTEGRATION OF POWER SERIES

Differentiation of Power Series

We begin with a simple but important result.

THEOREM 11.8.1

If

$$\sum_{k=0}^{\infty} a_k x^k = a_0 + a_1 x + a_2 x^2 + \cdots + a_n x^n + \cdots$$

converges on $(-c, c)$, then

$$\sum_{k=0}^{\infty} \frac{d}{dx}(a_k x^k) = \sum_{k=1}^{\infty} k a_k x^{k-1} = a_1 + 2a_2 x + \cdots + n a_n x^{n-1} + \cdots$$

also converges on $(-c, c)$.

PROOF Assume that

$$\sum_{k=0}^{\infty} a_k x^k \quad \text{converges on } (-c, c).$$

By Theorem 11.7.2, it converges absolutely on this interval.

Now let x be some fixed number in $(-c, c)$ and choose $\epsilon > 0$ such that

$$|x| < |x| + \epsilon < c.$$

Since $|x| + \epsilon$ lies within the interval of convergence,

$$\sum_{k=0}^{\infty} |a_k(|x| + \epsilon)^k| \quad \text{converges.}$$

As you are asked to show in Exercise 48, for all k sufficiently large,

$$|k\, x^{k-1}| \le (|x| + \epsilon)^k.$$

It follows that for all such k,

$$|k a_k x^{k-1}| \le |a_k(|x| + \epsilon)^k|.$$

Since

$$\sum_{k=0}^{\infty} |a_k(|x| + \epsilon)^k| \quad \text{converges,}$$

we can conclude that

$$\sum_{k=0}^{\infty} \left| \frac{d}{dx}(a_k x^k) \right| = \sum_{k=1}^{\infty} |k a_k x^{k-1}| \quad \text{converges,}$$

and thus that

$$\sum_{k=0}^{\infty} \frac{d}{dx}(a_k x^k) = \sum_{k=1}^{\infty} k a_k x^{k-1} \quad \text{converges.} \quad \square$$

Repeated application of the theorem shows that

$$\sum_{k=0}^{\infty} \frac{d^2}{dx^2}(a_k x^k), \quad \sum_{k=0}^{\infty} \frac{d^3}{dx^3}(a_k x^k), \quad \sum_{k=0}^{\infty} \frac{d^4}{dx^4}(a_k x^k), \qquad \text{and so on,}$$

all converge on $(-c, c)$.

Example 1 Since the geometric series

$$\sum_{k=0}^{\infty} x^k = 1 + x + x^2 + x^3 + x^4 + x^5 + x^6 + \cdots$$

converges on $(-1, 1)$, the series

$$\sum_{k=0}^{\infty} \frac{d}{dx}(x^k) = \sum_{k=1}^{\infty} kx^{k-1} = 1 + 2x + 3x^2 + 4x^3 + 5x^4 + 6x^5 + \cdots,$$

$$\sum_{k=0}^{\infty} \frac{d^2}{dx^2}(x^k) = \sum_{k=2}^{\infty} k(k-1)x^{k-2} = 2 + 6x + 12x^2 + 20x^3 + 30x^4 + \cdots,$$

$$\sum_{k=0}^{\infty} \frac{d^3}{dx^3}(x^k) = \sum_{k=3}^{\infty} k(k-1)(k-2)x^{k-3} = 6 + 24x + 60x^2 + 120x^3 + \cdots,$$

$$\vdots$$

all converge on $(-1, 1)$. $\square$

Remark Even though $\sum a_k x^k$ and its "derivative" $\sum k a_k x^{k-1}$ have the same radius of convergence, their intervals of convergence may be different due to possible differences in behavior at the endpoints. For example, the interval of convergence of the series

$$\sum_{k=1}^{\infty} \frac{1}{k^2} x^k$$

is $[-1, 1]$, whereas the interval of convergence of its derivative

$$\sum_{k=1}^{\infty} \frac{1}{k} x^{k-1}$$

is $[-1, 1)$. Endpoints must always be checked separately. $\square$

Suppose now that

$$\sum_{k=0}^{\infty} a_k x^k \quad \text{converges on } (-c, c).$$

Then, as we have seen,

$$\sum_{k=0}^{\infty} \frac{d}{dx}(a_k x^k) \quad \text{also converges on } (-c, c).$$

Using the first series, we can define a function f on $(-c, c)$ by setting

$$f(x) = \sum_{k=0}^{\infty} a_k x^k.$$

Using the second series, we can define a function g on $(-c, c)$ by setting

$$g(x) = \sum_{k=0}^{\infty} \frac{d}{dx}(a_k x^k).$$

The crucial point is that

$$f'(x) = g(x).$$

THEOREM 11.8.2 THE DIFFERENTIABILITY THEOREM

If

$$f(x) = \sum_{k=0}^{\infty} a_k x^k \quad \text{for all } x \text{ in } (-c, c),$$

then f is differentiable on $(-c, c)$ and

$$f'(x) = \sum_{k=0}^{\infty} \frac{d}{dx}(a_k x^k) \quad \text{for all } x \text{ in } (-c, c).$$

By applying this theorem to f', you can see that f' is itself differentiable. This in turn implies that f'' is differentiable, and so on. In short, f has derivatives of all orders.

The discussion up to this point can be summarized as follows:

In the interior of its interval of convergence a power series defines an infinitely differentiable function, the derivatives of which can be obtained by differentiating term by term:

$$\frac{d^n}{dx^n}\left(\sum_{k=0}^{\infty} a_k x^k\right) = \sum_{k=0}^{\infty} \frac{d^n}{dx^n}(a_k x^k) \quad \text{for all } n.$$

For a detailed proof of the differentiability theorem, see the supplement at the end of this section. We go on to examples.

Example 2 You know that $\frac{d}{dx}(e^x) = e^x$. You can see this directly by differentiating the exponential series:

$$\frac{d}{dx}(e^x) = \frac{d}{dx}\left(\sum_{k=0}^{\infty}\frac{x^k}{k!}\right) = \sum_{k=0}^{\infty}\frac{d}{dx}\left(\frac{x^k}{k!}\right) = \sum_{k=1}^{\infty}\frac{x^{k-1}}{(k-1)!} = \sum_{n=0}^{\infty}\frac{x^n}{n!} = e^x. \quad \square$$

set $n = k - 1$

Example 3 You have seen that $\sin x = x - \frac{x^3}{3!} + \frac{x^5}{5!} - \frac{x^7}{7!} + \frac{x^9}{9!} - \cdots$

and
$$\cos x = 1 - \frac{x^2}{2!} + \frac{x^4}{4!} - \frac{x^6}{6!} + \frac{x^8}{8!} - \cdots.$$

The relations

$$\frac{d}{dx}(\sin x) = \cos x, \qquad \frac{d}{dx}(\cos x) = -\sin x$$

can be confirmed by differentiating these series term by term:

$$\begin{aligned}\frac{d}{dx}(\sin x) &= 1 - \frac{3x^2}{3!} + \frac{5x^4}{5!} - \frac{7x^6}{7!} + \frac{9x^8}{9!} - \cdots\\ &= 1 - \frac{x^2}{2!} + \frac{x^4}{4!} - \frac{x^6}{6!} + \frac{x^8}{8!} - \cdots = \cos x,\\ \frac{d}{dx}(\cos x) &= -\frac{2x}{2!} + \frac{4x^3}{4!} - \frac{6x^5}{6!} + \frac{8x^7}{8!} - \cdots\\ &= -x + \frac{x^3}{3!} - \frac{x^5}{5!} + \frac{x^7}{7!} - \cdots\\ &= -\left(x - \frac{x^3}{3!} + \frac{x^5}{5!} - \frac{x^7}{7!} + \cdots\right) = -\sin x. \quad \square\end{aligned}$$

Example 4 We can sum the series $\sum_{k=1}^{\infty}\frac{x^k}{k}$ for all x in $(-1, 1)$ by setting

$$g(x) = \sum_{k=1}^{\infty}\frac{x^k}{k} \quad \text{for all } x \text{ in } (-1, 1)$$

and noting that

$$g'(x) = \sum_{k=1}^{\infty}\frac{kx^{k-1}}{k} = \sum_{k=1}^{\infty}x^{k-1} = \sum_{n=0}^{\infty}x^n = \frac{1}{1-x}.$$

the geometric series

With
$$g'(x) = \frac{1}{1-x} \quad \text{and} \quad g(0) = 0,$$

we can conclude that

$$g(x) = -\ln(1-x) = \ln\left(\frac{1}{1-x}\right).$$

It follows that

$$\sum_{k=1}^{\infty} \frac{x^k}{k} = \ln\left(\frac{1}{1-x}\right) \qquad \text{for all } x \text{ in } (-1, 1). \quad \square$$

Integration of Power Series

Power series can also be integrated term by term.

THEOREM 11.8.3 TERM-BY-TERM INTEGRATION

If $f(x) = \sum_{k=0}^{\infty} a_k x^k$ converges on $(-c, c)$, then

$$g(x) = \sum_{k=0}^{\infty} \frac{a_k}{k+1} x^{k+1} \text{ converges on } (-c, c) \text{ and } \int f(x)\,dx = g(x) + C.$$

PROOF If $\sum_{k=0}^{\infty} a_k x^k$ converges on $(-c, c)$, then $\sum_{k=0}^{\infty} |a_k x^k|$ converges on $(-c, c)$ (Theorem 11.7.2). Since

$$\left|\frac{a_k}{k+1} x^k\right| \le |a_k x^k| \qquad \text{for all } k,$$

we know by comparison that

$$\sum_{k=0}^{\infty} \left|\frac{a_k}{k+1} x^k\right| \qquad \text{also converges on } (-c, c).$$

It follows that

$$x \sum_{k=0}^{\infty} \frac{a_k}{k+1} x^k = \sum_{k=0}^{\infty} \frac{a_k}{k+1} x^{k+1} \qquad \text{converges on } (-c, c).$$

With

$$f(x) = \sum_{k=0}^{\infty} a_k x^k \quad \text{and} \quad g(x) = \sum_{k=0}^{\infty} \frac{a_k}{k+1} x^{k+1},$$

we know from the differentiability theorem that

$$g'(x) = f(x) \quad \text{and therefore} \quad \int f(x)\,dx = g(x) + C. \quad \square$$

Term-by-term integration can be expressed as follows:

(11.8.4)
$$\int \left(\sum_{k=0}^{\infty} a_k x^k\right) dx = \left(\sum_{k=0}^{\infty} \frac{a_k}{k+1} x^{k+1}\right) + C.$$

Remark If $\sum a_k x^k$ has radius of convergence r, then its "integral" $\sum [a_k/(k+1)]x^k$ also has radius of convergence r. As in the case of differentiating power series, convergence of the "integral" at the endpoints must be tested separately. □

If a power series converges at c and converges at d, then it converges at all numbers between c and d, and

$$\int_c^d \left(\sum_{k=0}^{\infty} a_k x^k\right) dx = \sum_{k=0}^{\infty}\left(\int_c^d a_k x^k dx\right) = \sum_{k=0}^{\infty} \frac{a_k}{k+1}(d^{k+1} - c^{k+1}). \tag{11.8.5}$$

Example 5 You are familiar with the series expansion

$$\frac{1}{1+x} = \frac{1}{1-(-x)} = \sum_{k=0}^{\infty} (-1)^k x^k.$$

It is valid for all x in $(-1, 1)$ and for no other x. Integrating term by term we have

$$\ln(1+x) = \int \left(\sum_{k=0}^{\infty} (-1)^k x^k\right) dx = \left(\sum_{k=0}^{\infty} \frac{(-1)^k}{k+1} x^{k+1}\right) + C$$

for all x in $(-1, 1)$. At $x = 0$ both $\ln(1+x)$ and the series on the right are 0. It follows that $C = 0$ and thus

$$\ln(1+x) = \sum_{k+1}^{\infty} \frac{(-1)^k}{k+1} x^{k+1} = x - \frac{x^2}{2} + \frac{x^3}{3} - \frac{x^4}{4} + \cdots$$

for all x in $(-1, 1)$. □

In Section 11.5 we were able to prove that this expansion for $\ln(1+x)$ was valid on the half-closed interval $(-1, 1]$; this gave us an expansion for ln 2. Term-by-term integration gives us only the open interval $(-1, 1)$. Well, you may say, it's easy to see that the logarithm series also converges at $x = 1$. † True enough, but why to ln 2? This takes us back to consideration of the remainder term, the method of Section 11.5.

There is, however, another way to proceed. The great Norwegian mathematician Niels Henrik Abel (1802–1829) proved the following result:

Suppose that $\sum_{k=0}^{\infty} a_k x^k$ converges on $(-c, c)$ and that $f(x) = \sum_{k=0}^{\infty} a_k x^k$ on this interval.

(1) If f is continuous at c and $\sum_{k=0}^{\infty} a_k c^k$ converges, then $f(c) = \sum_{k=0}^{\infty} a_k c^k$.

(2) If f is continuous at $-c$ and $\sum_{k=0}^{\infty} a_k(-c)^k$ converges, then $f(-c) = \sum_{k=0}^{\infty} a_k(-c)^k$.

From Abel's theorem it is evident that the series for $\ln(1+x)$ does represent the function at $x = 1$.

† An alternating series with $a_k \to 0$.

We come now to another important series expansion:

(11.8.6)
$$\tan^{-1} x = x - \frac{x^3}{3} + \frac{x^5}{5} - \frac{x^7}{7} + \cdots \qquad \text{for } -1 \le x \le 1.$$

PROOF For x in $(-1, 1)$

$$\frac{1}{1+x^2} = \frac{1}{1-(-x^2)} = \sum_{k=0}^{\infty}(-1)^k x^{2k},$$

so that, by integration,

$$\tan^{-1} x = \int \left(\sum_{k=0}^{\infty}(-1)^k x^{2k}\right) dx = \left(\sum_{k=0}^{\infty}\frac{(-1)^k}{2k+1}x^{2k+1}\right) + C.$$

The constant C is 0 because the series on the right and the inverse tangent are both 0 at $x = 0$. Thus, for all x in $(-1, 1)$, we have

$$\tan^{-1} x = \sum_{k=0}^{\infty}\frac{(-1)^k}{2k+1}x^{2k+1} = x - \frac{x^3}{3} + \frac{x^5}{5} - \frac{x^7}{7} + \cdots.$$

That the series also represents the function at $x = -1$ and $x = 1$ follows directly from Abel's theorem: at both these points $\tan^{-1} x$ is continuous, and at both of these points the series converges. ❑

Since $\tan^{-1} 1 = \frac{1}{4}\pi$, we have

$$\tfrac{1}{4}\pi = 1 - \tfrac{1}{3} + \tfrac{1}{5} - \tfrac{1}{7} + \tfrac{1}{9} - \cdots.$$

This series was known to the Scottish mathematician James Gregory in 1671. It is an elegant formula for π, but it converges too slowly for computational purposes. A much more effective way of computing π is outlined in the Project at the end of this section.

Term-by-term integration provides a method of calculating some (otherwise rather intractable) definite integrals. Suppose that you are trying to evaluate

$$\int_a^b f(x)\, dx,$$

but cannot find an antiderivative. If you can expand $f(x)$ in a convergent power series, then you can estimate the integral by forming the series and integrating term by term.

Example 6 We will estimate $\displaystyle\int_0^1 e^{-x^2}\, dx$ by expanding the integral in a power series and integrating term by term. Our starting point is the expansion

$$e^x = 1 + x + \frac{x^2}{2!} + \frac{x^3}{3!} + \frac{x^4}{4!} + \frac{x^5}{5!} + \frac{x^6}{6!} + \cdots \qquad \text{for all } x.$$

From this we see that

$$e^{-x^2} = 1 - x^2 + \frac{x^4}{2!} - \frac{x^6}{3!} + \frac{x^8}{4!} - \frac{x^{10}}{5!} + \frac{x^{12}}{6!} - \cdots \qquad \text{for all } x,$$

and therefore

$$\int_0^1 e^{-x^2}\,dx = \left[x - \frac{x^3}{3} + \frac{x^5}{5(2!)} - \frac{x^7}{7(3!)} + \frac{x^9}{9(4!)} - \frac{x^{11}}{11(5!)} + \frac{x^{13}}{13(6!)} - \cdots\right]_0^1$$

$$= 1 - \frac{1}{3} + \frac{1}{5(2!)} - \frac{1}{7(3!)} + \frac{1}{9(4!)} - \frac{1}{11(5!)} + \frac{1}{13(6!)} - \cdots .$$

This is an alternating series with decreasing terms. Therefore we know that the integral lies between consecutive partial sums. In particular it lies between

$$1 - \frac{1}{3} + \frac{1}{5(2!)} - \frac{1}{7(3!)} + \frac{1}{9(4!)} - \frac{1}{11(5!)}$$

and
$$\left[1 - \frac{1}{3} + \frac{1}{5(2!)} - \frac{1}{7(3!)} + \frac{1}{9(4!)} - \frac{1}{11(5!)}\right] + \frac{1}{13(6!)}.$$

As you can check, the first sum is greater than 0.74672 and the second one is less than 0.74684. It follows that

$$0.74672 < \int_0^1 e^{-x^2}\,dx < 0.74684.$$

The estimate 0.7468 approximates the integral to within 0.0001. ☐

The integral of Example 6 was easy to estimate numerically because it could be expressed as an alternating series with decreasing terms. The next example requires more subtlety and illustrates a method more general than that used in Example 6.

Example 7 We want to estimate $\int_0^1 e^{x^2}\,dx$. If we proceed exactly as in Example 6, we find that

$$\int_0^1 e^{x^2}\,dx = 1 + \frac{1}{3} + \frac{1}{5(2!)} + \frac{1}{7(3!)} + \frac{1}{9(4!)} + \frac{1}{11(5!)} + \frac{1}{13(6!)} + \cdots .$$

We now have a series expansion for the integral, but that expansion does not guide us directly to a numerical estimate for the integral. We know that s_n, the nth partial sum of the series, approximates the integral, but we don't know the accuracy of the approximation. We have no handle on the remainder left by s_n.

We start again, this time keeping track of the remainder. For $x \in [0, 1]$,

$$0 \le e^x - \left(1 + x + \frac{x^2}{2!} + \cdots + \frac{x^n}{n!}\right) = R_n(x) \overset{(11.5.3)}{\le} e\left[\frac{x^{n+1}}{(n+1)!}\right] \le \frac{3}{(n+1)!}.$$

If $x \in [0, 1]$, then $x^2 \in [0, 1]$, and therefore

$$0 \le e^{x^2} - \left(1 + x^2 + \frac{x^4}{2!} + \cdots + \frac{x^{2n}}{n!}\right) \le \frac{3}{(n+1)!}.$$

Integrating this inequality from $x = 0$ to $x = 1$, we have

$$0 \le \int_0^1 \left[e^{x^2} - \left(1 + x^2 + \frac{x^4}{2!} + \cdots + \frac{x^{2n}}{n!}\right)\right] dx \le \int_0^1 \frac{3}{(n+1)!}\,dx.$$

Carrying out the integration where possible, we see that

$$0 \le \int_0^1 e^{x^2}\,dx - \left[1 + \frac{1}{3} + \frac{1}{5(2!)} + \cdots + \frac{1}{(2n+1)(n!)}\right] \le \frac{3}{(n+1)!}.$$

We can use this inequality to estimate the integral as closely as we wish. Since

$$\frac{3}{7!} = \frac{1}{1680} < 0.0006,$$

we see that

$$\alpha = 1 + \frac{1}{3} + \frac{1}{5(2!)} + \frac{1}{7(3!)} + \frac{1}{9(4!)} + \frac{1}{11(5!)} + \frac{1}{13(6!)}$$

approximates the integral to within 0.0006. Arithmetical computation shows that

$$1.4626 \le \alpha \le 1.4627.$$

It follows that

$$1.4626 \le \int_0^1 e^{x^2}\,dx \le 1.4627 + 0.0006 = 1.4633.$$

The estimate 1.463 approximates the integral to within 0.0004. □

Power Series; Taylor Series

It is time to relate Taylor series

$$\sum_{k=0}^{\infty} \frac{f^{(k)}(0)}{k!} x^k$$

to power series in general. The relationship is very simple:

> On its interval of convergence a power series is the Taylor series of its sum.

To see this, all you have to do is differentiate

$$f(x) = a_0 + a_1 x + a_2 x^2 + \cdots + a_k x^k + \cdots$$

term by term. Do this and you will find that $f^{(k)}(0) = k!a_k$, and therefore

$$a_k = \frac{f^{(k)}(0)}{k!}.$$

The a_k are the Taylor coefficients of f.

We end this section by carrying out a few simple expansions.

Example 8 Expand $\cosh x$ and $\sinh x$ in powers of x.

SOLUTION There is no need to go through the labor of computing the Taylor coefficients

$$\frac{f^{(k)}(0)}{k!}$$

by differentiation. We know that

$$\cosh x = \tfrac{1}{2}(e^x + e^{-x}) \quad \text{and} \quad \sinh x = \tfrac{1}{2}(e^x - e^{-x}). \tag{7.8.1}$$

Since

$$e^x = 1 + x + \frac{x^2}{2!} + \frac{x^3}{3!} + \frac{x^4}{4!} + \frac{x^5}{5!} + \cdots,$$

we have

$$e^{-x} = 1 - x + \frac{x^2}{2!} - \frac{x^3}{3!} + \frac{x^4}{4!} - \frac{x^5}{5!} + \cdots.$$

Thus

$$\cosh x = \frac{1}{2}\left(2 + 2\frac{x^2}{2!} + 2\frac{x^4}{4!} + \cdots\right) = 1 + \frac{x^2}{2!} + \frac{x^4}{4!} + \cdots = \sum_{k=0}^{\infty} \frac{x^{2k}}{(2k)!}$$

and

$$\sinh x = \frac{1}{2}\left(2x + 2\frac{x^3}{3!} + 2\frac{x^5}{5!} + \cdots\right) = x + \frac{x^3}{3!} + \frac{x^5}{5!} + \cdots = \sum_{k=0}^{\infty} \frac{x^{2k+1}}{(2k+1)!}.$$

Both expansions are valid for all real x, since the exponential expansions are valid for all real x. □

Example 9 Expand $x^2 \cos x^3$ in powers of x.

SOLUTION

$$\cos x = 1 - \frac{x^2}{2!} + \frac{x^4}{4!} - \frac{x^6}{6!} + \cdots.$$

Thus

$$\cos x^3 = 1 - \frac{(x^3)^2}{2!} + \frac{(x^3)^4}{4!} - \frac{(x^3)^6}{6!} + \cdots = 1 - \frac{x^6}{2!} + \frac{x^{12}}{4!} - \frac{x^{18}}{6!} + \cdots,$$

and

$$x^2 \cos x^3 = x^2 - \frac{x^8}{2!} + \frac{x^{14}}{4!} - \frac{x^{20}}{6!} + \cdots.$$

This expansion is valid for all real x, since the expansion for $\cos x$ is valid for all real x.

ALTERNATIVE SOLUTION Since

$$x^2 \cos x^3 = \frac{d}{dx}\left(\tfrac{1}{3}\sin x^3\right),$$

we can derive the expansion for $x^2 \cos x^3$ by expanding $\frac{1}{3}\sin x^3$ and then differentiating term by term. □

EXERCISES 11.8

In Exercises 1–6, expand f in powers of x, basing your calculations on the geometric series

$$\frac{1}{1-x} = 1 + x + x^2 + \cdots + x^n + \cdots.$$

1. $f(x) = \dfrac{1}{(1-x)^2}$.
2. $f(x) = \dfrac{1}{(1-x)^3}$.
3. $f(x) = \dfrac{1}{(1-x)^k}$.
4. $f(x) = \ln(1-x)$.
5. $f(x) = \ln(1-x^2)$.
6. $f(x) = \ln(2-3x)$.

In Exercises 7 and 8, expand f in powers of x, basing your calculations on the tangent series:

$$\tan x = x + \tfrac{1}{3}x^3 + \tfrac{2}{15}x^5 + \tfrac{17}{315}x^7 + \cdots.$$

7. $f(x) = \sec^2 x$.
8. $f(x) = \ln \cos x$.

In Exercises 9 and 10, find $f^{(9)}(0)$.

9. $f(x) = x^2 \sin x$.
10. $f(x) = x \cos x^2$.

Expand f in powers of x.

11. $f(x) = \sin x^2$.
12. $f(x) = x^2 \tan^{-1} x$.
13. $f(x) = e^{3x^3}$.
14. $f(x) = \dfrac{1-x}{1+x}$.
15. $f(x) = \dfrac{2x}{1-x^2}$.
16. $f(x) = x \sinh x^2$.
17. $f(x) = \dfrac{1}{1-x} + e^x$.
18. $f(x) = \cosh x \sinh x$.
19. $f(x) = x \ln(1+x^3)$.
20. $f(x) = (x^2+x)\ln(1+x)$.
21. $f(x) = x^3 e^{-x^3}$.
22. $f(x) = x^5(\sin x + \cos 2x)$.

Evaluate the given limit in two ways: (a) using L 'Hôpital's rule, and (b) using power series.

23. $\displaystyle\lim_{x\to 0} \frac{1-\cos x}{x^2}$.
24. $\displaystyle\lim_{x\to 0} \frac{\sin x - x}{x^2}$.
25. $\displaystyle\lim_{x\to 0} \frac{\cos x - 1}{x \sin x}$.
26. $\displaystyle\lim_{x\to 0} \frac{e^x - 1 - x}{x \tan^{-1} x}$.

Find a powers series representation for the improper integral.

27. $\displaystyle\int_0^x \frac{\ln(1+t)}{t}\,dt$.
28. $\displaystyle\int_0^x \frac{1-\cos t}{t^2}\,dt$.
29. $\displaystyle\int_0^x \frac{\tan^{-1} t}{t}\,dt$.
30. $\displaystyle\int_0^x \frac{\sinh t}{t}\,dt$.

C Estimate to within 0.01.

31. $\displaystyle\int_0^1 e^{-x^3}\,dx$.
32. $\displaystyle\int_0^1 \sin x^2\,dx$.
33. $\displaystyle\int_0^1 \sin\sqrt{x}\,dx$.
34. $\displaystyle\int_0^1 x^4 e^{-x^2}\,dx$.
35. $\displaystyle\int_0^1 \tan^{-1} x^2\,dx$.
36. $\displaystyle\int_1^2 \frac{1-\cos x}{x}\,dx$.

C Use a power series to estimate the integral to within 0.0001.

37. $\displaystyle\int_0^1 \frac{\sin x}{x}\,dx$.
38. $\displaystyle\int_0^{0.5} \frac{1-\cos x}{x^2}\,dx$.
39. $\displaystyle\int_0^{0.5} \frac{\ln(1+x)}{x}\,dx$.
40. $\displaystyle\int_0^{0.2} x \sin x\,dx$.

Sum the series.

41. $\displaystyle\sum_{k=0}^{\infty} \frac{1}{k!}x^{3k}$.
42. $\displaystyle\sum_{k=0}^{\infty} \frac{1}{k!}x^{3k+1}$.
43. $\displaystyle\sum_{k=0}^{\infty} \frac{3k}{k!}x^{3k-1}$.
44. Let $f(x) = \dfrac{e^x - 1}{x}$.
    (a) Expand $f(x)$ in a power series.
    (b) Differentiate the power series in part (a) and show that
    $$\sum_{n=1}^{\infty} \frac{n}{(n+1)!} = 1.$$
45. Let $f(x) = xe^x$.
    (a) Expand $f(x)$ in a power series.
    (b) Integrate the power series in part (a) and show that
    $$\sum_{n=1}^{\infty} \frac{1}{n!(n+2)} = \frac{1}{2}.$$
46. Deduce the differentiation formulas
    $$\frac{d}{dx}(\sinh x) = \cosh x, \qquad \frac{d}{dx}(\cosh x) = \sinh x$$
    from the expansions of $\sinh x$ and $\cosh x$ in powers of x.
47. Show that, if $\sum a_k x^k$ and $\sum b_k x^k$ both converge to the same sum on some interval, then $a_k = b_k$ for each k.
48. Show that, if $\epsilon > 0$, then
    $$|kx^{k-1}| < (|x| + \epsilon)^k \quad \text{for all } k \text{ sufficiently large.}$$
    HINT: Take the kth root and let $k \to \infty$.
49. Suppose that the function f has the power series representation $f(x) = \sum_{k=0}^{\infty} a_k x^k$.
    (a) Show that if f is an even function, then $a_{2k+1} = 0$ for all k.
    (b) Show that if f is an odd function, then $a_{2k} = 0$ for all k.
50. Suppose that the function f is infinitely differentiable on an interval containing 0, and suppose that $f'(x) = -2f(x)$ and $f(0) = 1$. Use these properties to find the power series representation of f in powers of x. What function does the series represent?

51. Suppose that the function f is infinitely differentiable on an interval containing 0, that $f''(x) = -2f(x)$ for all x, and $f(0) = 0, f'(0) = 1$. Use these properties to find a power series representation of f. What is the sum of this series?

52. Use a CAS to expand $f(x), f'(x)$, and $\int f(x)\,dx$ in a power series.

(a) $f(x) = x2^{-x^2}$.

(b) $f(x) = x\tan^{-1} x$.

In Exercises 53–56, estimate to within 0.001 by the method of this section and check your result by carrying out the integration directly.

53. $\displaystyle\int_0^{1/2} x \ln(1+x)\,dx.$

54. $\displaystyle\int_0^1 x \sin x\,dx.$

55. $\displaystyle\int_0^1 x\,e^{-x} dx.$

56. Show that

$$0 \le \int_0^2 e^{x^2} dx - \left[2 + \frac{2^3}{3} + \frac{2^5}{5(2!)} + \cdots + \frac{2^{2n+1}}{(2n+1)n!}\right] < \frac{e^4 2^{2n+3}}{(n+1)!}.$$

■ PROJECT 11.8 Approximating π

The value of π to twenty decimal places is

$$\pi \cong 3.14159\ 26535\ 89793\ 23846.$$

In Exercises 8.7 we estimated

$$\int_0^1 \frac{4}{1+x^2}\,dx$$

by using the trapezoidal rule and Simpson's rule. The results provided approximations of π since

$$\int_0^1 \frac{4}{1+x^2}\,dx = \pi.$$

In this project we will approximate π by using the inverse tangent series

$$\tan^{-1} x = x - \frac{x^3}{3} + \frac{x^5}{5} - \frac{x^7}{7} + \cdots \quad \text{for} \quad -1 \le x \le 1$$

and the relation

$$\text{(1)} \qquad \frac{\pi}{4} = 4\tan^{-1}\frac{1}{5} - \tan^{-1}\frac{1}{239}.$$

(This relation was discovered in 1706 by John Machin, a Scotsman.)

Problem 1. Verify (1). HINT: Using the addition formula for the tangent function, calculate $\tan(2\tan^{-1}\frac{1}{5})$, then $\tan(4\tan^{-1}\frac{1}{5})$, and finally $\tan(4\tan^{-1}\frac{1}{5} - \tan^{-1}\frac{1}{239})$.

The inverse tangent series gives

$$\tan^{-1}\frac{1}{5} = \frac{1}{5} - \frac{1}{3}(\tfrac{1}{5})^3 + \frac{1}{5}(\tfrac{1}{5})^5 - \frac{1}{7}(\tfrac{1}{5})^7 + \cdots$$

and

$$\tan^{-1}\frac{1}{239} = \frac{1}{239} - \frac{1}{3}\left(\frac{1}{239}\right)^3 + \frac{1}{5}\left(\frac{1}{239}\right)^5 - \frac{1}{7}\left(\frac{1}{239}\right)^7 + \cdots.$$

These are convergent alternating series. Thus, we know

$$\frac{1}{5} - \frac{1}{3}\left(\frac{1}{5}\right)^3 \le \tan^{-1}\frac{1}{5} \le \frac{1}{5} - \frac{1}{3}\left(\frac{1}{5}\right)^3 + \frac{1}{5}(\tfrac{1}{5})^5$$

and

$$\frac{1}{239} - \frac{1}{3}\left(\frac{1}{239}\right)^3 \le \tan^{-1}\frac{1}{239} \le \frac{1}{239}.$$

Problem 2.

(a) Use these inequalities, together with (1), to show that $3.14 < \pi < 3.147$.

(b) Show that $3.14159262 < \pi < 3.14159267$ by using six terms of the series for $\tan^{-1}\frac{1}{5}$ and two terms of the series for $\tan^{-1}\frac{1}{239}$.

Greater accuracy can be obtained by using more terms. For example, 15 terms of the series for $\tan^{-1}\frac{1}{5}$ and just four terms of the series for $\tan^{-1}\frac{1}{239}$ determine π accurate to twenty decimal places.

Greater accuracy can be obtained by using more terms.

Problem 3.

(a) Use a CAS to obtain the sum of the first fifteen terms of the series for $\tan^{-1}\frac{1}{5}$ and the first four terms of the series for $\tan^{-1}\frac{1}{239}$.

(b) Use the results in part (a) to approximate π. Compare your approximation with the twenty-decimal place approximation given above.

*SUPPLEMENT TO SECTION 11.8

PROOF OF SECTION 11.8.2

Set $$f(x) = \sum_{k=0}^{\infty} a_k x^k \quad \text{and} \quad g(x) = \sum_{k=0}^{\infty} \frac{d}{dx}(a_k x^k) = \sum_{k=1}^{\infty} k a_k x^{k-1}.$$

Select x from $(-c, c)$. We want to show that

$$f'(x) = \lim_{h \to 0} \frac{f(x+h) - f(x)}{h} = g(x).$$

For $x + h$ in $(-c, c)$, $h \neq 0$, we have

$$\begin{aligned} \left| g(x) - \frac{f(x+h) - f(x)}{h} \right| &= \left| \sum_{k=1}^{\infty} k a_k x^{k-1} - \sum_{k=0}^{\infty} \frac{a_k (x+h)^k - a_k x^k}{h} \right| \\ &= \left| \sum_{k=1}^{\infty} k a_k x^{k-1} - \sum_{k=1}^{\infty} a_k \left[\frac{(x+h)^k - x^k}{h} \right] \right|. \end{aligned}$$

By the mean-value theorem,

$$\frac{(x+h)^k - x^k}{h} = k(t_k)^{k-1}$$

for some number t_k between x and $x + h$. Thus we can write

$$\begin{aligned} \left| g(x) - \frac{f(x+h) - f(x)}{h} \right| &= \left| \sum_{k=1}^{\infty} k a_k x^{k-1} - \sum_{k=1}^{\infty} k a_k (t_k)^{k-1} \right| \\ &= \left| \sum_{k=1}^{\infty} k a_k [x^{k-1} - (t_k)^{k-1}] \right| \\ &= \left| \sum_{k=2}^{\infty} k a_k [x^{k-1} - (t_k)^{k-1}] \right|. \end{aligned}$$

By the mean-value theorem,

$$\frac{x^{k-1} - (t_k)^{k-1}}{x - t_k} = (k-1)(p_{k-1})^{k-2}$$

for some number p_{k-1} between x and t_k. It follows that

$$|x^{k-1} - (t_k)^{k-1}| = |x - t_k| \, |(k-1)(p_{k-1})^{k-2}|.$$

Since $|x - t_k| < |h|$ and $|p_{k-1}| \leq |\alpha|$ where $|\alpha| = \max\{|x|, |x+h|\}$,

$$|x^{k-1} - (t_k)^{k-1}| \leq |h| \, |(k-1)\alpha^{k-2}|.$$

Thus $$\left| g(x) - \frac{f(x+h) - f(x)}{h} \right| \leq |h| \sum_{k=2}^{\infty} |k(k-1) a_k \alpha^{k-2}|.$$

Since this series converges,

$$\lim_{h\to 0}\left(|h|\sum_{k=2}^{\infty}|k(k-1)a_k\alpha^{k-2}|\right)=0.$$

This gives

$$\lim_{h\to 0}\left|g(x)-\frac{f(x+h)-f(x)}{h}\right|=0 \quad\text{and thus}\quad f'(x)=\lim_{h\to 0}\frac{f(x+h)-f(x)}{h}=g(x). \quad \square$$

■ *11.9 THE BINOMIAL SERIES

Through a collection of problems, we invite you to derive for yourself the basic properties of one of the most celebrated series of all — *the binomial series.*

Start with the binomial $1+x$. Choose a real number $\alpha \neq 0$ and form the function

$$f(x)=(1+x)^{\alpha}.$$

If α is a positive integer n, then

$$(1+x)^n = 1+nx+\frac{n(n-1)}{2}x^2+\cdots+n\,x^{n-1}+x^n.$$

This is the familiar binomial theorem. The binomial series is a generalization of the binomial theorem.

Problem 1. Show that

$$\frac{f^{(k)}(0)}{k!}=\frac{\alpha[\alpha-1][\alpha-2]\cdots[\alpha-(k-1)]}{k!}.$$

The number you just obtained is the coefficient of x^k in the expansion of $(1+x)^{\alpha}$. It is called *the kth binomial coefficient* and is usually denoted by $\binom{\alpha}{k}$:

(11.9.1)
$$\binom{\alpha}{k}=\frac{\alpha[\alpha-1][\alpha-2]\cdots[\alpha-(k-1)]}{k!}.$$

For example, if $\alpha=7$ and $k=3$, then

$$\binom{7}{3}=\frac{7\cdot 6\cdot 5}{3!}=35;$$

if $\alpha=3/2$ and $k=3$, then

$$\binom{3/2}{3}=\frac{(3/2)[(3/2)-1][(3/2)-2]}{3!}=\frac{(3/2)(1/2)(-1/2)}{6}=-\frac{1}{16}. \quad \square$$

Problem 2. Show that if α is not a positive integer, then the binomial series

$$\sum\binom{\alpha}{k}x^k$$

has radius of convergence 1. HINT: Use the ratio test. $\square$

From Problem 2 you know that the binomial series converges on the open interval $(-1, 1)$ and defines there an infinitely differentiable function. The next thing to show is that this function (the one defined by the series) is actually $(1+x)^\alpha$. To do this, you first need some other results.

Problem 3. Verify the identity

$$(k+1)\binom{\alpha}{k+1} + k\binom{\alpha}{k} = \alpha\binom{\alpha}{k}. \quad \square$$

Problem 4. Use the identity of Problem 3 to show that the sum of the binomial series

$$\phi(x) = \sum_{k=0}^{\infty}\binom{\alpha}{k}x^k$$

satisfies the differential equation

$$(1+x)\phi'(x) = \alpha\phi(x) \quad \text{for all } x \text{ in } (-1, 1),$$

together with the initial condition $\phi(0) = 1$. □

You are now in a position to prove the main result.

Problem 5. Show that

(11.9.2)
$$(1+x)^\alpha = \sum_{k=0}^{\infty}\binom{\alpha}{k}x^k \quad \text{for all } x \text{ in } (-1, 1).$$

You can probably get a better feeling for the series by writing out the first few terms:

(11.9.3)
$$(1+x)^\alpha = 1 + \alpha x + \frac{\alpha(\alpha-1)}{2!}x^2 + \frac{\alpha(\alpha-1)(\alpha-2)}{3!}x^3 + \cdots. \quad \square$$

EXERCISES 11.9

In Exercises 1–10, expand f in powers of x up to x^4.

1. $f(x) = \sqrt{1+x}$.
2. $f(x) = \sqrt{1-x}$.
3. $f(x) = \sqrt{1+x^2}$.
4. $f(x) = \sqrt{1-x^2}$.
5. $f(x) = \dfrac{1}{\sqrt{1+x}}$.
6. $f(x) = \dfrac{1}{\sqrt[3]{1+x}}$.
7. $f(x) = \sqrt[4]{1-x}$.
8. $f(x) = \dfrac{1}{\sqrt[4]{1+x}}$.
9. $f(x) = (4+x)^{3/2}$.
10. $f(x) = \sqrt{1+x^4}$.
11. (a) Use a binomial series to find the Taylor series of $f(x) = 1/\sqrt{1-x^2}$ in powers of x.
    (b) Use the series for f in part (a) to find the Taylor series for $F(x) = \sin^{-1} x$ and give the radius of convergence.
12. (a) Use a binomial series to find the Taylor series of $f(x) = 1/\sqrt{1+x^2}$ in powers of x.
    (b) Use the series for f in part (a) to find the Taylor series for $F(x) = \sinh^{-1} x$ and give the radius of convergence.

Ⓒ Estimate by using the first three terms of a binomial expansion, rounding off your answer to four decimal places.

13. $\sqrt{98}$. HINT: $\sqrt{98} = (100-2)^{1/2} = 10(1-\frac{1}{50})^{1/2}$.
14. $\sqrt[5]{36}$.
15. $\sqrt[3]{9}$.
16. $\sqrt[4]{620}$.
17. $17^{-1/4}$.
18. $9^{-1/3}$.

Ⓒ Approximate each integral to within 0.001.

19. $\displaystyle\int_0^{1/3}\sqrt{1+x^3}\,dx$.
20. $\displaystyle\int_0^{1/5}\sqrt{1+x^4}\,dx$.
21. $\displaystyle\int_0^{1/2}\frac{1}{\sqrt{1+x^2}}\,dx$.
22. $\displaystyle\int_0^{1/2}\frac{1}{\sqrt{1-x^3}}\,dx$.

■ CHAPTER HIGHLIGHTS

11.1 Infinite Series

sigma notation (p. 633) partial sums (p. 634)
convergence, divergence (p. 635)
sum of a series(p. 635)
a divergence test (p. 641)

geometric series: $\sum_{k=0}^{\infty} x^k = \begin{cases} \dfrac{1}{1-x}, & |x| < 1 \\ \text{diverges}, & |x| \geq 1 \end{cases}$

If $\sum_{k=0}^{\infty} a^k$ converges, then $a_k \to 0$. The converse is false.

11.2 The integral Test; Comparison Tests

integral test (p. 644) basic comparison (p. 647)
limit comparison (p. 649)

harmonic series: $\sum_{k=1}^{\infty} \dfrac{1}{k}$ diverges p-series: $\sum_{k=1}^{\infty} \dfrac{1}{k^p}$ converges iff $p > 1$

11.3 The Root Test; The Ratio Test

root test (p. 653) ratio test (p. 654)
summary on convergence tests (p. 656)

11.4 Absolute and Conditional Convergence; Alternating Series

absolutely convergent, conditionally convergent (p. 657)
convergence theorem for alternating series (p.659)
an estimate for alternating series (p.660)
rearrangements (p.662)

11.5 Taylor Polynomials in *x*; Taylor Series in *x*

Taylor polynomials in x (p. 665) remainder term $R_n(x)$ (p. 667).
remainder estimate (p. 668) Lagrange form of the remainder (p. 668)

Taylor series in x (Maclaurin series): $\sum_{k=1}^{\infty} \dfrac{f^{(k)}(0)}{k!} x^k$

$$e^x = \sum_{k=0}^{\infty} \frac{x^k}{k!} \quad \text{all real } x$$

$$\ln(1+x) = \sum_{k=1}^{\infty} \frac{(-1)^{k+1}}{k} x^k, -1 < x \leq 1$$

$$\sin x = \sum_{k=0}^{\infty} \frac{(-1)^k}{(2k+1)!} x^{2k+1}, \quad \text{all real } x$$

$$\cos x = \sum_{k=0}^{\infty} \frac{(-1)^k}{(2k)!} x^{2k}, \quad \text{all real } x$$

11.6 Taylor Polynomials and Taylor Series in $x - a$

Taylor series in $x - a$: $\sum_{k=0}^{\infty} \dfrac{g^{(k)}(a)}{k!} (x-a)^k$

11.7 Power Series

power series (p. 681) radius of convergence (p. 682)
interval of convergence (p. 683)

If a power series converges at $c \neq 0$, then it converges absolutely for $|x| < |c|$; if it diverges at d, then it diverges for $|x| > |d|$.

11.8 Differentiation and Integration of Power Series

$$\tan^{-1} x = \sum_{k=0}^{\infty} \frac{(-1)^k}{2k+1} x^{2k+1}, \quad -1 \leq x \leq 1$$

$$\cosh x = \sum_{k=0}^{\infty} \frac{x^{2k}}{(2k)!}, \quad \text{all real } x$$

$$\sinh x = \sum_{k=0}^{\infty} \frac{x^{2k+1}}{(2k+1)!}, \quad \text{all real } x$$

On the interior of its interval of convergence, a power series can be differentiated and integrated term by term.

On its interval of convergence, a power series is the Taylor series of its sum.

*11.9 The Binomial Series

$$(1+x)^\alpha = \sum_{k=0}^{\infty} \binom{\alpha}{k} x^k = 1 + \alpha x + \frac{\alpha(\alpha-1)}{2!} x^2 + \cdots,$$
$$-1 < x < 1$$

SKILL MASTERY REVIEW Polar Coordinates; Parametric Equations; More on Limits; Series

Sketch the graph of the polar curve and find the area it encloses.

1. $r = 2(1 - \cos\theta)$

2. $r^2 = 4\sin 2\theta$

3. $r = 3\sin 3\theta$

4. $r = 2 + \cos\theta$

5. Find the area of the region that is inside the circle $r = \sin\theta$ and outside the cardioid $r = 1 - \cos\theta$.

6. Find equations in x and y for the tangent and normal lines to the curve $C : x = 3e^t, y = 5e^{-t}$ at the point where $t = 0$.

7. Find an equation in x and y for the line tangent to the polar curve $C : r = 2\sin 2\theta$ at the point where $\theta = \pi/4$.

8. Find dy/dx and d^2y/dx^2 for the curve $C : x = 3t^2, y = 4t^3$.

9. An object moves in a plane so that dx/dt and d^2y/dt^2 are nonzero constants. Identify the path of the object.

Find the length of the plane curve.

10. $y = x^{3/2}$ from $x = 0$ to $x = \frac{5}{9}$.

11. $y = \ln(1 - x^2)$ from $x = 0$ to $x = \frac{1}{2}$.

12. $x(t) = \cos t,\ y(t) = \sin^2 t$ from $t = 0$ to $t = \pi$.

Find the area of the surface generated by revolving the curve about the x–axis.

13. $y^2 = 4x$, from $x = 0$ to $x = 24$.

14. $x(t) = \frac{2}{3}t^{3/2},\ y(t) = t$, from $t = 3$ to $t = 8$.

15. A particle moves from time $t = 0$ to $t = 1$ so that $x(t) = 4t - \sin\pi t,\ y(t) = 4t + \cos\pi t$.

(a) When does the particle have minimum speed? When does it have maximum speed?

(b) What is the slope of the tangent line at the point where $t = \frac{1}{4}$?

Determine the boundedness and monotonicity of the sequence.

16. $\dfrac{2n}{3n+1}$

17. $\dfrac{n^2 - 1}{n}$

18. $1 + \dfrac{(-1)^n}{n}$

19. $\dfrac{2^n}{n^2}$

State whether the sequence converges, and if it does, find the limit.

20. $n\,2^{1/n}$

21. $\dfrac{(n+1)(n+2)}{(n+3)(n+4)}$

22. $\left(\dfrac{n}{1+n}\right)^{1/n}$

23. $\dfrac{n^2 + 5n + 1}{n^3 + 1}$

24. $\cos\left(\dfrac{1}{n}\pi\right)\sin\left(\dfrac{1}{n}\pi\right)$

25. $\left(2 + \dfrac{1}{n}\right)^n$

26. $\left[\ln\left(1 + \dfrac{1}{n}\right)\right]^n$

27. $3\ln 2n - \ln(n^3 + 1)$

28. $\displaystyle\int_n^{n+1} e^{-x}dx$

Find the limit.

29. $\displaystyle\lim_{x\to\infty} \frac{5x + 2\ln x}{x + 3\ln x}$

30. $\displaystyle\lim_{x\to 0} \frac{e^x - 1}{\tan 2x}$

31. $\displaystyle\lim_{x\to 0} \frac{\ln(\cos x)}{x^2}$

32. $\displaystyle\lim_{x\to\infty} \left(1 + \frac{4}{x}\right)^{2x}$

33. $\displaystyle\lim_{x\to 0^+} x^2 \ln x$

34. $\displaystyle\lim_{x\to 0} \frac{e^x + e^{-x} - x^2 - 2}{\sin^2 x - x^2}$

35. Evaluate $\displaystyle\int_0^a \ln(1/x)dx$ for $a > 0$.

36. Calculate $\displaystyle\lim_{x\to\infty} xe^{-x^2}\int_0^x e^{t^2}dt$.

Evaluate the improper integrals that converge.

37. $\displaystyle\int_1^\infty \frac{e^{-\sqrt{x}}}{\sqrt{x}}dx$

38. $\displaystyle\int_0^1 \frac{x}{\sqrt{1 - x^2}}dx$

39. $\displaystyle\int_0^1 \frac{1}{x^2 - 1}dx$

40. $\displaystyle\int_1^\infty \frac{\sin(\pi/x)}{x^2}dx$

Find the sum of the series.

41. $\displaystyle\sum_{k=0}^\infty \left(\frac{3}{4}\right)^k$

42. $\displaystyle\sum_{k=0}^\infty (-1)^k \left(\frac{1}{2}\right)^k$

43. $\displaystyle\sum_{k=0}^\infty \frac{(\ln 2)^k}{k!}$

44. $\displaystyle\sum_{k=1}^\infty \frac{1}{k(k+1)}$

Determine convergence or divergence if the series has only nonnegative terms; determine whether the series is absolutely convergent, conditionally convergent, or divergent if it contains both positive and negative terms.

45. $\sum_{k=0}^{\infty} \frac{1}{2k+1}$

46. $\sum_{k=0}^{\infty} \frac{1}{(2k+1)(2k+3)}$

47. $\sum_{k=0}^{\infty} \frac{(-1)^k}{(k+1)(k+2)}$

48. $\sum_{k=2}^{\infty} \frac{1}{k \ln k}$

49. $\sum_{k=0}^{\infty} \frac{(-1)^k}{(2k+1)}$

50. $\sum_{k=0}^{\infty} \frac{k+1}{3^k}$

51. $\sum_{k=0}^{\infty} \frac{(-1)^k (100)^k}{k!}$

52. $\sum_{k=0}^{\infty} \frac{k + \cos k}{k^3+1}$

53. $\sum_{k=0}^{\infty} \frac{(-1)^k}{\sqrt{(k+1)(k+2)}}$

54. $\sum_{k=1}^{\infty} k \left(\frac{3}{4}\right)^k$

55. $\sum_{k=0}^{\infty} \frac{k^e}{e^k}$

56. $\sum_{k=1}^{\infty} (-1)^{k-1} \frac{\ln k}{\sqrt{k}}$

57. $\sum_{k=0}^{\infty} \frac{(2k)!}{2^k k!}$

58. $\sum_{k=0}^{\infty} \frac{(-1)^k}{\sqrt{k^3+1}}$

59. $\sum_{k=0}^{\infty} \frac{(\tan^{-1} k)^2}{1+k^2}$

60. $\sum_{k=0}^{\infty} \frac{2^k + k^4}{3^k}$

Find the interval of convergence for the power series.

61. $\sum \frac{5^k}{k} x^k$

62. $\sum \frac{(-1)^k}{3^k} x^{k+1}$

63. $\sum \frac{2^k}{(2k)!} (x-1)^{2k}$

64. $\sum \frac{1}{2^k} (x-2)^k$

65. $\sum \frac{(-1)^k k}{3^{2k}} x^k$

66. $\sum \frac{k}{2k+1} x^{2k+1}$

Find the Taylor series expansion in powers of x.

67. $f(x) = xe^{2x^2}$

68. $f(x) = \ln(1+x^2)$

69. $f(x) = \sqrt{x} \tan^{-1} \sqrt{x}$

70. $f(x) = a^x, a > 0$.

71. $f(x) = (1-x^2)^{-1/2}$ up to x^3

72. $f(x) = \sin^{-1} x$ up to x^4

Find the Taylor series expansion of f and give the radius of convergence.

73. $f(x) = e^{-2x}$ in powers of $(x+1)$

74. $f(x) = \sin 2x$ in powers of $(x - \pi/4)$

75. $f(x) = \ln x$ in powers of $(x-1)$

76. $f(x) = \sqrt{x+1}$ in powers of x.

77. Use the Lagrange form of the remainder to show that the approximation

$$\sin x \cong x - \tfrac{1}{6}x^3 + \tfrac{1}{120}x^5$$

is accurate to four decimal places for $0 \le x \le \pi/4$.

78. Use the Lagrange form of the remainder to show that the approximation

$$\cos x \cong 1 - \tfrac{1}{2}x^2 + \tfrac{1}{24}x^4 - \tfrac{1}{720}x^6$$

is accurate to five decimal places for $0 \le x \le \pi/4$.

CHAPTER 12

VECTORS

■ 12.1 CARTESIAN SPACE COORDINATES

To introduce a Cartesian coordinate system in three-dimensional space, we begin with a plane Cartesian coordinate O-xy. Through the point O, which we continue to call the origin, we pass a third line, perpendicular to the other two. This third line we call the z-axis. We assign coordinates to the z-axis using the same scale, assigning the z-coordinate 0 to the origin O.

For later convenience we orient the z-axis so that O-xyz forms a "right-handed" system. That is, if the index finger of the right hand points along the positive x-axis and the middle finger along the positive y-axis, then the thumb will point along the positive z-axis (see Figure 12.1.1).

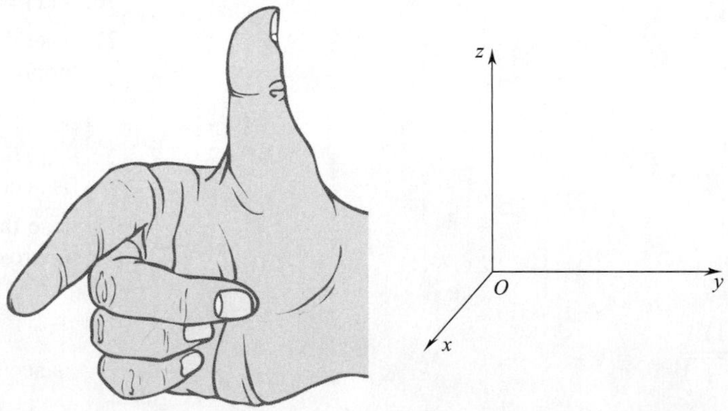

Figure 12.1.1

There are now three coordinate planes: the *xy-plane*, the *xz-plane*, and the *yz-plane*.

The point on the x-axis with x-coordinate x_0 is given space coordinates $(x_0, 0, 0)$; the point on the y-axis with y-coordinate y_0 is given space coordinates $(0, y_0, 0)$; the point on the z-axis with z-coordinate z_0 is given space coordinates $(0, 0, z_0)$.

An arbitrary point P in three-dimensional space (see Figure 12.1.2) is assigned coordinates (x_0, y_0, z_0) provided that

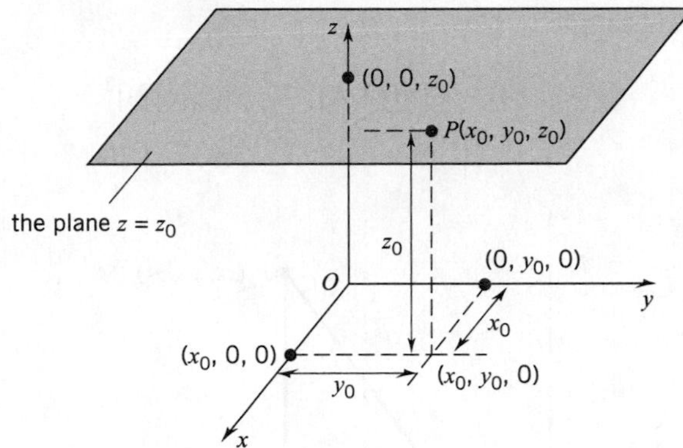

Figure 12.1.2

(1) the plane through P parallel to the yz-plane intersects the x-axis at $(x_0, 0, 0)$;

(2) the plane through P parallel to the xz-plane intersects the y-axis at $(0, y_0, 0)$;

(3) the plane through P parallel to the xy-plane intersects the z-axis at $(0, 0, z_0)$.

The space coordinates (x_0, y_0, z_0) are called the *Cartesian coordinates of P* or simply the *rectangular coordinates of P*.

A point is in the xy-plane iff it is of the form $(x, y, 0)$. Thus the equation $z = 0$ represents the xy-plane. The equation $z = z_0$ represents the set of all points (x, y, z_0), that is, the set of all points with z-coordinate z_0. This is a plane parallel to the xy-plane. Similarly, the equation $x = x_0$ represents a plane parallel to the yz-plane and the equation $y = y_0$ represents a plane parallel to the xz-plane. (See, for example, Figure 12.1.3. There we have drawn the planes $x = 1$ and $y = 3$.)

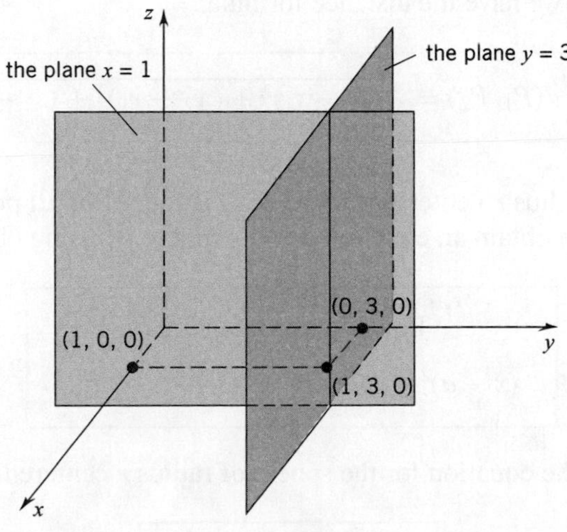

Figure 12.1.3

The Distance Formula

The distance $d(P_1, P_2)$ between two points $P_1(x_1, y_1, z_1)$ and $P_2(x_2, y_2, z_2)$ can be found by applying the Pythagorean theorem twice. With Q and R as in Figure 12.1.4, P_1P_2R

and P_1RQ are both right triangles. From the first triangle

$$[d(P_1, P_2)]^2 = [d(P_1, R)]^2 + [d(R, P_2)]^2,$$

and from the second triangle

$$[d(P_1, R)]^2 = [d(Q, R)]^2 + [d(P_1, Q)]^2.$$

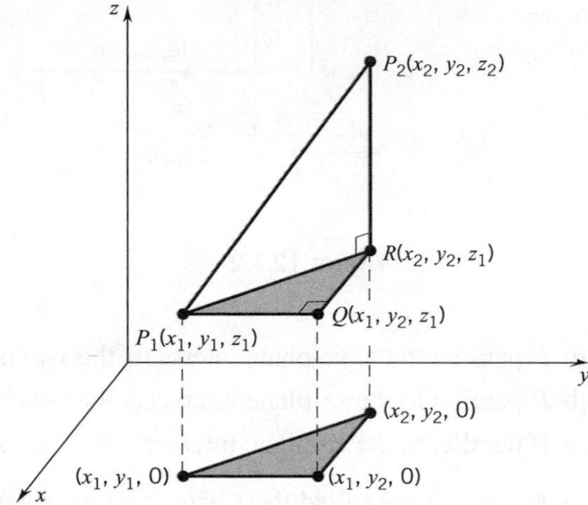

Figure 12.1.4

Combining equations,

$$\begin{aligned}[d(P_1, P_2)]^2 &= [d(Q, R)]^2 + [d(P_1, Q)]^2 + [d(R, P_2)]^2 \\ &= (x_2 - x_1)^2 + (y_2 - y_1)^2 + (z_2 - z_1)^2.\end{aligned}$$

Taking square roots, we have the distance formula:

(12.1.1)

$$d(P_1, P_2) = \sqrt{(x_2 - x_1)^2 + (y_2 - y_1)^2 + (z_2 - z_1)^2}.$$

The *sphere* of radius r centered at $P_0(a, b, c)$ is the set of all points $P(x, y, z)$ with $d(P, P_0) = r$. We can obtain an equation for this sphere by using (12.1.1):

(12.1.2)

Equation for a Sphere

$$(x - a)^2 + (y - b)^2 + (z - c)^2 = r^2.$$

See Figure 12.1.5. The equation for the sphere of radius r centered at the origin is

(12.1.3)

$$x^2 + y^2 + z^2 = r^2.$$

See Figure 12.1.6.

Example 1 The equation $(x - 5)^2 + (y + 2)^2 + z^2 = 9$ represents the sphere of radius 3 centered at the point $(5, -2, 0)$. □

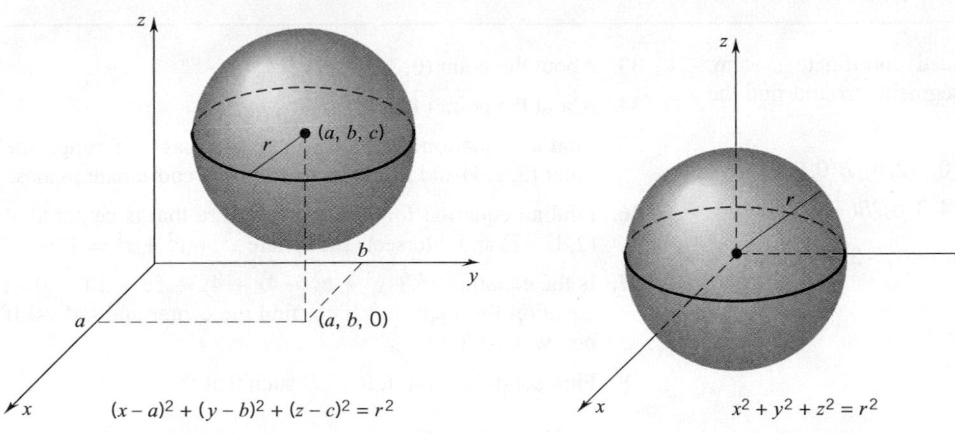

Figure 12.1.5

Figure 12.1.6

Example 2 Show that the equation $x^2+y^2+z^2+6x+2y-4z=11$ represents a sphere. Find the center of the sphere and the radius.

SOLUTION We write the equation as $(x^2+6x)+(y^2+2y)+(z^2-4z)=11$ and complete the squares. The result,

$$(x^2+6x+9)+(y^2+2y+1)+(z^2-4z+4)=11+9+1+4=25,$$

can be written

$$(x+3)^2+(y+1)^2+(z-2)^2=25.$$

This equation represents the sphere of radius 5 centered at $(-3,-1,2)$. □

Symmetry

You are already familiar with two kinds of symmetry: symmetry about a point and symmetry about a line. In space we can also speak of symmetry about a plane. These ideas are illustrated in Figure 12.1.7

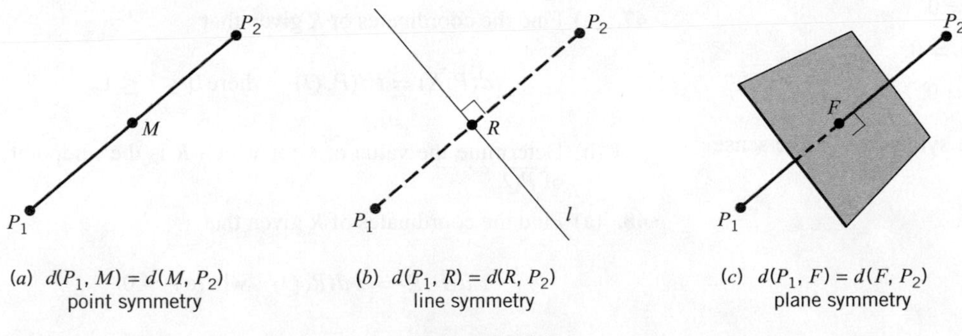

(a) $d(P_1, M) = d(M, P_2)$ point symmetry

(b) $d(P_1, R) = d(R, P_2)$ line symmetry

(c) $d(P_1, F) = d(F, P_2)$ plane symmetry

Figure 12.1.7

The endpoints $P_1(x_1, y_1, z_1)$ and $P_2(x_2, y_2, z_2)$ of the line segment $\overline{P_1P_2}$ are symmetric about the midpoint of the segment. This leads to the *midpoint formula*

(12.1.4)
$$\left(\frac{x_1+x_2}{2}, \frac{y_1+y_2}{2}, \frac{z_1+z_2}{2}\right).$$

EXERCISES 12.1

Plot points A and B on a right-handed coordinate system. Then calculate the length of the line segment $\overline{AB}$ and find the midpoint.

1. $A(2, 0, 0), B(0, 0, -4)$.
2. $A(0, -2, 0), B(0, 0, 6)$.
3. $A(0, -2, 5), B(4, 1, 0)$.
4. $A(4, 3, 0), B(-2, 0, 6)$.

Find an equation for the plane through $(3, 1, -2)$ that satisfies the given condition.

5. Parallel to the xy-plane.
6. Parallel to the xz-plane.
7. Perpendicular to the y-axis.
8. Perpendicular to the z-axis.
9. Parallel to the yz-plane.
10. Perpendicular to the x-axis.

Find an equation for the sphere that satisfies the given conditions.

11. Centered at $(0, 2, -1)$ with radius 3.
12. Centered at $(1, 0, -2)$ with radius 4.
13. Centered at $(2, 4, -4)$ and passes through the origin.
14. Centered at the origin and passes through $(1, -2, 2)$.
15. The line segment joining $(0, 4, 2)$ and $(6, 0, 2)$ is a diameter.
16. Centered at $(2, 3, -4)$ and tangent to the xy-plane.
17. Centered at $(2, 3, -4)$ and tangent to the plane $x = 7$.
18. Centered at $(2, 3, -4)$ and tangent to the plane $y = 1$.

Show that the equation represents a sphere; find the center and radius.

19. $x^2 + y^2 + z^2 + 4x - 8y - 2z + 5 = 0$.
20. $3x^2 + 3y^2 + 3z^2 - 12x - 6z + 3 = 0$.
21. $x^2 + y^2 + z^2 - 6x + 10y - 2z - 1 = 0$.
22. $4x^2 + 4y^2 + 4z^2 - 4x - 8y - 11 = 0$.

The points $P(a, b, c)$ and $Q(2, 3, 5)$ are symmetric in the sense given in Exercises 23–34. Find a, b, c.

23. About the xy-plane.
24. About the xz-plane.
25. About the yz-plane.
26. About the x-axis.
27. About the y-axis.
28. About the z-axis.
29. About the origin.
30. About the plane $x = 1$.
31. About the plane $y = -1$.
32. About the plane $z = 4$.
33. About the point $(0, 2, 1)$.
34. About the point $(4, 0, 1)$.
35. Find an equation for each sphere that passes through the point $(5, 1, 4)$ and is tangent to all three coordinate planes.
36. Find an equation for the largest sphere that is centered at $(2, 1, -2)$ and intersects the sphere $x^2 + y^2 + z^2 = 1$.
37. Is the equation $x^2 + y^2 + z^2 - 4x + 4y + 6z + 20 = 0$ an equation for a sphere? If so, find the center and radius. If not, why isn't it?
38. Find conditions on A, B, C, D such that the equation

$$x^2 + y^2 + z^2 + Ax + By + Cz + D = 0$$

represents a sphere.

39. Show that the points $P(1, 2, 3)$, $Q(4, -5, 2)$, $R(0, 0, 0)$ are the vertices of a right triangle.
40. The points $(5, -1, 3)$, $(4, 2, 1)$, $(2, 1, 0)$ are the midpoints of the sides of a triangle PQR. Find the vertices P, Q, R of the triangle.

Describe the region Ω

41. $\Omega = \{(x, y, z) : x^2 + y^2 + z^2 \leq 4\}$.
42. $\Omega = \{(x, y, z) : x^2 + y^2 + z^2 > 9\}$.
43. $\Omega = \{(x, y, z) : 0 \leq x \leq 1,\ 0 \leq y \leq 2,\ 0 \leq z \leq 3\}$.
44. $\Omega = \{(x, y, z) : |x| \leq 2,\ |y| \leq 2,\ |z| \leq 2\}$.
45. $\Omega = \{(x, y, z) : x^2 + y^2 \leq 4,\ 0 \leq z \leq 4\}$.
46. $\Omega = \{(x, y, z) : 4 < x^2 + y^2 + z^2 < 9\}$.

In Exercises 47 and 48, the point R lies on the line segment that joins $P(a_1, a_2, a_3)$ and $Q(b_1, b_2, b_3)$.

47. (a) Find the coordinates of R given that

$$d(P, R) = t\, d(P, Q) \quad \text{where } 0 \leq t \leq 1.$$

(b) Determine the value of t for which R is the midpoint of $\overline{PQ}$.

48. (a) Find the coordinates of R given that

$$d(P, R) = r\, d(R, Q) \quad \text{where } r > 0.$$

(b) Determine the value of r for which R is the midpoint of $\overline{PQ}$.

49. Use a CAS to find the equation of the sphere that has the line segment joining the points $P(3, -2, -2)$ and $Q(-1, 4, -3)$ as a diameter. Sketch the sphere.
50. Use a CAS to find the perimeter and the area of the triangle with vertices $P(4, -3, 2)$, $Q(-6, -2, 7)$, $R(5, -1, -2)$. Sketch the triangle.

■ 12.2 DISPLACEMENTS AND FORCES

Our purpose here is to motivate the notion of vector. A quick reading of this section will do.

Displacements

A displacement along a coordinate line can be specified by a real number and depicted by an arrow. For a displacement of a_1 units we can use the number a_1 and an arrow that begins at any number x and ends at $x + a_1$. By convention, if $a_1 > 0$, the arrow will point to the right, and if $a_1 < 0$, the arrow will point to the left (Figure 12.2.1). The *magnitude* of the displacement a_1 is defined to be the length of the arrow. Thus, the magnitude of the displacement a_1 is $|a_1|$.

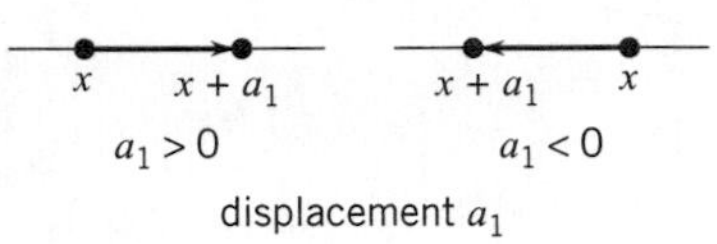

displacement a_1

Figure 12.2.1

Displacements in the plane are more interesting. Instead of having two possible directions, there are an infinite number of possible directions. Displacements in the plane are specified by ordered pairs of real numbers. Figure 12.2.2 shows a displacement (a_1, a_2) beginning at the point (x, y) and ending at the point $(x + a_1, y + a_2)$. The magnitude of the displacement (a_1, a_2) is given by $\sqrt{a_1^2 + a_2^2}$.

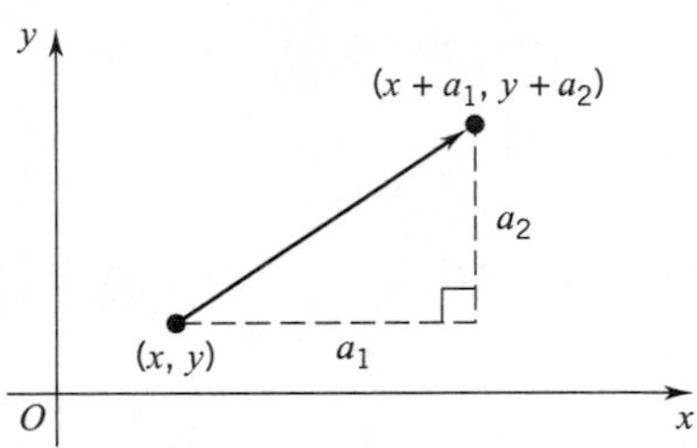

a displacement $(a_1; a_2)$ in the xy-plane

Figure 12.2.2

The most general displacements take place in space. Here ordered triples of numbers come into play. A displacement of a_1 units in x-coordinate, a_2 units in y-coordinate, and a_3 units in z-coordinate can be indicated by an arrow that begins at any point (x, y, z) and ends at the point $(x + a_1, y + a_2, z + a_3)$. This displacement is represented by the ordered triple (a_1, a_2, a_3). (Figure 12.2.3). The magnitude of the displacement (a_1, a_2, a_3) is $\sqrt{a_1^2 + a_2^2 + a_3^2}$.

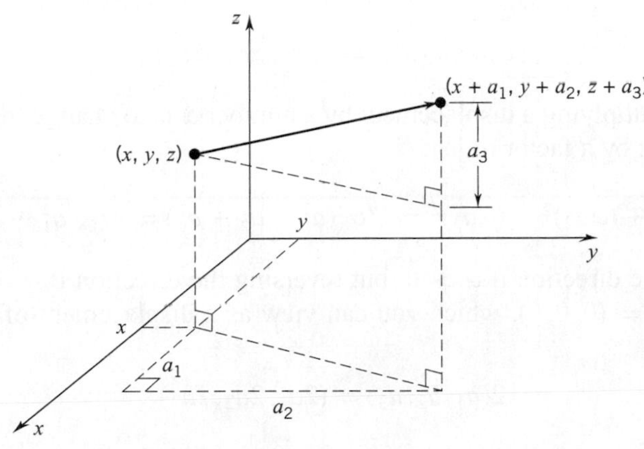

a displacement (a_1, a_2, a_3) in space

Figure 12.2.3

We can follow one displacement by another. A displacement (a_1, a_2, a_3) followed by (b_1, b_2, b_3) results in a total displacement $(a_1 + b_1, a_2 + b_2, a_3 + b_3)$. We can express this by writing

$$(a_1, a_2, a_3) + (b_1, b_2, b_3) = (a_1 + b_1, a_2 + b_2, a_3 + b_3).$$

We can picture the first displacement by an arrow from some point $P(x, y, z)$ to

$$Q(x + a_1, y + a_2, z + a_3),$$

the second displacement by an arrow from $Q(x + a_1, y + a_2, z + a_3)$ to

$$R(x + a_1 + b_1, y + a_2 + b_2, z + a_3 + b_3),$$

and then the resultant displacement by an arrow from $P(x, y, z)$ to

$$R(x + a_1 + b_1, y + a_2 + b_2, z + a_3 + b_3).$$

The three arrows then form a triangular pattern that is easy to remember (Figure 12.2.4).

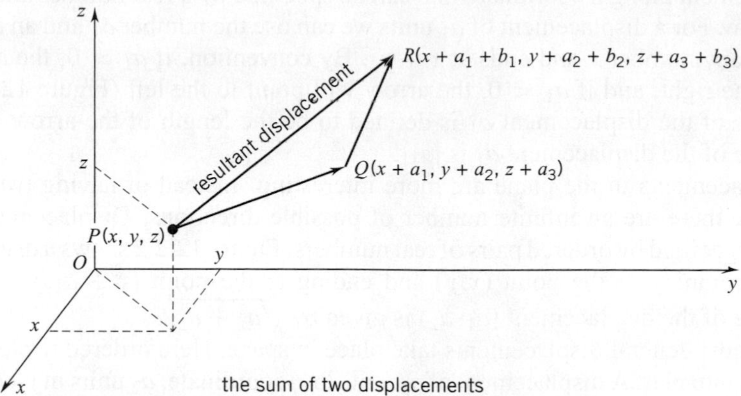

Figure 12.2.4

From a displacement (a_1, a_2, a_3) and a real number α, we can form a new displacement $(\alpha a_1, \alpha a_2, \alpha a_3)$. We view this new displacement as α times the initial displacement and write

$$\alpha(a_1, a_2, a_3) = (\alpha a_1, \alpha a_2, \alpha a_3).$$

The effect of multiplying a displacement by a number α is to change the magnitude of the displacement by a factor of $|\alpha|$:

$$\sqrt{(\alpha a_1)^2 + (\alpha a_2)^2 + (\alpha a_3)^2} = \sqrt{\alpha^2(a_1^2 + a_2^2 + a_3^2)} = |\alpha|\sqrt{a_1^2 + a_2^2 + a_3^2},$$

keeping the same direction if $\alpha > 0$, but reversing the direction if $\alpha < 0$. [If $\alpha = 0$, then $(a_1, a_2, a_3) = (0, 0, 0)$, which you can view as a displacement of length 0.] The displacement

$$2(a_1, a_2, a_3) = (2a_1, 2a_2, 2a_3)$$

is twice as long as (a_1, a_2, a_3) and has the same direction; the displacement

$$-(a_1, a_2, a_3) = (-a_1, -a_2, -a_3)$$

has the same length as (a_1, a_2, a_3) but the opposite direction; the displacement

$$-\tfrac{3}{2}(a_1, a_2, a_3) = (-\tfrac{3}{2}a_1, -\tfrac{3}{2}a_2, -\tfrac{3}{2}a_3)$$

is one and one-half times as long as (a_1, a_2, a_3) and has the opposite direction (Figure 12.2.5).

Forces

The algebraic patterns

$$(a_1, a_2, a_3) + (b_1, b_2, b_3) = (a_1 + b_1, a_2 + b_2, a_3 + b_3)$$
$$\alpha(a_1, a_2, a_3) = (\alpha a_1, \alpha a_2, \alpha a_3)$$

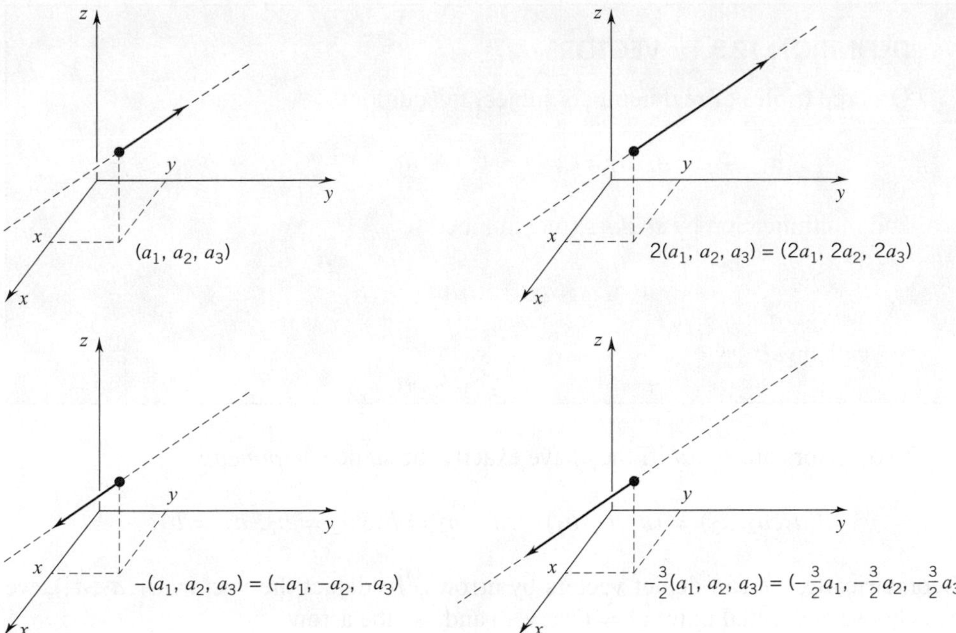

Figure 12.2.5

arise naturally in other settings; for example in the analysis of forces. A force **F** acting in three-dimensional space is completely determined by its components along the x, y, and z axes. If these components are a_1, a_2, a_3, respectively, then the force can be represented by the ordered triple (a_1, a_2, a_3). See Figure 12.2.6.

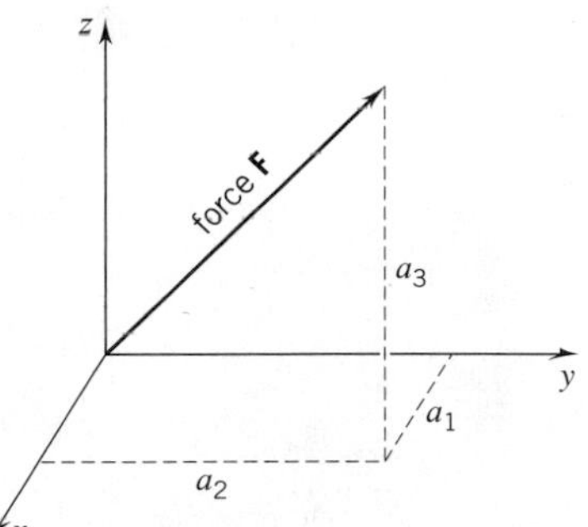

Figure 12.2.6

If two forces $\mathbf{F}_1 = (a_1, a_2, a_3)$ and $\mathbf{F}_2 = (b_1, b_2, b_3)$ are applied simultaneously at the same point, the effect is the same as that produced by the single force

$$\mathbf{F}_3 = (a_1 + b_1, a_2 + b_2, a_3 + b_3).$$

We call $\mathbf{F}_3$ the *resultant* or *total force* and write $\mathbf{F}_1 + \mathbf{F}_2 = \mathbf{F}_3$. For the ordered triples,

$$(a_1, a_2, a_3) + (b_1, b_2, b_3) = (a_1 + b_1, a_2 + b_2, a_3 + b_3).$$

Pictorially we have the usual force diagram, Figure 12.2.7. It is a parallelogram with the sides representing $\mathbf{F}_1 = (a_1, a_2, a_3)$ and $\mathbf{F}_2 = (b_1, b_2, b_3)$ and the diagonal representing

$$\mathbf{F}_3 = \mathbf{F}_1 + \mathbf{F}_2 = (a_1 + b_1, a_2 + b_2, a_3 + b_3).$$

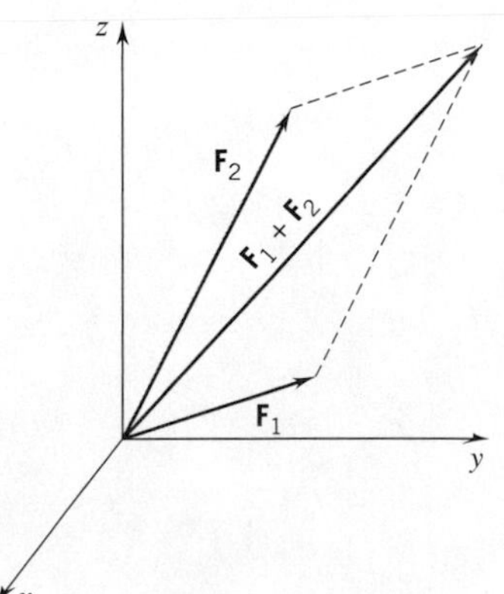

Figure 12.2.7

For any force $\mathbf{F} = (a_1, a_2, a_3)$ and any real number α, the force $\alpha\mathbf{F}$ is defined by the equation

$$\alpha\mathbf{F} = (\alpha a_1, \alpha a_2, \alpha a_3).$$

Thus, once again we have

$$\alpha(a_1, a_2, a_3) = (\alpha a_1, \alpha a_2, \alpha a_3).$$

■ 12.3 VECTORS

The algebra of number triples that we introduced in the last section to discuss displacements and forces is so prodigiously rich in applications that it has found a firm place in the world of science and engineering, and has generated much of mathematics. It is to this mathematics that we now turn.

DEFINITION 12.3.1 VECTORS

Ordered triples of real numbers subject to addition:

$$(a_1, a_2, a_3) + (b_1, b_2, b_3) = (a_1 + b_1,\ a_2 + b_2,\ a_3 + b_3),$$

and multiplication by *scalars* (real numbers):

$$\alpha(a_1, a_2, a_3) = (\alpha a_1, \alpha a_2, \alpha a_3)$$

are called *vectors*. †

Two vectors are *equal* iff they have exactly the same *components:*

$$(a_1, a_2, a_3) = (b_1, b_2, b_3) \quad \text{iff} \quad a_1 = b_1,\ a_2 = b_2,\ a_3 = b_3.$$

Geometrically, we can depict vectors by arrows. To depict the vector (a_1, a_2, a_3), we can choose any initial point $Q = Q(x, y, z)$ and use the arrow

$$\overrightarrow{QR} \quad \text{with} \quad R = R(x + a_1, y + a_2, z + a_3). \qquad \text{(Figure 12.3.1)}$$

Usually we choose the origin as the initial point and use the arrow

$$\overrightarrow{OP} \quad \text{with} \quad P = P(a_1, a_2, a_3).\ ††$$

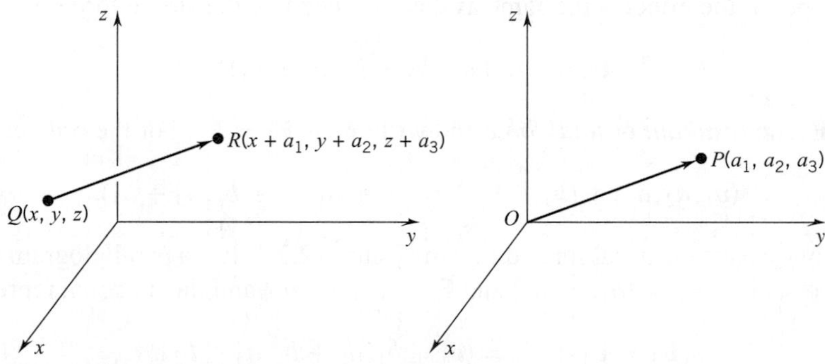

Figure 12.3.1

We use boldface letters to denote vectors. Thus for

$$\mathbf{a} = (a_1, a_2, a_3) \qquad \text{and} \qquad \mathbf{b} = (b_1, b_2, b_3)$$

we have

$$\mathbf{a} + \mathbf{b} = (a_1 + b_1, a_2 + b_2, a_3 + b_3),$$

† Strictly speaking, these are vectors in three-dimensional space. Two-dimensional vectors are represented by ordered pairs of real numbers. There are also four-dimensional vectors which are represented by ordered quadruples, five-dimensional vectors represented by ordered quintuples, and so on. We will be working in three dimensions.

†† The ordered triple $(0, 0, 0)$ will not be represented by an arrow but simply by the origin.

and, if α is a *scalar* (a real number),

$$\alpha\mathbf{a} = (\alpha a_1, \alpha a_2, \alpha a_3).$$

Addition of vectors satisfies the commutative and associative laws:

$$\mathbf{a} + \mathbf{b} = \mathbf{b} + \mathbf{a}$$
$$\mathbf{a} + (\mathbf{b} + \mathbf{c}) = (\mathbf{a} + \mathbf{b}) + \mathbf{c}.$$

These laws follow immediately from the definition of vector addition and the corresponding properties of real numbers. For the *zero vector* $(0, 0, 0)$ we will use the symbol $\mathbf{0}$. Obviously

$$0\mathbf{a} = \mathbf{0} \quad \text{for all vectors } \mathbf{a}.$$

By the vector $-\mathbf{b}$ we mean $(-1)\,\mathbf{b}$; that is,

$$-(b_1, b_2, b_3) = (-b_1, -b_2, -b_3).$$

By $\mathbf{a} - \mathbf{b}$ we mean $\mathbf{a} + (-\mathbf{b})$; that is,

$$\begin{aligned}(a_1, a_2, a_3) - (b_1, b_2, b_3) &= (a_1, a_2, a_3) + (-b_1, -b_2, -b_3)\\ &= (a_1 - b_1, a_2 - b_2, a_3 - b_3).\end{aligned}$$

Example 1 Given that $\mathbf{a} = (1, -1, 2)$, $\mathbf{b} = (2, 3, -1)$, $\mathbf{c} = (8, 7, 1)$, find **(a)** $\mathbf{a} - \mathbf{b}$. **(b)** $2\mathbf{a} + \mathbf{b}$. **(c)** $3\mathbf{a} - 7\mathbf{b}$. **(d)** $2\mathbf{a} + 3\mathbf{b} - \mathbf{c}$.

SOLUTION

(a) $\mathbf{a} - \mathbf{b} = (1, -1, 2) - (2, 3, -1) = (1 - 2, -1 - 3, 2 + 1) = (-1, -4, 3)$.

(b) $2\mathbf{a} + \mathbf{b} = 2(1, -1, 2) + (2, 3, -1) = (2, -2, 4) + (2, 3, -1) = (4, 1, 3)$.

(c) $3\mathbf{a} - 7\mathbf{b} = 3(1, -1, 2) - 7(2, 3, -1)$
$= (3, -3, 6) - (14, 21, -7) = (-11, -24, 13)$.

(d) $2\mathbf{a} + 3\mathbf{b} - \mathbf{c} = 2(1, -1, 2) + 3(2, 3, -1) - (8, 7, 1)$
$= (2, -2, 4) + (6, 9, -3) - (8, 7, 1) = (0, 0, 0) = \mathbf{0}$. □

The addition of vectors can be visualized as the "addition" of arrows by the parallelogram law (as in the case of forces). If you picture **a**, **b**, and **a** + **b** all as arrows emanating from the same point, say the origin, then **a** + **b** acts as the diagonal of the parallelogram generated by **a** and **b**. (See Figure 12.3.2.)

The addition of vectors can also be visualized as the "tail-to-head" addition of arrows (as in the case of displacements). If, instead of starting **b** at the origin, you start **b** at the tip of **a**, then **a** + **b** goes from the tail of **a** to the tip of **b** (Figure 12.3.3). These two pictorial representations of vector addition lead to the same result.

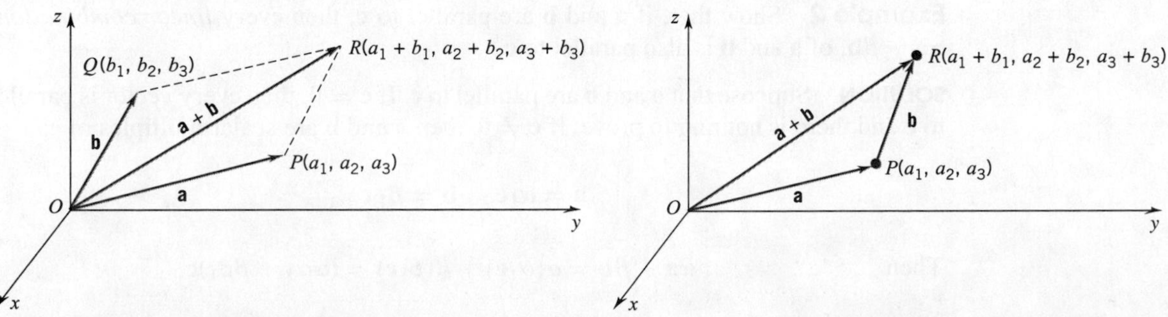

Figure 12.3.2 **Figure 12.3.3**

DEFINITION 12.3.2 PARALLEL VECTORS

Two nonzero vectors **a** and **b** are said to be *parallel* provided that

$$\mathbf{a} = \alpha\mathbf{b} \quad \text{for some real number } \alpha .$$

If $\alpha > 0$, **a** and **b** are said to have the *same direction;* if $\alpha < 0$, they are said to have *opposite directions*.

In the case of $\mathbf{a} = (2, -2, 6)$, $\mathbf{b} = (1, -1, 3)$, $\mathbf{c} = (-1, 1, -3)$ we have

$$\mathbf{a} = 2\mathbf{b} \quad \text{and} \quad \mathbf{a} = -2\mathbf{c}.$$

This tells us that **a** and **b** are parallel and have the same direction, whereas **a** and **c**, though parallel, have opposite directions. (See Figure 12.3.4.)

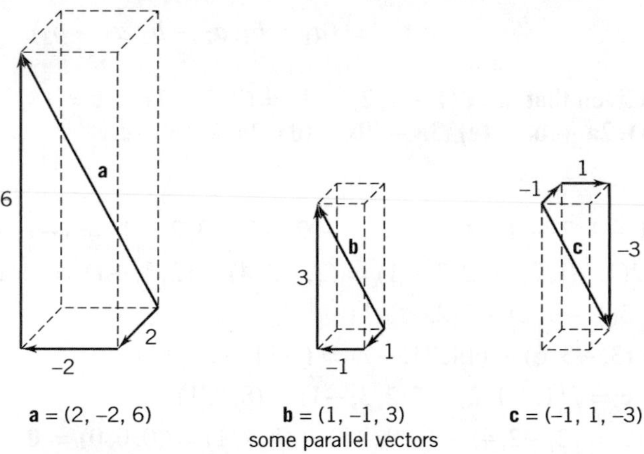

Figure 12.3.4

Definition 12.3.2 did not include the zero vector **0**. By special convention, **0** is said to be *parallel to every vector*.

(Since **0** is represented geometrically by a point, there is no geometric meaning to saying that **0** is parallel to another vector. However, it simplifies the statement of certain results to maintain that **0** is parallel to every vector. Algebraically this is warranted by the fact that **0** is a scalar multiple of every vector **b**: $\mathbf{0} = 0\mathbf{b}$.)

Example 2 Show that, if **a** and **b** are parallel to **c**, then every *linear combination*, $\alpha\mathbf{a} + \beta\mathbf{b}$, of **a** and **b** is also parallel to **c**.

SOLUTION Suppose that **a** and **b** are parallel to **c**. If $\mathbf{c} = \mathbf{0}$, then every vector is parallel to **c** and there is nothing to prove. If $\mathbf{c} \neq \mathbf{0}$, then **a** and **b** are scalar multiples of **c**:

$$\mathbf{a} = \alpha_1\mathbf{c}, \quad \mathbf{b} = \beta_1\mathbf{c}.$$

Then
$$\alpha\mathbf{a} + \beta\mathbf{b} = \alpha(\alpha_1\mathbf{c}) + \beta(\beta_1\mathbf{c}) = (\alpha\alpha_1 + \beta\beta_1)\mathbf{c}$$

is also parallel to **c**. □

DEFINITION 12.3.3 NORM

The *norm* of a vector $\mathbf{a} = (a_1, a_2, a_3)$, denoted by $\|\mathbf{a}\|$, is the number

$$\|\mathbf{a}\| = \sqrt{a_1^2 + a_2^2 + a_3^2}.$$

The norm of **a** is also called the *length* or *magnitude* of **a**: if we represent the vector **a** by an arrow, $\overrightarrow{QR}$, with $Q = Q(x, y, z)$ and $R(x + a_1, y + a_2, z + a_3)$, then $\|\mathbf{a}\|$ gives the length of $\overrightarrow{QR}$. (Figure 12.3.5)

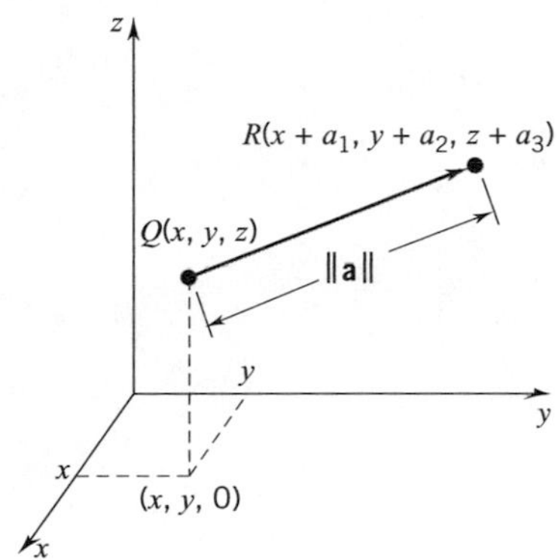

Figure 12.3.5

The norm properties of vectors are very similar to the absolute value properties of real numbers. In particular

(12.3.4)

(1) $\|\mathbf{a}\| \geq 0$ and $\|\mathbf{a}\| = 0$ iff $\mathbf{a} = \mathbf{0}$.

(2) $\|\alpha\mathbf{a}\| = |\alpha|\ \|\mathbf{a}\|$.

(3) $\|\mathbf{a} + \mathbf{b}\| \leq \|\mathbf{a}\| + \|\mathbf{b}\|$. (the triangle inequality)

Property (1) is obvious. Property (2) is easy to verify:

$$\|\alpha\mathbf{a}\| = \sqrt{(\alpha a_1)^2 + (\alpha a_2)^2 + (\alpha a_3)^2} = |\alpha|\sqrt{a_1^2 + a_2^2 + a_3^2} = |\alpha|\ \|\mathbf{a}\|.$$

We prove Property (3), the triangle inequality, in the next section, where we have "dot products" at our disposal. A proof at this time would be laborious.

You can interpret the triangle inequality as saying that the length of a side of a triangle cannot exceed the sum of the lengths of the other two sides. (See Figure 12.3.6.)

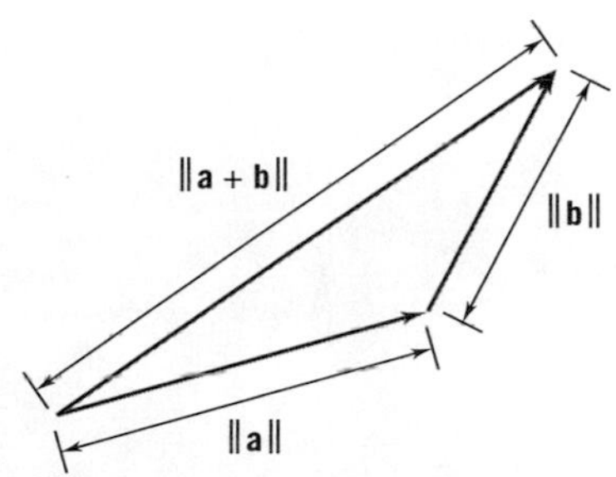

Figure 12.3.6

Example 3 Given that $\mathbf{a} = (1, -2, 3)$ and $\mathbf{b} = (-4, 1, 0)$, calculate

(a) $\|\mathbf{a}\|$. **(b)** $\|\mathbf{b}\|$. **(c)** $\|\mathbf{a} + \mathbf{b}\|$. **(d)** $\|\mathbf{a} - \mathbf{b}\|$. **(e)** $\|-7\mathbf{a}\|$.
(f) $\|2\mathbf{a} - 3\mathbf{b}\|$.

SOLUTION

(a) $\|\mathbf{a}\| = \sqrt{1^2 + (-2)^2 + 3^2} = \sqrt{1 + 4 + 9} = \sqrt{14}$.

(b) $\|\mathbf{b}\| = \sqrt{(-4)^2 + 1^2 + 0^2} = \sqrt{16 + 1 + 0} = \sqrt{17}$.

(c) $\|\mathbf{a} + \mathbf{b}\| = \|(-3, -1, 3)\| = \sqrt{(-3)^2 + (-1)^2 + 3^2} = \sqrt{9 + 1 + 9} = \sqrt{19}$.

(d) $\|\mathbf{a} - \mathbf{b}\| = \|(5, -3, 3)\| = \sqrt{5^2 + (-3)^2 + 3^2} = \sqrt{25 + 9 + 9} = \sqrt{43}$.

(e) $\|-7\mathbf{a}\| = |-7|\ \|\mathbf{a}\| = 7\sqrt{14}$.

(f) $\|2\mathbf{a} - 3\mathbf{b}\| = \|2(1, -2, 3) - 3(-4, 1, 0)\|$

$$= \|(14, -7, 6)\| = \sqrt{14^2 + (-7)^2 + 6^2}$$

$$= \sqrt{196 + 49 + 36} = \sqrt{281}. \quad \square$$

To multiply a nonzero vector by a nonzero scalar α is to change its length by a factor of $|\alpha|$,

$$\|\alpha\mathbf{a}\| = |\alpha|\ \|\mathbf{a}\|,$$

keeping the same direction if $\alpha > 0$ and reversing the direction if $\alpha < 0$. The vector obtained from $\mathbf{a}$ simply by reversing its direction is the vector $(-1)\mathbf{a} = -\mathbf{a}$. (See Figure 12.3.7.)

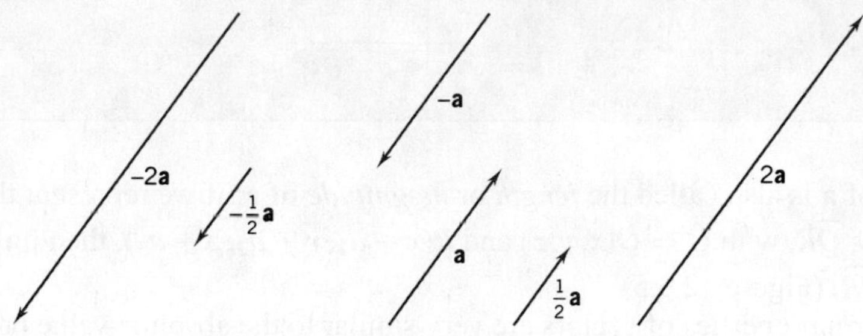

Figure 12.3.7

Since $\mathbf{a} - \mathbf{b} = \mathbf{a} + (-\mathbf{b})$, we can draw the vector $\mathbf{a} - \mathbf{b}$ by drawing $-\mathbf{b}$ and adding it to the vector $\mathbf{a}$ (Figure 12.3.8).

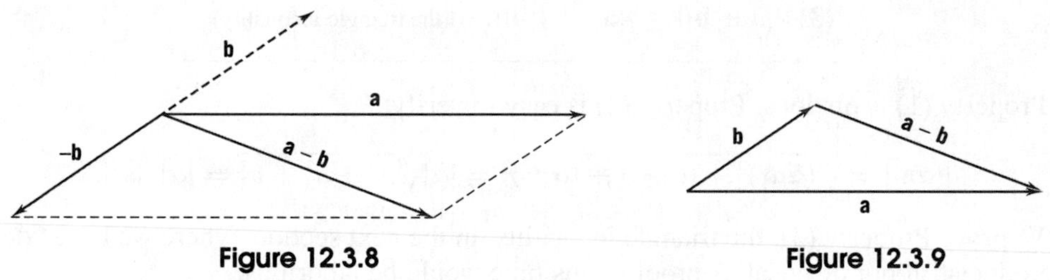

Figure 12.3.8 Figure 12.3.9

We can obtain the same result more easily by noting that $\mathbf{a} - \mathbf{b}$ is the vector that we must add to $\mathbf{b}$ to obtain $\mathbf{a}$ (Figure 12.3.9).

Vectors of norm 1 are called *unit vectors*. If $\mathbf{b}$ is a nonzero vector, then there is a unit vector $\mathbf{u_b}$ that has the direction of $\mathbf{b}$. To find $\mathbf{u_b}$, note that

$$||\mathbf{u_b}|| = 1 \quad \text{and} \quad \mathbf{u_b} = \alpha\mathbf{b} \quad \text{for some } \alpha > 0.$$

It follows that

$$1 = ||\mathbf{u_b}|| = ||\alpha\mathbf{b}|| = |\alpha|\,||\mathbf{b}|| = \alpha||\mathbf{b}||.$$

Thus

$$\alpha = \frac{1}{||\mathbf{b}||} \quad \text{and consequently} \quad \mathbf{u_b} = \frac{1}{||\mathbf{b}||}\mathbf{b} = \frac{\mathbf{b}}{||\mathbf{b}||}.$$

While

$$\mathbf{u_b} = \frac{\mathbf{b}}{||\mathbf{b}||}$$

is the unit vector in the direction of $\mathbf{b}$,

$$-\mathbf{u_b} = -\frac{\mathbf{b}}{||\mathbf{b}||}$$

is the unit vector in the opposite direction.

We single out for special attention the vectors

$$\mathbf{i} = (1, 0, 0), \quad \mathbf{j} = (0, 1, 0), \quad \mathbf{k} = (0, 0, 1).$$

These vectors all have norm 1 and, if pictured as emanating from the origin, lie along the positive coordinate axes. They are called *the unit coordinate vectors*. (See Figure 12.3.10)

Every vector can be expressed as a linear combination of the unit coordinate vectors:

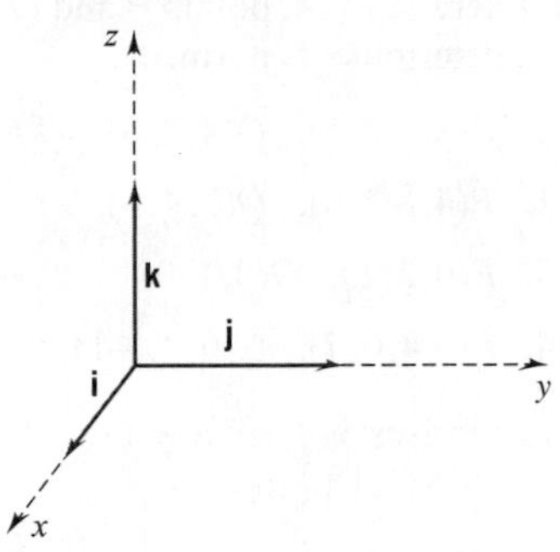

Figure 12.3.10

(12.3.5) $$\text{for } \mathbf{a} = (a_1, a_2, a_3) \quad \text{we have} \quad \mathbf{a} = a_1\,\mathbf{i} + a_2\,\mathbf{j} + a_3\,\mathbf{k}.$$

PROOF

$$\begin{aligned}(a_1, a_2, a_3) &= (a_1, 0, 0) + (0, a_2, 0) + (0, 0, a_3)\\ &= a_1(1, 0, 0) + a_2(0, 1, 0) + a_3(0, 0, 1)\\ &= a_1\,\mathbf{i} + a_2\,\mathbf{j} + a_3\,\mathbf{k}. \quad \square\end{aligned}$$

The numbers a_1, a_2, a_3 are called the **i**, **j**, **k** components of the vector **a**.

Example 4 Give that $\mathbf{a} = 3\mathbf{i} - \mathbf{j} + \mathbf{k}$ and $\mathbf{b} = 2\mathbf{i} + 3\mathbf{j} - \mathbf{k}$,

(1) Express $2\mathbf{a} - \mathbf{b}$ as a linear combination of **i**, **j**, **k**.

(2) Calculate $||2\mathbf{a} - \mathbf{b}||$.

(3) Find the unit vector $\mathbf{u}_\mathbf{c}$ in the direction of $\mathbf{c} = 2\mathbf{a} - \mathbf{b}$.

SOLUTION

(1) $$\begin{aligned}2\mathbf{a} - \mathbf{b} &= 2(3\mathbf{i} - \mathbf{j} + \mathbf{k}) - (2\mathbf{i} + 3\mathbf{j} - \mathbf{k})\\ &= 6\mathbf{i} - 2\mathbf{j} + 2\mathbf{k} - 2\mathbf{i} - 3\mathbf{j} + \mathbf{k} = 4\mathbf{i} - 5\mathbf{j} + 3\mathbf{k}.\end{aligned}$$

(2) $||2\mathbf{a} - \mathbf{b}|| = ||4\mathbf{i} - 5\mathbf{j} + 3\mathbf{k}|| = \sqrt{16 + 25 + 9} = \sqrt{50} = 5\sqrt{2}$.

(3) $\mathbf{u}_\mathbf{c} = \dfrac{2\mathbf{a} - \mathbf{b}}{||2\mathbf{a} - \mathbf{b}||} = \dfrac{1}{5\sqrt{2}}(4\mathbf{i} - 5\mathbf{j} + 3\mathbf{k})$. $\square$

Remark: Vectors in the Plane A vector $\mathbf{a} = a_1\,\mathbf{i} + a_2\,\mathbf{j} + a_3\,\mathbf{k}$ for which $a_3 = 0$ is a vector in the xy-plane. Such a vector can be written more simply as $\mathbf{a} = a_1\,\mathbf{i} + a_2\,\mathbf{j}$ and identified with the ordered pair (a_1, a_2).

The unit coordinate vectors $\mathbf{i} = (1, 0, 0)$ and $\mathbf{j} = (0, 1, 0)$ are vectors in the xy-plane and are identified with the ordered pairs $(1, 0)$ and $(0, 1)$, respectively. These are the unit coordinate vectors and we will continue to denote them by **i** and **j**. As in three-dimensional space, every vector $\mathbf{a} = (a_1, a_2)$ in the plane can be expressed as a linear combination of the unit coordinate vectors:

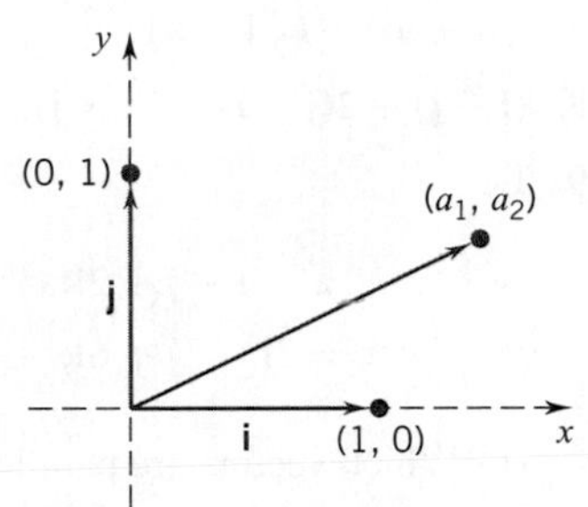

Figure 12.3.11

(12.3.6) $$\mathbf{a} = a_1(1, 0) + a_2(0, 1) = a_1\,\mathbf{i} + a_2\,\mathbf{j}.$$ (Figure 12.3.11)

The definitions and results stated for vectors in space hold for vectors in the plane as well. In particular, if $\mathbf{a} = (a_1, a_2)$ and $\mathbf{b} = (b_1, b_2)$ are vectors in the plane and α is a scalar, then

(1) $\mathbf{a} = \mathbf{b}$ iff $a_1 = b_1 \quad a_2 = b_2$.

(2) $\mathbf{a} + \mathbf{b} = (a_1 + b_1, a_2 + b_2) = (a_1 + b_1)\mathbf{i} + (a_2 + b_2)\mathbf{j}$.

(3) $\alpha\mathbf{a} = (\alpha a_1, \alpha a_2) = \alpha a_1\,\mathbf{i} + \alpha a_2\,\mathbf{j}$.

(4) $\mathbf{0} = (0, 0)$ is the zero vector.

(5) $||\mathbf{a}|| = \sqrt{a_1^2 + a_2^2}$ is the norm, or magnitude of **a**.

We will use ordered pairs (a_1, a_2) or the form (12.3.6) to treat two dimensional problems.

EXERCISES 12.3

In Exercises 1–4, points P and Q are given. Find the vector $\overrightarrow{PQ}$ and determine its norm.

1. $P(1, -2, 5)$, $Q(4, 2, 3)$.

2. $P(4, -2, 0)$, $Q(2, 4, 0)$.

3. $P(0, 3, 1)$, $Q(0, 1, 0)$.

4. $P(-4, 0, 7)$, $Q(0, 3, -1)$.

In Exercises 5–8, set $\mathbf{a} = (1, -2, 3)$, $\mathbf{b} = (3, 0, -1)$, $\mathbf{c} = (-4, 2, 1)$. Find:

5. $2\mathbf{a} - \mathbf{b}$.

6. $2\mathbf{b} + 3\mathbf{c}$.

7. $-2\mathbf{a} + \mathbf{b} - \mathbf{c}$.

8. $\mathbf{a} + 3\mathbf{b} - 2\mathbf{c}$.

Simplify the linear combinations.

9. $(2\mathbf{i} - \mathbf{j} + \mathbf{k}) + (\mathbf{i} - 3\mathbf{j} + 5\mathbf{k})$.

10. $(6\mathbf{j} - \mathbf{k}) + (3\mathbf{i} - \mathbf{j} + 2\mathbf{k})$.

11. $2(\mathbf{j} + \mathbf{k}) - 3(\mathbf{i} + \mathbf{j} - 2\mathbf{k})$.

12. $2(\mathbf{i} - \mathbf{j}) + 6(2\mathbf{i} + \mathbf{j} - 2\mathbf{k})$.

Calculate the norm of the vector.

13. $3\mathbf{i} + 4\mathbf{j}$.

14. $\mathbf{i} - \mathbf{j}$.

15. $2\mathbf{i} + \mathbf{j} - 2\mathbf{k}$.

16. $6\mathbf{i} + 2\mathbf{j} - \mathbf{k}$.

17. $\frac{1}{2}(\mathbf{i} + 4\mathbf{j}) - (\frac{3}{2}\,\mathbf{i} + \mathbf{k})$.

18. $(\mathbf{i} - \mathbf{j}) + 2(\mathbf{j} - \mathbf{i}) + (\mathbf{k} - \mathbf{j})$.

19. Let

$$\mathbf{a} = \mathbf{i} - \mathbf{j} + 2\mathbf{k}, \quad \mathbf{b} = 2\mathbf{i} - \mathbf{j} + 2\mathbf{k},$$
$$\mathbf{c} = 3\mathbf{i} - 3\mathbf{j} + 6\mathbf{k}, \quad \mathbf{d} = -2\mathbf{i} + 2\mathbf{j} - 4\mathbf{k}.$$

(a) Which vectors are parallel?

(b) Which vectors have the same direction?

(c) Which vectors have opposite directions?

20. (*Important*) Prove the following version of the triangle inequality:

$$\big|\, \|\mathbf{a}\| - \|\mathbf{b}\| \,\big| \le \|\mathbf{a} - \mathbf{b}\|.$$

HINT: $\mathbf{a} = (\mathbf{a} - \mathbf{b}) + \mathbf{b}$

Find the unit vector in the direction of $\mathbf{a}$.

21. $\mathbf{a} = (3, -4, 0)$.

22. $\mathbf{a} = -2\mathbf{i} + 3\mathbf{j}$.

23. $\mathbf{a} = \mathbf{i} - 2\mathbf{j} + 2\mathbf{k}$.

24. $\mathbf{a} = (2, 1, 2)$.

In Exercises 25 and 26, find the unit vector in the direction opposite to the direction of $\mathbf{a}$.

25. $\mathbf{a} = -\mathbf{i} + 3\mathbf{j} + 2\mathbf{k}$.

26. $\mathbf{a} = 2\mathbf{i} - \mathbf{k}$.

27. Label the vectors.

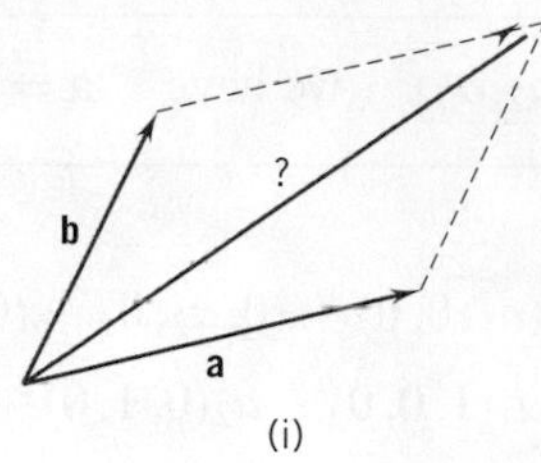

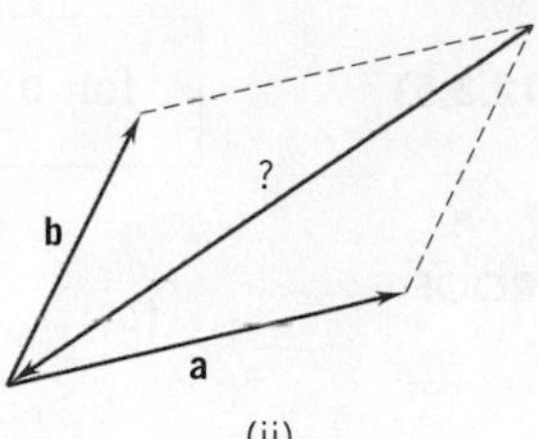

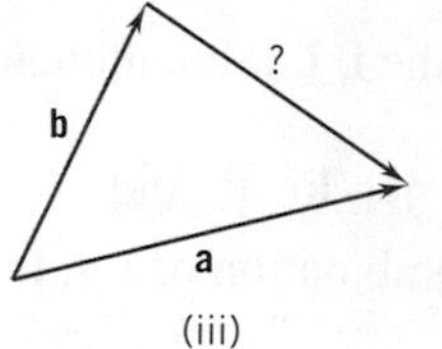

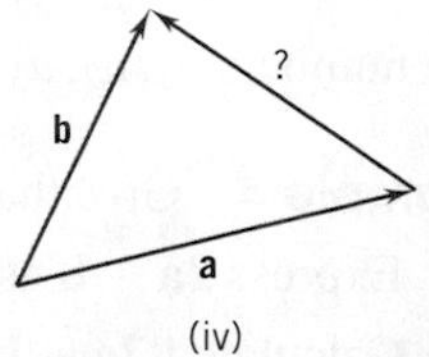

28. Let $\mathbf{a} = (1, 1, 1)$, $\mathbf{b} = (-1, 3, 2,)$, $\mathbf{c} = (-3, 0, 1)$, $\mathbf{d} = (4, -1, 1)$.

(a) Express $\mathbf{a} + 2\mathbf{b} + 3\mathbf{c} + 4\mathbf{d}$ as a linear combination of $\mathbf{i}, \mathbf{j}, \mathbf{k}$.

(b) Find scalars A, B, C such that $\mathbf{d} = A\mathbf{a} + B\mathbf{b} + C\mathbf{c}$.

29. Let $\mathbf{a} = (2, 0, -1)$, $\mathbf{b} = (1, 3, 5)$, $\mathbf{c} = (-1, 1, 1)$, $\mathbf{d} = (1, 1, 6)$.

(a) Express $\mathbf{a} - 3\mathbf{b} + 2\mathbf{c} + 4\mathbf{d}$ as a linear combination of $\mathbf{i}, \mathbf{j}, \mathbf{k}$.

(b) Find scalars A, B, C such that $\mathbf{d} = A\mathbf{a} + B\mathbf{b} + C\mathbf{c}$.

30. Find α given that $3\mathbf{i} + \mathbf{j} - \mathbf{k}$ and $\alpha\mathbf{i} - 4\mathbf{j} + 4\mathbf{k}$ are parallel.

31. Find α given that $3\mathbf{i} + \mathbf{j}$ and $\alpha\mathbf{j} - \mathbf{k}$ have the same length.

32. Find the unit vector in the direction of $\mathbf{i} - 2\mathbf{j} + 2\mathbf{k}$.

33. Find α given that $\|\alpha\mathbf{i} + (\alpha - 1)\mathbf{j} + (\alpha + 1)\mathbf{k}\| = 2$.

34. Find the vector of norm 2 in the direction of $\mathbf{i} + 2\mathbf{j} - \mathbf{k}$.

35. Find the vectors of norm 2 parallel to $3\mathbf{j} + 2\mathbf{k}$.

36. Express $\mathbf{c}$ in terms of $\mathbf{a}$ and $\mathbf{b}$, given that the tip of $\mathbf{c}$ bisects the line segment.

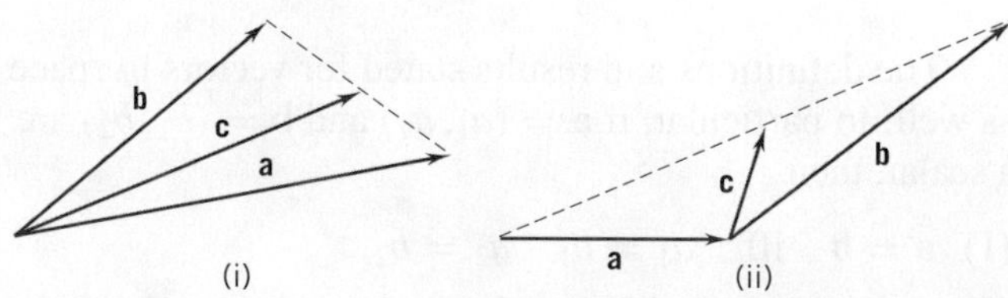

37. Let $\mathbf{a}$ and $\mathbf{b}$ be nonzero vectors such that

$$\|\mathbf{a} - \mathbf{b}\| = \|\mathbf{a} + \mathbf{b}\|.$$

(a) What can you conclude about the parallelogram generated by $\mathbf{a}$ and $\mathbf{b}$?

(b) Show that, if $\mathbf{a} = a_1\mathbf{i}+a_2\mathbf{j}+a_3\mathbf{k}$ and $\mathbf{b} = b_1\mathbf{i}+b_2\mathbf{j}+b_3\mathbf{k}$, then

$$a_1b_1 + a_2b_2 + a_3b_3 = 0.$$

38. (a) Show that, if **a** and **b** have the same direction, then

$$||\mathbf{a} + \mathbf{b}|| = ||\mathbf{a}|| + ||\mathbf{b}||.$$

(b) Does this equation necessarily hold if **a** and **b** are only parallel?

39. Let P and Q be two points in space and let M be the midpoint of the line segment $\overline{PQ}$. Let $\mathbf{p} = \overrightarrow{OP}$, $\mathbf{q} = \overrightarrow{OQ}$, and $\mathbf{m} = \overrightarrow{OM}$.

(a) Show that $\mathbf{m} = \mathbf{p} + \frac{1}{2}(\mathbf{q} - \mathbf{p})$.

(b) Derive the midpoint formula (12.1.4).

40. Let P and Q be two points in space and let R be the point on $\overline{PQ}$ which is twice as far from P as it is from Q. Let $\mathbf{p} = \overrightarrow{OP}$, $\mathbf{q} = \overrightarrow{OQ}$, and $\mathbf{r} = \overrightarrow{OR}$. Prove that $\mathbf{r} = \frac{1}{3}\mathbf{p} + \frac{2}{3}\mathbf{q}$.

A vector **r** emanating from the origin is called a *radius* vector. Each radius vector determines a unique point of space: the point at the tip of the vector. Conversely, each point of space determines a unique radius vector: the radius vector whose tip falls on that point. This one-to-one correspondence between the set of all radius vectors and the set of all points in three-dimensional space enables us to use radius vectors to specify sets in space. Thus, for example, the radius-vector equation $||\mathbf{r}|| = 3$ can be used to represent the sphere of radius 3 centered at the origin: the sphere consists of the tips of all the radius vectors **r** that satisfy that equation.

41. Write a radius-vector equation or inequality for each of the following sets.

(a) The sphere of radius 3 centered at $P(a_1, a_2, a_3)$.

(b) The set of all points on or inside the sphere of radius 2 centered at the origin. (This set is called the *ball* of radius 2 about the origin.)

(c) The ball of radius 1 about the point $P(a_1, a_2, a_3)$.

42. Write a radius-vector equation for each of the following sets.

(a) The set of all points equidistant from $P(a_1, a_2, a_3)$ and $Q(b_1, b_2, b_3)$. (This set forms a plane.)

(b) The set of all points the sum of whose distances from $P(a_1, a_2, a_3)$ and $Q(b_1, b_2, b_3)$ is a constant $k > d(P, Q)$. [Such a set is an example of an *ellipsoid*. An ellipsoid is a three-dimensional analogue of the ellipse.] What happens if $k = d(P, Q)$? If $k < d(P, Q)$?

■ 12.4 THE DOT PRODUCT

In this section we introduce the first of two products that we define for vectors.

Introduction

We begin with two nonzero vectors

$$\mathbf{a} = a_1\mathbf{i} + a_2\mathbf{j} + a_3\mathbf{k}, \quad \mathbf{b} = b_1\mathbf{i} + b_2\mathbf{j} + b_3\mathbf{k}.$$

How can we tell from the components of these vectors whether these vectors meet at right angles? To explore this question we draw Figure 12.4.1. By the Pythagorean theorem, **a** and **b** meet at right angles iff

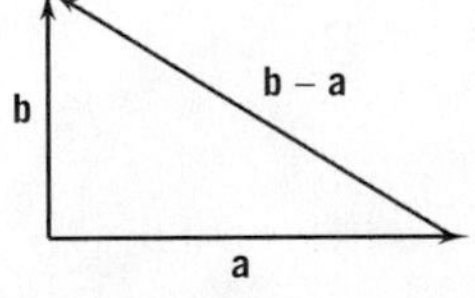

Figure 12.4.1

$$||\mathbf{a}||^2 + ||\mathbf{b}||^2 = ||\mathbf{b} - \mathbf{a}||^2.$$

In terms of components this equation reads

$$(a_1^2 + a_2^2 + a_3^2) + (b_1^2 + b_2^2 + b_3^2) = (b_1 - a_1)^2 + (b_2 - a_2)^2 + (b_3 - a_3)^2,$$

which, as you can readily check, simplifies to

$$a_1b_1 + a_2b_2 + a_3b_3 = 0.$$

The expression $a_1b_1 + a_2b_2 + a_3b_3$ is widely used in geometry and in physics. It has a name, the *dot product* of **a** and **b**, and there is a special notation for it, $\mathbf{a} \cdot \mathbf{b}$. The notion is so important that it deserves a formal definition.

Definition of the Dot Product

DEFINITION 12.4.1

For any two vectors

$$\mathbf{a} = a_1\,\mathbf{i} + a_2\,\mathbf{j} + a_3\,\mathbf{k} \qquad \text{and} \qquad \mathbf{b} = b_1\,\mathbf{i} + b_2\,\mathbf{j} + b_3\,\mathbf{k},$$

we define the *dot product* $\mathbf{a} \cdot \mathbf{b}$ by setting

$$\mathbf{a} \cdot \mathbf{b} = a_1 b_1 + a_2 b_2 + a_3 b_3.$$

Remark The dot product of two vectors in n-dimensional space, $n = 2, 4, 5, \ldots,$ is defined in the same way. In particular, for $\mathbf{a} = a_1\mathbf{i} + a_2\mathbf{j}$ and $\mathbf{b} = b_1\mathbf{i} + b_2\mathbf{j}$ in the xy-plane,

$$\mathbf{a} \cdot \mathbf{b} = a_1 b_1 + a_2 b_2.$$

The properties of the dot product presented in this section for three-dimensional vectors also hold for vectors in n-dimensional space. □

Example 1 For $\mathbf{a} = 2\mathbf{i} - \mathbf{j} + 3\mathbf{k}, \quad \mathbf{b} = -3\mathbf{i} + \mathbf{j} + 4\mathbf{k}, \quad \mathbf{c} = \mathbf{i} + 3\mathbf{j},$

we have

$$\begin{aligned} \mathbf{a} \cdot \mathbf{b} &= (2)(-3) + (-1)(1) + (3)(4) = -6 - 1 + 12 = 5, \\ \mathbf{a} \cdot \mathbf{c} &= (2)(1) + (-1)(3) + (3)(0) = 2 - 3 = -1, \\ \mathbf{b} \cdot \mathbf{c} &= (-3)(1) + (1)(3) + (4)(0) = -3 + 3 = 0. \end{aligned}$$

The last equation tells us that **b** and **c** meet at right angles. (Verify this by drawing a figure.) □

Because $\mathbf{a} \cdot \mathbf{b}$ is not a vector, but a scalar, it is sometimes called the *scalar product* of **a** and **b**. We will continue to call it the dot product and speak of "dotting **a** with **b**."

Properties of the Dot Product

If we dot a vector with itself, we obtain the square of its norm:

(12.4.2) $$\mathbf{a} \cdot \mathbf{a} = \|\mathbf{a}\|^2.$$

PROOF

$$\mathbf{a} \cdot \mathbf{a} = a_1 a_1 + a_2 a_2 + a_3 a_3 = a_1^2 + a_2^2 + a_3^2 = \|\mathbf{a}\|^2. \quad \square$$

The dot product of any vector with the zero vector is zero:

(12.4.3) $$\mathbf{a} \cdot \mathbf{0} = 0, \qquad \mathbf{0} \cdot \mathbf{a} = 0.$$

PROOF

$$(a_1)(0) + (a_2)(0) + (a_3)(0) = 0, \qquad (0)(a_1) + (0)(a_2) + (0)(a_3) = 0. \quad \square$$

The dot product is commutative:

(12.4.4) $$\mathbf{a}\cdot\mathbf{b} = \mathbf{b}\cdot\mathbf{a},$$

and scalars can be factored:

(12.4.5) $$\alpha\mathbf{a}\cdot\beta\mathbf{b} = \alpha\beta(\mathbf{a}\cdot\mathbf{b}).$$

PROOF

$$\mathbf{a}\cdot\mathbf{b} = a_1b_1 + a_2b_2 + a_3b_3 = b_1a_1 + b_2a_2 + b_3a_3 = \mathbf{b}\cdot\mathbf{a},$$

and
$$\begin{aligned}\alpha\mathbf{a}\cdot\beta\mathbf{b} &= (\alpha a_1)(\beta b_1) + (\alpha a_2)(\beta b_2) + (\alpha a_3)(\beta b_3)\\ &= \alpha\beta(a_1b_1 + a_2b_2 + a_3b_3) = \alpha\beta(\mathbf{a}\cdot\mathbf{b}). \quad \square\end{aligned}$$

The dot product satisfies the following distributive laws:

(12.4.6) $$\mathbf{a}\cdot(\mathbf{b}+\mathbf{c}) = \mathbf{a}\cdot\mathbf{b} + \mathbf{a}\cdot\mathbf{c}, \qquad (\mathbf{a}+\mathbf{b})\cdot\mathbf{c} = \mathbf{a}\cdot\mathbf{c} + \mathbf{b}\cdot\mathbf{c}.$$

PROOF

$$\begin{aligned}\mathbf{a}\cdot(\mathbf{b}+\mathbf{c}) &= a_1(b_1+c_1) + a_2(b_2+c_2) + a_3(b_3+c_3)\\ &= a_1b_1 + a_1c_1 + a_2b_2 + a_2c_2 + a_3b_3 + a_3c_3\\ &= (a_1b_1 + a_2b_2 + a_3b_3) + (a_1c_1 + a_2c_2 + a_3c_3)\\ &= \mathbf{a}\cdot\mathbf{b} + \mathbf{a}\cdot\mathbf{c}.\end{aligned}$$

The second equation can be verified in a similar manner. □

Example 2 Given that
$\|\mathbf{a}\| = 1,\quad \|\mathbf{b}\| = 3,\quad \|\mathbf{c}\| = 4,\quad \mathbf{a}\cdot\mathbf{b} = 0,\quad \mathbf{a}\cdot\mathbf{c} = 1,\quad \mathbf{b}\cdot\mathbf{c} = -2,$
find (a) $3\mathbf{a}\cdot(\mathbf{b}+4\mathbf{c})$. (b) $(\mathbf{a}-\mathbf{b})\cdot(2\mathbf{a}+\mathbf{b})$. (c) $[(\mathbf{b}\cdot\mathbf{c})\,\mathbf{a} - (\mathbf{a}\cdot\mathbf{c})\,\mathbf{b}]\cdot\mathbf{c}$.

SOLUTION

(a) $3\mathbf{a}\cdot(\mathbf{b}+4\mathbf{c}) = (3\mathbf{a}\cdot\mathbf{b}) + (3\mathbf{a}\cdot 4\mathbf{c}) = 3(\mathbf{a}\cdot\mathbf{b}) + 12(\mathbf{a}\cdot\mathbf{c}) = 12.$

(b)
$$\begin{aligned}(\mathbf{a}-\mathbf{b})\cdot(2\mathbf{a}+\mathbf{b}) &= (\mathbf{a}\cdot 2\mathbf{a}) + (\mathbf{a}\cdot\mathbf{b}) + (-\mathbf{b}\cdot 2\mathbf{a}) + (-\mathbf{b}\cdot\mathbf{b})\\ &= 2(\mathbf{a}\cdot\mathbf{a}) + (\mathbf{a}\cdot\mathbf{b}) - 2(\mathbf{b}\cdot\mathbf{a}) - (\mathbf{b}\cdot\mathbf{b})\\ &= 2\|\mathbf{a}\|^2 + (\mathbf{a}\cdot\mathbf{b}) - 2(\mathbf{a}\cdot\mathbf{b}) - \|\mathbf{b}\|^2\\ &= 2 + 0 - 2(0) - 9 = -7.\end{aligned}$$

(c)
$$\begin{aligned}[(\mathbf{b}\cdot\mathbf{c})\,\mathbf{a} - (\mathbf{a}\cdot\mathbf{c})\,\mathbf{b}]\cdot\mathbf{c} &= [(\mathbf{b}\cdot\mathbf{c})\,\mathbf{a}\cdot\mathbf{c}] - [(\mathbf{a}\cdot\mathbf{c})\,\mathbf{b}\cdot\mathbf{c}]\\ &= (\mathbf{b}\cdot\mathbf{c})(\mathbf{a}\cdot\mathbf{c}) - (\mathbf{a}\cdot\mathbf{c})(\mathbf{b}\cdot\mathbf{c}) = 0. \quad \square\end{aligned}$$

Geometric Interpretation of the Dot Product

We begin with a triangle with sides a, b, c (Figure 12.4.2). If θ were $\frac{1}{2}\pi$, the pythagorean theorem would tell us that $c^2 = a^2 + b^2$. The law of cosines,

$$c^2 = a^2 + b^2 - 2ab\cos\theta,$$

is a generalization of the Pythagorean theorem.

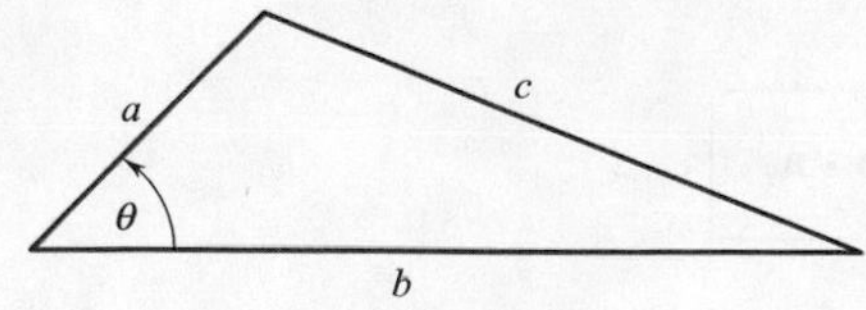

Figure 12.4.2

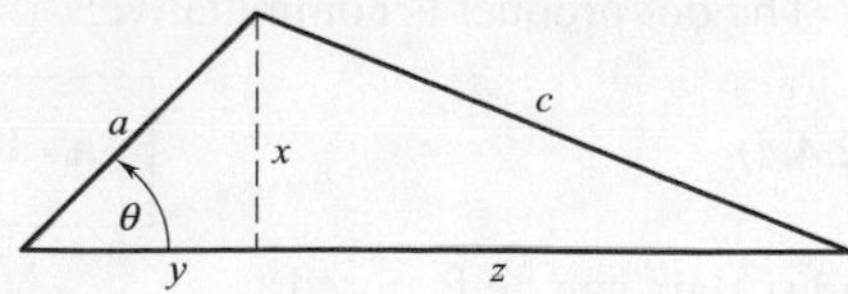

Figure 12.4.3

To derive the law of cosines, we drop a perpendicular to side b (Figure 12.4.3). We then have

$$c^2 = z^2 + x^2 = (b - y)^2 + x^2 = b^2 - 2by + y^2 + x^2.$$

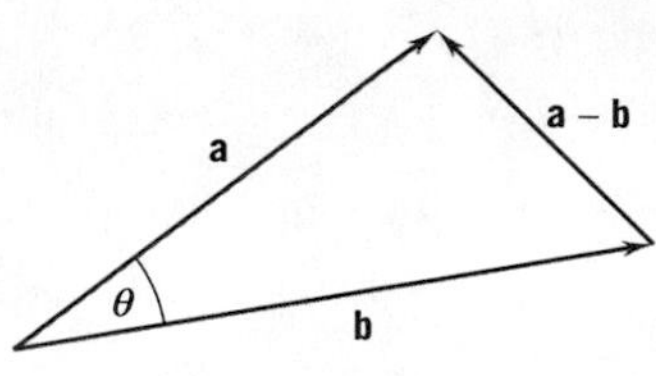

Figure 12.4.4

From the figure, $y^2 + x^2 = a^2$ and $y = a \cos\theta$. Therefore

$$c^2 = a^2 + b^2 - 2ab\cos\theta,$$

as asserted. (What if the angle θ is obtuse? We leave that case to you.)

Now back to dot products. If neither $\mathbf{a}$ nor $\mathbf{b}$ is zero, we can interpret $\mathbf{a} \cdot \mathbf{b}$ from the triangle of Figure 12.4.4. The lengths of the sides are $||\mathbf{a}||$, $||\mathbf{b}||$, $||\mathbf{a} - \mathbf{b}||$. By the law of cosines

$$||\mathbf{a} - \mathbf{b}||^2 = ||\mathbf{a}||^2 + ||\mathbf{b}||^2 - 2||\mathbf{a}||\,||\mathbf{b}||\cos\theta.$$

This gives

$$2||\mathbf{a}||\,||\mathbf{b}||\cos\theta = ||\mathbf{a}||^2 + ||\mathbf{b}||^2 - ||\mathbf{a} - \mathbf{b}||^2 = a_1^2 + a_2^2 + a_3^2 + b_1^2 + b_2^2 + b_3^2 - (a_1 - b_1)^2 - (a_2 - b_2)^2 - (a_3 - b_3)^2 = 2(a_1b_1 + a_2b_2 + a_3b_3) = 2(\mathbf{a} \cdot \mathbf{b}),$$

and thus

(12.4.7) $$\boxed{\mathbf{a} \cdot \mathbf{b} = ||\mathbf{a}||\,||\mathbf{b}||\cos\theta.}$$

By convention, θ, the angle between $\mathbf{a}$ and $\mathbf{b}$, is measured in radians and taken from 0 to π; no negative angles.

From (12.4.7) you can see that the dot product of two vectors depends on the norms of the vectors and on the angle between them. For vectors of a given norm, the dot product measures the extent to which the vectors agree in direction. As the difference in direction increases, the dot product decreases:

If $\mathbf{a}$ and $\mathbf{b}$ have the same direction, then $\theta = 0$ and

$$\mathbf{a} \cdot \mathbf{b} = ||\mathbf{a}||\,||\mathbf{b}||; \qquad (\cos 0 = 1)$$

this is the largest possible value for $\mathbf{a} \cdot \mathbf{b}$.

If $\mathbf{a}$ and $\mathbf{b}$ meet at right angles, then $\theta = \frac{1}{2}\pi$ and

$$\mathbf{a} \cdot \mathbf{b} = 0. \qquad (\cos \tfrac{1}{2}\pi = 0)$$

If $\mathbf{a}$ and $\mathbf{b}$ have opposite directions, then $\theta = \pi$ and

$$\mathbf{a} \cdot \mathbf{b} = -||\mathbf{a}||\,||\mathbf{b}||; \qquad (\cos\pi = -1)$$

this is the least possible value for $\mathbf{a} \cdot \mathbf{b}$.

Two vectors are said to be *perpendicular* if they lie at right angles or one of the vectors is the zero vector; in other words, two vectors are said to be perpendicular iff their dot product is zero.† In symbols

(12.4.8) $$\mathbf{a} \perp \mathbf{b} \quad \text{iff} \quad \mathbf{a} \cdot \mathbf{b} = 0.$$ (Figure 12.4.5)

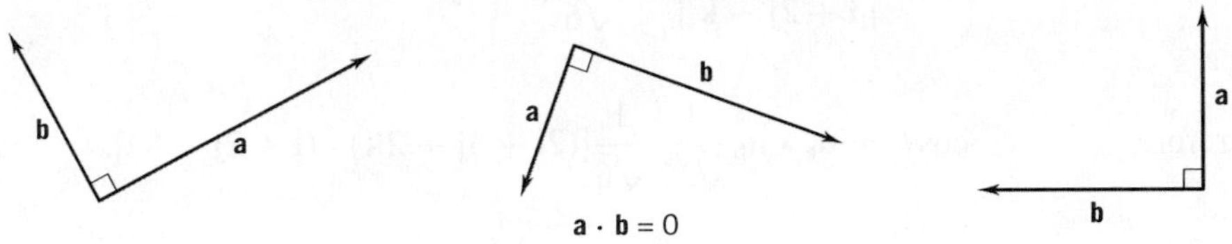

a · b = 0

Figure 12.4.5

The unit coordinate vectors are obviously mutually perpendicular:

$$\mathbf{i} \cdot \mathbf{j} = 0, \qquad \mathbf{i} \cdot \mathbf{k} = 0, \qquad \mathbf{j} \cdot \mathbf{k} = 0.$$

Example 3 Verify that the vectors $\mathbf{a} = 2\mathbf{i} + \mathbf{j} + \mathbf{k}$ and $\mathbf{b} = \mathbf{i} + \mathbf{j} - 3\mathbf{k}$ are perpendicular.

SOLUTION

$$\mathbf{a} \cdot \mathbf{b} = (2)(1) + (1)(1) + (1)(-3) = 2 + 1 - 3 = 0. \quad \square$$

Example 4 Find the value of α for which $(3\mathbf{i} - \alpha\mathbf{j} + \mathbf{k}) \perp (\mathbf{i} + 2\mathbf{j})$.

SOLUTION For the two vectors to be perpendicular, their dot product must be zero.

Since $$(3\mathbf{i} - \alpha\mathbf{j} + \mathbf{k}) \cdot (\mathbf{i} + 2\mathbf{j}) = (3)(1) + (-\alpha)(2) + (1)(0) = 3 - 2\alpha,$$

α must be $\frac{3}{2}$. $\square$

With **a** and **b** both different from zero, we can divide the equation

$$\mathbf{a} \cdot \mathbf{b} = ||\mathbf{a}||\,||\mathbf{b}|| \cos\theta$$

by $||\mathbf{a}||\,||\mathbf{b}||$ to obtain

$$\cos\theta = \frac{\mathbf{a} \cdot \mathbf{b}}{||\mathbf{a}||\,||\mathbf{b}||} = \frac{\mathbf{a}}{||\mathbf{a}||} \cdot \frac{\mathbf{b}}{||\mathbf{b}||}.$$

Writing $$\mathbf{u_a} = \frac{\mathbf{a}}{||\mathbf{a}||} \quad \text{and} \quad \mathbf{u_b} = \frac{\mathbf{b}}{||\mathbf{b}||},$$

we have

(12.4.9) $$\cos\theta = \mathbf{u_a} \cdot \mathbf{u_b}.$$

The cosine of the angle between the vectors is the dot product of the corresponding unit vectors.

† This makes the zero vector both parallel and perpendicular to every vector. There is, however, no contradiction since we do not apply the notion of "direction" to the zero vector.

Example 5 Calculate the angle between $\mathbf{a} = 2\mathbf{i}+3\mathbf{j}+2\mathbf{k}$ and $\mathbf{b} = \mathbf{i}+2\mathbf{j}-\mathbf{k}$.

SOLUTION

$$\mathbf{u_a} = \frac{2\mathbf{i}+3\mathbf{j}+2\mathbf{k}}{||2\mathbf{i}+3\mathbf{j}+2\mathbf{k}||} = \frac{1}{\sqrt{17}}(2\mathbf{i}+3\mathbf{j}+2\mathbf{k}),$$

$$\mathbf{u_b} = \frac{\mathbf{i}+2\mathbf{j}-\mathbf{k}}{||\mathbf{i}+2\mathbf{j}-\mathbf{k}||} = \frac{1}{\sqrt{6}}(\mathbf{i}+2\mathbf{j}-\mathbf{k}).$$

Therefore $$\cos\theta = \mathbf{u_a}\cdot\mathbf{u_b}\frac{1}{\sqrt{17}}\frac{1}{\sqrt{6}}[(2\mathbf{i}+3\mathbf{j}+2\mathbf{k})\cdot(\mathbf{i}+2\mathbf{j}-\mathbf{k})].$$

Since $$(2\mathbf{i}+3\mathbf{j}+2\mathbf{k})\cdot(\mathbf{i}+2\mathbf{j}-\mathbf{k}) = (2)(1)+(3)(2)+(2)(-1) = 6,$$

we have $$\cos\theta = \frac{6}{\sqrt{17}\sqrt{6}} = \frac{1}{17}\sqrt{102} \cong \frac{10.1}{17} \cong 0.594.$$

Thus, $\theta \cong 0.935$ radians, which is about 54 degrees. □

Example 6 Show that, if **a** and **b** are both perpendicular to **c**, then every linear combination $\alpha\mathbf{a}+\beta\mathbf{b}$ is also perpendicular to **c**.

SOLUTION Suppose that **a** and **b** are both perpendicular to **c**. Then

$$\mathbf{a}\cdot\mathbf{c} = 0 \qquad \text{and} \qquad \mathbf{b}\cdot\mathbf{c} = 0.$$

It follows that

$$(\alpha\mathbf{a}+\beta\mathbf{b})\cdot\mathbf{c} = \alpha\underbrace{(\mathbf{a}\cdot\mathbf{c})}_{0}+\beta\underbrace{(\mathbf{b}\cdot\mathbf{c})}_{0} = 0$$

and therefore $\alpha\mathbf{a}+\beta\mathbf{b}$ is perpendicular to **c**. □

Projections and Components

If $\mathbf{b} \neq \mathbf{0}$, then every vector **a** can be written in a unique manner as the sum of a vector $\mathbf{a}_{||}$ parallel to **b** and a vector $\mathbf{a}_{\perp}$ perpendicular to **b**:

$$\mathbf{a} = \mathbf{a}_{||} + \mathbf{a}_{\perp}. \qquad \text{(Exercise 53)}$$

The idea is illustrated in Figure 12.4.6. If **a** is parallel to **b**, then $\mathbf{a}_{||} = \mathbf{a}$ and $\mathbf{a}_{\perp} = \mathbf{0}$. If **a** is perpendicular to **b**, then $\mathbf{a}_{||} = \mathbf{0}$ and $\mathbf{a}_{\perp} = \mathbf{a}$.

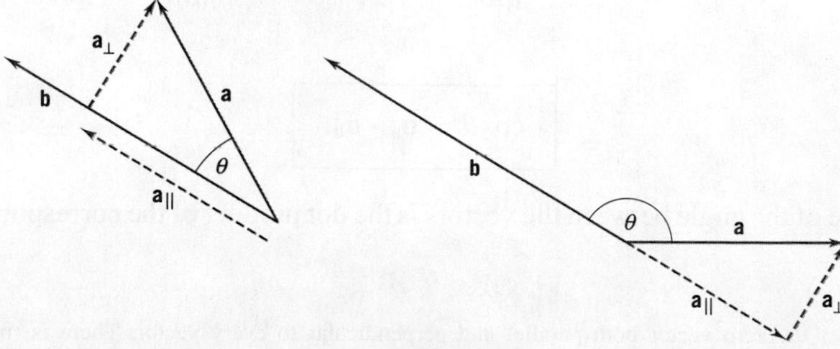

Figure 12.4.6

The vector $\mathbf{a}_{\parallel}$ is called *the projection of* **a** *on* **b** and is denoted by $\text{proj}_{\mathbf{b}}\mathbf{a}$. Since $\text{proj}_{\mathbf{b}}\mathbf{a}$ is parallel to **b**, it is a scalar multiple of the unit vector in the direction of **b**:

$$\text{proj}_{\mathbf{b}}\mathbf{a} = \lambda \mathbf{u}_{\mathbf{b}}.$$

The scalar λ is called *the component of* **a** *in the direction of* **b** (or, more briefly, the **b**-*component* of **a**) and is denoted by $\text{comp}_{\mathbf{b}}\mathbf{a}$. In symbols,

(12.4.10)

$$\text{proj}_{\mathbf{b}}\mathbf{a} = (\text{comp}_{\mathbf{b}}\mathbf{a})\,\mathbf{u}_{\mathbf{b}}.$$

The component of **a** in the direction of **b** measures the "advance" of **a** in the direction of **b**. In Figure 12.4.6 we used θ to indicate the angle between **a** and **b**. If $0 \le \theta < \frac{1}{2}\pi$, the projection and **b** have the same direction and the component is positive. If $\theta = \frac{1}{2}\pi$, the projection is **0** and the component is 0. If $\frac{1}{2}\pi < \theta \le \pi$, the projection and **b** have opposite directions and, consequently, the component is negative.

Projections and components are closely related to dot products: if $\mathbf{b} \neq \mathbf{0}$, then

(12.4.11)

$$\text{proj}_{\mathbf{b}}\mathbf{a} = (\mathbf{a}\cdot\mathbf{u}_{\mathbf{b}})\,\mathbf{u}_{\mathbf{b}} \quad \text{and} \quad \text{comp}_{\mathbf{b}}\mathbf{a} = \mathbf{a}\cdot\mathbf{u}_{\mathbf{b}}.$$

PROOF The second assertion follows immediately from the first. We will prove the first. We begin with the identity

$$\mathbf{a} = (\mathbf{a}\cdot\mathbf{u}_{\mathbf{b}})\,\mathbf{u}_{\mathbf{b}} + [\mathbf{a} - (\mathbf{a}\cdot\mathbf{u}_{\mathbf{b}})\,\mathbf{u}_{\mathbf{b}}].$$

Since the first vector $(\mathbf{a}\cdot\mathbf{u}_{\mathbf{b}})\,\mathbf{u}_{\mathbf{b}}$ is a scalar multiple of **b**, it is parallel to **b**. All we have to show now is that the second vector is perpendicular to **b**. We do this by showing that its dot product with $\mathbf{u}_{\mathbf{b}}$ is zero:

$$[\mathbf{a} - (\mathbf{a}\cdot\mathbf{u}_{\mathbf{b}})\,\mathbf{u}_{\mathbf{b}}]\cdot\mathbf{u}_{\mathbf{b}} = (\mathbf{a}\cdot\mathbf{u}_{\mathbf{b}}) - (\mathbf{a}\cdot\mathbf{u}_{\mathbf{b}})(\mathbf{u}_{\mathbf{b}}\cdot\mathbf{u}_{\mathbf{b}}) = 0. \quad \square$$

$\uparrow$ $\mathbf{u}_{\mathbf{b}}\cdot\mathbf{u}_{\mathbf{b}} = \|\mathbf{u}_{\mathbf{b}}\|^2 = 1$

Example 7 Find $\text{comp}_{\mathbf{b}}\mathbf{a}$ and $\text{proj}_{\mathbf{b}}\mathbf{a}$ given that

$$\mathbf{a} = -2\mathbf{i} + \mathbf{j} + \mathbf{k} \qquad \text{and} \qquad \mathbf{b} = 4\mathbf{i} - 3\mathbf{j} + \mathbf{k}.$$

SOLUTION Since $\|\mathbf{b}\| = \sqrt{4^2 + (-3)^2 + 1^2} = \sqrt{26}$,

we have

$$\mathbf{u}_{\mathbf{b}} = \frac{\mathbf{b}}{\|\mathbf{b}\|} = \frac{1}{\sqrt{26}}(4\mathbf{i} - 3\mathbf{j} + \mathbf{k}).$$

Thus

$$\text{comp}_{\mathbf{b}}\mathbf{a} = \mathbf{a}\cdot\mathbf{u}_{\mathbf{b}} = (-2\mathbf{i} + \mathbf{j} + \mathbf{k})\cdot\frac{1}{\sqrt{26}}(4\mathbf{i} - 3\mathbf{j} + \mathbf{k})$$

$$= \frac{1}{\sqrt{26}}[(-2)(4) + (1)(-3) + (1)(1)] = -\frac{10}{\sqrt{26}} = -\frac{5}{13}\sqrt{26}$$

and

$$\text{proj}_{\mathbf{b}}\mathbf{a} = (\text{comp}_{\mathbf{b}}\mathbf{a})\,\mathbf{u}_{\mathbf{b}} = -\tfrac{5}{13}(4\mathbf{i} - 3\mathbf{j} + \mathbf{k}). \quad \square$$

The following characterization of $\mathbf{a}\cdot\mathbf{b}$ is frequently used in physical applications:

(12.4.12)

$$\text{if } \mathbf{b} \neq \mathbf{0}, \quad \mathbf{a}\cdot\mathbf{b} = (\text{comp}_{\mathbf{b}}\mathbf{a})\|\mathbf{b}\|.$$

PROOF

$$\mathbf{a}\cdot\mathbf{b} = \left(\mathbf{a}\cdot\frac{\mathbf{b}}{||\mathbf{b}||}\right)||\mathbf{b}|| = (\mathbf{a}\cdot\mathbf{u_b})||\mathbf{b}|| = (\text{comp}_{\mathbf{b}}\mathbf{a})||\mathbf{b}||. \quad \square$$

For an arbitrary vector $\mathbf{a} = a_1\,\mathbf{i} + a_2\,\mathbf{j} + a_3\,\mathbf{k}$ we have

$$\text{comp}_{\mathbf{i}}\mathbf{a} = \mathbf{a}\cdot\mathbf{i} = a_1, \quad \text{comp}_{\mathbf{j}}\mathbf{a} = \mathbf{a}\cdot\mathbf{j} = a_2, \quad \text{comp}_{\mathbf{k}}\mathbf{a} = \mathbf{a}\cdot\mathbf{k} = a_3.$$

This agrees with our previous use of the term "component" (Section 12.3) and gives the identity

(12.4.13) $$\boxed{\mathbf{a} = (\mathbf{a}\cdot\mathbf{i})\,\mathbf{i} + (\mathbf{a}\cdot\mathbf{j})\,\mathbf{j} + (\mathbf{a}\cdot\mathbf{k})\,\mathbf{k}.}$$

Direction Angles, Direction Cosines

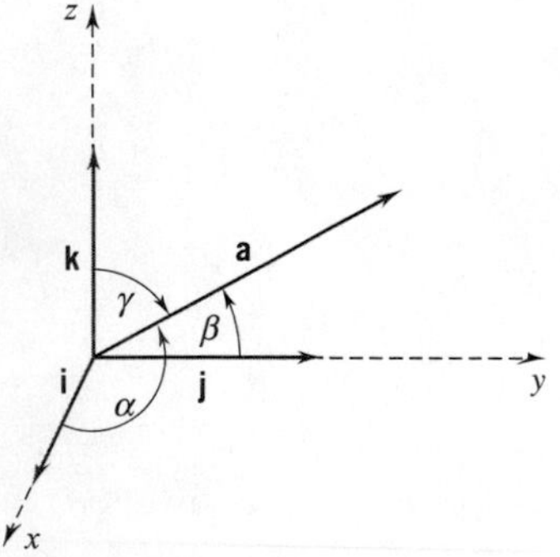

Figure 12.4.7

In Figure 12.4.7 we show a nonzero vector $\mathbf{a}$. The angles α, β, γ that the vector makes with the unit coordinate vectors are called the *direction angles* of $\mathbf{a}$, and $\cos\alpha, \cos\beta, \cos\gamma$ are called the *direction cosines*. By (12.4.7)

$$\mathbf{a}\cdot\mathbf{i} = ||\mathbf{a}||\cos\alpha, \quad \mathbf{a}\cdot\mathbf{j} = ||\mathbf{a}||\cos\beta, \quad \mathbf{a}\cdot\mathbf{k} = ||\mathbf{a}||\cos\gamma.$$

Thus by (12.4.13)

(12.4.14) $$\boxed{\mathbf{a} = ||\mathbf{a}||(\cos\alpha\,\mathbf{i} + \cos\beta\,\mathbf{j} + \cos\gamma\,\mathbf{k}).}$$

Taking the norm of both sides, we have

$$||\mathbf{a}|| = ||\mathbf{a}||\sqrt{\cos^2\alpha + \cos^2\beta + \cos^2\gamma}$$

and therefore

(12.4.15) $$\boxed{\cos^2\alpha + \cos^2\beta + \cos^2\gamma = 1.}$$

The sum of the squares of the direction cosines is always 1.

For a unit vector $\mathbf{u}$, Equation 12.4.14 takes the form

(12.4.16) $$\boxed{\mathbf{u} = \cos\alpha\,\mathbf{i} + \cos\beta\,\mathbf{j} + \cos\gamma\,\mathbf{k}.}$$

Thus, for a unit vector, the $\mathbf{i}$, $\mathbf{j}$, $\mathbf{k}$ components are simply the direction cosines.

Example 8 Find the unit vector with direction angles

$$\alpha = \tfrac{1}{4}\pi, \quad \beta = \tfrac{2}{3}\pi, \quad \gamma = \tfrac{1}{3}\pi.$$

What is the vector of norm 4 with these same direction angles?

SOLUTION The unit vector with these direction angles is

$$\cos\tfrac{1}{4}\pi\,\mathbf{i} + \cos\tfrac{2}{3}\pi\,\mathbf{j} + \cos\tfrac{1}{3}\pi\,\mathbf{k} = \tfrac{1}{2}\sqrt{2}\,\mathbf{i} - \tfrac{1}{2}\mathbf{j} + \tfrac{1}{2}\mathbf{k}.$$

The vector of norm 4 with these direction angles is

$$4(\tfrac{1}{2}\sqrt{2}\,\mathbf{i} - \tfrac{1}{2}\,\mathbf{j} + \tfrac{1}{2}\,\mathbf{k}) = 2\sqrt{2}\,\mathbf{i} - 2\,\mathbf{j} + 2\,\mathbf{k}. \quad \square$$

Example 9 Find the direction cosines of $\mathbf{a} = 2\mathbf{i} + 3\mathbf{j} - 6\mathbf{k}$. What are the direction angles?

SOLUTION Here $||\mathbf{a}|| = \sqrt{2^2 + 3^2 + (-6)^2} = 7$

so that
$$2 = 7\cos\alpha, \qquad 3 = 7\cos\beta, \qquad -6 = 7\cos\gamma$$

and
$$\cos\alpha = \tfrac{2}{7}, \qquad \cos\beta = \tfrac{3}{7}, \qquad \cos\gamma = -\tfrac{6}{7}.$$

Since angles between vectors are measured in radians and taken from 0 to π,

$$\alpha = \cos^{-1}\tfrac{2}{7} \cong 1.28 \text{ radians}, \qquad \beta = \cos^{-1}\tfrac{3}{7} \cong 1.13 \text{ radians},$$
$$\gamma = \cos^{-1}\left(-\tfrac{6}{7}\right) \cong 2.60 \text{ radians}. \quad \square$$

Proving the Triangle Inequality

By taking absolute values, the relation

$$\mathbf{a} \cdot \mathbf{b} = ||\mathbf{a}||\,||\mathbf{b}|| \cos\theta$$

gives

(12.4.17)
$$|\mathbf{a} \cdot \mathbf{b}| \le ||\mathbf{a}||\,||\mathbf{b}||. \qquad (\cos\theta \le 1)$$

This inequality, called *Schwarz's inequality*, enables us to give a simple proof of the triangle inequality

$$||\mathbf{a} + \mathbf{b}|| \le ||\mathbf{a}|| + ||\mathbf{b}||.$$

PROOF

$$\begin{aligned}
||\mathbf{a} + \mathbf{b}||^2 &= (\mathbf{a} + \mathbf{b}) \cdot (\mathbf{a} + \mathbf{b}) \\
&= (\mathbf{a} \cdot \mathbf{a}) + (\mathbf{b} \cdot \mathbf{a}) + (\mathbf{a} \cdot \mathbf{b}) + (\mathbf{b} \cdot \mathbf{b}) \\
&= ||\mathbf{a}||^2 + 2(\mathbf{a} \cdot \mathbf{b}) + ||\mathbf{b}||^2 \\
&\le ||\mathbf{a}||^2 + 2|\mathbf{a} \cdot \mathbf{b}| + ||\mathbf{b}||^2 \qquad (\mathbf{a} \cdot \mathbf{b} \le |\mathbf{a} \cdot \mathbf{b}|) \\
&\le ||\mathbf{a}||^2 + 2||\mathbf{a}||\,||\mathbf{b}|| + ||\mathbf{b}||^2 = (||\mathbf{a}|| + ||\mathbf{b}||)^2.
\end{aligned}$$

by Schwarz's inequality ↑

Taking square roots, we have

$$||\mathbf{a} + \mathbf{b}|| \le ||\mathbf{a}|| + ||\mathbf{b}||. \quad \square$$

It is worth remarking that Schwarz's inequality, and hence the triangle inequality, can be proved by purely algebraic methods. (See Exercise 56.)

EXERCISES 12.4

Find $\mathbf{a} \cdot \mathbf{b}$.

1. $\mathbf{a} = (2, -3, 1)$, $\mathbf{b} = (-2, 0, 3)$.
2. $\mathbf{a} = (4, 2, -1)$, $\mathbf{b} = (-2, 2, 1)$.
3. $\mathbf{a} = (2, -4, 0)$, $\mathbf{b} = (1, \frac{1}{2}, 0)$.
4. $\mathbf{a} = (-2, 0, 5)$, $\mathbf{b} = (3, 0, 1)$.
5. $\mathbf{a} = 2\mathbf{i} + \mathbf{j} - 2\mathbf{k}$, $\mathbf{b} = \mathbf{i} + \mathbf{j} + 2\mathbf{k}$.
6. $\mathbf{a} = 2\mathbf{i} + 3\mathbf{j} + \mathbf{k}$, $\mathbf{b} = \mathbf{i} + 4\mathbf{j}$.

Simplify.

7. $(3\mathbf{a} \cdot \mathbf{b}) - (\mathbf{a} \cdot 2\mathbf{b})$.
8. $\mathbf{a} \cdot (\mathbf{a} - \mathbf{b}) + \mathbf{b} \cdot (\mathbf{b} + \mathbf{a})$.
9. $(\mathbf{a} - \mathbf{b}) \cdot \mathbf{c} + \mathbf{b} \cdot (\mathbf{c} + \mathbf{a})$.
10. $\mathbf{a} \cdot (\mathbf{a} + 2\mathbf{c}) + (2\mathbf{b} - \mathbf{a}) \cdot (\mathbf{a} + 2\mathbf{c}) - 2\mathbf{b} \cdot (\mathbf{a} + 2\mathbf{c})$.
11. Taking

$$\mathbf{a} = 2\mathbf{i} + \mathbf{j}, \qquad \mathbf{b} = 3\mathbf{i} - \mathbf{j} + 2\mathbf{k}, \qquad \mathbf{c} = 4\mathbf{i} + 3\mathbf{k},$$

calculate:

(a) the three dot products $\mathbf{a} \cdot \mathbf{b}$, $\mathbf{a} \cdot \mathbf{c}$, $\mathbf{b} \cdot \mathbf{c}$;

(b) the cosines of the angles between these vectors;

(c) the component of $\mathbf{a}$ (i) in the $\mathbf{b}$ direction, (ii) in the $\mathbf{c}$ direction;

(d) the projection of $\mathbf{a}$ (i) in the $\mathbf{b}$ direction, (ii) in the $\mathbf{c}$ direction.

12. Repeat Exercise 11 with $\mathbf{a} = \mathbf{j} + 3\mathbf{k}$, $\mathbf{b} = 2\mathbf{i} - \mathbf{j} + 2\mathbf{k}$, $\mathbf{c} = 3\mathbf{i} - \mathbf{k}$.
13. Find the unit vector with direction angles $\frac{1}{3}\pi, \frac{1}{4}\pi, \frac{2}{3}\pi$.
14. Find the vector of norm 2 with direction angles $\frac{1}{4}\pi, \frac{1}{4}\pi, \frac{1}{2}\pi$.
15. Find the angle between the vectors $3\mathbf{i} - \mathbf{j} - 2\mathbf{k}$ and $\mathbf{i} + 2\mathbf{j} - 3\mathbf{k}$.
16. Find the angle between the vectors $2\mathbf{i} - 3\mathbf{j} + \mathbf{k}$ and $-3\mathbf{i} + \mathbf{j} + 9\mathbf{k}$.
17. Find the direction angles of the vector $\mathbf{i} - \mathbf{j} + \sqrt{2}\,\mathbf{k}$.
18. Find the direction angles of the vector $\mathbf{i} - \sqrt{3}\,\mathbf{k}$.

C Estimate the angle between the vectors. Express your answers in radians rounded to the nearest hundredth of a radian, and in degrees to the nearest tenth of a degree.

19. $\mathbf{a} = (3, 1, -1)$, $\mathbf{b} = (-2, 1, 4)$.
20. $\mathbf{a} = (-2, -3, 0)$, $\mathbf{b} = (-6, 0, 4)$.
21. $\mathbf{a} = -\mathbf{i} + 2\mathbf{k}$, $\mathbf{b} = 3\mathbf{i} + 4\mathbf{j} - 5\mathbf{k}$.
22. $\mathbf{a} = -3\mathbf{i} + \mathbf{j} - \mathbf{k}$, $\mathbf{b} = \mathbf{i} - \mathbf{j}$.
23. (C) Use a CAS to determine the angles and the perimeter of the triangle with vertices $P(1, 3, -2)$, $Q(3, 1, 2)$, $R(2, -3, 1)$.
24. (C) Use a CAS to find the direction cosines and the direction angles of the vector from $P(5, 7, -2)$ to $Q(-3, 4, 1)$.

Find the direction cosines and direction angles of the vector. Express the angles in degrees rounded to the nearest tenth of a degree.

25. $\mathbf{a} = (1, 2, 2)$.
26. $\mathbf{a} = (2, 6, -1)$.
27. $\mathbf{a} = 3\mathbf{i} + 12\mathbf{j} + 4\mathbf{k}$.
28. $\mathbf{a} = 3\mathbf{i} + 5\mathbf{j} - 4\mathbf{k}$.
29. Find all the numbers x for which

$$2\mathbf{i} + 5\mathbf{j} + 2x\,\mathbf{k} \perp 6\mathbf{i} + 4\mathbf{j} - x\,\mathbf{k}.$$

30. Find all the numbers x for which

$$(x\mathbf{i} + 11\mathbf{j} - 3\mathbf{k}) \perp (2x\mathbf{i} - x\mathbf{j} - 5\mathbf{k}).$$

31. Find all the numbers x for which the angle between $\mathbf{c} = x\mathbf{i} + \mathbf{j} + \mathbf{k}$ and $\mathbf{d} = \mathbf{i} + x\mathbf{j} + \mathbf{k}$ is $\frac{1}{3}\pi$.
32. Set $\mathbf{a} = \mathbf{i} + x\mathbf{j} + \mathbf{k}$ and $\mathbf{b} = 2\mathbf{i} - \mathbf{j} + y\mathbf{k}$. Compute all values of x and y for which $\mathbf{a} \perp \mathbf{b}$ and $||\mathbf{a}|| = ||\mathbf{b}||$.
33. (a) Show that $\frac{1}{4}\pi, \frac{1}{6}\pi, \frac{2}{3}\pi$ cannot be the direction angles of a vector.

(b) Show that, if $\mathbf{a} = a_1\mathbf{i} + a_2\mathbf{j} + a_3\mathbf{k}$ has direction angles $\alpha, \frac{1}{4}\pi, \frac{1}{4}\pi$, then $a_1 = 0$.

34. If a vector has direction angles $\alpha = \pi/3$, $\beta = \pi/4$, find the third direction angle γ.
35. What are the direction angles of $-\mathbf{a}$ if the direction angles of $\mathbf{a}$ are α, β, γ?
36. Suppose that the direction angles of a vector are equal. What are the angles?
37. Find the unit vectors $\mathbf{u}$ that are perpendicular to both $\mathbf{i} + 2\mathbf{j} + \mathbf{k}$ and $3\mathbf{i} - 4\mathbf{j} + 2\mathbf{k}$.
38. Find two mutually perpendicular unit vectors that are perpendicular to $2\mathbf{i} + 3\mathbf{j}$.
39. Find the angle between the diagonal of a cube and one of the edges.
40. Find the angle between the diagonal of a cube and the diagonal of one of the faces.
41. Show that

(a) $\text{proj}_{\mathbf{b}}\alpha\mathbf{a} = \alpha\, \text{proj}_{\mathbf{b}}\mathbf{a}$ for all real α, and

(b) $\text{proj}_{\mathbf{b}}(\mathbf{a} + \mathbf{c}) = \text{proj}_{\mathbf{b}}\mathbf{a} + \text{proj}_{\mathbf{b}}\mathbf{c}$.

42. Show that

(a) $\text{proj}_{\beta\mathbf{b}}\mathbf{a} = \text{proj}_{\mathbf{b}}\mathbf{a}$ for all real $\beta \neq 0$, but

(b) $\text{comp}_{\beta\mathbf{b}}\mathbf{a} = \begin{cases} \text{comp}_{\mathbf{b}}\mathbf{a}, & \text{for } \beta > 0 \\ -\,\text{comp}_{\mathbf{b}}\mathbf{a}, & \text{for } \beta < 0. \end{cases}$

43. (a) (*Important*) Let $\mathbf{a} \neq \mathbf{0}$. Show that $\mathbf{a} \cdot \mathbf{b} = \mathbf{a} \cdot \mathbf{c}$ does not necessarily imply that $\mathbf{b} = \mathbf{c}$, but only that $\mathbf{b}$ and $\mathbf{c}$ have the same projection on $\mathbf{a}$. Draw a figure illustrating this for $\mathbf{b}$ and $\mathbf{c}$ different from $\mathbf{0}$.

(b) Show that if $\mathbf{u} \cdot \mathbf{b} = \mathbf{u} \cdot \mathbf{c}$ for all unit vectors $\mathbf{u}$, then $\mathbf{b} = \mathbf{c}$. HINT: Consider the unit coordinate vectors.

44. What can you conclude about $\mathbf{a}$ and $\mathbf{b}$ given that

(a) $||\mathbf{a}||^2 + ||\mathbf{b}||^2 = ||\mathbf{a} + \mathbf{b}||^2$?

(b) $||\mathbf{a}||^2 + ||\mathbf{b}||^2 = ||\mathbf{a} - \mathbf{b}||^2$?

HINT: Draw figures.

45. (a) Show that

$$4(\mathbf{a} \cdot \mathbf{b}) = ||\mathbf{a} + \mathbf{b}||^2 - ||\mathbf{a} - \mathbf{b}||^2.$$

(b) Use part (a) to verify that

$$\mathbf{a} \perp \mathbf{b} \quad \text{iff} \quad \|\mathbf{a}+\mathbf{b}\| = \|\mathbf{a}-\mathbf{b}\|.$$

(c) Show that, if **a** and **b** are nonzero vectors such that

$$(\mathbf{a}+\mathbf{b}) \perp (\mathbf{a}-\mathbf{b}) \quad \text{and} \quad \|\mathbf{a}+\mathbf{b}\| = \|\mathbf{a}-\mathbf{b}\|,$$

then the parallelogram generated by **a** and **b** is a square.

46. Under what conditions does $|\mathbf{a}\cdot\mathbf{b}| = \|\mathbf{a}\|\,\|\mathbf{b}\|$?

47. Given two vectors **a** and **b**, prove the *parallelogram law*:

$$\|\mathbf{a}+\mathbf{b}\|^2 + \|\mathbf{a}-\mathbf{b}\|^2 = 2\|\mathbf{a}\|^2 + 2\|\mathbf{b}\|^2.$$

Geometric interpretation: The sum of the squares of the lengths of the diagonals of a parallelogram equals the sum of the squares of the lengths of the four sides. See the figure.

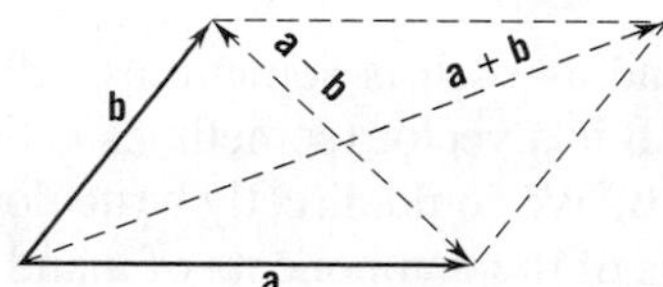

48. A *rhombus* is a parallelogram with sides of equal length. Show that the diagonals of a rhombus are perpendicular.

49. Let **a** and **b** be nonzero vectors. Show that the vector $\mathbf{c} = \|\mathbf{b}\|\,\mathbf{a} + \|\mathbf{a}\|\,\mathbf{b}$ bisects the angle between **a** and **b**.

50. Let θ be the angle between **a** and **b**, and let β be a negative number. Use the dot product to compute the angle between **a** and $\beta\mathbf{b}$ in terms of θ. Draw a figure to verify your answer geometrically.

51. Show that if $\mathbf{a} \perp \mathbf{b}$ and $\mathbf{a} \perp \mathbf{c}$, then $\mathbf{a} \perp (\alpha\mathbf{b} + \beta\mathbf{c})$ for all real α, β.

52. Show that, if $\mathbf{a} \parallel \mathbf{b}$ and $\mathbf{a} \parallel \mathbf{c}$, then $\mathbf{a} \parallel (\alpha\mathbf{b} + \beta\mathbf{c})$ for all real α, β. ($\mathbf{a} \parallel \mathbf{b}$ is used to indicate that **a** is parallel to **b**)

53. (*Important*) Show that, if **b** is a nonzero vector, then every vector **a** can be written in a unique manner as the sum of a vector $\mathbf{a}_{\parallel}$ parallel to **b** and a vector $\mathbf{a}_{\perp}$ perpendicular to **b**:

$$\mathbf{a} = \mathbf{a}_{\parallel} + \mathbf{a}_{\perp}.$$

54. let $r = f(\theta)$ be the polar equation of a curve in the plane and let

$$\mathbf{u}_r = (\cos\theta)\mathbf{i} + (\sin\theta)\mathbf{j} \qquad \mathbf{u}_\theta = (-\sin\theta)\mathbf{i} + (\cos\theta)\mathbf{j}.$$

(a) Show that $\mathbf{u}_r$ and $\mathbf{u}_\theta$ are unit vectors and that they are perpendicular.

(b) Let $P[r,\theta]$ be a point on the curve. Show that $\mathbf{u}_r$ has the same direction as the vector $\overrightarrow{OP}$ and that $\mathbf{u}_\theta$ is 90° counterclockwise from $\mathbf{u}_r$.

55. Two points on a sphere are called *antipodal* if they are opposite endpoints of a diameter. Show that, if P_1 and P_2 are antipodal points on a sphere and Q is any other point on the sphere, then $\overrightarrow{P_1Q} \perp \overrightarrow{P_2Q}$.

56. (*Important*) Give an algebraic proof of Schwarz's inequality $|\mathbf{a}\cdot\mathbf{b}| \le \|\mathbf{a}\|\,\|\mathbf{b}\|$. HINT: If $\mathbf{b} = 0$, the inequality is trivial, so assume $\mathbf{b} \neq \mathbf{0}$. Note that for any number λ we have $\|\mathbf{a} - \lambda\mathbf{b}\|^2 \ge 0$. First expand this inequality using the fact that $\|\mathbf{a} - \lambda\mathbf{b}\|^2 = (\mathbf{a} - \lambda\mathbf{b})\cdot(\mathbf{a} - \lambda\mathbf{b})$. After collecting terms, make the special choice $\lambda = (\mathbf{a}\cdot\mathbf{b})/\|\mathbf{b}\|^2$ and see what happens.

■ PROJECT 12.4 Work

If a constant force **F** is applied to an object moving in a straight line throughout a displacement **r** (see the figure), then the work done by **F** is defined by the equation

$$W = (\text{comp}_{\mathbf{r}}\,\mathbf{F})\|\mathbf{r}\|$$

This generalizes (6.5.1).

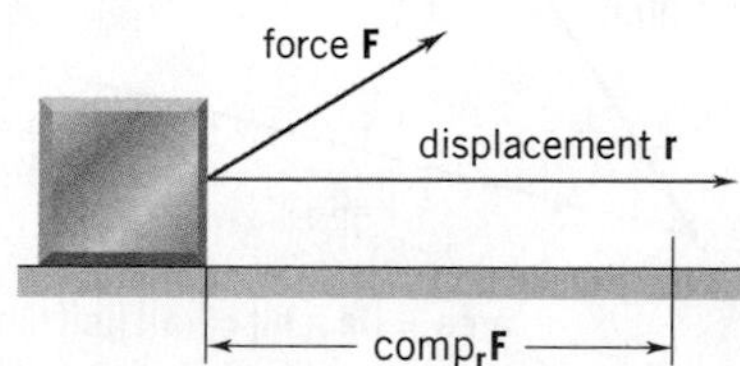

If **F** is measured in pounds and distance in feet, then work is measured in *foot-pounds*; if **F** is measured in newtons and distance in meters, then work is measured in *newton-meters* or *joules*.

Problem 1. Let the force **F** be applied throughout a displacement **r**.

a. Express the work done by **F** as a dot product.

b. What is the work done by **F** if $\mathbf{F} \perp \mathbf{r}$?

c. Show that the work done by $\mathbf{F} = \|\mathbf{F}\|\,\mathbf{i}$ applied throughout the displacement $\mathbf{r} = (b-a)\mathbf{i}$ reduces to (6.5.1).

Problem 2.

(a) A sled is pulled along level ground by a force of 15 Newtons along a rope that makes an angle of 35° with the ground. Find the work done by the force in pulling the sled 50 meters. See Figure A.

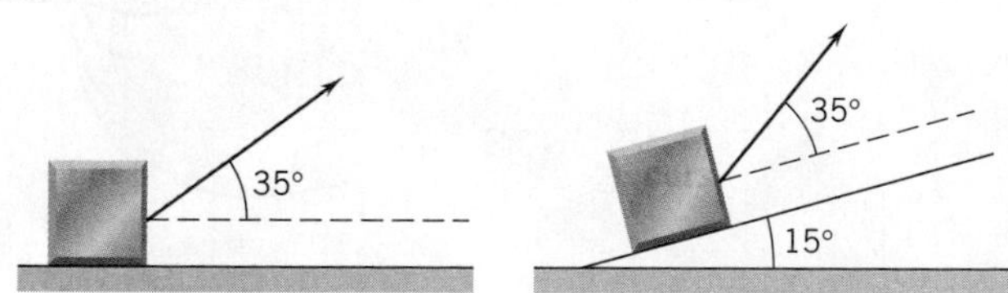

Figure A **Figure B**

(b) Suppose the same sled is pulled 50 meters up a hill that makes an angle of 15° with level ground. Find the work done by the force in this case. See Figure B.

Problem 3. A wooden crate is pulled along a level floor by a rope that makes an angle of 40° with the floor. If the force of friction (which acts in a direction opposite to motion) between the carton and the floor is 50 pounds, what is the minimum force that must be applied to the rope to move the crate?

Problem 4. Two forces of the same magnitude, $\mathbf{F}_1$ and $\mathbf{F}_2$, are applied throughout a displacement $\mathbf{r}$ at angles θ_1 and θ_2, respectively. Compare the work done by $\mathbf{F}_1$ to that done by $\mathbf{F}_2$ if

(a) $\theta_1 = -\theta_2$.

(b) $\theta_1 = \pi/3$ and $\theta_2 = \pi/6$.

Problem 5. What is the total work done by a constant force $\mathbf{F}$ if the object to which it is applied moves around a triangle? Justify your answer.

■ 12.5 THE CROSS PRODUCT

Everything that we have done with vectors so far (other than draw pictures) can be generalized to higher dimensions. We come now to a notion that is particular to three-dimensional space and cannot be generalized to higher dimensions.

Definition of the Cross Product

While the dot product $\mathbf{a} \cdot \mathbf{b}$ is a scalar (and as such is sometimes called the scalar product of $\mathbf{a}$ and $\mathbf{b}$), the cross product $\mathbf{a} \times \mathbf{b}$ is a vector (sometimes called the *vector product* of $\mathbf{a}$ and $\mathbf{b}$). What is the vector $\mathbf{a} \times \mathbf{b}$? We could directly write down a formula that gives the components of $\mathbf{a} \times \mathbf{b}$ in terms of the components of $\mathbf{a}$ and $\mathbf{b}$, but at this stage that would reveal little. Instead we will begin geometrically. We will define $\mathbf{a} \times \mathbf{b}$ by giving its direction and its magnitude.

The Direction of $\mathbf{a} \times \mathbf{b}$ If the vectors $\mathbf{a}$ and $\mathbf{b}$ are not parallel, they determine a plane. The vector $\mathbf{a} \times \mathbf{b}$ is perpendicular to this plane and is directed in such a way that (like $\mathbf{i}, \mathbf{j}, \mathbf{k}$) the vectors $\mathbf{a}, \mathbf{b}, \mathbf{a} \times \mathbf{b}$ form a right-handed triple. (See Figure 12.5.1.) If the index finger of the right hand points along $\mathbf{a}$ and the middle finger points along $\mathbf{b}$, then the thumb will point in the direction of $\mathbf{a} \times \mathbf{b}$. NOTE: In saying that $\mathbf{a}, \mathbf{b}, \mathbf{a} \times \mathbf{b}$ form a right-handed triple, we do not require the vectors $\mathbf{a}$ and $\mathbf{b}$ to be perpendicular (like $\mathbf{i}$ and $\mathbf{j}$); the general notion of a "right-handed triple" does not require the vectors involved to be perpendicular to each other.

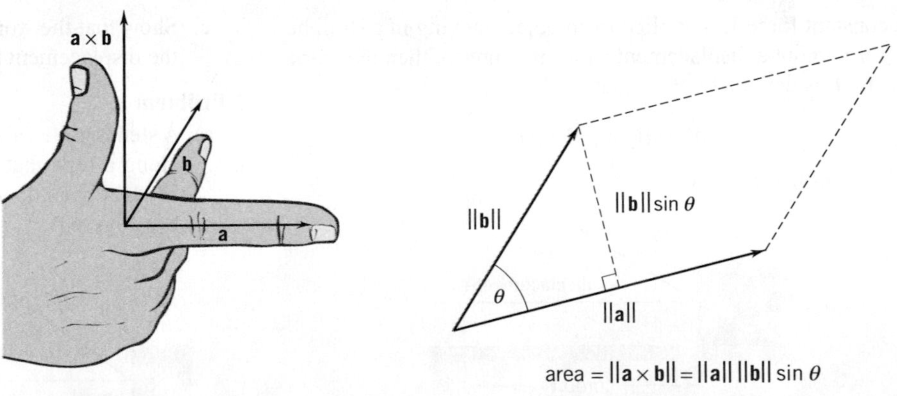

Figure 12.5.1

Figure 12.5.2

The Magnitude of $\mathbf{a} \times \mathbf{b}$ If $\mathbf{a}$ and $\mathbf{b}$ are not parallel, they form the sides of a parallelogram. (See Figure 12.5.2.) The magnitude of $\mathbf{a} \times \mathbf{b}$ is the area of this parallelogram: $\|\mathbf{a}\|\,\|\mathbf{b}\| \sin\theta$. (Recall that the area of a parallelogram with base b and height h is given by $A = bh$.)

One more point. What if $\mathbf{a}$ and $\mathbf{b}$ are parallel? Then there is no parallelogram and we define $\mathbf{a} \times \mathbf{b}$ to be $\mathbf{0}$.

We summarize all this below.

DEFINITION 12.5.1

If $\mathbf{a}$ and $\mathbf{b}$ are not parallel, then $\mathbf{a} \times \mathbf{b}$ is the vector with the following properties:

1. $\mathbf{a} \times \mathbf{b}$ is perpendicular to the plane of $\mathbf{a}$ and $\mathbf{b}$.
2. $\mathbf{a}, \mathbf{b}, \mathbf{a} \times \mathbf{b}$ form a right-handed triple.
3. $\|\mathbf{a} \times \mathbf{b}\| = \|\mathbf{a}\| \, \|\mathbf{b}\| \sin\theta$, where θ is the angle between $\mathbf{a}$ and $\mathbf{b}$.

If $\mathbf{a}$ and $\mathbf{b}$ are parallel, then $\mathbf{a} \times \mathbf{b} = \mathbf{0}$.

Properties of Right-Handed Triples

I. Note first of all that if $(\mathbf{a}, \mathbf{b}, \mathbf{c})$ forms a right-handed triple, then $(\mathbf{b}, \mathbf{c}, \mathbf{a})$ and $(\mathbf{c}, \mathbf{a}, \mathbf{b})$ also form right-handed triples (Figure 12.5.3). To maintain right-handedness we don't have to keep the vectors in the same order, but we do have to keep them in the *same cyclic order*:

$$\mathbf{a} \to \mathbf{b} \to \mathbf{c}$$

Alter the cyclic order and you reverse the orientation.

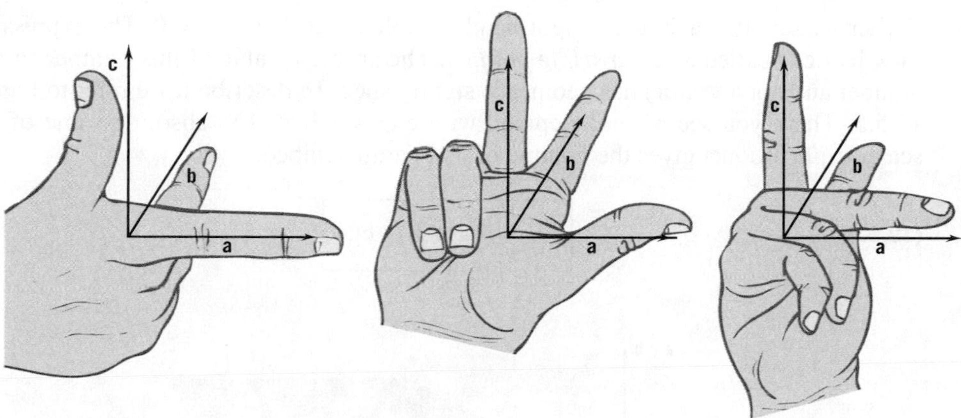

Figure 12.5.3

II. A triple $(\mathbf{a}, \mathbf{b}, \mathbf{c})$ is right-handed iff $\mathbf{c}$ and $\mathbf{a} \times \mathbf{b}$ lie on the same side of the plane determined by $\mathbf{a}$ and $\mathbf{b}$ (Figure 12.5.4). This means that $(\mathbf{a}, \mathbf{b}, \mathbf{c})$ is right-handed iff $(\mathbf{a} \times \mathbf{b}) \cdot \mathbf{c} > 0$. (Explain)

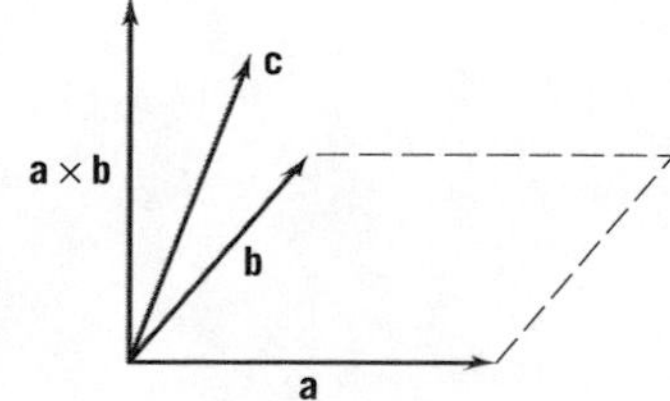

Figure 12.5.4

III. It follows from II that, if $(\mathbf{a}, \mathbf{b}, \mathbf{c})$ is right-handed, then $(\mathbf{a}, \mathbf{b}, -\mathbf{c})$ is not right-handed. Similarly, $(-\mathbf{a}, \mathbf{b}, \mathbf{c})$ and $(\mathbf{a}, -\mathbf{b}, \mathbf{c})$ are not right-handed. However, multiplication by positive scalars does maintain right-handedness: if $(\mathbf{a}, \mathbf{b}, \mathbf{c})$ is right-handed and α, β, γ are positive, then $(\alpha\mathbf{a}, \beta\mathbf{b}, \gamma\mathbf{c})$ is also right-handed.

Properties of the Cross Product

The cross product is *anticommutative*:

(12.5.2) $$\mathbf{b} \times \mathbf{a} = -(\mathbf{a} \times \mathbf{b}).$$

To see this, note that both vectors are perpendicular to the plane determined by **a** and **b** and both have the same norm. Thus $\mathbf{b} \times \mathbf{a} = \pm(\mathbf{a} \times \mathbf{b})$. That the minus sign holds, not the plus sign, follows from observing that $\mathbf{b}, \mathbf{a}, \mathbf{b} \times \mathbf{a}$ is a right-handed triple and that $\mathbf{b}, \mathbf{a}, \mathbf{a} \times \mathbf{b}$ is not right-handed (since $\mathbf{a}, \mathbf{b}, \mathbf{a} \times \mathbf{b}$ is right-handed).

Scalars can be factored:

(12.5.3)
$$\alpha\mathbf{a} \times \beta\mathbf{b} = \alpha\beta(\mathbf{a} \times \mathbf{b}).$$

If α or β is zero, the result is obvious. We will assume that α and β are both nonzero. In this case the two vectors are perpendicular to **a**, perpendicular to **b**, and have the same norm. Thus $\alpha\mathbf{a} \times \beta\mathbf{b} = \pm\alpha\beta(\mathbf{a} \times \mathbf{b})$. That the positive sign holds comes from noting that $\alpha\mathbf{a}, \beta\mathbf{b}, \alpha\beta(\mathbf{a} \times \mathbf{b})$ is a right-handed triple. This is obvious if α and β are both positive. If not, two of the three coefficients $\alpha, \beta, \alpha\beta$ are negative and the other is positive. In this case, the first minus sign reverses the orientation (that is, changes right-handed to left-handed) but the second one restores it.

Finally, there are two distributive laws, the verification of which we postpone for a moment.

(12.5.4)
$$\mathbf{a} \times (\mathbf{b} + \mathbf{c}) = (\mathbf{a} \times \mathbf{b}) + (\mathbf{a} \times \mathbf{c}),$$
$$(\mathbf{a} + \mathbf{b}) \times \mathbf{c} = (\mathbf{a} \times \mathbf{c}) + (\mathbf{b} \times \mathbf{c}).$$

The Scalar Triple Product

Earlier we saw that $\mathbf{a}, \mathbf{b}, \mathbf{c}$ is a right-handed triple iff $(\mathbf{a} \times \mathbf{b}) \cdot \mathbf{c} > 0$. The expression $(\mathbf{a} \times \mathbf{b}) \cdot \mathbf{c}$ is called a *scalar triple product*. The absolute value of this number (it is a number and not a vector) has geometric significance. To describe it we refer to Figure 12.5.5. There you see a parallelepiped with edges $\mathbf{a}, \mathbf{b}, \mathbf{c}$. The absolute value of the scalar triple product gives the volume of that parallelepiped:

(12.5.5)
$$V = |(\mathbf{a} \times \mathbf{b}) \cdot \mathbf{c}|.$$

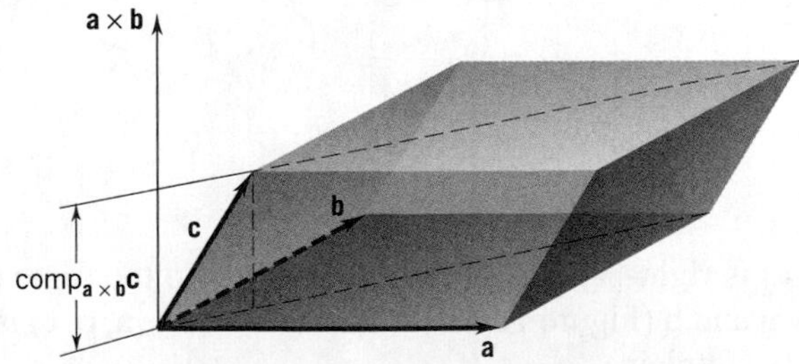

Figure 12.5.5

PROOF The area of the base is $\|\mathbf{a} \times \mathbf{b}\|$. The height is $|\text{comp}_{\mathbf{a}\times\mathbf{b}}\mathbf{c}|$. Therefore

$$V = |\text{comp}_{\mathbf{a}\times\mathbf{b}}\,\mathbf{c}|\,\|\mathbf{a} \times \mathbf{b}\| = |(\mathbf{a} \times \mathbf{b}) \cdot \mathbf{c}|. \quad \square$$

↑ (12.4.12)

Of course, we could have formed the same parallelogram using a different base (for example, using the vectors **c** and **a**) with a correspondingly different height (comp $_{\mathbf{c}\times\mathbf{a}}$ $\mathbf{b}$). Therefore

$$|(\mathbf{a}\times\mathbf{b})\cdot\mathbf{c}| = |(\mathbf{c}\times\mathbf{a})\cdot\mathbf{b}| = |(\mathbf{b}\times\mathbf{c})\cdot\mathbf{a}|.$$

Since the $\mathbf{a}, \mathbf{b}, \mathbf{c}$ appear in the same cyclic order, the expressions inside the absolute value signs all have the same sign (Property II of right-handed triples). Therefore

(12.5.6)
$$(\mathbf{a}\times\mathbf{b})\cdot\mathbf{c} = (\mathbf{c}\times\mathbf{a})\cdot\mathbf{b} = (\mathbf{b}\times\mathbf{c})\cdot\mathbf{a}.$$

Verification of the Distributive Laws

We will verify the first distributive law,

$$\mathbf{a}\times(\mathbf{b}+\mathbf{c}) = (\mathbf{a}\times\mathbf{b}) + (\mathbf{a}\times\mathbf{c}).$$

The second follows readily from this one. The argument is left to you as an exercise.

Take an arbitrary vector $\mathbf{r}$ and form the dot product $[\mathbf{a}\times(\mathbf{b}+\mathbf{c})]\cdot\mathbf{r}$. We can then write

$$\begin{aligned}
[\mathbf{a}\times(\mathbf{b}+\mathbf{c})]\cdot\mathbf{r} &= (\mathbf{r}\times\mathbf{a})\cdot(\mathbf{b}+\mathbf{c}) && (12.5.6)\\
&= [(\mathbf{r}\times\mathbf{a})\cdot\mathbf{b}] + [(\mathbf{r}\times\mathbf{a})\cdot\mathbf{c}] && (12.4.6)\\
&= [(\mathbf{a}\times\mathbf{b})\cdot\mathbf{r}] + [(\mathbf{a}\times\mathbf{c})\cdot\mathbf{r}] && (12.5.6)\\
&= [(\mathbf{a}\times\mathbf{b})] + (\mathbf{a}\times\mathbf{c})]\cdot\mathbf{r}. && (12.4.6)
\end{aligned}$$

Since this holds true for all vectors $\mathbf{r}$, it holds true for $\mathbf{i}, \mathbf{j}, \mathbf{k}$ and proves that

$$\mathbf{a}\times(\mathbf{b}+\mathbf{c}) = (\mathbf{a}\times\mathbf{b}) + (\mathbf{a}\times\mathbf{c}).$$

The Components of $\mathbf{a}\times\mathbf{b}$

You have learned a lot about cross products, but you still have not seen $\mathbf{a}\times\mathbf{b}$ expressed in terms of the components of $\mathbf{a}$ and $\mathbf{b}$. To derive the formula that does this, we need to observe one more fact, one which follows from the definition of cross product:

(12.5.7)
$$\mathbf{i}\times\mathbf{j} = \mathbf{k}, \quad \mathbf{j}\times\mathbf{k} = \mathbf{i}, \quad \mathbf{k}\times\mathbf{i} = \mathbf{j}. \qquad \text{(Figure 12.5.6)}$$

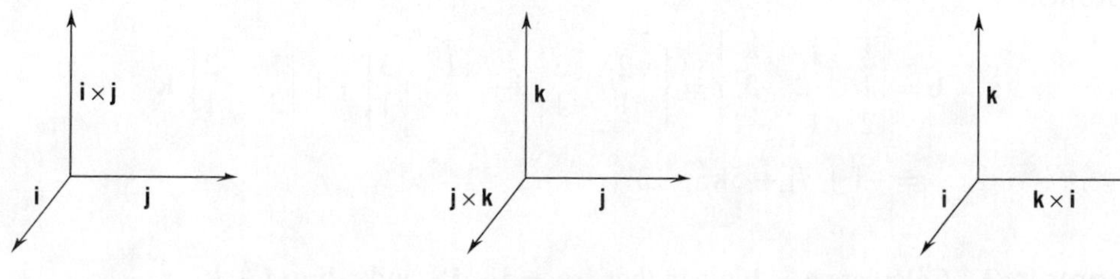

Figure 12.5.6

One way to remember these products is to arrange $\mathbf{i}, \mathbf{j}, \mathbf{k}$ in cyclic order, $\mathbf{i}\rightarrow\mathbf{j}\rightarrow\mathbf{k}$, and note that

[each coordinate unit vector] × [the next one] = [the third one].

THEOREM 12.5.8

For vectors $\mathbf{a} = a_1\mathbf{i} + a_2\mathbf{j} + a_3\mathbf{k}$ and $\mathbf{b} = b_1\mathbf{i} + b_2\mathbf{j} + b_3\mathbf{k}$,

$$\mathbf{a} \times \mathbf{b} = (a_2b_3 - a_3b_2)\mathbf{i} - (a_1b_3 - a_3b_1)\mathbf{j} + (a_1b_2 - a_2b_1)\mathbf{k}.$$

Those of you who have studied some linear algebra will recognize that the jumble of symbols we have just written down for $\mathbf{a} \times \mathbf{b}$ can be elegantly summarized by the use of determinants. Here is Theorem 12.5.8 stated in terms of determinants. (If you are not familiar with determinants, see Appendix A-2.)

THEOREM 12.5.8′

For vectors $\mathbf{a} = a_1\mathbf{i} + a_2\mathbf{j} + a_3\mathbf{k}$ and $\mathbf{b} = b_1\mathbf{i} + b_2\mathbf{j} + b_3\mathbf{k}$,

$$\mathbf{a} \times \mathbf{b} = \begin{vmatrix} \mathbf{i} & \mathbf{j} & \mathbf{k} \\ a_1 & a_2 & a_3 \\ b_1 & b_2 & b_3 \end{vmatrix} = \begin{vmatrix} a_2 & a_3 \\ b_2 & b_3 \end{vmatrix}\mathbf{i} - \begin{vmatrix} a_1 & a_3 \\ b_1 & b_3 \end{vmatrix}\mathbf{j} + \begin{vmatrix} a_1 & a_2 \\ b_1 & b_2 \end{vmatrix}\mathbf{k}.$$

(The 3×3 determinant with $\mathbf{i}, \mathbf{j}, \mathbf{k}$ in the top row is there only as a mnemonic device.)

PROOF The hard work has all been done. With what you know about cross products now, the proof is just a matter of algebraic manipulation:

$$\begin{aligned} \mathbf{a} \times \mathbf{b} &= (a_1\mathbf{i} + a_2\mathbf{j} + a_3\mathbf{k}) \times (b_1\mathbf{i} + b_2\mathbf{j} + b_3\mathbf{k}) \\ &= a_1b_2(\mathbf{i}\times\mathbf{j}) + a_1b_3(\mathbf{i}\times\mathbf{k}) + a_2b_1(\mathbf{j}\times\mathbf{i}) + a_2b_3(\mathbf{j}\times\mathbf{k}) + a_3b_1(\mathbf{k}\times\mathbf{i}) + a_3b_2(\mathbf{k}\times\mathbf{j}) \\ &\uparrow\!\!\text{———}\ \mathbf{i}\times\mathbf{i} = \mathbf{j}\times\mathbf{j} = \mathbf{k}\times\mathbf{k} = \mathbf{0} \\ &= a_1b_2\mathbf{k} - a_1b_3\mathbf{j} - a_2b_1\mathbf{k} + a_2b_3\mathbf{i} + a_3b_1\mathbf{j} - a_3b_2\mathbf{i} \\ &= (a_2b_3 - a_3b_2)\mathbf{i} - (a_1b_3 - a_3b_1)\mathbf{j} + (a_1b_2 - a_2b_1)\mathbf{k} \quad \text{(this proves Theorem 12.5.8)} \\ &= \begin{vmatrix} a_2 & a_3 \\ b_2 & b_3 \end{vmatrix}\mathbf{i} - \begin{vmatrix} a_1 & a_3 \\ b_1 & b_3 \end{vmatrix}\mathbf{j} + \begin{vmatrix} a_1 & a_2 \\ b_1 & b_2 \end{vmatrix}\mathbf{k}. \quad \square \end{aligned}$$

Example 1 Calculate $\mathbf{a} \times \mathbf{b}$ given that $\mathbf{a} = \mathbf{i} - 2\mathbf{j} + 3\mathbf{k}$ and $\mathbf{b} = 2\mathbf{i} + \mathbf{j} - \mathbf{k}$.

SOLUTION

$$\begin{aligned} \mathbf{a} \times \mathbf{b} &= \begin{vmatrix} \mathbf{i} & \mathbf{j} & \mathbf{k} \\ 1 & -2 & 3 \\ 2 & 1 & -1 \end{vmatrix} = \begin{vmatrix} -2 & 3 \\ 1 & -1 \end{vmatrix}\mathbf{i} - \begin{vmatrix} 1 & 3 \\ 2 & -1 \end{vmatrix}\mathbf{j} + \begin{vmatrix} 1 & -2 \\ 2 & 1 \end{vmatrix}\mathbf{k} \\ &= -\mathbf{i} + 7\mathbf{j} + 5\mathbf{k}. \quad \square \end{aligned}$$

Example 2 Calculate $\mathbf{a} \times \mathbf{b}$ given that $\mathbf{a} = \mathbf{i} - \mathbf{j}$ and $\mathbf{b} = \mathbf{i} + \mathbf{k}$.

SOLUTION

$$\begin{aligned} \mathbf{a} \times \mathbf{b} &= \begin{vmatrix} \mathbf{i} & \mathbf{j} & \mathbf{k} \\ 1 & -1 & 0 \\ 1 & 0 & 1 \end{vmatrix} = \begin{vmatrix} -1 & 0 \\ 0 & 1 \end{vmatrix}\mathbf{i} - \begin{vmatrix} 1 & 0 \\ 1 & 1 \end{vmatrix}\mathbf{j} + \begin{vmatrix} 1 & -1 \\ 1 & 0 \end{vmatrix}\mathbf{k} \\ &= -\mathbf{i} - \mathbf{j} + \mathbf{k}. \quad \square \end{aligned}$$

In Examples 1 and 2 we calculated some cross products using Theorem 12.5.8′. Of course, we can obtain the same results just by applying the distributive laws. For example, for $\mathbf{a} = \mathbf{i} - 2\mathbf{j} + 3\mathbf{k}$ and $\mathbf{b} = 2\mathbf{i} + \mathbf{j} - \mathbf{k}$, we have

$$\begin{aligned}\mathbf{a} \times \mathbf{b} &= (\mathbf{i} - 2\mathbf{j} + 3\mathbf{k}) \times (2\mathbf{i} + \mathbf{j} - \mathbf{k})\\ &= (\mathbf{i} \times \mathbf{j}) - (\mathbf{i} \times \mathbf{k}) - 4(\mathbf{j} \times \mathbf{i}) + 2(\mathbf{j} \times \mathbf{k}) + 6(\mathbf{k} \times \mathbf{i}) + 3(\mathbf{k} \times \mathbf{j})\\ &= \mathbf{k} + \mathbf{j} + 4\mathbf{k} + 2\mathbf{i} + 6\mathbf{j} - 3\mathbf{i} = -\mathbf{i} + 7\mathbf{j} + 5\mathbf{k}.\end{aligned}$$

Example 3 Show that the scalar triple product can be written

(12.5.9)
$$(\mathbf{a} \times \mathbf{b}) \cdot \mathbf{c} = \begin{vmatrix} a_1 & a_2 & a_3 \\ b_1 & b_2 & b_3 \\ c_1 & c_2 & c_3 \end{vmatrix}.$$

SOLUTION

$$\begin{aligned}(\mathbf{a} \times \mathbf{b}) \cdot \mathbf{c} &= \mathbf{c} \cdot (\mathbf{a} \times \mathbf{b})\\ &= (c_1\mathbf{i} + c_2\mathbf{j} + c_3\mathbf{k}) \cdot \left(\begin{vmatrix} a_2 & a_3 \\ b_2 & b_3 \end{vmatrix} \mathbf{i} - \begin{vmatrix} a_1 & a_3 \\ b_1 & b_3 \end{vmatrix} \mathbf{j} + \begin{vmatrix} a_1 & a_2 \\ b_1 & b_2 \end{vmatrix} \mathbf{k} \right)\\ &= c_1 \begin{vmatrix} a_2 & a_3 \\ b_2 & b_3 \end{vmatrix} - c_2 \begin{vmatrix} a_1 & a_3 \\ b_1 & b_3 \end{vmatrix} + c_3 \begin{vmatrix} a_1 & a_2 \\ b_1 & b_2 \end{vmatrix}.\end{aligned}$$

This is the expansion of

$$\begin{vmatrix} a_1 & a_2 & a_3 \\ b_1 & b_2 & b_3 \\ c_1 & c_2 & c_3 \end{vmatrix}$$

by the elements of the bottom row. □

A SUGGESTION: Vectors were defined as ordered triples, and many of the early proofs were done by "breaking up" vectors into their components. This may give you the impression that the method of "breakup" and working with the components is the first thing to try when confronted with a problem that involves vectors. If it is a *computational* problem, this method may give good results. But if you have to *analyze* a situation involving vectors, particularly one in which geometry plays a role, then the "breakup" strategy is seldom the best. Think instead of using the *operations* we have defined on vectors: addition, subtraction, scalar multiplication, dot product, cross product. Being geometrically motivated, these operations are likely to provide greater understanding than breaking up everything in sight into components. □

Example 4 Let $\mathbf{a}, \mathbf{b}, \mathbf{c}$ be nonzero vectors that do not lie in the same plane. Find all the vectors $\mathbf{d}$ for which

$$(*) \qquad \mathbf{d} \cdot \mathbf{a} = \mathbf{d} \cdot \mathbf{b} = \mathbf{d} \cdot \mathbf{c}.$$

SOLUTION We could begin by writing

$$\mathbf{d} = d_1\mathbf{i} + d_2\mathbf{j} + d_3\mathbf{k}, \quad \mathbf{a} = a_1\mathbf{i} + a_2\mathbf{j} + a_3\mathbf{k}, \quad \text{and so on.}$$

Equation $(*)$ would then take the form

$$d_1a_1 + d_2a_2 + d_3a_3 = d_1b_1 + d_2b_2 + d_3b_3 = d_1c_1 + d_2c_2 + d_3c_3,$$

and we would be faced with finding all d_1, d_2, d_3 that satisfy these equations. This is a messy task.

Here is a better approach. The vectors **d** that satisfy (∗) are the vectors **d** for which

$$\mathbf{d}\cdot(\mathbf{a}-\mathbf{b}) = 0 \qquad \text{and} \qquad \mathbf{d}\cdot(\mathbf{b}-\mathbf{c}) = 0.$$

These are the vectors **d** that are perpendicular to both $\mathbf{a}-\mathbf{b}$ and $\mathbf{b}-\mathbf{c}$. One such vector is $(\mathbf{a}-\mathbf{b})\times(\mathbf{b}-\mathbf{c})$. The vectors **d** that satisfy (∗) are the scalar multiples of that cross product. ❑

Example 5 Verify *Lagrange's identity*: $\|\mathbf{a}\times\mathbf{b}\|^2 + (\mathbf{a}\cdot\mathbf{b})^2 = \|\mathbf{a}\|^2\,\|\mathbf{b}\|^2$.

SOLUTION We could begin by writing

$$\begin{aligned}
\|\mathbf{a}\times\mathbf{b}\|^2 &= (a_2b_3 - a_3b_2)^2 + (a_1b_3 - a_3b_1)^2 + (a_1b_2 - a_2b_1)^2\\
(\mathbf{a}\cdot\mathbf{b})^2 &= (a_1b_1 + a_2b_2 + a_3b_3)^2\\
\|\mathbf{a}\|^2\,\|\mathbf{b}\|^2 &= (a_1^2 + a_2^2 + a_3^2)(b_1^2 + b_2^2 + b_3^2),
\end{aligned}$$

but this would take us into a morass of arithmetic. It is much more fruitful to proceed as follows:

$$\|\mathbf{a}\times\mathbf{b}\| = \|\mathbf{a}\|\,\|\mathbf{b}\|\sin\theta \qquad \text{and} \qquad \mathbf{a}\cdot\mathbf{b} = \|\mathbf{a}\|\,\|\mathbf{b}\|\cos\theta.$$

Therefore

$$\begin{aligned}
\|\mathbf{a}\times\mathbf{b}\|^2 + (\mathbf{a}\cdot\mathbf{b})^2 &= \|\mathbf{a}\|^2\|\mathbf{b}\|^2\sin^2\theta + \|\mathbf{a}\|^2\|\mathbf{b}\|^2\cos^2\theta\\
&= \|\mathbf{a}\|^2\,\|\mathbf{b}\|^2(\sin^2\theta + \cos^2\theta) = \|\mathbf{a}\|^2\|\mathbf{b}\|^2. \quad ❑
\end{aligned}$$

Three Important Identities

It may be tempting to think that $\mathbf{a}\times(\mathbf{b}\times\mathbf{c})$ and $(\mathbf{a}\times\mathbf{b})\times\mathbf{c}$ are equal, that is, that the cross product satisfies the associative law. In general, this is false. For example,

$$\mathbf{i}\times(\mathbf{i}\times\mathbf{j}) = \mathbf{i}\times\mathbf{k} = -\mathbf{j} \qquad \text{but} \qquad (\mathbf{i}\times\mathbf{i})\times\mathbf{j} = \mathbf{0}\times\mathbf{j} = \mathbf{0}.$$

What is true instead is that

(12.5.10)

$$\begin{aligned}
\mathbf{a}\times(\mathbf{b}\times\mathbf{c}) &= (\mathbf{a}\cdot\mathbf{c})\mathbf{b} - (\mathbf{a}\cdot\mathbf{b})\mathbf{c},\\
(\mathbf{a}\times\mathbf{b})\times\mathbf{c} &= (\mathbf{c}\cdot\mathbf{a})\mathbf{b} - (\mathbf{c}\cdot\mathbf{b})\mathbf{a}.
\end{aligned}$$

There is one more identity that we want to mention:

(12.5.11)

$$(\mathbf{a}\times\mathbf{b})\cdot(\mathbf{c}\times\mathbf{d}) = (\mathbf{a}\cdot\mathbf{c})(\mathbf{b}\cdot\mathbf{d}) - (\mathbf{a}\cdot\mathbf{d})(\mathbf{b}\cdot\mathbf{c}).$$

The proof of this, as well as the proof of (12.5.10), is left to you in the Exercises.

Remark Dot products and cross products appear frequently in physics and in engineering. Work is a dot product. So is the power expended by a force. Torque and angular momentum are cross products. Turn on a television set and watch the dots on the screen. How they move is determined by the laws of electromagnetism. It is all based on Maxwell's four equations. Two of them specify dot products; two of them specify cross products. ❑

EXERCISES 12.5

Calculate.

1. $(\mathbf{i}+\mathbf{j})\times(\mathbf{i}-\mathbf{j})$.
2. $(\mathbf{i}-\mathbf{j})\times(\mathbf{j}-\mathbf{i})$.
3. $(\mathbf{i}-\mathbf{j})\times\mathbf{j}-\mathbf{k})$.
4. $\mathbf{j}\times(2\mathbf{i}-\mathbf{k})$.
5. $(2\mathbf{j}-\mathbf{k})\times(\mathbf{i}-3\mathbf{j})$.
6. $\mathbf{i}\cdot(\mathbf{j}\times\mathbf{k})$.
7. $\mathbf{j}\cdot(\mathbf{i}\times\mathbf{k})$.
8. $(\mathbf{j}\times\mathbf{i})\cdot(\mathbf{i}\times\mathbf{k})$.
9. $(\mathbf{i}\times\mathbf{j})\times\mathbf{k}$.
10. $\mathbf{k}\cdot(\mathbf{j}\times\mathbf{i})$.
11. $\mathbf{j}\cdot(\mathbf{k}\times\mathbf{i})$.
12. $\mathbf{j}\times(\mathbf{k}\times\mathbf{i})$.

Calculate

13. $(\mathbf{i}+3\mathbf{j}-\mathbf{k})\times(\mathbf{i}+\mathbf{k})$.
14. $(3\mathbf{i}-2\mathbf{j}+\mathbf{k})\times(\mathbf{i}-\mathbf{j}+\mathbf{k})$.
15. $(\mathbf{i}+\mathbf{j}+\mathbf{k})\times(2\mathbf{i}+\mathbf{k})$.
16. $(2\mathbf{i}-\mathbf{k})\times(\mathbf{i}-2\mathbf{j}+2\mathbf{k})$.
17. $[2\mathbf{i}+\mathbf{j}]\cdot[(\mathbf{i}-3\mathbf{j}+\mathbf{k})\times(4\mathbf{i}+\mathbf{k})]$.
18. $[(-2\mathbf{i}+\mathbf{j}-3\mathbf{k})\times\mathbf{i}]\times[\mathbf{i}+\mathbf{j}]$.
19. $[(\mathbf{i}-\mathbf{j})\times(\mathbf{j}-\mathbf{k})]\times[\mathbf{i}+5\mathbf{k}]$.
20. $[\mathbf{i}-\mathbf{j}]\times[(\mathbf{j}-\mathbf{k})\times(\mathbf{j}+5\mathbf{k})]$.
21. Find two unit vectors which are perpendicular to the vectors $\mathbf{a}=(1,3,-1)$ and $\mathbf{b}=(2,0,1)$.
22. Repeat Exercise 21 for the vectors $\mathbf{a}=(1,2,3)$ and $\mathbf{b}=(2,1,1)$.

In Exercises 23–26, find a vector $\mathbf{N}$ that is perpendicular to the plane determined by the points P, Q, R, and find the area of triangle PQR.

23. $P(0,1,0),\ Q(-1,1,2),\ R(2,1,-1)$.
24. $P(1,2,3),\ Q(-1,3,2),\ R(3,-1,2)$.
25. $P(1,-1,4),\ Q(2,0,1),\ R(0,2,3)$.
26. $P(2,-1,3),\ Q(4,1,-1),\ R(-3,0,5)$.

In Exercises 27 and 28, find the volume of the parallelepiped with the given edges.

27. $\mathbf{i}+\mathbf{j},\quad 2\mathbf{i}-\mathbf{k},\quad 3\mathbf{j}+\mathbf{k}$.
28. $\mathbf{i}-3\mathbf{j}+\mathbf{k},\quad 2\mathbf{j}-\mathbf{k},\quad \mathbf{i}+\mathbf{j}-2\mathbf{k}$.
29. Given the points $O(0,0,0), P(1,2,3), Q(1,1,2), R(2,1,1)$, find the volume of the parallelepiped with edges $\overrightarrow{OP}, \overrightarrow{OQ}$, and $\overrightarrow{OR}$.
30. Given the points $P(1,-1,4), Q(2,0,1), R(0,2,3), S(3,5,7)$, find the volume of the parallelepiped with edges $\overrightarrow{PQ}, \overrightarrow{PR}, \overrightarrow{PS}$.
31. Express $(\mathbf{a}+\mathbf{b})\times(\mathbf{a}-\mathbf{b})$ as a scalar multiple of $\mathbf{a}\times\mathbf{b}$.
32. Earlier we verified that $\mathbf{a}\times(\mathbf{b}+\mathbf{c})=(\mathbf{a}\times\mathbf{b})+(\mathbf{a}\times\mathbf{c})$. Show now that
$$(\mathbf{a}+\mathbf{b})\times\mathbf{c}=(\mathbf{a}\times\mathbf{c})+(\mathbf{b}\times\mathbf{c}).$$
33. Suppose that $\mathbf{a}\times\mathbf{i}=\mathbf{0}$ and $\mathbf{a}\times\mathbf{j}=\mathbf{0}$. What can you conclude about $\mathbf{a}$?
34. Let $\mathbf{a}=a_1\mathbf{i}+a_2\mathbf{j}$ and $\mathbf{b}=b_1\mathbf{i}+b_2\mathbf{j}$ be nonzero vectors in the xy-plane. Show that $\mathbf{a}\times\mathbf{b}$ is parallel to $\mathbf{k}$.
35. Express $(\alpha\mathbf{a}+\beta\mathbf{b})\times(\gamma\mathbf{a}+\delta\mathbf{b})$ as a scalar multiple of $\mathbf{a}\times\mathbf{b}$.
36. (a) Let $\mathbf{a}, \mathbf{b}, \mathbf{c}$ be distinct nonzero vectors. Show that
$$\mathbf{a}\times\mathbf{b}=\mathbf{a}\times\mathbf{c}\quad\text{iff}\quad\mathbf{a}\quad\text{and}\quad\mathbf{b}-\mathbf{c}\ \text{are parallel}.$$
(b) Sketch a figure depicting all the vectors $\mathbf{c}$ that satisfy the relation $\mathbf{a}\times\mathbf{b}=\mathbf{a}\times\mathbf{c}$.
37. Which of the following dot products are equal?
$$\mathbf{a}\cdot(\mathbf{b}\times\mathbf{c}),\ \ \mathbf{a}\cdot(\mathbf{c}\times\mathbf{b}),\ \ (\mathbf{a}\times\mathbf{b})\cdot\mathbf{c},\ \ (\mathbf{c}\times\mathbf{a})\cdot\mathbf{b},$$
$$(\mathbf{b}\times\mathbf{c})\cdot\mathbf{a},\ \ \mathbf{c}\cdot(\mathbf{b}\times\mathbf{a}),\ \ (-\mathbf{a}\times\mathbf{b})\cdot\mathbf{c},\ \ (\mathbf{a}\times-\mathbf{c})\cdot\mathbf{b}.$$
38. Show that $(\mathbf{a}\times\mathbf{b})\cdot\mathbf{b}=0$ for all vectors $\mathbf{a}$ and $\mathbf{b}$.
39. Show that the vectors $\mathbf{a}, \mathbf{b}, \mathbf{c}$ are coplanar iff
$$(\mathbf{a}\times\mathbf{b})\cdot\mathbf{c}=0.$$
40. Given that $\mathbf{a}, \mathbf{b}, \mathbf{c}$ are mutually perpendicular, show that
$$\mathbf{a}\times(\mathbf{b}\times\mathbf{c})=\mathbf{0}.$$
41. Let $\mathbf{a}\neq\mathbf{0}$. Show that if
$$\mathbf{a}\times\mathbf{b}=\mathbf{a}\times\mathbf{c}\quad\text{and}\quad\mathbf{a}\cdot\mathbf{b}=\mathbf{a}\cdot\mathbf{c},\quad\text{then}\quad\mathbf{b}=\mathbf{c}.$$
42. (a) Show that
$$\mathbf{a}\times(\mathbf{b}\times\mathbf{c})=(\mathbf{a}\cdot\mathbf{c})\mathbf{b}-(\mathbf{a}\cdot\mathbf{b})\mathbf{c}.$$
HINT: Verify that the $\mathbf{i}$ components of the two sides agree. A similar verification can be carried out for the $\mathbf{j}$ and $\mathbf{k}$ components
(b) Show that
$$(\mathbf{a}\times\mathbf{b})\times\mathbf{c}=(\mathbf{c}\cdot\mathbf{a})\mathbf{b}-(\mathbf{c}\cdot\mathbf{b})\mathbf{a}.$$
HINT: $(\mathbf{a}\times\mathbf{b})\times\mathbf{c}=-[\mathbf{c}\times(\mathbf{a}\times\mathbf{b})]$.
(c) Finally, show that
$$(\mathbf{a}\times\mathbf{b})\cdot(\mathbf{c}\times\mathbf{d})=(\mathbf{a}\cdot\mathbf{c})(\mathbf{b}\cdot\mathbf{d})-(\mathbf{a}\cdot\mathbf{d})(\mathbf{b}\cdot\mathbf{c}).$$
HINT: Set $\mathbf{c}\times\mathbf{d}=\mathbf{r}$ and use (12.5.6).
43. Let $\mathbf{a}$ and $\mathbf{b}$ be nonzero vectors with $\mathbf{a}\perp\mathbf{b}$, and set $\mathbf{c}=\mathbf{a}\times\mathbf{b}$. Express $\mathbf{c}\times\mathbf{a}$ as a multiple of $\mathbf{b}$.
44. Given that $\mathbf{u}$ is a unit vector, show that each vector $\mathbf{a}$ can be decomposed as follows into a part parallel to $\mathbf{u}$ and a part perpendicular to $\mathbf{u}$:
$$\mathbf{a}=\underbrace{(\mathbf{a}\cdot\mathbf{u})\mathbf{u}}_{\substack{\text{parallel}\\ \text{to } \mathbf{u}}}+\underbrace{(\mathbf{u}\times\mathbf{a})\times\mathbf{u}}_{\substack{\text{perpendicular}\\ \text{to } \mathbf{u}}}.$$
HINT: Use Exercise 42(b).
45. Suppose $\mathbf{a}$ and $\mathbf{b}$ are vectors such that
$$\mathbf{a}\times\mathbf{b}=\mathbf{0}\quad\text{and}\quad\mathbf{a}\cdot\mathbf{b}=0.$$
What can you conclude about $\mathbf{a}$ or $\mathbf{b}$?

Exercises 46 and 47 refer to the tetrahedron with vertices $O(0,0,0), P(a,0,0), Q(0,b,0), R(0,0,c)$ shown in the figure.

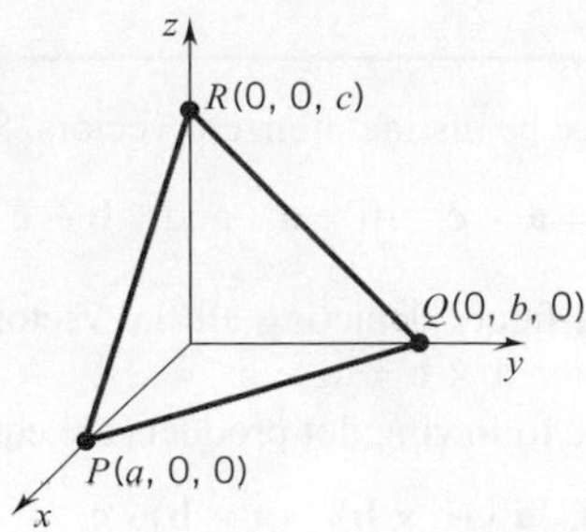

46. Use part 3 of the definition of cross product (Definition 12.5.1) to derive a formula for the area D of the face of the tetrahedron with vertices P, Q, R.

47. Let A be the area of the face opposite vertex P, let B be the area of the face opposite vertex Q, and let C be the area of the face opposite vertex R. Show that

$$A^2 + B^2 + C^2 = D^2.$$

This result is a three-dimensional version of the Pythagorean theorem.

48. Let **a**, **b** and **c** be vectors. Which of the following expressions make sense and which do not? Explain your answer in each case.

(a) $\mathbf{a} \cdot (\mathbf{b} \times \mathbf{c})$. (b) $\mathbf{a} \times (\mathbf{b} \cdot \mathbf{c})$.

(c) $\mathbf{a} \cdot (\mathbf{b} \cdot \mathbf{c})$. (d) $\mathbf{a} \times (\mathbf{b} \times \mathbf{c})$.

■ PROJECT 12.5 Torque

Suppose that a rigid body is free to rotate about a fixed point O. If a force **F** acts on the body at a point P, then the body tends to rotate about an axis through O. This effect is measured by the *torque* vector $\boldsymbol{\tau}$ which is given by

$$\boldsymbol{\tau} = \mathbf{r} \times \mathbf{F},$$

where **r** is the position vector $\overrightarrow{OP}$. The straight line through O determined by $\boldsymbol{\tau}$ is the axis of rotation. The vectors **r**, **F** and $\boldsymbol{\tau}$ form a right-handed system, and the magnitude of $\boldsymbol{\tau}$ is

$$||\boldsymbol{\tau}|| = ||\mathbf{r}||\ ||\mathbf{F}|| \sin\theta,$$

where θ is the angle between the position and force vectors (see the figure).

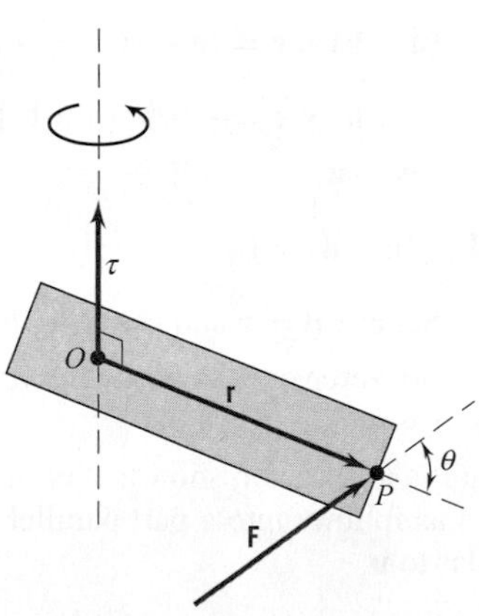

Problem 1. Find the magnitude of the torque exerted at the origin by the force $\mathbf{F} = \mathbf{i} + 2\mathbf{j} + \mathbf{k}$ applied at the point $P(1, 1, 1)$.

Problem 2. A bolt is being tightened by a 20-pound force applied to a 10-inch wrench as shown in the figure. Find the magnitude of the torque. Assuming that the wrench and the force are in the plane of the paper, in what direction will the bolt move?

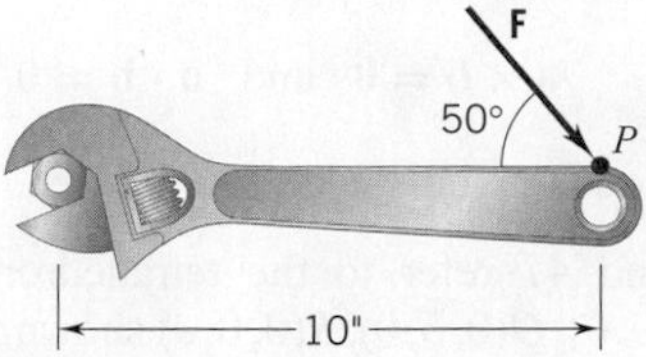

Problem 3. Repeat problem 2 if the 20-pound force is applied as shown in the figure.

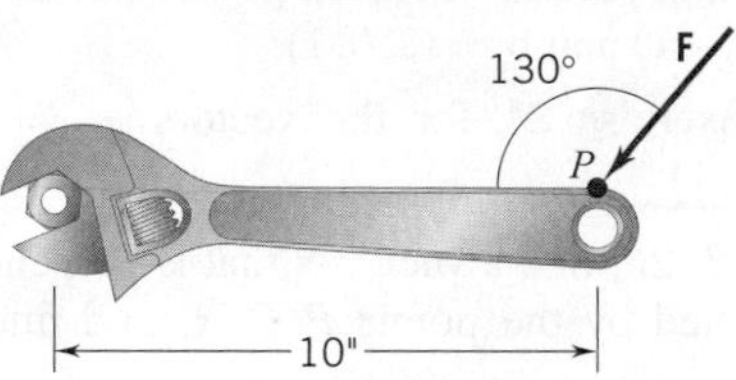

Problem 4. A bicycle with a front-wheel brake comes to a sudden stop. The horizontal braking force exerted by the brake on the front wheel is 650 newtons. The center of mass of the bicycle and its rider is 90 centimeters above the ground and 70 centimeters behind the point at which the front wheel touches the ground (see the figure).

a. What is the torque of this force about the center of mass?

b. What is the direction of the torque?

c. What rotation does it produce?

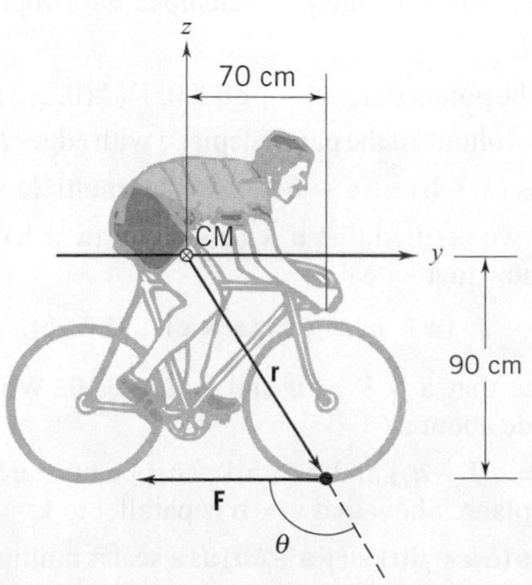

■ 12.6 LINES

Vectors used to specify position are called *position vectors*. Position vectors that emanate from the origin are known as *radius vectors*. In this section we use radius vectors to characterize lines.

Vector Parametrizations

We begin with the idea that two distinct points determine a line. In Figure 12.6.1 we have marked two points, P and Q, and the line l that they determine. To obtain a vector characterization of l, we choose the vectors $\mathbf{r}_0$ and $\mathbf{d} = \overrightarrow{PQ}$ as in Figure 12.6.2. Since we began with two distinct points, the vector $\mathbf{d}$ is nonzero.

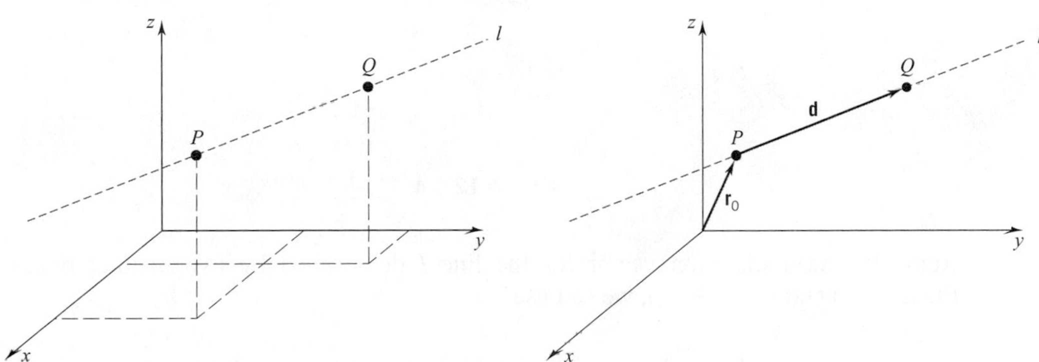

Figure 12.6.1 Figure 12.6.2

In Figure 12.6.3 we have drawn an additional vector $\mathbf{r}$. The vector that begins at the tip of $\mathbf{r}_0$ and ends at the tip of $\mathbf{r}$ is $\mathbf{r} - \mathbf{r}_0$. Therefore, the tip of $\mathbf{r}$ will fall on l iff

$$\mathbf{r} - \mathbf{r}_0 \quad \text{and} \quad \mathbf{d} \quad \text{are parallel;}$$

this in turn will happen iff

$$\mathbf{r} - \mathbf{r}_0 = t\mathbf{d} \qquad \text{for some real number } t,$$

or, equivalently, iff

$$\mathbf{r} = \mathbf{r}_0 + t\mathbf{d} \qquad \text{for some real number } t.$$

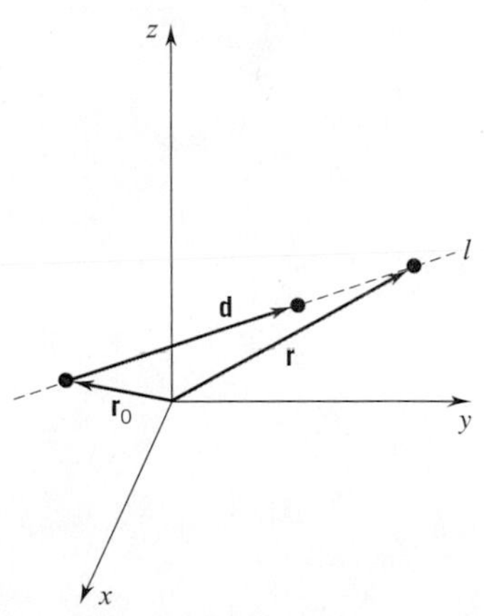

Figure 12.6.3

The vector equation

(12.6.1)
$$\boxed{\mathbf{r}(t) = \mathbf{r}_0 + t\mathbf{d}, \quad t \text{ real}}$$

parametrizes the line l: by varying t, we vary the vector $\mathbf{r}(t)$, but its tip remains on l; as t ranges over the set of real numbers, the tip of $\mathbf{r}(t)$ traces out the line l.

Now set

$$\mathbf{r}_0 = x_0\mathbf{i} + y_0\mathbf{j} + z_0\mathbf{k}, \quad \mathbf{d} = d_1\mathbf{i} + d_2\mathbf{j} + d_3\mathbf{k} \neq \mathbf{0}.$$

The tip of $\mathbf{r}_0$ is the point $P(x_0, y_0, z_0)$. The line l given by

(12.6.2)
$$\boxed{\mathbf{r}(t) = \mathbf{r}_0 + t\mathbf{d} = (x_0 + td_1)\mathbf{i} + (y_0 + td_2)\mathbf{j} + (z_0 + td_3)\mathbf{k}}$$

passes through the point $P(x_0, y_0, z_0)$ and is parallel to **d** (Figure 12.6.4). The vector **d**, which by assumption is not zero, is called a *direction vector* for l, and the components d_1, d_2, d_3 are called *direction numbers*.

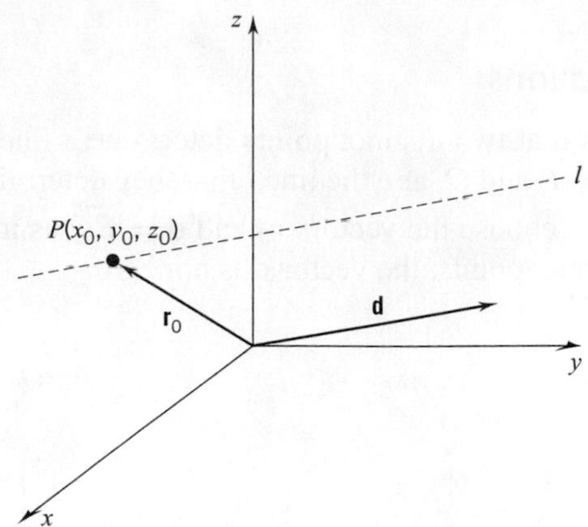

Figure 12.6.4

Remark As a direction vector for the line l determined by two distinct points $P(x_0, y_0, z_0)$ and $Q(x_1, y_2, z_3)$, we can use

$$\mathbf{d} = \overrightarrow{PQ} = (x_1 - x_0)\mathbf{i} + (y_1 - y_0)\mathbf{j} + (z_1 - z_0)\mathbf{k}.$$

The numbers $(x_1 - x_0)$, $(y_1 - y_0)$, $(z_1 - z_0)$ constitute a set of direction numbers for the line. □

Example 1 Find a vector equation that parametrizes the line that passes through the point $P(1, -1, 2)$ and is parallel to the vector $2\mathbf{i} - 3\mathbf{j} + \mathbf{k}$.

SOLUTION Here we can set

$$\mathbf{r}_0 = \mathbf{i} - \mathbf{j} + 2\mathbf{k} \qquad \text{and} \qquad \mathbf{d} = 2\mathbf{i} - 3\mathbf{j} + \mathbf{k}.$$

As a vector parametrization for the line we have

$$\mathbf{r}(t) = (\mathbf{i} - \mathbf{j} + 2\mathbf{k}) + t(2\mathbf{i} - 3\mathbf{j} + \mathbf{k}),$$

which we can rewrite as

$$\mathbf{r}(t) = (1 + 2t)\mathbf{i} - (1 + 3t)\mathbf{j} + (2 + t)\mathbf{k}. \quad □$$

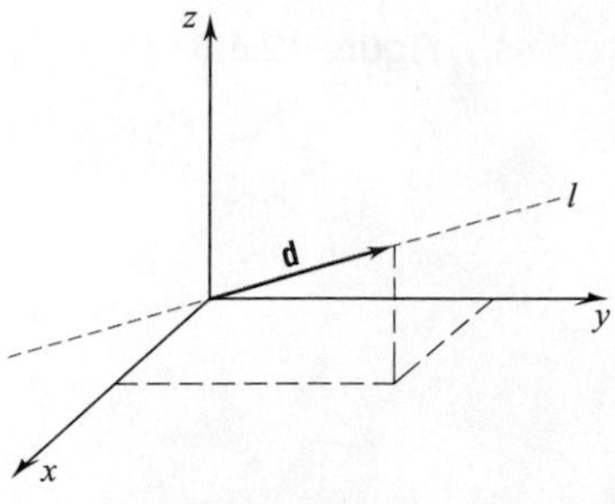

Figure 12.6.5

As a direction vector for a given line we can take any nonzero vector that is parallel to the line. Thus, if **d** is a direction vector for l, so is $\alpha\mathbf{d}$, provided that $\alpha \neq 0$. If d_1, d_2, d_3 are direction numbers for l, so are $\alpha d_1, \alpha d_2, \alpha d_3$, provided again that $\alpha \neq 0$.

The line that passes through the origin with direction vector **d** can be parametrized by the vector equation

(12.6.3) $$\boxed{\mathbf{r}(t) = t\mathbf{d} = td_1\mathbf{i} + td_2\mathbf{j} + td_3\mathbf{k}.}$$ (Figure 12.6.5)

There are, however, other ways to parametrize this line.

Example 2 Find all vector parametrizations (12.6.1) for the line through the origin with direction vector $\mathbf{d}$.

SOLUTION Since $\mathbf{d}$ is a direction vector, so is every vector $\alpha\mathbf{d}$ with $\alpha \neq 0$. We therefore write

$$(*) \qquad \mathbf{r}(t) = \mathbf{r}_0 + t\alpha\mathbf{d}.$$

Since the line passes through the origin, it may appear at first glance that $\mathbf{r}_0$ has to be $\mathbf{0}$, but that is not true. From the fact that the line passes through the origin we can conclude only that $\mathbf{r}_0 + t\alpha\mathbf{d} = \mathbf{0}$ is for some value of t. Call this value t_0. Then

$$\mathbf{r}_0 + t_0\alpha\mathbf{d} = \mathbf{0} \quad \text{and thus} \quad \mathbf{r}_0 = -t_0\alpha\mathbf{d}.$$

Substitution in $(*)$ gives

$$\mathbf{r}(t) = -t_0\alpha\mathbf{d} + t\alpha\mathbf{d} = (t - t_0)\alpha\mathbf{d}.$$

All the desired parametrizations can be written

$$\mathbf{r}(t) = (t - t_0)\alpha\mathbf{d} \quad \text{with } \alpha \text{ and } t_0 \text{ real, } \alpha \neq 0,$$

and all equations of this form parametrize that same line. ☐

Scalar Parametric Equations

It follows from (12.6.2) that the line that passes through the point $P(x_0, y_0, z_0)$ with direction numbers d_1, d_2, d_3 can be parametrized by three scalar equations:

(12.6.4)
$$x(t) = x_0 + d_1t, \quad y(t) = y_0 + d_2t, \quad z(t) = z_0 + d_3t.$$

These quantities are the $\mathbf{i}, \mathbf{j}, \mathbf{k}$ components of the vector $\mathbf{r}(t) = \mathbf{r}_0 + t\mathbf{d}$.

Example 3 Write scalar parametric equations for the line that passes through the point $P(-1, 4, 2)$ with direction numbers $1, 2, 3$.

SOLUTION In this case the scalar equations

$$x(t) = x_0 + d_1t, \quad y(t) = y_0 + d_2t, \quad z(t) = z_0 + d_3t$$

take the form

$$x(t) = -1 + t, \quad y(t) = 4 + 2t, \quad z(t) = 2 + 3t. \quad ☐$$

Example 4 What direction numbers are displayed by the parametric equations

$$x(t) = 3 - t, \quad y(t) = 2 + 4t, \quad z(t) = 1 - 5t?$$

What other direction numbers could be used for the same line?

SOLUTION The direction numbers displayed are $-1, 4, -5$. Any triple of the form

$$-\alpha, \ 4\alpha, \ -5\alpha \quad \text{with } \alpha \neq 0$$

could be used as a set of direction numbers for that same line. ☐

Symmetric Form

If the direction numbers are all nonzero, then each of the scalar parametric equations can be solved for t:

$$t = \frac{x(t) - x_0}{d_1}, \quad t = \frac{y(t) - y_0}{d_2}, \quad t = \frac{z(t) - z_0}{d_3}.$$

Eliminating the parameter t, we obtain three equations:

$$\frac{x - x_0}{d_1} = \frac{y - y_0}{d_2}, \quad \frac{y - y_0}{d_2} = \frac{z - z_0}{d_3}, \quad \frac{x - x_0}{d_1} = \frac{z - z_0}{d_3}.$$

Any two of these equations suffice; the third is redundant and can be discarded. Rather than decide which equation to discard, we simply write

(12.6.5)

$$\frac{x - x_0}{d_1} = \frac{y - y_0}{d_2} = \frac{z - z_0}{d_3}.$$

These are the equations of a line written in *symmetric form*. They can be used only if d_1, d_2, d_3 are all different from zero.

Example 5 Write equations in symmetric form for the line that passes through the point $P(x_0, y_0, z_0)$ and $Q(x_1, y_1, z_1)$. Under what conditions are the equations valid?

SOLUTION As direction numbers we can take the triple

$$x_1 - x_0, \quad y_1 - y_0, \quad z_1 - z_0.$$

We can base our calculations on $P(x_0, y_0, z_0)$ and write

$$\frac{x - x_0}{x_1 - x_0} = \frac{y - y_0}{y_1 - y_0} = \frac{z - z_0}{z_1 - z_0},$$

or we can base our calculations on $Q(x_1, y_1, z_1)$ and write

$$\frac{x - x_1}{x_1 - x_0} = \frac{y - y_1}{y_1 - y_0} = \frac{z - z_1}{z_1 - z_0}.$$

Both sets of equations are valid provided that $x_1 \neq x_0, y_1 \neq y_0, z_1 \neq z_0$. □

Equations (12.6.5) can be used only if the direction numbers are all different from zero. If one of the direction numbers is zero, then one of the coordinates is constant. As you will see, this simplifies the algebra. Geometrically, it means that the line lies on a plane that is parallel to one of the coordinate planes.

Suppose, for example, that $d_3 = 0$. Then the scalar parametric equations take the form

$$x(t) = x_0 + d_1 t, \quad y(t) = y_0 + d_2 t, \quad z(t) = z_0.$$

Eliminating t, we are left with two equations:

$$\frac{x - x_0}{d_1} = \frac{y - y_0}{d_2}, \qquad z = z_0.$$

The line lies on the horizontal plane $z = z_0$ and its projection onto the xy-plane (see Figure 12.6.6) is the line l' with equation

$$\frac{x - x_0}{d_1} = \frac{y - y_0}{d_2}.$$

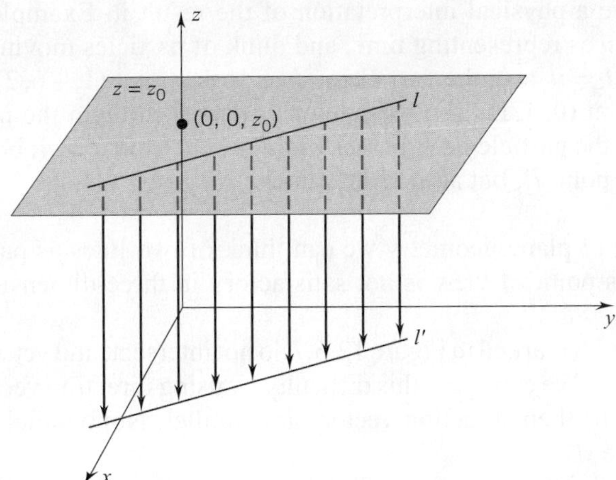

Figure 12.6.6

Intersecting Lines, Parallel Lines

Two distinct lines

$$l_1 : \mathbf{r}(t) = \mathbf{r}_0 + t\mathbf{d}, \quad l_2 : \mathbf{R}(u) = \mathbf{R}_0 + u\mathbf{D}$$

intersect iff there are numbers t and u at which

$$\mathbf{r}(t) = \mathbf{R}(u).$$

Example 6 Find the point at which the lines

$$l_1 : \mathbf{r}(t) = (\mathbf{i} - 6\mathbf{j} + 2\mathbf{k}) + t(\mathbf{i} + 2\mathbf{j} + \mathbf{k}), \quad l_2 : \mathbf{R}(u) = (4\mathbf{j} + \mathbf{k}) + u(2\mathbf{i} + \mathbf{j} + 2\mathbf{k})$$

intersect.

SOLUTION We set $\mathbf{r}(t) = \mathbf{R}(u)$

and solve for t and u:

$$(\mathbf{i} - 6\mathbf{j} + 2\mathbf{k}) + t(\mathbf{i} + 2\mathbf{j} + \mathbf{k}) = (4\mathbf{j} + \mathbf{k}) + u(2\mathbf{i} + \mathbf{j} + 2\mathbf{k}),$$
$$(1 + t)\mathbf{i} + (-6 + 2t)\,\mathbf{j} + (2 + t)\,\mathbf{k} = 2u\,\mathbf{i} + (4 + u)\,\mathbf{j} + (1 + 2u)\,\mathbf{k}$$

and therefore

$$(1 + t - 2u)\,\mathbf{i} + (-10 + 2t - u)\,\mathbf{j} + (1 + t - 2u)\,\mathbf{k} = \mathbf{0}.$$

This tells us that

$$1 + t - 2u = 0,$$
$$-10 + 2t - u = 0,$$
$$1 + t - 2u = 0.$$

Note that the first and third equations are the same. Solving the first two equations simultaneously, we obtain $t = 7, u = 4$. As you can verify,

$$\mathbf{r}(7) = 8\mathbf{i} + 8\mathbf{j} + 9\mathbf{k} = \mathbf{R}(4).$$

The two lines intersect at the tip of this vector, which is the point $P(8, 8, 9)$. □

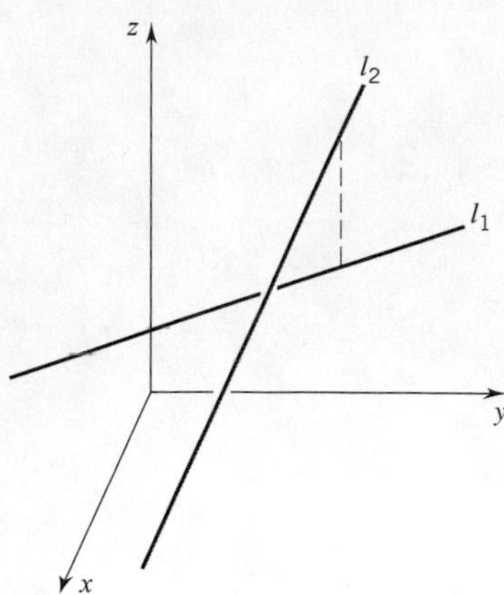

Figure 12.6.7

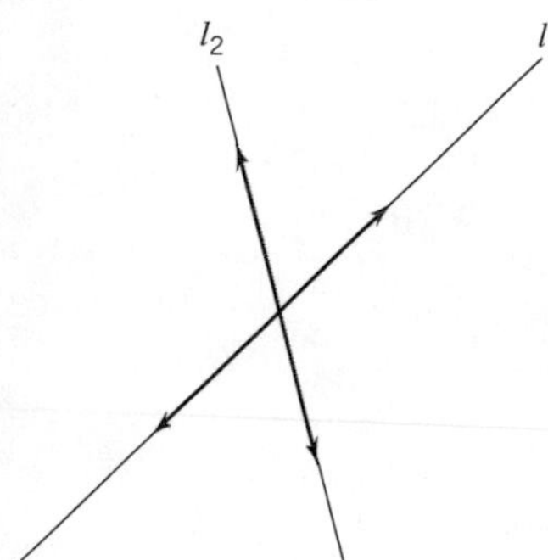

Figure 12.6.8

Remark To give a physical interpretation of the result in Example 6, think of the parameters t and u as representing time, and think of particles moving along the lines l_2 and l_2. At time $t = u = 0$, the particle on l_1 is at the point $(1, -6, 2)$ and the particle on l_2 is at the point $(0, 4, 1)$. The particle on l_1 passes through the point $P(8, 8, 9)$ at time $t = 7$, while the particle on l_2 passes through P at time $u = 4$; both particles pass through the same point P, but at different times. □

In the setting of plane geometry we can think of two lines as parallel iff they do not intersect. This point of view is not satisfactory in three-dimensional space. (See Figure 12.6.7.)

The lines l_1 and l_2 marked in Figure 12.6.7 do not intersect, and yet we would hesitate to call them parallel. We can avoid this difficulty by using direction vectors: two distinct lines are *parallel* iff their direction vectors are parallel. Nonparallel, nonintersecting lines are said to be *skew*.

If two lines l_1, l_2 intersect, we can find the angle between them by finding the angle between their direction vectors, $\mathbf{d}$ and $\mathbf{D}$. Depending on our choice of direction vectors, there are two such angles, each the supplement of the other (Figure 12.6.8). We choose the smaller of the two angles, the one with nonnegative cosine:

(12.6.6) $$\boxed{\cos\theta = |\mathbf{u_d} \cdot \mathbf{u_D}|.}$$

Example 7 Earlier we verified that the lines

$$l_1 : \mathbf{r}(t) = (\mathbf{i} - 6\mathbf{j} + 2\mathbf{k}) + t(\mathbf{i} + 2\mathbf{j} + \mathbf{k}), \quad l_2 : \mathbf{R}(u) = (4\mathbf{j} + \mathbf{k}) + u(2\mathbf{i} + \mathbf{j} + 2\mathbf{k})$$

intersect at $P(8, 8, 9)$. What is the angle between these lines?

SOLUTION As direction vectors we can take

$$\mathbf{d} = \mathbf{i} + 2\mathbf{j} + \mathbf{k} \qquad \text{and} \qquad \mathbf{D} = 2\mathbf{i} + \mathbf{j} + 2\mathbf{k}.$$

Then, as you can check,

$$\mathbf{u_d} = \tfrac{1}{6}\sqrt{6}(\mathbf{i} + 2\mathbf{j} + \mathbf{k}) \qquad \text{and} \qquad \mathbf{u_D} = \tfrac{1}{3}(2\mathbf{i} + \mathbf{j} + 2\mathbf{k}).$$

It follows that

$$\cos\theta = |\mathbf{u_d} \cdot \mathbf{u_D}| = \tfrac{1}{3}\sqrt{6} \qquad \text{and} \qquad \theta \cong 0.615 \text{ radians, about } 35.26^\circ \quad \square$$

Two intersecting lines are said to be *perpendicular* if their direction vectors are perpendicular.

Example 8 Let l_1 and l_2 be the lines of the last example. These lines intersect at $P(8, 8, 9)$. Find a vector parametrization for the line l_3 that passes through $P(8, 8, 9)$ and is perpendicular to both l_1 and l_2.

SOLUTION We are given that l_3 passes through $P(8, 8, 9)$. All we need to parametrize that line is a direction vector $\mathbf{c}$. We require that $\mathbf{c}$ be perpendicular to the direction vectors of l_1 and l_2; namely, we require that

$$\mathbf{c} \perp \mathbf{d} \quad \text{and} \quad \mathbf{c} \perp \mathbf{D}, \quad \text{where} \quad \mathbf{d} = \mathbf{i}+2\mathbf{j}+\mathbf{k} \quad \text{and} \quad \mathbf{D} = 2\mathbf{i}+\mathbf{j}+2\mathbf{k}.$$

Since $\mathbf{d} \times \mathbf{D}$ is perpendicular to both $\mathbf{d}$ and $\mathbf{D}$, we can set

$$\mathbf{c} = \mathbf{d} \times \mathbf{D} = \begin{vmatrix} \mathbf{i} & \mathbf{j} & \mathbf{k} \\ 1 & 2 & 1 \\ 2 & 1 & 2 \end{vmatrix} = \begin{vmatrix} 2 & 1 \\ 1 & 2 \end{vmatrix}\mathbf{i} - \begin{vmatrix} 1 & 1 \\ 2 & 2 \end{vmatrix}\mathbf{j} + \begin{vmatrix} 1 & 2 \\ 2 & 1 \end{vmatrix}\mathbf{k} = 3\mathbf{i} - 3\mathbf{k}.$$

As a parametrization for l_3 we can write

$$\mathbf{s}(t) = (8\mathbf{i} + 8\mathbf{j} + 9\mathbf{k}) + t(3\mathbf{i} - 3\mathbf{k}). \quad \square$$

Example 9

(a) Find a vector parametrization for the line

$$l : y = mx + b \quad \text{in the } xy\text{-plane.}$$

(b) Show by vector methods that

$$l_1 : y = m_1x + b_1 \perp l_2 : y = m_2x + b_2 \quad \text{iff} \quad m_1m_2 = -1.$$

SOLUTION

(a) We seek a parametrization of the form

$$\mathbf{r}(t) = \mathbf{r}_0 + t\mathbf{d}.$$

Since $P(0, b)$ lies on l, we can set

$$\mathbf{r}_0 = 0\mathbf{i} + b\mathbf{j} = b\mathbf{j}.$$

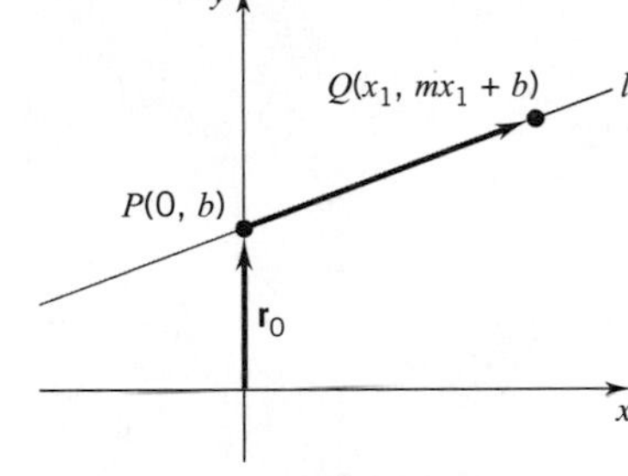

Figure 12.6.9

To find a direction vector, we take $x_1 \neq 0$ and note that the point $Q(x_1, mx_1 + b)$ also lies on l. (See Figure 12.6.9.) As direction numbers we can take

$$x_1 - 0 = x_1 \qquad \text{and} \qquad (mx_1 + b) - b = mx_1$$

or, more simply, 1 and m. This choice of direction numbers gives us the direction vector $\mathbf{d} = \mathbf{i} + m\mathbf{j}$. The vector equation

$$\mathbf{r}(t) = b\mathbf{j} + t(\mathbf{i} + m\mathbf{j})$$

parametrizes the line l.

(b) As direction vectors for l_1 and l_2 we have

$$\mathbf{d}_1 = \mathbf{i} + m_1\mathbf{j} \qquad \text{and} \qquad \mathbf{d}_2 = \mathbf{i} + m_2\mathbf{j}.$$

Since

$$\mathbf{d}_1 \cdot \mathbf{d}_2 = (\mathbf{i} + m_1\mathbf{j}) \cdot (\mathbf{i} + m_2\mathbf{j}) = 1 + m_1m_2,$$

you can see that

$$\mathbf{d}_1 \cdot \mathbf{d}_2 = 0 \qquad \text{iff} \qquad m_1m_2 = -1. \quad \square$$

Distance from a Point to a Line

In Figure 12.6.10 we have drawn a line l and a point P_1 not on l. We are interested in finding the distance $d(P_1, l)$ between P_1 and l.

Let P_0 be a point on l and let $\mathbf{d}$ be a direction vector for l. With P_0 and Q as shown in the figure, you can see that

$$d(P_1, l) = d(P_1, Q) = ||\overrightarrow{P_0P_1}|| \sin\theta$$

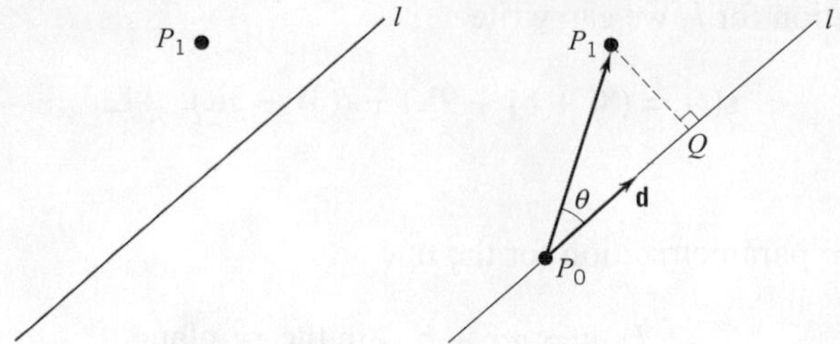

Figure 12.6.10

Since $\|\overrightarrow{P_0P_1} \times \mathbf{d}\| = \|\overrightarrow{P_0P_1}\|\,\|\mathbf{d}\| \sin\theta$, we have

(12.6.7)
$$d(P_1, l) = \frac{\|\overrightarrow{P_0P_1} \times \mathbf{d}\|}{\|\mathbf{d}\|}.$$

This elegant little formula gives the distance from a point P_1 to any line l in terms of any point P_0 on l and any direction vector $\mathbf{d}$ for l.

Computations based on this formula are left to the Exercises.

EXERCISES 12.6

1. Which of the points $P(1, 2, 0), Q(-5, 1, 5), R(-4, 2, 5)$ lie on the line

$$l : \mathbf{r}(t) = (\mathbf{i} + 2\mathbf{j}) + t(6\mathbf{i} + \mathbf{j} - 5\mathbf{k})?$$

2. Determine which of the lines are parallel:

$$l_1 : \mathbf{r}_1(t) = (\mathbf{i} + 2\mathbf{k}) + t(\mathbf{i} - 2\mathbf{j} + 3\mathbf{k}),$$
$$l_2 : \mathbf{r}_2(u) = (\mathbf{i} + 2\mathbf{k}) + u(\mathbf{i} + 2\mathbf{j} - 3\mathbf{k}),$$
$$l_3 : \mathbf{r}_3(v) = (6\mathbf{i} - \mathbf{j}) - v(2\mathbf{i} - 4\mathbf{j} + 6\mathbf{k}),$$
$$l_4 : \mathbf{r}_4(w) = (\tfrac{1}{2} + \tfrac{1}{2}w)\mathbf{i} - w\mathbf{j} + (1 + \tfrac{3}{2}w)\mathbf{k}.$$

Find a vector parametrization for the line that satisfies the given conditions.

3. Passes through $P(3, 1, 0)$ and is parallel to the line $\mathbf{r}(t) = (\mathbf{i} - \mathbf{j}) + t\mathbf{k}$.

4. Passes through $P(1, -1, 2)$ and is parallel to the line $\mathbf{r}(t) = t(3\mathbf{i} - \mathbf{j} + \mathbf{k})$.

5. Passes through the origin and $Q(x_1, y_1, z_1)$.

6. Passes through $P(x_0, y_0, z_0)$ and $Q(x_1, y_1, z_1)$.

Find a set of scalar parametric equations for the line that satisfies the given conditions.

7. Passes through $P(1, 0, 3)$ and $Q(2, -1, 4)$.

8. Passes through $P(x_0, y_0, z_0)$ and $Q(x_1, y_1, z_1)$.

9. Passes through $P(2, -2, 3)$ and is perpendicular to the xz-plane.

10. Passes through $P(1, 4, -3)$ and is perpendicular to the yz-plane.

11. Give a vector parametrization for the line that passes through $P(-1, 2, -3)$ and is parallel to the line $2(x + 1) = 4(y - 3) = z$.

12. Write equations in symmetric form for the line that passes through the origin and the point $P(x_0, y_0, z_0)$, $x_0, y_0, z_0 \neq 0$.

Determine whether the lines l_1 and l_2 are parallel, skew, or intersecting. If they intersect, find the point of intersection.

13. $l_1 : \mathbf{r}(t) = (3\mathbf{i} + \mathbf{j} + 5\mathbf{k}) + t(\mathbf{i} - \mathbf{j} + 2\mathbf{k})$,
$l_2 : \mathbf{R}(u) = (\mathbf{i} + 4\mathbf{j} + 2\mathbf{k}) + u(\mathbf{j} + \mathbf{k})$.

14. $l_1 : \mathbf{r}(t) = (-\mathbf{i} + 2\mathbf{j} + \mathbf{k}) + t(\mathbf{i} - 3\mathbf{j} + 2\mathbf{k})$,
$l_2 : \mathbf{R}(u) = (2\mathbf{i} - \mathbf{j}) + u(-2\mathbf{i} + 6\mathbf{j} - 4\mathbf{k})$.

15. $l_1 : x_1(t) = 3 + 2t,\ y_1(t) = -1 + 4t,\ z_1(t) = 2 - t$,
$l_2 : x_2(u) = 3 + 2u,\ y_2(u) = 2 + u,\ z_2(u) = -2 + 2u$.

16. $l_1 : x_1(t) = 1 + t,\ y_1(y) = -1 - t,\ z_1(t) = -4 + 2t$,
$l_2 : x_2(u) = 1 - u,\ y_2(u) = 1 + 3u,\ z_2(u) = 2u$.

17. $l_1 : x_1(t) = 1 - 6t, y_1(t) = 2 + 9t, z_1(t) = -3t$,
$l_2 : x_2(u) = 2 + 2u,\ y_2(u) = 3 - 3u,\ z_2(u) = u$.

18. $l_1 : x - 2 = \dfrac{y + 1}{2} = \dfrac{z - 1}{3}, \quad l_2 : \dfrac{x - 5}{3} = \dfrac{y - 1}{2} = z - 4.$

19. $l_1 : \dfrac{x - 4}{2} = \dfrac{y + 5}{4} = \dfrac{z - 1}{3}, \quad l_2 : x - 2 = \dfrac{y + 1}{3} = \dfrac{z}{2}.$

20. $l_1 : x_1(t) = 1 + t,\ y_1(t) = 2t,\ z_1(t) = 1 + 3t$,
$l_2 : x_2(u) = 3u,\ y_2(u) = 2u,\ z_2(u) = 2 + u$.

In Exercises 21 and 22, find the point where l_1 and l_2 intersect and find the angle between l_1 and l_2.

21. $l_1 : \mathbf{r}_1(t) = \mathbf{i} + t\mathbf{j}, \quad l_2 : \mathbf{r}_2(u) = \mathbf{j} + u(\mathbf{i} + \mathbf{j})$.

22. $l_1 : \mathbf{r}_1(t) = (\mathbf{i} - 4\sqrt{3}\,\mathbf{j}) + t(\mathbf{i} + \sqrt{3}\,\mathbf{j})$,
$l_2 : \mathbf{r}_2(u) = (4\,\mathbf{i} + 3\sqrt{3}\,\mathbf{j}) + u(\mathbf{i} - \sqrt{3}\,\mathbf{j})$.

23. Where does the line

$$\frac{x - x_0}{d_1} = \frac{y - y_0}{d_2} = \frac{z - z_0}{d_3}$$

intersect the xy-plane?

24. What can you conclude about the lines

$$\frac{x - x_0}{d_1} = \frac{y - y_0}{d_2} = \frac{z - z_0}{d_3}, \quad \frac{x - x_0}{D_1} = \frac{y - y_0}{D_2} = \frac{z - z_0}{D_3}$$

given that $d_1D_1 + d_2D_2 + d_3D_3 = 0$?

25. What can you conclude about the lines

$$\frac{x - x_0}{d_1} = \frac{y - y_0}{d_2} = \frac{z - z_0}{d_3}, \quad \frac{x - x_1}{D_1} = \frac{y - y_1}{D_2} = \frac{z - z_1}{D_3}$$

given that $d_1/D_1 = d_2/D_2 = d_3/D_3$?

26. (*Important*) Let P_0, P_1 be two distinct points and let $\mathbf{r}_0, \mathbf{r}_1$ be the radius vectors that they determine:

$$\mathbf{r}_0 = \overrightarrow{OP_0}, \quad \mathbf{r}_1 = \overrightarrow{OP_1}.$$

As t ranges over the set of real numbers, $\mathbf{r}(t) = \mathbf{r}_0 + t(\mathbf{r}_1 - \mathbf{r}_0)$ traces out the line determined by P_0 and P_1. Restrict t so that $\mathbf{r}(t)$ traces out only the line segment $\overline{P_0P_1}$.

27. Find a vector parametrization for the line segment that begins at $(2, 7, -1)$ and ends at $(4, 2, 3)$.

28. Restrict t so that the equations

$$x(t) = 7 - 5t, \quad y(t) = -3 + 2t, \quad z(t) = 4 - t$$

parametrize the line segment that begins at $(12, -5, 5)$ and ends at $(-3, 1, 2)$.

29. Determine a unit vector $\mathbf{u}$ and the values of t for which the equation

$$\mathbf{r}(t) = (6\mathbf{i} - 5\mathbf{j} + \mathbf{k}) + t\mathbf{u}$$

parametrizes the line segment that begins at $P(0, -2, 7)$ and ends at $Q(-4, 0, 11)$.

30. Suppose that the lines

$$l_1 : \mathbf{r}(t) = \mathbf{r}_0 + t\mathbf{d}, \qquad l_2 : \mathbf{R}(u) = \mathbf{R}_0 + u\mathbf{D}$$

intersect at right angles. Show that the point of intersection is the origin iff $\mathbf{r}(t) \perp \mathbf{R}(u)$ for all real numbers t and u.

31. Find scalar parametric equations for all the lines that are perpendicular to the line

$$x(t) = 1 + 2(t), \quad y(t) = 3 - 4t, \quad z(t) = 2 + 6t$$

and intersect that line at the point $P(3, -1, 8)$.

32. Suppose that $\mathbf{r}(t) = \mathbf{r}_0 + t\mathbf{d}$ and $\mathbf{R}(u) = \mathbf{R}_0 + u\mathbf{D}$ both parametrize the same line. (a) Show that $\mathbf{R}_0 = \mathbf{r}_0 + t_0\mathbf{d}$ for some real number t_0. (b) Then show that, for some real number α, $\mathbf{R}(u) = \mathbf{r}_0 + (t_0 + \alpha u)\,\mathbf{d}$ for all real u.

Find the distance from $P(1, 0, 2)$ to the indicated line.

33. The line through the origin parallel to $2\mathbf{i} - \mathbf{j} + 2\mathbf{k}$.

34. The line through $P_0(1, -1, 1)$ parallel to $\mathbf{i} - 2\mathbf{j} - 2\mathbf{k}$.

Find the distance from the point to the line.

35. $P(1, 2, 3), \quad l : \mathbf{r}(t) = \mathbf{i} + 2\mathbf{k} + t(\mathbf{i} - 2\mathbf{j} + 3\mathbf{k})$.

36. $P(0, 0, 0), \quad l : \mathbf{r}(t) = \mathbf{i} + t\mathbf{j}$.

37. $P(1, 0, 1), \quad l : \mathbf{r}(t) = 2\mathbf{i} - \mathbf{j} + t(\mathbf{i} + \mathbf{j})$.

38. Find the distance from the point $P(x_0, y_0, z_0)$ to the line $y = mx + b$ in the xy-plane.

39. What is the distance from the origin: (a) to the line that joins $P(1, 1, 1)$ and $Q(2, 2, 1)$? (b) to the line segment that joins these same points? [For part (b) find the point of the line segment $\overline{PQ}$ closest to the origin and calculate its distance from the origin.]

40. Let l be the line

$$\mathbf{r}(t) = \mathbf{r}_0 + t\mathbf{d}.$$

(a) Find the scalar t_0 for which $\mathbf{r}(t_0) \perp l$.

(b) Find the parametrizations $\mathbf{R}(t) = \mathbf{R}_0 + t\mathbf{D}$ for l in which $\mathbf{R}_0 \perp l$ and $\|\mathbf{D}\| = 1$. These are the *standard vector parametrizations* for l.

In Exercises 41 and 42, find the standard vector parametrizations (Exercise 40) for the specified line.

41. The line through $P(0, 1, -2)$ parallel to $\mathbf{i} - \mathbf{j} + 3\mathbf{k}$.

42. The line through $P(\sqrt{3}, 0, 0)$ parallel to $\mathbf{i} + \mathbf{j} + \mathbf{k}$.

43. Let A, B, C be the vertices of a triangle in the xy-plane. Given that $0 < s < 1$, determine the values of t for which the tip of the radius vector

$$\overrightarrow{OA} + s\overrightarrow{AB} + t\overrightarrow{BC}$$

lies inside the triangle. HINT: Draw a diagram.

44. (*Distance between skew lines*). The distance between two skew lines l_1 and l_2 is defined to be the minimum of the distances $d(P, Q)$ where P is a point on l_1 and Q is a point on l_2. This distance is the length of the line segment joining l_1 and l_2 and perpendicular to both lines. See the figure. Show that the distance between two skew lines is given by

$$d(l_1, l_2) = \frac{|\overrightarrow{PQ} \cdot (\mathbf{d}_1 \times \mathbf{d}_2)|}{\|\mathbf{d}_1 \times \mathbf{d}_2\|}.$$

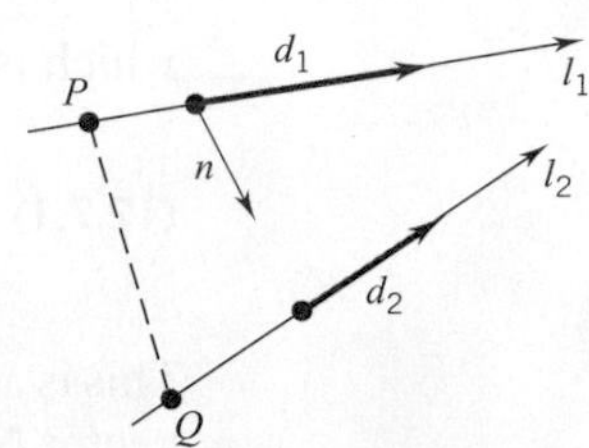

45. Show that the lines

$$l_1 : x_1(t) = 2 + t,\ y_1(t) = -1 + 3t,\ z_1(t) = 1 - 2t,$$
$$l_2 : x_2(u) = -1 + 4u,\ y_2(u) = 2 - u,\ z_2(u) = -3 + 2u.$$

are skew and find the distance between them.

46. Repeat Exercise 45 with

$$l_1 : x_1(t) = 1 + t,\ y_1(t) = -2 + 3t,\ z_1(t) = 4 - 2t,$$
$$l_2 : x_2(u) = 2u,\ y_2(u) = 3 + u,\ z_2(u) = -3 + 4u.$$

■ 12.7 PLANES

Ways of Specifying a Plane

How can we specify a plane? There are a number of ways of doing so. For example by giving three distinct points on it, so long as they are not all on the same line; by giving two distinct lines on it; or by giving a line on it and a point on it, so long as the point does not lie on the line. There is still another way to specify a plane, and that is to give a point on the plane and a nonzero vector perpendicular to the plane.

Scalar Equation of a Plane

Figure 12.7.1 shows a plane. On it we have marked a point $P(x_0, y_0, z_0)$ and, starting at that point, a nonzero vector $\mathbf{N} = A\mathbf{i} + B\mathbf{j} + C\mathbf{k}$ perpendicular to the plane. We call $\mathbf{N}$ a *normal vector*. We can obtain an equation for the plane in terms of the coordinates of P and the components of $\mathbf{N}$.

To find such an equation we take an arbitrary point $Q(x, y, z)$ in space and form the vector

$$\overrightarrow{PQ} = (x - x_0)\mathbf{i} + (y - y_0)\mathbf{j} + (z - z_0)\mathbf{k}.$$

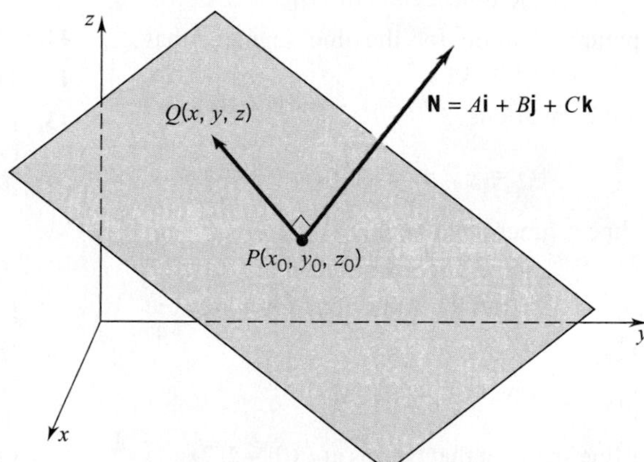

Figure 12.7.1

The point Q will lie on the given plane iff

$$\mathbf{N} \cdot \overrightarrow{PQ} = 0,$$

which is to say, iff

(12.7.1)
$$A(x - x_0) + B(y - y_0) + C(z - z_0) = 0.$$

This is an equation in x, y, z for the plane that passes through $P(x_0, y_0, z_0)$ and has normal vector $\mathbf{N} = A\mathbf{i} + B\mathbf{j} + C\mathbf{k}$.

Remark If **N** is normal to a given plane, then so is every nonzero scalar multiple of **N**. Suppose we had chosen $-2\mathbf{N}$ as our normal. Then (12.7.1) would have read

$$-2A(x-x_0)-2B(y-y_0)-2C(z-z_0)=0.$$

Canceling the -2, we would have the same equation we had before. It does not matter which normal we choose. All normals give equivalent equations. □

We can write (12.7.1) in the form

$$Ax+By+Cz+D=0$$

simply by setting $D=-Ax_0-By_0-Cz_0$.

Example 1 Write an equation for the plane that passes through the point $P(1,0,2)$ and has normal vector $\mathbf{N}=3\mathbf{i}-2\mathbf{j}+\mathbf{k}$.

SOLUTION The general equation

$$A(x-x_0)+B(y-y_0)+C(z-z_0)=0$$

becomes

$$3(x-1)+(-2)(y-0)+(z-2)=0,$$

which simplifies to

$$3x-2y+z-5=0. \quad \square$$

Example 2 Find an equation for the plane p that passes through $P(-2,3,5)$ and is perpendicular to the line l with scalar parametric equations: $x=-2+t$, $y=1+2t$, $z=4$.

SOLUTION We can take $\mathbf{N}=\mathbf{i}+2\mathbf{j}$ as a direction vector for l. Since p and l are perpendicular, **N** is a normal vector for p. Thus, as an equation for p, we can write

$$(x+2)+2(y-3)+0(z-5)=0,$$

which simplifies to

$$x+2y-4=0.$$

The last equation looks very much like the equation of a line in the xy-plane. If the context of our discussion were the xy-plane, then the equation $x+2y-4=0$ would represent a line. In this case, however, our context is three-dimensional space. Hence, the equation $x+2y-4=0$ represents the set of all points $Q(x,y,z)$, where $x+2y-4=0$ and z is unrestricted. This set forms a vertical plane that intersects the xy-plane in the indicated line (Figure 12.7.2). □

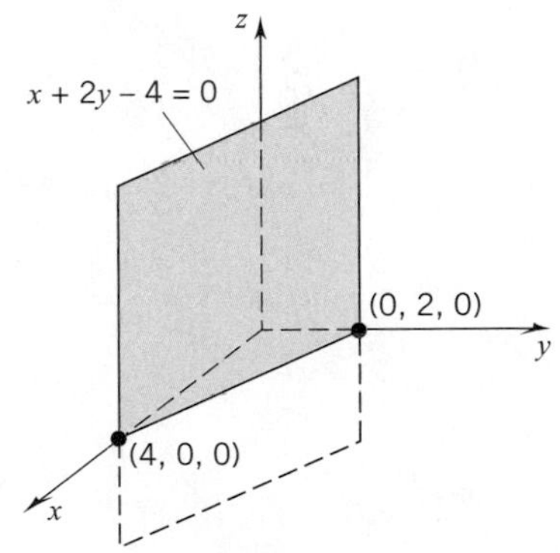

Figure 12.7.2

Example 3 Show that every equation

$$ax+by+cz+d=0 \quad \text{with} \quad \sqrt{a^2+b^2+c^2}\neq 0$$

represents a plane in space.

SOLUTION Since $\sqrt{a^2+b^2+c^2}\neq 0$, the numbers a, b, c are not all zero, and therefore there exist numbers x_0, y_0, z_0 such that

$$ax_0+by_0+cz_0+d=0.\dagger$$

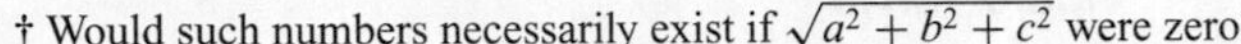

† Would such numbers necessarily exist if $\sqrt{a^2+b^2+c^2}$ were zero?

The equation

$$ax + by + cz + d = 0$$

can now be written

$$(ax + by + cz + d) - (ax_0 + by_0 + cz_0 + d) = 0,$$

and so, after factoring, we have

$$a(x - x_0) + b(y - y_0) + c(z - z_0) = 0.$$

This equation (and hence the initial equation) represents the plane through the point $P(x_0, y_0, z_0)$ with normal $\mathbf{N} = a\mathbf{i} + b\mathbf{j} + c\mathbf{k}$. The initial assumption that $\sqrt{a^2 + b^2 + c^2} \neq 0$ guarantees that $\mathbf{N} \neq \mathbf{0}$. □

Vector Equation of a Plane

We can write the equation of a plane entirely in vector notation. With

$$\mathbf{N} = A\mathbf{i} + B\mathbf{j} + C\mathbf{k}$$

and

$$\mathbf{r}_0 = x_0\mathbf{i} + y_0\mathbf{j} + z_0\mathbf{k}, \quad \mathbf{r} = x\mathbf{i} + y\mathbf{j} + z\mathbf{k},$$

Equation (12.7.1) reads

(12.7.2)

$$\boxed{\mathbf{N} \cdot (\mathbf{r} - \mathbf{r}_0) = 0.}$$

This vector equation represents the plane that passes through the tip of $\mathbf{r}_0$ and has normal $\mathbf{N}$. (See Figure 12.7.3.) If the plane passes through the origin, we can take $\mathbf{r}_0 = \mathbf{0}$. Equation (12.7.2) then takes the form

(12.7.3)

$$\boxed{\mathbf{N} \cdot \mathbf{r} = 0.}$$

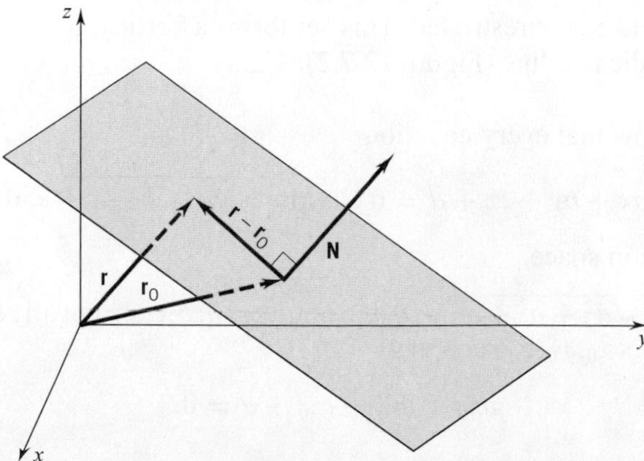

Figure 12.7.3

Collinear Vectors, Coplanar Vectors

Collinear points are points that lie on the same line; *coplanar points* are points that lie on the same plane. The terms "collinear" and "coplanar" are also applied to vectors: by definition, two vectors **a** and **b** are said to be *collinear* if there exist scalars s and t not both 0 such that

$$s\mathbf{a} + t\mathbf{b} = \mathbf{0}.$$

If $s \neq 0$, then $\mathbf{a} = -(t/s)\mathbf{b}$; if $t \neq 0$, then $\mathbf{b} = -(s/t)\mathbf{a}$. Collinear vectors are thus parallel. If we set

$$\mathbf{a} = \overrightarrow{PA} \qquad \text{and} \qquad \mathbf{b} = \overrightarrow{PB},$$

then the points P, A, B all fall on the same line, hence the term "collinear vectors."

Three vectors **a**, **b**, **c** are said to be *coplanar* if there exist scalars s, t, u not all zero such that

$$s\mathbf{a} + t\mathbf{b} + u\mathbf{c} = \mathbf{0}.$$

This term, too, is justified:

(12.7.4) $\mathbf{a} = \overrightarrow{PA}, \mathbf{b} = \overrightarrow{PB}, \mathbf{c} = \overrightarrow{PC}$ are coplanar vectors iff the points P, A, B, C all lie on the same plane .

PROOF Here we show that if the three vectors are coplanar, then the four points all lie on the same plane. We leave the converse to you (Exercise 45).

Suppose that the three vectors are coplanar. Then we can write

$$s\,\overrightarrow{PA} + t\,\overrightarrow{PB} + u\,\overrightarrow{PC} = \mathbf{0} \quad \text{with } s, t, u \text{ not all zero}.$$

Without loss of generality, we assume that $s \neq 0$. Then

$$\overrightarrow{PA} = -\frac{t}{s}\overrightarrow{PB} - \frac{u}{s}\overrightarrow{PC}.$$

Since $\overrightarrow{PB} \times \overrightarrow{PC}$ is perpendicular to both $\overrightarrow{PB}$ and $\overrightarrow{PC}$, we see that

$$\begin{aligned}(\overrightarrow{PB} \times \overrightarrow{PC}) \cdot \overrightarrow{PA} &= (\overrightarrow{PB} \times \overrightarrow{PC}) \cdot \left(-\frac{t}{s}\overrightarrow{PB} - \frac{u}{s}\overrightarrow{PC}\right)\\ &= -\frac{t}{s}(\overrightarrow{PB} \times \overrightarrow{PC}) \cdot \overrightarrow{PB} - \frac{u}{s}(\overrightarrow{PB} \times \overrightarrow{PC}) \cdot \overrightarrow{PC}\\ &= 0.\end{aligned}$$

But $|(\overrightarrow{PB} \times \overrightarrow{PC}) \cdot \overrightarrow{PA}|$ gives the volume of the parallelepiped with edges $\overrightarrow{PA}, \overrightarrow{PB}, \overrightarrow{PC}$. This volume can be zero only if P, A, B, C all lie on the same plane. ☐

Unit Normals

If **N** is normal to a given plane, then all other normals to that plane are parallel to **N** and hence scalar multiples of **N**. In particular there are only two normals of length 1:

$$\mathbf{u_N} = \frac{\mathbf{N}}{\|\mathbf{N}\|} \qquad \text{and} \qquad -\,\mathbf{u_N} = \frac{\mathbf{N}}{\|\mathbf{N}\|}.$$

These are called the *unit normals*.

Example 4 Find the unit normals for the plane $3x - 4y + 12z + 8 = 0$.

SOLUTION We can take $\mathbf{N} = 3\mathbf{i} - 4\mathbf{j} + 12\mathbf{k}$.

Since $$||\mathbf{N}|| = \sqrt{3^2 + (-4)^2 + 12^2} = \sqrt{169} = 13,$$

we have $\mathbf{u_N} = \frac{1}{13}(3\mathbf{i} - 4\mathbf{j} + 12\mathbf{k})$ and $-\mathbf{u_N} = -\frac{1}{13}(3\mathbf{i} - 4\mathbf{j} + 12\mathbf{k})$. □

Parallel Planes, Intersecting Planes

Two planes are called *parallel* iff their normals are parallel. If two planes, p_1 and p_2, are not parallel, we can find the angle between them by finding the angle between their normals, $\mathbf{N}_1, \mathbf{N}_2$. (See Figure 12.7.4.) Depending on our choice of normals, there are two such angles, each the supplement of the other. We will choose the smaller angle, the one with the nonnegative cosine:

(12.7.5) $$\cos\theta = |\mathbf{u}_{\mathbf{N}_1} \cdot \mathbf{u}_{\mathbf{N}_2}|.$$

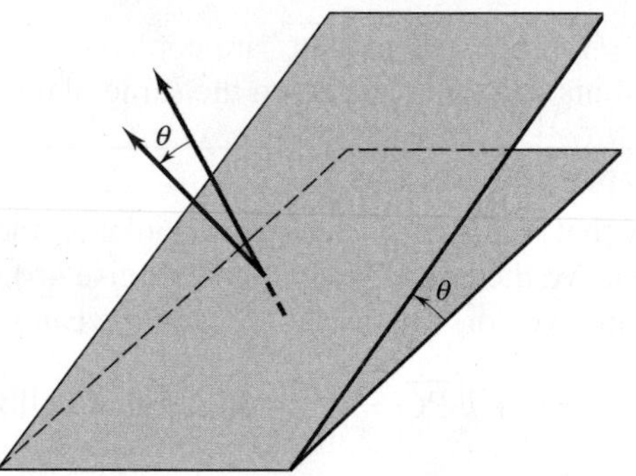

Figure 12.7.4

Example 5 Here are some planes:

$$p_1 : 2(x-1) - 3y + 5(z-2) = 0, \qquad p_2 : -4x + 6y + 10z + 24 = 0,$$
$$p_3 : 4x - 6y - 10z + 1 = 0, \qquad p_4 : 2x - 3y + 5z - 12 = 0.$$

(a) Indicate which planes are identical.
(b) Indicate which planes are distinct but parallel.
(c) Find the angle between p_1 and p_2.

SOLUTION

(a) p_1 and p_4 are identical, as you can verify by simplifying the equation of p_1.
(b) p_2 and p_3 are distinct but parallel. The planes are distinct since $P(0, 0, \frac{1}{10})$ lies on p_3 but not on p_2; they are parallel since the normals

$$-4\mathbf{i} + 6\mathbf{j} + 10\mathbf{k} \quad \text{and} \quad 4\mathbf{i} - 6\mathbf{j} - 10\mathbf{k}$$

are parallel.
(c) Taking

$$\mathbf{N}_1 = 2\mathbf{i} - 3\mathbf{j} + 5\mathbf{k} \quad \text{and} \quad \mathbf{N}_2 = -4\mathbf{i} + 6\mathbf{j} + 10\mathbf{k},$$

we have

$$\mathbf{u}_{\mathbf{N}_1} = \frac{1}{\sqrt{38}}(2\mathbf{i} - 3\mathbf{j} + 5\mathbf{k}) \quad \text{and} \quad \mathbf{u}_{\mathbf{N}_2} = \frac{1}{\sqrt{152}}(-4\mathbf{i} + 6\mathbf{j} + 10\mathbf{k}).$$

As you can check,

$$\cos\theta = |\mathbf{u}_{\mathbf{N}_1} \cdot \mathbf{u}_{\mathbf{N}_2}| = \tfrac{6}{19} \quad \text{and thus} \quad \theta \cong 1.25 \text{ radians, about } 71.59^\circ. \quad \square$$

Example 6 The planes

$$p_1 : A_1x + B_1y + C_1z + D_1 = 0, \qquad p_2 : A_2x + B_2y + C_2z + D_2 = 0$$

intersect to form a line l. Find a vector parametrization for l.

SOLUTION We need to find a point P_0 on l and a direction vector for l. Finding P_0 is a matter of finding numbers x, y, z that simultaneously satisfy the equations given for p_1 and p_2. In concrete cases this is not hard, and we will not try to give a general formula for such a P_0. We will just assume that P_0 has been found and focus on finding a direction vector for l.

Since p_1 and p_2 intersect in a line, the normals

$$\mathbf{N}_1 = A_1\mathbf{i} + B_1\mathbf{j} + C_1\mathbf{k}, \quad \mathbf{N}_2 = A_2\mathbf{i} + B_2\mathbf{j} + C_2\mathbf{k}$$

are not parallel. This guarantees that $\mathbf{N}_1 \times \mathbf{N}_2$ is not $\mathbf{0}$. Since l lies on both p_1 and p_2, the line l, like the vector $\mathbf{N}_1 \times \mathbf{N}_2$, is perpendicular to both $\mathbf{N}_1$ and $\mathbf{N}_2$. This makes l parallel to $\mathbf{N}_1 \times \mathbf{N}_2$. (See Figure 12.7.5.) We can therefore take $\mathbf{N}_1 \times \mathbf{N}_2$ as a

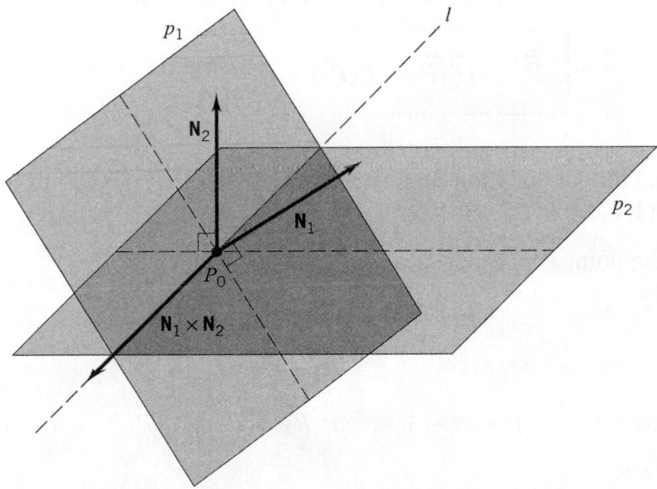

Figure 12.7.5

direction vector for l and write

$$l : \mathbf{r}(t) = \overrightarrow{OP_0} + t(\mathbf{N}_1 \times \mathbf{N}_2). \quad \square$$

Example 7 Given the planes $p_1 : 2x - 3y + 2z = 9$ and $p_2 : x + 2y - z = -4$, show that p_1 and p_2 are not parallel. Then find scalar parametric equations of the line l which is formed by the intersection of the two planes.

SOLUTION As normal vectors for p_1 and p_2, we can take $\mathbf{N}_1 = 2\mathbf{i} - 3\mathbf{j} + 2\mathbf{k}$ and $\mathbf{N}_2 = \mathbf{i} + 2\mathbf{j} - \mathbf{k}$. Since neither vector is a scalar multiple of the other, the vectors are not parallel. Therefore, p_1 and p_2 are not parallel.

As shown in Example 6, we can use

$$\mathbf{N}_1 \times \mathbf{N}_2 = \begin{vmatrix} \mathbf{i} & \mathbf{j} & \mathbf{k} \\ 2 & -3 & 2 \\ 1 & 2 & -1 \end{vmatrix} = -\mathbf{i} + 4\mathbf{j} + 7\mathbf{k}$$

as a direction vector for l. Now we need a point that lies on l. To find one, we solve the equations for p_1 and p_2 simultaneously. If, for example, we set $x = 0$ in the two equations, we get

$$-3y + 2z = 9$$
$$2y - z = -4.$$

Solving this pair of equations for y and z, we find that $y = 1$ and $z = 6$. Thus, the point $(0, 1, 6)$ is on l. As scalar parametric equations for l, we can write

$$x = -t, \quad y = 1 + 4t, \quad z = 6 + 7t. \quad \square$$

The Plane Determined by Three Noncollinear Points

Suppose now that we are given three noncollinear points P_1, P_2, P_3. These points determine a plane. How can we find an equation for this plane?

First we form the vectors $\overrightarrow{P_1P_2}, \overrightarrow{P_1P_3}$. Since P_1, P_2, P_3 are noncollinear,the vectors are not parallel. Therefore their cross product $\overrightarrow{P_1P_2} \times \overrightarrow{P_1P_3}$ can be used as a normal for the plane. We are back in a familiar situation. We have a point of the plane, say P_1, and we have a normal vector, $\overrightarrow{P_1P_2} \times \overrightarrow{P_1P_3}$. A point P will lie on the plane iff

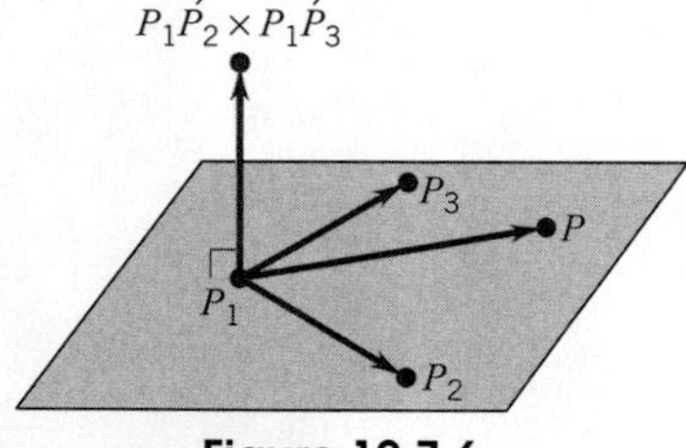

Figure 12.7.6

(12.7.6)
$$\overrightarrow{P_1P} \cdot (\overrightarrow{P_1P_2} \times \overrightarrow{P_1P_3}) = 0.$$
(Figure 12.7.6)

Example 8 Find an equation in x, y, z for the plane that passes through the points $P_1(0, 1, 1), P_2(1, 0, 1), P_3(1, -3, -1)$.

SOLUTION The point $P = P(x, y, z)$ will lie on this plane iff

$$\overrightarrow{P_1P} \cdot (\overrightarrow{P_1P_2} \times \overrightarrow{P_1P_3}) = 0.$$

Here

$$\overrightarrow{P_1P} = x\mathbf{i} + (y-1)\mathbf{j} + (z-1)\mathbf{k}, \quad \overrightarrow{P_1P_2} = \mathbf{i} - \mathbf{j}, \quad \overrightarrow{P_1P_3} = \mathbf{i} - 4\mathbf{j} - 2\mathbf{k}.$$

As you can check,

$$\overrightarrow{P_1P_2} \times \overrightarrow{P_1P_3} = 2\mathbf{i} + 2\mathbf{j} - 3\mathbf{k}.$$

Thus,

$$\begin{aligned}\overrightarrow{P_1P} \cdot (\overrightarrow{P_1P_2} \times \overrightarrow{P_1P_3}) &= [x\mathbf{i} + (y-1)\mathbf{j} + (z-1)\mathbf{k}] \cdot [2\mathbf{i} + 2\mathbf{j} - 3\mathbf{k}] \\ &= 2x + 2(y-1) - 3(z-1) = 2x + 2y - 3z + 1.\end{aligned}$$

As an equation for the plane, we can use

$$2x + 2y - 3z + 1 = 0. \quad \square$$

The Distance from a Point to a Plane

In Figure 12.7.7, we have drawn a plane $p : Ax + By + Cz + D = 0$ and a point $P(x_1, y_1, z_1)$ not on p. The distance between the point P_1 and the plane p is given by the formula

(12.7.7) $$d(P_1, p) = \frac{|Ax_1 + By_1 + Cz_1 + D|}{\sqrt{A^2 + B^2 + C^2}}.$$

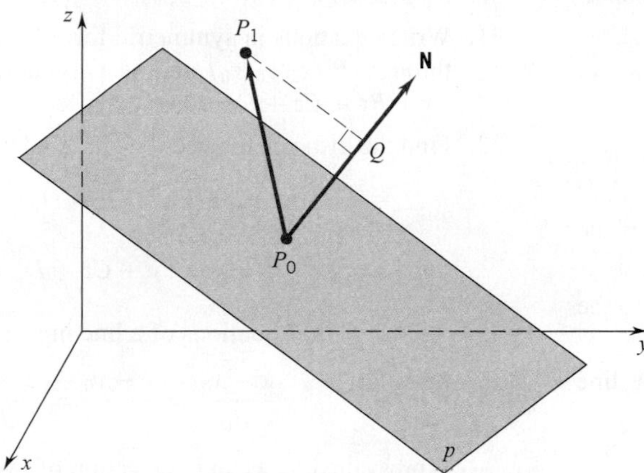

Figure 12.7.7

PROOF Pick any point $P_0(x_0, y_0, z_0)$ in the plane. As a normal to p we can take the vector

$$\mathbf{N} = A\mathbf{i} + B\mathbf{j} + C\mathbf{k}.$$

Then
$$\mathbf{u_N} = \frac{A\mathbf{i} + B\mathbf{j} + C\mathbf{k}}{\sqrt{A^2 + B^2 + C^2}}$$

is the corresponding unit normal. From Figure 12.7.7,

$$\begin{aligned} d(P_1, p) = d(P_0, Q) &= |\text{comp}_{\mathbf{N}}\, \overrightarrow{P_0P_1}| \\ &= |\overrightarrow{P_0P_1} \cdot \mathbf{u_N}| \\ &= \frac{|(x_1 - x_0)A + (y_1 - y_0)B + (z_1 - z_0)C|}{\sqrt{A^2 + B^2 + C^2}} \\ &= \frac{|Ax_1 + By_1 + Cz_1 - (Ax_0 + By_0 + Cz_0)|}{\sqrt{A^2 + B^2 + C^2}} \end{aligned}$$

Since $P_0(x_0, y_0, z_0)$ lies on p,

$$Ax_0 + By_0 + Cz_0 = -D,$$

and we have

$$d(P_1, p) = \frac{|Ax_1 + By_1 + Cz_1 + D|}{\sqrt{A^2 + B^2 + C^2}}. \quad \square$$

EXERCISES 12.7

1. Which of the points $P(3, 2, 1)$, $Q(2, 3, 1)$, $R(1, 4, 1)$ lie on the plane

$$3(x - 1) + 4y - 5(z + 2) = 0?$$

2. Which of the points $P(2, 1, -2)$, $Q(2, 0, 0)$, $R(4, 1, -1)$, $S(0, -1, -3)$ lie on the plane $\mathbf{N} \cdot (\mathbf{r} - \mathbf{r}_0) = 0$ if $\mathbf{N} = \mathbf{i} - 3\mathbf{j} + \mathbf{k}$ and $\mathbf{r}_0 = 4\mathbf{i} + \mathbf{j} - \mathbf{k}$?

Find an equation for the plane which satisfies the given conditions.

3. Passes through the point $P(2, 3, 4)$ and is perpendicular to $\mathbf{i} - 4\mathbf{j} + 3\mathbf{k}$.

4. Passes through the point $P(1, -2, 3)$ and is perpendicular to $\mathbf{j} + 2\mathbf{k}$.

5. Passes through the point $P(2, 1, 1)$ and is parallel to the plane $3x - 2y + 5z - 2 = 0$.

6. Passes through the point $P(3, -1, 5)$ and is parallel to the plane $4x + 2y - 7z + 5 = 0$.

7. Passes through the point $P(1, 3, 1)$ and contains the line $l : x = t,\ y = t,\ z = -2 + t$.

8. Passes through the point $P(2, 0, 1)$ and contains the line $l : x = 1 - 2t,\ y = 1 + 4t,\ z = 2 + t$.

9. Passes through the point $P_0(x_0, y_0, z_0)$ and is perpendicular to $\overrightarrow{OP_0}$.

Find the unit normals to the plane.

10. $2x - 3y + 7z - 3 = 0$.

11. $2x - y + 5z - 10 = 0$.

12. Show that the plane $x/a + y/b + z/c = 1$ intersects the coordinate axes at $x = a, y = b, z = c$. This is the equation of a plane in *intercept form*.

Write the equation of the plane in intercept form and find the points where it intersects the coordinate axes.

13. $4x + 5y - 6z = 60$.

14. $3x - y + 4z + 2 = 0$.

Find the angle between the planes.

15. $5(x - 1) - 3(y + 2) + 2z = 0$,
$x + 3(y - 1) + 2(z + 4) = 0$.

16. $2x - y + 3z = 5,\quad 5x + 5y - z = 1$.

17. $x - y + z - 1 = 0,\quad 2x + y + 3z + 5 = 0$.

18. $4x + 4y - 2z = 3,\quad 2x + y + z = -1$.

Determine whether the vectors are coplanar.

19. $4\mathbf{j} - \mathbf{k},\quad 3\mathbf{i} + \mathbf{j} + 2\mathbf{k},\quad \mathbf{0}$.

20. $\mathbf{i},\quad \mathbf{i} - 2\mathbf{j},\quad 3\mathbf{j} + \mathbf{k}$.

21. $\mathbf{i} + \mathbf{j} + \mathbf{k},\quad 2\mathbf{i} - \mathbf{j},\quad 3\mathbf{i} - \mathbf{j} - \mathbf{k}$.

22. $\mathbf{j} - \mathbf{k},\quad 3\mathbf{i} - \mathbf{j} + 2\mathbf{k},\quad 3\mathbf{i} - 2\mathbf{j} + 3\mathbf{k}$.

Find the distance from the point P to the given plane.

23. $P(2, -1, 3);\quad 2x + 4y - z + 1 = 0$.

24. $P(3, -5, 2);\quad 8x - 2y + z = 5$.

25. $P(1, -3, 5);\quad -3x + 4z + 5 = 0$.

26. $P(1, 3, 4);\quad x + y - 2z = 0$.

Find an equation in x, y, z for the plane that passes through the given points.

27. $P_1(1, 0, 1),\quad P_2(2, 1, 0),\quad P_3(1, 1, 1)$.

28. $P_1(1, 1, 1),\quad P_2(2, -2, -1),\quad P_3(0, 2, 1)$.

29. $P_1(3, -4, 1),\quad P_2(3, 2, 1),\quad P_3(-1, 1, -2)$.

30. $P_1(3, 2, -1),\quad P_2(3, -2, 4),\quad P_3(1, -1, 3)$.

31. Write equations in symmetric form for the line that passes through $P_0(x_0, y_0, z_0)$ and is perpendicular to the plane $Ax + By + Cz + D = 0$.

32. Find the distance between the parallel planes

$$Ax + By + Cz + D_1 = 0$$

and

$$Ax + By + Cz + D_2 = 0.$$

33. Show that the equations of a line in symmetric form

$$\frac{x - x_0}{d_1} = \frac{y - y_0}{d_2} = \frac{z - z_0}{d_3}$$

express the line as an intersection of two planes by finding equations for two such planes.

34. Find scalar parametric equations for the line formed by the two intersecting planes.
(a) $z = z_0,\quad y = y_0$. (b) $x = x_0,\quad z = z_0$.

In Exercises 35 and 36, find a set of scalar parametric equations for the line formed by the two intersecting planes.

35. $p_1 : x + 2y + 3z = 0,\quad p_2 : -3x + 4y + z = 0$.

36. $p_1 : x + y + z + 1 = 0,\quad p_2 : x - y + z + 2 = 0$.

In Exercises 37 and 38, let l be the line determined by P_1, P_2, and let p be the plane determined by Q_1, Q_2, Q_3. Where, if anywhere, does l intersect p?

37. $P_1(1, -1, 2),\quad P_2(-2, 3, 1)$;
$Q_1(2, 0, -4),\quad Q_2(1, 2, 3),\quad Q_3(-1, 2, 1)$.

38. $P_1(4, -3, 1),\quad P_2(2, -2, 3)$;
$Q_1(2, 0, -4),\quad Q_2(1, 2, 3),\quad Q_3(-1, 2, 1)$.

39. Let l_1, l_2 be lines that pass through the origin and have direction vectors

$$\mathbf{d} = \mathbf{i} + 2\mathbf{j} + 4\mathbf{k},\quad \mathbf{D} = -\mathbf{i} - \mathbf{j} + 3\mathbf{k}.$$

Find an equation for the plane that contains l_1 and l_2.

40. Show that two nonparallel lines $\mathbf{r}(t) = \mathbf{r}_0 + t\mathbf{d}$ and $\mathbf{R}(t) = \mathbf{R}_0 + t\mathbf{D}$ intersect iff the vectors $\mathbf{r}_0 - \mathbf{R}_0, \mathbf{d}$, and $\mathbf{D}$ are coplanar.

41. Given that a plane contains the point P and has normal $\mathbf{N}$, describe the set of points Q on the plane for which $(\mathbf{N} + \overrightarrow{PQ}) \perp (\mathbf{N} - \overrightarrow{PQ})$. HINT: Draw a figure.

42. Let $\mathbf{a}, \mathbf{b}, \mathbf{c}$ be three nonzero vectors such that the angle between any pair of them is $\frac{1}{2}\pi$. Can these vectors be coplanar?

43. Suppose that N $= A\mathbf{i} + B\mathbf{j} + C\mathbf{k}$ is a nonzero vector with its initial point on the plane $Ax + By + Cz + D = 0$. Take $P_1(x_1, y_1, z_1)$ as a point of space and set $\alpha = Ax_1 + By_1 + Cz_1 + D$. If $\alpha = 0$, then P_1 lies on the plane. What can you conclude about P_1 if α is positive? If α is negative?

44. Suppose that $\mathbf{a}, \mathbf{b}, \mathbf{c}$ are radius vectors the tips of which are not collinear. Give a geometric interpretation of the equation

$$\begin{vmatrix} x - a_1 & y - a_2 & z - a_3 \\ x - b_1 & y - b_2 & z - b_3 \\ x - c_1 & y - c_2 & z - c_3 \end{vmatrix} = 0.$$

45. Suppose that the points P, A, B, and C all lie on the same plane. Show that the vectors $\mathbf{a} = \overrightarrow{PA}, \mathbf{b} = \overrightarrow{PB}$, and $\mathbf{c} = \overrightarrow{PC}$ are coplanar vectors.

(*Sketching planes*). In Exercises 46–49, the equation of a plane p is given.

(a) Find the intercepts of p.

(b) Find the *traces* of p. (The traces of p are the lines of intersection of p with the coordinate planes.

(c) Find the unit normals.

(d) Sketch the plane.

46. $x + 2y + 3z - 6 = 0$.

47. $5x + 4y + 10z = 20$.

48. $3x + 2y - 6 = 0$.

49. $3x + 2z - 12 = 0$.

In Exercises 50–53, find an equation for the plane shown in the figure.

50.

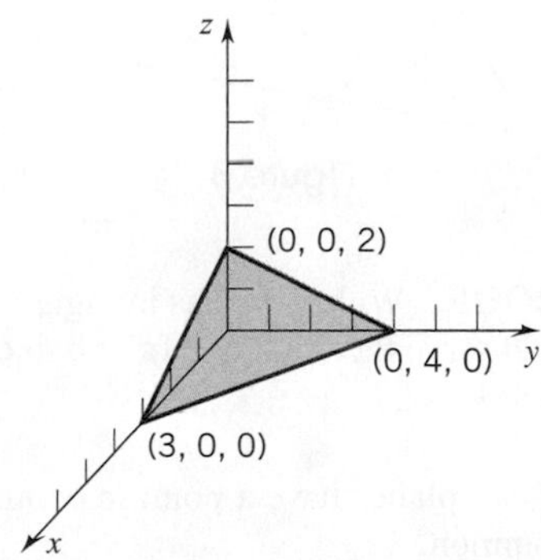

51.

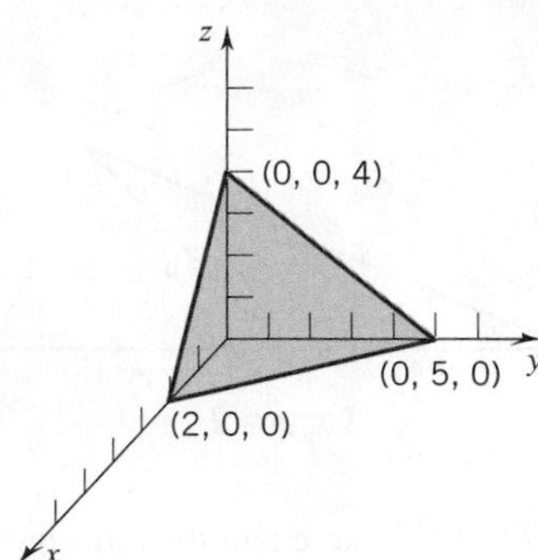

52.

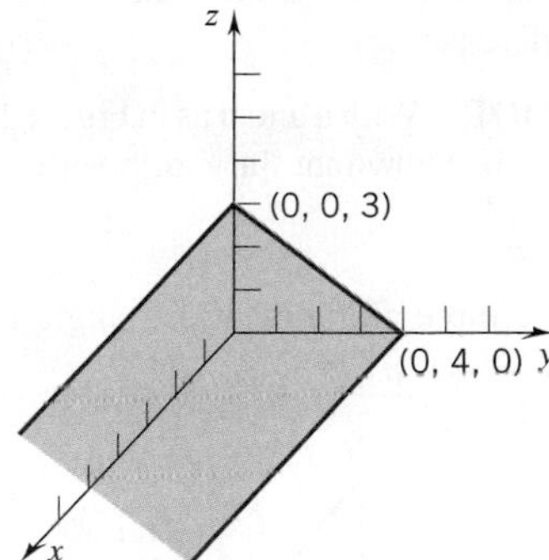

53.

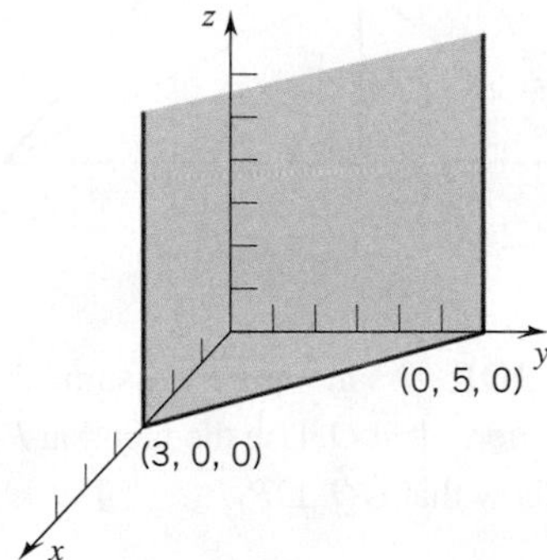

■ PROJECT 12.7 Some Geometry by Vector Methods

Try your hand at proving the following theorems by vector methods. Follow the hints if you like, but you may find it more interesting to disregard them and come up with proofs that are entirely you own.

Problem 1. The diagonals of a parallelogram are perpendicular iff the parallelogram is a rhombus.

HINT FOR PROOF With $\mathbf{a}$ and $\mathbf{b}$ as in Figure 1, the diagonals are $\mathbf{a} + \mathbf{b}$ and $\mathbf{a} - \mathbf{b}$.
Show that $(\mathbf{a} + \mathbf{b}) \cdot (\mathbf{a} - \mathbf{b}) = 0$ iff $\|\mathbf{a}\| = \|\mathbf{b}\|$. ☐

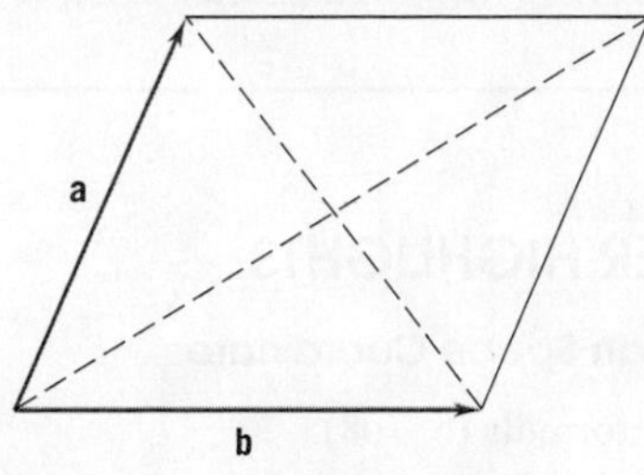

Figure 1

Problem 2. Every angle inscribed in a semicircle is a right angle.

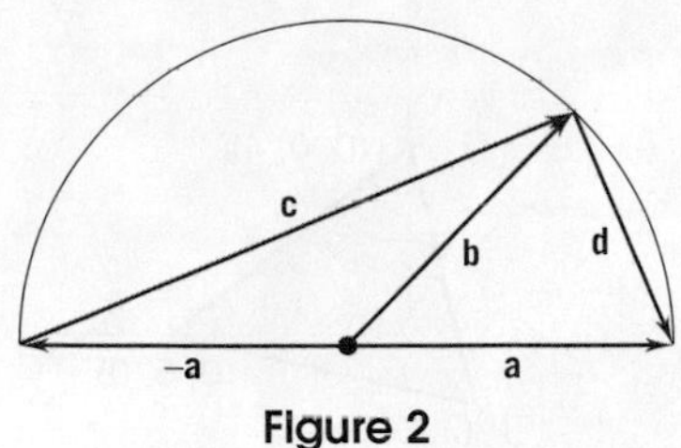

Figure 2

HINT FOR PROOF Take **c** and **d** as in Figure 2; express **c** and **d** in terms of **a** and **b**; then show that $\mathbf{c} \cdot \mathbf{d} = 0$. □

Problem 3. In a parallelogram the sum of the squares of the lengths of the diagonals equals the sum of the squares of the lengths of the sides.

HINT FOR PROOF With **a** and **b** as in Figure 1, the diagonals are $\mathbf{a} + \mathbf{b}$ and $\mathbf{a} - \mathbf{b}$. Show that $||\mathbf{a} + \mathbf{b}||^2 + ||\mathbf{a} - \mathbf{b}||^2 = 2||\mathbf{a}||^2 + 2||\mathbf{b}||^2$. □

Problem 4. The three altitudes of a triangle intersect at one point.

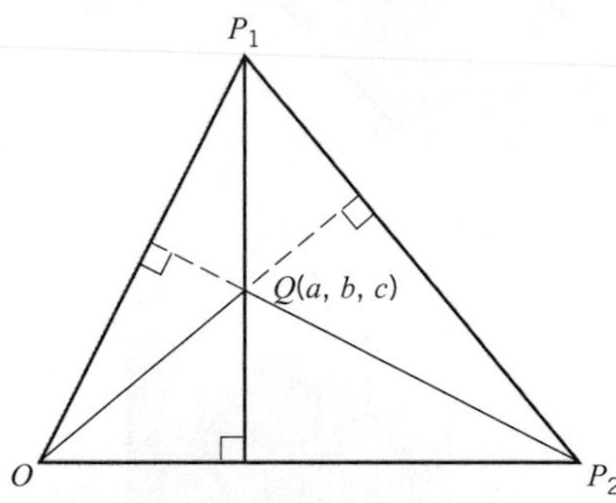

Figure 3

HINT FOR PROOF As in Figure 3 assume that the altitudes from P_1 and P_2 intersect at Q. Use the fact that $\overrightarrow{P_1Q} \perp \overrightarrow{OP_2}$ and $\overrightarrow{P_2Q} \perp \overrightarrow{OP_1}$ to show that $\overrightarrow{OQ} \perp \overrightarrow{P_1P_2}$. □

Problem 5. The three medians of a triangle intersect at one point.

HINT FOR PROOF With l_1, l_2, l_3 as in Figure 4,

$$l_1 : \mathbf{r}_1(t) = t(\mathbf{a} + \mathbf{b}),$$

$$l_2 : \mathbf{r}_2(u) = \tfrac{1}{2}\mathbf{b} + u(\mathbf{a} - \tfrac{1}{2}\mathbf{b}),$$

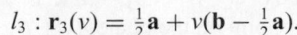

$$l_3 : \mathbf{r}_3(v) = \tfrac{1}{2}\mathbf{a} + v(\mathbf{b} - \tfrac{1}{2}\mathbf{a}).$$

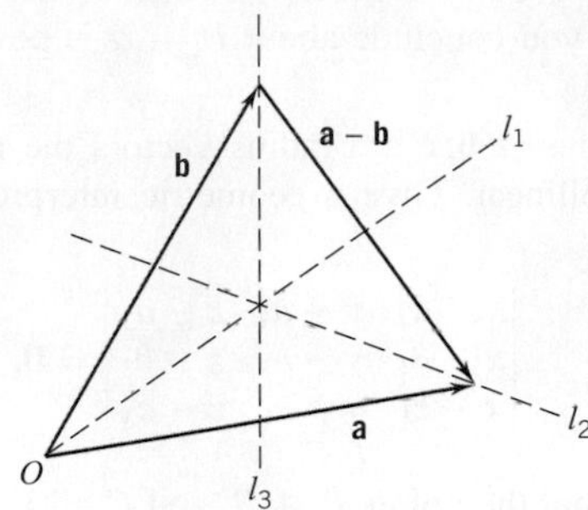

Figure 4

Show that l_1 intersects both l_2 and l_3 at the same point. □

Problem 6 (The Law of Sines). If a triangle has sides **a**, **b**, **c** and opposite angles A, B, C, then

$$\frac{\sin A}{||\mathbf{a}||} = \frac{\sin B}{||\mathbf{b}||} = \frac{\sin C}{||\mathbf{c}||}.$$

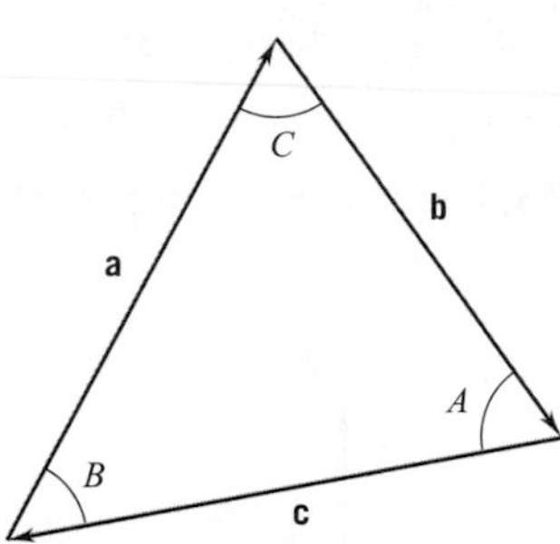

Figure 5

HINT FOR PROOF With **a**, **b**, **c** as in Figure 5, $\mathbf{a} + \mathbf{b} + \mathbf{c} = \mathbf{0}$. Observe then that $\mathbf{a} \times [\mathbf{a} + \mathbf{b} + \mathbf{c}] = \mathbf{0}$ and $\mathbf{b} \times [\mathbf{a} + \mathbf{b} + \mathbf{c}] = \mathbf{0}$. □

Problem 7. If two planes have a point in common, then they have a line in common.

HINT FOR PROOF As equations for the two planes, take $\mathbf{n} \cdot (\mathbf{r} - \mathbf{r}_0) = 0$ and $\mathbf{N} \cdot (\mathbf{R} - \mathbf{R}_0) = 0$. If the point $P(a_1, a_2, a_3)$ lies on both planes, then the vector $\mathbf{a} = a_1\mathbf{i} + a_2\mathbf{j} + a_3\mathbf{k}$ satisfies both equations. In that case we have $\mathbf{n} \cdot (\mathbf{a} - \mathbf{r}_0) = 0$ and $\mathbf{N} \cdot (\mathbf{a} - \mathbf{R}_0) = 0$. Consider the line $\mathbf{r}(t) = \mathbf{a} + t(\mathbf{n} \times \mathbf{N})$. □

■ CHAPTER HIGHLIGHTS

12.1 Cartesian Space Coordinates

distance formula (p. 708)
equation of sphere (p. 708)
midpoint formula (p. 709)

12.2 Displacements and Forces

12.3 Vectors

addition, multiplication by scalars (p. 714)
component (p.714)

parallel vectors (p. 716)
norm (p. 717)
unit vector (p. 718)
vectors in the plane (p. 719)
The zero vector **0** is parallel to every vector.

12.4 The Dot Product

$\mathbf{a} \cdot \mathbf{b} = a_1b_1 + a_2b_2 + a_3b_3 = ||\mathbf{a}||\,||\mathbf{b}|| \cos\theta$
$\mathbf{a} \perp \mathbf{b}$ iff $\mathbf{a} \cdot \mathbf{b} = 0$
projections and components (p. 726)
directions angles, direction cosines (p. 728)
Schwarz's inequality (p. 729)

12.5 The Cross product

definition of $\mathbf{a} \times \mathbf{b}$ (p. 732)
properties of right-handed triples (p. 733)
properties of the cross product (p. 733)
distributive laws (p. 734)
scalar triple product (p. 734)
components of $\mathbf{a} \times \mathbf{b}$ (p. 735)
identities (p. 738)

12.6 Lines

position vector, radius vector (p. 741)
vector parametrization: $\mathbf{r}(t) = \mathbf{r}_0 + t\mathbf{d}$
direction vector, direction numbers (p. 742)
scalar parametric equations : $x(t) = x_0 + d_1t$,
$y(t) = y_0 + d_2t, \quad z(t) = z_0 + d_3t$
symmetric form (p. 744)
distance from a point to a line (p. 747)

Two lines are parallel iff their direction vectors are parallel; two intersecting lines are perpendicular iff their direction vectors are perpendicular.

12.7 Planes

normal vector (p. 750)
scalar equation: $A(x - x_0) + B(y - y_0) + C(z - z_0) = 0$
vector equation of a plane (p. 752)
collinear vectors, coplanar vectors (p. 753)
parallel planes (p. 754)
angle between intersecting planes (p. 754)
plane determined by three noncollinear points (p. 756)
distance between a point and a plane (p. 757)

CHAPTER

13

VECTOR CALCULUS

■ 13.1 VECTOR FUNCTIONS

Introduction

If f_1, f_2, f_3 are real-valued functions defined on some interval I, then for each $t \in I$ we can form the vector

$$\mathbf{f}(t) = f_1(t)\,\mathbf{i} + f_2(t)\,\mathbf{j} + f_3(t)\,\mathbf{k}$$

and thereby create a *vector-valued function* **f**. For short we will call such a function a *vector function*. The real-valued functions f_1, f_2, f_3 are called the *components* of **f**. A point t is in the *domain* of a vector function **f** iff it is in the domain of each of its components.

For example, from the scalar functions

$$f_1(t) = x_0 + d_1 t, \quad f_2(t) = y_0 + d_2 t, \quad f_3(t) = z_0 + d_3 t,$$

we can form the vector function

$$\mathbf{f}(t) = (x_0 + d_1 t)\,\mathbf{i} + (y_0 + d_2 t)\,\mathbf{j} + (z_0 + d_3 t)\,\mathbf{k}.$$

The domain of **f** is the set of all real numbers. If d_1, d_2, d_3 are not all 0, then the radius vector $\mathbf{f}(t)$ traces out the line that passes through the point $P(x_0, y_0, z_0)$ and has direction numbers d_1, d_2, d_3. If d_1, d_2, d_3 are all 0, then we have the constant function

$$\mathbf{f}(t) = x_0\,\mathbf{i} + y_0\,\mathbf{j} + z_0\,\mathbf{k}.$$

Example 1 From the functions

$$f_1(t) = \cos t, \qquad f_2(t) = \sin t, \qquad f_3(t) = 0$$

we can form the vector function

$$\mathbf{f}(t) = \cos t\,\mathbf{i} + \sin t\,\mathbf{j}.$$

For each t

$$\|\mathbf{f}(t)\| = \sqrt{\cos^2 t + \sin^2 t} = 1.$$

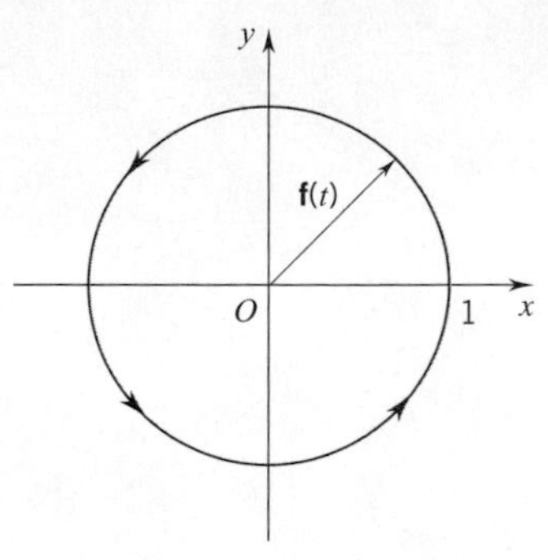

$\mathbf{f}(t) = \cos t\,\mathbf{i} + \sin t\,\mathbf{j}$

Figure 13.1.1

Since the third component is zero, the radius vector $\mathbf{f}(t)$ lies in the xy-plane. As t increases, the tip of $\mathbf{f}(t)$ traces out the unit circle in a counterclockwise manner, effecting a complete revolution as t increases by 2π (Figure 13.1.1). □

Example 2 Each real-valued function f defined on an interval $[a, b]$ gives rise to a vector-valued function $\mathbf{f}$ in a natural way. Setting

$$f_1(t) = t, \quad f_2(t) = f(t), \quad f_3(t) = 0,$$

we obtain the vector function

$$\mathbf{f}(t) = t\,\mathbf{i} + f(t)\,\mathbf{j}.$$

As t ranges from a to b, the radius vector $\mathbf{f}(t)$ traces out the graph of f from left to right. See Figure 13.1.2. □

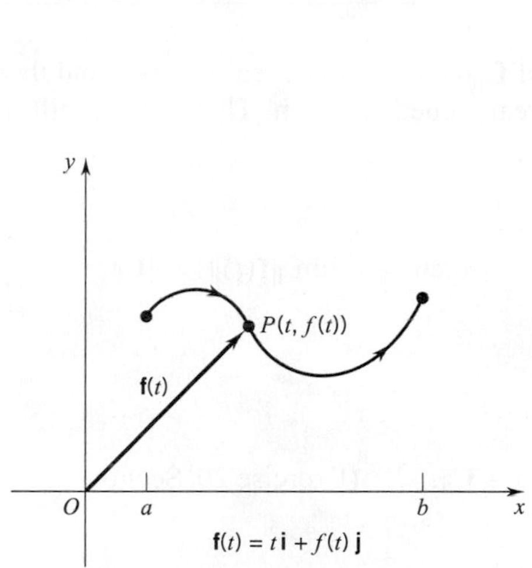

$\mathbf{f}(t) = t\,\mathbf{i} + f(t)\,\mathbf{j}$

Figure 13.1.2

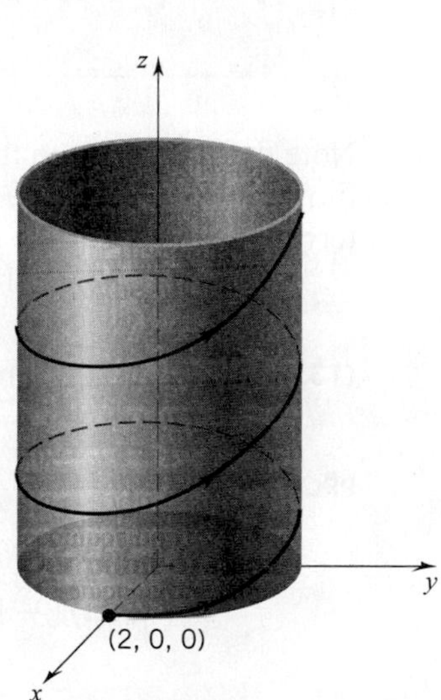

Figure 13.1.3

Example 3 From the functions

$$f_1(t) = 2\cos t, \qquad f_2(t) = 2\sin t, \qquad f_3(t) = t, \qquad t \geq 0,$$

we can form the vector function

$$\mathbf{f}(t) = 2\cos t\,\mathbf{i} + 2\sin t\,\mathbf{j} + t\,\mathbf{k}, \qquad t \geq 0.$$

At $t = 0$, the tip of the radius vector $\mathbf{f}(0)$ is at the point $(2, 0, 0)$. As t increases, the tip of $\mathbf{f}(t)$ spirals up the circular cylinder $x^2 + y^2 = 4$ (z arbitrary), effecting a complete turn on every t-interval of length 2π. This spiraling curve is called a *circular helix* (Figure 13.1.3). □

We return to the general case: a vector function

$$\mathbf{f}(t) = f_1(t)\,\mathbf{i} + f_2(t)\,\mathbf{j} + f_3(t)\,\mathbf{k}$$

defined on some t-interval I. As in the examples given, and under conditions to be spelled out later, as t ranges over the interval I, the tip of the radius vector $\mathbf{f}(t)$ traces out a curve C. The equations

$$x = f_1(t), \qquad y = f_2(t), \qquad z = f_3(t)$$

formed from the components of $\mathbf{f}$ serve as parametric equations for C. If one of the components is identically 0 on I, for example, if $\mathbf{f}$ has the form $\mathbf{f}(t) = f_1(t)\,\mathbf{i} + f_2(t)\,\mathbf{j}$, then C is a *plane curve*; otherwise C is a *space curve*.

The Limit Process

DEFINITION 13.1.1 LIMIT OF A VECTOR FUNCTION

$$\lim_{t\to t_0} \mathbf{f}(t) = \mathbf{L} \quad \text{if} \quad \lim_{t\to t_0} \|\mathbf{f}(t) - \mathbf{L}\| = 0.$$

Note that for each t in the domain of $\mathbf{f}$, $\|\mathbf{f}(t) - \mathbf{L}\|$ is a real number, and therefore the limit on the right is the limit of a real-valued function. Thus we are still in familiar territory.

The first thing we show is that

(13.1.2) $$\text{if} \quad \lim_{t\to t_0} \mathbf{f}(t) = \mathbf{L}, \qquad \text{then} \qquad \lim_{t\to t_0} \|\mathbf{f}(t)\| = \|\mathbf{L}\|.$$

PROOF We know that

$$0 \le \Big|\, \|\mathbf{f}(t)\| - \|\mathbf{L}\| \,\Big| \le \|\mathbf{f}(t) - \mathbf{L}\|. \qquad \text{(Exercise 20, Section 12.3)}$$

It follows from the pinching theorem that

$$\text{if} \quad \lim_{t\to t_0} \|\mathbf{f}(t) - \mathbf{L}\| = 0, \qquad \text{then} \qquad \lim_{t\to t_0} \Big|\, \|\mathbf{f}(t)\| - \|\mathbf{L}\| \,\Big| = 0. \quad \square$$

Remark The converse of (13.1.2) is false, as you can see by setting $\mathbf{f}(t) = \mathbf{k}$ and taking $\mathbf{L} = -\mathbf{k}$. ☐

We can indicate that $\lim_{t\to t_0} \mathbf{f}(t) = \mathbf{L}$ by writing

$$\text{as } t \to t_0, \quad \mathbf{f}(t) \to \mathbf{L}.$$

We will state the limit rules in this form. As you will see below, there are no surprises.

THEOREM 13.1.3 LIMIT RULES

Let $\mathbf{f}$ and $\mathbf{g}$ be vector functions and let u be a real-valued function. Suppose that, as $t \to t_0$,

$$\mathbf{f}(t) \to \mathbf{L}, \quad \mathbf{g}(t) \to \mathbf{M}, \quad u(t) \to A.$$

Then

$$\mathbf{f}(t) + \mathbf{g}(t) \to \mathbf{L} + \mathbf{M}, \quad \alpha\mathbf{f}(t) \to \alpha\mathbf{L},$$

$$u(t)\mathbf{f}(t) \to A\mathbf{L}, \quad \mathbf{f}(t) \cdot \mathbf{g}(t) \to \mathbf{L} \cdot \mathbf{M}, \quad \mathbf{f}(t) \times \mathbf{g}(t) \to \mathbf{L} \times \mathbf{M}.$$

Each of these limit rules is easy to verify. We will verify the last one. To do this, we have to show that

$$\text{as } t \to t_0, \quad \|[\mathbf{f}(t) \times \mathbf{g}(t)] - [\mathbf{L} \times \mathbf{M}]\| \to 0.$$

We do this as follows:

$$\begin{aligned}
\|[\mathbf{f}(t) \times \mathbf{g}(t)] - [\mathbf{L} \times \mathbf{M}]\| &= \|[\mathbf{f}(t) \times \mathbf{g}(t)] - [\mathbf{L} \times \mathbf{g}(t)] + [\mathbf{L} \times \mathbf{g}(t)] - [\mathbf{L} \times \mathbf{M}]\| \\
&= \|[(\mathbf{f}(t) - \mathbf{L}) \times \mathbf{g}(t)] + [\mathbf{L} \times (\mathbf{g}(t) - \mathbf{M})]\| \qquad (12.5.4) \\
&\leq \|(\mathbf{f}(t) - \mathbf{L}) \times \mathbf{g}(t)\| + \|\mathbf{L} \times (\mathbf{g}(t) - \mathbf{M})\| \\
&\uparrow \text{(triangle inequality)} \\
&\leq \|\mathbf{f}(t) - \mathbf{L}\| \, \|\mathbf{g}(t)\| + \|\mathbf{L}\| \, \|\mathbf{g}(t) - \mathbf{M}\|. \\
&\uparrow \text{explain}
\end{aligned}$$

As $t \to t_0$, $\|\mathbf{g}(t)\| \to \|\mathbf{M}\|$. (This follows from 13.1.2.) Therefore, as $t \to t_0$,

$$\|\mathbf{f}(t) - \mathbf{L}\| \, \|\mathbf{g}(t)\| + \|\mathbf{L}\| \, \|\mathbf{g}(t) - \mathbf{M}\| \to (0)\|\mathbf{M}\| + \|\mathbf{L}\|(0) = 0.$$

Since $0 \leq \| \mathbf{f}(t) \times \mathbf{g}(t)] - [\mathbf{L} \times \mathbf{M}]\|$, it follows from the pinching theorem that $\|[\mathbf{f}(t) \times \mathbf{g}(t)] - [\mathbf{L} \times \mathbf{M}]\| \to 0$.

The limit process can be carried out component by component. Let $\mathbf{f}(t) = f_1(t)\,\mathbf{i} + f_2(t)\,\mathbf{j} + f_3(t)\,\mathbf{k}$ and let $\mathbf{L} = L_1\,\mathbf{i} + L_2\,\mathbf{j} + L_3\,\mathbf{k}$. Then

(13.1.4)
$$\lim_{t \to t_0} \mathbf{f}(t) = \mathbf{L} \quad \text{iff}$$
$$\lim_{t \to t_0} f_1(t) = L_1, \quad \lim_{t \to t_0} f_2(t) = L_2, \quad \lim_{t \to t_0} f_3(t) = L_3.$$

PROOF

$$\begin{aligned}
\lim_{t \to t_0} \mathbf{f}(t) = \mathbf{L} \quad &\text{iff} \quad \lim_{t \to t_0} \|\mathbf{f}(t) - \mathbf{L}\| = 0 \\
&\text{iff} \quad \lim_{t \to t_0} \sqrt{[f_1(t) - L_1]^2 + [f_2(t) - L_2]^2 + [f_3(t) - L_3]^2} = 0 \\
&\text{iff} \quad \lim_{t \to t_0} f_1(t) = L_1, \quad \lim_{t \to t_0} f_2(t) = L_2, \quad \lim_{t \to t_0} f_3(t) = L_3. \quad \square
\end{aligned}$$

Example 4 Find $\lim_{t\to 0} \mathbf{f}(t)$ given that

$$\mathbf{f}(t) = \cos(t+\pi)\,\mathbf{i} + \sin(t+\pi)\,\mathbf{j} + e^{-t^2}\mathbf{k}.$$

SOLUTION

$$\begin{aligned}\lim_{t\to 0}\mathbf{f}(t) &= \lim_{t\to 0}[\cos(t+\pi)\,\mathbf{i} + \sin(t+\pi)\,\mathbf{j} + e^{-t^2}\mathbf{k}]\\ &= \left[\lim_{t\to 0}\cos(t+\pi)\right]\mathbf{i} + \left[\lim_{t\to 0}\sin(t+\pi)\right]\mathbf{j} + \left[\lim_{t\to 0}e^{-t^2}\right]\mathbf{k}\\ &= (-1)\,\mathbf{i} + (0)\,\mathbf{j} + (1)\,\mathbf{k} = -\mathbf{i} + \mathbf{k}. \quad \square\end{aligned}$$

Continuity and Differentiability

As you would expect, $\mathbf{f}$ is said to be *continuous* at t_0 if

$$\lim_{t\to t_0}\mathbf{f}(t) = \mathbf{f}(t_0).$$

Thus, by (13.1.4), $\mathbf{f}$ is continuous at t_0 iff each component of $\mathbf{f}$ is continuous at t_0.

The derivative of a vector function is defined as the limit of a *vector difference quotient*:

DEFINITION 13.1.5 DERIVATIVE OF A VECTOR FUNCTION

The vector function $\mathbf{f}$ is *differentiable* at t if

$$\lim_{h\to 0}\frac{\mathbf{f}(t+h) - \mathbf{f}(t)}{h} \quad \text{exists.}$$

If this limit exists, it is called the *derivative of* $\mathbf{f}$ at t and is denoted $\mathbf{f}'(t)$.

Differentiation can be carried out component by component; which is to say, if $\mathbf{f}(t) = f_1(t)\,\mathbf{i} + f_2(t)\,\mathbf{j} + f_3(t)\,\mathbf{k}$ is differentiable at t, then

$$\mathbf{f}'(t) = f_1'(t)\,\mathbf{i} + f_2'(t)\,\mathbf{j} + f_3'(t)\,\mathbf{k}.$$

PROOF

$$\begin{aligned}\mathbf{f}'(t) &= \lim_{h\to 0}\frac{\mathbf{f}(t+h) - \mathbf{f}(t)}{h}\\ &= \lim_{h\to 0}\left[\frac{f_1(t+h) - f_1(t)}{h}\,\mathbf{i} + \frac{f_2(t+h) - f_2(t)}{h}\,\mathbf{j} + \frac{f_3(t+h) - f_3(t)}{h}\,\mathbf{k}\right]\\ &= \left[\lim_{h\to 0}\frac{f_1(t+h) - f_1(t)}{h}\right]\mathbf{i} + \left[\lim_{h\to 0}\frac{f_2(t+h) - f_2(t)}{h}\right]\mathbf{j} +\\ &\qquad \left[\lim_{h\to 0}\frac{f_3(t+h) - f_3(t)}{h}\right]\mathbf{k}\\ &= f_1'(t)\,\mathbf{i} + f_2'(t)\,\mathbf{j} + f_3'(t)\,\mathbf{k}. \quad \square\end{aligned}$$

As with real-valued functions, if $\mathbf{f}$ is differentiable at t, then $\mathbf{f}$ is continuous at t. (Exercise 53.)

Interpretations of the vector derivative and applications of vector differentiation are introduced later in the chapter. Here we limit ourselves to computation.

Example 5 Given that $\mathbf{f}(t) = t\,\mathbf{i} + \sqrt{t+1}\,\mathbf{j} - e^t\,\mathbf{k}$, find:

(1) The domain of $\mathbf{f}$. (2) $\mathbf{f}(0)$.

(3) $\mathbf{f}'(t)$. (4) $\mathbf{f}'(0)$.

(5) $\|\mathbf{f}(t)\|$. (6) $\mathbf{f}(t)\cdot\mathbf{f}'(t)$.

SOLUTION

(1) For a number to be in the domain of $\mathbf{f}$, it is necessary only that it be in the domain of each of the components. The domain of $\mathbf{f}$ is $[-1, \infty)$.

(2) $\mathbf{f}(0) = 0\,\mathbf{i} + \sqrt{0+1}\,\mathbf{j} - e^0\,\mathbf{k} = \mathbf{j} - \mathbf{k}$.

(3) $\mathbf{f}'(t) = \mathbf{i} + \dfrac{1}{2\sqrt{t+1}}\mathbf{j} - e^t\,\mathbf{k}$.

(4) $\mathbf{f}'(0) = \mathbf{i} + \dfrac{1}{2\sqrt{0+1}}\mathbf{j} - e^0\,\mathbf{k} = i + \dfrac{1}{2}\mathbf{j} - \mathbf{k}$.

(5) $\|\mathbf{f}(t)\| = \sqrt{t^2 + (\sqrt{t+1})^2 + (-e^t)^2} = \sqrt{t^2 + t + 1 + e^{2t}}$.

(6) $$\mathbf{f}(t)\cdot\mathbf{f}'(t) = (t\,\mathbf{i} + \sqrt{t+1}\,\mathbf{j} - e^t\,\mathbf{k})\cdot\left(\mathbf{i} + \frac{1}{2\sqrt{t+1}}\mathbf{j} - e^t\,\mathbf{k}\right)$$
$$= (t)(1) + (\sqrt{t+1})\left(\frac{1}{2\sqrt{t+1}}\right) + (-e^t)(-e^t) = t + \frac{1}{2} + e^{2t}. \quad \square$$

If $\mathbf{f}'$ is itself differentiable, we can calculate the second derivative $\mathbf{f}''$ and so on.

Example 6 Find $\mathbf{f}''(t)$ for $\mathbf{f}(t) = t\sin t\,\mathbf{i} + e^{-t}\,\mathbf{j} + t\,\mathbf{k}$.

SOLUTION

$$\mathbf{f}'(t) = (t\cos t + \sin t)\,\mathbf{i} - e^{-t}\mathbf{j} + \mathbf{k}.$$
$$\mathbf{f}''(t) = (-t\sin t + \cos t + \cos t)\,\mathbf{i} + e^{-t}\mathbf{j} = (2\cos t - t\sin t)\,\mathbf{i} + e^{-t}\mathbf{j}. \quad \square$$

Integration

Just as we can differentiate vector functions component by component, we can integrate component by component. For $\mathbf{f}(t) = f_1(t)\,\mathbf{i} + f_2(t)\,\mathbf{j} + f_3(t)\,\mathbf{k}$ continuous on $[a, b]$, we set

(13.1.6) $$\int_a^b \mathbf{f}(t)\,dt = \left(\int_a^b f_1(t)\,dt\right)\mathbf{i} + \left(\int_a^b f_2(t)\,dt\right)\mathbf{j} + \left(\int_a^b f_3(t)\,dt\right)\mathbf{k}.$$

Example 7 Find $\displaystyle\int_0^1 \mathbf{f}(t)\,dt$ for $\mathbf{f}(t) = t\,\mathbf{i} + \sqrt{t+1}\,\mathbf{j} - e^t\,\mathbf{k}$.

SOLUTION

$$\int_0^1 \mathbf{f}(t)\,dt = \left(\int_0^1 t\,dt\right)\mathbf{i} + \left(\int_0^1 \sqrt{t+1}\,dt\right)\mathbf{j} + \left(\int_0^1 (-e^t)\,dt\right)\mathbf{k}$$

$$= \left[\tfrac{1}{2}t^2\right]_0^1\mathbf{i} + \left[\tfrac{2}{3}(t+1)^{3/2}\right]_0^1\mathbf{j} + \left[-e^t\right]_0^1\mathbf{k}$$

$$= \tfrac{1}{2}\mathbf{i} + \tfrac{2}{3}(2\sqrt{2}-1)\mathbf{j} + (1-e)\mathbf{k}. \quad \square$$

We can calculate indefinite integrals.

Example 8 Find $\mathbf{f}(t)$ given that

$$\mathbf{f}'(t) = 2\cos t\,\mathbf{i} - t\sin t^2\,\mathbf{j} + 2t\,\mathbf{k} \quad \text{and} \quad \mathbf{f}(0) = \mathbf{i} + 3\mathbf{k}.$$

SOLUTION By integrating $\mathbf{f}'(t)$, we find that

$$\mathbf{f}(t) = (2\sin t + C_1)\,\mathbf{i} + \left(\tfrac{1}{2}\cos t^2 + C_2\right)\mathbf{j} + (t^2 + C_3)\,\mathbf{k},$$

where C_1, C_2, C_3 are constants to be determined. Since

$$\mathbf{i} + 3\mathbf{k} = \mathbf{f}(0) = C_1\mathbf{i} + \left(\tfrac{1}{2} + C_2\right)\mathbf{j} + C_3\mathbf{k},$$

you can see that

$$C_1 = 1, \quad C_2 = -\tfrac{1}{2}, \quad C_3 = 3.$$

Thus $$\mathbf{f}(t) = (2\sin t + 1)\,\mathbf{i} + \left(\tfrac{1}{2}\cos t^2 - \tfrac{1}{2}\right)\mathbf{j} + (t^2 + 3)\,\mathbf{k}. \quad \square$$

(Integration can also be carried out without direct reference to components. See Exercise 54.)

Properties of the Integral

It is easy to see that

(13.1.7) $$\int_a^b [\mathbf{f}(t) + \mathbf{g}(t)]\,dt = \int_a^b \mathbf{f}(t)dt + \int_a^b \mathbf{g}(t)\,dt$$

and

(13.1.8) $$\int_a^b [\alpha\mathbf{f}(t)]\,dt = \alpha\int_a^b \mathbf{f}(t)\,dt \qquad \text{for every constant scalar } \alpha.$$

It is also true that

(13.1.9) $$\int_a^b [\mathbf{c}\cdot\mathbf{f}(t)]\,dt = \mathbf{c}\cdot\left(\int_a^b \mathbf{f}(t)\,dt\right) \qquad \text{for every constant vector } \mathbf{c}$$

and

$$\left\| \int_a^b \mathbf{f}(t)\, dt \right\| \le \int_a^b \|\mathbf{f}(t)\|\, dt. \tag{13.1.10}$$

The proof of (13.1.9) is left as an exercise (Exercise 56). Here we prove (13.1.10). It is an important inequality.

PROOF Set $\mathbf{r} = \displaystyle\int_a^b \mathbf{f}(t)\, dt$ and note that

$$\begin{aligned} \|\mathbf{r}\|^2 &= \mathbf{r} \cdot \mathbf{r} = \mathbf{r} \cdot \int_a^b \mathbf{f}(t)\, dt \\ &= \int_a^b [\mathbf{r} \cdot \mathbf{f}(t)]\, dt \le \int_a^b \|\mathbf{r}\|\, \|\mathbf{f}(t)\|\, dt = \|\mathbf{r}\| \int_a^b \|\mathbf{f}(t)\|\, dt. \end{aligned}$$

by (13.1.9) ↑ ↑ by Schwarz's inequality (12.4.17)

If $\mathbf{r} \neq \mathbf{0}$, we can divide by $\|\mathbf{r}\|$ and conclude that

$$\|\mathbf{r}\| \le \int_a^b \|\mathbf{f}(t)\|\, dt.$$

If $\mathbf{r} = \mathbf{0}$, the result is obvious in the first place. □

EXERCISES 13.1

Find the derivative.

1. $\mathbf{f}(t) = (1 + 2t)\,\mathbf{i} + (3 - t)\,\mathbf{j} + (2 + 3t)\,\mathbf{k}$.
2. $\mathbf{f}(t) = 2\,\mathbf{i} - \cos t\,\mathbf{k}$.
3. $\mathbf{f}(t) = \sqrt{1-t}\,\mathbf{i} + \sqrt{1+t}\,\mathbf{j} + (1-t)^{-1}\,\mathbf{k}$.
4. $\mathbf{f}(t) = e^t\,\mathbf{i} + \ln t\,\mathbf{j} + \tan^{-1} t\,\mathbf{k}$.
5. $\mathbf{f}(t) = \sin t\,\mathbf{i} + \cos t\,\mathbf{j} + \tan t\,\mathbf{k}$.
6. $\mathbf{f}(t) = e^t(\mathbf{i} + t\,\mathbf{j} + t^2\mathbf{k})$.
7. $\mathbf{f}(t) = \ln(1-t)\,\mathbf{i} + \cos t\,\mathbf{j} + t^2\,\mathbf{k}$.
8. $\mathbf{f}(t) = \dfrac{t+1}{t-1}\,\mathbf{i} + t\,e^{2t}\,\mathbf{j} + \sec t\,\mathbf{k}$.

Calculate the second derivative

9. $\mathbf{f}(t) = 4t\,\mathbf{i} + 2t^3\,\mathbf{j} + (t^2 + 2t)\,\mathbf{k}$.
10. $\mathbf{f}(t) = t\sin t\,\mathbf{i} + t\cos t\,\mathbf{k}$.
11. $\mathbf{f}(t) = \cos 2t\,\mathbf{i} + \sin 2t\,\mathbf{j} + 4t\,\mathbf{k}$.
12. $\mathbf{f}(t) = \sqrt{t}\,\mathbf{i} + t\sqrt{t}\,\mathbf{j} + \ln t\,\mathbf{k}$.

13. (C) Use a CAS to find $\mathbf{r}'(t_0)$.
    (a) $\mathbf{r}(t) = te^{-t^2}\,\mathbf{i} + t^2e^{-t}\,\mathbf{j}$, $t_0 = 0$.
    (b) $\mathbf{r}(t) = \ln(\sin t)\,\mathbf{i} + \ln(\cos t)\,\mathbf{j} + (2\sin t - 3\cot t)\,\mathbf{k}$, $t_0 = \pi/4$.

14. (C) Use a CAS to find $\mathbf{r}''(t_0)$.
    (a) $\mathbf{r}(t) = t^2e^{-t}\,\mathbf{i} + te^{-t}\,\mathbf{j}$, $t_0 = 1$.
    (b) $\mathbf{r}(t) = t\ln t\,\mathbf{i} + \dfrac{t}{\ln t}\,\mathbf{j} + \sqrt{\ln t}\,\mathbf{k}$, $t_0 = e$.

Carry out the intergration

15. $\displaystyle\int_1^2 \mathbf{f}(t)\, dt$ for $\mathbf{f}(t) = \mathbf{i} + 2t\,\mathbf{j}$.
16. $\displaystyle\int_0^\pi \mathbf{r}(t)\, dt$ for $\mathbf{r}(t) = \sin t\,\mathbf{i} + \cos t\,\mathbf{j} + t\,\mathbf{k}$.
17. $\displaystyle\int_0^1 \mathbf{g}(t)\, dt$ for $\mathbf{g}(t) = e^t\,\mathbf{i} + e^{-t}\,\mathbf{k}$.
18. $\displaystyle\int_0^1 \mathbf{h}(t)\, dt$ for $\mathbf{h}(t) = e^{-t}[t^2\,\mathbf{i} + \sqrt{2}t\,\mathbf{j} + \mathbf{k}]$.
19. $\displaystyle\int_0^1 \mathbf{f}(t)\, dt$ for $\mathbf{f}(t) = \dfrac{1}{1+t^2}\,\mathbf{i} + \sec^2 t\,\mathbf{j}$.
20. $\displaystyle\int_1^3 \mathbf{F}(t)\, dt$ for $\mathbf{F}(t) = \dfrac{1}{t}\,\mathbf{i} + \dfrac{\ln t}{t}\,\mathbf{j} + e^{-2t}\,\mathbf{k}$.

Find $\lim_{t\to 0} \mathbf{f}(t)$ if it exists.

21. $\mathbf{f}(t) = \dfrac{\sin t}{2t}\,\mathbf{i} + e^{2t}\,\mathbf{j} + \dfrac{t^2}{e^t}\,\mathbf{k}$.

22. $\mathbf{f}(t) = 3(t^2 - 1)\,\mathbf{i} + \cos t\,\mathbf{j} + \dfrac{t}{|t|}\,\mathbf{k}$.

23. $\mathbf{f}(t) = t^2\,\mathbf{i} + \dfrac{1 - \cos t}{3t}\,\mathbf{j} + \dfrac{t}{t+1}\,\mathbf{k}$.

24. $\mathbf{f}(t) = 3t\,\mathbf{i} + (t^2 + 1)\,\mathbf{j} + e^{2t}\,\mathbf{k}$.

C 25. Use a CAS to evaluate the definite integral.

(a) $\displaystyle\int_0^1 \mathbf{f}(t)\,dt$ for $\mathbf{f}(t) = te^t\,\mathbf{i} + te^{t^2}\,\mathbf{j}$.

(b) $\displaystyle\int_3^8 \mathbf{f}(t)\,dt$ for $\mathbf{f}(t) = \dfrac{t}{t+1}\,\mathbf{i} + \dfrac{t}{(t+1)^2}\,\mathbf{j} + \dfrac{t}{(t+1)^3}\,\mathbf{k}$.

C 26. Use a CAS to find the limit.

(a) $\displaystyle\lim_{t\to\pi/6} (\cos^2 t\,\mathbf{i} + \sin^2 t\,\mathbf{j} + \mathbf{k})$.

(b) $\displaystyle\lim_{t\to e^2} \left(t \ln t\,\mathbf{i} + \frac{\ln t^2}{t^2}\,\mathbf{j} + \sqrt{\ln t^2}\,\mathbf{k}\right)$.

Sketch the curve traced out by the vector-valued function and indicate the direction in which the curve is traversed as t increases.

27. $\mathbf{r}(t) = 2t\,\mathbf{i} + t^2\,\mathbf{j}$. $t \geq 0$.

28. $\mathbf{r}(t) = t^3\,\mathbf{i} + 2t\,\mathbf{j}$, $t \geq 0$.

29. $\mathbf{r}(t) = 2\sinh t\,\mathbf{i} + 2\cosh t\,\mathbf{j}$, $t \geq 0$.

30. $\mathbf{r}(t) = 3\cos t\,\mathbf{i} + 3\sin t\,\mathbf{k}$, $0 \leq t \leq 2\pi$.

31. $\mathbf{r}(t) = 2\cos t\,\mathbf{i} + 3\sin t\,\mathbf{j}$, $0 \leq t \leq 2\pi$.

32. $\mathbf{r}(t) = 2t\,\mathbf{i} + (5 - 2t)\,\mathbf{j} + 3t\,\mathbf{k}$, $t \geq 0$.

33. $\mathbf{r}(t) = (t^2 + 1)\,\mathbf{i} + t\,\mathbf{j} + 4\,\mathbf{k}$, $-2 \leq t \leq 2$.

34. $\mathbf{r}(t) = 2\cos t\,\mathbf{i} + 2\sin t\,\mathbf{j} + (2\pi - t)\,\mathbf{k}$, $0 \leq t \leq 2\pi$.

C Use a graphing utility to sketch the curve generated by the vector-valued function and indicate the direction in which the curve is traversed as t increases.

35. $\mathbf{r}(t) = 2\cos(t^2)\,\mathbf{i} + (2 - \sqrt{t})\,\mathbf{j}$.

36. $\mathbf{r}(t) = e^{\cos 2t}\,\mathbf{i} + e^{-\sin t}\,\mathbf{j}$.

37. $\mathbf{r}(t) = (2 - \sin 2t)\,\mathbf{i} + (3 + 2\cos t)\,\mathbf{j}$.

38. $\mathbf{r}(t) = (t - \sin t)\,\mathbf{i} + (1 - \cos t)\,\mathbf{j}$. (a cycloid)

Find a vector-valued function $\mathbf{f}$ that traces out the given curve in the indicated direction.

39. $4x^2 + 9y^2 = 36$ (a) Counterclockwise. (b) Clockwise.

40. $(x - 1)^2 + y^2 = 1$ (a) Counterclockwise. (b) Clockwise.

41. $y = x^2$ (a) From left to right. (b) From right to left.

42. $y = x^3$ (a) From left to right. (b) From right to left.

43. The directed line segment from $(1, 4, -2)$ to $(3, 9, 6)$.

44. The directed line segment from $(3, 2, -5)$ to $(7, 2, 9)$.

45. Set $\mathbf{f}(t) = t\,\mathbf{i} + f(t)\,\mathbf{j}$ and calculate

$$\mathbf{f}'(t_0), \quad \int_a^b \mathbf{f}(t)dt, \quad \int_a^b \mathbf{f}'(t)\,dt$$

given that

$$f'(t_0) = m, \quad f(a) = c, \quad f(b) = d, \quad \int_a^b f(t)dt = A.$$

Find $\mathbf{f}(t)$ from the following information.

46. $\mathbf{f}'(t) = t\,\mathbf{i} + t(1+t^2)^{-1/2}\,\mathbf{j} + te^t\,\mathbf{k}$ and $\mathbf{f}(0) = \mathbf{i} + 2\mathbf{j} + 3\mathbf{k}$.

47. $\mathbf{f}'(t) = \mathbf{i} + t^2\mathbf{j}$ and $\mathbf{f}(0) = \mathbf{j} - \mathbf{k}$.

48. $\mathbf{f}'(t) = 2\mathbf{f}(t)$ and $\mathbf{f}(0) = \mathbf{i} - \mathbf{k}$.

49. $\mathbf{f}'(t) = \alpha\mathbf{f}(t)$ with α a real number and $\mathbf{f}(0) = \mathbf{c}$.

50. No ϵ, δ's have surfaced so far, but they are still there at the heart of the limit process. Give an ϵ, δ characterization of

$$\lim_{t\to t_0} \mathbf{f}(t) = \mathbf{L}.$$

51. (a) Show that, if $\mathbf{f}'(t) = \mathbf{0}$ for all t in an interval I, then $\mathbf{f}$ is a constant vector on I.

(b) Show that, if $\mathbf{f}'(t) = \mathbf{g}'(t)$ for all t in an interval I, then $\mathbf{f}$ and $\mathbf{g}$ differ by a constant vector on I.

52. Assume that, as $t \to t_0$, $\mathbf{f}(t) \to \mathbf{L}$ and $\mathbf{g}(t) \to \mathbf{M}$. Show that

$$\mathbf{f}(t) \cdot \mathbf{g}(t) \to \mathbf{L} \cdot \mathbf{M}.$$

53. Show that, if $\mathbf{f}$ is differentiable at t, then $\mathbf{f}$ is continuous at t.

54. A vector-valued function $\mathbf{G}$ is called an *antiderivative* for $\mathbf{f}$ on $[a, b]$ iff (i) $\mathbf{G}$ is continuous on $[a, b]$ and (ii) $\mathbf{G}'(t) = \mathbf{f}(t)$ for all $t \in (a, b)$. Show that:

(a) If $\mathbf{f}$ is continuous on $[a, b]$ and $\mathbf{G}$ is an antiderivative for $\mathbf{f}$ on $[a, b]$, then

$$\int_a^b \mathbf{f}(t)\,dt = \mathbf{G}(b) - \mathbf{G}(a).$$

(This is the vector version of the fundamental theorem of integral calculus.)

(b) If $\mathbf{f}$ is continuous on an interval I and $\mathbf{F}$ and $\mathbf{G}$ are antiderivatives for $\mathbf{f}$, then

$$\mathbf{F} = \mathbf{G} + \mathbf{C}$$

for some constant vector $\mathbf{C}$.

55. Is it always true that

$$\int_a^b [\mathbf{f}(t) \cdot \mathbf{g}(t)]\,dt = \left[\int_a^b \mathbf{f}(t)\,dt\right] \cdot \left[\int_a^b \mathbf{g}(t)\,dt\right]?$$

56. Prove that, if $\mathbf{f}$ is continuous on $[a, b]$, then for each constant vector $\mathbf{c}$,

$$\int_a^b [\mathbf{c} \cdot \mathbf{f}(t)]\,dt = \mathbf{c} \cdot \int_a^b \mathbf{f}(t)\,dt \quad \text{and}$$

$$\int_a^b [c \times \mathbf{f}(t)]\,dt = c \times \int_a^b \mathbf{f}(t)\,dt.$$

57. Let $\mathbf{f}$ be a differentiable vector-valued function. Show that if $\|\mathbf{f}(t)\| \neq 0$, then

$$\frac{d}{dt}(\|\mathbf{f}(t)\|) = \frac{\mathbf{f}(t)\cdot\mathbf{f}'(t)}{\|\mathbf{f}(t)\|}.$$

58. Let $\mathbf{f}$ be a differentiable vector-valued function. Show that where $\|\mathbf{f}(t)\| \neq 0$,

$$\frac{d}{dt}\left(\frac{\mathbf{f}(t)}{\|\mathbf{f}(t)\|}\right) = \frac{\mathbf{f}'(t)}{\|\mathbf{f}(t)\|} - \frac{\mathbf{f}(t)\cdot\mathbf{f}'(t)}{\|\mathbf{f}(t)\|^3}\mathbf{f}(t).$$

C In Exercises 59–62, use a graphing utility that has three-dimensional capability.

59. The vector function

$$\mathbf{r}(t) = \cos at\,\mathbf{i} + \sin at\,\mathbf{j} + f(t)\,\mathbf{k}$$

describes the motion of an object on a circular cylinder.

(a) Set $f(t)=t$. Plot the helix for $a=1,2,4,5$ and $0\le t\le 4\pi$.

(b) Set $a=1$. Plot the helix for $f(t)=bt, b=1,2,4,5$ and $0\le t\le 4\pi$.

(c) Describe the effect that the constants a and b have on the curve.

(d) Set $a=1$ and experiment with other functions f. For example, try $f(t)=t^2$, $f(t)=e^t$, $f(t)=\ln(t+1)$ and describe the effect on the curve.

60. The path traced out by the vector function

$$\mathbf{r}(t) = A\cos at\,\mathbf{i} + B\sin at\,\mathbf{j} + f(t)\,\mathbf{k}$$

is also a curve on a cylinder.

(a) Set $f(t)=t$ and $a=1$. Plot the curve for the pairs $A=2, B=3; A=4, B=2; 0\le t\le 4\pi$. What name would you give to this type of curve?

(b) Experiment with other values for the constants a, A, B and other functions f

61. Let $\mathbf{r}(t) = A\cos t\,\mathbf{i} + B\sin t\,\mathbf{j} + f(t)\,\mathbf{k}$.

(a) Set $A=B=1$ and $f(t)=\sin bt$. Plot the curve for $b=1,2,3,4,5; 0\le t\le 2\pi$.

(b) Set $A=B=1$ and $f(t)=\cos bt$. Plot the curve for $b=1,2,3,4,5; 0\le t\le 2\pi$.

(c) Set $f(t)=\sin bt$. Plot the curves for the pairs $A=2$, $B=3; A=4, B=2$.

(d) Describe the effects that the constants A, B, and b have on the curve.

62. The vector function

$$\mathbf{r}(t) = (A\cos at + B\cos bt)\,\mathbf{i} + A\sin at\,\mathbf{j} + B\sin bt\,\mathbf{k}.$$

describes the motion of an object on a torus. Sketch the curve generated by letting:

(a) $A=2, B=1, a=1, b=1$.

(b) $A=1, B=2, a=2, b=1$.

(c) Experiment with other values of the constants to see the effect on the curve.

■ 13.2 DIFFERENTIATION FORMULAS

Vector functions with a common domain can be combined in many ways to form new functions. From $\mathbf{f}$ and $\mathbf{g}$ we can form the sum $\mathbf{f}+\mathbf{g}$:

$$(\mathbf{f}+\mathbf{g})(t) = \mathbf{f}(t) + \mathbf{g}(t).$$

We can form scalar multiples $\alpha\mathbf{f}$ and thus linear combinations $\alpha\,\mathbf{f} + \beta\,\mathbf{g}$:

$$(\alpha\mathbf{f})(t) = \alpha\mathbf{f}(t), \quad (\alpha\mathbf{f}+\beta\mathbf{g})(t) = \alpha\mathbf{f}(t) + \beta\mathbf{g}(t).$$

We can form the dot product $\mathbf{f}\cdot\mathbf{g}$:

$$(\mathbf{f}\cdot\mathbf{g})(t) = \mathbf{f}(t)\cdot\mathbf{g}(t).$$

We can also form the cross product $\mathbf{f}\times\mathbf{g}$:

$$(\mathbf{f}\times\mathbf{g})(t) = \mathbf{f}(t)\times\mathbf{g}(t).$$

These operations on vector functions are simply the pointwise application of the algebraic operations on vectors that we introduced in Chapter 12.

There are two ways of bringing *scalar functions* (real-valued functions) into play. If a scalar function u has the same domain as $\mathbf{f}$, we can form the product $u\,\mathbf{f}$:

$$(u\,\mathbf{f})(t) = u(t)\mathbf{f}(t).$$

If $u(t)$ is in the domain of $\mathbf{f}$ for each t in some interval, then we can form the composition $\mathbf{f} \circ u$:

$$(\mathbf{f} \circ u)(t) = \mathbf{f}(u(t)).$$

For example, with $u(t) = t^2$ and $\mathbf{f}(t) = e^t\,\mathbf{i} + \sin 2t\,\mathbf{j}$,

$$(u\,\mathbf{f})(t) = u(t)\mathbf{f}(t) = t^2 e^t\,\mathbf{i} + t^2 \sin 2t\,\mathbf{j} \text{ and } (\mathbf{f} \circ u)(t) = \mathbf{f}(u(t)) = e^{t^2}\mathbf{i} + \sin 2t^2\,\mathbf{j}.$$

It follows from Theorem 13.1.3 that if $\mathbf{f}, \mathbf{g}$ and u are continuous on a common domain, then $\mathbf{f}+\mathbf{g}$, $\alpha\mathbf{f}$, $\mathbf{f} \cdot \mathbf{g}$, $\mathbf{f} \times \mathbf{g}$, and $u\mathbf{f}$ are all continuous on that same set. We have yet to show the continuity of $\mathbf{f} \circ u$. The verification of that is left to you (Exercise 36). What interests us here is that, if $\mathbf{f}, \mathbf{g}$ and u are differentiable, then the newly constructed functions are also differentiable and their derivatives satisfy the following rules:

(13.2.1)

$$
\begin{aligned}
&(1)\ (\mathbf{f}+\mathbf{g})'(t) = \mathbf{f}'(t) + \mathbf{g}'(t).\\
&(2)\ (\alpha\mathbf{f})(t) = \alpha\mathbf{f}'(t) \quad (\alpha \text{ constant}).\\
&(3)\ (u\mathbf{f})'(t) = u(t)\mathbf{f}'(t) + u'(t)\mathbf{f}(t).\\
&(4)\ (\mathbf{f} \cdot \mathbf{g})'(t) = [\mathbf{f}(t) \cdot \mathbf{g}'(t)] + [\mathbf{f}'(t) \cdot \mathbf{g}(t)].\\
&(5)\ (\mathbf{f} \times \mathbf{g})'(t) = [\mathbf{f}(t) \times \mathbf{g}'(t)] + [\mathbf{f}'(t) \times \mathbf{g}(t)].\\
&(6)\ (\mathbf{f} \circ u)'(t) = \mathbf{f}'(u(t))u'(t) = u'(t)\mathbf{f}'(u(t)) \quad (\text{the chain rule}).
\end{aligned}
$$

Rules (3),(4),(5) are all "product" rules and should remind you of the rule for differentiating the product of ordinary functions. Keep in mind, however, that the cross product is not commutative and therefore the order in Rule (5) is important.

In Rule (6) we first wrote the scalar part $u'(t)$ on the right so that the formula would look like the chain rule for ordinary functions. In general, $\mathbf{a}\alpha$ has the same meaning as $\alpha\mathbf{a}$.

Example 1 Taking $\mathbf{f}(t) = 2t^2\,\mathbf{i} - 3\mathbf{j}$, $\quad g(t) = \mathbf{i} + t\,\mathbf{j} + t^2\,\mathbf{k}$, $\quad u(t) = \frac{1}{3}t^3$

we have $\quad \mathbf{f}'(t) = 4t\,\mathbf{i}, \quad \mathbf{g}'(t) = \mathbf{j} + 2t\,\mathbf{k}, \quad u'(t) = t^2.$

Therefore

(a) $(\mathbf{f}+\mathbf{g})'(t) = \mathbf{f}'(t) + \mathbf{g}'(t) = 4t\,\mathbf{i} + (\mathbf{j} + 2t\,\mathbf{k}) = 4t\,\mathbf{i} + \mathbf{j} + 2t\,\mathbf{k};$

(b) $(2\mathbf{f})'(t) = 2\mathbf{f}'(t) = 2(4t\,\mathbf{i}) = 8t\,\mathbf{i};$

(c) $(u\mathbf{f})'(t) = u(t)\mathbf{f}'(t) + u'(t)\mathbf{f}(t) = \frac{1}{3}t^3(4t\,\mathbf{i}) + t^2(2t^2\mathbf{i} - 3\mathbf{j}) = \frac{10}{3}t^4\mathbf{i} - 3t^2\mathbf{j};$

(d) $(\mathbf{f} \cdot \mathbf{g})'(t) = [\mathbf{f}(t) \cdot \mathbf{g}'(t)] + [\mathbf{f}'(t) \cdot \mathbf{g}(t)]$

$$= [(2t^2\,\mathbf{i} - 3\mathbf{j}) \cdot (\mathbf{j} + 2t\,\mathbf{k})] + [4t\,\mathbf{i} \cdot (i + t\,\mathbf{j} + t^2\,\mathbf{k})] = -3 + 4t;$$

(e)
$$
\begin{aligned}
(\mathbf{f} \times \mathbf{g})'(t) &= [\mathbf{f}(t) \times \mathbf{g}'(t)] + [\mathbf{f}'(t) \times \mathbf{g}(t)] \\
&= [(2t^2\,\mathbf{i} - 3\,\mathbf{j}) \times (\mathbf{j} + 2t\,\mathbf{k})] + [4t\,\mathbf{i} \times (\mathbf{i} + t\,\mathbf{j} + t^2\,\mathbf{k})] \\
&= (2t^2\,\mathbf{k} - 4t^3\,\mathbf{j} - 6t\,\mathbf{i}) + (4t^2\,\mathbf{k} - 4t^3\,\mathbf{j}) = -6t\,\mathbf{i} - 8t^3\,\mathbf{j} + 6t^2\,\mathbf{k}
\end{aligned}
$$

while

$$
\begin{aligned}
(\mathbf{g} \times \mathbf{f})'(t) &= [\mathbf{g}(t) \times \mathbf{f}'(t)] + [\mathbf{g}'(t) \times \mathbf{f}(t)] \\
&= [(\mathbf{i} + t\,\mathbf{j} + t^2\,\mathbf{k}) \times 4t\,\mathbf{i}] + [(\mathbf{j} + 2t\,\mathbf{k}) \times (2t^2\,\mathbf{i} - 3\mathbf{j})] \\
&= (-4t^2\,\mathbf{k} + 4t^3\,\mathbf{j}) + (-2t^2\,\mathbf{k} + 4t^3\,\mathbf{j} + 6t\,\mathbf{i}) \\
&= 6t\,\mathbf{i} + 8t^3\,\mathbf{j} - 6t^2\,\mathbf{k} = -(\mathbf{f} \times \mathbf{g})'(t);
\end{aligned}
$$

(f)
$$
\begin{aligned}
(\mathbf{f} \circ u)'(t) &= \mathbf{f}'(u(t))u'(t) \\
&= [4u(t)\,\mathbf{i}]u'(t) = [4(\tfrac{1}{3}t^3)\,\mathbf{i}]t^2 = \tfrac{4}{3}t^5\,\mathbf{i}. \quad \square
\end{aligned}
$$

The differentiation formulas that we have given can all be derived component by component, and they can all be derived in a component-free manner. Take, for example, formula (3):

$$
(u\mathbf{f})'(t) = u(t)\mathbf{f}'(t) + u'(t)\mathbf{f}(t).
$$

COMPONENT-BY-COMPONENT DERIVATION Set

$$
\mathbf{f}(t) = f_1(t)\,\mathbf{i} + f_2(t)\,\mathbf{j} + f_3(t)\,\mathbf{k}.
$$

Then
$$
(u\,\mathbf{f})(t) = u(t)\,\mathbf{f}(t) = u(t)\,f_1(t)\,\mathbf{i} + u(t)\,f_2(t)\,\mathbf{j} + u(t)\,f_3(t)\,\mathbf{k}
$$

and
$$
\begin{aligned}
(u\,\mathbf{f})'(t) &= [u(t)\,f_1'(t) + u'(t)\,f_1(t)]\,\mathbf{i} + [u(t)\,f_2'(t) + u'(t)\,f_2(t)]\,\mathbf{j} + \\
&\qquad [u(t)\,f_3'(t) + u'(t)\,f_3(t)]\,\mathbf{k} \\
&= u(t)[f_1'(t)\,\mathbf{i} + f_2'(t)\,\mathbf{j} + f_3'(t)\,\mathbf{k}] + u'(t)[f_1(t)\,\mathbf{i} + f_2(t)\,\mathbf{j} + f_3(t)\,\mathbf{k}] \\
&= u(t)\mathbf{f}'(t) + u'(t)\mathbf{f}(t). \quad \square
\end{aligned}
$$

COMPONENT-FREE DERIVATION We find $(u\,\mathbf{f})'(t)$ by taking the limit as $h \to 0$ of the difference quotient

$$
\frac{u(t+h)\mathbf{f}(t+h) - u(t)\mathbf{f}(t)}{h}.
$$

By adding and subtracting $u(t+h)\mathbf{f}(t)$, we can rewrite this quotient as

$$
\frac{u(t+h)\mathbf{f}(t+h) - u(t+h)\mathbf{f}(t) + u(t+h)\mathbf{f}(t) - u(t)\mathbf{f}(t)}{h},
$$

which is equal to

$$
u(t+h)\frac{\mathbf{f}(t+h) - \mathbf{f}(t)}{h} + \frac{u(t+h) - u(t)}{h}\mathbf{f}(t).
$$

As $h \to 0$.

$$u(t+h) \to u(t), \qquad \text{(differentiable functions are continuous)}$$

$$\frac{\mathbf{f}(t+h)-\mathbf{f}(t)}{h} \to \mathbf{f}'(t), \qquad \text{(definition of derivative for vector functions)}$$

$$\frac{u(t+h)-u(t)}{h} \to u'(t). \qquad \text{(definition of derivative for scalar functions)}$$

It follows from the limit rules (Theorem 13.1.3) that

$$u(t+h)\frac{\mathbf{f}(t+h)-\mathbf{f}(t)}{h} \to u(t)\mathbf{f}'(t), \qquad \frac{u(t+h)-u(t)}{h}\mathbf{f}(t) \to u'(t)\mathbf{f}(t)$$

and therefore

$$u(t+h)\frac{\mathbf{f}(t+h)-\mathbf{f}(t)}{h} + \frac{u(t+h)-u(t)}{h}\mathbf{f}(t) \to u(t)\mathbf{f}'(t) + u'(t)\mathbf{f}(t). \quad \square$$

In Leibniz's notation the formulas take the following form:

(13.2.2)

$$(1)\ \frac{d}{dt}(\mathbf{f}+\mathbf{g}) = \frac{d\mathbf{f}}{dt} + \frac{d\mathbf{g}}{dt}.$$

$$(2)\ \frac{d}{dt}(\alpha\mathbf{f}) = \alpha\frac{d\mathbf{f}}{dt}. \qquad (\alpha \text{ constant})$$

$$(3)\ \frac{d}{dt}(u\mathbf{f}) = u\frac{d\mathbf{f}}{dt} + \frac{du}{dt}\mathbf{f}. \qquad (u = u(t))$$

$$(4)\ \frac{d}{dt}(\mathbf{f}\cdot\mathbf{g}) = \left(\mathbf{f}\cdot\frac{d\mathbf{g}}{dt}\right) + \left(\frac{d\mathbf{f}}{dt}\cdot\mathbf{g}\right).$$

$$(5)\ \frac{d}{dt}(\mathbf{f}\times\mathbf{g}) = \left(\mathbf{f}\times\frac{d\mathbf{g}}{dt}\right) + \left(\frac{d\mathbf{f}}{dt}\times\mathbf{g}\right).$$

$$(6)\ \frac{d\mathbf{f}}{dt} = \frac{d\mathbf{f}}{du}\frac{du}{dt}. \qquad \text{(chain rule)}$$

We conclude this section with two results that will prove useful as we go on.

Example 2 Let $\mathbf{r}$ be a differentiable vector function of t and set $r = \|\mathbf{r}\|$. Show that r is differentiable wherever it is not 0 and

(13.2.3)
$$\mathbf{r}\cdot\frac{d\mathbf{r}}{dt} = r\frac{dr}{dt}.$$

SOLUTION If $\mathbf{r}$ is differentiable, then $\mathbf{r}\cdot\mathbf{r} = \|\mathbf{r}\|^2 = r^2$ is differentiable. Let's assume now that $r \neq 0$. Since the square-root function is differentiable at all positive numbers and r^2 is positive, we can apply the square-root function to r^2 and conclude by the chain rule that r is itself differentiable.

To obtain the formula we differentiate the identity $\mathbf{r}\cdot\mathbf{r} = r^2$:

$$\mathbf{r}\cdot\frac{d\mathbf{r}}{dt} + \frac{d\mathbf{r}}{dt}\cdot\mathbf{r} = 2r\frac{dr}{dt}$$

$$2\mathbf{r}\cdot\frac{d\mathbf{r}}{dt} = 2r\frac{dr}{dt}$$

$$\mathbf{r}\cdot\frac{d\mathbf{r}}{dt} = r\frac{dr}{dt}. \quad \square$$

Example 3 Let $\mathbf{r}$ be a differentiable vector function of t and set $r = \|\mathbf{r}\|$. Show that where $r \neq 0$

$$\frac{d}{dt}\left(\frac{\mathbf{r}}{r}\right) = \frac{1}{r^3}\left[\left(\mathbf{r} \times \frac{d\mathbf{r}}{dt}\right) \times \mathbf{r}\right]. \tag{13.2.4}$$

SOLUTION This is a little tricky:

$$\begin{aligned}\frac{d}{dt}\left(\frac{\mathbf{r}}{r}\right) &= \frac{1}{r}\frac{d\mathbf{r}}{dt} - \frac{1}{r^2}\frac{dr}{dt}\mathbf{r}\\ &= \frac{1}{r^3}\left[r^2\frac{d\mathbf{r}}{dt} - r\frac{dr}{dt}\mathbf{r}\right]\\ &= \frac{1}{r^3}\left[(\mathbf{r}\cdot\mathbf{r})\frac{d\mathbf{r}}{dt} - \left(\mathbf{r}\cdot\frac{d\mathbf{r}}{dt}\right)\mathbf{r}\right] = \frac{1}{r^3}\left[\left(\mathbf{r}\times\frac{d\mathbf{r}}{dt}\right)\times\mathbf{r}\right]. \quad \square\end{aligned}$$

$(\mathbf{a} \times \mathbf{b}) \times \mathbf{c} = (\mathbf{c} \cdot \mathbf{a})\mathbf{b} - (\mathbf{c} \cdot \mathbf{b})\mathbf{a}$

EXERCISES 13.2

Find $\mathbf{f}'(t)$ and $\mathbf{f}''(t)$.

1. $\mathbf{f}(t) = \mathbf{a} + t\,\mathbf{b}$.
2. $\mathbf{f}(t) = \mathbf{a} + t\,\mathbf{b} + t^2\,\mathbf{c}$.
3. $\mathbf{f}(t) = e^{2t}\,\mathbf{i} - \sin t\,\mathbf{j}$.
4. $\mathbf{f}(t) = [(t^2\,\mathbf{i} - \mathbf{j}) \cdot (\mathbf{i} - t^2\,\mathbf{j})]\,\mathbf{i}$.
5. $\mathbf{f}(t) = [(t^2\,\mathbf{i} - 2t\,\mathbf{j}) \cdot (t\,\mathbf{i} + t^3\,\mathbf{j})]\,\mathbf{j}$.
6. $\mathbf{f}(t) = [(3t\,\mathbf{i} - t^2\,\mathbf{j} + \mathbf{k}) \cdot (\mathbf{i} + t^3\,\mathbf{j} - 2t\,\mathbf{k})]\mathbf{k}$.
7. $\mathbf{f}(t) = (e^t\,\mathbf{i} + t\,\mathbf{k}) \times (t\,\mathbf{j} + e^{-t}\,\mathbf{k})$.
8. $\mathbf{f}(t) = (t\,\mathbf{i} - t^2\,\mathbf{j} + \mathbf{k}) \times (\mathbf{i} + t^3\,\mathbf{j} + 5t\,\mathbf{k})$.
9. $\mathbf{f}(t) = (\cos t\,\mathbf{i} + \sin t\,\mathbf{j} + \mathbf{k}) \times (\sin 2t\,\mathbf{i} + \cos 2t\,\mathbf{j} + t\,\mathbf{k})$.
10. $\mathbf{f}(t) = t\mathbf{g}(t^2)$.
11. $\mathbf{f}(t) = t\mathbf{g}(\sqrt{t})$.
12. $\mathbf{f}(t) = (e^{2t}\mathbf{i} + e^{-2t}\,\mathbf{j} + \mathbf{k}) \times (e^{2t}\,\mathbf{i} - e^{-2t}\,\mathbf{j} + \mathbf{k})$.

Find the indicated derivative.

13. $\dfrac{d}{dt}[e^{\cos t}\,\mathbf{i} + e^{\sin t}\,\mathbf{j}]$.
14. $\dfrac{d^2}{dt^2}[e^t\,\cos t\,\mathbf{i} + e^t\,\sin t\,\mathbf{j}]$.
15. $\dfrac{d^2}{dt^2}\left[(e^t\,\mathbf{i} + e^{-t}\,\mathbf{j}) \cdot (e^t\,\mathbf{i} - e^{-t}\,\mathbf{j})\right]$.
16. $\dfrac{d}{dt}\left[(\ln t\,\mathbf{i} + t\,\mathbf{j} - (t^2 + 1)\,\mathbf{k}) \times \left(\frac{1}{t}\mathbf{i} + t^2\,\mathbf{j} - t\,\mathbf{k}\right)\right]$.
17. $\dfrac{d}{dt}[(\mathbf{a} + t\,\mathbf{b}) \times (\mathbf{c} + t\,\mathbf{d})]$.
18. $\dfrac{d}{dt}[(\mathbf{a} + t\,\mathbf{b}) \times (\mathbf{a} + t\,\mathbf{b} + t^2\,\mathbf{c})]$.
19. $\dfrac{d}{dt}[(\mathbf{a} + t\,\mathbf{b}) \cdot (\mathbf{c} + t\,\mathbf{d})]$.
20. $\dfrac{d}{dt}[(\mathbf{a} + t\,\mathbf{b}) \cdot (\mathbf{a} + t\,\mathbf{b} + t^2\,\mathbf{c})]$.

Find $\mathbf{r}(t)$ given that:

21. $\mathbf{r}'(t) = \mathbf{b}$ for all real t, $\quad \mathbf{r}(0) = \mathbf{a}$.
22. $\mathbf{r}''(t) = \mathbf{c}$ for all real t, $\quad \mathbf{r}'(0) = \mathbf{b}$, $\quad \mathbf{r}(0) = \mathbf{a}$.
23. $\mathbf{r}''(t) = \mathbf{a} + t\mathbf{b}$ for all real t, $\quad \mathbf{r}'(0) = \mathbf{c}$, $\quad \mathbf{r}(0) = \mathbf{d}$.
24. $\mathbf{r}''(t) = \cos 2t\,\mathbf{i} + \sin 2t\,\mathbf{j}$ for all real t, $\mathbf{r}'(0) = 2\,\mathbf{i} - \frac{1}{2}\,\mathbf{j}$, $\quad \mathbf{r}(0) = \frac{3}{4}\,\mathbf{i} + \mathbf{j}$.
25. Show that, if $\mathbf{r}(t) = \sin t\,\mathbf{i} + \cos t\,\mathbf{j}$, then $\mathbf{r}(t)$ and $\mathbf{r}''(t)$ are parallel. Is there a value of t for which $\mathbf{r}(t)$ and $\mathbf{r}''(t)$ have the same direction?
26. Show that, if $\mathbf{r}(t) = e^{kt}\mathbf{i} + e^{-kt}\mathbf{j}$, then $\mathbf{r}(t)$ and $\mathbf{r}''(t)$ have the same direction.
27. Calculate $\mathbf{r}(t) \cdot \mathbf{r}'(t)$ and $\mathbf{r}(t) \times \mathbf{r}'(t)$ for $\mathbf{r}(t) = \cos t\,\mathbf{i} + \sin t\,\mathbf{j}$.

Assume the rule for differentiating a cross product and show the following.

28. $(\mathbf{g} \times \mathbf{f})'(t) = -(\mathbf{f} \times \mathbf{g})'(t)$.
29. $\dfrac{d}{dt}[\mathbf{f}(t) \times \mathbf{f}'(t)] = \mathbf{f}(t) \times \mathbf{f}''(t)$.
30. $\dfrac{d}{dt}[u_1(t)\mathbf{r}_1(t) \times u_2(t)\mathbf{r}_2(t)] = u_1(t)u_2(t)\dfrac{d}{dt}[\mathbf{r}_1(t) \times \mathbf{r}_2(t)] + [\mathbf{r}_1(t) \times \mathbf{r}_2(t)]\,\dfrac{d}{dt}[u_1(t)u_2(t)]$.
31. Set $\mathbf{E}(t) = \mathbf{f}(t) \cdot [\mathbf{g}(t) \times \mathbf{h}(t)]$ and show that $\mathbf{E}'(t) = \mathbf{f}'(t) \cdot [\mathbf{g}(t) \times \mathbf{h}(t)] + \mathbf{f}(t) \cdot [\mathbf{g}'(t) \times \mathbf{h}(t)] + \mathbf{f}(t) \cdot [\mathbf{g}(t) \times \mathbf{h}'(t)]$.
32. Suppose that $\mathbf{f}(t)$ is parallel to $\mathbf{f}''(t)$ for all t. Prove that $\mathbf{f} \times \mathbf{f}'$ is constant. HINT: See Exercise 29.
33. Assume the rule for differentiating a dot product and show that

$$\|\mathbf{r}(t)\| \text{ is constant} \qquad \text{iff} \quad \mathbf{r}(t) \cdot \mathbf{r}'(t) = 0 \text{ identically.}$$

34. Derive the formula

$$(\mathbf{f}\cdot\mathbf{g})'(t) = [\mathbf{f}(t)\cdot\mathbf{g}'(t)] + [\mathbf{f}'(t)\cdot\mathbf{g}(t)]$$

(a) by appealing to components;

(b) without appealing to components.

35. Derive the formula

$$(\mathbf{f}\times\mathbf{g})'(t) = [\mathbf{f}(t)\times\mathbf{g}'(t)] + [\mathbf{f}'(t)\times\mathbf{g}(t)]$$

without appealing to components.

36. (a) Show that, if u is continuous at t_0 and $\mathbf{f}$ is continuous at $u(t_0)$, then the composition $\mathbf{f}\circ u$ is continuous at t_0.

(b) Derive the chain rule for vector functions:

$$\frac{d\mathbf{f}}{dt} = \frac{d\mathbf{f}}{du}\frac{du}{dt}.$$

■ 13.3 CURVES

Introduction

Earlier we explained how every linear vector function

$$\mathbf{r}(t) = \mathbf{r}_0 + t\,\mathbf{d} \quad \text{with} \quad \mathbf{d} \neq \mathbf{0}$$

parametrizes a line. More generally, every differentiable vector function parametrizes a curve.

Suppose that

$$\mathbf{r}(t) = x(t)\,\mathbf{i} + y(t)\,\mathbf{j} + z(t)\,\mathbf{k}$$

is differentiable on some interval I. (At the endpoints, if there are any, we require only continuity.) For each number $t \in I$, the tip of the radius vector $\mathbf{r}(t)$ is the point $P(x(t), y(t), z(t))$. As t ranges over I, the point P traces out some path C (Figure 13.3.1). We call C a *differentiable curve* and say that C is *parametrized* by $\mathbf{r}$ with parameter t. It is important to understand that a parametrized curve C is an *oriented* curve in the sense that as t increases on I, the tip of the radius vector traces out the curve C in a certain direction. For example, the curve parametrized by

$$\mathbf{r}(t) = \cos t\,\mathbf{i} + \sin t\,\mathbf{j}, \quad t \in [0, 2\pi]$$

is the unit circle traversed in the counterclockwise direction starting at the point $(1, 0)$ (Figure 13.3.2). The orientation of the elliptical helix parametrized by

$$\mathbf{r}(t) = t\,\mathbf{i} + 2\cos t\,\mathbf{j} + 3\sin t\,\mathbf{k}, \quad t \geq 0,$$

is indicated by the arrows drawn in Figure 13.3.3.

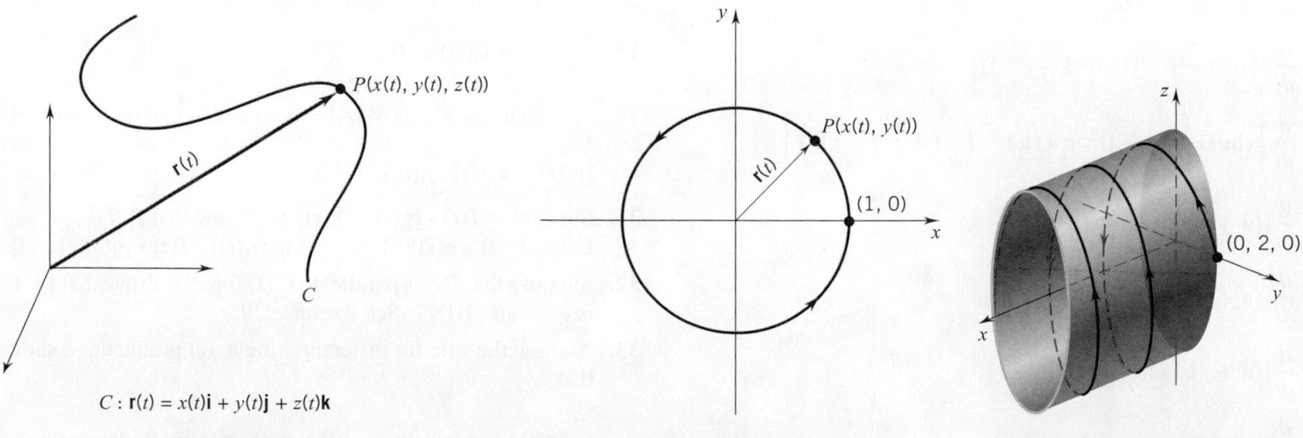

Figure 13.3.1 Figure 13.3.2 Figure 13.3.3

Tangent Vector, Tangent Line

Let's try to interpret the derivative

$$\mathbf{r}'(t) = x'(t)\,\mathbf{i} + y'(t)\,\mathbf{j} + z'(t)\,\mathbf{k}$$

geometrically. First of all,

$$\mathbf{r}'(t) = \lim_{h \to 0} \frac{\mathbf{r}(t+h) - \mathbf{r}(t)}{h}.$$

If $\mathbf{r}'(t) \neq \mathbf{0}$, then we can be sure that for $t + h$ close enough to t, the vector

$$\mathbf{r}(t+h) - \mathbf{r}(t)$$

will not be $\mathbf{0}$. (Explain.) Consequently, we can think of the vector $\mathbf{r}(t+h) - \mathbf{r}(t)$ as pictured in Figure 13.3.4.

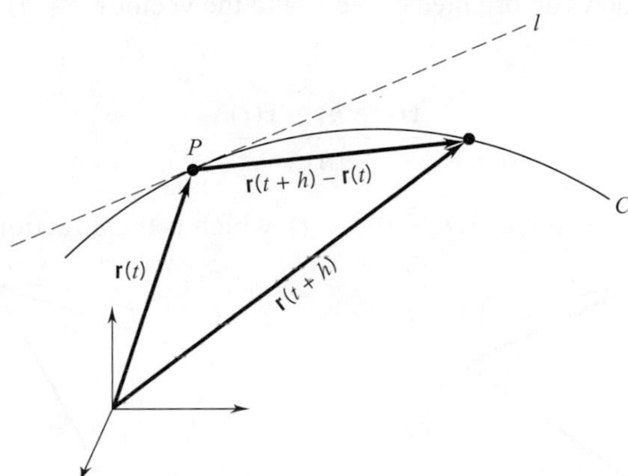

Figure 13.3.4

Let's agree that the line marked l in the figure corresponds to our intuitive notion of the tangent line at the point P. As h tends to zero, the vector

$$\mathbf{r}(t+h) - \mathbf{r}(t)$$

comes increasingly closer to serving as a direction vector for that tangent line. It may be tempting therefore to take the limiting case

$$\lim_{h \to 0} [\mathbf{r}(t+h) - \mathbf{r}(t)]$$

and call that a direction vector for the tangent line. The trouble is that this limit vector is $\mathbf{0}$ and $\mathbf{0}$ has no direction.

We can circumvent this difficulty by replacing $\mathbf{r}(t+h) - \mathbf{r}(t)$ by a vector which, for small h, has greater length: the difference quotient

$$\frac{\mathbf{r}(t+h) - \mathbf{r}(t)}{h}.$$

For each real number $h \neq 0$, the vector $[\mathbf{r}(t+h) - \mathbf{r}(t)]/h$ is parallel to $\mathbf{r}(t+h) - \mathbf{r}(t)$, and therefore its limit,

$$\mathbf{r}'(t) = \lim_{h \to 0} \frac{\mathbf{r}(t+h) - \mathbf{r}(t)}{h},$$

which by assumption is not $\mathbf{0}$, can be taken as a direction vector for the tangent line. Hence the following definition.

DEFINITION 13.3.1 TANGENT VECTOR

Let

$$C : \mathbf{r}(t) = x(t)\,\mathbf{i} + y(t)\,\mathbf{j} + z(t)\,\mathbf{k}$$

be a differentiable curve. The vector $\mathbf{r}'(t)$, if not $\mathbf{0}$, is *tangent* to the curve C at the point $P(x(t), y(t), z(t))$.

Now the following question arises: Assuming that $\mathbf{r}'(t) \neq \mathbf{0}$, in which of the two possible directions does $\mathbf{r}'(t)$ point?

Figure 13.3.5 shows an oriented curve C and the vector $\mathbf{r}(t+h) - \mathbf{r}(t)$ with $h > 0$. In this case,

$$\frac{\mathbf{r}(t+h) - \mathbf{r}(t)}{h}$$

points in the same direction as $\mathbf{r}(t+h) - \mathbf{r}(t)$, which is the direction of increasing t.

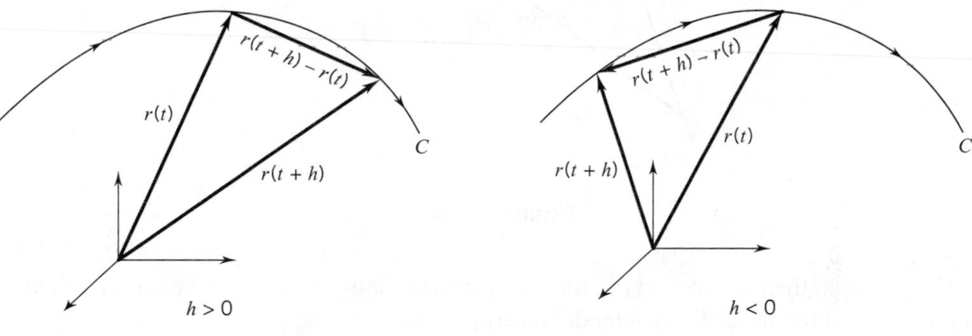

Figure 13.3.5 Figure 13.3.6

Figure 13.3.6 shows the vector $\mathbf{r}(t+h) - \mathbf{r}(t)$ for $h < 0$. In this case, dividing by h to form the difference quotient *reverses* the direction of the vector. Thus

$$\frac{\mathbf{r}(t+h) - \mathbf{r}(t)}{h}$$

also points in the direction of increasing t. We can now conclude that

$\mathbf{r}'(t)$ points in the direction of increasing t.

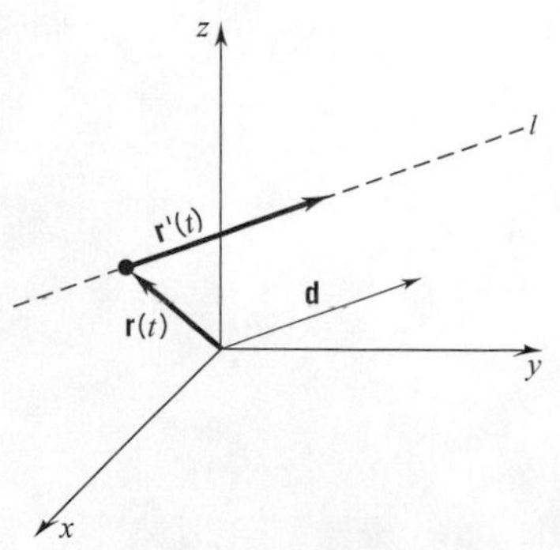

Figure 13.3.7

At each point of a line

$$l: \quad \mathbf{r}(t) = \mathbf{r}_0 + t\mathbf{d}$$

the tangent vector $\mathbf{r}'(t)$ is parallel to the line itself:

$$\mathbf{r}'(t) = \mathbf{d} \quad \text{and } \mathbf{d} \text{ is parallel to } l. \qquad \text{(Figure 13.3.7)}$$

In the case of a circle

$$C: \quad \mathbf{r}(t) = a\cos t\,\mathbf{i} + a\sin t\,\mathbf{j}, \quad (a > 0)$$

the tangent vector $\mathbf{r}'(t)$ is perpendicular to the radius vector $\mathbf{r}(t)$:

$$\mathbf{r}'(t)\cdot\mathbf{r}(t) = (-a\sin t\,\mathbf{i} + a\cos t\,\mathbf{j})\cdot(a\cos t\,\mathbf{i} + a\sin t\,\mathbf{j})$$
$$= -a^2\sin t\cos t + a^2\cos t\sin t = 0. \qquad \text{(Figure 13.3.8)}$$

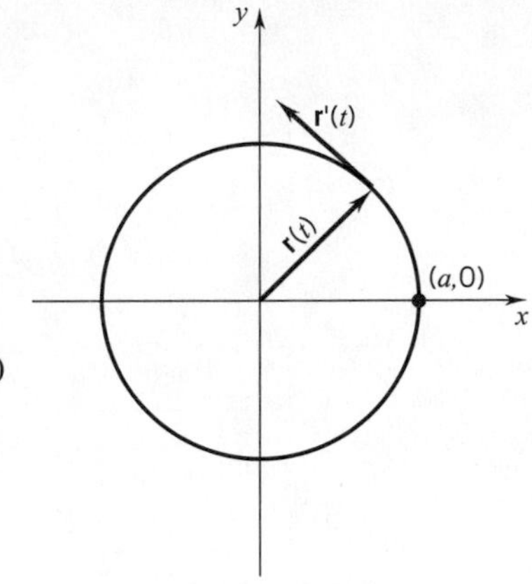

Figure 13.3.8

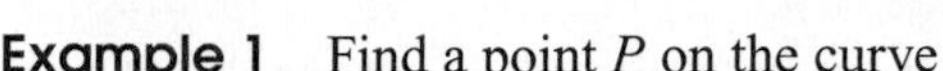

Example 1 Find a point P on the curve

$$\mathbf{r}(t) = (1-2t)\,\mathbf{i} + t^2\,\mathbf{j} + 2e^{2(t-1)}\,\mathbf{k}$$

at which the tangent vector $\mathbf{r}'(t)$ is parallel to the radius vector $\mathbf{r}(t)$.

SOLUTION $\mathbf{r}'(t) = -2\,\mathbf{i} + 2t\,\mathbf{j} + 4e^{2(t-1)}\,\mathbf{k}$.

For $\mathbf{r}'(t)$ to be parallel to $\mathbf{r}(t)$ there must exist a scalar α such that

$$\mathbf{r}(t) = \alpha\mathbf{r}'(t).$$

This vector equation holds iff

$$1-2t = -2\alpha, \quad t^2 = 2\alpha t, \quad 2e^{2(t-1)} = 4\alpha e^{2(t-1)}.$$

The last scalar equation requires that $\alpha = \frac{1}{2}$. The only value of t that satisfies all three equations with $\alpha = \frac{1}{2}$ is $t = 1$. Therefore the only point at which $\mathbf{r}'(t)$ is parallel to $\mathbf{r}(t)$ is the tip of $\mathbf{r}(1)$. This is the point $P(-1, 1, 2)$. ❑

If $\mathbf{r}'(t_0) \neq \mathbf{0}$, then $\mathbf{r}'(t_0)$ is tangent to the curve at the tip of $\mathbf{r}(t_0)$. The *tangent line* at this point can be parametrized by setting

(13.3.2) $$\boxed{\mathbf{R}(u) = \mathbf{r}(t_0) + u\,\mathbf{r}'(t_0).}$$

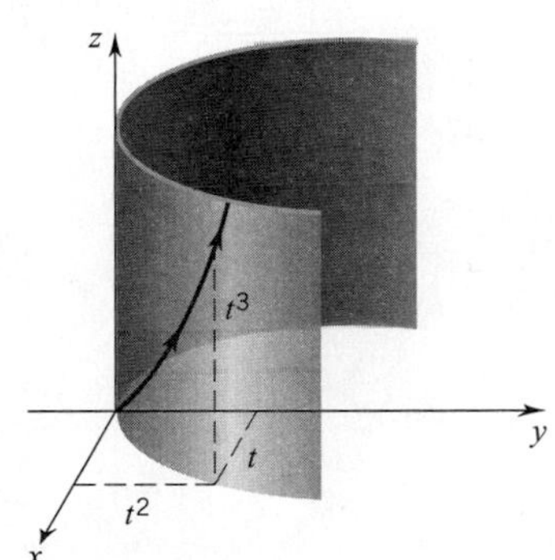

the twisted cubic, $t > 0$

Figure 13.3.9

Example 2 Find a vector tangent to the *twisted cubic*

$$\mathbf{r}(t) = t\,\mathbf{i} + t^2\mathbf{j} + t^3\mathbf{k} \qquad \text{(Figure 13.3.9)}$$

at the point $P(2, 4, 8)$, and then parametrize the tangent line.

SOLUTION Here $\mathbf{r}'(t) = \mathbf{i} + 2t\,\mathbf{j} + 3t^2\,\mathbf{k}$.

Since $P(2, 4, 8)$ is the tip of $\mathbf{r}(2)$, the vector

$$\mathbf{r}'(2) = \mathbf{i} + 4\mathbf{j} + 12\mathbf{k}$$

is tangent to the curve at the point $P(2, 4, 8)$. The vector function

$$\mathbf{R}(u) = (2\mathbf{i} + 4\mathbf{j} + 8\mathbf{k}) + u(\mathbf{i} + 4\mathbf{j} + 12\mathbf{k})$$

parametrizes the tangent line. ❑

Intersecting Curves

Two curves

$$C_1: \ \mathbf{r}_1(t) = x_1(t)\,\mathbf{i} + y_1(t)\,\mathbf{j} + z_1(t)\,\mathbf{k}, \quad C_2: \ \mathbf{r}_2(u) = x_2(u)\,\mathbf{i} + y_2(u)\,\mathbf{j} + z_2(u)\,\mathbf{k}$$

intersect iff there are numbers t and u for which

$$\mathbf{r}_1(t) = \mathbf{r}_2(u).$$

If two curves C_1 and C_2 intersect at a point $\mathbf{r}_1(t_1) = \mathbf{r}_2(u_2)$, we define the angle between the curves at this point to be the angle between the corresponding tangent vectors $\mathbf{r}'_1(t_1)$ and $\mathbf{r}'_2(u_2)$.

Example 3 Show that the circles

$$C_1: \ \mathbf{r}_1(t) = \cos t\,\mathbf{i} + \sin t\,\mathbf{j}, \quad C_2: \ \mathbf{r}_2(u) = \cos u\,\mathbf{j} + \sin u\,\mathbf{k}$$

intersect at right angles at $P(0, 1, 0)$ and $Q(0, -1, 0)$.

SOLUTION Since $\mathbf{r}_1(\pi/2) = \mathbf{j} = \mathbf{r}_2(0)$, the curves meet at the tip of $\mathbf{j}$, which is $P(0, 1, 0)$. Also, since $\mathbf{r}_1(3\pi/2) = -\mathbf{j} = \mathbf{r}_2(\pi)$, the curves meet at the tip of $-\mathbf{j}$, which is $Q(0, -1, 0)$. Differentiation gives

$$\mathbf{r}_1'(t) = -\sin t\,\mathbf{i} + \cos t\,\mathbf{j} \quad \text{and} \quad \mathbf{r}_2'(u) = -\sin u\,\mathbf{j} + \cos u\,\mathbf{k}.$$

Since $\mathbf{r}_1'(\pi/2) = -\mathbf{i}$ and $\mathbf{r}_2'(0) = \mathbf{k}$, we have

$$\mathbf{r}_1'(\pi/2) \cdot \mathbf{r}_2'(0) = 0.$$

This tells us that the curves are perpendicular at $P(0, 1, 0)$. Since $\mathbf{r}_1'(3\pi/2) = \mathbf{i}$ and $\mathbf{r}_2'(\pi) = -\mathbf{k}$, we have

$$\mathbf{r}_1'(3\pi/2) \cdot \mathbf{r}_2'(\pi) = 0.$$

This tells us that the curves are perpendicular at $Q(0, -1, 0)$. The curves appear in Figure 13.3.10. ❑

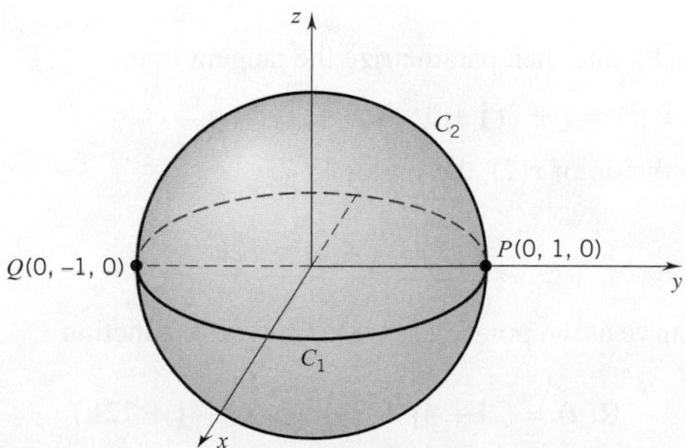

Figure 13.3.10

The Unit Tangent, the Principal Normal, the Osculating Plane

Suppose now that the curve

$$C:\ \mathbf{r}(t) = x(t)\,\mathbf{i} + y(t)\,\mathbf{j} + z(t)\,\mathbf{k}$$

is twice differentiable and $\mathbf{r}'(t)$ is never zero. Then at each point $P(x(t), y(t), z(t))$ of the curve, there is a *unit tangent vector*:

(13.3.3)
$$\boxed{\mathbf{T}(t) = \frac{\mathbf{r}'(t)}{\|\mathbf{r}'(t)\|}.}$$

Since $\|\mathbf{r}'(t)\| > 0$, $\mathbf{T}(t)$ points in the direction of $\mathbf{r}'(t)$ that is, in the direction of increasing t. Since $\|\mathbf{T}(t)\| = 1$, we have $\mathbf{T}(t)\cdot\mathbf{T}(t) = 1$. Differentiation gives

$$\mathbf{T}(t)\cdot\mathbf{T}'(t) + \mathbf{T}'(t)\cdot\mathbf{T}(t) = 0.$$

Since the dot product is commutative, we have

$$2[\mathbf{T}'(t)\cdot\mathbf{T}(t)] = 0 \qquad \text{and thus} \qquad \mathbf{T}'(t)\cdot\mathbf{T}(t) = 0.$$

At each point of the curve *the vector* $\mathbf{T}'(t)$ *is perpendicular to* $\mathbf{T}(t)$.

The vector $\mathbf{T}'(t)$ measures the rate of change of $\mathbf{T}(t)$ with respect to t. Since the norm of $\mathbf{T}(t)$ is constantly 1, $\mathbf{T}(t)$ can change only in direction. The vector $\mathbf{T}'(t)$ measures this change in direction.

If the unit tangent vector is not changing in direction (as in the case of a straight line), then $\mathbf{T}'(t) = \mathbf{0}$. If $\mathbf{T}'(t) \neq \mathbf{0}$, then we can form what is called the *principal normal vector*:

(13.3.4)
$$\boxed{\mathbf{N}(t) = \frac{\mathbf{T}'(t)}{\|\mathbf{T}'(t)\|}.}$$

This is the unit vector in the direction of $\mathbf{T}'(t)$. The *normal line* at P is the line through P parallel to the principal normal.

Figure 13.3.11 shows a curve on which we have marked several points. At each point we have drawn the unit tangent and the principal normal. The plane determined by these two vectors is called the *osculating plane* (literally, the "kissing plane"). This is the plane of greatest contact with the curve at the point in question. The principal normal points in the direction in which the curve is curving, that is, on the concave side of the curve.

If a curve is a plane curve but not a straight line, then the osculating plane is simply the plane of the curve. A straight line does not have an osculating plane. There is no principal normal vector $[\mathbf{T}'(t) = \mathbf{0}]$ and there is no single plane of greatest contact. Each straight line lies on an infinite number of planes.

Example 4 In Figure 13.3.12 you can see a curve spiraling up a circular cylinder. The curve is called a *circular helix* if the rate of climb is constant. The simplest parametrization for a circular helix takes the form

$$\mathbf{r}(t) = a\cos t\,\mathbf{i} + a\sin t\,\mathbf{j} + bt\,\mathbf{k} \quad \text{with } a > 0, \quad b > 0.$$

The first two components produce the rotational effect; the third component gives the rate of climb.

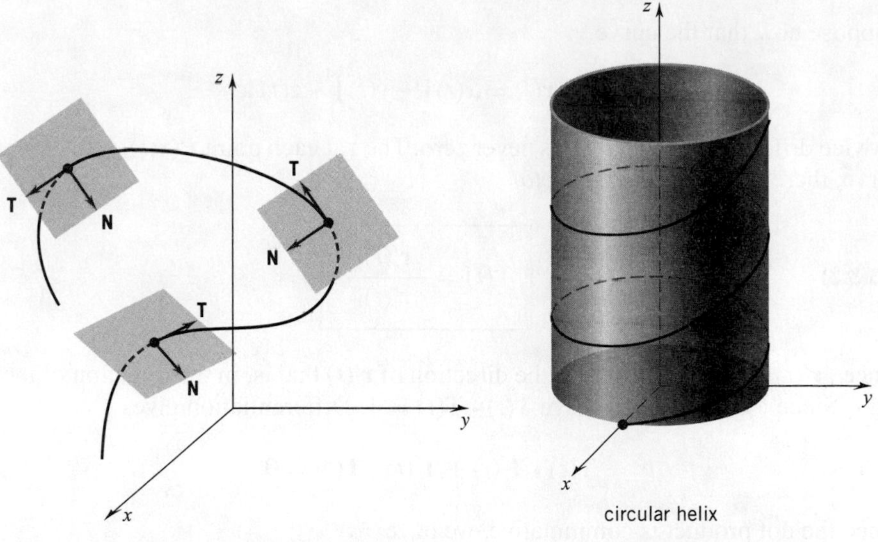

Figure 13.3.11

Figure 13.3.12

We will find the unit tangent, the principal normal, and then an equation for the osculating plane. Since

$$\mathbf{r}'(t) = -a\sin t\,\mathbf{i} + a\cos t\,\mathbf{j} + b\,\mathbf{k},$$

we have
$$\|\mathbf{r}'(t)\| = \sqrt{a^2\sin^2 t + a^2\cos^2 t + b^2} = \sqrt{a^2+b^2}$$

and therefore

$$\mathbf{T}(t) = \frac{\mathbf{r}'(t)}{\|\mathbf{r}'(t)\|} = \frac{1}{\sqrt{a^2+b^2}}(-a\sin t\,\mathbf{i} + a\cos t\,\mathbf{j} + b\,\mathbf{k}).$$

This is the unit tangent vector.

The principal normal vector is the unit vector in the direction of $\mathbf{T}'(t)$. Since

$$\frac{d}{dt}(-a\sin t\,\mathbf{i} + a\cos t\,\mathbf{j} + b\,\mathbf{k}) = -a\cos t\,\mathbf{i} - a\sin t\,\mathbf{j} \quad \text{and} \quad a > 0,$$

you can see that

$$\mathbf{N}(t) = -\cos t\,\mathbf{i} - \sin t\,\mathbf{j}.$$

The principal normal is horizontal and points directly toward the z-axis.

Now let's find an equation for the osculating plane p at an arbitrary point $P(a\cos t, a\sin t, bt)$. The cross product $\mathbf{T}(t)\times\mathbf{N}(t)$ is perpendicular to p. Therefore, as a normal for p, we can take any nonzero scalar multiple of $\mathbf{T}(t)\times\mathbf{N}(t)$. In particular, we can take

$$(a\sin t\,\mathbf{i} - a\cos t\,\mathbf{j} - b\,\mathbf{k}) \times (\cos t\,\mathbf{i} + \sin t\,\mathbf{j}).$$

As you can check, this simplifies to

$$b\sin t\,\mathbf{i} - b\cos t\,\mathbf{j} + a\,\mathbf{k}.$$

The equation for the osculating plane at the point $P(a\cos t, a\sin t, bt)$ thus takes the form

$$b\sin t(x - a\cos t) - b\cos t(y - a\sin t) + a(z - bt) = 0. \qquad (12.7.1)$$

This simplifies to

$$(b\sin t)x - (b\cos t)y + az = abt.$$

To visualize how this osculating plane changes from point to point, think of a playground spiral slide or the threaded surface on a bolt. ❑

Reversing the Orientation of a Curve

As noted at the beginning of this section, a parametrized curve is an *oriented* curve; it is not just a set of points, but rather a succession of points traversed in a certain order.

We make a distinction between the curve

$$\mathbf{r} = \mathbf{r}(t), \quad t \in [a, b]$$

and the curve

$$\mathbf{R}(u) = \mathbf{r}(a + b - u), \quad u \in [a, b].$$

Both vector functions trace out the same set of points (check that out), but the order has been reversed. Whereas the first curve starts at $\mathbf{r}(a)$ and ends at $\mathbf{r}(b)$, the second curve starts at $\mathbf{r}(b)$ and ends at $\mathbf{r}(a)$:

$$\mathbf{R}(a) = \mathbf{r}(a + b - a) = \mathbf{r}(b), \qquad \mathbf{R}(b) = \mathbf{r}(a + b - b) = \mathbf{r}(a).$$

The vector function

$$\mathbf{r}(t) = \cos t\,\mathbf{i} + \sin t\,\mathbf{j}, \quad t \in [0, 2\pi]$$

gives the unit circle traversed counterclockwise, while

$$\mathbf{R}(u) = \cos(2\pi - u)\,\mathbf{i} + \sin(2\pi - u)\,\mathbf{j}, \quad u \in [0, 2\pi]$$

gives the unit circle traversed clockwise.

What happens to the unit tangent $\mathbf{T}$, the principal normal $\mathbf{N}$, and the osculating plane when we reverse the orientation of a curve? As you are asked to show in Exercise 43, $\mathbf{T}$ is replaced by $-\mathbf{T}$, but $\mathbf{N}$ remains the same. The osculating plane also remains the same.

Other Changes of Parameter

Not all changes of parameter change the succession of points. Suppose that

$$\mathbf{r} = \mathbf{r}(t), \quad t \in I$$

is a differentiable curve, and let ϕ be a function that maps some interval J onto the interval I (domain J, range I). Now set

$$\mathbf{R}(u) = \mathbf{r}(\phi(u)) \quad \text{for all } u \in J.$$

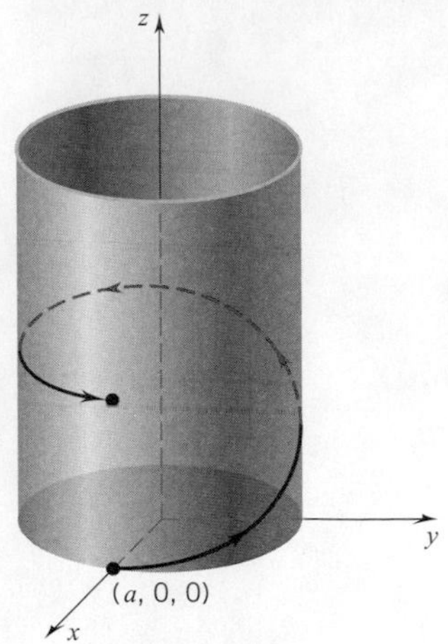

Figure 13.3.13

If $\phi'(u) > 0$ for all $u \in J$, then $\mathbf{r}$ and $\mathbf{R}$ are said to differ by an *orientation-preserving change of parameter*. In this case, $\mathbf{r}$ and $\mathbf{R}$ take on exactly the same values in exactly the same order. In other words, they produce exactly the same oriented curve (Exercise 44).

If, on the other hand, $\phi'(u) < 0$ for all $u \in J$. then the change in parameter is said to be *orientation-reversing*. In this case $\mathbf{r}$ and $\mathbf{R}$ still take on exactly the same values but in opposite order. The paths are the same, but they are traversed in opposite directions. You have already seen one example of this:

$$\mathbf{R}(u) = \mathbf{r}(a + b - u).$$

Example 5 The parametrization

$$\mathbf{r}(t) = a\cos t\,\mathbf{i} + a\sin t\,\mathbf{j} + bt\,\mathbf{k}. \quad t \in [0, 2\pi]$$

gives one "spiral" of the circular helix with the orientation indicated by the arrows (Figure 13.3.13). If we let $\phi(u) = \pi u$, $u \in [0, 2]$, then ϕ maps the interval $J = [0, 2]$ onto the interval $I = [0, 2\pi]$, and $\phi'(u) = \pi > 0$. Thus,

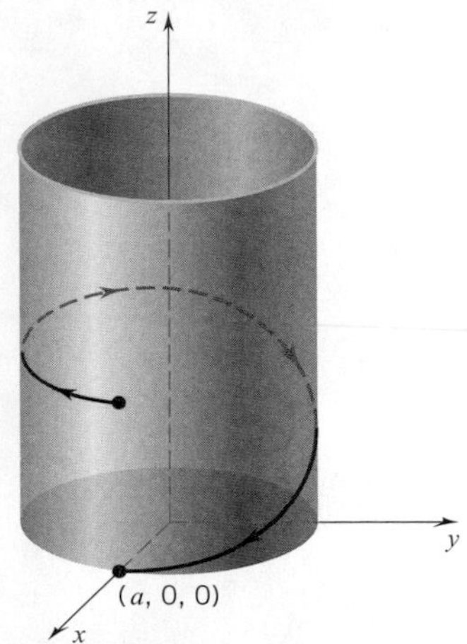

Figure 13.3.14

$$\mathbf{R}(u) = \mathbf{r}(\phi(u)) = a\cos(\pi u)\,\mathbf{i} + a\sin(\pi u)\,\mathbf{j} + b\pi u\,\mathbf{k}, \quad u \in [0, 2]$$

is precisely the same curve with the same orientation.

On the other hand, if we let $\psi(u) = (2 - u)\pi$, $u \in [0, 2]$, then ψ also maps the interval $J = [0, 2]$ onto the interval $I = [0, 2\pi]$, but $\psi'(u) = -\pi < 0$. Thus,

$$\mathbf{R}(u) = \mathbf{r}(\psi(u)) = a\cos(2 - u)\pi\,\mathbf{i} + a\sin(2 - u)\pi\,\mathbf{j} + b(2 - u)\pi\,\mathbf{k}$$

produces the same curve but with the opposite orientation. See Figure 13.3.14. □

EXERCISES 13.3

Find the tangent vector $\mathbf{r}'(t)$ at the indicated point and parametrize the tangent line.

1. $\mathbf{r}(t) = \cos\pi t\,\mathbf{i} + \sin\pi t\,\mathbf{j} + t\,\mathbf{k}$ at $t = 2$.
2. $\mathbf{r}(t) = e^t\,\mathbf{i} + e^{-t}\,\mathbf{j} - \ln t\,\mathbf{k}$ at $t = 1$.
3. $\mathbf{r}(t) = \mathbf{a} + t\,\mathbf{b} + t^2\,\mathbf{c}$ at $t = -1$.
4. $\mathbf{r}(t) = (t + 1)\,\mathbf{i} + (t^2 + 1)\,\mathbf{j} + (t^3 + 1)\,\mathbf{k}$ at $P(1, 1, 1)$.
5. $\mathbf{r}(t) = 2t^2\,\mathbf{i} + (1 - t)\,\mathbf{j} + (3 + 2t^2)\,\mathbf{k}$ at $P(2, 0, 5)$.
6. $\mathbf{r}(t) = 3t\,\mathbf{a} + \mathbf{b} - t^2\,\mathbf{c}$ at $t = 2$.
7. $\mathbf{r}(t) = 2\cos t\,\mathbf{i} + 3\sin t\,\mathbf{j} + t\,\mathbf{k}$; $t = \pi/4$.
8. $\mathbf{r}(t) = t\sin t\,\mathbf{i} + t\cos t\,\mathbf{j} + 2t\,\mathbf{k}$; $t = \pi/2$.
9. Show that $\mathbf{r}(t) = at\,\mathbf{i} + bt^2\,\mathbf{j}$ parametrizes a parabola. Find an equation in x and y for this parabola.
10. Show that $\mathbf{r}(t) = \frac{1}{2}a(e^{\omega t} + e^{-\omega t})\,\mathbf{i} + \frac{1}{2}a(e^{\omega t} - t^{-\omega t})\,\mathbf{j}$ parametrizes the right branch ($x > 0$) of the hyperbola $x^2 - y^2 = a^2$.
11. Find (a) the points on the curve $\mathbf{r}(t) = t\,\mathbf{i} + (1 + t^2)\,\mathbf{j}$ at which $\mathbf{r}(t)$ and $\mathbf{r}'(t)$ are perpendicular; (b) the points at which they have the same direction; (c) the points at which they have opposite directions.
12. Find the curve given that $\mathbf{r}'(t) = \alpha\mathbf{r}(t)$ for all real t and $\mathbf{r}(0) = \mathbf{i} + 2\mathbf{j} + 3\mathbf{k}$.
13. Suppose that $\mathbf{r}'(t)$ and $\mathbf{r}(t)$ are parallel for all t. Show that, if $\mathbf{r}'(t)$ is never $\mathbf{0}$, then the tangent line at each point passes through the origin.

C In Exercises 14–16, the given curves intersect at the indicated point. Find the angle of intersection. Express your answer in radians rounded to the nearest hundredth, and in degrees rounded to the nearest tenth.

14. $\mathbf{r}_1(t) = t\,\mathbf{i} + t^2\,\mathbf{j} + t^3\,\mathbf{k}$,
    $\mathbf{r}_2(u) = \sin 2u\,\mathbf{i} + u\cos u\,\mathbf{j} + u\,\mathbf{k}$; $P(0, 0, 0)$.
15. $\mathbf{r}_1(t) = (e^t - 1)\,\mathbf{i} + 2\sin t\,\mathbf{j} + \ln(t + 1)\,\mathbf{k}$,
    $\mathbf{r}_2(u) = (u + 1)\,\mathbf{i} + (u^2 - 1)\,\mathbf{j} + (u^3 + 1)\,\mathbf{k}$; $P(0, 0, 0)$.

16. $\mathbf{r}_1(t) = e^{-t}\,\mathbf{i} + \cos t\,\mathbf{j} + (t^2+4)\,\mathbf{k}$,
$\mathbf{r}_2(u) = (2+u)\,\mathbf{i} + u^4\,\mathbf{j} + 4u^2\,\mathbf{k}$; $P(1,1,4)$.

17. Find the point at which the curves

$$\mathbf{r}_1(t) = e^t\,\mathbf{i} + 2\sin(t+\tfrac{1}{2}\pi)\,\mathbf{j} + (t^2-2)\,\mathbf{k},$$
$$\mathbf{r}_2(u) = u\,\mathbf{i} + 2\,\mathbf{j} + (u^2-3)\,\mathbf{k}$$

intersect and find the angle of intersection.

18. Consider the vector function $\mathbf{f}(t) = t\,\mathbf{i} + f(t)\,\mathbf{j}$ formed from a differentiable real-valued function f. The vector function $\mathbf{f}$ parametrizes the graph of f.

(a) Parametrize the tangent line at $P(t_0, f(t_0))$.

(b) Show that the parametrization obtained in part (a) can be reduced to the usual equation for the tangent line:

$$y - f(t_0) = f'(t_0)(x - t_0) \quad \text{if} \quad f'(t_0) \neq 0;$$
$$y = f(t_0) \quad \text{if } f'(t_0) = 0.$$

19. Define a vector function $\mathbf{r}$ on the interval $[0, 2\pi]$ that satisfies the initial condition $\mathbf{r}(0) = a\mathbf{i}$ and, as t increases to 2π, traces out the ellipse $b^2x^2 + a^2y^2 = a^2b^2$:

(a) Once in a counterclockwise manner.

(b) Once in a clockwise manner.

(c) Twice in a counterclockwise manner.

(d) Three times in a clockwise manner.

20. Repeat Exercise 19 given that $\mathbf{r}(0) = b\,\mathbf{j}$.

In Exercise 21–26, sketch the plane curve determined by the given vector-valued function $\mathbf{r}$ and indicate the orientation. Find $\mathbf{r}'(t)$ and draw the position vector and the tangent vector for the indicated value of t, placing the tangent vector at the tip of the position vector.

21. $\mathbf{r}(t) = \frac{1}{4}t^4\,\mathbf{i} + t^2\,\mathbf{j}$; $t = 2$.

22. $\mathbf{r}(t) = 2t\,\mathbf{i} + (t^2+1)\,\mathbf{j}$; $t = 4$.

23. $\mathbf{r}(t) = e^{2t}\,\mathbf{i} + e^{-4t}\,\mathbf{j}$; $t = 0$.

24. $\mathbf{r}(t) = \sin t\,\mathbf{i} - 2\cos t\,\mathbf{j}$; $t = \pi/3$.

25. $\mathbf{r}(t) = 2\cos t\,\mathbf{i} + 3\sin t\,\mathbf{j}$; $t = \pi/6$.

26. $\mathbf{r}(t) = \sec t\,\mathbf{i} + \tan t\,\mathbf{j}$, $|t| < \pi/2$; $t = \pi/4$.

Find a vector parametrization for the curve.

27. $y^2 = x - 1$, $y \geq 1$.

28. $r = 1 - \cos\theta$, $\theta \in [0, 2\pi]$. (Polar coordinates)

29. $r = \sin 3\theta$, $\theta \in [0, \pi]$. (Polar coordinates)

30. $y^4 = x^3$, $y \leq 0$.

31. Find an equation in x and y for the curve $\mathbf{r}(t) = t^3\,\mathbf{i} + t^2\,\mathbf{j}$. Draw the curve. Does the curve have a tangent vector at the origin? If so, what is the unit tangent vector?

32. (a) Show that the curve

$$\mathbf{r}(t) = (t^2 - t + 1)\,\mathbf{i} + (t^3 - t + 2)\,\mathbf{j} + (\sin \pi t)\,\mathbf{k}$$

intersects itself at $P(1,2,0)$ by finding numbers $t_1 < t_2$ for which P is the tip of both $\mathbf{r}(t_1)$ and $\mathbf{r}(t_2)$.

(b) Find the unit tangents at $P(1,2,0)$, first taking $t = t_1$, then taking $t = t_2$.

33. Find the point(s) at which the twisted cubic

$$\mathbf{r}(t) = t\,\mathbf{i} + t^2\,\mathbf{j} + t^3\,\mathbf{k}$$

intersects the plane $4x + 2y + z = 24$. What is the angle of intersection between the curve and the normal to the plane?

34. (a) Find the unit tangent and the principal normal at an arbitrary point of the ellipse

$$\mathbf{r}(t) = a\cos t\,\mathbf{i} + b\sin t\,\mathbf{j}.$$

(b) Write vector equations for the tangent line and the normal line at the tip of $\mathbf{r}(\frac{1}{4}\pi)$.

Find the unit tangent vector, the principal normal vector, and an equation in x, y, z for the osculating plane at the point on the curve corresponding to the indicated value of t.

35. $\mathbf{r}(t) = \mathbf{i} + 2t\,\mathbf{j} + t^2\,\mathbf{k}$; $t = 1$.

36. $\mathbf{r}(t) = t\,\mathbf{i} + t^2\,\mathbf{j} + 2t^2\,\mathbf{k}$; $t = 1$.

37. $\mathbf{r}(t) = \cos 2t\,\mathbf{i} + \sin 2t\,\mathbf{j} + t\,\mathbf{k}$ at $t = \frac{1}{4}\pi$.

38. $\mathbf{r}(t) = t\,\mathbf{i} + 2t\,\mathbf{j} + t^2\,\mathbf{k}$ at $t = 2$.

39. $\mathbf{r}(t) = t\,\mathbf{i} + t^2\,\mathbf{j} + t^3\,\mathbf{k}$ at $t = 1$.

40. $\mathbf{r}(t) = \cos 3t\,\mathbf{i} + t\,\mathbf{j} - \sin 3t\,\mathbf{k}$ at $t = \frac{1}{3}\pi$.

41. $\mathbf{r}(t) = e^t \sin t\,\mathbf{i} + e^t \cos t\,\mathbf{j} + e^t\,\mathbf{k}$; $t = 0$.

42. $\mathbf{r}(t) = (\cos t + t\sin t)\,\mathbf{i} + (\sin t - t\cos t)\,\mathbf{j} + 2k$; $t = \frac{1}{4}\pi$.

43. Let $\mathbf{r} = \mathbf{r}(t)$, $t \in [a,b]$ and set

$$\mathbf{R}(u) = \mathbf{r}(a + b - u), \quad u \in [a,b].$$

Show that this change of parameter changes the sign of the unit tangent but does not alter the principal normal.

HINT: Let P be the tip of $\mathbf{R}(u) = \mathbf{r}(a+b-u)$. At that point, $\mathbf{R}$ produces a unit tangent $\mathbf{T}_1(u)$ and a principal normal $\mathbf{N}_1(u)$. At that same point, $\mathbf{r}$ produces a unit tangent $\mathbf{T}(a + b - u)$ and a principal normal $\mathbf{N}(a+b-u)$.

44. Show that two vector functions that differ by a orientation-preserving change of parameter take on exactly the same values in exactly the same order. That is, set

$$\mathbf{r} = \mathbf{r}(t), \quad t \in I.$$

Assume that ϕ maps an interval J onto the interval I and that $\phi'(u) > 0$ for all $u \in J$. Set

$$\mathbf{R}(u) = \mathbf{r}(\phi(u)),$$

and show that $\mathbf{R}$ and $\mathbf{r}$ take on exactly the same values in exactly the same order.

45. Show that the unit tangent vector, the principal normal vector, and the osculating plane are left invariant (left unchanged) by every orientation-preserving change of parameter.

In Exercises 46 and 47, let **r** be the vector-valued function defined by

$$\mathbf{r}(t) = 2\cos t\,\mathbf{i} + 2\sin t\,\mathbf{j} + 4t\,\mathbf{k} \quad \text{for} \quad 0 \le t \le 2\pi.$$

The curve generated is one turn along a circular helix, starting at the point (2, 0, 0) and ending at the point $(2, 0, 8\pi)$.

46. Let $\varphi(u) = u^2$ for $0 \le u \le \sqrt{2\pi}$, and let

$$\mathbf{R}(u) = \mathbf{r}[\varphi(u)] = 2\cos u^2\mathbf{i} + 2\sin u^2\,\mathbf{j} + 4u^2\,\mathbf{k}.$$

(a) Show that φ determines a orientation-preserving change of parameter on $[0, \sqrt{2\pi}]$.

(b) Show that the unit tangent and principal normal vectors for **r** at the point $t = \frac{1}{4}\pi$ are the same as the unit tangent and principal normal vectors for **R** at $u = \frac{1}{2}\sqrt{\pi}$.

47. Let $\psi(v) = 2\pi - v^2$ for $0 \le v \le \sqrt{2\pi}$, and let

$$\mathbf{R}(v) = \mathbf{r}[\psi(v)] = 2\cos(2\pi - v^2)\,\mathbf{i} + 2\sin(2\pi - v^2)\,\mathbf{j} + 4(2\pi - v^2)\,\mathbf{k}.$$

(a) Show that ψ determines a orientation-reversing change of parameter on $[0, \sqrt{2\pi}]$.

(b) Show that the principal normal vector for **r** at $t = \pi/4$ is the same as the principal normal for **R** at $v = \frac{1}{2}\sqrt{7\pi}$, and show that the unit tangent vector for **r** at $\pi/4$ is the negative of the unit tangent vector for **R** at $v = \frac{1}{2}\sqrt{7\pi}$.

48. Let $\mathbf{r}(t) = \sqrt{2}\cos t\,\mathbf{i} + \sqrt{2}\sin t\,\mathbf{j} + t\,\mathbf{k}, \quad 0 \le t \le 2\pi$.

(a) Find scalar parametric equations for the tangent line to the curve at the point $(1, 1, \pi/4)$.

(b) Use a CAS to draw the curve and the tangent line together. Experiment with the t-interval to find a good illustration of the curve and the tangent line.

(c) Are there points on the curve where the tangent line is parallel to the xy-plane? If so, find them.

49. Let $\mathbf{r}(t) = \sqrt{2}\cos t\,\mathbf{i} + \sqrt{2}\sin t\,\mathbf{j} + \sin 5t\,\mathbf{k}, \quad 0 \le t \le 2\pi$.

(a) Find scalar parametric equations for the tangent line to the curve at the point $(1, 1, -\sqrt{2}/2)$.

(b) Use a CAS to draw the curve and the tangent line together. Experiment with the t-interval to find a good illustration of the curve and the tangent line.

(c) Are there points on the curve where the tangent line is parallel to the xy-plane? If so, find them.

50. Let $\mathbf{r}(t) = \sqrt{2}\cos t\,\mathbf{i} + \sqrt{2}\sin t\,\mathbf{j} + \ln t\,\mathbf{k}, \quad 0 \le t \le 2\pi$.

(a) Find the scalar parametric equations for the tangent line to the curve at the point $(1, 1, \ln(\pi/4))$.

(b) Use a CAS to draw the curve and the tangent line together. Experiment with the t-interval to find a good illustration of the curve and the tangent line.

■ 13.4 ARC LENGTH

In Section 9.8 we considered arc length in an intuitive manner and decided that the length of the path C traced out by a pair of continuously differentiable functions

$$x = x(t), \qquad y = y(t), \qquad t \in [a, b]$$

is given by the formula

$$L(C) = \int_a^b \sqrt{[x'(t)]^2 + [y'(t)]^2}\,dt.$$

Applied to a path C in space traced out by

$$x = x(t), \qquad y = y(t), \qquad z = z(t), \qquad t \in [a, b]$$

the formula becomes

$$L(C) = \int_a^b \sqrt{[x'(t)]^2 + [y'(t)]^2 + [z'(t)]^2}\,dt.$$

In vector notation, both formulas can be written

$$L(C) = \int_a^b \|\mathbf{r}'(t)\|\,dt.$$

We will prove the result in this form, but first we give a precise definition of arc length.

In Figure 13.4.1 we have sketched the path C traced out by a continuously differentiable vector function

$$\mathbf{r} = \mathbf{r}(t), \quad t \in [a, b].$$

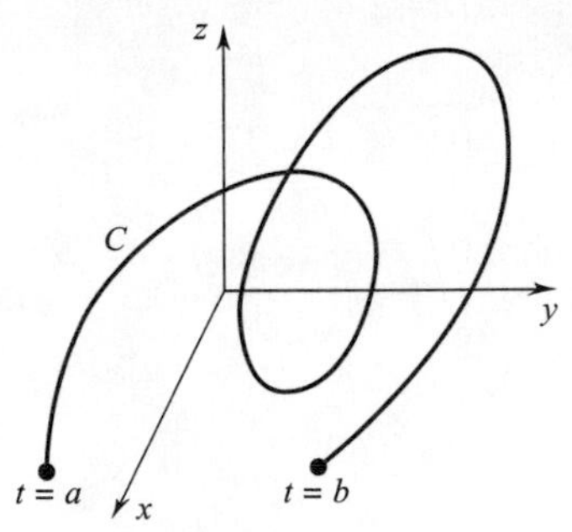

Figure 13.4.1

To decide what should be meant by the length of C, we approximate C by the union of a finite number of line segments.

Choosing a finite number of points in $[a, b]$,

$$a = t_0 < t_1 < \cdots < t_{i-1} < t_i < \cdots < t_{n-1} < t_n = b,$$

we obtain a finite number of points $P_0, P_1, \ldots, P_{i-1}, P_i, \ldots, P_{n-1}, P_n$, on C, where for each $k, 0 \le k \le n$, P_k denotes the point $P(x(t_k), y(t_k), z(t_k))$. Join these points consecutively by line segments and call the resulting path

$$\gamma = \overline{P_0P_1} \cup \cdots \cup \overline{P_{i-1}P_i} \cup \cdots \cup \overline{P_{n-1}P_n},$$

a *polygonal path* inscribed in C (Figure 13.4.2).

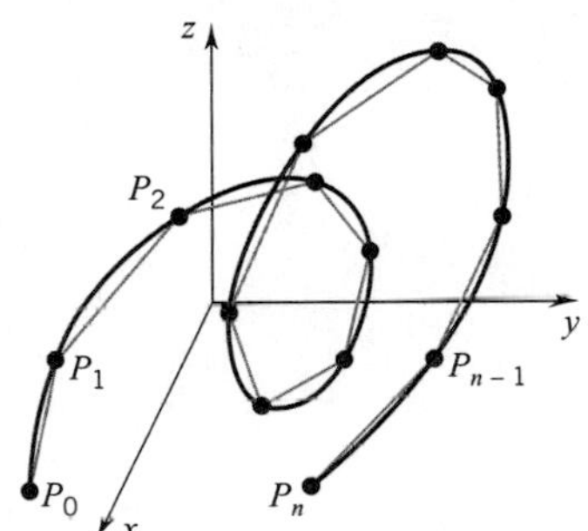

Figure 13.4.2

The length of such a polygonal path is the sum of the distances between consecutive vertices:

$$L(\gamma) = d(P_0, P_1) + \cdots + d(P_{i-1}, P_i) + \cdots + d(P_{n-1}, P_n).$$

The path γ serves as an approximation to the path C, and better approximations can be obtained by adding more vertices to γ. We now ask ourselves exactly what should we require of the number that we shall call the length of C. Certainly we require that

$$L(\gamma) \le L(C) \quad \text{for each } \gamma \text{ inscribed in } C.$$

But that is not enough. There is another requirement that seems reasonable. If we can choose γ to approximate C as closely as we wish, then we should be able to choose γ so that $L(\gamma)$ approximates the length of C as closely as we wish; namely, for each positive number ϵ there should exist a polygonal path γ such that

$$L(C) - \epsilon < L(\gamma) \le L(C).$$

In Section 10.1, we introduced the concept of least upper bound of a set of real numbers. Theorem 10.1.2 tells us that we can achieve the result we want by defining the length of C as the least upper bound of the set of all $L(\gamma)$. This is in fact what we do.

DEFINITION 13.4.1 ARC LENGTH

$$L(C) = \begin{cases} \text{the least upper bound of the set of all} \\ \text{lengths of polygonal paths inscribed in } C. \end{cases}$$

We are now ready to establish the arc length formula.

THEOREM 13.4.2 ARC LENGTH FORMULA

Let C be the path traced out by a continuously differentiable vector function

$$\mathbf{r} = \mathbf{r}(t), \qquad t \in [a.b].$$

The length of C is given by the formula

$$L(C) = \int_a^b \|\mathbf{r}'(t)\|\, dt.$$

PROOF First we show that

$$L(C) \leq \int_a^b \|\mathbf{r}'(t)\|\, dt.$$

To do this, we begin with an arbitrary partition P of $[a, b]$:

$$P = \{a = t_0, \ldots, t_{i-1}, t_i, \ldots, t_n = b\}.$$

Such a partition gives rise to a finite number of points of C:

$$\mathbf{r}(a) = \mathbf{r}(t_0), \ldots, \mathbf{r}(t_{i-1}), \mathbf{r}(t_i), \ldots, \mathbf{r}(t_n) = \mathbf{r}(b)$$

and thus to an inscribed polygonal path of total length

$$L_P = \sum_{i=1}^{n} \|\mathbf{r}(t_i) - \mathbf{r}(t_{i-1})\|. \qquad \text{(Figure 13.4.3)}$$

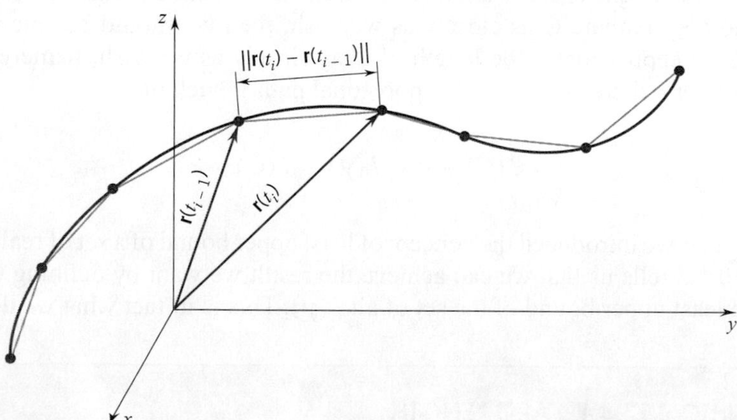

Figure 13.4.3

For each i,

$$\mathbf{r}(t_i) - \mathbf{r}(t_{i-1}) = \int_{t_{i-1}}^{t_i} \mathbf{r}'(t)\, dt.$$

This gives

$$\|\mathbf{r}(t_i) - \mathbf{r}(t_{i-1})\| = \left\| \int_{t_{i-1}}^{t_i} \mathbf{r}'(t)\,dt \right\| \overset{\text{by (13.1.10)}}{\leq} \int_{t_{i-1}}^{t_i} \|\mathbf{r}'(t)\|\,dt$$

and thus

$$L_P = \sum_{i=1}^{n} \|\mathbf{r}(t_i) - \mathbf{r}(t_{i-1})\| \leq \sum_{i=1}^{n} \int_{t_{i-1}}^{t_i} \|\mathbf{r}'(t)\|\,dt = \int_a^b \|\mathbf{r}'(t)\|\,dt.$$

Since the partition P is arbitrary, we know that the inequality

$$L_P \leq \int_a^b \|\mathbf{r}'(t)\|\,dt$$

must hold for all the L_P. This makes the integral on the right an upper bound for all the L_P. Since $L(C)$ is the *least* upper bound of all the L_P, we can conclude right now that

$$L(C) \leq \int_a^b \|\mathbf{r}'(t)\|\,dt.$$

The next step is to show that this inequality is actually an equation. To do this we need to know that arc length, *as we have defined it*, is additive. That is, with P, Q, and R as in Figure 13.4.4, we need to know that the arc length from P to Q plus the arc length from Q to R equals the arc length from P to R. It is clear arc length should have this property. We must prove that it does. A proof has been placed in a supplement to this section. For the moment we shall assume that arc length is additive and continue with the proof of the arc length formula

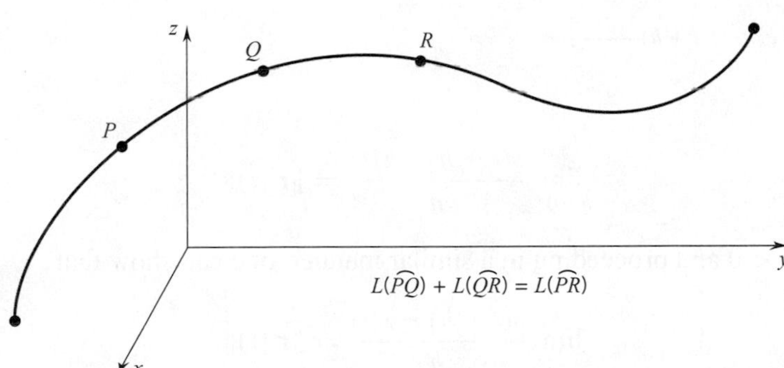

Figure 13.4.4

In Figure 13.4.5, we display the initial vector $\mathbf{r}(a)$, a general radius vector $\mathbf{r}(t)$, and a nearby vector $\mathbf{r}(t+h)$. Set

$$s(t) = \text{ length of the path from } \mathbf{r}(a) \text{ to } \mathbf{r}(t).$$

Then

$$s(a) = 0 \quad \text{and} \quad s(t+h) = \text{ length of the path from } \mathbf{r}(a) \text{ to } \mathbf{r}(t+h).$$

By the additivity of arc length (remember, we are assuming this for the moment),

$$s(t+h) - s(t) = \text{ length of the curve from } \mathbf{r}(t) \text{ to } \mathbf{r}(t+h).$$

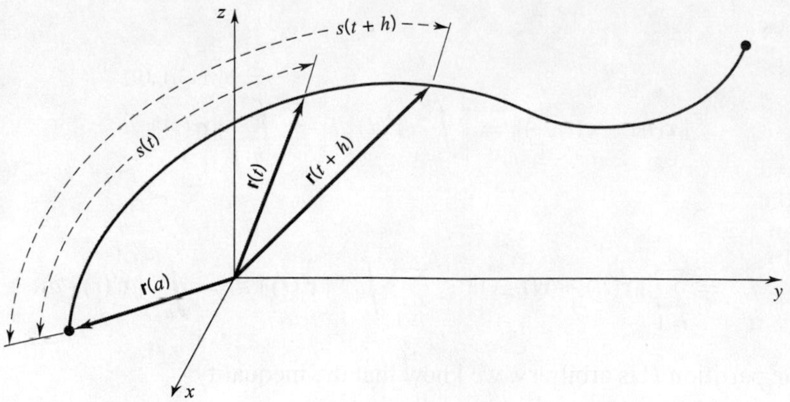

Figure 13.4.5

From that we have shown already, you can see that

$$\|\mathbf{r}(t+h) - \mathbf{r}(t)\| \leq s(t+h) - s(t) \leq \int_t^{t+h} \|\mathbf{r}'(u)\|\, du.$$

Dividing this inequality by h (which we are taking as positive), we get

$$\left\|\frac{\mathbf{r}(t+h) - \mathbf{r}(t)}{h}\right\| \leq \frac{s(t+h) - s(t)}{h} \leq \frac{1}{h}\int_t^{t+h} \|\mathbf{r}'(u)\|\, du.$$

As $h \to 0^+$, the left-hand side tends to $\|\mathbf{r}'(t)\|$ and, by the first mean-value theorem for integrals (Theorem 5.9.1), so does the right-hand side:

$$\frac{1}{h}\int_t^{t+h} \|\mathbf{r}'(u)\|\, du = \frac{1}{h}\|\mathbf{r}'(c_h)\|(t+h-t) = \|\mathbf{r}'(c_h)\| \to \|\mathbf{r}'(t)\|.$$

$c_h \in (t, t+h)$ ⟶ ↑

Therefore

$$\lim_{h\to 0^+} \frac{s(t+h) - s(t)}{h} = \|\mathbf{r}'(t)\|.$$

By taking $h < 0$ and proceeding in a similar manner, one can show that

$$\lim_{h\to 0^-} \frac{s(t+h) - s(t)}{h} = \|\mathbf{r}'(t)\|.$$

Therefore, we can conclude that

$$\lim_{h\to 0} \frac{s(t+h) - s(t)}{h} = \|\mathbf{r}'(t)\|$$

and so $s'(t) = \|\mathbf{r}'(t)\|$. Integrating this equation from a to t, we get

$$s(t) - s(a) = \int_a^t s'(u)\, du = \int_a^t \|\mathbf{r}'(u)\|\, du.$$

Since $s(a) = 0$, it follows that

$$s(t) = \int_a^t \|\mathbf{r}'(u)\|\, du.$$

The total length of C is therefore

$$s(b) = \int_a^b \|\mathbf{r}'(t)\|\, dt. \quad \square$$

For convenience, and to follow custom, we shall speak of the length of a parametrized curve. You can take this to mean the length of the path traced out by the parametrizing vector function.

Example 1 Find the length of the curve

$$\mathbf{r}(t) = 2t^{3/2}\,\mathbf{i} + 4t\,\mathbf{j} \quad \text{from} \quad t = 0 \text{ to } t = 1.$$

SOLUTION

$$\mathbf{r}'(t) = 3t^{1/2}\,\mathbf{i} + 4\,\mathbf{j}.$$

$$\|\mathbf{r}'(t)\| = \sqrt{(3t^{1/2})^2 + 4^2} = \sqrt{9t + 16}.$$

$$L(C) = \int_0^1 \|\mathbf{r}'(t)\|\, dt = \int_0^1 \sqrt{9t + 16}\, dt$$

$$= \left[\frac{1}{9}\left(\frac{2}{3}\right)(9t + 16)^{3/2}\right]_0^1 = \frac{250}{27} - \frac{128}{27} = \frac{122}{27}. \quad \square$$

Example 2 Find the length of the curve

$$\mathbf{r}(t) = 2\cos t\,\mathbf{i} + 2\sin t\,\mathbf{j} + t^2\,\mathbf{k} \qquad \text{from } t = 0 \text{ to } t = \pi/2$$

and compare it to the straight-line distance between the endpoints of the curve. (See Figure 13.4.6.)

SOLUTION

$$\mathbf{r}'(t) = -2\sin t\,\mathbf{i} + 2\cos t\,\mathbf{j} + 2t\,\mathbf{k}.$$

$$\|\mathbf{r}'(t)\| = \sqrt{4\sin^2 t + 4\cos^2 t + 4t^2} = 2\sqrt{\sin^2 t + \cos^2 t + t^2} = 2\sqrt{1 + t^2}.$$

$$L(C) = \int_0^{\pi/2} \|\mathbf{r}'(t)\|\, dt = \int_0^{\pi/2} 2\sqrt{1 + t^2}\, dt$$

$$= \left[t\sqrt{1 + t^2} + \ln\left(t + \sqrt{1 + t^2}\right)\right]_0^{\pi/2} \qquad \text{(Formula 78)}$$

$$= \frac{\pi}{2}\sqrt{1 + \frac{\pi^2}{4}} + \ln\left[\frac{\pi}{2} + \sqrt{1 + \frac{\pi^2}{4}}\right] \cong 4.158.$$

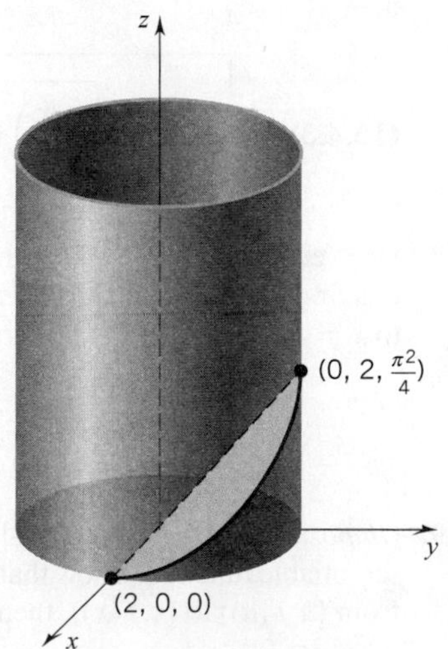

Figure 13.4.6

The curve begins at $\mathbf{r}(0) = 2\,\mathbf{i}$ and ends at $\mathbf{r}(\pi/2) = 2\,\mathbf{j} + (\pi^2/4)\,\mathbf{k}$. The straight-line distance between these two points is

$$\|\mathbf{r}(\pi/2) - \mathbf{r}(0)\| = \sqrt{2^2 + 2^2 + \frac{\pi^4}{16}} \cong 3.753.$$

The curve is about 11 percent longer than the straight-line distance between the endpoints of the curve. $\square$

EXERCISES 13.4

Find the length of the curve.

1. $\mathbf{r}(t) = t\,\mathbf{i} + \frac{2}{3}t^{3/2}\,\mathbf{j}$ from $t = 0$ to $t = 8$.

2. $\mathbf{r}(t) = (\frac{1}{3}t^3 - t)\,\mathbf{i} + t^2\,\mathbf{j}$ from $t = 0$ to $t = 2$.

3. $\mathbf{r}(t) = a\cos t\,\mathbf{i} + a\sin t\,\mathbf{j} + bt\,\mathbf{k}$ from $t = 0$ to $t = 2\pi$.

4. $\mathbf{r}(t) = t\,\mathbf{i} + \frac{2}{3}\sqrt{2}\,t^{3/2}\,\mathbf{j} + \frac{1}{2}t^2\,\mathbf{k}$ from $t = 0$ to $t = 2$.

5. $\mathbf{r}(t) = t\,\mathbf{i} + \ln(\sec t)\,\mathbf{j} + 3\,\mathbf{k}$ from $t = 0$ to $t = \frac{1}{4}\pi$.

6. $\mathbf{r}(t) = \tan^{-1} t\,\mathbf{i} + \frac{1}{2}\ln(1 + t^2)\,\mathbf{j}$ from $t = 0$ to $t = 1$.

7. $\mathbf{r}(t) = t^3\,\mathbf{i} + t^2\,\mathbf{j}$ from $t = 0$ to $t = 1$.

8. $\mathbf{r}(t) = t\,\mathbf{i} + \mathbf{j} + (\frac{1}{6}t^3 + \frac{1}{2}t^{-1})\,\mathbf{k}$ from $t = 1$ to $t = 3$.

9. $\mathbf{r}(t) = e^t[\cos t\,\mathbf{i} + \sin t\,\mathbf{j}]$ from $t = 0$ to $t = \pi$.

10. $\mathbf{r}(t) = 3t\cos t\,\mathbf{i} + 3t\sin t\,\mathbf{j} + 4t\,\mathbf{k}$ from $t = 0$ to $t = 4$.

11. $\mathbf{r}(t) = 2t\,\mathbf{i} + (t^2 - 2)\,\mathbf{j} + (1 - t^2)\,\mathbf{k}$ from $t = 0$ to $t = 2$.

12. $\mathbf{r}(t) = t^2\,\mathbf{i} + (t^2 - 2)\,\mathbf{j} + (1 - t^2)\,\mathbf{k}$ from $t = 0$ to $t = 2$.

13. $\mathbf{r}(t) = (\ln t)\,\mathbf{i} + 2t\,\mathbf{j} + t^2\,\mathbf{k}$ from $t = 1$ to $t = e$.

14. $\mathbf{r}(t) = (t\sin t + \cos t)\,\mathbf{i} + (t\cos t - \sin t)\,\mathbf{j} + 2\,\mathbf{k}$ from $t = 0$ to $t = 2$.

15. $\mathbf{r}(t) = (\cos t + t\sin t)\,\mathbf{i} + (\sin t - t\cos t)\,\mathbf{j} + \frac{1}{2}\sqrt{3}t^2\,\mathbf{k}$ from $t = 0$ to $t = 2\pi$.

16. $\mathbf{r}(t) = \frac{2}{3}(1 + t)^{3/2}\,\mathbf{i} + \frac{2}{3}(1 - t)^{3/2}\,\mathbf{j} + \sqrt{2}t\,\mathbf{k}$ from $t = -\frac{1}{2}$ to $t = \frac{1}{2}$.

17. (*Important*) Let $\mathbf{r}(t) = x(t)\,\mathbf{i} + y(t)\,\mathbf{j} + z(t)\,\mathbf{k}, t \in [a, b]$ be a continuously differentiable curve. Show that, if s is the length of the curve from the tip of $\mathbf{r}(a)$ to the tip of $\mathbf{r}(t)$, then

(13.4.3) $$\frac{ds}{dt} = \sqrt{\left(\frac{dx}{dt}\right)^2 + \left(\frac{dy}{dt}\right)^2 + \left(\frac{dz}{dt}\right)^2}.$$

18. Use vector methods to show that, if $y = f(x)$ has a continuous first derivative, then the length of the graph from $x = a$ to $x = b$ is given by the integral

$$\int_a^b \sqrt{1 + [f'(x)]^2}\,dx.$$

19. (*Important*) Let $y = f(x)$, $x \in [a, b]$, be a continuously differentiable function. Show that, if s is the length of the graph from $(a, f(a))$ to $(x, f(x))$, then

(13.4.4) $$\frac{ds}{dx} = \sqrt{1 + \left(\frac{dy}{dx}\right)^2}.$$

20. Let C_1 be the curve

$$\mathbf{r}(t) = (t - \ln t)\,\mathbf{i} + (t + \ln t)\,\mathbf{j}. \quad 1 \le t \le e$$

and let C_2 be the graph of

$$y = e^x, \quad 0 \le x \le 1.$$

Find a relation between the length of C_1, and the length of C_2.

21. Show that the length of a continuously differentiable curve is left invariant (left unchanged) by an orientation-preserving (or an orientation-reversing) change of parameter. That is, set

$$\mathbf{r} = \mathbf{r}(t), \quad t \in [a, b].$$

Assume that ϕ maps $[c, d]$ onto $[a, b]$ and that ϕ' is positive (or negative) and continuous on $[a, b]$. Set $\mathbf{R}(u) = r(\phi(u))$ and show that the length of the curve as computed from $\mathbf{R}$ is the length of the curve as computed from $\mathbf{r}$.

22. Let $\mathbf{r}(t) = x(t)\,\mathbf{i} + y(t)\,\mathbf{j} + z(t)\,\mathbf{k}$ be a differentiable vector-valued function such that $\mathbf{r}'(t) \ne 0$ for all $t \ge 0$.

(a) Show that the arc length function s defined by

$$s(t) = \int_0^t \sqrt{\left(\frac{dx}{dt}\right)^2 + \left(\frac{dy}{dt}\right)^2 + \left(\frac{dz}{dt}\right)^2}\,dt, \; t \ge 0$$

has an inverse, $t = \varphi(s)$.

(b) Let $\mathbf{R}(s) = \mathbf{r}[\varphi(s)]$. Show that $\|\mathbf{R}'(s)\| = 1$ for all s.

23. Consider the circular helix

$$\mathbf{r}(t) = 3\cos t\,\mathbf{i} + 3\sin t\,\mathbf{j} + 4t\,\mathbf{k} \text{ for } t \ge 0.$$

(a) Determine the arc length s as a function of t by evaluating the integral

$$s = \int_0^t \sqrt{\left(\frac{dx}{dt}\right)^2 + \left(\frac{dy}{dt}\right)^2 + \left(\frac{dz}{dt}\right)^2}\,dt.$$

(b) Use the relation found in (a) to express t as a function of $s, t = \varphi(s)$, and set

$$\mathbf{R}(s) = \mathbf{r}[\varphi(s)] = 3\cos\varphi(s)\,\mathbf{i} + 3\sin\varphi(s)\,\mathbf{j} + 4\varphi(s)\,\mathbf{k}.$$

(c) Find the coordinates of the point Q on the helix such that the arc length from $P(3, 0, 0)$ to Q is 5π.

(d) Show that $\|\mathbf{R}'(s)\| = 1$ for all s.

24. Repeat Exercise 23 (a), (b), (d) for the vector-valued function

$$\mathbf{r}(t) = (\sin t - t\cos t)\,\mathbf{i} + (\cos t + t\sin t)\,\mathbf{j} + \frac{1}{2}t^2\,\mathbf{k}$$

for $t \ge 0$.

C In Exercise 25–28, use a CAS to estimate the length of the given curve.

25. $\mathbf{r}(t) = \frac{2}{5}t^{5/2}\,\mathbf{j} + t\,\mathbf{k}$ from $t = 0$ to $t = \frac{1}{2}$.

26. $\mathbf{r}(t) = t\,\mathbf{i} + \frac{1}{3}t^3\,\mathbf{j}$ from $t = 0$ to $t = 2$.

27. $\mathbf{r}(t) = 3\cos t\,\mathbf{i} + 4\sin t\,\mathbf{j} + 2\,\mathbf{k}$ from $t = 0$ to $t = 2\pi$.

28. $\mathbf{r}(t) = t\,\mathbf{i} + t^2\,\mathbf{j} + (\ln t)\,\mathbf{k}$ from $t = 0$ to $t = 4$.

29. Let $\mathbf{r}(t) = \cos t\,\mathbf{i} + \sin t\,\mathbf{j} + \sin 4t\,\mathbf{k}$, $0 \le t \le 2\pi$.
(a) Use a graphing utility to draw the curve.
(b) Use a CAS to estimate the length of the curve.

30. Let $\mathbf{r}(t) = \cos t\,\mathbf{i} + \ln(1+t)\,\mathbf{j} + \sin t\,\mathbf{k}$, $0 \le t \le 2\pi$.
(a) Use a graphing utility to draw the curve.
(b) Use a CAS to estimate the length of the curve.

*SUPPLEMENT TO SECTION 13.4

The Additivity of Arc Length We wish to show that with P, Q, R as in Figure 14.4.3,

$$L(\widehat{PQ}) + L(\widehat{QR}) = L(\widehat{PR}).$$

Let γ_1 be an arbitrary polygonal path inscribed in $\widehat{PQ}$ and γ_2 an arbitrary polygonal path inscribed in $\widehat{QR}$. Then $\gamma_1 \cup \gamma_2$ is a polygonal path inscribed in $\widehat{PR}$. Since

$$L(\gamma_1) + L(\gamma_2) = L(\gamma_1 \cup \gamma_2) \quad \text{and} \quad L(\gamma_1 \cup \gamma_2) \le L(\widehat{PR}),$$

we have

$$L(\gamma_1) + L(\gamma_2) \le L(\widehat{PR}) \quad \text{and thus} \quad L(\gamma_1) \le L(\widehat{PR}) - L(\gamma_2).$$

Since γ_1 is arbitrary, we can conclude that $L(\widehat{PR}) - L(\gamma_2)$ is an upper bound for the set of all lengths of polygonal paths inscribed in $\widehat{PQ}$. Since $L(\widehat{PQ})$is the *least* upper bound of this set, we have

$$L(\widehat{PQ}) \le L(\widehat{PR}) - L(\gamma_2).$$

It follows that

$$L(\gamma_2) \le L(\widehat{PR}) - L(\widehat{PQ}).$$

Arguing as we did with γ_1, we can conclude that

$$L(\widehat{QR}) \le L(\widehat{PR}) - L(\widehat{PQ}).$$

This gives

$$L(\widehat{PQ}) + L(\widehat{QR}) \le L(\widehat{PR}).$$

We now set out to prove that $L(\widehat{PR}) \le L(\widehat{PQ}) + L(\widehat{QR})$. To do this, we need only take $\gamma = \overline{T_0T_1} \cup \cdots \cup \overline{T_{n-1}T_n}$ as an arbitrary polygonal path inscribed in $\widehat{PR}$ and show that

$$L(\gamma) \le L(\widehat{PQ}) + L(\widehat{QR}).$$

If Q is one of the T_i, say $Q = T_k$, then

$$\gamma_1 = \overline{T_0T_1} \cup \cdots \cup \overline{T_{k-1}T_k} \quad \text{is inscribed in } \widehat{PQ}$$

and

$$\gamma_2 = \overline{T_kT_{k+1}} \cup \cdots \cup \overline{T_{n-1}T_n} \quad \text{is inscribed in } \widehat{QR}.$$

Moreover, $L(\gamma) = L(\gamma_1) + L(\gamma_2)$, so that

$$L(\gamma) \le L(\widehat{PQ}) + L(\widehat{QR}).$$

If Q is none of the T_i, then Q lies between two consecutive points T_k and T_{k+1}. Set

$$\gamma' = \overline{T_0T_1} \cup \cdots \cup \overline{T_kQ} \cup \overline{QT_{k+1}} \cup \cdots \cup \overline{T_{n-1}T_n}.$$

Since $$d(T_k, T_{k+1}) \le d(T_k, Q) + d(Q, T_{k+1}),$$

we have $$L(\gamma) \le L(\gamma').$$

Proceed as before and you will see that

$$L(\gamma') \le L(\widehat{PQ}) + L(\widehat{QR}),$$

and once again

$$L(\gamma) \le L(\widehat{PQ}) + L(\widehat{QR}). \quad \square$$

■ 13.5 CURVILINEAR MOTION; CURVATURE

Curvilinear Motion from a Vector Viewpoint

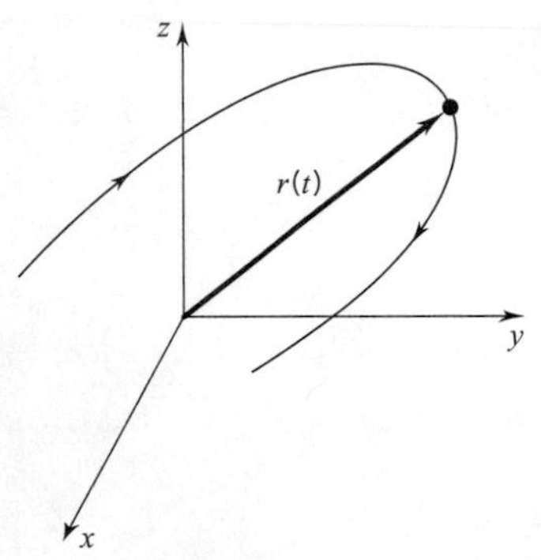

Figure 13.5.1

Here we use the theory we have developed for vector-valued functions to study the motion of an object moving in space. We can describe the position of a moving object at time t by a radius vector $\mathbf{r}(t)$. As t ranges over a time interval I, the object traces out some path

$$C : \mathbf{r}(t) = x(t)\,\mathbf{i} + y(t)\,\mathbf{j} + z(t)\,\mathbf{k}, \quad t \in I \qquad \text{(Figure 13.5.1)}$$

If $\mathbf{r}$ is twice differentiable, we can form $\mathbf{r}'(t)$ and $\mathbf{r}''(t)$. In this context these vectors have special names and special significance: $\mathbf{r}'(t)$ is called the *velocity* of the object at time t, and $\mathbf{r}''(t)$ is called the *acceleration*. In symbols, we have

(13.5.1) $$\mathbf{r}'(t) = \mathbf{v}(t) \quad \text{and} \quad \mathbf{r}''(t) = \mathbf{v}'(t) = \mathbf{a}(t).$$

There should be nothing surprising about this. As before, velocity is the time rate of change of position and acceleration the time rate of change of velocity.

Since $\mathbf{v}(t) = \mathbf{r}'(t)$, the velocity vector, when not $\mathbf{0}$, is tangent to the path of the motion at the tip of $\mathbf{r}(t)$. (See Section 13.3.) The direction of the velocity vector at time t thus gives the direction of the motion at time t (Figure 13.5.2).

The magnitude of the velocity vector is called the *speed* of the object:

(13.5.2) $$\|\mathbf{v}(t)\| = \text{ the speed at time } t.$$

The reasoning is as follows: during a time interval $[t_0, t]$ the object moves along its path from $\mathbf{r}(t_0)$ to $\mathbf{r}(t)$ for a total distance

$$s(t) = \int_{t_0}^{t} \|\mathbf{r}'(u)\|\, du. \qquad \text{(Section 13.4)}$$

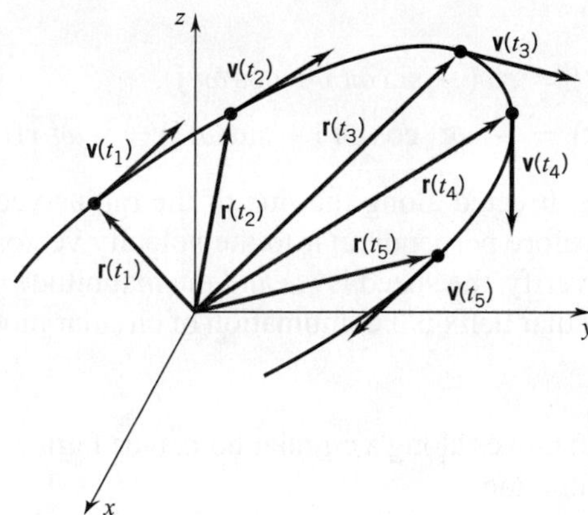

Figure 13.5.2

Differentiating with respect to t, we have

$$\frac{ds}{dt} = \|\mathbf{r}'(t)\|.$$

The magnitude of the velocity vector is thus *the rate of change of arc distance with respect to time*. This is why we call it the speed of the object.

Motion Along a Straight Line

The position at time t is given by a function of the form

$$\mathbf{r}(t) = \mathbf{r}_0 + f(t)\,\mathbf{d}, \quad \mathbf{d} \neq 0.$$

For convenience we take $\mathbf{d}$ as a unit vector.

The velocity and acceleration vectors are both directed along the line of the motion:

$$\mathbf{v}(t) = f'(t)\,\mathbf{d} \quad \text{and} \quad \mathbf{a}(t) = f''(t)\,\mathbf{d}.$$

The speed is $|f'(t)|$:

$$\|\mathbf{v}(t)\| = \|f'(t)\mathbf{d}\| = |f'(t)|\;\|\mathbf{d}\| = |f'(t)|,$$

and the magnitude of the acceleration is $|f''(t)|$:

$$\|\mathbf{a}(t)\| = \|f''(t)\mathbf{d}\| = |f''(t)|\;\|\mathbf{d}\| = |f''(t)|.$$

Circular Motion About the Origin

The position function can be written

$$\mathbf{r}(t) = r[\cos\theta(t)\,\mathbf{i} + \sin\theta(t)\,\mathbf{j}], \quad r > 0 \text{ constant}.$$

Here $\theta'(t)$ gives the time rate of change of the central angle θ. If $\theta'(t) > 0$, the motion is counterclockwise; if $\theta'(t) < 0$, the motion is clockwise. We call $\theta'(t)$ the *angular velocity* and $|\theta'(t)|$ the *angular speed*.

Uniform circular motion is circular motion with constant angular speed $\omega > 0$. The position function for uniform circular motion in the counterclockwise direction can be written

$$\mathbf{r}(t) = \mathbf{r}(\cos\omega t\,\mathbf{i} + \sin\omega t\,\mathbf{j}).$$

Differentiation gives

$$\mathbf{v}(t) = r\omega(-\sin\omega t\,\mathbf{i} + \cos\omega t\,\mathbf{j}),$$
$$\mathbf{a}(t) = -r\omega^2(\cos\omega t\,\mathbf{i} + \sin\omega t\,\mathbf{j}) = -\omega^2\,\mathbf{r}(t).$$

The acceleration is directed along the line of the radius vector toward the center of the circle and is therefore perpendicular to the velocity vector, which, as always, is tangential. As you can verify, the speed is $r\omega$ and the magnitude of acceleration is $r\omega^2$.

Motion along a circular helix is a combination of circular motion and motion along a straight line.

Example 1 An object moves along a circular helix (see Figure 13.3.12) with position at time t given by the function

$$\mathbf{r}(t) = a\cos\omega t\,\mathbf{i} + a\sin\omega t\,\mathbf{j} + b\omega t\,\mathbf{k} \quad (a > 0, b > 0, \omega > 0).$$

For each time t, find

(a) the velocity of the particle; **(b)** the speed; **(c)** the acceleration;
(d) the magnitude of the acceleration;
(e) the angle between the velocity vector and the acceleration vector.

SOLUTION

(a) Velocity: $\mathbf{v}(t) = \mathbf{r}'(t) = -a\omega\sin\omega t\,\mathbf{i} + a\omega\cos\omega t\,\mathbf{j} + b\omega\,\mathbf{k}.$

(b) Speed: $\|\mathbf{v}(t)\| = \sqrt{a^2\omega^2\sin^2\omega t + a^2\omega^2\cos^2\omega t + b^2\omega^2}$
$= \sqrt{a^2\omega^2 + b^2\omega^2} = \omega\sqrt{a^2+b^2}.$

(The speed is thus constant.)

(c) Acceleration : $\mathbf{a}(t) = \mathbf{v}'(t) = -a\omega^2\cos\omega t\,\mathbf{i} - a\omega^2\sin\omega t\,\mathbf{j}$
$= -a\omega^2(\cos\omega t\,\mathbf{i} + \sin\omega t\,\mathbf{j}).$

(Since the speed is constant, the acceleration comes entirely from the change in direction.)

(d) Magnitude of the acceleration: $\|\mathbf{a}(t)\| = a\omega^2.$

(e) Angle between $\mathbf{v}(t)$ and $\mathbf{a}(t)$:

$$\cos\theta = \frac{\mathbf{v}(t)}{\|\mathbf{v}(t)\|}\cdot\frac{\mathbf{a}(t)}{\|\mathbf{a}(t)\|}$$
$$= \left[\frac{-a\omega\sin\omega t\,\mathbf{i} + a\omega\cos\omega t\,\mathbf{j} + b\omega\,\mathbf{k}}{\omega\sqrt{a^2+b^2}}\right]\cdot\left[\frac{-a\omega^2(\cos\omega t\,\mathbf{i} + \sin\omega t\,\mathbf{j})}{a\omega^2}\right]$$
$$= \frac{a(\sin\omega t\cos\omega t - \cos\omega t\sin\omega t)}{\sqrt{a^2+b^2}} = 0.$$

Therefore $\theta = \frac{1}{2}\pi$. (At each point the acceleration vector is perpendicular to the velocity vector.) □

The Curvature of a Plane Curve

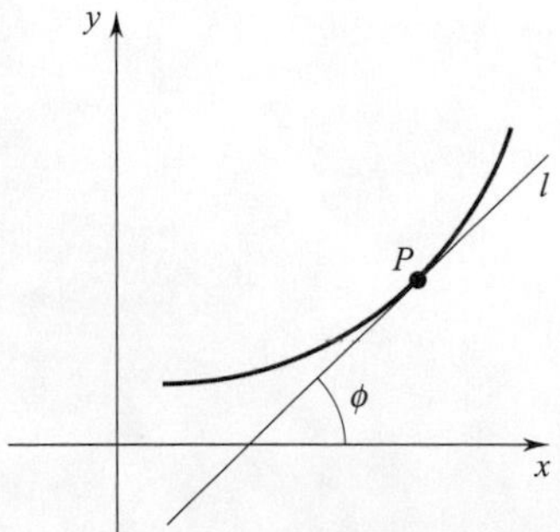

Figure 13.5.3

Figure 13.5.3 shows a plane curve which we assume to be twice differentiable. At the point P we have drawn the tangent line l. The angle that l makes with the x-axis has been labeled ϕ. As P moves along the curve, l changes and ϕ changes also. The magnitude κ† of the change in ϕ per unit of arc length is called the *curvature*:

† The symbol κ is the lower case Greek letter "kappa."

(13.5.3) $$\kappa = \left|\frac{d\phi}{ds}\right|.$$

Calculating Curvature

If the curve is the graph of a twice differentiable function

$$y = y(x),$$

then the curvature can be calculated from the formula

(13.5.4) $$\kappa = \frac{|y''|}{[1 + (y')^2]^{3/2}},$$

where the primes indicate differentiation with respect to x.

DERIVATION OF FORMULA 13.5.4 We know that $\tan\phi = y'$. Therefore,

$$\phi = \tan^{-1}(y').$$

Differentiating with respect to x, we have

$$\frac{d\phi}{dx} = \frac{1}{1 + (y')^2}\cdot\frac{d}{dx}(y') = \frac{y''}{1 + (y')^2}.$$

Since

$$\underbrace{\frac{d\phi}{dx} = \frac{d\phi}{ds}\frac{ds}{dx}}_{\text{chain rule}} = \underbrace{\frac{d\phi}{ds}\sqrt{1 + (y')^2}}_{(13.4.4)},$$

we have

$$\frac{d\phi}{ds}\sqrt{1 + (y')^2} = \frac{y''}{1 + (y')^2} \quad \text{and therefore} \quad \left|\frac{d\phi}{ds}\right| = \frac{|y''|}{[1 + (y')^2]^{3/2}}. \quad \square$$

If the curve is given parametrically by a twice differentiable vector function

$$\mathbf{r}(t) = x(t)\,\mathbf{i} + y(t)\,\mathbf{j},$$

then

(13.5.5) $$\kappa = \frac{|x'y'' - y'x''|}{[(x')^2 + (y')^2]^{3/2}},$$

where the primes now indicate differentiation with respect to t.

We will derive the formula under the assumption that $x' \neq 0$. Actually, the formula holds provided that $(x')^2 + (y')^2 \neq 0$.

DERIVATION OF FORMULA 13.5.5

$$\frac{dy}{dx} = \frac{dy/dt}{dx/dt} = \frac{y'}{x'}.$$

Therefore, as you can verify,

$$\frac{d^2y}{dx^2} \underset{(9.7.5)}{=} \frac{(dx/dt)(d^2y/dt^2) - (dy/dt)(d^2x/dt^2)}{(dx/dt)^3} = \frac{x'y'' - y'x''}{(x')^3}.$$

Thus

$$\kappa = \frac{|d^2y/dx^2|}{[1+(dy/dx)^2]^{3/2}} = \left|\frac{x'y''-y'x''}{(x')^3}\right| \frac{1}{[1+(y'/x')^2]^{3/2}} = \frac{|x'y''-y'x''|}{[(x')^2+(y')^2]^{3/2}}. \quad \square$$

↑ (13.5.4)

Example 2 Since a straight line has constant inclination, we have

$$\frac{d\phi}{ds} = 0 \quad \text{and thus} \quad \kappa = 0.$$

Along a straight line the curvature is constantly zero.

□

Example 3 For a circle of radius r,

$$\mathbf{r}(t) = r(\cos t\,\mathbf{i} + \sin t\,\mathbf{j}),$$

we have

$$x = r\cos t, \quad y = r\sin t.$$

Differentiation with respect to t gives

$$x' = -r\sin t, \quad x'' = -r\cos t; \quad y' = r\cos t, \quad y'' = -r\sin t.$$

Thus

$$\kappa = \frac{|x'y''-y'x''|}{[(x')^2+(y')^2]^{3/2}} = \frac{|(-r\sin t)(-r\sin t)-(r\cos t)(-r\cos t)|}{[(-r\sin t)^2+(r\cos t)^2]^{3/2}} = \frac{r^2}{r^3} = \frac{1}{r}.$$

Along a circle of radius r the curvature is constantly $1/r$.

Hardly surprising. It is geometrically evident that along a circular path the change in direction takes place at a constant rate. Since a complete revolution entails a change of direction of 2π radians and this change is effected on a path of length $2\pi r$, the change in direction per unit of arc length is $2\pi/2\pi r = 1/r$. □

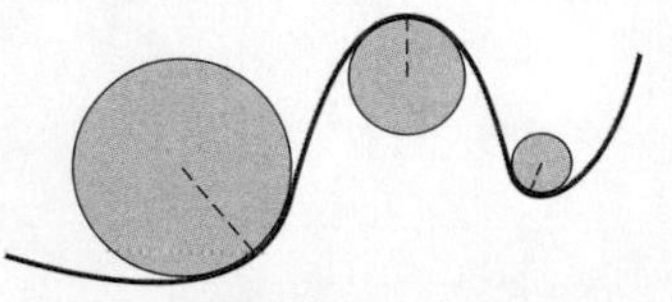

Figure 13.5.4

To say that a curve $y = f(x)$ has slope m at a point P is to say that at the point P the curve is rising or falling at the rate of a line of slope m. To say that a plane curve C has curvature $1/r$ at a point P is to say that at the point P the curve is turning at the rate of a circle of radius r. (Figure 13.5.4). The smaller the circle, the tighter the turn and, thus, the greater the curvature.

The reciprocal of the curvature,

$$\rho = \frac{1}{k}, \qquad \text{(for } k \neq 0\text{)}$$

is called the *radius of curvature*. The point at a distance ρ from the curve in the direction of the principal normal is called the *center of curvature*.

Example 4 Find the curvature at an arbitrary point (x, y) of the ellipse

$$\frac{x^2}{a^2} + \frac{y^2}{b^2} = 1. \qquad (a > b > 0). \qquad \text{(Figure 13.5.5)}$$

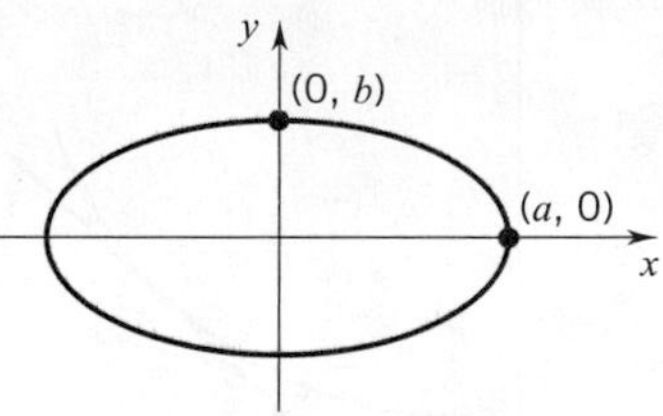

Figure 13.5.5

Determine the points of maximal curvature and the points of minimal curvature. What is the radius of curvature at each of these points?

SOLUTION We parametrize the ellipse by setting

$$\mathbf{r}(t) = a\cos t\,\mathbf{i} + b\sin t\,\mathbf{j}$$

and use the fact that

$$\kappa = \frac{|x'y'' - y'x''|}{[(x')^2 + (y')^2]^{3/2}}.$$

Here

$$x = a\cos t, \quad y = b\sin t,$$

and therefore

$$x' = -a\sin t, \quad x'' = -a\cos t; \quad y' = b\cos t, \quad y'' = -b\sin t.$$

Thus

$$k = \frac{|(-a\sin t)(-b\ \sin t) - (b\ \cos t)(-a\cos t)|}{[(-a\sin t)^2 + (b\ \cos t)^2]^{3/2}} = \frac{ab}{[a^2\sin^2 t + b^2\cos^2 t]^{3/2}}.$$

As you can check, the curvature at the point (x, y) can be written

$$\kappa = \frac{a^4b^4}{(b^4x^2 + a^4y^2)^{3/2}}.$$

To find the points of maximal and minimal curvature, we go back to the parameter t. Observe that

$$\begin{aligned} a^2\sin^2 t + b^2\cos^2 t &= (a^2 - b^2)\sin^2 t + b^2(\sin^2 t + \cos^2 t) \\ &= (a^2 - b^2)\sin^2 t + b^2. \end{aligned}$$

Thus we have

$$\kappa = \frac{ab}{[(a^2 - b^2)\sin^2 t + b^2]^{3/2}}.$$

Since we have assumed that $a > b > 0$, the curvature is maximal when $\sin^2 t = 0$; that is, when $t = 0$ and when $t = \pi$. Thus the points of maximal curvature are the points $P(\pm a, 0)$, the ends of the major axis. The curvature at these points is a/b^2 and the radius of curvature is b^2/a. The curvature is minimal when $\sin^2 t = 1$; that is, when $t = \frac{1}{2}\pi$ and when $t = \frac{3}{2}\pi$. The points of minimal curvature are the points $P(0, \pm b)$, the ends of the minor axis. The curvature at these points is b/a^2 and the radius of curvature is a^2/b. □

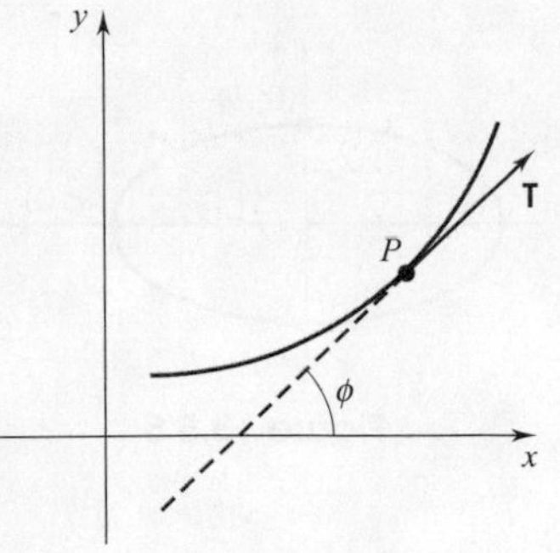

Figure 13.5.6

In Figure 13.5.6 you can see a plane curve. At the point P we have affixed the unit tangent vector $\mathbf{T}$. As P moves along the curve, $\mathbf{T}$ changes, not in length, but in direction. The curvature of the curve is the magnitude of the change in $\mathbf{T}$ per unit of arc length:

(13.5.6)
$$\kappa = \left\| \frac{d\mathbf{T}}{ds} \right\|.$$

PROOF Since $\mathbf{T}$ has length 1, we can write

$$\mathbf{T} = \cos\phi\,\mathbf{i} + \sin\phi\,\mathbf{j},$$

where ϕ is the angle between the tangent line and the x-axis. Differentiation with respect to s gives

$$\frac{d\mathbf{T}}{ds} = -\sin\phi\frac{d\phi}{ds}\,\mathbf{i} + \cos\phi\frac{d\phi}{ds}\,\mathbf{j} = \frac{d\phi}{ds}(-\sin\phi\mathbf{i} + \cos\phi\mathbf{j}).$$

Taking norms we have

$$\left\|\frac{d\mathbf{T}}{ds}\right\| = \left|\frac{d\phi}{ds}\right| \sqrt{\sin^2\phi + \cos^2\phi} = \left|\frac{d\phi}{ds}\right| = k$$

as asserted. □

This characterization of curvature generalizes to space curves.

The Curvature of a Space Curve

A space curve bends in two ways. It bends in the osculating plane (the plane of the unit tangent $\mathbf{T}$ and the principal normal $\mathbf{N}$) and it bends away from that plane. The first form of bending is measured by the rate at which the unit tangent $\mathbf{T}$ changes direction. The second form of bending is measured by the rate at which the vector $\mathbf{T} \times \mathbf{N}$ changes direction. We will concentrate here on the first form of bending, the bending in the osculating plane. The measure of this is called *curvature*.

What was a theorem on curvature in the case of a plane curve becomes a definition of curvature in the case of a space curve; namely, in the case of space curve, we *define* the curvature κ by setting

$$\kappa = \left\| \frac{d\mathbf{T}}{ds} \right\|.$$

If the space curve is given in terms of a parameter t, say

$$C: \quad \mathbf{r}(t) = x(t)\,\mathbf{i} + y(t)\,\mathbf{j} + z(t)\,\mathbf{k}, \qquad t \in [a, b],$$

then the curvature can be calculated from the formula

(13.5.7)
$$\kappa = \frac{\|d\mathbf{T}/dt\|}{ds/dt}.$$

We arrive at this by noting that

$$\frac{d\mathbf{T}}{ds}\frac{ds}{dt} = \frac{d\mathbf{T}}{dt},$$

then dividing through by ds/dt and taking norms. The assumption here is that $ds/dt = \|\mathbf{r}'(t)\|$ remains nonzero.

Example 5 Calculate the curvature of the circular helix

$$\mathbf{r}(t) = r\sin t\,\mathbf{i} + r\cos t\,\mathbf{j} + t\,\mathbf{k}. \qquad (r > 0)$$

SOLUTION We will use the Leibniz notation. We have:

$$\frac{d\mathbf{r}}{dt} = r\cos t\,\mathbf{i} - r\sin t\,\mathbf{j} + \mathbf{k}, \qquad \frac{ds}{dt} = \left\|\frac{d\mathbf{r}}{dt}\right\| = \sqrt{r^2+1},$$

$$\mathbf{T} = \frac{d\mathbf{r}/dt}{\|d\mathbf{r}/dt\|} = \frac{r\cos t\,\mathbf{i} - r\sin t\,\mathbf{j} + \mathbf{k}}{\sqrt{r^2+1}},$$

$$\frac{d\mathbf{T}}{dt} = \frac{-r\sin t\,\mathbf{i} - r\cos t\,\mathbf{j}}{\sqrt{r^2+1}}, \qquad \text{and} \qquad \left\|\frac{d\mathbf{T}}{dt}\right\| = \frac{r}{\sqrt{r^2+1}}.$$

Therefore,

$$\kappa = \frac{\|d\mathbf{T}/dt\|}{ds/dt} = \frac{r/\sqrt{r^2+1}}{\sqrt{r^2+1}} = \frac{r}{r^2+1}. \qquad \square$$

Components of Acceleration

In straight-line motion, acceleration is purely tangential; that is, the acceleration vector points along the line of the motion. In uniform circular motion, the acceleration is normal; the acceleration vector is perpendicular to the tangent vector and points along the line of the normal vector toward the center of the circle.

In general, acceleration has two components, a tangential component and a normal component. To see this, let's suppose that the position of an object at time t is given by the vector function

$$\mathbf{r}(t) = x(t)\,\mathbf{i} + y(t)\,\mathbf{j} + z(t)\,\mathbf{k}.$$

Since

$$\mathbf{T} = \frac{d\mathbf{r}/dt}{\|d\mathbf{r}/dt\|} = \frac{\mathbf{v}}{ds/dt},$$

we have

$$\mathbf{v} = \frac{ds}{dt}\mathbf{T}.$$

Differentiation gives

$$\mathbf{a} = \frac{d^2s}{dt^2}\mathbf{T} + \frac{ds}{dt}\frac{d\mathbf{T}}{dt}.$$

Observe now that

$$\frac{d\mathbf{T}}{dt} \underset{(13.3.4)}{=} \left\|\frac{d\mathbf{T}}{dt}\right\|\mathbf{N} \underset{(13.5.7)}{=} \kappa\frac{ds}{dt}\mathbf{N}.$$

Substitution in the previous display gives

(13.5.8)
$$\boxed{\mathbf{a} = \frac{d^2s}{dt^2}\mathbf{T} + \kappa\left(\frac{ds}{dt}\right)^2\mathbf{N}.}$$

The acceleration vector lies in the osculating plane, the plane of **T** and **N**. The tangential component of acceleration,

$$a_{\mathbf{T}} = \frac{d^2 s}{dt^2},$$

depends only on the change of speed of the object; if the speed is constant, the tangential component of acceleration is zero and the acceleration is directed entirely toward the center of curvature of the path. On the other hand, the normal component of acceleration,

$$a_{\mathbf{N}} = \kappa \left(\frac{ds}{dt}\right)^2,$$

depends both on the speed of the object and the curvature of the path. At a point where the curvature is zero, the normal component of acceleration is zero and the acceleration is directed entirely along the path of motion. If the curvature is not zero, then the normal component of acceleration is a multiple of the *square* of the speed. This means, for example, that if you are in a car going around a curve at 50 miles per hour, you will feel *four times* the normal component of acceleration that you would feel going around the same curve at 25 miles per hour.

We can use Equation (13.5.8) to obtain alternative formulas for $a_{\mathbf{T}}, a_{\mathbf{N}}$, and the curvature, κ.

If we take the dot product of **T** with **a**, we get

$$\mathbf{T} \cdot \mathbf{a} = a_{\mathbf{T}}(\mathbf{T} \cdot \mathbf{T}) + a_N(\mathbf{T} \cdot \mathbf{N}) = a_{\mathbf{T}}.$$

Therefore,

(13.5.9)
$$a_{\mathbf{T}} = \mathbf{T} \cdot \mathbf{a} = \frac{\mathbf{v} \cdot \mathbf{a}}{\|\mathbf{v}\|} = \frac{\mathbf{v} \cdot \mathbf{a}}{(ds/dt)}.$$

If we take the cross product of **T** with **a**, we get

$$\mathbf{T} \times \mathbf{a} = a_{\mathbf{T}}(\mathbf{T} \times \mathbf{T}) + a_{\mathbf{N}}(\mathbf{T} \times \mathbf{N}) = a_{\mathbf{N}}(\mathbf{T} \times \mathbf{N}),$$

and so

$$\|\mathbf{T} \times \mathbf{a}\| = a_{\mathbf{N}} \|\mathbf{T} \times \mathbf{N}\| = a_{\mathbf{N}} \|\mathbf{T}\| \|\mathbf{N}\| \sin(\pi/2) = a_{\mathbf{N}}.$$

Therefore

(13.5.10)
$$a_{\mathbf{N}} = \|\mathbf{T} \times \mathbf{a}\| = \frac{\|\mathbf{v} \times \mathbf{a}\|}{\|\mathbf{v}\|} = \frac{\|\mathbf{v} \times \mathbf{a}\|}{(ds/dt)}.$$

Since $a_{\mathbf{N}} = \kappa(ds/dt)^2$, it follows that

(13.5.11)
$$\kappa = \frac{\|\mathbf{v} \times \mathbf{a}\|}{(ds/dt)^3}.$$

Example 6 The position of a moving object at time t is given by

$$\mathbf{r}(t) = \ln t\,\mathbf{i} + 2t\,\mathbf{j} + t^2\,\mathbf{k}, \qquad t > 0.$$

Find the tangential and normal components of acceleration and the curvature of the path of the object at time $t = 1$.

SOLUTION $\mathbf{r}'(t) = \mathbf{v}(t) = \dfrac{1}{t}\,\mathbf{i} + 2\,\mathbf{j} + 2t\,\mathbf{k}$, and $\mathbf{r}''(t) = \mathbf{a}(t) = -\dfrac{1}{t^2}\,\mathbf{i} + 2\,\mathbf{k}$.

At $t = 1$, we have

$$\mathbf{v}(1) = \mathbf{i} + 2\mathbf{j} + 2\mathbf{k}, \qquad \|\mathbf{v}(1)\| = ds/dt = \sqrt{9} = 3, \qquad \text{and} \qquad \mathbf{a}(1) = -\mathbf{i} + 2\mathbf{k}.$$

Now,

$$a_{\mathbf{T}}(1) = \frac{\mathbf{v} \cdot \mathbf{a}}{(ds/dt)} = \frac{-1+4}{3} = 1,$$

$$\mathbf{a_N}(1) = \frac{\|\mathbf{v} \times \mathbf{a}\|}{(ds/dt)} = \tfrac{1}{3} \left\| \begin{vmatrix} \mathbf{i} & \mathbf{j} & \mathbf{k} \\ 1 & 2 & 2 \\ -1 & 0 & 2 \end{vmatrix} \right\| = \tfrac{1}{3}\,\|4\mathbf{i} - 4\mathbf{j} + 2\mathbf{k}\| = \frac{\sqrt{36}}{3} = 2,$$

and

$$k(1) = \frac{\|\mathbf{v} \times \mathbf{a}\|}{(ds/dt)^3} = \frac{\sqrt{36}}{27} = \frac{2}{9}. \quad \square$$

EXERCISES 13.5

1. A particle moves in a circle of radius r at constant speed v. Find the angular speed and the magnitude of the acceleration.
2. A particle moves so that
$$\mathbf{r}(t) = (a\cos \pi t + bt^2)\,\mathbf{i} + (a \sin \pi t - bt^2)\,\mathbf{j}.$$
Find the velocity, speed, acceleration, and the magnitude of the acceleration all at time $t = 1$.
3. A particle moves so that $\mathbf{r}(t) = at\,\mathbf{i} + b\sin at\,\mathbf{j}$. Show that the magnitude of the acceleration of the particle is proportional to its distance from the x-axis.
4. A particle moves so that $\mathbf{r}(t) = 2\,\mathbf{i} + t^2\,\mathbf{j} + (t-1)^2\,\mathbf{k}$. At what time is the speed a minimum?

Sketch the curve. Then compute and sketch the acceleration vector at the indicated points.

5. $\mathbf{r}(t) = (t/\pi)\,\mathbf{i} + \cos t\,\mathbf{j}$, $t \in [0, 2\pi]$; at $t = \frac{1}{4}\pi, \frac{1}{2}\pi, \pi$.
6. $\mathbf{r}(t) = t^3\,\mathbf{i} + t\,\mathbf{j}$, t real; at $t = -\frac{1}{2}, \frac{1}{2}, 1$.
7. $\mathbf{r}(t) = \sec t\,\mathbf{i} + \tan t\,\mathbf{j}$, $[-\frac{1}{4}\pi, \frac{1}{2}\pi)$; at $t = -\frac{1}{6}\pi, 0, \frac{1}{3}\pi$.
8. $\mathbf{r}(t) = \sin \pi t\,\mathbf{i} + t\,\mathbf{j}$, $t \in [0, 2]$; at $t = \frac{1}{2}, 1, \frac{5}{4}$.
9. An object moves so that
$$\mathbf{r}(t) = x_0\,\mathbf{i} + [\,y_0 + (\alpha\cos\theta)t]\,\mathbf{j} + [z_0 + (\alpha \sin\theta)t - 16t^2]\,\mathbf{k}, \quad t \ge 0.$$
Find (a) the initial position, (b) the initial velocity, (c) the initial speed, (d) the acceleration throughout the motion. Finally, (e) identify the curve.
10. A particle moves so that $\mathbf{r}(t) = 2\cos 2t\,\mathbf{i} + 3\cos t\,\mathbf{j}$. (a) Show that the particle oscillates on an arc of the parabola $4y^2 - 9x = 18$. (b) Draw the path. (c) What are the acceleration vectors at the points of zero velocity? (d) Draw these vectors at the points in question.
11. Let $\mathbf{r}(t)$ be the position vector of a moving particle. Show that $\|\mathbf{r}(t)\|$ is constant iff $\mathbf{r}(t) \perp \mathbf{r}'(t)$.
12. Let $\mathbf{r}(t)$ be the position vector of a moving particle. Show that if the speed of the particle is constant, then the velocity vector is perpendicular to the acceleration vector.

Find the curvature of the given curve.

13. $y = e^{-x}$.
14. $y = x^3$.
15. $y = \sqrt{x}$.
16. $y = x - x^2$.
17. $y = \ln \sec x$.
18. $y = \tan x$.
19. $y = \sin x$.
20. $x^2 - y^2 = a^2$.

Find the radius of curvature at the indicated point.

21. $6y = x^3$; $(2, \frac{4}{3})$.
22. $2y = x^2$; $(0, 0)$.
23. $y^2 = 2x$; $(2, 2)$.
24. $y = 2\sin 2x$. $(\frac{1}{4}\pi, 2)$.
25. $y = \ln(x+1)$; $(2, \ln 3)$.
26. $y = \sec x$; $(\frac{1}{4}\pi, \sqrt{2})$.
27. Find the point of maximal curvature on the curve $y = \ln x$.
28. Find the curvature of the graph of $y = 3x - x^3$ at the point where the function takes on its local maximum value.

Express the curvature in terms of t.

29. $\mathbf{r}(t) = t\,\mathbf{i} + \frac{1}{2}t^2\,\mathbf{j}$.
30. $\mathbf{r}(t) = e^t\,\mathbf{i} + e^{-t}\,\mathbf{j}$.

31. $\mathbf{r}(t) = 2t\,\mathbf{i} + t^3\,\mathbf{j}$.

32. $\mathbf{r}(t) = t^2\,\mathbf{i} + t^3\,\mathbf{j}$.

33. $\mathbf{r}(t) = e^t(\cos t\,\mathbf{i} + \sin t\,\mathbf{j})$.

34. $\mathbf{r}(t) = 2\cos t\,\mathbf{i} + 3\sin t\,\mathbf{j}$.

35. $\mathbf{r}(t) = (t\cos t)\,\mathbf{i} + (t\sin t)\,\mathbf{j}$.

36. $\mathbf{r}(t) = (\cos t + t\sin t)\,\mathbf{i} + (\sin t - t\cos t)\,\mathbf{j},\ t > 0$.

37. Find the radius of curvature of the hyperbola $xy = 1$ at the points $(1, 1)$ and $(-1, -1)$.

38. Find the radius of curvature at the vertices of the hyperbola $x^2 - y^2 = 1$.

39. Find the curvature at each point (x, y) on the hyperbola $b^2x^2 - a^2y^2 = a^2b^2$.

HINT: Parametrize the hyperbola by setting $\mathbf{r}(t) = a\cosh t\,\mathbf{i} + b\sinh t\,\mathbf{j}$.

40. Find the curvature at the highest point of an arch of the cycloid

$$x(t) = r(t - \sin)t, \quad y(t) = r(1 - \cos t).$$

In Exercises 41–47, interpret $\mathbf{r}(t)$ as the position of a moving object at time t. Find the curvature of the path and determine the tangential and normal components of acceleration.

41. $\mathbf{r}(t) = e^t\cos t\,\mathbf{i} + e^t\sin t\,\mathbf{j} + e^t\,\mathbf{k}$.

42. $\mathbf{r}(t) = 2\cos t\,\mathbf{i} + t\,\mathbf{j} + \sin t\,\mathbf{k}$.

43. $\mathbf{r}(t) = \cos 2t\,\mathbf{i} + \sin 2t\,\mathbf{j} + \mathbf{k}$.

44. $\mathbf{r}(t) = 2t\,\mathbf{i} + t^2\,\mathbf{j} + \ln t\,\mathbf{k}$.

45. $\mathbf{r}(t) = 2t\,\mathbf{i} + t^2\,\mathbf{j} + \frac{1}{3}t^3\mathbf{k}$.

46. $\mathbf{r}(t) = (\cos t + t\sin t)\,\mathbf{i} + (\sin t - t\cos t)\,\mathbf{j} + \frac{1}{2}\sqrt{3}t^2\,\mathbf{k}$, from $t = 0$ to $t = 2\pi$.

47. $\mathbf{r}(t) = \frac{2}{3}(1 + t)^{3/2}\mathbf{i} + \frac{2}{3}(1 - t)^{3/2}\,\mathbf{j} + \sqrt{2}t\,\mathbf{k}$.

C In Exercises 48 and 49, let $\mathbf{r}(t) = 6t\,\mathbf{i} + 3t^2\,\mathbf{j} + 2t^3\,\mathbf{k}$

48. (a) Use a graphing utility to draw the curve.

(b) Use a CAS to find the maximum curvature of the curve.

49. Use a CAS to find the tangential and normal components of the acceleration.

50. Show that the curvature of a polar curve $r = f(\theta)$ is given by

$$\kappa = \frac{|[f(\theta)]^2 + 2[f'(\theta)]^2 - f(\theta)f''(\theta)|}{([f(\theta)]^2 + [f'(\theta)]^2)^{3/2}}.$$

51. Find the curvature of the logarithmic spiral $r = e^{a\theta}, a > 0$.

52. Find the curvature of the spiral of Archimedes $r = a\theta$, $a > 0$.

53. Find the curvature of the cardioid $r = a(1 - \cos\theta)$ in terms of r.

54. Find the curvature of the petal curve $r = a\sin 2\theta, a > 0$.

55. Let $s(\theta)$ be the arc distance from the highest point of the cycloidal arch

$$x(\theta) = R(\theta - \sin\theta),\ y(\theta) = R(1 - \cos\theta),\ \theta \in [0, 2\pi]$$

to the point $(x(\theta), y(\theta))$ of that same arch. Let $\rho(\theta)$ be the radius of curvature at the point $(x(\theta), y(\theta))$. (a) Calculate $s(\theta)$. (b) Calculate $\rho(\theta)$. (c) Then find an equation in s and ρ for that arch. (Such an equation is called a *natural equation* for the curve.)

56. Let $s(\theta)$ be the arc distance from the origin to the point $(x(\theta), y(\theta))$ along the exponential spiral $r = ae^{c\theta}$. (Take $a > 0, c > 0$.) Let $\rho(\theta)$ be the radius of curvature at that same point. Find an equation in s and ρ for that curve.

57. Let $s(\theta)$ be the arc distance from the point $(-2a, 0)$ to the point $(x(\theta), y(\theta))$ along the cardioid $r = a(1 - \cos\theta)$. (Take $a > 0$.) Let $\rho(\theta)$ be the radius of curvature at that same point. Find an equation is s and ρ for that curve.

(*The Frenet formulas*) The figure shows a space curve. At a point of the curve we have drawn the unit tangent. $\mathbf{T}$, the principal normal $\mathbf{N}$, and the vector $\mathbf{B} = \mathbf{T} \times \mathbf{N}$, which, being normal to both $\mathbf{T}$ and $\mathbf{N}$, is called the *binormal*. At each point of the curve, the vectors $\mathbf{T}, \mathbf{N}, \mathbf{B}$ form what is called the *Frenet frame* a set of mutually perpendicular unit vectors that, in the order given, form a local right-handed coordinate system.

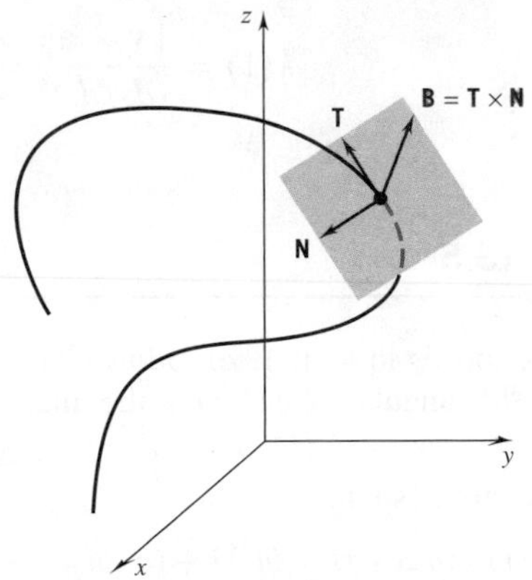

58. (a) Show that $d\mathbf{B}/ds$ is parallel to $\mathbf{N}$, and therefore there is a scalar τ for which

$$\frac{d\mathbf{B}}{ds} = \tau\mathbf{N}.$$

HINT: Since $\mathbf{B}$ has constant length one, $d\mathbf{B}/ds \perp \mathbf{B}$. Show that $d\,\mathbf{B}/ds \perp \mathbf{T}$ by carrying out the differentiation

$$\frac{d\,\mathbf{B}}{ds} = \frac{d}{ds}(\mathbf{T} \times \mathbf{N}).$$

(b) Now show that

$$\frac{d\,\mathbf{N}}{ds} = -\kappa\mathbf{T} - \tau\mathbf{B}.$$

HINT: Since $\mathbf{T}, \mathbf{N}, \mathbf{B}$ form a right-handed system of mutually perpendicular unit vectors, we can show that

$$\mathbf{N} \times \mathbf{B} = \mathbf{T} \quad \text{and} \quad \mathbf{B} \times \mathbf{T} = \mathbf{N}.$$

You can assume these relations.

(c) The scalar τ is called the *torsion* of the curve. Give a geometric interpretation to $|\tau|$.

■ PROJECT 13.5 Transition Curves

In the design of an automobile fender, engineers face the problem of connecting curved pieces in a smooth and elegant manner. In this context problems of the following sort arise.

The total curve of a fender is to be made up of two pieces each the graph of a cubic polynomial. The first piece p is to begin at the point $(1, 3)$ and end at the point $(3, 7)$. This requires that $p(1) = 3$ and $p(3) = 7$. The second piece q is to meet the first piece at the point $(3, 7)$ and end at the point $(9, -2)$. This requires that $q(3) = 7$ and $q(9) = -2$. The two pieces must meet smoothly. This requires that $p'(3) = q'(3)$ and $p''(3) = q''(3)$. Finally, the fender must be straight at the ends. This requires that $p''(1) = 0$ and $q''(9) = 0$. A curve that meets such requirements is called a *cubic spline*.

Problem 1. Let $p(x) = ax^3 + bx^2 + cx + d$ and $q(x) = \alpha x^3 + \beta x^2 + \gamma x + \delta$. Write the system of equations generated by the specified conditions and use a CAS to find the solution of the system.

Problem 2. Let

$$F(x) = \begin{cases} p(x), & 1 \le x \le 3 \\ q(x), & 3 \le x \le 9. \end{cases}$$

Show that F, F' and F'' are continuous on $[1, 9]$. Does F have continuous curvature? Sketch the graph of F using a graphing utility.

Problem 3. You are given the data in the table:

x	3	4	6
y	10	15	35

a. Define a cubic polynomial p on $[3, 4]$ and a cubic polynomial q on $[4, 6]$ using the data in the table.

b. Write the 8×8 system of equations as in Problem 1.

c. Use a CAS to solve the system.

When engineers lay railroad track, they cannot allow any abrupt changes in curvature. To join a straight away that ends at a point P to a curved track that begins at a point Q, they need to lay some *transitional track* that has zero curvature at P and the curvature of the second piece at Q (see the figure).

Problem 4. Find an arc of the form $y = Cx^n$, $x \in [0, 1]$ that joins the arcs C_1 and C_2 in the figure without any discontinuities in curvature.

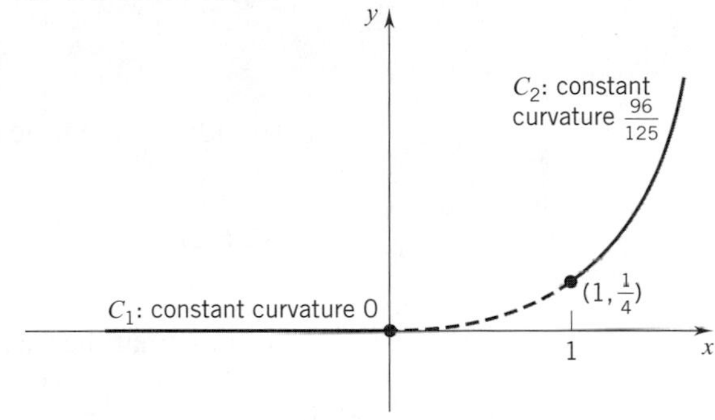

■ 13.6 VECTOR CALCULUS IN MECHANICS

The tools we have developed in the preceding sections have their premier application in Newtonian mechanics, the study of bodies in motion subject to Newton's laws. The heart of Newton's mechanics is his second law of motion:

$$\text{force} = \text{mass} \times \text{acceleration}.$$

We have worked with Newton's second law, but only in a very restricted context: motion along a coordinate line under the influence of a force directed along that same line. In that special setting, Newton's law was written as a scalar equation: $F = ma$. In general, objects do not move along straight lines (they move along curved paths) and the forces on them vary in direction. What happens to Newton's second law then? It becomes the vector equation

$$\mathbf{F} = m\,\mathbf{a}.$$

This is Newton's second law in its full glory.

An Introduction to Vector Mechanics

We are now ready to work with Newton's second law of motion in its vector form: $\mathbf{F} = m\,\mathbf{a}$. Since at each time t we have $\mathbf{a}(t) = \mathbf{r}''(t)$, Newton's law can be written

(13.6.1)
$$\mathbf{F}(t) = m\,\mathbf{a}(t) = m\,\mathbf{r}''(t).$$

This is a second-order differential equation in t. In Chapter 18 we give an introduction to the general theory of differential equations. Second-order differential equations of the type that we encounter here are treated in Sections 18.4 and 18.5. Our approach in this section is intuitive; we will search for solutions of (13.6.1) in particular situations.

When objects are moving, certain quantities (positions, velocities, and so on) are continually changing. This can make a situation difficult to grasp. In these circumstances it is particularly satisfying to find quantities that do not change. Such quantities are said to be *conserved*. (These conserved quantities are called the *constants of the motion*.) Mathematically we can determine whether a quantity is conserved by looking at its derivative with respect to time (the time derivative): *The quantity is conserved* (*is constant*) *iff its time derivative remains zero*. A *conservation law* is the assertion that in a given context a certain quantity does not change.

Momentum

We start with the idea of momentum. The *momentum* $\mathbf{p}$ of an object is the mass of the object times the velocity of the object:

$$\mathbf{p} = m\mathbf{v}.$$

To indicate the time dependence we write

(13.6.2)
$$\boxed{\mathbf{p}(t) = m\mathbf{v}(t) = m\mathbf{r}'(t).}$$

Assume that the mass of the object is constant. Then differentiation gives

$$\mathbf{p}'(t) = m\mathbf{r}''(t) = \mathbf{F}(t).$$

Thus, *the time derivative of the momentum of an object is the net force on the object*. If the net force on an object is continually zero, the momentum $\mathbf{p}(t)$ is constant. This is the law of *conservation of momentum*:

(13.6.3) If the net force on an object is continually zero, then the momentum of the object is conserved.

Angular Momentum

The angular momentum of an object about any given point is a vector quantity that is intended to measure the extent to which the object is circling about that point. If the position of the object at time t is given by the radius vector $\mathbf{r}(t)$, then the object's *angular momentum about the origin* is defined by the formula

(13.6.4)
$$\boxed{\mathbf{L}(t) = \mathbf{r}(t) \times \mathbf{p}(t) = \mathbf{r}(t) \times m\mathbf{v}(t).}$$

At each time t of a motion, $\mathbf{L}(t)$ is perpendicular to $\mathbf{r}(t)$, perpendicular to $\mathbf{v}(t)$, and oriented so that $\mathbf{r}(t), \mathbf{v}(t), \mathbf{L}(t)$ form a right-handed triple. The magnitude of $\mathbf{L}(t)$ is given by the relation

$$\|\mathbf{L}(t)\| = \|\mathbf{r}(t)\|\ \|m\mathbf{v}(t)\| \sin\theta(t),$$

where $\theta(t)$ is the angle between $\mathbf{r}(t)$ and $\mathbf{v}(t)$. (All this, of course, comes from the definition of the cross product.)

If $\mathbf{r}(t)$ and $\mathbf{v}(t)$ are not zero, then we can express $\mathbf{v}(t)$ as a vector parallel to $\mathbf{r}(t)$ plus a vector perpendicular to $\mathbf{r}(t)$:

$$\mathbf{v}(t) = \mathbf{v}_{\parallel}(t) + \mathbf{v}_{\perp}(t). \qquad \text{(see Figure 13.6.1)}$$

Thus

$$\begin{aligned} \mathbf{L}(t) &= \mathbf{r}(t) \times m\mathbf{v}(t) \\ &= \mathbf{r}(t) \times m\,[\mathbf{v}_{\parallel}(t) + \mathbf{v}_{\perp}(t)] \\ &= \underbrace{[\mathbf{r}(t) \times m\mathbf{v}_{\parallel}(t)]}_{\mathbf{0}} + [\mathbf{r}(t) \times m\mathbf{v}_{\perp}(t)] \\ &= \mathbf{r}(t) \times m\mathbf{v}_{\perp}(t). \end{aligned}$$

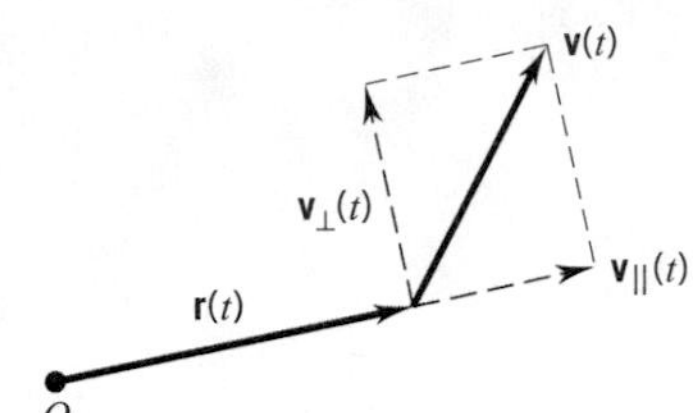

Figure 13.6.1

The component of velocity that is parallel to the radius vector contributes nothing to angular momentum. *The angular momentum comes entirely from the component of velocity that is perpendicular to the radius vector.*

Example 1 In uniform circular motion about the origin,

$$\mathbf{r}(t) = \mathbf{r}(\cos\omega t\,\mathbf{i} + \sin\omega t\,\mathbf{j}),$$

the velocity vector $\mathbf{v}(t)$ is always perpendicular to the radius vector $\mathbf{r}(t)$. In this case all of $\mathbf{v}(t)$ contributes to the angular momentum.

We can calculate $\mathbf{L}(t)$ as follows:

$$\begin{aligned} \mathbf{L}(t) &= \mathbf{r}(t) \times m\mathbf{v}(t) \\ &= [r(\cos\omega t\,\mathbf{i} + \sin\omega t\,\mathbf{j})] \times [mr(-\omega\sin\omega t\,\mathbf{i} + \omega\cos\omega t\,\mathbf{j})] \\ &= mr^2\omega(\cos^2\omega t + \sin^2\omega t)\,\mathbf{k} = mr^2\omega\,\mathbf{k}. \end{aligned}$$

The angular momentum is constant and is perpendicular to the xy-plane. If the motion is counterclockwise (if $\omega > 0$), then the angular momentum points up from the xy-plane. If the motion is clockwise (if $\omega < 0$), then the angular momentum points down from the xy-plane. (This is the right-handedness of the cross product coming in.) □

Example 2 In uniform straight-line motion with constant velocity $\mathbf{d}$,

$$\mathbf{r}(t) = \mathbf{r}_0 + t\,\mathbf{d},$$

we have

$$\mathbf{L}(t) = \mathbf{r}(t) \times m\mathbf{v}(t) = (\mathbf{r}_0 + t\,\mathbf{d}) \times m\mathbf{d} = m(\mathbf{r}_0 \times \mathbf{d}).$$

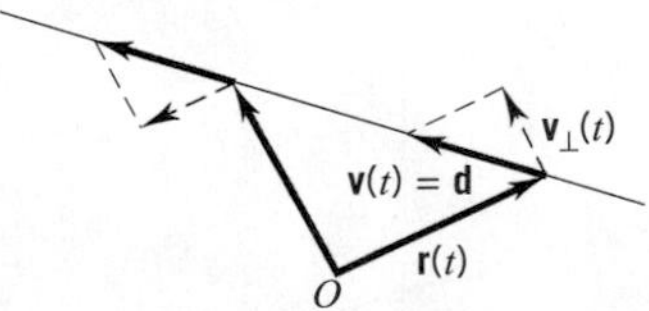

Figure 13.6.2

Here again the angular momentum is constant, but all is not quite so simple as it looks. In Figure 13.6.2 you can see the radius vector $\mathbf{r}(t)$, the velocity vector $\mathbf{v}(t) = \mathbf{d}$, and the part of the velocity vector that gives rise to angular momentum, the part perpendicular to $\mathbf{r}(t)$. As before, we have called this $\mathbf{v}_{\perp}(t)$. While $\mathbf{v}(t)$ is constant, $\mathbf{v}_{\perp}(t)$ is not constant. What happens here is that $\mathbf{r}(t)$ and $\mathbf{v}_{\perp}(t)$ vary in such a way that the cross product

$$\mathbf{L}(t) = \mathbf{r}(t) \times m\mathbf{v}_{\perp}(t)$$

remains constant. If the line of motion passes through the origin, then $\mathbf{v}(t)$ is parallel to $\mathbf{r}(t)$, $\mathbf{v}_{\perp}(t)$ is zero, and the angular momentum is zero. □

Torque

How the angular momentum of an object changes in time depends on the force acting on the object and on the position of the object relative to the origin: since $\mathbf{L}(t) = \mathbf{r}(t) \times m\mathbf{r}'(t)$,

$$\mathbf{L}'(t) = [\mathbf{r}(t) \times m\mathbf{r}''(t)] + \underbrace{[\mathbf{r}'(t) \times m\,\mathbf{r}'(t)]}_{\mathbf{0}} = \mathbf{r}(t) \times \mathbf{F}(t).$$

The cross product

(13.6.5)
$$\boxed{\boldsymbol{\tau}(t) = \mathbf{r}(t) \times \mathbf{F}(t)}$$

is called the *torque* about the origin.† See Project 12.5.

Since $\mathbf{L}'(t) = \boldsymbol{\tau}(t)$, we have the following conservation law:

(13.6.6) If the net torque on an object is continually zero, then the angular momentum of the object is conserved.

A force $\mathbf{F} = \mathbf{F}(t)$ is called a *central force* (*radial force*) if $\mathbf{F}(t)$ is always parallel to $\mathbf{r}(t)$. (Gravitational force, for example, is a central force.) For a central force, the cross product $\mathbf{r}(t) \times \mathbf{F}(t)$ is always zero. Thus a central force produces no torque about the origin. As you will see, this places severe restrictions on the kind of motion possible under a central force.

THEOREM 13.6.7

If an object moves under a central force and has constant angular momentum $\mathbf{L}$ different from zero, then:

1. The object is confined to the plane that passes through the origin and is perpendicular to $\mathbf{L}$.
2. The radius vector of the object sweeps out equal areas in equal times.

(This theorem plays an important role in astronomy and is embodied in Kepler's three celebrated laws of planetary motion. We will study Kepler's laws in Section *13.7.)

PROOF OF THEOREM 13.6.7 The first assertion is easy to verify: all the radius vectors pass through the origin and (by the very definition of angular momentum) they are all perpendicular to the constant vector $\mathbf{L}$.

To verify the second assertion, we introduce a right-handed coordinate system O–xyz setting the xy-plane as the plane of the motion, the positive z-axis pointing along $\mathbf{L}$. On the xy-plane we introduce polar coordinates r, θ. Thus, at time t, the object has some position $[r(t), \theta(t)]$.

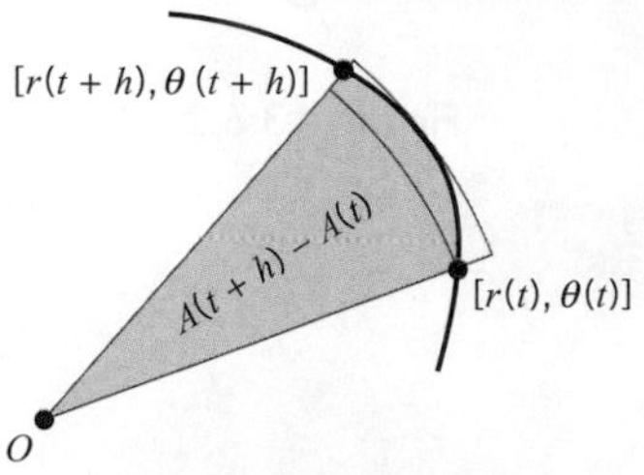

Let's denote by $A(t)$ the area swept out by the radius vector from some fixed time t_0 up to time t. Our task is to show that $A'(t)$ is constant.

The area swept out during the time interval $[t, t+h]$ is simply

$$A(t+h) - A(t).$$

Figure 13.6.3

Assuming that the motion takes place in the direction of increasing polar angle θ (see Figure 13.6.3), we have with obvious notation,

† The symbol τ is the Greek letter tau. The word *torque* comes from the Latin word *torquere*, "to twist."

$$\underbrace{\tfrac{1}{2}\min\,[r(t)]^2\cdot[\theta(t+h)-\theta(t)]}_{\text{area of inner sector}} \le A(t+h)-A(t) \le \underbrace{\tfrac{1}{2}\max\,[r(t)]^2\cdot[\theta(t+h)-\theta(t)]}_{\text{area of outer sector}}.$$

Divide through by h, take the limit as h tends to 0, and you will see that

$$(*)\qquad A'(t)=\tfrac{1}{2}[r(t)]^2\theta'(t).$$

Now

$$\begin{aligned}
\mathbf{r}(t) &= r(t)[\cos\theta(t)\,\mathbf{i}+\sin\theta(t)\,\mathbf{j}],\\
\mathbf{v}(t) &= r'(t)[\cos\theta(t)\,\mathbf{i}+\sin\theta(t)\,\mathbf{j}]+r(t)\theta'(t)[-\sin\theta(t)\,\mathbf{i}+\cos\theta(t)\,\mathbf{j}].\\
&= [r'(t)\cos\theta(t)-r(t)\theta'(t)\sin\theta(t)]\,\mathbf{i}+[r'(t)\sin\theta(t)+r(t)\theta'(t)\cos\theta(t)]\,\mathbf{j}.
\end{aligned}$$

A calculation that you can carry out yourself shows that

$$\mathbf{L}=\mathbf{r}(t)\times m\mathbf{v}(t)=mr^2(t)\theta'(t)\,\mathbf{k}.$$

Since $\mathbf{L}$ is constant, $r^2(t)\theta'(t)$ is constant. Thus, by $(*)$, $A'(t)$ is constant:

$$A'(t)=L/2m \qquad \text{where}\quad L=\|\mathbf{L}\|.\quad \square$$

Initial-Value Problems

In physics one tries to make predictions about the future on the basis of current information and a knowledge of the forces at work. In the case of an object in motion, the task can be to determine $\mathbf{r}(t)$ for all t given the force and some "initial conditions." Frequently the initial conditions give the position and velocity of the object at some time t_0. The problem then is to solve the differential equation

$$\mathbf{F}=m\mathbf{r}''$$

subject to conditions of the form

$$\mathbf{r}(t_0)=\mathbf{r}_0,\qquad \mathbf{v}(t_0)=\mathbf{v}_0.$$

Such problems are known as initial-value problems. We have considered initial-value problems in several different contexts earlier in the text.

By far the simplest problem of this sort concerns a *free particle*, an object on which there is no net force.

Example 3 At time t_0 a free particle has position $\mathbf{r}(t_0)=\mathbf{r}_0$ and velocity $\mathbf{v}(t_0)=\mathbf{v}_0$. Find $\mathbf{r}(t)$ for all t.

SOLUTION Since there is no net force on the object, the acceleration is zero and the velocity is constant. Since $\mathbf{v}(t_0)=\mathbf{v}_0$,

$$\mathbf{v}(t)=\mathbf{v}_0 \qquad \text{for all } t.$$

Integration with respect to t gives.

$$\mathbf{r}(t)=t\mathbf{v}_0+\mathbf{c},$$

where $\mathbf{c}$, the constant of integration, is a vector that we can determine from the initial position. The initial position $\mathbf{r}(t_0)=\mathbf{r}_0$ gives

$$\mathbf{r}_0=t_0\mathbf{v}_0+\mathbf{c} \qquad \text{and therefore} \qquad \mathbf{c}=\mathbf{r}_0-t_0\mathbf{v}_0.$$

Using this value for $\mathbf{c}$ in our equation for $\mathbf{r}(t)$, we have

$$\mathbf{r}(t) = t\mathbf{v}_0 + (\mathbf{r}_0 - t_0\mathbf{v}_0),$$

which we write as

$$\mathbf{r}(t) = \mathbf{r}_0 + (t - t_0)\mathbf{v}_0.$$

This is the equation of a straight line with direction vector $\mathbf{v}_0$. Free particles travel in straight lines with constant velocity. (We have tacitly assumed that $\mathbf{v}_0 \neq \mathbf{0}$. If $\mathbf{v}_0 = \mathbf{0}$, the particle remains at rest at $\mathbf{r}_0$.) □

Example 4 An object of mass m is subject to a force of the form

$$\mathbf{F}(t) = -m\omega^2\mathbf{r}(t) \quad \text{with} \quad \omega > 0.$$

Find $\mathbf{r}(t)$ for all t given that

$$\mathbf{r}(0) = a\,\mathbf{i} \quad \text{and} \quad \mathbf{v}(0) = \omega a\,\mathbf{j} \quad \text{with} \quad a > 0.$$

SOLUTION The force is a vector version of the restoring force exerted by a linear spring (Hooke's law). Since the force is central, the angular momentum of the object, $\mathbf{L}(t) = \mathbf{r}(t) \times m\mathbf{v}(t)$, is conserved. So $\mathbf{L}(t)$ is constantly equal to the value it had at time $t = 0$; for all t.

$$\mathbf{L}(t) = \mathbf{L}(0) = \mathbf{r}(0) \times m\mathbf{v}(0) = a\,\mathbf{i} \times m\omega a\,\mathbf{j} = ma^2\omega\,\mathbf{k}.$$

From our earlier discussion (Theorem 13.6.7) we can conclude that the motion takes place in the plane that passes through the origin and is perpendicular to $\mathbf{k}$. This is the xy-plane. Thus we can write

$$\mathbf{r}(t) = x(t)\,\mathbf{i} + y(t)\,\mathbf{j}.$$

Since $\mathbf{F}(t) = m\,\mathbf{r}''(t)$, the force equation can be written $\mathbf{r}''(t) = -\omega^2\mathbf{r}(t)$. In terms of components we have

$$x''(t) = -\omega^2 x(t), \qquad y''(t) = -\omega^2 y(t).$$

These are the equations of simple harmonic motion. We have already seen that functions of the form

$$x(t) = A_1 \sin(\omega t + \phi_1), \qquad y(t) = A_2 \sin(\omega t + \phi_2)$$

are solutions of these equations (Exercises 73 and 74, Section 3.6). In Section 18.4 we show that *all* solutions have this form. To evaluate the constants, we use the initial conditions. The condition $\mathbf{r}(0) = a\,\mathbf{i}$ means that $x(0) = a$ and $y(0) = 0$. So

$$(*) \qquad A_1 \sin\phi_1 = a \qquad A_2 \sin\phi_2 = 0.$$

The condition $\mathbf{v}(0) = \omega a\,\mathbf{j}$ means that $x'(0) = 0$ and $y'(0) = \omega a$. Since

$$x'(t) = \omega A_1 \cos(\omega t + \phi_1) \quad \text{and} \quad y'(t) = \omega A_2 \cos(\omega t + \phi_2),$$

we have

$$(**) \qquad \omega A_1 \cos\phi_1 = 0, \qquad \omega A_2 \cos\phi_2 = \omega a.$$

Conditions $(*)$ and $(**)$ are met by setting $A_1 = a, A_2 = a, \phi_1 = \frac{1}{2}\pi, \phi_2 = 0$. Thus

$$x(t) = a\sin(\omega t + \tfrac{1}{2}\pi) = a\cos\omega t, \quad y(t) = a\sin\omega t.$$

The vector equation reads

$$\mathbf{r}(t) = a\cos\omega t\,\mathbf{i} + a\sin\omega t\,\mathbf{j}.$$

The object moves in a circle of radius a about the origin with constant angular velocity ω. □

Example 5 A particle of charge q in a magnetic field $\mathbf{B}$ is subject to the force

$$\mathbf{F}(t) = \frac{q}{c}[\mathbf{v}(t) \times \mathbf{B}(t)],$$

where c is the speed of light and $\mathbf{v}$ is the velocity of the particle. Given that $\mathbf{r}(0) = \mathbf{r}_0$ and $\mathbf{v}(0) = \mathbf{v}_0$, find the path of the particle in the constant magnetic field $\mathbf{B}(t) = B_0\,\mathbf{k}$, $B_0 \neq 0$.

SOLUTION There is no conservation law that we can conveniently appeal to here. Neither momentum nor angular momentum is conserved: the force is not zero and it is not central. We start directly with Newton's $\mathbf{F} = m\mathbf{r}''$.

Since $\mathbf{r}'' = \mathbf{v}'$, we have

$$m\mathbf{v}'(t) = \frac{q}{c}[\mathbf{v}(t) \times B_0\mathbf{k}],$$

which we can write as

$$\mathbf{v}'(t) = \frac{qB_0}{mc}[\mathbf{v}(t) \times \mathbf{k}].$$

To simplify notation, we set $qB_0/mc = \omega$. We then have

$$\mathbf{v}'(t) = \omega[\mathbf{v}(t) \times \mathbf{k}].$$

Placing $\mathbf{v}(t) = v_1(t)\mathbf{i} + v_2(t)\mathbf{j} + v_3(t)\mathbf{k}$ in this last equation and working out the cross product, we find that

$$v_1'(t)\mathbf{i} + v_2'(t)\mathbf{j} + v_3'(t)\mathbf{k} = \omega[v_2(t)\mathbf{i} - v_1(t)\mathbf{j}].$$

This gives the scalar equations

$$v_1'(t) = \omega v_2(t), \qquad v_2'(t) = -\omega v_1(t), \qquad v_3'(t) = 0.$$

The last equation is trivial. It says that v_3 is constant:

$$v_3(t) = C.$$

The equations for v_1 and v_2 are linked together. We can get an equation that involves only v_1 by differentiating the first equation:

$$v_1''(t) = \omega v_2'(t) = -\omega^2 v_1(t).$$

As we know from our earlier work, this gives

$$v_1(t) = A_1\sin(\omega t + \phi_1).$$

Since $v_1'(t) = \omega v_2(t)$, we have

$$v_2(t) = \frac{v_1'(t)}{\omega} = \frac{A_1\omega}{\omega}\cos(\omega t + \phi_1) = A_1\cos(\omega t + \phi_1).$$

Therefore $$\mathbf{v}(t) = A_1\sin(\omega t + \phi_1)\,\mathbf{i} + A_1\cos(\omega t + \phi_1)\,\mathbf{j} + C\,\mathbf{k}.$$

A final integration with respect to t gives

$$\mathbf{r}(t) = \left[-\frac{A_1}{\omega}\cos(\omega t + \phi_1) + D_1\right]\mathbf{i} + \left[\frac{A_1}{\omega}\sin(\omega t + \phi_1) + D_2\right]\mathbf{j} + [Ct + D_3]\,\mathbf{k}$$

where $D_1,\ D_2,\ D_3$ are constants of integration. All six constants of integration — $A_1, \phi_1, C,\ D_1,\ D_2,\ D_3$— can be evaluated from the initial conditions. We will not pursue this. What is important here is that the path of the particle is a circular helix with axis parallel to $\mathbf{B}$, in this case parallel to $\mathbf{k}$. You should be able to see this from the equation for $\mathbf{r}(t)$: the z-component of $\mathbf{r}$ varies linearly with t from the value D_3, while the x and y components represent uniform motion with angular velocity ω in a circle of radius $|A_1/\omega|$ around the center (D_1, D_2). □

(Physicists express the behavior just found by saying that charged particles *spiral around* the magnetic field lines. Qualitatively, this behavior still holds even if the magnetic field lines are "bent," as is the case with the earth's magnetic field. Many charged particles become trapped by the earth's magnetic field. They keep spiraling around the magnetic field lines that run from pole to pole.)

EXERCISES 13.6

1. An object of mass m moves so that

$$\mathbf{r}(t) = \tfrac{1}{2}a(e^{\omega t} + e^{-\omega t})\,\mathbf{i} + \tfrac{1}{2}b(e^{\omega t} - e^{-\omega t})\,\mathbf{j}.$$

(a) What is the velocity at $t = 0$? (b) Show that the acceleration vector is a constant positive multiple of the radius vector. (This shows that the force is central and repelling.) (c) What does (b) imply about the angular momentum and the torque? Verify your answers by direct calculation.

2. (a) An object moves so that

$$\mathbf{r}(t) = a_1e^{bt}\,\mathbf{i} + a_2e^{bt}\,\mathbf{j} + a_3e^{bt}\,\mathbf{k}.$$

Show that, if $b > 0$, the object experiences a repelling central force.

(b) An object moves so that

$$\mathbf{r}(t) = \sin t\,\mathbf{i} + \cos t\,\mathbf{j} + (\sin t + \cos t)\,\mathbf{k}.$$

Show that the object experiences an attracting central force.

(c) Compute the angular momentum $\mathbf{L}(t)$ for the motion in part (*b*).

3. A constant force of magnitude α directed upward from the xy-plane is continually applied to an object of mass m. Given that the object starts at time 0 at the point $P(0, y_0, z_0)$ with initial velocity $2\mathbf{j}$, find: (a) the velocity of the object t seconds later; (b) the speed of the object t seconds later; (c) the momentum of the object t seconds later; (d) the path followed by the object, both in vector form and in Cartesian coordinates.

4. Show that, if the force on an object is always perpendicular to the velocity of the object, then the *speed* of the object is constant. (This tells us that the speed of a charged particle in a magnetic field is constant.)

5. Find the force required to propel a particle of mass m so that $\mathbf{r}(t) = t\mathbf{j} + t^2\mathbf{k}$.

6. Show that for an object of constant velocity, the angular momentum is constant.

7. At each point $P(x(t), y(t), z(t))$ of its motion, an object of mass m is subject to a force

$$\mathbf{F}(t) = m\pi^2[a\cos\pi t\,\mathbf{i} + b\sin\pi t\,\mathbf{j}]. \quad (a > 0, b > 0)$$

Given that $\mathbf{v}(0) = -\pi b\,\mathbf{j} + \mathbf{k}$ and $\mathbf{r}(0) = b\,\mathbf{j}$, find the following at time $t = 1$:

(a) The velocity. (b) The speed.
(c) The acceleration. (d) The momentum.
(e) The angular momentum. (f) The torque.

8. If an object of mass m moves with velocity $\mathbf{v}(t)$ subject to a force $\mathbf{F}(t)$, the scalar product

(13.6.8) $$\mathbf{F}(t) \cdot \mathbf{v}(t)$$

is called the *power* (expended by the force) and the number

(13.6.9) $$\tfrac{1}{2}m[v(t)]^2$$

is called the *kinetic energy* of the object. Show that the time rate of change of the kinetic energy of an object is the power expended on it:

(13.6.10) $$\frac{d}{dt}\left(\tfrac{1}{2}m[v(t)]^2\right) = \mathbf{F}(t) \cdot \mathbf{v}(t).$$

9. Two particles of equal mass m, one with constant velocity $\mathbf{v}$ and the other at rest, collide elastically (i.e., the kinetic energy of the system is preserved) and go off in different directions. Show that the two particles go off at right angles.

10. *(Elliptic harmonic motion)* Show that if the force on a particle of mass m is of the form

$$\mathbf{F}(t) = -m\omega^2\mathbf{r}(t),$$

then the path of the particle may be written

$$\mathbf{r}(t) = \cos\omega t\mathbf{A} + \sin\omega t\mathbf{B},$$

where $\mathbf{A}$ and $\mathbf{B}$ are constant vectors. Give the physical significance of $\mathbf{A}$ and $\mathbf{B}$ and specify conditions on $\mathbf{A}$ and $\mathbf{B}$ that restrict the particle to a circular path.

HINT: The solutions of the differential equation

$$x''(t) = -\omega^2 x(t)$$

can be written in the form

$$x(t) = A\cos\omega t + B\sin\omega t. \qquad \text{(Exercise 73, Section 3.6)}$$

11. A particle moves with constant acceleration $\mathbf{a}$. Show that the path of the particle lies entirely in some plane. Find a vector equation for this plane.

12. In Example 5 we stated that the path

$$\mathbf{r}(t) = \left[-\frac{A_1}{\omega}\cos(\omega t + \phi_1) + D_1\right]\mathbf{i} + \left[\frac{A_1}{\omega}\sin(\omega t + \phi_1) + D_2\right]\mathbf{j} + [Ct + D_3]\mathbf{k}$$

was a circular helix, Set $\omega = -1$ and show that, if $\mathbf{r}(0) = a\mathbf{i}$ and $\mathbf{v}(0) = a\,\mathbf{j} + b\,\mathbf{k}$, then the path takes the form

$$\mathbf{r}(t) = a\cos t\,\mathbf{i} + a\sin t\,\mathbf{j} + bt\,\mathbf{k},$$

the circular helix described in Section 13.3.

13. A charged particle in a time-independent electric field $\mathbf{E}$ experiences the force $q\mathbf{E}$, where q is the charge of the particle. Assume that the field has the constant value $\mathbf{E} = E_0\mathbf{k}$ and find the path of the particle given that $\mathbf{r}(0) = \mathbf{i}$ and $\mathbf{v}(0) = \mathbf{j}$.

14. *(Important)* A wheel is rotating about an axle with angular speed ω. Let $\boldsymbol{\omega}$ be the *angular velocity vector*, the vector of length ω that points along the axis of the wheel in such a direction that, observed from the tip of $\boldsymbol{\omega}$, the wheel rotates counterclockwise. Take the origin as the center of the wheel and let $\mathbf{r}$ be the vector from the origin to a point P on the rim of the wheel. Express the velocity $\mathbf{v}$ of P in terms of $\boldsymbol{\omega}$ and $\mathbf{r}$.

15. Solve the initial-value problem

$$\mathbf{F}(t) = m\mathbf{r}''(t) = t\mathbf{i} + t^2\mathbf{j}, \quad \mathbf{r}_0 = \mathbf{r}(0) = \mathbf{i},$$
$$\mathbf{v}_0 = \mathbf{v}(0) = \mathbf{k}.$$

16. Solve the initial-value problem

$$\mathbf{F}(t) = m\mathbf{r}''(t) = -m\beta^2 z(t)\,\mathbf{k}, \quad \mathbf{r}_0 = \mathbf{r}(0) = \mathbf{k},$$
$$\mathbf{v}_0 = \mathbf{v}(0) = 0.$$

[Here $z(t)$ is the z-component of $\mathbf{r}(t)$.]

17. An object of mass m moves subject to the force

$$\mathbf{F}(\mathbf{r}) = 4r^2\mathbf{r}$$

where $\mathbf{r} = \mathbf{r}(t)$ is the position of the object. Suppose $\mathbf{r}(0) = \mathbf{0}$ and $\mathbf{v}(0) = 2\mathbf{u}$, where $\mathbf{u}$ is a unit vector. Show that at each time t the speed v of the object satisfies the relation

$$v = \sqrt{4 + \frac{2}{m}r^4}.$$

HINT: Examine the quantity $\frac{1}{2}mv^2 - r^4$. (This is the *energy* of the object, a notion we will take up in Chapter 17.)

■ *13.7 PLANETARY MOTION

Tycho Brahe, Johannes Kepler

In the middle of the sixteenth century the arguments on planetary motion persisted. Was Copernicus right? Did the planets move in circles about the sun? Obviously not. Was not the earth the center of the universe?

In 1576, with the generous support of his king, the Danish astronomer Tycho Brahe built an elaborate astronomical observatory on the isle of Hveen and began his painstaking observations. For more than twenty years he looked through his telescopes and recorded what he saw. He was a meticulous observer, but he could draw no definite conclusions.

In 1599 the German astronomer-mathematician Johannes Kepler began his study of Brahe's voluminous tables. For a year and a half Brahe and Kepler worked together. Then Brahe died and Keper went on wrestling with the data. His persistence paid off. By 1619 Kepler had made three stupendous discoveries, known today as *Kepler's laws of planetary motion*:

I. Each planet moves in a plane, not in a circle, but in an elliptical orbit with the sun at one focus.

II. The radius vector from the sun to the planet sweeps out equal areas in equal times.

III. The square of the period of the motion varies directly as the cube of the major semiaxis, and the constant of proportionality is the same for all the planets.

What Kepler formulated empirically, Newton was able to explain. Each of these laws, Newton showed, was deducible from his laws of motion and his law of gravitation.

Newton's Second Law of Motion for Extended Three-Dimensional Objects

Imagine an object that consists of n particles with masses $m_1, m_2, \ldots, m_n$ located at $\mathbf{r}_1, \mathbf{r}_2, \ldots \mathbf{r}_n$.† The total mass M of the object is the sum of the masses of the constituent particles:

$$M = m_1 + \cdots + m_n.$$

The center of mass of the object is by definition the point $\mathbf{R}_M$ where

$$M\mathbf{R}_M = m_1\mathbf{r}_1 + \cdots + m_n\mathbf{r}_n.$$

The total force $\mathbf{F}_{\text{TOT}}$ on the object is by definition the sum of all the forces that act on the particles that constitute the object:

$$\mathbf{F}_{\text{TOT}} = \mathbf{F}_1 + \cdots + \mathbf{F}_n.$$

Since $\mathbf{F}_1 = m_1\mathbf{r}_1'', \cdots, \mathbf{F}_n = m_n\mathbf{r}_n''$, we have

$$\mathbf{F}_{\text{TOT}} = m_1\mathbf{r}_1'' + \cdots + m_n\mathbf{r}_n'',$$

which we can write as

$$\mathbf{F}_{\text{TOT}} = M\mathbf{R}_M''.$$

The total force on an extended object is thus the total mass of the object times the acceleration of the center of mass.

We can simplify this still further. The forces that act between the constituent particles, the so-called internal forces, cancel in pairs: if particle 23 tugs at particle 71 in

† The case of a continuously distributed mass is taken up in Chapter 16.

a certain direction with a certain strength, then particle 71 tugs at particle 23 in the opposite direction with the same strength. (Newton's third law: To every action there is an equal reaction.) Therefore, in calculating the total force on our object, we can disregard the internal forces and simply add up the external forces. $\mathbf{F}_{\text{TOT}} = \mathbf{F}_{\text{TOT}}^{(\text{Ext})}$ and Newton's second law takes the form

(13.7.1) $$\boxed{\mathbf{F}_{\text{TOT}}^{(\text{Ext})} = M\,\mathbf{R}_M''.}$$

The total external force on an extended three-dimensional object is thus the total mass of the object times the acceleration of the center of mass.

When an external force is applied to an extended object, we cannot predict the reaction of all the constituent particles, but we can predict the reaction of the center of mass. The center of mass will react to the force as if it were a particle with all the mass concentrated there. Suppose, for example, that a bomb is dropped from an airplane. The center of mass, "feeling" only the force of gravity (we are neglecting air resistance), falls in a parabolic arc toward the ground even if the bomb explodes at a thousand meters and individual pieces fly every which way. The forces of explosion are internal and do not affect the motion of the center of mass.

Some Preliminary Comments About the Planets

Roughly speaking, a planet is a massive object in the form of a ball with the center of mass at the center. In what follows, when we refer to the position of a planet, you are to understand that we really mean the position of the center of mass of the planet. Two other points require comment. First, we will write our equations as if the sun affected the motion of the planet, but not vice versa: we will assume that the sun stays put and that the planet moves. Really, each affects the other. Our viewpoint is justified by the immense difference in mass between the planets and the sun. In the case of the earth and the sun, for example, a reasonable analogy is to imagine a tug of war in space between someone who weighs three pounds and someone who weighs a million pounds: to a good approximation, the million-pound person does not move. The second point is that the planet is affected not only by the pull from the sun, but also by the gravitational pulls from the other planets and all the other celestial bodies. But these forces are much smaller, and they tend to cancel. We will ignore them. (Our results are only approximations, but they prove to be very good approximations.)

A Derivation of Kepler's Laws from Newton's Laws of Motion and His Law of Gravitation

The gravitational force exerted by the sun on a planet can be written in vector form as follows:

(∗) $$\mathbf{F}(\mathbf{r}) = -G\frac{mM}{r^3}\mathbf{r}.$$

Here m is the mass of the planet, M is the mass of the sun, G is the gravitational constant, $\mathbf{r}$ is the vector from the center of the sun to the center of the planet, and r is the magnitude of $\mathbf{r}$. (Thus we are placing the sun at the origin of our coordinate system; namely, we are using what is known as a *heliocentric* coordinate system, from *hēlios*, the Greek word for sun.)

Let's make sure that equation (∗) conforms to our earlier characterization of gravitational force. First of all, because of the minus sign, the direction is toward the origin, where the sun is located. Taking norms we have

$$\|\mathbf{F}(\mathbf{r})\| = \frac{GmM}{r^3}\|\mathbf{r}\| = \underset{\uparrow\ \|\mathbf{r}\| = r}{\frac{GmM}{r^2}}$$

Thus the magnitude of the force is as expected: it does vary directly as the product of the masses and inversely as the square of the distance between them. So we are back in familiar territory.

We will derive Kepler's laws in a somewhat piecemeal manner. Since the force on the planet is a central force, **L**, the angular momentum of the planet, is conserved (Section 13.6). If **L** were zero, we would have

$$\|\mathbf{L}\| = L = m\|\mathbf{r}\|\ \|\mathbf{v}\| \sin\theta = 0.$$

This would mean that either $\mathbf{r} = \mathbf{0}$, $\mathbf{v} = \mathbf{0}$, or $\sin\theta = 0$. The first equation would place the planet at the center of the sun. The second equation could hold only if the planet stopped. The third equation could hold only if the planet moved directly toward the sun or directly away from the sun. We can be thankful that none of those things happen.

Since the planet moves under a central force and **L** is not zero, *the planet stays on the plane that passes through the center of the sun and is perpendicular to* **L**, and the radius vector does sweep out equal areas in equal times. (We know all this from Theorem 13.6.7.)

Now we go on to show that the path is an ellipse with the sun at one focus.† The equation of motion of a planet of mass m can be written

$$m\,\mathbf{a}(t) = -GmM\frac{\mathbf{r}(t)}{[r(t)]^3}.$$

Clearly m drops out of the equation. Setting $GM = \rho$ and suppressing the explicit dependence on t, we have

$$\mathbf{a} = -\rho\frac{\mathbf{r}}{r^3},$$

which gives

$$(**) \qquad \frac{\mathbf{r}}{r^3} = -\frac{1}{\rho}\mathbf{a} = -\frac{1}{\rho}\frac{d\mathbf{v}}{dt}.$$

From (13.2.4) we know that in general

$$\frac{d}{dt}\left(\frac{\mathbf{r}}{r}\right) = (\mathbf{r}\times\mathbf{v})\times\frac{\mathbf{r}}{r^3}.$$

Since $\mathbf{L} = \mathbf{r}\times m\mathbf{v}$, we have

$$\frac{d}{dt}\left(\frac{\mathbf{r}}{r}\right) = \frac{\mathbf{L}}{m}\times\frac{\mathbf{r}}{r^3}.$$

† If some of the steps seem unanticipated to you, you should realize that we are discussing one of the most celebrated physical problems in history and there have been three hundred years to think of ingenious ways to deal with it. The argument we give here follows the lines of the excellent discussion that appears in Harry Pollard's *Celestial Mechanics*, Mathematical Association of America (1976).

Inserting (∗∗) we see that

$$\frac{d}{dt}\left(\frac{\mathbf{r}}{r}\right) = \frac{\mathbf{L}}{m} \times \left(-\frac{1}{\rho}\frac{d\mathbf{v}}{dt}\right) = \frac{d\mathbf{v}}{dt} \times \frac{\mathbf{L}}{m\rho} = \frac{d}{dt}\left(\mathbf{v} \times \frac{\mathbf{L}}{m\rho}\right)$$

$\mathbf{L}, m, \rho$, are all constant

and therefore

$$\frac{d}{dt}\left[\left(\mathbf{v} \times \frac{\mathbf{L}}{m\rho}\right) - \frac{\mathbf{r}}{r}\right] = \mathbf{0}.$$

Integration with respect to t gives

$$(\ast\ast\ast) \qquad \left(\mathbf{v} \times \frac{\mathbf{L}}{m\rho}\right) - \frac{\mathbf{r}}{r} = \mathbf{e}$$

where **e** is a constant vector that depends on the initial conditions. Dotting both sides with **r**, we have

$$\mathbf{r} \cdot \left(\mathbf{v} \times \frac{\mathbf{L}}{m\rho}\right) - \frac{\mathbf{r} \cdot \mathbf{r}}{r} = \mathbf{r} \cdot \mathbf{e}.$$

Since

$$\mathbf{r} \cdot \left(\mathbf{v} \times \frac{\mathbf{L}}{m\rho}\right) = \frac{\mathbf{L}}{m\rho} \cdot (\mathbf{r} \times \mathbf{v}) = \frac{\mathbf{L}}{m\rho} \cdot \frac{\mathbf{L}}{m} = \frac{L^2}{m^2\rho} \quad \text{and} \quad \frac{\mathbf{r} \cdot \mathbf{r}}{r} = \frac{r^2}{r} = r,$$

(12.5.6)

we find that

$$\frac{L^2}{m^2\rho} - r = \mathbf{r} \cdot \mathbf{e},$$

which we write as

$$r + (\mathbf{r} \cdot \mathbf{e}) = \frac{L^2}{m^2\rho}. \qquad \text{(orbit equation)}$$

If $\mathbf{e} = \mathbf{0}$, the orbit is a circle:

$$r = \frac{L^2}{m^2\rho}.$$

This is a possibility that requires very special initial conditions, conditions not met by any of the planets in our solar system. In our solar system at least, $\mathbf{e} \neq \mathbf{0}$.

Given that $\mathbf{e} \neq \mathbf{0}$, we can write the orbit equation as

$$\textbf{(13.7.2)} \qquad \boxed{r(1 + e\cos\theta) = \frac{L^2}{m^2\rho}}$$

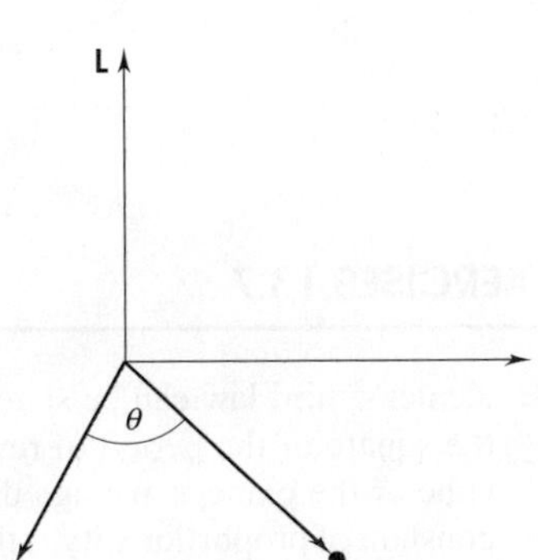

Figure 13.7.1

where $e = \|\mathbf{e}\|$ and θ is the angle between **r** and **e**. We are almost through. Since $\mathbf{v} \times \mathbf{L}$ and **r** are both perpendicular to **L**, we know from (∗∗∗) that **e** is perpendicular to **L**. Therefore, in the plane of **e** and **r** (see Figure 13.7.1) we can interpret r and θ as the usual polar coordinates. The orbit equation (13.7.2) is then a polar equation. We refer you to Project 9.4, where it is shown that an equation of the form (13.7.2) represents a conic section with focus at the origin, which is to say *focus at the sun*. Accordingly, the conic section can be a parabola, a hyperbola, or an ellipse. The repetitiveness of planetary motion rules out the parabola and the hyperbola. *The orbit is an ellipse.*

Finally we will verify Kepler's third law: The square of the period of the motion varies directly as the cube of the major semi-axis, and the constant of proportionality is the same for all planets.

The elliptic orbit has an equation of the form

$$r(1 + e\cos\theta) = ed \qquad \text{with} \qquad 0 < e < 1, \quad d > 0.$$

Rewriting this equation in rectangular coordinates ($x = r\cos\theta$, $y = r\sin\theta$), we get

$$\frac{(x+c)^2}{a^2} + \frac{y^2}{a^2 - c^2} = 1, \qquad \text{where} \qquad a = \frac{ed}{1-e^2} \quad \text{and} \quad c = \frac{e^2 d}{1-e^2}.$$

By Section 9.2, the lengths of the major and minor semi-axes of the ellipse are given by

$$a = \frac{ed}{1-e^2} \qquad \text{and} \qquad b = \sqrt{a^2 - c^2} = a\sqrt{1-e^2}.$$

Denote the period of revolution by T. Since the radius vector sweeps out area at the constant rate of $L/2m$ (we know this from the proof of Theorem 13.6.7)

$$\left(\frac{L}{2m}\right) T = \text{area of the ellipse} = \pi ab = \pi a^2\sqrt{1-e^2}.$$

Thus

$$T = \frac{2\pi m a^2\sqrt{1-e^2}}{L} \qquad \text{and} \qquad T^2 = \frac{4\pi^2 m^2 a^4(1-e^2)}{L^2}.$$

From (13.7.2) we know that

$$ed = \frac{L^2}{m^2\rho} = \frac{L^2}{m^2 GM} \qquad \text{and therefore} \qquad \frac{m^2}{L^2} = \frac{1}{edGM}.$$

It follows that

$$T^2 = \frac{4\pi^2 a^4(1-e^2)}{edGM} = \underset{\uparrow\, ed = a(1-e^2)}{\frac{4\pi^2 a^4(1-e^2)}{a(1-e^2)GM}} = \frac{4\pi^2}{GM}a^3.$$

T^2 does vary directly with a^3, and the constant of proportionality $4\pi^2/GM$ is the same for all planets.

EXERCISES 13.7

1. Kepler's third law can be stated as follows: For each planet the square of the period of revolution varies directly as the cube of the planet's average distance from the sun, and the constant of proportionality is the same for all planets. A *year* on a given planet is the time taken by the planet to make one circuit around the sun. Thus, a year on a planet is the period of revolution of that planet. Given that on average Venus is 0.72 times as far from the sun as the earth, how does the length of a "Venus year" compare with the length of an "earth year"?

2. Verify by differentiation with respect to time t that if the acceleration of a planet is given by

$$\mathbf{a} = -\rho\frac{\mathbf{r}}{r^3},$$

then the energy $E = \frac{1}{2}mv^2 - \dfrac{m\rho}{r}$ is constant.

3. Given that a planet moves in a plane, its motion can be described by rectangular coordinates (x, y) or polar coordinates $[r, \theta]$, with the origin at the sun. The kinetic energy of a planet is

$$\frac{1}{2}mv^2 = \frac{1}{2}m\left[\left(\frac{dx}{dt}\right)^2 + \left(\frac{dy}{dt}\right)^2\right].$$

Show that in polar coordinates

$$\frac{1}{2}mv^2 = \frac{1}{2}m\left[\left(\frac{dr}{dt}\right)^2 + r^2\left(\frac{d\theta}{dt}\right)^2\right].$$

4. We have seen that the energy of a planet

$$E = \frac{1}{2}mv^2 - \frac{m\rho}{r}$$

is constant (Exercise 2). Setting $dr/dt = \dot{r}, d\theta/dt = \dot{\theta}$, and using Exercise 3, we have

$$E = \frac{1}{2}m(\dot{r}^2 + r^2\dot{\theta}^2) - \frac{m\rho}{r}.$$

We also know that the angular momentum $\mathbf{L}$ is constant and that $L = mr^2\dot{\theta}$. Use this fact to verify that

$$E = \frac{L^2}{2m}\left\{\frac{1}{r^2} + \frac{1}{r^4}\left(\frac{dr}{d\theta}\right)^2\right\} - \frac{m\rho}{r}.$$

Since E is a constant, this is a differential equation for r as a function of θ.

5. Show that the function

$$r = \frac{a}{1 + e\cos\theta} \quad \text{with} \quad a = \frac{L^2}{m^2\rho} \quad \text{and} \quad e^2 = \frac{2Ea}{m\rho} + 1$$

satisfies the equation derived in Exercise 4.

■ CHAPTER HIGHLIGHTS

13.1 Vector Functions

basic definition: $\lim\limits_{y\to t_0} \mathbf{f}(t) = \mathbf{L}$ if $\lim\limits_{t\to t_0} \|\mathbf{f}(t)\| - \|\mathbf{L}\| = 0$

theorem: if $\lim\limits_{t\to t_0} \mathbf{f}(t) = \mathbf{L}$ then $\lim\limits_{t\to t_0} \|\mathbf{f}(t)\| = \|\mathbf{L}\|$

limit rules (p. 765)
limits can be taken component by component (p. 765)
continuity and differentiability (p. 766) integration (p. 767)
properties of the integral (p. 768)

13.2 Differentiation Formulas

$$\frac{d}{dt}(\mathbf{f} + \mathbf{g}) = \frac{d\mathbf{f}}{dt} + \frac{d\mathbf{g}}{dt}.$$

$$\frac{d}{dt}(\alpha\mathbf{f}) = \alpha\frac{d\mathbf{f}}{dt} (\alpha \text{ constant}).$$

$$\frac{d}{dt}(u\mathbf{f}) = u\frac{d\mathbf{f}}{dt} + \frac{du}{dt}\mathbf{f} \quad (u = u(t)).$$

$$\frac{d}{dt}(\mathbf{f}\cdot\mathbf{g}) = \left(\mathbf{f}\cdot\frac{d\mathbf{g}}{dt}\right) + \left(\frac{d\mathbf{f}}{dt}\cdot g\right).$$

$$\frac{d}{dt}(\mathbf{f}\times\mathbf{g}) = \left(\mathbf{f}\times\frac{d\mathbf{g}}{dt}\right) + \left(\frac{d\mathbf{f}}{dt}\times g\right).$$

$$\frac{d\mathbf{f}}{dt} = \frac{d\mathbf{f}}{du}\frac{du}{dt} \quad \text{(chain rule).}$$

13.3 Curves

tangent vector (p. 778) tangent line (p. 779)
unit tangent (p. 781) principal normal (p. 781)
normal line (p. 781) osculating plane (p. 781)
orientation-preserving (orientation-reversing) change of parameter (p. 784)

13.4 Arc Length

$$L(C) = \int_a^b \|\mathbf{r}'(t)\|\,dt \qquad \frac{ds}{dt} = \sqrt{\left(\frac{dx}{dt}\right)^2 + \left(\frac{dy}{dt}\right)^2 + \left(\frac{dz}{dt}\right)^2}$$

13.5 Curvature; Curvilinear Motion

velocity, acceleration (p. 794) speed (p. 794)
angular speed, uniform circular motion (p. 795)
curvature (p. 796) radius of curvature (p. 798)
components of acceleration (p. 801)
Frenet formulas (p. 804)

*13.6 Vector Calculus in Mechanics

conservation law (p. 806)
angular momentum (p. 806) torque (p. 808)
central force (radial force) (p. 808)
effect of a central force on an object with nonzero angular momentum (p. 806)
initial-value problems (p. 809) power, kinetic energy (p. 813)

*13.7 Planetary Motion

Kepler's three laws of planetary motion (p. 814)

CHAPTER
14
FUNCTIONS
OF SEVERAL
VARIABLES

■ 14.1 ELEMENTARY EXAMPLES

First a remark on notation. Points $P(x, y)$ of the xy-plane will be written (x, y) and points $P(x, y, z)$ of three-space will be written (x, y, z).

Let D be a nonempty subset of the xy-plane. A rule f that assigns a real number $f(x, y)$ to each point (x, y) in D is called a *real-valued function of two variables*. The set D is called the *domain of* f and the set of all values $f(x, y)$ is called the *range of* f.

Example 1 Take D as the entire xy-plane and to each point (x, y) assign the number

$$f(x, y) = xy. \quad \square$$

Example 2 Take D as the set of all points (x, y) with $y \neq 0$. To each such point assign the number

$$f(x, y) = \tan^{-1}\left(\frac{x}{y}\right). \quad \square$$

Example 3 Take D as the *open unit disc*: $D = \{(x, y) : x^2 + y^2 < 1\}$. This set consists of all points inside the *unit circle* $x^2 + y^2 = 1$; the circle itself is omitted.† To each point (x, y) in D assign the number

$$f(x, y) = \frac{1}{\sqrt{1 - (x^2 + y^2)}}. \quad \square$$

Now let D be a nonempty subset of three-space. A rule f that assigns a real number $f(x, y, z)$ to each point (x, y, z) in D is called a *real-valued function of three variables*. The set D is called the *domain of* f and the set of all values $f(x, y, z)$ is called the *range of* f.

† The *closed unit disc* $D = \{(x, y) : x^2 + y^2 \leq 1\}$ consists of all points on or inside the unit circle; the circle is included.

Example 4 Take D as all of three-space and to each point (x, y, z) assign the number

$$f(x, y, z) = xyz. \quad \square$$

Example 5 Take D as the set of all points (x, y, z) with $z \neq x + y$. (Thus D consists of all points not on the plane $x + y - z = 0$.) To each point of D assign the number

$$f(x, y, z) = \cos\left(\frac{1}{x + y - z}\right). \quad \square$$

Example 6 Take D as the *open unit ball*: $D = \{(x, y, z) : x^2 + y^2 + z^2 < 1\}$. This set consists of all points inside the *unit sphere* $x^2 + y^2 + z^2 = 1$; the sphere itself is omitted. † To each point (x, y, z) in D assign the number

$$f(x, y, z) = \frac{1}{\sqrt{1 - (x^2 + y^2 + z^2)}}. \quad \square$$

Functions of several variables arise naturally in very elementary settings.

$f(x, y) = \sqrt{x^2 + y^2}$ gives the distance between (x, y) and the origin;

$f(x, y) = xy$ gives the area of a rectangle of dimensions x, y; and

$f(x, y) = 2(x + y)$ gives the perimeter.

$f(x, y, z) = \sqrt{x^2 + y^2 + z^2}$ gives the distance between (x, y, z) and the origin;

$f(x, y, z) = xyz$ gives the volume of a rectangular solid of dimensions x, y, z; and

$f(x, y, z) = 2(xy + xz + yz)$ gives the total surface area.

In general, a *real-valued function of n variables* is a rule f that assigns a real number $f(x_1, x_2, \ldots x_n)$ to each ordered n-tuple $(x_1, x_2, \ldots, x_n)$ in a subset D of n-dimensional space. For example, the function

$$f(x_1, x_2, \ldots, x_n) = \sqrt{x_1^2 + x_2^2 + \cdots + x_n^2}$$

gives the distance between the point $(x_1, x_2, \ldots, x_n)$ and the origin in n-dimensional space. In our study of functions of several variables we will restrict our attention to functions of two and three variables because, as you will see, graphs and other visualizations are possible.

Functions of several variables arise in many problems in science, engineering, economics, and so on. Indeed, the mathematical models for "real" problems are much more likely to involve functions of several variables than functions of a single variable; large-scale problems often involve functions of hundreds or even thousands of variables. Here are some simple examples.

An initial investment A_o compounded continuously at interest rate r grows over time t to have value.

$$A(t, r) = A_0 e^{rt} \qquad \text{(Section 7.6)}$$

Thus the principal A is a function of two variables r and t.

The magnitude of the gravitational force exerted by a body of mass M situated at the origin on a body of mass m at the point (x, y, z) is given by the function

$$F(x, y, z) = \frac{GmM}{x^2 + y^2 + z^2}. \qquad \text{(Section 13.6)}$$

† The *closed unit ball* $D = \{(x, y, z) : x^2 + y^2 + z^2 \leq 1\}$ consists of all points on or inside the unit sphere; the sphere is included.

The *ideal gas law* states that the pressure P of a gas depends on the volume V and on the temperature T according to the equation

$$P = \frac{cT}{V}$$

where c is a constant.

If a metal sphere is being heated by a flame, then the temperature T at a point $P(x, y, z)$ on the sphere is a function f of four variables: the position variables x, y, z and the time t,

$$T = f(x, y, z, t).$$

If the domain of a function of several variables is not explicitly given, it is to be understood that the domain is the maximal set of points for which the definition generates a real number. Thus, in the case of

$$f(x, y) = \frac{1}{x - y},$$

the domain is understood to be all points (x, y) with $y \neq x$, that is, all points of the plane not on the line $y = x$. In the case of

$$g(x,y,z) = \sin^{-1}(x + y + z),$$

the domain is understood to be all points (x, y, z) with $-1 \leq x + y + z \leq 1$. This set is the slab bounded by the parallel planes

$$x + y + z = -1 \quad \text{and} \quad x + y + z = 1.$$

To say that a function is *bounded* is to say that its range is bounded. Since the function

$$f(x,y) = \frac{1}{x - y}$$

takes on all values other than 0, its range is $(-\infty, 0) \cup (0, \infty)$. The function is unbounded. In the case of

$$g(x, y, z) = \sin^{-1}(x + y + z),$$

the range is the closed interval $[-\frac{1}{2}\pi, \frac{1}{2}\pi]$. This function is bounded, below by $-\frac{1}{2}\pi$ and above by $\frac{1}{2}\pi$.

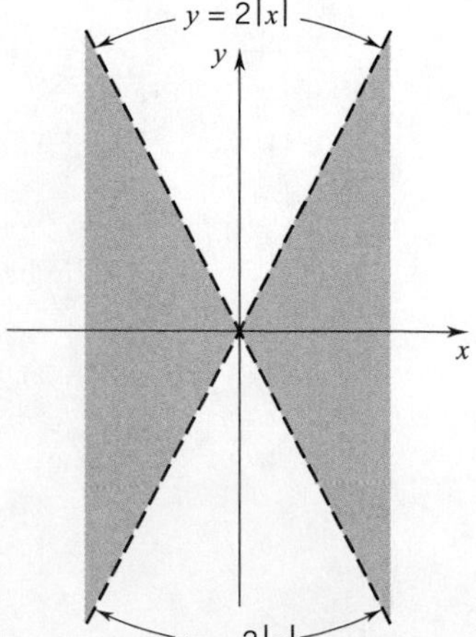

Figure 14.1.1

Example 7 Find the domain and range of the function $f(x,y) = \dfrac{1}{\sqrt{4x^2 - y^2}}$.

SOLUTION A point (x, y) is in the domain of f iff $4x^2 - y^2 > 0$. This occurs iff

$$y^2 < 4x^2$$

and thus iff

$$-2|x| < y < 2|x|.$$

The domain of f is the shaded region shown in Figure 14.1.1. It consists of all the points of the xy-plane that lie between the graph of $y = -2|x|$ and the graph of $y = 2|x|$. On this set $\sqrt{4x^2 - y^2}$ takes on all positive values, and so does its reciprocal $f(x,y)$. The range of f is $(0, \infty)$. □

EXERCISES 14.1

Find the domain and range of the function.

1. $f(x, y) = \sqrt{xy}$.
2. $f(x, y) = \sqrt{1 - xy}$.
3. $f(x, y) = \dfrac{1}{x + y}$.
4. $f(x, y) = \dfrac{1}{x^2 + y^2}$.
5. $f(x, y) = \dfrac{e^x - e^y}{e^x + e^y}$.
6. $f(x, y) = \dfrac{x^2}{x^2 + y^2}$.
7. $f(x, y) = \ln(xy)$.
8. $f(x, y) = \ln(1 - xy)$.
9. $f(x, y) = \dfrac{1}{\sqrt{y - x^2}}$.
10. $f(x,y) = \dfrac{\sqrt{9 - x^2}}{1 + \sqrt{1 - y^2}}$.
11. $f(x, y) = \sqrt{9 - x^2} - \sqrt{4 - y^2}$.
12. $f(x, y, z) = \cos x + \cos y + \cos z$.
13. $f(x, y, z) = \dfrac{x + y + z}{|x + y + z|}$.
14. $f(x, y, z) = \dfrac{z^2}{x^2 - y^2}$.
15. $f(x, y, z) = -\dfrac{z^2}{\sqrt{x^2 - y^2}}$.
16. $f(x, y, z) = \dfrac{z}{x - y}$.
17. $f(x, y) = \dfrac{2}{\sqrt{9 - (x^2 + y^2)}}$.
18. $f(x, y, z) = \ln(|x + 2y + 3z| + 1)$.
19. $f(x, y, z) = \ln(x + 2y + 3z)$.
20. $f(x, y, z) = e^{\sqrt{4-(x^2+y^2+z^2)}}$.
21. $f(x, y, z) = e^{-(x^2+y^2+z^2)}$.
22. $f(x, y, z) = \dfrac{\sqrt{1 - x^2} + \sqrt{4 - y^2}}{1 + \sqrt{9 - z^2}}$.
23. Let $f(x) = \sqrt{x}$, $g(x, y) = \sqrt{x}$, $h(x, y, z) = \sqrt{x}$. Determine the domain and range of each function and compare the results. Sketch the graphs of f and g.
24. Let $f(x, y) = \cos \pi x \sin \pi y$ and $g(x, y, z) = \cos \pi x \sin \pi y$. Determine the domain and range of each of these functions and compare the results.
25. (C) Let $f(x, y) = \sqrt{1 - 4xy}$. Use a graphing utility to draw the graph of $g(x,y) = 1 - 4xy = 0$. Use this graph to help determine the domain of f.
26. (C) Repeat Exercise 25 with $f(x,y) = \ln(x^2 + y - 2)$ and $g(x, y) = x^2 + y - 2$.

In Exercises 27–32, form the difference quotients

$$\frac{f(x + h,y) - f(x,y)}{h} \quad \text{and} \quad \frac{f(x,y + h) - f(x,y)}{h}, \quad (h \neq 0).$$

Then, assuming that x and y are fixed, calculate the limit as $h \to 0$.

27. $f(x,y) = 2x^2 - y$.
28. $f(x,y) = xy + 2y$.
29. $f(x,y) = 3x - xy + 2y^2$.
30. $f(x,y) = x \sin y$.
31. $f(x,y) = \cos(xy)$.
32. $f(x,y) = x^2 e^y$.
33. (C) Use a CAS to find the difference quotients for $f(x,y) = \dfrac{x - 2y}{x + y}$.
    (a) $\dfrac{f(x + h,y) - f(x,y)}{h}$.
    (b) $\dfrac{f(x,y + h) - f(x,y)}{h}$.
    (c) Calculate the limits as $h \to 0$.
34. (C) Repeat Exercise 33 for the function $f(x,y) = x^2 - 4xy + y^2$.
35. Express each of the following as a function of two variables.
    (a) The volume of a box with a square base of side length x and height y.
    (b) The volume of a right circular cylinder whose radius is x and whose height is y.
    (c) The area of the parallelogram whose sides are the vectors $2\mathbf{i}$ and $x\,\mathbf{i} + y\,\mathbf{j}$.
36. Express each of the following as a function of three variables.
    (a) The surface area of a box with no top whose sides have lengths x, y and z.
    (b) The angle between the vectors $\mathbf{i} + \mathbf{j}$ and $x\,\mathbf{i} + y\,\mathbf{j} + z\,\mathbf{k}$.
    (c) The volume of the parallelepiped whose sides are the vectors $\mathbf{i}, \mathbf{i} + \mathbf{j}$, and $x\,\mathbf{i} + y\,\mathbf{j} + z\,\mathbf{k}$.
37. A closed box is to have a total surface area of 20 square feet. Express the volume V of the box as a function of the length l and the height h.
38. An open box is to contain a volume of 12 cubic meters. Given that the material for the sides of the box costs \$2 per square meter and the material for the bottom costs \$4 per square meter, express the total cost C of the box as a function of the length l and width w.
39. A petrochemical company is designing a cylindrical tank with hemispherical ends to be used in the transportation of its products. (See the figure.) Express the volume of the tank as a function of the radius r and the length h of the cylindrical portion.

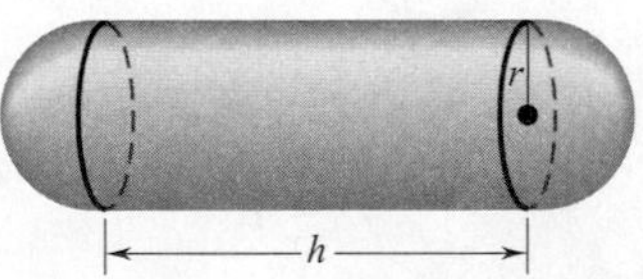

40. A 10-foot section of gutter is to be made from a 12-inch-wide strip of metal by folding up strips of length x on each side so that they make an angle θ with the bottom of the gutter. (See the figure.) Express the area of the trapezoidal cross section as a function of x and θ.

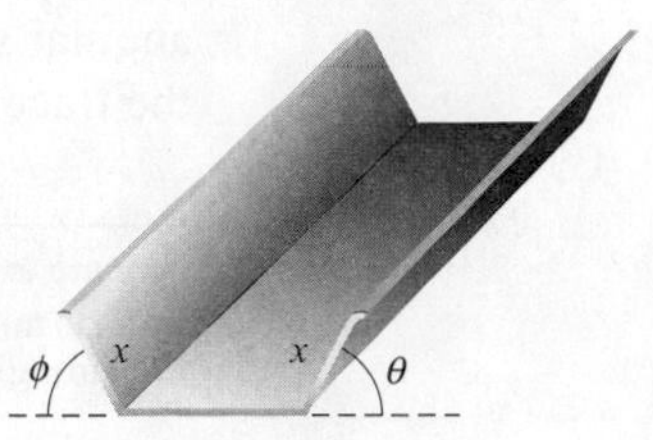

■ 14.2 A BRIEF CATALOG OF THE QUADRIC SURFACES; PROJECTIONS

As you know, the graph of an equation in x and y is typically a curve in the xy-plane. As you will see in this section, and in the sections which follow, the graph of an equation in three variables is, in general, a surface in three-space.

In this section we examine in a systematic manner the surfaces defined by equations of the form

$$Ax^2 + By^2 + Cz^2 + Dxy + Exz + Fyz + Hx + Iy + Jz + K = 0$$

where $A, B, C, \ldots, K$ are constants. Such surfaces are called *quadric surfaces.*

By suitable translations and rotations of the coordinate axes (see Section 9.1 and Appendix A.1), we can simplify such equations and thereby show that the nondegenerate† quadrics fall into nine distinct types:

1. The ellipsoid.
2. The hyperboloid of one sheet.
3. The hyperboloid of two sheets.
4. The elliptic cone.
5. The elliptic paraboloid.
6. The hyperbolic paraboloid.
7. The parabolic cylinder.
8. The elliptic cylinder.
9. The hyperbolic cylinder.

As you go on with calculus, you will encounter these surfaces time and time again. Here we give you a picture of each one, together with its equation in standard form and some information about its special properties. These are some of the things to look for:

(a) The *intercepts* (the points at which the surface intersects the coordinate axes).
(b) The *traces* (the intersections with the coordinate planes).
(c) The *sections* (the intersections with planes in general).
(d) The *center* (some quadrics have a center; some do not).
(e) *Symmetry.*
(f) *Boundedness, unboundedness.*

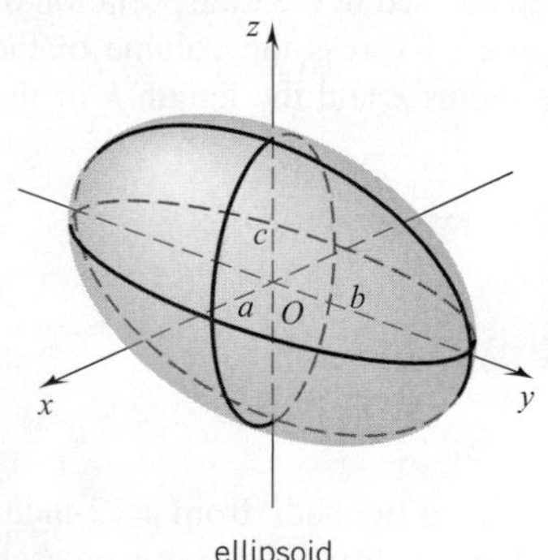

Figure 14.2.1

The Ellipsoid

$$\frac{x^2}{a^2} + \frac{y^2}{b^2} + \frac{z^2}{c^2} = 1.\dagger\dagger \qquad \text{(Figure 14.2.1)}$$

This ellipsoid is centered at the origin and is symmetric about the three coordinate planes. It intersects the coordinate axes at six points: $(\pm a, 0, 0), (0, \pm b, 0), (0, 0, \pm c)$. These points are called the *vertices*. The surface is bounded, being contained in the rectangular solid: $|x| \le a, \ |y| \le b, \ |z| \le c$. All three traces are ellipses; thus, for example, the trace in the xy-plane (set $z = 0$) is the ellipse

†We are excluding such degenerate quadrics as $x^2 + y^2 + z^2 + 1 = 0$ and $x^2 + y^2 + z^2 = 0$. The first one has no points and the second consists of only one point, the origin.
†† Throughout this section we take a, b, c as positive constants.

$$\frac{x^2}{a^2} + \frac{y^2}{b^2} = 1.$$

Sections parallel to the coordinate planes are also ellipses; for example, taking $y = y_0$ we have

$$\frac{x^2}{a^2} + \frac{z^2}{c^2} = 1 - \frac{y_0^2}{b^2}.$$

This ellipse is the intersection of the ellipsoid with the plane $y = y_0$. The numbers a, b, c are called the *semiaxes* of the ellipsoid. If two of the semiaxes are equal, then we have an *ellipsoid of revolution.* (If, for example, $a = c$, then all sections parallel to the xz-plane are circles and the surface can be obtained by revolving the trace in the xy-plane about the y-axis.) If all three semiaxes are equal, the surface is a *sphere*.

The Hyperboloid of One Sheet

$$\frac{x^2}{a^2} + \frac{y^2}{b^2} - \frac{z^2}{c^2} = 1. \qquad \text{(Figure 14.2.2)}$$

The surface is unbounded. It is centered at the origin and is symmetric about the three coordinate planes. The surface intersects the coordinate axes at four points: $(\pm a,\ 0,\ 0), (0,\ \pm b,\ 0)$. The trace in the xy-plane (set $z = 0$) is the ellipse

$$\frac{x^2}{a^2} + \frac{y^2}{b^2} = 1.$$

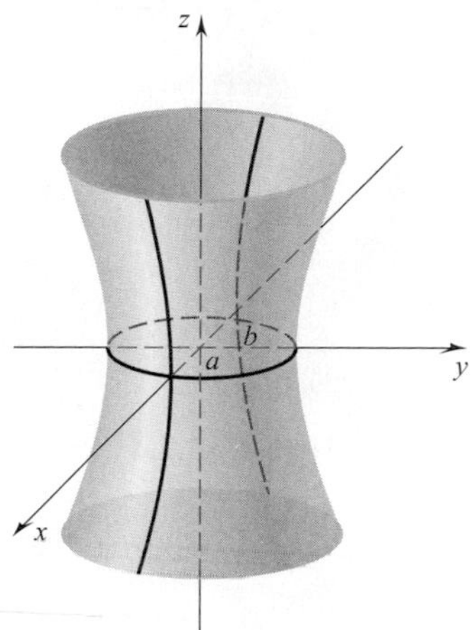

hyperboloid of one sheet

Figure 14.2.2

Sections parallel to the xy-plane are ellipses. The trace in the xz-plane (set $y = 0$) is the hyperbola

$$\frac{x^2}{b^2} - \frac{z^2}{c^2} = 1,$$

and the trace in the yz-plane (set $x = 0$) is the hyperbola

$$\frac{y^2}{b^2} - \frac{z^2}{c^2} = 1.$$

Sections parallel to the xz-plane or yz-plane are hyperbolas. If $a = b$, then sections parallel to the xy-plane are circles and we have a *hyperboloid of revolution*.

The Hyperboloid of Two Sheets

$$\frac{x^2}{a^2} + \frac{y^2}{b^2} - \frac{z^2}{c^2} = -1. \qquad \text{(Figure 14.2.3)}$$

The surface intersects the coordinate axes only at the two vertics $(0,\ 0,\ \pm c)$. The surface consists of two parts: one for which $z \geq c$, another for which $z \leq -c$. We can see this by rewriting the equation as

$$\frac{x^2}{a^2} + \frac{y^2}{b^2} = \frac{z^2}{c^2} - 1.$$

The equation requires

$$\frac{z^2}{c^2} - 1 \geq 0, \quad z^2 \geq c^2, \quad |z| \geq c.$$

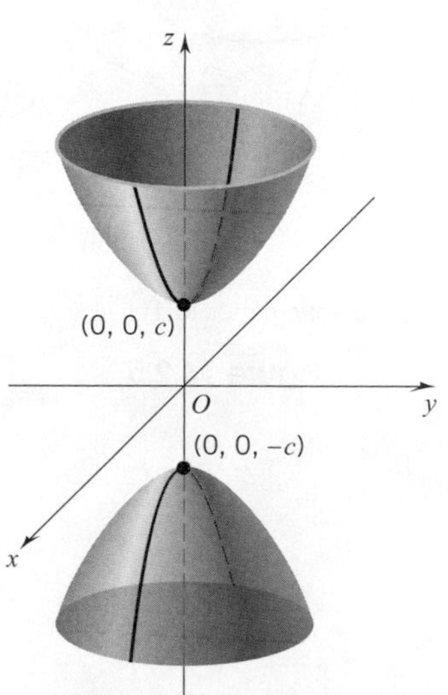

hyperboloid of two sheets

Figure 14.2.3

Each of the two parts is unbounded. Sections parallel to the xy-plane are ellipses: setting $z = z_0$ with $|z_0| \geq c$, we have

$$\frac{x^2}{a^2} + \frac{y^2}{b^2} = \frac{z_0^2}{c^2} - 1.$$

If $a = b$, then all sections parallel to the xy-plane are circles and we have a hyperboloid of revolution. Sections parallel to the other coordinate planes are hyperbolas; for example, setting $y = y_0$, we have

$$\frac{z^2}{c^2} - \frac{x^2}{a^2} = 1 + \frac{y_0^2}{b^2}.$$

The entire surface is symmetric about the three coordinate planes and is centered at the origin.

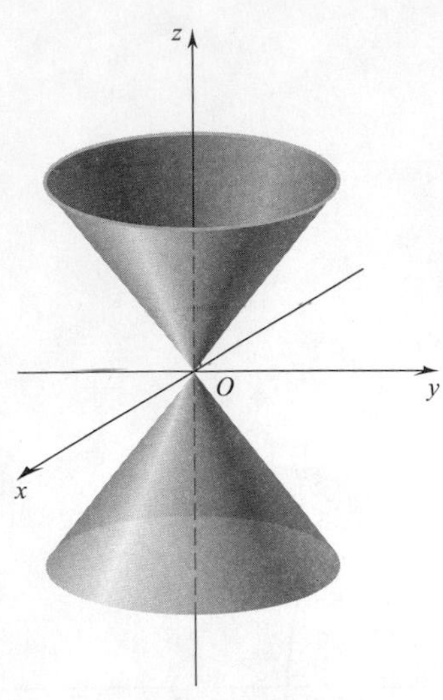

elliptic cone

Figure 14.2.4

The Elliptic Cone

$$\frac{x^2}{a^2} + \frac{y^2}{b^2} = z^2. \qquad \text{(Figure 14.2.4)}$$

The surface intersects the coordinate axes only at the origin. The surface is unbounded. Once again there is symmetry about the three coordinate planes. The trace in the xz-plane is a pair of intersecting lines: $z = \pm x/a$. The trace in the yz-plane is also a pair of intersecting lines: $z = \pm y/b$. The trace in the xy-plane is just the origin. Sections parallel to the xy-plane are ellipses. If $a = b$, these sections are circles and we have a surface of revolution, what is commonly called a *double circular cone* or simply a *cone*. The upper and lower portions of the cone are called *nappes.*

We come now to the *paraboloids*. The equations in standard form will involve x^2 and y^2, but then z instead of z^2.

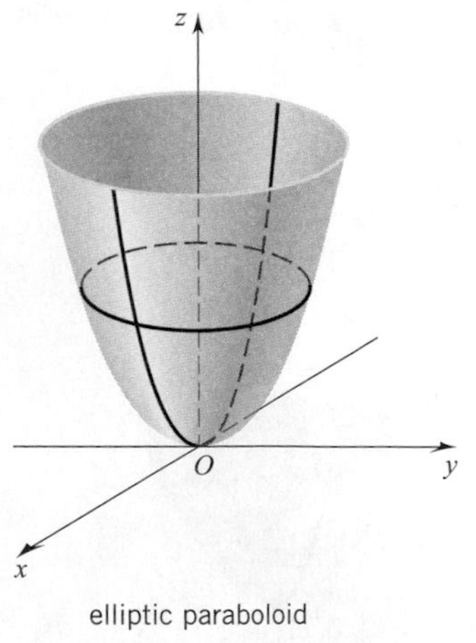

elliptic paraboloid

Figure 14.2.5

The Elliptic Paraboloid

$$\frac{x^2}{a^2} + \frac{y^2}{b^2} = z. \qquad \text{(Figure 14.2.5)}$$

The surface does not extend below the xy-plane; it is unbounded above. The origin is called the *vertex.* Sections parallel to the xy-plane are ellipses: sections parallel to the other coordinate planes are parabolas. Hence the term "elliptic paraboloid". The surface is symmetric about the xz-plane and about the yz-plane. It is also symmetric about the z-axis. If $a = b$, then the surface is a *paraboloid of revolution*.

The Hyperbolic Paraboloid

$$\frac{x^2}{a^2} - \frac{y^2}{b^2} = z. \qquad \text{(Figure 14.2.6)}$$

Here there is symmetry about the xz-plane and yz plane. Sections parallel to the xy-plane are hyperbolas; sections parallel to the other coordinate planes are parabolas. Hence the term "hyperbolic paraboloid." The origin is a minimum point for the trace in the xz-plane, but a maximum point for the trace in the yz-plane. The origin is called a *minimax* or *saddle point* of the surface. NOTE: The orientation of the coordinate axes was chosen to enhance our view of the surface.

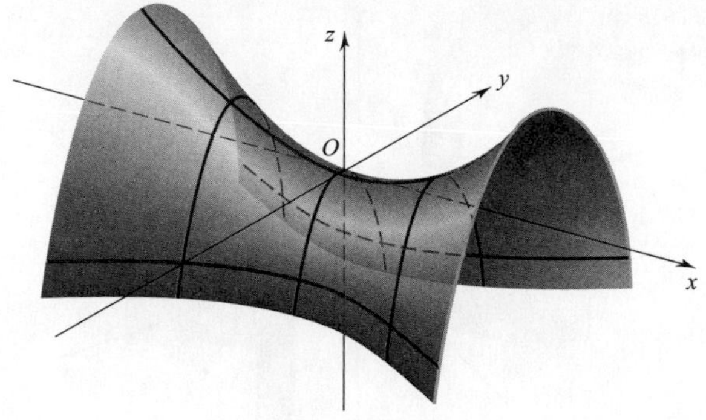

hyperbolic paraboloid

Figure 14.2.6

The rest of the quadric surfaces are *cylinders*. The term deserves definition. Take any plane curve C. All the lines through C that are perpendicular to the plane of C form a surface. Such a surface is called a *cylinder*, the cylinder with *base curve C*. The perpendicular lines are known as the *generators* of the cylinder.

If the base curve lies in the xy-plane (or in a plane parallel to the xy-plane), then the generators of the cylinder are parallel to the z-axis. In such a case the equation of the cylinder involves only x and y. The z-coordinate is left unrestricted; it can take on all values.

There are three basic types of quadric cylinders. We give you their equations in standard form: base curve in the xy-plane, generators parallel to the z-axis.

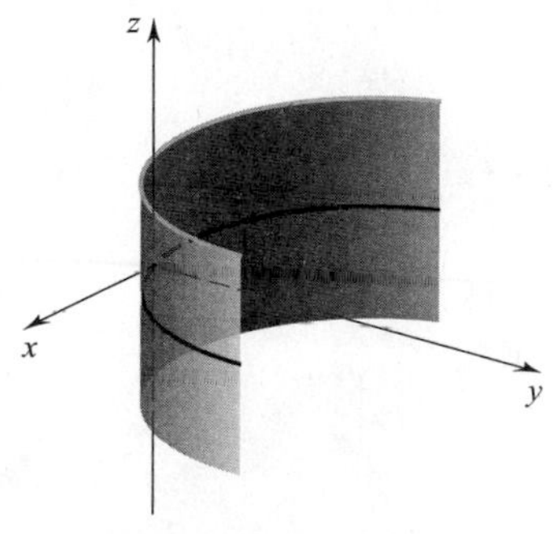

parabolic cylinder

Figure 14.2.7

The Parabolic Cylinder

$$x^2 = 4cy. \qquad \text{(Figure 14.2.7)}$$

This surface is formed by all lines that pass through the parabola $x^2 = 4cy$ and are perpendicular to the xy-plane.

The Elliptic Cylinder

$$\frac{x^2}{a^2} + \frac{y^2}{b^2} = 1. \qquad \text{(Figure 14.2.8)}$$

The surface is formed by all lines that pass through the ellipse

$$\frac{x^2}{a^2} + \frac{y^2}{b^2} = 1$$

and are perpendicular to the xy-plane. If $a = b$, we have the common *right circular cylinder*.

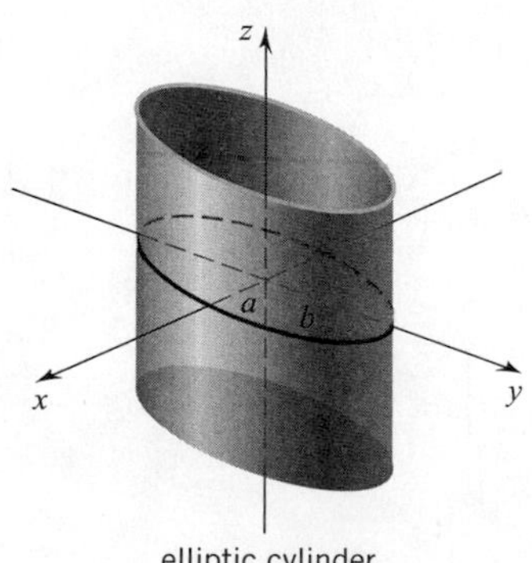

elliptic cylinder

Figure 14.2.8

The Hyperbolic Cylinder

$$\frac{x^2}{a^2} - \frac{y^2}{b^2} = 1. \qquad \text{(Figure 14.2.9)}$$

The surface has two parts, each generated by a branch of the hyperbola

$$\frac{x^2}{a^2} - \frac{y^2}{b^2} = 1.$$

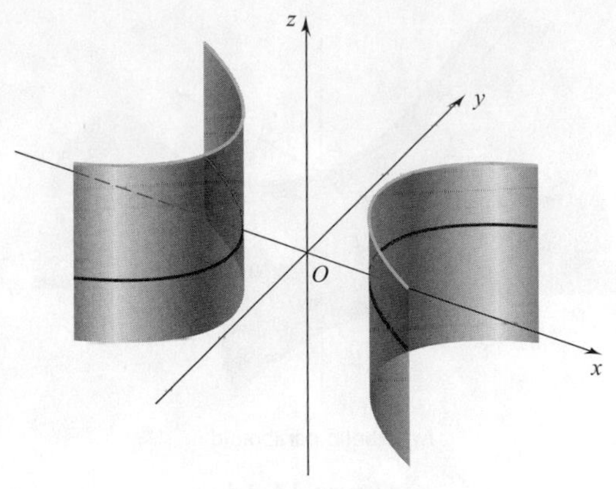

hyperbolic cylinder

Figure 14.2.9

Projections

Suppose that $S_1 : z = f(x, y)$ and $S_2 : z = g(x, y)$ are surfaces in three-space that intersect in a space curve C. (See Figure 14.2.10.)

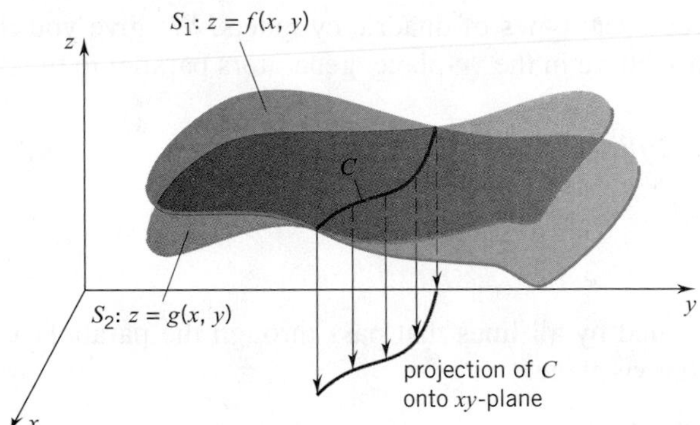

Figure 14.2.10

The curve C is the set of all points (x, y, z) with

$$z = f(x, y) \qquad \text{and} \qquad z = g(x, y).$$

The set of all points (x, y, z) with

$$f(x, y) = g(x, y) \qquad \text{(here z is unrestricted)}$$

is the vertical cylinder that passes through C.

The set of all points $(x, y, 0)$ with

$$f(x, y) = g(x, y) \qquad \text{(here } z = 0\text{)}$$

is called the *projection of C onto the xy-plane*. In Figure 14.2.10 it appears as the curve in the xy-plane that lies directly below C.

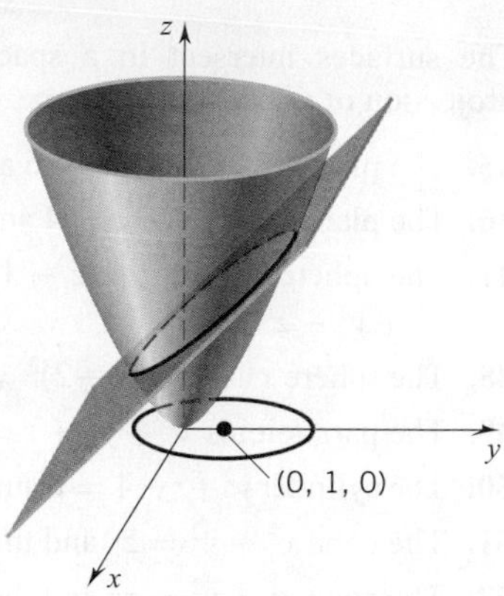

Figure 14.2.11

Example 1 The paraboloid of revolution $z = x^2 + y^2$ and the plane

$$z = 2y + 3$$

intersect in a curve C. See Figure 14.2.11. The projection of this curve onto the xy-plane is the set of all points $(x, y, 0)$ with

$$x^2 + y^2 = 2y + 3.$$

This equation can be written

$$x^2 + (y - 1)^2 = 4.$$

The projection of C onto the xy-plane is the circle of radius 2 centered at $(0, 1, 0)$. □

EXERCISES 14.2

Identify the surface.

1. $x^2 + 4y^2 - 16z^2 = 0.$
2. $x^2 + 4y^2 + 16z^2 - 12 = 0.$
3. $x - 4y^2 = 0.$
4. $x^2 - 4y^2 - 2z = 0.$
5. $5x^2 + 2y^2 - 6z^2 - 10 = 0.$
6. $2x^2 + 4y^2 - 1 = 0.$
7. $x^2 + y^2 + z^2 - 4 = 0.$
8. $5x^2 + 2y^2 - 6z^2 + 10 = 0.$
9. $x^2 + 2y^2 - 4z = 0.$
10. $2x^2 - 3y^2 - 6 = 0.$
11. $x - y^2 + 2z^2 = 0.$
12. $x - y^2 - 6z^2 = 0.$

Sketch the cylinder.

13. $25y^2 + 4z^2 - 100 = 0.$
14. $25x^2 + 4y^2 - 100 = 0.$
15. $y^2 - z = 0.$
16. $x^2 - y + 1 = 0.$
17. $y^2 + z = 0.$
18. $25x^2 - 9y^2 - 225 = 0.$
19. $x^2 + y^2 = 9.$
20. $\dfrac{x^2}{4} + \dfrac{y^2}{9} = 1.$
21. $y^2 - 4x^2 = 4.$
22. $z = x^2.$
23. $y = x^2 + 1.$
24. $(x - 1)^2 + (y - 1)^2 = 1.$

Identify the surface and find the traces. Then sketch the surface.

25. $9x^2 + 4y^2 - 36z = 0.$
26. $9x^2 + 4y^2 + 36z^2 - 36 = 0.$
27. $9x^2 + 4y^2 - 36z^2 = 0.$
28. $9x^2 + 4y^2 - 36z^2 - 36 = 0.$
29. $9x^2 + 4y^2 - 36z^2 + 36 = 0.$
30. $9x^2 - 4y^2 - 36z = 0.$
31. $9x^2 - 4y^2 - 36z^2 = 36.$
32. $9x^2 + 4z^2 - 36y^2 - 36 = 0.$
33. $4x^2 + 9z^2 - 36y = 0.$
34. $9x^2 + 4z^2 - 36y^2 = 0.$
35. $9y^2 - 4x^2 - 36z^2 - 36 = 0.$
36. $9y^2 + 4z^2 - 36x = 0.$
37. $x^2 + y^2 - 4z = 0.$
38. $36x^2 + 9y^2 + 4z^2 - 36 = 0.$
39. Identify all possibilities for the surface
$$z = Ax^2 + By^2$$
taking (a) $AB > 0.$ (b) $AB < 0.$ (c) $AB = 0.$
40. Find the planes of symmetry for the cylinder $x - 4y^2 = 0.$
41. Write an equation for the surface obtained by revolving the parabola $4z - y^2 = 0$ about the z-axis.
42. The hyperbola $c^2y^2 - b^2z^2 - b^2c^2 = 0$ is revolved about the z-axis. Find an equation for the resulting surface.
43. (a) The equation
$$\sqrt{x^2 + y^2} = kz \qquad \text{with } k > 0$$
represents the upper nappe of a cone, with vertex at the origin and the positive z-axis as the axis of symmetry. Describe the section in the plane $z = z_0$, $z_0 > 0$.
(b) Let S be one nappe of a cone, with vertex at the origin. Write an equation for S given that
(i) the negative z-axis is the axis of symmetry and the section in the plane $z = -2$ is a circle of radius 6,
(ii) the positive y-axis is the axis of symmetry and the section in the plane $y = 3$ is a circle of radius 1.
44. Form the elliptic paraboloid
$$x^2 + \frac{y^2}{b^2} = z.$$
(a) Describe the section in the plane $z = 1$.
(b) What happens to this section as b tends to infinity?
(c) What happens to the paraboloid as b tends to infinity?

The surfaces intersect in a space curve C. Determine the projection of C onto the xy-plane.

45. The planes $x + 2y + 3z = 6$ and $x + y - 2z = 6$.

46. The planes $x - 2y + z = 4$ and $3x + y - 2z = 1$.

47. The sphere $x^2 + y^2 + (z - 1)^2 = \frac{3}{2}$ and the hyperboloid $x^2 + y^2 - z^2 = 1$.

48. The sphere $x^2 + y^2 + (z - 2)^2 = 2$ and the cone $x^2 + y^2 = z^2$.

49. The paraboloids $x^2 + y^2 + z = 4$ and $x^2 + 3y^2 = z$.

50. The cylinder $y^2 + z - 4 = 0$ and the paraboloid $x^2 + 3y^2 = z$.

51. The cone $x^2 + y^2 = z^2$ and the plane $y + z = 2$.

52. The cone $x^2 + y^2 = z^2$ and the plane $y + 2z = 2$.

53. The ellipsoid $\dfrac{x^2}{a^2} + \dfrac{y^2}{b^2} + \dfrac{z^2}{c^2} = 1$ can be parametrized by the vector function of two variables

$$\mathbf{r}(u, v) = a \cos u \cos v\,\mathbf{i} + b \cos u \sin v\,\mathbf{j} + c \sin u\,\mathbf{k}.$$

(a) Verify that $\mathbf{r}$ parametrizes an ellipsoid.

(b) Use a graphing utility to draw the ellipsoid with $a = 3$, $b = 4$, $c = 2$.

(c) Experiment with other values of a, b, c to see how the ellipsoid changes shape. How would you choose a, b, c to obtain a sphere?

54. The hyperboloid of one sheet $\dfrac{x^2}{a^2} + \dfrac{y^2}{b^2} - \dfrac{z^2}{c^2} = 1$ can be parametrized by the vector function of two variables

$$\mathbf{r}(u, v) = a \sec u \cos v\,\mathbf{i} + b \sec u \sin v\,\mathbf{j} + c \tan u\,\mathbf{k}.$$

(a) Verify that $\mathbf{r}$ parametrizes a hyperboloid.

(b) Use a graphing utility to draw the hyperboloid with $a = 2, b = 3, c = 4$.

(c) Experiment with other values of a, b, c to see how the hyperboloid changes shape.

55. The elliptic cone $\dfrac{x^2}{a^2} + \dfrac{y^2}{b^2} = \dfrac{z^2}{c^2}$ can be parametrized by the vector function of two variables

$$\mathbf{r}(u, v) = a v \cos u\,\mathbf{i} + b v \sin u\,\mathbf{j} + c v\,\mathbf{k}$$

(a) Verify that $\mathbf{r}$ parametrizes an elliptic cone.

(b) Use a graphing utility to draw the elliptic cone with $a = 1$, $b = 2$, $c = 3$.

(c) Experiment with other values of a, b, c to see how the cone changes shape. In particular, what effect does c have on the cone?

■ 14.3 GRAPHS; LEVEL CURVES AND LEVEL SURFACES

We begin with a function f of two variables defined on a subset D of the xy-plane. By the *graph of f* we mean the graph of the equation

$$z = f(x, y) \qquad (x, y) \in D.$$

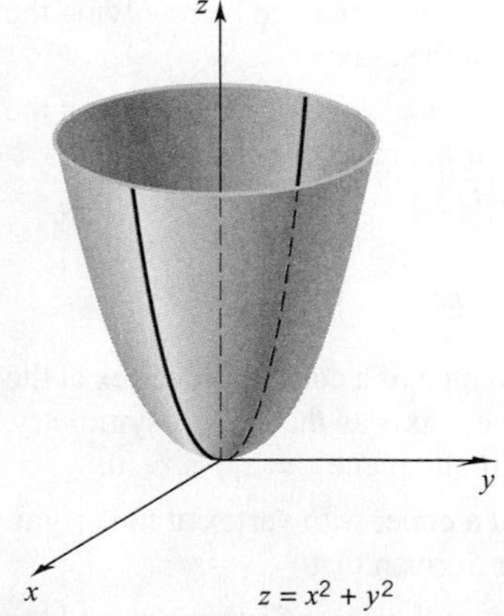

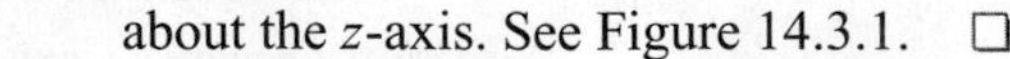

Figure 14.3.1

Example 1 In the case of $f(x, y) = x^2 + y^2$, the domain is the entire plane. The graph of f is a paraboloid of revolution:

$$z = x^2 + y^2.$$

This surface can be generated by revolving the parabola

$$z = x^2 \qquad \text{(in the xz-plane)}$$

about the z-axis. See Figure 14.3.1. □

Example 2 Let a, b, and c be positive constants. The domain of the function

$$g(x, y) = c - ax - by$$

is the entire xy-plane. The graph of g is the plane

$$z = c - ax - by \qquad \text{(Figure 14.3.2)}$$

with intercepts $x = c/a$, $y = c/b$, $z = c$. □

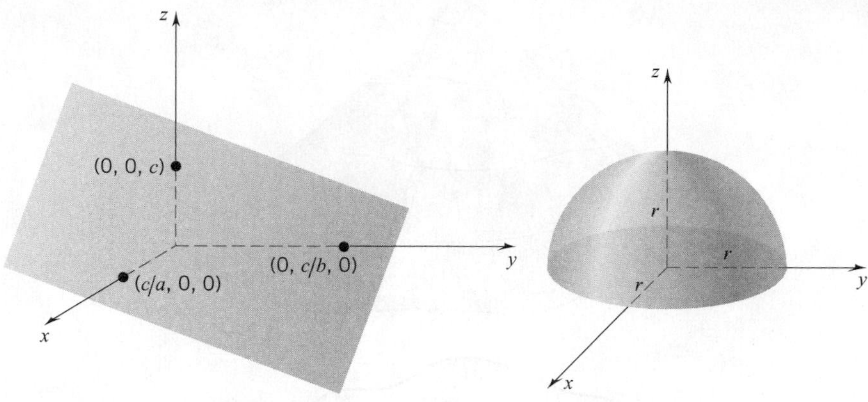

Figure 14.3.2

Figure 14.3.3

Example 3 The function $f(x,y) = \sqrt{r^2 - (x^2 + y^2)}$, $r > 0$ is defined only on the closed disc $x^2 + y^2 \leq r^2$. The graph is the surface

$$z = \sqrt{r^2 - (x^2 + y^2)}. \qquad \text{(Figure 14.3.3)}$$

This is the upper half of the sphere

$$x^2 + y^2 + z^2 = r^2. \quad \square$$

Example 4 The function $f(x,y) = xy$ is simple enough, but its graph, the surface $z = xy$, is quite difficult to draw. It is a "saddle-shaped" surface, a hyperbolic paraboloid: rotate the x and y axes by $\frac{1}{4}\pi$ radians and the equation takes the form

$$\frac{X^2}{2} - \frac{Y^2}{2} = z.\dagger$$

Try to visualize the surface in Figure 14.2.6 rotated $\frac{1}{4}\pi$ radians in the clockwise direction. $\square$

Level Curves

In practice, the graph of a function of two variables is difficult to visualize and difficult to draw. Moreover, if drawn, the drawing is often difficult to interpret. Computer-generated graphics can be very useful in resolving these matters. We will provide some illustration of this later in the section.

Here we discuss an approach which we take from the map maker. In mapping mountainous terrain, it is common practice to sketch curves joining points of constant elevation. The collection of such curves, called a topographic map, gives a good idea of the altitude variations in a region and suggests the shape of the mountains and valleys. (See Figure 14.3.5)

We can apply this technique to functions of two variables. Suppose that f is a nonconstant function defined on some portion of the xy-plane. If c is a value in the range of f, then we can sketch the curve $f(x, y) = c$. Such a curve is called a *level curve* for f. It can be obtained by intersecting the graph of f with the horizontal plane $z = c$ and then projecting that intersection onto the xy-plane. (See Figure 14.3.4)

The level curve $f(x,y) = c$ lies entirely in the domain of f, and on this curve f is constantly C.

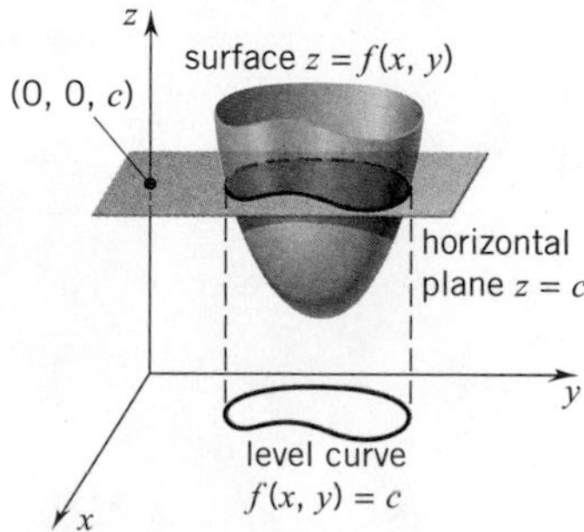

Figure 14.3.4

† To see this, set $\alpha = \frac{1}{4}\pi$ in Formula A.1.2, Appendix A-1.

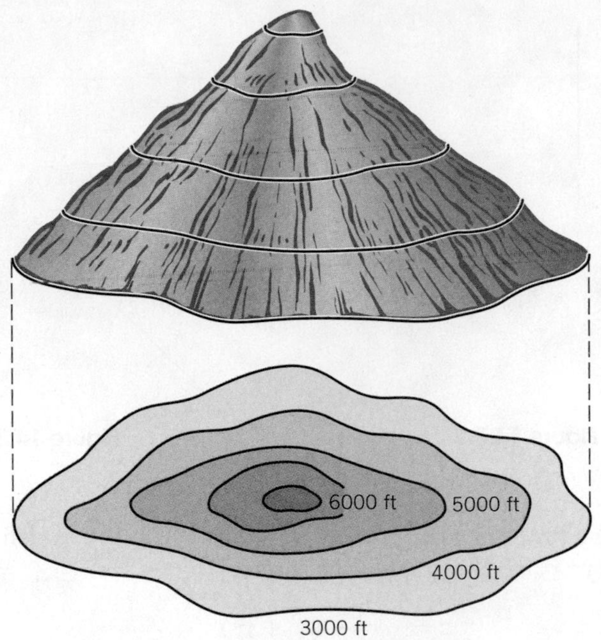

Figure 14.3.5

A collection of level curves, properly drawn and labeled, can lead to a good understanding of the overall behavior of a function.

Example 5 We begin with function $f(x, y) = x^2 + y^2$ (see Figure 14.3.1). The level curves are circles centered at the origin:

$$x^2 + y^2 = c, \qquad c \geq 0. \qquad \text{(Figure 14.3.6)}$$

The function has the value c on the circle of radius $\sqrt{c}$ centered at the origin. At the origin, the function has the value 0. ❑

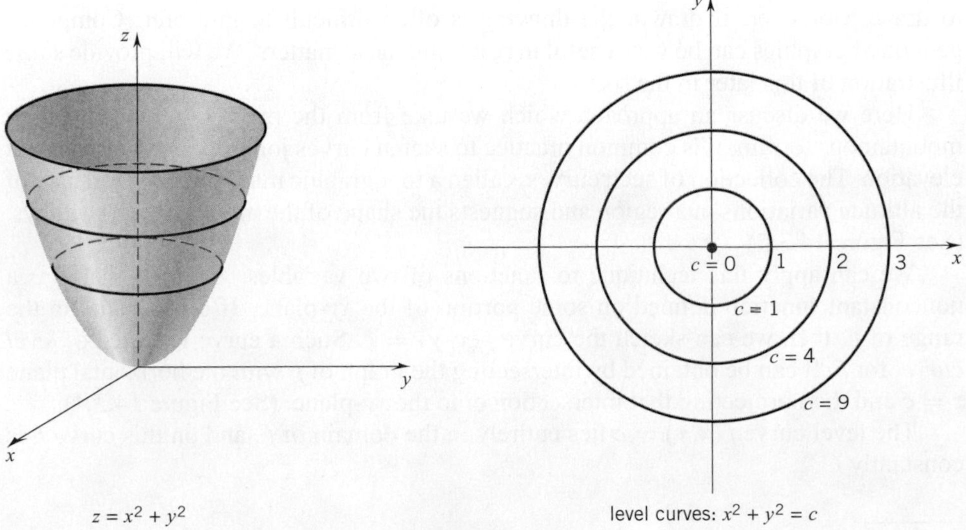

Figure 14.3.6

Example 6 The graph of the function $g(x, y) = 4 - x - y$ is a plane. The level curves are parallel lines of the form $4 - x - y = c$.

The surface and the level curves are indicated in Figure 14.3.7. □

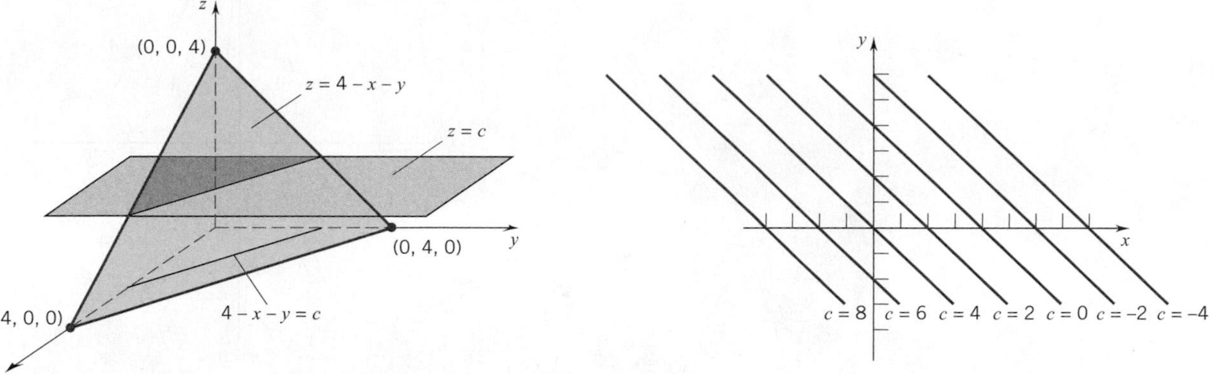

Figure 14.3.7

Example 7 Now we consider the function

$$h(x, y) = \begin{cases} \sqrt{x^2 + y^2}, & x \geq 0 \\ |y|, & x < 0. \end{cases}$$

For $x \geq 0, h(x, y)$ is the distance from (x, y) to the origin. For $x < 0, h(x, y)$ is the distance from (x, y) to the x-axis. The level curves are pictured in Figure 14.3.8. The 0-level curve is the nonpositive x-axis. The other level curves are horseshoe-shaped: pairs of horizontal rays capped on the right by semicircles. □

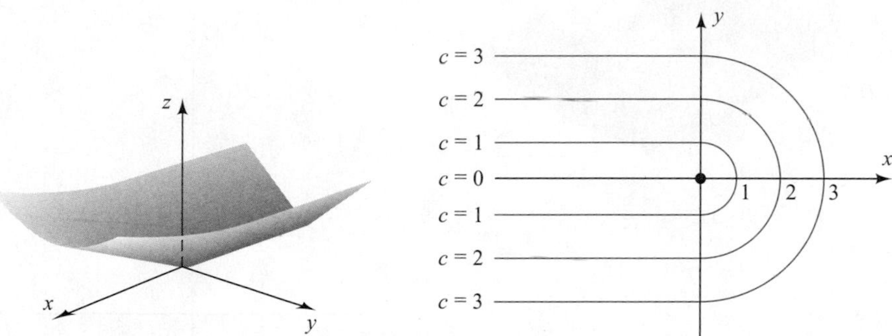

Figure 14.3.8

Example 8 Let's return to the function $f(x, y) = xy$. Earlier we noted that the graph is a "saddled-shaped" surface. You can visualize the surface from the few level curves sketched in Figure 14.3.9. The 0-level curve, $xy = 0$, consists of the two coordinate axes. The other level curves, $xy = c$ with $c \neq 0$, are hyperbolas.

Computer-Generated Graphs

The preceding examples illustrate how difficult it is to sketch an accurate graph of a function of two variables. But powerful help is at hand. Three-dimensional graphing programs for modern computers make it possible to visualize even quite complicated surfaces. These programs allow the user to view a surface from different perspectives, and they show level curves and the sections in various planes. Examples of computer-generated graphs are shown in Figure 14.3.10 and in the Exercises.

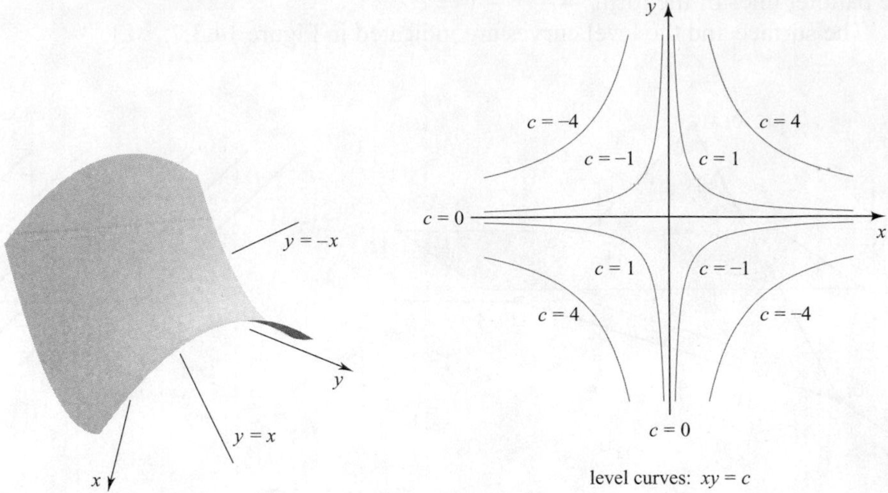

Figure 14.3.9

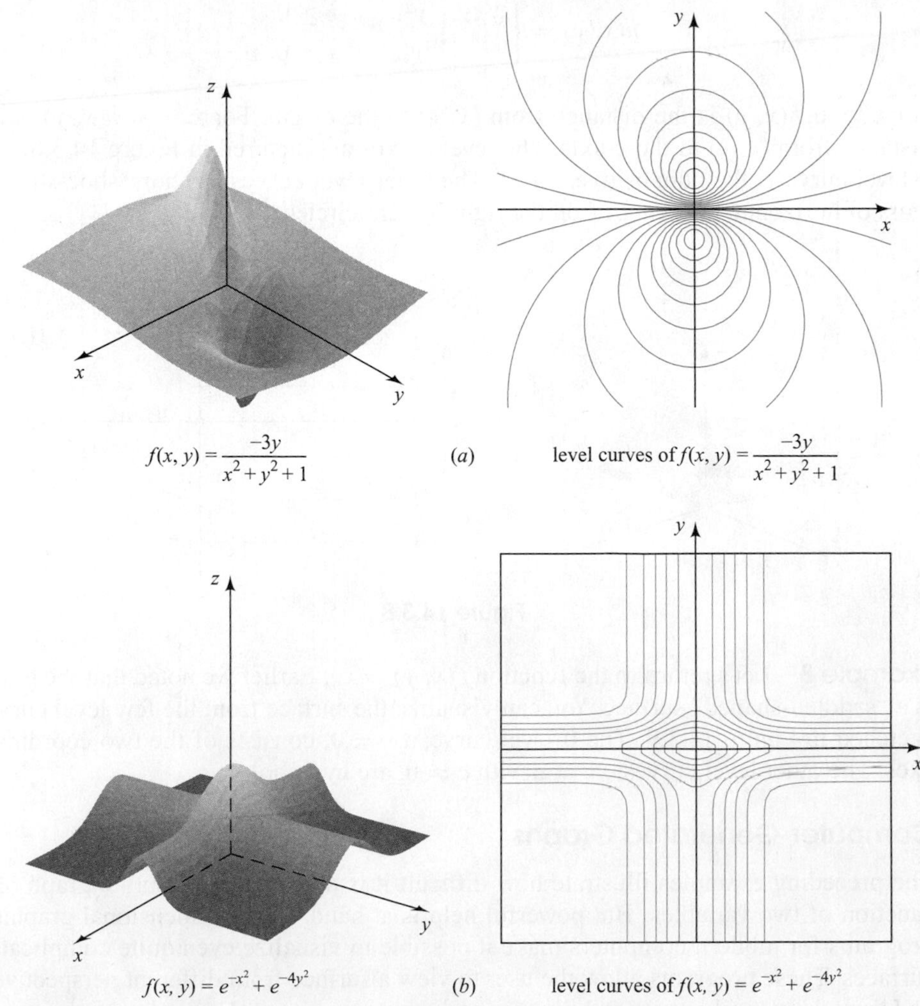

Figure 14.3.10

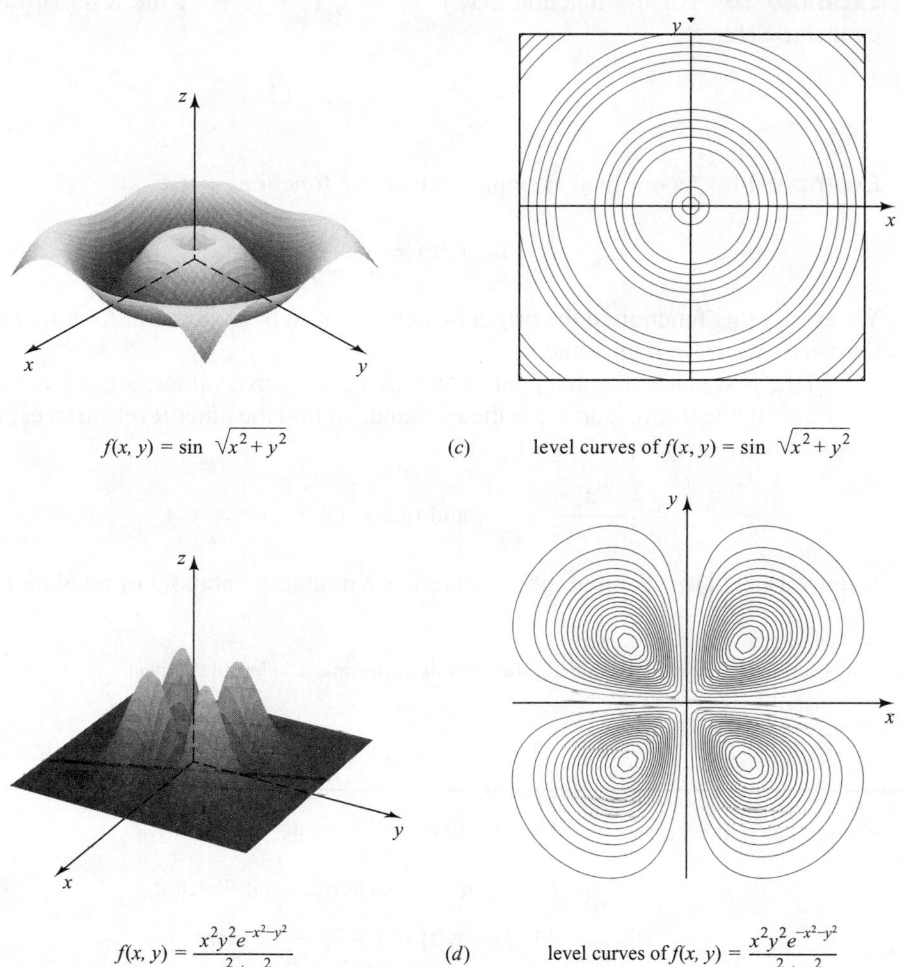

Figure 14.3.10 ***Continued***

Level Surfaces

While drawing graphs for functions of two variables is quite difficult, drawing graphs for functions of three variables is actually impossible. To draw such figures we would need four dimensions at our disposal; the domain itself is a portion of three-space.

One can try to visualize the behavior of a function of three variables, $w = f(x, y, z)$, by examining the *level surfaces* of f. These are the subsets of the domain of f with equations of the form

$$f(x,y,z) = c$$

where c is a value in the range of f.

Level surfaces are usually difficult to draw. Nevertheless, a knowledge of what they are can be helpful. Here we restrict ourselves to a few simple examples.

Example 9 For the function $f(x,y,z) = Ax + By + Cz$, the level surfaces are parallel planes

$$Ax + By + Cz = c. \quad \square$$

Example 10 For the function $g(x, y, z) = \sqrt{x^2 + y^2 + z^2}$, the level surfaces are concentric spheres

$$x^2 + y^2 + z^2 = c^2. \quad \square$$

Example 11 As our final example we take the function

$$f(x, y, z) = \frac{|z|}{x^2 + y^2}.$$

We extend this function to the origin by defining it to be zero there. At other points of the z-axis we leave f undefined.

In the first place note that f takes on only nonnegative values. Since f is zero only when $z = 0$, the 0-level surface is the xy-plane. To find the other level surfaces, we take $c > 0$ and set $f(x, y, z) = c$. This gives

$$\frac{|z|}{x^2 + y^2} = c \quad \text{and thus} \quad |z| = c(x^2 + y^2)$$

(Figure 14.3.11). Each of these surfaces is a double-paraboloid of revolution.† $\square$

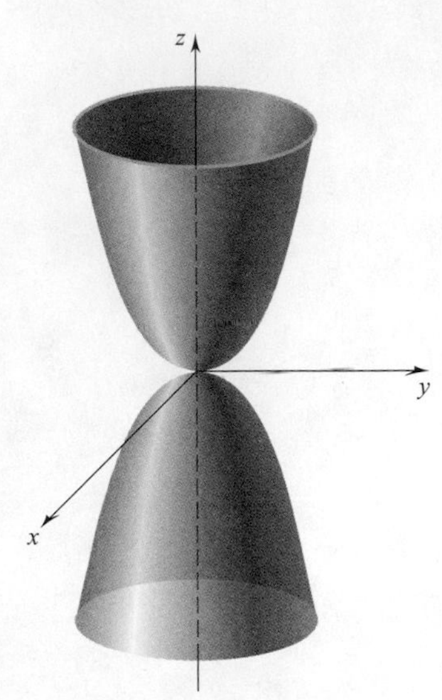

level surface: $|z| = c(x^2 + y^2)$, $(c > 0)$

Figure 14.3.11

† It is surface 5 of the last section together with its mirror image below the xy-plane.

EXERCISES 14.3

Identify the level curves $f(x, y) = c$ and sketch the curves corresponding to the indicated values of c.

1. $f(x, y) = x - y; \quad c = -2, 0, 2.$

2. $f(x, y) = 2x - y; \quad c = -2, 0, 2.$

3. $f(x, y) = x^2 - y; \quad c = -1, 0, 1, 2.$

4. $f(x, y) = \dfrac{1}{x - y^2}; \quad c = -2, -1, 1, 2.$

5. $f(x, y) = \dfrac{x}{x + y}; \quad c = -1, 0, 1, 2.$

6. $f(x, y) = \dfrac{y}{x^2}; \quad c = -1, 0, 1, 2.$

7. $f(x, y) = x^3 - y; \quad c = -1, 0, 1, 2.$

8. $f(x, y) = e^{xy}; \quad c = \frac{1}{2}, 1, 2, 3.$

9. $f(x, y) = x^2 - y^2; \quad c = -2, -1, 0, 1, 2.$

10. $f(x, y) = x^2; \quad c = 0, 1, 4, 9.$

11. $f(x, y) = y^2; \quad c = 0, 1, 4, 9.$

12. $f(x, y) = x(y - 1); \quad c = -2, -1, 0, 1, 2.$

13. $f(x, y) = \ln(x^2 + y^2); \quad c = -1, 0, 1.$

14. $f(x, y) = \ln\left(\dfrac{y}{x^2}\right); \quad c = -2, -1, 0, 1, 2.$

15. $f(x, y) = \dfrac{\ln y}{x^2}; \quad c = -2, -1, 0, 1, 2.$

16. $f(x, y) = x^2y^2; \quad c = -4, -1, 0, 1, 4.$

17. $f(x, y) = \dfrac{x^2}{x^2 + y^2}, \quad c = 0, \frac{1}{4}, \frac{1}{2}.$

18. $f(x, y) = \dfrac{\ln y}{x}; \quad c = -2, -1, 0, 1, 2.$

Identify the c-level surface and sketch it.

19. $f(x, y, z) = x + 2y + 3z, \quad c = 0.$

20. $f(x, y, z) = x^2 + y^2, \quad c = 4.$

21. $f(x, y, z) = z(x^2 + y^2)^{-1/2}, \quad c = 1.$

22. $f(x, y, z) = x^2/4 + y^2/6 + z^2/9, \quad c = 1.$

23. $f(x, y, z) = 4x^2 + 9y^2 - 72z, \quad c = 0.$

24. $f(x, y, z) = z^2 - 36x^2 - 9y^2, \quad c = 1.$

25. Identify the c-level surfaces of

$$f(x,y,z) = x^2 + y^2 - z^2$$

taking (i) $c < 0$, (ii) $c = 0$, (iii) $c > 0$.

26. Identify the c-level surfaces of

$$f(x,y,z) = 9x^2 - 4y^2 + 36z^2$$

taking (i) $c < 0$, (ii) $c = 0$, (iii) $c > 0$.

Find an equation for the the level curve of f that contains the point P.

27. $f(x,y) = 1 - 4x^2 - y^2; \quad P(0, 1).$

28. $f(x,y) = (x^2 + y^2)\,e^{xy}; \quad P(1, 0).$

29. $f(x,y) = y^2 \tan^{-1} x; \quad P(1, 2).$

30. $f(x,y) = (x^2 + y)\ln[2 - x + e^y]; \quad P(2, 1).$

Find an equation for the level surface of f that contains the point P.

31. $f(x,y,z) = x^2 + 2y^2 - 2xyz;\quad P(-1,2,1).$

32. $f(x,y,z) = \sqrt{x^2+y^2} - \ln z;\quad P(3,4,e).$

C▶33. Use a graphing utility to draw (a) the surfaces and (b) the default level curves.

(a) $f(x,y) = 3x + y^3.\qquad$ (b) $f(x,y) = \dfrac{x^2+1}{y^2+4}.$

C▶34. Use a graphing utility to draw the level surfaces corresponding to the values of c.

(a) $f(x,y,z) = x + 2y + 4z;\qquad c = 0, 4, 8.$

(b) $f(x,y,z) = \dfrac{x+y}{1+z^2};\qquad c = -2, 0, 2.$

C▶35. Use a CAS to find the level curve/surface at the point P.

(a) $f(x,y) = \dfrac{3x+2y+1}{4x^2+9};\qquad P(2,4).$

(b) $f(x,y,z) = x^2 + 2y^2 - z^2;\qquad P(2,-3,1).$

C▶36. Use a CAS to draw the surface and the level curves.

(a) $f(x,y) = (x^2 - y^2)\,e^{(-x^2-y^2)};\qquad -2 \le x \le 2,$
$-2 \le y \le 2.$

(b) $f(x,y) = xy^3 - yx^3;\qquad -5 \le x \le 5,\quad -5 \le y \le 5.$

37. The magnitude of the gravitational force exerted by a body of mass M situated at the origin on a body of mass m located at the point (x,y,z) is given by

$$F(x,y,z) = \frac{GmM}{x^2+y^2+z^2}$$

where G is the universal gravitational constant. If m and M are constants, describe the level surfaces of F. What is the physical significance of these surfaces?

38. The strength E of an electric field at a point (x,y,z) due to an infinitely long charged wire lying along the y-axis is given by

$$E(x,y,z) = \frac{k}{\sqrt{x^2+z^2}}$$

where k is a positive constant. Describe the level surfaces of E.

39. A thin metal plate is situated in the xy-plane. The temperature T(in °C) at the point (x,y) is inversely proportional to the square of its distance from the origin.

(a) Express T as a function of x and y.

(b) Describe the level curves and sketch a representative set. NOTE: The level curves of T are called *isothermals*; all points on an isothermal have the same temperature.

(c) Suppose the temperature at the point $(1,2)$ is 50°. What is the temperature at the point $(4,3)$?

40. The formula

$$V(x,y) = \frac{k}{\sqrt{r^2-x^2-y^2}},$$

where k and r are positive constants, gives the electric potential (in volts) at a point (x,y) in the xy-plane. Describe the level curves of V and sketch a representative set. NOTE: The level curves of V are called the *equipotential curves*; all points on an equipotential curve have the same electric potential.

In Exercises 41–46, a function f, together with a set of level curves for f, is given. Figures **A–F** are the surfaces $z = f(x, y)$ (in some order). Match f and its system of level curves with its graph $z = f(x, y)$.

41. $f(x, y) = y^2 - y^3.$

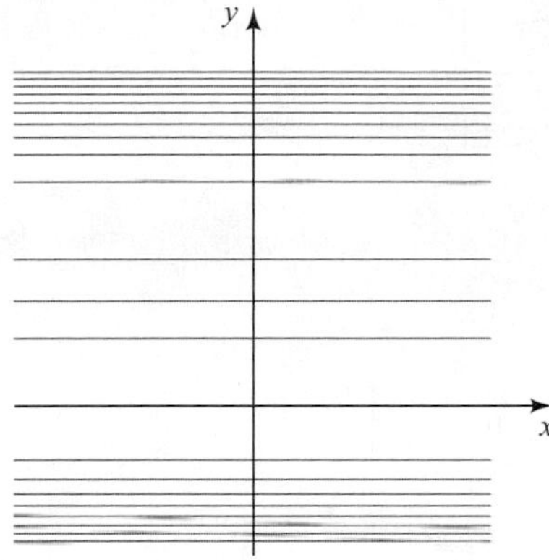

42. $f(x,y) = \sin x, 0 \le x \le 2\pi.$

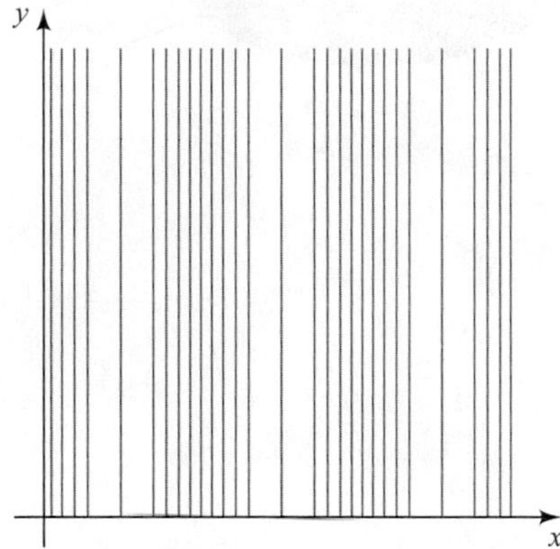

43. $f(x,y) = \cos\sqrt{x^2+y^2}, -10 \le x \le 10, -10 \le y \le 10.$

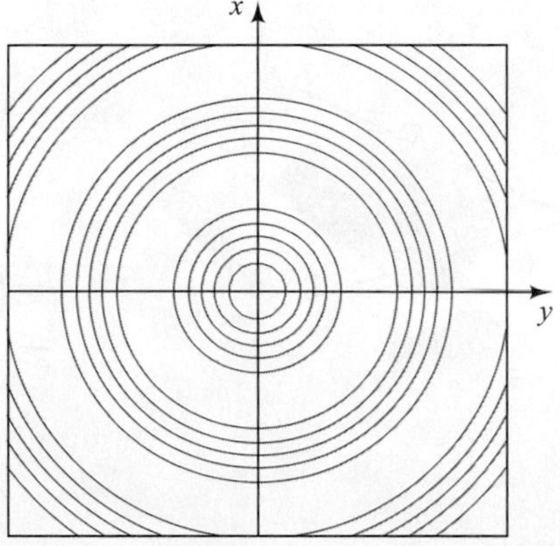

44. $f(x, y) = 2x^2 + 4y^2$.

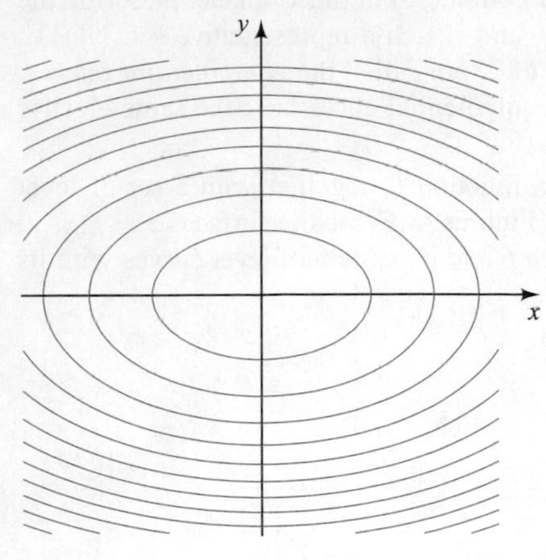

45. $f(x, y) = xye^{-(x^2+y^2)/2}$.

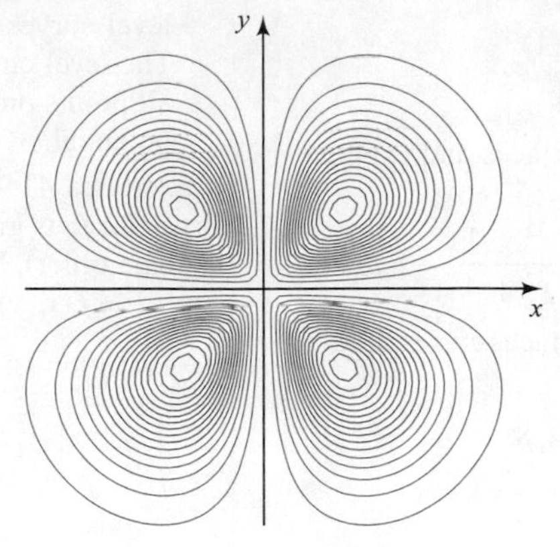

46. $f(x, y) = \sin x \sin y$.

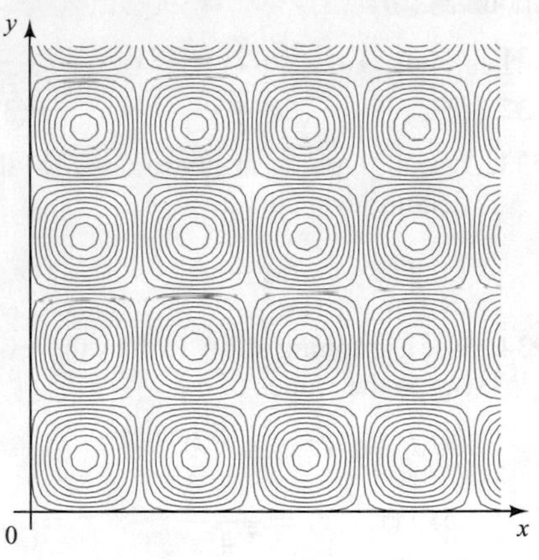

A

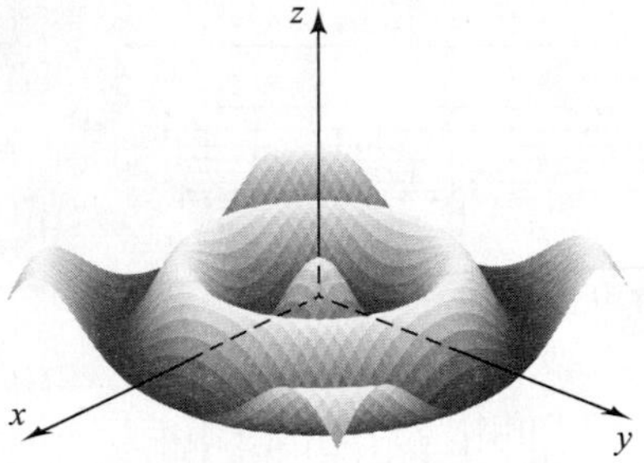

B

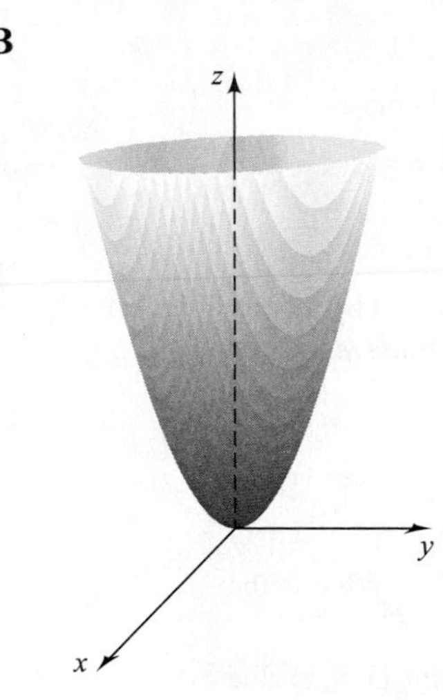

C

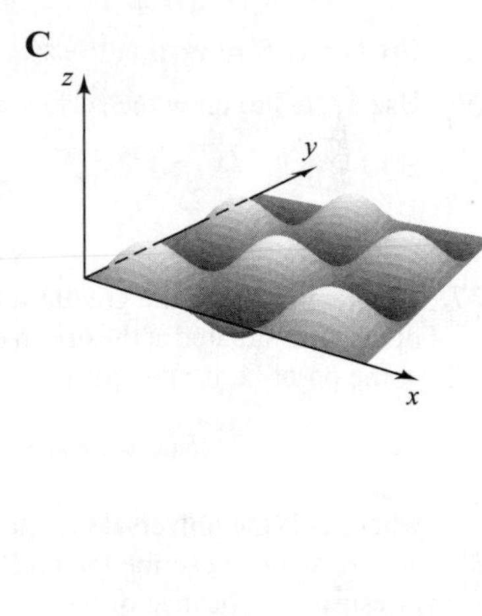

D

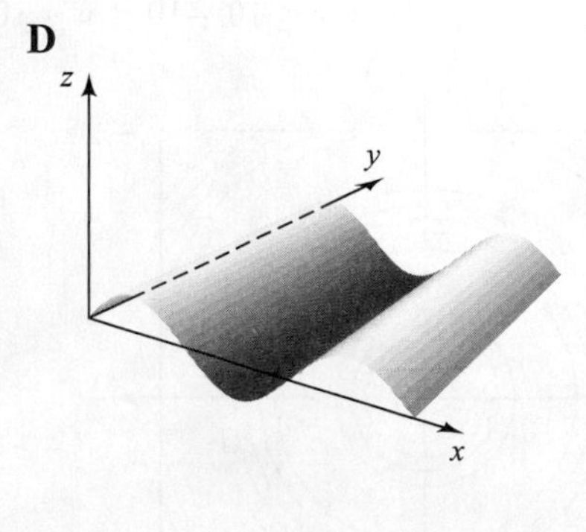

E

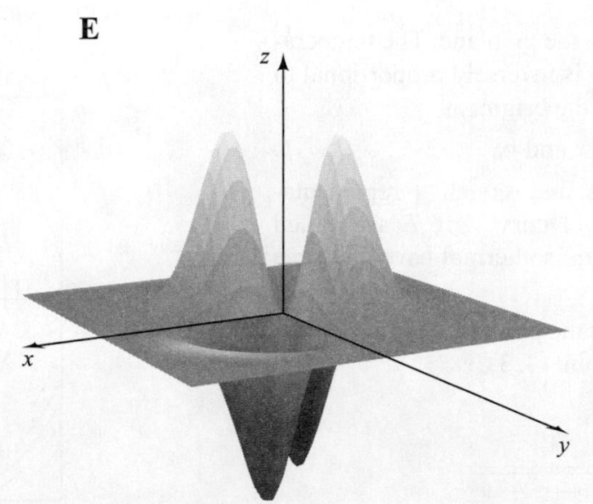

F

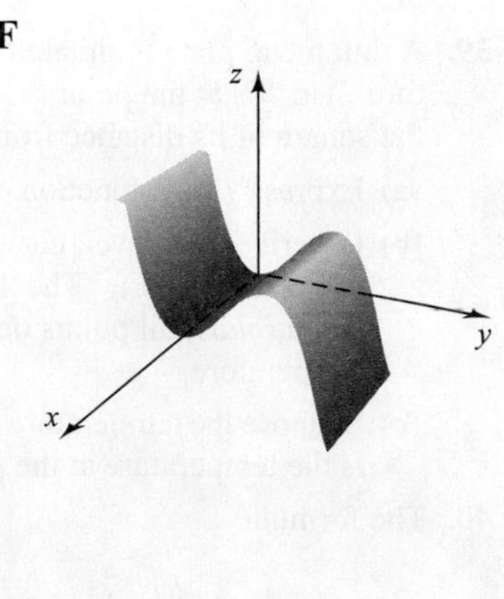

■ PROJECT 14.3 Level Curves and Surfaces

Computer systems such as Derive, Maple, and Mathematica are able to map the level curves of a function $f = f(x, y)$. In this project you are asked to map the level curves of a given function over a given rectangle and then you are asked to "visualize" the surface $z = f(x, y)$. For example, the level curves of

$$f(x,y) = x^2y^2 e^{-(x^2+y^2)}$$

on the rectangle: $-3 \leq x \leq 3, -3 \leq y \leq 3$ are

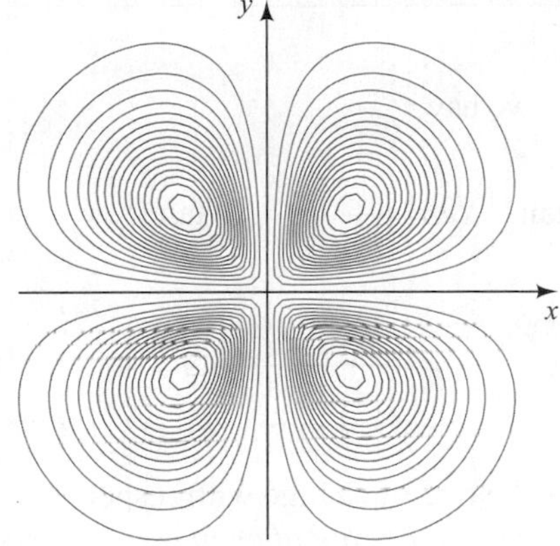

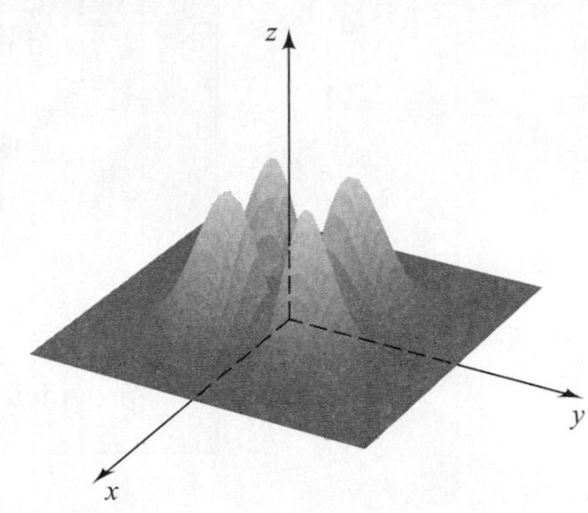

This map of level curves suggests that the surface $z = f(x, y)$ has either "peaks" or "pits" symmetrically placed in the four quadrants. The graph of the surface shown to the right confirms this conjecture.

Problem 1. Make a map of the level curves of $f(x,y) = \dfrac{1}{x^2+y^2}$ over the rectangle: $-3 \leq x \leq 3, -3 \leq y \leq 3$. Try to visualize the graph of the surface from your map of the level curves. Then graph the surface $z = f(x,y)$ to confirm your visualization.

Problem 2. Repeat Problem 1 for the function $f(x,y) = \dfrac{2y}{x^2+y^2+1}$ over the rectangle $-5 \leq x \leq 5, -5 \leq y \leq 5$.

Problem 3. Repeat Problem 1 for the function $f(x,y) = \cos x \cos y\, e^{-(1/4)\sqrt{x^2+y^2}}$ over the rectangle $-2\pi \leq x \leq 2\pi$, $-2\pi \leq y \leq 2\pi$.

Problem 4. Repeat Problem 1 for the function $f(x,y) = \dfrac{-xy}{e^{x^2+y^2}}$ over the rectangle: $-2 \leq x \leq 2, -2 \leq y \leq 2$.

■ 14.4 PARTIAL DERIVATIVES

Functions of Two Variables

Let f be a function of x and y; for example

$$f(x, y) = 3x^2y - 5x \cos \pi y.$$

The *partial derivative of f with respect to x* is the function f_x obtained by differentiating f with respect to x, treating y as a constant. In this case

$$f_x(x, y) = 6xy - 5 \cos \pi y.$$

The *partial derivative of f with respect to y* is the function f_y obtained by differentiating f with respect to y, treating x as a constant. In this case

$$f_y(x, y) = 3x^2 + 5\pi x \sin \pi y.$$

These partial derivatives are formally defined as limits:

DEFINITION 14.4.1 PARTIAL DERIVATIVES OF *f*(*x*, *y*)

Let f be a function of two variables. The partial derivatives of f with respect to x and y are the functions f_x and f_y defined by setting

$$f_x(x, y) = \lim_{h \to 0} \frac{f(x+h, y) - f(x, y)}{h},$$

$$f_y(x, y) = \lim_{h \to 0} \frac{f(x, y+h) - f(x, y)}{h},$$

provided these limits exist.

Example 1 For $f(x, y) = x \tan^{-1} xy$, we have

$$f_x(x, y) = x\frac{y}{1+(xy)^2} + \tan^{-1} xy = \frac{xy}{1+x^2y^2} + \tan^{-1} xy$$

and
$$f_y(x, y) = x\frac{x}{1+(xy)^2} = \frac{x^2}{1+x^2y^2}. \quad \square$$

In the one-variable case, $f'(x_0)$ gives the rate of change with respect to x of $f(x)$ at $x = x_0$. In the two-variable case, *$f_x(x_0, y_0)$ gives the rate of change with respect to x of $f(x, y_0)$ at $x = x_0$, and $f_y(x_0, y_0)$ gives the rate of change with respect to y of $f(x_0, y)$ at $y = y_0$.*

Example 2 For the function $f(x, y) = e^{xy} + \ln(x^2 + y)$, we have

$$f_x(x, y) = ye^{xy} + \frac{2x}{x^2+y} \quad \text{and} \quad f_y(x, y) = xe^{xy} + \frac{1}{x^2+y}.$$

The number

$$f_x(2, 1) = e^2 + \frac{4}{4+1} = e^2 + \frac{4}{5}$$

gives the rate of change with respect to x of the function

$$f(x, 1) = e^x + \ln(x^2 + 1) \quad \text{at } x = 2;$$

the number

$$f_y(2, 1) = 2\,e^2 + \frac{1}{4+1} = 2\,e^2 + \frac{1}{5}$$

gives the rate of change with respect to y of the function

$$f(2, y) = e^{2y} + \ln(4 + y) \quad \text{at } y = 1. \quad \square$$

A Geometric Interpretation

In Figure 14.4.1 we have sketched a surface $z = f(x, y)$, which you can take as everywhere defined. Through the surface we have passed a plane $y = y_0$ parallel to the

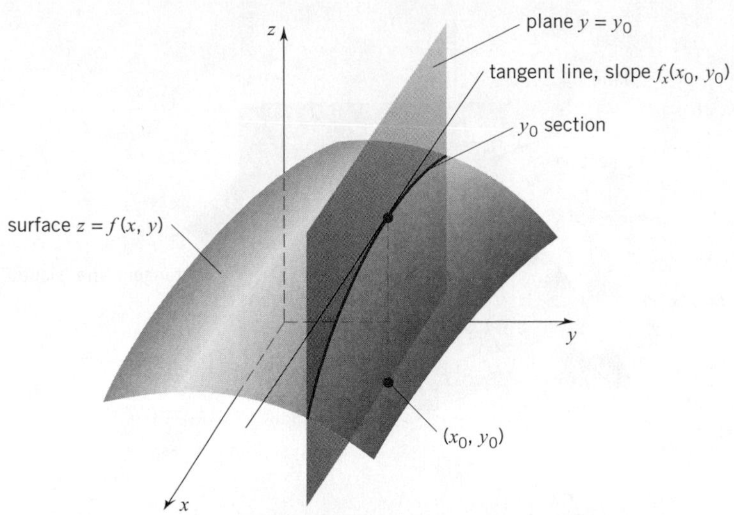

Figure 14.4.1

xz-plane. The plane $y = y_0$ intersects the surface in a curve, the y_0-section of the surface.

The y_0-section of the surface is the graph of the function

$$g(x) = f(x, y_0).$$

Differentiating with respect to x, we have

$$g'(x) = f_x(x, y_0)$$

and, in particular,

$$g'(x_0) = f_x(x_0, y_0).$$

The number $f_x(x_0, y_0)$ is thus the slope of the y_0-section of the surface $z = f(x, y)$ at the point $P(x_0, y_0, f(x_0, y_0))$.

The other partial derivative f_y can be given a similar interpretation. In Figure 14.4.2 you can see the same surface $z = f(x, y)$, this time sliced by a plane $x = x_0$ parallel to the yz-plane. The plane $x = x_0$ intersects the surface in a curve, the x_0-section of the surface.

The x_0-section of the surface is the graph of the function

$$h(y) = f(x_0, y).$$

Differentiating, this time with respect to y, we have

$$h'(y) = f_y(x_0, y)$$

and thus

$$h'(y_0) = f_y(x_0, y_0).$$

The number $f_y(x_0, y_0)$ is the slope of the x_0-section of the surface $z = f(x, y)$ at the point $P(x_0, y_0, f(x_0, y_0))$.

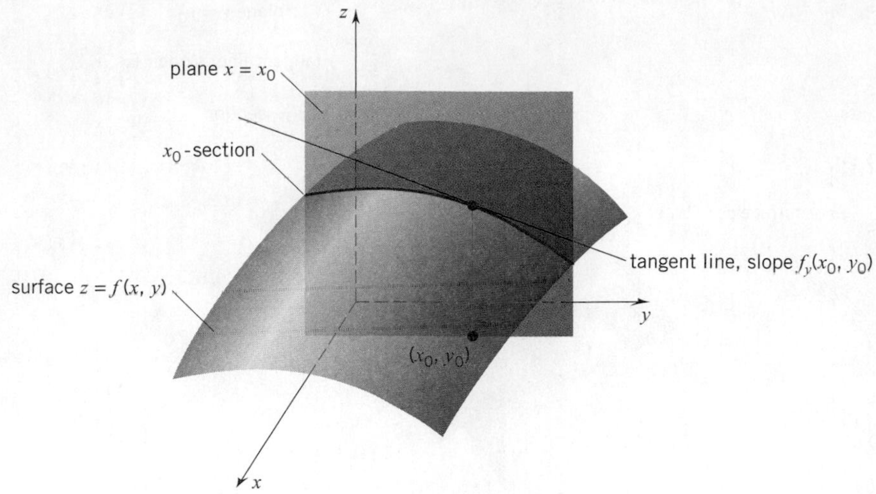

Figure 14.4.2

Functions of Three Variables

In the case of a function of three variables, you can look for three partial derivatives: the partial with respect to x, the partial with respect to y, and the partial with respect to z. These partials,

$$f_x(x, y, z), \quad f_y(x, y, z), \quad f_z(x, y, z),$$

are defined as follows.

DEFINITION 14.4.2 PARTIAL DERIVATIVES OF $f(x, y, z)$

Let f be a function of three variables. The partial derivatives of f with respect to x, y, and z are the functions f_x, f_y, and f_z defined by setting

$$f_x(x, y, z) = \lim_{h\to 0} \frac{f(x+h, y, z) - f(x, y, z)}{h},$$

$$f_y(x, y, z) = \lim_{h\to 0} \frac{f(x, y+h, z) - f(x, y, z)}{h},$$

$$f_z(x, y, z) = \lim_{h\to 0} \frac{f(x, y, z+h) - f(x, y, z)}{h},$$

provided these limits exist.

Each partial can be found by differentiating with respect to the subscript variable, treating the other two variables as constants.

Example 3 For the function $f(x, y, z) = xy^2z^3$ the partial derivatives are:

$$f_x(x, y, z) = y^2z^3, \quad f_y(x, y, z) = 2xyz^3, \quad f_z(x, y, z) = 3xy^2z^2.$$

In particular,

$$f_x(1, -2, -1) = -4, \quad f_y(1, -2, -1) = 4, \quad f_z(1, -2, -1) = 12. \quad \square$$

Example 4 For $g(x, y, z) = x^2 e^{y/z}$ we have

$$g_x(x,y,z) = 2x\, e^{y/z}, \quad g_y(x, y, z) = \frac{x^2}{z}\, e^{y/z}, \quad g_z(x, y, z) = -\frac{x^2 y}{z^2} e^{y/z}. \quad \square$$

Example 5 For a function of the form $f(x, y, z) = F(x,y)G(y,z)$ we can write

$$f_x(x, y, z) = F_x(x, y)G(y, z),$$
$$f_y(x, y, z) = F(x, y)G_y(y, z) + F_y(x, y)G(y, z),$$
$$f_z(x, y, z) = F(x, y)G_z(y, z). \quad \square$$

The number $f_x(x_0, y_0, z_0)$ gives the rate of change with respect to x of $f(x, y_0, z_0)$ at $x = x_0$; $f_y(x_0, y_0, z_0)$ gives the rate of change with respect to y of $f(x_0, y, z_0)$ at $y = y_0$; $f_z(x_0, y_0, z_0)$ gives the rate of change with respect to z of $f(x_0, y_0, z)$ at $z = z_0$.

Example 6 The function $f(x, y, z) = xy^2 - yz^2$ has partial derivatives

$$f_x(x, y, z) = y^2, \quad f_y(x, y, z) = 2xy - z^2, \quad f_z(x, y, z) = -2yz.$$

The number $f_x(1, 2, 3) = 4$ gives the rate of change with respect to x of the function

$$f(x, 2, 3) = 4x - 18 \qquad \text{at } x = 1;$$

$f_y(1, 2, 3) = -5$ gives the rate of change with respect to y of the function

$$f(1, y, 3) = y^2 - 9y \qquad \text{at } y = 2;$$

$f_z(1, 2, 3) = -12$ gives the rate of change with respect to z of the function

$$f(1, 2, z) = 4 - 2z^2 \qquad \text{at } z = 3. \quad \square$$

In general, if f is a function of n variables, $x_1, x_2, \ldots, x_n$, then the partial derivative of f with respect to the kth variable, x_k, is given by

$$f_{x_k}(x_1, x_2, \ldots, x_n) = \lim_{h \to 0} \frac{f(x_1, \ldots, x_{k-1}, x_k + h, x_{k+1}, \ldots x_n) - f(x_1, \ldots, x_k, \ldots, x_n)}{h}$$

provided the limit exists.

The geometric illustrations of the partial derivatives of a function of two variables given in Figures 14.4.1 and 14.4.2 require three-dimensional sketches. Corresponding illustrations for functions of three or more variables would require drawings in four or higher dimensions. Clearly, such drawings are not possible.

Other Notations

There is obviously no need to restrict ourselves to the variables x, y, z. Where more convenient we can use other letters.

Example 7 The volume of the frustum of a cone (Figure 14.4.3) is given by the function

$$V(R, r, h) = \tfrac{1}{3}\pi h(R^2 + Rr + r^2).$$

Find the rate of change of the volume with respect to each of its dimensions if the other dimensions are held constant. Determine the values of these rates of change when $R = 8, r = 4$, and $h = 6$.

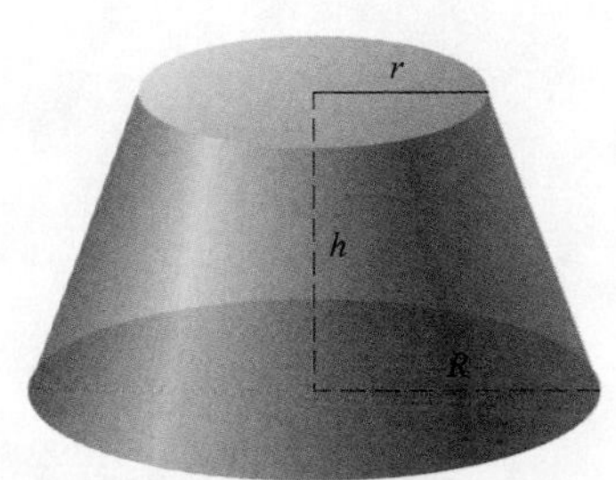

frustum of a cone

Figure 14.4.3

SOLUTION The partial derivatives of V with respect to R, r, and h are as follows:

$$V_R(R, r, h) = \tfrac{1}{3}\pi h(2R + r),$$
$$V_r(R, r, h) = \tfrac{1}{3}\pi h(R + 2r),$$
$$V_h(R, r, h) = \tfrac{1}{3}\pi(R^2 + Rr + r^2).$$

When $R = 8, r = 4$, and $h = 6$,

the rate of change of V with respect to R is $V_R(8, 4, 6) = 40\pi$,
the rate of change of V with respect to r is $V_r(8, 4, 6) = 32\pi$,
the rate of change of V with respect to h is $V_h(8, 4, 6) = \frac{112}{3}\pi$. □

The subscript notation is not the only one used in partial differentiation. A variant of Leibniz's double-d notation is also commonly used. In this notation the partials f_x, f_y, f_z are denoted by

$$\frac{\partial f}{\partial x}, \frac{\partial f}{\partial y}, \frac{\partial f}{\partial z}.$$

Thus, for

$$f(x, y, z) = x^3y^2z + \sin xy$$

we have

$$\frac{\partial f}{\partial x}(x, y, z) = 3x^2y^2z + y\cos xy, \quad \frac{\partial f}{\partial y}(x, y, z) = 2x^3yz + x\cos xy,$$
$$\frac{\partial f}{\partial z}(x, y, z) = x^3y^2,$$

or more simply,

$$\frac{\partial f}{\partial x} = 3x^2y^2z + y\cos xy, \quad \frac{\partial f}{\partial y} = 2x^3yz + x\cos xy, \quad \frac{\partial f}{\partial z} = x^3y^2.$$

We can also write

$$\frac{\partial}{\partial x}(x^3y^2z + \sin xy) = 3x^2y^2z + y\cos xy,$$

$$\frac{\partial}{\partial y}(x^3y^2z + \sin xy) = 2x^3yz + x\cos xy, \quad \frac{\partial}{\partial z}(x^3y^2z + \sin xy) = x^3y^2.$$

Of course, this notation is not restricted to the letters x, y, z. For instance, we can write

$$\frac{\partial}{\partial r}(r^2\cos\theta + e^{\theta r}) = 2r\cos\theta + \theta e^{\theta r},$$
$$\frac{\partial}{\partial \theta}(r^2\cos\theta + e^{\theta r}) = -r^2\sin\theta + r e^{\theta r}.$$

For the function

$$\rho = \sin 2\theta \cos 3\phi$$

we have

$$\frac{\partial \rho}{\partial \theta} = 2\cos 2\theta \cos 3\phi \quad \text{and} \quad \frac{\partial \rho}{\partial \phi} = -3\sin 2\theta \sin 3\phi.$$

EXERCISES 14.4

Calculate the partial derivatives.

1. $f(x, y) = 3x^2 - xy + y$.
2. $g(x, y) = x^2e^{-y}$.
3. $\rho = \sin\phi\cos\theta$.
4. $\rho = \sin^2(\theta - \phi)$.
5. $f(x, y) = e^{x-y} - e^{y-x}$.
6. $z = \sqrt{x^2 - 3y}$.
7. $g(x, y) = \dfrac{Ax + By}{Cx + Dy}$.
8. $u = \dfrac{e^z}{xy^2}$.
9. $u = xy + yz + zx$.
10. $z = Ax^2 + Bxy + Cy^2$.
11. $f(x, y, z) = z\sin(x - y)$.
12. $g(u, v, w) = \ln(u^2 + vw - w^2)$.
13. $\rho = e^{\theta+\phi}\cos(\theta - \phi)$.
14. $f(x, y) = (x + y)\sin(x - y)$.
15. $f(x, y) = x^2y\sec xy$.
16. $g(x, y) = \tan^{-1}(2x + y)$.
17. $h(x, y) = \dfrac{x}{x^2 + y^2}$.
18. $z = \ln\sqrt{x^2 + y^2}$.
19. $f(x, y) = \dfrac{x\sin y}{y\cos x}$.
20. $f(x, y, z) = e^{xy}\sin xz$.
21. $h(x, y) = [f(x)]^2g(y)$.
22. $h(x, y) = e^{f(x)g(y)}$.
23. $f(x, y, z) = z^{xy^2}$.
24. $h(x, y, z) = [f(x,y)]^3[g(x,z)]^2$.
25. $h(r,\theta,t) = r^2e^{2t}\cos(\theta - t)$.
26. $u = \ln(x/y) - ye^{xz}$.
27. $f(x, y, z) = z\tan^{-1}(y/x)$.
28. $w = xy\sin z - yz\sin x$.
29. Find $f_x(0, e)$ and $f_y(0, e)$ given that $f(x, y) = e^x\ln y$.
30. Find $g_x(0, \frac{1}{4}\pi)$ and $g_y(0, \frac{1}{4}\pi)$ given that $g(x, y) = e^{-x}\sin(x + 2y)$.
31. Find $f_x(1, 2)$ and $f_y(1, 2)$ given that $f(x, y) = \dfrac{x}{x + y}$.
32. Find $g_x(1, 2)$ and $g_y(1, 2)$ given that $g(x, y) = \dfrac{y}{x + y^2}$.

Find $f_x(x, y)$ and $f_y(x, y)$ by forming the appropriate difference quotient and taking the limit as h tends to zero.

33. $f(x, y) = x^2y$.
34. $f(x, y) = y^2$.
35. $f(x, y) = \ln(x^2y)$.
36. $f(x, y) = \dfrac{1}{x + 4y}$.
37. $f(x, y) = \dfrac{1}{x - y}$.
38. $f(x, y) = e^{2x+3y}$.

Find $f_x(x, y, z)$, $f_y(x, y, z)$, and $f_z(x, y, z)$ by forming the appropriate difference quotient and taking the limit as h tends to zero.

39. $f(x, y, z) = xy^2z$.
40. $f(x, y, z) = \dfrac{x^2y}{z}$.
41. The intersection of a surface $z = f(x, y)$ with a plane $y = y_0$ is a curve C in space. The slope of the line tangent to C at the point $P(x_0, y_0, f(x_0, y_0))$ is $f_x(x_0, y_0)$. (See Figure 14.4.1.)
    (a) Show that the equations for the tangent line can be written in the form
    $$y = y_0, \quad z - z_0 = f_x(x_0, y_0)(x - x_0).$$
    (b) Now let C be the curve formed by intersecting the surface $z = f(x, y)$ with the plane $x = x_0$. Derive equations for the line tangent to C at the point $P(x_0, y_0, f(x_0, y_0))$. (See Figure 14.4.2.)

In Exercises 42 and 43, let $z = x^2 + y^2$ and let C be the curve of intersection of the surface with the given plane. Find equations for the line tangent to C at the point P.

42. Plane $y = 3$; $P(1, 3, 10)$.
43. Plane $x = 2$; $P(2, 1, 5)$.

In Exercises 44 and 45, let

$$z = \frac{x^2}{y^2 - 3}$$

and let C be the curve of intersection of the surface with the given plane. Find equations for the line tangent to C at the point P.

44. Plane $x = 3$; $P(3, 2, 9)$.
45. Plane $y = 2$; $P(3, 2, 9)$.
46. The surface $z = \sqrt{4 - x^2 - y^2}$ is a hemisphere of radius 2 centered at the origin.
    (a) Find equations for the line l_1 tangent to the curve of intersection of the hemisphere with plane $x = 1$ at the point $(1, 1, \sqrt{2})$.
    (b) Find equations for the line l_2 tangent to the curve of intersection of the hemisphere with the plane $y = 1$ at the point $(1, 1, \sqrt{2})$.
    (c) The tangent lines l_1 and l_2 determine a plane. Find an equation for this plane. As you might expect, this plane may be viewed as tangent to the surface at the point $(1, 1, \sqrt{2})$.
47. (C) Let $f(x, y) = 3x^2 - 6xy + 2y^3$. Use a graphing utility to draw the graph of f in a region around the point $P(1, 2)$.
    (a) Use a CAS to find $m_x = \partial f/\partial x$ at P. Use a graphing utility to draw the graph of $z = f(x, 2)$ together with the line through P with slope m_x.
    (b) Use a CAS to find $m_y = \partial f/\partial y$ at P. Use a graphing utility to draw the graph of $z = f(1, y)$ together with the line through P with slope m_y.
48. (C) Repeat Exercise 47 with $f(x, y) = \dfrac{x - y}{x^2 + y^2}$ and $P(1, 2)$.

In Exercises 49–52, show that the functions u and v satisfy

$$u_x(x, y) = v_y(x, y) \quad \text{and} \quad u_y(x, y) = -v_x(x, y).$$

These equations are called the *Cauchy-Riemann equations*. They arise in the study of functions of a complex variable and are of fundamental importance in that setting.

49. $u(x, y) = x^2 - y^2$; $v(x, y) = 2xy$.

50. $u(x,y) = e^x \cos y;\quad v(x,y) = e^x \sin y.$

51. $u(x,y) = \frac{1}{2}\ln(x^2+y^2);\quad v(x,y) = \tan^{-1}\frac{y}{x}.$

52. $u(x,y) = \dfrac{x}{x^2+y^2};\quad v(x,y) = \dfrac{-y}{x^2+y^2}.$

53. Assume that f is a function defined on a set D in the xy-plane, and assume that the partial derivatives exist throughout D.

(a) Suppose that $f_x(x,y) = 0$ for all $(x,y) \in D$. What can you conclude about f?

(b) Suppose that $f_y(x,y) = 0$ for all $(x,y) \in D$. What can you conclude about f?

54. The law of cosines for a triangle can be written

$$a^2 = b^2 + c^2 - 2bc\cos\theta.$$

At time t_0 we have $b_0 = 10$ inches, $c_0 = 15$ inches, $\theta_0 = \frac{1}{3}\pi$ radians.

(a) Find a_0.

(b) Find the rate of change of a with respect to b at time t_0 given that c and θ remain constant.

(c) Using the rate found in part (b), calculate (by differentials) the approximate change in a if b is decreased by 1 inch.

(d) Find the rate of change of a with respect to θ at time t_0 given that b and c remain constant.

(e) Find the rate of change of c with respect to θ at time t_0 given that a and b remain constant.

55. The area of a triangle is given by the formula

$$A = \tfrac{1}{2}bc\sin\theta.$$

At time t_0 we have $b_0 = 10$ inches, $c_0 = 20$ inches, $\theta_0 = \frac{1}{3}\pi$ radians.

(a) Find the area of the triangle at time t_0.

(b) Find the rate of change of the area with respect to b at time t_0 given that c and θ remain constant.

(c) Find the rate of change of the area with respect to θ at time t_0 given that b and c remain constant.

(d) Using the rate found in part (c), calculate (by differentials) the approximate change in area if angle θ is increased by one degree.

(e) Find the rate of change of c with respect to b at time t_0 if the area and angle θ are to remain constant.

56. Let f be a function of x and y that satisfies a relation of the form

$$\frac{\partial f}{\partial x} = kf, \quad k \text{ a constant.}$$

Show that

$$f(x,y) = g(y)e^{kx},$$

where g is some function of y.

57. Let $z = f(x,y)$ be a function everywhere defined.

(a) Find a vector function that parametrizes the y_0-section of the graph. (See Figure 14.4.1.) Find a vector function that parametrizes the line tangent to this section at the point $P(x_0, y_0, f(x_0,y_0))$. See Exercise 41.

(b) Find a vector function that parametrizes the x_0-section of the graph. (See Figure 14.4.2.) Find a vector function that parametrizes the line tangent to this section at the point $P(x_0, y_0, f(x_0,y_0))$.

(c) Show that the equation of the plane determined by the tangent lines found in parts (a) and (b) can be written

$$z - f(x_0,y_0) = (x - x_0)\frac{\partial f}{\partial x}(x_0,y_0) + (y - y_0)\frac{\partial f}{\partial y}(x_0,y_0).$$

58. *(A chain rule)* Let f be a function of x and y, and g be a function of a single variable. Form the composition $h(x,y) = g(f(x,y))$ and show that

$$h_x(x,y) = g'(f(x,y))\,f_x(x,y) \quad \text{and}$$
$$h_y(x,y) = g'(f(x,y))\,f_y(x,y).$$

In Leibniz's notation, setting $u = f(x,y)$, we have

$$\frac{\partial h}{\partial x} = \frac{dg}{du}\frac{\partial u}{\partial x} \quad \text{and} \quad \frac{\partial h}{\partial y} = \frac{dg}{du}\frac{\partial u}{\partial y}.$$

59. Let g be a differentiable function of a single variable. Use Exercise 58 to verify the following results.

(a) If $w = g(ax + by)$, a, b constant, then

$$b\frac{\partial w}{\partial x} = a\frac{\partial w}{\partial y}.$$

(b) If m and n are nonzero integers and $w = g(x^m y^n)$, then

$$nx\frac{\partial w}{\partial x} = my\frac{\partial w}{\partial y}.$$

60. Given that $x = r\cos\theta$ and $y = r\sin\theta$, find

$$\frac{\partial x}{\partial r}\frac{\partial y}{\partial \theta} - \frac{\partial x}{\partial \theta}\frac{\partial y}{\partial r}.$$

61. For a gas confined in a container, the ideal gas law states that the pressure P is related to the volume V and the temperature T by an equation of the form

$$P = k\frac{T}{V}$$

where k is a positive constant. Show that

$$V\frac{\partial P}{\partial V} = -P \quad \text{and} \quad V\frac{\partial P}{\partial V} + T\frac{\partial P}{\partial T} = 0.$$

62. Three resistances R_1, R_2, R_3 connected in parallel in an electrical circuit produce a resistance R that is given by the formula

$$\frac{1}{R} = \frac{1}{R_1} + \frac{1}{R_2} + \frac{1}{R_3}.$$

Find $\partial R/\partial R_1$.

■ 14.5 OPEN AND CLOSED SETS

A *neighborhood* of a real number x_0 is by definition a set of the form $\{x : |x - x_0| < \delta\}$ where δ is a positive number. This is just an open interval centered at x_0:

$$(x_0 - \delta, x_0 + \delta).$$

If we remove x_0 from the set, we obtain the set

$$(x_0 - \delta, x_0) \cup (x_0, x_0 + \delta).$$

Such a set is called a *deleted neighborhood* of x_0.

From your study of one-variable calculus you know that for a function to have a limit at x_0 it must be defined at least on a deleted neighborhood of x_0, and for it to be continuous or differentiable at x_0 it must be defined at least on a full neighborhood of x_0.

To pave the way for the calculus of functions of several variables, we will extend the notions of neighborhood and deleted neighborhood to higher dimensions and, in so doing, obtain access to other fruitful ideas.

Points in the domain of a function of several variables can be written in vector notation. In the two-variable case, set

$$\mathbf{x} = (x, y),$$

and, in the three-variable case, set

$$\mathbf{x} = (x, y, z).$$

The vector notation enables us to treat the two cases together.

In this section we introduce five important notions:

(1) Neighborhood of a point.

(2) Interior of a set.

(3) Boundary of a set.

(4) Open set.

(5) Closed Set.

For our purposes, the fundamental notion here is "neighborhood of a point." The other four notions can be derived from it.

DEFINITION 14.5.1 NEIGHBORHOOD OF A POINT

A *neighborhood* of a point $\mathbf{x}_0$ is a set of the form

$$\{\mathbf{x} : ||\mathbf{x} - \mathbf{x}_0|| < \delta\}$$

where δ is some number greater than zero.

In the plane, a neighborhood of $\mathbf{x}_0 = (x_0, y_0)$ consists of all the points inside a disc centered at (x_0, y_0). In three-space, a neighborhood of $\mathbf{x}_0 = (x_0, y_0, z_0)$. consists of all the points inside a ball (sphere) centered at (x_0, y_0, z_0). See Figure 14.5.1.

Remark On the real line, the norm of a number x is its absolute value, $||x|| = |x|$. Thus a neighborhood of a point x_0 on the real line is a set of the form $\{x : |x - x_0| < \delta\}$ for some $\delta > 0$, which is simply the open interval $(x_0 - \delta, x_0 + \delta)$.

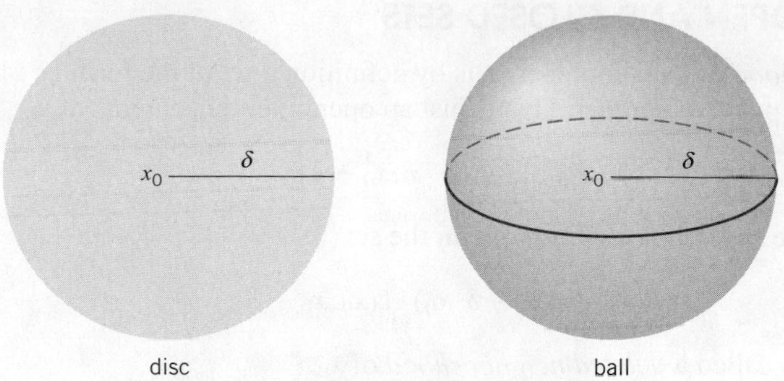

Figure 14.5.1

> **DEFINITION 14.5.2 THE INTERIOR OF A SET**
>
> A point $\mathbf{x}_0$ is said to be an *interior point* of the set S if the set S contains some neighborhood of $\mathbf{x}_0$. The set of all interior points of S is called the *interior* of S.

Example 1 Let Ω be the plane set shown in Figure 14.5.2. The point marked $\mathbf{x}_1$ is an interior point of Ω because Ω contains a neighborhood of $\mathbf{x}_1$. The point $\mathbf{x}_2$ is not an interior point of Ω because *no* neighborhood of $\mathbf{x}_2$ is completely contained in Ω. (Every neighborhood of $\mathbf{x}_2$ has points that lie outside of Ω.) □

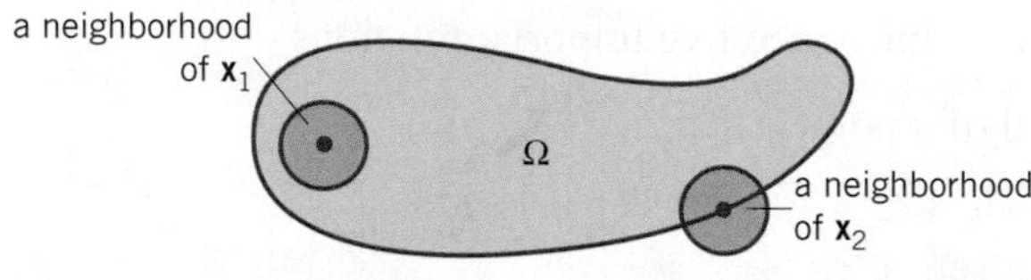

Figure 14.5.2

> **DEFINITION 14.5.3 THE BOUNDARY OF A SET**
>
> A point $\mathbf{x}_0$ is said to be a *boundary point* of the set S if every neighborhood of $\mathbf{x}_0$ contains points that are in S and points that are not in S. The set of all boundary points of S is called the *boundary* of S.

Example 2 The point marked $\mathbf{x}_2$ in Figure 14.5.2 is a boundary point of Ω: each neighborhood of $\mathbf{x}_2$ contains points in Ω and points not in Ω. □

> **DEFINITION 14.5.4 OPEN SET**
>
> A set S is said to be *open* if it contains a neighborhood of each of its points.

Equivalently:

(a) A set S is open iff each of its points is an interior point.

(b) A set S is open iff it contains no boundary points.

DEFINITION 14.5.5 CLOSED SET

A set S is said to be *closed* if it contains its boundary.

Here are some examples of sets that are open, sets that are closed, and sets that are neither open nor closed:

Two-Dimensional Examples

The sets

$$S_1 = \{(x, y) : 1 < x < 2, 1 < y < 2\},$$
$$S_2 = \{(x, y) : 3 \leq x \leq 4, 1 \leq y \leq 2\},$$
$$S_3 = \{(x, y) : 5 \leq x \leq 6, 1 < y < 2\}$$

are displayed if Figure 14.5.3. S_1 is the inside of the first square. S_1 is open because it contains a neighborhood of each of its points. S_2 is the inside of the second square together with the four bounding line segments. S_2 is closed because it contains its entire boundary. S_3 is the inside of the last square together with the two vertical bounding line segments. S_3 is not open because it contains part of its boundary, and it is not closed because it does not contain all of its boundary.

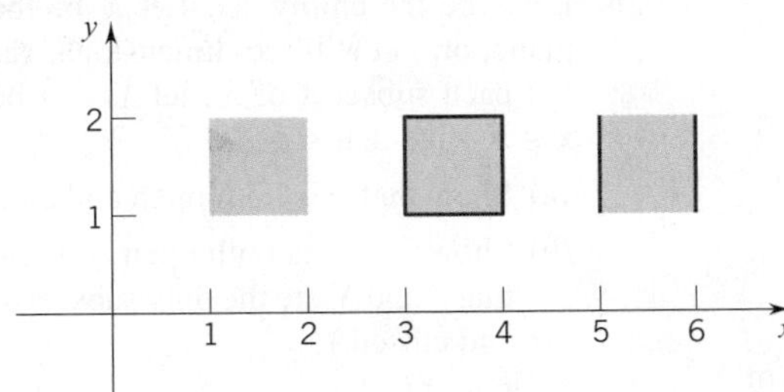

Figure 14.5.3

Three-Dimensional Examples

We now examine some three-dimensional sets:

$$S_1 = \{(x, y, z) : z > x^2 + y^2\},$$
$$S_2 = \{(x, y, z) : z \geq x^2 + y^2\},$$
$$S_3 = \left\{(x, y, z) : 1 \geq \frac{x^2 + y^2}{z}\right\}.$$

The boundary of each of these sets is the paraboloid of revolution

$$z = x^2 + y^2. \qquad \text{(Figure 14.5.4)}$$

The first set consists of all points above this surface. This set is open because, if a point is above this surface, then all points sufficiently close to it are also above this surface. Thus the set contains a neighborhood of each of its points. The second set is closed because it contains all of its boundary. The third set is neither open nor closed. It is not open because it contains some boundary points; for example, it contains the point (1, 1, 2). It is not closed because it fails to contain the boundary point (0, 0, 0).

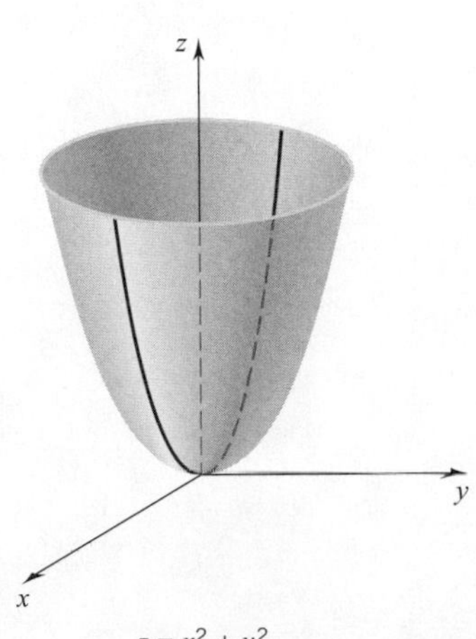

Figure 14.5.4

A Final Remark A neighborhood of $\mathbf{x}_0$ is a set of the form

$$\{\mathbf{x} : ||\mathbf{x} - \mathbf{x}_0|| < \delta\}.$$

If we remove $\mathbf{x}_0$ from the set, we have the set

$$\{\mathbf{x} : 0 < ||\mathbf{x} - \mathbf{x}_0|| < \delta\}.$$

Such a set is called a *deleted neighborhood* of $\mathbf{x}_0$. (We will use deleted neighborhoods in the next section to support the definition of limit.) □

EXERCISES 14.5

Specify the interior and the boundary of the set. State whether the set is open, closed, or neither. Then sketch the set.

1. $\{(x,y) : 2 \le x \le 4, 1 \le y \le 3\}$.

2. $\{(x,y) : 2 < x < 4, 1 < y < 3\}$.

3. $\{(x,y) : 1 < x^2 + y^2 < 4\}$.

4. $\{(x,y) : 1 \le x^2 \le 4\}$.

5. $\{(x,y) : 1 < x^2 \le 4\}$. **6.** $\{(x,y) : y < x^2\}$.

7. $\{(x,y) : y \le x^2\}$.

8. $\{(x,y,z) : 1 \le x \le 2, 1 \le y \le 2, 1 \le z < 2\}$.

9. $\{(x,y,z) : x^2 + y^2 \le 1, 0 \le z \le 4\}$.

10. $\{(x,y,z) : (x-1)^2 + (y-1)^2 + (z-1)^2 < \frac{1}{4}\}$.

11. Let $S = \{\mathbf{x}_1, \mathbf{x}_2, \ldots, \mathbf{x}_n\}$ be a nonempty, finite set of points. (a) What is the interior of S? (b) What is the boundary of S? (c) Is S open, closed, or neither?

All the notions introduced in this section can be applied to sets of real numbers: write x for $\mathbf{x}$ and $|x|$ for $||\mathbf{x}||$. As indicated earlier a *neighborhood* of a number x_0, a set of the form

$$\{x : |x - x_0| < \delta\} \quad \text{with } \delta > 0,$$

is just the open interval $(x_0 - \delta, x_0 + \delta)$. In Exercises 12–19, specify the interior and boundary of the given set of real numbers. State whether the set is open, closed, or neither.

12. $\{x : 1 < x < 3\}$. **13.** $\{x : 1 \le x \le 3\}$.

14. $\{x : 1 \le x < 3\}$. **15.** $\{x : x > 1\}$.

16. $\{x : x \le -1\}$. **17.** $\{x : x < -1 \text{ or } x \ge 1\}$.

18. The set of positive integers: $\{1, 2, 3, \ldots, n, \ldots\}$.

19. The set of reciprocals: $\{1, 1/2, 1/3, \ldots, 1/n \ldots\}$.

20. Let $\emptyset$ be the empty set. Let X be the real line, the entire plane, or, in the three-dimensional case, all of three-space. For each subset A of X, let $X - A$ be the set of all points $\mathbf{x} \in X$ such that $\mathbf{x} \notin A$.

(a) Show that $\emptyset$ is both open and closed.

(b) Show that X is both open and closed. (It can be shown that $\emptyset$ and X are the only subsets of X that are both open and closed.)

(c) Let U be a subset of X. Show that U is open iff $X - U$ is closed.

(d) Let F be a subset of X. Show that F is closed iff $X - F$ is open.

■ 14.6 LIMITS AND CONTINUITY; EQUALITY OF MIXED PARTIALS

The Basic Notions

The limit process used in taking partial derivatives involved nothing new because in each instance all but one of the variables remained fixed. In this section we take up limits of the form

$$\lim_{(x,y)\to(x_0,y_0)} f(x,y) \qquad \text{and} \qquad \lim_{(x,y,z)\to(x_0,y_0,z_0)} f(x,y,z).$$

To avoid having to treat the two-and three-variable cases separately, we will write instead

$$\lim_{\mathbf{x}\to\mathbf{x}_0} f(\mathbf{x}).$$

This gives us both the two-variable case [set $\mathbf{x} = (x, y)$ and $\mathbf{x}_0 = (x_0, y_0)$] and the three-variable case [set $\mathbf{x} = (x, y, z)$ and $\mathbf{x}_0 = (x_0, y_0, z_0)$].

To take the limit of $f(\mathbf{x})$ as $\mathbf{x}$ tends to $\mathbf{x}_0$, we do not need f to be defined at $\mathbf{x}_0$ itself, but we do need f to be defined at points $\mathbf{x}$ close to $\mathbf{x}_0$. At this stage, we will assume that f is defined at all points $\mathbf{x}$ in some deleted neighborhood of $\mathbf{x}_0$ (f may or may not be defined at $\mathbf{x}_0$). This will guarantee that we can form $f(\mathbf{x})$ for all $\mathbf{x} \neq \mathbf{x}_0$ that are "sufficiently close" to $\mathbf{x}_0$. This approach is consistent with our approach to limits of functions of one variable in Chapter 2.

To say that

$$\lim_{\mathbf{x}\to\mathbf{x}_0} f(\mathbf{x}) = L$$

is to say that for $\mathbf{x}$ sufficiently close to $\mathbf{x}_0$ but different from $\mathbf{x}_0$, the number $f(\mathbf{x})$ is close to L; or, to put it another way, as $||\mathbf{x} - \mathbf{x}_0||$ tends to zero but remains different from zero, $|f(\mathbf{x}) - L|$ tends to zero. The $\epsilon - \delta$ definition is a direct generalization of the $\epsilon - \delta$ definition in the single-variable case.

DEFINITION 14.6.1 THE LIMIT OF A FUNCTION OF SEVERAL VARIABLES

Let f be a function defined at least on some deleted neighborhood of $\mathbf{x}_0$.

$$\lim_{\mathbf{x}\to\mathbf{x}_0} f(\mathbf{x}) = L$$

if for each $\epsilon > 0$ there exists a $\delta > 0$ such that

$$\text{if} \quad 0 < ||\mathbf{x} - \mathbf{x}_0|| < \delta \quad \text{then} \quad |f(\mathbf{x}) - L| < \epsilon.$$

Example 1 We will show that the function $f(x,y) = \dfrac{xy + y^3}{x^2 + y^2}$ does not have a limit at $(0, 0)$. Note that f is not defined at $(0, 0)$, but is defined for all $(x, y) \neq (0, 0)$.

Along the obvious paths to $(0, 0)$, the coordinate axes, the limiting value is 0:

Along the x-axis, $y = 0$; thus, $f(x,y) = f(x,0) = 0$ and $\lim_{x\to 0} f(x,0) = \lim_{x\to 0} 0 = 0.$

Along the y-axis, $x = 0$; thus, $f(x,y) = f(0,y) = y$ and $\lim_{y\to 0} f(0,y) = \lim_{y\to 0} y = 0.$

Along the line $y = 2x$, however, the limiting value is $\frac{2}{5}$:

$$f(x,y) = f(x,2x) = \frac{2x^2 + 8x^3}{x^2 + 4x^2} = \frac{2 + 8x}{5} \to \frac{2}{5} \qquad \text{as } x \to 0.$$

There is nothing special about the line $y = 2x$ here. For example, as you can verify, $f(x,y) \to -\frac{1}{2}$ as $(x,y) \to (0,0)$ along the line $y = -x$. See Figure 14.6.1.

We have shown that not all paths to $(0, 0)$ yield the same limiting value. It follows that f does not have a limit at $(0, 0)$. □

Example 2 Show that the function $g(x,y) = \dfrac{x^2 y}{x^4 + y^2}$ has limiting value 0 as $(x,y) \to (0,0)$ along *any* line through the origin, but

$$\lim_{(x,y)\to(0,0)} g(x,y)$$

still does not exist. Note that the domain of g is all $(x,y) \neq (0,0)$.

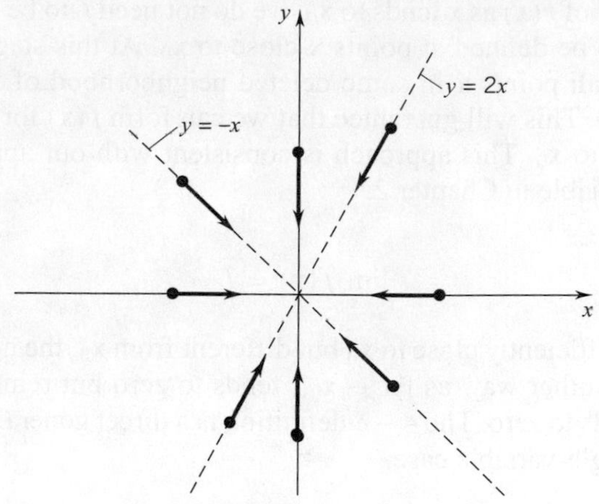

Figure 14.6.1

SOLUTION As in Example 1, it is easy to verify that $g(x, y) \to 0$ as $(x, y) \to (0, 0)$. along the coordinate axes. If we let $y = mx$, then

$$g(x, y) = g(x, mx) = \frac{mx^3}{x^4 + m^2x^2} = \frac{mx}{x^2 + m^2} \qquad (x \neq 0)$$

and
$$\lim_{x\to 0} g(x, mx) = \lim_{x\to 0} \frac{mx}{x^2 + m^2} = 0.$$

Therefore, $g(x, y) \to 0$ as $(x, y) \to (0, 0)$ along any line through the origin.

Now suppose that $(x, y) \to (0, 0)$ along the parabola $y = x^2$. Then we have

$$g(x, y) = g(x, x^2) = \frac{x^4}{x^4 + x^4} = \frac{1}{2}$$

and
$$\lim_{x\to 0} g(x, x^2) = \lim_{x\to 0} \frac{1}{2} = \frac{1}{2}.$$

Thus, $g(x, y) \to \dfrac{1}{2}$ as $(x, y) \to (0, 0)$ along $y = x^2$ (Figure 14.6.2).

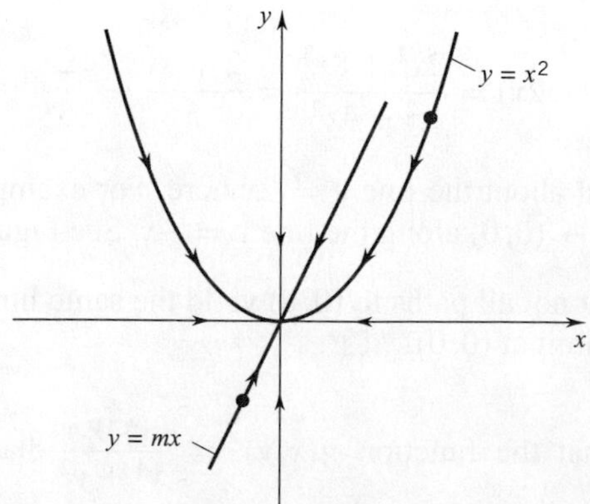

Figure 14.6.2

Since not all paths to $(0, 0)$ yield the same limiting value, we conclude that g does not have a limit at $(0, 0)$. □

As in the one-variable case, the limit (if it exists) is unique. Moreover, if

$$\lim_{\mathbf{x}\to\mathbf{x}_0} f(\mathbf{x}) = L \qquad \text{and} \qquad \lim_{\mathbf{x}\to\mathbf{x}_0} g(\mathbf{x}) = M,$$

then $\quad \lim_{\mathbf{x}\to\mathbf{x}_0} [f(\mathbf{x}) + g(\mathbf{x})] = L + M, \quad \lim_{\mathbf{x}\to\mathbf{x}_0} [\alpha f(\mathbf{x})] = \alpha L, \quad \alpha$ a real number,

$$\lim_{\mathbf{x}\to\mathbf{x}_0} [f(\mathbf{x})g(\mathbf{x})] = LM, \qquad \text{and} \qquad \lim_{\mathbf{x}\to\mathbf{x}_0} [f(\mathbf{x})/g(\mathbf{x})] = L/M \quad \text{provided } M \neq 0.$$

These results are not hard to derive. You can do it simply by imitating the corresponding arguments in the one-variable case.

Suppose now that $\mathbf{x}_0$ is an interior point of the domain of f. To say that f is *continuous* at $\mathbf{x}_0$ is to say that

(14.6.2)
$$\lim_{\mathbf{x}\to\mathbf{x}_0} f(\mathbf{x}) = f(\mathbf{x}_0).$$

For the two variable case we can write

$$\lim_{(x,y)\to(x_0,y_0)} f(x,y) = f(x_0,y_0)$$

and for three variables,

$$\lim_{(x,y,z)\to(x_0,y_0,z_0)} f(x,y,z) = f(x_0,y_0,z_0).$$

Another way to indicate that f is continuous at $\mathbf{x}_0$ is to write

(14.6.3)
$$\lim_{\mathbf{h}\to\mathbf{0}} f(\mathbf{x}_0 + \mathbf{h}) = f(\mathbf{x}_0).$$

To say that f is *continuous on an open set* S is to say that f is continuous at all points of S.

Some Examples of Continuous Functions

Polynomials in several variables, for example,

$$P(x,y) = x^2y + 3x^3y^4 - x + 2y \qquad \text{and} \qquad Q(x,y,z) = 6x^3z - yz^3 + 2xyz$$

are everywhere continuous. In the two-variable case, that means continuity at each point of the xy-plane, and in the three-variable case, continuity at each point of three-space.

Rational functions (quotients of polynomials) are continuous everywhere except where the denominator is zero. Thus

$$f(x,y) = \frac{2x - y}{x^2 + y^2}$$

is continuous at each point of the xy-plane other than the origin $(0,0)$;

$$g(x,y) = \frac{x^4}{x - y}$$

is continuous except on the line $y = x$;

$$h(x, y) = \frac{1}{x^2 - y}$$

is continuous except on the parabola $y = x^2$;

$$F(x, y, z) = \frac{2x}{x^2 + y^2 + z^2}$$

is continuous at each point of three-space other than the origin $(0, 0, 0)$;

$$G(x, y, z) = \frac{x^5 - y}{ax + by + cz},$$

where a, b, c are constants, is continuous except on the plane $ax + by + cz = 0$.

You can construct more elaborate continuous functions by forming composites: take, for example,

$$f(x, y, z) = \tan^{-1}\left(\frac{xz^2}{x + y}\right), \quad g(x, y, z) = \sqrt{x^2 + y^4 + z^6}, \quad h(x, y, z) = \sin xyz.$$

The first function is continuous except along the vertical plane $x + y = 0$. The other two functions are continuous at each point of space. The continuity of such composites follows from a simple theorem that we now state and prove. In the theorem, g is a function of several variables, but f is a function of a single variable.

THEOREM 14.6.4 THE CONTINUITY OF COMPOSITE FUNCTIONS

If g is continuous at the point $\mathbf{x_0}$ and f is continuous at the number $g(\mathbf{x_0})$, then the composition $f \circ g$ is continuous at the point $\mathbf{x_0}$.

PROOF We begin with $\epsilon > 0$. We must show that there exists a $\delta > 0$ such that

$$\text{if } \|\mathbf{x} - \mathbf{x_0}\| < \delta, \quad \text{then} \quad |f(g(\mathbf{x})) - f(g(\mathbf{x_0}))| < \epsilon.$$

From the continuity of f at $g(\mathbf{x_0})$, we know that there exists a $\delta_1 > 0$ such that

$$\text{if } |u - g(\mathbf{x_0})| < \delta_1, \quad \text{then} \quad |f(u) - f(g(\mathbf{x_0}))| < \epsilon.$$

From the continuity of g at $\mathbf{x_0}$, we know that there exists a $\delta > 0$ such that

$$\text{if } \|\mathbf{x} - \mathbf{x_0}\| < \delta, \quad \text{then} \quad |g(\mathbf{x}) - g(\mathbf{x_0})| < \delta_1.$$

This last δ obviously works; namely,

$$\text{if } \|\mathbf{x} - \mathbf{x_0}\| < \delta, \quad \text{then} \quad |g(\mathbf{x}) - g(\mathbf{x_0})| < \delta_1,$$

and so

$$|f(g(\mathbf{x})) - f(g(\mathbf{x_0}))| < \epsilon. \quad \square$$

Continuity in Each Variable Separately

A continuous function of several variables is continuous in each of its variables separately. In the two-variable case, this means that, if

$$\lim_{(x,y)\to(x_0,y_0)} f(x, y) = f(x_0, y_0),$$

then $\qquad \lim_{x\to x_0} f(x,y_0) = f(x_0,y_0) \quad \text{and} \quad \lim_{y\to y_0} f(x_0,y) = f(x_0,y_0).$

(This is not hard to prove.) The converse is false. *It is possible for a function to be continuous in each variable separately and yet fail to be continuous as a function of several variables.* You can see this in the next example.

Example 3 We set

$$f(x,y) = \begin{cases} \dfrac{2xy}{x^2+y^2}, & (x,y) \neq (0,0) \\ 0, & (x,y) = (0,0). \end{cases}$$

Since $\qquad f(x,0) = 0 \quad \text{for all } x \qquad \text{and} \qquad f(0,y) = 0 \quad \text{for all } y,$

we have $\qquad \lim_{x\to 0} f(x,0) = 0 = f(0,0) \qquad \text{and} \qquad \lim_{y\to 0} f(0,y) = 0 = f(0,0).$

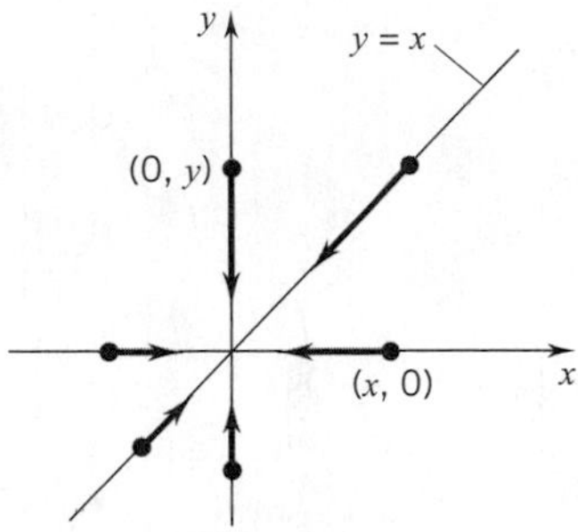

Figure 14.6.3

Thus, at the point $(0,0)$, f is continuous in x and continuous in y. As a function of two variables, however, f is not continuous at $(0,0)$. One way to see this is to note that we can approach $(0,0)$ as closely as we wish by points of the form (t,t) with $t \neq 0$ (that is, along the line $y = x$.) (See Figure 14.6.3.) At such points f takes on the value 1:

$$f(t,t) = \frac{2t^2}{t^2+t^2} = 1.$$

Hence, f cannot tend to $f(0,0) = 0$ as required. ❑

Continuity and Partial Differentiability

For functions of a single variable the existence of the derivative guarantees continuity (Theorem 3.1.4). *For functions of several variables the existence of partial derivatives fails to guarantee continuity.*†

To show this, we can use the same function

$$f(x,y) = \begin{cases} \dfrac{2xy}{x^2+y^2}, & (x,y) \neq (0,0) \\ 0, & (x,y) = (0,0). \end{cases}$$

Since both $f(x,0)$ and $f(0,y)$ are constantly zero, both partials exist (and are zero) at $(0,0)$, and yet, as you saw, the function is discontinuous at $(0,0)$.

It is not hard to understand how a function can have partial derivatives and yet fail to be continuous. The existence of $\partial f/\partial x$ at (x_0,y_0) depends on the behavior of f only at points of the form (x_0+h, y_0). Similarly, the existence of $\partial f/\partial y$ at (x_0,y_0) depends on the behavior of f only at points of the form (x_0, y_0+k). On the other hand, continuity at (x_0,y_0) depends on the behavior of f at points of the more general form (x_0+h, y_0+k). More briefly, we can put it this way: *the existence of a partial derivative depends on the behavior of the function along a line segment parallel to one of the axes, whereas continuity depends on the behavior of the function in all directions.*

† See. however, Exercise 32.

Derivatives of Higher Order; Equality of Mixed Partials

Suppose that f is a function of x and y with first partials f_x and f_y. These are again functions of x and y and may themselves possess partial derivatives: $(f_x)_x, (f_x)_y, (f_y)_x, (f_y)_y$. These functions are called the *second-order partials.* If $z = f(x, y)$, we use the following notations for second-order partials

$$(f_x)_x = f_{xx} = \frac{\partial}{\partial x}\left(\frac{\partial f}{\partial x}\right) = \frac{\partial^2 f}{\partial x^2} = \frac{\partial^2 z}{\partial x^2},$$

$$(f_x)_y = f_{xy} = \frac{\partial}{\partial y}\left(\frac{\partial f}{\partial x}\right) = \frac{\partial^2 f}{\partial y \partial x} = \frac{\partial^2 z}{\partial y \partial x},$$

$$(f_y)_x = f_{yx} = \frac{\partial}{\partial x}\left(\frac{\partial f}{\partial y}\right) = \frac{\partial^2 f}{\partial x \partial y} = \frac{\partial^2 z}{\partial x \partial y},$$

$$(f_y)_y = f_{yy} = \frac{\partial}{\partial y}\left(\frac{\partial f}{\partial y}\right) = \frac{\partial^2 f}{\partial y^2} = \frac{\partial^2 z}{\partial y^2}.$$

Note that there are two "mixed" partials: f_{xy} $(\partial^2 f/\partial y \partial x)$ and f_{yx} $(\partial^2 f/\partial x \partial y)$. The first of these is obtained by differentiating first respect to x and then with respect to y. The second is obtained by differentiating first with respect to y and then with respect to x.

Example 4 The function $f(x, y) = \sin x^2 y$ has first partials

$$\frac{\partial f}{\partial x} = 2xy \cos x^2 y \qquad \text{and} \qquad \frac{\partial f}{\partial y} = x^2 \cos x^2 y.$$

The second-order partials are

$$\frac{\partial^2 f}{\partial x^2} = -4x^2y^2 \sin x^2 y + 2y \cos x^2 y, \qquad \frac{\partial^2 f}{\partial y \partial x} = -2x^3 y \sin x^2 y + 2x \cos x^2 y,$$

$$\frac{\partial^2 f}{\partial x \partial y} = -2x^3 y \sin x^2 y + 2x \cos x^2 y, \qquad \frac{\partial^2 f}{\partial y^2} = -x^4 \sin x^2 y. \quad \square$$

Example 5 Setting $f(x, y) = \ln(x^2 + y^3)$, we have

$$\frac{\partial f}{\partial x} = \frac{2x}{x^2 + y^3} \qquad \text{and} \qquad \frac{\partial f}{\partial y} = \frac{3y^2}{x^2 + y^3}.$$

The second-order partials are

$$\frac{\partial^2 f}{\partial x^2} = \frac{(x^2 + y^3)2 - 2x(2x)}{(x^2 + y^3)^2} = \frac{2(y^3 - x^2)}{(x^2 + y^3)^2},$$

$$\frac{\partial^2 f}{\partial y \partial x} = \frac{-2x(3y^2)}{(x^2 + y^3)^2} = -\frac{6xy^2}{(x^2 + y^3)^2},$$

$$\frac{\partial^2 f}{\partial x \partial y} = \frac{-3y^2(2x)}{(x^2 + y^3)^2} = -\frac{6xy^2}{(x^2 + y^3)^2},$$

$$\frac{\partial^2 f}{\partial y^2} = \frac{(x^2 + y^3)6y - 3y^2(3y^2)}{(x^2 + y^3)^2} = \frac{3y(2x^2 - y^3)}{(x^2 + y^3)^2}. \quad \square$$

Perhaps you noticed that in both examples we had

$$\frac{\partial^2 f}{\partial y \partial x} = \frac{\partial^2 f}{\partial x \partial y}.$$

Since in neither case was f symmetric in x and y, this equality of the mixed partials was not due to symmetry. Actually it was due to continuity. It can be proved that

(14.6.5)

$$\frac{\partial^2 f}{\partial y \partial x} = \frac{\partial^2 f}{\partial x \partial y}$$

on every open set U on which f and its partials

$$\frac{\partial f}{\partial x}, \quad \frac{\partial f}{\partial y}, \quad \frac{\partial^2 f}{\partial y \partial x}, \quad \frac{\partial^2 f}{\partial x \partial y}$$

are all continuous. †

In the case of a function of three variables you can look for three first partials

$$\frac{\partial f}{\partial x}, \quad \frac{\partial f}{\partial y}, \quad \frac{\partial f}{\partial z}$$

and nine second partials

$$\frac{\partial^2 f}{\partial x^2}, \quad \frac{\partial^2 f}{\partial x \partial y}, \quad \frac{\partial^2 f}{\partial x \partial z}, \qquad \frac{\partial^2 f}{\partial y \partial x}, \quad \frac{\partial^2 f}{\partial y^2}, \quad \frac{\partial^2 f}{\partial y \partial z}, \qquad \frac{\partial^2 f}{\partial z \partial x}, \quad \frac{\partial^2 f}{\partial z \partial y}, \quad \frac{\partial^2 f}{\partial z^2}.$$

Here again, there is equality of the mixed partials

$$\frac{\partial^2 f}{\partial y \partial x} = \frac{\partial^2 f}{\partial x \partial y}, \qquad \frac{\partial^2 f}{\partial z \partial x} = \frac{\partial^2 f}{\partial x \partial z}, \qquad \frac{\partial^2 f}{\partial y \partial z} = \frac{\partial^2 f}{\partial z \partial y}$$

provided that f and its first and second partials are all continuous.

Example 6 For $f(x, y, z) = xe^y \sin \pi z$

we have

$$\frac{\partial f}{\partial x} = e^y \sin \pi z, \qquad \frac{\partial f}{\partial y} = xe^y \sin \pi z, \qquad \frac{\partial f}{\partial z} = \pi xe^y \cos \pi z.$$

$$\frac{\partial^2 f}{\partial x^2} = 0, \qquad \frac{\partial^2 f}{\partial y^2} = xe^y \sin \pi z, \qquad \frac{\partial^2 f}{\partial z^2} = -\pi^2 xe^y \sin \pi z.$$

$$\frac{\partial^2 f}{\partial y \partial x} = e^y \sin \pi z = \frac{\partial^2 f}{\partial x \partial y},$$

$$\frac{\partial^2 f}{\partial z \partial x} = \pi e^y \cos \pi z = \frac{\partial^2 f}{\partial x \partial z},$$

$$\frac{\partial^2 f}{\partial y \partial z} = \pi xe^y \cos \pi z = \frac{\partial^2 f}{\partial z \partial y}. \quad \square$$

† For a proof, consult a text on advanced calculus.

EXERCISES 14.6

Calculate the second-order partial derivatives. (Treat A, B, C, D as constants.)

1. $f(x, y) = Ax^2 + 2Bxy + Cy^2$.
2. $f(x, y) = Ax^3 + Bx^2y + Cxy^2$.
3. $f(x, y) = Ax + By + Ce^{xy}$.
4. $f(x, y) = x^2 \cos y + y^2 \sin x$.
5. $f(x, y, z) = (x + y^2 + z^3)^2$.
6. $f(x, y) = \sqrt{x + y^2}$.
7. $f(x, y) = \ln\left(\dfrac{x}{x+y}\right)$.
8. $f(x, y) = \dfrac{Ax + By}{Cx + Dy}$.
9. $f(x, y, z) = (x + y)(y + z)(z + x)$.
10. $f(x, y, z) = \tan^{-1} xyz$.
11. $f(x, y) = x^y$.
12. $f(x, y, z) = \sin(x + z^y)$.
13. $f(x, y) = xe^y + ye^x$.
14. $f(x, y) = \tan^{-1}(y/x)$.
15. $f(x, y) = \ln\sqrt{x^2 + y^2}$.
16. $f(x, y) = \sin(x^3y^2)$.
17. $f(x, y) = \cos^2(xy)$.
18. $f(x, y) = e^{xy^2}$.
19. $f(x, y, z) = xy \sin z - xz \sin y$.
20. $f(x, y, z) = xe^y + ye^z + ze^x$.
21. Show that

$$\text{if} \quad u = \frac{xy}{x+y},$$

$$\text{then} \quad x^2\frac{\partial^2 u}{\partial x^2} + 2xy\frac{\partial^2 u}{\partial x \partial y} + y^2\frac{\partial^2 u}{\partial y^2} = 0.$$

22. Verify that

$$\frac{\partial^2 f}{\partial y \partial x} = \frac{\partial^2 f}{\partial x \partial y}$$

for
(a) $f(x, y) = g(x) + h(y)$ with g and h differentiable.
(b) $f(x, y) = g(x)h(y)$ with g and h differentiable.
(c) $f(x, y)$ a polynomial in x and y. HINT: Check each term $x^m y^n$ separately.

23. Let f be a function of x and y with everywhere continuous second partials. Is it possible that

(a) $\dfrac{\partial f}{\partial x} = x + y$ and $\dfrac{\partial f}{\partial y} = y - x$?

(b) $\dfrac{\partial f}{\partial x} = xy$ and $\dfrac{\partial f}{\partial y} = xy$?

24. Let g be a twice-differentiable function of one variable and set

$$h(x, y) = g(x + y) + g(x - y).$$

Show that

$$\frac{\partial^2 h}{\partial x^2} = \frac{\partial^2 h}{\partial y^2}.$$

HINT: Use the chain rule of Exercise 58, Section 14.4.

25. Let f be a function of x and y with third-order partials

$$\frac{\partial^3 f}{\partial x^2 \partial y} = \frac{\partial}{\partial x}\left(\frac{\partial^2 f}{\partial x \partial y}\right) \quad \text{and} \quad \frac{\partial^3 f}{\partial y \partial x^2} = \frac{\partial}{\partial y}\left(\frac{\partial^2 f}{\partial x^2}\right).$$

Show that, if all the partials are continuous, then

$$\frac{\partial^3 f}{\partial x^2 \partial y} = \frac{\partial^2 f}{\partial y \partial x^2}.$$

26. Show that the following functions do not have a limit at $(0, 0)$:

(a) $f(x, y) = \dfrac{x^2 - y^2}{x^2 + y^2}$. (b) $f(x, y) = \dfrac{y^2}{x^2 + y^2}$.

In Exercises 27 and 28, evaluate the limit as (x, y) approaches the origin along:

(a) The x-axis. (b) The y-axis.
(c) The line $y = mx$. (d) The spiral $r = \theta, \theta > 0$.
(e) The differentiable curve $y = f(x)$, with $f(0) = 0$.
(f) The arc $r = \sin 3\theta$, $\frac{1}{6}\pi < \theta < \frac{1}{3}\pi$.
(g) The path $\mathbf{r}(t) = \dfrac{1}{t}\mathbf{i} + \dfrac{\sin t}{t}\mathbf{j}, t > 0$.

27. $\displaystyle\lim_{(x,y)\to(0,0)} \frac{xy}{x^2 + y^2}$.
28. $\displaystyle\lim_{(x,y)\to(0,0)} \frac{xy^2}{(x^2 + y^2)^{3/2}}$.

29. Set

$$g(x, y) = \begin{cases} \dfrac{x^2y^2}{x^4 + y^4}, & (x, y) \neq (0, 0) \\ 0, & (x, y) = (0, 0). \end{cases}$$

(a) Show that $\partial g/\partial x$ and $\partial g/\partial y$ both exist at $(0, 0)$. What are their values at $(0, 0)$?
(b) Show that $\displaystyle\lim_{(x,y)\to(0,0)} g(x, y)$ does not exist.

30. Set

$$f(x, y) = \frac{x - y^4}{x^3 - y^4}.$$

Determine whether or not f has a limit at $(1, 1)$.
HINT: Let (x, y) tend to $(1, 1)$ along the line $x = 1$ and along the line $y = 1$.

31. Set

$$f(x, y) = \begin{cases} \dfrac{xy(y^2 - x^2)}{x^2 + y^2}, & (x, y) \neq (0, 0) \\ 0, & (x, y) = (0, 0). \end{cases}$$

It can be shown that some of the second partials are discontinuous at $(0, 0)$. Show that

$$\frac{\partial^2 f}{\partial y \partial x}(0, 0) \neq \frac{\partial^2 f}{\partial x \partial y}(0, 0).$$

32. If a function of several variables has all first partials at a point, then it is continuous in each variable separately at

that point. Show, for example, that if f_x exists at (x_0, y_0), then f is continuous in x at (x_0, y_0).

33. Let f be a function of x and y which has continuous first and second partial derivatives throughout some set D in the plane. Suppose that $f_{xy}(x,y) = 0$ for all $(x,y) \in D$. What can you conclude about f?

34. Use a graphing utility to draw the graph of the function in Exercise 26(*a*) on the square $-2 \le x \le 2, -2 \le y \le 2$. Can you see that the limit as $(x,y) \to (0,0)$ along the x-axis is 1 and the limit as $(x,y) \to (0,0)$ along the y-axis is -1?

35. Use a graphing utility to draw the graph of the function in Exercise 27 on the square $-2 \le x \le 2, -2 \le y \le 2$. Can you see that the limit as $(x,y) \to (0,0)$ along the coordinate axes is 0 and the limit as $(x,y) \to (0,0)$ along the line $y = x$ is $\frac{1}{2}$?

36. Use a graphing utility to draw the graph of

$$f(x,y) = \frac{x^2 - y^2}{x^2 + y^2}$$

on the square $-2 \le x \le 2, -2 \le y \le 2$. From the graph, determine the limit of f as $(x,y) \to (0,0)$ along the x-axis and as $(x,y) \to (0,0)$ along the y-axis. Reverse the roles of x and y and see what happens.

■ PROJECT 14.6 Partial Differential Equations

The differential equations that we have studied so far are *ordinary differential equations.* They involve only ordinary derivatives, derivatives of functions of one variable. Here we examine some *partial differential equations,* the most prominent of which are equations that relate two or more of the partial derivatives of an unknown function of several variables.

Partial differential equations play an enormous role in science because the description of most natural phenomena is based on models that involve functions of several variables. For example, the partial differential equation known as the Schrödinger.† equation is viewed by many physicists as the cornerstone of quantum mechanics. Below we introduce two of the classical equations of physics having broad applications in science and engineering.

Problem 1. Show that the given function satisfies the corresponding partial differential equation

(a) $u = \dfrac{x^2y^2}{x+y}$; $\quad x\dfrac{\partial u}{\partial x} + y\dfrac{\partial u}{\partial y} = 3u.$

(b) $u = x^2y + y^2z + z^2x$; $\quad \dfrac{\partial u}{\partial x} + \dfrac{\partial u}{\partial y} + \dfrac{\partial u}{\partial z} = (x+y+z)^2.$

Laplace's Equation †† The partial differential equation

$$\frac{\partial^2 f}{\partial x^2} + \frac{\partial^2 f}{\partial y^2} = 0.$$

is known as *Laplace's equation in two dimensions.* It is used to describe potentials and steady-state temperature distributions in the plane. In three dimensions, Laplace's equation reads

$$\frac{\partial^2 f}{\partial x^2} + \frac{\partial^2 f}{\partial y^2} + \frac{\partial^2 f}{\partial z^2} = 0.$$

The equation is satisfied by gravitational and electrostatic potentials and by steady-state temperature distributions in space. Functions that satisfy Laplace's equation are called *harmonic* functions.

Problem 2.

a. Show that the given functions satisfy Laplace's equation in two dimensions:

(*i*) $f(x,y) = x^3 - 3xy^2$

(*ii*) $f(x,y) = \cos x \sinh y + \sin x \cosh y$

(*iii*) $f(x,y) = \ln\sqrt{x^2+y^2}$.

b. Show that the given functions satisfy Laplace's equation in three dimensions:

(*i*) $f(x,y,z) = \dfrac{1}{\sqrt{x^2+y^2+z^2}}$ $\quad$ (*ii*) $f(x,y,z) = e^{x+y}\cos\sqrt{2}\,z.$

The Wave Equation The partial differential equation

$$\frac{\partial^2 f}{\partial t^2} - c^2\frac{\partial^2 f}{\partial x^2} = 0,$$

where c is a positive constant, is known as the *wave equation.* It arises in the study of phenomena involving the propagation of waves in a continuous medium. For example, studies of water waves, sound waves, and light waves are all based on this equation. The wave equation is also used in the study of mechanical vibrations.

Problem 3. Show that the given functions satisfy the wave equation:

(i) $f(x,t) = (Ax+B)(Ct+D)$

(ii) $f(x,t) = \sin(x+ct)\cos(2x+2ct)$

(iii) $f(x,t) = \ln(x+ct)$

(iv) $f(x,t) = (Ae^{kx} + Be^{-kx})(Ce^{ckt} + De^{-ckt})$

Problem 4. Let $f(x,t) = g(x+ct) + h(x-ct)$, where g and h are any two, twice differentiable functions. Show that f is a solution of the wave equation. [This is the most general form of a solution of the wave equation.]

† Introduced in 1926 by the Austrian theoretical physicist Ervin Schrödinger (1881–1961).

†† Named after the French mathematician Pierre-Simon Laplace (1749–1827). Laplace wrote two monumental works: one on celestial mechanics, the other on probability theory. He also made major contributions to the theory of differential equations.

■ CHAPTER HIGHLIGHTS

14.1 Elementary Examples

domain, range (p. 820) open (closed) unit disc (p. 820)
open (closed) unit ball (p. 821)

14.2 A Brief Catalog of the Quadric Surface: Projections

traces, sections (p. 824) ellipsoid (p. 824)
hyperboloid of one sheet (p. 825)
hyperboloid of two sheets (p. 825)
elliptic cone, nappe (p. 826) elliptic paraboloid (p. 826)
hyperbolic paraboloid (p. 826) cylinder (p. 827)
parabolic cylinder (p. 827) elliptic cylinder (p. 827)
hyperbolic cylinder (p. 827) projection (p. 828)

14.3 Graphs; Level Curves and Level Surfaces

level curve (p. 831) level surface (p. 835)

14.4 Partial Derivatives

for a function of two variables:
  limit definition(p. 840)
  geometric interpretation (Figs. 14.4.1 and 14.4.2)
for a function of three variables:
  limit definitions (p. 842)

14.5 Open Sets and Closed Sets

neighborhood (p. 847) interior (p. 848)
boundary (p. 848)
open set (p. 848) closed set (p. 849)
deleted neighborhood (p. 850)
Some sets are neither open nor closed.

14.6 Limits and Continuity; Equality of Mixed Partials

limit of a function (p. 851) continuity at $\mathbf{x}_0$ (p. 853)

Points in the domain of a function of several variables can be written in vector notation. In the two-variable case, set $\mathbf{x}=(x, y)$, and, in the three-variable case, set $\mathbf{x} = (x, y, z)$. The vector notation enables us to treat the two cases together simply by writing $f(\mathbf{x})$.

A continuous function of several variables is continuous in each of its variables. The converse is false: it is possible for a function to be continuous in each variable separately and yet fail to be continuous as a function of several variables. (p. 855)

For functions of several variables the existence of partial derivatives fails to guarantee continuity. (p. 855)

The existence of a partial derivative depends on the behavior of the function along a line segment (two directions), whereas continuity depends on the behavior of the function in all directions. (p. 855)

equality of mixed partials (p. 857)

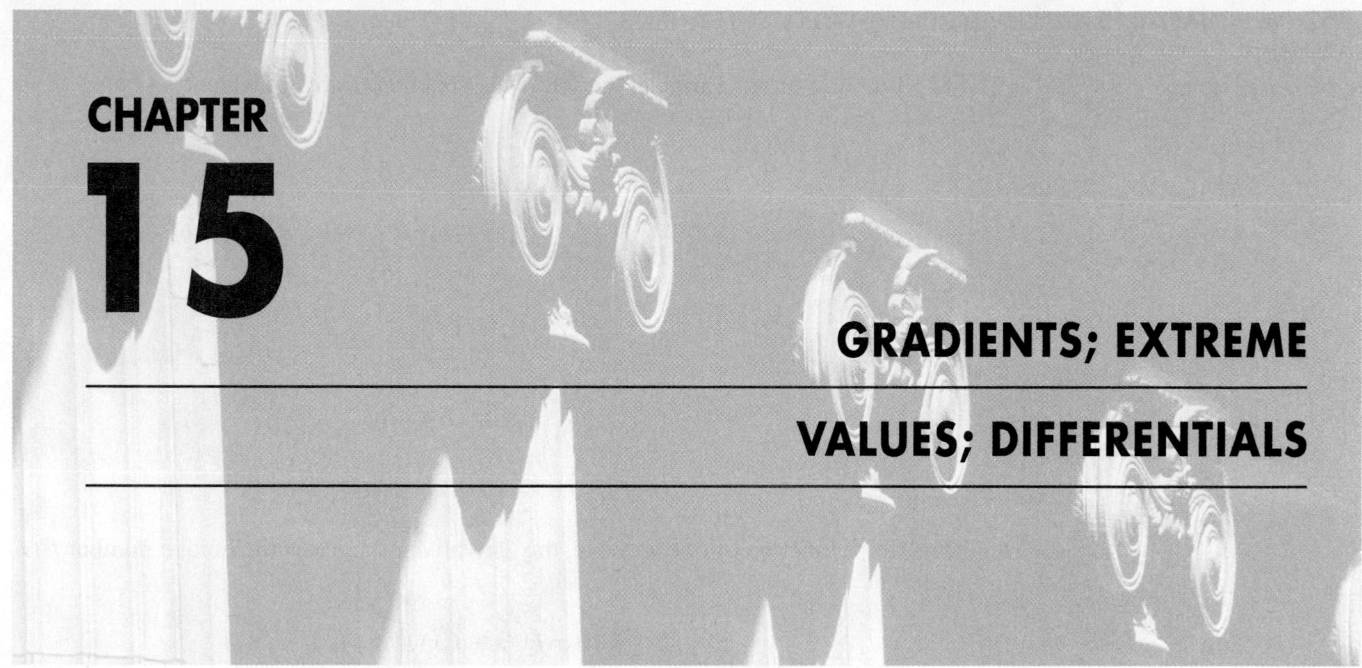

CHAPTER 15

GRADIENTS; EXTREME VALUES; DIFFERENTIALS

■ 15.1 DIFFERENTIABILITY AND GRADIENT

The Notion of Differentiability

Our object here is to extend the notion of differentiability from real-valued functions of one variable to real-valued functions of several variables. Partial derivatives alone do not fulfill this role because they reflect behavior only along paths parallel to the coordinate axes.

In the one-variable case we formed the difference quotient

$$\frac{f(x+h)-f(x)}{h}$$

and called f differentiable at x provided that this quotient had a limit as h tended to zero. In the multivariable case we can still form the difference

$$f(\mathbf{x}+\mathbf{h})-f(\mathbf{x}),$$

but the "quotient"

$$\frac{f(\mathbf{x}+\mathbf{h})-f(\mathbf{x})}{\mathbf{h}}$$

is not defined because it makes no sense to divide a real number $[f(\mathbf{x}+\mathbf{h})-f(\mathbf{x})]$ by a vector $\mathbf{h}$.

We can get around this difficulty by going back to an idea introduced in the exercises to Section 3.9. We review the idea here.

Let g be a real-valued function of a single variable which is defined on some open interval containing 0. We say that $g(h)$ is *little-o(h)* (read "little oh of h") and write $g(h) = o(h)$ iff

$$\lim_{h\to 0}\frac{g(h)}{|h|}=0.$$

For a function of one variable, the following statements are equivalent:

$$\lim_{h \to 0} \frac{f(x+h) - f(x)}{h} = f'(x),$$

$$\lim_{h \to 0} \frac{[f(x+h) - f(x)] - f'(x)h}{h} = 0,$$

$$\lim_{h \to 0} \frac{[f(x+h) - f(x)] - f'(x)h}{|h|} = 0,$$

$$[f(x+h) - f(x)] - f'(x)h = o(h),$$

$$[f(x+h) - f(x)] = f'(x)h + o(h).$$

Thus, for a function of one variable, the derivative of f at x is the unique number $f'(x)$ such that

$$f(x+h) - f(x) = f'(x)h + o(h).$$

It is this view of the derivative that inspires the notion of differentiability in the multivariable case.

Paralleling the definition for a function of a single variable, let g be a function of several variables which is defined in some neighborhood of $\mathbf{0}$. We will say that $g(\mathbf{h})$ is $o(\mathbf{h})$ if

$$\lim_{h \to 0} \frac{g(\mathbf{h})}{\|\mathbf{h}\|} = 0.$$

We will denote by $o(\mathbf{h})$ any expression $g(\mathbf{h})$ that has this property.

Now let f be a function of several variables *defined at least in some neighborhood of* $\mathbf{x}$. [In the three-variable case, $\mathbf{x} = (x, y, z)$; in the two-variable case, $\mathbf{x} = (x, y)$.]

DEFINITION 15.1.1 DIFFERENTIABILITY

We say that f is *differentiable at* $\mathbf{x}$ if there exists a vector $\mathbf{y}$ such that

$$f(\mathbf{x} + \mathbf{h}) - f(\mathbf{x}) = \mathbf{y} \cdot \mathbf{h} + o(\mathbf{h}).$$

It is not hard to show that, if such a vector $\mathbf{y}$ exists, it is unique (Exercise 41). We call this unique vector *the gradient of f at* $\mathbf{x}$ and denote it by $\nabla f(\mathbf{x})$: †

DEFINITION 15.1.2 GRADIENT

Let f be differentiable at $\mathbf{x}$. The *gradient of f at* $\mathbf{x}$ is the unique vector $\nabla f(\mathbf{x})$ such that

$$f(\mathbf{x} + \mathbf{h}) - f(\mathbf{x}) = \nabla f(\mathbf{x}) \cdot \mathbf{h} + o(\mathbf{h}).$$

† The symbol ∇, an inverted capital delta, is called a *nabla* and is read "del." The gradient of f is sometimes written grad f

The similarities between the one-variable case,

$$f(x+h)-f(x)=f'(x)h+o(h),$$

and the multivariable case,

$$f(\mathbf{x}+\mathbf{h})-f(\mathbf{x})=\nabla f(\mathbf{x})\cdot\mathbf{h}+o(\mathbf{h}),$$

are obvious. We point to the differences. There are essentially two of them:

(1) While the derivative, $f'(x)$, is a number, the gradient $\nabla f(\mathbf{x})$ is a vector.

(2) While $f'(x)h$ is the ordinary product of two real numbers, $\nabla f(\mathbf{x})\cdot\mathbf{h}$ is the dot product of two vectors.

Calculating Gradients

First we calculate some gradients by applying the definition directly. Then we give a theorem that makes such calculations much easier. Finally, we calculate some gradients with the aid of the theorem. As for notation, in the two-variable case we write

$$\nabla f(\mathbf{x})=\nabla f(x,y)\qquad\text{and}\qquad\mathbf{h}=(h_1,h_2),$$

and in the three-variable case,

$$\nabla f(\mathbf{x})=\nabla f(x,y,z)\qquad\text{and}\qquad\mathbf{h}=(h_1,h_2,h_3).$$

Example 1 For the function $f(x,y)=x^2+y^2$, we have

$$\begin{aligned}f(\mathbf{x}+\mathbf{h})-f(\mathbf{x})&=f(x+h_1,y+h_2)-f(x,y)\\&=\left[(x+h_1)^2+(y+h_2)^2\right]-\left[x^2+y^2\right]\\&=[2xh_1+2yh_2]+\left[h_1^2+h_2^2\right]\\&=[2x\,\mathbf{i}+2y\,\mathbf{j}]\cdot\mathbf{h}+\|\mathbf{h}\|^2.\end{aligned}$$

The remainder $\|\mathbf{h}\|^2$ is $o(\mathbf{h})$:

$$\frac{\|\mathbf{h}\|^2}{\|\mathbf{h}\|}=\|\mathbf{h}\|\to 0\quad\text{as}\quad\mathbf{h}\to\mathbf{0}.$$

Thus,
$$\nabla f(\mathbf{x})=\nabla f(x,y)=2x\,\mathbf{i}+2y\,\mathbf{j}.\quad\square$$

Example 2 For the function $f(x,y,z)=2xy-3z^2$, we have

$$\begin{aligned}f(\mathbf{x}+\mathbf{h})-f(\mathbf{x})&=f(x+h_1,y+h_2,z+h_3)-f(x,y,z)\\&=2(x+h_1)(y+h_2)-3(z+h_3)^2-\left[2xy-3z^2\right]\\&=2xh_2+2yh_1+2h_1h_2-6zh_3-3h_3^2\\&=(2y\,\mathbf{i}+2x\,\mathbf{j}-6z\,\mathbf{k})\cdot(h_1\,\mathbf{i}+h_2\,\mathbf{j}+h_3\,\mathbf{k})+2h_1h_2-3h_3^2\\&=(2y\,\mathbf{i}+2x\,\mathbf{j}-6z\,\mathbf{k})\cdot\mathbf{h}+2h_1h_2-3h_3^2.\end{aligned}$$

It remains to be shown that the remainder $g(\mathbf{h})=2h_1h_2-3h_3^2$ is $o(\mathbf{h})$. Since

$$g(\mathbf{h})=(2h_2\,\mathbf{i}-3h_3\,\mathbf{k})\cdot(h_1\,\mathbf{i}+h_2\,\mathbf{j}+h_3\,\mathbf{k})=(2h_2\,\mathbf{i}-3h_3\,\mathbf{k})\cdot\mathbf{h},$$

we can write

$$\frac{|g(\mathbf{h})|}{||\mathbf{h}||} = \frac{||2h_2\,\mathbf{i} - 3h_3\,\mathbf{k}||\;||\mathbf{h}||\;|\cos\theta|}{||\mathbf{h}||} \le \frac{||2h_2\,\mathbf{i} - 3h_3\,\mathbf{k}||\;||\mathbf{h}||}{||\mathbf{h}||} = ||2h_2\,\mathbf{i} - 3h_3\,\mathbf{k}||.$$

Since $\mathbf{h} \to \mathbf{0}$ iff $h_1 \to 0, h_2 \to 0$, and $h_3 \to 0$, it follows that $||2h_2\,\mathbf{i} - 3h_3\,\mathbf{k}|| \to 0$ as $\mathbf{h} \to 0$. Therefore, $g(\mathbf{h})/||\mathbf{h}|| \to 0$ as $\mathbf{h} \to \mathbf{0}$, and

$$\nabla f(\mathbf{x}) = 2y\,\mathbf{i} + 2x\,\mathbf{j} - 6z\,\mathbf{k}. \quad \square$$

Examples 1 and 2 illustrate the calculation of gradients directly from the definition of gradient. An easier way to calculate gradients is made possible by the theorem that follows

THEOREM 15.1.3

If f has continuous first partials in a neighborhood of $\mathbf{x}$, then f is differentiable at $\mathbf{x}$ and

$$\nabla f(\mathbf{x}) = \frac{\partial f}{\partial x}(\mathbf{x})\,\mathbf{i} + \frac{\partial f}{\partial y}(\mathbf{x})\,\mathbf{j} \qquad \text{(two variables)}$$

or

$$\nabla f(\mathbf{x}) = \frac{\partial f}{\partial x}(\mathbf{x})\,\mathbf{i} + \frac{\partial f}{\partial y}(\mathbf{x})\,\mathbf{j} + \frac{\partial f}{\partial z}(\mathbf{x})\,\mathbf{k} \qquad \text{(three variables)}$$

The proof is somewhat difficult. A proof of the two-variable case is given in a supplement at the end of this section.

Returning to Examples 1 and 2: for $f(x,y) = x^2+y^2$, $\partial f/\partial x = 2x$ and $\partial f/\partial y = 2y$. Since these functions are continuous,

$$\nabla f(\mathbf{x}) = 2x\,\mathbf{i} + 2y\,\mathbf{j}.$$

For $f(x,y,z) = 2xy - 3z^2$, $\partial f/\partial x = 2y$, $\partial f/\partial y = 2x$ and $\partial f/\partial z = -6z$. Since these functions are continuous,

$$\nabla f(\mathbf{x}) = 2y\,\mathbf{i} + 2x\,\mathbf{j} - 6z\,\mathbf{k}.$$

Example 3 For $f(x,y) = x\,e^y - y\,e^x$, we have

$$\frac{\partial f}{\partial x}(x,y) = e^y - y\,e^x, \quad \frac{\partial f}{\partial y}(x,y) = x\,e^y - e^x,$$

and therefore

$$\nabla f(x,y) = (e^y - y\,e^x)\,\mathbf{i} + (x\,e^y - e^x)\,\mathbf{j}. \quad \square$$

When there is no reason to emphasize the point of evaluation, we don't write

$$\nabla f(\mathbf{x}) \quad \text{or} \quad \nabla f(x,y) \quad \text{or} \quad \nabla f(x,y,z)$$

but simply ∇f. Thus for the function

$$f(x,y) = x\,e^y - y\,e^x$$

we write $$\frac{\partial f}{\partial x} = e^y - y\,e^x, \quad \frac{\partial f}{\partial x} = x\,e^y - e^x$$

and $$\nabla f = (e^y - y\,e^x)\,\mathbf{i} + (x\,e^y - e^x)\,\mathbf{j}.$$

Example 4 For $f(x, y, z) = \sin\,(xy^2z^3)$, we have

$$\frac{\partial f}{\partial x} = y^2z^3\cos\,(xy^2z^3), \quad \frac{\partial f}{\partial y} = 2xyz^3\cos\,(xy^2z^3), \quad \frac{\partial f}{\partial z} = 3xy^2z^2\cos\,(xy^2z^3)$$

and $$\nabla f = yz^2\cos\,(xy^2z^3)[\,yz\,\mathbf{i} + 2xz\,\mathbf{j} + 3xy\,\mathbf{k}]. \quad \square$$

Example 5 Let f be the function defined by

$$f(x, y, z) = x\sin\pi y + y\cos\pi z.$$

Evaluate ∇f at $(0, 1, 2)$.

SOLUTION Here

$$\frac{\partial f}{\partial x} = \sin\pi y, \quad \frac{\partial f}{\partial y} = \pi x\cos\pi y + \cos\pi z, \quad \frac{\partial f}{\partial z} = -\pi y\sin\pi z.$$

At $(0, 1, 2)$,

$$\frac{\partial f}{\partial x} = 0, \quad \frac{\partial f}{\partial y} = 1, \quad \frac{\partial f}{\partial z} = 0 \qquad \text{and thus} \qquad \nabla f(0, 1, 2) = \mathbf{j}. \quad \square$$

Of special interest for later work are the powers of r where, as usual,

$$r = \|\mathbf{r}\| \qquad \text{and} \qquad \mathbf{r} = x\,\mathbf{i} + y\,\mathbf{j} + z\,\mathbf{k}.$$

We begin by showing that, for $r \neq 0$,

(15.1.4) $$\boxed{\nabla r = \frac{\mathbf{r}}{r} \qquad \text{and} \qquad \nabla\left(\frac{1}{r}\right) = -\frac{\mathbf{r}}{r^3}.}$$

PROOF

$$\begin{aligned}
\nabla r &= \nabla(x^2 + y^2 + z^2)^{1/2} \\
&= \frac{\partial}{\partial x}(x^2 + y^2 + z^2)^{1/2}\,\mathbf{i} + \frac{\partial}{\partial y}(x^2 + y^2 + z^2)^{1/2}\,\mathbf{j} + \frac{\partial}{\partial z}(x^2 + y^2 + z^2)^{1/2}\,\mathbf{k} \\
&= \frac{x}{(x^2 + y^2 + z^2)^{1/2}}\,\mathbf{i} + \frac{y}{(x^2 + y^2 + z^2)^{1/2}}\,\mathbf{j} + \frac{z}{(x^2 + y^2 + z^2)^{1/2}}\,\mathbf{k} \\
&= \frac{1}{(x^2 + y^2 + z^2)^{1/2}}(x\,\mathbf{i} + y\,\mathbf{j} + z\,\mathbf{k}) = \frac{\mathbf{r}}{r}.
\end{aligned}$$

$$\begin{aligned}\nabla\left(\frac{1}{r}\right) &= \nabla(x^2+y^2+z^2)^{-1/2}\\ &= \frac{\partial}{\partial x}(x^2+y^2+z^2)^{-1/2}\,\mathbf{i}+\frac{\partial}{\partial y}(x^2+y^2+z^2)^{-1/2}\,\mathbf{j}+\frac{\partial}{\partial z}(x^2+y^2+z^2)^{-1/2}\,\mathbf{k}\\ &= -\frac{x}{(x^2+y^2+z^2)^{3/2}}\,\mathbf{i}-\frac{y}{(x^2+y^2+z^2)^{3/2}}\,\mathbf{j}-\frac{z}{(x^2+y^2+z^2)^{3/2}}\,\mathbf{k}\\ &= -\frac{1}{(x^2+y^2+z^2)^{3/2}}(x\,\mathbf{i}+y\,\mathbf{j}+z\,\mathbf{k}) = -\frac{\mathbf{r}}{r^3}.\quad \square\end{aligned}$$

The formulas we just derived can be generalized. As you are asked to show in the Exercises, for each integer n and all $\mathbf{r} \neq \mathbf{0}$,

(15.1.5) $$\nabla r^n = nr^{n-2}\,\mathbf{r}.$$

(If n is positive and even, the result also holds at $\mathbf{r} = \mathbf{0}$.)

Differentiability Implies Continuity

As in the one-variable case, differentiability implies continuity; namely,

(15.1.6) if f is differentiable at $\mathbf{x}$, then f is continuous at $\mathbf{x}$.

To see this, write

$$f(\mathbf{x}+\mathbf{h}) - f(\mathbf{x}) = \nabla f(\mathbf{x}) \cdot \mathbf{h} + o(\mathbf{h})$$

and note that

$$|f(\mathbf{x}+\mathbf{h}) - f(\mathbf{x})| = |\nabla f(\mathbf{x}) \cdot \mathbf{h} + o(\mathbf{h})| \leq |\nabla f(\mathbf{x}) \cdot \mathbf{h}| + |o(\mathbf{h})|. \quad \text{(triangle inequality)}$$

As $\mathbf{h} \to \mathbf{0}$,

$$|\nabla f(\mathbf{x}) \cdot \mathbf{h}| \leq \|\nabla f(\mathbf{x})\|\,\|\mathbf{h}\| \to 0 \quad \text{and} \quad |o(\mathbf{h})| \to 0.$$

Schwarz's inequality (for the first inequality); Exercise 42 (for $|o(\mathbf{h})| \to 0$)

It follows that

$$f(\mathbf{x}+\mathbf{h}) - f(\mathbf{x}) \to 0 \quad \text{and therefore} \quad f(\mathbf{x}+\mathbf{h}) \to f(\mathbf{x}). \quad \square$$

EXERCISES 15.1

Find the gradient.

1. $f(x,y) = 3x^2 - xy + y$.
2. $f(x,y) = Ax^2 + Bxy + Cy^2$.
3. $f(x,y) = xe^{xy}$.
4. $f(x,y) = \dfrac{x-y}{x^2+y^2}$.
5. $f(x,y) = 2xy^2 \sin(x^2+1)$.
6. $f(x,y) = \ln(x^2+y^2)$.
7. $f(x,y) = e^{x-y} - e^{y-x}$.
8. $f(x,y) = \dfrac{Ax+By}{Cx+Dy}$.
9. $f(x,y,z) = x^2y + y^2z + z^2x$.
10. $f(x,y,z) = \sqrt{x^2+y^2+z^2}$.
11. $f(x,y,z) = x^2ye^{-z}$.
12. $f(x,y,z) = xyz \ln(x+y+z)$.
13. $f(x,y,z) = e^{x+2y}\cos(z^2+1)$.
14. $f(x,y,z) = e^{yz^2/x^3}$.
15. $f(x,y,z) = \sin(2xy) + \ln(x^2z)$.
16. $f(x,y,z) = x^2y/z - 3xz^4$.

Find the gradient vector at the point P.

17. $f(x,y) = 2x^2 - 3xy + 4y^2$ at $P(2,3)$.
18. $f(x,y) = 2x(x-y)^{-1}$ at $P(3,1)$.
19. $f(x,y) = \ln(x^2+y^2)$ at $P(2,1)$.

20. $f(x,y) = x\tan^{-1}(y/x)$ at $P(1,1)$.

21. $f(x,y) = x\sin(xy)$ at $P(1,\pi/2)$.

22. $f(x,y) = xye^{-(x^2+y^2)}$ at $P(1,-1)$.

23. $f(x,y,z) = e^{-x}\sin(z+2y)$ at $P(0,\frac{1}{4}\pi,\frac{1}{4}\pi)$.

24. $f(x,y,z) = (x-y)\cos\pi z$ at $P(1,0,\frac{1}{2})$.

25. $f(x,y,z) = x - \sqrt{y^2+z^2}$ at $P(2,-3,4)$.

26. $f(x,y,z) = \cos(xyz^2)$ at $P(\pi,\frac{1}{4},-1)$.

In Exercises 27 and 28, use a CAS to find the gradient of f at the point P.

27. (a) $f(x,y) = xy^2e^{-xy}$; $P(0,2)$.
(b) $f(x,y) = \sin(2x+y) - \cos(x-2y)$; $P(\pi/4,\pi/6)$.
(c) $f(x,y) = x - y\ln(x^2y)$; $P(1,e)$.

28. (a) $f(x,y,z) = \sqrt{x+y^2-z^3}$; $P(1,2,-3)$.
(b) $f(x,y,z) = \dfrac{xy}{x-y+z}$; $P(1,-2,3)$
(c) $f(x,y,z) = x\sin(z\ln y)$; $P(1,e^2,\pi/6)$.

Obtain the gradient directly from Definition 15.1.2

29. $f(x,y) = 3x^2 - xy + y$.

30. $f(x,y) = \frac{1}{2}x^2 + 2xy + y^2$.

31. $f(x,y,z) = x^2y + y^2z + z^2x$.

32. $f(x,y,z) = 2x^2y - \dfrac{1}{z}$.

Find a function f with the gradient $\mathbf{F}$.

33. $\mathbf{F}(x,y) = 2xy\,\mathbf{i} + (1+x^2)\,\mathbf{j}$.

34. $\mathbf{F}(x,y) = (2xy+x)\,\mathbf{i} + (x^2+y)\,\mathbf{j}$.

35. $\mathbf{F}(x,y) = (x+\sin y)\,\mathbf{i} + (x\cos y - 2y)\,\mathbf{j}$.

36. $\mathbf{F}(x,y,z) = yz\,\mathbf{i} + (xz+2yz)\,\mathbf{j} + (xy+y^2)\,\mathbf{k}$.

37. Find (a) $\nabla(\ln r)$, (b) $\nabla(\sin r)$, (c) $\nabla(e^r)$, where $r = \sqrt{x^2+y^2+z^2}$.

38. Derive (15.1.5).

39. Let $f(x,y) = 1 + x^2 + y^2$.
(a) Find the points (x,y), if any, at which $\nabla f(x,y) = \mathbf{0}$.
(b) Sketch the graph of the surface $z = f(x,y)$.
(c) What can you say about the surface at the point(s) found in part (a)?

40. Repeat Exercise 39 for $f(x,y) = \sqrt{4-x^2-y^2}$.

41. (a) Show that, if $\mathbf{c}\cdot\mathbf{h}$ is $o(\mathbf{h})$, then $\mathbf{c} = \mathbf{0}$. HINT: First set $\mathbf{h} = h\,\mathbf{i}$, then set $\mathbf{h} = h\,\mathbf{j}$, then $\mathbf{h} = h\,\mathbf{k}$.
(b) Show that, if

$$f(\mathbf{x}+\mathbf{h}) - f(\mathbf{x}) = \mathbf{y}\cdot\mathbf{h} + o(\mathbf{h})$$

and

$$f(\mathbf{x}+\mathbf{h}) - f(\mathbf{x}) = \mathbf{z}\cdot\mathbf{h} + o(\mathbf{h}),$$

then

$$\mathbf{y} = \mathbf{z}.$$

42. Show that, if g is $o(\mathbf{h})$, then

$$\lim_{\mathbf{h}\to\mathbf{0}} g(\mathbf{h}) = 0.$$

43. Set

$$f(x,y) = \begin{cases} \dfrac{2xy}{x^2+y^2}, & (x,y) \neq (0,0) \\ 0, & (x,y) = (0,0). \end{cases}$$

(a) Show that f is not differentiable at $(0,0)$.
(b) In Section 14.6 you saw that the first partials $\partial f/\partial x$ and $\partial f/\partial y$ exist at $(0,0)$. Since these partials obviously exist at every other point of the plane, we can conclude from Theorem 15.1.3 that at least one of these partials is not continuous in a neighborhood of (0,0). Show that $\partial f/\partial x$ is discontinuous at $(0,0)$.

■ PROJECT 15.1 Points Where $\nabla f = 0$

In this project we find points (x,y) where $\nabla f = 0$. Then we investigate the behavior of the surface $z = f(x,y)$ at these points. (Recall the significance of $f'(x) = 0$ for a function f of a single variable.)

For example, let

$$f(x,y) = \frac{-2y}{x^2+y^2+1}.$$

Then,

$$\frac{\partial f}{\partial x} = \frac{4xy}{(x^2+y^2+1)^2}, \quad \frac{\partial f}{\partial y} = \frac{2y^2-2x^2-2}{(x^2+y^2+1)^2}$$

and

$$\nabla f(x,y) = \frac{4xy}{(x^2+y^2+1)^2}\,\mathbf{i} + \frac{2y^2-2x^2-2}{(x^2+y^2+1)^2}\,\mathbf{j}.$$

Setting $\nabla f(x,y) = 0$, we get the pair of equations

$$\frac{4xy}{(x^2+y^2+1)^2} = 0. \quad \frac{2y^2-2x^2-2}{(x^2+y^2+1)^2} = 0$$

which are satisfied iff

$$4xy = 0 \quad \text{and} \quad 2y^2 - 2x^2 - 2 = 0$$

From the first equation, we get $x = 0$ or $y = 0$. Setting $x = 0$ in the second equation yields $y = \pm 1$; setting $y = 0$ in the second equation yields $x^2 = -1$, which has no solutions. Thus $\nabla f(x,y) = 0$ only at $(0,1)$ and $(0,-1)$. The figure shows the graph of the surface $z = f(x,y)$. Note that f has a maximum at $(0,-1)$ and a minimum at $(0,1)$.

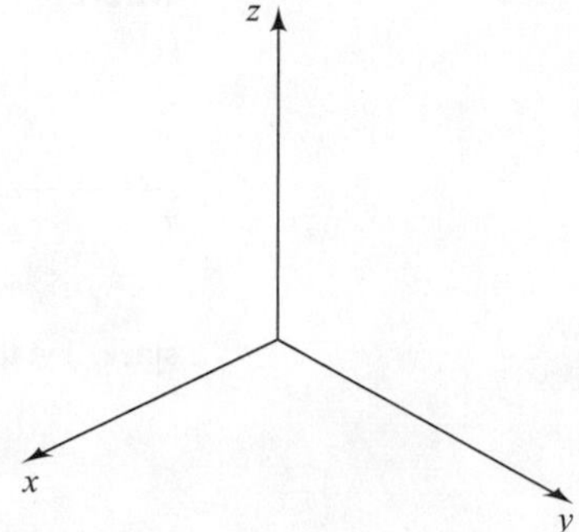

Problem 1. Let $f(x,y)=\dfrac{5x}{x^2+y^2+1}$, $-5\le x\le 5,\quad -5\le y\le 5.$

(a) Find the points (x,y), if any, at which $\nabla f(x,y)=0$.

(b) Use a CAS to draw the surface $z=f(x,y)$ and the level curves. Compare with the results of part (a).

(c) Investigate the behavior of f at the points found in part (a).

Problem 2. Repeat Problem 1 with $f(x,y)=(\sin x)(\sin y)$, $0\le x\le 4\pi, 0\le y\le 4\pi$.

Problem 3. Repeat Problem 1 with $f(x,y)=4xye^{-(x^2+y^2)}$, $-2\le x\le 2, -2\le y\le 2$.

Problem 4. Repeat Problem 1 with $f(x,y)=(x^2+4y^2)e^{1-(x^2+y^2)}$, $-2\le x\le 2, -2\le y\le 2$.

*SUPPLEMENT TO SECTION 15.1

PROOF OF THEOREM 15.1.3

We prove the theorem in the two-variable case. A similar argument yields a proof in the three-variable case, but there the details are more burdensome.

In the first place,

$$f(\mathbf{x}+\mathbf{h})-f(\mathbf{x})=f(x+h_1,y+h_2)-f(x,y).$$

Adding and subtracting $f(x,y+h_2)$, we have

$$(1)\qquad f(\mathbf{x}+\mathbf{h})-f(\mathbf{x})=[f(x+h_1,y+h_2)-f(x,y+h_2)]+[f(x,y+h_2)-f(x,y)].$$

By the mean-value theorem for functions of one variable, we know that there are numbers

$$0<\theta_1<1 \quad\text{and}\quad 0<\theta_2<1$$

such that

$$f(x+h_1,y+h_2)-f(x,y+h_2)=\frac{\partial f}{\partial x}(x+\theta_1h_1,y+h_2)h_1$$

and

$$f(x,y+h_2)-f(x,y)=\frac{\partial f}{\partial y}(x,y+\theta_2h_2)h_2. \qquad \text{(Exercise 40, Section 4.1)}$$

By the continuity of $\partial f/\partial x$,

$$\frac{\partial f}{\partial x}(x+\theta_1h_1,y+h_2)=\frac{\partial f}{\partial x}(x,y)+\epsilon_1(\mathbf{h})$$

where $\quad \epsilon_1(\mathbf{h})\to 0 \quad\text{as}\quad \mathbf{h}\to 0.$†

† $$\epsilon_1(\mathbf{h})=\frac{\partial f}{\partial x}(x+\theta_1h_1,y+h_2)-\frac{\partial f}{\partial x}(x,y)\to 0$$

since, by the continuity of $\partial f/\partial x$,

$$\frac{\partial f}{\partial x}(x+\theta_1h_1,y+h_2)\to\frac{\partial f}{\partial x}(x,y).$$

By the continuity of $\partial f/\partial y$,

$$\frac{\partial f}{\partial x}(x, y+\theta_2 h_2) = \frac{\partial f}{\partial x}(x, y) + \epsilon_2(\mathbf{h})$$

where $\qquad \epsilon_2(\mathbf{h}) \to 0 \quad \text{as} \quad \mathbf{h} \to \mathbf{0}.$

Substituting these expressions in equation (1), we find that

$$\begin{aligned} f(\mathbf{x}+\mathbf{h}) - f(\mathbf{x}) &= \left[\frac{\partial f}{\partial x}(x,y) + \epsilon_1(\mathbf{h})\right] h_1 + \left[\frac{\partial f}{\partial y}(x,y) + \epsilon_2(\mathbf{h})\right] h_2 \\ &= \left[\frac{\partial f}{\partial x}(x,y) + \epsilon_1(\mathbf{h})\right](\mathbf{i}\cdot\mathbf{h}) + \left[\frac{\partial f}{\partial y}(x,y) + \epsilon_2(\mathbf{h})\right](\mathbf{j}\cdot\mathbf{h}) \\ &= \left[\frac{\partial f}{\partial x}(x,y)\,\mathbf{i} + \epsilon_1(\mathbf{h})\,\mathbf{i}\right]\cdot\mathbf{h} + \left[\frac{\partial f}{\partial y}(x,y)\,\mathbf{j} + \epsilon_2(\mathbf{h})\,\mathbf{j}\right]\cdot\mathbf{h} \\ &= \left[\frac{\partial f}{\partial x}(x,y)\,\mathbf{i} + \frac{\partial f}{\partial y}(x,y)\,\mathbf{j}\right]\cdot\mathbf{h} + \left[\epsilon_1(\mathbf{h})\,\mathbf{i} + \epsilon_2(\mathbf{h})\,\mathbf{j}\right]\cdot\mathbf{h}. \end{aligned}$$

To complete the proof of the theorem we need only show that

$$\text{(2)} \qquad [\epsilon_1(\mathbf{h})\,\mathbf{i} + \epsilon_2(\mathbf{h})\,\mathbf{j}]\cdot\mathbf{h} = o(\mathbf{h}).$$

From Schwarz's inequality, $|\mathbf{a}\cdot\mathbf{b}| \le \|\mathbf{a}\|\,\|\mathbf{b}\|$, we know that

$$|[\epsilon_1(\mathbf{h})\,\mathbf{i} + \epsilon_2(\mathbf{h})\,\mathbf{j}]\cdot\mathbf{h}| \le \|\epsilon_1(\mathbf{h})\,\mathbf{i} + \epsilon_2(\mathbf{h})\,\mathbf{j}\|\,\|\mathbf{h}\|.$$

It follows that

$$\frac{|[\epsilon_1(\mathbf{h})\,\mathbf{i} + \epsilon_2(\mathbf{h})\,\mathbf{j}]\cdot\mathbf{h}|}{\|\mathbf{h}\|} \le \|\epsilon_1(\mathbf{h})\,\mathbf{i} + \epsilon_2(\mathbf{h})\,\mathbf{j}\| \underset{\uparrow}{\le} \|\epsilon_1(\mathbf{h})\,\mathbf{i}\| + \|\epsilon_2(\mathbf{h})\,\mathbf{j}\| = |\epsilon_1(\mathbf{h})| + |\epsilon_2(\mathbf{h})|.$$

by the triangle inequality

As $\mathbf{h} \to \mathbf{0}$, the expression on the right tends to 0. This shows that (2) holds and completes the proof of the theorem. ❑

■ 15.2 GRADIENTS AND DIRECTIONAL DERIVATIVES

Some Elementary Formulas

In many respects gradients behave just as derivatives do in the one-variable case. In particular, if $\nabla f(\mathbf{x})$ and $\nabla g(\mathbf{x})$ exist, then $\nabla[f(\mathbf{x}) + g(\mathbf{x})]$, $\nabla[\alpha f(\mathbf{x})]$, and $\nabla[f(\mathbf{x})g(\mathbf{x})]$ all exist, and

(15.2.1)
$$\begin{gathered} \nabla[f(\mathbf{x}) + g(\mathbf{x})] = \nabla f(\mathbf{x}) + \nabla g(\mathbf{x}), \\ \nabla[\alpha f(\mathbf{x})] = \alpha\nabla f(\mathbf{x}), \\ \nabla[f(\mathbf{x})g(\mathbf{x})] = f(\mathbf{x})\,\nabla g(\mathbf{x}) + g(\mathbf{x})\,\nabla f(\mathbf{x}). \end{gathered}$$

The first two formulas are easy to derive. To derive the third formula, let's assume that $\nabla f(\mathbf{x})$ and $\nabla g(\mathbf{x})$ both exist. Our task is to show that

$$f(\mathbf{x}+\mathbf{h})g(\mathbf{x}+\mathbf{h}) - f(\mathbf{x})g(\mathbf{x}) = [f(\mathbf{x})\,\nabla g(\mathbf{x}) + g(\mathbf{x})\,\nabla f(\mathbf{x})]\cdot\mathbf{h} + o(\mathbf{h}).$$

We now sketch how this can be done. We leave it to you to justify each step. The key to proving a product rule is to add and subtract an appropriate expression (see, for instance, the proof of product rule Theorem 3.2.6). Starting from

$$f(\mathbf{x}+\mathbf{h})g(\mathbf{x}+\mathbf{h})-f(\mathbf{x})g(\mathbf{x}),$$

we add and subtract the term $f(\mathbf{x})g(\mathbf{x}+\mathbf{h})$. This gives

$$\begin{aligned}
&[f(\mathbf{x}+\mathbf{h})g(\mathbf{x}+\mathbf{h})-f(\mathbf{x})g(\mathbf{x}+\mathbf{h})]+[f(\mathbf{x})g(\mathbf{x}+\mathbf{h})-f(\mathbf{x})g(\mathbf{x})]\\
&=[f(\mathbf{x}+\mathbf{h})-f(\mathbf{x})]g(\mathbf{x}+\mathbf{h})+f(\mathbf{x})[g(\mathbf{x}+\mathbf{h})-g(\mathbf{x})]\\
&=[\nabla f(\mathbf{x})\cdot\mathbf{h}+o(\mathbf{h})]g(\mathbf{x}+\mathbf{h})+f(\mathbf{x})[\nabla g(\mathbf{x})\cdot\mathbf{h}+o(\mathbf{h})]\\
&=g(\mathbf{x}+\mathbf{h})\nabla f(\mathbf{x})\cdot\mathbf{h}+f(\mathbf{x})\nabla g(\mathbf{x})\cdot\mathbf{h}+o(\mathbf{h}) \qquad \text{(Exercise 30)}\\
&=g(\mathbf{x})\nabla f(\mathbf{x})\cdot\mathbf{h}+f(\mathbf{x})\nabla g(\mathbf{x})\cdot\mathbf{h}+[g(\mathbf{x}+\mathbf{h})-g(\mathbf{x})]\nabla f(\mathbf{x})\cdot\mathbf{h}+o(\mathbf{h})\\
&=[g(\mathbf{x})\nabla f(\mathbf{x})+f(\mathbf{x})\nabla g(\mathbf{x})]\cdot\mathbf{h}+o(\mathbf{h}). \qquad \text{(Exercise 30)} \quad \square
\end{aligned}$$

In Exercise 43, you are asked to derive this formula from Theorem 15.1.3.

Directional Derivatives

Here we take up an idea that generalizes the notion of partial derivative. Its connection with gradients will be made clear as we go on. We begin by recalling the definitions of the first partial derivatives:

(two variables)

$$\frac{\partial f}{\partial x}(x,y)=\lim_{h\to 0}\frac{f(x+h,y)-f(x,y)}{h},$$

$$\frac{\partial f}{\partial y}(x,y)=\lim_{h\to 0}\frac{f(x,y+h)-f(x,y)}{h},$$

(three variables)

$$\frac{\partial f}{\partial x}(x,y,z)=\lim_{h\to 0}\frac{f(x+h,y,z)-f(x,y,z)}{h},$$

$$\frac{\partial f}{\partial y}(x,y,z)=\lim_{h\to 0}\frac{f(x,y+h,z)-f(x,y,z)}{h},$$

$$\frac{\partial f}{\partial z}(x,y,z)=\lim_{h\to 0}\frac{f(x,y,z+h)-f(x,y,z)}{h}.$$

Expressed in vector notation, these definitions take the form

$(\mathbf{x}=(x,y))$

$$\frac{\partial f}{\partial x}(\mathbf{x})=\lim_{h\to 0}\frac{f(\mathbf{x}+h\,\mathbf{i})-f(\mathbf{x})}{h},$$

$$\frac{\partial f}{\partial y}(\mathbf{x})=\lim_{h\to 0}\frac{f(\mathbf{x}+h\,\mathbf{j})-f(\mathbf{x})}{h},$$

$(\mathbf{x}=(x,y,z))$

$$\frac{\partial f}{\partial x}(\mathbf{x})=\lim_{h\to 0}\frac{f(\mathbf{x}+h\,\mathbf{i})-f(\mathbf{x})}{h},$$

$$\frac{\partial f}{\partial y}(\mathbf{x})=\lim_{h\to 0}\frac{f(\mathbf{x}+h\,\mathbf{j})-f(\mathbf{x})}{h},$$

$$\frac{\partial f}{\partial z}(\mathbf{x})=\lim_{h\to 0}\frac{f(\mathbf{x}+h\,\mathbf{k})-f(\mathbf{x})}{h}.$$

Each partial is thus the limit of a quotient

$$\frac{f(\mathbf{x}+h\,\mathbf{u})-f(\mathbf{x})}{h}$$

where $\mathbf{u}$ is one of the unit coordinate vectors, $\mathbf{i}, \mathbf{j}, \mathbf{k}$. There is no reason to be so restrictive on $\mathbf{u}$. If f is defined in a neighborhood of $\mathbf{x}$, then, for small h, the difference quotient

$$\frac{f(\mathbf{x}+h\,\mathbf{u})-f(\mathbf{x})}{h}$$

makes sense for any unit vector $\mathbf{u}$.

DEFINITION 15.2.2 DIRECTIONAL DERIVATIVE

For each unit vector **u**, the limit

$$f'_u(\mathbf{x}) = \lim_{h\to 0} \frac{f(\mathbf{x}+h\mathbf{u}) - f(\mathbf{x})}{h},$$

if it exists, is called the *directional derivative of f at* **x** *in the direction* **u**.

It is important to recognize that each partial derivative $\partial f/\partial x, \partial f/\partial y, \partial f/\partial z$ is itself a directional derivative:

(15.2.3)

$$\frac{\partial f}{\partial x}(\mathbf{x}) = f'_{\mathbf{i}}(\mathbf{x}), \quad \frac{\partial f}{\partial y}(\mathbf{x}) = f'_{\mathbf{j}}(\mathbf{x}), \quad \frac{\partial f}{\partial z}(\mathbf{x}) = f'_{\mathbf{k}}(\mathbf{x}).$$

As you know, the partials of f give the rates of change of f in the **i**, **j**, **k** directions. The directional derivative $f'_{\mathbf{u}}$ gives *the rate of change of f in the direction* **u**.

A geomentric interpretation of the directional derivative for a function f of two variables can be obtained from Figure 15.2.1. Fix a point (x, y) in the domain of f and let **u** be a unit vector with initial point (x, y) in the xy-plane. Let C be the curve of intersection of the surface $z = f(x, y)$ and the plane p which contains **u** and is perpendicular to the xy-plane. Then $f'_{\mathbf{u}}(\mathbf{x})$ is the slope of the line tangent to C at the point $(x, y, f(x, y))$.

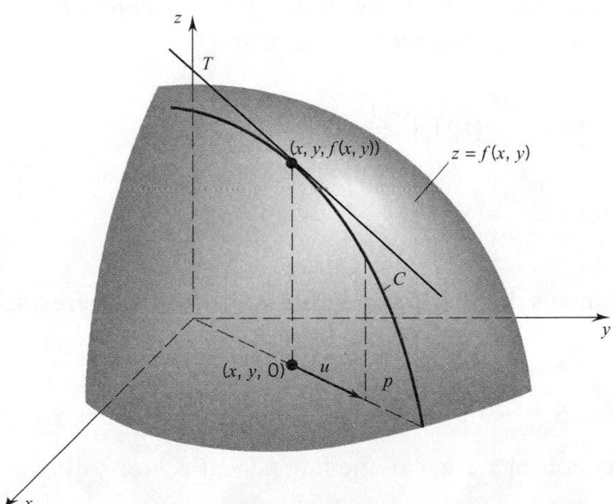

Figure 15.2.1

Remark The definition of the directional derivative in the direction **u** requires **u** to be a unit vector. However, we can extend the definition to arbitrary nonzero vectors by adopting the following convention: the directional derivative of f at **x** in the direction of a nonzero vector **a** is by definition $f'_{\mathbf{u}}(\mathbf{x})$ where $\mathbf{u} = \mathbf{a}/||\mathbf{a}||$ is the unit vector which has the same direction as **a**. □

There is an important connection between the gradient at **x** and the directional derivatives at **x**.

THEOREM 15.2.4

If f is differentiable at $\mathbf{x}$, then f has a directional derivative at $\mathbf{x}$ in every direction $\mathbf{u}$, where $\mathbf{u}$ is a unit vector, and

$$f'_{\mathbf{u}}(\mathbf{x}) = \nabla f(\mathbf{x}) \cdot \mathbf{u}.$$

PROOF We take $\mathbf{u}$ as a unit vector and assume that f is differentiable at $\mathbf{x}$. The differentiability at $\mathbf{x}$ tells us that $\nabla f(\mathbf{x})$ exists and

$$f(\mathbf{x} + h\,\mathbf{u}) - f(\mathbf{x}) = \nabla f\,(\mathbf{x}) \cdot h\,\mathbf{u} + o(h\,\mathbf{u}).$$

Division by h gives

$$\frac{f(\mathbf{x} + h\,\mathbf{u}) - f(\mathbf{x})}{h} = \nabla f(\mathbf{x}) \cdot \mathbf{u} + \frac{o(h\,\mathbf{u})}{h}.$$

Since $\left|\dfrac{o(h\,\mathbf{u})}{h}\right| = \dfrac{|o(h\,\mathbf{u})|}{|h|} = \dfrac{|o(h\,\mathbf{u})|}{\|h\,\mathbf{u}\|} \to 0,$

you can see that

$$\frac{o(h\,\mathbf{u})}{h} \to 0 \quad \text{and thus} \quad \frac{f(\mathbf{x} + h\,\mathbf{u}) - f(\mathbf{x})}{h} \to \nabla f(\mathbf{x}) \cdot \mathbf{u}. \quad \square$$

Earlier (Theorem 15.1.3) you saw that, *if f has continuous first partials in a neighborhood of* $\mathbf{x}$, *then f is differentiable at* $\mathbf{x}$ *and*

$$\nabla f(\mathbf{x}) = \frac{\partial f}{\partial x}(\mathbf{x})\,\mathbf{i} + \frac{\partial f}{\partial y}(\mathbf{x})\,\mathbf{j}, \qquad [\mathbf{x} = (x, y)]$$

$$\nabla f(\mathbf{x}) = \frac{\partial f}{\partial x}(\mathbf{x})\,\mathbf{i} + \frac{\partial f}{\partial y}(\mathbf{x})\,\mathbf{j} + \frac{\partial f}{\partial z}(\mathbf{x})\,\mathbf{k}. \qquad [\mathbf{x} = (x, y, z)]$$

The next theorem shows that this formula for $\nabla f(\mathbf{x})$ holds wherever f is differentiable.

THEOREM 15.2.5

If f is differentiable at $\mathbf{x}$, then all the first partial derivatives of f exist at $\mathbf{x}$ and

$$\nabla f(\mathbf{x}) = \frac{\partial f}{\partial x}(\mathbf{x})\,\mathbf{i} + \frac{\partial f}{\partial y}(\mathbf{x})\,\mathbf{j} \qquad (\mathbf{x} = (x, y)),$$

$$\nabla f(\mathbf{x}) = \frac{\partial f}{\partial x}(\mathbf{x})\,\mathbf{i} + \frac{\partial f}{\partial y}(\mathbf{x})\,\mathbf{j} + \frac{\partial f}{\partial z}(\mathbf{x})\,\mathbf{k} \qquad (\mathbf{x} = (x, y, z)).$$

PROOF It is sufficient to prove the theorem for the case $\mathbf{x} = (x, y, z)$. Assume that f is differentiable at $\mathbf{x}$. Then $\nabla f(\mathbf{x})$ exists and we can write

$$\nabla f(\mathbf{x}) = [\nabla f(\mathbf{x}) \cdot \mathbf{i}]\,\mathbf{i} + [\nabla f(\mathbf{x}) \cdot \mathbf{j}]\,\mathbf{j} + [\nabla f(\mathbf{x}) \cdot \mathbf{k}]\,\mathbf{k}. \qquad (12.4.13)$$

The result follows from observing that

$$\begin{aligned}
&\qquad\quad (15.2.4)\quad (15.2.3)\\
&\qquad\qquad\downarrow\qquad\quad\downarrow\\
\nabla f(\mathbf{x})\cdot\mathbf{i} &= f'_{\mathbf{i}}(\mathbf{x}) = \frac{\partial f}{\partial x}(\mathbf{x}),\\
\nabla f(\mathbf{x})\cdot\mathbf{j} &= f'_{\mathbf{j}}(\mathbf{x}) = \frac{\partial f}{\partial y}(\mathbf{x}),\\
\nabla f(\mathbf{x})\cdot\mathbf{k} &= f'_{\mathbf{k}}(\mathbf{x}) = \frac{\partial f}{\partial z}(\mathbf{x}). \quad \square
\end{aligned}$$

Example 1 Find the directional derivative of the function $f(x, y) = x^2 + y^2$ at the point $(1, 2)$ in the direction of the vector $2\,\mathbf{i} - 3\,\mathbf{j}$.

SOLUTION In the first place, $2\,\mathbf{i} - 3\,\mathbf{j}$ is not a unit vector; its norm is $\sqrt{13}$. The unit vector in the direction of $2\,\mathbf{i} - 3\,\mathbf{j}$ is the vector

$$\mathbf{u} = \frac{1}{\sqrt{13}}[2\,\mathbf{i} - 3\,\mathbf{j}].$$

Next, $$\nabla f = 2x\,\mathbf{i} + 2y\,\mathbf{j},$$

and therefore $$\nabla f(1, 2) = 2\,\mathbf{i} + 4\,\mathbf{j}.$$

By Theorem 15.2.4 we have

$$\begin{aligned}
f'_{\mathbf{u}}(1, 2) &= \nabla f(1, 2)\cdot\mathbf{u}\\
&= (2\mathbf{i} + 4\mathbf{j})\cdot\frac{1}{\sqrt{13}}[2\,\mathbf{i} - 3\,\mathbf{j}] = \frac{-8}{\sqrt{13}} \cong -2.219. \quad \square
\end{aligned}$$

Example 2 Find the directional derivative of the function

$$f(x, y, z) = 2xz^2 \cos \pi y$$

at the point $P(1, 2, -1)$ toward the point $Q(2, 1, 3)$.

SOLUTION The vector from P to Q is given by $\overrightarrow{PQ} = \mathbf{i} - \mathbf{j} + 4\,\mathbf{k}$. The unit vector in this direction is the vector

$$\mathbf{u} = \frac{1}{3\sqrt{2}}[\mathbf{i} - \mathbf{j} + 4\,\mathbf{k}].$$

Here, $$\frac{\partial f}{\partial x} = 2z^2 \cos \pi y, \qquad \frac{\partial f}{\partial y} = -2\pi x z^2 \sin \pi y, \qquad \frac{\partial f}{\partial z} = 4xz \cos \pi y,$$

so that $$\frac{\partial f}{\partial x}(1, 2, -1) = 2, \qquad \frac{\partial f}{\partial y}(1, 2, -1) = 0, \qquad \frac{\partial f}{\partial z}(1, 2, -1) = -4.$$

Therefore, $$\nabla f(1, 2, -1) = 2\,\mathbf{i} - 4\,\mathbf{k}$$

and $$f'_{\mathbf{u}}(1,2,-1) = \nabla f(1,2,-1) \cdot \mathbf{u} = (2\,\mathbf{i} - 4\,\mathbf{k}) \cdot \frac{1}{3\sqrt{2}}[\mathbf{i} - \mathbf{j} + 4\mathbf{k}]$$

$$= -\frac{14}{3\sqrt{2}} \cong -3.30. \quad \square$$

You know that for each unit vector $\mathbf{u}$

$$f'_{\mathbf{u}}(\mathbf{x}) = \nabla f(\mathbf{x}) \cdot \mathbf{u}$$

Since $$\nabla f(\mathbf{x}) \cdot \mathbf{u} = \text{comp}_{\mathbf{u}} \nabla f(\mathbf{x}), \qquad (12.4.11)$$

we have

(15.2.6) $$\boxed{f'_{\mathbf{u}}(\mathbf{x}) = \text{comp}_{\mathbf{u}} \nabla f(\mathbf{x}).}$$

Namely, the directional derivative in a direction $\mathbf{u}$ *is the component of the gradient vector in that direction.* (See Figure 15.2.2.)

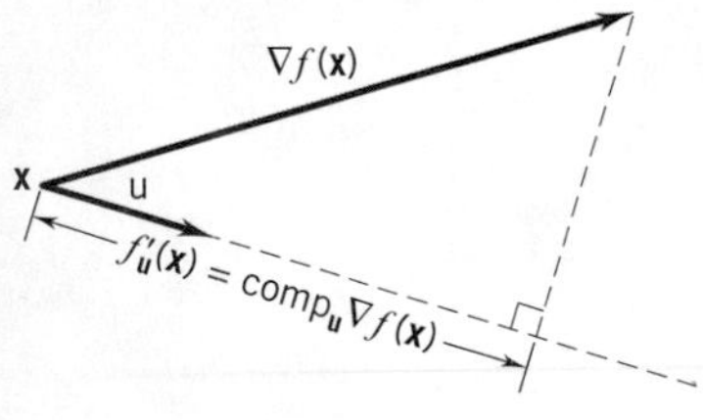

Figure 15.2.2

If $\nabla f(\mathbf{x}) \neq \mathbf{0}$, then

$$f'_{\mathbf{u}}(\mathbf{x}) = \nabla f(\mathbf{x}) \cdot \mathbf{u} = \underset{\uparrow\,(12.4.7)}{=} ||\nabla f(\mathbf{x})||\;||\mathbf{u}|| \cos\theta \underset{\uparrow\,||\mathbf{u}||=1}{=} ||\nabla f(\mathbf{x})|| \cos\theta$$

where θ is the angle between $\nabla f(\mathbf{x})$ and $\mathbf{u}$. Since $-1 \leq \cos\theta \leq 1$, we have

$$-||\nabla f(\mathbf{x})|| \leq f'_{\mathbf{u}}(\mathbf{x}) \leq ||\nabla f(\mathbf{x})|| \quad \text{for all directions } \mathbf{u}.$$

If $\mathbf{u}$ points in the direction of $\nabla f(\mathbf{x})$, then

$$f'_{\mathbf{u}}(\mathbf{x}) = ||\nabla f(\mathbf{x})||, \qquad (\theta = 0, \cos\theta = 1)$$

and if $\mathbf{u}$ points in the direction of $-\nabla f(\mathbf{x})$, then

$$f'_{\mathbf{u}}(\mathbf{x}) = -||\nabla f(\mathbf{x})||. \qquad (\theta = \pi, \cos\theta = -1)$$

Since the directional derivative gives the rate of change of the function in that direction, it is clear that

(15.2.7) from each point $\mathbf{x}$ of the domain, a differentiable function f increases most rapidly in the direction of the gradient (the rate of change at $\mathbf{x}$ being $||\nabla f(\mathbf{x})||$); the function decreases most rapidly in the opposite direction (the rate of change at $\mathbf{x}$ being $-||\nabla f(\mathbf{x})||$).

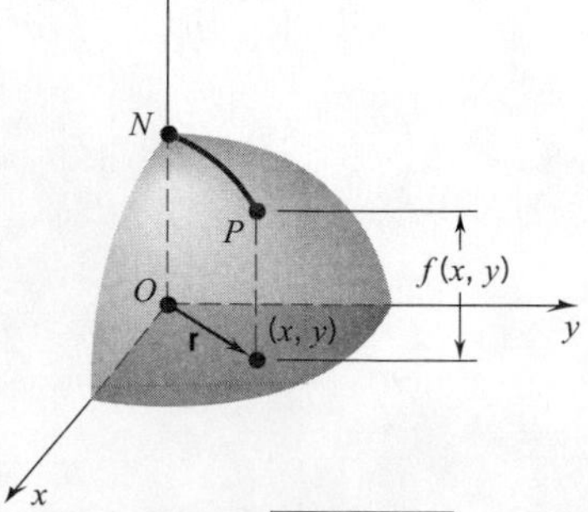

$f(x, y) = \sqrt{1 - (x^2 + y^2)}$

Figure 15.2.3

Example 3 The graph of the function $f(x, y) = \sqrt{1 - (x^2 + y^2)}$ is the upper half of the unit sphere $x^2 + y^2 + z^2 = 1$. The function is defined on the closed unit disc $x^2 + y^2 \leq 1$, but differentiable only on the open unit disc.

In Figure 15.2.3 we have marked a point (x, y) and drawn the corresponding radius vector $\mathbf{r} = x\,\mathbf{i} + y\,\mathbf{j}$. The gradient

$$\nabla f(x, y) = \frac{-x}{\sqrt{1 - (x^2 + y^2)}}\,\mathbf{i} + \frac{-y}{\sqrt{1 - (x^2 + y^2)}}\,\mathbf{j}$$

is a negative multiple of $\mathbf{r}$:

$$\nabla f(x,y) = -\frac{1}{\sqrt{1-(x^2+y^2)}}(x\,\mathbf{i}+y\,\mathbf{j}) = -\frac{1}{\sqrt{1-(x^2+y^2)}}\,\mathbf{r}.$$

Since $\mathbf{r}$ points from the origin to (x,y), the gradient points from (x,y) to the origin. This means that f increases most rapidly toward the origin. This is borne out by the observation that along the hemispherical surface the path of steepest ascent from the point $P(x,y,f(x,y))$ is the "great circle route to the North Pole." ☐

Example 4 Suppose that the temperature at each point of a metal plate is given by the function

$$T(x,y) = e^x \cos\, y + e^y \cos x.$$

(a) In what direction does the temperature increase most rapidly at the point $(0,0)$? What is this rate of increase?

(b) In what direction does the temperature decrease most rapidly at $(0,0)$?

SOLUTION

$$\begin{aligned}\nabla T(x,y) &= \frac{\partial T}{\partial x}(x,y)\,\mathbf{i} + \frac{\partial T}{\partial y}(x,y)\,\mathbf{j}\\ &= (e^x\cos y - e^y\sin x)\,\mathbf{i} + (e^y\cos x - e^x\sin y)\,\mathbf{j}.\end{aligned}$$

(a) At $(0,0)$ the temperature increases most rapidly in the direction of the gradient

$$\nabla T(0,0) = \mathbf{i}+\mathbf{j}.$$

This rate of increase is

$$||\nabla T(0,0)|| = ||\mathbf{i}+\mathbf{j}|| = \sqrt{2}.$$

(b) The temperature decreases most rapidly in the direction of

$$-\nabla T(0,0) = -\mathbf{i}-\mathbf{j}. \quad ☐$$

Example 5 Suppose that the mass density (mass per unit volume) of a metal ball centered at the origin is given by the function

$$\lambda(x,y,z) = k\,e^{-(x^2+y^2+z^2)}, \quad k \text{ a positive constant.}$$

(a) In what direction does the density increase most rapidly at the point (x,y,z)? What is this rate of density increase?

(b) In what direction does the density decrease most rapidly?

(c) What are the rates of density change at (x,y,z) in the $\mathbf{i}, \mathbf{j}, \mathbf{k}$ directions?

SOLUTION

$$\begin{aligned}\nabla\lambda(x,y,z) &= \frac{\partial\lambda}{\partial x}(x,y,z)\,\mathbf{i} + \frac{\partial\lambda}{\partial y}(x,y,z)\,\mathbf{j} + \frac{\partial\lambda}{\partial z}(x,y,z)\,\mathbf{k}\\ &= -2k\,e^{-(x^2+y^2+z^2)}[x\,\mathbf{i}+y\,\mathbf{j}+z\,\mathbf{k}].\end{aligned}$$

Since $\lambda(x,y,z) = k\,e^{-(x^2+y^2+z^2)}$, we have

$$\nabla\lambda(x,y,z) = -2\lambda(x,y,z)\mathbf{r}.$$

From this, we see that the gradient points from (x,y,z) in the direction opposite to that of the radius vector.

(a) The density increases most rapidly toward the origin. The rate of increase is

$$\|\nabla\lambda(x,y,z)\| = 2\lambda(x,y,z)\|\mathbf{r}\| = 2\lambda(x,y,z)\sqrt{x^2+y^2+z^2}.$$

(b) The density decreases most rapidly directly away from the origin.

(c) The rates of density change in the $\mathbf{i},\mathbf{j},\mathbf{k}$ directions are given by the directional derivatives

$$\lambda'_{\mathbf{i}}(x,y,z) = \nabla\lambda(x,y,z)\cdot\mathbf{i} = -2x\,\lambda(x,y,z),$$
$$\lambda'_{\mathbf{j}}(x,y,z) = \nabla\lambda(x,y,z)\cdot\mathbf{j} = -2y\,\lambda(x,y,z),$$
$$\lambda'_{\mathbf{k}}(x,y,z) = \nabla\lambda(x,y,z)\cdot\mathbf{k} = -2z\,\lambda(x,y,z).$$

These are just the first partials of λ. □

Example 6 Suppose that the temperature at each point of a metal plate is given by the function

$$T(x,y) = 1 + x^2 - y^2.$$

Find the path followed by a heat-seeking particle that originates at $(-2, 1)$.

SOLUTION The particle moves in the direction of the gradient vector

$$\nabla T = 2x\,\mathbf{i} - 2y\,\mathbf{j}.$$

We want the curve

$$C : \mathbf{r}(t) = x(t)\,\mathbf{i} + y(t)\,\mathbf{j}$$

which begins at $(-2, 1)$ and at each point has its tangent vector in the direction of ∇T. We can satisfy the first condition by setting

$$x(0) = -2, \qquad y(0) = 1.$$

We can satisfy the second condition by setting

$$x'(t) = 2x(t), \qquad y'(t) = -2y(t). \qquad \text{(explain)}$$

These differential equations, together with initial conditions at $t = 0$, imply that

$$x(t) = -2\,e^{2t}, \qquad y(t) = e^{-2t}. \qquad \text{(Section 7.6)}$$

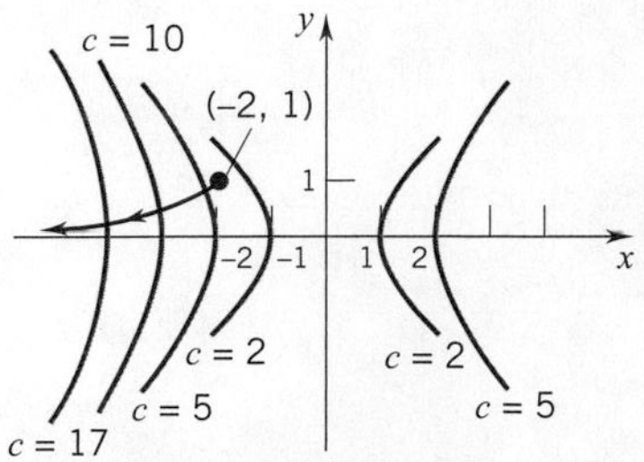

Figure 15.2.4

We can eliminate the parameter t by noting that

$$x(t)y(t) = (-2\,e^{2t})(e^{-2t}) = -2.$$

In terms of just x and y we have

$$xy = -2.$$

The particle moves from the point $(-2, 1)$ along the left branch of the hyperbola $xy = -2$ in the direction of decreasing x (see Figure 15.2.4).

The level curves, *isothermals* of the temperature distribution T, are also hyperbolas. As you can verify, the path of the object is perpendicular to each of the isothermals $x^2 - y^2 = c - 1$. □

Remark The pair of differential equations

$$x'(t) = 2x(t), \qquad y'(t) = -2y(t)$$

can be set as a single differential equation in x and y: the relation

$$\frac{y'(t)}{x'(t)} = -\frac{y(t)}{x(t)} \quad \text{gives} \quad \frac{dy}{dx} = -\frac{y}{x}.$$

This equation is readily solved directly (Section 8.9):

$$\begin{aligned} \frac{1}{y}\frac{dy}{dx} &= -\frac{1}{x} \\ \ln|y| &= -\ln|x| + C \\ \ln|x| + \ln|y| &= C \\ \ln|xy| &= C \end{aligned}$$

Thus xy is constant:

$$xy = k.$$

Since the curve passes through the point $(-2, 1)$, $k = -2$ and once again we have the curve

$$xy = -2.$$

You will be called on to use this method of solution in some of the Exercises. □

EXERCISES 15.2

Find the directional derivative at the point P in the direction indicated.

1. $f(x,y) = x^2 + 3y^2$ at $P(1,1)$ in the direction of $\mathbf{i} - \mathbf{j}$.

2. $f(x,y) = x + \sin(x+y)$ at $P(0,0)$ in the direction of $2\mathbf{i} + \mathbf{j}$.

3. $f(x,y) = x\,e^y - y\,e^x$ at $P(1,0)$ in the direction of $3\mathbf{i} + 4\mathbf{j}$.

4. $f(x,y) = \dfrac{2x}{x-y}$ at $P(1,0)$ in the direction of $\mathbf{i} - \sqrt{3}\,\mathbf{j}$.

5. $f(x,y) = \dfrac{ax+by}{x+y}$ at $P(1,1)$ in the direction of $\mathbf{i} - \mathbf{j}$.

6. $f(x,y) = \dfrac{x+y}{cx+dy}$ at $P(1,1)$ in the direction of $c\mathbf{i} - d\mathbf{j}$.

7. $f(x,y) = \ln(x^2 + y^2)$ at $P(0,1)$ in the direction of $8\mathbf{i} + \mathbf{j}$.

8. $f(x,y) = x^2 y + \tan y$ at $P(-1, \pi/4)$ in the direction of $\mathbf{i} - 2\mathbf{j}$.

9. $f(x,y,z) = xy + yz + zx$ at $P(1,-1,1)$ in the direction of $\mathbf{i} + 2\mathbf{j} + \mathbf{k}$.

10. $f(x,y,z) = x^2 y + y^2 z + z^2 x$ at $P(1,0,1)$ in the direction of $3\mathbf{j} - \mathbf{k}$.

11. $f(x,y,z) = (x + y^2 + z^3)^2$ at $P(1,-1,1)$ in the direction of $\mathbf{i} + \mathbf{j}$.

12. $f(x,y,z) = Ax^2 + Bxyz + Cy^2$ at $P(1,2,1)$ in the direction of $A\mathbf{i} + B\mathbf{j} + C\mathbf{k}$.

13. $f(x,y,z) = x\tan^{-1}(y+z)$ at $P(1,0,1)$ in the direction of $\mathbf{i} + \mathbf{j} - \mathbf{k}$.

14. $f(x,y,z) = xy^2\cos z - 2yz^2\sin \pi x + 3zx^2$ at $P(0,-1,\pi)$ in the direction of $2\mathbf{i} - \mathbf{j} + 2\mathbf{k}$.

15. Find the directional derivative of $f(x,y) = \ln\sqrt{x^2+y^2}$ at $(x,y) \neq (0,0)$ toward the origin.

16. Find the directional derivative of $f(x,y) = (x-1)y^2 e^{xy}$ at $(0,1)$ toward the point $(-1,3)$.

17. Find the directional derivative of $f(x,y) = Ax^2 + 2Bxy + Cy^2$ at (a,b) toward (b,a) (a) if $a > b$; (b) if $a < b$.

18. Find the directional derivative of $f(x,y,z) = z\ln(x/y)$ at $(1,1,2)$ toward the point $(2,2,1)$.

19. Find the directional derivative of $f(x,y,z) = x\,e^{y^2 - z^2}$ at $(1,2,-2)$ in the direction of the path $\mathbf{r}(t) = t\,\mathbf{i} + 2\cos(t-1)\,\mathbf{j} - 2e^{t-1}\,\mathbf{k}$.

20. Find the directional derivative of $f(x,y,z) = x^2 + yz$ at $(1,-3,2)$ in the direction of the path $\mathbf{r}(t) = t^2\mathbf{i} + 3t\mathbf{j} + (1-t^3)\mathbf{k}$.

21. Find a directional derivative of $f(x,y,z) = x^2 + 2xyz - yz^2$ at $(1,1,2)$ in a direction parallel to the straight line

$$\frac{x-1}{2} = y - 1 = \frac{z-2}{-3}.$$

22. Find a directional derivative of $f(x,y,z) = e^x \cos \pi yz$ at $(0, 1, \frac{1}{2})$ in a direction parallel to the line of intersection of the planes $x + y - z - 5 = 0$ and $4x - y - z + 2 = 0$.

In Exercises 23–26, find a unit vector in the direction in which f increases most rapidly at P and give the rate of change of f in that direction; find a unit vector in the direction in which f decreases most rapidly at P and give the rate of change of f in that direction.

23. $f(x,y) = y^2 e^{2x}$; $P(0,1)$.

24. $f(x,y) = x + \sin(x + 2y)$; $P(0,0)$.

25. $f(x,y,z) = \sqrt{x^2+y^2+z^2}$; $P(1,-2,1)$.

26. $f(x,y,z) = x^2 z e^y + xz^2$; $P(1, \ln 2, 2)$.

27. Let $f = f(x)$ be a differentiable function of one variable. What is the gradient of f at x_0? What is the geometric significance of the direction of this gradient?

28. Suppose that f is differentiable at (x_0, y_0) and $\nabla f(x_o, y_0) \neq 0$. Compute the rate of change of f in the direction of the vector

$$\frac{\partial f}{\partial y}(x_0,y_0)\,\mathbf{i} - \frac{\partial f}{\partial x}(x_0,y_0)\,\mathbf{j}.$$

Give a geometric interpretation to your answer.

29. Let

$$f(x,y) = \sqrt{x^2+y^2}.$$

(a) Show that $\partial f/\partial x$ is not defined at $(0,0)$.

(b) Is f differentiable at $(0,0)$?

30. Verify that, if g is continuous at $\mathbf{x}$, then

(a) $g(\mathbf{x}+\mathbf{h})\,o(\mathbf{h}) = o(\mathbf{h})$ and

(b) $[g(\mathbf{x}+\mathbf{h}) - g(\mathbf{x})]\nabla f(\mathbf{x}) \cdot \mathbf{h} = o(\mathbf{h})$.

31. Given the density function $\lambda(x,y) = 48 - \frac{4}{3}x^2 - 3y^2$, find the rate of density change (a) at $(1,-1)$ in the direction of the most rapid density decrease; (b) at $(1,2)$ in the $\mathbf{i}$ direction; (c) at (2,2) away from the origin.

32. The intensity of light in a neighborhood of the point $(-2,1)$ is given by a function of the form $I(x,y) = A - 2x^2 - y^2$. Find the path followed by a light-seeking particle that originates at the center of the neighborhood.

33. Determine the path of steepest descent along the surface $z = x^2 + 3y^2$ from each of the following points: (a) $(1,1,4)$; (b) $(1,-2,13)$.

34. Determine the path of steepest ascent along the hyperbolic paraboloid $z = \frac{1}{2}x^2 - y^2$ from each of the following points: (a) $(-1, 1, -\frac{1}{2})$; (b) $(1, 0, \frac{1}{2})$.

35. Determine the path of steepest descent along the surface $z = a^2x^2 + b^2y^2$ from the point: $(a, b, a^4 + b^4)$.

36. The temperature in a neighborhood of the origin is given by a function of the form

$$T(x,y) = T_0 + e^y \sin x.$$

Find the path followed by a heat-fleeing particle that originates at the origin.

37. The temperature in a neighborhood of the point $(\frac{1}{4}\pi, 0)$ is given by the function

$$T(x,y) = \sqrt{2}\,e^{-y}\cos x.$$

Find the path followed by a heat-seeking particle that originates at the center of the neighborhood.

38. Determine the path of steepest descent along the surface $z = A + x + 2y - x^2 - 3y^2$ from the point $(0,0,A)$.

39. Set $f(x,y) = 3x^2 + y$.

(a) Find

$$\lim_{h\to 0} \frac{f(x(2+h), y(2+h)) - f(2,4)}{h}$$

given that $x(t) = t$ and $y(t) = t^2$. (These functions parametrize the parabola $y = x^2$.)

(b) Find

$$\lim_{h\to 0} \frac{f(x(4+h), y(4+h)) - f(2,4)}{h}$$

given that $x(t) = \frac{1}{4}(t+4)$ and $y(t) = t$. (These functions parametrize the line $y = 4x - 4$.)

(c) Compute the directional derivative of f at $(2,4)$ in the direction of $\mathbf{i} + 4\mathbf{j}$.

(d) Observe that $\mathbf{i} + 4\mathbf{j}$ is a direction vector for the line $y = 4x - 4$ and that this line is tangent to the parabola $y = x^2$ at $(2,4)$. Explain then why the computations in (a),(b), and (c) yield different values.

40. According to Newton's law of gravitation, the force exerted on a particle of mass m located at the point (x,y,z) by a particle of mass M located at the origin is given by

$$\mathbf{F}(x,y,z) = \frac{-GMm}{r^3}\mathbf{r}$$

where $\mathbf{r} = x\,\mathbf{i} + y\,\mathbf{j} + z\,\mathbf{k}$, $r = ||\mathbf{r}||$, and G is the gravitational constant. Show that $\mathbf{F}$ is the gradient of

$$f(x,y,z) = \frac{GMm}{r}.$$

41. Let $\mathbf{u}$ be a unit position vector (initial point at the origin) in the plane and let θ be the angle measured in the counterclockwise direction from the positive x-axis to $\mathbf{u}$. Let f be a differentiable function of two variables.

(a) Show that $f'_{\mathbf{u}}(x,y) = \dfrac{\partial f}{\partial x}\cos\theta + \dfrac{\partial f}{\partial y}\sin\theta$.

(b) Let $f(x,y) = x^3 + 2xy - xy^2$ and $\theta = 2\pi/3$. Find $f'_u(-1,2)$.

42. Refer to Exercise 41. Let $f(x,y) = x^2e^{2y}$ and $\theta = 5\pi/4$. Find $f'_{\mathbf{u}}(x,y)$ and $f'_{\mathbf{u}}(2, \ln 2)$.

43. Assume that $\nabla f(\mathbf{x})$ and $\nabla g(\mathbf{x})$ exist. Use Theorem 15.1.3 to derive the product rule

$$\nabla[f(\mathbf{x})g(\mathbf{x})] = f(\mathbf{x})\nabla g(\mathbf{x}) + g(\mathbf{x})\nabla f(\mathbf{x}).$$

44. Assume that $\nabla f(\mathbf{x})$ and $\nabla g(\mathbf{x})$ exist, and that $g(\mathbf{x}) \neq 0$. Derive the quotient rule

$$\nabla\left[\frac{f(\mathbf{x})}{g(\mathbf{x})}\right] = \frac{g(\mathbf{x})\nabla f(\mathbf{x}) - f(\mathbf{x})\nabla g(\mathbf{x})}{g^2(\mathbf{x})}.$$

45. Assume that $\nabla f(\mathbf{x})$ exists. Prove that, for each integer n,

$$\nabla f^n(\mathbf{x}) = nf^{n-1}(\mathbf{x})\nabla f(\mathbf{x}).$$

Does this result hold if n is replaced by an arbitrary real number?

■ 15.3 THE MEAN-VALUE THEOREM; CHAIN RULES

The Mean-Value Theorem

You have seen the important role played by the mean-value theorem in the calculus of functions of one variable. Here we take up the analogous result for functions of several variables. Let **a** and **b** be points (either in the plane or in three-space); by $\overline{\mathbf{ab}}$ we mean the line segment that joins point **a** to point **b**.

THEOREM 15.3.1 THE MEAN-VALUE THEOREM (SEVERAL VARIABLES)

If f is differentiable at each point of the line segment $\overline{\mathbf{ab}}$, then there exists on that line segment a point **c** between **a** and **b** such that

$$f(\mathbf{b}) - f(\mathbf{a}) = \nabla f(\mathbf{c}) \cdot (\mathbf{b} - \mathbf{a}).$$

PROOF As t ranges from 0 to 1, $\mathbf{a} + t(\mathbf{b} - \mathbf{a})$ traces out the line segment $\overline{\mathbf{ab}}$. The idea of the proof is to apply the one-variable mean-value theorem to the function

$$g(t) = f(\mathbf{a} + t[\mathbf{b} - \mathbf{a}]), \quad t \in [0, 1].$$

To show that g is differentiable on the open interval $(0, 1)$ we take $t \in (0, 1)$ and form

$$\begin{aligned} g(t+h) - g(t) &= f(\mathbf{a} + (t+h)[\mathbf{b} - \mathbf{a}]) - f(\mathbf{a} + t[\mathbf{b} - \mathbf{a}]) \\ &= f(\mathbf{a} + t[\mathbf{b} - \mathbf{a}] + h[\mathbf{b} - \mathbf{a}]) - f(\mathbf{a} + t[\mathbf{b} - \mathbf{a}]) \\ &= \nabla f(\mathbf{a} + t[\mathbf{b} - \mathbf{a}]) \cdot h[\mathbf{b} - \mathbf{a}] + o(h[\mathbf{b} - \mathbf{a}]). \end{aligned}$$

Since $\qquad \nabla f(\mathbf{a} + t[\mathbf{b} + \mathbf{a}]) \cdot h(\mathbf{b} - \mathbf{a}) = [\nabla f(\mathbf{a} + t[\mathbf{b} - \mathbf{a}]) \cdot (\mathbf{b} - \mathbf{a})]h$

and the $o(h[\mathbf{b} - \mathbf{a}])$ term is obviously $o(h)$, we can write

$$g(t+h) - g(t) = [\nabla f(\mathbf{a} + t[\mathbf{b} - \mathbf{a}]) \cdot (\mathbf{b} - \mathbf{a})]h + o(h).$$

Dividing both sides by h, we see that g is differentiable and

$$g'(t) = \nabla f(\mathbf{a} + t[\mathbf{b} - \mathbf{a}]) \cdot (\mathbf{b} - \mathbf{a}).$$

The function g is clearly continuous at 0 and at 1. Applying the one-variable mean-value theorem to g, we can conclude that there exists a number t_0 between 0 and 1 such that

$$(*) \qquad g(1) - g(0) = g'(t_0)(1 - 0).$$

Since $g(1) = f(\mathbf{b}), g(0) = f(\mathbf{a})$, and $g'(t_0) = \nabla f(\mathbf{a} + t_0[\mathbf{b} - \mathbf{a}]) \cdot (\mathbf{b} - \mathbf{a})$, condition $(*)$ gives

$$f(\mathbf{b}) - f(\mathbf{a}) = \nabla f(\mathbf{a} + t_0[\mathbf{b} - \mathbf{a}]) \cdot (\mathbf{b} - \mathbf{a}).$$

Setting $\mathbf{c} = \mathbf{a} + t_0[\mathbf{b} - \mathbf{a}]$, we have

$$f(\mathbf{b}) - f(\mathbf{a}) = \nabla f(\mathbf{c}) \cdot (\mathbf{b} - \mathbf{a}). \quad \square$$

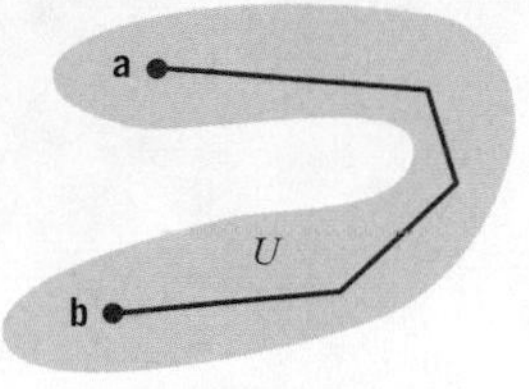

U is connected

Figure 15.3.1

A nonempty open set U (in the plane or in three-space) is said to be *connected* if any two points of U can be joined by a polygonal path that lies entirely in U. You can see such a set in Figure 15.3.1.

The set shown in Figure 15.3.2 is the union of two disjoint open sets. The set is open but not connected: it is impossible to join **a** and **b** by a polygonal path that lies within the set.

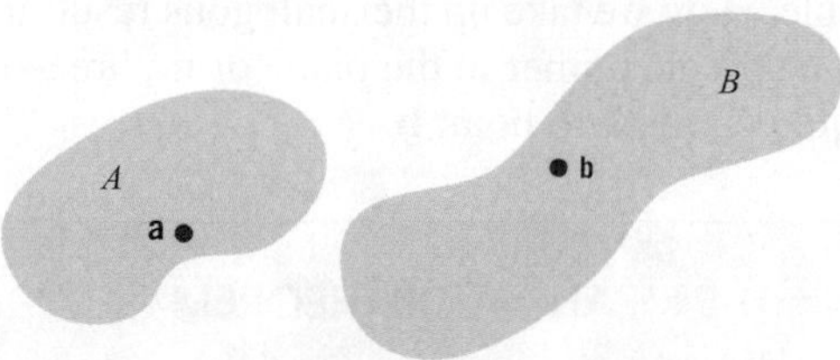

$U = A \cup B$ is not connected

Figure 15.3.2

In Chapter 4 you saw that, if $f'(x) = 0$ for all x in an open interval I, then f is constant on I. We have a similar result for functions of several variables.

THEOREM 15.3.2

Let U be an open connected set and let f be a differentiable function on U.

If $\nabla f(\mathbf{x}) = \mathbf{0}$ for all $\mathbf{x}$ in U, then f is constant on U.

PROOF Let **a** and **b** be any two points in U. Since U is connected, we can join these points by a polygonal path with vertices $\mathbf{a} = \mathbf{c}_0, \mathbf{c}_1, \mathbf{c}_2, \ldots, \mathbf{c}_{n-1}, \mathbf{c}_n = \mathbf{b}$ (see Figure 15.3.3). By the mean-value theorem (Theorem 15.3.1) there exist points

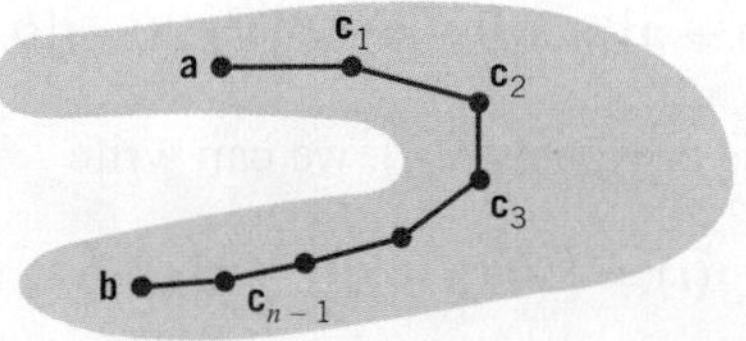

Figure 15.3.3

$\mathbf{c}_1^*$ between $\mathbf{c}_0$ and $\mathbf{c}_1$ such that $f(\mathbf{c}_1) - f(\mathbf{c}_0) = \nabla f(\mathbf{c}_1^*) \cdot (\mathbf{c}_1 - \mathbf{c}_0)$,

$\mathbf{c}_2^*$ between $\mathbf{c}_1$ and $\mathbf{c}_2$ such that $f(\mathbf{c}_2) - f(\mathbf{c}_1) = \nabla f(\mathbf{c}_2^*) \cdot (\mathbf{c}_2 - \mathbf{c}_1)$,

$\vdots$

$\mathbf{c}_n^*$ between $\mathbf{c}_{n-1}$ and $\mathbf{c}_n$ such that $f(\mathbf{c}_n) - f(\mathbf{c}_{n-1}) = \nabla f(\mathbf{c}_n^*) \cdot (\mathbf{c}_n - \mathbf{c}_{n-1})$.

If $\nabla f(\mathbf{x}) = \mathbf{0}$ for all $\mathbf{x}$ in U, then

$$f(\mathbf{c}_1) - f(\mathbf{c}_0) = 0,\ f(\mathbf{c}_2) - f(\mathbf{c}_1) = 0,\ \cdots, f(\mathbf{c}_n) - f(\mathbf{c}_{n-1}) = 0.$$

This shows that

$$f(\mathbf{a}) = f(\mathbf{c}_0) = f(\mathbf{c}_1) = f(\mathbf{c}_2) = \cdots = f(\mathbf{c}_{n-1}) = f(\mathbf{c}_n) = f(\mathbf{b}).$$

Since $\mathbf{a}$ and $\mathbf{b}$ are arbitrary points of U, f must be constant on U. ❑

THEOREM 15.3.3

Let U be an open connected set and let f and g be differentiable functions on U.

If $\nabla f(\mathbf{x}) = \nabla g(\mathbf{x})$ for all $\mathbf{x}$ in U, then f and g differ by a constant on U.

PROOF If $\nabla f(\mathbf{x}) = \nabla g(\mathbf{x})$ for all $\mathbf{x}$ in U, then

$$\nabla[f(\mathbf{x}) - g(\mathbf{x})] = \nabla f(\mathbf{x}) - \nabla g(\mathbf{x}) = 0 \quad \text{for all } \mathbf{x} \text{ in } U.$$

By Theorem 15.3.2, $f - g$ must be constant on U. ❑

The Chain Rule

You are by now thoroughly familiar with the chain rule for functions of a single variable (Theorem 3.5.7): If g is differentiable at x and f is differentiable at $g(x)$, then

$$\frac{d}{dx}(f[g(x)]) = f'[g(x)]g'(x).$$

Here we obtain generalizations of the chain rule for functions of several variables. As you will see, there are many versions of the chain rule in this setting.

A vector-valued function is said to be *continuous* provided that its components are continuous. If $f = f(x, y, z)$ is a scalar-valued function (a real-valued function), then its gradient ∇f is a vector-valued function. We say that f is *continuously differentiable* on an *open set U* if f is differentiable on U and ∇f is continuous on U.

If a curve $\mathbf{r}$ lies in the domain of f, then we can form the composition

$$(f \circ \mathbf{r})(t) = f(\mathbf{r}(t)).$$

The composition $f \circ \mathbf{r}$ is a real-valued function of the real variable t. The numbers $f(\mathbf{r}(t))$ are the values taken on by f *along the curve* $\mathbf{r}$. For example, let

$$f(x, y) = \tfrac{1}{3}(x^3 + y^3) \quad \text{and} \quad \mathbf{r}(t) = a\cos t\,\mathbf{i} + b\sin t\,\mathbf{j} \quad \text{(an ellipse)}.$$

Then, with $x(t) = a\cos t$ and $y(t) = b\sin t$, we have

$$f(\mathbf{r}(t)) = \tfrac{1}{3}(a^3\cos^3 t + b^3\sin^3 t).$$

The chain rule is a formula for calculating the derivative of the composition $f \circ \mathbf{r}$.

THEOREM 15.3.4 CHAIN RULE (ALONG A CURVE)

If f is continuously differentiable on an open set U and $\mathbf{r} = \mathbf{r}(t)$ is a differentiable curve that lies in U, then the composition $f \circ \mathbf{r}$ is differentiable and

$$\frac{d}{dt}[f(\mathbf{r}(t))] = \nabla f(\mathbf{r}(t)) \cdot \mathbf{r}'(t).$$

PROOF We will show that

$$\lim_{h\to 0} \frac{f(\mathbf{r}(t+h)) - f(\mathbf{r}(t))}{h} = \nabla f(\mathbf{r}(t)) \cdot \mathbf{r}'(t).$$

For $h \neq 0$ and sufficiently small, the line segment that joins $\mathbf{r}(t)$ to $\mathbf{r}(t+h)$ lies entirely in U. This we know because U is open and $\mathbf{r}$ is continuous. (See Figure 15.3.4.) For such h, the mean-value theorem assures us that there exists a point $\mathbf{c}(h)$ between $\mathbf{r}(t)$ and $\mathbf{r}(t+h)$ such that

$$f(\mathbf{r}(t+h)) - f(\mathbf{r}(t)) = \nabla f(\mathbf{c}(h)) \cdot [\mathbf{r}(t+h) - \mathbf{r}(t)].$$

Dividing both sides by h, we have

$$\frac{f(\mathbf{r}(t+h)) - f(\mathbf{r}(t))}{h} = \nabla f(\mathbf{c}(h)) \cdot \left[\frac{\mathbf{r}(t+h) - \mathbf{r}(t)}{h}\right].$$

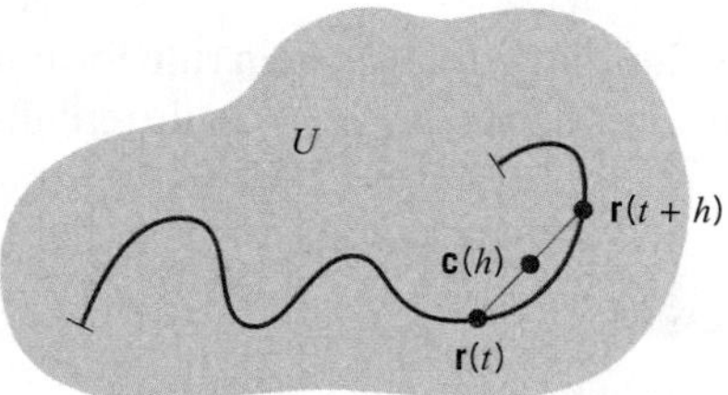

Figure 15.3.4

As h tends to zero, $\mathbf{c}(h)$ tends to $\mathbf{r}(t)$ and, by the continuity of ∇f,

$$\nabla f(\mathbf{c}(h)) \to \nabla f(\mathbf{r}(t)).$$

Since

$$\frac{\mathbf{r}(t+h) - \mathbf{r}(t)}{h} \to \mathbf{r}'(t),$$

the result follows. □

Example 1 Use the chain rule to find the rate of change of

$$f(x, y) = \tfrac{1}{3}(x^3 + y^3)$$

with respect to t along the ellipse $\mathbf{r}(t) = a\cos t\,\mathbf{i} + b\sin t\,\mathbf{j}$.

SOLUTION The rate of change of f with respect to t along the curve $\mathbf{r}$ is the derivative

$$\frac{d}{dt}[f(\mathbf{r}(t))].$$

By the chain rule (Theorem 15.3.4)

$$\frac{d}{dt}[f(\mathbf{r}(t))] = \nabla f(\mathbf{r}(t)) \cdot \mathbf{r}'(t).$$

Here
$$\nabla f = x^2\,\mathbf{i} + y^2\,\mathbf{j}.$$

With $x(t) = a\cos t$ and $y(t) = b\sin t$, we have

$$\nabla f(\mathbf{r}(t)) = a^2\cos^2 t\,\mathbf{i} + b^2\sin^2 t\,\mathbf{j}.$$

Since $\mathbf{r}'(t) = -a\sin t\,\mathbf{i} + b\cos t\,\mathbf{j}$, you can see that

$$\begin{aligned}\frac{d}{dt}[f(\mathbf{r}(t))] &= \nabla f(\mathbf{r}(t)) \cdot \mathbf{r}'(t)\\ &= (a^2\cos^2 t\,\mathbf{i} + b^2\sin^2 t\,\mathbf{j}) \cdot (-\,a\sin t\,\mathbf{i} + b\cos t\,\mathbf{j})\\ &= -a^3\sin t\cos^2 t + b^3\sin^2 t\cos t\\ &= \sin t\cos t(b^3\sin t - a^3\cos t). \quad \square\end{aligned}$$

Remark Note that we could have obtained the same result without invoking Theorem 15.3.4 by first forming $f(\mathbf{r}(t))$ and then differentiating with respect to t. As you saw in the calculations carried out before the statement of the theorem,

$$f(\mathbf{r}(t)) = \tfrac{1}{3}(a^3\cos^3 t + b^3\sin^3 t).$$

This gives

$$\begin{aligned}\frac{d}{dt}[f(\mathbf{r}(t))] &= \tfrac{1}{3}[3a^3(\cos^2 t)(-\sin t) + 3b^3\sin^2 t\cos t]\\ &= \sin t\cos t(b^3\sin t - a^3\cos t). \quad \square\end{aligned}$$

Example 2 Use the chain rule to find the rate of change of

$$f(x,y,z) = x^2y + z\cos x$$

with respect to t along the twisted cubic $\mathbf{r}(t) = t\,\mathbf{i} + t^2\,\mathbf{j} + t^3\,\mathbf{k}$.

SOLUTION Once again we use the relation

$$\frac{d}{dt}[f(\mathbf{r}(t))] = \nabla f(\mathbf{r}(t)) \cdot \mathbf{r}'(t).$$

This time
$$\nabla f = (2xy - z\sin x)\,\mathbf{i} + x^2\,\mathbf{j} + \cos x\,\mathbf{k}.$$

With $x(t) = t, y(t) = t^2, z(t) = t^3$, we have

$$\nabla f(\mathbf{r}(t)) = (2t^3 - t^3\sin t)\,\mathbf{i} + t^2\,\mathbf{j} + \cos t\,\mathbf{k}.$$

Since $\mathbf{r}'(t) = \mathbf{i} + 2t\,\mathbf{j} + 3t^2\,\mathbf{k}$, we have

$$\begin{aligned}\frac{d}{dt}[(\mathbf{r}(t))] &= \nabla f(\mathbf{r}(t)) \cdot \mathbf{r}'(t)\\ &= [(2t^3 - t^3\sin t)\,\mathbf{i} + t^2\,\mathbf{j} + \cos t\,\mathbf{k}] \cdot [\mathbf{i} + 2t\,\mathbf{j} + 3t^2\,\mathbf{k}]\\ &= 2t^3 - t^3\sin t + 2t^3 + 3t^2\cos t\\ &= 4t^3 - t^3\sin t + 3t^2\cos t.\end{aligned}$$

You can check this answer by first forming $f(\mathbf{r}(t))$ and then differentiating. $\square$

Another Formulation of Theorem 15.3.4

The chain rule for functions of one variable,

$$\frac{d}{dt}[u(x(t))] = u'(x(t))x'(t),$$

can be written

$$\frac{du}{dt} = \frac{du}{dx}\frac{dx}{dt}.$$

In a similar manner, the relation

$$\frac{d}{dt}[u(\mathbf{r}(t))] = \nabla u(\mathbf{r}(t)) \cdot \mathbf{r}'(t)$$

can be written

(1) $$\frac{du}{dt} = \nabla u \cdot \frac{d\mathbf{r}}{dt}.$$

With $\nabla u = \frac{\partial u}{\partial x}\mathbf{i} + \frac{\partial u}{\partial y}\mathbf{j} + \frac{\partial u}{\partial z}\mathbf{k}$ and $\frac{d\mathbf{r}}{dt} = \frac{dx}{dt}\mathbf{i} + \frac{dy}{dt}\mathbf{j} + \frac{dz}{dt}\mathbf{k}$,

equation (1) takes the form

(15.3.5) $$\frac{du}{dt} = \frac{\partial u}{\partial x}\frac{dx}{dt} + \frac{\partial u}{\partial y}\frac{dy}{dt} + \frac{\partial u}{\partial z}\frac{dz}{dt}.$$

In the two-variable case, the z-term drops out and we have

(15.3.6) $$\frac{du}{dt} = \frac{\partial u}{\partial x}\frac{dx}{dt} + \frac{\partial u}{\partial y}\frac{dy}{dt}.$$

Example 3 Find du/dt given that $u = x^2 - y^2$ and $x = t^2 - 1, \quad y = 3\sin \pi t$.

SOLUTION Here we are in the two-variable case

$$\frac{du}{dt} = \frac{\partial u}{\partial x}\frac{dx}{dt} + \frac{\partial u}{\partial y}\frac{dy}{dt}.$$

Since $\frac{\partial u}{\partial x} = 2x, \quad \frac{\partial u}{\partial y} = -2y$ and $\frac{dx}{dt} = 2t, \quad \frac{dy}{dt} = 3\pi \cos \pi t,$

we have
$$\begin{aligned}\frac{du}{dt} &= (2x)(2t) + (-\,2y)(3\pi \cos \pi t)\\ &= 2(t^2-1)(2t) + (-\,2)(3\sin \pi t)(3\pi \cos \pi t)\\ &= 4t^3 - 4t - 18\pi \sin \pi t \cos \pi t.\end{aligned}$$

You can obtain this same result by first writing u directly as a function of t and then differentiating:

$$u = x^2 - y^2 = (t^2-1)^2 - (3\sin \pi t)^2,$$

so that

$$\frac{du}{dt} = 2(t^2 - 1)2t - 2(3\sin\pi t)3\pi\cos\pi t = 4t^3 - 4t - 18\pi\sin\pi t\cos\pi t. \quad \square$$

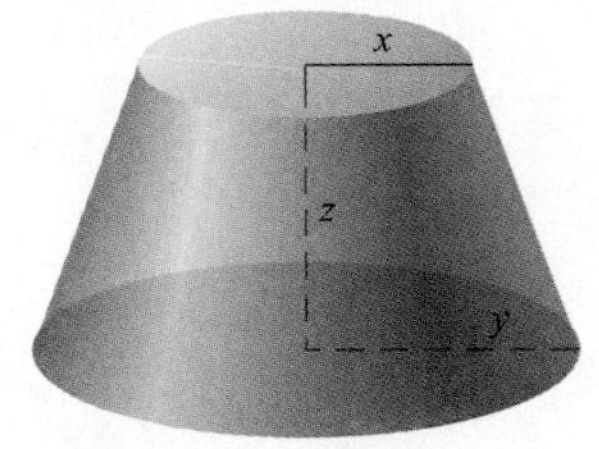

Figure 15.3.5

Example 4 A solid is in the shape of a frustum of a right circular cone (see Figure 15.3.5). If the upper radius x decreases at the rate of 2 inches per minute, the lower radius y increases at the rate of 3 inches per minute, and the height z decreases at the rate of 4 inches per minute, at what rate is the volume V changing at the instant the upper radius is 10 inches, the lower radius is 12 inches, and the height is 18 inches.

SOLUTION In Exercise 39 of Section 6.2, you were asked to derive the formula for the volume of a frustum of a cone.

With x, y, z as given,

$$V = \frac{1}{3}\pi z(x^2 + xy + y^2).$$

Here $\quad \dfrac{\partial V}{\partial x} = \dfrac{1}{3}\pi z(2x + y), \quad \dfrac{\partial V}{\partial y} = \dfrac{1}{3}\pi z(x + 2y), \quad \dfrac{\partial V}{\partial z} = \dfrac{1}{3}\pi(x^2 + xy + y^2).$

Since $$\frac{dV}{dt} = \frac{\partial V}{\partial x}\frac{dx}{dt} + \frac{\partial V}{\partial y}\frac{dy}{dt} + \frac{\partial V}{\partial z}\frac{dz}{dt},$$

we have $$\frac{dV}{dt} = \frac{1}{3}\pi z(2x + y)\frac{dx}{dt} + \frac{1}{3}\pi z(x + 2y)\frac{dy}{dt} + \frac{1}{3}\pi(x^2 + xy + y^2)\frac{dz}{dt}.$$

Set $\quad x = 10, \quad y = 12, \quad z = 18, \quad \dfrac{dx}{dt} = -2, \quad \dfrac{dy}{dt} = 3, \quad \dfrac{dz}{dt} = -4,$

and you will find that

$$\frac{dV}{dt} = -\frac{772}{3}\pi \cong -808.4.$$

The volume decreases at the rate of approximately 808 cubic inches per minute. $\square$

Other Chain Rules

In the setting of functions of several variables there are numerous versions of the chain rule. Some are stated here, others in the Exercises. They can all be deduced from Theorem 15.3.4 and its corollaries, (15.3.5) and (15.3.6).

If, for example,

$$u = u(x, y) \quad \text{where} \quad x = x(s, t) \quad \text{and} \quad y = y(s, t),$$

then

(15.3.7)
$$\frac{\partial u}{\partial s} = \frac{\partial u}{\partial x}\frac{\partial x}{\partial s} + \frac{\partial u}{\partial y}\frac{\partial y}{\partial s} \quad \text{and} \quad \frac{\partial u}{\partial t} = \frac{\partial u}{\partial x}\frac{\partial x}{\partial t} + \frac{\partial u}{\partial y}\frac{\partial y}{\partial t}.$$

To obtain the first equation, keep t fixed and differentiate u with respect to s according to Formula (15.3.6); to obtain the second equation, keep s fixed and differentiate u with respect to t.

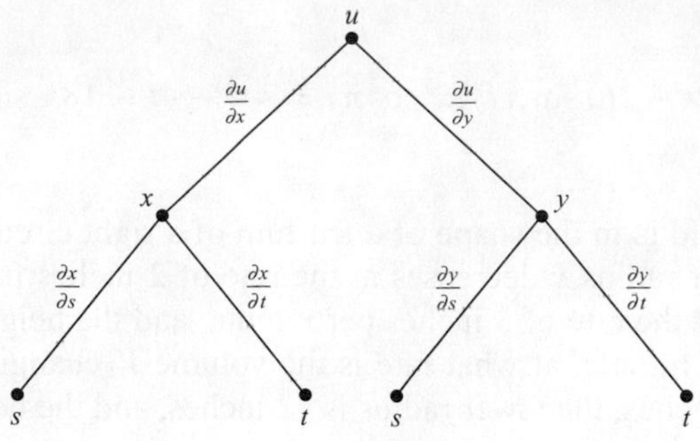

Figure 15.3.6

In Figure 15.3.6 we have drawn a *tree diagram* for Formula (15.3.7). We construct such a tree by branching at each stage from a function to all the variables that directly determine it. Each path starting at u and ending at a variable determines a product of (partial) derivatives. The partial derivative of u with respect to each variable is the sum of the products generated by all the direct paths to that variable.

Example 5 Let $u = x^2 - 2xy + 2y^3$, where $x = s^2 \ln t$ and $y = 2st^3$. Find $\partial u/\partial s$ and $\partial u/\partial t$.

SOLUTION Here u is a function of two variables, x and y, each of which is itself a function of two variables, s and t Thus (15.3.7) applies. Since

$$\frac{\partial u}{\partial x} = 2x - 2y, \qquad \frac{\partial u}{\partial y} = -2x + 6y^2$$

and

$$\frac{\partial x}{\partial s} = 2s \ln t, \qquad \frac{\partial y}{\partial s} = 2t^3, \qquad \frac{\partial x}{\partial t} = \frac{s^2}{t}, \qquad \frac{\partial y}{\partial t} = 6st^2,$$

we have

$$\frac{\partial u}{\partial s} = (2x - 2y)(2s \ln t) + (-2x + 6y^2)(2t^3)$$

and

$$\frac{\partial u}{\partial t} = (2x - 2y)\left(\frac{s^2}{t}\right) + (-2x + 6y^2)(6st^2).$$

These results can be expressed entirely in terms of s and t by replacing x by $s^2 \ln t$ and y by $2st^3$:

$$\frac{\partial u}{\partial s} = (2s^2 \ln t - 4st^3)(2s \ln t) + (-2s^2 \ln t + 24s^2t^6)(2t^3),$$

$$\frac{\partial u}{\partial t} = (2s^2 \ln t - 4st^3)\left(\frac{s^2}{t}\right) + (-2s^2 \ln t + 24s^2t^6)(6st^2). \quad \square$$

Now suppose that u is a function of three variables:

$$u = u(x, y, z) \quad \text{where} \quad x = x(s, t) \quad y = y(s, t), \quad z = z(s, t).$$

A tree diagram for the partials of u appears in Figure 15.3.7. The partials of u with respect to s and t can be read from the diagram:

(15.3.8)
$$\frac{\partial u}{\partial s} = \frac{\partial u}{\partial x}\frac{\partial x}{\partial s} + \frac{\partial u}{\partial y}\frac{\partial y}{\partial s} + \frac{\partial u}{\partial z}\frac{\partial z}{\partial s}, \qquad \frac{\partial u}{\partial t} = \frac{\partial u}{\partial x}\frac{\partial x}{\partial t} + \frac{\partial u}{\partial y}\frac{\partial y}{\partial t} + \frac{\partial u}{\partial z}\frac{\partial z}{\partial t}.$$

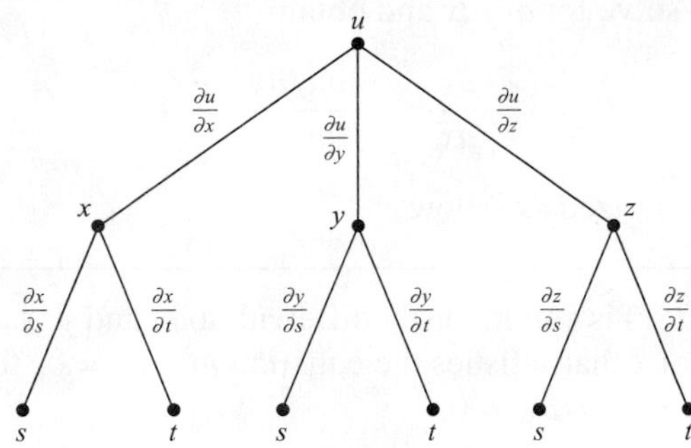

Figure 15.3.7

Example 6 Let $u = x^2y^3e^{xz}$, where $x = s^2 + t^2$, $y = 2st$, and $z = s \ln t$. Find $\partial u/\partial s$.

SOLUTION In this case, u is a function of three variables, x, y, z, each of which is a function of two variables, s and t. Thus (15.3.8) applies. To find $\partial u/\partial s$ we use

$$\frac{\partial u}{\partial s} = \frac{\partial u}{\partial x}\frac{\partial x}{\partial s} + \frac{\partial u}{\partial y}\frac{\partial y}{\partial s} + \frac{\partial u}{\partial z}\frac{\partial z}{\partial s}.$$

Here $$\frac{\partial u}{\partial x} = 2xy^3e^{xz} + x^2y^3ze^{xz}, \qquad \frac{\partial u}{\partial y} = 3x^2y^2e^{xz}, \qquad \frac{\partial u}{\partial z} = x^3y^3e^{xz}$$

and $$\frac{\partial x}{\partial s} = 2s, \qquad \frac{\partial y}{\partial s} = 2t, \qquad \frac{\partial z}{\partial s} = \ln t.$$

Therefore,

$$\frac{\partial u}{\partial s} = (2xy^3e^{xz} + x^2y^3ze^{xz})\,2s + (3x^2y^2e^{xz})\,2t + (x^3y^3e^{xz})\ln t.$$

The result can be expressed entirely in terms of s and t by substituting $s^2 + t^2$ for x, $2st$ for y, and $s \ln t$ for z. ❑

Implicit Differentiation

We return to implicit differentiation, a technique we introduced in Section 3.7.

Suppose that $u = u(x, y)$ is a continuously differentiable function. If y is a differentiable function of x that satisfies the equation $u(x, y) = 0$, then we can find the derivative of y without having to express y *explicitly* in terms of x. The process by which we do this is called *implicit differentiation*.

The process is based on (15.3.6.). To be able to apply that formula, we introduce a new variable t by setting $x = t$. We then have

$$u = u(x, y) \qquad \text{with} \quad x = t \qquad \text{and} \qquad y = y(t).$$

Formula (15.3.6) states that

$$\frac{du}{dt} = \frac{\partial u}{\partial x}\frac{dx}{dt} + \frac{\partial u}{\partial y}\frac{dy}{dt}.$$

Since $u(x(t), y(t)) = 0$ for all t under consideration, $du/dt = 0$ for such t. Since $x = t$, we have $dx/dt = 1$ and $dy/dt = dy/dx$. Therefore,

$$0 = \frac{\partial u}{\partial x} + \frac{\partial u}{\partial y}\frac{dy}{dx}.$$

If $\partial u/\partial y \neq 0$, we can solve for dy/dx and obtain

$$\frac{dy}{dx} = -\frac{\partial u/\partial x}{\partial u/\partial y}.$$

The result can be summarized as follows:

(15.3.9) If $u = u(x, y)$ is continuously differentiable, and y is a differentiable function of x that satisfies the equation $u(x, y) = 0$, then at all points (x, y) where $\partial u/\partial y \neq 0$,

$$\frac{dy}{dx} = -\frac{\partial u/\partial x}{\partial u/\partial y}.$$

Example 7 Suppose that y is a differentiable function of x that satisfies the equation

$$u(x, y) = 2x^2y - y^3 + 1 - x - 2y = 0.$$

Since

$$\frac{\partial u}{\partial x} = 4xy - 1 \quad \text{and} \quad \frac{\partial u}{\partial y} = 2x^2 - 3y^2 - 2,$$

we know that

$$\frac{dy}{dx} = -\frac{4xy - 1}{2x^2 - 3y^2 - 2}.$$

We obtained this result by a slightly different method in Section 3.7, Example 2 (a). □

This method of implicit differentiation can be applied to expressions involving more than two variables. For example,

(15.3.10) If $u = u(x, y, z)$ is continuously differentiable, and $z = z(x, y)$ is a differentiable function that satisfies the equation $u(x, y, z) = 0$, then at the points (x, y, z) where $\partial u/\partial z \neq 0$,

$$\frac{\partial z}{\partial x} = -\frac{\partial u/\partial x}{\partial u/\partial z} \quad \text{and} \quad \frac{\partial z}{\partial y} = -\frac{\partial u/\partial y}{\partial u/\partial z}.$$

PROOF To apply (15.3.8), we write

$$u = u(x, y, z) \quad \text{with} \quad x = s, \quad y = t, \quad z = z(s, t).$$

Then

$$\frac{\partial u}{\partial s} = \frac{\partial u}{\partial x}\frac{\partial x}{\partial s} + \frac{\partial u}{\partial y}\frac{\partial y}{\partial s} + \frac{\partial u}{\partial z}\frac{\partial z}{\partial s}.$$

Since $u(s, t, z(s, t)) = 0$, $\partial u/\partial s = 0$. Also, since

$$\frac{\partial x}{\partial s} = 1 \quad \text{and} \quad \frac{\partial y}{\partial s} = 0,$$

we have

$$0 = \frac{\partial u}{\partial x}(1) + \frac{\partial u}{\partial y}(0) + \frac{\partial u}{\partial z}\frac{\partial z}{\partial s} = \frac{\partial u}{\partial x} + \frac{\partial u}{\partial z}\frac{\partial z}{\partial s} \underset{x = s}{=} \frac{\partial u}{\partial x} + \frac{\partial u}{\partial z}\frac{\partial z}{\partial x}$$

At those points (x, y, z) where $\partial u/\partial z \neq 0$,

$$\frac{\partial z}{\partial x} = -\frac{\partial u/\partial x}{\partial u/\partial z}.$$

The formula for $\partial z/\partial y$ can be obtained in a similar manner. ❑

EXERCISES 15.3

1. Let $f(x, y) = x^3 - xy$. Set $\mathbf{a} = (0, 1)$ and $\mathbf{b} = (1, 3)$. Find a point $\mathbf{c}$ on the line segment $\overline{\mathbf{ab}}$ for which

$$f(\mathbf{b}) - f(\mathbf{a}) = \nabla f(\mathbf{c}) \cdot (\mathbf{b} - \mathbf{a}).$$

2. Let $f(x, y, z) = 4xz - y^2 + z^2$. Set $\mathbf{a} = (0, 1, 1)$ and $\mathbf{b} = (1, 3, 2)$. Find a point $\mathbf{c}$ on the line segment $\overline{\mathbf{ab}}$ for which

$$f(\mathbf{b}) - f(\mathbf{a}) = \nabla f(\mathbf{c}) \cdot (\mathbf{b} - \mathbf{a}).$$

3. (a) Find f if $\nabla f(x, y, z) = a_1\mathbf{i} + a_2\mathbf{j} + a_3\mathbf{k}$ for all (x, y, z).

(b) What can you conclude about f and g if $\nabla f(x, y, z) - \nabla g(x, y, z) = a_1\mathbf{i} + a_2\mathbf{j} + a_3\mathbf{k}$ for all (x, y, z)?

4. (*Rolle's theorem for functions of several variables*) Show that, if f is differentiable at each point of the line segment $\overline{\mathbf{ab}}$ and $f(\mathbf{a}) = f(\mathbf{b})$, then there exists a point $\mathbf{c}$ between $\mathbf{a}$ and $\mathbf{b}$ for which $\nabla f(\mathbf{c}) \perp (\mathbf{b} - \mathbf{a})$.

5. Let $U = \{\mathbf{x} : ||\mathbf{x}|| \neq 1\}$. Define f on U by setting

$$f(\mathbf{x}) = \begin{cases} 0, & ||\mathbf{x}|| < 1 \\ 1, & ||\mathbf{x}|| > 1. \end{cases}$$

(a) Note that $\nabla f(\mathbf{x}) = 0$ for all $\mathbf{x}$ in U, but f is not constant on U. Explain how this does not contradict Theorem 15.3.2.

(b) Define a function g on U different from f such that $\nabla f(\mathbf{x}) = \nabla g(\mathbf{x})$ for all $\mathbf{x}$ in U and $f - g$ is (i) constant on U, (ii) not constant on U.

6. A set of points is said to be *convex* provided that every pair of points in the set can be joined by a line segment that lies entirely within the set. Show that, if $||\nabla f(\mathbf{x})|| \leq M$ for all $\mathbf{x}$ in some convex set Ω, then

$$|f(\mathbf{x}_1) - f(\mathbf{x}_2)| \leq M||\mathbf{x}_1 - \mathbf{x}_2|| \qquad \text{for all } \mathbf{x}_1 \text{ and } \mathbf{x}_2 \text{ in } \Omega.$$

Find the rate of change of f with respect to t along the given curve.

7. $f(x, y) = x^2y, \quad \mathbf{r}(t) = e^t\,\mathbf{i} + e^{-t}\,\mathbf{j}$.

8. $f(x, y) = x - y, \quad \mathbf{r}(t) = at\,\mathbf{i} + b\cos at\,\mathbf{j}$.

9. $f(x, y) = \tan^{-1}(y^2 - x^2), \quad \mathbf{r}(t) = \sin t\,\mathbf{i} + \cos t\,\mathbf{j}$.

10. $f(x, y) = \ln(2x^2 + y^3), \quad \mathbf{r}(t) = e^{2t}\,\mathbf{i} + t^{1/3}\,\mathbf{j}$.

11. $f(x, y) = x\,e^y + y\,e^{-x}, \quad \mathbf{r}(t) = (\ln t)\,\mathbf{i} + t(\ln \mathrm{t})\,\mathbf{j}$.

12. $f(x, y, z) = \ln(x^2 + y^2 + z^2)$,
$\mathbf{r}(t) = \sin t\,\mathbf{i} + \cos t\,\mathbf{j} + e^{2t}\,\mathbf{k}$.

13. $f(x, y, z) = xy - yz, \quad \mathbf{r}(t) = t\,\mathbf{i} + t^2\,\mathbf{j} + t^3\,\mathbf{k}$.

14. $f(x, y, z) = x^2 + y^2$,
$\mathbf{r}(t) = a\cos\omega t\,\mathbf{i} + b\sin\omega t\,\mathbf{j} + b\omega t\,\mathbf{k}$.

15. $f(x, y, z) = x^2 + y^2 + z$,
$\mathbf{r}(t) = a\cos\omega t\,\mathbf{i} + b\sin\omega t\,\mathbf{j} + b\omega t\,\mathbf{k}$.

16. $f(x, y, z) = y^2\sin(x + z), \quad \mathbf{r}(t) = 2t\,\mathbf{i} + \cos t\,\mathbf{j} + t^3\,\mathbf{k}$.

Find du/dt by applying (15.3.5) or (15.3.6).

17. $u = x^2 - 3xy + 2y^2; \quad x = \cos t, \quad y = \sin t$.

18. $u = x + 4\sqrt{xy} - 3y; \quad x = t^3, \quad y = t^{-1} \quad (t > 0)$.

19. $u = e^x\sin y + e^y\sin x; \quad x = \frac{1}{2}t, \quad y = 2t$.

20. $u = 2x^2 - xy + y^2; \quad x = \cos 2t, \quad y = \sin t$.

21. $u = e^x\sin y; \quad x = t^2, \quad y = \pi t$.

22. $u = z\ln\left(\frac{y}{x}\right); \quad x = t^2 + 1, \quad y = \sqrt{t}, \quad z = t\,e^t$

23. $u = xy + yz + zx; \quad x = t^2, \quad y = t(1 - t), \quad z = (1 - t)^2$.

24. $u = x\sin\pi y - z\cos\pi x; \quad x = t^2, \quad y = 1 - t, \quad z = 1 - t^2$.

25. The radius of a right circular cone is increasing at the rate of 3 inches per second and the height is decreasing at the rate of 2 inches per second. At what rate is the volume of the cone changing at the instant the height is 20 inches and the radius is 14 inches?

26. The radius of a right circular cylinder is decreasing at the rate of 2 centimeters per second and the height is increasing at the rate of 3 centimeters per second. At what rate is the volume of the cylinder changing at the instant the radius is 13 centimeters and the height is 18 centimeters?

27. If the lengths of two sides of a triangle are x and y, and θ is the angle between the two sides, then the area A of the triangle is given by $A = \frac{1}{2}xy\sin\theta$. See the figure. If the sides are each increasing at the rate of 3 inches per second and θ is decreasing at the rate of 0.10 radian per second, how fast is the area changing at the instant $x = 1.5$ feet, $y = 2$ feet, and $\theta = 1$ radian?

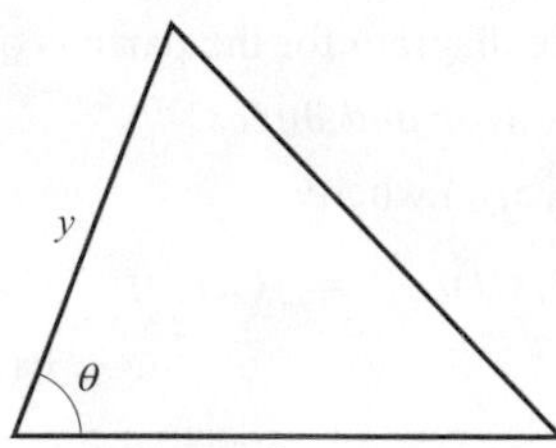

28. An object is moving along the curve of intersection of the paraboloid $z = x^2 + \frac{1}{4}y^2$ and the circular cylinder $x^2 + y^2 = 13$. If the x-coordinate is increasing at the rate of 5 centimeters per second, how fast is the z-coordinate changing at the instant when $x = 2$ centimeters and $y = 3$ centimeters?

Find $\partial u/\partial s$ and $\partial u/\partial t$.

29. $u = x^2 - xy$ where $x = s\cos t$, $y = t\sin s$.

30. $u = \sin(x - y) + \cos(x + y)$ where $x = st$, $y = s^2 - t^2$.

31. $u = x^2 \tan y$ where $x = s^2 t$, $y = s + t^2$.

32. $u = z^2 \sec xy$ where $x = 2st$, $y = s - t^2$, $z = s^2 t$.

33. $u = x^2 - xy + z^2$ where $x = s\cos t$, $y = \sin(t - s)$, $z = t\sin s$.

34. $u = x\,e^{yz^2}$ where $x = \ln st$, $y = t^3$, $z = s^2 + t^2$.

35. An object moves so that at time t it has position $\mathbf{r}(t) = x(t)\,\mathbf{i} + y(t)\,\mathbf{j} + z(t)\,\mathbf{k}$. Show that

$$\frac{d}{dt}[f(\mathbf{r}(t))]$$

is the directional derivative of f in the direction of the motion times the speed of the object.

36. (*Important*) Set $r = \|\mathbf{r}\|$ where $\mathbf{r} = x\,\mathbf{i} + y\,\mathbf{j} + z\,\mathbf{k}$. If f is a continuously differentiable function of r, then

(15.3.11)
$$\nabla[f(r)] = f'(r)\frac{\mathbf{r}}{r}.$$

Derive this formula.

Calculate the following gradients taking $r = \|\mathbf{r}\|$.

37. (a) $\nabla(\sin r)$. (b) $\nabla(r\sin r)$.

38. (a) $\nabla(r\ln r)$. (b) $\nabla(e^{1-r^2})$.

39. (a) $\nabla\left(\dfrac{\sin r}{r}\right)$. (b) $\nabla\left(\dfrac{r}{\sin r}\right)$.

40. (a) Draw a tree diagram for du/dt given that $u = u(x,y)$ where $x = x(s)$, $y = y(s)$, and $s = s(t)$.
(b) Calculate du/dt.

41. Set $u = u(x,y,z)$ where

$$x = x(w,t),\quad y = y(w,t),\quad z = z(w,t)$$
$$w = w(r,s),\quad t = t(r,s).$$

(a) Draw a tree diagram for the partials of u.
(b) Calculate $\partial u/\partial r$ *and* $\partial u/\partial s$.

42. Set $u = u(x,y,z,w)$ where

$$x = x(r,s,t),\quad y = y(s,t,v),\quad z = z(r,t),$$
$$w = w(r,s,t,v).$$

(a) Draw a tree diagram for the partials of u.
(b) Calculate $\partial u/\partial r$ *and* $\partial u/\partial v$.

Higher Derivatives

43. Let $u = u(x,y)$, where $x = x(t)$ and $y = y(t)$, and assume that these functions have continuous second derivatives. Show that

$$\frac{d^2u}{dt^2} = \frac{\partial^2 u}{\partial x^2}\left(\frac{dx}{dt}\right)^2 + 2\frac{\partial^2 u}{\partial x\partial y}\frac{dx}{dt}\frac{dy}{dt} + \frac{\partial^2 u}{\partial y^2}\left(\frac{dy}{dt}\right)^2 + \frac{\partial u}{\partial x}\frac{d^2x}{dt^2} + \frac{\partial u}{\partial y}\frac{d^2y}{dt^2}.$$

44. Let $u = u(x,y)$, where $x = x(s,t)$ and $y = y(s,t)$, and assume that all these functions have continuous second derivatives. Show that

$$\frac{\partial^2 u}{\partial s^2} = \frac{\partial^2 u}{\partial x^2}\left(\frac{\partial x}{\partial s}\right)^2 + 2\frac{\partial^2 u}{\partial x\partial y}\frac{\partial x}{\partial s}\frac{\partial y}{\partial s} + \frac{\partial^2 u}{\partial y^2}\left(\frac{\partial y}{\partial s}\right)^2 + \frac{\partial u}{\partial x}\frac{\partial^2 x}{\partial s^2} + \frac{\partial u}{\partial y}\frac{\partial^2 y}{\partial s^2}.$$

Polar Coordinates

45. Assume that $u = u(x,y)$ is differentiable.
(a) Show that the change of variables to polar coordinates $x = r\cos\theta$ and $y = r\sin\theta$ gives

$$\frac{\partial u}{\partial r} = \frac{\partial u}{\partial x}\cos\theta + \frac{\partial u}{\partial y}\sin\theta,$$
$$\frac{\partial u}{\partial \theta} = -\frac{\partial u}{\partial x}r\sin\theta + \frac{\partial u}{\partial y}r\cos\theta.$$

(b) Express

$$\left(\frac{\partial u}{\partial r}\right)^2 + \frac{1}{r^2}\left(\frac{\partial u}{\partial \theta}\right)^2$$

entirely in terms of $\partial u/\partial x$ and $\partial u/\partial y$.

46. Let w be a function of polar coordinates r and θ. Then w can be expressed in rectangular coordinates by using

$$x = r\cos\theta \quad\text{and}\quad y = r\sin\theta.$$

(a) Using the first part of Exercise 45, verify that

$$\frac{\partial w}{\partial x} = \frac{\partial w}{\partial r}\cos\theta - \frac{1}{r}\frac{\partial w}{\partial \theta}\sin\theta,$$
$$\frac{\partial w}{\partial y} = \frac{\partial w}{\partial r}\sin\theta + \frac{1}{r}\frac{\partial w}{\partial \theta}\cos\theta.$$

(b) Deduce from part (a) that

$$\frac{\partial r}{\partial x} = \cos\theta,\qquad \frac{\partial r}{\partial y} = \sin\theta;$$
$$\frac{\partial \theta}{\partial x} = -\frac{1}{r}\sin\theta,\qquad \frac{\partial \theta}{\partial y} = \frac{1}{r}\cos\theta.$$

(c) Find the fallacy in the following argument:

$$x = r\cos\theta,\qquad r = \frac{x}{\cos\theta},\qquad \frac{\partial r}{\partial x} = \frac{1}{\cos\theta}.$$

47. (*The gradient in polar coordinates*) Let $u = u(x,y)$ be differentiable. Show that if u is written in terms of polar coordinates, then

$$\nabla u = \frac{\partial u}{\partial r}\,\mathbf{e}_r + \frac{1}{r}\frac{\partial u}{\partial \theta}\,\mathbf{e}_\theta$$

where

$$\mathbf{e}_r = \cos\theta\,\mathbf{i} + \sin\theta\,\mathbf{j} \quad\text{and}\quad \mathbf{e}_\theta = -\sin\theta\,\mathbf{i} + \cos\theta\,\mathbf{j}.$$

In Exercises 48 and 49, use the formula in Exercise 47 to express the gradient of the given function in polar coordinates.

48. $u(x, y) = x^2 + y^2$.

49. $u(x, y) = x^2 - xy + y^2$.

50. Let $u = u(x, y)$, where $x = r\cos\theta$ and $y = r\sin\theta$, and assume that u has continuous second partial derivatives. Derive a formula for $\partial^2 u/\partial r\partial\theta$.

51. (*The Laplace operator in polar coordinates*). Assume that $u = u(x, y)$ has continuous second partial derivatives. Show that

$$\frac{\partial^2 u}{\partial x^2} + \frac{\partial^2 u}{\partial y^2} = \frac{\partial^2 u}{\partial r^2} + \frac{1}{r^2}\frac{\partial^2 u}{\partial \theta^2} + \frac{1}{r}\frac{\partial u}{\partial r}.$$

Assume that y is a differentiable function of x that satisfies the given equation. Find dy/dx by implicit differentiation.

52. $x^2 - 2xy + y^4 = 4$.

53. $x\,e^y + y\,e^x - 2x^2 y = 0$.

54. $x^{2/3} + y^{2/3} = a^{2/3}$ $\qquad$ (a a constant).

55. $x\cos xy + y\cos x = 2$.

Assume that z is a differentiable function of (x, y) that satisfies the given equation. Find $\partial z/\partial x$ and $\partial z/\partial y$ by implicit differentiation.

56. $z^4 + x^2 z^3 + y^2 + xy = 2$.

57. $\cos xyz + \ln(x^2 + y^2 + z^2) = 0$.

58. (*A chain rule for vector-valued functions*) Suppose that

$$\mathbf{u}(x, y) = u_1(x, y)\,\mathbf{i} + u_2(x, y)\,\mathbf{j}$$
where $x = x(t), \quad y = y(t)$.

(a) Show that

$$\text{(15.3.12)} \qquad \frac{d\mathbf{u}}{dt} = \frac{\partial \mathbf{u}}{\partial x}\frac{dx}{dt} + \frac{\partial \mathbf{u}}{\partial y}\frac{dy}{dt},$$

where

$$\frac{\partial \mathbf{u}}{\partial x} = \frac{\partial u_1}{\partial x}\mathbf{i} + \frac{\partial u_2}{\partial x}\mathbf{j} \quad\text{and}\quad \frac{\partial \mathbf{u}}{\partial y} = \frac{\partial u_1}{\partial y}\mathbf{i} + \frac{\partial u_2}{\partial y}\mathbf{j}.$$

(b) Let

$$\mathbf{u} = e^x \cos y\,\mathbf{i} + e^x \sin y\,\mathbf{j}$$
where $x = \frac{1}{2}t^2, \quad y = \pi t$.

Calculate $d\mathbf{u}/dt$ (i) by applying (15.3.12), (ii) by forming $\mathbf{u}(t)$ directly.

59. Set

$$\mathbf{u}(x, y) = u_1(x, y)\,\mathbf{i} + u_2(x, y)\,\mathbf{j}$$
where $x = x(s, t), \quad y = y(s, t)$.

Find $\partial\mathbf{u}/\partial s$ and $\partial\mathbf{u}/\partial t$.

60. Set

$$\mathbf{u}(x, y, z) = u_1(x, y, z)\,\mathbf{i} + u_2(x, y, z)\,\mathbf{j} + u_3(x, y, z)\,\mathbf{k},$$

where

$$x = x(t), \quad y = y(t), \quad z = z(t).$$

Derive a formula for $d\mathbf{u}/dt$ analogous to (15.3.12).

*SUPPLEMENT TO SECTION 15.3

AN INTERMEDIATE-VALUE THEOREM

You have seen that a function of single variable that is continuous on an interval skips no values (the intermediate-value theorem, Theorem 2.6.1). There is an analogous result for functions of several variables. First, some terminology.

Open Regions, Closed Regions, Continuity

An open connected set is called an *open region*. If we start with an open region and adjoin to it the boundary, then we have what is called a *closed region*. (A closed region is therefore a closed set, the interior of which is an open region.)

Continuity on a closed region Ω requires continuity at the interior points *and* continuity at the boundary points. The meaning of this second requirement has to be explained.

Let the function f be defined on a closed region Ω, and let $\mathbf{x}_0 \in \Omega$. If $\mathbf{x}_0$ is an interior point of Ω, then all points $\mathbf{x}$ sufficiently close to $\mathbf{x}_0$ are in Ω and, by definition f is continuous at $\mathbf{x}_0$ iff

$$\text{as } \mathbf{x} \text{ approaches } \mathbf{x}_0, \quad f(\mathbf{x}) \text{ approaches } f(\mathbf{x}_0). \qquad \text{(Figure 15.3.8)}$$

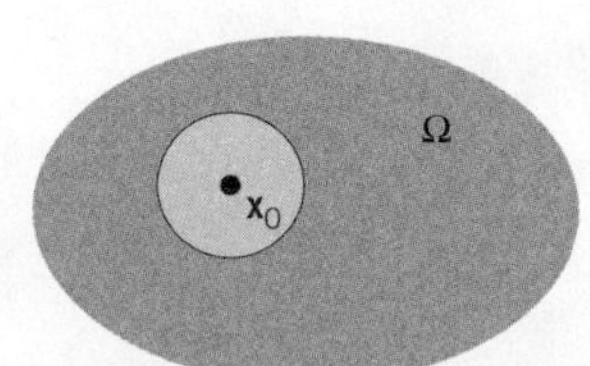

Figure 15.3.8

If $\mathbf{x}_0$ is a boundary point of Ω, then we have to modify the definition and say: f is continuous at $\mathbf{x}_0$ if

as $\mathbf{x}$ approaches $\mathbf{x}_0$ from *within* Ω, $f(\mathbf{x})$ approaches $f(\mathbf{x}_0)$. (Figure 15.3.9)

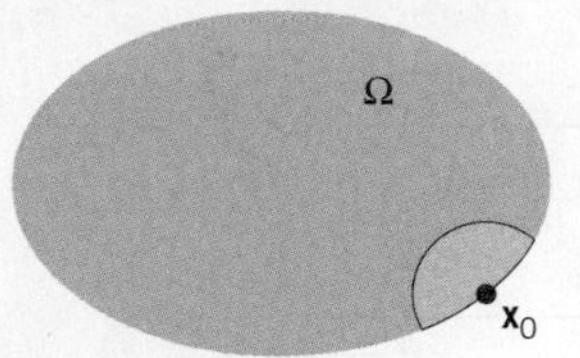

Figure 15.3.9

> **THEOREM 15.3.13 AN INTERMEDIATE-VALUE THEOREM**
>
> Suppose that f is continuous on a closed region Ω and A, B, C are real numbers such that $A < C < B$. If, somewhere in Ω, f takes on the value A and, somewhere in Ω, f takes on the value B, then, somewhere in Ω, f takes on the value C.

PROOF Let $\mathbf{a}$ and $\mathbf{b}$ be points in Ω for which

$$f(\mathbf{a}) = A \qquad \text{and} \qquad f(\mathbf{b}) = B.$$

We must show that there exists a point $\mathbf{c}$ in Ω for which $f(\mathbf{c}) = C$.

Let U be the interior of Ω and assume first that $\mathbf{a}$ and $\mathbf{b}$ are in U. Since U is connected, there is a polygonal path γ in U that joins $\mathbf{a}$ to $\mathbf{b}$. Let $\mathbf{r} = \mathbf{r}(t), a \leq t \leq b$, be a continuous parametrization of the path γ with $\mathbf{r}(a) = \mathbf{a}$ and $\mathbf{r}(b) = \mathbf{b}$. Since $\mathbf{r}$ is continuous, the composition

$$g(t) = f(\mathbf{r}(t))$$

is also continuous on $[a, b]$. Since

$$g(a) = f(\mathbf{r}(a)) = f(\mathbf{a}) = A \qquad \text{and} \qquad g(b) = f(\mathbf{r}(b)) = f(\mathbf{b}) = B,$$

we know from Theorem 2.6.1 that there is a number c in $[a, b]$ for which $g(c) = C$. Setting $\mathbf{c} = \mathbf{r}(c)$ we have $f(\mathbf{c}) = C$.

Now let $\mathbf{a}$ and $\mathbf{b}$ be *any* two points in Ω for which

$$f(\mathbf{a}) = A \qquad \text{and} \qquad f(\mathbf{b}) = B.$$

To take care of the possibility that one or both of these points lie on the boundary of Ω, we proceed as follows. Take ϵ small enough that

$$A + \epsilon < C < B - \epsilon.$$

By continuity there exist points $\mathbf{x}_1, \mathbf{x}_2$ in U, the interior of Ω, for which

$$f(\mathbf{x}_1) < A + \epsilon \qquad \text{and} \qquad B - \epsilon < f(\mathbf{x}_2).$$

Then

$$f(\mathbf{x}_1) < C < f(\mathbf{x}_2)$$

and the result follows by our argument above. □

■ 15.4 THE GRADIENT AS A NORMAL; TANGENT LINES AND TANGENT PLANES

Functions of Two Variables

We begin with a nonconstant function $f = f(x, y)$ that is continuously differentiable. (*Remember:* That means f is differentiable and its gradient ∇f is continuous.) You have seen that at each point of the domain, the gradient vector, if not $\mathbf{0}$, points in the direction of the most rapid increase of f. Here we show that

(15.4.1) at each point of the domain, the gradient vector ∇f, if not $\mathbf{0}$, is perpendicular to the level curve of f that passes through that point.

PROOF We choose a point (x_0, y_0) in the domain and assume that $\nabla f(x_0, y_0) \neq \mathbf{0}$. The level curve through this point has equation

$$f(x, y) = c \quad \text{where} \quad c = f(x_0, y_0).$$

Under our assumptions on f, this curve can be parametrized in a neighborhood of (x_0, y_0) by a continuously differentiable vector function

$$\mathbf{r}(t) = x(t)\,\mathbf{i} + y(t)\,\mathbf{j}, \quad t \in I$$

with nonzero tangent vector $\mathbf{r}'(t)$. †

Now take t_0 such that

$$\mathbf{r}(t_0) = x_0\,\mathbf{i} + y_0\,\mathbf{j} = (x_0, y_0).$$

We will show that

$$\nabla f(\mathbf{r}(t_0)) \perp \mathbf{r}'(t_0).$$

Since f is constantly c on the curve, we have

$$f(\mathbf{r}(t)) = c \quad \text{for all } t \in I.$$

For such t

$$\frac{d}{dt}[f(\mathbf{r}(t))] = \nabla f(\mathbf{r}(t)) \cdot \mathbf{r}'(t) = 0.$$

↑ (Theorem 15.3.4)

In particular

$$\nabla f(\mathbf{r}(t_0)) \cdot \mathbf{r}'(t_0) = 0,$$

and thus

$$\nabla f(\mathbf{r}(t_0)) \perp \mathbf{r}'(t_0). \quad \square$$

Figure 15.4.1 illustrates the result.

† This follows from a result of advanced calculus known as the *implicit function theorem.*

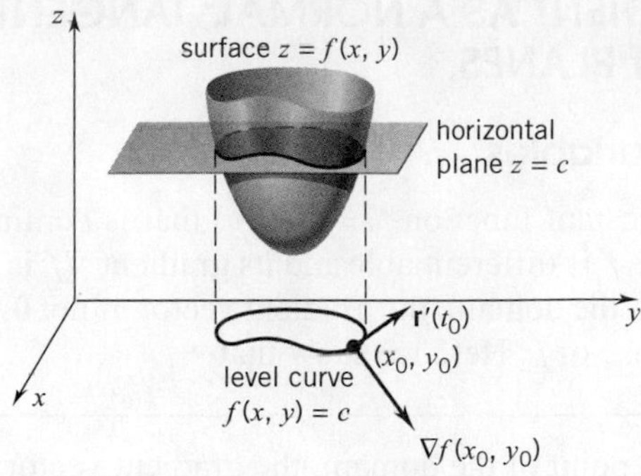

Figure 15.4.1

Example 1 For the function $f(x, y) = x^2 + y^2$ the level curves are concentric circles:

$$x^2 + y^2 = c.$$

At each point $(x, y) \neq (0, 0)$ the gradient vector

$$\nabla f(x, y) = 2x\,\mathbf{i} + 2y\,\mathbf{j} = 2\mathbf{r}$$

points away from the origin along the line of the radius vector and is thus perpendicular to the circle in question. At the origin the level curve is reduced to a point and the gradient is simply $\mathbf{0}$. See Figure 15.4.2. □

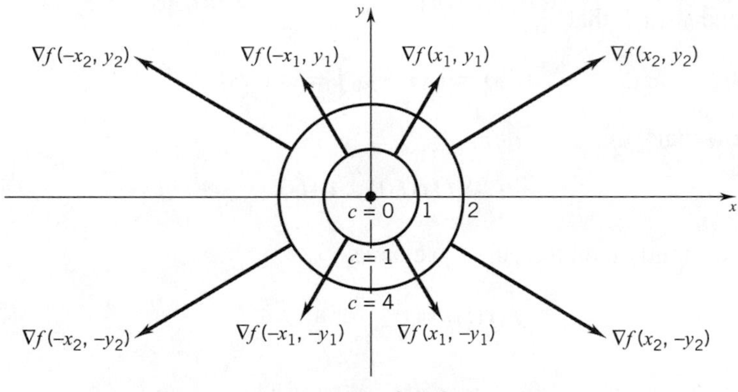

Figure 15.4.2

Consider now a curve in the xy-plane

$$C\colon\ f(x, y) = c.$$

As before, we assume that f is nonconstant and continuously differentiable. Let's suppose that (x_0, y_0) lies on the curve and $\nabla f(x_0, y_0) \neq \mathbf{0}$.

We can view C as the c-level curve of f and conclude from (15.4.1) that the gradient

(15.4.2)
$$\boxed{\nabla f(x_0, y_0) = \frac{\partial f}{\partial y}(x_0, y_0)\,\mathbf{i} + \frac{\partial f}{\partial y}(x_0, y_0)\,\mathbf{j}}$$

is perpendicular to C at (x_0, y_0). We call it a *normal vector.*

The vector

(15.4.3)
$$\mathbf{t}(x_0, y_0) = \frac{\partial f}{\partial y}(x_0, y_0)\,\mathbf{i} - \frac{\partial f}{\partial x}(x_0, y_0)\,\mathbf{j}$$

is perpendicular to the gradient:

$$\nabla f(x_0, y_0) \cdot \mathbf{t}(x_0, y_0) = \frac{\partial f}{\partial x}(x_0, y_0)\frac{\partial f}{\partial y}(x_0, y_0) - \frac{\partial f}{\partial y}(x_0, y_0)\frac{\partial f}{\partial x}(x_0, y_0) = 0.$$

It is therefore a *tangent vector*.

The line through (x_0, y_0) perpendicular to the gradient is the tangent line. To obtain an equation for the tangent line, we refer to Figure 15.4.3. A point (x, y) will lie on the tangent line iff

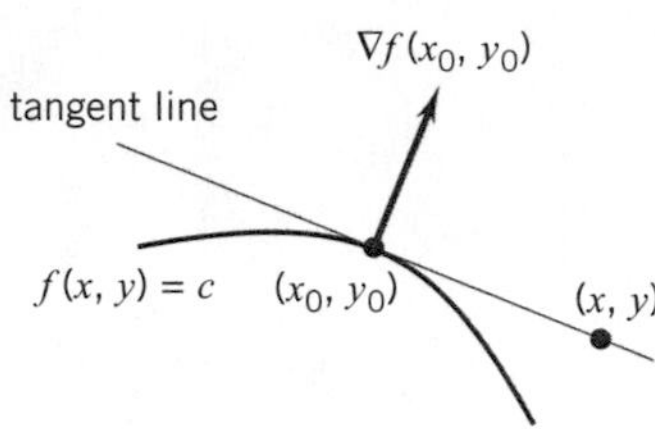

Figure 15.4.3

$$[(x - x_0)\,\mathbf{i} + (y - y_0)\,\mathbf{j}] \cdot \nabla f(x_0, y_0) = 0,$$

that is, iff

(15.4.4)
$$\frac{\partial f}{\partial x}(x_0, y_0)(x - x_0) + \frac{\partial f}{\partial y}(x_0, y_0)(y - y_0) = 0.$$

This is an equation for the *tangent line.*

The line through (x_0, y_0) perpendicular to the tangent vector $\mathbf{t}(x_0, y_0)$ is the normal line (Figure 15.4.4). A point (x, y) will lie on the normal line iff

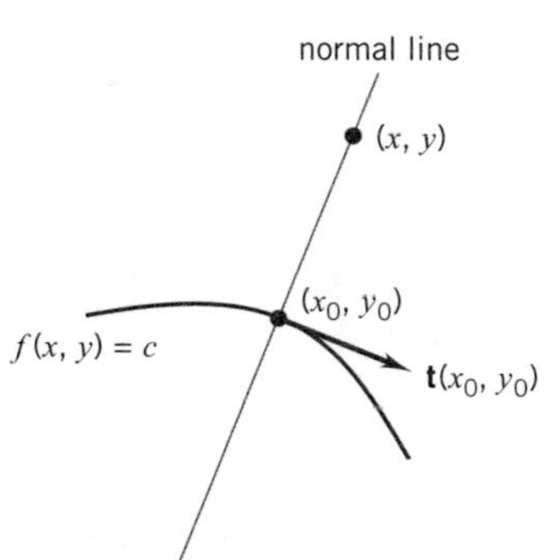

Figure 15.4.4

$$[(x - x_0)\,\mathbf{i} + (y - y_0)\,\mathbf{j}] \cdot \mathbf{t}(x_0, y_0) = 0,$$

that is, iff

(15.4.5)
$$\frac{\partial f}{\partial y}(x_0, y_0)(x - x_0) - \frac{\partial f}{\partial x}(x_0, y_0)(y - y_0) = 0.$$

This is an equation for the *normal line.*

Example 2 Find a normal vector and a tangent vector to the plane curve

$$x^2 + 2y^3 = xy + 4$$

at the point (2,1). Then write equations for the tangent line and normal line to the curve at this point.

SOLUTION Set $f(x, y) = x^2 + 2y^3 - xy$. Then the given curve is the level curve $f(x, y) = 4$. Now the gradient of f is

$$\nabla f = (2x - y)\,\mathbf{i} + (6y^2 - x)\,\mathbf{j} \quad \text{and} \quad \nabla f(2, 1) = 3\,\mathbf{i} + 4\,\mathbf{j}.$$

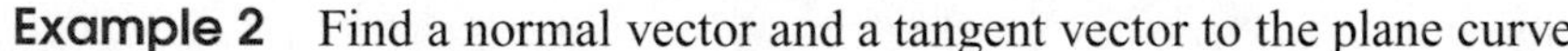

Therefore, we have

normal vector: $\nabla f(2, 1) = 3\,\mathbf{i} + 4\,\mathbf{j}$, tangent vector: $\mathbf{t}(2, 1) = 4\,\mathbf{i} - 3\,\mathbf{j}$,

equation of tangent line: $3(x - 2) + 4(y - 1) = 0$ or $y = -\frac{3}{4}x + \frac{5}{2}$,

equation of normal line: $4(x - 2) - 3(y - 1) = 0$ or $y = \frac{4}{3}x - \frac{5}{3}$. □

Functions of Three Variables

Here, instead of level curves, we have level surfaces, but the results are similar. If $f = f(x, y, z)$ is nonconstant and continuously differentiable, then

(15.4.6) at each point of the domain, the gradient vector, if not $\mathbf{0}$, is perpendicular to the level surface that passes through that point.

PROOF We choose a point $\mathbf{x}_0 = (x_0, y_0, z_0)$ in the domain and assume that $\nabla f(x_0, y_0, z_0) \neq \mathbf{0}$. The level surface through this point has equation

$$f(x, y, z) = c \qquad \text{where} \qquad c = f(x_0, y_0, z_0).$$

We suppose now that

$$\mathbf{r}(t) = x(t)\,\mathbf{i} + y(t)\,\mathbf{j} + z(t)\,\mathbf{k}, \quad t \in I,$$

is a differentiable curve that lies on this surface and passes through the point $\mathbf{x}_0 = (x_0, y_0, z_0)$. We choose t_0 so that

$$\mathbf{r}(t_0) = \mathbf{x_0} = (x_0, y_0, z_0)$$

and suppose that $\mathbf{r}'(t_0) \neq \mathbf{0}$.

Since the curve lies on the given surface, we have

$$f(\mathbf{r}(t)) = c \quad \text{for all } t \in I.$$

For such t

$$\frac{d}{dt}[f(\mathbf{r}(t))] = \nabla f(\mathbf{r}(t)) \cdot \mathbf{r}'(t) = 0.$$

In particular,

$$\nabla f(\mathbf{r}(t_0)) \cdot \mathbf{r}'(t_0) = 0.$$

The gradient vector

$$\nabla f(\mathbf{r}(t_0)) = \nabla f(\mathbf{x}_0) = \nabla f(x_0, y_0, z_0)$$

is thus perpendicular to the curve in question.

This same argument applies to *every* differentiable curve that lies on this level surface and passes through the point $\mathbf{x}_0 = (x_0, y_0, z_0)$ with nonzero tangent vector. (See Figure 15.4.5.) Consequently, $\nabla f(\mathbf{x}_0)$ must be perpendicular to the surface itself. □

Example 3 For the function $f(x, y, z) = x^2 + y^2 + z^2$ the level surfaces are concentric spheres:

$$x^2 + y^2 + z^2 = c.$$

At each point $(x, y, z) \neq (0, 0, 0)$, the gradient vector

$$\nabla f(x, y, z) = 2x\,\mathbf{i} + 2y\,\mathbf{j} + 2z\,\mathbf{k} = 2\mathbf{r}$$

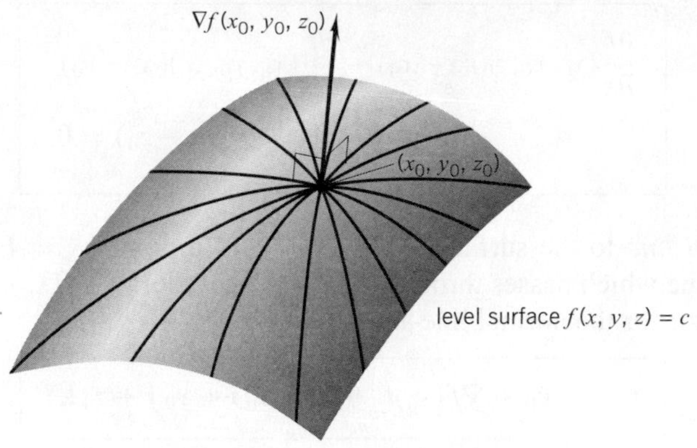

The gradient vector $\nabla f(x_0, y_0, z_0)$ is perpendicular to the level surface at (x_0, y_0, z_0)

Figure 15.4.5

points away from the origin along the line of the radius vector and is thus perpendicular to the sphere in question. At the origin the level surface is reduced to a point and the gradient is $\mathbf{0}$. □

The *tangent plane* for a surface

$$f(x,y,z) = c$$

at a point $\mathbf{x}_0 = (x_0, y_0, z_0)$ is the plane through $\mathbf{x}_0$ with normal $\nabla f(\mathbf{x}_0)$. See Figure 15.4.6.

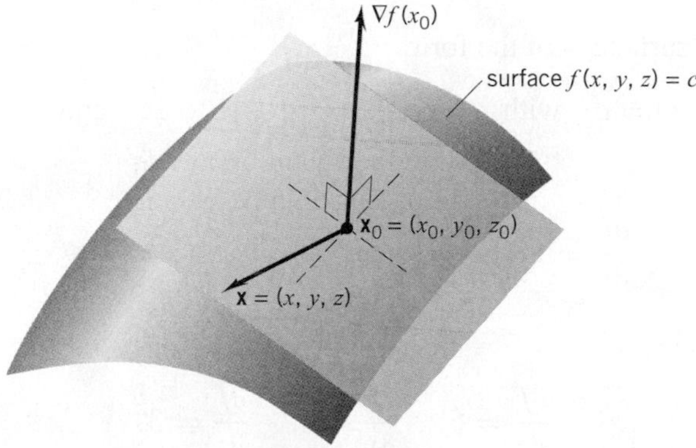

tangent plane to a surface $f(x, y, z) = c$

Figure 15.4.6

The tangent plane at a point $\mathbf{x}_0$ is the plane through $\mathbf{x}_0$ that best approximates the surface in a neighborhood of $\mathbf{x}_0$. (We return to this later.)

A point $\mathbf{x}$ lies on the tangent plane through $\mathbf{x}_0$ iff

(15.4.7) $$\boxed{\nabla f(\mathbf{x}_0) \cdot (\mathbf{x} - \mathbf{x}_0) = 0.}$$ (Figure 15.4.6)

This is an equation for the tangent plane in vector notation. In Cartesian coordinates the equation takes the form

(15.4.8)
$$\frac{\partial f}{\partial x}(x_0, y_0, z_0)(x - x_0) + \frac{\partial f}{\partial y}(x_0, y_0, z_0)(y - y_0) + \frac{\partial f}{\partial z}(x_0, y_0, z_0)(z - z_0) = 0.$$

The *normal line* to the surface $f(x, y, z) = c$ at a point $\mathbf{x}_0 = (x_0, y_0, z_0)$ on the surface is the line which passes through (x_0, y_0, z_0) parallel to $\nabla f(\mathbf{x}_0)$. Thus, $\nabla f(\mathbf{x}_0)$ is a direction vector for the normal line and

(15.4.9)
$$\mathbf{r}(t) = \mathbf{r}_0 + \nabla f(\mathbf{x}_0)t \quad (\mathbf{r}_0 = x_0\,\mathbf{i} + y_0\,\mathbf{j} + z_0\,\mathbf{k})$$

is a vector equation for the line. In scalar parametric form, equations for the normal line can be written

(15.4.10)
$$x = x_0 + \frac{\partial f}{\partial x}(x_0, y_0, z_0)\,t,$$
$$y = y_0 + \frac{\partial f}{\partial y}(x_0, y_0, z_0)\,t,$$
$$z = z_0 + \frac{\partial f}{\partial z}(x_0, y_0, z_0)\,t.$$

Example 4 Find an equation for the plane tangent to the surface

$$xy + yz + zx = 11 \quad \text{at the point } (1, 2, 3).$$

SOLUTION The surface is of the form

$$f(x, y, z) = c \quad \text{with} \quad f(x, y, z) = xy + yz + zx \quad \text{and} \quad c = 11.$$

Observe that

$$\frac{\partial f}{\partial x} = y + z, \quad \frac{\partial f}{\partial y} = x + z, \quad \frac{\partial f}{\partial z} = x + y.$$

At the point (1,2,3)

$$\frac{\partial f}{\partial x} = 5, \quad \frac{\partial f}{\partial y} = 4, \quad \frac{\partial f}{\partial z} = 3.$$

The equation for the tangent plane can therefore be written

$$5(x - 1) + 4(y - 2) + 3(z - 3) = 0.$$

This simplifies to

$$5x + 4y + 3z - 22 = 0. \quad \square$$

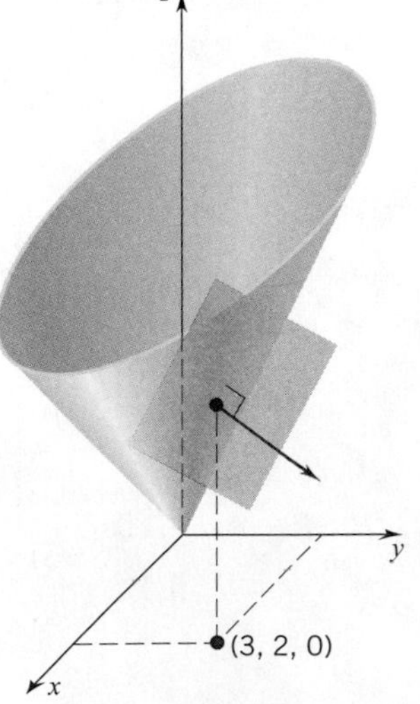

Figure 15.4.7

Example 5 Find an equation for the tangent plane and find scalar parametric equations for the normal line to the elliptic cone

$$x^2 + 4y^2 = z^2$$

at the point (3,2,5) on the cone (Figure 15.4.7).

SOLUTION The surface is of the form $f(x, y, z) = c$ with

$$f(x, y, z) = x^2 + 4y^2 - z^2 \qquad \text{and} \qquad c = 0.$$

The partial derivatives of f are

$$\frac{\partial f}{\partial x} = 2x, \quad \frac{\partial f}{\partial y} = 8y, \quad \frac{\partial f}{\partial z} = -2z$$

and
$$\nabla f = 2x\,\mathbf{i} + 8y\,\mathbf{j} - 2z\,\mathbf{k}.$$

Now, $\nabla f(3, 2, 5) = 6\mathbf{i} + 16\mathbf{j} - 10\mathbf{k}$ is normal to the cone at the point (3,2,5). Note that $\frac{1}{2}\nabla f(3, 2, 5) = 3\mathbf{i} + 8\mathbf{j} - 5\mathbf{k}$ is also normal to the cone and is a little easier to work with.

The equation for the tangent plane can be written

$$3(x - 3) + 8(y - 2) - 5(z - 5) = 0,$$

which simplifies to

$$3x + 8y - 5z = 0.$$

Note that this plane passes through the origin, as we would expect. The following are scalar parametric equations for the normal line:

$$x = 3 + 3t, \quad y = 2 + 8t, \quad z = 5 - 5t. \quad \square$$

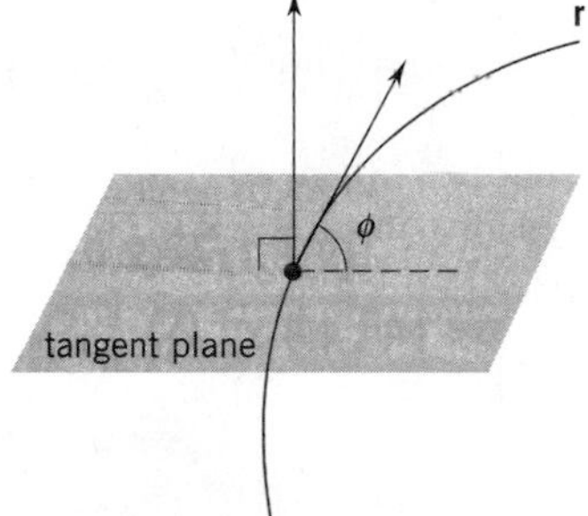

Figure 15.4.8

Example 6 The curve $\mathbf{r}(t) = \frac{1}{2}t^2\,\mathbf{i} + 4t^{-1}\,\mathbf{j} + (\frac{1}{2}t - t^2)\,\mathbf{k}$ intersects the hyperbolic paraboloid $x^2 - 4y^2 - 4z = 0$ at the point $(2, 2, -3)$. What is the angle of intersection?

SOLUTION We want the angle ϕ between the tangent vector of the curve and the tangent plane of the surface at the point of intersection (Figure 15.4.8).

A simple calculation shows that the curve passes through the point $(2, 2, -3)$ at $t = 2$. Since

$$\mathbf{r}'(t) = t\,\mathbf{i} - 4t^{-2}\,\mathbf{j} + (\tfrac{1}{2} - 2t)\,\mathbf{k},$$

we have
$$\mathbf{r}'(2) = 2\,\mathbf{i} - \mathbf{j} - \tfrac{7}{2}\,\mathbf{k}.$$

Now set
$$f(x, y, z) = x^2 - 4y^2 - 4z.$$

This function has gradient $2x\,\mathbf{i} - 8y\,\mathbf{j} - 4\,\mathbf{k}$. At the point $(2, 2, -3)$,

$$\nabla f = 4\,\mathbf{i} - 16\,\mathbf{j} - 4\,\mathbf{k}.$$

Now let θ be the angle between $\mathbf{r}'(2)$ and this gradient. By (12.4.9),

$$\cos\theta = \frac{\mathbf{r}'(2)}{||\mathbf{r}'(2)||} \cdot \frac{\nabla f}{||\nabla f||} = \frac{19}{414}\sqrt{138} \cong 0.539,$$

so that $\theta \cong 1.00$ radian. Since the gradient is normal to the tangent plane, the angle ϕ we want is

$$\phi = \tfrac{1}{2}\pi - \theta \cong 1.57 - 1.00 = 0.57 \text{ radian.} \quad \square$$

A surface of the form

$$z = g(x, y)$$

can be written in the form

$$f(x, y, z) = 0$$

by setting

$$f(x, y, z) = g(x, y) - z.$$

If g is differentiable, so is f. Moreover,

$$\frac{\partial f}{\partial x}(x, y, z) = \frac{\partial g}{\partial x}(x, y), \quad \frac{\partial f}{\partial y}(x, y, z) = \frac{\partial g}{\partial y}(x, y), \quad \frac{\partial f}{\partial z}(x, y, z) = -1.$$

By (15.4.8), the tangent plane at (x_0, y_0, z_0) has equation

$$\frac{\partial g}{\partial x}(x_0, y_0)(x - x_0) + \frac{\partial g}{\partial y}(x_0, y_0)(y - y_0) + (-1)(z - z_0) = 0$$

which we can rewrite as

(15.4.11)
$$z - z_0 = \frac{\partial g}{\partial x}(x_0, y_0)(x - x_0) + \frac{\partial g}{\partial y}(x_0, y_0)(y - y_0).$$

If $\nabla g(x_0, y_0) = \mathbf{0}$, then both partials of g are zero at (x_0, y_0) and the equation reduces to

$$z = z_0.$$

In this case the tangent plane is *horizontal*.

Scalar parametric equations for the normal line to the surface $z = g(x, y)$ at the point (x_0, y_0, z_0) take the form

(15.4.12)
$$x = x_0 + \frac{\partial g}{\partial x}(x_0, y_0)\,t, \quad y = y_0 + \frac{\partial g}{\partial y}(x_0, y_0)\,t, \quad z = z_0 + (-1)\,t.$$

Example 7 Find an equation for the tangent plane and symmetric equations for the line normal to the surface

$$z = \ln(x^2 + y^2)$$

at the point $(-2, 1, \ln 5)$ on the surface.

SOLUTION Set $g(x, y) = \ln(x^2 + y^2)$. The partial derivatives of g are

$$\frac{\partial g}{\partial x}(x, y) = \frac{2x}{x^2 + y^2}, \quad \frac{\partial g}{\partial y}(x, y) = \frac{2y}{x^2 + y^2}.$$

Where $x = -2$ and $y = 1$,

$$\frac{\partial g}{\partial x} = -\frac{4}{5}, \quad \frac{\partial g}{\partial y} = \frac{2}{5}.$$

Therefore, at the point $(-2, 1, \ln 5)$, the tangent plane has equation

$$z - \ln 5 = -\frac{4}{5}(x+2) + \frac{2}{5}(y-1).$$

The symmetric equations for the normal line are

$$\frac{x+2}{-\frac{4}{5}} = \frac{y-1}{\frac{2}{5}} = \frac{z - \ln 5}{-1}. \quad \square$$

Example 8 At what points on the surface $z = 3xy - x^3 - y^3$, is the tangent plane horizontal?

SOLUTION The function $g(x,y) = 3xy - x^3 - y^3$ has first partials

$$\frac{\partial g}{\partial x}(x,y) = 3y - 3x^2, \quad \frac{\partial g}{\partial y}(x,y) = 3x - 3y^2.$$

We set these partial derivatives equal to 0 and solve the resulting system of equations for x and y:

$$\begin{aligned} 3y - 3x^2 &= 0 \quad \text{simplifies to} \quad & y - x^2 &= 0 \\ 3x - 3y^2 &= 0 & x - y^2 &= 0. \end{aligned}$$

From the first equation, we get $y = x^2$. Substituting this into the second equation, we have

$$x - x^4 = 0,$$

and thus

$$x(1 - x^3) = 0.$$

Therefore, $x = 0$ (which implies that y=0) or $x = 1$ (which implies that $y = 1$). Thus, the partials are both zero only at $(0,0)$ and $(1,1)$. The surface has a horizontal tangent plane only at the points $(0,0,0)$ and $(1,1,1)$. A graph of the surface and the level curves are shown in Figure 15.4.9 $\square$

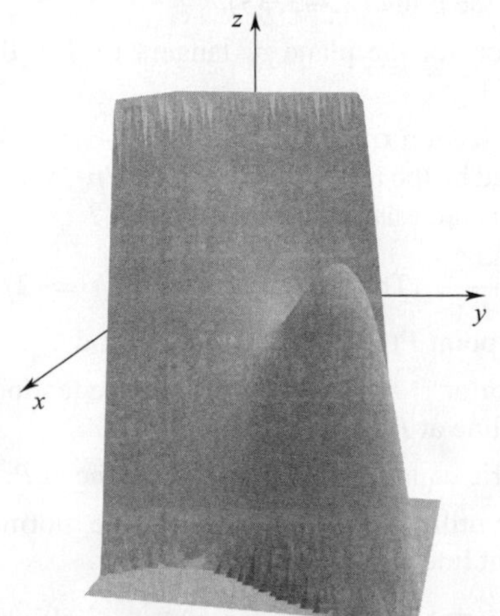

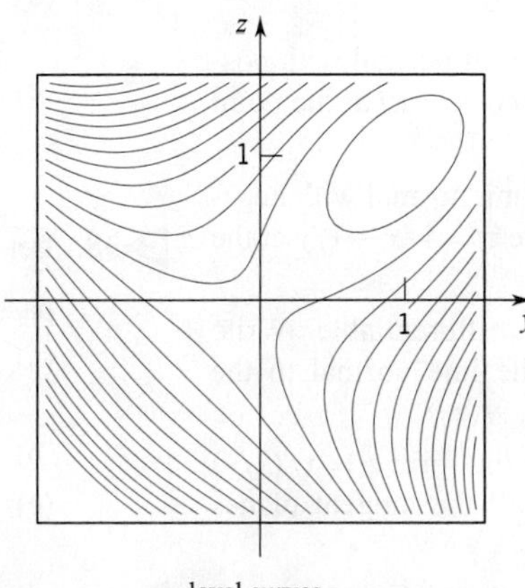

Figure 15.4.9

EXERCISES 15.4

Find a normal vector and a tangent vector at the point P. Write an equation for the tangent line and an equation for the normal line.

1. $x^2 + xy + y^2 = 3$; $P(-1, -1)$.
2. $(y - x)^2 = 2x$; $P(2, 4)$.
3. $(x^2 + y^2)^2 = 9(x^2 - y^2)$; $P(\sqrt{2}, 1)$.
4. $x^3 + y^3 = 9$; $P(1, 2)$.
5. $xy^2 - 2x^2 + y + 5x = 6$; $P(4, 2)$.
6. $x^5 + y^5 = 2x^3$; $P(1, 1)$.
7. $2x^3 - x^2y^2 = 3x - y - 7$; $P(1, -2)$.
8. $x^3 + y^2 + 2x = 6$; $P(-1, 3)$.
9. Find the slope of the curve $x^2y = a^2(a - y)$ at the point $(0, a)$.

Find an equation for the tangent plane and scalar parametric equations for the normal line at the point P.

10. $z = (x^2 + y^2)^2$; $P(1, 1, 4)$.
11. $x^3 + y^3 = 3xyz$; $P(1, 2, \frac{3}{2})$.
12. $xy^2 + 2z^2 = 12$; $P(1, 2, 2)$.
13. $z = axy$; $P(1, 1/a, 1)$.
14. $\sqrt{x} + \sqrt{y} + \sqrt{z} = 4$; $P(1, 4, 1)$.
15. $z = \sin x + \sin y + \sin(x + y)$; $P(0, 0, 0)$.
16. $z = x^2 + xy + y^2 - 6x + 2$; $P(4, -2, -10)$.
17. $b^2c^2x^2 - a^2c^2y^2 - a^2b^2z^2 = a^2b^2c^2$; $P(x_0, y_0, z_0)$.
18. $z = \sin(x \cos y)$; $P(1, \frac{1}{2}\pi, 0)$.

Find the point(s) on the surface at which the tangent plane is horizontal.

19. $xy + a^3x^{-1} + b^3y^{-1} - z = 0$.
20. $z = 4x + 2y - x^2 + xy - y^2$.
21. $z = xy$.
22. $x + y + z + xy - x^2 - y^2 = 0$.
23. $z - 2x^2 - 2xy + y^2 + 5x - 3y + 2 = 0$.
24. (a) Find the *upper unit normal* (the unit normal with positive $\mathbf{k}$ component) for the surface $z = xy$ at the point $(1, 1, 1)$.
    (b) Find the *lower unit normal* (the unit normal with negative $\mathbf{k}$ component) for the surface $z = 1/x - 1/y$ at the point $(1, 1, 0)$.
25. Let $f = f(x, y, z)$ be continuously differentiable. Write equations in symmetric form for the line normal to the surface $f(x, y, z) = c$ at the point (x_0, y_0, z_0).
26. Show that in the case of a surface of the form $z = xf(x/y)$ with f continuously differentiable, all the tangent planes have a point in common.
27. Given that the surfaces $F(x, y, z) = 0$ and $G(x, y, z) = 0$ intersect at right angles in a curve γ, what condition must be satisfied by the partial derivatives of F and G on γ?
28. Show that, for all planes tangent to the surface $\sqrt{x} + \sqrt{y} + \sqrt{z} = \sqrt{a}$, the sum of the intercepts is the same.
29. Show that all pyramids formed by the coordinate planes and a plane tangent to the surface $xyz = a^3$ have the same volume. What is this volume?
30. Show that, for all planes tangent to the surface $x^{2/3} + y^{2/3} + z^{2/3} = a^{2/3}$, the sum of the squares of the intercepts is the same.
31. The curve $\mathbf{r}(t) = 2t\,\mathbf{i} + 3t^{-1}\,\mathbf{j} - 2t^2\,\mathbf{k}$ and the ellipsoid $x^2 + y^2 + 3z^2 = 25$ intersect at $(2, 3, -2)$. What is the angle of intersection?
32. Show that the curve $\mathbf{r}(t) = \frac{3}{2}(t^2 + 1)\,\mathbf{i} + (t^4 + 1)\,\mathbf{j} + t^3\mathbf{k}$ is perpendicular to the ellipsoid $x^2 + 2y^2 + 3z^2 = 20$ at the point $(3, 2, 1)$.
33. The surfaces $x^2y^2 + 2x + z^3 = 16$ and $3x^2 + y^2 - 2z = 9$ intersect in a curve that passes through the point $(2, 1, 2)$. What are the equations of the respective tangent planes for the two surfaces at this point?
34. Show that the sphere $x^2 + y^2 + z^2 - 8x - 8y - 6z + 24 = 0$ is tangent to the ellipsoid $x^2 + 3y^2 + 2z^2 = 9$ at the point $(2, 1, 1)$.
35. Show that the sphere $x^2 + y^2 + z^2 - 4y - 2z + 2 = 0$ is perpendicular to the paraboloid $3x^2 + 2y^2 - 2z = 1$ at the point $(1, 1, 2)$.
36. Show that the following surfaces are mutually perpendicular:
$$xy = az^2, \quad x^2 + y^2 + z^2 = b, \quad z^2 + 2x^2 = c(z^2 + 2y^2).$$
37. The surface $S: z = x^2 + 3y^2 + 2$ intersects the vertical plane $p: 3x + 4y + 6 = 0$ in a space curve C.
    (a) Let C_1 be the projection of C onto the xy-plane. Find an equation for C_1.
    (b) Find a parametrization $\mathbf{r}(t) = x(t)\,\mathbf{i} + y(t)\,\mathbf{j} + z(t)\,\mathbf{k}$ for C setting $x(t) = 4t - 2$.
    (c) Find a parametrization $\mathbf{R}(s) = \mathbf{R}_0 + s\mathbf{d}$ for the line l tangent to C at the point $(2, -3, 33)$.
    (d) Find an equation for the plane p_1 tangent to S at the point $(2, -3, 33)$.
    (e) Find a parametrization $\mathbf{r}(t) = x(t)\,\mathbf{i} + y(t)\,\mathbf{j} + z(t)\,\mathbf{k}$ for the line l' formed by the intersection of p with p_1, taking $x(t) = t$. What is the relation between l and l'?
38. (C) Let $f(x, y) = \dfrac{x^2 - 2y}{x^2 + y^2}$. The level curve $f(x, y) = 2/5$ passes through the point $P(2, 1)$. Use a CAS to find:
    (a) A normal vector at P and scalar parametric equations for the normal line at P.
    (b) Scalar parametric equations for the tangent line at P.
    (c) Use a graphing utility to draw the level curve, normal line, and tangent line together.
39. (C) Let $f(x, y, z) = x^2 + (y - 1)^2 + z^2$. The level surface $f(x, y, z) = 6$ passes through the point $P(1, 2, 2)$. Use a CAS to find:
    (a) A normal vector at P and scalar parametric equations for the normal line at P.

(b) The equation of the tangent plane at P.

(c) Use a graphing utility to draw the level surface, normal line, and tangent plane together.

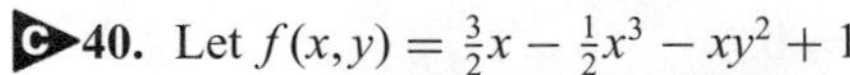

40. Let $f(x,y) = \frac{3}{2}x - \frac{1}{2}x^3 - xy^2 + 1$.

(a) Use a graphing utility to draw the surface $z = f(x,y)$. Choose a viewing window that reveals the features of the surface.

(b) Draw the level curves of the surface and use these to estimate the points where the surface may have a horizontal tangent plane.

(c) Calculate ∇f and find the points where $\nabla f(x,y) = \mathbf{0}$. Compare your answers with your estimates from part(b).

41. Repeat Exercise 40 with $f(x,y) = x^4 - y^4 - 2x^2 + 2y^2 + 2$.

42. Repeat Exercise 40 with $f(x,y) = 8xy\, e^{-(x^2+y^2)}$.

■ 15.5 LOCAL EXTREME VALUES

In Chapter 4 we discussed local extreme values for a function of one variable. Here we take up the same subject for functions of several variables. The ideas are very similar.

DEFINITION 15.5.1 LOCAL MAXIMUM AND LOCAL MINIMUM

Let f be a function of several variables and let $\mathbf{x}_0$ be an interior point of the domain:

f is said to take on a *local maximum* at $\mathbf{x}_0$ if

$$f(\mathbf{x}_0) \geq f(\mathbf{x}) \quad \text{for all } \mathbf{x} \text{ in some neighborhood of } \mathbf{x}_0;$$

f is said to take on a *local minimum* at $\mathbf{x}_0$ if

$$f(\mathbf{x}_0) \leq f(\mathbf{x}) \quad \text{for all } \mathbf{x} \text{ in some neighborhood of } \mathbf{x}_0.$$

As in the one-variable case, the local maxima and local minima together comprise the *local extreme values*.

In the one-variable case we know that if f takes on a local extreme value at x_0, then

$$\text{either} \quad f'(x_0) = 0 \quad \text{or} \quad f'(x_0) \text{ does not exist.}$$

We have a similar result for functions of several variables.

THEOREM 15.5.2

If f takes on a local extreme value at $\mathbf{x}_0$, then

$$\text{either} \qquad \nabla f(\mathbf{x}_0) = \mathbf{0} \quad \text{or} \quad \nabla f(\mathbf{x}_0) \text{ does not exist.}$$

PROOF We assume that f takes on a local extreme value at $\mathbf{x}_0$ and that f is differentiable at $\mathbf{x}_0$ [namely, that $\nabla f(\mathbf{x}_0)$ exists]. We need to show that $\nabla f(\mathbf{x}_0) = \mathbf{0}$. For simplicity we set $\mathbf{x}_0 = (x_0, y_0)$. The three-variable case is similar.

Since f takes on a local extreme value at (x_0, y_0), the function $g(x) = f(x, y_0)$ takes on a local a extreme value at x_0. Since f is differentiable at (x_0, y_0), g is differentiable at x_0 and therefore

$$g'(x_0) = \frac{\partial f}{\partial x}(x_0, y_0) = 0.$$

Similarly, the function $h(y)=f(x_0,y)$ takes on extreme value at y_0 and, being differentiable there, satisfies the relation

$$h'(y_0) = \frac{\partial f}{\partial y}(x_0, y_0) = 0.$$

The gradient is $\mathbf{0}$ since both partials are 0. □

Interior points of the domain at which the gradient is zero or the gradient does not exist are called *critical points*. By Theorem 15.5.2 these are the only points that can give rise to local extreme values.

Although the ideas introduced so far are completely general, their application to functions of more than two variables is generally laborious. We restrict ourselves mostly to functions of two variables. Not only are the computations less formidable, but also we can make use of our geometric intuition.

Two Variables

We suppose for the moment that $f = f(x,y)$ is defined on an open set and is continuously differentiable there. The graph of f is a surface

$$z = f(x,y).$$

Where f takes on a local maximum, the surface has a local high point. Where f takes on a local minimum, the surface has a local low point. Where f has either a local maximum or a local minimum, the gradient is $\mathbf{0}$ and therefore the tangent plane is horizontal. See Figure 15.5.1.

A zero gradient signals the possibility of a local extreme value; it does not guarantee it. For example, in the case of the saddle-shaped surface of Figure 15.5.2, there is a horizontal tangent plane at the origin and therefore the gradient is zero there, yet the origin gives neither a local maximum nor a local minimum.

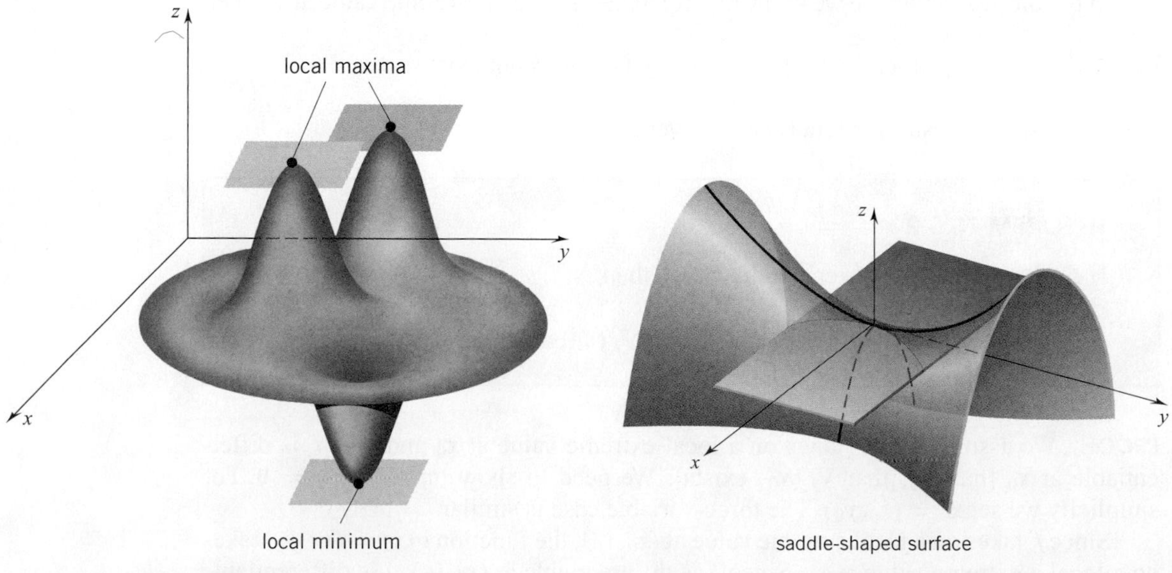

Figure 15.5.1

Figure 15.5.2

Critical points at which the gradient is zero are called *stationary points*. The stationary points that do not give rise to local extreme values are called *saddle points*.

Below we test some differentiable functions for extreme values. In each case, our first step is to find the stationary points.

Example 1 For the function $f(x,y) = 2x^2 + y^2 - xy - 7y$,

we have $$\nabla f(x,y) = (4x - y)\,\mathbf{i} + (2y - x - 7)\,\mathbf{j}.$$

To find the stationary points, we set $\nabla f(x,y) = \mathbf{0}$. This gives

$$4x - y = 0 \quad \text{and} \quad 2y - x - 7 = 0.$$

The only simultaneous solution to these equations is $x = 1, y = 4$. The point $(1,4)$ is therefore the only stationary point.

We now compare the value of f at $(1,4)$ with the values of f at nearby points $(1+h, 4+k)$:

$$\begin{aligned} f(1,4) &= 2 + 16 - 4 - 28 = -14, \\ f(1+h, 4+k) &= 2(1+h)^2 + (4+k)^2 - (1+h)(4+k) - 7(4+k) \\ &= 2 + 4h + 2h^2 + 16 + 8k + k^2 - 4 - 4h - k - hk - 28 - 7k \\ &= 2h^2 + k^2 - hk - 14. \end{aligned}$$

The difference

$$\begin{aligned} f(1+h, 4+k) - f(1,4) &= 2h^2 + k^2 - hk \\ &= h^2 + \left(h^2 - hk + k^2\right) \\ &\geq h^2 + \left(h^2 - 2|h||k| + k^2\right) \\ &= h^2 + (|h| - |k|)^2 \geq 0. \end{aligned}$$

Thus, $f(1+h, 4+k) \geq f(1,4)$ for all small h and k (in fact for all real h and k). †

It follows that f takes on local minimum at $(1,4)$. This local minimum is -14. □

Example 2 In the case of $f(x,y) = y^2 - xy + 2x + y + 1$

we have $$\nabla f(x,y) = (2 - y)\,\mathbf{i} + (2y - x + 1)\,\mathbf{j}.$$

The gradient is $\mathbf{0}$ where

$$2 - y = 0 \quad \text{and} \quad 2y - x + 1 = 0.$$

The only simultaneous solution to these equations is $x = 5, y = 2$. The point $(5,2)$ is the only stationary point.

We now compare the value of f at $(5,2)$ with the values of f at nearby points $(5+h, 2+k)$:

$$\begin{aligned} f(5,2) &= 4 - 10 + 10 + 2 + 1 = 7, \\ f(5+h, 2+k) &= (2+k)^2 - (5+h)(2+k) + 2(5+h) + (2+k) + 1 \\ &= 4 + 4k + k^2 - 10 - 2h - 5k - hk + 10 + 2h + 2 + k + 1 \\ &= k^2 - hk + 7. \end{aligned}$$

---

† Another way to see that $2h^2 + k^2 - hk$ is nonnegative is to complete the square:

$$2h^2 + k^2 - hk = \tfrac{1}{4}h^2 - hk + k^2 + \tfrac{7}{4}h^2 = (\tfrac{1}{2}h - k)^2 + \tfrac{7}{4}h^2 \geq 0.$$

The difference

$$d = f(5+h, 2+k) - f(5,2) = k^2 - hk = k(k-h)$$

does not keep a constant sign for small h and k. See Figure 15.5.3. Therefore, $(5, 2)$ is a saddle point. ❑

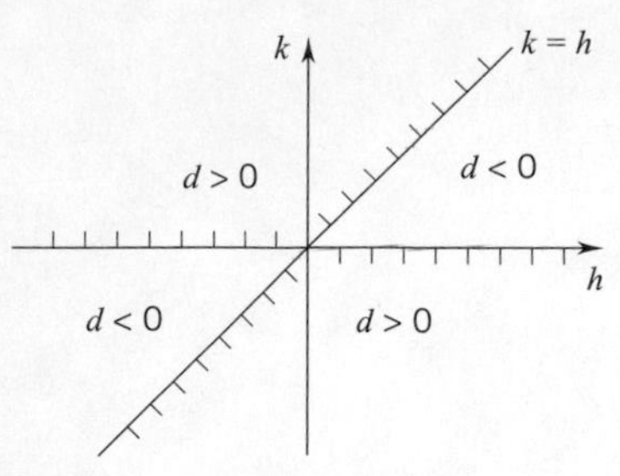

Figure 15.5.3

Example 3 The function $f(x,y) = 1 + \sqrt{x^2+y^2}$ is everywhere defined and everywhere continuous. The graph is the upper nappe of a right circular cone. (See Figure 15.5.4). The number $f(0,0) = 1$ is obviously a local minimum.

Since the partials

$$\frac{\partial f}{\partial x} = \frac{x}{\sqrt{x^2+y^2}}, \qquad \frac{\partial f}{\partial y} = \frac{y}{\sqrt{x^2+y^2}}$$

are not defined at $(0, 0)$, the gradient is not defined at $(0, 0)$ (Theorem 15.2.5). The point $(0, 0)$ is thus a critical point, but not a stationary point. At $(0, 0, 1)$ the surface comes to a sharp point and there is no tangent plane. ❑

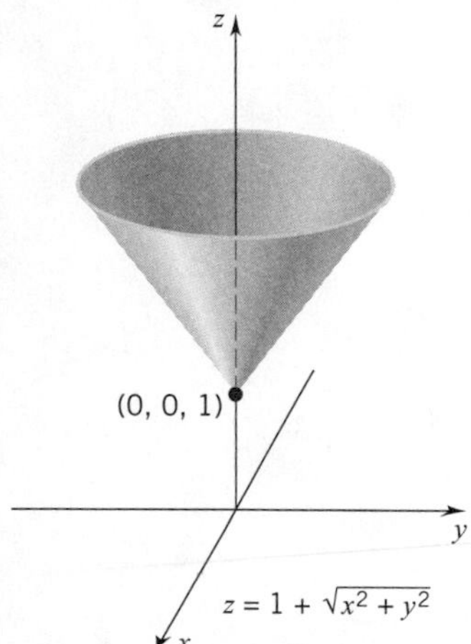

Figure 15.5.4

Second-Partials Test

Suppose that g is a function of one variable and $g'(x_0) = 0$. Then, according to the second-derivative test (Theorem 4.3.5), g takes on

$$\text{a local minimum at } x_0 \quad \text{if} \quad g''(x_0) > 0,$$
$$\text{a local maximum at } x_0 \quad \text{if} \quad g''(x_0) < 0.$$

We have a similar test for functions of two variables. As one might expect, the test is somewhat more complicated to state and definitely more difficult to prove. We will omit the proof. †

THEOREM 15.5.3 THE SECOND-PARTIALS TEST

Suppose that f has continuous second-order partial derivatives in a neighborhood of (x_0, y_0) and that $\nabla f(x_0, y_0) = \mathbf{0}$. Set

$$A = \frac{\partial^2 f}{\partial x^2}(x_0, y_0), \quad B = \frac{\partial^2 f}{\partial y \partial x}(x_0, y_0), \quad C = \frac{\partial^2 f}{\partial y^2}(x_0, y_0)$$

and form the *discriminant* $D = AC - B^2$

1. If $D < 0$, then (x_0, y_0) is a saddle point.
2. If $D > 0$, then f takes on

$$\text{a local minimum at } (x_0, y_0) \quad \text{if } A > 0,$$
$$\text{a local maximum at } (x_0, y_0) \quad \text{if } A < 0.$$

The test is geometrically evident for functions of the form

$$f(x,y) = \tfrac{1}{2}ax^2 + \tfrac{1}{2}cy^2. \qquad (a \neq 0, c \neq 0)$$

† You can find a proof in most texts on advanced calculus.

The graph of such a function is a paraboloid:

$$z = \tfrac{1}{2}ax^2 + \tfrac{1}{2}cy^2.$$

The gradient is $\mathbf{0}$ at the origin $(0, 0)$. Moreover

$$A = \frac{\partial^2 f}{\partial x^2}(0,0) = a, \quad B = \frac{\partial^2 f}{\partial y \partial x}(0,0) = 0, \quad C = \frac{\partial^2 f}{\partial y^2}(0,0) = c.$$

and $D = AC - B^2 = ac$. If $D < 0$, then a and c have opposite signs and we have a saddle point. (Figure 15.5.5).

Suppose now that $D > 0$. If $a > 0$, then $c > 0$ and the surface has a minimum point; if $a < 0$, then $c < 0$ and the surface has a maximum point.

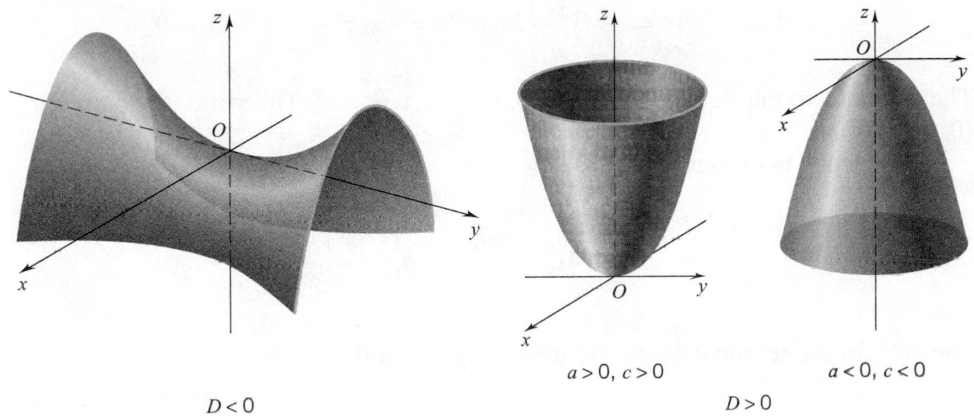

Figure 15.5.5

In the examples that follow we apply the second-partials test to a variety of functions.

Example 4 In Example 1, we saw that the point $(1, 4)$ is the only stationary point of the function

$$f(x, y) = 2x^2 + y^2 - xy - 7y.$$

The first partials of f are

$$\frac{\partial f}{\partial x} = 4x - y \quad \text{and} \quad \frac{\partial f}{\partial y} = 2y - x - 7,$$

and the second partials are constant:

$$\frac{\partial^2 f}{\partial x^2} = 4, \quad \frac{\partial^2 f}{\partial y \partial x} = -1, \quad \frac{\partial^2 f}{\partial y^2} = 2.$$

Thus, $A = 4, B = -1, C = 2$, and $D = AC - B^2 = 7 > 0$. Since $A > 0$, it follows from the second-partials test that

$$f(1, 4) = 2 + 16 - 4 - 28 = -14$$

is a local minimum. □

Example 5 The function $f(x,y) = -\frac{1}{4}x^4 + \frac{2}{3}x^3 + 4xy - y^2$ has partial derivatives

$$\frac{\partial f}{\partial x} = -x^3 + 2x^2 + 4y, \qquad \frac{\partial f}{\partial y} = 4x - 2y.$$

Setting both partials equal to zero, we have

$$-x^3 + 2x^2 + 4y = 0, \quad 4x - 2y = 0.$$

We get $y = 2x$ from the second equation. Substituting $y = 2x$ into the first equation, we have

$$-x^3 + 2x^2 + 8x = -x(x^2 - 2x - 8) = -x(x-4)(x+2) = 0.$$

The solutions to this equation are $x = 0$, $x = 4$, $x = -2$. The stationary points are $(0,0)$, $(4,8)$, $(-2,-4)$.

The second partials are

$$\frac{\partial^2 f}{\partial x^2} = -3x^2 + 4x, \qquad \frac{\partial^2 f}{\partial x \partial y} = 4, \qquad \frac{\partial^2 f}{\partial y^2} = -2.$$

The data for the second partials test are given in the following table.

Point	A	B	C	D	Result
$(0,0)$	0	4	-2	-16	Saddle point
$(4,8)$	-32	4	-2	48	Loc. max.
$(-2,-4)$	-20	4	-2	24	Loc. max.

A graph of the surface and the level curves are shown in Figure 15.5.6. □

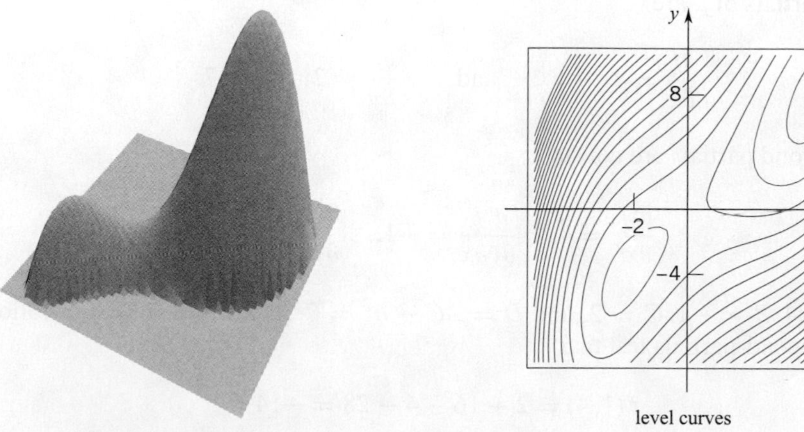

Figure 15.5.6

Example 6 For the function $f(x,y) = -xy\,e^{-(x^2+y^2)/2}$

we have $$\frac{\partial f}{\partial x} = -y\,e^{-(x^2+y^2)/2} + x^2y\,e^{-(x^2+y^2)/2} = y(x^2-1)\,e^{-(x^2+y^2)/2},$$

$$\frac{\partial f}{\partial y} = -x\,e^{-(x^2+y^2)/2} + xy^2\,e^{-(x^2+y^2)/2} = x(y^2-1)\,e^{-(x^2+y^2)/2},$$

and $$\nabla f(x,y) = e^{-(x^2+y^2)/2}[y(x^2-1)\,\mathbf{i} + x(y^2-1)\,\mathbf{j}].$$

Since $e^{-(x^2+y^2)/2} \neq 0$ for all (x,y), $\nabla f(x,y) = \mathbf{0}$ iff

$$y(1-x^2) = 0 \quad \text{and} \quad x(y^2-1) = 0.$$

The simultaneous solutions to these equations are $x = y = 0$; $x = 1, y = \pm 1$; $x = -1$, $y = \pm 1$. Thus, $(0,0), (1,1), (1,-1), (-1,1)$, and $(-1,-1)$ are the stationary points.

You can verify that the second partial derivatives of f are

$$\frac{\partial^2 f}{\partial x^2} = xy(3-x^2)\,e^{-(x^2+y^2)/2}, \qquad \frac{\partial^2 f}{\partial y^2} = xy(3-y^2)\,e^{-(x^2+y^2)/2},$$

and $$\frac{\partial^2 f}{\partial y \partial x} = (x^2-1)(1-y^2)e^{-(x^2+y^2)/2}.$$

The data for the second-partials test are recorded in the following table.

Point	A	B	C	D	Result
(0,0)	0	-1	0	-1	Saddle point
(1,1)	$2\,e^{-1}$	0	$2\,e^{-1}$	$4\,e^{-2}$	Loc. min.
(1,−1)	$-2\,e^{-1}$	0	$-2\,e^{-1}$	$4\,e^{-2}$	Loc. max.
(− 1,1)	$-2\,e^{-1}$	0	$-2\,e^{-1}$	$4\,e^{-2}$	Loc. max.
(− 1,−1)	$2\,e^{-1}$	0	$2\,e^{-1}$	$4\,e^{-2}$	Loc. min.

A computer-generated graph of this function is shown in Figure 15.5.7. □

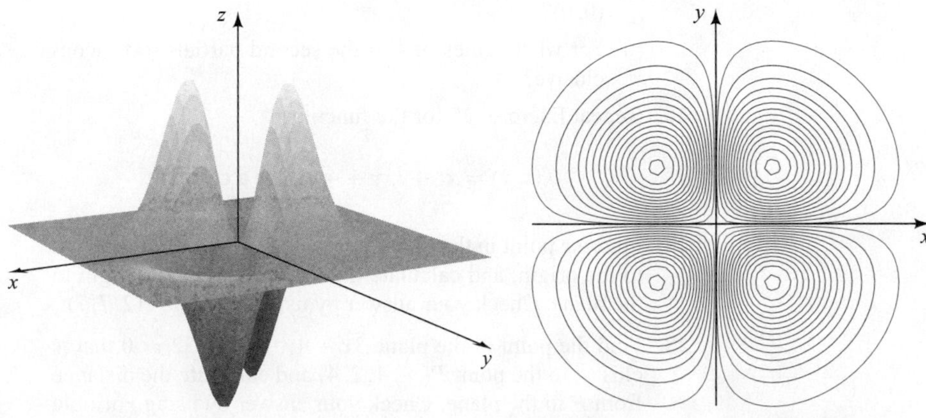

Figure 15.5.7

The second-derivative test for a function of one variable applies to points x_0 where $g'(x_0) = 0$ but $g''(x_0) \neq 0$. If $g''(x_0) = 0$, the second-derivative test provides no conclusive information. The second-partials test suffers from a similar limitation. It applies to points (x_0, y_0) where $\nabla f(x_0, y_0) = \mathbf{0}$ but $D \neq 0$. If $D = 0$, the second-partials test provides no information.

Consider, for example, the functions

$$f(x,y) = x^4 + y^4, \quad g(x,y) = -(x^4 + y^4), \quad h(x,y) = x^4 - y^4.$$

Each of these functions has zero gradient at the origin, and, as you can check, in each case $D = 0$. Yet,

(1) for f, $(0,0)$ gives a local minimum;
(2) for g, $(0,0)$ gives a local maximum;
(3) for h, $(0,0)$ is a saddle point.

Statements (1) and (2) are obvious. To confirm (3), note that $h(0,0) = 0$, but in every neighborhood of $(0,0)$ the function h takes on both positive and negative values:

$$h(x,0) > 0 \quad \text{for } x \neq 0, \quad \text{while} \quad h(0,y) < 0 \quad \text{for } y \neq 0. \quad \square$$

EXERCISES 15.5

Find the stationary points and use the method illustrated in Examples 1 and 2 to determine the local extreme values.

1. $f(x,y) = 2x - x^2 - y^2$.

2. $f(x,y) = 2x + 2y - x^2 + y^2 + 5$.

3. $f(x,y) = x^2 + xy + y^2 + 3x + 1$.

4. $f(x,y) = x^3 - 3x + y$.

Find the stationary points and the local extreme values.

5. $f(x,y) = x^2 + xy + y^2 - 6x + 2$

6. $f(x,y) = x^2 + 2xy + 3y^2 + 2x + 10y + 1$.

7. $f(x,y) = x^3 - 6xy + y^3$.

8. $f(x,y) = 3x^2 + xy - y^2 + 5x - 5y + 4$.

9. $f(x,y) = x^3 + y^2 - 6xy + 6x + 3y - 2$.

10. $f(x,y) = x^2 - 2xy + 2y^2 - 3x + 5y$.

11. $f(x,y) = x \sin y$.

12. $f(x,y) = y + x \sin y$.

13. $f(x,y) = (x + y)(xy + 1)$.

14. $f(x,y) = xy^{-1} - yx^{-1}$.

15. $f(x,y) = xy + x^{-1} + 8y^{-1}$.

16. $f(x,y) = x^2 - 2xy - y^2 + 1$.

17. $f(x,y) = xy + x^{-1} + y^{-1}$.

18. $f(x,y) = (x - y)(xy - 1)$.

19. $f(x,y) = \dfrac{-2x}{x^2 + y^2 + 1}$.

20. $f(x,y) = (x - 3) \ln xy$.

21. $f(x,y) = x^4 - 2x^2 + y^2 - 2$.

22. $f(x,y) = (x^2 + y^2)e^{x^2 - y^2}$.

23. $f(x,y) = \sin x \, \sin y, \quad 0 < x < 2\pi, \quad 0 < y < 2\pi$.

24. $f(x,y) = \cos x \, \cosh y, \quad -2\pi < x < 2\pi$.

25. Let $f(x,y) = x^2 + kxy + y^2$, k a constant.

(a) Show that f has a stationary point at $(0,0)$ no matter what value is assigned to k.

(b) For what values of k will f have a saddle point at $(0,0)$?

(c) For what values of k will f have a local minimum at $(0,0)$?

(d) For what values of k is the second-partials test inconclusive?

26. Repeat Exercise 25 for the function

$$f(x,y) = x^2 + kxy + 4y^2, \quad k \text{ a constant.}$$

27. Find the point in the plane $2x - y + 2z = 16$ that is closest to the origin, and calculate the distance from the origin to the plane. Check your answer by using Formula (12.7.7).

28. Find the point in the plane $3x - 4y + 2z + 32 = 0$ that is closest to the point $P(-1, 2, 4)$ and calculate the distance from P to the plane. Check your answer by using Formula (12.7.7).

29. Find the shortest distance from the point $(1, 2, 0)$ to the elliptic cone $z = \sqrt{x^2 + 2y^2}$. HINT: Minimize the square of the distance.

30. Find the maximum volume for a rectangular solid inscribed in the sphere.

$$x^2 + y^2 + z^2 = a^2.$$

If a continuous function of one variable has local maxima at $x = a$ and $x = b, a < b$, then it must have at least one local minimum at some number $c \in (a, b)$. The corresponding result holds with the roles of the local extrema interchanged. Exercises 31 and 32 illustrate that these properties do not hold in higher dimensions.

31. Let $f(x, y) = 4xy - x^4 - y^4 + 1$.

(a) Use a graphing utility to draw graphs of f and the level curves.

(b) Your graphs should show that f has local maxima at $(1, 1)$ and $(-1, -1)$, a saddle point at $(0, 0)$, and no local minima.

(c) Verify the conclusions of part (b) using the methods of this section.

32. Let $f(x, y) = x^4 - 2x^2 + y^2 + 1$.

(a) Use a graphing utility to draw graphs of f and the level curves.

(b) Your graphs should show that f has local minima at $(1, 0)$ and $(-1, 0)$, a saddle point at $(0, 0)$, and no local maxima.

(c) Verify the conclusions of part (b) using the methods of this section.

Use a graphing utility to draw graphs of f and the level curves. Locate the stationary points, if any, and at each stationary point state whether f has a local maximum, a local minimum or a saddle point.

33. $f(x, y) = 3xy - x^3 - y^3 + 2$.

34. $f(x, y) = (x^2 + 2y^2)e^{-(x^2+y^2)}$.

35. $f(x, y) = \dfrac{-2x}{x^2 + y^2 + 1}$.

36. $f(x, y) = \sin x + \sin y - \cos(x + y)$, $0 \le x \le 3\pi, 0 \le y \le 3\pi$.

■ 15.6 ABSOLUTE EXTREME VALUES

In this section we consider absolute maxima and minima. As in the preceding section, the ideas here are similar to those discussed in Chapter 4.

DEFINITION 15.6.1 ABSOLUTE MAXIMUM AND ABSOLUTE MINIMUM

Let f be a function of several variables with domain D:

f takes on an *absolute maximum* at $\mathbf{x}_0$ if

$$f(\mathbf{x}_0) \ge f(\mathbf{x}) \qquad \text{for all } \mathbf{x} \in D;$$

f takes on an *absolute minimum* at $\mathbf{x}_0$ if

$$f(\mathbf{x}_0) \le f(\mathbf{x}) \qquad \text{for all } \mathbf{x} \in D.$$

In Chapter 2 we saw that a function of one variable that is continuous on a bounded closed interval takes on both an absolute maximum and an absolute minimum on that interval (Theorem 2.6.2). Before stating the general theorem, we need to extend the notion of bounded set of real numbers to include sets in higher dimensions.

DEFINITION 15.6.2 BOUNDED SET

A set S (of the real line, the plane, or three-space) is *bounded* if there exists a positive number R such that

$$\|\mathbf{x}\| \le R \qquad \text{for all} \quad \mathbf{x} \in S.$$

Thus a set S in the plane or three-space is bounded iff it can be contained in a ball of radius R.

THEOREM 15.6.3 EXTREME-VALUE THEOREM

If f is continuous on a bounded closed set D, then f takes on an absolute maximum value and an absolute minimum value.

Two Variables

In the search for local extreme values, the critical points are interior points of the domain: the stationary points and the points at which the gradient does not exist. In the search for absolute extreme values, we must also test the boundary points. This usually requires special methods. One approach is to try to parametrize the boundary by some vector function $\mathbf{r} = \mathbf{r}(t)$ and then work with the function of one variable $f(\mathbf{r}(t))$. This is the approach we take in this section. A more sophisticated approach, one which generalizes to functions of three or more variables, is given in the next section.

The general procedure for finding the absolute extrema of a function f continuous on a bounded closed set D is an extension of the procedure given in Section 4.4, namely:

1. Find the critical points in the interior of D.
2. Find the extreme points on the boundary of D.
3. Evaluate f at the points found in Steps 1 and 2.
4. The largest of the numbers found in Step 3 is the absolute maximum value of f and the smallest is the absolute minimum.

Example 1 Find the absolute extreme values of the function

$$f(x, y) = x^2 + y^2 - 2x - 2y + 4$$

on the closed disc $D = \{(x, y) : x^2 + y^2 \leq 9\}$. (Figure 15.6.1)

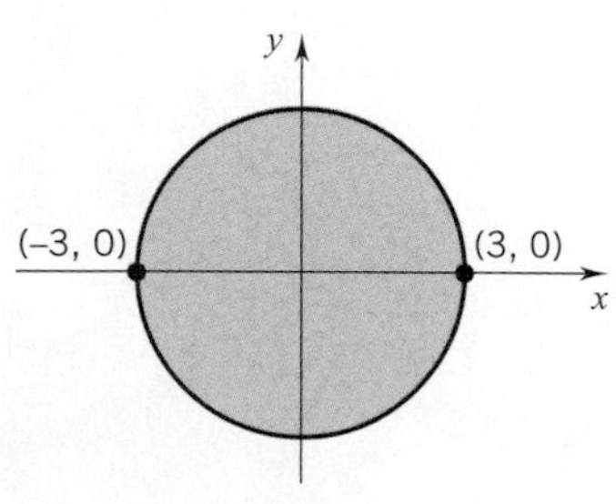

Figure 15.6.1

SOLUTION The region D is a bounded closed set and the function f, being continuous everywhere, is continuous on D. Therefore, we know that f takes on an absolute maximum and an absolute minimum on D.

First we find the critical points in the interior of D. The gradient of f,

$$\nabla f(x, y) = (2x - 2)\,\mathbf{i} + (2y - 2)\,\mathbf{j},$$

is defined everywhere. The gradient is $\mathbf{0}$ where

$$2x - 2 = 0 \quad \text{and} \quad 2y - 2 = 0.$$

The only simultaneous solution to these equations is $x = 1, y = 1$. The point $(1, 1)$ is the only stationary point of f and it is in the interior of D.

Now we look for extreme values on the boundary of D. The boundary can be parametrized by the equations

$$x = 3\cos t, \qquad y = 3\sin t, \qquad 0 \leq t \leq 2\pi.$$

The values of f on the boundary are given by the function

$$\begin{aligned} F(t) = f(\mathbf{r}(t)) &= 9\cos^2 t + 9\sin^2 t - 6\cos t - 6\sin t + 4 \\ &= 13 - 6\cos t - 6\sin t, \qquad 0 \leq t \leq 2\pi. \end{aligned}$$

Since F is a continuous function on a bounded closed interval, it has an absolute maximum and an absolute minimum. We find the absolute extreme values of F by the method of Section 4.4. To find the critical numbers, we differentiate:

$$F'(t) = 6\sin t - 6\cos t.$$

Setting $F'(t) = 0$, we get

$$\sin t = \cos t.$$

The solutions of this equation are $t = \pi/4$ and $t = 5\pi/4$.

The extreme values of f on the boundary are

$$F(0) = F(2\pi) = f(3,0) = 7,$$
$$F(\pi/4) = f(\tfrac{3}{2}\sqrt{2}, \tfrac{3}{2}\sqrt{2}) = 13 - 6\sqrt{2} \cong 4.51,$$
$$F(5\pi/4) = f(-\tfrac{3}{2}\sqrt{2}, -\tfrac{3}{2}\sqrt{2}) = 13 + 6\sqrt{2} \cong 21.49.$$

The value of f at the stationary point is $f(1,1) = 2$.

Thus, the absolute maximum of f on D is $13 + 6\sqrt{2}$ and the absolute minimum value of f is 2. □

Example 2 Find the absolute extreme values of the function

$$f(x,y) = 4xy - x^2 - y^2 - 6x$$

on the triangular region $D = \{(x,y) : 0 \leq x \leq 2, 0 \leq y \leq 3x\}$. (Figure 15.6.2)

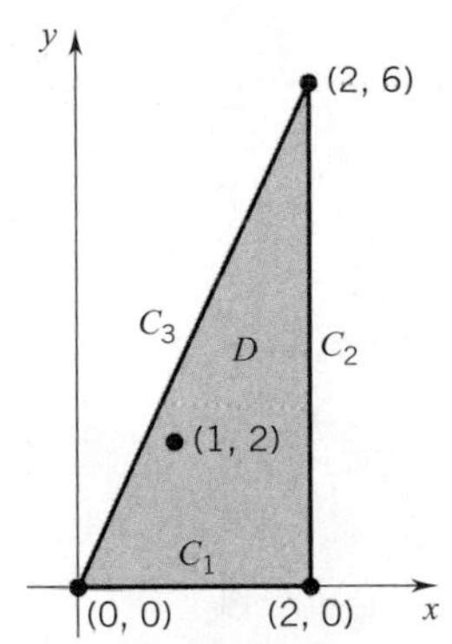

Figure 15.6.2

SOLUTION Since f is continuous and D is a bounded closed set, we know that f takes on an absolute maximum and an absolute minimum on D.

First we find the critical points of f in the interior of D. The gradient of f,

$$\nabla f = (4y - 2x - 6)\,\mathbf{i} + (4x - 2y)\,\mathbf{j},$$

is defined everywhere. The gradient is $\mathbf{0}$ iff

$$4y - 2x - 6 = 0 \quad \text{and} \quad 4x - 2y = 0.$$

Solving these equations simultaneously, we get $x = 1, y = 2$. The stationary point $(1,2)$ is in D.

Now we look for extreme values on the boundary by writing each side of the triangle in the form $\mathbf{r} = \mathbf{r}(t)$ and then analyzing $f(\mathbf{r}(t))$. With C_1, C_2, C_3 as in the figure, we have

$$C_1 : \mathbf{r}_1(t) = t\,\mathbf{i}, \qquad t \in [0,2];$$
$$C_2 : \mathbf{r}_2(t) = 2\,\mathbf{i} + t\,\mathbf{j}, \qquad t \in [0,6];$$
$$C_3 : \mathbf{r}_3(t) = t\,\mathbf{i} + 3t\,\mathbf{j}, \qquad t \in [0,2].$$

The values of f on these line segments are given by the functions

$$f_1(t) = f(\mathbf{r}_1(t)) = -t^2 - 6t, \qquad t \in [0,2];$$
$$f_2(t) = f(\mathbf{r}_2(t)) = -(t-4)^2, \qquad t \in [0,6];$$
$$f_3(t) = f(\mathbf{r}_3(t)) = 2t^2 - 6t, \qquad t \in [0,2].$$

As you can check, f_1 has no critical numbers in $(0, 2)$, f_2 has no critical numbers in $(0, 6)$, and f_3 has a critical number at $t = 3/2$. Evaluating these functions at the endpoints of their domains and at the critical number for f_3, we find that

$$\begin{aligned} f_1(0) &= f_3(0) = f(0,0) = 0, \\ f_1(2) &= f_2(0) = f(2,0) = -16, \\ f_2(6) &= f_3(2) = f(2,6) = -4, \\ f_3(\tfrac{3}{2}) &= f(\tfrac{3}{2}, \tfrac{9}{2}) = -\tfrac{9}{2}. \end{aligned}$$

The value of f at the stationary point is $f(1, 2) = -3$.

Thus the absolute maximum of f is 0 taken on at $(0, 0)$, and the absolute minimum is -16 taken on at $(2, 0)$. ☐

In some cases, physical or geometric considerations may allow us to conclude that an absolute maximum or an absolute minimum exists even if the function is not continuous, or the domain is not bounded or not closed.

Example 3 The rectangle $\{(x, y) : 0 \le x \le a, -b \le y \le b\}$ is a bounded closed subset of the plane. The function

$$f(x, y) = 1 + \sqrt{x^2 + y^2},$$

being everywhere continuous, is continuous on this rectangle. Thus we can be sure that f takes on both an absolute maximum and an absolute minimum on this set. The absolute maximum is taken on at the points $(a, -b)$ and (a, b), the points of the rectangle farthest away from the origin. (This should be clear from Figure 15.5.4.) The value at these points is $1 + \sqrt{a^2 + b^2}$. The absolute minimum is taken on at the origin $(0, 0)$. The value there is 1.

Now let's continue with the same function but apply it instead to the rectangle

$$\{(x, y) : 0 < x \le a, -b \le y \le b\}.$$

This rectangle is bounded but not closed. On this set f takes on an absolute maximum (the same maximum as before and at the same points), but it takes on no absolute minimum (the origin is not in the set).

Finally, on the entire plane (which is closed but not bounded), f takes on an absolute minimum (1 at the origin) but no maximum. ☐

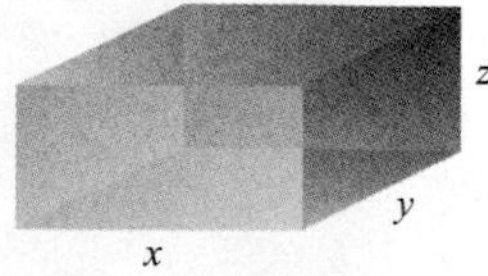

Figure 15.6.3

Example 4 A rectangular box without a top is to have a volume of 12 cubic feet. Find the dimensions of the box that will have minimum surface area.

SOLUTION Let the dimensions of the box be length x, width y, height z (Figure 15.6.3). Then the surface area is given by the expression

$$S = xy + 2xz + 2yz.$$

We can write S as a function of x and y by using the fact that the volume of the box is given by $xyz = 12$, so that

$$z = \frac{12}{xy}.$$

Thus, the expression for S can be written

$$S(x, y) = xy + \frac{24}{x} + \frac{24}{y}, \qquad \text{where} \quad x > 0, \ y > 0.$$

Note that the domain $D = \{(x, y) : x > 0, y > 0\}$ of S is open and unbounded.

The gradient of S,

$$\nabla S = \left(y - \frac{24}{x^2}\right)\mathbf{i} + \left(x - \frac{24}{y^2}\right)\mathbf{j},$$

is $\mathbf{0}$ iff

$$y - \frac{24}{x^2} = 0 \quad \text{and} \quad x - \frac{24}{y^2} = 0.$$

Solving these equations simultaneously, we get $x = y = 2\sqrt[3]{3}$. The point $(2\sqrt[3]{3}, 2\sqrt[3]{3})$ is the only stationary point.

The second partials are

$$\frac{\partial^2 S}{\partial x^2} = \frac{48}{x^3}, \quad \frac{\partial^2 S}{\partial x \partial y} = 1, \quad \frac{\partial^2 S}{\partial y^2} = \frac{48}{y^3}.$$

Evaluating these partials at $(2\sqrt[3]{3}, 2\sqrt[3]{3})$, we get

$$A = C = \tfrac{48}{24} = 2 \quad \text{and} \quad B = 1.$$

Thus $D = AC - B^2 = 4 - 1 = 3 > 0$. Since $A = 2 > 0$, it follows from the second-partials test that S has a local minimum at $(2\sqrt[3]{3}, 2\sqrt[3]{3})$. We can conclude that S actually has an absolute minimum at this point because $S(x, y) \to \infty$ as either $x \to 0$ or $y \to 0$, and $S(x, y) \to \infty$ as $x \to \infty$ or $y \to \infty$. Finally, the relation

$$xyz = 12$$

implies that $z = \sqrt[3]{3}$ when $x = y = 2\sqrt[3]{3}$, and so the dimensions that yield the minimum surface area are

$$\text{length} = 2\sqrt[3]{3}, \quad \text{width} = 2\sqrt[3]{3}, \quad \text{and} \quad \text{height} = \sqrt[3]{3}. \quad \square$$

EXERCISES 15.6

Find the absolute extreme values taken on by f on the set D.

1. $f(x, y) = 2x^2 + y^2 - 4x - 2y + 2, \quad D = \{(x, y) : 0 \le x \le 2, 0 \le y \le 2x\}$.

2. $f(x, y) = 2 - 3x + 2y$, D the closed region enclosed by the triangle with vertices $(0, 0), (4, 0), (0, 6)$.

3. $f(x, y) = x^2 + xy + y^2 - 6x - 1, \quad D = \{(x, y) : 0 \le x \le 5, -3 \le y \le 0\}$.

4. $f(x, y) = x^2 + 2xy + 3y^2, \quad D = \{(x, y) : |x| \le 2, |y| \le 2\}$.

5. $f(x, y) = x^2 + y^2 + 3xy + 2, \quad D = \{(x, y) : x^2 + y^2 \le 4\}$.

6. $f(x, y) = y(x - 3), \quad D = \{(x, y) : x^2 + y^2 \le 9\}$.

7. $f(x, y) = (x - 1)^2 + (y - 1)^2, \quad D = \{(x, y) : x^2 + y^2 \le 4\}$.

8. $f(x, y) = 3 + x - y + xy$, D the closed region enclosed by $y = x^2$ and $y = 4$.

9. $f(x, y) = \dfrac{-2x}{x^2 + y^2 + 1}, \quad D = \{(x, y) : |x| \le 2, |y| \le 2\}$.

10. $f(x, y) = \dfrac{-2x}{x^2 + y^2 + 1}, \quad D = \{(x, y) : 0 \le x \le 2, -x \le y \le x\}$.

11. $f(x, y) = (4x - 2x^2)\cos y, \quad D = \{(x, y) : 0 \le x \le 2, -\pi/4 \le y \le \pi/4\}$.

12. $f(x, y) = (x - 3)^2 + y^2, \quad D = \{(x, y) : 0 \le x \le 4, x^2 \le y \le 4x\}$.

13. $f(x, y) = x^3 - 3xy - y^3, \quad D = \{(x, y) : -2 \le x \le 2, x \le y \le 2\}$.

14. $f(x, y) = (x - 4)^2 + y^2, \quad D = \{(x, y) : 0 \le x \le 2, x^3 \le y \le 4x\}$.

15. $f(x, y) = \dfrac{-2y}{x^2 + y^2 + 1}, \quad D = \{(x, y) : x^2 + y^2 \le 4\}$.

16. $f(x, y) = x^2 + 4y^2 + x - 2y$, D the closed region enclosed by the ellipse $\frac{1}{4}x^2 + y^2 = 1$.

17. $f(x, y) = x^2 - 2xy + y^2, \quad D = \{(x, y) : 0 \le x \le 6, 0 \le y \le 12 - 2x\}$.

18. $f(x, y) = \dfrac{1}{\sqrt{x^2 + y^2}}, \quad D = \{(x, y) : 1 \le x \le 3, 1 \le y \le 4\}$.

19. Find positive numbers x, y, z such that $x + y + z = 18$ and xyz is a maximum. HINT: Maximize the function $f(x, y) = 18xy - x^2y - xy^2$ on the triangular region bounded by the positive x- and y-axes and the line $x + y = 18$.

20. Find positive numbers x, y, z such that $x + y + z = 30$ and xyz^2 is a maximum. HINT: Maximize the function $f(y, z) = 30yz^2 - y^2z^2 - yz^3$ on the triangular region bounded by the positive y- and z-axes and the line $y+z = 30$.

21. Find the maximum volume for a rectangular solid in the first octant ($x \geq 0, y \geq 0, z \geq 0$) with one vertex at the origin and opposite vertex on the plane $x + y + z = 1$.

22. Find the maximum volume for a rectangular solid in the first octant with one vertex at the origin and the opposite vertex on the plane

$$\frac{x}{a} + \frac{y}{b} + \frac{z}{c} = 1.$$

23. Define $f(x, y) = \frac{1}{4}x^2 - \frac{1}{9}y^2$ on the closed unit disc. Find

(a) the stationary points,

(b) the local extreme values,

(c) the absolute extreme values.

24. Suppose that the material to be used to construct the box of Example 4 costs \$3 per square foot for the sides and \$4 per square foot for the bottom. What dimensions will yield the minimum cost?

25. Describe the behavior of the function at the origin.

(a) $f(x, y) = x^2 - y^3$.

(b) $f(x, y) = 2\cos(x + y) + e^{xy}$.

26. Let n be an integer greater than 2 and set

$$f(x, y) = ax^n + cy^n, \quad \text{taking } ac \neq 0.$$

(a) Find the stationary points.

(b) Find the discriminant at each stationary point.

(c) Find the local and absolute extreme values given that

(i) $a > 0, c > 0$.

(ii) $a < 0, c < 0$.

(iii) $a > 0, c < 0$.

27. Find the point with the property that the sum of the squares of its distances from $P_1(x_1, y_1), P_2(x_2, y_2), P_3(x_3, y_3)$ is an absolute minimum.

28. Given that $0 < a < b$, find the absolute maximum value taken on by the function

$$f(x, y) = \frac{xy}{(a + x)(x + y)(b + y)}$$

on the open square $\{(x, y) : a < x < b, a < y < b\}$.

29. A pentagon is composed of a rectangle surmounted by an isosceles triangle (see the figure). Given that the perimeter of the pentagon has a fixed value P, find the dimensions for maximum area.

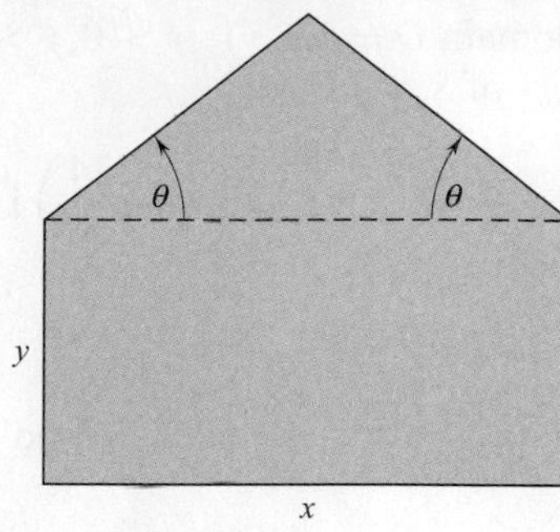

30. A bakery produces two types of bread, one at a cost of 50 cents per loaf, the other at a cost of 60 cents per loaf. Assume that if the first bread is sold at x cents a loaf and the second at y cents a loaf, then the number of loaves that can be sold each week is given by the formulas

$$N_1 = 250(y - x), \quad N_2 = 32{,}000 + 250(x - 2y).$$

Determine x and y for maximum profit.

31. Find the distance between the lines $x = \frac{1}{2}y = \frac{1}{3}z$ and $x = y - 2 = z$.

32. Find the absolute maximum value of the function

$$f(x, y) = \frac{(ax + by + c)^2}{x^2 + y^2 + 1}.$$

33. Find the dimensions of the most economical open-top rectangular crate 96 cubic meters in volume given that the base costs 30 cents per square meter and the sides cost 10 cents per square meter.

34. Let $f(x, y) = ax^2 + bxy + cy^2$, taking $abc \neq 0$.

(a) Find the discriminant D.

(b) Find the stationary points and local extreme values if $D \neq 0$.

(c) Suppose that $D = 0$. Find the stationary points and the local and absolute extreme values given that

(i) $a > 0, c > 0$.

(ii) $a < 0, c < 0$.

35. Show that a closed rectangular box of maximum volume having a prescribed surface area S is a cube.

36. If an open rectangular box has a prescribed surface area S, what dimensions yield the maximum volume?

37. (*The method of least squares*) In this exercise we illustrate an important method of fitting a curve to a collection of points. Consider three points

$$(x_1, y_1) = (0, 2), \quad (x_2, y_2) = (1, -5), \quad (x_3, y_3) = (2, 4).$$

(a) Find the line $y = mx + b$ that minimizes the sum of the squares of the vertical distances $d_i = |y_i - (mx_i + b)|$ from these points to the line.

(b) Find the parabola $y = \alpha x^2 + \beta$ that minimizes the sum of the squares of the vertical distances $d_i = |y_i - (\alpha x_i^2 + \beta)|$ from the points to the parabola.

38. Exercise 37 taking

$$(x_1, y_1) = (-1, 2),\ (x_2, y_2) = (0, -1),\ (x_3, y_3) = (1, 1).$$

39. According to U.S. Postal Service regulations, the length plus the girth (the perimeter of a cross section) of a package cannot exceed 108 inches. (See the figure.)

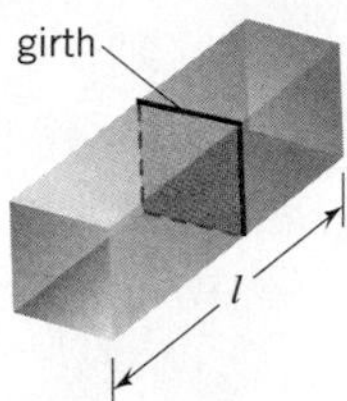

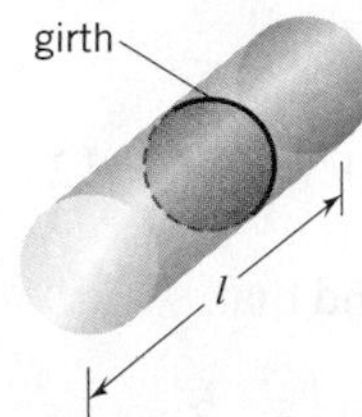

(a) Find the dimensions of the rectangular box of maximum volume that is acceptable for mailing.

(b) Find the dimensions of the cylindrical tube of maximum volume that is acceptable for mailing.

40. A petrochemical company is designing a cylindrical tank with hemispherical ends to be used in transporting its products. If the volume of the tank is to be 10,000 cubic meters, what dimensions should be used to minimize the amount of metal required?

41. A 10-foot section of gutter is to be made from a 12-inch wide strip of metal by folding up strips of length x on each side so that they make an angle θ with the bottom of the gutter. (See the figure.) Determine values for x and θ that will maximize the carrying capacity of the gutter.

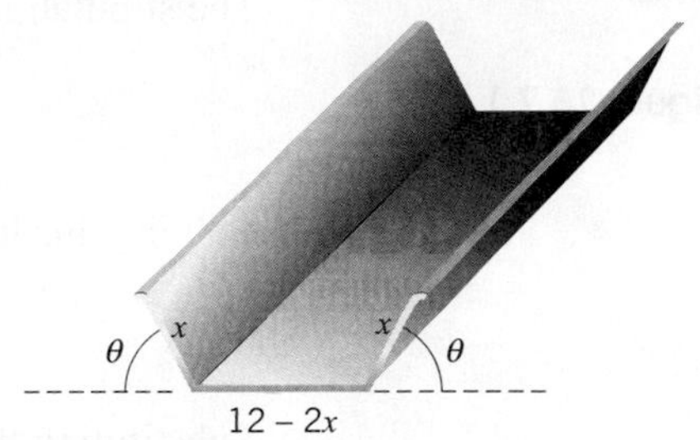

42. Find the volume of the largest rectangular box with edges parallel to the coordinate axes that can be inscribed in the ellipsoid

$$\frac{x^2}{a^2} + \frac{y^2}{b^2} + \frac{z^2}{c^2} = 1.$$

■ 15.7 MAXIMA AND MINIMA WITH SIDE CONDITIONS

When we ask for the distance from a point $P(x_0, y_0)$ to a line $l\colon Ax + By + C = 0$, we are asking for the minimum value of

$$f(x, y) = \sqrt{(x - x_0)^2 + (y - y_0)^2}$$

with (x, y) subject to the side condition† $Ax + By + C = 0$. When we ask for the distance from a point $P(x_0, y_0, z_0)$ to a plane $p : Ax + By + Cz + D = 0$, we are asking for a minimum value of

$$f(x, y, z) = \sqrt{(x - x_0)^2 + (y - y_0)^2 + (z - z_0)^2}$$

with (x, y, z) subject to the side condition $Ax + By + Cz + D = 0$.

We have already treated these particular problems by special techniques. Our interest here is to present techniques for handling problems of this sort in general. In the two-variable case, the problems will take the form of maximizing (or minimizing) some expression $f(x, y)$ subject to a side condition $g(x, y) = 0$. In the three-variable case, we will seek to maximize (or minimize) some expression $f(x, y, z)$ subject to a side condition $g(x, y, z) = 0$. We begin with two simple examples.

Example 1 Maximize the product xy subject to the side condition $x + y - 1 = 0$.

SOLUTION The condition $x + y - 1 = 0$ gives $y = 1 - x$. The original problem can therefore be solved simply by maximizing the product $h(x) = x(1 - x)$. The derivative $h'(x) = 1 - 2x$ is 0 only at $x = \frac{1}{2}$. Since $h''(x) = -2 < 0$, we know from the second-derivative test that $h(\frac{1}{2}) = \frac{1}{2}(1 - \frac{1}{2}) = \frac{1}{4}$ is the desired maximum. □

† Side conditions are often called *constraints*.

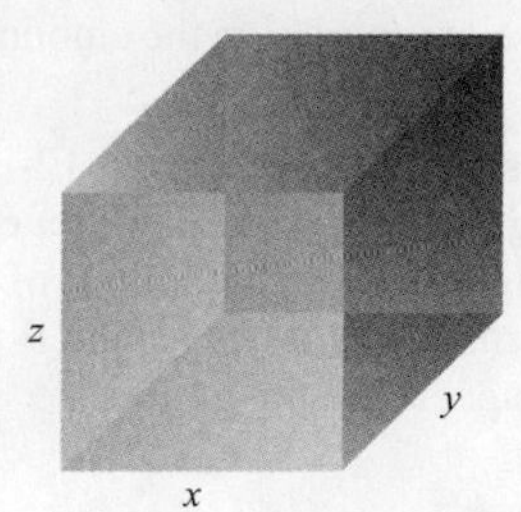

Figure 15.7.1

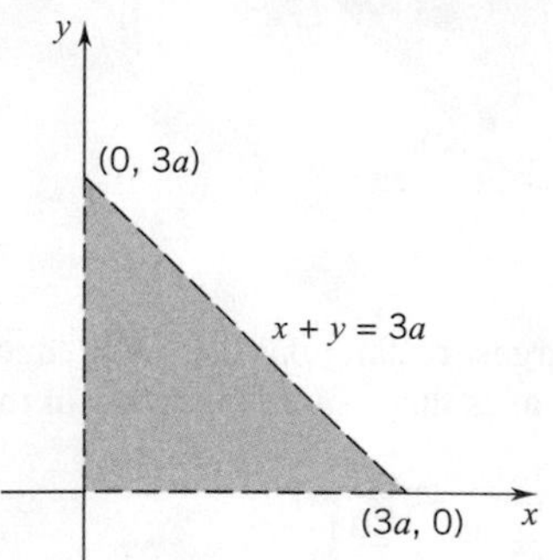

Figure 15.7.2

Example 2 Find the maximum volume of a rectangular solid given that the sum of the lengths of its edges is 12a.

SOLUTION We denote the dimensions of the solid by x, y, z. (Figure 15.7.1.) The volume is given by

$$V = xyz.$$

The stipulation on the edges requires that

$$4(x + y + z) = 12a.$$

Solving this last equation for z, we find that

$$z = 3a - (x + y).$$

Substituting this expression for z in the volume formula, we have

$$V = xy[3a - (x + y)].$$

Since $V = 0$ on the sides $(x = 0, y = 0, z = 3a - (x + y) = 0)$ of the triangle shown in Figure 15.7.2, we can conclude that the maximum value of V must be attained in the interior of the triangle.

The first partials are:

$$\frac{\partial V}{\partial x} = xy(-1) + y[3a - (x + y)] = 3ay - 2xy - y^2,$$
$$\frac{\partial V}{\partial y} = xy(-1) + x[3a - (x + y)] = 3ax - x^2 - 2xy.$$

Setting both partials equal to zero, we have

$$(3a - 2x - y)y = 0 \qquad \text{and} \qquad (3a - x - 2y)x = 0.$$

Since x and y are assumed positive, we can divide by x and y and get

$$3a - 2x - y = 0 \qquad \text{and} \qquad 3a - x - 2y = 0.$$

Solving these equations simultaneously, we find that $x = y = a$. The point (a, a), which does lie within the triangle, is the only stationary point. The value of V at that point is a^3. The conditions of the problem make it clear that this is a maximum. (If you are skeptical, you can confirm this by appealing to the second-partials test.) □

The last two problems were easy. They were easy in part because the side conditions were such that we could solve for one of the variables in terms of the other (s). In general this is not possible and a more sophisticated approach is required.

The Method of Lagrange

We begin with what looks like a detour. To avoid having to make separate statements for the two-and three-variable cases, we will use vector notation.

Throughout the discussion f will be a function of two or three variables which is continuously differentiable on some open set U. We take

$$C: \mathbf{r} = \mathbf{r}(t), \quad t \in I$$

to be a curve that lies entirely in U and has at each point a nonzero tangent vector $\mathbf{r}'(t)$. The basic result is this:

(15.7.1) if $\mathbf{x}_0$ maximizes (or minimizes) $f(\mathbf{x})$ on C, then $\nabla f(\mathbf{x}_0)$ is perpendicular to C at $\mathbf{x}_0$.

PROOF Assume that $\mathbf{x}_0$ maximizes (or minimizes) $f(\mathbf{x})$ on C. Choose t_0, so that

$$\mathbf{r}(t_0) = \mathbf{x}_0.$$

The composition $f(\mathbf{r}(t))$ has a maximum (or minimum) at t_0. Consequently, its derivative,

$$\frac{d}{dt}[f(\mathbf{r}(t))] = \nabla f(\mathbf{r}(t)) \cdot \mathbf{r}'(t),$$

must be zero at t_0 :

$$0 = \nabla f(\mathbf{r}(t_0)) \cdot \mathbf{r}'(t_0) = \nabla f(\mathbf{x}_0) \cdot \mathbf{r}'(t_0).$$

This shows that

$$\nabla f(\mathbf{x}_0) \perp \mathbf{r}'(t_0).$$

Since $\mathbf{r}'(t_0)$ is tangent to C at $\mathbf{x}_0$, $\nabla f(\mathbf{x}_0)$ is perpendicular to C at $\mathbf{x}_0$. ❑

We are now ready for side-condition problems. Suppose that g is a continuously differentiable function of two or three variables defined on a subset of the domain of f. Lagrange made the following observation: †

(15.7.2) if $\mathbf{x}_0$ maximizes (or minimizes) $f(\mathbf{x})$ subject to the side condition $g(\mathbf{x}) = 0$, then $\nabla f(\mathbf{x}_0)$ and $\nabla g(\mathbf{x}_0)$ are parallel. Thus, if $\nabla g(\mathbf{x}_0) \neq \mathbf{0}$, then there exists a scalar λ such that

$$\nabla f(\mathbf{x}_0) = \lambda \nabla g(\mathbf{x}_0).$$

Such a scalar λ has come to be called a *Lagrange multiplier.*

PROOF OF (15.7.2) Let's suppose that $\mathbf{x}_0$ maximizes (or minimizes) $f(\mathbf{x})$ subject to the side condition $g(\mathbf{x}) = 0$. If $\nabla g(\mathbf{x}_0) = \mathbf{0}$, the result is trivially true: every vector is parallel to the zero vector. We suppose therefore that $\nabla g(\mathbf{x}_0) \neq \mathbf{0}$.

In the two-variable case we have

$$\mathbf{x}_0 = (x_0, y_0) \quad \text{and} \quad \text{the side condition} \quad g(x, y) = 0.$$

The side condition defines a curve C that has a nonzero tangent vector at (x_0, y_0). †† Since (x_0, y_0) maximizes (or minimizes) $f(x, y)$ on C, we know from (15.7.1) that $\nabla f(x_0, y_0)$ is perpendicular to C at (x_0, y_0). By (15.4.2), $\nabla g(x_0, y_0)$ is also perpendicular to C at (x_0, y_0). The two gradients are therefore parallel. See Figure 15.7.3.

† Another contribution of the French mathematician Joseph Louis Lagrange.

†† $\mathbf{t}(x_0, y_0) = \dfrac{\partial g}{\partial y}(x_0, y_0)\,\mathbf{i} - \dfrac{\partial g}{\partial x}(x_0, y_0)\,\mathbf{j} \neq \mathbf{0}.$

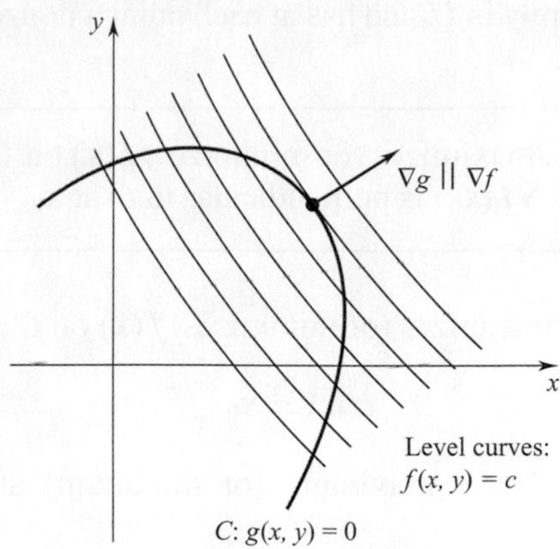

Figure 15.7.3

In the three-variable case we have

$$\mathbf{x}_0 = (x_0, y_0, z_0) \qquad \text{and} \qquad \text{the side condition} \quad g(x, y, z) = 0.$$

The side condition defines a surface Γ that lies in the domain of f. Now let C be a curve that lies on Γ and passes through (x_0, y_0, z_0) with nonzero tangent vector. We know that (x_0, y_0, z_0) maximizes (or minimizes) $f(x, y, z)$ on C. Consequently, $\nabla f(x_0, y_0, z_0)$ is perpendicular to C at (x_0, y_0, z_0). Since this is true for each such curve C, $\nabla f(x_0, y_0, z_0)$ must be perpendicular to Γ itself. But $\nabla g(x_0, y_0, z_0)$ is also perpendicular to Γ at (x_0, y_0, z_0) by (15.4.6). It follows that $\nabla f(x_0, y_0, z_0)$ and $\nabla g(x_0, y_0, z_0)$ are parallel. □

We come now to some problems that are susceptible to Lagrange's method. In each case ∇g is not $\mathbf{0}$ where g is 0 and therefore we can focus entirely on those points $\mathbf{x}$ that satisfy the Lagrange condition

$$\nabla f(\mathbf{x}) = \lambda \nabla g(\mathbf{x}). \tag{15.7.2}$$

Example 3 Maximize and minimize

$$f(x, y) = xy \quad \text{on the unit circle} \quad x^2 + y^2 = 1.$$

SOLUTION Since f is continuous and the unit circle is closed and bounded, it is clear that both a maximum and a minimum exist (see Section 15.6).

To apply Lagrange's method we set

$$g(x, y) = x^2 + y^2 - 1.$$

We want to maximize and minimize

$$f(x, y) = xy \quad \text{subject to the side condition } g(x, y) = 0.$$

The gradients are

$$\nabla f(x, y) = y\,\mathbf{i} + x\,\mathbf{j}, \quad \nabla g(x, y) = 2x\,\mathbf{i} + 2y\,\mathbf{j}.$$

Setting $\qquad \nabla f(x, y) = \lambda \nabla g(x, y),$

we obtain $\qquad y = 2\lambda x, \quad x = 2\lambda y.$

Multiplying the first equation by y and the second equation by x, we find that

$$y^2 = 2\lambda xy, \quad x^2 = 2\lambda xy$$

and thus

$$y^2 = x^2.$$

The side condition $x^2 + y^2 = 1$ now implies that $2x^2 = 1$ and therefore that $x = \pm\frac{1}{2}\sqrt{2}$. The only points that can give rise to an extreme value are

$$(\tfrac{1}{2}\sqrt{2}, \tfrac{1}{2}\sqrt{2}), \quad (\tfrac{1}{2}\sqrt{2}, -\tfrac{1}{2}\sqrt{2}), \quad (-\tfrac{1}{2}\sqrt{2}, \tfrac{1}{2}\sqrt{2}), \quad (-\tfrac{1}{2}\sqrt{2}, -\tfrac{1}{2}\sqrt{2}).$$

At the first and fourth points f takes on the value $\frac{1}{2}$. At the second and third points f takes on the value $-\frac{1}{2}$. Clearly then, $\frac{1}{2}$ is the maximum value and $-\frac{1}{2}$ the minimum value. ❑

Example 4 Find the minimum value taken on by the function

$$f(x, y) = x^2 + (y-2)^2 \quad \text{on the hyperbola} \quad x^2 - y^2 = 1.$$

SOLUTION Note that the expression $f(x, y) = x^2 + (y-2)^2$ gives the square of the distance between the points $(0, 2)$ and (x, y). Therefore, the problem asks us to minimize the square of the distance from the point $(0, 2)$ to the hyperbola, and this minimum value clearly exists. Note, also, that there is no maximum value. See Figure 15.7.4.

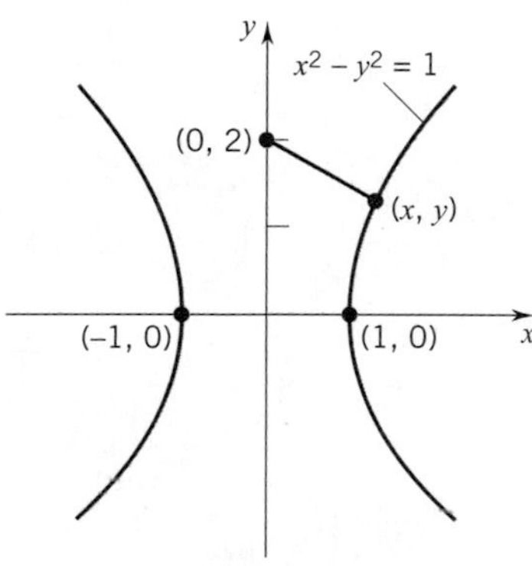

Figure 15.7.4

Now set

$$g(x, y) = x^2 - y^2 - 1.$$

We want to minimize

$$f(x, y) = x^2 + (y-2)^2 \quad \text{subject to the side condition} \quad g(x, y) = 0.$$

Here

$$\nabla f(x, y) = 2x\,\mathbf{i} + 2(y-2)\,\mathbf{j}, \quad \nabla g(x, y) = 2x\,\mathbf{i} - 2y\,\mathbf{j}.$$

The Lagrange condition $\nabla f(x, y) = \lambda \nabla g(x, y)$ gives

$$2x = 2\lambda x, \quad 2(y-2) = -2\lambda y,$$

which we can simplify to

$$x = \lambda x, \quad y - 2 = -\lambda y.$$

The side condition $x^2 - y^2 = 1$ shows that x cannot be zero. Dividing $x = \lambda x$ by x, we get $\lambda = 1$. This means that $y - 2 = -y$ and therefore $y = 1$. With $y = 1$, the side condition gives $x = \pm\sqrt{2}$. The points to be checked are therefore $(-\sqrt{2}, 1)$ and $(\sqrt{2}, 1)$. At each of these points f takes on the value 3. This is the desired minimum. ❑

REMARK The last problem could have been solved more simply by rewriting the side condition as $x^2 = 1 + y^2$ and eliminating x from $f(x, y)$ by substitution. Then it would have been simply a matter of minimizing the function $h(y) = 1 + y^2 + (y-2)^2 = 2y^2 - 4y + 5$. ❑

In the next example we use Lagrange's method to solve the problem of Example 4 of the previous section.

Example 5 A rectangular box without a top is to have a volume of 12 cubic feet. Find the dimensions of the box that will have minimum surface area.

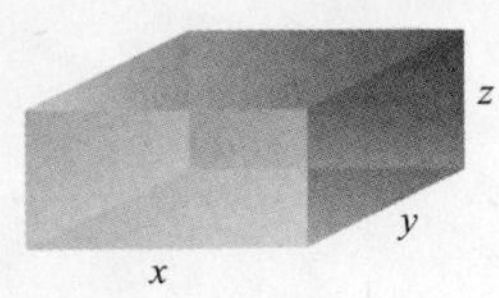

Figure 15.7.5

SOLUTION With the dimensions indicated in Figure 15.7.5, the surface area is given by the expression

$$S = xy + 2xz + 2yz.$$

We want to minimize S subject to the side condition $xyz = 12$ with $x > 0, y > 0, z > 0$.

We begin by setting

$$g(x,y,z) = xyz - 12$$

so that the side condition becomes $g(x,y,z) = 0$. We seek those triples (x,y,z) that simultaneously satisfy the Lagrange condition

$$\nabla f(x,y,z) = \lambda \nabla g(x,y,z) \qquad \text{and the side condition} \qquad g(x,y,z) = 0.$$

The gradients are

$$\nabla f = (y + 2z)\,\mathbf{i} + (x + 2z)\,\mathbf{j} + (2x + 2y)\,\mathbf{k}, \qquad \nabla g = yz\,\mathbf{i} + xz\,\mathbf{j} + xy\,\mathbf{k}.$$

The Lagrange condition gives

$$y + 2z = \lambda yz, \quad x + 2z = \lambda xz, \quad 2x + 2y = \lambda xy.$$

Multiplying the first equation by x, the second by $-y$, and adding the resulting equations, we get

$$2z(x - y) = 0.$$

Since $z \neq 0$, it follows that $y = x$. Replacing y by x in the third equation yields the equation

$$4x = \lambda x^2.$$

Since $x \neq 0$, we conclude that $x = 4/\lambda$, and since $y = x$, $y = 4/\lambda$. We can now solve either the first or second equation for z in terms of λ. The result is $z = 2/\lambda$.

Finally, substituting $x = y = 4/\lambda$ and $z = 2/\lambda$ into the side condition, we get

$$\left(\frac{4}{\lambda}\right)\left(\frac{4}{\lambda}\right)\left(\frac{2}{\lambda}\right) = 12 \qquad \text{which implies} \qquad \lambda^3 = \frac{8}{3} \qquad \text{so that} \qquad \lambda = \frac{2}{\sqrt[3]{3}}.$$

With $\lambda = 2/\sqrt[3]{3}$, we find that $x = y = 2\sqrt[3]{3}$ and $z = \sqrt[3]{3}$. The dimensions that minimize the surface area are

$$\text{length} = 2\sqrt[3]{3}, \qquad \text{width} = 2\sqrt[3]{3} \qquad \text{and} \qquad \text{height} = \sqrt[3]{3},$$

which is the result we obtained before. □

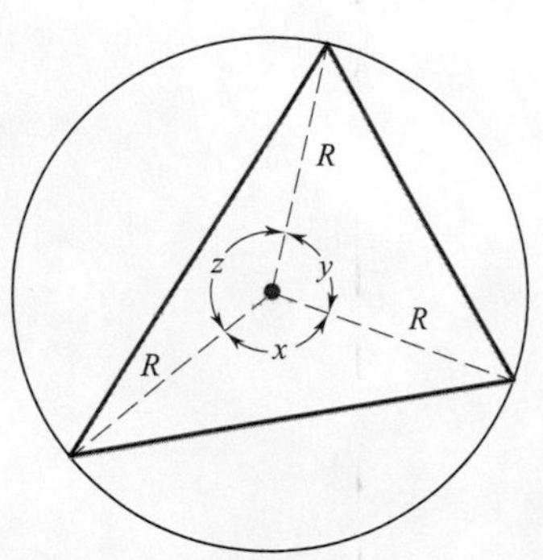

Figure 15.7.6

Example 6 Show that, of all the triangles inscribed in a fixed circle of radius R, the equilateral triangle has the largest perimeter.

SOLUTION It is intuitively clear that this maximum exists and geometrically clear that the triangle that offers this maximum contains the center of the circle in its interior or on its boundary. As in Figure 15.7.6, we denote by x, y, z the central angles that subtend the three sides. As you can verify by trigonometry, the perimeter of the triangle is given by the function.

$$f(x, y, z) = 2R(\sin \tfrac{1}{2}x + \sin \tfrac{1}{2}y + \sin \tfrac{1}{2}z).$$

As a side condition we have

$$g(x, y, z) = x + y + z - 2\pi = 0.$$

To maximize the perimeter we form the gradients

$$\nabla f(x, y, z) = R\left(\cos \tfrac{1}{2}x\,\mathbf{i} + \cos \tfrac{1}{2}y\,\mathbf{j} + \cos \tfrac{1}{2}z\,\mathbf{k}\right), \quad \nabla g(x, y, z) = \mathbf{i} + \mathbf{j} + \mathbf{k}.$$

The Lagrange condition $\nabla f(x, y, z) = \lambda \nabla g(x, y, z)$ gives

$$\lambda = R\cos \tfrac{1}{2}x, \quad \lambda = R\cos \tfrac{1}{2}y, \quad \lambda = R\cos \tfrac{1}{2}z,$$

and therefore

$$\cos \tfrac{1}{2}x = \cos \tfrac{1}{2}y = \cos \tfrac{1}{2}z.$$

With x, y, z all in $(0, \pi]$, we can conclude that $x = y = z$. Since the central angles are equal, the sides are equal. The triangle is therefore equilateral. ❑

An Application of the Cross Product

The Lagrange condition can be replaced by a cross-product equation: points that satisfy $\nabla f = \lambda \nabla g$ satisfy

(15.7.3)
$$\boxed{\nabla f \times \nabla g = \mathbf{0}.}$$

If f and g are functions of two variables,

$$\nabla f \times \nabla g = \begin{vmatrix} \mathbf{i} & \mathbf{j} & \mathbf{k} \\ f_x & f_y & 0 \\ g_x & g_y & 0 \end{vmatrix} = (f_x g_y - f_y g_x)\,\mathbf{k}.$$

Thus, in two variables, the condition $\nabla f \times \nabla g = \mathbf{0}$ gives

(15.7.4)
$$\boxed{\frac{\partial f}{\partial x}\frac{\partial g}{\partial y} - \frac{\partial f}{\partial y}\frac{\partial g}{\partial x} = 0.}$$

Example 7 Maximize and minimize $f(x, y) = xy$ on the unit circle $x^2 + y^2 = 1$.

SOLUTION This problem was solved earlier by means of the Lagrange equation (see Example 3). This time we will use (15.7.4) instead.

As before, we set

$$g(x, y) = x^2 + y^2 - 1,$$

so that the side condition takes the form $g(x, y) = 0$. Since

$$\frac{\partial f}{\partial x} = y, \quad \frac{\partial f}{\partial y} = x \quad \text{and} \quad \frac{\partial g}{\partial x} = 2x, \quad \frac{\partial g}{\partial y} = 2y,$$

(15.7.4) takes the form

$$y(2y) - x(2x) = 0.$$

This gives $x^2 = y^2$.

As before, the side condition $x^2 + y^2 = 1$ implies that $2x^2 = 1$ and therefore $x = \pm\frac{1}{2}\sqrt{2}$. The points under consideration are

$$(\tfrac{1}{2}\sqrt{2}, \tfrac{1}{2}\sqrt{2}), \quad (\tfrac{1}{2}\sqrt{2}, -\tfrac{1}{2}\sqrt{2}), \quad (-\tfrac{1}{2}\sqrt{2}, \tfrac{1}{2}\sqrt{2}), \quad (-\tfrac{1}{2}\sqrt{2}, -\tfrac{1}{2}\sqrt{2}).$$

As we saw in Example 3, f takes on its maximum value $\frac{1}{2}$ at the first and fourth points, and its minimum value $-\frac{1}{2}$ at the second and third points. □

In three variables the computations demanded by the cross-product equation are often quite complicated, and it is usually easier to follow the method of Lagrange.

EXERCISES 15.7

1. Minimize $x^2 + y^2$ on the hyperbola $xy = 1$.

2. Maximize xy on the ellipse $b^2x^2 + a^2y^2 = a^2b^2$.

3. Minimize xy on the ellipse $b^2x^2 + a^2y^2 = a^2b^2$.

4. Minimize xy^2 on the unit circle $x^2 + y^2 = 1$.

5. Maximize xy^2 on the ellipse $b^2x^2 + a^2y^2 = a^2b^2$.

6. Maximize $x + y$ on the curve $x^4 + y^4 = 1$.

7. Maximize $x^2 + y^2$ on the curve $x^4 + 7x^2y^2 + y^4 = 1$.

8. Minimize xyz on the unit sphere $x^2 + y^2 + z^2 = 1$.

9. Maximize xyz on the ellipsoid $x^2/a^2 + y^2/b^2 + z^2/c^2 = 1$.

10. Minimize $x + 2y + 4z$ on the sphere $x^2 + y^2 + z^2 = 7$.

11. Maximize $2x + 3y + 5z$ on the sphere $x^2 + y^2 + z^2 = 19$.

12. Minimize $x^4 + y^4 + z^4$ on the plane $x + y + z = 1$.

13. Maximize the volume of a rectangular solid in the first octant with one vertex at the origin and opposite vertex on the plane $x/a + y/b + z/c = 1$. (Take $a > 0, b > 0, c > 0$.)

14. Show that the square has the largest area of all the rectangles with a given perimeter.

15. Find the distance from the point $(0, 1)$ to the parabola $x^2 = 4y$.

16. Find the distance from the point $(p, 4p)$ to the parabola $y^2 = 2px$.

17. Find the points on the sphere $x^2 + y^2 + z^2 = 1$ that are closest to and farthest from the point $(2, 1, 2)$.

18. Let x, y, and z be the angles of a triangle. Determine the maximum value of $f(x, y, z) = \sin x \sin y \sin z$.

19. Maximize $f(x, y, z) = 3x - 2y + z$ on the sphere $x^2 + y^2 + z^2 = 14$.

20. A rectangular box has three of its faces on the coordinate planes and one vertex in the first octant on the paraboloid $z = 4 - x^2 - y^2$. Determine the maximum volume of the box.

21. Use the method of Lagrange to find the distance from the origin to the plane with equation $Ax + By + Cz + D = 0$.

22. Maximize the volume of a rectangular solid given that the sum of the areas of the six faces is $6a^2$.

23. Within a triangle there is a point P such that the sum of the squares of the distances from P to the sides of the triangle is a minimum. Find this minimum.

24. Show that of all the triangles inscribed in a fixed circle the equilateral one has the largest: (a) product of the lengths of the sides; (b) sum of the squares of the lengths of the sides.

25. The curve $x^3 - y^3 = 1$ is asymptotic to the line $y = x$. Find the point(s) on the curve $x^3 - y^3 = 1$ farthest from the line $y = x$.

26. A plane passes through the point $(a, b, c.)$ Find its intercepts with the coordinate axes if the volume of the solid bounded by the plane and the coordinate planes is to be a minimum.

27. Show that, of all the triangles with a given perimeter, the equilateral triangle has the largest area. HINT: Area$=\sqrt{s(s-a)(s-b)(s-c)}$, where s represents the semiperimeter $s = \frac{1}{2}(a + b + c)$.

28. Show that the rectangular box of maximum volume that can be inscribed in the sphere $x^2 + y^2 + z^2 = a^2$ is a cube.

29. Determine the maximum value of $f(x,y) = (xy)^{1/2}$ given that x and y are nonnegative numbers and $x + y = k$, k a constant. This result shows that if x and y are nonnegative numbers, then

$$(xy)^{1/2} \le \frac{x+y}{2}.$$

(See Exercise 58, Section 1.3.)

30. (a) Determine the maximum value of $f(x,y,z) = (xyz)^{1/3}$ given that x, y, and z are nonnegative numbers and $x + y + z = k$, k a constant.

(b) Use the result in part (a) to show that if x, y, and z are nonnegative numbers, then

$$(xyz)^{1/3} \le \frac{x+y+z}{3}.$$

NOTE: $(xyz)^{1/3}$ is the *geometric mean* of x, y, z.

31. Let $x_1, x_2, \dots, x_n$ be nonnegative numbers such that $x_1 + x_2 + \cdots + x_n = k$, k a constant. Prove that

$$(x_1 x_2 \cdots x_n)^{1/n} \le \frac{x_1 + x_2 + \cdots + x_n}{n}.$$

Thus, for n nonnegative numbers, *the geometric mean is less than or equal to the arithmetic mean.*

32. Assume that the Celsius temperature T at a point (x, y, z) on the sphere $x^2 + y^2 + z^2 = 1$ is given by

$$T(x,y,z) = 10xy^2z.$$

Find the point(s) on the sphere at which the temperature is greatest and the point(s) at which it is least. Give the temperature at each of these points.

33. A soft drink manufacturer wants to design an aluminum can in the shape of a right circular cylinder to hold a given volume V (measured in cubic inches.) If the objective is to minimize the amount of aluminum needed (top, sides, and bottom), what dimensions should be used?

Use the Lagrange method to give alternative solutions to the indicated exercises in Section 15.6.

34. Exercise 19.

35. Exercise 22.

36. Exercise 20.

37. Exercise 33.

38. Exercise 35.

39. Exercise 36.

40. Exercise 40.

41. Exercise 39.

42. Exercise 42.

43. A manufacturer can produce three distinct products in quantities Q_1, Q_2, Q_3, releate, and thereby derive a profit $p(Q_1, Q_2, Q_3) = 2Q_1, +8Q_2 + 24Q_3$. Find the values of Q_1, Q_2, Q_3 that maximize profit if production is subject to the constraint $Q_1^2 + 2Q_2^2 + 4Q_3^2 = 4.5 \times 10^9$.

44. Find the volume of the largest rectangular box that can be inscribed in the ellipsoid

$$4x^2 + 9y^2 + 36z^2 = 36$$

if the edges of the box are parallel to the coordinate axes.

■ PROJECT 15.7 Maxima and Minima with Two Side Conditions

The Lagrange method can be extended to problems with two side conditions. If $\mathbf{x}_0$ gives rise to a maximum (or minimum) of $f(\mathbf{x})$ subject to the two side conditions $g(\mathbf{x}) = 0$ and $h(\mathbf{x}) = 0$, and if $\nabla g(\mathbf{x_0}), \nabla h(\mathbf{x_0})$ are nonzero and nonparallel, then there exist scalars λ and μ such that

$$\nabla f(\mathbf{x}_0) = \lambda \nabla g(\mathbf{x}_0) + \mu \nabla h(\mathbf{x}_0).$$

Assume this result and solve the following problems.

Problem 1 Find the extreme values of

$$f(x,y,z) = xy + z^2$$

subject to the side conditions:

$$x^2 + y^2 + z^2 = 4 \qquad \text{and} \qquad y - x = 0.$$

Problem 2. The planes $x + 2y + 3z = 0$ and $2x + 3y + z = 4$ intersect in a straight line. Find the point on that line that is closest to the origin.

Problem 3. The plane $x + y - z + 1 = 0$ intersects the upper nappe of the cone $z^2 = x^2 + y^2$ in an ellipse. Find the points on this ellipse that are closest to and farthest from the origin.

■ 15.8 DIFFERENTIALS

We begin by reviewing the one-variable case. If f is differentiable at x, then for small h, the increment

$$\nabla f = f(x+h) - f(x)$$

can be approximated by the differential

$$df = f'(x)h.$$

The geometric interpretations of Δf and df are shown in Figure 15.8.1. We write

$$\Delta f \cong df.$$

How good is this approximation? It is good enough that the ratio

$$\frac{\Delta f - df}{|h|}$$

tends to 0 as h tends to 0.

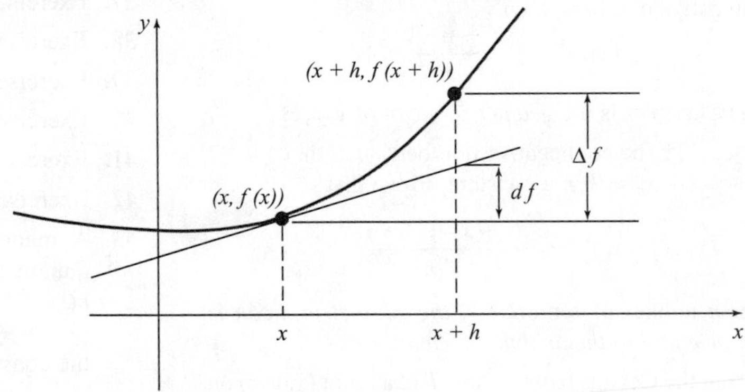

Figure 15.8.1

The differential of a function of several variables, defined in an analogous manner, plays a similar approximating role. Let's suppose that f, now a function of several variables, is differentiable at $\mathbf{x}$. The difference

(15.8.1) $$\Delta f = f(\mathbf{x} + \mathbf{h}) - f(\mathbf{x})$$

is called the *increment* of f, and the dot product

(15.8.2) $$df = \nabla f(\mathbf{x}) \cdot \mathbf{h}$$

is called the *differential* (more formally, the *total differential*). As in the one-variable case, for small $\mathbf{h}$, the differential and the increment are approximately equal:

(15.8.3) $$\Delta f \cong df.$$

How approximately equal are they? Enough so that the ratio

$$\frac{\Delta f - df}{\|\mathbf{h}\|}$$

tends to 0 as $\mathbf{h}$ tends to $\mathbf{0}$. How do we know this? We know that

$$f(\mathbf{x} + \mathbf{h}) - f(\mathbf{x}) = \nabla f(\mathbf{x}) \cdot \mathbf{h} + o(\mathbf{h}).$$

Therefore, $$[f(\mathbf{x}+\mathbf{h})-f(\mathbf{x})]-\nabla f(\mathbf{x})\cdot\mathbf{h}=o(\mathbf{h}),$$

and so

$$\frac{\overbrace{[f(\mathbf{x}+\mathbf{h})-f(\mathbf{x})]}^{\Delta f}-\overbrace{\nabla f(\mathbf{x})\cdot\mathbf{h}}^{df}}{\|\mathbf{h}\|}\to 0 \text{ as } \mathbf{h}\to\mathbf{0}. \qquad \text{(Section 15.1)}$$

In the two-variable case we set $\mathbf{x}=(x,y)$ and $\mathbf{h}=(\Delta x,\Delta y)$. The increment $\Delta f=f(\mathbf{x}+\mathbf{h})-f(\mathbf{x})$ then takes the form

$$\Delta f=f(x+\Delta x,y+\Delta y)-f(x,y),$$

and the differential $df=\nabla f(\mathbf{x})\cdot\mathbf{h}$ takes the form

$$df=\frac{\partial f}{\partial x}(x,y)\Delta x+\frac{\partial f}{\partial y}(x,y)\Delta y.$$

By suppressing the point of evaluation, we can write

(15.8.4) $$\boxed{df=\frac{\partial f}{\partial x}\Delta x+\frac{\partial f}{\partial y}\Delta y.}$$

The approximation $\Delta f\cong df$ is illustrated in Figure 15.8.2. There we have represented f as a surface $z=f(x,y)$. Through a point $P(x_0,y_0,f(x_0,y_0))$ we have drawn the tangent plane. *The difference* $df-\Delta f$ *is the vertical separation between this tangent plane and the surface as measured at the point* $(x_0+\Delta x,y_0+\Delta y)$.

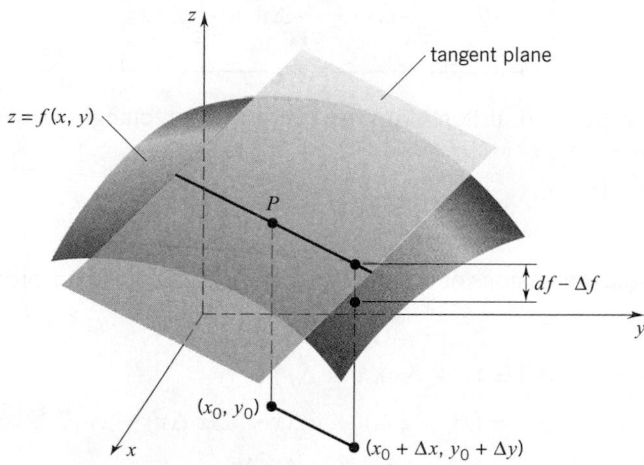

Figure 15.8.2

PROOF The tangent plane at P has equation

$$z-f(x_0,y_0)=\frac{\partial f}{\partial x}(x_0,y_0)(x-x_0)+\frac{\partial f}{\partial y}(x_0,y_0)(y-y_0).$$

The z-coordinate of this plane at the point $(x_0+\Delta x,y_0+\Delta y)$ is

$$f(x_0,y_0)+\frac{\partial f}{\partial x}(x_0,y_0)\Delta x+\frac{\partial f}{\partial y}(x_0,y_0)\Delta y. \qquad \text{(check this)}$$

The z-coordinate of the surface at this same point is

$$f(x_0 + \Delta x, y_0 + \Delta y).$$

The difference between these two,

$$\left[f(x_0, y_0) + \frac{\partial f}{\partial x}(x_0, y_0)\Delta x + \frac{\partial f}{\partial y}(x_0, y_0)\Delta y\right] - \left[f(x_0 + \Delta x, y_0 + \Delta y)\right],$$

can be written as

$$\left[\frac{\partial f}{\partial x}(x_0, y_0)\Delta x + \frac{\partial f}{\partial y}(x_0, y_0)\Delta y\right] - \left[f(x_0 + \Delta x, y_0 + \Delta y) - f(x_0, y_0)\right].$$

This is just $df - \Delta f$. □†

For the three-variable case we set $\mathbf{x} = (x, y, z)$ and $\mathbf{h} = (\Delta x, \Delta y, \Delta z)$. The increment becomes

$$\Delta f = f(x + \Delta x, y + \Delta y, z + \Delta z) - f(x, y, z),$$

and the approximating differential becomes

$$df = \frac{\partial f}{\partial x}(x, y, z)\Delta x + \frac{\partial f}{\partial y}(x, y, z)\Delta y + \frac{\partial f}{\partial z}(x, y, z)\,\Delta z.$$

Suppressing the point of evaluation we have

(15.8.5)
$$\boxed{df = \frac{\partial f}{\partial x}\Delta x + \frac{\partial f}{\partial y}\Delta y + \frac{\partial f}{\partial z}\Delta z.}$$

To illustrate the use of differentials, we begin with a rectangle of sides x and y. The area is given by

$$A(x, y) = xy.$$

An increase in the dimensions of the rectangle to $x + \Delta x$ and $y + \Delta y$ produces a change in area

$$\begin{aligned}\Delta A &= (x + \Delta x)(y + \Delta y) - xy\\ &= (xy + x\,\Delta y + y\,\Delta x + \Delta x\,\Delta y) - xy\\ &= x\,\Delta y + y\,\Delta x + \Delta x\,\Delta y.\end{aligned}$$

The differential estimate for this change in area is

$$dA = \frac{\partial A}{\partial x}\Delta x + \frac{\partial A}{\partial y}\Delta y = y\,\Delta x + x\,\Delta y. \qquad \text{(Figure 15.8.3)}$$

† As in the figure, we have been assuming that the tangent plane lies above the surface. If the tangent plane lies below the surface, then $df - \Delta f$ is negative. The vertical separation between the tangent plane and the surface is then $\Delta f - df$.

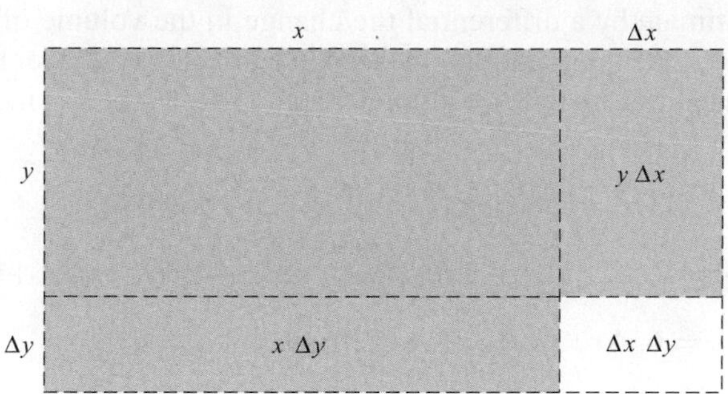

Figure 15.8.3

The error of our estimate, the difference between the actual change and the estimated change, is the difference $\Delta A - dA = \Delta x\, \Delta y$.

Example 1 Given that $f(x,y) = yx^{2/5} + x\sqrt{y}$, estimate by a differential the change in f from (32, 16) to (35, 18).

SOLUTION Since

$$\frac{\partial f}{\partial x} = \frac{2y}{5}\left(\frac{1}{x}\right)^{3/5} + \sqrt{y} \quad \text{and} \quad \frac{\partial f}{\partial y} = x^{2/5} + \frac{x}{2\sqrt{y}},$$

we have
$$df = \left[\frac{2y}{5}\left(\frac{1}{x}\right)^{3/5} + \sqrt{y}\right]\Delta x + \left[x^{2/5} + \frac{x}{2\sqrt{y}}\right]\Delta y.$$

At $x = 32, y = 16, \Delta x = 3, \Delta y = 2$, and

$$df = \left[\frac{32}{5}\left(\frac{1}{32}\right)^{3/5} + \sqrt{16}\right]3 + \left[32^{2/5} + \frac{32}{2\sqrt{16}}\right]2 = 30.4.$$

The change increases the value of f by approximately 30.4. □

Example 2 Use differentials to estimate $\sqrt{27}\,\sqrt[3]{1021}$.

SOLUTION We know $\sqrt{25}$ and $\sqrt[3]{1000}$. What we need is an estimate for the increase of

$$f(x,y) = \sqrt{x}\,\sqrt[3]{y} = x^{1/2}y^{1/3}$$

from $x = 25, y = 1000$ to $x = 27, y = 1021$. The differential is

$$df = \tfrac{1}{2}x^{-1/2}y^{1/3}\Delta x + \tfrac{1}{3}x^{1/2}y^{-2/3}\Delta y.$$

With $x = 25, y = 1000, \Delta x = 2, \Delta y = 21, df$ becomes

$$(\tfrac{1}{2} \cdot 25^{-1/2} \cdot 1000^{1/3})2 + (\tfrac{1}{3} \cdot 25^{1/2} \cdot 1000^{-2/3})21 = 2.35.$$

The change increases the value of the function by about 2.35. It follows that

$$\sqrt{27}\,\sqrt[3]{1021} \cong \sqrt{25}\,\sqrt[3]{1000} + 2.35 = 52.35.$$

(Our calculator gives $\sqrt{27}\,\sqrt[3]{1021} \cong 52.323$.) □

Example 3 Estimate by a differential the change in the volume of the frustum of a right circular cone if the upper radius r is decreased from 3 to 2.7 centimeters, the base radius R is increased from 8 to 8.1 centimeters, and the height h is increased from 6 to 6.3 centimeters.

SOLUTION Since $V(r, R, h) = \frac{1}{3}\pi h(R^2 + Rr + r^2)$, we have

$$dV = \tfrac{1}{3}\pi h(R + 2r)\Delta r + \tfrac{1}{3}\pi h(2R + r)\Delta R + \tfrac{1}{3}\pi(R^2 + Rr + r^2)\Delta h.$$

At $r = 3, R = 8, h = 6, \Delta r = -0.3, \Delta R = 0.1$, and $\Delta h = 0.3$,

$$dV = (28\pi)(-0.3) + (38\pi)(0.1) + \tfrac{1}{3}(97\pi)(0.3) = 5.1\,\pi \cong 16.02.$$

According to our differential estimate, the volume increases by about 16 cubic centimeters. ❑

EXERCISES 15.8

Find the differential df.

1. $f(x,y) = x^3y - x^2y^2$.
2. $f(x,y,z) = xy + yz + xz$.
3. $f(x,y) = x\cos y - y\cos x$.
4. $f(x,y,z) = x^2y\,e^{2z}$.
5. $f(x,y,z) = x - y\,\tan z$.
6. $f(x,y) = (x - y)\ln(x + y)$.
7. $f(x,y,z) = \dfrac{xy}{x^2 + y^2 + z^2}$.
8. $f(x,y) = \ln(x^2 + y^2) + x\,e^{xy}$.
9. $f(x,y) = \sin(x + y) + \sin(x - y)$.
10. $f(x,y) = x\ln\left[\dfrac{1+y}{1-y}\right]$.
11. $f(x,y,z) = y^2e^{xz} + x\ln z$.
12. $f(x,y) = xy\,e^{-(x^2+y^2)}$.
13. Calculate Δu and du for $u = x^2 - 3xy + 2y^2$ at $x = 2$, $y = -3$, $\Delta x = -0.3$, $\Delta y = 0.2$.
14. Calculate du for $u = (x + y)\sqrt{x - y}$ at $x = 6, y = 2$, $\Delta x = \frac{1}{4}, \Delta y = -\frac{1}{2}$.
15. Calculate Δu and du for $u = x^2z - 2yz^2 + 3xyz$ at $x = 2$, $y = 1, z = 3, \Delta x = 0.1, \Delta y = 0.3, \Delta z = -0.2$.
16. Calculate du for
$$u = \frac{xy}{\sqrt{x^2 + y^2 + z^2}}$$
at $x = 1, y = 3, z = -2, \Delta x = \frac{1}{2}, \Delta y = \frac{1}{4}, \Delta z = -\frac{1}{4}$.

Use differentials to find the approximate value.

17. $\sqrt{125}\,\sqrt[4]{17}$.
18. $(1 - \sqrt{10})(1 + \sqrt{24})$.
19. $\sin\frac{6}{7}\pi\,\cos\frac{1}{5}\pi$.
20. $\sqrt{8}\tan\frac{5}{16}\pi$.

Use differentials to approximate the value of f at the point P.

21. $f(x,y) = x^2e^{xy}$; $P(2.9, 0.01)$.
22. $f(x,y,z) = x^2y\cos\pi z$; $P(2.12, 2.92, 3.02)$.
23. $f(x,y,z) = x\tan^{-1} yz$; $P(2.94, 1.1, 0.92)$.
24. $f(x,y) = \sqrt{x^2 + y^2}$; $P(3.06, 3.88)$.
25. Given that $z = (x - y)(x + y)^{-1}$, use dz to find the approximate change in z if x is increased from 4 to $4\frac{1}{10}$ and y is increased from 2 to $2\frac{1}{10.}$ What is the exact change?
26. Estimate by a differential the change in the volume of a right circular cylinder if the height is increased from 12 to 12.2 inches and the radius is decreased from 8 to 7.7 inches.
27. Estimate the change in the total surface area for the cylinder of Exercise 26.
28. Use a differential to estimate the change in $T = x^2\cos\pi z - y^2\sin\pi z$ from $x = 2, y = 2, z = 2$ to $x = 2.1, y = 1.9, z = 2.2$.
29. Estimate the surface area of a closed rectangular box whose dimensions are: length $= 9.98$ inches, width $= 5.88$ inches, height $= 4.08$ inches.
30. Estimate the volume of a right circular cone of base radius 7.2 centimeters and height 10.15 centimeters.
31. The dimensions of a closed rectangular box change from length $= 12$, width $= 8$, height $= 6$ to length $= 12.02$, width $= 7.95$, height $= 6.03$.
    (a) Use a differential to approximate the change in volume.
    (b) Calculate the exact change in volume.
32. Use the dimensions of the rectangular box in Exercise 31.
    (a) Approximate the change in the surface area using a differential.
    (b) Calculate the exact change in the surface area.
33. The function $T(x,y,z) = 100 - x^2 - y^2 - z^2 + 2xyz$ is defined at all points in space. Use differentials to approximate the difference between $T(1,3,4)$ and $T(1.15, 2.90, 4.10)$.
34. According to the ideal gas law, the relation between the pressure P, the temperature T, and the volume V of a confined gas is given by the equation $PV = kT$, where k is a

constant. If $P = 4$ pounds per square inch when $V = 81$ cubic inches and $T = 300$ K, approximate the change in pressure if the volume is decreased to 75 cubic inches and the temperature is increased to 325 K.

35. As illustrated in the following figure, the side x in the right triangle is increased by Δx and the angle θ is increased by $\Delta\theta$. Use a differential to approximate the change in the area of the triangle. Is the area more sensitive to a change in x or to a change in θ?

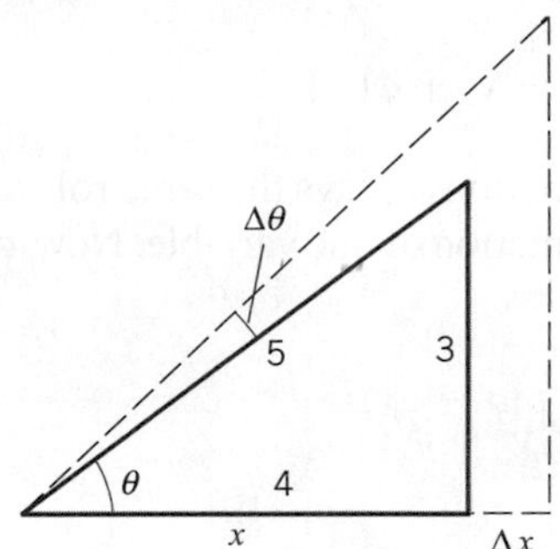

36. Use a differential to approximate the change in the area of the isosceles triangle shown in the figure if x changes by Δx and θ changes by $\Delta\theta$. Is the area more sensitive to changes in x or to changes in θ? HINT: The area of a triangle with sides a and b and included angle θ is given by $A = \frac{1}{2}ab\sin\theta$.

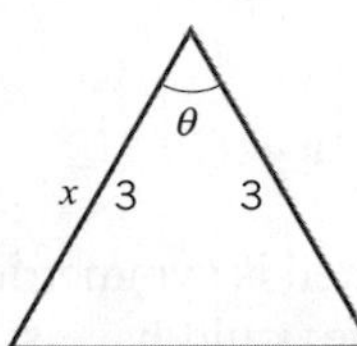

37. The radius of a right circular cylinder of height h is increased from r to $r + \Delta r$.

(a) Determine the exact change in h that will keep the volume constant. Then estimate this change in h by using a differential.

(b) Determine the exact change in h that will keep the total surface area constant. Then estimate this change in h by using a differential.

38. The dimensions of a rectangular box with a top are length $= 4$ feet, width $= 2$ feet, height $= 3$ feet. It has a coat of paint $\frac{1}{16}$ inch thick. Estimate the amount of paint (cubic inches) on the box.

Error Estimates

Let $u = u(x, y)$ be differentiable. If the variables x and y are known to be $x_0 \pm \Delta x$ and $y_0 \pm \Delta y$, then the maximum possible error in the calculated value of $u(x_0, y_0)$ is

$$\frac{\partial u}{\partial x}(x_0, y_0)(\pm\Delta x) + \frac{\partial u}{\partial y}(x_0, y_0)(\pm\Delta y).$$

39. The legs of a right triangle are measured to be 5 and 12 centimeters with a possible error of ± 15 millimeters in each measurement. What is the maximum possible error in the calculated value of (a) the hypotenuse and (b) the area of the triangle?

40. The radius of a right circular cone is measured to be 5 inches with a possible error of ± 0.2 inch and the height is measured to be 12 inches with a possible error of ± 0.3 inch. What is the maximum possible error in the calculated values of (a) the volume and (b) the lateral surface area of the cone?

41. The specific gravity of a solid is given by the formula $s = A(A - W)^{-1}$ where A is the weight in air and W is the weight in water. What is the maximum possible error in the calculated value of s if A is measured to be 9 pounds (within a tolerance of 0.01 pound) and W is measured to be 5 pounds (within a tolerance of 0.02 pound)?

42. The measurements of a closed rectangular box are length $=$ 5 feet, width $= 3$ feet, and height $= 3.5$ feet, with a possible error of $\pm\frac{1}{12}$ inch in each measurement. What is the maximum possible error in the calculated value of (a) the volume and (b) the surface area of the box? HINT: Extend the error estimate to three dimensions.

■ 15.9 RECONSTRUCTING A FUNCTION FROM ITS GRADIENT

This section has three parts. In Part 1 we show how to find $f(x, y)$ given its gradient

$$\nabla f(x, y) = \frac{\partial f}{\partial x}(x, y)\,\mathbf{i} + \frac{\partial f}{\partial y}(x, y)\,\mathbf{j}.$$

In Part 2 we show that, although all gradients $\nabla f(x, y)$ are expressions of the form

$$P(x, y)\,\mathbf{i} + Q(x, y)\,\mathbf{j}$$

(set $P = \partial f/\partial x$ and $Q = \partial f/\partial y$), not all such expressions are gradients. In Part 3 we tackle the problem of recognizing which expressions $P(x, y)\,\mathbf{i} + Q(x, y)\,\mathbf{j}$ are actually gradients.

Part 1

Example 1 Find f given that $\nabla f(x, y) = (4x^3y^3 - 3x^2)\,\mathbf{i} + (3x^4y^2 + \cos 2y)\,\mathbf{j}$.

SOLUTION The first partial derivatives of f are

$$\frac{\partial f}{\partial x}(x, y) = 4x^3y^3 - 3x^2, \qquad \frac{\partial f}{\partial y}(x, y) = 3x^4y^2 + \cos 2y.$$

Integrating $\partial f/\partial x$ with respect to x, treating y as a constant, we find that

$$f(x, y) = x^4y^3 - x^3 + \phi(y)$$

where ϕ is an unknown function of y. The function ϕ plays the same role as the arbitrary constant C that arises when you integrate a function of one variable. Now, differentiation with respect to y gives

$$\frac{\partial f}{\partial y}(x, y) = 3x^4y^2 + \phi'(y).$$

Equating the two expressions for $\partial f/\partial y$, we have

$$3x^4y^2 + \phi'(y) = 3x^4y^2 + \cos 2y,$$

which implies that

$$\phi'(y) = \cos 2y \qquad \text{and thus} \qquad \phi(y) = \tfrac{1}{2}\sin 2y + C. \qquad (C \text{ a constant})$$

This means that

$$f(x, y) = x^4y^3 - x^3 + \tfrac{1}{2}\sin 2y + C. \quad \square$$

Remark The procedure for finding f just illustrated is symmetric in x and y. That is, rather than integrating $\partial f/\partial x$ with respect to x, we could have started by integrating $\partial f/\partial y$ with respect to y, with x held constant, followed by differentiating the result with respect to x:

$$\text{if} \quad \frac{\partial f}{\partial y}(x, y) = 3x^4y^2 + \cos 2y \qquad \text{then} \qquad f(x, y) = x^4y^3 + \tfrac{1}{2}\sin 2y + \psi(x)$$

where ψ is an unknown function of x. Now, differentiating with respect to x, we have

$$\frac{\partial f}{\partial x}(x, y) = 4x^3y^3 + \psi'(x).$$

Equating the two expressions for $\partial f/\partial x$, we have

$$4x^3y^3 + \psi'(x) = 4x^3y^3 - 3x^2.$$

Therefore,

$$\psi'(x) = -3x^2 \quad \text{which implies that} \quad \psi(x) = -x^3 + C. \qquad (C \text{ a constant})$$

Thus, $$f(x, y) = x^4y^3 + \tfrac{1}{2}\sin 2y - x^3 + C,$$

as we found before. $\square$

Example 2 Find f given that

$$\nabla f(x,y) = \left(\sqrt{y} - \frac{y}{2\sqrt{x}} + 2x\right)\mathbf{i} + \left(\frac{x}{2\sqrt{y}} - \sqrt{x} + 1\right)\mathbf{j}.$$

SOLUTION Here we have

$$\frac{\partial f}{\partial x}(x,y) = \sqrt{y} - \frac{y}{2\sqrt{x}} + 2x, \qquad \frac{\partial f}{\partial y}(x,y) = \frac{x}{2\sqrt{y}} - \sqrt{x} + 1.$$

Integrating $\partial f/\partial x$ with respect to x, we have

$$f(x,y) = x\sqrt{y} - y\sqrt{x} + x^2 + \phi(y)$$

with $\phi(y)$ independent of x. Differentiation with respect to y gives

$$\frac{\partial f}{\partial y}(x,y) = \frac{x}{2\sqrt{y}} - \sqrt{x} + \phi'(y).$$

The two equations for $\partial f/\partial y$ can be reconciled only by having

$$\phi'(y) = 1 \quad \text{and thus} \quad \phi(y) = y + C.$$

This means that

$$f(x,y) = x\sqrt{y} - y\sqrt{x} + x^2 + y + C. \quad \square$$

The function

$$f(x,y) = x\sqrt{y} - y\sqrt{x} + x^2 + y + C$$

is the *general solution* of the vector differential equation

$$\nabla f(x,y) = \left(\sqrt{y} - \frac{y}{2\sqrt{x}} + 2x\right)\mathbf{i} + \left(\frac{x}{2\sqrt{y}} - \sqrt{x} + 1\right)\mathbf{j}.$$

Each *particular solution* can be obtained by assigning a particular value to the constant C.

Part 2

The next example shows that not all linear combinations $P(x,y)\,\mathbf{i} + Q(x,y)\,\mathbf{j}$ are gradients.

Example 3 Show that $y\,\mathbf{i} - x\,\mathbf{j}$ is not a gradient.

SOLUTION Suppose on the contrary that it is a gradient. Then there exists a function f such that

$$\nabla f(x,y) = y\,\mathbf{i} - x\,\mathbf{j}.$$

This implies that

$$\frac{\partial f}{\partial x}(x,y) = y, \qquad \frac{\partial f}{\partial y}(x,y) = -x$$

$$\frac{\partial^2 f}{\partial y \partial x}(x,y) = 1, \qquad \frac{\partial^2 f}{\partial x \partial y}(x,y) = -1$$

and thus
$$\frac{\partial^2 f}{\partial y \partial x}(x, y) \neq \frac{\partial^2 f}{\partial x \partial y}(x, y).$$

This contradicts (14.6.5): the four partial derivatives under consideration are everywhere continuous and thus, according to (14.6.5), we must have

$$\frac{\partial^2 f}{\partial y \partial x}(x, y) = \frac{\partial^2 f}{\partial x \partial y}(x, y).$$

This contradiction shows that $y\,\mathbf{i} - x\,\mathbf{j}$ is not a gradient. □

Part 3

We come now to the problem of recognizing which linear combinations

$$P(x, y)\,\mathbf{i} + Q(x, y)\,\mathbf{j}$$

are actually gradients. But first we need to review some ideas and establish some terminology.

As indicated earlier (Section 15.3), an open set (in the plane or in three-space) is said to be *connected* if each pair of points of the set can be joined by a polygonal path that lies entirely within the set. An open connected set is called an *open region*. A curve

$$C: \quad \mathbf{r} = \mathbf{r}(t), \quad t \in [a, b]$$

is said to be *closed* if it begins and ends at the same point:

$$\mathbf{r}(a) = \mathbf{r}(b).$$

It is said to be *simple* if it does not intersect itself:

$$a < t_1 < t_2 < b \quad \text{implies} \quad \mathbf{r}(t_1) \neq \mathbf{r}(t_2).$$

These notions are illustrated in Figure 15.9.1.

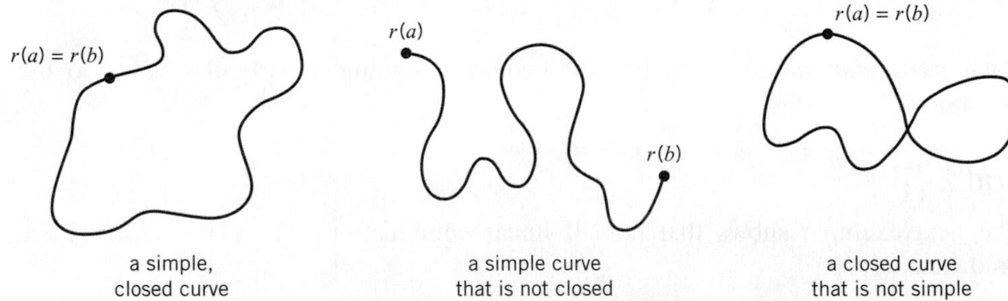

Figure 15.9.1

As is intuitively clear (Figure 15.9.2), a simple closed curve in the plane separates the plane into two disjoint open connected sets: a bounded inner region consisting of all points surrounded by the curve and an unbounded outer region consisting of all points not surrounded by the curve.†

†Add to this the assertion that the curve in question constitutes the total boundary of both regions and you have what is called the Jordan curve theorem, named after the French mathematician Camille Jordan (1838–1922). Jordan was the first to point out that, although all this is apparently obvious, it nevertheless requires proof. In recognition of Jordan, simple closed plane curves are now commonly known as *Jordan curves*.

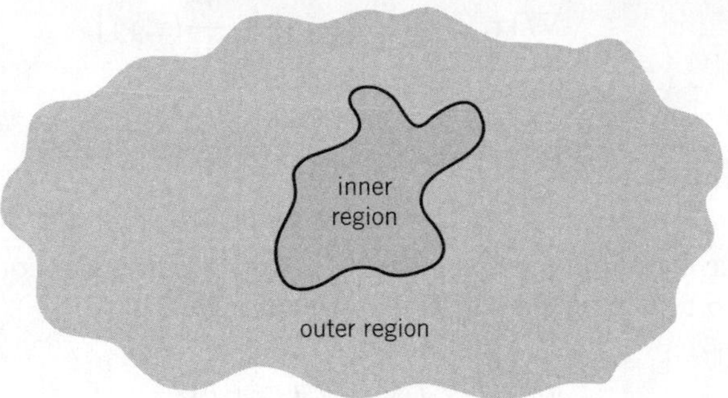

Figure 15.9.2

Finally, we come to a notion we will need in our work with gradients:

(15.9.1) Let Ω be an open region of the plane. Ω is said to be *simply connected* if, for every simple closed curve C in Ω, the inner region of C is contained in Ω.

The first two regions in Figure 15.9.3 are simply connected. The annular region is not; the annular region contains the simple closed curve drawn there, but it does not contain all of the inner region of that curve.

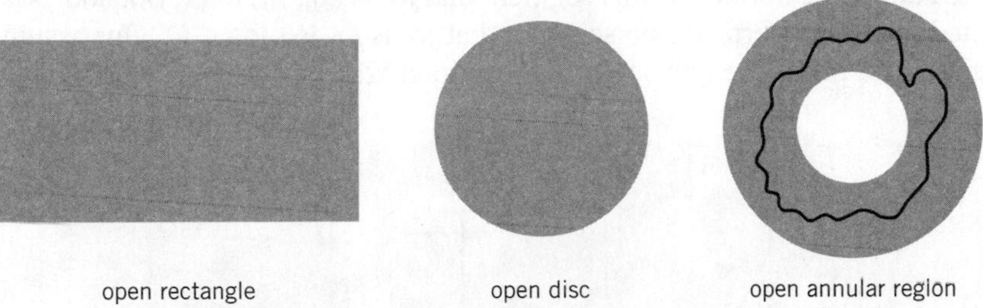

Figure 15.9.3

THEOREM 15.9.2

Let P and Q be functions of two variables, each continuously differentiable on a simply connected open region Ω. The linear combination $P(x,y)\,\mathbf{i} + Q(x,y)\,\mathbf{j}$ is a gradient on Ω iff

$$\frac{\partial P}{\partial y}(x,y) = \frac{\partial Q}{\partial x}(x,y) \qquad \text{for all } (x,y) \in \Omega.$$

A complete proof of this theorem for a general region Ω is complicated. We will prove the result under the additional assumption that Ω has the form of an open rectangle with sides parallel to the coordinate axes.

Suppose that $P(x,y)\,\mathbf{i} + Q(x,y)\,\mathbf{j}$ is a gradient on this open rectangle Ω, say,

$$\nabla f(x,y) = P(x,y)\,\mathbf{i} + Q(x,y)\,\mathbf{j}.$$

Since
$$\nabla f(x,y) = \frac{\partial f}{\partial x}(x,y)\,\mathbf{i} + \frac{\partial f}{\partial y}(x,y)\,\mathbf{j},$$

we have
$$P = \frac{\partial f}{\partial x} \quad \text{and} \quad Q = \frac{\partial f}{\partial y}.$$

Since P and Q have continuous first partials, f has continuous second partials. Thus, according to (14.6.5), the mixed partials are equal and we have

$$\frac{\partial P}{\partial y} = \frac{\partial^2 f}{\partial y \partial x} = \frac{\partial^2 f}{\partial x \partial y} = \frac{\partial Q}{\partial x}.$$

Conversely, suppose that

$$\frac{\partial P}{\partial y}(x,y) = \frac{\partial Q}{\partial x}(x,y) \quad \text{for all} \quad (x,y) \in \Omega.$$

We must show that $P(x,y)\,\mathbf{i} + Q(x,y)\,\mathbf{j}$ is a gradient on Ω. To do this, we choose a point (x_0, y_0) in Ω and form the function

$$f(x,y) = \int_{x_0}^{x} P(u, y_0)\,du + \int_{y_0}^{y} Q(x, v)\,dv, \quad (x,y) \in \Omega.$$

[If you want to visualize f, you can refer to Figure 15.9.4. The function P is being integrated along the horizontal line segment that joins (x_0, y_0) to (x, y_0), and Q is being integrated along the vertical line segment that joins (x, y_0) to (x, y). Our assumptions on Ω guarantee that these line segments remain in Ω.]

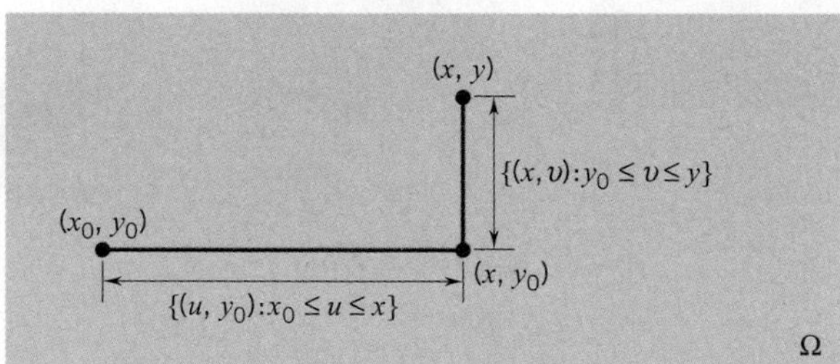

Figure 15.9.4

The first integral is independent of y. Hence

$$\frac{\partial f}{\partial y}(x,y) = \frac{\partial}{\partial y}\left(\int_{y_0}^{y} Q(x,v)\,dv\right) = Q(x,y).$$

The last equality holds because we are differentiating an integral with respect to its upper limit (Theorem 5.2.5). Differentiating f with respect to x we have

$$\frac{\partial f}{\partial x}(x,y) = \frac{\partial}{\partial x}\left(\int_{x_0}^{x} P(u,y_0)\,du\right) + \frac{\partial}{\partial x}\left(\int_{y_0}^{y} Q(x,v)\,dv\right).$$

The first term is $P(x, y_0)$, since once again we are differentiating with respect to the upper limit. In the second term the variable x appears in the integrand. It can be shown that, since Q and $\partial Q/\partial x$ are continuous,

$$\frac{\partial}{\partial x}\left(\int_{y_0}^{y} Q(x,v)\,dv\right) = \int_{y_0}^{y} \frac{\partial Q}{\partial x}(x,v)\,dv.\dagger$$

Anticipating this result, we have

$$\frac{\partial f}{\partial x}(x,y) = P(x,y_0) + \int_{y_0}^{y} \frac{\partial Q}{\partial x}(x,v)\,dv = P(x,y_0) + \int_{y_0}^{y} \frac{\partial P}{\partial y}(x,v)\,dv$$

explain (pointing to the second equality)

$$= P(x,y_0) + P(x,y) - P(x,y_0) = P(x,y).$$

We have now shown that

$$P(x,y) = \frac{\partial f}{\partial x}(x,y) \quad \text{and} \quad Q(x,y) = \frac{\partial f}{\partial y}(x,y) \quad \text{for all } (x,y) \in \Omega.$$

It follows that $P(x,y)\,\mathbf{i} + Q(x,y)\,\mathbf{j}$ is the gradient of f on Ω. □

Example 4 The vector functions

$$\mathbf{F}(x,y) = 2x\sin y\,\mathbf{i} + x^2\cos y\,\mathbf{j} \quad \text{and} \quad \mathbf{G}(x,y) = xy\,\mathbf{i} + \tfrac{1}{2}(x+1)^2y^2\,\mathbf{j}$$

are both defined everywhere. The first vector function is the gradient of a function that is defined everywhere, since for $P(x,y) = 2x\sin y$ and $Q(x,y) = x^2\cos y$, we have

$$\frac{\partial P}{\partial y}(x,y) = 2x\cos y = \frac{\partial Q}{\partial x} \quad \text{for all } (x,y).$$

It is easy to verify that if $f(x,y) = x^2\sin y + C$ where C is a constant, then $\nabla f(x,y) = \mathbf{F}(x,y)$.

The vector function $\mathbf{G}$ is not a gradient: $P(x,y) = xy$, $Q(x,y) = \frac{1}{2}(x+1)^2y^2$, and

$$\frac{\partial P}{\partial y}(x,y) = x \quad \text{and} \quad \frac{\partial Q}{\partial x}(x,y) = (x+1)y^2.$$

Thus,

$$\frac{\partial P}{\partial y}(x,y) \neq \frac{\partial Q}{\partial x}(x,y). \quad \square$$

Example 5 The vector function $\mathbf{F}$ defined on the *punctured disc* $0 < x^2 + y^2 < 1$ by setting

$$\mathbf{F}(x,y) = \frac{y}{x^2+y^2}\,\mathbf{i} - \frac{x}{x^2+y^2}\,\mathbf{j}$$

satisfies

$$\frac{\partial P}{\partial y}(x,y) = \frac{\partial Q}{\partial x}(x,y) \quad \text{on the punctured disc.}$$

(Check this out.) Nevertheless, as you will see in Exercise 28, Section 17.2, $\mathbf{F}$ is not a gradient on that set. The punctured disc is not simply connected and therefore Theorem 15.9.2 does not apply. □

†The validity of this equality is the subject of Exercise 60, Section 16.3.

EXERCISES 15.9

Determine whether or not the vector function is the gradient $\nabla f(x,y)$ of a function everywhere defined. If so, find all the functions with that gradient.

1. $xy^2\,\mathbf{i} + x^2y\,\mathbf{j}$.
2. $x\,\mathbf{i} + y\,\mathbf{j}$.
3. $y\,\mathbf{i} + x\,\mathbf{j}$.
4. $(x^2 + y)\,\mathbf{i} + (y^3 + x)\,\mathbf{j}$.
5. $(y^3 + x)\,\mathbf{i} + (x^2 + y)\,\mathbf{j}$.
6. $(y^2e^x - y)\,\mathbf{i} + (2y\,e^x - x)\,\mathbf{j}$.
7. $(\cos x - y\sin x)\,\mathbf{i} + \cos x\,\mathbf{j}$.
8. $(1 + e^y)\,\mathbf{i} + (x\,e^y + y^2)\,\mathbf{j}$.
9. $e^x\cos y^2\,\mathbf{i} - 2y\,e^x\sin y^2\mathbf{j}$.
10. $e^x\cos y\,\mathbf{i} + e^x\sin y\,\mathbf{j}$.
11. $y\,e^x(1 + x)\,\mathbf{i} + (x\,e^x - e^{-y})\,\mathbf{j}$.
12. $(e^x + 2xy)\,\mathbf{i} + (x^2 + \sin y)\,\mathbf{j}$.
13. $(x\,e^{xy} + x^2)\,\mathbf{i} + (y\,e^{xy} - 2y)\,\mathbf{j}$.
14. $(y\sin x + xy\cos x)\,\mathbf{i} + (x\sin x + 2y + 1)\,\mathbf{j}$.
15. $(1 + y^2 + xy^2)\,\mathbf{i} + (x^2y + y + 2xy + 1)\,\mathbf{j}$.
16. $\left[2\ln(3y) + \dfrac{1}{x}\right]\mathbf{i} + \left[\dfrac{2x}{y} + y^2\right]\mathbf{j}$.

In Exercises 17–20, find the most general function with the given gradient.

17. $\dfrac{x}{\sqrt{x^2+y^2}}\,\mathbf{i} + \dfrac{y}{\sqrt{x^2+y^2}}\,\mathbf{j}$.
18. $(x\tan y + \sec^2 x)\,\mathbf{i} + (\tfrac{1}{2}x^2\sec^2 y + \pi y)\,\mathbf{j}$.
19. $(x^2\sin^{-1}y)\,\mathbf{i} + \left(\dfrac{x^3}{3\sqrt{1-y^2}} - \ln y\right)\mathbf{j}$.
20. $\left(\dfrac{\tan^{-1}y}{\sqrt{1-x^2}} + \dfrac{x}{y}\right)\mathbf{i} + \left(\dfrac{\sin^{-1}x}{1+y^2} - \dfrac{x^2}{2y^2} + 1\right)\mathbf{j}$.

21. Use a CAS to determine whether **F** is a gradient.

    (a) $\mathbf{F}(x,y) = (y - 2xy + y^2)\,\mathbf{i} + (x - x^2 + 2xy)\,\mathbf{j}$.

    (b) $\mathbf{F}(x,y) = [2xy^2\cos(x^2 - y)]\,\mathbf{i} + [-y^2\cos(x^2 - y) + 2y\sin(x^2 - y)]\,\mathbf{j}$.

    (c) $\mathbf{F}(x,y) = 2xy(y - x)e^{-x^2y}\,\mathbf{i} + x^2(x - y)e^{-x^2y}\,\mathbf{j}$.

22. (a) Use a CAS to find f within an additive constant if

$$\nabla f = [(1-2xy[x-y])e^{-x^2y}]\,\mathbf{i} + [-(1+x^2[x-y])e^{-x^2y}]\,\mathbf{j}.$$

    (b) Use a CAS to find f if $f\left(\frac{\pi}{3}, \frac{\pi}{4}\right) = 6$ and

$$\nabla f = [\cos(x+y) + \sin(x-y)]\,\mathbf{i} + [\cos(x+y) - \sin(x-y)]\,\mathbf{j}.$$

23. Find the general solution of the differential equation $\nabla f(x,y) = f(x,y)\,\mathbf{i} + f(x,y)\,\mathbf{j}$.

24. Given that g and its first and second partials are everywhere continuous, find the general solution of the differential equation $\nabla f(x,y) = e^{g(x,y)}[g_x(x,y)\,\mathbf{i} + g_y(x,y)\,\mathbf{j}]$.

Theorem 15.9.2 has a three-dimensional analog. In particular we can show that, if P, Q, R are continuously differentiable on an open rectangular box S, then the vector function

$$P(x,y,z)\,\mathbf{i} + Q(x,y,z)\,\mathbf{j} + R(x,y,z)\,\mathbf{k}$$

is a gradient on S iff

$$\frac{\partial P}{\partial y} = \frac{\partial Q}{\partial x}, \quad \frac{\partial P}{\partial z} = \frac{\partial R}{\partial x}, \quad \frac{\partial Q}{\partial z} = \frac{\partial R}{\partial y} \quad \text{throughout S.}$$

25. (a) Verify that $2x\,\mathbf{i} + z\,\mathbf{j} + y\,\mathbf{k}$ is the gradient of a function f that is everywhere defined.

    (b) Deduce from the relation $\partial f/\partial x = 2x$ that $f(x,y,z) = x^2 + g(y,z)$.

    (c) Verify then that $\partial f/\partial y = z$ gives $g(y,z) = zy + h(z)$ and finally that $\partial f/\partial z = y$ gives $h(z) = C$.

    (d) What is $f(x,y,z)$?

Determine whether the vector function is a gradient $\nabla f(x,y,z)$ and, if so, find all functions f with that gradient.

26. $yz\,\mathbf{i} + xz\,\mathbf{j} + xy\,\mathbf{k}$.
27. $(2x + y)\,\mathbf{i} + (2y + x + z)\,\mathbf{j} + (y - 2z)\,\mathbf{k}$.
28. $(2x\sin 2y\cos z)\,\mathbf{i} + (2x^2\cos 2y\cos z)\,\mathbf{j} - (x^2\sin 2y\sin z)\,\mathbf{k}$.
29. $(y^2z^3 + 1)\,\mathbf{i} + (2xyz^3 + y)\,\mathbf{j} + (3xy^2z^2 + 1)\,\mathbf{k}$.
30. $\left[\dfrac{y}{z} - e^z\right]\mathbf{i} + \left[\dfrac{x}{z} + 1\right]\mathbf{j} - \left[xe^z + \dfrac{xy}{z^2}\right]\mathbf{k}$.
31. Verify that the gravitational force function

$$\mathbf{F}(\mathbf{r}) = -G\frac{mM}{r^3}\mathbf{r} \quad (\mathbf{r} = x\,\mathbf{i} + y\,\mathbf{j} + z\,\mathbf{k})$$

    is a gradient.

32. Verify that every vector function of the form

$$\mathbf{h}(\mathbf{r}) = kr^n\mathbf{r} \qquad (k \text{ constant, } n \text{ an integer})$$

    is a gradient.

■ CHAPTER HIGHLIGHTS

15.1 Differentiability and Gradient

gradient of f at $\mathbf{x}$: $\nabla f(\mathbf{x})$ (p. 862)

$$\nabla f(x,y) = \frac{\partial f}{\partial x}\mathbf{i} + \frac{\partial f}{\partial y}\mathbf{j}$$

$$\nabla f(x,y,z) = \frac{\partial f}{\partial x}\mathbf{i} + \frac{\partial f}{\partial y}\mathbf{j} + \frac{\partial f}{\partial z}\mathbf{k}$$

If f is differentiable at $\mathbf{x}$, then f is continuous at $\mathbf{x}$.

$$\nabla r^n = nr^{n-2}\mathbf{r}.$$

15.2 Gradients and Directional Derivatives

directional derivative: $f'_{\mathbf{u}}(\mathbf{x}) = \nabla f(\mathbf{x}) \cdot \mathbf{u}$ (p. 872)

The directional derivative $f'_{\mathbf{u}}$ gives the rate of change of f in the direction of the unit vector $\mathbf{u}$.

The directional derivative in a direction $\mathbf{u}$ is the component of the gradient vector in that direction. (p. 874)

A differentiable function f increases most rapidly in the direction of the gradient (the rate of change is then $\|\nabla f(\mathbf{x})\|$) and it decreases most rapidly in the opposite direction (the rate of change is then $-\|\nabla f(\mathbf{x})\|$).

15.3 The Mean-Value Theorem; Chain Rules

The mean-value theorem (p. 879)
open, connected set (p. 880)
chain rule along a curve (p. 882)
tree diagram (p. 886)
intermediate-value theorem (p. 891)
open, closed regions (p. 891)

In the setting of functions of several variables, there are numerous versions of the chain rule. They can all be deduced from the chain rule along a curve.

If f is a continuously differentiable function of $r = \|\mathbf{r}\|$, then

$$\nabla[f(r)] = f'(r)\frac{\mathbf{r}}{r}.$$

15.4 The Gradient as a Normal; Tangent Lines and Tangent Planes

tangent and normal lines to a curve $f(x,y) = c$ (p. 895)
tangent plane to a surface $f(x,y,z) = c$ (p. 897)
upper and lower unit normals (p. 902)

At each point of the domain of a function, the gradient vector, if not $\mathbf{0}$, is perpendicular to the level curve (level surface) that passes through that point.

15.5 Local Extreme Values

local maximum and local minimum (p. 903)
critical points (p. 904)
stationary points, saddle points (p. 904)

If f has a local extreme value at $\mathbf{x}_0$, then either $\nabla f(\mathbf{x}_0) = \mathbf{0}$ or $\nabla f(\mathbf{x}_0)$ does not exist.

second-partials test, discriminant (p. 906)

15.6 Absolute Extreme Values

absolute maximum and absolute minimum (p. 911)
bounded set (p. 911)
extreme-value theorem (p. 912)
procedure for finding extreme values (p. 912)
method of least squares (p. 916)

15.7 Maxima and Minima with side Conditions

If $\mathbf{x}_0$ maximizes (or minimizes) $f(\mathbf{x})$ subject to the side condition $g(\mathbf{x}) = 0$, then $\nabla f(\mathbf{x}_0)$ and $\nabla g(\mathbf{x}_0)$ are parallel. Thus, if $\nabla g(\mathbf{x}_0) \neq \mathbf{0}$, then there exists a scalar λ, called a Lagrange multiplier, such that $\nabla f(\mathbf{x}_0) = \lambda \nabla g(\mathbf{x}_0)$.

15.8 Differentials

increment : $\Delta f = f(\mathbf{x}+\mathbf{h}) - f(\mathbf{x})$
differential : $df = \nabla f(\mathbf{x}) \cdot \mathbf{h}$

$$\Delta f \cong df \quad \text{in the sense that} \quad \frac{\Delta f - df}{\|\mathbf{h}\|} \to 0 \text{ as } \mathbf{h} \to \mathbf{0}$$

$$\text{two variables:} \quad df = \frac{\partial f}{\partial x}\Delta x + \frac{\partial f}{\partial y}\Delta y$$

$$\text{three variables:} \quad df = \frac{\partial f}{\partial x}\Delta x + \frac{\partial f}{\partial y}\Delta y + \frac{\partial f}{\partial z}\Delta z$$

15.9 Reconstructing a Function from its Gradient

finding f from ∇f (p. 932)
closed curve, simple curve (p. 934)
simply connected open region (p. 935)

necessary and sufficient conditions for a vector-valued function to be a gradient:

two variables (p. 935)
three variables (p. 938)

Let $\mathbf{a} = 3\,\mathbf{i} + 2\,\mathbf{j} - \mathbf{k}$, $\mathbf{b} = 5\,\mathbf{i} + 3\,\mathbf{j}$, $\mathbf{c} = -2\,\mathbf{i} + 4\,\mathbf{j} + \mathbf{k}$. Find the indicated vector or scalar.

1. $2\,\mathbf{a} - 3\,\mathbf{b}$

2. $\mathbf{a} \cdot (\mathbf{b} + \mathbf{c})$

3. $\|\mathbf{a} + \mathbf{b}\|$

4. A unit vector in the same direction as **a**.

5. The cosine of the angle between **b** and **c**.

6. $\mathbf{a} \times \mathbf{b}$

7. A unit vector perpendicular to **a** and **c**.

8. The volume of the parallelpiped with **a**, **b**, and **c** as sides.

9. Let $P(3, 2, -1)$, $Q(7, -5, 4)$, $R(5, 6, -3)$ be points in space.

(a) Find scalar parametric equations for the line that passes through R parallel to the line determined by P and Q.

(b) Find scalar parametric equations for the line that passes through R perpendicular to the line determined by P and Q.

(c) The lines found in parts (a) and (b) determine a plane. Find an equation for it.

10. Repeat Exercise 9 with $P(4, 2, 3)$, $Q(-2, 1, 4)$, $R(1, -1, -6)$.

11. (a) Are the points $P(3, 2, -1)$, $Q(7, -5, 4)$, $R(5, -1, 1)$ collinear?

(b) Are the points $P(3, 2, -1)$, $Q(7, -5, 4)$, $R(5, -1, 1)$, $S(1, 2, 0)$ coplanar?

Write an equation for the plane that satisfies the given conditions.

12. Contains the points $P(1, -2, 1)$, $Q(2, 0, 3)$, $R(0, 1, -1)$.

13. Contains the point $P(2, 1, -3)$ and is perpendicular to the line

$$\frac{x+1}{2} = \frac{y-1}{3} = -\frac{z}{4}$$

14. Contains the point $P(1, -2, -1)$ and is parallel to the plane $3x + 2y - z = 4$.

15. Contains the point $P(3, -1, 2)$ and the line $x = 2 + 2t$, $y = -1 + 3t$, $z = -2t$.

16. Find scalar parametric equations for the line of intersection of the planes $2x + y - 3z + 6 = 0$ and $x + 4y + 5z - 7 = 0$.

Find $\mathbf{f}'$ and $\mathbf{f}''$ for the vector-valued function **f**.

17. $\mathbf{f}(t) = e^{2t}\,\mathbf{i} + \ln(t^2 + 1)\,\mathbf{j}$

18. $\mathbf{f}(t) = e^t \cos t\,\mathbf{i} - \cos 2t\,\mathbf{j} + 3\,\mathbf{k}$

19. $\mathbf{f}(t) = \sinh 2t\,\mathbf{i} - te^{-t}\,\mathbf{j} + \cosh\,t\,\mathbf{k}$

Find the velocity, speed, and acceleration of the object with position **r**.

20. $\mathbf{r}(t) = \cos 2t\,\mathbf{i} + \sin 2t\,\mathbf{j} - t^2\mathbf{k}$

21. $\mathbf{r}(t) = 2t\,\mathbf{i} + \ln t\,\mathbf{j} - t^2\,\mathbf{k}$

22. $\mathbf{r}(t) = \cosh t\,\mathbf{i} + \sinh t\,\mathbf{j} + t\,\mathbf{k}$

23. Find $\mathbf{f}(t)$ if $\mathbf{f}'(t) = t^2\,\mathbf{i} + (e^{2t} + 1)\,\mathbf{j} + \sqrt{2t+1}\,\mathbf{k}$ and $\mathbf{f}(0) = \mathbf{i} - 3\,\mathbf{j} + 3\,\mathbf{k}$.

24. Find the position, velocity and speed of an object with initial velocity $\mathbf{v}_0 = \mathbf{k}$, initial position $\mathbf{r}_0 = \mathbf{i}$ and acceleration vector $\mathbf{a}(t) = -\cos\,t\,\mathbf{i} - \sin\,t\,\mathbf{j}$.

Sketch the curve represented by the vector-valued function and indicate the orientation.

25. $\mathbf{r}(t) = 2t^2\,\mathbf{i} - t\,\mathbf{j};\; t \geq 0$

26. $\mathbf{r}(t) = e^{-t}\,\mathbf{i} + 2e^{2t}\,\mathbf{j}$; for all real t

27. $\mathbf{r}(t) = t\,\mathbf{i} + t\,\mathbf{j} + \sin t\,\mathbf{k};\; t \geq 0$

Find the tangent vector and scalar parametric equations for the tangent line at the indicated point.

28. $\mathbf{r}(t) = t^2\,\mathbf{i} + (t+1)\,\mathbf{j} - t^3\,\mathbf{k};\; P(1, 2, -1)$

29. $\mathbf{r}(t) = \cos\,2t\,\mathbf{i} + \sin\,2t\,\mathbf{j} + t\,\mathbf{k};\; t = \pi/3$

Find the unit tangent vector and the principal unit normal vector for the curve **r**.

30. $\mathbf{r}(t) = 2t\,\mathbf{i} + \ln t\,\mathbf{j} - t^2\,\mathbf{k}$

31. $\mathbf{r}(t) = \cos\,t\,\mathbf{i} + \cos\,t\,\mathbf{j} - \sqrt{2}\sin\,t\,\mathbf{k}$

32. $\mathbf{r}(t) = e^t\,\mathbf{i} + e^{-t}\,\mathbf{j} - t\sqrt{2}\,\mathbf{k}$

Find the length of the curve **r**.

33. $\mathbf{r}(t) = 2t\,\mathbf{i} + \frac{2}{3}t^{3/2}\,\mathbf{j}$; from $t = 0$ to $t = 5$.

34. $\mathbf{r}(t) = e^t\,\mathbf{i} + e^{-t}\,\mathbf{j} - t\sqrt{2}\,\mathbf{k}$; from $t = 0$ to $t = \ln 3$.

35. $\mathbf{r}(t) = \sinh t\,\mathbf{i} + \cosh t\,\mathbf{j} + t\,\mathbf{k}$; from $t = 0$ to $t = 1$.

36. Find the curvature of the plane curves.

(a) $y = x^{3/2}$ (b) $y = \cos\,2x$

37. Find the curvature of the plane curves.

(a) $x(t) = 2e^{-t}, y(t) = e^{-2t}$ (b) $\mathbf{r}(t) = \frac{1}{3}t^3\,\mathbf{i} + \frac{1}{2}t^2\,\mathbf{j}$

Interpret $\mathbf{r}(t)$ as the position of a moving object at time t. Find the curvature of the path and the tangential and normal components of acceleration.

38. $\mathbf{r}(t) = e^t\,\mathbf{i} + e^{-t}\,\mathbf{j} - t\sqrt{2}\,\mathbf{k}$

39. $\mathbf{r}(t) = \frac{4}{5}\cos\,t\,\mathbf{i} - \frac{3}{5}\cos\,t\,\mathbf{j} + (1 + \sin\,t)\,\mathbf{k}$

40. $\mathbf{r}(t) = \sinh t\,\mathbf{i} + \cosh t\,\mathbf{j} + t\,\mathbf{k}$

41. Find the domain and range of the function f.

(a) $f(x, y) = \ln(x^2 - y^2 - 1)$

(b) $f(x, y, z) = \sqrt{z - x^2 - y^2}$

42. Determine a function f whose value at (x, y) is:

(a) The volume of a circular cone of radius x and height y.

(b) The volume of a box whose length x is twice its width, and whose height is y.

(c) The cosine of the angle between the vectors $y\,\mathbf{i} + 2xy\,\mathbf{j}$ and $x\,\mathbf{i} + y\,\mathbf{j}$.

Identify the level curves $f(x,y) = c$ and sketch the curves corresponding to the given values of c.

43. $f(x,y) = \sqrt{x^2+y^2-4}$; $c = 0, \sqrt{5}$

44. $f(x,y) = \dfrac{y}{x^2} = c$; $c = -4, -1, 1, 4$

Identify the level surface $F(x,y,z) = c$ and sketch it.

45. $F(x,y,z) = 2x+y+3z$; $c = 6$

46. $F(x,y,z) = 4x^2+9y^2+36z^2$; $c = 36$

Calculate the first-order partial derivatives

47. $z = x^2 \sin xy^2$

48. $f(x,y,z) = \dfrac{2xy}{x+y+z}$

49. $g(x,y,z) = \ln\sqrt{x^2+y^2+z^2}$

Calculate the indicated second-order partial derivatives of f.

50. $f(x,y) = x^3y^2 - 4xy^3 + 2x - y$; f_{xx}, f_{yx}.

51. $f(x,y,z) = 2x^2yz^3 + e^{xyz}$; f_{xx}, f_{zx}, f_{yz}.

52. The surface $z = \sqrt{20-2x^2-3y^2}$ is the top half of an ellipsoid centered at the origin.

(a) Find scalar parametric equations for the line l_1 tangent to the curve of intersection of the ellipsoid and the plane $x = 2$ at the point (2, 1, 3) on the surface.

(b) Find scalar parametric equations for the line l_2 tangent to the curve of intersection of the ellipsoid and the plane $y = 1$ at the point (2, 1, 3) on the surface.

(c) The tangent lines l_1 and l_2 determine a plane. Find an equation for this plane.

53. Find the gradient of the function f.

(a) $f(x,y) = 2x^2 - 4xy + y^3$ (b) $f(x,y) = \dfrac{xy}{x^2+y^2}$

54. Find the gradient of the function F

(a) $F(x,y,z) = \ln\sqrt{x^2+y^2+z^2}$

(b) $F(x,y,z) = x^2e^{-yz}\cos 2z$

Find the directional derivative at the point P in the direction indicated.

55. $f(x,y) = x^2 - 2xy$ at $(1, -2)$ in the direction of $\mathbf{a} = \mathbf{i} + 2\,\mathbf{j}$.

56. $f(x,y,z) = e^{x^2+y^2+z^2}$ at (0, 0, 0) in the direction of the line $\dfrac{x}{2} = \dfrac{y}{-1} = \dfrac{z}{4}$.

57. Find the directional derivative of $f(x,y) = e^{2x}(\cos y - \sin y)$ at $(\frac{1}{2}, -\frac{1}{2}\pi)$ in the direction of the greatest increase of f.

58. Find the directional derivative of $F(x,y,z) = \sin(xyz)$ at $(\frac{1}{2}, \frac{1}{3}, \pi)$ in the direction of the greatest decrease of F.

Find an equation for the tangent plane and scalar parametric equations for the normal line to the surface at the point indicated.

59. $z = \sqrt{4-x^2-y^2}$ at $(1, -1, \sqrt{2})$.

60. $z^3 + xyz - 2 = 0$ at (1, 1, 1).

61. $ye^{xy} + 2z^2 = 1$ at $(0, -1, 1)$.

62. $f(x,y) = \ln(x^2+y^2)$ at $(1, -1, \ln 2)$.

Find the stationary points; determine the local maxima and minima and the saddle points.

63. $f(x,y) = x^2y - 2xy + 2y^2 - 15y - 2$

64. $f(x,y) = 3x^2 - 3xy^2 + y^3 + 3y^2$

65. $f(x,y) = (x-3)\ln xy$

Find the absolute maximum and minimum of f on the set D.

66. $f(x,y) = x^2+y^2-2x+2y+2$; $D = \{(x,y): x^2+y^2 \le 4\}$.

67. $f(x,y) = 2x^2 - 4x + y^2 - 4y + 3$; D the closed triangular region bounded by the lines $x = 0$, $y = 3$, $y = x$.

68. Use differentials to calculate the approximate value.

(a) $e^{0.02}\sqrt{15.2 + (1.01)^3}$ (b) $\sqrt[3]{64.5}\cos^2(28°)$

69. Determine whether the given vector function is the gradient of a function f. If it is, find f.

(a) $\mathbf{R}(x,y) = (y^2e^{2x} + 4x + 4 - 2y)\,\mathbf{i} + (ye^{2x} - 2x^3 + 2y)\,\mathbf{j}$

(b) $\mathbf{S}(x,y) = \left(\dfrac{y}{x^2} + 4x^3 - 1 + 3y\sin 3x\right)\mathbf{i} + \left(3y^2 + 2 - \dfrac{1}{x} - \cos 3x\right)\mathbf{j}$

Find the extreme values of f subject to the side condition.

70. $f(x,y) = x^2y^2$; $\frac{1}{9}x^2 + \frac{1}{4}y^2 = 1$

71. $f(x,y,z) = xz + 2y$; $x^2+y^2+z^2 = 36$

72. Find the volume of the largest rectangular box that can be inscribed in the ellipsoid

$$4x^2 + 9y^2 + 36z^2 = 36$$

if the edges are parallel to the coordinate planes.

73. A closed rectangular box having a volume of 16 cubic feet will be constructed from three types of metal. The cost of the metal for the bottom is \$0. 50 per square foot, the metal for the sides costs \$0. 25 per square foot, and the cost for the top is \$0. 10 per square foot. Find the dimensions of the box that will minimize the cost of construction.

74. A metal silo in the shape of a right circular cylinder 22 feet high and 10 feet in diameter will be given a coat of paint 0.01 inches thick. The contractor for the job wants to estimate the number of gallons of paint that will be needed. Use differentials to obtain an estimate (there are 231 cubic inches in a gallon).

75. Find the absolute maximum value and the absolute minimum value of

$$T(x,y) = 2x^2 + y^2 - y$$

on the closed disc $x^2 + y^2 \le 1$.

CHAPTER 16

DOUBLE AND TRIPLE INTEGRALS

We began with ordinary integrals

$$\int_a^b f(x)\,dx$$

which, with $b > a$, we can write as

$$\int_{[a,b]} f(x)\,dx.$$

Here we will study double integrals

$$\iint_{\Omega} f(x,y)\,dxdy$$

where Ω is a region in the xy-plane and, a little later, triple integrals

$$\iiint_{T} f(x,y,z)\,dxdydz$$

where T is a solid in three-dimensional space. Our first step is to introduce some new notation.

■ 16.1 MULTIPLE-SIGMA NOTATION

In an ordinary sequence $\{a_n\}$, each term a_i is indexed by a single integer. The sum of all the a_i from $i = 1$ to $i = m$ is then denoted by

$$\sum_{i=1}^{m} a_i.$$

When two indices are involved, say,

$$a_{ij} = 2^i 5^j, \qquad a_{ij} = \frac{2i}{5+j} \qquad \text{or} \qquad a_{ij} = (1+i)^j,$$

then we use double-sigma notation to denote the sum of all the doubly indexed terms. By

(16.1.1)
$$\boxed{\sum_{i=1}^{m}\sum_{j=1}^{n} a_{ij}}$$

we mean *the sum of all the* a_{ij} *where* i *ranges from* 1 *to* m *and* j *ranges from* 1 *to* n. For example,

$$\sum_{i=1}^{3}\sum_{j=1}^{2} 2^i 5^j = 2\cdot 5 + 2\cdot 5^2 + 2^2\cdot 5 + 2^2\cdot 5^2 + 2^3\cdot 5 + 2^3\cdot 5^2 = 420.$$

Since addition is associative and commutative, we can add the terms of (16.1.1) in any order we choose. Usually we set

(16.1.2)
$$\boxed{\sum_{i=1}^{m}\sum_{j=1}^{n} a_{ij} = \sum_{i=1}^{m}\left(\sum_{j=1}^{n} a_{ij}\right).}$$

We can expand the expression on the right by expanding first with respect to i and then with respect to j:

$$\begin{aligned}\sum_{i=1}^{m}\left(\sum_{j=1}^{n} a_{ij}\right) &= \sum_{j=1}^{n} a_{1j} + \sum_{j=1}^{n} a_{2j} + \cdots + \sum_{j=1}^{n} a_{mj}\\ &= (a_{11} + a_{12} + \cdots + a_{1n}) + (a_{21} + a_{22} + \cdots + a_{2n}) +\\ &\quad \cdots + (a_{m1} + a_{m2} + \cdots + a_{mn}),\end{aligned}$$

or we can expand first with respect to j and then with respect to i:

$$\begin{aligned}\sum_{i=1}^{m}\left(\sum_{j=1}^{n} a_{ij}\right) &= \sum_{i=1}^{m} (a_{i1} + a_{i2} + \cdots + a_{in})\\ &= (a_{11} + a_{12} + \cdots + a_{1n}) + (a_{21} + a_{22} + \cdots + a_{2n}) +\\ &\quad \cdots + (a_{m1} + a_{m2} + \cdots + a_{mn}).\end{aligned}$$

The results are the same.

For example, we can write

$$\begin{aligned}\sum_{i=1}^{3}\left(\sum_{j=1}^{2} a_{ij}\right) &= \sum_{j=1}^{2} a_{1j} + \sum_{j=1}^{2} a_{2j} + \sum_{j=1}^{2} a_{3j}\\ &= (a_{11} + a_{12}) + (a_{21} + a_{22}) + (a_{31} + a_{32}),\end{aligned}$$

or we can write

$$\sum_{i=1}^{3}\left(\sum_{j=1}^{2} a_{ij}\right) = \sum_{i=1}^{3}(a_{i1} + a_{i2}) = (a_{11} + a_{12}) + (a_{21} + a_{22}) + (a_{31} + a_{32}).$$

Since constants can be factored through single sums, they can also be factored through double sums; namely,

(16.1.3)
$$\sum_{i=1}^{m}\sum_{j=1}^{n} \alpha a_{ij} = \alpha \sum_{i=1}^{m}\sum_{j=1}^{n} a_{ij}.$$

Also,

(16.1.4)
$$\sum_{i=1}^{m}\sum_{j=1}^{n} (a_{ij} + b_{ij}) = \sum_{i=1}^{m}\sum_{j=1}^{n} a_{ij} + \sum_{i=1}^{m}\sum_{j=1}^{n} b_{ij}.$$

The easiest double sums to evaluate are those where each term a_{ij} appears as a product $b_i c_j$ in which each of the factors bears only one index. In that case, we can express the double sum as the product of two single sums:

(16.1.5)
$$\sum_{i=1}^{m}\sum_{j=1}^{n} b_i c_j = \left(\sum_{i=1}^{m} b_i\right)\left(\sum_{j=1}^{n} c_j\right).$$

PROOF Set

$$B = \sum_{i=1}^{m} b_i, \quad C = \sum_{j=1}^{n} c_j.$$

Then

$$\sum_{i=1}^{m}\sum_{j=1}^{n} b_i c_j = \sum_{i=1}^{m}\left(\sum_{j=1}^{n} b_i c_j\right) \overset{\dagger}{=} \sum_{i=1}^{m} b_i\left(\sum_{j=1}^{n} c_j\right) = \sum_{i=1}^{m} b_i C$$
$$= C\sum_{i=1}^{m} b_i = CB = BC. \quad \square$$

For example,

$$\sum_{i=1}^{3}\sum_{j=1}^{2} 2^i 5^j = \left(\sum_{i=1}^{3} 2^i\right)\left(\sum_{j=1}^{2} 5^j\right)$$
$$= (2 + 2^2 + 2^3)(5 + 5^2) = (14)(30) = 420.$$

† Since b_i is independent of j, it can be factored through the j-summation.

Triple-sigma notation is used when three indices are involved. The sum of all the a_{ijk} where i ranges from 1 to m, j from 1 to n, and k from 1 to q can be written

(16.1.6)
$$\sum_{i=1}^{m}\sum_{j=1}^{n}\sum_{k=1}^{q} a_{ijk}.$$

Multiple sums appear in the following sections. We introduced them here so as to avoid lengthy asides later.

EXERCISES 16.1

Evaluate the sum.

1. $\sum_{i=1}^{3}\sum_{j=1}^{3} 2^{i-1}3^{j+1}.$

2. $\sum_{i=1}^{4}\sum_{j=1}^{2} (1+i)^{j}.$

3. $\sum_{i=1}^{4}\sum_{j=1}^{3} (i^2+3i)(j-2).$

4. $\sum_{i=1}^{3}\sum_{j=1}^{2}\sum_{k=1}^{3} \dfrac{2i}{k+j^2}.$

For Exercises 5–16, let

$$P_1 = \{x_0, x_1, \ldots, x_m\} \text{ be a partition of } [a_1, a_2],$$
$$P_2 = \{y_0, y_1, \ldots, y_n\} \text{ be a partition of } [b_1, b_2],$$
$$P_3 = \{z_0, z_1, \ldots, z_q\} \text{ be a partition of } [c_1, c_2],$$

and let

$$\Delta x_i = x_i - x_{i-1}, \quad \Delta y_j = y_j - y_{j-1}, \quad \Delta z_k = z_k - z_{k-1}.$$

Evaluate the sum.

5. $\sum_{i=1}^{m} \Delta x_i.$

6. $\sum_{j=1}^{n} \Delta y_j.$

7. $\sum_{i=1}^{m}\sum_{j=1}^{n} \Delta x_i \Delta y_j.$

8. $\sum_{j=1}^{n}\sum_{k=1}^{q} \Delta y_j \Delta z_k.$

9. $\sum_{i=1}^{m} (x_i + x_{i-1})\Delta x_i.$

10. $\sum_{j=1}^{n} \frac{1}{2}(y_j^2 + y_j y_{j-1} + y_{j-1}^2)\Delta y_j.$

11. $\sum_{i=1}^{m}\sum_{j=1}^{n} (x_i + x_{i-1})\Delta x_i \Delta y_j.$

12. $\sum_{i=1}^{m}\sum_{j=1}^{n} (y_j + y_{j-1})\Delta x_i \Delta y_j.$

13. $\sum_{i=1}^{m}\sum_{j=1}^{n} (2\Delta x_i - 3\Delta y_j).$

14. $\sum_{i=1}^{m}\sum_{j=1}^{n} (3\Delta x_i - 2\Delta y_j).$

15. $\sum_{i=1}^{m}\sum_{j=1}^{n}\sum_{k=1}^{q} \Delta x_i \Delta y_j \Delta z_k.$

16. $\sum_{i=1}^{m}\sum_{j=1}^{n}\sum_{k=1}^{q} (x_i + x_{i-1})\Delta x_i \Delta y_j \Delta z_k.$

17. Evaluate

$$\sum_{i=1}^{n}\sum_{j=1}^{n}\sum_{k=1}^{n} \delta_{ijk}\, a_{ijk} \quad \text{where} \quad \delta_{ijk} = \begin{cases} 1, & \text{if } i = j = k \\ 0, & \text{otherwise.} \end{cases}$$

18. Show that any sum written in double-sigma notation can be expressed in ordinary (that is, in single) sigma notation.

■ 16.2 DOUBLE INTEGRALS

The Double Integral over a Rectangle

We start with a function f continuous on a rectangle.

$$R: \quad a \le x \le b, \quad c \le y \le d.$$

See Figure 16.2.1. Our object is to define the double integral of f over R:

$$\iint_R f(x, y)\, dxdy.$$

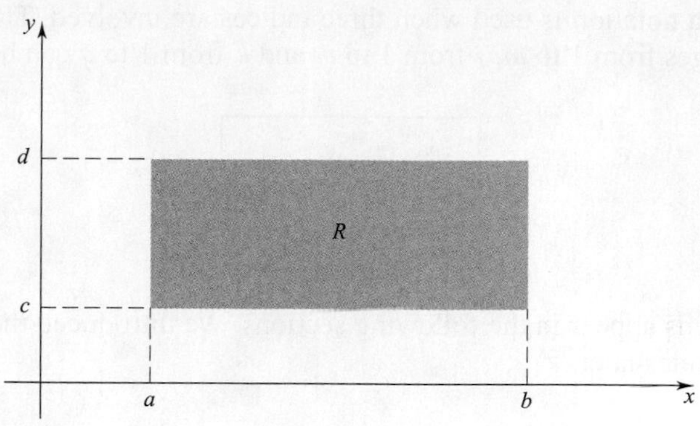

Figure 16.2.1

To define the integral

$$\int_a^b f(x)\,dx,$$

we introduced some auxiliary notions, namely: partition P of $[a, b]$, upper sum $U_f(P)$, and lower sum $L_f(P)$. We were then able to define

$$\int_a^b f(x)\,dx$$

as the unique number I that satisfies the inequality

$$L_f(P) \leq I \leq U_f(P) \quad \text{for all partitions } P \text{ of } [a, b].$$

We will follow exactly the same procedure to define the double integral

$$\iint_R f(x, y)\,dxdy.$$

First we explain what we mean by a partition of the rectangle R. To do this, we begin with a partition

$$P_1 = \{x_0, x_1, \ldots, x_m\} \quad \text{of} \quad [a, b]$$

and a partition

$$P_2 = \{y_0, y_1, \ldots, y_n\} \quad \text{of} \quad [c, d].$$

The set

$$P = P_1 \times P_2 = \{(x_i, y_j) : x_i \in P_1, y_j \in P_2\}\dagger$$

is called a *partition of R* (see Figure 16.2.2); P consists of all the grid points (x_i, y_j).

† $P_1 \times P_2$ is the Cartesian product of P_1 and P_2; if A and B are sets, then the *Cartesian product* of A and B is by definition the set $A \times B = \{(a, b) : a \in A \text{ and } b \in B\}$.

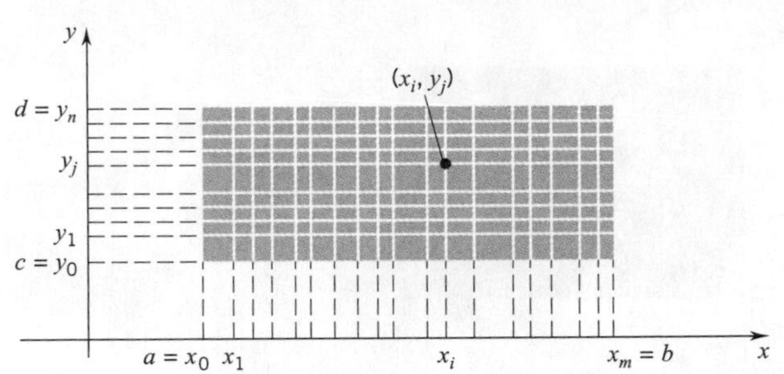

Figure 16.2.2

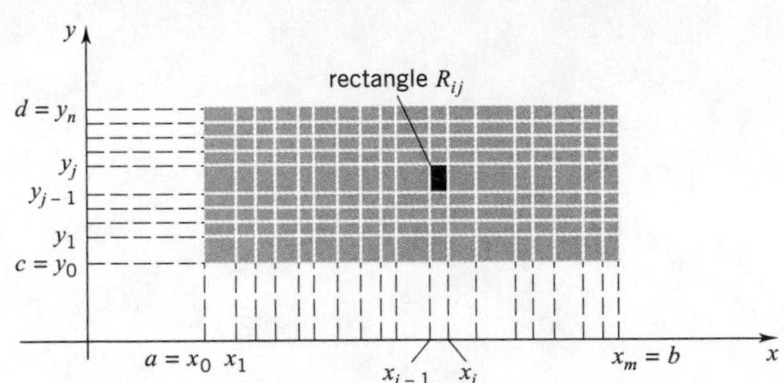

Figure 16.2.3

Using the partition P, we break up R into $m \times n$ nonoverlapping rectangles

$$R_{ij}: \quad x_{i-1} \leq x \leq x_i, \quad y_{j-1} \leq y \leq y_j, \qquad \text{(Figure 16.2.3)}$$

where $1 \leq i \leq m$, $1 \leq j \leq n$. On each rectangle R_{ij}, the function f takes on a maximum value M_{ij} and a minimum value m_{ij}. We know this because f is continuous and R_{ij} is closed and bounded (Section 15.6). The sum of all the products

$$M_{ij}(\text{ area of } R_{ij}) = M_{ij}(x_i - x_{i-1})(y_j - y_{j-1}) = M_{ij}\,\Delta x_i \Delta y_j$$

is called the P *upper sum* for f:

(16.2.1) $$U_f(P) = \sum_{i=1}^{m}\sum_{j=1}^{n} M_{ij}(\text{ area of } R_{ij}) = \sum_{i=1}^{m}\sum_{j=1}^{n} M_{ij}\,\Delta x_i\,\Delta y_j.$$

The sum of all the products

$$m_{ij}(\text{ area of } R_{ij}) = m_{ij}(x_i - x_{i-1})(y_j - y_{j-1}) = m_{ij}\,\Delta x_i\,\Delta y_j$$

is called the P *lower sum* for f:

(16.2.2) $$L_f(P) = \sum_{i=1}^{m}\sum_{j=1}^{n} m_{ij}(\text{area of } R_{ij}) = \sum_{i=1}^{m}\sum_{j=1}^{n} m_{ij}\,\Delta x_i\,\Delta y_j.$$

Example 1 Consider the function $f(x, y) = x + y - 2$ on the rectangle

$$R: \quad 1 \leq x \leq 4, \quad 1 \leq y \leq 3.$$

As a partition of [1, 4] take

$$P_1 = \{1, 2, 3, 4\},$$

and as a partition of [1, 3] take

$$P_2 = \{1, \tfrac{3}{2}, 3\}.$$

The partition $P = P_1 \times P_2$ then breaks up the initial rectangle into the six rectangles marked in Figure 16.2.4. On each rectangle R_{ij}, the function f takes on its maximum value M_{ij} at the point (x_i, y_j), the corner farthest from the origin:

$$M_{ij} = f(x_i, y_j) = x_i + y_j - 2.$$

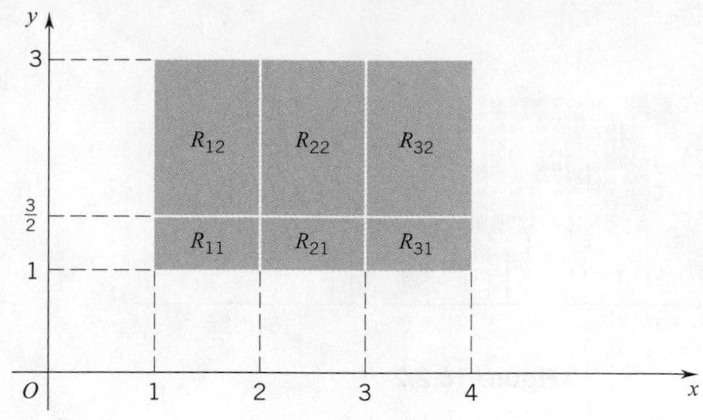

Figure 16.2.4

Thus
$$\begin{aligned}U_f(P) &= M_{11}(\text{area of } R_{11}) + M_{12}(\text{area of } R_{12}) + M_{21}(\text{area of } R_{21})\\ &\quad + M_{22}(\text{area of } R_{22}) + M_{31}(\text{area of } R_{31}) + M_{32}(\text{area of } R_{32})\\ &= \tfrac{3}{2}(\tfrac{1}{2}) + 3(\tfrac{3}{2}) + \tfrac{5}{2}(\tfrac{1}{2}) + 4(\tfrac{3}{2}) + \tfrac{7}{2}(\tfrac{1}{2}) + 5(\tfrac{3}{2}) = \tfrac{87}{4}.\end{aligned}$$

On each rectangle R_{ij}, f takes on its minimum value m_{ij} at the point (x_{i-1}, y_{j-1}), the corner closest to the origin:

$$m_{ij} = f(x_{i-1}, y_{j-1}) = x_{i-1} + y_{j-1} - 2.$$

Thus
$$\begin{aligned}L_f(P) &= m_{11}(\text{ area of } R_{11}) + m_{12}(\text{ area of } R_{12}) + m_{21}(\text{ area of } R_{21})\\ &\quad + m_{22}(\text{ area of } R_{22}) + m_{31}(\text{ area of } R_{31}) + m_{32}(\text{ area of } R_{32})\\ &= 0(\tfrac{1}{2}) + \tfrac{1}{2}(\tfrac{3}{2}) + 1(\tfrac{1}{2}) + \tfrac{3}{2}(\tfrac{3}{2}) + 2(\tfrac{1}{2}) + \tfrac{5}{2}(\tfrac{3}{2}) = \tfrac{33}{4}. \quad \square\end{aligned}$$

We return now to the general situation. As in the one-variable case, it can be shown that if f is continuous, then there exists one and only one number I that satisfies the inequality

$$L_f(P) \le I \le U_f(P) \quad \text{for all partitions } P \text{ of } R.$$

DEFINITION 16.2.3 THE DOUBLE INTEGRAL OVER A RECTANGLE *R*

Let f be continuous on a closed rectangle R. The unique number I that satisfies the inequality

$$L_f(P) \le I \le U_f(P) \quad \text{for all partitions } P \text{ of } R$$

is called the *double integral* of f over R, and is denoted by

$$\iint_R f(x, y)\, dxdy.\dagger$$

† The double integral $\iint_R f(x, y)\, dxdy$ can also be written $\iint_R f(x, y)\, dA$.

The Double Integral as a Volume

If f is continuous and nonnegative on the rectangle R, the equation

$$z = f(x, y)$$

represents a surface that lies above R. (See Figure 16.2.5.) In this case the double integral

$$\iint_R f(x, y)\, dxdy$$

gives the volume of the solid that is bounded below by R and bounded above by the surface $z = f(x, y)$.

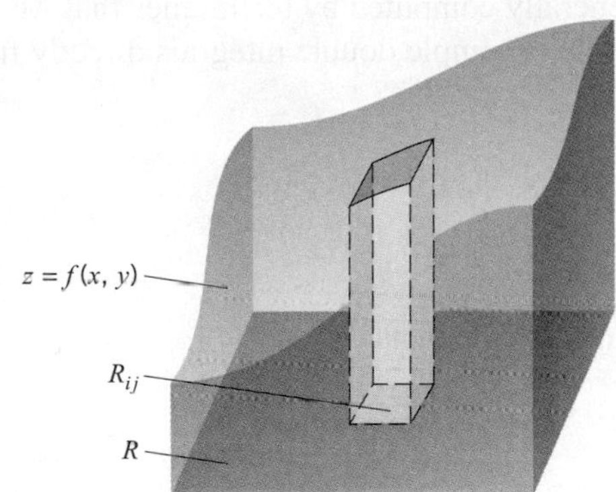

Figure 16.2.5

To see this, consider a partition P of R. P breaks up R into subrectangles R_{ij} and thus the entire solid T into parts T_{ij}. Since T_{ij} contains a rectangular solid with base R_{ij} and height m_{ij} (Figure 16.2.6), we must have

$$m_{ij}(\text{area of } R_{ij}) \leq \text{volume of } T_{ij}.$$

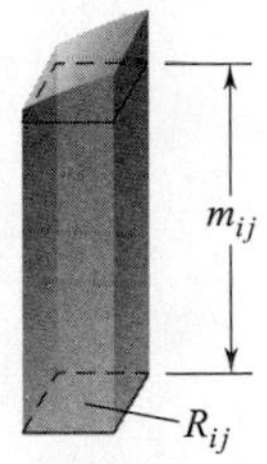

Figure 16.2.6

Since T_{ij} is contained in a rectangular solid with base R_{ij} and height M_{ij} (Figure 16.2.7), we must have

$$\text{volume of } T_{ij} \leq M_{ij}(\text{area of } R_{ij}).$$

In short, for each pair of indices i and j, we must have

$$m_{ij}(\text{area of } R_{ij}) \leq \text{ volume of } T_{ij} \leq M_{ij}(\text{area of } R_{ij}).$$

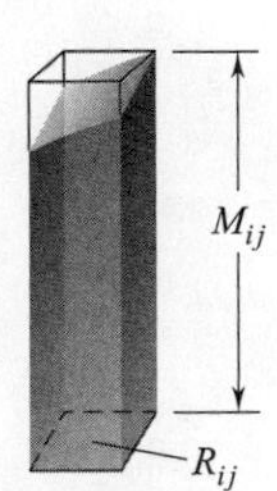

Figure 16.2.7

Adding up these inequalities, we can conclude that

$$L_f(P) \leq \text{ volume of } T \leq U_f(P).$$

Since P is arbitrary, the volume of T must be the double integral:

(16.2.4)

$$\boxed{\text{volume of } T = \iint_R f(x, y)\, dxdy.}$$

The double integral

$$\iint_R 1 \; dxdy = \iint_R dxdy$$

gives the volume of a solid of constant height 1 erected over R. In square units this is just the area of R:

(16.2.5)
$$\boxed{\text{area of } R = \iint_R dxdy.}$$

Some Computations

Double integrals are generally computed by techniques that we will take up later. It is possible, however, to evaluate simple double integrals directly from the definition.

Example 2 Evaluate

$$\iint_R \alpha \; dxdy$$

where α is a constant and R is the rectangle

$$R: \quad a \le x \le b, \quad c \le y \le d.$$

SOLUTION Here $f(x, y) = \alpha$ for all $(x, y) \in R$.

We begin with $P_1 = \{x_0, x_1, \ldots, x_m\}$ as an arbitrary partition of the interval $[a, b]$ and $P_2 = \{y_0, y_1, \ldots, y_n\}$ as an arbitrary partition of $[c, d]$. This gives

$$P = P_1 \times P_2 = \{(x_i, y_j) : x_i \in P_1, \; y_j \in P_2\}$$

as an arbitrary partition of R. On each rectangle R_{ij}, f has constant value α. Therefore we have $M_{ij} = \alpha$ and $m_{ij} = \alpha$ throughout. Thus

$$U_f(P) = \sum_{i=1}^{m}\sum_{j=1}^{n} \alpha \, \Delta x_i \, \Delta y_j = \alpha \left(\sum_{i=1}^{m} \Delta x_i\right)\left(\sum_{j=1}^{n} \Delta y_j\right) = \alpha(b-a)(d-c).$$

Similarly, $$L_f(P) = \alpha(b-a)(d-c).$$

The inequality $$L_f(P) \le I \le U_f(P) \qquad \text{for all } P$$

forces $$\alpha(b-a)(d-c) \le I \le \alpha(b-a)(d-c).$$

The only number I that can satisfy this inequality is

$$I = \alpha(b-a)(d-c).$$

Therefore $$\iint_R f(x, y) \; dxdy = \alpha(b-a)(d-c). \quad \square$$

Remark If $\alpha > 0$,

$$\iint_R \alpha \; dxdy = \alpha(b-a)(d-c)$$

gives the volume of the rectangular solid of constant height α erected over the rectangle R. (Figure 16.2.8) ☐

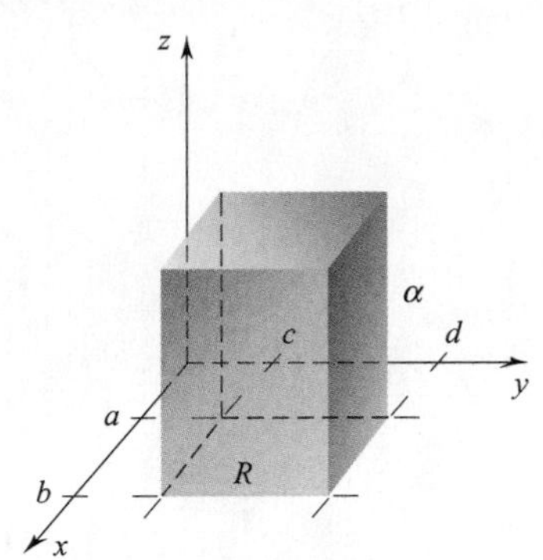

Figure 16.2.8

Example 3 Evaluate

$$\iint_R (x+y-2)\; dxdy,$$

where R is the rectangle: $1 \le x \le 4, 1 \le y \le 3$. (This is a continuation of Example 1.)

SOLUTION With $P_1 = \{x_0, x_1, \ldots, x_m\}$ as an arbitrary partition of the interval $[1, 4]$ and $P_2 = \{y_0, y_1, \ldots, y_n\}$ as an arbitrary partition of $[1, 3]$, we have

$$P_1 \times P_2 = \{(x_i, y_j) : x_i \in P_1, y_j \in P_2\}$$

as an arbitrary partition of R. On each rectangle $R_{ij} : x_{i-1} \le x \le x_i, y_{j-1} \le y \le y_j$, the function

$$f(x,y) = x + y - 2$$

has a maximum $M_{ij} = x_i + y_j - 2$ and a minimum $m_{ij} = x_{i-1} + y_{j-1} - 2$. Thus

$$L_f(P) = \sum_{i=1}^{m}\sum_{j=1}^{n}(x_{i-1} + y_{j-1} - 2)\Delta x_i\; \Delta y_j$$

$$U_f(P) = \sum_{i=1}^{m}\sum_{j=1}^{n}(x_i + y_j - 2)\; \Delta x_i\; \Delta y_j.$$

Now, for each pair of indices i and j

$$x_{i-1} + y_{j-1} - 2 \le \tfrac{1}{2}(x_i + x_{i-1}) + \tfrac{1}{2}(y_j + y_{j-1}) - 2 \le x_i + y_j - 2. \qquad \text{(explain)}$$

Adding up these inequalities, we have

$$L_f(P) \le \sum_{i=1}^{m}\sum_{j=1}^{n}[\tfrac{1}{2}(x_i + x_{i-1}) + \tfrac{1}{2}(y_j + y_{j-1}) - 2]\; \Delta x_i\; \Delta y_j \le U_f(P).$$

The double sum in the middle can be written

$$\sum_{i=1}^{m}\sum_{j=1}^{n}\tfrac{1}{2}(x_i + x_{i-1})\; \Delta x_i\; \Delta y_j + \sum_{i=1}^{m}\sum_{j=1}^{n}\tfrac{1}{2}(y_j + y_{j-1})\; \Delta x_i\; \Delta y_j - \sum_{i=1}^{m}\sum_{j=1}^{n}2\Delta x_i\; \Delta y_j.$$

The first double sum reduces to

$$\begin{aligned}\sum_{i=1}^{m}\sum_{j=1}^{n}\tfrac{1}{2}(x_i^2 - x_{i-1}^2)\Delta y_j &= \tfrac{1}{2}\left(\sum_{i=1}^{m}(x_i^2 - x_{i-1}^2)\right)\left(\sum_{j=1}^{n}\Delta y_j\right)\\ &= \tfrac{1}{2}(16-1)(3-1) = 15.\end{aligned}$$

The second double sum reduces to

$$\sum_{i=1}^{m}\sum_{j=1}^{n}\tfrac{1}{2}\Delta x_i(y_j^2 - y_{j-1}^2) = \tfrac{1}{2}\left(\sum_{i=1}^{m}\Delta x_i\right)\left(\sum_{j=1}^{n}(y_j^2 - y_{j-1}^2)\right)$$

$$= \tfrac{1}{2}(4-1)(9-1) = 12.$$

The third double sum reduces to

$$-\sum_{i=1}^{m}\sum_{j=1}^{n}2\Delta x_i\,\Delta y_j = -2\left(\sum_{i=1}^{m}\Delta x_i\right)\left(\sum_{j=1}^{n}\Delta y_j\right) = -2(4-1)(3-1) = -12.$$

The sum of these numbers, $15 + 12 - 12 = 15$, satisfies the inequality

$$L_f(P) \leq 15 \leq U_f(P) \qquad \text{for arbitrary } P.$$

Thus,
$$\iint_R (x + y - 2)\; dxdy = 15. \quad \square$$

Remark Since $f(x, y) = x + y - 2 \geq 0$ on the rectangle R, the double integral gives the volume of the prism bounded above by the plane $z = x + y - 2$ and below by R. See Figure 16.2.9. $\square$

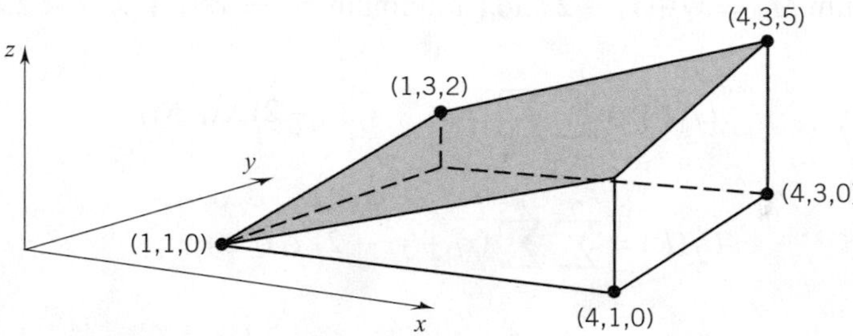

Figure 16.2.9

The Double Integral over a Region

We start with a closed and bounded set Ω in the xy-plane such as that depicted in Figure 16.2.10. We assume that Ω is a *basic region*; that is, we assume that Ω is a connected set (see Section 15.3) the total boundary of which consists of a finite number of continuous arcs of the form $y = \phi(x), x = \psi(y)$. See, for example, Figure 16.2.11.

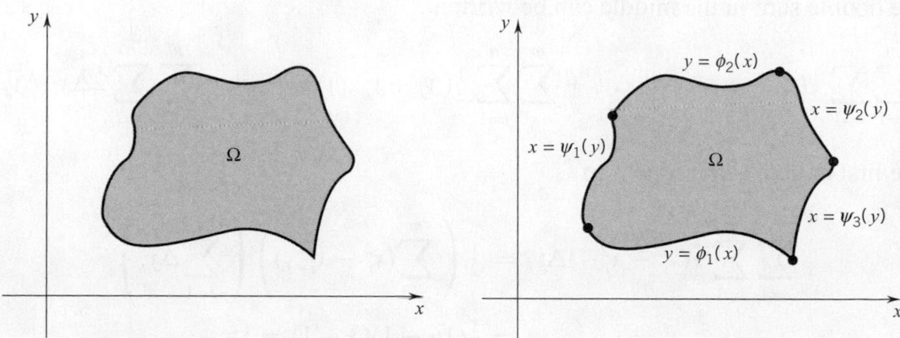

Figure 16.2.10

Figure 16.2.11

Now let f be a function continuous on Ω. We want to define the double integral

$$\iint_{\Omega} f(x,y)\ dxdy.$$

To do this, we enclose Ω by a rectangle R with sides parallel to the coordinate axes as in Figure 16.2.12. We extend f to all of R by setting f equal to 0 outside Ω. This extended function, which we continue to call f, is bounded on R, and it is continuous on all of R except possibly at the boundary of Ω. In spite of these possible discontinuities, it can be shown that f is still integrable on R; that is, there still exists a unique number I such that

$$L_f(P) \leq I \leq U_f(P) \qquad \text{for all partitions } P \text{ of } R.$$

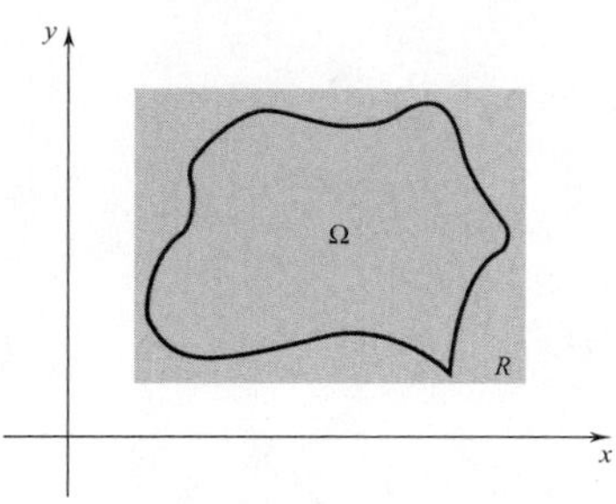

Figure 16.2.12

(We will not attempt to prove this.) This number I is by definition the double integral

$$\iint_{R} f(x,y)\ dxdy.$$

As you have probably guessed by now, we define the double integral over Ω by setting

(16.2.6)
$$\iint_{\Omega} f(x,y)\ dxdy = \iint_{R} f(x,y)\ dxdy.$$

If f is continuous and nonnegative over Ω, the extended f is nonnegative on all of R (Figure 16.2.13). The double integral gives the volume of the solid bounded above by the surface $z = f(x,y)$ and bounded below by the rectangle R. But since the surface has height 0 outside of Ω, the volume outside Ω is 0. It follows that

$$\iint_{\Omega} f(x,y)\ dxdy$$

gives the volume of the solid T bounded above by $z = f(x,y)$ and bounded below by Ω:

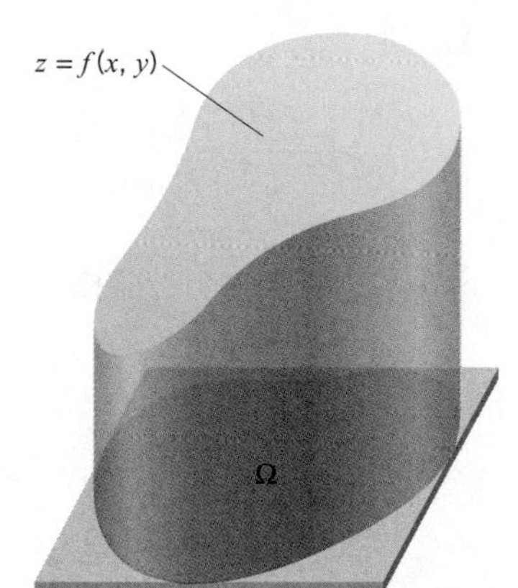

Figure 16.2.13

(16.2.7)
$$\text{volume of } T = \iint_{\Omega} f(x,y)\ dxdy.$$

The double integral

$$\iint_{\Omega} 1\ dxdy = \iint_{\Omega} dxdy$$

gives the volume of a solid of constant height 1 erected over Ω. In square units this is the area of Ω:

(16.2.8)
$$\text{area of } \Omega = \iint_{\Omega} dxdy.$$

Below we list four elementary properties of the double integral. They are all analogous to what you saw in the one-variable case. As specified above, the Ω referred to is a basic region. The functions f and g are assumed to be continuous on Ω.

I. Linearity: The double integral of a linear combination is the linear combination of the double integrals:

$$\iint_{\Omega} [\alpha f(x,y) + \beta g(x,y)]\, dxdy = \alpha \iint_{\Omega} f(x,y)\, dxdy + \beta \iint_{\Omega} g(x,y)\, dxdy.$$

II. Order: The double integral preserves order:

$$\text{if } f \geq 0 \text{ on } \Omega, \text{ then } \iint_{\Omega} f(x,y)\, dxdy \geq 0;$$

$$\text{if } f \leq g \text{ on } \Omega, \text{ then } \iint_{\Omega} f(x,y)\, dxdy \leq \iint_{\Omega} g(x,y)\, dxdy.$$

III. Additivity: If Ω is broken up into a finite number of nonoverlapping basic regions $\Omega_1, \ldots, \Omega_n$, then

$$\iint_{\Omega} f(x,y)\, dxdy = \iint_{\Omega_1} f(x,y)\, dxdy + \cdots + \iint_{\Omega_n} f(x,y)\, dxdy.$$

See, for example, Figure 16.2.14.

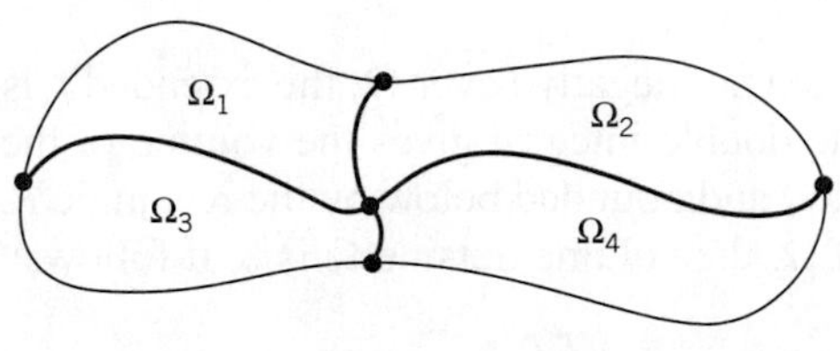

$$\iint_{\Omega} f(x,y)\, dxdy = \iint_{\Omega_1} f(x,y)\, dxdy + \iint_{\Omega_2} f(x,y)\, dxdy + \iint_{\Omega_3} f(x,y)\, dxdy + \iint_{\Omega_4} f(x,y)\, dxdy$$

Figure 16.2.14

IV. Mean-value condition: There is a point (x_0, y_0) in Ω for which

$$\iint_{\Omega} f(x,y)\, dxdy = f(x_0, y_0) \cdot (\text{ area of } \Omega).$$

We call $f(x_0, y_0)$ *the average value of f on Ω*.

This notion of average given in Property IV enables us to write

(16.2.9)
$$\iint_{\Omega} f(x,y)\, dxdy = \begin{pmatrix}\text{the average value} \\ \text{of } f \text{ on } \Omega\end{pmatrix} \cdot (\text{area of } \Omega).$$

This is a powerful, intuitive way of viewing the double integral. We will capitalize on it as we go on.

THEOREM 16.2.10 MEAN-VALUE THEOREM FOR DOUBLE INTEGRALS

Let f and g be functions continuous on the basic region Ω. If g is nonnegative on Ω, then there exists a point (x_0, y_0) in Ω for which

$$\iint_\Omega f(x,y)\, g(x,y)\; dxdy = f(x_0, y_0) \iint_\Omega g(x,y)\; dxdy.\dagger$$

We call $f(x_0, y_0)$ the *g-weighted average of f on Ω.*

PROOF Since f is continuous on Ω, and Ω is closed and bounded, we know that f takes on a minimum value m and a maximum value M. Since g is nonnegative on Ω,

$$mg(x,y) \le f(x,y)g(x,y) \le Mg(x,y) \quad \text{for all } (x,y) \text{ in } \Omega.$$

Therefore (by Property II)

$$\iint_\Omega m\, g(x,y)\; dxdy \le \iint_\Omega f(x,y)\, g(x,y)\; dxdy \le \iint_\Omega M\, g(x,y)\; dxdy,$$

and (by Property I)

$$(*) \qquad m \iint_\Omega g(x,y)\; dxdy \le \iint_\Omega f(x,y)g(x,y)\; dxdy \le M \iint_\Omega g(x,y)\; dxdy.$$

We know that $\iint_\Omega g(x,y)\; dxdy \ge 0$ (again, by Property II). If $\iint_\Omega g(x,y)\; dxdy = 0$, then by $(*)$ we have $\iint_\Omega f(x,y)\, g(x,y)\; dxdy = 0$ and the theorem holds for all choices of (x_0, y_0) in Ω. If $\iint_\Omega g(x,y)\; dxdy > 0$, then

$$m \le \frac{\iint_\Omega f(x,y)\, g(x,y)\; dxdy}{\iint_\Omega g(x,y)\; dxdy} \le M,$$

and, by the intermediate-value theorem (given in the supplement to Section 15.3), there exists (x_0, y_0) in Ω for which

$$f(x_0, y_0) = \frac{\iint_\Omega f(x,y)\, g(x,y)\; dxdy}{\iint_\Omega g(x,y)\; dxdy}.$$

Obviously, then,

$$f(x_0, y_0) \iint_\Omega g(x,y)\; dxdy = \iint_\Omega f(x,y)\, g(x,y)\; dxdy. \quad \square$$

† Property IV is this equation with g constantly 1.

EXERCISES 16.2

For Exercises 1–3, take

$$f(x, y) = x + 2y \quad \text{on} \quad R: 0 \le x \le 2, \quad 0 \le y \le 1.$$

and let P be the partition $P = P_1 \times P_2$.

1. Find $L_f(P)$ and $U_f(P)$ given that $P_1 = \{0, 1, \frac{3}{2}, 2\}$ and $P_2 = \{0, \frac{1}{2}, 1\}$.

2. Find $L_f(P)$ and $U_f(P)$ given that $P_1 = \{0, \frac{1}{2}, 1, \frac{3}{2}, 2\}$ and $P_2 = \{0, \frac{1}{4}, \frac{1}{2}, \frac{3}{4}, 1\}$.

3. (a) Find $L_f(P)$ and $U_f(P)$ given that

$$P_1 = \{x_0, x_1, \dots, x_m\}$$

is an arbitrary partition of $[0, 2]$, and

$$P_2 = \{y_0, y_1, \dots, y_n\}$$

is an arbitrary partition of $[0, 1]$.

(b) Use your answer to part (a) to evaluate the double integral

$$\iint_R (x + 2y)\, dxdy,$$

and give a geometric interpretation to your answer.

For Exercises 4–6, take

$$f(x, y) = x - y \quad \text{on} \quad R: 0 \le x \le 1, \quad 0 \le y \le 1.$$

and let P be the partition $P = P_1 \times P_2$.

4. Find $L_f(P)$ and $U_f(P)$ given that $P_1 = \{0, \frac{1}{2}, \frac{3}{4}, 1\}$ and $P_2 = \{0, \frac{1}{2}, 1\}$.

5. Find $L_f(P)$ and $U_f(P)$ given that $P_1 = \{0, \frac{1}{4}, \frac{1}{2}, \frac{3}{4}, 1\}$ and $P_2 = \{0, \frac{1}{3}, \frac{2}{3}, 1\}$.

6. (a) Find $L_f(P)$ and $U_f(P)$ given that

$$P_1 = \{x_0, x_1, \dots, x_m\} \text{ and } P_2 = \{y_0, y_1, \dots, y_n\}$$

are arbitrary partitions of $[0, 1]$.

(b) Use your answer to part (a) to evaluate the double integral

$$\iint_R (x - y)\, dxdy.$$

For Exercises 7–9, take $R: 0 \le x \le b, \ 0 \le y \le d$ and let $P = P_1 \times P_2$, where

$P_1 = \{x_0, x_1, \dots, x_m\}$ is an arbitrary partition of $[0, b]$,
$P_2 = \{y_0, y_1, \dots, y_n\}$ is an arbitrary partition of $[0, d]$.

7. (a) Find $L_f(P)$ and $U_f(P)$ for $f(x, y) = 4xy$.

(b) Calculate

$$\iint_R 4xy\, dxdy.$$

HINT: $4x_{i-1}y_{j-1} \le (x_i + x_{i-1})(y_j + y_{j-1}) \le 4x_iy_j$.

8. (a) Find $L_f(P)$ and $U_f(P)$ for $f(x, y) = 3(x^2 + y^2)$.

(b) Calculate

$$\iint_R 3(x^2 + y^2)\, dxdy.$$

HINT: If $0 \le s \le t$, then $3s^2 \le t^2 + ts + s^2 \le 3t^2$.

9. (a) Find $L_f(P)$ and $U_f(P)$ for $f(x, y) = 3(x^2 - y^2)$.

(b) Calculate

$$\iint_R 3(x^2 - y^2)\, dxdy.$$

10. Let $f = f(x, y)$ be continuous on the rectangle R: $a \le x \le b, c \le y \le d$. Suppose that $L_f(P) = U_f(P)$ for some partition P of R. What can you conclude about f? What is

$$\iint_R f(x, y)\, dxdy?$$

11. Let $\phi = \phi(x)$ be continuous and nonnegative on the interval $[a, b]$, and set

$$\Omega = \{(x, y) : a \le x \le b, 0 \le y \le \phi(x)\}.$$

Compare

$$\iint_\Omega dxdy \quad \text{to} \quad \int_a^b \phi(x)\, dx.$$

12. Begin with a function f that is continuous on a closed bounded region Ω. Now surround Ω by a rectangle R as in Figure 16.2.12 and extend f to all of R by defining f to be 0 outside of Ω. Explain how the extended f can fail to be continuous on the boundary of Ω although the original function f, being continuous on all of Ω, was continuous on the boundary of Ω.

13. Suppose that f is continuous on a disc Ω centered at (x_0, y_0) and assume that

$$\iint_R f(x, y)\, dxdy = 0$$

for every rectangle R contained in Ω. Show that $f(x_0, y_0) = 0$.

14. Calculate the average value of $f(x,y) = x + 2y$ on the rectangle R: $0 \leq x \leq 2$, $0 \leq y \leq 1$. HINT: See Exercise 3.

15. Calculate the average value of $f(x,y) = 4xy$ on the rectangle $R: 0 \leq x \leq 2$, $0 \leq y \leq 3$. HINT: See Exercise 7.

16. Calculate the average value of $f(x,y) = x^2 + y^2$ on the rectangle R: $0 \leq x \leq b, 0 \leq y \leq d$. HINT: See Exercise 8.

17. Let f be continuous on a closed bounded region Ω and let (x_0, y_0) be a point in the interior of Ω. Let D_r be the closed disc with center (x_0, y_0) and radius r. Show that

$$\lim_{r\to 0} \frac{1}{\pi r^2} \iint_{D_r} f(x,y)\ dxdy = f(x_0, y_0).$$

18. Let $f(x,y) = \sin(x+y)$ on $R: 0 \leq x \leq 1$, $0 \leq y \leq 1$. Show that

$$0 \leq \iint_R \sin(x+y)\ dxdy \leq 1.$$

Sometimes it is possible to evaluate an integral by identifying it as the volume of an elementary solid that is known from geometry. Use this approach to evaluate the integrals in Exercises 19–21.

19. $\iint_\Omega \sqrt{4 - x^2 - y^2}\ dxdy$ where Ω is the quarter disk $x^2 + y^2 \leq 4, x \geq 0, y \geq 0$.

20. $\iint_\Omega 8 - 4\sqrt{x^2 + y^2}\ dxdy$ where Ω is the disk $x^2 + y^2 \leq 4$.

21. $\iint_\Omega (6 - 2x - 3y)\ dxdy$ where Ω is the triangular region bounded by the coordinate axes and the line $2x + 3y = 6$.

C **22.** Let $f(x,y) = 3y^2 - 2x$ on the rectangle R: $2 \leq x \leq 5$, $1 \leq y \leq 3$. Let P_1 be a regular partition of $[2,5]$ with $n = 100$ subintervals, let P_2 be a regular partition of $[1,3]$ with $m = 200$ subintervals, and let $P = P_1 \times P_2$.

(a) Use a CAS to find $L_f(P)$ and $U_f(P)$.

(b) Find $L_f(P)$ and $U_f(P)$ for several values of $n > 100$, $m > 200$.

(c) Estimate $\iint_R f(x,y)\ dxdy$.

■ 16.3 THE EVALUATION OF DOUBLE INTEGRALS BY REPEATED INTEGRALS

The Reduction Formulas

If an integral

$$\int_a^b f(x)\ dx$$

proves difficult to evaluate, it is not because of the interval $[a,b]$ but because of the integrand f. Difficulty in evaluating a double integral

$$\iint_\Omega f(x,y)\ dxdy$$

can come from two sources: from the integrand f or from the base region Ω. Even such a simple-looking integral as $\iint_\Omega 1\, dxdy$ is difficult to evaluate if Ω is complicated.

In this section we introduce a technique for evaluating double integrals of continuous functions over regions of Type I or Type II as depicted in Figure 16.3.1. In each case the region Ω is a basic region and so we know that the double integral exists. The fundamental idea of this section is that double integrals over such regions can each be reduced to a pair of ordinary integrals.

Type I Region. The *projection* of Ω onto the x-axis is a closed interval $[a,b]$ and Ω consists of all points (x,y) with

$$a \leq x \leq b \quad \text{and} \quad \phi_1(x) \leq y \leq \phi_2(x).$$

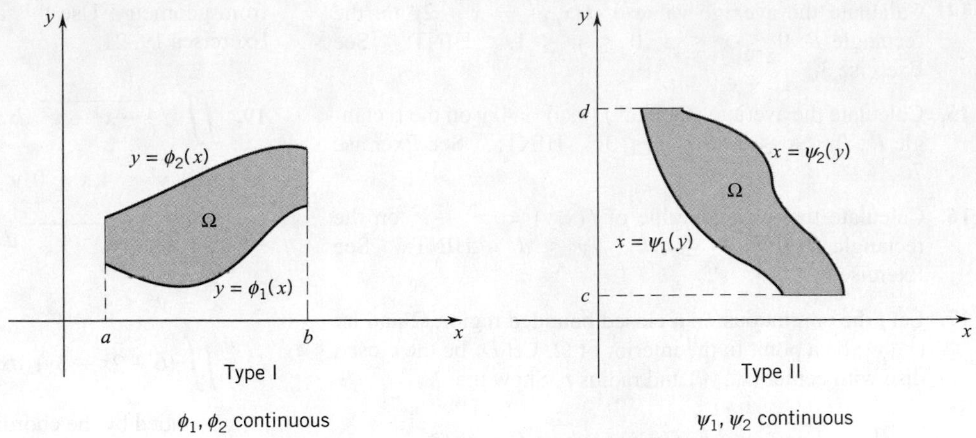

Figure 16.3.1

Then

(16.3.1) $$\iint_{\Omega} f(x,y)\, dxdy = \int_a^b \left(\int_{\phi_1(x)}^{\phi_2(x)} f(x,y)\, dy \right) dx.$$

Here we first calculate

$$\int_{\phi_1(x)}^{\phi_2(x)} f(x,y)\, dy$$

by integrating $f(x,y)$ with respect to y from $y = \phi_1(x)$ to $y = \phi_2(x)$. The resulting expression is a function of x alone, which we then integrate with respect to x from $x = a$ to $x = b$.

Type II Region. The *projection* of Ω onto the y-axis is a closed interval $[c, d]$ and Ω consists of all points (x, y) with

$$c \le y \le d \qquad \text{and} \qquad \psi_1(y) \le x \le \psi_2(y).$$

Then

(16.3.2) $$\iint_{\Omega} f(x,y)\, dxdy = \int_c^d \left(\int_{\psi_1(y)}^{\psi_2(y)} f(x,y)\, dx \right) dy.$$

This time we first calculate

$$\int_{\psi_1(y)}^{\psi_2(y)} f(x,y)\, dx$$

by integrating $f(x,y)$ with respect to x from $x = \psi_1(y)$ to $x = \psi_2(y)$. The resulting expression is a function of y alone, which we can then integrate with respect to y from $y = c$ to $y = d$.

The integrals on the right-hand sides of formulas (16.3.1) and (16.3.2) are called *repeated integrals*. These formulas are easy to understand geometrically.

The Reduction Formulas Viewed Geometrically

Suppose that f is nonnegative and Ω is a region of Type I. The double integral over Ω gives the volume of the solid T bounded above by the surface $z = f(x, y)$ and bounded below by the region Ω:

$$(1) \qquad \iint_{\Omega} f(x, y)\, dxdy = \text{ volume of } T. \qquad \text{(Figure 16.3.2)}$$

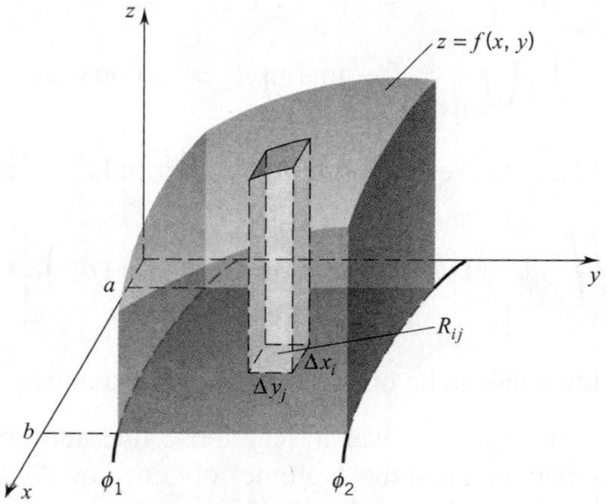

volume by double integration

$$V = \iint_{\Omega} f(x,y)\, dxdy$$

Figure 16.3.2

We can also compute the volume of T by the method of parallel cross sections (see Section 6.2). As in Figure 16.3.3, let $A(x)$ be the area of the cross section of T all points of which have first coordinate x. Then by (6.2.1)

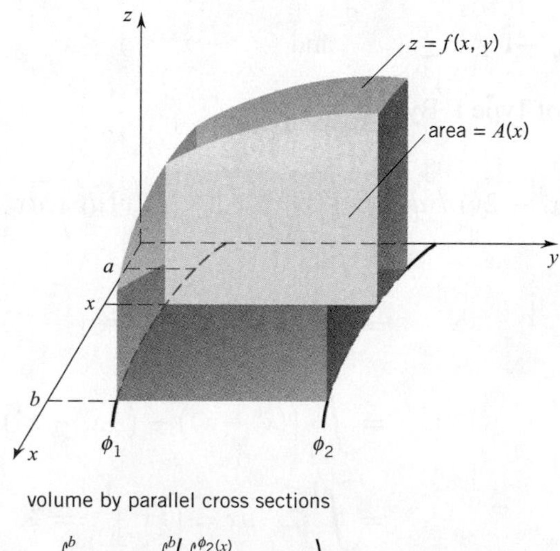

volume by parallel cross sections

$$V = \int_a^b A(x)\, dx = \int_a^b \left(\int_{\phi_1(x)}^{\phi_2(x)} f(x,y)\, dy \right) dx$$

Figure 16.3.3

$$\int_a^b A(x)\,dx = \text{volume of } T.$$

Since
$$A(x) = \int_{\phi_1(x)}^{\phi_2(x)} f(x,y)\,dy,$$
we have

$$(2) \qquad \int_a^b \left(\int_{\phi_1(x)}^{\phi_2(x)} f(x,y)\,dy \right) dx = \text{volume of } T.$$

Combining (1) with (2), we have the first reduction formula

$$\iint_\Omega f(x,y)\,dxdy = \int_a^b \left(\int_{\phi_1(x)}^{\phi_2(x)} f(x,y)\,dy \right) dx.$$

The other reduction formula can be obtained in a similar manner.

Remark Note that our argument was a very loose one and certainly not a proof. How do we know, for example, that the "volume" obtained by double integration is the same as the "volume" obtained by the method of parallel cross sections? Intuitively it seems evident but, actually it is quite difficult to prove. The result is a special case of Fubini's theorem (after the Italian mathematician Guido Fubini (1879–1943) who proved a general version of the result in 1907). ❑

Computations

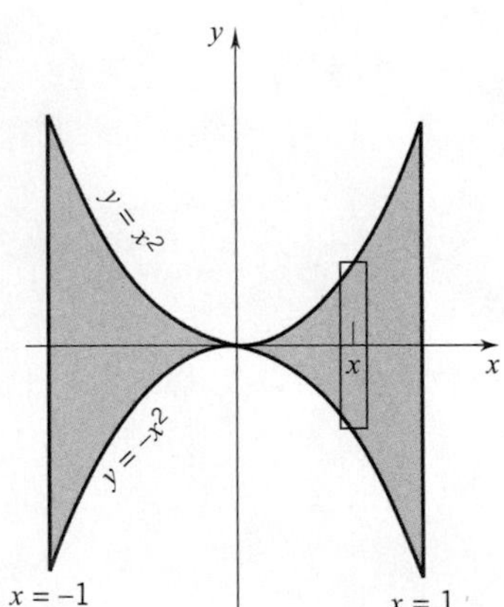

Figure 16.3.4

Example 1 Evaluate $\iint_\Omega (x^4 - 2y)\,dxdy$ with Ω as in Figure 16.3.4.

SOLUTION By projecting Ω onto the x-axis, we obtain the interval $[-1, 1]$. The region Ω consists of all points (x, y) with

$$-1 \le x \le 1 \qquad \text{and} \qquad -x^2 \le y \le x^2.$$

This is a region of Type I. By (16.3.1)

$$\begin{aligned}\iint_\Omega (x^4 - 2y)\,dxdy &= \int_{-1}^{1} \left(\int_{-x^2}^{x^2} [x^4 - 2y]\,dy \right) dx \\ &= \int_{-1}^{1} \left[x^4 y - y^2 \right]_{-x^2}^{x^2} dx \\ &= \int_{-1}^{1} [(x^6 - x^4) - (-x^6 - x^4)]\,dx \\ &= \int_{-1}^{1} 2x^6\,dx = \left[\tfrac{2}{7}x^7 \right]_{-1}^{1} = \tfrac{4}{7}. \quad ❑\end{aligned}$$

Example 2 Evaluate $\iint_\Omega (xy - y^3)\,dxdy$ with Ω as in Figure 16.3.5.

Figure 16.3.5

SOLUTION By projecting Ω onto the y-axis, we obtain the interval $[0, 1]$. The region Ω consists of all points (x, y) with

$$0 \le y \le 1 \quad \text{and} \quad -1 \le x \le y.$$

This is a region of Type II. By (16.3.2)

$$\begin{aligned}\iint_{\Omega} (xy - y^3)\, dxdy &= \int_0^1 \left(\int_{-1}^{y} (xy - y^3)\, dx \right) dy \\ &= \int_0^1 \left[\tfrac{1}{2}x^2y - xy^3 \right]_{-1}^{y} dy \\ &= \int_0^1 \left[\left(\tfrac{1}{2}y^3 - y^4\right) - \left(\tfrac{1}{2}y + y^3\right) \right] dy \\ &= \int_0^1 \left(-\tfrac{1}{2}y^3 - y^4 - \tfrac{1}{2}y\right) dy \\ &= \left[-\tfrac{1}{8}y^4 - \tfrac{1}{5}y^5 - \tfrac{1}{4}y^2 \right]_0^1 = -\tfrac{23}{40}.\end{aligned}$$

We can also project Ω onto the x-axis and express Ω as a region of Type I, but then the lower boundary is defined piecewise (see the figure) and the calculations are somewhat more complicated: setting

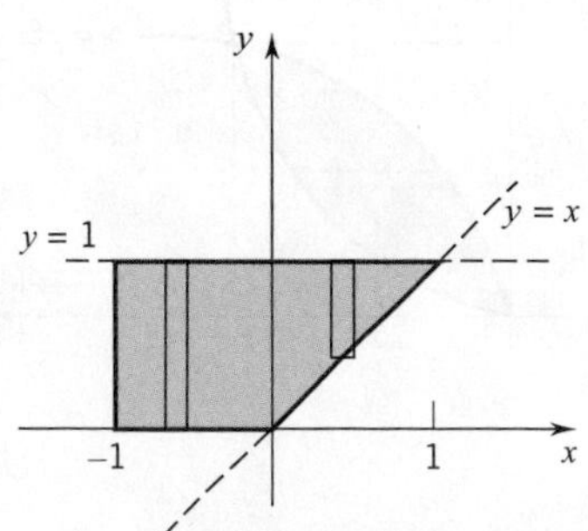

$$\phi(x) = \begin{cases} 0, & -1 \le x \le 0 \\ x, & 0 \le x \le 1, \end{cases}$$

we have Ω as the set of all points (x, y) with

$$-1 \le x \le 1 \quad \text{and} \quad \phi(x) \le y \le 1;$$

thus

$$\begin{aligned}\iint_{\Omega} (xy - y^3)\, dxdy &= \int_{-1}^{1} \left(\int_{\phi(x)}^{1} (xy - y^3)\, dy \right) dx \\ &= \int_{-1}^{0} \left(\int_{\phi(x)}^{1} (xy - y^3)\, dy \right) dx + \int_{0}^{1} \left(\int_{\phi(x)}^{1} (xy - y^3)\, dy \right) dx\end{aligned}$$

$$= \int_{-1}^{0}\left(\int_{0}^{1}(xy - y^3)\,dy\right)dx + \int_{0}^{1}\left(\int_{x}^{1}(xy - y^3)\,dy\right)dx$$

as you can check

$$= (-\tfrac{1}{2}) + (-\tfrac{3}{40}) = -\tfrac{23}{40}. \quad \square$$

Repeated integrals

$$\int_a^b\left(\int_{\phi_1(x)}^{\phi_2(x)} f(x,y)\,dy\right)dx \qquad \text{and} \qquad \int_c^d\left(\int_{\psi_1(y)}^{\psi_2(y)} f(x,y)\,dx\right)dy$$

can be written in more compact form by omitting the large parentheses. From now on we will simply write

$$\int_a^b\int_{\phi_1(x)}^{\phi_2(x)} f(x,y)\,dy\,dx \qquad \text{and} \qquad \int_c^d\int_{\psi_1(y)}^{\psi_2(y)} f(x,y)\,dx\,dy$$

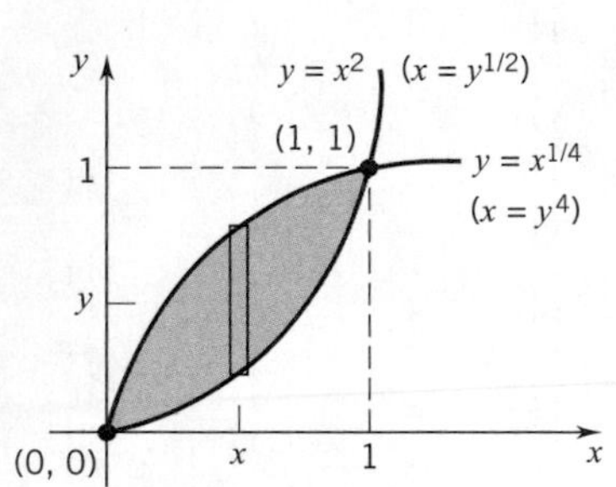

Figure 16.3.6

Example 3 Evaluate $\iint_{\Omega} (x^{1/2} - y^2)\,dxdy$ with Ω as in Figure 16.3.6.

SOLUTION The projection of Ω onto the x-axis is the closed interval $[0, 1]$ and Ω can be characterized as the set of all (x, y) with

$$0 \le x \le 1 \qquad \text{and} \qquad x^2 \le y \le x^{1/4}.$$

Thus
$$\iint_{\Omega}(x^{1/2} - y^2)\,dxdy = \int_0^1\int_{x^2}^{x^{1/4}}(x^{1/2} - y^2)\,dy\,dx$$

$$= \int_0^1\left[x^{1/2}y - \tfrac{1}{3}y^3\right]_{x^2}^{x^{1/4}} dx$$

$$= \int_0^1\left(\tfrac{2}{3}x^{3/4} - x^{5/2} + \tfrac{1}{3}x^6\right)dx$$

$$= \left[\tfrac{8}{21}x^{7/4} - \tfrac{2}{7}x^{7/2} + \tfrac{1}{21}x^7\right]_0^1 = \tfrac{8}{21} - \tfrac{2}{7} + \tfrac{1}{21} = \tfrac{1}{7}.$$

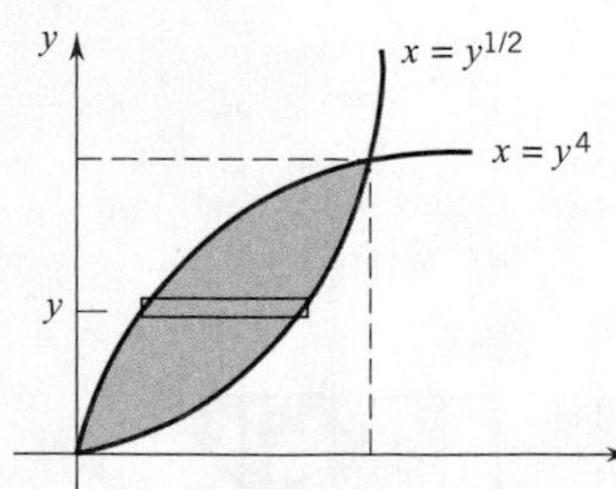

We can also integrate in the other order. The projection of Ω onto the y-axis is the closed interval $[0, 1]$, and Ω can be characterized as the set of all (x, y) with

$$0 \le y \le 1 \qquad \text{and} \qquad y^4 \le x \le y^{1/2}.$$

This gives the same result:

$$\iint_{\Omega}(x^{1/2} - y^2)\,dxdy = \int_0^1\int_{y^4}^{y^{1/2}}(x^{1/2} - y^2)\,dx\,dy$$

$$= \int_0^1\left[\tfrac{2}{3}x^{3/2} - y^2x\right]_{y^4}^{y^{1/2}} dy$$

$$= \int_0^1\left(\tfrac{2}{3}y^{3/4} - y^{5/2} + \tfrac{1}{3}y^6\right)dy$$

$$= \left[\tfrac{8}{21}y^{7/4} - \tfrac{2}{7}y^{7/2} + \tfrac{1}{21}y^7\right]_0^1 = \tfrac{8}{21} - \tfrac{2}{7} + \tfrac{1}{21} = \tfrac{1}{7}. \quad \square$$

Example 4 Use double integration to calculate the area of the region Ω enclosed by

$$y = x^2 \qquad \text{and} \qquad x + y = 2.$$

SOLUTION The region Ω is pictured in Figure 16.3.7. Its area is given by the double integral

$$\iint_{\Omega} dxdy.$$

We project Ω onto the x-axis and write the boundaries as functions of x,

$$y = x^2 \qquad \text{and} \qquad y = 2 - x,$$

and view Ω as the set of all (x, y) with $-2 \leq x \leq 1$ and $x^2 \leq y \leq 2 - x$. This gives

$$\iint_{\Omega} dxdy = \int_{-2}^{1} \int_{x^2}^{2-x} dy\,dx = \int_{-2}^{1} (2 - x - x^2)\ dx = \left[2x - \tfrac{1}{2}x^2 - \tfrac{1}{3}x^3\right]_{-2}^{1}$$

$$= (2 - \tfrac{1}{2} - \tfrac{1}{3}) - (-4 - 2 + \tfrac{8}{3}) = \tfrac{9}{2}.$$

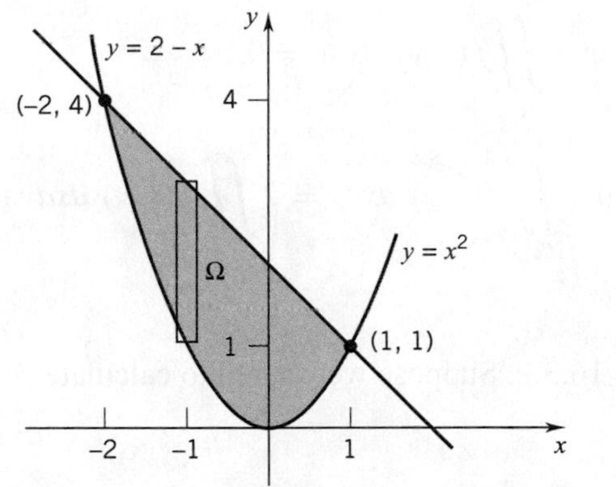

Figure 16.3.7

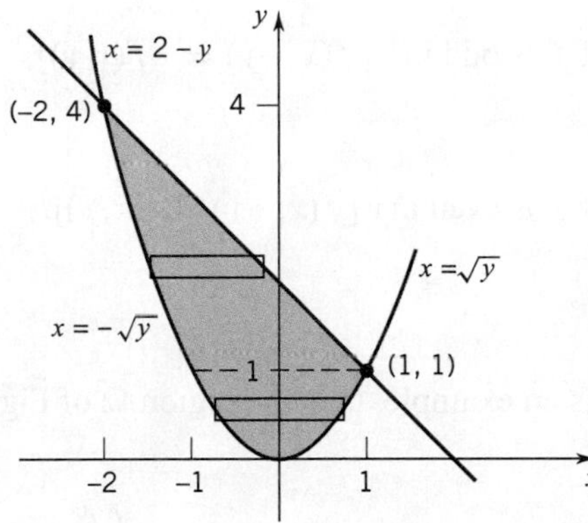

Figure 16.3.8

We can also project Ω onto the y-axis and write the boundaries as functions of y, but then the calculations become more complicated. As illustrated in Figure 16.3.8, Ω is the set of all (x, y) with

$$0 \leq y \leq 4 \qquad \text{and} \qquad -\sqrt{y} \leq x \leq \psi(y)$$

where

$$\psi(y) = \begin{cases} \sqrt{y}, & 0 \leq y \leq 1 \\ 2 - y, & 1 \leq y \leq 4. \end{cases}$$

Thus

$$\iint_{\Omega} dxdy = \int_{0}^{1} \int_{-\sqrt{y}}^{\sqrt{y}} dx\,dy + \int_{1}^{4} \int_{-\sqrt{y}}^{2-y} dx\,dy.$$

Carry out the calculations and you will get the same result as above. □

Symmetry in Double Integration

First we go back to the one-variable case (Section 5.8). Let's suppose that g is continuous on an interval that is symmetric about the origin, say $[-a, a]$.

$$\text{If } g \text{ is odd,} \quad \text{then} \quad \int_{-a}^{a} g(x)\, dx = 0.$$

$$\text{If } g \text{ is even,} \quad \text{then} \quad \int_{-a}^{a} g(x)\, dx = 2\int_{0}^{a} g(x)\, dx.$$

We have similar results for double integrals.

Suppose that Ω is symmetric about the y-axis.

$$\text{If } f \text{ is odd in } x\, [f(-x,y) = -f(x,y)], \quad \text{then} \quad \iint_{\Omega} f(x,y)\, dxdy = 0.$$

$$\text{If } f \text{ is even in } x\, [f(-x,y) = f(x,y)], \quad \text{then} \quad \iint_{\Omega} f(x,y)\, dxdy = 2\iint_{\substack{\text{right half}\\ \text{of } \Omega}} f(x,y)\, dxdy.$$

Suppose that Ω is symmetric about the x-axis.

$$\text{If } f \text{ is odd in } y\, [f(x,-y) = -f(x,y)], \quad \text{then} \quad \iint_{\Omega} f(x,y)\, dxdy = 0.$$

$$\text{If } f \text{ is even in } y\, [f(x,-y) = f(x,y)], \quad \text{then} \quad \iint_{\Omega} f(x,y)\, dxdy = 2\iint_{\substack{\text{upper half}\\ \text{of } \Omega}} f(x,y)\, dxdy.$$

As an example, take the region Ω of Figure 16.3.9. Suppose we wanted to calculate

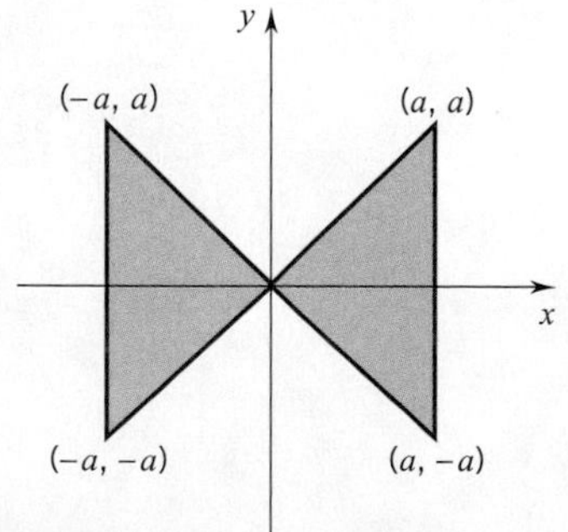

Figure 16.3.9

$$\iint_{\Omega} (2x - \sin x^2 y)\, dxdy.$$

First,

$$\iint_{\Omega} (2x - \sin x^2 y)\, dxdy = \iint_{\Omega} 2x\, dxdy - \iint_{\Omega} \sin x^2 y\, dxdy.$$

The symmetry of Ω about the y-axis gives

$$\iint_{\Omega} 2x\, dxdy = 0. \qquad \text{(the integrand is odd in } x\text{)}$$

The symmetry of Ω about the x-axis gives

$$\iint_{\Omega} \sin x^2 y\, dxdy = 0. \qquad \text{(the integrand is odd in } y\text{)}$$

Therefore $$\iint_{\Omega} (2x - \sin x^2 y)\, dxdy = 0.$$

Let's go back to Example 1 and reevaluate

$$\iint_{\Omega} (x^4 - 2y)\, dxdy,$$

this time capitalizing on the symmetry of Ω. Note that

$$\underbrace{\iint_{\Omega} 2y\, dxdy = 0}_{\text{symmetry about } x\text{-axis}} \quad \text{and} \quad \iint_{\Omega} x^4\, dxdy \overset{\text{symmetry about } y\text{-axis}}{=} 2 \iint_{\substack{\text{right half}\\ \text{of } \Omega}} x^4\, dxdy \overset{\text{symmetry about } x\text{-axis}}{=} 4 \iint_{\substack{\text{upper part}\\ \text{of right half}\\ \text{of } \Omega}} x^4\, dxdy.$$

Therefore $$\iint_{\Omega} (x^4 - 2y)\, dxdy = 4 \int_0^1 \int_0^{x^2} x^4\, dy\, dx = 4 \int_0^1 x^6\, dx = \tfrac{4}{7}.$$

Example 5 Calculate the volume within the cylinder $x^2 + y^2 = b^2$ between the planes $y + z = a$ and $z = 0$ given that $a \geq b > 0$.

SOLUTION See Figure 16.3.10. The solid in question is bounded below by the disc

$$\Omega:\quad 0 \leq x^2 + y^2 \leq b^2$$

and above by the plane

$$z = a - y.$$

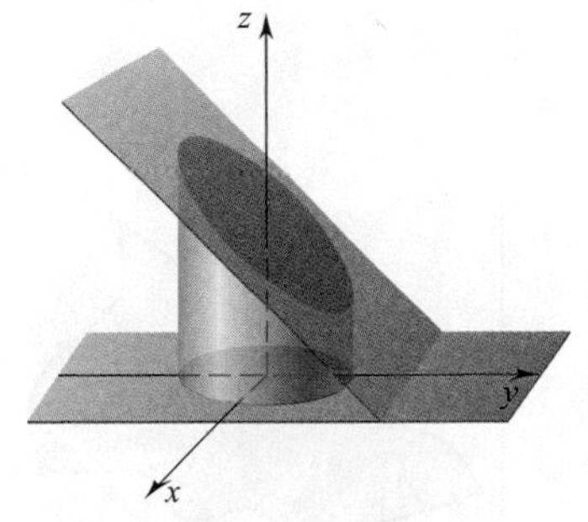

Figure 16.3.10

The volume is given by the double integral

$$\iint_{\Omega} (a - y)\, dxdy.$$

Since Ω is symmetric about the x-axis,

$$\iint_{\Omega} y\, dxdy = 0.$$

Thus

$$\iint_{\Omega} (a - y)\, dxdy = \iint_{\Omega} a\, dxdy = a \iint_{\Omega} dxdy = a\,(\text{area of } \Omega) = \pi a b^2. \quad \square$$

Concluding Remarks

When two orders of integration are possible, one order may be easy to carry out, while the other may present serious difficulties. Take as an example the double integral

$$\iint_{\Omega} \cos \tfrac{1}{2}\pi x^2 \, dxdy$$

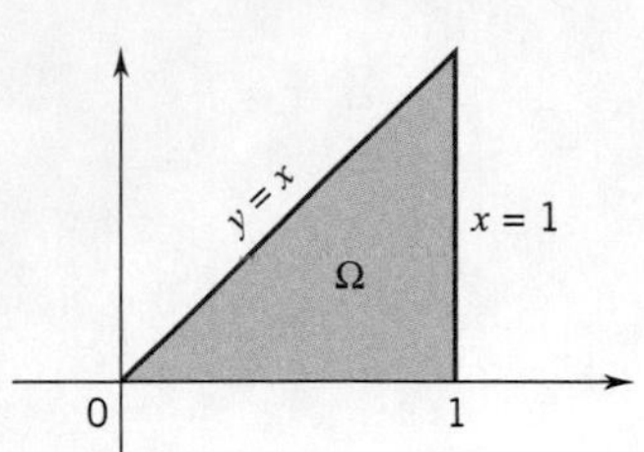

Figure 16.3.11

with Ω as in Figure 16.3.11. Projection onto the x-axis leads to

$$\int_0^1 \int_0^x \cos \tfrac{1}{2}\pi x^2 \, dy \, dx.$$

Projection onto the y-axis leads to

$$\int_0^1 \int_y^1 \cos \tfrac{1}{2}\pi x^2 \, dx \, dy.$$

The first expression is easy to evaluate:

$$\begin{aligned} \int_0^1 \int_0^x \cos \tfrac{1}{2}\pi x^2 \, dy \, dx &= \int_0^1 \left(\int_0^x \cos \tfrac{1}{2}\pi x^2 \, dy \right) dx \\ &= \int_0^1 \left[y \cos \tfrac{1}{2}\pi x^2 \right]_0^x dx = \int_0^1 x \cos \tfrac{1}{2}\pi x^2 \, dx \\ &= \left[\tfrac{1}{\pi} \sin \tfrac{1}{2}\pi x^2 \right]_0^1 = \tfrac{1}{\pi}. \end{aligned}$$

The second expression is not easy to evaluate:

$$\int_0^1 \int_y^1 \cos \tfrac{1}{2}\pi x^2 \, dx \, dy = \int_0^1 \left(\int_y^1 \cos \tfrac{1}{2}\pi x^2 \, dx \right) dy,$$

and $\cos \frac{1}{2}\pi x^2$ does not have an elementary antiderivative.

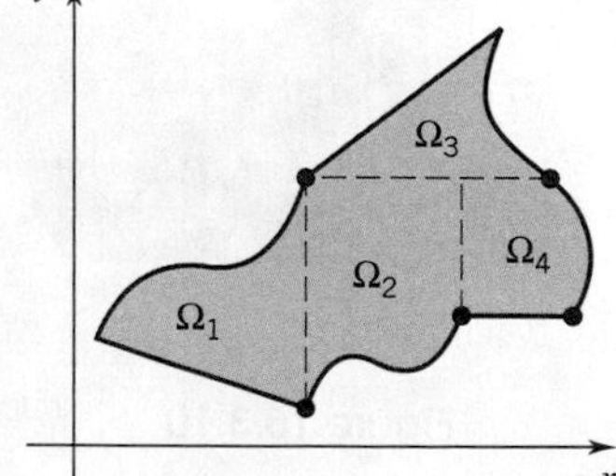

Figure 16.3.12

Finally, if Ω, the region of integration, is neither of Type I nor of Type II, it may be possible to break it up into a finite number of regions $\Omega_1, \ldots, \Omega_n$, each of which is of Type I or Type II. (See Figure 16.3.12.) Since the double integral is additive,

$$\iint_{\Omega_1} f(x,y) \, dxdy + \cdots + \iint_{\Omega_4} f(x,y) \, dxdy = \iint_{\Omega} f(x,y) \, dxdy.$$

Each of the integrals on the left can be evaluated by the methods of this section.

EXERCISES 16.3

Evaluate the integral taking $\Omega: 0 \le x \le 1, 0 \le y \le 3$.

1. $\iint_{\Omega} x^2 \, dxdy$. **2.** $\iint_{\Omega} e^{x+y} \, dxdy$.

3. $\iint_{\Omega} xy^2 \, dxdy$.

Evaluate the integral taking $\Omega: 0 \le x \le 1, 0 \le y \le x$.

4. $\iint_{\Omega} x^3 y \, dxdy$. **5.** $\iint_{\Omega} xy^3 \, dxdy$.

6. $\iint_{\Omega} x^2 y^2 \, dxdy$.

Evaluate the integral taking $\Omega: 0 \le x \le \frac{1}{2}\pi, 0 \le y \le \frac{1}{2}\pi$.

7. $\iint_{\Omega} \sin(x+y)\, dxdy$.

8. $\iint_{\Omega} \cos(x+y)\, dxdy$.

9. $\iint_{\Omega} (1+xy)\, dxdy$.

Evaluate the double integral.

10. $\iint_{\Omega} (x+3y^3)\, dxdy, \quad \Omega: 0 \le x^2+y^2 \le 1$.

11. $\iint_{\Omega} \sqrt{xy}\, dxdy, \quad \Omega: 0 \le y \le 1, \; y^2 \le x \le y$.

12. $\iint_{\Omega} y\,e^x\, dxdy, \quad \Omega: 0 \le y \le 1, \quad 0 \le x \le y^2$.

13. $\iint_{\Omega} (4-y^2)\, dxdy$, Ω the bounded region between $y^2 = 2x$ and $y^2 = 8-2x$.

14. $\iint_{\Omega} (x^4+y^2)\, dxdy$, Ω the bounded region between $y = x^3$ and $y = x^2$.

15. $\iint_{\Omega} (3xy^3 - y)\, dxdy$, Ω the region between $y = |x|$ and $y = -|x|$, $x \in [-1, 1]$.

16. $\iint_{\Omega} e^{-y^2/2}\, dxdy$, Ω the triangle formed by the y-axis, $2y = x$, $y = 1$.

17. $\iint_{\Omega} e^{x^2}\, dxdy$, Ω the triangle formed by the x-axis, $2y = x, x = 2$.

18. $\iint_{\Omega} (x+y)\, dxdy$, Ω the region between $y = x^3$ and $y = x^4$, $x \in [-1, 1]$.

Sketch the region Ω that gives rise to the repeated integral and change the order of integration.

19. $\int_0^1 \int_{x^4}^{x^2} f(x,y)\, dy\, dx$.

20. $\int_0^1 \int_0^{y^2} f(x,y)\, dx\, dy$.

21. $\int_0^1 \int_{-y}^{y} f(x,y)\, dx\, dy$.

22. $\int_{1/2}^1 \int_{x^3}^{x} f(x,y)\, dy\, dx$.

23. $\int_1^4 \int_x^{2x} f(x,y)\, dy\, dx$.

24. $\int_1^3 \int_{-x}^{x^2} f(x,y)\, dy\, dx$.

Calculate by double integration the area of the bounded region determined by the given pairs of curves.

25. $x^2 = 4y, \quad 2y - x - 4 = 0$.

26. $y = x, \quad x = 4y - y^2$.

27. $y = x, \quad 4y^3 = x^2$.

28. $x + y = 5, \quad xy = 6$.

Sketch the region Ω that gives rise to the repeated integral, change the order of integration, and then evaluate.

29. $\int_0^1 \int_{\sqrt{x}}^1 \sin\left(\frac{y^3+1}{2}\right) dy\, dx$.

30. $\int_{-1}^0 \int_{-\sqrt{y+1}}^{\sqrt{y+1}} x^2\, dx\, dy$.

31. $\int_1^2 \int_0^{\ln y} e^{-x}\, dx\, dy$.

32. $\int_0^1 \int_{x^2}^1 \frac{x^3}{\sqrt{x^4+y^2}}\, dy\, dx$.

33. Find the area of the first quadrant region bounded by $xy = 2, \; y = 1, \; y = x+1$.

34. Find the volume of the solid bounded above by $z = x+y$ and below by the triangle with vertices $(0,0,0)$, $(0,1,0)$, $(1,0,0)$.

35. Find the volume of the solid bounded by $\frac{1}{2}x + \frac{1}{3}y + \frac{1}{4}z = 1$ and the coordinate planes.

36. Find the volume of the solid bounded above by the plane $z = 2x + 3y$ and below by the unit square $0 \le x \le 1$, $0 \le y \le 1$.

37. Find the volume of the solid bounded above by $z = x^3y$ and below by the triangle with vertices $(0,0,0)$, $(2,0,0)$, $(0,1,0)$.

38. Find the volume under the paraboloid $z = x^2 + y^2$ within the cylinder $x^2 + y^2 \le 1, z \ge 0$.

39. Find the volume of the solid bounded above by the plane $z = 2x + 1$ and below by the disc $(x-1)^2 + y^2 \le 1$.

40. Find the volume of the solid bounded above by $z = 4 - y^2 - \frac{1}{4}x^2$ and below by the disc $(y-1)^2 + x^2 \le 1$.

41. Find the volume of the solid in the first octant ($x \ge 0, y \ge 0, z \ge 0$) bounded by $z = x^2 + y^2$, the plane $x + y = 1$, and the coordinate planes.

42. Find the volume of the solid bounded by the circular cylinder $x^2 + y^2 = 1$, the plane $z = 0$, and the plane $x + z = 1$.

43. Find the volume of the solid in the first octant bounded above by $z = x^2 + 3y^2$, below by the xy-plane, and on the sides by the cylinder $y = x^2$ and the plane $y = x$.

44. Find the volume of the solid bounded above by the surface $z = 1 + xy$ and below by the triangle with vertices $(1,1), (4,1)$, and $(3,2)$.

45. Find the volume of the solid in the first octant bounded by the two cylinders $x^2 + y^2 = a^2, \; x^2 + z^2 = a^2$.

46. Find the volume of the tetrahedron bounded by the coordinate planes and the plane

$$\frac{x}{a} + \frac{y}{b} + \frac{z}{c} = 1, \qquad a, b, c > 0.$$

Evaluate.

47. $\int_0^1 \int_y^1 e^{y/x}\, dx\, dy$.

48. $\int_0^1 \int_0^{\cos^{-1} y} e^{\sin x}\, dx\, dy$.

49. $\int_0^1 \int_x^1 x^2 e^{y^4}\, dy\, dx$.

50. $\int_0^1 \int_x^1 e^{y^2}\, dy\, dx$.

Calculate the average value of f over the region Ω.

51. $f(x,y)=x^2y$; $\quad\Omega$: $-1\le x\le 1,\quad 0\le y\le 4$.

52. $f(x,y)=xy$; $\quad\Omega$: $0\le x\le 1,\quad 0\le y\le\sqrt{1-x^2}$.

53. $f(x,y)=\dfrac{1}{xy}$; $\quad\Omega$: $\ln 2\le x\le 2\ln 2$, $\ln 2\le y\le 2\ln 2$.

54. $f(x,y)=e^{x+y}$; $\quad\Omega$: $0\le x\le 1,\quad x-1\le y\le x+1$.

55. (*Separated variables over a rectangle*) Let R be the rectangle $a\le x\le b,\ c\le y\le d$. Show that, if f is continuous on $[a,b]$ and g is continuous on $[c,d]$, then

(16.3.3)
$$\iint_R f(x)g(y)\,dxdy = \left[\int_a^b f(x)\,dx\right]\cdot\left[\int_c^b g(y)\,dy\right].$$

56. Let R be a rectangle symmetric about the x-axis, sides parallel to the coordinate axes. Show that, if f is odd with respect to y, then the double integral of f over R is 0.

57. Let R be a rectangle symmetric about the y-axis, sides parallel to the coordinate axes. Show that, if f is odd with respect to x, then the double integral of f over R is 0.

58. Given that $f(-x,-y)=-f(x,y)$ for all (x,y) in Ω, what form of symmetry in Ω will ensure that the double integral of f over Ω is zero?

59. Let Ω be the triangle with vertices $(0,0),(0,1),(1,1)$. Show that

$$\text{if}\quad \int_0^1 f(x)\,dx=0\quad\text{then}\quad\iint_\Omega f(x)f(y)\,dxdy=0.$$

60. (*Differentiation under the integral sign*) If f and $\partial f/\partial x$ are continuous, then the function

$$H(t)=\int_a^b\frac{\partial f}{\partial x}(t,y)\,dy$$

can be shown to be continuous. Use the identity

$$\int_0^x\int_a^b\frac{\partial f}{\partial x}(t,y)\,dy\,dt=\int_a^b\int_0^x\frac{\partial f}{\partial x}(t,y)\,dt\,dy$$

to verify that

$$\frac{d}{dx}\left[\int_a^b f(x,y)\,dy\right]=\int_a^b\frac{\partial f}{\partial x}(x,y)\,dy.$$

61. We integrate over regions of Type I by means of the formula

$$\iint_\Omega f(x,y)\ dx\,dy=\int_a^b\int_{\phi_1(x)}^{\phi_2(x)} f(x,y)\ dy\,dx.$$

Here f is assumed to be continuous on Ω and ϕ_1,ϕ_2 are assumed to be continuous on $[a,b]$. Show that the function

$$F(x)=\int_{\phi_1(x)}^{\phi_2(x)} f(x,y)\,dy$$

is continuous on $[a,b]$.

C 62. Use a CAS to evaluate the expression

(a) $\displaystyle\int_{-1}^{3}\int_{2}^{5} x\,e^{-xy}\,dy\,dx.$ $\qquad$ (b) $\displaystyle\int_{3}^{7}\int_{1}^{4}\frac{xy}{x^2+y^2}\,dy\,dx.$

C 63. Use a graphing utility to draw the region Ω bounded by $y=x^2-2x+2$ and $y=1+\sqrt{x-1}$. Find the area of Ω by evaluating the following integrals.

(a) $\displaystyle\int_{x=?}^{x=?}\int_{y=?}^{y=?} 1\,dy\,dx.$ $\qquad$ (b) $\displaystyle\int_{y=?}^{y=?}\int_{x=?}^{x=?} 1\,dx\,dy.$

C 64. Use a graphing utility to draw the region Ω bounded by $x-2y=0,\ x+y=3$ and $y=0$. Use a CAS to calculate

(a) $\displaystyle\int_{y=?}^{y=?}\int_{x=?}^{x=?}\sqrt{2x+y}\ dx\,dy.$

(b) $\displaystyle\int_{x=?}^{x=?}\int_{y=?}^{y=?}\sqrt{2x+y}\ dy\,dx.$

■ PROJECT 16.3 Numerical Methods for Double Integrals

The methods used to approximate ordinary definite integrals (see Section 8.7) can be extended to multiple integrals, but due the large number of calculations involved, numerical integration methods for multiple integrals are much more time consuming. In this project, we explore numerical methods for the double integral

$$I=\iint_R f(x,y)\ dxdy$$

where R is the rectangle: $a\le x\le b,\ c\le y\le d$. Let P_1 be a partition of $[a,b]$ into m subintervals of equal length $(b-a)/m$, and let P_2 be a partition of $[c,d]$ into n subintervals each of length $(d-c)/n$. Then $P=P_1\times P_2$ partitions R into $m\times n$ rectangles $R_{ij},\ 1\le i\le m,\ 1\le j\le n$.

Midpoint Approximation. Let u_i be the midpoint of the x-subinterval $[x_{i-1},x_i]$ and let v_j be the midpoint of the y-subinterval $[y_{j-1},y_j]$. Then the midpoint approximation of I is given by the double sum.

$$I\cong M_{mn}=\frac{b-a}{m}\cdot\frac{d-c}{n}\sum_{i=1}^{m}\sum_{j=1}^{n} f(u_i,v_j).$$

The figure illustrates the case $m = 3, n = 2$.

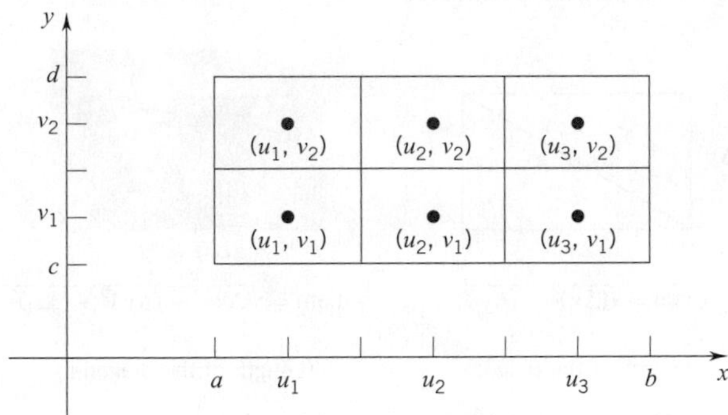

$$M_{32} = \frac{b-a}{3} \cdot \frac{d-c}{2} \, [f(u_1, v_1) + f(u_2, v_1) + f(u_3, v_1) + f(u_1, v_2) + f(u_2, v_2) + f(u_3, v_2)].$$

Problem 1. Let $I = \int_0^3 \int_0^2 (x + 2y) \, dx \, dy$.

a. Calculate M_{32}.

b. Use technology to calculate M_{mn} for larger values of m and n.

c. Calculate the exact value of I and compare it with your numerical approximations.

Problem 2. Let $I = \int_0^2 \int_0^2 4xy \, dx \, dy$.

a. Calculate M_{22}.

b. Use technology to calculate M_{mn} for larger values of m and n.

c. Calculate the exact value of I and compare it with your numerical approximations.

Problem 3. Let $I = \int_0^{\pi/2} \int_0^{\pi/2} \sin x \sin y \, dx \, dy$.

a. Calculate M_{22}.

b. Use technology to calculate M_{mn} for larger values of m and n.

c. Calculate the exact value of I and compare it with your numerical approximations.

Problem 4. Trapezoidal Rule: Derive a formula which extends the trapezoidal rule to the double integral over a rectangle. Denote your trapezoidal approximation of I by T_{mn}.

Problem 5. Let $I = \int_0^3 \int_0^2 (x + 2y) \, dx \, dy$.

a. Calculate T_{32}.

b. Use technology to calculate T_{mn} for larger values of m and n.

c. Compare your results here with your approximations in Problem 1 and with the exact value of I.

■ 16.4 THE DOUBLE INTEGRAL AS A LIMIT OF RIEMANN SUMS; POLAR COORDINATES

In the one-variable case we can write the integral as the limit of Riemann sums:

$$\int_a^b f(x) \, dx = \lim_{\max \Delta x \to 0} \sum_{i=1}^{n} f(x_i^*) \, \Delta x_i.$$

The same approach works with double integrals. To explain it, we need to explain what we mean by the *diameter of a set*.

Suppose that S is a bounded closed set (on the line, in the plane, or in three-space). For any two points P and Q of S, we can measure their separation, $d(P, Q)$. The maximal separation between points of S is called the *diameter of S*:

$$\text{diam } S = \max_{P, Q \in S} d(P, Q).\dagger$$

† If a set S is bounded, it is contained in some ball, which in turn has some finite diameter D. The set of all distances between points of S, being bounded above by D, has a least upper bound. This least upper bound is called the diameter of S:

$$\text{diam } S = \underset{P, Q \in S}{\text{lub}} \; d(P, Q).$$

It can be shown that, if S is bounded and closed, then this least upper bound is attained, in which case we can define

$$\text{diam } S = \max_{P, Q \in S} d(P, Q).$$

For a circle, a circular disc, a sphere, or a ball, this sense of diameter agrees with the usual one. Figure 16.4.1 gives some other examples.

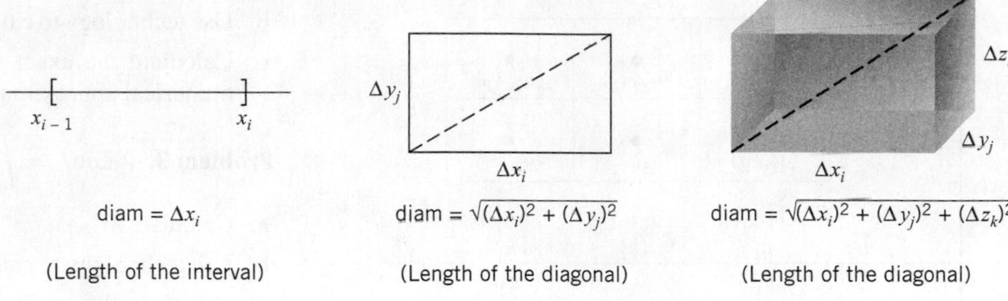

Figure 16.4.1

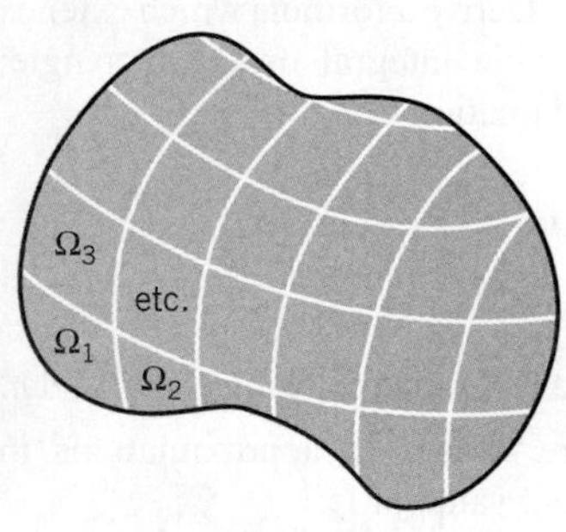

Figure 16.4.2

Now let's start with a basic region Ω and decompose it into a finite number of basic subregions $\Omega_1, \ldots, \Omega_N$. (See Figure 16.4.2.) If f is continuous on Ω, then f is continuous on each Ω_i. Now from each Ω_i we pick an arbitrary point (x_i^*, y_i^*) and form the *Riemann sum*

$$\sum_{i=1}^{N} f(x_i^*, y_i^*)(\text{area of } \Omega_i).$$

As you would expect, the double integral over Ω can be obtained as the limit of such sums; namely, given any $\epsilon > 0$, there exists $\delta > 0$ such that, if the diameters of the Ω_i are all less than δ, then

$$\left| \sum_{i=1}^{N} f(x_i^*, y_i^*)(\text{area of } \Omega_i) - \iint_{\Omega} f(x, y)\, dxdy \right| < \epsilon$$

no matter how the (x_i^*, y_i^*) are chosen within the Ω_i. We express this by writing

(16.4.1)
$$\iint_{\Omega} f(x, y)\, dxdy = \lim_{\text{diam } \Omega_i \to 0} \sum_{i=1}^{N} f(x_i^*, y_i^*)(\text{area of } \Omega_i).$$

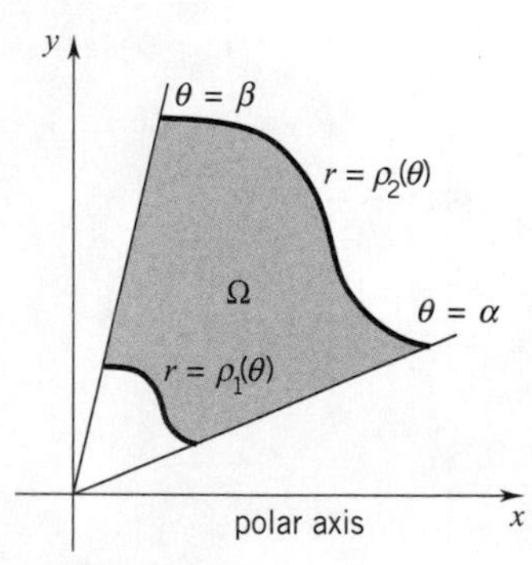

Figure 16.4.3

Evaluating Double Integrals Using Polar Coordinates

Here we explain how to calculate double integrals

$$\iint_{\Omega} f(x, y)\, dxdy$$

using polar coordinates $[r, \theta]$. Throughout we take $r \geq 0$.

We will work with type of region shown in Figure 16.4.3. The region Ω is then the set of all points (x, y) that have polar coordinates $[r, \theta]$ in the set

$$\Gamma: \quad \alpha \leq \theta \leq \beta, \quad \rho_1(\theta) \leq r \leq \rho_2(\theta),$$

where $\beta \leq \alpha + 2\pi$.

You already know how to calculate the area of Ω. By (9.5.2),

$$\text{area of } \Omega = \int_{\alpha}^{\beta} \tfrac{1}{2}\left([\rho_2(\theta)]^2 - [\rho_1(\theta)]^2 \right) d\theta.$$

We can write this as a double integral over Γ:

(16.4.2)
$$\text{area of } \Omega = \iint_{\Gamma} r\, dr d\theta.$$

PROOF Simply note that

$$\tfrac{1}{2}\left([\rho_2(\theta)]^2 - [\rho_1(\theta)]^2\right) = \int_{\rho_1(\theta)}^{\rho_2(\theta)} r\, dr,$$

and therefore

$$\text{area of } \Omega = \int_{\alpha}^{\beta} \int_{\rho_1(\theta)}^{\rho_2(\theta)} r\, dr d\theta = \iint_{\Gamma} r\, dr d\theta. \quad \square$$

Now let's suppose that f is some function continuous at each point (x, y) of Ω. Then the composition

$$F(r, \theta) = f(r\cos\theta, r\sin\theta)$$

is continuous at each point $[r, \theta]$ of Γ. We will show that

(16.4.3)
$$\iint_{\Omega} f(x, y)\, dx dy = \iint_{\Gamma} f(r\cos\theta, r\sin\theta)\, r\, dr d\theta. \qquad \text{(note the extra } r\text{)}$$

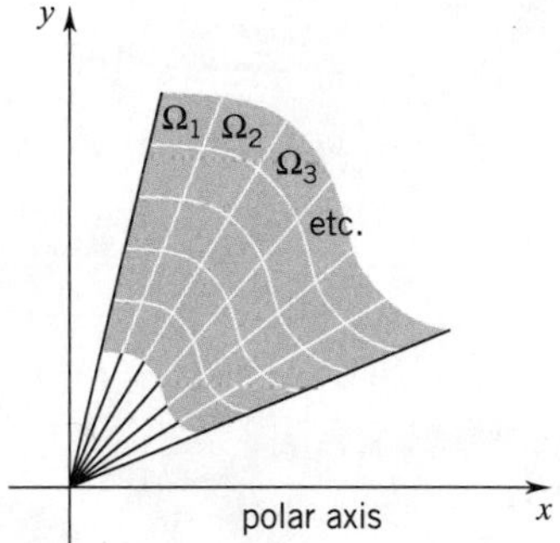

Figure 16.4.4

PROOF Our first step is to place a grid on Ω by using a finite number of rays $\theta = \theta_j$ and a finite number of continuous curves $r = \rho_k(\theta)$ in the manner of Figure 16.4.4. This grid decomposes Ω into a finite number of regions

$$\Omega_1, \ldots, \Omega_N$$

with polar coordinates in sets $\Gamma_1, \ldots, \Gamma_N$. Note that by (16.4.2)

$$\text{area of each } \Omega_i = \iint_{\Gamma_i} r\, dr d\theta.$$

Writing $F(r, \theta)$ for $f(r\cos\theta, r\sin\theta)$, we have

$$\iint_{\Gamma} F(r, \theta)\, r\, dr d\theta = \sum_{i=1}^{N} \iint_{\Gamma_i} F(r, \theta)\, r\, dr d\theta$$

additivity

$$= \sum_{i=1}^{N} F(r_i^*, \theta_i^*) \iint_{\Gamma_i} r\, dr d\theta$$

for some $[r_i^*, \theta_i^*] \in \Gamma_i$ (Theorem 16.2.10)

$$= \sum_{i=1}^{N} F(r_i^*, \theta_i^*)(\text{area of } \Omega_i)$$

with $x_i^* = r_i^* \cos\theta_i^*, y_i^* = r_i^* \sin\theta_i^*$

$$\overset{\downarrow}{=} \sum_{i=1}^{N} f(x_i^*, y_i^*)(\text{area of } \Omega_i).$$

This last expression is a Riemann sum for the double integral

$$\iint_{\Omega} f(x,y)\ dxdy.$$

As such, by (16.4.1), it differs from that integral by less than any preassigned positive ϵ provided only that the diameters of all the Ω_i are sufficiently small. This we can guarantee by making our grid sufficiently fine. □

Example 1 Use polar coordinates to evaluate $\iint_{\Omega} xy\,dxdy$, where Ω is the portion of the unit disc that lies in the first quadrant.

SOLUTION Here $\Gamma: 0 \le \theta \le \frac{1}{2}\pi, \quad 0 \le r \le 1.$ Therefore

$$\iint_{\Omega} xy\ dxdy = \iint_{\Gamma} (r\cos\theta)(r\sin\theta)r\ drd\theta$$

$$= \int_0^{\pi/2}\int_0^1 r^3\cos\theta\sin\theta\ dr\,d\theta = \tfrac{1}{8}. \quad \square$$

check this ⟶

Example 2 Use polar coordinates to calculate the volume of a sphere of radius R.

SOLUTION In rectangular coordinates,

$$V = 2\iint_{\Omega} \sqrt{R^2 - (x^2+y^2)}\ dxdy$$

where Ω is the disc of radius R centered at the origin. (Verify.) In polar coordinates, the disc of radius R centered at the origin is given by

$$\Gamma: \quad 0 \le \theta \le 2\pi, \quad 0 \le r \le R.$$

Therefore

$$V = 2\iint_{\Gamma} \sqrt{R^2 - r^2}\ r\,drd\theta = 2\int_0^{2\pi}\int_0^R \sqrt{R^2 - r^2}\ r\,dr\,d\theta = \tfrac{4}{3}\pi R^3. \quad \square$$

check this ⟶

Example 3 Calculate the volume of the solid bounded above by the cone $z = 2 - \sqrt{x^2+y^2}$ and bounded below by the disc $\Omega: (x-1)^2 + y^2 \le 1$. (See Figure 16.4.5.)

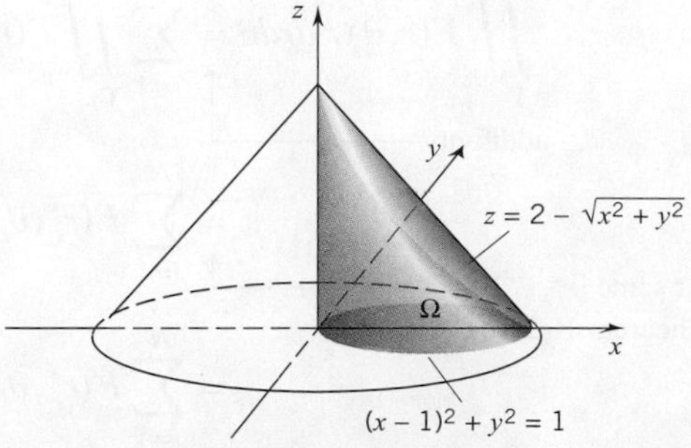

Figure 16.4.5

SOLUTION

$$V = \iint_{\Omega} (2 - \sqrt{x^2 - y^2})\, dxdy$$

$$= 2\iint_{\Omega} dxdy - \iint_{\Omega} \sqrt{x^2 + y^2}\, dxdy.$$

The first integral is $2 \times (\text{area of } \Omega) = 2\pi$. We evaluate the second integral by changing to polar coordinates.

The equation $(x-1)^2 + y^2 = 1$ simplifies to $x^2 + y^2 = 2x$. In polar coordinates this becomes $r^2 = 2r\cos\theta$, which simplifies to $r = 2\cos\theta$. The disc Ω is the set of all points with polar coordinates in the set

$$\Gamma\colon -\tfrac{1}{2}\pi \le \theta \le \tfrac{1}{2}\pi, \quad 0 \le r \le 2\cos\theta.$$

Therefore $$\iint_{\Omega} \sqrt{x^2+y^2}\, dxdy = \iint_{\Gamma} r^2 dr d\theta = \int_{-\pi/2}^{\pi/2} \int_0^{2\cos\theta} r^2\, dr\, d\theta = \tfrac{32}{9}.$$

check this

It follows that $$V = 2\pi - \tfrac{32}{9} \cong 2.73. \quad \square$$

Example 4 Evaluate $\displaystyle\iint_{\Omega} \frac{1}{(1+x^2+y^2)^{3/2}}\, dxdy$ where Ω is the triangle of Figure 16.4.6.

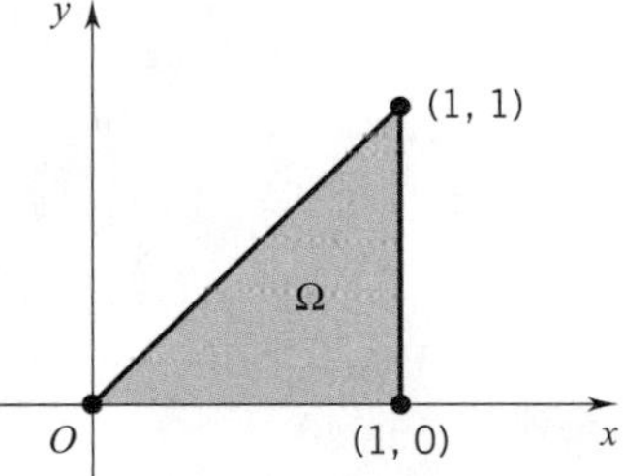

Figure 16.4.6

SOLUTION The vertical side of the triangle is part of the line $x = 1$. In polar coordinates this is $r\cos\theta = 1$, which can be written $r = \sec\theta$. Therefore

$$\iint_{\Omega} \frac{1}{(1+x^2+y^2)^{3/2}}\, dxdy = \iint_{\Gamma} \frac{r}{(1+r^2)^{3/2}}\, drd\theta$$

where $\quad \Gamma\colon 0 \le \theta \le \pi/4, \quad 0 \le r \le \sec\theta.$ (Figure 16.4.7)

The double integral over Γ reduces to

$$\int_0^{\pi/4}\int_0^{\sec\theta} \frac{r}{(1+r^2)^{3/2}}\, dr\, d\theta = \int_0^{\pi/4} \left[\frac{-1}{\sqrt{1+r^2}}\right]_0^{\sec\theta} d\theta$$

$$= \int_0^{\pi/4} \left(1 - \frac{1}{\sqrt{1+\sec^2\theta}}\right) d\theta.$$

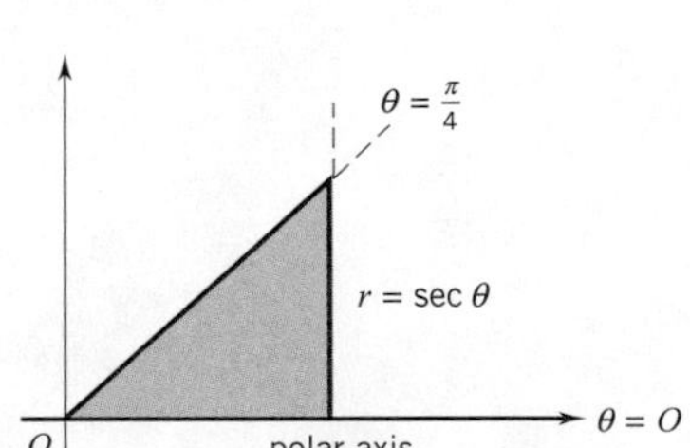

Figure 16.4.7

For $\theta \in [0, \pi/4]$,

$$\frac{1}{\sqrt{1+\sec^2\theta}} = \frac{\cos\theta}{\sqrt{\cos^2\theta + 1}} = \frac{\cos\theta}{\sqrt{2 - \sin^2\theta}}.$$

Therefore the integral can be written

$$\int_0^{\pi/4} \left(1 - \frac{\cos\theta}{\sqrt{2-\sin^2\theta}}\right) d\theta = \left[\theta - \sin^{-1}\left(\frac{\sin\theta}{\sqrt{2}}\right)\right]_0^{\pi/4} = \frac{\pi}{4} - \frac{\pi}{6} = \frac{\pi}{12}. \quad \square$$

(7.7.4)

The function $f(x) = e^{-x^2}$ has no elementary antiderivative. Nevertheless, by taking a circuitous route and then using polar coordinates, we can show that

(16.4.4)
$$\int_{-\infty}^{\infty} e^{-x^2}\,dx = \sqrt{\pi}.\ \dagger$$

PROOF The circular disc D_b: $x^2+y^2 \le b^2$ is the set of all (x, y) with polar coordinates $[r, \theta]$ in the set Γ: $0 \le \theta \le 2\pi, 0 \le r \le b$. Therefore

$$\iint_{D_b} e^{-(x^2+y^2)}\,dxdy = \iint_{\Gamma} e^{-r^2} r\,drd\theta = \int_0^{2\pi}\int_0^b e^{-r^2} r\,dr\,d\theta$$
$$= \int_0^{2\pi} \tfrac{1}{2}(1 - e^{-b^2})\,d\theta = \pi(1 - e^{-b^2}).$$

Let S_a be the square $-a \le x \le a, -a \le y \le a$. Since $D_a \subseteq S_a \subseteq D_{2a}$ and $e^{-(x^2+y^2)}$ is positive,

$$\iint_{D_a} e^{-(x^2+y^2)}\,dxdy \le \iint_{S_a} e^{-(x^2+y^2)}\,dxdy \le \iint_{D_{2a}} e^{-(x^2+y^2)}\,dxdy.$$

It follows that

$$\pi(1 - e^{-a^2}) \le \iint_{S_a} e^{-(x^2+y^2)}\,dxdy \le \pi(1 - e^{-4a^2}).$$

As $a \to \infty$, $\pi(1 - e^{-a^2}) \to \pi$ and $\pi(1 - e^{-4a^2}) \to \pi$. Therefore

$$\lim_{a\to\infty} \iint_{S_a} e^{-(x^2+y^2)}\,dxdy = \pi.$$

But
$$\iint_{S_a} e^{-(x^2+y^2)}\,dxdy = \int_{-a}^{a}\int_{-a}^{a} e^{-(x^2+y^2)}\,dx\,dy$$
$$= \int_{-a}^{a}\int_{-a}^{a} e^{-x^2}\cdot e^{-y^2}\,dx\,dy$$
$$= \left(\int_{-a}^{a} e^{-x^2}\,dx\right)\left(\int_{-a}^{a} e^{-y^2}\,dy\right) = \left(\int_{-a}^{a} e^{-x^2}\,dx\right)^2.$$

Therefore
$$\lim_{a\to\infty}\int_{-a}^{a} e^{-x^2}\,dx = \lim_{a\to\infty}\left(\iint_{S_a} e^{-(x^2+y^2)}\,dxdy\right)^{1/2} = \sqrt{\pi}.\quad \square$$

† This integral comes up frequently in probability theory and plays an important role in the branch of physics called "statistical mechanics." This integral was evaluated by numerical methods in Project 8.7

EXERCISES 16.4

Evaluate the repeated integral.

1. $\int_0^{\pi/2}\int_0^{\sin\theta} r\cos\theta\, dr\, d\theta.$

2. $\int_0^{\pi/4}\int_0^{\cos 2\theta} r\, dr\, d\theta.$

3. $\int_0^{\pi/2}\int_0^{3\sin\theta} r^2 dr\, d\theta.$

4. $\int_{-\pi/3}^{2\pi/3}\int_0^{2\cos\theta} r\sin\theta\, dr\, d\theta.$

5. Integrate $f(x,y) = \cos(x^2+y^2)$ over:
(a) the closed unit disc;
(b) the annular region $1 \le x^2+y^2 \le 4$.

6. Integrate $f(x,y) = \sin(\sqrt{x^2+y^2})$ over:
(a) the closed unit disc;
(b) the annular region $1 \le x^2+y^2 \le 4$.

7. Integrate $f(x,y) = x+y$ over:
(a) $0 \le x^2+y^2 \le 1,\ x \ge 0, y \ge 0$;
(b) $1 \le x^2+y^2 \le 4,\ x \ge 0, y \ge 0$.

8. Integrate $f(x,y) = \sqrt{x^2+y^2}$ over the triangle with vertices $(0,0), (1,0), (1,\sqrt{3})$.

Calculate by changing to polar coordinates.

9. $\int_{-1}^{1}\int_0^{\sqrt{1-y^2}} \sqrt{x^2+y^2}\, dx\, dy.$

10. $\int_0^{2}\int_0^{\sqrt{4-x^2}} \sqrt{x^2+y^2}\, dy\, dx.$

11. $\int_{1/2}^{1}\int_0^{\sqrt{1-x^2}} dy\, dx.$

12. $\int_0^{1/2}\int_0^{\sqrt{1-x^2}} xy\sqrt{x^2+y^2}\, dy\, dx.$

13. $\int_0^{1}\int_0^{\sqrt{1-x^2}} \sin\sqrt{x^2+y^2}\, dy\, dx.$

14. $\int_{-1}^{1}\int_{-\sqrt{1-y^2}}^{\sqrt{1-y^2}} e^{-(x^2+y^2)}\, dx\, dy.$

15. $\int_0^{2}\int_0^{\sqrt{2x-x^2}} x\, dy\, dx.$

16. $\int_0^{1}\int_{-\sqrt{x-x^2}}^{\sqrt{x-x^2}} (x^2+y^2)\, dy\, dx.$

In Exercises 17–22, use a double integral to find the area of the given region.

17. One leaf of the petal curve $r = 3\sin 3\theta$.

18. The region enclosed by the cardioid $r = 2(1-\cos\theta)$.

19. The region inside the circle $r = 4\cos\theta$ but outside the circle $r = 2$.

20. The region inside the large loop but outside the small loop of the limaçon $r = 1 + 2\cos\theta$.

21. The region enclosed by the lemniscate $r^2 = 4\cos 2\theta$.

22. The region inside the circle $r = 3\cos\theta$ but outside the cardioid $r = 1 + \cos\theta$.

23. Find the volume of the solid bounded above by the plane $z = y + b$, below by the xy-plane, and on the sides by the circular cylinder $x^2 + y^2 = b^2$.

24. Find the volume of the solid bounded below by the xy-plane and above by the paraboloid $z = 1 - (x^2+y^2)$.

25. Find the volume of the ellipsoid

$$x^2/4 + y^2/4 + z^2/3 = 1.$$

26. Find the volume of the solid bounded below by the xy-plane and above by the surface $x^2+y^2+z^6 = 5$.

27. Find the volume of the solid bounded below by the xy-plane, above by the spherical surface $x^2+y^2+z^2 = 4$, and on the sides by the cylinder $x^2+y^2 = 1$.

28. Find the volume of the solid bounded above by the surface $z = 1-(x^2+y^2)$, below by the xy-plane, and on the sides by the cylinder $x^2+y^2-x = 0$.

29. Find the volume of the solid bounded above by the plane $z = 2x$ and below by the disc $(x-1)^2+y^2 \le 1$.

30. Find the volume of the solid bounded above by the cone $z^2 = x^2+y^2$ and below by the region Ω which lies inside the curve $x^2+y^2 = 2ax$.

31. Find the volume of the solid bounded below by the xy-plane, above by the ellipsoid of revolution $b^2x^2 + b^2y^2 + a^2z^2 = a^2b^2$, and on the sides by the cylinder $x^2+y^2-ay = 0$.

32. A cylindrical hole of radius r is drilled through the center of a sphere of radius R.
(a) Determine the volume of the material that has been removed from the sphere.
(b) Determine the volume of the ring-shaped solid that remains.

▶C 33. Use a graphing utility to draw the petal curve $r = 2\cos 2\theta$. Then use a CAS to find the area of one petal by evaluating a double integral in polar coordinates.

▶C 34. Let $I = \iint_\Omega e^{x^2+y^2}\, dxdy$ where Ω is the annular region between the circles $x^2+y^2 = 4$ and $x^2+y^2 = 16$. Use a CAS to evaluate this integral after transforming it into a double integral in polar coordinates.

■ 16.5 SOME APPLICATIONS OF DOUBLE INTEGRATION

The Mass of a Plate

Suppose that a thin distribution of matter, called a *plate*, is laid out in the xy-plane in the form of a basic region Ω. If the mass density of the plate (the mass per unit area) is a constant λ, then the total mass M of the plate is simply the density λ times the area of the plate:

$$M = \lambda \times \text{the area of } \Omega.$$

If the density varies continuously from point to point, say $\lambda = \lambda(x, y)$, then the mass of the plate is the average density of the plate times the area of the plate:

$$M = \text{average density} \times \text{the area of } \Omega.$$

This is a double integral:

(16.5.1)
$$M = \iint_{\Omega} \lambda(x, y)\ dxdy.$$

The Center of Mass of a Plate

The center of mass x_M of a rod is a density-weighted average of position taken over the interval occupied by the rod:

$$x_M M = \int_a^b x\lambda(x)\ dx.$$

[This you have seen: (5.9.5).]

The coordinates of the center of mass of a plate (x_M, y_M) are determined by two density-weighted averages of position, each taken over the region occupied by the plate:

(16.5.2)
$$x_M M = \iint_{\Omega} x\lambda(x, y)\ dxdy, \quad y_M M = \iint_{\Omega} y\lambda(x, y)\ dxdy.$$

Example 1 A plate is in the form of a half-disc of radius a. Find the mass of the plate and the center of mass given that the mass density of the plate is directly proportional to the distance from the midpoint of the straight edge of the plate.

SOLUTION Place the plate over the region Ω: $-a \le x \le a,\ 0 \le y \le \sqrt{a^2 - x^2}$. See Figure 16.5.1. The mass density can then be written $\lambda(x, y) = k\sqrt{x^2 + y^2}$, where $k > 0$ is the constant of proportionality. Now

$$M = \iint_{\Omega} k\sqrt{x^2 + y^2}\ dxdy = \int_0^{\pi}\int_0^a (kr)\, r\, dr\, d\theta = k\left(\int_0^{\pi} 1\, d\theta\right)\left(\int_0^a r^2\, dr\right)$$

change to polar coordinates (at the second equality)

$$= k(\pi)(\tfrac{1}{3}a^3) = \tfrac{1}{3}ka^3\pi,$$

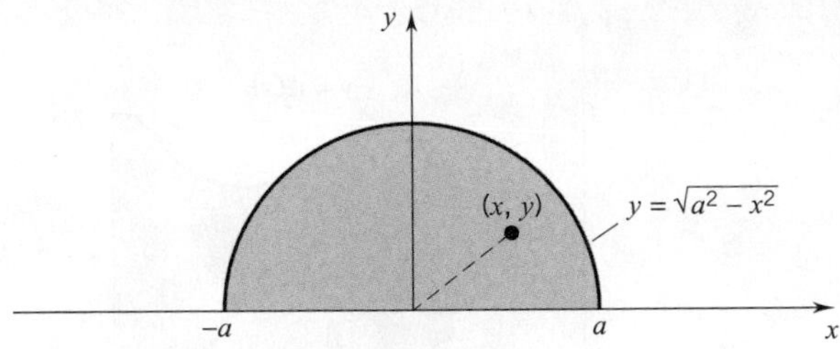

Figure 16.5.1

and

$$x_M M = \iint_{\Omega} x(k\sqrt{x^2+y^2})\, dxdy = 0.$$

(Ω is symmetric with respect to the y-axis and the integrand is odd with respect to x.) Thus $x_M = 0$. Also,

$$\begin{aligned} y_M M = \iint_{\Omega} y(k\sqrt{x^2+y^2})\, dxdy &= \int_0^{\pi}\int_0^a (r\sin\theta)(kr)\, r\, dr\, d\theta \\ &= k\left(\int_0^{\pi} \sin\theta\, d\theta\right)\left(\int_0^a r^3\, dr\right) \\ &= k(2)(\tfrac{1}{4}a^4) = \tfrac{1}{2}ka^4. \end{aligned}$$

Since $M = \frac{1}{3}ka^3\pi$, we have $y_M = (\frac{1}{2}ka^4)/(\frac{1}{3}ka^3\pi) = 3a/2\pi$. The center of mass of the plate is the point $(0, 3a/2\pi) \cong (0, 0.48a)$. □

Centroids

If the plate is homogeneous, then the mass density λ is constantly M/A where A is the area of the base region Ω. In this case the center of mass of the plate falls on the *centroid* of the base region (a notion with which you are already familiar). The centroid $(\bar{x}, \bar{y})$ depends only on the geometry of Ω:

$$\bar{x}M = \iint_{\Omega} x(M/A)\, dxdy = (M/A)\iint_{\Omega} x\, dxdy,$$

$$\bar{y}M = \iint_{\Omega} y(M/A)\, dxdy = (M/A)\iint_{\Omega} y\, dxdy.$$

Dividing by M and multiplying through by A, we have

(16.5.3)
$$\bar{x}A = \iint_{\Omega} x\, dxdy, \quad \bar{y}A = \iint_{\Omega} y\, dxdy.$$

Thus, in the sense of (16.2.9), $\bar{x}$ is the average x-coordinate on Ω and $\bar{y}$ is the average y-coordinate on Ω. The mass of the plate does not enter into this at all.

Example 2 Find the centroid of the region

$$\Omega: \quad a \le x \le b, \quad \phi_1(x) \le y \le \phi_2(x). \qquad \text{(Figure 16.5.2)}$$

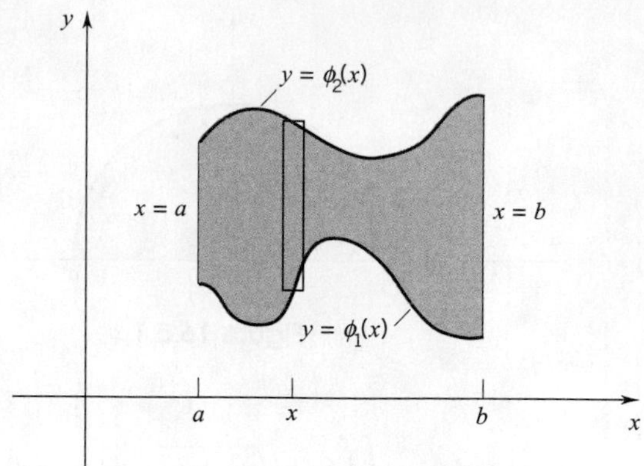

Figure 16.5.2

SOLUTION

$$\bar{x}A = \iint_{\Omega} x \; dxdy = \int_a^b \int_{\phi_1(x)}^{\phi_2(x)} x \, dy \, dx = \int_a^b x[\phi_2(x) - \phi_1(x)] \, dx;$$

$$\bar{y}A = \iint_{\Omega} y \; dxdy = \int_a^b \int_{\phi_1(x)}^{\phi_2(x)} y \, dy \, dx = \int_a^b \tfrac{1}{2}([\phi_2(x)]^2 - [\phi_1(x)]^2) \, dx.$$

These are the formulas for the centroid that we developed in Section 6.4. Having calculated many centroids there, we won't do so here. ☐

Kinetic Energy and Moment of Inertia

A particle of mass m at a distance r from a given line rotates about that line (called the *axis of rotation*) with angular speed ω. The speed v of the particle is then $r\omega$, and the kinetic energy is given by the formula

$$\text{KE} = \tfrac{1}{2}mv^2 = \tfrac{1}{2}mr^2\omega^2.$$

Imagine now a rigid body composed of a finite number of point masses m_i located at distances r_i from some fixed line. If the rigid body rotates about that line with angular speed ω, then all the point masses rotate about that same line with that same angular speed ω. The kinetic energy of the body can be obtained by adding up the kinetic energies of all the individual particles:

$$\text{KE} = \sum_i \tfrac{1}{2}m_i r_i^2 \omega^2 = \tfrac{1}{2}\left(\sum_i m_i r_i^2\right)\omega^2.$$

The expression in parentheses is called the *moment of inertia* (or *rotational inertia*) of the body and is denoted by the letter I:

(16.5.4)
$$\boxed{I = \sum_i m_i r_i^2.}$$

For a rigid body in straight-line motion

$$\text{KE} = \tfrac{1}{2}Mv^2, \qquad \text{where } v \text{ is the speed of the body and } M = \sum m_i.$$

For a rigid body in rotational motion

$$\text{KE} = \tfrac{1}{2}I\omega^2, \qquad \text{where } \omega \text{ is the angular speed of the body.}$$

The Moment of Inertia of a Plate

Suppose that a plate in the shape of a basic region Ω rotates about a line. The moment of inertia of the plate about that axis of rotation is given by the formula

(16.5.5)
$$I = \iint_{\Omega} \lambda(x,y)[r(x,y)]^2 \, dxdy$$

where $\lambda = \lambda(x,y)$ is the mass density function and $r(x,y)$ is the distance from the axis to the point (x,y).

DERIVATION OF (16.5.5) Decompose the plate into N pieces in the form of basic regions $\Omega_1, \ldots, \Omega_N$. From each Ω_i choose a point (x_i^*, y_i^*) and view all the mass of the ith piece as concentrated there. The moment of inertia of this piece is then approximately

$$\underbrace{[\lambda(x_i^*,y_i^*)(\text{area of }\Omega_i)]}_{\text{approx. mass of piece}}\underbrace{[r(x_i^*,y_i^*)]^2}_{(\text{approx. distance})^2} = \lambda(x_i^*,y_i^*)[r(x_i^*,y_i^*)]^2(\text{area of }\Omega_i).$$

The sum of these approximations,

$$\sum_{i=1}^{N} \lambda(x_i^*,y_i^*)[r(x_i^*,y_i^*)]^2(\text{area of }\Omega_i),$$

is a Riemann sum for the double integral

$$\iint_{\Omega} \lambda(x,y)[r(x,y)]^2 \, dxdy.$$

As the maximum diameter of the Ω_i tends to zero, the Riemann sum tends to this integral. □

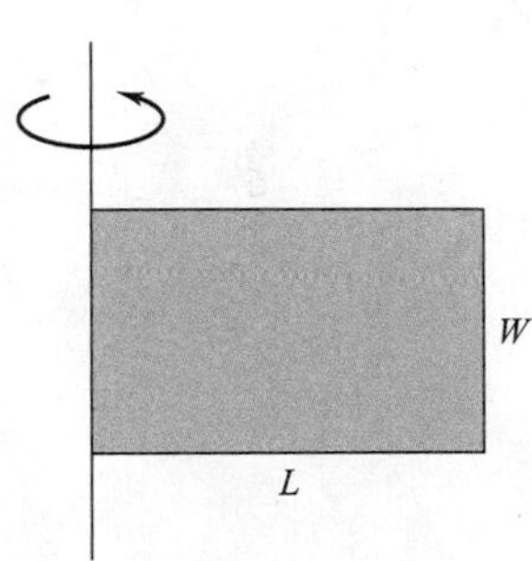

Figure 16.5.3

Example 3 A rectangular plate of mass M, length L, and width W rotates about the line shown in Figure 16.5.3. Find the moment of inertia of the plate about that line: **(a)** given that the plate has uniform mass density; **(b)** given that the mass density of the plate is directly proportional to the square of the distance from the rightmost side.

SOLUTION Coordinatize the plate as in Figure 16.5.4 and call the base region R.

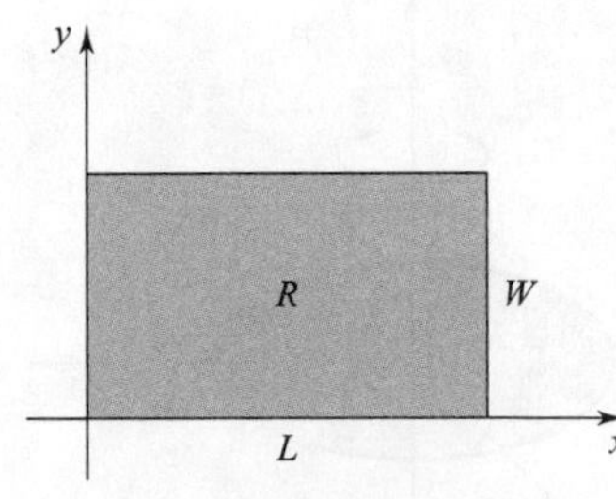

Figure 16.5.4

(a) Here $\lambda(x,y) = M/LW$ and $r(x,y) = x$. Thus

$$I = \iint_R \frac{M}{LW}x^2 \, dxdy = \frac{M}{LW}\int_0^W \int_0^L x^2 \, dx \, dy$$

$$= \frac{M}{LW}W\int_0^L x^2 \, dx = \frac{M}{L}\left(\tfrac{1}{3}L^3\right) = \tfrac{1}{3}ML^2.$$

(b) In this case $\lambda(x, y) = k(L - x)^2$, but we still have $r(x, y) = x$. Therefore

$$I = \iint_R k(L-x)^2 x^2 \, dxdy = k \int_0^W \int_0^L (L-x)^2 x^2 \, dx\, dy$$

$$= kW \int_0^L (L^2x^2 - 2Lx^3 + x^4) \, dx = \tfrac{1}{30} kL^5 W.$$

We can eliminate the constant of proportionality k by noting that

$$M = \iint_R k(L-x)^2 \, dxdy = k \int_0^W \int_0^L (L-x)^2 \, dx\, dy$$

$$= kW\Big[-\tfrac{1}{3}(L-x)^3\Big]_0^L = \tfrac{1}{3}kWL^3.$$

Therefore,

$$k = \frac{3M}{WL^3} \quad \text{and} \quad I = \tfrac{1}{30}\left(\frac{3M}{WL^3}\right) L^5 W = \tfrac{1}{10} ML^2. \quad \square$$

Radius of Gyration

If the mass M of an object is all concentrated at a distance r from a given line, then the moment of inertia about that line is given by the product Mr^2.

Suppose now that we have a plate of mass M (actually any object of mass M will do here), and suppose that l is some line. The object has some moment of inertia I about l. Its *radius of gyration* about l is the distance K for which

$$I = MK^2.$$

Namely, the radius of gyration about l is the distance from l at which all the mass of the object would have to be concentrated to effect the same moment of inertia. The formula for radius of gyration is usually written

(16.5.6)
$$\boxed{K = \sqrt{I/M}.}$$

Example 4 A homogeneous circular plate of mass M and radius R rotates about an axle that passes through the center of the plate and is perpendicular to the plate. Calculate the moment of inertia and the radius of gyration.

SOLUTION Take the axle as the z-axis and let the plate rest on the circular region $\Omega : x^2 + y^2 \le R^2$ (Figure 16.5.5). The density of the plate is $M/A = M/\pi R^2$ and $r(x, y) = \sqrt{x^2 + y^2}$. Hence

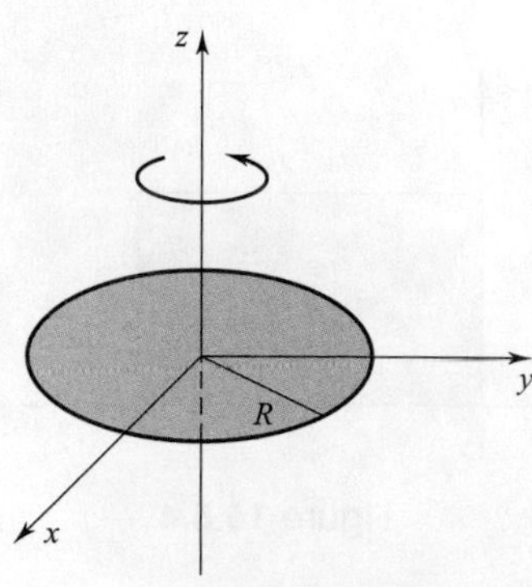

Figure 16.5.5

$$I = \iint_\Omega \frac{M}{\pi R^2}(x^2 + y^2) \, dxdy = \frac{M}{\pi R^2} \int_0^{2\pi} \int_0^R r^3 \, dr\, d\theta = \tfrac{1}{2} MR^2.$$

The radius of gyration K is $K = \sqrt{I/M} = R/\sqrt{2}$.

The circular plate of radius R has the same moment of inertia about the central axle as a circular wire of the same mass with radius $R/\sqrt{2}$. $\square$

The Parallel Axis Theorem

Suppose we have an object of mass M and a line l_M that passes through the center of mass (x_M, y_M) of the object. The object has some moment of inertia about that line; call it I_M. If l is any line parallel to l_M, then the object has a certain moment of inertia about l; call that I. The parallel axis theorem states that

(16.5.7) $$\boxed{I = I_M + d^2 M}$$

where d is the distance between the axes.

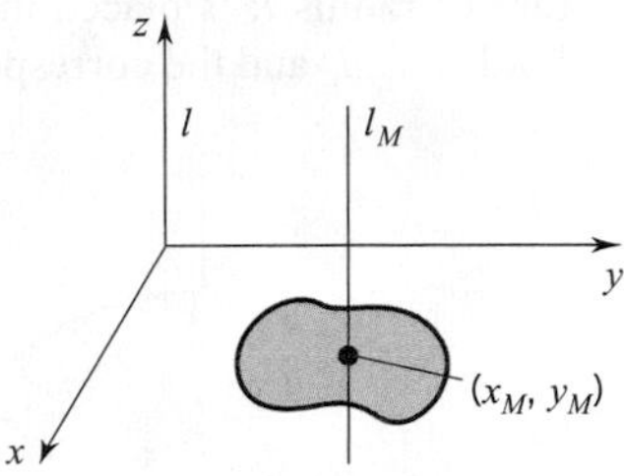

Figure 16.5.6

We prove the theorem under somewhat restrictive assumptions. Assume that the object is a plate of mass M in the shape of a basic region Ω, and assume that l_M is perpendicular to the plate. Call l the z-axis. Call the plane of the plate the xy-plane. (See Figure 16.5.6.) Denoting the points of Ω by (x, y) we have

$$\begin{aligned}
I - I_M &= \iint_\Omega \lambda(x,y)(x^2+y^2)\,dxdy - \iint_\Omega \lambda(x,y)[(x-x_M)^2+(y-y_M)^2]\,dxdy \\
&= \iint_\Omega \lambda(x,y)[2x_M x + 2y_M y - (x_M^2+y_M^2)]\,dxdy \\
&= 2x_M \iint_\Omega x\lambda(x,y)\,dxdy + 2y_M \iint_\Omega y\lambda(x,y)\,dxdy \\
&\quad -(x_M^2+y_M^2)\iint_\Omega \lambda(x,y)\,dxdy \\
&= 2x_M^2 M + 2y_M^2 M - (x_M^2+y_M^2)M = (x_M^2+y_M^2)M = d^2M. \quad \square
\end{aligned}$$

An obvious consequence of the parallel axis theorem is that $I_M \leq I$ for all lines l parallel to l_M. To minimize the moment of inertia we must pass the axis of rotation through the center of mass.

EXERCISES 16.5

Find the mass and center of mass of the plate that occupies the region Ω and has the density function λ.

1. $\Omega: \quad -1 \leq x \leq 1, \quad 0 \leq y \leq 1, \quad \lambda(x,y) = x^2.$
2. $\Omega: \quad 0 \leq x \leq 1, \quad 0 \leq y \leq \sqrt{x}, \quad \lambda(x,y) = x+y.$
3. $\Omega: \quad 0 \leq x \leq 1, \quad x^2 \leq y \leq 1, \quad \lambda(x,y) = xy.$
4. $\Omega: \quad 0 \leq x \leq \pi, \quad 0 \leq y \leq \sin x, \quad \lambda(x,y) = y.$
5. $\Omega: \quad 0 \leq x \leq 8, \quad 0 \leq y \leq \sqrt[3]{x}, \quad \lambda(x,y) = y^2.$
6. $\Omega: \quad 0 \leq x \leq a, \quad 0 \leq y \leq \sqrt{a^2-x^2}, \quad \lambda(x,y) = xy.$
7. $\Omega:$ the triangle with vertices $(0,0)$, $(1,2)$, and $(1,3)$; $\lambda(x,y) = xy$.
8. $\Omega:$ the triangular region in the first quadrant bounded by the lines $x = 0$, $y = 0$, and $3x + 2y = 6$; $\lambda(x,y) = x + y$.
9. $\Omega:$ the region bounded by the cardioid $r = 1 + \cos\theta$; λ is the distance to the pole.
10. Ω: the region inside the circle $r = 2\sin\theta$ but outside the circle $r = 1$; $\lambda(x,y) = y$.

In the exercises that follow, I_x, I_y, I_z denote the moments of inertia about the x, y, z axes.

11. A rectangular plate of mass M, length L, and width W is placed on the xy-plane with center at the origin, long sides parallel to the x-axis. (We assume here that $L \geq W$.) Find I_x, I_y, I_z if the plate is homogeneous. Determine the corresponding radii of gyration K_x, K_y, K_z.
12. Verify that I_x, I_y, I_z are unchanged if the mass density of the plate of Exercise 11 varies directly as the distance from the leftmost side.
13. Determine the center of mass of the plate of Exercise 11 if the mass density varies as in Exercise 12.

14. Show that for any plate in the xy-plane

$$I_z = I_x + I_y.$$

How are the corresponding radii of gyration K_x, K_y, K_z related?

15. A homogeneous plate of mass M in the form of a quarter disc of radius R is placed in the xy-plane as in the figure. Find I_x, I_y, I_z and the corresponding radii of gyration.

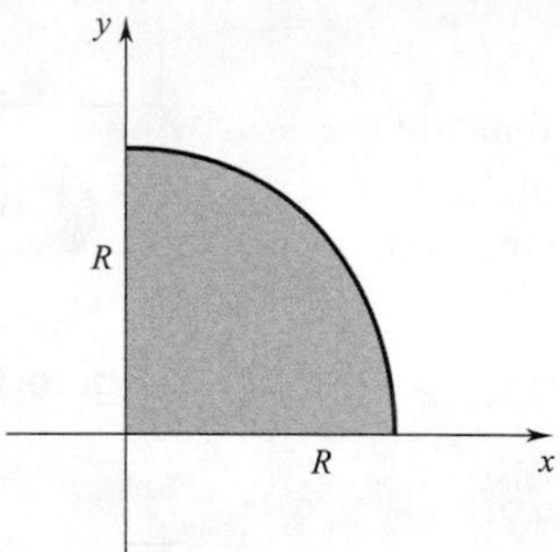

16. A plate in the xy-plane undergoes a rotation in that plane about its center of mass. Show that I_z remains unchanged.

17. A homogeneous disc of mass M and radius R is to be placed on the xy-plane so that it has moment of inertia I_0 about the z-axis. Where should the disc be placed?

18. A homogeneous plate of mass density λ occupies the region under the curve $y = f(x)$ from $x = a$ to $x = b$. Show that

$$I_x = \tfrac{1}{3}\lambda \int_a^b [f(x)]^3\, dx \qquad \text{and} \qquad I_y = \lambda \int_a^b x^2 f(x)\, dx.$$

19. A homogeneous plate of mass M in the form of an elliptical quadrant is placed on the xy-plane. (See the figure.) Find I_x, I_y, I_z.

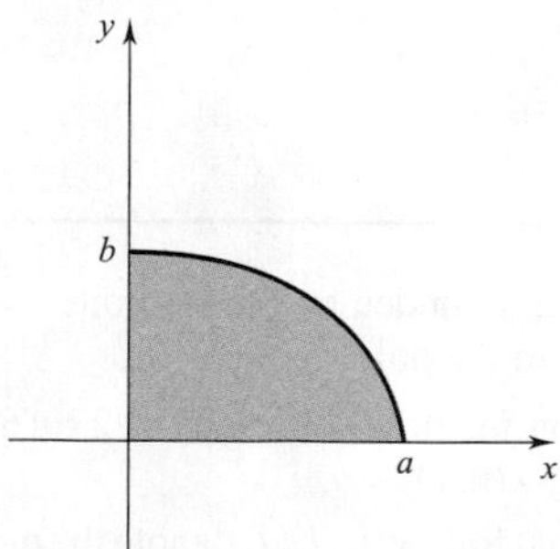

20. Find I_x, I_y, I_z for the plate in Exercise 2.

21. Find I_x, I_y, I_z for the plate in Exercise 3.

22. Find I_x, I_y, I_z for the plate in Exercise 5.

23. Find I_x, I_y, I_z for the plate in Exercise 9.

24. A plate of varying density occupies the region $\Omega = \Omega_1 \cup \Omega_2$ shown in the figure. Find the center of mass of the plate given that the Ω_1 piece has mass M_1 and center of mass (x_1, y_1), and the Ω_2 piece has mass M_2 and center of mass (x_2, y_2).

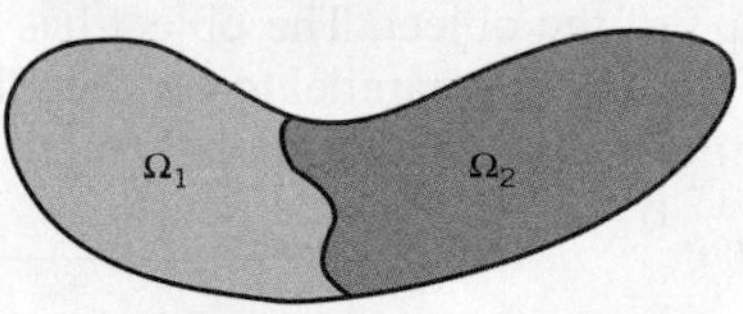

25. A homogeneous plate of mass M is in the form of a ring. (See the figure.) Calculate the moment of inertia of the plate:

(a) about a diameter;

(b) about a tangent to the inner circle;

(c) about a tangent to the outer circle.

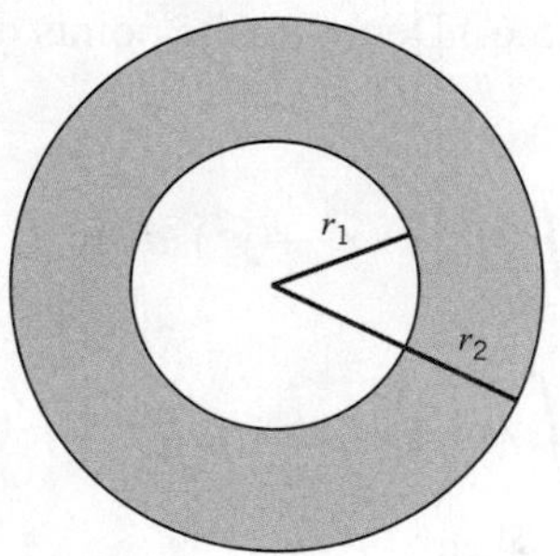

26. Find the moment of inertia of a homogeneous circular wire of mass M and radius r:

(a) about a diameter;

(b) about a tangent. HINT: Use the previous exercise.

27. The plate of Exercise 25 rotates about the axis that is perpendicular to the plate and passes through the center. Find the moment of inertia.

28. Prove the parallel axis theorem for the case where the line through the center of mass lies in the plane of the plate.

29. A plate of mass M has the form of a half disc Ω, $-R \le x \le R, 0 \le y \le \sqrt{R^2 - x^2}$. Find the center of mass given that the mass density varies directly as the distance from the curved boundary.

30. Find I_x, I_y, I_z for the plate of Exercise 29.

31. A plate of mass M is in the form of a disc of radius R. Given that the mass density of the plate varies directly as the distance from a point P on the boundary of the plate, locate the center of mass.

32. A plate of mass M is in the form of a right triangle of base b and height h. Given that the mass density of the plate varies directly as the square of the distance from the vertex of the right angle, locate the center of mass of the plate.

33. Use double integrals to justify an assumption we made about centroids in Chapter 6: Formula (6.4.1).

C 34. A plate is in the form of a triangle with vertices $(0,0)$, $(0,1)$, $(2,1)$. The mass density of the plate is given by $\lambda(x,y) = x + y$. Use a CAS to find: (a) the center of mass of the plate, and (b) the moments of inertia I_x and I_y.

■ 16.6 TRIPLE INTEGRALS

Now that you are familiar with double integrals

$$\iint_{\Omega} f(x,y)\, dxdy,$$

you will find it easy to understand triple integrals

$$\iiint_{T} f(x,y,z)\, dxdydz.$$

Basically the only difference is that, instead of working with functions of two variables continuous on a plane region Ω, we will be working with functions of three variables continuous on some portion T of three-space.

The Triple Integral over a Box

For double integration we began with a rectangle

$$R: a_1 \leq x \leq a_2, \quad b_1 \leq y \leq b_2.$$

For triple integration we begin with a *box* (a rectangular solid)

$$\Pi: a_1 \leq x \leq a_2, \quad b_1 \leq y \leq b_2, \quad c_1 \leq z \leq c_2. \qquad \text{(Figure 16.6.1)}$$

To partition this box, we first partition the edges. Taking

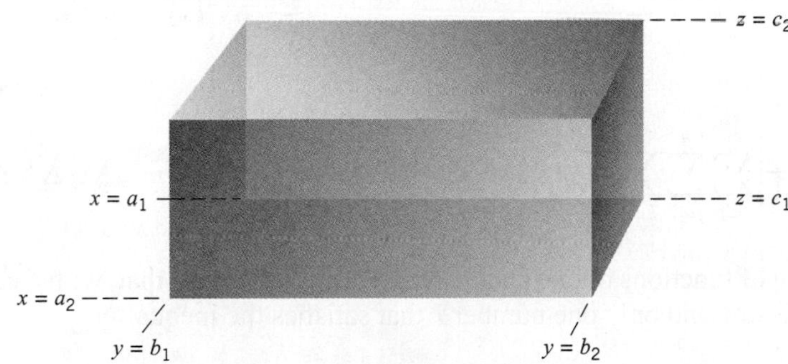

Figure 16.6.1

$$\begin{aligned} P_1 &= \{x_0, \ldots, x_m\} && \text{as a partition of } [a_1, a_2], \\ P_2 &= \{y_0, \ldots, y_n\} && \text{as a partition of } [b_1, b_2], \\ P_3 &= \{z_0, \ldots, z_q\} && \text{as a partition of } [c_1, c_2], \end{aligned}$$

we form the set

$$P = P_1 \times P_2 \times P_3 = \{(x_i, y_j, z_k) : x_i \in P_1, y_j \in P_2, z_k \in P_3\}\dagger$$

and call this a *partition of* Π. The partition P breaks up Π into $m \times n \times q$ nonoverlapping boxes

$$\Pi_{ijk}: \quad x_{i-1} \leq x \leq x_i, \quad y_{j-1} \leq y \leq y_j, \quad z_{k-1} \leq z \leq z_k.$$

---

†$P_1 \times P_2 \times P_3$ is called the *Cartesian product* of P_1, P_2, P_3.

A typical such box is pictured in Figured 16.6.2.

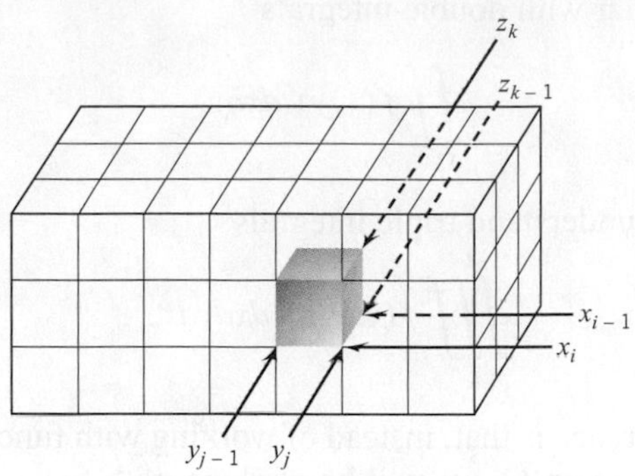

Figure 16.6.2

Let f be continuous on Π. Then, taking

M_{ijk} as the maximum value of f on Π_{ijk}

and m_{ijk} as the minimum value of f on Π_{ijk},

we form the *upper sum*

$$U_f(P) = \sum_{i=1}^{m}\sum_{j=1}^{n}\sum_{k=1}^{q} M_{ijk}(\text{ volume of } \Pi_{ijk}) = \sum_{i=1}^{m}\sum_{j=1}^{n}\sum_{k=1}^{q} M_{ijk}\,\Delta x_i\,\Delta y_j\,\Delta z_k$$

and the *lower sum*

$$Ł_f(P) = \sum_{i=1}^{m}\sum_{j=1}^{n}\sum_{k=1}^{q} m_{ijk}(\text{ volume of } \Pi_{ijk}) = \sum_{i=1}^{m}\sum_{j=1}^{n}\sum_{k=1}^{q} m_{ijk}\,\Delta x_i\,\Delta y_j\,\Delta z_k.$$

As in the case of functions of one and two variables, it turns out that, with f continuous on Π, there is one and only one number I that satisfies the inequality

$$L_f(P) \le I \le u_f(P) \qquad \text{for all partitions } P \text{ of } \Pi.$$

DEFINITION 16.6.1 THE TRIPLE INTEGRAL OVER A BOX $\prod$

Let f be continuous on a closed box Π. The unique number I that satisfies the inequality

$$L_f(P) \le I \le U_f(P) \qquad \text{for all partitions } P \text{ of } \Pi$$

is called the *triple integral* of f over Π and is denoted by

$$\iiint\limits_{\Pi} f(x, y, z)\; dxdydz.\dagger$$

---

† The triple integral can also be written $\int\int\limits_{\Pi}\int f(x, y, z)\,dV$

The Triple Integral over a More General Solid

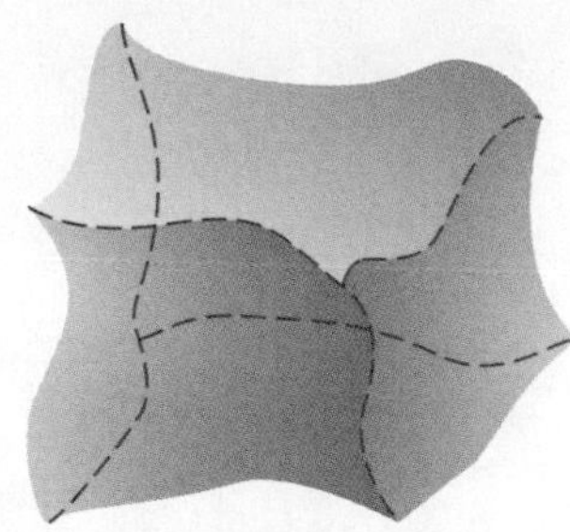
Figure 16.6.3

We start with a three-dimensional, bounded, closed, connected set T. We assume that T is a *basic solid;* that is, we assume that the boundary of T consists of a finite number of continuous surfaces $z = \alpha(x, y)$, $y = \beta(x, z)$, $x = \gamma(y, z)$. See, for example, Figure 16.6.3.

Now let's suppose that f is some function continuous on T. To define the triple integral of f over T, we first encase T in a rectangular box Π with sides parallel to the coordinate planes (Figure 16.6.4). We then extend f to all of Π by defining f to be zero outside of T. This extended function f is bounded on Π, and it is continuous on all of Π except possibly at the boundary of T. In spite of these possible discontinuities, f is still integrable over Π; that is, there still exists a unique number I such that

$$L_f(P) \leq I \leq U_f(P) \qquad \text{for all partitions } P \text{ of } \Pi$$

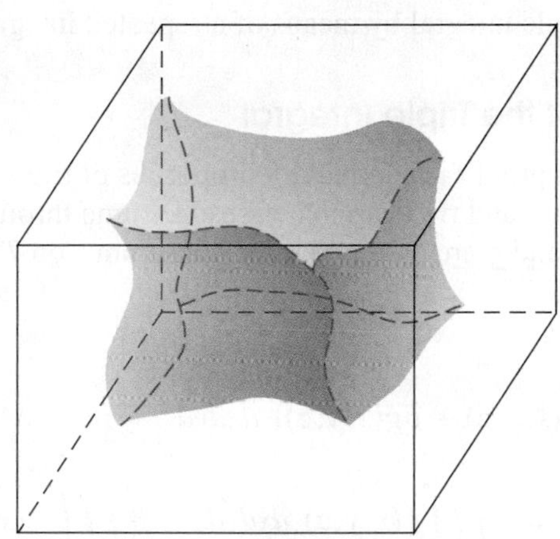
Figure 16.6.4

(We will not attempt to prove this.) The number I is by definition the triple integral

$$\iiint\limits_{\Pi} f(x, y, z)\, dxdydz.$$

We define the triple integral over T by setting

(16.6.2)
$$\iiint\limits_{T} f(x, y, z)\, dxdydz = \iiint\limits_{\Pi} f(x, y, z)\, dxdydz.$$

Volume as a Triple Integral

The simplest triple integral of interest is the triple integral of the function that is constantly 1 on T. This gives the volume of T:

(16.6.3)
$$\text{volume of } T = \iiint\limits_{T} dxdydz.$$

PROOF Set $f(x, y, z) = 1$ for all (x, y, z) in T. Encase T in a box Π. Define f to be zero outside of T. An arbitrary partition P of Π breaks up T into little boxes Π_{ijk}. Note that

$$L_f(P) = \text{ the sum of the volumes of all the } \Pi_{ijk} \text{ that are contained in } T$$

$$U_f(P) = \text{ the sum of the volumes of all the } \Pi_{ijk} \text{ that intersect } T.$$

It follows that

$$L_f(P) \leq \text{ the volume of } T \leq U_f(P).$$

The arbitrariness of P gives the formula. □

Remark As in the case of double integrals, there is no implied order in writing *dxdydz* in (16.6.2) and (16.6.3). The *dxdydz* represents an element of volume and could just as well have been written in any other order. An order of integration will be introduced when we evaluate a triple integral by means of a repeated integral. □

Some Properties of the Triple Integral

Below we give without proof the elementary properties of triple integrals analogous to what you saw in the one- and two-variable cases. Assume throughout that T is a basic solid. The functions f and g are assumed to be continuous on T.

I. Linearity:

$$\iiint_T [\alpha f(x, y, z) + \beta g(x, y, z)]\ dxdydz$$
$$= \alpha \iiint_T f(x, y, z)\ dxdydz + \beta \iiint_T g(x, y, z)\ dxdydz.$$

II. Order:

$$\text{if } f \geq 0 \text{ on } T, \text{ then } \iiint_T f(x, y, z)\ dxdydz \geq 0;$$

$$\text{if } f \leq g \text{ on } T, \text{ then } \iiint_T f(x, y, z)\ dxdydz \leq \iiint_T g(x, y, z)\ dxdydz.$$

III. Additivity: If T is broken up into a finite number of basic solids $T_1, \ldots, T_n$, then

$$\iiint_T f(x, y, z)\ dxdydz = \iiint_{T_1} f(x, y, z)\ dxdydz +$$
$$\cdots + \iiint_{T_n} f(x, y, z)\ dxdydz.$$

IV. Mean-value condition: There is a point (x_0, y_0, z_0) in T for which

$$\iiint_T f(x, y, z)\ dxdydz = f(x_0, y_0, z_0) \cdot (\text{ volume of } T).$$

We call $f(x_0, y_0, z_0)$ *the average value of f on T*.

The notion of average enables us to write

(16.6.4)
$$\iiint_T f(x,y,z)\,dxdydz = \left(\begin{array}{c}\text{the average value}\\ \text{of } f \text{ on } T\end{array}\right)\cdot(\text{ volume of } T).$$

We can also take weighted averages: if f and g are continuous and g is nonnegative on T, then there is a point (x_0, y_0, z_0) in T for which

(16.6.5)
$$\iiint_T f(x,y,z)\,g(x,y,z)\,dxdydz = f(x_0,y_0,z_0)\iiint_T g(x,y,z)\,dxdydz.$$

As you would expect, we call $f(x_0, y_0, z_0)$ *the g-weighted average of f on T*.

The formulas for mass, center of mass, and moments of inertia derived in the previous section for two-dimensional plates are easily extended to three-dimensional objects.

Suppose that T is an object in the form of a basic solid. If T has constant mass density λ (here density is mass per unit volume), then the mass of T is the density λ times the volume of T:

$$M = \lambda V.$$

If the mass density varies continuously over T, say, $\lambda = \lambda(x, y, z)$, then *the mass of T is the average density of T times the volume of T*. This is a triple integral

(16.6.6)
$$M = \iiint_T \lambda(x,y,z)\,dxdydz.$$

The coordinates of the center of mass (x_M, y_M, z_M) are density-weighted averages of position, each taken over the portion of space occupied by the solid.

(16.6.7)
$$x_M M = \iiint_T x\lambda(x,y,z)\,dxdydz, \qquad \text{etc.}$$

If the object T is homogeneous (constant mass density M/V), then the center of mass of T depends only on the geometry of T and falls on the centroid $(\bar{x}, \bar{y}, \bar{z})$ of the space occupied by T. The density is irrelevant. The coordinates of the centroid are simple averages over T:

(16.6.8)
$$\bar{x}V = \iiint_T x\,dxdydz, \qquad \text{etc.}$$

The moment of inertia of T about a line is given by the formula

(16.6.9)
$$I = \iiint_T \lambda(x,y,z)[r(x,y,z)]^2\,dxdydz.$$

Here $\lambda(x,y,z)$ is the mass density of T at (x,y,z) and $r(x,y,z)$ is the distance of (x,y,z) from the line in question. The moments of inertia about the x,y,z axes are again denoted by I_x, I_y, I_z.

All of this should be readily understandable. Techniques for evaluating triple integrals are introduced in the next three sections.

EXERCISES 16.6

1. Let $f(x,y)$ be a function continuous and nonnegative on a basic solid Ω and set

$$T = \{(x,y,z) : (x,y) \in \Omega, \quad 0 \le z \le f(x,y)\}.$$

Compare

$$\iiint_T dxdydz \quad \text{to} \quad \iint_\Omega f(x,y)\, dxdy.$$

2. Set $f(x,y,z) = xyz$ on $\Pi : 0 \le x \le 1, \quad 0 \le y \le 1, \quad 0 \le z \le 1$ and take P as the partition $P_1 \times P_2 \times P_3$.

(a) Find $L_f(P)$ and $U_f(P)$ given that

$$P_1 = \{x_0, \dots, x_m\}, \quad P_2 = \{y_0, \dots, y_n\},$$

$$P_3 = \{z_0, \dots, z_q\}$$

are all arbitrary partitions of [0, 1].

(b) Use your answer to (a) to calculate

$$\iiint_\Pi xyz\, dxdydz.$$

3. Let $f(x,y,z) = \alpha$, constant, over the rectangular solid Π : $a_1 \le x \le a_2, \quad b_1 \le y \le b_2, \quad c_1 \le z \le c_2$. Show that

$$\iiint_\Pi \alpha\, dxdydz = \alpha(a_2 - a_1)(b_2 - b_1)(c_2 - c_1).$$

4. Find the average value of $f(x,y,z) = xyz$ over the solid Π in Exercise 2.

5. Calculate

$$\iiint_\Pi xy\, dxdydz \quad \text{where} \quad \Pi : 0 \le x \le a, \quad 0 \le y \le b, \quad 0 \le z \le c.$$

6. Let T be a basic solid of varying mass density $\lambda = \lambda(x,y,z)$. The moment of inertia of T about the xy-plane is defined by setting

$$I_{xy} = \iiint_T \lambda(x,y,z)z^2\, dxdydz.$$

The other plane moments of inertia, I_{xz} and I_{yz}, have comparable definitions. Express I_x, I_y, I_z in terms of the plane moments of inertia.

7. A box $\Pi_1 : \ 0 \le x \le 2a, \ 0 \le y \le 2b, \ 0 \le z \le 2c$ is cut away from a larger box $\Pi_0 : 0 \le x \le 2A, \ 0 \le y \le 2B, \ 0 \le z \le 2C$. Locate the centroid of the remaining solid.

8. Show that, if f is continuous and nonnegative on a basic solid T, then the triple integral of f over T is nonnegative.

9. Calculate the mass M of the cube $\Pi : \ 0 \le x \le a, \ 0 \le y \le a, 0 \le z \le a$ given that the mass density varies directly with the distance from the face on the xy-plane.

10. Locate the center of mass of the cube of Exercise 9.

11. Find the moment of inertia I_z of the cube of Exercise 9.

12. Let $f(x,y,z) = 3y^2 - 2x + z$ on the box $B : 2 \le x \le 5, \ 1 \le y \le 3, \ 3 \le z \le 4$. Let P_1 be a regular partition of $[2,5]$ with $k = 100$ subintervals, let P_2 be a regular partition of $[1,3]$ with $m = 200$ subintervals, let P_3 be a regular partition of $[3,4]$ with $n = 150$ subintervals, and let $P = P_1 \times P_2 \times P_3$.

(a) Use a CAS to find $L_f(P)$ and $U_f(P)$.

(b) Investigate $L_f(P)$ and $U_f(P)$ for values of $k > 100$, $m > 200$, and $n > 150$.

(c) Estimate $\int\int\int_B f(x,y,z)\, dxdydz$.

■ 16.7 REDUCTION TO REPEATED INTEGRALS

In this section we give no proofs. You can assume that all the solids that appear are basic solids and all the functions that you encounter are continuous.

In Figure 16.7.1 we have sketched a solid T. The projection of T onto the xy-plane has been labeled Ω_{xy}. The solid T is then the set of all (x,y,z) with

$$(x,y) \text{ in } \Omega_{xy} \qquad \text{and} \qquad \psi_1(x,y) \le z \le \psi_2(x,y).$$

The triple integral over T can be evaluated by setting

$$(*)\qquad \iiint_T f(x,y,z)\;dxdydz = \iint_{\Omega_{xy}} \left(\int_{\psi_1(x,y)}^{\psi_2(x,y)} f(x,y,z)\,dz\right) dxdy.$$

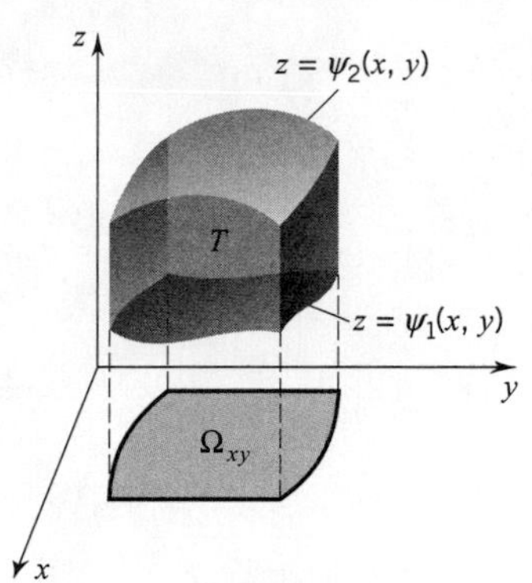

Figure 16.7.1

Moving to Figure 16.7.2, we see that in this case Ω_{xy} is the region

$$a_1 \le x \le a_2, \qquad \phi_1(x) \le y \le \phi_2(x),$$

and T itself is the set of all (x, y, z) with

$$a_1 \le x \le a_2, \qquad \phi_1(x) \le y \le \phi_2(x), \qquad \psi_1(x,y) \le z \le \psi_2(x,y).$$

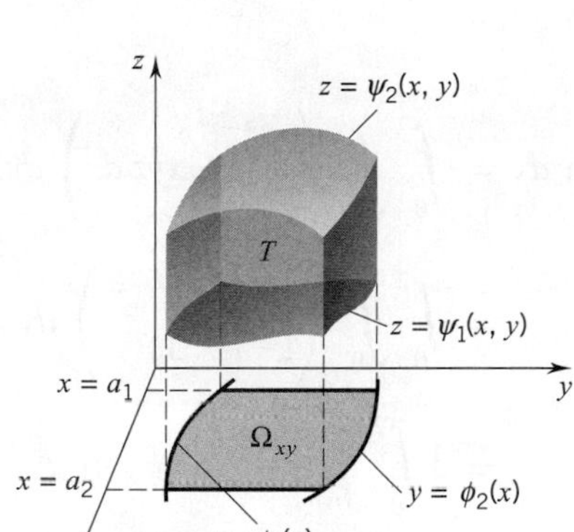

Figure 16.7.2

The triple integral over T can then be expressed by three ordinary integrals:

$$\iiint_T f(x,y,z)\;dxdydz = \int_{a_1}^{a_2}\left[\int_{\phi_1(x)}^{\phi_2(x)}\left(\int_{\psi_1(x,y)}^{\psi_2(x,y)} f(x,y,z)\,dz\right) dy\right] dx.$$

It is customary to omit the brackets and parentheses and write

(16.7.1)
$$\iiint_T f(x,y,z)\;dxdydz = \int_{a_1}^{a_2}\int_{\phi_1(x)}^{\phi_2(x)}\int_{\psi_1(x,y)}^{\psi_2(x,y)} f(x,y,z)\,dz\,dy\,dx. \;\dagger$$

Here we first integrate with respect to z [from $z = \psi_1(x,y)$ to $z = \psi_2(x,y)$], then with respect to y [from $y = \phi_1(x)$ to $y = \phi_2(x)$], and finally with respect to x [from $x = a_1$ to $x = a_2$].

There is nothing special about this order of integration. Other orders of integration are possible and in some cases more convenient. Suppose, for example, that the projection of T onto the xz-plane is a region of the form

$$\Omega_{xz}\colon a_1 \le z \le a_2, \qquad \phi_1(z) \le x \le \phi_2(z).$$

† This formula is formula $(*)$ taken one step further. Usually we skip the double-integral stage and go directly to three integrals.

If T is the set of all (x, y, z) with

$$a_1 \le z \le a_2, \quad \phi_1(z) \le x \le \phi_2(z), \quad \psi_1(x,z) \le y \le \psi_2(x,z),$$

then $$\iiint_T f(x,y,z)\, dxdydz = \int_{a_1}^{a_2} \int_{\phi_1(z)}^{\phi_2(z)} \int_{\psi_1(x,z)}^{\psi_2(x,z)} f(x,y,z)\, dy\, dx\, dz.$$

In this case we integrate first with respect to y, then with respect to x, and finally with respect to z. Still four other orders of integration are possible.

Example 1 Evaluate the expression $\displaystyle\int_0^2 \int_0^x \int_0^{4-x^2} xyz\, dz\, dy\, dx.$

SOLUTION

$$\begin{aligned}
\int_0^2 \int_0^x \int_0^{4-x^2} xyz\, dz\, dy\, dx &= \int_0^2 \int_0^x \left(\int_0^{4-x^2} xyz\, dz \right) dy\, dx \\
&= \int_0^2 \int_0^x \left(\left[\tfrac{1}{2} xyz^2 \right]_0^{4-x^2} \right) dy\, dx \\
&= \tfrac{1}{2} \int_0^2 \int_0^x x(4-x^2)^2 y\, dy\, dx \\
&= \tfrac{1}{2} \int_0^2 \left(\int_0^x x(4-x^2)^2 y\, dy \right) dx \\
&= \tfrac{1}{2} \int_0^2 \left(\left[\tfrac{1}{2} x(4-x^2)^2 y^2 \right]_0^x \right) dx \\
&= \tfrac{1}{4} \int_0^2 x^3(4-x^2)^2\, dx = \tfrac{1}{4} \int_0^2 x^3(16 - 8x^2 + x^4)\, dx \\
&= \tfrac{1}{4} \left[4x^4 - \tfrac{8}{6} x^6 + \tfrac{1}{8} x^8 \right]_0^2 = \tfrac{8}{3}. \quad \square
\end{aligned}$$

Remark The solid determined by the limits of integration in Example 1 is the solid T in the first octant bounded by the parabolic cylinder $z = 4 - x^2$, the plane $z = 0$, the plane $y = x$, and the plane $y = 0$. This solid is shown in Figure 16.7.3. □

Example 2 Use triple integration to find the volume of the tetrahedron T shown in Figure 16.7.4, and find the coordinates of the centroid.

SOLUTION The volume of T is given by the triple integral

$$V = \iiint_T dxdydz.$$

To evaluate this triple integral, we can project T onto any one of the three coordinate planes. We will project onto the xy-plane. The base region is then the triangle

$$\Omega_{xy}: \quad 0 \le x \le 1, \quad 0 \le y \le 1 - x. \qquad \text{(Figure 16.7.5)}$$

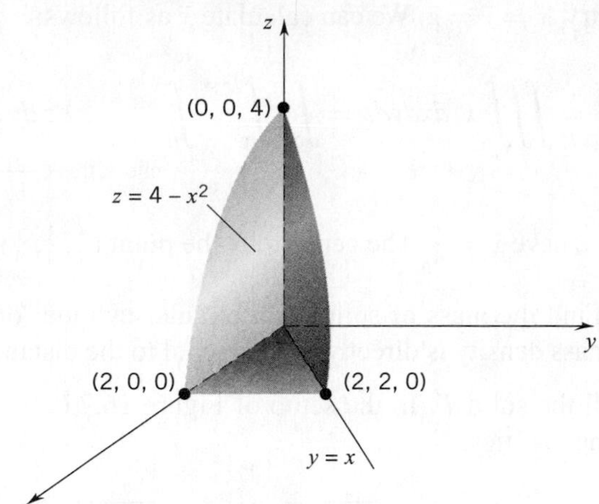

Figure 16.7.3

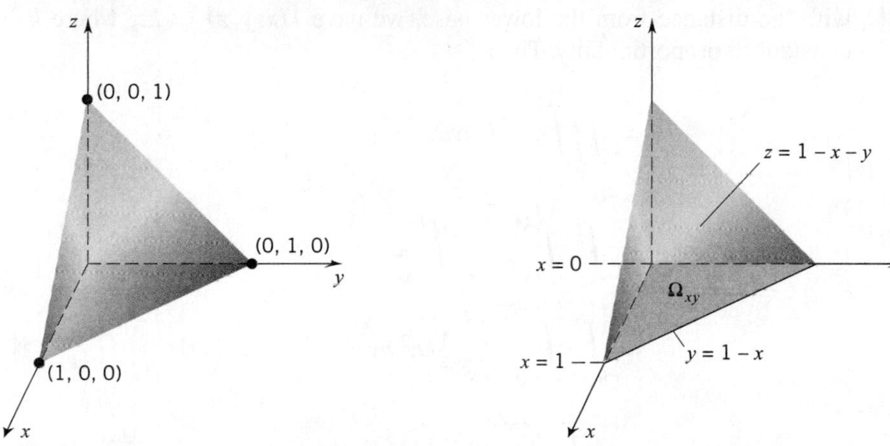

Figure 16.7.4

Figure 16.7.5

Since the inclined face is part of the plane $z = 1 - x - y$, we have T as the set of all (x, y, z) with

$$0 \leq x \leq 1, \quad 0 \leq y \leq 1 - x, \quad 0 \leq z \leq 1 - x - y.$$

It follows that

$$\begin{aligned} V = \iiint_T dxdydz &= \int_0^1 \int_0^{1-x} \int_0^{1-x-y} dz\,dy\,dx \\ &= \int_0^1 \int_0^{1-x} (1 - x - y)\,dy\,dx \\ &= \int_0^1 \left[(1 - x)y - \tfrac{1}{2}y^2\right]_0^{1-x} dx \\ &= \int_0^1 \tfrac{1}{2}(1 - x)^2\,dx = \left[-\tfrac{1}{6}(1 - x)^3\right]_0^1 = \tfrac{1}{6}. \end{aligned}$$

By symmetry, $\bar{x} = \bar{y} = \bar{z}$. We can calculate $\bar{x}$ as follows:

$$\bar{x}\,V = \iiint_T x \; dxdydz = \int_0^1 \int_0^{1-x} \int_0^{1-x-y} x \, dz \, dy \, dx = \tfrac{1}{24}.$$

check this

Since $V = \frac{1}{6}$, we have $\bar{x} = \frac{1}{4}$. The centroid is the point $(\frac{1}{4}, \frac{1}{4}, \frac{1}{4})$. □

Example 3 Find the mass of solid right circular cylinder of radius r and height h given that the mass density is directly proportional to the distance from the lower base.

SOLUTION Call the solid T. In the setup of Figure 16.7.6, we can characterize T by the following inequalities:

$$-r \le x \le r, \quad -\sqrt{r^2 - x^2} \le y \le \sqrt{r^2 - x^2}, \quad 0 \le z \le h.$$

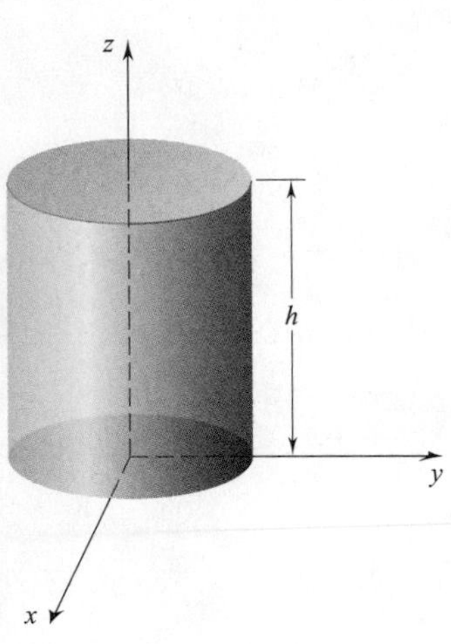

Figure 16.7.6

The first two inequalities define the base region Ω_{xy}. Since the density varies directly with the distance from the lower base, we have $\lambda(x, y, z) = kz$, where $k > 0$ is the constant of proportionality. Then

$$\begin{aligned} M &= \iiint_T kz \; dxdydz \\ &= \int_{-r}^{r} \int_{-\sqrt{r^2-x^2}}^{\sqrt{r^2-x^2}} \int_0^h kz \, dz \, dy \, dx \\ &= \int_{-r}^{r} \int_{-\sqrt{r^2-x^2}}^{\sqrt{r^2-x^2}} \tfrac{1}{2}kh^2 \, dy \, dx \\ &= 4 \int_0^r \int_0^{\sqrt{r^2-x^2}} \tfrac{1}{2}kh^2 \, dy \, dx \qquad \text{(using the symmetry)} \\ &= 2kh^2 \int_0^r \sqrt{r^2 - x^2} \, dx. \end{aligned}$$

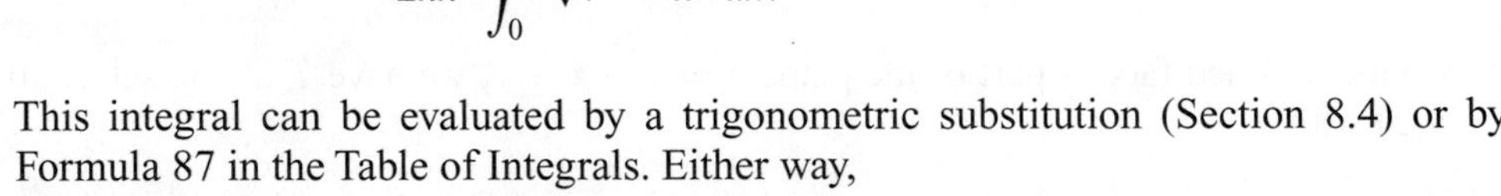
This integral can be evaluated by a trigonometric substitution (Section 8.4) or by Formula 87 in the Table of Integrals. Either way,

$$\int_0^r \sqrt{r^2 - x^2} \, dx = \tfrac{1}{4}\pi r^2.$$

It follows that

$$M = 2kh^2(\tfrac{1}{4}\pi r^2) = \tfrac{1}{2}kh^2 r^2 \pi. \quad \square$$

Remark In Example 3 we would have profited by not skipping the double integral stage; namely, we could have written

$$\begin{aligned} M &= \iint_{\Omega_{xy}} \left(\int_0^h kz \, dz \right) dxdy = \iint_{\Omega_{xy}} \tfrac{1}{2}kh^2 \; dxdy \\ &= \tfrac{1}{2}kh^2(\text{area of } \Omega_{xy}) = \tfrac{1}{2}kh^2 r^2 \pi. \quad \square \end{aligned}$$

Example 4 Integrate $f(x,y,z) = yz$ over that part of the first octant $x \geq 0,\ y \geq 0,\ z \geq 0$ that is contained in the ellipsoid

$$\frac{x^2}{a^2} + \frac{y^2}{b^2} + \frac{z^2}{c^2} = 1.$$

SOLUTION Call the solid T. The upper boundary of T has equation

$$z = \psi(x,y) = \frac{c}{ab}\sqrt{a^2b^2 - b^2x^2 - a^2y^2}.$$

This surface intersects the xy-plane in the curve

$$y = \phi(x) = \frac{b}{a}\sqrt{a^2 - x^2}.$$

We can take

$$\Omega_{xy}: \quad 0 \leq x \leq a, \quad 0 \leq y \leq \phi(x)$$

as the base region (see Figure 16.7.7) and characterize T as the set of all (x,y,z) with

$$0 \leq x \leq a, \quad 0 \leq y \leq \phi(x), \quad 0 \leq z \leq \psi(x,y).$$

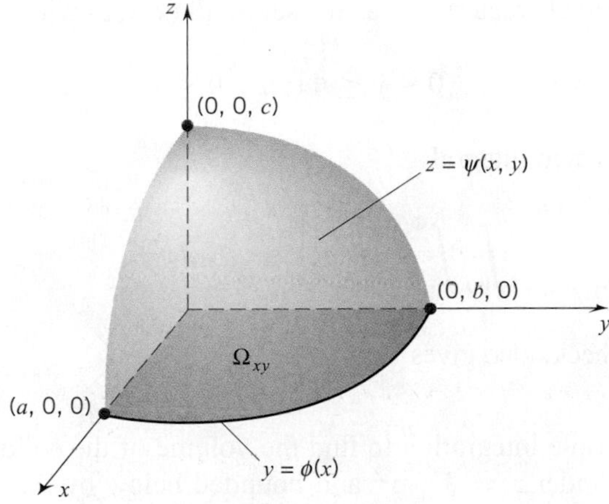

Figure 16.7.7

We can therefore calculate the triple integral by evaluating

$$\int_0^a \int_0^{\phi(x)} \int_0^{\psi(x,y)} yz\, dz\, dy\, dx.$$

A straightforward (but somewhat lengthy) computation that you can verify gives an answer of $\frac{1}{15}ab^2c^2$.

ANOTHER SOLUTION This time we carry out the integration in a different order. In Figure 16.7.8 you can see the same solid projected this time onto the yz-plane. In terms of y and z, the curved surface has equation

$$x = \Psi(y,z) = \frac{a}{bc}\sqrt{b^2c^2 - c^2y^2 - b^2z^2}.$$

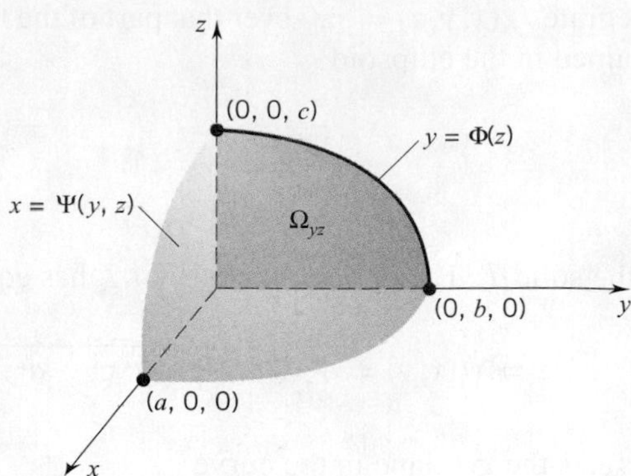

Figure 16.7.8

This surface intersects the yz-plane in the curve

$$y = \Phi(z) = \frac{b}{c}\sqrt{c^2 - z^2}.$$

We can take

$$\Omega_{yz}: \quad 0 \le z \le c, \quad 0 \le y \le \Phi(z)$$

as the base region and characterize T as the set of all (x, y, z) with

$$0 \le z \le c, \quad 0 \le y \le \Phi(z), \quad 0 \le x \le \Psi(y, z).$$

This leads to the repeated integral

$$\int_0^c \int_0^{\Phi(z)} \int_0^{\Psi(y,z)} yz \; dx \, dy \, dz,$$

which, as you can check, also gives $\frac{1}{15}ab^2c^2$. □

Example 5 Use triple integration to find the volume of the solid T bounded above by the parabolic cylinder $z = 4 - y^2$ and bounded below by the elliptic paraboloid $z = x^2 + 3y^2$.

SOLUTION Solving the two equations simultaneously, we have

$$4 - y^2 = x^2 + 3y^2 \qquad \text{and thus} \qquad x^2 + 4y^2 = 4.$$

This tells us that the two surfaces intersect in a space curve that lies on the elliptic cylinder $x^2 + 4y^2 = 4$. The projection of this intersection onto the xy-plane is the ellipse $x^2 + 4y^2 = 4$. (See Figure 16.7.9.)

The projection of T onto the xy-plane is the region

$$\Omega_{xy}: \quad -2 \le x \le 2, \quad -\tfrac{1}{2}\sqrt{4 - x^2} \le y \le \tfrac{1}{2}\sqrt{4 - x^2}.$$

The solid T is then the set of all (x, y, z) with

$$-2 \le x \le 2, \quad -\tfrac{1}{2}\sqrt{4 - x^2} \le y \le \tfrac{1}{2}\sqrt{4 - x^2}, \quad x^2 + 3y^2 \le z \le 4 - y^2.$$

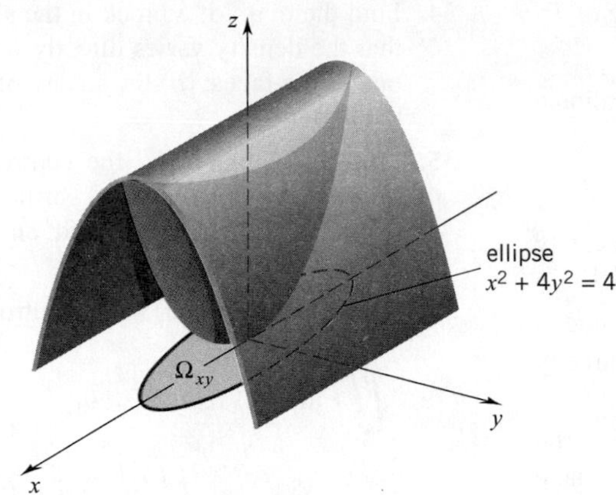

Figure 16.7.9

Its volume is given by

$$V = \int_{-2}^{2}\int_{-\frac{1}{2}\sqrt{4-x^2}}^{\frac{1}{2}\sqrt{4-x^2}}\int_{x^2+3y^2}^{4-y^2} dz\,dy\,dx$$

$$= 4\int_0^2\int_0^{\frac{1}{2}\sqrt{4-x^2}}\int_{x^2+3y^2}^{4-y^2} dz\,dy\,dx = 4\pi. \quad \square$$

explain (arrow pointing to the 4) — check this (arrow pointing to 4π)

EXERCISES 16.7

Evaluate the repeated integral.

1. $\displaystyle\int_0^a\int_0^b\int_0^c dx\,dy\,dz.$

2. $\displaystyle\int_0^1\int_0^x\int_0^y y\,dz\,dy\,dx.$

3. $\displaystyle\int_0^1\int_1^{2y}\int_0^x (x+2z)\,dz\,dx\,dy.$

4. $\displaystyle\int_0^1\int_{1-x}^{1+x}\int_0^{xy} 4z\,dz\,dy\,dx.$

5. $\displaystyle\int_0^2\int_{-1}^1\int_1^3 (z-xy)\,dz\,dy\,dx.$

6. $\displaystyle\int_0^2\int_{-1}^1\int_1^3 (z-xz)\,dy\,dx\,dz.$

7. $\displaystyle\int_0^{\pi/2}\int_0^1\int_0^{\sqrt{1-x^2}} x\cos z\,dy\,dx\,dz.$

8. $\displaystyle\int_{-1}^2\int_1^{y-2}\int_e^{e^2} \frac{x+y}{z}\,dz\,dx\,dy.$

9. $\displaystyle\int_1^2\int_y^{y^2}\int_0^{\ln x} y\,e^z\,dz\,dx\,dy.$

10. $\displaystyle\int_0^{\pi/2}\int_0^{\pi/2}\int_0^1 e^z\cos x\sin y\,dz\,dy\,dx.$

11. (*Separated variables over a box*) Set $\Pi: a_1 \le x \le a_2$, $b_1 \le y \le b_2, c_1 \le z \le c_2$. Show that, if f is continuous on $[a_1, a_2]$, g is continuous on $[b_1, b_2]$, and h is continuous on $[c_1, c_2]$, then

(16.7.2)
$$\iiint_{\Pi} f(x)\,g(y)\,h(z)\,dxdydz = \left(\int_{a_1}^{a_2} f(x)\,dx\right)\left(\int_{b_1}^{b_2} g(y)\,dy\right)\left(\int_{c_1}^{c_2} h(z)\,dz\right).$$

In Exercises 12 and 13, evaluate the triple integral, taking Π : $0 \le x \le 1$, $0 \le y \le 2$, $0 \le z \le 3$.

12. $\displaystyle\iiint_{\Pi} x^3y^2z\ dxdydz.$

13. $\displaystyle\iiint_{\Pi} x^2y^2z^2\ dxdydz.$

In Exercises 14–16, the mass density of a box $\Pi: 0 \le x \le a$, $0 \le y \le b$, $0 \le z \le c$ varies directly with the product xyz.

14. Calculate the mass of Π.

15. Locate the center of mass.

16. Determine the moment of inertia of Π about: (a) the vertical line that passes through the point (a, b, c); (b) the vertical line that passes through the center of mass.

In Exercises 17–20, a homogeneous solid T of mass M consists of all points (x, y, z) with $0 \le x \le 1$, $0 \le y \le 1$, $0 \le z \le 1 - y$.

17. Sketch T.

18. Find the volume of T.

19. Locate the center of mass.

20. Find the moments of inertia of T about the coordinate axes.

Express by repeated integrals. Do not evaluate.

21. The mass of a ball $x^2+y^2+z^2 \le r^2$ given that the density varies directly with the distance from the outer shell.

22. The mass of the solid bounded above by $z=1$ and bounded below by $z=\sqrt{x^2+y^2}$ given that the density varies directly with the distance from the origin. Identify the solid.

23. The volume of the solid bounded above by the parabolic cylinder $z=1-y^2$, below by the plane $2x+3y+z+10=0$, and on the sides by the circular cylinder $x^2+y^2-x=0$.

24. The volume of the solid bounded above by the paraboloid $z=4-x^2-y^2$ and bounded below by the parabolic cylinder $z=2+y^2$.

25. The mass of the solid bounded by the elliptic paraboloids $z=4-x^2-\frac{1}{4}y^2$ and $z=3x^2+\frac{1}{4}y^2$ given that the density varies directly with the vertical distance from the lower surface.

26. The mass of the solid bounded by the paraboloid $x=z^2+2y^2$ and the parabolic cylinder $x=4-z^2$ given that the density varies directly with the distance from the z-axis.

Evaluate the triple integral.

27. $\iiint_T (x^2z+y)\, dxdydz$, where T is the solid bounded by the planes $x=0, x=1, y=1, y=3, z=0, z=2$.

28. $\iiint_T 2y\, e^x\, dxdydz$, where T is the solid given by $0\le y\le 1,\ 0\le x\le y,\ 0\le z\le x+y$.

29. $\iiint_T x^2y^2z^2\, dxdydz$, where T is the solid bounded by the planes $z=y+1,\ y+z=1,\ x=0,\ x=1,\ z=0$.

30. $\iiint_T xy\, dxdydz$, where T is the solid in the first octant bounded by the coordinate planes and the hemisphere $z=\sqrt{4-x^2-y^2}$.

31. $\iiint_T y^2\, dxdydz$, where T is the tetrahedron in the first octant bounded by the coordinate planes and the plane $2x+3y+z=6$.

32. $\iiint_T y^2\, dxdydz$, where T is the solid in the first octant bounded by the cylinders $x^2+y=1,\ z^2+y=1$.

33. Find the volume of the portion of the first octant bounded by the planes $z=x,\ y-x=2$, and the cylinder $y=x^2$. Where is the centroid?

34. Find the mass of a block in the shape of a unit cube given that the density varies directly with: (a) the distance from one of the faces; (b) the square of the distance from one of the vertices.

35. Find the volume and the centroid of the solid bounded above by the cylindrical surface $x^2+z=4$, below by plane $x+z=2$, and on the sides by the planes $y=0$ and $y=3$.

36. Show that, if $(\bar{x},\bar{y},\bar{z})$ is the centroid of a solid T, then

$$\iiint_T (x-\bar{x})\, dxdydz = 0,$$

$$\iiint_T (y-\bar{y})\, dxdydz = 0,$$

$$\iiint_T (z-\bar{z})\, dxdydz = 0.$$

37. Taking a, b, c as positive, find the volume of the tetrahedron with vertices $(0,0,0)$, $(a,0,0)$, $(0,b,0)$, $(0,0,c)$. Where is the centroid?

38. A homogeneous solid of mass M in the form and position of the tetrahedron of Figure 16.7.4 rotates about the z-axis. Find the moment of inertia I_z.

39. A homogeneous box of mass M has edges a, b, c. Calculate the moment of inertia

(a) about the edge c;

(b) about the line that passes through the center of the box and is parallel to the edge c;

(c) about the line that passes through the center of the face bc and is parallel to the edge c.

40. Where is the centroid of the solid bounded above by the plane $z=1+x+y$, below by the plane $z=-2$, and on the sides by the planes $x=1,\ x=2,\ y=1,\ y=2$?

41. Let T be the solid bounded above by the plane $z=y$, below by the xy-plane, and on the sides by the planes $x=0,\ x=1,\ y=1$. Find the mass of T given that the density varies directly with the square of the distance from the origin. Where is the center of mass?

42. What can you conclude about T given that

$$\iiint_T f(x,y,z)\, dxdydz = 0$$

(a) for every continuous function f that is odd in x?

(b) for every continuous function f that is odd in y?

(c) for every continuous function f that is odd in z?

(d) for every continuous function f that satisfies the relation $f(-x,-y,-z)=-f(x,y,z)$?

43. (a) Integrate $f(x,y,z)=x+y^3+z$ over the unit ball centered at the origin.

(b) Integrate $f(x,y,z)=a_1x+a_2y+a_3z+a_4$ over the unit ball centered at the origin.

44. Integrate $f(x, y, z) = x^2y^2$ over the solid bounded above by the cylinder $y^2 + z = 4$, below by the plane $y + z = 2$, and on the sides by the planes $x = 0$ and $x = 2$.

45. Use triple integrals to find the volume enclosed by the sphere $x^2 + y^2 + z^2 = a^2$.

46. Use triple integrals to find the volume enclosed by the ellipsoid

$$\frac{x^2}{a^2} + \frac{y^2}{b^2} + \frac{z^2}{c^2} = 1.$$

47. Find the mass of the solid in Example 5 given that the density varies directly with $|x|$.

48. Find the volume of the solid bounded by the paraboloids $z = 2 - x^2 - y^2$ and $z = x^2 + y^2$.

49. Find the mass and the center of mass of the solid of Exercise 35 given that the density varies directly with $1 + y$.

50. Let T be a solid with volume

$$V = \iiint_T dxdydz = \int_0^2 \int_0^{9-x^2} \int_0^{2-x} dz\,dy\,dx.$$

Sketch T and fill in the blanks.

(a) $V = \int_\square^\square \int_\square^\square \int_\square^\square dy\,dx\,dz.$

(b) $V = \int_\square^\square \int_\square^\square \int_\square^\square dy\,dz\,dx.$

(c) $V = \int_0^5 \int_\square^\square \int_\square^\square dz\,dx\,dy + \int_5^9 \int_\square^\square \int_\square^\square dz\,dx\,dy.$

51. Let T be a solid with volume

$$V = \iiint_T dxdydz = \int_0^3 \int_0^{6-x} \int_0^{2x} dz\,dy\,dx.$$

Sketch T and fill in the blanks.

(a) $V = \int_\square^\square \int_\square^\square \int_\square^\square dy\,dx\,dz.$

(b) $V = \int_\square^\square \int_\square^\square \int_\square^\square dy\,dz\,dx.$

(c) $V = \int_0^6 \int_\square^\square \int_\square^\square dx\,dy\,dz + \int_\square^\square \int_\square^\square \int_\square^\square dx\,dy\,dz.$

For the remaining exercises, let V be the volume of the solid T enclosed by the parabolic cylinder $y = 4 - z^2$ and the V-shaped cylinder $y = |x|$. Let $\Omega_{xy}, \Omega_{yz}, \Omega_{xz}$ be the projections of T onto the xy-, yz-, and xz-planes, respectively. Fill in the blanks.

52. (a) $V = \iint_{\Omega_{xy}} \square\, dxdy.$

(b) $V = \iint_{\Omega_{xy}} \left(\int_\square^\square dz \right) dxdy.$

(c) $V = \int_\square^\square \int_\square^\square \int_\square^\square dz\,dy\,dx.$

(d) $V = \int_\square^\square \int_\square^\square \int_\square^\square dz\,dx\,dy.$

53. (a) $V = \iint_{\Omega_{yz}} \square\, dydz.$

(b) $V = \iint_{\Omega_{yz}} \left(\int_\square^\square dx \right) dydz.$

(c) $V = \int_\square^\square \int_\square^\square \int_\square^\square dx\,dz\,dy.$

(d) $V = \int_\square^\square \int_\square^\square \int_\square^\square dx\,dy\,dz.$

54. (a) $V = \iint_{\Omega_{xz}} \square\, dxdz.$

(b) $V = \iint_{\Omega_{xz}} \left(\int_\square^\square dy \right) dxdz.$

(c) $V = \int_\square^\square \int_\square^\square \int_\square^\square dy\,dx\,dz.$

(d) $V = \int_{-2}^0 \int_\square^\square \int_\square^\square dy\,dz\,dx + \int_0^2 \int_\square^\square \int_\square^\square dy\,dz\,dx.$

55. Use a CAS to evaluate the triple integrals.

(a) $\int_2^4 \int_3^5 \int_1^2 \frac{\ln xy}{z}\,dz\,dy\,dx.$

(b) $\int_0^4 \int_1^2 \int_0^3 x\sqrt{yz}\,dz\,dy\,dx.$

56. Use a CAS to find the volume of the solid bounded by the plane $3x + 6y + 2z = 6$, the elliptic paraboloid $y = 36 - 9x^2 - 4z^2$, the yz-plane, and the xy-plane.

■ 16.8 CYLINDRICAL COORDINATES

Introduction to Cylindrical Coordinates

The cylindrical coordinates (r, θ, z) of a point P in xyz-space are shown geometrically in Figure 16.8.1. The first two coordinates, r and θ, are the usual plane polar coordinates, except that r is taken to be nonnegative and θ is restricted to the interval $[0, 2\pi]$. † The third coordinate is the third rectangular coordinate z.

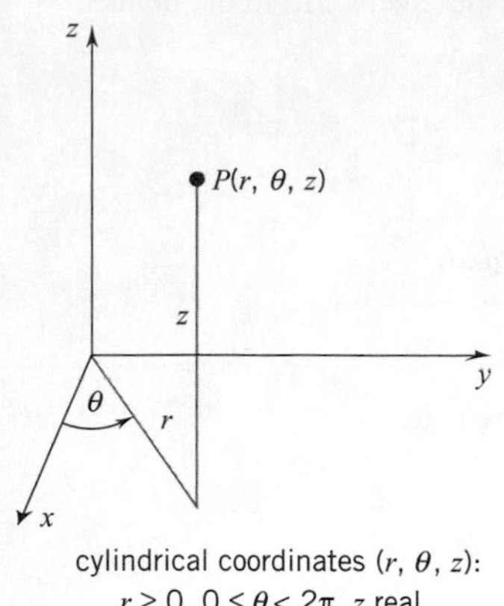

cylindrical coordinates (r, θ, z): $r \geq 0$, $0 \leq \theta < 2\pi$, z real

Figure 16.8.1

In rectangular coordinates, the coordinate surfaces

$$x = x_0, \qquad y = y_0, \qquad z = z_0$$

are three mutually perpendicular planes. In cylindrical coordinates, the coordinate surfaces take the form

$$r = r_0, \qquad \theta = \theta_0, \qquad z = z_0. \qquad \text{(Figure 16.8.2)}$$

The surface $r = r_0$ is a right circular cylinder of radius r_0. The central axis of the cylinder is the z-axis. The surface $\theta = \theta_0$ is a vertical half-plane hinged at the z-axis. The plane stands at an angle of θ_0 radians from the positive x-axis. The last coordinate surface is the plane $z = z_0$.

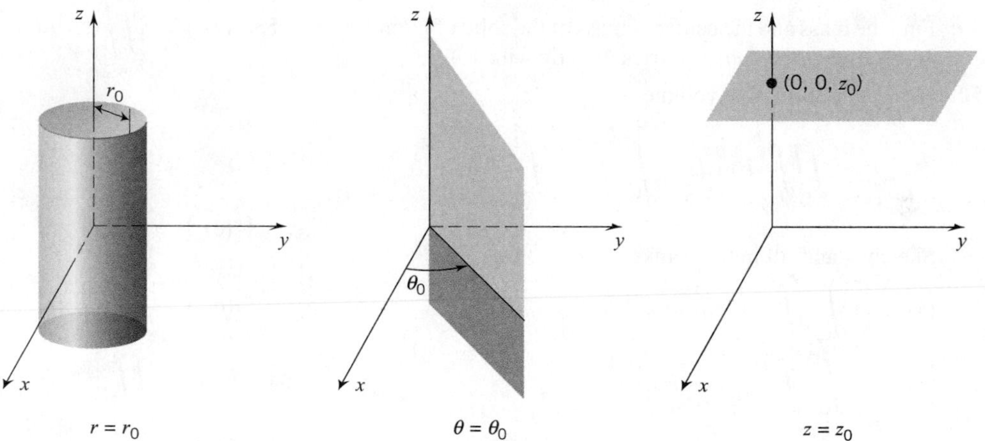

Figure 16.8.2

The point P with rectangular coordinates (x_0, y_0, z_0) lies on the plane $x = x_0$, on the plane $y = y_0$, and on the plane $z = z_0$. P is at the intersection of these three planes.

The point P with cylindrical coordinates (r_0, θ_0, z_0) lies on the cylinder $r = r_0$, on the vertical half-plane $\theta = \theta_0$, and on the horizontal plane $z = z_0$. P is at the intersection of these three surfaces.

Rectangular coordinates (x, y, z) can be obtained from cylindrical coordinates (r, θ, z) by means of the equations

$$x = r\cos\theta, \qquad y = r\sin\theta, \qquad z = z.$$

Conversely, with the obvious exclusions, cylindrical coordinates (r, θ, z) can be obtained from rectangular coordinates (x, y, z) by means of the equations

$$r = \sqrt{x^2 + y^2}, \qquad \tan\theta = \frac{y}{x}, \qquad z = z.$$

The solids in xyz-space easiest to describe in cylindrical coordinates are the *cylindrical wedges*. Such a wedge is pictured in Figure 16.8.3. The wedge consists of all points (x, y, z) that have cylindrical coordinates (r, θ, z) in the box

$$\Pi: \quad a_1 \leq r \leq a_2, \quad b_1 \leq \theta \leq b_2, \quad c_1 \leq z \leq c_2.$$

† By allowing θ to take on both 0 and 2π, we lose uniqueness but we gain flexibility and convenience.

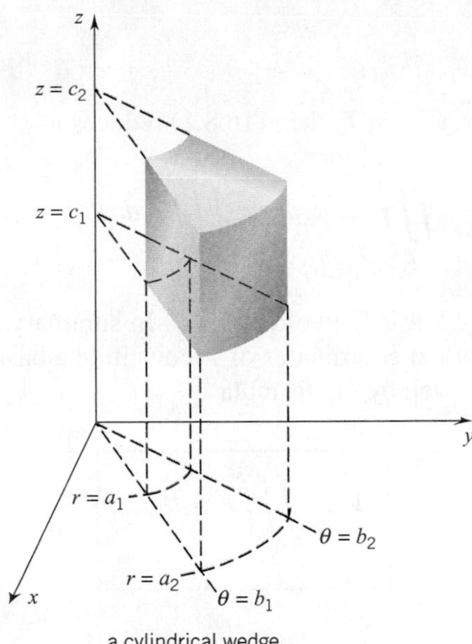

a cylindrical wedge

Figure 16.8.3

Evaluating Triple Integrals Using Cylindrical Coordinates

Suppose that T is some basic solid in xyz-space, not necessarily a wedge. If T is the set of all (x, y, z) with cylindrical coordinates in some basic solid S in $r\theta z$-space, then

(16.8.1)
$$\iiint_T f(x,y,z)\; dxdydz = \iiint_S f(r\cos\theta, r\sin\theta, z)\, r\, drd\theta dz.$$

DERIVATION OF (16.8.1) We will carry out the argument on the assumption that T is projectable onto some basic region Ω_{xy} of the xy-plane. (It is for such solids that the formula is most useful.) T has some lower boundary $z = \psi_1(x, y)$ and some upper boundary $z = \psi_2(x, y)$. T is then the set of all (x, y, z) with

$$(x,y) \in \Omega_{xy} \qquad \text{and} \qquad \psi_1(x,y) \le z \le \psi_2(x,y).$$

The region Ω_{xy} has polar coordinates in some set $\Omega_{r\theta}$ (which we assume is a basic region). Then S is the set of all (r, θ, z) with

$$[r,\theta] \in \Omega_{r\theta} \qquad \text{and} \qquad \psi_1(r\cos\theta,\ r\sin\theta) \le z \le \psi_2(r\cos\theta,\ r\sin\theta).$$

Therefore

$$\iiint_T f(x,y,z)\; dxdydz = \iint_{\Omega_{xy}} \left(\int_{\psi_1(x,y)}^{\psi_2(x,y)} f(x,y,z)\, dz\right) dxdy$$

$$\underset{(16.4.3)}{=} \iint_{\Omega_{r\theta}} \left(\int_{\psi_1(r\cos\theta,\ r\sin\theta)}^{\psi_2(r\cos\theta,\ r\sin\theta)} f(r\cos\theta,\ r\sin\theta, z)\, dz\right) r\, drd\theta$$

$$= \iiint_S f(r\cos\theta, r\sin\theta, z)\, r\, drd\theta dz. \quad \square$$

Volume Formula

If $f(x, y, z) = 1$ for all (x, y, z) in T, then (16.8.1) reduces to

$$\iiint_T dxdydz = \iiint_S r\, drd\theta dz.$$

The triple integral on the left is the volume of T. In summary, if T is a basic solid in xyz-space and the cylindrical coordinates of T constitute a basic solid S in $r\theta z$-space, then the volume of T is given by the formula

(16.8.2)
$$V = \iiint_S r\, drd\theta dz.$$

Calculations

Cylindrical coordinates are particularly useful in cases where there is an axis of symmetry. The axis of symmetry is then taken as the z-axis.

Example 1 Use cylindrical coordinates to calculate

$$\iiint_T (x^2 + y^2)\, dxdydz$$

for $\quad T\colon -2 \le x \le 2, \quad -\sqrt{4 - x^2} \le y \le \sqrt{4 - x^2}, \quad 0 \le z \le 4 - x^2 - y^2.$

SOLUTION The solid is bounded above by the paraboloid of revolution $z = 4 - x^2 - y^2$ and below by the xy-plane (see Figure 16.8.4). Since the solid is symmetric about the z-axis, the integral has a simpler representation in cylindrical coordinates:

$$S\colon 0 \le r \le 2, \quad 0 \le \theta \le 2\pi, \quad 0 \le z \le 4 - r^2.$$

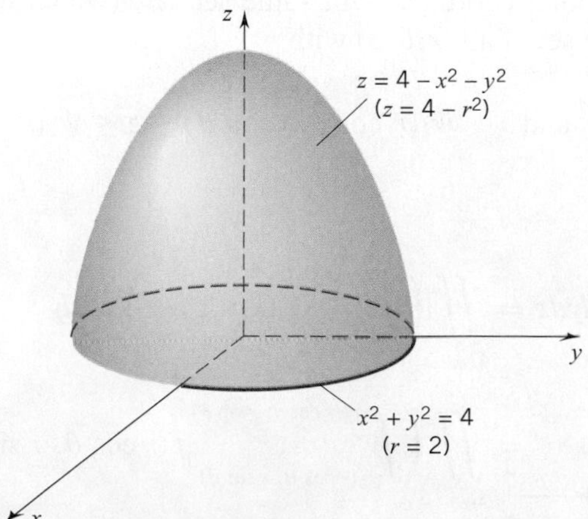

Figure 16.8.4

Now $\displaystyle\iiint_T (x^2+y^2)\,dxdydz = \iiint_S r^2\, r\,drd\theta dz = \int_0^{2\pi}\int_0^2\int_0^{4-r^2} r^3\,dz\,dr\,d\theta$

$$= \int_0^{2\pi}\int_0^2 \Big[r^3 z\Big]_0^{4-r^2} dr\,d\theta = \int_0^{2\pi}\int_0^2 (4r^3 - r^5)\,dr\,d\theta$$

$$= \int_0^{2\pi}\Big[r^4 - \tfrac{1}{6}r^6\Big]_0^2 d\theta$$

$$= \tfrac{16}{3}\int_0^{2\pi} d\theta = \frac{32\pi}{3}. \quad \square$$

Example 2 Find the mass of a solid right circular cylinder T of radius R and height h given that the density varies directly with the distance from the axis of the cylinder.

SOLUTION Place the cylinder T on the xy-plane so that the axis of T coincides with the z-axis. The density function then takes the form $\lambda(x,y,z) = k\sqrt{x^2+y^2}$, and T consists of all points (x,y,z) with cylindrical coordinates (r,θ,z) in the set

$$S:\quad 0 \le r \le R,\quad 0 \le \theta \le 2\pi,\quad 0 \le z \le h.$$

Therefore $\displaystyle M = \iiint_T k\sqrt{x^2+y^2}\,dxdydz = \iiint_S (kr)\,r\,drd\theta dz$

$$= k\int_0^R\int_0^{2\pi}\int_0^h r^2\,dz\,d\theta\,dr = \tfrac{2}{3}k\pi R^3 h. \quad \square$$

↑ check this

Example 3 Use cylindrical coordinates to find the volume of the solid T bounded above by the plane $z = y$ and below by the paraboloid $z = x^2 + y^2$.

SOLUTION In cylindrical coordinates the plane has equation $z = r\sin\theta$ and the paraboloid has equation $z = r^2$. Solving these two equations simultaneously, we have $r = \sin\theta$. This tells us that the two surfaces intersect in a space curve that lies along the circular cylinder $r = \sin\theta$. The projection of this intersection onto the xy-plane is the circle with polar equation $r = \sin\theta$. (See Figure 16.8.5.) The base region Ω_{xy} is thus the set of all (x,y) with polar coordinates in the set

$$0 \le \theta \le \pi,\quad 0 \le r \le \sin\theta.$$

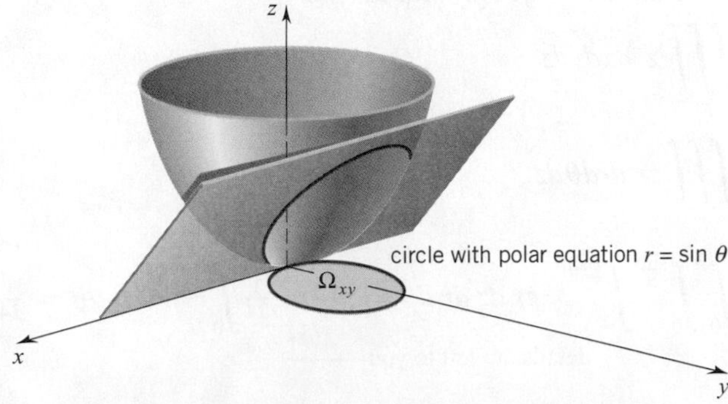

Figure 16.8.5

T itself is the set of all (x, y, z) with cylindrical coordinates in the set

$$S:\quad 0 \le \theta \le \pi, \quad 0 \le r \le \sin\theta, \quad r^2 \le z \le r\sin\theta. \qquad \text{(check this)}$$

Therefore,

$$\begin{aligned} V &= \iiint_T dxdydz = \iiint_S r\,drd\theta dz \\ &= \int_0^{\pi}\int_0^{\sin\theta}\int_{r^2}^{r\sin\theta} r\,dz\,dr\,d\theta \\ &= \int_0^{\pi}\int_0^{\sin\theta} (r^2\sin\theta - r^3)\,dr\,d\theta \\ &= \int_0^{\pi}\left[\tfrac{1}{3}r^3\sin\theta - \tfrac{1}{4}r^4\right]_0^{\sin\theta} d\theta \\ &= \tfrac{1}{12}\int_0^{\pi}\sin^4\theta\,d\theta = \tfrac{1}{12}\left(\tfrac{3}{8}\pi\right) = \tfrac{1}{32}\pi. \quad \square \end{aligned}$$

(verify this; Exercise 13, Section 8.3)

Example 4 Locate the centroid of the solid in Example 3.

SOLUTION Since T is symmetric about the yz-plane, we see that $\bar{x} = 0$. To get $\bar{y}$ we begin as usual:

$$\begin{aligned} \bar{y}V &= \iiint_T y\,dxdydz = \iiint_S (r\sin\theta)\,r\,drd\theta dz \\ &= \int_0^{\pi}\int_0^{\sin\theta}\int_{r^2}^{r\sin\theta} r^2\sin\theta dz\,dr\,d\theta \\ &= \int_0^{\pi}\int_0^{\sin\theta} (r^3\sin^2\theta - r^4\sin\theta)\,dr\,d\theta \\ &= \int_0^{\pi}\left[\tfrac{1}{4}r^4\sin^2\theta - \tfrac{1}{5}r^5\sin\theta\right]_0^{\sin\theta} d\theta \\ &= \tfrac{1}{20}\int_0^{\pi}\sin^6\theta\,d\theta = \tfrac{1}{20}\left(\tfrac{5}{16}\pi\right) = \tfrac{1}{64}\pi. \end{aligned}$$

(verify this; Exercise 13, Section 8.3)

Since $V = \frac{1}{32}\pi$, we have $\bar{y} = \frac{1}{2}$. Now for $\bar{z}$:

$$\begin{aligned} \bar{z}V &= \iiint_T z\;dxdydz \\ &= \iiint_S zr\,drd\theta dz \\ &= \int_0^{\pi}\int_0^{\sin\theta}\int_{r^2}^{r\sin\theta} zr\,dz\,dr\,d\theta = \cdots = \tfrac{1}{24}\int_0^{\pi}\sin^6\theta\,d\theta = \tfrac{1}{24}\left(\tfrac{5}{16}\pi\right) = \tfrac{5}{384}\pi. \end{aligned}$$

details are left to you

Division by $V = \frac{1}{32}\pi$ gives $\bar{z} = \frac{5}{12}$. The centroid is thus the point $(0, \frac{1}{2}, \frac{5}{12})$. $\square$

EXERCISE 16.8

Write the given equation in cylindrical coordinates and sketch the graph of the surface.

1. $x^2 + y^2 + z^2 = 9$.

2. $x^2 + y^2 = 4$.

3. $z = 2\sqrt{x^2 + y^2}$.

4. $x = 4z$.

5. $4x^2 + 4y^2 - z^2 = 0$.

6. $y^2 + z^2 = 8$.

The volume of a solid T is given in cylindrical coordinates. Sketch T and evaluate the repeated integral.

7. $\displaystyle\int_0^{\pi/2}\int_0^2\int_0^{4-r^2} r\,dz\,dr\,d\theta$.

8. $\displaystyle\int_0^{\pi/4}\int_0^1\int_0^{\sqrt{1-r^2}} r\,dz\,dr\,d\theta$.

9. $\displaystyle\int_0^{2\pi}\int_0^2\int_0^{r^2} r\,dz\,dr\,d\theta$.

10. $\displaystyle\int_0^3\int_0^{2\pi}\int_r^3 r\,dz\,d\theta\,dr$.

Evaluate using cylindrical coordinates.

11. $\displaystyle\iiint_T dxdydz;\quad T: 0 \le x \le 1,\ 0 \le y \le \sqrt{1-x^2},\ 0 \le z \le \sqrt{4-(x^2+y^2)}$.

12. $\displaystyle\iiint_T z^3\,dxdydz;\quad T: -1 \le x \le 1,\ 0 \le y \le \sqrt{1-x^2},\ \sqrt{x^2+y^2} \le z \le 1$.

13. $\displaystyle\iiint_T \frac{1}{\sqrt{x^2+y^2}}\,dxdydz;\quad T: 0 \le x \le \sqrt{9-y^2}, 0 \le y \le 3,\ 0 \le z \le \sqrt{9-(x^2+y^2)}$.

14. $\displaystyle\iiint_T z\,dxdydz;\quad T: 0 \le x \le 1,\ 0 \le y \le \sqrt{1-x^2},\ 0 \le z \le \sqrt{1-x^2-y^2}$.

15. $\displaystyle\iiint_T \sin\,(x^2+y^2)\,dxdydz;\quad T: 0 \le x \le 1,\ 0 \le y \le \sqrt{1-x^2},\ 0 \le z \le 2$.

16. $\displaystyle\iiint_T \sqrt{x^2+y^2}\,dxdydz;\quad T: -1 \le x \le 1,\ -\sqrt{1-x^2} \le y \le \sqrt{1-x^2},\ x^2+y^2 \le z \le 2-(x^2+y^2)$.

17. Find the volume of the solid bounded above by the cone $z^2 = x^2 + y^2$, below by the xy-plane, and on the sides by the cylinder $x^2 + y^2 = 2ax$.

18. Find the volume of the solid bounded by the paraboloid of revolution $x^2 + y^2 = az$, the xy-plane, and the cylinder $x^2 + y^2 = 2ax$.

19. Find the volume of the solid bounded above by $z = a - \sqrt{x^2 + y^2}$, below by the xy-plane, and on the sides by the cylinder $x^2 + y^2 = ax$.

20. Find the volume of the solid bounded above by the plane $2z = 4 + x$, below by the xy-plane, and on the sides by the cylinder $x^2 + y^2 = 2x$.

21. Find the volume of the solid bounded by the paraboloid $z = x^2 + y^2$ and the plane $z = x$.

22. Find the volume of the solid that is bounded above by $x^2 + y^2 + z^2 = 25$ and below by $z = \sqrt{x^2 + y^2} + 1$.

23. Find the volume of the "ice cream cone" bounded below by the half-cone $z = \sqrt{3(x^2 + y^2)}$ and above by the unit sphere $x^2 + y^2 + z^2 = 1$.

24. Find the volume of the solid bounded by the hyperboloid $z^2 = a^2 + x^2 + y^2$ and the upper nappe of the cone $z^2 = 2(x^2 + y^2)$.

25. Find the volume of the solid that is bounded below by the xy-plane and lies inside the sphere $x^2 + y^2 + z^2 = 9$ but outside the cylinder $x^2 + y^2 = 1$.

26. Find the volume of the solid that lies between the cylinders $x^2 + y^2 = 1$ and $x^2 + y^2 = 4$, and is bounded above by the ellipsoid $x^2 + y^2 + 4z^2 = 36$ and below by the xy-plane.

In Exercises 27–29, let T be a solid right circular cylinder of base radius R and height h. Assume that the mass density varies directly with the distance from one of the bases.

27. Use cylindrical coordinates to find the mass M of T.

28. Locate the center of mass of T.

29. Find the moment of inertia of T about the axis of the cylinder.

30. Let T be a homogeneous right circular cylinder of mass M, base radius R, and height h. Find the moment of inertia of the cylinder about: (a) the central axis; (b) a line that lies in the plane of one of the bases and passes through the center of that base; (c) a line that passes through the center of the cylinder and is parallel to the bases.

In Exercises 31–34, let T be a homogeneous solid right circular cone of mass M, base radius R, and height h.

31. Use cylindrical coordinates to verify that the volume of the cone is given by the formula $V = \frac{1}{3}\pi R^2 h$.

32. Locate the center of mass.

33. Find the moment of inertia about the axis of the cone.

34. Find the moment of inertia about a line that passes through the vertex and is parallel to the base.

In Exercises 35–37, let T be the solid bounded above by the paraboloid $z = 1 - (x^2 + y^2)$ and bounded below by the xy-plane.

35. Use cylindrical coordinates to find the volume of T.

36. Find the mass of T if the density varies directly with the distance to the xy-plane.

37. Find the mass of T if the density varies directly with the square of the distance from the origin.

■ 16.9 THE TRIPLE INTEGRAL AS THE LIMIT OF RIEMANN SUMS; SPHERICAL COORDINATES

The Triple Integral as the Limit of Riemann Sums

You have seen how single integrals and double integrals can be obtained as limits of Riemann sums. The same holds true for triple integrals.

Start with a basic solid T in xyz-space and decompose it into a finite number of basic solids $T_1 \ldots, T_N$. If f is continuous on T, then f is continuous on each T_i. From each T_i pick an arbitrary point (x_i^*, y_i^*, z_i^*) and form the *Riemann sum*

$$\sum_{i=1}^{N} f(x_i^*, y_i^*, z_i^*)(\text{ volume of } T_i).$$

As you would expect, the triple integral over T is the limit of such sums; namely, given any $\epsilon > 0$, there exists a $\delta > 0$ such that, if the diameters of the T_i are all less than δ, then

$$\left| \sum_{i=1}^{N} f(x_i^*, y_i^*, z_i^*)(\text{ volume of } T_i) - \iiint_T f(x,y,z)\, dxdydz \right| < \epsilon$$

no matter how the (x_i^*, y_i^*, z_i^*) are chosen within the T_i. We express this by writing

(16.9.1) $$\iiint_T f(x,y,z)\, dxdydz = \lim_{\operatorname{diam} T_i \to 0} \sum_{i=1}^{N} f(x_i^*, y_i^*, z_i^*)(\text{ volume of } T_i).$$

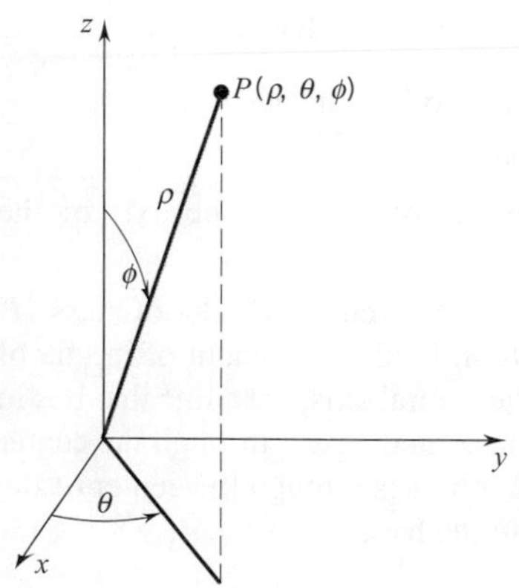

spherical coordinates (ρ, θ, ϕ): $\rho \geq 0,\ 0 \leq \theta < 2\pi,\ 0 \leq \phi \leq \pi$

Figure 16.9.1

Introduction to Spherical Coordinates

The spherical coordinates (ρ, θ, ϕ) of a point P in xyz-space are shown geometrically in Figure 16.9.1. The first coordinate, ρ, is the distance from P to the origin; thus $\rho \geq 0$. The second coordinate, the angle marked θ, is the second coordinate of cylindrical coordinates; θ ranges from 0 to 2π. We call θ the *longitude*. The third coordinate, the angle marked ϕ, ranges only from 0 to π. We call ϕ the *colatitude*, or more simply the *polar angle*. (The complement of ϕ would be the *latitude* on a globe.)

The coordinate surfaces

$$\rho = \rho_0, \quad \theta = \theta_0, \quad \phi = \phi_0$$

are shown in Figure 16.9.2. The surface $\rho = \rho_0$ is a sphere; the radius is ρ_0 and the center is the origin. The second surface, $\theta = \theta_0$, is the same as in cylindrical coordinates: the vertical half-plane hinged at the z-axis and standing at an angle of θ_0 radians from the positive x-axis. The surface $\phi = \phi_0$ requires detailed explanation. If $0 \leq \phi_0 \leq \frac{1}{2}\pi$ or $\frac{1}{2}\pi \leq \phi_0 < \pi$, the surface is a nappe of a cone; it is generated by revolving about the z-axis any ray that emerges from the origin at an angle of ϕ_0 radians from the positive z-axis. The surface $\phi = \frac{1}{2}\pi$ is the xy-plane. (The nappe of the cone has opened up completely.) The equation $\phi = 0$ gives the nonnegative z-axis, and the equation $\phi = \pi$ gives the nonpositive z-axis. (When $\phi = 0$ or $\phi = \pi$, the nappe of the cone has closed up completely.)

The point P with spherical coordinates $(\rho_0, \theta_0, \phi_0)$ is located at the intersection of the three surfaces $\rho = \rho_0$, $\theta = \theta_0$, $\phi = \phi_0$.

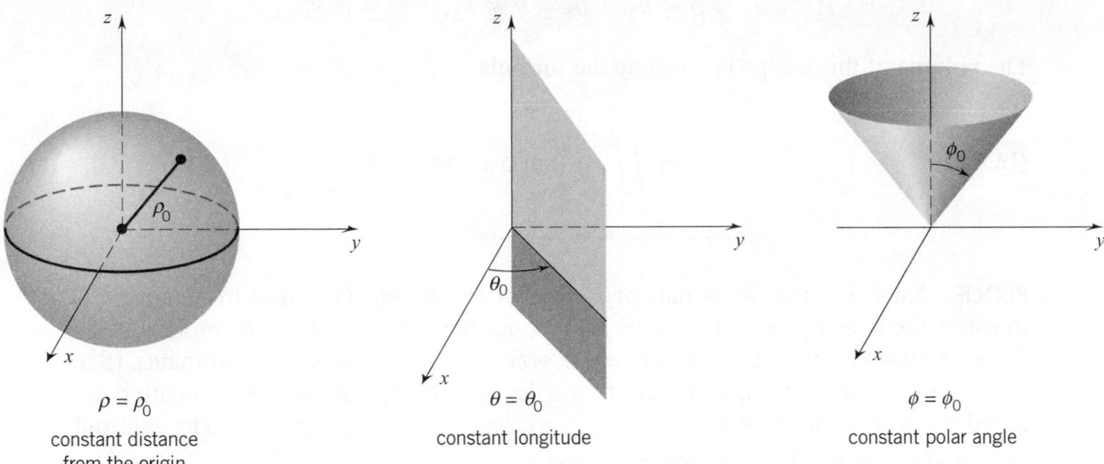

Figure 16.9.2

Rectangular coordinates (x, y, z) are related to spherical coordinates (ρ, θ, ϕ) by the following equations:

$$x = \rho \sin\phi \cos\theta, \quad y = \rho \sin\phi \sin\theta, \quad z = \rho \cos\phi.$$

You can verify these relations by referring to Figure 16.9.3. (Note that the factor $\rho \sin\phi$ appearing in the first two equations is the r of cylindrical coordinates: $r = \rho \sin\phi$.) Conversely, with obvious exclusions, we have

$$\rho = \sqrt{x^2 + y^2 + z^2}, \quad \tan\theta = \frac{y}{x}, \quad \cos\phi = \frac{z}{\sqrt{x^2 + y^2 + z^2}}.$$

The Volume of a Spherical Wedge

Figure 16.9.4 shows a *spherical wedge* W in *xyz*-space. The wedge W consists of all points (x, y, z) that have spherical coordinates in the box

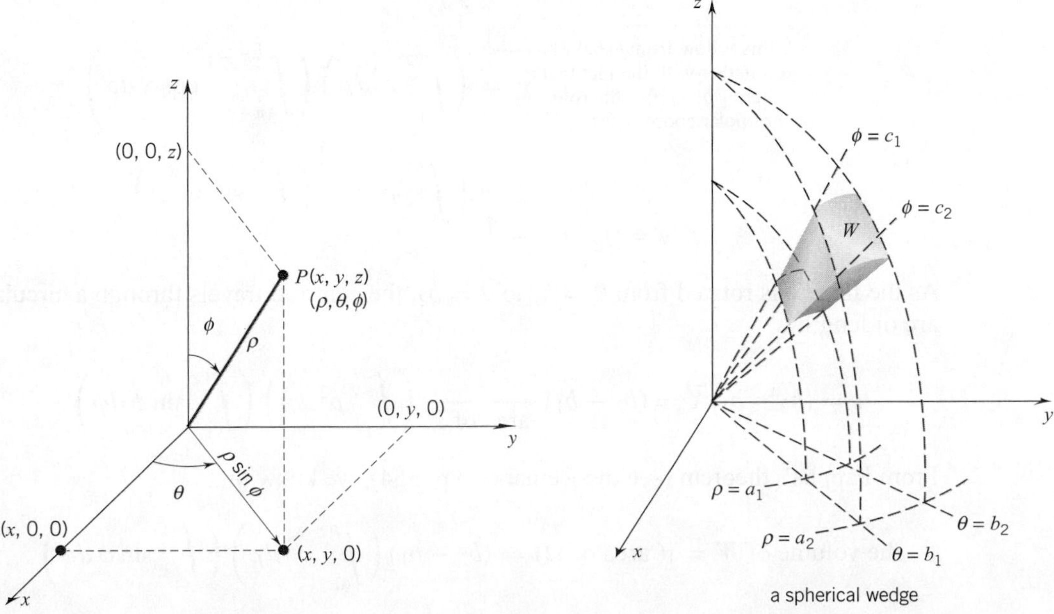

Figure 16.9.3

Figure 16.9.4

$$\Pi: \quad a_1 \le \rho \le a_2, \quad b_1 \le \theta \le b_2, \quad c_1 \le \phi \le c_2.$$

The volume of this wedge is given by the formula

(16.9.2)
$$V = \iiint_{\Pi} \rho^2 \sin\phi \, d\rho d\theta d\phi.$$

PROOF Note first that W is part of a solid of revolution. One way to obtain W is to rotate the $\theta = b_1$ face of W, call it Ω, about the z-axis for $b_2 - b_1$ radians. (See Figure 16.9.4.) On that face ρ and $\alpha = \frac{1}{2}\pi - \phi$ play the role of polar coordinates. (See Figure 16.9.5.) In the setup of Figure 16.9.5 the face Ω is the set of all (X, z) with polar coordinates $[\rho, \alpha]$ in the set $\Gamma: a_1 \le \rho \le a_2, \frac{1}{2}\pi - c_2 \le \alpha \le \frac{1}{2}\pi - c_1$. The centroid of Ω is at a distance $\overline{X}$ from the z-axis where

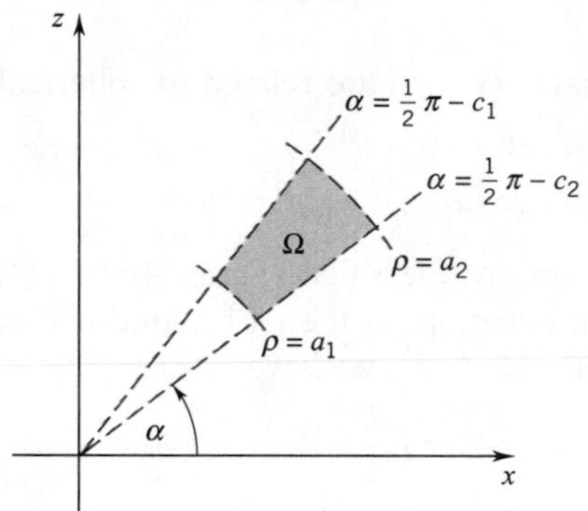

Figure 16.9.5

$$\overline{X}(\text{ area of } \Omega) = \iint_{\Omega} X \, dX \, dz = \iint_{\Gamma} \rho^2 \cos\alpha \, d\rho d\alpha$$

this follow from (16.4.3) together with the fact that here $[\rho, \alpha]$ play the role of polar coordinates

$$= \left(\int_{a_1}^{a_2} \rho^2 d\rho\right)\left(\int_{\frac{1}{2}\pi - c_2}^{\frac{1}{2}\pi - c_1} \cos\alpha \, d\alpha\right)$$

$$= \left(\int_{a_1}^{a_2} \rho^2 d\rho\right)\left(\int_{c_1}^{c_2} \sin\phi \, d\phi\right).$$

$\phi = \frac{1}{2}\pi - \alpha$

As the face Ω is rotated from $\theta = b_1$ to $\theta = b_2$, the centroid travels through a circular arc of length

$$s = (b_2 - b_1)\overline{X} = (b_2 - b_1)\frac{1}{\text{area of } \Omega}\left(\int_{a_1}^{a_2} \rho^2 \, d\rho\right)\left(\int_{c_1}^{c_2} \sin\phi \, d\phi\right).$$

From Pappus's theorem (see the Remark on p. 354), we know that

$$\text{the volume of } W = s(\text{ area of } \Omega) = (b_2 - b_1)\left(\int_{a_1}^{a_2} \rho^2 \, d\rho\right)\left(\int_{c_1}^{c_2} \sin\phi \, d\phi\right)$$

$$= \left(\int_{b_1}^{b_2} d\theta\right)\left(\int_{a_1}^{a_2} \rho^2 \, d\rho\right)\left(\int_{c_1}^{c_2} \sin\phi \, d\phi\right)$$

$$= \int_{a_1}^{a_2}\int_{b_1}^{b_2}\int_{c_1}^{c_2} \rho^2 \sin\phi \, d\phi \, d\theta \, d\rho$$

$$= \iiint_{\Pi} \rho^2 \sin\phi \, d\rho d\theta d\phi. \quad \square$$

Evaluating Triple Integrals Using Spherical Coordinates

Suppose that T is a basic solid in xyz-space with spherical coordinates in some basic solid S of $\rho\theta\phi$-space. Then

(16.9.3)
$$\iiint_T f(x,y,z)\, dxdydz = \iiint_S f(\rho\sin\phi\cos\theta,\ \rho\sin\phi\sin\theta,\ \rho\cos\phi)\,\rho^2\sin\phi\, d\rho d\theta d\phi.$$

DERIVATION OF (16.9.3) Assume first that T is a spherical wedge W. The solid S is then a box Π. Now decompose Π into N boxes $\Pi_1, \ldots, \Pi_N$. This induces a subdivision of W into N spherical wedges $W_1, \ldots, W_N$.

Writing $F(\rho, \theta, \phi)$ for $f(\rho\sin\phi\cos\theta,\ \rho\sin\phi\sin\theta,\ \rho\cos\phi)$ to save space, we have

$$\iiint_{\Pi} F(\rho,\theta,\phi)\,\rho^2\sin\phi\, d\rho d\theta d\phi \underset{\text{additivity}}{=} \sum_{i=1}^{N}\iiint_{\Pi_i} F(\rho,\theta,\phi)\,\rho^2\sin\phi\, d\rho d\theta d\phi$$

$$\underset{\substack{\text{for some } (\rho_i^*,\theta_i^*,\phi_i^*)\in\Pi_i \\ \text{(by 16.6.5)}}}{=} \sum_{i=1}^{N} F(\rho_i^*,\theta_i^*,\phi_i^*)\iiint_{\Pi_i}\rho^2\sin\phi\, d\rho d\theta d\phi$$

$$\underset{\text{(16.9.2) applied to } \Pi_i}{=} \sum_{i=1}^{N} F(\rho_i^*,\theta_i^*,\phi_i^*)(\text{volume of } W_i)$$

$$\underset{\substack{x_i^* = \rho_i^*\sin\phi_i^*\cos\theta_i^*,\\ y_i^* = \rho_i^*\sin\phi_i^*\sin\theta_i^*,\\ z_i^* = \rho_i^*\cos\phi_i^*}}{=} \sum_{i=1}^{N} f(x_i^*,y_i^*,z_i^*)(\text{volume of } W_i).$$

This last expression is a Riemann sum for

$$\iiint_W f(x,y,z)\, dxdydz$$

and, as such, by (16.9.1), will differ from that integral by less than any preassigned positive number ϵ provided only that the diameters of all the W_i are sufficiently small. This we can guarantee by making the diameters of all the Π_i sufficiently small.

This verifies the formula for the case where T is a spherical wedge. The more general case is left to you. HINT: Encase T in a wedge W and define f to be zero outside of T. $\square$

Volume Formula

If $f(x, y, z) = 1$ for all (x, y, z) in T, then the change of variables formula reduces to

$$\iiint_T dxdydz = \iiint_S \rho^2 \sin\phi \, d\rho d\theta d\phi.$$

The integral on the left is the volume of T. It follows that the volume of T is given by the formula

(16.9.4) $$V = \iiint_S \rho^2 \sin\phi \, d\rho d\theta d\phi.$$

Calculations

Spherical coordinates are commonly used in applications where there is a center of symmetry. The center of symmetry is then taken as the origin.

Example 1 Calculate the mass M of a solid ball of radius 1 given that the density varies directly with the square of the distance from the center of the ball.

SOLUTION Center the ball at the origin. The ball, call it T, is now the set of all (x, y, z) with spherical coordinates (ρ, θ, ϕ) in the box

$$S: \quad 0 \le \rho \le 1, \quad 0 \le \theta \le 2\pi, \quad 0 \le \phi \le \pi. \qquad \text{(Figure 16.9.6)}$$

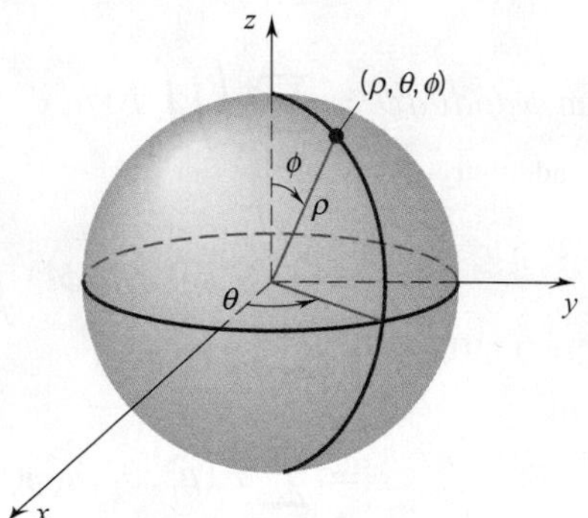

Figure 16.9.6

$$\text{Therefore } M = \iiint_T k(x^2 + y^2 + z^2) \, dxdydz$$

$$= \iiint_S (k\rho^2)\rho^2 \sin\phi \, d\rho d\theta d\phi = k \int_0^{\pi}\int_0^{2\pi}\int_0^1 \rho^4 \sin\phi \, d\rho \, d\theta \, d\phi$$

$$= k\left(\int_0^{\pi} \sin\phi \, d\phi\right)\left(\int_0^{2\pi} d\theta\right)\left(\int_0^1 \rho^4 \, d\rho\right) = k(2)(2\pi)(\tfrac{1}{5}) = \tfrac{4}{5}k\pi. \quad \square$$

Example 2 Find the volume of the solid T that is bounded above by the cone $z^2 = x^2 + y^2$, below by the xy-plane, and on the sides by the hemisphere $z = \sqrt{4 - x^2 - y^2}$ (see Figure 16.9.7).

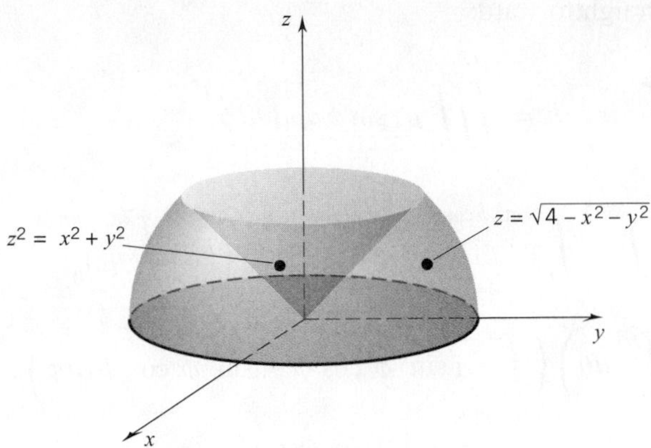

Figure16.9.7

SOLUTION In terms of spherical coordinates, the hemisphere is given by $\rho = 2$, $0 \leq \phi \leq \pi/2$. As you can verify, the hemisphere and the cone intersect in a circle which lies in the plane $z = \sqrt{2}$ and has its center on the z-axis. For points on this circle, the angle ϕ is $\pi/4$ (verify this). It follows that the solid T is the set of all (x, y, z) with spherical coordinates (ρ, θ, ϕ) in the set

$$S: \quad 0 \leq \rho \leq 2, \quad 0 \leq \theta \leq 2\pi, \quad \pi/4 \leq \phi \leq \pi/2.$$

Thus,

$$V = \iiint_T dxdydz = \iiint_S \rho^2 \sin\phi \, d\rho d\phi d\theta$$

$$= \int_{\pi/4}^{\pi/2} \int_0^{2\pi} \int_0^2 \rho^2 \sin\phi \, d\rho \, d\theta \, d\phi = \left(\int_{\pi/4}^{\pi/2} \sin\phi \, d\phi\right)\left(\int_0^{2\pi} d\theta\right)\left(\int_0^2 \rho^2 \, d\rho\right)$$

$$= (\sqrt{2}/2)(2\pi)(\tfrac{8}{3}) = \frac{8\pi\sqrt{2}}{3} \cong 11.85. \quad \square$$

Example 3 Find the volume of the solid T enclosed by the surface

$$(x^2 + y^2 + z^2)^2 = 2z(x^2 + y^2).$$

SOLUTION In spherical coordinates the bounding surface takes the form

$$\rho = 2\sin^2\phi\cos\phi. \qquad \text{(check this out)}$$

This equation places no restriction on θ; thus θ can range from 0 to 2π. Since ρ remains nonnegative, ϕ can range only from 0 to $\frac{1}{2}\pi$. Thus the solid T is the set of all (x, y, z) with spherical coordinates (ρ, θ, ϕ) in the set

$$S: \quad 0 \leq \theta \leq 2\pi, \quad 0 \leq \phi \leq \tfrac{1}{2}\pi, \quad 0 \leq \rho \leq 2\sin^2\phi\cos\phi.$$

The rest is straightforward:

$$V = \iiint_T dxdydz = \iiint_S \rho^2 \sin\phi \, d\rho d\theta d\phi$$

$$= \int_0^{2\pi} \int_0^{\pi/2} \int_0^{2\sin^2\phi\cos\phi} \rho^2 \sin\phi \, d\rho \, d\phi \, d\theta = \int_0^{2\pi} \int_0^{\pi/2} \tfrac{8}{3} \sin^7\phi \cos^3\phi \, d\phi \, d\theta$$

$$= \tfrac{8}{3} \left(\int_0^{2\pi} d\theta \right) \left(\int_0^{\pi/2} (\sin^7\phi \cos\phi - \sin^9\phi \cos\phi) \, d\phi \right)$$

$$= \tfrac{8}{3}(2\pi)(\tfrac{1}{40}) = \tfrac{2}{15}\pi \cong 0.42. \quad \square$$

EXERCISES 16.9

1. Find the spherical coordinates (ρ, θ, ϕ) of the point with rectangular coordinates $(1, 1, 1)$.
2. Find the rectangular coordinates of the point with spherical coordinates $(2, \frac{1}{6}\pi, \frac{1}{4}\pi)$.
3. Find the rectangular coordinates of the point with spherical coordinates $(3, \frac{1}{3}\pi, \frac{1}{6}\pi)$.
4. Find the spherical coordinates of the point with cylindrical coordinates $(2, \frac{2}{3}\pi, 6)$.
5. Find the spherical coordinates of the point with rectangular coordinates $(2, 2, \frac{2}{3}\sqrt{6})$.
6. Find the spherical coordinates of the point with rectangular coordinates $(2\sqrt{2}, -2\sqrt{2}, -4\sqrt{3})$.
7. Find the rectangular coordinates of the point with spherical coordinates $(3, \pi/2, 0)$.
8. Find the spherical coordinates of the point with cylindrical coordinates $(1/\sqrt{2}, \pi/4, 1/\sqrt{2})$.

Equations are given in spherical coordinates. Interpret each one geometrically.

9. $\rho \sin\phi = 1$.
10. $\sin\phi = 1$.
11. $\cos\phi = -\frac{1}{2}\sqrt{2}$.
12. $\tan\theta = 1$.
13. $\rho\cos\phi = 1$.
14. $\rho = \cos\phi$.

The volume of a solid T is given in spherical coordinates. Sketch T and evaluate the repeated integral.

15. $\int_0^{2\pi} \int_0^{\pi} \int_0^{2} \rho^2 \sin\phi \, d\rho \, d\phi \, d\theta.$

16. $\int_0^{\pi/4} \int_0^{\pi/2} \int_0^{1} \rho^2 \sin\phi \, d\rho \, d\phi \, d\theta.$

17. $\int_{\pi/6}^{\pi/2} \int_0^{\pi/2} \int_0^{3} \rho^2 \sin\phi \, d\rho \, d\theta \, d\phi.$

18. $\int_0^{\pi/4} \int_0^{2\pi} \int_0^{\sec\varphi} \rho^2 \sin\phi \, d\rho \, d\theta \, d\phi.$

Evaluate using spherical coordinates.

19. $\iiint_T dxdydz$; $T: 0 \le x \le 1, 0 \le y \le \sqrt{1-x^2}$, $\sqrt{x^2+y^2} \le z \le \sqrt{2-(x^2+y^2)}$.

20. $\iiint_T (x^2+y^2+z^2) \, dxdydz$; $T: 0 \le x \le \sqrt{4-y^2}, 0 \le y \le 2$, $\sqrt{x^2+y^2} \le z \le \sqrt{4-x^2-y^2}$.

21. $\iiint_T z\sqrt{x^2+y^2+z^2} \, dxdydz$; $T: 0 \le x \le \sqrt{9-y^2}, 0 \le y \le 3$, $0 \le z \le \sqrt{9-(x^2+y^2)}$.

22. $\iiint_T \frac{1}{(x^2+y^2+z^2)} \, dxdydz$; $T: 0 \le x \le 1, 0 \le y \le \sqrt{1-x^2}$, $0 \le z \le \sqrt{1-x^2-y^2}$.

23. Derive the formula for the volume of a sphere of radius R using spherical coordinates.
24. Express cylindrical coordinates in terms of spherical coordinates.
25. A wedge is cut from a ball of radius R by two planes that meet in a diameter. Find the volume of the wedge if the angle between the planes is α radians.
26. Find the mass of a ball of radius R given that the density varies directly with the distance from the boundary.
27. Find the mass of a right circular cone of base radius r and height h given that the density varies directly with the distance from the vertex.
28. Use spherical coordinates to derive the formula for the volume of a right circular cone of base radius r and height h.

In Exercises 29 and 30, let T be a homogeneous ball of mass M and radius R.

29. Calculate the moment of inertia about: (a) a diameter; (b) a tangent line.

30. Locate the center of mass of the upper half given that the center of the ball is at the origin.

In Exercises 31 and 32, let T be a homogeneous solid bounded by two concentric spherical shells, an outer shell of radius R_2 and an inner shell of radius R_1.

31. (a) Calculate the moment of inertia about a diameter.
(b) Use your result in part (a) to determine the moment of inertia of a spherical shell of radius R and mass M about a diameter.
(c) What is the moment of inertia of that same shell about a tangent line?

32. (a) Locate the center of mass of the upper half of T given that the center of T is at the origin.
(b) Use your result in part (a) to locate the center of mass of a homogeneous hemispherical shell of radius R.

33. Find the volume of the solid common to the sphere $\rho = a$ and the cone $\phi = \alpha$. Take $\alpha \in (0, \frac{1}{2}\pi)$.

34. Let T be the solid bounded below by the half-cone $z = \sqrt{x^2 + y^2}$ and above by the spherical surface $x^2 + y^2 + z^2 = 1$. Use spherical coordinates to evaluate

$$\iiint_T e^{(x^2+y^2+z^2)^{3/2}}\, dxdydz.$$

35. (a) Find an equation in spherical coordinates for the sphere $x^2 + y^2 + (z - R)^2 = R^2$.
(b) Express the upper half of the ball $x^2 + y^2 + (z - R)^2 \le R^2$ by inequalities in spherical coordinates.

36. Find the mass of the ball $\rho \le 2R\cos\phi$ given that the density varies directly with (a) ρ; (b) $\rho\sin\phi$; (c) $\rho\cos^2\theta\sin\phi$.

37. Find the volume of the solid common to the spheres $\rho = 2\sqrt{2}\cos\phi$ and $\rho = 2$.

38. Find the volume of the solid enclosed by the surface $\rho = 1 - \cos\phi$.

39. Finish the argument in (16.9.3).

40. (*Gravitational attraction*) Let T be a basic solid and let (a, b, c) be a point not in T. Show that, if T has continuously varying mass density $\lambda = \lambda(x, y, z)$, then T attracts a point mass m at (a, b, c) with a force

$$\mathbf{F} = \iiint_T Gm\,\lambda(x, y, z)\,\mathbf{f}(x, y, z)\; dxdydz, \text{ where}$$

$$\mathbf{f}(x, y, z) = \frac{[(x - a)\,\mathbf{i} + (y - b)\,\mathbf{j} + (z - c)\,\mathbf{k}]}{\left[(x - a)^2 + (y - b)^2 + (z - c)^2\right]^{3/2}}.$$

[Assume that a point mass m_1 at P_1 attracts a point mass m_2 at P_2 with a force $\mathbf{F} = -(Gm_1m_2/r^3)\mathbf{r}$, where $\mathbf{r}$ is the vector $\overrightarrow{P_1P_2}$. Interpret the triple integral component by component.]

41. Let T be the upper half of the ball $x^2 + y^2 + (z - R)^2 \le R^2$. Given that T is homogeneous and has mass M, find the gravitational force exerted by T on a point mass m located at the origin. (Note Exercise 40.)

42. A point mass m is placed on the axis of a homogeneous solid right circular cylinder at a distance α from the nearest base of the cylinder. Find the gravitational force exerted by the cylinder on the point mass given that the cylinder has base radius R, height h, and mass M. (Note Exercise 40.)

■ 16.10 JACOBIANS; CHANGING VARIABLES IN MULTIPLE INTEGRATION

During the course of the last few sections you have met several formulas for changing variables in multiple integration: to polar coordinates, to cylindrical coordinates, to spherical coordinates. The purpose of this section is to bring some unity into that material and provide a general description for other changes of variable.

We begin with a consideration of area. Figure 16.10.1 shows a basic region Γ in a plane that we are calling the uv-plane. (In this plane we denote the abscissa of a point by u and the ordinate by v.) Suppose that

$$x = x(u, v), \quad y = y(u, v)$$

are continuously differentiable functions on Γ. As (u, v) ranges over Γ, the point $(x, y) = (x(u, v),\ y(u, v))$ generates a region Ω in the xy-plane. If the mapping

$$(u, v) \to (x, y)$$

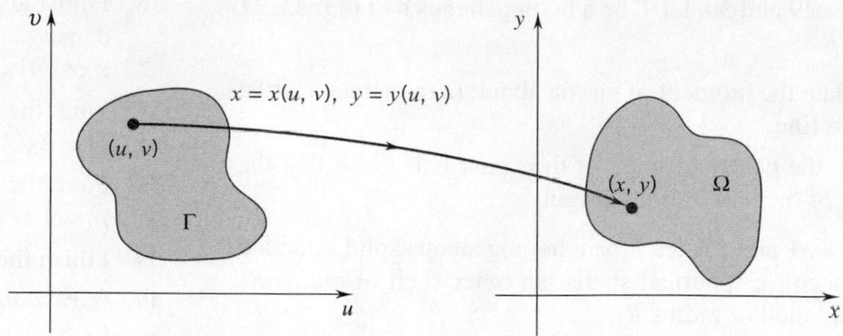

Figure 16.10.1

is one-to-one on the interior of Γ, and the *Jacobian J* given by the 2×2 determinant

$$J(u,v) = \begin{vmatrix} \dfrac{\partial x}{\partial u} & \dfrac{\partial y}{\partial u} \\ \dfrac{\partial x}{\partial v} & \dfrac{\partial y}{\partial v} \end{vmatrix} = \frac{\partial x}{\partial u}\frac{\partial y}{\partial v} - \frac{\partial x}{\partial v}\frac{\partial y}{\partial u},$$

is never zero on the interior of Γ, then

(16.10.1)
$$\text{area of } \Omega = \iint_{\Gamma} |J(u,v)|\, dudv.$$

It is very difficult to prove this assertion without making additional assumptions. A proof valid for all cases of practical interest is given in the supplement to Section 17.5. At this point we simply assume this area formula and go on from there.

Suppose now that we want to integrate some continuous function $f = f(x,y)$ over Ω. If this proves difficult to do directly, then we can change variables to u, v and try to integrate over Γ instead. It follows from (16.10.1) that

(16.10.2)
$$\iint_{\Omega} f(x,y)\; dxdy = \iint_{\Gamma} f(x(u,v), y(u,v))|J(u,v)|\, dudv.$$

The derivation of this formula from (16.10.1) follows the usual lines. Break up Γ into N little basic regions $\Gamma_1, \ldots, \Gamma_N$. These induce a decomposition of Ω into N little basic regions $\Omega_1, \ldots, \Omega_N$. We can then write

$$\iint_{\Gamma} f(x(u,v), y(u,v))|J(u,v)|\, dudv = \sum_{i=1}^{N} \iint_{\Gamma_i} f(x(u,v), y(u,v))|J(u,v)|\, dudv$$

additivity

$$= \sum_{i=1}^{N} f(x(u_i^*, v_i^*), y(u_i^*, v_i^*)) \iint_{\Gamma_i} |J(u,v)|\, dudv$$

Theorem 16.2.10 applied to Γ_i

$$= \sum_{i=1}^{N} f(x_i^*, y_i^*) \iint_{\Gamma_i} |J(u,v)|\, dudv$$

set $x_i^* = x(u_i^*, v_i^*), y_i^* = y(u_i^*, v_i^*)$

$$= \sum_{i=1}^{N} f(x_i^*, y_i^*)(\text{area of } \Omega_i).$$

(16.10.1) applied to Γ_i

This last expression is a Riemann sum for

$$\iint_{\Omega} f(x,y)\,dxdy$$

and tends to that integral as the maximum diameter of the Ω_i tends to zero. We can ensure this by letting the maximum diameter of the Γ_i tend to zero. ☐

Example 1 Evaluate

$$\iint_{\Omega} (x+y)^2\,dxdy$$

where Ω is the parallelogram bounded by the lines

$$x+y=0, \qquad x+y=1, \qquad 2x-y=0, \qquad 2x-y=3.$$

SOLUTION The parallelogram is shown in Figure 16.10.2. The boundaries suggest that we set

$$u = x+y, \qquad v = 2x-y.$$

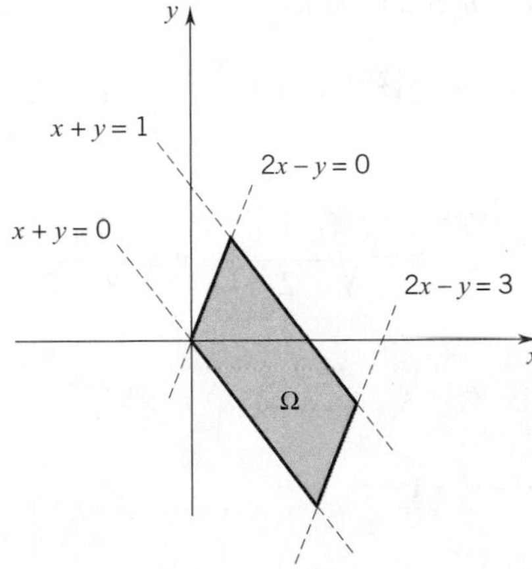

Figure 16.10.2

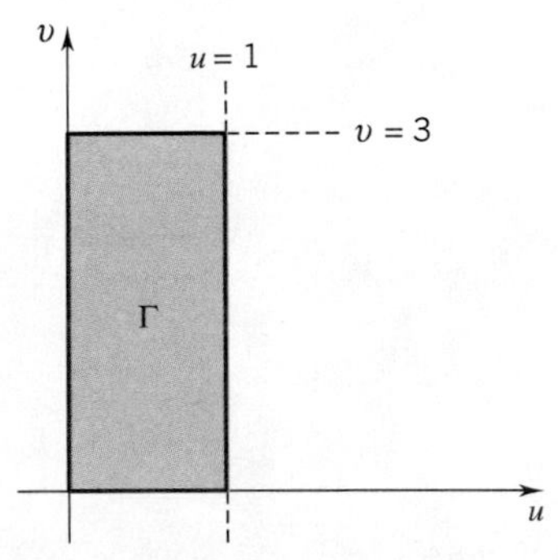

Figure 16.10.3

We want x and y in terms of u and v. Since

$$u+v = (x+y)+(2x-y) = 3x \quad \text{and} \quad 2u-v = (2x+2y)-(2x-y) = 3y,$$

we have
$$x = \frac{u+v}{3}, \quad y = \frac{2u-v}{3}.$$

This transformation maps the rectangle Γ of Figure 16.10.3 onto Ω with Jacobian

$$J(u,v) = \begin{vmatrix} \dfrac{\partial}{\partial u}\left(\dfrac{u+v}{3}\right) & \dfrac{\partial}{\partial u}\left(\dfrac{2u-v}{3}\right) \\[2ex] \dfrac{\partial}{\partial v}\left(\dfrac{u+v}{3}\right) & \dfrac{\partial}{\partial v}\left(\dfrac{2u-v}{3}\right) \end{vmatrix} = \begin{vmatrix} \frac{1}{3} & \frac{2}{3} \\ \frac{1}{3} & -\frac{1}{3} \end{vmatrix} = -\tfrac{1}{3}.$$

Therefore $\displaystyle\iint_{\Omega} (x+y)^2 \, dxdy = \iint_{\Omega} u^2 |J(u,v)| \, dudv$

$$= \tfrac{1}{3}\int_0^3 \int_0^1 u^2 \, dudv$$

$$= \tfrac{1}{3}\left(\int_0^3 dv\right)\left(\int_0^1 u^2 \, du\right) = \tfrac{1}{3}(3)\tfrac{1}{3} = \tfrac{1}{3}. \quad \square$$

Example 2 Evaluate

$$\iint_{\Omega} xy \, dxdy$$

where Ω is the first-quadrant region bounded by the curves

$$x^2+y^2=4, \qquad x^2+y^2=9, \qquad x^2-y^2=1, \qquad x^2-y^2=4.$$

SOLUTION The region is shown in Figure 16.10.4. The boundaries suggest that we set

$$u = x^2+y^2, \qquad v = x^2-y^2.$$

We want x and y in terms of u and v. Since

$$u+v = 2x^2 \quad \text{and} \quad u-v = 2y^2,$$

we have

$$x = \sqrt{\frac{u+v}{2}}, \quad y = \sqrt{\frac{u-v}{2}}.$$

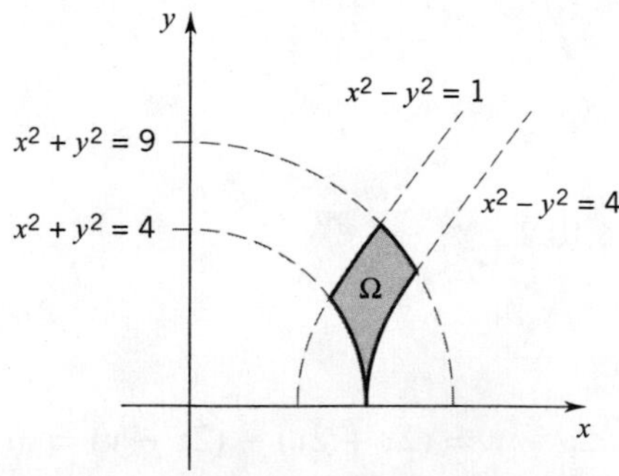

Figure 16.10.4

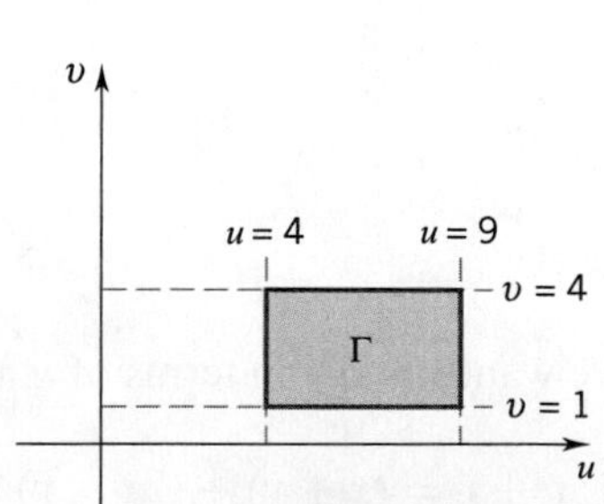

Figure 16.10.5

The transformation maps the rectangle Γ of Figure 16.10.5 onto Ω with Jacobian

$$J(u,v) = \begin{vmatrix} \dfrac{\partial}{\partial u}\left(\sqrt{\dfrac{u+v}{2}}\right) & \dfrac{\partial}{\partial u}\left(\sqrt{\dfrac{u-v}{2}}\right) \\ \dfrac{\partial}{\partial v}\left(\sqrt{\dfrac{u+v}{2}}\right) & \dfrac{\partial}{\partial v}\left(\sqrt{\dfrac{u-v}{2}}\right) \end{vmatrix} = -\frac{1}{4\sqrt{u^2-v^2}}.$$

check this

Therefore
$$\iint_{\Omega} xy\;dxdy = \iint_{\Gamma} \left(\sqrt{\frac{u+v}{2}}\right)\left(\sqrt{\frac{u-v}{2}}\right)\left(\frac{1}{4\sqrt{u^2-v^2}}\right) dudv$$
$$= \iint_{\Gamma} \tfrac{1}{8}\,du\,dv = \tfrac{1}{8}(\text{area of }\Gamma) = \tfrac{15}{8}. \quad \square$$

In Section 16.4 you saw the formula for changing variables from rectangular coordinates (x, y) to polar coordinates $[r, \theta]$. The formula reads

$$\iint_{\Omega} f(x,y)\;dxdy = \iint_{\Gamma} f(r\cos\theta, r\sin\theta)\,r\,drd\theta. \tag{16.4.3}$$

The factor r in the double integral over Γ is the Jacobian of the transformation $x = r\cos\theta,\ y = r\sin\theta$:

$$J(r,\theta) = \begin{vmatrix} \dfrac{\partial}{\partial r}(r\cos\theta) & \dfrac{\partial}{\partial r}(r\sin\theta) \\ \dfrac{\partial}{\partial\theta}(r\cos\theta) & \dfrac{\partial}{\partial\theta}(r\sin\theta) \end{vmatrix} = \begin{vmatrix} \cos\theta & \sin\theta \\ -r\sin\theta & r\cos\theta \end{vmatrix} = r(\cos^2\theta + \sin^2\theta) = r.$$

As you can see, (16.4.3) is a special case of (16.10.2).

When changing variables in a triple integral we make three coordinate changes:

$$x = x(u, v, w), \qquad y = y(u, v, w), \qquad z = z(u, v, w).$$

If these functions carry a basic solid Γ onto a solid T, then, under conditions analogous to the two-dimensional case,

$$\text{volume of } T = \iiint_{\Gamma} |J(u, v, w)|\;dudvdw$$

where now the Jacobian† is a three-by-three determinant:

$$J(u, v, w) = \begin{vmatrix} \dfrac{\partial x}{\partial u} & \dfrac{\partial y}{\partial u} & \dfrac{\partial z}{\partial u} \\ \dfrac{\partial x}{\partial v} & \dfrac{\partial y}{\partial v} & \dfrac{\partial z}{\partial v} \\ \dfrac{\partial x}{\partial w} & \dfrac{\partial y}{\partial w} & \dfrac{\partial z}{\partial w} \end{vmatrix}.$$

In this case the change of variables formula reads

$$\iiint_{T} f(x,y,z)\;dxdydz = \iiint_{\Gamma} f(x(u,v,w),\ y(u,v,w),\ z(u,v,w))|\,J(u,v,w)|\;dudvdw.$$

† The study of these functional determinants goes back to a memoir by the German mathematician C. G. Jacobi, 1804–1851.

EXERCISES 16.10

Find the Jacobian of the transformation.

1. $x = au + bv, \quad y = cu + dv.$ (linear transformation)

2. $x = u\cos\theta - v\sin\theta, \quad y = u\sin\theta + v\cos\theta.$ (rotation by θ)

3. $x = uv, \quad y = u^2 + v^2.$

4. $x = u\ln v, \quad y = uv.$

5. $x = uv^2, \quad y = u^2v.$

6. $x = u - \ln v, \quad y = \ln u + v.$

7. $x = au, \quad y = bv, \quad z = cw.$

8. $x = v + w, \quad y = u + w, \quad z = u + v.$

9. $x = r\cos\theta, \quad y = r\sin\theta, \quad z = z.$ (cylindrical coordinates)

10. $x = \rho\sin\phi\cos\theta, \quad y = \rho\sin\phi\sin\theta, \quad z = \rho\cos\phi.$ (spherical coordinates)

11. $x = (1 + w\cos v)\cos u, \quad y = (1 + w\cos v)\sin u,$ $z = w\sin v.$

12. Every linear transformation

$$x = au + bv, \quad y = cu + dv \quad \text{with} \quad ad - bc \neq 0$$

maps lines of the uv-plane onto lines of the xy-plane. Find the image of (a) a vertical line $u = u_0$; (b) a horizontal line $v = v_0$.

For Exercises 13–15, take Ω as the parallelogram bounded by

$$x + y = 0, \quad x + y = 1, \quad x - y = 0, \quad x - y = 2.$$

Evaluate.

13. $\displaystyle\iint_\Omega (x^2 - y^2)\,dx\,dy.$

14. $\displaystyle\iint_\Omega 4xy\,dx\,dy.$

15. $\displaystyle\iint_\Omega (x - y)\cos[\pi(x - y)]\,dx\,dy.$

For Exercises 16–18, take Ω as the parallelogram bounded by

$$x - y = 0, \quad x - y = \pi, \quad x + 2y = 0, \quad x + 2y = \tfrac{1}{2}\pi.$$

Evaluate.

16. $\displaystyle\iint_\Omega (x + y)\,dx\,dy.$

17. $\displaystyle\iint_\Omega \sin(x - y)\cos(x + 2y)\,dx\,dy.$

18. $\displaystyle\iint_\Omega \sin 3x\,dx\,dy.$

19. Let Ω be the first-quadrant region bounded by the curves $xy = 1$, $xy = 4$, $y = x$, $y = 4x$. (a) Determine the area of Ω and (b) locate the centroid.

20. Show that the ellipse $b^2x^2 + a^2y^2 = a^2b^2$ has area πab by setting $x = ar\cos\theta$, $y = br\sin\theta$.

21. A homogeneous plate in the xy-plane is in the form of a parallelogram. The parallelogram is bounded by the lines $x + y = 0, x + y = 1, 3x - 2y = 0, 3x - 2y = 2$. Calculate the moments of inertia of the plate about the three coordinate axes. Express your answers in terms of the mass of the plate.

22. Calculate the area of the region Ω bounded by the curves

$$x^2 - 2xy + y^2 + x + y = 0, \qquad x + y + 4 = 0.$$

HINT: Set $u = x - y, v = x + y$.

23. Calculate the area of the region Ω bounded by the curves

$$x^2 - 4xy + 4y^2 - 2x - y - 1 = 0, \qquad y = \tfrac{2}{5}.$$

24. Locate the centroid of the region in Exercise 22.

25. Calculate the area of the region Ω enclosed by the curve

$$11x^2 + 4\sqrt{3}xy + 7y^2 - 1 = 0.$$

HINT: Use a rotation $x = u\cos\theta - v\sin\theta$, $y = u\sin\theta + v\cos\theta$ such that the resulting uv-equation has no uv-term.

26. Evaluate

$$\int_{-\infty}^{\infty}\int_{-\infty}^{\infty} \frac{e^{-(x-y)^2}}{1 + (x + y)^2}\,dx\,dy$$

by integrating over the square $S_a: -a \le x \le a, -a \le y \le a$ and taking the limit as $a \to \infty$.
HINT: Set $u = x - y, v = x + y$ and see (16.4.4).

For Exercises 27–30, let T be the solid ellipsoid $x^2/a^2 + y^2/b^2 + z^2/c^2 \le 1$.

27. Calculate the volume of T by setting

$$x = a\rho\sin\phi\cos\theta, \quad y = b\rho\sin\phi\sin\theta, \quad z = c\rho\cos\phi.$$

28. Locate the centroid of the upper half of T.

29. View the upper half of T as a homogeneous solid of mass M. Find the moments of inertia of this solid about the coordinate axes.

30. Evaluate

$$\iiint_T \left(\frac{x^2}{a^2} + \frac{y^2}{b^2} + \frac{z^2}{c^2}\right)dx\,dy\,dz.$$

■ PROJECT 16.10 Generalized Polar Coordinates

Recall the equations that transform polar coordinates to rectangular coordinates:

$$x = r\cos\theta, \quad y = r\sin\theta.$$

In this project, we investigate a generalization of these equations, and we apply the generalized polar coordinates to the problem of finding the area of regions enclosed by curves given in rectangular coordinates.

Let a, b, and α be fixed positive numbers, and let (x, y) be related to (r, θ) by the equations

$$(1) \qquad x = ar(\cos\theta)^{\alpha}, \quad y = br(\sin\theta)^{\alpha}.$$

Problem 1.
a. Show that the mapping defined by (1) carries the polar region $\Gamma : 0 \le r < \infty, 0 \le \theta \le \pi/2$ onto the first quadrant in the xy-plane. HINT: Find a point $[r, \theta]$ that maps onto (x, y) given that $x \ge 0$ and $y \ge 0$.

b. Show that the mapping is one-to-one on the interior of Γ.

Problem 2. Determine the Jacobian of the mapping defined by (1).

Problem 3. The curve given by the equation $x^{2/3} + y^{2/3} = a^{2/3}, a > 0$, is called an *astroid*.

a. Use a graphing utility to graph the astroid for several values of a.

b. Calculate the area enclosed by the astroid in the first quadrant by setting $x = ar\cos^3\theta, y = ar\sin^3\theta$.

c. What is the entire area enclosed by the astroid?

Problem 4. Consider the curve defined by the equation

$$\left(\frac{x}{a}\right)^{1/4} + \left(\frac{y}{b}\right)^{1/4} = 1.$$

a. Use a graphing utility to graph this curve in the cases $a = 3, b = 2$ and $a = 2, b = 3$.

b. Calculate the area enclosed by the curve in the first quadrant by setting $x = ar\cos^8\theta, y = br\sin^8\theta$.

■ CHAPTER HIGHLIGHTS

16.1 Multiple-Sigma Notation

16.2 Double Integrals

The Double Integral over a Rectangle:
- partition (p. 946) upper sum, lower sum (p. 947)
- double integral over a rectangle (p. 948)

The Double Integral over a Region:
- basic region (p. 952)
- double integral over a region (p. 953)
- double integral as a volume (p. 953)
- area as a double integral (p. 953)
- properties of the double integral (p. 954)
- average value over a region (p. 954)
- weighted average over a region (p. 955)

16.3 The Evaluation of Double Integrals by Repeated Integrals

- evaluation formulas (p. 958)
- symmetry (p. 964)
- separated variables over a rectangle (p. 968)

16.4 The Double Integral as the Limit of Riemann Sums; Polar Coordinates

- diameter of set (p. 969) Riemann sum (p. 970)

$$\iint_{\Omega} f(x, y)\,dx\,dy = \lim_{\text{diam}\,\Omega_i \to 0} \sum_{i=1}^{N} f(x_i^*, y_i^*)(\text{ area of } \Omega_i)$$

- area formula (p. 971)
- evaluating double integrals using polar coordinates (p. 971)

16.5 Some Applications of Double Integration

- mass of a plate (p. 976)
- center of mass of a plate (p. 976)
- centroid of a region (p. 977)
- moment of inertia of a plate (p. 979)
- radius of gyration (p. 980)
- parallel axis theorem (p. 981)

16.6 Triple Integrals

- box (p. 983) partition of a box (p. 983)
- upper sum, lower sum (p. 984)
- triple integral over a box (p. 984)
- basic solid (p. 985)
- triple integral over a basic solid (p. 985)
- volume as a triple integral (p. 985)
- properties of the triple integral (p. 986)
- average value (p. 987) weighted average (p. 987)
- mass (p. 987) center of mass (p. 987)
- centroid (p. 987) moment of inertia (p. 987)

16.7 Reduction to Repeated Integrals

- possible orders of integration (p. 989–990)
- separated variables over a box (p. 995)

16.8 Cylindrical Coordinates

cylindrical coordinates (r, θ, z) (p. 997)
cylindrical wedge (p. 998)
evaluating triple integrals using cylindrical coordinates (p. 999)
volume in cylindrical coordinates (p. 1000)

16.9 The Triple Integral as the Limit of Riemann Sums; Spherical Coordinates

$$\iiint_T f(x,y,z)\,dx\,dy\,dz = \lim_{\operatorname{diam} T_i \to 0} \sum_{i=1}^{N} f(x_i^*, y_i^*, z_i^*) \text{ (volume of } T_i)$$

spherical coordinates (ρ, θ, ϕ) (p. 1004)
spherical wedge (p. 1005)
volume of a spherical wedge (p. 1005)
evaluating triple integrals using spherical coordinates (p. 1007)
volume in spherical coordinates (p. 1008)

16.10 Jacobians; Changing Variables in Multiple Integration

Jacobian: two-dimensional case (p. 1012)
change of variables in double integration (p. 1012)
Jacobian: three-dimensional case (p. 1015)
change of variables in triple integration (p. 1015)

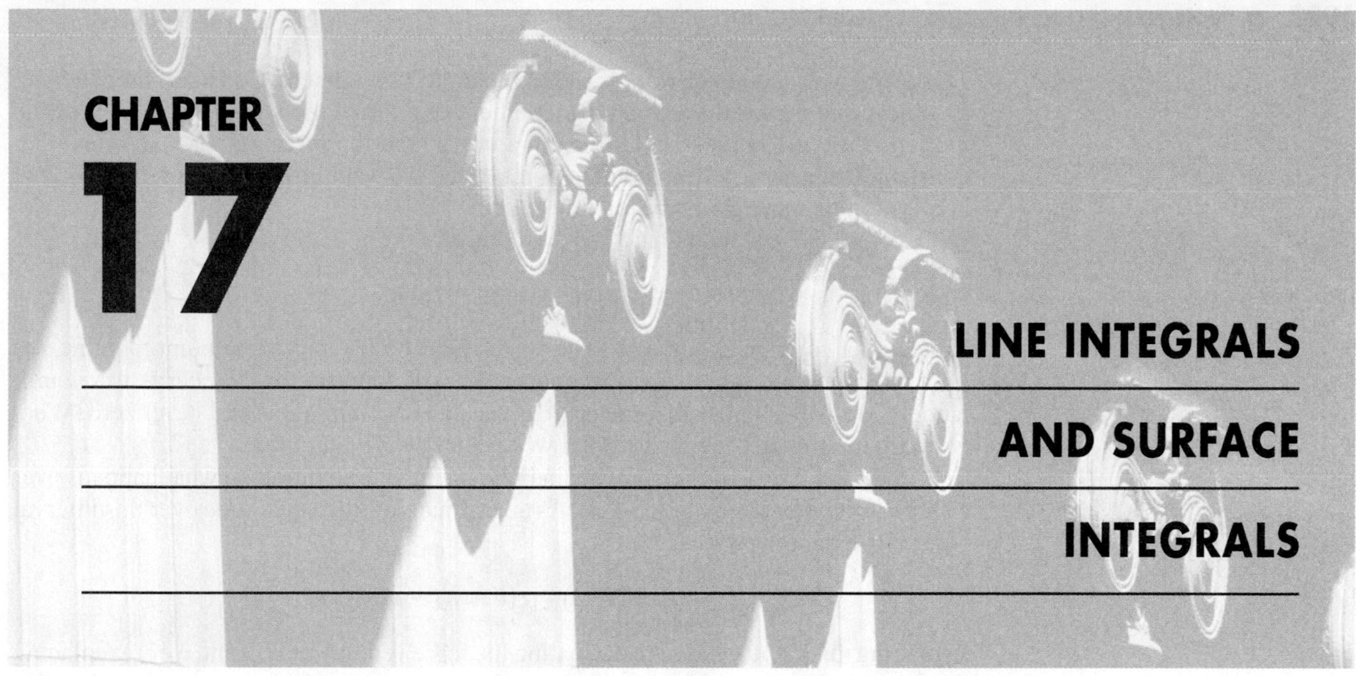

CHAPTER 17 LINE INTEGRALS AND SURFACE INTEGRALS

In this chapter we will study integration along curves and integration over surfaces. At the heart of this subject lie three great integration theorems: *Green's theorem, Gauss's theorem* (commonly known as the *divergence theorem*), and *Stokes's theorem*.

All three theorems are ultimately based on *The Fundamental Theorem of Integral Calculus*, and all can be cast in the same general form:

an integral over a set S = a related integral over the boundary of S.

A word about terminology. Suppose that S is some subset of the plane or of three-dimensional space. A function that assigns a scalar to each point of S (say, the temperature at that point or the mass density at that point) is known in science as a *scalar field*. A function that assigns a vector to each point of S (say, the wind velocity at that point or the gradient of a function f evaluated at that point) is called a *vector field*. We will be using this "field" language throughout.

■ 17.1 LINE INTEGRALS

We are led to the definition of *line integral* by the notion of work.

The Work Done by a Varying Force over a Curved Path

The work done by a constant force $\mathbf{F}$ on an object that moves along a straight line is, by definition, the component of $\mathbf{F}$ in the direction of the displacement multiplied by the length of the displacement vector $\mathbf{d}$ (Project 12.4):

$$W = (\text{comp}_{\mathbf{d}}\,\mathbf{F})\,||\mathbf{d}||$$

We can write this more briefly as a dot product:

(17.1.1) $$\boxed{W = \mathbf{F} \cdot \mathbf{d}}$$

This elementary notion of work is useful, but it is not sufficient. Consider, for example, an object that moves through a magnetic field or a gravitational field. The path of the motion is usually not a straight line but a curve, and the force, rather than remaining constant, tends to vary from point to point. What we want now is a notion of work that applies to this more general situation.

Let's suppose that an object moves along a curve

$$C:\ \mathbf{r}(u) = x(u)\,\mathbf{i} + y(u)\,\mathbf{j} + z(u)\,\mathbf{k}, \qquad u \in [a, b]$$

subject to a continuous force **F**. (The vector field **F** may vary from point to point, not only in magnitude but also in direction.) We will suppose that the curve is *smooth*; namely, we will suppose that the tangent vector $\mathbf{r}'$ is continuous and never zero. What we want to do here is define the total work done by **F** along the curve C.

To decide how to do this, we begin by focusing our attention on what happens over a short parameter interval $[u, u+h]$. As an estimate for the work done over this interval we can use the dot product

$$\mathbf{F}(\mathbf{r}(u)) \cdot [\mathbf{r}(u+h) - \mathbf{r}(u)].$$

In making this estimate, we are evaluating the force vector **F** at $\mathbf{r}(u)$ and we are replacing the curved path from $\mathbf{r}(u)$ to $\mathbf{r}(u+h)$ by the line segment from $\mathbf{r}(u)$ to $\mathbf{r}(u+h)$. (See Figure 17.1.1.) If we set

$$W(u) = \text{ total work done by } \mathbf{F} \text{ from } \mathbf{r}(a) \text{ to } \mathbf{r}(u), \text{ and}$$
$$W(u+h) = \text{ total work done by } \mathbf{F} \text{ from } \mathbf{r}(a) \text{ to } \mathbf{r}(u+h),$$

then the work done by **F** from $\mathbf{r}(u)$ to $\mathbf{r}(u+h)$ must be the difference

$$W(u+h) - W(u).$$

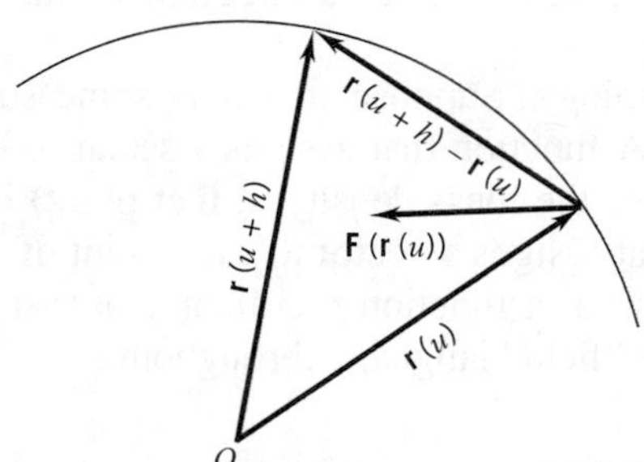

Figure 17.1.1

Bringing our estimate into play, we are led to the approximate equation

$$W(u+h) - W(u) \cong \mathbf{F}(\mathbf{r}(u)) \cdot [\mathbf{r}(u+h) - \mathbf{r}(u)],$$

which, upon division by h, becomes

$$\frac{W(u+h) - W(u)}{h} \cong \mathbf{F}(\mathbf{r}(u)) \cdot \frac{[\mathbf{r}(u+h) - \mathbf{r}(u)]}{h}.$$

The quotients here are average rates of change, and the equation is only an approximate one. The notion of work is made precise by requiring that both sides have exactly the same limit as h tends to zero; in other words, by requiring that

$$W'(u) = \mathbf{F}(\mathbf{r}(u)) \cdot \mathbf{r}'(u).$$

The rest is now determined. Since

$$W(a) = 0 \qquad \text{and} \qquad W(b) = \text{total work done by } \mathbf{F} \text{ on } C,$$

we have:

$$\text{total work done by } \mathbf{F} \text{ on } C = W(b) - W(a) = \int_a^b W'(u)\,du = \int_a^b [\mathbf{F}(\mathbf{r}(u)) \cdot \mathbf{r}'(u)]\,du.$$

In short, we have arrived at the following definition of work:

(17.1.2)
$$W = \int_a^b [\mathbf{F}(\mathbf{r}(u)) \cdot \mathbf{r}'(u)]\,du.$$

Example 1 Determine the work done by the force

$$\mathbf{F}(x, y, z) = xy\,\mathbf{i} + 2\,\mathbf{j} + 4z\,\mathbf{k}$$

along the circular helix C: $\mathbf{r}(u) = \cos u\,\mathbf{i} + \sin u\,\mathbf{j} + u\,\mathbf{k}$, from $u = 0$ to $u = 2\pi$.

SOLUTION We have: $x(u) = \cos u, y(u) = \sin u, z(u) = u$, and

$$\mathbf{F}(\mathbf{r}(u)) = \cos u \sin u\,\mathbf{i} + 2\,\mathbf{j} + 4u\,\mathbf{k},$$
$$\mathbf{r}'(u) = -\sin u\,\mathbf{i} + \cos u\,\mathbf{j} + \mathbf{k}.$$

Now,
$$\mathbf{F}(\mathbf{r}(u)) \cdot \mathbf{r}'(u) = -\sin^2 u \cos u + 2\cos u + 4u$$

and
$$W = \int_0^{2\pi} \mathbf{F}(\mathbf{r}(u)) \cdot \mathbf{r}'(u)\,du = \int_0^{2\pi} (-\sin^2 u \cos u + 2\cos u + 4u)\,du$$
$$= \left[-\frac{\sin^3 u}{3} + 2\sin u + 2u^2\right]_0^{2\pi} = 8\pi^2. \quad \square$$

Line Integrals

The integral on the right of (17.1.2) can be calculated not only for a force function **F** but for any vector field **h** continuous on C.

DEFINITION 17.1.3 LINE INTEGRAL

Let $\mathbf{h}(x, y, z) = h_1(x, y, z)\,\mathbf{i} + h_2(x, y, z)\,\mathbf{j} + h_3(x, y, z)\,\mathbf{k}$ be a vector field that is continuous on a smooth curve

$$C\colon\ \mathbf{r}(u) = x(u)\,\mathbf{i} + y(u)\,\mathbf{j} + z(u)\,\mathbf{k}, \qquad u \in [a, b].$$

The *line integral* of **h** over C is the number

$$\int_C \mathbf{h}(\mathbf{r}) \cdot \mathbf{dr} = \int_a^b [\mathbf{h}(\mathbf{r}(u)) \cdot \mathbf{r}'(u)]\,du.$$

Note also that, while we speak of integrating over C, we actually carry out the calculations over the parameter set $[a, b]$. If our definition of line integral is to make sense, the line integral as defined must be independent of the particular parametrization chosen for C. Within the limitations spelled out as follows, this is indeed the case:

THEOREM 17.1.4

Let $\mathbf{h}$ be a vector field that is continuous on a smooth curve C. The line integral

$$\int_C \mathbf{h}(\mathbf{r}) \cdot \mathbf{dr} = \int_a^b [\mathbf{h}(\mathbf{r}(u)) \cdot \mathbf{r}'(u)]\, du$$

is left invariant by every *orientation-preserving* change of parameter.†

PROOF Suppose that ϕ maps $[c, d]$ onto $[a, b]$ and that ϕ' is positive and continuous on $[c, d]$. We must show that the line integral over C as parametrized by

$$\mathbf{R}(w) = \mathbf{r}(\phi(w)), \qquad w \in [c, d]$$

equals the line integral over C as parametrized by $\mathbf{r}$. The argument is straightforward:

$$\begin{aligned}
\int_C \mathbf{h}(\mathbf{R}) \cdot \mathbf{dR} &= \int_c^d [\mathbf{h}(\mathbf{R}(w)) \cdot \mathbf{R}'(w)]\, dw \\
&= \int_c^d [\mathbf{h}(\mathbf{r}(\phi(w))) \cdot \mathbf{r}'(\phi(w))\phi'(w)]\, dw \\
&= \int_c^d [\mathbf{h}(\mathbf{r}(\phi(w))) \cdot \mathbf{r}'(\phi(w))]\phi'(w)\, dw \\
&= \int_a^b [\mathbf{h}(\mathbf{r}(u)) \cdot \mathbf{r}'(u)]\, du = \int_C \mathbf{h}(\mathbf{r}) \cdot \mathbf{dr}. \quad \square
\end{aligned}$$

Set $u = \phi(w)$, $du = \phi'(w)\, dw$.
At $w = c, u = a$; at $w = d, u = b$.

Example 2 Calculate $\int_C \mathbf{h}(\mathbf{r}) \cdot \mathbf{dr}$ given that

$$\mathbf{h}(x, y) = xy\,\mathbf{i} + y^2\,\mathbf{j} \quad \text{and} \quad C\colon\ \mathbf{r}(u) = u\,\mathbf{i} + u^2\,\mathbf{j}, \quad u \in [0, 1].$$

SOLUTION Here $x(u) = u$, $y(u) = u^2$ and

$$\begin{aligned}
\mathbf{h}(\mathbf{r}(u)) \cdot \mathbf{r}'(u) &= [x(u)y(u)\,\mathbf{i} + [y(u)]^2\,\mathbf{j}] \cdot [x'(u)\,\mathbf{i} + y'(u)\,\mathbf{j}] \\
&= x(u)y(u)x'(u) + [y(u)]^2 y'(u) \\
&= u(u^2)(1) + u^4(2u) = u^3 + 2u^5.
\end{aligned}$$

It follows that $\displaystyle\int_C \mathbf{h}(\mathbf{r}) \cdot \mathbf{dr} = \int_0^1 (u^3 + 2u^5)\, du = \left[\tfrac{1}{4}u^4 + \tfrac{1}{3}u^6\right]_0^1 = \tfrac{7}{12}.$ $\square$

† Changes of parameter were explained in Section 13.3.

Example 3 Integrate the vector field $\mathbf{h}(x,y,z) = xy\,\mathbf{i} + yz\,\mathbf{j} + xz\,\mathbf{k}$ over the twisted cubic $\mathbf{r}(u) = u\,\mathbf{i} + u^2\,\mathbf{j} + u^3\,\mathbf{k}$ from $(-1, 1, -1)$ to $(1, 1, 1)$.

SOLUTION The path of integration begins at $u = -1$ and ends at $u = 1$. In this case

$$x(u) = u, \qquad y(u) = u^2, \qquad z(u) = u^3.$$

Therefore

$$\begin{aligned}\mathbf{h}(\mathbf{r}(u)) \cdot \mathbf{r}'(u) &= [x(u)y(u)\,\mathbf{i} + y(u)z(u)\,\mathbf{j} + x(u)z(u)\,\mathbf{k}] \cdot [x'(u)\,\mathbf{i} + y'(u)\,\mathbf{j} + z'(u)\,\mathbf{k}]\\ &= x(u)y(u)x'(u) + y(u)z(u)y'(u) + x(u)z(u)z'(u)\\ &= u(u^2)(1) + u^2(u^3)2u + u(u^3)3u^2\\ &= u^3 + 5u^6\end{aligned}$$

and $$\int_C \mathbf{h}(\mathbf{r}) \cdot \mathbf{dr} = \int_{-1}^{1} (u^3 + 5u^6)\,du = \left[\tfrac{1}{4}u^4 + \tfrac{5}{7}u^7\right]_{-1}^{1} = \tfrac{10}{7}. \quad \square$$

If a curve C is not smooth but is made up of a finite number of adjoining smooth pieces $C_1, C_2, \ldots, C_n$, then we define the integral over C as the sum of the integrals over the C_i:

(17.1.5) $$\int_C = \int_{C_1} + \int_{C_2} + \cdots + \int_{C_n}$$

A curve of this type is said to be *piecewise smooth*. See Figure 17.1.2.

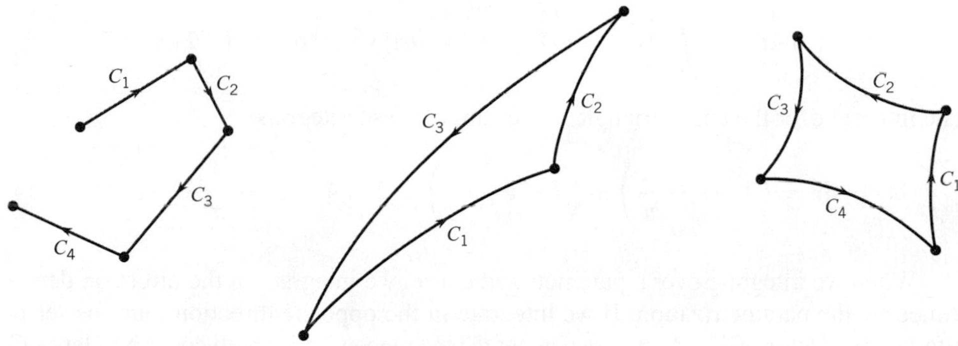

Figure 17.1.2

All polygonal paths are piecewise-smooth curves. In the next example we integrate over a triangle. We do this by integrating over each of the sides and then adding up the results. Observe that the directed line segment that begins at $\mathbf{a} = (a_1, a_2)$ and ends at $\mathbf{b} = (b_1, b_2)$ can be parametrized by setting

$$\mathbf{r}(u) = (1 - u)\mathbf{a} + u\mathbf{b}, \qquad u \in [0, 1].$$

Example 4 Evaluate the line integral $\int_C \mathbf{h}(\mathbf{r}) \cdot \mathbf{dr}$ if $\mathbf{h}(x,y) = e^y\,\mathbf{i} - \sin \pi x\,\mathbf{j}$ and C is the triangle with vertices $(1, 0), (0, 1), (-1, 0)$ traversed counterclockwise.

SOLUTION The path C is made up of three line segments:

$$\begin{aligned}C_1 &: \quad \mathbf{r}(u) = (1 - u)\,\mathbf{i} + u\,\mathbf{j}, \quad u \in [0, 1],\\ C_2 &: \quad \mathbf{r}(u) = (1 - u)\,\mathbf{j} + u(-\mathbf{i}) = -u\,\mathbf{i} + (1 - u)\,\mathbf{j}, \; u \in [0, 1],\\ C_3 &: \quad \mathbf{r}(u) = (1 - u)(-\mathbf{i}) + u\,\mathbf{i} = (2u - 1)\,\mathbf{i}, \; u \in [0, 1].\end{aligned}$$

See Figure 17.1.3.

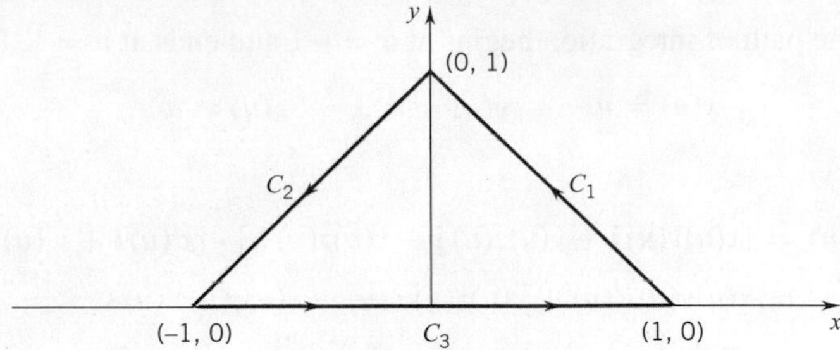

Figure 17.1.3

As you can verify,

$$\int_{C_1} \mathbf{h}(\mathbf{r}) \cdot \mathbf{dr} = \int_0^1 \left[e^{y(u)}x'(u) - \sin\left[\pi x(u)\right] y'(u)\right] du$$

$$= \int_0^1 \left[-e^u - \sin\left[\pi(1-u)\right]\right] du = 1 - e - \frac{2}{\pi};$$

$$\int_{C_2} \mathbf{h}(\mathbf{r}) \cdot \mathbf{dr} = \int_0^1 \left[e^{y(u)}x'(u) - \sin\left[\pi x(u)\right] y'(u)\right] du$$

$$= \int_0^1 \left[-e^{1-u} + \sin\left(-\pi u\right)\right] du = 1 - e - \frac{2}{\pi};$$

$$\int_{C_3} \mathbf{h}(\mathbf{r}) \cdot \mathbf{dr} = \int_0^1 \left[e^{y(u)}x'(u) - \sin\left[\pi x(u)\right] y'(u)\right] du = \int_0^1 2\, du = 2.$$

The integral over the entire triangle is the sum of these integrals:

$$\int_C \mathbf{h}(\mathbf{r}) \cdot \mathbf{dr} = \left(1 - e - \frac{2}{\pi}\right) + \left(1 - e - \frac{2}{\pi}\right) + 2 = 4 - 2e - \frac{4}{\pi} \cong -2.71. \quad \square$$

When we integrate over a parametrized curve, we integrate in the direction determined by the parametrization. If we integrate in the opposite direction, our answer is altered by a factor of -1. To be precise, let C be a piecewise-smooth curve and let $-C$ denote the same path traversed in the *opposite orientation*. (See Figure 17.1.4.) If C is parametrized by a vector function $\mathbf{r}$ defined on $[a, b]$, then $-C$ can be parametrized by setting

$$\mathbf{R}(w) = \mathbf{r}(a + b - w), \qquad w \in [a, b]. \qquad \text{(Section 13.3)}$$

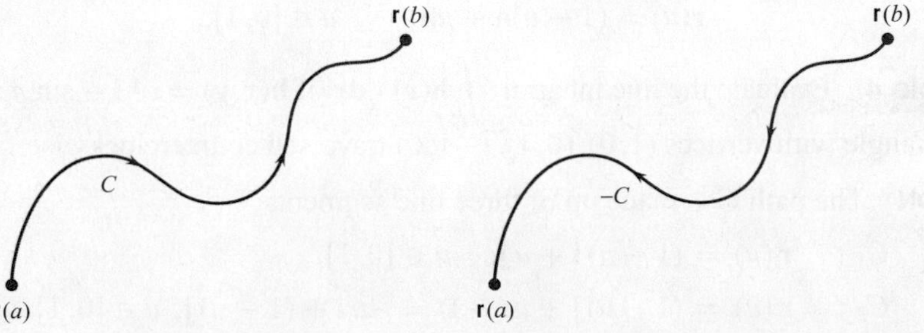

Figure 17.1.4

Our assertion is that

(17.1.6)
$$\int_{-C} \mathbf{h}(\mathbf{R}) \cdot d\mathbf{R} = -\int_{C} \mathbf{h}(\mathbf{r}) \cdot d\mathbf{r},$$

or, more briefly, that

(17.1.7)
$$\int_{-C} = -\int_{C}.$$

We leave the proof of this to you.

We were led to the definition of line integral by the notion of work. It follows from (17.1.2) that if a force $\mathbf{F}$ is continually applied to an object that moves over a piecewise-smooth curve C, then the work done by $\mathbf{F}$ is the line integral of $\mathbf{F}$ over C:

(17.1.8)
$$W = \int_{C} \mathbf{F}(\mathbf{r}) \cdot d\mathbf{r}.$$

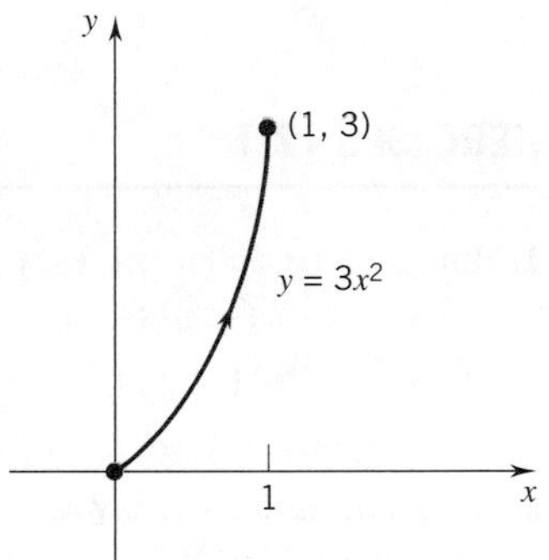

Figure 17.1.5

Example 5 An object, acted on by various forces, moves along the parabola $y = 3x^2$ from the origin to the point $(1, 3)$. (See Figure 17.1.5.) One of the forces acting on the object is $\mathbf{F}(x, y) = x^3\,\mathbf{i} + y\,\mathbf{j}$. Calculate the work done by $\mathbf{F}$.

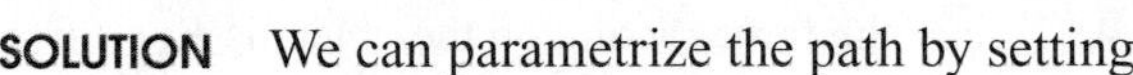
SOLUTION We can parametrize the path by setting

$$C\colon\ \mathbf{r}(u) = u\,\mathbf{i} + 3u^2\,\mathbf{j}, \quad u \in [0, 1].$$

Here
$$x(u) = u, \quad y(u) = 3u^2$$

and
$$\mathbf{F}(\mathbf{r}(u)) \cdot \mathbf{r}'(u) = [x(u)]^3 x'(u) + y(u)y'(u) = u^3(1) + 3u^2(6u) = 19u^3.$$

It follows that

$$W = \int_{C} \mathbf{F}(\mathbf{r}) \cdot d\mathbf{r} = \int_0^1 19u^3\,du = \tfrac{19}{4}. \quad \square$$

If an object of mass m moves so that at time t it has position $\mathbf{r}(t)$, then, from Newton's second law, $\mathbf{F} = m\mathbf{a}$, the total force acting on the object at time t must be $m\mathbf{r}''(t)$.

Example 6 An object of mass m moves from time $t = 0$ to $t = 1$ so that its position a time t is given by the vector function

$$\mathbf{r}(t) = \alpha t^2\,\mathbf{i} + \sin\beta t\,\mathbf{j} + \cos\beta t\,\mathbf{k}, \qquad \alpha, \beta \text{ constant.}$$

Find the total force acting on the object at time t and calculate the total work done by this force.

SOLUTION Differentiation gives

$$\mathbf{r}'(t) = 2\alpha t\,\mathbf{i} + \beta\cos\beta t\,\mathbf{j} - \beta\sin\beta t\,\mathbf{k}, \qquad \mathbf{r}''(t) = 2\alpha\,\mathbf{i} - \beta^2\sin\beta t\,\mathbf{j} - \beta^2\cos\beta t\,\mathbf{k}.$$

The total force $\mathbf{F}(t)$ on the object at time t is therefore

$$\mathbf{F}(t) = m\,\mathbf{r}''(t) = m(2\alpha\,\mathbf{i} - \beta^2 \sin \beta t\,\mathbf{j} - \beta^2 \cos \beta t\,\mathbf{k}).$$

We can calculate the total work done by this force by integrating the force over the curve

$$C: \quad \mathbf{r}(t) = \alpha t^2\,\mathbf{i} + \sin \beta t\,\mathbf{j} + \cos \beta t\,\mathbf{k}, \qquad t \in [0, 1].$$

We leave it to you to verify that

$$W = \int_0^1 [m\,\mathbf{r}''(t) \cdot \mathbf{r}'(t)]\,dt = m \int_0^1 4\alpha^2 t\,dt = 2\alpha^2 m. \quad \square$$

EXERCISES 17.1

1. Integrate $\mathbf{h}(x, y) = y\,\mathbf{i} + x\,\mathbf{j}$ over the indicated path:
   (a) $\mathbf{r}(u) = u\,\mathbf{i} + u^2\,\mathbf{j}, \quad u \in [0, 1]$.
   (b) $\mathbf{r}(u) = u^3\,\mathbf{i} - 2u\,\mathbf{j}, \quad u \in [0, 1]$.
2. Integrate $\mathbf{h}(x, y) = x\,\mathbf{i} + y\,\mathbf{j}$ over the paths of Exercise 1.
3. Integrate $\mathbf{h}(x, y) = y\,\mathbf{i} + x\,\mathbf{j}$ over the unit circle traversed clockwise.
4. Integrate $\mathbf{h}(x, y) = xy^2\,\mathbf{i} + 2\,\mathbf{j}$ over the indicated path:
   (a) $\mathbf{r}(u) = e^u\,\mathbf{i} + e^{-u}\,\mathbf{j}, \quad u \in [0, 1]$.
   (b) $\mathbf{r}(u) = (1 - u)\,\mathbf{i}, \quad u \in [0, 2]$.
5. Integrate $\mathbf{h}(x, y) = (x - y)\,\mathbf{i} + xy\,\mathbf{j}$ over the indicated path:
   (a) The line segment from (2, 3) to (1, 2).
   (b) The line segment from (1, 2) to (2, 3).
6. Integrate $\mathbf{h}(x, y) = x^{-1}y^{-2}\,\mathbf{i} + x^{-2}y^{-1}\,\mathbf{j}$ over the indicated path:
   (a) $\mathbf{r}(u) = \sqrt{u}\,\mathbf{i} + \sqrt{1 + u}\,\mathbf{j}, \quad u \in [1, 4]$.
   (b) The line segment from (1, 1) to (2, 2).
7. Integrate $\mathbf{h}(x, y) = y\,\mathbf{i} - x\,\mathbf{j}$ over the triangle with vertices $(-2, 0), (2, 0), (0, 2)$ traversed counterclockwise.
8. Integrate $\mathbf{h}(x, y) = e^{x-y}\,\mathbf{i} + e^{x+y}\,\mathbf{j}$ over the line segment from $(-1, 1)$ to $(1, 2)$.
9. Integrate $\mathbf{h}(x, y) = (x + y)\,\mathbf{i} + (y^2 - x)\,\mathbf{j}$ over the closed curve that begins at $(-1, 0)$, goes along the x-axis to (1,0), and returns to $(-1, 0)$ by the upper part of the unit circle.
10. Integrate $\mathbf{h}(x, y) = 3x^2y\,\mathbf{i} + (x^3 + 2y)\,\mathbf{j}$ over the square with vertices $(0, 0), (1, 0), (1, 1), (0, 1)$ traversed counterclockwise.
11. Integrate $\mathbf{h}(x, y, z) = yz\,\mathbf{i} + x^2\,\mathbf{j} + xz\,\mathbf{k}$ over the indicated path:
    (a) The line segment from (0, 0, 0) to (1, 1, 1).
    (b) $\mathbf{r}(u) = u\,\mathbf{i} + u^2\,\mathbf{j} + u^3\,\mathbf{k}, \quad u \in [0, 1]$.
12. Integrate $\mathbf{h}(x, y, z) = e^x\,\mathbf{i} + e^y\,\mathbf{j} + e^z\,\mathbf{k}$ over the paths of Exercise 11.
13. Integrate $\mathbf{h}(x, y, z) = \cos x\,\mathbf{i} + \sin y\,\mathbf{j} + yz\,\mathbf{k}$ over the indicated path:
    (a) The line segment from $(0, 0, 0)$ to $(2, 3, -1)$.
    (b) $\mathbf{r}(u) = u^2\,\mathbf{i} - u^3\,\mathbf{j} + u\,\mathbf{k}, \quad u \in [0, 1]$.
14. Integrate $\mathbf{h}(x, y, z) = xy\,\mathbf{i} + x^2z\,\mathbf{j} + xyz\,\mathbf{k}$ over the indicated path:
    (a) The line segment from $(0, 0, 0)$ to $(2, -1, 1)$.
    (b) $\mathbf{r}(u) = e^u\,\mathbf{i} + e^{-u}\,\mathbf{j} + u\,\mathbf{k}, \quad u \in [0, 1]$.
15. An object moves along the parabola $y = x^2$ from $(0, 0)$ to $(2, 4)$. One of the forces acting on the object is $\mathbf{F}(x, y) = (x + 2y)\,\mathbf{i} + (2x + y)\,\mathbf{j}$. Calculate the work done by $\mathbf{F}$.
16. An object moves along the polygonal path that connects $(0, 0), (1, 0), (1, 1), (0, 1)$ in the order indicated. One of the forces acting on the object is $\mathbf{F}(x, y) = x \cos y\,\mathbf{i} - y \sin x\,\mathbf{j}$. Calculate the work done by $\mathbf{F}$.
17. An object moves along the straight line from $(0, 1, 4)$ to $(1, 0, -4)$. One of the forces acting on the object is $\mathbf{F}(x, y, z) = x\,\mathbf{i} + xy\,\mathbf{j} + xyz\,\mathbf{k}$. Calculate the work done by $\mathbf{F}$.
18. An object moves along the polygonal path that connects $(0, 0, 0), (1, 0, 0), (1, 1, 0), (1, 1, 1)$ in the order indicated. One of the forces acting on the object is $\mathbf{F}(x, y, z) = yz\,\mathbf{i} + xz\,\mathbf{j} + xy\,\mathbf{k}$. Calculate the work done by $\mathbf{F}$.
19. An object moves along the circular helix $\mathbf{r}(u) = \cos u\,\mathbf{i} + \sin u\,\mathbf{j} + u\,\mathbf{k}$ from $(1, 0, 0)$ to $(1, 0, 2\pi)$. One of the forces acting on the object is $\mathbf{F}(x, y, z) = x^2\,\mathbf{i} + xy\,\mathbf{j} + z^2\,\mathbf{k}$. Calculate the work done by $\mathbf{F}$.
20. A mass m, moving in a force field, traces out a circular arc at constant speed. Show that the force field does no work. Give a physical explanation for this.
21. Let $C: \mathbf{r} = \mathbf{r}(u), u \in [a, b]$ be a smooth curve and $\mathbf{q}$ a fixed vector. Show that

$$\int_C \mathbf{q} \cdot d\mathbf{r} = \mathbf{q} \cdot [\mathbf{r}(b) - \mathbf{r}(a)] \quad \text{and}$$

$$\int_C \mathbf{r} \cdot d\mathbf{r} = \frac{\|\mathbf{r}(b)\|^2 - \|\mathbf{r}(a)\|^2}{2}.$$

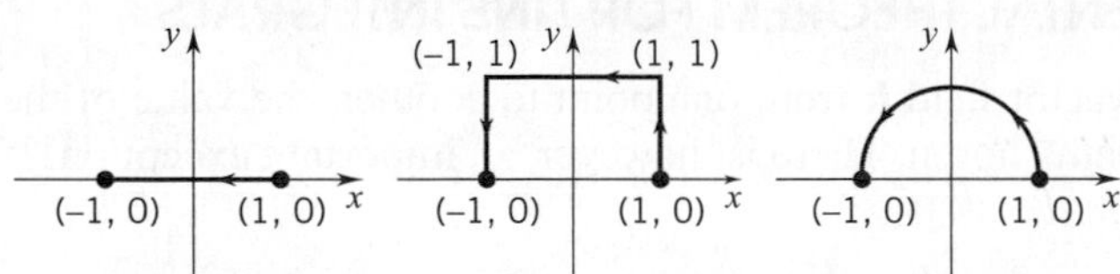

C_1: straight-line path C_2: rectangular path C_3: semicircular path

22. The preceding figure shows three paths from $(1, 0)$ to $(-1, 0)$. Calculate the line integral of

$$\mathbf{h}(x, y) = x^2\,\mathbf{i} + y\,\mathbf{j}$$

over (a) the straight-line path; (b) the rectangular path; (c) the semicircular path.

23. Let f be a continuous real-valued function of a real variable. Show that, if

$$\mathbf{f}(x, y, z) = f(x)\,\mathbf{i} \quad \text{and} \quad C\colon \mathbf{r}(u) = u\,\mathbf{i}, \quad u \in [a, b],$$

then
$$\int_C \mathbf{f}(\mathbf{r}) \cdot \mathbf{dr} = \int_a^b f(u)\,du.$$

24. (*Linearity*) Show that, if **f** and **g** are continuous vector fields and C is piecewise smooth, then

(17.1.9)
$$\int_C [\alpha\,\mathbf{f}(\mathbf{r}) + \beta\,\mathbf{g}(\mathbf{r})] \cdot \mathbf{dr} = \alpha \int_C \mathbf{f}(\mathbf{r}) \cdot \mathbf{dr} + \beta \int_C \mathbf{g}(\mathbf{r}) \cdot \mathbf{dr}$$

for all real α, β.

25. The force $\mathbf{F}(x, y) = -\frac{1}{2}[y\,\mathbf{i} - x\,\mathbf{j}]$ is continually applied to an object that orbits an ellipse in standard position. Find a relation between the work done during each orbit and the area of the ellipse.

26. An object of mass m moves from time $t = 0$ to $t = 1$ so that its position at time t is given by the vector function

$$\mathbf{r}(t) = \alpha t\,\mathbf{i} + \beta t^2\,\mathbf{j}, \quad \alpha, \beta \quad \text{constant}.$$

Find the total force acting on the object at time t and calculate the work done by that force during the time interval $[0, 1]$.

27. Repeat Exercise 26 with $\mathbf{r}(t) = \alpha t\,\mathbf{i} + \beta t^2\,\mathbf{j} + \gamma t^3\,\mathbf{k}$.

28. (*Important*) The *circulation* of a vector field **v** around an oriented closed curve C is by definition the line integral

$$\int_C \mathbf{v}(\mathbf{r}) \cdot \mathbf{dr}.$$

Let **v** be the velocity field of a fluid in counterclockwise circular motion about the z-axis with constant angular speed ω.

(a) Verify that $\mathbf{v}(\mathbf{r}) = \omega\,\mathbf{k} \times \mathbf{r}$.

(b) Show that the circulation of **v** around any circle C in the xy-plane with center at the origin is $\pm 2\omega$ times the area of the circle.

29. Let **v** be the velocity field of a fluid that moves radially from the origin: $\mathbf{v} = f(x, y)\,\mathbf{r}$. What is the circulation of **v** around a circle C centered at the origin?

30. The force exerted on a charged particle at the point $(x, y) \neq (0, 0)$ in the xy-plane by an infinitely long uniformly charged wire lying along the z-axis is given by

$$\mathbf{F}(x, y) = \frac{c(x\,\mathbf{i} + y\,\mathbf{j})}{x^2 + y^2}, \qquad c > 0, \quad \text{constant}.$$

Find the work done by **F** in moving the particle along the indicated path:

(a) The line segment from $(1, 0)$ to $(1, 2)$.

(b) The line segment from $(0, 1)$ to $(1, 1)$.

31. An inverse-square force field is given by

$$\mathbf{F}(x, y, z) = \frac{c\,\mathbf{r}}{||\mathbf{r}||^3} = \frac{c(x\,\mathbf{i} + y\,\mathbf{j} + z\,\mathbf{k})}{(x^2 + y^2 + z^2)^{3/2}}, \qquad c > 0, \ \text{constant}.$$

Find the work done by **F** in moving an object along the indicated path:

(a) The line segment from $(1, 0, 2)$ to $(1, 3, 2)$.

(b) From $(1, 0, 0)$ to $(0, \frac{5}{2}\sqrt{2}, \frac{5}{2}\sqrt{2})$ by the line segment from $(1, 0, 0)$ to $(5, 0, 0)$ and then to $(0, \frac{5}{2}\sqrt{2}, \frac{5}{2}\sqrt{2})$ along a path on the surface of the sphere $x^2 + y^2 + z^2 = 25$.

32. Suppose that $\mathbf{F}(\mathbf{r}) = c\,\mathbf{r}/||\mathbf{r}||^3$ is an inverse-square force field, where $\mathbf{r} = x\,\mathbf{i} + y\,\mathbf{j} + z\,\mathbf{k}$. Express the work done by **F** in moving an object from a point P_0 to a point P_1 in terms of the distances of these points to the origin.

33. An object moves along the curve $y = \alpha x(1 - x)$ from $(0, 0)$ to $(1, 0)$. One of the forces acting on the object is $\mathbf{F}(x, y) = (y^2 + 1)\,\mathbf{i} + (x + y)\,\mathbf{j}$. What value of α minimizes the work done by **F**?

34. Assume that the earth is located at the origin of a rectangular coordinate system. The gravitational force on an object at the point (x, y) is given by

$$\mathbf{F}(x, y) = \frac{-c\,\mathbf{r}}{||\mathbf{r}||^3} = \frac{-c(x\,\mathbf{i} + y\,\mathbf{j})}{(x^2 + y^2)^{3/2}}, \qquad c > 0, \quad \text{constant}.$$

Find the work done by **F** in moving an object from $(3, 0)$ to $(0, 4,)$ along the indicated path:

(a) The first quadrant part of the ellipse $\mathbf{r}(u) = 3\cos u\,\mathbf{i} + 4\sin u\,\mathbf{j}, \quad u \in [0, \pi/2]$.

(b) The line segment connecting the two points.

■ 17.2 THE FUNDAMENTAL THEOREM FOR LINE INTEGRALS

In general, if we integrate a vector field **h** from one point to another, the value of the line integral depends on the path chosen. There is, however, an important exception. If the vector field is a *gradient field*,

$$\mathbf{h} = \nabla f,$$

then the value of the line integral depends only on the endpoints of the path and not on the path itself. The details are spelled out in the following theorem.

THEOREM 17.2.1 THE FUNDAMENTAL THEOREM FOR LINE INTEGRALS

Let $C\colon \mathbf{r} = \mathbf{r}(u)$, $u \in [a, b]$, be a piecewise-smooth curve that begins at $\mathbf{a} = \mathbf{r}(a)$ and ends at $\mathbf{b} = \mathbf{r}(b)$. If the scalar field f is continuously differentiable on an open set that contains the curve C, then

$$\int_C \nabla f(\mathbf{r}) \cdot \mathbf{dr} = f(\mathbf{b}) - f(\mathbf{a}).$$

PROOF If C is smooth,

$$\begin{aligned}\int_C \nabla f(\mathbf{r}) \cdot \mathbf{dr} &= \int_a^b [\nabla f(\mathbf{r}(u)) \cdot \mathbf{r}'(u)]\, du \\ &= \int_a^b \frac{d}{du}[f(\mathbf{r}(u))]\, du \quad \text{(chain rule (15.3.4))} \\ &= f(\mathbf{r}(b)) - f(\mathbf{r}(a)) \\ &= f(\mathbf{b}) - f(\mathbf{a}).\end{aligned}$$

If C is not smooth but only piecewise smooth, then we break up C into smooth pieces

$$C = C_1 \cup C_2 \cup \cdots \cup C_n.$$

With obvious notation,

$$\begin{aligned}\int_C \nabla f(\mathbf{r}) \cdot \mathbf{dr} &= \int_{C_1} \nabla f(\mathbf{r}) \cdot \mathbf{dr} + \int_{C_2} \nabla f(\mathbf{r}) \cdot \mathbf{dr} + \cdots + \int_{C_n} \nabla f(\mathbf{r}) \cdot \mathbf{dr} \\ &= [f(\mathbf{a}_1) - f(\mathbf{a}_0)] + [f(\mathbf{a}_2) - f(\mathbf{a}_1)] + \cdots + [f(\mathbf{a}_n) - f(\mathbf{a}_{n-1})] \\ &= f(\mathbf{a}_n) - f(\mathbf{a}_0) \\ &= f(\mathbf{b}) - f(\mathbf{a}). \quad \square\end{aligned}$$

The theorem we just proved has an important corollary:

(17.2.2) If the curve C is closed [that is, if $\mathbf{b} = a$], then

$$\int_C \nabla f(\mathbf{r}) \cdot \mathbf{dr} = 0.$$

Example 1 Integrate the vector field $\mathbf{h}(x, y) = y^2\,\mathbf{i} + (2xy - e^{2y})\,\mathbf{j}$ over the circular arc

$$C\colon \quad \mathbf{r}(u) = \cos u\,\mathbf{i} + \sin u\,\mathbf{j}, \qquad u \in [0, \tfrac{1}{2}\pi].$$

SOLUTION First we try to determine whether **h** is a gradient. We do this by applying Theorem 15.9.2.

Note that $\mathbf{h}(x, y)$ has the form $P(x, y)\,\mathbf{i} + Q(x, y)\,\mathbf{j}$ with

$$P(x, y) = y^2 \quad \text{and} \quad Q(x, y) = 2xy - e^{2y}.$$

Since P and Q are continuously differentiable everywhere and

$$\frac{\partial P}{\partial y} = 2y = \frac{\partial Q}{\partial x},$$

we can conclude that **h** is a gradient. Therefore, since the integral depends only on the endpoints of C and not on C itself, we can simplify the computations by integrating over the line segment C' that joins these same endpoints. (See Figure 17.2.1.)

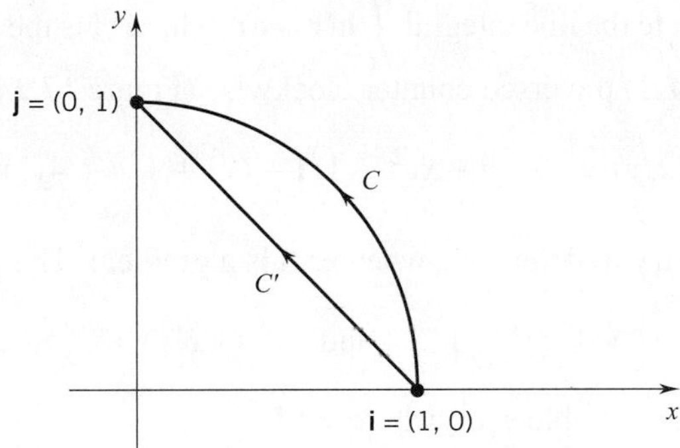

Figure 17.2.1

We parametrize C' by setting

$$\mathbf{r}(u) = (1 - u)\,\mathbf{i} + u\,\mathbf{j}, \quad u \in [0, 1].$$

We then have

$$\begin{aligned}
\int_C \mathbf{h}(\mathbf{r}) \cdot \mathbf{dr} &= \int_{C'} \mathbf{h}(\mathbf{r}) \cdot \mathbf{dr} \\
&= \int_0^1 [\mathbf{h}(\mathbf{r}(u)) \cdot \mathbf{r}'(u)]\, du \\
&= \int_0^1 \Big[[\,y(u)]^2 x'(u) + [2x(u)y(u) - e^{2y(u)}]y'(u)\Big]\, du \\
&= \int_0^1 \Big[u^2(-1) + [2(1 - u)u - e^{2u}](1)\Big]\, du \\
&= \int_0^1 [2u - 3u^2 - e^{2u}]\, du = \Big[u^2 - u^3 - \tfrac{1}{2}e^{2u}\Big]_0^1 \\
&= \tfrac{1}{2} - \tfrac{1}{2}e^2.
\end{aligned}$$

ALTERNATIVE SOLUTION Once we recognize that $\mathbf{h}(x, y) = y^2\,\mathbf{i} + (2xy - e^{2y})\,\mathbf{j}$ is a gradient ∇f, we can try to determine $f(x, y)$ by the methods of Section 15.9. Since

$$\frac{\partial f}{\partial x} = y^2 \quad \text{and} \quad \frac{\partial f}{\partial y} = 2xy - e^{2y},$$

we have $\quad f(x,y) = xy^2 + \phi(y) \quad$ and therefore $\quad \dfrac{\partial f}{\partial y} = 2xy + \phi'(y).$

The two expressions for $\partial f/\partial y$ can be reconciled only if

$$\phi'(y) = -e^{2y} \quad \text{and thus} \quad \phi(y) = -\tfrac{1}{2}e^{2y} + K \qquad (K \text{ an arbitrary constant}).$$

Each function

$$f(x,y) = xy^2 - \tfrac{1}{2}e^{2y} + K$$

has gradient **h**. No matter what value we assign to K,

$$\int_C \mathbf{h}(\mathbf{r}) \cdot \mathbf{dr} = f(0,1) - f(1,0) = (-\tfrac{1}{2}e^2) - (-\tfrac{1}{2}) = \tfrac{1}{2} - \tfrac{1}{2}e^2. \quad \square$$

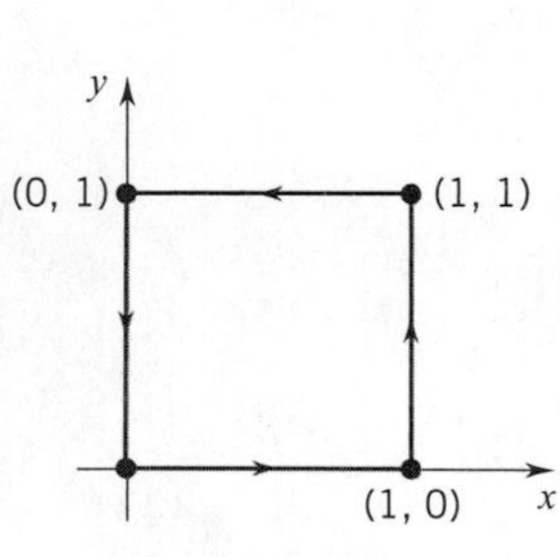

Figure 17.2.2

Example 2 Evaluate the line integral $\int_C \mathbf{h}(\mathbf{r}) \cdot \mathbf{dr}$, where C is the square with vertices $(0,0), (1,0), (1,1), (0,1)$ traversed counter clockwise (Figure 17.2.2), and

$$\mathbf{h}(x,y) = (3x^2y + xy^2 - 1)\,\mathbf{i} + (x^3 + x^2y + 4y^3)\,\mathbf{j}.$$

SOLUTION First we try to determine whether **h** is a gradient. The functions

$$P(x,y) = 3x^2y + xy^2 - 1 \quad \text{and} \quad Q(x,y) = x^3 + x^2y + 4y^3$$

are continuously differentiable everywhere, and

$$\frac{\partial P}{\partial y} = 3x^2 + 2xy = \frac{\partial Q}{\partial x}.$$

Therefore **h** is the gradient of a function f. By (17.2.2),

$$\int_C \mathbf{h}(\mathbf{r}) \cdot \mathbf{dr} = \int_C \nabla f(\mathbf{r}) \cdot \mathbf{dr} = 0. \quad \square$$

Example 3 Evaluate the line integral $\int_C \mathbf{h}(\mathbf{r}) \cdot \mathbf{d\,r}$, where C is the unit circle

$$C\colon\ \mathbf{r}(u) = \cos u\,\mathbf{i} + \sin u\,\mathbf{j}, \quad u \in [0, 2\pi]$$

and

$$\mathbf{h}(x,y) = (y^2 + y)\,\mathbf{i} + (2xy - e^{2y})\,\mathbf{j}.$$

SOLUTION Although $(y^2 + y)\,\mathbf{i} + (2xy - e^{2y})\,\mathbf{j}$ is not a gradient $[\partial P/\partial y \neq \partial Q/\partial x]$, part of it,

$$y^2\,\mathbf{i} + (2xy - e^{2y})\,\mathbf{j},$$

is a gradient, as shown in Example 1. Therefore, we can write **h** as

$$\begin{aligned} \mathbf{h}(x,y) = (y^2 + y)\,\mathbf{i} + (2xy - e^{2y})\,\mathbf{j} &= [y^2\,\mathbf{i} + (2xy - e^{2y})\,\mathbf{j}] + y\,\mathbf{i} \\ &= \nabla f(x,y) + \mathbf{g}(x,y) \end{aligned}$$

where $\mathbf{g}(x,y) = y\,\mathbf{i}$. Now

$$\int_C \mathbf{h}(\mathbf{r}) \cdot \mathbf{dr} = \int_C \nabla f(\mathbf{r}) \cdot \mathbf{dr} + \int_C \mathbf{g}(\mathbf{r}) \cdot \mathbf{dr}.$$

Since we are integrating over a closed curve, the contribution of the gradient part is 0. The contribution of the remaining part is

$$\int_C \mathbf{g}(\mathbf{r}) \cdot d\mathbf{r} = \int_0^{2\pi} [\mathbf{g}(\mathbf{r}(u)) \cdot \mathbf{r}'(u)]\, du = \int_0^{2\pi} y(u)\, x'(u)\, du$$

$$= \int_0^{2\pi} -\sin^2 u\, du = -\pi.$$

Therefore $\int_C \mathbf{h}(\mathbf{r}) \cdot d\mathbf{r} = -\pi.$ □

EXERCISES 17.2

Determine whether $\mathbf{h}$ is a gradient, then calculate the line integral of $\mathbf{h}$ over the indicated curve.

1. $\mathbf{h}(x,y) = x\,\mathbf{i} + y\,\mathbf{j}$; $\mathbf{r}(u) = a\cos u\,\mathbf{i} + b\sin u\,\mathbf{j}$, $u \in [0, 2\pi]$.

2. $\mathbf{h}(x,y) = (x+y)\,\mathbf{i} + y\,\mathbf{j}$; the curve of Exercise 1.

3. $\mathbf{h}(x,y) = \cos\pi y\,\mathbf{i} - \pi x\sin\pi y\,\mathbf{j}$; $\mathbf{r}(u) = u^2\,\mathbf{i} - u^3\,\mathbf{j}$, $u \in [0, 1]$.

4. $\mathbf{h}(x,y) = (x^2 - y)\,\mathbf{i} + (y^2 - x)\,\mathbf{j}$; the curve of Exercise 1.

5. $\mathbf{h}(x,y) = xy^2\,\mathbf{i} + x^2y\,\mathbf{j}$; $\mathbf{r}(u) = u\sin\pi u\,\mathbf{i} + \cos\pi u^2\,\mathbf{j}$, $u \in [0, 1]$.

6. $\mathbf{h}(x,y) = (1+e^y)\,\mathbf{i} + (x e^y - x)\,\mathbf{j}$; the square with vertices $(-1,-1), (1,-1), (1,1), (-1,1)$ traversed counterclockwise.

7. $\mathbf{h}(x,y) = (2xy - y^2)\,\mathbf{i} + (x^2 - 2xy)\,\mathbf{j}$; $\mathbf{r}(u) = \cos u\,\mathbf{i} + \sin u\,\mathbf{j}$, $u \in [0, \pi]$.

8. $\mathbf{h}(x,y) = 3x(x^2+y^4)^{1/2}\,\mathbf{i} + 6y^3(x^2+y^4)^{1/2}\,\mathbf{j}$; the circular arc $y = (1-x^2)^{1/2}$ from $(1,0)$ to $(-1,0)$.

9. $\mathbf{h}(x,y) = 3x(x^2+y^4)^{1/2}\,\mathbf{i} + 6y^3(x^2+y^4)^{1/2}\,\mathbf{j}$; the arc $y = -(1-x^2)^{1/2}$ from $(-1,0)$ to $(1,0)$.

10. $\mathbf{h}(x,y) = 2xy\sinh x^2y\,\mathbf{i} + x^2\sinh x^2y\,\mathbf{j}$; the curve of Exercise 1.

11. $\mathbf{h}(x,y) = (2x\cosh y - y)\,\mathbf{i} + (x^2\sinh y - y)\,\mathbf{j}$; the square of Exercise 6.

Verify that the given vector field $\mathbf{h}$ is a gradient. Then calculate the line integral of $\mathbf{h}$ over the indicated curve C in two ways: (a) by calculating $\int_C \mathbf{h}(\mathbf{r}) \cdot d\mathbf{r}$ and (b) by finding f such that $\nabla f = \mathbf{h}$ and evaluating f at the endpoints of C.

12. $\mathbf{h}(x,y) = xy^2\,\mathbf{i} + yx^2\,\mathbf{j}$; $\mathbf{r}(u) = u\,\mathbf{i} + u^2\,\mathbf{j}$, $u \in [0, 2]$.

13. $\mathbf{h}(x,y) = (3x^2y^3 + 2x)\,\mathbf{i} + (3x^3y^2 - 4y)\,\mathbf{j}$; $\mathbf{r}(u) = u\,\mathbf{i} + e^u\,\mathbf{j}$, $u \in [0, 1]$.

14. $\mathbf{h}(x,y) = (2x\sin y - e^x)\,\mathbf{i} + (x^2\cos y)\,\mathbf{j}$; $\mathbf{r}(u) = \cos u\,\mathbf{i} + u\,\mathbf{j}$, $u \in [0, \pi]$.

15. $\mathbf{h}(x,y) = (e^{2y} - 2xy)\,\mathbf{i} + (2x\,e^{2y} - x^2 + 1)\,\mathbf{j}$; $\mathbf{r}(u) = u e^u\,\mathbf{i} + (1+u)\,\mathbf{j}$, $u \in [0, 1]$.

Use the three-dimensional analog of Theorem 15.9.2 given in Exercises 15.9 to show that the vector function $\mathbf{h}$ is a gradient. Then evaluate the line integral of $\mathbf{h}$ over the indicated curve.

16. $\mathbf{h}(x,y,z) = y^2z^3\,\mathbf{i} + 2xyz^3\,\mathbf{j} + 3xy^2z^2\mathbf{k}$; $\mathbf{r}(u) = u^2\,\mathbf{i} + u^4\,\mathbf{j} + u^6\,\mathbf{k}$, $u \in [0, 1]$.

17. $\mathbf{h}(x,y,z) = (2xz + \sin y)\,\mathbf{i} + x\cos y\,\mathbf{j} + x^2\,\mathbf{k}$; $\mathbf{r}(u) = \cos u\,\mathbf{i} + \sin u\,\mathbf{j} + u\,\mathbf{k}$, $u \in [0, 2\pi]$.

18. $\mathbf{h}(x,y,z) = \pi yz\cos\pi x\,\mathbf{i} + z\sin\pi x\,\mathbf{j} + y\sin\pi x\,\mathbf{k}$; $\mathbf{r}(u) = \cos u\,\mathbf{i} + \sin u\,\mathbf{j} + u\,\mathbf{k}$, $u \in [0, \pi/3]$.

19. $\mathbf{h}(x,y,z) = (2xy + z^2)\,\mathbf{i} + x^2\,\mathbf{j} + 2xz\,\mathbf{k}$; $\mathbf{r}(u) = 2u\,\mathbf{i} + (u^2+2)\,\mathbf{j} - u\,\mathbf{k}$, $u \in [0, 1]$.

20. $\mathbf{h}(x,y,z) = e^{-x}\ln y\,\mathbf{i} - \dfrac{e^{-x}}{y}\,\mathbf{j} + 3z^2\,\mathbf{k}$; $\mathbf{r}(u) = (u+1)\,\mathbf{i} + e^{2u}\,\mathbf{j} + (u^2+1)\,\mathbf{k}$, $u \in [0, 1]$.

21. Calculate the work done by the force $\mathbf{F}(x,y) = (x + e^{2y})\,\mathbf{i} + (2y + 2x\,e^{2y})\,\mathbf{j}$ applied to an object that traverses the curve $\mathbf{r}(u) = 3\cos u\,\mathbf{i} + 4\sin u\,\mathbf{j}, u \in [0, 2\pi]$.

22. Calculate the work done by the force $\mathbf{F}(x,y,z) = (2x\ln y - yz)\,\mathbf{i} + [(x^2/y) - xz]\,\mathbf{j} - xy\,\mathbf{k}$ applied to an object that moves from the point $(1,2,1)$ to the point $(3,2,2)$.

23. If g is a continuously differentiable real-valued function defined on $[a,b]$, then by the fundamental theorem of integral calculus

$$\int_a^b g'(u)\, du = g(b) - g(a).$$

Show how this result is included in Theorem 17.2.1.

24. Let $\mathbf{r} = x\,\mathbf{i} + y\,\mathbf{j} + z\,\mathbf{k}$ and set $r = \|\mathbf{r}\|$. The central force field

$$\mathbf{F}(\mathbf{r}) = \frac{K}{r^n}\mathbf{r}, \qquad n \text{ a positive integer}$$

is a gradient field. Find f such that $\nabla f(\mathbf{r}) = \mathbf{F}(\mathbf{r})$ if: (a) $n = 2$; (b) $n \neq 2$.

25. Let $\mathbf{r} = x\,\mathbf{i} + y\,\mathbf{j} + z\,\mathbf{k}$ and set $r = \|\mathbf{r}\|$. Suppose that $\mathbf{F}$ is a vector field that is directed away from the origin with magnitude proportional to the square of the distance to the origin.

(a) Show that $\mathbf{F}(\mathbf{r}) = cr(x\,\mathbf{i} + y\,\mathbf{j} + z\,\mathbf{k}), c > 0$, constant.

(b) Show that $\mathbf{F}$ is a gradient field and find f such that $\mathbf{F}(\mathbf{r}) = \nabla f(\mathbf{r})$.

26. Let $\mathbf{r} = x\,\mathbf{i} + y\,\mathbf{j} + z\,\mathbf{k}$ and set $r = \|\mathbf{r}\|$. Suppose that $\mathbf{F}(\mathbf{r}) = g(r^2)\,\mathbf{r}$ where g is a continuous, real-valued function defined on $[0, \infty)$.

(a) Show that $\mathbf{F}$ is a gradient field

(b) Find f such that $\mathbf{F}(\mathbf{r}) = \nabla f(\mathbf{r})$.

27. Let $\mathbf{r} = x\,\mathbf{i} + y\,\mathbf{j} + z\,\mathbf{k}$ and set $r = \|\mathbf{r}\|$. The function

$$\mathbf{F}(\mathbf{r}) = -\frac{mG}{r^3}\,\mathbf{r} \quad (G \text{ is the gravitational constant})$$

gives the gravitational force exerted by a unit mass at the origin on a mass m located at $\mathbf{r}$. What is the work done by $\mathbf{F}$ if m moves from $\mathbf{r}_1$ to $\mathbf{r}_2$?

28. Set

$$P(x, y) = \frac{y}{x^2 + y^2} \quad \text{and} \quad Q(x, y) = -\frac{x}{x^2 + y^2}$$

on *the punctured unit disc* $\Omega : 0 < x^2 + y^2 < 1$.

(a) Verify that P and Q are continuously differentiable on Ω and that

$$\frac{\partial P}{\partial y}(x, y) = \frac{\partial Q}{\partial x}(x, y) \quad \text{for all } (x, y) \in \Omega.$$

(b) Verify that, in spite of (a), the vector field $\mathbf{h}(x, y) = P(x, y)\,\mathbf{i} + Q(x, y)\,\mathbf{j}$ is not a gradient on Ω. HINT: Integrate $\mathbf{h}$ over a circle of radius less than 1 centered at the origin.

(c) Show that part (b) does not contradict Theorem 15.9.2.

29. The gravitational force acting on an object of mass m at a height z above the surface of the earth is given by

$$\mathbf{F}(x, y, z) = \frac{-mG\,r_0^2}{(r_0 + z)^2}\,\mathbf{k},$$

where G is the gravitational constant and r_0 is the radius of the earth. Show that $\mathbf{F}$ is a gradient field and find f such that $\nabla f = \mathbf{F}$.

30. A rocket of mass m falls to the earth from a height of 300 miles. How much work is done by the gravitational force? Use Exercise 29 and assume that the radius of the earth is 4000 miles.

■ 17.3 WORK–ENERGY FORMULA; CONSERVATION OF MECHANICAL ENERGY

Suppose that a continuous force field $\mathbf{F} = \mathbf{F}(\mathbf{r})$ accelerates a mass m from $\mathbf{r} = \mathbf{a}$ to $\mathbf{r} = \mathbf{b}$ along some smooth curve C. The object undergoes a change in kinetic energy:

$$\tfrac{1}{2}m[v(\beta)]^2 - \tfrac{1}{2}m[v(\alpha)]^2.$$

The force does a certain amount of work W. How are these quantities related? They are equal:

(17.3.1)
$$\boxed{W = \tfrac{1}{2}m[v(\beta)]^2 - \tfrac{1}{2}m[v(\alpha)]^2.}$$

This relation is called the *work–energy formula*.

DERIVATION OF THE WORK-ENERGY FORMULA We parametrize the path of the motion by the time parameter t:

$$C: \quad \mathbf{r} = \mathbf{r}(t), \quad t \in [\alpha, \beta]$$

where $\mathbf{r}(\alpha) = \mathbf{a}$ and $\mathbf{r}(\beta) = \mathbf{b}$. The work done by $\mathbf{F}$ is given by the formula

$$W = \int_C \mathbf{F}(\mathbf{r}) \cdot d\mathbf{r} = \int_\alpha^\beta [\mathbf{F}(\mathbf{r}(t)) \cdot \mathbf{r}'(t)]\,dt.$$

From Newton's second law of motion, we know that at time t,

$$\mathbf{F}(\mathbf{r}(t)) = m\mathbf{a}(t) = m\mathbf{r}''(t).$$

It follows that

$$\begin{aligned}\mathbf{F}(\mathbf{r}(t)) \cdot \mathbf{r}'(t) &= m\,\mathbf{r}''(t) \cdot \mathbf{r}'(t) \\ &= \frac{d}{dt}\left[\frac{1}{2}m[\mathbf{r}'(t) \cdot \mathbf{r}'(t)]\right] = \frac{d}{dt}\left[\frac{1}{2}m\|\mathbf{r}'(t)\|^2\right] = \frac{d}{dt}\left[\frac{1}{2}m[v(t)]^2\right].\end{aligned}$$

Substituting this last expression into the work integral, we see that

$$W = \int_{\alpha}^{\beta} \frac{d}{dt}\left(\frac{1}{2}m[v(t)]^2\right) dt = \frac{1}{2}m[v(\beta)]^2 - \frac{1}{2}m[v(\alpha)]^2$$

as asserted. □

Conservative Force Fields

In general, if an object moves from one point to another, the work done (and hence the change in kinetic energy) depends on the path of the motion. There is, however, an important exception: if the force field is a gradient field,

$$\mathbf{F} = \nabla f,$$

then the work done (and hence the change in kinetic energy) depends only on the endpoints of the path and not on the path itself. (This follows directly from the fundamental theorem for line integrals.) A force field that is a gradient field is called a *conservative field.*

Since the line integral over a closed path is zero, *the work done by a conservative field over a closed path is always zero. An object that passes through a given point with a certain kinetic energy returns to that same point with exactly the same kinetic energy.*

Potential Energy Functions

Suppose that $\mathbf{F}$ is a conservative force field. It is then a gradient field. Then $-\mathbf{F}$ is also a gradient field. The functions U for which $\nabla U = -\mathbf{F}$ are called *potential energy functions* for $\mathbf{F}$.

The Conservation of Mechanical Energy

Suppose that $\mathbf{F}$ is a conservative force field: $\mathbf{F} = -\nabla U$. In our derivation of the work–energy formula we showed that

$$\frac{d}{dt}\left(\tfrac{1}{2}m[v(t)]^2\right) = \mathbf{F}(\mathbf{r}(t)) \cdot \mathbf{r}'(t).$$

Since
$$\frac{d}{dt}[U(\mathbf{r}(t))] = \nabla U(\mathbf{r}(t)) \cdot \mathbf{r}'(t) = -\mathbf{F}(\mathbf{r}(t)) \cdot \mathbf{r}'(t),$$

we have
$$\frac{d}{dt}[\tfrac{1}{2}m[v(t)]^2 + U(\mathbf{r}(t))] = 0,$$

and therefore

$$\boxed{\underbrace{\tfrac{1}{2}m[v(t)]^2}_{\text{KE}} + \underbrace{U(\mathbf{r}(t))}_{\text{PE}} = \text{a constant.}}$$

As an object moves in a conservative force field, its kinetic energy can vary and its potential energy can vary, but the sum of these two quantities remains constant. We call this constant *the total mechanical energy.*

The total mechanical energy is usually denoted by the letter E. The law of conservation of mechanical energy can then be written

(17.3.2) $$\boxed{\tfrac{1}{2}mv^2 + U = E.}$$

The conservation of energy is one of the cornerstones of physics. Here we have been talking about mechanical energy. There are other forms of energy and other energy conservation laws.

Differences in Potential Energy

Potential energy at a particular point has no physical significance. Only differences in potential energy are significant:

$$U(\mathbf{b}) - U(\mathbf{a}) = \int_C -\mathbf{F}(\mathbf{r}) \cdot \mathbf{dr}$$

is the work required to move from $\mathbf{r} = \mathbf{a}$ to $\mathbf{r} = \mathbf{b}$ *against* the force field $\mathbf{F}$.

Example 1 A planet moves in the gravitational field of the sun,

$$\mathbf{F}(\mathbf{r}) = -\rho m \frac{\mathbf{r}}{r^3}$$

where ρ is a constant and m is the mass of the planet. Show that the force field is conservative, find a potential energy function, and determine the total energy of the planet. How does the planet's speed vary with the planet's distance from the sun?

SOLUTION The field is conservative since

$$\mathbf{F}(\mathbf{r}) = -\rho m \frac{\mathbf{r}}{r^3} = \nabla\left(\frac{\rho m}{r}\right). \qquad \text{(check this out)}$$

As a potential energy function we can use

$$U(\mathbf{r}) = -\frac{\rho m}{r}.$$

The total energy of the planet is the constant

$$E = \tfrac{1}{2}mv^2 - \frac{\rho m}{r}.$$

(You met this quantity before: Exercises 2 and 4 of Section 13.7.)

Solving the energy equation for v, we have

$$v = \sqrt{\frac{2E}{m} + \frac{2\rho}{r}}.$$

As r decreases, $2\rho/r$ increases, and v increases; as r increases, $2\rho/r$ decreases, and v decreases. Thus every planet speeds up as it comes near the sun and slows down as it moves away. The same holds true for Halley's comet. The fact that it slows down as it gets farther away helps explain why it comes back. The simplicity of all this is a testimony to the power of the principle of energy conservation. ☐

EXERCISES 17.3

1. Let f be a continuous real-valued function of the real variable x. Show that the force field $\mathbf{F}(x, y, z) = f(x)\mathbf{i}$ is conservative and the potential functions for $\mathbf{F}$ are (except for notation) the antiderivatives of $-f$.

2. A particle with electric charge e and velocity $\mathbf{v}$ moves in a magnetic field $\mathbf{B}$, experiencing the force

$$\mathbf{F} = \frac{e}{c}[\mathbf{v} \times \mathbf{B}]. \qquad (c \text{ is the velocity of light})$$

$\mathbf{F}$ is not a gradient — it can't be, depending as it does on the *velocity* of the particle. Still, we can find a conserved quantity; the *kinetic energy* $\frac{1}{2}mv^2$. Show by differentiation with respect to t that this quantity is constant. (Assume Newton's second law.)

3. An object is subject to a constant force in the direction of $-\mathbf{k}$: $\mathbf{F} = -c\mathbf{k}$ with $c > 0$. Find a potential energy function for $\mathbf{F}$, and use energy conservation to show that the speed of the object at time t_2 is related to that at time t_1 by the equation

$$v(t_2) = \sqrt{[v(t_1)]^2 + \frac{2c}{m}[z(t_1) - z(t_2)]}$$

where $z(t_1)$ and $z(t_2)$ are the z-coordinates of the object at times t_1 and t_2. (This analysis is sometimes used to model the behavior of an object in the gravitational field near the surface of the earth.)

4. (*Escape velocity*) An object is to be fired straight up from the surface of the earth. Assume that the only force acting on the object is the gravitational pull of the earth and determine the initial speed v_0 necessary to send the object off to infinity.

HINT: Appeal to conservation of energy and use the idea that the object is to arrive at infinity with zero speed.

5. (a) Justify the statement that a conservative force field $\mathbf{F}$ always acts so as to encourage motion toward regions of lower potential energy U.

(b) Evaluate $\mathbf{F}$ at a point where U has a minimum.

6. A harmonic oscillator has a restoring force $\mathbf{F} = -\lambda x\,\mathbf{i}$. The associated potential is $U(x, y, z) = -\frac{1}{2}\lambda x^2$, and the constant total energy is

$$E = \tfrac{1}{2}mv^2 + U(x, y, z) = \tfrac{1}{2}mv^2 + \tfrac{1}{2}\lambda x^2.$$

Given that $x(0) = 2$ and $x'(0) = 1$, calculate the maximum speed of the oscillator and the maximum value of x.

7. The *equipotential surfaces* of a conservative field $\mathbf{F}$ are the surfaces where the potential energy is constant. Show that: (a) the speed of an object in such a field is constant on every equipotential surface; and (b) at each point of such a surface the force field is perpendicular to the surface.

8. Suppose a force field $\mathbf{F}$ is directed away from the origin with a magnitude that is inversely proportional to the distance from the origin. Show that $\mathbf{F}$ is a conservative field.

9. Let $\mathbf{F}$ be the inverse-square force field:

$$\mathbf{F}(x, y, z) = \frac{k(x\,\mathbf{i} + y\,\mathbf{j} + z\,\mathbf{k})}{(x^2 + y^2 + z^2)^{3/2}}$$

and let C be any curve on the unit sphere $x^2 + y^2 + z^2 = 1$. Show that the work done by $\mathbf{F}$ in moving an object along C is 0. Explain this result.

■ 17.4 ANOTHER NOTATION FOR LINE INTEGRALS; LINE INTEGRALS WITH RESPECT TO ARC LENGTH

If $\mathbf{h}(x, y, z) = P(x, y, z)\,\mathbf{i} + Q(x, y, z)\,\mathbf{j} + R(x, y, z)\,\mathbf{k}$, the line integral

$$\int_C \mathbf{h}(\mathbf{r}) \cdot d\mathbf{r} \qquad \text{can be written} \qquad \int_C P(x, y, z)\,dx + Q(x, y, z)\,dy + R(x, y, z)\,dz.$$

The notation arises as follows. With

$$C: \quad \mathbf{r}(u) = x(u)\,\mathbf{i} + y(u)\,\mathbf{j} + z(u)\,\mathbf{k}, \quad u \in [a, b],$$

the line integral

$$\int_C \mathbf{h}(\mathbf{r}) \cdot d\mathbf{r} = \int_a^b [\mathbf{h}(\mathbf{r}(u)) \cdot \mathbf{r}'(u)]\,du$$

expands to

$$\int_a^b \{P[x(u), y(u), z(u)]x'(u) + Q[x(u), y(u), z(u)]\,y'(u) + R[x(u), y(u), z(u)]z'(u)\}\,du.$$

Now set $$\int_C P(x,y,z)\,dx = \int_a^b P[x(u),y(u),z(u)]\,x'(u)\,du,$$

$$\int_C Q(x,y,z)\,dy = \int_a^b Q[x(u),y(u),z(u)]\,y'(u)\,du,$$

$$\int_C R(x,y,z)\,dz = \int_a^b R[x(u),y(u),z(u)]\,z'(u)\,du.$$

Writing the sum of these integrals as

$$\int_C P(x,y,z)\,dx + Q(x,y,z)\,dy + R(x,y,z)\,dz,$$

we have $$\int_C P(x,y,z)\,dx + Q(x,y,z)\,dy + R(x,y,z)\,dz = \int_C \mathbf{h}(\mathbf{r})\cdot d\mathbf{r}.$$

If C lies in the xy-plane and $\mathbf{h}(x,y) = P(x,y)\,\mathbf{i} + Q(x,y)\,\mathbf{j}$, then the line integral reduces to

$$\int_C P(x,y)\,dx + Q(x,y)\,dy.$$

Example 1 Evaluate $\int_C x^2y\,dx + xy\,dy$, where C is

(a) The straight-line path connecting (1, 0) to (0, 1).

(b) The circular path $y = \sqrt{1-x^2}$ connecting $(1,0)$ to $(0,1)$.

(c) The polygonal path $(1,0)$, $(1,1)$, $(0,1)$.

SOLUTION The given integral is the line integral $\int_C \mathbf{h}(\mathbf{r})\cdot d\mathbf{r}$, where

$$\mathbf{h}(x,y) = x^2y\,\mathbf{i} + xy\,\mathbf{j} \quad \text{and} \quad \mathbf{r} = \mathbf{r}(u) \text{ is a parametrization of } C.$$

(a) The straight-line path connecting $(1,0)$ to $(0,1)$ can be parametrized by

$$\mathbf{r}(u) = (1-u)\,\mathbf{i} + u\,\mathbf{j}, \quad 0 \le u \le 1.$$

Thus, $$\begin{aligned}\int_C x^2y\,dx + xy\,dy &= \int_0^1 [x^2(u)y(u)x'(u) + x(u)y(u)y'(u)]\,du\\ &= \int_0^1 [(1-u)^2u(-1) + (1-u)u]\,du\\ &= \int_0^1 (u^2-u^3)\,du = \left[\tfrac{1}{3}u^3 - \tfrac{1}{4}u^4\right]_0^1 = \tfrac{1}{12}.\end{aligned}$$

(b) We parametrize the circular arc $y = \sqrt{1-x^2}$, $0 \le x \le 1$ by setting

$$\mathbf{r}(u) = \cos u\,\mathbf{i} + \sin u\,\mathbf{j}, \quad 0 \le u \le \pi/2.$$

Thus,
$$\begin{aligned}\int_C x^2y\,dx + xy\,dy &= \int_0^{\pi/2} [x^2(u)y(u)x'(u) + x(u)y(u)y'(u)]\,du\\
&= \int_0^{\pi/2} [\cos^2 u \sin u\,(-\sin u) + \cos u \sin u\,(\cos u)]\,du\\
&= \int_0^{\pi/2} [-\cos^2 u \sin^2 u + \cos^2 u \sin u]\,du\\
&= -\int_0^{\pi/2} \tfrac{1}{4}\sin^2 2u\,du + \int_0^{\pi/2} \cos^2 u \sin u\,du\\
&= -\tfrac{1}{8}\int_0^{\pi/2} (1-\cos 4u)\,du + \Big[-\tfrac{1}{3}\cos^3 u\Big]_0^{\pi/2}\\
&= -\tfrac{1}{8}\Big[u - \tfrac{1}{4}\sin 4u\Big]_0^{\pi/2} + \tfrac{1}{3} = \tfrac{1}{3} - \tfrac{\pi}{16}.\end{aligned}$$

(c) The polygonal path $(1, 0), (1, 1), (0, 1)$ is made up of the two segments

$$C_1\colon\ \mathbf{r}(u) = \mathbf{i} + u\,\mathbf{j},\ 0 \le u \le 1, \quad\text{and}\quad C_2\colon\ \mathbf{r}(u) = (1-u)\,\mathbf{i} + \mathbf{j},\ 0 \le u \le 1.$$

Now
$$\int_{C_1} x^2y\,dx + xy\,dy = \int_{C_1} xy\,dy = \int_0^1 x(u)\,y(u)\,y'(u)\,du = \int_0^1 u\,du = \tfrac{1}{2}$$

and
$$\begin{aligned}\int_{C_2} x^2y\,dx + xy\,dy = \int_{C_2} x^2y\,dx &= \int_0^1 x^2(u)\,y(u)\,x'(u)\,du\\
&= \int_0^1 -(1-u)^2\,du = -\tfrac{1}{3}.\end{aligned}$$

Therefore,

$$\int_C x^2y\,dx + xy\,dy = \int_{C_1} x^2y\,dx + xy\,dy + \int_{C_2} x^2y\,dx + xy\,dy = \tfrac{1}{2} - \tfrac{1}{3} = \tfrac{1}{6}. \quad \square$$

Line Integrals with Respect to Arc Length

Suppose that f is a scalar field continuous on a piecewise-smooth curve

$$C\colon\quad \mathbf{r}(u) = x(u)\,\mathbf{i} + y(u)\,\mathbf{j} + z(u)\,\mathbf{k}, \quad u \in [a, b].$$

If $s(u)$ is the length of the curve from the tip of $\mathbf{r}(a)$ to the tip of $\mathbf{r}(u)$, then, as you have seen,

$$s'(u) = \|\mathbf{r}'(u)\| = \sqrt{[x'(u)]^2 + [y'(u)]^2 + [z'(u)]^2}.$$

The line integral of f over C *with respect to arc length s* is defined by setting

(17.4.1)
$$\boxed{\int_C f(\mathbf{r})\,ds = \int_a^b f(\mathbf{r}(u))\,s'(u)\,du.}$$

In the xyz-notation we have

$$\int_C f(x, y, z)\,ds = \int_a^b f(x(u), y(u), z(u))\,s'(u)\,du,$$

which, in the two-dimensional case, becomes

$$\int_C f(x,y)\,ds = \int_a^b f(x(u),y(u))\,s'(u)\,du.$$

Suppose now that C represents a thin wire (a material curve) of varying mass density $\lambda = \lambda(\mathbf{r})$. (Here mass density is mass per unit length.) The *length* of the wire can be written

(17.4.2) $$L = \int_C ds.$$

The *mass* of the wire is given by

(17.4.3) $$M = \int_C \lambda(\mathbf{r})\,ds,$$

and the *center of mass* $\mathbf{r}_M$ can be obtained from the vector equation

(17.4.4) $$\mathbf{r}_M M = \int_C \mathbf{r}\,\lambda(\mathbf{r})\,ds.$$

The equivalent scalar equations read

$$x_M M = \int_C x\,\lambda(\mathbf{r})\,ds, \quad y_M M = \int_C y\,\lambda(\mathbf{r})\,ds, \quad z_M M = \int_C z\,\lambda(\mathbf{r})\,ds.$$

Finally, the *moment of inertia* about an axis is given by the formula

(17.4.5) $$I = \int_C \lambda(\mathbf{r})[R(\mathbf{r})]^2\,ds$$

where $R(\mathbf{r})$ is the distance from the axis to the tip of $\mathbf{r}$.

Example 2 The mass density of a semicircular wire of radius a varies directly as the distance from the diameter that joins the two endpoints of the wire. (a) Find the mass of the wire. (b) Locate the center of mass. (c) Determine the moment of inertia of the wire about the diameter.

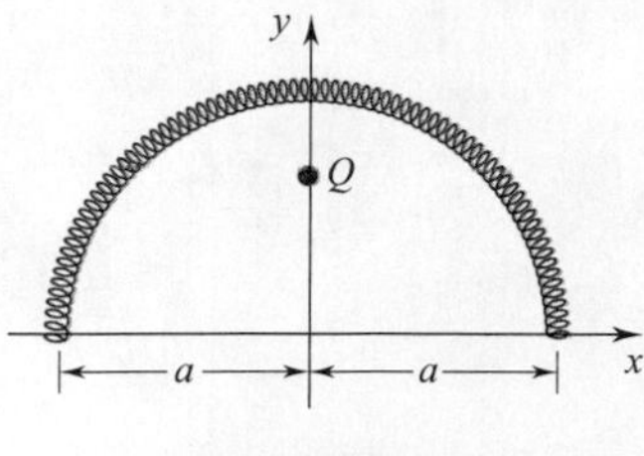

Figure 17.4.1

SOLUTION Placed as in Figure 17.4.1, the wire can be parametrized by

$$\mathbf{r}(u) = a\cos u\,\mathbf{i} + a\sin u\,\mathbf{j}, \quad u \in [0,\pi]$$

and the mass density function can be written $\lambda(x,y) = ky$. Since $\mathbf{r}'(u) = -a\sin u\,\mathbf{i} + a\cos u\,\mathbf{j}$, we have

$$s'(u) = \|\mathbf{r}'(u)\| = a.$$

Therefore $\quad M = \int_C \lambda(x,y)\,ds = \int_C ky\,ds$

$$= \int_0^{\pi} ky(u)\,s'(u)\,du$$

$$= \int_0^{\pi} k(a\sin u)\,a\,du = ka^2 \int_0^{\pi} \sin u\,du = 2ka^2.$$

By the symmetry of the configuration, $x_M = 0$. To find y_M we have to integrate:

$$y_M M = \int_C y\lambda(x,y)\,ds = \int_C ky^2\,ds$$

$$= \int_0^{\pi} k[\,y(u)]^2\,s'(u)\,du$$

$$= \int_0^{\pi} k(a\sin u)^2 a\,du = ka^3 \int_0^{\pi} \sin^2 u\,du = \tfrac{1}{2}ka^3\pi.$$

Since $M = 2ka^2$, we have $y_M = (\frac{1}{2}ka^3\pi)/(2ka^2) = \frac{1}{4}a\pi$. The center of mass Q lies on the perpendicular bisector of the wire at a distance $\frac{1}{4}a\pi$ from the diameter. (See Figure 17.4.1.) Note that in this instance, the center of mass does not lie on the wire.

Now let's find the moment of inertia about the diameter:

$$I = \int_C \lambda(x,y)[R(x,y)]^2\,ds = \int_C (ky)y^2\,ds$$

$$= \int_C k[\,y(u)]^3\,s'(u)\,du$$

$$= \int_0^{\pi} k(a\sin u)^3 a\,du$$

$$= ka^4 \int_0^{\pi} \sin^3 u\,du = \tfrac{4}{3}ka^4.$$

It is customary to express I in terms of M. With $M = 2ka^2$, we have

$$I = \tfrac{2}{3}(2ka^2)a^2 = \tfrac{2}{3}Ma^2. \quad \square$$

EXERCISES 17.4

In Exercises 1–4, evaluate

$$\int_C (x - 2y)\,dx + 2x\,dy$$

along the given path C from $(0,0)$ to $(1,2)$.

1. The straight-line path.
2. The parabolic path $y = 2x^2$.
3. The polygonal path $(0,0), (1,0), (1,2)$.
4. The polygonal path $(0,0), (0,2), (1,2)$.

In Exercises 5–8, evaluate

$$\int_C y\,dx + xy\,dy$$

along the given path C from $(0,0)$ to $(2,1)$.

5. The parabolic path $x = 2y^2$.
6. The straight-line path.
7. The polygonal path $(0,0), (0,1), (2,1)$.
8. The cubic path $x = 2y^3$.

In Exercises 9–12, evaluate

$$\int_C y^2\,dx + (xy - x^2)\,dy$$

along the given path C from $(0, 0)$ to $(2, 4)$.

9. The straight-line path.

10. The parabolic path $y = x^2$.

11. The parabolic path $y^2 = 8x$.

12. The polygonal path $(0, 0), (2, 0), (2, 4)$.

In Exercises 13–16, evaluate

$$\int_C (y^2 + 2x + 1)\,dx + (2xy + 4y - 1)\,dy$$

along the given path C from $(0, 0)$ to $(1, 1)$.

13. The straight-line path.

14. The parabolic path $y = x^2$.

15. The cubic path $y = x^3$.

16. The polygonal path $(0, 0), (4, 0), (4, 2), (1, 1)$.

In Exercises 17–20, evaluate

$$\int_C y\,dx + 2z\,dy + x\,dz$$

along the given path C from $(0, 0, 0)$ to $(1, 1, 1)$.

17. The straight-line path.

18. $\mathbf{r}(u) = u\,\mathbf{i} + u^2\,\mathbf{j} + u^3\,\mathbf{k}$.

19. The polygonal path $(0, 0, 0), (0, 0, 1), (0, 1, 1), (1, 1, 1)$.

20. The polygonal path $(0, 0, 0), (1, 0, 0), (1, 1, 0), (1, 1, 1)$.

In Exercises 21–24, evaluate

$$\int_C xy\,dx + 2z\,dy + (y + z)\,dz$$

along the given path C from $(0, 0, 0)$ to $(2, 2, 8)$.

21. The straight-line path.

22. The polygonal path $(0, 0, 0), (2, 0, 0), (2, 2, 0), (2, 2, 8)$.

23. The parabolic path $\mathbf{r}(u) = u\,\mathbf{i} + u\,\mathbf{j} + 2u^2\,\mathbf{k}$.

24. The polygonal path $(0, 0, 0), (2, 2, 2), (2, 2, 8)$.

25. Evaluate $\int_C x^2 y\,dx + y\,dy + xz\,dz$ where C is the curve of intersection of the cylinder $y - 2z^2 = 1$ and the plane $z = x + 1$ from $(0, 3, 1)$ to $(1, 9, 2)$.

26. Evaluate $\int_C y\,dx + yz\,dy + z(x - 1)\,dz$ where C is the curve of intersection of the sphere $x^2 + y^2 + z^2 = 4$ and the cylinder $(x - 1)^2 + y^2 = 1$ from $(2, 0, 0)$ to $(0, 0, 2)$.

27. Let $\mathbf{h}$ be the vector field

$$\mathbf{h}(x, y) = (x^2 + 6xy - 2y^2)\,\mathbf{i} + (3x^2 - 4xy + 2y)\,\mathbf{j}.$$

(a) Show that $\mathbf{h}$ is a gradient field.

(b) What is the value of

$$\int_C (x^2 + 6xy - 2y^2)\,dx + (3x^2 - 4xy + 2y)\,dy$$

along any piecewise-smooth curve from $(3, 0)$ to $(0, 4)$?

(c) What is the value of

$$\int_C (x^2 + 6xy - 2y^2)\,dx + (3x^2 - 4xy + 2y)\,dy$$

where C is any piecewise-smooth curve from $(4, 0)$ to $(0, 3)$?

28. Let $\mathbf{h}$ be the vector field

$$\mathbf{h}(x, y, z) = (2xy + z^2)\,\mathbf{i} + (x^2 - 2yz)\,\mathbf{j} + (2xz - y^2)\,\mathbf{k}.$$

(a) Show that $\mathbf{h}$ is a gradient field.

(b) What is the value of

$$\int_C (2xy + z^2)\,dx + (x^2 - 2yz)\,dy + (2xz - y^2)\,dz$$

along any piecewise-smooth curve from $(1, 0, 1)$ to $(3, 2, -1)$?

(c) What is the value of

$$\int_C (2xy + z^2)\,dx + (x^2 - 2yz)\,dy + (2xz - y^2)\,dz$$

where C is any piecewise-smooth curve from $(3, 2, -1)$ to $(1, 0, 1)$?

29. A wire in the shape of the quarter-circle

$$C\colon\ \mathbf{r}(u) = a(\cos u\,\mathbf{i} + \sin u\,\mathbf{j}),\quad u \in [0, \tfrac{1}{2}\pi]$$

has varying mass density $\lambda(x, y) = k(x + y)$ where k is a positive constant.

(a) Find the total mass of the wire and locate the center of mass.

(b) What is the moment of inertia of the wire about the x-axis?

30. Find the moment of inertia of a homogeneous circular wire of radius a and mass M about (a) a diameter; (b) the axis through the center that is perpendicular to the plane of the wire.

31. Find the moment of inertia of the wire of Exercise 29 about

(a) the z-axis; (b) the line $y = x$.

32. A wire of constant mass density k has the form

$$\mathbf{r}(u) = (1 - \cos u)\,\mathbf{i} + (u - \sin u)\,\mathbf{j},\quad u \in [0, 2\pi].$$

(a) Determine the mass of the wire.

(b) Locate the center of mass.

33. A homogeneous wire of mass M winds around the z-axis as

$$C: \quad \mathbf{r}(u) = a\cos u\,\mathbf{i} + a\sin u\,\mathbf{j} + bu\,\mathbf{k}, \quad u \in [0, 2\pi].$$

(a) Find the length of the wire.

(b) Locate the center of mass.

(c) Determine the moments of inertia of the wire about the coordinate axes.

34. A homogeneous wire of mass M is of the form

$$C: \quad \mathbf{r}(u) = u\,\mathbf{i} + u^2\,\mathbf{j} + \tfrac{2}{3}u^3\,\mathbf{k}, \quad u \in [0, a].$$

(a) Find the length of the wire.

(b) Locate the center of mass.

(c) Determine the moment of inertia of the wire about the z-axis.

35. Calculate the mass of the wire of Exercise 33 given that the mass density varies directly as the square of the distance from the origin.

36. Show that

(17.4.6) $$\int_C \mathbf{h}(\mathbf{r}) \cdot \mathbf{dr} = \int_C [\mathbf{h}(\mathbf{r}) \cdot \mathbf{T}(\mathbf{r})]\, ds$$

where $\mathbf{T}$ is the unit tangent vector.

■ 17.5 GREEN'S THEOREM

Green's theorem is the first of the three integration theorems heralded at the beginning of this chapter.

A *Jordan curve*, as you may recall from a footnote in Section 15.9, is a plane curve that is both closed and simple. Thus circles, ellipses, and triangles are Jordan curves; figure eights are not.

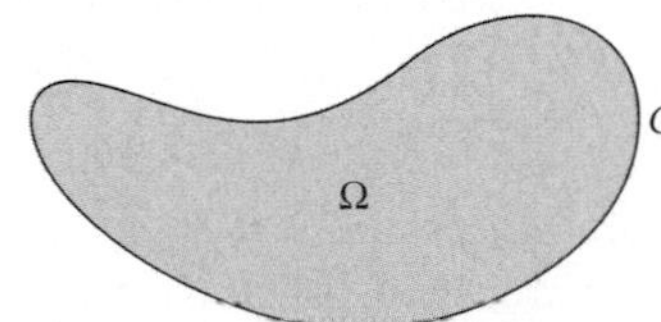

Figure 17.5.1

Figure 17.5.1 depicts a closed region Ω, the total boundary of which is a Jordan curve C. Such a region is called a *Jordan region*. We know how to integrate over Ω, and if the boundary C is piecewise smooth, we know how to integrate over C. Green's theorem expresses a double integral over Ω as a line integral over C.

THEOREM 17.5.1 GREEN'S THEOREM †

Let Ω be a Jordan region with a piecewise-smooth boundary C. If P and Q are scalar fields that are continuously differentiable on an open set that contains Ω, then

$$\iint_{\Omega} \left[\frac{\partial Q}{\partial x}(x, y) - \frac{\partial P}{\partial y}(x, y)\right] dxdy = \oint_C P(x, y)\, dx + Q(x, y)\, dy$$

where the integral on the right is the line integral taken over C in the counterclockwise direction.††

We will prove the theorem only for special cases. First of all let's assume that Ω is an *elementary region*, a region that is both of Type I and Type II as defined in Section 16.3. For simplicity we take Ω as in Figure 17.5.2.

Ω being of Type I, we can show that

$$(1) \qquad \oint_C P(x, y)\, dx = \iint_{\Omega} -\frac{\partial P}{\partial y}(x, y)\, dxdy.$$

† The result was established in 1828 by the English mathematician George Green (1793–1841).

†† Counterclockwise as viewed from $z > 0$ in a right-handed coordinate system. The integral over C in the clockwise direction is denoted $\oint\!\!\!\!\circlearrowright$.

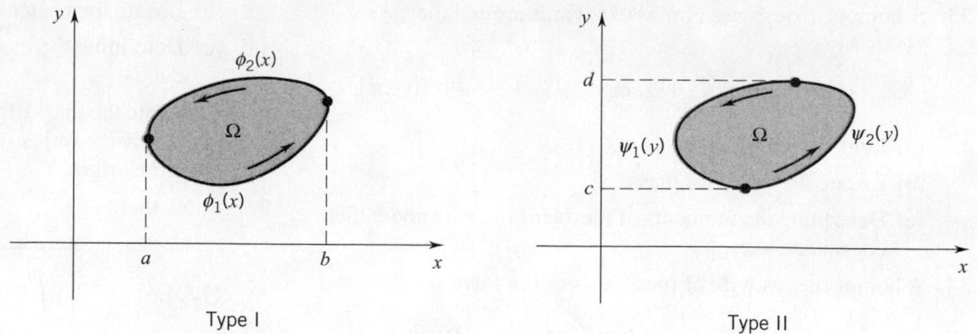

Ω is an elementary region: it is both of Type I and Type II

Figure 17.5.2

In the first place

$$\iint_{\Omega} -\frac{\partial P}{\partial y}(x,y)\,dxdy = -\int_a^b \int_{\phi_1(x)}^{\phi_2(x)} \frac{\partial P}{\partial y}(x,y)\,dy\,dx$$

$$= -\int_a^b \{P[x,\phi_2(x)] - P[x,\phi_1(x)]\}\,dx$$

by the fundamental theorem of integral calculus ↑

$$(*) \qquad = \int_a^b P[x,\phi_1(x)]\,dx - \int_a^b P[x,\phi_2(x)]\,dx.$$

The graph of ϕ_1 parametrized from left to right is the curve

$$C_1: \quad \mathbf{r}_1(u) = u\,\mathbf{i} + \phi_1(u)\,\mathbf{j}, \quad u \in [a,b];$$

the graph of ϕ_2, also parametrized from left to right, is the curve

$$C_2: \quad \mathbf{r}_2(u) = u\,\mathbf{i} + \phi_2(u)\,\mathbf{j}, \quad u \in [a,b].$$

Since C traversed counterclockwise consists of C_1 followed by $-C_2$ (C_2 traversed from right to left), you can see that

$$\oint_C P(x,y)\,dx = \int_{C_1} P(x,y)\,dx - \int_{C_2} P(x,y)\,dx$$

$$= \int_a^b P[u,\phi_1(u)]\,du - \int_a^b P[u,\phi_2(u)]\,du.$$

Since u is a dummy variable, it can be replaced by x. Comparison with $(*)$ proves (1).

We leave it to you to verify that

$$(2) \qquad \oint_C Q(x,y)\,dy = \iint_{\Omega} \frac{\partial Q}{\partial x}(x,y)\,dxdy$$

by using the fact that Ω is of Type II. This completes the proof of the theorem for Ω as in Figure 17.5.2.

A slight modification of this argument applies to elementary regions which are bordered entirely or in part by line segments parallel to the coordinate axes.

Figure 17.5.3 shows a Jordan region that is not elementary but can be broken up into two elementary regions. (See Figure 17.5.4.) Green's theorem applied to the elementary parts tells us that

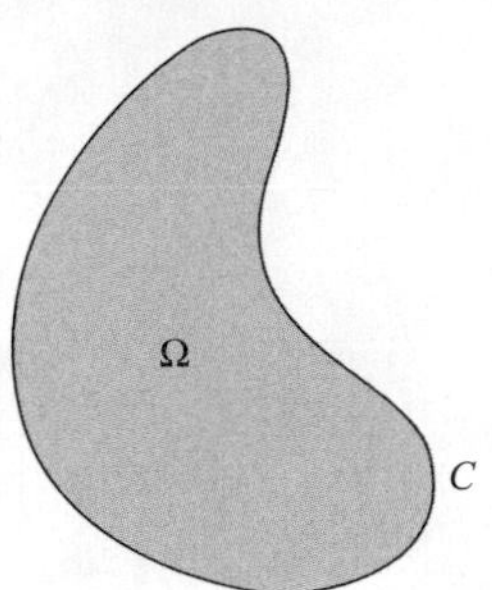

Figure 17.5.3

$$\iint_{\Omega_1}\left(\frac{\partial Q}{\partial x}-\frac{\partial P}{\partial y}\right)dxdy=\oint_{\text{bdry of }\Omega_1}P\,dx+Q\,dy,$$

$$\iint_{\Omega_2}\left(\frac{\partial Q}{\partial x}-\frac{\partial P}{\partial y}\right)dxdy=\oint_{\text{bdry of }\Omega_2}P\,dx+Q\,dy.$$

We now add these equations. The sum of the double integrals is, by additivity, the double integral over Ω. The sum of the line integrals is the integral over C (see the figure) plus the integrals over the crosscut. Since the crosscut is traversed twice and in opposite directions, the total contribution of the crosscut is zero and therefore Green's theorem holds:

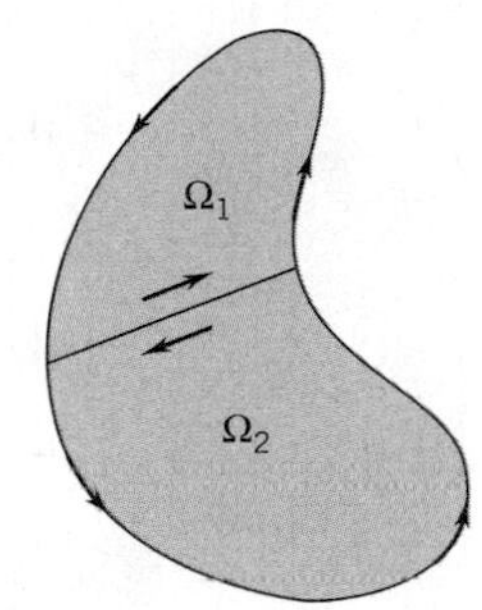

Figure 17.5.4

$$\iint_{\Omega}\left(\frac{\partial Q}{\partial x}-\frac{\partial P}{\partial y}\right)dxdy=\oint_C P\,dx+Q\,dy.$$

This same argument can be extended to a Jordan region Ω that breaks up into n elementary regions $\Omega_1,\ldots,\Omega_n$. (Figure 17.5.5 gives an example with $n=4$.) The double integrals over the Ω_i add up to the double integral over Ω, and, since the line integrals over the crosscuts cancel, the line integrals over the boundaries of the Ω_i add up to the line integral over C. (This is as far as we will carry the proof of Green's theorem. It is far enough to cover all the Jordan regions encountered in practice.)

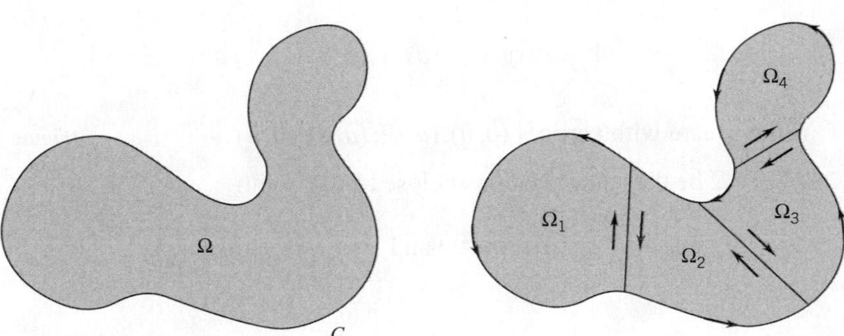

Figure 17.5.5

Example 1 Use Green's theorem to evaluate

$$\oint_C (3x^2+y)\,dx+(2x+y^3)\,dy$$

where C is the circle $x^2+y^2=a^2$. (Figure 17.5.6)

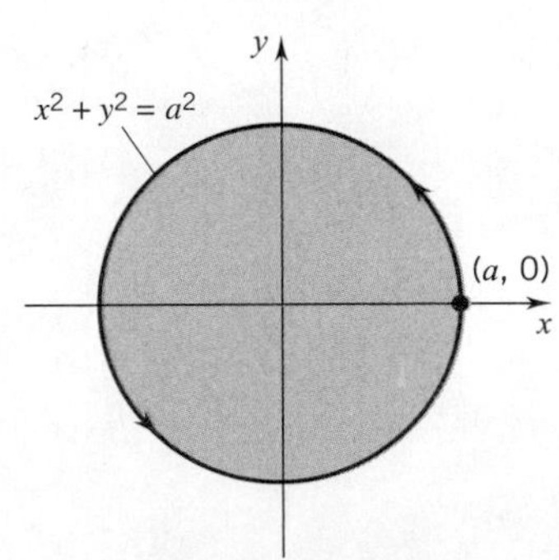

Figure 17.5.6

SOLUTION Let Ω be the closed disc $0\le x^2+y^2\le a^2$. With

$$P(x,y)=3x^2+y \quad\text{and}\quad Q(x,y)=2x+y^3,$$

we have $\dfrac{\partial P}{\partial y} = 1, \quad \dfrac{\partial Q}{\partial x} = 2, \quad$ and $\quad \dfrac{\partial Q}{\partial x} - \dfrac{\partial P}{\partial y} = 2 - 1 = 1.$

By Green's theorem

$$\oint_C (3x^2 + y)\,dx + (2x + y^3)\,dy = \iint_\Omega 1\,dxdy = \text{area of } \Omega = \pi a^2. \quad \square$$

Remark The line integral in Example 1 could have been calculated directly as follows: The circle $x^2 + y^2 = a^2$ can be parametrized counterclockwise by setting

$$x = a\cos u, \quad y = a\sin u, \quad 0 \le u \le 2\pi.$$

Thus

$$\oint_C (3x^2 + y)\,dx + (2x + y^3)\,dy$$

$$= \int_0^{2\pi} [(3a^2\cos^2 u + a\sin u)(-a\sin u) + (2a\cos u + a^3\sin^3 u)(a\cos u)]\,du$$

$$= \int_0^{2\pi} [-3a^3\cos^2 u\sin u - a^2\sin^2 u + 2a^2\cos^2 u + a^4\sin^3 u\,\cos u]\,du,$$

which, as you can verify, also yields πa^2. In this case, at least, Green's theorem gives us a more direct route to the answer. $\square$

Example 2 Use Green's theorem to evaluate

$$\oint_C (1 + 10xy + y^2)\,dx + (6xy + 5x^2)\,dy$$

where C is the square with vertices $(0, 0), (a, 0), (a, a), (0, a)$. (Figure 17.5.7)

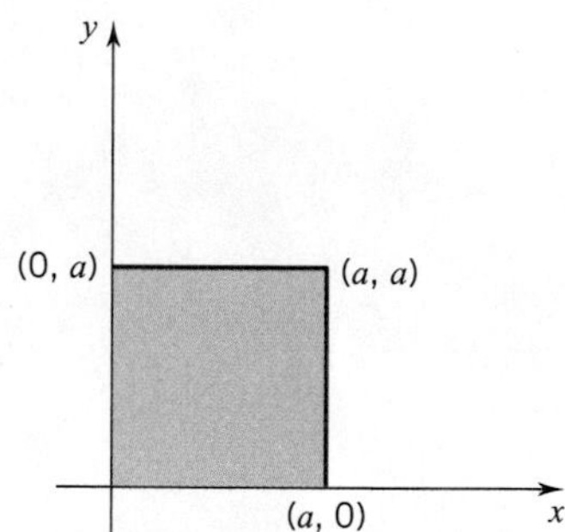

Figure 17.5.7

SOLUTION Let Ω be the square region enclosed by C. With

$$P(x,y) = 1 + 10xy + y^2 \quad \text{and} \quad Q(x,y) = 6xy + 5x^2,$$

we have $\dfrac{\partial P}{\partial y} = 10x + 2y, \quad \dfrac{\partial Q}{\partial x} = 6y + 10x, \quad \dfrac{\partial Q}{\partial x} - \dfrac{\partial P}{\partial y} = 4y.$

By Green's theorem,

$$\oint_C (1 + 10xy + y^2)\,dx + (6xy + 5x^2)\,dy = \iint_\Omega 4y\,dxdy$$

$$= \int_0^a \int_0^a 4y\,dx\,dy$$

$$= \left(\int_0^a dx\right)\left(\int_0^a 4y\,dy\right)$$

$$= (a)(2a^2) = 2a^3.$$

ALTERNATIVE SOLUTION By (16.5.3),

$$\iint_{\Omega} y\,dxdy = \bar{y}\,(\text{area of }\Omega)$$

where $\bar{y}$ is the y-coordinate of the centroid of Ω. Since $\bar{y} = \frac{1}{2}a$, it is evident that

$$\iint_{\Omega} 4y\,dxdy = 4\bar{y}\,(\text{area of }\Omega) = 4(\tfrac{1}{2}a)a^2 = 2a^3. \quad \square$$

Example 3 Use Green's theorem to evaluate

$$\oint_C e^x \sin y\,dx + e^x \cos y\,dy$$

where C is the closed curve consisting of the semicircle $y = \sqrt{1-x^2}$ and the interval $[-1, 1]$. (See Figure 17.5.8.)

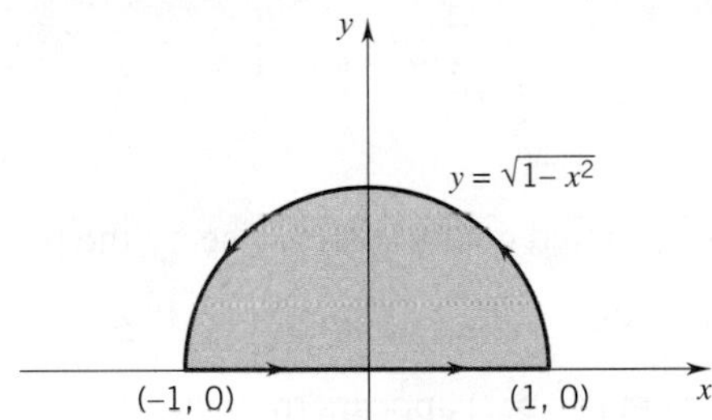

Figure 17.5.8

SOLUTION The curve bounds the closed semicircular disc $\Omega : x^2 + y^2 \leq 1,\ y \geq 0$.

Here $\quad \dfrac{\partial P}{\partial y} = e^x \cos y, \quad \dfrac{\partial Q}{\partial x} = e^x \cos y \quad$ and $\quad \dfrac{\partial Q}{\partial x} - \dfrac{\partial P}{\partial y} = 0.$

By Green's theorem,

$$\oint_C e^x \sin y\,dx + e^x \cos y\,dy = \iint_{\Omega} 0\,dxdy = 0.$$

To see the power of Green's theorem, try to evaluate this line integral directly. $\square$

The preceding examples illustrate the use of Green's theorem to convert a line integral over the boundary of a Jordan region Ω into a double integral over Ω. In some instances, the theorem can be used in reverse. That is, it may be possible to find the value of a double integral over a Jordan region by evaluating a line integral over its boundary. For example, Green's theorem enables us to calculate the area of a Jordan region by integrating over the boundary of the region.

(17.5.2)

> The area of a Jordan region with boundary C is given by each of the following integrals :
>
> $$\oint_C -y\,dx, \quad \oint_C x\,dy, \quad \tfrac{1}{2}\oint_C -y\,dx + x\,dy.$$

PROOF Let Ω be the region enclosed by C. In the first integral

$$P(x,y) = -y, \quad Q(x,y) = 0.$$

Therefore $\quad \dfrac{\partial P}{\partial y} = -1, \quad \dfrac{\partial Q}{\partial x} = 0, \quad$ and $\quad \dfrac{\partial Q}{\partial x} - \dfrac{\partial P}{\partial y} = 1.$

Thus by Green's theorem

$$\oint_C -y\,dx = \iint_\Omega 1\,dxdy = \text{area of } \Omega.$$

That the second integral also gives the area of Ω can be verified in a similar manner.

We can see the validity of the third formula by observing that

$$\oint_C -y\,dx + \oint_C x\,dy = \text{twice the area of } \Omega. \quad \square$$

Example 4 Show that the area of the region Ω enclosed by the ellipse

$$\frac{x^2}{a^2} + \frac{y^2}{b^2} = 1 \qquad \text{(Figure 17.5.9)}$$

is πab.

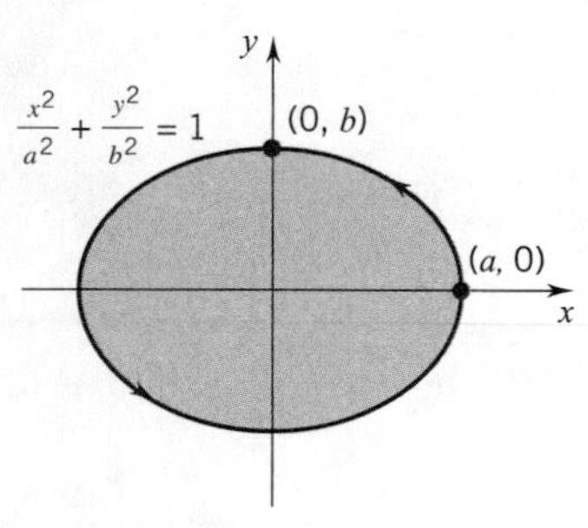

Figure 17.5.9

SOLUTION The ellipse is oriented counterclockwise by the parametrization

$$x = a\cos u, \qquad y = b\sin u, \qquad 0 \le u \le 2\pi.$$

Although the third integral in (17.5.2) appears to be the most complicated of the three, it is in this case the simplest to use.

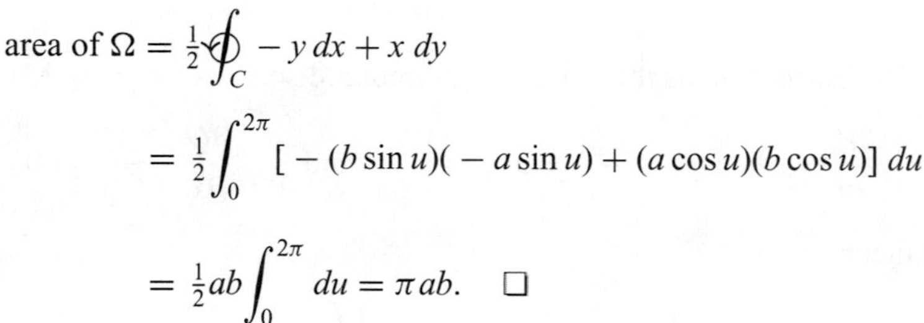

$$\begin{aligned} \text{area of } \Omega &= \tfrac{1}{2}\oint_C -y\,dx + x\,dy \\ &= \tfrac{1}{2}\int_0^{2\pi} [-(b\sin u)(-a\sin u) + (a\cos u)(b\cos u)]\,du \\ &= \tfrac{1}{2}ab\int_0^{2\pi} du = \pi ab. \quad \square \end{aligned}$$

Example 5 Let Ω be a Jordan region of area A with a piecewise-smooth boundary C. Show that the coordinates of the centroid of Ω are given by the formulas

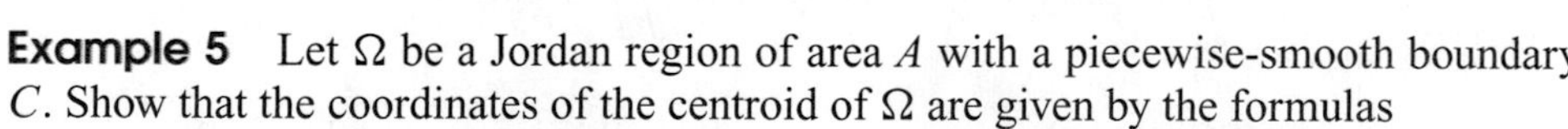

$$\bar{x}A = \tfrac{1}{2}\oint_C x^2\,dy, \qquad \bar{y}A = -\tfrac{1}{2}\oint_C y^2\,dx.$$

SOLUTION

$$\tfrac{1}{2}\oint_C x^2\,dy = \tfrac{1}{2}\iint_\Omega 2x\,dxdy = \iint_\Omega x\,dxdy = \bar{x}A,$$

by Green's theorem

$$-\tfrac{1}{2}\oint_C y^2\,dx = -\tfrac{1}{2}\iint_\Omega (-2y)\,dxdy = \iint_\Omega y\,dxdy = \bar{y}A. \quad \square$$

Regions Bounded by Two or More Jordan Curves

(All the curves that appear here are assumed to be piecewise smooth.)

Figure 17.5.10 shows an annular region Ω. The region is not a Jordan region: the boundary consists of two Jordan curves C_1 and C_2. We cannot apply Green's theorem to Ω directly, but we can break up Ω into two Jordan regions as in Figure 17.5.11 and then apply Green's theorem to each piece. With Ω_1 and Ω_2 as in Figure 17.5.11,

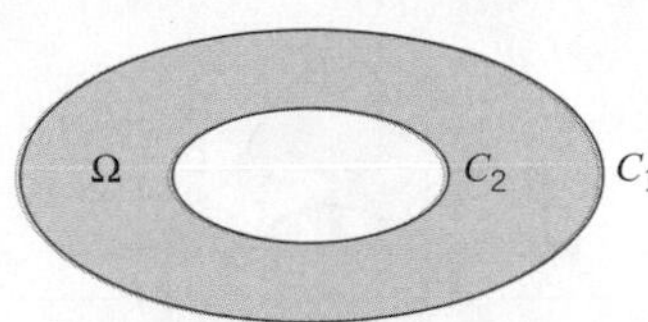

Figure 17.5.10

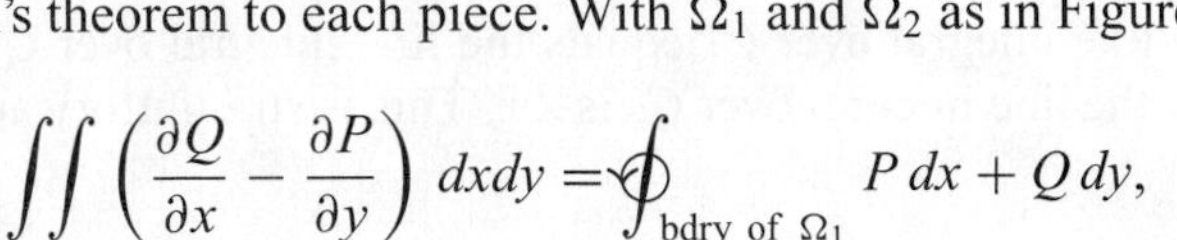

$$\iint_{\Omega_1} \left(\frac{\partial Q}{\partial x} - \frac{\partial P}{\partial y}\right) dxdy = \oint_{\text{bdry of } \Omega_1} P\,dx + Q\,dy,$$

$$\iint_{\Omega_2} \left(\frac{\partial Q}{\partial x} - \frac{\partial P}{\partial y}\right) dxdy = \oint_{\text{bdry of } \Omega_2} P\,dx + Q\,dy.$$

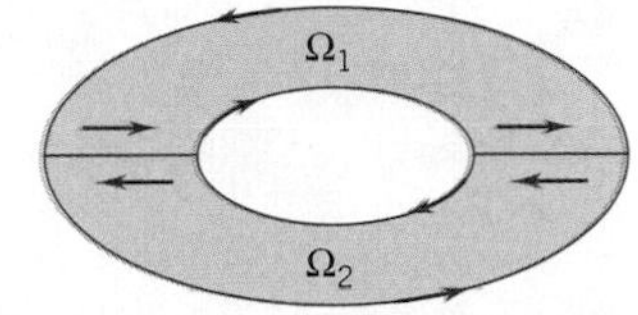

Figure 17.5.11

When we add the double integrals, we get the double integral over Ω. When we add the line integrals, the integrals over the crosscuts cancel and we are left with the *counterclockwise integral* over C_1 and the *clockwise integral* over C_2. (See the figure.) Thus, for the annular region,

(17.5.3)
$$\iint_{\Omega} \left(\frac{\partial Q}{\partial x} - \frac{\partial P}{\partial y}\right) dxdy = \oint_{C_1} P\,dx + Q\,dy + \oint_{C_2} P\,dx + Q\,dy.$$

If $\partial Q/\partial x = \partial P/\partial y$ throughout Ω, then the double integral on the left is 0, and the sum of the integrals on the right is also 0. Therefore we see that

(17.5.4)
if $\partial Q/\partial x = \partial P/\partial y$ throughout Ω, then
$$\oint_{C_1} P\,dx + Q\,dy = \oint_{C_2} P\,dx + Q\,dy.$$

Example 6 Let C_1 be a Jordan curve that does not pass through the origin $(0, 0)$. Show that

$$\oint_{C_1} -\frac{y}{x^2+y^2}\,dx + \frac{x}{x^2+y^2}\,dy = \begin{cases} 0 & \text{if } C_1 \text{ does not enclose the origin} \\ 2\pi & \text{if } C_1 \text{ does enclose the origin}. \end{cases}$$

SOLUTION In this case

$$\frac{\partial P}{\partial y} = \frac{\partial}{\partial y}\left(-\frac{y}{x^2+y^2}\right) = -\left[\frac{(x^2+y^2)1 - 2y^2}{(x^2+y^2)^2}\right] = \frac{y^2-x^2}{(x^2+y^2)^2},$$

$$\frac{\partial Q}{\partial x} = \frac{\partial}{\partial x}\left(\frac{x}{x^2+y^2}\right) = \frac{(x^2+y^2)1 - 2x^2}{(x^2+y^2)^2} = \frac{y^2-x^2}{(x^2+y^2)^2}.$$

Thus
$$\frac{\partial Q}{\partial x} = \frac{\partial P}{\partial y} \quad \text{except at the origin.}$$

If C_1 does not enclose the origin, then $\partial Q/\partial x - \partial P/\partial y = 0$ throughout the region enclosed by C_1, and, by Green's theorem, the line integral is 0.

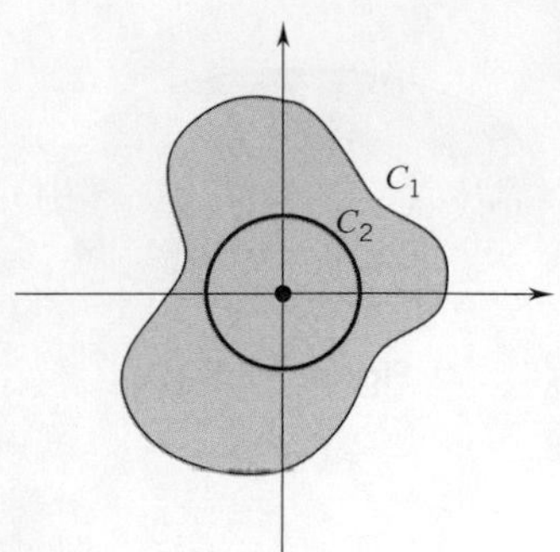

Figure 17.5.12

If C_1 does enclose the origin, we draw within the inner region of C_1 a small circle centered at the origin

$$C_2: \quad x^2 + y^2 = a^2. \qquad \text{(Figure 17.5.12)}$$

Since $\partial Q/\partial x - \partial P/\partial y = 0$ on the annular region bounded by C_1 and C_2, we know from (17.5.4) that the line integral over C_1 equals the line integral over C_2. All we have to show now is that the line integral over C_2 is 2π. This is straightforward. Parametrizing the circle by

$$\mathbf{r}(u) = a\cos u\,\mathbf{i} + a\sin u\,\mathbf{j} \qquad \text{with} \qquad u \in [0, 2\pi],$$

we have

$$\oint_{C_2} -\frac{y}{x^2+y^2}\,dx + \frac{x}{x^2+y^2}\,dy = \int_0^{2\pi} (\sin^2 u + \cos^2 u)\,du = \int_0^{2\pi} du = 2\pi. \quad \square$$

check this

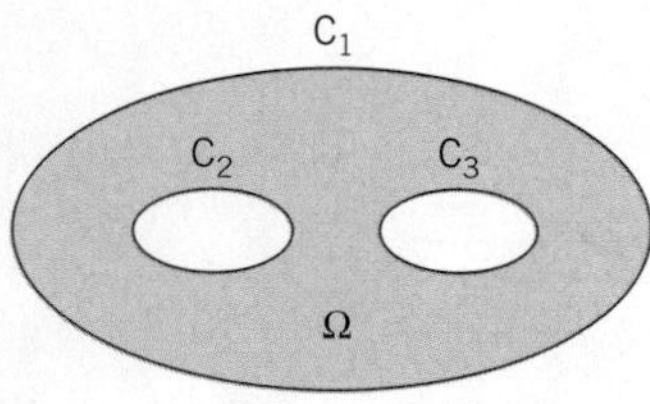

Figure 17.5.13

Figure 17.5.13 shows a region bounded by three Jordan curves: C_2 and C_3, each exterior to the other, both within C_1. For such a region Green's theorem gives

$$\iint_\Omega \left(\frac{\partial Q}{\partial x} - \frac{\partial P}{\partial y}\right) dx\,dy =$$

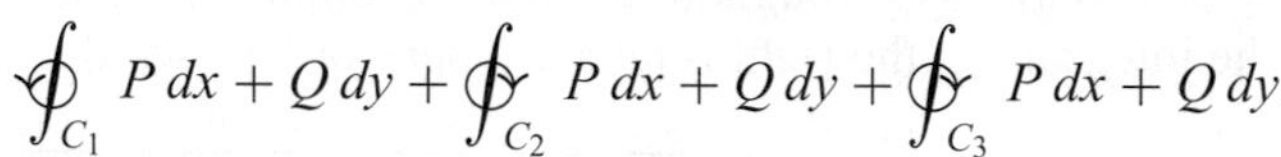

$$\oint_{C_1} P\,dx + Q\,dy + \oint_{C_2} P\,dx + Q\,dy + \oint_{C_3} P\,dx + Q\,dy.$$

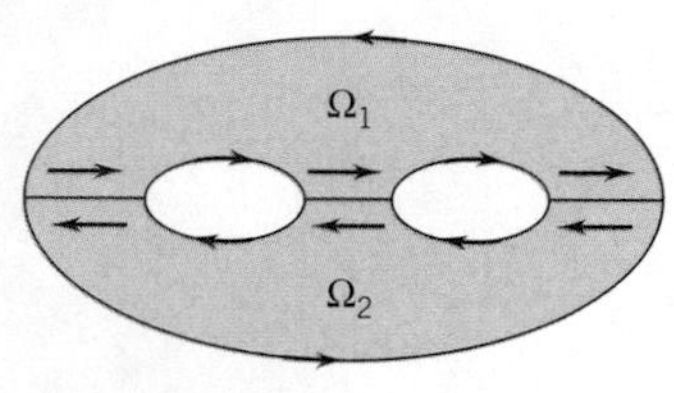

Figure 17.5.14

To see this, break up Ω into two regions by making the crosscuts shown in Figure 17.5.14.

The general formula for configurations of this type reads

$$\iint_\Omega \left(\frac{\partial Q}{\partial x} - \frac{\partial P}{\partial y}\right) dx\,dy = \oint_{C_1} P\,dx + Q\,dy + \sum_{i=2}^{n} \oint_{C_i} P\,dx + Q\,dy.$$

EXERCISES 17.5

Evaluate the line integral (a) directly; and (b) by applying Green's theorem.

1. $\oint_C xy\,dx + x^2\,dy$; where C is the triangle with vertices $(0, 0), (0, 1), (1, 1)$.

2. $\oint_C x^2 y\,dx + 2y^2\,dy$; where C is the square with vertices $(0, 0), (1, 0), (1, 1), (0, 1)$.

3. $\oint_C (3x^2 + y)\,dx + (2x + y^3)\,dy; \quad C: 9x^2 + 4y^2 = 36.$

4. $\oint_C y^2\,dx + x^2\,dy$; where C is the boundary of the region that lies between the curves $y = x$ and $y = x^2$.

Evaluate by Green's theorem.

5. $\oint_C 3y\,dx + 5x\,dy; \quad C: x^2 + y^2 = 1.$

6. $\oint_C 5x\,dx + 3y\,dy; \quad C: (x-1)^2 + (y+1)^2 = 1.$

7. $\oint_C x^2 dy$; where C is the rectangle with vertices $(0, 0), (a, 0), (a, b), (0, b)$.

8. $\oint_C y^2\,dx$; where C is the rectangle of Exercise 7.

9. $\oint_C (3xy + y^2)\,dx + (2xy + 5x^2)dy$; $C: (x-1)^2 + (y+2)^2 = 1.$

10. $\oint_C (xy + 3y^2)\,dx + (5xy + 2x^2)\,dy;$
$C\colon (x-1)^2 + (y+2)^2 = 1.$

11. $\oint_C (2x^2 + xy - y^2)\,dx + (3x^2 - xy + 2y^2)\,dy;$
$C\colon (x-a)^2 + y^2 = r^2.$

12. $\oint_C (x^2 - 2xy + 3y^2)\,dx + (5x + 1)\,dy;$
$C\colon x^2 + (y-b)^2 = r^2.$

13. $\oint_C e^x \sin y\,dx + e^x \cos y\,dy;$
$C : (x-a)^2 + (y-b)^2 = r^2.$

14. $\oint_C e^x \cos y\,dx + e^x \sin y\,dy$ where C is the rectangle with vertices $(0,0), (1,0), (1,\pi), (0,\pi)$.

15. $\oint_C 2xy\,dx + x^2 dy$ where C is the cardioid $r = 1 - \cos\theta, \theta \in [0, 2\pi]$.

16. $\oint_C y^2\,dx + 2xy\,dy$ where C is the first quadrant loop of the petal curve $r = 2\sin 2\theta$.

In Exercises 17 and 18, find the area enclosed by the curve by integrating over the curve.

17. The circle $x^2 + y^2 = a^2$.

18. The astroid $x^{2/3} + y^{2/3} = a^{2/3}$.

19. Sketch the region Ω bounded by the curves $xy = 4$ and $x + y = 5$. Then use Green's theorem to find the area of Ω.

20. Sketch the region Ω bounded by the curves $y^2 - x^2 = 5$ and $y = 3$. Then use Green's theorem to find the area of Ω.

21. Let C be a piecewise-smooth Jordan curve. Calculate

$$\oint_C (ay + b)\,dx + (cx + d)\,dy$$

given that C encloses a region of area A.

22. Calculate

$$\oint_C \mathbf{F(r)} \cdot d\mathbf{r}$$

given that $\mathbf{F}(x, y) = 2y\,\mathbf{i} - 3x\,\mathbf{j}$ and C is the astroid $x^{2/3} + y^{2/3} = a^{2/3}$.

23. Use Green's theorem to find the area under one arch of the cycloid

$$x(\theta) = R(\theta - \sin\theta), \quad y(\theta) = R(1 - \cos\theta).$$

24. Find the Jordan curve C that maximizes the line integral

$$\oint_C y^3\,dx + (3x - x^3)\,dy.$$

25. Complete the proof of Green's theorem for the elementary region of Figure 17.5.2 by showing that

$$\oint_C Q(x,y)dy = \iint_\Omega \frac{\partial Q}{\partial x}(x,y)\,dxdy.$$

26. Suppose that f and g have continuous first-partial derivatives in a simply connected open region Ω. Show that if C is any piecewise-smooth simple closed curve in Ω, then

$$\oint_C [f(\mathbf{r})\nabla g(\mathbf{r}) + g(\mathbf{r})\nabla f(\mathbf{r})] \cdot d\mathbf{r} = 0.$$

27. Let $(\bar{x}, \bar{y})$ be the centroid of a Jordan region with piecewise-smooth boundary C and area A. Show that

$$\bar{x}A = \tfrac{1}{2}\oint_C x^2 dy \qquad \text{and} \qquad \bar{y}A = -\tfrac{1}{2}\oint_C y^2\,dx.$$

28. Let Ω be a plate of constant mass density λ in the form of a Jordan region with a piecewise-smooth boundary C. Show that the moments of inertia of the plate about the coordinate axes are given by the formulas

(17.5.5) $$I_x = -\frac{\lambda}{3}\oint_C y^3\,dx, \quad I_y = \frac{\lambda}{3}\oint_C x^3 dy.$$

29. Let P and Q be continuously differentiable functions on the region Ω of Figure 17.5.13. Given that $\partial P/\partial y = \partial Q/\partial x$ on Ω, find a relation between the line integrals

$$\oint_{C_1} P\,dx + Q\,dy, \quad \oint_{C_2} P\,dx + Q\,dy,$$

$$\oint_{C_3} P\,dx + Q\,dy.$$

30. Show that, if $f = f(x)$ and $g = g(y)$ are everywhere continuously differentiable, then

$$\int_C f(x)\,dx + g(y)\,dy = 0$$

for all piecewise-smooth Jordan curves C.

31. Let C be a piecewise-smooth Jordan curve that does not pass through the origin. Evaluate

$$\oint_C \frac{x}{x^2 + y^2}\,dx + \frac{y}{x^2 + y^2}\,dy$$

(a) if C does not enclose the origin.
(b) if C does enclose the origin.

32. Let C be a piecewise-smooth Jordan curve that does not pass through the origin. Evaluate

$$\oint_C -\frac{y^3}{(x^2 + y^2)^2}\,dx + \frac{xy^2}{(x^2 + y^2)^2}\,dy$$

(a) if C does not enclose the origin.
(b) if C does enclose the origin.

33. Let $\mathbf{v}$ be a vector field continuously differentiable on the entire plane. Use Green's theorem to verify that if $\mathbf{v}$ is a gradient field $[\mathbf{v} = \nabla\phi]$, then

$$\oint_C \mathbf{v} \cdot d\mathbf{r} = 0$$

for every piecewise-smooth Jordan curve C.

34. Let C be the line segment from the point (x_1, y_1) to the point (x_2, y_2). Show that

$$\int_C -y\,dx + x\,dy = x_1y_2 - x_2y_1.$$

35. Let $(x_1, y_1), (x_2, y_2), \ldots, (x_n, y_n)$ be the vertices of a polygon in counterclockwise order. Show that the area of the polygon is

$$A = \tfrac{1}{2}[(x_1y_2 - x_2y_1) + (x_2y_3 - x_3y_2) + \cdots$$
$$+ (x_{n-1}y_n - x_ny_{n-1}) + (x_ny_1 - x_1y_n)].$$

36. Use the formula of Exercise 35 to find the area of:

(a) The triangle with vertices $(0, 0), (2, 1), (1, 4)$.

(b) The pentagon with vertices $(0, 0), (3, 1), (2, 4), (0, 6), (-1, 2)$.

■ PROJECT 17.5 The Folium of Descartes

The word *folium* means "leaf" in Latin. The *folium of Descartes*, shown in the figure, was introduced by Descartes in 1638. It is the graph of the equation:

$$x^3 + y^3 = 3axy, \ a > 0.$$

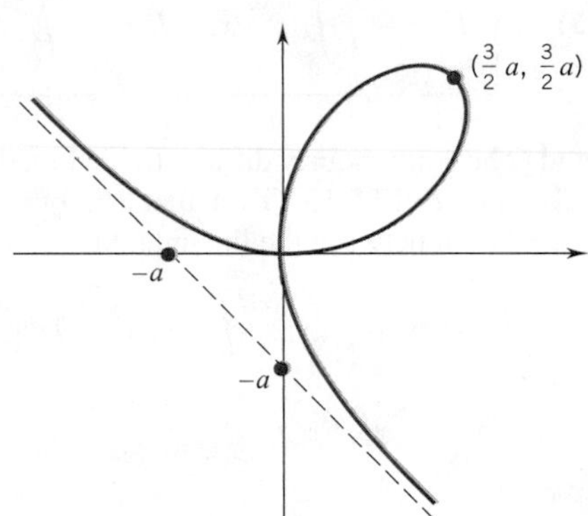

Problem 1. Show that if the parameter t is defined by setting $t = y/x$, then

$$x = \frac{3at}{1+t^3}, \quad y = \frac{3at^2}{1+t^3}, \ t \neq -1,$$

are parametric equations for the folium.

Problem 2. Use this parametric representation of the folium to show that the line $x + y + a = 0$ is an asymptote of the curve. HINT: Show that $x + y \to -a$ as $t \to -1$.

Problem 3. Show that the folium is symmetric with respect to the line $y = x$. Use this fact and the results obtained in Problems 1 and 2 to describe the orientation of the curve as t varies from $-\infty$ to ∞ $(t \neq -1)$.

Problem 4. Express the area of the loop as an integral. Use the third equation in (17.5.2) to find the area of the loop. HINT: You may want to double the area of the bottom half of the loop.

Problem 5. Show that the area of the loop equals the area of the region between the curve and its asymptote.

The curve defined by

$$x^{2n+1} + y^{2n+1} = (2n+1)ax^ny^n, \quad a > 0,$$

where n is a positive integer greater than 1, is called the *generalized folium of Descartes*. The following is a parametric representation of the generalized folium:

$$x = \frac{(2n+1)at^n}{1+t^{2n+1}}, \quad y = \frac{(2n+1)at^{n+1}}{1+t^{2n+1}}, \ t \neq -1.$$

Problem 6.

a. Use a graphing utility to graph several of these curves.

b. Calculate the area of the loop of the generalized folium.

---

*SUPPLEMENT TO SECTION 17.5

A JUSTIFICATION OF THE JACOBIAN AREA FORMULA

We based the change of variables for double integrals on the Jacobian area formula [Formula (16.10.1)]. Green's theorem enables us to derive this formula under the conditions spelled out as follows:

Let Γ be a Jordan region in the uv-plane with a piecewise-smooth boundary C_Γ. A vector function $\mathbf{r}(u, v) = x(u, v)\,\mathbf{i} + y(u, v)\,\mathbf{j}$ with continuous second partials maps Γ onto a region Ω of the xy-plane. If $\mathbf{r}$ is one-to-one on the interior of Γ and the Jacobian J of the components of $\mathbf{r}$ is different from zero on the interior of Γ, then

$$\text{area of } \Omega = \iint_\Gamma |J(u, v)|\, dudv.$$

PROOF Suppose that C_Γ is parametrized by $u = u(t), v = v(t)$ with $t \in [a, b]$. Then the boundary of Ω is a piecewise-smooth curve C given by

$$\mathbf{r}[u(t), v(t)] = x[u(t), v(t)]\,\mathbf{i} + y[u(t), v(t)]\,\mathbf{j}, \quad t \in [a, b].$$

By Green's theorem, the area of Ω is

$$\begin{aligned}
\oint_C x\, dy &= \left| \int_a^b x[u(t), v(t)]\, \frac{d}{dt}\,(y[u(t), v(t)])\, dt \right| \\
&= \left| \int_a^b x[u(t), v(t)] \left(\frac{\partial y}{\partial u}\,[u(t), v(t)]\, u'(t) + \frac{\partial y}{\partial v}\,[u(t), v(t)]\, v'(t) \right) dt \right| \\
&= \left| \int_a^b \left(x[u(t), v(t)] \frac{\partial y}{\partial u}[u(t), v(t)]\, u'(t) + x[u(t), v(t)] \frac{\partial y}{\partial v}[u(t), v(t)]\, v'(t) \right) dt \right| \\
&= \left| \int_{C_\Gamma} x \frac{\partial y}{\partial u}\, du + x \frac{\partial y}{\partial v} dv \right| \\
&\overset{\text{again by Green's theorem}}{=} \left| \iint_\Gamma \left[\frac{\partial}{\partial u}\left(x \frac{\partial y}{\partial v} \right) - \frac{\partial}{\partial v}\left(x \frac{\partial y}{\partial u} \right) \right] dudv \right|.
\end{aligned}$$

Now
$$\begin{aligned}
\frac{\partial}{\partial u}\left(x \frac{\partial y}{\partial v} \right) - \frac{\partial}{\partial v}\left(x \frac{\partial y}{\partial u} \right) &= \frac{\partial x}{\partial u}\frac{\partial y}{\partial v} + x \frac{\partial^2 y}{\partial u \partial v} - \frac{\partial x}{\partial v}\frac{\partial y}{\partial u} - x \frac{\partial^2 y}{\partial v \partial u} \\
&= \frac{\partial x}{\partial u}\frac{\partial y}{\partial v} - \frac{\partial x}{\partial v}\frac{\partial y}{\partial u} = J(u, v).
\end{aligned}$$

Therefore
$$\text{area of } \Omega = \left| \iint_\Gamma J(u, v)\, dudv \right| = \iint_\Gamma |J(u, v)|\, dudv,$$

the final equality holding because $J(u, v)$ cannot change sign on Γ. ☐

■ 17.6 PARAMETRIZED SURFACES; SURFACE AREA

You have seen that a space curve can be parametrized by a vector function $\mathbf{r} = \mathbf{r}(u)$ where u ranges over some interval I of the u-axis (Figure 17.6.1). In an analogous manner, we can parametrize a surface S in space by a vector function $\mathbf{r} = \mathbf{r}(u, v)$ where (u, v) ranges over some region Ω of the uv-plane (Figure 17.6.2).

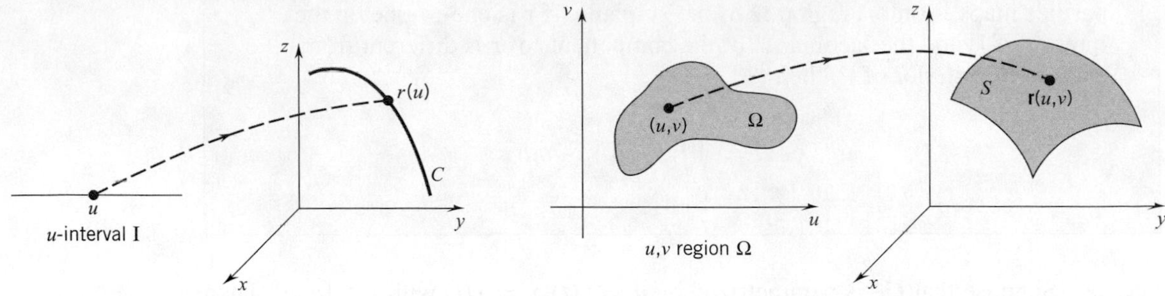

Figure 17.6.1 Figure 17.6.2

Example 1 (*The graph of a function*) Just as the graph of a function

$$y = f(x), \qquad x \in [a, b]$$

can be parametrized by setting

$$\mathbf{r}(u) = u\,\mathbf{i} + f(u)\,\mathbf{j}, \qquad u \in [a, b],$$

the graph of a function

$$z = f(x, y), \qquad (x, y) \in \Omega$$

can be parametrized by setting

$$\mathbf{r}(u, v) = u\,\mathbf{i} + v\,\mathbf{j} + f(u, v)\,\mathbf{k}, \qquad (u, v) \in \Omega.$$

As (u, v) ranges over Ω, the tip of $\mathbf{r}(u, v)$ traces out the surface, which is the graph of f. □

Example 2 (*A plane*) If two vectors $\mathbf{a}$ and $\mathbf{b}$ are not parallel, then the set of all linear combinations $u\,\mathbf{a} + v\,\mathbf{b}$ generate a plane p_0 that passes through the origin. We can parametrize this plane by setting

$$\mathbf{r}(u, v) = u\,\mathbf{a} + v\,\mathbf{b}, \quad u, v \text{ real numbers.}$$

The plane p that is parallel to p_0 and passes through the tip of $\mathbf{c}$ can be parametrized by setting

$$\mathbf{r}(u, v) = u\,\mathbf{a} + v\,\mathbf{b} + \mathbf{c}, \quad u, v \text{ real numbers.}$$

Note that the plane contains the lines

$$l_1\colon \mathbf{r}(u, 0) = u\,\mathbf{a} + \mathbf{c} \qquad \text{and} \qquad l_2\colon \mathbf{r}(0, v) = v\,\mathbf{b} + \mathbf{c}. \quad \square$$

Example 3 (*A sphere*) The sphere of radius a centered at the origin can be parametrized by

$$\mathbf{r}(u, v) = a\cos u\cos v\,\mathbf{i} + a\sin u\cos v\,\mathbf{j} + a\sin v\,\mathbf{k}$$

with (u, v) ranging over the rectangle $R\colon 0 \le u \le 2\pi, -\frac{1}{2}\pi \le v \le \frac{1}{2}\pi$.

To derive this parametrization, we refer to Figure 17.6.3. The points of latitude v (see the figure) form a circle of radius $a\cos v$ on the horizontal plane $z = a\sin v$.

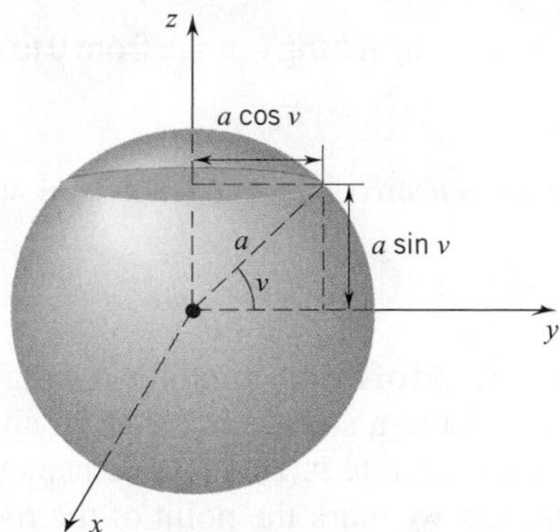

Figure 17.6.3

This circle can be parametrized by

$$\mathbf{R}(u) = a\cos v\,(\cos u\,\mathbf{i} + \sin u\,\mathbf{j}) + a\sin v\,\mathbf{k}, \quad u \in [0, 2\pi].$$

This expands to give

$$\mathbf{R}(u) = a\cos u\cos v\,\mathbf{i} + a\sin u\cos v\,\mathbf{j} + a\sin v\,\mathbf{k}, \quad u \in [0, 2\pi].$$

Letting v range from $-\frac{1}{2}\pi$ to $\frac{1}{2}\pi$, we obtain the entire sphere.

The xyz-equation for this same sphere is $x^2 + y^2 + z^2 = a^2$. It is easy to verify that the parametrization satisfies this equation:

$$\begin{aligned} x^2 + y^2 + z^2 &= a^2\cos^2 u\cos^2 v + a^2\sin^2 u\cos^2 v + a^2\sin^2 v \\ &= a^2(\cos^2 u + \sin^2 u)\cos^2 v + a^2\sin^2 v \\ &= a^2(\cos^2 v + \sin^2 v) = a^2. \quad \square \end{aligned}$$

Example 4 (*A cone*) Figure 17.6.4 shows a right circular cone with vertex semiangle α and slant height s. The points of slant height v (see the figure) form a circle of radius $v\sin\alpha$ on the horizontal plane $z = v\cos\alpha$. This circle can be parametrized by

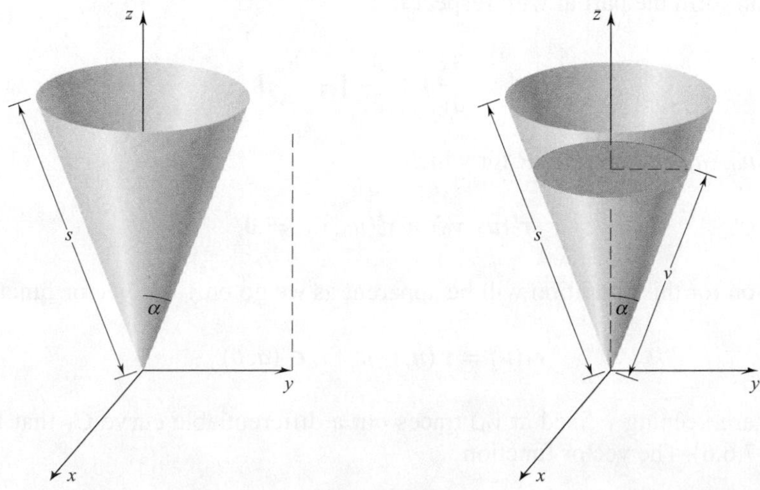

Figure 17.6.4

$$\begin{aligned}\mathbf{R}(u) &= v\sin\alpha(\cos u\,\mathbf{i} + \sin u\,\mathbf{j}) + v\cos\alpha\,\mathbf{k}\\ &= v\cos u\sin\alpha\,\mathbf{i} + v\sin u\sin\alpha\,\mathbf{j} + v\cos\alpha\,\mathbf{k}, \qquad u \in [0, 2\pi].\end{aligned}$$

Since we can obtain the entire cone by letting v range from 0 to s, the cone is parametrized by setting

$$\mathbf{r}(u, v) = v\cos u\sin\alpha\,\mathbf{i} + v\sin u\sin\alpha\,\mathbf{j} + v\cos\alpha\,\mathbf{k},$$

with $0 \le u \le 2\pi, 0 \le v \le s$. □

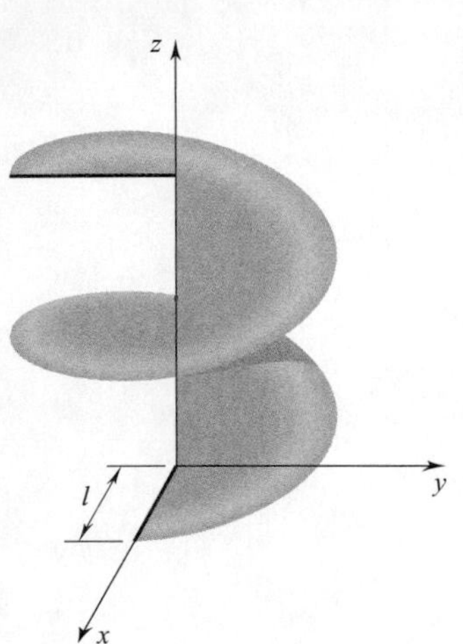

Figure 17.6.5

Example 5 (*A spiral ramp*) A rod of length l initially resting on the x-axis and attached at one end to the z-axis sweeps out a surface by rotating about the z-axis at constant rate ω while climbing at a constant rate b. The surface is pictured in Figure 17.6.5.

To parametrize this surface, we mark the point of the rod at a distance u from the z-axis ($0 \le u \le l$) and ask for the position of this point at time v. At time v the rod will have climbed a distance bv and rotated through an angle ωv. Thus the point will be found at the tip of the vector

$$u(\cos\omega v\,\mathbf{i} + \sin\omega v\,\mathbf{j}) + bv\,\mathbf{k} = u\cos\omega v\,\mathbf{i} + u\sin\omega v\,\mathbf{j} + bv\,\mathbf{k}.$$

The entire surface can be parametrized by setting

$$\mathbf{r}(u, v) = u\cos\omega v\,\mathbf{i} + u\sin\omega v\,\mathbf{j} + bv\,\mathbf{k} \quad \text{with } 0 \le u \le l,\ v \ge 0. \ \square$$

The Fundamental Vector Product

Let S be a surface parametrized by a differentiable vector function

$$\mathbf{r} = \mathbf{r}(u, v) = x(u, v)\,\mathbf{i} + y(u, v)\,\mathbf{j} + z(u, v)\,\mathbf{k}.$$

For simplicity, let us suppose that (u, v) varies over the open rectangle R: $a < u < b$, $c < v < d$. Since $\mathbf{r}$ is a function of u and v, we can form the partial with respect to u,

$$\mathbf{r}'_u = \frac{\partial x}{\partial u}\mathbf{i} + \frac{\partial y}{\partial u}\mathbf{j} + \frac{\partial z}{\partial u}\mathbf{k},$$

and we can form the partial with respect to v,

$$\mathbf{r}'_v = \frac{\partial x}{\partial v}\mathbf{i} + \frac{\partial y}{\partial v}\mathbf{j} + \frac{\partial z}{\partial v}\mathbf{k}.$$

Now let (u_0, v_0) be a point of R for which

$$\mathbf{r}'_u(u_0, v_0) \times \mathbf{r}'_v(u_0, v_0) \ne \mathbf{0}.$$

(The reason for this condition will be apparent as we go on.) The vector function

$$\mathbf{r}_1(u) = \mathbf{r}(u, v_0), \quad u \in (a, b)$$

(here we are keeping v fixed at v_0) traces out a differentiable curve C_1 that lies on S (Figure 17.6.6). The vector function

$$\mathbf{r}_2(v) = \mathbf{r}(u_0, v), \quad v \in (c, d)$$

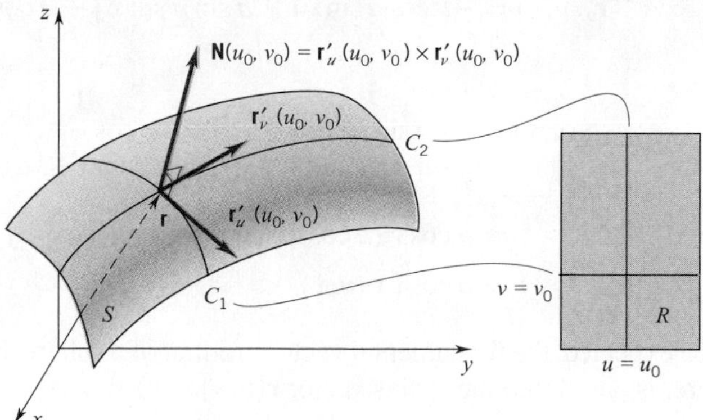

Figure 17.6.6

(this time we are keeping u fixed at u_0) traces out a differentiable curve C_2 that also lies on S. Both curves pass through the tip of $\mathbf{r}(u_0, v_0)$:

$$\text{at } C_1 \text{ with tangent vector } \mathbf{r}_1'(u_0) = \mathbf{r}_u'(u_0, v_0),$$
$$\text{at } C_2 \text{ with tangent vector } \mathbf{r}_2'(v_0) = \mathbf{r}_v'(u_0, v_0).$$

The cross product $\mathbf{N}(u_0, v_0) = \mathbf{r}_u'(u_0, v_0) \times \mathbf{r}_v'(u_0, v_0)$, which we have assumed to be different from zero, is perpendicular to both curves at the tip of $\mathbf{r}(u_0, v_0)$ and can be taken as a normal to the surface at that point. We record the result as follows:

(17.6.1)

> If S is the surface given by a differentiable function $\mathbf{r} = \mathbf{r}(u, v)$, then the vector $\mathbf{N}(u, v) = \mathbf{r}_u'(u, v) \times \mathbf{r}_v'(u, v)$ is perpendicular to the surface at the tip of $\mathbf{r}(u, v)$ and, if different from zero, can be taken as a normal to the surface at this point.

The cross product

$$\mathbf{N} = \mathbf{r}_u' \times \mathbf{r}_v' = \begin{vmatrix} \mathbf{i} & \mathbf{j} & \mathbf{k} \\ \dfrac{\partial x}{\partial u} & \dfrac{\partial y}{\partial u} & \dfrac{\partial z}{\partial u} \\ \dfrac{\partial x}{\partial v} & \dfrac{\partial y}{\partial v} & \dfrac{\partial z}{\partial v} \end{vmatrix}$$

is called the *fundamental vector product* of the surface.

Example 6 For the plane $\mathbf{r}(u, v) = u\,\mathbf{a} + v\,\mathbf{b} + \mathbf{c}$ we have

$$\mathbf{r}_u'(u, v) = \mathbf{a}, \quad \mathbf{r}_v'(u, v) = \mathbf{b} \quad \text{and therefore} \quad \mathbf{N}(u, v) = \mathbf{a} \times \mathbf{b}.$$

The vector $\mathbf{a} \times \mathbf{b}$ is normal to the plane. □

Example 7 We parametrized the sphere $x^2 + y^2 + z^2 = a^2$ by setting

$$\mathbf{r}(u, v) = a\cos u \cos v\,\mathbf{i} + a\sin u \cos v\,\mathbf{j} + a\sin v\,\mathbf{k}$$

with $0 \leq u \leq 2\pi, -\frac{1}{2}\pi \leq v \leq \frac{1}{2}\pi$. In this case

$$\mathbf{r}_u'(u, v) = -a\sin u \cos v\,\mathbf{i} + a\cos u \cos v\,\mathbf{j}$$

and $\qquad \mathbf{r}'_v(u,v) = -a\cos u\sin v\,\mathbf{i} - a\sin u\sin v\,\mathbf{j} + a\cos v\,\mathbf{k}.$

Thus

$$\mathbf{N}(u,v) = \begin{vmatrix} \mathbf{i} & \mathbf{j} & \mathbf{k} \\ -a\sin u\cos v & a\cos u\cos v & 0 \\ -a\cos u\sin v & -a\sin u\sin v & a\cos v \end{vmatrix}$$

check this ↓

$$= a\cos v\,(a\cos u\cos v\,\mathbf{i} + a\sin u\cos v\,\mathbf{j} + a\sin v\,\mathbf{k})$$

$$= a\cos v\,\mathbf{r}(u,v).$$

As was to be expected, the fundamental vector product of a sphere, being perpendicular to the sphere, is parallel to the radius vector $\mathbf{r}(u,v)$. ❑

The Area of a Parametrized Surface

A linear function

$$\mathbf{r}(u,v) = u\,\mathbf{a} + v\,\mathbf{b} + \mathbf{c} \qquad (\mathbf{a} \text{ and } \mathbf{b} \text{ not parallel})$$

parametrizes a plane p. Horizontal lines from the uv-plane, lines with equations of the form $v = v_0$, are mapped onto lines parallel to **a**, and vertical lines, $u = u_0$, are mapped onto lines parallel to **b**:

$$\mathbf{r}(u,v_0) = \underset{\uparrow\ \text{direction vector}}{u\,\mathbf{a}} + \underbrace{v_0\,\mathbf{b} + \mathbf{c}}_{\text{constant}}, \qquad \mathbf{r}(u_0,v) = \underset{\uparrow\ \text{direction vector}}{v\,\mathbf{b}} + \underbrace{u_0\,\mathbf{a} + \mathbf{c}}_{\text{constant}}.$$

Thus a rectangle R in the uv-plane with sides parallel to the u and v axes,

$$R: \quad u_1 \le u \le u_2, \quad v_1 \le v \le v_2, \qquad \text{(see Figure 17.6.7)}$$

is mapped onto a parallelogram on p with sides parallel to **a** and **b**. What is important to us here is that

$$\text{the area of the parallelogram} = \|\mathbf{a} \times \mathbf{b}\| \cdot \text{(the area of } R).$$

The parallelogram is generated by the vectors

$$\mathbf{r}(u_2,v_1) - \mathbf{r}(u_1,v_1) = (u_2\,\mathbf{a} + v_1\,\mathbf{b} + \mathbf{c}) - (u_1\,\mathbf{a} + v_1\,\mathbf{b} + \mathbf{c}) = (u_2 - u_1)\,\mathbf{a},$$

$$\mathbf{r}(u_1,v_2) - \mathbf{r}(u_1,v_1) = (u_1\,\mathbf{a} + v_2\,\mathbf{b} + \mathbf{c}) - (u_1\,\mathbf{a} + v_1\,\mathbf{b} + \mathbf{c}) = (v_2 - v_1)\,\mathbf{b}.$$

The area of the parallelogram is thus

$$\|(u_2 - u_1)\,\mathbf{a} \times (v_2 - v_1)\,\mathbf{b}\| = \|\mathbf{a} \times \mathbf{b}\|(u_2 - u_1)(v_2 - v_1)$$

$$= \|\mathbf{a} \times \mathbf{b}\| \cdot (\text{area of } R).$$

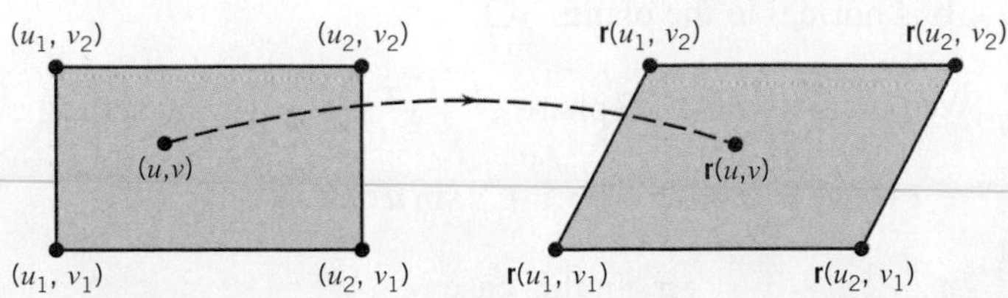

Figure 17.6.7

We can summarize as follows:

(17.6.2) Let R be a rectangle in the uv-plane with sides parallel to the coordinate axes. If $\mathbf{a}$ and $\mathbf{b}$ are not parallel, the linear function

$$\mathbf{r}(u, v) = u\,\mathbf{a} + v\,\mathbf{b} + \mathbf{c}, \quad (u, v) \in R$$

parametrizes a parallelogram with sides parallel to $\mathbf{a}$ and $\mathbf{b}$, and

$$\text{the area of the parallelogram } = \|\mathbf{a} \times \mathbf{b}\| \cdot (\text{the area of } R.)$$

More generally, let's suppose that we have a surface S parametrized by a continuously differentiable function

$$\mathbf{r} = \mathbf{r}(u, v), \qquad (u, v) \in \Omega.$$

We assume that Ω is a basic region in the uv-plane and that $\mathbf{r}$ is one-to-one on the interior of Ω. (We don't want $\mathbf{r}$ to cover parts of S more than once.) Also we assume that the fundamental vector product $\mathbf{N} = \mathbf{r}'_u \times \mathbf{r}'_v$ is never zero on the interior of Ω. (We can then use it as a normal.) Under these conditions we call S a *smooth surface* and define

(17.6.3)
$$\text{area of } S = \iint_{\Omega} \|\mathbf{N}(u, v)\|\, du dv.$$

We show the reasoning behind this definition in the case where Ω is a rectangle R with sides parallel to the coordinate axes. We begin by breaking up R into N little rectangles $R_1, \ldots, R_N$. This induces a decomposition of S into little pieces $S_1, \ldots, S_N$. Taking (u_i^*, v_i^*) as the center of R_i, we have the tip of $\mathbf{r}(u_i^*, v_i^*)$ in S_i. Since the vector $\mathbf{r}'_u(u_i^*, v_i^*) \times \mathbf{r}'_v(u_i^*, v_i^*)$ is normal to the surface at the tip of $\mathbf{r}(u_i^*, v_i^*)$, we can parametrize the tangent plane at this point by the linear function

$$\mathbf{f}(u, v) = u\,\mathbf{r}'_u(u_i^*, v_i^*) + v\,\mathbf{r}'_v(u_i^*, v_i^*) + [\mathbf{r}(u_i^*, v_i^*) - u_i^*\,\mathbf{r}'_u(u_i^*, v_i^*) - v_i^*\,\mathbf{r}'_v(u_i^*, v_i^*)].$$

(Check that this linear function gives the right plane.) S_i is the portion of S that corresponds to R_i. The portion of the tangent plane that corresponds to this same R_i is a parallelogram with area

$$\|\mathbf{r}'_u(u_i^*, v_i^*) \times \mathbf{r}'_v(u_i^*, v_i^*)\| \cdot (\text{area of } R_i) = \|\mathbf{N}(u_i^*, v_i^*)\| \cdot (\text{area of } R_i). \qquad \text{[by (17.6.2)]}$$

Taking this as our estimate for the area of S_i, we have

$$\text{area of } S = \sum_{i=1}^{N} \text{area of } S_i \cong \sum_{i=1}^{N} \|\mathbf{N}(u_i^*, v_i^*)\| \cdot (\text{area of } R_i).$$

This is a Riemann sum for

$$\iint_{R} \|\mathbf{N}(u, v)\|\, du dv.$$

and tends to this integral as the maximal diameter of the R_i tends to zero.

To make sure that Formula (17.6.3) does not violate our previously established notion of area, we must verify that it gives the expected result both for plane regions and for surfaces of revolution. This is done in Examples 9 and 10. By way of introduction we begin with the sphere.

Example 8 (*The surface area of a sphere*) The function

$$\mathbf{r}(u, v) = a\cos u\cos v\,\mathbf{i} + a\sin u\cos v\,\mathbf{j} + a\sin v\,\mathbf{k},$$

with (u, v) ranging over the set $\Omega: 0 \le u \le 2\pi, -\frac{1}{2}\pi \le v \le \frac{1}{2}\pi$, parametrizes a sphere of radius a. For this parametrization

$$\mathbf{N}(u, v) = a\cos v\,\mathbf{r}(u, v) \qquad \text{and} \qquad \|\mathbf{N}(u, v)\| = a^2|\cos v| = a^2\cos v.$$

Example 7 ↑ $\qquad$ $-\frac{1}{2}\pi \le v \le \frac{1}{2}\pi$ ↑

According to the new formula,

$$\text{area of the sphere} = \iint_{\Omega} a^2\cos v\,dudv$$

$$= \int_0^{2\pi}\left(\int_{-\frac{1}{2}\pi}^{\frac{1}{2}\pi} a^2\cos v\,dv\right)du = 2\pi a^2\int_{-\frac{1}{2}\pi}^{\frac{1}{2}\pi}\cos v\,dv = 4\pi a^2,$$

which, as you know, is correct. □

Example 9 (*The area of a plane region*) If S is a plane region Ω, then S can be parametrized by setting

$$\mathbf{r}(u, v) = u\,\mathbf{i} + v\,\mathbf{j}, \qquad (u, v) \in \Omega.$$

Here $\mathbf{N}(u, v) = \mathbf{r}'_u(u, v) \times \mathbf{r}'_v(u, v) = \mathbf{i} \times \mathbf{j} = \mathbf{k}$ and $\|\mathbf{N}(u, v)\| = 1$. In this case (17.6.3) reduces to the familiar formulas

$$A = \iint_{\Omega} dudv. \quad □$$

Example 10 (*The area of a surface of revolution*) Let S be the surface generated by revolving the graph of a function

$$y = f(x), \quad x \in [a, b], \qquad \text{(Figure 17.6.8)}$$

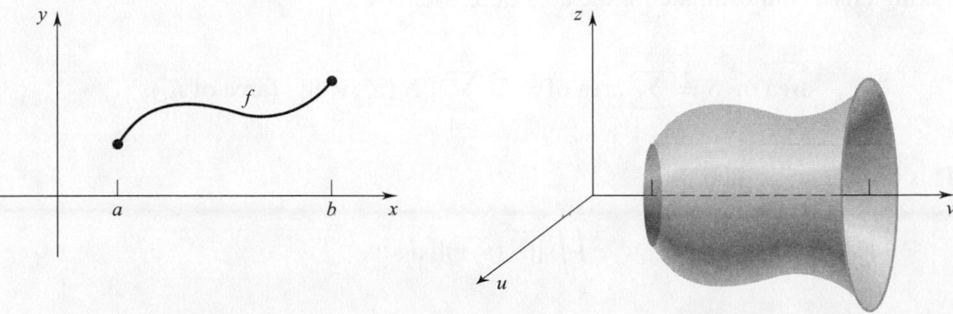

Figure 17.6.8

about the x-axis. We assume that f is positive and continuously differentiable. We can parametrize S by setting.

$$\mathbf{r}(u,v) = v\,\mathbf{i} + f(v)\cos u\,\mathbf{j} + f(v)\sin u\,\mathbf{k}$$

with (u, v) ranging over the set $\Omega: 0 \le u \le 2\pi, a \le v \le b$. (We leave it to you to verify that this is right.) In this case

$$\mathbf{N}(u,v) = \mathbf{r}'_u(u,v) \times \mathbf{r}'_v(u,v) = \begin{vmatrix} \mathbf{i} & \mathbf{j} & \mathbf{k} \\ 0 & -f(v)\sin u & f(v)\cos u \\ 1 & f'(v)\cos u & f'(v)\sin u \end{vmatrix}$$

$$= -f(v)f'(v)\,\mathbf{i} + f(v)\cos u\,\mathbf{j} + f(v)\sin u\,\mathbf{k}.$$

Therefore $\|\mathbf{N}(u,v)\| = f(v)\sqrt{[f'(v)]^2+1}$ and

$$\text{area of } S = \iint_\Omega f(v)\sqrt{[f'(v)]^2+1}\,du\,dv$$

$$= \int_0^{2\pi}\left(\int_a^b f(v)\sqrt{[f'(v)]^2+1}\,dv\right)du = \int_a^b 2\pi f(v)\sqrt{[f'(v)]^2+1}\,dv.$$

This is in agreement with Formula (9.9.3). ☐

Example 11 (*Spiral ramp*) One turn of the spiral ramp of Example 5 is the surface

$$S: \quad \mathbf{r}(u,v) = u\cos\omega v\,\mathbf{i} + u\sin\omega v\,\mathbf{j} + bv\,\mathbf{k}$$

with (u, v) ranging over the set $\Omega: 0 \le u \le l, 0 \le v \le 2\pi/\omega$. In this case

$$\mathbf{r}'_u(u,v) = \cos\omega v\,\mathbf{i} + \sin\omega v\,\mathbf{j}, \quad \mathbf{r}'_v(u,v) = -\omega u\sin\omega v\,\mathbf{i} + \omega u\cos\omega v\,\mathbf{j} + b\,\mathbf{k}.$$

Therefore

$$\mathbf{N}(u,v) = \begin{vmatrix} \mathbf{i} & \mathbf{j} & \mathbf{k} \\ \cos\omega v & \sin\omega v & 0 \\ -\omega u\sin\omega v & \omega u\cos\omega v & b \end{vmatrix} = b\sin\omega v\,\mathbf{i} - b\cos\omega v\,\mathbf{j} + \omega u\,\mathbf{k},$$

and

$$\|\mathbf{N}(u,v)\| = \sqrt{b^2+\omega^2u^2}.$$

Thus

$$\text{area of } S = \iint_\Omega \sqrt{b^2+\omega^2u^2}\,du\,dv$$

$$= \int_0^{2\pi/\omega}\left(\int_0^l \sqrt{b^2+\omega^2u^2}\,du\right)dv = \frac{2\pi}{\omega}\int_0^l \sqrt{b^2+\omega^2u^2}\,du.$$

The integral can be evaluated by setting $u = (b/\omega)\tan x$. ☐

The Area of a Surface $z = f(x, y)$

Figure 17.6.9 shows a surface that projects onto a basic region Ω of the xy-plane. Above each point (x, y) of Ω there is one and only one point of S. The surface S is then the graph of a function

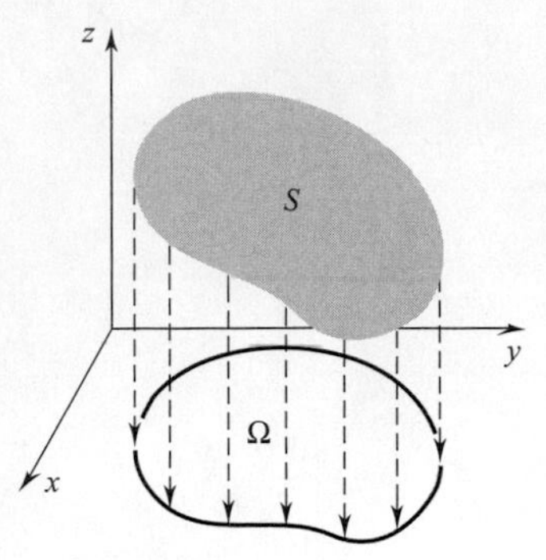

Figure 17.6.9

$$z = f(x, y), \quad (x, y) \in \Omega.$$

As we show, if f is continuously differentiable, then

(17.6.4)
$$\text{area of } S = \iint_\Omega \sqrt{[f_x(x, y)]^2 + [f_y(x, y)]^2 + 1}\, dxdy.$$

DERIVATION OF FORMULA (17.6.4) We can parametrize S by setting

$$\mathbf{r}(u, v) = u\,\mathbf{i} + v\,\mathbf{j} + f(u, v)\,\mathbf{k}, \quad (u, v) \in \Omega.$$

We may just as well use x and y and write

$$\mathbf{r}(x, y) = x\,\mathbf{i} + y\,\mathbf{j} + f(x, y)\,\mathbf{k}, \quad (x, y) \in \Omega.$$

Clearly $\quad \mathbf{r}'_x(x, y) = \mathbf{i} + f_x(x, y)\,\mathbf{k} \quad \text{and} \quad \mathbf{r}'_y(x, y) = \mathbf{j} + f_y(x, y)\,\mathbf{k}.$

Thus
$$\mathbf{N}(x, y) = \begin{vmatrix} \mathbf{i} & \mathbf{j} & \mathbf{k} \\ 1 & 0 & f_x(x, y) \\ 0 & 1 & f_y(x, y) \end{vmatrix} = -f_x(x, y)\,\mathbf{i} - f_y(x, y)\,\mathbf{j} + \mathbf{k}.$$

Therefore $\|\mathbf{N}(x, y)\| = \sqrt{[f_x(x, y)]^2 + [f_y(x, y)]^2 + 1}$ and the formula is verified. □

Example 12 Find the surface area of that part of the parabolic cylinder $z = y^2$ that lies over the triangle with vertices $(0, 0), (0, 1), (1, 1)$ in the xy-plane.

SOLUTION Here $f(x, y) = y^2$ so that

$$f_x(x, y) = 0, \quad f_y(x, y) = 2y.$$

The base triangle can be expressed by writing

$$\Omega\colon\ 0 \le y \le 1, \quad 0 \le x \le y.$$

The surface has area

$$\begin{aligned} A &= \iint_\Omega \sqrt{[f_x(x, y)]^2 + [f_y(x, y)]^2 + 1}\, dxdy \\ &= \int_0^1 \int_0^y \sqrt{4y^2 + 1}\, dx\, dy \\ &= \int_0^1 y\sqrt{4y^2 + 1}\, dy = \left[\tfrac{1}{12}(4y^2 + 1)^{3/2}\right]_0^1 = \tfrac{1}{12}(5\sqrt{5} - 1). \end{aligned}$$ □

Example 13 Find the surface area of that part of the hyperbolic paraboloid $z = xy$ that lies inside the cylinder $x^2 + y^2 = a^2$. See Figure 17.6.10.

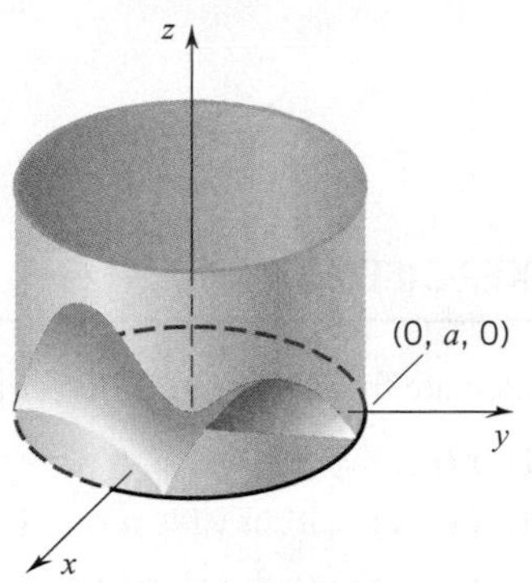

Figure 17.6.10

SOLUTION Here $f(x,y) = xy$, so that $f_x(x,y) = y, \quad f_y(x,y) = x$.
The formula gives

$$A = \iint_{\Omega} \sqrt{y^2 + x^2 + 1}\, dxdy.$$

In polar coordinates the base region takes the form

$$\Gamma\colon\ 0 \le r \le a, \quad 0 \le \theta \le 2\pi.$$

Thus we have

$$A = \iint_{\Gamma} \sqrt{r^2 + 1}\, r\, drd\theta = \int_0^{2\pi}\int_0^a \sqrt{r^2 + 1}\, r\, dr\, d\theta = \tfrac{2}{3}\pi[(a^2 + 1)^{3/2} - 1]. \quad \square$$

There is an elegant version of this last area formula [Formula (17.6.4)] that is geometrically vivid. We know that the vector

$$\mathbf{r}'_x(x,y) \times \mathbf{r}'_y(x,y) = -f_x(x,y)\,\mathbf{i} - f_y(x,y)\,\mathbf{j} + \mathbf{k}$$

is normal to the surface at the point $(x, y, f(x,y))$. The unit vector in that direction, the vector

$$\mathbf{n}(x,y) = \frac{-f_x(x,y)\,\mathbf{i} - f_y(x,y)\,\mathbf{j} + \mathbf{k}}{\sqrt{[f_x(x,y)]^2 + [f_y(x,y)]^2 + 1}},$$

is called the *upper unit normal*. (It is the unit normal with a nonnegative **k**-component.)

Now let $\gamma(x,y)$ be the angle between $\mathbf{n}(x,y)$ and $\mathbf{k}$ (Figure 17.6.11). Since $\mathbf{n}(x,y)$ and $\mathbf{k}$ are both unit vectors,

$$\cos[\gamma(x,y)] = \mathbf{n}(x,y) \cdot \mathbf{k} = \frac{1}{\sqrt{[f_x(x,y)]^2 + [f_y(x,y)]^2 + 1}}.$$

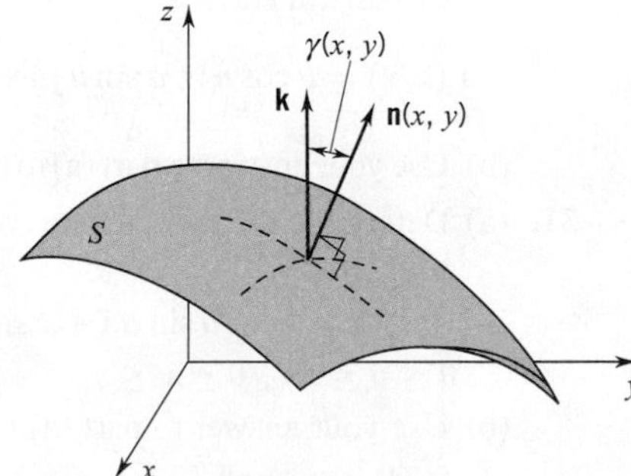

Figure 17.6.11

Taking reciprocals we have

$$\sec[\gamma(x,y)] = \sqrt{[f_x(x,y)]^2 + [f_y(x,y)]^2 + 1}.$$

The area formula can therefore be written

(17.6.5) $$A = \iint_{\Omega} \sec[\gamma(x,y)]\,dxdy.$$

EXERCISES 17.6

Calculate the fundamental vector product.

1. $\mathbf{r}(u,v) = (u^2 - v^2)\mathbf{i} + (u^2 + v^2)\mathbf{j} + 2uv\,\mathbf{k}$.

2. $\mathbf{r}(u,v) = u\cos v\,\mathbf{i} + u\sin v\,\mathbf{j} + \mathbf{k}$.

3. $\mathbf{r}(u,v) = (u+v)\mathbf{i} + (u+v)\mathbf{j} + (u-v)\mathbf{k}$.

4. $\mathbf{r}(u,v) = \cos u\sin v\,\mathbf{i} + \sin u\cos v\,\mathbf{j} + u\,\mathbf{k}$.

Find a parametric representation for the surface.

5. The upper half of the ellipsoid $4x^2 + 9y^2 + z^2 = 36$.

6. The part of the cylinder $x^2 + y^2 = 4$ that lies between the planes $z = 1$ and $z = 4$.

7. The part of the sphere $x^2 + y^2 + z^2 = 4$ that lies above the plane $z = \sqrt{2}$.

8. The part of the plane $z = x + 2$ that lies inside the cylinder $x^2 + y^2 = 1$.

9. $y = g(x,z), (x,z) \in \Omega$. **10.** $x = h(y,z), (y,z) \in \Gamma$.

Find an equation in x, y, z for the given surface and identify the surface.

11. $\mathbf{r}(u,v) = a\cos u\cos v\,\mathbf{i} + b\sin u\cos v\,\mathbf{j} + c\sin v\,\mathbf{k}$; $0 \le u \le 2\pi, -\frac{1}{2}\pi \le v \le \frac{1}{2}\pi$.

12. $\mathbf{r}(u,v) = au\cos v\,\mathbf{i} + bu\sin v\,\mathbf{j} + u^2\,\mathbf{k}$; $0 \le u, 0 \le v \le 2\pi$.

13. $\mathbf{r}(u,v) = au\cosh v\,\mathbf{i} + bu\sinh v\,\mathbf{j} + u^2\,\mathbf{k}$; u real, v real.

C 14. Use a graphing utility to draw the surface.

(a) Exercise 11 with $a = 4, b = 3, c = 2$; experiment with other values of a, b, c.

(b) Exercise 12 with $a = 3, b = 2$; experiment with other values of a and b.

(c) Exercise 13 with $a = 3, b = 2$; experiment with viewpoints to obtain a good view of the surface and try other values of a and b.

15. The graph of a continuously differentiable function $y = f(x), x \in [a,b]$ is revolved about the y-axis. Parametrize the surface given that $a \ge 0$.

16. Show that the area of the surface of Exercise 15 is given by the formula

$$A = \int_a^b 2\pi x\sqrt{1 + [f'(x)]^2}\,dx.$$

17. A plane p intersects the xy-plane at an angle γ. (Draw a figure.) Find the area of the region Γ on p given that the projection of Γ onto the xy-plane is a region Ω of area A_Ω.

18. Determine the area of the portion of the plane $x + y + z = a$ that lies within the cylinder $x^2 + y^2 = b^2$.

19. Find the area of the part of the plane $bcx + acy + abz = abc$ that lies within the first octant.

20. Find the area of the surface $z^2 = x^2 + y^2$ from $z = 0$ to $z = 1$.

21. Find the area of the surface $z = x^2 + y^2$ from $z = 0$ to $z = 4$.

Calculate the area of the surface

22. $z^2 = 2xy$ with $0 \le x \le a,\ 0 \le y \le b,\ z \ge 0$.

23. $z = a^2 - (x^2 + y^2)$ with $\frac{1}{4}a^2 \le x^2 + y^2 \le a^2$.

24. $3z^2 = (x+y)^3$ with $x + y \le 2,\ x \ge 0,\ y \ge 0$.

25. $3z = x^{3/2} + y^{3/2}$ with $0 \le x \le 1,\ 0 \le y \le x$.

26. $z = y^2$ with $0 \le x \le 1,\ 0 \le y \le 1$.

27. $x^2 + y^2 + z^2 - 4z = 0$ with $0 \le 3(x^2 + y^2) \le z^2,\ z \ge 2$.

28. $x^2 + y^2 + z^2 - 2az = 0$ with $0 \le x^2 + y^2 \le bz$. Assume $a > b > 0$.

29. (a) Find a formula for the area of a surface that is projectable onto a region Ω of the yz- plane; say,

$$S:\ x = g(y,z), \quad (y,z) \in \Omega.$$

Assume that g is continuously differentiable.

(b) Find a formula for the area of a surface that is projectable onto a region Ω of the xz-plane; say,

$$S:\ y = h(x,z), \quad (x,z) \in \Omega.$$

Assume that h is continuously differentiable.

30. (a) Determine the fundamental vector product for the cylindrical surface

$$\mathbf{r}(u,v) = a\cos u\,\mathbf{i} + a\sin u\,\mathbf{j} + v\,\mathbf{k};\quad 0 \le u \le 2\pi, 0 \le v \le l.$$

(b) Use your answer to part (a) to find the area of the surface.

31. (a) Determine the fundamental vector product for the cone of Example 4:

$\mathbf{r}(u,v) = v\cos u\sin\alpha\,\mathbf{i} + v\sin u\sin\alpha\,\mathbf{j} + v\cos\alpha\,\mathbf{k}$; $0 \le u \le 2\pi,\ 0 \le v \le s$.

(b) Use your answer to part (a) to calculate the area of the cone.

C 32. (a) Show that $\mathbf{r}(u) = a\cos u\sin v\mathbf{i} + a\sin u\sin v\mathbf{j} + b\cos v\mathbf{k}$, $0 \le u \le 2\pi, 0 \le v \le \pi$, parametrizes the ellipsoid of revolution

$$\frac{x^2}{a^2} + \frac{y^2}{a^2} + \frac{z^2}{b^2} = 1.$$

(b) Use a graphing utility to draw the surface with $a = 3, b = 4$; experiment with other values of a and b.

(c) Show that the surface area of the ellipsoid is given by the formula

$$A = 2\pi a \int_0^{\pi} \sin v \sqrt{b^2 \sin^2 v + a^2 \cos^2 v}\, dv.$$

33. (a) Show that $\mathbf{r}(u) = a \cos u \cosh v\, \mathbf{i} + b \sin u \cosh v\, \mathbf{j} + c \sinh v\, \mathbf{k}, 0 \le u \le 2\pi, v$ real, parametrizes the hyperboloid of one sheet

$$\frac{x^2}{a^2} + \frac{y^2}{b^2} - \frac{z^2}{c^2} = 1.$$

(b) Use a graphing utility to draw the surface with $a = 3, b = 2, c = 4$; experiment with other values of a, b, c.

(c) Set up a double integral for the surface area of the part of the hyperboloid in part (b) that lies between the planes $z = -3$ and $z = 3$.

34. (a) Show that $\mathbf{r}(u) = a \cos u \sinh v\, \mathbf{i} + b \sin u \sinh v\, \mathbf{j} + c \cosh v\, \mathbf{k}, \quad 0 \le u \le 2\pi, v$ real, parametrizes the hyperboloid of two sheets

$$\frac{x^2}{a^2} + \frac{y^2}{a^2} - \frac{z^2}{b^2} = -1.$$

(b) Use a graphing utility to draw the surface with $a = 3$, $b = 2, c = 4$; experiment with other values of a, b, c.

(c) Explain why the your graph shows only the upper surface of the hyperboloid. Change the parametrization so that your graph displays the lower half of the surface.

35. Let Ω be a plane region in space and let A_1, A_2, A_3 be the areas of the projections of Ω onto the three coordinate planes. Express the area of Ω in terms of A_1, A_2, A_3.

36. Let S be a surface given in cylindrical coordinates by an equation of the form $z = f(r, \theta), (r, \theta) \in \Omega$. Show that if f is continuously differentiable, then

$$\text{area of } S = \iint_{\Omega} \sqrt{r^2[f_r(r,\theta)]^2 + [(f_\theta(r,\theta)]^2 + r^2}\, dr d\theta$$

provided the integrand is never zero on the interior of Ω.

37. The following surfaces are given in cylindrical coordinates. Find the surface area.

(a) $z = r + \theta; \quad 0 \le r \le 1, \quad 0 \le \theta \le \pi.$

(b) $z = re^{\theta}; \quad 0 \le r \le a, \quad 0 \le \theta \le 2\pi.$

38. Show that, for a flat surface S that is part of the xy-plane, (17.6.3) gives

$$\text{area of } S = \iint_{\Omega} |J(u, v)|\, du\, dv$$

where J is the Jacobian of the components of a vector function, which is defined on some region Ω and parametrizes S. Except for notation this is Formula (16.10.1).

■ 17.7 SURFACE INTEGRALS

The Mass of a Material Surface

Imagine a thin distribution of matter spread out over a surface S. We call this a *material surface*.

If the mass density (the mass per unit area) is a constant λ throughout, then the total mass of the material surface is the density λ times the area of S:

$$M = \lambda\,(\text{area of } S).$$

If, however, the mass density varies continuously from point to point, $\lambda = \lambda(x, y, z)$, then the total mass must be calculated by integration.

To develop the appropriate integral, we suppose that

$$S\colon\ \mathbf{r} = \mathbf{r}(u, v) = x(u, v)\,\mathbf{i} + y(u, v)\,\mathbf{j} + z(u, v)\,\mathbf{k}, \qquad (u, v) \in \Omega,$$

is a smooth surface, a surface that meets the conditions for area formula (17.6.3).† Our first step is to break up Ω into N little basic regions $\Omega_1, \ldots, \Omega_N$. This decomposes the surface into N little pieces $S_1, \ldots, S_N$. The area of S_i is given by the integral

$$\iint_{\Omega_i} ||\mathbf{N}(u, v)||\, dudv. \qquad \text{[Formula (17.6.3)]}$$

† We repeat the conditions here: $\mathbf{r}$ is continuously differentiable; Ω is a basic region in the uv-plane; $\mathbf{r}$ is one-to-one on the interior of Ω; $\mathbf{N} = \mathbf{r}'_u \times \mathbf{r}'_v$ is never zero on the interior of Ω.

By the mean-value theorem for double integrals, there exists a point (u_i^*, v_i^*) in Ω_i for which

$$\iint_{\Omega_i} \|\mathbf{N}(u,v)\|\, du dv = \|\mathbf{N}(u_i^*, v_i^*)\|(\text{area of } \Omega_i).$$

It follows that

$$\text{area of } S_i = \|\mathbf{N}(u_i^*, v_i^*)\|(\text{area of } \Omega_i).$$

Since the point (u_i^*, v_i^*) is in Ω_i, the tip of $\mathbf{r}(u_i^*, v_i^*)$ is on S_i. The mass density at this point is $\lambda[\mathbf{r}(u_i^*, v_i^*)]$. If S_i is small (which we can guarantee by choosing Ω_i small), then the mass density on S_i is approximately the same throughout. Thus we can estimate M_i, the mass contribution of S_i, by writing

$$M_i \cong \lambda[\mathbf{r}(u_i^*, v_i^*)](\text{area of } S_i) = \lambda[\mathbf{r}(u_i^*, v_i^*)]\, \|\mathbf{N}(u_i^*, v_i^*)\|(\text{area of } \Omega_i).$$

Adding up these estimates, we have an estimate for the total mass of the surface:

$$M \cong \sum_{i=1}^{N} \lambda[\mathbf{r}(u_i^*, v_i^*)]\, \|\mathbf{N}(u_i^*, v_i^*)\|(\text{area of } \Omega_i)$$

$$= \sum_{i=1}^{N} \lambda[x(u_i^*, v_i^*), y(u_i^*, v_i^*), z(u_i^*, v_i^*)]\, \|\mathbf{N}(u_i^*, v_i^*)\|(\text{area of } \Omega_i).$$

This last expression is a Riemann sum for

$$\iint_{\Omega} \lambda[x(u,v), y(u,v), z(u,v)]\, \|\mathbf{N}(u,v)\|\, du dv$$

and tends to this integral as the maximal diameter of the Ω_i tends to zero. We can therefore conclude that

(17.7.1) $$M = \iint_{\Omega} \lambda[x(u,v), y(u,v), z(u,v)]\, \|\mathbf{N}(u,v)\|\, du dv.$$

Surface Integrals

The double integral in (17.7.1) can be calculated not only for a mass density function λ but for any scalar field H continuous over S. We call this integral *the surface integral of H over S* and write

(17.7.2) $$\iint_S H(x,y,z)\, d\sigma = \iint_{\Omega} H[x(u,v), y(u,v)\, z(u,v)]\, \|\mathbf{N}(u,v)\|\, du dv$$

Note that, if $H(x,y,z)$ is identically 1, then the right-hand side of (17.7.2) gives the area of S. Thus

(17.7.3) $$\iint_S d\sigma = \text{area of } S.$$

Example 1 Let $\mathbf{a} = a_1\,\mathbf{i}+a_2\,\mathbf{j}+a_3\,\mathbf{k}$ and $\mathbf{b} = b_1\,\mathbf{i}+b_2\,\mathbf{j}+b_3\,\mathbf{k}$ be nonzero vectors. Calculate

$$\iint_S xy\,d\sigma \quad \text{where} \quad S\colon \ \mathbf{r}(u,v) = u\,\mathbf{a} + v\,\mathbf{b};\ \ 0 \le u \le 1,\ \ 0 \le v \le 1.$$

SOLUTION Call the parameter set Ω. Then

$$\iint_S xy\,d\sigma = \iint_\Omega x(u,v)\,y(u,v)||\mathbf{N}(u,v)||\,dudv.$$

A simple calculation shows that $||\mathbf{N}(u,v)|| = ||\mathbf{a}\times\mathbf{b}||$. Thus

$$\iint_S xy\,d\sigma = ||\mathbf{a}\times\mathbf{b}|| \iint_\Omega x(u,v)\,y(u,v),\ dudv.$$

To find $x(u,v)$ and $y(u,v)$, we need the $\mathbf{i}$ and $\mathbf{j}$ components of $\mathbf{r}(u,v)$. We can get these as follows:

$$\begin{aligned}\mathbf{r}(u,v) = u\,\mathbf{a} + v\,\mathbf{b} &= u(a_1\,\mathbf{i} + a_2\,\mathbf{j} + a_3\,\mathbf{k}) + v(b_1\,\mathbf{i} + b_2\,\mathbf{j} + b_3\,\mathbf{k})\\ &= (a_1u + b_1v)\,\mathbf{i} + (a_2u + b_2v)\,\mathbf{j} + (a_3u + b_3v)\,\mathbf{k}.\end{aligned}$$

Therefore $x(u,v) = a_1u + b_1v$ and $y(u,v) = a_2u + b_2v$. We can now write

$$\begin{aligned}\iint_S xy\,d\sigma &= ||\mathbf{a}\times\mathbf{b}|| \iint_\Omega (a_1u + b_1v)(a_2u + b_2v)\,du\,dv\\ &= ||\mathbf{a}\times\mathbf{b}|| \int_0^1 \left(\int_0^1 \left[a_1a_2u^2 + (a_1b_2 + b_1a_2)uv + b_1b_2v^2\right] du\right) dv\\ &= ||\mathbf{a}\times\mathbf{b}||\ \left[\tfrac{1}{3}a_1a_2 + \tfrac{1}{4}(a_1b_2 + b_1a_2) + \tfrac{1}{3}b_1b_2\right]. \quad \square\end{aligned}$$

check this ⟶↑

Example 2 Calculate

$$\iint_S \sqrt{x^2 + y^2}\,d\sigma$$

where S is the spiral ramp of Example 11, Section 17.6:

$$S\colon \quad \mathbf{r}(u,v) = u\cos\omega v\,\mathbf{i} + u\sin\omega v\,\mathbf{j} + bv\,\mathbf{k};\quad 0 \le u \le l,\quad 0 \le v \le 2\pi/\omega.$$

SOLUTION Call the parameter set Ω. As you saw in Example 11, Section 17.6,

$$||\mathbf{N}(u,v)|| = \sqrt{b^2 + \omega^2u^2}.$$

Therefore

$$\begin{aligned}
\iint_S \sqrt{x^2+y^2}\,d\sigma &= \iint_\Omega \sqrt{[x(u,v)]^2+[y(u,v)]^2}\;||\mathbf{N}(u,v)||\,dudv \\
&= \iint_\Omega \sqrt{u^2\cos^2\omega v + u^2\sin^2\omega v}\;\sqrt{b^2+\omega^2u^2}\,dudv \\
&= \iint_\Omega u\sqrt{b^2+\omega^2u^2}\,dudv \qquad (u \geq 0 \text{ on } \Omega) \\
&= \int_0^{2\pi/\omega}\left(\int_0^l u\sqrt{b^2+\omega^2u^2}\,du\right)dv \\
&= \frac{2\pi}{\omega}\int_0^l u\sqrt{b^2+\omega^2u^2}\,du = \frac{2\pi}{3\omega^3}\left[(b^2+\omega^2l^2)^{3/2}-b^3\right]. \quad \square
\end{aligned}$$

Like the other integrals you have studied, the surface integral satisfies a mean-value condition; namely, if the scalar field H is continuous, then there is a point (x_0, y_0, z_0) on S for which

$$\iint_S H(x,y,z)\,d\sigma = H(x_0,y_0,z_0)(\text{ area of } S).$$

we call $H(x_0, y_0, z_0)$ *the average value of H on S*. Thus we can write

(17.7.4)
$$\iint_S H(x,y,z)\,d\sigma = \begin{pmatrix}\text{average value}\\ \text{of } H \text{ on } S\end{pmatrix}\cdot(\text{area of } S).$$

We can also take weighted averages: if H and G are continuous on S and G is nonnegative on S, then there is a point (x_0, y_0, z_0) on S for which

(17.7.5)
$$\iint_S H(x,y,z)G(x,y,z)\,d\sigma = H(x_0,y_0,z_0)\iint_S G(x,y,z)\,d\sigma.$$

As you would expect, we call $H(x_0, y_0, z_0)$ *the G-weighted average of H on S*.

The coordinates of the centroid $(\overline{x}, \overline{y}, \overline{z})$ of a surface are simply averages taken over the surface: for a surface S of area A,

$$\overline{x}A = \iint_S x\,d\sigma, \quad \overline{y}A = \iint_S y\,d\sigma, \quad \overline{z}A = \iint_S z\,d\sigma.$$

In the case of a material surface of mass density $\lambda = \lambda(x, y, z)$, the coordinates of the center of mass (x_M, y_M, z_M) are density-weighted averages: for a surface S of total mass M,

$$x_M M = \iint_S x\lambda(x,y,z)\,d\sigma, \quad y_M M = \iint_S y\lambda(x,y,z)\,d\sigma, \quad z_M M = \iint_S z\lambda(x,y,z)\,d\sigma.$$

Example 3 Locate the center of mass of a material surface in the form of a hemispherical shell $x^2 + y^2 + z^2 = a^2$ with $z \geq 0$ given that the mass density is directly proportional to the distance from the xy-plane. See Figure 17.7.1.

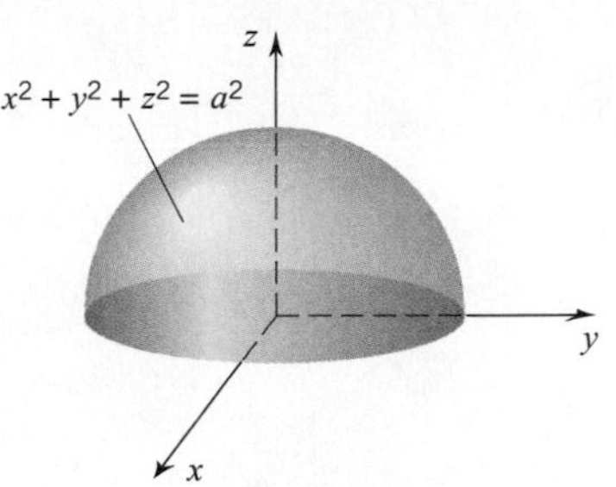

Figure 17.7.1

SOLUTION The surface S can be parametrized by the function

$$\mathbf{r}(u, v) = a \cos u \cos v\,\mathbf{i} + a \sin u \cos v\,\mathbf{j} + a \sin v\,\mathbf{k}; \quad 0 \leq u \leq 2\pi, 0 \leq v \leq \tfrac{1}{2}\pi.$$

Call the parameter set Ω and recall that $||\mathbf{N}(u, v)|| = a^2 \cos v$ (Example 8, Section 17.6). The density function can be written $\lambda(x, y, z) = kz$, where k is the constant of proportionality. We can calculate the mass as follows:

$$\begin{aligned} M = \iint_S \lambda(x, y, z)\, d\sigma = k \iint_S z\, d\sigma &= k \iint_\Omega z(u, v)||\mathbf{N}(u, v)||\, du dv \\ &= k \int_0^{2\pi} \left(\int_0^{\pi/2} (a \sin v)(a^2 \cos v)\, dv \right) du \\ &= 2\pi k a^3 \int_0^{\pi/2} \sin v \cos v\, dv = \pi k a^3. \end{aligned}$$

By symmetry $x_M = 0$ and $y_M = 0$. To find z_M we write

$$\begin{aligned} z_M M = \iint_S z\lambda(x, y, z)\, d\sigma &= k \iint_S z^2 d\sigma \\ &= k \iint_\Omega [z(u, v)]^2 ||\mathbf{N}(u, v)||\, du dv \\ &= k \int_0^{2\pi} \left(\int_0^{\pi/2} (a^2 \sin^2 v)(a^2 \cos v)\, dv \right) du \\ &= 2\pi k a^4 \int_0^{\pi/2} \sin^2 v \cos v\, dv = \tfrac{2}{3}\pi k a^4. \end{aligned}$$

Since $M = \pi k a^3$, we see that $z_M = \frac{2}{3}\pi k a^4/M = \frac{2}{3}a$. The center of mass is the point $(0, 0, \frac{2}{3}a)$. □

Suppose that a material surface S rotates about an axis. The moment of inertia of the surface about that axis is given by the formula

(17.7.6)
$$I = \iint_S \lambda(x, y, z)[R(x, y, z)]^2\, d\sigma$$

where $\lambda = \lambda(x, y, z)$ is the mass density function and $R(x, y, z)$ is the distance from the axis to the point (x, y, z). (As usual, the moments of inertia about the x, y, z axes are denoted by I_x, I_y, I_z.)

Example 4 A homogeneous material surface with mass density 1 is the shape of a spherical shell

$$S\colon\ x^2 + y^2 + z^2 = a^2. \qquad \text{(Figure 17.7.2)}$$

Calculate the moment of inertia about the z-axis.

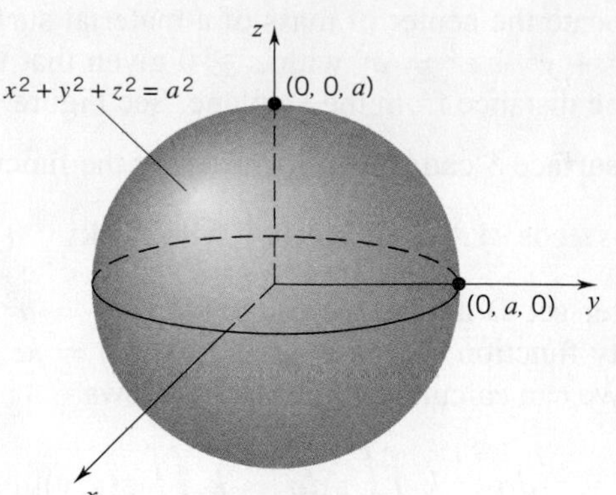

Figure 17.7.2

SOLUTION We parametrize S by setting

$$\mathbf{r}(u,v) = a\cos u\cos v\,\mathbf{i} + a\sin u\cos v\,\mathbf{j} + a\sin v\,\mathbf{k}; \quad 0 \le u \le 2\pi, -\tfrac{1}{2}\pi \le v \le \tfrac{1}{2}\pi.$$

Call the parameter set Ω and recall that $||\mathbf{N}(u,v)|| = a^2\cos v$. We can calculate the moment of inertia as follows:

$$\begin{aligned}
I_z = \iint_S (1)(x^2+y^2)\,d\sigma &= \iint_\Omega \left([x(u,v)]^2 + [y(u,v)]^2\right) ||\mathbf{N}(u,v)||\,d\sigma \\
&= \iint_\Omega (a^2\cos^2 v)(a^2\cos v)\,du dv \\
&= a^4 \int_0^{2\pi} \left(\int_{-\pi/2}^{\pi/2} \cos^3 v\,dv \right) du \\
&= 2\pi a^4 \int_{-\pi/2}^{\pi/2} \cos^3 v\,dv = \tfrac{8}{3}\pi a^4.
\end{aligned}$$

↑ check this

Since the surface has mass $M = A = 4\pi a^2$, it follows that $I_z = \frac{2}{3}Ma^2$. ❑

A surface

$$S:\ z = f(x,y),\ (x,y) \in \Omega$$

can be parametrized by the function

$$\mathbf{r}(x,y) = x\,\mathbf{i} + y\,\mathbf{j} + f(x,y)\,\mathbf{k}, \quad (x,y) \in \Omega.$$

As you saw in Section 17.6, $||\mathbf{N}(x,y)|| = \sec[\gamma(x,y)]$ where $\gamma(x,y)$ is the angle between $\mathbf{k}$ and the upper unit normal. Therefore, for any continuous scalar field H on S,

(17.7.7)
$$\iint_S H(x,y,z)\,d\sigma = \iint_\Omega H(x,y,z)\sec[\gamma(x,y)]\,dxdy.$$

In evaluating this last integral we use the fact that

$$\sec[\gamma(x,y)] = \sqrt{[f_x(x,y)]^2 + [f_y(x,y)]^2 + 1}. \qquad \text{(Section 17.6)}$$

Example 5 Calculate

$$\iint_S \sqrt{x^2+y^2}\, d\sigma \quad \text{with} \quad S: z = xy,\ 0 \le x^2+y^2 \le 1.$$

SOLUTION The base region Ω is the unit disc. The function $z = f(x,y) = xy$ has partial derivatives $f_x(x,y) = y$, $f_y(x,y) = x$. Therefore

$$\sec[\gamma(x,y)] = \sqrt{y^2+x^2+1} = \sqrt{x^2+y^2+1}$$

and
$$\iint_S \sqrt{x^2+y^2}\, d\sigma = \iint_\Omega \sqrt{x^2+y^2}\sqrt{x^2+y^2+1}\, dxdy.$$

We evaluate this last integral by changing to polar coordinates. The region Ω is the set of all (x,y) with polar coordinates $[r,\theta]$ in the set

$$\Gamma: \quad 0 \le \theta \le 2\pi, \quad 0 \le r \le 1.$$

Therefore
$$\iint_S \sqrt{x^2+y^2}\, d\sigma = \iint_\Gamma r\sqrt{r^2+1}\, r\, drd\theta = \int_0^{2\pi}\left(\int_0^1 r^2\sqrt{r^2+1}\, dr\right) d\theta$$

$$= 2\pi \int_0^1 r^2\sqrt{r^2+1}\, dr$$

$$= 2\pi \int_0^{\pi/4} \tan^2\phi \sec^3\phi\, d\phi$$

$r = \tan\phi$

$$= 2\pi \int_0^{\pi/4} [\sec^5\phi - \sec^3\phi]\, d\phi$$

$$= \tfrac{1}{4}\pi[3\sqrt{2} - \ln(\sqrt{2}+1)]. \quad \square$$

The Flux of a Vector Field

Suppose that
$$S: \mathbf{r} = \mathbf{r}(u,v), \quad (u,v) \in \Omega$$

is a smooth surface with a unit normal $\mathbf{n} = \mathbf{n}(x,y,z)$ that is continuous on all of S. Such a surface is called an *oriented surface*. Note that an oriented surface has two sides: the side with normal $\mathbf{n}$ and the side with normal $-\mathbf{n}$.† If $\mathbf{v} = \mathbf{v}(x,y,z)$ is a vector field continuous on S, then we can form the surface integral

(17.7.8)
$$\iint_S (\mathbf{v}\cdot\mathbf{n})\, d\sigma = \iint_S [\mathbf{v}(x,y,z)\cdot\mathbf{n}(x,y,z)]\, d\sigma.$$

This surface integral is called *the flux of* $\mathbf{v}$ *across* S *in the direction of* $\mathbf{n}$.

† Not all surfaces have two sides. In Exercise 41 we exhibit a surface (the Möbius band) which has only one side.

Note that the flux across a surface depends on the choice of unit normal. If $-\mathbf{n}$ is chosen instead of $\mathbf{n}$, the sign of the flux is reversed:

$$\iint_S (\mathbf{v}\cdot[-\mathbf{n}])\,d\sigma = \iint_S -(\mathbf{v}\cdot\mathbf{n})\,d\sigma = -\iint_S (\mathbf{v}\cdot\mathbf{n})\,d\sigma.$$

Example 6 Calculate the flux of the vector field $\mathbf{v}(x,y,z) = x\,\mathbf{i}+y\,\mathbf{j}$ across the sphere $S: x^2+y^2+z^2=a^2$ in the outward direction (Figure 17.7.3).

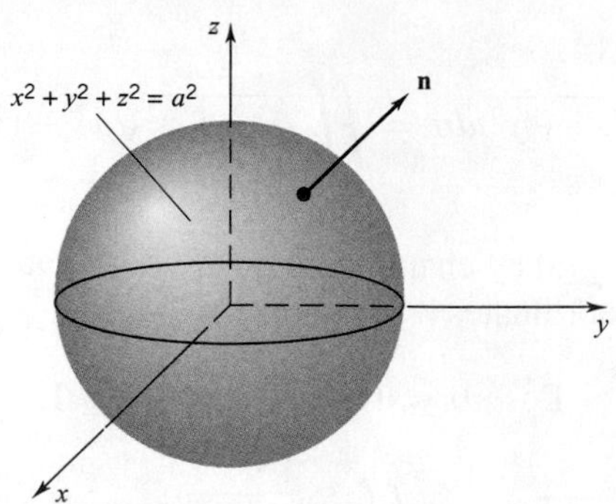

Figure 17.7.3

SOLUTION The outward unit normal is the vector

$$\mathbf{n}(x,y,z) = \frac{1}{a}(x\,\mathbf{i}+y\,\mathbf{j}+z\,\mathbf{k}).$$

Here
$$\mathbf{v}\cdot\mathbf{n} = (x\,\mathbf{i}+y\,\mathbf{j})\cdot\frac{1}{a}(x\,\mathbf{i}+y\,\mathbf{j}+z\,\mathbf{k}) = \frac{1}{a}(x^2+y^2).$$

Therefore
$$\text{flux out of } S = \frac{1}{a}\iint_S (x^2+y^2)\,d\sigma.$$

To evaluate this integral we use the usual parametrization

$$\mathbf{r}(u,v) = a\cos u\cos v\,\mathbf{i} + a\sin u\cos v\,\mathbf{j} + a\sin v\,\mathbf{k};\quad 0\le u\le 2\pi,\ -\tfrac{1}{2}\pi\le v\le \tfrac{1}{2}\pi.$$

Recall that $||\mathbf{N}(u,v)|| = a^2\cos v$. Thus

$$\begin{aligned}\text{flux out of } S &= \frac{1}{a}\iint_\Omega ([x(u,v)]^2+[y(u,v)]^2||\mathbf{N}(u,v)||\,dudv\\ &= \frac{1}{a}\iint_\Omega (a^2\cos^2 u\cos^2 v + a^2\sin^2 u\cos^2 v)(a^2\cos v)\,dudv\\ &= a^3\iint_\Omega \cos^3 v\,dudv\\ &= a^3\int_0^{2\pi}\left(\int_{-\pi/2}^{\pi/2}\cos^3 v\,dv\right)du = 2\pi a^3\int_{-\pi/2}^{\pi/2}\cos^3 v\,dv = \tfrac{8}{3}\pi a^3.\quad\square\end{aligned}$$

If S is the graph of a function $z = f(x,y)$, $(x,y) \in \Omega$ and $\mathbf{n}$ is the upper unit normal, then the flux of the vector field $\mathbf{v} = v_1\mathbf{i} + v_2\mathbf{j} + v_3\mathbf{k}$ across S in the direction of $\mathbf{n}$ is

(17.7.9)
$$\iint_S (\mathbf{v} \cdot \mathbf{n})\, d\sigma = \iint_\Omega (-v_1 f_x - v_2 f_y + v_3)\, dxdy.$$

PROOF From Section 17.6 we know that

$$\mathbf{n} = \frac{-f_x\mathbf{i} - f_y\mathbf{j} + \mathbf{k}}{\sqrt{(f_x)^2 + (f_y)^2 + 1}} = (-f_x\mathbf{i} - f_y\mathbf{j} + \mathbf{k}) \cos\gamma$$

where γ is the angle between $\mathbf{n}$ and $\mathbf{k}$. Thus $\mathbf{v} \cdot \mathbf{n} = (-v_1 f_x - v_2 f_y + v_3)\cos\gamma$ and

$$\iint_S (\mathbf{v} \cdot \mathbf{n})\, d\sigma = \iint_\Omega (\mathbf{v} \cdot \mathbf{n}) \sec\gamma\, dxdy = \iint_\Omega (-v_1 f_x - v_2 f_y + v_3)\, dxdy. \quad \square$$

Example 7 Let S be the part of the paraboloid $z = 1 - (x^2 + y^2)$ that lies above the unit disc Ω. Calculate the flux of $\mathbf{v} = x\,\mathbf{i} + y\,\mathbf{j} + z\,\mathbf{k}$ across this surface in the direction of the upper unit normal. (See Figure 17.7.4.)

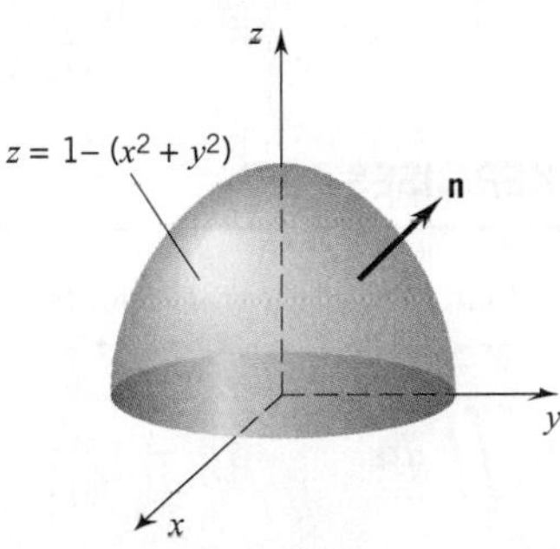

Figure 17.7.4

SOLUTION

$$f(x,y) = 1 - (x^2 + y^2); \qquad f_x = -2x, \qquad f_y = -2y.$$

$$\begin{aligned} \text{Flux} &= \iint_\Omega (-v_1 f_x - v_2 f_y + v_3)\, dxdy \\ &= \iint_\Omega [(-x)(-2x) - y(-2y) + 1 - (x^2 + y^2)]\, dxdy \\ &= \iint_\Omega (1 + x^2 + y^2)\, dxdy = \int_0^{2\pi} \left(\int_0^1 (1 + r^2)\, r\, dr \right) d\theta = \tfrac{3}{2}\pi. \quad \square \end{aligned}$$

in polar coordinates (referring to the last integral)

The flux through a closed *piecewise-smooth oriented surface* (a closed surface that consists of a finite number of smooth oriented pieces joined together at the boundaries) can be evaluated by integrating over each smooth piece and adding up the results.

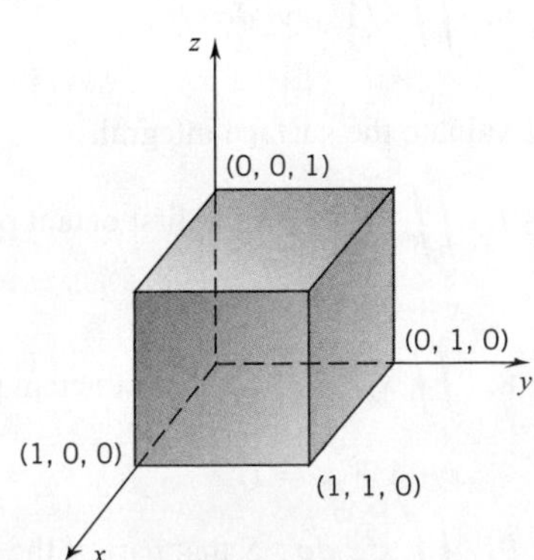

Figure 17.7.5

Example 8 Calculate the total flux of $\mathbf{v}(x,y,z) = xy\,\mathbf{i} + 4yz^2\,\mathbf{j} + yz\,\mathbf{k}$ out of the unit cube: $0 \le x \le 1$, $0 \le y \le 1$, $0 \le z \le 1$ (Figure 17.7.5).

SOLUTION The total flux is the sum of the fluxes across the faces of the cube:

Face	$\mathbf{n}$	$\mathbf{v} \cdot \mathbf{n}$	Flux
$x = 0$	$-\mathbf{i}$	$-xy = 0$	0
$x = 1$	$\mathbf{i}$	$xy = y$	$\frac{1}{2}$
$y = 0$	$-\mathbf{j}$	$-4yz^2 = 0$	0
$y = 1$	$\mathbf{j}$	$4yz^2 = 4z^2$	$\frac{4}{3}$
$z = 0$	$-\mathbf{k}$	$-yz = 0$	0
$z = 1$	$\mathbf{k}$	$yz = y$	$\frac{1}{2}$

The total flux is $\frac{1}{2} + \frac{4}{3} + \frac{1}{2} = \frac{7}{3}$. □

Flux plays an important role in the study of fluid motion. Imagine a surface S within a fluid and choose a unit normal $\mathbf{n}$. Take $\mathbf{v}$ as the velocity of the fluid that passes through S. You can expect $\mathbf{v}$ to vary from point to point, but, for simplicity, let's assume that $\mathbf{v}$ does not change with time. (Such a time-independent flow is said to be *steady state*.) The flux of $\mathbf{v}$ across S is the average component of $\mathbf{v}$ in the direction of $\mathbf{n}$ times the area of the surface S. This is just the volume of fluid that passes through S in unit time, from the $-\mathbf{n}$ side of S to the $\mathbf{n}$ side of S. If S is a closed surface (such as a cube, a sphere, or an ellipsoid) and $\mathbf{n}$ is chosen as the *outer unit normal*, then the flux across S gives the volume of fluid that flows *out* through S in unit time; if $\mathbf{n}$ is chosen as the *inner unit normal*, then the flux gives the volume of liquid that flows *in* through S in unit time.

EXERCISES 17.7

Evaluate these integrals over the surface S: $z = \frac{1}{2}y^2$; $0 \le x \le 1$, $0 \le y \le 1$.

1. $\iint_S d\sigma$.

2. $\iint_S x^2 d\sigma$.

3. $\iint_S 3y\, d\sigma$.

4. $\iint_S (x - y)\, d\sigma$.

5. $\iint_S \sqrt{2z}\, d\sigma$.

6. $\iint_S \sqrt{1 + y^2}\, d\sigma$.

Evaluate the surface integral.

7. $\iint_S xy\, d\sigma$; S the first octant part of the plane $x + 2y + 3z = 6$.

8. $\iint_S xyz\, d\sigma$; S the first octant part of the plane $x + y + z = 1$.

9. $\iint_S x^2 z\, d\sigma$; S that part of the cylinder $x^2 + z^2 = 1$ which lies between the planes $y = 0$ and $y = 2$, and is above the xy-plane.

10. $\iint_S (x^2 + y^2 + z^2)\, d\sigma$; S that part of the plane $z = x + 2$ which lies inside the cylinder $x^2 + y^2 = 1$.

11. $\iint_S (x^2 + y^2)\, d\sigma$; S the hemisphere $z = \sqrt{1 - (x^2 + y^2)}$.

12. $\iint_S (x^2 + y^2)\, d\sigma$; S that part of the paraboloid $z = 1 - x^2 - y^2$ which lies above the xy-plane.

In Exercises 13–15, find the mass of a material surface in the shape of a triangle $(a, 0, 0), (0, a, 0), (0, 0, a)$ given that the mass density varies as indicated. Take $a > 0$.

13. $\lambda(x, y, z) = k$.

14. $\lambda(x, y, z) = k(x + y)$.

15. $\lambda(x, y, z) = kx^2$.

16. Locate the centroid of the triangle $(a, 0, 0), (0, a, 0), (0, 0, a)$. Take $a > 0$.

17. Locate the centroid of the hemisphere $x^2 + y^2 + z^2 = a^2$, $z \ge 0$.

In Exercises 18 and 19, let S be the parallelogram given by $\mathbf{r}(u, v) = (u + v)\,\mathbf{i} + (u - v)\,\mathbf{j} + 2u\,\mathbf{k}$; $0 \le u \le 1, 0 \le v \le 1$.

18. Find the area of S.

19. Determine the flux of $\mathbf{v} = x\,\mathbf{i} - y\,\mathbf{j}$ across S in the direction of the fundamental vector product.

20. Find the mass of the material surface $S: z = 1 - \frac{1}{2}(x^2 + y^2)$ with $0 \le x \le 1$, $0 \le y \le 1$ given that the density at each point (x, y, z) is proportional to xy.

In Exercises 21–23, calculate the flux out of the sphere $x^2 + y^2 + z^2 = a^2$.

21. $\mathbf{v} = z\,\mathbf{k}$. **22.** $\mathbf{v} = x\,\mathbf{i} + y\,\mathbf{j} + z\,\mathbf{k}$. **23.** $\mathbf{v} = y\,\mathbf{i} - x\,\mathbf{j}$.

24. A homogeneous plate of mass density 1 is in the form of the parallelogram of Exercises 18 and 19. Determine the moments of inertia about the coordinate axes: (a) I_x. (b) I_y. (c) I_z.

In Exercises 25–27, determine the flux across the triangle $(a, 0, 0), (0, a, 0), (0, 0, a), a > 0$, in the direction of the upper unit normal.

25. $\mathbf{v} = x\,\mathbf{i} + y\,\mathbf{j} + z\,\mathbf{k}$.

26. $\mathbf{v} = (x + z)\,\mathbf{k}$.

27. $\mathbf{v} = x^2\,\mathbf{i} - y^2\,\mathbf{j}$.

In Exercises 28–30, determine the flux across $S: z = xy$ with $0 \le x \le 1, 0 \le y \le 2$ in the direction of the upper unit normal.

28. $\mathbf{v} = -xy^2\,\mathbf{i} + z\,\mathbf{j}$.

29. $\mathbf{v} = xz\,\mathbf{j} - xy\,\mathbf{k}$.

30. $\mathbf{v} = x^2y\,\mathbf{i} + z^2\,\mathbf{k}$.

31. Calculate the flux of $\mathbf{v} = x\,\mathbf{i} + y\,\mathbf{j} + z\,\mathbf{k}$ out of the cylindrical surface

$$S: \mathbf{r}(u, v) = a\cos u\,\mathbf{i} + a\sin u\,\mathbf{j} + v\,\mathbf{k};$$
$$0 \le u \le 2\pi, 0 \le v \le l.$$

32. (*The gravitational field*) A mass M at the origin exerts a force

$$\mathbf{F}(\mathbf{r}) = -G\frac{mM}{r^3}\mathbf{r}$$

on a mass m located at the tip of the radius vector $\mathbf{r} = x\mathbf{i} + y\mathbf{j} + z\mathbf{k}$. Find the flux of $\mathbf{F}$ into the sphere $x^2 + y^2 + z^2 = a^2$.

In Exercises 33-36, find the flux across $S: z = \frac{2}{3}(x^{3/2} + y^{3/2})$ with $0 \le x \le 1, 0 \le y \le 1 - x$ in the direction of the upper unit normal.

33. $\mathbf{v} = x\,\mathbf{i} - y\,\mathbf{j} + \frac{3}{2}z\,\mathbf{k}$.

34. $\mathbf{v} = x^2\,\mathbf{i}$.

35. $\mathbf{v} = y^2\,\mathbf{j}$.

36. $\mathbf{v} = y\,\mathbf{i} - \sqrt{xy}\,\mathbf{j}$.

37. The cone

$$\mathbf{r}(u, v) = v\cos u \sin\alpha\,\mathbf{i} + v\sin u \sin\alpha\,\mathbf{j} + v\cos\alpha\,\mathbf{k},$$
$$0 \le u \le 2\pi, \quad 0 \le v \le s,$$

has area $A = \pi s^2 \sin\alpha$. Locate the centroid.

In Exercises 38–40, the mass density of a material cone $z = \sqrt{x^2 + y^2}$ with $0 \le z \le 1$ varies directly as the distance from the z-axis.

38. Find the mass of the cone.

39. Locate the center of mass.

40. Determine the moments of inertia about the coordinate axes: (*a*) I_x. (*b*) I_y. (*c*) I_z.

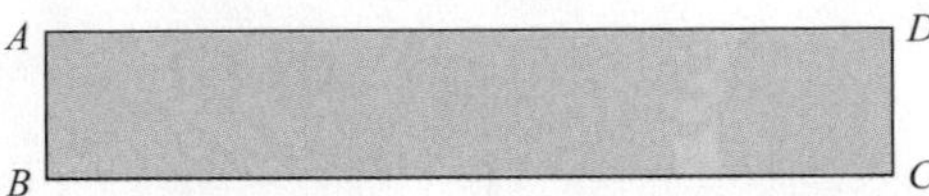

41. You have seen that, if S is a smooth oriented surface immersed in a fluid of velocity $\mathbf{v}$, then the flux

$$\iint_S (\mathbf{v} \cdot \mathbf{n})\, d\sigma$$

is the volume of fluid that passes through S in unit time from the $-\mathbf{n}$ side of S to the $\mathbf{n}$ side of S. This requires that S be a two-sided surface. There are, however, one-sided surfaces; for example, the Möbius band. To construct a material Möbius band, start with a piece of paper in the form of the rectangle in the figure. Now give the piece of paper a single twist and join the two far edges together so that C coincides with A and D coincides with B. (a) Convince yourself that this surface is one-sided and therefore the notion of flux cannot be applied to it. (b) The surface is not smooth because it is impossible to erect a unit normal $\mathbf{n}$ that varies continuously over the entire surface. Convince yourself of this as follows: erect a unit normal $\mathbf{n}$ at some point P and make a circuit of the surface with $\mathbf{n}$. Note that, as $\mathbf{n}$ returns to P, the direction of $\mathbf{n}$ has been reversed.

In Exercises 42–44, assume that the parallelogram of Exercises 18 and 19 is a material surface with a mass density that varies directly as the square of the distance from the x-axis.

42. Determine the mass.

43. Find the x-coordinate of the center of mass.

44. Find the moment of inertia about the z-axis.

45. Calculate the total flux of $\mathbf{v}(x, y, z) = y\,\mathbf{i} - x\,\mathbf{j}$ out of the solid bounded on the sides by the cylinder $x^2 + y^2 = 1$ and above and below by the planes $z = 1$ and $z = 0$. HINT: Draw a figure.

46. Calculate the total flux of $\mathbf{v}(x, y, z) = y\,\mathbf{i} - x\,\mathbf{j}$ out of the solid bounded above by $z = 4$ and below by $z = x^2 + y^2$.

47. Calculate the total flux of $\mathbf{v}(x, y, z) = x\,\mathbf{i} + y\,\mathbf{j} + z\,\mathbf{k}$ out of the solid bounded above by $z = \sqrt{2 - (x^2 + y^2)}$ and below by $z = x^2 + y^2$.

48. Calculate the total flux of $\mathbf{v}(x, y, z) = xz\,\mathbf{i} + 4xyz^2\,\mathbf{j} + 2z\,\mathbf{k}$ out of the unit cube: $0 \le x \le 1, 0 \le y \le 1, 0 \le z \le 1$.

■ 17.8 THE VECTOR DIFFERENTIAL OPERATOR ∇

Divergence $\nabla \cdot \mathbf{v}$, Curl $\nabla \times \mathbf{v}$

The *vector differential operator* ∇ is defined formally by setting

(17.8.1)
$$\nabla = \frac{\partial}{\partial x}\mathbf{i} + \frac{\partial}{\partial y}\mathbf{j} + \frac{\partial}{\partial z}\mathbf{k}.$$

By "formally," we mean that this is not an ordinary vector. Its "components" are differentiation symbols. As the term "operator" suggests, ∇ is to be thought of as something that "operates" on things. What sorts of things? Scalar fields and vector fields.

Suppose that f is a differentiable scalar field. Then ∇ operates on f as follows:

$$\nabla f = \left(\frac{\partial}{\partial x}\mathbf{i} + \frac{\partial}{\partial y}\mathbf{j} + \frac{\partial}{\partial z}\mathbf{k}\right) f = \frac{\partial f}{\partial x}\mathbf{i} + \frac{\partial f}{\partial y}\mathbf{j} + \frac{\partial f}{\partial z}\mathbf{k}.$$

This is just the *gradient* of f, with which you are already familiar.

How does ∇ operate on vector fields? In two ways. If $\mathbf{v} = v_1\,\mathbf{i} + v_2\,\mathbf{j} + v_3\,\mathbf{k}$ is a differentiable vector field, then, by definition,

(17.8.2)
$$\nabla \cdot \mathbf{v} = \frac{\partial v_1}{\partial x} + \frac{\partial v_2}{\partial y} + \frac{\partial v_3}{\partial z}$$

and

(17.8.3)
$$\nabla \times \mathbf{v} = \begin{vmatrix} \mathbf{i} & \mathbf{j} & \mathbf{k} \\ \dfrac{\partial}{\partial x} & \dfrac{\partial}{\partial y} & \dfrac{\partial}{\partial z} \\ v_1 & v_2 & v_3 \end{vmatrix}$$
$$= \left(\frac{\partial v_3}{\partial y} - \frac{\partial v_2}{\partial z}\right)\mathbf{i} + \left(\frac{\partial v_1}{\partial z} - \frac{\partial v_3}{\partial x}\right)\mathbf{j} + \left(\frac{\partial v_2}{\partial x} - \frac{\partial v_1}{\partial y}\right)\mathbf{k}.$$

The first "product," $\nabla \cdot \mathbf{v}$, defined in imitation of the ordinary dot product, is called the *divergence* of $\mathbf{v}$:

$$\nabla \cdot \mathbf{v} = \text{div}\,\mathbf{v}.$$

The second "product," $\nabla \times \mathbf{v}$, defined in imitation of the ordinary cross product, is called the *curl* of $\mathbf{v}$:

$$\nabla \times \mathbf{v} = \text{curl}\,\mathbf{v}.$$

Interpretation of Divergence and Curl

Suppose we know the divergence of a field and also the curl. What does that tell us? For definitive answers we must wait for the divergence theorem and Stokes's theorem, but, in a preliminary way, we can give you some rough answers right now.

View $\mathbf{v}$ as the velocity field of some fluid. The divergence of $\mathbf{v}$ at a point P gives us an indication of whether the fluid tends to accumulate near P (negative divergence) or tends to move away from P (positive divergence). In the first case, P is sometimes called a *sink*, and in the second case, it is called a *source*. The curl at P measures the rotational tendency of the fluid.

Example 1 Set

$$\mathbf{v}(x,y,z) = \alpha\,\mathbf{r} = \alpha x\,\mathbf{i} + \alpha y\,\mathbf{j} + \alpha z\,\mathbf{k}. \qquad (\alpha \text{ a constant})$$

The divergence is

$$\nabla \cdot \mathbf{v} = \alpha\frac{\partial x}{\partial x} + \alpha\frac{\partial y}{\partial y} + \alpha\frac{\partial z}{\partial z} = 3\alpha.$$

The curl is

$$\nabla \times \mathbf{v} = \begin{vmatrix} \mathbf{i} & \mathbf{j} & \mathbf{k} \\ \dfrac{\partial}{\partial x} & \dfrac{\partial}{\partial y} & \dfrac{\partial}{\partial z} \\ \alpha x & \alpha y & \alpha z \end{vmatrix} = \alpha \begin{vmatrix} \mathbf{i} & \mathbf{j} & \mathbf{k} \\ \dfrac{\partial}{\partial x} & \dfrac{\partial}{\partial y} & \dfrac{\partial}{\partial z} \\ x & y & z \end{vmatrix} = \mathbf{0}$$

because the partial derivatives that appear in the expanded determinant

$$\frac{\partial y}{\partial x}, \quad \frac{\partial x}{\partial y}, \quad \text{etc.},$$

are all zero. □

The field of Example 1

$$\mathbf{v}(x,y,z) = \alpha(x\,\mathbf{i} + y\,\mathbf{j} + z\,\mathbf{k}) = \alpha\,\mathbf{r}$$

can be viewed as the velocity field of a fluid in radial motion—toward the origin if $\alpha < 0$, away from the origin if $\alpha > 0$. Figure 17.8.1 shows a point (x, y, z), a spherical neighborhood of that point, and a cone emanating from the origin that is tangent to the boundary of the neighborhood.

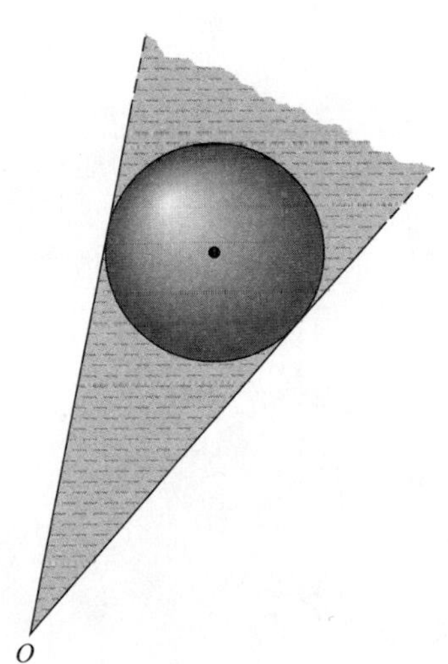

Figure 17.8.1

Note two things: all the fluid in the cone stays in the cone, and the speed of the fluid is proportional to its distance from the origin. Therefore, if the divergence 3α is negative, then α is negative, the motion is toward the origin, and the neighborhood *gains fluid* because the fluid coming in is moving more quickly than the fluid going out. (Also, the entry area is larger than the exit area.) If, however, the divergence 3α is positive, then α is positive, the motion is away from the origin, and the neighborhood *loses fluid* because the fluid coming in is moving more slowly than the fluid going out. (Also, the entry area is smaller than the exit area.)

Since the motion is radial, the fluid has no rotational tendency whatsoever, and we would expect the curl to be identically zero. It is.

Example 2 Set

$$\mathbf{v}(x,y,z) = -\omega y\,\mathbf{i} + \omega x\,\mathbf{j}. \qquad (\omega \text{ a positive constant})$$

The divergence is

$$\nabla \cdot \mathbf{v} = -\omega\frac{\partial y}{\partial x} + \omega\frac{\partial x}{\partial y} = 0 + 0 = 0.$$

The curl is

$$\nabla \times \mathbf{v} = \begin{vmatrix} \mathbf{i} & \mathbf{j} & \mathbf{k} \\ \dfrac{\partial}{\partial x} & \dfrac{\partial}{\partial y} & \dfrac{\partial}{\partial z} \\ -\omega y & \omega x & 0 \end{vmatrix} = \left(\omega\frac{\partial x}{\partial x} + \omega\frac{\partial y}{\partial y}\right)\mathbf{k} = 2\omega\,\mathbf{k}. \quad \square$$

The field of Example 2,

$$\mathbf{v}(x, y, z) = -\omega y\,\mathbf{i} + \omega x\,\mathbf{j},$$

is the velocity field of uniform counterclockwise rotation about the z-axis with angular speed ω. You can see this by noting that $\mathbf{v}$ is perpendicular to $\mathbf{r} = x\,\mathbf{i} + y\,\mathbf{j} + z\,\mathbf{k}$:

$$\mathbf{v} \cdot \mathbf{r} = (-\omega y\,\mathbf{i} + x\,\mathbf{j}) \cdot (x\,\mathbf{i} + y\,\mathbf{j} + z\,\mathbf{k}) = -\omega yx + \omega xy = 0$$

and the speed at each point is ωR where R is the radius of rotation:

$$v = \sqrt{\omega^2 y^2 + \omega^2 x^2} = \omega\sqrt{x^2 + y^2} = \omega R.$$

How is the curl, $2\omega\,\mathbf{k}$, related to the rotation? The angular velocity vector (see Exercise 14, Section 13.6) is the vector $\boldsymbol{\omega} = \omega\,\mathbf{k}$. In this case, then, the curl of $\mathbf{v}$ is twice the angular velocity vector.

With this rotation no neighborhood gains any fluid and no neighborhood loses any fluid. As we saw, the divergence is identically zero.

Basic Identities

For vectors we have $\mathbf{a} \times \mathbf{a} = \mathbf{0}$. Is it true that $\nabla \times \nabla = \mathbf{0}$? Define $(\nabla \times \nabla)f$ by

$$(\nabla \times \nabla)f = \nabla \times (\nabla f).$$

THEOREM 17.8.4 THE CURL OF A GRADIENT IS ZERO

If f is a scalar field with continuous second partials, then

$$\nabla \times (\nabla f) = \mathbf{0}.$$

PROOF

$$\nabla \times (\nabla f) = \begin{vmatrix} \mathbf{i} & \mathbf{j} & \mathbf{k} \\ \dfrac{\partial}{\partial x} & \dfrac{\partial}{\partial y} & \dfrac{\partial}{\partial z} \\ \dfrac{\partial f}{\partial x} & \dfrac{\partial f}{\partial y} & \dfrac{\partial f}{\partial z} \end{vmatrix} =$$

$$\left(\frac{\partial^2 f}{\partial y \partial z} - \frac{\partial^2 f}{\partial z \partial y}\right)\mathbf{i} + \left(\frac{\partial^2 f}{\partial z \partial x} - \frac{\partial^2 f}{\partial x \partial z}\right)\mathbf{j} + \left(\frac{\partial^2 f}{\partial x \partial y} - \frac{\partial^2 f}{\partial y \partial x}\right)\mathbf{k} = \mathbf{0}. \quad \square$$

For vectors we have $\mathbf{a}\cdot(\mathbf{a}\times\mathbf{c})=0$. The analogous operator formula, $\nabla\cdot(\nabla\times\mathbf{v})=0$, is also valid.

THEOREM 17.8.5 THE DIVERGENCE OF A CURL IS ZERO

If the components of the vector field $\mathbf{v}=v_1\mathbf{i}+v_2\mathbf{j}+v_3\mathbf{k}$ have continuous second partials, then

$$\nabla\cdot(\nabla\times\mathbf{v})=0.$$

PROOF Again the key is the equality of the mixed partials:

$$\nabla\cdot(\nabla\times\mathbf{v})=\frac{\partial}{\partial x}\left(\frac{\partial v_3}{\partial y}-\frac{\partial v_2}{\partial z}\right)+\frac{\partial}{\partial y}\left(\frac{\partial v_1}{\partial z}-\frac{\partial v_3}{\partial x}\right)+\frac{\partial}{\partial z}\left(\frac{\partial v_2}{\partial x}-\frac{\partial v_1}{\partial y}\right)=0,$$

since for each component v_i the mixed partials cancel. Try it for v_1. ❑

The next two identities are product rules. Here f is a scalar field and $\mathbf{v}$ is a vector field.

(17.8.6) $$\nabla\cdot(f\mathbf{v})=(\nabla f)\cdot\mathbf{v}+f(\nabla\cdot\mathbf{v}).\quad[\text{div}\,(f\mathbf{v})=(\text{grad}\,f)\cdot\mathbf{v}+f(\text{div}\,\mathbf{v})]$$

(17.8.7) $$\nabla\times(f\mathbf{v})=(\nabla f)\times\mathbf{v}+f(\nabla\times\mathbf{v}).\quad[\text{curl}\,(f\mathbf{v})=(\text{grad}\,f)\times\mathbf{v}+f(\text{curl}\,\mathbf{v})]$$

The verification of these identities is left to you in the Exercises.

We know from Example 1 that $\nabla\cdot\mathbf{r}=3$ and $\nabla\times\mathbf{r}=\mathbf{0}$ at all points of space. Now we can show that if n is an integer, then, for all $\mathbf{r}\neq\mathbf{0}$,

(17.8.8) $$\nabla\cdot(r^n\mathbf{r})=(n+3)r^n\quad\text{and}\quad\nabla\times(r^n\mathbf{r})=\mathbf{0}.$$ †

PROOF Recall that $\nabla r^n=nr^{n-2}\mathbf{r}$. [(15.1.5.)] Using (17.8.6), we have

$$\begin{aligned}\nabla\cdot(r^n\mathbf{r})&=(\nabla r^n)\cdot\mathbf{r}+r^n(\nabla\cdot\mathbf{r})\\&=(nr^{n-2}\mathbf{r})\cdot\mathbf{r}+r^n(3)\\&=nr^{n-2}(\mathbf{r}\cdot\mathbf{r})+3r^n=nr^n+3r^n=(n+3)r^n.\end{aligned}$$

From (15.1.5) you can see that $r^n\mathbf{r}$ is a gradient. Its curl is therefore $\mathbf{0}$. (17.8.4) ❑

The Laplacian

From the operator ∇ we can construct other operators, the most important of which is the Laplacian $\nabla^2=\nabla\cdot\nabla$. The Laplacian (named after the French mathematician Pierre-Simon Laplace) operates on scalar fields according to the following rule:

(17.8.9) $$\nabla^2 f=\nabla\cdot(\nabla f)=\frac{\partial^2 f}{\partial x^2}+\frac{\partial^2 f}{\partial y^2}+\frac{\partial^2 f}{\partial z^2}.$$ ††

† If n is positive and even, these formulas also hold at $\mathbf{r}=\mathbf{0}$.

†† In some texts, you will see the Laplacian of f written Δf. Unfortunately this can be misread as the increment of f.

Example 3 If $f(x,y,z) = x^2 + y^2 + z^2$, then

$$\begin{aligned}\nabla^2 f &= \frac{\partial^2}{\partial x^2}(x^2+y^2+z^2) + \frac{\partial^2}{\partial y^2}(x^2+y^2+z^2) + \frac{\partial^2}{\partial z^2}(x^2+y^2+z^2)\\ &= 2+2+2 = 6. \quad \square\end{aligned}$$

Example 4 If $f(x,y,z) = e^{xyz}$, then

$$\begin{aligned}\nabla^2 f &= \frac{\partial}{\partial x^2}(e^{xyz}) + \frac{\partial^2}{\partial y^2}(e^{xyz}) + \frac{\partial^2}{\partial z^2}(e^{xyz})\\ &= \frac{\partial}{\partial x}(yz\,e^{xyz}) + \frac{\partial}{\partial y}(xz\,e^{xyz}) + \frac{\partial}{\partial z}(xy\,e^{xyz})\\ &= y^2z^2\,e^{xyz} + x^2z^2\,e^{xyz} + x^2y^2\,e^{xyz}\\ &= (y^2z^2 + x^2z^2 + x^2y^2)\,e^{xyz}. \quad \square\end{aligned}$$

Example 5 To calculate $\nabla^2(\sin r) = \nabla^2(\sin\sqrt{x^2+y^2+z^2})$, we could write

$$\frac{\partial^2}{\partial x^2}(\sin\sqrt{x^2+y^2+z^2}) + \frac{\partial^2}{\partial y^2}(\sin\sqrt{x^2+y^2+z^2}) + \frac{\partial^2}{\partial z^2}(\sin\sqrt{x^2+y^2+z^2})$$

and proceed from there. The calculations are straightforward but lengthy. We will proceed in a different way.

Recall that

$$\nabla^2 f = \nabla\cdot\nabla f \quad (17.8.9) \qquad \nabla\cdot(f\mathbf{v}) = (\nabla f)\cdot\mathbf{v} + f(\nabla\cdot\mathbf{v}) \quad (17.8.6)$$

$$\nabla f(r) = f'(r)r^{-1}\mathbf{r} \quad (15.3.11) \qquad \nabla\cdot(r^n\mathbf{r}) = (n+3)r^n \quad (17.8.8).$$

Using these relations, we have

$$\begin{aligned}\nabla^2(\sin r) = \nabla\cdot(\nabla\sin r) &= \nabla\cdot[(\cos r)\,r^{-1}\mathbf{r}]\\ &= [(\nabla\cos r)\cdot r^{-1}\mathbf{r}] + \cos r\,(\nabla\cdot r^{-1}\mathbf{r})\\ &= \{(-\sin r)\,r^{-1}\mathbf{r}]\cdot r^{-1}\mathbf{r}\} + (\cos r)(2r^{-1})\\ &= -\sin r + 2r^{-1}\cos r.\end{aligned}$$

We leave it to you to verify each step. $\square$

EXERCISES 17.8

Calculate $\nabla\cdot\mathbf{v}$ and $\nabla\times\mathbf{v}$.

1. $\mathbf{v}(x,y) = x\,\mathbf{i} + y\,\mathbf{j}$.
2. $\mathbf{v}(x,y) = y\,\mathbf{i} + x\,\mathbf{j}$.
3. $\mathbf{v}(x,y) = \dfrac{x}{x^2+y^2}\mathbf{i} + \dfrac{y}{x^2+y^2}\mathbf{j}$.
4. $\mathbf{v}(x,y) = \dfrac{y}{x^2+y^2}\mathbf{i} + \dfrac{x}{x^2+y^2}\mathbf{j}$.
5. $\mathbf{v}(x,y,z) = x\,\mathbf{i} + 2y\,\mathbf{j} + 3z\,\mathbf{k}$.
6. $\mathbf{v}(x,y,z) = yz\,\mathbf{i} + xz\,\mathbf{j} + xy\,\mathbf{k}$.
7. $\mathbf{v}(x,y,z) = xyz\,\mathbf{i} + xz\,\mathbf{j} + z\,\mathbf{k}$.
8. $\mathbf{v}(x,y,z) = x^2y\,\mathbf{i} + y^2z\,\mathbf{j} + xy^2\,\mathbf{k}$.
9. $\mathbf{v}(\mathbf{r}) = r^{-2}\,\mathbf{r}$
10. $\mathbf{v}(\mathbf{r}) = e^r\,\mathbf{r}$.
11. $\mathbf{v}(\mathbf{r}) = e^{r^2}\,(\mathbf{i}+\mathbf{j}+\mathbf{k})$.
12. $\mathbf{v}(\mathbf{r}) = e^{y^2}\,\mathbf{i} + e^{z^2}\,\mathbf{j} + e^{x^2}\mathbf{k}$.
13. Suppose that f is a differentiable function of one variable and $\mathbf{v}(x,y,z) = f(x)\,\mathbf{i}$. Determine $\nabla\cdot\mathbf{v}$ and $\nabla\times\mathbf{v}$.
14. Show that, if $\mathbf{v}$ is a differentiable vector field of the form $\mathbf{v}(\mathbf{r}) = f(x)\,\mathbf{i} + g(y)\,\mathbf{j} + h(z)\,\mathbf{k}$, then $\nabla\times\mathbf{v} = \mathbf{0}$.

15. Show that divergence and curl are *linear* operators:

$$\nabla \cdot (\alpha\mathbf{u} + \beta\mathbf{v}) = \alpha(\nabla \cdot \mathbf{u}) + \beta(\nabla \cdot \mathbf{v}) \quad \text{and}$$

$$\nabla \times (\alpha\mathbf{u} + \beta\mathbf{v}) = \alpha(\nabla \times \mathbf{u}) + \beta(\nabla \times \mathbf{v}).$$

16. (*Important*) Show that the gravitational field

$$\mathbf{F}(\mathbf{r}) = -\frac{GmM}{r^3}\mathbf{r}$$

has zero divergence and zero curl at each $\mathbf{r} \neq \mathbf{0}$.

A vector field $\mathbf{v}$ with the property that $\nabla \cdot \mathbf{v} = 0$ is said to be *solenoidal*, from the Greek word for "tubular." Exercise 16 shows that the gravitational field is solenoidal. If $\mathbf{v}$ is the velocity field of some fluid and $\nabla \cdot \mathbf{v} = 0$ in a solid T in three-dimensional space, then $\mathbf{v}$ has no sources or sinks within T.

17. Show that the vector field $\mathbf{v}(x, y, z) = (2x + y + 2z)\,\mathbf{i} + (x + 4y - 3z)\,\mathbf{j} + (2x - 3y - 6z)\,\mathbf{k}$ is solenoidal.

18. Show that $\mathbf{v}(x, y, z) = 3x^2\,\mathbf{i} - y^2\,\mathbf{j} + (2yz - 6xz)\,\mathbf{k}$ is solenoidal.

A vector field $\mathbf{v}$ with the property that $\nabla \times \mathbf{v} = \mathbf{0}$ is said to be *irrotational*. Exercise 16 shows that the gravitational field is irrotational. If $\mathbf{v}$ is the velocity field of some fluid, then $\nabla \times \mathbf{v}$ measures the tendency of the fluid to "curl" or rotate about an axis. Thus $\nabla \times \mathbf{v} = \mathbf{0}$ in some solid T in three-dimensional space can be interpreted to mean that the fluid is tending to move in a straight line.

19. Show that the vector field $\mathbf{v}(x, y, z) = x\,\mathbf{i} + y\,\mathbf{j} - 2z\,\mathbf{k}$ is irrotational.

20. Show that the vector field of Exercise 17 is irrotational.

In Exercises 21–26, calculate the Laplacian $\nabla^2 f$.

21. $f(x, y, z) = x^4 + y^4 + z^4$.

22. $f(x, y, z) = xyz$.

23. $f(x, y, z) = x^2y^3z^4$.

24. $f(\mathbf{r}) = \cos r$.

25. $f(\mathbf{r}) = e^r$.

26. $f(\mathbf{r}) = \ln r$.

27. Given a vector field $\mathbf{u}$, the operator $\mathbf{u} \cdot \nabla$ is defined by setting

$$(\mathbf{u} \cdot \nabla)f = \mathbf{u} \cdot \nabla f = u_1\frac{\partial f}{\partial x} + u_2\frac{\partial f}{\partial y} + u_3\frac{\partial f}{\partial z}.$$

Calculate $(\mathbf{r} \cdot \nabla)f$: (a) for $f(\mathbf{r}) = r^2$; (b) for $f(\mathbf{r}) = 1/r$.

28. (Based on Exercise 27) We can also apply $\mathbf{u} \cdot \nabla$ to a vector field $\mathbf{v}$ by applying it to each component. By definition

$$(\mathbf{u} \cdot \nabla)\mathbf{v} = (\mathbf{u} \cdot \nabla v_1)\,\mathbf{i} + (\mathbf{u} \cdot \nabla v_2)\,\mathbf{j} + (\mathbf{u} \cdot \nabla v_3)\,\mathbf{k}.$$

(a) Calculate $(\mathbf{u} \cdot \nabla)\,\mathbf{r}$ for an arbitrary vector field $\mathbf{u}$.

(b) Calculate $(\mathbf{r} \cdot \nabla)\,\mathbf{u}$ given that $\mathbf{u} = yz\,\mathbf{i} + xz\,\mathbf{j} + xy\,\mathbf{k}$.

29. Show that, if $f(\mathbf{r}) = g(r)$ and g is twice differentiable, then

$$\nabla^2 f = g''(r) + 2r^{-1}g'(r).$$

30. Verify the following identities.

(a) $\nabla \cdot (f\mathbf{v}) = (\nabla f) \cdot \mathbf{v} + f(\nabla \cdot \mathbf{v})$.

(b) $\nabla \times (f\mathbf{v}) = (\nabla f) \times \mathbf{v} + f(\nabla \times \mathbf{v})$.

(c) $\nabla \times (\nabla \times \mathbf{v}) = \nabla(\nabla \cdot \mathbf{v}) - \nabla^2\mathbf{v}$, where $\nabla^2\mathbf{v} = (\nabla^2 v_1)\,\mathbf{i} + (\nabla^2 v_2)\,\mathbf{j} + (\nabla^2 v_3)\,\mathbf{k}$.

HINT: Begin part (c) by writing out the ith component of each side.

As you saw in Project 14.6, the equation

$$\nabla^2 f = \frac{\partial^2 f}{\partial x^2} + \frac{\partial^2 f}{\partial y^2} + \frac{\partial^2 f}{\partial z^2} = 0$$

is called *Laplace's equation in three dimensions*. A scalar field $f = f(x, y, z)$ with continuous second partial derivatives is said to be *harmonic* on a solid T if it is a solution of Laplace's equation.

31. Show that the scalar field

$$f(x, y, z) = x^2 + 2y^2 - 3z^2 + xy + 2xz - 3yz$$

is harmonic.

32. Show that the scalar field

$$f(x, y, z) = \frac{1}{\sqrt{x^2 + y^2 + z^2}}$$

is harmonic on every solid T that excludes the origin. Except for a constant factor, f is a potential function for the gravitational field.

33. For what nonzero integers n is $f(\mathbf{r}) = r^n$ harmonic on every solid T that excludes the origin?

34. Show that if $f = f(x, y, z)$ satisfies Laplace's equation, then its gradient field is both solenoidal and irrotational.

■ 17.9 THE DIVERGENCE THEOREM

Let Ω be a Jordan region with a piecewise-smooth boundary C, and let P and Q be continuously differentiable scalar fields on an open set containing Ω. Green's theorem allows us to express a double integral over Ω as a line integral over C:

$$\iint_{\Omega} \left(\frac{\partial Q}{\partial x} - \frac{\partial P}{\partial y}\right) dxdy = \oint_C P\,dx + Q\,dy.$$

In vector terms Green's theorem can be written:

(17.9.1)
$$\iint_{\Omega} (\nabla \cdot \mathbf{v})\, dxdy = \oint_C (\mathbf{v} \cdot \mathbf{n})\, ds.$$

Here **n** is the outer unit normal and the integral on the right is taken with respect to arc length. (Section 17.4)

PROOF Set $\mathbf{v} = Q\,\mathbf{i} - P\,\mathbf{j}$. Then

$$\iint_{\Omega} (\nabla \cdot \mathbf{v})\, dxdy = \iint_{\Omega} \left(\frac{\partial Q}{\partial x} - \frac{\partial P}{\partial y}\right) dxdy.$$

All we have to show then is that

$$\oint_C (\mathbf{v} \cdot \mathbf{n})\, ds = \oint_C P\, dx + Q\, dy.$$

For C traversed counterclockwise, $\mathbf{n} = \mathbf{T} \times \mathbf{k}$ where **T** is the unit tangent vector. (Draw a figure.) Thus

$$\mathbf{v} \cdot \mathbf{n} = \mathbf{v} \cdot (\mathbf{T} \times \mathbf{k}) = (-\mathbf{v}) \cdot (\mathbf{k} \times \mathbf{T}) \underset{\uparrow\ \mathbf{a} \cdot (\mathbf{b} \times \mathbf{c}) = (\mathbf{a} \times \mathbf{b}) \cdot \mathbf{c}}{=} (-\mathbf{v} \times \mathbf{k}) \cdot \mathbf{T}.$$

Since $-\mathbf{v} \times \mathbf{k} = (P\,\mathbf{j} - Q\,\mathbf{i}) \times \mathbf{k} = P\,\mathbf{i} + Q\,\mathbf{j}$, we have $\mathbf{v} \cdot \mathbf{n} = (P\,\mathbf{i} + Q\,\mathbf{j}) \cdot \mathbf{T}$.

Therefore

$$\oint_C (\mathbf{v} \cdot \mathbf{n})\, ds = \oint_C [(P\,\mathbf{i} + Q\,\mathbf{j}) \cdot \mathbf{T}]\, ds \underset{(17.4.6)}{=} \oint_C (P\,\mathbf{i} + Q\,\mathbf{j}) \cdot d\mathbf{r} = \oint_C P\, dx + Q\, dy. \quad \square$$

Green's theorem expressed as (17.9.1) has a higher dimensional analog that is known as the divergence theorem.†

THEOREM 17.9.2 THE DIVERGENCE THEOREM

Let T be a solid bounded by a closed oriented surface S which, if not smooth, is piecewise smooth. If the vector field $\mathbf{v} = \mathbf{v}(x, y, z)$ is continuously differentiable throughout T, then

$$\iiint_T (\nabla \cdot \mathbf{v})\, dxdydz = \iint_S (\mathbf{v} \cdot \mathbf{n})\, d\sigma$$

where **n** is the outer unit normal.

PROOF We will carry out the proof under the assumption that S is smooth and that any line parallel to a coordinate axis intersects S at most twice.

† This is also called Gauss's theorem after the German mathematician Carl Friedrich Gauss (1777–1855). Often referred to as "The Prince of Mathematicians," Gauss is regarded by many as one of the greatest geniuses of all time.

Our first step is to express the outer unit normal $\mathbf{n}$ in terms of its direction cosines:

$$\mathbf{n} = \cos\alpha_1\,\mathbf{i} + \cos\alpha_2\,\mathbf{j} + \cos\alpha_3\,\mathbf{k}.$$

Then, for $\mathbf{v} = v_1\,\mathbf{i} + v_2\,\mathbf{j} + v_3\,\mathbf{k}$,

$$\mathbf{v} \cdot \mathbf{n} = v_1 \cos\alpha_1 + v_2 \cos\alpha_2 + v_3 \cos\alpha_3.$$

The idea of the proof is to show that

(1) $$\iint_S v_1 \cos\alpha_1 \, d\sigma = \iiint_T \frac{\partial v_1}{\partial x}\, dxdydz,$$

(2) $$\iint_S v_2 \cos\alpha_2 \, d\sigma = \iiint_T \frac{\partial v_2}{\partial y}\, dxdydz,$$

(3) $$\iint_S v_3 \cos\alpha_3 \, d\sigma = \iiint_T \frac{\partial v_3}{\partial z}\, dxdydz.$$

All three equations can be verified in much the same manner. We will carry out the details only for the third equation.

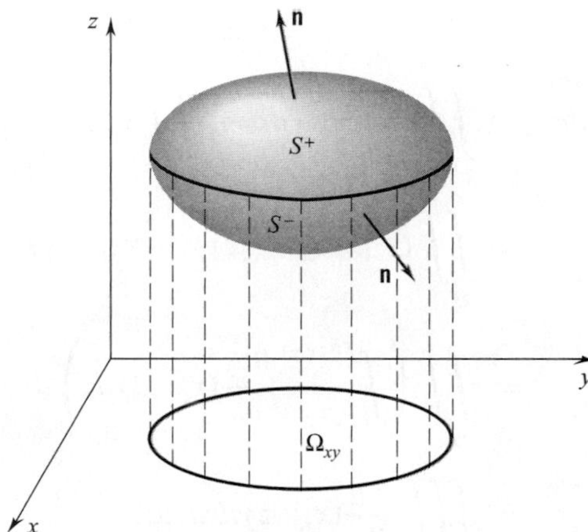

Figure 17.9.1

Let Ω_{xy} be the projection of T onto the xy-plane. (See Figure 17.9.1.) If $(x, y) \in \Omega_{xy}$, then, by assumption, the vertical line through (x, y) intersects S in at most two points, an upper point P^+ and a lower point P^-. (If the vertical line intersects S at only one point P, we set $P = P^+ = P^-$.) As (x, y) ranges over Ω_{xy}, the upper point P^+ describes a surface

$$S^+ : \quad z = f^+(x, y), \quad (x, y) \in \Omega_{xy} \qquad \text{(see the figure)}$$

and the lower point describes a surface

$$S^- : \quad z = f^-(x, y), \quad (x, y) \in \Omega_{xy}.$$

By our assumptions, f^+ and f^- are continuously differentiable, $S = S^+ \cup S^-$, and the solid T is the set of all points (x, y, z) with

$$f^-(x,y) \le z \le f^+(x,y), \quad (x,y) \in \Omega_{xy}.$$

Now let λ be the angle between the positive z-axis and the upper unit normal. On S^+ the outer unit normal $\mathbf{n}$ is the upper unit normal. Thus on S^+,

$$\lambda = \alpha_3 \qquad \text{and} \qquad \cos\alpha_3 \sec\gamma = 1.$$

On S^- the outer unit normal $\mathbf{n}$ is the lower unit normal. In this case,

$$\lambda = \pi - \alpha_3 \qquad \text{and} \qquad \cos\alpha_3 \sec\gamma = -1.$$

Thus,

$$\iint_{S^+} v_3 \cos\alpha_3 \, d\sigma = \underset{\uparrow (17.7.7)}{\iint_{\Omega_{xy}}} v_3 \cos\alpha_3 \sec\gamma \, dxdy = \iint_{\Omega_{xy}} v_3[x, y, f^+(x,y)]\, dxdy$$

and

$$\iint_{S^-} v_3 \cos\alpha_3 \, d\sigma = \iint_{\Omega_{xy}} v_3 \cos\alpha_3 \sec\gamma \, dxdy = -\iint_{\Omega_{xy}} v_3[x, y, f^-(x,y)]\, dxdy.$$

It follows that

$$\begin{aligned}
\iint_S v_3 \cos\alpha_3 \, d\sigma &= \iint_{S^+} v_3 \cos\alpha_3 \, d\sigma + \iint_{S^-} v_3 \cos\alpha_3 \, d\sigma \\
&= \iint_{\Omega xy} (v_3[x, y, f^+(x,y)] - v_3[x, y, f^-(x,y)])\, dxdy \\
&= \iint_{\Omega xy} \left(\int_{f^-(x,y)}^{f^+(x,y)} \frac{\partial v_3}{\partial z}(x, y, z)\, dz \right) dxdy \\
&= \iiint_T \frac{\partial v_3}{\partial z}(x, y, z)\, dxdydz.
\end{aligned}$$

This confirms Equation (3). Equation (2) can be confirmed by projection onto the xz-plane; Equation (1) can be confirmed by projection onto the yz-plane. ❑

Divergence as Outward Flux per Unit Volume

Choose a point P and surround it by a closed ball N_ϵ of radius ϵ. According to the divergence theorem,

$$\iiint_{N_\epsilon} (\nabla \cdot \mathbf{v})\, dxdydz = \text{flux of } \mathbf{v} \text{ out of } N_\epsilon.$$

Thus $\qquad$ (average divergence of $\mathbf{v}$ on N_ϵ) (volume of N_ϵ) = flux of $\mathbf{v}$ out of N_ϵ

and $\qquad \text{average divergence of } \mathbf{v} \text{ on } N_\epsilon = \dfrac{\text{flux of } \mathbf{v} \text{ out of } N_\epsilon}{\text{volume of } N_\epsilon}.$

Taking the limit of both sides as ϵ shrinks to 0, we have

$$\text{divergence of } \mathbf{v} \text{ at } P = \lim_{\epsilon \to 0^+} \frac{\text{flux of } \mathbf{v} \text{ out of } N_\epsilon}{\text{volume of } N_\epsilon}.$$

In this sense *divergence is outward flux per unit volume.*

Think of $\mathbf{v}$ as the velocity of a fluid. As suggested in Section 17.8, negative divergence at P signals an accumulation of fluid near P:

$$\nabla \cdot \mathbf{v} < 0 \text{ at } P \Longrightarrow \text{ flux out of } N_\epsilon < 0 \Longrightarrow \text{ net flow into } N_\epsilon.$$

Positive divergence at P signals a flow of liquid away from P:

$$\nabla \cdot \mathbf{v} > 0 \text{ at } P \Longrightarrow \text{ flux out of } N_\epsilon > 0 \Longrightarrow \text{ net flow out of } N_\epsilon.$$

Points at which the divergence is negative are called *sinks;* points at which the divergence is positive are called *sources*. If the divergence of $\mathbf{v}$ is 0 throughout, then the flow has no sinks and no sources and $\mathbf{v}$ is called *solenoidal*. (See Exercises 17.8)

Solids Bounded by Two or More Closed Surfaces

The divergence theorem, stated for solids bounded by a single closed oriented surface, can be extended to solids bounded by several closed surfaces. Suppose, for example, that we start with a solid bounded by a closed oriented surface S_1 and extract from the interior of that solid a solid bounded by a closed oriented surface S_2. The remaining solid, call it T and see Figure 17.9.2, has a boundary S that consists of two pieces: an outer piece S_1 and an inner piece S_2. The key here is to note that the *outer* normal for T points *out* of S_1 but *into* S_2. The divergence theorem can be proved for T by slicing T into two pieces T_1 and T_2 as in Figure 17.9.3 and applying the divergence theorem to each piece:

$$\iiint_{T_1} (\nabla \cdot \mathbf{v})\, dxdydz = \iint_{\text{bdry of } T_1} (\mathbf{v} \cdot \mathbf{n})\, d\sigma,$$

$$\iiint_{T_2} (\nabla \cdot \mathbf{v})\, dxdydz = \iint_{\text{bdry of } T_2} (\mathbf{v} \cdot \mathbf{n})\, d\sigma,$$

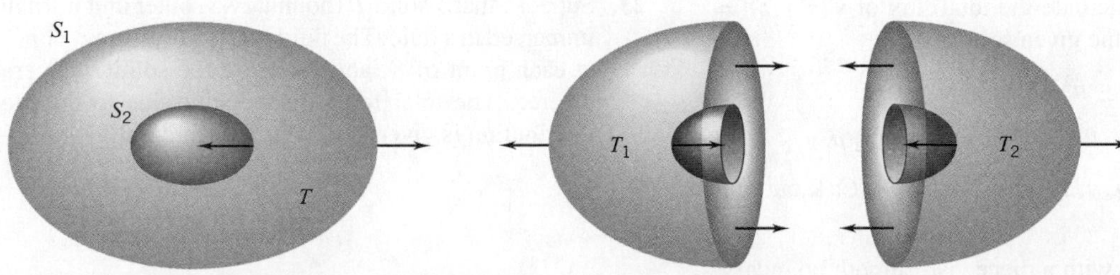

Figure 17.9.2 $\qquad$ Figure 17.9.3

The triple integrals over T_1 and T_2 add up to the triple integral over T. When the surface integrals are added together, the integrals along the common cut cancel (because the normals are in opposite directions), and therefore only the integrals over S_1 and S_2 remain. Thus the surface integrals add up to the surface integral over $S = S_1 \cup S_2$ and the divergence theorem still holds:

$$\iiint_T (\nabla \cdot \mathbf{v})\, dxdydz = \iint_S (\mathbf{v} \cdot \mathbf{n})\, d\sigma.$$

EXERCISES 17.9

Calculate the flux $\mathbf{v}$ out of the unit ball $x^2 + y^2 + z^2 \leq 1$ by applying the divergence theorem.

1. $\mathbf{v}(x, y, z) = x\,\mathbf{i} + y\,\mathbf{j} + z\,\mathbf{k}$.

2. $\mathbf{v}(x, y, z) = (1 - x)\,\mathbf{i} + (2 - y)\,\mathbf{j} + (3 - z)\,\mathbf{k}$.

3. $\mathbf{v}(x, y, z) = x^2\,\mathbf{i} + y^2\,\mathbf{j} + z^2\,\mathbf{k}$.

4. $\mathbf{v}(x, y, z) = (1 - x^2)\,\mathbf{i} + y^2\,\mathbf{j} + z\,\mathbf{k}$.

Verify the divergence theorem on the unit cube $0 \leq x \leq 1$, $0 \leq y \leq 1, 0 \leq z \leq 1$ for the following vector fields.

5. $\mathbf{v}(x, y, z) = x\,\mathbf{i} + y\,\mathbf{j} + z\,\mathbf{k}$.

6. $\mathbf{v}(x, y, z) = xy\,\mathbf{i} + yz\,\mathbf{j} + xz\,\mathbf{k}$.

7. $\mathbf{v}(x, y, z) = x^2\,\mathbf{i} - xz\,\mathbf{j} + z^2\,\mathbf{k}$.

8. $\mathbf{v}(x, y, z) = x\,\mathbf{i} + xy\,\mathbf{j} + xyz\,\mathbf{k}$.

Use the divergence theorem to find the total flux out of the given solid.

9. $\mathbf{v}(x, y, z) = x\,\mathbf{i} + 2y^2\,\mathbf{j} + 3z^2\,\mathbf{k}$; $\quad x^2 + y^2 \leq 9, \quad 0 \leq z \leq 1$.

10. $\mathbf{v}(x, y, z) = xy\,\mathbf{i} + yz\,\mathbf{j} + xz\,\mathbf{k}$; $\quad 0 \leq x \leq 1$, $0 \leq y \leq 1 - x, \quad 0 \leq z \leq 1 - x - y$.

11. $\mathbf{v}(x, y, z) = x^2\,\mathbf{i} + xy\,\mathbf{j} - 2xz\,\mathbf{k}$; $\quad 0 \leq x \leq 1$, $0 \leq y \leq 1 - x, \quad 0 \leq z \leq 1 - x - y$.

12. $\mathbf{v}(x, y, z) = (2xy + 2z)\,\mathbf{i} + (y^2 + 1)\,\mathbf{j} - (x + y)\,\mathbf{k}$; $0 \leq x \leq 4, \quad 0 \leq y \leq 4 - x, \quad 0 \leq z \leq 4 - x - y$.

13. $\mathbf{v}(x, y, z) = x^2\,\mathbf{i} + y^2\,\mathbf{j} + z^2\,\mathbf{k}$; the cylinder $x^2 + y^2 \leq 4, \quad 0 \leq z \leq 4$, including the top and base.

14. $\mathbf{v}(x, y, z) = 2x\,\mathbf{i} + xy\,\mathbf{j} + xz\,\mathbf{k}$; the ball $x^2 + y^2 + z^2 \leq 4$.

In Exercises 15 and 16, calculate the total flux of $\mathbf{v}(x, y, z) = 2xy\,\mathbf{i} + y^2\,\mathbf{j} + 3yz\,\mathbf{k}$ out of the given solid.

15. The ball: $x^2 + y^2 + z^2 \leq a^2$.

16. The cube: $0 \leq x \leq a, \quad 0 \leq y \leq a, \quad 0 \leq z \leq a$.

17. What is the flux of $\mathbf{v}(x, y, z) = Ax\,\mathbf{i} + By\,\mathbf{j} + Cz\,\mathbf{k}$ out of a solid of volume V?

18. Let T be a basic solid with a piecewise-smooth boundary. Show that if f is harmonic on T (defined in Exercises 17.8), then the flux of ∇f out of T is zero.

19. Let S be a closed smooth surface with continuous unit normal $\mathbf{n} = \mathbf{n}(x, y, z)$. Show that

$$\iint_S \mathbf{n}\, d\sigma = \left(\iint_S n_1\, d\sigma\right)\mathbf{i} + \left(\iint_S n_2\, d\sigma\right)\mathbf{j} + \left(\iint_S n_3\, d\sigma\right)\mathbf{k} = \mathbf{0}.$$

20. Let T be a solid with a piecewise-smooth boundary S and let $\mathbf{n}$ be the outer unit normal.

(a) Verify the identity $\nabla \cdot (f \nabla f) = ||\nabla f||^2 + f(\nabla^2 f)$ and show that, if f is harmonic on T, then

$$\iint_S (f f'_{\mathbf{n}})\, d\sigma = \iiint_T ||\nabla f||^2\, dxdydz$$

where $f'_{\mathbf{n}}$ is the directional derivative $\nabla f \cdot \mathbf{n}$.

(b) Show that, if g is continuously differentiable on T, then

$$\iint_S (g f'_{\mathbf{n}})\, d\sigma = \iiint_T [(\nabla g \cdot \nabla f) + g(\nabla^2 f)]\, dxdydz.$$

21. Let T be a solid with a piecewise-smooth boundary. Show that if f and g have continuous second partials, then the flux of $\nabla f \times \nabla g$ out of T is zero.

22. Let T be a solid with a piecewise-smooth boundary S. Express the volume of T as a surface integral over S.

23. Suppose that a solid T (boundary S, outer unit normal $\mathbf{n}$) is immersed in a fluid. The fluid exerts a pressure $p = p(x, y, z)$ at each point of S, and therefore the solid T experiences a force. The total force on the solid due to the pressure distribution is given by the surface integral

$$\mathbf{F} = -\iint_S p\,\mathbf{n}\, d\sigma.$$

(The formula says that the force on the solid is the average pressure against S times the area of S.) Now choose a coordinate system with the z-axis vertical and assume that the fluid fills a region of space to the level $z = c$. The depth of a point (x, y, z) is then $c - z$ and we have $p(x, y, z) = \rho(c - z)$, where ρ is the weight density of the fluid (the weight per unit volume). Apply the divergence theorem to each component of $\mathbf{F}$ to show that $\mathbf{F} = W\mathbf{k}$ where W is the weight of the fluid displaced by the solid. We call this the *buoyant force* on the solid. (This shows that the object is not pushed from side to side by the pressure and verifies the *principle of Archimedes*: that the buoyant force on an object at rest in a fluid equals the weight of the fluid displaced.)

24. If $\mathbf{F}$ is a force applied at the tip of a radius vector $\mathbf{r}$, then the *torque*, or twisting strength, of $\mathbf{F}$ about the origin is given by the cross product $\boldsymbol{\tau} = \mathbf{r} \times \mathbf{F}$. From physics we learn that the total torque on the solid T of Exercise 23 is given by the formula

$$\boldsymbol{\tau}_{\text{Tot}} = -\iint_S [\mathbf{r} \times \rho(c - z)\,\mathbf{n}]\,d\sigma.$$

Use the divergence theorem to find the components of $\boldsymbol{\tau}_{\text{Tot}}$ and show that $\boldsymbol{\tau}_{\text{Tot}} = \bar{\mathbf{r}} \times \mathbf{F}$ where $\mathbf{F}$ is the buoyant force $W\mathbf{k}$ and $\bar{\mathbf{r}}$ is the centroid of T. (This indicates that for calculating the twisting effect of the buoyant force we can view this force as being applied at the centroid. This is very important in ship design. Imagine, for example, a totally submerged submarine. While the buoyant force acts upward through the centroid of the submarine, gravity acts downward through the center of mass. Suppose the submarine should tilt a bit to one side as depicted in the figure. If the centroid lies above the center of mass, then the buoyant force acts to restore the submarine to an upright position. If, however, the centroid lies below the center of mass, then disaster. Once the submarine has tilted a bit, the buoyant force will make it tilt further. This kind of

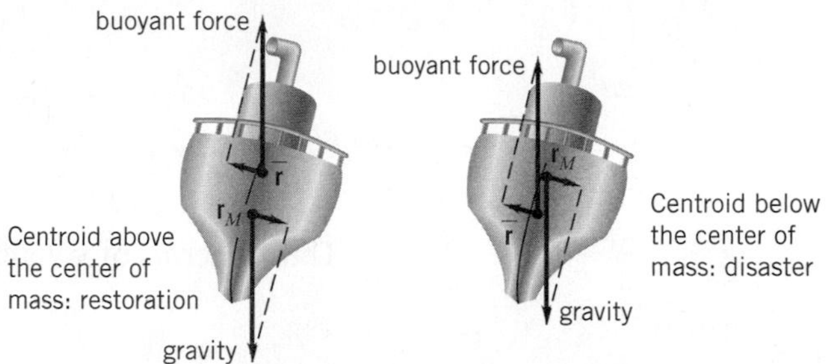

analysis also applies to surface ships, but in this case the buoyant force acts upward through *the centroid of the portion of the ship that is submerged*. This point is called the *center of flotation*. One must design and load a ship to keep the center of flotation above the center of mass. Putting a lot of heavy cargo on the deck, for instance, tends to raise the center of mass and destabilize the ship.)

■ PROJECT 17.9 Static Charges

Consider a point charge q somewhere in space. This charge creates around itself an *electric field* $\mathbf{E}$, which in turn exerts an electric force on every other nearby charge. If we center our coordinate system at q, then the electric field at the point $\mathbf{r}$ can be written

$$\mathbf{E}(\mathbf{r}) = q\frac{\mathbf{r}}{r^3}.$$

This result is found experimentally. Note that this field has the same form as a gravitational field.

Problem 1. Show that $\nabla \cdot \mathbf{E} = 0$ for all $\mathbf{r} \neq \mathbf{0}$.]

We are interested in the flux of $\mathbf{E}$ out of a closed surface S. We assume that S does not pass through q.

Problem 2. Suppose that the charge q is outside S. Then $\mathbf{E}$ is continuously differentiable on the region T bounded by S. Use the divergence theorem to show that

$$\text{flux of } \mathbf{E} \text{ out of } S = 0.$$

If q is inside S, then the divergence theorem does not apply to T directly because $\mathbf{E}$ is not differentiable on all of T. We can circumvent this difficulty by surrounding q by a small sphere S_a of radius a and applying the divergence theorem to the region T' bounded on the outside by S and on the inside by S_a. See the figure.

Since $\mathbf{E}$ is continuously differentiable on T',

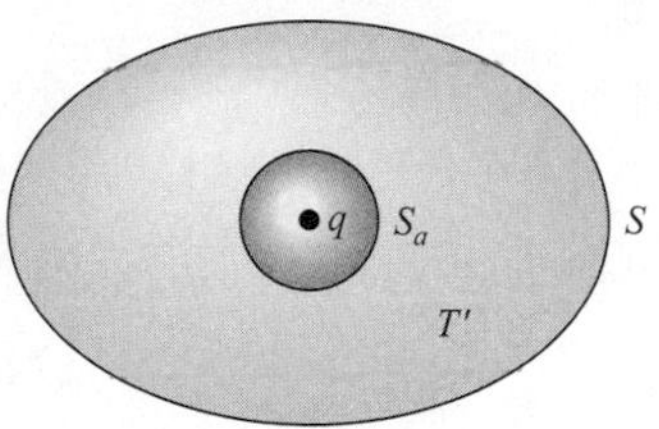

$$\iiint_T (\nabla \cdot \mathbf{E})\,dxdydz = \text{flux of } \mathbf{E} \text{ out of } S + \text{flux of } \mathbf{E} \text{ into } S_a$$
$$= \text{flux of } \mathbf{E} \text{ out of } S - \text{flux of } \mathbf{E} \text{ out of } S_a.$$

Since $\nabla \cdot \mathbf{E} = 0$ on T', it follows that

$$\text{flux of } \mathbf{E} \text{ out of } S = \text{flux of } \mathbf{E} \text{ out of } S_a.$$

Problem 3. Show that: flux out of $S_a = 4\pi q$.

Thus, if $\mathbf{E}$ is the electric field of a point charge q and S is a closed surface that does not pass through q, then

$$\text{flux of } \mathbf{E} \text{ out of } S = \begin{cases} 0 & \text{if } q \text{ is outside } S \\ 4\pi q & \text{if } q \text{ is inside } S. \end{cases}$$

■ 17.10 STOKES'S THEOREM

We return to Green's theorem

$$\iint_{\Omega} \left(\frac{\partial Q}{\partial x} - \frac{\partial P}{\partial y}\right) dxdy = \oint P\,dx + Q\,dy.$$

Setting $\mathbf{v} = P\,\mathbf{i} + Q\,\mathbf{j} + R\,\mathbf{k}$, we have

$$(\nabla \times \mathbf{v}) \cdot \mathbf{k} = \begin{vmatrix} \mathbf{i} & \mathbf{j} & \mathbf{k} \\ \dfrac{\partial}{\partial x} & \dfrac{\partial}{\partial y} & \dfrac{\partial}{\partial z} \\ P & Q & R \end{vmatrix} \cdot \mathbf{k} = \frac{\partial Q}{\partial x} - \frac{\partial P}{\partial y}.$$

Thus in terms of $\mathbf{v}$, Green's theorem can be written

$$\iint_{\Omega} [(\nabla \times \mathbf{v}) \cdot \mathbf{k}]\, dxdy = \oint_C \mathbf{v}(\mathbf{r}) \cdot d\mathbf{r}.$$

Since any plane can be coordinatized as the xy-plane, this result can be phrased as follows: Let S be a flat surface in space bounded by a Jordan curve C. If $\mathbf{v}$ is continuously differentiable on S, then

$$\iint_S [(\nabla \times \mathbf{v}) \cdot \mathbf{n}]\, d\sigma = \oint_C \mathbf{v}(\mathbf{r}) \cdot d\mathbf{r}$$

where $\mathbf{n}$ is a unit normal for S and the line integral is taken in the *positive sense*, meaning in the direction of the unit tangent $\mathbf{T}$ for which $\mathbf{T} \times \mathbf{n}$ points away from the surface. See Figure 17.10.1. (An observer marching along C with the same orientation as $\mathbf{n}$ keeps the surface to his left.) The symbol $\oint_C$ denotes this line integral.

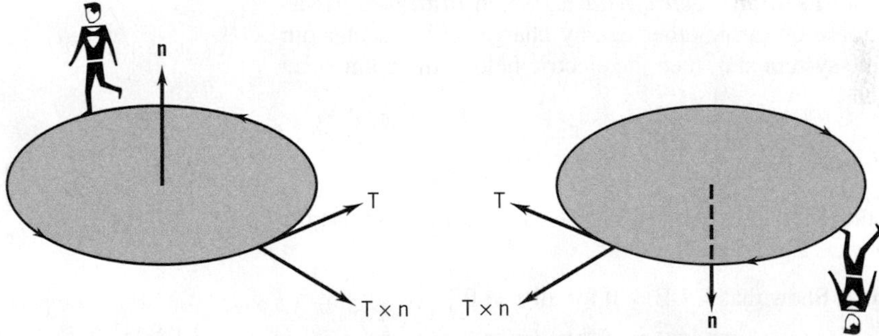

Figure 17.10.1

Figure 17.10.2 shows a *polyhedral surface* S bounded by a closed polygonal path C. The surface S consists of a finite number of flat faces $S_1, \ldots, S_N$ with polygonal boundaries $C_1, \ldots, C_N$ and unit normals $\mathbf{n}_1, \ldots, \mathbf{n}_N$. We choose these unit normals in a consistent manner; that is, they emanate from the same side of the surface. Now let $\mathbf{n} = \mathbf{n}(x, y, z)$ be a vector function of norm 1 which is $\mathbf{n}_1$ on S_1, $\mathbf{n}_2$ on S_2, $\mathbf{n}_3$ on S_3, etc. It is immaterial how $\mathbf{n}$ is defined on the line segments that join the different faces. Suppose now that $\mathbf{v} = \mathbf{v}(x, y, z)$ is a vector function continuously differentiable on an open set that contains S. Then

$$\iint_S [(\nabla \times \mathbf{v}) \cdot \mathbf{n}]\, d\sigma = \sum_{i=1}^{N} \iint_{S_i} [(\nabla \times \mathbf{v}) \cdot \mathbf{n}_i]\, d\sigma = \sum_{i=1}^{N} \oint_{C_i} \mathbf{v}(\mathbf{r}) \cdot d\mathbf{r},$$

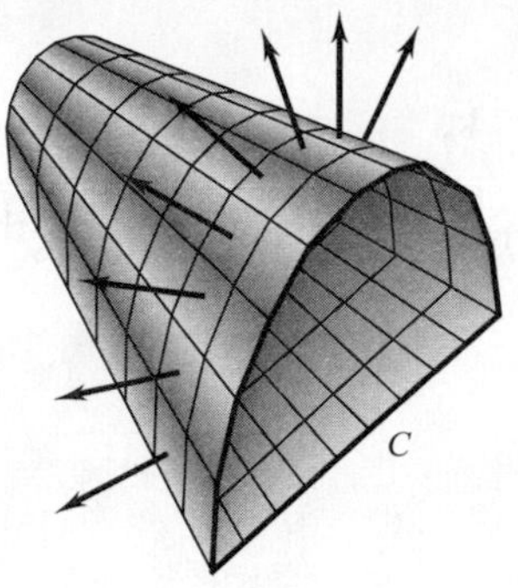

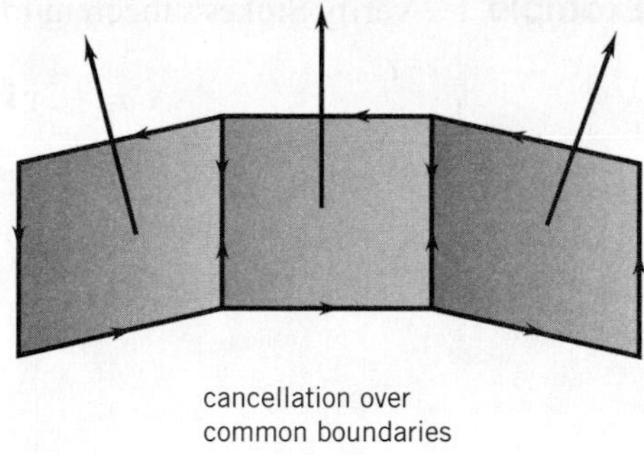

Figure 17.10.2

the integral over C_i being taken in the positive sense with respect to $\mathbf{n}_i$. Now when we add these line integrals, we find that all the line segments that make up the C_i but are not part of C are traversed twice and in opposite directions. (See the figure.) Thus these line segments contribute nothing to the sum of the line integrals and we are left with the integral around C. It follows that for a polyhedral surface S with boundary C

$$\iint_S [(\nabla \times \mathbf{v}) \cdot \mathbf{n}]\, d\sigma = \oint_C \mathbf{v}(\mathbf{r}) \cdot d\mathbf{r}.$$

This result can be extended to smooth oriented surfaces with smooth bounding curves (see Figure 17.10.3) by approximating these configurations by polyhedral configurations of the type considered and using a limit process. In an admittedly informal way we have arrived at Stokes's theorem.

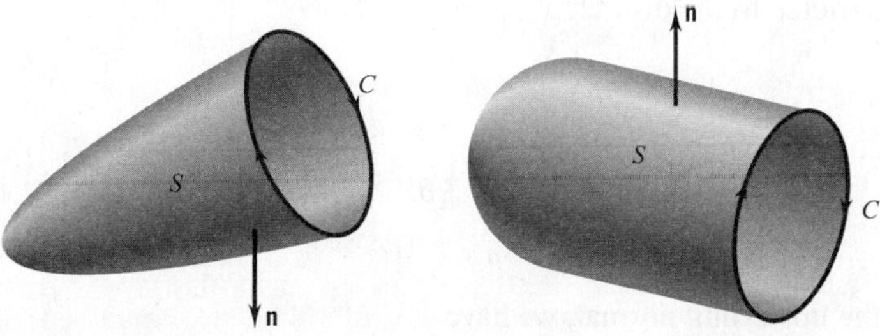

Figure 17.10.3

THEOREM 17.10.1 STOKES'S THEOREM †

Let S be a smooth oriented surface with a smooth bounding curve C. If $\mathbf{v} = \mathbf{v}(x, y, z)$ is a continuously differentiable vector field on an open set that contains S, then

$$\iint_S [(\nabla \times \mathbf{v}) \cdot \mathbf{n}]\, d\sigma = \oint_C \mathbf{v}(\mathbf{r}) \cdot d\mathbf{r},$$

where $\mathbf{n} = \mathbf{n}(x, y, z)$ is a unit normal that varies continuously on S and the line integral is taken in the positive sense with respect to $\mathbf{n}$.

† The result was announced publicly for the first time by George Gabriel Stokes (1819–1903), an Irish mathematician and physicist who, like Green, was a Cambridge professor.

Example 1 Verify Stokes's theorem for

$$\mathbf{v} = -3y\,\mathbf{i} + 3x\,\mathbf{j} + z^4\,\mathbf{k},$$

taking S as the portion of the ellipsoid $2x^2 + 2y^2 + z^2 = 1$ that lies above the plane $z = 1/\sqrt{2}$. (Figure 17.10.4)

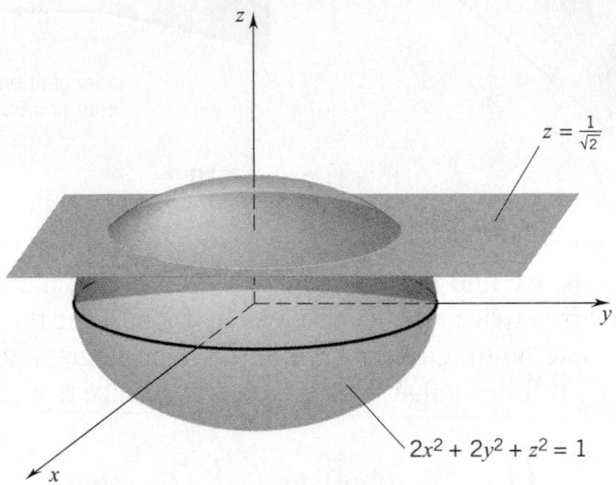

Figure 17.10.4

SOLUTION A little algebra shows that S is the graph of

$$f(x, y) = \sqrt{1 - 2(x^2 + y^2)}$$

with (x, y) restricted to the disc $\Omega: x^2 + y^2 \le \frac{1}{4}$. Now

$$\nabla \times \mathbf{v} = \begin{vmatrix} \mathbf{i} & \mathbf{j} & \mathbf{k} \\ \dfrac{\partial}{\partial x} & \dfrac{\partial}{\partial y} & \dfrac{\partial}{\partial z} \\ -3y & 3x & z^4 \end{vmatrix} = \left[\frac{\partial}{\partial x}(3x) + \frac{\partial}{\partial y}(3y)\right]\mathbf{k} = 6\,\mathbf{k}.$$

Taking $\mathbf{n}$ as the upper unit normal, we have

$$\begin{aligned} \iint_S [(\nabla \times \mathbf{v}) \cdot \mathbf{n}]\, d\sigma &= \iint_S (6\mathbf{k} \cdot \mathbf{n})\, d\sigma \\ &\underset{(17.7.9)}{=} \iint_\Omega (-(0) f_x - (0) f_y + 6)\, dxdy \\ &= \iint_\Omega 6\, dxdy = 6\,(\text{area of } \Omega) = 6(\tfrac{1}{4}\pi) = \tfrac{3}{2}\pi. \end{aligned}$$

The bounding curve C is the set of all (x, y, z) with $x^2 + y^2 = \frac{1}{4}$ and $z = 1/\sqrt{2}$. We can parametrize C by setting

$$\mathbf{r}(u) = \tfrac{1}{2}\cos u\,\mathbf{i} + \tfrac{1}{2}\sin u\,\mathbf{j} + \frac{1}{\sqrt{2}}\mathbf{k}, \qquad u \in [0, 2\pi].$$

Since $\mathbf{n}$ is the upper unit normal, this parametrization gives C in the positive sense.

Thus

$$\oint_C \mathbf{v}(\mathbf{r}) \cdot d\mathbf{r} = \int_0^{2\pi} (-\tfrac{3}{2}\sin u\,\mathbf{i} + \tfrac{3}{2}\cos u\,\mathbf{j} + \tfrac{1}{4}\mathbf{k}) \cdot (-\tfrac{1}{2}\sin u\,\mathbf{i} + \tfrac{1}{2}\cos u\,\mathbf{j})\,du$$

$$= \int_0^{2\pi} (\tfrac{3}{4}\sin^2 u + \tfrac{3}{4}\cos^2 u)\,du = \int_0^{2\pi} \tfrac{3}{4}\,du = \tfrac{3}{2}\pi.$$

This is the value we obtained for the surface integral. ☐

Example 2 Verify Stokes's theorem for

$$\mathbf{v} = z^2\,\mathbf{i} - 2x\,\mathbf{j} + y^3\,\mathbf{k},$$

taking S as the upper half of the unit sphere $x^2 + y^2 + z^2 = 1$.

SOLUTION We use the upper unit normal $\mathbf{n} = x\,\mathbf{i} + y\,\mathbf{j} + z\,\mathbf{k}$. Now

$$\nabla \times \mathbf{v} = \begin{vmatrix} \mathbf{i} & \mathbf{j} & \mathbf{k} \\ \dfrac{\partial}{\partial x} & \dfrac{\partial}{\partial y} & \dfrac{\partial}{\partial z} \\ z^2 & -2x & y^3 \end{vmatrix} = 3y^2\,\mathbf{i} + 2z\,\mathbf{j} - 2\,\mathbf{k}.$$

Therefore

$$\iint_S [(\nabla \times \mathbf{v}) \cdot \mathbf{n}]\,d\sigma = \iint_S [(3y^2\,\mathbf{i} + 2z\,\mathbf{j} - 2\,\mathbf{k}) \cdot (x\,\mathbf{i} + y\,\mathbf{j} + z\,\mathbf{k})]\,d\sigma$$

$$= \iint_S (3xy^2 + 2yz - 2z)\,d\sigma$$

$$= \iint_S 3xy^2\,d\sigma + \iint_S 2yz\,d\sigma - \iint_S 2z\,d\sigma.$$

The first integral is zero because S is symmetric about the yz-plane and the integrand is odd with respect to x. The second integral is zero because S is symmetric about the xz-plane and the integrand is odd with respect to y. Thus

$$\iint_S [(\nabla \times \mathbf{v}) \cdot \mathbf{n}]\,d\sigma = -\iint_S 2z\,d\sigma = -2\bar{z}(\text{area of } S) = -2(\tfrac{1}{2})2\pi = -2\pi.$$

Exercise 17, Section 17.7 ↑

This is also the value of the integral along the bounding base circle taken in the positive sense: $\mathbf{r}(u) = \cos u\,\mathbf{i} + \sin u\,\mathbf{j},\ u \in [0, 2\pi]$, and

$$\oint_C \mathbf{v}(\mathbf{r}) \cdot d\mathbf{r} = \oint_C z^2\,dx - 2x\,dy = -2\oint_C x\,dy$$

$$= -2\int_0^{2\pi} \cos^2 u\,du = -2\pi. \quad ☐$$

Earlier you saw that the curl of a gradient is zero. Using Stokes's theorem we can prove a partial converse.

(17.10.2) If a vector field $\mathbf{v} = \mathbf{v}(x, y, z)$ is continuously differentiable on an open convex† set U and $\nabla \times \mathbf{v} = \mathbf{0}$ on all of U, then $\mathbf{v}$ is the gradient of some scalar field ϕ defined on U.

---

† *A* set U is said to be *convex* 0 if for each pair of points $p, q \in U$, the line segment $\overline{pq}$ lies entirely in U.

PROOF Choose a point $\mathbf{a}$ in U, and for each point $\mathbf{x}$ in U, define

$$\phi(\mathbf{x}) = \int_{\mathbf{a}}^{\mathbf{x}} \mathbf{v}(\mathbf{r}) \cdot \mathbf{dr}.$$

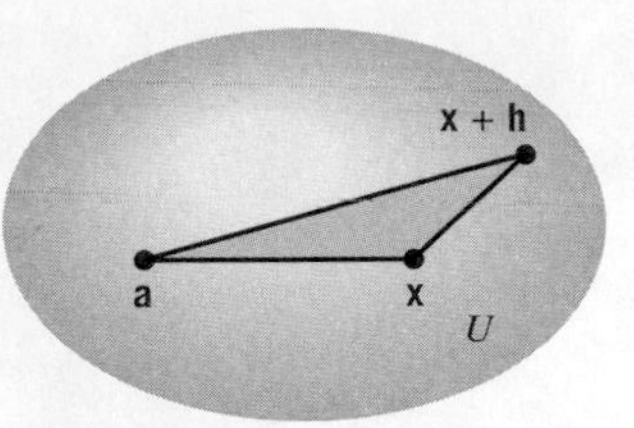

Figure 17.10.5

(This is the line integral from $\mathbf{a}$ to $\mathbf{x}$ taken along the line segment that joins these two points. We know that this line segment lies in U because U is convex.)

Since U is open, $\mathbf{x} + \mathbf{h}$ is in U for all $\mathbf{h}$ sufficiently small. Assume then that $\mathbf{h}$ is sufficiently small for $\mathbf{x} + \mathbf{h}$ to be in U. Since U is convex, the triangular region with vertices at $\mathbf{a}, \mathbf{x}, \mathbf{x} + \mathbf{h}$ lies in U. (See Figure 17.10.5.) Since $\nabla \times \mathbf{v} = \mathbf{0}$ on U, we can conclude from Stokes's theorem that

$$\int_{\mathbf{a}}^{\mathbf{x}} \mathbf{v}(\mathbf{r}) \cdot \mathbf{dr} + \int_{\mathbf{x}}^{\mathbf{x}+\mathbf{h}} \mathbf{v}(\mathbf{r}) \cdot \mathbf{dr} + \int_{\mathbf{x}+\mathbf{h}}^{\mathbf{a}} \mathbf{v}(\mathbf{r}) \cdot \mathbf{dr} = 0.$$

Therefore

$$\begin{aligned} \int_{\mathbf{x}}^{\mathbf{x}+\mathbf{h}} \mathbf{v}(\mathbf{r}) \cdot \mathbf{dr} &= -\int_{\mathbf{x}+\mathbf{h}}^{\mathbf{a}} \mathbf{v}(\mathbf{r}) \cdot \mathbf{dr} - \int_{\mathbf{a}}^{\mathbf{x}} \mathbf{v}(\mathbf{r}) \cdot \mathbf{dr} \\ &= \int_{\mathbf{a}}^{\mathbf{x}+\mathbf{h}} \mathbf{v}(\mathbf{r}) \cdot \mathbf{dr} - \int_{\mathbf{a}}^{\mathbf{x}} \mathbf{v}(\mathbf{r}) \cdot \mathbf{dr}. \end{aligned}$$

By our definition of ϕ, we have

$$\phi(\mathbf{x} + \mathbf{h}) - \phi(\mathbf{x}) = \int_{\mathbf{x}}^{\mathbf{x}+\mathbf{h}} \mathbf{v}(\mathbf{r}) \cdot \mathbf{dr}.$$

We can parametrize the line segment from $\mathbf{x}$ to $\mathbf{x} + \mathbf{h}$ by $\mathbf{r}(u) = \mathbf{x} + u\mathbf{h}$ with $u \in [0, 1]$. Therefore

$$\begin{aligned} \phi(\mathbf{x} + \mathbf{h}) - \phi(\mathbf{x}) &= \int_0^1 [\mathbf{v}(\mathbf{r}(u)) \cdot \mathbf{r}'(u)]\, du \\ &= \int_0^1 [\mathbf{v}(\mathbf{r}(u)) \cdot \mathbf{h}]\, du \\ \text{Theorem 5.8.1} \longrightarrow \quad &= \mathbf{v}(\mathbf{r}(u_0)) \cdot \mathbf{h} \qquad \text{for some } u_0 \text{ in } [0, 1] \\ = \mathbf{v}(\mathbf{x} + u_0\mathbf{h}) \cdot \mathbf{h} &= \mathbf{v}(\mathbf{x}) \cdot \mathbf{h} + [\mathbf{v}(\mathbf{x} + u_0\mathbf{h}) - \mathbf{v}(\mathbf{x})] \cdot \mathbf{h}. \end{aligned}$$

The fact that $\mathbf{v} = \nabla\phi$ follows from observing that $[\mathbf{v}(\mathbf{x} + u_o\mathbf{h}) - \mathbf{v}(\mathbf{x})] \cdot \mathbf{h}$ is $o(\mathbf{h})$:

$$\begin{aligned} \frac{|[\mathbf{v}(\mathbf{x} + u_0\,\mathbf{h}) - \mathbf{v}(\mathbf{x})] \cdot \mathbf{h}|}{||\mathbf{h}||} &\le \frac{||\mathbf{v}(\mathbf{x} + u_0\mathbf{h}) - \mathbf{v}\,(\mathbf{x})||\;||\mathbf{h}||}{||\mathbf{h}||} \\ &= ||\mathbf{v}(\mathbf{x} + u_0\mathbf{h}) - \mathbf{v}(\mathbf{x})|| \to 0 \end{aligned}$$

as $\mathbf{h} \to \mathbf{0}$. □

The Normal Component of $\nabla\times$ v as Circulation per Unit Area; Irrotational Flow

Figure 17.10.6

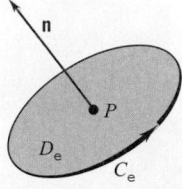

Interpret $\mathbf{v} = \mathbf{v}(x, y, z)$ as the velocity of a fluid flow. In Section 17.8 we stated that $\nabla \times \mathbf{v}$ measures the rotational tendency of the fluid. Now we can be more precise.

Take a point P within the flow and choose a unit vector $\mathbf{n}$. Let D_ϵ be the ϵ-disc that is centered at P and is perpendicular to $\mathbf{n}$. Let C_ϵ be the circular boundary of D_ϵ directed in the positive sense with respect to $\mathbf{n}$. (See Figure 17.10.6.) By Stokes's theorem,

$$\iint_{D_\epsilon} [(\nabla \times \mathbf{v}) \cdot \mathbf{n}]\, d\sigma = \oint_{C_\epsilon} \mathbf{v}(\mathbf{r}) \cdot \mathbf{dr}.$$

The line integral on the right is called *the circulation* of $\mathbf{v}$ around C_ϵ. Thus we can say that

$$\begin{pmatrix}\text{the average } \mathbf{n}\text{-component of} \\ \nabla \times \mathbf{v} \text{ on } D_\epsilon\end{pmatrix} (\text{ the area of } D_\epsilon) = \text{ the circulation of } \mathbf{v} \text{ around } C_\epsilon.$$

It follows that

$$\text{the average } \mathbf{n}\text{-component of } \nabla \times \mathbf{v} \text{ on } D_\epsilon = \frac{\text{the circulation of } \mathbf{v} \text{ around } C_\epsilon}{\text{the area of } D_\epsilon}.$$

Taking the limit as ϵ shrinks to 0, you can see that

$$\text{the } \mathbf{n}\text{-component of } \nabla \times \mathbf{v} \text{ at } P = \lim_{\epsilon \to 0^-} \frac{\text{the circulation of } \mathbf{v} \text{ around } C_\epsilon}{\text{the area of } D_\epsilon}.$$

At each point P the component of $\nabla \times \mathbf{v}$ *in any direction* $\mathbf{n}$ *is the circulation of* $\mathbf{v}$ *per unit area in the plane normal to* $\mathbf{n}$. *If* $\nabla \times \mathbf{v} = \mathbf{0}$ *identically, the fluid has no rotational tendency, and the flow is called irrotational.*

Remark Flux and circulation apply to vector fields where no material substance is flowing. Electromagnetic phenomena result from the action and interaction of two vector fields: the electric field E and the magnetic field B. The four fundamental laws of electromagnetism can be stated as equations that give the flux and circulation of these two fields.

EXERCISES 17.10

In Exercises 1–4, let S be the upper half of the unit sphere $x^2 + y^2 + z^2 = 1$ and take $\mathbf{n}$ as the upper unit normal. Find

$$\iint_S [(\nabla \times \mathbf{v}) \cdot \mathbf{n}]\, d\sigma$$

(a) by direct calculation. (b) by Stokes's theorem.

1. $\mathbf{v}(x, y, z) = x\,\mathbf{i} + y\,\mathbf{j} + z\,\mathbf{k}$.

2. $\mathbf{v}(x, y, z) = y\,\mathbf{i} - x\,\mathbf{j} + z\,\mathbf{k}$.

3. $\mathbf{v}(x, y, z) = z^2\,\mathbf{i} + 2x\,\mathbf{j} - y^3\,\mathbf{k}$.

4. $\mathbf{v}(x, y, z) = 6xz\,\mathbf{i} - x^2\,\mathbf{j} - 3y^2\,\mathbf{k}$.

In Exercises 5–7, let S be the triangular surface with vertices $(2, 0, 0), (0, 2, 0), (0, 0, 2)$ and take $\mathbf{n}$ as the upper unit normal. Find

$$\iint_S [(\nabla \times \mathbf{v}) \cdot \mathbf{n}]\, d\sigma$$

(a) by direct calculation. (b) by Stokes's theorem.

5. $\mathbf{v}\,(x, y, z) = 2z\,\mathbf{i} - y\,\mathbf{j} + x\,\mathbf{k}$.

6. $\mathbf{v}\,(x, y, z) = (x^2 + y^2)\,\mathbf{i} + y^2\,\mathbf{j} + (x^2 + z^2)\,\mathbf{k}$.

7. $\mathbf{v}\,(x, y, z) = x^4\,\mathbf{i} + xy\,\mathbf{j} + z^4\,\mathbf{k}$.

8. Show that if $\mathbf{v} = \mathbf{v}(x, y, z)$ is continuously differentiable everywhere and its curl is identically zero, then

$$\int_C \mathbf{v}(\mathbf{r}) \cdot \mathbf{dr} = 0 \quad \text{for every smooth closed curve } C.$$

9. Let $\mathbf{v} = y\,\mathbf{i} + z\,\mathbf{j} + x^2y^2\,\mathbf{k}$ and let S be the surface $z = x^2 + y^2$ from $z = 0$ to $z = 4$. Calculate the flux of $\nabla \times \mathbf{v}$ in the direction of the lower unit normal $\mathbf{n}$.

10. Let $\mathbf{v} = \frac{1}{2}y\,\mathbf{i} + 2xz\,\mathbf{j} - 3x\,\mathbf{k}$ and let S be the surface $y = 1 - (x^2 + z^2)$ from $y = -8$ to $y = 1$. Calculate the flux of $\nabla \times \mathbf{v}$ in the direction of the unit normal $\mathbf{n}$ with positive $\mathbf{j}$-component.

11. Let $\mathbf{v} = 2x\,\mathbf{i} + 2y\,\mathbf{j} + x^2y^2z^2\,\mathbf{k}$ and let S be the lower half of the ellipsoid.

$$\frac{x^2}{4} + \frac{y^2}{9} + \frac{z^2}{27} = 1.$$

Calculate the flux of $\nabla \times \mathbf{v}$ in the direction of the upper unit normal $\mathbf{n}$.

12. Let S be a smooth closed surface and let $\mathbf{v} = \mathbf{v}(x, y, z)$ be a vector field with second partials continuous on an open convex set that contains S. Show that

$$\iint_S [(\nabla \times \mathbf{v}) \cdot \mathbf{n}]\, d\sigma = 0$$

where $\mathbf{n} = \mathbf{n}(x, y, z)$ is the outer unit normal.

13. The upper half of the ellipsoid $\frac{1}{2}x^2 + \frac{1}{2}y^2 + z^2 = 1$ intersects the cylinder $x^2 + y^2 - y = 0$ in a curve C. Calculate the circulation of $\mathbf{v} = y^3\,\mathbf{i} + (xy + 3xy^2)\,\mathbf{j} + z^4\,\mathbf{k}$ around C by using Stokes's theorem.

14. The sphere $x^2 + y^2 + z^2 = a^2$ intersects the plane $x + 2y + z = 0$ in a curve C. Calculate the circulation of $\mathbf{v} = 2y\,\mathbf{i} - z\,\mathbf{j} + 2x\,\mathbf{k}$ about C by using Stokes's theorem.

15. The paraboloid $z = x^2 + y^2$ intersects the plane $z = y$ in a curve C. Calculate the circulation of $\mathbf{v} = 2z\,\mathbf{i} + x\,\mathbf{j} + y\,\mathbf{k}$ about C using Stokes's theorem.

16. The cylinder $x^2 + y^2 = b^2$ intersects the plane $y + z = a^2$ in a curve C. Assume $a^2 \geq b > 0$. Calculate the circulation of $\mathbf{v} = xy\,\mathbf{i} + yz\,\mathbf{j} + xz\,\mathbf{k}$ about C using Stokes's theorem.

17. Let S be a smooth oriented surface with a smooth bounding curve C and let $\mathbf{a}$ be a fixed vector. Show that

$$\iint_S (2\mathbf{a} \cdot \mathbf{n})\, d\sigma = \oint_C (\mathbf{a} \times \mathbf{r}) \cdot d\mathbf{r}$$

where $\mathbf{n} = \mathbf{n}(x, y, z)$ is a unit normal vector that varies continuously over S and the line integral is taken in the positive sense with respect to $\mathbf{n}$.

18. Let S be a smooth oriented surface with smooth bounding curve C. Show that, if ϕ and ψ are sufficiently differentiable scalar fields, then

$$\iint_S [(\nabla\phi \times \nabla\psi) \cdot \mathbf{n}]\, d\sigma = \oint_C (\phi\nabla\psi) \cdot d\mathbf{r}$$

where $\mathbf{n} = \mathbf{n}(x, y, z)$ is a unit normal that varies continuously on S and the line integral is taken in the positive sense with respect to $\mathbf{n}$.

19. Let S be a smooth oriented surface with a smooth plane bounding curve C and let $\mathbf{v} = \mathbf{v}(x, y, z)$ be a vector field with second partials continuous on an open convex set that contains S. If S does not cross the plane of C, then Stokes's theorem for S follows readily from the divergence theorem and Green's theorem. Carry out the argument.

20. Our derivation of Stokes's theorem was admittedly nonrigorous. The following version (17.10.3) of Stokes's theorem lends itself more readily to rigorous proof.

Give a detailed proof of the theorem. HINT: Set $\mathbf{v} = v_1\mathbf{i} + v_2\mathbf{j} + v_3\mathbf{k}$. Then

$$\iint_S [(\nabla \times \mathbf{v}) \cdot \mathbf{n}]\, d\sigma = \iint_S [(\nabla \times v_1\mathbf{i}) \cdot \mathbf{n}]\, d\sigma$$
$$+ \iint_S [(\nabla \times v_2\mathbf{j}) \cdot \mathbf{n}]\, d\sigma + \iint_S [(\nabla \times v_3\mathbf{k}) \cdot \mathbf{n}]\, d\sigma$$

and $$\oint_C \mathbf{v}(\mathbf{r}) \cdot d\mathbf{r} = \int_C v_1\, dx + \int_C v_2\, dy + \int_C v_3\, dz.$$

Show that $$\iint_S [(\nabla \times v_1\mathbf{i}) \cdot \mathbf{n}]\, d\sigma = \int_C v_1\, dx$$

by showing that both integrals can be written

$$\iint_\Gamma \left[\frac{\partial v_1}{\partial u}\frac{\partial x}{\partial v} - \frac{\partial v_1}{\partial v}\frac{\partial x}{\partial u}\right] du\, dv.$$

A similar argument (no need to carry it out) equates the integrals for v_2 and v_3 and proves the theorem.

THEOREM 17.10.3

Let Γ be a Jordan region in the uv-plane with a piecewise-smooth boundary C_Γ given in a counterclockwise sense by a pair of functions $u = u(t)$, $v = v(t)$ with $t \in [a, b]$. Let $\mathbf{R}\,(u, v) = x(u, v)\,\mathbf{i} + y(u, v)\,\mathbf{j} + z(u, v)\,\mathbf{k}$ be a vector function with continuous second partials on Γ. Assume that $\mathbf{R}$ is one-to-one on Γ and that the fundamental vector product $\mathbf{N} = \mathbf{R}'_u \times \mathbf{R}'_v$ is never zero. The surface $S\colon \mathbf{R} = \mathbf{R}\,(u, v), (u, v) \in \Gamma$ is a smooth oriented surface bounded by the oriented space curve $C : \mathbf{r}\,(t) = \mathbf{R}\,(u(t), v(t))$, $t \in [a, b]$. If $\mathbf{v} = \mathbf{v}\,(x, y, z)$ is a vector field continuously differentiable on S, then

$$\iint_S [(\nabla \times \mathbf{v}) \cdot \mathbf{n}]\, d\sigma = \oint_C \mathbf{v}(\mathbf{r}) \cdot d\mathbf{r}$$

where $\mathbf{n}$ is the unit normal in the direction of the fundamental vector product.

■ CHAPTER HIGHLIGHTS

17.1 Line Integrals

smooth curve (p. 1020) work along a curve (p. 1021)
line integral (p. 1021)
invariance under changes in parameter (p. 1022)
piecewise-smooth curve (p. 1023)
work as a line integral (p. 1025)
circulation of a vector field (p. 1027)

17.2 The Fundamental Theorem for Line Integrals

gradient field (p. 1028)
the fundamental theorem (p. 1028)

17.3 Work-Energy Formula; Conservation of Mechanical Energy

work-energy formula (p. 1032)
conservative force fields (p. 1033)
potential energy functions (p. 1033)
conservation of mechanical energy (p. 1033)
differences in potential energy (p. 1034)
escape velocity (p. 1035)
equipotential surfaces (p. 1035)

17.4 Another Notation for Line Integrals; Line Integrals with Respect to Arc Length

$$\int_C P\,dx + Q\,dy + R\,dz \qquad \int_C f(\mathbf{r})\,ds$$

mass, center of mass, moment of inertia of a wire (p. 1038)

17.5 Green's Theorem

Jordan region (p. 1041)
Green's theorem for a Jordan region (p. 1041)
area enclosed by a plane curve (p. 1045)
Green's theorem for regions bounded by several Jordan curves (p. 1047)
a justification of the Jacobian area formula (supplement) (p. 1050)

17.6 Parametrized Surfaces; Surface Area

parametrized surface: $\mathbf{r} = \mathbf{r}(u, v),\ (u, v) \in \Omega$.
fundamental vector product: $\mathbf{N} = \mathbf{r}'_u \times \mathbf{r}'_v$
$\mathbf{N}$ as a normal (p. 1055)
area of a parametrized surface (p. 1056)
area of a surface $z = f(x, y,)$ (p. 1060)
upper unit normal (p. 1061)
secant area formula (p. 1062)

17.7 Surface Integrals

mass of a material surface (p. 1064)
surface integral (p. 1064)
average value on a surface, weighted average (p. 1066)
centroid (p. 1066)
center of mass of a material surface (p. 1066)
moment of inertia (p. 1067)
oriented surface (p. 1069)
flux of a vector field (p. 1069)
closed, piecewise-smooth surface (p. 1071)
one-sided surface, Möbius band (p. 1073)

17.8 The Vector Differential Operator ∇

the operator del: $\nabla = \dfrac{\partial}{\partial x}\mathbf{i} + \dfrac{\partial}{\partial y}\mathbf{j} + \dfrac{\partial}{\partial z}\mathbf{k}$

gradient of f : ∇f
divergence of $\mathbf{v}$: $\nabla \cdot \mathbf{v}$ (p. 1074)
curl of $\mathbf{v}$: $\nabla \times \mathbf{v}$ (p. 1074)
curl of a gradient is zero (p. 1076)
divergence of a curl is zero (p. 1077)

$$\nabla \cdot (f\,\mathbf{v}) = (\nabla f) \cdot \mathbf{v} + f\,(\nabla \cdot \mathbf{v})$$
$$\nabla \times (f\mathbf{v}) = (\nabla f) \times \mathbf{v} + f(\nabla \times \mathbf{v})$$
$$\nabla \cdot (r^n\,\mathbf{r}) = (n+3)\,r^n$$
$$\nabla \times (r^n\mathbf{r}) = 0$$

Laplacian: $\nabla^2 f = \nabla \cdot \nabla f = \dfrac{\partial^2 f}{\partial x^2} + \dfrac{\partial^2 f}{\partial y^2} + \dfrac{\partial^2 f}{\partial z^2}$

17.9 The Divergence Theorem

divergence theorem for a solid bounded by a single closed surface (p. 1080)
divergence as outward flux per unit volume (p. 1082)
sinks and sources (p. 1083)
divergence theorem for solids bounded by two or more closed surfaces (p. 1083)
buoyant force, principle of Archimedes (p. 1085)
an application to static charges (p. 1085)

17.10 Stokes's Theorem

positive sense along a curve bounding an open surface (p. 1086)
Stokes's theorem (p. 1087)
conditions under which $\nabla \times \mathbf{v} = \mathbf{0}$ implies $\mathbf{v} = \nabla f$
normal component of $\nabla \times \mathbf{v}$ as circulation per unit area
irrotational flow (p. 1090)

Evaluate the repeated integral.

1. $\int_0^1 \int_y^{\sqrt{y}} xy^2\, dx\, dy$

2. $\int_0^1 \int_{-y}^{y} e^{x+y}\, dx\, dy$

3. $\int_0^1 \int_x^{3x} 2ye^{x^3}\, dy\, dx$

4. $\int_{-1}^{2} \int_0^4 \int_0^1 xyz\, dx\, dy\, dz$

5. $\int_0^2 \int_0^{2-3x} \int_0^{x+y} x\, dz\, dy\, dx$

6. $\int_0^{\pi/2} \int_z^{\pi/2} \int_0^{\sin z} 3x^2 \sin y\, dx\, dy\, dz$

7. $\int_{-\pi/2}^{0} \int_0^{2\sin\theta} \int_0^{r^2} r^2 \cos\theta\, dz\, dr\, d\theta$

8. $\int_{-\pi/6}^{\pi/2} \int_0^{\pi/2} \int_0^1 \rho^3 \sin\varphi\, \cos\varphi\, d\rho\, d\theta\, d\varphi$

Sketch the region Ω corresponding to the repeated integral. Then change the order of integration and evaluate.

9. $\int_0^1 \int_0^{\sqrt{1-x^2}} \frac{1}{\sqrt{1-y^2}}\, dy\, dx$

10. $\int_0^2 \int_{\frac{1}{2}x}^{1} \cos\,(y^2)\, dy\, dx$

Evaluate the integral.

11. $\iint_\Omega xy\, dxdy;\ \Omega : 0 \le x^2 + y^2 \le 1, x, y > 0$

12. $\iint_\Omega (x-y) dxdy$; Ω is the region between the curves $y^2 = 3x$ and $y^2 = 4 - x$

13. $\iint_\Omega (x^2 - xy) dxdy$; Ω is the region between the curves $y = x$ and $y = 3x - x^2$

14. $\iiint_T z\, dxdydz$; T is the region bounded by the planes $x = 0,\ y = 0,\ z = 0, y + z = 1,\ x + z = 1$

15. $\iiint_T xy\, dxdydz$; T is the region in the first octant bounded by the coordinate planes and the hemisphere $z = \sqrt{4 - x^2 - y^2}$

16. $\iiint_T (x^2 + 2z)\, dxdydz$; T is the region bounded by the planes $z = 0$ and $y + z = 4$, and the cylinder $y = x^2$

17. Find the volume of the solid in the first octant bounded by $z = x^2 + y^2$, $x + y = 1$, and the coordinate planes.

18. Find the volume of the solid in the first octant bounded by the cylinder $x^2 + y^2 = 9$ and the planes $z = y$ and $z = 0$.

Use polar coordinates to evaluate the repeated integral.

19. $\int_0^2 \int_0^{\sqrt{4-y^2}} e^{\sqrt{x^2+y^2}}\, dx\, dy$

20. $\int_{-1}^{1} \int_0^{\sqrt{1-x^2}} \tan^{-1}(y/x)\, dy\, dx$

Find the mass and center of mass of the plate that occupies the region Ω and has the density function λ.

21. Ω is the region between the curves $y = x$ and $y = \sqrt{x}$; $\lambda(x, y) = 2x$.

22. Ω is the region bounded below by the x-axis and above by the cardioid $r = 2(1 + \cos\theta)$; λ is the distance to the pole.

23. An isosceles triangle of base b and height h forms the boundary of a homogeneous plate.
   (a) Find the centroid of the plate.
   (b) Find the moment of inertia about the base.
   (c) Find the moment of inertia about the axis of symmetry of the triangle.

Use a triple integral to find the volume of the solid. Use either rectangular, cylindrical, or spherical coordinates, whichever seem appropriate.

24. The solid bounded above by the paraboloid $z = 4(x^2 + y^2)$, below by the plane $z = -1$, and on the sides by the cylinders $y = x^2$ and $y = x$.

25. The solid bounded above by the elliptic paraboloid $z = 12 - x^2 - 2y^2$ and below by the elliptic paraboloid $z = 2x^2 + y^2$. HINT: Find the curve of intersection of the two surfaces.

26. The solid in the first octant bounded by the plane $2x + y + z = 2$ and inside the cylinder $y^2 + z^2 = 1$.

27. The solid bounded above by the sphere $x^2 + y^2 + z^2 = 4$ and below by the plane $z = 1$.

28. The solid bounded above by the sphere $x^2 + y^2 + z^2 = 4$, on the sides by the cylinder $x^2 + y^2 = 1$, and below by the x, y-plane.

29. A homogeneous solid in the first octant bounded is by the cylinders $x^2 + z^2 = 1$ and $y^2 + z^2 = 1$.
   (a) Find the centroid of the solid. (b) Find I_z.

30. Find the Jacobian of the transformation.
   (a) $x = u^2 - v^2,\ y = 2uv$
   (b) $x = u^2 + 2vw,\ y = v^2 + 2uw,\ z = uvw$

Evaluate the integral using the suggested transformation.

31. $\iint_\Omega \sin\left(\frac{y-x}{y+x}\right) dxdy$; Ω is the region in the first quadrant bounded by the two lines $x + y = 1$ and $x + y = 2$. Let $x = \frac{1}{2}(v - u), y = \frac{1}{2}(v + u)$.

32. $\iiint_T dxdydz$; T is the solid that lies between the paraboloids $z = x^2 + y^2$ and $z = 4x^2 + 4y^2$, and between the planes $z = 1$ and $z = 4$. Let $x = (r/u)\cos\theta$, $y = (r/u)\sin\theta, z = r^2$.

33. Integrate $\mathbf{h}(x, y) = x^2y\,\mathbf{i} - xy\,\mathbf{j}$ over the path:
   (a) The straight-line segment from $(0, 0)$ to $(1, 1)$.
   (b) $\mathbf{r}(u) = u^2\,\mathbf{i} + u^3\,\mathbf{j}, 0 \le u \le 1$.

34. Integrate $\mathbf{h}(x, y) = (2xy^2 + x)\,\mathbf{i} + (2x^2y - 1)\,\mathbf{j}$ over the path:
(a) The straight-line segment from $(-1, 2)$ to $(2, 4)$.
(b) The polygonal path from $(-1, 2)$ to $(0, 0)$ to $(2, 4)$.
(c) The straight-line segment from $(-1, 2)$ to $(0, 0)$, then the parabolic path $y = x^2$ from $(0, 0)$ to $(2, 4)$.

35. Integrate $\mathbf{h}(x, y, z) = \sin y\,\mathbf{i} + xe^{xy}\,\mathbf{j} + \sin z\,\mathbf{k}$ over the curve $\mathbf{r}(u) = u^2\,\mathbf{i} + u\,\mathbf{j} + u^3\,\mathbf{k}, 0 \le u \le 3$.

36. Integrate $\mathbf{h}(x, y, z) = x^2\,\mathbf{i} + xy\,\mathbf{j} + z^2\,\mathbf{k}$ over the curve $\mathbf{r}(u) = \cos u\,\mathbf{i} + \sin u\,\mathbf{j} + u^2\,\mathbf{k}, 0 \le u \le \pi/2$.

37. The force exerted by a charged particle at the origin on a charged particle at a point (x, y, z) is of the form

$$\mathbf{F}(x, y, z) = \frac{C(x\,\mathbf{i} + y\,\mathbf{j} + z\,\mathbf{k})}{\sqrt{x^2 + y^2 + z^2}}, \ C \text{ constant.}$$

Find the work done by $\mathbf{F}$ applied to a particle that moves in a straight line from $(1, 0, 0)$ to $(3, 0, 4)$.

38. An object is moving through a force field $\mathbf{F}$ in such a way that its velocity vector at each point (x, y, z) is orthogonal to $\mathbf{F}(x, y, z)$. Show that the work done by $\mathbf{F}$ on the object is 0.

Verify that the vector field $\mathbf{h}$ is a gradient field. Calculate the line integral of $\mathbf{h}$ over C (a) directly, and (b) by applying the fundamental theorem for line integrals.

39. $\mathbf{h}(x, y) = (ye^{xy} + 2x)\,\mathbf{i} + (xe^{xy} - 2y)\,\mathbf{j}$; C: $\mathbf{r}(u) = u\,\mathbf{i} + u^2\,\mathbf{j}$; $0 \le u \le 2$.

40. $\mathbf{h}(x, y, z = 4x^3y^3z^2\,\mathbf{i} + 3x^4y^2z^2\,\mathbf{j} + 2x^4y^3z\,\mathbf{k}$. C: $\mathbf{r}(u) = u\,\mathbf{i} + u^2\,\mathbf{j} + u^3\,\mathbf{k}, 0 \le u \le 1$.

41. Evaluate $\int_C y^2\,dx + (x^2 - xy)\,dy$ given that c is the following path from $(0, 0)$ to $(2, 8)$:
(a) the straight-line segment.
(b) the polygonal path $(0, 0)$ to $(2, 0)$ to $(2, 8)$.
(c) the cubic $y = x^3$.

42. Evaluate $\int_C z\,dx + x\,dy + y\,dz$ given that C is the circular helix: $\mathbf{r}(u) = a\cos u\,\mathbf{i} + a\sin u\,\mathbf{j} + u\,\mathbf{k}, 0 \le u \le 2\pi$.

43. Evaluate $\int_C ye^{xy}\,dx + \cos x\,dy + (xy/z)\,dz$ given that C is the twisted cubic: $\mathbf{r}(u) = u\,\mathbf{i} + u^2\,\mathbf{j} + u^3\,\mathbf{k}, 0 \le u \le 2$.

44. A wire winds around the z-axis in the shape of the circular helix C : $\mathbf{r}(u) = \sin u\,\mathbf{i} - \cos u\,\mathbf{j} + 4u\,\mathbf{k}, \pi \le u \le 2\pi$. The mass density at the point $P(x, y, z)$ on the wire is equal to the square of the distance from P to the x-axis. Find the mass of the wire.

Verify Green's theorem (a) by calculating the line integral over the simple closed curve C directly, and (b) by calculating the corresponding double integral of the region enclosed by C.

45. $\oint_C xy^2 dx - x^2 y dy$; C: the closed curve in the first quadrant determined by the parabolas $y = x^2$ and $y^2 = x$.

46. $\oint_C (x^2 + y^2)dx + (x^2 - y^2)dy$; C : the unit circle $x^2 + y^2 = 1$.

Evaluate the line integral using Green's theorem.

47. $\oint_C (x - 2y^2)\,dx + 2xy\,dy$; C : the rectangle with vertices $(0, 0), (2, 0), (2, 1), (0, 1)$.

48. $\oint_C xy\,dx + (\frac{1}{2}x^2 + xy)dy$; C : the upper semi-ellipse $x^2 + 4y^2 = 1$ together with the interval $[-1, 1]$.

49. $\oint_C \ln(x^2 + y^2)dx + \ln(x^2 + y^2)dy$; C : the boundary of the semi-circular ring determined by $x^2 + y^2 = 1$ and $x^2 + y^2 = 4, y \ge 0$.

50. $\oint_C y^2 dx$; C: the cardioid $r = 1 + \sin\theta$.

Find the area of the surface.

51. The part of the sphere $x^2 + y^2 + z^2 = 4$ that is inside the cylinder $x^2 + y^2 = 2x$.

52. The part of the plane $x + y + 2z = 4$ that lies inside the cylinder $x^2 + y^2 = 4$.

53. The part of the cone $z = \sqrt{x^2 + y^2}$ that lies between the planes $z = 0$ and $z = 3$.

Evaluate the surface integral.

54. $\iint_S xz\,d\sigma$; S is the first octant part of the plane $x + y + z = 1$

55. $\iint_S (x^2 + y^2 + z^2)d\sigma$; S is the cylinder $y^2 + z^2 = 4$, $0 \le x \le 2$, together with the circular discs at each end.

56. Verify the divergence theorem on the surface $S : y^2 + z^2 = 1, 0 \le x \le 4$, for the vector field $\mathbf{v}(x, y, z) = (x + z)\,\mathbf{i} + (y + z)\,\mathbf{j} + (x + z)\,\mathbf{k}$.

57. Calculate the total flux of $\mathbf{v}(x, y, z) = 2x\,\mathbf{i} + xz\,\mathbf{j} + z^2\,\mathbf{k}$ out of the solid bounded by the paraboloid $z = 9 - x^2 - y^2$ and the xy-plane.

58. Calculate the total flux of $\mathbf{v}(x, y, z) = x^2\,\mathbf{i} - xz\,\mathbf{j} + z^2\,\mathbf{k}$ out of the cube $0 \le x \le a, 0 \le y \le a, 0 \le z \le a$.

59. Let S be the hemisphere $z = \sqrt{4 - x^2 - y^2}$ and take $\mathbf{n}$ as the upper unit normal. Let $\mathbf{v}(x, y, z) = z\,\mathbf{i} + x\,\mathbf{j} + y\,\mathbf{k}$ and find

$$\iint_S [(\nabla \times \mathbf{v}) \cdot \mathbf{n}]\,d\sigma$$

(a) by direct calculation; and (b) by Stokes's theorem.

60. Let S be that part of the paraboloid $z = 9 - x^2 - y^2$ for which $z \ge 0$ and take $\mathbf{n}$ as the upper unit normal. Let $\mathbf{v}(x, y, z) = z^3\,\mathbf{i} + x\,\mathbf{j} + y^2\,\mathbf{k}$ and find

$$\iint_S [(\nabla \times \mathbf{v}) \cdot \mathbf{n}]\,d\sigma$$

(a) by direct calculation; and (b) by Stokes's theorem.

CHAPTER

18 ELEMENTARY DIFFERENTIAL EQUATIONS

■ 18.1 INTRODUCTION; REVIEW OF EQUATIONS ALREADY CONSIDERED

Suppose that y is an unknown function of x. If we know y', then we can recover y up to an additive constant, by integrating y' :

$$y(x) = \int y'(x)\,dx + C = F(x) + C. \qquad \text{(where } F \text{ is an antiderivative for } y'\text{)}$$

If we don't know y' but we know y'', then we can still recover y, but now we have to integrate twice and there will be two arbitrary constants:

$$\begin{aligned} y'(x) &= \int y''(x)\,dx + C_1 = G(x) + C_1 && \text{(where } G \text{ is an antiderivative for } y''\text{)} \\ y(x) &= \int [G(x) + C_1]\,dx + C_2 \\ &= \int G(x)\,dx + \int C_1\,dx + C_2 \\ &= F(x) + C_1 x + C_2. && \text{(where } F \text{ is an antiderivative for } G\text{)} \end{aligned}$$

It often happens in mathematics and in applications to other fields that we don't know y', we don't know y'', we don't know any of the derivatives explicitly, but we do have an equation that relates y to one or more of its derivatives. An equation that relates an unknown function to one or more of its derivatives is called an *ordinary differential equation*.† The recovery of a function y from a differential equation is called *solving* the differential equation. From some differential equations we can recover y completely and describe its action explicitly as a function of x (up to one or more arbitrary constants).

† In contrast to partial differential equations, which arise in the study of functions of several variables.

More frequently, we cannot recover y completely, but we can obtain an equation in x and y which is satisfied by y and involves none of the derivatives of y. Such an equation, carrying one or more arbitrary constants, represents a family of curves called *integral curves* (*solution curves*) of the differential equation.

Finally, there are differential equations from which we can extract no explicit solutions and no integral curves. Such equations have to be approached by other methods.

The *order* of a differential equation is the order of the highest derivative that appears in the equation. The equations used by scientists to model the processes of nature are almost all equations of order one or order two. We can be thankful for that. It would be rather burdensome to have to solve differential equations of order, say, 100.

This is not your first encounter with differential equations. They have appeared off and on throughout the text. We review here the equations with which you are expected to be familiar before you begin the study of this chapter.

First-Order Linear Differential Equations

An equation of the form

$$(1) \qquad y' + p(x)\,y = q(x)$$

where p and q are known functions of x defined and continuous on some interval I is called a *first-order linear differential equation with continuous coefficients.* Such an equation is intimately related to the exponential function.

To solve the equation, we multiply it by $e^{H(x)}$ where $H(x)$ is an antiderivative for p. That makes the left side a derivative with respect to x and, as shown in Section *8.8, eventually leads to the conclusion that

$$(2) \qquad y(x) = e^{-H(x)}\left\{\int e^{H(x)}q(x)\,dx + C\right\}$$

where C is an arbitrary constant.

There is no point memorizing this expression. We are led to it naturally once we have multiplied by $e^{H(x)}$. What's important here is that (2) gives us explicitly *every* function that satisfies the differential equation. Because of this, we call (2) the *general solution* of the differential equation (1). By assigning particular values to the constant C, we obtain what are called *particular solutions* of the differential equation.

Separable First-Order Equations

In Section *8.9 we discussed *separable equations.* These are equations that can be written in the form

$$(3) \qquad p(x) + q(y)\,y' = 0$$

with p and q continuous where defined.

Since y is assumed to be a function of x, we can write

$$p(x) + q(y(x))\,y'(x) = 0.$$

Integration with respect to x gives

$$\int p(x)\,dx + \int q(y(x))\,y'(x)\,dx = C,$$

which leads to

$$\int p(x)\,dx + \int q(y)\,dy = C.$$

We have separated the variables. If P is an antiderivative for p and Q is an antiderivative for q, we have

$$P(x) + Q(y) = C,$$

an equation that we can write as

$$F(x, y) = C. \tag{4}$$

If y is related to y' by equation (3), then y is related to x by equation (4). The curves in the xy-plane generated by (4) are *integral curves* (*solution curves*) of the differential equation. Different values of C give different integral curves.

Each integral curve is a solution of the differential equation in the sense that along the curve the numbers x, y, y' are related as prescribed by the differential equation.

In some cases equation (4) can be solved for y in terms of x, giving us functions that satisfy the differential equation. Usually equation (4) cannot be solved for y in terms of x, and we have to be satisfied with equation (4) and the integral curves that it generates. We can, however, assert the following: any function $y = y(x)$ whose graph lies on one of the integral curves satisfies the differential equation. In Section *8.9 we showed that the integral curves of the differential equation

$$x + yy' = 0$$

are of the form

$$x^2 + y^2 = C, \qquad C \geq 0.$$

The graphs of the functions

$$y = \sqrt{1 - x^2} \qquad \text{and} \qquad y = -\sqrt{1 - x^2}$$

both lie on the integral curve

$$x^2 + y^2 = 1.$$

As you can readily verify, both functions satisfy the differential equation.

The Equation of Harmonic Motion

In the Exercises for Section 3.6 and Section 4.4 we introduced the *second-order linear differential equation*

$$\frac{d^2y}{dt^2} + \omega^2 y = 0. \qquad (\omega \text{ a constant}) \tag{5}$$

This is the equation of *simple harmonic motion*, motion during which the acceleration remains a constant negative multiple of the displacement from a point of equilibrium. This equation models the motion of a simple pendulum and the up-and-down oscillations of a mass suspended from a spring.

We cited three (completely equivalent) ways to express all the solutions of (5). Here we remind you of one of them:

$$y(t) = A\cos\omega t + B\sin\omega t$$

where A and B are arbitrary constants. Particular solutions of the differential equation are obtained by assigning numerical values to these constants.

EXERCISES 18.1

Determine whether the differential equation has the indicated functions as solutions.

1. $2y' - y = 0$; $y_1(x) = e^{x/2}$, $y_2(x) = x^2 + 2e^{x/2}$.

2. $y' + xy = x$; $y_1(x) = e^{-x^2/2}$, $y_2(x) = 1 + Ce^{-x^2/2}$, C any constant

3. $y' + y = y^2$; $y_1(x) = \dfrac{1}{e^x + 1}$, $y_2(x) = \dfrac{1}{Ce^x + 1}$, C any constant.

4. $y'' + 4y = 0$; $y_1(x) = 2\sin 2x$, $y_2(x) = 2\cos x$.

5. $y'' - 4y = 0$; $y_1(x) = e^{2x}$, $y_2(x) = C\sinh 2x$, C any constant.

6. $y'' - 2y' - 3y = 7e^{3x}$; $y_1(x) = e^{-x} + 2e^{3x}$, $y_2(x) = \frac{7}{4}xe^{3x}$.

Show that the members of the given family of functions are solutions of the differential equation. Then find a member of the family that satisfies the initial condition(s).

7. $y = Ce^{5x}$; $y' = 5y$, $y(0) = 2$.

8. $y = \dfrac{x^2}{3} + \dfrac{C}{x}$; $xy' + y = x^2$, $y(3) = 2$.

9. $y = \dfrac{1}{Ce^x + 1}$; $y' + y = y^2$, $y(1) = -1$.

10. $y = x\ln\dfrac{C}{x}$; $y' = \dfrac{y - x}{x}$, $y(2) = 4$.

11. $y = C_1x + C_2x^{1/2}$; $2x^2y'' - xy' + y = 0$, $y(4) = 1$, $y'(4) = -2$.

12. $y = C_1\sin 3x + C_2\cos 3x$; $y'' + 9y = 0$, $y(\pi/2) = y'(\pi/2) = 1$.

13. $y = C_1x^2 + C_2x^2\ln x$; $x^2y'' - 3xy' + 4y = 0$, $y(1) = 0$, $y'(1) = 1$.

14. $y = C_1 + C_2e^x + C_3e^{2x} + \frac{1}{4}x^2 + \frac{3}{4}x - xe^x$; $y''' - 3y'' + 2y' = x + e^x$, $y(0) = 1$, $y'(0) = -\frac{1}{4}$, $y''(0) = -\frac{3}{2}$.

Identify the differential equation as linear or separable and then solve the equation.

15. $y' + y = 2e^{-2x}$.

16. $y^2 + 1 = yy'\sec^2 x$.

17. $y' = \dfrac{x^2y - y}{y + 1}$.

18. $\dfrac{dy}{dx} = \dfrac{x^3 - 2y}{x}$.

19. $xy' + 2y = \dfrac{\cos x}{x}$.

20. $y' = \dfrac{\ln x}{xy + xy^3}$.

21. $yy' = 4x\sqrt{y^2 + 1}$.

22. $xy' = x^2 + 2y$.

Determine the values of r, if any, such that $y = e^{rx}$ is a solution of the given differential equation.

23. $y' + 3y = 0$.

24. $y'' - 5y' + 6y = 0$.

25. $y'' + 6y' + 9y = 0$.

26. $y''' - 3y' + 2y = 0$.

Determine values of r, if any, such that $y = x^r$ is a solution of the given differential equation.

27. $xy'' + y' = 0$.

28. $x^2y'' + xy' - y = 0$.

29. $4x^2y'' - 4xy' + 3y = 0$.

30. $x^3y''' - 2x^2y'' - 2xy' + 8y = 0$.

An nth-order differential equation together with conditions that are specified at two, or more, points on an interval I is called a *boundary-value problem*. In particular, if conditions are specified at two points, the problem is called a *two-point boundary-value problem.*

31. Each member of the family of functions $y = C_1\sin 4x + C_2\cos 4x$ is a solution of the differential equation $y'' + 16y = 0$.

(a) Find all the members of this family that satisfy the boundary conditions:

$$y(0) = 0, \quad y(\pi/2) = 0.$$

(b) Find all the members of this family that satisfy the boundary conditions:

$$y(0) = 0, \quad y(\pi/8) = 0.$$

32. For each real number r, each member of the family $y = C_1\sin rx + C_2\cos rx$ is a solution of the differential equation $y'' + r^2y = 0$.

(a) Determine the numbers r such that the two-point boundary-value problem

$$y'' + r^2y = 0, \quad y(0) = 0, \quad y(\pi) = 0$$

has a nonzero solution.

(b) Determine the numbers r such that the two-point boundary-value problem

$$y'' + r^2y = 0, \quad y(0) = 0, \quad y(\pi/2) = 0$$

has a nonzero solution.

■ 18.2 BERNOULLI EQUATIONS; HOMOGENEOUS EQUATIONS; NUMERICAL METHODS

Bernoulli Equations

A first-order equation of the form

$$y' + p(x)y = q(x)y^r \tag{18.2.1}$$

where p and q are functions defined and continuous on some interval I, and r is a real number different from 0 and 1, is called a *Bernoulli equation.*† We exclude 0 and 1 because in each of these cases the equation is simply a linear equation.

METHOD OF SOLUTION To solve (18.2.1), we multiply the equation by y^{-r} and obtain

$$y^{-r}y' + p(x)\,y^{1-r} = q(x).$$

We can transform this equation into a linear equation by setting $v = y^{1-r}$. For then

$$v' = (1-r)\,y^{-r}y'$$

and our differential equation becomes

$$\frac{1}{1-r}v' + p(x)\,v = q(x),$$

which we can write as

$$v' + (1-r)p(x)\,v = (1-r)\,q(x).$$

This equation is linear in v, and we can solve for v by the method introduced in Section *8. 8 and reviewed in the introduction to this chapter. Replacing v by y^{1-r}, we have the integral curves in terms of x and y. □

Example 1 Find the integral curves of the equation $y' + 4y = 3e^{2x}y^2$.

SOLUTION The equation is a Bernoulli equation with $p(x) = 4,\ q(x) = 3e^{2x},\ r = 2$. We will solve the equation by the method just outlined. Our first step is to multiply the equation by y^{-2} (thereby excluding, at least for the moment, $y = 0$):

$$y^{-2}y' + 4y^{-1} = 3e^{2x}.$$

We set $v = y^{-1}$. Differentiation gives $v' = -y^{-2}y'$ and transforms the equation into

$$-v' + 4v = 3e^{2x},$$

which we write as

$$v' - 4v = -3e^{2x}.$$

We solve this last equation by setting $H(x) = \int(-4)\,dx = -4x$ and multiplying through by $e^{H(x)} = e^{-4x}$. This gives us

$$e^{-4x}v' - 4e^{-4x}v = -3e^{-2x},$$

which we recognize as stating that

$$\frac{d}{dx}(e^{-4x}v) = -3e^{-2x}.$$

Integration gives

$$e^{-4x}\,v = \tfrac{3}{2}e^{-2x} + C,$$

which we write as

$$v = \tfrac{3}{2}e^{2x} + Ce^{4x}.$$

† These equations were introduced by Jacob Bernoulli (1654—1705), who, along with his brother Johann, made many contributions to the development of calculus and its applications.

Replacing v by y^{-1}, we have

$$\frac{1}{y} = \frac{3}{2}e^{2x} + Ce^{4x}.$$

These are the integral curves of our Bernoulli equation. As you can verify, this family of solutions can also be written

$$y = \frac{2}{3e^{2x} + Ke^{4x}} \qquad (K = 2C). \quad \square$$

Remark In applying our method of solution, we had to exclude $y = 0$. Note that the constant function $y = 0$ is a solution of the differential equation, but this function is not a member of the family of integral curves (there is no value that can be assigned to C to produce $y = 0$). A solution of a differential equation that is not included in the family of integral curves is called a *singular solution*. □

Homogeneous Equations

A first-order differential equation

(18.2.2) $$y' = f(x, y) \qquad (f \text{ continuous})$$

is said to be *homogeneous* if

$$f(tx, ty) = f(x, y) \qquad \text{for all} \quad t \neq 0.$$

Remark The term "homogeneous" requires some explanation. As you may have noted in your study of algebra, a function $H = H(x, y)$ is said to be *homogeneous to degree n* if

$$H(tx, ty) = t^n H(x, y) \qquad \text{for all } t \neq 0.$$

Thus the $f = f(x, y)$ of the homogeneous differential equation (18.2.2) can be viewed as a function *homogeneous to degree* 0:

$$f(tx, ty) = t^0 f(x, y) = f(x, y) \qquad \text{for all } t \neq 0.$$

To construct such a function, simply divide two functions homogeneous to the same degree:

$$\frac{M(tx, ty)}{N(tx, ty)} = \frac{t^n M(x, y)}{t^n N(x, y)} = \frac{M(x, y)}{N(x, y)} \qquad \text{for all } t \neq 0. \quad \square$$

Now back to our differential equation (18.2.2).

METHOD OF SOLUTION The first step is to write the equation in the form

(1) $$y' = g\left(\frac{y}{x}\right)$$

where g is a continuous function of only one variable. That this can be done is shown as follows: observe that for $x \neq 0$

$$f(x, y) = f\left(x, x\left(\frac{y}{x}\right)\right) = f\left(1, \frac{y}{x}\right)$$

and define $$g\left(\frac{y}{x}\right) = f\left(1, \frac{y}{x}\right).$$

We now set $y/x = v$ and transform (1) into an equation in v and x:

$$y = vx \qquad \text{gives} \qquad y' = v + v'x$$

and $\qquad g\left(\frac{y}{x}\right) \qquad$ becomes $\qquad g(v)$.

Thus equation (1) can be written

$$v + v'x = g(v),$$

which can be rearranged to give

$$\frac{1}{x} + \left[\frac{1}{v - g(v)}\right] v' = 0.$$

This equation is separable. We can solve it by the method introduced in Section *8.9 (reviewed in the introduction to this chapter) and write the resulting integral curves in terms of x and y by substituting y/x for v. □

Example 2 Show that the differential equation $y' = \dfrac{3x^2 + y^2}{xy}$ is homogeneous and find the integral curves.

SOLUTION The equation is homogeneous: for $t \neq 0$

$$f(tx, ty) = \frac{3(tx)^2 + (ty)^2}{(tx)(ty)} = \frac{t^2(3x^2 + y^2)}{t^2(xy)} = \frac{3x^2 + y^2}{xy} = f(x, y).$$

Setting $y/x = v$, we have

$$\frac{3x^2 + y^2}{xy} = \frac{3 + (y/x)^2}{y/x} = \frac{3 + v^2}{v},$$

and, since $y = vx$,

$$y' = v + v'x.$$

Our differential equation now reads

$$v + v'x = \frac{3 + v^2}{v}.$$

Multiplying both sides by v and simplifying, we get

$$v'xv = 3,$$

which can be rearranged to give

$$-\frac{1}{x} + \frac{1}{3} v v' = 0.$$

This is a separable equation which we can readily solve:

$$\int -\frac{1}{x}\, dx + \frac{1}{3}\int v v' dx = 0$$

$$-\int \frac{1}{x}\, dx + \frac{1}{3}\int v\, dv = 0$$

$$-\ln|x| + \tfrac{1}{6}v^2 = C.$$

This gives

$$v^2 = 6(\ln|x| + C).$$

Substituting y/x back in for v, we have

$$\frac{y^2}{x^2} = 6(\ln|x| + C).$$

The integral curves take the form

$$y^2 = 6x^2(\ln|x| + C). \quad \square$$

Numerical Methods

As indicated earlier, the differential equations that occur most frequently in applications are of order one or order two. We have presented techniques for solving some first-order equations; methods for solving certain second-order equations are presented in Sections 18.4 and 18.5. However, many of the differential equations that arise in the study of physical phenomena cannot be solved exactly. Therefore, we need other methods to describe solutions or to approximate them. So-called *qualitative methods* which are used to determine the behavior of solutions are studied in more advanced courses on differential equations. Here we present two numerical methods for approximating solutions. We will apply these methods to first-order initial value problems: problems in which we are given a first-order differential equation and some point on the solution curve from which we can start the process.

So that you can keep track of the accuracy (or inaccuracy) of the approximations, we will work with an initial-value problem for which an exact solution is available. As you can check, the initial-value problem

(18.2.3) $$y' = x + 2y, \qquad y(0) = 1$$

has the exact solution

(18.2.4) $$y = \tfrac{1}{4}(5\,e^{2x} - 2x - 1).$$

We will use our numerical methods to estimate $y(1)$; the actual value is

$$y(1) = \tfrac{1}{4}(5e^2 - 3) \cong 8.4863.$$

The first method we consider is called the *Euler method*. Figure 18.2.1 shows the graph of a differentiable function f. Given $f(x_0)$, we can estimate $f(x_0 + h)$ by proceeding along the tangent line, the line with slope $f'(x_0)$:

$$f(x_0 + h) \cong f(x_0) + hf'(x_0).$$

Using this estimate for $f(x_0 + h)$, we can go on to estimate $f(x_0 + 2h)$ by proceeding along the line with slope $f'(x_0 + h)$. We can go on to estimate $f(x_0 + 3h), f(x_0 + 4h)$, etc. We illustrate the process in Figure 18.2.2. Presumably, if h is taken small enough, the path of line segments stays fairly close to the graph of f. This way of approximating a curve by line segments is the basis of the Euler method for solving initial-value problems.

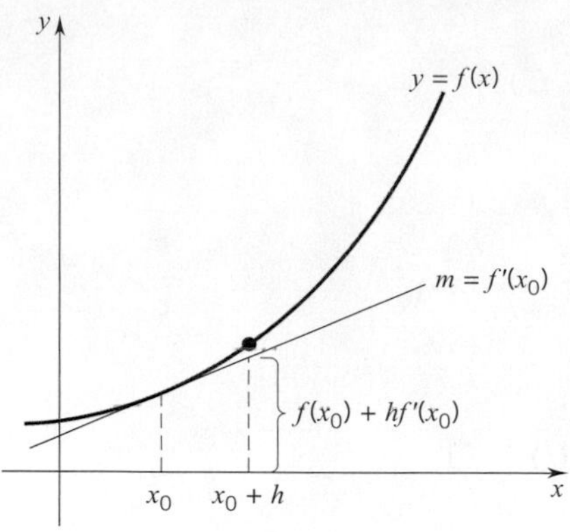

Figure 18.2.1

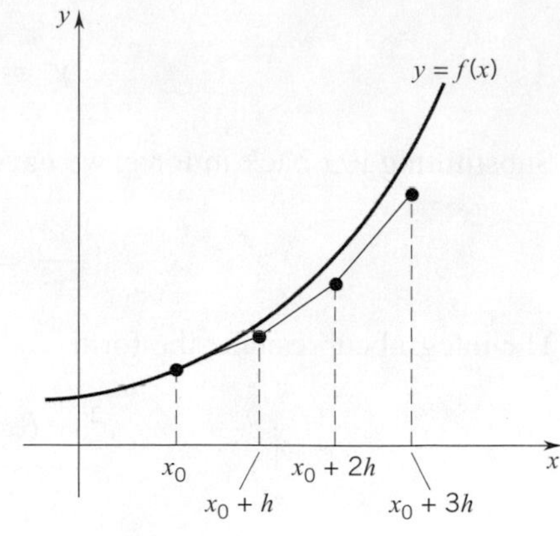

Figure 18.2.2

Consider the initial-value problem

(18.2.5) $$y' = f(x, y), \qquad y(x_0) = y_0.$$

The point (x_0, y_0) lies on the solution curve. The point on the curve with x-coordinate $x_1 = x_0 + h$ has approximate y-coordinate

$$y_1 = y_0 + hy'(x_0) = y_0 + hf(x_0, y_0).$$

The point on the curve with x-coordinate $x_2 = x_0 + 2h$ has approximate y-coordinate

$$y_2 = y_1 + hy'(x_1) = y_1 + hf(x_1, y_1).$$

Successive steps of h units (to the right if $h > 0$ and to the left if $h < 0$) produce a succession of points near the graph of the solution curve:

(18.2.6) $$(x_n, y_n): \quad x_n = x_0 + nh, \quad y_n = y_{n-1} + hf(x_{n-1}, y_{n-1}), \qquad (n \geq 1).$$

Example 3 Now let's return to the initial-value problem (18.2.3):

$$y' = x + 2y, \quad y(0) = 1.$$

In the notation of (18.2.5)

$$f(x, y) = x + 2y, \quad x_0 = 0, \quad y_0 = 1.$$

We want to estimate $y(1)$. Setting $h = 0.1$, we will need 10 iterations.

By (18.2.6)

$$x_n = (0.1)n \quad \text{and} \quad y_n = y_{n-1} + 0.1(x_{n-1} + 2y_{n-1}), \qquad 1 \leq n \leq 10.$$

The computations are given in the chart that follows. The numbers in the last column represent the actual values (rounded off to four decimal places) as calculated from the exact solution (18.2.4).

n	x_n	y_n	$y(x_n)$
0	0	1	1.0000
1	0.1	1.2000	1.2268
2	0.2	1.4500	1.5148
3	0.3	1.7600	1 .8776
4	0.4	2.1420	2.3319
5	0.5	2.6104	2.8979
6	0.6	3.1825	3.6001
7	0.7	3.8790	4.4690
8	0.8	4.7248	5.5413
9	0.9	5.7497	6.8621
10	1.0	6.9897	8.4863

The error in this estimate for $y(1)$ is about 17.6%. In this case the Euler method has not given us a very accurate estimate. Presumably we could improve on the accuracy of the estimate by using a smaller h, but as we explain below, that's not necessarily the case. □

Computational errors arise in several ways. With each iteration of the procedure we introduce the error inherent in the approximation method, and this error is compounded by the errors introduced in the previous steps. To reduce this contribution to the error, we can use smaller values of h. However, this improvement is made at the expense of increased round-off error. Each calculation produces an error in the last decimal place carried. Increasing the number of times the procedure is iterated by using smaller h increases the number of times a round-off error is made and thus tends to increase the total error.

The second numerical method that we consider is called the *Runge-Kutta method.* This is one of the oldest methods for generating numerical solutions of differential equations. Yet it remains one of the most accurate methods known. In the Euler method we estimated $y(x_n)$ from our estimate for $y(x_{n-1})$ by using the slope of y at x_{n-1}. The idea behind the Runge-Kutta method is to select a slope more representative of the derivative of y on the interval $[x_{n-1}, x_n]$. This is done by selecting intermediate points in the interval and then forming a weighted average of the slopes at these points. Details of the development of the formulas used below can be found in any text on numerical analysis or differential equations.

In the Runge-Kutta method successive points of approximation for the initial-value problem

(1) $$y' = f(x, y), \qquad y(x_0) = y_0$$

are chosen as follows:

(2) $$(x_n, y_n) \quad \text{with } x_n = x_0 + nh, \quad y_n = y_{n-1} + hK$$

where
$$\begin{aligned} K_1 &= f(x_{n-1}, y_{n-1}), \\ K_2 &= f(x_{n-1} + \tfrac{1}{2}h, y_{n-1} + \tfrac{1}{2}hK_1), \\ K_3 &= f(x_{n-1} + \tfrac{1}{2}h, y_{n-1} + \tfrac{1}{2}hK_2), \\ K_4 &= f(x_{n-1} + h, y_{n-1} + hK_3), \end{aligned}$$

and
$$K = \tfrac{1}{6}(K_1 + 2K_2 + 2K_3 + K_4).$$

Note that as a consequence of (1), the numbers K_1, K_2, K_3, K_4 give slopes, and K is a weighted average of these slopes. As you can see from (2), K gives the slope of the line segment that joins the approximation (x_{n-1}, y_{n-1}) to the approximation (x_n, y_n).

Example 4 We reconsider the initial-value problem

$$y' = x + 2y, \qquad y(0) = 1,$$

but this time we estimate $y(1)$ by the Runge-Kutta method. Again we have $f(x,y) = x + 2y$, and take $h = 0.1$.

The results of the computations (these can be verified rather quickly on a computer or programmable calculator) are tabulated below. Again, (18.2.4) was used to calculate the actual values of $y(x_n)$ (to four decimal places.)

n	x_n	y_n	$y(x_n)$
0	0	1	1.0000
1	0.1	1.2267	1.2268
2	0.2	1.5148	1.5148
3	0.3	1.8776	1.8776
4	0.4	2.3319	2.3319
5	0.5	2.8978	2.8979
6	0.6	3.6001	3.6001
7	0.7	4.4689	4.4690
8	0.8	5.5412	5.5413
9	0.9	6.8618	6.8621
10	1.0	8.4861	8.4863

The error in this approximation of $y(1)$ is about 0.002%. Runge-Kutta has given us a much better approximation than we obtained by the Euler method. □

EXERCISES 18.2

Find the integral curves.

1. $y' + xy = xy^3$.

2. $y' + y^2(x^2 + x + 1) = y$.

3. $y' = 4y + 2e^x\sqrt{y}$.

4. $2xy\,y' = 1 + y^2$.

5. $(x-2)\,y' + y = 5(x-2)^2 y^{1/2}$.

6. $y\,y' - xy^2 + x = 0$.

Find a solution to the initial value problem.

7. $y' + xy - y^3 e^{x^2} = 0; \quad y(0) = \frac{1}{2}$.

8. $xy' + y - y^2 \ln x = 0; \quad y(1) = 1$.

9. $2x^3 y' = y(y^2 + 3x^2); \quad y(1) = 1$.

10. $y' + y\tan x - y^2 \sec^3 x = 0; \quad y(0) = 3$.

11. Show that the change of variable $\mu = \ln y$ transforms the equation

$$y' - \left(\frac{y}{x}\right)\ln y = xy$$

into a linear equation. Find the integral curves.

12. (a) Show that the change of variable indicated in Exercise 11 transforms

$$y' + yf(x)\ln y = g(x)y$$

into a first-order linear equation.

(b) Find a change of variable which transforms

$$y'\cos y + g(x)\sin y = f(x)$$

into a linear equation:

Verify that the equation is homogeneous and find the integral curves.

13. $y' = \dfrac{x^2 + y^2}{2xy}$.

14. $y' = \dfrac{y^2}{xy + x^2}$.

15. $y' = \dfrac{x - y}{x + y}$.

16. $y' = \dfrac{x + y}{x - y}$.

17. $y' = \dfrac{x^2(e^y)^{1/x} + y^2}{xy}$.

18. $y' = \dfrac{x^2 + 3y^2}{4xy}$.

19. $y' = \dfrac{y}{x} + \sin\left(\dfrac{y}{x}\right)$.

20. $x\,dy = y\left[1 + \ln\left(\frac{y}{x}\right)\right] dx.$

Find the integral curve that-satisfies the initial condition.

21. $y' = \dfrac{y^3 - x^3}{xy^2}, \quad y(1) = 2.$

22. $x \sin\left(\frac{y}{x}\right) dy = \left[x + y \sin\left(\frac{y}{x}\right)\right] dx, \quad y(1) = 0.$

In Exercises 23–32, solve the initial value problem (a) by the Euler method, (b) by the Runge-Kutta method. Then solve the initial-value problem and estimate the accuracy of your numerical estimate using

$$\text{relative percentage error} = \frac{y_{\text{actual}} - y_{\text{approx}}}{y_{\text{actual}}} \times 100\%.$$

23. Estimate $y(1)$ if $y' = y$ and $y(0) = 1$, setting $h = 0.2$.

24. Estimate $y(1)$ if $y' = x + y$ and $y(0) = 2$, setting $h = 0.2$.

25. Exercise 23 setting $h = 0.1$.

26. Exercise 24 setting $h = 0.1$.

27. Estimate $y(1)$ if $y' = 2x$ and $y(2) = 5$, setting $h = 0.1$.

28. Estimate $y(0)$ if $y' = 3x^2$ and $y(1) = 2$, setting $h = 0.1$.

29. Estimate $y(2)$ if $y' = 1/(2y)$ and $y(1) = 1$, setting $h = 0.1$.

30. Estimate $y(2)$ if $y' = 1/(3y^2)$ and $y(1) = 1$, setting $h = 0.1$.

31. Exercise 23 setting $h = 0.05$.

32. Exercise 24 setting $h = 0.05$.

■ PROJECT 18.2 Direction Fields

Here we introduce a geometric approach to differential equations of the form $y' = f(x, y)$ that enables us to produce sketches of solution curves without requiring us to calculate any integrals. The approach does not produce equations in x and y; it produces pictures, pictures from which we can gather useful information. Do the curves slant up or do they slant down? Are there any maxima or minima? What is the concavity of the curves? We usually do not get precise answers to such questions, but we can get a good qualitative sense of what the curves look like.

The geometric approach to which we have been alluding is based on the construction of what is called a *direction field* (a *slope field*) for the differential equation. What this means is described below.

If a solution curve for the equation $y' = f(x, y)$ passes through the point (x, y), then it does so with slope $f(x, y)$. We construct a direction field for the differential equation by selecting a grid of points $(x_i, y_i), i = 1, 2 \cdots, n$, and drawing at each point a short line segment with slope $f(x_i, y_i)$. We can then use these little line segments to sketch the solution curve for the initial-value problem

$$y' = f(x, y), \quad y(a) = b$$

by starting at the point (a, b) and following the line segments in both directions. Figure A shows a direction field for the differential equation

$$y' = x - y$$

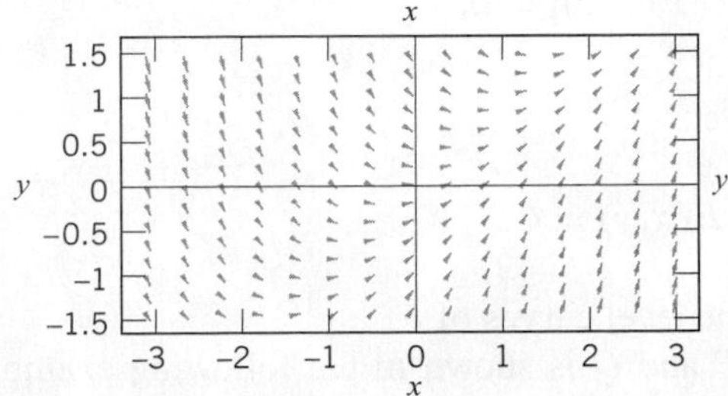

Figure A

drawn within the rectangle $R: -3 \le x \le 3, -1.5 \le y \le 1.5$. A sketch of the solution curve that satisfies the initial condition $y(0) = 0$ is shown in Figure B.

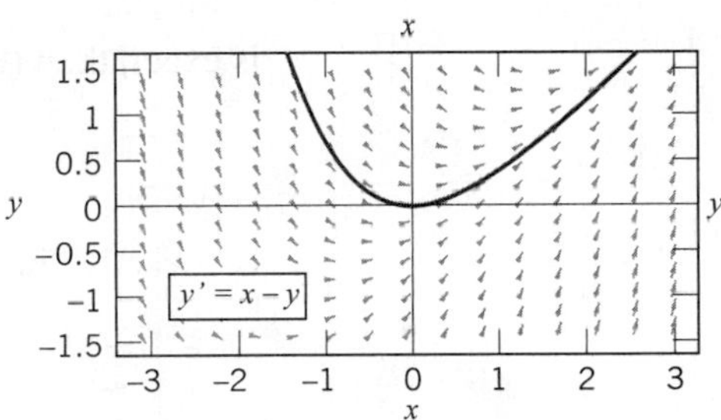

Figure B

REMARK Computer algebra systems usually include a feature for sketching direction fields.

Problem 1. In this problem we consider the initial-value problem: $y' = y,\ y(0) = 1$.

(a) Use a CAS to draw a direction field in the rectangle R: $-3 \le x \le 1.5, -1 \le y \le 3$.

(b) Use this direction field to sketch the solution curve that satisfies the initial condition. Experiment with other rectangles to obtain additional views of the solution curve.

(c) Solve the initial-value problem by other means and then compare the graph of your solution to the curve you obtained in part (b).

Problem 2. Repeat Problem 1 with $y' = x + 2y,\ y(0) = 1$ and $R: -1 \le x \le 2, -1 \le y \le 9$.

Problem 3. Repeat Problem 1 with $y' = 2xy,\ y(0) = 1$ and R: $-1.5 \le x \le 3, -1 \le y \le 8$.

Problem 4. In this problem we consider the initial-value problem: $y' = -4x/y,\ y(1) = 1$.

(a) Use a CAS to draw a direction field in the rectangle R: $-2 \le x \le 2, -3 \le y \le 3$.

(b) Use this direction field to sketch the solution curve that satisfies the initial condition. Experiment with other rectangles to obtain additional views of the solution curve.

(c) Solve the initial value problem by other means and then compare the graph of your solution to the curve you obtained in part (b).

■ 18.3 EXACT DIFFERENTIAL EQUATIONS; INTEGRATING FACTORS

We begin with two functions $P = P(x, y)$ and $Q = Q(x, y)$, each continuously differentiable on a simply connected region Ω. The differential equation

(18.3.1) $$P(x, y) + Q(x, y)\, y' = 0$$

is said to be *exact* on Ω if

$$\frac{\partial P}{\partial y}(x, y) = \frac{\partial Q}{\partial x}(x, y) \quad \text{for all } (x, y) \in \Omega.$$

The reason for this terminology is as follows. If the equation

$$P(x, y) + Q(x, y)\, y' = 0$$

is exact, then (by Theorem 15.9.2) the vector-valued function

$$P(x, y)\,\mathbf{i} + Q(x, y)\,\mathbf{j}$$

is "exactly" in the form of a gradient and there is a function F defined on Ω such that

$$\frac{\partial F}{\partial x} = P \quad \text{and} \quad \frac{\partial F}{\partial y} = Q.$$

Therefore, we can write (18.3.1) as

(1) $$\frac{\partial F}{\partial x}(x, y) + \frac{\partial F}{\partial y}(x, y)\, y' = 0.$$

Since, by the chain rule [(15.3.6)],

$$\frac{d}{dx}[F(x, y)] = \frac{\partial F}{\partial x}(x, y) + \frac{\partial F}{\partial y}(x, y)\, y',$$

equation (1) can be written

$$\frac{d}{dx}[F(x, y)] = 0.$$

Integrating with respect to x, we have

$$F(x, y) = C.$$

The integral curves of (18.3.1) are the level curves of F.

How F can be obtained from P and Q is shown in the following example. The process was explained in Section 15.9.

Example 1 The differential equation $(xy^2 - x^3) + (x^2y - y)y' = 0$ is everywhere exact: the coefficients

$$P(x, y) = xy^2 - x^3 \qquad \text{and} \qquad Q(x, y) = x^2y - y$$

are everywhere continuously differentiable, and at all points

$$\frac{\partial P}{\partial y} = 2xy = \frac{\partial Q}{\partial x}.$$

To find the integral curves

$$F(x, y) = C,$$

we set

$$\frac{\partial F}{\partial x}(x, y) = xy^2 - x^3 \qquad \text{and} \qquad \frac{\partial F}{\partial y}(x, y) = x^2y - y.$$

Integrating $\partial F/\partial x$ with respect to x, we have

$$F(x, y) = \tfrac{1}{2}x^2y^2 - \tfrac{1}{4}x^4 + \phi(y)$$

where $\phi(y)$ is independent of x but may depend on y. Differentiation with respect to y gives

$$\frac{\partial F}{\partial y} = x^2y - \phi'(y).$$

The two equations for $\partial F/\partial y$ can be reconciled by having

$$\phi'(y) = -y$$

and setting

$$\phi(y) = -\tfrac{1}{2}y^2.$$

The integral curves of the differential equation can be written

$$\tfrac{1}{2}x^2y^2 - \tfrac{1}{4}x^4 - \tfrac{1}{2}y^2 = C.$$

Checking: Differentiation with respect to x gives

$$\tfrac{1}{2}x^2(2yy') + xy^2 - x^3 - yy' = 0$$
$$(xy^2 - x^3) + (x^2y - y)y' = 0.$$

This is the original equation. □

If the equation

$$P(x, y) + Q(x, y)y' = 0, \qquad (x, y) \in \Omega$$

is not exact on Ω, it may be possible to find a function $\mu = \mu(x, y)$ not identically zero such that the equation

$$\mu(x, y)P(x, y) + \mu(x, y)Q(x, y)y' = 0$$

is exact. If μ is never zero on Ω, then any solution of this second equation gives a solution of the first equation. We call $\mu(x, y)$ an *integrating factor.*

Example 2 Consider the differential equation

$$(*) \qquad \left(2y^2 + 3x + \frac{2}{x^2}\right) + \left(2xy - \frac{y}{x}\right) y' = 0$$

on the right half-plane $\Omega = \{(x, y) : x > 0\}$.

The coefficients are continuously differentiable on Ω, but the equation is not exact there:

$$\frac{\partial}{\partial y}\left(2y^2 + 3x + \frac{2}{x^2}\right) = 4y \qquad \text{but} \qquad \frac{\partial}{\partial x}\left(2xy - \frac{y}{x}\right) = 2y + \frac{y}{x^2}.$$

However, multiplication by x gives

$$(**) \qquad (2xy^2 + 3x^2 + 2x^{-1}) + (2x^2y - y)\, y' = 0,$$

and this equation is exact:

$$\frac{\partial}{\partial y}(2xy^2 + 3x + 2x^{-1}) = 4xy = \frac{\partial}{\partial x}(2x^2y - y).$$

Thus we can solve $(*)$ by solving $(**)$. As you can check, the integral curves on Ω (where x remains positive) are of the form

$$x^2y^2 + x^3 + 2 \ln x - \tfrac{1}{2}y^2 = C. \quad \square$$

Let's return to the general situation and write our equation in the form

$$(2) \qquad P + Qy' = 0.$$

If this equation is not exact, how can we find an integrating factor?

Observe first of all that the equation

$$\mu P + \mu Q y' = 0$$

is exact iff

$$\frac{\partial}{\partial y}(\mu P) = \frac{\partial}{\partial x}(\mu Q),$$

and this occurs iff

$$(3) \qquad \mu\frac{\partial P}{\partial y} + P\frac{\partial \mu}{\partial y} = \mu\frac{\partial}{\partial x}Q + Q\frac{\partial \mu}{\partial x}.$$

Thus μ is an integrating factor for (2) iff it satisfies equation (3).

In theory all we have to do to find an integrating factor for (2) is solve equation (3) for μ. Unfortunately equation (3) is a partial differential equation that is usually more difficult to solve than equation (2). To get anywhere, we will have to make assumptions on the nature of μ.

The assumption that μ depends not on both x and y but only on one of these variables simplifies matters considerably. We will *assume* that μ is independent of y. Then equation (3) reduces to

$$\mu\frac{\partial P}{\partial y} = \mu\frac{\partial Q}{\partial x} + \frac{d\mu}{dx}Q$$

and gives

(4)
$$\frac{1}{\mu}\frac{d\mu}{dx} = \frac{1}{Q}\left(\frac{\partial P}{\partial y} - \frac{\partial Q}{\partial x}\right).$$

Since the left side of (4) is independent of y, the right side is independent of y. As you can check, this equation is satisfied by setting

$$\mu = e^{\int r(x)dx} \qquad \text{where} \qquad r(x) = \frac{1}{Q}\left(\frac{\partial P}{\partial y} - \frac{\partial Q}{\partial x}\right).$$

What does all this mean? It means that

(18.3.2)

> if
> $$r = \frac{1}{Q}\left(\frac{\partial P}{\partial y} - \frac{\partial Q}{\partial x}\right)$$
> is independent of y, then the function
> $$\mu = e^{\int r(x)\,dx}$$
> is an integrating factor for the equation
> $$P + Qy' = 0.$$

PROOF We assume that $r = \frac{1}{Q}\left(\frac{\partial P}{\partial y} - \frac{\partial Q}{\partial x}\right)$ is independent of y and write

$$P\,e^{\int r(x)\,dx} + Qe^{\int r(x)\,dx}y' = 0.$$

All we have to show is that

$$\frac{\partial}{\partial y}\left(P\,e^{\int r(x)\,dx}\right) = \frac{\partial}{\partial x}\left(Q\,e^{\int r(x)\,dx}\right).$$

This can be seen as follows:

$$\begin{aligned}
\frac{\partial}{\partial x}\left(Q\,e^{\int r(x)\,dx}\right) &= Q\,r(x)e^{\int r(x)\,dx} + \frac{\partial Q}{\partial x}e^{\int r(x)\,dx} \\
&= \left[Q\,r(x) + \frac{\partial Q}{\partial x}\right]e^{\int r(x)\,dx} \\
&= \left[\left(\frac{\partial P}{\partial y} - \frac{\partial Q}{\partial x}\right) + \frac{\partial Q}{\partial x}\right]e^{\int r(x)\,dx} \\
&= \frac{\partial P}{\partial y}\,e^{\int r(x)\,dx} = \frac{\partial}{\partial y}\left(P\,e^{\int r(x)\,dx}\right). \quad \square
\end{aligned}$$

Example 3 Earlier we considered the equation

$$\left(2y^2 + 3x + \frac{2}{x^2}\right) + \left(2xy - \frac{y}{x}\right)y' = 0$$

on the right half-plane and found that the equation was not exact there. We made it exact by multiplying through by x. We can obtain this integrating factor by using (18.3.2). In this case

$$r(x) = \frac{1}{Q}\left(\frac{\partial P}{\partial y} - \frac{\partial Q}{\partial x}\right) = \frac{1}{2xy - (y/x)}\left[4y - \left(2y + \frac{y}{x^2}\right)\right] = \frac{2y - (y/x^2)}{2xy - (y/x)} = \frac{1}{x}$$

so that $$e^{\int r(x)\,dx} = e^{\int (1/x)\,dx} = e^{\ln x} = x. \quad \square$$

In the Exercises you will be asked to show that

(18.3.3)

if
$$R = -\frac{1}{P}\left(\frac{\partial P}{\partial y} - \frac{\partial Q}{\partial x}\right)$$
is independent of x, then the function
$$\mu = e^{\int R(y)\,dy}$$
is an integrating factor for the equation
$$P + Qy' = 0.$$

One final remark. We have been working with equations
$$P(x,y) + Q(x,y)\,y' = 0.$$
Such equations are often written
$$P(x,y)\,dx + Q(x,y)\,dy = 0.$$
To accustom you to this notation, we will use it in some of the Exercises.

EXERCISES 18.3

Find the maximal simply connected region on which the equation is exact (in each of these cases there is one) and find the integral curves.

1. $(xy^2 - y) + (x^2y - x)\,y' = 0.$

2. $e^x \sin y + (e^x \cos y)\,y' = 0.$

3. $(e^y - y\,e^x) + (x\,e^y - e^x)\,y' = 0.$

4. $\sin y + (x\cos y + 1)\,y' = 0.$

5. $\ln y + 2xy + \left(\dfrac{x}{y} + x^2\right)\,y' = 0.$

6. $2x\tan^{-1} y + \left(\dfrac{x^2}{1+y^2}\right)\,y' = 0.$

7. $\left(\dfrac{y}{x} + 6x\right)dx + (\ln x - 2)\,dy = 0.$

8. $e^x + \ln y + \dfrac{y}{x} + \left(\dfrac{x}{y} + \ln x + \sin y\right)\,y' = 0.$

9. $(y^3 - y^2 \sin x - x)\,dx + (3xy^2 + 2y\cos x + e^{2y})\,dy = 0.$

10. $(e^{2y} - y\cos xy) + (2x\,e^{2y} - x\cos xy + 2y)y' = 0.$

11. Let p and q be functions of one variable everywhere continuously differentiable.

(a) Is the equation $p(x) + q(y)\,y' = 0$ necessarily exact?

(b) Show that the equation $p(y) + q(x)y' = 0$ is not necessarily exact. Then find an integrating factor.

12. Prove (18.3.3).

Solve the equation using an integrating factor if necessary.

13. $(e^{y-x} - y) + (x\,e^{y-x} - 1)\,y' = 0.$

14. $(x + e^y) - \frac{1}{2}x^2y' = 0.$

15. $(3x^2y^2 + x + e^y) + (2x^3y + y + x\,e^y)\,y' = 0.$

16. $\sin 2x\cos y - (\sin^2 x \sin y)\,y' = 0.$

17. $(y^3 + x + 1) + (3y^2)\,y' = 0.$

18. $(e^{2x+y} - 2y) + (x\,e^{2x+y} + 1)\,y' = 0.$

Find an integral curve that passes through the point (x_0, y_0). Use an integrating factor if necessary.

19. $(x^2 + y) + (x + e^y)\,y' = 0;\quad (x_0, y_0) = (1, 0).$

20. $(3x^2 - 2xy + y^3) + (3xy^2 - x^2)\,y' = 0;\quad (x_0, y_0) = (1, -1).$

21. $(2y^2 + x^2 + 2) + (2xy)\,y' = 0;\quad (x_0, y_0) = (1, 0).$

22. $(x^2 + y) + (3x^2y^2 - x)\,y' = 0;\quad (x_0, y_0) = (1, 1).$

23. $y^3 + (1 + xy^2)\,y' = 0;\quad (x_0, y_0) = (-2, -1).$

24. $(x + y)^2 + (2xy + x^2 - 1)\,y' = 0;\quad (x_0, y_0) = (1, 1).$

25. $[\cosh(x - y^2) + e^{2x}]\,dx + y[1 - 2\cosh(x - y^2)]\,dy = 0;$ $(x_0, y_0) = (2, \sqrt{2}).$

26. In section 8.8 we solved the linear differential equation
$$y' + p(x)\,y = q(x)$$
by using the integrating factor
$$e^{\int p(x)\,dx}.$$
Show that this integrating factor is obtainable by the methods of this section.

27. (a) Find a value of k, if possible, such that the differential equation

$$(xy^2 + kx^2y + x^3)\,dx + (x^3 + x^2y + y^2)\,dy = 0$$

is everywhere exact.

(b) Find a value of k, if possible, such that the differential equation

$$y\,e^{2xy} + 2x + (kx\,e^{2xy} - 2y)\,y' = 0$$

is everywhere exact.

28. (a) Find functions f and g, not both identically zero, such that the differential equation

$$g(y)\sin x\,dx + y^2 f(x)\,dy = 0$$

is everywhere exact.

(b) Find all functions g such that the differential equation $g(y)\,e^y + xy\,y' = 0$ is everywhere exact.

In Exercises 29–34, solve the given differential equation by any means at your disposal.

29. $y' = y^2x^3$.

30. $y\,y' = 4x\,e^{2x+y}$.

31. $y' + \dfrac{4y}{x} = x^4$.

32. $y' + 2xy - 2x^3 = 0$.

33. $(y\,e^{xy} - 2x)\,dx + \left(\dfrac{2}{y} + x\,e^{xy}\right)dy = 0$.

34. $y\,dx + (2xy - e^{-2y})\,dy = 0$.

■ 18.4 THE EQUATION $y'' + ay' + by = 0$

A differential equation of the form

$$y'' + ay' + by = \phi(x)$$

where a and b are real numbers and ϕ is a continuous function on some interval I is a *second-order linear differential equation with constant coefficients.* By a solution of such an equation we mean a function $y = y(x)$ that satisfies the equation for all x in I. In this section we set $\phi = 0$ and consider the *reduced equation*†

(18.4.1) $$y'' + ay' + by = 0.$$

As you will see, the solutions of the reduced equation are defined for all real x.

The Characteristic Equation

Earlier you saw that the function $y = e^{-ax}$ satisfies the first-order linear equation

$$y' + ay = 0.$$

This suggests that the differential equation

$$y'' + ay' + by = 0$$

may have a solution of the form $y = e^{rx}$.

If $y = e^{rx}$, then

$$y' = r\,e^{rx} \qquad \text{and} \qquad y'' = r^2\,e^{rx}.$$

Substitution into the differential equation gives

$$r^2\,e^{rx} + ar\,e^{rx} + b\,e^{rx} = e^{rx}(r^2 + ar + b) = 0,$$

and since $e^{rx} \neq 0$,

$$r^2 + ar + b = 0.$$

This shows that the function $y = e^{rx}$ satisfies the differential equation iff

$$r^2 + ar + b = 0.$$

†Sometimes referred to as the *homogeneous equation.*

This quadratic equation in r is called the *characteristic equation.*†

The nature of the solutions of the differential equation

$$y'' + ay' + by = 0$$

depends on the nature of the roots of the characteristic equation. There are three cases to be considered.

Case 1: *The characteristic equation has two distinct real roots r_1 and r_2.* In this case both

$$y_1(x) = e^{r_1 x} \qquad \text{and} \qquad y_2(x) = e^{r_2 x}$$

are solutions of the reduced equation.

Case 2: *The characteristic equation has only one real root $r_1 = r_2 = \alpha$.* In this case the characteristic equation $(r - \alpha)^2 = 0$ can be written

$$r^2 - 2\alpha r + \alpha^2 = 0.$$

As you are asked to show in Exercise 33, the substitution $y = u\,e^{\alpha x}$ gives

$$u'' = 0.$$

This equation is satisfied by

the constant function $u_1 = 1$ and the identity function $u_2 = x$.

Thus the reduced equation is satisfied by the products

$$y_1(x) = e^{\alpha x} \qquad \text{and} \qquad y_2(x) = x\,e^{\alpha x}.$$

Case 3: *The characteristic equation has two complex roots $r_1 = \alpha + i\beta$ and $r_2 = \alpha - i\beta$.* In this case the characteristic equation $[r - (\alpha + i\beta)][r - (\alpha - i\beta)] = 0$ can be written

$$r^2 - 2\alpha r + (\alpha^2 + \beta^2) = 0.$$

As you are asked to show in Exercise 33, the substitution $y = u\,e^{\alpha x}$ eliminates α and gives

$$u'' + \beta^2 u = 0.$$

This equation, the equation of harmonic motion, is satisfied by the functions

$$u_1(x) = \cos\beta x \qquad \text{and} \qquad u_2(x) = \sin\beta x.\,\dagger\dagger$$

Thus, the reduced equation is satisfied by the products

$$y_1(x) = e^{\alpha x}\cos\beta x \qquad \text{and} \qquad y_2(x) = e^{\alpha x}\sin\beta x. \quad \square$$

---

†Sometimes called the *auxiliary equation.*

†† We reviewed this in the introduction to this chapter. In any case, the statements are easy to verify.

Linear Combinations of Solutions; Existence and Uniqueness of Solutions; Wronskians

Observe that *if y_1 and y_2 are both solutions of the reduced equation, then every linear combination*

$$u(x) = C_1 y_1(x) + C_2 y_2(x)$$

is also a solution.

PROOF Set

$$u = C_1 y_1 + C_2 y_2$$

and observe that

$$u' = C_1 y_1' + C_2 y_2' \qquad \text{and} \qquad u'' = C_1 y_1'' + C_2 y_2''.$$

Since y_1 and y_2 are solutions of (18.4.1),

$$y_1'' + ay_1' + by_1 = 0 \qquad \text{and} \qquad y_2'' + ay_2' + by_2 = 0.$$

Therefore,

$$\begin{aligned} u'' + au' + bu &= (C_1 y_1'' + C_2 y_2'') + a(C_1 y_1' + C_2 y_2') + b(C_1 y_1 + C_2 y_2) \\ &= C_1(y_1'' + ay_1' + by_1) + C_2(y_2'' + ay_2' + by_2) \\ &= C_1(0) + C_2(0) = 0 \quad \square \end{aligned}$$

You have seen how to obtain solutions of the differential equation

$$y'' + ay' + by = 0$$

from the characteristic equation

$$r^2 + ar + b = 0.$$

We can form more solutions by taking linear combinations of these solutions. Question: Are there still other solutions or do all solutions arise in this manner? Answer: All solutions of the reduced equation are linear combinations of the solutions that we have already found.

To show this, we have to go a little deeper into the theory. Our point of departure is a result that we prove in a supplement to this section.

THEOREM 18.4.2 EXISTENCE AND UNIQUENESS THEOREM

Let x_0, α_0, α_1 be arbitrary real numbers. The reduced equation

$$y'' + ay' + by = 0$$

has a unique solution $y = y(x)$ that satisfies the initial conditions

$$y(x_0) = \alpha_0, \qquad y'(x_0) = \alpha_1.$$

Geometrically, the theorem says that there is one and only one solution the graph of which passes through a prescribed point (x_0, α_0) with a prescribed slope α_1. We assume the result and go on from there.

DEFINITION 18.4.3

Let y_1 and y_2 be two solutions of

$$y'' + ay' + by = 0.$$

The *Wronskian*† of y_1 and y_2 is the function W defined for all real x by

$$W(x) = y_1(x)\,y_2'(x) - y_2(x)\,y_1'(x).$$

Note that the Wronskian can be written as the 2×2 determinant

$$W(x) = \begin{vmatrix} y_1(x) & y_2(x) \\ y_1'(x) & y_2'(x) \end{vmatrix}.$$

Wronskians have a very special property.

THEOREM 18.4.4

If both y_1 and y_2 are solutions of

$$y'' + ay' + by = 0,$$

then their Wronskian W is either identically zero or never zero.

PROOF Assume that both y_1 and y_2 are solutions of the equation and set

$$W = y_1y_2' - y_2y_1'.$$

Differentiation gives

$$W' = y_1y_2'' + y_1'y_2' - y_1'y_2' - y_1''y_2 = y_1y_2'' - y_1''y_2.$$

Since y_1 and y_2 are solutions, we know that

$$y_1'' + ay_1' + by_1 = 0$$

and

$$y_2'' + ay_2' + by_2 = 0.$$

Multiplying the first equation by $-y_2$ and the second equation by y_1, we have

$$\begin{aligned} -y_1''y_2 - ay_1'y_2 - by_1y_2 &= 0 \\ y_2''y_1 + ay_2'y_1 + by_2y_1 &= 0. \end{aligned}$$

We now add these two equations and obtain

$$(y_1y_2'' - y_2y_1'') + a(y_1y_2' - y_2y_1') = 0.$$

† Named after Count Hoëné Wronski, a Polish mathematician (1776–1853).

In terms of the Wronskian, we have

$$W' + aW = 0.$$

This is a first-order linear differential equation with general solution

$$W(x) = Ce^{-ax}.$$

If $C = 0$, then W is identically 0; if $C \neq 0$, then W is never zero. ❑

Remarks There is a simple way to determine whether a Wronskian is zero, a way that does not require that we calculate a determinant.

As you are asked to show in Exercise 37,
The Wronskian

$$W(x) = \begin{vmatrix} y_1(x) & y_2(x) \\ y_1'(x) & y_1'(x) \end{vmatrix}$$

of two solutions y_1, y_2 is zero iff one of these functions is a scalar multiple of the other.

Part of the proof is straightforward; part of it is somewhat delicate. The straightforward part is showing that

$$\text{if } y_2 = \alpha y_1 \quad \text{or} \quad y_1 = \alpha y_2 \quad \text{for some constant } \alpha,$$

then

$$W = 0.$$

Showing that

$$\text{if} \quad W = 0,$$

then

$$y_2 = \alpha y_1 \quad \text{or} \quad y_1 = \alpha y_2 \quad \text{for some constant } \alpha$$

is the delicate part. ❑

THEOREM 18.4.5

Every solution of the equation

$$y'' + ay' + by = 0$$

can be expressed in a unique manner as the linear combination of any two solutions with a nonzero Wronskian.

PROOF Let u be any solution of the equation and let y_1, y_2 be any two solutions with nonzero Wronskian. Choose a number x_0 and form the equations

$$(1) \qquad \begin{aligned} C_1 y_1(x_0) + C_2 y_2(x_0) &= u(x_0) \\ C_1 y_1'(x_0) + C_2 y_2'(x_0) &= u'(x_0). \end{aligned}$$

The Wronskian of y_1 and y_2 at x_0,

$$W(x_0) = y_1(x_0)\,y_2'(x_0) - y_2(x_0)\,y_1'(x_0),$$

is different from zero. This guarantees that the system of equations (1) has a unique solution given by

$$C_1 = \frac{u(x_0)\,y_2'(x_0) - y_2(x_0)u'(x_0)}{y_1(x_0)\,y_2'(x_0) - y_2(x_0)\,y_1'(x_0)}, \qquad C_2 = \frac{y_1(x_0)u'(x_0) - u(x_0)\,y_1'(x_0)}{y_1(x_0)\,y_2'(x_0) - y_2(x_0)\,y_1'(x_0)}.$$

Our work is finished. The function $C_1y_1 + C_2y_2$ is a solution of the equation which by (1) has the same value as u at x_0 and the same derivative. Thus, by Theorem 18.4.2, $C_1y_1 + C_2y_2$ and u cannot be different functions; that is,

$$u = C_1y_1 + C_2y_2.$$

This proves the theorem. □

The General Solution

The arbitrary linear combination $y = C_1y_1 + C_2y_2$ of any two solutions with nonzero Wronskian is the *general solution*. By Theorem 18.4.5, we can obtain any *particular solution* by adjusting C_1 and C_2.

We now return to the solutions obtained earlier and prove the final result.

THEOREM 18.4.6

Given the equation

$$y'' + ay' + by = 0,$$

we form the characteristic equation

$$r^2 + ar + b = 0.$$

I. If the characteristic equation has two distinct real roots r_1 and r_2, then the general solution takes the form

$$y = C_1e^{r_1x} + C_2e^{r_2x}.$$

II. If the characteristic equation has only one real root, α, then the general solution takes the form

$$y = C_1e^{\alpha x} + C_2xe^{\alpha x} = (C_1 + C_2x)e^{\alpha x}.$$

III. If the characteristic equation has two complex roots

$$r_1 = \alpha + i\beta \quad \text{and} \quad r_2 = \alpha - i\beta,$$

then the general solution takes the form

$$y = C_1\,e^{\alpha x}\cos\beta x + C_2\,e^{\alpha x}\sin\beta x = e^{\alpha x}(C_1\cos\beta x + C_2\sin\beta x).$$

PROOF To prove this theorem, it is enough to show that the three solution pairs

$$e^{r_1 x},\ e^{r_2 x} \qquad e^{\alpha x},\ x\,e^{\alpha x} \qquad e^{\alpha x}\cos\beta x,\ e^{\alpha x}\sin\beta x$$

all have nonzero Wronskians. The Wronskian of the first pair is the function

$$\begin{aligned} W(x) &= e^{r_1 x}\frac{d}{dx}(e^{r_2 x}) - \frac{d}{dx}(e^{r_1 x})\,e^{r_2 x} \\ &= e^{r_1 x} r_2\, e^{r_2 x} - r_1\, e^{r_1 x}\, e^{r_2 x} = (r_2 - r_1)\,e^{(r_1+r_2)x}. \end{aligned}$$

$W(x)$ is different from zero since, by assumption, $r_2 \neq r_1$.

We leave it to you to verify that the other two pairs also have nonzero Wronskians. ❑

It's time to look at specific examples.

Example 1 Find the general solution of the equation $y'' + 2y' - 15y = 0$. Then find the particular solution that satisfies the initial conditions

$$y(0) = 0, \qquad y'(0) = -1.$$

SOLUTION The characteristic equation is the quadratic $r^2 + 2r - 15 = 0$. Factoring the left side, we have

$$(r+5)(r-3) = 0.$$

There are two real roots: -5 and 3. The general solution takes the form

$$y = C_1\,e^{-5x} + C_2\,e^{3x}.$$

Differentiating the general solution, we have

$$y' = -5C_1\,e^{-5x} + 3C_2\,e^{3x}.$$

The conditions

$$y(0) = 0, \qquad y'(0) = -1$$

are satisfied iff

$$C_1 + C_2 = 0 \qquad \text{and} \qquad -5C_1 + 3C_2 = -1.$$

Solving these two equations simultaneously, we find that

$$C_1 = \tfrac{1}{8}, \quad C_2 = -\tfrac{1}{8}.$$

The solution that satisfies the prescribed side conditions is the function

$$y = \tfrac{1}{8}\,e^{-5x} - \tfrac{1}{8}\,e^{3x}. \quad ❑$$

Example 2 Find the general solution of the equation $y'' + 4y' + 4y = 0$.

SOLUTION The characteristic equation takes the form $r^2 + 4r + 4 = 0$, which can be written

$$(r+2)^2 = 0.$$

The number -2 is the only root and

$$y = C_1 e^{-2x} + C_2 x e^{-2x}.$$

is the general solution. ❑

Example 3 Find the general solution of the equation $y'' + y' + 3y = 0$.

SOLUTION The characteristic equation is $r^2 + r + 3 = 0$. The quadratic formula shows that there are two complex roots:

$$r_1 = -\tfrac{1}{2} + i\tfrac{1}{2}\sqrt{11}, \quad r_2 = -\tfrac{1}{2} - i\tfrac{1}{2}\sqrt{11}.$$

The general solution takes the form

$$y = e^{-x/2}[C_1 \cos(\tfrac{1}{2}\sqrt{11}\,x) + C_2 \sin(\tfrac{1}{2}\sqrt{11}\,x)]. \quad \square$$

In our final example we revisit the equation of simple harmonic motion.

Example 4 Find the general solution of the equation

$$y'' + \omega^2 y = 0. \qquad (\omega \neq 0)$$

SOLUTION The characteristic equation is $r^2 + \omega^2 = 0$ and the roots are

$$r_1 = \omega i, \quad r_2 = -\omega i.$$

Thus the general solution is

$$y = C_1 \cos \omega x + C_2 \sin \omega x. \quad \square$$

Remark As you probably recall, the equation in Example 4 describes the oscillatory motion of an object suspended by a spring under the assumption that there are no forces acting on the spring-mass system other than the restoring force of the spring. This spring-mass problem and some generalizations of it are studied in Section 18.6. In the Exercises you are asked to show that the general solution that we gave above can be written

$$y = A \sin(\omega x + \phi_0)$$

where A and ϕ_0 are constants with $A > 0$ and $\phi_0 \in [0, 2\pi)$. $\square$

EXERCISES 18.4

Find the general solution.

1. $y'' + 2y' - 8y = 0$.

2. $y'' - 13y' + 42y = 0$.

3. $y'' + 8y' + 16y = 0$.

4. $y'' + 7y' + 3y = 0$.

5. $y'' + 2y' + 5y = 0$.

6. $y'' - 3y' + 8y = 0$.

7. $2y'' + 5y' - 3y = 0$.

8. $y'' - 12y = 0$.

9. $y'' + 12y = 0$.

10. $y'' - 3y' + \frac{9}{4}y = 0$.

11. $5y'' + \frac{11}{4}y' - \frac{3}{4}y = 0$.

12. $2y'' + 3y' = 0$.

13. $y'' + 9y = 0$.

14. $y'' - y' - 30y = 0$.

15. $2y'' + 2y' + y = 0$.

16. $y'' - 4y' + 4y = 0$.

17. $8y'' + 2y' - y = 0$.

18. $5y'' - 2y' + y = 0$.

Solve the initial-value problem.

19. $y'' - 5y' + 6y = 0$, $y(0) = 1$, $y'(0) = 1$.

20. $y'' + 2y' + y = 0$, $y(2) = 1$, $y'(2) = 2$.

21. $y'' + \frac{1}{4}y = 0$, $y(\pi) = 1$, $y'(\pi) = -1$.

22. $y'' - 2y' + 2y = 0$, $y(0) = -1$, $y'(0) = -1$.

23. $y'' + 4y' + 4y = 0$, $y(-1) = 2$, $y'(-1) = 1$.

24. $y'' - 2y' + 5y = 0$, $y(\pi/2) = 0$, $y'(\pi/2) = 2$.

25. Find all solutions of the equation $y'' - y' - 2y = 0$ that satisfy the given initial conditions:
(a) $y(0) = 1$. (b) $y'(0) = 1$.
(c) $y(0) = 1$, $y'(0) = 1$.

26. Prove that the general solution of the differential equation

$$y'' - \omega^2 y = 0 \qquad (\omega > 0)$$

can be written

$$y = C_1 \cosh \omega x + C_2 \sinh \omega x.$$

27. Suppose that the roots r_1 and r_2 of the characteristic equation of (18.4.1) are real and distinct. Then they can be written as $r_1 = \alpha + \beta$ and $r_2 = \alpha - \beta$, where α and β are real. Show that the general solution of the equation (18.4.1) can be expressed in the form

$$y = e^{\alpha x}(C_1 \cosh \beta x + C_2 \sinh \beta x).$$

28. Show that the general solution of the differential equation

$$y'' + \omega^2 y = 0$$

can be written

$$y = A \sin(\omega x + \phi_0),$$

where A and ϕ_0 are constants with $A > 0$ and $\phi_0 \in [0, 2\pi)$.

29. Complete the proof of Theorem 18.4.6 by showing that the following solutions have nonzero Wronskians.

(a) $y_1 = e^{\alpha x}, \quad y_2 = x\,e^{\alpha x}$. (one root case)

(b) $y_1 = e^{\alpha x} \cos \beta x, \quad y_2 = e^{\alpha x} \sin \beta x$. (complex root case)

30. In the absence of any external electromotive force, the current i in a simple electrical circuit varies with time t according to the formula

$$L\frac{d^2 i}{dt^2} + R\frac{di}{dt} + \frac{1}{C} i = 0. \quad (L, R, C \text{ constants})\dagger$$

Find the general solution of this equation given that $L = 1, R = 10^3$, and

(a) $C = 5 \times 10^{-6}$.

(b) $C = 4 \times 10^{-6}$.

(c) $C = 2 \times 10^{-6}$.

31. Find a differential equation $y'' + ay' + by = 0$ that is satisfied by both functions.

(a) $y_1 = e^{2x}, \quad y_2 = e^{-4x}$.

(b) $y_1 = 3\,e^{-x}, \quad y_2 = 4\,e^{5x}$.

(c) $y_1 = 2\,e^{3x}, \quad y_2 = x\,e^{3x}$.

32. Find a differential equation $y'' + ay' + by = 0$ that is satisfied by both functions.

(a) $y_1 = 2 \cos 2x, \quad y_2 = -\sin 2x$.

(b) $y_1 = e^{-2x} \cos 3x, \quad y_2 = 2\,e^{-2x} \sin 3x$.

33. (a) Show that the substitution $y = e^{\alpha x} u$ transforms

$$y'' - 2\alpha y' + \alpha^2 y = 0 \quad \text{into} \quad u'' = 0.$$

(b) Show that the substitution $y = e^{\alpha x} u$ transforms

$$y'' - 2\alpha y' + (\alpha^2 + \beta^2) y = 0 \quad \text{into} \quad u'' + \beta^2 u = 0.$$

† L is inductance, R is resistance, and C is capacitance. If L is given in henrys, R in ohms, C in farads, and t in seconds, then the current is given in amperes.

Exercises 34 and 35 relate to the differential equation $y'' + ay' + by = 0$ where a and b are nonnegative constants.

34. Prove that if a and b are both positive, then $y(x) \to 0$ as $x \to \infty$ for all solutions y of the equation.

35. (a) Prove that if $a = 0$ and $b > 0$, then all solutions of the equation are bounded.

(b) Suppose that $a > 0$, $b = 0$, and $y = y(x)$ is a solution of the equation. Prove that

$$\lim_{x \to \infty} y(x) = k$$

for some constant k. Determine k for the solution that satisfies the initial conditions: $y(0) = y_0$, $y'(0) = y_1$.

36. Let y_1, y_2 be solutions of the reduced equation defined for all real x. Show that if $y_1(a) = y_2(a) = 0$ at some number a, then y_1 and y_2 are scalar multiples of each other.

37. Let y_1, y_2 be solutions of the reduced equation defined for all real x. Show that the Wronskian of y_1, y_2 is zero iff one of these functions is a scalar multiple of the other.

Euler Equations An equation of the form

$$(*) \qquad x^2 y'' + \alpha x y' + \beta y = 0,$$

where α and β are real numbers, is called an *Euler equation.*

38. Show that the Euler equation $(*)$ can be transformed into an equation of the form

$$\frac{d^2 y}{dz^2} + a\frac{dy}{dz} + by = 0$$

where a and b are real numbers, by means of the change of variable $z = \ln x$. HINT: If $z = \ln x$, then by the chain rule,

$$\frac{dy}{dx} = \frac{dy}{dz}\frac{dz}{dx} = \frac{dy}{dz}\frac{1}{x}.$$

Now calculate d^2y/dx^2 and substitute the result into the differential equation.

In Exercises 39–42, use the change of variable indicated in Exercise 38 to transform the given equation into an equation with constant coefficients. Find the general solution of that equation, and then express it in terms of x.

39. $x^2 y'' - xy' - 8y = 0$.

40. $x^2 y'' - 2xy' + 2y = 0$.

41. $x^2 y'' - 3xy' + 4y = 0$.

42. $x^2 y'' - xy' + 5y = 0$.

*SUPPLEMENT TO SECTION 18.4

PROOF OF THEOREM 18.4.2

Existence: Take two solutions y_1, y_2 with nonzero Wronskian

$$W(x) = y_1(x)\, y_2'(x) - y_2(x)\, y_1'(x).$$

For any numbers x_0, α_0, α_1, the equations

$$C_1 y_1(x_0) + C_2 y_2(x_0) = \alpha_0,$$
$$C_1 y_1'(x_0) + C_2 y_2'(x_0) = \alpha_1$$

can be solved for C_1 and C_2. For those values of C_1 and C_2, the function

$$y = C_1 y_1 + C_2 y_2$$

is a solution of (18.4.1) that satisfies the prescribed initial conditions.

Uniqueness: Let us assume that there are two distinct solutions y_1, y_2 that satisfy the same prescribed initial conditions

$$y_1(x_0) = \alpha_0 = y_2(x_0) \quad \text{and} \quad y_1'(x_0) = \alpha_1 = y_2'(x_0).$$

Then the solution $y = y_1 - y_2$ satisfies the initial conditions

$$y(x_0) = 0,\ y'(x_0) = 0.$$

Since y_1 and y_2 are, by assumption, distinct functions, there is at least one number x at which y is not zero. Therefore, by the continuity of y there exists an interval I on which y is does not take on the value zero.

Now let u be *any* solution of (18.4.1). The Wronskian of y and u is zero at x_0:

$$W(x_0) = y(x_0)\, u'(x_0) - u(x_0)\, y'(x_0) = (0)u'(x_0) - u(x_0)(0) = 0.$$

Therefore the Wronskian of y and u is everywhere zero. Since $y(x) \neq 0$ for all $x \in I$, the quotient u/y is defined on I, and on that interval

$$\frac{d}{dx}\left(\frac{u}{y}\right) = \frac{yu' - uy'}{y^2} = \frac{W}{y^2} = 0, \quad \frac{u}{y} = C, \quad \text{and} \quad u = Cy.$$

We have shown that on the interval I every solution is some scalar multiple of y.

Now let u_1 and u_2 be any two solutions with a nonzero Wronskian W. From what we have just shown, there are constants C_1 and C_2 such that on I

$$u_1 = C_1 y \quad \text{and} \quad u_2 = C_2 y.$$

Then on I

$$W = u_1 u_2' - u_2 u_1' = (C_1 y)(C_2 y') - (C_2 y)(C_1 y') = C_1 C_2 (yy' - yy') = 0.$$

This contradicts the statement that $W \neq 0$.

The assumption that there are two distinct solutions that satisfy the same prescribed initial conditions has led to a contradiction. This proves uniqueness. □

■ 18.5 THE EQUATION $y'' + ay' + by = \phi(x)$

In Section 18.4 we solved the reduced equation

$$y'' + ay' + by = 0.$$

Here we go on to the *complete equation*

(18.5.1) $$y'' + ay' + by = \phi(x).\dagger$$

The function $\phi = \phi(x)$ that appears on the right will be called the *forcing function*.†† We assume that ϕ is continuous on some interval I and we will solve the differential equation on that interval.

We begin by proving two simple but important results.

(18.5.2) If both y_1 and y_2 are solutions of the complete equation, then their difference $u = y_1 - y_2$ is a solution of the reduced equation.

PROOF If

$$y_1'' + ay_1' + by_1 = \phi(x) \qquad \text{and} \qquad y_2'' + ay_2' + by_2 = \phi(x),$$

then

$$\begin{aligned} u'' + au' + bu &= (y_1'' - y_2'') + a(y_1' - y_2') + b(y_1 - y_2) \\ &= (y_1'' + ay_1' + by_1) - (y_2'' + ay_2' + by_2) \\ &= \phi(x) - \phi(x) = 0 \quad \square \end{aligned}$$

(18.5.3) If y_p is a particular solution of the complete equation, then every solution of the complete equation can be written as a solution of the reduced equation plus y_p.

PROOF Let y_p be a solution of the complete equation. If y is another solution of the complete equation, then, by (18.5.2), $y - y_p$ is a solution of the reduced equation. Obviously

$$y = (y - y_p) + y_p. \quad \square$$

It follows from (18.5.3) that we can obtain the *general solution* of the complete equation by starting with the general solution of the reduced equation and then adding to it a particular solution of the complete equation. The general solution of the complete equation can thus be written

(18.5.4) $$y = C_1u_1 + C_2u_2 + y_p$$

where u_1, u_2 are any two solutions of the reduced equation which have a nonzero Wronskian and y_p is any particular solution of the complete equation.

The main task before us is to search for functions which can serve as y_p. In this search the following result can sometimes be used to advantage. It is called the

† Sometimes referred to as the *nonhomogeneous equation*.

†† Because this is the role played by ϕ in the study of vibrations.

superposition principle.

(18.5.5)

> If y_1 is a solution of
> $$y'' + ay' + by = \phi_1(x)$$
> and y_2 is a solution of
> $$y'' + ay' + by = \phi_2(x),$$
> then $y_1 + y_2$ is a solution of
> $$y'' + ay' + by = \phi_1(x) + \phi_2(x).$$

Using the superposition principle, we can find solutions to an equation in which the forcing function has several terms by finding solutions to equations in which the forcing function has only one term and then adding up the results. Verification of the superposition principle is left to you as an exercise.

We are now ready to describe two methods by which we can find some solution of the complete equation, some function that can play the role of y_p.

Variation of Parameters

The method that we outline here gives particular solutions to all equations of the form

(1) $$y'' + ay' + by = \phi(x).$$

The general solution of the reduced equation

$$y'' + ay' + by = 0$$

can be written

$$y = C_1u_1 + C_2u_2$$

where u_1, u_2 are any two solutions with a nonzero Wronskian and the coefficients C_1, C_2 are arbitrary constants. In the method called *variation of parameters,* we let the coefficients vary. That is, we replace the constants C_1, C_2 by functions

$$z_1 = z_1(x), \qquad z_2 = z_2(x)$$

and seek solutions of the form

(2) $$y_p = z_1u_1 + z_2u_2.$$

Differentiating (2), we have

$$y_p = z_1u_1' + z_1'u_1 + z_2u_2' + z_2'u_2 = (z_1u_1' + z_2u_2') + (z_1'u_1 + z_2'u_2).$$

We now impose a restriction on z_1, z_2 : we require that

(3) $$z_1'u_1 + z_2'u_2 = 0.$$

Having imposed this restriction, we have

$$y_p' = z_1u_1' + z_2u_2'$$

and, differentiating once more,

$$y_p'' = z_1 u_1'' + z_1' u_1' + z_2 u_2'' + z_2' u_2'.$$

A straightforward calculation that we leave to you shows that y_p will satisfy equation (1) iff

$$(4) \qquad z_1' u_1' + z_2' u_2' = \phi(x).$$

Equations (3) and (4) can now be solved simultaneously for z_1' and z_2'. As you can verify yourself, the unique solutions are

$$(5) \qquad z_1' = -\frac{u_2\phi}{W} \quad \text{and} \quad z_2' = \frac{u_1\phi}{W}$$

where the denominator $W = u_1u_2' - u_2u_1'$ is the Wronskian of u_1 and u_2. The functions z_1, z_2 are now found by integration:

$$z_1 = -\int \frac{u_2\phi}{W}\,dx, \qquad z_2 = \int \frac{u_1\phi}{W}\,dx.$$

The function

$$\textbf{(18.5.6)} \qquad \boxed{y_p = \left(-\int \frac{u_2(x)\phi(x)}{W}\,dx\right) u_1(x) + \left(\int \frac{u_1(x)\phi(x)}{W}\right) u_2(x)}$$

is a particular solution of the equation

$$y'' + ay' + by = \phi(x).$$

Example 1 Use variation of parameters to find a solution of the equation

$$y'' + y = \tan x, \qquad -\frac{\pi}{2} < x < \frac{\pi}{2}.$$

Then give the general solution.

SOLUTION The reduced equation $y'' + y = 0$ has solutions

$$u_1 = \cos x, \quad u_2 = \sin x.$$

The Wronskian of these solutions is identically 1:

$$W = u_1u_2' - u_2u_1' = (\cos x)(\cos x) - (-\sin x)\sin x = \cos^2 x + \sin^2 x = 1.$$

Since $\phi(x) = \tan x$, we can set

$$\begin{aligned} z_1 &= -\int \frac{u_2\phi}{W}\,dx \\ &= -\int \frac{\sin x \tan x}{1}\,dx \\ &= -\int \frac{\sin^2 x}{\cos x}\,dx \\ &= \int \frac{\cos^2 x - 1}{\cos x}\,dx = \int (\cos x - \sec x)\,dx = \sin x - \ln|\sec x + \tan x| \end{aligned}$$

and $$z_2 = \int \frac{u_1\phi}{W}\,dx = \int \frac{\cos x \tan x}{1}\,dx = \int \sin x\,dx = -\cos x.$$

We didn't include any arbitrary constants here. At this stage we are looking for only one solution of the complete equation, not a family of solutions.

By (18.5.6), the function

$$\begin{aligned} y_p &= (\sin x - \ln|\sec x + \tan x|)\cos x + (-\cos x)\sin x \\ &= -(\ln|\sec x + \tan x|)\cos x \end{aligned}$$

is a solution of the complete equation.

The general solution can be written

$$y = C_1 \cos x + C_2 \sin x - (\ln|\sec x + \tan x|)\cos x. \quad \square$$

Example 2 Find the general solution of $y'' = 5y' + 6y = 4e^{2x}$.

SOLUTION The equation $y'' - 5y' + 6y = 0$ has characteristic equation

$$r^2 - 5r + 6 = (r-2)(r-3) = 0.$$

Thus, $u_1(x) = e^{2x}, u_2(x) = e^{3x}$ are solutions. Their Wronskian W is e^{5x}:

$$W = u_1u_2' - u_2u_1' = e^{2x}\,3e^{3x} - e^{3x}\,2e^{2x} = e^{5x}.$$

Since $\phi(x) = 4e^{2x}$, we have

$$z_1 = -\int \frac{u_2\phi}{W}\,dx = -\int \frac{e^{3x}\,4e^{2x}}{e^{5x}}\,dx = -\int 4\,dx = -4x$$

and

$$z_2 = \int \frac{u_1\phi}{W}\,dx = \int \frac{e^{2x}\,4e^{2x}}{e^{5x}}\,dx = \int 4e^{-x}\,dx = -4e^{-x}.$$

Therefore, by (18.5.6),

$$y_p = -4x\,e^{2x} - 4\,e^{-x}e^{3x} = -4x\,e^{2x} - 4\,e^{2x}$$

is a solution of the complete equation.

The general solution can be written

$$\begin{aligned} y &= A_1u_1 + A_2u_2 + y_p = A_1e^{2x} + A_2\,e^{3x} - 4x\,e^{2x} - 4\,e^{2x} \\ &= (A_1 - 4)\,e^{2x} + A_2\,e^{3x} - 4x\,e^{2x} \\ &= C_1\,e^{2x} + C_2\,e^{3x} - 4x\,e^{2x}. \qquad (C_1 = A_1 - 4, C_2 = A_2) \quad \square \end{aligned}$$

Undetermined Coefficients

In equations

$$y'' + ay' + by = \phi(x)$$

that arise in the study of physical phenomena, the forcing function is often a polynomial, a sine, a cosine, an exponential, or a simple combination thereof. Particular solutions of such equations can usually be found by what is formally called the method of *undetermined coefficients,* but is probably just as aptly called the method of "informed guessing."

Instead of trying to lay out formal rules of procedure, we take a practical approach and proceed directly to examples. In the first few examples we won't be looking for the general solution of the equation. We will be looking for any solution that we can get hold of, any function that can act as y_p.

Example 3 Suppose that we are faced with the equation

$$(*) \qquad y'' + 2y' + 5y = 10e^{-2x}.$$

From what we know about exponentials, it seems reasonable to guess a solution of the form $y = Ae^{-2x}$. Proceeding with our guess we have

$$y = Ae^{-2x}, \qquad y' = -2Ae^{-2x}, \qquad y'' = 4Ae^{-2x}.$$

Therefore $\qquad y'' + 2y' + 5y = 4Ae^{-2x} + 2(-2Ae^{-2x}) + 5Ae^{-2x} = 5Ae^{-2x}.$

Our exponential function satisfies $(*)$ provided

$$5Ae^{-2x} = 10e^{-2x}.$$

It follows from this equation that $5A = 10$ and so $A = 2$. The function $y_p = 2e^{-2x}$ is a particular solution. □

Example 4 This time we seek a solution to the equation

$$(*) \qquad y'' + 2y' + y = 10\cos 3x.$$

Following the lead of Example 3, we may be tempted to try a solution of the form $y = A\cos 3x$. For this function

$$y'' + 2y' + y = -8A\cos 3x - 6A\sin 3x.$$

Verify this. Let us see what happens with $y = B\sin 3x$. For this function

$$y'' + 2y' + y = 6B\cos 3x - 8B\sin 3x.$$

Verify this. Combining these two calculations, we find that the function

$$y = A\cos 3x + B\sin 3x$$

satisfies the equation

$$y'' + 2y' + y = (-8A + 6B)\cos 3x + (-6A - 8B)\sin 3x.$$

We can satisfy the equation $(*)$ by having

$$-8A + 6B = 10 \qquad \text{and} \qquad -6A - 8B = 0.$$

As you can check, these relations lead to $A = -\frac{4}{5}, B = \frac{3}{5}$. Therefore, the function

$$y_p = -\tfrac{4}{5}\cos 3x + \tfrac{3}{5}\sin 3x$$

is a particular solution of $(*)$. □

Example 5 For the equation

$$(*) \qquad y'' - 5y' + 6y = 4e^{2x}$$

you may be tempted to try to find a solution of the form $y = Ae^{2x}$. This won't work. The characteristic equation of the reduced equation reads $r^2 - 5r + 6 = 0$, which factors into $(r-2)(r-3) = 0$. This tells us that all functions of the form $y = Ae^{2x}$ are solutions of the reduced equation. For such functions the left side of $(*)$ is zero and cannot be $4e^{2x}$. What can we do? We need an e^{2x} and we don't want any sines, cosines, or any other exponentials around. We try a function of the form $y = Axe^{2x}$. Perhaps the left side of the equation, $y'' - 5y' + 6y$, will eliminate the x and leave us with a constant multiple of e^{2x}, which is what we want. Substituting y and its derivatives

$$y' = Ae^{2x} + 2Axe^{2x}, \qquad y'' = 4Ae^{2x} + 4Axe^{2x}$$

into equation $(*)$, we get

$$(4Ae^{2x} + 4Axe^{2x}) - 5(Ae^{2x} + 2Axe^{2x}) + 6(Axe^{2x}) = 4e^{2x},$$

which simplifies to

$$-Ae^{2x} = 4e^{2x}.$$

Thus, $y = Axe^{2x}$ satisfies $(*)$ provided that $A = -4$. The function

$$y_p = -4xe^{2x}$$

is a particular solution of equation $(*)$. Now go back and look at Example 2. □

In the next example we will use the superposition principle: the fact that if y_1 is a solution of

$$y'' + ay' + by = \phi_1(x)$$

and y_2 is a solution of

$$y'' + ay' + by = \phi_2(x),$$

then $y_1 + y_2$ is a solution of

$$y'' + ay' + by = \phi_1(x) + \phi_2(x).$$

Example 6 We consider the equation $y'' - 2y' + y = e^x + e^{-x}\sin x$. This time we want the general solution.

First we calculate the general solution of the reduced equation

$$y'' - 2y' + y = 0.$$

Since the characteristic equation $r^2 - 2r + 1 = 0$ has the factored form $(r-1)^2 = 0$, the general solution takes the form

$$y_g = C_1e^x + C_2xe^x.$$

Now we look for some solution of

$$(*) \qquad y'' - 2y' + y = e^x.$$

Functions of the form $y = Ae^x$ or $y = Axe^x$ won't work because they are solutions of the reduced equation. So we go one step further and try $y = Ax^2e^x$. As you can check, substituting this guess into (∗) leads to the conclusion that this y is a solution provided $A = \frac{1}{2}$. The function

$$y_1 = \tfrac{1}{2}x^2e^x$$

is a particular solution of (∗).

Next we look for a solution of

$$(**)\qquad y'' - 2y' + y = e^{-x}\sin x.$$

The result in Example 4 suggests that we try to find solution of the form

$$y = Ae^{-x}\sin x + Be^{-x}\cos x.$$

For this function, you can verify that

$$y'' - 2y' + y = (3A + 4B)e^{-x}\sin x + (-4A + 3B)e^{-x}\cos x.$$

We can satisfy equation (∗∗) by having

$$3A + 4B = 1 \qquad \text{and} \qquad -4A + 3B = 0.$$

These relations lead to $A = \frac{3}{25}$, $B = \frac{4}{25}$. The function

$$y_2 = \tfrac{3}{25}e^{-x}\sin x + \tfrac{4}{25}e^{-x}\cos x$$

is a particular solution of (∗∗).

Therefore, the general solution of the equation

$$y'' - 2y' + y = e^x + e^{-x}\sin x$$

can be written

$$y = y_g + y_1 + y_2 = C_1e^x + C_2xe^x + \tfrac{1}{2}x^2e^x + \tfrac{3}{25}e^{-x}\sin x + \tfrac{4}{25}e^{-x}\cos x. \quad \square$$

We have approached these problems as tests of ingenuity. There are detailed recipes that give suggested trial solutions for a multitude of forcing functions ϕ. We won't attempt to give them here.

EXERCISES 18.5

Find a particular solution.

1. $y'' + 5y' + 6y = 3x + 4$.

2. $y'' - 3y' - 10y = 5$.

3. $y'' + 2y' + 5y = x^2 - 1$.

4. $y'' + y' - 2y = x^3 + x$.

5. $y'' + 6y' + 9y = e^{3x}$.

6. $y'' + 6y' + 9y = e^{-3x}$.

7. $y'' + 2y' + 2y = e^x$.

8. $y'' + 4y' + 4y = x\,e^{-x}$.

9. $y'' - y' - 12y = \cos x$.

10. $y'' - y' - 12y = \sin x$.

11. $y'' + 7y' + 6y = 3\cos 2x$.

12. $y'' + y' + 3y = \sin 3x$.

13. $y'' - 2y' + 5y = e^{-x}\sin 2x$.

14. $y'' + 4y' + 5y = e^{2x}\cos x$.

15. $y'' + 6y' + 8y = 3\,e^{-2x}$.

16. $y'' - 2y' + 5y = e^x\sin x$.

Find the general solution.

17. $y'' + y = e^x$.

18. $y'' - 2y' + y = -25\sin 2x$.

19. $y'' - 3y' - 10y = -x - 1$.

20. $y'' + 4y = x\cos 2x$.

21. $y'' + 3y' - 4y = e^{-4x}$.

22. $y'' + 2y' = 4\sin 2x$.

23. $y'' + y' - 2y = 3x\,e^x$.

24. $y'' + 4y' + 4y = x\,e^{-2x}$.

25. Prove the superposition principle (18.5.5).

26. Use (18.5.5) to find a particular solution.

(a) $y'' + 2y' - 15y = x + e^{2x}$.

(b) $y'' - 7y' - 12y = e^{-x} + \sin 2x$.

27. Find the general solution of the equation

$$y'' - 4y' + 3y = \cosh x.$$

Use variation of parameters to find a particular solution.

28. $y'' + y = 3 \sin x \sin 2x$.

29. $y'' - 2y' + y = x e^x \cos x$.

30. $y'' + y = \csc x, \quad 0 < x < \pi$.

31. $y'' - 4y' + 4y = \frac{1}{3}x^{-1}e^{2x}, \quad x > 0$.

32. $y'' + 4y = \sec^2 2x$.

33. $y'' + 4y' + 4y = \dfrac{e^{-2x}}{x^2}$.

34. $y'' + 2y' + y = e^{-x} \ln x$.

35. $y'' - 2y' + 2y = e^x \sec x$.

36. Show that the change of variable $y = ve^{kx}$ transforms the equation

$$y'' + ay' + by = (c_n x^n + \cdots + c_1 x + c_0)e^{kx}$$

into

$$v'' + (2k + a)v' + (k^2 + ak + b)v = c_n x^n + \cdots + c_1 x + c_0.$$

37. In Exercise 30 of Section 18.4 we introduced a differential equation for the electrical current in a simple circuit. In the presence of an external electromotive force $F(t)$, the equation takes the form

$$L\frac{d^2 i}{dt^2} + R\frac{di}{dt} + \frac{1}{C}i = F(t).$$

Find the current i given that $F(t) = F_0, i(0) = 0, i'(0) = F_0/L$ and (a) $CR^2 = 4L$; (b) $CR^2 < 4L$.

38. (a) Show that $y_1 = x,\ y_2 = x \ln x$ are solutions of the Euler equation

$$x^2 y'' - xy' + y = 0$$

and that their Wronskian is nonzero on $(0, \infty)$.

(b) Use variation of parameters to find a particular solution of the equation

$$x^2 y'' - xy' + y = 4x \ln x.$$

39. (a) Show that $y_1 = \sin(\ln x^2)$ and $y_2 = \cos(\ln x^2)$ are solutions of the Euler equation

$$x^2 y'' + xy' + 4y = 0.$$

Verify that their Wronskian is nonzero on $(0, \infty)$.

(b) Use variation of parameters to find a particular solution of the equation.

$$x^2 y'' + xy' + 4y = \sin(\ln x).$$

■ 18.6 MECHANICAL VIBRATIONS

Simple Harmonic Motion

An object moves along a straight line. Instead of continuing in one direction, it moves back and forth, oscillating about a central point. Call the central point $x = 0$ and denote by $x(t)$ the displacement of the object at time t. If the acceleration is a constant negative multiple of the displacement,

$$a(t) = -kx(t), \quad k > 0,$$

then the object is said to be in *simple harmonic motion*.

Since, by definition,

$$a(t) = x''(t),$$

we have

$$x''(t) = -kx(t),$$

and therefore

$$x''(t) + kx(t) = 0.$$

To emphasize that k is positive, we set $k = \omega^2$ where $\omega = \sqrt{k} > 0$. The equation of motion then takes the form

(18.6.1)
$$\boxed{x''(t) + \omega^2 x(t) = 0.}$$

This is a second-order, linear differential equation with constant coefficients. The characteristic equation reads

$$r^2 + \omega^2 = 0,$$

and the roots are $\pm\omega i$. Therefore, the general solution of (18.6.1) has the form

$$x(t) = C_1 \cos \omega t + C_2 \sin \omega t.$$

A routine calculation shows that the general solution can be written

(Exercise 28, Section 18.4)

(18.6.2)
$$\boxed{x(t) = A \sin(\omega t + \phi_0)}$$

where A and ϕ_0 are constants with $A > 0$ and $\phi_0 \in [0, 2\pi)$.

Now let's analyze the motion measuring t in seconds. By adding $2\pi/\omega$ to t, we increase $\omega t + \phi_0$ by 2π:

$$\omega\left(t + \frac{2\pi}{\omega}\right) + \phi_0 = \omega t + \phi_0 + 2\pi.$$

Therefore the motion is *periodic* with *period*

$$T = \frac{2\pi}{\omega}.$$

A complete oscillation takes $2\pi/\omega$ seconds. The reciprocal of the period gives the number of complete oscillations per second. This is called the *frequency*:

$$f = \frac{\omega}{2\pi}.$$

The number ω is called the *angular frequency*. Since $\sin(\omega t + \phi_0)$ oscillates between -1 and 1,

$$x(t) = A \sin(\omega t + \phi_0)$$

oscillates between $-A$ and A. The number A is called the *amplitude* of the motion.

In Figure 18.6.1 we have plotted x against t. The oscillations along the x-axis are now waves in the tx-plane. The period of the motion, $2\pi/\omega$, is the t distance (the time separation) between consecutive wave crests. The amplitude of the motion, A, is the height of the waves measured in x units from $x = 0$. The number ϕ_0 is known as the *phase constant*, or *phase shift*. The phase constant determines the initial displacement (the height of the wave at time $t = 0$). If $\phi_0 = 0$, the object starts at the center of the interval of motion (the wave starts at the origin of the tx-plane).

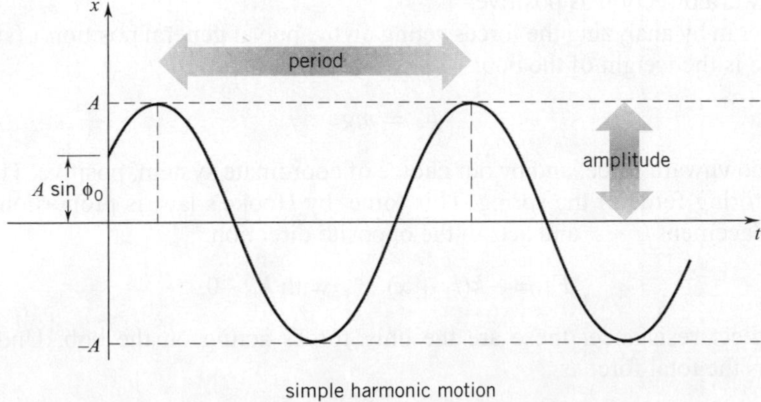

simple harmonic motion

Figure 18.6.1

Example 1 Find an equation for the oscillatory motion of an object, given that the period is $2\pi/3$ and, at time $t = 0, x = 1, x' = 3$.

SOLUTION We begin by setting $x(t) = A\sin(\omega t + \phi_0)$. In general the period is $2\pi/\omega$, so that here

$$\frac{2\pi}{\omega} = \frac{2\pi}{3} \quad \text{and thus} \quad \omega = 3.$$

The equation of motion takes the form

$$x(t) = A\sin(3t + \phi_0).$$

By differentiation

$$x'(t) = 3A\cos(3t + \phi_0).$$

The conditions at $t = 0$ give

$$1 = x(0) = A\sin\phi_0, \quad 3 = x'(0) = 3A\cos\phi_0$$

and therefore

$$1 = A\sin\phi_0, \quad 1 = A\cos\phi_0.$$

Adding the squares of these equations, we have

$$2 = A^2\sin^2\phi_0 + A^2\cos^2\phi_0 = A^2.$$

Since $A > 0, A = \sqrt{2}$.

To find ϕ_0 we note that

$$1 = \sqrt{2}\sin\phi_0 \quad \text{and} \quad 1 = \sqrt{2}\cos\phi_0.$$

These equations are satisfied by setting $\phi_0 = \frac{1}{4}\pi$. The equation of motion can be written

$$x(t) = \sqrt{2}\sin(3t + \tfrac{1}{4}\pi). \quad \square$$

Undamped Vibrations

A coil spring hangs naturally to a length l_0. When a bob of mass m is attached to it, the spring stretches l_1 inches. The bob is later pulled down an additional x_0 inches and then released. What is the resulting motion? Throughout we refer to Figure 18.6.2, taking the downward direction as positive.

We begin by analyzing the forces acting on the bob at general position x (stage IV). First there is the weight of the bob:

$$F_1 = mg.$$

This is a downward force, and by our choice of coordinate system, positive. Then there is the restoring force of the spring. This force, by Hooke's law, is proportional to the total displacement $l_1 + x$ and acts in the opposite direction:

$$F_2 = -k(l_1 + x) \qquad \text{with } k > 0.$$

If we neglect resistance, these are the only forces acting on the bob. Under these conditions the total force is

$$F = F_1 + F_2 = mg - k(l_1 + x),$$

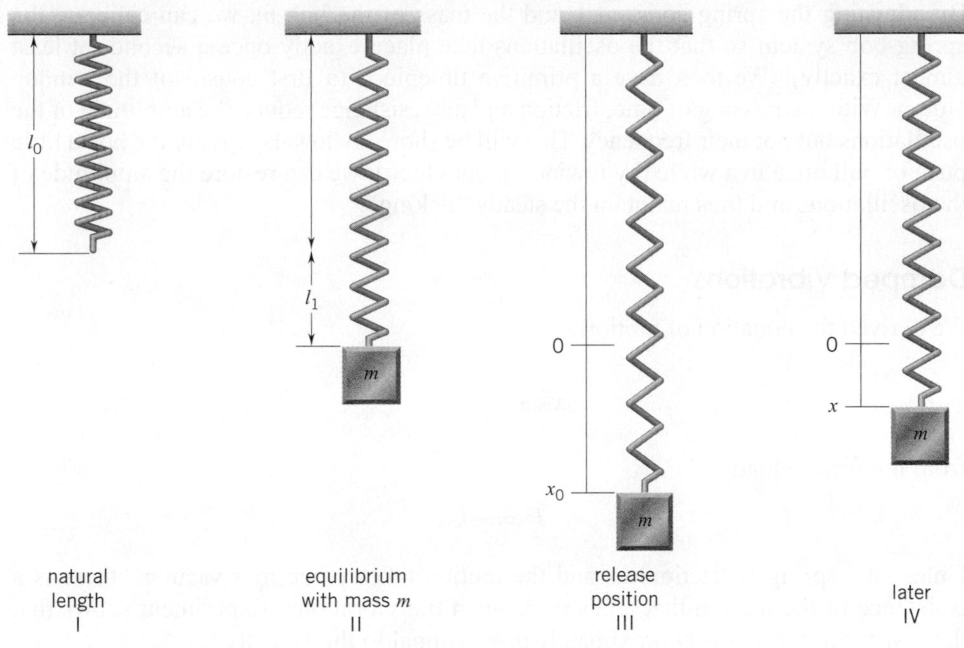

Figure 18.6.2

which we rewrite as

$$(1) \qquad F = (mg - kl_1) - kx.$$

At stage II (Figure 18.6.2) there was equilibrium. The force of gravity, mg, plus the force of the spring, $-kl_1$, must have been 0:

$$mg - kl_1 = 0.$$

Equation (1) can therefore be simplified to read

$$F = -kx.$$

Using Newton's second law,

$$F = ma, \qquad \text{(force = mass × accelration)}$$

we have $\qquad ma = -kx \quad \text{and} \quad \text{thus} \quad a = -\frac{k}{m}x.$

At each time t we have

$$x''(t) = -\frac{k}{m}x(t) \quad \text{and therefore} \quad x''(t) + \frac{k}{m}x(t) = 0.$$

Since $k/m > 0$, we can set $\omega = \sqrt{k/m}$ and write

$$x''(t) + \omega^2 x(t) = 0.$$

The motion of the bob is simple harmonic motion with period $T = 2\pi/\omega$. □

There is something remarkable about simple harmonic motion that we have not yet specifically pointed out; namely, that the frequency $f = \omega/2\pi$ is completely independent of the amplitude of the motion. The oscillations of the bob occur with frequency

$$f = \frac{\sqrt{k/m}}{2\pi}. \qquad (\text{here } \omega = \sqrt{k/m})$$

By adjusting the spring constant k and the mass of the bob m, we can calibrate the spring-bob system so that the oscillations take place exactly once a second (at least almost exactly). We then have a primitive timepiece (a first cousin of the windup clock). With the passing of time, friction and air resistance reduce the amplitude of the oscillations but not their frequency. This will be shown below. By giving the bob a little push or pull once in a while (by rewinding our clock), we can restore the amplitude of the oscillations and thus maintain the steady "ticking".

Damped Vibrations

We derived the equation of motion

$$x'' + \frac{k}{m}x = 0$$

from the force equation

$$F = -kx.$$

Unless the spring is frictionless and the motion takes place in a vacuum, there is a resistance to the motion that tends to dampen the vibrations. Experiment shows that the resistance force R is approximately proportional to the velocity x':

$$R = -cx'. \qquad (c > 0)$$

Taking this resistance term into account, the force equation reads

$$F = -kx - cx'.$$

Newton's law $F = ma = mx''$ then gives

$$mx'' = -cx' - kx,$$

which we can write as

(18.6.3)
$$\boxed{x'' + \frac{c}{m}x' + \frac{k}{m}x = 0.}$$

This is the equation of motion in the presence of a *damping factor.* To study the motion, we analyze this equation.

The characteristic equation

$$r^2 + \frac{c}{m}r + \frac{k}{m} = 0$$

has roots

$$r = \frac{-c \pm \sqrt{c^2 - 4km}}{2m}.$$

There are three possibilities:

$$c^2 - 4km < 0, \quad c^2 - 4km > 0, \quad c^2 - 4km = 0.$$

Case 1: $c^2 - 4km < 0$. In this case the characteristic equation has two complex roots:

$$r_1 = \frac{c}{2m} + i\omega, \quad r_2 = -\frac{c}{2m} - i\omega \quad \text{where} \quad \omega = \frac{\sqrt{4km - c^2}}{2m}.$$

The general solution of (18.6.3),

$$x = e^{-(c/2m)t}(C_1 \cos \omega t + C_2 \sin \omega t),$$

can be written

(18.6.4) $$\boxed{x(t) = A\, e^{(-c/2m)t} \sin\, (\omega t + \phi_0),}$$

where, as before, A and ϕ_0 are constants, $A > 0, \phi_0 \in [0, 2\pi)$. This is called the *underdamped case*. The motion is similar to simple harmonic motion, except that the damping term $e^{(-c/2m)t}$ ensures that $x \to 0$ as $t \to \infty$. The vibrations continue indefinitely with constant frequency $2\pi/\omega$ but with diminishing amplitude $A\, e^{(-c/2m)t}$. As $t \to \infty$, the amplitude of the vibrations tends to zero; the vibrations die down. The motion is illustrated in Figure 18.6.3. □

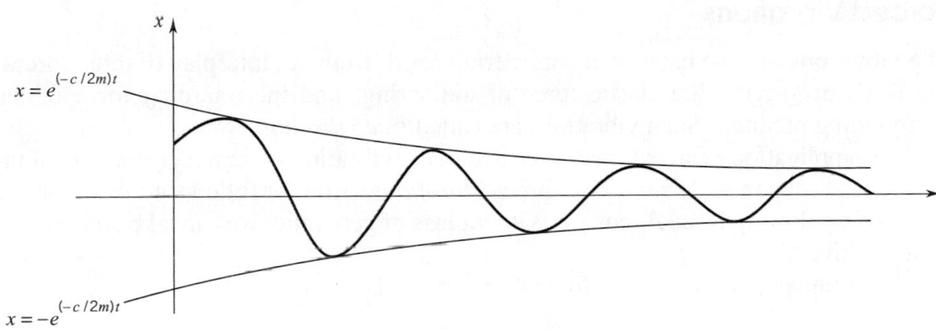

damped harmonic motion

Figure 18.6.3

Case 2: $c^2 - 4km > 0$. In this case the characteristic equation has two distinct real roots:

$$r_1 = \frac{-c + \sqrt{c^2 - 4km}}{2m}, \quad r_2 = \frac{-c - \sqrt{c^2 - 4km}}{2m}.$$

The general solution takes the form

(18.6.5) $$\boxed{x = C_1\, e^{r_1 t} + C_2\, e^{r_2 t}.}$$

This is called the *overdamped case*. The motion is nonoscillatory. Since $\sqrt{c^2 - 4km} < \sqrt{c^2} = c$, both r_1 and r_2 are negative. As $t \to \infty$, $x \to 0$. □

Case 3: $c^2 - 4km = 0$. In this case the characteristic equation has only one root

$$r_1 = -\frac{c}{2m},$$

and the general solution takes the form

(18.6.6) $$\boxed{x = C_1\, e^{-(c/2m)t} + C_2\, te^{-(c/2m)t}.}$$

This is called the *critically damped case*. Once again the motion is nonoscillatory. Moreover, as $t \to \infty$, $x \to 0$. □

In both the overdamped and critically damped cases, the mass moves slowly back to its equilibrium position ($x \to 0$ as $t \to \infty$). Depending on the initial conditions, the mass may move through the equilibrium once, but only once; there is no oscillatory motion. Two typical examples of the motion are shown in Figure 18.6.4.

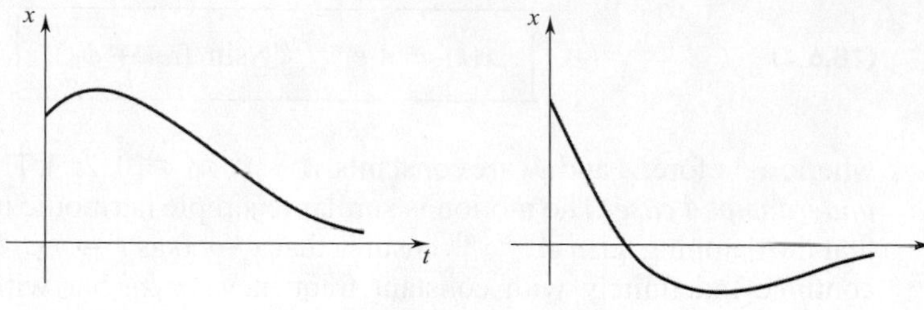

Figure 18.6.4

Forced Vibrations

The vibrations that we have been considering result from the interplay of three forces: the force of gravity, the elastic force of the spring, and the retarding force of the surrounding medium. Such vibrations are called *free vibrations.*

The application of an external force to a freely vibrating system modifies the vibrations and results in what are called *forced vibrations* . In what follows we examine the effect of a pulsating force $F_0 \cos \gamma t$. Without loss of generality we can take both F_0 and γ as positive.

In an undamped system the force equation reads

$$F = -kx + F_0 \cos \gamma t,$$

and the equation of motion takes the form

(18.6.7)
$$\boxed{x'' + \frac{k}{m} x = \frac{F_0}{m} \cos \gamma t.}$$

We set $\omega = \sqrt{k/m}$ and write

(18.6.8)
$$\boxed{x'' + \omega^2 x = \frac{F_0}{m} \cos \gamma t.}$$

As you'll see, the nature of the vibrations depends on the relation between the *applied frequency,* $\gamma/2\pi$, and the *natural frequency* of the system, $\omega/2\pi$.

Case 1: $\gamma \neq \omega$. In this case the method of undetermined coefficients gives the particular solution

$$x_p = \frac{F_0/m}{\omega^2 - \gamma^2} \cos \gamma t.$$

The general equation of motion can thus be written

(18.6.9)
$$\boxed{x = A \sin(\omega t + \phi_0) + \frac{F_0/m}{\omega^2 - \gamma^2} \cos \gamma t.}$$

If ω/γ is rational, the vibrations are periodic. If, on other hand, ω/γ is not rational, then the vibrations are not periodic and the motion, though bounded by

$$|A| + \left|\frac{F_0/m}{\omega^2 - \gamma^2}\right|,$$

can be highly irregular. These motions are illustrated in Figures 18.6.5 and 18.6.6. Each graph is the solution of the initial value problem:

$$x'' + 4x = 3\cos\gamma t,\ x(0) = 0,\ x'(0) = 1. \quad \square$$

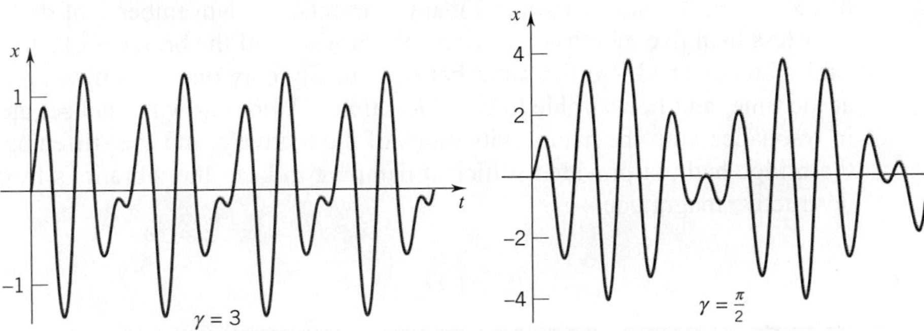

Figure 18.6.5 **Figure 18.6.6**

Case 2: $\gamma = \omega$. In this case the method of undetermined coefficients gives

$$x_p = \frac{F_0}{2\omega m}\, t\sin\omega t$$

and the general solution takes the form

(18.6.10)
$$x = A\sin(\omega t + \phi_0) + \frac{F_0}{2\omega m}\, t\sin\omega t.$$

The undamped system is said to be in *resonance.* The motion is oscillatory but, because of the extra t present in the second-term, it is far from periodic. As $t \to \infty$, the amplitude of vibration increases without bound. The motion is illustrated in Figure 18.6.7. The graph is the solution of the initial value problem

$$x'' + 4x = 3\cos 2t,\ x(0) = 0,\ x'(0) = 1. \quad \square$$

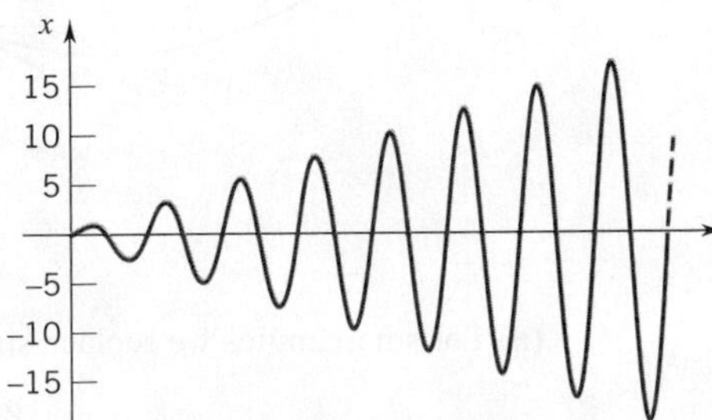

Figure 18.6.7

Undamped systems and unbounded vibrations are mathematical fictions. No real mechanical system is totally undamped, and unbounded vibrations do not occur in nature. Nevertheless, a form of resonance can occur in a real mechanical system. (See Exercises 24–28.) A periodic external force applied to a mechanical system that is insufficiently damped can set up vibrations of very large amplitude. Such vibrations have caused the destruction of some formidable man-made structures. In 1850 the suspension bridge at Angers, France, was destroyed by vibrations set up by the unified step of a column of marching soldiers. More than two hundred French soldiers were killed in that catastrophe. (Soldiers today are told to break ranks before crossing a bridge.) The collapse of the bridge at Tacoma, Washington, is a more recent event. Slender in construction and graceful in design, the Tacoma bridge was opened to traffic on July 1, 1940. The third longest suspension bridge in the world, with a main span of 2800 feet, the bridge attracted many admirers. On November 1 of that same year, after less than five months of service, the main span of the bridge broke loose from its cables and crashed into the water below. (Luckily only one person was on the bridge at the time, and he was able to crawl to safety.) A driving wind had set up vibrations in resonance with the natural vibrations of the roadway, and the stiffening girders of the bridge had not provided sufficient damping to keep the vibrations from reaching destructive magnitude.

EXERCISES 18.6

1. An object is in simple harmonic motion. Find an equation for the motion given that the period is $\frac{1}{4}\pi$ and, at time $t = 0, x = 1$ and $x' = 0$. What is the amplitude? What is the frequency?

2. An object is in simple harmonic motion. Find an equation for the motion given that the frequency is $1/\pi$ and, at time $t = 0, x = 0$ and $x' = -2$. What is the amplitude? What is the period?

3. An object is in simple harmonic motion with period T and amplitude A. What is the velocity at the central point $x = 0$?

4. An object is in simple harmonic motion with period T. Find the amplitude given that $x' = \pm v_0$ at $x = x_0$.

5. An object in simple harmonic motion passes through the central point $x = 0$ at time $t = 0$ and every 3 seconds thereafter. Find the equation of motion given that $x'(0) = 5$.

6. Show that simple harmonic motion $x(t) = A \sin(\omega t + \phi_0)$ can just as well be written: (a) $x(t) = A\cos(\omega t + \phi_1)$; (b) $x(t) = B \sin \omega t + C \cos \omega t$.

Exercises 7–12 relate to the motion of the bob depicted in Figure 18.6.2.

7. What is $x(t)$ for the bob of mass m?

8. Find the positions of the bob where it attains: (a) maximum speed; (b) zero speed; (c) maximum acceleration; (d) zero acceleration.

9. Where does the bob take on half of its maximum speed?

10. Find the maximal kinetic energy obtained by the bob. (Remember: $\text{KE} = \frac{1}{2}mv^2$, where m is the mass of the object and v is the speed.)

11. Find the time average of the kinetic energy of the bob during one period T.

12. Express the velocity of the bob in terms of k, m, x_0, and $x(t)$.

13. Given that $x''(t) = 8 - 4x(t)$ with $x(0) = 0$ and $x'(0) = 0$, show that the motion is simple harmonic motion centered at $x = 2$. Find the amplitude and the period.

14. The figure shows a pendulum of mass m swinging on an arm of length L. The angle θ is measured counterclockwise. Neglecting friction and the weight of the arm, we can describe the motion by the equation

$$mL\theta''(t) = -mg \sin \theta(t),$$

which reduces to

$$\theta''(t) = -\frac{g}{L} \sin \theta(t).$$

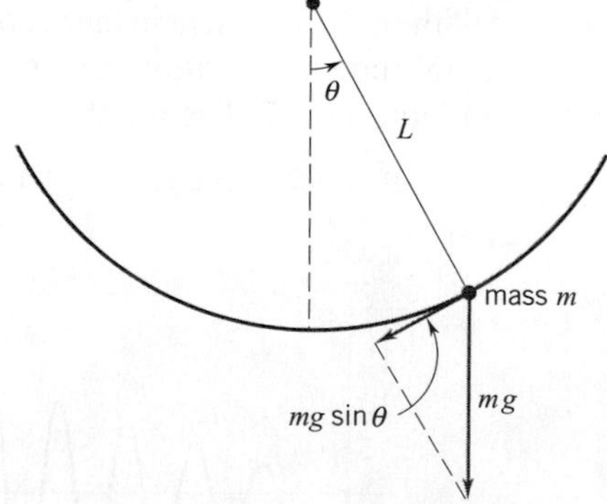

(a) For small angles we replace $\sin\theta$ by θ and write

$$\theta''(t) \cong -\frac{g}{L}\theta(t).$$

Justify this step.

(b) Solve the approximate equation in part (a) part given that the pendulum

(i) is held at an angle $\theta_0 > 0$ and released at time $t = 0$.
(ii) passes through the vertical position at time $t = 0$ and $\theta'(0) = -\sqrt{g/L}\,\theta_1$.

(c) Find L given that the motion repeats itself every 2 seconds.

15. A cylindrical buoy of mass m and radius r centimeters floats with its axis vertical in a liquid of density ρ kilograms per cubic centimeter. Suppose that the buoy is pushed x_0 centimeters down into the liquid. See the figure.

(a) Neglecting friction and given that the buoyancy force is equal to the weight of the liquid displaced, show that the buoy bobs up and down in simple harmonic motion by finding the equation of motion.

(b) Solve the equation obtained in part (a). Specify the amplitude and the period.

16. Explain in detail the connection between uniform circular motion and simple harmonic motion.

17. What is the effect of an increase in the resistance constant c on the amplitude and frequency of the vibrations given by Equation (18.6.4)?

18. Prove that the motion given by (18. 6. 5) can pass through the equilibrium point at most once. How many times can the motion change directions?

19. Prove that the motion given by (18. 6. 6) can pass through the equilibrium point at most once. How many times can the motion change directions?

20. Show that, if $\gamma \neq \omega$, then the method of undetermined coefficients applied to (18.6.8) gives

$$x_p = \frac{F_0/m}{\omega^2 - \gamma^2}\cos\gamma t.$$

21. Show that if ω/γ is rational, then the vibrations given by (18.6.9) are periodic.

22. Show that, if $\gamma = \omega$, then the method of undetermined coefficients applied to (18.6.8) gives

$$x_p = \frac{F_0}{2\omega m}t\sin\omega t.$$

Forced Vibrations in a Damped System

Write the equation

$$x'' + \frac{c}{m}x' + \frac{k}{m}x = \frac{F_0}{m}\cos\gamma t$$

as

$$(*) \qquad x'' + 2\alpha x' + \omega^2 x = \frac{F_0}{m}\cos\gamma t.$$

We will assume throughout that $0 < \alpha < \omega$. (For large α the resistance is large and the motion is not as interesting.)

23. Find the general solution of the reduced equation $x'' + 2\alpha x' + \omega^2 x = 0$.

24. Verify that the function

$$x_p = \frac{F_0/m}{(\omega^2-\gamma^2)^2 + 4\alpha^2\gamma^2}[(\omega^2-\gamma^2)\cos\gamma t + 2\alpha\gamma\sin\gamma t]$$

is a particular solution of $(*)$.

25. Determine x_p if $\omega = \gamma$. Show that the amplitude of the vibrations is very large if the resistance constant c is very small.

26. Show that the solution x_p in Exercise 24 can be written

$$x_p = \frac{F_0/m}{\sqrt{(\omega^2-\gamma^2)^2 + 4\alpha^2\gamma^2}}\sin(\gamma t + \phi).$$

27. Show that, if $2\alpha^2 \geq \omega^2$, then the amplitude of vibration of the solution x_p in Exercise 26 decreases as γ increases.

28. Suppose now that $2\alpha^2 \leq \omega^2$.

(a) Find the value of γ that maximizes the amplitude of the solution x_p in Exercise 26.

(b) Determine the frequency that corresponds to this value of γ. (This is called the *resonant frequency* of the damped system).

(c) What is the *resonant amplitude* of the system? (In other words, what is the amplitude of the vibrations if the applied force is at resonant frequency?)

(d) Show that, if c, the constant of resistance, is very small, then the resonant amplitude is very large.

■ CHAPTER HIGHLIGHTS

18.1 Introduction; Review of Equations Already Considered

ordinary differential equation (p. 1096)
order (p. 1097)
first-order linear differential equations (p. 1097)
separable first-order equations (p. 1097)
equations of harmonic motion (p. 1098)

18.2 Bernoulli Equation; Homogeneous Equations; Numerical Methods

Bernoulli equation: $y' + p(x)\,y = q(x)\,y^r$ (p. 1099)
method of solution — transformation into a linear equation (p. 1100)
homogeneous function (p. 1101)

homogeneous equation: $y' = f(x, y)$ where f is a homogeneous function to degree 0 (p. 1101)
method of solution — transformation into a separable equation (p. 1101)
Euler method (p. 1103)
Runge-Kutta method (p. 1105)

18.3 Exact Differential Equations; Integrating Factors

exact differential equation: $P(x, y) + Q(x, y)\, y' = 0$ (p. 1108)
integrating factor (p. 1109)
transformation into an exact differential equation (p. 1111, p. 1112)

18.4 The Equation $y'' + ay' + by = 0$

the reduced equation (homogeneous equation) (p. 1113)
the characteristic equation (p. 1113)
linear combinations of solutions (p. 1115)
existence and uniqueness of solutions (p. 1115)
Wronskians (p. 1116)
the general solution of $y'' + ay' + by = 0$ (p. 1118)

18.5 The Equation $y'' + ay' + by = \phi(x)$

the complete equation (nonhomogeneous equation) (p. 1123)
forcing function (p. 1123)
particular solution of $y'' + ay' + by = \phi(x)$ (p. 1123)
the general solution of $y'' + ay' + by = \phi(x)$ (p. 1123)
superposition principle (p. 1124)
variation of parameters (p. 1124)
undetermined coefficients (p. 1125)

18.6 Mechanical Vibrations

simple harmonic motion: $x'' + \omega^2 x = 0$ (p. 1130)
undamped vibrations (p. 1132)
damped vibrations: $x'' + \dfrac{c}{m}x' + \dfrac{k}{m}x = 0$ (p. 1132)
damping factor (p. 1134)
underdamped (p. 1135) overdamped (p. 1135)
critically damped (p. 1135)
forced vibrations: $x'' + \omega^2 x = \dfrac{F_0}{m}\cos\gamma t$ (p. 1136)
applied frequency (p. 1136) natural frequency (p. 1136)

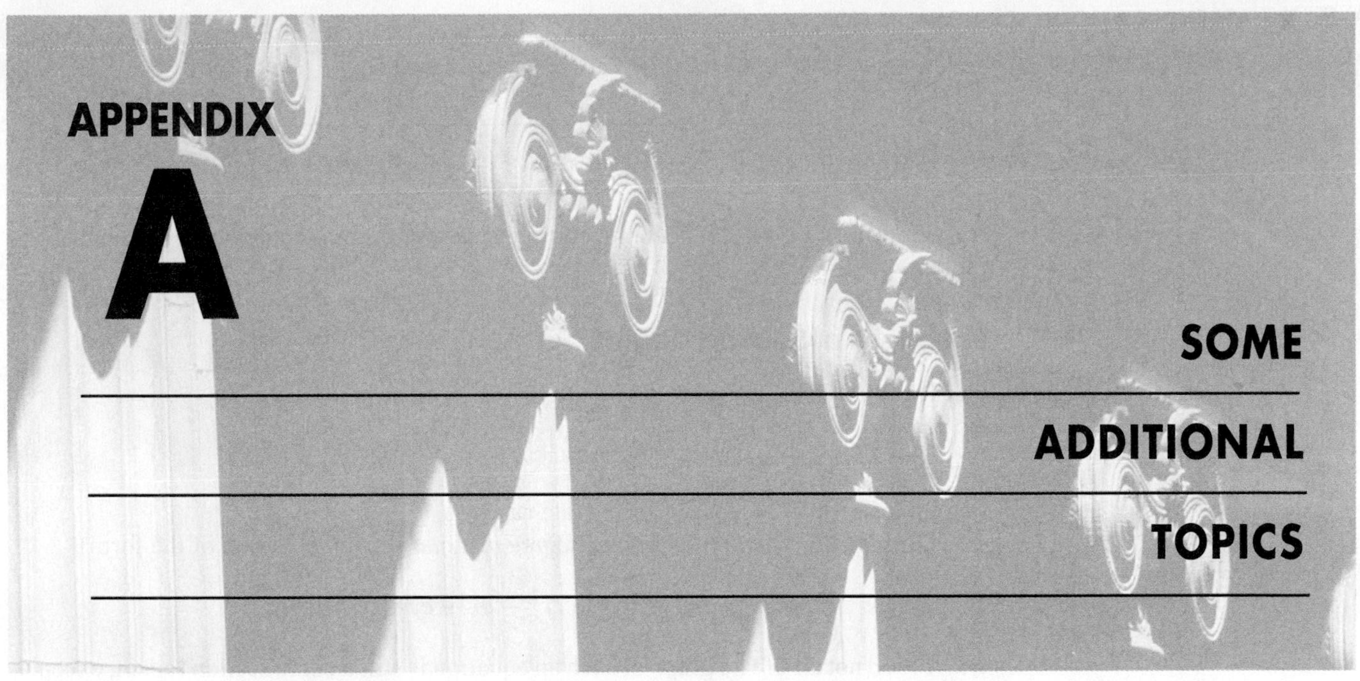

■ A.1 ROTATION OF AXES; EQUATIONS OF SECOND DEGREE

Rotation of Axes

We begin by referring to Figure A.1.1. From the figure,

$$\cos\theta = \frac{x}{r}, \quad \sin\theta = \frac{y}{r}.$$

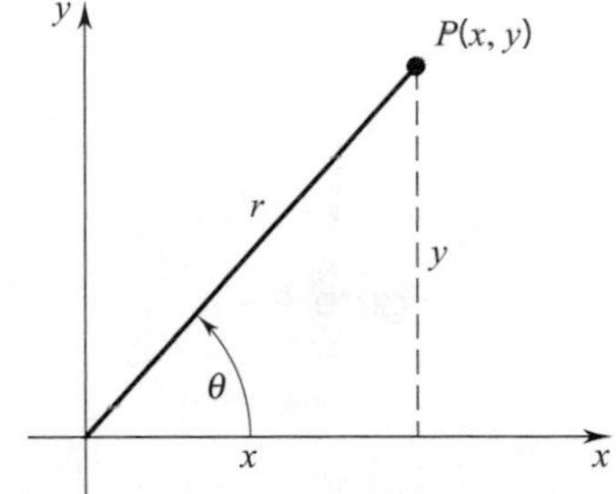

Figure A.1.1

Therefore

(A.1.1) $$x = r\cos\theta, \quad y = r\sin\theta.$$

These equations come up repeatedly in calculus. In particular, these are the equations that we use to convert polar coordinates to rectangular coordinates.

Consider now a rectangular coordinate system Oxy. By rotating this system about the origin counterclockwise through an angle of α radians, we obtain a new coordinate system OXY. See Figure A.1.2.

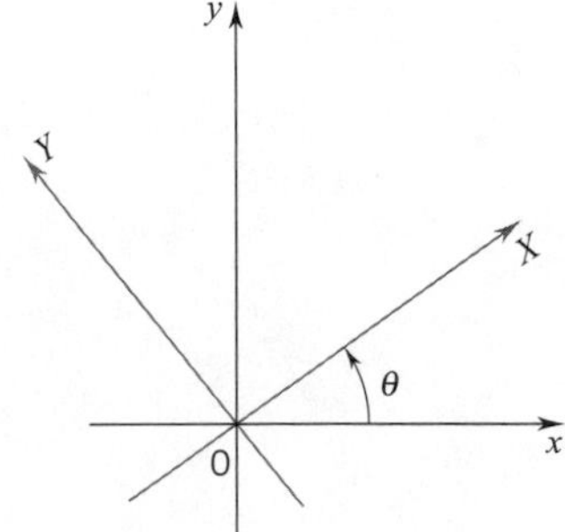

Figure A.1.2

A point P now has two pairs of rectangular coordinates:

(x, y) in the Oxy system and (X, Y) in the OXY system.

Here we investigate the relation between (x, y) and (X, Y). With P as in Figure A.1.3,

$$x = r\cos(\alpha+\beta), \quad y = r\sin(\alpha+\beta)$$

and

$$X = r\cos\beta, \quad Y = r\sin\beta.$$

Since

$$\cos(\alpha+\beta) = \cos\alpha\cos\beta - \sin\alpha\sin\beta,$$
$$\sin(\alpha+\beta) = \sin\alpha\cos\beta + \cos\alpha\sin\beta,$$

we have

$$x = r\cos(\alpha+\beta) = (\cos\alpha)\, r\cos\beta - (\sin\alpha)\, r\sin\beta,$$
$$y = r\sin(\alpha+\beta) = (\sin\alpha)\, r\cos\beta + (\cos\alpha)\, r\sin\beta,$$

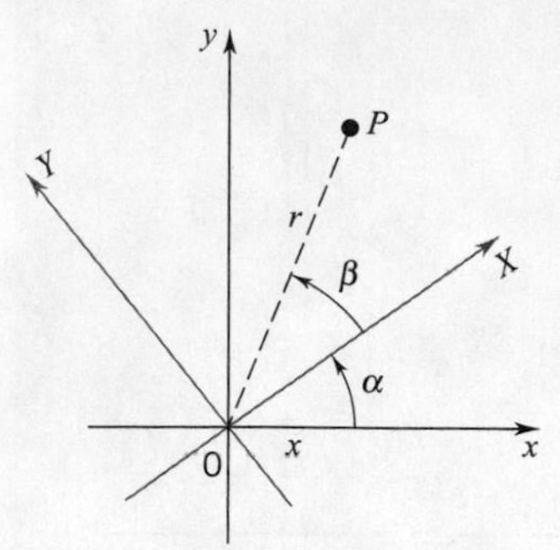

Figure A.1.3

and therefore

(A.1.2) $$x = (\cos\alpha)X - (\sin\alpha)Y, \quad y = (\sin\alpha)X + (\cos\alpha)Y.$$

These formulas give the algebraic consequences of a counterclockwise rotation of α radians.

Equations of Second Degree

As you know, the graph of an equation of the form

$$ax^2 + cy^2 + dx + ey + f = 0 \qquad (a, c \text{ not both } 0)$$

is a conic section (except for degenerate cases).

The *general equation of second degree in x and y* is an equation of the form

(A.1.3) $$ax^2 + bxy + cy^2 + dx + ey + f = 0$$

with a, b, c not all 0. The graph of such an equation is still a conic section (again, except for degenerate cases). For example, the graph of

$$xy - 2 = 0$$

is the hyperbola shown in Figure A.1.4.

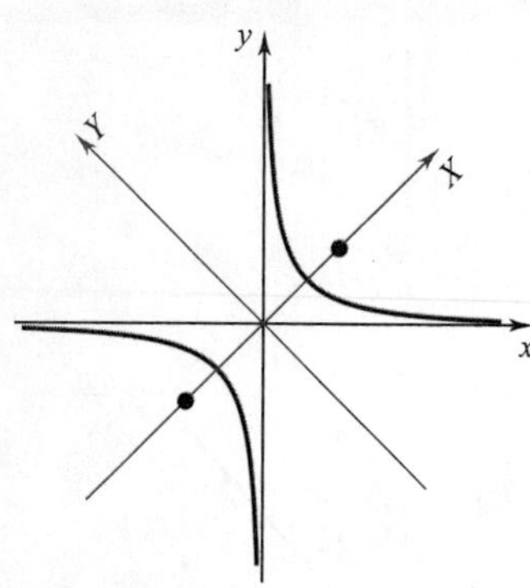

Figure A.1.4

Eliminating the *xy*-Term

Rotations of the coordinate system enable us to simplify equations of the second degree by eliminating the xy-term; that is, if in the Oxy coordinate system, a curve S has an equation of the form

(1) $$ax^2 + bxy + cy^2 + dx + ey + f = 0 \qquad \text{with} \qquad b \neq 0,$$

then there exists a coordinate system OXY, differing from Oxy by a rotation α, with $0 < \alpha < \pi/2$, such that in the OXY system S has an equation of the form

(2) $$AX^2 + CY^2 + DX + EY + F = 0,$$

with A and C not both zero. To see this, substitute

$$x = (\cos\alpha)X - (\sin\alpha)Y, \quad y = (\sin\alpha)X + (\cos\alpha)Y$$

in equation (1). This will give you a second-degree equation in X and Y in which the coefficient of XY is

$$-2a\cos\alpha\sin\alpha + b(\cos^2\alpha - \sin^2\alpha) + 2c\cos\alpha\sin\alpha.$$

This can be simplified to

$$(c - a)\sin 2\alpha + b\cos 2\alpha.$$

To eliminate the XY term, we must have this coefficient equal to zero, that is, we must have

$$b\cos 2\alpha = (a - c)\sin 2\alpha,$$

which, with $b \neq 0$, gives

$$\cot 2\alpha = \frac{a-c}{b}.$$

With $0 < \alpha < \frac{\pi}{2}$, we have $0 < 2\alpha < \pi$. Therefore

$$2\alpha = \cot^{-1}\left(\frac{a-c}{b}\right)$$

and

$$\alpha = \frac{1}{2}\cot^{-1}\left(\frac{a-c}{b}\right).$$

We have shown that an equation of the form (1) can be transformed into an equation of the form (2) by rotating the axes through the angle α given by

(A.1.4)

$$\boxed{\alpha = \frac{1}{2}\cot^{-1}\left(\frac{a-c}{b}\right)}$$

We leave it as an exercise to show that the coefficients A and C in (2) are not both zero.

Example 1 In the case of $xy - 2 = 0$, we have

$a = c = 0, b = 1$, and $\alpha = \frac{1}{2}\cot^{-1}(0) = \frac{1}{2}(\frac{\pi}{2}) = \frac{1}{4}\pi$.

Setting

$$x = (\cos\tfrac{1}{4}\pi)X - (\sin\tfrac{1}{4}\pi)Y = \tfrac{1}{2}\sqrt{2}(X - Y),$$

$$y = (\sin\tfrac{1}{4}\pi)X + (\cos\tfrac{1}{4}\pi)Y = \tfrac{1}{2}\sqrt{2}(X + Y),$$

we find that $xy - 2 = 0$ becomes

$$\tfrac{1}{2}(X^2 - Y^2) - 2 = 0,$$

which can be written

$$\frac{X^2}{4} - \frac{Y^2}{4} = 1.$$

This is the equation of a hyperbola in standard position in the OXY system. The hyperbola is shown in Figure A.1.4. □

Example 2 In the case of $11x^2 + 4\sqrt{3}xy + 7y^2 - 1 = 0$, we have

$a = 11, b = 4\sqrt{3}$, and $c = 7$.

Thus we choose

$$\alpha = \tfrac{1}{2}\cot^{-1}\left(\frac{11-7}{4\sqrt{3}}\right) = \tfrac{1}{2}\cot^{-1}\left(\frac{1}{\sqrt{3}}\right) = \tfrac{1}{6}\pi.$$

Setting

$$x = (\cos\tfrac{1}{6}\pi)X - (\sin\tfrac{1}{6}\pi)Y = \tfrac{1}{2}(\sqrt{3}X - Y),$$

$$y = (\sin\tfrac{1}{6}\pi)X + (\cos\tfrac{1}{6}\pi)Y = \tfrac{1}{2}(X + \sqrt{3}Y),$$

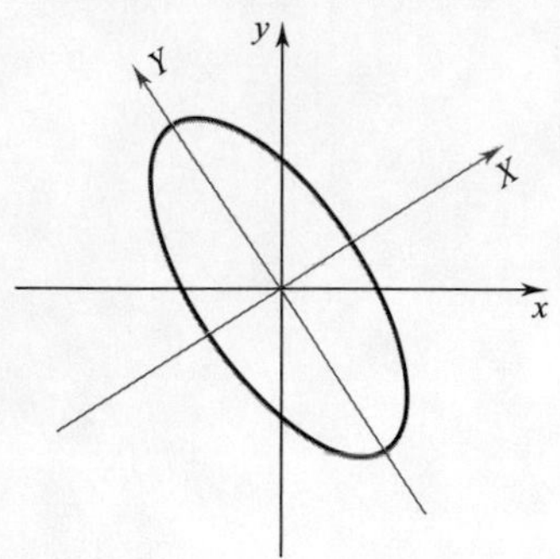

Figure A.1.5

we find that our initial equation simplifies to $13X^2 + 5Y^2 - 1 = 0$, which we can write as

$$\frac{X^2}{(1/\sqrt{13})^2} + \frac{Y^2}{(1/\sqrt{5})^2} = 1.$$

This is the equation of an ellipse. The ellipse is pictured in Figure A.1.5. □

The Discriminant

It is possible to draw general conclusions about the graph of a second-degree equation

$$ax^2 + bxy + cy^2 + dx + ey + f = 0, \quad a, b, c \text{ not all } 0,$$

just from the *discriminant* $\Delta = b^2 - 4ac$. There are three cases:

Case 1. If $\Delta < 0$, the graph is an ellipse, a circle, a point, or empty.

Case 2. If $\Delta > 0$, the graph is a hyperbola or a pair of intersecting lines.

Case 3. If $\Delta = 0$, the graph is a parabola, a line, a pair of lines, or empty.

Below we outline how these assertions can be verified. A useful first step is to rotate the coordinate system so that the equation takes the form

$$(3) \qquad AX^2 + CY^2 + DX + EY + F = 0.$$

An elementary but time-consuming computation shows that the discriminant is unchanged by a rotation, so that in this instance we have

$$\Delta = b^2 - 4ac = -4AC.$$

Moreover, A and C cannot both be zero. If $\Delta < 0$, then $AC > 0$, and we can rewrite (3) as

$$\frac{X^2}{C} + \frac{D}{AC}X + \frac{Y^2}{A} + \frac{E}{AC}Y + \frac{F}{AC} = 0.$$

By completing the squares, we obtain an equation of the form

$$\frac{(X-\alpha)^2}{(\sqrt{|C|})^2} + \frac{(Y-\beta)^2}{(\sqrt{|A|})^2} = K.$$

If $K > 0$, we have an ellipse or a circle. If $K = 0$, we have the point (α, β). If $K < 0$, the set is empty.

If $\Delta > 0$, then $AC < 0$. Proceeding as before, we obtain an equation of the form

$$\frac{(X-\alpha)^2}{(\sqrt{|C|})^2} - \frac{(Y-\beta)^2}{(\sqrt{|A|})^2} = K.$$

If $K \neq 0$, we have a hyperbola. If $K = 0$, the equation becomes

$$\left(\frac{X-\alpha}{\sqrt{|C|}} - \frac{Y-\beta}{\sqrt{|A|}}\right)\left(\frac{X-\alpha}{\sqrt{|C|}} + \frac{Y-\beta}{\sqrt{|A|}}\right) = 0,$$

so that we have a pair of lines intersecting at the point (α, β).

If $\Delta = 0$, then $AC = 0$, so that either $A = 0$ or $C = 0$. Since A and C are not both zero, there is no loss in generality in assuming that $A \neq 0$ and $C = 0$. In this case equation (3) reduces to

$$AX^2 + DX + EY + F = 0.$$

Dividing by A and completing the square, we have an equation of the form

$$(X - \alpha)^2 = \beta Y + K.$$

If $\beta \neq 0$, we have a parabola. If $\beta = 0$ and $K = 0$, we have a line. If $\beta = 0$ and $K > 0$, we have a pair of parallel lines. If $\beta = 0$ and $K < 0$, the set is empty.

EXERCISES A.1

In Exercises 1–8, (a) use the discriminant to identify the curve; (b) find a rotation $\alpha \in (0, \pi/2)$ that eliminates the xy-term; (c) rewrite the equation in terms of the new coordinate system; (d) sketch the curve displaying both coordinate systems.

1. $xy = 1$.

2. $xy - y + x = 1$.

3. $11x^2 + 10\sqrt{3}xy + y^2 - 4 = 0$.

4. $52x^2 - 72xy + 73y^2 - 100 = 0$.

5. $x^2 - 2xy + y^2 + x + y = 0$.

6. $3x^2 + 2\sqrt{3}xy + y^2 - 2x + 2\sqrt{3}y = 0$.

7. $x^2 + 2\sqrt{3}xy + 3y^2 + 2\sqrt{3}x - 2y = 0$.

8. $2x^2 + 4\sqrt{3}xy + 6y^2 + (8 - \sqrt{3})x + (8\sqrt{3} + 1)y + 8 = 0$.

In Exercises 9 and 10, find a rotation $\alpha \in (0, \pi/2)$ that eliminates the xy-term. Then find $\cos\alpha$ and $\sin\alpha$.

9. $x^2 + xy + Kx + Ly + M = 0$.

10. $5x^2 + 24xy + 12y^2 + Kx + Ly + M = 0$.

11. Show that after a rotation of axes through an angle α, the coefficients in the equation

$$AX^2 + BXY + CY^2 + DX + EY + F = 0$$

are related to the coefficients in the equation

$$ax^2 + bxy + cy^2 + dx + ey + f = 0, \quad a, b, c \text{ not all } 0,$$

as follows:

$$\begin{aligned} A &= a\cos^2\alpha + b\cos\alpha\,\sin\alpha + c\sin^2\alpha, \\ B &= 2(c - a)\cos\alpha\,\sin\alpha + b(\cos^2\alpha - \sin^2\alpha), \\ C &= a\sin^2\alpha - b\cos\alpha\,\sin\alpha + c\cos^2\alpha, \\ D &= d\cos\alpha + e\sin\alpha, \\ E &= e\cos\alpha - d\sin\alpha, \\ F &= f. \end{aligned}$$

12. Use the results of Exercise 11 to show the following:

(a) $B^2 - 4AC = b^2 - 4ac$.

(b) If $B = 0$, then A and C cannot both be 0.

■ A.2 DETERMINANTS

By a *matrix* we mean a rectangular arrangement of numbers enclosed in parentheses. For example,

$$\begin{pmatrix} 2 & 4 \\ 3 & 1 \end{pmatrix} \quad \begin{pmatrix} 1 & 6 & 3 \\ 5 & 2 & 2 \end{pmatrix} \quad \begin{pmatrix} 2 & 4 & 0 \\ 4 & 7 & 1 \\ 0 & 1 & 1 \end{pmatrix}$$

are all matrices. The numbers that appear in a matrix are called the *entries*.

Each matrix has a certain number of rows and a certain number of columns. A matrix with m rows and n columns is called an $m \times n$ *matrix*. Thus the first matrix above is a 2×2 matrix, the second a 2×3 matrix, the third a 3×3 matrix. The first and third matrices are called *square*; they have the same number of rows as columns. Here we will be working with square matrices as these are the only ones that have determinants.

We could give a definition of determinant that is applicable to all square matrices, but the definition is complicated and would serve little purpose at this point. Our interest here is in the 2×2 case and in the 3×3 case. We begin with the 2×2 case.

(A.2.1)

The *determinant* of the matrix

$$\begin{pmatrix} a_1 & a_2 \\ b_1 & b_2 \end{pmatrix}$$

is the number $a_1b_2 - a_2b_1$.

We have a special notation for the determinant. We change the parentheses of the matrix to vertical bars:

$$\text{Determinant of} \begin{pmatrix} a_1 & a_2 \\ b_1 & b_2 \end{pmatrix} = \begin{vmatrix} a_1 & a_2 \\ b_1 & b_2 \end{vmatrix} = a_1b_2 - a_2b_1.$$

Thus, for example,

$$\begin{vmatrix} 5 & 8 \\ 4 & 2 \end{vmatrix} = (5 \cdot 2) - (8 \cdot 4) = 10 - 32 = -22$$

$$\text{and} \quad \begin{vmatrix} 4 & 0 \\ 0 & \frac{1}{4} \end{vmatrix} = (4 \cdot \tfrac{1}{4}) - (0 \cdot 0) = 1.$$

We remark on three properties of 2×2 determinants:

1. If the rows or columns of a 2×2 determinant are interchanged, the determinant changes sign:

$$\begin{vmatrix} b_1 & b_2 \\ a_1 & a_2 \end{vmatrix} = -\begin{vmatrix} a_1 & a_2 \\ b_1 & b_2 \end{vmatrix}, \qquad \begin{vmatrix} a_2 & a_1 \\ b_2 & b_1 \end{vmatrix} = -\begin{vmatrix} a_1 & a_2 \\ b_1 & b_2 \end{vmatrix}.$$

PROOF Just note that

$$b_1a_2 - b_2a_1 = -(a_1b_2 - a_2b_1) \quad \text{and} \quad a_2b_1 - a_1b_2 = -(a_1b_2 - a_2b_1). \quad \square$$

2. A common factor can be removed from any row or column and placed as a factor in front of the determinant:

$$\begin{vmatrix} \lambda a_1 & \lambda a_2 \\ b_1 & b_2 \end{vmatrix} = \lambda \begin{vmatrix} a_1 & a_2 \\ b_1 & b_2 \end{vmatrix}, \qquad \begin{vmatrix} \lambda a_1 & a_2 \\ \lambda b_1 & b_2 \end{vmatrix} = \lambda \begin{vmatrix} a_1 & a_2 \\ b_1 & b_2 \end{vmatrix}.$$

PROOF Just note that

$$(\lambda a_1)b_2 - (\lambda a_2)b_1 = \lambda(a_1b_2 - a_2b_1)$$

and

$$(\lambda a_1)b_2 - a_2(\lambda b_1) = \lambda(a_1b_2 - a_2b_1). \quad \square$$

3. If the rows or columns of a 2×2 determinant are the same, the determinant is 0.

PROOF

$$\begin{vmatrix} a_1 & a_2 \\ a_1 & a_2 \end{vmatrix} = a_1a_2 - a_2a_1 = 0, \qquad \begin{vmatrix} a_1 & a_1 \\ b_1 & b_1 \end{vmatrix} = a_1b_1 - a_1b_1 = 0. \quad \square$$

The determinant of a 3×3 matrix is harder to define. One definition is this:

$$\begin{vmatrix} a_1 & a_2 & a_3 \\ b_1 & b_2 & b_3 \\ c_1 & c_2 & c_3 \end{vmatrix} = a_1b_2c_3 - a_1b_3c_2 + a_2b_3c_1 - a_2b_1c_3 + a_3b_1c_2 - a_3b_2c_1.$$

The problem with this definition is that it is hard to remember. What saves us is that the expansion on the right can be conveniently written in terms of 2×2 determinants; namely, the expression on the right can be written

$$a_1(b_2c_3 - b_3c_2) - a_2(b_1c_3 - b_3c_1) + a_3(b_1c_2 - b_2c_1),$$

which turns into

$$a_1\begin{vmatrix} b_2 & b_3 \\ c_2 & c_3 \end{vmatrix} - a_2\begin{vmatrix} b_1 & b_3 \\ c_1 & c_3 \end{vmatrix} + a_3\begin{vmatrix} b_1 & b_2 \\ c_1 & c_2 \end{vmatrix}.$$

We then have

(A.2.2)
$$\begin{vmatrix} a_1 & a_2 & a_3 \\ b_1 & b_2 & b_3 \\ c_1 & c_2 & c_3 \end{vmatrix} = a_1\begin{vmatrix} b_2 & b_3 \\ c_2 & c_3 \end{vmatrix} - a_2\begin{vmatrix} b_1 & b_3 \\ c_1 & c_3 \end{vmatrix} + a_3\begin{vmatrix} b_1 & b_2 \\ c_1 & c_2 \end{vmatrix}.$$

We will take this as our definition. It is called the *expansion of the determinant by elements of the first row.* Note that the coefficients are the entries a_1, a_2, a_3 of the first row, that they occur alternately with + and −signs, and that each is multiplied by a determinant. You can remember which determinant goes with which entry a_i as follows: in the original matrix, mentally cross out the row and column in which the entry a_i is found, and take the determinant of the remaining 2×2 matrix. For example, the determinant that goes with a_3 is

$$\begin{vmatrix} \not{a_1} & \not{a_2} & \not{a_3} \\ b_1 & b_2 & \not{b_3} \\ c_1 & c_2 & \not{c_3} \end{vmatrix} = \begin{vmatrix} b_1 & b_2 \\ c_1 & c_2 \end{vmatrix}.$$

When first starting to work with specific 3×3 determinants, it is a good idea to set up the formula with blank 2×2 determinants:

$$\begin{vmatrix} a_1 & a_2 & a_3 \\ b_1 & b_2 & b_3 \\ c_1 & c_2 & c_3 \end{vmatrix} = a_1\begin{vmatrix} & \\ & \end{vmatrix} - a_2\begin{vmatrix} & \\ & \end{vmatrix} + a_3\begin{vmatrix} & \\ & \end{vmatrix}$$

and then fill in the 2×2 determinants by using the "crossing out" rule explained above.

Example 1

$$\begin{aligned} \begin{vmatrix} 1 & 2 & 1 \\ 0 & 3 & 4 \\ 6 & 2 & 5 \end{vmatrix} &= 1\begin{vmatrix} 3 & 4 \\ 2 & 5 \end{vmatrix} - 2\begin{vmatrix} 0 & 4 \\ 6 & 5 \end{vmatrix} + 1\begin{vmatrix} 0 & 3 \\ 6 & 2 \end{vmatrix} \\ &= 1(15 - 8) - 2(0 - 24) + 1(0 - 18) \\ &= 7 + 48 - 18 = 37. \quad \square \end{aligned}$$

A straightforward (but somewhat laborious) calculation shows that 3×3 determinants have the three properties we proved earlier for 2×2 determinants.

1. If two rows or columns are interchanged, the determinant changes sign.
2. A common factor can be removed from any row or column and placed as a factor in front of the determinant.
3. If two rows or columns are the same, the determinant is 0.

EXERCISES A.2

Evaluate the following determinants.

1. $\begin{vmatrix} 1 & 2 \\ 3 & 4 \end{vmatrix}$.

2. $\begin{vmatrix} 1 & -1 \\ -1 & 1 \end{vmatrix}$.

3. $\begin{vmatrix} 1 & 1 \\ a & a \end{vmatrix}$.

4. $\begin{vmatrix} a & b \\ b & d \end{vmatrix}$.

5. $\begin{vmatrix} 1 & 0 & 3 \\ 2 & 4 & 1 \\ 0 & 1 & 0 \end{vmatrix}$.

6. $\begin{vmatrix} 1 & 0 & 0 \\ 0 & 2 & 0 \\ 0 & 0 & 3 \end{vmatrix}$.

7. $\begin{vmatrix} 0 & 0 & 1 \\ 0 & 2 & 0 \\ 3 & 0 & 0 \end{vmatrix}$.

8. $\begin{vmatrix} a & 0 & 0 \\ b & c & 0 \\ d & e & f \end{vmatrix}$.

9. If A is a matrix, its *transpose* A^T is obtained by interchanging the rows and columns. Thus

$$\begin{pmatrix} a_1 & a_2 \\ b_1 & b_2 \end{pmatrix}^T = \begin{pmatrix} a_1 & b_1 \\ a_2 & b_2 \end{pmatrix}$$

and

$$\begin{pmatrix} a_1 & a_2 & a_3 \\ b_1 & b_2 & b_3 \\ c_1 & c_2 & c_3 \end{pmatrix}^T = \begin{pmatrix} a_1 & b_1 & c_1 \\ a_2 & b_2 & c_2 \\ a_3 & b_3 & c_3 \end{pmatrix}.$$

Show that the determinant of a matrix equals the determinant of its transpose: (a) for the 2×2 case; (b) for the 3×3 case.

Justify the assertions made in Exercises 10–14 by invoking the relevant properties of determinants.

10. $\begin{vmatrix} 1 & 2 & 3 \\ 4 & 5 & 6 \\ 7 & 8 & 9 \end{vmatrix} + \begin{vmatrix} 4 & 5 & 6 \\ 1 & 2 & 3 \\ 7 & 8 & 9 \end{vmatrix} = 0.$

11. $\begin{vmatrix} 1 & 2 & 3 \\ 4 & 5 & 6 \\ 7 & 8 & 9 \end{vmatrix} = \begin{vmatrix} 4 & 5 & 6 \\ 7 & 8 & 9 \\ 1 & 2 & 3 \end{vmatrix}.$

12. $\begin{vmatrix} 1 & 2 & 3 \\ 4 & 5 & 6 \\ 7 & 8 & 9 \end{vmatrix} + \begin{vmatrix} 1 & 2 & 3 \\ 1 & 2 & 3 \\ 7 & 8 & 9 \end{vmatrix} = \begin{vmatrix} 1 & 2 & 3 \\ 4 & 5 & 6 \\ 7 & 8 & 9 \end{vmatrix}.$

13. $\frac{1}{2}\begin{vmatrix} 1 & 0 & 7 \\ 3 & 4 & 5 \\ 2 & 4 & 6 \end{vmatrix} = \begin{vmatrix} 1 & 0 & 7 \\ 3 & 4 & 5 \\ 1 & 2 & 3 \end{vmatrix}.$

14. $\begin{vmatrix} 1 & 2 & 3 \\ x & 2x & 3x \\ 4 & 5 & 6 \end{vmatrix} = 0.$

15. (a) Verify that the equations

$$\begin{aligned} 3x + 4y &= 6 \\ 2x - 3y &= 7 \end{aligned}$$

can be solved by the prescription

$$x = \frac{\begin{vmatrix} 6 & 4 \\ 7 & -3 \end{vmatrix}}{\begin{vmatrix} 3 & 4 \\ 2 & -3 \end{vmatrix}}, \quad y = \frac{\begin{vmatrix} 3 & 6 \\ 2 & 7 \end{vmatrix}}{\begin{vmatrix} 3 & 4 \\ 2 & -3 \end{vmatrix}}.$$

(b) More generally, verify that the equations

$$\begin{aligned} a_1x + a_2y &= d \\ b_1x + b_2y &= e \end{aligned}$$

can be solved by the prescription

$$x = \frac{\begin{vmatrix} d & a_2 \\ e & b_2 \end{vmatrix}}{\begin{vmatrix} a_1 & a_2 \\ b_1 & b_2 \end{vmatrix}}, \quad y = \frac{\begin{vmatrix} a_1 & d \\ b_1 & e \end{vmatrix}}{\begin{vmatrix} a_1 & a_2 \\ b_1 & b_2 \end{vmatrix}}$$

provided that the determinant in the denominator is different from 0.

(c) Devise an analogous rule for solving three linear equations in three unknowns.

16. Show that a 3×3 determinant can be " expanded by the elements of the bottom row" as follows:

$$\begin{vmatrix} a_1 & a_2 & a_3 \\ b_1 & b_2 & b_3 \\ c_1 & c_2 & c_3 \end{vmatrix} = c_1\begin{vmatrix} a_2 & a_3 \\ b_2 & b_3 \end{vmatrix} - c_2\begin{vmatrix} a_1 & a_3 \\ b_1 & b_3 \end{vmatrix} + c_3\begin{vmatrix} a_1 & a_2 \\ b_1 & b_2 \end{vmatrix}.$$

HINT: You can check this directly by writing out the values of the determinants on the right, or you can interchange rows twice to bring the bottom row to the top and then expand by the elements of the top row.

APPENDIX
B
SOME
ADDITIONAL
PROOFS

In this appendix we present some proofs that many would consider too advanced for the main body of the text. Some details are omitted. These are left to you.

The arguments presented in Sections B.1, B.2, and B.4 require some familiarity with the *least upper bound axiom.* This is discussed in Section 10.1. In addition, Section B.4 requires some understanding of *sequences*, for which we refer you to Sections 10.2 and 10.3.

■ B.1 THE INTERMEDIATE-VALUE THEOREM

LEMMA B.1.1

Let f be continuous on $[a, b]$. If $f(a) < 0 < f(b)$ or $f(b) < 0 < f(a)$, then there is a number c between a and b for which $f(c) = 0$.

PROOF Suppose that $f(a) < 0 < f(b)$. (The other case can be treated in a similar manner.) Since $f(a) < 0$, we know from the continuity of f that there exists a number ξ such that f is negative on $[a, \xi)$. Let

$$c = \text{lub}\,\{\xi : f \text{ is negative on } [a, \xi)\}.$$

Clearly, $c \leq b$. We cannot have $f(c) > 0$, for then f would be positive on some interval extending to the left of c, and we know that, to the left of c, f is negative. Incidentally, this argument excludes the possibility that $c = b$ and means that $c < b$. We cannot have $f(c) < 0$, for then there would be an interval $[a, t)$, with $t > c$, on which f is negative, and this would contradict the definition of c. It follows that $f(c) = 0$. ❑

THEOREM B.1.2 THE INTERMEDIATE-VALUE THEOREM

If f is continuous on $[a, b]$ and K is a number between $f(a)$ and $f(b)$, then there is at least one number c between a and b for which $f(c) = K$.

PROOF Suppose, for example, that

$$f(a) < K < f(b).$$

(The other possibility can be handled in a similar manner.) The function

$$g(x) = f(x) - K$$

is continuous on $[a, b]$. Since

$$g(a) = f(a) - K < 0 \qquad \text{and} \qquad g(b) = f(b) - K > 0,$$

we know from the lemma that there is a number c between a and b for which $g(c) = 0$. Obviously, then, $f(c) = K$. □

■ B.2 BOUNDEDNESS; EXTREME-VALUE THEOREM

LEMMA B.2.1

If f is continuous on $[a, b]$ then f is bounded on $[a, b]$.

PROOF Consider

$$\{x : x \in [a, b] \text{ and } f \text{ is bounded on } [a, x]\}.$$

It is easy to see that this set is nonempty and bounded above by b. Thus we can set

$$c = \text{lub}\,\{x : f \text{ is bounded on } [a, x]\}.$$

Now we argue that $c = b$. To do so, we suppose that $c < b$. From the continuity of f at c, it is easy to see that f is bounded on $[c - \epsilon, c + \epsilon]$ for some $\epsilon > 0$. Being bounded on $[a, c - \epsilon]$ and on $[c - \epsilon, c + \epsilon]$, it is obviously bounded on $[a, c + \epsilon]$. This contradicts our choice of c. We can therefore conclude that $c = b$. This tells us that f is bounded on $[a, x]$ for all $x < b$. We are now almost through. From the continuity of f, we know that f is bounded on some interval of the form $[b - \epsilon, b]$. Since $b - \epsilon < b$, we know from what we have just proved that f is bounded on $[a, b - \epsilon]$. Being bounded on $[a, b - \epsilon]$ and bounded on $[b - \epsilon, b]$, it is bounded on $[a, b]$. □

THEOREM B.2.2 THE EXTREME-VALUE THEOREM

If f is continuous on $[a, b]$, then f takes on both a maximum value M and a minimum value m on $[a, b]$.

PROOF By the lemma, f is bounded on $[a, b]$. Set

$$M = \text{lub}\,\{f(x) : x \in [a, b]\}.$$

We must show that there exists c in $[a, b]$ such that $f(c) = M$. To do this, we set

$$g(x) = \frac{1}{M - f(x)}.$$

If f does not take on the value M, then g is continuous on $[a, b]$ and thus, by the lemma, bounded on $[a, b]$. A look at the definition of g makes it clear that g cannot be bounded on $[a, b]$. The assumption that f does not take on the value M has led to a contradiction. (That f takes a minimum value m can be proved in a similar manner.) □

■ B.3 INVERSES

THEOREM B.3.1 CONTINUITY OF THE INVERSE

Let f be a one-to-one function defined on an interval (a, b). If f is continuous, then its inverse f^{-1} is also continuous.

PROOF If f is continuous, then, being one-to-one, f either increases throughout (a, b) or it decreases throughout (a, b). The proof of this assertion we leave to you.

Suppose now that f increases throughout (a, b). Let's take c in the domain of f^{-1} and show that f^{-1} is continuous at c.

We first observe that $f^{-1}(c)$ lies in (a, b) and choose $\epsilon > 0$ sufficiently small so that $f^{-1}(c) - \epsilon$ and $f^{-1}(c) + \epsilon$ also lie in (a, b). We seek a $\delta > 0$ such that

$$\text{if } c - \delta < x < c + \delta, \qquad \text{then} \qquad f^{-1}(c) - \epsilon < f^{-1}(x) < f^{-1}(c) + \epsilon.$$

This condition can be met by choosing δ to satisfy

$$f(f^{-1}(c) - \epsilon) < c - \delta \qquad \text{and} \qquad c + \delta < f(f^{-1}(c) + \epsilon),$$

for then, if $c - \delta < x < c + \delta$,

$$f(f^{-1}(c) - \epsilon) < x < f(f^{-1}(c) + \epsilon),$$

and, since f^{-1} also increases,

$$f^{-1}(c) - \epsilon < f^{-1}(x) < f^{-1}(c) + \epsilon.$$

The case where f decreases throughout (a, b) can be handled in a similar manner. □

THEOREM B.3.2 DIFFERENTIABILITY OF THE INVERSE

Let f be a one-to-one function differentiable on an open interval I. Let a be a point of I and let $f(a) = b$. If $f'(a) \neq 0$, then f^{-1} is differentiable at b and

$$(f^{-1})'(b) = \frac{1}{f'(a)}.$$

PROOF (Here we use the characterization of derivative spelled out in Theorem 3.5.8.) We take $\epsilon > 0$ and show that there exists a $\delta > 0$ such that

$$\text{if} \quad 0 < |t - b| < \delta, \qquad \text{then} \qquad \left|\frac{f^{-1}(t) - f^{-1}(b)}{t - b} - \frac{1}{f'(a)}\right| < \epsilon.$$

Since f is differentiable at a, there exists a $\delta_1 > 0$ such that

$$\text{if} \quad 0 < |x - a| < \delta_1, \qquad \text{then} \qquad \left|\frac{1}{\dfrac{f(x) - f(a)}{x - a}} - \frac{1}{f'(a)}\right| < \epsilon,$$

and therefore $$\left|\frac{x-a}{f(x)-f(a)} - \frac{1}{f'(a)}\right| < \epsilon.$$

By the previous theorem, f^{-1} is continuous at b. Hence there exists a $\delta > 0$ such that

$$\text{if} \quad 0 < |t-b| < \delta, \qquad \text{then} \quad 0 < |f^{-1}(t) - f^{-1}(b)| < \delta_1,$$

and therefore $$\left|\frac{f^{-1}(t)-f^{-1}(b)}{t-b} - \frac{1}{f'(a)}\right| < \epsilon. \quad \square$$

■ B.4 THE INTEGRABILITY OF CONTINUOUS FUNCTIONS

The aim here is to prove that, if f is continuous on $[a, b]$, then there is one and only one number I that satisfies the inequality

$$L_f(P) \leq I \leq U_f(P) \qquad \text{for all partitions } P \text{ of } [a, b].$$

DEFINITION B.4.1

A function f is said to be *uniformly continuous* on $[a, b]$, if for each $\epsilon > 0$ there exists $\delta > 0$ such that

$$\text{if } x, y \in [a, b] \quad \text{and} \quad |x - y| < \delta, \quad \text{then} \quad |f(x) - f(y)| < \epsilon.$$

For convenience, let's agree to say that *the interval* $[a, b]$ *has the property* P_ϵ if there exist sequences $\{x_n\}$, $\{y_n\}$ satisfying

$$x_n, y_n \in [a, b], \quad |x_n - y_n| < 1/n, \quad |f(x_n) - f(y_n)| \geq \epsilon.$$

LEMMA B.4.2

If f is not uniformly continuous on $[a, b]$, then $[a, b]$ has the property P_ϵ for some $\epsilon > 0$.

PROOF If f is not uniformly continuous on $[a, b]$, then there is at least one $\epsilon > 0$ for which there is no $\delta > 0$ such that

$$\text{if } x, y \in [a, b] \quad \text{and} \quad |x - y| < \delta, \quad \text{then} \quad |f(x) - f(y)| < \epsilon.$$

The interval $[a, b]$ has the property P_ϵ for that choice of ϵ. The details of the argument are left to you. $\square$

LEMMA B.4.3

Let f be continuous on $[a, b]$. If $[a, b]$ has the property P_ϵ, then at least one of the subintervals $[a, \frac{1}{2}(a + b)]$, $[\frac{1}{2}(a + b), b]$ has the property P_ϵ.

PROOF Let's suppose that the lemma is false. For convenience, we let $c = \frac{1}{2}(a + b)$, so that the halves become $[a, c]$ and $[c, b]$. Since $[a, c]$ fails to have the property P_ϵ, there exists an integer p such that

$$\text{if } x, y \in [a, c] \quad \text{and} \quad |x - y| < 1/p, \quad \text{then} \quad |f(x) - f(y)| < \epsilon.$$

Since $[c, b]$ fails to have the property P_ϵ, there exists an integer q such that

$$\text{if } x, y \in [c, b] \quad \text{and} \quad |x - y| < 1/q, \quad \text{then} \quad |f(x) - f(y)| < \epsilon.$$

Since f is continuous at c, there exists an integer r such that, if $|x - c| < 1/r$, then $|f(x) - f(c)| < \frac{1}{2}\epsilon$. Set $s = \max\{p, q, r\}$ and suppose that

$$x, y \in [a, b], \quad |x - y| < 1/s.$$

If x, y are both in $[a, c]$ or both in $[c, b]$, then

$$|f(x) - f(y)| < \epsilon.$$

The only other possibility is that $x \in [a, c]$ and $y \in [c, b]$. In this case we have

$$|x - c| < 1/r, \quad |y - c| < 1/r,$$

and thus

$$|f(x) - f(c)| < \tfrac{1}{2}\epsilon, \quad |f(y) - f(c)| < \tfrac{1}{2}\epsilon.$$

By the triangle inequality, we again have

$$|f(x) - f(y)| < \epsilon.$$

In summary, we have obtained the existence of an integer s with the property that

$$x, y \in [a, b], \quad |x - y| < 1/s \quad \text{implies} \quad |f(x) - f(y)| < \epsilon.$$

Hence $[a, b]$ does not have the property P_ϵ. This is a contradiction and proves the lemma. □

THEOREM B.4.4

If f is continuous on $[a, b]$, then f is uniformly continuous on $[a, b]$.

PROOF We suppose that f is not uniformly continuous on $[a, b]$ and base our argument on a mathematical version of the classical maxim "Divide and conquer".

By the first lemma of this section, we know that $[a, b]$ has the property P_ϵ for some $\epsilon > 0$. We bisect $[a, b]$ and note by the second lemma that one of the halves, say $[a_1, b_1]$, has the property P_ϵ. We then bisect $[a_1, b_1]$ and note that one of the halves, say $[a_2, b_2]$, has the property P_ϵ. Continuing in this manner, we obtain a sequence of intervals $[a_n, b_n]$, each with the property P_ϵ. Then for each n, we can choose $x_n, y_n \in [a_n, b_n]$ such that

$$|x_n - y_n| < 1/n \quad \text{and} \quad |f(x_n) - f(y_n)| \geq \epsilon.$$

Since

$$a \leq a_n \leq a_{n+1} < b_{n+1} \leq b_n \leq b,$$

we see that sequences $\{a_n\}$ and $\{b_n\}$ are both bounded and monotonic. Thus they are convergent. Since $b_n - a_n \to 0$, we see that $\{a_n\}$ and $\{b_n\}$ both converge to the same limit, say L. From the inequality

$$a_n \leq x_n \leq y_n \leq b_n,$$

we conclude that

$$x_n \to L \quad \text{and} \quad y_n \to L.$$

This tells us that

$$|f(x_n) - f(y_n)| \to |f(L) - f(L)| = 0,$$

which contradicts the statement that $|f(x_n) - f(y_n)| \geq \epsilon$ for all n. ❑

LEMMA B.4.5

If P and Q are partitions of $[a, b]$, then $L_f(P) \leq U_f(Q)$.

PROOF $P \cup Q$ is a partition of $[a, b]$ that contains both P and Q. It is obvious then that

$$L_f(P) \leq L_f(P \cup Q) \leq U_f(P \cup Q) \leq U_f(Q). \quad ❑$$

From this lemma it follows that the set of all lower sums is bounded above and has a least upper bound L. The number L satisfies the inequality

$$L_f(P) \leq L \leq U_f(P) \quad \text{for all partitions } P$$

and is clearly the least of such numbers. Similarly, we find that the set of all upper sums is bounded below and has a greatest lower bound U. The number U satisfies the inequality

$$L_f(P) \leq U \leq U_f(P) \quad \text{for all partitions } P$$

and is clearly the largest of such numbers.

We are now ready to prove the basic theorem.

THEOREM B.4.6 THE INTEGRABILITY THEOREM

If f is continuous on $[a, b]$, then there exists one and only one number I that satisfies the inequality

$$L_f(P) \leq I \leq U_f(P) \quad \text{for all partitions } P \text{ of } [a, b].$$

PROOF We know that

$$L_f(P) \leq L \leq U \leq U_f(P) \quad \text{for all } P,$$

so that existence is no problem. We will have uniqueness if we can prove that

$$L = U.$$

To do this, we take $\epsilon > 0$ and note that f, being continuous on $[a, b]$, is uniformly continuous on $[a, b]$. Thus there exists a $\delta > 0$ such that, if

$$x, y \in [a, b] \quad \text{and} \quad |x - y| < \delta, \quad \text{then} \quad |f(x) - f(y)| < \frac{\epsilon}{b - a}.$$

We now choose a partition $P = \{x_0, x_1, \ldots, x_n\}$ for which $\max \Delta x_i < \delta$. For this partition P, we have

$$\begin{aligned} U_f(P) - L_f(P) &= \sum_{i=1}^{n} M_i \Delta x_i - \sum_{i=1}^{n} m_i \Delta x_i \\ &= \sum_{i=1}^{n} (M_i - m_i) \Delta x_i \\ &< \sum_{i=1}^{n} \frac{\epsilon}{b - a} \Delta x_i = \frac{\epsilon}{b - a} \sum_{i=1}^{n} \Delta x_i = \frac{\epsilon}{b - a}(b - a) = \epsilon. \end{aligned}$$

Since $\quad U_f(P) - L_f(P) < \epsilon \quad$ and $\quad 0 \le U - L \le U_f(P) - L_f(P),$

you can see that

$$0 \le U - L < \epsilon.$$

Since ϵ was chosen arbitrarily, we must have $U - L = 0$ and $L = U$. □

■ B.5 THE INTEGRAL AS THE LIMIT OF RIEMANN SUMS

For the notation we refer to Section 5.2.

THEOREM B.5.1

If f is continuous on $[a, b]$, then

$$\int_a^b f(x)\,dx = \lim_{||P|| \to 0} S^*(P).$$

PROOF Let $\epsilon > 0$. We must show that there exists a $\delta > 0$ such that

$$\text{if} \quad ||P|| < \delta, \quad \text{then} \quad \left| S^*(P) - \int_a^b f(x)\,dx \right| < \epsilon.$$

From the proof of Theorem B.4.6 we know that there exists a $\delta > 0$ such that

$$\text{if} \quad ||P|| < \delta, \quad \text{then} \quad U_f(P) - L_f(P) < \epsilon.$$

For such P we have

$$U_f(P) - \epsilon < L_f(P) \le S^*(P) \le U_f(P) < L_f(P) + \epsilon.$$

This gives

$$\int_a^b f(x)\,dx - \epsilon < S^*(P) < \int_a^b f(x)\,dx + \epsilon,$$

and therefore

$$\left| S^*(P) - \int_a^b f(x)\,dx \right| < \epsilon. \quad \square$$

ANSWERS TO ODD-NUMBERED EXERCISES

CHAPTER 1

SECTION 1.2

1. rational **3.** rational **5.** integer, rational **7.** integer, rational **9.** integer, rational **11.** =

13. > **15.** < **17.** 6 **19.** 4 **21.** 13 **23.** $5-\sqrt{5}$

25. $-\frac{3}{2}$ 0 x **27.** -4 0 4 x **29.** 0 x

31. 0 2 6 x **33.** -3 0 x **35.** 0 2 x

37. -2 0 5 x **39.** 0 x

41. bounded; lower bound 0, upper bound 4 **43.** not bounded

45. not bounded **47.** bounded above; $\sqrt{2}$ is an upper bound

49. $x_0 = 2$, $x_1 \cong 2.75$, $x_2 \cong 2.5864$, $x_3 \cong 2.57133$, $x_4 \cong 2.57128$, $x_5 \cong 2.57128$; bounded; lower bound 2, upper bound 3 (the smallest upper bound $\cong 2.57128\ldots$); $x_n \cong 2.5712815907$ (10 decimal places).

51. $(x-5)^2$ **53.** $8(x^2+2)(x^4-2x^2+4)$ **55.** $(2x+3)^2$

57. $2, -1$ **59.** 3 **61.** none **63.** no real zeros **65.** 120 **67.** 56 **69.** 1

71. If r and $r+s$ are rational, then $s = (r+s) - r$ is rational.

73. The product could be either rational or irrational; $0 \cdot \sqrt{2} = 0$ is rational, $1 \cdot \sqrt{2} = \sqrt{2}$ is irrational.

75. Suppose $\sqrt{2} = \dfrac{p}{q}$, where p, q are integers with no common divisor (other than ± 1). Then $p^2 = 2q^2$, which implies p^2 is even, which, in turn, implies $p = 2r$ is even. Thus $4r^2 = 2q^2$, which implies q^2 and hence q are even. This is a contradiction.

77. If the length of a rectangle with perimeter P is x, $0 < x < \frac{1}{2}P$, then the width is $\frac{1}{2}P - x$ and the area A is $A(x) = x(\frac{1}{2}P - x) = \left(\dfrac{P}{4}\right)^2 - \left(x - \dfrac{P}{4}\right)^2$. Clearly A is a maximum when $x = P/4$. The width in this case is also $P/4$, and the rectangle is a square.

SECTION 1.3

1. $(-\infty, 1)$ **3.** $(-\infty, -3]$ **5.** $\left(-\infty, -\frac{1}{5}\right)$ **7.** $(-1, 1)$ **9.** $(-\infty, -2] \cup [3, \infty)$ **11.** $\left[-1, \frac{1}{2}\right]$

13. $(0,1) \cup (2,\infty)$ **15.** $[0, \infty)$ **17.** $(0, 2)$ **19.** $(-\infty, -6) \cup (2, \infty)$ **21.** $(-2, 2)$ **23.** $(-\infty, -3) \cup (3, \infty)$

25. $\left(\frac{3}{2}, \frac{5}{2}\right)$ **27.** $(-1, 0) \cup (0, 1)$ **29.** $\left(\frac{3}{2}, 2\right) \cup \left(2, \frac{5}{2}\right)$ **31.** $(-5, 3) \cup (3, 11)$ **33.** $\left(-\frac{5}{8}, -\frac{3}{8}\right)$

35. $(-\infty, -4) \cup (-1, \infty)$ **37.** $|x| < 3$ **39.** $|x-2| < 5$ **41.** $|x+2| < 5$

43. $A \geq 2$ **45.** $A \leq \frac{4}{3}$ **47.** (a) $\dfrac{1}{x} < \dfrac{1}{\sqrt{x}} < 1 < \sqrt{x} < x$ (b) $x < \sqrt{x} < 1 < \dfrac{1}{\sqrt{x}} < \dfrac{1}{x}$

49. $a < b$ and $ab > 0$ implies

$$\frac{1}{b} - \frac{1}{a} = \frac{a-b}{ab} < 0.$$

The result follows.

51. $b - a = (\sqrt{b} + \sqrt{a})(\sqrt{b} - \sqrt{a})$; since $\sqrt{b} + \sqrt{a} \geq 0$, $b - a$ and $\sqrt{b} - \sqrt{a}$ have the same sign.

53. $\big||a| - |b|\big|^2 = (|a| - |b|)^2 = |a|^2 - 2|a|\,|b| + |b|^2 = a^2 - 2|ab| + b^2 \leq a^2 - 2ab + b^2 = (a-b)^2$

$\uparrow$ — $(ab \leq |ab|)$

Thus, $\big||a| - |b|\big| \leq \sqrt{(a-b)^2} = |a - b|$.

55. $0 \le a \le b$ implies

$$\frac{a}{1+a} - \frac{b}{1+b} = \frac{a-b}{(1+a)(1+b)} \le 0.$$

The result follows.

57. $a < b$ implies.

$$a - \frac{a+b}{2} = \frac{a-b}{2} < 0$$

and

$$\frac{a+b}{2} - b = \frac{a-b}{2} < 0.$$

Thus

$$a < \frac{a+b}{2} < b;$$

$(a+b)/2$ is the midpoint of the line segment $\overline{ab}$.

SECTION 1.4

1. 10 **3.** $4\sqrt{5}$ **5.** $(4,6)$ **7.** $\left(\frac{9}{2},-3\right)$ **9.** $-\frac{2}{3}$ **11.** -1 **13.** $-y_0/x_0$

15. slope 2, y-intercept -4 **17.** slope $\frac{1}{3}$, y-intercept 2 **19.** slope $\frac{7}{3}$, y-intercept $\frac{4}{3}$ **21.** $y = 5x + 2$ **23.** $y = -5x + 2$ **25.** $y = 3$

27. $x = -3$ **29.** $y = 7$ **31.** $3y - 2x - 17 = 0$ **33.** $2y + 3x - 20 = 0$ **35.** $\left(\frac{1}{2}\sqrt{2}, \frac{1}{2}\sqrt{2}\right), \left(-\frac{1}{2}\sqrt{2}, -\frac{1}{2}\sqrt{2}\right)$ **37.** $(3,4), \left(\frac{117}{25}, \frac{44}{25}\right)$

39. $(1,1)$ **41.** $\left(-\frac{2}{23}, \frac{38}{23}\right)$ **43.** $\frac{17}{2}$ **45.** parabola, vertex $(-1,-1)$ **47.** ellipse, center $(2,-1)$ **49.** hyperbola, center $(1,2)$

51. hyperbola, center $(3,-2)$ **53.** $-\frac{5}{12}$ **55.** $x - 2y - 3 = 0$ **57.** $(2.36, -0.21)$ **59.** $(0.61, 2.94), (2.64, 1.42)$

61. $3x + 13y - 40 = 0$ **63.** isosceles right triangle **65.** isosceles right triangle

67. The midpoint of the hypotenuse is $M(\frac{a}{2}, \frac{b}{2})$. The distance between M and the origin, M and $(0,b)$, and M and $(a,0)$ is $\frac{1}{2}\sqrt{a^2+b^2}$.

69. $(1, \frac{10}{3})$ **71.** If $A(0,0)$ and $B(a,0)$ are two vertices of a parallelogram, and $C(b,c)$ is the vertex opposite B, then $D(a+b,c)$ is the vertex opposite A. The midpoint of the diagonal AD = midpoint of the diagonal $BC = \left(\frac{1}{2}(a+b), \frac{1}{2}c\right)$.

73. $F = \frac{9}{5}C + 32; -40°$

SECTION 1.5

1. $f(0) = 2, f(1) = 1, f(-2) = 16, f\left(\frac{3}{2}\right) = 2$ **3.** $f(0) = 0, f(1) = \sqrt{3}, f(-2) = 0, f\left(\frac{3}{2}\right) = \dfrac{\sqrt{21}}{2}$

5. $f(0) = 0, f(1) = \frac{1}{2}, f(-2) = -1, f(\frac{3}{2}) = \frac{12}{23}$ **7.** $f(-x) = x^2 + 2x, f\left(\dfrac{1}{x}\right) = \dfrac{1-2x}{x^2}, f(a+b) = a^2 + 2ab + b^2 - 2a - 2b$

9. $f(-x) = \sqrt{1+x^2}, f\left(\dfrac{1}{x}\right) = \dfrac{\sqrt{x^2+1}}{|x|}, f(a+b) = \sqrt{a^2+2ab+b^2+1}$ **11.** $2a^2 + 4ah + 2h^2 - 3a - 3h; 4a - 3 + 2h$

13. 1, 3 **15.** -2 **17.** $3, -3$ **19.** dom $(f) = (-\infty, \infty)$; range $(f) = [0, \infty)$ **21.** dom $(f) = (-\infty, \infty)$; range $(f) = (-\infty, \infty)$

23. dom $(f) = (-\infty, 0) \cup (0, \infty)$; range $(f) = (0, \infty)$ **25.** dom $(f) = (-\infty, 1]$; range $(f) = [0, \infty)$

27. dom $(f) = (-\infty, 7]$; range $(f) = [-1, \infty)$ **29.** dom $(f) = (-\infty, 2)$; range $(f) = (0, \infty)$

31. horizontal line one unit above x-axis **33.** line through the origin with slope 2 **35.** line through $(0,2)$ with slope $\frac{1}{2}$

37. upper semicircle of radius 2 centered at the origin **39.**

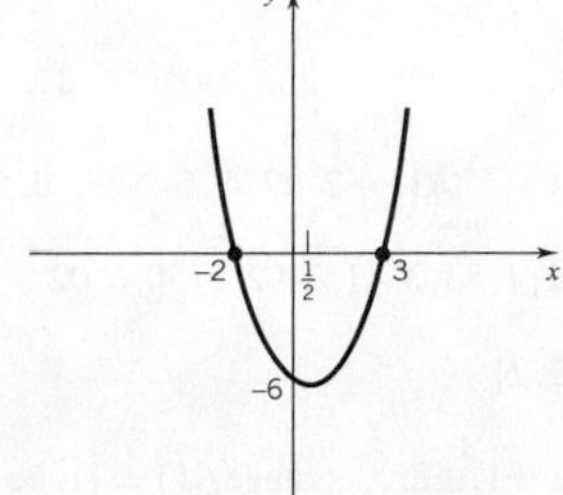

41. dom $(f) = (-\infty, 0) \cup (0, \infty)$; range $(f) = \{-1, 1\}$

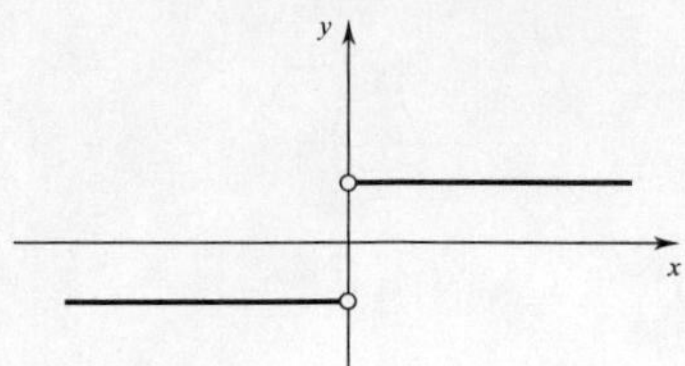

43. dom $(f) = [0, \infty)$; range $(f) = [1, \infty)$

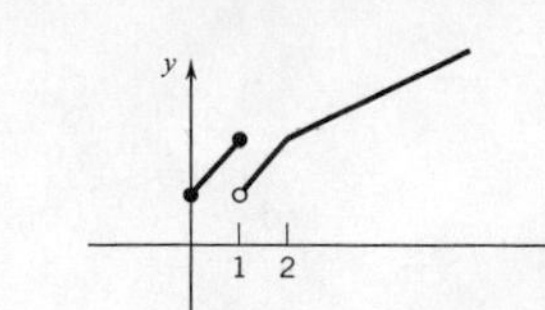

45. yes, dom $(f) = [-2, 2]$; range $(f) = [-2, 2]$ **47.** no **49.** odd **51.** neither **53.** even **55.** (a) even (b) odd (c) neither

57. (a)

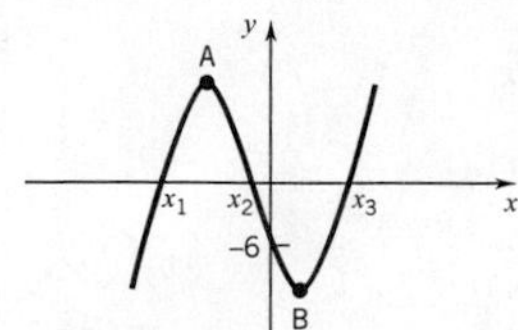

(b) $x_1 = -6.566, x_2 = -0.493, x_3 = 5.559$
(c) $A(-4, 28.667), B(3, -28.500)$

59.

-3.7 0 0.2 6.6

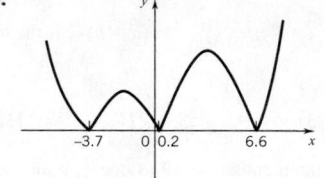

$-5 \le x \le 8,\ 0 \le y \le 100$

61. range: $[-9, \infty)$

63. $A = \dfrac{C^2}{4\pi}$, where C is the circumference; dom $(A) = [0, \infty)$

65. $V = s^{3/2}$, where s is the area of a face; dom $V = [0, \infty)$

67. $S = 3d^2$, where d is the diagonal of a face; dom $(S) = [0, \infty)$

69. $A = \dfrac{\sqrt{3}}{4}x^2$, where x is the length of a side; dom $(A) = [0, \infty)$

71. $A = \dfrac{15x}{2} - \dfrac{x^2}{2} - \dfrac{\pi x^2}{8}, 0 \le x \le \dfrac{30}{\pi + 2}$ **73.** $A = bx - \dfrac{b}{a}x^2, 0 \le x \le a$

75. $A = \dfrac{P^2}{16} + \dfrac{(28 - P)^2}{4\pi}, 0 \le P \le 28$ **77.** $V = \pi r^2(108 - 2\pi r)$

SECTION 1.6

1. polynomial, degree 0 **3.** rational function **5.** neither **7.** neither **9.** neither

11. dom $(f) = (-\infty, \infty)$

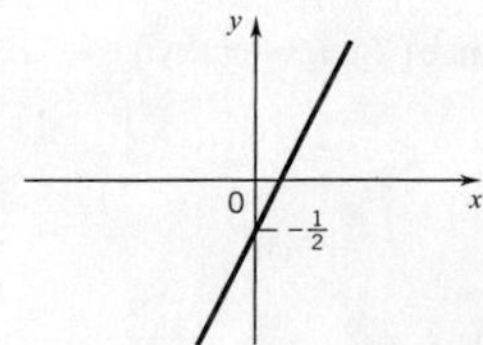

13. dom $(f) = (-\infty, \infty)$

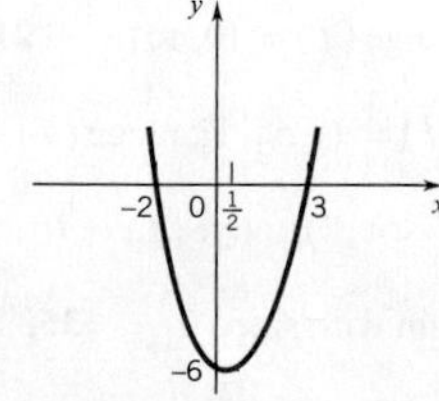

15. dom $(f) = (-\infty, -2) \cup (-2, 2) \cup (2, \infty)$

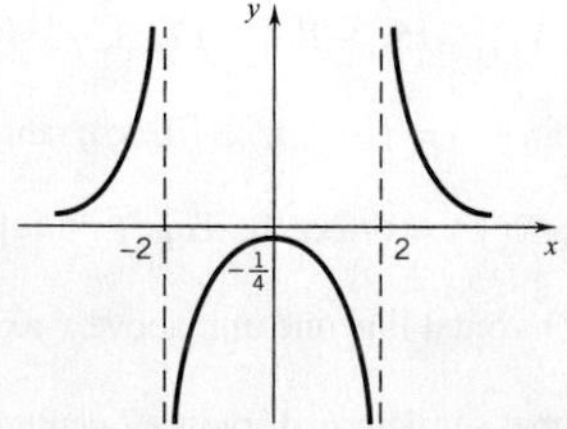

17. $\dfrac{5\pi}{4}$ **19.** $-\dfrac{5\pi}{3}$ **21.** $\dfrac{\pi}{12}$ **23.** $-270°$ **25.** $300°$ **27.** $114.59°$ **29.** $\dfrac{\pi}{6}, \dfrac{5\pi}{6}$ **31.** $\dfrac{\pi}{2}$ **33.** $\dfrac{\pi}{4}, \dfrac{7\pi}{4}$ **35.** $\dfrac{\pi}{4}, \dfrac{3\pi}{4}, \dfrac{5\pi}{4}, \dfrac{7\pi}{4}$

37. 0.7772 **39.** 0.7101 **41.** 3.1524 **43.** -2.8974 **45.** $0.5505, \pi - 0.5505$ **47.** $1.4231, \pi + 1.4231$

49. $1.7997, 2\pi - 1.7997$ **51.** $x \cong 1.31, 1.83, 3.41, 3.92, 5.50, 6.02$ **53.** dom $(f) = (-\infty, \infty)$; range $(f) = [0, 1]$

55. dom $(f) = (-\infty, \infty)$; range $(f) = [-2, 2]$

57. dom $(f) = \left(k\pi - \dfrac{\pi}{2}, k\pi + \dfrac{\pi}{2}\right), k = 0, \pm 1, \pm 2, \ldots$; range $(f) = [1, \infty)$

59.

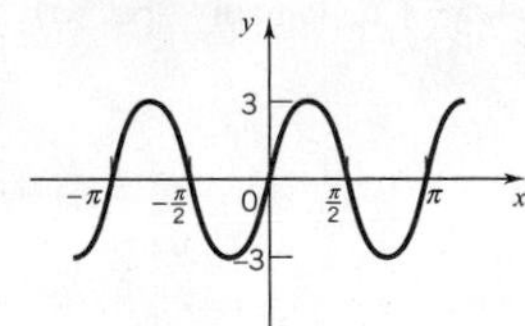

61.

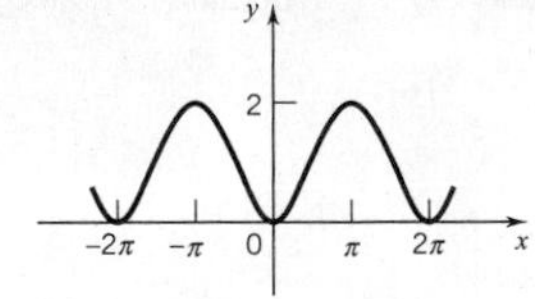

63.

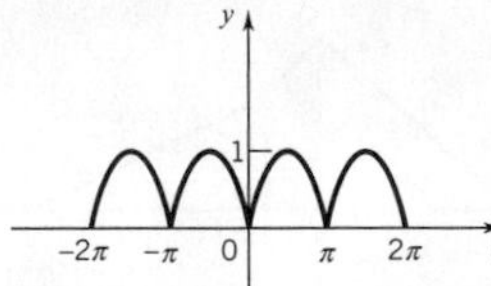

65. odd **67.** even **69.** odd **71.** Let $m_1 = \tan\theta_1, m_2 = \tan\theta_2, \alpha = |\theta_2 - \theta_1|$

$$\tan\alpha = |\tan(\theta_2 - \theta_1)| = \left|\frac{\tan\theta_2 - \tan\theta_1}{1 + \tan\theta_2\tan\theta_1}\right| = \left|\frac{m_2 - m_1}{1 + m_2 m_1}\right|$$

73. $\left(\frac{23}{37}, \frac{116}{37}\right)$; approx 73° **75.** $\left(\frac{-17}{73}, \frac{-2}{73}\right)$; approx 82°

77. $b\sin A = a\sin B = h$, so $\frac{\sin A}{a} = \frac{\sin B}{b}$; similarly, $\frac{\sin A}{a} = \frac{\sin C}{c}$

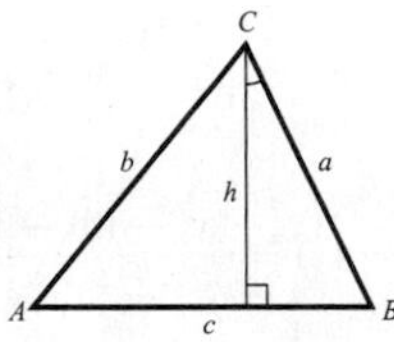

79. $A = \frac{1}{2}ah = \frac{1}{2}a^2\sin\theta$

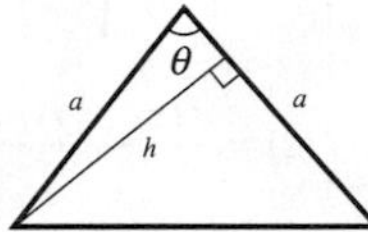

81.

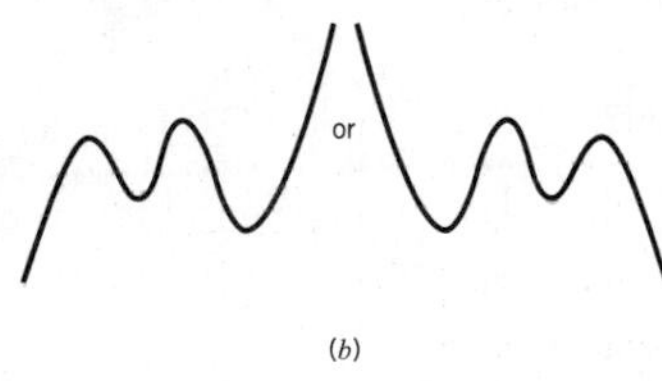

83.

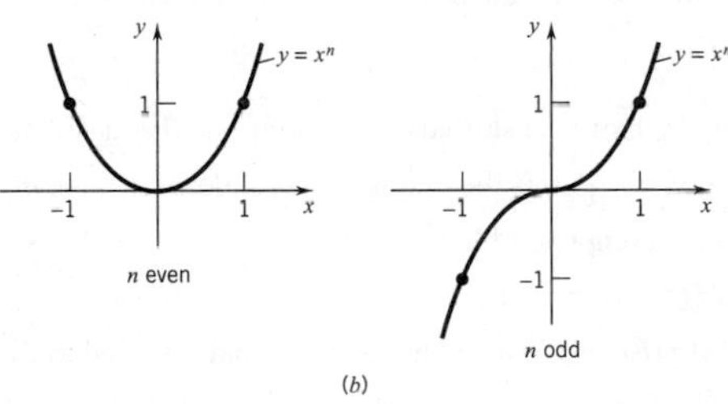

(c) $f_k(x) \geq f_{k+1}(x)$ on $[0, 1]$; $f_{k+1}(x) \geq f_k(x)$ on $[1, \infty)$

85. $x = 10 : 0.097656$, $x = 20 : 0.000381$, $x = 30 : 8.381903 \times 10^{-7}$, $x = 40 : 1.455192 \times 10^{-9}$, $x = 50 : 2.220446 \times 10^{-12}$; f/g approaches 0 as x gets "large."

SECTION 1.7

1. $\frac{15}{2}$ **3.** $\frac{105}{2}$ **5.** $\frac{-27}{4}$ **7.** 3

9. $(f + g)(x) = x - 1$; domain $(-\infty, \infty)$

$(f - g)(x) = 3x - 5$; domain $(-\infty, \infty)$

$(f \cdot g)(x) = -2x^2 + 7x - 6$; domain $(-\infty, \infty)$

$\left(\frac{f}{g}\right)(x) = \frac{2x - 3}{2 - x}$; domain: all real numbers except $x = 2$

11. $(f + g)(x) = x + \sqrt{x - 1} - \sqrt{x + 1}$; domain $[1, \infty)$

$(f - g)(x) = \sqrt{x - 1} + \sqrt{x + 1} - x$; domain $[1, \infty)$

$(f \cdot g)(x) = \sqrt{x - 1}(x - \sqrt{x + 1}) = x\sqrt{x - 1} - \sqrt{x^2 - 1}$; domain $[1, \infty)$

$\left(\frac{f}{g}\right)(x) = \frac{\sqrt{x - 1}}{x - \sqrt{x + 1}}$; domain $\left[1, \frac{1 + \sqrt{5}}{2}\right) \cup \left(\frac{1 + \sqrt{5}}{2}, \infty\right)$

13. (a) $(6f + 3g)(x) = 6x + 3\sqrt{x}$, $x > 0$ (b) $(f - g)(x) = x + \frac{3}{\sqrt{x}} - \sqrt{x}$, $x > 0$ (c) $(f/g)(x) = \frac{x\sqrt{x} + 1}{x - 2}$, $x > 0$, $x \neq 2$

15.

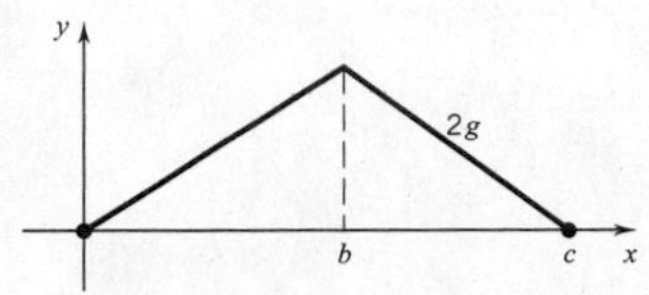

17.

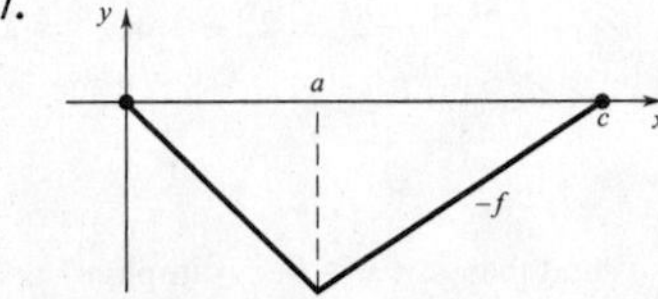

19.

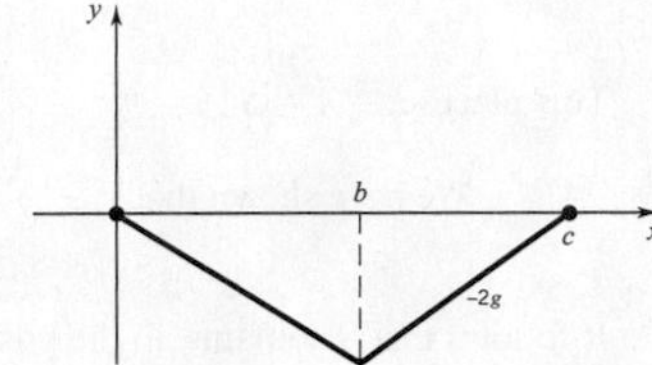

21.

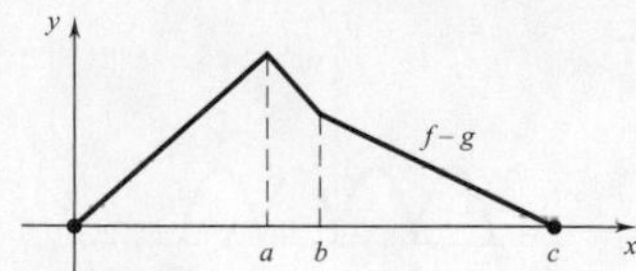

23. $(f \circ g)(x) = 2x^2 + 5$; domain $(-\infty, \infty)$ **25.** $(f \circ g)(x) = \sqrt{x^2 + 5}$; domain $(-\infty, \infty)$

27. $(f \circ g)(x) = \dfrac{x}{x-2}$; domain: all real numbers except $x = 0, x = 2$ **29.** $(f \circ g)(x) = |\sin 2x|$; domain $(-\infty, \infty)$

31. $(f \circ g \circ h)(x) = 4(x^2 - 1)$; domain $(-\infty, \infty)$ **33.** $(f \circ g \circ h)(x) = 2x^2 + 1$; domain $(-\infty, \infty)$ **35.** $f(x) = \dfrac{1}{x}$ **37.** $f(x) = 2 \sin x$

39. $g(x) = \left(1 - \dfrac{1}{x^4}\right)^{2/3}$ **41.** $g(x) = 2x^3 - 1$ **43.** $(f \circ g)(x) = |x|; (g \circ f)(x) = x$ **45.** $(f \circ g)(x) = \cos^2 x; (g \circ f)(x) = \sin(1 - x^2)$

47. $(f \circ g)(x) = x; (g \circ f)(x) = x$ **49.** fg is an even function since $(fg)(-x) = f(-x)g(-x) = f(x)g(x) = (fg)(x)$

51. (a) $f(x) = \begin{cases} -x, & -1 \le x < 0 \\ 1, & x < -1 \end{cases}$ (b) $f(x) = \begin{cases} x, & -1 \le x < 0 \\ -1, & x < -1 \end{cases}$ **53.** $g(-x) = f(-x) + f(x) = f(x) + f(-x) = g(x)$

55. $f(x) = \dfrac{f(x) + f(-x)}{2} + \dfrac{f(x) - f(-x)}{2}$ **57.** (a) $(f \circ g)(x) = \dfrac{5x^2 + 16x - 16}{(2-x)^2}$ (b) $(g \circ k)(x) = x$ (c) $(f \circ k \circ g)(x) = x^2 - 4$

59. (a) For fixed a, varying b varies the y-coordinate of the vertex of the parabola.
(b) For fixed b, varying a varies the x-coordinate of the vertex of the parabola.
(c) The graph of $-F$ is the reflection of the graph of F in the x-axis.

61. (a) The graph of $f(x) + c$ is the graph of $f(x)$ shifted c units up if $c > 0$ and shifted $|c|$ units down if $c < 0$.
(b) For $a > 0$ $(a < 0)$, the graph of $f(x - a)$ is the graph of $f(x)$ shifted horizontally $|a|$ units to the right (left).
(c) For $b > 1$, the graph of $f(bx)$ is compressed horizontally.
For $0 < b < 1$, the graph of $f(bx)$ is stretched horizontally.
For $-1 < b < 0$, the graph of $f(bx)$ is stretched horizontally and reflected in the y-axis.
For $b < -1$, the graph of $f(bx)$ is compressed horizontally and reflected in the y-axis.

63. (a) For $A > 0$, the graph of Af is the graph of f scaled vertically by the factor A.
For $A < 0$, the graph of Af is the graph of f scaled vertically by the factor $|A|$ and then reflected in the x-axis.
(b) See Answer 61(c)

SECTION 1.8

1. Let S be the set of integers for which the statement is true. Since $2(1) \le 2^1$, S contains 1. Assume now that $k \in S$. This tells us that $2k \le 2^k$, and thus

$$2(k+1) = 2k + 2 \le 2^k + 2 \le 2^k + 2^k = 2(2^k) = 2^{k+1}.$$
$$\uparrow \underline{\qquad} (k \ge 1)$$

This places $k + 1$ in S.

We have shown that

$$1 \in S \quad \text{and that} \quad k \in S \quad \text{implies} \quad k + 1 \in S.$$

It follows that S contains all the positive integers.

3. Let S be the set of integers for which the statement is true. Since $2^0 = 2^1 - 1 = 1$, S contains 1. Assume now that $k \in S$. This tells us that

$$2^0 + 2^1 + 2^2 + 2^3 + \cdots + 2^{k-1} = 2^k - 1.$$

Therefore,

$$2^0 + 2^1 + 2^2 + 2^3 + \cdots + 2^{k-1} + 2^k = 2^k - 1 + 2^k = 2(2^k) - 1 = 2^{k+1} - 1.$$

This places $k + 1 \in S$.

We have shown that

$$1 \in S \quad \text{and that} \quad k \in S \quad \text{implies} \quad k + 1 \in S.$$

It follows that S contains all the positive integers.

5. Use $1^2 + 2^2 + \cdots + k^2 + (k+1)^2 = \frac{1}{6}k(k+1)(2k+1) + (k+1)^2$

$$= \tfrac{1}{6}(k+1)[(k(2k+1) + 6(k+1)]$$
$$= \tfrac{1}{6}(k+1)(2k^2 + 7k + 6)$$
$$= \tfrac{1}{6}(k+1)(k+2)(2k+3) = \tfrac{1}{6}(k+1)[(k+1)+1][2(k+1)+1].$$

7. By Exercise 6 and Example 1.

$$1^3 + 2^3 + \cdots + (n-1)^3 = \left[\tfrac{1}{2}(n-1)n\right]^2 = \tfrac{1}{4}(n-1)^2n^2 < \tfrac{1}{4}n^4$$

and

$$1^3 + 2^3 + \cdots + n^3 = [\tfrac{1}{2}n(n+1)]^2 = \tfrac{1}{4}n^2(n+1)^2 > \tfrac{1}{4}n^4.$$

9. Use $\dfrac{1}{\sqrt{1}} + \dfrac{1}{\sqrt{2}} + \dfrac{1}{\sqrt{3}} + \cdots + \dfrac{1}{\sqrt{n}} + \dfrac{1}{\sqrt{n+1}} > \sqrt{n} + \dfrac{1}{\sqrt{n+1}} > \sqrt{n} + \dfrac{1}{\sqrt{n+1}+\sqrt{n}}\left(\dfrac{\sqrt{n+1}-\sqrt{n}}{\sqrt{n+1}-\sqrt{n}}\right) = \sqrt{n+1}.$

11. Let S be the set of integers for which the statement is true. Since

$$3^{2(1)+1} + 2^{1+2} = 27 + 8 = 35$$

is divisible by 7, $1 \in S$.

Assume now that $k \in S$. This tells us that

$$3^{2k+1} + 2^{k+2} \text{ is divisible by 7.}$$

It follows that

$$3^{2(k+1)+1} + 2^{(k+1)+2} = 3^2 \cdot 3^{2k+1} + 2 \cdot 2^{k+2} = 9 \cdot 3^{2k+1} + 2 \cdot 2^{k+2} = 7 \cdot 3^{2k+1} + 2(3^{2k+1} + 2^{k+2})$$

is also divisible by 7. This places $k+1 \in S$.

We have shown that

$$1 \in S \quad \text{and that} \quad k \in S \quad \text{implies} \quad k+1 \in S.$$

It follows that S contains all the positive integers.

13. For all positive integers $n \geq 2$,

$$\left(1 - \frac{1}{2}\right)\left(1 - \frac{1}{3}\right)\cdots\left(1 - \frac{1}{n}\right) = \frac{1}{n}.$$

To see this let S be the set of integers n for which the formula holds. Since $1 - \frac{1}{2} = \frac{1}{2}$, $2 \in S$. Suppose now that $k \in S$. This tells us that

$$\left(1 - \frac{1}{2}\right)\left(1 - \frac{1}{3}\right)\cdots\left(1 - \frac{1}{k}\right) = \frac{1}{k}$$

and therefore that

$$\left(1 - \frac{1}{2}\right)\left(1 - \frac{1}{3}\right)\cdots\left(1 - \frac{1}{k}\right)\left(1 - \frac{1}{k+1}\right) = \frac{1}{k}\left(1 - \frac{1}{k+1}\right) = \frac{1}{k}\left(\frac{k}{k+1}\right) = \frac{1}{k+1}.$$

This places $k+1$ in S and verifies the formula for $n \geq 2$.

15. From the figure, observe that adding a vertex V_{N+1} to an N-sided polygon increases the number of diagonals by $(N-2) + 1 = N - 1$.

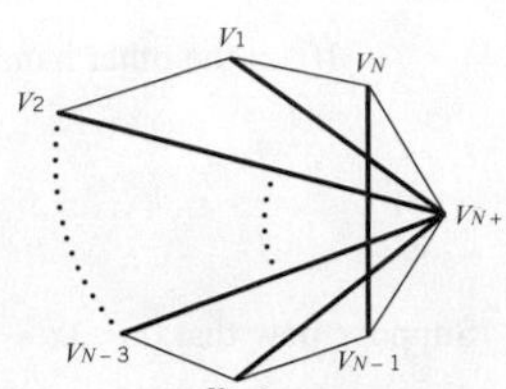

Then use the identity

$$\tfrac{1}{2}N(N-3) + (N-1) = \tfrac{1}{2}(N+1)(N+1-3).$$

17. To go from k to $k+1$, take $A = \{a_1, \ldots, a_{k+1}\}$ and $B = \{a_1, \ldots, a_k\}$. Assume that B has 2^k subsets: $B_1, B_2, \ldots, B_{2^k}$. The subsets of A are then $B_1, B_2, \ldots, B_{2^k}$ together with

$$B_1 \cup \{a_{k+1}\}, B_2 \cup \{a_{k+1}\}, \ldots, B_{2^k} \cup \{a_{k+1}\}.$$

This gives $2(2^k) = 2^{k+1}$ subsets for A.

CHAPTER 2

SECTION 2.1

1. (a) 2 (b) -1 (c) does not exist (d) -3 **3.** (a) does not exist (b) -3 (c) does not exist (d) -3

5. (a) does not exist (b) does not exist (c) does not exist (d) 1 **7.** (a) 2 (b) 2 (c) 2 (d) -1 **9.** (a) 0 (b) 0 (c) 0 (d) 0

11. $c = 0, 6$ **13.** -1 **15.** 12 **17.** 1 **19.** $\frac{3}{2}$ **21.** does not exist **23.** 2 **25.** does not exist **27.** 1 **29.** does not exist

31. 2 **33.** 2 **35.** 0 **37.** 1 **39.** 16 **41.** does not exist **43.** 4 **45.** does not exist **47.** $1/\sqrt{2}$ **49.** 4

51. (b) the limits do not exist (c)

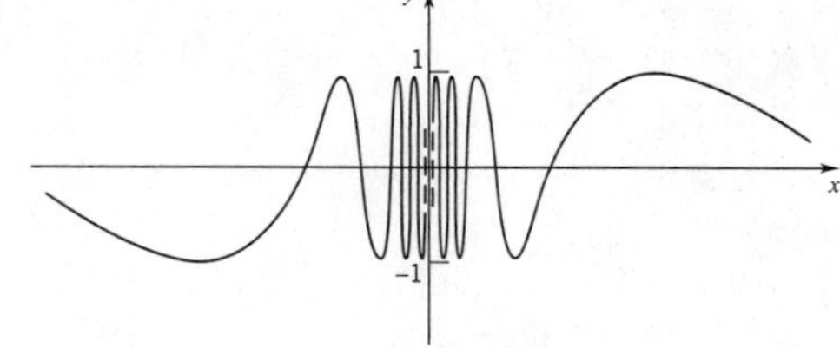

53. 2 **55.** $\frac{3}{2}$

57. (a) (i) 5 (ii) does not exist **59.** (a) (i) $-5/4$ (ii) 0 **61.** $c = -1$ **63.** $c = -2$

SECTION 2.2

1. $\frac{1}{2}$ **3.** does not exist **5.** 4 **7.** does not exist **9.** -1 **11.** does not exist **13.** 0 **15.** 2 **17.** 1 **19.** 1 **21.** δ_1

23. $\frac{1}{2}\epsilon$ **25.** 2ϵ **27.** $\delta = 1.75$ **29.** for $\epsilon = 0.5$, take $\delta = 0.24$; for $\epsilon = 0.25$, take $\delta = 0.1$

31. for $\epsilon = 0.5$, take $\delta = 0.75$; for $\epsilon = 0.25$, take $\delta = 0.43$

33. for $\epsilon = 0.25$, take $\delta = 0.23$; for $\epsilon = 0.1$, take $\delta = 0.14$

35. Take $\delta = \frac{1}{2}\epsilon$. If $0 < |x-4| < \frac{1}{2}\epsilon$, then $|(2x-5)-3| = 2|x-4| < \epsilon$.

37. Take $\delta = \frac{1}{6}\epsilon$. If $0 < |x-3| < \frac{1}{6}\epsilon$, then $|(6x-7)-11| = 6|x-3| < \epsilon$.

39. Take $\delta = \frac{1}{3}\epsilon$. If $0 < |x-2| < \frac{1}{3}\epsilon$, then $||1-3x|-5| \le 3|x-2| < \epsilon$. **41.** Statements (b), (e), (g), and (i) are necessarily true.

43. (i) $\lim_{x\to 3} \frac{1}{x-1} = \frac{1}{2}$ (ii) $\lim_{h\to 0} \frac{1}{(3+h)-1} = \frac{1}{2}$ (iii) $\lim_{x\to 3} \left(\frac{1}{x-1} - \frac{1}{2}\right) = 0$ (iv) $\lim_{x\to 3} \left|\frac{1}{x-1} - \frac{1}{2}\right| = 0$

45. (i) and (iv) of (2.2.6) with $L = 0$

47. Let $\epsilon > 0$. If $\lim_{x\to c} f(x) = L$, then there must exist $\delta > 0$ such that

$$(*) \qquad \text{if} \quad 0 < |x-c| < \delta \quad \text{then} \quad |f(x) - L| < \epsilon.$$

Suppose now that $0 < |h| < \delta$. Then $0 < |(c+h) - c| < \delta$, and thus by $(*)$, $|f(c+h) - L| < \epsilon$. This proves that, if $\lim_{x\to c} f(x) = L$, then $\lim_{h\to 0} f(c+h) = L$.

If, on the other hand, $\lim_{h\to 0} f(c+h) = L$, then there must exist $\delta > 0$ such that

$$(**) \qquad \text{if} \quad 0 < |h| < \delta \quad \text{then} \quad |f(c+h) - L| < \epsilon.$$

Suppose now that $0 < |x-c| < \delta$. Then by $(**)$, $|f(c + (x-c)) - L| < \epsilon$. More simply stated, $|f(x) - L| < \epsilon$. This proves that, if $\lim_{h\to 0} f(c+h) = L$, then $\lim_{x\to c} f(x) = L$.

49. (a) Set $\delta = \epsilon\sqrt{c}$. By the hint

$$\text{if} \quad 0 < |x - c| < \epsilon\sqrt{c}, \quad \text{then} \quad |\sqrt{x} - \sqrt{c}| < \frac{1}{\sqrt{c}}|x - c| < \epsilon.$$

(b) Set $\delta = \epsilon^2$. If $0 < x < \epsilon^2$, then $|\sqrt{x} - 0| = \sqrt{x} < \epsilon$.

51. Take $\delta =$ minimum of 1 and $\epsilon/7$. If $0 < |x-1| < \delta$, then $0 < x < 2$ and $|x-1| < \epsilon/7$. Therefore $|x^3 - 1| = |x^2 + x + 1||x - 1| < 7|x - 1| < 7(\epsilon/7) = \epsilon$.

53. Set $\delta = \epsilon^2$. If $3 - \epsilon^2 < x < 3$, then $-\epsilon^2 < x - 3, 0 < 3 - x < \epsilon^2$ and therefore $|\sqrt{3 - x} - 0| < \epsilon$.

55. Suppose on the contrary, that $\lim_{x\to c} f(x) = L$ for some particular c. Taking $\epsilon = \frac{1}{2}$, there must exist $\delta > 0$ such that,

$$\text{if} \quad 0 < |x - c| < \delta, \quad \text{then} \quad |f(x) - L| < \tfrac{1}{2}.$$

Let x_1 be a rational number satisfying $0 < |x_1 - c| < \delta$, and x_2 an irrational number satisfying $0 < |x_2 - c| < \delta$. (That such numbers exist follows from the fact that every interval contains both rational and irrational numbers.) Now $f(x_1) = 1$ and $f(x_2) = 0$. Thus we must have both $|1 - L| < \frac{1}{2}$ and $|0 - L| < \frac{1}{2}$. From the first inequality we conclude that $L > \frac{1}{2}$. From the second, we conclude that $L < \frac{1}{2}$. Clearly no such number L exists.

57. We begin by assuming that $\lim_{x\to c^+} f(x) = L$ and showing that $\lim_{h\to 0} f(c + |h|) = L$.

Let $\epsilon > 0$. Since $\lim_{x\to c^+} f(x) = L$, there exists $\delta > 0$ such that

$$(*) \qquad \text{if} \quad c < x < c + \delta \quad \text{then} \quad |f(x) - L| < \epsilon.$$

Suppose now that $0 < |h| < \delta$. Then $c < c + |h| < c + \delta$ and, by $(*)$, $|f(c + |h|) - L| < \epsilon$ Thus $\lim_{h\to 0} f(c + |h|) = L$.

Conversely let's assume that $\lim_{h\to 0} f(c + |h|) = L$ and again take $\epsilon > 0$. Then there exists $\delta > 0$ such that

$$(**) \qquad \text{if} \quad 0 < |h| < \delta \quad \text{then} \quad |f(c + |h|) - L| < \epsilon.$$

Suppose now that $c < x < c + \delta$. Then $0 < x - c < \delta$ so that, by $(**)$,

$$|f(x) - L| = |f(c + (x - c)) - L| < \epsilon.$$

Thus $\lim_{h\to c^+} f(x) = L$.

59. (a) Let $\epsilon = L$. There exist $\gamma > 0$ such that if $0 < |x - c| < \gamma$, then $|f(x) - L| < L$, which is equivalent to $0 < f(x) < 2L$. Thus $f(x) > 0$ for all $x \in (c - \gamma, c + \gamma), x \neq c$.

(b) Use the same argument.

61. (a) Suppose $\lim_{x\to c} f(x) = L$ and $\lim_{x\to c} g(x) = M$. Then $\lim_{x\to c} [g(x) - f(x)] = M - L$ exists. If $M - L < 0$, then by Exercise 59(b), there exists $\gamma > 0$ such that $g(x) - f(x) < 0$ on $(c - \gamma, c + \gamma), x \neq c$, which contradicts the fact that $f(x) \leq g(x)$ on $(c - p, c + p), x \neq c$. Thus $M - L \geq 0$.

(b) No. Consider $f(x) = 1 - x^2$ and $g(x) = 1 + x^2$ on $(-1, 1)$, and let $c = 0$.

SECTION 2.3

1. (a) 3 (b) 4 (c) -2 (d) 0 (e) does not exit (f) $\frac{1}{3}$

3. $\lim_{x\to 4}\left[\left(\frac{1}{x} - \frac{1}{4}\right)\left(\frac{1}{x-4}\right)\right] = \lim_{x\to 4}\left[\left(\frac{4-x}{4x}\right)\left(\frac{1}{x-4}\right)\right] = \lim_{x\to 4}\frac{-1}{4x} = -\frac{1}{16}$; Theorem 2.3.2 does not apply since $\lim_{x\to 4}\frac{1}{x-4}$ does not exist.

5. 3 **7.** -3 **9.** 5 **11.** does not exist **13.** -1 **15.** does not exist **17.** 1 **19.** 4 **21.** $\frac{1}{4}$ **23.** $-\frac{2}{3}$

25. does not exist **27.** -1 **29.** 4 **31.** a/b **33.** 5/4 **35.** does not exist **37.** 2

39. (a) 0 (b) $-\frac{1}{16}$ (c) 0 (d) does not exist **41.** (a) 4 (b) -2 (c) 2 (d) does not exist **43.** $f(x) = 1/x, g(x) = -1/x, c = 0$

45. True. Let $\lim_{x\to c} [f(x) + g(x)] = L$. If $\lim_{x\to c} g(x) = M$ exists, then $\lim_{x\to c} f(x) = \lim_{x\to c} [f(x) + g(x) - g(x)] = L - M$ also exists. Thus, $\lim_{x\to c} g(x)$ cannot exist.

47. True. If $\lim_{x\to c}\sqrt{f(x)} = L$ exists, then $\lim_{x\to c} f(x) = \lim_{x\to c}\sqrt{f(x)} \bullet \sqrt{f(x)} = L^2$ exists. 49. False. Let $f(x) = x, c = 0$

51. False. Let $f(x) = 1 - x^2, g(x) = 1 + x^2$, and $c = 0$

53. If $\lim_{x\to c} f(x) = L$ and $\lim_{x\to c} g(x) = L$, then

$$\lim_{x\to c} h(x) = \lim_{x\to c} \tfrac{1}{2}\{[f(x) + g(x)] - |f(x) - g(x)|\}$$

$$= \lim_{x\to c} \tfrac{1}{2}[f(x) + g(x)] - \lim_{x\to c} |f(x) - g(x)| = \tfrac{1}{2}(L + L) - \tfrac{1}{2}|L - L| = L$$

A similar argument works for H.

55. (a) Suppose $\lim_{x\to c} g(x) = k$ exists. Then $\lim_{x\to c} f(x) \bullet g(x) = 0 \bullet k = 0$. Thus $\lim_{x\to c} g(x)$ cannot exist. (b) $\lim_{x\to c} g(x) = \dfrac{1}{L}$

57. 5 **59.** $\frac{1}{4}$ **61.** (a) 1 (b) $2x$ (c) $3x^2$ (d) $4x^3$ (e) nx^{n-1}

SECTION 2.4

1. (a) $x = -3, x = 0, x = 2, x = 6$ (b) At -3, neither; at 0, continuous from the right; at 2, neither; at 6, neither

(c) removable discontinuity at $x = 2$; jump discontinuity at $x = 0$

3. continuous **5.** continuous **7.** continuous **9.** removable discontinuity **11.** jump discontinuity

13. continuous **15.** infinite discontinuity **17.** no discontinuities

19. no discontinuities **21.** infinite discontinuity at $x = 3$ **23.** no discontinuities

25. jump discontinuities at 0 and 2 **27.** removable discontinuity at -2; jump discontinuity at 3 **29.**

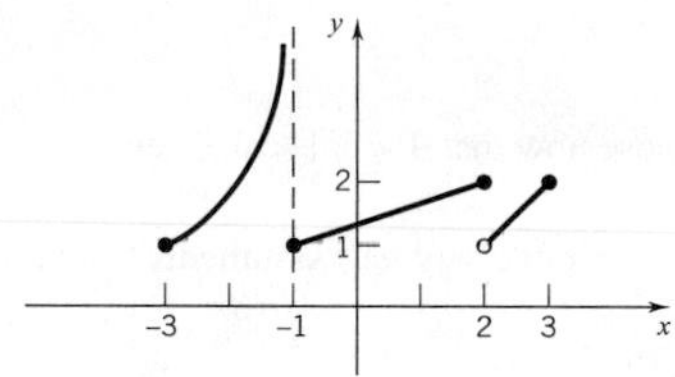

31. $f(1) = 2$ **33.** impossible **35.** 4 **37.** $A - B = 3$ with $B \neq 3$ **39.** $c = -3$ **41.** $f(5) = \frac{1}{6}$ **43.** $f(5) = \frac{1}{3}$

45. nowhere **47.** $x = 0, x = 2$, and all nonintegral values of x **49.** Refer to (2.2.6). Use the equivalence of (i) to (ii) setting $L = f(c)$.

51. Let $A = \{x_1, x_2, \ldots, x_n\}$ and let $\epsilon > 0$. In $A - \{c\}$ there is one point closest to c. Call it d and set $\delta_1 = |c - d|$. Note that if $0 < |x - c| < \delta_1$, then $f(x) = g(x)$.

Suppose now that g is continuous at c. Then there exists a positive number δ less than δ_1 such that

$$\text{if} \quad 0 < |x - c| < \delta, \quad \text{then} \quad |g(x) - g(c)| < \epsilon.$$

But for such $x, f(x) = g(x)$. Therefore, we see that

$$\text{if} \quad 0 < |x - c| < \delta, \quad \text{then} \quad |f(x) - g(c)| < \epsilon.$$

This means that

$$\lim_{x\to c} f(x) = g(c)$$

and contradicts the assumption that f has a nonremovable discontinuity at c.

53. The hypothesis implies that f is defined at c and $B > 0$. Let $\epsilon > 0$ and choose $\delta = \min(p, \epsilon/B)$. Then $|x - c| < \delta$ implies

$$|f(x) - f(c)| \leq B|x - c| < B(\epsilon/B) = \epsilon.$$

It now follows that f is continuous at c.

55. $\lim_{h\to 0} f(c + h) - f(c) = \lim_{h\to 0}\left[\dfrac{f(c + h) - f(c)}{h} \bullet h\right] = \lim_{h\to 0}\left[\dfrac{f(c + h) - f(c)}{h}\right] \bullet \lim_{h\to 0} h = L \bullet 0 = 0.$ Therefore f is continuous at c by Exercise 49.

57. $k \cong 1.61$ **59.** $\lim_{x\to 0^-} f(x) = -1$, $\lim_{x\to 0^+} f(x) = 1$; F is discontinuous at 0 for all k.

SECTION 2.5

1. 3 **3.** $\frac{3}{5}$ **5.** 2 **7.** 0 **9.** does not exist **11.** $\frac{9}{5}$ **13.** $\frac{2}{3}$ **15.** 1 **17.** $\frac{1}{2}$ **19.** -4 **21.** 1 **23.** $\frac{3}{5}$

25. 0 **27.** $\frac{2}{\pi}\sqrt{2}$ **29.** -1 **31.** 0

33. We will show that

$$\lim_{x\to c} \cos x = \cos c \quad \text{by showing that} \quad \lim_{h\to 0} \cos(c+h) = \cos c.$$

Note that $\cos(c + h) = \cos c \cos h - \sin c \sin h$. We know that

$$\lim_{h\to 0} \cos h = 1 \qquad \text{and} \qquad \lim_{h\to 0} \sin h = 0.$$

Therefore

$$\lim_{h\to 0} \cos(c + h) = (\cos c)(\lim_{h\to 0} \cos h) - (\sin c)(\lim_{h\to 0} \sin h) = (\cos c)(1) - (\sin c)(0) = \cos c.$$

Here is a different proof. The addition formula for the sine gives $\cos x = \sin(\frac{1}{2}\pi + x)$. Being the composition of functions that are everywhere continuous, the cosine function is itself everywhere continuous. Therefore

$$\lim_{x\to c} \cos x = \cos c \qquad \text{for all real } c.$$

35. 0 **37.** 1 **39.** $\frac{\sqrt{2}}{2}$; **41.** $-\sqrt{3}$; **43.** $0 \le |x \sin(1/x)| \le |x|$ for $x \ne 0$

45. $0 \le |x - 1|\,|\sin x| \le |x - 1|$ for all x **47.** $0 \le |xf(x)| \le B\,|x|$ for $x \ne 0$ **49.** $0 \le |f(x) - L| \le B|x - c|$ for $x \ne c$

51. $\lim_{x\to 0} g(x) = 1$ **53.** does not exist **55.** 1/3

SECTION 2.6

1. $f(1) = -1 < 0, f(2) = 6 > 0$ **3.** $f(0) = 2 > 0, f(\pi/2) = 1 - \pi^2/4 < 0$ **5.** $f(\frac{1}{4}) = \frac{1}{16} > 0, f(1) = -\frac{1}{2} < 0$

7. Let $f(x) = x^3 - \sqrt{x+2}$; $f(1) = 1 - \sqrt{3} < 0, f(2) = 6 > 0$ **9.** $(-\infty, 2) \cup (2, 5)$ **11.** $(-\infty, 0] \cup \{1\}$ **13.** $(1, 2) \cup (6, \infty)$

15. $(-\infty, -\frac{4}{3}) \cup [2, \infty)$ **17.** $(-\sqrt{6}, -\frac{3}{2}) \cup (-1, \frac{1}{2}) \cup (\sqrt{6}, \infty)$

19. Let $F(x) = x^5 - 2x^2 + 5x - 1$; $F(0) = -1 < 0$, $F(1) = 3 > 0$. Therefore there is a number $c \in (0, 1)$ such that $F(c) = 0$ which implies $f(c) = 1$.

21. $f(-3) = -13 < 0, f(-2) = 2 > 0$; $f(0) = 2 > 0, f(1) = -1 < 0, f(2) = 2 > 0$; f has a root in $(-3, -2)$, $(0, 1)$, and $(1, 2)$

23.
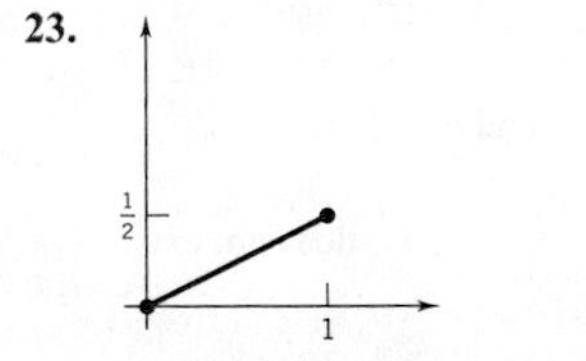

25.
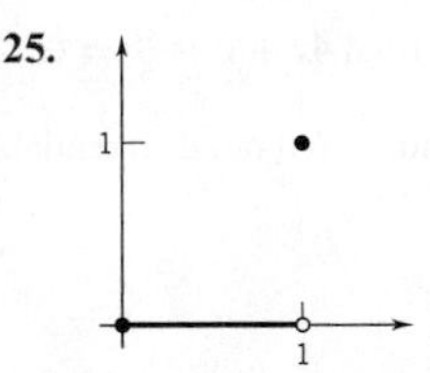

27.
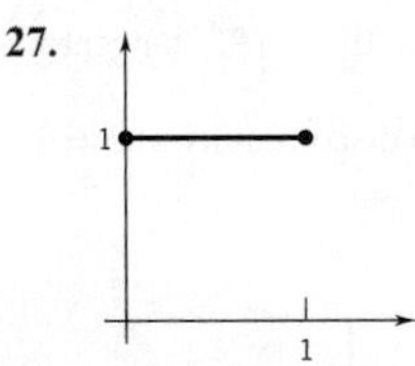

29.
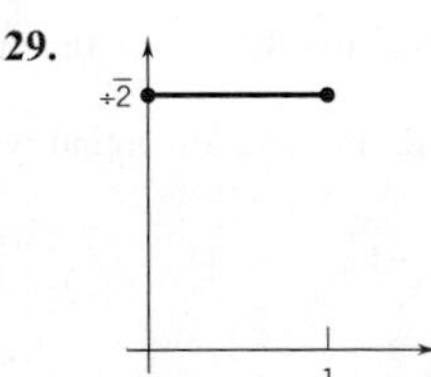

31. impossible by the intermediate-value theorem

33.
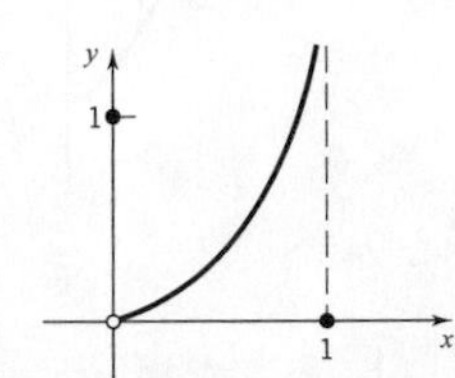

35. Set $g(x) = x - f(x)$. Since g is continuous on [0, 1] and $g(0) \le 0 \le g(1)$, there exists c in [0, 1] such that $g(c) = c - f(c) = 0$.

37. Let $f(x) = x^2$. Then f is continuous on $[0, b]$ for all $b > 0$. Since $x^2 \to \infty$ as $x \to \infty$, we can choose a number b such that $b^2 > a$. Now, $f(0) = 0 < a$ and $f(b) = b^2 > a$. Therefore, by the intermediate-value theorem there is a number c in $(0, b)$ such that $f(c) = a$.

39. The cubic polynomial $P(x) = x^3 + ax^2 + bx + c$ is continuous on $(-\infty, \infty)$. Writing P as

$$P(x) = x^3\left(1 + \frac{a}{x} + \frac{b}{x^2} + \frac{c}{x^3}\right), x \neq 0,$$

it follows that $P(x) < 0$ for large negative values of x and $P(x) > 0$ for large positive values of x. Thus, there exists a negative number N such that $P(x) < 0$ for $x < N$, and a positive number M such that $P(x) > 0$ for $x > M$. By the intermediate-value theorem, P has a zero on $[N, M]$.

41. The function T is continuous on $[4000, 4500]$; $T(4000) \cong 98.0995$ and $T(4500) \cong 97.9478$. Thus, by the intermediate-value theorem, there is an elevation h between 4000 and 4500 meters such that $T(h) = 98$.

43. A circle of radius r has area πr^2. Let $A(r) = \pi r^2, r \in [0, 10]$. Then $A(0) = 0$ and $A(10) = 100\pi \cong 314$. Since $0 < 250 < 314$, it follows from the intermediate-value theorem that there exists a number $c \in (0, 10)$ such that $A(c) = 250$

45. Inscribe a rectangle in a circle of radius R and then introduce a coordinate system as shown in the figure. Then the area of the rectangle is given by

$$A(x) = 4x\sqrt{R^2 - x^2}, x \in [0, R]$$

Since A is continuous on $[0, R]$, A has a maximum value. (A also has a minimum value, namely, 0.)

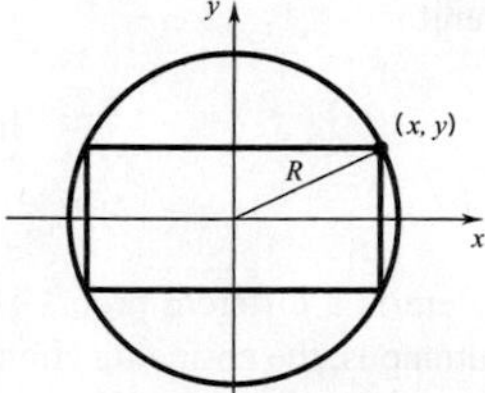

47. f has a zero on $(-3, -2)$, $(0, 1)$, $(1, 2)$; the zeros of f are $r_1 = -2.4909, r_2 = 0.6566, r_3 = 1.8343$

49. f has a zero on $(-2, -1)$, $(0, 1)$, $(1, 2)$; the zeros of f are $r_1 = -1.3482, r_2 = 0.2620, r_3 = 1.0816$

51. f is not continuous at $x = 1$. Therefore f does not satisfy the hypothesis of the intermediate-value theorem. $\dfrac{f(-3) + f(2)}{2} = \dfrac{167}{32} = f(c)$ for $c \cong 0.163$.

53. f satisfies the hypothesis of the intermediate-value theorem.

$$\frac{f(\pi/2) + f(2\pi)}{2} = \frac{1}{2} = f(c) \text{ for } c \cong 2.38, 4.16, 5.25.$$

55. f is bounded; the maximum value of f is 1, the minimum value is -1

57. f is bounded; no maximum value, the minimum value is approximately 0.3540

59. f satisfies the hypothesis of the extreme-value theorem on every interval of the form $[a, b]$ where $-2 \leq a < b \leq 2$. (a) $[-1, 1]$ (b) $[\frac{3}{2}, 2]$ (c) $[-2, 2]$

CHAPTER 3

SECTION 3.1

1. 2 **3.** 6 **5.** -2 **7.** -3 **9.** $5 - 2x$ **11.** $4x^3$ **13.** $\frac{1}{2}(x - 1)^{-1/2}$ **15.** $-2x^{-3}$

17. tangent $y + 3x - 16 = 0$; normal $3y - x - 8 = 0$ **19.** tangent $x - 4y + 3 = 0$; normal $4x + y + \frac{31}{4} = 0$

21. (a) Removable discontinuity at $c = -1$; jump discontinuity at $c = 1$ (b) f is continuous but not differentiable at $c = 0$ and $c = 3$

23. $x = -1$

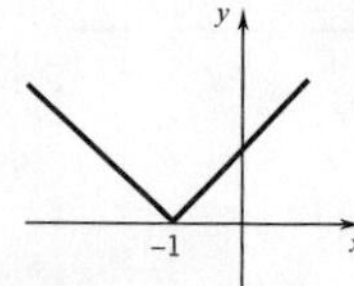

25. $x = 0$

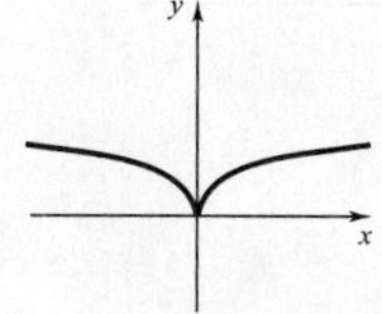

27. $x = 1$

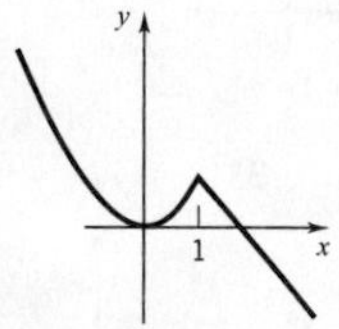

29. 4 **31.** does not exist

33.

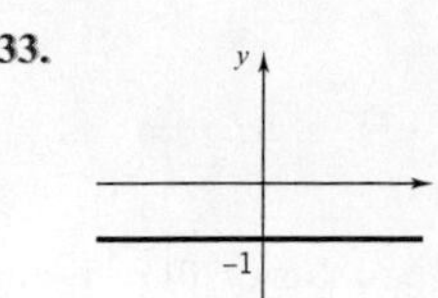

35.

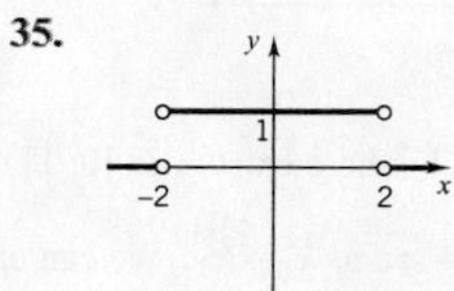

37.

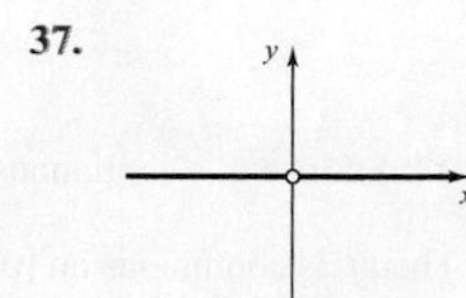

39. $f(x) = x^2; c = 1$ **41.** $f(x) = \sqrt{x}; c = 4$ **43.** $f(x) = \cos x; c = \pi$

45. Since $f(1) = 1$ and $\lim_{x\to 1^+} f(x) = 2$, f is not continuous at 1 and thus, by (3.1.5), is not differentiable at 1.

47. $A = 3,\ B = -2$

In 49–53, there are many possible answers. Here are some.

49. $f(x) = c$, c any constant **51.** $f(x) = |x+1|; f(x) = \begin{cases} 0, & x \neq -1 \\ 1, & x = -1 \end{cases}$ **53.** $f(x) = 2x + 5$

55. (a) $\lim_{x\to 2^-}(x^2 - x) = \lim_{h\to 2^+}(2x-2) = 2 = f(2)$ (b) $f'_-(2) = 3, f'_+(2) = 2$ (c) f is not differentiable at 2

57. (a) $f'(x) = \dfrac{-1}{2\sqrt{1-x}}$ (b) $f'_+(0) = -\frac{1}{2}$ (c) $f'_-(1)$ does not exist

59. No; $f(x) = |x|$ is differentiable on $[-1, 0]$ and on $[0, 1]$, but f is not differentiable on $[-1, 1]$ since it is not differentiable at $x = 0$.

61. Let $L = f'_-(c) = f'_+(c)$ and let $\epsilon > 0$. There exists $\delta_1 > 0$ such that

$$\left|\frac{f(c+h) - f(c)}{h} - L\right| < \epsilon$$

whenever $h \in (-\delta_1, 0)$. There exists $\delta_2 > 0$ such that

$$\left|\frac{f(c+h) - f(c)}{h} - L\right| < \epsilon$$

whenever $h \in (0, \delta_2)$. Let $\delta = \min(\delta_1, \delta_2)$. Then

$$\left|\frac{f(c+h) - f(c)}{h} - L\right| < \epsilon$$

whenever $|h| < \delta, h \neq 0$. Thus f is differentiable at c and $f'(c) = L$.

63. (a) $\lim_{x\to 0} x \sin\dfrac{1}{x} = 0 = f(0); \lim_{x\to 0} x^2 \sin\dfrac{1}{x} = 0 = g(0)$

(b) $\lim_{h\to 0} \dfrac{h\sin(1/h) - 0}{h} = \lim_{h\to 0} \sin\dfrac{1}{h}$ does not exist

(c) $\lim_{h\to 0} \dfrac{h^2\sin(1/h) - 0}{h} = \lim_{h\to 0} h \sin\dfrac{1}{h} = 0; g'(0) = 0$

65. $f'(1) = -1$ **67.** $f'(-1) = \frac{1}{3}$ **69.** (b) 7.071 (c) $D(0.001) \cong 7.074; D(-0.001) \cong 7.068$

71. (a) $f'(x) = \dfrac{5}{2\sqrt{5x-4}}; f'(3) = \dfrac{5}{2\sqrt{11}}$

(b) $f'(x) = -2x + 16x^3 - 6x^5; f'(-2) = 68$

(c) $f'(x) = -\dfrac{3(3-2x)}{(2+3x)^2} - \dfrac{2}{(2+3x)}; f'(-1) = -13$

73. (c) $f'(x) = 10x - 21x^2$

(d) $f'(x) = 0$ at $x = 0,\ 10/21$

75. (a) $f'(\frac{3}{2}) = -\frac{11}{4}$; $T: y - \frac{21}{8} = -\frac{11}{4}(x - \frac{3}{2}); N: y - \frac{21}{8} = \frac{4}{11}(x - \frac{3}{2})$

(c) (1.453, 1.547)

77. (a) $-3, 3$

(b) -2

(c) f is not differentiable at $x = -1$ since it is not defined there.

SECTION 3.2

1. -1 **3.** $55x^4 - 18x^2$ **5.** $2ax + b$ **7.** $\dfrac{2}{x^3}$ **9.** $3x^2 - 6x - 1$ **11.** $\dfrac{3x^2 - 2x^3}{(1-x)^2}$ **13.** $\dfrac{2(x^2+3x+1)}{(2x+3)^2}$ **15.** $8x^3 + 15x^2 - 8x - 10$

17. $-\dfrac{2(3x^2 - x + 1)}{x^2(x-2)^2}$ **19.** $-80x^9 + 81x^8 - 64x^7 + 63x^6$ **21.** $f'(0) = -\frac{1}{4}; f'(1) = -1$ **23.** $f'(0) = 0; f'(1) = -1$

25. $f'(0) = \dfrac{ad - bc}{d^2}; f'(1) = \dfrac{ad - bc}{(c+d)^2}$ **27.** $f'(0) = 3$ **29.** $f'(0) = \frac{20}{9}$ **31.** $2y - x - 8 = 0$ **33.** $y - 4x + 12 = 0$

35. $(-1, 27), (3, -5)$ **37.** $(-1, -\frac{5}{3}), (1, \frac{5}{3})$

39. (a) $-2, 0, 2$

(b) $(-2, 0) \cup (2, \infty)$

(c) $(-\infty, -2) \cup (0, 2)$

41. (a) 2

(b) $(-\infty, 0) \cup (2, \infty)$

(c) $(0, 2)$

43. $(-2, -10)$

45. $f(x) = x^3 + x^2 + x + C$, C any constant **47.** $f(x) = \frac{2}{3}x^3 - \frac{3}{2}x^2 + \frac{1}{x} + C$, C any constant **49.** $A = -2, B = -8$ **51.** $\frac{425}{8}$

53. $A = -1, B = 0, C = 4$ **55.** $x = -\dfrac{b}{2a}$ **57.** $c = -1, 1$

59. Let $a > 0$. An equation for the tangent line to the graph of $f(x) = 1/x$ at $x = a$ is $y = (-1/a^2)x + 2/a$. The y-intercept is $2/a$ and the x-intercept is $2a$. Thus, the area of the triangle formed by this line and the coordinate axes is $A = \frac{1}{2}(2/a)(2a) = 2$ square units.

61. $y = -x, y + 24 = 26(x + 3)$

63. Since f and $f + g$ are differentiable, $g = (f + g) - f$ is differentiable. The functions $f(x) = |x|$ and $g(x) = -|x|$ are not differentiable at 0; their sum $h(x) = 0$ is differentiable everywhere.

65. Since

$$\left(\frac{f}{g}\right)(x) = \frac{f(x)}{g(x)} = f(x) \cdot \frac{1}{g(x)},$$

it follows from the product and reciprocal rules that

$$\left(\frac{f}{g}\right)'(x) = \left(f \cdot \frac{1}{g}\right)'(x) = f(x)\left(-\frac{g'(x)}{[g(x)]^2}\right) + f'(x) \cdot \frac{1}{g(x)} = \frac{g(x)f'(x) - f(x)g'(x)}{[g(x)]^2}$$

67. $F'(x) = 2x\left(1 + \dfrac{1}{x}\right)(2x^3 - x + 1) + (x^2 + 1)\left(-\dfrac{1}{x^2}\right)(2x^3 - x + 1) + (x^2 + 1)\left(1 + \dfrac{1}{x}\right)(6x^2 - 1)$

69. $g(x) = [f(x)]^2 = f(x) \cdot f(x); g'(x) = f(x)f'(x) + f(x)f'(x) = 2f(x)f'(x)$ **71.** $g'(x) = 3(x^3 - 2x^2 + x + 2)^2(3x^2 - 4x + 1)$

73. (a) $0, -2$

(b) $(-\infty, -2) \cup (0, \infty)$

(c) $(-2, -1) \cup (-1, 0)$

75. (a) $f'(x) \neq 0$ for all $x \neq 0$

(b) $(0, \infty)$

(c) $(-\infty, 0)$

77. (a) Let $D(h) = \dfrac{\sin(x + h) - \sin x}{h}$.

At $x = 0, D(0.001) \cong 0.99999, D(-0.001) \cong 0.99999$; at $x = \pi/6, D(0.001) \cong 0.86578, D(-0.001) \cong 0.86628$; at $x = \pi/4, D(0.001) \cong 0.70675, D(-0.001) \cong 0.70746$; at $x = \pi/3, D(0.001) \cong 0.49957, D(-0.001) \cong 0.50043$; at $x = \pi/2, D(0.001) \cong -0.0005, D(-0.001) \cong 0.0005$

(b) $\cos(0) = 1, \cos(\pi/6) \cong 0.866025, \cos(\pi/4) \cong 0.707107, \cos(\pi/3) = 0.5, \cos(\pi/2) = 0$ (c) $f'(x) = \cos x$.

79. (a) Let $D(h) = \dfrac{2^{x+h} - 2^x}{h}$

At $x = 0, D(0.001) \cong 0.69339, D(-0.001) \cong 0.69291$; at $x = 1, D(0.001) \cong 1.38678, D(-0.001) \cong 1.38581$; at $x = 2, D(0.001) \cong 2.77355, D(-0.001) \cong 2.77163$; at $x = 3, D(0.001) \cong 5.54710, D(-0.001) \cong 5.54326$

(b) $\dfrac{f'(x)}{f(x)} \cong 0.693$ (c) $f'(x) \cong (0.693)\,2^x$

SECTION 3.3

1. $\dfrac{dy}{dx} = 12x^3 - 2x$ **3.** $\dfrac{dy}{dx} = 1 + \dfrac{1}{x^2}$ **5.** $\dfrac{dy}{dx} = \dfrac{1 - x^2}{(1 + x^2)^2}$ **7.** $\dfrac{dy}{dx} = \dfrac{2x - x^2}{(1 - x)^2}$ **9.** $\dfrac{dy}{dx} = \dfrac{-6x^2}{(x^3 - 1)^2}$ **11.** 2

13. $18x^2 + 30x + 5x^{-2}$ **15.** $\dfrac{2t^3(t^3 - 2)}{(2t^3 - 1)^2}$ **17.** $\dfrac{2}{(1 - 2u)^2}$ **19.** $-\left[\dfrac{1}{(u - 1)^2} + \dfrac{1}{(u + 1)^2}\right] = -2\left[\dfrac{u^2 + 1}{(u^2 - 1)^2}\right]$

21. $\dfrac{-2}{(x - 1)^2}$ **23.** 47 **25.** $\frac{1}{4}$ **27.** $42x - 120x^3$ **29.** $-6x^{-3}$ **31.** $4 - 12x^{-4}$ **33.** 2 **35.** 0 **37.** $6 + 60x^{-6}$

39. $1 - 4x$ **41.** -24 **43.** -24 **45.** $y = x^4 - \frac{1}{3}x^3 + 2x^2 + C$, C any constant **47.** $y = x^5 - x^{-4} + C$, C any constant

49. $p(x) = 2x^2 - 6x + 7$ **51.** (a) $n!$ (b) 0 (c) $f^{(k)}(x) = n(n-1)\cdots(n-k+1)x^{n-k}$

53. (a) $f'_+(0) = 0$ and $f'_-(0) = 0$. Thus $f'(0) = 0$.

(b) $f'(x) = \begin{cases} 2x, & x \geq 0 \\ 0, & x < 0 \end{cases}$

(c) $f''_+(0) = 2$ and $f''_-(0) = 0$. Thus $f''(0)$ does not exist.

(d)

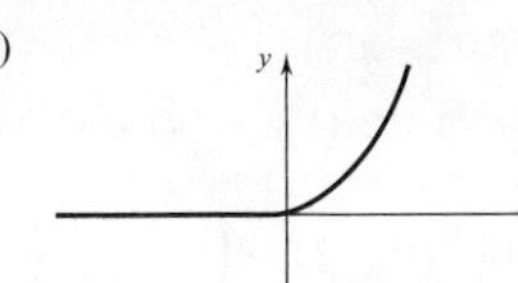

55. If $f(x) = g(x) = x$, then $(fg)(x) = x^2$, so that $(fg)''(x) = 2$, but $f(x)g''(x) + f''(x)g(x) = x \bullet 0 + 0 \bullet x = 0$.

57. (a) $x = 0$ (b) $x > 0$ (c) $x < 0$ **59.** (a) $x = -2, x = 1$ (b) $x < -2, x > 1$ (c) $-2 < x < 1$

61. The result is true for $n = 1: d^1y/dx^1 = dy/dx = -x^{-2} = (-1)^1 1!x^{-1-1}$. If the result is true for $n = k: d^ky/dx^k = (-1)^k k!x^{-k-1}$, then the result is true for $n = k + 1$:

$$\frac{d^{k+1}y}{dx^{k+1}} = \frac{d}{dx}\left[\frac{d^ky}{dx^k}\right] = (-1)^k k!(-k-1)x^{-k-2} = (-1)^{k+1}(k+1)!x^{-(k-1)-1}$$

63. $\dfrac{d}{dx}(uvw) = vw\dfrac{du}{dx} + uw\dfrac{dv}{dx} + uv\dfrac{dw}{dx}$ **65.** $\dfrac{d^{n+1}}{dx^{n+1}}[x^n] = 0$ for all positive integers n.

67. (b) $\left(-\dfrac{1}{\sqrt{2}}, \dfrac{1}{2\sqrt{2}}\right), \left(\dfrac{1}{\sqrt{2}}, -\dfrac{1}{2\sqrt{2}}\right)$

69. (a) $f'(x) = 3x^2 + 2x - 4$ (c) The graph is "falling" when $f'(x) < 0$; the graph is rising when $f'(x) > 0$. **71.** (a) $y = 4x + 1$ (c) $(6, 25)$

SECTION 3.4

1. $\dfrac{dA}{dr} = 2\pi r, 4\pi$ **3.** $\dfrac{dA}{dz} = z, 4$ **5.** $-\frac{5}{36}$ **7.** $\dfrac{dV}{dr} = 4\pi r^2$ **9.** $x_0 = \frac{3}{4}$ **11.** (a) $\dfrac{3\sqrt{2}}{4}w^2$ (b) $\dfrac{\sqrt{3}}{3}z^2$

13. (a) $\frac{1}{2}r^2$ (b) $r\theta$ (c) $-4Ar^{-3} = -2\theta/r$ **15.** $x = \frac{1}{2}$ **17.** $x(5) = -6$, $v(5) = -7$, $a(5) = -2$, speed $= 7$

19. $x(1) = 6$, $v(1) = -2$, $a(1) = \frac{4}{3}$, speed $= 2$ **21.** $x(1) = 0$, $v(1) = 18$, $a(1) = 54$, speed $= 18$ **23.** never **25.** at $-2 + \sqrt{5}$ **27.** A

29. A **31.** A and B **33.** A **35.** A and C **37.** $(0, 2)$, $(7, \infty)$ **39.** $(0, 3)$ $(4, \infty)$ **41.** $(2, 5)$

43. $(0, 2 - \frac{2}{3}\sqrt{3}), (4, \infty)$ **45.** 576 ft **47.** $v_0^2/2g$ meters

49. $y(t) = -\frac{1}{2}gt^2 + v_0t + y_0$. If $y(t_1) = y(t_2), t_1 \neq t_2$, then $v(t_1) = -gt_1 + v_0 = -(-g\,t_2 + v_0) = -v(t_2)$.

51. 9 ft/sec **53.** (a) 2 sec (b) 16 ft (c) 48 ft/sec **55.** (a) $\frac{1625}{16}$ ft (b) $\frac{6475}{64}$ ft (c) 100 ft **57.** 984 ft

59. $C'(100) = 0.04$

$C(101) - C(100) = 0.0401$

61. $C'(100) = 0$

$C(101) - C(100) = \frac{1}{10{,}100}$

63. $C'(10) = 23$

$C(11) - C(10) = 22.90$

65. (a) profit function: $P(x) = 16x - 1400 - x^2/50$; break-even points: $x_1 = 100, x_2 = 700$

(b) $P'(x) = 16 - x/25$; $P'(x) = 0$ at $x = 400$

(c)

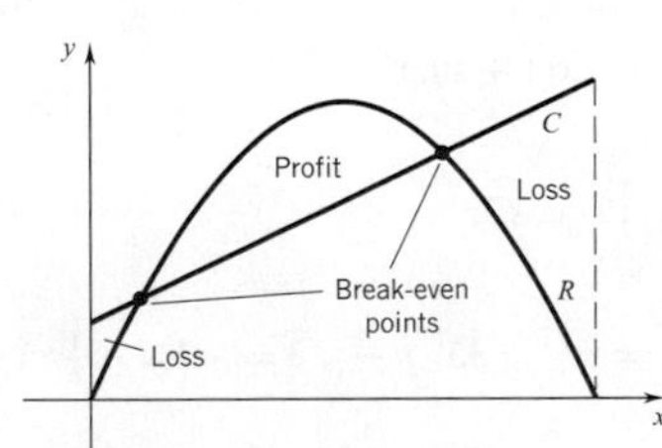

67. (a) 202.551 meters

(b) 11.531 seconds

69. (a) $v(t) = 3t^2 - 14t + 10$

(b) moving to the right when $0 < t < 0.88$ and when $3.79 < t < 5$; moving to the left when $0.88 < t < 3.79$

(c) the object stops at $t \cong 0.88$ and $t \cong 3.79$; the maximum speed is approximately 6.33

(d) $a(t) = 6t - 14$; slowing down when $0 < t < 0.88$ and when $2.33 < t < 3.79$; speeding up when $0.88 < t < 2.33$ and when $3.79 < t < 5$

SECTION 3.5

1. $f(x) = x^4 + 2x^2 + 1; f'(x) = 4x^3 + 4x = 4x(x^2 + 1)$

$f(x) = (x^2 + 1)^2; f'(x) = 2(x^2 + 1)(2x) = 4x(x^2 + 1)$

3. $f(x) = 8x^3 + 12x^2 + 6x + 1; f'(x) = 24x^2 + 24x + 6 = 6(2x + 1)^2$

$f(x) = (2x + 1)^3; f'(x) = 3(2x + 1)^2(2) = 6(2x + 1)^2$

5. $f(x) = x^2 + 2 + x^{-2}; f'(x) = 2x - 2x^{-3} = 2x(1 - x^{-4})$
$f(x) = (x + x^{-1})^2; f'(x) = 2(x + x^{-1})(1 - x^{-2}) = 2x(1 + x^{-2})(1 - x^{-2}) = 2x(1 - x^{-4})$

7. $2(1 - 2x)^{-2}$ **9.** $20(x^5 - x^{10})^{19}(5x^4 - 10x^9)$ **11.** $4\left(x - \frac{1}{x}\right)^3\left(1 + \frac{1}{x^2}\right)$ **13.** $4(x - x^3 - x^5)^3(1 - 3x^2 - 5x^4)$

15. $-4(t^{-1} + t^{-2})^3(t^{-2} + 2t^{-3})$ **17.** $324x^3\left[\frac{1 - x^2}{(x^2 + 1)^5}\right]$ **19.** $-\left(\frac{x^3}{3} + \frac{x^2}{2} + \frac{x}{1}\right)^{-2}(x^2 + x + 1)$

21. -1 **23.** 0 **25.** $\frac{dy}{dt} = \frac{dy}{du}\frac{du}{dx}\frac{dx}{dt} = \frac{7(2t - 5)^4 + 12(2t + 5)^2 - 2}{[(2t - 5)^4 + 2(2t - 5)^2 + 2]^2}[4(2t - 5)]$ **27.** 16 **29.** 1 **31.** 1 **33.** 1 **35.** 2

37. 0 **39.** $12(x^3 + x)^2\left[(3x^2 + 1)^2 + 2x(x^3 + x)\right]$ **41.** $\frac{6x(1 + x)}{(1 - x)^5}$ **43.** $2xf'(x^2 + 1)$ **45.** $2f(x)f'(x)$

47. (a) $x = 0$ (b) $x < 0$ (c) $x > 0$ **49.** (a) $x = -1, x = 1$ (b) $-1 < x < 1$ (c) $x < -1, x > 1$ **51.** at 3 **53.** at 2 and $2\sqrt{3}$

55. $y^{(n)} = \frac{n!}{(1 - x)^{n+1}}$ **57.** $y^{(n)} = n!b^n$ **59.** $y = (x^2 + 1)^3 + C$, C any constant **61.** $y = (x^3 - 2)^2 + C$, C any constant

63. $L'(x^2 + 1) = \frac{2x}{x^2 + 1}$ **65.** $T'(x) = 0$

67. If $p(x) = (x - a)^2q(x)$, where $q(a) \neq 0$, then $p(a) = p'(a) = 0$ and $p''(a) \neq 0$. Now suppose that $p(a) = p'(a) = 0$ and $p''(a) \neq 0$. Then $p(x) = (x - a)g(x)$ for some polynomial g. Since $p'(x) = (x - a)g'(x) + g(x)$ and $p'(a) = 0$, it follows that $g(a) = 0$. Therefore $g(x) = (x - a)q(x)$ for some polynomial q, and $p(x) = (x - a)^2q(x)$. Finally $p''(a) \neq 0$ implies $q(a) \neq 0$.

69. Let $p(x)$ be a polynomial function of degree n. The number a is a zero of multiplicity k for $p, k < n$, iff $p(a) = p'(a) = \cdots = p^{(k-1)}(a) = 0$ and $p^{(k)}(a) \neq 0$.

71. 800π cm³/sec **73.** $\frac{d(KE)}{dt} = mv\frac{dv}{dt}$ **77.** (a) $-\frac{f'(1/x)}{x^2}$ (b) $\frac{4xf'[(x^2 - 1)/(x^2 + 1)]}{(1 + x^2)^2}$ (c) $\frac{f'(x)}{[1 + f(x)]^2}$

79. $[g'(x)]^2f''[g(x)] + f'[g(x)]g''(x)$

SECTION 3.6

1. $\frac{dy}{dx} = -3\sin x - 4\sec x\tan x$ **3.** $\frac{dy}{dx} = 3x^2\csc x - x^3\csc x\cot x$ **5.** $\frac{dy}{dt} = -2\cos t\sin t$ **7.** $\frac{dy}{du} = 2u^{-1/2}\sin^3\sqrt{u}\cos\sqrt{u}$

9. $\frac{dy}{dx} = 2x\sec^2 x^2$ **11.** $\frac{dy}{dx} = 4(1 - \pi\csc^2\pi x)(x + \cot\pi x)^3$ **13.** $\frac{d^2y}{dx^2} = -\sin x$ **15.** $\frac{d^2y}{dx^2} = \cos x(1 + \sin x)^{-2}$

17. $\frac{d^2y}{du^2} = 12\cos 2u(2\sin^2 2u - \cos^2 2u)$ **19.** $\frac{d^2y}{dt^2} = 8\sec^2 2t\tan 2t$ **21.** $\frac{d^2y}{dx^2} = (2 - 9x^2)\sin 3x + 12x\cos 3x$

23. $\frac{d^2y}{dx^2} = 0$ **25.** $\sin x$ **27.** $(27t^3 - 12t)\sin 3t - 45t^2\cos 3t$ **29.** $3\cos 3xf'(\sin 3x)$ **31.** $y = x$ **33.** $y - \sqrt{3} = -4(x - \frac{1}{6}\pi)$

35. $y - \sqrt{2} = \sqrt{2}(x - \frac{1}{4}\pi)$ **37.** at π **39.** at $\frac{1}{6}\pi, \frac{7}{6}\pi$ **41.** at $\frac{1}{2}\pi, \pi, \frac{3}{2}\pi$ **43.** at $\frac{1}{4}\pi, \frac{3}{4}\pi, \frac{5}{4}\pi, \frac{7}{4}\pi$ **45.** at $\frac{7}{6}\pi, \frac{11}{6}\pi$

47. $(\frac{1}{3}\pi, \frac{2}{3}\pi), (\frac{7}{6}\pi, \frac{4}{3}\pi), (\frac{11}{6}\pi, 2\pi)$ **49.** $(0, \frac{1}{4}\pi), (\frac{7}{4}\pi, 2\pi)$ **51.** $(\frac{5}{6}\pi, \frac{3}{2}\pi)$

53. (a) $\frac{dy}{dt} = \frac{dy}{du}\frac{du}{dx}\frac{dx}{dt} = (2u)(\sec x\tan x)(\pi) = 2\pi\sec^2\pi t\tan\pi t$ (b) $y = \sec^2\pi t - 1$; $\frac{dy}{dt} = 2\sec\pi t(\sec\pi t\tan\pi t)\pi = 2\pi\sec^2\pi t\tan\pi t$

55. (a) $\frac{dy}{dt} = \frac{dy}{du}\frac{du}{dx}\frac{dx}{dt} = 4[\frac{1}{2}(1 - u)]^3(-\frac{1}{2}) \cdot (-\sin x) \cdot 2 = 4[\frac{1}{2}(1 - \cos 2t)]^3\sin 2t = (4\sin^6 t)(2\sin t\cos t) = 8\sin^7 t\cos t$
(b) $y = [\frac{1}{2}(1 - \cos 2t)]^4 = \sin^8 t$; $\frac{dy}{dt} = 8\sin^7 t\cos t$

57. $\dfrac{d^n}{dx^n}(\cos x) = \begin{cases} (-1)^{(n+1)/2}\sin x, & n \text{ odd} \\ (-1)^{n/2}\cos x, & n \text{ even} \end{cases}$

59. $\dfrac{d}{dx}(\cos x) = \dfrac{d}{dx}[\sin(\frac{\pi}{2} - x)] = \cos(\frac{\pi}{2} - x)(-1) = -\sin x$ **61.** $f'(0) = \lim\limits_{x\to 0} \dfrac{f(x) - f(0)}{x - 0} = \lim\limits_{x\to 0} \dfrac{\sin x}{x}$

63. $f(x) = 2\sin x + 3\cos x + C$, C any constant **65.** $f(x) = \sin 2x + \sec x + C$, C any constant

67. $f(x) = \sin(x^2) + \cos 2x + C$, C any constant

69. (a) $f'(x) = \sin\dfrac{1}{x} - \dfrac{1}{x}\cos\dfrac{1}{x}$; $g'(x) = 2x\sin\dfrac{1}{x} - \cos\dfrac{1}{x}$ (b) $\lim\limits_{x\to 0} g'(x) = \lim\limits_{x\to 0}\left(2x\sin\dfrac{1}{x} - \cos\dfrac{1}{x}\right) = \lim\limits_{x\to 0}\cos\dfrac{1}{x}$ does not exist

71. (a) $a = -\frac{1}{2}$, $b = \frac{\sqrt{3}}{2} + \frac{\pi}{3}$ (b)

73. $y'' = -A\omega^2 \sin\omega t - B\omega^2\cos\omega t = -\omega^2 y$. Thus $y'' + \omega^2 y = 0$. **75.** $A(x) = \frac{1}{2}c^2\sin x$; $A'(x) = \frac{1}{2}c^2\cos x$

77. (a) $\lim\limits_{\theta\to 0}\dfrac{\sin\theta}{\theta} \cong 0.01745$ (b) $\dfrac{\pi}{180} \cong 0.01745$

79. $A = 2$, $B = -\dfrac{3\sqrt{2}}{2}$

81. horizontal tangents at $\left(\dfrac{\pi}{6}, \dfrac{1}{4}\right), \left(\dfrac{\pi}{2}, 0\right), \left(\dfrac{5\pi}{6}, \dfrac{1}{4}\right), \left(\dfrac{3\pi}{2}, -2\right)$

83.

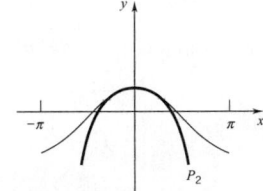

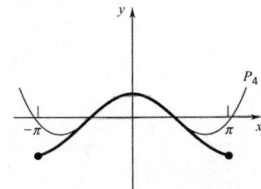

SECTION 3.7

1. $-\dfrac{x}{y}$ **3.** $-\dfrac{4x}{9y}$ **5.** $-\dfrac{x^2(x+3y)}{x^3+y^3}$ **7.** $\dfrac{2(x-y)}{2(x-y)+1}$ **9.** $\dfrac{y - \cos(x+y)}{\cos(x+y) - x}$ **11.** $\dfrac{16}{(x+y)^3}$ **13.** $\dfrac{90}{(2y+x)^3}$

15. $\dfrac{d^2y}{dx^2} = \frac{3}{2}x\cos^2 y - \frac{9}{8}x^4\sin y\cos^3 y$ **17.** $\dfrac{dy}{dx} = \dfrac{5}{8}, \dfrac{d^2y}{dx^2} = -\dfrac{9}{128}$ **19.** $\dfrac{dy}{dx} = -\dfrac{1}{2}, \dfrac{d^2y}{dx^2} = 0$

21. tangent $2x + 3y - 5 = 0$; normal $3x - 2y + 12 = 0$ **23.** tangent $x + 2y + 8 = 0$; normal $2x - y + 1 = 0$

25. tangent: $y = \dfrac{-2}{\sqrt{3}}x + \left(\dfrac{1}{\sqrt{3}} + \dfrac{\pi}{3}\right)$; normal: $y = \dfrac{\sqrt{3}}{2}x + \left(\dfrac{\pi}{3} - \dfrac{\sqrt{3}}{4}\right)$ **27.** $\frac{3}{2}x^2(x^3+1)^{-1/2}$ **29.** $\dfrac{x}{(\sqrt[4]{2x^2+1})^3}$

31. $\dfrac{x(2x^2-5)}{\sqrt{2-x^2}\sqrt{3-x^2}}$ **33.** $\dfrac{1}{2}\left(\dfrac{1}{\sqrt{x}} - \dfrac{1}{x\sqrt{x}}\right)$ **35.** $\dfrac{1}{(\sqrt{x^2+1})^3}$

37. (a) (b) (c) **39.** $-\dfrac{2b^2}{9\left(\sqrt[3]{a+bx}\right)^5}$

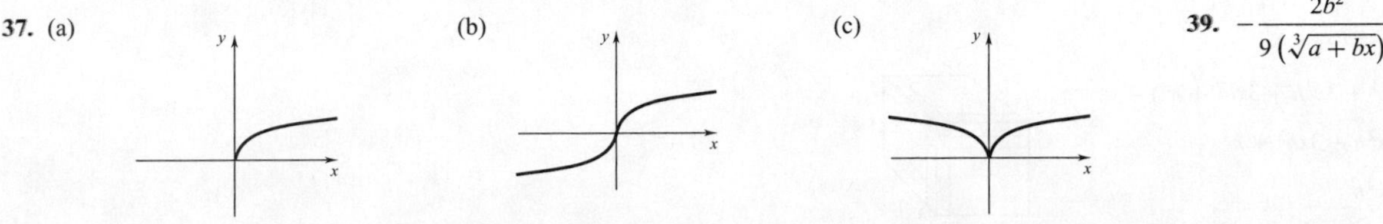

41. $\dfrac{\sqrt{x}\sec^2\sqrt{x} - \tan\sqrt{x} + 2x\sec^2\sqrt{x}\tan\sqrt{x}}{4x\sqrt{x}}$

43. Let (x_0, y_0) be a point of the circle. If $x_0 = 0$, the normal line is the y-axis; if $y_0 = 0$, the normal line is the x-axis. For all other choices of (x_0, y_0) the normal line takes the form

$$y - y_0 = \frac{y_0}{x_0}(x - x_0) \qquad \text{which reduces to} \qquad y = \frac{y_0}{x_0}x.$$

In each case the normal line passes through the origin.

45. at right angles **47.** at $(1,1), \alpha = \pi/4$; at $(0,0), \alpha = \pi/2$

49. The hyperbola and the ellipse intersect at the four points $(\pm 3, \pm 2)$. For the hyperbola, $\frac{dy}{dx} = \frac{x}{y}$. For the ellipse, $\frac{dy}{dx} = -\frac{4x}{9y}$. The product of these slopes is therefore $-\frac{4x^2}{9y^2}$. At each of the points of intersection this product is -1.

51. For the circles, $\frac{dy}{dx} = -\frac{x}{y}, y \neq 0$. For the straight lines $\frac{dy}{dx} = m = \frac{y}{x}, x \neq 0$. Thus, at a point of intersection of a circle $x^2 + y^2 = r^2$ and a line $y = mx$, we have $-\frac{x}{y} \cdot \frac{y}{x} = -1 (x \neq 0, y \neq 0)$.

53. $y - 2x + 12 = 0, y - 2x - 12 = 0$ **55.** $\left(-\frac{\sqrt{6}}{4}, \pm\frac{\sqrt{2}}{4}\right), \left(\frac{\sqrt{6}}{4}, \pm\frac{\sqrt{2}}{4}\right)$

57. Let (x_0, y_0) be a point on the graph. An equation for the tangent line at (x_0, y_0) is

$$y - y_0 = -\left(\frac{y_0}{x_0}\right)^{1/2}(x - x_0)$$

The y-intercept is $(x_0y_0)^{1/2} + y_0$; the x-intercept is $(x_0y_0)^{1/2} + x_0$; and $(x_0y_0)^{1/2} + y_0 + (x_0y_0)^{1/2} + x_0 = (x_0^{1/2} + y_0^{1/2})^2 = (c^{1/2})^2 = c$.

59. (b) $d(h) \to \infty$ as $h \to 0^-$ and as $h \to 0^+$ **61.** $f'(x) > 0$ on $(-\infty, \infty)$

(c) The graph of f has a vertical tangent at $(0, 0)$.

63.

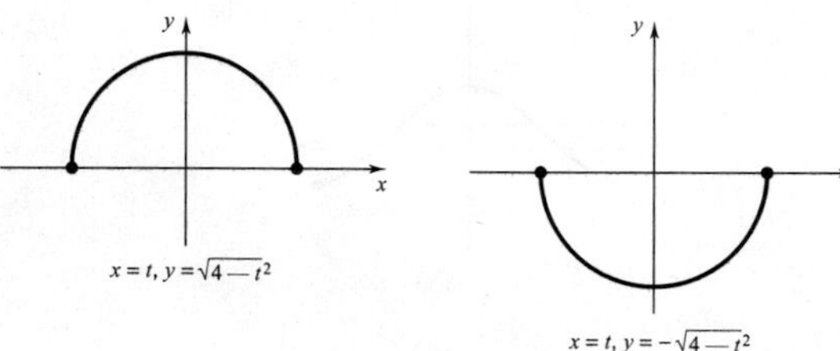

65. $y'\big|_{(3,4)} = 3$ **67.** (a)

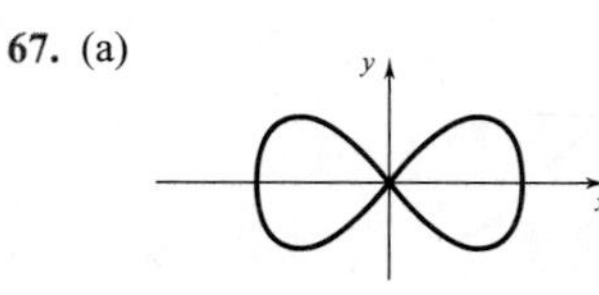

(b) $x = \pm\frac{\sqrt{2}}{2}$

SECTION 3.8

1. (a) -2 units/sec (b) 4 units/sec **3.** 9 units/sec **5.** $\frac{-12 \sin t \cos t}{\sqrt{16\cos^2 t + 4\sin^2 t}}; -\frac{3}{5}\sqrt{10}$ **7.** $-\frac{2}{27}$ m/min, $-\frac{8}{3}$ m²/min

9. (a) $A(\theta) = 50 \sin\theta$ (b) $\frac{25\pi}{18} \cong 4.36\text{cm}^2/\text{min}$ (c) $\theta = \frac{\pi}{2}$ **11.** 6 cm **13.** decreasing 7 in.²/sec **15.** boat A **17.** $-\frac{119}{5}$ ft²/sec

19. 10 ft³/hr **21.** $\frac{1600}{3}$ ft/min, $\frac{2800}{3}$ ft/min **23.** 0.5634 lb/sec **25.** dropping $1/2\pi$ in./min **27.** $10/\pi$ cm³/min

29. decreasing 0.04 rad / min **31.** 5π mi / min **33.** $\frac{4}{130} \cong 0.031$ rad / sec **35.** 0.1 ft / min **37.** decreasing 0.12 rad / sec

39. increasing $\frac{4}{101}$ rad/min **41.** $-\frac{15}{\sqrt{82}}$ ft/sec **43.** $\frac{23}{4}$ **45.** $-\frac{1}{2\sqrt{3}}$

SECTION 3.9

1.

$$\begin{aligned}\Delta V &= (x+h)^3 - x^3 \\ &= (x^3 + 3x^2h + 3xh^2 + h^3) - x^3 \\ &= 3x^2h + 3xh^2 + h^3 \\ dV &= 3x^2h \\ \Delta V - dV &= 3xh^2 + h^2 \qquad \text{(see figure)}\end{aligned}$$

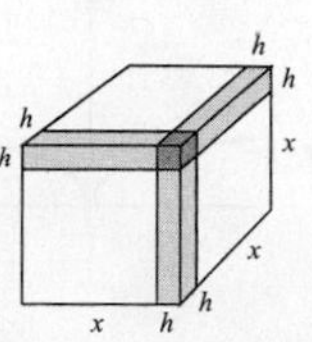

3. $10 + \frac{1}{150}$ taking $x = 1000$; 10.0067

5. $2 - \frac{1}{64}$ taking $x = 16$; 1.9844

7. 8.15 taking $x = 32$; 8.1491 **9.** 0.719; 0.7193 **11.** 0.531; 0.5317 **13.** 1.6 **15.** $2\pi rht$ **17.** error ≤ 0.01 ft **19.** 98 gallons

21. $P^2 = 4\pi^2 \frac{L}{g}$ **23.** 0.00307 sec **25.** within $\frac{1}{2}\%$ **27.** (a) $x_{n+1} = \frac{1}{2}x_n + 12\left(\frac{1}{x_n}\right)$ (b) $x_4 \cong 4.89898$

$$2P\frac{dP}{dL} = \frac{4\pi^2}{g} = \frac{P^2}{L}$$

$$\text{Thus}\ \frac{dP}{P} = \frac{1}{2}\frac{dL}{L}$$

29. (a) $x_{n+1} = \frac{2}{3}x_n + \frac{25}{3}\left(\frac{1}{x_n}\right)^2$ (b) $x_4 \cong 2.92402$ **31.** (a) $x_{n+1} = \dfrac{x_n \sin x_n + \cos x_n}{\sin x_n + 1}$ (b) $x_4 \cong 0.73909$

33. $x_2 = -2x_1, x_3 = 4x_1, x_4 = -8x_1, \ldots, x_n = (-2)^{n-1}x_1, \ldots$

35. (a) $x_1 = \frac{1}{2}, x_2 = -\frac{1}{2}, x_3 = \frac{1}{2}, \ldots, x_n = (-1)^{n-1}\frac{1}{2}, \ldots$ (b) $x_4 = 1.56165$ **37.** (b) $x_4 \cong 2.84382$; $f(x_4) \cong -0.00114$

39. (a) and (b) **41.** $\lim\limits_{h\to 0} \dfrac{g_1(h) + g_2(h)}{h} = \lim\limits_{h\to 0} \dfrac{g_1(h)}{h} + \lim\limits_{h\to 0} \dfrac{g_2(h)}{h} = 0 + 0 = 0$

$$\lim_{h\to 0} \frac{g_1(h)g_2(h)}{h} = \lim_{h\to 0} h\frac{g_1(h)g_2(h)}{h^2} = \left(\lim_{h\to 0} h\right)\left(\lim_{h\to 0} \frac{g_1(h)}{h}\right)\left(\lim_{h\to 0} \frac{g_2(h)}{h}\right) = (0)(0)(0) = 0$$

43. (b) finite (2)
(c) 0.8241
(d) by symmetry, -0.8241

45. (b) infinite
(c) 4.4934
(d) 0

CHAPTER 4

SECTION 4.1

1. $c = \dfrac{\sqrt{3}}{3} \cong 0.577$ **3.** $c = \dfrac{\pi}{4}, \dfrac{3\pi}{4}, \dfrac{5\pi}{4}, \dfrac{7\pi}{4}$ **5.** $c = \frac{3}{2}$ **7.** $c = \frac{1}{3}\sqrt{39}$ **9.** $c = \frac{1}{2}\sqrt{2}$ **11.** $c = 0$

13. No. By mean-value theorem there exists at least one number $c \in (0, 2)$ such that $f'(c) = \dfrac{f(2) - f(0)}{2 - 0} = \dfrac{3}{2}$.

15. $f'(x) = \begin{cases} 2, & x \leq -1 \\ 3x^2 - 1, & x > -1; \end{cases}$ $-3 < c \leq -1$ and $c = 1$

17. $\dfrac{f(b) - f(a)}{b - a} = A(b + a) + B$; $f'(c) = 2Ac + B$. Equating and solving for c gives $c = \dfrac{a + b}{2}$.

19. $\dfrac{f(1) - f(-1)}{(1) - (-1)} = 0$ and $f'(x)$ is never zero; f is not differentiable at 0.

21. Set $P(x) = 6x^4 - 7x + 1$. If there existed three numbers $a < b < c$ at which $P(x) = 0$, then by Rolle's theorem $P'(x)$ would have to be zero for some x in (a, b) and also for some x in (b, c). This is not the case: $P'(x) = 24x^3 - 7$ is zero only at $x = (\frac{7}{24})^{1/3}$.

23. Set $P(x) = x^3 + 9x^2 + 33x - 8$. Note that $P(0) < 0$ and $P(1) > 0$. Thus by the intermediate-value theorem there exists some number c between 0 and 1 at which $P(x) = 0$. If the equation $P(x) = 0$ had an additional real root, then by Rolle's theorem there would have to be some real number at which $P'(x) = 0$. This is not the case: $P'(x) = 3x^2 + 18x + 33$ is never 0 since the discriminant $b^2 - 4ac = (18)^2 - 12(33) < 0$.

25. Let c and d be two consecutive roots of the equation $P'(x) = 0$. The equation $P(x) = 0$ cannot have two more roots between c and d, for then, by Rolle's theorem, $P'(x)$ would have to be zero somewhere between these two roots and thus between c and d. In this case, c and d would no longer be consecutive roots of $P'(x) = 0$.

27. If $x_1, x_2 \in I$ are fixed points of f, then $g(x_1) = g(x_2) = 0$. By Rolle's theorem there exists a number $c \in (x_1, x_2)$ such that $g'(c) = 1 - f'(c) = 0$, which implies $f'(c) = 1$, contradicting the hypothesis.

29. (a) $f'(x) = 3x^2 - 3 = 3(x^2 - 1) \neq 0$ on $(-1, 1)$
(b) $-2 < b < 2$

31. $p'(x) = nx^{n-1} + a$ has at most one real zero.

33. If $x_1 = x_2$, then $|f(x_1) - f(x_2)|$ and $|x_1 - x_2|$ are both 0 and the inequality holds. If $x_1 \neq x_2$, then you know by the mean-value theorem that

$$\frac{f(x_1) - f(x_2)}{x_1 - x_2} = f'(c)$$

for some number c between x_1 and x_2. Since $|f'(c)| \leq 1$, you can conclude that

$$\left|\frac{f(x_1) - f(x_2)}{x_1 - x_2}\right| \leq 1 \quad \text{and thus that} \quad |f(x_1) - f(x_2)| \leq |x_1 - x_2|.$$

35. Set, for instance, $f(x) = \begin{cases} 1, & a < x < b \\ 0, & x = a, b \end{cases}$

37. (a) By the mean-value theorem, there exists $c \in (a, b)$ such that $f(b) - f(a) = f'(c)(b - a)$. Since $f'(x) \leq M$ for all $x \in (a, b)$, it follows that $f(b) \leq f(a) + M(b - a)$. (b) Same argument as (a). (c) $|f'(x)| \leq L$ implies $-L \leq f'(x) \leq L$.

39. $f(x)g'(x) - g(x)f'(x) = \cos^2 x + \sin^2 x = 1$ for all $x \in I$. The result follows from Exercise 38.

41. $$f'(x_0) = \lim_{h\to 0} \frac{f(x_0 + h) - f(x_0)}{h} \underset{\uparrow\ \text{by the hint}}{=} \lim_{h\to 0} \frac{f'(x_0 + \theta h)h}{h} = \lim_{h\to 0} f'(x_0 + \theta h) \underset{\uparrow\ \text{by (2.2.6)}}{=} \lim_{x\to x_0} f'(x) = L$$

43. Using the hint, F is continuous on $[a, b]$, differentiable on (a, b), and $F(a) = F(b)$. Therefore, by Exercise 42, there is at least one number $c \in (a, b)$ such that $F'(c) = 0$ and the result follows.

45. Let T be the time the horses finish the race. Then $f(t) = f_1(t) - f_2(t)$ satisfies the hypotheses in Exercise 42. Therefore, there exists at least one time $c \in (0, T)$ such that $f'(c) = 0$ and the result follows.

47. The driver must have exceeded the speed limit at some time during the trip. Let $s(t)$ denote the car's position at time t, with $s(0) = 0$ and $s(1.67) = 120$. Then, by the mean-value theorem, there exists at least one number (time) c such that

$$v(c) = s'(c) = \frac{s(1.67) - s(0)}{1.67 - 0} = \frac{120}{1.67} \cong 71.86 \text{ mi/hr}.$$

49. Yes. Since $\dfrac{f(6) - f(0)}{6 - 0} = \dfrac{280}{6} \cong 46.67$ ft/sec, the driver's speed at some time $c \in (0, 6)$ was 46.67 ft/sec $\cong 32$ mph, by the mean-value theorem. The driver's speed must have been greater than 32 mph at the instant the brakes were applied.

51. $\sqrt{65} \cong 8.0625$ **53.** $c = 0.676$ **55.** $c = 0.3045$ **57.** $c = 0$ **59.** $c = 2.205$

SECTION 4.2

1. increases on $(-\infty, -1]$ and $[1, \infty]$, decreases on $[-1, 1]$ **3.** increases on $(-\infty, -1]$ and $[1, \infty)$, decreases on $[-1, 0)$ and $(0, 1]$

5. increases on $[-\frac{3}{4}, \infty)$, decreases on $(-\infty, -\frac{3}{4}]$ **7.** increases on $[-1, \infty)$, decreases on $(-\infty, -1]$

9. increases on $(-\infty, 2)$, decreases on $(2, \infty)$ **11.** increases on $(-\infty, -1)$ and $(-1, 0]$, decreases on $[0, 1)$ and $(1, \infty)$

13. increases on $[-\sqrt{5}, 0]$ and $[\sqrt{5}, \infty)$, decreases on $(-\infty, -\sqrt{5}]$ and $[0, \sqrt{5}]$ **15.** increases on $(-\infty, -1)$ and $(-1, \infty)$

17. increases on $[0, \infty)$, decreases on $(-\infty, 0]$ **19.** increases on $[0, 2\pi]$

21. increases on $\left[\frac{2}{3}\pi, \pi\right]$, decreases on $\left[0, \frac{2}{3}\pi\right]$ **23.** increases on $\left[0, \frac{2}{3}\pi\right]$ and $\left[\frac{5}{6}\pi, \pi\right]$, decreases on $\left[\frac{2}{3}\pi, \frac{5}{6}\pi\right]$

25. $f(x) = \frac{1}{3}x^3 - x + \frac{8}{3}$ **27.** $f(x) = x^5 + x^4 + x^3 + x^2 + x + 5$ **29.** $f(x) = \frac{3}{4}x^{4/3} - \frac{2}{3}x^{3/2} + 1, x \geq 0$ **31.** $f(x) = 2x - \cos x + 4$

33. increases on $(-\infty, -3)$ and $[-1, 1]$, decreases on $[-3, -1]$ and $[1, \infty)$ **35.** increases on $(-\infty, 0]$ and $[3, \infty)$, decreases on $[0, 1)$ and $[1, 3]$

37.

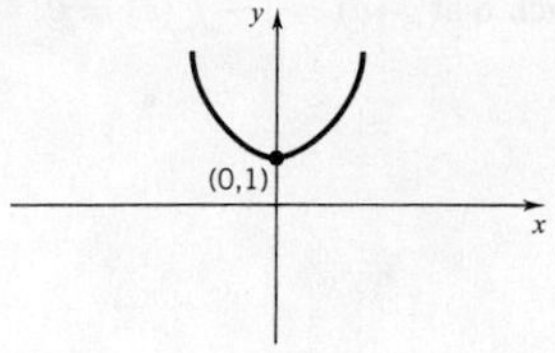

39.

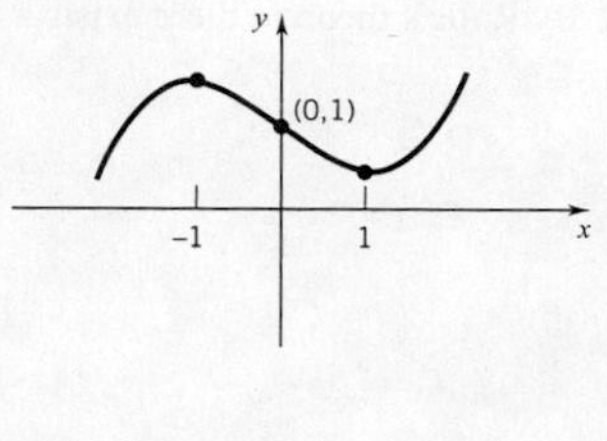

41.

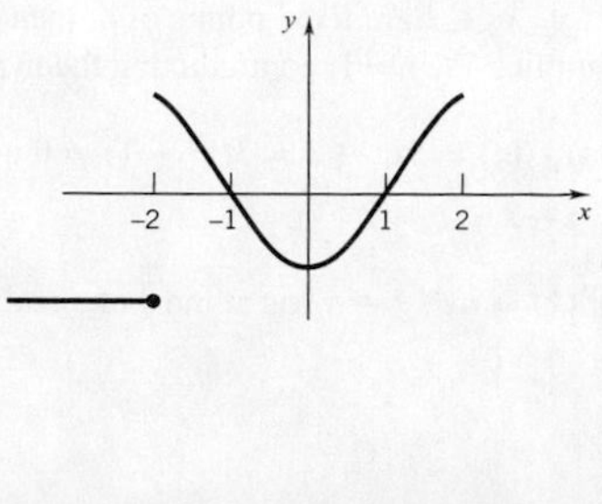

43.

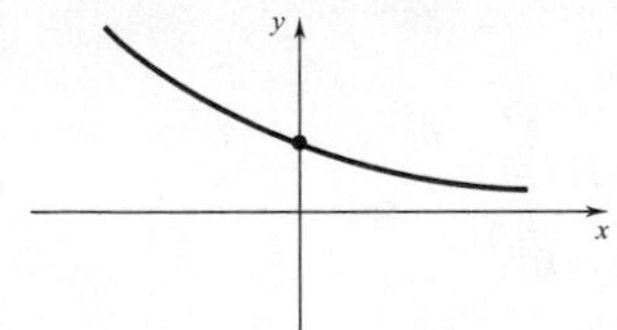

45. Not possible; f is increasing, so $f(2)$ must be greater than $f(-1)$.

47. $v(t) = 3t^2 - 12t + 9 = 3(t-1)(t-3)$

sign of v:

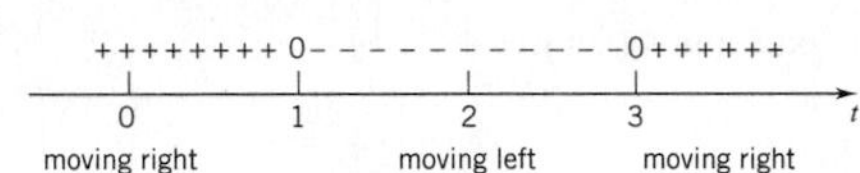

$a(t) = 6t - 12$

sign of a:

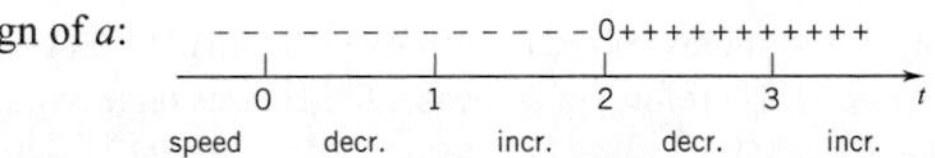

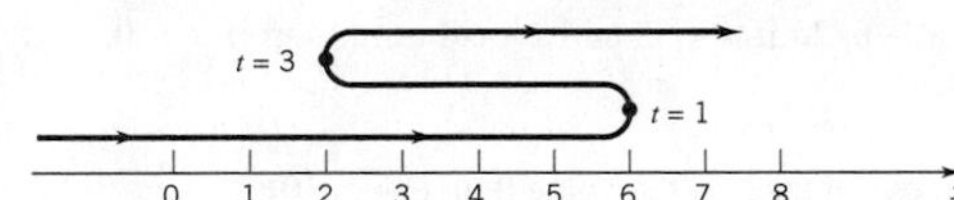

49. $v(t) = 6\cos 3t$

sign of v:

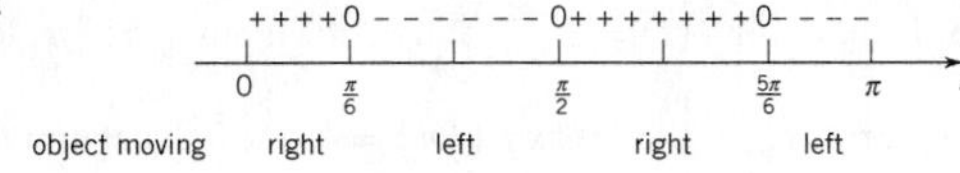

$a(t) = -18\sin 3t$

sign of a:

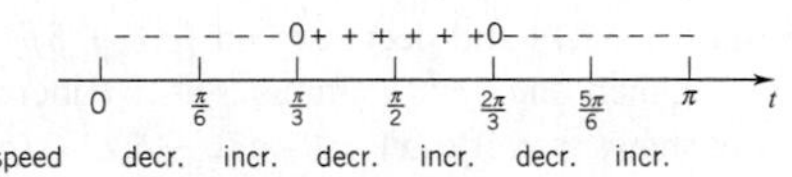

speed decr. incr. decr. incr. decr. incr.

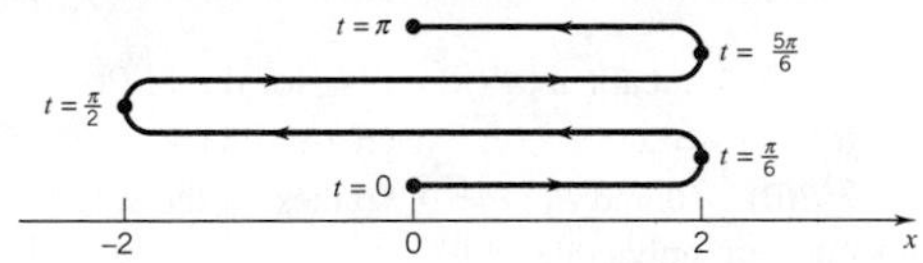

51. (a) $M \le L \le N$ (b) none (c) $M = L = N$

53. If there exists $c \in (a, b)$ such that $f'(c) < 0$, then for sufficiently small $h > 0, f(c-h) > f(c) > f(c+h)$ (Theorem 4.1.2) which contradicts the fact that f is increasing.

55. (a) $f'(x) = 1 - \cos x > 0$ except for $x = \dfrac{\pi}{2} + 2n\pi, n = 0, \pm 1, \pm 2, \ldots$. Thus f is increasing on $(-\infty, \infty)$. (b) Since f is increasing on $(-\infty, \infty)$ and $f(0) = 0, f(x) < 0$ on $(-\infty, 0)$ and $f(x) > 0$ on $(0, \infty)$. Thus $\sin x < x$ on $(-\infty, 0)$ and $\sin x > x$ on $(0, \infty)$.

57. (a) $f'(x) = 2\sec x(\sec x\tan x) = 2\sec^2 x\tan x$ $g'(x) = 2\tan x(\sec^2 x) = 2\sec^2 x\tan x$ (b) $C = 1$

59. (a) Let $h(x) = f^2(x) + g^2(x)$. Then $h'(x) = 0$; thus $h(x) = C$, constant.
(b) $f^2(a) + g^2(a) = C$ implies $C = 1; f(x) = \cos(x-a), g(x) = \sin(x-a)$.

61. Let $f(x) = \tan x$ and $g(x) = x$. Then $f(0) = g(0) = 0$ and $f'(x) = \sec^2 x > g'(x) = 1$ on $\left(0, \dfrac{\pi}{2}\right)$. The result follows from Exercise 60.

63. Choose an integer $n > 1$, and let $f(x) = (1+x)^n, g(x) = 1 + nx$. Then $f(0) = g(0) = 1$ and $f'(x) = n(1+x)^{n-1} > g'(x) = n$ since $(1+x)^{n-1} > 1$. The result follows from Exercise 60.

65. $0.06975 < \sin 4° < 0.06981$

67. $f'(x) = 0$ at $x = -0.633, 0.5, 2.633$
f is decreasing on $[-2, -0.633]$ and $[0.5, 2.633]$
f is increasing on $[-0.633, 0.5]$ and $[2.633, 5]$

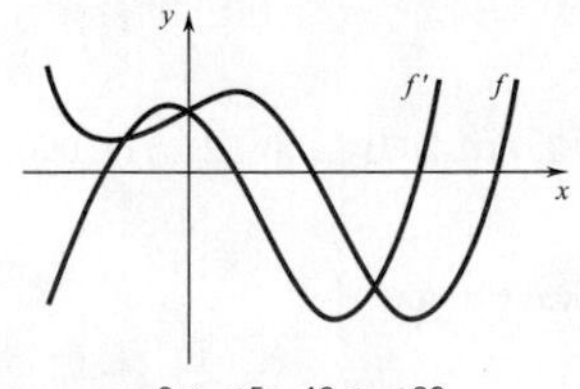

69. $f'(x) = 0$ at $x = 0.770, 2.155, 3.798, 5.812$
f is decreasing on $[0, 0.770]$, $[2.155, 3.798]$, and $[5.812, 6]$
f is increasing on $[0.770, 2.155]$ and $[3.798, 5.812]$

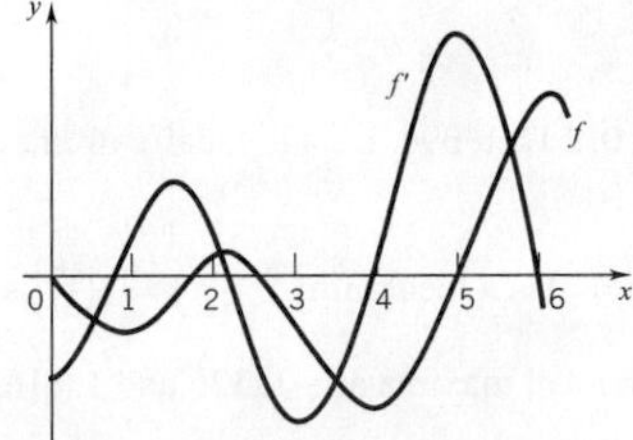

71. (a) $0, \frac{\pi}{2}, \pi, \frac{3\pi}{2}, 2\pi$
(b) $(\pi, \frac{3\pi}{2}) \cup (\frac{3\pi}{2}, 2\pi)$
(c) $(0, \frac{\pi}{2}) \cup (\frac{\pi}{2}, \pi)$

73. (a) 0
(b) $(0, \infty)$
(c) $(-\infty, 0)$

75. $f = C$ constant; $f'(x) = 0$

SECTION 4.3

1. no critical nos.; no local extreme values **3.** critical nos. ± 1; local max $f(-1) = -2$, local min $f(1) = 2$

5. critical nos. $0, \frac{2}{3}$; $f(0) = 0$ local min, $f(\frac{2}{3}) = \frac{4}{27}$ local max **7.** no critical nos.; no local extreme values

9. critical no. $-\frac{1}{2}$; local max $f\left(-\frac{1}{2}\right) = -8$ **11.** critical nos. $0, \frac{3}{5}, 1$; local max $f(\frac{3}{5}) = 2^2 3^3/5^5$, local min $f(1) = 0$

13. critical nos. $\frac{5}{8}, 1$; local max $f(\frac{5}{8}) = \frac{27}{2048}$ **15.** critical nos. $-2, 0$; local max $f(-2) = -4$, local min $f(0) = 0$

17. critical nos. $-2, -\frac{12}{7}, 0$; local max $f\left(-\frac{12}{7}\right) = \frac{144}{49}\left(\frac{2}{7}\right)^{1/3}$, local min $f(0) = 0$ **19.** critical nos. $-\frac{1}{2}, 3$; local min $f\left(-\frac{1}{2}\right) = \frac{7}{2}$

21. critical no. 1; local min $f(1) = 3$ **23.** critical nos. $\frac{1}{4}\pi, \frac{5}{4}\pi$, local max $f\left(\frac{1}{4}\pi\right) = \sqrt{2}$, local min $f\left(\frac{5}{4}\pi\right) = -\sqrt{2}$

25. critical nos. $\frac{1}{3}\pi, \frac{1}{2}\pi, \frac{2}{3}\pi$ local max $f\left(\frac{1}{2}\pi\right) = 1 - \sqrt{3}$, local min $f\left(\frac{1}{3}\pi\right) = f\left(\frac{2}{3}\pi\right) = -\frac{3}{4}$

27. critical nos. $\frac{1}{3}\pi, \frac{5}{3}\pi$, local max $f(\frac{5}{3}\pi) = \frac{5}{4}\sqrt{3} + \frac{10}{3}\pi$, local min $f(\frac{1}{3}\pi) = -\frac{5}{4}\sqrt{3} + \frac{2}{3}\pi$

29. (i) f increases on $(c - \delta, c]$ and decreases on $[c, c + \delta)$. (ii) f decreases on $(c - \delta, c]$ and increases on $[c, c + \delta)$. (iii) If $f'(x) > 0$ on $(c - \delta, c) \cup (c, c + \delta)$, then, since f is continuous at c, f increases on $(c - \delta, c]$ and also on $[c, c + \delta)$. Therefore, in this case, f increases on $(c - \delta, c + \delta)$. A similar argument shows that, if $f'(x) < 0$ on $(c - \delta, c) \cup (c, c + \delta)$, then f decreases on $(c - \delta, c + \delta)$.

31. $x = -b/2a$ is a critical number of f and $f''(x) = 2a$. Thus, f has a local maximum at $-b/2a$ if $a < 0$, and a local minimum if $a > 0$.

33. critical numbers 1, 2, 3; local max $P(2) = -4$, local min $P(1) = P(3) = -5$

Since $P'(x) < 0$ for $x < 0$, P decreases on $(-\infty, 0]$. Since $P(0) = 4$, P does not take on the value 0 on $(-\infty, 0]$.

Since $P(0) > 0$ and $P(1) < 0$, P takes on the value 0 at least once on $(0, 1)$. Since $P'(x) < 0$ on $(0, 1)$, P decreases on $[0, 1]$. It follows that P takes on the value zero only once on $[0, 1]$.

Since $P'(x) > 0$ on $(1, 2)$ and $P'(x) < 0$ on $(2, 3)$, P increases on $[1, 2]$ and decreases on $[2, 3]$. Since $P(1), P(2), P(3)$ are all negative, P cannot take on the value 0 between 1 and 3.

Since $P(3) < 0$ and $P(100) > 0$, P takes on the value 0 at least once on $(3, 100)$. Since $P'(x) > 0$ on $(3, 100)$, P increases on $[3, 100]$. It follows that P takes on the value zero only once on $[3, 100]$.

Since $P'(x) > 0$ on $(100, \infty)$, P increases on $[100, \infty)$. Since $P(100) > 0$, P does not take on the value 0 on $[100, \infty)$.

35.

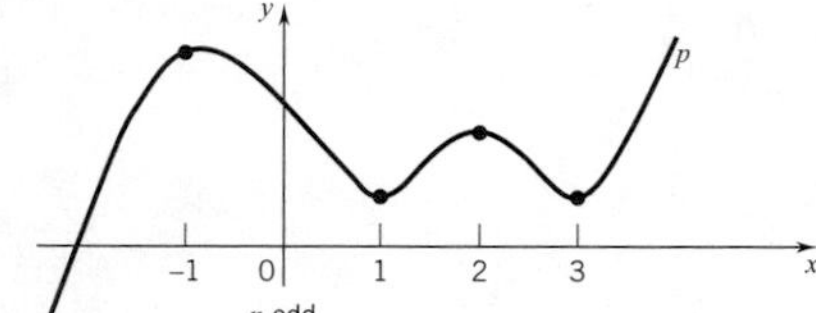

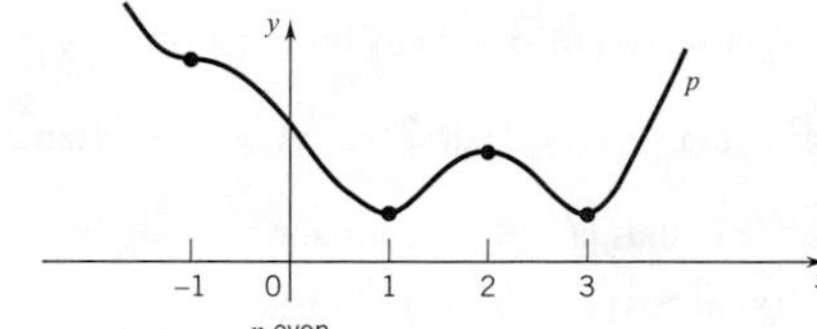

37. $a = 4, b = \pm 2$. **39.** The function f takes on both positive and negative values in every open interval containing $x = 0$.

41. The line through $(0, 0)$ and $(c, f(c))$ has equation $y = \dfrac{f(c)}{c}x$; the tangent line to the graph of f at c has equation $y - f(c) = f'(c)(x - c)$. If D has an extreme value at c, then $\dfrac{f(c)}{c} = -\dfrac{1}{f'(c)}$.

43. (a) $f'(x) = 4x^3 - 14x - 8$; $f'(2) = -4$, $f'(3) = 58$. Therefore f' has a zero in $(2, 3)$. Since $f''(x) = 12x^2 - 14 > 0$ on $[2, 3]$, f' has exactly one zero in $(2, 3)$. (b) $c \cong 2.1091$; f has a local minimum at c.

45. $f'(x) = \dfrac{2\,(ad - bc)x}{(cx^2 + d)^2}$; $x = 0$ is a critical number; $f''(0) = \dfrac{2\,(ad - bc)}{d^2}$; $ad - bc > 0$ implies $f(0)$ is a local minimum, $ad - bc < 0$ implies $f(0)$ is a local maximum.

47. (a) critical nos. $-2.085, -1, 0.207, 1.096, 1.544$; local extreme values: $f(-2.085) \cong -6.255$, $f(-1) = 7$, $f(0.207) \cong 0.621$, $f(1.096) \cong 7.097$, $f(1.544) \cong 4.635$

49. $f'(x) = 0$ at $x = 2, \frac{12}{5}, 3$; $f(2) = 0$ is a local min; $f\left(\frac{12}{5}\right) = \frac{108}{3125}$ is a local max. **51.** $f'(x) > 0$ on $(\frac{2}{3}, \infty)$; no local extrema.

53. critical nos. $-1.326, 0, 1.816$; local maxima at -1.326 and 1.816, local minimum at 0.

SECTION 4.4

1. $f(-2) = 0$ endpt min and absolute min

3. critical no. 2; $f(0) = 1$ endpt max and absolute max, $f(2) = -3$ local min and absolute min, $f(3) = -2$ endpt max.

5. critical no. $2^{-1/3}$; $f(2^{-2/3}) = 3 \cdot 2^{-2/3}$ local min

7. critical no. $2^{-1/3}$; $f\left(\frac{1}{10}\right) = 10\frac{1}{100}$ endpt max and absolute max; $f(2^{-1/3}) = 3 \cdot 2^{-2/3}$ local min and absolute min, $f(2) = 4\frac{1}{2}$ endpt max

9. critical no. $\frac{3}{2}$; $f(0) = 2$ endpt max and absolute max, $f\left(\frac{3}{2}\right) = -\frac{1}{4}$ local min and absolute min, $f(2) = 0$ endpt max

11. critical no. -2; $f(-3) = -\frac{3}{13}$ endpt max, $f(-2) = -\frac{1}{4}$ local min and absolute min, $f(1) = \frac{1}{5}$ endpt max and absolute max

13. critical nos. $\frac{1}{4}, 1$; $f(0) = 0$ endpt min and absolute min, $f\left(\frac{1}{4}\right) = \frac{1}{16}$ local max, $f(1) = 0$ local min and absolute min

15. critical no. 2; $f(2) = 2$ local max and absolute max, $f(3) = 0$ endpt min

17. critical no. 1; no extreme values

19. critical no. $\frac{5}{6}\pi$; $f(0) = -\sqrt{3}$ endpt min and absolute min, $f\left(\frac{5}{6}\pi\right) = \frac{7}{4}$ local max and absolute max, $f(\pi) = \sqrt{3}$ endpt min

21. $f(0) = 5$ endpt max and absolute max, $f(\pi) = -5$ endpt min and absolute min

23. critical no. 0; $f(-\frac{1}{3}\pi) = \frac{1}{3}\pi - \sqrt{3}$ endpt min and absolute min, no absolute max

25. critical nos. 1, 4; $f(0) = 0$ endpt, min, $f(1) = -2$ local min and absolute min, $f(4) = 1$ local max and absolute max, $f(7) = -2$ endpt min and absolute min

27. critical nos. $-1, 1, 3$; $f(-2) = 5$ endpt max, $f(-1) = 2$ local min and absolute min, $f(1) = 6$ local max and absolute max, $f(3) = 2$ local min and absolute min

29. critical nos. $-1, 0, 2$; $f(-3) = 2$ endpt max and absolute max, $f(-1) = 0$ local min, $f(0) = 2$ local max and absolute max, $f(2) = -2$ local min and absolute min

31.

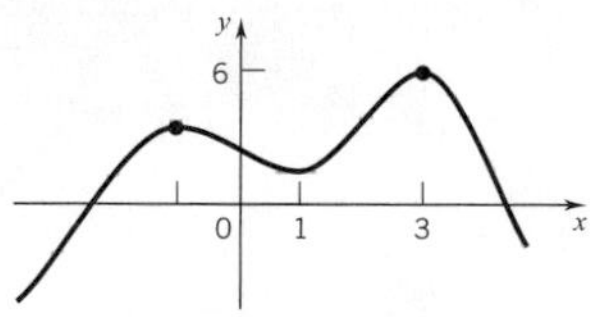

33. Not possible: $f(1) > 0$ and f is increasing on $(1, 3)$. Therefore $f(3)$ cannot be 0.

35. The discriminant of the quadratic polynomial $p'(x) = 3x^2 + 2ax + b$ is $4a^2 - 12b = 4(a^2 - 3b)$. If $a^2 \leq 3b$, then p' does not change sign.

37. By contradiction. If f is continuous at c, then $f(c)$ is not a local maximum by the first-derivative test (4.3.4).

39. If f is not differentiable on (a, b), then f has a critical number at each point c in (a, b) where $f'(c)$ does not exist. If f is differentiable on (a, b), then there exists c in (a, b) where $f'(c) = (f(b) - f(a))/(b - a)$ (mean-value theorem). With $f(b) = f(a)$, we have $f'(c) = 0$ and thus c is a critical number of f.

41. $f(x) = \begin{cases} 1 & \text{if } x \text{ is a rational number} \\ 0 & \text{if } x \text{ is an irrational number.} \end{cases}$

43. $P(x) - M \geq a_n x^n - (|a_{n-1}|x^{n-1} + \cdots + |a_1|x + |a_0| + M)$

for $x > 0$ ⟶ ↑

$\geq a_n x^n - (|a_{n-1}| + \cdots + |a_1| + |a_0| + M) \geq 0$ for $x \geq \left(\dfrac{|a_{n-1}| + \cdots + |a_1| + |a_0| + M}{a_n}\right)^{1/n}$

for $x > 1$ ⟶ ↑

45. Let x be the length and y the width of a rectangle with diagonal of length c. Then $y = \sqrt{c^2 - x^2}$ and the area $A = x\sqrt{c^2 - x^2}$. The maximum value of A occurs when $x = y = \dfrac{c}{\sqrt{2}}$.

47. $x = \pi/4$ **49.** critical nos.: -1.452, local max; 0.727, local min; $f(-2.5)$ absolute min; $f(3)$ absolute max;

51. critical nos.: -1.683, local max; -0.284, local min; 0.645, local max; 1.760, local min; $f(-\pi)$ absolute min; $f(\pi)$ absolute max.

53. Yes; $M = f(2) = 1$; $m = f(1) = f(3) = 0$ **55.** Yes; $M = f(6) = 2 + \sqrt{3}$; $m = f(1) = \frac{3}{2}$

57. $f(-2) = \frac{3}{4}$ absolute max.; $f(2[\sqrt{5} - 2]) \cong -0.05902$ absolute min. **59.** critical no. 0; $f(2) = 52$ absolute max.; $f(0) = 0$ absolute min.

SECTION 4.5

1. 400 **3.** 20 by 10 ft **5.** 32 **7.** 100 by 150 ft with the divider 100 ft long

9. radius of semi-circle $\dfrac{3p}{12 + 5\pi}$ **11.** $x = 2, y = \frac{3}{2}$ **13.** $-\frac{5}{2}$ **15.** $\frac{10}{3}\sqrt{3}$ by $\frac{5}{3}\sqrt{3}$ in.

17. equilateral triangle with side 4 **19.** (1, 1) **21.** height of rectangle $\frac{15}{11}(5-\sqrt{3}) \cong 4.46$ in; side of triangle $\frac{10}{11}(6+\sqrt{3}) \cong 7.03$ in.

23. $\frac{5}{3} \times \frac{5}{3}$ **25.** $(0, \sqrt{3})$ **27.** $5\sqrt{5}$ ft **29.** 54 by 72 in. **31.** (a) use it all for the circle
(b) use $28\pi/(4+\pi) \cong 12.32$ in. for the circle

33. base radius $\frac{10}{3}$ and height $\frac{8}{3}$ **35.** 10 by 10 by 12.5 ft **37.** equilateral triangle with side $2r\sqrt{3}$

39. base radius $\frac{1}{3}R\sqrt{6}$ and height $\frac{2}{3}R\sqrt{3}$

41. base radius $\frac{2}{3}R\sqrt{2}$ and height $\frac{4}{3}R$ **43.** \$160,000 **45.** $\tan\theta = m$ **47.** $x = \dfrac{a^{1/3}s}{a^{1/3}+b^{1/3}}$

49. The slope of the line through (a, b) and $(x, f(x))$ is $\dfrac{f(x)-b}{x-a}$. Let $D(x) = [x-a]^2 + [b-f(x)]^2$. Then $dD/dx = 0$ implies $[x-a]-[b-f(x)]f'(x) = 0$ which, in turn, implies $f'(x) = \dfrac{x-a}{b-f(x)}$.

51. $6\sqrt{6}$ ft **53.** 125 customers **55.** $\dfrac{dA}{dx} = \dfrac{xC'(x)-C(x)}{x^2}$; $\dfrac{dA}{dx} = 0$ implies $C'(x) = \dfrac{C(x)}{x}$. **57.** $m = 1$

59. $x = 55$ mph **61.** (b) $(1+\sqrt{2}, 2+\sqrt{2})$ **63.** $(\frac{21}{10}, \frac{7}{10})$ **65.** $x = \frac{3}{\sqrt{2}}$; max. area $A = 24$.

(c) $y-(2+\sqrt{2}) = \dfrac{1-\sqrt{2}}{3-\sqrt{2}}(x-[1+\sqrt{2}])$

(e) $l_{PQ} = l_N$

SECTION 4.6

1. (a) increasing on $[a, b]$, $[d, n]$, decreasing on $[b, d]$, $[n, p]$
(b) concave up on (c, k), (l, m), concave down on (a, c), (k, l), (m, p); points of inflection at $x = c, k, l$, and m.

3. concave down on $(-\infty, 0)$, concave up on $(0, \infty)$

5. concave down on $(-\infty, 0)$, concave up on $(0, \infty)$; pt of inflection $(0, 2)$

7. concave up on $(-\infty, -\frac{1}{3}\sqrt{3})$, concave down on $(-\frac{1}{3}\sqrt{3}, \frac{1}{3}\sqrt{3})$, concave up on $(\frac{1}{3}\sqrt{3}, \infty)$; pts of inflection $(-\frac{1}{3}\sqrt{3}, -\frac{5}{36})$, $(\frac{1}{3}\sqrt{3}, -\frac{5}{36})$

9. concave down on $(-\infty, -1)$ and on $(0, 1)$, concave up on $(-1, 0)$ and on $(1, \infty)$; pt of inflection $(0, 0)$

11. concave up on $(-\infty, -\frac{1}{3}\sqrt{3})$, concave down on $(-\frac{1}{3}\sqrt{3}, \frac{1}{3}\sqrt{3})$, concave up on $(\frac{1}{3}\sqrt{3}, \infty)$; pts of inflection $(-\frac{1}{3}\sqrt{3}, \frac{4}{9})$, $(\frac{1}{3}\sqrt{3}, \frac{4}{9})$

13. concave up on $(0, \infty)$ **15.** concave down on $(-\infty, -2)$, concave up on $(-2, \infty)$; pt of inflection $(-2, 0)$

17. concave up on $(0, \frac{1}{4}\pi)$, concave down on $(\frac{1}{4}\pi, \frac{3}{4}\pi)$, concave up on $(\frac{3}{4}\pi, \pi)$; pts of inflection $(\frac{1}{4}\pi, \frac{1}{2})$ and $(\frac{3}{4}\pi, \frac{1}{2})$

19. concave up on $(0, \frac{1}{12}\pi)$, concave down on $(\frac{1}{12}\pi, \frac{5}{12}\pi)$, concave up on $(\frac{5}{12}\pi, \pi)$; pts of inflection $(\frac{1}{12}\pi, \frac{1}{2}+\frac{1}{144}\pi^2)$ and $(\frac{5}{12}\pi, \frac{1}{2}+\frac{25}{144}\pi^2)$

21. $(\pm 3.94822, 10.39228)$

23. $(-3, 0)$, $(-2.11652, 2.39953)$, $(-0.28349, -18.43523)$

25. (a) f increases on $(-\infty, -\sqrt{3}]$ and $[\sqrt{3}, \infty)$, decreases on $[-\sqrt{3}, \sqrt{3}]$
(b) $f(-\sqrt{3}) \cong 10.39$ local max; $f(\sqrt{3}) \cong -10.39$ local min.
(c) concave down on $(-\infty, 0)$, concave up on $(0, \infty)$
(d) $(0, 0)$ is a point of inflection

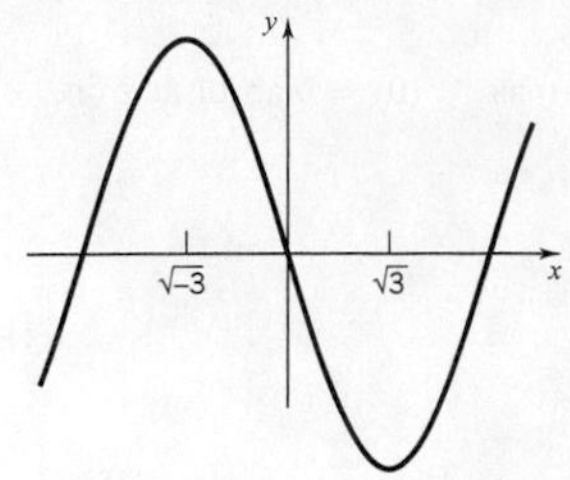

27. (a) f decreases on $(-\infty, -1]$ and $[1, \infty)$, increases on $[-1, 1]$
(b) $f(-1) = -1$ local min; $f(1) = 1$ local max
(c) concave down on $(-\infty, -\sqrt{3})$ and $(0, \sqrt{3})$, concave up on $(-\sqrt{3}, 0)$ and $(\sqrt{3}, \infty)$
(d) points of inflection $\left(-\sqrt{3}, -\dfrac{\sqrt{3}}{2}\right)$, $(0, 0)$, $\left(\sqrt{3}, \dfrac{\sqrt{3}}{2}\right)$

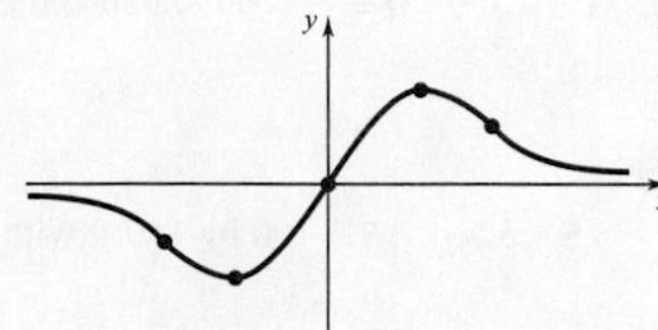

29. (a) increasing on $[-\pi, \pi]$

(b) no local max or min

(c) concave up on $(-\pi, 0)$, concave down on $(0, \pi)$

(d) point of inflection $(0, 0)$

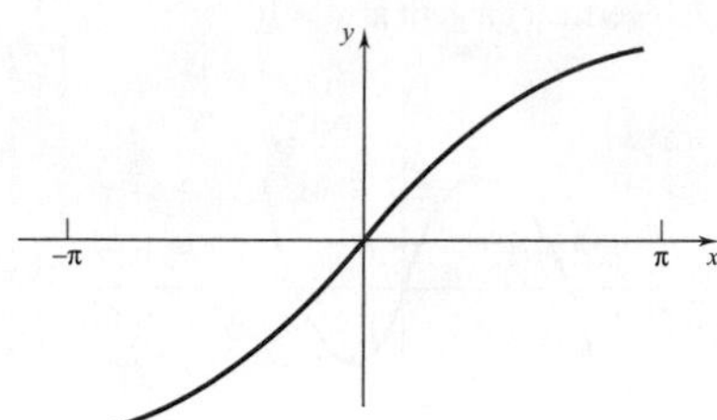

31. (a) increasing on $(-\infty, \infty)$

(b) no local max or min

(c) concave down on $(-\infty, 0)$, concave up on $(0, 1)$, no concavity on $(1, \infty)$

(d) point of inflection $(0, 0)$

$y = 3x - 2$

$y = x^3$

33.

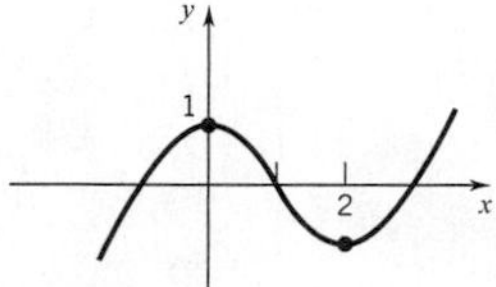

35.

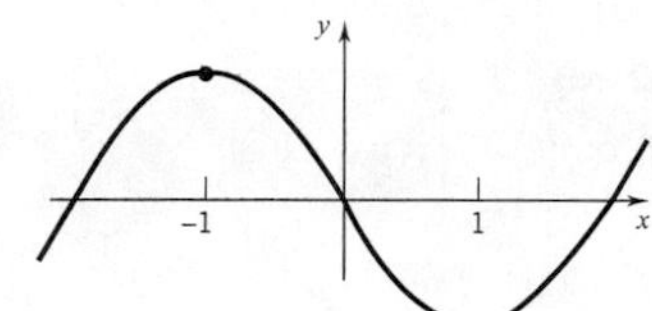

37. $d = \frac{1}{3}(a + b + c)$ **39.** $a = -\frac{1}{2}, b = \frac{1}{2}$ **41.** $A = 18, B = -4$ **43.** $f(x) = x^3 - 3x^2 + 3x - 3$

45. (a) $p''(x) = 6x - 2a$ has exactly one zero

(b) $p'(x) = 3x^2 + 2ax + b$ has real zeros iff $a^2 > 3b$.

47. (a)

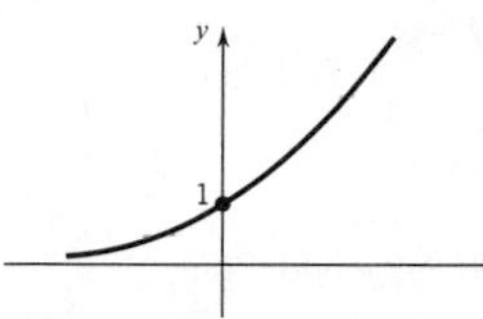

(b) No. If $f''(x) < 0$ and $f'(x) < 0$ for all x, then $f(x) < f'(0)x + f(0)$ on $(0, \infty)$ which implies $f(x) \to -\infty$ as $x \to \infty$.

49. (a) concave up on $(-4, -0.913)$ and $(0.913, 4)$; concave down on $(-0.913, 0.913)$

(b) points of inflection at $x \cong -0.913, 0.913$

51. (a) concave up on $(-\pi, -1.996)$ and $(-0.345, 2.550)$; concave down on $(-1.996, -0.345)$ and $(2.550, \pi)$

(b) points of inflection at $x \cong -1.996, -0.345, 2.550$

53. (a) $0.68824, 2.27492, 4.00827, 5.59494$

(b) $(0.68824, 2.27492) \cup (4.00827, 5.59494)$

(c) $(0, 0.68824) \cup (2.27492, 4.00827) \cup (5.59494, 2\pi)$

55. (a) $-1, \pm 0.71523, \pm 0.32654, 0, 1$

(b) $(-0.71523, -0.32654) \cup (0, 0.32654) \cup (0.71523, 1), (1, \infty)$

(c) $(-\infty, -1) \cup (-1, -0.71523) \cup (-0.32654, 0) \cup (0.32654, 0.71523)$

SECTION 4.7

1. (a) ∞ (b) $-\infty$ (c) ∞ (d) 1 (e) 0 (f) $x = -1, x = 1$ (g) $y = 0, y = 1$ **3.** vertical: $x = \frac{1}{3}$; horizontal: $y = \frac{1}{3}$

5. vertical: $x = 2$; horizontal: none **7.** vertical: $x = \pm 3$; horizontal: $y = 0$ **9.** vertical: $x = -\frac{4}{3}$; horizontal: $y = \frac{4}{9}$

11. vertical: $x = \frac{5}{2}$; horizontal; $y = 0$ **13.** vertical: none; horizontal: $y = \pm\frac{3}{2}$ **15.** vertical: $x = 1$; horizontal: $y = 0$

17. vertical: none; horizontal: $y = 0$ **19.** vertical: $x = (2n + \frac{1}{2})\pi$; horizontal: none **21.** neither **23.** cusp

25. tangent **27.** neither **29.** cusp **31.** cusp **33.** neither; f not continuous at $x = 0$

35.

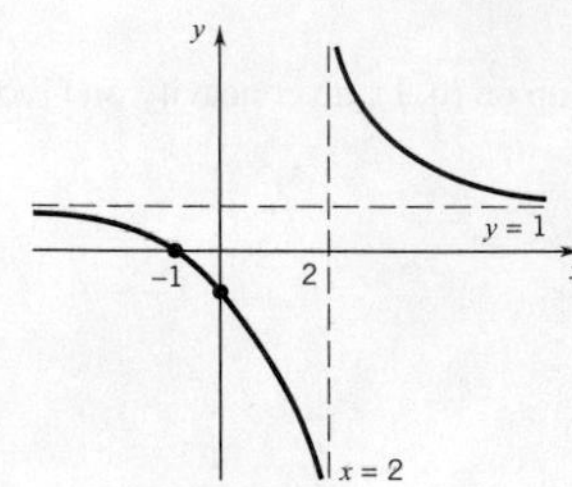

37.

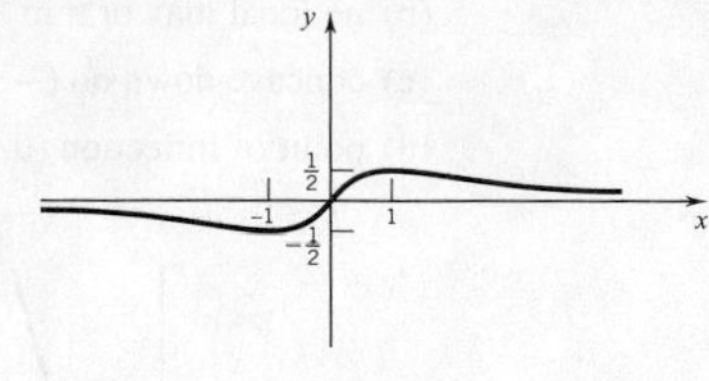

39. (a) increasing on $(-\infty, -1], [1, \infty)$; decreasing on $[-1, 1]$
(b) concave down on $(-\infty, 0)$; concave up on $(0, \infty)$

vertical tangent at $x = 0$

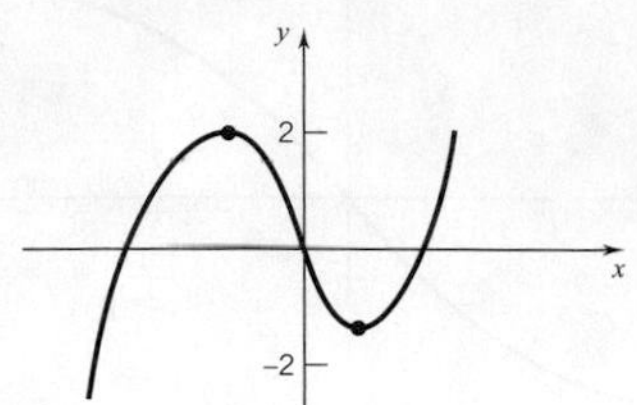

41. (a) decreasing on $[0, 2]$; increasing on $(-\infty, 0], [2, \infty)$
(b) concave up on $(-1, 0), (0, \infty)$; concave down on $(-\infty, -1)$ vertical cusp at $x = 0$

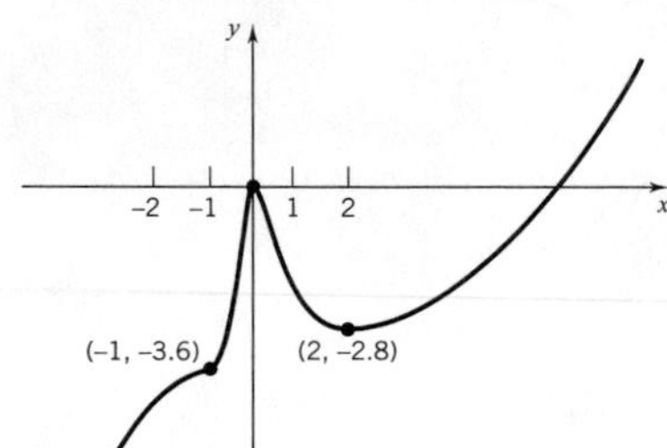

43. $x = 1$ vertical asymptote
$y = 0$ and $y = 2$ horizontal asymptotes

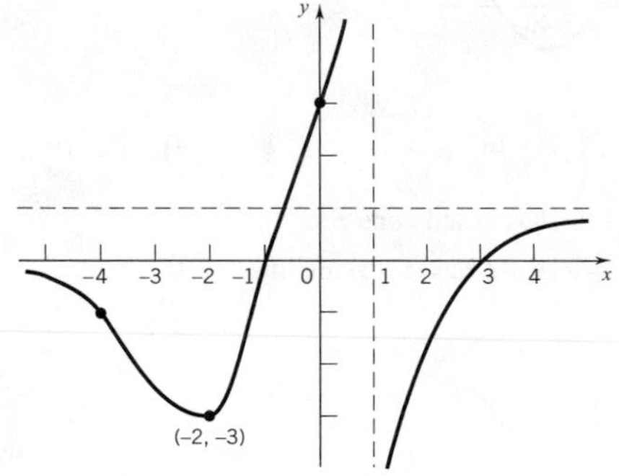

45. vertical cusp at $x = 0$

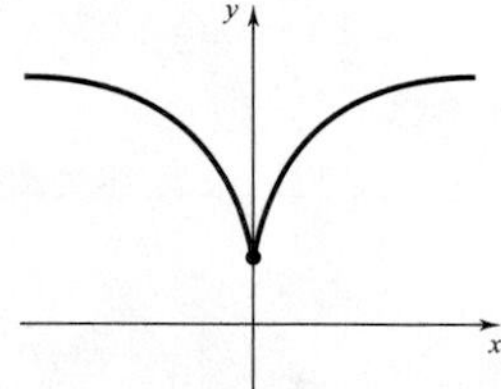

47. vertical tangent line at $x = 0$
$y = 1$ and $y = -1$ horizontal asymptotes

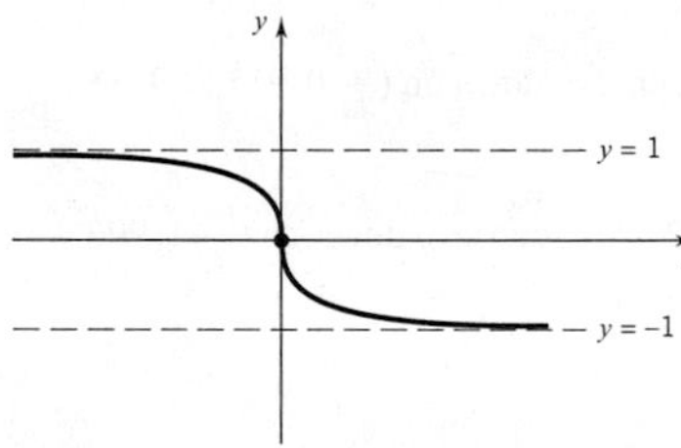

49. (a) p odd
(b) p even

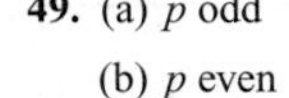

51.

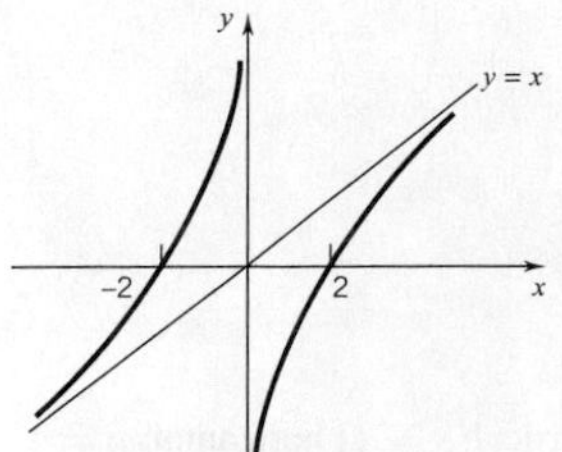

53.

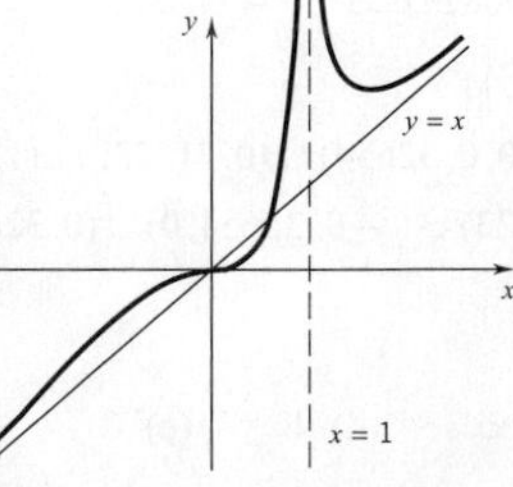

55. $-\frac{1}{2}$ **57.** $y = 3x - 4$

59. (a) $y = 0$
(b) no horizontal asymptotes
(c) no horizontal asymptotes

61. $N = 252$

SECTION 4.8

1.
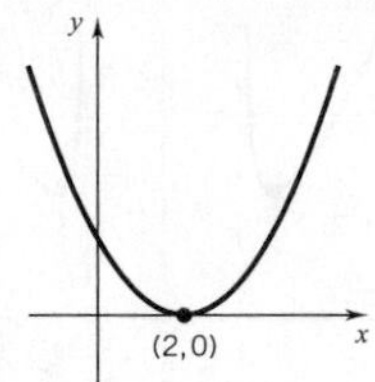

3.
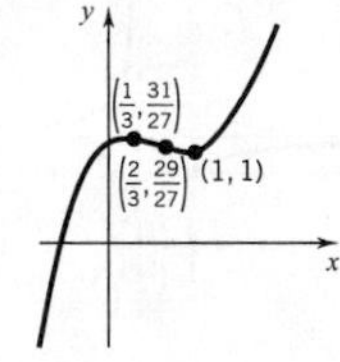

5.
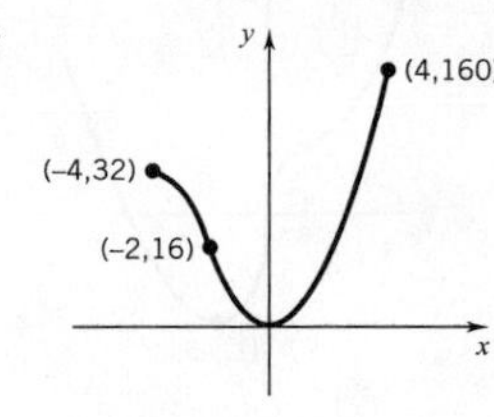

7.
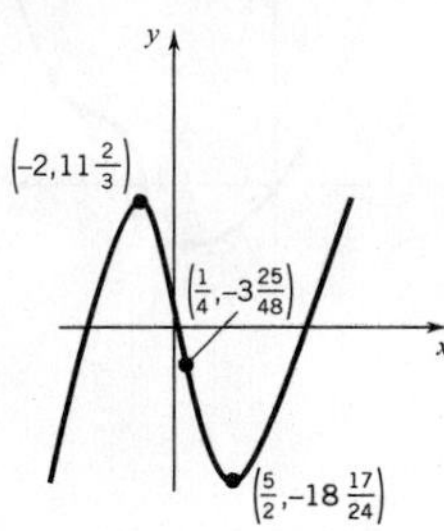

9.
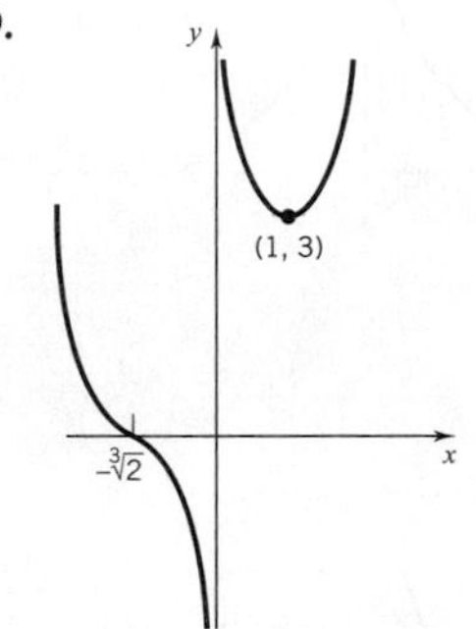

11.
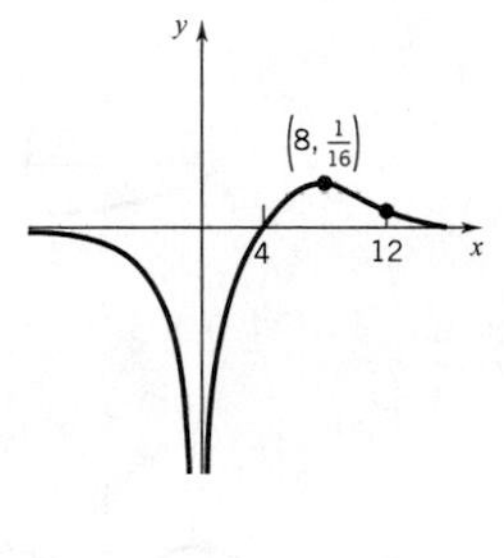

13.
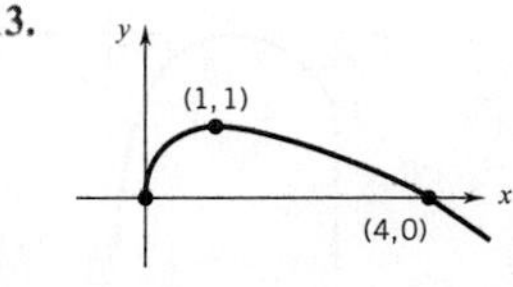

15.
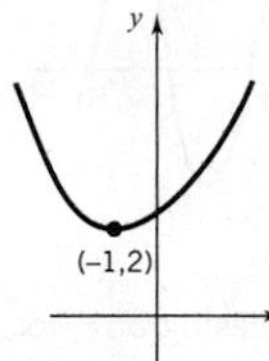

17.
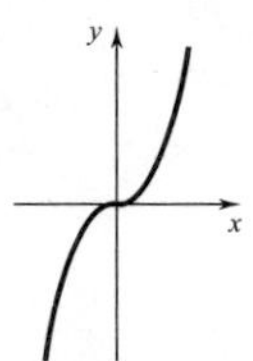

19.
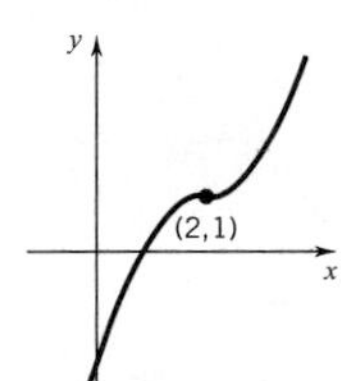

21.
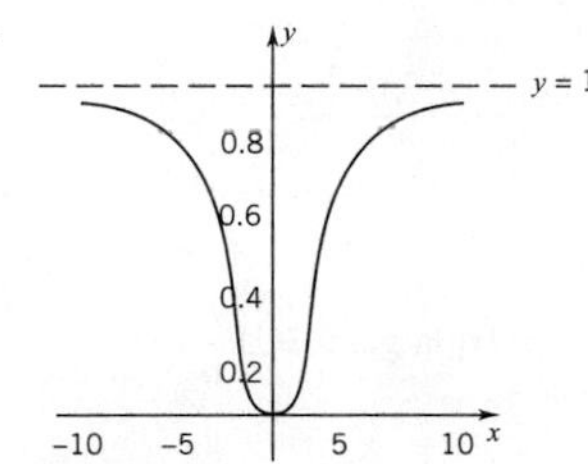

23.
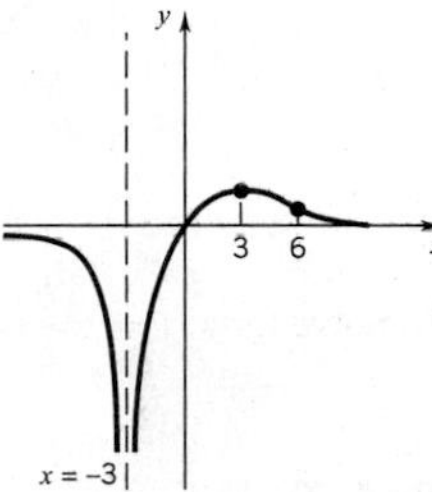

25.
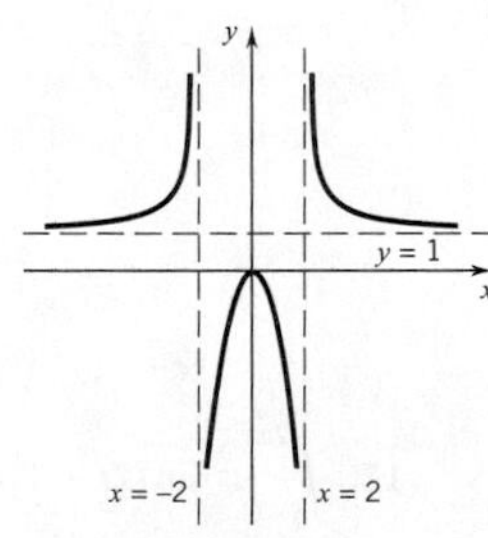

27.
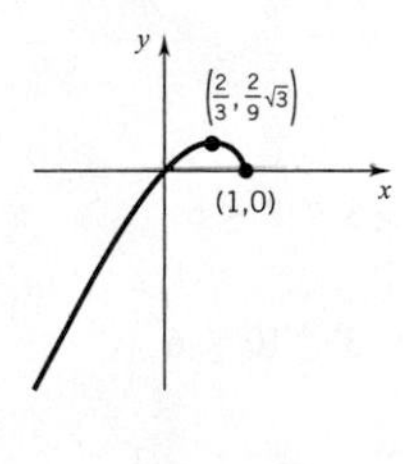

29.
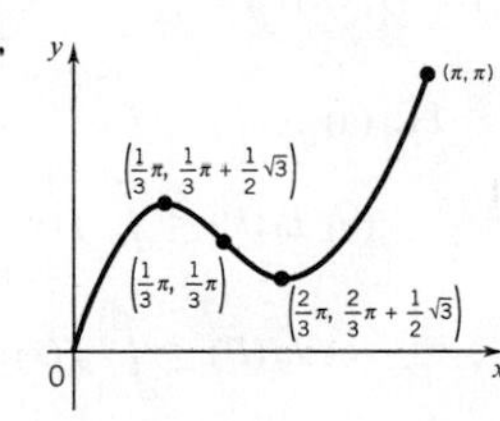

31.
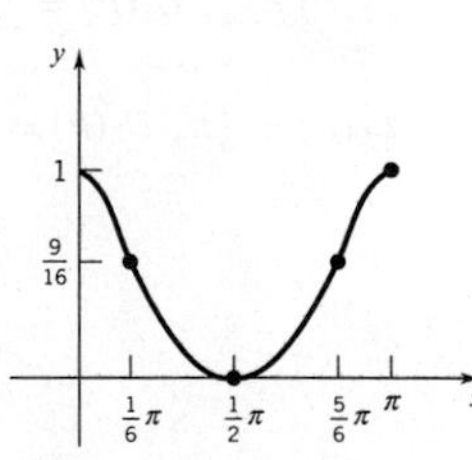

33.
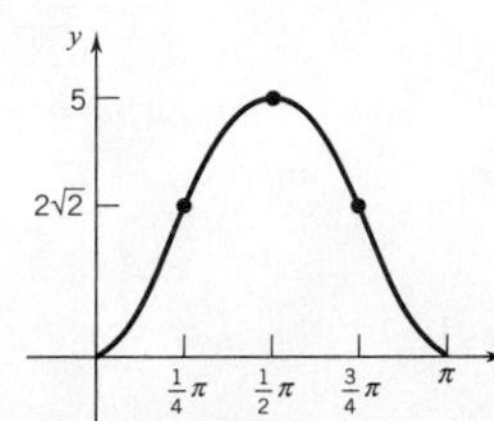

35.
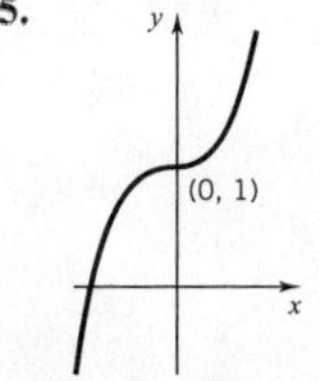

37.
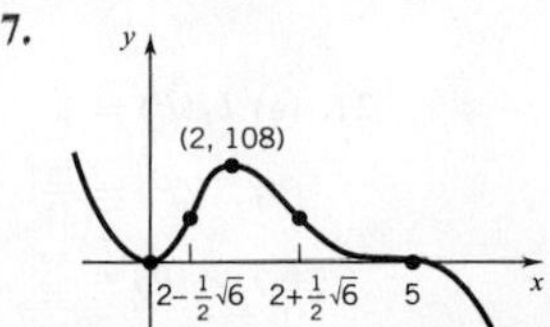

39.
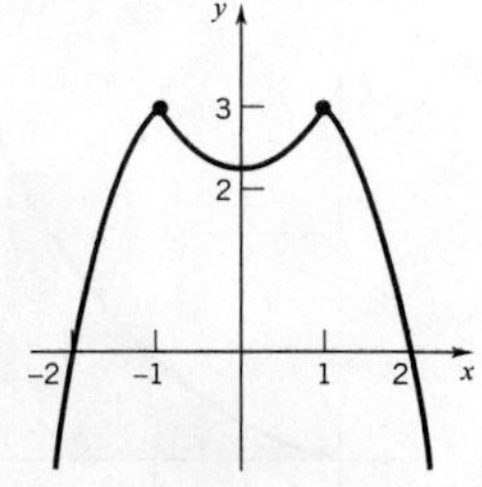

41.
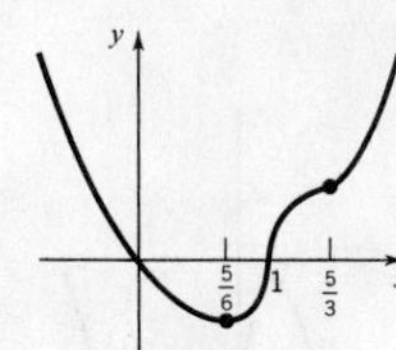

43.
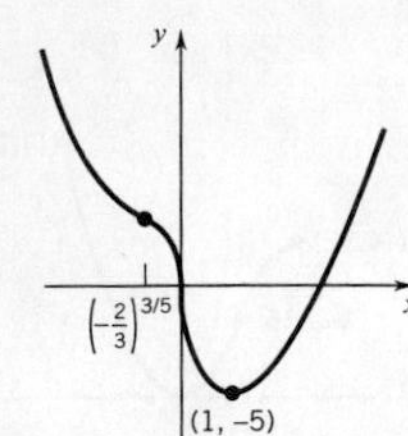

45.
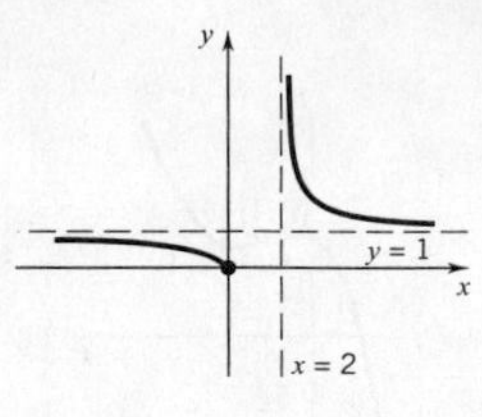

47.
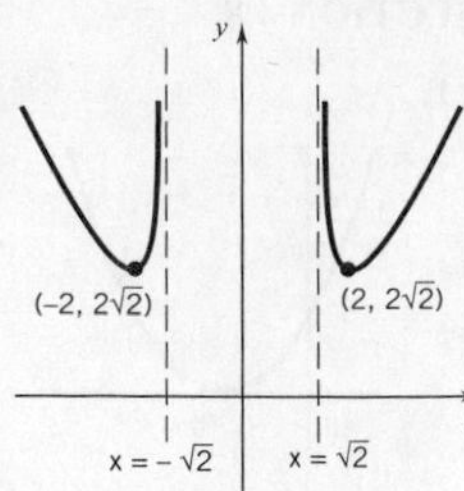

49.
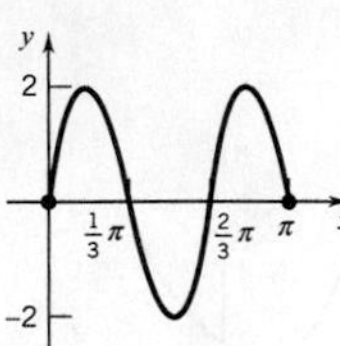

51.
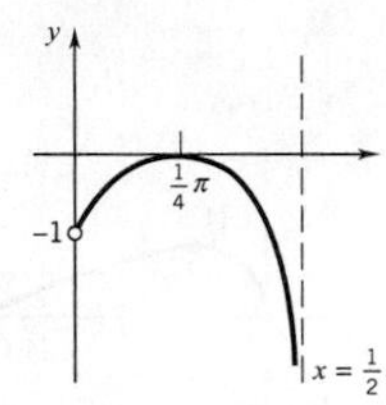

53.
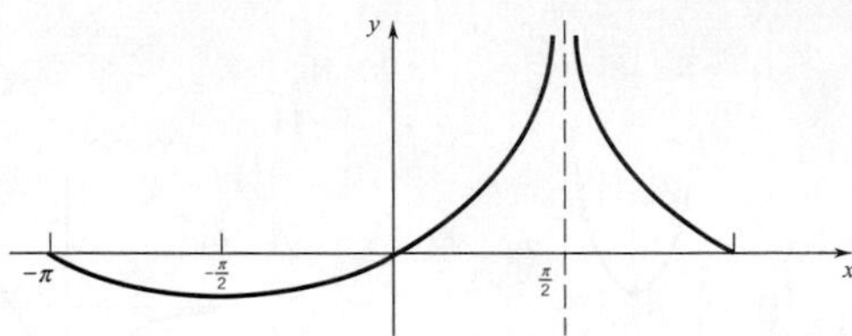

55. (a) increasing on $(-\infty, -1], (0, 1], [3, \infty)$; decreasing on $[-1, 0), [1, 3]$ critical nos. $x = -1, 0, 1, 3$
(b) concave up on $(-\infty, -3), (2, \infty)$ concave down on $(-3, 0), (0, 2)$ (c)

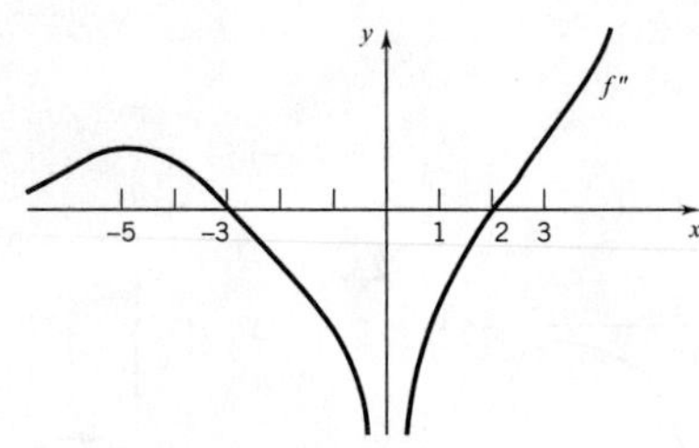

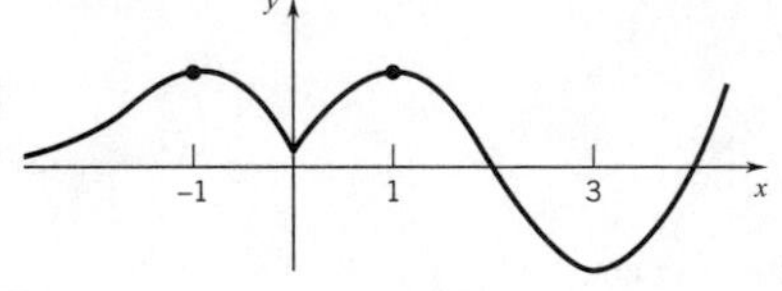

57. Solve for y: $y = \pm\frac{b}{a}x\sqrt{1 - \frac{a^2}{x^2}}$. For $|x|$ large, $y \cong \pm\frac{b}{a}x$.

CHAPTER 5

SECTION 5.2

1. $L_f(P) = \frac{5}{8},\ U_f(P) = \frac{11}{8}$ **3.** $L_f(P) = \frac{9}{64},\ U_f(P) = \frac{37}{64}$ **5.** $L_f(P) = \frac{17}{16},\ U_f(P) = \frac{25}{16}$ **7.** $L_f(P) = \frac{3}{16},\ U_f(P) = \frac{43}{32}$

9. $L_f(P) = \frac{1}{6}\pi,\ U_f(P) = \frac{11}{12}\pi$ **11.** (a) $L_f(P) \le U_f(P)$ but $3 \not\le 2$
(b) $L_f(P) \le \int_{-1}^{1} f(x)\,dx \le U_f(P)$ but $3 \not\le 2 \le 6$
(c) $L_f(P) \le \int_{-1}^{1} f(x)\,dx \le U_f(P)$ but $3 \le 10 \not\le 6$

13. (a) $L_f(P) = -3x_1(x_1 - x_0) - 3x_2(x_2 - x_1) - \cdots - 3x_n(x_n - x_{n-1})$ (b) $-\frac{3}{2}(b^2 - a^2)$ **15.** $\int_{-1}^{2} (x^2 + 2x - 3)\,dx$ **17.** $\int_{0}^{2\pi} t^2 \sin(2t + 1)\,dt$

$U_f(P) = -3x_0(x_1 - x_0) - 3x_1(x_2 - x_1) - \cdots - 3x_{n-1}(x_n - x_{n-1})$

19.
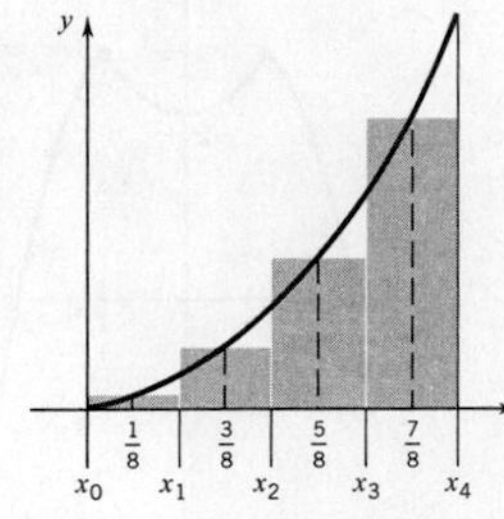

21. (a) $L_f(P) = \frac{25}{32}$
(b) $S^*(P) = \frac{15}{16}$
(c) $U_f(P) = \frac{39}{32}$
(d) $\int_a^b f(x)\,dx = 1$

23. $L_f(P) = x_0^3(x_1 - x_0) + x_1^3(x_2 - x_1) + \cdots + x_{n-1}^3(x_n - x_{n-1})$

$U_f(P) = x_1^3(x_1 - x_0) + x_2^3(x_2 - x_1) + \cdots + x_n^3(x_n - x_{n-1})$

For each index, $i, x_{i-1}^3 \le \frac{1}{4}(x_i^3 + x_i^2 x_{i-1} + x_i x_{i-1}^2 + x_{i-1}^3) \le x_i^3$ and thus by the hint $x_{i-1}^3(x_i - x_{i-1}) \le \frac{1}{4}(x_i^4 - x_{i-1}^4) \le x_i^3(x_i - x_{i-1})$. Adding up these inequalities, we find that $L_f(P) \le \frac{1}{4}(x_n^4 - x_0^4) \le U_f(P)$. Since $x_n = 1$ and $x_0 = 0$, the middle term is $\frac{1}{4}$. Thus the integral is $\frac{1}{4}$.

25. Let f be continuous and increasing on $[a, b]$, and let P be a regular partition. Then

$$\begin{aligned} U_f(P) &= [f(x_1) + f(x_2) + \cdots + f(x_n)]\Delta x \\ L_f(P) &= [f(x_0) + f(x_1) + \cdots + f(x_{n-1})]\Delta x \end{aligned}$$

and $U_f(P) - L_f(P) = [f(x_n) - f(x_0)]\Delta x = [f(b) - f(a)]\Delta x$.

27. necessarily holds: $L_g(P) \le \int_a^b g(x)\,dx < \int_a^b f(x)\,dx \le U_f(P)$

29. necessarily holds: $L_g(P) \le \int_a^b g(x)\,dx < \int_a^b f(x)\,dx$

31. necessarily holds: $U_f(P) \ge \int_a^b f(x)\,dx > \int_a^b g(x)\,dx$

33. (a) From Definition 5.2.3, $0 \le I - L_f(P) \le U_f(P) - L_f(P)$

(b) Use Definition 5.2.3

35. (a) Let $x_1, x_2 \in [0, 2], x_2 \ge x_1$. Then

$$f(x_2) - f(x_1) = \frac{(x_2 + x_1)(x_2 - x_1)}{\sqrt{1 + x_2^2} + \sqrt{1 + x_1^2}} \ge 0.$$

(b) $n = 25$ (c) 3.0

37. Let S be the set of positive integers for which the statement is true. Since $1 = \frac{1(2)}{2}$, $1 \in S$. Assume $k \in S$. Then

$$\begin{aligned} 1 + 2 + \cdots + k + k + 1 &= (1 + 2 + \cdots + k) + k + 1 \\ &= \frac{k(k+1)}{2} + k + 1 = \frac{(k+1)(k+2)}{2}. \end{aligned}$$

Thus $k + 1 \in S$ and so S is the set of positive integers.

39. Let $f(x) = x$ on $[0, b]$ and let $P = \{x_0, x_1, \ldots, x_n\}$ be a regular partition.

(a) $L_f(P) = \left[0 \cdot \frac{b}{n} + 1 \cdot \frac{b}{n} + \cdots + (n-1) \cdot \frac{b}{n}\right] \frac{b}{n}$

$= \frac{b^2}{n^2}[1 + 2 + \cdots + (n-1)]$

(b) $U_f(P) = \left[1 \cdot \frac{b}{n} + 2 \cdot \frac{b}{n} + \cdots + n \cdot \frac{b}{n}\right] \frac{b}{n}$

$= \frac{b^2}{n^2}[1 + 2 + \cdots + n]$

(c) $L_f(P) = \frac{b^2}{n^2} \cdot \frac{(n-1)n}{2} \to \frac{b^2}{2}$ as $n \to \infty$

$U_f(P) = \frac{b^2}{n^2} \cdot \frac{n(n+1)}{2} \to \frac{b^2}{2}$ as $n \to \infty$

Thus $\int_0^b x\,dx = \frac{b^2}{2}$.

41. (a) $\frac{1}{n^2}(1 + 2 + \cdots + n) = \frac{1}{n^2}\left[\frac{n(n+1)}{2}\right] = \frac{1}{2} + \frac{1}{2n}$

(b) $S_n^* = \frac{1}{2} + \frac{1}{2n}$, $\int_0^1 x\,dx = \left[\frac{1}{2}x^2\right]_0^1 = \frac{1}{2}$

$\left|S_n^* - \int_0^1 x\,dx\right| = \frac{1}{2n} < \frac{1}{n} < \epsilon$ if $n > \frac{1}{\epsilon}$

43. (a) $\frac{1}{n^4}(1^3 + 2^3 + \cdots + n^3) = \frac{1}{n^4}\left[\frac{n^2(n+1)^2}{4}\right] = \frac{1}{4} + \frac{1}{2n} + \frac{1}{4n^2}$

(b) $S_n^* = \frac{1}{4} + \frac{1}{2n} + \frac{1}{4n^2}$, $\int_0^1 x^3\,dx = \left[\frac{1}{4}x^4\right]_0^1 = \frac{1}{4}$

$\left|S_n^* - \int_0^1 x^3\,dx\right| = \frac{1}{2n} + \frac{1}{4n^2} < \frac{1}{n} < \epsilon$ if $n > \frac{1}{\epsilon}$

45. $\frac{5}{4}$ units

47. Let P be an arbitrary partition of $[0, 4]$. Since each $m_i = 2$ and each $M_i \ge 2$.

$$\begin{aligned} L_g(P) &= 2\Delta x_i + \cdots + 2\Delta x_n = 2(\Delta x_1 + \cdots + \Delta x_n) = 2 \cdot 4 = 8, \\ U_g(P) &\ge 2\Delta x_1 + \cdots + 2\Delta x_n = 2(\Delta x_1 + \cdots + \Delta x_n) = 2 \cdot 4 = 8. \end{aligned}$$

Thus $L_g(P) \le 8 \le U_g(P)$ for all partitions P of $[0, 4]$.

Uniqueness: Suppose that

$$(*) \qquad L_g(P) \leq I \leq U_g(P) \quad \text{for all partitions } P \text{ of } [0, 4].$$

Since $L_g(P) = 8$ for all P, I is at least 8. Suppose now that $I > 8$ and choose a partition P of $[0, 4]$ with max $\Delta x_i < \frac{1}{5}(I - 8)$ and $0 = x_1 < \cdots < x_{i-1} < 3 < x_i < \cdots < x_n = 4$. Then

$$\begin{aligned} U_g(P) &= 2\Delta x_1 + \cdots + 2\Delta x_{i-1} + 7\Delta x_i + 2\Delta x_{i+1} + \cdots + 2\Delta x_n \\ &= 2(\Delta x_1 + \cdots + \Delta x_n) + 5\Delta x_i = 8 + 5\Delta x_i < 8 + \tfrac{5}{5}(I - 8) = I \end{aligned}$$

and I does not satisfy $(*)$. This contradiction proves that I is not greater than 8 and therefore $I = 8$.

49. Let $P = \{x_0, x_1, \ldots, x_n\}$ be any partition of $[2, 10]$.

(a) Since each subinterval of $[2, 10]$ contains both rational and irrational numbers,

$$\begin{aligned} L_f(P) &= 4\Delta x_1 + 4\Delta x_2 + \cdots + 4\Delta x_n = 4(10 - 2) = 32 \\ U_f(P) &= 7\Delta x_1 + 7\Delta x_2 + \cdots + 7\Delta x_n = 7(10 - 2) = 56. \end{aligned}$$

(b) There is more than one number I that satisfies $L_f(P) \leq I \leq U_f(P)$ for all partitions P.

(c) See (a).

51. (a) $L_f(P) \cong 0.6105, U_f(P) \cong 0.7105$ (b) $\frac{1}{2}[L_f(P) + U_f(P)] = 0.6605$ (c) $S^*(P) \cong 0.6684$

53. (a) $L_f(P) \cong 0.53138 \quad U_f(P) \cong 0.73138$ (b) $\dfrac{1}{2}[L_f(P) + U_f(P)] = 0.63138$ (c) $S^*(P) \cong 0.63926$

SECTION 5.3

1. (a) 5 (b) -2 (c) -1 (d) 0 (e) -4 (f) 1

3. With $P = \{1, \frac{3}{2}, 2\}$ and $f(x) = \dfrac{1}{x}$, we have $0.5 < \frac{7}{12} = L_f(P) \leq \displaystyle\int_1^2 \frac{dx}{x} \leq U_f(P) = \frac{5}{6} < 1$.

5. (a) $F(0) = 0$ (b) $F'(x) = x\sqrt{x+1}$ (c) $F'(2) = 2\sqrt{3}$ (d) $F(2) = \displaystyle\int_0^2 t\sqrt{t+1}\,dt$ (e) $-F(x) = \displaystyle\int_x^0 t\sqrt{t+1}\,dt$

7. (a) $\frac{1}{10}$ (b) $\frac{1}{9}$ (c) $\frac{4}{37}$ (d) $\dfrac{-2x}{(x^2+9)^2}$

9. (a) $\sqrt{2}$ (b) 0 (c) $-\frac{1}{4}\sqrt{5}$ (d) $-(\sqrt{x^2+1} + x^2/\sqrt{x^2+1})$ **11.** (a) -1 (b) 1 (c) 0 (d) $-\pi \sin \pi x$

13. (a) Since $P_1 \subseteq P_2, U_f(P_2) \leq U_f(P_1)$ but $5 \nleq 4$. (b) Since $P_1 \subseteq P_2, L_f(P_1) \leq L_f(P_2)$ but $5 \nleq 4$.

15. $x = 1$ is a critical number; F has a local minimum at $x = 1$.

17. (a) $F'(x) = \dfrac{1}{x}, x > 0; F$ is increasing on $(0, \infty)$. (b) $F''(x) = \dfrac{-1}{x^2}, x > 0$; the graph of F is concave down on $(0,\infty)$. (c)

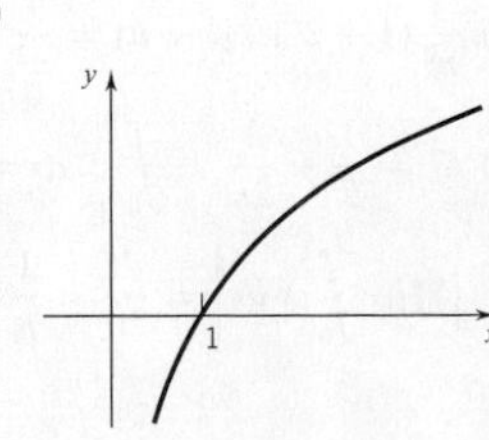

19. (a) f is continuous; Theorem 5.3.5

(b) $F'(x) = f(x)$ and f is differentiable; $F''(x) = f'(x)$

(c) $F'(1) = f(1) = 0$

(d) $F''(1) = f'(1) > 0$

(e) $F(0) = 0,\ F'(x) < 0$ on $(0, 1),\ F''(x) > 0$ on $(0, \infty)$

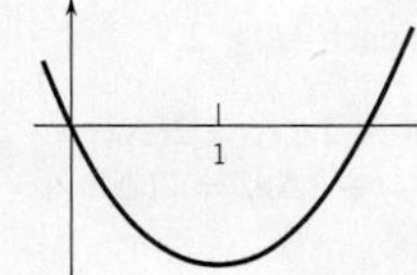

21. (a)

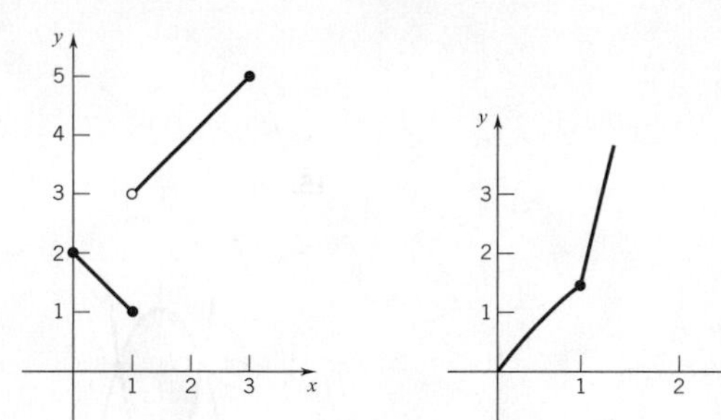

(b) $F(x) = \begin{cases} 2x - \frac{1}{2}x^2, & 0 \le x \le 1 \\ 2x + \frac{1}{2}x^2 - 1, & 1 < x \le 3 \end{cases}$

(c) f is discontinuous at $x = 1$; F is continuous but not differentiable at $x = 1$.

23. $F'(x) = 3x^5 \cos(x^3)$. **25.** $F'(x) = 2x[\sin^2(x^2) - x^2]$. **27.** (a) 0 (b) 2 (c) 2

29. (a) $f(0) = \frac{1}{2}$ (b) $2, -2$

31. Apply the mean-value theorem to the function $F(x) = \int_a^x f(t)\,dt$ and observe that

$$F(b) = \int_a^b f(t)\,dt, \quad F(a) = 0, \quad \text{and} \quad F'(c) = f(c).$$

33. Set $G(x) = \int_a^x f(t)\,dt$. Then $F(x) = \int_c^a f(t)\,dt + G(x)$. By (5.3.5) G, and thus F, is continuous on $[a, b]$, is differentiable on (a, b), and $F'(x) = G'(x) = f(x)$ for all x in (a, b).

35. (a) $F'(x) = f(x) = G'(x)$ so $F(x) = G(x) + C$ by Theorem 4.2.5. (b) Set $x = d$ in $F(x) = G(x) + C$.

37. (a) $F'(x) = 0$ at $x = -1, 4$;
F increasing on $(-\infty, -1], [4, \infty)$;
F decreasing on $[-1, 4]$
(b) $F''(x) = 0$ at $x = \frac{3}{2}$;
the graph of F is concave up on $(\frac{3}{2}, \infty)$; concave down on $(-\infty, \frac{3}{2})$

39. (a) $F'(x) = 0$ at $x = 0, \frac{\pi}{2}, \pi, \frac{3\pi}{2}, 2\pi$;
F increasing on $[\frac{\pi}{2}, \pi], [\frac{3\pi}{2}, 2\pi]$;
F decreasing on $[0, \frac{\pi}{2}], [\pi, \frac{3\pi}{2}]$
(b) $F''(x) = 0$ at $x = \frac{\pi}{4}, \frac{3\pi}{4}, \frac{5\pi}{4}, \frac{7\pi}{4}$;
the graph of F is concave up on $(\frac{\pi}{4}, \frac{3\pi}{4}), (\frac{5\pi}{4}, \frac{7\pi}{4})$;
the graph of F is concave down on $(0, \frac{\pi}{4}), (\frac{3\pi}{4}, \frac{5\pi}{4}), (\frac{7\pi}{4}, 2\pi)$

SECTION 5.4

1. -2 **3.** 1 **5.** $\frac{28}{3}$ **7.** $\frac{32}{3}$ **9.** $\frac{2}{3}$ **11.** $\frac{13}{2}$ **13.** $-\frac{4}{15}$ **15.** $\frac{1}{18}(2^{18} - 1)$ **17.** $\frac{1}{6}a^2$ **19.** $\frac{7}{4}$

21. $\frac{21}{2}$ **23.** 1 **25.** 2 **27.** $2 - \sqrt{2}$ **29.** 0 **31.** $\frac{\pi}{9} - 2\sqrt{3}$ **33.** $\sqrt{13} - 2$ **35.** $F'(x) = (x + 2)^2$

37. $F'(x) = \sec(2x + 1)\tan(2x + 1)$ **39.** (a) $\int_2^x \frac{dt}{t}$ (b) $-3 + \int_2^x \frac{dt}{t}$ **41.** $\frac{32}{3}$ **43.** $2 + \sqrt{2}$ **45.** (a) 3/2 (b) 5/2

47. (a) 4/3 (b) 4 **49.** valid **51.** not valid; $1/x^3$ is not defined at $x = 0$.

53. (a) $x(t) = 5t^2 - \frac{1}{3}t^3, 0 \le t \le 10$ (b) At $t = 5$; $x(5) = \frac{250}{3}$ **55.** $\frac{13}{2}$ **57.** $\frac{2 + \sqrt{3}}{2} + 2\pi$

59. (a) $g(x) = \begin{cases} \frac{1}{2}x^2 + 2x + 2, & -2 \le x \le 0 \\ 2x + 2, & 0 < x \le 1 \\ -x^2 + 4x + 1, & 1 < x \le 2 \end{cases}$

(b)

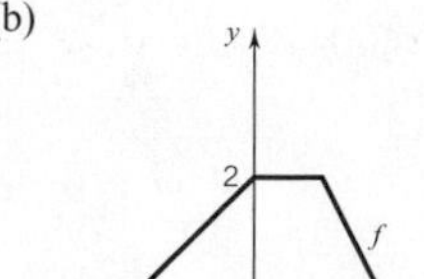

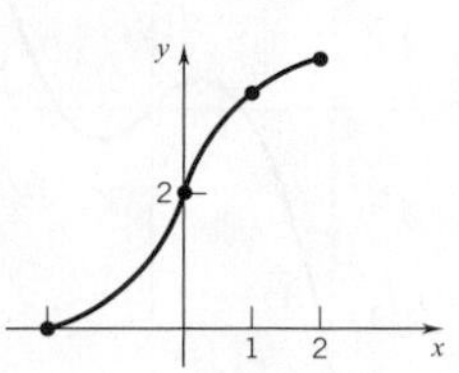

(c) f is continuous on $[-2, 2]$; f is differentiable on $(-2, 0), (0, 1), (1, 2)$; g is differentiable on $(-2, 2)$.

61. Since f is an antiderivative of f', it follows that $\int_a^b f'(x)\,dx = \Big[f(x)\Big]_a^b = f(b) - f(a)$.

63. $f(x)$ and $f(x) - f(a)$, respectively **65.** $h \simeq 9.44892$

SECTION 5.5

1. $\frac{9}{4}$ **3.** $\frac{38}{3}$ **5.** $\frac{47}{15}$ **7.** $\frac{5}{3}$ **9.** $\frac{1}{2}$

11. area $= \frac{1}{3}$

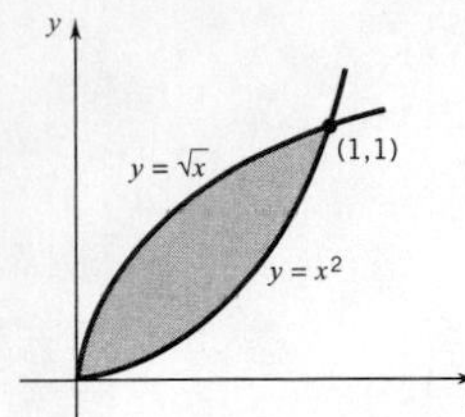

13. area $\frac{9}{2}$

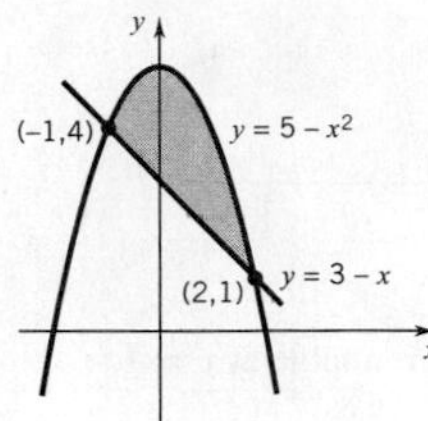

15. area $= \frac{64}{3}$

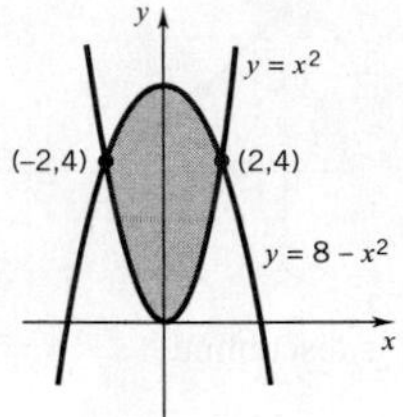

17. area $= 10$

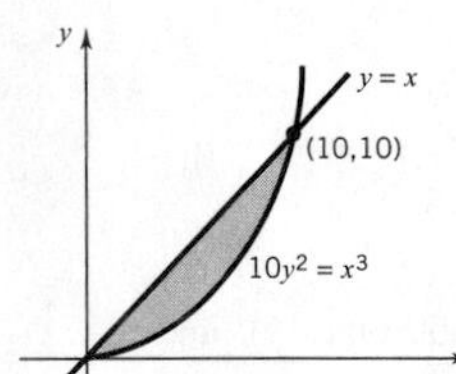

19. area $= \frac{32}{3}$

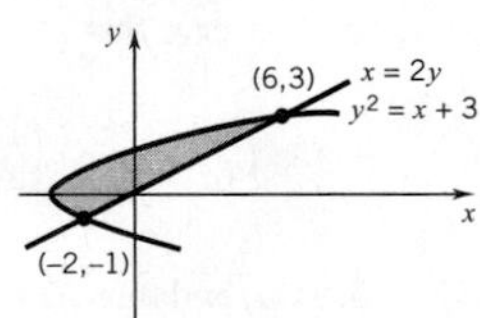

21. area $= 4$

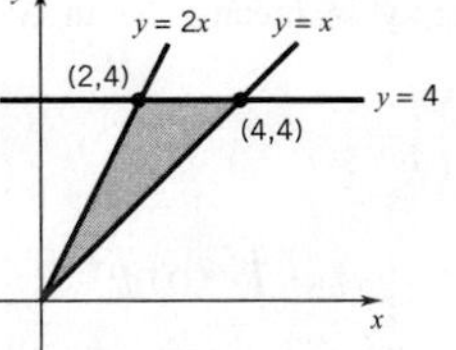

23. area $= 2 + \frac{2}{3}\pi^3$

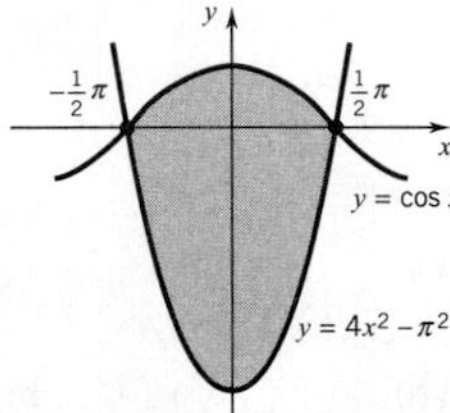

25. area $= \frac{1}{8}\pi^2 - 1$

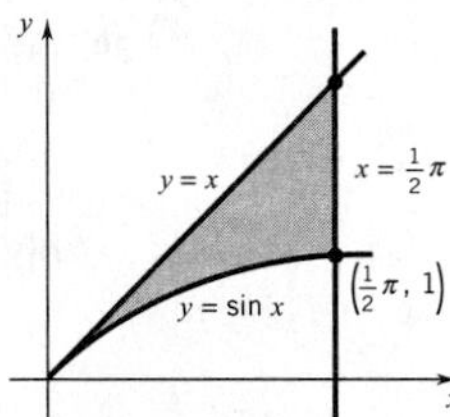

27. (a) $-\frac{91}{6}$, the area of the region bounded by f and the x-axis for $x \in [-3, -2] \cup [3, 4]$ minus the area of the region bounded by f and the x-axis for $x \in [-2, 3]$.

(b) $\dfrac{53}{2}$ (c) $\dfrac{125}{6}$

29. (a) 0 (b) 5

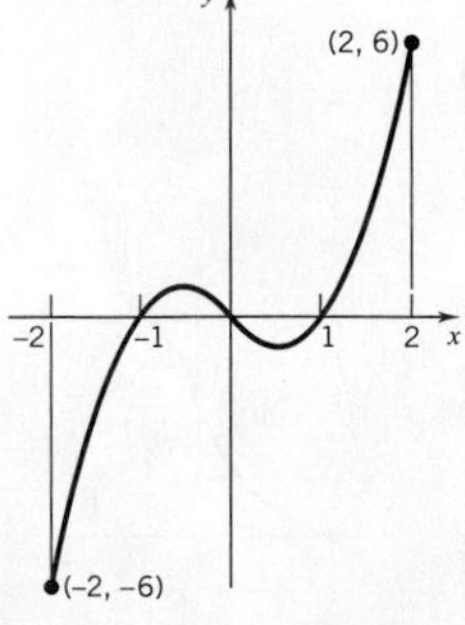

31. (a) $\dfrac{65}{4}$ (b) 17.87 **33.** $\dfrac{10}{3}$ **35.** area $= 2 - \sqrt{2}$ **37.** 2.86

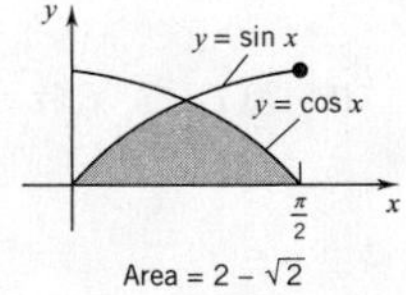

39. (b) ,(d) $\frac{13}{2}$

(c) $f(x) = \begin{cases} -x + 1 & -2 \le x < 1 \\ x - 1 & 1 \le x \le 3 \end{cases}$

SECTION 5.6

1. $-\frac{1}{3x^3} + C$ 3. $\frac{1}{2}ax^2 + bx + C$ 5. $2\sqrt{1+x} + C$ 7. $\frac{1}{2}x^2 + \frac{1}{x} + C$ 9. $\frac{1}{3}t^3 - \frac{1}{2}(a+b)t^2 + abt + C$

11. $\frac{2}{9}t^{9/2} - \frac{2}{5}(a+b)t^{5/2} + 2abt^{1/2} + C$ 13. $\frac{1}{2}[g(x)]^2 + C$ 15. $\frac{1}{2}\sec^2 x + C$ 17. $-\frac{1}{4x+1} + C$ 19. $x^2 - x - 2$

21. $\frac{1}{2}ax^2 + bx - 2a - 2b$ 23. $3 - \cos x$ 25. $x^3 - x^2 + x + 2$ 27. $\frac{1}{12}(x^4 - 2x^3 + 2x + 23)$ 29. $x - \cos x + 3$

31. $\frac{1}{3}x^3 - \frac{3}{2}x^2 - \frac{1}{3}x + 3$ 33. $\frac{d}{dx}\left(\int f(x)\,dx\right) = f(x); \int \frac{d}{dx}[f(x)]\,dx = f(x) + C$

35. (a) 34 units to the right of the origin (b) 44 units 37. (a) $v(t) = 2(t+1)^{1/2} - 1$ (b) $x(t) = \frac{4}{3}(t+1)^{3/2} - t - \frac{4}{3}$

39. (a) 4.4 sec (b) 193.6 ft 41. $[v(t)]^2 = (at + v_0)^2 = a^2t^2 + 2av_0t + v_0^2 = v_0^2 + 2a(\frac{1}{2}at^2 + v_0t) = v_0^2 + 2a[x(t) - x_0]$

$x(t) = \frac{1}{2}at^2 + v_0t + x_0$ ↑

43. 42 sec 45. $x(t) = x_0 + v_0t + At^2 + Bt^3$ 47. at $(\frac{160}{3}, 50)$ 49. $A = -\frac{5}{2}, B = 2$ 51. (a) at $t = \frac{11}{6}\pi$ sec (b) at $t = \frac{13}{6}\pi$ sec

53. mean-value theorem 55. $v(t) = v_0(1 - 2tv_0)^{-1}$

57. $\frac{d}{dx}\left[\int(\cos x - 2\sin x)\,dx\right] = \cos x - 2\sin x; \int \frac{d}{dx}[f(x)]\,dx = \cos x - 2\sin x + C$

59. $f(x) = \sin x + 2\cos x + 1$ 61. $f(x) = \frac{1}{12}x^4 - \frac{1}{2}x^3 + \frac{5}{2}x^2 + 4x - 3$

SECTION 5.7

1. $\frac{1}{3(2-3x)} + C$ 3. $\frac{1}{3}(2x+1)^{3/2} + C$ 5. $\frac{4}{7a}(ax+b)^{7/4} + C$ 7. $-\frac{1}{8(4t^2+9)} + C$ 9. $\frac{4}{15}(1+x^3)^{5/4} + C$

11. $-\frac{1}{4(2+s^2)^2} + C$ 13. $\sqrt{x^2+1} + C$ 15. $-\frac{5}{4}(x^2+1)^{-2} + C$ 17. $-4(x^{1/4}+1)^{-1} + C$ 19. $-\frac{b^3}{2a^4}\sqrt{1 - a^4x^4} + C$

21. $\frac{15}{8}$ 23. 0 25. $\frac{1}{3}|a|^3$ 27. $\frac{13}{3}$ 29. $\frac{39}{400}$ 31. $\frac{2}{5}(x+1)^{5/2} - \frac{2}{3}(x+1)^{3/2} + C$ 33. $\frac{1}{10}(2x-1)^{5/2} + \frac{1}{6}(2x-1)^{3/2} + C$

35. $\frac{16}{3}\sqrt{2} - \frac{14}{3}$ 37. $y = \frac{1}{3}(x^2+1)^{3/2} + \frac{2}{3}$ 39. $\frac{1}{3}\sin(3x+1) + C$

41. $-(\cot \pi x)/\pi + C$ 43. $\frac{1}{2}\cos(3 - 2x) + C$ 45. $-\frac{1}{5}\cos^5 x + C$ 47. $-2\cos x^{1/2} + C$ 49. $\frac{2}{3}(1 + \sin x)^{3/2} + C$ 51. $\tan x + C$

53. $\frac{1}{8}\sin^4 x^2 + C$ 55. $2(1 + \tan x)^{1/2} + C$ 57. 0 59. $(\sqrt{3} - 1)/\pi$ 61. $\frac{1}{4}$

63. $\int \sin^2 x\,dx = \int \frac{1 - \cos 2x}{2}\,dx = \frac{1}{2}\int(1 - \cos 2x)\,dx = \frac{1}{2}x - \frac{1}{4}\sin 2x + C$ 65. $\frac{1}{2}x + \frac{1}{20}\sin 10x + C$ 67. $\frac{\pi}{4}$ 69. 2 71. $1/2\pi$

73. $(4\sqrt{3} - 6)/3\pi$ 75. (a) $\frac{1}{2}\sec^2 x + C$ (b) $\frac{1}{2}\tan^2 x + C'$

(c) $\frac{1}{2}\sec^2 x + C = \frac{1}{2}(1 + \tan^2 x) + C = \frac{1}{2}\tan^2 x + (C + \frac{1}{2}) = \frac{1}{2}\tan^2 x + C'$; $C + \frac{1}{2}$ and C' each represent an arbitrary constant

77. πab 79. $\frac{4(1+\sqrt{x})\sqrt{x}}{\sqrt{x + x^{3/2}}}$ 81. $\frac{267}{5440}$ 83. $\frac{1076}{15}, u = x + 1$

SECTION 5.8

1. yes; $\int_a^b [f(x) - g(x)]\,dx = \int_a^b f(x)\,dx - \int_a^b g(x)dx > 0$ 3. yes; otherwise we would have $f(x) \le g(x)$ for all $x \in [a, b]$, and it would follow that $\int_a^b f(x)\,dx \le \int_a^b g(x)\,dx$

5. no; take $f(x) = 0, g(x) = -1$ on $[0, 1]$ 7. no; take, for example, any odd function on an interval of the form $[-c, c]$

9. no; $\int_{-1}^1 x\,dx = 0$ but $\int_{-1}^1 |x|\,dx \ne 0$ 11. yes; $U_f(P) \ge \int_a^b f(x)\,dx = 0$ 13. no; $L_f(P) \le \int_a^b f(x)\,dx = 0$

15. yes; $\int_a^b [f(x) + 1]\,dx = \int_a^b f(x)\,dx + \int_a^b 1\,dx = 0 + b - a = b - a$ 17. $\frac{2x}{\sqrt{2x^2+7}}$ 19. $-f(x)$ 21. $-\frac{2\sin(x^2)}{x}$

23. $\dfrac{\sqrt{x}}{2(1+x)}$ **25.** $\dfrac{1}{x}$ **27.** $4x\sqrt{1+4x^2} - \tan x \sec^2 x|\sec x|$

29. (a) With P a partition of $[a, b]$

$$L_f(P) \le \int_a^b f(x)\,dx.$$

If f is nonnegative on $[a, b]$, then $L_f(P)$ is nonnegative and, consequently, so is the integral. If f is positive on $[a, b]$, then $L_f(P)$ is positive and, consequently, so is the integral.

(b) Take F as an antiderivative of f on $[a, b]$. Observe that

$$F'(x) = f(x) \text{ on } (a, b) \quad \text{and} \quad \int_a^b f(x)\,dx = F(b) - F(a).$$

If $f(x) \ge 0$ on $[a, b]$, then F is nondecreasing on $[a, b]$ and $F(b) - F(a) \ge 0$.

If $f(x) > 0$ on $[a, b]$, then F is increasing on $[a, b]$ and $F(b) - F(a) > 0$.

(c) same argument as (a)

31. Since $m \le f(x) \le M$ for all $x \in [a, b]$, it follows from (5.8.3) that

$$\int_a^b m\,dx \le \int_a^b f(x)\,dx \le \int_a^b M\,dx$$

and so $m(b-a) \le \displaystyle\int_a^b f(x)\,dx \le M(b-a)$

33. $H(x) = \displaystyle\int_{2x}^{x^3-4} \frac{x\,dt}{1+\sqrt{t}} = x\int_{2x}^{x^3-4} \frac{dt}{1+\sqrt{t}}$

$$H'(x) = x\left[\frac{3x^2}{1+\sqrt{x^3-4}} - \frac{2}{1+\sqrt{2x}}\right] + \int_{2x}^{x^3-4} \frac{dt}{1+\sqrt{t}}$$

$$H'(2) = 2\left[\frac{12}{3} - \frac{2}{3}\right] + \underbrace{\int_4^4 \frac{dt}{1+\sqrt{t}}}_{=0} = \frac{20}{3}$$

35. (a) Let $u = -x$ (b) $\displaystyle\int_{-a}^a f(x)\,dx = \int_{-a}^a f(x)\,dx + \int_0^a f(x)\,dx = \int_0^a f(-x)\,dx + \int_0^a f(x)\,dx = \int_0^a [f(x) + f(-x)]\,dx$

37. 0 **39.** $\frac{2}{3}\pi + \frac{2}{81}\pi^3 - \sqrt{3}$ **41.** $\displaystyle\int_3^5 f(x)\,dx = 18$

43. $\displaystyle\int_3^6 f(x)\,dx = -\frac{147}{4}; \int_3^6 |f(x)|\,dx \cong 44.9738$ **45.** $f(-x) = 2(-x) - \sin(-x) = -2x + \sin x = -f(x)$

SECTION 5.9

1. $A.V. = \frac{1}{2}mc + b, \quad x = \frac{1}{2}c$ **3.** $A.V. = 0, \quad x = 0$ **5.** $A.V. = 1, \quad x = \pm 1$ **7.** $A.V. = \frac{2}{3}, \quad x = 1 \pm \frac{1}{3}\sqrt{3}$

9. $A.V. = 2, \quad x = 4$ **11.** $A.V. = 0, \quad x = 0, \pi, 2\pi$ **13.** $A.V. = \dfrac{b^{n+1} - a^{n+1}}{(n+1)(b-a)}$

15. average of f' on $[a, b] = \dfrac{1}{b-a}\displaystyle\int_a^b f'(x)\,dx = \dfrac{1}{b-a}\Big[f(x)\Big]_a^b = \dfrac{f(b) - f(a)}{b-a}$

17. $D = \sqrt{x^2 + x^4}; A.V. = \frac{7}{9}\sqrt{3}$ **19.** (a) The terminal velocity is twice the average velocity.
(b) The average velocity during the first $\frac{1}{2}t$ seconds is one-third of the average velocity during the next $\frac{1}{2}t$ seconds.

21. Suppose $f(x) \ne 0$ for all x. Then, since f is continuous, either $f(x) > 0$ on (a, b) or $f(x) < 0$ on (a, b). In either case, $\displaystyle\int_a^b f(x)\,dx \ne 0$.

23. (a) $v(t) = at, x(t) = \frac{1}{2}at^2 + x_0$ (b) $V_{\text{avg}} = \dfrac{1}{t_2 - t_1}\displaystyle\int_{t_1}^{t_2} at\;dt = \dfrac{at_1 + at_2}{2} = \dfrac{v(t_1) + v(t_2)}{2}$

25. (a) $M = 24\left(\sqrt{7} - 1\right); x_M = \dfrac{4\sqrt{7} + 2}{3\sqrt{7} - 3}$

(b) $A.V. = 4\left(\sqrt{7} - 1\right)$

27. (a) $M = \frac{2}{3}kL^{3/2}, x_M = \frac{3}{5}L$ (b) $M = \frac{1}{3}kL^3, x_M = \frac{1}{4}L$ **29.** $x_{M_2} = (2M - M_1)L/8M_2$ **31.** $x = \dfrac{2M \pm kL^2}{2kL}$

33. see answer to Exercise 31, Section 5.3

35. If f and g take on the same average on every interval $[a,x]$, then

$$\frac{1}{x-a}\int_a^x f(t)dt = \frac{1}{x-a}\int_a^x g(t)\,dt.$$

Multiplication by $(x-a)$ gives

$$\int_a^x f(t)\,dt = \int_a^x g(t)\,dt.$$

Differentiation with respect to x gives $f(x) = g(x)$. This shows that, if the averages are the same on every interval, then the functions are everywhere the same.

37. $\int_a^b f(x)\,dx = \int_{x_0}^{x_1} f(x)\,dx + \int_{x_1}^{x^2} f(x)\,dx + \cdots + \int_{x_{n-1}}^{x_n} f(x)\,dx$

By the mean-value theorem for integrals, there exists a number $x_i^* \in (x_{x_{i-1}}, x_i)$ such that

$$\int_{x_{i-1}}^{x_i} f(x)\,dx = f(x_i^*)\,(x_i - x_{i-1}) = f(x_i^*)\Delta x_i, \quad i = 1, 2, \ldots, n$$

39. (a) A rod is lying on the x-axis from $x = a$ to $x = b$. The mass density of the rod is given by the continuous function $\lambda = \lambda(x)$. Let $P = \{a = x_0, x_1, x_2, \ldots, x_{n-1}, x_n = b\}$ be a partition of the interval $[a,b]$. Let x_i^* be a point in the ith interval, and let $\Delta x_i = x_i - x_{i-1}, i = 1, 2, \ldots, n$. The mass contributed by $[x_{i-1}, x_i]$ is approximately $\lambda(x_i^*)\Delta x_i$. The sum of these contributions,

$$\lambda(x_1^*)\Delta x_1 + \cdots + \lambda(x_n^*)\,\Delta x_n,$$

is a Riemann sum, which as $||P|| \to 0$, gives the mass of the rod

$$M = \int_a^b \lambda(x)\,dx. \qquad (5.9.4)$$

Take M_i as the mass contributed by $[x_{i-1}, x_i]$. Then $x_{M_i} M_i \cong x_i^* \lambda(x_i^*)\,\Delta x_i$. Therefore

$$x_M M = x_{M_1} M_1 + \cdots + x_{M_n} M_n \cong x_1^* \lambda(x_1^*)\,\Delta x_1 + \cdots + x_n^* \lambda(x_n^*)\Delta x_n.$$

As $||P|| \to 0$, the sum on the right converges to

$$x_M M = \int_a^b x\,\lambda(x)\,dx. \qquad (5.9.5)$$

(b) The mass moment about the center of mass is

$$m_1(x_1 - x_M) + m_2(x_2 - x_M) + \cdots + m_n(x_n - x_M)$$

which can be written

$$m_1 x_1 + m_2 x_2 + \cdots + m_n x_n - x_M M.$$

The value of this expression is 0.

41. $x = -\frac{3}{5}L$

43. (a) $A.V. = \frac{2}{\pi}$
(b) $c \cong 0.691$

45. (a) $h \cong 14.3333$
(c) $d \cong -4.0551, 2.0551$
(d) $\frac{430}{3}$
(e) $\frac{43}{3} = 14.3333$

47. $a \cong -3.4743, b \cong 3.4743$

(a) $f(c) = A.V. \cong 36.0948 \quad C \cong \pm 2.9545$ or $C \cong \pm 1.1274$,

CHAPTER 6

SECTION 6.1

1.

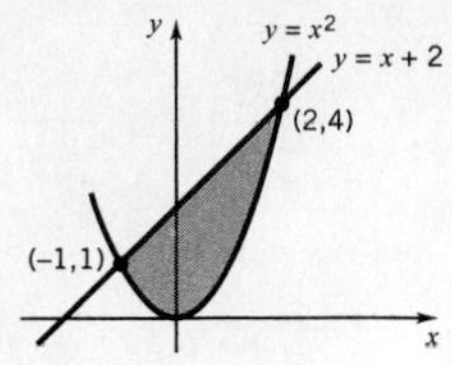

(a) $\int_{-1}^{2}[(x+2)-x^2]\,dx$

(b) $\int_{0}^{1}[\sqrt{y}-(-\sqrt{y})]\,dy+\int_{1}^{4}[\sqrt{y}-(y-2)]\,dy$

3.

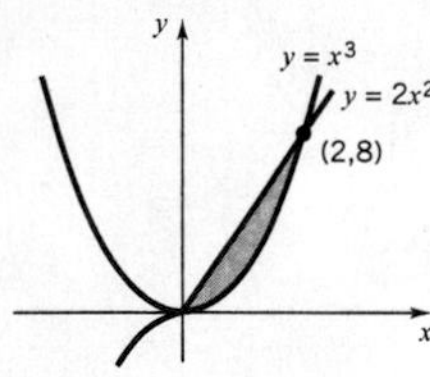

(a) $\int_{0}^{2}[2x^2-x^3]\,dx$

(b) $\int_{0}^{8}\left[y^{1/3}-\left(\frac{1}{2}y\right)^{1/2}\right]dy$

5.

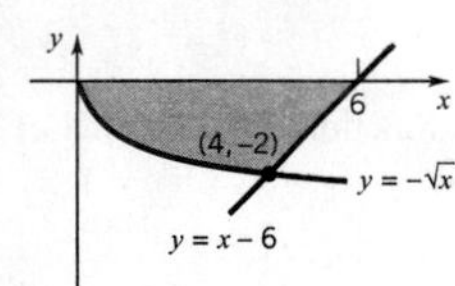

(a) $\int_{0}^{4}[0-(-\sqrt{x})]\,dx+\int_{4}^{6}[0-(x-6)]\,dx$

(b) $\int_{-2}^{0}[(y+6)-y^2]\,dy$

7.

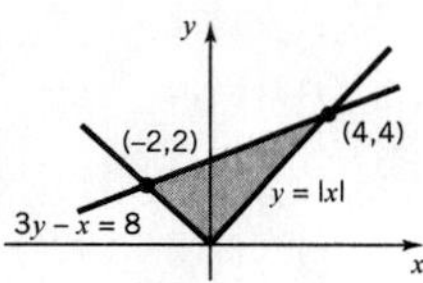

(a) $\int_{-2}^{0}\left[\frac{8+x}{3}-(-x)\right]dx+\int_{0}^{4}\left[\frac{8+x}{3}-x\right]dx$

(b) $\int_{0}^{2}[y-(-y)]\,dy+\int_{2}^{4}[y-(3y-8)]\,dy$

9.

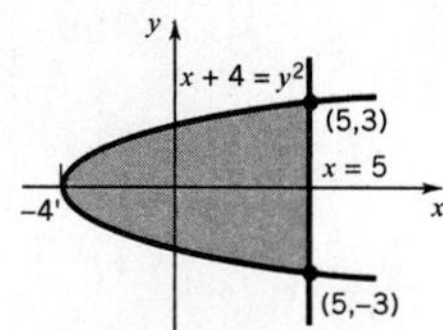

(a) $\int_{-4}^{5}[\sqrt{4+x}-(-\sqrt{4+x})]\,dx$

(b) $\int_{-3}^{3}[5-(y^2-4)]\,dy$

11.

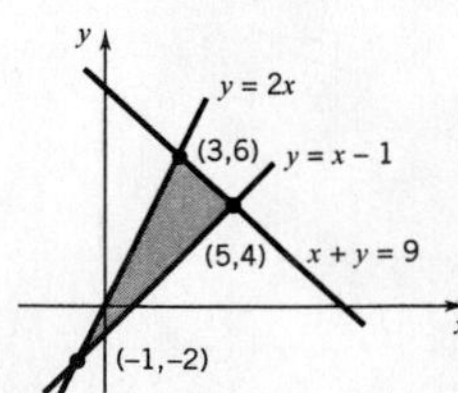

(a) $\int_{-1}^{3}[2x-(x-1)]\,dx+\int_{3}^{5}[(9-x)-(x-1)]\,dx$

(b) $\int_{-2}^{4}\left[(y+1)-\frac{1}{2}y\right]dy+\int_{4}^{6}\left[(9-y)-\frac{1}{2}y\right]dy$

13.

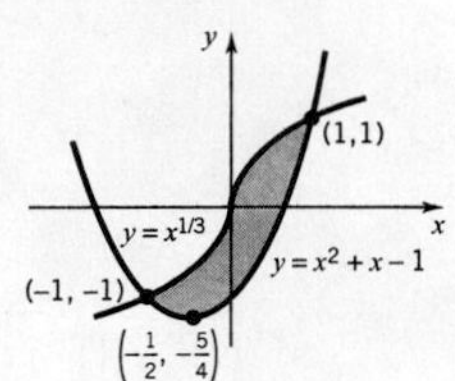

(a) $\int_{-1}^{1}[x^{1/3}-(x^2+x-1)]\,dx$

(b) $\int_{-5/4}^{-1}\left[\left(-\frac{1}{2}+\frac{1}{2}\sqrt{4y+5}\right)-\left(-\frac{1}{2}-\frac{1}{2}\sqrt{4y+5}\right)\right]dy+\int_{-1}^{1}\left[\left(-\frac{1}{2}+\frac{1}{2}\sqrt{4y+5}\right)-y^3\right]dy$

15. area $= \frac{9}{8}$

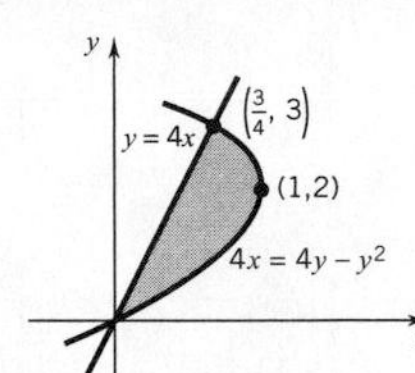

17. area $= 32$

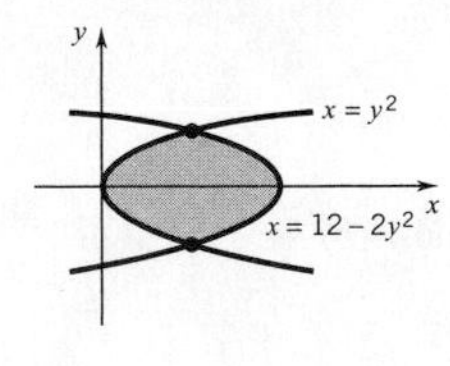

19. area $= \frac{37}{12}$

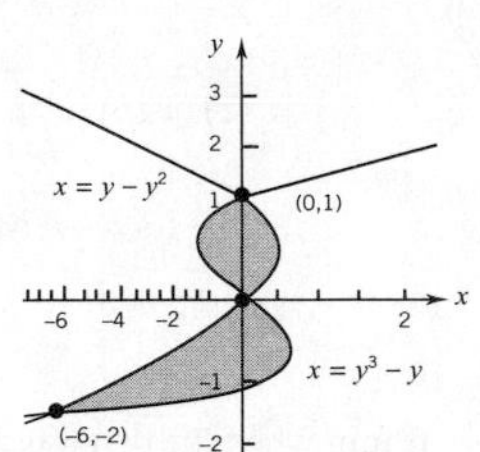

21. area $= 2 - \sqrt{2}$

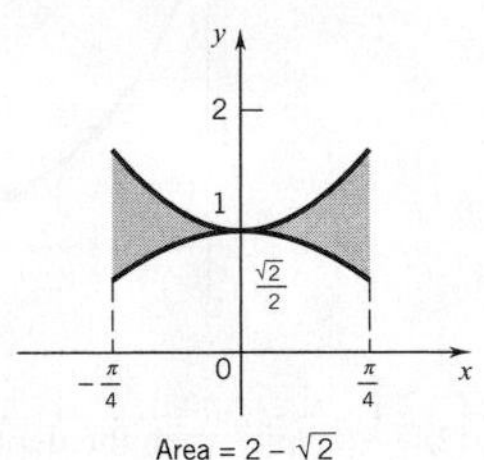

23. area $= 8$

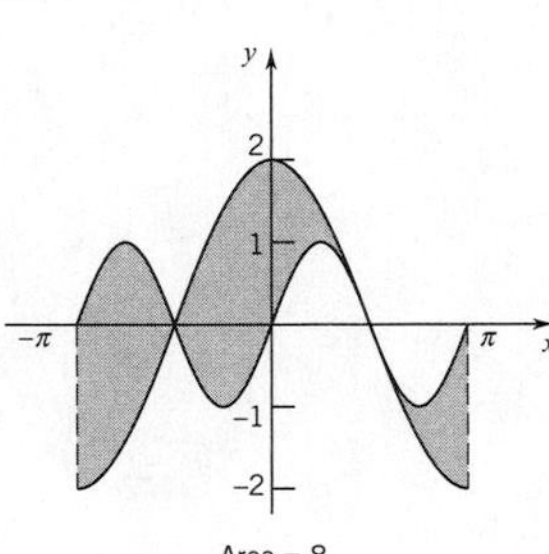

25. area $= \frac{1}{5}$

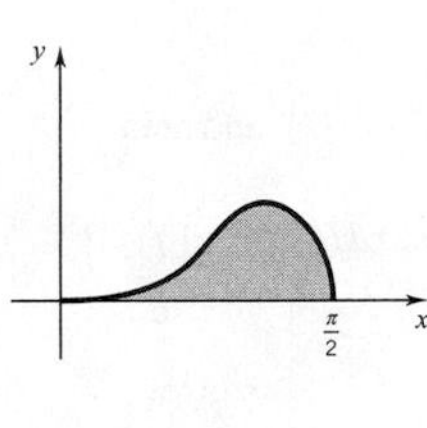

27. 4 **29.** $\frac{39}{2}$

31. area $= 27$

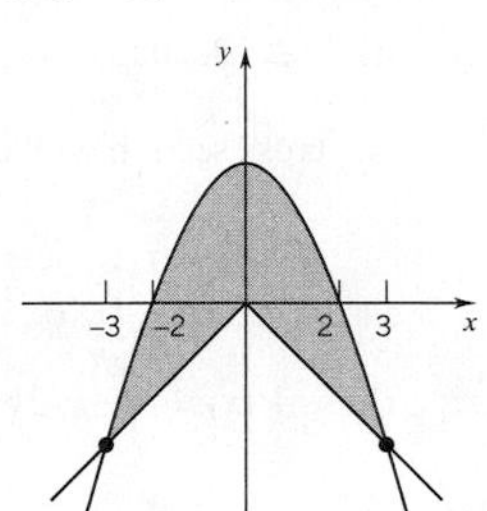

33.

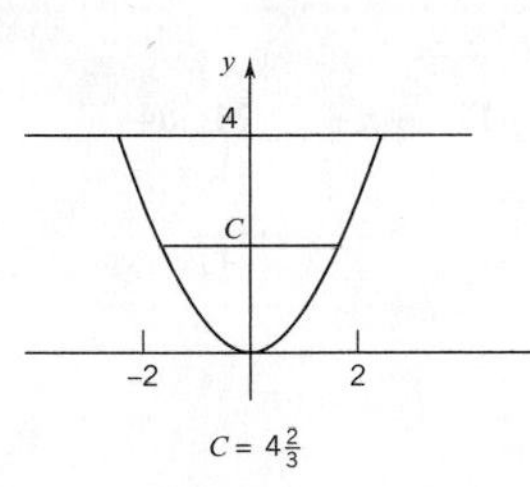

35. $A = \int_0^{\sqrt{3}} \left[\sqrt{4 - y^2} - \frac{1}{\sqrt{3}}y\right] dy$

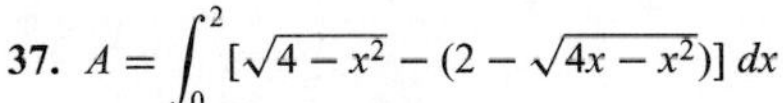

37. $A = \int_0^2 [\sqrt{4 - x^2} - (2 - \sqrt{4x - x^2})]\, dx$

39. The ratio is $\frac{1}{n+1}$.

41.

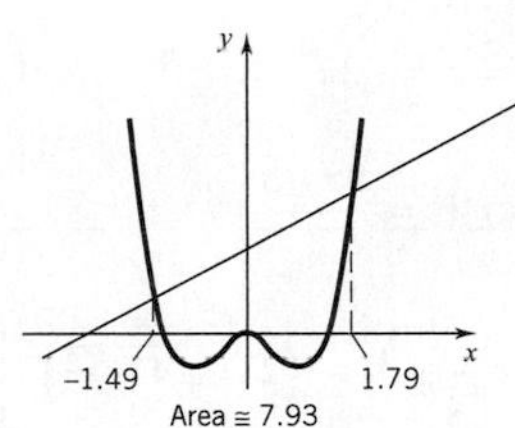

43. (b) x-coordinates of points of intersection: $x = -1.41421, 1.41421, 3$

(c) 152.068

45. 1536 cu in. $\cong 0.89$ cu ft **47.** (a) $A(b) = 2\sqrt{b} - 2$

(b) $A(b) \to \infty$ as $x \to \infty$; the region under the graph of f on $[1, \infty)$ has "infinite" area.

SECTION 6.2

1. $\frac{1}{3}\pi$ **3.** $\frac{1944}{5}\pi$ **5.** $\frac{5}{14}\pi$ **7.** $\frac{3790}{21}\pi$ **9.** $\frac{72}{5}\pi$ **11.** $\frac{32}{3}\pi$ **13.** π **15.** $\frac{\pi^2}{24}(\pi^2 + 6\pi + 6)$

17. $\frac{16}{3}\pi$ **19.** $\frac{768}{7}\pi$ **21.** $\frac{2}{5}\pi$ **23.** $\frac{128}{3}\pi$ **25.** $\frac{16}{3}\pi$ **27.** (a) $\frac{16}{3}r^3$ (b) $\frac{4}{3}\sqrt{3}r^3$ **29.** (a) $\frac{512}{15}$ (b) $\frac{64}{15}\pi$ (c) $\frac{128}{15}\sqrt{3}$

31. (a) 32 (b) 4π (c) $8\sqrt{3}$ **33.** (a) $\frac{64}{3}$ (b) $\frac{16}{3}$ **35.** (a) $\sqrt{3}$ (b) 4 **37.** $\frac{4}{3}\pi ab^2$ **39.** $\frac{1}{3}\pi h(R^2 + rR + r^2)$

41. (a) $31\frac{1}{4}\%$ (b) $14\frac{22}{27}\%$ **43.** $V = \frac{\pi}{3}(2r^3 - 3r^2h + h^3)$

45. (a)

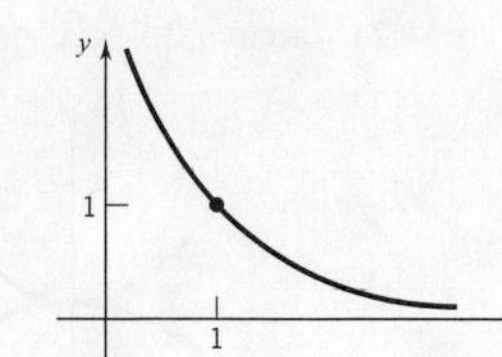

(b) $A(b) = \int_1^b x^{-2/3}\,dx = 3(b^{1/3} - 1)$

(c) $V(b) = \int_1^b \pi(x^{-2/3})^2\,dx = 3\pi(1 - b^{-1/3})$

(d) As $b \to \infty$, $A(b) \to \infty$ and $V(b) \to 3\pi$

47. $\dfrac{1}{\pi}$ ft/min when the depth is 1 foot; $\dfrac{2}{3\pi}$ ft/min when the depth is 2 feet.

49. (b) x-coordinates of points of intersection: $x = 1, 3.57474$

(c) $A \cong 3.11484$

(d) $V \cong 22.2025$

51. The cross section with coordinate x is a washer with outer radius k, inner radius $k - f(x)$, and area

$$A(x) = \pi k^2 - \pi[k - f(x)]^2 = 2\pi k f(x) - \pi[f(x)]^2.$$

Thus, $V(x) = \int_a^b \pi(2kf(x) - [f(x)]^2)dx.$

53. $\frac{40}{3}\pi$ **55.** $4\pi - \frac{1}{2}\pi^2$ **57.** 250π **59.** (a) $\frac{32}{3}\pi$ (b) $\frac{64}{5}\pi$ **61.** (a) 64π (b) $\frac{1024}{35}\pi$ (c) $\frac{704}{5}\pi$ (d) $\frac{512}{7}\pi$

SECTION 6.3

1. $\frac{2}{3}\pi$ **3.** $\frac{128}{5}\pi$ **5.** $\frac{2}{5}\pi$ **7.** 16π **9.** $\frac{72}{5}\pi$ **11.** 36π **13.** 8π **15.** $\frac{1944}{5}\pi$ **17.** $\frac{5}{14}\pi$ **19.** $\frac{72}{5}\pi$ **21.** 64π

23. $\frac{1}{3}\pi$ **25.** (a) $V = \int_0^1 2\pi x(1 - \sqrt{x})\,dx$ (b) $V = \int_0^1 \pi y^4\,dy;\quad V = \frac{1}{5}\pi$

27. (a) $V = \int_0^1 \pi(x - x^4)\,dx$ (b) $V = \int_0^1 2\pi y(\sqrt{y} - y^2)\,dy;\quad V = \frac{3}{10}\pi$

29. (a) $V = \int_0^1 2\pi x^3\,dx$ (b) $V = \int_0^1 \pi(1 - y)\,dy;\quad V = \dfrac{\pi}{2}$ **31.** (a) $\frac{4}{3}\pi ba^2$ **33.** $\frac{1}{4}\pi a^3\sqrt{3}$

35. (a) 64π (b) $\frac{1024}{35}\pi$ (c) $\frac{704}{5}\pi$ (d) $\frac{512}{7}\pi$ **37.** (a) $F'(x) = x\cos x$ (b) $V = \pi^2 - 2\pi$

39. (a) $V = \int_0^1 2\sqrt{3}\pi x^2\,dx + \int_1^2 2\pi x\sqrt{4 - x^2}\,dx$ (b) $V = \int_0^{\sqrt{3}} \pi\left(4 - \frac{4}{3}y^2\right)\,dy$ (c) $V = \dfrac{8\pi\sqrt{3}}{3}$

41. (a) $V = \int_0^1 2\sqrt{3}\pi x(2 - x)\,dx + \int_1^2 2\pi(2 - x)\sqrt{4 - x^2}\,dx$ (b) $V = \int_0^{\sqrt{3}} \pi\left[\left(2 - \dfrac{y}{\sqrt{3}}\right)^2 - \left(2 - \sqrt{4 - y^2}\right)^2\right]\,dy$

43. (a) $V = 2\int_{b-a}^{b+a} 2\pi x\sqrt{a^2 - (x - b)^2}\,dx$ (b) $V = \int_{-a}^{a} \pi\left[\left(b + \sqrt{a^2 - y^2}\right)^2 - \left(b - \sqrt{a^2 - y^2}\right)^2\right]\,dy$

45. $V = \int_0^r 2\pi x\left(-\frac{h}{r}x + h\right)\,dx = \frac{1}{3}\pi r^2 h$

47. (b) $V = \int_0^1 \pi\sin^2(\pi x^2)\,dx \cong 1.18732$; $\sin^2(\pi x^2)$ does not have an "elementary" antiderivative (c) 2

49. (b) x-coordinates of points of intersection: $x = 1, 3.57474$

(c) $A \cong 3.11484$

(d) $V \cong 44.5806$

SECTION 6.4

1. $(\frac{12}{5}, \frac{3}{4}), V_x = 8\pi, V_y = \frac{128}{5}\pi$ **3.** $(\frac{3}{7}, \frac{12}{25}), V_x = \frac{2}{5}\pi, V_y = \frac{5}{14}\pi$ **5.** $(\frac{7}{3}, \frac{10}{3}), V_x = \frac{80}{3}\pi, V_y = \frac{56}{3}\pi$ **7.** $(\frac{3}{4}, \frac{22}{5}), V_x = \frac{704}{15}\pi, V_y = 8\pi$

9. $(\frac{2}{5}, \frac{2}{5}), V_x = \frac{4}{15}\pi, V_y = \frac{4}{15}\pi$ **11.** $(\frac{45}{28}, \frac{93}{70}), V_x = \frac{31}{5}\pi, V_y = \frac{15}{2}\pi$ **13.** $(3, \frac{5}{3}), V_x = \frac{40}{3}\pi, V_y = 24\pi$ **15.** $(\frac{5}{2}, 5)$ **17.** $(1, \frac{8}{5})$

19. $(\frac{10}{3}, \frac{40}{21})$ **21.** $(2, 4)$ **23.** $(-\frac{3}{5}, 0)$ **25.** (a) $(0, 0)$ (b) $\left(\frac{14}{5\pi}, \frac{14}{5\pi}\right)$ (c) $\left(0, \frac{14}{5\pi}\right)$ **27.** $V = \pi ab(2c + \sqrt{a^2 + b^2})$

29. (a) $(\frac{2}{3}a, \frac{1}{3}h)$ (b) $(\frac{2}{3}a + \frac{1}{3}b, \frac{1}{3}h)$ (c) $(\frac{1}{3}a + \frac{1}{3}b, \frac{1}{3}h)$ **31.** (a) $\frac{1}{3}\pi R^3 \sin^2\theta(2\sin\theta + \cos\theta)$ (b) $\dfrac{2R\sin\theta(2\sin\theta + \cos\theta)}{3(\pi\sin\theta + 2\cos\theta)}$

33. An annular region; see Exercise 25(a). **35.** (a) $A = \frac{1}{2}$ (b) $(\frac{16}{35}, \frac{16}{35})$ (c) $V = \frac{16}{35}\pi$ (d) $V = \frac{16}{35}\pi$

37. (a) $A = \frac{250}{3}$ (b) $(-\frac{9}{8}, \frac{290}{21}) \cong (-1.125, 13.8095)$

SECTION 6.5

1. 817.5 ft-lb **3.** $\frac{1}{3}(64 - 7^{3/2})$ ft-lb **5.** $\dfrac{35\pi^2}{72} - \dfrac{1}{4}$ newton-meters **7.** 625 ft-lb **9.** (a) 25-ft-lb (b) $\frac{225}{4}$ ft-lb **11.** 1.95 ft

13. (a) $(6480\pi + 8640)$ ft-lb (b) $(15{,}120\pi + 8640)$ ft-lb **15.** (a) $\frac{11}{192}\pi r^2 h^2 \sigma$ ft-lb (b) $(\frac{11}{192}\pi r^2 h^2 \sigma + \frac{7}{24}\pi r^2 hk\sigma)$ ft-lb

17. (a) $384\pi\sigma$ newton-meters (b) $480\pi\sigma$ newton-meters **19.** 48,000 ft-lb **21.** (a) 20,000 ft-lb (b) 30,000 ft-lb **23.** 788 ft-lb

25. (a) $\frac{1}{2}\sigma l^2$ ft-lb (b) $\frac{3}{2}l^2\sigma$ ft-lb **27.** 20,800 ft-lb

29. Let $\lambda(x)$ be the mass density of the chain at the point x units above the ground. Let g be the gravitational constant. The work done to pull the chain to the top of the building is given by

$$\begin{aligned} W &= \int_0^H (H - x)g\lambda(x)\,dx = Hg\int_0^H \lambda(x)\,dx - g\int_0^H x\lambda(x)\,dx \\ &= HgM - g\bar{x}M = (H - \bar{x})gM \\ &= (\text{weight of chain}) \times (\text{distance from center of mass to top of building}). \end{aligned}$$

31. Total weight 400 lbs.; center of mass (chain plus bucket): $\bar{x} = 25$ ft from ground;

$$W = 400(100 - 25) = 400(75) = 30{,}000 \text{ ft-lbs}.$$

33. $W = \displaystyle\int_a^b ma\,dx = \int_a^b mv\,dv = \frac{1}{2}mv_b^2 - \frac{1}{2}mv_a^2$ **35.** 94.8 ft-lb **37.** 9.714×10^9 ft-lb

39. (a) 670 sec or 11 min, 10 sec (b) 1116 sec or 18 min, 36 sec

SECTION 6.6

1. 9000 lb **3.** 1.437×10^8 newtons **5.** 1.7052×10^6 newtons **7.** 2160 lb **9.** $\frac{8000}{3}\sqrt{2}$ lb **11.** 333.33 lb **13.** 2560 lb

15. (a) 41,250 lb (b) 41,250 lb **17.** (a) 297,267 newtons (b) 39,200 newtons at the shallow end; 352,800 newtons at the deep end

19. 2.21749×10^6 newtons **21.** $F = \sigma\bar{x}A$ where A is the area of the submerged surface and $\bar{x}$ is the depth of its centroid.

CHAPTER 7

SECTION 7.1

1. $f^{-1}(x) = \frac{1}{5}(x - 3)$ **3.** not one-to-one **5.** $f^{-1}(x) = (x - 1)^{1/5}$ **7.** $f^{-1}(x) = [\frac{1}{3}(x - 1)]^{1/3}$ **9.** $f^{-1}(x) = 1 - x^{1/3}$

11. $f^{-1}(x) = (x - 2)^{1/3} - 1$ **13.** $f^{-1}(x) = x^{5/3}$ **15.** $f^{-1}(x) = \frac{1}{3}(2 - x^{1/3})$ **17.** $f^{-1}(x) = \sin^{-1}(x)$ (to be studied in Section 7.7)

19. $f^{-1}(x) = 1/x$ **21.** not one-to-one **23.** $f^{-1}(x) = \left(\dfrac{1 - x}{x}\right)^{1/3}$ **25.** $f^{-1}(x) = (2 - x)/(x - 1)$ **27.** they are equal

29. **31.**

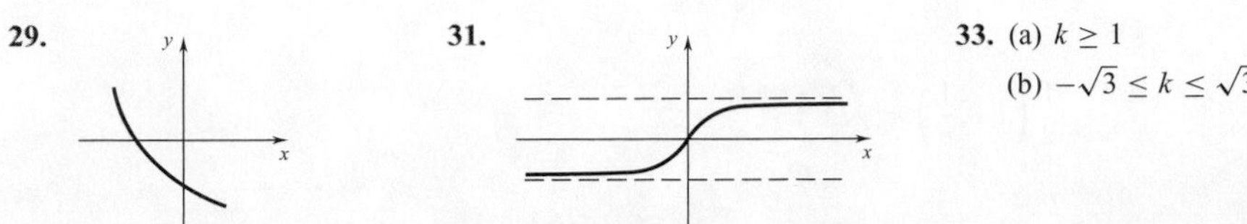

33. (a) $k \geq 1$
(b) $-\sqrt{3} \leq k \leq \sqrt{3}$

35. $f'(x) = 3x^2 \geq 0$ on $(-\infty, \infty)$, $f'(x) = 0$ only at $x = 0$; $(f^{-1})'(9) = \frac{1}{12}$ **37.** $f'(x) = 1 + \dfrac{1}{\sqrt{x}} > 0$ on $(0, \infty)$; $(f^{-1})'(8) = \frac{2}{3}$

39. $f'(x) = 2 - \sin x > 0$ on $(-\infty, \infty)$; $(f^{-1})'(\pi) = 1$ **41.** $f'(x) = \sec^2 x > 0$ on $(-\pi/2, \pi/2)$; $(f^{-1})'(\sqrt{3}) = \frac{1}{4}$

43. $f'(x) = 3x^2 + \dfrac{3}{x^4} > 0$ on $(0, \infty)$; $(f^{-1})'(2) = \frac{1}{6}$ **45.** $(f^{-1})'(x) = \dfrac{1}{x}$ **47.** $(f^{-1})'(x) = \dfrac{1}{\sqrt{1-x^2}}$

49. (a) $f'(x) = \dfrac{ad - bc}{(cx+d)^2} \neq 0$ iff $ad - bc \neq 0$ (b) $f^{-1}(x) = \dfrac{dx - b}{a - cx}$ **51.** (a) $f'(x) = \sqrt{1 + x^2} > 0$ (b) $(f^{-1})'(0) = \dfrac{1}{\sqrt{5}}$

53. (a) $g(t) \geq 0$ or $g(t) \leq 0$ on $[a, b]$, with $g(t) = 0$ only at "isolated" points. (b) $g(x) \neq 0$ (c) $(f^{-1})'(x) = \dfrac{1}{g(x)}$

55. (a) $g'(x) = \dfrac{1}{f'(g(x))}$; $g''(x) = -\dfrac{f''(g(x))g'(x)}{[f'(g(x))]^2} = -\dfrac{f''(g(x))}{[f'(g(x))]^3}$

(b) If f is increasing, then the graphs of f and g have opposite concavity; if f is decreasing, then the graphs of f and g have the same concavity.

57. $(f^{-1})'(x) = \dfrac{1}{\sqrt{1-x^2}}$. **59.** $f^{-1}(x) = -1 + (2-x)^{1/3}$ **61.** $f^{-1}(x) = \dfrac{x^2 - 8x + 25}{9}$, $x \geq 4$ **63.** $f^{-1}(x) = \dfrac{1-x}{1+x}$

65.

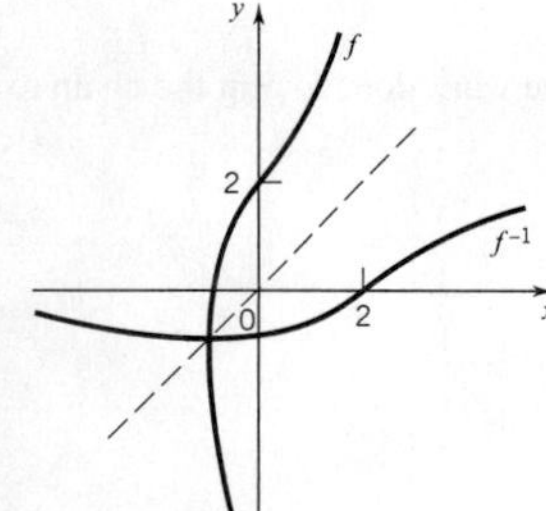

67.

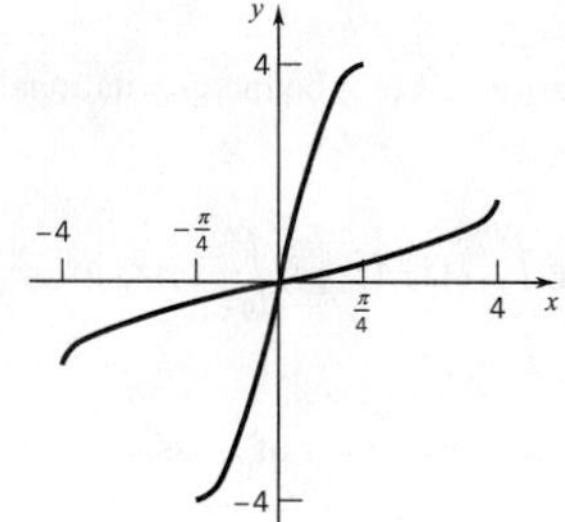

SECTION 7.2

1. $\ln 2 + \ln 10 \cong 2.99$

3. $2\ln 4 - \ln 10 \cong 0.48$

5. $-\ln 10 \cong -2.30$

7. $\ln 8 + \ln 9 - \ln 10 \cong 1.98$

9. $\frac{1}{2}\ln 2 \cong 0.35$

11.

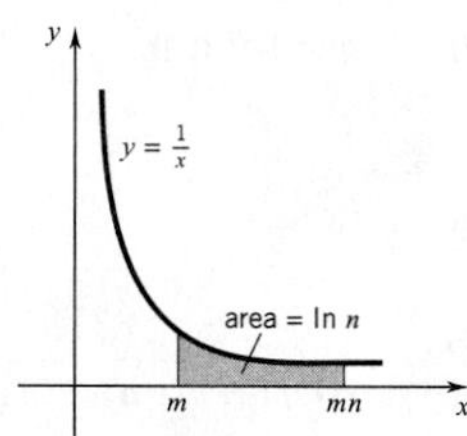

13. 0.406 **15.** (a) 1.65 (b) 1.57 (c) 1.71 **17.** $x = e^2$ **19.** $x = 1, e^2$ **21.** $x = 1$ **23.** $\displaystyle\lim_{x\to 1}\frac{\ln x}{x-1} = \left.\frac{d(\ln x)}{dx}\right|_{x=1} = 1$

25. (a) $P = \{1, 2, \ldots, n\}$ is a regular partition of $[1, n]$; $L_f(P) = \dfrac{1}{2} + \dfrac{1}{3} + \cdots + \dfrac{1}{n} < \displaystyle\int_1^n \frac{1}{t}\,dt = \ln n < 1 + \frac{1}{2} + \cdots + \frac{1}{n-1} = U_f(P)$.

(b) Sum of shaded areas $= V_f(P) - \displaystyle\int_1^n \frac{1}{t}\,dt = 1 + \frac{1}{2} + \cdots + \frac{1}{n-1} - \ln n$.

(c) Connect the points $(1,1), (2, \frac{1}{2}), \ldots, \left(n, \dfrac{1}{n}\right)$ by straight-line segments. The sum of the areas of the triangles that are formed is

$$\frac{1}{2}\cdot 1\left[\left(1 - \frac{1}{2}\right) + \left(\frac{1}{2} - \frac{1}{3}\right) + \cdots + \left(\frac{1}{n-1} - \frac{1}{n}\right)\right] = \frac{1}{2}\left(1 - \frac{1}{n}\right),$$

so

$$\frac{1}{2}\left(1 - \frac{1}{n}\right) < \gamma.$$

The sum of the areas of the indicated rectangles is

$$1\left[\left(1 - \frac{1}{2}\right) + \left(\frac{1}{2} - \frac{1}{3}\right) + \cdots + \left(\frac{1}{n-1} - \frac{1}{n}\right)\right] = 1 - \frac{1}{n},$$

so

$$\gamma < 1 - \frac{1}{n}.$$

Letting $n \to \infty$, we have $\frac{1}{2} < \gamma < 1$.

27. (a) $\ln 3 - \sin 3 \cong 0.96 > 0$; $\ln 2 - \sin 2 \cong -0.22 < 0$ (b) $r \cong 2.2191$ **29.** 1

31. (b) $e^{\pi/2} \cong 4.81048$; $e^{3\pi/2} \cong 111.318$ (c) $e^{\pi/2+2n\pi}$; $e^{3\pi/2+2n\pi}$

SECTION 7.3

1. domain $(0, \infty)$, $f'(x) = \dfrac{1}{x}$ **3.** domain $(-1, \infty)$, $f'(x) = \dfrac{3x^2}{x^3+1}$ **5.** domain $(-\infty, \infty)$, $f'(x) = \dfrac{x}{1+x^2}$

7. domain all $x \neq \pm 1$, $f'(x) = \dfrac{4x^3}{x^4-1}$ **9.** domain $(-\frac{1}{2}, \infty)$, $f'(x) = 2(2x+1)\left[1 + 2\ln(2x+1)\right]$

11. domain $(0,1) \cup (1,\infty)$, $f'(x) = -\dfrac{1}{x(\ln x)^2}$ **13.** domain $(0,\infty)$; $f'(x) = \dfrac{1}{x}\cos(\ln x)$ **15.** $\ln|x+1|+C$ **17.** $-\frac{1}{2}\ln|3-x^2| + C$

19. $\frac{1}{3}\ln|\sec 3x| + C$ **21.** $\frac{1}{2}\ln|\sec x^2 + \tan x^2| + C$ **23.** $\dfrac{1}{2(3-x^2)} + C$ **25.** $-\ln|2+\cos x| + C$ **27.** $\ln|\ln x| + C$

29. $\dfrac{-1}{\ln x} + C$ **31.** $-\ln|\sin x + \cos x| + C$ **33.** $\frac{2}{3}\ln|1 + x\sqrt{x}| + C$ **35.** $x + 2\ln|\sec x + \tan x| + \tan x + C$

37. 1 **39.** 1 **41.** $\frac{1}{2}\ln\frac{8}{5}$ **43.** $\ln\frac{4}{3}$ **45.** $\frac{1}{2}\ln 2$

47. The integrand is not defined at $x = 2$. **49.** $g'(x) = (x^2+1)^2(x-1)^5x^3\left(\dfrac{4x}{x^2+1} + \dfrac{5}{x-1} + \dfrac{3}{x}\right)$ **51.** $g'(x) = \dfrac{x^4(x-1)}{(x+2)(x^2+1)}\left(\dfrac{4}{x} + \dfrac{1}{x-1} - \dfrac{1}{x+2} - \dfrac{2x}{x^2+1}\right)$

53. $\frac{1}{3}\pi - \frac{1}{2}\ln 3$ **55.** $\frac{1}{4}\pi - \frac{1}{2}\ln 2$ **57.** $\frac{15}{8} - \ln 4$ **59.** $\pi\ln 9$ **61.** $2\pi\ln(2+\sqrt{3})$ **63.** $\ln 5$ ft **65.** $(-1)^{n-1}\dfrac{(n-1)!}{x^n}$

67.
$$\int \csc x\,dx = \int \frac{\csc x(\csc x - \cot x)}{\csc x - \cot x}\,dx$$

Let $u = \csc x - \cot x$, $du = \csc x(\csc x - \cot x)dx$. Then

$$\int \csc x\,dx = \int \frac{1}{u}du = \ln|u| + C = \ln|\csc x - \cot x| + C.$$

69. (i) domain $(-\infty, 4)$
(ii) decreases throughout
(iii) no extreme values
(iv) concave down throughout; no pts of inflection
(v)

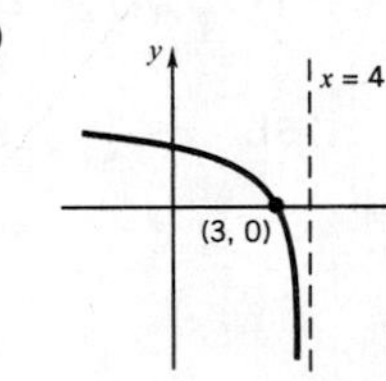

71. (i) domain $(0,\infty)$
(ii) decreases on $(0, e^{-1/2})$, increases on $[e^{-1/2}, \infty]$
(iii) $f(e^{-1/2}) = -\frac{1}{2e}$ local and absolute min.
(iv) concave down on $(0, e^{-3/2})$; concave up on $(e^{-3/2}, \infty)$ pt of inflection at $(e^{-3/2}, -\frac{3}{2}e^{-3})$
(v)

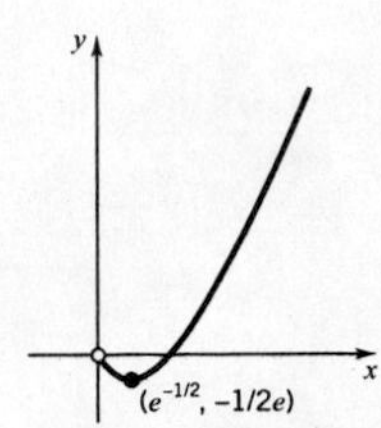

73. (i) domain $(0, \infty)$
(ii) increases on $(0, 1]$; decreases on $[1, \infty)$
(iii) $f(1) = \ln \frac{1}{2}$ local and absolute max
(iv) concave down on $(0, 2.0582)$; concave up on $(2.0582, \infty)$; point of inflection $(2.0582, -0.9338)$ (approx.)
(v)

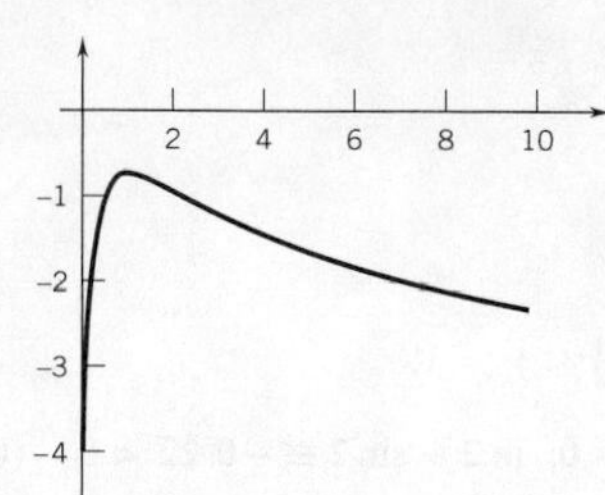

75. average slope $= \dfrac{1}{b-a}\displaystyle\int_a^b \frac{1}{x}\,dx = \frac{1}{b-a}\ln(b/a)$ **77.** x-intercept: 1; absolute min at $x = e^{-2}$; absolute max at $x = 10$

79. x-intercepts: 1,23.1407; absolute max at $x \cong 4.8105$, absolute min at $x = 100$

81. (a) $v(t) = 2 + 2t - t^2 + 3\ln(t+1)$ (c) max velocity at $t \cong 1.5811$; min velocity at $t = 0$

83. (b) x-coordinates of points of intersection: $x = 1, 3.30278$
(c) $A \cong 2.34042$

85. (a) $f(x) = \ln x$
(b) $f(x) = x\ln x$
(c) $f(x) = x^2 \ln x$

87. (a) $f'(x) = \dfrac{1 - 2\ln x}{x^3}$; $f''(x) = \dfrac{-5 + 6\ln x}{x^4}$
(b) $f(1) = 0$; $f'(e^{1/2}) = 0$; $f''(e^{5/6}) = 0$
(c) $f(x) > 0$ on $(1, \infty)$; $f'(x) > 0$ on $(0, e^{1/2})$; $f''(x) > 0$ on $(e^{5/6}, \infty)$;
$f(x) < 0$ on $(0, 1)$; $f'(x) < 0$ on $(e^{1/2}, \infty)$; $f''(x) < 0$ on $(0, e^{5/6})$

SECTION 7.4

1. $\dfrac{dy}{dx} = -2e^{-2x}$ **3.** $\dfrac{dy}{dx} = 2xe^{x^2-1}$ **5.** $\dfrac{dy}{dx} = e^x\left(\dfrac{1}{x} + \ln x\right)$ **7.** $\dfrac{dy}{dx} = -(x^{-1} + x^{-2})e^{-x}$ **9.** $\dfrac{dy}{dx} = \frac{1}{2}(e^x - e^{-x})$

11. $\dfrac{dy}{dx} = \dfrac{1}{2}e^{\sqrt{x}}\left(\dfrac{1}{x} + \dfrac{\ln\sqrt{x}}{\sqrt{x}}\right)$ **13.** $\dfrac{dy}{dx} = 4xe^{x^2}(e^{x^2} + 1)$ **15.** $\dfrac{dy}{dx} = x^2e^x$ **17.** $\dfrac{dy}{dx} = \dfrac{2e^x}{(e^x + 1)^2}$ **19.** $\dfrac{dy}{dx} = 4x^3$

21. $f'(x) = 2\cos(e^{2x})e^{2x}$ **23.** $f'(x) = -e^{-2x}(2\cos x + \sin x)$ **25.** $\frac{1}{2}e^{2x} + C$ **27.** $\dfrac{1}{k}e^{kx} + C$ **29.** $\frac{1}{2}e^{x^2} + C$ **31.** $-e^{1/x} + C$

33. $\frac{1}{2}x^2 + C$ **35.** $-8e^{-x/2} + C$ **37.** $2\sqrt{e^x + 1} + C$ **39.** $\frac{1}{4}\ln(2e^{2x} + 3) + C$ **41.** $e^{\sin x} + C$

43. $e - 1$ **45.** $\frac{1}{6}(1 - \pi^{-6})$ **47.** $2 - \dfrac{1}{e}$ **49.** $\ln\frac{3}{2}$ **51.** $\frac{1}{2}e + \frac{1}{2}$ **53.** (a) $f^{(n)}(x) = a^n e^{ax}$ (b) $f^{(n)}(x) = (-1)^n a^n e^{-ax}$

55. at $\left(\pm\dfrac{1}{\sqrt{2}}, \dfrac{1}{\sqrt{e}}\right)$

57. (a) f is an even function; symmetric with respect to the y-axis.
(b) f increases on $(-\infty, 0]$; f decreases on $[0, \infty)$.
(c) $f(0) = 1$ is a local and absolute maximum.
(d) the graph is concave up on $(-\infty, -1/\sqrt{2}) \cup (1/\sqrt{2}, \infty)$; the graph is concave down on $(-1/\sqrt{2}, 1/\sqrt{2})$; points of inflection at $(-1/\sqrt{2}, e^{-1/2})$ and $(1/\sqrt{2}, e^{-1/2})$
(e) the x-axis (f)

59. (a) $\pi\left(1 - \frac{1}{e}\right)$
(b) $\displaystyle\int_0^1 \pi e^{-2x^2}\,dx$

61. $\frac{1}{2}(3e^4 + 1)$ **63.** $e^2 - e - 2$

65. (a) domain $(-\infty, 0) \cup (0, \infty)$
(b) increases on $(-\infty, 0)$, decreases on $(0, \infty)$
(c) no extreme values
(d) concave up on $(-\infty, 0)$ and on $(0, \infty)$
(e)

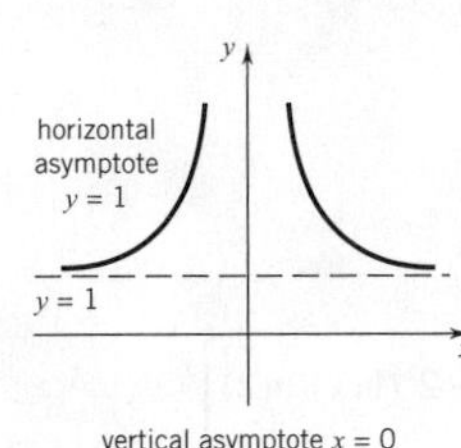

67. (a) domain $(0, \infty)$
(b) f increases on $(e^{-1/2}, \infty)$; f decreases on $(0, e^{-1/2})$.
(c) $f(e^{-1/2}) = -1/2e$ is a local and absolute minimum.
(d) the graph is concave down on $(0, e^{-3/2})$; the graph is concave up on $(e^{-3/2}, \infty)$; point of inflection at $(e^{-3/2}, -3/2e^3)$
(e)

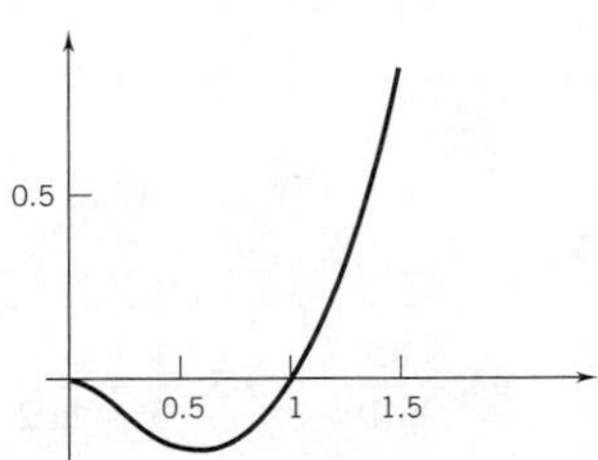

69. (a) $\lim_{x\to 0^+} f(x) = 0$ for all $k > 0$
(b) $e^{-1/k}$

71. (a) $\left(\pm\frac{1}{a}, e\right)$ (b) $\frac{1}{a}(e-2)$ (c) $\frac{1+2a^2e}{a^3e}$ **73.** for $x > (n+1)!$, $e^x > 1 + x + \cdots + \frac{x^{n+1}}{(n+1)!} > \frac{x^{n+1}}{(n+1)!} = x^n\left[\frac{x}{(n+1)!}\right] > x^n$

75. (a)

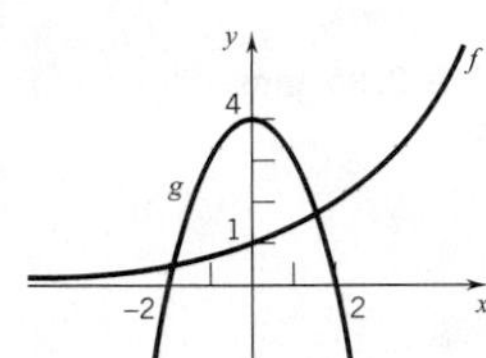

(b) $x = -1.9646$; $x = 1.0580$
(c) 6.4240

77.

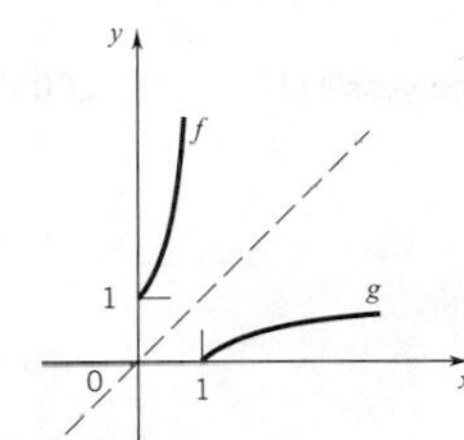

79. (a) $x \cong 1.14477, 1.83788$
(b) $x = \ln(|n\pi|), n = \pm 1, \pm 2, \ldots$

81. (b) $x \cong 1.3098$
(c) $f'(1.3098) \cong -0.26987$; $g'(1.3098) \cong 0.76348$
(d) no

83. (a) $x - \ln|e^x - 1| + C$
(b) $-\frac{1}{5}e^{-5x} + e^{-4x} - 2e^{-3x} + 2e^{-2x} - e^{-x} + C$
(c) $e^{\tan x} + C$

SECTION 7.5

1. 6 **3.** $-\frac{1}{6}$ **5.** 0 **7.** 3 **9.** $\log_p xy = \frac{\ln xy}{\ln p} = \frac{\ln x + \ln y}{\ln p} = \frac{\ln x}{\ln p} + \frac{\ln y}{\ln p} = \log_p x + \log_p y$

11. $\log_p x^y = \frac{\ln x^y}{\ln p} = y\frac{\ln x}{\ln p} = y \log_p x$ **13.** 0 **15.** 2 **17.** $t_1 < \ln a < t_2$ **19.** $f'(x) = 2(\ln 3)3^{2x}$

21. $f'(x) = \left(5\ln 2 + \frac{\ln 3}{x}\right)2^{5x}3^{\ln x}$ **23.** $g'(x) = \frac{1}{2\ln 3} \cdot \frac{1}{x\sqrt{\log_3 x}}$ **25.** $f'(x) = \frac{\sec^2(\log_5 x)}{x \ln 5}$

27. $F'(x) = \ln 2(2^{-x} - 2^x)\sin(2^x + 2^{-x})$ **29.** $\frac{3^x}{\ln 3} + C$ **31.** $\frac{1}{4}x^4 - \frac{3^{-x}}{\ln 3} + C$ **33.** $\log_5|x| + C$ **35.** $\frac{3}{\ln 4}(\ln x)^2 + C$ **37.** $\frac{1}{e\ln 3}$

39. $\frac{1}{e}$ **41.** $f(x) = p^x$

$\ln f(x) = x \ln p$

$\frac{f'(x)}{f(x)} = \ln p$

$f'(x) = p^x \ln p$

43. $(x+1)^x \left[\frac{x}{x+1} + \ln(x+1)\right]$ **45.** $(\ln x)^{\ln x}\left[\frac{1+\ln(\ln x)}{x}\right]$ **47.** $x^{\sin x}\left(\cos x \ln x + \frac{\sin x}{x}\right)$

49. $(\sin x)^{\cos x}\left[\frac{\cos^2 x}{\sin x} - \sin x \ln(\sin x)\right]$ **51.** $x^{2^x}\left[\frac{2^x}{x} + 2^x(\ln x)(\ln 2)\right]$

53.

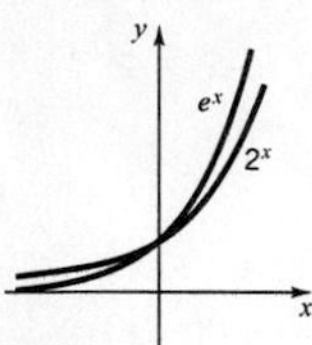

55.

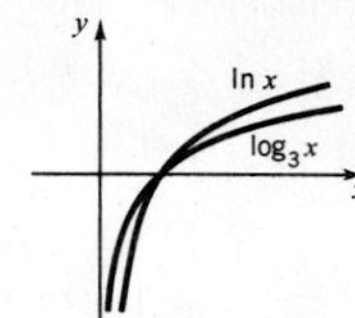

57.

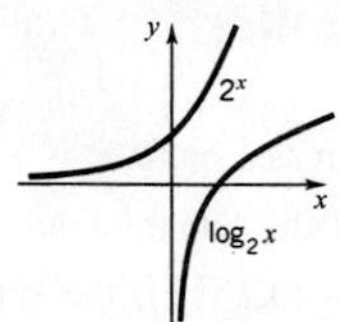

59. $\frac{1}{4 \ln 2}$ **61.** 2 **63.** $\frac{45}{\ln 10}$ **65.** $\frac{1}{3} + \frac{1}{\ln 2}$ **67.** approx. 16.999999; $5^{(\ln 17)/(\ln 5)} = (e^{\ln 5})^{(\ln 17)/(\ln 5)} = e^{\ln 17} = 17$

69. (b) the x-coordinates of the points of intersection are: $x \cong -1.198, x = 3$ and $x \cong 3.408$.
(c) for the interval $[-1.198, 3], A \cong 5.5376$; for the interval $[3, 3.408], A \cong 0.1373$

SECTION 7.6

1. (a) \$411.06 (b) \$612.77 (c) \$859.14 **3.** about $5\frac{1}{2}\%$: $(\ln 3)/20 \cong 0.0549$ **5.** 16,000

7. (a) $P(t) = 10{,}000\, e^{t \ln 2} = 10{,}000(2)^t$ (b) $P(26) = 10{,}000(2)^{26}, P(52) = 10{,}000(2)^{52}$ **9.** (a) $e^{0.35}$ (b) $k = \frac{\ln 2}{15}$

11. $P(20) \cong 317.1$ million; $P(11) \cong 284.4$ million **13.** in the year 2112 **15.** $200\left(\frac{4}{5}\right)^{t/5}$ liters **17.** $5\left(\frac{4}{5}\right)^{5/2} \cong 2.86$ gms

19. $100[1 - (\frac{1}{2})^{1/n}]\%$ **21.** 80.7%, 3240 yrs

23. (a) $x_1(t) = 10^6 t, x_2(t) = e^t - 1$
(b) $\frac{d}{dt}[x_1(t) - x_2(t)] = \frac{d}{dt}[10^6 t - (e^t - 1)] = 10^6 - e^t$.
This derivative is zero at $t = 6 \ln 10 \cong 13.8$. After that the derivative is negative
(c) $x_2(15) < e^{15} = (e^3)^5 \cong 20^5 = 2^5(10^5) = 3.2(10^6) < 15(10^6) = x_1(15)$
$x_2(18) = e^{18} - 1 = (e^3)^6 - 1 \cong 20^6 - 1 = 64(10^6) - 1 > 18(10^6) = x_1(18)$
$x_2(18) - x_1(18) \cong 64(10^6) - 1 - 18(10^6) \cong 46(10^6)$
(d) If by time t_1 EXP has passed LIN, then $t_1 > 6 \ln 10$. For all $t \geq t_1$ the speed of EXP is greater than the speed of LIN: for $t \geq t_1 > 6 \ln 10$, $v_2(t) = e^t > 10^6 = v_1(t)$.

25. (a) $15(\frac{2}{3})^{1/2} \cong 12.25$ lb/in.^{2} (b) $15(\frac{2}{3})^{3/2} \cong 8.16$ lb/in.^{2} **27.** 6.4% **29.** (a) \$18,589.35 (b) \$20,339.99 (c) \$22,933.27

31. $176/\ln 2 \cong 254$ ft **33.** 11,400 years **35.** $f(t) = Ce^{t^2/2}$ **37.** $f(t) = Ce^{\sin t}$

SECTION 7.7

1. (a) 0 (b) $-\frac{\pi}{3}$ **3.** (a) $\frac{2\pi}{3}$ (b) $\frac{3\pi}{4}$ **5.** (a) $\frac{1}{2}$ (b) $\frac{\pi}{4}$ **7.** (a) does not exist (b) does not exist **9.** (a) $\frac{\sqrt{3}}{2}$ (b) $-\frac{7}{25}$

11. $\frac{1}{x^2 + 2x + 2}$ **13.** $\frac{2}{x\sqrt{4x^4 - 1}}$ **15.** $\frac{2x}{\sqrt{1 - 4x^2}} + \sin^{-1} 2x$ **17.** $\frac{2 \sin^{-1} x}{\sqrt{1 - x^2}}$ **19.** $\frac{x - (1 + x^2)\tan^{-1} x}{x^2(1 + x^2)}$ **21.** $\frac{1}{(1 + 4x^2)\sqrt{\tan^{-1} 2x}}$

23. $\frac{1}{x[1 + (\ln x)^2]}$ **25.** $-\frac{r}{|r|\sqrt{1 - r^2}}$ **27.** $2x \sec^{-1}\left(\frac{1}{x}\right) - \frac{x^2}{\sqrt{1 - x^2}}$ **29.** $\cos[\sec^{-1}(\ln x)] \cdot \frac{1}{x|\ln x|\sqrt{(\ln x)^2 - 1}}$ **31.** $\sqrt{\frac{c - x}{c + x}}$

33. Set $au = x + b$, $a\,du = dx$.

$$\int \frac{dx}{\sqrt{a^2 - (x+b)^2}} = \int \frac{a\,du}{\sqrt{a^2 - a^2u^2}} = \int \frac{du}{\sqrt{1 - u^2}} = \sin^{-1} u + C = \sin^{-1}\left(\frac{x+b}{a}\right) + C$$

35. Set $au = x + b,\ a\,du = dx$.

$$\int \frac{dx}{(x+b)\sqrt{(x+b)^2 - a^2}} = \int \frac{a\,du}{au\sqrt{a^2u^2 - a^2}} = \frac{1}{a}\int \frac{du}{u\sqrt{u^2-1}} = \frac{1}{a}\sec^{-1}\frac{|x+b|}{a} + C.$$

37. domain $(-\infty, \infty)$, range $(0, \pi)$ **39.** $\frac{1}{4}\pi$ **41.** $\frac{1}{4}\pi$ **43.** $\frac{1}{20}\pi$ **45.** $\frac{1}{24}\pi$ **47.** $\frac{1}{3}\sec^{-1}4 - \frac{\pi}{9}$ **49.** $\frac{1}{6}\pi$

51. $\tan^{-1}2 - \frac{1}{4}\pi \cong 0.322$ **53.** $\frac{1}{2}\sin^{-1}x^2 + C$ **55.** $\frac{1}{2}\tan^{-1}x^2 + C$ **57.** $\frac{1}{3}\tan^{-1}(\frac{1}{3}\tan x) + C$ **59.** $\frac{1}{2}(\sin^{-1}x)^2 + C$

61. $\sin^{-1}(\ln x) + C$ **63.** $\frac{1}{\sqrt{1-x^2}}$ is not defined for $x \geq 1$. **65.** $\frac{\pi}{3}$ **67.** $2\pi - \frac{4}{3}$ **69.** $4\pi(\sqrt{2} - 1)$

71. $\sqrt{s^2 + sk}$ feet from the point where the line of the sign intersects the road.

73. (b) $\frac{1}{2}\pi a^2$; area of semicircle of radius a

75. (a) There exist constants C_1, C_2 such that

$$f(x) + g(x) = C_1 \quad \text{for } x < 0; \quad f(x) + g(x) = C_2 \text{ for } x > 0.$$

(b) $\lim_{x\to 0^+} f(x) = \frac{\pi}{2}$; $\lim_{x\to 0^-} f(x) = -\frac{\pi}{2}$ (d) This is clear from the graphs in (a). (e) $C_1 = \frac{\pi}{2}$; $C_2 = -\frac{\pi}{2}$

77. estimate $\cong 0.523$, $\sin 0.523 \cong 0.499$ explanation: the integral $= \sin^{-1}0.5$; therefore sin (integral)$= 0.5$

79. (a) $\frac{16}{87}$
(b) 0
(c) $-\frac{120}{169}$

81. (a) $(0.78615, 0.66624)$
(b) $A \cong 0.37743$

SECTION 7.8

1. $2x\cosh x^2$ **3.** $\dfrac{a\sinh ax}{2\sqrt{\cosh ax}}$ **5.** $\dfrac{1}{1-\cosh x}$ **7.** $ab(\cosh bx - \sinh ax)$ **9.** $\dfrac{a\cosh ax}{\sinh ax}$ **11.** $2\,e^{2x}\cosh(e^{2x})$

13. $-e^{-x}\cosh 2x + 2\,e^{-x}\sinh 2x$ **15.** $\tanh x$ **17.** $(\sinh x)^x\,[\ln(\sinh x) + x\coth x]$

19. $\cosh^2 t - \sinh^2 t = \left(\frac{e^t + e^{-t}}{2}\right)^2 - \left(\frac{e^t - e^{-t}}{2}\right)^2 = \frac{e^{2t} - 2 + e^{-2t}}{4} - \frac{e^{2t} - 2 + e^{-2t}}{4} = 1$

21. $\cosh t\cosh s + \sinh t\sinh s = \left(\frac{e^t + e^{-t}}{2}\right)\left(\frac{e^s + e^{-s}}{2}\right) + \left(\frac{e^t - e^{-t}}{2}\right)\left(\frac{e^s - e^{-s}}{2}\right)$

$= \frac{1}{4}(e^{t-s} + e^{s-t} + e^{t-s} + e^{-t-s} + e^{t+s} - e^{s-t} - e^{t-s} + e^{-t-s})$

$= \frac{1}{2}(e^{t-s} + e^{-(t+s)}) = \cosh(t+s)$

23. $\cosh^2 t + \sinh^2 t = \left(\frac{e^t + e^{-t}}{2}\right)^2 + \left(\frac{e^t + e^{-t}}{2}\right)^2 = \frac{1}{4}(e^{2t} + 2 + e^{-2t} + e^{2t} - 2 - e^{-2t}) = \frac{e^{2t} + e^{-2t}}{2} = \cosh 2t$

25. $\sinh(-t) = \frac{e^{-t} - e^{-(-t)}}{2} = \frac{e^{-t} - e^{t}}{2} = -\sinh t$ **27.** absolute max -3

29. $[\cosh x + \sinh x]^n = \left[\frac{e^x + e^{-x}}{2} + \frac{e^x - e^{-x}}{2}\right]^n = [e^x]^n = e^{nx} = \frac{e^{nx} + e^{-nx}}{2} + \frac{e^{nx} - e^{-nx}}{2} = \cosh nx + \sinh nx$

31. $A = 2, B = \frac{1}{3}, C = 3$ **33.** $\frac{1}{a}\sinh ax + C$ **35.** $\frac{1}{3a}\sinh^3 ax + C$ **37.** $\frac{1}{a}\ln(\cosh ax) + C$ **39.** $-\frac{1}{a\cosh ax} + C$

41. $\frac{1}{2}(\sinh x\cosh x + x) + C$ **43.** $2\cosh\sqrt{x} + C$ **45.** $\sinh 1 \cong 1.175$ **47.** $\frac{81}{20}$ **49.** π

51. $\pi[\ln 5 + \frac{1}{4}\sinh(4\ln 5)] \cong 250.492$ 53. (a) $(0.69315, 1.25)$
(b) $A \cong 0.38629$

SECTION 7.9

1. $2\tanh x\,\mathrm{sech}^2 x$ **3.** $\mathrm{sech}\,x\,\mathrm{csch}\,x$ **5.** $\dfrac{2e^{2x}\cosh(\tan^{-1}e^{2x})}{1-e^{4x}}$ **7.** $\dfrac{-x\,\mathrm{csch}^2(\sqrt{x^2+1})}{\sqrt{x^2+1}}$ **9.** $\dfrac{-\mathrm{sech}\,x(\tanh x + 2\sinh x)}{(1+\cosh x)^2}$

11. $\frac{d}{dx}(\coth x) = \frac{d}{dx}\left(\frac{\cosh x}{\sinh x}\right) = \frac{\sinh^2 x - \cosh^2 x}{\sinh^2 x} = \frac{-1}{\sinh^2 x} = -\operatorname{csch}^2 x$

13. $\frac{d}{dx}(\operatorname{csch} x) = \frac{d}{dx}\left(\frac{1}{\sinh x}\right) = -\frac{\cosh x}{\sinh^2 x} = -\operatorname{csch} x \coth x$ **15.** (a) $\frac{3}{5}$ (b) $\frac{5}{3}$ (c) $\frac{4}{3}$ (d) $\frac{5}{4}$ (e) $\frac{3}{4}$

17. If $x \leq 0$, the result is obvious. Suppose then that $x > 0$. Since $x^2 \geq 1$, we have $x \geq 1$. consequently,

$$x - 1 = \sqrt{x-1}\sqrt{x-1} \leq \sqrt{x-1}\sqrt{x+1} = \sqrt{x^2-1} \text{ and therefore } x - \sqrt{x^2-1} \leq 1.$$

19. $\frac{d}{dx}(\sinh^{-1} x) = \frac{d}{dx}[\ln(x + \sqrt{x^2+1})] = \frac{1 + \frac{x}{\sqrt{x^2+1}}}{x + \sqrt{x^2+1}} = \frac{1}{\sqrt{x^2+1}}$

21. $\frac{d}{dx}(\tanh^{-1} x) = \frac{1}{2}\frac{d}{dx}\left[\ln\left(\frac{1+x}{1-x}\right)\right] = \frac{1}{2} \cdot \frac{1}{\left(\frac{1+x}{1-x}\right)} \cdot \frac{2}{(1-x)^2} = \frac{1}{1-x^2}$

23. Let $y = \operatorname{csch}^{-1} x$. Then $\operatorname{csch} y = x$ and $\sinh y = \frac{1}{x}$. Thus $\cosh y \cdot y' = -\frac{1}{x^2}$ and

$$y' = -\frac{1}{x^2 \cosh y} = -\frac{1}{x^2\sqrt{1 + \left(\frac{1}{x}\right)^2}} = -\frac{1}{|x|\sqrt{1+x^2}}.$$

25. (a) absolute max $(0, 1)$

(b) points of inflection at $x = \ln(1 + \sqrt{2}) \cong 0.881$, $x = -\ln(1 + \sqrt{2}) \cong -0.881$

(c) concave up on $(-\infty, -\ln(1+\sqrt{2})) \cup (\ln(1+\sqrt{2}), \infty)$; concave down on $(-\ln(1+\sqrt{2}), \ln(1+\sqrt{2}))$

(d)

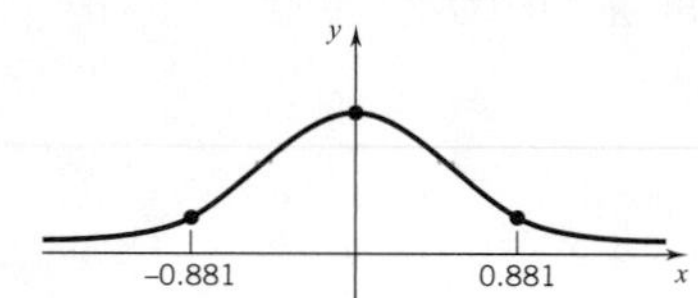

27.

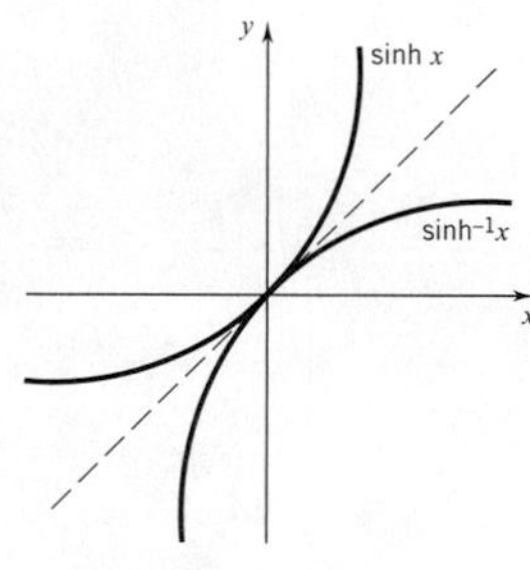

29. (a) $\tan\phi = \sinh x$

$$\phi = \tan^{-1}(\sinh x)$$

$$\frac{d\phi}{dx} = \frac{\cosh x}{1 + \sinh^2 x} = \frac{\cosh x}{\cosh^2 x} = \frac{1}{\cosh x} = \operatorname{sech} x$$

(b) $\sinh x = \tan\phi$

$$x = \sinh^{-1}(\tan\phi)$$
$$= \ln(\tan\phi + \sqrt{\tan^2\phi + 1})$$
$$= \ln(\tan\phi + \sec\phi) = \ln(\sec\phi + \tan\phi)$$

(c) $x = \ln(\sec\phi + \tan\phi)$

$$\frac{dx}{d\phi} = \frac{\sec\phi\tan\phi + \sec^2\phi}{\tan\phi + \sec\phi} = \sec\phi$$

31. $\ln(\cosh x) + C$ **33.** $2\tan^{-1}(e^x) + C$ **35.** $-\frac{1}{3}\operatorname{sech}^3 x + C$ **37.** $\frac{1}{2}[\ln(\cosh x)]^2 + C$ **39.** $\ln|1 + \tanh x| + C$

41. Let $x = a\sinh u$, $dx = a\cosh u\, du$. Then

$$\int \frac{dx}{\sqrt{a^2+x^2}} = \int \frac{a\cosh u}{\sqrt{a^2 + a^2\sinh^2 u}}\,du = \int du = \sinh^{-1}\left(\frac{x}{a}\right) + C.$$

43. Suppose $|x| < a$. Let $x = a\tanh u$, $dx = a\,\operatorname{sech}^2 u\, du$. Then

$$\int \frac{dx}{a^2 - x^2} = \int \frac{a\,\operatorname{sech}^2 u}{a^2 - a^2\tanh^2 u}\,du = \frac{1}{a}\int du = \frac{1}{a}\tanh^{-1}\left(\frac{x}{a}\right) + C.$$

The other case is done in the same way.

45. (a) $f(x) = \frac{\operatorname{sech} x}{(1 + \tanh x)^2}$

(b) $f(x) = x\tanh^{-1}(x^2)$

CHAPTER 8

SECTION 8.1

1. $-e^{2-x}+C$ **3.** $2/\pi$ **5.** $-\tan(1-x)+C$ **7.** $\frac{1}{2}\ln 3$ **9.** $-\sqrt{1-x^2}+C$ **11.** 0 **13.** $e-\sqrt{e}$ **15.** $\pi/4c$

17. $\frac{2}{3}\sqrt{3\tan\theta+1}+C$ **19.** $(1/a)\ln|a\,e^x-b|+C$ **21.** $\frac{1}{2}\ln[(x+1)^2+4]-\frac{1}{2}\tan^{-1}(\frac{1}{2}[x+1])+C$ **23.** $\frac{1}{2}\sin^{-1}x^2+C$

25. $\tan^{-1}(x+3)+C$ **27.** $-\frac{1}{2}\cos x^2+C$ **29.** $\tan x-x+C$ **31.** $3/2$ **33.** $\frac{1}{2}(\sin^{-1}x)^2+C$ **35.** $\ln|\ln x|+C$ **37.** $\sqrt{2}$

39. (Formula 99) $\frac{x}{2}\sqrt{x^2-4}-2\ln|x+\sqrt{x^2-4}|+C$ **41.** (Formula 18) $\frac{1}{2}[\sin 2t-\frac{1}{3}\sin^3 2t]+C$ **43.** (Formula 108) $\frac{1}{3}\ln\left|\frac{x}{2x+3}\right|+C$

45. (Formula 81) $-\frac{\sqrt{x^2+9}}{x}+\ln|x+\sqrt{x^2+9}|+C$ **47.** (Formula 11) $x^4\left[\frac{\ln x}{4}-\frac{1}{16}\right]+C$ **49.** $2\sqrt{2}$

51. (a) $\int_0^\pi \sin^2 nx\,dx=\int_0^\pi\left(\frac{1}{2}-\frac{\cos 2nx}{2}\right)dx=\left[\frac{1}{2}x-\frac{\sin 2nx}{4n}\right]_0^\pi=\frac{\pi}{2}$

(b) $\int_0^\pi \sin nx\cos nx\,dx=\frac{1}{n}\int_0^0 u\,du=0$ $\left(u=\sin nx,\ du=\frac{1}{n}\cos nx\,dx\right)$ (c) $\int_0^{\pi/n}\sin nx\cos nx\,dx=\frac{1}{n}\int_0^0 u\,du=0.$

53. (a) $\frac{1}{2}\tan^2 x-\ln|\sec x|+C$ (b) $\frac{1}{4}\tan^4 x-\frac{1}{2}\tan^2 x+\ln|\sec x|+C$ (c) $\frac{1}{6}\tan^6 x-\frac{1}{4}\tan^4 x+\frac{1}{2}\tan^2 x-\ln|\sec x|+C$

(d) $\int\tan^{2k+1}x\,dx=\frac{1}{2k}\tan^{2k}x-\int\tan^{2k-1}x\,dx$

55. (b) $A=\sqrt{2},\quad B=\frac{\pi}{4}$ (c) $\frac{\sqrt{2}}{2}\ln\left(\frac{\sqrt{2}+1}{\sqrt{2}-1}\right)$ **57.** (b) $-0.80,\ 5.80$ (c) 27.60

SECTION 8.2

1. $-xe^{-x}-e^{-x}+C$ **3.** $-\frac{1}{3}e^{-x^3}+C$ **5.** $2-5e^{-1}$ **7.** $-2x^2(1-x)^{1/2}-\frac{8}{3}x(1-x)^{3/2}-\frac{16}{15}(1-x)^{5/2}+C$

9. $\frac{3}{8}e^4+\frac{1}{8}$ **11.** $2\sqrt{x+1}\ln(x+1)-4\sqrt{x+1}+C$ **13.** $x(\ln x)^2-2x\ln x+2x+C$

15. $3^x\left(\frac{x^3}{\ln 3}-\frac{3x^2}{(\ln 3)^2}+\frac{6x}{(\ln 3)^3}-\frac{6}{(\ln 3)^4}\right)+C$ **17.** $\frac{1}{15}x(x+5)^{15}-\frac{1}{240}(x+5)^{16}+C$ **19.** $\frac{1}{2\pi}-\frac{1}{\pi^2}$

21. $\frac{1}{10}x^2(x+1)^{10}-\frac{1}{55}x(x+1)^{11}+\frac{1}{660}(x+1)^{12}+C$ **23.** $\frac{1}{2}e^x(\sin x-\cos x)+C$ **25.** $\ln 2+\frac{\pi}{2}-2$

27. $\frac{x^{n+1}}{n+1}\ln x-\frac{x^{n+1}}{(n+1)^2}+C$ **29.** $-\frac{1}{2}x^2\cos x^2+\frac{1}{2}\sin x^2+C$ **31.** $\frac{\pi}{24}+\frac{\sqrt{3}-2}{4}$ **33.** $\frac{\pi}{8}-\frac{1}{4}\ln 2$

35. $\frac{1}{2}x^2\sinh 2x-\frac{1}{2}x\cosh 2x+\frac{1}{4}\sinh 2x+C$ **37.** $\ln x\sin^{-1}(\ln x)+\sqrt{1-(\ln x)^2}+C$ **39.** $\frac{x}{2}[\sin(\ln x)-\cos(\ln x)]+C$

41. Set $u=\ln x$, $dv=dx$ and integrate by parts. **43.** Set $u=\ln x$, $dv=x^k\,dx$ and integrate by parts.

45. Integrate by parts twice and solve for $\int e^{ax}\sin bx\,dx$. **47.** π **49.** $\frac{\pi}{12}+\frac{\sqrt{3}-2}{2}$

51. (a) 1 (b) $\bar{x}=\frac{e^2}{4}+\frac{1}{4},\ \bar{y}=\frac{e}{2}-1$ (c) x-axis: $\pi(e-2)$, y-axis: $\frac{\pi}{2}(e^2+1)$ **53.** $\bar{x}=1/(e-1),\ \bar{y}=(e+1)/4$

55. $\bar{x}=\frac{1}{2}\pi,\ \bar{y}=\frac{1}{8}\pi$ **57.** (a) $M=(e^k-1)/k$ (b) $x_M=[(k-1)e^k+1]/[k(e^k-1)]$ **59.** $V=4-8/\pi$ **61.** $V=2\pi(e-2)$

63. $\bar{x}=(e^2+1)/[2(e^2-1)]$ **65.** area $=\sinh 1=\frac{e^2-1}{2e}$; $\bar{x}=\frac{2}{e+1}$, $\bar{y}=\frac{e^4+4e^2-1}{8e(e^2-1)}$ **67.** Let $u=x^n$, $dv=e^{ax}\,dx$. Then $du=nx^{n-1}\,dx$, $v=\frac{1}{a}e^{ax}$.

69. $(\frac{1}{2}x^3-\frac{3}{4}x^2+\frac{3}{4}x-\frac{3}{8})e^{2x}+C$ **71.** $x[(\ln x)^3-3(\ln x)^2+6\ln x-6]+C$ **73.** $e^x[x^3-3x^2+6x-6]+C$

75. (a) $(x^2-5x+6)e^x+C$ (b) $(x^3-3x^2+4x-4)e^x+C$

77. Let $u=f(x)$, $dv=g''(x)\,dx$. Then $du=f'(x)\,dx$, $v=g'(x)$, and

$$\int_a^b f(x)g''(x)\,dx=[f(x)g'(x)]_a^b-\int_a^b f'(x)g'(x)\,dx.$$

Now let $u=f'(x)$, $dv=g'(x)\,dx$ and integrate by parts again. The result follows.

79. (a) π
(b) 3π
(c) 5π
(d) $(2n+1)\pi,\ n = 0, 1, 2, \ldots$

81. (a) $\pi - 2 \cong 1.1416$
(b) $\pi^3 - 2\pi^2 \cong 11.2671$
(c) $(\frac{1}{2}\pi, 0.31202)$

SECTION 8.3

1. $\frac{1}{3}\cos^3 x - \cos x + C$ **3.** $\dfrac{\pi}{12}$ **5.** $-\frac{1}{5}\cos^5 x + \frac{1}{7}\cos^7 x + C$ **7.** $\frac{1}{4}\sin^4 x - \frac{1}{6}\sin^6 x + C$ **9.** $(1/\pi)\tan \pi x + C$

11. $\frac{1}{2}\tan^2 x + \ln|\cos x| + C$ **13.** $\frac{3}{8}\pi$ **15.** $\frac{1}{2}\cos x - \frac{1}{10}\cos 5x + C$ **17.** $\frac{1}{3}\tan^3 x + C$

19. $\frac{1}{2}\sin^4 x + C$ **21.** $\frac{5}{16}x - \frac{1}{4}\sin 2x + \frac{3}{64}\sin 4x + \frac{1}{48}\sin^3 2x + C$ **23.** $\sqrt{3} - \dfrac{\pi}{3}$ **25.** $-\frac{1}{5}\csc^5 x + \frac{1}{3}\csc^3 x + C$

27. $\frac{1}{6}\sin 3x - \frac{1}{14}\sin 7x + C$ **29.** $\frac{2}{7}\sin^{7/2} x - \frac{2}{11}\sin^{11/2} x + C$ **31.** $\frac{1}{12}\tan^4 3x - \frac{1}{6}\tan^2 3x + \frac{1}{3}\ln|\sec 3x| + C$ **33.** $\dfrac{2}{105\pi}$ **35.** $-1/6$

37. $\frac{1}{7}\tan^7 x + \frac{1}{5}\tan^5 x + C$ **39.** $\frac{1}{3}\cos(\frac{3}{2}x) - \frac{1}{5}\cos(\frac{5}{2}x) + C$ **41.** $\dfrac{1}{4}$

43. $\dfrac{\sqrt{3}}{2} - \dfrac{\pi}{6}$ **45.** $\pi/2$ **47.** $\dfrac{3\pi^2}{8}$ **49.** $\dfrac{\pi^2}{2} - \pi$ **51.** $\pi\left[1 - \dfrac{\pi}{4} + \ln 2\right]$

53. $\sin mx \sin nx = \frac{1}{2}[\cos(m-n)x - \cos(m+n)x],\ m \neq n$

$\sin mx \sin nx = \sin^2 mx = \dfrac{1 - \cos 2mx}{2},\ m = n$

57. Let $u = \cos^{n-1} x,\ dv = \cos x\, dx$. Then $du = (n-1)\cos^{n-2} x(-\sin x)\, dx,\ v = \sin x$.

$$\int \cos^n x\, dx = \int \cos^{n-1} x \cos x\, dx = \cos^{n-1} x \sin x + (n-1)\int \cos^{n-2} x \sin^2 x\, dx$$
$$= \cos^{n-1} x \sin x + (n-1)\int (\cos^{n-2} x - \cos^n x)\, dx$$

Now solve for $\int \cos^n x\, dx$.

59. $\frac{16}{35}$ **61.** $\displaystyle\int \cot^n x\, dx = \int \cot^{n-2} x(\csc^2 x - 1)\, dx = -\frac{\cot^{n-1} x}{n-1} - \int \cot^{n-2} x\, dx$

63. $\int \csc^n x\, dx = \int \csc^{n-2} x \csc^2 x\, dx$. Now let $u = \csc^{n-2} x,\ dv = \csc^2 x\, dx$ and use integration by parts.

65. (a) $\frac{1}{2}\sin^2 x + C$
(b) $-\frac{1}{2}\cos^2 x + C$
(c) $-\frac{1}{4}\cos 2x + C$
(d) The results differ by a constant.

67. (a) $A = \frac{1}{2}\pi^2 \cong 4.9348$

SECTION 8.4

1. $\sin^{-1}\left(\dfrac{x}{a}\right) + C$ **3.** $\frac{1}{2}x\sqrt{x^2-1} - \frac{1}{2}\ln|x + \sqrt{x^2-1}| + C$ **5.** $2\sin^{-1}\left(\dfrac{x}{2}\right) - \dfrac{1}{2}x\sqrt{4-x^2} + C$

7. $\dfrac{1}{\sqrt{1-x^2}} + C$ **9.** $\dfrac{2\sqrt{3}-\pi}{6}$ **11.** $-\frac{1}{3}(4-x^2)^{3/2} + C$ **13.** $\dfrac{625\pi}{16}$ **15.** $\ln(\sqrt{8+x^2} + x) - \dfrac{x}{\sqrt{8+x^2}} + C$

17. $\dfrac{1}{a}\ln\left|\dfrac{a - \sqrt{a^2-x^2}}{x}\right| + C$ **19.** $18 - 9\sqrt{2}$ **21.** $-\dfrac{1}{a^2 x}\sqrt{a^2+x^2} + C$ **23.** $\frac{1}{10}$ **25.** $\dfrac{1}{a^2 x}\sqrt{x^2-a^2} + C$

27. $\frac{1}{9}e^{-x}\sqrt{e^{2x}-9} + C$ **29.** $\begin{cases} -\dfrac{1}{2(x-2)^2} + C, & x > 2 \\ \dfrac{1}{2(2-x)^2} + C, & x < 2 \end{cases}$ **31.** $-\frac{1}{3}(6x-x^2-8)^{3/2} + \frac{3}{2}\sin^{-1}(x-3) + \frac{3}{2}(x-3)\sqrt{6x-x^2-8} + C$

33. $\dfrac{x^2+x}{8(x^2+2x+5)} - \dfrac{1}{16}\tan^{-1}\left(\dfrac{x+1}{2}\right) + C$ **35.** Let $u = \sec^{-1} x,\ dv = dx$ and integrate by parts.

37. Let $x = a\tan u$, $dx = a\sec^2 u\,du$; $\sqrt{x^2+a^2} = a\sec u$. Then $(x^2+a^2)^n = a^{2n}\sec^{2n} u$ and the result follows by substitution.

39. $\frac{3}{8}\tan^{-1} x + \frac{3x}{8(x^2+1)} + \frac{x}{4(x^2+1)^2} + C$ **41.** $\frac{1}{4}(2x^2-1)\sin^{-1} x + \frac{x}{4}\sqrt{1-x^2} + C$ **43.** $\frac{\pi^2}{8} + \frac{\pi}{4}$

45. $A = \frac{1}{2}r^2\sin\theta\cos\theta + \int_{r\cos\theta}^{r}\sqrt{r^2-x^2}\,dx = \frac{1}{2}r^2\theta$ **47.** $\frac{8}{3}[10 - \frac{9}{2}\ln 3]$ **49.** $M = \ln(1+\sqrt{2})$, $x_M = \frac{(\sqrt{2}-1)a}{\ln(1+\sqrt{2})}$

51. $A = \frac{1}{2}a^2[\sqrt{2} - \ln(\sqrt{2}+1)]$; $\bar{x} = \frac{2a}{3[\sqrt{2}-\ln(\sqrt{2}+1)]}$, $\bar{y} = \frac{(2-\sqrt{2})a}{3[\sqrt{2}-\ln(\sqrt{2}+1)]}$ **53.** $V_y = \frac{2}{3}\pi a^3$, $\bar{y} = \frac{3}{8}a$

55. (a) Let $x = a\sec u$, $dx = a\sec u\tan u\,du$, $\sqrt{x^2-a^2} = a\tan u$.

(b) Let $x = a\cosh u$, $dx = a\sinh u$, $\sqrt{x^2-a^2} = a\sinh u$.

57. (b) $\ln(2+\sqrt{3}) - \frac{\sqrt{3}}{2}$ (c) $\bar{x} = \frac{2(3\sqrt{3}-\pi)}{2\ln(2+\sqrt{3})-\sqrt{3}}$, $\bar{y} = \frac{5}{72[2\ln(2+\sqrt{3})-\sqrt{3}]}$

SECTION 8.5

1. $\frac{1/5}{x+1} - \frac{1/5}{x+6}$ **3.** $\frac{1/4}{x-1} + \frac{1/4}{x+1} - \frac{x/2}{x^2+1}$ **5.** $\frac{1/2}{x} + \frac{3/2}{x+2} - \frac{1}{x-1}$ **7.** $\frac{3/2}{x-1} - \frac{9}{x-2} + \frac{19/2}{x-3}$ **9.** $\ln\left|\frac{x-2}{x+5}\right| + C$

11. $x^2 - 2x + \frac{3}{x} + 5\ln|x-1| - 3\ln|x| + C$ **13.** $\frac{1}{4}x^4 + \frac{4}{3}x^3 + 6x^2 + 32x - \frac{32}{x-2} + 80\ln|x-2| + C$ **15.** $5\ln|x-2| - 4\ln|x-1| + C$

17. $\frac{-1}{2(x-1)^2} + C$ **19.** $\frac{3}{4}\ln|x-1| - \frac{1}{2(x-1)} + \frac{1}{4}\ln|x+1| + C$ **21.** $\frac{1}{32}\ln\left|\frac{x-2}{x+2}\right| - \frac{1}{16}\tan^{-1}\frac{x}{2} + C$

23. $\frac{1}{2}\ln(x^2+1) + \frac{3}{2}\tan^{-1}x + \frac{5(1-x)}{2(x^2+1)} + C$ **25.** $\frac{1}{16}\ln\left[\frac{x^2+2x+2}{x^2-2x+2}\right] + \frac{1}{8}\tan^{-1}(x+1) + \frac{1}{8}\tan^{-1}(x-1) + C$

27. $\frac{3}{x} + 4\ln\left|\frac{x}{x+1}\right| + C$ **29.** $-\frac{1}{6}\ln|x| + \frac{3}{10}\ln|x-2| - \frac{2}{15}\ln|x+3| + C$ **31.** $\ln(\frac{125}{108})$ **33.** $\ln(\frac{27}{4}) - 2$

35. $\frac{1}{6}\ln\left|\frac{\sin\theta-4}{\sin\theta+2}\right| + C$ **37.** $\frac{1}{4}\ln\left|\frac{\ln t-2}{\ln t+2}\right| + C$ **39.** $\frac{u}{a+bu} = \frac{1}{b}\left[1 - \frac{a}{a+bu}\right]$

41. $\frac{1}{u^2(a+bu)} = \frac{(-b/a^2)}{u} + \frac{(1/a)}{u^2} + \frac{(b^2/a^2)}{a+bu}$ **43.** $\frac{1}{(a+bu)(c+du)} = \frac{-b/(ad-bc)}{a+bu} + \frac{d/(ad-bc)}{c+du}$

45. $\int\frac{u}{a^2-u^2}du = -\frac{1}{2}\int v^{-1}\,dv$, where $v = a^2-u^2$ **47.** (a) Exercise 44 (b) $x = a\sin u$, $dx = a\cos u\,du$

49. (a) $\frac{\pi}{4}\ln 7$ (b) $\pi(4-\sqrt{7})$ **51.** $\bar{x} = (2\ln 2)/\pi$, $\bar{y} = (\pi+2)/4\pi$ **53.** (a) $\frac{1}{x} - \frac{2}{x^2} + \frac{5}{x+1} - \frac{4}{(x+1)^3}$ (b) $\frac{1}{x^2+4} + \frac{3}{x+3} - \frac{4}{x-3}$ (c) $\frac{2x-1}{x^2+2x+4} - \frac{3}{x}$ **55.** (b) $3\ln 7 - 5\ln 3$ **57.** (b) $11 - \ln 12$

SECTION 8.6

1. $-2(\sqrt{x} + \ln|1-\sqrt{x}|) + C$ **3.** $2\ln(\sqrt{1+e^x}-1) - x + 2\sqrt{1+e^x} + C$ **5.** $\frac{2}{5}(1+x)^{5/2} - \frac{2}{3}(1+x)^{3/2} + C$

7. $\frac{2}{5}(x-1)^{5/2} + 2(x-1)^{3/2} + C$ **9.** $-\frac{1+2x^2}{4(1+x^2)^2} + C$ **11.** $x + 2\sqrt{x} + 2\ln|\sqrt{x}-1| + C$ **13.** $x + 4\sqrt{x-1} + 4\ln|\sqrt{x-1}-1| + C$

15. $2\ln(\sqrt{1+e^x}-1) - x + C$ **17.** $\frac{2}{3}(x-8)\sqrt{x+4} + C$ **19.** $\frac{1}{16}(4x+1)^{1/2} + \frac{1}{8}(4x+1)^{-1/2} - \frac{1}{48}(4x+1)^{-3/2} + C$ **21.** $\frac{4b+2ax}{a^2\sqrt{ax+b}} + C$

23. $-\ln\left|1 - \tan\frac{x}{2}\right| + C$ **25.** $\frac{2}{\sqrt{3}}\tan^{-1}\left[\frac{1}{\sqrt{3}}(2\tan\frac{x}{2} + 1)\right] + C$ **27.** $\frac{1}{2}\ln\left|\tan\frac{x}{2}\right| - \frac{1}{4}\tan^2\frac{x}{2} + C$

29. $\ln\left|\frac{1}{1+\sin x}\right| - \frac{2}{1+\tan(x/2)} + C$ **31.** $\frac{4}{5} + 2\tan^{-1}2$ **33.** $2 + 4\ln\frac{2}{3}$ **35.** $\ln\left(\frac{\sqrt{3}-1}{\sqrt{3}}\right)$

37. Let $u = \tan\frac{x}{2}$. Then $\int \frac{1}{\cos x}\,dx = 2\int \frac{du}{1-u^2}$ and the result follows.

39. $\int \csc x\,dx = \int \frac{\sin x}{\sin^2 x}\,dx = \int \frac{\sin x}{1-\cos^2 x}\,dx = -\int \frac{du}{1-u^2}$ where $u = \cos x$. The result follows.

41. $2\tan^{-1}\left(\tanh\frac{x}{2}\right) + C$ **43.** $\frac{-2}{1+\tanh(x/2)} + C$

SECTION 8.7

1. (a) 506 (b) 650 (c) 572 (d) 578 (e) 576 **3.** (a) 1.394 (b) 0.9122 (c) 1.1776 (d) 1.1533 (e) 1.1614

5. (a) $\pi \cong 3.1312$ (b) $\pi \cong 3.1416$ **7.** (a) 1.8440 (b) 1.7915 (c) 1.8090 **9.** (a) 0.8818 (b) 0.8821

11. Such a curve passes through the three points $(a_1, b_1), (a_2, b_2), (a_3, b_3)$ iff

$$b_1 = a_1^2 A + a_1 B + C,\quad b_2 = a_2^2 A + a_2 B + C,\quad b_3 = a_3^2 A + a_3 B + C,$$

which happens iff

$$A = \frac{b_1(a_2-a_3) - b_2(a_1-a_3) + b_3(a_1-a_2)}{(a_1-a_3)(a_1-a_2)(a_2-a_3)},\quad B = -\frac{b_1(a_2^2-a_3^2) - b_2(a_1^2-a_3^2) + b_3(a_1^2-a_2^2)}{(a_1-a_3)(a_1-a_2)(a_2-a_3)},$$

$$C = \frac{a_1^2(a_2b_3 - a_3b_2) - a_2^2(a_1b_3 - a_3b_1) + a_3^2(a_1b_2 - a_2b_1)}{(a_1-a_3)(a_1-a_2)(a_2-a_3)}.$$

13. (a) $n \ge 8$ (b) $n \ge 2$ **15.** (a) $n \ge 238$ (b) $n \ge 10$ **17.** (a) $n \ge 51$ (b) $n \ge 4$ **19.** (a) $n \ge 37$ (b) $n \ge 3$

21. (a) 78 (b) 7 **23.** $f^{(4)}(x) = 0$ for all x; therefore by (8.7.3) the theoretical error is zero

25. (a) $\left|T_2 - \int_0^1 x^2\,dx\right| = \frac{3}{8} - \frac{1}{3} = \frac{1}{24} = E_2^T$ (b) $\left|S_1 - \int_0^1 x^4\,dx\right| = \frac{5}{24} - \frac{1}{5} = \frac{1}{120} = E_1^S$

27. Using the hint, $M_n =$ area $ABCD$ = area $AEFD \le \int_a^b f(x)\,dx \le T_n$. **29.** (a) 49.4578 (b) 1280.56 **31.** error $\le 4.01 \times 10^{-7}$

33. $\int_0^1 \frac{4}{1+x^2}\,dx = 4\tan^{-1}x\Big|_0^1 = 4\left(\frac{\pi}{4} - 0\right) = \pi \simeq 3.14159$

(a) 3.14141 (b) 3.14159

SECTION 8.8

1. y_1 is; y_2 is not **3.** y_1 and y_2 are solutions **5.** y_1 and y_2 are solutions **7.** $y = -\frac{1}{2} + Ce^{2x}$ **9.** $y = \frac{2}{5} + Ce^{-(5/2)x}$

11. $y = x + Ce^{2x}$ **13.** $y = \frac{2}{3}nx + Cx^4$ **15.** $y = Ce^{e^x}$ **17.** $y = 1 + C(e^{-x} + 1)$ **19.** $y = e^{-x^2}\left(\frac{1}{2}x^2 + C\right)$ **21.** $y = C(x+1)^{-2}$

23. $y = 2e^{-x} + x - 1$ **25.** $y = e^{-x}\left[\ln(1+e^x) + e - \ln 2\right]$ **27.** $y = x^2(e^x - e)$ **29.** $y = C_1e^x + C_2xe^x$ **35.** $T(1) \cong 40.10°$; 1.62 min

37. (a) $v(t) = \frac{32}{k}(1 - e^{-kt})$
(b) $1 - e^{-kt} < 1$; $e^{-kt} \to 0$ as $t \to \infty$

39. (a) $i(t) = \frac{E}{R}[1 - e^{-(R/L)t}]$
(b) $i(t) \to \frac{E}{R}$ (amps) as $t \to \infty$
(c) $t = \frac{L}{R}\ln 10$ seconds

41. (a) $200\left(\frac{4}{5}\right)^{t/5}$
(b) $200\left(\frac{4}{5}\right)^{t^2/25}$ liters

43. (a) $\frac{dP}{dt} = k(M - P)$
(b) $P(t) = M(1 - e^{-0.0357t})$
(c) 65 days

45. (a) $P(t) = 1000\,e^{(\sin 2\pi t)/\pi}$
(b) $P(t) = 2000\,e^{(\sin 2\pi t)/\pi} - 1000$

SECTION 8.9

1. $y = Ce^{-(1/2)\cos(2x+3)}$ **3.** $x^4 + \frac{2}{y^2} = C$ **5.** $y\sin y + \cos y = -\cos\left(\frac{1}{x}\right) + C$ **7.** $e^{-y} = e^x - xe^x + C$

9. $\ln|y+1| + \frac{1}{y+1} = \ln|\ln x| + C$ **11.** $y^2 = C(\ln x)^2 - 1$ **13.** $\sin^{-1}y = 1 - \sqrt{1-x^2}$ **15.** $y + \ln|y| = \frac{x^3}{3} - x - 5$

17. $\frac{x^2}{2} + x + \frac{1}{2}\ln(y^2+1) - \tan^{-1} y = 4$ **19.** $y = \ln\left[3e^{2x} - 2\right]$

21. $y = \frac{3}{2}x + C$ **23.** $x^2 - y^2 = C$ **25.** $y^2 = -2x + C$

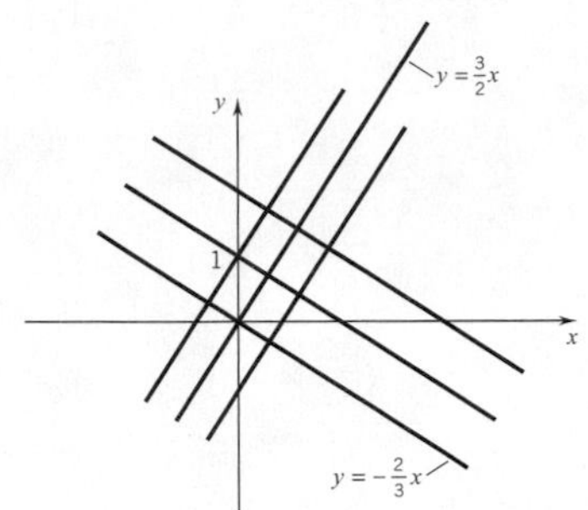

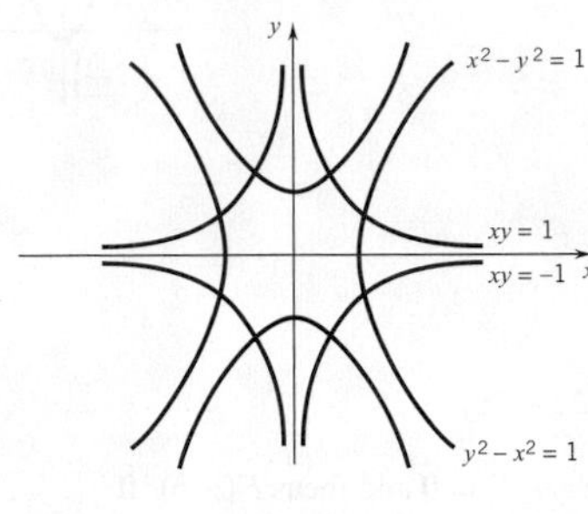

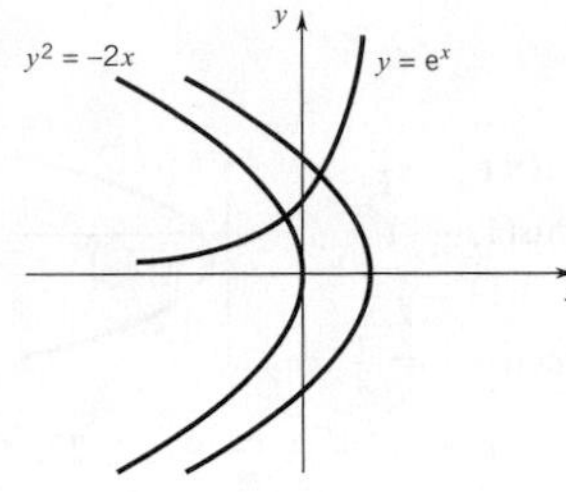

27. A differential equation for the given family is $y^2 = 2xyy' + y^2(y')^2$. Now replace y' by $-\frac{1}{y'}$. Then the resulting differential equation is $y^2 = -\frac{2xy}{y'} + \frac{y^2}{(y')^2}$, which simplifies to $y^2 = 2xyy' + y^2(y')^2$. Thus the given family is self-orthogonal.

29. (a) $C(t) = \frac{kA_0^2 t}{1 + kA_0 t}$ (b) $C(t) = \frac{A_0 B_0 (e^{kA_0 t} - e^{kB_0 t})}{A_0\, e^{kA_0 t} - B_0\, e^{kB_0 t}}$

31. (a) $v(t) = \frac{\alpha}{c\, e^{(\alpha/m)t} - \beta}$, where C is an arbitrary constant. (b) $v(t) = \frac{\alpha v_0}{(\alpha + \beta v_0)\, e^{(\alpha/m)t} - \beta v_0} = \frac{\alpha v_0 e^{-(\alpha/m)t}}{\alpha + \beta v_0 - \beta v_0\, e^{-(\alpha/m)t}}$ (c) $\lim_{t\to\infty} v(t) = 0$

33. (a) $y(t) = \frac{25,000}{1 + 249\, e^{-0.1398t}}$, $y(20) \cong 1544$ (b) 40 days **35.** (a) 88.82 m/sec (b) $v(t) = \frac{15.65(1 + 0.70\, e^{-1.25t})}{1 - 0.70\, e^{-1.25t}}$ (c) 15.65 m/sec

CHAPTER 9

SECTION 9.1

1. (a) $\frac{2}{13}$ (b) $\frac{29}{13}$

3. (0, 1) is the closest point (− 1, 1) the farthest away **5.** $\frac{17}{2}$

7. Adjust the sign of A and B so that the equation reads $Ax + By = |C|$. Then we have

$$x\frac{A}{\sqrt{A^2+B^2}} + y\frac{B}{\sqrt{A^2+B^2}} = \frac{|C|}{\sqrt{A^2+B^2}}.$$

Now set

$$\frac{A}{\sqrt{A^2+B^2}} = \cos\alpha, \qquad \frac{B}{\sqrt{A^2+B^2}} = \sin\alpha, \qquad \frac{|C|}{\sqrt{A^2+B^2}} = p.$$

p is the length of $\overline{OQ}$, the distance between the line and the origin; α is the angle from the positive x-axis to the line segment $\overline{OQ}$.

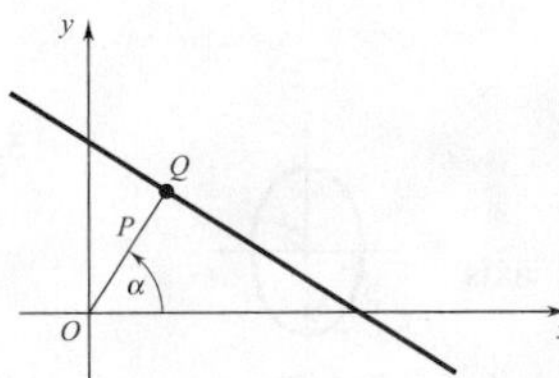

9. $y^2 = 8x$ **11.** $(x+1)^2 = -12(y-3)$ **13.** $4y = (x-1)^2$ **15.** $(y-1)^2 = -2(x - \frac{3}{2})$

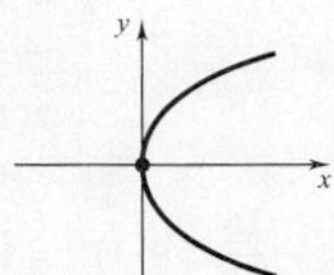

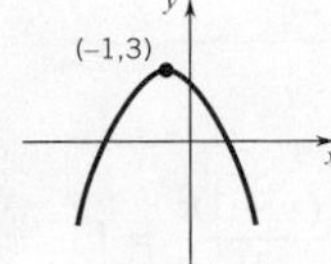

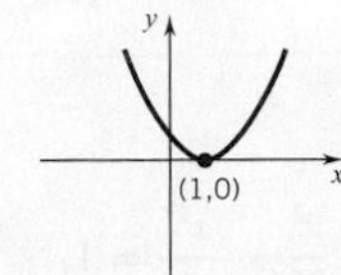

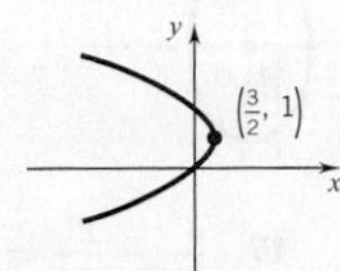

17. vertex $(0, 0)$
focus $(\frac{1}{2}, 0)$
axis $y = 0$
directrix $x = -\frac{1}{2}$

19. vertex $(0, -\frac{1}{2})$
focus $(0, -\frac{3}{8})$
axis $x = 0$
directrix $y = -\frac{5}{8}$

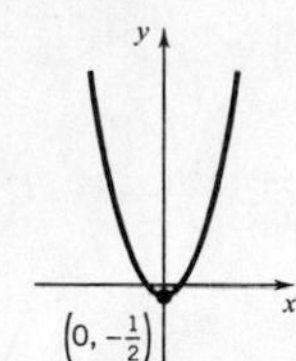

21. vertex $(-2, \frac{3}{2})$
focus $(-2, -\frac{1}{2})$
axis $x = -2$
directrix $y = \frac{7}{2}$

23. vertex $(\frac{3}{4}, -\frac{1}{2})$
focus$(1, -\frac{1}{2})$
axis $y = -\frac{1}{2}$
directrix $x = \frac{1}{2}$

25. $(x - y)^2 = 6x + 10y - 9$ **27.** $(x + y)^2 = -12x + 20y + 28$

29. $P(x, y)$ is on the parabola with directrix $\ell : Ax + By + C = 0$ and focus $F(a, b)$ iff

$$d(P, l) = d(P, F) \text{ which happens iff } \frac{|(Ax + By + C)|}{\sqrt{A^2 + B^2}} = \sqrt{(x - a)^2 + (y - b)^2}.$$

Square this last equation and simplify.

31. We can choose the coordinate system so that the parabola has an equation of the form $y = \alpha x^2, \alpha > 0$. One of the points of intersection is then the origin and the other is of the form $(c, \alpha c^2)$. We will assume that $c > 0$.

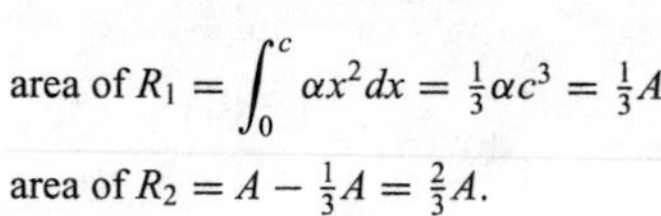

$$\text{area of } R_1 = \int_0^c \alpha x^2 dx = \tfrac{1}{3}\alpha c^3 = \tfrac{1}{3}A$$

$$\text{area of } R_2 = A - \tfrac{1}{3}A = \tfrac{2}{3}A.$$

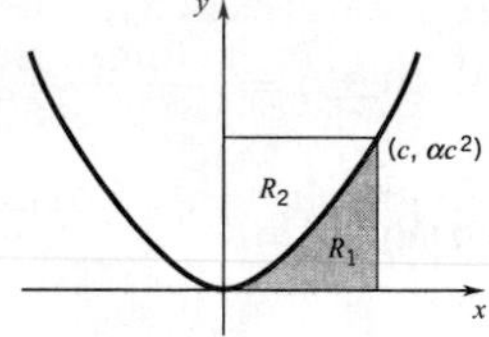

33. $2y = x^2 - 4x + 7, 18y = x^2 - 4x + 103$ **35.** $4c$ **37.** $A = \frac{8}{3}c^3;\ \bar{x} = 0,\ \bar{y} = \frac{3}{5}c$

39. $\dfrac{kx}{p(0)} = \tan\theta = \dfrac{dy}{dx},\ y = \dfrac{k}{2p(0)}x^2 + C$

In our figure, $C = y(0) = 0$. Thus the equation of the cable is $y = kx^2/2p(0)$, the equation of a parabola.

41. Start with any two parabolas, γ_1, γ_2. By moving then we can see to it that they have equations of the following form:

$$\gamma_1 : x^2 = 4c_1 y,\ c_1 > 0; \quad \gamma_2 : x^2 = 4c_2 y,\ c_2 > 0.$$

Now we change the scale for γ_2 so that the equation for γ_2 will look exactly like the equation for γ_1. Set $X = (c_1/c_2)x, y = (c_1/c_2)y$. Then

$$x^2 = 4c_2 y \Longrightarrow (c_2/c_1)^2 X^2 = 4c_2(c_2/c_1)Y \Longrightarrow X^2 = 4c_1 Y.$$

Now γ_2 has exactly the same equation as γ_1; only the scale, the units by which we measure distance, has changed.

SECTION 9.2

1. center $(0, 0)$
foci $(\pm\sqrt{5}, 0)$
length of major axis 6
length of minor axis 4

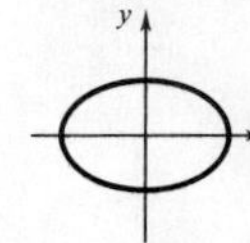

3. center $(0, 0)$
foci $(0, \pm\sqrt{2})$
length of major axis $2\sqrt{6}$
length of minor axis 4

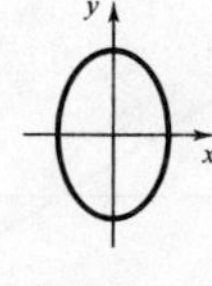

5. center $(0, 1)$
foci $(\pm\sqrt{5}, 1)$
length of major axis 6
length of minor axis 4

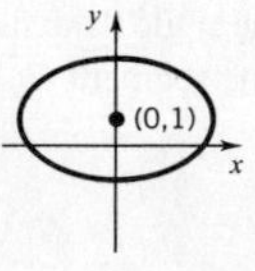

7. center $(1, 0)$
foci $(1, \pm 4\sqrt{3})$
length of major axis 16
length of minor axis 8

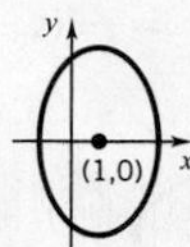

9. $\dfrac{x^2}{9} + \dfrac{y^2}{8} = 1$ **11.** $\dfrac{(x-1)^2}{16} + \dfrac{(y-6)^2}{25} = 1$ **13.** $\dfrac{(x-1)^2}{21} + \dfrac{(y-3)^2}{25} = 1$

15. $\dfrac{(x-3)^2}{25} + \dfrac{(y+1)^2}{9} = 1$ **17.** $\dfrac{x^2}{9} - \dfrac{y^2}{16} = 1$ **19.** $\dfrac{y^2}{25} - \dfrac{x^2}{144} = 1$ **21.** $\dfrac{x^2}{9} - \dfrac{(y-1)^2}{16} = 1$ **23.** $16y^2 - \dfrac{16}{15}(x+1)^2 = 1$

25. center $(0,0)$
transverse axis 2
vertices $(\pm 1, 0)$
foci $(\pm\sqrt{2}, 0)$
asymptotes $y = \pm x$

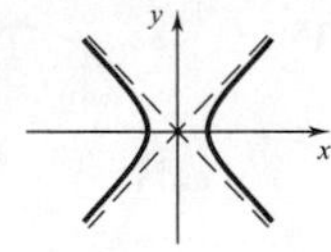

27. center $(0,0)$
transverse axis 6
vertices $(\pm 3, 0)$
foci $(\pm 5, 0)$
asymptotes $y = \pm\frac{4}{3}x$

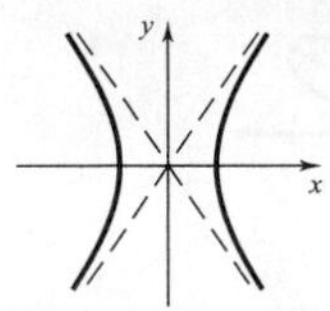

29. center $(0,0)$
transverse axis 8
vertices $(0, \pm 4)$
foci $(0, \pm 5)$
asymptotes $y = \pm\frac{4}{3}x$

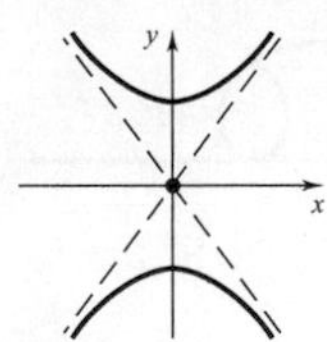

31. center $(1,3)$
transverse axis 6
vertices $(4,3)$ and $(-2,3)$
foci $(6,3)$ and $(-4,3)$
asymptotes $y = \pm\frac{4}{3}(x-1)+3$

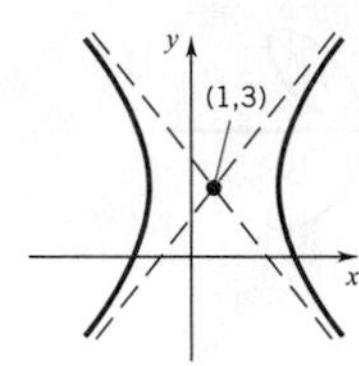

33. center $(1,3)$
transverse axis 4
vertices $(1,5)$ and $(1,1)$
foci $(1, 3\pm\sqrt{5})$
asymptotes $y = 2x+1, y = -2x+5$

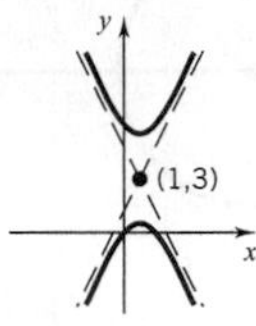

35. $d(F_1, F_2) + k = 2(c+a)$ **37.** $2\sqrt{\pi^2 a^4 - A^2}/\pi a$ **39.** $(5 \pm \frac{5}{21}\sqrt{5}, 0)$

41. center $(0,0)$, vertices $(1,1)$ and $(-1,-1)$, foci $(\sqrt{2},\sqrt{2})$ and $(-\sqrt{2},-\sqrt{2})$, asymptotes $x=0$ and $y=0$, transverse axis $2\sqrt{2}$

43. $[2\sqrt{3} - \ln(2+\sqrt{3})]ab$ **45.** $\frac{3}{5}$ **47.** $\frac{4}{5}$. **49.** E_1 is fatter than E_2, more like a circle

51. The ellipse tends to a line segment of length $2a$. **53.** $x^2/9 + y^2 = 1$ **55.** $\frac{5}{3}$ **57.** $\sqrt{2}$

59. The branches of H_1 open up less quickly than the branches of H_2.

61. The hyperbola tends to a pair of parallel lines separated by the transverse axis.

63. about 0.25 mile west and 1.46 miles north of point A

SECTION 9.3

1. -7. See figure to the right. **9.** $(0,3)$ **11.** $(1,0)$ **13.** $(-\frac{3}{2}, \frac{3}{2}\sqrt{3})$ **15.** $(0,-3)$

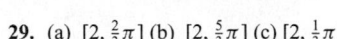

17. $[1, \frac{1}{2}\pi + 2n\pi], [-1, \frac{3}{2}\pi + 2n\pi]$ **19.** $[3, \pi + 2n\pi], [-3, 2n\pi]$

21. $[2\sqrt{2}, \frac{7}{4}\pi + 2n\pi], [-2\sqrt{2}, \frac{3}{4}\pi + 2n\pi]$ **23.** $[8, \frac{1}{6}\pi + 2n\pi], [-8, \frac{7}{6}\pi + 2n\pi]$

25. $\sqrt{r_1^2 + r_2^2 - 2r_1 r_2 \cos(\theta_1 - \theta_2)}$ **27.** (a) $[\frac{1}{2}, \frac{11}{6}\pi]$ (b) $[\frac{1}{2}, \frac{5}{6}\pi]$ (c) $[\frac{1}{2}, \frac{7}{6}\pi]$

29. (a) $[2, \frac{2}{3}\pi]$ (b) $[2, \frac{5}{3}\pi]$ (c) $[2, \frac{1}{3}\pi]$ **31.** symmetry about the x-axis

33. no symmetry about the coordinate axes; no symmetry about the origin

35. symmetry about the origin **37.** $r\cos\theta = 2$ **39.** $r^2 \sin 2\theta = 1$ **41.** $r = 4\sin\theta$

43. $\theta = \pi/4$ **45.** $r = 1 - \cos\theta$ **47.** $r^2 = \sin 2\theta$ **49.** the horizontal line $y = 4$

51. the line $y = \sqrt{3}x$ **53.** the parabola $y^2 = 4(x+1)$ **55.** the circle $x^2 + y^2 = 3x$

57. the line $y = 2x$ **59.** $3x^2 + 4y^2 - 8x = 16$, ellipse **61.** $y^2 = 8x + 16$, parabola

63. $(x - \frac{b}{2})^2 + (y - \frac{a}{2})^2 = \frac{a^2+b^2}{4}$; center: $\left(\frac{b}{2}, \frac{a}{2}\right)$, radius: $\frac{\sqrt{a^2+b^2}}{2}$ **65.** $r = \frac{d}{2 - \cos\theta}$

SECTION 9.4

1.

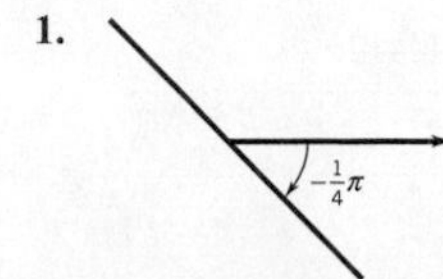

3.

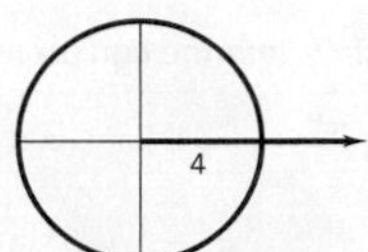

5.

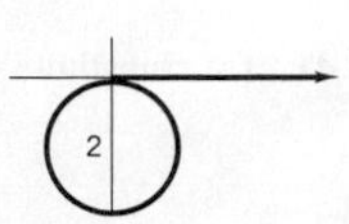

7.

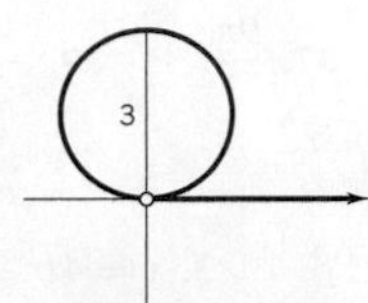

9. 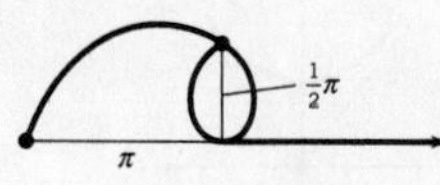

11. 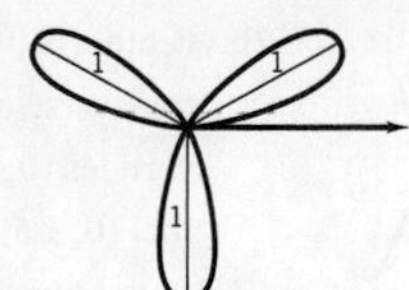

13. 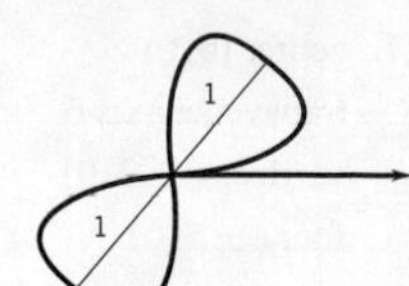

15. 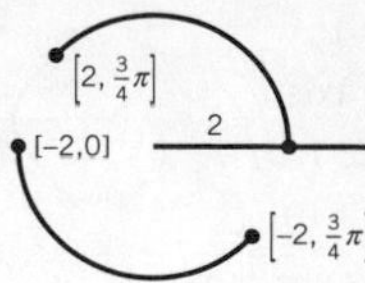

17. 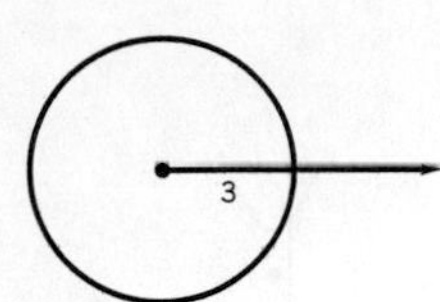

19. 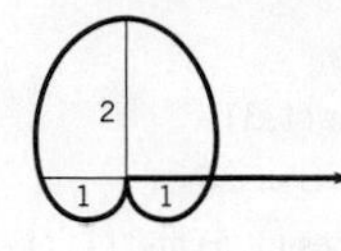

21. 

23. 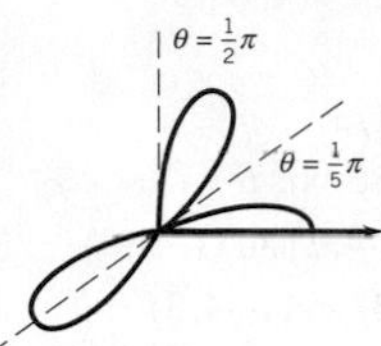

25. 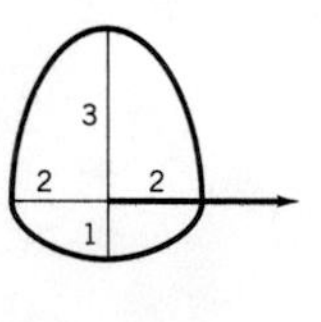

27. 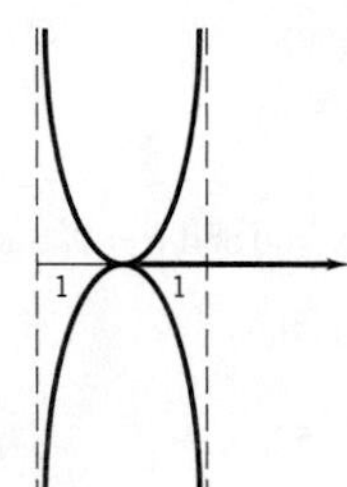

29. 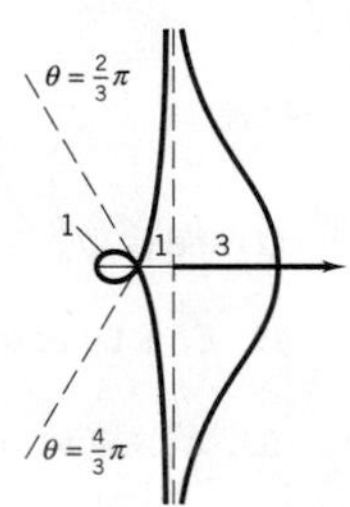

31. 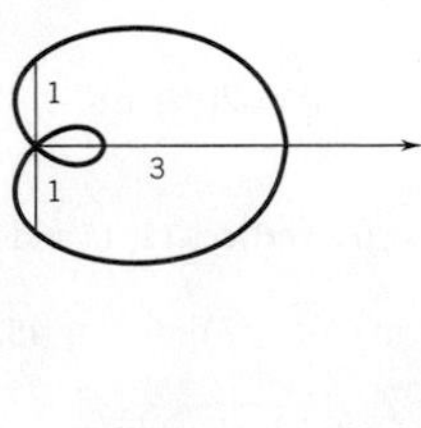

33. yes; $[1, \pi] = [-1, 0]$ and the pair $r = -1, \theta = 0$ satisfies the equation **35.** yes: the pair $r = \frac{1}{2}, \theta = \frac{1}{2}\pi$ satisfies equation

37. $[2, \pi] = [-2, 0]$. The coordinates of $[-2, 0]$ satisfy the equation $r^2 = 4\cos\theta$, and the coordinates of $[2, \pi]$ satisfy the equation $r = 3 + \cos\theta$.

39. $(0,0), (-\frac{1}{2}, \frac{1}{2})$ **41.** $(-1,0), (1,0)$ **43.** $(0,0), (\frac{1}{4}, \pm\frac{1}{4}\sqrt{3}rt)$ **45.** $(0,0), (\pm\frac{\sqrt{3}}{4}, \frac{3}{4})$ **47.** center: (b, a); radius: $\sqrt{a^2 + b^2}$

49. (b) The curves intersect at the pole and at $[1.175, 0.176], [1.86, 1.036], [0.90, 3.243]$. **51.** $\theta = \frac{\pi}{6}, \frac{\pi}{2}, \frac{5\pi}{6}$

53. (b) The curves intersect at the pole and at: **55.** butterfly **57.** a petal curve with $2m$ petals

$r = 1 - 3\cos\theta$	$r = 2 - 5\sin\theta$
$[-2, 0]$	$[2, \pi]$
$[3.800, 3.510]$	$[3.800, 3.510]$
$[2.412, 4.223]$	$[-2.412, 1.081]$
$[-1.267, 0.713]$	$[-1.267, 0.713]$

SECTION 9.5

1. $\frac{1}{4}\pi a^2$ **3.** $\frac{1}{2}a^2$ **5.** $\frac{1}{2}\pi a^2$ **7.** $\frac{1}{4} - \frac{1}{16}\pi$ **9.** $\frac{3}{16}\pi + \frac{3}{8}$ **11.** $\frac{5}{2}a^2$ **13.** $\frac{1}{12}(3e^{2\pi} - 3 - 2\pi^3)$ **15.** $\frac{1}{4}(e^{2\pi} + 1 - 2e^{\pi})$

17. $\int_{\pi/6}^{5\pi/6} \frac{1}{2}([4\sin\theta]^2 - [2]^2)\,d\theta$ **19.** $\int_{-\pi/3}^{\pi/3} \frac{1}{2}([4]^2 - [2\sec\theta]^2)\,d\theta$ **21.** $2\left[\int_0^{\pi/3} \frac{1}{2}(2\sec\theta)^2 d\theta + \int_{\pi/3}^{\pi/2} \frac{1}{2}(4)^2 d\theta\right]$

23. $\int_0^{\pi/3} \frac{1}{2}(2\sin 3\theta)^2 d\theta$ **25.** $2\left[\int_0^{\pi/6} \frac{1}{2}(\sin\theta)^2 d\theta + \int_{\pi/6}^{\pi/2} \frac{1}{2}(1 - \sin\theta)^2 d\theta\right]$ **27.** $\pi - 8\int_0^{\pi/4} \frac{1}{2}(\cos 2\theta)^2 d\theta$ **29.** $\frac{\pi}{6} - \frac{\sqrt{3}}{16}$

31. For $r = a\cos 2n\theta$, area of one petal is $a^2 \int_0^{\pi/4n} \cos^2 2n\theta\, d\theta = \dfrac{\pi a^2}{8n}$. For $r = a\sin 2n\theta$, area of one petal is $a^2 \int_0^{\pi/4n} \sin^2 2n\theta\, d\theta = \dfrac{\pi a^2}{8n}$.
Total area $= \dfrac{\pi a^2}{2}$.

35. $(5/6, 0)$ **37.** $\dfrac{9\pi}{2}$ **39.** $\dfrac{4\pi}{3} + 2\sqrt{3}$ **41.** (a) Substitute $x = r\cos\theta, y = r\sin\theta$ into the equation and solve for r. (c) $8 - 2\pi$

SECTION 9.6

1. $4x = (y-1)^2$ **3.** $y = 4x^2 + 1,\ x \ge 0$ **5.** $9x^2 + 4y^2 = 36$ **7.** $1 + x^2 = y^2$ **9.** $y = 2 - x^2,\ -1 \le x \le 1$

11. $2y - 6 = x,\ -4 \le x \le 4$

13. $y = x - 1$

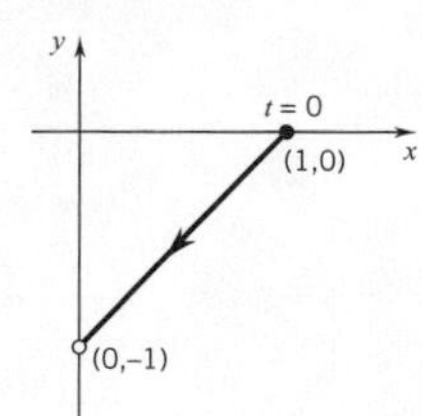

15. $xy = 1$

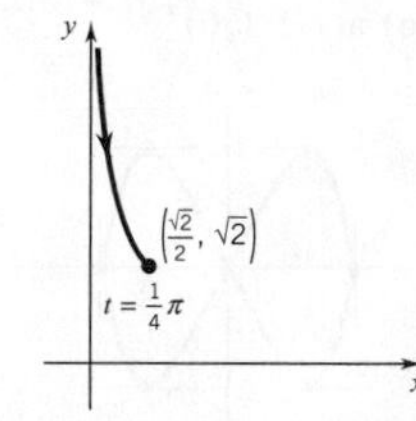

17. $y + 2x = 11$

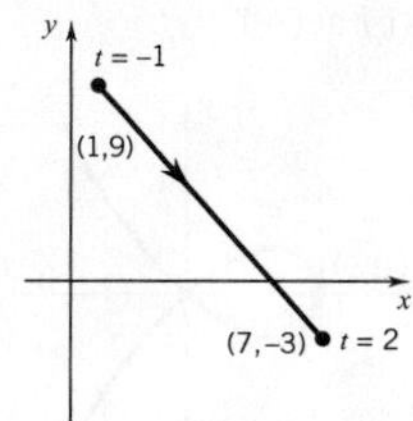

19. $x = \sin \frac{1}{2}\pi y$

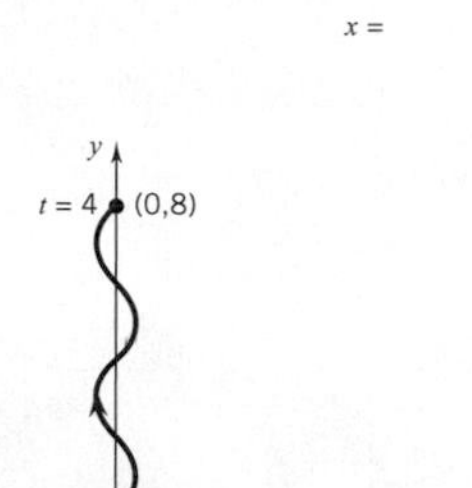

21. $y^2 = x^2 + 1$

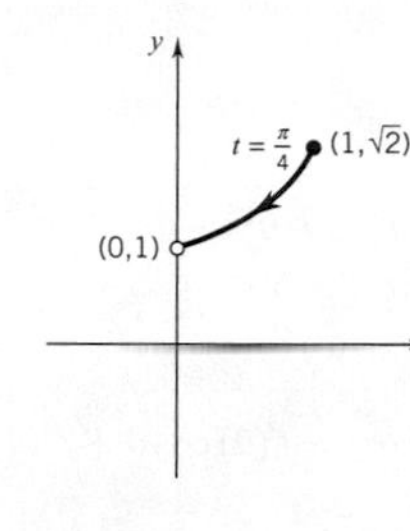

23. (a) $x(t) = -\sin 2\pi t, y(t) = \cos 2\pi t$ (b) $x(t) = \sin 4\pi t,\ y(t) = \cos 4\pi t$ (c) $x(t) = \cos \frac{1}{2}\pi t,\ y(t) = \sin \frac{1}{2}\pi t$
(d) $x(t) = \cos \frac{3}{2}\pi t,\ y(t) = -\sin \frac{3}{2}\pi t$

25. $x(t) = \tan \frac{1}{2}\pi t, \quad y(t) = 2$ **27.** $x(t) = 3 + 5t, \quad y(t) = 7 - 2t$ **29.** $x(t) = \sin^2 \pi t, \quad y(t) = -\cos \pi t$

31. $x(t) = (2 - t)^2, \quad y(t) = (2 - t)^3$ **33.** $\int_c^d y(t)x'(t)dt = \int_c^d f(x(t))x'(t)dt = \int_a^b f(x)dx =$ area below C

35. $\int_c^d \pi[y(t)]^2 x'(t)\,dt = \int_c^d \pi[f(x(t))]^2 x'(t)\,dt = \int_a^b \pi[f(x)]^2 dx = V_x;\ \int_c^d 2\pi x(t)y(t)x'(t)\,dt = \int_c^d 2\pi x(t)f(x(t))x'(t)\,dt = \int_a^b 2\pi x f(x)\,dx = V_y$

37. $A = 2\pi a^2$

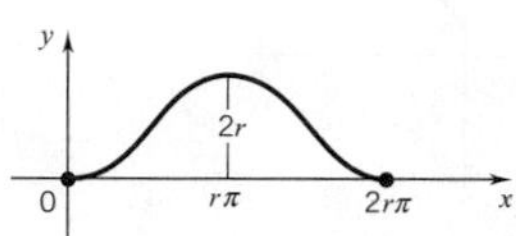

39. (a) $V_x = 3\pi^2 r^3$ (b) $V_y = 4\pi^3 r^3$

41. $x(t) = -a\cos t,\ y(t) = b\sin t;\ t \in [0, \pi]$ **43.** (a) paths intersect at (6, 5) and (8, 1) (b) particles collide at (8, 1)

45. curve intersects itself at (0, 0) **47.** curve intersects itself at (0, 0) and $(0, \frac{3}{4})$ **49.** $x^2 - y^2 = 1$

51. The particle moves along the parabola $y = 2x - \dfrac{x^2}{4}$ from (0, 0) to (12, –12).

53. The particle moves around the unit circle in the counterclockwise direction starting from the point (1, 0).

55. (a) The coefficient a determines the amplitude and the period. (b) $\dfrac{dy}{dx} \to -\infty$ as $\theta \to 2\pi^-$; $\dfrac{dy}{dx} \to \infty$ as $\theta \to 2\pi^+$ (c) cusp

57. (c) $|x| \le 1$ and $|y| \le 1$ for all t.

SECTION 9.7

1. $3x - y - 3 = 0$ **3.** $y = 1$ **5.** $3x + y - 3 = 0$ **7.** $2x + 2y - \sqrt{2} = 0$ **9.** $2x + y - 8 = 0$ **11.** $x - 5y + 4 = 0$

13. $x + 2y + 1 = 0$ **15.** $x(t) = t,\ y(t) = t^3$; tangent line $y = 0$ **17.** $x(t) = t^{5/3},\ y(t) = t$; tangent line $x = 0$ **19.** (a) none;
(b) at (2, 2) and (–2, 0)

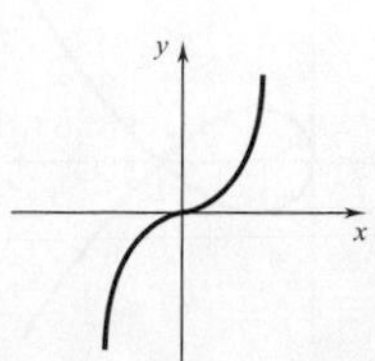

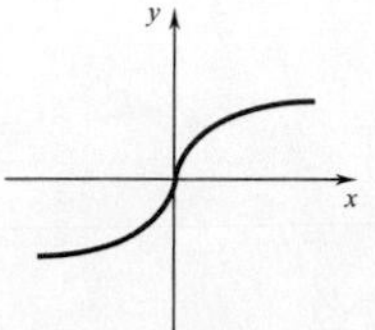

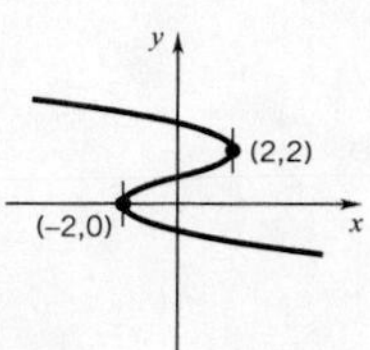

21. (a) at $(3, 7)$ and $(3, 1)$;
(b) at $(-1, 4)$ and $(7, 4)$

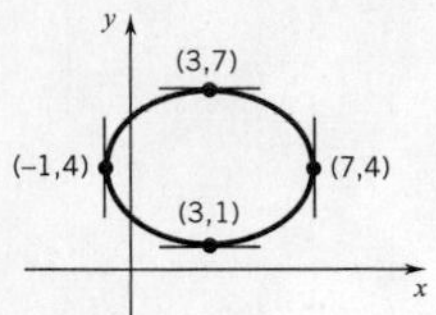

23. (a) at $(-\frac{2}{3}, \pm\frac{2}{9}\sqrt{3})$;
(b) at $(-1, 0)$

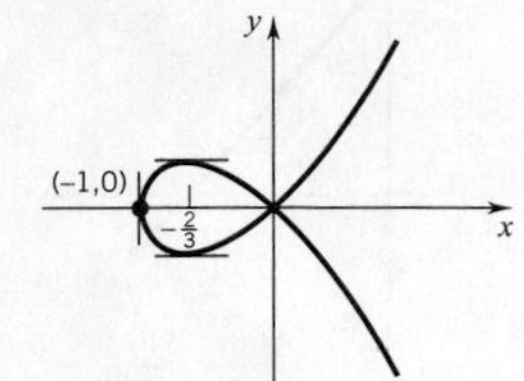

25. (a) at $(\pm\frac{1}{2}\sqrt{2}, \pm 1)$;
(b) at $(\pm 1, 0)$

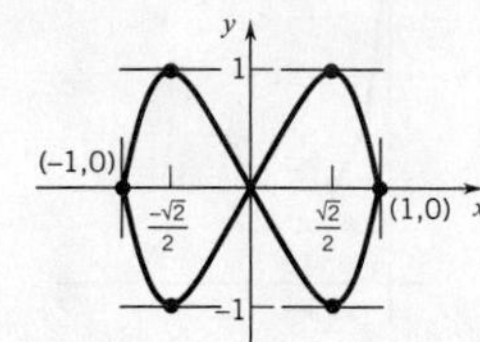

27. $y = 0, \quad (\pi - 2)y + 32x - 64 = 0$

29. The slope of $\overline{OP}$ is $\tan\theta_1$. The curve $r = f(\theta)$ can be parametrized by setting

$$x(\theta) = f(\theta)\cos\theta, \quad y(\theta) = f(\theta)\sin\theta.$$

Differentiation gives

$$x'(\theta) = -f(\theta)\sin\theta + f'(\theta)\cos\theta, \quad y'(\theta) = f(\theta)\cos\theta + f'(\theta)\sin\theta.$$

If $f'(\theta_1) = 0$, then

$$x'(\theta_1) = -f(\theta_1)\sin\theta_1, \quad y'(\theta_1) = f(\theta_1)\cos\theta_1.$$

Since $f(\theta_1) \neq 0$, we have

$$m = \frac{y'(\theta_1)}{x'(\theta_1)} = -\cot\theta_1 = -\frac{1}{\text{slope of}\overline{OP}}.$$

31.

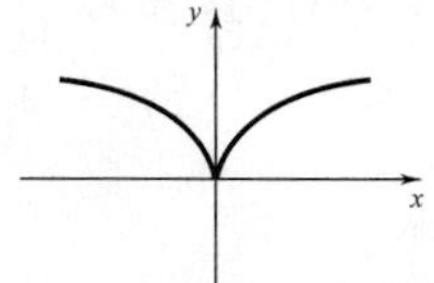

33.

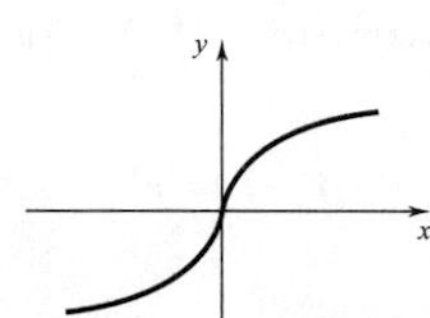

35.

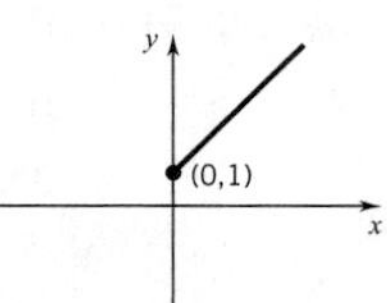

37. -8 **39.** 2 **41.** $\dfrac{1}{\sin^3 t}$ **43.** $y - 2 = -\frac{16}{3}(x - \frac{1}{8})$

SECTION 9.8

1. $\sqrt{5}$ **3.** 7 **5.** $2\sqrt{3}$ **7.** $\frac{4}{3}$ **9.** $6 + \frac{1}{2}\ln 5$ **11.** $\frac{63}{8}$ **13.** $\ln(1 + \sqrt{2})$ **15.** $\frac{3}{2}$ **17.** $\frac{1}{3}\pi + \frac{1}{2}\sqrt{3}$

19. initial speed 2, terminal speed 4; $s = 2\sqrt{3} + \ln(2 + \sqrt{3})$ **21.** initial speed 0, terminal speed $\sqrt{13}$; $x = \frac{1}{27}(13\sqrt{13} - 8)$

23. initial speed $\sqrt{2}$, terminal speed $\sqrt{2}\,e^{\pi}$; $s = \sqrt{2}(e^{\pi} - 1)$ **25.** $8a$

27. (a) $24a$ (b) use the identities $\cos 3\theta = 4\cos^3\theta - 3\cos\theta$, $\sin 3\theta = 3\sin\theta - 4\sin^3\theta$ **29.** 2π **31.** $\sqrt{2}(e^{4\pi} - 1)$

33. $\frac{1}{2}\sqrt{5}(e^{4\pi} - 1)$ **35.** $4 - 2\sqrt{2}$ **37.** $\ln(1 + \sqrt{2})$ **39.** $c = 1$ **41.** (a) $(\frac{1}{2}, -\frac{7}{2})$

43. $L = \int_a^b \sqrt{1 + \sinh^2 x}\, dx = \int_a^b \sqrt{\cosh^2 x}\, dx = \int_a^b \cosh x\, dx = A$ **45.** $L \cong 4.6984$ **47.** (a) (b) 2.7156

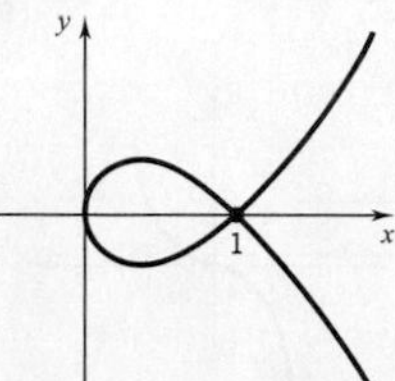

49. 4 **51.** (b) 28.3617 **53.** $\sqrt{1 + [f'(x)]^2} = \sqrt{1 + \tan^2[\alpha(x)]} = |\sec[\alpha(x)]|$

SECTION 9.9

1. $L = 1, (\bar{x}, \bar{y}) = (\frac{1}{2}, 4),\ A_x = 8\pi$ **3.** $L = 5, (\bar{x}, \bar{y}) = (\frac{3}{2}, 2), A_x = 20\pi$ **5.** $L = 10,\ (\bar{x}, \bar{y}) = (3, 4),\ A_x = 80\pi$

7. $L = \frac{1}{3}\pi,\quad \bar{x} = 6/\pi,\quad \bar{y} = 6(2 - \sqrt{3})/\pi,\quad A_x = 4\pi(2 - \sqrt{3})$ **9.** $L = \frac{1}{3}\pi a;\quad \bar{x} = 0,\quad \bar{y} = 3a/\pi,\quad A_x = 2\pi a^2$ **11.** $\frac{1}{9}\pi(17\sqrt{17} - 1)$

13. $\frac{61}{432}\pi$ **15.** $\pi[\sqrt{2} + \ln(1 + \sqrt{2})]$ **17.** $\frac{2}{5}\sqrt{2}\pi(2e^{\pi} + 1)$ **19.** (a) $3\pi a^2$ (b) $\dfrac{64\pi a^2}{3}$

21. See the figure.

$$A = \tfrac{1}{2}\theta s_2^2 - \tfrac{1}{2}\theta s_1^2$$
$$= \tfrac{1}{2}(\theta s_2 + \theta s_1)(s_2 - s_1)$$
$$= \tfrac{1}{2}(2\pi R - 2\pi r)s = \pi(R + r)s$$

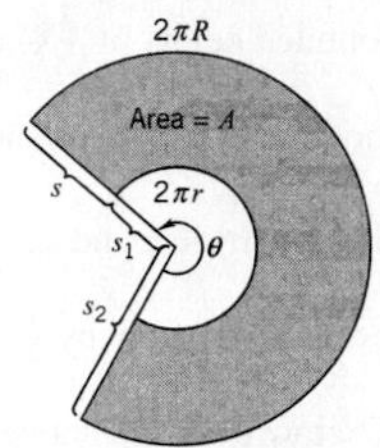

23. (a) the 3, 4, 5 sides have centroids

$(\frac{3}{2}, 0),\ (4, 2),\ (\frac{3}{2}, 2)$

(b) $\bar{x} = 2,\ \bar{y} = \frac{3}{2}$
(c) $\bar{x} = 2,\ \bar{y} = \frac{4}{3}$
(d) $\bar{x} = \frac{13}{6}, \bar{y} = 2$
(e) $A = 20\pi$
(f) $A = 36\pi$

25. $4\pi^2 ab$

27. The band can be obtained by revolving about the x-axis the graph of a function

$$f(x) = \sqrt{r^2 - x^2}, x \in [a, b].$$

A straightforward calculation shows that the surface area of the band is $2\pi r(b - a)$.

29. (a) $2\pi b^2 + \dfrac{2\pi ab}{e}\sin^{-1} e$ (b) $2\pi a^2 + \dfrac{\pi b^2}{e}\ln\left|\dfrac{1 + e}{1 - e}\right|$, where e is the eccentricity $c/a = \sqrt{a^2 - b^2}/a$

31. at the midpoint of the axis of the hemisphere **33.** on the axis of the cone $\left(\dfrac{2R + r}{R + r}\right)\dfrac{h}{3}$ units from the base of radius r

CHAPTER 10

SECTION 10.1

1. lub = 2; glb = 0 **3.** no lub; glb = 0 **5.** lub = 2; glb = −2 **7.** no lub; glb = 2 **9.** lub = $2\frac{1}{2}$; glb = 2

11. lub = 1; glb = 0.9 **13.** lub = e; glb = 0 **15.** lub = $\frac{1}{2}(-1 + \sqrt{5})$; glb = $\frac{1}{2}(-1 - \sqrt{5})$ **17.** no lub; no glb

19. no lub; no glb **21.** glb $S = 0$, $\quad 0 \le \left(\frac{1}{11}\right)^3 < 0 + 0.001$ **23.** glb $S = 0$, $\quad 0 \le \left(\frac{1}{10}\right)^{2n-1} < 0 + \left(\frac{1}{10}\right)^k,\quad n > \frac{1}{2}(k + 1)$

25. Let $\epsilon > 0$. The condition $m \le s$ is satisfied by all numbers s in S. All we have to show therefore is that there is some number s in S such that $s < m + \epsilon$. Suppose on the contrary that there is no such number in S. We then have $m + \epsilon \le x$ for all $x \in S$, so that $m + \epsilon$ becomes a lower bound for S. But this cannot happen, for it makes $m + \epsilon$ a lower bound that is *greater* than m, and by assumption, m is the *greatest* lower bound.

27. Let $c =$ lub S. Since $b \in S, b \le c$. Since b is an upper bound for $S, c \le b$. Thus $b = c$.

29. (a) Any upper bound for S is an upper bound for T; any lower bound for S is a lower bound for T.

(b) Let $a =$ glb S. Then $a \le t$ for all $t \in T$. Therefore $a \le$ glb T. Similarly, if $b =$ lub S, then $t \le b$ for all $t \in T$, so lub $T \le b$. It now follows that glb $S \le$ glb $T \le$ lub $T \le$ lub S.

31. Let M be any positive number and consider M/c. Since the set of positive integers is not bounded above, there exists a positive integer k such that $k \ge M/c$. This implies $kc \ge M$. Since $Kc \in S$, it follows that S is not bounded above.

35. (a) 0.5, 0.38554, 0.36971, 0.36806, 0.36970

(b) lub = 0.5, glb = $\frac{1}{e}$

37. (a)

a_1	a_2	a_3	a_4	a_5	a_6	a_7	a_8	a_9	a_{10}
1.4142	1.6818	1.8340	1.9152	1.9571	1.9785	1.9892	1.9946	1.9973	1.9986

(b) Let S be the set of positive integers for which $a_n < 2$. Then $1 \in S$ since $a_1 = \sqrt{2} \cong 1.4142 < 2$. Assume that $k \in S$. Now $a_{k+1}^2 = 2a_k < 4$, which implies $a_{k+1} < 2$. Thus $k + 1 \in S$ and S is the set of positive integers.

(c) yes (d) The number you chose is the least upper bound of the corresponding set S.

SECTION 10.2

1. $a_n = 2 + 3(n-1), \quad n = 1, 2, 3, \ldots$ **3.** $a_n = \dfrac{(-1)^{n-1}}{2n-1}, \quad n = 1, 2, 3, \ldots$ **5.** $a_n = \dfrac{n^2+1}{n}, \quad n = 1, 2, 3, \ldots$

7. $a_n = \begin{cases} n & \text{if } n = 2k-1, \\ 1/n & \text{if } n = 2k, \end{cases}$ where $k = 1, 2, 3, \ldots$ **9.** decreasing; bounded below by 0 and above by 2

11. not monotonic; bounded below by 0 and above by $\frac{3}{2}$ **13.** decreasing; bounded below by 0 and above by 0.9

15. increasing; bounded below by $\frac{1}{2}$ but not bounded above **17.** increasing; bounded below by $\frac{4}{5}\sqrt{5}$ and above by 2

19. increasing; bounded below by $\frac{2}{51}$ but not bounded above **21.** increasing; bounded below by 0 and above by ln 2

23. decreasing; bounded below by 1 and above by 4 **25.** increasing; bounded below by $\sqrt{3}$ and above by 2

27. decreasing; bounded above by -1 but not bounded below **29.** increasing; bounded below by $\frac{1}{2}$ and above by 1

31. decreasing; bounded below by 0 and above by 1 **33.** decreasing: bounded below by 0 and above by $\frac{5}{6}$

35. decreasing; bounded below by 0 and above by $\frac{1}{2}$ **37.** decreasing; bounded below by 0 and above by $\frac{1}{3}\ln 3$

39. increasing; bounded below by $\frac{3}{4}$ but not bounded above **41.** for $n \geq 5$,

$$\frac{a_{n+1}}{a_n} = \frac{5^{n+1}}{(n+1)!} \bullet \frac{n!}{5^n} = \frac{5}{n+1} < 1 \quad \text{and} \quad a_{n+1} < a_n.$$

Sequence is not nonincreasing : $a_1 = 5 < \frac{25}{2} = a_2$.

43. boundedness : $0 < (c^n + d^n)^{1/n} < (2d^n)^{1/n} = 2^{1/n} d \leq 2d$.

monotonicity : $a_{n+1}^{n+1} = c^{n+1} + d^{n+1} = cc^n + dd^n < (c^n + d^n)^{1/n} c^n + (c^n + d^n)^{1/n} d^n$
$= (c^n + d^n)^{1+(1/n)} = (c^n + d^n)^{(n+1)/n} = a_n^{n+1}$.

Taking the $n+1$-th root of each side we have $a_{n+1} < a_n$. The sequence is monotonic decreasing.

45. $a_1 = 1, a_2 = \frac{1}{2}, a_3 = \frac{1}{6}, a_4 = \frac{1}{24}, a_5 = \frac{1}{120}, a_6 = \frac{1}{720}; \; a_n = 1/n!$

47. $a_1 = a_2 = a_3 = a_4 = a_5 = a_6 = 1; \; a_n = 1$

49. $a_1 = 1, a_2 = 3, a_3 = 5, a_4 = 7, a_5 = 9, a_6 = 11; \; a_n = 2n - 1$

51. $a_1 = 1, a_2 = 4, a_3 = 9, a_4 = 16, a_5 = 25, a_6 = 36; \; a_n = n^2$

53. $a_1 = 1, a_2 = 1, a_3 = 2, a_4 = 4, a_5 = 8, a_6 = 16; \; a_n = 2^{n-2}$ for $n \geq 3$

55. $a_1 = 1, a_2 = 3, a_3 = 5, a_4 = 7, a_5 = 9, a_6 = 11; \; a_n = 2n - 1$

57. First $a_1 = 2^1 - 1 = 1$. Next suppose $a_k = 2^k - 1$ for some $k \geq 1$. Then $a_{k+1} = 2a_k + 1 = 2(2^k - 1) + 1 = 2^{k+1} - 1$.

59. First $a_1 = \dfrac{1}{2^0} = 1$. Next suppose $a_k = \dfrac{k}{2^{k-1}}$ for some $k \geq 1$. Then $a_{k+1} = \dfrac{k+1}{2k} a_k = \dfrac{k+1}{2k} \dfrac{k}{2^{k-1}} = \dfrac{k+1}{2^k}$.

61. (a) n (b) $\dfrac{1-r^n}{1-r}$ **63.** (a) $150\left(\frac{3}{4}\right)^{n-1}$ (b) $\dfrac{5\sqrt{3}}{2}\left(\dfrac{3}{4}\right)^{\frac{n-1}{2}}$ **65.** $\{a_n\}$ is increasing; limit $\frac{1}{2}$

67. (a) $a_2 = 1 + \sqrt{a_1} = 2 > 1 = a_1$. Assume $a_k = 1 + \sqrt{a_{k-1}} > a_{k-1}$. Then $a_{k+1} = 1 + \sqrt{a_k} > 1 + \sqrt{a_{k-1}} = a_k$.
Thus $\{a_n\}$ is an increasing sequence.

(b) $a_n = 1 + \sqrt{a_{n-1}} < 1 + \sqrt{a_n}$, since $a_{n-1} < a_n$.

$a_n - \sqrt{a_n} - 1 < 0$, or $(\sqrt{a_n})^2 - \sqrt{a_n} - 1 < 0$, which implies (solve the inequality) that $\sqrt{a_n} < \dfrac{1+\sqrt{5}}{2}$, hence $a_n < \dfrac{3+\sqrt{5}}{2}$ for all n.

(c) lub $\{a_n\} \cong 2.6180$.

69. (b), (c) It appears that $\lim_{n\to\infty} a_n = 1$; the sequence is bounded.

SECTION 10.3

1. diverges **3.** converges to 0 **5.** converges to 1 **7.** converges to 0 **9.** converges to 0 **11.** diverges **13.** converges to 0

15. converges to 1 **17.** converges to $\frac{4}{9}$ **19.** converges to $\frac{1}{2}\sqrt{2}$ **21.** diverges **23.** converges to 1 **25.** converges to 0

27. converges to $\frac{1}{2}$ **29.** converges to e^2 **31.** diverges **33.** (a) 0 (b) $\frac{\pi}{4}$ (c) $\frac{1}{2}$

35. $b < \sqrt[n]{a^n + b^n} = b\sqrt[n]{(a/b)^n + 1} < b\sqrt[n]{2}$. Since $2^{1/n} \to 1$ as $n \to \infty$, it follows that $\sqrt[n]{a^n + b^n} \to b$ by the pinching theorem.

37. Use $|(a_n + b_n) - (L + M)| \le |a_n - L| + |b_n - M|$. **39.** $\left(1 + \frac{1}{n}\right)^{n+1} = \left(1 + \frac{1}{n}\right)^n \left(1 + \frac{1}{n}\right)$. Note that $\left(1 + \frac{1}{n}\right)^n \to e$ and $\left(1 + \frac{1}{n}\right) \to 1$.

41. Imitate the proof given for the nondecreasing case in Theorem 10.3.6. **43.** Let $\epsilon > 0$. Choose k so that, for $n \ge k$,

$$L - \epsilon < a_n < L + \epsilon, \quad L - \epsilon < c_n < L + \epsilon \quad \text{and} \quad a_n \le b_n \le c_n.$$

For such n,

$$L - \epsilon < b_n < L + \epsilon.$$

45. Let $\epsilon > 0$. Since $a_n \to L$, there exists a positive integer N such that $L - \epsilon < a_n < L + \epsilon$ for all $n \ge N$. Now $a_n \le M$ for all n, so $L - \epsilon < M$, or $L < M + \epsilon$. Since ϵ is arbitrary, $L \le M$.

47. Assume $a_n \to 0$. Let $\epsilon > 0$. There exists a positive integer N such that $|a_n - 0| < \epsilon$ for all $n \ge N$. Since $||a_n| - 0| \le |a_n - 0|$, it follows that $|a_n| \to 0$. Now assume $|a_n| \to 0$. Since $-|a_n| \le a_n \le |a_n|$, $a_n \to 0$ by the pinching theorem.

49. By the continuity of f, $f(L) = f(\lim_{n\to\infty} a_n) = \lim_{n\to\infty} f(a_n) = \lim_{n\to\infty} a_{n+1} = L$. **51.** Use Theorem 10.3.12 with $f(x) = x^{1/p}$. **53.** converges to 0

55. converges to 0 **57.** diverges **59.** $L = 0,\ n = 32$ **61.** $L = 0,\ n = 4$ **63.** $L = 0,\ n = 7$ **65.** $L = 0,\ n = 65$

67. (a) $\dfrac{3 + \sqrt{5}}{2}$ (b) 3

69. (a)

a_2	a_3	a_4	a_5	a_6	a_7	a_8	a_9	a_{10}
0.540302	0.857553	0.654290	0.793480	0.701369	0.763960	0.722102	0.750418	0.731404

(b) 0.739085; it is the fixed point of $f(x) = \cos x$.

SECTION 10.4

1. converges to 1 **3.** converges to 0 **5.** converges to 0 **7.** converges to 0 **9.** converges to 1 **11.** converges to 0 **13.** converges to 1

15. converges to 1 **17.** converges to π **19.** converges to 1 **21.** converges to 0 **23.** diverges **25.** converges to 0

27. converges to e^{-1} **29.** converges to 0 **31.** converges to 0 **33.** converges to e^x **35.** converges to 0 **37.** (a) 2 (b) 0 (c) 1

39. $\sqrt{n+1} - \sqrt{n} = \dfrac{\sqrt{n+1} - \sqrt{n}}{\sqrt{n+1} + \sqrt{n}}(\sqrt{n+1} + \sqrt{n}) = \dfrac{1}{\sqrt{n+1} + \sqrt{n}} \to 0$

41. (b) $2\pi r$. As $n \to \infty$, the perimeter of the polygon tends to the circumference of the circle. **43.** $\frac{1}{2}$ **45.** $\frac{1}{8}$

47. (a) $m_{n+1} - m_n = \dfrac{1}{n+1}(a_1 + \cdots + a_n + a_{n+1}) - \dfrac{1}{n}(a_1 + \cdots + a_n)$

$$= \frac{1}{n(n+1)}\Big[na_{n+1} - (a_1 + \cdots + a_n)\Big] > 0 \text{ since } \{a_n\} \text{ is increasing.}$$

(b) We begin with the hint $m_n < \dfrac{|a_1 + \cdots + a_j|}{n} + \dfrac{\epsilon}{2}\left(\dfrac{n-j}{n}\right)$. Since j is fixed, $\dfrac{|a_1 + \cdots + a_j|}{n} \to 0$, and therefore for n sufficiently large $\dfrac{|a_1 + \cdots + a_j|}{n} < \dfrac{\epsilon}{2}$. Since $\dfrac{\epsilon}{2}\left(\dfrac{n-j}{n}\right) < \dfrac{\epsilon}{2}$, we see that, for n sufficiently large, $|m_n| < \epsilon$. This shows that $m_n \to 0$.

49. (a) Let S be the set of positive integers $n (n \ge 2)$ for which the inequalities hold. Since $(\sqrt{b})^2 - 2\sqrt{ab} + (\sqrt{a})^2 = (\sqrt{b} - \sqrt{a})^2 > 0$, it follows that $\dfrac{a+b}{2} > \sqrt{ab}$ and $a_1 > b_1$. Now $a_2 = \dfrac{a_1 + b_1}{2} < a_1$ and $b_2 = \sqrt{a_1 b_1} > b_1$. Also, by the argument above, $a_2 = \dfrac{a_1 + b_1}{2} > \sqrt{a_1 b_1} = b_2$, and so $a_1 > a_2 > b_2 > b_1$. Thus $2 \in S$. Assume that $k \in S$. Then $a_{k+1} = \dfrac{a_k + b_k}{2} < \dfrac{a_k + a_k}{2} = a_k$, $b_{k+1} = \sqrt{a_k b_k} > \sqrt{b_k^2} = b_k$, and $a_{k+1} = \dfrac{a_k + b_k}{2} > \sqrt{a_k b_k} = b_{k+1}$. Thus $k + 1 \in S$. Therefore the inequalities hold for all $n \ge 2$.

(b) $\{a_n\}$ is a decreasing sequence which is bounded below.

$\{b_n\}$ is an increasing sequence which is bounded above.

Let $L_a = \lim_{n\to\infty} a_n$, $L_b = \lim_{n\to\infty} b_n$. Then $a_n = \dfrac{a_{n-1} + b_{n-1}}{2}$ implies $L_a = \dfrac{L_a + L_b}{2}$ and $L_a = L_b$.

51. The numerical work suggests $L \cong 1$. Justification: Set $f(x) = \sin x - x^2$. Note that $f(0) = 0$ and for x close to 0, $f'(x) = \cos x - 2x > 0$. Therefore $\sin x - x^2 > 0$ for x close to 0 and $\sin(1/n) - 1/n^2 > 0$ for n large. Thus, for n large,

$$\frac{1}{n^2} < \sin\frac{1}{n} < \frac{1}{n} \qquad |\sin x| \le |x| \quad \text{for all } x$$

$$\left(\frac{1}{n^2}\right)^{1/n} < \left(\sin\frac{1}{n}\right)^{1/n} < \left(\frac{1}{n}\right)^{1/n}$$

$$\left(\frac{1}{n^{1/n}}\right)^2 < \left(\sin\frac{1}{n}\right)^{1/n} < \frac{1}{n^{1/n}}.$$

As $n \to \infty$ both bounds tend to 1 and therefore the middle term also tends to 1.

53. (a)

a_3	a_4	a_5	a_6	a_7	a_8	a_9	a_{10}
2	3	5	8	13	21	34	55

(b)

r_1	r_2	r_3	r_4	r_5	r_6
1	2	1.5	1.6667	1.6000	1.625

(c) $L = \dfrac{1+\sqrt{5}}{2} \cong 1.618033989$

SECTION 10.5

1. 0 **3.** 1 **5.** $\frac{1}{2}$ **7.** $\ln 2$ **9.** $\frac{1}{4}$ **11.** 2 **13.** $\dfrac{1+\pi}{1-\pi}$ **15.** $\frac{1}{2}$ **17.** π **19.** $-\frac{1}{2}$ **21.** -2

23. $\frac{1}{3}\sqrt{6}$ **25.** $-\frac{1}{8}$ **27.** 4 **29.** $\frac{1}{2}$ **31.** $\frac{1}{2}$ **33.** 1 **35.** 1 **37.** 0 **39.** 1 **41.** 1

43. $\lim_{x\to 0}(2 + x + \sin x) \neq 0$, $\lim_{x\to 0}(x^3 + x - \cos x) \neq 0$ **45.** $a = \pm 4$, $b = 1$ **47.** $-\dfrac{e}{2}$

49. $f(0)$ **51.** (a) 1 (b) $-\frac{1}{3}$ **53.** $\frac{3}{4}$ **55.** (a) $f(x) \to \infty$ as $x \to \pm\infty$ (b) 10 **57.** (b) $\ln 2 \cong 0.6931$

SECTION 10.6

1. ∞ **3.** -1 **5.** ∞ **7.** $\frac{1}{5}$ **9.** 1 **11.** 0 **13.** ∞ **15.** $\frac{1}{3}$ **17.** e **19.** 1 **21.** $\frac{1}{2}$ **23.** 0 **25.** 1

27. e^3 **29.** e **31.** 0 **33.** $-\frac{1}{2}$ **35.** 0 **37.** 1 **39.** 1 **41.** 0 **43.** 1 **45.** 0

47. y-axis vertical asymptote

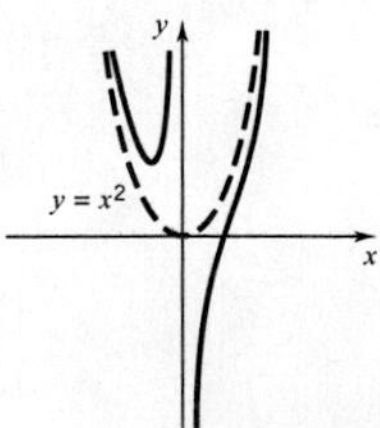

49. x-axis horizontal asymptote

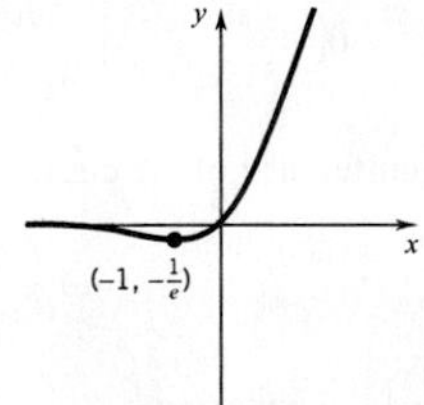

51. x-axis horizontal asymptote

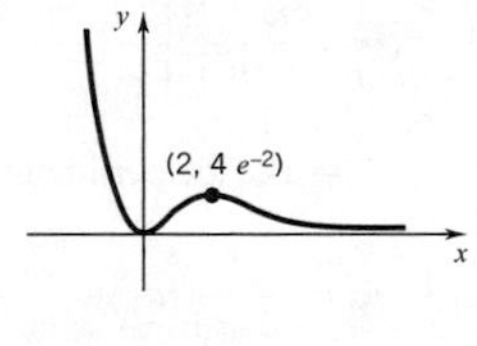

53. $\dfrac{b}{a}\sqrt{x^2-a^2} - \dfrac{b}{a}x = \dfrac{\sqrt{x^2-a^2}+x}{\sqrt{x^2-a^2}+x}\left(\dfrac{b}{a}\right)\left(\sqrt{x^2-a^2}-x\right) = \dfrac{-ab}{\sqrt{x^2-a^2}+x} \to 0$ as $x \to \infty$ **55.** example : $f(x) = x^2 + \dfrac{(x-1)(x-2)}{x^3}$

57. $\lim_{x\to 0^+} \cos x \neq 0$

59. (a) Let S be the set of positive integers for which the statement is true. Since $\lim_{x\to\infty}\dfrac{\ln x}{x} = 0$, $1 \in S$. Assume that $k \in S$. By L'Hôpital's rule,

$$\lim_{x\to\infty}\frac{(\ln x)^{k+1}}{x} \overset{*}{=} \lim_{x\to\infty}\frac{(k+1)(\ln x)^k}{x} = 0 \quad (\text{since } k \in S).$$

Thus $k + 1 \in S$, and S is the set of positive integers.

(b) Choose any positive number α. Let $k - 1$ and k be positive integers such that $k - 1 \le \alpha \le k$. Then, for $x > e$,

$$\frac{(\ln x)^{k-1}}{x} \le \frac{(\ln x)^{\alpha}}{x} \le \frac{(\ln x)^k}{x}$$

and the result follows by the pinching theorem.

61. The limit has the form 1^∞; use L'Hôpital's rule

63. (a) $A_b = 1 - (1+b)\,e^{-b}$

(b) $\bar{x}_b = \dfrac{2-(2+2b+b^2)e^{-b}}{1-(1+b)e^{-b}}, \quad \bar{y}_b = \dfrac{\frac{1}{4}-\frac{1}{4}(1+2b+2b^2)e^{-2b}}{2[1-(1+b)e^{-b}]}$

(c) $\lim_{b\to\infty} A_b = 1; \quad \lim_{b\to\infty} \bar{x}_b = 2, \quad \lim_{b\to\infty} \bar{y}_b = \frac{1}{8}$

65. $\lim_{x\to 0^+} (1+x^2)^{1/x} = 1.$ **67.** $\lim_{x\to\infty} g(x) = -5/3.$

SECTION 10.7

1. 1 **3.** $\frac{1}{4}\pi$ **5.** diverges **7.** 6 **9.** $\frac{1}{2}\pi$ **11.** 2 **13.** diverges **15.** $-\frac{1}{4}$ **17.** π **19.** diverges **21.** $\ln 2$

23. 4 **25.** diverges **27.** diverges **29.** diverges **31.** $\frac{1}{2}$ **33.** $2e-2$ **35.** (a) converges: $\frac{1}{32}$ (b) converges: $\frac{\pi}{16}$ (c) converges: $\frac{\pi}{16}$ (d) diverges **37.** $\frac{\pi}{2}-1$ **39.** π

41. surface area $= \displaystyle\int_1^\infty 2\pi\left(\frac{1}{x}\right)\sqrt{1+\frac{1}{x^4}}\,dx = 2\pi\int_1^\infty \frac{\sqrt{x^4+1}}{x^3}dx > 2\pi\int_1^\infty \frac{1}{x}\,dx = \infty$

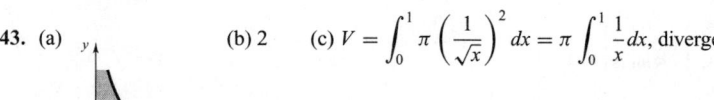

43. (a) (b) 2 (c) $V = \displaystyle\int_0^1 \pi\left(\frac{1}{\sqrt{x}}\right)^2 dx = \pi\int_0^1 \frac{1}{x}dx$, diverges

45. (a)

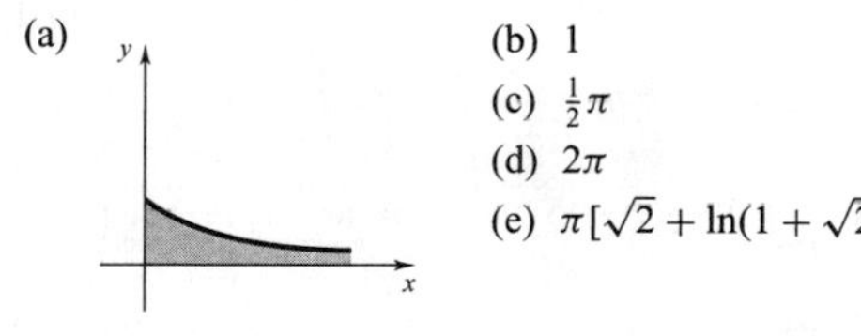

(b) 1 (c) $\frac{1}{2}\pi$ (d) 2π (e) $\pi[\sqrt{2}+\ln(1+\sqrt{2})]$

47. (a) The interval [0, 1] causes no problem. For $x \geq 1, e^{-x^2} \leq e^{-x}$and $\displaystyle\int_1^\infty e^{-x}dx$ is finite.

(b) $V_y = \displaystyle\int_0^\infty 2\pi x\, e^{-x^2}\,dx = \pi$

49. (a) (b) $\frac{4}{3}$ (c) 2π (d) $\frac{8}{7}\pi$

51. converges by comparison with $\displaystyle\int_0^\infty \frac{dx}{x^{3/2}}$

53. diverges since for x large the integrand is greater than $\dfrac{1}{x}$ and $\displaystyle\int_1^\infty \frac{1}{x}\,dx$ diverges **55.** converges by comparison with $\displaystyle\int_1^\infty \frac{dx}{x^{3/2}}$

57. (a) $\displaystyle\int_0^\infty \frac{2x}{1+x^2}\,dx = \lim_{b\to\infty}\int_0^b \frac{2x}{1+x^2}\,dx = \lim_{b\to\infty}\left[\ln(1+x^2)\right]_0^b = \infty$

(b) $\displaystyle\lim_{b\to\infty}\int_{-b}^b \frac{2x}{1+x^2}\,dx = \lim_{b\to\infty}\left[\ln(1+x^2)\right]_{-b}^b$

$\displaystyle = \lim_{b\to\infty}\left[\ln(1+b^2) - \ln(1+b^2)\right] = \lim_{b\to\infty} 0 = 0$

59. $L = (a\sqrt{1+c^2}\,/\,c)\,e^{c\theta_1}$ **61.** $\dfrac{1}{s}$; dom $(F) = (0,\infty)$ **63.** $\dfrac{s}{s^2+4}$; dom $(F) = (0,\infty)$

65. $f(x) \geq 0$ for all x and $\displaystyle\int_{-\infty}^\infty f(x)\,dx = \int_0^\infty \frac{6x}{(1+3x^2)^2}dx = 1$ **67.** $\dfrac{1}{k}$

69. Observe that $F(t) = \displaystyle\int_1^t f(x)\,dx$ is continuous and increasing, that $a_n = \displaystyle\int_1^n f(x)\,dx$ is increasing, and that $a_n \leq \displaystyle\int_1^t f(x)\,dx \leq a_{n+1}$ for $t \in [n, n+1]$. The result follows.

CHAPTER 11

SECTION 11.1

1. 12 **3.** 15 **5.** $-\frac{2}{15}$ **7.** $\sum_{k=1}^{11}(2k-1)$ **9.** $\sum_{k=1}^{35}k(k+1)$ **11.** $\sum_{k=1}^{n}M_k\Delta x_k$ **13.** $\sum_{k=3}^{10}\frac{1}{2^k},\ \sum_{i=0}^{7}\frac{1}{2^{i+3}}$

15. $\sum_{k=3}^{10}(-1)^{k-1}\frac{k}{k+1},\ \sum_{i=0}^{7}(-1)^i\frac{i+3}{i+4}$ **17.** let $k=n+3$ **19.** let $k=n-3$ **21.** 140 **23.** 680 **25.** $\frac{1}{2}$ **27.** $\frac{11}{18}$

29. $\frac{10}{3}$ **31.** $-\frac{3}{2}$ **33.** 24 **35.** $\sum_{k=1}^{\infty}\frac{7}{10^k}=\frac{7}{9}$ **37.** $\sum_{k=1}^{\infty}\frac{24}{100^k}=\frac{8}{33}$ **39.** $\frac{62}{100}+\frac{1}{100}\sum_{k=1}^{\infty}\frac{45}{100^k}=\frac{687}{1100}$

41. Let $x=\overbrace{a_1a_2\cdots a_n}\,\overbrace{a_1a_2\cdots a_n}\cdots$. Then

$$x=\sum_{k=1}^{\infty}\frac{a_1a_2\cdots a_n}{(10^n)^k}=a_1a_2\cdots a_n\sum_{k=1}^{\infty}\left(\frac{1}{10^n}\right)^k=a_1a_2\cdots a_n\left[\frac{1}{1-\frac{1}{10^n}}-1\right]=\frac{a_1a_2\cdots a_n}{10^n-1}.$$

43. $\frac{1}{1+x}=\frac{1}{1-(-x)}=\sum_{k=0}^{\infty}(-x)^k=\sum_{k=0}^{\infty}(-1)^kx^k$ **45.** $\sum_{k=0}^{\infty}x^{k+1}$ **47.** $\sum_{k=0}^{\infty}(-1)^kx^{2k+1}$ **49.** $\sum_{k=0}^{\infty}\left(\frac{3}{2}\right)^k$; geometric series with $r=\frac{3}{2}>1$

51. $\lim_{k\to\infty}\left(\frac{k+1}{k}\right)^k=e\neq 0$ **53.** (a) 87.9935; diverges
(b) 8.17837; diverges
(c) 1.64443; converges, $S\cong 1.645$
(d) 1.71828; converges, the sum is $e-1$

55. 18 **57.** $\sum_{k=1}^{\infty}n_k\left(1+\frac{r}{100}\right)^{-k}$ **59.** \$9 **61.** 32

63. $\lim_{n\to\infty}S_n=L=\sum_{k=0}^{\infty}a_k$. Thus $\lim_{n\to\infty}R_n=\lim_{n\to\infty}\left(S_n-\sum_{k=0}^{\infty}a_k\right)=\lim_{n\to\infty}S_n-L=0.$ **65.** $\frac{1+(-1)^n}{2},\quad n=0,1,2,\ldots$

67. $S_n=\sum_{k=1}^{\infty}\ln\left(\frac{k+1}{k}\right)=\sum_{k=1}^{\infty}[\ln(k+1)-\ln k]=\ln(n+1)\to\infty$

69. (a) $S_n=\sum_{k=1}^{n}(d_k-d_{k+1})=d_1-d_{n+1}\to d_1$

(b) (i) $\sum_{k=1}^{\infty}\frac{\sqrt{k+1}-\sqrt{k}}{\sqrt{k(k+1)}}=\sum_{k=1}^{\infty}\left(\frac{1}{\sqrt{k}}-\frac{1}{\sqrt{k+1}}\right)=1$ (ii) $\sum_{k=1}^{\infty}\frac{2k+1}{2k^2(k+1)^2}=\sum_{k=1}^{\infty}\frac{1}{2}\left(\frac{1}{k^2}-\frac{1}{(k+1)^2}\right)=\frac{1}{2}$

71. $N=6$ **73.** $N=9999$ **75.** $N=\left[\!\left[\frac{\ln(\epsilon[1-x])}{\ln|x|}\right]\!\right]$, where [[ ]] denotes the greatest integer function.

SECTION 11.2

1. converges; comparison $\sum 1/k^2$ **3.** converges; comparison $\sum 1/k^2$ **5.** diverges; comparison $\sum 1/(k+1)$

7. diverges; limit comparison $\sum 1/k$ **9.** converges; integral test **11.** diverges; p-series with $p=\frac{2}{3}\le 1$ **13.** diverges; $a_k\not\to 0$

15. diverges; comparison $\sum 1/k$ **17.** diverges; $a_k\not\to 0$ **19.** converges; limit comparison $\sum 1/k^2$ **21.** diverges; integral test

23. converges; limit comparison with $\sum\frac{2^k}{5^k}$ **25.** diverges; limit comparison $\sum 1/k$ **27.** converges; limit comparison $\sum 1/k^{3/2}$

29. converges; integral test **31.** converges; comparison $\sum 3/k^2$ **33.** converges; comparison $\sum 2/k^2$ **35.** (a) converges; $S\cong 3.18459$
(b) converges; $S\cong 1.1752$
(c) converges; $S=2$

37. $p>1$

39. (a) The improper integral $\int_0^{\infty}e^{-\alpha x}dx=\frac{1}{\alpha}$ converges. (b) The improper integral $\int_0^{\infty}x\,e^{-\alpha x}\,dx=\frac{1}{\alpha^2}$ converges.

(c) The improper integral $\int_0^{\infty}x^ne^{-\alpha x}dx=\frac{n!}{\alpha^{n+1}}$ converges.

41. (a) 1.1777 (b) $0.02 < R_4 < 0.0313$ (c) $1.1977 < \sum_{k=1}^{\infty} \frac{1}{k^3} < 1.209$ **43.** (a) $1/101 < R_{100} < 1/100$ (b) 10,001 **45.** (a) 15 (b) 1.082

47. (a) If $a_k/b_k \to 0$, then $a_k/b_k < 1$ for all $k \geq K$ for some K. But then $a_k < b_k$ for all $k \geq K$ and, since $\sum b_k$ converges, $\sum a_k$ converges. [The basic comparison test, 11.2.5]

(b) Similar to (a) except that this time we appeal to part (ii) of Theorem 11.2.5.

(c) $\sum a_k = \sum \frac{1}{k^2}$ converges, $\sum b_k = \sum \frac{1}{k^{3/2}}$ converges, $\frac{1/k^2}{1/k^{3/2}} = \frac{1}{\sqrt{k}} \to 0$

$\sum a_k = \sum \frac{1}{k^2}$ converges, $\sum b_k = \sum \frac{1}{\sqrt{k}}$ diverges, $\frac{1/k^2}{1/\sqrt{k}} = \frac{1}{k^{3/2}} \to 0$

(d) $\sum b_k = \sum \frac{1}{\sqrt{k}}$ diverges, $\sum a_k = \sum \frac{1}{k^2}$ converges, $\frac{1/k^2}{1/\sqrt{k}} = \frac{1}{k^{3/2}} \to 0$

$\sum b_k = \sum \frac{1}{\sqrt{k}}$ diverges, $\sum a_k = \sum \frac{1}{k}$ diverges, $\frac{1/k}{1/\sqrt{k}} = \frac{1}{\sqrt{k}} \to 0$

49. (a) Since $\sum a_k$ converges, $\lim_{k\to\infty} a_k = 0$. Therefore there exists a positive integer N such that $0 < a_k < 1$ for $k \geq N$.

Thus, for $k \geq N, a_k^2 < a_k$ and so $\sum a_k^2$ converges by the comparison test.

(b) $\sum a_k$ may either converge or diverge.

$\sum 1/k^4$ converges and $\sum 1/k^2$ converges; $\sum 1/k^2$ converges and $\sum 1/k$ diverges.

51. $0 < L - \sum_{k=1}^{n} f(k) = L - S_n = \sum_{k=n+1}^{\infty} f(k) < \int_n^{\infty} f(x)\,dx$ **53.** $N = 3$.

55. (a) Set $f(x) = x^{1/4} - \ln x$. Then $f'(x) = \frac{1}{4}x^{-3/4} - \frac{1}{x} = \frac{1}{4x}(x^{1/4} - 4)$. Since $f(e^{12}) = e^3 - 12 > 0$ and $f'(x) > 0$ for $x > e^{12}$, we know that $k^{1/4} > \ln k$ and therefore $\frac{1}{k^{5/4}} > \frac{\ln k}{k^{3/2}}$ for sufficiently large k. Since $\sum \frac{1}{k^{5/4}}$ is a convergent p-series, $\sum \frac{\ln k}{k^{3/2}}$ converges by the basic comparison test.

(b) By L'Hôpital's rule $\lim_{x\to\infty} \left[\left(\frac{\ln x}{x^{3/2}}\right) / \left(\frac{1}{x^{5/4}}\right)\right] = 0$

SECTION 11.3

1. converges; ratio test **3.** converges; root test **5.** diverges; ratio test **7.** diverges; limit comparison $\sum 1/k$ **9.** converges; root test

11. diverges; limit comparison $\sum 1/\sqrt{k}$ **13.** diverges; ratio test **15.** converges; comparison $\sum 1/k^{3/2}$

17. converges; comparison $\sum 1/k^2$ **19.** diverges; integral test **21.** diverges; $a_k \to e^{-100} \neq 0$ **23.** diverges; limit comparison $\sum 1/k$

25. converges; ratio test **27.** converges; comparison $\sum 1/k^{3/2}$ **29.** converges; ratio test **31.** converges; ratio test: $a_{k+1}/a_k \to \frac{4}{27}$

33. converges; ratio test **35.** converges; root test **37.** converges; root test **39.** converges; ratio test

41. (a) converges; $a_{k+1}/a_k \to 0$ **43.** $\frac{10}{81}$ **45.** The series $\sum \frac{k!}{k^k}$ converges. Therefore $\lim_{k\to\infty} \frac{k!}{k^k} = 0$ by Theorem 11.1.5. **47.** $p \geq 2$

(b) diverges; $a_{k+1}/a_k \to 2$

49. Set $b_k = a_k r^k$. If $(a_k)^{1/k} \to \rho$ and $\rho < 1/r$, then

$(b_k)^{1/k} = (a_k r^k)^{1/k} = (a_k)^{1/k} r \to \rho r < 1$

and thus, by the root test, $\sum b_k = \sum a_k r^k$ converges.

SECTION 11.4

1. diverges; $a_k \not\to 0$ **3.** diverges; $a_k \not\to 0$ **5.** (a) does not converge absolutely; integral test (b) converges conditionally; Theorem 11.4.3

7. diverges; limit comparison $\sum 1/k$ **9.** (a) does not converge absolutely; limit comparison $\sum 1/k$ (b) converges conditionally; Theorem 11.4.3

11. diverges; $a_k \not\to 0$ **13.** (a) does not converge absolutely; comparison $\sum 1/\sqrt{k+1}$ (b) converges conditionally; Theorem 11.4.3

15. converges absolutely (terms already positive); $\sum \sin\left(\frac{\pi}{4k^2}\right) \leq \sum \frac{\pi}{4k^2} = \frac{\pi}{4}\sum \frac{1}{k^2}$ $(|\sin x| \leq |x|)$ **17.** converges absolutely; ratio test

19. (a) does not converge absolutely; limit comparison $\sum 1/k$ (b) converges conditionally; Theorem 11.4.3 **21.** diverges; $a_k \not\to 0$

23. diverges; $a_k \not\to 0$ **25.** converges absolutely; ratio test **27.** diverges; $a_k = \frac{1}{k}$ for all k **29.** converges absolutely; comparison $\sum 1/k^2$

31. diverges; $a_k \not\to 0$ **33.** 0.1104 **35.** 0.001 **37.** $\frac{10}{11}$ **39.** $N = 39{,}998$ **41.** $n = 999$ **43.** (a) 4 (b) 6

45. No. For instance, set $a_{2k} = 2/k$ and $a_{2k+1} = 1/k$.

47. (a) Since $\sum |a_k|$ converges, $\sum |a_k|^2 = \sum a_k^2$ converges (Exercise 49, Section 11.2).

(b) $\sum 1/k^2$ is convergent, $\sum \frac{(-1)^k}{k}$ is not absolutely convergent.

49. See the proof of Theorem 11.7.2.

51. (a) $\sum_{k=1}^{2n} \frac{(-1)^{k+1}}{k} = S_{2n} - S_n$

(b) Let $P = \{n, n+1, n+2, \ldots, 2n\}$ be the regular partition of the interval $[n, 2n]$ into n subintervals; form the Riemann sum by evaluating $f(x) = 1/x$ at the right endpoints (= the lower sum).

(c) $\int_n^{2n} (1/x)\,dx = \ln(2n) - \ln(n) = \ln 2$.

SECTION 11.5

1. $-1 + x + \frac{1}{2}x^2 - \frac{1}{24}x^4$ **3.** $-\frac{1}{2}x^2 - \frac{1}{12}x^4$ **5.** $1 - x + x^2 - x^3 + x^4 - x^5$ **7.** $x + \frac{1}{3}x^3 + \frac{2}{15}x^5$

9. $P_0(x) = 1,\quad P_1(x) = 1 - x,\quad P_2(x) = 1 - x + 3x^2,\quad P_3(x) = 1 - x + 3x^2 + 5x^3$ **11.** $\sum_{k=0}^{n} (-1)^k \frac{x^k}{k!}$

13. $\sum_{k=0}^{m} \frac{x^{2k}}{(2k!)}$ where $m = \frac{n}{2}$ and n is even **15.** $\sum_{k=0}^{n} \frac{r^k}{k!}x^k$ **17.** 0.00002 **19.** $n = 9$ **21.** $n = 6$ **23.** $|x| < 1.513$

25. 79/48 $(79/48 \cong 1.646)$ **27.** 5/6 $(5/6 \cong 0.833)$ **29.** 13/24 $(13/24 \cong 0.542)$ **31.** 0.17

33. $\frac{4e^{2c}}{15}x^5,\quad |c| < |x|$ **35.** $\frac{-4\sin 2c}{15}x^5,\quad |c| < |x|$ **37.** $\frac{3\sec^4 c - 2\sec^2 c}{3}x^3,\quad |c| < |x|$

39. $\frac{3c^2 - 1}{3(1 + c^2)^3}x^3,\quad |c| < |x|$ **41.** $\frac{(-1)^{n+1}e^{-c}}{(n+1)!}x^{n+1},\quad |c| < |x|$ **43.** $\frac{1}{(1-c)^{n+2}}x^{n+1},\quad |c| < |x|$

45. (a) 4 (b) 2 (c) 999 **47.** (a) 1.649 (b) 0.368

49. For $0 \le k \le n$, $P^{(k)}(0) = k!a_k$; for $k > n$, $P^{(k)}(0) = 0$. Thus $P(x) = \sum_{k=0}^{\infty} P^{(k)}(0)\frac{x^k}{k!}$.

51. $\left.\frac{d^{2k}(\sinh x)}{dx^{2k}}\right|_{x=0} = \sinh(0) = 0$: $\left.\frac{d^{2k+1}(\sinh x)}{dx^{2k+1}}\right|_{x=0} = \cosh(0) = 1$

Therefore $\sinh x = x + \frac{x^3}{3!} + \frac{x^5}{5!} + \cdots = \sum_{k=0}^{\infty} \frac{1}{(2k+1)!}x^{2k+1}$

53. $\sum_{k=0}^{\infty} \frac{a^k}{k!}x^k,\quad (-\infty, \infty)$ **55.** $\sum_{k=0}^{\infty} \frac{(-1)^k a^{2k}}{(2k)!}x^{2k},\quad (-\infty, \infty)$ **57.** $\ln a + \sum_{k=1}^{\infty} \frac{(-1)^{k-1}}{ka^k}x^k,\quad (-a, a]$

59. $\ln 2 = \ln\left(\frac{1 + \frac{1}{3}}{1 - \frac{1}{3}}\right) \cong 2\left[\frac{1}{3} + \frac{1}{3}\left(\frac{1}{3}\right)^3 + \frac{1}{5}\left(\frac{1}{3}\right)^5\right] = \frac{842}{1215}\quad \left(\frac{842}{1215} \cong 0.693\right)$ **61.** routine; use $u = (x - t)^k$ and $dv = f^{(k+1)}(t)\,dt$

63. (b) $f(x) = \frac{x^{-n}}{e^{1/x^2}}$ and $\lim_{x\to 0} f(x)$ has the form ∞/∞. Successive applications of L'Hôpital's rule will finally produce a quotient of the form $\frac{cx^k}{e^{1/x^2}}$, where k is a nonnegative integer and c is a constant. It follows that $\lim_{x\to 0} f(x) = 0$.

(c) $f'(0) = \lim_{x\to 0} \frac{e^{-1/x^2} - 0}{x} = 0$ by part (b). Assume that $f^{(k)}(0) = 0$. Then

$$f^{(k+1)}(0) = \lim_{x\to 0} \frac{f^{(k)}(x) - 0}{x} = \lim_{x\to 0} \frac{f^{(k)}(x)}{x}.$$

Now, $\dfrac{f^{(k)}(x)}{x}$ is a sum of terms of the form $\dfrac{c\,e^{-1/x^2}}{x^n}$, where n is a positive integer and c is a constant. By part (b), $f^{(k+1)}(0)=0$. Therefore $f^{(n)}(0)=0$ for all n.

(d) 0 (e) $x=0$

65. $P_2(x)=x-\frac{1}{2}x^2, P_3(x)=x-\frac{1}{2}x^2+\frac{1}{3}x^3, P_4(x)=x-\frac{1}{2}x^2+\frac{1}{3}x^3-\frac{1}{4}x^4, P_5(x)=x-\frac{1}{2}x^2+\frac{1}{3}x^3-\frac{1}{4}x^4+\frac{1}{5}x^5$ **67.** $\sum_{k=0}^{\infty}\frac{(\ln 3)^k}{k!}x^k;\ R=\infty$

SECTION 11.6

1. $P_3(x)=2+\frac{1}{4}(x-4)-\frac{1}{64}(x-4)^2+\frac{1}{512}(x-4)^3$

$R_3(x)=\dfrac{-5}{128c^{7/2}}(x-4)^4,\quad |c-4|<|x-4|$

3. $P_4(x)=\dfrac{\sqrt{2}}{2}+\dfrac{\sqrt{2}}{2}\left(x-\dfrac{\pi}{4}\right)-\dfrac{\sqrt{2}}{4}\left(x-\dfrac{\pi}{4}\right)^2-\dfrac{\sqrt{2}}{12}\left(x-\dfrac{\pi}{4}\right)^3+\dfrac{\sqrt{2}}{48}\left(x-\dfrac{\pi}{4}\right)^4$

$R_4(x)=\dfrac{\cos c}{120}\left(x-\dfrac{\pi}{4}\right)^5, \left|c-\dfrac{\pi}{4}\right|<\left|x-\dfrac{\pi}{4}\right|$

5. $P_3(x)=\dfrac{\pi}{4}+\dfrac{1}{2}(x-1)-\dfrac{1}{4}(x-1)^2+\dfrac{1}{12}(x-1)^3\quad R_3(x)=\dfrac{c(1-c^2)}{(1+c^2)^4}(x-1)^4,\quad |c-1|<|x-1|$

7. $6+9(x-1)+7(x-1)^2+3(x-1)^3,\quad (-\infty,\infty)$ **9.** $-3+5(x+1)-19(x+1)^2+20(x+1)^3-10(x+1)^4+2(x+1)^5,\quad (-\infty,\infty)$

11. $\sum_{k=0}^{\infty}(-1)^k\left(\dfrac{1}{2}\right)^{k+1}(x-1)^k,\quad (-1,3)$ **13.** $\dfrac{1}{5}\sum_{k=0}^{\infty}\left(\dfrac{2}{5}\right)^k(x+2)^k,\ \left(-\dfrac{9}{2},\dfrac{1}{2}\right)$ **15.** $\sum_{k=0}^{\infty}\dfrac{(-1)^{k+1}}{(2k+1)!}(x-\pi)^{2k+1},\quad (-\infty,\infty)$

17. $\sum_{k=0}^{\infty}\dfrac{(-1)^{k+1}}{(2k)!}(x-\pi)^{2k},\quad (-\infty,\infty)$ **19.** $\sum_{k=0}^{\infty}\dfrac{(-1)^k}{(2k)!}\left(\dfrac{\pi}{2}\right)^{2k}(x-1)^{2k},\quad (-\infty,\infty)$ **21.** $\ln 3+\sum_{k=1}^{\infty}\dfrac{(-1)^{k+1}}{k}\left(\dfrac{2}{3}\right)^k(x-1)^k,\left(-\dfrac{1}{2},\dfrac{5}{2}\right]$

23. $2\ln 2+(1+\ln 2)(x-2)+\sum_{k=2}^{\infty}\dfrac{(-1)^k}{k(k-1)2^{k-1}}(x-2)^k$ **25.** $\sum_{k=0}^{\infty}\dfrac{(-1)^k}{(2k+1)!}x^{2k+2}$ **27.** $\sum_{k=0}^{\infty}(k+2)(k+1)\dfrac{2^{k-1}}{5^{k+3}}(x+2)^k$

29. $1+\sum_{k=1}^{\infty}\dfrac{(-1)^k 2^{2k-1}}{(2k)!}(x-\pi)^{2k}$ **31.** $\sum_{k=0}^{\infty}\dfrac{n!}{(n-k)!k!}(x-1)^k$

33. (a) $\dfrac{e^x}{e^a}=e^{x-a}=\sum_{k=0}^{\infty}\dfrac{(x-a)^k}{k!},\quad e^x=e^a\sum_{k=0}^{\infty}\dfrac{(x-a)^k}{k!}$ (b) $e^{a+(x-a)}=e^x=e^a\sum_{k=0}^{\infty}\dfrac{(x-a)^k}{k!},\quad e^{x_1+x_2}=e^{x_1}\sum_{k=0}^{\infty}\dfrac{x_2^k}{k!}$

(c) $e^{-a}\sum_{k=0}^{\infty}(-1)^k\dfrac{(x-a)^k}{k!}$

35. (a) $P_3(x)=\dfrac{1}{2}+\dfrac{\sqrt{3}}{2}\left(x-\dfrac{\pi}{6}\right)-\dfrac{1}{4}\left(x-\dfrac{\pi}{6}\right)^2-\dfrac{\sqrt{3}}{12}\left(x-\dfrac{\pi}{6}\right)^3$ (b) 0.5736 **37.** $P_2(x)=6+\frac{1}{12}(x-36)-\frac{1}{1728}(x-36)^2;\quad \sqrt{38}\cong 6.164$

39. $P_6(x)=\dfrac{\pi}{4}+\dfrac{1}{2}(x-1)-\dfrac{1}{4}(x-1)^2+\dfrac{1}{12}(x-1)^3-\dfrac{1}{40}(x-1)^5+\dfrac{1}{48}(x-1)^6$

SECTION 11.7

1. (a) absolutely convergent (b) absolutely convergent (c) ? (d) ? **3.** $(-1,1)$ **5.** $(-\infty,\infty)$ **7.** $\{0\}$ **9.** $[-2,2)$

11. $\{0\}$ **13.** $[-\frac{1}{2},\frac{1}{2})$ **15.** $(-1,1)$ **17.** $(-10,10)$ **19.** $(-\infty,\infty)$ **21.** $(-\infty,\infty)$ **23.** $(-3/2,3/2)$

25. converges only at $x=1$ **27.** $(-4,0)$ **29.** $(-\infty,\infty)$ **31.** $(-1,1)$ **33.** $(0,4)$

35. $(-\frac{5}{2},\frac{1}{2})$ **37.** $(-2,2)$ **39.** $\left[-\dfrac{1}{\sqrt{3}},\dfrac{1}{\sqrt{3}}\right]$ **41.** (a) absolutely convergent (b) absolutely convergent (c) ?

43. (a) $\sum|a_k r^k|=\sum|a_k(-r)^k|$

(b) If $\sum|a_k(-r)^k|$ converges, then $\sum a_k(-r)^k$ converges.

45. (a) $\sum a_k x^k=\sum(a_0+a_1x+a_2x^2)x^{3k}$, a geometric series with $a=a_0+a_1x+a_2x^2$ and $r=x^3$; $|x^3|<1$ implies $|x|<1$

(b) $\sum a_k x^k=\dfrac{a_0+a_1x+a_2x^2}{1-x^3}$

47. Examine the convergence of $\sum|a_k x^k|$; for (a) use the root test and for (b) use the ratio test.

49. (a) $R=\infty$ (b) $R=\frac{1}{2}$ (c) $R=2$ **51.** (c) $(-1,1)$

SECTION 11.8

1. $1 + 2x + 3x^2 + \cdots + nx^{n-1} + \cdots$ **3.** $1 + kx + \frac{(k+1)k}{2!}x^2 + \cdots + \frac{(n+k-1)!}{n!(k-1)!}x^n + \cdots$

5. $\ln(1-x^2) = -x^2 - \frac{1}{2}x^4 - \frac{1}{3}x^6 - \cdots - \frac{1}{n+1}x^{2n+2} - \cdots$ **7.** $1 + x^2 + \frac{2}{3}x^4 + \frac{17}{45}x^6 + \cdots$ **9.** $-72.$ **11.** $\sum_{k=0}^{\infty} \frac{(-1)^k}{(2k+1)!}x^{4k+2}$

13. $\sum_{k=0}^{\infty} \frac{3^k}{k!}x^{3k}$ **15.** $2\sum_{k=0}^{\infty} x^{2k+1}$ **17.** $\sum_{k=0}^{\infty} \frac{(k!+1)}{k!}x^k$ **19.** $\sum_{k=1}^{\infty} \frac{(-1)^{k+1}}{k}x^{3k+1}$ **21.** $\sum_{k=0}^{\infty} \frac{(-1)^k}{k!}x^{3k+3}$ **23.** $\frac{1}{2}$

25. $-\frac{1}{2}$ **27.** $\sum_{k=1}^{\infty} \frac{(-1)^{k-1}}{k^2}x^k, -1 \le x \le 1$ **29.** $\sum_{k=0}^{\infty} \frac{(-1)^k}{(2k+1)^2}x^{2k+1}$ **31.** $0.804 \le I \le 0.808$ **33.** $0.600 \le I \le 0.603$

35. $0.294 \le I \le 0.304$ **37.** 0.9461 **39.** 0.4485 **41.** e^{x^3} **43.** $3x^2e^{x^3}$

45. (a) $\sum_{k=0}^{\infty} \frac{1}{k!}x^{k+1}$ (b) $\int_0^1 xe^x\,dx = 1 = \int_0^1 \left(\sum_{k=0}^{\infty} \frac{1}{k!}x^{k+1}\right) dx = \sum_{k=0}^{\infty} \frac{1}{k!(k+2)} = \frac{1}{2} + \sum_{k=1}^{\infty} \frac{1}{k!(k+2)}$

47. Let $f(x)$ be the sum of these series; a_k and b_k are both $f^{(k)}(0)/k!$.

49. (a) If f is even, then $f^{(2k-1)}$ is odd for $k = 1, 2, \ldots$ This implies that $f^{(2k-1)}(0) = 0$, and so $a_{2k-1} = \frac{f^{(2k-1)}(0)}{(2k-1)!} = 0$ for all k.

(b) If f is odd, then $f^{(2k)}$ is odd for $k = 1, 2, \ldots$, which implies $a_{2k} = 0$ for all k.

51. $f(x) = x - \frac{2}{3!}x^3 + \frac{4}{5!}x^5 - \frac{8}{7!}x^7 + \cdots = \sum_{k=0}^{\infty} \frac{(-1)^k 2^k}{(2k+1)!}x^{2k+1}$; $\frac{1}{\sqrt{2}}\sin(x\sqrt{2})$

53. $0.0352 \le I \le 0.0359; I = \frac{3}{16} - \frac{3}{8}\ln 1.5 \cong 0.0354505$ **55.** $0.2640 \le I \le 0.2643; I = 1 - 2/e \cong 0.2642411$

SECTION 11.9

1. $1 + \frac{1}{2}x - \frac{1}{8}x^2 + \frac{1}{16}x^3 - \frac{5}{128}x^4$ **3.** $1 + \frac{1}{2}x^2 - \frac{1}{8}x^4$ **5.** $1 - \frac{1}{2}x + \frac{3}{8}x^2 - \frac{5}{16}x^3 + \frac{35}{128}x^4$ **7.** $1 - \frac{1}{4}x - \frac{3}{32}x^2 - \frac{7}{128}x^3 - \frac{77}{2048}x^4$

9. $8 + 3x + \frac{3}{16}x^2 - \frac{1}{128}x^3 + \frac{3}{4096}x^4$ **11.** (a) $\sum_{k=0}^{\infty}(-1)^k\binom{-1/2}{k}x^{2k}$ (b) $\sum_{k=0}^{\infty}(-1)^k\binom{-1/2}{k}\frac{1}{2k+1}x^{2k+1}$, $R = 1$ **13.** 9.8995

15. 2.0799 **17.** 0.4925 **19.** 0.3349 **21.** 0.4815

CHAPTER 12

SECTION 12.1

1. **3.**

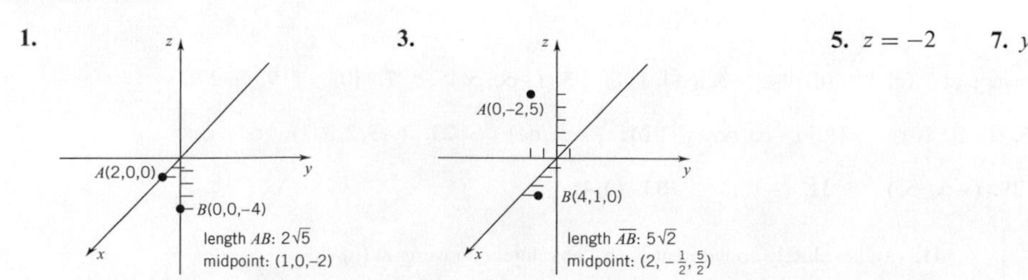

5. $z = -2$ **7.** $y = 1$ **9.** $x = 3$

11. $x^2 + (y-2)^2 + (z+1)^2 = 9$ **13.** $(x-2)^2 + (y-4)^2 + (z+4)^2 = 36$ **15.** $(x-3)^2 + (y-2)^2 + (z-2)^2 = 13$

17. $(x-2)^2 + (y-3)^2 + (z+4)^2 = 25$ **19.** center $(-2, 4, 1)$, radius 4 **21.** center: $(3, -5, 1)$; radius: 6 **23.** $(2, 3, -5)$ **25.** $(-2, 3, 5)$

27. $(-2, 3, -5)$ **29.** $(-2, -3, -5)$ **31.** $(2, -5, 5)$ **33.** $(-2, 1, -3)$

35. $(x-3)^2 + (y-3)^2 + (z-3)^2 = 9$, $(x-7)^2 + (y-7)^2 + (z-7)^2 = 49$

37. not a sphere; the equation is equivalent to $(x-2)^2 + (y+2)^2 + (z+3)^2 = -3$

39. $d(P,R)=\sqrt{14}, \quad d(Q,R)=\sqrt{45}, \quad d(P,Q)=\sqrt{59}, \quad [d(P,R)]^2+[d(Q,R)]^2=[d(P,Q)]^2$

41. The sphere of radius 2 centered at the origin, together with its interior

43. A rectangular box with sides on the coordinate planes and dimensions $1 \times 2 \times 3$, together with its interior

45. A circular cylinder with base the circle $x^2+y^2=4$ and height 4, together with its interior

47. (a) $x=a_1+t(b_1-a_1), \quad y=a_2+t(b_2-a_2), \quad z=a_3+t(b_3-a_3)$ (b) $t=\frac{1}{2}$ **49.** $(x-1)^2+(y-1)^2+(z+\frac{5}{2})^2=\frac{53}{4}$

SECTION 12.3

1. $(3,4,-2), \sqrt{29}$ **3.** $(0,-2,-1), \sqrt{5}$ **5.** $(-1,-4,7)$ **7.** $(5,2,-8)$ **9.** $3\mathbf{i}-4\mathbf{j}+6\mathbf{k}$ **11.** $-3\mathbf{i}-\mathbf{j}+8\mathbf{k}$ **13.** 5 **15.** 3

17. $\sqrt{6}$ **19.** (a) **a, c, d** (b) **a, c** (c) **a** and **c** both have direction opposite to **d** **21.** $(\frac{3}{5},-\frac{4}{5},0)$ **23.** $\frac{1}{3}\mathbf{i}-\frac{2}{3}\mathbf{j}+\frac{2}{3}\mathbf{k}$

25. $\dfrac{1}{\sqrt{14}}\mathbf{i}-\dfrac{3}{\sqrt{14}}\mathbf{j}-\dfrac{2}{\sqrt{14}}\mathbf{k}$ **27.** (i) $\mathbf{a}+\mathbf{b}$ (ii) $-(\mathbf{a}+\mathbf{b})$ (iii) $\mathbf{a}-\mathbf{b}$ (iv) $\mathbf{b}-\mathbf{a}$

29. (a) $\mathbf{i}-3\mathbf{j}+10\mathbf{k}$ (b) $A=-2, B=\frac{3}{2}, C=-\frac{7}{2}$ **31.** $\alpha=\pm 3$ 33. $\alpha=\pm\frac{1}{3}\sqrt{6}$ **35.** $\pm\frac{2}{13}\sqrt{13}(3\mathbf{j}+2\mathbf{k})$

37. (a) the parallelogram is a rectangle
(b) simplify $\sqrt{(a_1-b_1)^2+(a_2-b_2)^2+(a_3-b_3)^2}=\sqrt{(a_1+b_1)^2+(a_2+b_2)^2+(a_3+b_3)^2}$

39. (a)

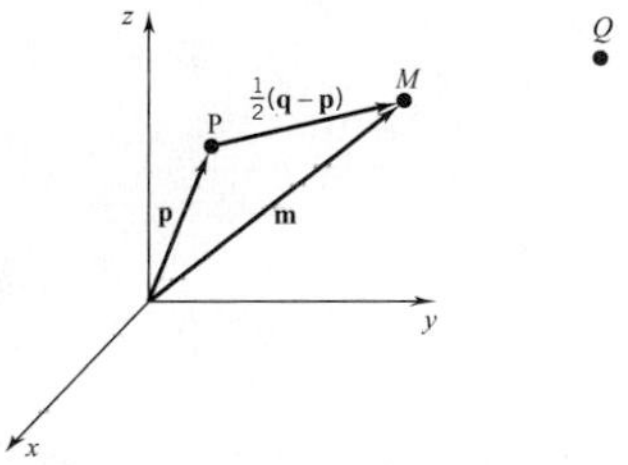

$\mathbf{m}=\mathbf{p}+\frac{1}{2}(\mathbf{q}-\mathbf{p})$

(b) Let $P(x_1,y_1,z_1), Q=(x_2,y_2,z_2)$, and $M=(x_m,y_m,z_m)$. Then

$$(x_m,y_m,z_m)=(x_1,y_1,z_1)+\tfrac{1}{2}(x_2-x_1,y_2-y_1,z_2-z_1)$$
$$=\left(\frac{x_1+x_2}{2},\frac{y_1+y_2}{2},\frac{z_1+z_2}{2}\right).$$

41. (a) $\|\mathbf{r}-\mathbf{a}\|=3$ where $\mathbf{a}=a_1\mathbf{i}+a_2\mathbf{j}+a_3\mathbf{k}$ (b) $\|\mathbf{r}\|\le 2$ (c) $\|\mathbf{r}-\mathbf{a}\|\le 1$ where $\mathbf{a}=a_1\mathbf{i}+a_2\mathbf{j}+a_3\mathbf{k}$

SECTION 12.4

1. -1 **3.** 0 **5.** -1 **7.** $\mathbf{a}\cdot\mathbf{b}$ **9.** $\mathbf{a}\cdot(\mathbf{b}+\mathbf{c})$

11. (a) $\mathbf{a}\cdot\mathbf{b}=5, \quad \mathbf{a}\cdot\mathbf{c}=8, \quad \mathbf{b}\cdot\mathbf{c}=18$ (b) $\cos\sphericalangle(\mathbf{a},\mathbf{b})=\frac{1}{14}\sqrt{70}, \quad \cos\sphericalangle(\mathbf{a},\mathbf{c})=\frac{8}{25}\sqrt{5}, \quad \cos\sphericalangle(\mathbf{b},\mathbf{c})=\frac{9}{35}\sqrt{14}$

(c) $\text{comp}_\mathbf{b}\mathbf{a}=\frac{5}{14}\sqrt{14}$, $\text{comp}_\mathbf{c}\mathbf{a}=\frac{8}{5}$ (d) $\mathbf{proj_b a}=\frac{5}{14}(3\mathbf{i}-\mathbf{j}+2\mathbf{k})$, $\mathbf{proj_c a}=\frac{8}{25}(4\mathbf{i}+3\mathbf{k})$ **13.** $\frac{1}{2}\mathbf{i}+\frac{1}{2}\sqrt{2}\mathbf{j}-\frac{1}{2}\mathbf{k}$ **15.** $\frac{1}{3}\pi$

17. $\frac{1}{3}\pi, \frac{2}{3}\pi, \frac{1}{4}\pi$ **19.** 2.2 radians, or 126.3° **21.** 2.5 radians, or 145.3°

23. angles: 38.51°, 95.52°, 45.97°; perimeter: $P\cong 15.924$ **25.** $\cos\alpha=\frac{1}{3}, \cos\beta=\frac{2}{3}, \cos\gamma=\frac{2}{3}$; $\alpha\cong 70.5°, \beta\cong 48.2°, \gamma\cong 48.2°$

27. $\cos\alpha=\frac{3}{13}, \cos\beta=\frac{12}{13}, \cos\gamma=\frac{4}{13}$; $\alpha\cong 76.7°, \beta\cong 22.6°, \gamma\cong 72.1°$ **29.** $x=\pm 4$ **31.** $x=0, x=4$

33. (a) The direction angles of a vector always satisfy $\cos^2\alpha+\cos^2\beta+\cos^2\gamma=1$, and, as you can check, $\cos^2\frac{1}{4}\pi+\cos^2\frac{1}{6}\pi+\cos^2\frac{2}{3}\pi\neq 1$.
(b) The relation $\cos^2\alpha+\cos^2\frac{1}{4}\pi+\cos^2\frac{1}{4}\pi=1$ gives

$$\cos^2\alpha+\tfrac{1}{2}+\tfrac{1}{2}=1, \quad \cos\alpha=0, \quad a_1=\|\mathbf{a}\|\cos\alpha=0.$$

35. $\pi-\alpha, \pi-\beta, \pi-\gamma$ **37.** $\mathbf{u}=\pm\frac{1}{165}\sqrt{165}\,(8\mathbf{i}+\mathbf{j}-10\mathbf{k})$ **39.** $\theta=\cos^{-1}(\frac{1}{3}\sqrt{3})\cong 0.96$ radians

41. (a) $\mathbf{proj_b}\,\alpha\mathbf{a}=(\alpha\mathbf{a}\cdot\mathbf{u_b})\mathbf{u_b}=\alpha(\mathbf{a}\cdot\mathbf{u_b})\mathbf{u_b}=\alpha\mathbf{proj_b\,a}$
(b) $\mathbf{proj_b}(\mathbf{a}+\mathbf{c})=[(\mathbf{a}+\mathbf{c})\cdot\mathbf{u_b}]\mathbf{u_b}$
$=(\mathbf{a}\cdot\mathbf{u_b}+\mathbf{c}\cdot\mathbf{u_b})\mathbf{u_b}$
$=(\mathbf{a}\cdot\mathbf{u_b})\mathbf{u_b}+(\mathbf{c}\cdot\mathbf{u_b})\mathbf{u_b}=\mathbf{proj_b\,a}+\mathbf{proj_b\,c}$

43. (a) for $\mathbf{a} \neq 0$ the following statements are equivalent:

$$\mathbf{a} \cdot \mathbf{b} = \mathbf{a} \cdot \mathbf{c}, \quad \mathbf{b} \cdot \mathbf{a} = \mathbf{c} \cdot \mathbf{a},$$

$$\mathbf{b} \cdot \frac{\mathbf{a}}{\|\mathbf{a}\|} = \mathbf{c} \cdot \frac{\mathbf{a}}{\|\mathbf{a}\|}, \quad \mathbf{b} \cdot \mathbf{u_a} = \mathbf{c} \cdot \mathbf{u_a},$$

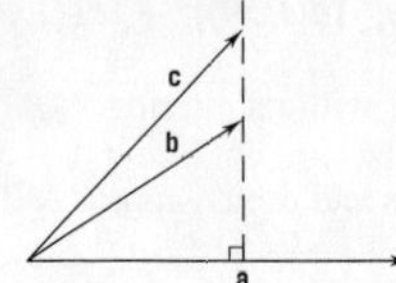

$$(\mathbf{b} \cdot \mathbf{u_a})\mathbf{u_a} = (\mathbf{c} \cdot \mathbf{u_a})\mathbf{u_a}, \quad \mathbf{proj_a b} = \mathbf{proj_a c}. \quad \mathbf{a} \cdot \mathbf{b} = \mathbf{a} \cdot \mathbf{c} \text{ but } \mathbf{b} \neq \mathbf{c}.$$

(b) $\mathbf{b} = (\mathbf{b} \cdot \mathbf{i})\mathbf{i} + (\mathbf{b} \cdot \mathbf{j})\mathbf{j} + (\mathbf{b} \cdot \mathbf{k})\mathbf{k} = (\mathbf{c} \cdot \mathbf{i})\mathbf{i} + (\mathbf{c} \cdot \mathbf{j})\mathbf{j} + (\mathbf{c} \cdot \mathbf{k})\mathbf{k} = \mathbf{c}.$

45. (a) Express the norms as dot products.

(b) The following statements are equivalent:

$$\mathbf{a} \perp \mathbf{b}, \quad \mathbf{a} \cdot \mathbf{b} = 0, \quad \|\mathbf{a}+\mathbf{b}\|^2 - \|\mathbf{a}-\mathbf{b}\|^2 = 0, \quad \|\mathbf{a}+\mathbf{b}\| = \|\mathbf{a}-\mathbf{b}\|.$$

(c) By (b), the relation $\|\mathbf{a}+\mathbf{b}\| = \|\mathbf{a}-\mathbf{b}\|$ gives $\mathbf{a} \perp \mathbf{b}$. The relation $\mathbf{a}+\mathbf{b} \perp \mathbf{a}-\mathbf{b}$ gives

$$0 = (\mathbf{a}+\mathbf{b}) \cdot (\mathbf{a}-\mathbf{b}) = \|\mathbf{a}\|^2 - \|\mathbf{b}\|^2 \quad \text{and thus} \quad \|\mathbf{a}\| = \|\mathbf{b}\|.$$

The parallelogram is a square since it has two adjacent sides of equal length that meet at right angles.

47. $\|\mathbf{a}+\mathbf{b}\|^2 = (\mathbf{a}+\mathbf{b}) \cdot (\mathbf{a}+\mathbf{b}) = \mathbf{a} \cdot \mathbf{a} + 2\mathbf{a} \cdot \mathbf{b} + \mathbf{b} \cdot \mathbf{b}$

$\|\mathbf{a}-\mathbf{b}\|^2 = (\mathbf{a}-\mathbf{b}) \cdot (\mathbf{a}-\mathbf{b}) = \mathbf{a} \cdot \mathbf{a} - 2\mathbf{a} \cdot \mathbf{b} + \mathbf{b} \cdot \mathbf{b}$

and the result follows.

49. $\dfrac{\mathbf{a} \cdot \mathbf{c}}{\|\mathbf{a}\|\,\|\mathbf{c}\|} = \|\mathbf{a}\|\,\|\mathbf{b}\| + \mathbf{a} \cdot \mathbf{b} = \dfrac{\mathbf{b} \cdot \mathbf{c}}{\|\mathbf{b}\|\,\|\mathbf{c}\|}$

51. If $\mathbf{a} \perp \mathbf{b}$ and $\mathbf{a} \perp \mathbf{c}$, then $\mathbf{a} \cdot \mathbf{b} = 0$ and $\mathbf{a} \cdot \mathbf{c} = 0$, so that

$$\mathbf{a}, \cdot\ (\alpha\mathbf{b} + \beta\mathbf{c}) = \alpha(\mathbf{a} \cdot \mathbf{b}) + \beta(\mathbf{a} \cdot \mathbf{c}) = 0.$$

Thus $\mathbf{a} \perp (\alpha\mathbf{b} + \beta\mathbf{c})$.

53. Existence of decomposition: $\mathbf{a} = (\mathbf{a} \cdot \mathbf{u_b})\mathbf{u_b} + [\mathbf{a} - (\mathbf{a} \cdot \mathbf{u_b})\mathbf{u_b}]$. Uniqueness of decomposition: suppose that $\mathbf{a} = \mathbf{a}_{\|} + \mathbf{a}_{\perp} = \mathbf{A}_{\|} + \mathbf{A}_{\perp}$. Then the vector $\mathbf{a}_{\|} - \mathbf{A}_{\|} = \mathbf{A}_{\perp} - \mathbf{a}_{\perp}$ is both parallel to $\mathbf{b}$ and perpendicular to $\mathbf{b}$. (Exercises 37 and 38.) Therefore it is zero. Consequently $\mathbf{A}_{\|} = \mathbf{a}_{\|}$ and $\mathbf{A}_{\perp} = \mathbf{a}_{\perp}$.

55. Place center of sphere at the origin.

$$\overrightarrow{P_1Q} \cdot \overrightarrow{P_2Q} = (-\mathbf{a}+\mathbf{b}) \cdot (\mathbf{a}+\mathbf{b}) = -\|\mathbf{a}\|^2 + \|\mathbf{b}\|^2 = 0.$$

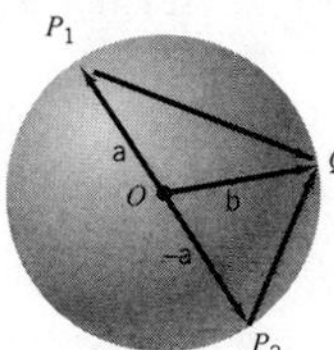

SECTION 12.5

1. $-2\mathbf{k}$ **3.** $\mathbf{i}+\mathbf{j}+\mathbf{k}$ **5.** $-3\mathbf{i}-\mathbf{j}-2\mathbf{k}$ **7.** -1 **9.** 0 **11.** 1 **13.** $3\mathbf{i}-2\mathbf{j}-3\mathbf{k}$ **15.** $\mathbf{i}+\mathbf{j}-2\mathbf{k}$ **17.** -3

19. $5\mathbf{i}-4\mathbf{j}-\mathbf{k}$ **21.** $\left(\frac{1}{\sqrt{6}}, \frac{-1}{\sqrt{6}}, \frac{-2}{\sqrt{6}}\right), \left(\frac{-1}{\sqrt{6}}, \frac{1}{\sqrt{6}}, \frac{2}{\sqrt{6}}\right)$ **23.** $\mathbf{N} = 3\mathbf{j}$; area $= \frac{3}{2}$ **25.** $\mathbf{N} = 8\mathbf{i}+4\mathbf{j}+4\mathbf{k}$; area $= 2\sqrt{6}$ **27.** 1

29. 2 **31.** $-2(\mathbf{a} \times \mathbf{b})$ **33.** $\mathbf{a} = \mathbf{0}$ **35.** $\begin{vmatrix} \alpha & \beta \\ \gamma & \delta \end{vmatrix}(\mathbf{a} \times \mathbf{b})$ **37.** $\mathbf{a} \cdot (\mathbf{b} \times \mathbf{c}) = (\mathbf{a} \times \mathbf{b}) \cdot \mathbf{c} = (\mathbf{c} \times \mathbf{a}) \cdot \mathbf{b} = (\mathbf{b} \times \mathbf{c}) \cdot \mathbf{a} = (\mathbf{a} \times -\mathbf{c}) \cdot \mathbf{b}$,
$\mathbf{a} \cdot (\mathbf{c} \times \mathbf{b}) = \mathbf{c} \cdot (\mathbf{b} \times \mathbf{a}) = (-\mathbf{a} \times \mathbf{b}) \cdot \mathbf{c}$

39. $\mathbf{a} \times \mathbf{b}$ is perpendicular to the plane determined by $\mathbf{a}$ and $\mathbf{b}$;
$\mathbf{c}$ is in this plane iff $\mathbf{a} \times \mathbf{b} \cdot \mathbf{c} = 0$.

41. $\mathbf{a} \cdot \mathbf{b} = \mathbf{a} \cdot \mathbf{c}$ implies $\mathbf{a} \cdot (\mathbf{b}-\mathbf{c}) = 0$; $\mathbf{a}$ is perpendicular to $\mathbf{b}-\mathbf{c}$.
$\mathbf{a} \times \mathbf{b} = \mathbf{a} \times \mathbf{c}$ implies $\mathbf{a} \times (\mathbf{b}-\mathbf{c}) = 0$; $\mathbf{a}$ is parallel to $\mathbf{b}-\mathbf{c}$.
Since $\mathbf{a} \neq \mathbf{0}$, it follows that $\mathbf{b}-\mathbf{c} = \mathbf{0}$, or $\mathbf{b} = \mathbf{c}$.

43. $\mathbf{c} \times \mathbf{a} = \|\mathbf{a}\|^2\mathbf{b}$ **45.** either $\mathbf{a} = \mathbf{0}$ or $\mathbf{b} = \mathbf{0}$

47. The result follows from Exercise 46.

SECTION 12.6

1. P and Q **3.** $\mathbf{r}(t) = (3\mathbf{i} + \mathbf{j}) + t\mathbf{k}$ **5.** $\mathbf{r}(t) = t(x_1\mathbf{i} + y_1\mathbf{j} + z_1\mathbf{k})$ **7.** $x(t) = 1 + t, \quad y(t) = -t, \quad z(t) = 3 + t$

9. $x(t) = 2, \quad y(t) = t, \quad z(t) = 3$ **11.** $\mathbf{r}(t) = (-\mathbf{i} + 2\mathbf{j} - 3\mathbf{k}) + t(2\mathbf{i} + \mathbf{j} + 4\mathbf{k})$ **13.** intersect at $(1, 3, 1)$

15. skew **17.** parallel **19.** skew **21.** $P(1, 2, 0), \quad \frac{1}{4}\pi$ rad **23.** $(x_0 - [d_1/d_3]z_0, y_0 - [d_2/d_3]z_0, 0)$

25. The lines are parallel. **27.** $\mathbf{r}(t) = (2\mathbf{i} + 7\mathbf{j} - \mathbf{k}) + t(2\mathbf{i} - 5\mathbf{j} + 4\mathbf{k}), \quad 0 \le t \le 1$ **29.** $\mathbf{u} = -\frac{2}{3}\mathbf{i} + \frac{1}{3}\mathbf{j} + \frac{2}{3}\mathbf{k}, \quad 9 \le t \le 15$

31. triples of the form $X(u) = 3 + au, \ Y(u) = -1 + bu, \ Z(u) = 8 + cu$ with $2a - 4b + 6c = 0$ **33.** 1

35. $\sqrt{69/14} \cong 2.22$ **37.** $\sqrt{3} \cong 1.73$ **39.** (a) 1 (b) $\sqrt{3}$ **41.** $\mathbf{r}(t) = \frac{1}{11}(7\mathbf{i} + 4\mathbf{j} - \mathbf{k}) \pm t[\frac{1}{11}\sqrt{11}(\mathbf{i} - \mathbf{j} + 3\mathbf{k})]$

43. $0 < t < s$ **45.** $d(l_1, l_2) = \dfrac{10}{\sqrt{285}}$

SECTION 12.7

1. Q **3.** $x - 4y + 3z - 2 = 0$ **5.** $3x - 2y + 5z - 9 = 0$ **7.** $y - z - 2 = 0$ **9.** $x_0(x - x_0) + y_0(y - y_0) + z_0(z - z_0) = 0$

11. $\dfrac{1}{\sqrt{30}}(2\mathbf{i} - \mathbf{j} + 5\mathbf{k}), \quad -\dfrac{1}{\sqrt{30}}(2\mathbf{i} - \mathbf{j} + 5\mathbf{k})$ **13.** $\dfrac{1}{15}x + \dfrac{1}{12}y - \dfrac{1}{10}z = 1$ **15.** $\frac{1}{2}\pi$ **17.** $\cos\theta = \frac{2}{21}\sqrt{42} \cong 0.617, \quad \theta \cong 0.91$ rad

19. coplanar **21.** not coplanar **23.** $\dfrac{2}{\sqrt{21}}$ **25.** $\frac{22}{5}$ **27.** $x + z = 2$ **29.** $3x - 4z - 5 = 0$

31. $\dfrac{x - x_0}{A} = \dfrac{y - y_0}{B} = \dfrac{z - z_0}{C}$ **33.** $(x - x_0)/d_1 = (y - y_0)/d_2, \quad (y - y_0)/d_2 = (z - z_0)/d_3$ **35.** $x(t) = t, \quad y(t) = t, \quad z(t) = -t$

37. $P(-\frac{19}{14}, \frac{15}{7}, \frac{17}{14})$ **39.** $10x - 7y + z = 0$ **41.** circle centered at P with radius $\|\mathbf{N}\|$

43. If $\alpha > 0$, then P_1 lies on the same side of the plane as the tip of $\mathbf{N}$;
if $\alpha < 0$, then P_1 and the tip of $\mathbf{N}$ lie on opposite sides of the plane

45. $\mathbf{a} \cdot \mathbf{b} \times \mathbf{c} = 0$

47. (a) $(4, 0, 0), (0, 5, 0), (0, 0, 2)$
(b) $5x + 4y = 20, x + 2z = 4, 2y + 5z = 10$
(c) $\pm\frac{1}{\sqrt{141}}(5\mathbf{i} + 4\mathbf{j} + 10\mathbf{k})$
(d)

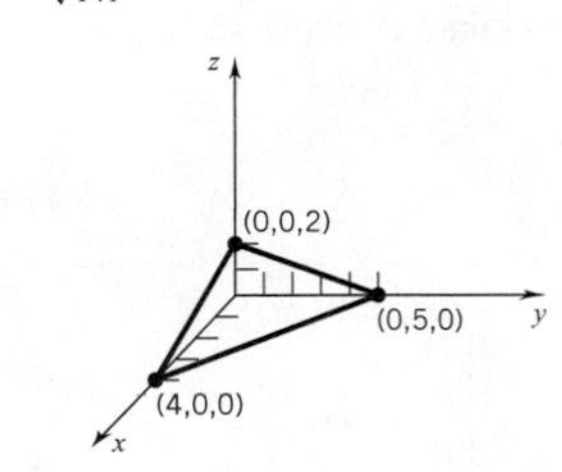

49. (a) (4,0,0), (0,0,6), no y-intercept
(b) $x = 4, \ 3x + 2z = 12, \ z = 6$
(c) $\pm\frac{1}{\sqrt{13}}(3\mathbf{i} + 2\mathbf{k})$
(d)

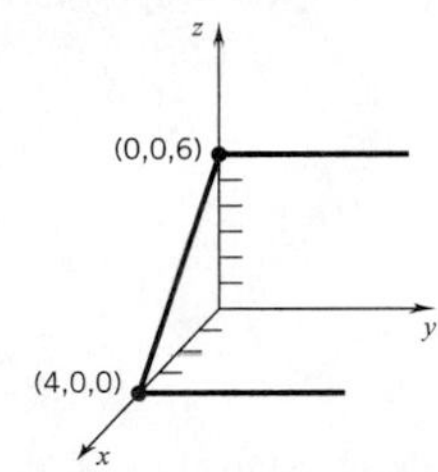

51. $10x + 4y + 5z = 20$ **53.** $5x + 3y = 15$

CHAPTER 13

SECTION 13.1

1. $\mathbf{f}'(t) = 2\mathbf{i} - \mathbf{j} + 3\mathbf{k}$ **3.** $\mathbf{f}'(t) = -\dfrac{1}{2\sqrt{1-t}}\mathbf{i} + \dfrac{1}{2\sqrt{1+t}}\mathbf{j} + \dfrac{1}{(1-t)^2}\mathbf{k}$ **5.** $\mathbf{f}'(t) = \cos t\,\mathbf{i} - \sin t\,\mathbf{j} + \sec^2 t\mathbf{k}$.

7. $-\dfrac{1}{1-t}\mathbf{i} - \sin t\,\mathbf{j} + 2t\,\mathbf{k}$ **9.** $12t\,\mathbf{j} + 2\mathbf{k}$ **11.** $-4\cos 2t\,\mathbf{i} - 4\sin 2t\,\mathbf{j}$

13. (a) $\mathbf{i}$
(b) $\mathbf{i} - \mathbf{j} + \frac{5}{\sqrt{2}}\mathbf{k}$

15. $\mathbf{i} + 3\mathbf{j}$ **17.** $(e - 1)\mathbf{i} + (1 - 1/e)\mathbf{k}$

19. $\dfrac{\pi}{4}\mathbf{i} + \tan(1)\mathbf{j}$

21. $\frac{1}{2}\mathbf{i} + \mathbf{j}$ **23.** $0\,\mathbf{i} + 0\,\mathbf{j} + 0\,\mathbf{k} = \mathbf{0}$ **25.** (a) $\mathbf{i} + \frac{e-1}{2}\,\mathbf{j}$
(b) $\left[5 + \ln\left(\frac{4}{9}\right)\right]\mathbf{i} + \left[\frac{-5}{36} + \ln\left(\frac{9}{4}\right)\right]\mathbf{j} + \frac{295}{2592}\mathbf{k}$

27.

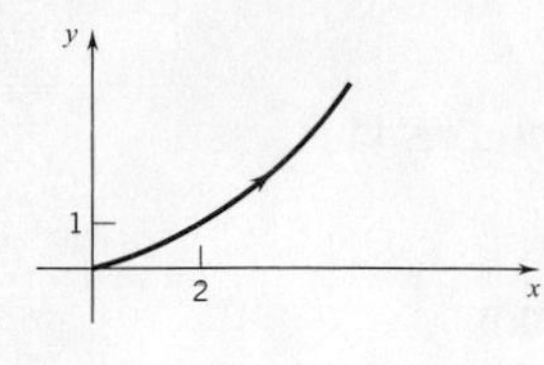

29.

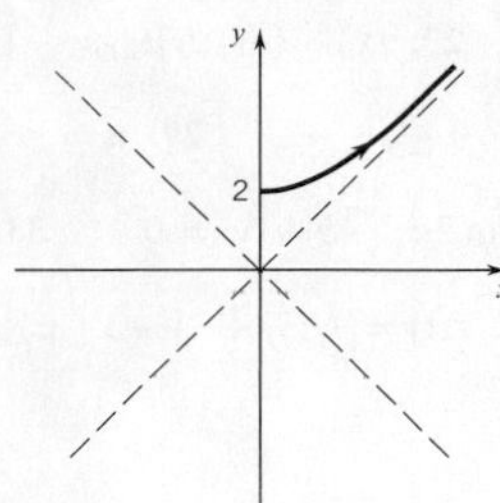

31.

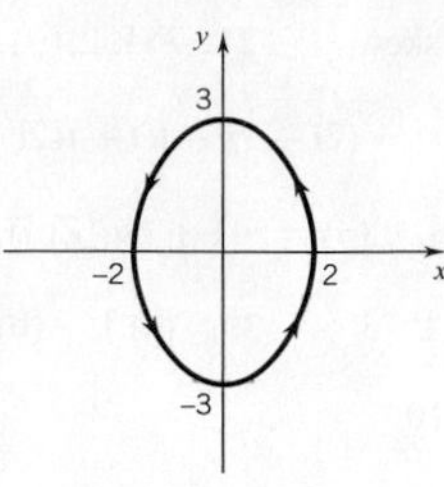

33.

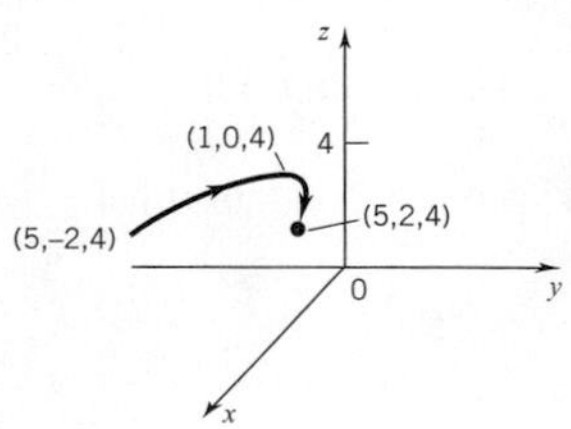

35.

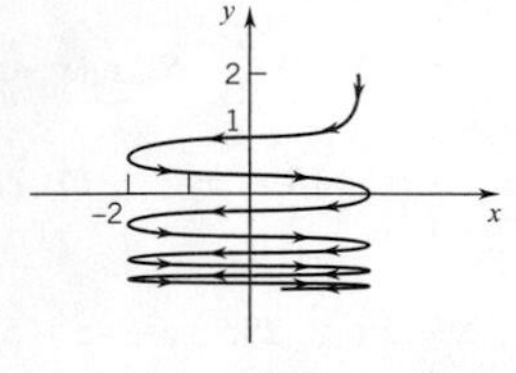

37.

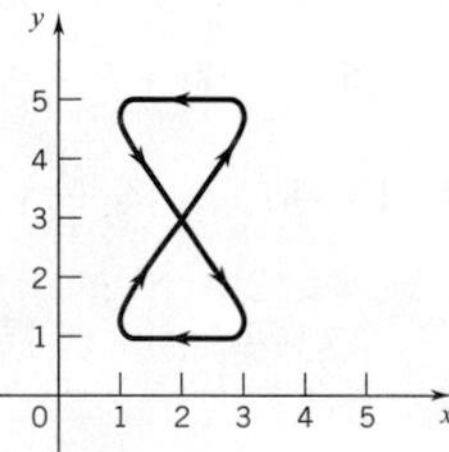

39. (a) $\mathbf{f}(t) = 3\cos t\,\mathbf{i} + 2\sin t\,\mathbf{j}$ (b) $\mathbf{f}(t) = 3\cos t\,\mathbf{i} - 2\sin t\,\mathbf{j}$ **41.** (a) $\mathbf{f}(t) = t\,\mathbf{i} + t^2\mathbf{j}$ (b) $\mathbf{f}(t) = -t\,\mathbf{i} + t^2\,\mathbf{j}$

43. $\mathbf{f}(t) = (1+2t)\mathbf{i} + (4+5t)\mathbf{j} + (-2+8t)\mathbf{k}, 0 \le t \le 1$

45. $\mathbf{f}'(t_0) = \mathbf{i} + m\,\mathbf{j}$ $\displaystyle\int_a^b \mathbf{f}(t)\,dt = \tfrac{1}{2}(b^2 - a^2)\mathbf{i} + A\mathbf{j},\quad \int_a^b \mathbf{f}'(t)\,dt = (b-a)\,\mathbf{i} + (d-c)\,\mathbf{j}$ **47.** $\mathbf{f}(t) = t\,\mathbf{i} + (\frac{1}{3}t^3 + 1)\,\mathbf{j} - \mathbf{k}$ **49.** $\mathbf{f}(t) = e^{\alpha t}\mathbf{c}$

51. (a) if $\mathbf{f}'(t_0) = \mathbf{0}$ on an interval, then the derivative of each component is 0 on that interval, each component is constant on that interval, and therefore $\mathbf{f}$ itself is constant on that interval
(b) set $\mathbf{h}(t) = \mathbf{f}(t) - \mathbf{g}(t)$ and apply part (a)

53. set $\mathbf{f}(t) = f_1(t)\,\mathbf{i} + f_2(t)\,\mathbf{j} + f_3(t)\mathbf{k}$, and apply (3. 1. 4) to f_1, f_2, f_3 **55.** no; as a counterexample set $\mathbf{f}(t) = \mathbf{i} = \mathbf{g}(t)$

57. $\|\mathbf{f}(t)\|^2 = \mathbf{f}(t) \cdot \mathbf{f}(t)$

$$2\|\mathbf{f}(t)\|\frac{d\|\mathbf{f}(t)\|}{dt} = 2\mathbf{f}(t) \cdot \mathbf{f}'(t)$$

$$\frac{d\|\mathbf{f}(t)\|}{dt} = \frac{\mathbf{f}(t) \cdot \mathbf{f}'(t)}{\|\mathbf{f}(t)\|}$$

59. (c) Increasing a causes the object to move faster around the cylinder; increasing b increases the rate at which the object rises.

61. (d) Increasing b increases the number of "peaks;" if $A = B$, overall shape is circular; if $A \ne B$, overall shape is elliptical.

SECTION 13.2

1. $\mathbf{f}'(t) = \mathbf{b}$, $\mathbf{f}''(t) = \mathbf{0}$ **3.** $\mathbf{f}'(t) = 2\,e^{2t}\,\mathbf{i} - \cos\,t\,\mathbf{j}$, $\mathbf{f}''(t) = 4\,e^{2t}\,\mathbf{i} + \sin t\,\mathbf{j}$ **5.** $\mathbf{f}'(t) = (3t^2 - 8t^3)\,\mathbf{j}, \mathbf{f}''(t) = (6t - 24t^2)\,\mathbf{j}$

7. $\mathbf{f}'(t) = -2t\,\mathbf{i} + e^t(t+1)\mathbf{k}$, $\mathbf{f}''(t) = -2\,\mathbf{i} + e^t(t+2)\mathbf{k}$

9. $\mathbf{f}'(t) = (\sin t + t\cos t + 2\sin 2t)\,\mathbf{i} + (2\cos 2t - \cos t + t\sin t)\,\mathbf{j} - 3\sin 3t\,\mathbf{k}$

$\mathbf{f}''(t) = (2\cos t - t\sin t + 4\cos 2t)\,\mathbf{i} + (-4\sin 2t + 2\sin t + t\cos t)\,\mathbf{j} - 9\cos 3t\,\mathbf{k}$

11. $\mathbf{f}'(t) = \frac{1}{2}\sqrt{t}\,\mathbf{g}'(\sqrt{t}) + \mathbf{g}(\sqrt{t})$, $\mathbf{f}''(t) = \frac{1}{4}\mathbf{g}''(\sqrt{t}) + \frac{3}{4}(1/\sqrt{t})\mathbf{g}'(\sqrt{t})$ **13.** $-\sin t\,e^{\cos t}\,\mathbf{i} + \cos t\,e^{\sin t}\,\mathbf{j}$ **15.** $4\,e^{2t} - 4\,e^{-2t}$

17. $(\mathbf{a} \times \mathbf{d}) + (\mathbf{b} \times \mathbf{c}) + 2t(\mathbf{b} \times \mathbf{d})$ **19.** $(\mathbf{a} \cdot \mathbf{d}) + (\mathbf{b} \cdot \mathbf{c}) + 2t(\mathbf{b} \cdot \mathbf{d})$ **21.** $\mathbf{r}(t) = \mathbf{a} + t\mathbf{b}$ **23.** $\mathbf{r}(t) = \frac{1}{2}t^2\mathbf{a} + \frac{1}{6}t^3\mathbf{b} + t\mathbf{c} + \mathbf{d}$

25. $\mathbf{r}''(t) = -\sin t\,\mathbf{i} - \cos t\,\mathbf{j} = -\mathbf{r}(t)$; no. **27.** $\mathbf{r}(t) \cdot \mathbf{r}'(t) = 0$, $\mathbf{r}(t) \times \mathbf{r}'(t) = \mathbf{k}$

29. $\dfrac{d}{dt}[\mathbf{f}(t)\times\mathbf{f}'(t)] = [\mathbf{f}(t)\times\mathbf{f}''(t)] + \underbrace{[\mathbf{f}'(t)\times\mathbf{f}'(t)]}_{0} = \mathbf{f}(t)\times\mathbf{f}''(t)$

31. $[\mathbf{f}\bullet\mathbf{g}\times\mathbf{h}]' = \mathbf{f}'\bullet(\mathbf{g}\times\mathbf{h}) + \mathbf{f}\bullet(\mathbf{g}\times\mathbf{h})' = \mathbf{f}'\bullet(\mathbf{g}\times\mathbf{h}) + \mathbf{f}\bullet[\mathbf{g}'\times\mathbf{h}+\mathbf{g}\times\mathbf{h}']$ and the result follows.

33. The following four statements are equivalent: $||\mathbf{r}(t)|| = \sqrt{\mathbf{r}(t)\bullet\mathbf{r}(t)}$ is constant, $\mathbf{r}(t)\bullet\mathbf{r}(t)$ is constant, $d/dt\,[\mathbf{r}(t)\bullet\mathbf{r}(t)] = 2[\mathbf{r}(t)\bullet\mathbf{r}'(t)] = 0$ identically, $\mathbf{r}(t)\bullet\mathbf{r}'(t) = 0$ identically.

35. $\dfrac{[\mathbf{f}(t+h)\times\mathbf{g}(t+h)] - [\mathbf{f}(t)\times\mathbf{g}(t)]}{h} = \left(\mathbf{f}(t+h)\times\left[\dfrac{\mathbf{g}(t+h)-\mathbf{g}(t)}{h}\right]\right) + \left(\left[\dfrac{\mathbf{f}(t+h)-\mathbf{f}(t)}{h}\right]\times\mathbf{g}(t)\right).$

Now take the limit as $h \to 0$.

(Appeal to Theorem 13.1.3.)

SECTION 13.3

1. $\pi\,\mathbf{j}+\mathbf{k},\quad R(u) = (\mathbf{i}+2\mathbf{k}) + u(\pi\mathbf{j}+\mathbf{k})$ **3.** $\mathbf{b}-2\mathbf{c},\quad \mathbf{R}(u) = (\mathbf{a}-\mathbf{b}+\mathbf{c}) + u(\mathbf{b}-2\mathbf{c})$

5. $4\mathbf{i}-\mathbf{j}+4\mathbf{k},\quad \mathbf{R}(u) = (2\mathbf{i}+5\mathbf{k}) + u(4\mathbf{i}-\mathbf{j}+4\mathbf{k})$

7. $-\sqrt{2}\,\mathbf{i} + \dfrac{3\sqrt{2}}{2}\mathbf{j}+\mathbf{k},\quad \mathbf{R}(u) = \left(\sqrt{2}\,\mathbf{i} + \dfrac{3\sqrt{2}}{2}\mathbf{j} + \dfrac{\pi}{4}\mathbf{k}\right) + u\left(-\sqrt{2}\,\mathbf{i} + \dfrac{3\sqrt{2}}{2}\mathbf{j}+\mathbf{k}\right)$

9. The scalar components $x(t) = at$ and $y(t) = bt^2$ satisfy the equation $a^2y(t) = a^2(bt^2) = b(at)^2 = b[x(t)]^2$ and generate the parabola $a^2y = bx^2$.

11. (a) $P(0,1)$ (b) $P(1,2)$ (c) $P(-1,2)$

13. The tangent line at $t = t_0$ has the form $\mathbf{R}(u) = \mathbf{r}(t_0) + u\mathbf{r}'(t_0)$. If $\mathbf{r}'(t_0) = \alpha\mathbf{r}(t_0)$, then

$\mathbf{R}(u) = \mathbf{r}(t_0) + u\,\alpha\,\mathbf{r}(t_0) = (1+u\alpha)\mathbf{r}(t_0).$

The tangent line passes through the origin at $u = -1/\alpha$.

15. $\pi/2 \cong 1.57$, or 90°

17. $P(1,2,-2);\quad \cos^{-1}\left(\tfrac{1}{5}\sqrt{5}\right) \cong 1.11\text{rad}$ **19.** (a) $\mathbf{r}(t) = a\cos t\,\mathbf{i} + b\sin t\,\mathbf{j}$ (b) $\mathbf{r}(t) = a\cos t\,\mathbf{i} - b\sin t\,\mathbf{j}$ (c) $\mathbf{r}(t) = a\cos 2t\,\mathbf{i} + b\sin 2t\,\mathbf{j}$ (d) $\mathbf{r}(t) = a\cos 3t\,\mathbf{i} - b\sin 3t\,\mathbf{j}$

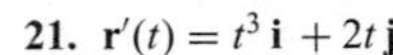

21. $\mathbf{r}'(t) = t^3\,\mathbf{i} + 2t\,\mathbf{j}$

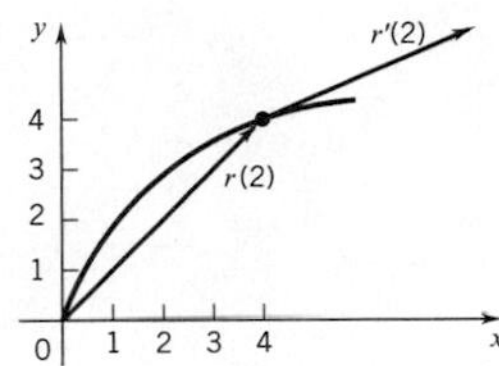

23. $\mathbf{r}'(t) = 2e^{2t}\,\mathbf{i} - 4e^{-4t}\,\mathbf{j}$

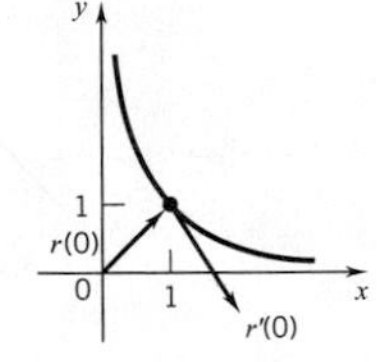

25. $\mathbf{r}'(t) = -2\sin t\,\mathbf{i} + 3\cos t\,\mathbf{j}$

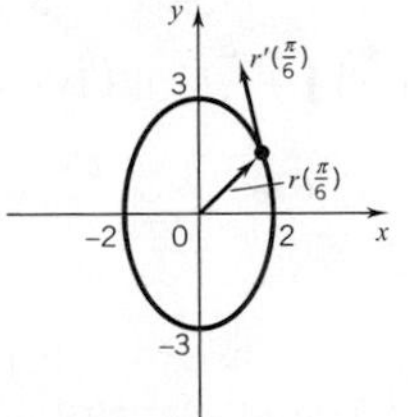

27. $\mathbf{r}(t) = (t^2+1)\,\mathbf{i} + t\,\mathbf{j},\quad t \ge 1;$ or, $\mathbf{r}(t) = \sec^2 t\,\mathbf{i} + \tan t\,\mathbf{j},\quad t \in [\tfrac{1}{4}\pi, \tfrac{1}{2}\pi)$ **29.** $\mathbf{r}(t) = \cos t\sin 3t\,\mathbf{i} + \sin t\,\sin 3t\,\mathbf{j},\quad t \in [0,\pi]$

31.

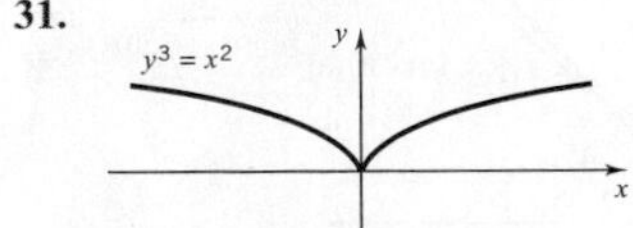

There is no tangent vector at the origin.

33. $(2,4,8);\quad \cos^{-1}\left(\dfrac{24}{\sqrt{21}\sqrt{161}}\right) \cong 1.15$ rad

35. $\mathbf{T}(1) = \dfrac{1}{\sqrt{2}}\mathbf{j} + \dfrac{1}{\sqrt{2}}\mathbf{k},\quad +\mathbf{N}(1) = \dfrac{-1}{\sqrt{2}}\mathbf{j} + \dfrac{1}{\sqrt{2}}\mathbf{k},\quad x-1=0$ **37.** $\tfrac{1}{5}\sqrt{5}(-2\mathbf{i}+\mathbf{k}),\quad -\mathbf{j},\quad x+2z = \tfrac{1}{2}\pi$

39. $\tfrac{1}{14}\sqrt{14}(\mathbf{i}+2\mathbf{j}+3\mathbf{k}),\quad \tfrac{1}{266}\sqrt{266}(-11\,\mathbf{i} - 8\,\mathbf{j} + 9\mathbf{k}),\quad 3x-3y+z=1$

41. $\mathbf{T}(0) = \dfrac{1}{\sqrt{3}}\mathbf{i} + \dfrac{1}{\sqrt{3}}\mathbf{j} + \dfrac{1}{\sqrt{3}}\mathbf{k},\quad \mathbf{N}(0) = \dfrac{1}{\sqrt{2}}\mathbf{i} - \dfrac{1}{\sqrt{2}}\mathbf{j},\quad x+y-2z+1=0$

43. $\mathbf{T}_1 = \dfrac{\mathbf{R}'(u)}{||\mathbf{R}'(u)||} = -\dfrac{\mathbf{r}'(a+b-u)}{||\mathbf{r}'(a+b-u)||} = -\mathbf{T}$. Then $\mathbf{T}_1'(u) = \mathbf{T}'(a+b-u)$ and thus $\mathbf{N}_1 = \mathbf{N}$.

45. Let $\mathbf{T}$ be the unit tangent at the tip of $\mathbf{R}(u) = \mathbf{r}(\phi(u))$ as calculated from the parametrization $\mathbf{r}$, and let $\mathbf{T}_1$ be the unit tangent at the same point as calculated from the parametrization $\mathbf{R}$. Then

$$\mathbf{T}_1 = \frac{\mathbf{R}'(u)}{||\mathbf{R}'(u)||} = \frac{\mathbf{r}'(\phi(u))\phi'(u)}{||\mathbf{r}'(\phi(u))\phi'(u)||} \underset{\phi'(u)>0}{=} \frac{\mathbf{r}'(\phi(u))}{||\mathbf{r}'(\phi(u))||} = \mathbf{T}.$$

This shows the invariance of the unit tangent. The invariance of the principal normal and the osculating plane follows directly from the invariance of the unit tangent.

47. (a) Let $t = \Psi(v) = 2\pi - v^2$. When t increases from 0 to 2π, v decreases from $\sqrt{2\pi}$ to 0.

(b) $\mathbf{T}_r\left(\frac{\pi}{4}\right) = -\frac{1}{\sqrt{10}}\mathbf{i} + \frac{1}{\sqrt{10}}\mathbf{j} + \frac{2}{\sqrt{5}}\mathbf{k}$, $\quad \mathbf{T}_R\left(\frac{\sqrt{7\pi}}{2}\right) = \frac{1}{\sqrt{10}}\mathbf{i} - \frac{1}{\sqrt{10}}\mathbf{j} - \frac{2}{\sqrt{5}}\mathbf{k}$

$\mathbf{N}_r\left(\frac{\pi}{4}\right) = -\frac{\sqrt{2}}{2}\mathbf{i} - \frac{\sqrt{2}}{2}\mathbf{j}$; $\quad \mathbf{N}_R\left(\frac{\sqrt{7\pi}}{2}\right) = -\frac{\sqrt{2}}{2}\mathbf{i} - \frac{\sqrt{2}}{2}\mathbf{j}$

49. (a) $x = 1 - t, y = 1 + t, z = -\frac{\sqrt{2}}{2} - \frac{5\sqrt{2}}{2}t$

(c) the tangent line is parallel to the xy-plane at the points where $t = \frac{(2n+1)\pi}{10}$, $n = 0, 1, 2, \ldots, 9$.

SECTION 13.4

1. $\frac{52}{3}$ **3.** $2\pi\sqrt{a^2+b^2}$ **5.** $\ln(1+\sqrt{2})$ **7.** $\frac{1}{27}(13\sqrt{13}-8)$ **9.** $\sqrt{2}(e^{\pi}-1)$ **11.** $6 + \frac{1}{2}\sqrt{2}\ln(2\sqrt{2}+3)$ **13.** e^2 **15.** $4\pi^2$

17. differentiate $s(t) = \int_a^t \sqrt{[x'(u)]^2 + [y'(u)]^2 + [z'(u)]^2}\, du$ **19.** see Exercise 16, differentiate $s(x) = \int_a^x \sqrt{1 + [f'(t)]^2}dt$

21. Let L be the length as computed from $\mathbf{r}$ and L^* the length as computed from $\mathbf{R}$. Then

$$L^* = \int_c^d ||\mathbf{R}'(u)||du = \int_c^d ||\mathbf{r}'(\phi(u))||\phi'(u)du) \underset{t=\phi(u)}{=} \int_a^b ||\mathbf{r}'(t)||dt = L.$$

23. (a) $s = 5t$ (b) $\mathbf{R}(s) = 3\cos\left(\frac{s}{5}\right)\mathbf{i} + 3\sin\left(\frac{s}{5}\right)\mathbf{j} + \frac{4s}{5}\mathbf{k}$ (c) $Q(-3, 0, 4\pi)$

(d) $\mathbf{R}'(s) = \frac{-3}{5}\sin\left(\frac{s}{5}\right)\mathbf{i} + \frac{3}{5}\cos\left(\frac{s}{5}\right)\mathbf{j} + \frac{4}{5}\mathbf{k}$, $||\mathbf{R}'(s)|| = 1$

25. 0.5077 **27.** 22.0939 **29.** $L \cong 17.6286$

SECTION 13.5

1. v/r, v^2/r **3.** $||\mathbf{r}''(t)|| = a^2|b\sin at| = a^2|y(t)|$ **5.** $y = \cos\pi x$, $0 \le x \le 2$ **7.** $x = \sqrt{1+y^2}$, $y \ge -1$

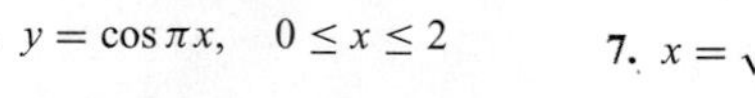

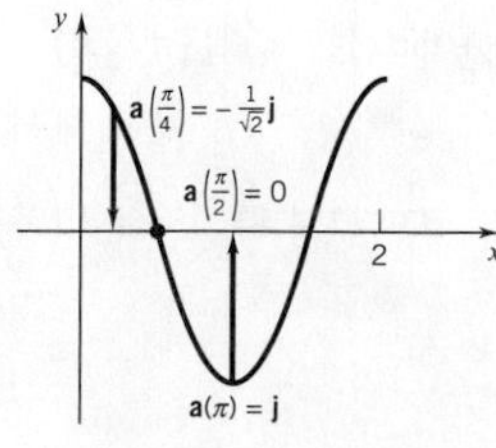

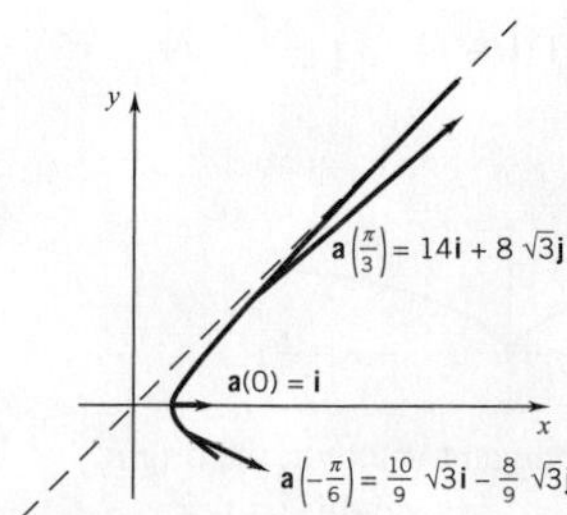

9. (a) (x_0, y_0, z_0) (b) $\alpha\cos\theta\,\mathbf{j} + \alpha\sin\theta\,\mathbf{k}$ (c) $|\alpha|$ (d) $-32\mathbf{k}$

(e) arc from parabola $z = z_0 + (\tan\theta)(y - y_0) - 16(y-y_0)^2/(\alpha^2\cos^2\theta)$ in the plane $x = x_0$

11. $||\mathbf{r}(t)||$ is constant iff $||\mathbf{r}(t)||^2 = \mathbf{r}(t)\cdot\mathbf{r}(t)$ is constant;

$\mathbf{r}(t)\cdot\mathbf{r}(t)$ is constant iff $\frac{d}{dt}[\mathbf{r}(t)\cdot\mathbf{r}(t)] = 2\mathbf{r}(t)\cdot\mathbf{r}'(t) = 0$.

13. $\dfrac{e^{-x}}{(1+e^{-2x})^{3/2}}$ **15.** $\dfrac{2}{(1+4x)^{3/2}}$ **17.** $|\cos x|$ **19.** $\dfrac{|\sin x|}{(1+\cos^2 x)^{3/2}}$ **21.** $\frac{5}{2}\sqrt{5}$ **23.** $5\sqrt{5}$ **25.** $\dfrac{10\sqrt{10}}{3}$

27. $(\frac{1}{2}\sqrt{2}, \frac{1}{2}\ln\frac{1}{2})$ **29.** $\dfrac{1}{(1+t^2)^{3/2}}$ **31.** $\dfrac{12|t|}{(4+9t^4)^{3/2}}$ **33.** $\frac{1}{2}\sqrt{2}\,e^{-t}$ **35.** $\dfrac{2+t^2}{(1+t^2)^{3/2}}$ **37.** $\sqrt{2}$ **39.** $\dfrac{a^4b^4}{(b^4x^2+a^4y^2)^{3/2}}$

41. $\kappa = \frac{1}{3}\sqrt{2}e^{-t}$, $a_{\mathbf{T}} = \sqrt{3}e^t$, $a_{\mathbf{N}} = \sqrt{2}e^t$ **43.** $\kappa = 1$, $a_{\mathbf{T}} = 0$, $a_{\mathbf{N}} = 4$ **45.** $\kappa = \dfrac{2}{2+t^2}$ $a_{\mathbf{T}} = 2t$, $a_{\mathbf{N}} = 4+2t^2$

47. $\kappa = \dfrac{1}{8}\sqrt{\dfrac{2}{1-t^2}}$; $a_{\mathbf{T}} = 0$; $a_{\mathbf{N}} = \dfrac{1}{2}\sqrt{\dfrac{2}{1-t^2}}$ **49.** $a_{\mathbf{T}} = \dfrac{6t+12t^3}{\sqrt{1+t^2+t^4}}$; $a_{\mathbf{N}} = 6\sqrt{\dfrac{1+4t^2+t^4}{1+t^2+t^4}}$ **51.** $\dfrac{e^{-a\theta}}{\sqrt{1+a^2}}$

53. $\dfrac{3}{2\sqrt{2a^2(1-\cos\theta)}} = \dfrac{3}{2\sqrt{2ar}}$ **55.** (a) $s(\theta) = 4R|\cos\frac{1}{2}\theta|$ (b) $\rho(\theta) = 4R\sin\frac{1}{2}\theta$ (c) $\rho^2+s^2 = 16R^2$ **57.** $9\rho^2+s^2 = 16a^2$

SECTION 13.6

1. (a) $\mathbf{r}'(0) = b\omega\,\mathbf{j}$ (b) $\mathbf{r}''(t) = \omega^2\mathbf{r}(t)$ (c) The torque is **0** and the angular momentum is constant.

3. (a) $\mathbf{v}(t) = 2\mathbf{j} + (\alpha/m)t\mathbf{k}$ (b) $v(t) = 1/m\sqrt{4m^2+\alpha^2t^2}$ (c) $\mathbf{p}(t) = 2m\,\mathbf{j} + \alpha t\mathbf{k}$
(d) $\mathbf{r}(t_1) = [2t+y_0]\,\mathbf{j} + [(\alpha/2m)t^2 + z_0]\mathbf{k}$, $t \geq 0$, $z = (\alpha/8m)(y-y_0)^2 + z_0$, $y \geq y_0$, $x = 0$

5. $\mathbf{F}(t) = 2m\mathbf{k}$ **7.** (a) $\pi b\,\mathbf{j} + \mathbf{k}$ (b) $\sqrt{\pi^2b^2+1}$ (c) $-\pi^2a\,\mathbf{i}$ (d) $m(\pi b\mathbf{j} + \mathbf{k})$
(e) $m[b(1-\pi)\mathbf{i} - 2a\,\mathbf{j} + 2\pi ab\mathbf{k}]$ (f) $-m\pi^2a[\mathbf{j} - b\mathbf{k}]$

9. We have $m\mathbf{v} = m\mathbf{v}_1 + m\mathbf{v}_2$ and $\frac{1}{2}mv^2 = \frac{1}{2}mv_1^2 + \frac{1}{2}mv_2^2$. Therefore $\mathbf{v} = \mathbf{v}_1 + \mathbf{v}_2$ and $v^2 = v_1^2 + v_2^2$. Since

$$v^2 = \mathbf{v}\cdot\mathbf{v} = (\mathbf{v}_1+\mathbf{v}_2)\cdot(\mathbf{v}_1+\mathbf{v}_2) = v_1^2 + v_2^2 + 2(\mathbf{v}_1\cdot\mathbf{v}_2),$$

we have $\mathbf{v}_1\cdot\mathbf{v}_2 = 0$ and $\mathbf{v}_1 \perp \mathbf{v}_2$.

11. Here $\mathbf{r}''(t) = \mathbf{a}, \mathbf{r}'(t) = \mathbf{v}(0) + t\mathbf{a}, \mathbf{r}(0) + t\mathbf{v}(0) + \frac{1}{2}t^2\mathbf{a}$. If neither $\mathbf{v}(0)$ nor $\mathbf{a}$ is zero, the displacement $\mathbf{r}(t) - \mathbf{r}(0)$ is a linear combination of $\mathbf{v}(0)$ and $\mathbf{a}$ and thus remains on the plane determined by these vectors. The equation of this plane can be written

$$[\mathbf{a}\times\mathbf{v}(0)]\cdot[\mathbf{r} - \mathbf{r}(0)] = 0.$$

[If either $\mathbf{v}(0)$ or $\mathbf{a}$ is zero, the motion is restricted to a straight line; if both of these vectors are zero, the particle remains at its initial position $\mathbf{r}(0)$.]

13. $\mathbf{r}(t) = \mathbf{i} + t\,\mathbf{i} + (qE_0/2m)t^2\mathbf{k}$ **15.** $\mathbf{r}(t) = (1+t^3/6m)\,\mathbf{i} + (t^4/12m)\,\mathbf{j} + t\,\mathbf{k}$.

17. $\dfrac{d}{dt}(\frac{1}{2}mv^2) = mv\dfrac{dv}{dt} = m\left(\mathbf{v}\cdot\dfrac{d\mathbf{v}}{dt}\right) = m\dfrac{d\mathbf{v}}{dt}\cdot\mathbf{v} = \mathbf{F}\cdot\dfrac{d\mathbf{r}}{dt} = 4r^2\left(\mathbf{r}\cdot\dfrac{d\mathbf{r}}{dt}\right) = 4r^2\left(r\dfrac{dr}{dt}\right) = 4r^3\dfrac{dr}{dt} = \dfrac{d}{dt}(r^4)$. Therefore $d/dt(\frac{1}{2}mv^2 - r^4) = 0$ and $\frac{1}{2}mv^2 - r^4$ is a constant E. Evaluating E from $t = 0$, we find that $E = 2m$. Thus $\frac{1}{2}mv^2 - r^4 = 2m$ and $v = \sqrt{4+(2/m)r^4}$.

SECTION 13.7

1. about 61.1% of an earth year **3.** set $x = r\cos\theta$, $y = r\sin\theta$

5. Substitute

$$r = \frac{a}{1+\,e\cos\theta}, \quad \left(\frac{dr}{d\theta}\right)^2 = \frac{(a\,e\sin\theta)^2}{(1+e\cos\theta)^4}$$

into the right side of the equation and you will see that, with a and e^2 as given, the expression reduces to E.

CHAPTER 14

SECTION 14.1

1. dom (f) = the first and third quadrants, including the axes; range $(f) = [0,\infty)$

3. dom (f) = the set of all points (x,y) not on the line $y = -x$; range $(f) = (-\infty,0)\cup(0,\infty)$ **5.** dom (f) = the entire plane; range $(f) = (-1,1)$

7. dom (f) = the first and third quadrants, excluding the axes; range $(f) = (-\infty,\infty)$

9. dom (f) = the set of all points (x,y) with $x^2 < y$; in other words, the set of all points of the plane above the parabola $y = x^2$; range $(f) = (0,\infty)$

11. dom (f) = the set of all points (x,y) with $-3 \leq x \leq 3, -2 \leq y \leq 2$ (a rectangle); range $(f) = [-2,3]$

13. dom (f) = the set of all points (x, y, z) not on the plane $x + y + z = 0$; range $(f) = \{-1, 1\}$

15. dom (f) = the set of all points (x, y, z) with $|y| < |x|$; range $(f) = (-\infty, 0]$

17. dom (f) = the set of all points (x, y) such that $x^2 + y^2 < 9$; in other words, the set of all points of the plane inside the circle $x^2 + y^2 = 9$; range $(f) = [\frac{2}{3}, \infty)$

19. dom (f) = the set of all points (x, y, z) with $x + 2y + 3z > 0$; in other words, the set of all points in space that lie on the same side of the plane $x + 2y + 3z = 0$ as the point $(1, 1, 1)$; range $(f) = (-\infty, \infty)$

21. dom (f) = all of space; range $(f) = (0, 1]$

23. dom $(f) = \{x : x \geq 0\}$; range $(f) = [0, \infty)$
dom $(g) = \{(x, y) : x \geq 0, y \text{ real}\}$; range $(g) = [0, \infty)$
dom $(h) = \{x, y, z) : x \geq 0, y, z \text{ real}\}$; range $(h) = [0, \infty)$

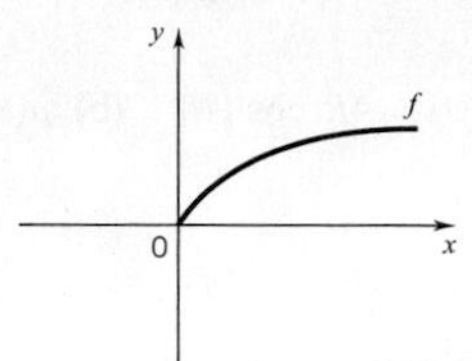

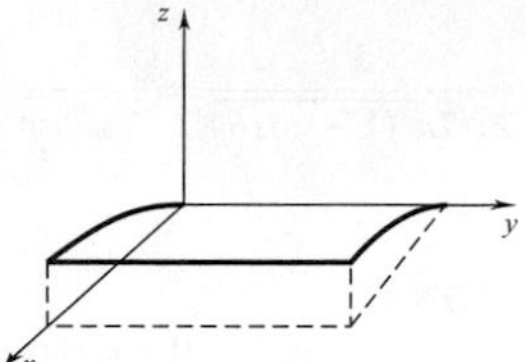

25. dom $(f) = \{(x, y) : 1 - 4xy \geq 0\}$ **27.** $\lim_{h\to0} \dfrac{f(x+h, y) - f(x, y)}{h} = 4x$; $\lim_{h\to0} \dfrac{f(x, y+h) - f(x, y)}{h} = -1$

29. $\lim_{h\to0} \dfrac{f(x+h, y) - f(x, y)}{h} = 3 - y$; $\lim_{h\to0} \dfrac{f(x, y+h) - f(x, y)}{h} = -x + 4y$

31. $\lim_{h\to0} \dfrac{f(x+h, y) - f(x, y)}{h} = -y \sin(xy)$; $\lim_{h\to0} \dfrac{f(x, y+h) - f(x, y)}{h} = -x \sin(xy)$

33. (a) $\dfrac{3y}{(x+y)(x+h+y)}$ (b) $-\dfrac{3x}{(x+y)(x+y+h)}$ (c) $\dfrac{3y}{(x+y)^2}$; $-\dfrac{3x}{(x+y)^2}$

35. (a) $f(x, y) = x^2 y$ (b) $f(x, y) = \pi x^2 y$ (c) $f(x, y) = 2|y|$ **37.** $V = \dfrac{lh(10 - lh)}{l + h}$ **39.** $V = \pi r^2 h + \frac{4}{3}\pi r^3$

SECTION 14.2

1. an elliptic cone **3.** a parabolic cylinder **5.** a hyperboloid of one sheet **7.** sphere of radius 2 centered at the origin

9. an elliptic paraboloid **11.** a hyperbolic paraboloid

13.

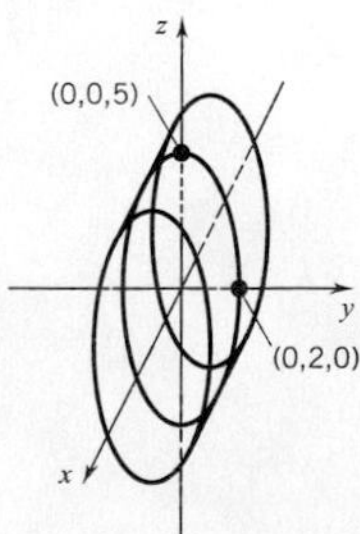

15.

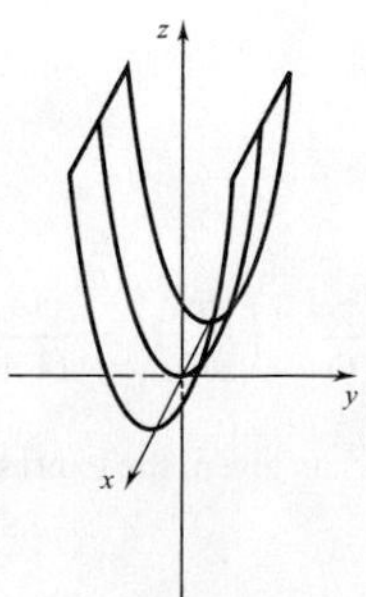

17.

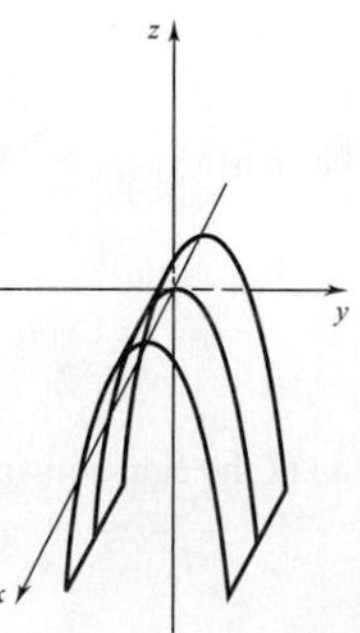

19.

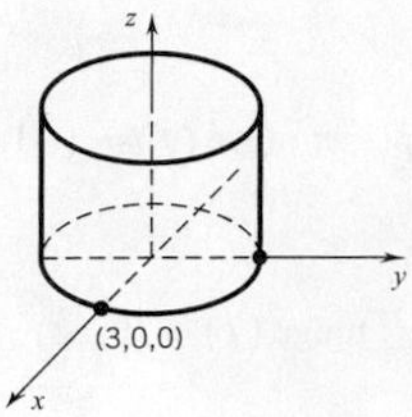

21.

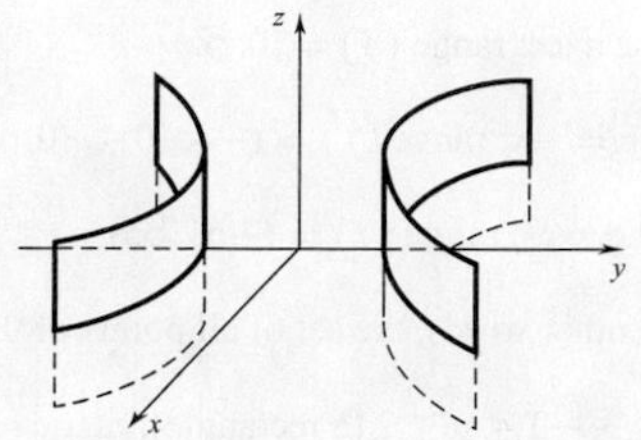

23.

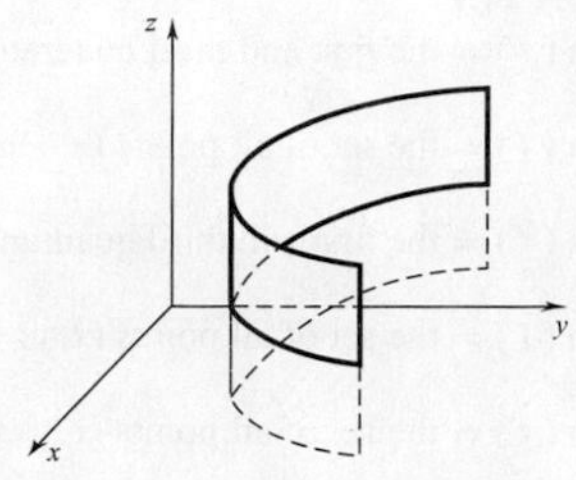

25. elliptic paraboloid, xy-trace: the origin, xz-trace: the parabola $x^2 = 4z$, yz-trace: the parabola $y^2 = 9z$, surface has the form of Figure 14.2.5

27. an elliptic cone, xy-trace: the origin, xz-trace: the lines $x = \pm 2z$, yz-trace: the lines $y = \pm 3z$, surface has the form of Figure 14.2.4

29. a hyperboloid of two sheets, xy-trace: none, xz-trace: the hyperbola $4z^2 - x^2 = 4$, yz-trace: the hyperbola $9z^2 - y^2 = 9$, surface has the form of Figure 14.2.3

31. hyperboloid of two sheets, xy-trace: the hyperbola $\dfrac{x^2}{4} - \dfrac{y^2}{9} = 1$, xz-trace: the hyperbola $\dfrac{x^2}{4} - z^2 = 1$, yz-trace: none, see Figure 14.2.3 for an example

33. elliptic paraboloid, xz-trace: the origin, yz-trace: the parabola $z^2 = 4y$, xy-trace: the parabola $x^2 = 9y$, surface has the form of Figure 14.2.5

35. hyperboloid of two sheets, xy-trace: the hyperbola $\dfrac{y^2}{4} - \dfrac{x^2}{9} = 1$, xz–trace: none, yz-trace: the hyperbola $\dfrac{y^2}{4} - z^2 = 1$, see Figure 14.2.3 for an example

37. paraboloid of revolution, xy-trace: the origin, xz-trace: the parabola $x^2 = 4z$, yz-trace: the parabola $y^2 = 4z$, surface has the form of figure 14.2.5.

39. (a) an elliptic paraboloid (opening up if A and B are both positive, opening down if A and B are both negative)
(b) a hyperbolic paraboloid (c) the xy-plane if A and B are both zero; otherwise, a parabolic cylinder

41. $x^2 + y^2 - 4z = 0$ (paraboloid of revolution) **43.** (a) a circle (b) (i) $\sqrt{x^2 + y^2} = -3z$ (ii) $\sqrt{x^2 + z^2} = \frac{1}{3}y$

45. the line $5x + 7y = 30$ **47.** the circle $x^2 + y^2 = \frac{5}{4}$ **49.** the ellipse $x^2 + 2y^2 = 2$ **51.** the parabola $x^2 = -4(y - 1)$

53. set $\dfrac{x}{a} = \cos u \cos v$, $\dfrac{y}{b} = \cos u \sin v$, $\dfrac{z}{c} = \sin u$. **55.** set $\dfrac{x}{a} = v \cos u$, $\dfrac{y}{b} = v \sin u$, $\dfrac{z}{c} = v$

SECTION 14.3

1. lines of slope 1 : $y = x - c$

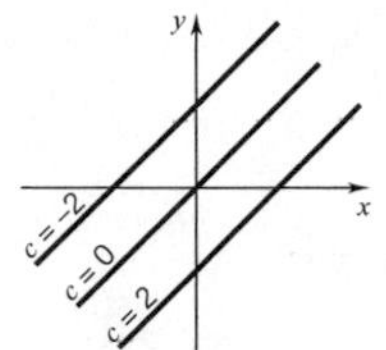

3. parabolas , $y = x^2 - c$

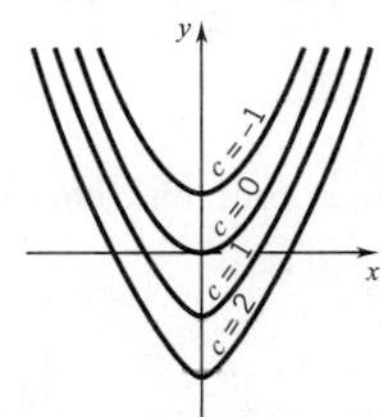

5. the y-axis and the lines $y = \left(\dfrac{1-c}{c}\right)x$,

the origin omitted throughout

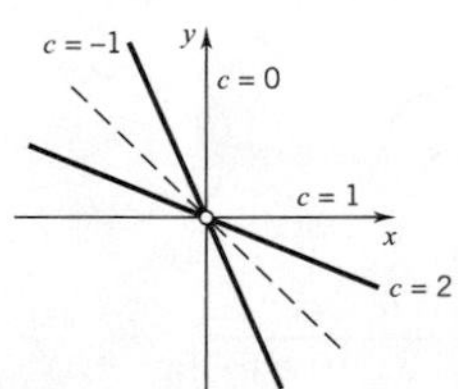

7. the cubics $y = x^3 - c$

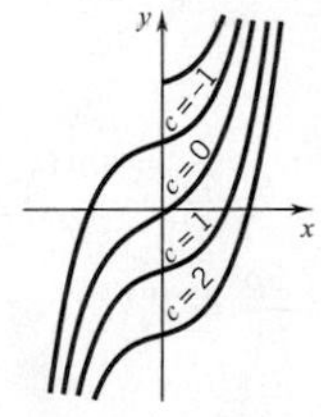

9. the lines $y = \pm x$ and the hyperbolas $x^2 - y^2 = c$

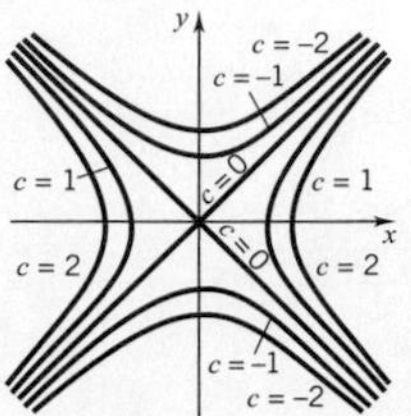

11. pairs of horizontal lines $y = \pm\sqrt{c}$ and the x-axis

13. the circle $x^2 + y^2 = e^c$ with c real

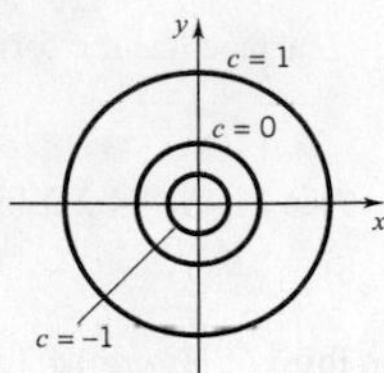

15. the curves $y = e^{cx^2}$ with the point $(0, 1)$ omitted

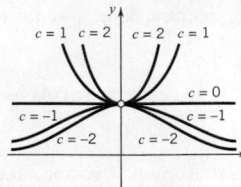

17. the coordinate axes and pairs of lines $y = \pm(\sqrt{1-c}/\sqrt{c})x$, the origin omitted throughout

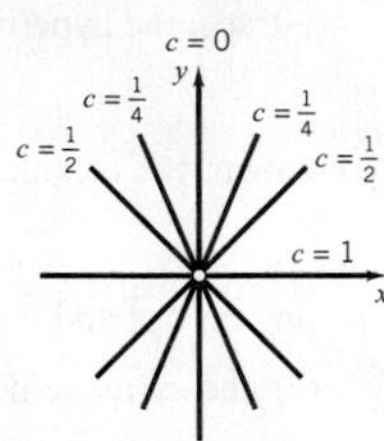

19. $x + 2y + 3z = 0$, plane through the origin **21.** $z = \sqrt{x^2+y^2}$, the upper nappe of the circular cone $z^2 = x^2 + y^2$; Figure 14.2.4

23. the elliptic paraboloid $\dfrac{x^2}{(3\sqrt{2})^2} + \dfrac{y^2}{(2\sqrt{2})^2} = z$; Figure 14.2.5

25. (i) hyperboloid of two sheets; Figure 14.2.3 (ii) circular cone; Figure 14.2.4 (iii) hyperboloid of one sheet; Figure 14.2.2

27. $4x^2 + y^2 = 1$ **29.** $y^2 \tan^{-1} x = \pi$ **31.** $x^2 + 2y^2 - 2xyz = 13$

35. (a) $\dfrac{3x + 2y + 1}{4x^2 + 9} = \dfrac{3}{5}$
(b) $x^2 + 2y^2 - z^2 = 21$

37. $x^2 + y^2 + z^2 = \dfrac{GmM}{c}$. The surfaces of constant gravitational force are concentric spheres.

39. (a) $T(x, y) = \dfrac{k}{x^2 + y^2}$ (b) $x^2 + y^2 = \dfrac{k}{c}$, concentric circles (c) 10° **41.** F **43.** A **45.** E

SECTION 14.4

1. $\dfrac{\partial f}{\partial x} = 6x - y, \quad \dfrac{\partial f}{\partial y} = 1 - x$ **3.** $\dfrac{\partial \rho}{\partial \phi} = \cos\phi\cos\theta, \quad \dfrac{\partial \rho}{\partial \theta} = -\sin\phi\sin\theta$ **5.** $\dfrac{\partial f}{\partial x} = e^{x-y} + e^{y-x}, \quad \dfrac{\partial f}{\partial y} = -e^{x-y} - e^{y-x}$

7. $\dfrac{\partial g}{\partial x} = \dfrac{(AD - BC)y}{(Cx + Dy)^2}, \quad \dfrac{\partial g}{\partial y} = \dfrac{(BC - AD)x}{(Cx + Dy)^2}$ **9.** $\dfrac{\partial u}{\partial x} = y + z, \quad \dfrac{\partial u}{\partial y} = x + z, \quad \dfrac{\partial u}{\partial z} = x + y$

11. $\dfrac{\partial f}{\partial x} = z\cos(x - y), \quad \dfrac{\partial f}{\partial y} = -z\cos(x - y), \quad \dfrac{\partial f}{\partial z} = \sin(x - y)$

13. $\dfrac{\partial \rho}{\partial \theta} = e^{\theta+\phi}\,[\cos(\theta - \phi) - \sin(\theta - \phi)], \quad \dfrac{\partial \rho}{\partial \phi} = e^{\theta+\phi}\,[\cos(\theta - \phi) + \sin(\theta - \phi)]$

15. $\dfrac{\partial f}{\partial x} = 2xy\sec(xy) + x^2y^2\sec(xy)\tan(xy), \quad \dfrac{\partial f}{\partial y} = x^2\sec(xy) + x^3y\sec(xy)\tan(xy)$ **17.** $\dfrac{\partial h}{\partial x} = \dfrac{y^2 - x^2}{(x^2 + y^2)^2}, \quad \dfrac{\partial h}{\partial y} = -\dfrac{2xy}{(x^2 + y^2)^2}$

19. $\dfrac{\partial f}{\partial x} = \dfrac{\sin y\,(\cos x + x\sin x)}{y\cos^2 x}, \quad \dfrac{\partial f}{\partial y} = \dfrac{x\,(y\cos y - \sin y)}{y^2\cos x}$ **21.** $\dfrac{\partial h}{\partial x} = 2f(x)f'(x)g(y), \quad \dfrac{\partial h}{\partial y} = [f(x)]^2\,g'(y)$

23. $\dfrac{\partial f}{\partial x} = (y^2\ln z)z^{xy^2}, \quad \dfrac{\partial f}{\partial y} = (2xy\ln z)z^{xy^2}, \quad \dfrac{\partial f}{\partial z} = xy^2z^{xy^2-1}$

25. $\dfrac{\partial h}{\partial r} = 2r\,e^{2t}\cos(\theta - t), \quad \dfrac{\partial h}{\partial \theta} = -r^2\,e^{2t}\sin(\theta - t), \quad \dfrac{\partial h}{\partial t} = r^2\,e^{2t}[2\cos(\theta - t) + \sin(\theta - t)]$

27. $\dfrac{\partial f}{\partial x} = -\dfrac{yz}{x^2 + y^2}, \quad \dfrac{\partial f}{\partial y} = \dfrac{xz}{x^2 + y^2}, \quad \dfrac{\partial f}{\partial z} = \tan^{-1}(y/x)$ **29.** $f_x(0, e) = 1, \quad f_y(0, e) = e^{-1}$ **31.** $f_x(1, 2) = \frac{2}{9}, \quad f_y(1, 2) = -\frac{1}{9}$

33. $f_x(x, y) = 2xy, \quad f_y(x, y) = x^2$ **35.** $f_x(x, y) = \dfrac{2}{x}, \quad f_y(x, y) = \dfrac{1}{y}$

37. $f_x(x,y) = -\dfrac{1}{(x-y)^2}, \quad f_y(x,y) = \dfrac{1}{(x-y)^2}$ **39.** $f_x(x,y,z) = y^2 z, \quad f_y(x,y,z) = 2xyz, \quad f_z(x,y,z) = xy^2$

41. (b) $x = x_0, \quad z - z_0 = f_y(x_0, y_0)(y - y_0)$ **43.** $x = 2, \quad z - 5 = 2(y-1)$ **45.** $y = 2, \quad z - 9 = 6(x-3)$

47. (a) $m_x = -6$; tangent line: $y = 2, \quad z = -6x + 13$
(b) $m_y = 18$; tangent line: $x = 1, z = 18y - 29$

49. $u_x = v_y = 2x, \quad u_y = -v_x = -2y$ **51.** $u_x = v_y = \dfrac{x}{x^2+y^2}, \quad u_y = -v_x = \dfrac{y}{x^2+y^2}$

53. (a) f depends only on y (b) f depends only on x

55. (a) $50\sqrt{3}$ in.^{2} (b) $5\sqrt{3}$ in.^{2} (c) 50 in.^{2} (d) $\frac{5}{18}\pi$ in.^{2} (e) -2

57. (a) y_0-section: $\mathbf{r}(x) = x\mathbf{i} + y_0\mathbf{j} + f(x_0, y_0)\mathbf{k}$

tangent line: $\mathbf{R}(t) = [x_0\mathbf{i} + y_0\mathbf{j} + f(x_0,y_0)\mathbf{k}] + t\left[\mathbf{i} + \dfrac{\partial f}{\partial x}(x_0,y_0)\mathbf{k}\right]$

(b) x_0-section: $\mathbf{r}(y) = x_0\mathbf{i} + y\mathbf{j} + f(x_0, y)\mathbf{k}$

tangent line: $\mathbf{R}(t) = [x_0\mathbf{i} + y_0\mathbf{j} + f(x_0,y_0)\mathbf{k}] + t\left[\mathbf{j} + \dfrac{\partial f}{\partial y}(x_0,y_0)\mathbf{k}\right]$

(c) For (x,y,z) in the plane

$$[(x-x_0)\mathbf{i} + (y-y_0)\mathbf{j} + (z - f(x_0,y_0))\mathbf{k}] \bullet \left[\left(\mathbf{i} + \frac{\partial f}{\partial x}(x_0,y_0)\mathbf{k}\right) \times \left(\mathbf{j} + \frac{\partial f}{\partial y}(x_0,y_0)\mathbf{k}\right)\right] = 0.$$

From this it follows that

$$z - f(x_0,y_0) = (x-x_0)\frac{\partial f}{\partial x}(x_0,y_0) + (y-y_0)\frac{\partial f}{\partial y}(x_0,y_0).$$

59. (a) Set $u = ax + by$. Then $\dfrac{\partial w}{\partial x} = ag'(u)$ and $\dfrac{\partial w}{\partial y} = bg'(u)$. (b) Set $u = x^m y^n$. Then $\dfrac{\partial w}{\partial x} = mx^{m-1}y^n g'(u)$ and $\dfrac{\partial w}{\partial y} = nx^m y^{n-1} g'(u)$.

61. $V\dfrac{\partial P}{\partial V} = V\left(-\dfrac{kT}{V^2}\right) = -k\dfrac{T}{V} = -P;\ V\dfrac{\partial P}{\partial V} + T\dfrac{\partial P}{\partial T} = -k\dfrac{T}{V} + T\left(\dfrac{k}{V}\right) = 0$

SECTION 14.5

1. interior $= \{(x,y) : 2 < x < 4, 1 < y < 3\}$. (the inside of the rectangle)
boundary = the union of the four line segments that bound the rectangle
set is closed

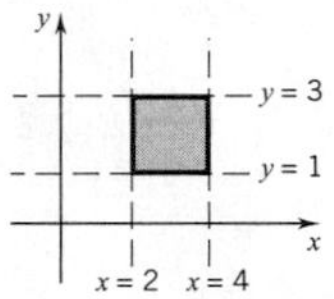

3. interior = the entire set (region between two concentric circles)
boundary $= \{(x,y) : x^2 + y^2 = 1 \text{ or } x^2 + y^2 = 4\}$ (the two circles)
set is open)

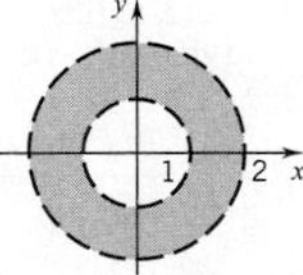

5. interior $= \{(x,y) : 1 < x^2 < 4\} = \{(x,y) : -2 < x < 1\} \cup \{(x,y) : 1 < x < 2\}$
(two vertical strips without the boundary lines)
boundary $= \{(x,y) : x = -2, x = -1, x = 1, \text{ or } x = 2\}$ (four vertical lines)
set is neither open nor closed

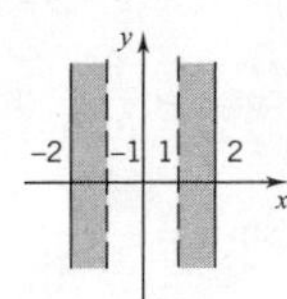

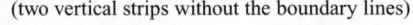

7. interior $= \{(x,y) : y < x^2\}$ (region below the parabola)
boundary $= \{(x,y) : y = x^2\}$ (the parabola)
set is closed

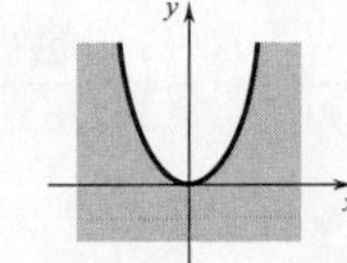

9. interior $= \{(x, y, z) : x^2 + y^2 < 1, 0 < z < 4\}$
(the inside of a cylinder)
boundary = the total surface of the cylinder
(the curved part, the top, the bottom)
set is closed

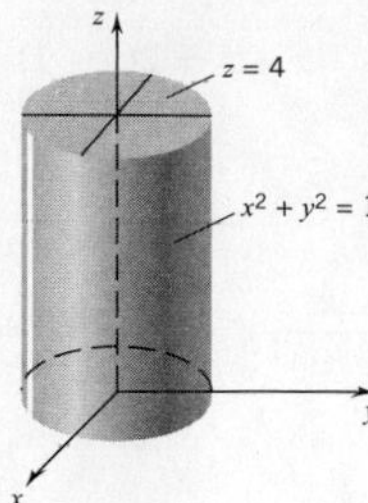

11. (a) ϕ (b) S (c) closed **13.** interior $= \{x : 1 < x < 3\}$, boundary $= \{1, 3\}$; set is closed

15. interior = the entire set, boundary $= \{1\}$; set is open **17.** interior $= \{x : |x| > 1\}$, boundary $= \{1, -1\}$; set is neither open nor closed

19. interior $= \phi$, boundary $=$ {the entire set} $\cup \{0\}$; the set is neither open nor closed

SECTION 14.6

1. $\dfrac{\partial^2 f}{\partial x^2} = 2A, \quad \dfrac{\partial^2 f}{\partial y^2} = 2C, \quad \dfrac{\partial^2 f}{\partial y \partial x} = \dfrac{\partial^2 f}{\partial x \partial y} = 2B$

3. $\dfrac{\partial^2 f}{\partial x^2} = Cy^2 e^{xy}, \quad \dfrac{\partial^2 f}{\partial y^2} = Cx^2 e^{xy}, \quad \dfrac{\partial^2 f}{\partial y \partial x} = \dfrac{\partial^2 f}{\partial x \partial y} = Ce^{xy}(xy + 1)$

5. $\dfrac{\partial^2 f}{\partial x^2} = 2, \quad \dfrac{\partial^2 f}{\partial y^2} = 4(x + 3y^2 + z^3), \quad \dfrac{\partial^2 f}{\partial z^2} = 6z(2x + 2y^2 + 5z^3)$

$\dfrac{\partial^2 f}{\partial x \partial y} = \dfrac{\partial^2 f}{\partial y \partial x} = 4y, \quad \dfrac{\partial^2 f}{\partial z \partial x} = \dfrac{\partial^2 f}{\partial x \partial z} = 6z^2, \quad \dfrac{\partial^2 f}{\partial z \partial y} = \dfrac{\partial^2 f}{\partial y \partial z} = 12yz^2$

7. $\dfrac{\partial^2 f}{\partial x^2} = \dfrac{1}{(x+y)^2} - \dfrac{1}{x^2}, \quad \dfrac{\partial^2 f}{\partial y^2} = \dfrac{1}{(x+y)^2}, \quad \dfrac{\partial^2 f}{\partial y \partial x} = \dfrac{\partial^2 f}{\partial x \partial y} = \dfrac{1}{(x+y)^2}$

9. $\dfrac{\partial^2 f}{\partial x^2} = 2(y + z), \quad \dfrac{\partial^2 f}{\partial y^2} = 2(x + z), \quad \dfrac{\partial^2 f}{\partial z^2} = 2(x + y)$; the second mixed partials are all $2(x + y + z)$

11. $\dfrac{\partial^2 f}{\partial x^2} = y(y-1)x^{y-2}, \quad \dfrac{\partial^2 f}{\partial y^2} = (\ln x)^2 x^y, \quad \dfrac{\partial^2 f}{\partial y \partial x} = \dfrac{\partial^2 f}{\partial x \partial y} = x^{y-1}(1 + y \ln x)$

13. $\dfrac{\partial^2 f}{\partial x^2} = y e^x, \quad \dfrac{\partial^2 f}{\partial y^2} = x e^y, \quad \dfrac{\partial^2 f}{\partial y \partial x} = e^y + e^x = \dfrac{\partial^2 f}{\partial x \partial y}$

15. $\dfrac{\partial^2 f}{\partial x^2} = \dfrac{y^2 - x^2}{(x^2 + y^2)^2}, \quad \dfrac{\partial^2 f}{\partial y^2} = \dfrac{x^2 - y^2}{(x^2 + y^2)^2}, \quad \dfrac{\partial^2 f}{\partial y \partial x} = -\dfrac{2xy}{(x^2 + y^2)^2} = \dfrac{\partial^2 f}{\partial x \partial y}$

17. $\dfrac{\partial^2 f}{\partial x^2} = -2y^2 \cos 2xy, \quad \dfrac{\partial^2 f}{\partial y^2} = -2x^2 \cos 2xy, \quad \dfrac{\partial^2 f}{\partial y \partial x} = -[\sin 2xy + 2xy \cos 2xy] = \dfrac{\partial^2 f}{\partial x \partial y}$

19. $\dfrac{\partial^2 f}{\partial x^2} = 0, \quad \dfrac{\partial^2 f}{\partial y^2} = xz \sin y, \quad \dfrac{\partial^2 f}{\partial z^2} = -xy \sin z$

$\dfrac{\partial^2 f}{\partial y \partial x} = \sin z - z \cos y = \dfrac{\partial^2 f}{\partial x \partial y}$

$\dfrac{\partial^2 f}{\partial z \partial x} = y \cos z - \sin y = \dfrac{\partial^2 f}{\partial x \partial z}$

$\dfrac{\partial^2 f}{\partial z \partial y} = x \cos z - x \cos y = \dfrac{\partial^2 f}{\partial y \partial z}$

21. $x^2 \dfrac{\partial^2 u}{\partial x^2} + 2xy \dfrac{\partial^2 u}{\partial x \partial y} + y^2 \dfrac{\partial^2 u}{\partial y^2} = x^2 \left(\dfrac{-2y^2}{(x+y)^3} \right) + 2xy \left(\dfrac{2xy}{(x+y)^3} \right) + y^2 \left(\dfrac{-2x^2}{(x+y)^3} \right) = 0$

23. (a) no, since $\dfrac{\partial^2 f}{\partial x \partial y} \neq \dfrac{\partial^2 f}{\partial y \partial x}$ (b) no, since $\dfrac{\partial^2 f}{\partial x \partial y} \neq \dfrac{\partial^2 f}{\partial y \partial x}$ for $x \neq y$

25.

$$\frac{\partial^3 f}{\partial x^2 \partial y} = \frac{\partial}{\partial x}\left(\frac{\partial^2 f}{\partial x \partial y}\right) \quad \text{(by defnition)}$$

$$= \frac{\partial}{\partial x}\left(\frac{\partial^2 f}{\partial y \partial x}\right) = \frac{\partial^2}{\partial x \partial y}\left(\frac{\partial f}{\partial x}\right) = \frac{\partial^2}{\partial y \partial x}\left(\frac{\partial f}{\partial x}\right) = \frac{\partial}{\partial y}\left(\frac{\partial^2 f}{\partial x^2}\right) = \frac{\partial^3 f}{\partial y \partial x^2}$$

(14.6.5) by definition (14.6.5) by definition

27. (a) 0 (b) 0 (c) $\dfrac{m}{1+m^2}$ (d) 0 (e) $\dfrac{f'(0)}{1+[f'(0)]^2}$ (f) $\frac{1}{4}\sqrt{3}$ (g) does not exist

29. (a) $\dfrac{\partial g}{\partial x}(0,0) = \lim_{h\to 0} \dfrac{g(h,0)-g(0,0)}{h} = \lim_{h\to 0} 0 = 0, \quad \dfrac{\partial g}{\partial y}(0,0) = \lim_{h\to 0} \dfrac{g(0,h)-g(0,0)}{h} - \lim_{h\to 0} = 0$

(b) as (x,y) tends to $(0,0)$ along the x-axis, $g(x,y) = g(x,0) = 0$ tends to 0;
as (x,y) tends to $(0,0)$ along the line $y = x$, $g(x,y) = g(x,x) = \frac{1}{2}$ tends to $\frac{1}{2}$

31. For $y \neq 0$, $\dfrac{\partial f}{\partial x}(0,y) = \lim_{h\to 0} \dfrac{f(h,y)-f(0,y)}{h} = \lim_{h\to 0} \dfrac{y(y^2-h^2)}{h^2+y^2} = y.$ Since $\dfrac{\partial f}{\partial x}(0,0) = \lim_{h\to 0} \dfrac{f(h,0)-f(0,0)}{h} = \lim_{h\to 0} 0 = 0,$

we have $\dfrac{\partial f}{\partial x}(0,y) = y$ for all y. For $x \neq 0$, $\dfrac{\partial f}{\partial y}(x,0) = \lim_{h\to 0} \dfrac{f(x,h)-f(x,0)}{h} = \lim_{h\to 0} \dfrac{x(h^2-x^2)}{x^2+h^2} = -x.$

Since $\dfrac{\partial f}{\partial y}(0,0) = \lim_{h\to 0} \dfrac{f(0,h)-f(0,0)}{h} = \lim_{h\to 0} 0 = 0$, we have $\dfrac{\partial f}{\partial y}(x,0) = -x$ for all x.

Therefore $\dfrac{\partial^2 f}{\partial y \partial x}(0,y) = 1$ for all y and $\dfrac{\partial^2 f}{\partial x \partial y}(x,0) = -1$ for all x. In particular, $\dfrac{\partial^2 f}{\partial y \partial x}(0,0) = 1$, while $\dfrac{\partial^2 f}{\partial x \partial y}(0,0) = -1$.

33. f must have the form: $f(x,y) = g(x) + h(y)$

CHAPTER 15

SECTION 15.1

1. $(6x-y)\,\mathbf{i} + (1-x)\,\mathbf{j}$ **3.** $e^{xy}[(xy+1)\,\mathbf{i} + x^2\,\mathbf{j}]$ **5.** $[2y^2 \sin(x^2+1) + 4x^2y^2\cos(x^2+1)]\,\mathbf{i} + 4xy\sin(x^2+1)\,\mathbf{j}$

7. $(e^{x-y}+e^{y-x})(\mathbf{i} - \mathbf{j})$ **9.** $(z^2+2xy)\,\mathbf{i} + (x^2+2yz)\,\mathbf{j} + (y^2+2xz)\,\mathbf{k}$ **11.** $e^{-z}(2xy\,\mathbf{i} + x^2\,\mathbf{j} - x^2y\,\mathbf{k})$

13. $e^{x+2y}\cos(z^2+1)\,\mathbf{i} + 2e^{x+2y}\cos(z^2+1)\,\mathbf{j} - 2ze^{x+2y}\sin(z^2+1)\,\mathbf{k}$ **15.** $\left[2y\cos(2xy) + \dfrac{2}{x}\right]\mathbf{i} + 2x\cos(2xy)\,\mathbf{j} + \dfrac{1}{z}\mathbf{k}$

17. $\nabla f = -\,\mathbf{i} + 18\,\mathbf{j}$ **19.** $\frac{4}{5}\,\mathbf{i} + \frac{2}{5}\,\mathbf{j}$ **21.** $\mathbf{i}$ **23.** $\nabla f = -\frac{1}{2}\sqrt{2}(\mathbf{i} + 2\,\mathbf{j} + \mathbf{k})$ **25.** $\mathbf{i} + \frac{3}{5}\,\mathbf{j} - \frac{4}{5}\mathbf{k}$

27. (a) $\nabla f(0,2) = 4\,\mathbf{i}$

(b) $\nabla f(\pi/4, \pi/6) = \left(-1 - \dfrac{-1+\sqrt{3}}{2\sqrt{2}}\right)\mathbf{i} + \left(-\dfrac{1}{2} + \dfrac{-1+\sqrt{3}}{\sqrt{2}}\right)\mathbf{j}$

(c) $\nabla f(1,e) = (1-2e)\,\mathbf{i} - 2\,\mathbf{j}$

29. $(6x-y)\,\mathbf{i} + (1-x)\,\mathbf{j}$ **31.** $(2xy+z^2)\,\mathbf{i} + (2yz+x^2)\,\mathbf{j} + (2xz+y^2)\,\mathbf{k}$

33. $f(x,y) = x^2y + y$ **35.** $f(x,y) = \dfrac{x^2}{2} + x\sin y - y^2$ **37.** (a) $(1/r^2)\,\mathbf{r}$ (b) $[(\cos r)/r]\mathbf{r}$ (c) $(e^r/r)\mathbf{r}$

39. (a) $(0,0)$ (b)

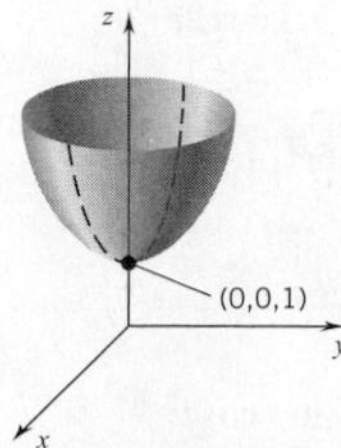

(c) f has an absolute minimum at $(0,0)$

41. (a) Let $\mathbf{c} = c_1\mathbf{i} + c_2\mathbf{j} + c_3\mathbf{k}$. First, we take $\mathbf{h} = h\mathbf{i}$. Since $\mathbf{c} \cdot \mathbf{h}$ is $o(\mathbf{h})$.

$$0 = \lim_{h\to 0} \frac{\mathbf{c} \cdot \mathbf{h}}{||\mathbf{h}||} = \lim_{h\to 0} \frac{c_1 h}{h} = c_1.$$

Similarly, $c_2 = 0$ and $c_3 = 0$.

(b) $(\mathbf{y} - \mathbf{z}) \cdot \mathbf{h} = [f(\mathbf{x}+\mathbf{h}) - f(\mathbf{x}) - \mathbf{z} \cdot \mathbf{h}] + [\mathbf{y} \cdot \mathbf{h} - f(\mathbf{x}+\mathbf{h}) + f(\mathbf{x})] = o(\mathbf{h}) + o(\mathbf{h}) = o(\mathbf{h})$, so that, by part (a), $\mathbf{y} - \mathbf{z} = 0$.

43. (a) In Section 14.6 we showed that f was not continuous at $(0, 0)$. It is therefore not differentiable at $(0, 0)$.

(b) For $(x, y) \neq (0, 0)$, $\dfrac{\partial f}{\partial x} = \dfrac{2y(y^2 - x^2)}{(x^2 + y^2)^2}$. As (x, y) tends to $(0, 0)$ along the y-axis, $\partial f/\partial x = 2/y$ tends to ∞.

SECTION 15.2

1. $-2\sqrt{2}$ **3.** $\frac{1}{5}(7 - 4e)$ **5.** $\frac{1}{4}\sqrt{2}(a - b)$ **7.** $\dfrac{2}{\sqrt{65}}$ **9.** $\frac{2}{3}\sqrt{6}$ **11.** $-3\sqrt{2}$ **13.** $\dfrac{\sqrt{3}\pi}{12}$ **15.** $-(x^2 + y^2)^{-1/2}$

17. (a) $\sqrt{2}[a(B - A) + b(C - B)]$ (b) $\sqrt{2}[a(A - B) + b(B - C)]$ **19.** $-\frac{7}{5}\sqrt{5}$ **21.** $\dfrac{18}{\sqrt{14}}$ or $\dfrac{-18}{\sqrt{14}}$

23. increases most rapidly in the direction of $\dfrac{1}{\sqrt{2}}\mathbf{i} + \dfrac{1}{\sqrt{2}}\mathbf{j}$, rate of change $2\sqrt{2}$; decreases most rapidly in the direction of $-\dfrac{1}{\sqrt{2}}\mathbf{i} - \dfrac{1}{\sqrt{2}}\mathbf{j}$, rate of change $-2\sqrt{2}$

25. increases most rapidly in the direction of $\dfrac{1}{\sqrt{6}}\mathbf{i} - \dfrac{2}{\sqrt{6}}\mathbf{j} + \dfrac{1}{\sqrt{6}}\mathbf{k}$, rate of change 1; decreases most rapidly in the direction of $-\dfrac{1}{\sqrt{6}}\mathbf{i} + \dfrac{2}{\sqrt{6}}\mathbf{j} - \dfrac{1}{\sqrt{6}}\mathbf{k}$, rate of change -1

27. $\nabla f = f'(x_0)\mathbf{i}$. If $f'(x_0) \neq 0$, the gradient points in the direction in which f increases: to the right if $f'(x_0) > 0$, to the left if $f'(x_0) < 0$.

29. (a) $\displaystyle\lim_{h\to 0} \frac{f(h, 0) - f(0, 0)}{h} = \lim_{h\to 0} \frac{\sqrt{h^2}}{h} = \lim_{h\to 0} \frac{|h|}{h}$ does not exist (b) no; by Theorem 15.2.5 f cannot be differentiable at $(0, 0)$

31. (a) $-\frac{2}{3}\sqrt{97}$ (b) $-\frac{8}{3}$ (c) $-\frac{26}{3}\sqrt{2}$ **33.** (a) its projection onto the xy-plane is the curve $y = x^3$ from $(1, 1)$ to $(0, 0)$

(b) its projection onto the xy-plane is the curve $y = -2x^3$ from $(1, -2)$ to $(0, 0)$

35. its projection onto the xy-plane is the curve $(b^2)^{a^2} x^{b^2} = (a^2)^{b^2} y^{a^2}$ from (a^2, b^2) to $(0, 0)$

37. the curve $y = \ln|\sqrt{2}\sin x|$ in the direction of decreasing x

39. (a) 16 (b) 4 (c) $\frac{16}{17}\sqrt{17}$

(d) The limits computed in (a) and (b) are not directional derivatives. In (a) and (b) we have, in essence, computed $\nabla f(2, 4) \cdot \mathbf{r}_0$ taking $\mathbf{r}_0 = \mathbf{i} + 4\mathbf{j}$ in (a) and $\mathbf{r}_0 = \frac{1}{4}\mathbf{i} + \mathbf{j}$ in (b). In neither case is $\mathbf{r}_0$ a unit vector.

41. (b) $\dfrac{2\sqrt{3} - 3}{2}$ **43.** $\nabla(fg) = \left(f\dfrac{\partial g}{\partial x} + g\dfrac{\partial f}{\partial x}\right)\mathbf{i} + \left(f\dfrac{\partial g}{\partial y} + g\dfrac{\partial f}{\partial y}\right)\mathbf{j} = f\nabla g + g\nabla f$

45. $\nabla f^n = \dfrac{\partial f^n}{\partial x}\mathbf{i} + \dfrac{\partial f^n}{\partial y}\mathbf{j} = nf^{n-1}\dfrac{\partial f}{\partial x}\mathbf{i} + nf^{n-1}\dfrac{\partial f}{\partial y}\mathbf{j} = nf^{n-1}\nabla f$

SECTION 15.3

1. $C = (\frac{1}{3}, \frac{5}{3})$ **3.** (a) $f(x, y, z) = a_1x + a_2y + a_3z + C$ (b) $f(x, y, z) = g(x, y, z) + a_1x + a_2y + a_3z + C$

5. (a) U is not connected (b) (i) $g(\mathbf{x}) = f(\mathbf{x}) - 1$ (ii) $g(\mathbf{x}) = -f(\mathbf{x})$ **7.** e^t **9.** $\dfrac{-2\sin 2t}{1 + \cos^2 2t}$

11. $t^t\left[\dfrac{1}{t} + \ln t + (\ln t)^2\right] + \dfrac{1}{t}$ **13.** $3t^2 - 5t^4$ **15.** $2\omega(b^2 - a^2)\sin\omega t\cos\omega t + b\omega$ **17.** $\sin 2t - 3\cos 2t$

19. $e^{t/2}(\frac{1}{2}\sin 2t + 2\cos 2t) + e^{2t}(2\sin\frac{1}{2}t + \frac{1}{2}\cos\frac{1}{2}t)$ **21.** $e^{t^2}\left[2t\sin\pi t + \pi\cos\pi t\right]$ **23.** $1 - 4t + 6t^2 - 4t^3$

25. increasing $\frac{1288}{3}\pi$ in.^{3}/ sec **27.** 41.34 sq in. / sec **29.** $\dfrac{\partial u}{\partial s} = 2s\cos^2 t - t\sin s\cos t - st\cos s\cos t$

$$\frac{\partial u}{\partial t} = -2s^2\sin t\cos t + st\sin s\sin t - s\sin s\cos t$$

31. $\dfrac{\partial u}{\partial s} = 4s^3t^2 \tan(s+t^2) + s^4t^2 \sec^2(s+t^2);$

$\dfrac{\partial u}{\partial t} = 2s^4 t \tan(s+t^2) + 2s^4t^3 \sec^2(s+t^2)$

33. $\dfrac{\partial u}{\partial s} = 2s\cos^2 t - \sin(t-s)\cos t + s\cos t\cos(t-s) + 2t^2 \sin s\cos s$

$\dfrac{\partial u}{\partial t} = -2s^2 \sin t\cos t + s\sin(t-s)\sin t - s\cos t\cos(t-s) + 2t\sin^2 s$

35. $\dfrac{d}{dt}[f(\mathbf{r}(t))] = \left[\nabla f(\mathbf{r}(t)) \cdot \dfrac{\mathbf{r}'(t)}{\|\mathbf{r}'(t)\|}\right] \|\mathbf{r}'(t)\| = f'_{(\mathbf{u})}(\mathbf{r}(t))\,\|\mathbf{r}'(t)\|$ where $\mathbf{u}(t) = \dfrac{\mathbf{r}'(t)}{\|\mathbf{r}'(t)\|}$

37. (a) $(\cos r)\dfrac{\mathbf{r}}{r}$ (b) $(r\cos r + \sin r)\dfrac{\mathbf{r}}{r}$ **39.** (a) $(r\cos r - \sin r)\dfrac{\mathbf{r}}{r^3}$ (b) $\left(\dfrac{\sin r - r\cos r}{\sin^2 r}\right)\dfrac{\mathbf{r}}{r}$

41. (a) See the figure

(b) $\dfrac{\partial u}{\partial r} = \dfrac{\partial u}{\partial x}\left(\dfrac{\partial x}{\partial w}\dfrac{\partial w}{\partial r} + \dfrac{\partial x}{\partial t}\dfrac{\partial t}{\partial r}\right) + \dfrac{\partial u}{\partial y}\left(\dfrac{\partial y}{\partial w}\dfrac{\partial w}{\partial r} + \dfrac{\partial y}{\partial t}\dfrac{\partial t}{\partial r}\right) + \dfrac{\partial u}{\partial z}\left(\dfrac{\partial z}{\partial w}\dfrac{\partial w}{\partial r} + \dfrac{\partial z}{\partial t}\dfrac{\partial t}{\partial r}\right)$

To obtain $\dfrac{\partial u}{\partial s}$, replace each r by s.

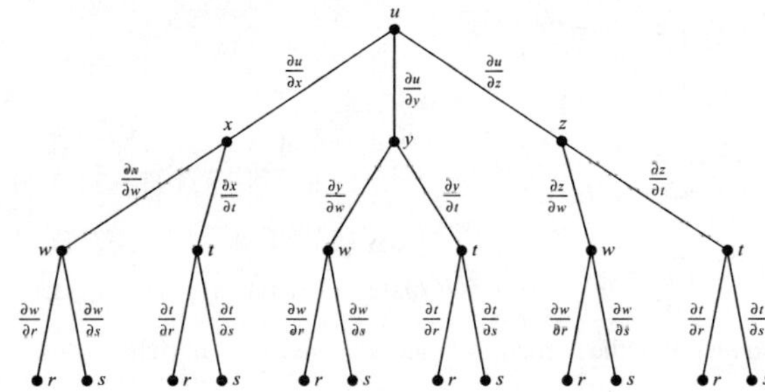

43. $\dfrac{du}{dt} = \dfrac{\partial u}{\partial x}\dfrac{dx}{dt} + \dfrac{\partial u}{\partial y}\dfrac{dy}{dt}$

$\dfrac{d^2u}{dt^2} = \dfrac{\partial u}{\partial x}\dfrac{d^2x}{dt^2} + \dfrac{dx}{dt}\left[\dfrac{\partial^2 u}{\partial x^2}\dfrac{dx}{dt} + \dfrac{\partial^2 u}{\partial y\partial x}\dfrac{dy}{dt}\right] + \dfrac{\partial u}{\partial y}\dfrac{d^2y}{dt^2} + \dfrac{dy}{dt}\left[\dfrac{\partial^2 u}{\partial x\partial y}\dfrac{dx}{dt} + \dfrac{\partial^2 u}{\partial y^2}\dfrac{dy}{dt}\right]$ and the result follows.

45. (a) $\dfrac{\partial u}{\partial r} = \dfrac{\partial u}{\partial x}\dfrac{\partial x}{\partial r} + \dfrac{\partial u}{\partial y}\dfrac{\partial y}{\partial r} = \dfrac{\partial u}{\partial x}\cos\theta + \dfrac{\partial u}{\partial y}\sin\theta,\quad \dfrac{\partial u}{\partial \theta} = \dfrac{\partial u}{\partial x}\dfrac{\partial x}{\partial \theta} + \dfrac{\partial u}{\partial y}\dfrac{\partial y}{\partial \theta} = \dfrac{\partial u}{\partial x}(-r\sin\theta) + \dfrac{\partial u}{\partial y}(r\cos\theta)$

(b) $\left(\dfrac{\partial u}{\partial r}\right)^2 = \left(\dfrac{\partial u}{\partial x}\right)^2\cos^2\theta + 2\dfrac{\partial u}{\partial x}\dfrac{\partial u}{\partial y}\cos\theta\sin\theta + \left(\dfrac{\partial u}{\partial y}\right)^2\sin^2\theta,\ \dfrac{1}{r^2}\left(\dfrac{\partial u}{\partial \theta}\right)^2 = \left(\dfrac{\partial u}{\partial x}\right)^2\sin^2\theta - 2\dfrac{\partial u}{\partial x}\dfrac{\partial u}{\partial y}\cos\theta\sin\theta + \left(\dfrac{\partial u}{\partial y}\right)^2\cos^2\theta,$

$\left(\dfrac{\partial u}{\partial r}\right)^2 + \dfrac{1}{r^2}\left(\dfrac{\partial u}{\partial \theta}\right)^2 = \left(\dfrac{\partial u}{\partial x}\right)^2(\cos^2\theta + \sin^2\theta) + \left(\dfrac{\partial u}{\partial y}\right)^2(\sin^2\theta + \cos^2\theta) = \left(\dfrac{\partial u}{\partial x}\right)^2 + \left(\dfrac{\partial u}{\partial y}\right)^2$

47. Solve the equations in Exercise 45 (a) for $\dfrac{\partial u}{\partial x}, \dfrac{\partial u}{\partial y}$:

$\dfrac{\partial u}{\partial x} = \dfrac{\partial u}{\partial r}\cos\theta - \dfrac{1}{r}\dfrac{\partial u}{\partial \theta}\sin\theta;\quad \dfrac{\partial u}{\partial y} = \dfrac{\partial u}{\partial r}\sin\theta + \dfrac{1}{r}\dfrac{\partial u}{\partial \theta}\cos\theta$

Then $\nabla u = \dfrac{\partial u}{\partial x}\mathbf{i} + \dfrac{\partial u}{\partial y}\mathbf{j} = \dfrac{\partial u}{\partial r}(\cos\theta\,\mathbf{i} + \sin\theta\,\mathbf{j}) + \dfrac{1}{r}\dfrac{\partial u}{\partial \theta}(-\sin\theta\,\mathbf{i} + \cos\theta\,\mathbf{j})$

49. $\nabla u = r(2 - \sin 2\theta)\,\mathbf{e}_r - r\cos 2\theta\,\mathbf{e}_\theta$

51. From Exercise 45(a);

$\dfrac{\partial^2 u}{\partial r^2} = \dfrac{\partial^2 u}{\partial x^2}\cos^2\theta + 2\dfrac{\partial^2 u}{\partial y\partial x}\sin\theta\cos\theta + \dfrac{\partial^2 u}{\partial y^2}\sin^2\theta$

$\dfrac{\partial^2 u}{\partial \theta^2} = \dfrac{\partial^2 u}{\partial x^2}r^2\sin^2\theta - 2\dfrac{\partial^2 u}{\partial y\partial x}r^2\sin\theta\cos\theta + \dfrac{\partial^2 u}{\partial y^2}r^2\cos^2\theta - r\left(\dfrac{\partial u}{\partial x}\cos\theta + \dfrac{\partial u}{\partial y}\sin\theta\right).$

53. $\dfrac{dy}{dx} = -\dfrac{e^y + y\,e^x - 4xy}{x\,e^y + e^x - 2x^2}$

The term in parentheses is just $\frac{\partial u}{\partial r}$, and the result follows.

55. $\frac{dy}{dx} = \frac{\cos xy - xy \sin xy - y \sin x}{x^2 \sin xy - \cos x}$ **57.** $\frac{\partial z}{\partial x} = -\frac{2x - yz(x^2+y^2+z^2)\sin xyz}{2z - xy(x^2+y^2+z^2)\sin xyz}$; $\frac{\partial z}{\partial y} = -\frac{2y - xz(x^2+y^2+z^2)\sin xyz}{2z - xy(x^2+y^2+z^2)\sin xyz}$

59. $\frac{\partial \mathbf{u}}{\partial s} = \frac{\partial \mathbf{u}}{\partial x}\frac{\partial x}{\partial s} + \frac{\partial \mathbf{u}}{\partial y}\frac{\partial y}{\partial s}$, $\frac{\partial \mathbf{u}}{\partial t} = \frac{\partial \mathbf{u}}{\partial x}\frac{\partial x}{\partial t} + \frac{\partial \mathbf{u}}{\partial y}\frac{\partial y}{\partial t}$

where

$\frac{\partial \mathbf{u}}{\partial x} = \frac{\partial u_1}{\partial x}\mathbf{i} + \frac{\partial u_2}{\partial x}\mathbf{j}$, $\frac{\partial \mathbf{u}}{\partial y} = \frac{\partial u_1}{\partial y}\mathbf{i} + \frac{\partial u_2}{\partial y}\mathbf{j}$

SECTION 15.4

1. normal vector $\mathbf{i}+\mathbf{j}$; tangent vector $\mathbf{i}-\mathbf{j}$
tangent line $x+y+2=0$; normal line $x-y=0$

3. normal vector $\sqrt{2}\,\mathbf{i} - 5\,\mathbf{j}$; tangent vector $5\,\mathbf{i} + \sqrt{2}\,\mathbf{j}$
tangent line $\sqrt{2}x - 5y + 3 = 0$; normal line $5x + \sqrt{2}y - 6\sqrt{2} = 0$

5. normal vector $7\,\mathbf{i} - 17\,\mathbf{j}$; tangent vector $17\,\mathbf{i} + 7\,\mathbf{j}$
tangent line $7x - 17y + 6 = 0$; normal line $17x + 7y - 82 = 0$

7. normal vector $\mathbf{i}-\mathbf{j}$; tangent vector $\mathbf{i}+\mathbf{j}$
tangent line $x - y - 3 = 0$; normal line $x + y + 1 = 0$

9 0.

11. $4x - 5y + 4z = 0$; $x = 1 + 4t,\ y = 2 - 5t,\ z = \frac{3}{2} + 4t$

13. $x + ay - z - 1 = 0$; $x = 1 + t,\ y = \frac{1}{a} + at,\ z = 1 - t$

15. $2x + 2y - z = 0$; $x = 2t,\ y = 2t,\ z = -t$

17. $b^2c^2x_0x - a^2c^2y_0y - a^2b^2z_0z - a^2 - b^2 - c^2 = 0$;
$x = x_0 + 2b^2c^2x_0t,\ y = y_0 - 2a^2c^2y_0t,\ z = z_0 - 2a^2bc^2z_0t$

19. $(a^2/b, b^2/a, 3ab)$ **21.** $(0,0,0)$ **23.** $(\frac{1}{3}, \frac{11}{6}, -\frac{1}{12})$ **25.** $\frac{x - x_0}{\partial f/\partial x(x_0,y_0,z_0)} = \frac{y - y_0}{\partial f/\partial y(x_0,y_0,z_0)} = \frac{z - z_0}{\partial f/\partial z(x_0,y_0,z_0)}$

27. the tangent planes meet at right angles and therefore the normals ∇F and ∇G must meet at right angles:

$$\frac{\partial F}{\partial x}\frac{\partial G}{\partial x} + \frac{\partial F}{\partial y}\frac{\partial G}{\partial y} + \frac{\partial F}{\partial z}\frac{\partial G}{\partial z} = 0$$

29. $\frac{9}{2}a^3\ (V = \frac{1}{3}Bh)$ **31.** approx. 0.528 rad **33.** $3x + 4y + 6z = 22,\quad 6x + y - z = 11$

35. $(1, 1, 2)$ lies on both surfaces and the normals at this point are perpendicular.

37. (a) $3x + 4y + 6 = 0$ (b) $\mathbf{r}(t) = (4t-2)\mathbf{i} - 3t\mathbf{j} + (43t^2 - 16t + 6)\mathbf{k}$ (c) $\mathbf{R}(s) = (2\mathbf{i} - 3\mathbf{j} + 33\mathbf{k}) + s(4\mathbf{i} - 3\mathbf{j} + 70\mathbf{k})$
(d) $4x - 18y - z = 29$ (e) $\mathbf{r}(t) = t\mathbf{i} - (\frac{3}{4}t + \frac{3}{2})\mathbf{j} + (\frac{35}{2}t - 2)\,\mathbf{k}$; $l = l'$

39. (a) $2\mathbf{i} + 2\mathbf{j} + 4\mathbf{k}$; $x = 1 + 2t,\ y = 2 + 2t,\ z = 2 + 4t$
(b) $x + y + 2z - 7 = 0$

41. (c) $\nabla f(x,y) = \mathbf{0}$ at $(0,0), (\pm 1, 0), (0, \pm 1), (1, \pm 1), (-1, \pm 1)$

SECTION 15.5

1. $(1, 0)$ gives a local max of 1 **3.** $(-2, 1)$ gives a local min of -2 **5.** $(4, -2)$ gives a local min of -10

7. $(0, 0)$ is a saddle point; $(2, 2)$ gives a local min of -8 **9.** $(1, \frac{3}{2})$ is a saddle point; $(5, \frac{27}{2})$ gives a local min of $-\frac{117}{4}$

11. $(0, n\pi)$ for integral n are saddle points; no local extreme values **13.** $(1, -1)$ and $(-1, 1)$ are saddle points; no local extreme values

15. $(\frac{1}{2}, 4)$ gives a local min of 6 **17.** $(1, 1)$ gives a local min of 3 **19.** $(1, 0)$ gives a local min of -1; $(-1, 0)$ gives a local max of 1

21. $(0, 0)$ is a saddle point; $(1, 0)$ and $(-1, 0)$ give a local min of -3

23. (π, π) is a saddle point; $\left(\frac{\pi}{2}, \frac{\pi}{2}\right)$ and $\left(\frac{3\pi}{2}, \frac{3\pi}{2}\right)$ give a local maximum of 1;
$\left(\frac{\pi}{2}, \frac{3\pi}{2}\right)$ and $\left(\frac{3\pi}{2}, \frac{\pi}{2}\right)$ give a local minimum of -1.

25. (a) $f_x = 2x + ky, f_y = 2y + kx$; $f_x(0,0) = f_y(0,0) = 0$ independent of k (b) $|k| > 2$ (c) $|k| < 2$ (d) $|k| = 2$

27. $(\frac{32}{9}, -\frac{16}{9}, \frac{32}{9})$; $\frac{16}{3}$ **29.** $\dfrac{\sqrt{114}}{6}$

33. $(0, 0)$ is a saddle point; $(1, 1)$ gives a local maximum of 3.

35. $(-1, 0)$ gives a local maximum of 1; $(1, 0)$ gives a local minimum of -1.

SECTION 15.6

1. $(1, 1)$ gives absolute min of -1; $(2, 4)$ gives absolute max of 10

3. $(4, -2)$ gives absolute min of -13 ; $(0, -3)$ gives absolute max of 8

5. $(\sqrt{2}, -\sqrt{2})$ and $(-\sqrt{2}, \sqrt{2})$ give absolute min of 0; $(\sqrt{2}, \sqrt{2})$ and $(-\sqrt{2}, -\sqrt{2})$ give absolute max of 12

7. $(1, 1)$ gives absolute min of 0; $\left(-\sqrt{2}, -\sqrt{2}\right)$ gives absolute max of $6 + 4\sqrt{2}$

9. $(1, 0)$ gives absolute min of -1; $(-1, 0)$ gives absolute max of 1

11. absolute min of 0 along the lines $x = 0$ and $x = 2$; $(1, 0)$ gives absolute max of 2

13. $(\sqrt{2}, 2)$ and $(-\sqrt{2}, -2)$ give absolute min of $-8 - 4\sqrt{2}$; $(-1, 1)$ gives absolute max of 1

15. $(0, 1)$ gives absolute min of -1; $(0, -1)$ gives absolute max of 1

17. absolute min of 0 along the line $y = x, 0 \le x \le 4$; $(0, 12)$ gives absolute max of 144

19. $x = 6, y = 6, z = 6$; maximum $= 216$ **21.** $\frac{1}{27}$

23. (a) $(0, 0)$ (b) no local extremes as $(0, 0)$ is a saddle point
(c) $(1, 0)$ and $(-1, 0)$ give absolute max of $\frac{1}{4}$; $(0, 1)$ and $(0, -1)$ give absolute min of $-\frac{1}{9}$

25. (a) saddle point, $f(x, y) = 0$ along the plane curve $y = x^{2/3}$ (see figure)
(b) $(0, 0)$ gives a local max of 3

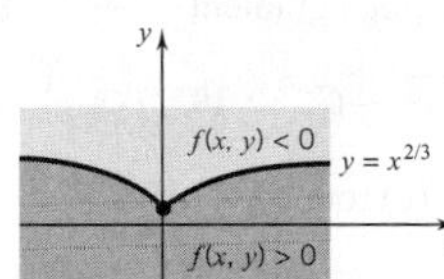

27. $\left(\dfrac{x_1 + x_2 + x_3}{3}, \dfrac{y_1 + y_2 + y_3}{3}\right)$ **29.** $\theta = \frac{1}{6}\pi, x = (2 - \sqrt{3})P, y = \frac{1}{6}(3 - \sqrt{3})P$ **31.** $\frac{2}{3}\sqrt{6}$ **33.** $4 \times 4 \times 6$ m

35. $V = \dfrac{xy(S - 2xy)}{2(x + y)}$ has a maximum value when $x = y = z = \sqrt{\dfrac{S}{6}}$. **37.** (a) $y = x - \frac{2}{3}$ (b) $y = \frac{14}{13}x^2 - \frac{19}{13}$

39. (a) cross section 18×18 inches; length 36 inches
(b) radius of cross section $36/\pi$ inches; length 36 inches

41. $x = 4$ in., $\theta = \dfrac{\pi}{3}$

SECTION 15.7

1. 2 **3.** $-\frac{1}{2}ab$ **5.** $\frac{2}{9}\sqrt{3}ab^2$ **7.** 1 **9.** $\frac{1}{9}\sqrt{3}abc$ **11.** $19\sqrt{2}$ **13.** $\frac{1}{27}abc$ **15.** 1

17. closest point $(\frac{2}{3}, \frac{1}{3}, \frac{2}{3})$; furthest point $(-\frac{2}{3}, -\frac{1}{3}, -\frac{2}{3})$ **19.** $f(3, -2, 1) = 14$ **21.** $|D|(A^2 + B^2 + C^2)^{-1/2}$

23. $4A^2(a^2 + b^2 + c^2)^{-1}$, where A is the area of the triangle and a, b, c, are the sides **25.** $(2^{-1/3}, -2^{-1/3})$ **27.** hint is given

29. (a) $f(\frac{k}{2}, \frac{k}{2}) = \frac{k}{2}$ is the maximum value (b) $(xy)^{1/2} = f(x, y) \le f(\frac{k}{2}, \frac{k}{2}) = \frac{k}{2} = \dfrac{x + y}{2}$

31. Same argument as Exercises 29 and 30: $f\left(\dfrac{k}{n}, \dfrac{k}{n}, \dots, \dfrac{k}{n}\right) = \dfrac{k}{n}$ is the maximum value of $f(x_1, x_2, \dots, x_n) = (x_1 x_2 \cdots x_n)^{1/n}$

33. radius $\sqrt[3]{\dfrac{V}{2\pi}}$; height $2\sqrt[3]{\dfrac{V}{2\pi}}$ **35.** $\dfrac{abc}{27}$ **37.** $4 \times 4 \times 6$ **39.** $\sqrt{\dfrac{S}{3}} \times \sqrt{\dfrac{S}{3}} \times \dfrac{1}{2}\sqrt{\dfrac{S}{3}}$

41. (a) cross section 18×18 inches; length 36 inches (b) radius of cross section $36/\pi$ inches; length 36 inches

43. $Q_1 = 10{,}000, Q_2 = 20{,}000, Q_3 = 30{,}000$

SECTION 15.8

1. $df = (3x^2y - 2xy^2)\Delta x + (x^3 - 2x^2y)\Delta y$ **3.** $df = (\cos y + y\sin x)\Delta x - (x\sin y + \cos x)\Delta y$

5. $df = \Delta x - (\tan z)\Delta y - (y\sec^2 z)\Delta z$

7. $df = \dfrac{y(y^2+z^2-x^2)}{(x^2+y^2+z^2)^2}\Delta x + \dfrac{x(x^2+z^2-y^2)}{(x^2+y^2+z^2)^2}\Delta y - \dfrac{2xyz}{(x^2+y^2+z^2)^2}\Delta z$

9. $df = [\cos(x+y) + \cos(x-y)]\Delta x + [\cos(x+y) - \cos(x-y)]\Delta y$

11. $df = (y^2 z\, e^{xz} + \ln z)\Delta x + 2y\, e^{xz}\Delta y + \left(xy^2 e^{xz} + \dfrac{x}{z}\right)\Delta z$ **13.** $\Delta u = -7.15, du = -7.50$ **15.** $\Delta u = 2.896; du = 2.5$

17. $22\frac{249}{352}$ taking $u = x^{1/2}y^{1/4}$, $x = 121$, $y = 16$, $\Delta x = 4$, $\Delta y = 1$

19. $\frac{\pi}{14}\sqrt{2}$ taking $u = \sin x\cos y$, $x = \pi$, $y = \frac{1}{4}\pi$, $\Delta x = -\frac{1}{7}\pi$, $\Delta y = -\frac{1}{20}\pi$ **21.** $f(2.9, 0.01) \cong 8.67$

23. $f(2.94, 1.1, 0.92) \cong 2.3391$ **25.** $dz = -\frac{1}{90}$, $\Delta z = -\frac{1}{93}$ **27.** decreases about 13.6π in.^{2} **29.** $S \cong 246.8$

31. (a) $dv = 0.24$ (b) $\Delta V = 0.22077$ **33.** $dT = 2.9$ **35.** $dA = 3\,\Delta x + 12.5\Delta\theta$; more sensitive to a change in θ

37. (a) $\Delta h = -\dfrac{(2r+\Delta r)h}{(r+\Delta r)^2}\Delta r$, $\Delta h \cong -\left(\dfrac{2h}{r}\right)\Delta r$ (b) $\Delta h = -\dfrac{(2r+h+\Delta r)}{r+\Delta r}\Delta r$, $\Delta h \cong -\left(\dfrac{2r+h}{r}\right)\Delta r$

39. (a) 1.962cm (b) 12.75cm^2 **41.** $2.23 \le s \pm |\Delta s| \le 2.27$

SECTION 15.9

1. $f(x,y) = \frac{1}{2}x^2y^2 + C$ **3.** $f(x,y) = xy + C$ **5.** not a gradient **7.** $f(x,y) = \sin x + y\cos x + C$

9. $f(x,y) = e^x\cos y^2 + C$ **11.** $f(x,y) = xy\,e^x + e^{-y} + C$ **13.** not a gradient

15. $f(x,y) = x + xy^2 + \frac{1}{2}x^2y^2 + \frac{1}{2}y^2 + y + C$ **17.** $f(x,y) = \sqrt{x^2+y^2} + C$ **19.** $f(x,y) = \frac{1}{3}x^3\sin^{-1}y + y - y\ln y + C$

21. (a) yes (b) yes (c) no **23.** $f(x,y) = Ce^{x+y}$ **25.** (a), (b), (c) routine; (d) $f(x,y,z) = x^2 + yz + C$

27. $f(x,y,z) = x^2 + y^2 - z^2 + xy + yz + C$ **29.** $f(x,y,z) = xy^2z^3 + x + \frac{1}{2}y^2 + z + C$ **31.** $\mathbf{F}(\mathbf{r}) = \nabla\left(G\dfrac{mM}{r}\right)$

CHAPTER 16

SECTION 16.1

1. 819 **3.** 0 **5.** $a_2 - a_1$ **7.** $(a_2-a_1)(b_2-b_1)$ **9.** $a_2^2 - a_1^2$ **11.** $(a_2^2 - a_1^2)(b_2 - b_1)$

13. $2n(a_2-a_1) - 3m(b_2-b_1)$ **15.** $(a_2-a_1)(b_2-b_1)(c_2-c_1)$ **17.** $a_{111} + a_{222} + \cdots + a_{nnn} = \sum_{p=1}^{n} a_{ppp}$

SECTION 16.2

1. $L_f(P) = 2\frac{1}{4}$, $U_f(P) = 5\frac{3}{4}$ **3.** (a) $L_f(P) = \sum_{i=1}^{m}\sum_{j=1}^{n}(x_{i-1} + 2y_{i-1})\,\Delta x_i\,\Delta y_j$, $U_f(P) = \sum_{i=1}^{m}\sum_{j=1}^{n}(x_i + 2y_j)\,\Delta x_i\,\Delta y_j$
(b) $I = 4$; the volume of the prism bounded above by the plane $z = x + 2y$ and below by R

5. $L_f(P) = -\frac{7}{24}$, $U_f(P) = \frac{7}{24}$ **7.** (a) $L_f(P) = \sum_{i=1}^{m}\sum_{j=1}^{n} 4x_{i-1}y_{j-1}\Delta x_i\,\Delta y_j$, $U_f(P) = \sum_{i=1}^{m}\sum_{j=1}^{n} 4x_iy_j\Delta x_i\,\Delta y_j$ (b) $I = b^2d^2$

9. (a) $L_f(P) = \sum_{i=1}^{m}\sum_{j=1}^{n} 3(x_{i-1}^2 - y_j^2)\,\Delta x_i\,\Delta y_j$, $U_fP = \sum_{i=1}^{m}\sum_{j=1}^{n} 3(x_i^2 - y_{j-1}^2)\,\Delta x_i\,\Delta y_j$

(b) $I = bd(b^2 - d^2)$

11. $\displaystyle\iint_\Omega dxdy = \int_a^b \phi(x)\,dx$

13. Suppose $f(x_0, y_0) \neq 0$. Assume $f(x_0, y_0) > 0$. Since f is continuous, there exists a disc Ω_ϵ with radius ϵ centered at (x_0, y_0) such that $f(x, y) > 0$ on Ω_ϵ. Let R be a rectangle contained in Ω_ϵ. Then $\iint\limits_R f(x, y)\,dx\,dy > 0$, which contradicts the hypothesis. **15.** 6

17. By Theorem 16.2.10, there exists a point (x_1, y_1) in D_r such that

$$\iint\limits_{D_r} f(x,y)\,dxdy = f(x_1,y_1)\iint\limits_{D_r} dxdy = f(x_1,y_1)\pi r^2$$

Thus $f(x_1, y_1) = \dfrac{1}{\pi r^2}\iint\limits_{D_r} f(x,y)\,dxdy$. As $r \to 0, (x_1, y_1) \to (x_0, y_0)$ and $f(x_1, y_1) \to f(x_0, y_0)$ since f is continuous. The result follows.

19. $\frac{1}{8}$ of a sphere of radius 2; $\frac{4}{3}\pi$

21. tetrahedron bounded by the coordinate planes and the plane $\dfrac{x}{3} + \dfrac{y}{2} + \dfrac{z}{6} = 1$; 6

SECTION 16.3

1. 1 **3.** $\frac{9}{2}$ **5.** $\frac{1}{24}$ **7.** 2 **9.** $\frac{1}{4}\pi^2 + \frac{1}{64}\pi^4$ **11.** $\frac{2}{27}$ **13.** $\frac{512}{15}$ **15.** 0 **17.** $\frac{1}{4}(e^4 - 1)$

19. $\displaystyle\int_0^1 \int_{y^{1/2}}^{y^{1/4}} f(x,y)\,dx\,dy$

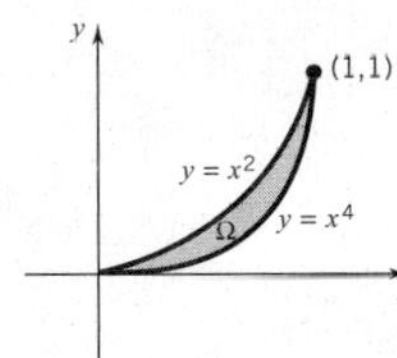

21. 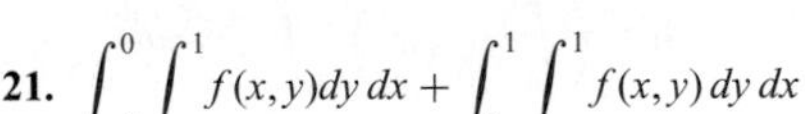
$\displaystyle\int_{-1}^{0} \int_{-x}^{1} f(x,y)dy\,dx + \int_0^1 \int_x^1 f(x,y)\,dy\,dx$

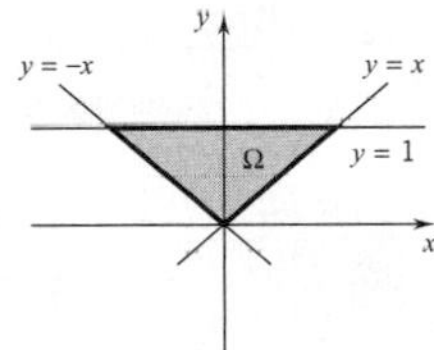

23. $\displaystyle\int_1^2 \int_1^y f(x,y)\,dx\,dy + \int_2^4 \int_{y/2}^{y} f(x,y)\,dx\,dy + \int_4^8 \int_{y/2}^{4} f(x,y)\,dx\,dy$ **25.** 9 **27.** $\frac{1}{160}$

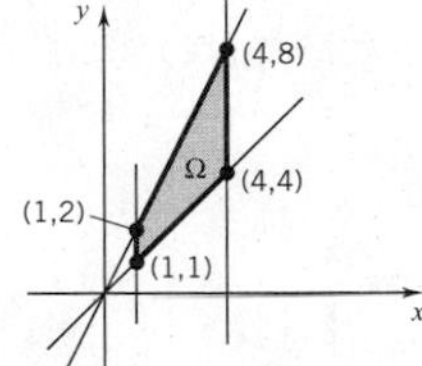

29. $\frac{2}{3}(\cos\frac{1}{2} - \cos 1)$ **31.** $1 - \ln 2$ 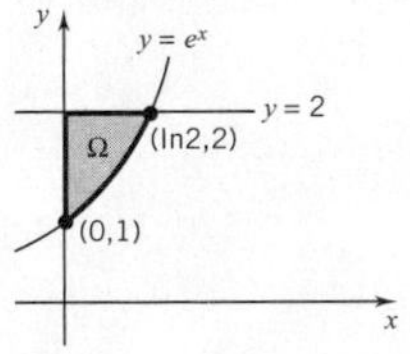 **33.** $\ln 4 - \frac{1}{2}$ **35.** 4 **37.** $\frac{2}{15}$

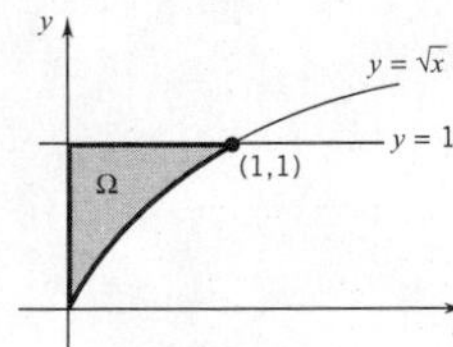

39. 3π **41.** $\frac{1}{6}$ **43.** $\frac{11}{70}$ **45.** $\frac{2}{3}a^3$ **47.** $\frac{1}{2}(e - 1)$ **49.** $\frac{1}{12}(e - 1)$ **51.** $\frac{2}{3}$ **53.** 1

55. $$\iint\limits_R f(x)g(y)\,dx\,dy = \int_c^d \int_a^b f(x)g(y)\,dx\,dy = \int_c^d \left(\int_a^b f(x)g(y)\,dx\right) dy = \int_c^d g(y)\left(\int_a^b f(x)\,dx\right) dy$$
$$= \left(\int_a^b f(x)\,dx\right)\left(\int_c^d g(y)\,dy\right)$$

57. We have $R: -a \le x \le a, c \le y \le d$. Set $f(x,y) = g_y(x)$. For each fixed $y \in [c,d]$, g_y is an odd function. Thus

$$\int_{-a}^{a} g_y(x)\,dx = 0.$$

Therefore

$$\iint_R f(x,y)\,dx\,dy = \int_c^d \int_{-a}^{a} f(x,y)\,dx\,dy = \int_c^d \int_{-a}^{a} g_y(x)\,dx\,dy = \int_c^d 0\,dy = 0.$$

59. Note that $\Omega = \{(x,y) : 0 \le x \le y, \quad 0 \le y \le 1\}$. Set $\Omega' = \{(x,y) : 0 \le y \le x, \quad 0 \le x \le 1\}$.

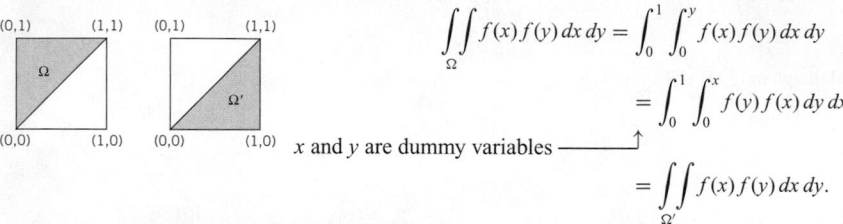

(0,1) (1,1) Ω (0,0) (1,0)   (0,1) (1,1) Ω′ (0,0) (1,0)

$$\iint_\Omega f(x)f(y)\,dx\,dy = \int_0^1 \int_0^y f(x)f(y)\,dx\,dy$$

$$= \int_0^1 \int_0^x f(y)f(x)\,dy\,dx \quad \text{(} x \text{ and } y \text{ are dummy variables)}$$

$$= \iint_{\Omega'} f(x)f(y)\,dx\,dy.$$

Note that Ω and Ω' overlap and their union is the unit square $R: \{(x,y) : 0 \le x \le 1, \quad 0 \le y \le 1\}$. If $\int_0^1 f(x)\,dx = 0$, then

$$0 = \left(\int_0^1 f(x)\,dx\right)\left(\int_0^1 f(y)\,dy\right) = \iint_R f(x)f(y)\,dx\,dy \quad \text{(by Exercise 55)}$$

$$= \iint_\Omega f(x)f(y)\,dx\,dy + \iint_\Omega f(x)f(y)\,dx\,dy$$

$$= 2\iint_\Omega f(x)f(y)\,dx\,dy$$

and therefore $\displaystyle\iint_\Omega f(x)f(y)\,dx\,dy = 0.$

61. Let M be the maximum value of $|f(x,y)|$ on Ω.

$$\int_{\phi_1(x+h)}^{\phi_2(x+h)} = \int_{\phi_1(x+h)}^{\phi_1(x)} + \int_{\phi_1(x)}^{\phi_2(x)} + \int_{\phi_2(x)}^{\phi_2(x+h)}$$

$$|F(x+h) - F(x)| = \left|\int_{\phi_1(x+h)}^{\phi_2(x+h)} f(x,y)\,dy - \int_{\phi_1(x)}^{\phi_2(x)} f(x,y)\,dy\right|$$

$$= \left|\int_{\phi_1(x+h)}^{\phi_1(x)} f(x,y)\,dy + \int_{\phi_2(x)}^{\phi_2(x+h)} f(x,y)\,dy\right| \le \left|\int_{\phi_1(x+h)}^{\phi_1(x)} f(x,y)\,dy\right| + \left|\int_{\phi_2(x)}^{\phi_2(x+h)} f(x,y)\,dy\right|$$

$$\le |\phi_1(x) - \phi_1(x+h)|M + |\phi_2(x+h) - \phi_2(x)|M.$$

The expression on the right tends to 0 as h tends to 0 since ϕ_1 and ϕ_2 are continuous.

63. (a) $\displaystyle\int_1^2 \int_{x^2-2x+2}^{1+\sqrt{x-1}} 1\,dy\,dx = \frac{1}{3}$

(b) $\displaystyle\int_1^2 \int_{y^2-2y+2}^{1+\sqrt{y-1}} 1\,dx\,dy = \frac{1}{3}$

SECTION 16.4

1. $\frac{1}{6}$ **3.** 6 **5.** (a) $\pi \sin 1$ (b) $\pi(\sin 4 - \sin 1)$ **7.** (a) $\frac{2}{3}$ (b) $\frac{14}{3}$ **9.** $\frac{1}{3}\pi$ **11.** $\frac{\pi}{6} - \frac{\sqrt{3}}{8}$ **13.** $\frac{\pi}{2}(\sin 1 - \cos 1)$

15. $\dfrac{\pi}{2}$ **17.** $\dfrac{3\pi}{4}$ **19.** $\dfrac{4\pi}{3} + 2\sqrt{3}$ **21.** 4 **23.** $b^3\pi$ **25.** $\frac{16}{3}\sqrt{3}\pi$ **27.** $\frac{2}{3}(8 - 3\sqrt{3})\pi$ **29.** 2π **31.** $\frac{1}{3}\pi a^2 b$ **33.** $\frac{\pi}{2}$

SECTION 16.5

1. $M = \frac{2}{3}, \quad x_M = 0, \quad y_M = \frac{1}{2}$ **3.** $M = \frac{1}{6}, \quad x_M = \frac{4}{7}, \quad y_M = \frac{3}{4}$ **5.** $M = \frac{32}{3}, \quad x_M = \frac{16}{3}, \quad y_M = \frac{9}{7}$

7. $M = \frac{5}{8}, \quad x_M = \frac{4}{5}, \quad y_m = \frac{152}{75}$ **9.** $M = \dfrac{5\pi}{3}, \quad x_M = \dfrac{21}{20}, \quad y_M = 0$

11. $I_x = \frac{1}{12}MW^2, \quad I_y = \frac{1}{12}ML^2, \quad I_z = \frac{1}{12}M(L^2 + W^2); \quad K_x = \frac{1}{6}\sqrt{3}W, \quad K_y = \frac{1}{6}\sqrt{3}L, \quad K_z = \frac{1}{6}\sqrt{3}\sqrt{L^2 + W^2}$

13. $x_M = \frac{1}{6}L, \quad y_M = 0$ **15.** $I_x = I_y = \frac{1}{4}MR^2, \quad I_z = \frac{1}{2}MR^2; \quad K_x = K_y = \frac{1}{2}R, \; K_z = \frac{1}{2}\sqrt{2}R$

17. center the disc at a distance $\sqrt{I_0 - \frac{1}{2}MR^2} / \sqrt{M}$ from the origin **19.** $I_x = \frac{1}{4}Mb^2, \; I_y = \frac{1}{4}Ma^2, \; I_z = \frac{1}{4}M(a^2 + b^2)$

21. $I_x = \frac{1}{10}, \quad I_y = \frac{1}{16}, \quad I_z = \frac{13}{80}$ **23.** $I_x = \dfrac{33\pi}{40}, \quad I_y = \dfrac{93\pi}{40}, \quad I_z = \dfrac{63\pi}{20}$

25. (a) $\frac{1}{4}M(r_2^2 + r_1^2)$ (b) $\frac{1}{4}M(r_2^2 + 5r_1^2)$ (c) $\frac{1}{4}M(5r_2^2 + r_1^2)$ **27.** $\frac{1}{2}M(r_2^2 + r_1^2)$ **29.** $x_M = 0, \quad y_M = R/\pi$

31. on the diameter through P at a distance $\frac{6}{5}R$ from P

33. Suppose Ω, a basic region of area A, is broken up into n basic regions $\Omega_1, \cdots, \Omega_n$ with areas $A_1, \cdots, A_n$. Then

$$\bar{x}A = \iint_{\Omega} x\,dx\,dy = \sum_{i=1}^{n}\left(\iint_{\Omega_i} x\,dx\,dy\right) = \sum_{i=1}^{n}\bar{x}_i A_i = \bar{x}_1 A_1 + \cdots + \bar{x}_n A_n.$$

The second formula follows in the same manner.

SECTION 16.6

1. they are equal **3.** $\displaystyle\iiint_{\Pi} \alpha\,dx\,dy\,dz = \alpha \iiint_{\Pi} dx\,dy\,dz = \alpha(\text{volume of }\Pi) = \alpha(a_2 - a_1)(b_2 - b_1)(c_2 - c_1)$ **5.** $\frac{1}{4}a^2b^2c$

7. $\bar{x} = \dfrac{A^2BC - a^2bc}{ABC - abc}, \; \bar{y} = \dfrac{AB^2C - ab^2c}{ABC - abc}, \; \bar{z} = \dfrac{ABC^2 - abc^2}{ABC - abc}$

9. $M = \frac{1}{2}Ka^4$ where K is the constant of proportionality for the density function **11.** $I_z = \frac{2}{3}Ma^2$

SECTION 16.7

1. abc **3.** $\frac{2}{3}$ **5.** 16 **7.** $\frac{1}{3}$ **9.** $\frac{47}{24}$

11.
$$\begin{aligned}
\iiint_{\Pi} f(x)g(y)h(z)\,dx\,dy\,dz &= \int_{c1}^{c2}\left[\int_{b1}^{b2}\left(\int_{a1}^{a2} f(x)g(y)h(z)\,dx\right)dy\right]dz \\
&= \int_{c_1}^{c_2}\left[\int_{b_1}^{b_2} g(y)h(z)\left(\int_{a_1}^{a_2} f(x)\,dx\right)dy\right]dz \\
&= \int_{c_1}^{c_2}\left[h(z)\left(\int_{a_1}^{a_2} f(x)\,dx\right)\left(\int_{b_1}^{b_2} g(y)\,dy\right)dz\right] \\
&= \left(\int_{a_1}^{a_2} f(x)\,dx\right)\left(\int_{b_1}^{b_2} g(y)\,dy\right)\left(\int_{c_1}^{c_2} h(z)\,dz\right)
\end{aligned}$$

13. 8 **15.** $(\frac{2}{3}a, \frac{2}{3}b, \frac{2}{3}c)$ **17.**

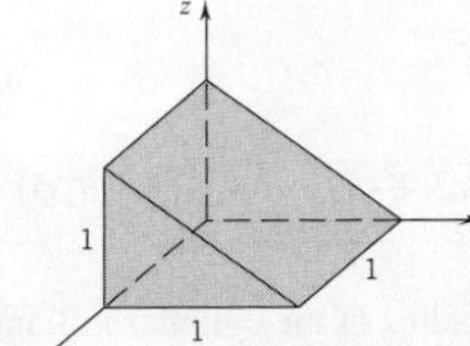

19. $(\frac{1}{2}, \frac{1}{3}, \frac{1}{3})$

21. $\displaystyle\int_{-r}^{r}\int_{-\sqrt{r^2-x^2}}^{\sqrt{r^2-x^2}}\int_{-\sqrt{r^2-(x^2+y^2)}}^{\sqrt{r^2-(x^2+y^2)}} k(r-\sqrt{x^2+y^2+z^2})\,dz\,dy\,dx$ **23.** $\displaystyle\int_{0}^{1}\int_{-\sqrt{x-x^2}}^{\sqrt{x-x^2}}\int_{-2x-3y-10}^{1-y^2} dz\,dy\,dx$

25. $\displaystyle\int_{-1}^{1}\int_{-2\sqrt{2-2x^2}}^{2\sqrt{2-2x^2}}\int_{3x^2+y^2/4}^{4-x^2-y^2/4} k(z-3x^2-\tfrac{1}{4}y^2)\,dz\,dy\,dx$ **27.** $\frac{28}{3}$ **29.** $\frac{1}{270}$ **31.** $\frac{12}{5}$ **33.** $V=\frac{8}{3}$, $(\frac{11}{10},\frac{9}{4},\frac{11}{20})$

35. $V=\frac{27}{2}$, $(\frac{1}{2},\frac{3}{2},\frac{12}{5})$ **37.** $V=\frac{1}{6}abc$, $(\frac{1}{4}a,\frac{1}{4}b,\frac{1}{4}c)$ **39.** (a) $\frac{1}{3}M(a^2+b^2)$ (b) $\frac{1}{12}M(a^2+b^2)$ (c) $\frac{1}{3}Ma^2+\frac{1}{12}Mb^2$

41. $M=\frac{1}{2}k$, $(\frac{7}{12},\frac{34}{45},\frac{37}{90})$ **43.** (a) 0 by symmetry (b) $\frac{4}{3}\pi a_4$ **45.** $\displaystyle 8\int_0^a\int_0^{\sqrt{a^2-x^2}}\int_0^{\sqrt{a^2-x^2-y^2}} dz\,dy\,dz=\frac{4}{3}\pi a^3$

47. $M=\frac{128}{15}k$

49. $M=\frac{135}{4}k$, $(\frac{1}{2},\frac{9}{5},\frac{12}{5})$ **51.** (a) $\displaystyle V=\int_0^6\int_{z/2}^3\int_x^{6-x} dy\,dx\,dz$

(b) $\displaystyle V=\int_0^3\int_0^{2x}\int_x^{6-x} dy\,dz\,dx$

(c) $\displaystyle V=\int_0^6\int_{z/2}^3\int_{z/2}^{y} dx\,dy\,dz+\int_0^6\int_3^{(12-z)/2}\int_{z/2}^{6-y} dx\,dy\,dz$

53. (a) $\displaystyle V=\iint_{\Omega_{yz}} 2y\,dy\,dz$ (b) $\displaystyle V=\iint_{\Omega_{yz}}\left(\int_{-y}^{y} dx\right) dy\,dz$ (c) $\displaystyle V=\int_0^4\int_{-\sqrt{4-y}}^{\sqrt{4-y}}\int_{-y}^{y} dx\,dz\,dy$ (d) $\displaystyle V=\int_{-2}^{2}\int_0^{4-z^2}\int_{-y}^{y} dx\,dy\,dz$

55. (a) 6.80703 (b) $\frac{16\sqrt{3}}{3}(4\sqrt{2}-2)\cong 33.7801$

SECTION 16.8

1. $r^2+z^2=9$ **3.** $z=2r$ **5.** $4r^2=z^2$ **7.** 2π **9.** 8π

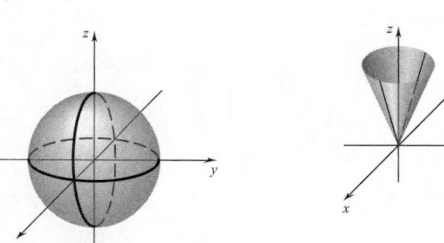

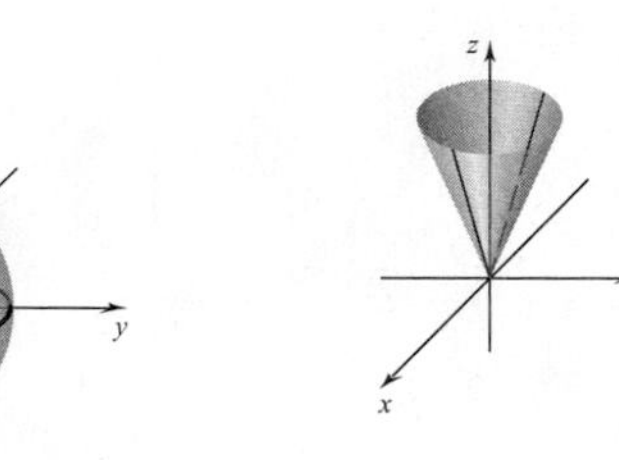

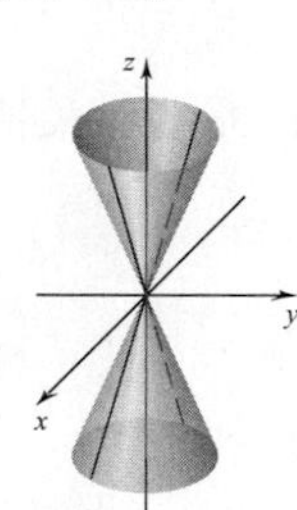

11. $\frac{1}{6}(8-3\sqrt{3})\pi$ **13.** $\dfrac{9\pi^2}{8}$ **15.** $\dfrac{\pi}{2}(1-\cos 1)\cong 0.7221$ **17.** $\frac{32}{9}a^3$ **19.** $\frac{1}{36}a^3(9\pi-16)$ **21.** $\frac{1}{32}\pi$ **23.** $\frac{1}{3}\pi(2-\sqrt{3})$

25. $\frac{32}{3}\pi\sqrt{2}$ **27.** $M=\frac{1}{2}k\pi R^2h^2$ **29.** $\frac{1}{2}MR^2$ **31.** Inverting the cone and placing the vertex at the origin, we have

$$V=\int_0^h\int_0^{2\pi}\int_0^{(R/h)z} r\,dr\,d\theta\,dz=\frac{1}{3}\pi R^2h.$$

33. $\frac{3}{10}MR^2$ **35.** $\frac{1}{2}\pi$ **37.** $\frac{1}{4}k\pi$

SECTION 16.9

1. $(\sqrt{3},\frac{1}{4}\pi,\cos^{-1}[\frac{1}{3}\sqrt{3}])$ **3.** $\left(\frac{3}{4},\frac{3}{4}\sqrt{3},\frac{3}{2}\sqrt{3}\right)$ **5.** $(\rho,\theta,\phi)=\left(\dfrac{4\sqrt{6}}{3},\dfrac{\pi}{4},\dfrac{\pi}{3}\right)$ 7. $(x,y,z)=(0,0,3)$

9. the circular cylinder $x^2+y^2=1$; the radius of the cylinder is 1 and the axis is the z-axis **11.** the lower nappe of the cone $z^2=x^2+y^2$

13. horizontal plane one unit above the xy-plane **15.** T: sphere centered at the origin, radius 2; $\frac{32\pi}{3}$

17. T: the portion of the sphere $x^2+y^2+z^2=9$ that lies between the planes $z=0$ and $z=\frac{3}{2}\sqrt{3}$; $\frac{9}{4}\pi\sqrt{3}$ **19.** $\frac{\pi}{3}(\sqrt{2}-1)$

21. $\frac{243\pi}{20}$ **23.** $V=\frac{4}{3}\pi R^3$ **25.** $V=\frac{2}{3}\alpha R^3$

27. $M=\frac{1}{6}k\pi h[(r^2+h^2)^{3/2}-h^3]$ **29.** (a) $\frac{2}{3}MR^2$ (b) $\frac{7}{5}MR^2$ **31.** (a) $\frac{2}{5}M\left(\frac{R_2^5-R_1^5}{R_2^3-R_1^3}\right)$ (b) $\frac{2}{3}MR^2$ (c) $\frac{5}{3}MR^2$ **33.** $V=\frac{2}{3}\pi(1-\cos\alpha)a^3$

35. (a) $\rho=2R\cos\phi$ (b) $0\le\theta\le 2\pi,\quad 0\le\phi\le\frac{1}{4}\pi,\quad R\sec\theta\le\rho\le 2R\cos\phi$ **37.** $V=\frac{1}{3}(16-6\sqrt{2})\pi$

39. Encase T in a spherical wedge W. W has spherical coordinates in a box Π that contains S. Define f to be zero outside of T. Then $F(\rho,\theta,\phi)=f(\rho\sin\phi\cos\theta,\rho\sin\phi\sin\theta,\rho\cos\phi)$ is zero outside of S and

$$\iiint_T f(x,y,z)\,dx\,dy\,dz = \iiint_W f(x,y,z)\,dx\,dy\,dz$$
$$= \iiint_\Pi F(\rho,\theta,\phi)\rho^2\sin\phi\,d\rho\,d\theta\,d\phi = \iiint_S F(\rho,\theta,\phi)\rho^2\sin\phi\,d\rho\,d\theta\,d\phi.$$

41. $\mathbf{F}=\frac{GmM}{R^2}(\sqrt{2}-1)\mathbf{k}$

SECTION 16.10

1. $ad-bc$ **3.** $2(v^2-u^2)$ **5.** $-3u^2v^2$ **7.** abc **9.** r **11.** $w(1+w\cos v)$ **13.** $\frac{1}{2}$ **15.** 0 **17.** $\frac{2}{3}$

19. (a) $A=3\ln 2$ (b) $\bar{x}=\frac{7}{9\ln 2}$, $\bar{y}=\frac{14}{9\ln 2}$ **21.** $I_x=\frac{4}{75}M$, $I_y=\frac{14}{75}M$, $I_z=\frac{18}{75}M$ **23.** $A=\frac{32}{15}$ **25.** $A=\pi/\sqrt{65}$

27. $V=\frac{4}{3}\pi abc$ **29.** $I_x=\frac{1}{5}M(b^2+c^2)$, $I_y=\frac{1}{5}M(a^2+c^2)$, $I_z=\frac{1}{5}M(a^2+b^2)$

CHAPTER 17

SECTION 17.1

1. (a) 1 (b) -2 **3.** 0 **5.** (a) $-\frac{17}{6}$ (b) $\frac{17}{6}$ **7.** -8 **9.** $-\pi$ **11.** (a) 1 (b) $\frac{23}{21}$

13. (a) $2+\sin 2-\cos 3$ (b) $\frac{4}{5}+\sin 1-\cos 1$ **15.** 26 **17.** $\frac{1}{3}$ **19.** $\frac{8\pi^3}{3}$

21. $\int_C \mathbf{q}\cdot d\mathbf{r}=\int_a^b[\mathbf{q}\cdot\mathbf{r}'(u)]\,du+\int_a^b\frac{d}{du}[\mathbf{q}\cdot\mathbf{r}(u)]\,du=[\mathbf{q}\cdot\mathbf{r}(b)]-[\mathbf{q}\cdot\mathbf{r}(a)]=\mathbf{q}\cdot[\mathbf{r}(b)-\mathbf{r}(a)]$

$\int_C \mathbf{r}\cdot d\mathbf{r}=\int_a^b[\mathbf{r}(u)\cdot\mathbf{r}'(u)]\,du=\frac{1}{2}\int_a^b\frac{d}{du}[\mathbf{r}(u)\cdot\mathbf{r}(u)]\,du=\frac{1}{2}\int_a^b\frac{d}{du}(\|\mathbf{r}(u)\|^2)\,du=\frac{1}{2}(\|\mathbf{r}(b)\|^2-\|\mathbf{r}(a)\|^2)$

23. $\int_C \mathbf{f}(\mathbf{r})\cdot d\mathbf{r}=\int_a^b[\,\mathbf{f}(\mathbf{r}(u))\cdot\mathbf{r}'(u)]\,du=\int_a^b[f(u)\mathbf{i}\cdot\mathbf{i}\,]\,du=\int_a^b f(u)\,du$ **25.** $|W|=$ area of ellipse

27. force at time $t=m\mathbf{r}''(t)=m(2\beta\,\mathbf{j}+6\gamma t\mathbf{k})$; $W=(2\beta^2+\frac{9}{2}\gamma^2)m$ **29.** 0 **31.** (a) $\left(\frac{1}{\sqrt{5}}-\frac{1}{\sqrt{14}}\right)c$ (b) $\frac{4}{5}c$ **33.** $\alpha=\frac{5}{2}$

SECTION 17.2

1. 0 **3.** -1 **5.** 0 **7.** 0 **9.** 0 **11.** 4 **13.** e^3-2e^2+3 **15.** e^5-2e^2+1 **17.** 2π **19.** 14 **21.** 0

23. Set $f(x,y,z)=g(x)$ and $C:\mathbf{r}(u)=u\,\mathbf{i}$, $u\in[a,b]$. In this case, $\nabla f(\mathbf{r}(u))=g'(x(u))\mathbf{i}=g'(u)\,\mathbf{i}$ and $\mathbf{r}'(u)=\mathbf{i}$, so that

$$\int_C\nabla f(\mathbf{r})\cdot d\mathbf{r}=\int_a^b[\nabla f(\mathbf{r}(u))\cdot\mathbf{r}'(u)]\,du=\int_a^b g'(u)\,du \quad\text{and}\quad f(\mathbf{r}(b))-f(\mathbf{r}(a))=g(b)-g(a).$$

The statement $\int_C\nabla f(\mathbf{r})\cdot d\mathbf{r}=f(\mathbf{r}(b))-f(\mathbf{r}(a))$ reduces to $\int_a^b g'(u)\,du=g(b)-g(a)$.

25. (a) $\mathbf{F}(\mathbf{r}) = cx\sqrt{x^2+y^2-z^2}\ \mathbf{i} + cy\sqrt{x^2+y^2+z^2}\ \mathbf{j} + cz\sqrt{x^2+y^2+z^2}\ \mathbf{k}$; $\|\mathbf{F}(\mathbf{r})\| = cr^2$
(b) $f(x,y,z) = \frac{c}{3}(x^2+y^2+z^2)^{3/2}$

27. $W = mG\left(\frac{1}{r_2} - \frac{1}{r_1}\right)$ **29.** $f(x,y,z) = \frac{mGr_0^2}{r_0+z}$

SECTION 17.3

1. If f is continuous, then $-f$ is continuous and has antiderivatives u. The scalar fields $U(x,y,z) = u(x)$ are potential functions for $\mathbf{F}$:

$$\nabla U = \frac{\partial U}{\partial x}\mathbf{i} + \frac{\partial U}{\partial y}\mathbf{j} + \frac{\partial U}{\partial z}\mathbf{k} = \frac{du}{dx}\mathbf{i} = -f\,\mathbf{i} = -\mathbf{F}.$$

3. The scalar field $U(x,y,z) = cz + d$ is a potential energy function for $\mathbf{F}$. We know that the total mechanical energy remains constant. Thus, for any times t_1 and t_2,

$$\tfrac{1}{2}m[v(t_1)]^2 + U(\mathbf{r}(t_1)) = \tfrac{1}{2}m[v(t_2)]^2 + U(\mathbf{r}(t_2)).$$

This gives

$$\tfrac{1}{2}m[v(t_1)]^2 + cz(t_1) + d = \tfrac{1}{2}m[v(t_2)]^2 + cz(t_2) + d.$$

Solve this equation for $v(t_2)$ and you have the desired formula.

5. (a) We know that $-\nabla U$ points in the direction of maximum decrease of U. Thus $\mathbf{F} = -\nabla U$ attempts to drive objects toward a region where U has lower values. (b) At a point where U has a minimum, $\nabla U = \mathbf{0}$ and therefore $\mathbf{F} = \mathbf{0}$.

7. (a) By conservation of energy $\frac{1}{2}mv^2 + U = E$. Since E is constant and U is constant, v is constant.
(b) ∇U is perpendicular to any surface where U is constant. Obviously so is $\mathbf{F} = -\nabla U$.

9. $f(x,y,z) = -\frac{k}{\sqrt{x^2+y^2+z^2}}$ is a potential function for $\mathbf{F}$. The work done by $\mathbf{F}$ moving an object along C is $W = \int_C \mathbf{F}(\mathbf{r}) \cdot d\mathbf{r} = \int_a^b \nabla f \cdot d\mathbf{r} =$ $f(\mathbf{r}(b)) - f(\mathbf{r}(a))$. Since $\mathbf{r}(a) = (x_0,y_0,z_0)$ and $\mathbf{r}(b) = (x_1,y_1,z_1)$ are points on the unit sphere, $f(\mathbf{r}(b)) = f(\mathbf{r}(a)) = -k$, and so $W = 0$.

SECTION 17.4

1. $\frac{1}{2}$ **3.** $\frac{9}{2}$ **5.** $\frac{11}{6}$ **7.** 2 **9.** 16 **11.** $\frac{104}{5}$ **13.** 4 **15.** 4 **17.** 2 **19.** 3 **21.** $\frac{176}{3}$ **23.** 56 **25.** $\frac{1177}{30}$

27. (a) $\frac{\partial P}{\partial y} = 6x - 4y = \frac{\partial Q}{\partial x}$ (b) 7 (c) $\frac{-37}{3}$ **29.** (a) $M = 2ka^2, x_M = y_M = \frac{1}{8}a(\pi+2)$ (b) $I_x = ka^4 = \frac{1}{2}Ma^2$

31. (a) $I_z = 2ka^4 = Ma^2$ (b) $I = \frac{1}{3}ka^4 = \frac{1}{6}Ma^2$

33. (a) $L = 2\pi\sqrt{a^2+b^2}$ (b) $x_M = y_M = 0, z_M = \pi b$ (c) $I_x = I_y = \frac{1}{6}M(3a^2 + 8b^2\pi^2), I_z = Ma^2$ **35.** $M = \frac{2}{3}\pi k\sqrt{a^2+b^2}(3a^2+4\pi^2b^2)$

SECTION 17.5

1. $\frac{1}{6}$ **3.** 6π **5.** 2π **7.** a^2b **9.** 7π **11.** $5a\pi r^2$ **13.** 0 **15.** 0 **17.** πa^2

19. $\frac{15}{2} - 4\ln 4$ **21.** $(c-a)A$ **23.** $3\pi R^2$

25. Taking Ω to be of type II, we have

$$\iint_\Omega \frac{\partial Q}{\partial x}(x,y)\,dx\,dy = \int_c^d \int_{\psi_1(y)}^{\psi_2(y)} \frac{\partial Q}{\partial x}(x,y)\,dx\,dy = \int_c^d \{Q[\psi_2(y),y] - Q[\psi_1(y),y]\}\,dy$$

$$* = \int_c^d Q[\psi_2(y),y]\,dy - \int_c^d Q[\psi_1(y),y]\,dy.$$

Set $C_3 : \mathbf{r}_3(u) = \psi_1(u)\mathbf{i} + u\mathbf{j},\ u \in [c,d]$ and $C_4 : \mathbf{r}_4(u) = \psi_2(u)\mathbf{i} + u\mathbf{j},\ u \in [c,d]$. Then

$$\oint_C Q(x,y)\,dy = \int_{C_4} Q(x,y)\,dy - \int_{C_3} Q(x,y)\,dy = \int_c^d Q[\psi_2(u),u]\,du - \int_c^d Q[\psi_1(u),u]\,du.$$

Comparison with $*$ proves the result.

27. by (16.5.3), $\bar{x}A = \iint_\Omega x\,dxdy$ and $\bar{y}A = \iint_\Omega y\,dxdy$;

by Green's theorem, $\iint_\Omega x\,dxdy = \oint_C \frac{1}{2}x^2 dy$ and $\iint_\Omega y\,dxdy = -\oint_C \frac{1}{2}y^2 dx$

29. $\oint_{C_1} = \oint_{C_2} + \oint_{C_3}$ **31.** (a) 0 (b) 0 **33.** If Ω is the region bounded by C, then

$$\oint_C \mathbf{v}\cdot d\mathbf{r} = \oint_C \frac{\partial\phi}{\partial x}dx + \frac{\partial\phi}{\partial y}dy = \iint_\Omega \left\{\frac{\partial}{\partial x}\left(\frac{\partial\phi}{\partial y}\right) - \frac{\partial}{\partial y}\left(\frac{\partial\phi}{\partial x}\right)\right\}dx\,dy$$

is zero by equality of mixed partials.

35. $A = \frac{1}{2}\oint_C(-y\,dx + x\,dy)$

$$= \frac{1}{2}\left[\int_{C_1} + \int_{C_2} + \cdots + \int_{C_n}\right]$$

Now

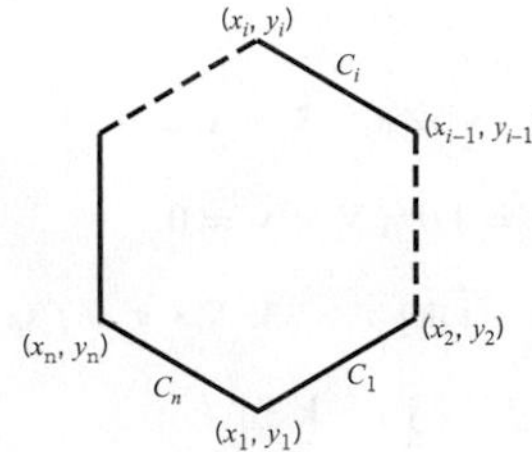

$$\int_{C_i}(-y\,dx + x\,dy) = \int_0^1 [-(y_i + u(y_{i+1} - y_i))(x_{i+1} - x_i) + (x_i + u(x_{i+1} - x_i))(y_{i+1} - y_i)]\,du$$

$$= x_i y_{i+1} - x_{i+1}y_i, \quad i = 1, 2, \ldots, n\ (x_{n+1} = x_1, y_{n+1} = y_1).$$

Thus $A = \frac{1}{2}[(x_1y_2 - x_2y_1) + (x_2y_3 - x_3y_2) + \cdots + (x_ny_1 - x_1y_n)]$.

SECTION 17.6

1. $4[(u^2 - v^2)\mathbf{i} - (u^2 + v^2)\mathbf{j} + 2uv\mathbf{k}]$ **3.** $2(\mathbf{j} - \mathbf{i})$ **5.** $\mathbf{r}(u, v) = 3\cos u\cos v\,\mathbf{i} + 2\sin u\cos v\,\mathbf{j} + 6\sin v\,\mathbf{k}$, $u \in [0, 2\pi]$, $v \in [0, \pi/2]$

7. $\mathbf{r}(u, v) = 2\cos u\cos v\,\mathbf{i} + 2\sin u\cos v\,\mathbf{j} + 2\sin v\,\mathbf{k}$, $u \in [0, 2\pi]$, $v \in (\pi/4, \pi/2]$ **9.** $\mathbf{r}(u, v) = u\,\mathbf{i} + g(u, v)\,\mathbf{j} + v\,\mathbf{k}$, $(u, v) \in \Omega$

11. $x^2/a^2 + y^2/b^2 + z^2/c^2 = 1$; ellipsoid **13.** $x^2/a^2 - y^2/b^2 = z$; hyperbolic paraboloid

15. $\mathbf{r}(u, v) = v\cos u\,\mathbf{i} + v\sin u\,\mathbf{j} + f(v)\,\mathbf{k}$; $0 \le u \le 2\pi, a \le v \le b$ **17.** area of $\Gamma = A_\Omega \sec\gamma$ **19.** $\frac{1}{2}\sqrt{a^2b^2 + a^2c^2 + b^2c^2}$

21. $\frac{1}{6}\pi(17\sqrt{17} - 1)$ **23.** $\frac{2}{3}\pi\left[(4a^2 + 1)^{3/2} - (a^2 + 1)^{3/2}\right]$ **25.** $\frac{1}{15}(36\sqrt{6} - 50\sqrt{5} + 32)$ **27.** 4π

29. (a) $\iint_\Omega \sqrt{\left[\frac{\partial g}{\partial y}(y, z)\right]^2 + \left[\frac{\partial g}{\partial z}(y, z)\right]^2 + 1}\,dydz = \iint_\Omega \sec[\alpha(y, z)]\,dydz$

where α is the angle between the unit normal with positive $\mathbf{i}$ component and the x-axis

(b) $\iint_\Omega \sqrt{\left[\frac{\partial h}{\partial x}(x, z)\right]^2 + \left[\frac{\partial h}{\partial z}(x, z)\right]^2 + 1}\,dxdz = \iint_\Omega \sec[\beta(x, z)]\,dxdz$

where β is the angle between the unit normal with positive $\mathbf{j}$ component and the y-axis

31. (a) $\mathbf{N}(u, v) = v\cos u\sin\alpha\cos\alpha\,\mathbf{i} + v\sin u\sin\alpha\cos\alpha\,\mathbf{j} - v\sin^2\alpha\mathbf{k}$ (b) $A = \pi s^2\sin\alpha$

33. (c) $A = \int_0^{2\pi}\int_{-\ln 2}^{\ln 2} ||\mathbf{N}(u, v)||\,dudv$

$$= \int_0^{2\pi}\int_{-\ln 2}^{\ln 2} \sqrt{64\cos^2 u\cosh^2 v + 144\sin^2 u\cosh^2 v + 36\cosh^2 v\sinh^2 v}\,dudv$$

35. $A = \sqrt{A_1^2 + A_2^2 + A_3^2}$

37. (a) $\frac{1}{4}\sqrt{2\pi}[\sqrt{6} + \ln(\sqrt{2} + \sqrt{3})]$

(b) $\frac{1}{2}a^2[\sqrt{2e^{4\pi} + 1} - \sqrt{3} + 2\pi + \ln(1 + \sqrt{3}) - \ln(1 + \sqrt{2\,e^{4\pi} + 1})]$

SECTION 17.7

1. $\frac{1}{2}[\sqrt{2}+\ln(1+\sqrt{2})]$ **3.** $2\sqrt{2}-1$ **5.** $\frac{1}{3}[2\sqrt{2}-1]$ **7.** $\dfrac{9\sqrt{14}}{2}$ **9.** $\frac{4}{3}$ **11.** $\dfrac{4\pi}{3}$ **13.** $\frac{1}{2}\sqrt{3}a^2k$

15. $\frac{1}{12}\sqrt{3}a^4k$ **17.** $(0,0,\frac{1}{2}a)$ **19.** 2 **21.** $\frac{4}{3}\pi a^3$ **23.** 0 **25.** $\dfrac{1}{2}\sqrt{3}a^3$ **27.** 0 **29.** $-\frac{3}{2}$ **31.** $2\pi a^2 l$

33. $\frac{8}{35}$ **35.** $-\frac{4}{63}$ **37.** $\bar{x}=\bar{y}=0,\quad \bar{z}=\frac{2}{3}s\cos\alpha$ **39.** $x_M=y_M=0,\quad z_M=\frac{3}{4}$ **41.** no answer required **43.** $x_M=\frac{11}{9}$

45. Total flux out of the solid is 0. It is clear from a diagram that the outer unit normal to the cylindrical side of the solid is given by $\mathbf{n}=x\,\mathbf{i}+y\,\mathbf{j}$, in which case $\mathbf{v}\cdot\mathbf{n}=0$. The outer unit normals to the top and bottom of the solid are $\mathbf{k}$ and $-\mathbf{k}$, respectively. So, here as well, $\mathbf{v}\cdot\mathbf{n}=0$ and the total flux is 0.

47. $(4\sqrt{2}-\frac{7}{2})\pi$

SECTION 17.8

1. $\nabla\cdot\mathbf{v}=2,\ \nabla\times\mathbf{v}=\mathbf{0}$ **3.** $\nabla\cdot\mathbf{v}=0,\ \nabla\times\mathbf{v}=\mathbf{0}$ **5.** $\nabla\cdot\mathbf{v}=6,\quad \nabla\times\mathbf{v}=\mathbf{0}$

7. $\nabla\cdot\mathbf{v}=yz+1,\ \nabla\times\mathbf{v}=-x\,\mathbf{i}+xy\,\mathbf{j}+(1-x)z\mathbf{k}$ **9.** $\nabla\cdot\mathbf{v}=1/r^2,\ \nabla\times\mathbf{v}=\mathbf{0}$

11. $\nabla\cdot\mathbf{v}=2(x+y+z)e^{r^2},\ \nabla\times\mathbf{v}=2e^{r^2}[(y-z)\,\mathbf{i}-(x-z)\,\mathbf{j}+(x-y)\mathbf{k}]$ **13.** $\nabla\cdot\mathbf{v}=f'(x),\quad \nabla\times\mathbf{v}=\mathbf{0}$ **15.** use components

17. $\nabla\cdot\mathbf{F}=\dfrac{\partial P}{\partial x}+\dfrac{\partial Q}{\partial y}+\dfrac{\partial R}{\partial z}=2+4-6=0$ **19.** $\nabla\times\mathbf{F}=\begin{vmatrix}\mathbf{i}&\mathbf{j}&\mathbf{k}\\ \dfrac{\partial}{\partial x}&\dfrac{\partial}{\partial y}&\dfrac{\partial}{\partial z}\\ x&y&-2z\end{vmatrix}=\mathbf{0}$ **21.** $\nabla^2 f=12(x^2+y^2+z^2)$

23. $\nabla^2 f=2y^3z^4+6x^2yz^4+12x^2y^3z^2$ **25.** $\nabla^2 f=e^r(1+2r^{-1})$ **27.** (a) $2r^2$ (b) $-1/r$

29.
$$\begin{aligned}\nabla^2 f&=\nabla^2 g(r)=\nabla\cdot(\nabla g(r))=\nabla\cdot(g'(r)r^{-1}\mathbf{r})\\ &=[(\nabla g'(r))\cdot r^{-1}\mathbf{r}]+g'(r)(\nabla\cdot r^{-1}\mathbf{r})\\ &=\{[g''(r)r^{-1}\mathbf{r}]\cdot r^{-1}\mathbf{r}\}+g'(r)(2r^{-1})=g''(r)+2r^{-1}g'(r)\end{aligned}$$

31. no answer required **33.** $n=-1$

SECTION 17.9

1. $\displaystyle\iint_S(\mathbf{v}\cdot\mathbf{n})\,d\sigma=\iiint_T(\nabla\cdot\mathbf{v})\,dxdydz=\iiint_T 3\,dxdydz=3V=4\pi$

3. $\displaystyle\iint_S(\mathbf{v}\cdot\mathbf{n})\,d\sigma=\iiint_T(\nabla\cdot\mathbf{v})\,dxdydz=\iiint_T 2(x+y+z)\,dxdydz.$

The flux is zero since the function $f(x,y,z)=2(x+y+z)$ satisfies the relation $f(-x,-y,-z)=-f(x,y,z)$ and T is symmetric about the origin.

5.

Face	n	v • n	Flux
$x=0$	$-\mathbf{i}$	0	0
$x=1$	$\mathbf{i}$	1	1
$y=0$	$-\mathbf{j}$	0	0
$y=1$	$\mathbf{j}$	1	1
$z=0$	$-\mathbf{k}$	0	0
$z=1$	$\mathbf{k}$	1	1

total flux = 3

$\displaystyle\iiint_T(\nabla\cdot\mathbf{v})\,dxdydz=\iiint_T 3\,dxdydz=3V=3$

7.

Face	n	v • n	Flux
$x=0$	$-\mathbf{i}$	0	0
$x=1$	$\mathbf{i}$	1	1
$y=0$	$-\mathbf{j}$	xz	fluxes added up to 0
$y=1$	$\mathbf{j}$	$-xz$	
$z=0$	$-\mathbf{k}$	0	0
$z=1$	$\mathbf{k}$	1	1

total flux = 2

$\displaystyle\iiint_T(\nabla\cdot\mathbf{v})\,dxdydz=\iiint_T 2(x+z)\,dxdydz=2(\bar{x}+\bar{z})V=2\left(\tfrac{1}{2}+\tfrac{1}{2}\right)1=2$

9. flux $=\displaystyle\iiint_T(1+4y+6z)\,dxdydz=(1+4\bar{y}+6\bar{z})V=(1+0+3)9\pi=36\pi$ **11.** $\frac{1}{24}$ **13.** $64\,\pi$ **15.** 0 **17.** $(A+B+C)V$

19. Let T be the solid enclosed by S and set $n = n_1\,\mathbf{i} + n_2\,\mathbf{j} + n_3\,\mathbf{k}$.

$$\iint_S n_1\,d\sigma = \iint_S (\mathbf{i}\cdot\mathbf{n})\,d\sigma = \iiint_T (\nabla\cdot\mathbf{i})\,dxdydz = \iiint_T 0\,dxdydz = 0.$$

Similarly $\iint_S n_2\,d\sigma = 0$ and $\iint_S n_3\,d\sigma = 0$.

21. A routine computation shows that $\nabla\cdot(\nabla f\times\nabla g) = 0$. Therefore

$$\iint_S [(\nabla f\times\nabla g)\cdot\mathbf{n}]d\sigma = \iiint_T [\nabla\cdot(\nabla f\times\nabla g)]\,dxdydz = 0.$$

23. Set $\mathbf{F} = F_1\,\mathbf{i} + F_2\,\mathbf{j} + F_3\mathbf{k}$.

$$F_1 = \iint_S [\rho(z-c)\mathbf{i}\cdot\mathbf{n}]\,d\sigma = \iiint_T [\nabla\cdot\rho(z-c)\,\mathbf{i}\,]\,dxdydz = \iiint_T \underbrace{\frac{\partial}{\partial x}[\rho(z-c)]}_{0}\,dxdydz = 0.$$

Similarly, $F_2 = 0$.

$$F_3 = \iint_S [\rho(z-c)\mathbf{k}\cdot\mathbf{n}]d\sigma = \iiint_T [\nabla\cdot\rho(z-c)\mathbf{k}]\,dxdydz = \iiint_T \frac{\partial}{\partial z}[(\rho(z-c)]\,dxdydz = \iiint_T \rho\,dxdydz = W.$$

SECTION 17.10

For Exercises 1 and 3 : $\mathbf{n} = x\,\mathbf{i} + y\,\mathbf{j} + z\,\mathbf{k}$ and $C : \mathbf{r}(u) = \cos u\,\mathbf{i} + \sin u\,\mathbf{j}$, $u\in[0,2\pi]$.

1. (a) $\iint_S [(\nabla\times\mathbf{v})\cdot\mathbf{n}]d\sigma = \iint_S (\mathbf{0}\cdot\mathbf{n})\,d\sigma = 0.$

(b) S is bounded by the unit circle $C : \mathbf{r}(u) = \cos u\,\mathbf{i} + \sin u\,\mathbf{j}, u\in[0,2\pi]$.

$\oint_c \mathbf{v}(\mathbf{r})\cdot d\mathbf{r} = 0$ since $\mathbf{v}$ is a gradient.

3. (a) $\iint_S [(\nabla\times\mathbf{v})\cdot\mathbf{n}]\,d\sigma = \iint_S [(-3y^2\,\mathbf{i} + 2z\,\mathbf{j} + 2\,\mathbf{k})\cdot(x\mathbf{i} - y\,\mathbf{j} + z\,\mathbf{k})]\,d\sigma$

$$= \iint_S (-3xy^2 + 2yz + 2z)\,d\sigma$$

$$= \underbrace{\iint_S (-3xy^2)\,d\sigma}_{0\text{ by symmetry}} + \underbrace{\iint_S 2yz\,d\sigma}_{0\text{ by symmetry}} + \underbrace{\iint_S 2z\,d\sigma}_{2\bar{z}A = 2\pi} = 2\pi$$

by Ex.17, Section 17.7

(b) $\oint_C \mathbf{v}(\mathbf{r})\cdot d\mathbf{r} = \oint_C z^2\,dx + 2x\,dy = \oint_C 2x\,dy = \int_0^{2\pi} 2\cos^2 u\,du = 2\pi$

For Exercises 5 and 7 take $S : z = 2 - x - y$ with $0\le x\le 2, 0\le y\le 2-x$ and C as the triangle $(2,0,0), (0,2,0), (0,0,2)$. Then $C = C_1\cup C_2\cup C_3$ with

$$\begin{aligned} C_2 : \mathbf{r}_1(u) &= 2(1-u)\mathbf{i} + 2u\,\mathbf{j}, u\in[0,1],\\ C_2 : \mathbf{r}_2(u) &= 2(1-u)\mathbf{j} + 2u\,\mathbf{k}, u\in[0,1],\\ C_3 : \mathbf{r}_3(u) &= 2(1-u)\mathbf{k} + 2u\,\mathbf{i}, u\in[0,1]\end{aligned}$$

$\mathbf{n} = \frac{1}{3}\sqrt{3}(\mathbf{i} + \mathbf{j} + \mathbf{k}).$

5. (a) $\iint_S [(\nabla\times\mathbf{v})\cdot\mathbf{n}]\,d\sigma = \iint_S \frac{1}{3}\sqrt{3}\,d\sigma = \frac{1}{3}\sqrt{3}A = 2$

(b) $\oint_C \mathbf{v}(r)\cdot \mathbf{dr} = \left(\int_{C_1} + \int_{C_2} + \int_{C_3}\right) \mathbf{v}(\mathbf{r})\cdot d\mathbf{r} = -2+2+2 = 2$

7. (a) $\iint_S [(\nabla\times\mathbf{v})\cdot\mathbf{n}]\,d\sigma \iint_S [y\mathbf{k}\cdot\frac{1}{3}\sqrt{3}(\mathbf{i}+\mathbf{j}+\mathbf{k})]\,d\sigma = \frac{1}{3}\sqrt{3}\iint_S y\,d\sigma = \frac{1}{3}\sqrt{3}\bar{y}A = \frac{4}{3}$ **9.** 4π **11.** 0 **13.** $\pm\frac{1}{8}\pi$ **15.** $\pm\frac{1}{4}\pi$

(b) $\oint_C \mathbf{v}(\mathbf{r})\cdot \mathbf{dr} = \left(\int_{C_1} + \int_{C_2} + \int_{C_3}\right) \mathbf{v}(\mathbf{r})\cdot \mathbf{dr} = (\frac{4}{3}-\frac{32}{5}) + \frac{32}{5} + 0 = \frac{4}{3}$

17. Straightforward calculation shows that

$$\nabla\times(\mathbf{a}\times\mathbf{r}) = \nabla\times[(a_2z-a_3y)\,\mathbf{i} + (a_3x-a_1z)\,\mathbf{j} + (a_1y-a_2x)\,\mathbf{k}] = 2\mathbf{a}.$$

19. In the plane of C, the curve C bounds some Jordan region that we call Ω. The surface $S\cup\Omega$ is a piecewise-smooth surface that bounds a solid T. Note that $\nabla\times\mathbf{v}$ is continuously differentiable on T. Thus, by the divergence theorem,

$$\iiint_T [\nabla\cdot(\nabla\times\mathbf{v})]\,dxdydz = \iint_{S\cup\Omega} [(\nabla\times\mathbf{v})\cdot\mathbf{n}]\,d\sigma$$

where $\mathbf{n}$ is the outer unit normal. Since the divergence of a curl is identically zero, we have

$$\iint_{S\cup\Omega} [(\nabla\times\mathbf{v})\cdot\mathbf{n}]d\sigma = 0.$$

Now $\mathbf{n}$ is $\mathbf{n}_1$ on S and $\mathbf{n}_2$ on Ω. Thus

$$\iint_S [(\nabla\times\mathbf{v})\cdot\mathbf{n}_1]\,d\sigma + \iint_\Omega [(\nabla\times\mathbf{v})\cdot\mathbf{n}_2]\,d\sigma = 0.$$

This gives

$$\iint_S [(\nabla\times\mathbf{v})\cdot\mathbf{n}_1]\,d\sigma = \iint_\Omega [(\nabla\times\mathbf{v})\cdot(-\mathbf{n}_2)]\,d\sigma = \oint_c \mathbf{v}(\mathbf{r})\cdot\mathbf{dr}$$

where C is traversed in a positive sense with respect to $-\mathbf{n}_2$ and therefore in a positive sense with respect to $\mathbf{n}_1$($-\mathbf{n}_2$ points toward S).

CHAPTER 18

SECTION 18.1

1. y_1 is; y_2 is not **3.** both y_1 and y_2 are solutions **5.** both y_1 and y_2 are solutions

7. $y = 2\,e^{5x}$ **9.** $y = \dfrac{1}{-2\,e^{x-1}+1}$ **11.** $y = -\dfrac{17}{4}x + 9x^{1/2}$ **13.** $y = x^2\ln x$ **15.** linear; $y = Ce^{-x} - 2e^{-2x}$

17. separable; $y + \ln|y| = \frac{1}{3}x^3 - x + C$ **19.** linear; $y = \dfrac{\sin x}{x^2} + \dfrac{C}{x^2}$ **21.** separable; $y^2 + 1 = (2x^2 + C)^2$

23. $r = -3$ **25.** $r = -3$ **27.** $r = 0$ **29.** $r = \frac{1}{2}$ or $r = \frac{3}{2}$ **31.** (a) $y = C_1\sin 4x$ (b) $y(0) = 0$

SECTION 18.2

1. $y^2 = \dfrac{1}{1 + C\,e^{x^2}}$ **3.** $y = (C\,e^{2x} - e^x)^2$ **5.** $y = \left[(x-2)^2 + \dfrac{C}{\sqrt{x-2}}\right]^2$ **7.** $y^{-2} = 4e^{x^2} - 2x\,e^{x^2}$ **9.** $y^2 = \dfrac{x^3}{2-x}$

11. $\ln y = x^2 + Cx$ **13.** $y^2 - x^2 = Cx$ **15.** $x^2 - 2xy - y^2 = C$ **17.** $y + x = x\,e^{y/x}[C - \ln x]$ **19.** $1 - \cos[y/x] = c\,x\sin[y/x]$

21. $y^3 + 3x^3\ln|x| = 8x^3$ **23.** (a) 2.48832, rel error 8.46% (b) 2.71825, rel error 0.001%

25. (a) 2.59374, rel error 4.58% (b) 2.71828, rel error 0% **27.** (a) 1.9, rel error 5.0% (b) 2.0 rel error 0%

29. (a) 1.42052, rel error −0.45% (b) 1.41421, rel error 0% **31.** (a) 2.65330, rel error 2.39% (b) 2.71828, rel error 0%

SECTION 18.3

1. the whole plane; $\dfrac{x^2y^2}{2} - xy = C$

3. the whole plane; $x\,e^y - y\,e^x = C$

5. the upper half-plane; $x \ln y + x^2 y = C$

7. the right half-plane; $y \ln x + 3x^2 - 2y = C$

9. the whole plane; $xy^3 + y^2 \cos x - \frac{1}{2}x^2 + \frac{1}{2}e^{2y} = C$

11. (a) yes (b) $\dfrac{1}{p(y)q(x)}$ $(p(y)q(x) \neq 0)$ **13.** $x\,e^y - y\,e^x = C$ **15.** $x^3y^2 + \frac{1}{2}x^2 + x\,e^y + \frac{1}{2}y^2 = C$ **17.** $y^3\,e^x + x\,e^x = C$

19. $x^3 + 3xy + 3\,e^y = 4$ **21.** $4x^2y^2 + x^4 + 4x^2 = 5$ **23.** $xy - \dfrac{1}{y} = 3$ **25.** $\sinh(x - y^2) + \frac{1}{2}\,e^{2x} + \frac{1}{2}y^2 = \frac{1}{2}e^4 + 1$

27. (a) $k = 3$ (b) $k = 1$ **29.** $y = \dfrac{-4}{x^4 + C}$ **31.** $y = \frac{1}{9}x^5 + Cx^{-4}$ **33.** $e^{xy} - x^2 + 2\ln|y| = C$

SECTION 18.4

1. $y = C_1\,e^{-4x} + C_2\,e^{2x}$ **3.** $y = C_1\,e^{-4x} + C_2\,x\,e^{-4x}$ **5.** $y = e^{-x}(C_1 \cos 2x + C_2 \sin 2x)$ **7.** $y = C_1\,e^{(1/2)x} + C_2\,e^{-3x}$

9. $y = C_1 \cos 2\sqrt{3}x + C_2 \sin 2\sqrt{3}x$ **11.** $y = C_1\,e^{(1/5)x} + C_2\,e^{-(3/4)x}$ **13.** $y = C_1 \cos 3x + C_2 \sin 3x$

15. $y = e^{-(1/2)x}(C_1 \cos \frac{1}{2}x + C_2 \sin \frac{1}{2}x)$ **17.** $y = C_1\,e^{(1/4)x} + C_2\,e^{-(1/2)x}$ **19.** $y = 2\,e^{2x} - e^{3x}$ **21.** $y = 2\cos\dfrac{x}{2} + \sin\dfrac{x}{2}$

23. $y = 7\,e^{-2(x+1)} + 5x\,e^{-2(x-1)}$ **25.** (a) $y = C\,e^{2x} + (1 - C)\,e^{-x}$ (b) $y = C\,e^{2x} + (2C - 1)\,e^{-x}$ (c) $y = \frac{2}{3}\,e^{2x} + \frac{1}{3}\,e^{-x}$

27. $\alpha = \dfrac{r_1 + r_2}{2}; \beta = \dfrac{r_1 - r_2}{2}$

$y = k_1\,e^{r_1 x} + k_2\,e^{r_2 x} = e^{\alpha x}(C_1 \cosh \beta x + C_2 \sinh \beta x)$, where $k_1 = \dfrac{C_1 + C_2}{2}$, $k_2 = \dfrac{C_1 - C_2}{2}$.

29. (a) The Wronskian of $y_1 = e^{\alpha x}$, $y_2 = x\,e^{\alpha x}$ is:

$W(x) = e^{\alpha x}\,[e^{\alpha x} + \alpha\,x\,e^{\alpha x}] - x\,e^{\alpha x}\,[\alpha\,e^{\alpha x}] = e^{2\alpha x} \neq 0.$

(b) The Wronskian of $y_1 = e^{\alpha x} \cos \beta x$, $y_2 = e^{\alpha x} \sin \beta x$, $\beta \neq 0$, is

$W(x) = e^{\alpha x} \cos \beta x[\alpha\,e^{\alpha x} \sin \beta x + \beta\,e^{\alpha x} \cos \beta x] - e^{\alpha x} \sin \beta x[\alpha\,e^{\alpha x} \cos \beta x - \beta\,e^{\alpha x} \sin \beta x] = \beta\,e^{2\alpha x} \neq 0.$

31. (a) $y'' + 2y' - 8y = 0$ (b) $y'' - 4y' - 5y = 0$ (c) $y'' - 6y' + 9y = 0$

33. Set $y = e^{\alpha x}u$. Then $y' = \alpha e^{\alpha x}u + e^{\alpha x}u'$ and $y'' = \alpha^2 e^{\alpha x}u + 2\alpha e^{\alpha x}u' + e^{\alpha x}u''$.

(a) Substituting into the differential equation yields $e^{\alpha x}u'' = 0$, which implies $u'' = 0$.

(b) Substituting into the differential equation yields $e^{\alpha x}(u'' + \beta^2 u) = 0$, which implies $u'' + \beta^2 u = 0$.

35. (a) If $a = 0$, $b > 0$, then the general solution is $y = C_1 \cos \sqrt{b}\,x + C_2 \sin \sqrt{b}\,x = A \cos(\sqrt{b}\,x + \phi)$, where A and ϕ are constants. Clearly $|y(x)| \le |A|$ for all x.

(b) If $a > 0$, $b = 0$, then the general solution is $y = C_1 + C_2\,e^{-ax}$ and $\lim_{x\to\infty} y(x) = C_1$. The solution which satisfies the conditions $y(0) = y_0, y'(0) = y_1$ is $y = y_0 + \dfrac{y_1}{a} - \dfrac{y_1}{a}e^{-ax}$; $k = y_0 + \dfrac{y_1}{a}$.

37. If $y_2 = k\,y_1$, then $W(y_1, y_2) = \begin{vmatrix} y_1 & ky_1 \\ y_1' & ky_1' \end{vmatrix} = 0$. Suppose that $W(y_1, y_2) = 0$. Let I be an interval on which y_1 is nonzero. Then

$$\left(\frac{y_2}{y_1}\right)' = \frac{y_1 y_2' - y_2 y_1'}{y_1^2} = \frac{W(y_1, y_2)}{y_1^2} = 0,$$

which implies $\dfrac{y_2}{y_1} = k$ constant.

39. $y = C_1\,x^4 + C_2\,x^{-2}$ **41.** $y = c_1\,x^2 + C_2\,x^2 \ln x$

SECTION 18.5

1. $y = \frac{1}{2}x + \frac{1}{4}$ **3.** $y = \frac{1}{5}x^2 - \frac{4}{25}x - \frac{27}{125}$ **5.** $y = \frac{1}{36}e^{3x}$ **7.** $y = \frac{1}{5}e^x$ **9.** $y = -\frac{13}{170}\cos x - \frac{1}{170}\sin x$

11. $y = \frac{3}{100}\cos 2x + \frac{21}{100}\sin 2x$ **13.** $y = \frac{1}{20}e^{-x}\sin 2x + \frac{1}{10}e^{-x}\cos 2x$ **15.** $y = \frac{3}{2}x\,e^{-2x}$ **17.** $y = C_1\cos x + C_2\sin x + \frac{1}{2}e^x$

19. $y = C_1 e^{5x} + C_2 e^{-2x} + \frac{1}{10}x + \frac{7}{100}$ **21.** $y = C_1 e^x + C_2 e^{-4x} - \frac{1}{5}x\,e^{-4x}$ **23.** $y = C_1 e^{-2x} + C_2 e^x + \frac{1}{2}x^2 e^x - \frac{1}{3}x\,e^x$

25. Let $z = y_1 + y_2$. Then

$$z'' + az' + bz = (y_1'' + y_2'') + a(y_1' + y_2') + b(y_1 + y_2)$$
$$= (y_1'' + ay_1' + by_1) + (y_2'' + ay_2' + by_2) = \phi_1 + \phi_2$$

27. $y = C_1 e^{-3x} + C_2 e^{-x} + \frac{1}{4}x\,e^{-x} + \frac{1}{16}e^x$

29. $y = 2e^x\sin x - x\,e^x\cos x$ **31.** $y = \frac{1}{3}x\ln|x|\,e^{2x}$ **33.** $y = -\ln|x|e^{-2x}$

35. $y = e^x(x\sin x + \cos x\ln|\cos x|)$

37. (a) $i(t) = -CF_0\,e^{-(R/2L)t} + \dfrac{F_0}{2L}(2 - RC)\,t\,e^{-(R/2L)t} + CF_0$

(b) $i(t) = e^{-(R/2L)t}\left[\dfrac{F_0(2 - RC)}{2L\beta}\sin\beta t - CF_0\cos\beta t\right] + CF_0,$ where $\beta = \sqrt{\dfrac{4L - CR^2}{4L^2C}}$

39. (a) $y_1y_2' - y_2y_1' = -\dfrac{2}{x} \neq 0$ (b) $y = \frac{1}{3}\sin(\ln x)$

SECTION 18.6

1. $x(t) = \sin(8t + \frac{1}{2}\pi);\quad A = 1,\quad f = 4/\pi$ **3.** $\pm 2\pi A/T$ **5.** $x(t) = (15/\pi)\sin\frac{1}{3}\pi t$ **7.** $x(t) = x_0\sin(t\sqrt{k/m} + \frac{1}{2}\pi)$

9. at $x = \pm\frac{1}{2}\sqrt{3}\,x_0$ **11.** $\frac{1}{4}k\,x_0^2$

13. Set $y(t) = x(t) - 2$. Equation $x''(t) = 8 - 4x(t)$ can be written $y''(t) + 4y(t) = 0$. This is simple harmonic motion centered at $y = 0$, which is $x = 2$.

$$y(t) = A\sin(2t + \phi_0).$$

The condition $x(0) = 0$ gives $y(0) = -2$ and thus

$$A\sin\phi_0 = -2. \qquad (*)$$

Since $y'(t) = x'(t)$ and $y'(t) = 2A\cos(2t + \phi_0)$, the condition $x'(0) = 0$ gives $y'(0) = 0$, and thus

$$2A\cos\phi_0 = 0. \qquad (**)$$

Equations $(*)$ and $(**)$ are satisfied by $A = 2$, $\phi_0 = \frac{3}{2}\pi$. The equation of motion can therefore be written

$$y(t) = 2\sin(2t + \tfrac{3}{2}\pi).$$

The amplitude is 2 and the period is π.

15. (a) $x''(t) + \omega^2 x(t) = 0$ with $\omega = r\sqrt{\pi\rho/m}$ (b) $x(t) = x_0\sin(r\sqrt{\pi p/m}t + \frac{1}{2}\pi)$, taking downward as positive; $A = x_0,\quad T = (2/r)\sqrt{m\pi/\rho}$

17. amplitude and frequency both decrease

19. at most once; at most once

21. if $\omega/\gamma = m/n$, then $m/\omega = n/\gamma$ is a period

23. $x = e^{-\alpha t}[c_1\cos\sqrt{\alpha^2 - \omega^2}\,t) + c_2\sin(\sqrt{\alpha^2 - \omega^2}\,t)]$ or equivalently $x = A\,e^{-\alpha t}\sin(\sqrt{\alpha^2 - \omega^2}\,t) + \phi_0$

25. $x_p = \dfrac{F_0}{2\alpha\gamma m}\sin\gamma t;$ as $c = 2\alpha m \to 0^+$, the amplitude $\left|\dfrac{F_0}{2\alpha\gamma m}\right| \to \infty$

27. $(\omega^2 - \gamma^2)^2 + 4\alpha^2\gamma^2 = \omega^4 + \gamma^4 + 2\gamma^2(2\alpha^2 - \omega^2)$ increases as γ increases

SKILL MASTERY REVIEW

ANSWERS, SKILL MASTERY REVIEW ONE (p. 260)

1. $(-\infty,-2]\cup[3,\infty)$ **3.** $(-2,-1)\cup(2,\infty)$ **5.** -5 **7.** $\frac{8}{9}$ **9.** -1 **11.** 0 **13.** $\frac{3}{5}$ **15.** 0 **17.** $\frac{4}{3}$ **19.** -1

21. does not exist **23.** (b) (*i*) 1 (*ii*) 0 (*iii*) does not exist (*iv*) -6 (*v*) 4 (*vi*) does not exist
(c) (*i*) yes, no (*ii*) no, yes **25.** $f(2)=\frac{1}{4}$

27. (a) 0

(b) Let $\lim_{x\to 0}\dfrac{f(x)}{x}=L$. Let $\epsilon>0$. There exists $\delta_1>0$ such that $\left|\dfrac{f(x)}{x}-L\right|<1$ whenever $0<|x|<\delta_1$. Let $\delta=\min\left(\delta_1,\dfrac{\epsilon}{|L|+1}\right)$. Then

$$|f(x)|=|f(x)-Lx+Lx|\le|f(x)-Lx|+|L|\,|x|=\left|\frac{f(x)}{x}-L\right||x|+|L|\,|x|$$
$$<(|L|+1)|x|<\epsilon$$

whenever $0<|x|<\delta$.

29. $\dfrac{-1}{(x-2)^2}$ **31.** $\dfrac{ax^2-c}{x^2}$ **33.** $\dfrac{-6b}{x^3}\left(a+\dfrac{b}{x^2}\right)^2$ **35.** $\dfrac{2(x^2+4x-4)}{(2x-x^2)^2}$ **37.** $2x\sec(x^2+1)\tan(x^2+1)+\dfrac{1}{2\sqrt{x}}\sec^2(\sqrt{x})$

39. $\dfrac{4}{(2-3t)^2}\left(\dfrac{2+3t}{2-3t}\right)^{-2/3}$ **41.** tangent line: $y-4=4(x-1)$; normal line: $y-4=-\frac{1}{4}(x-1)$ **43.** $\dfrac{6(x^2+2)}{\sqrt{x^2+4}}$ **45.** $n!\,b^n$

47. $\dfrac{dy}{dx}=-\dfrac{3x^2y+y^3}{3xy^2+x^3}$ **49.** (a) $x=1$ (b) $x=3$ (c) $x=\frac{1}{3}$ **51.** $\frac{1}{6}$ **53.** (a) 7.96875 (b) $4.2\overline{3}$

55. $x=\frac{\pi}{4},\frac{5\pi}{4}$ **57.** f is not differentiable on $(0,2)$ **59.** absolute min $f(3\pi/2)=-1$, absolute max $f(\pi/6)=f(5\pi/6)=\frac{5}{4}$

61.

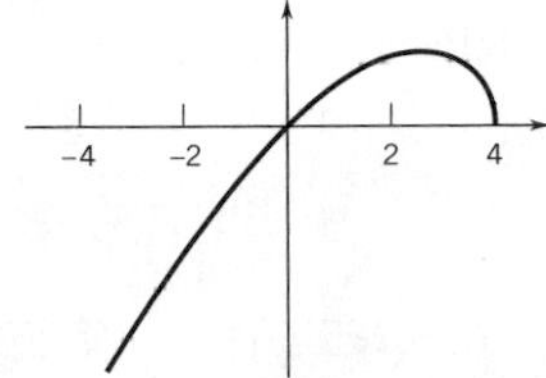

63.

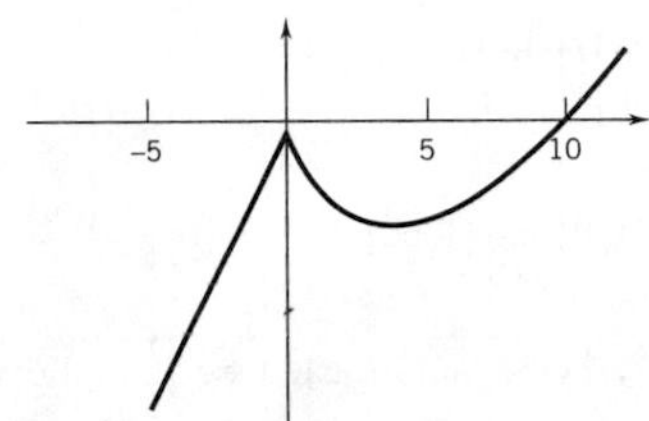

65.

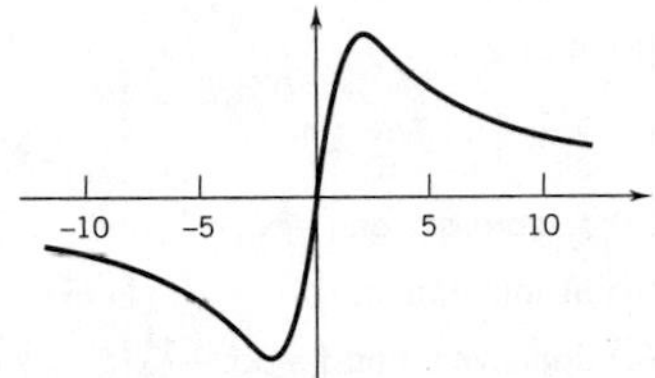

67.

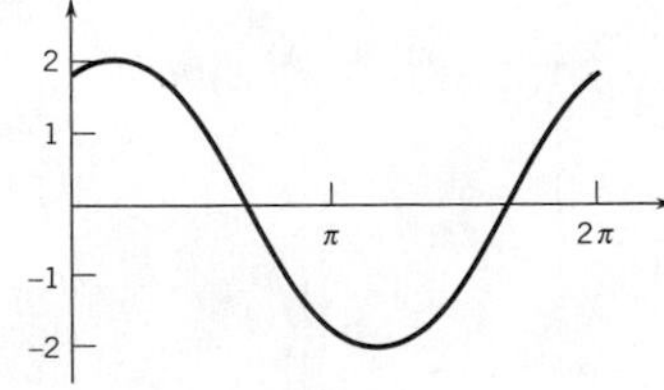

69.

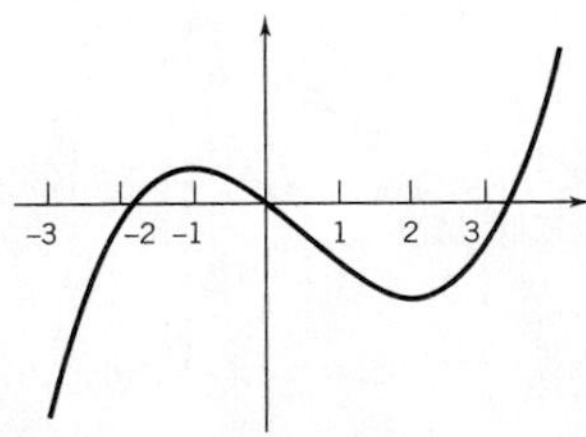

71. -2.25 inches per minute

73. (a) $a(t)=-\frac{1}{4}(t+1)^{-3/2}=-2v^3(t)$
(b) $x(17)\simeq 4.25,\ v(17)\simeq 0.1172,\ a(17)\simeq .0032$

75. (a) maximum height: 256 ft at $t=4$ seconds
(b) $t=8$ seconds, $v(8)=-128$ ft/sec

77. $A(2,0)$, $B(0,4)$

ANSWERS, SKILL MASTERY REVIEW TWO (p. 369)

1. $\frac{2}{7}x^{7/2}-\frac{4}{3}x^{3/2}+2x^{1/2}+C$ **3.** $\frac{1}{33}(1+t^3)^{11}+C$ **5.** $\frac{1}{2}(t^{2/3}-1)^3+C$ **7.** $\frac{2}{5}(2-x)^{5/2}-\frac{4}{3}(2-x)^{3/2}+C$ **9.** $\frac{1}{3}(1+\sqrt{x})^6+C$

11. $2\sqrt{1+\sin x}+C$ **13.** $\frac{1}{3}\tan 3\theta-\frac{1}{3}\cot 3\theta-4\theta+C$ **15.** $\dfrac{1}{2}\tan x+C$ **17.** $\frac{1}{3\pi}\sec^3\pi x+C$ **19.** $\dfrac{2a}{5b^2}(1+bx)^{5/2}-\dfrac{2a}{3b^2}(1+bx)^{3/2}+C$

21. $\sqrt{1+g^2(x)}+C$ **23.** 9 **25.** $\frac{1}{8}$ **27.** $4-\dfrac{6^{4/3}}{4}$ **29.** 19 **31.** $\frac{9}{2}$ **33.** 1 **35.** $\frac{3}{4}$

37. (a) $\int_{-1}^{2}(2+x-x^2)dx=\frac{9}{2}$

(b) $\int_{-2}^{1}\left(\sqrt{2-y}+y\right)dy+\int_{1}^{2}2\sqrt{2-y}\,dy$

39. (a) $\int_{1}^{3}2\sqrt{2(x-1)}\,dx+\int_{3}^{9}\left(\sqrt{2(x-1)}-x+5\right)dx$

(b) $\int_{-2}^{4}(y+4-\frac{1}{2}y^2)dy=18$

41. $\dfrac{1}{1+x^2}$ **43.** $\dfrac{2x}{1+x^4}-\dfrac{1}{1+x^2}$ **45.** (a) -1 (b) $f'(x)=-\sin x$ **47.** $\dfrac{8+\pi^2}{2\pi}$ **49.** $\frac{4}{15}\pi$ **51.** $\frac{6}{7}\pi$

53. $\pi-\frac{1}{4}\pi^2$ **55.** 2π **57.** (a) $\frac{2}{3}\pi r^3$ (b) $\frac{4}{3}r^3$ **59.** $6\sqrt{3}$ **61.** $\bar{x}=\frac{16}{15},\ \bar{y}=\frac{64}{21}$ **63.** $\bar{x}=0,\ \bar{y}=\frac{1}{8}\pi$

65. $\bar{x}=\bar{y}=\frac{16}{35}$; the two volumes are equal: $V=\frac{16}{35}\pi$ **67.** $\frac{13}{2}$ inches **69.** 1100 ft-lbs

ANSWERS, SKILL MASTERY REVIEW THREE (p. 512)

1. $f^{-1}(x)=(x-2)^3$ **3.** f is not one-to-one **5.** $f'(x)=\dfrac{-e^x}{(1+e^x)^2}<0$ on $(-\infty,\infty)$; $(f^{-1})'\left(\frac{1}{2}\right)=-4$

7. $f'(x)=\sqrt{4+x^2}>0$ on $(-\infty,\infty)$; $(f^{-1})'(0)=\frac{1}{2}$ **9.** $\dfrac{24}{x}[\ln x]^2$ **11.** $\dfrac{e^x-e^{3x}}{(1+e^{2x})^2}$ **13.** $\dfrac{3x^2+3^x\ln 3}{x^3+3^x}$

15. $(\cosh x)^{1/x}\left[\dfrac{x\tanh x-\ln(\cosh x)}{x^2}\right]$ **17.** $\frac{2}{3}$ **19.** $2[x\cosh x-\sinh x]+C$ **21.** $4\ln|x|+\dfrac{3}{x}-4\ln|x+1|+C$ **23.** $-\frac{2}{3}\cos^3 x+C$

25. $\frac{1}{2}[x\ln|x+1|-x+\ln|x+1|]+C$ **27.** $\ln|\sec x|-\frac{1}{2}\sin^2 x+C$ **29.** $\frac{3}{4}-\frac{1}{4}e^{-2}$ **31.** $-\frac{1}{2}[\ln(\cos x)]^2+C$ **33.** $\frac{\pi}{12}$

35. $\dfrac{x\,2^x}{\ln 2}-\dfrac{2^x}{(\ln 2)^2}+C$ **37.** $-\dfrac{\sqrt{a^2-x^2}}{x}-\sin^{-1}\left(\dfrac{x}{a}\right)+C$ **39.** $\frac{1}{6}\tan^6 x+C$ **41.** $\frac{1}{2}\ln\left(2+\sqrt{3}\right)-\frac{1}{4}\ln 3$

43. $\frac{1}{2}\sin^{-1}x-\frac{1}{2}x\sqrt{1-x^2}-\sqrt{1-x^2}+C$ **45.** $\frac{1}{2}$ **47.** $a^2\ln 2$ **49.** $\bar{x}=\dfrac{6-3\sqrt{3}}{\pi},\ \bar{y}=\dfrac{3\ln 3}{2\pi}$

51. (b) $A=2$

(c) $\bar{x}=\frac{1}{2}\pi,\ \bar{y}=\frac{5}{8}\pi$

53. $\frac{1}{2}\pi\,[1-\ln 4+e^2(1+\ln 4)]$

55. (a) f increases on $(-\infty,-1]$ and $[0,1]$; f decreases on $[-1,0]$ and $[1,\infty]$;

(b) absolute max: $f(-1)=f(1)=e^{-1}$; absolute min: $f(0)=0$

(c) concave up on $\left(-\infty,-\frac{1}{2}\sqrt{5+\sqrt{17}}\right)$, $\left(-\frac{1}{2}\sqrt{5-\sqrt{17}},\ \frac{1}{2}\sqrt{5-\sqrt{17}}\right)$, and $\left(\frac{1}{2}\sqrt{5+\sqrt{17}},\infty\right)$; concave down on $\left(-\frac{1}{2}\sqrt{5+\sqrt{17}},\ -\frac{1}{2}\sqrt{5-\sqrt{17}}\right)$ and $\left(\frac{1}{2}\sqrt{5-\sqrt{17}},\ \frac{1}{2}\sqrt{5+\sqrt{17}}\right)$

(d)

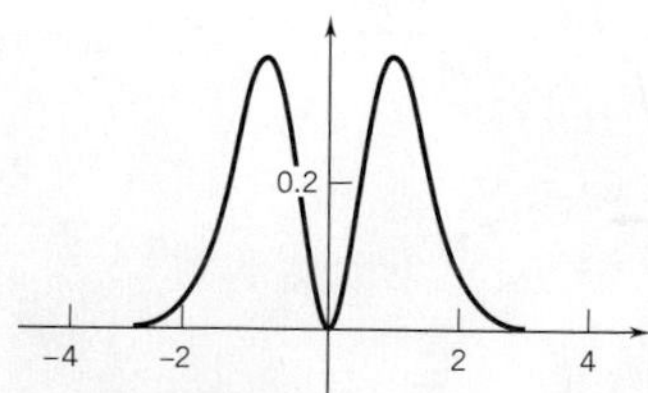

57. $T=-\dfrac{\ln 2}{\ln 0.8}\cong 3.1$ years **59.** (a) 6.94 grams (b) 18,935 years ago

ANSWERS, SKILL MASTERY REVIEW FOUR (p. 704)

1.

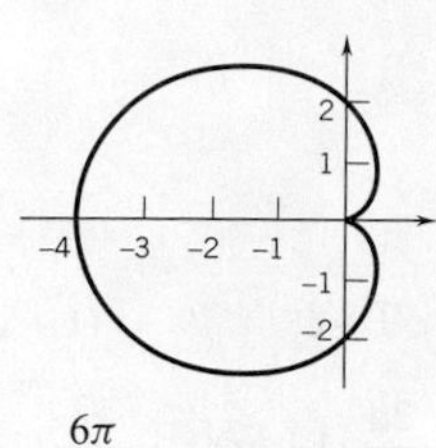

6π

3.

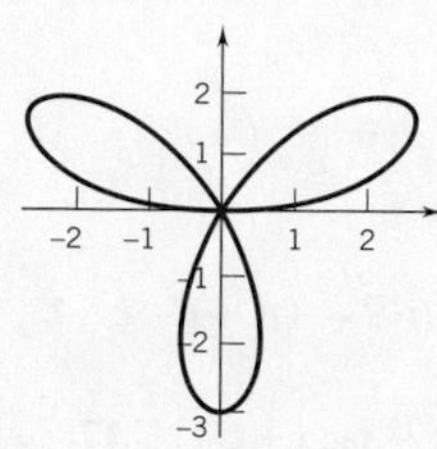

$\frac{9}{4}\pi$

5. $1 - \frac{1}{4}\pi$ **7.** $x + y = 2\sqrt{2}$ **9.** parabola **11.** $\ln 3 - \frac{1}{2}$ **13.** $\frac{992}{3}\pi$

15. (a) min speed at $t = \frac{1}{4}$, max speed at $t = 1$ (b) 1 **17.** unbounded, increasing **19.** unbounded, increasing from a_3 on **21.** converges to 1

23. converges to 0 **25.** diverges **27.** converges to ln 8 **29.** 5 **31.** $-\frac{1}{2}$ **33.** 0 **35.** $a - a\ln a$ **37.** $\frac{2}{e}$ **39.** diverges

41. 4 **43.** 2 **45.** divergent **47.** absolutely convergent **49.** conditionally convergent **51.** absolutely convergent

53. conditionally convergent **55.** convergent **57.** divergent **59.** convergent **61.** $[-\frac{1}{5}, \frac{1}{5})$ **63.** $(-\infty, \infty)$ **65.** $(-9, 9)$

67. $\sum_{k=0}^{\infty} \frac{2^k}{k!} x^{2k+1}$ **69.** $\sum_{k=1}^{\infty} \frac{(-1)^{k-1}}{2k-1} x^k$ **71.** $1 + \frac{1}{2}x^2$ **73.** $e^2 \sum_{k=0}^{\infty} \frac{(-1)^k 2^k}{k!}(x+1)^k,\ R = \infty$ **75.** $\sum_{k=1}^{\infty} \frac{(-1)^{k+1}}{k}(x-1)^k,\ R = 1$

ANSWERS, SKILL MASTERY REVIEW FIVE (p. 940)

1. $-9\mathbf{i} - 5\mathbf{j} - 2\mathbf{k}$ **3.** $\sqrt{90}$ **5.** $\cos\theta = \frac{2}{\sqrt{714}}$ **7.** $\pm\frac{1}{\sqrt{293}}(6\mathbf{i} - \mathbf{j} + 16\mathbf{k})$ **9.** (a) $x = 5 + 4t,\ y = 6 - 7t,\ z = -3 + 5t$ (b) $x = 5 - 10t,\ y = 6 - 5t,\ z = -3 + t$ (c) $x - 3y - 5z - 2 = 0$

11. (a) no (b) yes **13.** $2x + 3y - 4z - 19 = 0$ **15.** $2x - 2y - z - 6 = 0$ **17.** $\mathbf{f}'(t) = 2e^{2t}\,\mathbf{i} + \frac{2t}{t^2+1}\,\mathbf{j};\ \mathbf{f}''(t) = 4e^{2t}\,\mathbf{i} + \frac{2 - 2t^2}{(t^2+1)^2}\,\mathbf{j}$

19. $\mathbf{f}'(t) = 2\cosh 2t\,\mathbf{i} + (te^{-t} - e^{-t})\,\mathbf{j} + \sinh t\,\mathbf{k}$;
$\mathbf{f}''(t) = 4\sinh 2t\,\mathbf{i} + (2e^{-t} - te^{-t})\,\mathbf{j} + \cosh t\,\mathbf{k}$

21. $\mathbf{v}(t) = 2\mathbf{i} + \frac{1}{t}\mathbf{j} - 2t\,\mathbf{k};\ \|\mathbf{v}(t)\| = 2t + \frac{1}{t}$;
$\mathbf{a}(t) = -\frac{1}{t^2}\,\mathbf{j} - 2\,\mathbf{k}$

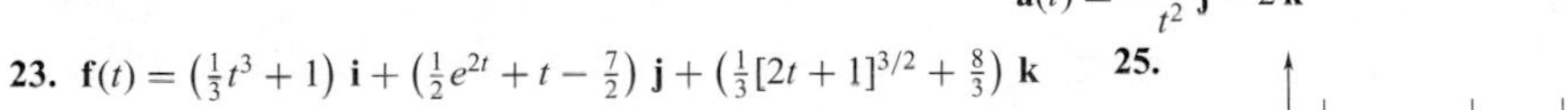
23. $\mathbf{f}(t) = \left(\frac{1}{3}t^3 + 1\right)\mathbf{i} + \left(\frac{1}{2}e^{2t} + t - \frac{7}{2}\right)\mathbf{j} + \left(\frac{1}{3}[2t+1]^{3/2} + \frac{8}{3}\right)\mathbf{k}$

25.

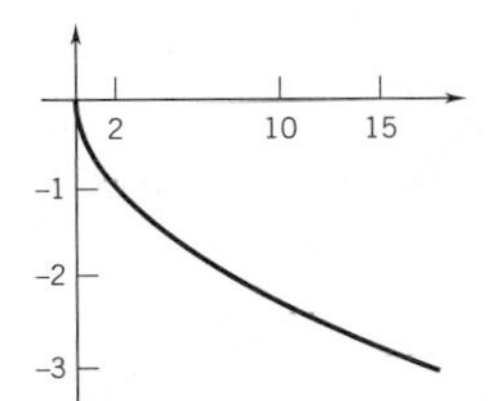

27.

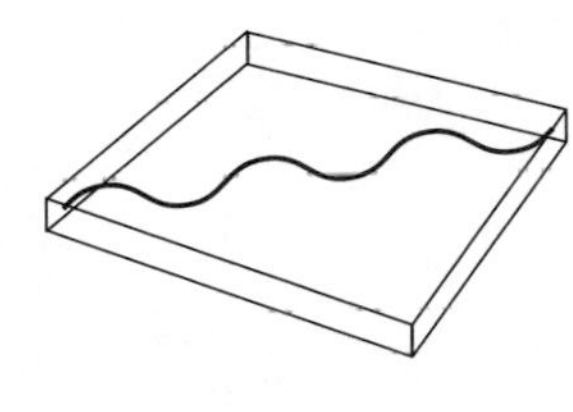

29. $\mathbf{r}'\left(\frac{\pi}{3}\right) = -\sqrt{3}\,\mathbf{i} - \mathbf{j} + \mathbf{k};\ x = -\frac{1}{2} - \sqrt{3}t,\ y = \frac{\sqrt{3}}{2} - t,\ z = \frac{\pi}{3} + t$

31. $\mathbf{T}(t) = -\frac{1}{\sqrt{2}}\sin t\,\mathbf{i} - \frac{1}{\sqrt{2}}\sin t\,\mathbf{j} - \cos t\,\mathbf{k}$
$\mathbf{N}(t) = -\frac{1}{\sqrt{2}}\cos t\,\mathbf{i} - \frac{1}{\sqrt{2}}\cos t\,\mathbf{j} + \sin t\,\mathbf{k}$

33. $\frac{38}{3}$

35. $\sqrt{2}\sinh 1$ **37.** (a) $\frac{1}{2(1 + e^{-2t})^{3/2}}$ (b) $\frac{1}{|t|(1 + t^2)^{3/2}}$ **39.** $\kappa = 1,\ a_T = 0,\ a_N = 1$

41. (a) $\text{dom}(f) = \{(x, y) : x^2 - y^2 > 1\}$; $\text{range}(f) = (-\infty, \infty)$
(b) $\text{dom}(f) = \{(x, y, z) : x^2 + y^2 < z\}$; $\text{range}(f) = [0, \infty)$

43. circles

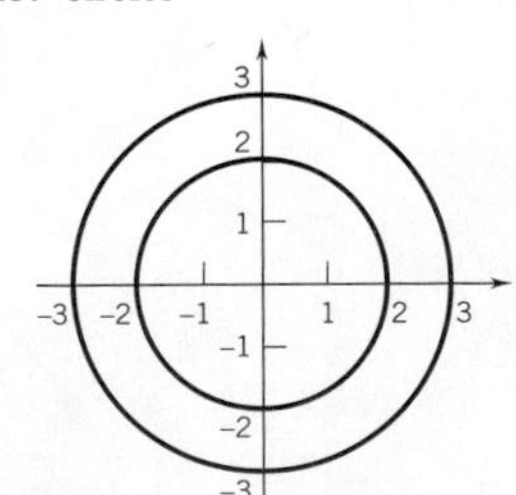

45. plane

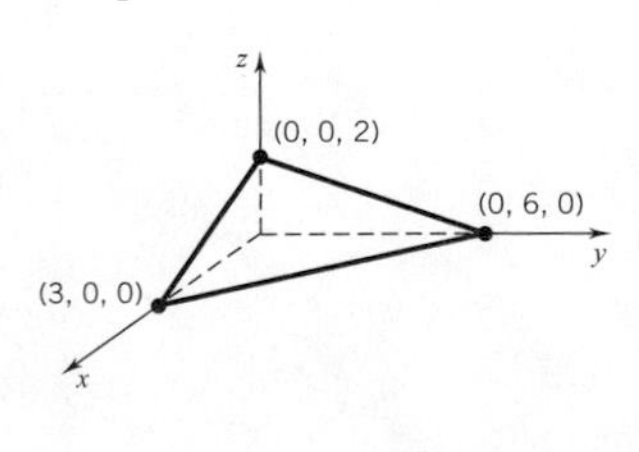

47. $\frac{\partial z}{\partial x} = x^2y^2\cos(xy^2) + 2x\sin(xy^2);\ \frac{\partial z}{\partial y} = 2x^3y\cos(xy^2)$ **49.** $g_x = \frac{x}{x^2 + y^2 + z^2};\ g_y = \frac{y}{x^2 + y^2 + z^2};\ g_z = \frac{z}{x^2 + y^2 + z^2}$

51. $f_{xx} = 4yz^3 + y^2z^2e^{xyz};\ f_{zx} = 12xyz^2 + ye^{xyz} + xy^2ze^{xyz}$
$f_{yz} = 6x^2z^2 + xe^{xyz} + x^2yze^{xyz}$

53. (a) $\nabla f = (4x - 4y)\,\mathbf{i} + (-4x + 3y^2)\,\mathbf{j}$
(b) $\nabla f = \frac{y^3 - x^2y}{(x^2 + y^2)^2}\,\mathbf{i} + \frac{x^3 - xy^2}{(x^2 + y^2)^2}\,\mathbf{j}$

55. $\frac{2}{\sqrt{5}}$ **57.** $e\sqrt{5}$ **59.** tangent plane: $-\frac{1}{\sqrt{2}}(x - 1) + \frac{1}{\sqrt{2}}(y + 1) - (z - \sqrt{2}) = 0$;
normal line: $x = 1 - \frac{1}{\sqrt{2}}t,\ y = -1 + \frac{1}{\sqrt{2}}t,\ z = \sqrt{2} - t$

61. tangent plane: $x + y + 4z - 3 = 0$; normal line: $x = t,\ y = -1 + t,\ z = 1 + 4t$

63. saddle points: $(5, 0)$, $(-3, 0)$, local min: $(1, 4)$

65. saddle point: $\left(3, \frac{1}{3}\right)$

67. absolute max: $f(3, 3) = 6$; absolute min: $f(1, 2) = -3$

69. (a) not a gradient (b) $f(x, y) = -\dfrac{y}{x} + x^4 - x - y\cos 3x + y^3 + 2y + C$

71. maximum: $f(4, 2, 4) = f(-4, 2, -4) = 20$; minimum: $f(-4, -2, 4) = f(4, -2, -4) = -20$

73. length $=$ width $= \dfrac{10}{\sqrt[3]{75}} \simeq 2.37$, height $= \dfrac{12}{\sqrt[3]{75}} \cong 2.85$

75. hottest: $T\left(\pm\frac{\sqrt{3}}{2}, -\frac{1}{2}\right) = \frac{9}{4}$; coldest: $T\left(0, \frac{1}{2}\right) = -\frac{1}{4}$

ANSWERS, SKILL MASTERY REVIEW SIX (p. 1094)

1. $\frac{1}{40}$ **3.** $\frac{8}{3}(e - 1)$ **5.** $-\frac{2}{3}$ **7.** $-\frac{16}{15}$ **9.** 1 **11.** $\frac{1}{8}$ **13.** $-\frac{8}{15}$ **15.** $\frac{32}{15}$ **17.** $\frac{1}{6}$ **19.** $\frac{1}{2}(1 + e^2)\pi$

21. $M = \frac{2}{15}$; $\bar{x} = \frac{15}{28}, \bar{y} = \frac{15}{24}$

23. Introduce a coordinate system with the y-axis the axis of symmetry and the base of the triangle on the x-axis.

(a) $\bar{x} = 0, \bar{y} = \frac{1}{3}h$ (b) $\frac{1}{6}Mh^2$ (c) $\frac{1}{24}Mb^2$

25. 24π **27.** $\frac{5}{3}\pi$ **29.** (a) $\bar{x} = \bar{y} = \frac{9}{64}\pi, \bar{z} = \frac{3}{8}$ (b) $\frac{8}{15}M$ **31.** 0 **33.** (a) $-\frac{1}{12}$ (b) $-\frac{11}{72}$ **35.** $\frac{2}{3} - 6\cos 3 + 2\sin 3 + \frac{1}{3}e^{27} - \cos 27$

37. 4C **39.** $e^8 - 13$ **41.** (a) $\frac{32}{3}$ (b) -32 (c) $-\frac{608}{35}$ **43.** $\frac{1}{3}e^8 + 4\sin 2 + 2\cos 2 + \frac{17}{3}$ **45.** $-\frac{1}{3}$ **47.** 6 **49.** -4

51. 8π **53.** $9\pi\sqrt{2}$ **55.** $\frac{224}{3}\pi$ **57.** 324π **59.** 4π

■ INDEX

D

E

F

G

H

I

J

K

L

M

N

Q

R

S

(*continued from the front*)

INVERSE TRIGONOMETRIC FUNCTIONS

64. $\int \sin^{-1} u\, du = \sin^{-1} u + \sqrt{1-u^2} + C$

65. $\int \cos^{-1} u\, du = u\cos^{-1} u - \sqrt{1-u^2} + C$

66. $\int \tan^{-1} u\, du = u\tan^{-1} u - \frac{1}{2}\ln(1+u^2) + C$

67. $\int \cot^{-1} u\, du = u\cot^{-1} u + \frac{1}{2}\ln(1+u^2) + C$

68. $\int \sec^{-1} u\, du = u\sec^{-1} u - \ln|u + \sqrt{u^2-1}| + C$

69. $\int \csc^{-1} u\, du = u\csc^{-1} u + \ln|u + \sqrt{u^2-1}| + C$

70. $\int u\sin^{-1} u\, du = \frac{1}{4}(2u^2-1)\sin^{-1} u + u\sqrt{1-u^2} + C$

71. $\int u\tan^{-1} u\, du = \frac{1}{2}(u^2+1)\tan^{-1} u - \frac{1}{2}u + C$

72. $\int u\cos^{-1} u\, du = \frac{1}{4}(2u^2-1)\cos^{-1} u - u\sqrt{1-u^2} + C$

73. $\int u^n \sin^{-1} u\, du = \frac{1}{n+1}\left[u^{n+1}\sin^{-1} u - \int \frac{u^{n+1}du}{\sqrt{1-u^2}}\right], n \neq -1$

74. $\int u^n \cos n^{-1} u\, du = \frac{1}{n+1}\left[u^{n+1}\cos^{-1} u - \int \frac{u^{n+1}\, du}{\sqrt{1-u^2}}\right], n \neq -1$

75. $\int u^n \tan^{-1} u\, du = \frac{1}{n+1}\left[u^{n+1}\tan^{-1} u - \int \frac{u^{n+1}du}{\sqrt{1-u^2}}\right], n \neq -1$

$\sqrt{a^2+u^2},\ a > 0$

76. $\int \frac{du}{a^2+u^2} = \frac{1}{a}\tan^{-1}\frac{u}{a} + C$

77. $\int \frac{du}{\sqrt{a^2+u^2}} = \ln|u + \sqrt{a^2+u^2}| + C$

78. $\int \sqrt{a^2+u^2}\, du = \frac{u}{2}\sqrt{a^2+u^2} + \frac{a^2}{2}\ln|u + \sqrt{a^2+u^2}| + C$

79. $\int u^2\sqrt{a^2+u^2}\, du = \frac{u}{8}(a^2+2u^2)\sqrt{a^2+u^2} - \frac{a^4}{8}\ln|u + \sqrt{a^2+u^2}| + C$

80. $\int \frac{\sqrt{a^2+u^2}}{u}\, du = \sqrt{a^2+u^2} - a\ln\left|\frac{a+\sqrt{a^2+u^2}}{u}\right| + C$

81. $\int \frac{\sqrt{a^2+u^2}}{u^2}\, du = -\frac{\sqrt{a^2+u^2}}{u} + \ln|u + \sqrt{a^2+u^2}| + C$

82. $\int \frac{u^2\, du}{\sqrt{a^2+u^2}} = \frac{u}{2}\sqrt{a^2+u^2} - \frac{a^2}{2}\ln|u + \sqrt{a^2+u^2}| + C$

83. $\int \frac{du}{u\sqrt{a^2+u^2}} = -\frac{1}{a}\ln\left|\frac{a+\sqrt{a^2+u^2}}{u}\right| + C$

84. $\int \frac{du}{u^2\sqrt{a^2+u^2}} = -\frac{\sqrt{a^2+u^2}}{a^2 u} + C$

85. $\int \frac{du}{(a^2+u^2)^{3/2}} = \frac{u}{a^2\sqrt{a^2+u^2}} + C$

$\sqrt{a^2-u^2},\ a > 0$

86. $\int \frac{du}{\sqrt{a^2-u^2}} = \sin^{-1}\frac{u}{a} + C$

87. $\int \sqrt{a^2-u^2}\, du = \frac{u}{2}\sqrt{a^2-u^2} + \frac{a^2}{2}\sin^{-1}\frac{u}{a} + C$

88. $\int u^2\sqrt{a^2-u^2}\, du = \frac{u}{8}(2u^2-a^2)\sqrt{a^2-u^2} + \frac{a^4}{8}\sin^{-1}\frac{u}{a} + C$